# Table of Integrals, Series, and Products

## Eighth Edition

# Table of
# Integrals, Series, and Products
## *Eighth Edition*

I. S. Gradshteyn and I. M. Ryzhik

Daniel Zwillinger, Editor
*Rensselaer Polytechnic Institute, USA*

Victor Moll (Scientific Editor)
*Tulane University, Department of Mathematics*

TRANSLATED FROM THE RUSSIAN BY SCRIPTA TECHNICA, INC

AMSTERDAM • BOSTON • HEIDELBERG • LONDON
NEW YORK • OXFORD • PARIS • SAN DIEGO
SAN FRANCISCO • SINGAPORE • SYDNEY • TOKYO

Academic Press is an imprint of Elsevier

Academic Press is an imprint of Elsevier
225 Wyman Street, Waltham, MA 02451, USA
525 B Street, Suite 1800, San Diego, CA 92101-4495, USA
32 Jamestown Road, London NW1 7BY, UK
The Boulevard, Langford Lane, Kidlington, Oxford OX5 1GB, UK

**ISBN: 978-0-12-384933-5**

**Library of Congress Cataloging-in-Publication Data**
Gradshtein, I. S. (Izrail? Solomonovich)
    [Tablitsy integralov, summ, riadov i proizvedenii. English]
    Table of integrals, series, and products. – Eighth edition / Daniel Zwillinger.
        pages cm
    Includes bibliographical references and index.
    ISBN 978-0-12-384933-5
    1. Mathematics–Tables. I. Zwillinger, Daniel, 1957- II. Title.
    QA55.G6613 2014
    510.2′1–dc23

                                                                              2014010276

**British Library Cataloguing-in-Publication Data**
A catalogue record for this book is available from the British Library.

For information on all Academic Press publications
visit our web site at store.elsevier.com

# Contents

# Preface to the Eighth Edition

Gradshteyn and Ryzhik continues to be a resource greatly used by mathematicians, scientists, and engineers in the theoretical, applied, and computational sciences. Since the publication in 2007 of the revised seventh edition, users have continued to submit corrections and new results that improve the book, and make suggestions for changes that improve the presentation of the material. We regret that the structure of the book makes it impossible to acknowledge these users by their individual contributions so, as usual, their names have been added to the acknowledgment list at the front of the book.

This eighth edition includes corrections received since the publication of the seventh edition, together with a substantial amount of new material acquired from isolated sources. From among the many contributions we have included just those integrals that appear commonly in different contexts.

Following our previous conventions, an amended entry has a superscript "12" added to its entry reference number, where the equivalent superscript number for the seventh edition was "11". Similarly, an asterisk on an entry's reference number indicates a new result. When, for technical reasons, an entry in a previous edition has been removed, the entry numbers will jump. This preserves the continuity of numbering between the new and older editions. This edition has also removed chapters present in the 7th edition that were not aligned with integrals, series, and products (e.g., the chapters on matrices and norms).

We wish to express our gratitude to all who have been in contact with us with the object of improving and extending the book. Special thanks are extended to both Dr. Francis J. O'Brien, Jr. of the Naval Station in Newport, Rhode Island and Dr. Andrej Tenne-Sens of Ottowa, Ontario, Canada. They have each spent an unimaginably large number of hours helping with this edition.

Experience over many years has shown that each new edition of Gradshteyn and Ryzhik generates many suggestions for new entries and new errata. Hence, we do not expect this new edition to be free from errors. All users who identify errors, or who wish to propose new entries, are invited to contact the authors whose email addresses are listed below. Corrections will be posted on the web site www.mathtable.com/gr/errata.

Finally, we mourn the passing of Professor Alan Jeffrey who, as Editor, guided this book from its 4th edition in 1965 to the 7th edition in 2007. Alan's dedication to this book was evident both by his longevity—well over 40 years of improvements—and in the consideration he gave to all of our correspondents for whom he evaluated a huge number of integrals.

Daniel Zwillinger
ZwillingerBooks@gmail.com

Victor Moll
vhm@tulane.edu

# Acknowledgments

The publisher and editors would like to take this opportunity to express their gratitude to the following users of the *Table of Integrals, Series, and Products* who either directly or through errata published in *Mathematics of Computation* have generously contributed corrections and addenda to the original printing.

Anonymous
Dr. Artem G. Abanov
Dr. A. Abbas
Dr. S. M. Abrarov
Dr. P. B. Abraham
Dr. Ari Abramson
Dr. Jose Adachi
Dr. R. J. Adler
Dr. N. Agmon
Dr. M. Ahmad
Dr. S. A. Ahmad
Dr. Rajai S. Alassar
Dr. Luis Alvarez-Ruso
Dr. Maarten H P Ambaum
Dr. R. K. Amiet
Dr. L. U. Ancarani
Dr. M. Antoine
Dr. Marian Apostol
Dr. C. R. Appledorn
Dr. D. R. Appleton
Dr. Mitsuhiro Arikawa
Dr. Ir. Luk R. Arnaut
Dr. Peter Arnold
Dr. P. Ashoshauvati
Dr. C. L. Axness
Dr. Scott Baalrud
Dr. E. Badralexe
Dr. S. B. Bagchi
Dr. L. J. Baker
Dr. R. Ball
Dr. Ingo Barth

Dr. M. P. Barnett
Dr. Fabio Bernardoni
Dr. Florian Baumann
Dr. Norman C. Beaulieu
Dr. Jerome Benoit
Mr. V. Bentley
Dr. Laurent Berger
Dr. M. van den Berg
Dr. N. F. Berk
Dr. C. A. Bertulani
Dr. J. Betancort-Rijo
Dr. P. Bickerstaff
Dr. Iwo Bialynicki-Birula
Dr. Chris Bidinosti
Dr. G. R. Bigg
Dr. Ian Bindloss
Dr. L. Blanchet
Dr. Mike Blaskiewicz
Dr. R. D. Blevins
Dr. Anders Blom
Dr. L. M. Blumberg
Dr. R. Blumel
Dr. S. E. Bodner
Dr. Simone Boi
Dr. M. Bonsager
Dr. George Boros
Dr. S. Bosanac
Dr. Ruben Van Boxem
Dr. Christoph Bruder
Dr. Patrick Bruno
Dr. B. Van den Bossche

Dr. A. Boström
Dr. J. E. Bowcock
Dr. T. H. Boyer
Dr. K. M. Briggs
Dr. D. J. Broadhurst
Dr. Chris Van Den Broeck
Dr. W. B. Brower
Dr. H. N. Browne
Dr. Christoph Bruegger
Dr. William J. Bruno
Dr. Vladimir Bubanja
Dr. D. J. Buch
Dr. D. J. Bukman
Dr. F. M. Burrows
Dr. R. Caboz
Dr. T. Calloway
Dr. F. Calogero
Dr. D. Dal Cappello
Dr. David Cardon
Dr. J. A. Carlson Gallos
Dr. B. Carrascal
Dr. A. R. Carr
Dr. Neal Carron
Dr. Florian Cartarius
Dr. S. Carter
Dr. Miguel Carvajal
Dr. G. Cavalleri
Mr. W. H. L. Cawthorne
Dr. Alexandre Cazé
Dr. A. Cecchini
Dr. B. Chan

Dr. M. A. Chaudhry
Dr. Sabino Chavez-Cerda
Dr. Julian Cheng
Dr. H. W. Chew
Dr. D. Chin
Dr. Young-seek Chung
Dr. S. Ciccariello
Dr. N. S. Clarke
Dr. R. W. Cleary
Dr. A. Clement
Dr. Alain Cochard
Dr. P. Cochrane
Dr. D. K. Cohoon
Dr. Howard S. Cohl
Dr. L. Cole
Dr. Filippo Colomo
Donal Connon
Dr. J. R. D. Copley
Henry Corback, Esq.
Dr. Daniel Benevides da Costa
Dr. Barry J. Cox
Dr. D. Cox
Dr. J. Cox
Dr. J. W. Criss
Dr. A. E. Curzon
Dr. D. Dadyburjor
Dr. D. Dajaputra
Dr. C. Dal Cappello
Dr. P. Daly
Dr. S. Dasgupta
Dr. Charles E. Davidson
Dr. John Davies
Dr. C. L. Davis
Dr. A. Degasperis
Dr. Gustav Delius
Dr. B. C. Denardo
Dr. R. W. Dent
Dr. E. Deutsch
Dr. D. deVries
Dr. Eran Dgani
Dr. Enno Diekema
Dr. P. Dita
Dr. P. J. de Doelder
Dr. Mischa Dohler
Dr. G. Dôme
Dr. Shi-Hai Dong
Dr. Balazs Dora

Dr. M. R. D'Orsogna
Dr. Forrest Doss
Dr. Adrian A. Dragulescu
Dr. Zvi Drezner
Dr. Eduardo Duenez
Mr. Tommi J. Dufva
Dr. Duc V. Duong
Dr. E. B. Dussan, V
Dr. Percy Dusek
Dr. C. A. Ebner
Dr. M. van der Ende
Dr. Jonathan Engle
Dr. G. Eng
Dr. E. S. Erck
Dr. Grant Erdmann
Dr. Jan Erkelens
Dr. Olivier Espinosa
Dr. G. A. Estévez
Dr. K. Evans
Dr. G. Evendon
Dr. Valery I. Fabrikant
Dr. L. A. Falkovsky
Dr. Kambiz Farahmand
Dr. Richard J. Fateman
Dr. G. Fedele
Dr. A. R. Ferchmin
Dr. P. Ferrant
Dr. André Ferrari
Dr. H. E. Fettis
Dr. W. B. Fichter
Dr. George Fikioris
Mr. J. C. S. S. Filho
Dr. L. Ford
Dr. Nicolao Fornengo
Dr. J. France
Dr. B. Frank
Dr. S. Frasier
Dr. Stefan Fredenhagen
Dr. A. J. Freeman
Dr. A. Frink
Dr. Jason M. Gallaspy
Dr. J. A. C. Gallas
Dr. J. A. Carlson Gallas
Dr. G. R. Gamertsfelder
Dr. Jianliang Gao
Dr. T. Garavaglia
Dr. Jaime Zaratiegui Garcia

Dr. C. G. Gardner
Dr. D. Garfinkle
Dr. P. N. Garner
Dr. F. Gasser
Dr. E. Gath
Dr. P. Gatt
Dr. D. Gay
Dr. M. P. Gelfand
Dr. M. R. Geller
Dr. Ali I. Genc
Dr. Vincent Genot
Dr. M. F. George
Dr. Teschl Gerald
Dr. P. Germain
Dr. Ing. Christoph Gierull
Dr. S. P. Gill
Dr. Federico Girosi
Dr. E. A. Gislason
Dr. M. I. Glasser
Dr. P. A. Glendinning
Dr. L. I. Goldfischer
Dr. Denis Golosov
Dr. I. J. Good
Dr. J. Good
Mr. L. Gorin
Dr. Martin Götz
Dr. R. Govindaraj
Dr. M. De Grauf
Dr. Gabriele Gradoni
Dr. L. Green
Mr. Leslie O. Green
Dr. R. Greenwell
Dr. K. D. Grimsley
Dr. Albert Groenenboom
Dr. V. Gudmundsson
Dr. J. Guillera
Dr. K. Gunn
Dr. D. L. Gunter
Dr. Julio C. Gutiérrez-Vega
Dr. Roger Haagmans
Dr. Howard Haber
Dr. H. van Haeringen
Dr. B. Hafizi
Dr. Bahman Hafizi
Dr. T. Hagfors
Dr. M. J. Haggerty
Dr. Timo Hakulinen

Dr. S. E. Hammel
Dr. E. Hansen
Dr. Wes Harker
Dr. T. Harrett
Dr. D. O. Harris
Dr. Frank Harris
Mr. Mazen D. Hasna
Dr. Peter Hawkins
Dr. Joel G. Heinrich
Dr. Adam Dade Henderson
Dr. Franck Hersant
Dr. Sten Herlitz
Dr. Chris Herzog
Dr. A. Higuchi
Dr. R. E. Hise
Dr. Henrik Holm
Dr. N. Holte
Dr. R. W. Hopper
Dr. P. N. Houle
Dr. C. J. Howard
Jie Hu
Dr. Ben Yu-Kuang Hu
Dr. J. H. Hubbell
Dr. Felix Huber
Dr. J. R. Hull
Dr. W. Humphries
Dr. Jean-Marc Huré
Dr. Jamal A. Hussein
Dr. Y. Iksbe
Dr. Philip Ingenhoven
Mr. L. Iossif
Dr. Sean A. Irvine
Dr. Óttar Ísberg
Dr. Kazuhiro Ishida
Dr. Cyril-Daniel Iskander
Dr. S. A. Jackson
Dr. John David Jackson
Dr. Francois Jaclot
Dr. B. Jacobs
Dr. Pierre Jacobs
Dr. E. C. James
Dr. B. Jancovici
Dr. D. J. Jeffrey
Dr. H. J. Jensen
Dr. Bin Jiang
Dr. Edwin F. Johnson
Dr. I. R. Johnson

Dr. Steven Johnson
Dr. Joel T. Johnson
Dr. Fredrik Johansson
Dr. I. Johnstone
Dr. Y. P. Joshi
Dr. Jae-Hun Jung
Dr. Damir Juric
Dr. Florian Kaempfer
Dr. S. Kanmani
Dr. Z. Kapal
Dr. Peter Karadimas
Dr. Dave Kasper
Dr. M. Kaufman
Dr. Eduardo Kausel
Dr. B. Kay
Louis Kempeneers
Dr. Jack Kerlin
Dr. Avinash Khare
Dr. Karen T. Kohl
Dr. Ilki Kim
Dr. Youngsun Kim
Dr. S. Klama
Dr. L. Klingen
Dr. C. Knessl
Dr. M. J. Knight
Dr. Yannis Kohninos
Dr. D. Koks
Dr. L. P. Kok
Dr. K. S. Kölbig
Dr. Yannis Komninos
Dr. D. D. Konowalow
Dr. Z. Kopal
Dr. I. Kostyukov
Dr. R. A. Krajcik
Dr. Vincent Krakoviack
Dr. Stefan Krämer
Dr. Tobias Kramer
Dr. Hermann Krebs
Dr. Chethan Krishnan
Dr. J. W. Krozel
Dr. E. D. Krupnikov
Dr. Kun-Lin Kuo
Dr. E. A. Kuraev
Dr. Heinrich Kuttler
Dr. Konstantinos Kyritsis
Dr. Velimir Labinac
Dr. Javier Navarro Laboulais

Dr. A. D. J. Lambert
Dr. A. Lambert
Dr. A. Larraza
Dr. K. D. Lee
Dr. M. Howard Lee
Dr. M. K. Lee
Dr. P. A. Lee
Dr. Todd Lee
Dr. J. Legg
Dr. Xianfu Lei
Dr. Yefim Leifman
Dr. Remigijus Leipus
Dr. Armando Lemus
Dr. Elvio Leonardo
Dr. Eugene Lepelaars
Dr. S. L. Levie
Dr. D. Levi
Dr. Michael Lexa
Dr. Liang Li
Dr. Kuo Kan Liang
Dr. Sergey Liflandsky
Dr. B. Linet
Dr. Andrej Likar
Dr. M. A. Lisa
Dr. Donald Livesay
Dr. H. Li
Dr. Xian-Fang Li
Dr. Georg Lohoefer
Dr. I. M. Longman
Dr. D. Long
Dr. Sylvie Lorthois
Dr. Wenzhou Lu
Dr. Phil Lucht
Dr. Stephan Ludwig
Dr. Arpad Lukacs
Dr. Y. L. Luke
Dr. W. Lukosz
Dr. T. Lundgren
Dr. E. A. Luraev
Dr. R. Lynch
Dr. K. B. Ma
Dr. Ilari Maasilta
Dr. R. Mahurin
Dr. R. Mallier
Dr. G. A. Mamon
Dr. A. Mangiarotti
Dr. I. Manning

Dr. Greg Marks
Dr. J. Marmur
Dr. A. Martin
Dr. Carmelo P. Martin
Sr. Yuzo Maruyama
Dr. Richard Marthar
Dr. David J. Masiello
Dr. Richard J. Mathar
Dr. H. A. Mavromatis
Dr. M. Mazzoni
Dr. P. McCullagh
Dr. J. H. McDonnell
Dr. J. R. McGregor
Dr. Kim McInturff
Dr. N. McKinney
Dr. David McA McKirdy
Dr. Andrew J. McHutchon
Dr. Rami Mehrem
Dr. W. N. Mei
Dr. Angelo Melino
Mr. José Ricardo Mendes
Dr. Zvi Mendlowitz
Dr. Andy Mennim
Dr. Sarah Messer
Dr. J. P. Meunier
Dr. Haixing Miao
Dr. Krys A. Michalski
Dr. Gerard P. Michon
Dr. D. F. R. Mildner
Dr. D. L. Miller
Dr. Steve Miller
Dr. P. C. D. Milly
Dr. S. P. Mitra
Dr. K. Miura
Dr. N. Mohankumar
Dr. M. Moll
Dr. Victor H. Moll
Dr. D. Monowalow
Mr. Tony Montagnese
Dr. Thierry Montagu
Dr. Jim Morehead
Dr. J. Morice
Dr. J. Guy Morgan
Dr. W. Mueck
Dr. C. Muhlhausen
Dr. S. Mukherjee
Dr. R. R. Müller

Dr. Pablo Parmezani Munhoz
Dr. Frank Namin
Dr. Paul Nanninga
Dr. A. Natarajan
Dr. Sven Peter Näsholm
Dr. Javier Navarro
Dr. Christian Netzel
Dr. Stefan Neumeier
Dr. C. T. Nguyen
Dr. A. C. Nicol
Dr. M. M. Nieto
Dr. P. Noerdlinger
Dr. Andrew N. Norris
Dr. K. H. Norwich
Dr. A. H. Nuttall
Dr. F. O'Brien
Dr. R. P. O'Keeffe
Dr. A. Ojo
Dr. O. Olendski
Dr. Oleg Olendski
Dr. P. Olsson
Dr. Gilad Oren
Dr. M. Ortner
Dr. Matthew Orton
Dr. S. Ostlund
Dr. J. Overduin
Dr. J. Pachner
Mr. Robert A. Padgug
Dr. D. Papadopoulos
Dr. F. J. Papp
Mr. Man Sik Park
Dr. Jong-Do Park
Dr. B. Patterson
Dr. R. F. Pawula
Dr. D. W. Peaceman
Dr. Vittorio Peano
Dr. Vincent Pegoraro
Dr. D. Pelat
Dr. L. Peliti
Dr. Y. P. Pellegrini
Dr. Thiago S. Pereira
Dr. G. J. Pert
Dr. Nicola Pessina
Dr. J. B. Peterson
Dr. Rickard Petersson
Dr. Arnaud Pierens
Dr. Emilio Pisanty

Dr. Ralph Pixley
Dr. Andrew Plumb
Dr. Dror Porat
Dr. E. A. Power
Dr. E. Predazzi
Dr. William S. Price
Dr. Gunnar Pruessner
Dr. Paul Radmore
Dr. Carl E. Rasmussen
Dr. F. Raynal
Dr. X. R. Resende
Dr. Guy A. Reynolds
Dr. J. M. Riedler
Dr. Sjoerd Rienstra
Dr. Thomas Richard
Dr. E. Ringel
Dr. T. M. Roberts
Dr. N. I. Robinson
Dr. P. A. Robinson
Dr. Elvira Romera
Dr. D. M. Rosenblum
Dr. R. A. Rosthal
Dr. J. R. Roth
Dr. Klaus Rottbrand
Dr. Bastien Roucaries
Dr. D. Roy
Dr. E. Royer
Dr. D. Rudermann
Dr. Ahmad Rushdi
Dr. Ali Rushdi
Dr. Niall Ryan
Dr. Sanjib Sabhapandit
Dr. C. T. Sachradja
Dr. J. Sadiku
Dr. A. Sadiq
Dr. Motohiko Saitoh
Dr. Naoki Saito
Dr. A. Salim
Dr. Sherwood Samn
Dr. J. H. Samson
Dr. Miguel A. Sanchis-Lozano
Dr. J. A. Sanders
Dr. M. A. F. Sanjun
Dr. P. Sarquiz
Dr. Avadh Saxena
Dr. Vito Scarola
Dr. O. Schärpf

Dr. A. Scherzinger
Dr. B. Schizer
Dr. Martin Schmid
Dr. J. Scholes
Dr. Mel Schopper
Dr. H. J. Schulz
Dr. Andreas Schulz
Dr. Markus Schwarz
Dr. G. J. Sears
Dr. Kazuhiko Seki
Dr. B. Seshadri
Dr. Roger Sewell
Dr. A. Shapiro
Dr. Sihong Shao
Dr. Masaki Shigemori
Dr. J. S. Sheng
Dr. Kenneth Ing Shing
Dr. Tomohiro Shirai
Dr. S. Shlomo
Dr. D. Siegel
Dr. Matthew Stapleton
Dr. Leo Stein
Dr. Michael L. Stein
Dr. Steven H. Simon
Dr. Ashok Kumar Singal
Dr. Constantin Siriteanu
Dr. C. Smith
Dr. Richard Smith
Dr. G. C. C. Smith
Dr. Stefan Llewellyn Smith
Dr. S. Smith
Dr. Sasha Sodin
Dr. G. Solt
Dr. J. Sondow
Dr. A. Sørenssen
Dr. Marcus Spradlin
Dr. Andrzej Staruszkiewicz
Dr. Philip C. L. Stephenson
Dr. Edgardo Stockmeyer
Dr. J. C. Straton
Mr. H. Suraweera
Dr. N. F. Svaiter
Dr. V. Svaiter
Dr. R. Szmytkowski
Dr. Sebastian S. Szyszkowicz
Dr. S. Tabachnik

Dr. Erik Talvila
Dr. G. Tanaka
Dr. C. Tanguy
Dr. G. K. Tannahill
Dr. B. T. Tan
Dr. C. Tavard
Dr. Gonçalo Tavares
Dr. Aba Teleki
Andrej Tenne-Sens
Dr. Gerald Teschl
Dr. Arash Dahi Taleghani
Dr. D. Temperley
Dr. A. J. Tervoort
Dr. Theodoros Theodoulidis
Dr. D. J. Thomas
Dr. Michael Thorwart
Dr. S. T. Thynell
Dr. Tamer Tlas
Dr. D. C. Torney
Dr. R. Tough
Dr. Marwan Toutounji
Dr. Dennis Trede
Dr. B. F. Treadway
Dr. Ming Tsai
Dr. N. Turkkan
Dr. Sandeep Tyagi
Dr. J. J. Tyson
Dr. Takahiro Ueda
Dr. S. Uehara
Dr. M. Vadacchino
Dr. Stathis Vagenas
Dr. O. T. Valls
Dr. D. Vandeth
Dr. Klaas Vantournhout
Mr. Andras Vanyolos
Dr. D. Veitch
Mr. Jose Lopez Vicario
Dr. Hari Vishnu
Dr. K. Vogel
Dr. J. M. M. J. Vogels
Dr. Alexis De Vos
Dr. Emanuel Voto
Dr. Stuart Walsh
Dr. Haiming Wang
Dr. J. J. Wang
Dr. Reinhold Wannemacher

Dr. S. Wanzura
Dr. J. Ward
Dr. S. I. Warshaw
Dr. Alex Watson
Dr. R. Weber
Dr. Steffen Weissmann
Dr. Wei Qian
Dr. Detmar Welz
Dr. Kyle Wendt
Dr. D. H. Werner
Dr. E. Wetzel
Dr. Robert Whittaker
Dr. Peter Widerin
Dr. D. T. Wilton
Dr. C. Wiuf
Dr. K. T. Wong
Dr. Rogério Nunes Wolff
Mr. J. N. Wright
Dr. J. D. Wright
Dr. D. Wright
Dr. Chong-shi Wu
Dr. D. Wu
Dr. Roahn Wynar
Dr. Takashi Yanagisawa
Dr. Michel Daoud Yacoub
Dr. Yu S. Yakovlev
Dr. H.-C. Yang
Dr. J. J. Yang
Dr. Z. J. Yang
Dr. Mingwu Yao
Dr. Yu-Min Yen
Mr. Chun Kin Au Yeung
Dr. Steve Young
Dr. Kazuya Yuasa
Dr. S. P. Yukon
Dr. Zhuo-Quan Zeng
Dr. B. Zhang
Dr. Peng Zhang
Dr. Y. C. Zhang
Dr. Y. Zhao
Dr. Ralf Zimmer
Dr. Chunhui Zhu
Dr. Zeng Zhuo-Quan

# The Order of Presentation of the Formulas

The question of the most expedient order in which to give the formulas, in particular, in what division to include particular formulas such as the definite integrals, turned out to be quite complicated. The thought naturally occurs to set up an order analogous to that of a dictionary. However, it is almost impossible to create such a system for the formulas of integral calculus. Indeed, in an arbitrary formula of the form

$$\int_a^b f(x)\,dx = A$$

one may make a large number of substitutions of the form $x = \varphi(t)$ and thus obtain a number of "synonyms" of the given formula. We must point out that the table of definite integrals by Bierens de Haan and the earlier editions of the present reference both sin in the plethora of such "synonyms" and formulas of complicated form. In the present edition, we have tried to keep only the simplest of the "synonym" formulas. Basically, we judged the simplicity of a formula from the standpoint of the simplicity of the arguments of the "outer" functions that appear in the integrand. Where possible, we have replaced a complicated formula with a simpler one. Sometimes, several complicated formulas were thereby reduced to a single simpler one. We then kept only the simplest formula. As a result of such substitutions, we sometimes obtained an integral that could be evaluated by use of the formulas of chapter two and the Newton–Leibniz formula, or to an integral of the form

$$\int_{-a}^a f(x)\,dx,$$

where $f(x)$ is an odd function. In such cases the complicated integrals have been omitted.

Let us give an example using the expression

$$\int_0^{\pi/4} \frac{(\cot x - 1)^{p-1}}{\sin^2 x}\,\ln\tan x\,dx = -\frac{\pi}{p}\,\operatorname{cosec} p\pi. \tag{0.1}$$

By making the natural substitution $u = \cot x - 1$, we obtain

$$\int_0^\infty u^{p-1}\ln(1+u)\,du = \frac{\pi}{p}\,\operatorname{cosec} p\pi. \tag{0.2}$$

Integrals similar to formula (0.1) are omitted in this new edition. Instead, we have formula (0.2).

As a second example, let us take

$$I = \int_0^{\pi/2} \ln\left(\tan^p x + \cot^p x\right) \ln\tan x \, \mathrm{d}x = 0.$$

The substitution $u = \tan x$ yields

$$I = \int_0^\infty \frac{\ln\left(u^p + u^{-p}\right)\ln u}{1 + u^2} \, \mathrm{d}u.$$

If we now set $v = \ln u$, we obtain

$$I = \int_{-\infty}^\infty \frac{v e^v}{1 + e^{2v}} \ln\left(e^{pv} + e^{-pv}\right) \mathrm{d}v = \int_{-\infty}^\infty v \, \frac{\ln\left(2\cosh pv\right)}{2\cosh v} \, \mathrm{d}v.$$

The integrand is odd and, consequently, the integral is equal to 0.

Thus, before looking for an integral in the tables, the user should simplify as much as possible the arguments (the "inner" functions) of the functions in the integrand.

The functions are ordered as follows: First we have the elementary functions:

1.    The function $f(x) = x$.
2.    The exponential function.
3.    The hyperbolic functions.
4.    The trigonometric functions.
5.    The logarithmic function.
6.    The inverse hyperbolic functions. (These are replaced with the corresponding logarithms in the formulas containing definite integrals.)
7.    The inverse trigonometric functions.

Then follow the special functions:

8.    Elliptic integrals.
9.    Elliptic functions.
10.   The logarithm integral, the exponential integral, the sine integral, and the cosine integral functions.
11.   Probability integrals and Fresnel's integrals.
12.   The gamma function and related functions.
13.   Bessel functions.
14.   Mathieu functions.
15.   Legendre functions.
16.   Orthogonal polynomials.
17.   Hypergeometric functions.
18.   Degenerate hypergeometric functions.
19.   Parabolic cylinder functions.
20.   Meijer's and MacRobert's functions.
21.   Riemann's zeta function.

The integrals are arranged in order of outer function according to the above scheme: the farther down in the list a function occurs, (i.e., the more complex it is) the later will the corresponding formula appear in the tables. Suppose that several expressions have the same outer function. For example, consider $\sin e^x$, $\sin x$, $\sin \ln x$. Here, the outer function is the sine function in all three cases. Such expressions are then arranged in order of the inner function. In the present work, these functions are therefore arranged in the following order: $\sin x$, $\sin e^x$, $\sin \ln x$.

Our list does not include polynomials, rational functions, powers, or other algebraic functions. An algebraic function that is included in tables of definite integrals can usually be reduced to a finite combination of roots of rational power. Therefore, for classifying our formulas, we can conditionally treat a power function as a generalization of an algebraic and, consequently, of a rational function.* We shall distinguish between all these functions and those listed above and we shall treat them as operators. Thus, in the expression $\sin^2 e^x$, we shall think of the squaring operator as applied to the outer function, namely, the sine. In the expression $\frac{\sin x + \cos x}{\sin x - \cos x}$, we shall think of the rational operator as applied to the trigonometric functions sine and cosine. We shall arrange the operators according to the following order:

1.  Polynomials (listed in order of their degree).
2.  Rational operators.
3.  Algebraic operators (expressions of the form $A^{p/q}$, where $q$ and $p$ are rational, and $q > 0$; these are listed according to the size of $q$).
4.  Power operators.

Expressions with the same outer and inner functions are arranged in the order of complexity of the operators. For example, the following functions (whose outer functions are all trigonometric, and whose inner functions are all $f(x) = x$) are arranged in the order shown:

$$\sin x, \quad \sin x \cos x, \quad \frac{1}{\sin x} = \operatorname{cosec} x, \quad \frac{\sin x}{\cos x} = \tan x, \quad \frac{\sin x + \cos x}{\sin x - \cos x}, \quad \sin^m x, \quad \sin^m x \cos x.$$

Furthermore, if two outer functions $\varphi_1(x)$ and $\varphi_2(x)$, where $\varphi_1(x)$ is more complex than $\varphi_2(x)$, appear in an integrand and if any of the operations mentioned are performed on them, the corresponding integral will appear (in the order determined by the position of $\varphi_2(x)$ in the list) after all integrals containing only the function $\varphi_1(x)$. Thus, following the trigonometric functions are the trigonometric and power functions (that is, $\varphi_2(x) = x$). Then come

- combinations of trigonometric and exponential functions,
- combinations of trigonometric functions, exponential functions, and powers, etc.,
- combinations of trigonometric and hyperbolic functions, etc.

Integrals containing two functions $\varphi_1(x)$ and $\varphi_2(x)$ are located in the division and order corresponding to the more complicated function of the two. However, if the positions of several integrals coincide because they contain the same complicated function, these integrals are put in the position defined by the complexity of the second function.

To these rules of a general nature, we need to add certain particular considerations that will be easily understood from the tables. For example, according to the above remarks, the function $e^{\frac{1}{x}}$ comes after $e^x$ as regards complexity, but $\ln x$ and $\ln \frac{1}{x}$ are equally complex since $\ln \frac{1}{x} = -\ln x$. In the section on "powers and algebraic functions", polynomials, rational functions, and powers of powers are formed from power functions of the form $(a + bx)^n$ and $(\alpha + \beta x)^\nu$.

---

*For any natural number $n$, the involution $(a + bx)^n$ of the binomial $a + bx$ is a polynomial. If $n$ is a negative integer, $(a + bx)^n$ is a rational function. If $n$ is irrational, the function $(a + bx)^n$ is not even an algebraic function.

# Use of the Tables*

For the effective use of the tables contained in this book it is necessary that the user should first become familiar with the classification system for integrals devised by the authors Gradshteyn and Ryzhik. This classification is described in detail in the section entitled *The Order of Presentation of the Formulas* (see page xxv) and essentially involves the separation of the integrand into *inner* and *outer* functions. The principal function involved in the integrand is called the *outer* function and its argument, which is itself usually another function, is called the *inner* function. Thus, if the integrand comprised the expression $\ln \sin x$, the *outer* function would be the logarithmic function while its argument, the *inner* function, would be the trigonometric function $\sin x$. The desired integral would then be found in the section dealing with logarithmic functions, its position within that section being determined by the position of the *inner* function (here a trigonometric function) in Gradshteyn and Ryzhik's list of functional forms.

It is inevitable that some duplication of symbols will occur within such a large collection of integrals and this happens most frequently in the first part of the book dealing with algebraic and trigonometric integrands. The symbols most frequently involved are $\alpha$, $\beta$, $\gamma$, $\delta$, $t$, $u$, $z$, $z_k$, and $\Delta$. The expressions associated with these symbols are used consistently within each section and are defined at the start of each new section in which they occur. Consequently, reference should be made to the beginning of the section being used in order to verify the meaning of the substitutions involved.

Integrals of algebraic functions are expressed as combinations of roots with rational power indices, and definite integrals of such functions are frequently expressed in terms of the Legendre elliptic integrals $F(\phi, k)$, $E(\phi, k)$ and $\Pi(\phi, n, k)$, respectively, of the first, second and third kinds.

The four inverse hyperbolic functions $\operatorname{arcsinh} z$, $\operatorname{arccosh} z$, $\operatorname{arctanh} z$ and $\operatorname{arccoth} z$ are introduced through the definitions

$$\arcsin z = \frac{1}{i} \operatorname{arcsinh}(iz)$$

$$\arccos z = \frac{1}{i} \operatorname{arccosh}(z)$$

$$\arctan z = \frac{1}{i} \operatorname{arctanh}(iz)$$

$$\operatorname{arccot} z = i \operatorname{arccoth}(iz)$$

---

*Prepared by Alan Jeffrey for the English language edition.

or

$$\operatorname{arcsinh} z = \frac{1}{i} \arcsin(iz)$$

$$\operatorname{arccosh} z = i \arccos z$$

$$\operatorname{arctanh} z = \frac{1}{i} \arctan(iz)$$

$$\operatorname{arccoth} z = \frac{1}{i} \operatorname{arccot}(-iz)$$

The numerical constants $\boldsymbol{C}$ and $\boldsymbol{G}$ which often appear in the definite integrals denote Euler's constant and Catalan's constant, respectively. Euler's constant $\boldsymbol{C}$ is defined by the limit

$$\boldsymbol{C} = \lim_{s \to \infty} \left( \sum_{m=1}^{s} \frac{1}{m} - \ln s \right) = 0.577215\ldots.$$

On occasions other writers denote Euler's constant by the symbol $\gamma$, but this is also often used instead to denote the constant

$$\gamma = e^{\boldsymbol{C}} = 1.781072\ldots.$$

Catalan's constant $\boldsymbol{G}$ is related to the complete elliptic integral

$$\mathbf{K} \equiv \mathbf{K}(k) \equiv \int_0^{\pi/2} \frac{\mathrm{d}x}{\sqrt{1 - k^2 \sin^2 x}}$$

by the expression

$$\boldsymbol{G} = \frac{1}{2} \int_0^1 \mathbf{K} \, \mathrm{d}k = \sum_{m=0}^{\infty} \frac{(-1)^m}{(2m+1)^2} = 0.915965\ldots.$$

Since the notations and definitions for higher transcendental functions that are used by different authors are by no means uniform, it is advisable to check the definitions of the functions that occur in these tables. This can be done by identifying the required function by symbol and name in the *Index of Special Functions and Notation* on page xxxvii, and by then referring to the defining formula or section number listed there. We now present a brief discussion of some of the most commonly used alternative notations and definitions for higher transcendental functions.

## Bernoulli and Euler Polynomials and Numbers

Extensive use is made throughout the book of the Bernoulli and Euler numbers $B_n$ and $E_n$ that are defined in terms of the Bernoulli and Euler polynomials of order $n$, $B_n(x)$ and $E_n(x)$, respectively. These polynomials are defined by the generating functions

$$\frac{te^{xt}}{e^t - 1} = \sum_{n=0}^{\infty} B_n(x) \frac{t^n}{n!} \qquad \text{for } |t| < 2\pi$$

and

$$\frac{2e^{xt}}{e^t + 1} = \sum_{n=0}^{\infty} E_n(x) \frac{t^n}{n!} \qquad \text{for } |t| < \pi.$$

The Bernoulli numbers are always denoted by $B_n$ and are defined by the relation

$$B_n = B_n(0) \qquad \text{for } n = 0, 1, \ldots,$$

when

$$B_0 = 1, \quad B_1 = -\frac{1}{2}, \quad B_2 = \frac{1}{6}, \quad B_4 = -\frac{1}{30}, \ldots.$$

The Euler numbers $E_n$ are defined by setting

$$E_n = 2^n \, E_n \left( \frac{1}{2} \right) \qquad \text{for } n = 0, 1, \ldots$$

The $E_n$ are all integral and $E_0 = 1$, $E_2 = -1$, $E_4 = 5$, $E_6 = -61$, ....

An alternative definition of Bernoulli numbers, which we shall denote by the symbol $B_n^*$, uses the same generating function but identifies the $B_n^*$ differently in the following manner:

$$\frac{t}{e^t - 1} = 1 - \frac{1}{2}t + B_1^* \frac{t^2}{2!} - B_2^* \frac{t^4}{4!} + \ldots.$$

This definition then gives rise to the alternative set of Bernoulli numbers

$$B_1^* = 1/6, \quad B_2^* = 1/30, \quad B_3^* = 1/42, \quad B_4^* = 1/30, \quad B_5^* = 5/66,$$
$$B_6^* = 691/2730, \quad B_7^* = 7/6, \quad B_8^* = 3617/510, \quad \ldots.$$

These differences in notation must also be taken into account when using the following relationships that exist between the Bernoulli and Euler polynomials:

$$B_n(x) = \frac{1}{2^n} \sum_{k=0}^{n} \binom{n}{k} B_{n-k} \, E_k(2x) \qquad n = 0, 1, \ldots$$

$$E_{n-1}(x) = \frac{2^n}{n} \left\{ B_n \left( \frac{x+1}{2} \right) - B_n \left( \frac{x}{2} \right) \right\}$$

or

$$E_{n-1}(x) = \frac{2}{n} \left\{ B_n(x) - 2^n B_n \left( \frac{x}{2} \right) \right\} \qquad n = 1, 2, \ldots$$

and

$$E_{n-2}(x) = 2 \binom{n}{2}^{-1} \sum_{k=0}^{n-2} \binom{n}{k} \left( 2^{n-k} - 1 \right) B_{n-k} \, B_k(x) \qquad n = 2, 3, \ldots$$

There are also alternative definitions of the Euler polynomial of order $n$, and it should be noted that some authors, using a modification of the third expression above, call

$$\left( \frac{2}{n+1} \right) \left\{ B_n(x) - 2^n B_n \left( \frac{x}{2} \right) \right\}$$

the Euler polynomial of order $n$.

## Elliptic Functions and Elliptic Integrals

The following notations are often used in connection with the inverse elliptic functions $\text{sn}\,u$, $\text{cn}\,u$, and $\text{dn}\,u$:

$$\text{ns}\,u = \frac{1}{\text{sn}u} \qquad\qquad \text{nc}\,u = \frac{1}{\text{cn}\,u} \qquad\qquad \text{nd}\,u = \frac{1}{\text{dn}\,u}$$

$$\text{sc}\,u = \frac{\text{sn}u}{\text{cn}\,u} \qquad\qquad \text{cs}\,u = \frac{\text{cn}\,u}{\text{sn}u} \qquad\qquad \text{ds}\,u = \frac{\text{dn}\,u}{\text{sn}u}$$

$$\text{sd}\,u = \frac{\text{sn}u}{\text{dn}\,u} \qquad\qquad \text{cd}\,u = \frac{\text{cn}\,u}{\text{dn}\,u} \qquad\qquad \text{dc}\,u = \frac{\text{dn}\,u}{\text{cn}\,u}$$

The elliptic integral of the third kind is defined by Gradshteyn and Ryzhik to be

$$\Pi\left(\varphi, n^2, k\right) = \int_0^\varphi \frac{da}{\left(1 - n^2 \sin^2 a\right)\sqrt{1 - k^2 \sin^2 a}}$$
$$= \int_0^{\sin \varphi} \frac{dx}{\left(1 - n^2 x^2\right)\sqrt{\left(1 - x^2\right)\left(1 - k^2 x^2\right)}} \qquad \left(-\infty < n^2 < \infty\right)$$

## The Jacobi Zeta Function and Theta Functions

The Jacobi zeta function $zn(u, k)$, frequently written $Z(u)$, is defined by the relation

$$zn(u, k) = Z(u) = \int_0^u \left\{ dn^2 v - \frac{\boldsymbol{E}}{\boldsymbol{K}} \right\} dv = \boldsymbol{E}(u) - \frac{\boldsymbol{E}}{\boldsymbol{K}} u.$$

This is related to the theta functions by the relationship

$$zn(u, k) = \frac{\partial}{\partial u} \ln \Theta(u)$$

giving

(i). $\quad zn(u, k) = \dfrac{\pi}{2\boldsymbol{K}} \dfrac{\vartheta_1'\left(\dfrac{\pi u}{2\boldsymbol{K}}\right)}{\vartheta_1\left(\dfrac{\pi u}{2\boldsymbol{K}}\right)} - \dfrac{cn\,u\,dn\,u}{sn\,u}$

(ii). $\quad zn(u, k) = \dfrac{\pi}{2\boldsymbol{K}} \dfrac{\vartheta_2'\left(\dfrac{\pi u}{2\boldsymbol{K}}\right)}{\vartheta_2\left(\dfrac{\pi u}{2\boldsymbol{K}}\right)} + \dfrac{dn\,u\,sn\,u}{cn\,u}$

(iii). $\quad zn(u, k) = \dfrac{\pi}{2\boldsymbol{K}} \dfrac{\vartheta_3'\left(\dfrac{\pi u}{2\boldsymbol{K}}\right)}{\vartheta_3\left(\dfrac{\pi u}{2\boldsymbol{K}}\right)} - k^2 \dfrac{sn\,u\,cn\,u}{dn\,u}$

(iv). $\quad zn(u, k) = \dfrac{\pi}{2\boldsymbol{K}} \dfrac{\vartheta_4'\left(\dfrac{\pi u}{2\boldsymbol{K}}\right)}{\vartheta_4\left(\dfrac{\pi u}{2\boldsymbol{K}}\right)}$

Many different notations for the theta function are in current use. The most common variants are the replacement of the argument $u$ by the argument $u/\pi$ and, occasionally, a permutation of the identification of the functions $\vartheta_1$ to $\vartheta_4$ with the function $\vartheta_4$ replaced by $\vartheta$.

## The Factorial (Gamma) Function

In older reference texts the gamma function $\Gamma(z)$, defined by the Euler integral

$$\Gamma(z) = \int_0^\infty t^{z-1} e^{-t} \, dt,$$

is sometimes expressed in the alternative notation

$$\Gamma(1 + z) = z! = \Pi(z).$$

On occasions the related derivative of the logarithmic factorial function $\Psi(z)$ is used where

$$\frac{d(\ln z!)}{dz} = \frac{(z!)'}{z!} = \Psi(z + 1).$$

This function satisfies the recurrence relation

$$\Psi(z) = \Psi(z-1) + \frac{1}{z-1}$$

and is defined by the series

$$\Psi(z) = -C + \sum_{n=0}^{\infty} \left( \frac{1}{n+1} - \frac{1}{z+n} \right).$$

The derivative $\Psi'(z)$ satisfies the recurrence relation

$$\Psi'(z+1) = \Psi'(z) - \frac{1}{z^2}$$

and is defined by the series

$$\Psi'(z) = \sum_{n=0}^{\infty} \frac{1}{(z+n)^2}.$$

## Exponential and Related Integrals

The exponential integrals $E_n(z)$ have been defined by Schloemilch using the integral

$$E_n(z) = \int_1^\infty e^{-zt} t^{-n} \, dt \qquad (n = 0, 1, \ldots, \quad \mathrm{Re}\, z > 0)$$

They should not be confused with the Euler polynomials already mentioned. The function $E_1(z)$ is related to the exponential integral $\mathrm{Ei}(z)$ through the expressions

$$E_1(z) = -\mathrm{Ei}(-z) = \int_z^\infty e^{-t} t^{-1} \, dt$$

and

$$\mathrm{li}(z) = \int_0^z \frac{dt}{\ln t} = \mathrm{Ei}(\ln z) \qquad [z > 1]$$

The functions $E_n(z)$ satisfy the recurrence relations

$$E_n(z) = \frac{1}{n-1} \left\{ e^{-z} - z\, E_{n-1}(z) \right\} \qquad [n > 1]$$

and

$$E_n'(z) = -E_{n-1}(z)$$

with

$$E_0(z) = e^{-z}/z.$$

The function $E_n(z)$ has the asymptotic expansion

$$E_n(z) \sim \frac{e^{-z}}{z} \left\{ 1 - \frac{n}{z} + \frac{n(n+1)}{z^2} - \frac{n(n+1)(n+2)}{z^3} + \cdots \right\} \qquad \left[ |\arg z| < \frac{3\pi}{2} \right]$$

while for large $n$,

$$E_n(x) = \frac{e^{-x}}{x+n} \left\{ 1 + \frac{n}{(x+n)^2} + \frac{n(n-2x)}{(x+n)^4} + \frac{n\left(6x^2 - 8nx + n^2\right)}{(x+n)^6} + R(n,x) \right\},$$

where

$$-0.36n^{-4} \le R(n,x) \le \left( 1 + \frac{1}{x+n-1} \right) n^{-4} \qquad [x > 0]$$

The sine and cosine integrals $\mathrm{si}(x)$ and $\mathrm{ci}(x)$ are related to the functions $\mathrm{Si}(x)$ and $\mathrm{Ci}(x)$ by the integrals

$$\mathrm{Si}(x) = \int_0^x \frac{\sin t}{t} \, dt = \mathrm{si}(x) + \frac{\pi}{2}$$

and

$$\mathrm{Ci}(x) = \boldsymbol{C} + \ln x + \int_0^x \frac{(\cos t - 1)}{t}\, dt.$$

The hyperbolic sine and cosine integrals $\mathrm{shi}(x)$ and $\mathrm{chi}(x)$ are defined by the relations

$$\mathrm{shi}(x) = \int_0^x \frac{\sinh t}{t}\, dt$$

and

$$\mathrm{chi}(x) = \boldsymbol{C} + \ln x + \int_0^x \frac{(\cosh t - 1)}{t}\, dt.$$

Some authors write

$$\mathrm{Cin}(x) = \int_0^x \frac{(1 - \cos t)}{t}\, dt$$

so that

$$\mathrm{Cin}(x) = -\,\mathrm{Ci}(x) + \ln x + \boldsymbol{C}.$$

The error function $\mathrm{erf}(x)$ is defined by the relation

$$\mathrm{erf}(x) = \Phi(x) = \frac{2}{\sqrt{\pi}} \int_0^x e^{-t^2}\, dt$$

and the complementary error function $\mathrm{erfc}(x)$ is related to the error function $\mathrm{erfc}(x)$ and to $\Phi(x)$ by the expression

$$\mathrm{erfc}(x) = 1 - \mathrm{erf}(x).$$

The Fresnel integrals $S(x)$ and $C(x)$ are defined by Gradshteyn and Ryzhik as

$$S(x) = \frac{2}{\sqrt{2\pi}} \int_0^x \sin t^2\, dt$$

and

$$C(x) = \frac{2}{\sqrt{2\pi}} \int_0^x \cos t^2\, dt.$$

Other definitions that are in use are

$$S_1(x) = \int_0^x \sin \frac{\pi t^2}{2}\, dt, \qquad C_1(x) = \int_0^x \cos \frac{\pi t^2}{2}\, dt$$

and

$$S_2(x) = \frac{1}{\sqrt{2\pi}} \int_0^x \frac{\sin t}{\sqrt{t}}\, dt, \qquad C_2(x) = \frac{1}{\sqrt{2\pi}} \int_0^x \frac{\cos t}{\sqrt{t}}\, dt$$

These are related by the expressions

$$S(x) = S_1\left(x\sqrt{\frac{2}{\pi}}\right) = S_2\left(x^2\right)$$

and

$$C(x) = C_1\left(x\sqrt{\frac{2}{\pi}}\right) = C_2\left(x^2\right)$$

## Hermite and Chebyshev Orthogonal Polynomials

The Hermite polynomials $H_n(x)$ are related to the Hermite polynomials $He_n(x)$ by the relations

$$He_n(x) = 2^{-n/2}\, H_n\left(\frac{x}{\sqrt{2}}\right)$$

and

$$H_n(x) = 2^{n/2}\, He_n\left(x\sqrt{2}\right).$$

These functions satisfy the differential equations

$$\frac{\mathrm{d}^2 H_n}{\mathrm{d}x^2} - 2x \frac{\mathrm{d} H_n}{\mathrm{d}x} + 2n\, H_n = 0$$

and

$$\frac{\mathrm{d}^2 He_n}{\mathrm{d}x^2} - x \frac{\mathrm{d} He_n}{\mathrm{d}x} + n\, He_n = 0.$$

They obey the recurrence relations

$$H_{n+1} = 2x\, H_n - 2n\, H_{n-1}$$

and

$$He_{n+1} = x\, He_n - n\, He_{n-1}$$

The first six orthogonal polynomials $He_n$ are

$$He_0 = 1, \quad He_1 = x, \quad He_2 = x^2 - 1, \quad He_3 = x^3 - 3x, \quad He_4 = x^4 - 6x^2 + 3, \quad He_5 = x^5 - 10x^3 + 15x.$$

Sometimes the Chebyshev polynomial $U_n(x)$ of the second kind is defined as a solution of the equation

$$(1 - x^2)\frac{\mathrm{d}^2 y}{\mathrm{d}x^2} - 3x \frac{\mathrm{d}y}{\mathrm{d}x} + n(n+2)y = 0.$$

## Bessel Functions

A variety of different notations for Bessel functions are in use. Some common ones involve the replacement of $Y_n(z)$ by $N_n(z)$ and the introduction of the symbol

$$\Lambda_n(z) = \left(\frac{1}{2}z\right)^{-n} \Gamma(n+1)\, J_n(z).$$

In the book by Gray, Mathews and MacRobert the symbol $Y_n(z)$ is used to denote $\frac{1}{2}\pi\, Y_n(z) + (\ln 2 - \mathbf{C})\, J_n(z)$ while Neumann uses the symbol $Y^{(n)}(z)$ for the identical quantity.

The Hankel functions $H_\nu^{(1)}(z)$ and $H_\nu^{(2)}(z)$ are sometimes denoted by $Hs_\nu(z)$ and $Hi_\nu(z)$ and some authors write $G_\nu(z) = \left(\frac{1}{2}\right)\pi i\, H_\nu^{(1)}(z)$.

The Neumann polynomial $O_n(t)$ is a polynomial of degree $n+1$ in $1/t$, with $O_0(t) = 1/t$. The polynomials $O_n(t)$ are defined by the generating function

$$\frac{1}{t-z} = J_0(z)\, O_0(t) + 2 \sum_{k=1}^{\infty} J_k(z)\, O_k(t),$$

giving

$$O_n(t) = \frac{1}{4} \sum_{k=0}^{[n/2]} \frac{n(n-k-1)!}{k!} \left(\frac{2}{t}\right)^{n-2k+1} \qquad \text{for } n = 1, 2, \ldots,$$

where $\left[\frac{1}{2}n\right]$ signifies the integral part of $\frac{1}{2}n$. The following relationship holds between three successive polynomials:

$$(n-1)\, O_{n+1}(t) + (n+1)\, O_{n-1}(t) - \frac{2\left(n^2 - 1\right)}{t}\, O_n(t) = \frac{2n}{t}\sin^2\frac{n\pi}{2}.$$

The Airy functions $\text{Ai}(z)$ and $\text{Bi}(z)$ are independent solutions of the equation

$$\frac{\mathrm{d}^2 u}{\mathrm{d}z^2} - zu = 0.$$

The solutions can be represented in terms of Bessel functions by the expressions

$$\text{Ai}(z) = \frac{1}{3}\sqrt{z}\left\{ I_{-1/3}\left(\frac{2}{3}z^{3/2}\right) - I_{1/3}\left(\frac{2}{3}z^{3/2}\right) \right\} = \frac{1}{\pi}\sqrt{\frac{z}{3}}K_{1/3}\left(\frac{2}{3}z^{3/2}\right)$$

$$\text{Ai}(-z) = \frac{1}{3}\sqrt{z}\left\{ J_{1/3}\left(\frac{2}{3}z^{3/2}\right) + J_{-1/3}\left(\frac{2}{3}z^{3/2}\right) \right\}$$

and by

$$\text{Bi}(z) = \sqrt{\frac{z}{3}}\left\{ I_{-1/3}\left(\frac{2}{3}z^{3/2}\right) + I_{1/3}\left(\frac{2}{3}z^{3/2}\right) \right\},$$

$$\text{Bi}(-z) = \sqrt{\frac{z}{3}}\left\{ J_{-1/3}\left(\frac{2}{3}z^{3/2}\right) - J_{1/3}\left(\frac{2}{3}z^{3/2}\right) \right\}.$$

## Parabolic Cylinder Functions and Whittaker Functions

The differential equation

$$\frac{\mathrm{d}^2 y}{\mathrm{d}z^2} + (az^2 + bz + c)y = 0$$

has associated with it the two equations

$$\frac{\mathrm{d}^2 y}{\mathrm{d}z^2} + \left(\frac{1}{4}z^2 + a\right)y = 0 \quad \text{and} \quad \frac{\mathrm{d}^2 y}{\mathrm{d}z^2} - \left(\frac{1}{4}z^2 + a\right)y = 0$$

the solutions of which are parabolic cylinder functions. The first equation can be derived from the second by replacing $z$ by $ze^{i\pi/4}$ and $a$ by $-ia$.

The solutions of the equation

$$\frac{\mathrm{d}^2 y}{\mathrm{d}z^2} - \left(\frac{1}{4}z^2 + a\right)y = 0$$

are sometimes written $U(a, z)$ and $V(a, z)$. These solutions are related to Whittaker's function $D_p(z)$ by the expressions

$$U(a, z) = D_{-a-\frac{1}{2}}(z)$$

and

$$V(a, z) = \frac{1}{\pi}\Gamma\left(\frac{1}{2} + a\right)\left\{ D_{-a-\frac{1}{2}}(-z) + (\sin \pi a)\, D_{-a-\frac{1}{2}}(z) \right\}.$$

## Mathieu Functions

There are several accepted notations for Mathieu functions and for their associated parameters. The defining equation used by Gradshteyn and Ryzhik is

$$\frac{\mathrm{d}^2 y}{\mathrm{d}z^2} + \left(a - 2k^2 \cos 2z\right)y = 0 \qquad \text{with } k^2 = q.$$

Different notations involve the replacement of $a$ and $q$ in this equation by $h$ and $\theta$, $\lambda$ and $h^2$ and $b$ and $c = 2\sqrt{q}$, respectively. The periodic solutions $\text{se}_n(z, q)$ and $\text{ce}_n(z, q)$ and the modified periodic solutions $\text{Se}_n(z, q)$ and $\text{Ce}_n(z, q)$ are suitably altered and, sometimes, re-normalized. A description of these relationships together with the normalizing factors is contained in: Tables relating to Mathieu functions. National Bureau of Standards, Columbia University Press, New York, 1951.

# Index of Special Functions

| Notation | Name of the function and the number of the formula containing its definition | |
|---|---|---|
| $\beta(x)$ | | 8.37 |
| $\Gamma(z)$ | Gamma function | 8.31–8.33 |
| $\gamma(a,x), \quad \Gamma(a,x)$ | Incomplete gamma functions | 8.35 |
| $\Delta(n-k)$ | Unit integer pulse function | 18.1 |
| $\xi(s)$ | | 9.56 |
| $\lambda(x,y)$ | | 9.640 |
| $\mu(x,\beta), \quad \mu(x,\beta,\alpha)$ | | 9.640 |
| $\nu(x), \quad \nu(x,\alpha)$ | | 9.640 |
| $\Pi(x)$ | Lobachevskiy angle of parallelism | 1.48 |
| $\Pi(\varphi,n,k)$ | Elliptic integral of the third kind | 8.11 |
| $\zeta(u)$ | Weierstrass zeta function | 8.17 |
| $\zeta(z,q), \quad \zeta(z)$ | Riemann zeta functions | 9.51–9.54 |
| $\Theta(u) = \vartheta_4\left(\frac{\pi u}{2\mathbf{K}}\right), \quad \Theta_1(u) = \vartheta_3\left(\frac{\pi u}{2\mathbf{K}}\right)$ | Jacobi theta function | 8.191–8.196 |
| $\begin{cases} \vartheta_0(v\mid\tau) = \vartheta_4(v\mid\tau), \\ \vartheta_1(v\mid\tau), \quad \vartheta_2(v\mid\tau), \\ \vartheta_3(v\mid\tau) \end{cases}$ | Elliptic theta functions | 8.18, 8.19 |
| $\sigma(u)$ | Weierstrass sigma function | 8.17 |
| $\Phi(x)$ | Probability integral | 8.25 |
| $\Phi(z,s,v)$ | Lerch function | 9.55 |
| $\Phi(\alpha,\gamma;x) = {}_1F_1(\alpha;\ \gamma;\ x)$ | Confluent hypergeometric function | 9.21 |
| $\begin{cases} \Phi_1(\alpha,\beta,\gamma,x,y) \\ \Phi_2(\beta,\beta',\gamma,x,y) \\ \Phi_3(\beta,\gamma,x,y) \end{cases}$ | Degenerate hypergeometric series in two variables | 9.26 |
| $\psi(x)$ | Euler psi function | 8.36 |
| $\wp(u)$ | Weierstrass elliptic function | 8.16 |
| $\mathrm{Ai}(x)$ | Airy function | page xxxvi |
| $\mathrm{am}(u,k)$ | Amplitude (of an elliptic function) | 8.141 |
| $\mathrm{Bi}(x)$ | Bairy function | page xxxvi |
| $B_n$ | Bernoulli numbers | 9.61, 9.71 |
| $B_n(x)$ | Bernoulli polynomials | 9.620 |
| $\mathrm{B}(x,y)$ | Beta functions | 8.38 |

*continued on next page*

| | continued from previous page | |
|---|---|---|

| Notation | Name of the function and the number of the formula containing its definition | |
|---|---|---|
| $\mathrm{B}_x(p,q)$ | Incomplete beta functions | 8.39 |
| $\mathrm{bei}(z),\quad \mathrm{ber}(z)$ | Thomson functions | 8.56 |
| $\boldsymbol{C}$ | Euler constant | 9.73, 8.367 |
| $C(x)$ | Fresnel cosine integral | 8.25 |
| $C_\nu(a)$ | Young functions | 3.76 |
| $C_n^\lambda(t)$ | Gegenbauer polynomials | 8.93 |
| $C_n^\lambda(x)$ | Gegenbauer functions | 8.932 1 |
| $\mathrm{ce}_{2n}(z,q),\quad \mathrm{ce}_{2n+1}(z,q)$ | Periodic Mathieu functions (Mathieu functions of the first kind) | 8.61 |
| $\mathrm{Ce}_{2n}(z,q),\quad \mathrm{Ce}_{2n+1}(z,q)$ | Associated (modified) Mathieu functions of the first kind | 8.63 |
| $\mathrm{chi}(x)$ | Hyperbolic cosine integral function | 8.22 |
| $\mathrm{ci}(x)$ | Cosine integral | 8.23 |
| $\mathrm{cn}(u)$ | Cosine amplitude | 8.14 |
| $D(k)\equiv \boldsymbol{D}$ | Elliptic integral | 8.112 |
| $D(\varphi,k)$ | Elliptic integral | 8.111 |
| $D_n(z),\quad D_p(z)$ | Parabolic cylinder functions | 9.24–9.25 |
| $\mathrm{dn}\,u$ | Delta amplitude | 8.14 |
| $e_1,e_2,e_3$ | (used with the Weierstrass function) | 8.162 |
| $E_n$ | Euler numbers | 9.63, 9.72 |
| $E(\varphi,k)$ | Elliptic integral of the second kind | 8.11–8.12 |
| $\left\{\begin{array}{l}\boldsymbol{E}(k)=\boldsymbol{E}\\ \boldsymbol{E}(k')=\boldsymbol{E}'\end{array}\right\}$ | Complete elliptic integral of the second kind | 8.11-8.12 |
| $E(p;\alpha_r:q,\varphi_s:x)$ | MacRobert function | 9.4 |
| $\mathbf{E}_\nu(z)$ | Weber function | 8.58 |
| $\mathrm{Ei}(z)$ | Exponential integral function | 8.21 |
| $\mathrm{erf}(x)$ | Error function | 8.25 |
| $\mathrm{erfc}(x)=1-\mathrm{erf}(x)$ | Complementary error function | 8.25 |
| $F(\varphi,k)$ | Elliptic integral of the first kind | 8.11–8.12 |
| $_pF_q\left(\alpha_1,\dots,\alpha_p;\beta_1,\dots,\beta_q;z\right)$ | Generalized hypergeometric series | 9.14 |
| $_2F_1\left(\alpha,\beta;\gamma;z\right)=F(\alpha,\beta;\gamma;z)$ | Gauss hypergeometric function | 9.10–9.13 |
| $_1F_1\left(\alpha;\gamma;z\right)=\Phi(\alpha,\gamma;z)$ | Degenerate hypergeometric function | 9.21 |
| $F_A(\alpha:\beta_1,\dots,\beta_n;$ $\gamma_1,\dots\dots,\gamma_n:z_1,\dots,z_n)$ | Hypergeometric function of several variables | 9.19 |
| $F_1,F_2,F_3,F_4$ | Hypergeometric functions of two variables | 9.18 |
| $\left\{\begin{array}{l}\mathrm{fe}_n(z,q),\mathrm{Fe}_n(z,q)\dots\\ \mathrm{Fey}_n(z,q),\mathrm{Fek}_n(z,q)\dots\end{array}\right\}$ | Other nonperiodic solutions of Mathieu's equation | 8.64, 8.663 |
| $\boldsymbol{G}$ | Catalan constant | 9.73 |
| $g_2,g_3$ | Invariants of the $\wp(u)$-function | 8.161 |
| $\mathrm{gd}\,x$ | Gudermannian | 1.49 |

continued on next page

*continued from previous page*

| Notation | Name of the function and the number of the formula containing its definition | |
|---|---|---|
| $\left.\begin{array}{l}\mathrm{ge}_n(z,q), \mathrm{Ge}_n(z,q) \\ \mathrm{Gey}_n(z,q), \mathrm{Gek}_n(z,q)\end{array}\right\}$ | Other nonperiodic solutions of Mathieu's equation | 8.64, 8.663 |
| $G_{p,q}^{m,n}\left(x \left\vert \begin{array}{c} a_1,\dots,a_p \\ b_1,\dots,b_q \end{array}\right.\right)$ | Meijer functions | 9.3 |
| $h(n)$ | Unit integer function | 18.1 |
| $\mathrm{hei}_\nu(z), \quad \mathrm{her}_\nu(z)$ | Thomson functions | 8.56 |
| $H_\nu^{(1)}(z), \quad H_\nu^{(2)}(z)$ | Hankel functions of the first and second kinds | 8.405, 8.42 |
| $H(u) = \vartheta_1\left(\frac{\pi u}{2K}\right)$ | Theta function | 8.192 |
| $H_1(u) = \vartheta_2\left(\frac{\pi u}{2K}\right)$ | Theta function | 8.192 |
| $H_n(z)$ | Hermite polynomials | 8.95 |
| $\mathbf{H}_\nu(z)$ | Struve functions | 8.55 |
| $I_\nu(z)$ | Bessel functions of an imaginary argument | 8.406, 8.43 |
| $I_x(p,q)$ | Normalized incomplete beta function | 8.39 |
| $J_\nu(z)$ | Bessel function | 8.402, 8.41 |
| $\mathbf{J}_\nu(z)$ | Anger function | 8.58 |
| $\mathrm{k}_\nu(x)$ | Bateman function | 9.210 3 |
| $K(k) = K, \quad K(k') = K'$ | Complete elliptic integral of the first kind | 8.11–8.12 |
| $K_\nu(z)$ | Bessel functions of imaginary argument | 8.407, 8.43 |
| $\mathrm{kei}(z), \quad \mathrm{ker}(z)$ | Thomson functions | 8.56 |
| $L(x)$ | Lobachevskiy function | 8.26 |
| $\mathbf{L}_\nu(z)$ | Modified Struve function | 8.55 |
| $L_n^\alpha(z)$ | Laguerre polynomials | 8.97 |
| $\mathrm{li}(x)$ | Logarithm integral | 8.24 |
| $M_{\lambda,\mu}(z)$ | Whittaker functions | 9.22, 9.23 |
| $O_n(x)$ | Neumann polynomials | 8.59 |
| $P_\nu^\mu(z), \quad P_\nu^\mu(x)$ | Associated Legendre functions of the first kind | 8.7, 8.8 |
| $P_\nu(z), \quad P_\nu(x)$ | Legendre functions and polynomials | 8.82, 8.83, 8.91 |
| $P\left\{\begin{array}{ccc} a & b & c \\ \alpha & \beta & \gamma & z \\ \alpha' & \beta' & \gamma' \end{array}\right\}$ | Riemann's differential equation | 9.160 |
| $P_n^{(\alpha,\beta)}(x)$ | Jacobi polynomials | 8.96 |
| $Q_\nu^\mu(z), \quad Q_\nu^\mu(x)$ | Associated Legendre functions of the second kind | 8.7, 8.8 |
| $Q_\nu(z), \quad Q_\nu(x)$ | Legendre functions of the second kind | 8.82, 8.83 |
| $R_C(x,y)$ | Elliptic Function | 8.111 |
| $R_D(x,y,z)$ | Elliptic Function | 8.111 |
| $R_F(x,y,z)$ | Elliptic Function | 8.111 |
| $R_J(x,y,z,p)$ | Elliptic Function | 8.111 |
| $S(x)$ | Fresnel sine integral | 8.25 |
| $S_n(x)$ | Schläfli polynomials | 8.59 |
| $s_{\mu,\nu}(z), \quad S_{\mu,\nu}(z)$ | Lommel functions | 8.57 |

*continued on next page*

| Notation | Name of the function and the number of the formula containing its definition | |
|---|---|---|
| $\mathrm{se}_{2n+1}(z,q)$,   $\mathrm{se}_{2n+2}(z,q)$ | Periodic Mathieu functions | 8.61 |
| $\mathrm{Se}_{2n+1}(z,q)$,   $\mathrm{Se}_{2n+2}(z,q)$ | Mathieu functions of an imaginary argument | 8.63 |
| $\mathrm{shi}(x)$ | Hyperbolic sine integral | 8.22 |
| $\mathrm{si}(x)$ | Sine integral | 8.23 |
| $\mathrm{sn}\,u$ | Sine amplitude | 8.14 |
| $T_n(x)$ | Chebyshev polynomial of the 1st kind | 8.94 |
| $U_n(x)$ | Chebyshev polynomials of the 2nd kind | 8.94 |
| $U_\nu(w,z)$,   $V_\nu(w,z)$ | Lommel functions of two variables | 8.578 |
| $W_{\lambda,\mu}(z)$ | Whittaker functions | 9.22, 9.23 |
| $Y_\nu(z)$ | Neumann functions | 8.403, 8.41 |
| $Z_\nu(z)$ | Bessel functions | 8.401 |
| $\mathfrak{Z}_\nu(z)$ | Bessel functions | 5.5 |

*continued from previous page*

# Notation

| Symbol | Meaning |
|---|---|
| $\lfloor x \rfloor$ | The integral part of the real number $x$ (also denoted by $[x]$). |
| $\displaystyle\int_a^{(b+)} \quad \int_a^{(b-)}$ | Contour integrals; the path of integration starting at the point $a$ extends to the point $b$ (along a straight line unless there is an indication to the contrary), encircles the point $b$ along a small circle in the positive (negative) direction, and returns to the point $a$, proceeding along the original path in the opposite direction. |
| $\int_C$ | Line integral along the curve $C$. |
| PV $\int$ | Principal value integral |
| $\bar{z} = x - iy$ | The complex conjugate of $z = x + iy$. |
| $n!$ | $= 1 \cdot 2 \cdot 3 \ldots n, \qquad 0! = 1.$ |
| $(2n+1)!!$ | $= 1 \cdot 3 \ldots (2n+1).$ (double factorial notation) |
| $(2n)!!$ | $= 2 \cdot 4 \ldots (2n).$ (double factorial notation) |
| $0!! = 1$ and $(-1)!! = 1$ | (cf. 3.372 for $n = 0$.) |
| $0^0 = 1$ | (cf. 0.112 and 0.113 for $q = 0$.) |
| $\binom{p}{n}$ | $= \dfrac{p(p-1)\ldots(p-n+1)}{1 \cdot 2 \ldots n} = \dfrac{p!}{n!(p-n)!}, \quad \binom{p}{0} = 1, \quad \binom{p}{n} = \dfrac{p!}{n!(p-n)!}$ $[n = 1, 2, \ldots, p \geq n].$ |
| $(a)_n$ | $= a(a+1)\ldots(a+n-1) = \frac{\Gamma(a+n)}{\Gamma(a)}$ (Pochhammer symbol). |
| $\displaystyle\sum_{k=m}^{n} u_k$ | $= u_m + u_{m+1} + \ldots + u_n.$ If $n < m$, we define $\displaystyle\sum_{k=m}^{n} u_k = 0.$ |
| $\displaystyle\sum_{n}{}', \quad \sum_{m,n}{}'$ | Summation over all integral values of $n$ excluding $n = 0$, and summation over all integral values of $n$ and $m$ excluding $m = n = 0$, respectively. |
| $\sum, \quad \prod$ | An empty $\sum$ has value 0 and an empty $\prod$ has value 1. |

*continued on next page*

*continued from previous page*

| Symbol | Meaning |
|---|---|
| $\delta_{ij} = \begin{cases} 1 & i = j \\ 0 & i \neq j \end{cases}$ | Kronecker delta |
| $\tau$ | Theta function parameter (cf. 8.18) |
| $\times$ and $\wedge$ | Vector product (cf. 10.11) |
| $\cdot$ | Scalar product (cf. 10.11) |
| $\nabla$ or "del" | Vector operator (cf. 10.21) |
| $\nabla^2$ | Laplacian (cf. 10.31) |
| $\sim$ | asymptotically equal to |
| $\arg z$ | The argument of the complex number $z = x + iy$. |
| curl or rot | Vector operator (cf. 10.21) |
| div | Vector operator (divergence) (cf. 10.21) |
| $\mathcal{F}$ | Fourier transform (cf. 17.21) |
| $\mathcal{F}_c$ | Fourier cosine transform (cf. 17.31) |
| $\mathcal{F}_s$ | Fourier sine transform (cf. 17.31) |
| grad | Vector operator (gradient) (cf. 10.21) |
| $h_i$ and $g_{ij}$ | Metric coefficients (cf. 10.51) |
| H | Hermitian transpose of a vector or matrix (cf. 13.123) |
| $H(x) = \begin{cases} 0 & x < 0 \\ 1 & x \geq 0 \end{cases}$ | Heaviside step function |
| $\operatorname{Im} z \equiv y$ | The imaginary part of the complex number $z = x + iy$. |
| $k$ | The letter $k$ (when not used as an index of summation) denotes a number in the interval $[0, 1]$. This notation is used in integrals that lead to elliptic integrals. In such a connection, the number $\sqrt{1 - k^2}$ is denoted by $k'$. |
| $\mathcal{L}$ | Laplace transform (cf. 17.11) |
| $\mathcal{M}$ | Mellin transform (cf. 17.41) |
| $\mathbb{N}$ | The natural numbers $(0, 1, 2, \dots)$ |
| $O(f(z))$ | The order of the function $f(z)$. Suppose that the point $z$ approaches $z_0$. If there exists an $M > 0$ such that $|g(z)| \leq M|f(z)|$ in some sufficiently small neighborhood of the point $z_0$, we write $g(z) = O(f(z))$. |

*continued on next page*

| Symbol | Meaning |
|---|---|
| *continued from previous page* | |
| $q$ | The nome, a theta function parameter (cf. 8.18) |
| $\mathbb{R}$ | The real numbers |
| $R(x)$ | A rational function |
| $\mathrm{Re}\, z \equiv x$ | The real part of the complex number $z = x + iy$. |
| $S_n^{(m)}$ | Stirling number of the first kind (cf. 9.74) |
| $\mathfrak{S}_n^{(m)}$ | Stirling number of the second kind (cf. 9.74) |
| $\operatorname{sign} x = \begin{cases} +1 & x > 0 \\ 0 & x = 0 \\ -1 & x < 0 \end{cases}$ | The sign (signum) of the real number $x$. |
| $\mathrm{T}$ | Transpose of a vector or matrix (cf. 13.115) |
| $\mathbb{Z}$ | The integers $(0, \pm 1, \pm 2, \ldots)$ |
| $Z_b$ | Bilateral $z$ transform (cf. 18.1) |
| $Z_u$ | Unilateral $z$ transform (cf. 18.1) |
| $f \sim g$ | $\frac{f}{g} \to 1$ in some appropriate limit |
| $f = o(g)$ | $\frac{f}{g} \to 0$ in some appropriate limit |
| $f = O(g)$ | $|f| < Ag$ for constant $A$, in some appropriate limit |
| $f \approx g$ | $f$ is approximately equal to $g$ |

# Note on the Bibliographic References

The letters and numbers following equations refer to the sources used by Russian editors. The key to the letters will be found preceding each entry in the Bibliography beginning on page 1105. Roman numerals indicate the volume number of a multivolume work. Numbers without parentheses indicate page numbers, numbers in single parentheses refer to equation numbers in the original sources.

Some formulas were changed from their form in the source material. In such cases, the letter $a$ appears at the end of the bibliographic references.

As an example we may use the reference to equation 3.354–5:

$$\text{ET I } 118 \ (1) \ a$$

The key on page 1105 indicates that the book referred to is:

$$\text{Erdélyi, A. et al., } \textit{Tables of Integral Transforms}$$

The Roman numeral denotes volume one of the work, 118 is the page on which the formula will be found, (1) refers to the number of the formula in this source, and the $a$ indicates that the expression appearing in the source differs in some respect from the formula in this book.

In several cases the editors have used Russian editions of works published in other languages. Under such circumstances, because the pagination and numbering of equations may be altered, we have referred the reader only to the original sources and dispensed with page and equation numbers.

Table of Integrals, Series, and Products. http://dx.doi.org/10.1016/B978-0-12-384933-5.00013-8

# 0 Introduction

## 0.1 Finite sums

## 0.11 Progressions

**0.111**[12]   Arithmetic progression.

$$\sum_{k=0}^{n-1}(a + kr) = \frac{n}{2}[2a + (n-1)r] = \frac{n}{2}(a + l) = \frac{1}{2r}(a + l)(l - a + r)$$

$$[l = a + (n-1)r \text{ is the last term}]$$

**0.112**   Geometric progression.

$$\sum_{k=1}^{n} aq^{k-1} = \frac{a(q^n - 1)}{q - 1}$$

$$[q \neq 1]$$

**0.113**   Arithmetic-geometric progression.

$$\sum_{k=0}^{n-1}(a + kr)q^k = \frac{a - [a + (n-1)r]q^n}{1 - q} + \frac{rq(1 - q^{n-1})}{(1 - q)^2}$$

$$[q \neq 1, \quad n > 1] \qquad \text{JO (5)}$$

**0.114**[8]   $$\sum_{k=1}^{n-1} k^2 x^k = \frac{(-n^2 + 2n - 1)x^{n+2} + (2n^2 - 2n - 1)x^{n+1} - n^2 x^n + x^2 + x}{(1 - x)^3}$$

## 0.12 Sums of powers of natural numbers

**0.121**   $$\sum_{k=1}^{n} k^q = \frac{n^{q+1}}{q + 1} + \frac{n^q}{2} + \frac{1}{2}\binom{q}{1} B_2 n^{q-1} + \frac{1}{4}\binom{q}{3} B_4 n^{q-3} + \frac{1}{6}\binom{q}{5} B_6 n^{q-5} + \cdots$$

$$= \frac{n^{q+1}}{q + 1} + \frac{n^q}{2} + \frac{qn^{q-1}}{12} - \frac{q(q-1)(q-2)}{720} n^{q-3} + \frac{q(q-1)(q-2)(q-3)(q-4)}{30,240} n^{q-5} - \cdots$$

$$[\text{last term contains either } n \text{ or } n^2] \qquad \text{CE 332}$$

1.   $$\sum_{k=1}^{n} k = \frac{n(n+1)}{2}$$

CE 333

2.   $$\sum_{k=1}^{n} k^2 = \frac{n(n+1)(2n+1)}{6}$$

CE 333

3. $\quad \sum_{k=1}^{n} k^3 = \left[\dfrac{n(n+1)}{2}\right]^2$      CE 333

4. $\quad \sum_{k=1}^{n} k^4 = \dfrac{1}{30}n(n+1)(2n+1)(3n^2+3n-1)$      CE 333

5. $\quad \sum_{k=1}^{n} k^5 = \dfrac{1}{12}n^2(n+1)^2(2n^2+2n-1)$      CE 333

6. $\quad \sum_{k=1}^{n} k^6 = \dfrac{1}{42}n(n+1)(2n+1)(3n^4+6n^3-3n+1)$      CE 333

7. $\quad \sum_{k=1}^{n} k^7 = \dfrac{1}{24}n^2(n+1)^2(3n^4+6n^3-n^2-4n+2)$      CE 333

**0.122** $\quad \sum_{k=1}^{n}(2k-1)^q = \dfrac{2^q}{q+1}n^{q+1} - \dfrac{1}{2}\binom{q}{1}2^{q-1}B_2n^{q-1} - \dfrac{1}{4}\binom{q}{3}2^{q-3}\left(2^3-1\right)B_4n^{q-3} - \cdots$

[last term contains either $n$ or $n^2$.]

1. $\quad \sum_{k=1}^{n}(2k-1) = n^2$

2. $\quad \sum_{k=1}^{n}(2k-1)^2 = \dfrac{1}{3}n(4n^2-1)$      JO (32a)

3. $\quad \sum_{k=1}^{n}(2k-1)^3 = n^2(2n^2-1)$      JO (32b)

4.[11] $\quad \sum_{k=1}^{n}(mk-1) = \dfrac{n}{2}[m(n+1)-2]$

5.[10] $\quad \sum_{k=1}^{n}(mk-1)^2 = \dfrac{1}{6}n[m^2(n+1)(2n+1)-6m(n+1)+6]$

6.[10] $\quad \sum_{k=1}^{n}(mk-1)^3 = \dfrac{1}{4}n[m^3n(n+1)^2-2m^2(n+1)(2n+1)+6m(n+1)-4]$

**0.123** $\quad \sum_{k=1}^{n}k(k+1)^2 = \dfrac{1}{12}n(n+1)(n+2)(3n+5)$

**0.124**

1. $\quad \sum_{k=1}^{q}k\left(n^2-k^2\right) = \dfrac{1}{4}q(q+1)\left(2n^2-q^2-q\right)$      $[q=1,2,\ldots]$

2.[10] $\quad \sum_{k=1}^{n}k(k+1)^3 = \dfrac{1}{60}n(n+1)\left(12n^3+63n^2+107n+58\right)$

**0.125**   $\displaystyle\sum_{k=1}^{n} k! \cdot k = (n+1)! - 1$          AD (188.1)

**0.126**   $\displaystyle\sum_{k=1}^{n} \frac{(n+k)!}{k!(n-k)!} = \sqrt{\frac{e}{\pi}} \, K_{n+\frac{1}{2}}\left(\frac{1}{2}\right)$          WA 94

## 0.13   Sums of reciprocals of natural numbers

**0.131**[12]      $\displaystyle\sum_{k=1}^{n} \frac{1}{k} = C + \ln n + \frac{1}{2n} - \sum_{k=2}^{\infty} \frac{A_k}{n(n+1)\ldots(n+k-1)},$       JO (59), AD (1876)

where

$$A_k = \frac{1}{k}\int_0^1 x(1-x)(2-x)(3-x)\cdots(k-1-x)\,\mathrm{d}x$$

$$A_2 = \frac{1}{12}, \qquad A_3 = \frac{1}{12},$$

$$A_4 = \frac{19}{120}, \qquad A_5 = \frac{9}{20},$$

$$A_6 = \frac{863}{504}, \qquad A_7 = \frac{1375}{168}$$

**0.132**[7]   $\displaystyle\sum_{k=1}^{n} \frac{1}{2k-1} = \frac{1}{2}\left(C + \ln n\right) + \ln 2 + \frac{B_2}{8n^2} + \frac{\left(2^3-1\right)B_4}{64n^4} + \ldots$      JO (71a)a

**0.133**   $\displaystyle\sum_{k=2}^{n} \frac{1}{k^2-1} = \frac{3}{4} - \frac{2n+1}{2n(n+1)}$      JO (184f)

**0.134***   $\displaystyle\sum_{n=2}^{\infty} (-1)^n \frac{\ln n}{n} = \frac{\ln 2}{2} - \frac{\ln 3}{3} + \frac{\ln 4}{4} - \cdots = C\ln 2 - \frac{(\ln 2)^2}{2}$

## 0.14   Sums of products of reciprocals of natural numbers

1.   $\displaystyle\sum_{k=1}^{n} \frac{1}{[p+(k-1)q](p+kq)} = \frac{n}{p(p+nq)}$      GI III (64)a

2.   $\displaystyle\sum_{k=1}^{n} \frac{1}{[p+(k-1)q](p+kq)[p+(k+1)q]} = \frac{n(2p+nq+q)}{2p(p+q)(p+nq)[p+(n+1)q]}$      GI III (65)a

3.   $\displaystyle\sum_{k=1}^{n} \frac{1}{[p+(k-1)q](p+kq)\ldots[p+(k+l)q]}$

$$= \frac{1}{(l+1)q}\left\{\frac{1}{p(p+q)\ldots(p+lq)} - \frac{1}{(p+nq)[p+(n+1)q]\ldots[p+(n+l)q]}\right\}$$

     AD (1856)a

4.[12]   $\displaystyle\sum_{k=1}^{n} \frac{1}{[1+(k-1)q][1+(k-1)q+p]} = \frac{1}{p}\left[\sum_{k=1}^{n} \frac{1}{1+(k-1)q} - \sum_{k=1}^{n} \frac{1}{1+(k-1)q+p}\right]$

     GI III (66)a

5.* $\displaystyle\sum_{k=1}^{n} \frac{1}{(k+x)(k+x+p)\dots(k+x+mp)}$

$$= \frac{1}{mp}\left\{\sum_{k=1}^{p}\frac{1}{[k+x]\dots[k+x+(m-1)p]} - \sum_{k=n+1}^{n+p}\frac{1}{[k+x]\dots[k+x+(m-1)p]}\right\}$$

$$[m>0, \quad p>0, \quad n>0, \quad x\neq -n-mp,\dots,-2,-1]$$

6.* $\displaystyle\sum_{k=1}^{\infty}\frac{1}{(k+x)(k+x+p)\dots(k+x+mp)} = \frac{1}{mp}\sum_{k=1}^{p}\frac{1}{[k+x]\dots[k+x+(m-1)p]}$

$$[m>0, \quad p>0, \quad x\neq -1,-2,\dots]$$

**0.142**  $\displaystyle\sum_{k=1}^{n}\frac{k^2+k-1}{(k+2)!} = \frac{1}{2} - \frac{n+1}{(n+2)!}$                                    JO (157)

## 0.15  Sums of the binomial coefficients

**Notation:** $n$ is a natural number

1. $\displaystyle\sum_{k=0}^{m}\binom{n+k}{n} = \binom{n+m+1}{n+1}$                                    KR 64 (70.1)

2. $1 + \dbinom{n}{2} + \dbinom{n}{4} + \dots = 2^{n-1}$                                    KR 62 (58.1)

3. $\dbinom{n}{1} + \dbinom{n}{3} + \dbinom{n}{5} + \dots = 2^{n-1}$                                    KR 62 (58.1)

4. $\displaystyle\sum_{k=0}^{m}(-1)^k\binom{n}{k} = (-1)^m\binom{n-1}{m}$                    $[n\geq 1]$                    KR 64 (70.2)

5.* $\displaystyle\sum_{j=k}^{n}2^{-2j}\binom{2j}{j} = 2^{-(2n+1)}\binom{2n+2}{n+1}\left[n+1-\frac{\Gamma(k+\frac{1}{2})\Gamma(n+2)}{\Gamma(k)\Gamma(n+\frac{3}{2})}\right]$

**0.152**

1. $\dbinom{n}{0} + \dbinom{n}{3} + \dbinom{n}{6} + \dots = \frac{1}{3}\left(2^n + 2\cos\frac{n\pi}{3}\right)$                                    KR 62 (59.1)

2. $\dbinom{n}{1} + \dbinom{n}{4} + \dbinom{n}{7} + \dots = \frac{1}{3}\left(2^n + 2\cos\frac{(n-2)\pi}{3}\right)$                                    KR 62 (59.2)

3. $\dbinom{n}{2} + \dbinom{n}{5} + \dbinom{n}{8} + \dots = \frac{1}{3}\left(2^n + 2\cos\frac{(n-4)\pi}{3}\right)$                                    KR 62 (59.3)

**0.153**

1. $\dbinom{n}{0} + \dbinom{n}{4} + \dbinom{n}{8} + \dots = \frac{1}{2}\left(2^{n-1} + 2^{\frac{n}{2}}\cos\frac{n\pi}{4}\right)$                                    KR 63 (60.1)

2. $\dbinom{n}{1} + \dbinom{n}{5} + \dbinom{n}{9} + \dots = \frac{1}{2}\left(2^{n-1} + 2^{\frac{n}{2}}\sin\frac{n\pi}{4}\right)$                                    KR 63 (60.2)

3. $\dbinom{n}{2} + \dbinom{n}{6} + \dbinom{n}{10} + \dots = \frac{1}{2}\left(2^{n-1} - 2^{\frac{n}{2}}\cos\frac{n\pi}{4}\right)$                                    KR 63 (60.3)

4. $\dbinom{n}{3} + \dbinom{n}{7} + \dbinom{n}{11} + \ldots = \dfrac{1}{2}\left(2^{n-1} - 2^{\frac{n}{2}} \sin\dfrac{n\pi}{4}\right)$ 
 KR 63 (60.4)

**0.154**

1. $\displaystyle\sum_{k=0}^{n}(k+1)\binom{n}{k} = 2^{n-1}(n+2)$ $[n \geq 0]$ KR 63 (66.1)

2. $\displaystyle\sum_{k=1}^{n}(-1)^{k+1}k\binom{n}{k} = 0$ $[n \geq 2]$ KR 63 (66.2)

3. $\displaystyle\sum_{k=0}^{n}(-1)^{k}\binom{N}{k}k^{n-1} = 0$ $[N \geq n \geq 1, \quad 0^0 \equiv 1]$

4. $\displaystyle\sum_{k=0}^{n}(-1)^{k}\binom{n}{k}k^{n} = (-1)^{n}n!$ $[n \geq 0, \quad 0^0 \equiv 1]$

5. $\displaystyle\sum_{k=0}^{n}(-1)^{k}\binom{n}{k}(\alpha+k)^{n} = (-1)^{n}n!$ $[n \geq 0, \quad 0^0 \equiv 1]$

6. $\displaystyle\sum_{k=0}^{n}(-1)^{k}\binom{N}{k}(\alpha+k)^{n-1} = 0$ $[N \geq n \geq 1, \quad 0^0 \equiv 1, \quad N, n \in N^+]$

**0.155**

1. $\displaystyle\sum_{k=1}^{n}\frac{(-1)^{k+1}}{k+1}\binom{n}{k} = \frac{n}{n+1}$ KR 63 (67)

2. $\displaystyle\sum_{k=0}^{n}\frac{1}{k+1}\binom{n}{k} = \frac{2^{n+1}-1}{n+1}$ KR 63 (68.1)

3. $\displaystyle\sum_{k=0}^{n}\frac{\alpha^{k+1}}{k+1}\binom{n}{k} = \frac{(\alpha+1)^{n+1}-1}{n+1}$ KR 63 (68.2)

4. $\displaystyle\sum_{k=1}^{n}\frac{(-1)^{k+1}}{k}\binom{n}{k} = \sum_{m=1}^{n}\frac{1}{m}$ KR 64 (69)

**0.156**

1. $\displaystyle\sum_{k=0}^{p}\binom{n}{k}\binom{m}{p-k} = \binom{n+m}{p}$ $[m \text{ is a natural number}]$ KR 64 (71.1)

2. $\displaystyle\sum_{k=0}^{n-p}\binom{n}{k}\binom{n}{p+k} = \frac{(2n)!}{(n-p)!(n+p)!}$ KR 64 (71.2)

**0.157**

1. $\displaystyle\sum_{k=0}^{n}\binom{n}{k}^{2} = \binom{2n}{n}$ KR 64 (72.1)

2. $\displaystyle\sum_{k=0}^{2n}(-1)^k \binom{2n}{k}^2 = (-1)^n \binom{2n}{n}$   KR 64 (72.2)

3.[12] $\displaystyle\sum_{k=0}^{2n+1}(-1)^k \binom{2n+1}{k}^m = 0$   KR 64 (72.3)

4. $\displaystyle\sum_{k=1}^{n} k \binom{n}{k}^2 = \frac{(2n-1)!}{[(n-1)!]^2}$   KR 64 (72.4)

**0.158**[10]

1. $\displaystyle\sum_{k=1}^{n}\left[2^k \binom{2n-k}{n-k} - 2^{k+1}\binom{2n-k-1}{n-k-1}\right]k = 4^n - \binom{2n}{n}$

2. $\displaystyle\sum_{k=1}^{n}\left[2^k \binom{2n-k}{n-k} - 2^{k+1}\binom{2n-k-1}{n-k-1}\right]k^2 = 4^n - \binom{2n}{n}3 \cdot 4^n$

3. $\displaystyle\sum_{k=1}^{n}\left[2^k \binom{2n-k}{n-k} - 2^{k+1}\binom{2n-k-1}{n-k-1}\right]k^3 = (6n+13)4^n - 18n\binom{2n}{n}$

4. $\displaystyle\sum_{k=1}^{n}\left[2^k \binom{2n-k}{n-k} - 2^{k+1}\binom{2n-k-1}{n-k-1}\right]k^4 = (32n^2 + 104n)\binom{2n}{n} - (60n+75)4^n$

**0.159**[10]

1. $\displaystyle\sum_{k=0}^{n}\left[\binom{2n}{n-k} - \binom{2n}{n-k-1}\right]k = \frac{1}{2}\left[4^n - \binom{2n}{n}\right]$

2. $\displaystyle\sum_{k=0}^{n}\left[\binom{2n}{n-k} - \binom{2n}{n-k-1}\right]k^2 = \frac{1}{2}\left[(2n+1)\binom{2n}{n} - 4^n\right]$

3. $\displaystyle\sum_{k=0}^{n}\left[\binom{2n}{n-k} - \binom{2n}{n-k-1}\right]k^3 = \frac{(3n+2)}{4}\cdot 4^n - \frac{1}{2}\binom{2n}{n}(3n+1)$

**0.160**[10]

1. $\displaystyle\sum_{k=n+1}^{2n}\binom{2n}{k}\alpha^k + \frac{1}{2}\binom{2n}{n}\alpha^n + \frac{(1+\alpha)^{2n-1}(1-\alpha)}{2}\sum_{k=0}^{n-1}\binom{2k}{k}\left[\frac{\alpha}{(1+\alpha)^2}\right]^k = \frac{1}{2}(1+\alpha)^{2n}$

2. $\displaystyle\sum_{r=0}^{n}(-1)^r \binom{n}{r}\frac{\Gamma(r+b)}{\Gamma(r+a)} = \frac{\mathrm{B}(n+a-b,b)}{\Gamma(a-b)}$

# 0.2 Numerical series and infinite products

## 0.21 The convergence of numerical series

The series

**0.211** $\quad \sum_{k=1}^{\infty} u_k = u_1 + u_2 + u_3 + \ldots$

is said to *converge absolutely* if the series

**0.212** $\quad \sum_{k=1}^{\infty} |u_k| = |u_1| + |u_2| + |u_3| + \cdots,$

composed of the absolute values of its terms converges. If the series **0.211** converges and the series **0.212** diverges, the series **0.211** is said to *converge conditionally*. Every absolutely convergent series converges.

## 0.22 Convergence tests

Suppose that

$$\lim_{k \to \infty} |u_k|^{1/k} = q$$

If $q < 1$, the series **0.211** converges absolutely. On the other hand, if $q > 1$, the series **0.211** diverges. (Cauchy)

**0.222** Suppose that

$$\lim_{k \to \infty} \left| \frac{u_{k+1}}{u_k} \right| = q$$

Here, if $q < 1$, the series **0.211** converges absolutely. If $q > 1$, the series **0.211** diverges. If $\left| \frac{u_{k+1}}{u_k} \right|$ approaches 1 but remains greater than unity, then the series **0.211** diverges. (d'Alembert)

**0.223** Suppose that

$$\lim_{k \to \infty} k \left\{ \left| \frac{u_k}{u_{k+1}} \right| - 1 \right\} = q$$

Here, if $q > 1$, the series **0.211** converges absolutely. If $q < 1$, the series **0.211** diverges. (Raabe)

**0.224** Suppose that $f(x)$ is a positive decreasing function and that

$$\lim_{k \to \infty} \frac{e^k f(e^k)}{f(k)} = q$$

for natural $k$. If $q < 1$, the series $\sum_{k=1}^{\infty} f(k)$ converges. If $q > 1$, this series diverges. (Ermakov)

**0.225** Suppose that

$$\left| \frac{u_k}{u_{k+1}} \right| = 1 + \frac{q}{k} + \frac{|v_k|}{k^p},$$

where $p > 1$ and the $|v_k|$ are bounded, that is, the $|v_k|$ are all less than some $M$, which is independent of $k$. Here, if $q > 1$, the series **0.211** converges absolutely. If $q \leq 1$, this series diverges. (Gauss)

**0.226**   Suppose that a function $f(x)$ defined for $x \geq q \geq 1$ is continuous, positive, and decreasing. Under these conditions, the series

$$\sum_{k=1}^{\infty} f(k)$$

converges or diverges according as the integral

$$\int_{q}^{\infty} f(x)\,\mathrm{d}x$$

converges or diverges (the Cauchy integral test.)

**0.227**   Suppose that all terms of a sequence $u_1, u_2, \ldots, u_n$ are positive. In such a case, the series

1.   $\displaystyle\sum_{k=1}^{\infty}(-1)^{k+1}u_k = u_1 - u_2 + u_3 - \ldots$

is called an *alternating series.*

If the terms of an alternating series decrease monotonically in absolute value and approach zero, that is, if

2.   $u_{k+1} < u_k$ and $\displaystyle\lim_{k\to\infty} u_k = 0,$

the series **0.227** 1 converges. Here, the remainder of the series is

3.[12]   $\displaystyle\sum_{k=n+1}^{\infty}(-1)^{k-n+1}u_k = \left|\sum_{k=1}^{\infty}(-1)^{k+1}u_k - \sum_{k=1}^{n}(-1)^{k+1}u_k\right| < u_{n+1}$                    (Leibniz)

**0.228**   If the series

1.   $\displaystyle\sum_{k=1}^{\infty} v_k = v_1 + v_2 + \ldots + v_k + \ldots$

converges and the numbers $u_k$ form a monotonic bounded sequence, that is, if $|u_k| < M$ for some number $M$ and for all $k$, the series

2.   $\displaystyle\sum_{k=1}^{\infty} u_k v_k = u_1 v_1 + u_2 v_2 + \ldots + u_k v_k + \ldots$                    FI II 354

converges. (Abel)

**0.229**   If the partial sums of the series **0.228** 1 are bounded and if the numbers $u_k$ constitute a monotonic sequence that approaches zero, that is, if

$$\left|\sum_{k=1}^{n} v_k\right| < M \qquad [n = 1, 2, \ldots] \qquad \text{and } \lim_{k\to\infty} u_k = 0,$$                    FI II 355

then the series **0.228** 2 converges. (Dirichlet)

## 0.23–0.24   Examples of numerical series

**0.231**   Progressions

1. $$\sum_{k=0}^{\infty} aq^k = \frac{a}{1-q} \qquad\qquad\qquad [|q| < 1]$$

2. $$\sum_{k=0}^{\infty} (a+kr)q^k = \frac{a}{1-q} + \frac{rq}{(1-q)^2} \qquad\qquad [|q| < 1] \qquad (\text{cf. } \mathbf{0.113})$$

**0.232**

1. $$\sum_{k=1}^{\infty} (-1)^{k+1} \frac{1}{k} = \ln 2 \qquad\qquad\qquad (\text{cf. } \mathbf{1.511})$$

2. $$\sum_{k=1}^{\infty} (-1)^{k+1} \frac{1}{2k-1} = 1 - 2\sum_{k=1}^{\infty} \frac{1}{(4k-1)(4k+1)} = \frac{\pi}{4}$$

$$(\text{cf. } \mathbf{1.643})$$

3. $$\sum_{k=1}^{\infty} \frac{k^a}{b^k} = \frac{1}{(b-1)^{a+1}} \sum_{i=1}^{a} \left[ \frac{1}{b^{a-i}} \sum_{j=0}^{i} \frac{(-1)^j (a+1)!(i-j)^a}{j!(a+1-j)!} \right]$$

$$[a = 1, 2, 3, \ldots, \quad b \neq 1]$$

**0.233**

1. $$\sum_{k=1}^{\infty} \frac{1}{k^p} = 1 + \frac{1}{2^p} + \frac{1}{3^p} + \ldots = \zeta(p) \qquad\qquad [\operatorname{Re} p > 1] \qquad\qquad \text{WH}$$

2. $$\sum_{k=1}^{\infty} (-1)^{k+1} \frac{1}{k^p} = (1 - 2^{1-p})\zeta(p) \qquad\qquad [\operatorname{Re} p > 0] \qquad\qquad \text{WH}$$

3.[10] $$\sum_{k=1}^{\infty} \frac{1}{k^{2n}} = \frac{2^{2n-1}\pi^{2n}}{(2n)!}|B_{2n}|, \qquad \sum_{k=1}^{\infty} \frac{1}{k^2} = \frac{\pi^2}{6} \qquad\qquad \text{FI II 721}$$

4. $$\sum_{k=1}^{\infty} (-1)^{k+1} \frac{1}{k^{2n}} = \frac{(2^{2n-1}-1)\pi^{2n}}{(2n)!}|B_{2n}| \qquad\qquad \text{JO (165)}$$

5. $$\sum_{k=1}^{\infty} \frac{1}{(2k-1)^{2n}} = \frac{(2^{2n}-1)\pi^{2n}}{2 \cdot (2n)!}|B_{2n}| \qquad\qquad \text{JO (184b)}$$

6. $$\sum_{k=1}^{\infty} (-1)^{k+1} \frac{1}{(2k-1)^{2n+1}} = \frac{\pi^{2n+1}}{2^{2n+2}(2n)!}|E_{2n}| \qquad\qquad \text{JO (184d)}$$

**0.234**

1. $$\sum_{k=1}^{\infty} (-1)^{k+1} \frac{1}{k^2} = \frac{\pi^2}{12} \qquad\qquad\qquad\qquad\qquad \text{EU}$$

2.    $\displaystyle\sum_{k=1}^{\infty} \frac{1}{(2k-1)^2} = \frac{\pi^2}{8}$                          EU

3.[12]    $\displaystyle\sum_{k=1}^{\infty} \frac{(-1)^k}{(2k+1)^2} = \boldsymbol{G} - 1$                  FI II 482

4.    $\displaystyle\sum_{k=1}^{\infty} \frac{(-1)^{k+1}}{(2k-1)^3} = \frac{\pi^3}{32}$                      EU

5.    $\displaystyle\sum_{k=1}^{\infty} \frac{1}{(2k-1)^4} = \frac{\pi^4}{96}$                        EU

6.    $\displaystyle\sum_{k=1}^{\infty} \frac{(-1)^{k+1}}{(2k-1)^5} = \frac{5\pi^5}{1536}$                 EU

7.    $\displaystyle\sum_{k=1}^{\infty} (-1)^{k+1} \frac{k}{(k+1)^2} = \frac{\pi^2}{12} - \ln 2$

8.[6]    $\displaystyle\sum_{k=1}^{\infty} \frac{1}{k(2k+1)} = 2 - 2\ln 2$

9.    $\displaystyle\sum_{n=1}^{\infty} \frac{\Gamma\left(n+\frac{1}{2}\right)}{n^2\,\Gamma(n)} = \sqrt{\pi}\,\ln 4$

**0.235**    $\displaystyle S_n = \sum_{k=1}^{\infty} \frac{1}{(4k^2-1)^n}$

$$S_1 = \frac{1}{2}, \quad S_2 = \frac{\pi^2-8}{16}, \quad S_3 = \frac{32-3\pi^2}{64}, \quad S_4 = \frac{\pi^4+30\pi^2-384}{768}$$

                                                        JO (186)

**0.236**

1.    $\displaystyle\sum_{k=1}^{\infty} \frac{1}{k(4k^2-1)} = 2\ln 2 - 1$                  BR 51a

2.    $\displaystyle\sum_{k=1}^{\infty} \frac{1}{k(9k^2-1)} = \frac{3}{2}(\ln 3 - 1)$              BR 51a

3.    $\displaystyle\sum_{k=1}^{\infty} \frac{1}{k(36k^2-1)} = -3 + \frac{3}{2}\ln 3 + 2\ln 2$      BR 52, AD (6913.3)

4.    $\displaystyle\sum_{k=1}^{\infty} \frac{k}{(4k^2-1)^2} = \frac{1}{8}$                       BR 52

5.    $\displaystyle\sum_{k=1}^{\infty} \frac{1}{k(4k^2-1)^2} = \frac{3}{2} - 2\ln 2$              BR 52

6.    $\displaystyle\sum_{k=1}^{\infty} \frac{12k^2-1}{k(4k^2-1)^2} = 2\ln 2$            AD (6917.3), BR 52

$7.^6 \quad \displaystyle\sum_{k=1}^{\infty} \frac{1}{k(2k+1)^2} = 4 - \frac{\pi^2}{4} - 2\ln 2$

**0.237**

1. $\displaystyle\sum_{k=1}^{\infty} \frac{1}{(2k-1)(2k+1)} = \frac{1}{2}$ 
   
   AD (6917.2), BR 52

2. $\displaystyle\sum_{k=1}^{\infty} \frac{1}{(4k-1)(4k+1)} = \frac{1}{2} - \frac{\pi}{8}$

3. $\displaystyle\sum_{k=2}^{\infty} \frac{1}{(k-1)(k+1)} = \frac{3}{4}$
   
   [see **0.133**],

4. $\displaystyle\sum_{k=1, k\neq m}^{\infty}{}' \frac{1}{(m+k)(m-k)} = -\frac{3}{4m^2}$
   
   [$m$ is an integer]     AD (6916.1)

5. $\displaystyle\sum_{k=1, k\neq m}^{\infty}{}' \frac{(-1)^{k-1}}{(m-k)(m+k)} = \frac{3}{4m^2}$
   
   [$m$ is an even number]     AD (6916.2)

**0.238**

1. $\displaystyle\sum_{k=1}^{\infty} \frac{1}{(2k-1)2k(2k+1)} = \ln 2 - \frac{1}{2}$
   
   GI III (93)

2. $\displaystyle\sum_{k=1}^{\infty} \frac{(-1)^{k+1}}{(2k-1)2k(2k+1)} = \frac{1}{2}(1 - \ln 2)$
   
   GI III (94)a

3. $\displaystyle\sum_{k=0}^{\infty} \frac{1}{(3k+1)(3k+2)(3k+3)(3k+4)} = \frac{1}{6} - \frac{1}{4}\ln 3 + \frac{\pi}{12\sqrt{3}}$
   
   GI III (95)

**0.239**

$1.^{11} \quad \displaystyle\sum_{k=1}^{\infty} (-1)^{k+1} \frac{1}{3k-2} = \frac{1}{3}\left(\frac{\pi}{\sqrt{3}} + \ln 2\right)$
   
   GI III (85), BR* 161 (1)

$2.^7 \quad \displaystyle\sum_{k=1}^{\infty} (-1)^{k+1} \frac{1}{3k-1} = \frac{1}{3}\left(\frac{\pi}{\sqrt{3}} - \ln 2\right)$
   
   BR* 161 (1)

3. $\displaystyle\sum_{k=1}^{\infty} (-1)^{k+1} \frac{1}{4k-3} = \frac{1}{4\sqrt{2}}\left[\pi + 2\ln\left(\sqrt{2}+1\right)\right]$
   
   BR* 161 (1)

4. $\displaystyle\sum_{k=1}^{\infty} (-1)^{\left[\frac{k+3}{2}\right]} \frac{1}{k} = \frac{\pi}{4} + \frac{1}{2}\ln 2$
   
   GI III (87)

5. $\displaystyle\sum_{k=1}^{\infty} (-1)^{\left[\frac{k+3}{2}\right]} \frac{1}{2k-1} = \frac{\pi}{2\sqrt{2}}$

6.    $\displaystyle\sum_{k=1}^{\infty}(-1)^{\left[\frac{k+5}{3}\right]}\frac{1}{2k-1}=\frac{5\pi}{12}$           GI III (88)

7.    $\displaystyle\sum_{k=1}^{\infty}\frac{1}{(8k-1)(8k+1)}=\frac{1}{2}-\frac{\pi}{16}\left(\sqrt{2}+1\right)$

**0.241**

1.    $\displaystyle\sum_{k=1}^{\infty}\frac{1}{2^k k}=\ln 2$           JO (172g)

2.    $\displaystyle\sum_{k=1}^{\infty}\frac{1}{2^k k^2}=\frac{\pi^2}{12}-\frac{1}{2}\left(\ln 2\right)^2$           JO (174)

3.[12]    $\displaystyle\sum_{k=0}^{\infty}\binom{2k}{k}p^k=\frac{1}{\sqrt{1-4p}}$           $\left[-\frac{1}{4}<p<\frac{1}{4}\right]$

4.[12]    $\displaystyle\sum_{k=1}^{\infty}\frac{p^k}{k^2}=\frac{\pi^2}{6}-\int_1^p\frac{\ln(1-x)}{x}\,\mathrm{d}x$           $[0\leq p\leq 1]$

5.[12]    $\displaystyle\sum_{k=1}^{n}\left[2^k\binom{2n-k}{n-k}-2^{k+1}\binom{2n-(k+1)}{n-(k+1)}\right]k=4^n-\binom{2n}{n}$

6.[12]    $\displaystyle\sum_{k=1}^{n}\left[2^k\binom{2n-k}{n-k}-2^{k+1}\binom{2n-(k+1)}{n-(k+1)}\right]k^2=4n\binom{2n}{n}-3\cdot 4^n$

7.[12]    $\displaystyle\sum_{k=1}^{n}\left[2^k\binom{2n-k}{n-k}-2^{k+1}\binom{2n-(k+1)}{n-(k+1)}\right]k^3=(6n+13)4^n-18n\binom{2n}{n}$

8.[12]    $\displaystyle\sum_{k=1}^{n}\left[2^k\binom{2n-k}{n-k}-2^{k+1}\binom{2n-(k+1)}{n-(k+1)}\right]k^4=(32n^2+104n)\binom{2n}{n}-(60n+75)4^n$

9.[12]    $\displaystyle\sum_{k=n+1}^{2n}\binom{2n}{k}k^k+\frac{1}{2}\binom{2n}{n}k^n+\frac{(1+k)^{2n-1}(1-k)}{2}\sum_{n=0}^{n-1}\binom{2n}{n}\left[\frac{k}{(1+k)^2}\right]^n=\frac{1}{2}(1+k)^{2n}$

10.[12]    $\displaystyle\sum_{k=0}^{n}\binom{n+k}{k}2^{n-k}=4^n$

11.[12]    $\displaystyle\sum_{k=0}^{n}\binom{n+k}{k}^{n-k}k=(n+1)4^n-(2n+1)\binom{2n}{n}$

12.[12]    $\displaystyle\sum_{k=0}^{n}\binom{2n}{k}=\frac{1}{2}\left[4^n+\binom{2n}{n}\right]$

13.[12]    $\displaystyle\sum_{k=0}^{n}\binom{2n}{k}k=\frac{n}{2}4^n$

**14.**[12] $\displaystyle\sum_{k=0}^{n} \binom{2n}{k} k^2 = (2n+1)n4^{n-1} - \frac{n^2}{2}\binom{2n}{n}$

**15.*** $\displaystyle\sum_{k=0}^{n} (-1)^k \frac{n!(x-k)^n}{k!(n-k)!} = n!$

**16.*** $\displaystyle\sum_{k=0}^{n} (-1)^k \binom{n}{k} k^{n+1} = (-1)^n n! \frac{n(n+1)}{2}$          $[n = 1, 2, \ldots]$

**17.*** $\displaystyle\sum_{k=0}^{n} (-1)^k \binom{n}{k} k^{n+m} = (-1)^n n! \sum_{k_m=0}^{n} k_m \ldots \sum_{k_1=0}^{k_2} k_1$      $[m, n = 1, 2, \ldots]$

**0.242**[12] $\displaystyle\sum_{k=0}^{\infty} (-1)^k \frac{1}{n^{2k}} = \frac{n^2}{n^2+1}$          $[|n| > 1]$

**0.243**

**1.** $\displaystyle\sum_{k=1}^{\infty} \frac{1}{[p+(k-1)q](p+kq)\ldots[p+(k+l)q]} = \frac{1}{(l+1)q}\frac{1}{p(p+q)\ldots(p+lq)}$

                                              (see also **0.141** 3)

**2.**[12] $\displaystyle\sum_{k=1}^{\infty} \frac{x^{k-1}}{[p+(k-1)q][p+(k-1)q+1][p+(k-1)q+2]\ldots[p+(k-1)q+l]} = \frac{1}{l!}\int_0^1 \frac{t^{p-1}(1-t)^t}{1-xt^q}\,dt$

                                           $[p > 0, \quad x^2 < 1]$    BR* 161 (2), AD (6.704)

**3.** $\displaystyle\sum_{k=0}^{\infty} \frac{1}{(2k+1)^3}\left(\frac{1}{x}\tanh\left[\frac{(2k+1)\pi x}{2}\right] + x\tanh\left[\frac{(2k+1)\pi}{2x}\right]\right) = \frac{\pi^3}{16}$

**0.244**

**1.** $\displaystyle\sum_{k=1}^{\infty} \frac{1}{(k+p)(k+q)} = \frac{1}{q-p}\int_0^1 \frac{x^p - x^q}{1-x}\,dx$      $[p > -1, \quad q > -1, \quad p \neq q]$    GI III (90)

**2.** $\displaystyle\sum_{k=1}^{\infty} (-1)^{k+1}\frac{1}{p+(k-1)q} = \int_0^1 \frac{t^{p-1}}{1+t^q}\,dt$      $[p > 0, \quad q > 0]$           BR* 161 (1)

**3.**[10] $\displaystyle\sum_{k=1}^{\infty} \frac{1}{(k+p)(k+q)} = \frac{1}{q-p}\sum_{m=p+1}^{q}\frac{1}{m}$      $[q > p > -1, \quad p \text{ and } q \text{ integers}]$

## Summations of reciprocals of factorials

**0.245**

**1.** $\displaystyle\sum_{k=0}^{\infty} \frac{1}{k!} = e$

**2.**[12] $\displaystyle\sum_{k=0}^{\infty} \frac{(-1)^k}{k!} = \frac{1}{e}$

3.[12]  $\displaystyle\sum_{k=1}^{\infty} \frac{k}{(2k+1)!} = \frac{1}{2e}$

4.  $\displaystyle\sum_{k=1}^{\infty} \frac{k}{(k+1)!} = 1$

5.  $\displaystyle\sum_{k=0}^{\infty} \frac{1}{(2k)!} = \frac{1}{2}\left(e + \frac{1}{e}\right)$

6.  $\displaystyle\sum_{k=0}^{\infty} \frac{1}{(2k+1)!} = \frac{1}{2}\left(e - \frac{1}{e}\right)$

7.  $\displaystyle\sum_{k=0}^{\infty} \frac{(-1)^k}{(2k)!} = \cos 1$

8.[12]  $\displaystyle\sum_{k=1}^{\infty} \frac{(-1)^{k-1}}{(2k-1)!} = \sin 1$

**0.246**

1.  $\displaystyle\sum_{k=0}^{\infty} \frac{1}{(k!)^2} = I_0(2) \approx 2.27958530$

2.  $\displaystyle\sum_{k=0}^{\infty} \frac{1}{k!(k+1)!} = I_1(2) \approx 1.590636855$

3.  $\displaystyle\sum_{k=0}^{\infty} \frac{1}{k!(k+n)!} = I_n(2)$

4.  $\displaystyle\sum_{k=0}^{\infty} \frac{(-1)^k}{(k!)^2} = J_0(2) \approx 0.22389078$

5.  $\displaystyle\sum_{k=0}^{\infty} \frac{(-1)^k}{k!(k+1)!} = J_1(2) \approx 0.57672481$

6.  $\displaystyle\sum_{k=0}^{\infty} \frac{(-1)^k}{k!(k+n)!} = J_n(2)$

**0.247**  $\displaystyle\sum_{k=1}^{\infty} \frac{k!}{(n+k-1)!} = \frac{1}{(n-2)\cdot(n-1)!}$

**0.248**  $\displaystyle\sum_{k=1}^{\infty} \frac{k^n}{k!} = S_n,$

$$S_1 = e, \qquad S_2 = 2e, \qquad S_3 = 5e, \qquad S_4 = 15e$$
$$S_5 = 52e, \qquad S_6 = 203e, \qquad S_7 = 877e, \qquad S_8 = 4140e$$

**0.249**[7]  $\displaystyle\sum_{k=0}^{\infty} \frac{(k+1)^3}{k!} = 15e$

## 0.25 Infinite products

**0.250** Suppose that a sequence of numbers $a_1, a_2, \ldots, a_k, \ldots$ is given. If the limit $\lim\limits_{n \to \infty} \prod\limits_{k=1}^{n} (1 + a_k)$ exists, whether finite or infinite (but of definite sign), this limit is called the value of the *infinite product* $\prod\limits_{k=1}^{\infty} (1 + a_k)$ and we write

1. $$\lim_{n \to \infty} \prod_{k=1}^{n} (1 + a_k) = \prod_{k=1}^{\infty} (1 + a_k)$$

If an infinite product has a finite *nonzero* value, it is said to converge. Otherwise, the infinite product is said to diverge. We assume that no $a_k$ is equal to $-1$.  FI II 400

**0.251** For the infinite product **0.250** 1. to converge, it is necessary that $\lim\limits_{k \to \infty} a_k = 0$.  FI II 403

**0.252** If $a_k > 0$ or $a_k < 0$ for all values of the index $k$ starting with some particular value, then, for the product **0.250** 1 to converge, it is necessary and sufficient that the series $\sum_{k=1}^{\infty} a_k$ converges.

**0.253** The product $\prod\limits_{k=1}^{\infty} (1 + a_k)$ is said to converge absolutely if the product $\prod\limits_{k=1}^{\infty} (1 + |a_k|)$ converges.  
FI II 403

**0.254** Absolute convergence of an infinite product implies its convergence.

**0.255** The product $\prod\limits_{k=1}^{\infty} (1 + a_k)$ converges absolutely if, and only if, the series $\sum\limits_{k=1}^{\infty} a_k$ converges absolutely.  FI II 406

## 0.26 Examples of infinite products

**0.261** $\displaystyle\prod_{k=1}^{\infty} \left(1 + \frac{(-1)^{k+1}}{2k-1}\right) = \sqrt{2}$  EU

**0.262**

1. $$\prod_{k=2}^{\infty} \left(1 - \frac{1}{k^2}\right) = \frac{1}{2}$$  FI II 401

2. $$\prod_{k=1}^{\infty} \left(1 - \frac{1}{(2k)^2}\right) = \frac{2}{\pi}$$  FI II 401

3. $$\prod_{k=1}^{\infty} \left(1 - \frac{1}{(2k+1)^2}\right) = \frac{\pi}{4}$$  FI II 401

**0.263**

1. $$e = \frac{2}{1} \cdot \left(\frac{4}{3}\right)^{1/2} \left(\frac{6 \cdot 8}{5 \cdot 7}\right)^{1/4} \left(\frac{10 \cdot 12 \cdot 14 \cdot 16}{9 \cdot 11 \cdot 13 \cdot 15}\right)^{1/8} \cdots$$

3.    $$\frac{\pi}{2} = \left(\frac{1}{2}\right)^{1/2} \left(\frac{2^2}{1 \cdot 3}\right)^{1/4} \left(\frac{2^3 \cdot 4}{1 \cdot 3^3}\right)^{1/8} \left(\frac{2^4 \cdot 4^4}{1 \cdot 3^6 \cdot 5}\right)^{1/16} \cdots$$

where the $n^{\text{th}}$ factor is the $(n+1)^{\text{th}}$ root of the product $\prod_{k=0}^{n}(k+1)^{(-1)^{k+1}\binom{n}{k}}$.

**0.264**

1.    $$e^{C} = \prod_{k=1}^{\infty} \frac{\sqrt[k]{e}}{1 + \dfrac{1}{k}}$$                          FI II 402

2.    $$e^{C} = \left(\frac{2}{1}\right)^{1/2} \left(\frac{2^2}{1 \cdot 3}\right)^{1/3} \left(\frac{2^3 \cdot 4}{1 \cdot 3^3}\right)^{1/4} \left(\frac{2^4 \cdot 4^4}{1 \cdot 3^6 \cdot 5}\right)^{1/5} \cdots$$

where the $n^{\text{th}}$ factor is the $(n+1)^{\text{th}}$ root of the product $\prod_{k=0}^{n}(k+1)^{(-1)^{k+1}\binom{n}{k}}$. Here $C$ is the Euler constant, denoted in other works by $\gamma$.

**0.265**    $$\frac{2}{\pi} = \sqrt{\frac{1}{2}} \cdot \sqrt{\frac{1}{2} + \frac{1}{2}\sqrt{\frac{1}{2}}} \cdot \sqrt{\frac{1}{2} + \frac{1}{2}\sqrt{\frac{1}{2} + \frac{1}{2}\sqrt{\frac{1}{2}}}} \cdots$$        FI II 402

**0.266**[8]    $$\prod_{k=0}^{\infty}\left(1 + x^{2^k}\right) = \frac{1}{1-x}$$             $[0 < x < 1]$        FI II 401

# 0.3 Functional series

## 0.30   Definitions and theorems

**0.301**   The series

1.    $$\sum_{k=1}^{\infty} f_k(x),$$

the terms of which are functions, is called a *functional series*. The set of values of the independent variable $x$ for which the series **0.301** 1 converges constitutes what is called the *region of convergence* of that series.

**0.302**   A series that converges for all values of $x$ in a region $M$ is said to *converge uniformly* in that region if, for every $\varepsilon \geq 0$, there exists a number $N$ such that, for $n > N$, the inequality

$$\left| \sum_{k=n+1}^{\infty} f_k(x) \right| < \varepsilon$$

holds for *all* $x$ in $M$.

**0.303**   If the terms of the functional series **0.301** 1 satisfy the inequalities:

$$|f_k(x)| < u_k \quad (k = 1, 2, 3, \ldots),$$

throughout the region $M$, where the $u_k$ are the terms of some *convergent* numerical series

$$\sum_{k=1}^{\infty} u_k = u_1 + u_2 + \ldots + u_k + \ldots,$$

the series **0.301** 1 converges uniformly in $M$. (Weierstrass)

**0.304**   Suppose that the series **0.301** 1 converges uniformly in a region $M$ and that a set of functions $g_k(x)$ constitutes (for each $x$) a monotonic sequence, and that these functions are uniformly bounded, that is, suppose that a number $L$ exists such that the inequalities

1.      $|g_n(x)| \leq L$

hold for all $n$ and $x$. Then, the series

2.      $\displaystyle\sum_{k=1}^{\infty} f_k(x) g_k(x)$

converges uniformly in the region $M$. (Abel)          FI II 451

**0.305**   Suppose that the partial sums of the series **0.301** 1 are uniformly bounded; that is, suppose that, for some $L$ and for all $n$ and $x$ in $M$, the inequalities

$$\left| \sum_{k=1}^{n} f_k(x) \right| < L$$

hold. Suppose also that for each $x$ the functions $g_n(x)$ constitute a monotonic sequence that approaches zero uniformly in the region $M$. Then, the series **0.304** 2 converges uniformly in the region $M$. (Dirichlet)          FI II 451

**0.306**[6]   If the functions $f_k(x)$ (for $k = 1, 2, 3, \ldots$) are integrable on the interval $[a, b]$ and if the series **0.301** 1 made up of these functions converges uniformly on that interval, this series may be integrated *termwise*; that is,

$$\int_a^b \left( \sum_{k=1}^{\infty} f_k(x) \right) dx = \sum_{k=1}^{\infty} \int_a^b f_k(x)\, dx \qquad [a \leq x \leq b] \qquad \text{FI II 459}$$

**0.307**   Suppose that the functions $f_k(x)$ (for $k = 1, 2, 3, \ldots$) have continuous derivatives $f_k'(x)$ on the interval $[a, b]$. If the series **0.301** 1 converges on this interval and if the series $\sum_{k=1}^{\infty} f_k'(x)$ of these derivatives converges uniformly, the series **0.301** 1 may be differentiated termwise; that is,

$$\left\{ \sum_{k=1}^{\infty} f_k(x) \right\}' = \sum_{k=1}^{\infty} f_k'(x) \qquad \text{FI II 460}$$

## 0.31   Power series

**0.311**   A functional series of the form

1.      $\displaystyle\sum_{k=0}^{\infty} a_k (x - \xi)^k = a_0 + a_1 (x - \xi) + a_2 (x - \xi)^2 + \ldots$

is called a *power series*. The following is true of any power series: if it is not everywhere convergent, the region of convergence is a circle with its center at the point $\xi$ and a radius equal to $R$; at every interior point of this circle, the power series **0.311** 1 converges absolutely and outside this circle, it diverges. This circle is called the *circle of convergence* and its radius is called the *radius of convergence*. If the series converges at all points of the complex plane, we say that the radius of convergence is infinite ($R = +\infty$).

**0.312**   Power series may be integrated and differentiated termwise inside the circle of convergence; that is,

$$\int_{\xi}^{x} \left\{ \sum_{k=0}^{\infty} a_k (x - \xi)^k \right\} dx = \sum_{k=0}^{\infty} \frac{a_k}{k+1} (x - \xi)^{k+1},$$

$$\frac{d}{dx} \left\{ \sum_{k=0}^{\infty} a_k (x - \xi)^k \right\} = \sum_{k=1}^{\infty} k a_k (x - \xi)^{k-1}.$$

The radius of convergence of a series that is obtained from termwise integration or differentiation of another power series coincides with the radius of convergence of the original series.

**Operations on power series**

**0.313**   Division of power series.

$$\frac{\sum\limits_{k=0}^{\infty} b_k x^k}{\sum\limits_{k=0}^{\infty} a_k x^k} = \frac{1}{a_0} \sum_{k=0}^{\infty} c_k x^k,$$

where

$$c_n + \frac{1}{a_0} \sum_{k=1}^{n} c_{n-k} a_k - b_n = 0,$$

or

$$c_n = \frac{(-1)^n}{a_0^n} \begin{bmatrix} a_1 b_0 - a_0 b_1 & a_0 & 0 & \cdots & 0 \\ a_2 b_0 - a_0 b_2 & a_1 & a_0 & & 0 \\ a_3 b_0 - a_0 b_3 & a_2 & a_1 & & 0 \\ \vdots & \vdots & \vdots & \ddots & \\ a_{n-1} b_0 - a_0 b_{n-1} & a_{n-2} & a_{n-3} & \cdots & a_0 \\ a_n b_0 - a_0 b_n & a_{n-1} & a_{n-2} & \cdots & a_1 \end{bmatrix}$$
     AD (6360)

**0.314**   Power series raised to powers.

$$\left( \sum_{k=0}^{\infty} a_k x^k \right)^n = \sum_{k=0}^{\infty} c_k x^k,$$

where

$$c_0 = a_0^n, \quad c_m = \frac{1}{m a_0} \sum_{k=1}^{m} (kn - m + k) a_k c_{m-k} \qquad \text{for } m \geq 1 \qquad [n \text{ is a natural number}] \qquad \text{AD (6361)}$$

**0.315**   The substitution of one series into another.

$$\sum_{k=1}^{\infty} b_k y^k = \sum_{k=1}^{\infty} c_k x^k \qquad y = \sum_{k=1}^{\infty} a_k x^k;$$

$$c_1 = a_1 b_1, \quad c_2 = a_2 b_1 + a_1^2 b_2, \quad c_3 = a_3 b_1 + 2 a_1 a_2 b_2 + a_1^3 b_3,$$

$$c_4 = a_4 b_1 + a_2^2 b_2 + 2 a_1 a_3 b_2 + 3 a_1^2 a_2 b_3 + a_1^4 b_4, \quad \ldots$$

     AD (6362)

**0.316**    Multiplication of power series

$$\sum_{k=0}^{\infty} a_k x^k \sum_{k=0}^{\infty} b_k x^k = \sum_{k=0}^{\infty} c_k x^k \qquad c_n = \sum_{k=0}^{n} a_k b_{n-k} \qquad\qquad \text{FI II 372}$$

## Taylor series

**0.317**    If a function $f(x)$ has derivatives of all orders throughout a neighborhood of a point $\xi$, then we may write the series

1.      $$f(\xi) + \frac{(x-\xi)}{1!} f'(\xi) + \frac{(x-\xi)^2}{2!} f''(\xi) + \frac{(x-\xi)^3}{3!} f'''(\xi) + \cdots,$$

which is known as the *Taylor series* of the function $f(x)$.

The Taylor series converges to the function $f(x)$ if the remainder

2.      $$R_n(x) = f(x) - f(\xi) - \sum_{k=1}^{n} \frac{(x-\xi)^k}{k!} f^{(k)}(\xi)$$

approaches zero as $n \to \infty$.

The following are different forms for the remainder of a Taylor series.

3.      $$R_n(x) = \frac{(x-\xi)^{n+1}}{(n+1)!} f^{(n+1)}(\xi + \theta(x-\xi)) \qquad\qquad [0 < \theta < 1] \qquad\qquad \text{(Lagrange)}$$

4.      $$R_n(x) = \frac{(x-\xi)^{n+1}}{n!} (1-\theta)^n f^{(n+1)}(\xi + \theta(x-\xi)) \qquad [0 < \theta < 1] \qquad\qquad \text{(Cauchy)}$$

5.      $$R_n(x) = \frac{\psi(x-\xi) - \psi(0)}{\psi'[(x-\xi)(1-\theta)]} \frac{(x-\xi)^n (1-\theta)^n}{n!} f^{(n+1)}(\xi + \theta(x-\xi))$$
$$[0 < \theta < 1], \qquad\qquad \text{(Schlömilch)}$$

where $\psi(x)$ is an arbitrary function satisfying the following two conditions: (1) It and its derivative $\psi'(x)$ are continuous in the interval $(0, x - \xi)$; and (2) the derivative $\psi'(x)$ does not change sign in that interval. If we set $\psi(x) = x^{p+1}$, we obtain the following form for the remainder:

$$R_n(x) = \frac{(x-\xi)^{n+1}(1-\theta)^{n-p-1}}{(p+1)n!} f^{(n+1)}(\xi + \theta(x-\xi)) \qquad [0 < p \le n, \quad 0 < \theta < 1] \quad \text{(Rouché)}$$

6.      $$R_n(x) = \frac{1}{n!} \int_{\xi}^{x} f^{(n+1)}(t)(x-t)^n \, dt$$

**0.318**    Other forms in which a Taylor series may be written:

1.[11]    $$f(a + x) = \sum_{k=0}^{\infty} \frac{x^k}{k!} f^{(k)}(a) = f(a) + \frac{x}{1!} f'(a) + \frac{x^2}{2!} f''(a) + \cdots$$

2.[12]    $$f(x) = \sum_{k=0}^{\infty} \frac{x^k}{k!} f^{(k)}(0) = f(0) + \frac{x}{1!} f'(0) + \frac{x^2}{2!} f''(0) + \cdots \qquad \text{(Maclaurin series)}$$

**0.319**    The Taylor series of functions of several variables:

$$f(x, y) = f(\xi, \eta) + (x - \xi) \frac{\partial f(\xi, \eta)}{\partial x} + (y - \eta) \frac{\partial f(\xi, \eta)}{\partial y}$$

$$+ \frac{1}{2!} \left\{ (x-\xi)^2 \frac{\partial^2 f(\xi, \eta)}{\partial x^2} + 2(x-\xi)(y-\eta) \frac{\partial^2 f(\xi, \eta)}{\partial x \, \partial y} + (y-\eta)^2 \frac{\partial^2 f(\xi, \eta)}{\partial y^2} \right\} + \cdots$$

## 0.32 Fourier series

**0.320** Suppose that $f(x)$ is a *periodic* function of period $2l$ and that it is absolutely integrable (possibly improperly) over the interval $(-l, l)$. The following trigonometric series is called the *Fourier series* of $f(x)$:

1.
$$\frac{a_0}{2} + \sum_{k=1}^{\infty} \left( a_k \cos \frac{k\pi x}{l} + b_k \sin \frac{k\pi x}{l}, \right)$$

the coefficients of which (the Fourier coefficients) are given by the formulas

2.
$$a_k = \frac{1}{l} \int_{-l}^{l} f(t) \cos \frac{k\pi t}{l}\, dt = \frac{1}{l} \int_{\alpha}^{\alpha+2l} f(t) \cos \frac{k\pi t}{l}\, dt \quad (k = 0, 1, 2, \ldots)$$

3.[11]
$$b_k = \frac{1}{l} \int_{-l}^{l} f(t) \sin \frac{k\pi t}{l}\, dt = \frac{1}{l} \int_{\alpha}^{\alpha+2l} f(t) \sin \frac{k\pi t}{l}\, dt \quad (k = 1, 2, \ldots)$$

### Convergence tests

**0.321** The Fourier series of a function $f(x)$ at a point $x_0$ converges to the number
$$\frac{f(x_0 + 0) + f(x_0 - 0)}{2},$$
if, for some $h > 0$, the integral
$$\int_0^h \frac{|f(x_0 + t) + f(x_0 - t) - f(x_0 + 0) - f(x_0 - 0)|}{t}\, dt$$
exists. Here, it is assumed that the function $f(x)$ either is continuous at the point $x_0$ or has a discontinuity of the first kind (a *saltus*) at that point and that both one-sided limits $f(x_0 + 0)$ and $f(x_0 - 0)$ exist. (Dini)                                                      FI III 524

**0.322** The Fourier series of a periodic function $f(x)$ that satisfies the Dirichlet conditions on the interval $[a, b]$ converges at every point $x_0$ to the value $\frac{1}{2}[f(x_0 + 0) + f(x_0 - 0)]$. (Dirichlet)

We say that a function $f(x)$ satisfies the Dirichlet conditions on the interval $[a, b]$ if it is bounded on that interval and if the interval $[a, b]$ can be partitioned into a finite number of subintervals inside each of which the function $f(x)$ is continuous and monotonic.

**0.323** The Fourier series of a function $f(x)$ at a point $x_0$ converges to $\frac{1}{2}[f(x_0 + 0) + f(x_0 - 0)]$ if $f(x)$ is of bounded variation in some interval $(x_0 - h, x_0 + h)$ with center at $x_0$. (Jordan–Dirichlet)       FI III 528

The definition of a function of bounded variation. Suppose that a function $f(x)$ is defined on some interval $[a, b]$, where $a < b$. Let us partition this interval in an arbitrary manner into subintervals with the dividing points
$$a = x_0 < x_1 < x_2 < \ldots < x_{n-1} < x_n = b$$
and let us form the sum
$$\sum_{k=1}^{n} |f(x_k) - f(x_{k-1})|$$
Different partitions of the interval $[a, b]$ (that is, different choices of points of division $x_i$) yield, generally speaking, different sums. If the set of these sums is bounded above, we say that the function $f(x)$ is *of bounded variation* on the interval $[a, b]$. The least upper bound of these sums is called the *total variation* of the function $f(x)$ on the interval $[a, b]$.

**0.324** Suppose that a function $f(x)$ is piecewise-continuous on the interval $[a, b]$ and that in each interval of continuity it has a piecewise-continuous derivative. Then, at every point $x_0$ of the interval $[a, b]$, the Fourier series of the function $f(x)$ converges to $\frac{1}{2}\left[f(x_0 + 0) + f(x_0 - 0)\right]$.

**0.325** A function $f(x)$ defined in the interval $(0, l)$ can be expanded in a cosine series of the form

1. $\quad \dfrac{a_0}{2} + \displaystyle\sum_{k=1}^{\infty} a_k \cos \dfrac{k\pi x}{l}$,

where

2. $\quad a_k = \dfrac{2}{l} \displaystyle\int_0^l f(t) \cos \dfrac{k\pi t}{l}\, dt$

**0.326** A function $f(x)$ defined in the interval $(0, l)$ can be expanded in a sine series of the form

1. $\quad \displaystyle\sum_{k=1}^{\infty} b_k \sin \dfrac{k\pi x}{l}$,

where

2. $\quad b_k = \dfrac{2}{l} \displaystyle\int_0^l f(t) \sin \dfrac{k\pi t}{l}\, dt$

The convergence tests for the series **0.325** 1 and **0.326** 1 are analogous to the convergence tests for the series **0.320** 1 (see **0.321–0.324**).

**0.327** The Fourier coefficients $a_k$ and $b_k$ (given by formulas **0.320** 2 and **0.320** 3) of an absolutely integrable function approach zero as $k \to \infty$.

If a function $f(x)$ is square-integrable on the interval $(-l, l)$, the equation of closure is satisfied:

$$\frac{a_0^2}{2} + \sum_{k=1}^{\infty}\left(a_k^2 + b_k^2\right) = \frac{1}{l}\int_{-l}^{l} f^2(x)\, dx \qquad \text{(A. M. Lyapunov)} \qquad \text{FI III 705}$$

**0.328** Suppose that $f(x)$ and $\varphi(x)$ are two functions that are square-integrable on the interval $(-l, l)$ and that $a_k, b_k$ and $\alpha_k, \beta_k$ are their Fourier coefficients. For such functions, the generalized equation of closure (Parseval's equation) holds:

$$\frac{a_0 \alpha_0}{2} + \sum_{k=1}^{\infty}\left(a_k \alpha_k + b_k \beta_k\right) = \frac{1}{l}\int_{-l}^{l} f(x)\varphi(x)\, dx \qquad \text{FI III 709}$$

For examples of Fourier series, see **1.44** and **1.45**.

## 0.33 Asymptotic series

**0.330** Included in the collection of all divergent series is the broad class of series known as *asymptotic* or *semiconvergent* series. *Despite the fact that these series diverge*, the values of the functions that they represent can be calculated with a high degree of accuracy if we take the sum of a suitable number of terms of such series. In the case of alternating asymptotic series, we obtain greatest accuracy if we break off the series in question at whatever term is of lowest absolute value. In this case, the error (in absolute value) does not exceed the absolute value of the first of the discarded terms (cf. **0.227** 3).

Asymptotic series have many properties that are analogous to the properties of convergent series and, for that reason, they play a significant role in analysis.

The asymptotic expansion of a function is denoted as follows:

$$f(z) \sim \sum_{n=0}^{\infty} \frac{A_n}{z^n}$$

The definition of an asymptotic expansion. The divergent series $\sum_{n=0}^{\infty} \frac{A_n}{z^n}$ is called the *asymptotic expansion* of a function $f(z)$ in a given region of values of $\arg z$ if the expression $R_k(z) = z^k [f(z) - S_k(z)]$, where $S_k(z) = \sum_{n=0}^{k} \frac{A_n}{z^n}$, satisfies the condition $\lim_{|z| \to \infty} R_k(z) = 0$ for fixed $k$.        FI II 820

A divergent series that represents the asymptotic expansion of some function is called an *asymptotic series*.

**0.331**    Properties of asymptotic series

1.     The operations of addition, subtraction, multiplication, and raising to a power can be performed on asymptotic series just as on absolutely convergent series. The series obtained as a result of these operations will also be asymptotic.

2.     One asymptotic series can be divided by another provided that the first term $A_0$ of the divisor is not equal to zero. The series obtained as a result of division will also be asymptotic.

                                                          FI II 823-825

3.     An asymptotic series can be integrated termwise, and the resultant series will also be asymptotic. In contrast, differentiation of an asymptotic series is, in general, not permissible.      FI II 824

4.     A single asymptotic expansion can represent different functions. On the other hand, a given function can be expanded in an asymptotic series in only one manner.

# 0.4 Certain formulas from differential calculus

## 0.41   Differentiation of a definite integral with respect to a parameter

**0.410**    $\dfrac{d}{da} \displaystyle\int_{\psi(a)}^{\varphi(a)} f(x,a)\,dx = f(\varphi(a),a)\dfrac{d\varphi(a)}{da} - f(\psi(a),a)\dfrac{d\psi(a)}{da} + \displaystyle\int_{\psi(a)}^{\varphi(a)} \dfrac{d}{da} f(x,a)\,dx$      FI II 680

**0.411**    In particular,

1.     $\dfrac{d}{da} \displaystyle\int_{b}^{a} f(x)\,dx = f(a)$

2.     $\dfrac{d}{db} \displaystyle\int_{b}^{a} f(x)\,dx = -f(b)$

## 0.42   The $n^{\text{th}}$ derivative of a product (Leibniz's rule)

Suppose that $u$ and $v$ are $n$-times-differentiable functions of $x$. Then,

$$\frac{d^n(uv)}{dx^n} = u\frac{d^n v}{dx^n} + \binom{n}{1}\frac{du}{dx}\frac{d^{n-1}v}{dx^{n-1}} + \binom{n}{2}\frac{d^2 u}{dx^2}\frac{d^{n-2}v}{dx^{n-2}} + \binom{n}{3}\frac{d^3 u}{dx^3}\frac{d^{n-3}v}{dx^{n-3}} + \cdots + v\frac{d^n u}{dx^n}$$

or, symbolically,

$$\frac{\mathrm{d}^n(uv)}{\mathrm{d}x^n} = (u+v)^n$$

FI I 272

## 0.43   The $n^{\text{th}}$ derivative of a composite function

**0.430**   If $f(x) = F(y)$ and $y = \varphi(x)$, then

1.   $$\frac{\mathrm{d}^n}{\mathrm{d}x^n} f(x) = \frac{U_1}{1!} F'(y) + \frac{U_2}{2!} F''(y) + \frac{U_3}{3!} F'''(y) + \ldots + \frac{U_n}{n!} F^{(n)}(y),$$

     where

$$U_k = \frac{\mathrm{d}^n}{\mathrm{d}x^n} y^k - \frac{k}{1!} y \frac{\mathrm{d}^n}{\mathrm{d}x^n} y^{k-1} + \frac{k(k-1)}{2!} y^2 \frac{\mathrm{d}^n}{\mathrm{d}x^n} y^{k-2} - \ldots + (-1)^{k-1} k y^{k-1} \frac{\mathrm{d}^n y}{\mathrm{d}x^n}$$

AD (7361) GO

2.   $$\frac{\mathrm{d}^n}{\mathrm{d}x^n} f(x) = \sum \frac{n!}{i!j!h!\ldots k!} \frac{\mathrm{d}^m F}{\mathrm{d}y^m} \left(\frac{y'}{1!}\right)^i \left(\frac{y''}{2!}\right)^j \left(\frac{y'''}{3!}\right)^h \ldots \left(\frac{y^{(l)}}{l!}\right)^k,$$

     Here, the symbol $\sum$ indicates summation over all solutions in non negative integers of the equation $i + 2j + 3h + \ldots + lk = n$ and $m = i + j + h + \ldots + k$.

**0.431**

1.   $$(-1)^n \frac{\mathrm{d}^n}{\mathrm{d}x^n} F\left(\frac{1}{x}\right) = \frac{1}{x^{2n}} F^{(n)}\left(\frac{1}{x}\right) + \frac{n-1}{x^{2n-1}} \frac{n}{1!} F^{(n-1)}\left(\frac{1}{x}\right)$$
$$+ \frac{(n-1)(n-2)}{x^{2n-2}} \frac{n(n-1)}{2!} F^{(n-2)}\left(\frac{1}{x}\right) + \ldots$$

AD (7362.1)

2.   $$(-1)^n \frac{\mathrm{d}^n}{\mathrm{d}x^n} e^{\frac{a}{x}} = \frac{1}{x^n} e^{\frac{a}{x}} \left[ \left(\frac{a}{x}\right)^n + (n-1) \binom{n}{1} \left(\frac{a}{x}\right)^{n-1} + (n-1)(n-2) \binom{n}{2} \left(\frac{a}{x}\right)^{n-2} \right.$$
$$\left. + (n-1)(n-2)(n-3) \binom{n}{3} \left(\frac{a}{x}\right)^{n-3} + \ldots \right]$$

AD (7362.2)

**0.432**

1.   $$\frac{\mathrm{d}^n}{\mathrm{d}x^n} F\left(x^2\right) = (2x)^n F^{(n)}\left(x^2\right) + \frac{n(n-1)}{1!} (2x)^{n-2} F^{(n-1)}\left(x^2\right)$$
$$+ \frac{n(n-1)(n-2)(n-3)}{2!} (2x)^{n-4} F^{(n-2)}\left(x^2\right) +$$
$$+ \frac{n(n-1)(n-2)(n-3)(n-4)(n-5)}{3!} (2x)^{n-6} F^{(n-3)}\left(x^2\right) + \ldots$$

AD (7363.1)

2.$^{12}$   $$\frac{\mathrm{d}^n}{\mathrm{d}x^n} e^{ax^2} = (2ax)^n e^{ax^2} \left[ 1 + \frac{n(n-1)}{1!(4ax^2)} + \frac{n(n-1)(n-2)(n-3)}{2!(4ax^2)^2} \right.$$
$$\left. + \frac{n(n-1)(n-2)(n-3)(n-4)(n-5)}{3!(4ax^2)^3} + \ldots \right] \quad \text{[there are} \lfloor \tfrac{n}{2} \rfloor \text{ terms]}$$

AD (7363.2)

3.    $\dfrac{\mathrm{d}^n}{\mathrm{d}x^n}\left(1+ax^2\right)^p = \dfrac{p(p-1)(p-2)\ldots(p-n+1)(2ax)^n}{\left(1+ax^2\right)^{n-p}}$

$$\times\left\{1+\frac{n(n-1)}{1!(p-n+1)}\frac{1+ax^2}{4ax^2}+\frac{n(n-1)(n-2)(n-3)}{2!(p-n+1)(p-n+2)}\left(\frac{1+ax^2}{4ax^2}\right)^2+\ldots\right\},$$

AD (7363.3)

4.    $\dfrac{\mathrm{d}^{m-1}}{\mathrm{d}x^{m-1}}\left(1-x^2\right)^{m-\frac{1}{2}} = (-1)^{m-1}\dfrac{(2m-1)!!}{m}\sin\left(m\arccos x\right)$      AD (7363.4)

5.    $(-1)^n\dfrac{\partial^n}{\partial a^n}\left(\dfrac{a}{a^2+b^2}\right) = n!\left(\dfrac{a}{a^2+b^2}\right)^{n+1}\displaystyle\sum_{0\le 2k\le n+1}(-1)^k\binom{n+1}{2k}\left(\dfrac{b}{a}\right)^{2k}$      (3.944.12)

6.    $(-1)^n\dfrac{\partial^n}{\partial a^n}\left(\dfrac{b}{a^2+b^2}\right) = n!\left(\dfrac{a}{a^2+b^2}\right)^{n+1}\displaystyle\sum_{0\le 2k\le n}(-1)^k\binom{n+1}{2k+1}\left(\dfrac{b}{a}\right)^{2k+1}$      (3.944.11)

**0.433**

1.    $\dfrac{\mathrm{d}^n}{\mathrm{d}x^n}F\left(\sqrt{x}\right) = \dfrac{F^{(n)}\left(\sqrt{x}\right)}{\left(2\sqrt{x}\right)^n}-\dfrac{n(n-1)}{1!}\dfrac{F^{(n-1)}\left(\sqrt{x}\right)}{\left(2\sqrt{x}\right)^{n+1}}+\dfrac{(n+1)n(n-1)(n-2)}{2!}\dfrac{F^{(n-2)}\left(\sqrt{x}\right)}{\left(2\sqrt{x}\right)^{n+2}}-\ldots$

AD (7364.1)

2.    $\dfrac{\mathrm{d}^n}{\mathrm{d}x^n}\left(1+a\sqrt{x}\right)^{2n-1} = \dfrac{(2n-1)!!}{2^n}\dfrac{a}{\sqrt{x}}\left(a^2-\dfrac{1}{x}\right)^{n-1}$      AD (7364.2)

**0.434**[12]   $\dfrac{\mathrm{d}^n}{\mathrm{d}x^n}y^p = p\binom{n-p}{n}\left\{-\binom{n}{1}\dfrac{1}{p-1}y^{p-1}\dfrac{\mathrm{d}^n y}{\mathrm{d}x^n}+\binom{n}{2}\dfrac{1}{p-2}y^{p-2}\dfrac{\mathrm{d}^n(y^2)}{\mathrm{d}x^n}-\ldots\right\}$    $[n<p]$

AD (737.1)

**0.435**[12]   $\dfrac{\mathrm{d}^n}{\mathrm{d}x^n}\ln y = \left\{\binom{n}{1}\dfrac{1}{1\cdot y}\dfrac{\mathrm{d}^n y}{\mathrm{d}x^n}-\binom{n}{2}\dfrac{1}{2\cdot y^2}\dfrac{\mathrm{d}^n(y^2)}{\mathrm{d}x^n}+\binom{n}{3}\dfrac{1}{3\cdot y^3}\dfrac{\mathrm{d}^n(y^3)}{\mathrm{d}x^n}x^n-\ldots\right\}$    AD (737.2)

## 0.44 Integration by substitution

**0.440**[11]   Let $f(g(x))$ and $g(x)$ be continuous in $[a,b]$. Further, let $g'(x)$ exist and be continuous there.
Then $\displaystyle\int_a^b f[g(x)]g'(x)\,\mathrm{d}x = \int_{g(a)}^{g(b)} f(u)\,\mathrm{d}u.$

Table of Integrals, Series, and Products. http://dx.doi.org/10.1016/B978-0-12-384933-5.00001-1

# 1 Elementary Functions

## 1.1 Power of Binomials

### 1.11 Power series

**1.110**  $(1+x)^q = 1 + qx + \dfrac{q(q-1)}{2!}x^2 + \cdots + \dfrac{q(q-1)\ldots(q-k+1)}{k!}x^k + \cdots = \displaystyle\sum_{k=0}^{\infty}\binom{q}{k}x^k$

If $q$ is neither a natural number nor zero, the series converges absolutely for $|x| < 1$ and diverges for $|x| > 1$. For $x = 1$, the series converges for $q > -1$ and diverges for $q \leq -1$. For $x = 1$, the series converges absolutely for $q > 0$. For $x = -1$, it converges absolutely for $q > 0$ and diverges for $q < 0$. If $q = n$ is a natural number, the series **1.110** is reduced to the finite sum **1.111**.                FI II 425

**1.111**  $(a+x)^n = \displaystyle\sum_{k=0}^{n}\binom{n}{k}x^k a^{n-k}$

**1.112**

1.  $(1+x)^{-1} = 1 - x + x^2 - x^3 + \cdots = \displaystyle\sum_{k=1}^{\infty}(-1)^{k-1}x^{k-1}$

(see also **1.121** 2)

2.  $(1+x)^{-2} = 1 - 2x + 3x^2 - 4x^3 + \cdots = \displaystyle\sum_{k=1}^{\infty}(-1)^{k-1}kx^{k-1}$

3.[11]  $(1+x)^{1/2} = 1 + \dfrac{1}{2}x - \dfrac{1\cdot1}{2\cdot4}x^2 + \dfrac{1\cdot1\cdot3}{2\cdot4\cdot6}x^3 - \dfrac{1\cdot1\cdot3\cdot5}{2\cdot4\cdot6\cdot8}x^4 + \cdots$

4.  $(1+x)^{-1/2} = 1 - \dfrac{1}{2}x + \dfrac{1\cdot3}{2\cdot4}x^2 - \dfrac{1\cdot3\cdot5}{2\cdot4\cdot6}x^3 + \cdots$

**1.113**  $\dfrac{x}{(1-x)^2} = \displaystyle\sum_{k=1}^{\infty}kx^k$           $\left[x^2 < 1\right]$

**1.114**

1.  $\left(1+\sqrt{1+x}\right)^q = 2^q\left[1 + \dfrac{q}{1!}\left(\dfrac{x}{4}\right) + \dfrac{q(q-3)}{2!}\left(\dfrac{x}{4}\right)^2 + \dfrac{q(q-4)(q-5)}{3!}\left(\dfrac{x}{4}\right)^3 + \cdots\right]$

$\left[x^2 < 1, \quad q \text{ is a real number}\right]$

AD (6351.1)

2. $$\left(x + \sqrt{1 + x^2}\right)^q = 1 + \sum_{k=0}^{\infty} \frac{q^2 \left(q^2 - 2^2\right)\left(q^2 - 4^2\right) \dots \left[q^2 - (2k)^2\right] x^{2k+2}}{(2k+2)!}$$
$$+ qx + q\sum_{k=1}^{\infty} \frac{\left(q^2 - 1^2\right)\left(q^2 - 3^2\right) \dots \left[q^2 - (2k-1)^2\right]}{(2k+1)!} x^{2k+1}$$

$$\left[x^2 < 1, \quad q \text{ is a real number}\right] \quad \text{AD}(6351.2)$$

## 1.12 Series of rational fractions

**1.121**

1. $$\frac{x}{1-x} = \sum_{k=1}^{\infty} \frac{2^{k-1} x^{2^{k-1}}}{1 + x^{2^{k-1}}} = \sum_{k=1}^{\infty} \frac{x^{2^{k-1}}}{1 - x^{2^k}} \qquad \left[x^2 < 1\right] \qquad \text{AD (6350.3)}$$

2. $$\frac{1}{x-1} = \sum_{k=1}^{\infty} \frac{2^{k-1}}{x^{2^{k-1}} + 1} \qquad \left[x^2 > 1\right] \qquad \text{AD (6350.3)}$$

# 1.2 The Exponential Function

## 1.21 Series representation

**1.211**

1.$^{11}$ $$e^x = \sum_{k=0}^{\infty} \frac{x^k}{k!}$$

2. $$a^x = \sum_{k=0}^{\infty} \frac{(x \ln a)^k}{k!}$$

3. $$e^{-x^2} = \sum_{k=0}^{\infty} (-1)^k \frac{x^{2k}}{k!}$$

4.* $$e^x = \lim_{n \to \infty} \left(1 + \frac{x}{n}\right)^n$$

**1.212** $$e^x(1+x) = \sum_{k=0}^{\infty} \frac{x^k(k+1)}{k!}$$

**1.213** $$\frac{x}{e^x - 1} = 1 - \frac{x}{2} + \sum_{k=1}^{\infty} \frac{B_{2k} x^{2k}}{(2k)!} \qquad [x < 2\pi] \qquad \text{FI II 520}$$

**1.214** $$e^{e^x} = e\left(1 + x + \frac{2x^2}{2!} + \frac{5x^3}{3!} + \frac{15x^4}{4!} + \dots\right) \qquad \text{AD (6460.3)}$$

**1.215**

1. $$e^{\sin x} = 1 + x + \frac{x^2}{2!} - \frac{3x^4}{4!} - \frac{8x^5}{5!} - \frac{3x^6}{6!} + \frac{56x^7}{7!} + \dots \qquad \text{AD (6460.4)}$$

2. $$e^{\cos x} = e\left(1 - \frac{x^2}{2!} + \frac{4x^4}{4!} - \frac{31x^6}{6!} + \dots\right) \qquad \text{AD (6460.5)}$$

3.    $e^{\tan x} = 1 + x + \dfrac{x^2}{2!} + \dfrac{3x^3}{3!} + \dfrac{9x^4}{4!} + \dfrac{37x^5}{5!} + \ldots$           AD (6460.6)

**1.216**

1.    $e^{\arcsin x} = 1 + x + \dfrac{x^2}{2!} + \dfrac{2x^3}{3!} + \dfrac{5x^4}{4!} + \ldots$          AD (6460.7)

2.    $e^{\arctan x} = 1 + x + \dfrac{x^2}{2!} - \dfrac{x^3}{3!} - \dfrac{7x^4}{4!} + \ldots$          AD (6460.8)

**1.217**

1.    $\pi \dfrac{e^{\pi x} + e^{-\pi x}}{e^{\pi x} - e^{-\pi x}} = x \displaystyle\sum_{k=-\infty}^{\infty} \dfrac{1}{x^2 + k^2}$     (cf. **1.421** 3)     AD (6707.1)

2.    $\dfrac{2\pi}{e^{\pi x} - e^{-\pi x}} = x \displaystyle\sum_{k=-\infty}^{\infty} \dfrac{(-1)^k}{x^2 + k^2}$     (cf. **1.422** 3)     AD (6707.2)

## 1.22 Functional relations

**1.221**

1.    $a^x = e^{x \ln a}$

2.    $a^{\log_a x} = a^{\frac{1}{\log_x a}} = x$

**1.222**

1.    $e^x = \cosh x + \sinh x$

2.    $e^{ix} = \cos x + i \sin x$

**1.223**[12]   $e^{ax} - e^{bx} = (a - b)x \exp\left[\dfrac{1}{2}(a + b)x\right] \displaystyle\prod_{k=1}^{\infty}\left[1 + \dfrac{(a - b)^2 x^2}{4k^2 \pi^2}\right]$       MO 216

## 1.23 Series of exponentials

**1.231**   $\displaystyle\sum_{k=0}^{\infty} a^{kx} = \dfrac{1}{1 - a^x}$     $[a > 1 \text{ and } x < 0 \text{ or } 0 < a < 1 \text{ and } x > 0]$

**1.232**

1.    $\tanh x = 1 + 2\displaystyle\sum_{k=1}^{\infty}(-1)^k e^{-2kx}$             $[x > 0]$

2.    $\operatorname{sech} x = 2\displaystyle\sum_{k=0}^{\infty}(-1)^k e^{-(2k+1)x}$        $[x > 0]$

3.    $\operatorname{cosech} x = 2\displaystyle\sum_{k=0}^{\infty} e^{-(2k+1)x}$            $[x > 0]$

4.    $\sin x = \exp\left[-\displaystyle\sum_{n=1}^{\infty}\dfrac{\cos^{2n} x}{2n}\right]$           $[0 \le x \le \pi]$

# 1.3–1.4 Trigonometric and Hyperbolic Functions

## 1.30  Introduction

The trigonometric and hyperbolic sines are related by the identities

$$\sinh x = \frac{1}{i}\sin(ix), \qquad \sin x = \frac{1}{i}\sinh(ix).$$

The trigonometric and hyperbolic cosines are related by the identities

$$\cosh x = \cos(ix), \qquad \cos x = \cosh(ix).$$

Because of this duality, every relation involving trigonometric functions has its formal counterpart involving the corresponding hyperbolic functions, and vice-versa. In many (though not all) cases, both pairs of relationships are meaningful.

   The idea of matching the relationships is carried out in the list of formulas given below. However, not all the meaningful "pairs" are included in the list.

## 1.31  The basic functional relations

**1.311**

1.  $\quad \sin x = \dfrac{1}{2i}\left(e^{ix} - e^{-ix}\right)$

$\qquad\qquad = -i\sinh(ix)$

2.  $\quad \sinh x = \dfrac{1}{2}\left(e^{x} - e^{-x}\right)$

$\qquad\qquad = -i\sin(ix)$

3.  $\quad \cos x = \dfrac{1}{2}\left(e^{ix} + e^{-ix}\right)$

$\qquad\qquad = \cosh(ix)$

4.  $\quad \cosh x = \dfrac{1}{2}\left(e^{x} + e^{-x}\right)$

$\qquad\qquad = \cos(ix)$

5.  $\quad \tan x \; = \dfrac{\sin x}{\cos x} = \dfrac{1}{i}\tanh(ix)$

6.  $\quad \tanh x \; = \dfrac{\sinh x}{\cosh x} = \dfrac{1}{i}\tan(ix)$

7.  $\quad \cot x \; = \dfrac{\cos x}{\sin x} = \dfrac{1}{\tan x} = i\coth(ix)$

8.  $\quad \coth x \; = \dfrac{\cosh x}{\sinh x} = \dfrac{1}{\tanh x} = i\cot(ix)$

**1.312**

1.  $\quad \cos^2 x + \sin^2 x = 1$

2.      $\cosh^2 x - \sinh^2 x = 1$

**1.313**

1.      $\sin(x \pm y) = \sin x \cos y \pm \sin y \cos x$

2.      $\sinh(x \pm y) = \sinh x \cosh y \pm \sinh y \cosh x$

3.      $\sin(x \pm iy) = \sin x \cosh y \pm i \sinh y \cos x$

4.      $\sinh(x \pm iy) = \sinh x \cos y \pm i \sin y \cosh x$

5.      $\cos(x \pm y) = \cos x \cos y \mp \sin x \sin y$

6.      $\cosh(x \pm y) = \cosh x \cosh y \pm \sinh x \sinh y$

7.      $\cos(x \pm iy) = \cos x \cosh y \mp i \sin x \sinh y$

8.      $\cosh(x \pm iy) = \cosh x \cos y \pm i \sinh x \sin y$

9.      $\tan(x \pm y) = \dfrac{\tan x \pm \tan y}{1 \mp \tan x \tan y}$

10.     $\tanh(x \pm y) = \dfrac{\tanh x \pm \tanh y}{1 \pm \tanh x \tanh y}$

11.     $\tan(x \pm iy) = \dfrac{\tan x \pm i \tanh y}{1 \mp i \tan x \tanh y}$

12.     $\tanh(x \pm iy) = \dfrac{\tanh x \pm i \tan y}{1 \pm i \tanh x \tan y}$

**1.314**

1.      $\sin x \pm \sin y = 2 \sin \dfrac{1}{2}(x \pm y) \cos \dfrac{1}{2}(x \mp y)$

2.      $\sinh x \pm \sinh y = 2 \sinh \dfrac{1}{2}(x \pm y) \cosh \dfrac{1}{2}(x \mp y)$

3.      $\cos x + \cos y = 2 \cos \dfrac{1}{2}(x + y) \cos \dfrac{1}{2}(x - y)$

4.      $\cosh x + \cosh y = 2 \cosh \dfrac{1}{2}(x + y) \cosh \dfrac{1}{2}(x - y)$

5.      $\cos x - \cos y = 2 \sin \dfrac{1}{2}(x + y) \sin \dfrac{1}{2}(y - x)$

6.      $\cosh x - \cosh y = 2 \sinh \dfrac{1}{2}(x + y) \sinh \dfrac{1}{2}(x - y)$

7.      $\tan x \pm \tan y = \dfrac{\sin(x \pm y)}{\cos x \cos y}$

8.      $\tanh x \pm \tanh y = \dfrac{\sinh(x \pm y)}{\cosh x \cosh y}$

9.      $\sin x \pm \cos y = \pm 2 \sin \left[ \dfrac{1}{2}(x + y) \pm \dfrac{\pi}{4} \right] \sin \left[ \dfrac{1}{2}(x - y) \pm \dfrac{\pi}{4} \right]$

$$= \pm 2 \cos \left[ \dfrac{1}{2}(x + y) \mp \dfrac{\pi}{4} \right] \cos \left[ \dfrac{1}{2}(x - y) \mp \dfrac{\pi}{4} \right]$$

$$= 2 \sin \left[ \dfrac{1}{2}(x \pm y) \pm \dfrac{\pi}{4} \right] \cos \left[ \dfrac{1}{2}(x \mp y) \mp \dfrac{\pi}{4} \right]$$

10.    $a \sin x \pm b \cos x = a\sqrt{1 + \left(\dfrac{b}{a}\right)^2} \sin\left[x \pm \arctan\left(\dfrac{b}{a}\right)\right]$

$$[a \neq 0]$$

11.    $\pm a \sin x + b \cos x = b\sqrt{1 + \left(\dfrac{a}{b}\right)^2} \cos\left[x \mp \arctan\left(\dfrac{a}{b}\right)\right]$

$$[b \neq 0]$$

12.[12]   $a \sin x \pm b \sin y = q\sqrt{1 + \left(\dfrac{r}{q}\right)^2} \sin\left[\dfrac{1}{2}(x \pm y) + \arctan\left(\dfrac{r}{q}\right)\right]$

$$q = (a+b)\cos\left[\frac{1}{2}(x \mp y)\right], \quad r = (a-b)\sin\left[\frac{1}{2}(x \mp y)\right] \qquad [q \neq 0]$$

13.[12]   $a \cos x + b \cos y = t\sqrt{1 + \left(\dfrac{s}{t}\right)^2} \cos\left[\dfrac{1}{2}(x \mp y) + \arctan\left(\dfrac{s}{t}\right)\right] \qquad [t \neq 0]$

$$= -s\sqrt{1 + \left(\frac{t}{s}\right)^2} \sin\left[\frac{1}{2}(x \mp y) - \arctan\left(\frac{t}{s}\right)\right] \qquad [s \neq 0]$$

$$s = (a-b)\sin\left[\frac{1}{2}(x \pm y)\right], \quad t = (a+b)\cos\left[\frac{1}{2}(x \pm y)\right]$$

## 1.315

1.    $\sin^2 x - \sin^2 y = \sin(x+y)\sin(x-y) = \cos^2 y - \cos^2 x$

2.    $\sinh^2 x - \sinh^2 y = \sinh(x+y)\sinh(x-y) = \cosh^2 x - \cosh^2 y$

3.    $\cos^2 x - \sin^2 y = \cos(x+y)\cos(x-y) = \cos^2 y - \sin^2 x$

4.    $\sinh^2 x + \cosh^2 y = \cosh(x+y)\cosh(x-y) = \cosh^2 x + \sinh^2 y$

## 1.316

1.    $(\cos x + i \sin x)^n = \cos nx + i \sin nx$           [n is an integer]

2.    $(\cosh x + \sinh x)^n = \sinh nx + \cosh nx$        [n is an integer]

## 1.317

1.    $\sin \dfrac{x}{2} = \pm\sqrt{\dfrac{1}{2}(1 - \cos x)}$

2.    $\sinh \dfrac{x}{2} = \pm\sqrt{\dfrac{1}{2}(\cosh x - 1)}$

3.    $\cos \dfrac{x}{2} = \pm\sqrt{\dfrac{1}{2}(1 + \cos x)}$

4.    $\cosh \dfrac{x}{2} = \sqrt{\dfrac{1}{2}(\cosh x + 1)}$

5.    $\tan \dfrac{x}{2} = \dfrac{1 - \cos x}{\sin x} = \dfrac{\sin x}{1 + \cos x}$

6. $\tanh \dfrac{x}{2} = \dfrac{\cosh x - 1}{\sinh x} = \dfrac{\sinh x}{\cosh x + 1}$

The signs in front of the radical in formulas **1.317** 1, **1.317** 2, and **1.317** 3 are taken so as to agree with the signs of the left hand members. The sign of the left hand members depends in turn on the value of $x$.

## 1.32 The representation of powers of trigonometric and hyperbolic functions in terms of functions of multiples of the argument (angle)

**1.320**

1. $\sin^{2n} x = \dfrac{1}{2^{2n}} \left\{ \displaystyle\sum_{k=0}^{n-1} (-1)^{n-k} 2 \binom{2n}{k} \cos 2(n-k)x + \binom{2n}{n} \right\}$    KR 56 (10, 2)

2. $\sinh^{2n} x = \dfrac{(-1)^n}{2^{2n}} \left\{ \displaystyle\sum_{k=0}^{n-1} (-1)^{n-k} 2 \binom{2n}{k} \cosh 2(n-k)x + \binom{2n}{n} \right\}$

3. $\sin^{2n-1} x = \dfrac{1}{2^{2n-2}} \displaystyle\sum_{k=0}^{n-1} (-1)^{n+k-1} \binom{2n-1}{k} \sin(2n-2k-1)x$    KR 56 (10, 4)

4. $\sinh^{2n-1} x = \dfrac{(-1)^{n-1}}{2^{2n-2}} \displaystyle\sum_{k=0}^{n-1} (-1)^{n+k-1} \binom{2n-1}{k} \sinh(2n-2k-1)x$

5. $\cos^{2n} x = \dfrac{1}{2^{2n}} \left\{ \displaystyle\sum_{k=0}^{n-1} 2 \binom{2n}{k} \cos 2(n-k)x + \binom{2n}{n} \right\}$    KR 56 (10, 1)

6. $\cosh^{2n} x = \dfrac{1}{2^{2n}} \left\{ \displaystyle\sum_{k=0}^{n-1} 2 \binom{2n}{k} \cosh 2(n-k)x + \binom{2n}{n} \right\}$

7. $\cos^{2n-1} x = \dfrac{1}{2^{2n-2}} \displaystyle\sum_{k=0}^{n-1} \binom{2n-1}{k} \cos(2n-2k-1)x$    KR 56 (10, 3)

8. $\cosh^{2n-1} x = \dfrac{1}{2^{2n-2}} \displaystyle\sum_{k=0}^{n-1} \binom{2n-1}{k} \cosh(2n-2k-1)x$

**Special cases**

**1.321**

1. $\sin^2 x = \dfrac{1}{2}(-\cos 2x + 1)$

2. $\sin^3 x = \dfrac{1}{4}(-\sin 3x + 3\sin x)$

3. $\sin^4 x = \dfrac{1}{8}(\cos 4x - 4\cos 2x + 3)$

4. $\sin^5 x = \dfrac{1}{16}(\sin 5x - 5\sin 3x + 10\sin x)$

5.      $\sin^6 x = \dfrac{1}{32} \left( -\cos 6x + 6\cos 4x - 15\cos 2x + 10 \right)$

6.      $\sin^7 x = \dfrac{1}{64} \left( -\sin 7x + 7\sin 5x - 21\sin 3x + 35\sin x \right)$

**1.322**

1.      $\sinh^2 x = \dfrac{1}{2} \left( \cosh 2x - 1 \right)$

2.      $\sinh^3 x = \dfrac{1}{4} \left( \sinh 3x - 3\sinh x \right)$

3.      $\sinh^4 x = \dfrac{1}{8} \left( \cosh 4x - 4\cosh 2x + 3 \right)$

4.      $\sinh^5 x = \dfrac{1}{16} \left( \sinh 5x - 5\sinh 3x + 10\sinh x \right)$

5.[12]    $\sinh^6 x = \dfrac{1}{32} \left( \cosh 6x - 6\cosh 4x + 15\cosh 2x - 10 \right)$

6.[12]    $\sinh^7 x = \dfrac{1}{64} \left( \sinh 7x - 7\sinh 5x + 21\sinh 3x - 35\sinh x \right)$

**1.323**

1.      $\cos^2 x = \dfrac{1}{2} \left( \cos 2x + 1 \right)$

2.      $\cos^3 x = \dfrac{1}{4} \left( \cos 3x + 3\cos x \right)$

3.      $\cos^4 x = \dfrac{1}{8} \left( \cos 4x + 4\cos 2x + 3 \right)$

4.      $\cos^5 x = \dfrac{1}{16} \left( \cos 5x + 5\cos 3x + 10\cos x \right)$

5.      $\cos^6 x = \dfrac{1}{32} \left( \cos 6x + 6\cos 4x + 15\cos 2x + 10 \right)$

6.      $\cos^7 x = \dfrac{1}{64} \left( \cos 7x + 7\cos 5x + 21\cos 3x + 35\cosh x \right)$

**1.324**

1.      $\cosh^2 x = \dfrac{1}{2} \left( \cosh 2x + 1 \right)$

2.      $\cosh^3 x = \dfrac{1}{4} \left( \cosh 3x + 3\cosh x \right)$

3.      $\cosh^4 x = \dfrac{1}{8} \left( \cosh 4x + 4\cosh 2x + 3 \right)$

4.      $\cosh^5 x = \dfrac{1}{16} \left( \cosh 5x + 5\cosh 3x + 10\cosh x \right)$

5.      $\cosh^6 x = \dfrac{1}{32} \left( \cosh 6x + 6\cosh 4x + 15\cosh 2x + 10 \right)$

6.      $\cosh^7 x = \dfrac{1}{64} \left( \cosh 7x + 7\cosh 5x + 21\cosh 3x + 35\cosh x \right)$

## 1.33 The representation of trigonometric and hyperbolic functions of multiples of the argument (angle) in terms of powers of these functions

**1.331**

1.[7]    $\sin nx = n \cos^{n-1} x \sin x - \binom{n}{3} \cos^{n-3} x \sin^3 x + \binom{n}{5} \cos^{n-5} x \sin^5 x - \ldots;$

$$= \sin x \left\{ 2^{n-1} \cos^{n-1} x - \binom{n-2}{1} 2^{n-3} \cos^{n-3} x \right.$$

$$\left. + \binom{n-3}{2} 2^{n-5} \cos^{n-5} x - \binom{n-4}{3} 2^{n-7} \cos^{n-7} x + \ldots \right\}$$

AD (3.175)

2.[12]    $\sinh nx = \sinh x \displaystyle\sum_{k=1}^{\lfloor (n+1)/2 \rfloor} \binom{n}{2k-1} \sinh^{2k-2} x \cosh^{n-2k+1} x$

$$= \sinh x \sum_{k=0}^{\lfloor (n-1)/2 \rfloor} (-1)^k \binom{n-k-1}{k} 2^{n-2k-1} \cosh^{n-2k-1} x$$

3.    $\cos nx = \cos^n x - \binom{n}{2} \cos^{n-2} x \sin^2 x + \binom{n}{4} \cos^{n-4} x \sin^4 x - \ldots;$

$$= 2^{n-1} \cos^n x - \frac{n}{1} 2^{n-3} \cos^{n-2} x + \frac{n}{2} \binom{n-3}{1} 2^{n-5} \cos^{n-4} x$$

$$- \frac{n}{3} \binom{n-4}{2} 2^{n-7} \cos^{n-6} x + \ldots$$

AD (3.175)

4.[3]    $\cosh nx = \displaystyle\sum_{k=0}^{\lfloor n/2 \rfloor} \binom{n}{2k} \sinh^{2k} x \cosh^{n-2k} x$

$$= 2^{n-1} \cosh^n x + n \sum_{k=1}^{\lfloor n/2 \rfloor} (-1)^k \frac{1}{k} \binom{n-k-1}{k-1} 2^{n-2k-1} \cosh^{n-2k} x$$

**1.332**

1.    $\sin 2nx = 2n \cos x \left\{ \sin x - \dfrac{4n^2 - 2^2}{3!} \sin^3 x + \dfrac{(4n^2 - 2^2)(4n^2 - 4^2)}{5!} \sin^5 x - \ldots \right\}$    AD (3.171)

$$= (-1)^{n-1} \cos x \left\{ 2^{2n-1} \sin^{2n-1} x - \frac{2n-2}{1!} 2^{2n-3} \sin^{2n-3} x \right.$$

$$+ \frac{(2n-3)(2n-4)}{2!} 2^{2n-5} \sin^{2n-5} x$$

$$\left. - \frac{(2n-4)(2n-5)(2n-6)}{3!} 2^{2n-7} \sin^{2n-7} x + \ldots \right\}$$

AD (3.173)

2. $\quad \sin(2n-1)x = (2n-1)\Bigg\{ \sin x - \dfrac{(2n-1)^2 - 1^2}{3!} \sin^3 x$

$\qquad\qquad + \dfrac{\left[(2n-1)^2 - 1^2\right]\left[(2n-1)^2 - 3^2\right]}{5!} \sin^5 x - \ldots \Bigg\}$    AD (3.172)

$\qquad\qquad = (-1)^{n-1}\Bigg\{ 2^{2n-2}\sin^{2n-1} x - \dfrac{2n-1}{1!} 2^{2n-4} \sin^{2n-3} x$

$\qquad\qquad + \dfrac{(2n-1)(2n-4)}{2!} 2^{2n-6} \sin^{2n-5} x$

$\qquad\qquad - \dfrac{(2n-1)(2n-5)(2n-6)}{3!} 2^{2n-8} \sin^{2n-7} x + \ldots \Bigg\}$    AD (3.174)

3.[12] $\quad \cos 2nx = 1 - \dfrac{4n^2}{2!}\sin^2 x + \dfrac{4n^2\left(4n^2 - 2^2\right)}{4!}\sin^4 x - \dfrac{4n^2\left(4n^2 - 2^2\right)\left(4n^2 - 4^2\right)}{6!}\sin^6 x + \ldots$

AD (3.171)

$\qquad\qquad = (-1)^n\Bigg\{ 2^{2n-1}\sin^{2n} x - \dfrac{2n}{1!} 2^{2n-3}\sin^{2n-2} x$

$\qquad\qquad + \dfrac{2n(2n-3)}{2!} 2^{2n-5}\sin^{2n-4} x - \dfrac{2n(2n-4)(2n-5)}{3!} 2^{2n-7}\sin^{2n-6} x + \ldots \Bigg\}$

AD (3.173)a

4. $\quad \cos(2n-1)x = \cos x\Bigg\{ 1 - \dfrac{(2n-1)^2 - 1^2}{2!}\sin^2 x$

$\qquad\qquad + \dfrac{\left[(2n-1)^2 - 1^2\right]\left[(2n-1)^2 - 3^2\right]}{4!} \sin^4 x - \ldots \Bigg\}$    AD (3.172)

$\qquad\qquad = (-1)^{n-1}\cos x\Bigg\{ 2^{2n-2}\sin^{2n-2} x - \dfrac{2n-3}{1!} 2^{2n-4}\sin^{2n-4} x$

$\qquad\qquad + \dfrac{(2n-4)(2n-5)}{2!} 2^{2n-6}\sin^{2n-6} x$

$\qquad\qquad - \dfrac{(2n-5)(2n-6)(2n-7)}{3!} 2^{2n-8}\sin^{2n-8} x + \ldots \Bigg\}$    AD (3.174)

5.* $\quad \sinh nx = \dfrac{\sinh(n-1)x + \sinh(n+1)x}{2\cosh x}$    $[n = 1, 2, \ldots]$

6.* $\quad \cosh nx = \dfrac{\cosh(n-1)x + \cosh(n+1)x}{2\cosh x}$    $[n = 1, 2, \ldots]$

**Special cases**

**1.333**

1. $\quad \sin 2x = 2\sin x \cos x$

2. $\quad \sin 3x = 3\sin x - 4\sin^3 x$

3. $\quad \sin 4x = \cos x\left(4\sin x - 8\sin^3 x\right)$

4. $\quad \sin 5x = 5\sin x - 20\sin^3 x + 16\sin^5 x$

5. $\quad \sin 6x = \cos x\left(6\sin x - 32\sin^3 x + 32\sin^5 x\right)$

6.      $\sin 7x = 7 \sin x - 56 \sin^3 x + 112 \sin^5 x - 64 \sin^7 x$

**1.334**

1.      $\sinh 2x = 2 \sinh x \cosh x$

2.      $\sinh 3x = 3 \sinh x + 4 \sinh^3 x$

3.[11]   $\sinh 4x = \cosh x \left( 4 \sinh x + 8 \sinh^3 x \right)$

4.      $\sinh 5x = 5 \sinh x + 20 \sinh^3 x + 16 \sinh^5 x$

5.[11]   $\sinh 6x = \cosh x \left( 6 \sinh x + 32 \sinh^3 x + 32 \sinh^5 x \right)$

6.      $\sinh 7x = 7 \sinh x + 56 \sinh^3 x + 112 \sinh^5 x + 64 \sinh^7 x$

**1.335**

1.      $\cos 2x = 2 \cos^2 x - 1$

2.      $\cos 3x = 4 \cos^3 x - 3 \cos x$

3.      $\cos 4x = 8 \cos^4 x - 8 \cos^2 x + 1$

4.      $\cos 5x = 16 \cos^5 x - 20 \cos^3 x + 5 \cos x$

5.      $\cos 6x = 32 \cos^6 x - 48 \cos^4 x + 18 \cos^2 x - 1$

6.      $\cos 7x = 64 \cos^7 x - 112 \cos^5 x + 56 \cos^3 x - 7 \cos x$

**1.336**

1.      $\cosh 2x = 2 \cosh^2 x - 1$

2.      $\cosh 3x = 4 \cosh^3 x - 3 \cosh x$

3.      $\cosh 4x = 8 \cosh^4 x - 8 \cosh^2 x + 1$

4.      $\cosh 5x = 16 \cosh^5 x - 20 \cosh^3 x + 5 \cosh x$

5.      $\cosh 6x = 32 \cosh^6 x - 48 \cosh^4 x + 18 \cosh^2 x - 1$

6.      $\cosh 7x = 64 \cosh^7 x - 112 \cosh^5 x + 56 \cosh^3 x - 7 \cosh x$

**1.337**

1.      $\dfrac{\cos 3x}{\cos^3 x} = 1 - 3 \tan^2 x$

2.      $\dfrac{\cos 4x}{\cos^4 x} = 1 - 6 \tan^2 x + \tan^4 x$

3.      $\dfrac{\cos 5x}{\cos^5 x} = 1 - 10 \tan^2 x + 5 \tan^4 x$

4.      $\dfrac{\cos 6x}{\cos^6 x} = 1 - 15 \tan^2 x + 15 \tan^4 x - \tan^6 x$

5.      $\dfrac{\sin 3x}{\cos^3 x} = 3 \tan x - \tan^3 x$

6.      $\dfrac{\sin 4x}{\cos^4 x} = 4 \tan x - 4 \tan^3 x$

7. $\dfrac{\sin 5x}{\cos^5 x} = 5\tan x - 10\tan^3 x + \tan^5 x$

8. $\dfrac{\sin 6x}{\cos^6 x} = 6\tan x - 20\tan^3 x + 6\tan^5 x$

9. $\dfrac{\cos 3x}{\sin^3 x} = \cot^3 x - 3\cot x$

10. $\dfrac{\cos 4x}{\sin^4 x} = \cot^4 x - 6\cot^2 x + 1$

11. $\dfrac{\cos 5x}{\sin^5 x} = \cot^5 x - 10\cot^3 x + 5\cot x$

12. $\dfrac{\cos 6x}{\sin^6 x} = \cot^6 x - 15\cot^4 x + 15\cot^2 x - 1$

13. $\dfrac{\sin 3x}{\sin^3 x} = 3\cot^2 x - 1$

14. $\dfrac{\sin 4x}{\sin^4 x} = 4\cot^3 x - 4\cot x$

15. $\dfrac{\sin 5x}{\sin^5 x} = 5\cot^4 x - 10\cot^2 x + 1$

16. $\dfrac{\sin 6x}{\sin^6 x} = 6\cot^5 x - 20\cot^3 x + 6\cot x$

## 1.34 Certain sums of trigonometric and hyperbolic functions

**1.341**

1. $\displaystyle\sum_{k=0}^{n-1} \sin(x+ky) = \sin\left(x + \frac{n-1}{2}y\right)\sin\frac{ny}{2}\operatorname{cosec}\frac{y}{2}$ $\hspace{2em}$ AD (361.8)

2. $\displaystyle\sum_{k=0}^{n-1} \sinh(x+ky) = \sinh\left(x + \frac{n-1}{2}y\right)\sinh\frac{ny}{2}\frac{1}{\sinh\frac{y}{2}}$

3. $\displaystyle\sum_{k=0}^{n-1} \cos(x+ky) = \cos\left(x + \frac{n-1}{2}y\right)\sin\frac{ny}{2}\operatorname{cosec}\frac{y}{2}$ $\hspace{2em}$ AD (361.9)

4. $\displaystyle\sum_{k=0}^{n-1} \cosh(x+ky) = \cosh\left(x + \frac{n-1}{2}y\right)\sinh\frac{ny}{2}\frac{1}{\sinh\frac{y}{2}}$

5. $\displaystyle\sum_{k=0}^{2n-1} (-1)^k \cos(x+ky) = \sin\left(x + \frac{2n-1}{2}y\right)\sin ny\sec\frac{y}{2}$ $\hspace{2em}$ JO (202)

6. $\displaystyle\sum_{k=0}^{n-1} (-1)^k \sin(x+ky) = \sin\left(x + \frac{n-1}{2}(y+\pi)\right)\sin\frac{n(y+\pi)}{2}\sec\frac{y}{2}$ $\hspace{2em}$ AD (202a)

## Special cases

**1.342**

1. $$\sum_{k=1}^{n} \sin kx = \sin \frac{n+1}{2} x \sin \frac{nx}{2} \operatorname{cosec} \frac{x}{2}$$      AD (361.1)

$2.^{10}$ $$\sum_{k=0}^{n} \cos kx = \cos \frac{n+1}{2} x \sin \frac{nx}{2} \operatorname{cosec} \frac{x}{2} + 1$$

$$= \cos \frac{nx}{2} \sin \frac{n+1}{2} x \operatorname{cosec} \frac{x}{2} = \frac{1}{2} \left( 1 + \frac{\sin \left( n + \frac{1}{2} \right) x}{\sin \frac{x}{2}} \right)$$

     AD (361.2)

3. $$\sum_{k=1}^{n} \sin(2k-1)x = \sin^2 nx \operatorname{cosec} x$$      AD (361.7)

4. $$\sum_{k=1}^{n} \cos(2k-1)x = \frac{1}{2} \sin 2nx \operatorname{cosec} x$$      JO (207)

**1.343**

1. $$\sum_{k=1}^{n} (-1)^k \cos kx = -\frac{1}{2} + \frac{(-1)^n \cos \left( \frac{2n+1}{2} x \right)}{2 \cos \frac{x}{2}}$$      AD (361.11)

2. $$\sum_{k=1}^{n} (-1)^{k+1} \sin(2k-1)x = (-1)^{n+1} \frac{\sin 2nx}{2 \cos x}$$      AD (361.10)

3. $$\sum_{k=1}^{n} \cos(4k-3)x + \sum_{k=1}^{n} \sin(4k-1)x = \sin 2nx \left( \cos 2nx + \sin 2nx \right) \left( \cos x + \sin x \right) \operatorname{cosec} 2x$$

     JO (208)

**1.344**

1. $$\sum_{k=1}^{n-1} \sin \frac{\pi k}{n} = \cot \frac{\pi}{2n}$$      AD (361.19)

2. $$\sum_{k=1}^{n-1} \sin \frac{2\pi k^2}{n} = \frac{\sqrt{n}}{2} \left( 1 + \cos \frac{n\pi}{2} - \sin \frac{n\pi}{2} \right)$$      AD (361.18)

3. $$\sum_{k=0}^{n-1} \cos \frac{2\pi k^2}{n} = \frac{\sqrt{n}}{2} \left( 1 + \cos \frac{n\pi}{2} + \sin \frac{n\pi}{2} \right)$$      AD (361.17)

## 1.35 Sums of powers of trigonometric functions of multiple angles

**1.351**

1. $$\sum_{k=1}^{n} \sin^2 kx = \frac{1}{4} \left[ (2n+1) \sin x - \sin(2n+1)x \right] \operatorname{cosec} x$$

$$= \frac{n}{2} - \frac{\cos(n+1)x \sin nx}{2 \sin x}$$

     AD (361.3)

2.    $\displaystyle\sum_{k=1}^{n} \cos^2 kx = \frac{n-1}{2} + \frac{1}{2}\cos nx \sin(n+1)x \operatorname{cosec} x$

$\qquad\qquad\qquad = \dfrac{n}{2} + \dfrac{\cos(n+1)x \sin nx}{2\sin x}$

                                                                 AD (361.4)a

3.    $\displaystyle\sum_{k=1}^{n} \sin^3 kx = \frac{3}{4}\sin\frac{n+1}{2}x \sin\frac{nx}{2}\operatorname{cosec}\frac{x}{2} - \frac{1}{4}\sin\frac{3(n+1)x}{2}\sin\frac{3nx}{2}\operatorname{cosec}\frac{3x}{2}$      JO (210)

4.    $\displaystyle\sum_{k=1}^{n} \cos^3 kx = \frac{3}{4}\cos\frac{n+1}{2}x \sin\frac{nx}{2}\operatorname{cosec}\frac{x}{2} + \frac{1}{4}\cos\frac{3(n+1)}{2}x \sin\frac{3nx}{2}\operatorname{cosec}\frac{3x}{2}$      JO (211)a

5.    $\displaystyle\sum_{k=1}^{n} \sin^4 kx = \frac{1}{8}\left[3n - 4\cos(n+1)x \sin nx \operatorname{cosec} x + \cos 2(n+1)x \sin 2nx \operatorname{cosec} 2x\right]$      JO (212)

6.    $\displaystyle\sum_{k=1}^{n} \cos^4 kx = \frac{1}{8}\left[3n + 4\cos(n+1)x \sin nx \operatorname{cosec} x + \cos 2(n+1)x \sin 2nx \operatorname{cosec} 2x\right]$      JO (213)

**1.352**

1.[11]    $\displaystyle\sum_{k=1}^{n-1} k \sin kx = \frac{\sin nx}{4\sin^2\frac{x}{2}} - \frac{n\cos\left(\frac{2n-1}{2}x\right)}{2\sin\frac{x}{2}}$      AD (361.5)

2.[11]    $\displaystyle\sum_{k=1}^{n-1} k \cos kx = \frac{n\sin\left(\frac{2n-1}{2}x\right)}{2\sin\frac{x}{2}} - \frac{1-\cos nx}{4\sin^2\frac{x}{2}}$      AD (361.6)

**1.353**

1.    $\displaystyle\sum_{k=1}^{n-1} p^k \sin kx = \frac{p\sin x - p^n \sin nx + p^{n+1}\sin(n-1)x}{1 - 2p\cos x + p^2}$      AD (361.12)a

2.    $\displaystyle\sum_{k=1}^{n-1} p^k \sinh kx = \frac{p\sinh x - p^n \sinh nx + p^{n+1}\sinh(n-1)x}{1 - 2p\cosh x + p^2}$

3.    $\displaystyle\sum_{k=0}^{n-1} p^k \cos kx = \frac{1 - p\cos x - p^n \cos nx + p^{n+1}\cos(n-1)x}{1 - 2p\cos x + p^2}$      AD (361.13)a₁

4.    $\displaystyle\sum_{k=0}^{n-1} p^k \cosh kx = \frac{1 - p\cosh x - p^n \cosh nx + p^{n+1}\cosh(n-1)x}{1 - 2p\cosh x + p^2}$      JO (396)

## 1.36  Sums of products of trigonometric functions of multiple angles

**1.361**

1.    $\displaystyle\sum_{k=1}^{n} \sin kx \sin(k+1)x = \frac{1}{4}\left[(n+1)\sin 2x - \sin 2(n+1)x\right]\operatorname{cosec} x$      JO (214)

2.    $\displaystyle\sum_{k=1}^{n} \sin kx \sin(k+2)x = \frac{n}{2}\cos 2x - \frac{1}{2}\cos(n+3)x \sin nx \operatorname{cosec} x$       JO (216)

3.    $\displaystyle 2\sum_{k=1}^{n} \sin kx \cos(2k-1)y = \sin\left(ny + \frac{n+1}{2}x\right)\sin\frac{n(x+2y)}{2}\operatorname{cosec}\frac{x+2y}{2}$

$$- \sin\left(ny - \frac{n+1}{2}x\right)\sin\frac{n(2y-x)}{2}\operatorname{cosec}\frac{2y-x}{2}$$

      JO (217)

**1.362**

1.    $\displaystyle\sum_{k=1}^{n}\left(2^k \sin^2\frac{x}{2^k}\right)^2 = \left(2^n \sin\frac{x}{2^n}\right)^2 - \sin^2 x$       AD (361.15)

2.    $\displaystyle\sum_{k=1}^{n}\left(\frac{1}{2^k}\sec\frac{x}{2^k}\right)^2 = \operatorname{cosec}^2 x - \left(\frac{1}{2^n}\operatorname{cosec}\frac{x}{2^n}\right)^2$       AD (361.14)

## 1.37 Sums of tangents of multiple angles

**1.371**

1.    $\displaystyle\sum_{k=0}^{n}\frac{1}{2^k}\tan\frac{x}{2^k} = \frac{1}{2^n}\cot\frac{x}{2^n} - 2\cot 2x$       AD (361.16)

2.    $\displaystyle\sum_{k=0}^{n}\frac{1}{2^{2k}}\tan^2\frac{x}{2^k} = \frac{2^{2n+2}-1}{3\cdot 2^{2n-1}} + 4\cot^2 2x - \frac{1}{2^{2n}}\cot^2\frac{x}{2^n}$       AD (361.20)

## 1.38 Sums leading to hyperbolic tangents and cotangents

**1.381**

1.[12]    $\displaystyle\sum_{k=0}^{n-1}\frac{\tanh x\dfrac{1}{n\sin^2\left(\frac{2k+1}{4n}\pi\right)}}{1 + \dfrac{\tanh^2 x}{\tan^2\left(\frac{2k+1}{4n}\pi\right)}} = \tanh(2nx)$       JO (402)a

2.[12]    $\displaystyle\sum_{k=1}^{n-1}\frac{\tanh x\dfrac{1}{n\sin^2\left(\frac{k\pi}{2n}\right)}}{1 + \dfrac{\tanh^2 x}{\tan^2\left(\frac{k\pi}{2n}\right)}} = \coth(2nx) - \frac{1}{2n}(\tanh x + \coth x)$       JO (403)

3.[12]    $\displaystyle\sum_{k=0}^{n-1}\frac{2\tanh x\dfrac{1}{(2n+1)\sin^2\left(\frac{2k+1}{2(2n+1)}\pi\right)}}{1 + \dfrac{\tanh^2 x}{\tan^2\left(\frac{2k+1}{2(2n+1)}\pi\right)}} = \tanh(2n+1)x - \frac{\tanh x}{2n+1}$       JO (404)

$$4.^{12} \quad \sum_{k=1}^{n} \frac{2 \tanh x \dfrac{1}{(2n+1) \sin^2 \left(\frac{k\pi}{2n+1}\right)}}{1 + \dfrac{\tanh^2 x}{\tan^2 \left(\frac{k\pi}{2n+1}\right)}} = \coth (2n+1) x - \frac{\coth x}{2n+1} \qquad \text{JO (405)}$$

**1.382**

$$1. \quad \sum_{k=0}^{n-1} \frac{1}{\left(\dfrac{\sin^2 \left(\frac{2k+1}{4n}\pi\right)}{\sinh x} + \dfrac{1}{2} \tanh \left(\frac{x}{2}\right)\right)} = 2n \tanh (nx) \qquad \text{JO (406)}$$

$$2. \quad \sum_{k=1}^{n-1} \frac{1}{\left(\dfrac{\sin^2 \left(\frac{k\pi}{2n}\right)}{\sinh x} + \dfrac{1}{2} \tanh \left(\frac{x}{2}\right)\right)} = 2n \coth (nx) - 2 \coth x \qquad \text{JO (407)}$$

$$3. \quad \sum_{k=0}^{n-1} \frac{1}{\left(\dfrac{\sin^2 \left(\frac{2k+1}{2(2n+1)}\pi\right)}{\sinh x} + \dfrac{1}{2} \tanh \left(\frac{x}{2}\right)\right)} = (2n+1) \tanh \left(\frac{(2n+1)x}{2}\right) - \tanh \frac{x}{2} \qquad \text{JO (408)}$$

$$4. \quad \sum_{k=1}^{n} \frac{1}{\left(\dfrac{\sin^2 \left(\frac{k\pi}{2n+1}\right)}{\sinh x} + \dfrac{1}{2} \tanh \left(\frac{x}{2}\right)\right)} = (2n+1) \coth \left(\frac{(2n+1)x}{2}\right) - \coth \frac{x}{2} \qquad \text{JO (409)}$$

## 1.39 The representation of cosines and sines of multiples of the angle as finite products

**1.391**

$$1. \quad \sin nx = n \sin x \cos x \prod_{k=1}^{\frac{n-2}{2}} \left(1 - \frac{\sin^2 x}{\sin^2 \frac{k\pi}{n}}\right) \qquad [n \text{ is even}] \qquad \text{JO (568)}$$

$$2. \quad \cos nx = \prod_{k=1}^{\frac{n}{2}} \left(1 - \frac{\sin^2 x}{\sin^2 \frac{(2k-1)\pi}{2n}}\right) \qquad [n \text{ is even}] \qquad \text{JO (569)}$$

$$3. \quad \sin nx = n \sin x \prod_{k=1}^{\frac{n-1}{2}} \left(1 - \frac{\sin^2 x}{\sin^2 \frac{k\pi}{n}}\right) \qquad [n \text{ is odd}] \qquad \text{JO (570)}$$

$$4. \quad \cos nx = \cos x \prod_{k=1}^{\frac{n-1}{2}} \left(1 - \frac{\sin^2 x}{\sin^2 \frac{(2k-1)\pi}{2n}}\right) \qquad [n \text{ is odd}] \qquad \text{JO (571)a}$$

**1.392**

1.     $\sin nx = 2^{n-1} \prod_{k=0}^{n-1} \sin\left(x + \frac{k\pi}{n}\right)$                        JO (548)

2.     $\cos nx = 2^{n-1} \prod_{k=1}^{n} \sin\left(x + \frac{2k-1}{2n}\pi\right)$               JO (549)

**1.393**

1.     $\prod_{k=0}^{n-1} \cos\left(x + \frac{2k}{n}\pi\right) = \dfrac{1}{2^{n-1}} \cos nx$         $[n \text{ odd}]$

                               $= \dfrac{1}{2^{n-1}} \left[(-1)^{\frac{n}{2}} - \cos nx\right]$     $[n \text{ even}]$

                                                                  JO (543)

2.[11]     $\prod_{k=0}^{n-1} \sin\left(x + \frac{2k}{n}\pi\right) = \dfrac{(-1)^{\frac{n-1}{2}}}{2^{n-1}} \sin nx$       $[n \text{ odd}]$

                               $= \dfrac{(-1)^{\frac{n}{2}}}{2^{n-1}} (1 - \cos nx)$     $[n \text{ even}]$

                                                                 JO (544)

**1.394**    $\prod_{k=0}^{n-1} \left\{x^2 - 2xy\cos\left(\alpha + \frac{2k\pi}{n}\right) + y^2\right\} = x^{2n} - 2x^n y^n \cos n\alpha + y^{2n}$       JO (573)

**1.395**

1.     $\cos nx - \cos ny = 2^{n-1} \prod_{k=0}^{n-1} \left\{\cos x - \cos\left(y + \frac{2k\pi}{n}\right)\right\}$     JO (573)

2.     $\cosh nx - \cos ny = 2^{n-1} \prod_{k=0}^{n-1} \left\{\cosh x - \cos\left(y + \frac{2k\pi}{n}\right)\right\}$     JO (538)

**1.396**

1.     $\prod_{k=1}^{n-1} \left(x^2 - 2x\cos\frac{k\pi}{n} + 1\right) = \dfrac{x^{2n} - 1}{x^2 - 1}$             KR 58 (28.1)

2.     $\prod_{k=1}^{n} \left(x^2 - 2x\cos\frac{2k\pi}{2n+1} + 1\right) = \dfrac{x^{2n+1} - 1}{x - 1}$       KR 58 (28.2)

3.     $\prod_{k=1}^{n} \left(x^2 + 2x\cos\frac{2k\pi}{2n+1} + 1\right) = \dfrac{x^{2n+1} - 1}{x + 1}$       KR 58 (28.3)

4.     $\prod_{k=0}^{n-1} \left(x^2 - 2x\cos\frac{(2k+1)\pi}{2n} + 1\right) = x^{2n} + 1$           KR 58 (28.4)

# 1.41 The expansion of trigonometric and hyperbolic functions in power series

**1.411**

1. $\quad \sin x = \sum\limits_{k=0}^{\infty} (-1)^k \dfrac{x^{2k+1}}{(2k+1)!}$

2. $\quad \sinh x = \sum\limits_{k=0}^{\infty} \dfrac{x^{2k+1}}{(2k+1)!}$

3. $\quad \cos x = \sum\limits_{k=0}^{\infty} (-1)^k \dfrac{x^{2k}}{(2k)!}$

4. $\quad \cosh x = \sum\limits_{k=0}^{\infty} \dfrac{x^{2k}}{(2k)!}$

5. $\quad \tan x = \sum\limits_{k=1}^{\infty} \dfrac{2^{2k} \left(2^{2k} - 1\right)}{(2k)!} |B_{2k}| x^{2k-1} \qquad \left[ x^2 < \dfrac{\pi^2}{4} \right]$ 　　　FI II 523

6.[11] $\quad \tanh x = x - \dfrac{x^3}{3} + \dfrac{2x^5}{15} - \dfrac{17}{315} x^7 + \cdots = \sum\limits_{k=1}^{\infty} \dfrac{2^{2k} \left(2^{2k} - 1\right)}{(2k)!} B_{2k} x^{2k-1}$

$$\left[ x^2 < \dfrac{\pi^2}{4} \right]$$

7. $\quad \cot x = \dfrac{1}{x} - \sum\limits_{k=1}^{\infty} \dfrac{2^{2k} |B_{2k}|}{(2k)!} x^{2k-1} \qquad \left[ x^2 < \pi^2 \right]$ 　　　FI II 523a

8. $\quad \coth x = \dfrac{1}{x} + \dfrac{x}{3} - \dfrac{x^3}{45} + \dfrac{2x^5}{945} - \cdots = \dfrac{1}{x} + \sum\limits_{k=1}^{\infty} \dfrac{2^{2k} B_{2k}}{(2k)!} x^{2k-1}$

$$\left[ x^2 < \pi^2 \right]$$ 　　　FI II 522a

9. $\quad \sec x = \sum\limits_{k=0}^{\infty} \dfrac{|E_{2k}|}{(2k)!} x^{2k} \qquad \left[ x^2 < \dfrac{\pi^2}{4} \right]$ 　　　CE 330a

10. $\quad \operatorname{sech} x = 1 - \dfrac{x^2}{2} + \dfrac{5x^4}{24} - \dfrac{61x^6}{720} + \cdots = 1 + \sum\limits_{k=1}^{\infty} \dfrac{E_{2k}}{(2k)!} x^{2k}$

$$\left[ x^2 < \dfrac{\pi^2}{4} \right]$$ 　　　CE 330

11. $\quad \operatorname{cosec} x = \dfrac{1}{x} + \sum\limits_{k=1}^{\infty} \dfrac{2 \left(2^{2k-1} - 1\right) |B_{2k}| x^{2k-1}}{(2k)!} \qquad \left[ x^2 < \pi^2 \right]$ 　　　CE 329a

12. $\quad \operatorname{cosech} x = \dfrac{1}{x} - \dfrac{1}{6} x + \dfrac{7x^3}{360} - \dfrac{31x^5}{15120} + \cdots = \dfrac{1}{x} - \sum\limits_{k=1}^{\infty} \dfrac{2 \left(2^{2k-1} - 1\right) B_{2k}}{(2k)!} x^{2k-1}$

$$\left[ x^2 < \pi^2 \right]$$ 　　　JO (418)

**1.412**

1.     $\sin^2 x = \sum_{k=1}^{\infty} (-1)^{k+1} \dfrac{2^{2k-1} x^{2k}}{(2k)!}$              JO (452)a

2.     $\cos^2 x = 1 - \sum_{k=1}^{\infty} (-1)^{k+1} \dfrac{2^{2k-1} x^{2k}}{(2k)!}$             JO (443)

3.     $\sin^3 x = \dfrac{1}{4} \sum_{k=1}^{\infty} (-1)^{k+1} \dfrac{3^{2k+1} - 3}{(2k+1)!} x^{2k+1}$       JO (452a)a

4.     $\cos^3 x = \dfrac{1}{4} \sum_{k=0}^{\infty} (-1)^k \dfrac{\left(3^{2k} + 3\right) x^{2k}}{(2k)!}$         JO (443a)

**1.413**

1.[12]     $\sinh x = \operatorname{cosec} x \sum_{k=1}^{\infty} (-1)^{k+1} \dfrac{2^{2k-1} x^{4k-2}}{(4k-2)!}$       JO (508)

2.     $\cosh x = \sec x + \sec x \sum_{k=1}^{\infty} (-1)^k \dfrac{2^{2k} x^{4k}}{(4k)!}$        JO (507)

3.     $\sinh x = \sec x \sum_{k=1}^{\infty} (-1)^{[k/2]} \dfrac{2^{k-1} x^{2k-1}}{(2k-1)!}$        JO (510)

4.     $\cosh x = \operatorname{cosec} x \sum_{k=1}^{\infty} (-1)^{[(k-1)/2]} \dfrac{2^{k-1} x^{2k-1}}{(2k-1)!}$       JO (509)

**1.414**

1.     $\cos\left[n \ln\left(x + \sqrt{1+x^2}\right)\right] = 1 - \sum_{k=0}^{\infty} (-1)^k \dfrac{\left(n^2 + 0^2\right)\left(n^2 + 2^2\right) \ldots \left[n^2 + (2k)^2\right]}{(2k+2)!} x^{2k+2}$

                                           $\left[x^2 < 1\right]$         AD (6456.1)

2.     $\sin\left[n \ln\left(x + \sqrt{1+x^2}\right)\right] = nx - n \sum_{k=1}^{\infty} (-1)^{k+1} \dfrac{\left(n^2 + 1^2\right)\left(n^2 + 3^2\right) \ldots \left[n^2 + (2k-1)^2\right] x^{2k+1}}{(2k+1)!}$

                                           $\left[x^2 < 1\right]$         AD (6456.2)

Power series for $\ln \sin x$, $\ln \cos x$, and $\ln \tan x$ see **1.518**.

## 1.42 Expansion in series of simple fractions

**1.421**

1.     $\tan \dfrac{\pi x}{2} = \dfrac{4x}{\pi} \sum_{k=1}^{\infty} \dfrac{1}{(2k-1)^2 - x^2}$        BR* (191), AD (6495.1)

2.[10]     $\tanh \dfrac{\pi x}{2} = \dfrac{4x}{\pi} \sum_{k=1}^{\infty} \dfrac{1}{(2k-1)^2 + x^2}$

3. $\cot \pi x = \dfrac{1}{\pi x} + \dfrac{2x}{\pi} \displaystyle\sum_{k=1}^{\infty} \dfrac{1}{x^2 - k^2} = \dfrac{1}{\pi x} + \dfrac{x}{\pi} \displaystyle\sum_{\substack{k=-\infty \\ k \neq 0}}^{\infty} \dfrac{1}{k(x-k)}$  AD (6495.2), JO (450a)

4. $\coth \pi x = \dfrac{1}{\pi x} + \dfrac{2x}{\pi} \displaystyle\sum_{k=1}^{\infty} \dfrac{1}{x^2 + k^2}$  (cf. **1.217** 1)

5. $\tan^2 \dfrac{\pi x}{2} = x^2 \displaystyle\sum_{k=1}^{\infty} \dfrac{2(2k-1)^2 - x^2}{(1^2 - x^2)^2 (3^2 - x^2)^2 \dots [(2k-1)^2 - x^2]^2}$  JO (450)

**1.422**

1. $\sec \dfrac{\pi x}{2} = \dfrac{4}{\pi} \displaystyle\sum_{k=1}^{\infty} (-1)^{k+1} \dfrac{2k-1}{(2k-1)^2 - x^2}$  AD (6495.3)a

2. $\sec^2 \dfrac{\pi x}{2} = \dfrac{4}{\pi^2} \displaystyle\sum_{k=1}^{\infty} \left\{ \dfrac{1}{(2k-1-x)^2} + \dfrac{1}{(2k-1+x)^2} \right\}$  JO (451)a

3. $\operatorname{cosec} \pi x = \dfrac{1}{\pi x} + \dfrac{2x}{\pi} \displaystyle\sum_{k=1}^{\infty} \dfrac{(-1)^k}{x^2 - k^2}$  (see also **1.217** 2)  AD (6495.4)a

4. $\operatorname{cosec}^2 \pi x = \dfrac{1}{\pi^2} \displaystyle\sum_{k=-\infty}^{\infty} \dfrac{1}{(x-k)^2} = \dfrac{1}{\pi^2 x^2} + \dfrac{2}{\pi^2} \displaystyle\sum_{k=1}^{\infty} \dfrac{x^2 + k^2}{(x^2 - k^2)^2}$  JO (446)

5. $\dfrac{1 + x \operatorname{cosec} x}{2x^2} = \dfrac{1}{x^2} - \displaystyle\sum_{k=1}^{\infty} \dfrac{(-1)^{k+1}}{(x^2 - k^2 \pi^2)}$  JO (449)

6. $\operatorname{cosec} \pi x = = \dfrac{2}{\pi} \displaystyle\sum_{k=-\infty}^{\infty} \dfrac{(-1)^k}{x^2 - k^2}$  JO (450b)

**1.423** $\dfrac{\pi^2}{4m^2} \operatorname{cosec}^2 \dfrac{\pi}{m} + \dfrac{\pi}{4m} \cot \dfrac{\pi}{m} - \dfrac{1}{2} = \displaystyle\sum_{k=1}^{\infty} \dfrac{1}{(1 - k^2 m^2)^2}$  JO (477)

## 1.43 Representation in the form of an infinite product

**1.431**

1. $\sin x = x \displaystyle\prod_{k=1}^{\infty} \left( 1 - \dfrac{x^2}{k^2 \pi^2} \right)$  EU

2. $\sinh x = x \displaystyle\prod_{k=1}^{\infty} \left( 1 + \dfrac{x^2}{k^2 \pi^2} \right)$  EU

3. $\cos x = \displaystyle\prod_{k=0}^{\infty} \left( 1 - \dfrac{4x^2}{(2k+1)^2 \pi^2} \right)$  EU

4. $\cosh x = \displaystyle\prod_{k=0}^{\infty} \left( 1 + \dfrac{4x^2}{(2k+1)^2 \pi^2} \right)$  EU

**1.432**

$1.^{11}$  $\quad \cos x - \cos y = 2\left(1 - \dfrac{x^2}{y^2}\right)\sin^2\dfrac{y}{2}\displaystyle\prod_{k=1}^{\infty}\left(1 - \dfrac{x^2}{(2k\pi + y)^2}\right)\left(1 - \dfrac{x^2}{(2k\pi - y)^2}\right)$ $\qquad$ AD (653.2)

2.  $\quad \cosh x - \cos y = 2\left(1 + \dfrac{x^2}{y^2}\right)\sin^2\dfrac{y}{2}\displaystyle\prod_{k=1}^{\infty}\left(1 + \dfrac{x^2}{(2k\pi + y)^2}\right)\left(1 + \dfrac{x^2}{(2k\pi - y)^2}\right)$ $\qquad$ AD (653.1)

**1.433** $\quad \cos\dfrac{\pi x}{4} - \sin\dfrac{\pi x}{4} = \displaystyle\prod_{k=1}^{\infty}\left[1 + \dfrac{(-1)^k x}{2k - 1}\right]$ $\qquad$ BR* 189

**1.434** $\quad \cos^2 x = \dfrac{1}{4}(\pi + 2x)^2\displaystyle\prod_{k=1}^{\infty}\left[1 - \left(\dfrac{\pi + 2x}{2k\pi}\right)^2\right]^2$ $\qquad$ MO 216

**1.435** $\quad \dfrac{\sin\pi(x + a)}{\sin\pi a} = \dfrac{x + a}{a}\displaystyle\prod_{k=1}^{\infty}\left(1 - \dfrac{x}{k - a}\right)\left(1 + \dfrac{x}{k + a}\right)$ $\qquad$ MO 216

**1.436** $\quad 1 - \dfrac{\sin^2\pi x}{\sin^2\pi a} = \displaystyle\prod_{k=-\infty}^{\infty}\left[1 - \left(\dfrac{x}{k - a}\right)^2\right]$ $\qquad$ MO 216

**1.437** $\quad \dfrac{\sin 3x}{\sin x} = -\displaystyle\prod_{k=-\infty}^{\infty}\left[1 - \left(\dfrac{2x}{x + k\pi}\right)^2\right]$ $\qquad$ MO 216

**1.438** $\quad \dfrac{\cosh x - \cos a}{1 - \cos a} = \displaystyle\prod_{k=-\infty}^{\infty}\left[1 + \left(\dfrac{x}{2k\pi + a}\right)^2\right]$ $\qquad$ MO 216

**1.439**

1.  $\quad \sin x = x\displaystyle\prod_{k=1}^{\infty}\cos\dfrac{x}{2^k}$ $\qquad\qquad [|x| < 1]$ $\qquad$ AD (615), MO 216

2.  $\quad \dfrac{\sin x}{x} = \displaystyle\prod_{k=1}^{\infty}\left[1 - \dfrac{4}{3}\sin^2\left(\dfrac{x}{3^k}\right)\right]$ $\qquad$ MO 216

# 1.44–1.45  Trigonometric (Fourier) series

**1.441**

1.  $\quad \displaystyle\sum_{k=1}^{\infty}\dfrac{\sin kx}{k} = \dfrac{\pi - x}{2}$ $\qquad\qquad [0 < x < 2\pi]$ $\qquad$ FI III 539

2.  $\quad \displaystyle\sum_{k=1}^{\infty}\dfrac{\cos kx}{k} = -\dfrac{1}{2}\ln\left[2\left(1 - \cos x\right)\right]$ $\qquad\qquad [0 < x < 2\pi]$ $\qquad$ FI III 530a, AD (6814)

3.  $\quad \displaystyle\sum_{k=1}^{\infty}\dfrac{(-1)^{k-1}\sin kx}{k} = \dfrac{x}{2}$ $\qquad\qquad [-\pi < x < \pi]$ $\qquad$ FI III 542

4.  $\quad \displaystyle\sum_{k=1}^{\infty}(-1)^{k-1}\dfrac{\cos kx}{k} = \ln\left(2\cos\dfrac{x}{2}\right)$ $\qquad\qquad [-\pi < x < \pi]$ $\qquad$ FI III 550

**1.442**

1.[11] $\displaystyle\sum_{k=1}^{\infty}\frac{\sin(2k-1)x}{2k-1}=\frac{\pi}{4}\operatorname{sign}x$ $\qquad[-\pi<x<\pi]$ FI III 541

2. $\displaystyle\sum_{k=1}^{\infty}\frac{\cos(2k-1)x}{2k-1}=\frac{1}{2}\ln\cot\frac{x}{2}$ $\qquad[0<x<\pi]$

BR* 168, JO (266), GI III(195)

3. $\displaystyle\sum_{k=1}^{\infty}(-1)^{k-1}\frac{\sin(2k-1)x}{2k-1}=\frac{1}{2}\ln\tan\left(\frac{\pi}{4}+\frac{x}{2}\right)$ $\qquad\left[-\frac{\pi}{2}<x<\frac{\pi}{2}\right]$ BR* 168, JO (268)a

4.[10] $\displaystyle\sum_{k=1}^{\infty}(-1)^{k-1}\frac{\cos(2k-1)x}{2k-1}=\frac{\pi}{4}$ $\qquad\left[-\frac{\pi}{2}<x<\frac{\pi}{2}\right]$

$\qquad\qquad\qquad\qquad\qquad\qquad\quad=-\frac{\pi}{4}$ $\qquad\left[\frac{\pi}{2}<x<\frac{3\pi}{2}\right]$

BR* 168, JO (269)

**1.443**

1.[12] $\displaystyle\sum_{k=1}^{\infty}\frac{\cos k\pi x}{k^{2n}}=(-1)^{n-1}2^{2n-1}\frac{\pi^{2n}}{(2n)!}\sum_{k=0}^{2n}(-1)^{k}\binom{2n}{k}B_{2n-k}\rho^{k}$

$\qquad\qquad\qquad=(-1)^{n-1}\frac{1}{2}\frac{(2\pi)^{2n}}{(2n)!}B_{2n}\left(\frac{x}{2}\right)$

$\qquad\qquad\qquad\left[0\leq x\leq1,\quad\rho=\frac{x}{2}-\left\lfloor\frac{x}{2}\right\rfloor\right]$ CE 340, GE 71

2. $\displaystyle\sum_{k=1}^{\infty}\frac{\sin k\pi x}{k^{2n+1}}=(-1)^{n-1}2^{2n}\frac{\pi^{2n+1}}{(2n+1)!}\sum_{k=0}^{2n+1}\binom{2n+1}{k}B_{2n-k+1}\rho^{k}$

$\qquad\qquad\qquad=(-1)^{n-1}\frac{1}{2}\frac{(2\pi)^{2n+1}}{(2n+1)!}B_{2n+1}\left(\frac{x}{2}\right)$

$\qquad\qquad\qquad\left[0<x<1,\quad\rho=\frac{x}{2}-\left\lfloor\frac{x}{2}\right\rfloor\right]$ CE 340

3. $\displaystyle\sum_{k=1}^{\infty}\frac{\cos kx}{k^{2}}=\frac{\pi^{2}}{6}-\frac{\pi x}{2}+\frac{x^{2}}{4}$ $\qquad[0\leq x\leq2\pi]$ FI III 547

4. $\displaystyle\sum_{k=1}^{\infty}(-1)^{k-1}\frac{\cos kx}{k^{2}}=\frac{\pi^{2}}{12}-\frac{x^{2}}{4}$ $\qquad[-\pi\leq x\leq\pi]$ FI III 544

5. $\displaystyle\sum_{k=1}^{\infty}\frac{\sin kx}{k^{3}}=\frac{\pi^{2}x}{6}-\frac{\pi x^{2}}{4}+\frac{x^{3}}{12}$ $\qquad[0\leq x\leq2\pi]$

6. $\displaystyle\sum_{k=1}^{\infty}\frac{\cos kx}{k^{4}}=\frac{\pi^{4}}{90}-\frac{\pi^{2}x^{2}}{12}+\frac{\pi x^{3}}{12}-\frac{x^{4}}{48}$ $\qquad[0\leq x\leq2\pi]$ AD (6617)

7. $\displaystyle\sum_{k=1}^{\infty}\frac{\sin kx}{k^{5}}=\frac{\pi^{4}x}{90}-\frac{\pi^{2}x^{3}}{36}+\frac{\pi x^{4}}{48}-\frac{x^{5}}{240}$ $\qquad[0\leq x\leq2\pi]$ AD (6818)

**1.444**

1. $$\sum_{k=1}^{\infty} \frac{\sin 2(k+1)x}{k(k+1)} = \sin 2x - (\pi - 2x)\sin^2 x - \sin x \cos x \ln\left(4\sin^2 x\right)$$

$$[0 \le x \le \pi] \qquad \text{BR* 168, GI III (190)}$$

2. $$\sum_{k=1}^{\infty} \frac{\cos 2(k+1)x}{k(k+1)} = \cos 2x - \left(\frac{\pi}{2} - x\right)\sin 2x + \sin^2 x \ln\left(4\sin^2 x\right)$$

$$[0 \le x \le \pi] \qquad \text{BR* 168}$$

3. $$\sum_{k=1}^{\infty} (-1)^k \frac{\sin(k+1)x}{k(k+1)} = \sin x - \frac{x}{2}(1 + \cos x) - \sin x \ln\left|2\cos\frac{x}{2}\right| \qquad \text{MO 213}$$

4. $$\sum_{k=1}^{\infty} (-1)^k \frac{\cos(k+1)x}{k(k+1)} = \cos x - \frac{x}{2}\sin x - (1 + \cos x)\ln\left|2\cos\frac{x}{2}\right| \qquad \text{MO 213}$$

5. $$\sum_{k=0}^{\infty} (-1)^k \frac{\sin(2k+1)x}{(2k+1)^2} = \frac{\pi}{4}x \qquad \qquad \left[-\frac{\pi}{2} \le x \le \frac{\pi}{2}\right]$$

$$= \frac{\pi}{4}(\pi - x) \qquad \qquad \left[\frac{\pi}{2} \le x \le \frac{3}{2}\pi\right]$$

$$\text{MO 213}$$

6.[6] $$\sum_{k=1}^{\infty} \frac{\cos(2k-1)x}{(2k-1)^2} = \frac{\pi}{4}\left(\frac{\pi}{2} - |x|\right) \qquad [-\pi \le x \le \pi] \qquad \text{FI III 546}$$

7. $$\sum_{k=1}^{\infty} \frac{\cos 2kx}{(2k-1)(2k+1)} = \frac{1}{2} - \frac{\pi}{4}\sin x \qquad \left[0 \le x \le \frac{\pi}{2}\right] \qquad \text{JO (591)}$$

**1.445**

1. $$\sum_{k=1}^{\infty} \frac{k\sin kx}{k^2 + \alpha^2} = \frac{\pi}{2}\frac{\sinh\alpha(\pi - x)}{\sinh\alpha\pi} \qquad [0 < x < 2\pi] \qquad \text{BR* 157, JO (411)}$$

2. $$\sum_{k=1}^{\infty} \frac{\cos kx}{k^2 + \alpha^2} = \frac{\pi}{2\alpha}\frac{\cosh\alpha(\pi - x)}{\sinh\alpha\pi} - \frac{1}{2\alpha^2} \qquad [0 \le x \le 2\pi] \qquad \text{BR* 257, JO (410)}$$

3. $$\sum_{k=1}^{\infty} \frac{(-1)^k \cos kx}{k^2 + \alpha^2} = \frac{\pi}{2\alpha}\frac{\cosh\alpha x}{\sinh\alpha\pi} - \frac{1}{2\alpha^2} \qquad [-\pi \le x \le \pi] \qquad \text{FI III 546}$$

4. $$\sum_{k=1}^{\infty} (-1)^{k-1} \frac{k\sin kx}{k^2 + \alpha^2} = \frac{\pi}{2}\frac{\sinh\alpha x}{\sinh\alpha\pi} \qquad [-\pi < x < \pi] \qquad \text{FI III, 546}$$

5. $$\sum_{k=1}^{\infty} \frac{k\sin kx}{k^2 - \alpha^2} = \pi\frac{\sin\{\alpha[(2m+1)\pi - x]\}}{2\sin\alpha\pi} \qquad \left[\text{if } x = 2m\pi, \text{ then } \sum\cdots = 0\right]$$

$$[2m\pi < x < (2m+2)\pi, \quad \alpha \text{ not an integer}] \qquad \text{MO 213}$$

6. $$\sum_{k=1}^{\infty} \frac{\cos kx}{k^2 - \alpha^2} = \frac{1}{2\alpha^2} - \frac{\pi}{2}\frac{\cos\left[\alpha\{(2m+1)\pi - x\}\right]}{\alpha\sin\alpha\pi}$$

$$[2m\pi \le x \le (2m+2)\pi, \quad \alpha \text{ not an integer}] \qquad \text{MO 213}$$

$7.^{12}$ $\displaystyle\sum_{k=1}^{\infty}(-1)^k\frac{k\sin kx}{k^2-\alpha^2}=\pi\frac{\sin[\alpha(2m\pi-x)]}{2\sin\alpha\pi}$ $\qquad\left[\text{if }x=(2m+1)\pi,\text{ then }\sum\cdots=0\right],$

$\qquad\qquad\qquad\qquad\qquad\qquad[(2m-1)\pi\le x\le(2m+1)\pi,\alpha\text{ not an integer}]$ FI III 545a

8. $\displaystyle\sum_{k=1}^{\infty}(-1)^k\frac{\cos kx}{k^2-\alpha^2}=\frac{1}{2\alpha^2}-\frac{\pi}{2}\frac{\cos[\alpha(2m\pi-x)]}{\alpha\sin\alpha\pi}$

$\qquad\qquad\qquad\qquad\qquad\qquad[(2m-1)\pi\le x\le(2m+1)\pi,\alpha\text{ not an integer}]$ FI III 545a

9. $\displaystyle\sum_{n=-\infty}^{\infty}\frac{e^{in\alpha}}{(n-\beta)^2+\gamma^2}=\frac{\pi}{\gamma}\frac{e^{i\beta(\alpha-2\pi)}\sinh(\gamma\alpha)+e^{i\beta\alpha}\sinh[\gamma(2\pi-\alpha)]}{\cosh(2\pi\gamma)-\cos(2\pi\beta)}$

$\qquad\qquad\qquad\qquad\qquad\qquad[0\le\alpha\le2\pi]$

**1.446** $\displaystyle\sum_{k=1}^{\infty}\frac{(-1)^{k+1}\cos(2k+1)x}{(2k-1)(2k+1)(2k+3)}=\frac{\pi}{8}\cos^2 x-\frac{1}{3}\cos x$

$\qquad\qquad\qquad\qquad\qquad\qquad\left[-\frac{\pi}{2}\le x\le\frac{\pi}{2}\right]$ BR* 256, GI III (189)

**1.447**

1. $\displaystyle\sum_{k=1}^{\infty}p^k\sin kx=\frac{p\sin x}{1-2p\cos x+p^2}$

$\qquad\qquad\qquad\qquad\qquad\qquad[|p|<1]$ FI II 559

2. $\displaystyle\sum_{k=0}^{\infty}p^k\cos kx=\frac{1-p\cos x}{1-2p\cos x+p^2}$

$\qquad\qquad\qquad\qquad\qquad\qquad[|p|<1]$ FI II 559

3. $1+2\displaystyle\sum_{k=1}^{\infty}p^k\cos kx=\frac{1-p^2}{1-2p\cos x+p^2}$

$\qquad\qquad\qquad\qquad\qquad\qquad[|p|<1]$ FI II 559a, MO 213

**1.448**

1. $\displaystyle\sum_{k=1}^{\infty}\frac{p^k\sin kx}{k}=\arctan\frac{p\sin x}{1-p\cos x}$

$\qquad\qquad\qquad\qquad\qquad\qquad\left[0<x<2\pi,\quad p^2\le1\right]$ FI II 559

2. $\displaystyle\sum_{k=1}^{\infty}\frac{p^k\cos kx}{k}=-\frac{1}{2}\ln\left(1-2p\cos x+p^2\right)$

$\qquad\qquad\qquad\qquad\qquad\qquad\left[0<x<2\pi,\quad p^2\le1\right]$ FI II 559

3. $\displaystyle\sum_{k=1}^{\infty}\frac{p^{2k-1}\sin(2k-1)x}{2k-1}=\frac{1}{2}\arctan\frac{2p\sin x}{1-p^2}$

$\qquad\qquad\qquad\qquad\qquad\qquad\left[0<x<2\pi,\quad p^2\le1\right]$ JO (594)

4. $\displaystyle\sum_{k=1}^{\infty}\frac{p^{2k-1}\cos(2k-1)x}{2k-1}=\frac{1}{4}\ln\frac{1+2p\cos x+p^2}{1-2p\cos x+p^2}$

$\qquad\qquad\qquad\qquad\qquad\qquad\left[0<x<2\pi,\quad p^2\le1\right]$ JO (259)

**5.** $\displaystyle\sum_{k=1}^{\infty} \frac{(-1)^{k-1}p^{2k-1}\sin(2k-1)x}{2k-1} = \frac{1}{4}\ln\frac{1+2p\sin x+p^2}{1-2p\sin x+p^2}$

$$\left[0 < x < \pi, \quad p^2 \le 1\right] \qquad\qquad \text{JO (261)}$$

**6.** $\displaystyle\sum_{k=1}^{\infty} \frac{(-1)^{k-1}p^{2k-1}\cos(2k-1)x}{2k-1} = \frac{1}{2}\arctan\frac{2p\cos x}{1-p^2}$

$$\left[0 < x < \pi, \quad p^2 \le 1\right] \qquad\qquad \text{JO (597)}$$

**1.449**

**1.** $\displaystyle\sum_{k=1}^{\infty} \frac{p^k\sin kx}{k!} = e^{p\cos x}\sin\left(p\sin x\right)$

$$\left[p^2 \le 1\right] \qquad\qquad \text{JO (486)}$$

**2.** $\displaystyle\sum_{k=0}^{\infty} \frac{p^k\cos kx}{k!} = e^{p\cos x}\cos\left(p\sin x\right)$

$$\left[p^2 \le 1\right] \qquad\qquad \text{JO (485)}$$

Let $S(x) = -\frac{1}{x}\cos x + \frac{1}{x}$ and $C(x) = \frac{1}{x}\sin x$.

**3.** $\displaystyle\sum_{k=1}^{\infty} \frac{k}{k^2-a^2}S(kx) = \frac{\pi}{2}\left[C(ax) - \cot(\pi a)S(ax)\right]$      $\left[0 < x < 2\pi, \quad a \ne 0, \pm 1, \pm 2, \ldots\right]$

**4.** $\displaystyle\sum_{k=1}^{\infty} \frac{1}{k^2-a^2}C(kx) = \frac{1}{2a^2} - \frac{\pi}{2a}\left[S(ax) - \cot(\pi a)C(ax)\right]$

$$\left[0 \le x \le 2\pi, \quad a \ne 0, \pm 1, \pm 2, \ldots\right]$$

**5.** $\displaystyle\sum_{k=1}^{\infty} \frac{(-1)^{k-1}k}{k^2-a^2}S(kx) = \frac{\pi}{2}\operatorname{cosec}(\pi a)S(ax)$      $\left[-\pi < x < \pi, \quad a \ne 0, \pm 1, \pm 2, \ldots\right]$

**6.** $\displaystyle\sum_{k=1}^{\infty} \frac{(-1)^{k-1}}{k^2-a^2}C(kx) = -\frac{1}{2a^2} + \frac{\pi}{2a}\operatorname{cosec}(\pi a)C(ax)$      $\left[-\pi < x < \pi, \quad a \ne 0, \pm 1, \pm 2, \ldots\right]$

**7.** $\displaystyle\sum_{k=1}^{\infty} \frac{2k-1}{(2k-1)^2-a^2}S(kx) = \frac{\pi}{4}\left[C(ax) + \tan\left(\frac{\pi a}{2}\right)S(ax)\right]$

$$\left[0 < x < \pi, \quad a \ne 0, \pm 1, \pm 2, \ldots\right]$$

**8.** $\displaystyle\sum_{k=1}^{\infty} \frac{1}{(2k-1)^2-a^2}C(kx) = -\frac{\pi}{4a}\left[S(ax) - \tan\left(\frac{\pi a}{2}\right)C(ax)\right]$

$$\left[0 \le x \le \pi, \quad a \ne 0, \pm 1, \pm 2, \ldots\right]$$

**9.** $\displaystyle\sum_{k=1}^{\infty} \frac{(-1)^{k-1}}{(2k-1)^2-a^2}S(kx) = \frac{\pi}{4a}\sec\left(\frac{\pi a}{2}\right)S(ax)$      $\left[-\frac{\pi}{2} \le x \le \frac{\pi}{2}, \quad a \ne 0, \pm 1, \pm 2, \ldots\right]$

**10.** $\displaystyle\sum_{k=1}^{\infty} \frac{(-1)^{k-1}(2k-1)}{(2k-1)^2-a^2}C(kx) = \frac{\pi}{4}\sec\left(\frac{\pi a}{2}\right)C(ax)$      $\left[-\frac{\pi}{2} \le x \le \frac{\pi}{2}, \quad a \ne 0, \pm 1, \pm 2, \ldots\right]$

**Fourier expansions of hyperbolic functions**

**1.451**

1. $\quad \sinh x = \cos x \sum_{k=0}^{\infty} \frac{\left(1+0^2\right)\left(1+2^2\right)\cdots\left[1+(2k)^2\right]}{(2k+1)!} \sin^{2k+1} x$ $\qquad$ JO (504)

2. $\quad \cosh x = \cos x + \cos x \sum_{k=1}^{\infty} \frac{\left(1+1\right)\left(1+3^2\right)\cdots\left[1+(2k-1)^2\right]}{(2k)!} \sin^{2k} x$ $\qquad$ JO (503)

**1.452**

1. $\quad \sinh\left(x\cos\theta\right) = \sec\left(x\sin\theta\right) \sum_{k=0}^{\infty} \frac{x^{2k+1}\cos(2k+1)\theta}{(2k+1)!}$

$\qquad\qquad\qquad\qquad\qquad\qquad\qquad \left[x^2 < 1\right]$ $\qquad$ JO (391)

2. $\quad \cosh\left(x\cos\theta\right) = \sec\left(x\sin\theta\right) \sum_{k=0}^{\infty} \frac{x^{2k}\cos 2k\theta}{(2k)!}$

$\qquad\qquad\qquad\qquad\qquad\qquad\qquad \left[x^2 < 1\right]$ $\qquad$ JO (390)

3. $\quad \sinh\left(x\cos\theta\right) = \operatorname{cosec}\left(x\sin\theta\right) \sum_{k=1}^{\infty} \frac{x^{2k}\sin 2k\theta}{(2k)!}$

$\qquad\qquad\qquad\qquad\qquad\qquad\qquad \left[x^2 < 1, \quad x\sin\theta \neq 0\right]$ $\qquad$ JO (393)

4. $\quad \cosh\left(x\cos\theta\right) = \operatorname{cosec}\left(x\sin\theta\right) \sum_{k=0}^{\infty} \frac{x^{2k+1}\sin(2k+1)\theta}{(2k+1)!}$

$\qquad\qquad\qquad\qquad\qquad\qquad\qquad \left[x^2 < 1, \quad x\sin\theta \neq 0\right]$ $\qquad$ JO (392)

## 1.46 Series of products of exponential and trigonometric functions

**1.461**

1. $\quad \sum_{k=0}^{\infty} e^{-kt}\sin kx = \frac{1}{2}\frac{\sin x}{\cosh t - \cos x}$ $\qquad [t > 0]$ $\qquad$ MO 213

2. $\quad 1 + 2\sum_{k=1}^{\infty} e^{-kt}\cos kx = \frac{\sinh t}{\cosh t - \cos x}$ $\qquad [t > 0]$ $\qquad$ MO 213

**1.462**[9] $\quad \sum_{k=1}^{\infty} \frac{\sin kx \sin ky}{k} e^{-2k|t|} = \frac{1}{4}\ln\left[\frac{\sin^2\dfrac{x+y}{2} + \sinh^2 t}{\sin^2\dfrac{x-y}{2} + \sinh^2 t}\right]$ $\qquad$ MO 214

**1.463**

1. $\quad e^{x\cos\varphi}\cos\left(x\sin\varphi\right) = \sum_{n=0}^{\infty} \frac{x^n\cos n\varphi}{n!}$ $\qquad \left[x^2 < 1\right]$ $\qquad$ AD (6476.1)

2. $\quad e^{x\cos\varphi}\sin\left(x\sin\varphi\right) = \sum_{n=1}^{\infty} \frac{x^n\sin n\varphi}{n!}$ $\qquad \left[x^2 < 1\right]$ $\qquad$ AD (6476.2)

## 1.47   Series of hyperbolic functions

**1.471**

1.   $\displaystyle\sum_{k=1}^{\infty} \frac{\sinh kx}{k!} = e^{\cosh x} \sinh(\sinh x)$.            JO (395)

2.   $\displaystyle\sum_{k=0}^{\infty} \frac{\cosh kx}{k!} = e^{\cosh x} \cosh(\sinh x)$.            JO (394)

3.   $\displaystyle\sum_{k=0}^{\infty} \frac{1}{(2k+1)^3}\left[\frac{1}{x}\tanh\frac{(2m+1)\pi x}{2} + x\tanh\frac{(2m+1)\pi}{2x}\right] = \frac{\pi^3}{16}$

**1.472**

1.   $\displaystyle\sum_{k=1}^{\infty} p^k \sinh kx = \frac{p\sinh x}{1 - 2p\cosh x + p^2}$      $\left[p^2 < 1\right]$        JO (396)

2.   $\displaystyle\sum_{k=0}^{\infty} p^k \cosh kx = \frac{1 - p\cosh x}{1 - 2p\cosh x + p^2}$      $\left[p^2 < 1\right]$        JO (397)a

## 1.48 Lobachevskiy's "Angle of parallelism" $\Pi(x)$

**1.480**   Definition.

1.   $\Pi(x) = 2\operatorname{arccot} e^x = 2\arctan e^{-x}$      $[x \geq 0]$      LO III 297, LOI 120

2.   $\Pi(x) = \pi - \Pi(-x)$      $[x < 0]$      LO III 183, LOI 193

**1.481**   Functional relations

1.   $\sin\Pi(x) = \dfrac{1}{\cosh x}$           LO III 297

2.   $\cos\Pi(x) = \tanh x$           LO III 297

3.   $\tan\Pi(x) = \dfrac{1}{\sinh x}$           LO III 297

4.   $\cot\Pi(x) = \sinh x$           LO III 297

5.   $\sin\Pi(x+y) = \dfrac{\sin\Pi(x)\sin\Pi(y)}{1 + \cos\Pi(x)\cos\Pi(y)}$           LO III 297

6.   $\cos\Pi(x+y) = \dfrac{\cos\Pi(x) + \cos\Pi(y)}{1 + \cos\Pi(x)\cos\Pi(y)}$           LO III 183

**1.482**   Connection with the Gudermannian.

$$\operatorname{gd}(-x) = \Pi(x) - \frac{\pi}{2}$$

(Definite) integral of the angle of parallelism; cf. **4.581** and **4.561**.

## 1.49 The hyperbolic amplitude (the Gudermannian) $\operatorname{gd} x$

**1.490**   Definition.

1.   $\operatorname{gd} x = \displaystyle\int_0^x \frac{dt}{\cosh t} = 2 \arctan e^x - \frac{\pi}{2}$                                   JA

2.   $x = \displaystyle\int_0^{\operatorname{gd} x} \frac{dt}{\cos t} = \ln \tan \left( \frac{\operatorname{gd} x}{2} + \frac{\pi}{4} \right)$                                   JA

**1.491**   Functional relations.

1.   $\cosh x = \sec(\operatorname{gd} x)$                                   AD (343.1), JA

2.   $\sinh x = \tan(\operatorname{gd} x)$                                   AD (343.2), JA

3.   $e^x = \sec(\operatorname{gd} x) + \tan(\operatorname{gd} x) = \tan \left( \frac{\pi}{4} + \frac{\operatorname{gd} x}{2} \right) = \dfrac{1 + \sin(\operatorname{gd} x)}{\cos(\operatorname{gd} x)}$                                   AD (343.5), JA

4.   $\tanh x = \sin(\operatorname{gd} x)$                                   AD (343.3), JA

5.   $\tanh \dfrac{x}{2} = \tan \left( \dfrac{1}{2} \operatorname{gd} x \right)$                                   AD (343.4), JA

6.   $\arctan (\tanh x) = \dfrac{1}{2} \operatorname{gd} 2x$                                   AD (343.6a)

**1.492**   If $\gamma = \operatorname{gd} x$, then $ix = \operatorname{gd} i\gamma$                                   JA

**1.493**   Series expansion.

1.   $\dfrac{\operatorname{gd} x}{2} = \displaystyle\sum_{k=0}^{\infty} \frac{(-1)^k}{2k+1} \tanh^{2k+1} \frac{x}{2}$                                   JA

2.   $\dfrac{x}{2} = \displaystyle\sum_{k=0}^{\infty} \frac{1}{2k+1} \tan^{2k+1} \left( \frac{1}{2} \operatorname{gd} x \right)$                                   JA

3.   $\operatorname{gd} x = x - \dfrac{x^3}{6} + \dfrac{x^5}{24} - \dfrac{61x^7}{5040} + \cdots$                                   JA

4.   $x = \operatorname{gd} x + \dfrac{(\operatorname{gd} x)^3}{6} + \dfrac{(\operatorname{gd} x)^5}{24} + \dfrac{61(\operatorname{gd} x)^7}{5040} + \cdots$        $\left[ \operatorname{gd} x < \dfrac{\pi}{2} \right]$                                   JA

## 1.5 The Logarithm

## 1.51  Series representation

**1.511**[12]   $\ln(1 + x) = x - \dfrac{1}{2}x^2 + \dfrac{1}{3}x^3 - \dfrac{1}{4}x^4 + \cdots = \displaystyle\sum_{k=1}^{\infty} (-1)^{k+1} \frac{x^k}{k} = x\, F(1, 1; 2; -x)$

$$[-1 < x \leq 1]$$

**1.512**

1.     $\ln x = (x-1) - \dfrac{1}{2}(x-1)^2 + \dfrac{1}{3}(x-1)^3 - \cdots = \displaystyle\sum_{k=1}^{\infty} (-1)^{k+1} \dfrac{(x-1)^k}{k}$

$$[0 < x \le 2]$$

2.     $\ln x = 2\left[ \dfrac{x-1}{x+1} + \dfrac{1}{3}\left(\dfrac{x-1}{x+1}\right)^3 + \dfrac{1}{5}\left(\dfrac{x-1}{x+1}\right)^5 + \ldots \right] = 2\displaystyle\sum_{k=1}^{\infty} \dfrac{1}{2k-1}\left(\dfrac{x-1}{x+1}\right)^{2k-1}$

$$[0 < x]$$

3.     $\ln x = \dfrac{x-1}{x} + \dfrac{1}{2}\left(\dfrac{x-1}{x}\right)^2 + \dfrac{1}{3}\left(\dfrac{x-1}{x}\right)^3 + \cdots = \displaystyle\sum_{k=1}^{\infty} \dfrac{1}{k}\left(\dfrac{x-1}{x}\right)^k$

$$\left[x \ge \tfrac{1}{2}\right] \qquad\qquad \text{AD (644.6)}$$

4.*     $\ln x = \displaystyle\lim_{\epsilon \to 0}\left(\dfrac{x^\epsilon - 1}{\epsilon}\right)$

**1.513**

1.[12]    $\ln\dfrac{1+x}{1-x} = 2\displaystyle\sum_{k=1}^{\infty} \dfrac{1}{2k-1} x^{2k-1} = 2x\, F\left(\dfrac{1}{2}, 1; \dfrac{3}{2}; x^2\right)$     $\left[x^2 < 1\right]$      FI II 421

2.     $\ln\dfrac{x+1}{x-1} = 2\displaystyle\sum_{k=1}^{\infty} \dfrac{1}{(2k-1)x^{2k-1}}$        $\left[x^2 > 1\right]$      AD (644.9)

3.     $\ln\dfrac{x}{x-1} = \displaystyle\sum_{k=1}^{\infty} \dfrac{1}{kx^k}$        $[x \le -1 \text{ or } x > 1]$      JO (88a)

4.     $\ln\dfrac{1}{1-x} = \displaystyle\sum_{k=1}^{\infty} \dfrac{x^k}{k}$        $[-1 \le x < 1]$      JO (88b)

5.     $\dfrac{1-x}{x}\ln\dfrac{1}{1-x} = 1 - \displaystyle\sum_{k=1}^{\infty} \dfrac{x^k}{k(k+1)}$        $[-1 \le x < 1]$      JO (102)

6.     $\dfrac{1}{1-x}\ln\dfrac{1}{1-x} = \displaystyle\sum_{k=1}^{\infty} x^k \sum_{n=1}^{k} \dfrac{1}{n}$        $\left[x^2 < 1\right]$      JO (88e)

7.     $\dfrac{(1-x)^2}{2x^3}\ln\dfrac{1}{1-x} = \dfrac{1}{2x^2} - \dfrac{3}{4x} + \displaystyle\sum_{k=1}^{\infty} \dfrac{x^{k-1}}{k(k+1)(k+2)}$    $[-1 \le x < 1]$    AD (6445.1)

**1.514**    $\ln\left(1 - 2x\cos\varphi + x^2\right) = -2\displaystyle\sum_{k=1}^{\infty} \dfrac{\cos k\varphi}{k} x^k; \quad \ln\left(x + \sqrt{1+x^2}\right) = \operatorname{arcsinh} x$

           (see **1.631, 1.641, 1.642, 1.646**)     $\left[x^2 \le 1, \quad x\cos\varphi \ne 1\right]$    MO 98, FI II 485

**1.515**

1.[11] $\ln\left(1+\sqrt{1+x^2}\right) = \ln 2 + \dfrac{1\cdot 1}{2\cdot 2}x^2 - \dfrac{1\cdot 1\cdot 3}{2\cdot 4\cdot 4}x^4 + \dfrac{1\cdot 1\cdot 3\cdot 5}{2\cdot 4\cdot 6\cdot 6}x^6 - \cdots$

$$= \ln 2 - \sum_{k=1}^{\infty}(-1)^k\frac{(2k-1)!}{2^{2k}\,(k!)^2}x^{2k}$$

$$\left[x^2 \le 1\right] \qquad\qquad \text{JO (91)}$$

2. $\ln\left(1+\sqrt{1+x^2}\right) = \ln x + \dfrac{1}{x} - \dfrac{1}{2\cdot 3x^3} + \dfrac{1\cdot 3}{2\cdot 4\cdot 5x^5} - \cdots$

$$= \ln x + \frac{1}{x} + \sum_{k=1}^{\infty}(-1)^k\frac{(2k-1)!}{2^{2k-1}\cdot k!(k-1)!(2k+1)x^{2k+1}}$$

$$\left[x^2 \ge 1\right] \qquad\qquad \text{AD (644.4)}$$

3.[12] $\sqrt{1+x^2}\ln\left(x+\sqrt{1+x^2}\right) = x - \sum_{k=1}^{\infty}(-1)^k\dfrac{2^{2k-1}(k-1)!k!}{(2k+1)!}x^{2k+1}$

$$\left[x^2 < 1\right] \qquad\qquad \text{JO (93)}$$

4.[12] $\dfrac{\ln\left(x+\sqrt{1+x^2}\right)}{\sqrt{1+x^2}} = \sum_{k=0}^{\infty}(-1)^k\dfrac{2^{2k}\,(k!)^2}{(2k+1)!}x^{2k+1}$ $\qquad \left[x^2 < 1\right] \qquad\qquad \text{JO (94)}$

**1.516**

1. $\dfrac{1}{2}\left\{\ln(1\pm x)\right\}^2 = \sum_{k=1}^{\infty}\dfrac{(\mp 1)^{k+1}\,x^{k+1}}{k+1}\sum_{n=1}^{k}\dfrac{1}{n}$ $\qquad \left[x^2 < 1\right] \qquad \text{JO (86), JO (85)}$

2. $\dfrac{1}{6}\left\{\ln(1+x)\right\}^3 = \sum_{k=1}^{\infty}\dfrac{(-1)^{k+1}x^{k+2}}{k+2}\sum_{n=1}^{k}\dfrac{1}{n+1}\sum_{m=1}^{n}\dfrac{1}{m}$ $\qquad \left[x^2 < 1\right] \qquad \text{AD (644.14)}$

3. $-\ln(1+x)\cdot\ln(1-x) = \sum_{k=1}^{\infty}\dfrac{x^{2k}}{k}\sum_{n=1}^{2k-1}\dfrac{(-1)^{n+1}}{n}$ $\qquad \left[x^2 < 1\right] \qquad \text{JO (87)}$

4. $\dfrac{1}{4x}\left\{\dfrac{1+x}{\sqrt{x}}\ln\dfrac{1+\sqrt{x}}{1-\sqrt{x}} + 2\ln(1-x)\right\} = \dfrac{1}{2x} + \sum_{k=1}^{\infty}\dfrac{x^{k-1}}{(2k-1)2k(2k+1)}$

$$\left[0 < x < 1\right] \qquad\qquad \text{AD (6445.2)}$$

**1.517**

1.[6] $\dfrac{1}{2x}\left\{1 - \ln(1+x) - \dfrac{1-x}{\sqrt{x}}\arctan\sqrt{x}\right\} = \sum_{k=1}^{\infty}\dfrac{(-1)^{k+1}x^{k-1}}{(2k-1)2k(2k+1)}$

$$\left[0 < x \le 1\right] \qquad\qquad \text{AD (6445.3)}$$

2. $\dfrac{1}{2}\arctan x\ln\dfrac{1+x}{1-x} = \sum_{k=1}^{\infty}\dfrac{x^{4k-2}}{2k-1}\sum_{n=1}^{2k-1}\dfrac{(-1)^{n-1}}{2n-1}$ $\qquad \left[x^2 < 1\right] \qquad \text{BR* 163}$

3.[12] $\dfrac{1}{2}\arctan x\ln\left(1+x^2\right) = \sum_{k=1}^{\infty}\dfrac{(-1)^{k+1}x^{2k+1}}{2k+1}\sum_{n=1}^{2k}\dfrac{1}{n}$ $\qquad \left[x^2 < 1\right] \qquad \text{AD (6455.3)}$

**1.518**

1.  $\ln \sin x = \ln x - \dfrac{x^2}{6} - \dfrac{x^4}{180} - \dfrac{x^6}{2835} - \cdots$

    $\quad = \ln x + \displaystyle\sum_{k=1}^{\infty} \dfrac{(-1)^k 2^{2k-1} B_{2k} x^{2k}}{k(2k)!}$

    $$[0 < x < \pi] \qquad\qquad \text{AD (643.1)a}$$

2.³  $\ln \cos x = -\dfrac{x^2}{2} - \dfrac{x^4}{12} - \dfrac{x^6}{45} - \dfrac{17x^8}{2520} - \cdots$

    $\quad = -\displaystyle\sum_{k=1}^{\infty} \dfrac{2^{2k-1}\left(2^{2k}-1\right)|B_{2k}|}{k(2k)!} x^{2k} = -\dfrac{1}{2}\displaystyle\sum_{k=1}^{\infty} \dfrac{\sin^{2k} x}{k}$

    $$\left[ x^2 < \dfrac{\pi^2}{4} \right] \qquad\qquad \text{FI II 524}$$

3.  $\ln \tan x = \ln x + \dfrac{x^2}{3} + \dfrac{7}{90}x^4 + \dfrac{62}{2835}x^6 + \dfrac{127}{18,900}x^8 + \cdots$

    $\quad = \ln x + \displaystyle\sum_{k=1}^{\infty}(-1)^{k+1}\dfrac{\left(2^{2k-1}-1\right)2^{2k}B_{2k}x^{2k}}{k(2k)!}$

    $$\left[0 < x < \dfrac{\pi}{2}\right] \qquad\qquad \text{AD (643.3)a}$$

## 1.52  Series of logarithms (cf. 1.431)

**1.521**

1.  $\displaystyle\sum_{k=1}^{\infty} \ln\left(1 - \dfrac{4x^2}{(2k-1)^2\pi^2}\right) = \ln \cos x \qquad\qquad \left[-\dfrac{\pi}{2} < x < \dfrac{\pi}{2}\right]$

2.  $\displaystyle\sum_{k=1}^{\infty} \ln\left(1 - \dfrac{x^2}{k^2\pi^2}\right) = \ln \sin x - \ln x \qquad\qquad [0 < x < \pi]$

3.*  $\displaystyle\sum_{k=2}^{\infty}(-1)^k \dfrac{\ln k}{k} = C \ln 2 - \dfrac{1}{2}(\ln 2)^2$

# 1.6 The Inverse Trigonometric and Hyperbolic Functions

## 1.61  The domain of definition

The principal values of the inverse trigonometric functions are defined by the inequalities:

1.  $-\dfrac{\pi}{2} \le \arcsin x \le \dfrac{\pi}{2};\quad 0 \le \arccos x \le \pi \qquad\qquad [-1 \le x \le 1] \qquad\qquad \text{FI II 553}$

2.  $-\dfrac{\pi}{2} < \arctan x < \dfrac{\pi}{2};\quad 0 < \text{arccot}\, x < \pi \qquad\qquad [-\infty < x < +\infty] \qquad\qquad \text{FI II 552}$

## 1.62–1.63  Functional relations

**1.621**   The relationship between the inverse and the direct trigonometric functions.

1.   $\arcsin(\sin x) = x - 2n\pi$

$$= -x + (2n+1)\pi$$

$$\left[2n\pi - \frac{\pi}{2} \le x \le 2n\pi + \frac{\pi}{2}\right]$$

$$\left[(2n+1)\pi - \frac{\pi}{2} \le x \le (2n+1)\pi + \frac{\pi}{2}\right]$$

2.   $\arccos(\cos x) = x - 2n\pi$

$$= -x + 2(n+1)\pi$$

$$[2n\pi \le x \le (2n+1)\pi]$$

$$[(2n+1)\pi \le x \le 2(n+1)\pi]$$

3.   $\arctan(\tan x) = x - n\pi$

$$\left[n\pi - \frac{\pi}{2} < x < n\pi + \frac{\pi}{2}\right]$$

4.   $\text{arccot}(\cot x) = x - n\pi$

$$[n\pi < x < (n+1)\pi]$$

**1.622**   The relationship between the inverse trigonometric functions, the inverse hyperbolic functions, and the logarithm.

1.   $\arcsin z = \dfrac{1}{i} \ln\left(iz + \sqrt{1 - z^2}\right) = \dfrac{1}{i} \text{arcsinh}(iz)$

2.   $\arccos z = \dfrac{1}{i} \ln\left(z + \sqrt{z^2 - 1}\right) = \dfrac{1}{i} \text{arccosh}\, z$

3.   $\arctan z = \dfrac{1}{2i} \ln\dfrac{1 + iz}{1 - iz} = \dfrac{1}{i} \text{arctanh}(iz)$

4.   $\text{arccot}\, z = \dfrac{1}{2i} \ln\dfrac{iz - 1}{iz + 1} = i\,\text{arccoth}(iz)$

5.   $\text{arcsinh}\, z = \ln\left(z + \sqrt{z^2 + 1}\right) = \dfrac{1}{i} \arcsin(iz)$

6.   $\text{arccosh}\, z = \ln\left(z + \sqrt{z^2 - 1}\right) = i\arccos z$

7.   $\text{arctanh}\, z = \dfrac{1}{2} \ln\dfrac{1 + z}{1 - z} = \dfrac{1}{i} \arctan(iz)$

8.   $\text{arccoth}\, z = \dfrac{1}{2} \ln\dfrac{z + 1}{z - 1} = \dfrac{1}{i} \text{arccot}(-iz)$

### Relations between different inverse trigonometric functions

**1.623**

1.   $\arcsin x + \arccos x = \dfrac{\pi}{2}$                    NV 43

2.   $\arctan x + \text{arccot}\, x = \dfrac{\pi}{2}$                    NV 43

**1.624**

1.   $\arcsin x = \arccos \sqrt{1 - x^2} \qquad [0 \le x \le 1]$                    NV 47 (5)

$$= -\arccos \sqrt{1 - x^2} \qquad [-1 \le x \le 0]$$                    NV 46 (2)

2.      $\arcsin x = \arctan \dfrac{x}{\sqrt{1-x^2}}$          $\left[x^2 < 1\right]$

3.      $\arcsin x = \operatorname{arccot} \dfrac{\sqrt{1-x^2}}{x}$     $[0 < x \le 1]$

             $= \operatorname{arccot} \dfrac{\sqrt{1-x^2}}{x} - \pi$     $[-1 \le x < 0]$          NV 49 (10)

4.      $\arccos x = \arcsin \sqrt{1-x^2}$     $[0 \le x \le 1]$

             $= \pi - \arcsin \sqrt{1-x^2}$     $[-1 \le x \le 0]$          NV 48 (6)

5.      $\arccos x = \arctan \dfrac{\sqrt{1-x^2}}{x}$     $[0 < x \le 1]$

             $= \pi + \arctan \dfrac{\sqrt{1-x^2}}{x}$     $[-1 \le x < 0]$          NV 48 (8)

6.      $\arccos x = \operatorname{arccot} \dfrac{x}{\sqrt{1-x^2}}$          $[-1 \le x < 1]$          NV 46 (4)

7.      $\arctan x = \arcsin \dfrac{x}{\sqrt{1+x^2}}$                      NV 6 (3)

8.      $\arctan x = \arccos \dfrac{1}{\sqrt{1+x^2}}$     $[x \ge 0]$

             $= -\arccos \dfrac{1}{\sqrt{1+x^2}}$     $[x \le 0]$          NV 48 (7)

9.[12]    $\arctan x = \operatorname{arccot} \dfrac{1}{x}$                      NV 49 (9)

10.[11]   $\operatorname{arccot} x = \arcsin \dfrac{1}{\sqrt{1+x^2}}$     $[x > 0]$

             $= \pi - \arcsin \dfrac{1}{\sqrt{1+x^2}}$     $[x < 0]$          NV 49 (11)

11.     $\operatorname{arccot} x = \arccos \dfrac{x}{\sqrt{1+x^2}}$                    NV 46 (4)

12.     $\operatorname{arccot} x = \arctan \dfrac{1}{x}$     $[x > 0]$

             $= \pi + \arctan \dfrac{1}{x}$     $[x < 0]$          NV 49 (12)

## 1.625

1.      $\arcsin x + \arcsin y = \arcsin\left(x\sqrt{1-y^2} + y\sqrt{1-x^2}\right)$      $\left[xy \le 0 \text{ or } x^2 + y^2 \le 1\right]$

                        $= \pi - \arcsin\left(x\sqrt{1-y^2} + y\sqrt{1-x^2}\right)$      $\left[x > 0, \quad y > 0 \text{ and } x^2 + y^2 > 1\right]$

                        $= -\pi - \arcsin\left(x\sqrt{1-y^2} + y\sqrt{1-x^2}\right)$      $\left[x < 0, \quad y < 0 \text{ and } x^2 + y^2 > 1\right]$

                                                      NV 54(1), GI I (880)

2.     $\arcsin x + \arcsin y = \arccos\left(\sqrt{1-x^2}\sqrt{1-y^2} - xy\right)$     $[x \geq 0, \quad y \geq 0]$

$\qquad = -\arccos\left(\sqrt{1-x^2}\sqrt{1-y^2} - xy\right)$     $[x < 0, \quad y < 0]$     NV 55

3.     $\arcsin x + \arcsin y = \arctan \dfrac{x\sqrt{1-y^2} + y\sqrt{1-x^2}}{\sqrt{1-x^2}\sqrt{1-y^2} - xy}$     $[xy \leq 0 \text{ or } x^2 + y^2 < 1]$

$\qquad = \arctan \dfrac{x\sqrt{1-y^2} + y\sqrt{1-x^2}}{\sqrt{1-x^2}\sqrt{1-y^2} - xy} + \pi$     $[x > 0, \quad y > 0 \text{ and } x^2 + y^2 > 1]$

$\qquad = \arctan \dfrac{x\sqrt{1-y^2} + y\sqrt{1-x^2}}{\sqrt{1-x^2}\sqrt{1-y^2} - xy} - \pi$     $[x < 0, \quad y < 0 \text{ and } x^2 + y^2 > 1]$

$\qquad\qquad\qquad\qquad\qquad\qquad\qquad\qquad\qquad\qquad\qquad\qquad$ NV 56

4.     $\arcsin x - \arcsin y = \arcsin\left(x\sqrt{1-y^2} - y\sqrt{1-x^2}\right)$     $[xy \geq 0 \text{ or } x^2 + y^2 \leq 1]$

$\qquad = \pi - \arcsin\left(x\sqrt{1-y^2} - y\sqrt{1-x^2}\right)$     $[x > 0, \quad y < 0 \text{ and } x^2 + y^2 > 1]$

$\qquad = -\pi - \arcsin\left(x\sqrt{1-y^2} - y\sqrt{1-x^2}\right)$     $[x < 0, \quad y > 0 \text{ and } x^2 + y^2 > 1]$

$\qquad\qquad\qquad\qquad\qquad\qquad\qquad\qquad\qquad\qquad\qquad\qquad$ NV 55(2)

5.[12]     $\arcsin x - \arcsin y = \arccos\left(\sqrt{1-x^2}\sqrt{1-y^2} + xy\right)$     $[x > y]$

$\qquad = -\arccos\left(\sqrt{1-x^2}\sqrt{1-y^2} + xy\right)$     $[x < y]$     NV 56

6.     $\arccos x + \arccos y = \arccos\left(xy - \sqrt{1-x^2}\sqrt{1-y^2}\right)$     $[x + y \geq 0]$

$\qquad = 2\pi - \arccos\left(xy - \sqrt{1-x^2}\sqrt{1-y^2}\right)$     $[x + y < 0]$     NV 57 (3)

7.[11]     $\arccos x - \arccos y = -\arccos\left(xy + \sqrt{1-x^2}\sqrt{1-y^2}\right)$     $[x \geq y]$

$\qquad = \arccos\left(xy + \sqrt{1-x^2}\sqrt{1-y^2}\right)$     $[x < y]$     NV 57 (4)

8.     $\arctan x + \arctan y = \arctan \dfrac{x+y}{1-xy}$     $[xy < 1]$

$\qquad = \pi + \arctan \dfrac{x+y}{1-xy}$     $[x > 0, \quad xy > 1]$

$\qquad = -\pi + \arctan \dfrac{x+y}{1-xy}$     $[x < 0, \quad xy > 1]$

$\qquad\qquad\qquad\qquad\qquad\qquad\qquad\qquad\qquad\qquad\qquad$ NV 59(5), GI I (879)

9.     $\arctan x - \arctan y = \arctan \dfrac{x-y}{1+xy}$     $[xy > -1]$

$\qquad = \pi + \arctan \dfrac{x-y}{1+xy}$     $[x > 0, \quad xy < -1]$

$\qquad = -\pi + \arctan \dfrac{x-y}{1+xy}$     $[x < 0, \quad xy < -1]$

$\qquad\qquad\qquad\qquad\qquad\qquad\qquad\qquad\qquad\qquad\qquad\qquad$ NV 59(6)

**1.626**

1.     $2\arcsin x = \arcsin\left(2x\sqrt{1-x^2}\right)$                                   $\left[|x| \le \dfrac{1}{\sqrt{2}}\right]$

   $= \pi - \arcsin\left(2x\sqrt{1-x^2}\right)$                                   $\left[\dfrac{1}{\sqrt{2}} < x \le 1\right]$

   $= -\pi - \arcsin\left(2x\sqrt{1-x^2}\right)$                                   $\left[-1 \le x < -\dfrac{1}{\sqrt{2}}\right]$

                                                                                 NV 61 (7)

2.     $2\arccos x = \arccos\left(2x^2 - 1\right)$                                   $[0 \le x \le 1]$

   $= 2\pi - \arccos\left(2x^2 - 1\right)$                                   $[-1 \le x < 0]$                                   NV 61 (8)

3.     $2\arctan x = \arctan\dfrac{2x}{1-x^2}$                                   $[|x| < 1]$

   $= \arctan\dfrac{2x}{1-x^2} + \pi$                                   $[x > 1]$

   $= \arctan\dfrac{2x}{1-x^2} - \pi$                                   $[x < -1]$

                                                                                 NV 61 (9)

4.*    $\arctan\left(x \pm \sqrt{x^2 + 1}\right) = 2\arctan(x) \pm \dfrac{\pi}{2}$

**1.627**

1.     $\arctan x + \arctan\dfrac{1}{x} = \dfrac{\pi}{2}$                                   $[x > 0]$

   $= -\dfrac{\pi}{2}$                                   $[x < 0]$                                   GI I (878)

2.     $\arctan x + \arctan\dfrac{1-x}{1+x} = \dfrac{\pi}{4}$                                   $[x > -1]$

   $= -\dfrac{3}{4}\pi$                                   $[x < -1]$                                   NV 62, GI I (881)

**1.628**

1.[12]    $\arcsin\dfrac{2x}{1+x^2} = -\pi - 2\arctan x$                                   $[x < -1]$

   $= 2\arctan x$                                   $[-1 \le x < 1]$

   $= \pi - 2\arctan x$                                   $[x \ge 1]$

                                                                                 NV 65

2.     $\arccos\dfrac{1-x^2}{1+x^2} = 2\arctan x$                                   $[x \ge 0]$

   $= -2\arctan x$                                   $[x \le 0]$                                   NV 66

**1.629**    $\dfrac{2x-1}{2} - \dfrac{1}{\pi}\arctan\left(\tan\dfrac{2x-1}{2}\pi\right) = E(x)$                                   GI (886)

**1.631** Relations between the inverse hyperbolic functions.

1. $\operatorname{arcsinh} x = \operatorname{arccosh} \sqrt{x^2 + 1} = \operatorname{arctanh} \dfrac{x}{\sqrt{x^2 + 1}}$  JA

2. $\operatorname{arccosh} x = \operatorname{arcsinh} \sqrt{x^2 - 1} = \operatorname{arctanh} \dfrac{\sqrt{x^2 - 1}}{x}$  JA

3. $\operatorname{arctanh} x = \operatorname{arcsinh} \dfrac{x}{\sqrt{1 - x^2}} = \operatorname{arccosh} \dfrac{1}{\sqrt{1 - x^2}} = \operatorname{arccoth} \dfrac{1}{x}$  JA

4. $\operatorname{arcsinh} x \pm \operatorname{arcsinh} y = \operatorname{arcsinh} \left( x\sqrt{1 + y^2} \pm y\sqrt{1 + x^2} \right)$  JA

5. $\operatorname{arccosh} x \pm \operatorname{arccosh} y = \operatorname{arccosh} \left( xy \pm \sqrt{(x^2 - 1)(y^2 - 1)} \right)$  JA

6. $\operatorname{arctanh} x \pm \operatorname{arctanh} y = \operatorname{arctanh} \dfrac{x \pm y}{1 \pm xy}$  JA

## 1.64 Series representations

**1.641**

1.[12] $\arcsin x = \dfrac{\pi}{2} - \arccos x = x + \dfrac{1}{2 \cdot 3} x^3 + \dfrac{1 \cdot 3}{2 \cdot 4 \cdot 5} x^5 + \dfrac{1 \cdot 3 \cdot 5}{2 \cdot 4 \cdot 6 \cdot 7} x^7 + \ldots$

$$= \sum_{k=0}^{\infty} \dfrac{(2k)!}{2^{2k} (k!)^2 (2k+1)} x^{2k+1} = x\, F\left( \dfrac{1}{2}, \dfrac{1}{2}; \dfrac{3}{2}; x^2 \right)$$

$\left[ x^2 < 1 \right]$  FI II 479

2.[12] $\operatorname{arcsinh} x = x - \dfrac{1}{2 \cdot 3} x^3 + \dfrac{1 \cdot 3}{2 \cdot 4 \cdot 5} x^5 - \ldots ;$

$$= \sum_{k=0}^{\infty} (-1)^k \dfrac{(2k)!}{2^{2k} (k!)^2 (2k+1)} x^{2k+1}$$

$$= x\, F\left( \tfrac{1}{2}, \tfrac{1}{2}; \tfrac{3}{2}; -x^2 \right)$$

$\left[ x^2 < 1 \right]$  FI II 480

**1.642**

1.[12] $\operatorname{arcsinh} x = \ln 2x + \dfrac{1}{2} \dfrac{1}{2x^2} - \dfrac{1 \cdot 3}{2 \cdot 4} \dfrac{1}{4x^4} + \ldots$

$$= \ln 2x + \sum_{k=1}^{\infty} (-1)^{k+1} \dfrac{(2k)! x^{-2k}}{2^{2k} (k!)^2 2k}$$  $\left[ x^2 > 1 \right]$

AD (6480.2)a

2.[12] $\operatorname{arccosh} x = \ln 2x - \sum_{k=1}^{\infty} \dfrac{(2k)! x^{-2k}}{2^{2k} (k!)^2 2k}$  $\left[ x^2 > 1 \right]$  AD (6480.3)a

**1.643**

1.[12] $\arctan x = x - \dfrac{x^3}{3} + \dfrac{x^5}{5} - \dfrac{x^7}{7} + \ldots$

$$= \sum_{k=0}^{\infty} \dfrac{(-1)^k x^{2k+1}}{2k+1} = x\, F\left( \dfrac{1}{2}, 1; \dfrac{3}{2}; -x^2 \right)$$

$\left[ x^2 \leq 1 \right]$  FI II 479

$2.^{12}$    $\operatorname{arctanh} x = x + \dfrac{x^3}{3} + \dfrac{x^5}{5} + \cdots = \displaystyle\sum_{k=0}^{\infty} \dfrac{x^{2k+1}}{2k+1} = x\, F\left(\dfrac{1}{2}, 1; \dfrac{3}{2}; x^2\right)$      $\left[x^2 < 1\right]$

AD (6480.4)

**1.644**

$1.^{12}$    $\operatorname{arctan} x = \dfrac{x}{\sqrt{1+x^2}} \displaystyle\sum_{k=0}^{\infty} \dfrac{(2k)!}{2^{2k}\,(k!)^2\,(2k+1)} \left(\dfrac{x^2}{1+x^2}\right)^k$

$\qquad\qquad = \dfrac{x}{\sqrt{1+x^2}}\, F\left(\dfrac{1}{2}, \dfrac{1}{2}; \dfrac{3}{2}; \dfrac{x^2}{1+x^2}\right)$

AD (641.3)

$2.^{12}$    $\operatorname{arctan} x = \dfrac{\pi}{2} - \dfrac{1}{x} + \dfrac{1}{3x^3} - \dfrac{1}{5x^5} + \dfrac{1}{7x^7} - \cdots = \dfrac{\pi}{2} - \displaystyle\sum_{k=0}^{\infty} (-1)^k \dfrac{1}{(2k+1)x^{2k+1}}$    $[x > 0]$

$\qquad\qquad = -\dfrac{\pi}{2} - \dfrac{1}{x} + \dfrac{1}{3x^3} - \dfrac{1}{5x^5} + \dfrac{1}{7x^7} - \cdots = -\dfrac{\pi}{2} - \displaystyle\sum_{k=0}^{\infty} (-1)^k \dfrac{1}{(2k+1)x^{2k+1}}$    $[x < 0]$

AD (641.4)

**1.645**

1.    $\operatorname{arcsec} x = \dfrac{\pi}{2} - \dfrac{1}{x} - \dfrac{1}{2 \cdot 3x^3} - \dfrac{1 \cdot 3}{2 \cdot 4 \cdot 5x^5} - \cdots = \dfrac{\pi}{2} - \displaystyle\sum_{k=0}^{\infty} \dfrac{(2k)!\,x^{-(2k+1)}}{(k!)^2\,2^{2k}(2k+1)}$

$\qquad\qquad = \dfrac{\pi}{2} - \dfrac{1}{x}\, F\left(\dfrac{1}{2}, \dfrac{1}{2}; \dfrac{3}{2}; \dfrac{1}{x^2}\right)$      $\left[x^2 > 1\right]$

AD (641.5)

2.    $(\operatorname{arcsin} x)^2 = \displaystyle\sum_{k=0}^{\infty} \dfrac{2^{2k}\,(k!)^2\,x^{2k+2}}{(2k+1)!\,(k+1)}$      $\left[x^2 \le 1\right]$      AD (642.2), GI III (152)a

3.    $(\operatorname{arcsin} x)^3 = x^3 + \dfrac{3!}{5!}3^2\left(1 + \dfrac{1}{3^2}\right)x^5 + \dfrac{3!}{7!}3^2 \cdot 5^2\left(1 + \dfrac{1}{3^2} + \dfrac{1}{5^2}\right)x^7 + \cdots$

$\qquad\qquad\qquad\qquad\qquad\qquad\qquad\qquad\qquad$ $\left[x^2 \le 1\right]$

$\qquad\qquad\qquad\qquad\qquad\qquad\qquad\qquad$ BR* 188, AD (642.2), GI III (153)a

**1.646**

$1.^{12}$    $\operatorname{arcsinh} \dfrac{1}{x} = \operatorname{arcosech} x = \displaystyle\sum_{k=0}^{\infty} \dfrac{(-1)^k (2k)!}{2^{2k}\,(k!)^2\,(2k+1)} x^{-2k-1}$

$\qquad\qquad\qquad\qquad\qquad\qquad$ $\left[x^2 > 1\right]$      AD (6480.5)

$2.^{12}$    $\operatorname{arccosh} \dfrac{1}{x} = \operatorname{arcsech} x = \ln\dfrac{2}{x} - \displaystyle\sum_{k=1}^{\infty} \dfrac{(2k)!}{2^{2k}\,(k!)^2\,2k} x^{2k}$    $[0 < x < 1]$      AD (6480.6)

$3.^{12}$    $\operatorname{arcsinh} \dfrac{1}{x} = \operatorname{arcosech} x = \ln\dfrac{2}{x} + \displaystyle\sum_{k=1}^{\infty} \dfrac{(-1)^{k+1}(2k)!}{2^{2k}\,(k!)^2\,2k} x^{2k}$

$\qquad\qquad\qquad\qquad\qquad\qquad$ $[0 < x < 1]$      AD (6480.7)a

4.    $\operatorname{arctanh} \dfrac{1}{x} = \operatorname{arccoth} x = \displaystyle\sum_{k=0}^{\infty} \dfrac{x^{-(2k+1)}}{2k+1}$      $\left[x^2 > 1\right]$      AD (6480.8)

**1.647**

1.$^{12}$ $$\sum_{k=1}^{\infty} \frac{\tanh(2k-1)\,(\pi/2)}{(2k-1)^{4n+3}} = \frac{\pi^{4n+3}}{2}\left(\sum_{j=1}^{n} \frac{(-1)^{j-1}\left(2^{2j}-1\right)\left(2^{4n-2j+4}-1\right)B_{2j-1}^{*}B_{4n-2j+3}^{*}}{(2j)!(4n-2j+4)!}\right.$$

$$\left. + \frac{(-1)^{n}\left(2^{2n+2}-1\right)^{2}B_{2n+1}^{*}{}^{2}}{\left[(2n+2)!\right]^{2}}\right)$$

$$[n=0,1,2,\ldots]$$

2.$^{12}$ $$\sum_{k=1}^{\infty} \frac{(-1)^{k-1}\operatorname{sech}(2k-1)\,(\pi/2)}{(2k-1)^{4n+1}} = \frac{\pi^{4n+1}}{2^{4n+3}}\left(\sum_{j=1}^{n-1} \frac{(-1)^{j}B_{2j}^{*}B_{4n-2j}^{*}}{(2j)!(4n-2j)!} + \frac{2B_{4n}^{*}}{(4n)!} + \frac{(-1)^{n}B_{2n}^{*}{}^{2}}{\left[(2n)\right]!^{2}}\right),$$

$$[n=1,2,\ldots]$$

(The summation term on the right is to be omitted for $n=1$.) (See page xxxi for the definition of $B_{r}^{*}$.)

Table of Integrals, Series, and Products. http://dx.doi.org/10.1016/B978-0-12-384933-5.00002-3

# 2 Indefinite Integrals of Elementary Functions

## 2.0 Introduction

### 2.00 General remarks

We omit the constant of integration in all the formulas of this chapter. Therefore, the equality sign ($=$) means that the functions on the left and right of this symbol differ by a constant. For example (see integral 2.01 15), we write

$$\int \frac{\mathrm{d}x}{1+x^2} = \arctan x = -\operatorname{arccot} x$$

although

$$\arctan x = -\operatorname{arccot} x \pm \frac{\pi}{2}$$

When we integrate certain functions, we obtain the logarithm of the absolute value (for example, $\int \frac{\mathrm{d}x}{\sqrt{1+x^2}} = \ln\left|x + \sqrt{1+x^2}\right|$). In such formulas, the absolute-value bars in the argument of the logarithm are omitted for simplicity in writing.

In certain cases, it is important to give the complete form of the primitive function. Such primitive functions, written in the form of definite integrals, are given in Chapter 2 and in other chapters.

Closely related to these formulas are formulas in which the limits of integration and the integrand depend on the same parameter.

A number of formulas lose their meaning for certain values of the constants (parameters) or for certain relationships between these constants (for example, formula **2.02** 8 for $n = -1$ or formula **2.02** 15 for $a = b$). These values of the constants and the relationships between them are for the most part completely clear from the very structure of the right hand member of the formula (the one not containing an integral sign). Therefore, throughout the chapter, we omit remarks to this effect. However, if the value of the integral is given by means of some other formula for those values of the parameters for which the formula in question loses meaning, we accompany this second formula with the appropriate explanation.

The letters $x$, $y$, $t$, ... denote independent variables; $f$, $g$, $\varphi$, ... denote functions of $x$, $y$, $t$, ...; $f'$, $g'$, $\varphi'$, ..., $f''$, $g''$, $\varphi''$, ... denote their first, second, etc., derivatives; $a$, $b$, $m$, $p$, ... denote constants, by which we generally mean arbitrary real numbers. If a particular formula is valid only for certain values of the constants (for example, only for positive numbers or only for integers), an appropriate remark is made provided the restriction that we make does not follow from the form of the formula itself. Thus, in formulas **2.148** 4 and **2.424** 6, we make no remark since it is clear from the form of these formulas themselves that $n$ must be a natural number (that is, a positive integer).

## 2.01 The basic integrals

1. $\displaystyle\int x^n \, dx = \frac{x^{n+1}}{n+1} \qquad (n \neq -1)$

2. $\displaystyle\int \frac{dx}{x} = \ln x$

3. $\displaystyle\int e^x \, dx = e^x$

4. $\displaystyle\int a^x \, dx = \frac{a^x}{\ln a}$

5. $\displaystyle\int \sin x \, dx = -\cos x$

6.[11] $\displaystyle\int \cos x \, dx = \sin x$

7. $\displaystyle\int \frac{dx}{\sin^2 x} = -\cot x$

8.[11] $\displaystyle\int \frac{dx}{\cos^2 x} = \tan x$

9. $\displaystyle\int \frac{\sin x}{\cos^2 x} \, dx = \sec x$

10. $\displaystyle\int \frac{\cos x}{\sin^2 x} \, dx = -\operatorname{cosec} x$

11. $\displaystyle\int \tan x \, dx = -\ln \cos x$

12. $\displaystyle\int \cot x \, dx = \ln \sin x$

13. $\displaystyle\int \frac{dx}{\sin x} = \ln \tan \frac{x}{2}$

14. $\displaystyle\int \frac{dx}{\cos x} = \ln \tan \left(\frac{\pi}{4} + \frac{x}{2}\right) = \ln \left(\sec x + \tan x\right)$

15.[12] $\displaystyle\int \frac{dx}{1+x^2} = \arctan x = -\operatorname{arccot} x$

16. $\displaystyle\int \frac{dx}{1-x^2} = \operatorname{arctanh} x = \frac{1}{2} \ln \frac{1+x}{1-x}$

17. $\displaystyle\int \frac{dx}{\sqrt{1-x^2}} = \arcsin x = -\arccos x$

18. $\displaystyle\int \frac{dx}{\sqrt{x^2+1}} = \operatorname{arcsinh} x = \ln \left(x + \sqrt{x^2+1}\right)$

19. $\displaystyle\int \frac{dx}{\sqrt{x^2-1}} = \operatorname{arccosh} x = \ln \left(x + \sqrt{x^2-1}\right)$

20. $\displaystyle\int \sinh x \, dx = \cosh x$

21. $\displaystyle\int \cosh x \, dx = \sinh x$

22.[11] $\displaystyle\int \frac{dx}{\sinh^2 x} = -\coth x$

23. $\displaystyle\int \frac{dx}{\cosh^2 x} = \tanh x$

24. $\displaystyle\int \tanh x \, dx = \ln \cosh x$

25. $\displaystyle\int \coth x \, dx = \ln \sinh x$

26. $\displaystyle\int \frac{dx}{\sinh x} = \ln \tanh \frac{x}{2}$

## 2.02 General formulas

1. $$\int af\,dx = a\int f\,dx$$

2. $$\int [af \pm b\varphi \pm c\psi \pm \ldots]\,dx = a\int f\,dx \pm b\int \varphi\,dx \pm c\int \psi\,dx \pm \ldots$$

3. $$\frac{d}{dx}\int f\,dx = f$$

4. $$\int f'\,dx = f$$

5. $$\int f'\varphi\,dx = f\varphi - \int f\varphi'\,dx \qquad\qquad\qquad\qquad \text{[integration by parts]}$$

6. $$\int f^{(n+1)}\varphi\,dx = \varphi f^{(n)} - \varphi' f^{(n-1)} + \varphi'' f^{(n-2)} - \ldots + (-1)^n \varphi^{(n)} f + (-1)^{n+1}\int \varphi^{(n+1)} f\,dx$$

7. $$\int f(x)\,dx = \int f[\varphi(y)]\varphi'(y)\,dy \qquad\qquad [x = \varphi(y)] \qquad \text{[change of variable]}$$

8.[11] $$\int (f)^n f'\,dx = \frac{(f)^{n+1}}{n+1} \qquad\qquad\qquad [n \neq -1]$$

For $n = -1$

$$\int \frac{f'\,dx}{f} = \ln f$$

9. $$\int (af + b)^n f'\,dx = \frac{(af + b)^{n+1}}{a(n+1)}$$

10. $$\int \frac{f'\,dx}{\sqrt{af+b}} = \frac{2\sqrt{af+b}}{a}$$

11. $$\int \frac{f'\varphi - \varphi' f}{\varphi^2}\,dx = \frac{f}{\varphi}$$

12. $$\int \frac{f'\varphi - \varphi' f}{f\varphi}\,dx = \ln \frac{f}{\varphi}$$

13. $$\int \frac{dx}{f(f \pm \varphi)} = \pm \int \frac{dx}{f\varphi} \mp \int \frac{dx}{\varphi(f \pm \varphi)}$$

14. $$\int \frac{f'\,dx}{\sqrt{f^2+a}} = \ln\left(f + \sqrt{f^2+a}\right)$$

15. $$\int \frac{f\,dx}{(f+a)(f+b)} = \frac{a}{a-b}\int \frac{dx}{(f+a)} - \frac{b}{a-b}\int \frac{dx}{(f+b)}$$

For $a = b$

$$\int \frac{f\,dx}{(f+a)^2} = \int \frac{dx}{f+a} - a\int \frac{dx}{(f+a)^2}$$

16. $$\int \frac{f\,dx}{(f+\varphi)^n} = \int \frac{dx}{(f+\varphi)^{n-1}} - \int \frac{\varphi\,dx}{(f+\varphi)^n}$$

17. $$\int \frac{f'\,dx}{p^2 + q^2 f^2} = \frac{1}{pq}\arctan \frac{qf}{p}$$

18. $\displaystyle\int \frac{f' \, dx}{q^2 f^2 - p^2} = \frac{1}{2pq} \ln \frac{qf - p}{qf + p}$

19. $\displaystyle\int \frac{f \, dx}{1 - f} = -x + \int \frac{dx}{1 - f}$

20. $\displaystyle\int \frac{f^2 \, dx}{f^2 - a^2} = \frac{1}{2} \int \frac{f \, dx}{f - a} + \frac{1}{2} \int \frac{f \, dx}{f + a}$

21. $\displaystyle\int \frac{f' \, dx}{\sqrt{a^2 - f^2}} = \arcsin \frac{f}{a}$

22. $\displaystyle\int \frac{f' \, dx}{af^2 + bf} = \frac{1}{b} \ln \frac{f}{af + b}$

23. $\displaystyle\int \frac{f' \, dx}{f \sqrt{f^2 - a^2}} = \frac{1}{a} \operatorname{arcsec} \frac{f}{a}$

24. $\displaystyle\int \frac{(f'\varphi - f\varphi') \, dx}{f^2 + \varphi^2} = \arctan \frac{f}{\varphi}$

25. $\displaystyle\int \frac{(f'\varphi - f\varphi') \, dx}{f^2 - \varphi^2} = \frac{1}{2} \ln \frac{f - \varphi}{f + \varphi}$

## 2.1 Rational Functions

### 2.10 General integration rules

**2.101** To integrate an arbitrary rational function $\dfrac{F(x)}{f(x)}$, where $F(x)$ and $f(x)$ are polynomials with no common factors, we first need to separate out the integral part $E(x)$ (where $E(x)$ is a polynomial), if there is an integral part, and then to integrate separately the integral part and the remainder, thus:
$$\int \frac{F(x) \, dx}{f(x)} = \int E(x) \, dx + \int \frac{\varphi(x)}{f(x)} \, dx.$$
Integration of the remainder, which is then a proper rational function (that is, one in which the degree of the numerator is less than the degree of the denominator) is based on the decomposition of the fraction into elementary fractions, the so-called *partial fractions*.

**2.102** If $a$, $b$, $c$, ..., $m$ are roots of the equation $f(x) = 0$ and if $\alpha$, $\beta$, $\gamma$, ..., $\mu$ are their corresponding multiplicities, so that $f(x) = (x-a)^\alpha (x-b)^\beta \ldots (x-m)^\mu$ then, $\dfrac{\varphi(x)}{f(x)}$ can be decomposed into the following partial fractions:
$$\frac{\varphi(x)}{f(x)} = \frac{A_\alpha}{(x-a)^\alpha} + \frac{A_{\alpha-1}}{(x-a)^{\alpha-1}} + \ldots + \frac{A_1}{x-a} + \frac{B_\beta}{(x-b)^\beta} + \frac{B_{\beta-1}}{(x-b)^{\beta-1}} + \ldots + \frac{B_1}{x-b} + \ldots$$
$$+ \frac{M_\mu}{(x-m)^\mu} + \frac{M_{\mu-1}}{(x-m)^{\mu-1}} + \ldots + \frac{M_1}{x-m},$$
where the numerators of the individual fractions are determined by the following formulas:
$$A_{\alpha-k+1} = \frac{\psi_1^{(k-1)}(a)}{(k-1)!}, \qquad B_{\beta-k+1} = \frac{\psi_2^{(k-1)}(b)}{(k-1)!}, \qquad \ldots, \qquad M_{\mu-k+1} = \frac{\psi_m^{(k-1)}(m)}{(k-1)!},$$
$$\psi_1(x) = \frac{\varphi(x)(x-a)^\alpha}{f(x)}, \qquad \psi_2(x) = \frac{\varphi(x)(x-b)^\beta}{f(x)}, \qquad \ldots, \qquad \psi_m(x) = \frac{\varphi(x)(x-m)^\mu}{f(x)}$$

If $a, b, \ldots, m$ are simple roots, that is, if $\alpha = \beta = \ldots = \mu = 1$, then

$$\frac{\varphi(x)}{f(x)} = \frac{A}{x-a} + \frac{B}{x-b} + \cdots + \frac{M}{x-m},$$

where

$$A = \frac{\varphi(a)}{f'(a)}, \qquad B = \frac{\varphi(b)}{f'(b)}, \qquad \ldots, \qquad M = \frac{\varphi(m)}{f'(m)}.$$

If some of the roots of the equation $f(x) = 0$ are imaginary, we group together the fractions that represent conjugate roots of the equation. Then, after certain manipulations, we represent the corresponding pairs of fractions in the form of real fractions of the form

$$\frac{M_1 x + N_1}{x^2 + 2Bx + C} + \frac{M_2 x + N_2}{(x^2 + 2Bx + C)^2} + \cdots + \frac{M_p x + N_p}{(x^2 + 2Bx + C)^p}.$$

**2.103** Thus, the integration of a proper rational fraction $\dfrac{\varphi(x)}{f(x)}$ reduces to integrals of the form $\displaystyle\int \frac{g \, dx}{(x-a)^\alpha}$

or $\displaystyle\int \frac{Mx + N}{(A + 2Bx + Cx^2)^p} \, dx$. Fractions of the first form yield rational functions for $\alpha > 1$ and logarithms for $\alpha = 1$. Fractions of the second form yield rational functions and logarithms or arctangents:

1.  $$\int \frac{g \, dx}{(x-a)^\alpha} = g \int \frac{d(x-a)}{(x-a)^\alpha} = -\frac{g}{(\alpha-1)(x-a)^{\alpha-1}}$$

2.  $$\int \frac{g \, dx}{x-a} = g \int \frac{d(x-a)}{x-a} = g \ln|x-a|$$

3.  $$\int \frac{Mx + N}{(A + 2Bx + Cx^2)^p} \, dx = \frac{NB - MA + (NC - MB)x}{2(p-1)(AC - B^2)(A + 2Bx + Cx^2)^{p-1}}$$
    $$+ \frac{(2p-3)(NC - MB)}{2(p-1)(AC - B^2)} \int \frac{dx}{(A + 2Bx + Cx^2)^{p-1}}$$

4.  $$\int \frac{dx}{A + 2Bx + Cx^2} = \frac{1}{\sqrt{AC - B^2}} \arctan \frac{Cx + B}{\sqrt{AC - B^2}} \qquad \text{for } [AC > B^2]$$
    $$= \frac{1}{2\sqrt{B^2 - AC}} \ln \left| \frac{Cx + B - \sqrt{B^2 - AC}}{Cx + B + \sqrt{B^2 - AC}} \right| \qquad \text{for } [AC < B^2]$$

5.  $$\int \frac{(Mx + N) \, dx}{A + 2Bx + Cx^2}$$
    $$= \frac{M}{2C} \ln|A + 2Bx + Cx^2| + \frac{NC - MB}{C\sqrt{AC - B^2}} \arctan \frac{Cx + B}{\sqrt{AC - B^2}} \qquad \text{for } [AC > B^2]$$
    $$= \frac{M}{2C} \ln|A + 2Bx + Cx^2| + \frac{NC - MB}{2C\sqrt{B^2 - AC}} \ln \left| \frac{Cx + B - \sqrt{B^2 - AC}}{Cx + B + \sqrt{B^2 - AC}} \right| \qquad \text{for } [AC < B^2]$$

## The Ostrogradskiy–Hermite method

**2.104** By means of the Ostrogradskiy–Hermite method, we can find the rational part of $\displaystyle\int \frac{\varphi(x)}{f(x)} \, dx$ without finding the roots of the equation $f(x) = 0$ and without decomposing the integrand into partial fractions:

$$\int \frac{\varphi(x)}{f(x)}\,dx = \frac{M}{D} + \int \frac{N\,dx}{Q}$$

FI II 49

Here, $M$, $N$, $D$, and $Q$ are rational functions of $x$. Specifically, $D$ is the greatest common divisor of the function $f(x)$ and its derivative $f'(x)$; $Q = \dfrac{f(x)}{D}$; $M$ is a polynomial of degree no higher than $m - 1$, where $m$ is the degree of the polynomial $D$; $N$ is a polynomial of degree no higher than $n - 1$, where $n$ is the degree of the polynomial $Q$. The coefficients of the polynomials $M$ and $N$ are determined by equating the coefficients of like powers of $x$ in the following identity:

$$\varphi(x) = M'Q - M\,(T - Q') + ND$$

where $T = \dfrac{f'(x)}{D}$ and $M'$ and $Q'$ are the derivatives of the polynomials $M$ and $Q$.

## 2.11–2.13  Forms containing the binomial $a + bx^k$

**2.110**    Reduction formulas for $z_k = a + bx^k$ and an explicit expression for the general case.

1.    $$\int x^n z_k^m\,dx = \frac{x^{n+1} z_k^m}{km + n + 1} + \frac{amk}{km + n + 1} \int x^n z_k^{m-1}\,dx$$

$$= \frac{x^{n+1}}{m+1} \sum_{s=0}^{p} \frac{(ak)^s (m+1)m(m-1)\ldots(m-s+1) z_k^{m-s}}{[mk+n+1][(m-1)k+n+1]\ldots[(m-s)k+n+1]}$$

$$+ \frac{(ak)^{p+1} m(m-1)\ldots(m-p+1)(m-p)}{[mk+n+1][(m-1)k+n+1]\ldots[(m-p)k+n+1]} \int x^n z_k^{m-p-1}\,dx$$

LA 126(4)

2.    $$\int x^n z_k^m\,dx = \frac{-x^{n+1} z_k^{m+1}}{ak(m+1)} + \frac{km+k+n+1}{ak(m+1)} \int x^n z_k^{m+1}\,dx$$

LA 126 (6)

3.    $$\int x^n z_k^m\,dx = \frac{x^{n+1} z_k^m}{n+1} - \frac{bkm}{n+1} \int x^{n+k} z_k^{m-1}\,dx$$

4.    $$\int x^n z_k^m\,dx = \frac{x^{n+1-k} z_k^{m+1}}{bk(m+1)} - \frac{n+1-k}{bk(m+1)} \int x^{n-k} z_k^{m+1}\,dx$$

LA 125 (2)

5.    $$\int x^n z_k^m\,dx = \frac{x^{n+1-k} z_k^{m+1}}{b(km+n+1)} - \frac{a(n+1-k)}{b(km+n+1)} \int x^{n-k} z_k^m\,dx$$

LA 126 (3)

6.    $$\int x^n z_k^m\,dx = \frac{x^{n+1} z_k^{m+1}}{a(n+1)} - \frac{b(km+k+n+1)}{a(n+1)} \int x^{n+k} z_k^m\,dx$$

LA 126 (5)

7.*    $$\int x^n \left(nx^b + c\right)^k\,dx = \frac{n^k}{b} \sum_{i=0}^{k} \frac{(-1)^i k!\,\Gamma\left(\frac{a+1}{b}\right)\left(n^b + \frac{c}{n}\right)^{k-i}}{(k-i)!\,\Gamma\left(\frac{a+1}{b} + i + 1\right)} x^{a+1+ib}$$

[$a$, $b$, $k \geq 0$ are all integers]

8.    $$\int x^n z_k^m\,dx = \frac{b^m}{k} \sum_{i=0}^{m} \frac{(-1)^i m!\,J!\left(x^k + \frac{a}{b}\right)^{m-i} x^{k(J+i+1)}}{(m-i)!(J+i+1)!}$$

$$J = \frac{n+1}{k} - 1 \qquad [a, b, k, m, n \text{ real}, \quad k \neq 0, \quad m \geq 0 \text{ an integer}]$$

**Forms containing the binomial $z_1 = a + bx$**

**2.111**

1. $$\int z_1^m \, dx = \frac{z_1^{m+1}}{b(m+1)}$$

   For $m = -1$

   $$\int \frac{dx}{z_1} = \frac{1}{b} \ln z_1$$

2. $$\int \frac{x^n \, dx}{z_1^m} = \frac{x^n}{z_1^{m-1}(n+1-m)b} - \frac{na}{(n+1-m)b} \int \frac{x^{n-1} \, dx}{z_1^m}$$

   For $n = m - 1$, we may use the formula

3.[8] $$\int \frac{x^{m-1} \, dx}{z_1^m} = -\frac{x^{m-1}}{z_1^{m-1}(m-1)b} + \frac{1}{b} \int \frac{x^{m-2} \, dx}{z_1^{m-1}}$$

   For $m = 1$

   $$\int \frac{x^n \, dx}{z_1} = \frac{x^n}{nb} - \frac{ax^{n-1}}{(n-1)b^2} + \frac{a^2 x^{n-2}}{(n-2)b^3} - \ldots + (-1)^{n-1} \frac{a^{n-1} x}{1 \cdot b^n} + \frac{(-1)^n a^n}{b^{n+1}} \ln z_1$$

4. $$\int \frac{x^n \, dx}{z_1^2} = \sum_{k=1}^{n-1} (-1)^{k-1} \frac{k a^{k-1} x^{n-k}}{(n-k) b^{k+1}} + (-1)^{n-1} \frac{a^n}{b^{n+1} z_1} + (-1)^{n+1} \frac{na^{n-1}}{b^{n+1}} \ln z_1$$

5. $$\int \frac{x \, dx}{z_1} = \frac{x}{b} - \frac{a}{b^2} \ln z_1$$

6. $$\int \frac{x^2 \, dx}{z_1} = \frac{x^2}{2b} - \frac{ax}{b^2} + \frac{a^2}{b^3} \ln z_1$$

**2.113**

1. $$\int \frac{dx}{z_1^2} = -\frac{1}{bz_1}$$

2. $$\int \frac{x \, dx}{z_1^2} = -\frac{x}{bz_1} + \frac{1}{b^2} \ln z_1 = \frac{a}{b^2 z_1} + \frac{1}{b^2} \ln z_1$$

3. $$\int \frac{x^2 \, dx}{z_1^2} = \frac{x}{b^2} - \frac{a^2}{b^3 z_1} - \frac{2a}{b^3} \ln z_1$$

**2.114**

1. $$\int \frac{dx}{z_1^3} = -\frac{1}{2bz_1^2}$$

2. $$\int \frac{x \, dx}{z_1^3} = -\left[ \frac{x}{b} + \frac{a}{2b^2} \right] \frac{1}{z_1^2}$$

3. $$\int \frac{x^2 \, dx}{z_1^3} = \left[ \frac{2ax}{b^2} + \frac{3a^2}{2b^3} \right] \frac{1}{z_1^2} + \frac{1}{b^3} \ln z_1$$

4.[6] $$\int \frac{x^3 \, dx}{z_1^3} = \left[ \frac{x^3}{b} + 2\frac{a}{b^2} x^2 - 2\frac{a^2}{b^3} x - \frac{5}{2} \frac{a^3}{b^4} \right] \frac{1}{z_1^2} - 3\frac{a}{b^4} \ln z_1$$

**2.115**

1.  $$\int \frac{dx}{z_1^4} = -\frac{1}{3bz_1^3}$$

2.  $$\int \frac{x\,dx}{z_1^4} = -\left[\frac{x}{2b} + \frac{a}{6b^2}\right]\frac{1}{z_1^3}$$

3.  $$\int \frac{x^2\,dx}{z_1^4} = -\left[\frac{x^2}{b} + \frac{ax}{b^2} + \frac{a^2}{3b^3}\right]\frac{1}{z_1^3}$$

4.  $$\int \frac{x^3\,dx}{z_1^4} = \left[\frac{3ax^2}{b^2} + \frac{9a^2x}{2b^2} + \frac{11a^3}{6b^4}\right]\frac{1}{z_1^3} + \frac{1}{b^4}\ln z_1$$

**2.116**

1.  $$\int \frac{dx}{z_1^5} = -\frac{1}{4bz_1^4}$$

2.  $$\int \frac{x\,dx}{z_1^5} = -\left[\frac{x}{3b} + \frac{a}{12b^2}\right]\frac{1}{z_1^4}$$

3.  $$\int \frac{x^2\,dx}{z_1^5} = -\left[\frac{x^2}{2b} + \frac{ax}{3b^2} + \frac{a^2}{12b^3}\right]\frac{1}{z_1^4}$$

4.  $$\int \frac{x^3\,dx}{z_1^5} = -\left[\frac{x^3}{b} + \frac{3ax^2}{2b^2} + \frac{a^2x}{b^3} + \frac{a^3}{4b^4}\right]\frac{1}{z_1^4}$$

**2.117**

1.  $$\int \frac{dx}{x^n z_1^m} = \frac{-1}{(n-1)ax^{n-1}z_1^{m-1}} + \frac{b(2-n-m)}{a(n-1)}\int \frac{dx}{x^{n-1}z_1^m}$$

2.  $$\int \frac{dx}{z_1^m} = -\frac{1}{(m-1)bz_1^{m-1}}$$

3.  $$\int \frac{dx}{xz_1^m} = \frac{1}{z_1^{m-1}a(m-1)} + \frac{1}{a}\int \frac{dx}{xz_1^{m-1}}$$

4.  $$\int \frac{dx}{x^n z_1} = \sum_{k=1}^{n-1}\frac{(-1)^k b^{k-1}}{(n-k)a^k x^{n-k}} + \frac{(-1)^n b^{n-1}}{a^n}\ln\frac{z_1}{x}$$

**2.118**

1.  $$\int \frac{dx}{xz_1} = -\frac{1}{a}\ln\frac{z_1}{x},$$

2.  $$\int \frac{dx}{x^2 z_1} = -\frac{1}{ax} + \frac{b}{a^2}\ln\frac{z_1}{x}$$

3.  $$\int \frac{dx}{x^3 z_1} = -\frac{1}{2ax^2} + \frac{b}{a^2 x} - \frac{b^2}{a^3}\ln\frac{z_1}{x}$$

**2.119**

1.  $$\int \frac{dx}{xz_1^2} = \frac{1}{az_1} - \frac{1}{a^2}\ln\frac{z_1}{x}$$

2. $\displaystyle \int \frac{dx}{x^2 z_1^2} = -\left[\frac{1}{ax} + \frac{2b}{a^2}\right]\frac{1}{z_1} + \frac{2b}{a^3}\ln\frac{z_1}{x}$

3. $\displaystyle \int \frac{dx}{x^3 z_1^2} = \left[-\frac{1}{2ax^2} + \frac{3b}{2a^2 x} + \frac{3b^2}{a^3}\right]\frac{1}{z_1} - \frac{3b^2}{a^4}\ln\frac{z_1}{x}$

**2.121**

1. $\displaystyle \int \frac{dx}{xz_1^3} = \left[\frac{3}{2a} + \frac{bx}{a^2}\right]\frac{1}{z_1^2} - \frac{1}{a^3}\ln\frac{z_1}{x}$

2. $\displaystyle \int \frac{dx}{x^2 z_1^3} = -\left[\frac{1}{ax} + \frac{9b}{2a^2} + \frac{3b^2 x}{a^3}\right]\frac{1}{z_1^2} + \frac{3b}{a^4}\ln\frac{z_1}{x}$

3. $\displaystyle \int \frac{dx}{x^3 z_1^3} = \left[-\frac{1}{2ax^2} + \frac{2b}{a^2 x} + \frac{9b^2}{a^3} + \frac{6b^3 x}{a^4}\right]\frac{1}{z_1^2} - \frac{6b^2}{a^5}\ln\frac{z_1}{x}$

**2.122**

1. $\displaystyle \int \frac{dx}{xz_1^4} = \left[\frac{11}{6a} + \frac{5bx}{2a^2} + \frac{b^2 x^2}{a^3}\right]\frac{1}{z_1^3} - \frac{1}{a^4}\ln\frac{z_1}{x}$

2. $\displaystyle \int \frac{dx}{x^2 z_1^4} = -\left[\frac{1}{ax} + \frac{22b}{3a^2} + \frac{10b^2 x}{a^3} + \frac{4b^3 x^2}{a^4}\right]\frac{1}{z_1^3} + \frac{4b}{a^5}\ln\frac{z_1}{x}$

3. $\displaystyle \int \frac{dx}{x^3 z_1^4} = \left[-\frac{1}{2ax^2} + \frac{5b}{2a^2 x} + \frac{55b^2}{3a^3} + \frac{25b^3 x}{a^4} + \frac{10b^4 x^2}{a^5}\right]\frac{1}{z_1^3} - \frac{10b^2}{a^6}\ln\frac{z_1}{x}$

**2.123**

1.[11] $\displaystyle \int \frac{dx}{xz_1^5} = \left[\frac{25}{12a} + \frac{13bx}{3a^2} + \frac{7b^2 x^2}{2a^3} + \frac{b^3 x^3}{a^4}\right]\frac{1}{z_1^4} - \frac{1}{a^5}\ln\frac{z_1}{x}$

2. $\displaystyle \int \frac{dx}{x^2 z_1^5} = \left[-\frac{1}{ax} - \frac{125b}{12a^2} - \frac{65b^2 x}{3a^3} - \frac{35b^3 x^2}{2a^4} - \frac{5b^4 x^3}{a^5}\right]\frac{1}{z_1^4} + \frac{5b}{a^6}\ln\frac{z_1}{x}$

3. $\displaystyle \int \frac{dx}{x^3 z_1^5} = \left[-\frac{1}{2ax^2} + \frac{3b}{a^2 x} + \frac{125b^2}{4a^3} + \frac{65b^3 x}{a^4} + \frac{105b^4 x^2}{2a^5} + \frac{15b^5 x^3}{a^6}\right]\frac{1}{z_1^4} - \frac{15b^2}{a^7}\ln\frac{z_1}{x}$

**2.124**　Forms containing the binomial $z_2 = a + bx^2$.

1.[12] $\displaystyle \int \frac{dx}{z_2} = \frac{1}{\sqrt{ab}}\arctan\left(x\sqrt{\frac{ab}{a}}\right)$　　[if $ab > 0$]　　(see also **2.141** 2)

$\displaystyle \qquad\quad = \frac{1}{2i\sqrt{ab}}\ln\frac{a + xi\sqrt{ab}}{a - xi\sqrt{ab}}$　　[if $ab < 0$]　　(see also **2.143** 2 and **2.143** 3)

2. $\displaystyle \int \frac{x\,dx}{z_2^m} = -\frac{1}{2b(m-1)z_2^{m-1}}$　　(see also **2.145** 2, **2.145** 6, and **2.18**)

**2.134**

1.  $$\int \frac{dx}{z_4^2} = \frac{x}{4az_4} + \frac{3}{4a}\int \frac{dx}{z_4}$$    (see **2.132** 1)

2.  $$\int \frac{x\,dx}{z_4^2} = \frac{x^2}{4az_4} + \frac{1}{2a}\int \frac{x\,dx}{z_4}$$    (see **2.132** 2)

3.  $$\int \frac{x^2\,dx}{z_4^2} = \frac{x^3}{4az_4} + \frac{1}{4a}\int \frac{x^2\,dx}{z_4}$$    (see **2.132** 3)

4.  $$\int \frac{x^3\,dx}{z_4^2} = \frac{x^4}{4az_4} = -\frac{1}{4bz_4}$$

**2.135**  $$\int \frac{dx}{x^n z_4^m} = -\frac{1}{(n-1)ax^{n-1}z_4^{m-1}} - \frac{b(4m+n-5)}{(n-1)a}\int \frac{dx}{x^{n-4}z_4^m}$$

For $n = 1$

$$\int \frac{dx}{xz_4^m} = \frac{1}{a}\int \frac{dx}{xz_4^{m-1}} - \frac{b}{a}\int \frac{dx}{x^{-3}z_4^m}$$

**2.136**

1.  $$\int \frac{dx}{xz_4} = \frac{\ln x}{a} - \frac{\ln z_4}{4a} = \frac{1}{4a}\ln \frac{x^4}{z_4}$$

2.  $$\int \frac{dx}{x^2 z_4} = -\frac{1}{ax} - \frac{b}{a}\int \frac{x^2\,dx}{z_4}$$    (see **2.132** 3)

## 2.14  Forms containing the binomial $1 \pm x^n$

**2.141**

1.  $$\int \frac{dx}{1+x} = \ln(1+x)$$

2.[11]  $$\int \frac{dx}{1+x^2} = \arctan x = -\arctan\left(\frac{1}{x}\right)$$    (see also **2.124** 1)

3.  $$\int \frac{dx}{1+x^3} = \frac{1}{3}\ln \frac{1+x}{\sqrt{1-x+x^2}} + \frac{1}{\sqrt{3}}\arctan \frac{x\sqrt{3}}{2-x}$$    (see also **2.126** 1)

4.  $$\int \frac{dx}{1+x^4} = \frac{1}{4\sqrt{2}}\ln \frac{1+x\sqrt{2}+x^2}{1-x\sqrt{2}+x^2} + \frac{1}{2\sqrt{2}}\arctan \frac{x\sqrt{2}}{1-x^2}$$

(see also **2.132** 1)

**2.142**  $$\int \frac{dx}{1+x^n} = -\frac{2}{n}\sum_{k=0}^{\frac{n}{2}-1} P_k \cos\left(\frac{2k+1}{n}\pi\right) + \frac{2}{n}\sum_{k=0}^{\frac{n}{2}-1} Q_k \sin\left(\frac{2k+1}{n}\pi\right)$$

for $n$ a positive even number
TI (43)a

$$= \frac{1}{n}\ln(1+x) - \frac{2}{n}\sum_{k=0}^{\frac{n-3}{2}} P_k \cos\left(\frac{2k+1}{n}\pi\right) + \frac{2}{n}\sum_{k=0}^{\frac{n-3}{2}} Q_k \sin\left(\frac{2k+1}{n}\pi\right)$$

for $n$ a positive odd number
TI (45)

where

$$P_k = \frac{1}{2} \ln \left( x^2 - 2x \cos \left( \frac{2k+1}{n} \pi \right) + 1 \right)$$

$$Q_k = \arctan \frac{x \sin \left( \frac{2k+1}{n} \pi \right)}{1 - x \cos \left( \frac{2k+1}{n} \pi \right)} = \arctan \frac{x - \cos \left( \frac{2k+1}{n} \pi \right)}{\sin \left( \frac{2k+1}{n} \pi \right)}$$

**2.143**

1.    $\displaystyle \int \frac{dx}{1-x} = -\ln(1-x)$

2.    $\displaystyle \int \frac{dx}{1-x^2} = \frac{1}{2} \ln \frac{1+x}{1-x} = \operatorname{arctanh} x$        $[-1 < x < 1]$      (see also **2.141** 1)

3.    $\displaystyle \int \frac{dx}{x^2-1} = \frac{1}{2} \ln \frac{x-1}{x+1} = -\operatorname{arccoth} x$        $[x > 1, \quad x < -1]$

4.    $\displaystyle \int \frac{dx}{1-x^3} = \frac{1}{3} \ln \frac{\sqrt{1+x+x^2}}{1-x} + \frac{1}{\sqrt{3}} \arctan \frac{x\sqrt{3}}{2+x}$      (see also **2.126** 1)

5.    $\displaystyle \int \frac{dx}{1-x^4} = \frac{1}{4} \ln \frac{1+x}{1-x} + \frac{1}{2} \arctan x = \frac{1}{2} \left( \operatorname{arctanh} x + \arctan x \right)$

                                          (see also **2.132** 1)

**2.144**

1.    $\displaystyle \int \frac{dx}{1-x^n} = \frac{1}{n} \ln \frac{1+x}{1-x} - \frac{2}{n} \sum_{k=1}^{\frac{n}{2}-1} P_k \cos \frac{2k}{n} \pi + \frac{2}{n} \sum_{k=1}^{\frac{n}{2}-1} Q_k \sin \frac{2k}{n} \pi$

                        for $n$ a positive even number      TI (47)

    where   $\displaystyle P_k = \frac{1}{2} \ln \left( x^2 + 2x \cos \frac{2k+1}{n} \pi + 1 \right), \qquad Q_k = \arctan \frac{x + \cos \frac{2k+1}{n} \pi}{\sin \frac{2k+1}{n} \pi}$

2.    $\displaystyle \int \frac{dx}{1-x^n} = -\frac{1}{n} \ln(1-x) + \frac{2}{n} \sum_{k=0}^{\frac{n-3}{2}} P_k \cos \frac{2k+1}{n} \pi + \frac{2}{n} \sum_{k=0}^{\frac{n-3}{2}} Q_k \sin \frac{2k+1}{n} \pi$

                        for $n$ a positive odd number      TI (49)

    where   $\displaystyle P_k = \frac{1}{2} \ln \left( x^2 - 2x \cos \frac{2k}{n} \pi + 1 \right), \qquad Q_k = \arctan \frac{x - \cos \frac{2k}{n} \pi}{\sin \frac{2k}{n} \pi}$

**2.145**

1.    $\displaystyle \int \frac{x \, dx}{1+x} = x - \ln(1+x)$

2.    $\displaystyle \int \frac{x \, dx}{1+x^2} = \frac{1}{2} \ln \left( 1 + x^2 \right)$

3.    $\displaystyle \int \frac{x \, dx}{1+x^3} = -\frac{1}{6} \ln \frac{(1+x)^2}{1-x+x^2} + \frac{1}{\sqrt{3}} \arctan \frac{2x-1}{\sqrt{3}}$      (see also **2.126** 2)

4. $\displaystyle\int \frac{x\,dx}{1+x^4} = \frac{1}{2}\arctan x^2$

5. $\displaystyle\int \frac{x\,dx}{1-x} = -\ln(1-x) - x$

6. $\displaystyle\int \frac{x\,dx}{1-x^2} = -\frac{1}{2}\ln\left(1-x^2\right)$

7. $\displaystyle\int \frac{x\,dx}{1-x^3} = -\frac{1}{6}\ln\frac{(1-x)^2}{1+x+x^2} - \frac{1}{\sqrt{3}}\arctan\frac{2x+1}{\sqrt{3}}$   (see also **2.126** 2)

8. $\displaystyle\int \frac{x\,dx}{1-x^4} = \frac{1}{4}\ln\frac{1+x^2}{1-x^2}$   (see also **2.132** 2)

**2.146**  For $m$ and $n$ natural numbers.

1. $\displaystyle\int \frac{x^{m-1}\,dx}{1+x^{2n}} = -\frac{1}{2n}\sum_{k=1}^{n}\cos\frac{m\pi(2k-1)}{2n}\ln\left(1-2x\cos\frac{2k-1}{2n}\pi + x^2\right)$
$$+\frac{1}{n}\sum_{k=1}^{n}\sin\frac{m\pi(2k-1)}{2n}\arctan\left(\frac{x-\cos\frac{2k-1}{2n}\pi}{\sin\frac{2k-1}{2n}\pi}\right)$$
$$[m<2n] \qquad\qquad\qquad \text{TI (44)a}$$

2. $\displaystyle\int \frac{x^{m-1}\,dx}{1+x^{2n+1}} = (-1)^{m+1}\frac{\ln(1+x)}{2n+1} - \frac{1}{2n+1}\sum_{k=1}^{n}\cos\frac{m\pi(2k-1)}{2n+1}\ln\left(1-2x\cos\frac{2k-1}{2n+1}\pi + x^2\right)$
$$+\frac{2}{2n+1}\sum_{k=1}^{n}\sin\frac{m\pi(2k-1)}{2n+1}\arctan\left(\frac{x-\cos\frac{2k-1}{2n+1}\pi}{\sin\frac{2k-1}{2n+1}\pi}\right)$$
$$[m\le 2n] \qquad\qquad\qquad \text{TI (46)a}$$

3.[11] $\displaystyle\int \frac{x^{m-1}\,dx}{1-x^{2n}} = \frac{1}{2n}\left\{(-1)^{m+1}\ln(1+x) - \ln(1-x)\right\} - \frac{1}{2n}\sum_{k=1}^{n-1}\cos\frac{km\pi}{n}\ln\left(1-2x\cos\frac{k\pi}{n} + x^2\right)$
$$+\frac{1}{n}\sum_{k=1}^{n-1}\sin\frac{km\pi}{n}\arctan\left(\frac{x-\cos\frac{k\pi}{n}}{\sin\frac{k\pi}{n}}\right)$$
$$[m<2n] \qquad\qquad\qquad \text{TI (48)}$$

4. $\displaystyle\int \frac{x^{m-1}\,dx}{1-x^{2n+1}} = -\frac{1}{2n+1}\ln(1-x)$
$$+(-1)^{m+1}\frac{1}{2n+1}\sum_{k=1}^{n}\cos\frac{m\pi(2k-1)}{2n+1}\ln\left(1+2x\cos\frac{2k-1}{2n+1}\pi + x^2\right)$$
$$+(-1)^{m+1}\frac{2}{2n+1}\sum_{k=1}^{n}\sin\frac{m\pi(2k-1)}{2n+1}\arctan\left(\frac{x+\cos\frac{2k-1}{2n+1}\pi}{\sin\frac{2k-1}{2n+1}\pi}\right)$$
$$[m\le 2n] \qquad\qquad\qquad \text{TI (50)}$$

**2.147**

1. $\displaystyle\int \frac{x^m\,dx}{1-x^{2n}} = \frac{1}{2}\int \frac{x^m\,dx}{1-x^n} + \frac{1}{2}\int \frac{x^m\,dx}{1+x^n}$

2. $\displaystyle\int \frac{x^m\,dx}{(1+x^2)^n} = -\frac{1}{2n-m-1}\cdot\frac{x^{m-1}}{(1+x^2)^{n-1}} + \frac{m-1}{2n-m-1}\int \frac{x^{m-2}\,dx}{(1+x^2)^n}$   LA 139 (28)

3.    $\displaystyle \int \frac{x^m}{1+x^2}\,dx = \frac{x^{m-1}}{m-1} - \int \frac{x^{m-2}}{1+x^2}\,dx$

4.    $\displaystyle \int \frac{x^m\,dx}{(1-x^2)^n} = \frac{1}{2n-m-1}\frac{x^{m-1}}{(1-x^2)^{n-1}} - \frac{m-1}{2n-m-1}\int \frac{x^{m-2}\,dx}{(1-x^2)^n}$

$\displaystyle \qquad\qquad = \frac{1}{2n-2}\frac{x^{m-1}}{(1-x^2)^{n-1}} - \frac{m-1}{2n-2}\int \frac{x^{m-2}\,dx}{(1-x^2)^{n-1}}$

LA 139 (33)

5.    $\displaystyle \int \frac{x^m\,dx}{1-x^2} = -\frac{x^{m-1}}{m-1} + \int \frac{x^{m-2}\,dx}{1-x^2}$

**2.148**

1.    $\displaystyle \int \frac{dx}{x^m(1+x^2)^n} = -\frac{1}{m-1}\frac{1}{x^{m-1}(1+x^2)^{n-1}} - \frac{2n+m-3}{m-1}\int \frac{dx}{x^{m-2}(1+x^2)^n}$    LA 139 (29)

For $m=1$

$\displaystyle \int \frac{dx}{x(1+x^2)^n} = \frac{1}{2n-2}\frac{1}{(1+x^2)^{n-1}} + \int \frac{dx}{x(1+x^2)^{n-1}}$    LA 139 (31)

For $m=1$ and $n=1$

$\displaystyle \int \frac{dx}{x(1+x^2)} = \ln\frac{x}{\sqrt{1+x^2}}$

2.    $\displaystyle \int \frac{dx}{x^m(1+x^2)} = -\frac{1}{(m-1)x^{m-1}} - \int \frac{dx}{x^{m-2}(1+x^2)}$

3.    $\displaystyle \int \frac{dx}{(1+x^2)^n} = \frac{1}{2n-2}\frac{x}{(1+x^2)^{n-1}} + \frac{2n-3}{2n-2}\int \frac{dx}{(1+x^2)^{n-1}}$    FI II 40

4.    $\displaystyle \int \frac{dx}{(1+x^2)^n} = \frac{x}{2n-1}\sum_{k=1}^{n-1}\frac{(2n-1)(2n-3)(2n-5)\cdots(2n-2k+1)}{2^k(n-1)(n-2)\ldots(n-k)(1+x^2)^{n-k}} + \frac{(2n-3)!!}{2^{n-1}(n-1)!}\arctan x$

TI (91)

**2.149**

1.    $\displaystyle \int \frac{dx}{x^m(1-x^2)^n} = -\frac{1}{(m-1)x^{m-1}(1-x^2)^{n-1}} + \frac{2n+m-3}{m-1}\int \frac{dx}{x^{m-2}(1-x^2)^n}$    LA 139 (34)

For $m=1$

$\displaystyle \int \frac{dx}{x(1-x^2)^n} = \frac{1}{2(n-1)(1-x^2)^{n-1}} + \int \frac{dx}{x(1-x^2)^{n-1}}$    LA 139 (36)

For $m=1$ and $n=1$

$\displaystyle \int \frac{dx}{x(1-x^2)} = \ln\frac{x}{\sqrt{1-x^2}}$

2.    $\displaystyle \int \frac{dx}{(1-x^2)^n} = \frac{1}{2n-2}\frac{x}{(1-x^2)^{n-1}} + \frac{2n-3}{2n-2}\int \frac{dx}{(1-x^2)^{n-1}}$    LA 139 (35)

3.    $\displaystyle \int \frac{dx}{(1-x^2)^n} = \frac{x}{2n-1}\sum_{k=1}^{n-1}\frac{(2n-1)(2n-3)(2n-5)\ldots(2n-2k+1)}{2^k(n-1)(n-2)\ldots(n-k)(1-x^2)^{n-k}} + \frac{(2n-3)!!}{2^n\cdot(n-1)!}\ln\frac{1+x}{1-x}$

TI (91)

## 2.15 Forms containing pairs of binomials: $a + bx$ and $\alpha + \beta x$

**Notation:** $z = a + bx;$    $t = \alpha + \beta x;$    $\Delta = a\beta - \alpha b$

**2.151** $\displaystyle \int z^n t^m \, dx = \frac{z^{n+1} t^m}{(m+n+1)b} - \frac{m\Delta}{(m+n+1)b} \int z^n t^{m-1} \, dx$

**2.152**

1.    $\displaystyle \int \frac{z}{t} \, dx = \frac{bx}{\beta} + \frac{\Delta}{\beta^2} \ln t$

2.    $\displaystyle \int \frac{t}{z} \, dx = \frac{\beta x}{b} - \frac{\Delta}{b^2} \ln z$

**2.153** $\displaystyle \int \frac{t^m \, dx}{z^n} = \frac{1}{(m-n+1)b} \frac{t^m}{z^{n-1}} - \frac{m\Delta}{(m-n+1)b} \int \frac{t^{m-1} \, dx}{z^n}$

$\displaystyle = \frac{1}{(n-1)\Delta} \frac{t^{m+1}}{z^{n-1}} - \frac{(m-n+2)\beta}{(n-1)\Delta} \int \frac{t^m \, dx}{z^{n-1}}$

$\displaystyle = -\frac{1}{(n-1)b} \frac{t^m}{z^{n-1}} + \frac{m\beta}{(n-1)b} \int \frac{t^{m-1}}{z^{n-1}} \, dx$

**2.154** $\displaystyle \int \frac{dx}{zt} = \frac{1}{\Delta} \ln \frac{t}{z}$

**2.155** $\displaystyle \int \frac{dx}{z^n t^m} = -\frac{1}{(m-1)\Delta} \frac{1}{t^{m-1}z^{n-1}} - \frac{(m+n-2)b}{(m-1)\Delta} \int \frac{dx}{t^{m-1}z^n}$

$\displaystyle = \frac{1}{(n-1)\Delta} \frac{1}{t^{m-1}z^{n-1}} + \frac{(m+n-2)\beta}{(n-1)\Delta} \int \frac{dx}{t^m z^{n-1}}$

**2.156** $\displaystyle \int \frac{x \, dx}{zt} = \frac{1}{\Delta} \left( \frac{a}{b} \ln z - \frac{\alpha}{\beta} \ln t \right)$

## 2.16 Forms containing the trinomial $a + bx^k + cx^{2k}$

**2.160**    Reduction formulas for $R_k = a + bx^k + cx^{2k}$.

1.    $\displaystyle \int x^{m-1} R_k^n \, dx = \frac{x^m R_k^{n+1}}{ma} - \frac{(m+k+nk)b}{ma} \int x^{m+k-1} R_k^n \, dx - \frac{(m+2k+2kn)c}{ma} \int x^{m+2k-1} R_k^n \, dx$

2.    $\displaystyle \int x^{m-1} R_k^n \, dx = \frac{x^m R_k^n}{m} - \frac{bkn}{m} \int x^{m+k-1} R_k^{n-1} \, dx - \frac{2ckn}{m} \int x^{m+2k-1} R_k^{n-1} \, dx$

3.    $\displaystyle \int x^{m-1} R_k^n \, dx = \frac{x^{m-2k} R_k^{n+1}}{(m+2kn)c} - \frac{(m-2k)a}{(m+2kn)c} \int x^{m-2k-1} R_k^n \, dx - \frac{(m-k+kn)b}{(m+2kn)c} \int x^{m-k-1} R_k^n \, dx$

$\displaystyle = \frac{x^m R_k^n}{m+2kn} + \frac{2kna}{m+2kn} \int x^{m-1} R_k^{n-1} \, dx + \frac{bkn}{m+2kn} \int x^{m+k-1} R_k^{n-1} \, dx$

**2.161**    Forms containing the trinomial $R_2 = a + bx^2 + cx^4$.

**Notation:** $f = \dfrac{b}{2} - \dfrac{1}{2}\sqrt{b^2 - 4ac}, \quad g = \dfrac{b}{2} + \dfrac{1}{2}\sqrt{b^2 - 4ac},$

$h = \sqrt{b^2 - 4ac}, \quad q = \sqrt[4]{\dfrac{a}{c}}, \quad l = 2a(n-1)\left(b^2 - 4ac\right), \quad \cos\alpha = -\dfrac{b}{2\sqrt{ac}}$

1.  $\displaystyle\int \frac{\mathrm{d}x}{R_2}$

$$= \frac{c}{h}\left\{\int \frac{\mathrm{d}x}{cx^2+f} - \int \frac{\mathrm{d}x}{cx^2+g}\right\} \qquad \left[h^2 > 0\right] \quad \text{LA 146 (5)}$$

$$= \frac{1}{4cq^3\sin\alpha}\left\{\sin\frac{\alpha}{2}\ln\frac{x^2+2qx\cos\frac{\alpha}{2}+q^2}{x^2-2qx\cos\frac{\alpha}{2}+q^2} + 2\cos\frac{\alpha}{2}\arctan\frac{x^2-q^2}{2qx\sin\frac{\alpha}{2}}\right\} \qquad \left[h^2 < 0\right] \quad \text{LA 146 (8)a}$$

2.  $\displaystyle\int \frac{x\,\mathrm{d}x}{R_2} = \frac{1}{2h}\ln\frac{cx^2+f}{cx^2+g} \qquad \left[h^2 > 0\right]$　　　　LA 146 (6)

$$= \frac{1}{2cq^2\sin\alpha}\arctan\frac{x^2-q^2\cos\alpha}{q^2\sin\alpha} \qquad \left[h^2 < 0\right] \qquad \text{LA 146 (9)a}$$

3.  $\displaystyle\int \frac{x^2\,\mathrm{d}x}{R_2} = \frac{g}{h}\int\frac{\mathrm{d}x}{cx^2+g} - \frac{f}{h}\int\frac{\mathrm{d}x}{cx^2+f} \qquad \left[h^2 > 0\right]$　　　　LA 146 (7)

4.  $\displaystyle\int \frac{\mathrm{d}x}{R_2^2} = \frac{bcx^3 + (b^2-2ac)\,x}{lR_2} + \frac{b^2-6ac}{l}\int\frac{\mathrm{d}x}{R_2} + \frac{bc}{l}\int\frac{x^2\,\mathrm{d}x}{R_2}$

5.[12]  $\displaystyle\int \frac{\mathrm{d}x}{R_2^n} = \frac{bcx^3 + (b^2-2ac)\,x}{lR_2^{n-1}} + \frac{(4n-7)bc}{l}\int\frac{x^2\,\mathrm{d}x}{R_2^{n-1}} + \frac{2(n-1)h^2 + 2ac - b^2}{l}\int\frac{\mathrm{d}x}{R_2^{n-1}}$

$$\left[n > 1\right] \qquad \text{LA 146}$$

6.[12]  $\displaystyle\int \frac{\mathrm{d}x}{x^m R_2^n} = -\frac{1}{(m-1)ax^{m-1}R_2^{n-1}} - \frac{(m+2n-3)b}{(m-1)a}\int\frac{\mathrm{d}x}{x^{m-2}R_2^n} - \frac{(m+4n-5)b}{(m-1)a}\int\frac{\mathrm{d}x}{x^{m-4}R_2^n}$

$$\text{LA 147 (12)a}$$

## 2.17 Forms containing the quadratic trinomial $a + bx + cx^2$ and powers of $x$

**Notation:** $R = a + bx + cx^2$;　$\Delta = 4ac - b^2$

**2.171**

1.  $\displaystyle\int x^{m+1}R^n\,\mathrm{d}x = \frac{x^m R^{n+1}}{c(m+2n+2)} - \frac{am}{c(m+2n+2)}\int x^{m-1}R^n\,\mathrm{d}x - \frac{b(m+n+1)}{c(m+2n+2)}\int x^m R^n\,\mathrm{d}x$

$$\text{TI (97)}$$

2.  $\displaystyle\int \frac{R^n\,\mathrm{d}x}{x^{m+1}} = -\frac{R^{n+1}}{amx^m} + \frac{b(n-m+1)}{am}\int\frac{R^n\,\mathrm{d}x}{x^m} + \frac{c(2n-m+2)}{am}\int\frac{R^n\,\mathrm{d}x}{x^{m-1}}$　　LA 142(3), TI (96)a

3.  $\displaystyle\int \frac{\mathrm{d}x}{R^{n+1}} = \frac{b+2cx}{n\Delta R^n} + \frac{(4n-2)c}{n\Delta}\int\frac{\mathrm{d}x}{R^n}$　　　　TI (94)a

4.[12]  $\displaystyle\int \frac{\mathrm{d}x}{R^{n+1}} = \frac{(2cx+b)}{2n+1}\sum_{k=0}^{n-1}\frac{2^k(2n+1)(2n-1)(2n-3)\dots(2n-2k+1)c^k}{n(n-1)\cdots(n-k)\Delta^{k+1}R^{n-k}} + 2^n\frac{(2n-1)!!c^n}{n!\Delta^n}\int\frac{\mathrm{d}x}{R}$

$$\text{TI (96)a}$$

**2.172**[12] $\displaystyle\int \frac{\mathrm{d}x}{R} = \frac{1}{\sqrt{-\Delta}}\ln\frac{(b+2cx)-\sqrt{-\Delta}}{(b+2cx)+\sqrt{-\Delta}} = \frac{-2}{\sqrt{-\Delta}}\operatorname{arctanh}\frac{b+2cx}{\sqrt{-\Delta}} \qquad \left[\text{for } \Delta < 0\right]$

$$= \frac{-2}{b+2cx} \qquad \left[\text{for } \Delta = 0,\ b \text{ and } c \text{ non-zero}\right]$$

$$= \frac{2}{\sqrt{\Delta}}\arctan\frac{b+2cx}{\sqrt{\Delta}} \qquad \left[\text{for } \Delta > 0\right]$$

**2.173**

1. $\int \dfrac{\mathrm{d}x}{R^2} = \dfrac{b+2cx}{\Delta R} + \dfrac{2c}{\Delta}\int \dfrac{\mathrm{d}x}{R}$  (see **2.172**)

2. $\int \dfrac{\mathrm{d}x}{R^3} = \dfrac{b+2cx}{\Delta}\left\{\dfrac{1}{2R^2} + \dfrac{3c}{\Delta R}\right\} + \dfrac{6c^2}{\Delta^2}\int \dfrac{\mathrm{d}x}{R}$  (see **2.172**)

**2.174**

1. $\int \dfrac{x^m\,\mathrm{d}x}{R^n} = -\dfrac{x^{m-1}}{(2n-m-1)cR^{n-1}} - \dfrac{(n-m)b}{(2n-m-1)c}\int \dfrac{x^{m-1}\,\mathrm{d}x}{R^n} + \dfrac{(m-1)a}{(2n-m-1)c}\int \dfrac{x^{m-2}\,\mathrm{d}x}{R^n}$

   For $m = 2n-1$, this formula is inapplicable. Instead, we may use

2. $\int \dfrac{x^{2n-1}\,\mathrm{d}x}{R^n} = \dfrac{1}{c}\int \dfrac{x^{2n-3}\,\mathrm{d}x}{R^{n-1}} - \dfrac{a}{c}\int \dfrac{x^{2n-3}\,\mathrm{d}x}{R^n} - \dfrac{b}{c}\int \dfrac{x^{2n-2}\,\mathrm{d}x}{R^n}$

**2.175**

1. $\int \dfrac{x\,\mathrm{d}x}{R} = \dfrac{1}{2c}\ln R - \dfrac{b}{2c}\int \dfrac{\mathrm{d}x}{R}$  (see **2.172**)

2. $\int \dfrac{x\,\mathrm{d}x}{R^2} = -\dfrac{2a+bx}{\Delta R} - \dfrac{b}{\Delta}\int \dfrac{\mathrm{d}x}{R}$  (see **2.172**)

3. $\int \dfrac{x\,\mathrm{d}x}{R^3} = -\dfrac{2a+bx}{2\Delta R^2} - \dfrac{3b(b+2cx)}{2\Delta^2 R} - \dfrac{3bc}{\Delta^2}\int \dfrac{\mathrm{d}x}{R}$  (see **2.172**)

4. $\int \dfrac{x^2\,\mathrm{d}x}{R} = \dfrac{x}{c} - \dfrac{b}{2c^2}\ln R + \dfrac{b^2-2ac}{2c^2}\int \dfrac{\mathrm{d}x}{R}$  (see **2.172**)

5. $\int \dfrac{x^2\,\mathrm{d}x}{R^2} = \dfrac{ab+(b^2-2ac)\,x}{c\Delta R} + \dfrac{2a}{\Delta}\int \dfrac{\mathrm{d}x}{R}$  (see **2.172**)

6. $\int \dfrac{x^2\,\mathrm{d}x}{R^3} = \dfrac{ab+(b^2-2ac)\,x}{2c\Delta R^2} + \dfrac{(2ac+b^2)(b+2cx)}{2c\Delta^2 R} + \dfrac{2ac+b^2}{\Delta^2}\int \dfrac{\mathrm{d}x}{R}$

   (see **2.172**)

7. $\int \dfrac{x^3\,\mathrm{d}x}{R} = \dfrac{x^2}{2c} - \dfrac{bx}{c^2} + \dfrac{b^2-ac}{2c^3}\ln R - \dfrac{b\left(b^2-3ac\right)}{2c^3}\int \dfrac{\mathrm{d}x}{R}$

   (see **2.172**)

8. $\int \dfrac{x^3\,\mathrm{d}x}{R^2} = \dfrac{1}{2c^2}\ln R + \dfrac{a\left(2ac-b^2\right)+b\left(3ac-b^2\right)x}{c^2\Delta R} - \dfrac{b\left(6ac-b^2\right)}{2c^2\Delta}\int \dfrac{\mathrm{d}x}{R}$

   (see **2.172**)

9. $\int \dfrac{x^3\,\mathrm{d}x}{R^3} = -\left(\dfrac{x^2}{c} + \dfrac{abx}{c\Delta} + \dfrac{2a^2}{c\Delta}\right)\dfrac{1}{2R^2} - \dfrac{3ab}{2c\Delta}\int \dfrac{\mathrm{d}x}{R^2}$  (see **2.173** 1)

**2.176** $\int \dfrac{\mathrm{d}x}{x^m R^n} = \dfrac{-1}{(m-1)ax^{m-1}R^{n-1}} - \dfrac{b(m+n-2)}{a(m-1)}\int \dfrac{\mathrm{d}x}{x^{m-1}R^n} - \dfrac{c(m+2n-3)}{a(m-1)}\int \dfrac{\mathrm{d}x}{x^{m-2}R^n}$

**2.177**

1. $\int \dfrac{\mathrm{d}x}{xR} = \dfrac{1}{2a}\ln \dfrac{x^2}{R} - \dfrac{b}{2a}\int \dfrac{\mathrm{d}x}{R}$  (see **2.172**)

2. $$\int \frac{\mathrm{d}x}{xR^2} = \frac{1}{2a^2} \ln \frac{x^2}{R} + \frac{1}{2aR}\left\{1 - \frac{b(b+2cx)}{\Delta}\right\} - \frac{b}{2a^2}\left(1 + \frac{2ac}{\Delta}\right)\int \frac{\mathrm{d}x}{R}$$

(see **2.172**)

3. $$\int \frac{\mathrm{d}x}{xR^3} = \frac{1}{4aR^2} + \frac{1}{2a^2R} + \frac{1}{2a^3}\ln \frac{x^2}{R} - \frac{b}{2a}\int \frac{\mathrm{d}x}{R^3} - \frac{b}{2a^2}\int \frac{\mathrm{d}x}{R^2} - \frac{b}{2a^3}\int \frac{\mathrm{d}x}{R}$$

(see **2.172, 2.173**)

4. $$\int \frac{\mathrm{d}x}{x^2R} = -\frac{b}{2a^2}\ln \frac{x^2}{R} - \frac{1}{ax} + \frac{b^2 - 2ac}{2a^2}\int \frac{\mathrm{d}x}{R}$$    (see **2.172**)

5. $$\int \frac{\mathrm{d}x}{x^2R^2} = -\frac{b}{a^3}\ln \frac{x^2}{R} - \frac{a+bx}{a^2xR} + \frac{(b^2-3ac)(b+2cx)}{a^2\Delta R} - \frac{1}{\Delta}\left(\frac{b^4}{a^3} - \frac{6b^2c}{a^2} + \frac{6c^2}{a}\right)\int \frac{\mathrm{d}x}{R}$$

(see **2.172**)

6. $$\int \frac{\mathrm{d}x}{x^2R^3} = -\frac{1}{axR^2} - \frac{3b}{a}\int \frac{\mathrm{d}x}{xR^3} - \frac{5c}{a}\int \frac{\mathrm{d}x}{R^3}$$    (see **2.173** and **2.177** 3)

7. $$\int \frac{\mathrm{d}x}{x^3R} = -\frac{ac-b^2}{2a^3}\ln \frac{x^2}{R} + \frac{b}{a^2x} - \frac{1}{2ax^2} + \frac{b(3ac-b^2)}{2a^3}\int \frac{\mathrm{d}x}{R}$$

(see **2.172**)

8. $$\int \frac{\mathrm{d}x}{x^3R^2} = \left(-\frac{1}{2ax^2} + \frac{3b}{2a^2x}\right)\frac{1}{R} + \left(\frac{3b^2}{a^2} - \frac{2c}{a}\right)\int \frac{\mathrm{d}x}{xR^2} + \frac{9bc}{2a^2}\int \frac{\mathrm{d}x}{R^2}$$

(see **2.173** 1 and **2.177** 2)

9. $$\int \frac{\mathrm{d}x}{x^3R^3} = \left(\frac{-1}{2ax^2} + \frac{2b}{a^2x}\right)\frac{1}{R^2} + \left(\frac{6b^2}{a^2} - \frac{3c}{a}\right)\int \frac{\mathrm{d}x}{xR^3} + \frac{10bc}{a^2}\int \frac{\mathrm{d}x}{R^3}$$

(see **2.173** 2 and **2.177** 3)

## 2.18 Forms containing the quadratic trinomial $a+bx+cx^2$ and the binomial $\alpha+\beta x$

**Notation:** $R = a + bx + cx^2$;  $z = \alpha + \beta x$;  $A = a\beta^2 - \alpha b\beta + c\alpha^2$;
$B = b\beta - 2c\alpha$;  $\Delta = 4ac - b^2$

1. $$\int z^m R^n \, \mathrm{d}x = \frac{\beta z^{m-1}R^{n+1}}{(m+2n+1)c} - \frac{(m+n)B}{(m+2n+1)c}\int z^{m-1}R^n \, \mathrm{d}x - \frac{(m-1)A}{(m+2n+1)c}\int z^{m-2}R^n \, \mathrm{d}x$$

2. $$\int \frac{R^n \, \mathrm{d}x}{z^m} = -\frac{1}{(m-2n-1)\beta}\frac{R^n}{z^{m-1}} - \frac{2nA}{(m-2n-1)\beta^2}\int \frac{R^{n-1} \, \mathrm{d}x}{z^m}$$
$$- \frac{nB}{(m-2n-1)\beta^2}\int \frac{R^{n-1} \, \mathrm{d}x}{z^{m-1}};$$    LA 184 (4)a

$$= \frac{-\beta}{(m-1)A}\frac{R^{n+1}}{z^{m-1}} - \frac{(m-n-2)B}{(m-1)A}\int \frac{R^n \, \mathrm{d}x}{z^{m-1}} - \frac{(m-2n-3)c}{(m-1)A}\int \frac{R^n \, \mathrm{d}x}{z^{m-2}}$$    LA 148 (5)

$$= -\frac{1}{(m-1)\beta}\frac{R^n}{z^{m-1}} + \frac{nB}{(m-1)\beta^2}\int \frac{R^{n-1} \, \mathrm{d}x}{z^{m-1}} + \frac{2nc}{(m-1)\beta^2}\int \frac{R^{n-1} \, \mathrm{d}x}{z^{m-2}}$$    LA 418 (6)

3.     $\displaystyle \int \frac{z^m \, dx}{R^n} = \frac{\beta}{(m-2n+1)c} \frac{z^{m-1}}{R^{n-1}} - \frac{(m-n)B}{(m-2n+1)c} \int \frac{z^{m-1} \, dx}{R^n} - \frac{(m-1)A}{(m-2n+1)c} \int \frac{z^{m-2} \, dx}{R^n}$

LA 147 (1)

$$= \frac{b+2cx}{(n-1)\Delta} \frac{z^m}{R^{n-1}} - \frac{2(m-2n+3)c}{(n-1)\Delta} \int \frac{z^m \, dx}{R^{n-1}} - \frac{Bm}{(n-1)\Delta} \int \frac{z^{m-1} \, dx}{R^{n-1}}$$

LA 148 (3)

4.[3]     $\displaystyle \int \frac{dx}{z^m R^n} = -\frac{\beta}{(m-1)A} \frac{1}{z^{m-1} R^{n-1}} - \frac{(m+n-2)B}{(m-1)A} \int \frac{dx}{z^{m-1} R^n} - \frac{(m+2n-3)c}{(m-1)A} \int \frac{dx}{z^{m-2} R^n}$

LA 148 (7)

$$= \frac{\beta}{2(n-1)A} \frac{1}{z^{m-1} R^{n-1}} - \frac{B}{2A} \int \frac{dx}{z^{m-1} R^n} + \frac{(m+2n-3)\beta^2}{2(n-1)A} \int \frac{dx}{z^m R^{n-1}}$$

LA 148 (8)

For $m=1$ and $n=1$

$$\int \frac{dx}{zR} = \frac{\beta}{2A} \ln \frac{z^2}{R} - \frac{B}{2A} \int \frac{dx}{R}$$

For $A=0$

$$\int \frac{dx}{z^m R^n} = -\frac{\beta}{(m+n-1)B} \frac{1}{z^m R^{n-1}} - \frac{(m+2n-2)c}{(m+n-1)B} \int \frac{dx}{z^{m-1} R^n}$$

LA 148 (9)

## 2.2 Algebraic functions

### 2.20 Introduction

**2.201**    The integrals $\displaystyle \int R\left(x, \left(\frac{\alpha x + \beta}{\gamma x + \delta}\right)^r, \left(\frac{\alpha x + \beta}{\gamma x + \delta}\right)^s, \ldots\right) dx$, where $r, s, \ldots$ are rational numbers, can be reduced to integrals of rational functions by means of the substitution

$$\frac{\alpha x + \beta}{\gamma x + \delta} = t^m,$$

FI II 57

where $m$ is the common denominator of the fractions $r, s, \ldots$.

**2.202**    Integrals of the form $\displaystyle \int x^m (a + bx^n)^p \, dx$,* where $m$, $n$, and $p$ are rational numbers, can be expressed in terms of elementary functions only in the following cases:

(a)     When $p$ is an integer; then, this integral takes the form of a sum of the integrals shown in **2.201**;

(b)     When $\frac{m+1}{n}$ is an integer: by means of the substitution $x^n = z$, this integral can be transformed to the form $\displaystyle \frac{1}{n} \int (a + bz)^p z^{\frac{m+1}{n} - 1} \, dz$, which we considered in **2.201**;

(c)     When $\frac{m+1}{n} + p$ is an integer; by means of the same substitution $x^n = z$, this integral can be reduced to an integral of the form $\displaystyle \frac{1}{n} \int \left(\frac{a + bz}{z}\right)^p z^{\frac{m+1}{n} + p - 1} \, dz$, considered in **2.201**;

For reduction formulas for integrals of binomial differentials, see **2.110**.

---

*Translator: The authors term such integrals "integrals of binomial differentials".

## 2.21 Forms containing the binomial $a + bx^k$ and $\sqrt{x}$

**Notation:** $z_1 = a + bx$.

**2.211**
$$\int \frac{dx}{z_1 \sqrt{x}} = \frac{2}{\sqrt{ab}} \arctan \sqrt{\frac{bx}{a}} \qquad\qquad [ab > 0]$$

$$= \frac{1}{i\sqrt{ab}} \ln \frac{a - bx + 2i\sqrt{xab}}{z_1} \qquad\qquad [ab < 0]$$

**2.212**
$$\int \frac{x^m \sqrt{x}}{z_1} \, dx = 2\sqrt{x} \sum_{k=0}^{m} \frac{(-1)^k a^k x^{m-k}}{(2m - 2k + 1)b^{k+1}} + (-1)^{m+1} \frac{a^{m+1}}{b^{m+1}} \int \frac{dx}{z_1 \sqrt{x}}$$

$$\text{(see **2.211**)}$$

**2.213**

1. $$\int \frac{\sqrt{x}\,dx}{z_1} = \frac{2\sqrt{x}}{b} - \frac{a}{b} \int \frac{dx}{z_1 \sqrt{x}} \qquad\qquad \text{(see **2.211**)}$$

2. $$\int \frac{x\sqrt{x}\,dx}{z_1} = \left( \frac{x}{3b} - \frac{a}{b^2} \right) 2\sqrt{x} + \frac{a^2}{b^2} \int \frac{dx}{z_1 \sqrt{x}} \qquad\qquad \text{(see **2.211**)}$$

3. $$\int \frac{x^2 \sqrt{x}\,dx}{z_1} = \left( \frac{x^2}{5b} - \frac{xa}{3b^2} + \frac{a^2}{b^3} \right) 2\sqrt{x} - \frac{a^3}{b^3} \int \frac{dx}{z_1 \sqrt{x}} \qquad\qquad \text{(see **2.211**)}$$

4. $$\int \frac{dx}{z_1^2 \sqrt{x}} = \frac{\sqrt{x}}{a z_1} + \frac{1}{2a} \int \frac{dx}{z_1 \sqrt{x}} \qquad\qquad \text{(see **2.211**)}$$

5. $$\int \frac{\sqrt{x}\,dx}{z_1^2} = -\frac{\sqrt{x}}{b z_1} + \frac{1}{2b} \int \frac{dx}{z_1 \sqrt{x}} \qquad\qquad \text{(see **2.211**)}$$

6. $$\int \frac{x\sqrt{x}\,dx}{z_1^2} = \frac{2x\sqrt{x}}{b z_1} - \frac{3a}{b} \int \frac{\sqrt{x}\,dx}{z_1^2} \qquad\qquad \text{(see **2.213** 5)}$$

7. $$\int \frac{x^2 \sqrt{x}\,dx}{z_1^2} = \left( \frac{x^2}{3b} - \frac{5ax}{3b^2} \right) \frac{2\sqrt{x}}{z_1} + \frac{5a^2}{b^2} \int \frac{\sqrt{x}\,dx}{z_1^2} \qquad\qquad \text{(see **2.213** 5)}$$

8. $$\int \frac{dx}{z_1^3 \sqrt{x}} = \left( \frac{1}{2a z_1^2} + \frac{3}{4a^2 z_1} \right) \sqrt{x} + \frac{3}{8a^2} \int \frac{dx}{z_1 \sqrt{x}} \qquad\qquad \text{(see **2.211**)}$$

9. $$\int \frac{\sqrt{x}\,dx}{z_1^3} = \left( -\frac{1}{2b z_1^2} + \frac{1}{4ab z_1} \right) \sqrt{x} + \frac{1}{8ab} \int \frac{dx}{z_1 \sqrt{x}} \qquad\qquad \text{(see **2.211**)}$$

10. $$\int \frac{x\sqrt{x}\,dx}{z_1^3} = -\frac{2x\sqrt{x}}{b z_1^2} + \frac{3a}{b} \int \frac{\sqrt{x}\,dx}{z_1^3} \qquad\qquad \text{(see **2.213** 9)}$$

11. $$\int \frac{x^2 \sqrt{x}\,dx}{z_1^3} = \left( \frac{x^2}{b} + \frac{5ax}{b^2} \right) \frac{2\sqrt{x}}{z_1^2} - \frac{15a^2}{b^2} \int \frac{\sqrt{x}\,dx}{z_1^3} \qquad\qquad \text{(see **2.213** 9)}$$

**Notation:** $z_2 = a + bx^2$, $\quad \alpha = \sqrt[4]{\dfrac{a}{b}}$, $\quad \alpha' = \sqrt[4]{-\dfrac{a}{b}}$.

**2.214**
$$\int \frac{dx}{z_2 \sqrt{x}} = \frac{1}{ba^3 \sqrt{2}} \left[ \ln \frac{x + \alpha\sqrt{2x} + \alpha^2}{\sqrt{z_2}} + \arctan \frac{\alpha\sqrt{2x}}{\alpha^2 - x} \right] \qquad \left[ \frac{a}{b} > 0 \right]$$

$$= \frac{1}{2b\alpha'^3} \left( \ln \frac{\alpha' - \sqrt{x}}{\alpha' + \sqrt{x}} - 2\arctan \frac{\sqrt{x}}{\alpha'} \right) \qquad \left[ \frac{a}{b} < 0 \right]$$

**2.215** $\displaystyle\int \frac{\sqrt{x}\,\mathrm{d}x}{z_2} = \frac{1}{b\alpha\sqrt{2}}\left[-\ln\frac{x + \alpha\sqrt{2x} + \alpha^2}{\sqrt{z_2}} + \arctan\frac{\alpha\sqrt{2x}}{\alpha^2 - x}\right]$  $\left[\dfrac{a}{b} > 0\right]$

$\displaystyle\qquad\qquad = \frac{1}{2b\alpha'}\left[\ln\frac{\alpha' - \sqrt{x}}{\alpha' + \sqrt{x}} + 2\arctan\frac{\sqrt{x}}{\alpha'}\right]$  $\left[\dfrac{a}{b} < 0\right]$

**2.216**

1.  $\displaystyle\int \frac{x\sqrt{x}\,\mathrm{d}x}{z_2} = \frac{2\sqrt{x}}{b} - \frac{a}{b}\int\frac{\mathrm{d}x}{z_2\sqrt{x}}$  (see **2.214**)

2.  $\displaystyle\int \frac{x^2\sqrt{x}\,\mathrm{d}x}{z_2} = \frac{2x\sqrt{x}}{3b} - \frac{a}{b}\int\frac{\sqrt{x}\,\mathrm{d}x}{z_2}$  (see **2.215**)

3.  $\displaystyle\int \frac{\mathrm{d}x}{z_2^2\sqrt{x}} = \frac{\sqrt{x}}{2az_2} + \frac{3}{4a}\int\frac{\mathrm{d}x}{z_2\sqrt{x}}$  (see **2.214**)

4.  $\displaystyle\int \frac{\sqrt{x}\,\mathrm{d}x}{z_2^2} = \frac{x\sqrt{x}}{2az_2} + \frac{1}{4a}\int\frac{\sqrt{x}\,\mathrm{d}x}{z_2}$  (see **2.215**)

5.  $\displaystyle\int \frac{x\sqrt{x}\,\mathrm{d}x}{z_2^2} = -\frac{\sqrt{x}}{2bz_2} + \frac{1}{4b}\int\frac{\mathrm{d}x}{z_2\sqrt{x}}$  (see **2.214**)

6.  $\displaystyle\int \frac{x^2\sqrt{x}\,\mathrm{d}x}{z_2^2} = -\frac{x\sqrt{x}}{2bz_2} + \frac{3}{4b}\int\frac{\sqrt{x}\,\mathrm{d}x}{z_2}$  (see **2.215**)

7.  $\displaystyle\int \frac{\mathrm{d}x}{z_2^3\sqrt{x}} = \left(\frac{1}{4az_2^2} + \frac{7}{16a^2z_2}\right)\sqrt{x} + \frac{21}{32a^2}\int\frac{\mathrm{d}x}{z_2\sqrt{x}}$  (see **2.214**)

8.  $\displaystyle\int \frac{\sqrt{x}\,\mathrm{d}x}{z_2^3} = \left(\frac{1}{4az_2^2} + \frac{5}{16a^2z_2}\right)x\sqrt{x} + \frac{5}{32a^2}\int\frac{\sqrt{x}\,\mathrm{d}x}{z_2}$  (see **2.215**)

9.  $\displaystyle\int \frac{x\sqrt{x}\,\mathrm{d}x}{z_2^3} = \frac{(bx^2 - 3a)\sqrt{x}}{16abz_2^2} + \frac{3}{32ab}\int\frac{\mathrm{d}x}{z_2\sqrt{x}}$  (see **2.214**)

10. $\displaystyle\int \frac{x^2\sqrt{x}\,\mathrm{d}x}{z_2^3} = -\frac{2x\sqrt{x}}{5bz_2^2} + \frac{3a}{5b}\int\frac{\sqrt{x}\,\mathrm{d}x}{z_2^3}$  (see **2.216** 8)

## 2.22–2.23 Forms containing $\sqrt[n]{(a + bx)^k}$

**Notation:** $z = a + bx$.

**2.220** $\displaystyle\int x^n \sqrt[l]{z^{lm+f}}\,\mathrm{d}x = \left\{\sum_{k=0}^{n}\frac{(-1)^k\binom{n}{k}z^{n-k}a^k}{ln - lk + l(m+1) + f}\right\}\frac{l\sqrt[l]{z^{l(m+1)+f}}}{b^{n+1}}$

**The square root**

**2.221** $\displaystyle\int x^n \sqrt{z^{2m-1}}\,\mathrm{d}x = \left\{\sum_{k=0}^{n}\frac{(-1)^k\binom{n}{k}z^{n-k}a^k}{2n - 2k + 2m + 1}\right\}\frac{2\sqrt{z^{2m+1}}}{b^{n+1}}$

**2.222**

1.  $\displaystyle\int \frac{\mathrm{d}x}{\sqrt{z}} = \frac{2}{b}\sqrt{z}$

2.  $\displaystyle\int \frac{x\,\mathrm{d}x}{\sqrt{z}} = \left(\frac{1}{3}z - a\right)\frac{2\sqrt{z}}{b^2}$

3.  $\displaystyle\int \frac{x^2\,\mathrm{d}x}{\sqrt{z}} = \left(\frac{1}{5}z^2 - \frac{2}{3}az + a^2\right)\frac{2\sqrt{z}}{b^3}$

**2.223**

1.  $\displaystyle\int \frac{\mathrm{d}x}{\sqrt{z^3}} = -\frac{2}{b\sqrt{z}}$

2.  $\displaystyle\int \frac{x\,\mathrm{d}x}{\sqrt{z^3}} = (z+a)\frac{2}{b^2\sqrt{z}}$

3.  $\displaystyle\int \frac{x^2\,\mathrm{d}x}{\sqrt{z^3}} = \left(\frac{z^2}{3} - 2az - a^2\right)\frac{2}{b^3\sqrt{z}}$

**2.224**

1.  $\displaystyle\int \frac{z^m\,\mathrm{d}x}{x^n\sqrt{z}} = -\frac{z^m\sqrt{z}}{(n-1)ax^{n-1}} + \frac{2m-2n+3}{2(n-1)}\frac{b}{a}\int \frac{z^m\,\mathrm{d}x}{x^{n-1}\sqrt{z}}$

2.  $\displaystyle\int \frac{z^m\,\mathrm{d}x}{x^n\sqrt{z}} = -z^m\sqrt{z}\left\{\frac{1}{(n-1)ax^{n-1}}\right.$

    $\displaystyle + \sum_{k=1}^{n-2} \frac{(2m-2n+3)(2m-2n+5)\ldots(2m-2n+2k+1)}{2^k(n-1)(n-2)\ldots(n-k-1)x^{n-k-1}}\frac{b^k}{a^{k+1}}\bigg\}$

    $\displaystyle + \frac{(2m-2n+3)(2m-2n+5)\ldots(2m-3)(2m-1)}{2^{n-1}(n-1)!x}\frac{b^{n-1}}{a^{n-1}}\int \frac{z^m\,\mathrm{d}x}{x\sqrt{z}}$

    For $n=1$

3.  $\displaystyle\int \frac{z^m}{x\sqrt{z}}\,\mathrm{d}x = \frac{2z^m}{(2m-1)\sqrt{z}} + a\int \frac{z^{m-1}}{x\sqrt{z}}\,\mathrm{d}x$

4.  $\displaystyle\int \frac{z^m}{x\sqrt{z}}\,\mathrm{d}x = \sum_{k=1}^{m} \frac{2a^{m-k}z^k}{(2k-1)\sqrt{z}} + a^m\int \frac{\mathrm{d}x}{x\sqrt{z}}$

5.[6]  $\displaystyle\int \frac{\mathrm{d}x}{x\sqrt{z}} = \frac{1}{\sqrt{a}}\ln\left|\frac{\sqrt{z}-\sqrt{a}}{\sqrt{z}+\sqrt{a}}\right|$ $\qquad\qquad\qquad [a>0]$

    $\displaystyle\phantom{\int \frac{\mathrm{d}x}{x\sqrt{z}}} = \frac{2}{\sqrt{-a}}\arctan\frac{\sqrt{z}}{\sqrt{-a}}$ $\qquad\qquad\qquad [a<0]$

**2.225**

1.  $\displaystyle\int \frac{\sqrt{z}\,\mathrm{d}x}{x} = 2\sqrt{z} + a\int \frac{\mathrm{d}x}{x\sqrt{z}}$ $\qquad\qquad$ (see **2.224** 4)

2.  $\displaystyle\int \frac{\sqrt{z}\,\mathrm{d}x}{x^2} = -\frac{\sqrt{z}}{x} + \frac{b}{2}\int \frac{\mathrm{d}x}{x\sqrt{z}}$ $\qquad\qquad$ (see **2.224** 4)

3.  $\displaystyle\int \frac{\sqrt{z}\,\mathrm{d}x}{x^3} = -\frac{\sqrt{z^3}}{2ax^2} + \frac{b\sqrt{z}}{4ax} - \frac{b^2}{8a}\int \frac{\mathrm{d}x}{x\sqrt{z}}$ $\qquad\qquad$ (see **2.224** 4)

**2.226**

1. $\quad \int \dfrac{\sqrt{z^3}\,dx}{x} = \left(\dfrac{z}{3} + a\right) 2\sqrt{z} + a^2 \int \dfrac{dx}{x\sqrt{z}}$ (see **2.224** 4)

2. $\quad \int \dfrac{\sqrt{z^3}\,dx}{x^2} = -\dfrac{\sqrt{z^5}}{ax} + \dfrac{3b}{2a} \int \dfrac{\sqrt{z^3}\,dx}{x}$ (see **2.226** 1)

3. $\quad \int \dfrac{\sqrt{z^3}\,dx}{x^3} = -\left(\dfrac{1}{2ax^2} + \dfrac{b}{4a^2x}\right)\sqrt{z^5} + \dfrac{3b^2}{8a^2} \int \dfrac{\sqrt{z^3}\,dx}{x}$

(see **2.226** 1)

**2.227** $\quad \int \dfrac{dx}{xz^m\sqrt{z}} = \sum_{k=0}^{m-1} \dfrac{2}{(2k+1)a^{m-k}z^k\sqrt{z}} + \dfrac{1}{a^m} \int \dfrac{dx}{x\sqrt{z}}$ (see **2.224** 4)

**2.228**

1. $\quad \int \dfrac{dx}{x^2\sqrt{z}} = -\dfrac{\sqrt{z}}{ax} - \dfrac{b}{2a} \int \dfrac{dx}{x\sqrt{z}}$ (see **2.224** 4)

2. $\quad \int \dfrac{dx}{x^3\sqrt{z}} = \left(-\dfrac{1}{2ax^2} + \dfrac{3b}{4a^2x}\right)\sqrt{z} + \dfrac{3b^2}{8a^2} \int \dfrac{dx}{x\sqrt{z}}$ (see **2.224** 4)

**2.229**

1. $\quad \int \dfrac{dx}{x\sqrt{z^3}} = \dfrac{2}{a\sqrt{z}} + \dfrac{1}{a} \int \dfrac{dx}{x\sqrt{z}}$ (see **2.224** 4)

2. $\quad \int \dfrac{dx}{x^2\sqrt{z^3}} = \left(-\dfrac{1}{ax} - \dfrac{3b}{a^2}\right)\dfrac{1}{\sqrt{z}} - \dfrac{3b}{2a^2} \int \dfrac{dx}{x\sqrt{z}}$ (see **2.224** 4)

3. $\quad \int \dfrac{dx}{x^3\sqrt{z^3}} = \left(-\dfrac{1}{2ax^2} + \dfrac{5b}{4a^2x} + \dfrac{15b^2}{4a^3}\right)\dfrac{1}{\sqrt{z}} + \dfrac{15b^2}{8a^3} \int \dfrac{dx}{x\sqrt{z}}$

(see **2.224** 4)

**Cube root**

**2.231**

1. $\quad \int \sqrt[3]{z^{3m+1}}\,x^n\,dx = \left\{ \sum_{k=0}^{n} \dfrac{(-1)^k \binom{n}{k} z^{n-k}a^k}{3n - 3k + 3(m+1) + 1} \right\} \dfrac{3\sqrt[3]{z^{3(m+1)+1}}}{b^{n+1}}$

2. $\quad \int \dfrac{x^n\,dx}{\sqrt[3]{z^{3m+2}}} = \left\{ \sum_{k=0}^{n} \dfrac{(-1)^k \binom{n}{k} z^{n-k}a^k}{3n - 3k - 3(m-1) - 2} \right\} \dfrac{3}{b^{n+1}\sqrt[3]{z^{3(m-1)+2}}}$

3. $\quad \int \sqrt[3]{z^{3m+2}}\,x^n\,dx = \left\{ \sum_{k=0}^{n} \dfrac{(-1)^k \binom{n}{k} z^{n-k}a^k}{3n - 3k + 3(m+1) + 2} \right\} \dfrac{3\sqrt[3]{z^{3(m+1)+2}}}{b^{n+1}}$

4. $\quad \int \dfrac{x^n\,dx}{\sqrt[3]{z^{3m+1}}} = \left\{ \sum_{k=0}^{n} \dfrac{(-1)^k \binom{n}{k} z^{n-k}a^k}{3n - 3k - 3(m-1) - 1} \right\} \dfrac{3}{b^{n+1}\sqrt[3]{z^{3(m-1)+1}}}$

5. $\quad \displaystyle\int \frac{z^n\,dx}{x^m\sqrt[3]{x^2}} = -\frac{z^{n+\frac{1}{3}}}{(m-1)ax^{m-1}} + \frac{3n-3m+4}{3(m-1)}\frac{b}{a}\int \frac{z^n\,dx}{x^{m-1}\sqrt[3]{z^2}}$

For $m=1$

$\displaystyle\int \frac{z^n\,dx}{x\sqrt[3]{z^2}} = \frac{3z^n}{(3n-2)\sqrt[3]{z^2}} + a\int \frac{z^{n-1}\,dx}{x\sqrt[3]{z^2}}$

6. $\quad \displaystyle\int \frac{dx}{xz^n\sqrt[3]{z^2}} = \frac{3\sqrt[3]{z}}{(3n-1)az^n} + \frac{1}{a}\int \frac{\sqrt[3]{z}\,dx}{xz^n}$

**2.232** $\quad \displaystyle\int \frac{dx}{x\sqrt[3]{z^2}} = \frac{1}{\sqrt[3]{a^2}}\left\{\frac{3}{2}\ln \frac{\sqrt[3]{z}-\sqrt[3]{a}}{\sqrt[3]{x}} - \sqrt{3}\arctan \frac{\sqrt{3}\sqrt[3]{z}}{\sqrt[3]{z}+2\sqrt[3]{a}}\right\}$

**2.233**

1. $\quad \displaystyle\int \frac{\sqrt[3]{z}\,dx}{x} = 3\sqrt[3]{z} + a\int \frac{dx}{x\sqrt[3]{z^2}}$ $\qquad\qquad$ (see **2.232**)

2. $\quad \displaystyle\int \frac{\sqrt[3]{z}\,dx}{x^2} = -\frac{z\sqrt[3]{z}}{ax} + \frac{b}{a}\sqrt[3]{z} + \frac{b}{3}\int \frac{dx}{x\sqrt[3]{z^2}}$ $\qquad$ (see **2.232**)

3. $\quad \displaystyle\int \frac{\sqrt[3]{z}\,dx}{x^3} = \left(-\frac{1}{2ax^2} + \frac{b}{3a^2x}\right)z\sqrt[3]{z} - \frac{b^2}{3a^2}\sqrt[3]{z} - \frac{b^2}{9a}\int \frac{dx}{x\sqrt[3]{z^2}}$

$\qquad\qquad\qquad\qquad\qquad\qquad\qquad\qquad\qquad\qquad\qquad\qquad$ (see **2.232**)

4. $\quad \displaystyle\int \frac{dx}{x^2\sqrt[3]{z^2}} = -\frac{\sqrt[3]{z}}{ax} - \frac{2b}{3a}\int \frac{dx}{x\sqrt[3]{z^2}}$ $\qquad\qquad$ (see **2.232**)

5. $\quad \displaystyle\int \frac{dx}{x^3\sqrt[3]{z^2}} = \left[-\frac{1}{2ax^2} + \frac{5b}{6a^2x}\right]\sqrt[3]{z} + \frac{5b^2}{9a^2}\int \frac{dx}{x\sqrt[3]{z^2}}$ $\qquad$ (see **2.232**)

**2.234**

1.[12] $\quad \displaystyle\int \frac{z^n\,dx}{x^m\sqrt[3]{z}} = -\frac{z^n\sqrt[3]{z}}{(m-1)ax^{m-1}} + \frac{3n-3m+5}{3(m-1)}\frac{b}{a}\int \frac{z^n\,dx}{x^{m-1}\sqrt[3]{z}}$

For $m=1$:

2. $\quad \displaystyle\int \frac{z^n\,dx}{x\sqrt[3]{z}} = \frac{3z^n}{(3n-1)\sqrt[3]{z}} + a\int \frac{z^{n-1}\,dx}{x\sqrt[3]{z}}$

3. $\quad \displaystyle\int \frac{dx}{xz^n\sqrt[3]{z}} = \frac{3\sqrt[3]{z}}{(3n-2)az^n} + \frac{1}{a}\int \frac{\sqrt[3]{z}\,dx}{xz^n}$

**2.235** $\quad \displaystyle\int \frac{dx}{x\sqrt[3]{z}} = \frac{1}{\sqrt[3]{a^2}}\left\{\frac{3}{2}\ln \frac{\sqrt[3]{z}-\sqrt[3]{a}}{\sqrt[3]{x}} + \sqrt{3}\arctan \frac{\sqrt{3}\sqrt[3]{z}}{\sqrt[3]{z}+2\sqrt[3]{a}}\right\}$

**2.236**

1. $\quad \displaystyle\int \frac{\sqrt[3]{z^2}\,dx}{x} = \frac{3}{2}\sqrt[3]{z^2} + a\int \frac{dx}{x\sqrt[3]{z}}$ $\qquad\qquad$ (see **2.235**)

2. $\quad \displaystyle\int \frac{\sqrt[3]{z^2}\,dx}{x^2} = -\frac{\sqrt[3]{z^5}}{ax} + \frac{b}{a}\sqrt[3]{z^2} + \frac{2b}{3}\int \frac{dx}{x\sqrt[3]{z}}$ $\qquad$ (see **2.235**)

3.
$$\int \frac{\sqrt[3]{z^2}\,dx}{x^3} = \left[-\frac{1}{2ax^2} + \frac{b}{6a^2x}\right] z^{5/3} - \frac{b^2}{6a^2}\sqrt[3]{z^2} - \frac{b^2}{9a}\int \frac{dx}{x\sqrt[3]{z}}$$

(see **2.235**)

4.
$$\int \frac{dx}{x^2\sqrt[3]{z}} = -\frac{\sqrt[3]{z^2}}{ax} - \frac{b}{3a}\int \frac{dx}{x\sqrt[3]{z}}$$   (see **2.235**)

5.
$$\int \frac{dx}{x^3\sqrt[3]{z}} = \left[-\frac{1}{2ax^2} + \frac{2b}{3a^2x}\right]\sqrt[3]{z} + \frac{2b^2}{9a^2}\int \frac{dx}{x\sqrt[3]{z}}$$   (see **2.235**)

## 2.24 Forms containing $\sqrt{a+bx}$ and the binomial $\alpha + \beta x$

**Notation**: $z = a + bx$,    $t = \alpha + \beta x$,    $\Delta = a\beta - b\alpha$.

**2.241**

1.
$$\int \frac{z^m t^n\,dx}{\sqrt{z}} = \frac{2}{(2n + 2m + 1)\beta} t^{n+1} z^{m-1}\sqrt{z} + \frac{(2m-1)\Delta}{(2n+2m+1)\beta}\int \frac{z^{m-1}t^n\,dx}{\sqrt{z}}$$   LA 176 (1)

2.
$$\int \frac{t^n z^m\,dx}{\sqrt{z}} = 2\sqrt{z^{2m+1}}\sum_{k=0}^{n}\binom{n}{k}\frac{\alpha^{n-k}\beta^k}{b^{k+1}}\sum_{p=0}^{k}(-1)^p\binom{k}{p}\frac{z^{k-p}a^p}{2k-2p+2m+1}$$

**2.242**

1.[11]
$$\int \frac{t\,dx}{\sqrt{z}} = \frac{2\alpha\sqrt{z}}{b} + \beta\left(\frac{z}{3} - a\right)\frac{2\sqrt{z}}{b^2}$$

2.
$$\int \frac{t^2\,dx}{\sqrt{z}} = \frac{2\alpha^2\sqrt{z}}{b} + 2\alpha\beta\left(\frac{z}{3} - a\right)\frac{2\sqrt{z}}{b^2} + \beta^2\left(\frac{z^2}{5} - \frac{2}{3}za + a^2\right)\frac{2\sqrt{z}}{b^3}$$

3.[12]
$$\int \frac{t^3\,dx}{\sqrt{z}} = \frac{2\alpha^3\sqrt{z}}{b} + 3\alpha^2\beta\left(\frac{z}{3} - a\right)\frac{2\sqrt{z}}{b^2} + 3\alpha\beta^2\left(\frac{z^2}{5} - \frac{2}{3}za + a^2\right)\frac{2\sqrt{z}}{b^3}$$
$$+\beta^3\left(\frac{z^3}{7} - \frac{3z^2a}{5} + za^2 - a^3\right)\frac{2\sqrt{z}}{b^4}$$

4.
$$\int \frac{tz\,dx}{\sqrt{z}} = \frac{2\alpha\sqrt{z^3}}{3b} + \beta\left(\frac{z}{5} - \frac{a}{3}\right)\frac{2\sqrt{z^3}}{b^2}$$

5.
$$\int \frac{t^2 z\,dx}{\sqrt{z}} = \frac{2\alpha^2\sqrt{z^3}}{3b} + 2\alpha\beta\left(\frac{z}{5} - \frac{a}{3}\right)\frac{2\sqrt{z^3}}{b^2} + \beta^2\left(\frac{z^2}{7} - \frac{2za}{5} + \frac{a^2}{3}\right)\frac{2\sqrt{z^3}}{b^3}$$

6.
$$\int \frac{t^3 z\,dx}{\sqrt{z}} = \frac{2\alpha^3\sqrt{z^3}}{3b} + 3\alpha^2\beta\left(\frac{z}{5} - \frac{a}{3}\right)\frac{2\sqrt{z^3}}{b^2} + 3\alpha\beta^2\left(\frac{z^2}{7} - \frac{2za}{5} + \frac{a^2}{3}\right)\frac{2\sqrt{z^3}}{b^3}$$
$$+\beta^3\left(\frac{z^3}{9} - \frac{3z^2a}{7} + \frac{3za^2}{5} - \frac{a^3}{3}\right)\frac{2\sqrt{z^3}}{b^4}$$

7.
$$\int \frac{tz^2\,dx}{\sqrt{z}} = \frac{2\alpha\sqrt{z^5}}{5b} + \beta\left(\frac{z}{7} - \frac{a}{5}\right)\frac{2\sqrt{z^5}}{b^2}$$

8.
$$\int \frac{t^2 z^2\,dx}{\sqrt{z}} = \frac{2\alpha^2\sqrt{z^5}}{5b} + 2\alpha\beta\left(\frac{z}{7} - \frac{a}{5}\right)\frac{2\sqrt{z^5}}{b^2} + \beta^2\left(\frac{z^2}{9} - \frac{2za}{7} + \frac{a^2}{5}\right)\frac{2\sqrt{z^5}}{b^3}$$

9.   $\displaystyle\int \frac{t^3 z^2\,dx}{\sqrt{z}} = \frac{2\alpha^3\sqrt{z^5}}{5b} + 3\alpha^2\beta\left(\frac{z}{7}-\frac{a}{5}\right)\frac{2\sqrt{z^5}}{b^2} + 3\alpha\beta^2\left(\frac{z^2}{9}-\frac{2za}{7}+\frac{a^2}{5}\right)\frac{2\sqrt{z^5}}{b^3}$

       $+\beta^3\left(\dfrac{z^3}{11}-\dfrac{3z^2a}{9}+\dfrac{3za^2}{7}-\dfrac{a^3}{5}\right)\dfrac{2\sqrt{z^5}}{b^4}$

10.   $\displaystyle\int \frac{tz^3\,dx}{\sqrt{z}} = \frac{2\alpha\sqrt{z^7}}{7b} + \beta\left(\frac{z}{9}-\frac{a}{7}\right)\frac{2\sqrt{z^7}}{b^2}$

11.   $\displaystyle\int \frac{t^2 z^3\,dx}{\sqrt{z}} = \frac{2\alpha^2\sqrt{z^7}}{7b} + 2\alpha\beta\left(\frac{z}{9}-\frac{a}{7}\right)\frac{2\sqrt{z^7}}{b^2} + \beta^2\left(\frac{z^2}{11}-\frac{2za}{9}+\frac{a^2}{7}\right)\frac{2\sqrt{z^7}}{b^3}$

12.   $\displaystyle\int \frac{t^3 z^3\,dx}{\sqrt{z}} = \frac{2\alpha^3\sqrt{z^7}}{7b} + 3\alpha^2\beta\left(\frac{z}{9}-\frac{a}{7}\right)\frac{2\sqrt{z^7}}{b^2} + 3\alpha\beta^2\left(\frac{z^2}{11}-\frac{2za}{9}+\frac{a^2}{7}\right)\frac{2\sqrt{z^7}}{b^3}$

       $+\beta^3\left(\dfrac{z^3}{13}-\dfrac{3z^2a}{11}+\dfrac{3za^2}{9}-\dfrac{a^3}{7}\right)\dfrac{2\sqrt{z^7}}{b^4}$

## 2.243

1.   $\displaystyle\int \frac{t^n\,dx}{z^m\sqrt{z}} = \frac{2}{(2m-1)\Delta}\frac{t^{n+1}}{z^m}\sqrt{z} - \frac{(2n-2m+3)\beta}{(2m-1)\Delta}\int \frac{t^n\,dx}{z^{m-1}\sqrt{z}}$

      $= -\dfrac{2}{(2m-1)b}\dfrac{t^n}{z^m}\sqrt{z} + \dfrac{2n\beta}{(2m-1)b}\displaystyle\int \frac{t^{n-1}\,dx}{z^{m-1}\sqrt{z}}$

                                                          LA 176 (2)

2.   $\displaystyle\int \frac{t^n\,dx}{z^m\sqrt{z}} = \frac{2}{\sqrt{z^{2m-1}}}\sum_{k=0}^{n}\binom{n}{k}\frac{\alpha^{n-k}\beta^k}{b^{k+1}}\sum_{p=0}^{k}(-1)^p\binom{k}{p}\frac{z^{k-p}a^p}{2k-2p-2m+1}$

## 2.244

1.   $\displaystyle\int \frac{t\,dx}{z\sqrt{z}} = -\frac{2\alpha}{b\sqrt{z}} + \frac{2\beta(z+a)}{b^2\sqrt{z}}$

2.   $\displaystyle\int \frac{t^2\,dx}{z\sqrt{z}} = -\frac{2\alpha^2}{b\sqrt{z}} + \frac{4\alpha\beta(z+a)}{b^2\sqrt{z}} + \frac{2\beta^2\left(\frac{z^2}{3}-2za-a^2\right)}{b^3\sqrt{z}}$

3.   $\displaystyle\int \frac{t^3\,dx}{z\sqrt{z}} = -\frac{2\alpha^3}{b\sqrt{z}} + \frac{6\alpha^2\beta(z+a)}{b^2\sqrt{z}} + \frac{6\alpha\beta^2\left(\frac{z^2}{3}-2za-a^2\right)}{b^3\sqrt{z}} + \frac{2\beta^3\left(\frac{z^3}{5}-z^2a+3za^2+a^3\right)}{b^4\sqrt{z}}$

4.   $\displaystyle\int \frac{t\,dx}{z^2\sqrt{z}} = -\frac{2\alpha}{3b\sqrt{z^3}} - \frac{2\beta\left(z-\frac{a}{3}\right)}{b^2\sqrt{z^3}}$

5.   $\displaystyle\int \frac{t^2\,dx}{z^2\sqrt{z}} = -\frac{2\alpha^2}{3b\sqrt{z^3}} - \frac{4\alpha\beta\left(z-\frac{a}{3}\right)}{b^2\sqrt{z^3}} + \frac{2\beta^2\left(z^2+2az-\frac{a^2}{3}\right)}{b^3\sqrt{z^3}}$

6.   $\displaystyle\int \frac{t^3\,dx}{z^2\sqrt{z}} = -\frac{2\alpha^3}{3b\sqrt{z^3}} - \frac{6\alpha^2\beta\left(z-\frac{a}{3}\right)}{b^2\sqrt{z^3}} + \frac{6\alpha\beta^2\left(z^2+2za-\frac{a^2}{3}\right)}{b^3\sqrt{z^3}} + \frac{2\beta^3\left(\frac{z^3}{3}-3z^2a-3za^2+\frac{a^3}{3}\right)}{b^4\sqrt{z^3}}$

7.   $\displaystyle\int \frac{t\,dx}{z^3\sqrt{z}} = -\frac{2\alpha}{5b\sqrt{z^5}} - \frac{2\beta\left(\frac{z}{3}-\frac{a}{5}\right)}{b^2\sqrt{z^5}}$

8.
$$\int \frac{t^2\,dx}{z^3\sqrt{z}} = -\frac{2\alpha^2}{5b\sqrt{z^5}} - \frac{4\alpha\beta\left(\frac{z}{3}-\frac{a}{5}\right)}{b^2\sqrt{z^5}} - \frac{2\beta^2\left(z^2-\frac{2za}{3}+\frac{a^2}{5}\right)}{b^3\sqrt{z^5}}$$

9.
$$\int \frac{t^3\,dx}{z^3\sqrt{z}} = -\frac{2\alpha^3}{5b\sqrt{z^5}} - \frac{6\alpha^2\beta\left(\frac{z}{3}-\frac{a}{5}\right)}{b^2\sqrt{z^5}} - \frac{6\alpha\beta^2\left(z^2-\frac{2za}{3}+\frac{a^2}{5}\right)}{b^3\sqrt{z^5}}$$
$$+ \frac{2\beta^3\left(z^3+3z^2a-za^2+\frac{a^3}{5}\right)}{b^4\sqrt{z^5}}$$

**2.245**

1.
$$\int \frac{z^m\,dx}{t^n\sqrt{z}} = -\frac{2}{(2n-2m-1)\beta}\frac{z^{m-1}}{t^{n-1}}\sqrt{z} - \frac{(2m-1)\Delta}{(2n-2m-1)\beta}\int \frac{z^{m-1}\,dx}{t^n\sqrt{z}}$$     LA 176 (3)
$$= -\frac{1}{(n-1)\beta}\frac{z^{m-1}}{t^{n-1}}\sqrt{z} + \frac{(2m-1)b}{2(n-1)\beta}\int \frac{z^{m-1}}{t^{n-1}\sqrt{z}}\,dx$$
$$= -\frac{1}{(n-1)\Delta}\frac{z^m}{t^{n-1}}\sqrt{z} - \frac{(2n-2m-3)b}{2(n-1)\Delta}\int \frac{z^m\,dx}{t^{n-1}\sqrt{z}}$$

2.[12]
$$\int \frac{z^m\,dz}{t^n\sqrt{z}} = -z^m\sqrt{z}\left[\frac{1}{(n-1)\Delta}\frac{1}{t^{n-1}}\right.$$
$$\left. + \sum_{k=2}^{n-1}\frac{(2n-2m-3)(2n-2m-5)\ldots(2n-2m-2k+1)(-b)^{k-1}}{2^{k-1}(n-1)(n-2)\ldots(n-k)\Delta^k}\frac{1}{t^{n-k}}\right]$$
$$- \frac{(2n-2m-3)(2n-2m-5)\ldots(-2m+3)(-2m+1)(-b)^{n-1}}{2^{n-1}\cdot(n-1)!\Delta^n}\int \frac{z^m\,dx}{t\sqrt{z}}$$

For $n = 1$

3.
$$\int \frac{z^m\,dx}{t\sqrt{z}} = \frac{2}{(2m-1)\beta}\frac{z^m}{\sqrt{z}} + \frac{\Delta}{\beta}\int \frac{z^{m-1}\,dx}{t\sqrt{z}}$$

4.
$$\int \frac{z^m\,dx}{t\sqrt{z}} = 2\sum_{k=0}^{m-1}\frac{\Delta^k}{(2m-2k-1)\beta^{k+1}}\frac{z^{m-k}}{\sqrt{z}} + \frac{\Delta^m}{\beta^m}\int \frac{dx}{t\sqrt{z}}$$

**2.246**[12]
$$\int \frac{dx}{t\sqrt{z}} = \frac{1}{\sqrt{\beta\Delta}}\ln\frac{\beta\sqrt{z}-\sqrt{\beta\Delta}}{\beta\sqrt{z}+\sqrt{\beta\Delta}}$$       $[\beta\Delta > 0]$
$$= \frac{2}{\sqrt{-\beta\Delta}}\arctan\frac{\beta\sqrt{z}}{\sqrt{-\beta\Delta}}$$       $[\beta\Delta < 0]$
$$= -\frac{2\sqrt{z}}{bt}$$       $[\Delta = 0]$

**2.247**
$$\int \frac{dx}{tz^m\sqrt{z}} = \frac{2}{z^{m-1}\sqrt{z}} + \sum_{k=1}^{m}\frac{\beta^{k-1}z^k}{\Delta^k(2m-2k+1)} + \frac{\beta^m}{\Delta^m}\int \frac{dx}{t\sqrt{z}}$$
$$\text{(see } \mathbf{2.246)}$$

**2.248**

1.
$$\int \frac{dx}{tz\sqrt{z}} = \frac{2}{\Delta\sqrt{z}} + \frac{\beta}{\Delta}\int \frac{dx}{t\sqrt{z}}$$       (see **2.246**)

2.    $\displaystyle\int \frac{dx}{tz^2\sqrt{z}} = \frac{2}{3\Delta z\sqrt{z}} + \frac{2\beta}{\Delta^2\sqrt{z}} + \frac{\beta^2}{\Delta^2}\int \frac{dx}{t\sqrt{z}}$      (see **2.246**)

3.    $\displaystyle\int \frac{dx}{tz^3\sqrt{z}} = \frac{2}{5\Delta z^2\sqrt{z}} + \frac{2\beta}{3\Delta^2 z\sqrt{z}} + \frac{2\beta^2}{\Delta^3\sqrt{z}} + \frac{\beta^3}{\Delta^3}\int \frac{dx}{t\sqrt{z}}$

                                                    (see **2.246**)

4.    $\displaystyle\int \frac{dx}{t^2\sqrt{z}} = -\frac{\sqrt{z}}{\Delta t} - \frac{b}{2\Delta}\int \frac{dx}{t\sqrt{z}}$      (see **2.246**)

5.    $\displaystyle\int \frac{dx}{t^2 z\sqrt{z}} = -\frac{1}{\Delta t\sqrt{z}} - \frac{3b}{\Delta^2\sqrt{z}} - \frac{3b\beta}{2\Delta^2}\int \frac{dx}{t\sqrt{z}}$      (see **2.246**)

6.    $\displaystyle\int \frac{dx}{t^2 z^2\sqrt{z}} = -\frac{1}{\Delta t z^2\sqrt{z}} - \frac{5b}{3\Delta^2 z\sqrt{z}} - \frac{5b\beta}{\Delta^3\sqrt{z}} - \frac{5b\beta^2}{2\Delta^3}\int \frac{dx}{t\sqrt{z}}$

                                                    (see **2.246**)

7.    $\displaystyle\int \frac{dx}{t^2 z^3\sqrt{z}} = -\frac{1}{\Delta t z^2\sqrt{z}} - \frac{7b}{5\Delta^2 z^2\sqrt{z}} - \frac{7b\beta}{3\Delta^3 z\sqrt{z}} - \frac{7b\beta^2}{\Delta^4\sqrt{z}} - \frac{7b\beta^3}{2\Delta^4}\int \frac{dx}{t\sqrt{z}}$

                                                    (see **2.246**)

8.    $\displaystyle\int \frac{dx}{t^3\sqrt{z}} = -\frac{\sqrt{z}}{2\Delta t^2} + \frac{3b\sqrt{z}}{4\Delta^2 t} + \frac{3b^2}{8\Delta^2}\int \frac{dx}{t\sqrt{z}}$      (see **2.246**)

9.    $\displaystyle\int \frac{dx}{t^3 z\sqrt{z}} = -\frac{1}{2\Delta t^2\sqrt{z}} + \frac{5b}{4\Delta^2 t\sqrt{z}} + \frac{15b^2}{4\Delta^3\sqrt{z}} + \frac{15b^2\beta}{8\Delta^3}\int \frac{dx}{t\sqrt{z}}$

                                                    (see **2.246**)

10.[12]    $\displaystyle\int \frac{dx}{t^3 z^2\sqrt{z}} = -\frac{1}{2\Delta t^2 z\sqrt{z}} + \frac{7b\sqrt{z}}{4\Delta^2 t z\sqrt{z}} + \frac{35b^2}{12\Delta^3 z\sqrt{z}} + \frac{35b^2\beta}{4\Delta^4\sqrt{z}} + \frac{35b^2\beta^2}{8\Delta^4}\int \frac{dx}{t\sqrt{z}}$

                                                   (see **2.246**)

11.    $\displaystyle\int \frac{dx}{t^3 z^3\sqrt{z}} = -\frac{1}{2\Delta t^2 z^2\sqrt{z}} + \frac{9b}{4\Delta^2 t z^2\sqrt{z}} + \frac{63b^2}{20\Delta^3 z^2\sqrt{z}} + \frac{21b^2\beta}{4\Delta^4 z\sqrt{z}} + \frac{63b^2\beta^2}{4\Delta^5\sqrt{z}} + \frac{63b^2\beta^3}{8\Delta^5}\int \frac{dx}{t\sqrt{z}}$

                                                   (see **2.246**)

12.    $\displaystyle\int \frac{z\,dx}{t\sqrt{z}} = \frac{2\sqrt{z}}{\beta} + \frac{\Delta}{\beta}\int \frac{dx}{t\sqrt{z}}$      (see **2.246**)

13.    $\displaystyle\int \frac{z^2\,dx}{t\sqrt{z}} = \frac{2z\sqrt{z}}{3\beta} + \frac{2\Delta\sqrt{z}}{\beta^2} + \frac{\Delta^2}{\beta^2}\int \frac{dx}{t\sqrt{z}}$      (see **2.246**)

14.    $\displaystyle\int \frac{z^3\,dx}{t\sqrt{z}} = \frac{2z^2\sqrt{z}}{5\beta} + \frac{2\Delta z\sqrt{z}}{3\beta^2} + \frac{2\Delta^2\sqrt{z}}{\beta^3} + \frac{\Delta^3}{\beta^3}\int \frac{dx}{t\sqrt{z}}$      (see **2.246**)

15.    $\displaystyle\int \frac{z\,dx}{t^2\sqrt{z}} = -\frac{z\sqrt{z}}{\Delta t} + \frac{b\sqrt{z}}{\beta\Delta} + \frac{b}{2\beta}\int \frac{dx}{t\sqrt{z}}$      (see **2.246**)

16.    $\displaystyle\int \frac{z^2\,dx}{t^2\sqrt{z}} = -\frac{z^2\sqrt{z}}{\Delta t} + \frac{bz\sqrt{z}}{\beta\Delta} + \frac{3b\sqrt{z}}{\beta^2} + \frac{3b\Delta}{2\beta^2}\int \frac{dx}{t\sqrt{z}}$      (see **2.246**)

17.    $\displaystyle \int \frac{z^3\,\mathrm{d}x}{t^2\sqrt{z}} = -\frac{z^3\sqrt{z}}{\Delta t} + \frac{bz^2\sqrt{z}}{\beta\Delta} + \frac{5bz\sqrt{z}}{3\beta^2} + \frac{5b\Delta\sqrt{z}}{\beta^3} + \frac{5\Delta^2 b}{2\beta^3}\int \frac{\mathrm{d}x}{t\sqrt{z}}$

(see 2.246)

18.[12]    $\displaystyle \int \frac{z\,\mathrm{d}x}{t^3\sqrt{z}} = -\frac{z\sqrt{z}}{2\Delta t^2} - \frac{bz\sqrt{z}}{4\Delta^2 t} + \frac{b^2\sqrt{z}}{4\beta\Delta^2} + \frac{b^2}{8\beta\Delta}\int \frac{\mathrm{d}x}{t\sqrt{z}}$    (see 2.246)

19.    $\displaystyle \int \frac{z^2\,\mathrm{d}x}{t^3\sqrt{z}} = -\frac{z^2\sqrt{z}}{2\Delta t^2} + \frac{bz^2\sqrt{z}}{4\Delta^2 t} + \frac{b^2 z\sqrt{z}}{4\beta\Delta^2} + \frac{3b^2\sqrt{z}}{4\beta^2\Delta} + \frac{3b^2}{8\beta^2}\int \frac{\mathrm{d}x}{t\sqrt{z}}$

(see 2.246)

20.    $\displaystyle \int \frac{z^3\,\mathrm{d}x}{t^3\sqrt{z}} = -\frac{z^3\sqrt{z}}{2\Delta t^2} + \frac{3bz^3\sqrt{z}}{\Delta^2 t} + \frac{3b^2 z^2\sqrt{z}}{4\beta\Delta^2} + \frac{5b^2 z\sqrt{z}}{4\beta^2\Delta} + \frac{15b^2\sqrt{z}}{4\beta^3} + \frac{15b^2\Delta}{8\beta^3}\int \frac{\mathrm{d}x}{t\sqrt{z}}$

(see 2.246)

**2.249**

1.    $\displaystyle \int \frac{\mathrm{d}x}{z^m t^n \sqrt{z}} = \frac{2}{(2m-1)\Delta}\frac{\sqrt{z}}{t^{n-1}z^m} + \frac{(2n+2m-3)\beta}{(2m-1)\Delta}\int \frac{\mathrm{d}x}{t^n z^{m-1}\sqrt{z}}$     LA 177 (4)

       $\displaystyle = -\frac{1}{(n-1)\Delta}\frac{\sqrt{z}}{z^m t^{n-1}} - \frac{(2n+2m-3)b}{2(n-1)\Delta}\int \frac{\mathrm{d}x}{t^{n-1}z^m\sqrt{z}}$

2.    $\displaystyle \int \frac{\mathrm{d}x}{z^m t^n \sqrt{z}} = \frac{\sqrt{z}}{z^m}\left[ \frac{-1}{(n-1)\Delta}\frac{1}{t^{n-1}} \right.$

       $\displaystyle \left. + \sum_{k=2}^{n-1}(-1)^k \frac{(2n+2m-3)(2n+2m-5)\ldots(2n+2m-2k+1)b^{k-1}}{2^{k-1}(n-1)(n-2)\ldots(n-k)\Delta^k}\cdot\frac{1}{t^{n-k}} \right]$

       $\displaystyle + (-1)^{n-1}\frac{(2n+2m-3)(2n+2m-5)\ldots(-2m+3)(-2m+1)b^{n-1}}{2^{n-1}(n-1)!\Delta^{n-1}}\int \frac{\mathrm{d}x}{tz^m\sqrt{z}}$

For $n=1$

$\displaystyle \int \frac{\mathrm{d}x}{z^m t\sqrt{z}} = \frac{2}{(2m-1)\Delta}\frac{1}{z^{m-1}\sqrt{z}} + \frac{\beta}{\Delta}\int \frac{\mathrm{d}x}{tz^{m-1}\sqrt{z}}$

## 2.25 Forms containing $\sqrt{a+bx+cx^2}$

### Integration techniques

**2.251**    It is possible to rationalize the integrand in integrals of the form $\int R\left(x, \sqrt{a+bx+cx^2}\right)\,\mathrm{d}x$ by using one or more of the following three substitutions, known as the "Euler substitutions".

1.    $\sqrt{a+bx+cx^2} = xt \pm \sqrt{a}$    for $a>0$;

2.    $\sqrt{a+bx+cx^2} = t \pm x\sqrt{c}$    for $c>0$;

3.    $\sqrt{c\left(x-x_1\right)\left(x-x_2\right)} = t\left(x-x_1\right)$ when $x_1$ and $x_2$ are real roots of the equation $a+bx+cx^2 = 0$.

**2.252**   Besides the Euler substitutions, there is also the following method of calculating integrals of the form $\int R\left(x, \sqrt{a + bx + cx^2}\right) dx$. By removing the irrational expressions in the denominator and performing simple algebraic operations, we can reduce the integrand to the sum of some rational function of $x$ and an expression of the form $\dfrac{P_1(x)}{P_2(x)\sqrt{a + bx + cx^2}}$, where $P_1(x)$ and $P_2(x)$ are both polynomials.

By separating the integral portion of the rational function $\dfrac{P_1(x)}{P_2(x)}$ from the remainder and decomposing the latter into partial fractions, we can reduce the integral of these partial fractions to the sum of integrals each of which is in one of the following three forms:

1.     $\displaystyle\int \frac{P(x)\,dx}{\sqrt{a + bx + cx^2}}$, where $P(x)$ is a polynomial of some degree $r$;

2.     $\displaystyle\int \frac{dx}{(x + p)^k \sqrt{a + bx + cx^2}}$;

3.     $\displaystyle\int \frac{(Mx + N)\,dx}{(a + bx + x^2)^m \sqrt{c\,(a_1 + b_1 x + x^2)}}$,     $\left(a_1 = \dfrac{a}{c}, \quad b_1 = \dfrac{b}{c}\right)$.

In more detail:

1.     $\displaystyle\int \frac{P(x)\,dx}{\sqrt{a + bx + cx^2}} = Q(x)\sqrt{a + bx + cx^2} + \lambda \int \frac{dx}{\sqrt{a + bx + cx^2}}$, where $Q(x)$ is a polynomial of degree $(r - 1)$. Its coefficients, and also the number $\lambda$, can be calculated by the method of undetermined coefficients from the identity

$$P(x) = Q'(x)\left(a + bx + cx^2\right) + \frac{1}{2}Q(x)(b + 2cx) + \lambda \qquad\qquad \text{LI II 77}$$

Integrals of the form $\displaystyle\int \frac{P(x)\,dx}{\sqrt{a + bx + cx^2}}$ (where $r \leq 3$) can also be calculated by use of formulas **2.26**.

2.     Integrals of the form $\displaystyle\int \frac{P(x)\,dx}{(x + p)^k \sqrt{a + bx + cx^2}}$, where the degree $n$ of the polynomial $P(x)$ is lower than $k$ can, by means of the substitution $t = \dfrac{1}{x + p}$, be reduced to an integral of the form $\displaystyle\int \frac{P(t)\,dt}{\sqrt{\alpha + \beta t + \gamma t^2}}$. (See also **2.281**).

3.[12]     Integrals of the form $\displaystyle\int \frac{(Mx + N)\,dx}{(\alpha + \beta x + x^2)^m \sqrt{c\,(a_1 + b_1 x + x^2)}}$ can be calculated by the following procedure.

     • If $b_1 \neq \beta$, by using the substitution

$$x = \frac{a_1 - \alpha}{\beta - b_1} + \frac{t - 1}{t + 1} \frac{\sqrt{(a_1 - \alpha)^2 - (\alpha b_1 - a_1 \beta)(\beta - b_1)}}{\beta - b_1}$$

we can reduce this integral to an integral of the form $\int \dfrac{P(t)\,dt}{(t^2+p)^m\,\sqrt{c\,(t^2+q)}}$, where $P(t)$

is a polynomial of degree no higher than $2m-1$. The integral $\int \dfrac{P(t)\,dt}{(t^2+p)^m\,\sqrt{t^2+q}}$ can be

reduced to the sum of integrals of the forms $\int \dfrac{t\,dt}{(t^2+p)^k\,\sqrt{t^2+q}}$ and $\int \dfrac{dt}{(t^2+p)^k\,\sqrt{t^2+q}}$.

- If $b_1=\beta$, we can reduce it to integrals of the form $\int \dfrac{P(t)\,dt}{(t^2+p)^m\,\sqrt{c\,(t^2+q)}}$ by means of the

substitution $t=x+\dfrac{b_1}{2}$.

The integral $\int \dfrac{t\,dt}{(t^2+p)^k\,\sqrt{c\,(t^2+q)}}$ can be evaluated by means of the substitution $t^2+q=u^2$.

The integral $\int \dfrac{dt}{(t^2+p)^k\,\sqrt{c\,(t^2+q)}}$ can be evaluated by means of the substitution $\dfrac{t}{\sqrt{t^2+q}}=v$ (see also **2.283**).

FI II 78-82

## 2.26 Forms containing $\sqrt{a+bx+cx^2}$ and integral powers of $x$

**Notation:** $R=a+bx+cx^2$, $\quad \Delta=4ac-b^2$

For simplified formulas for the case $b=0$, see **2.27**.

**2.260**

1. $\displaystyle\int x^m\sqrt{R^{2n+1}}\,dx=\frac{x^{m-1}\sqrt{R^{2n+3}}}{(m+2n+2)c}-\frac{(2m+2n+1)b}{2(m+2n+2)c}\int x^{m-1}\sqrt{R^{2n+1}}\,dx$
$\displaystyle\qquad\qquad\qquad\quad -\frac{(m-1)a}{(m+2n+2)c}\int x^{m-2}\sqrt{R^{2n+1}}\,dx$

TI (192)a

2. $\displaystyle\int \sqrt{R^{2n+1}}\,dx=\frac{2cx+b}{4(n+1)c}\sqrt{R^{2n+1}}+\frac{2n+1}{8(n+1)}\frac{\Delta}{c}\int \sqrt{R^{2n-1}}\,dx$      TI (188)

3. $\displaystyle\int \sqrt{R^{2n+1}}\,dx=\frac{(2cx+b)\sqrt{R}}{4(n+1)c}\left\{R^n+\sum_{k=0}^{n-1}\frac{(2n+1)(2n-1)\ldots(2n-2k+1)}{8^{k+1}n(n-1)\ldots(n-k)}\left(\frac{\Delta}{c}\right)^{k+1}R^{n-k-1}\right\}$
$\displaystyle\qquad\qquad\qquad +\frac{(2n+1)!!}{8^{n+1}(n+1)!}\left(\frac{\Delta}{c}\right)^{n+1}\int \frac{dx}{\sqrt{R}}$

TI (190)

**2.261**[12]  For $n=-1$

$\displaystyle\int \frac{dx}{\sqrt{R}}=\frac{1}{\sqrt{c}}\ln\left(2\sqrt{cR}+2cx+b\right)$     $\left[c>0,\quad 2cx+b>\sqrt{-\Delta},\quad \Delta<0\right]$     TI (127)

$\displaystyle\qquad\quad =-\frac{1}{\sqrt{c}}\ln\left(2\sqrt{cR}-2cx-b\right)$     $\left[c>0,\quad 2cx+b<-\sqrt{-\Delta},\quad \Delta<0\right]$

$\displaystyle\qquad\quad =\frac{1}{\sqrt{c}}\operatorname{arcsinh}\left(\frac{2cx+b}{\sqrt{\Delta}}\right)$     $\left[c>0,\quad \Delta>0\right]$     DW

$\displaystyle\qquad\quad =\frac{1}{\sqrt{c}}\ln(2cx+b)$     $\left[c>0,\quad \Delta=0\right]$     DW

$\displaystyle\qquad\quad =\frac{-1}{\sqrt{-c}}\arcsin\left(\frac{2cx+b}{\sqrt{-\Delta}}\right)$     $\left[c<0,\quad \Delta<0\right]$     TI (128)

**2.262**

1. $\displaystyle\int \sqrt{R}\,\mathrm{d}x = \frac{(2cx+b)\sqrt{R}}{4c} + \frac{\Delta}{8c}\int \frac{\mathrm{d}x}{\sqrt{R}}$ (see **2.261**)

2. $\displaystyle\int x\sqrt{R}\,\mathrm{d}x = \frac{\sqrt{R^3}}{3c} - \frac{(2cx+b)b}{8c^2}\sqrt{R} - \frac{b\Delta}{16c^2}\int \frac{\mathrm{d}x}{\sqrt{R}}$ (see **2.261**)

3. $\displaystyle\int x^2\sqrt{R}\,\mathrm{d}x = \left(\frac{x}{4c} - \frac{5b}{24c^2}\right)\sqrt{R^3} + \left(\frac{5b^2}{16c^2} - \frac{a}{4c}\right)\frac{(2cx+b)\sqrt{R}}{4c} + \left(\frac{5b^2}{16c^2} - \frac{a}{4c}\right)\frac{\Delta}{8c}\int \frac{\mathrm{d}x}{\sqrt{R}}$

(see **2.261**)

4. $\displaystyle\int x^3\sqrt{R}\,\mathrm{d}x = \left(\frac{x^2}{5c} - \frac{7bx}{40c^2} + \frac{7b^2}{48c^3} - \frac{2a}{15c^2}\right)\sqrt{R^3} - \left(\frac{7b^3}{32c^3} - \frac{3ab}{8c^2}\right)\frac{(2cx+b)\sqrt{R}}{4c}$

$\displaystyle\qquad\qquad - \left(\frac{7b^3}{32c^3} - \frac{3ab}{8c^2}\right)\frac{\Delta}{8c}\int \frac{\mathrm{d}x}{\sqrt{R}}$

(see **2.261**)

5. $\displaystyle\int \sqrt{R^3}\,\mathrm{d}x = \left(\frac{R}{8c} + \frac{3\Delta}{64c^2}\right)(2cx+b)\sqrt{R} + \frac{3\Delta^2}{128c^2}\int \frac{\mathrm{d}x}{\sqrt{R}}$

(see **2.261**)

6. $\displaystyle\int x\sqrt{R^3}\,\mathrm{d}x = \frac{\sqrt{R^5}}{5c} - (2cx+b)\left(\frac{b}{16c^2}\sqrt{R^3} + \frac{3\Delta b}{128c^3}\sqrt{R}\right) - \frac{3\Delta^2 b}{256c^3}\int \frac{\mathrm{d}x}{\sqrt{R}}$

(see **2.261**)

7. $\displaystyle\int x^2\sqrt{R^3}\,\mathrm{d}x = \left(\frac{x}{6c} - \frac{7b}{60c^2}\right)\sqrt{R^5} + \left(\frac{7b^2}{24c^2} - \frac{a}{6c}\right)\left(2x + \frac{b}{c}\right)\left(\frac{\sqrt{R^3}}{8} + \frac{3\Delta}{64c}\sqrt{R}\right)$

$\displaystyle\qquad\qquad + \left(\frac{7b^2}{4c} - a\right)\frac{\Delta^2}{256c^3}\int \frac{\mathrm{d}x}{\sqrt{R}}$

(see **2.261**)

8. $\displaystyle\int x^3\sqrt{R^3}\,\mathrm{d}x = \left(\frac{x^2}{7c} - \frac{3bx}{28c^2} + \frac{3b^2}{40c^3} - \frac{2a}{35c^2}\right)\sqrt{R^5}$

$\displaystyle\qquad\qquad - \left(\frac{3b^3}{16c^3} - \frac{ab}{4c^2}\right)\left(2x + \frac{b}{c}\right)\left(\frac{\sqrt{R^3}}{8} + \frac{3\Delta}{64c}\sqrt{R}\right)$

$\displaystyle\qquad\qquad - \left(\frac{3b^2}{4c} - a\right)\frac{3\Delta^2 b}{512c^4}\int \frac{\mathrm{d}x}{\sqrt{R}}$

(see **2.261**)

**2.263**

1. $\displaystyle\int \frac{x^m\,\mathrm{d}x}{\sqrt{R^{2n+1}}} = \frac{x^{m-1}}{(m-2n)c\sqrt{R^{2n-1}}} - \frac{(2m-2n-1)b}{2(m-2n)c}\int \frac{x^{m-1}\,\mathrm{d}x}{\sqrt{R^{2n+1}}} - \frac{(m-1)a}{(m-2n)c}\int \frac{x^{m-2}\,\mathrm{d}x}{\sqrt{R^{2n+1}}}$

TI (193)a

For $m = 2n$

2. $\displaystyle\int \frac{x^{2n}\,\mathrm{d}x}{\sqrt{R^{2n+1}}} = -\frac{x^{2n-1}}{(2n-1)c\sqrt{R^{2n-1}}} - \frac{b}{2c}\int \frac{x^{2n-1}}{\sqrt{R^{2n+1}}}\,\mathrm{d}x + \frac{1}{c}\int \frac{x^{2n-2}}{\sqrt{R^{2n-1}}}\,\mathrm{d}x$

TI (194)a

3. $\quad \displaystyle\int \frac{dx}{\sqrt{R^{2n+1}}} = \frac{2(2cx+b)}{(2n-1)\Delta\sqrt{R^{2n-1}}} + \frac{8(n-1)c}{(2n-1)\Delta}\int \frac{dx}{\sqrt{R^{2n-1}}}$                    TI (189)

4. $\quad \displaystyle\int \frac{dx}{\sqrt{R^{2n+1}}} = \frac{2(2cx+b)}{(2n-1)\Delta\sqrt{R^{2n-1}}}\left\{1 + \sum_{k=1}^{n-1} \frac{8^k(n-1)(n-2)\ldots(n-k)}{(2n-3)(2n-5)\ldots(2n-2k-1)}\frac{c^k}{\Delta^k}R^k\right\}$

$$[n \geq 1].$$                    TI (191)

**2.264**

1. $\quad \displaystyle\int \frac{dx}{\sqrt{R}}$                    (see **2.261**)

2. $\quad \displaystyle\int \frac{x\,dx}{\sqrt{R}} = \frac{\sqrt{R}}{c} - \frac{b}{2c}\int \frac{dx}{\sqrt{R}}$                    (see **2.261**)

3. $\quad \displaystyle\int \frac{x^2\,dx}{\sqrt{R}} = \left(\frac{x}{2c} - \frac{3b}{4c^2}\right)\sqrt{R} + \left(\frac{3b^2}{8c^2} - \frac{a}{2c}\right)\int \frac{dx}{\sqrt{R}}$                    (see **2.261**)

4. $\quad \displaystyle\int \frac{x^3\,dx}{\sqrt{R}} = \left(\frac{x^2}{3c} - \frac{5bx}{12c^2} + \frac{5b^2}{8c^3} - \frac{2a}{3c^2}\right)\sqrt{R} - \left(\frac{5b^3}{16c^3} - \frac{3ab}{4c^2}\right)\int \frac{dx}{\sqrt{R}}$

$$\text{(see \textbf{2.261})}$$

5. $\quad \displaystyle\int \frac{dx}{\sqrt{R^3}} = \frac{2(2cx+b)}{\Delta\sqrt{R}}$

6. $\quad \displaystyle\int \frac{x\,dx}{\sqrt{R^3}} = -\frac{2(2a+bx)}{\Delta\sqrt{R}}$

7. $\quad \displaystyle\int \frac{x^2\,dx}{\sqrt{R^3}} = -\frac{\left(\Delta - b^2\right)x - 2ab}{c\Delta\sqrt{R}} + \frac{1}{c}\int \frac{dx}{\sqrt{R}}$                    (see **2.261**)

8. $\quad \displaystyle\int \frac{x^3\,dx}{\sqrt{R^3}} = \frac{c\Delta x^2 + b\left(10ac - 3b^2\right)x + a\left(8ac - 3b^2\right)}{c^2\Delta\sqrt{R}} - \frac{3b}{2c^2}\int \frac{dx}{\sqrt{R}}$

$$\text{(see \textbf{2.261})}$$

**2.265** $\quad \displaystyle\int \frac{\sqrt{R^{2n+1}}}{x^m}\,dx = -\frac{\sqrt{R^{2n+3}}}{(m-1)ax^{m-1}} + \frac{(2n-2m+5)b}{2(m-1)a}\int \frac{\sqrt{R^{2n+1}}}{x^{m-1}}\,dx$

$$+ \frac{(2n-m+4)c}{(m-1)a}\int \frac{\sqrt{R^{2n+1}}}{x^{m-2}}\,dx$$

TI (195)

For $m = 1$

$\displaystyle\int \frac{\sqrt{R^{2n+1}}}{x}\,dx = \frac{\sqrt{R^{2n+1}}}{2n+1} + \frac{b}{2}\int \sqrt{R^{2n-1}}\,dx + a\int \frac{\sqrt{R^{2n-1}}}{x}\,dx$                    TI (198)

For $a = 0$

$\displaystyle\int \frac{\sqrt{(bx+cx^2)^{2n+1}}}{x^m}\,dx = \frac{2\sqrt{(bx+cx^2)^{2n+3}}}{(2n-2m+3)bx^m} + \frac{2(m-2n-3)c}{(2n-2m+3)b}\int \frac{\sqrt{(bx+cx^2)^{2n+1}}}{x^{m-1}}$                    LA 169 (3)

For $m = 0$ see **2.260** 2 and **2.260** 3.

For $n = -1$ and $m = 1$:

**2.266**[12] $\displaystyle\int \frac{dx}{x\sqrt{R}} = -\frac{1}{\sqrt{a}} \ln\left( \frac{2\sqrt{aR} + 2a + bx}{|x|} \right)$    $\left[ a > 0, \quad 2a + bx > \sqrt{-\Delta}|x|, \quad \Delta < 0 \right]$     TI (137)

$\displaystyle = -\frac{1}{\sqrt{a}} \ln\left( \frac{2\sqrt{aR} - 2a - bx}{|x|} \right)$    $\left[ a > 0, \quad 2a + bx < -\sqrt{-\Delta}|x|, \quad \Delta < 0 \right]$

$\displaystyle = \frac{1}{\sqrt{-a}} \arcsin \frac{2a + bx}{x\sqrt{-\Delta}}$    $[a < 0, \quad \Delta < 0]$     TI (138)

$\displaystyle = \frac{1}{\sqrt{-a}} \arctan \frac{2a + bx}{2\sqrt{-a}\sqrt{R}}$    $[a < 0]$     LA 178 (6)a

$\displaystyle = -\frac{1}{\sqrt{a}} \operatorname{arcsinh} \frac{2a + bx}{x\sqrt{\Delta}}$    $[a > 0, \quad \Delta > 0]$     DW

$\displaystyle = -\frac{1}{\sqrt{a}} \operatorname{arctanh} \frac{2a + bx}{2\sqrt{a}\sqrt{R}}$    $[a > 0]$

$\displaystyle = \frac{1}{\sqrt{a}} \ln \frac{x}{2a + bx}$    $[a > 0, \quad \Delta = 0]$

$\displaystyle = -\frac{2\sqrt{bx + cx^2}}{bx}$    $[a = 0, \quad b \neq 0]$     LA 170 (16)

$\displaystyle = \frac{1}{\sqrt{a}} \operatorname{arccosh}\left( \frac{2a + bx}{x\sqrt{-\Delta}} \right)$    $[a > 0, \quad \Delta < 0]$

**2.267**

1.   $\displaystyle\int \frac{\sqrt{R}\,dx}{x} = \sqrt{R} + a\int \frac{dx}{x\sqrt{R}} + \frac{b}{2}\int \frac{dx}{\sqrt{R}}$     (see **2.261** and **2.266**)

2.   $\displaystyle\int \frac{\sqrt{R}\,dx}{x^2} = -\frac{\sqrt{R}}{x} + \frac{b}{2}\int \frac{dx}{x\sqrt{R}} + c\int \frac{dx}{\sqrt{R}}$     (see **2.261** and **2.266**)

For $a = 0$

$\displaystyle\int \frac{\sqrt{bx + cx^2}}{x^2}\,dx = -\frac{2\sqrt{bx + cx^2}}{x} + c\int \frac{dx}{\sqrt{bx + cx^2}}$     (see **2.261**)

3.   $\displaystyle\int \frac{\sqrt{R}\,dx}{x^3} = -\left( \frac{1}{2x^2} + \frac{b}{4ax} \right)\sqrt{R} - \left( \frac{b^2}{8a} - \frac{c}{2} \right)\int \frac{dx}{x\sqrt{R}}$

                                            (see **2.266**)

For $a = 0$

$\displaystyle\int \frac{\sqrt{bx + cx^2}}{x^3}\,dx = -\frac{2\sqrt{(bx + cx^2)^3}}{3bx^3}$

4.   $\displaystyle\int \frac{\sqrt{R^3}}{x}\,dx = \frac{\sqrt{R^3}}{3} + \frac{2bcx + b^2 + 8ac}{8c}\sqrt{R} + a^2\int \frac{dx}{x\sqrt{R}} + \frac{b\left(12ac - b^2\right)}{16c}\int \frac{dx}{\sqrt{R}}$

                                            (see **2.261** and **2.266**)

5.   $\displaystyle\int \frac{\sqrt{R^3}}{x^2}\,dx = -\frac{\sqrt{R^5}}{ax} + \frac{cx + b}{a}\sqrt{R^3} + \frac{3}{4}(2cx + 3b)\sqrt{R} + \frac{3}{2}ab\int \frac{dx}{x\sqrt{R}} + \frac{3\left(4ac + b^2\right)}{8}\int \frac{dx}{\sqrt{R}}$

                                            (see **2.261** and **2.266**)

For $a = 0$

$\displaystyle\int \frac{\sqrt{(bx + cx^2)^3}}{x^2} = \frac{\sqrt{(bx + cx^2)^3}}{2x} + \frac{3b}{4}\sqrt{bx + cx^2} + \frac{3b^2}{8}\int \frac{dx}{\sqrt{bx + cx^2}}$

                                            (see **2.261**)

6. $\int \dfrac{\sqrt{R^3}}{x^3}\,dx = -\left(\dfrac{1}{2ax^2} + \dfrac{b}{4a^2x}\right)\sqrt{R^5} + \dfrac{bcx + 2ac + b^2}{4a^2}\sqrt{R^3} + \dfrac{3\left(bcx + 2ac + b^2\right)}{4a}\sqrt{R}$

$\qquad\qquad + \dfrac{3}{8}\left(4ac + b^2\right)\int\dfrac{dx}{x\sqrt{R}} + \dfrac{3}{2}bc\int\dfrac{dx}{\sqrt{R}}$

$\qquad\qquad\qquad\qquad\qquad\qquad\qquad\qquad\qquad$ (see **2.261** and **2.266**)

For $a = 0$

$\int \dfrac{\sqrt{\left(bx + cx^2\right)^3}}{x^3}\,dx = \left(c - \dfrac{2b}{x}\right)\sqrt{bx + cx^2} + \dfrac{3bc}{2}\int\dfrac{dx}{\sqrt{bx + cx^2}}$

$\qquad\qquad\qquad\qquad\qquad\qquad\qquad\qquad\qquad$ (see **2.261**)

**2.268** $\int\dfrac{dx}{x^m\sqrt{R^{2n+1}}} = -\dfrac{1}{(m-1)ax^{m-1}\sqrt{R^{2n-1}}}$

$\qquad\qquad\qquad\qquad - \dfrac{(2n + 2m - 3)b}{2(m-1)a}\int\dfrac{dx}{x^{m-1}\sqrt{R^{2n+1}}} - \dfrac{(2n + m - 2)c}{(m-1)a}\int\dfrac{dx}{x^{m-2}\sqrt{R^{2n+1}}}$

$\qquad\qquad\qquad\qquad\qquad\qquad\qquad\qquad\qquad\qquad\qquad\qquad\qquad\qquad\qquad\qquad$ TI (196)

For $m = 1$

$\qquad\int\dfrac{dx}{x\sqrt{R^{2n+1}}} = \dfrac{1}{(2n-1)a\sqrt{R^{2n-1}}} - \dfrac{b}{2a}\int\dfrac{dx}{\sqrt{R^{2n+1}}} + \dfrac{1}{a}\int\dfrac{dx}{x\sqrt{R^{2n-1}}}$ $\qquad$ TI (199)

For $a = 0$

$\int\dfrac{dx}{x^m\sqrt{\left(bx + cx^2\right)^{2n+1}}} = -\dfrac{2}{(2n + 2m - 1)bx^m\sqrt{\left(bx + cx^2\right)^{2n-1}}}$

$\qquad\qquad\qquad\qquad\qquad - \dfrac{(4n + 2m - 2)c}{(2n + 2m - 1)b}\int\dfrac{dx}{x^{m-1}\sqrt{\left(bx + cx^2\right)^{2n+1}}}$

$\qquad\qquad\qquad\qquad\qquad\qquad\qquad\qquad\qquad$ (see **2.265**)

**2.269**

1. $\int\dfrac{dx}{x\sqrt{R}}$ $\qquad\qquad\qquad\qquad\qquad\qquad\qquad$ (see **2.266**)

2. $\int\dfrac{dx}{x^2\sqrt{R}} = -\dfrac{\sqrt{R}}{ax} - \dfrac{b}{2a}\int\dfrac{dx}{x\sqrt{R}}$ $\qquad\qquad$ (see **2.266**)

   For $a = 0$

   $\int\dfrac{dx}{x^2\sqrt{bx + cx^2}} = \dfrac{2}{3}\left(-\dfrac{1}{bx^2} + \dfrac{2c}{b^2x}\right)\sqrt{bx + cx^2}$

3. $\int\dfrac{dx}{x^3\sqrt{R}} = \left(-\dfrac{1}{2ax^2} + \dfrac{3b}{4a^2x}\right)\sqrt{R} + \left(\dfrac{3b^2}{8a^2} - \dfrac{c}{2a}\right)\int\dfrac{dx}{x\sqrt{R}}$

   $\qquad\qquad\qquad\qquad\qquad\qquad\qquad\qquad\qquad$ (see **2.266**)

   For $a = 0$

   $\int\dfrac{dx}{x^3\sqrt{bx + cx^2}} = \dfrac{2}{5}\left(-\dfrac{1}{bx^3} + \dfrac{4c}{3b^2x^2} - \dfrac{8c^2}{3b^3x}\right)\sqrt{bx + cx^2}$

4. $\int\dfrac{dx}{x\sqrt{R^3}} = -\dfrac{2\left(bcx - 2ac + b^2\right)}{a\Delta\sqrt{R}} + \dfrac{1}{a}\int\dfrac{dx}{x\sqrt{R}}$ $\qquad$ (see **2.266**)

   For $a = 0$

$$\int \frac{dx}{x\sqrt{(bx+cx^2)^3}} = \frac{2}{3}\left(-\frac{1}{bx} + \frac{4c}{b^2} + \frac{8c^2x}{b^3}\right)\frac{1}{\sqrt{bx+cx^2}}$$

5.[11]
$$\int \frac{dx}{x^2\sqrt{R^3}} = -\frac{A}{\sqrt{R}} - \frac{3b}{2a^2}\int \frac{dx}{x\sqrt{R}}$$

$$\text{where } A = \left(-\frac{1}{ax} - \frac{b\left(10ac - 3b^2\right)}{a^2\Delta} - \frac{c\left(8ac - 3b^2\right)x}{a^2\Delta}\right) \qquad \text{(see } \mathbf{2.266})$$

For $a = 0$

$$\int \frac{dx}{x^2\sqrt{(bx+cx^2)^3}} = \frac{2}{5}\left(-\frac{1}{bx^2} + \frac{2c}{b^2x} - \frac{8c^2}{b^3} - \frac{16c^3x}{b^4}\right)\frac{1}{\sqrt{bx+cx^2}}$$

6.
$$\int \frac{dx}{x^3\sqrt{R^3}}$$

$$= \left(-\frac{1}{ax^2} + \frac{5b}{2a^2x} - \frac{15b^4 - 62acb^2 + 24a^2c^2}{2a^3\Delta} - \frac{bc\left(15b^2 - 52ac\right)x}{2a^3\Delta}\right)\frac{1}{2\sqrt{R}} + \frac{15b^2 - 12ac}{8a^3}\int \frac{dx}{x\sqrt{R}}$$

$$\text{(see } \mathbf{2.266})$$

For $a = 0$

$$\int \frac{dx}{x^3\sqrt{(bx+cx^2)^3}} = \frac{2}{7}\left(-\frac{1}{bx^3} + \frac{8c}{5b^2x^2} - \frac{16c^2}{5b^3x} + \frac{64c^3}{5b^4} + \frac{128c^4x}{5b^5}\right)\frac{1}{\sqrt{bx+cx^2}}$$

## 2.27[12] Forms containing $\sqrt{a+cx^2}$ and integral powers of $x$

**Notation:** $u = \sqrt{a+cx^2}$.

$$I_1 = \frac{1}{\sqrt{c}}\ln\left(x\sqrt{c} + u\right) \qquad\qquad \left[c > 0, \quad x > \sqrt{-\frac{a}{c}}, \quad a < 0\right]$$

$$= -\frac{1}{\sqrt{c}}\ln\left(-x\sqrt{c} + u\right) \qquad \left[c > 0, \quad x < -\sqrt{-\frac{a}{c}}, \quad a < 0\right]$$

$$= \frac{1}{\sqrt{-c}}\arcsin x\sqrt{-\frac{c}{a}} \qquad\qquad [c < 0 \text{ and } a > 0]$$

$$I_2 = \frac{1}{2\sqrt{a}}\ln\frac{u - \sqrt{a}}{u + \sqrt{a}} \qquad\qquad [a > 0 \text{ and } c > 0]$$

$$= \frac{1}{2\sqrt{a}}\ln\frac{\sqrt{a} - u}{\sqrt{a} + u} \qquad\qquad [a > 0 \text{ and } c < 0]$$

$$= \frac{1}{\sqrt{-a}}\operatorname{arcsec} x\sqrt{-\frac{c}{a}} = \frac{1}{\sqrt{-a}}\arccos\frac{1}{x}\sqrt{-\frac{a}{c}} \qquad [a < 0 \text{ and } c > 0]$$

**2.271**

1.  $\displaystyle\int u^5\,dx = \frac{1}{6}xu^5 + \frac{5}{24}axu^3 + \frac{5}{16}a^2xu + \frac{5}{16}a^3I_1$ 　　　　　　　　　　DW

2.  $\displaystyle\int u^3\,dx = \frac{1}{4}xu^3 + \frac{3}{8}axu + \frac{3}{8}a^2I_1$ 　　　　　　　　　　DW

3.  $\displaystyle\int u\,dx = \frac{1}{2}xu + \frac{1}{2}aI_1$ 　　　　　　　　　　DW

4.  $\displaystyle\int \frac{dx}{u} = I_1$ 　　　　　　　　　　DW

5.    $\displaystyle\int \frac{dx}{u^3} = \frac{1}{a}\frac{x}{u}$        DW

6.    $\displaystyle\int \frac{dx}{u^{2n+1}} = \frac{1}{a^n}\sum_{k=0}^{n-1}\frac{(-1)^k}{2k+1}\binom{n-1}{k}\frac{c^k x^{2k+1}}{u^{2k+1}}$

7.    $\displaystyle\int \frac{x\,dx}{u^{2n+1}} = -\frac{1}{(2n-1)cu^{2n-1}}$        DW

**2.272**

1.    $\displaystyle\int x^2 u^3 \,dx = \frac{1}{6}\frac{xu^5}{c} - \frac{1}{24}\frac{axu^3}{c} - \frac{1}{16}\frac{a^2xu}{c} - \frac{1}{16}\frac{a^3}{c}I_1$        DW

2.    $\displaystyle\int x^2 u \,dx = \frac{1}{4}\frac{xu^3}{c} - \frac{1}{8}\frac{axu}{c} - \frac{1}{8}\frac{a^2}{c}I_1$        DW

3.    $\displaystyle\int \frac{x^2}{u}\,dx = \frac{1}{2}\frac{xu}{c} - \frac{1}{2}\frac{a}{c}I_1$        DW

4.    $\displaystyle\int \frac{x^2}{u^3}\,dx = -\frac{x}{cu} + \frac{1}{c}I_1$        DW

5.    $\displaystyle\int \frac{x^2}{u^5}\,dx = \frac{1}{3}\frac{x^3}{au^3}$        DW

6.    $\displaystyle\int \frac{x^2\,dx}{u^{2n+1}} = \frac{1}{a^{n-1}}\sum_{k=0}^{n-2}\frac{(-1)^k}{2k+3}\binom{n-2}{k}\frac{c^k x^{2k+3}}{u^{2k+3}}$

7.    $\displaystyle\int \frac{x^3\,dx}{u^{2n+1}} = -\frac{1}{(2n-3)c^2u^{2n-3}} + \frac{a}{(2n-1)c^2u^{2n-1}}$        DW

**2.273**

1.    $\displaystyle\int x^4 u^3 \,dx = \frac{1}{8}\frac{x^3u^5}{c} - \frac{axu^5}{16c^2} + \frac{a^2xu^3}{64c^2} + \frac{3a^3xu}{128c^2} + \frac{3a^4}{128c^2}I_1$        DW

2.    $\displaystyle\int x^4 u \,dx = \frac{1}{6}\frac{x^3u^3}{c} - \frac{axu^3}{8c^2} + \frac{a^2xu}{16c^2} + \frac{a^3}{16c^2}I_1$        DW

3.    $\displaystyle\int \frac{x^4}{u}\,dx = \frac{1}{4}\frac{x^3u}{c} - \frac{3}{8}\frac{axu}{c^2} + \frac{3}{8}\frac{a^2}{c^2}I_1$        DW

4.    $\displaystyle\int \frac{x^4}{u^3}\,dx = \frac{1}{2}\frac{xu}{c^2} + \frac{ax}{c^2u} - \frac{3}{2}\frac{a}{c^2}I_1$        DW

5.    $\displaystyle\int \frac{x^4}{u^5}\,dx = -\frac{x}{c^2u} - \frac{1}{3}\frac{x^3}{cu^3} + \frac{1}{c^2}I_1$        DW

6.    $\displaystyle\int \frac{x^4}{u^7}\,dx = \frac{1}{5}\frac{x^5}{au^5}$        DW

7.    $\displaystyle\int \frac{x^4\,dx}{u^{2n+1}} = \frac{1}{a^{n-2}}\sum_{k=0}^{n-3}\frac{(-1)^k}{2k+5}\binom{n-3}{k}\frac{c^k x^{2k+5}}{u^{2k+5}}$

8.$^{12}$ $\displaystyle\int \frac{x^5\,dx}{u^{2n+1}} = -\frac{1}{(2n-5)c^3 u^{2n-5}} + \frac{2a}{(2n-3)c^3 u^{2n-3}} - \frac{a^2}{(2n-1)c^3 u^{2n-1}}$    DW

**2.274**

1.   $\displaystyle\int x^6 u^3\,dx = \frac{1}{10}\frac{x^5 u^5}{c} - \frac{ax^3 u^5}{16c^2} + \frac{a^2 x u^5}{32c^3} - \frac{a^3 x u^3}{128c^3} - \frac{3a^4 x u}{256c^3} - \frac{3}{256}\frac{a^5}{c^3}I_1$

2.   $\displaystyle\int x^6 u\,dx = \frac{1}{8}\frac{x^5 u^3}{c} - \frac{5}{48}\frac{ax^3 u^3}{c^2} + \frac{5a^2 x u^3}{64c^3} - \frac{5a^3 x u}{128c^3} - \frac{5}{128}\frac{a^4}{c^3}I_1$

3.   $\displaystyle\int \frac{x^6}{u}\,dx = \frac{1}{6}\frac{x^5 u}{c} - \frac{5}{24}\frac{ax^3 u}{c^2} + \frac{5}{16}\frac{a^2 x u}{c^3} - \frac{5}{16}\frac{a^3}{c^3}I_1$    DW

4.   $\displaystyle\int \frac{x^6}{u^3}\,dx = \frac{1}{4}\frac{x^5}{cu} - \frac{5}{8}\frac{ax^3}{c^2 u} - \frac{15}{8}\frac{a^2 x}{c^3 u} + \frac{15}{8}\frac{a^2}{c^3}I_1$    DW

5.   $\displaystyle\int \frac{x^6}{u^5}\,dx = \frac{1}{2}\frac{x^5}{cu^3} + \frac{10}{3}\frac{ax^3}{c^2 u^3} + \frac{5}{2}\frac{a^2 x}{c^3 u^3} - \frac{5}{2}\frac{a}{c^3}I_1$    DW

6.   $\displaystyle\int \frac{x^6}{u^7}\,dx = -\frac{23}{15}\frac{x^5}{cu^5} - \frac{7}{3}\frac{ax^3}{c^2 u^5} - \frac{a^2 x}{c^3 u^5} + \frac{1}{c^3}I_1$    DW

7.   $\displaystyle\int \frac{x^6}{u^9}\,dx = \frac{1}{7}\frac{x^7}{au^7}$    DW

8.   $\displaystyle\int \frac{x^6\,dx}{u^{2n+1}} = \frac{1}{a^{n-3}}\sum_{k=0}^{n-4}\frac{(-1)^k}{2k+7}\binom{n-4}{k}\frac{c^k x^{2k+7}}{u^{2k+7}}$

9.   $\displaystyle\int \frac{x^7\,dx}{u^{2n+1}} = -\frac{1}{(2n-7)c^4 u^{2n-7}} + \frac{3a}{(2n-5)c^4 u^{2n-5}} - \frac{3a^2}{(2n-3)c^4 u^{2n-3}} + \frac{a^3}{(2n-1)c^4 u^{2n-1}}$    DW

**2.275**

1.   $\displaystyle\int \frac{u^5}{x}\,dx = \frac{u^5}{5} + \frac{1}{3}au^3 + a^2 u + a^3 I_2$    DW

2.   $\displaystyle\int \frac{u^3}{x}\,dx = \frac{u^3}{3} + au + a^2 I_2$    DW

3.   $\displaystyle\int \frac{u}{x}\,dx = u + a I_2$    DW

4.   $\displaystyle\int \frac{dx}{xu} = I_2$    DW

5.   $\displaystyle\int \frac{dx}{xu^{2n+1}} = \frac{1}{a^n}I_2 + \sum_{k=0}^{n-1}\frac{1}{(2k+1)a^{n-k}u^{2k+1}}$

6.   $\displaystyle\int \frac{u^5}{x^2}\,dx = -\frac{u^5}{x} + \frac{5}{4}cxu^3 + \frac{15}{8}acxu + \frac{15}{8}a^2 I_1$    DW

7.   $\displaystyle\int \frac{u^3}{x^2}\,dx = -\frac{u^3}{x} + \frac{3}{2}cxu + \frac{3}{2}aI_1$    DW

8.$^{12}$ $\displaystyle\int \frac{u}{x^2}\,dx = -\frac{u}{x} + I_1$    DW

9. $$\int \frac{dx}{x^2 u^{2n+1}} = -\frac{1}{a^{n+1}} \left\{ \frac{u}{x} + \sum_{k=1}^{n} \frac{(-1)^{k+1}}{2k-1} \binom{n}{k} c^k \left(\frac{x}{u}\right)^{2k-1} \right\}$$

**2.276**

1. $$\int \frac{u^5}{x^3} \, dx = -\frac{u^5}{2x^2} + \frac{5}{6}cu^3 + \frac{5}{2}acu + \frac{5}{2}a^2 c I_2 \qquad \text{DW}$$

2. $$\int \frac{u^3}{x^3} \, dx = -\frac{u^3}{2x^2} + \frac{3}{2}cu + \frac{3}{2}ac I_2 \qquad \text{DW}$$

3. $$\int \frac{u}{x^3} \, dx = -\frac{u}{2x^2} + \frac{c}{2}I_2 \qquad \text{DW}$$

4. $$\int \frac{dx}{x^3 u} = -\frac{u}{2ax^2} - \frac{c}{2a}I_2 \qquad \text{DW}$$

5. $$\int \frac{dx}{x^3 u^3} = -\frac{1}{2ax^2 u} - \frac{3c}{2a^2 u} - \frac{3c}{2a^2}I_2 \qquad \text{DW}$$

6. $$\int \frac{dx}{x^3 u^5} = -\frac{1}{2ax^2 u^3} - \frac{5}{6}\frac{c}{a^2 u^3} - \frac{5}{2}\frac{c}{a^3 u} - \frac{5}{2}\frac{c}{a^3}I_2 \qquad \text{DW}$$

7. $$\int \frac{u^5}{x^4} \, dx = -\frac{au^3}{3x^3} - \frac{2acu}{x} + \frac{c^2 xu}{2} + \frac{5}{2}ac I_1 \qquad \text{DW}$$

8. $$\int \frac{u^3}{x^4} \, dx = -\frac{u^3}{3x^3} - \frac{cu}{x} + c I_1 \qquad \text{DW}$$

9. $$\int \frac{u}{x^4} \, dx = -\frac{u^3}{3ax^3} \qquad \text{DW}$$

10. $$\int \frac{dx}{x^4 u^{2n+1}} = \frac{1}{a^{n+2}} \left\{ -\frac{u^3}{3x^3} + (n+1)\frac{cu}{x} + \sum_{k=2}^{n+1} \frac{(-1)^k}{2k-3} \binom{n+1}{k} c^k \left(\frac{x}{u}\right)^{2k-3} \right\}$$

**2.277**

1. $$\int \frac{u^3}{x^5} \, dx = -\frac{u^3}{4x^4} - \frac{3}{8}\frac{cu^3}{ax^2} + \frac{3}{8}\frac{c^2 u}{a} + \frac{3}{8}c^2 I_2 \qquad \text{DW}$$

2. $$\int \frac{u}{x^5} \, dx = -\frac{u}{4x^4} - \frac{1}{8}\frac{cu}{ax^2} - \frac{1}{8}\frac{c^2}{a}I_2 \qquad \text{DW}$$

3. $$\int \frac{dx}{x^5 u} = -\frac{u}{4ax^4} + \frac{3}{8}\frac{cu}{a^2 x^2} + \frac{3}{8}\frac{c^2}{a^2}I_2 \qquad \text{DW}$$

4. $$\int \frac{dx}{x^5 u^3} = -\frac{1}{4ax^4 u} + \frac{5}{8}\frac{c}{a^2 x^2 u} + \frac{15}{8}\frac{c^2}{a^3 u} + \frac{15}{8}\frac{c^2}{a^3}I_2 \qquad \text{DW}$$

$$\text{DW}$$

**2.278**

1. $$\int \frac{u^3}{x^6} \, dx = -\frac{u^5}{5ax^5} \qquad \text{DW}$$

2. $$\int \frac{u}{x^6} \, dx = -\frac{u^3}{5ax^5} + \frac{2}{15}\frac{cu^3}{a^2 x^3} \qquad \text{DW}$$

3. $\displaystyle \int \frac{dx}{x^6 u} = \frac{1}{a^3}\left(-\frac{u^5}{5x^5} + \frac{2}{3}\frac{cu^3}{x^3} - \frac{c^2 u}{x}\right)$     DW

4. $\displaystyle \int \frac{dx}{x^6 u^{2n+1}} = \frac{1}{a^{n+3}}\left\{-\frac{u^5}{5x^5} + \frac{1}{3}\binom{n+2}{1}\frac{cu^3}{x^3} - \binom{n+2}{2}\frac{c^2 u}{x} + \sum_{k=3}^{n+2}\frac{(-1)^k}{2k-5}\binom{n+2}{k}c^k\left(\frac{x}{u}\right)^{2k-5}\right\}$

## 2.28 Forms containing $\sqrt{a+bx+cx^2}$ and first-and second-degree polynomials

**Notation:** $R = a + bx + cx^2$

**See also 2.252**

**2.281**[3]     $\displaystyle \int \frac{dx}{(x+p)^n \sqrt{R}} = -\int \frac{t^{n-1}\,dt}{\sqrt{c + (b-2pc)t + (a-bp+cp^2)t^2}}$

$$\left[t = \frac{1}{x+p} > 0\right]$$

**2.282**

1.[3]   $\displaystyle \int \frac{\sqrt{R}\,dx}{x+p} = c\int \frac{x\,dx}{\sqrt{R}} + (b-cp)\int \frac{dx}{\sqrt{R}} + \left(a - bp + cp^2\right)\int \frac{dx}{(x+p)\sqrt{R}}$

$$[x+p > 0]$$

2.   $\displaystyle \int \frac{dx}{(x+p)(x+q)\sqrt{R}} = \frac{1}{q-p}\int \frac{dx}{(x+p)\sqrt{R}} + \frac{1}{p-q}\int \frac{dx}{(x+q)\sqrt{R}}$

3.   $\displaystyle \int \frac{\sqrt{R}\,dx}{(x+p)(x+q)} = \frac{1}{q-p}\int \frac{\sqrt{R}\,dx}{x+p} + \frac{1}{p-q}\int \frac{\sqrt{R}\,dx}{x+q}$

4.   $\displaystyle \int \frac{(x+p)\sqrt{R}\,dx}{x+q} = \int \sqrt{R}\,dx + (p-q)\int \frac{\sqrt{R}\,dx}{x+q}$

5.   $\displaystyle \int \frac{(rx+s)\,dx}{(x+p)(x+q)\sqrt{R}} = \frac{s-pr}{q-p}\int \frac{dx}{(x+p)\sqrt{R}} + \frac{s-qr}{p-q}\int \frac{dx}{(x+q)\sqrt{R}}$

**2.283**   $\displaystyle \int \frac{(Ax+B)\,dx}{(p+R)^n \sqrt{R}} = \frac{A}{c}\int \frac{du}{(p+u^2)^n} + \frac{2Bc-Ab}{2c}\int \frac{\left(1-cv^2\right)^{n-1}dv}{\left[p+a-\dfrac{b^2}{4c}-cpv^2\right]^n},$

where $u = \sqrt{R}$ and $v = \dfrac{b+2cx}{2c\sqrt{R}}$.

**2.284**[12]   $\displaystyle \int \frac{Ax+B}{(p+R)\sqrt{R}}\,dx = \frac{A}{c}I_1 + \frac{2Bc-Ab}{c\sqrt{p\left[b^2-4(a+p)c\right]}}I_2,$

where

$I_1 = \dfrac{1}{\sqrt{p}}\arctan\sqrt{\dfrac{R}{p}}$     $[p > 0]$

$\phantom{I_1} = \dfrac{1}{2\sqrt{-p}}\ln\dfrac{\sqrt{-p}-\sqrt{R}}{\sqrt{-p}+\sqrt{R}}$     $[p < 0]$

$$I_2 = \arctan \sqrt{\frac{p}{b^2 - 4(a+p)c}} \frac{b+2cx}{\sqrt{R}} \qquad \left[p\left\{b^2 - 4(a+p)c\right\} > 0, \quad p < 0\right]$$

$$= -\arctan \frac{p}{\sqrt{p\left[b^2 - 4(a+p)c\right]}} \frac{b+2cx}{\sqrt{R}} \qquad \left[p\left\{b^2 - 4(a+p)c\right\} > 0, \quad p > 0\right]$$

$$= \frac{i}{2} \ln\left(\frac{\sqrt{p\left[4(a+p)c - b^2\right]}\sqrt{R} + p(b+2cx)}{\sqrt{p\left[4(a+p)c - b^2\right]}\sqrt{R} - p(b+2cx)}\right) \qquad \left[p\left\{b^2 - 4(a+p)c\right\} < 0\right]$$

## 2.29 Integrals that can be reduced to elliptic or pseudo-elliptic integrals

**2.290**　Integrals of the form $\int R\left(x, \sqrt{P(x)}\right)\,dx$, where $P(x)$ is a third-or fourth-degree polynomial can, by means of algebraic transformations, be reduced to a sum of integrals expressed in terms of elementary functions and elliptic integrals (see **8.11**). Since the substitutions that transform the given integral into an elliptic integral in the normal Legendre form are different for different intervals of integration, the corresponding formulas are given in the chapter on definite integrals (see **3.13**, **3.17**).

**2.291**　Certain integrals of the form $\int R\left(x, \sqrt{P(x)}\right)\,dx$, where $P_n(x)$ is a polynomial of not more than fourth degree, can be reduced to integrals of the form $\int R\left(x, \sqrt[k]{P_n(x)}\right)\,dx$ with $k \geq 2$. Below are examples of this procedure.

1.　$$\int \frac{dx}{\sqrt{1-x^6}} = -\int \frac{dz}{\sqrt{3+3z^2+z^4}} \qquad \left[x^2 = \frac{1}{1+z^2}\right]$$

2.　$$\int \frac{dx}{\sqrt{a+bx^2+cx^4+dx^6}} = \frac{1}{2}\int \frac{dz}{\sqrt{az+bz^2+cz^3+dz^4}}$$

$$\left[x^2 = z\right]$$

3.　$$\int \left(a+2bx+cx^2+gx^3\right)^{\pm 1/3}\,dx = \frac{3}{2}\int \frac{z^2 A^{\pm\frac{1}{3}}\,dz}{B}$$

$$\left[a+2bx+cx^2 = z^3, \quad A = g\left(\frac{-b+\sqrt{b^2+(z^3-a)c}}{c}\right)^3 + z^3, \quad B = \sqrt{b^2+(z^3-a)c}\,\right]$$

This integral involving a one-third power cannot be reduced to an elliptic integral.

4.[12]　$$\int \frac{dx}{\sqrt{a+bx+cx^2+dx^3+cx^4+bx^5+ax^6}}$$

$$= -\frac{1}{\sqrt{2}}\int \frac{dx}{\sqrt{(z+1)p}} - \frac{1}{\sqrt{2}}\int \frac{dz}{\sqrt{(z-1)p}} \qquad \left[x = z + \sqrt{z^2-1}\right]$$

$$= -\frac{1}{\sqrt{2}}\int \frac{dx}{\sqrt{(z+1)p}} + \frac{1}{\sqrt{2}}\int \frac{dz}{\sqrt{(z-1)p}} \qquad \left[x = z - \sqrt{z^2-1}\right]$$

where $p = 2a\left(4z^3 - 3z\right) + 2b\left(2z^2 - 1\right) + 2cz + d$.

5.[12]
$$\int \frac{dx}{\sqrt{a + bx^2 + cx^4 + bx^6 + ax^8}} = \frac{1}{2} \int \frac{dy}{\sqrt{y}\sqrt{a + by + cy^2 + by^3 + ay^4}} \qquad [x = \sqrt{y}]$$

$$= -\frac{1}{2\sqrt{2}} \int \frac{dz}{\sqrt{(z+1)p}} - \frac{1}{2\sqrt{2}} \int \frac{dz}{\sqrt{(z-1)p}} \qquad \left[ y = z + \sqrt{z^2 - 1} \right]$$

$$= -\frac{1}{2\sqrt{2}} \int \frac{dz}{\sqrt{(z+1)p}} + \frac{1}{2\sqrt{2}} \int \frac{dz}{\sqrt{(z-1)p}} \qquad \left[ y = z - \sqrt{z^2 - 1} \right]$$

where $p = 2a \left(2z^2 - 1\right) + 2bz + c$.

6.[12]
$$\int \frac{dx}{\sqrt{a + bx^4 + cx^8}} = \frac{1}{2} \sqrt[8]{\frac{a}{c}} \int \frac{dt}{\sqrt{t}\sqrt{a + b_1 t^2 + at^4}} \qquad \left[ x = \sqrt[8]{\frac{a}{c}}\sqrt{t} \right];$$

$$= -\frac{1}{2\sqrt{2}} \sqrt[8]{\frac{a}{c}} \left\{ \int \frac{dz}{\sqrt{(z+1)p}} - \int \frac{dz}{\sqrt{(z-1)p}} \right\} \qquad \left[ t = z + \sqrt{z^2 - 1} \right]$$

$$= -\frac{1}{2\sqrt{2}} \sqrt[8]{\frac{a}{c}} \left\{ \int \frac{dz}{\sqrt{(z+1)p}} + \int \frac{dz}{\sqrt{(z-1)p}} \right\} \qquad \left[ t = z - \sqrt{z^2 - 1} \right]$$

where $p = 2a \left(2z^2 - 1\right) + b_1; \ b_1 = b\sqrt{\frac{a}{c}}$.

7.
$$\int \frac{x\,dx}{\sqrt[4]{a + bx^2 + cx^4}} = 2 \int \frac{z^2\,dz}{\sqrt{A + Bz^4}} \qquad \left[ a + bx^2 + cx^4 = z^4, \quad A = b^2 - 4ac, \quad B = 4c \right]$$

8.
$$\int \frac{dx}{\sqrt[4]{a + 2bx^2 + cx^4}} = \int \frac{\sqrt{b^2 - a\left(c - z^4\right)} + b}{\left(c - z^4\right)\sqrt{b^2 - a\left(c - z^4\right)}} z^2\,dz = \int R_1\left(z^4\right) z^2\,dz + \int \frac{R_2\left(z^4\right) z^2\,dz}{\sqrt{b^2 - a\left(c - z^4\right)}},$$

where $R_1\left(z^4\right)$ and $R_2\left(z^4\right)$ are rational functions of $z^4$ and $a + 2bx^2 + cx^4 = x^4 z^4$.

**2.292**[12]   In certain cases, integrals of the form $\int R\left(x, \sqrt{P(x)}\right) dx$, where $P(x)$ is a third-or fourth-degree polynomial, can be expressed in terms of elementary functions. Such integrals are called *pseudo-elliptic* integrals.

Thus, if the relations

$$f_1(x) = -f_1\left(\frac{1}{k^2 x}\right), \qquad f_2(x) = -f_2\left(\frac{1 - k^2 x}{k^2(1 - x)}\right), \qquad f_3(x) = -f_3\left(\frac{1 - x}{1 - k^2 x}\right),$$

hold, then

1.
$$\int \frac{f_1(x)\,dx}{\sqrt{x(1 - x)\left(1 - k^2 x\right)}} = \int R_1(z)\,dz \qquad \left[ zx = \sqrt{x(1 - x)\left(1 - k^2 x\right)} \right]$$

2.
$$\int \frac{f_2(x)\,dx}{\sqrt{x(1 - x)\left(1 - k^2 x\right)}} = \int R_2(z)\,dz \qquad \left[ z = \frac{\sqrt{x\left(1 - k^2 x\right)}}{\sqrt{1 - x}} \right]$$

3.
$$\int \frac{f_3(x)\,dx}{\sqrt{x(1 - x)\left(1 - k^2 x\right)}} = \int R_3(z)\,dz \qquad \left[ z = \frac{\sqrt{x(1 - x)}}{\sqrt{1 - k^2 x}} \right]$$

where $R_1(z)$, $R_2(z)$, and $R_3(z)$ are rational functions of $z$.

## 2.3 The Exponential Function

### 2.31 Forms containing $e^{ax}$

**2.311** $\displaystyle\int e^{ax}\,dx = \frac{e^{ax}}{a}$

**2.312** $a^x$ in an integrand should be replaced with $e^{x\ln a} = a^x$

**2.313**

1. $\displaystyle\int \frac{dx}{a + be^{mx}} = \frac{1}{am}\left[mx - \ln\left(a + be^{mx}\right)\right]$ $\qquad$ PE (410)

2. $\displaystyle\int \frac{dx}{1 + e^x} = \ln\frac{e^x}{1 + e^x} = x - \ln\left(1 + e^x\right)$ $\qquad$ PE (409)

**2.314** $\displaystyle\int \frac{dx}{ae^{mx} + be^{-mx}} = \frac{1}{m\sqrt{ab}}\arctan\left(e^{mx}\sqrt{\frac{a}{b}}\right)$ $\qquad [ab > 0]$ $\qquad$ PE (411)

$$= \frac{1}{2m\sqrt{-ab}}\ln\left|\frac{b + e^{mx}\sqrt{-ab}}{b - e^{mx}\sqrt{-ab}}\right| \qquad [ab < 0]$$

**2.315** $\displaystyle\int \frac{dx}{\sqrt{a + be^{mx}}} = \frac{1}{m\sqrt{a}}\ln\frac{\sqrt{a + be^{mx}} - \sqrt{a}}{\sqrt{a + be^{mx}} + \sqrt{a}}$ $\qquad [a > 0]$

$$= \frac{2}{m\sqrt{-a}}\arctan\frac{\sqrt{a + be^{mx}}}{\sqrt{-a}} \qquad [a < 0]$$

**2.316\*** $\displaystyle\int \exp\left(a^x\right)dx = \frac{\mathrm{Ei}\left(a^x\right)}{\ln a}$

## 2.32 The exponential combined with rational functions of $x$

**2.321**

1.[12] $\displaystyle\int x^n e^{ax}\,dx = \frac{x^n e^{ax}}{a} - \frac{n}{a}\int x^{n-1}e^{ax}\,dx$ $\qquad [a \neq 0]$

2.[11] $\displaystyle\int x^n e^{ax}\,dx = e^{ax}\left(\sum_{k=0}^{n} \frac{(-1)^k k!\binom{n}{k}}{a^{k+1}}x^{n-k}\right)$

**2.322**

1. $\displaystyle\int xe^{ax}\,dx = e^{ax}\left(\frac{x}{a} - \frac{1}{a^2}\right)$

2. $\displaystyle\int x^2 e^{ax}\,dx = e^{ax}\left(\frac{x^2}{a} - \frac{2x}{a^2} + \frac{2}{a^3}\right)$

3. $\displaystyle\int x^3 e^{ax}\,dx = e^{ax}\left(\frac{x^3}{a} - \frac{3x^2}{a^2} + \frac{6x}{a^3} - \frac{6}{a^4}\right)$

4.[10] $\displaystyle\int x^4 e^{ax}\,dx = e^{ax}\left(\frac{x^4}{a} - \frac{4x^3}{a^2} + \frac{12x^2}{a^3} - \frac{24x}{a^4} + \frac{24}{a^5}\right)$

**2.323** $\displaystyle\int P_m(x)e^{ax}\,dx = \frac{e^{ax}}{a}\sum_{k=0}^{m}(-1)^k\frac{P^{(k)}(x)}{a^k}$,

where $P_m(x)$ is a polynomial in $x$ of degree $m$ and $P^{(k)}(x)$ is the $k$-th derivative of $P_m(x)$ with respect to $x$.

**2.324**

1.[12]   $$\int \frac{e^{ax}\, dx}{x^m} = \frac{1}{m-1}\left[-\frac{e^{ax}}{x^{m-1}} + a\int \frac{e^{ax}\, dx}{x^{m-1}}\right] \qquad [m \neq 1]$$

2.   $$\int \frac{e^{ax}}{x^n}\, dx = -e^{ax}\sum_{k=1}^{n-1} \frac{a^{k-1}}{(n-1)(n-2)\ldots(n-k)x^{n-k}} + \frac{a^{n-1}}{(n-1)!}\,\mathrm{Ei}(ax)$$

**2.325**

1.   $$\int \frac{e^{ax}}{x}\, dx = \mathrm{Ei}(ax)$$

2.   $$\int \frac{e^{ax}}{x^2}\, dx = -\frac{e^{ax}}{x} + a\,\mathrm{Ei}(ax)$$

3.   $$\int \frac{e^{ax}}{x^3}\, dx = -\frac{e^{ax}}{2x^2} - \frac{ae^{ax}}{2x} + \frac{a^2}{2}\,\mathrm{Ei}(ax)$$

4.   $$\int \frac{e^{ax}}{x^4}\, dx = -\frac{e^{ax}}{3x^3} - \frac{ae^{ax}}{6x^2} - \frac{a^2e^{ax}}{6x} + \frac{a^3}{6}\,\mathrm{Ei}(ax)$$

5.   $$\int \frac{e^{\pm ax^n}}{x^m}\, dx = \frac{1}{m-1}\left[-\frac{e^{\pm ax^n}}{x^{m-1}} \pm na\int \frac{e^{\pm ax^n}}{x^{m-n}}\, dx\right] \qquad [m \neq 1]$$

6.[12]   $$\int \frac{e^{ax^n}}{x^m}\, dx = \frac{(-1)^{z+1}\gamma\, a^z\, \Gamma\left(-z, -ax^n\right)}{n}$$
$$= \frac{(-1)^{z+1} a^z}{n}\int_{-ax^n}^{\infty} \frac{e^{-t}}{t^{z+1}}\, dt$$

$$z = \frac{m-1}{n}, \quad \text{for } \Gamma(\alpha, x) \text{ see } 8.350.2 \quad [a \neq 0, \quad n \neq 0]$$

7.   $$\int \frac{e^{ax^n}}{x}\, dx = \frac{\mathrm{Ei}\left(ax^n\right)}{n} \qquad\qquad [a \neq 0, \quad n \neq 0]$$

8.[12]   $$\int \frac{e^{ax^n}}{x^m}\, dx = -e^{ax^n}\frac{\sum_{k=0}^{z-1} k!\, \frac{a^{z-k-1}}{x^{n(k+1)}}}{nz!} + \frac{a^z\,\mathrm{Ei}\left(ax^n\right)}{nz!}$$

$$\left[a \neq 0, \quad n \neq 0, \quad z = \frac{m-1}{n} = 1, 2, \ldots, \quad m = 2, 3, \ldots\right]$$

9.   $$\int \frac{e^{ax^n}}{x^m}\, dx = -\frac{e^{ax^n}}{nx^n} + \frac{a\,\mathrm{Ei}\left(ax^n\right)}{n} \qquad\qquad \left[a \neq 0, \quad z = \frac{m-1}{n} = 1\right]$$

10.   $$\int \frac{e^{ax^n}}{x^m}\, dx = -\frac{e^{ax^n}}{2nx^{2n}} - \frac{ae^{ax^n}}{2nx^n} + \frac{a^2\,\mathrm{Ei}\left(ax^n\right)}{2n} \qquad \left[a \neq 0, \quad z = \frac{m-1}{n} = 2\right]$$

11.   $$\int \frac{e^{ax^n}}{x^m}\, dx = -\frac{e^{ax^n}}{3nx^{3n}} - \frac{e^{ax^n}}{6nx^{2n}} - \frac{a^2e^{ax^n}}{6nx^n} + \frac{a^3\,\mathrm{Ei}\left(ax^n\right)}{6n}$$

$$\left[a \neq 0, \quad z = \frac{m-1}{n} = 3\right]$$

12.   $$\int \frac{e^{ax^2}}{x^2}\, dx = -\frac{e^{ax^2}}{x} + \sqrt{a\pi}\,\mathrm{erfi}\left(\sqrt{a}x\right) \qquad \text{where } \mathrm{erfi}(z) = \frac{\mathrm{erf}(iz)}{i}$$

**2.326**  $\displaystyle\int \frac{x e^{ax}\,\mathrm{d}x}{(1+ax)^2} = \frac{e^{ax}}{a^2(1+ax)}$                                    $[a \neq 0]$

**2.33**

1.[8]  $\displaystyle\int e^{-(ax^2+2bx+c)}\,\mathrm{d}x = \frac{1}{2}\sqrt{\frac{\pi}{a}}\exp\left(\frac{b^2-ac}{a}\right)\operatorname{erf}\left(\sqrt{a}x + \frac{b}{\sqrt{a}}\right)$

$$[a \neq 0]$$

2.  $\displaystyle\int e^{ax^2}\,\mathrm{d}x = \frac{1}{2}\sqrt{\frac{\pi}{a}}\operatorname{erfi}\left(\sqrt{a}x\right)$                  where $\operatorname{erfi}(z) = \dfrac{\operatorname{erf}(iz)}{i}$      $[a \neq 0]$

3.[12]  $\displaystyle\int e^{ax^2+bx+c}\,\mathrm{d}x = \frac{1}{2}\sqrt{\frac{\pi}{a}}\exp\left(\frac{4ac-b^2}{4a}\right)\operatorname{erfi}\left(\sqrt{a}x + \frac{b}{2\sqrt{a}}\right)$

where $\operatorname{erfi}(z) = \dfrac{\operatorname{erf}(iz)}{i}$      $[a \neq 0]$

4.[12]  $\displaystyle\int x^m e^{\pm ax^n}\,\mathrm{d}x = \pm\frac{x^{m+1-n}}{na}e^{\pm ax^n} \mp \frac{m+1-n}{na}\int x^{m-n}e^{\pm ax^n}\,\mathrm{d}x$

$$[a \neq 0, \quad n \neq 0]$$

5.[12]  $\displaystyle\int x^m e^{ax^n}\,\mathrm{d}x = \frac{e^{ax^n}}{n}\left[(\gamma-1)!\sum_{k=0}^{\gamma-1}(-1)^{k+1-\gamma}\frac{x^{nk}}{k!a^{\gamma-k}}\right]$

$$\left[a \neq 0, \quad n \neq 0, \quad \gamma = \frac{m+1}{n} = 1,2,\ldots\right]$$

6.  $\displaystyle\int x^m e^{ax^n}\,\mathrm{d}x = \frac{e^{ax^n}}{na}$                  $\left[a \neq 0, \text{ if } \dfrac{m+1}{n} = 1\right]$

7.  $\displaystyle\int x^m e^{ax^n}\,\mathrm{d}x = \frac{e^{ax^n}}{n}\left(\frac{x^n}{a} - \frac{1}{a^2}\right)$                  $\left[a \neq 0, \text{ if } \dfrac{m+1}{n} = 2\right]$

8.  $\displaystyle\int x^m e^{ax^n}\,\mathrm{d}x = \frac{e^{ax^n}}{n}\left(\frac{x^{2n}}{a} - \frac{2x^n}{a^2} + \frac{2}{a^3}\right)$                  $\left[a \neq 0, \text{ if } \dfrac{m+1}{n} = 3\right]$

9.  $\displaystyle\int x^m e^{ax^n}\,\mathrm{d}x = \frac{e^{ax^n}}{n}\left(\frac{x^{3n}}{a} - \frac{3x^{2n}}{a^2} + \frac{6x^n}{a^3} - \frac{6}{a^4}\right)$                  $\left[a \neq 0, \text{ if } \dfrac{m+1}{n} = 4\right]$

10.  $\displaystyle\int x^m e^{-\beta x^n}\,\mathrm{d}x = -\frac{\Gamma\left(\gamma,\beta x^n\right)}{n\beta^\gamma}$                  for $\Gamma(\alpha,x)$ see 8.350.2

$$= -\frac{1}{n\beta^\gamma}\int_{\beta x^n}^{\infty} t^{\gamma-1}e^{-t}\,\mathrm{d}t \qquad \left[\gamma = \frac{m+1}{n}, \quad \beta \neq 0, \quad n \neq 0\right]$$

11.  $\displaystyle\int x^m \exp\left(-\beta x^n\right)\,\mathrm{d}x = -\frac{(\gamma-1)!}{n}\exp\left(-\beta x^n\right)\left[\sum_{k=0}^{\gamma-1}\frac{x^{nk}}{k!\beta^{\gamma-k}}\right]$

$$\left[\gamma = \frac{m+1}{n} = 1,2,\ldots\right]$$

12.  $\displaystyle\int x^m \exp\left(-\beta x^n\right)\,\mathrm{d}x = -\frac{\exp\left(-\beta x^n\right)}{n\beta}$                  $\left[\text{if } \dfrac{m+1}{n} = 1\right]$

13.    $\displaystyle \int x^m \exp\left(-\beta x^n\right) dx = -\frac{\exp\left(-\beta x^n\right)}{n} \left(\frac{x^n}{\beta} + \frac{1}{\beta^2}\right)$     $\left[\text{if } \dfrac{m+1}{n} = 2\right]$

14.    $\displaystyle \int x^m \exp\left(-\beta x^n\right) dx = -\frac{\exp\left(-\beta x^n\right)}{n} \left(\frac{x^{2n}}{\beta} + \frac{2x^n}{\beta^2} + \frac{2}{\beta^3}\right)$

$$\left[\text{if } \frac{m+1}{n} = 3\right]$$

15.    $\displaystyle \int x^m \exp\left(-\beta x^n\right) dx = -\frac{\exp\left(-\beta x^n\right)}{n} \left(\frac{x^{3n}}{\beta} + \frac{3x^{2n}}{\beta^2} + \frac{6x^n}{\beta^3} + \frac{6}{\beta^4}\right)$

$$\left[\text{if } \frac{m+1}{n} = 4\right]$$

16.[12]    $\displaystyle \int e^{-\beta x^2} dx = \frac{1}{2}\sqrt{\frac{\pi}{\beta}} \exp\left(\sqrt{\beta}x\right)$        $[\beta \neq 0]$

17.[12]    $\displaystyle \int \frac{\exp\left(-\beta x^n\right)}{x^m} dx = -\frac{\beta^z \, \Gamma\left(-z, \beta x^n\right)}{n}$

$$= -\frac{\beta^z}{n} \int_{\beta x^n}^{\infty} \frac{e^{-t}}{t^{z+1}} \, dt$$

$$\left[n \neq 0, \quad \beta \neq 0, \quad z = \frac{m-1}{n}\right]$$

18.    $\displaystyle \int \frac{\exp\left(-\beta x^n\right)}{x} dx = \frac{\text{Ei}\left(-\beta x^n\right)}{n}$

19.[12]    $\displaystyle \int \frac{\exp\left(-\beta x^n\right)}{x^m} dx = (-1)^z \frac{\text{Ei}\left(-\beta x^n\right)}{nz!} \sum_{k=0}^{z-1} (-1)^k k! \frac{\beta^{z-k-1}}{x^{n(k+1)}} + (-1)^z \frac{\beta^z}{nz!} \text{Ei}\left(-\beta x^n\right)$

$$\left[n \neq 0, \quad \beta \neq 0, \quad z = \frac{m-1}{n} = 1, 2, \ldots, \quad m = 2, 3, \ldots\right]$$

20.    $\displaystyle \int \frac{\exp\left(-\beta x^n\right)}{x^m} dx = -\frac{\exp\left(-\beta x^n\right)}{nx^n} - \frac{\beta \, \text{Ei}\left(-\beta x^n\right)}{n}$     $\left[z = \dfrac{m-1}{n} = 1\right]$

21.    $\displaystyle \int \frac{\exp\left(-\beta x^n\right)}{x^m} dx = -\frac{\exp\left(-\beta x^n\right)}{2nx^{2n}} + \frac{\beta \exp\left(-\beta x^n\right)}{2nx^n} + \frac{\beta^2 \, \text{Ei}\left(-\beta x^n\right)}{2n}$

$$\left[z = \frac{m-1}{n} = 2\right]$$

22.    $\displaystyle \int \frac{\exp\left(-\beta x^n\right)}{x^m} dx = -\frac{\exp\left(-\beta x^n\right)}{3nx^{3n}} + \frac{\beta \exp\left(-\beta x^n\right)}{6nx^{2n}} - \frac{\beta^2 \exp\left(-\beta x^n\right)}{6nx^n} - \frac{\beta^3 \, \text{Ei}\left(-\beta x^n\right)}{6n}$

$$\left[z = \frac{m-1}{n} = 3\right]$$

23.    $\displaystyle \int \frac{\exp\left(-\beta x^n\right)}{x^2} dx = -\frac{\beta \exp\left(-\beta x^2\right)}{x} - \sqrt{\beta\pi} \, \text{erf}\left(\sqrt{\beta}x\right)$

## 2.4 Hyperbolic Functions

### 2.41–2.43 Powers of $\sinh x$, $\cosh x$, $\tanh x$, and $\coth x$

**2.411**
$$\int \sinh^p x \cosh^q x \,\mathrm{d}x = \frac{\sinh^{p+1} x \cosh^{q-1} x}{p+q} + \frac{q-1}{p+q} \int \sinh^p x \cosh^{q-2} x \,\mathrm{d}x$$
$$= \frac{\sinh^{p-1} x \cosh^{q+1} x}{p+q} - \frac{p-1}{p+q} \int \sinh^{p-2} x \cosh^q x \,\mathrm{d}x$$
$$= \frac{\sinh^{p-1} x \cosh^{q+1} x}{q+1} - \frac{p-1}{q+1} \int \sinh^{p-2} x \cosh^{q+2} x \,\mathrm{d}x$$
$$= \frac{\sinh^{p+1} x \cosh^{q-1} x}{p+1} - \frac{q-1}{p+1} \int \sinh^{p+2} x \cosh^{q-2} x \,\mathrm{d}x$$
$$= \frac{\sinh^{p+1} x \cosh^{q+1} x}{p+1} - \frac{p+q+2}{p+1} \int \sinh^{p+2} x \cosh^q x \,\mathrm{d}x$$
$$= -\frac{\sinh^{p+1} x \cosh^{q+1} x}{q+1} + \frac{p+q+2}{q+1} \int \sinh^p x \cosh^{q+2} x \,\mathrm{d}x$$

**2.412**

1. $$\int \sinh^p x \cosh^{2n} x \,\mathrm{d}x = \frac{\sinh^{p+1} x}{2n+p}\left[\cosh^{2n-1} x\right.$$
$$+ \sum_{k=1}^{n-1} \frac{(2n-1)(2n-3)\ldots(2n-2k+1)}{(2n+p-2)(2n+p-4)\ldots(2n+p-2k)} \cosh^{2n-2k-1} x\left.\right]$$
$$+ \frac{(2n-1)!!}{(2n+p)(2n+p-2)\ldots(p+2)} \int \sinh^p x \,\mathrm{d}x$$
This formula is applicable for arbitrary real $p$ except for the following negative even integers: $-2$, $-4$, ..., $-2n$. If $p$ is a natural number and $n=0$, we have

2. $$\int \sinh^{2m} x \,\mathrm{d}x = (-1)^m \binom{2m}{m} \frac{x}{2^{2m}} + \frac{1}{2^{2m-1}} \sum_{k=0}^{m-1} (-1)^k \binom{2m}{k} \frac{\sinh(2m-2k)x}{2m-2k}$$    TI (543)

3. $$\int \sinh^{2m+1} x \,\mathrm{d}x = \frac{1}{2^{2m}} \sum_{k=0}^{m} (-1)^k \binom{2m+1}{k} \frac{\cosh(2m-2k+1)x}{2m-2k+1};$$    TI (544)
$$= (-1)^n \sum_{k=0}^{m} (-1)^k \binom{m}{k} \frac{\cosh^{2k+1} x}{2k+1}$$    GU (351) (5)

4. $$\int \sinh^p x \cosh^{2n+1} x \,\mathrm{d}x$$
$$= \frac{\sinh^{p+1} x}{2n+p+1}\left\{\cosh^{2n} x + \sum_{k=1}^{n} \frac{2^k n(n-1)\ldots(n-k+1)\cosh^{2n-2k} x}{(2n+p-1)(2n+p-3)\ldots(2n+p-2k+1)}\right\}$$
This formula is applicable for arbitrary real $p$ except for the following negative odd integers: $-1$, $-3$, ..., $-(2n+1)$.

**2.413**

1. $$\int \cosh^p x \sinh^{2n} x \, dx = \frac{\cosh^{p+1} x}{2n+p} \left[ \sinh^{2n-1} x \right.$$

$$+ \sum_{k=1}^{n-1}(-1)^k \frac{(2n-1)(2n-3)\ldots(2n-2k+1)\sinh^{2n-2k-1} x}{(2n+p-2)(2n+p-4)\ldots(2n+p-2k)} \left. \right]$$

$$+(-1)^n \frac{(2n-1)!!}{(2n+p)(2n+p-2)\ldots(p+2)} \int \cosh^p x \, dx$$

This formula is applicable for arbitrary real $p$ except for the following negative even integers: $-2$, $-4, \ldots, -2n$. If $p$ is a natural number and $n = 0$, we have

2. $$\int \cosh^{2m} x \, dx = \binom{2m}{m} \frac{x}{2^{2m}} + \frac{1}{2^{2m-1}} \sum_{k=0}^{m-1} \binom{2m}{k} \frac{\sinh(2m-2k)x}{2m-2k} \qquad \text{TI (541)}$$

3. $$\int \cosh^{2m+1} x \, dx = \frac{1}{2^{2m}} \sum_{k=0}^{m} \binom{2m+1}{k} \frac{\sinh(2m-2k+1)x}{2m-2k+1} \qquad \text{TI (542)}$$

$$= \sum_{k=0}^{m} \binom{m}{k} \frac{\sinh^{2k+1} x}{2k+1} \qquad \text{GU (351) (8)}$$

4. $$\int \cosh^p x \sinh^{2n+1} x \, dx = \frac{\cosh^{p+1} x}{2n+p+1} \left[ \sinh^{2n} x \right.$$

$$+ \sum_{k=1}^{n}(-1)^k \frac{2^k n(n-1)\ldots(n-k+1)\sinh^{2n-2k} x}{(2n+p-1)(2n+p-3)\ldots(2n+p-2k+1)} \left. \right]$$

This formula is applicable for arbitrary real $p$ except for the following negative odd integers: $-1$, $-3, \ldots, -(2n+1)$.

**2.414**

1. $$\int \sinh ax \, dx = \frac{1}{a} \cosh ax$$

2. $$\int \sinh^2 ax \, dx = \frac{1}{4a} \sinh 2ax - \frac{x}{2}$$

3. $$\int \sinh^3 x \, dx = -\frac{3}{4} \cosh x + \frac{1}{12} \cosh 3x = \frac{1}{3} \cosh^3 x - \cosh x$$

4. $$\int \sinh^4 x \, dx = \frac{3}{8} x - \frac{1}{4} \sinh 2x + \frac{1}{32} \sinh 4x = \frac{3}{8} x - \frac{3}{8} \sinh x \cosh x + \frac{1}{4} \sinh^3 x \cosh x$$

5. $$\int \sinh^5 x \, dx = \frac{5}{8} \cosh x - \frac{5}{48} \cosh 3x + \frac{1}{80} \cosh 5x$$
$$= \frac{4}{5} \cosh x + \frac{1}{5} \sinh^4 x \cosh x - \frac{4}{15} \cosh^3 x$$

6. $$\int \sinh^6 x \, dx = -\frac{5}{16} x + \frac{15}{64} \sinh 2x - \frac{3}{64} \sinh 4x + \frac{1}{192} \sinh 6x$$
$$= -\frac{5}{16} x + \frac{1}{6} \sinh^5 x \cosh x - \frac{5}{24} \sinh^3 x \cosh x + \frac{5}{16} \sinh x \cosh x$$

7. $\int \sinh^7 x \, dx = -\frac{35}{64} \cosh x + \frac{7}{64} \cosh 3x - \frac{7}{320} \cosh 5x + \frac{1}{448} \cosh 7x$

$$= -\frac{24}{35} \cosh x + \frac{8}{35} \cosh^3 x - \frac{6}{35} \cosh x \sinh^4 x + \frac{1}{7} \cosh x \sinh^6 x$$

8. $\int \cosh ax \, dx = \frac{1}{a} \sinh ax$

9. $\int \cosh^2 ax \, dx = \frac{x}{2} + \frac{1}{4a} \sinh 2ax$

10. $\int \cosh^3 x \, dx = \frac{3}{4} \sinh x + \frac{1}{12} \sinh 3x = \sinh x + \frac{1}{3} \sinh^3 x$

11. $\int \cosh^4 x \, dx = \frac{3}{8}x + \frac{1}{4} \sinh 2x + \frac{1}{32} \sinh 4x = \frac{3}{8}x + \frac{3}{8} \sinh x \cosh x + \frac{1}{4} \sinh x \cosh^3 x$

12. $\int \cosh^5 x \, dx = \frac{5}{8} \sinh x + \frac{5}{48} \sinh 3x + \frac{1}{80} \sinh 5x$

$$= \frac{4}{5} \sinh x + \frac{1}{5} \cosh^4 x \sinh x + \frac{4}{15} \sinh^3 x$$

13. $\int \cosh^6 x \, dx = \frac{5}{16}x + \frac{15}{64} \sinh 2x + \frac{3}{64} \sinh 4x + \frac{1}{192} \sinh 6x$

$$= \frac{5}{16}x + \frac{5}{16} \sinh x \cosh x + \frac{5}{24} \sinh x \cosh^3 x + \frac{1}{6} \sinh x \cosh^5 x$$

14. $\int \cosh^7 x \, dx = \frac{35}{64} \sinh x + \frac{7}{64} \sinh 3x + \frac{7}{320} \sinh 5x + \frac{1}{448} \sinh 7x$

$$= \frac{24}{35} \sinh x + \frac{8}{35} \sinh^3 x + \frac{6}{35} \sinh x \cosh^4 x + \frac{1}{7} \sinh x \cosh^6 x$$

**2.415**

1. $\int \sinh ax \cosh bx \, dx = \frac{\cosh(a+b)x}{2(a+b)} + \frac{\cosh(a-b)x}{2(a-b)}$

2. $\int \sinh ax \cosh ax \, dx = \frac{1}{4a} \cosh 2ax$

3. $\int \sinh^2 x \cosh x \, dx = \frac{1}{3} \sinh^3 x$

4. $\int \sinh^3 x \cosh x \, dx = \frac{1}{4} \sinh^4 x$

5. $\int \sinh^4 x \cosh x \, dx = \frac{1}{5} \sinh^5 x$

6. $\int \sinh x \cosh^2 x \, dx = \frac{1}{3} \cosh^3 x$

7. $\int \sinh^2 x \cosh^2 x \, dx = -\frac{1}{8}x + \frac{1}{32} \sinh 4x$

8. $\int \sinh^3 x \cosh^2 x \, dx = \frac{1}{5} \left( \sinh^2 x - \frac{2}{3} \right) \cosh^3 x$

9. $\int \sinh^4 x \cosh^2 x \, dx = \frac{1}{16}x - \frac{1}{64} \sinh 2x - \frac{1}{64} \sinh 4x + \frac{1}{192} \sinh 6x$

10.   $\int \sinh x \cosh^3 x \, dx = \frac{1}{4} \cosh^4 x$

11.   $\int \sinh^2 x \cosh^3 x \, dx = \frac{1}{5} \left( \cosh^2 x + \frac{2}{3} \right) \sinh^3 x$

12.   $\int \sinh^3 x \cosh^3 x \, dx = -\frac{3}{64} \cosh 2x + \frac{1}{192} \cosh 6x = \frac{1}{48} \cosh^3 2x - \frac{1}{16} \cosh 2x$

$$= \frac{\sinh^6 x}{6} + \frac{\sinh^4 x}{4} = \frac{\cosh^6 x}{6} - \frac{\cosh^4 x}{4}$$

13.   $\int \sinh^4 x \cosh^3 x \, dx = \frac{1}{7} \sinh^3 x \left( \cosh^4 x - \frac{3}{5} \cosh^2 x - \frac{2}{5} \right) = \frac{1}{7} \left( \cosh^2 x + \frac{2}{5} \right) \sinh^5 x$

14.   $\int \sinh x \cosh^4 x \, dx = \frac{1}{5} \cosh^5 x$

15.   $\int \sinh^2 x \cosh^4 x \, dx = -\frac{1}{16} x - \frac{1}{64} \sinh 2x + \frac{1}{64} \sinh 4x + \frac{1}{192} \sinh 6x$

16.   $\int \sinh^3 x \cosh^4 x \, dx = \frac{1}{7} \cosh^3 x \left( \sinh^4 x + \frac{3}{5} \sinh^2 x - \frac{2}{5} \right) = \frac{1}{7} \left( \sinh^2 x - \frac{2}{5} \right) \cosh^5 x$

17.   $\int \sinh^4 x \cosh^4 x \, dx = \frac{3}{128} x - \frac{1}{128} \sinh 4x + \frac{1}{1024} \sinh 8x$

**2.416**

1.[10]   $\int \dfrac{\sinh^p x}{\cosh^{2n} x} \, dx = \dfrac{\sinh^{p+1} x}{2n-1} \Bigg[ \operatorname{sech}^{2n-1} x$

$$+ \sum_{k=1}^{n-1} \frac{(2n-p-2)(2n-p-4)\ldots(2n-p-2k)}{(2n-3)(2n-5)\ldots(2n-2k-1)} \operatorname{sech}^{2n-2k-1} x$$

$$+ \frac{(2n-p-2)(2n-p-4)\ldots(-p+2)(-p)}{(2n-1)!!} \int \sinh^p x \, dx \Bigg]$$

This formula is applicable for arbitrary real $p$. For $\int \sinh^p x \, dx$, where $p$ is a natural number, see **2.412** 2 and **2.412** 3. For $n = 0$ and $p$ a negative integer, we have for this integral:

2.   $\int \dfrac{dx}{\sinh^{2m} x} = \dfrac{\cosh x}{2m-1} \Bigg[ -\operatorname{cosech}^{2m-1} x$

$$+ \sum_{k=1}^{m-1} (-1)^{k-1} \cdot \frac{2^k (m-1)(m-2)\ldots(m-k)}{(2m-3)(2m-5)\ldots(2m-2k-1)} \operatorname{cosech}^{2m-2k-1} x \Bigg]$$

3.   $\int \dfrac{dx}{\sinh^{2m+1} x} = \dfrac{\cosh x}{2m} \Bigg[ -\operatorname{cosech}^{2m} x$

$$+ \sum_{k=1}^{m-1} (-1)^{k-1} \cdot \frac{(2m-1)(2m-3)\ldots(2m-2k+1)}{2^k (m-1)(m-2)\ldots(m-k)} \operatorname{cosech}^{2m-2k} x \Bigg]$$

$$+ (-1)^m \frac{(2m-1)!!}{(2m)!!} \ln \tanh \frac{x}{2}$$

**2.417**

1. $\displaystyle\int\frac{\sinh^p x}{\cosh^{2n+1} x}\,dx = \frac{\sinh^{p+1} x}{2n}\left[\vphantom{\sum}\sech^{2n} x\right.$

$\displaystyle\left. + \sum_{k=1}^{n-1}\frac{(2n-p-1)(2n-p-3)\ldots(2n-p-2k+1)}{2^k(n-1)(n-2)\ldots(n-k)}\sech^{2n-2k} x\right]$

$\displaystyle + \frac{(2n-p-1)(2n-p-3)\ldots(3-p)(1-p)}{2^n n!}\int\frac{\sinh^p x}{\cosh x}\,dx$

This formula is applicable for arbitrary real $p$. For $n=0$ and $p$ integral, we have

2. $\displaystyle\int\frac{\sinh^{2m+1} x}{\cosh x}\,dx = \sum_{k=1}^{m}\frac{(-1)^{m+k}}{2k}\sinh^{2k} x + (-1)^m \ln\cosh x$

$\displaystyle = \sum_{k=1}^{m}\frac{(-1)^{m+k}}{2k}\binom{m}{k}\cosh^{2k} x + (-1)^m \ln\cosh x \qquad [m\geq 1]$

3. $\displaystyle\int\frac{\sinh^{2m} x}{\cosh x}\,dx = \sum_{k=1}^{m}\frac{(-1)^{m+k}}{2k-1}\sinh^{2k-1} x + (-1)^m \arctan(\sinh x)$

$$[m\geq 1]$$

4. $\displaystyle\int\frac{dx}{\sinh^{2m+1} x\cosh x} = \sum_{k=1}^{m}\frac{(-1)^k\cosech^{2m-2k+2} x}{2m-2k+2} + (-1)^m \ln\tanh x$

5. $\displaystyle\int\frac{dx}{\sinh^{2m} x\cosh x} = \sum_{k=1}^{m}\frac{(-1)^k\cosech^{2m-2k+2} x}{2m-2k+1} + (-1)^m \arctan\sinh x$

**2.418**

1. $\displaystyle\int\frac{\cosh^p x}{\sinh^{2n} x}\,dx = -\frac{\cosh^{p+1} x}{2n-1}\left[\vphantom{\sum}\cosech^{2n-1} x\right.$

$\displaystyle\left. + \sum_{k=1}^{n-1}\frac{(-1)^k(2n-p-2)(2n-p-4)\ldots(2n-p-2k)}{(2n-3)(2n-5)\ldots(2n-2k-1)}\cosech^{2n-2k-1} x\right]$

$\displaystyle + \frac{(-1)^n(2n-p-2)(2n-p-4)\ldots(-p+2)(-p)}{(2n-1)!!}\int\cosh^p x\,dx$

This formula is applicable for arbitrary real $p$. For the integral $\int\cosh^p x\,dx$, where $p$ is a natural number, see **2.413** 2 and **2.413** 3. If $p$ is a negative integer, we have for this integral:

2. $\displaystyle\int\frac{dx}{\cosh^{2m} x} = \frac{\sinh x}{2m-1}\left\{\sech^{2m-1} x + \sum_{k=1}^{m-1}\frac{2^k(m-1)(m-2)\ldots(m-k)}{(2m-3)(2m-5)\ldots(2m-2k-1)}\sech^{2m-2k-1} x\right\}$

3. $\displaystyle\int\frac{dx}{\cosh^{2m+1} x} = \frac{\sinh x}{2m}\left\{\sech^{2m} x + \sum_{k=1}^{m-1}\frac{(2m-1)(2m-3)\ldots(2m-2k+1)}{2^k(m-1)(m-2)\ldots(m-k)}\sech^{2m-2k} x\right\}$

$\displaystyle + \frac{(2m-1)!!}{(2m)!!}\arctan\sinh x$

**2.419**

1.   $\displaystyle \int \frac{\cosh^p x}{\sinh^{2n+1} x}\, dx = -\frac{\cosh^{p+1} x}{2n}\Bigg[ \operatorname{cosech}^{2n} x$

$\displaystyle + \sum_{k=1}^{n-1} \frac{(-1)^k (2n-p-1)(2n-p-3)\ldots(2n-p-2k+1)}{2^k(n-1)(n-2)\ldots(n-k)} \operatorname{cosech}^{2n-2k} x \Bigg]$

$\displaystyle + \frac{(-1)^n (2n-p-1)(2n-p-3)\ldots(3-p)(1-p)}{2^n n!} \int \frac{\cosh^p x}{\sinh x}\, dx$

This formula is applicable for arbitrary real $p$. For $n=0$ and $p$ an integer

2.   $\displaystyle \int \frac{\cosh^{2m} x}{\sinh x}\, dx = \sum_{k=1}^{m} \frac{\cosh^{2k-1} x}{2k-1} + \ln \tanh \frac{x}{2}$

3.   $\displaystyle \int \frac{\cosh^{2m+1} x}{\sinh x}\, dx = \sum_{k=1}^{m} \frac{\cosh^{2k} x}{2k} + \ln \sinh x$

$\displaystyle = \sum_{k=1}^{m} \binom{m}{k} \frac{\sinh^{2k} x}{2k} + \ln \sinh x$

4.   $\displaystyle \int \frac{dx}{\sinh x \cosh^{2m} x} = \sum_{k=1}^{m} \frac{\operatorname{sech}^{2m-2k+1} x}{2m-2k+1} + \ln \tanh \frac{x}{2}$

5.   $\displaystyle \int \frac{dx}{\sinh x \cosh^{2m+1} x} = \sum_{k=1}^{m} \frac{\operatorname{sech}^{2m-2k+2} x}{2m-2k+2} + \ln \tanh x$

**2.421**  In formulas **2.421** 1 and **2.421** 2, $s=1$ for $m$ odd and $m < 2n+1$; in all other cases, $s=0$.

GI (351)(11, 13)

1.$^{10}$   $\displaystyle \int \frac{\sinh^{2n+1} x}{\cosh^m x}\, dx = \sum_{\substack{k=0 \\ k \ne \frac{m-1}{2}}}^{n} (-1)^{n+k} \binom{n}{k} \frac{\cosh^{2k-m+1} x}{2k-m+1} + s(-1)^{n+\frac{m-1}{2}} \binom{n}{\frac{m-1}{2}} \ln \cosh x$

2.   $\displaystyle \int \frac{\cosh^{2n+1} x}{\sinh^m x}\, dx = \sum_{\substack{k=0 \\ k \ne \frac{m-1}{2}}}^{n} \binom{n}{k} \frac{\sinh^{2k-m+1} x}{2k-m+1} + s\binom{n}{\frac{m-1}{2}} \ln \sinh x$

**2.422**

1.   $\displaystyle \int \frac{dx}{\sinh^{2m} x \cosh^{2n} x} = \sum_{k=0}^{m+n-1} \frac{(-1)^{k+1}}{2m-2k-1} \binom{m+n-1}{k} \tanh^{2k-2m+1} x$

2.   $\displaystyle \int \frac{dx}{\sinh^{2m+1} x \cosh^{2n+1} x} = \sum_{\substack{k=0 \\ k \ne m}}^{m+n} \frac{(-1)^{k+1}}{2m-2k} \binom{m+n}{k} \tanh^{2k-2m} x + (-1)^m \binom{m+n}{m} \ln \tanh x$

GI (351)(15)

**2.423**

1.   $\displaystyle \int \frac{dx}{\sinh x} = \ln \tanh \frac{x}{2} = \frac{1}{2} \ln \frac{\cosh x - 1}{\cosh x + 1}$

2. $\int \dfrac{dx}{\sinh^2 x} = -\coth x$

3. $\int \dfrac{dx}{\sinh^3 x} = -\dfrac{\cosh x}{2\sinh^2 x} - \dfrac{1}{2}\ln\tanh\dfrac{x}{2}$

4. $\int \dfrac{dx}{\sinh^4 x} = -\dfrac{\cosh x}{3\sinh^3 x} + \dfrac{2}{3}\coth x = -\dfrac{1}{3}\coth^3 x + \coth x$

5. $\int \dfrac{dx}{\sinh^5 x} = -\dfrac{\cosh x}{4\sinh^4 x} + \dfrac{3}{8}\dfrac{\cosh x}{\sinh^2 x} + \dfrac{3}{8}\ln\tanh\dfrac{x}{2}$

6. $\int \dfrac{dx}{\sinh^6 x} = -\dfrac{\cosh x}{5\sinh^5 x} + \dfrac{4}{15}\coth^3 x - \dfrac{4}{5}\coth x$
$\qquad\qquad = -\dfrac{1}{5}\coth^5 x + \dfrac{2}{3}\coth^3 x - \coth x$

7. $\int \dfrac{dx}{\sinh^7 x} = -\dfrac{\cosh x}{6\sinh^2 x}\left(\dfrac{1}{\sinh^4 x} - \dfrac{5}{4\sinh^2 x} + \dfrac{15}{8}\right) - \dfrac{5}{16}\ln\tanh\dfrac{x}{2}$

8. $\int \dfrac{dx}{\sinh^8 x} = \coth x - \coth^3 x + \dfrac{3}{5}\coth^5 x - \dfrac{1}{7}\coth^7 x$

9. $\int \dfrac{dx}{\cosh x} = \arctan(\sinh x)$
$\qquad\qquad = \arcsin(\tanh x)$
$\qquad\qquad = 2\arctan(e^x)$
$\qquad\qquad = \operatorname{gd} x$

10. $\int \dfrac{dx}{\cosh^2 x} = \tanh x$

11. $\int \dfrac{dx}{\cosh^3 x} = \dfrac{\sinh x}{2\cosh^2 x} + \dfrac{1}{2}\arctan(\sinh x)$

12. $\int \dfrac{dx}{\cosh^4 x} = \dfrac{\sinh x}{3\cosh^3 x} + \dfrac{2}{3}\tanh x$
$\qquad\qquad = -\dfrac{1}{3}\tanh^3 x + \tanh x$

13. $\int \dfrac{dx}{\cosh^5 x} = \dfrac{\sinh x}{4\cosh^4 x} + \dfrac{3}{8}\dfrac{\sinh x}{\cosh^2 x} + \dfrac{3}{8}\arctan(\sinh x)$

14. $\int \dfrac{dx}{\cosh^6 x} = \dfrac{\sinh x}{5\cosh^5 x} - \dfrac{4}{15}\tanh^3 x + \dfrac{4}{5}\tanh x$
$\qquad\qquad = \dfrac{1}{5}\tanh^5 x - \dfrac{2}{3}\tanh^3 x + \tanh x$

15.[12] $\int \dfrac{dx}{\cosh^7 x} = \dfrac{\sinh x}{6\cosh^2 x}\left(\dfrac{1}{\cosh^4 x} + \dfrac{5}{4\cosh^6 x} + \dfrac{15}{8}\right) + \dfrac{5}{16}\arctan(\sinh x)$

16. $\int \dfrac{dx}{\cosh^8 x} = -\dfrac{1}{7}\tanh^7 x + \dfrac{3}{5}\tanh^5 x - \tanh^3 x + \tanh x$

17. $\int \dfrac{\sinh x}{\cosh x}\,dx = \ln\cosh x$

18. $\displaystyle\int \frac{\sinh^2 x}{\cosh x}\,dx = \sinh x - \arctan(\sinh x)$

19. $\displaystyle\int \frac{\sinh^3 x}{\cosh x}\,dx = \frac{1}{2}\sinh^2 x - \ln\cosh x$

$\qquad\qquad\qquad = \frac{1}{2}\cosh^2 x - \ln\cosh x$

20. $\displaystyle\int \frac{\sinh^4 x}{\cosh x}\,dx = \frac{1}{3}\sinh^3 x - \sinh x + \arctan(\sinh x)$

21. $\displaystyle\int \frac{\sinh x}{\cosh^2 x}\,dx = -\frac{1}{\cosh x}$

22. $\displaystyle\int \frac{\sinh^2 x}{\cosh^2 x}\,dx = x - \tanh x$

23. $\displaystyle\int \frac{\sinh^3 x}{\cosh^2 x}\,dx = \cosh x + \frac{1}{\cosh x}$

24. $\displaystyle\int \frac{\sinh^4 x}{\cosh^2 x}\,dx = -\frac{3}{2}x + \frac{1}{4}\sinh 2x + \tanh x$

25. $\displaystyle\int \frac{\sinh x}{\cosh^3 x}\,dx = -\frac{1}{2\cosh^2 x}$

$\qquad\qquad\qquad = \frac{1}{2}\tanh^2 x$

26. $\displaystyle\int \frac{\sinh^2 x}{\cosh^3 x}\,dx = -\frac{\sinh x}{2\cosh^2 x} + \frac{1}{2}\arctan(\sinh x)$

27. $\displaystyle\int \frac{\sinh^3 x}{\cosh^3 x}\,dx = -\frac{1}{2}\tanh^2 x + \ln\cosh x$

$\qquad\qquad\qquad = \frac{1}{2\cosh^2 x} + \ln\cosh x$

28. $\displaystyle\int \frac{\sinh^4 x}{\cosh^3 x}\,dx = \frac{\sinh x}{2\cosh x} + \sinh x - \frac{3}{2}\arctan(\sinh x)$

29. $\displaystyle\int \frac{\sinh x}{\cosh^4 x}\,dx = -\frac{1}{3\cosh^3 x}$

30. $\displaystyle\int \frac{\sinh^2 x}{\cosh^4 x}\,dx = \frac{1}{3}\tanh^3 x$

31. $\displaystyle\int \frac{\sinh^3 x}{\cosh^4 x}\,dx = -\frac{1}{\cosh x} + \frac{1}{3\cosh^3 x}$

32. $\displaystyle\int \frac{\sinh^4 x}{\cosh^4 x}\,dx = -\frac{1}{3}\tanh^3 x - \tanh x + x$

33. $\displaystyle\int \frac{\cosh x}{\sinh x}\,dx = \ln\sinh x$

34. $\displaystyle\int \frac{\cosh^2 x}{\sinh x}\,dx = \cosh x + \ln\tanh\frac{x}{2}$

35.  $\displaystyle\int \frac{\cosh^3 x}{\sinh x}\,\mathrm{d}x = \frac{1}{2}\cosh^2 x + \ln\sinh x$

36.  $\displaystyle\int \frac{\cosh^4 x}{\sinh x}\,\mathrm{d}x = \frac{1}{3}\cosh^3 x + \cosh x + \ln\tanh\frac{x}{2}$

37.  $\displaystyle\int \frac{\cosh x}{\sinh^2 x}\,\mathrm{d}x = -\frac{1}{\sinh x}$

38.  $\displaystyle\int \frac{\cosh^2 x}{\sinh^2 x}\,\mathrm{d}x = x - \coth x$

39.  $\displaystyle\int \frac{\cosh^3 x}{\sinh^2 x}\,\mathrm{d}x = \sinh x - \frac{1}{\sinh x}$

40.  $\displaystyle\int \frac{\cosh^4 x}{\sinh^2 x}\,\mathrm{d}x = \frac{3}{2}x + \frac{1}{4}\sinh 2x - \coth x$

41.  $\displaystyle\int \frac{\cosh x}{\sinh^3 x}\,\mathrm{d}x = -\frac{1}{2\sinh^2 x}$
$$= -\frac{1}{2}\coth^2 x$$

42.  $\displaystyle\int \frac{\cosh^2 x}{\sinh^3 x}\,\mathrm{d}x = -\frac{\cosh x}{2\sinh^2 x} + \ln\tanh\frac{x}{2}$

43.  $\displaystyle\int \frac{\cosh^3 x}{\sinh^3 x}\,\mathrm{d}x = -\frac{1}{2\sinh^2 x} + \ln\sinh x$
$$= -\frac{1}{2}\coth^2 x + \ln\sinh x$$

44.  $\displaystyle\int \frac{\cosh^4 x}{\sinh^3 x}\,\mathrm{d}x = -\frac{\cosh x}{2\sinh^2 x} + \cosh x + \frac{3}{2}\ln\tanh\frac{x}{2}$

45.  $\displaystyle\int \frac{\cosh x}{\sinh^4 x}\,\mathrm{d}x = -\frac{1}{3\sinh^3 x}$

46.  $\displaystyle\int \frac{\cosh^2 x}{\sinh^4 x}\,\mathrm{d}x = -\frac{1}{3}\coth^3 x$

47.  $\displaystyle\int \frac{\cosh^3 x}{\sinh^4 x}\,\mathrm{d}x = -\frac{1}{\sinh x} - \frac{1}{3\sinh^3 x}$

48.  $\displaystyle\int \frac{\cosh^4 x}{\sinh^4 x}\,\mathrm{d}x = -\frac{1}{3}\coth^3 x - \coth x + x$

49.  $\displaystyle\int \frac{\mathrm{d}x}{\sinh x \cosh x} = \ln\tanh x$

50.  $\displaystyle\int \frac{\mathrm{d}x}{\sinh x \cosh^2 x} = \frac{1}{\cosh x} + \ln\tanh\frac{x}{2}$

51.  $\displaystyle\int \frac{\mathrm{d}x}{\sinh x \cosh^3 x} = \frac{1}{2\cosh^2 x} + \ln\tanh x$
$$= -\frac{1}{2}\tanh^2 x + \ln\tanh x$$

52. $$\int \frac{dx}{\sinh x \cosh^4 x} = \frac{1}{\cosh x} + \frac{1}{3\cosh^3 x} + \ln\tanh\frac{x}{2}$$

53. $$\int \frac{dx}{\sinh^2 x \cosh x} = -\frac{1}{\sinh x} - \arctan\sinh x$$

54. $$\int \frac{dx}{\sinh^2 x \cosh^2 x} = -2\coth 2x$$

55. $$\int \frac{dx}{\sinh^2 x \cosh^3 x} = -\frac{\sinh x}{2\cosh^2 x} - \frac{1}{\sinh x} - \frac{3}{2}\arctan\sinh x$$

56. $$\int \frac{dx}{\sinh^2 x \cosh^4 x} = \frac{1}{3\sinh x \cosh^3 x} - \frac{8}{3}\coth 2x$$

57. $$\int \frac{dx}{\sinh^3 x \cosh x} = -\frac{1}{2\sinh^2 x} - \ln\tanh x$$
$$= -\frac{1}{2}\coth^2 x + \ln\coth x$$

58. $$\int \frac{dx}{\sinh^3 x \cosh^2 x} = -\frac{1}{\cosh x} - \frac{\cosh x}{2\sinh^2 x} - \frac{3}{2}\ln\tanh\frac{x}{2}$$

59. $$\int \frac{dx}{\sinh^3 x \cosh^3 x} = -\frac{2\cosh 2x}{\sinh^2 2x} - 2\ln\tanh x$$
$$= \frac{1}{2}\tanh^2 x - \frac{1}{2}\coth^2 x - 2\ln\tanh x$$

60. $$\int \frac{dx}{\sinh^3 x \cosh^4 x} = -\frac{2}{\cosh x} - \frac{1}{3\cosh^2 x} - \frac{\cosh x}{2\sinh^2 x} - \frac{5}{2}\ln\tanh\frac{x}{2}$$

61. $$\int \frac{dx}{\sinh^4 x \cosh x} = \frac{1}{\sinh x} - \frac{1}{3\sinh^3 x} + \arctan\sinh x$$

62. $$\int \frac{dx}{\sinh^4 x \cosh^2 x} = -\frac{1}{3\cosh x \sinh^3 x} + \frac{8}{3}\coth 2x$$

63. $$\int \frac{dx}{\sinh^4 x \cosh^3 x} = \frac{2}{\sinh x} - \frac{1}{3\sinh^3 x} + \frac{\sinh x}{2\cosh^2 x} + \frac{5}{2}\arctan\sinh x$$

64. $$\int \frac{dx}{\sinh^4 x \cosh^4 x} = 8\coth 2x - \frac{8}{3}\coth^3 2x$$

## 2.424

1. $$\int \tanh^p x\, dx = -\frac{\tanh^{p-1} x}{p-1} + \int \tanh^{p-2} x\, dx \qquad [p \neq 1]$$

2. $$\int \tanh^{2n+1} x\, dx = \sum_{k=1}^{n} \frac{(-1)^{k-1}}{2k}\binom{n}{k}\frac{1}{\cosh^{2k} x} + \ln\cosh x$$
$$= -\sum_{k=1}^{n} \frac{\tanh^{2n-2k+2} x}{2n-2k+2} + \ln\cosh x$$

3. $$\int \tanh^{2n} x\, dx = -\sum_{k=1}^{n} \frac{\tanh^{2n-2k+1} x}{2n-2k+1} + x \qquad\qquad\qquad \text{GU (351)(12)}$$

4. $$\int \coth^p x\, dx = -\frac{\coth^{p-1} x}{p-1} + \int \coth^{p-2} x\, dx \qquad [p \neq 1]$$

5. $\quad \displaystyle\int \coth^{2n+1} x \, dx = -\sum_{k=1}^{n} \frac{1}{2n} \binom{n}{k} \frac{1}{\sinh^{2k} x} + \ln \sinh x$

$\qquad\qquad\qquad = -\displaystyle\sum_{k=1}^{n} \frac{\coth^{2n-2k+2} x}{2n - 2k + 2} + \ln \sinh x$

6. $\quad \displaystyle\int \coth^{2n} x \, dx = -\sum_{k=1}^{n} \frac{\coth^{2n-2k+1} x}{2n - 2k + 1} + x$          GU (351)(14)

For formulas containing powers of $\tanh x$ and $\coth x$ equal to $n = 1,\ 2,\ 3,\ 4$, see **2.423** 17, **2.423** 22, **2.423** 27, **2.423** 32, **2.423** 33, **2.423** 38, **2.423** 43, **2.423** 48.

**Powers of hyperbolic functions and hyperbolic functions of linear functions of the argument**

**2.425**

1. $\quad \displaystyle\int \sinh(ax + b) \sinh(cx + d) \, dx = \frac{1}{2(a + c)} \sinh[(a + c)x + b + d]$

$\qquad\qquad\qquad\qquad\qquad - \dfrac{1}{2(a - c)} \sinh[(a - c)x + b - d]$

$\qquad\qquad\qquad\qquad\qquad\qquad\qquad \left[ a^2 \neq c^2 \right]$      GU (352)(2a)

2. $\quad \displaystyle\int \sinh(ax + b) \cosh(cx + d) \, dx = \frac{1}{2(a + c)} \cosh[(a + c)x + b + d]$

$\qquad\qquad\qquad\qquad\qquad + \dfrac{1}{2(a - c)} \cosh[(a - c)x + b - d]$

$\qquad\qquad\qquad\qquad\qquad\qquad\qquad \left[ a^2 \neq c^2 \right]$      GU (352)(2c)

3. $\quad \displaystyle\int \cosh(ax + b) \cosh(cx + d) \, dx = \frac{1}{2(a + c)} \sinh[(a + c)x + b + d]$

$\qquad\qquad\qquad\qquad\qquad + \dfrac{1}{2(a - c)} \sinh[(a - c)x + b - d]$

$\qquad\qquad\qquad\qquad\qquad\qquad\qquad \left[ a^2 \neq c^2 \right]$      GU (352)(2b)

When $a = c$:

4. $\quad \displaystyle\int \sinh(ax + b) \sinh(ax + d) \, dx = -\frac{x}{2} \cosh(b - d) + \frac{1}{4a} \sinh(2ax + b + d)$      GU (352)(3a)

5. $\quad \displaystyle\int \sinh(ax + b) \cosh(ax + d) \, dx = \frac{x}{2} \sinh(b - d) + \frac{1}{4a} \cosh(2ax + b + d)$      GU (352)(3c)

6. $\quad \displaystyle\int \cosh(ax + b) \cosh(ax + d) \, dx = \frac{x}{2} \cosh(b - d) + \frac{1}{4a} \sinh(2ax + b + d)$      GU (352)(3b)

**2.426**

1. $\quad \displaystyle\int \sinh ax \sinh bx \sinh cx \, dx = \frac{\cosh(a + b + c)x}{4(a + b + c)} - \frac{\cosh(-a + b + c)x}{4(-a + b + c)}$

$\qquad\qquad\qquad\qquad\qquad - \dfrac{\cosh(a - b + c)x}{4(a - b + c)} - \dfrac{\cosh(a + b - c)x}{4(a + b - c)}$

         GU (352)(4a)

**2.**

$$\int \sinh ax \sinh bx \cosh cx \, dx = \frac{\sinh(a+b+c)x}{4(a+b+c)} - \frac{\sinh(-a+b+c)x}{4(-a+b+c)}$$
$$- \frac{\sinh(a-b+c)x}{4(a-b+c)} + \frac{\sinh(a+b-c)x}{4(a+b-c)}$$

GU (352)(4b)

**3.**

$$\int \sinh ax \cosh bx \cosh cx \, dx = \frac{\cosh(a+b+c)x}{4(a+b+c)} - \frac{\cosh(-a+b+c)x}{4(-a+b+c)}$$
$$+ \frac{\cosh(a-b+c)x}{4(a-b+c)} + \frac{\cosh(a+b-c)x}{4(a+b-c)}$$

GU (352)(4c)

**4.**

$$\int \cosh ax \cosh bx \cosh cx \, dx = \frac{\sinh(a+b+c)x}{4(a+b+c)} + \frac{\sinh(-a+b+c)x}{4(-a+b+c)}$$
$$+ \frac{\sinh(a-b+c)x}{4(a-b+c)} + \frac{\sinh(a+b-c)x}{4(a+b-c)}$$

GU (352)(4d)

**2.427**

**1.**[12]   $\displaystyle \int \sinh^p x \sinh ax \, dx = \frac{1}{p+a} \left\{ \sinh^p x \cosh ax - p \int \sinh^{p-1} x \cosh(a-1)x \, dx \right\}$

**2.**   $\displaystyle \int \sinh^p x \sinh(2n+1)x \, dx = \frac{\Gamma(p+1)}{\Gamma\left(\frac{p+3}{2}+n\right)}$

$$\times \left[ \sum_{k=0}^{n-1} \frac{\Gamma\left(\frac{p+1}{2}+n-2k\right)}{2^{2k+1}\Gamma(p-2k+1)} \sinh^{p-2k} x \cosh(2n-2k+1)x \right.$$
$$\left. - \frac{\Gamma\left(\frac{p-1}{2}+n-2k\right)}{2^{2k+2}\Gamma(p-2k)} \sinh^{p-2k-1} x \sinh(2n-2k)x \right]$$
$$+ \frac{\Gamma\left(\frac{p+3}{2}-n\right)}{2^{2n}\Gamma(p+1-2n)} \int \sinh^{p-2n} x \sinh x \, dx$$

[p is not a negative integer]

**3.**   $\displaystyle \int \sinh^p x \sinh 2nx \, dx = \frac{\Gamma(p+1)}{\Gamma\left(\frac{p}{2}+n+1\right)}$

$$\times \sum_{k=0}^{n-1} \left[ \frac{\Gamma\left(\frac{p}{2}+n-2k\right)}{2^{2k+1}\Gamma(p-2k+1)} \sinh^{p-2k} x \cosh(2n-2k)x \right.$$
$$\left. - \frac{\Gamma\left(\frac{p}{2}+n-2k-1\right)}{2^{2k+2}\Gamma(p-2k)} \sinh^{p-2k-1} x \sinh(2n-2k-1)x \right]$$

[p is not a negative integer]    GU (352)(5)a

**2.428**

**1.**   $\displaystyle \int \sinh^p x \cosh ax \, dx = \frac{1}{p+a} \left\{ \sinh^p x \sinh ax - p \int \sinh^{p-1} x \sinh(a-1)x \, dx \right\}$

2. $\displaystyle\int \sinh^p x \cosh(2n+1)x\,dx = \frac{\Gamma(p+1)}{\Gamma\left(\frac{p+3}{2}+n\right)}$

$$\times \left\{\left[\sum_{k=0}^{n-1} \frac{\Gamma\left(\frac{p+1}{2}+n-2k\right)}{2^{2k+1}\,\Gamma(p-2k+1)}\sinh^{p-2k}x\sinh(2n-2k+1)x\right.\right.$$

$$-\frac{\Gamma\left(\frac{p-1}{2}+n-2k\right)}{2^{2k+2}\,\Gamma(p-2k)}\sinh^{p-2k-1}x\cosh(2n-2k)x\Bigg]$$

$$\left.+\frac{\Gamma\left(\frac{p+3}{2}-n\right)}{2^{2n}\,\Gamma(p+1-2n)}\int \sinh^{p-2n}x\cosh x\,dx\right\}$$

$$[p \text{ is not a negative integer}]$$

3. $\displaystyle\int \sinh^p x \cosh 2nx\,dx = \frac{\Gamma(p+1)}{\Gamma\left(\frac{p}{2}+n+1\right)}$

$$\times \left\{\sum_{k=0}^{n-1}\left[\frac{\Gamma\left(\frac{p}{2}+n-2k\right)}{2^{2k+1}\,\Gamma(p-2k+1)}\sinh^{p-2k}x\sinh(2n-2k)x\right.\right.$$

$$-\frac{\Gamma\left(\frac{p}{2}+n-2k-1\right)}{2^{2k+2}\,\Gamma(p-2k)}\sinh^{p-2k-1}x\cosh(2n-2k-1)x\Bigg]$$

$$\left.+\frac{\Gamma\left(\frac{p}{2}-n+1\right)}{2^{2n}\,\Gamma(p+1-2n)}\int \sinh^{p-2n}x\,dx\right\}$$

$$[p \text{ is not a negative integer}] \qquad \text{GU (352)(6)a}$$

**2.429**

1. $\displaystyle\int \cosh^p x \sinh ax\,dx = \frac{1}{p+a}\left\{\cosh^p x\cosh ax + p\int \cosh^{p-1}x\sinh(a-1)x\,dx\right\}$

2. $\displaystyle\int \cosh^p x \sinh(2n+1)x\,dx = \frac{\Gamma(p+1)}{\Gamma\left(\frac{p+3}{2}+n\right)}\left[\sum_{k=0}^{n-1}\frac{\Gamma\left(\frac{p+1}{2}+n-k\right)}{2^{k+1}\,\Gamma(p-k+1)}\cosh^{p-k}x\cosh(2n-k+1)x\right.$

$$\left.+\frac{\Gamma\left(\frac{p+3}{2}\right)}{2^n\,\Gamma(p-n+1)}\int \cosh^{p-n}x\sinh(n+1)x\,dx\right]$$

$$[p \text{ is not a negative integer}]$$

3. $\displaystyle\int \cosh^p x \sinh 2nx\,dx = \frac{\Gamma(p+1)}{\Gamma\left(\frac{p}{2}+n+1\right)}\left[\sum_{k=0}^{n-1}\frac{\Gamma\left(\frac{p}{2}+n-k\right)}{2^{k+1}\,\Gamma(p-k+1)}\cosh^{p-k}x\cosh(2n-k)x\right.$

$$\left.+\frac{\Gamma\left(\frac{p}{2}+1\right)}{2^n\,\Gamma(p-n+1)}\int \cosh^{p-n}x\sinh nx\,dx\right]$$

$$[p \text{ is not a negative integer}] \qquad \text{GU (352)(7)a}$$

**2.431**

1. $$\int \cosh^p x \cosh ax \, dx = \frac{1}{p+a} \left\{ \cosh^p x \sinh ax + p \int \cosh^{p-1} x \cosh(a-1)x \, dx \right\}$$

2. $$\int \cosh^p x \cosh(2n+1)x \, dx = \frac{\Gamma(p+1)}{\Gamma\left(\frac{p+3}{2}+n\right)} \left[ \sum_{k=0}^{n-1} \frac{\Gamma\left(\frac{p+1}{2}+n-k\right)}{2^{k+1}\Gamma(p-k+1)} \cosh^{p-k} x \sinh(2n-k+1)x \right.$$
$$\left. + \frac{\Gamma\left(\frac{p+3}{2}\right)}{2^n \Gamma(p-n+1)} \int \cosh^{p-n} x \cosh(n+1)x \, dx \right]$$

$$[p \text{ is not a negative integer}]$$

3. $$\int \cosh^p x \cosh 2nx \, dx = \frac{\Gamma(p+1)}{\Gamma\left(\frac{p}{2}+n+1\right)} \left[ \sum_{k=0}^{n-1} \frac{\Gamma\left(\frac{p}{2}+n-k\right)}{2^{k+1}\Gamma(p-k+1)} \cosh^{p-k} x \sinh(2n-k)x \right.$$
$$\left. + \frac{\Gamma\left(\frac{p}{2}+1\right)}{2^n \Gamma(p-n+1)} \cosh^{p-n} x \cosh nx \, dx \right]$$

$$[p \text{ is not a negative integer}] \quad \text{GU (352)(8)a}$$

**2.432**

1. $$\int \sinh(n+1)x \sinh^{n-1} x \, dx = \frac{1}{n} \sinh^n x \sinh nx$$

2. $$\int \sinh(n+1)x \cosh^{n-1} x \, dx = \frac{1}{n} \cosh^n x \cosh nx$$

3. $$\int \cosh(n+1)x \sinh^{n-1} x \, dx = \frac{1}{n} \sinh^n x \cosh nx$$

4. $$\int \cosh(n+1)x \cosh^{n-1} x \, dx = \frac{1}{n} \cosh^n x \sinh nx$$

**2.433**

1. $$\int \frac{\sinh(2n+1)x}{\sinh x} \, dx = 2 \sum_{k=0}^{n-1} \frac{\sinh(2n-2k)x}{2n-2k} + x$$

2. $$\int \frac{\sinh 2nx}{\sinh x} \, dx = 2 \sum_{k=0}^{n-1} \frac{\sinh(2n-2k-1)x}{2n-2k-1} \qquad \text{GU (352)(5d)}$$

3. $$\int \frac{\cosh(2n+1)x}{\sinh x} \, dx = 2 \sum_{k=0}^{n-1} \frac{\cosh(2n-2k)x}{2n-2k} + \ln \sinh x$$

4. $$\int \frac{\cosh 2nx}{\sinh x} \, dx = 2 \sum_{k=0}^{n-1} \frac{\cosh(2n-2k-1)x}{2n-2k-1} + \ln \tanh \frac{x}{2} \qquad \text{GU (352)(6d)}$$

5. $$\int \frac{\sinh(2n+1)x}{\cosh x} \, dx = 2 \sum_{k=0}^{n-1} (-1)^k \frac{\cosh(2n-2k)x}{2n-2k} + (-1)^n \ln \cosh x$$

6. $\displaystyle \int \frac{\sinh 2nx}{\cosh x}\, dx = 2 \sum_{k=0}^{n-1} (-1)^k \frac{\cosh(2n - 2k - 1)x}{2n - 2k - 1}$                                    GU (352)(7d)

7. $\displaystyle \int \frac{\cosh(2n + 1)x}{\cosh x}\, dx = 2 \sum_{k=0}^{n-1} (-1)^k \frac{\sinh(2n - 2k)x}{2n - 2k} + (-1)^n x$

8. $\displaystyle \int \frac{\cosh 2nx}{\cosh x}\, dx = 2 \sum_{k=0}^{n-1} (-1)^k \frac{\sinh(2n - 2k - 1)x}{2n - 2k - 1} + (-1)^n \arcsin(\tanh x)$                                    GU (352)(8d)

9. $\displaystyle \int \frac{\sinh 2x}{\sinh^n x}\, dx = -\frac{2}{(n - 2)\sinh^{n-2} x}$

For $n = 2$:

10. $\displaystyle \int \frac{\sinh 2x}{\sinh^2 x}\, dx = 2 \ln \sinh x$

11. $\displaystyle \int \frac{\sinh 2x\, dx}{\cosh^n x} = \frac{2}{(2 - n)\cosh^{n-2} x}$

For $n = 2$:

12. $\displaystyle \int \frac{\sinh 2x}{\cosh^2 x}\, dx = 2 \ln \cosh x$

13. $\displaystyle \int \frac{\cosh 2x}{\sinh x}\, dx = 2 \cosh x + \ln \tanh \frac{x}{2}$

14. $\displaystyle \int \frac{\cosh 2x}{\sinh^2 x}\, dx = -\coth x + 2x$

15. $\displaystyle \int \frac{\cosh 2x}{\sinh^3 x}\, dx = -\frac{\cosh x}{2 \sinh^2 x} + \frac{3}{2} \ln \tanh \frac{x}{2}$

16. $\displaystyle \int \frac{\cosh 2x}{\cosh x}\, dx = 2 \sinh x - \arcsin(\tanh x)$

17. $\displaystyle \int \frac{\cosh 2x}{\cosh^2 x}\, dx = -\tanh x + 2x$

18. $\displaystyle \int \frac{\cosh 2x}{\cosh^3 x}\, dx = -\frac{\sinh x}{2 \cosh^2 x} + \frac{3}{2} \arcsin(\tanh x)$

19. $\displaystyle \int \frac{\sinh 3x}{\sinh x}\, dx = x + \sinh 2x$

20. $\displaystyle \int \frac{\sinh 3x}{\sinh^2 x}\, dx = 3 \ln \tanh \frac{x}{2} + 4 \cosh x$

21. $\displaystyle \int \frac{\sinh 3x}{\sinh^3 x}\, dx = -3 \coth x + 4x$

22. $\displaystyle \int \frac{\sinh 3x}{\cosh^n x}\, dx = \frac{4}{(3 - n)\cosh^{n-3} x} - \frac{1}{(1 - n)\cosh^{n-1} x}$

For $n = 1$ and $n = 3$:

23. $\displaystyle \int \frac{\sinh 3x}{\cosh x}\, dx = 2 \sinh^2 x - \ln \cosh x$

24. $$\int \frac{\sinh 3x}{\cosh^3 x}\, dx = \frac{1}{2\cosh^2 x} + 4\ln\cosh x$$

25. $$\int \frac{\cosh 3x}{\sinh^n x}\, dx = \frac{4}{(3-n)\sinh^{n-3} x} + \frac{1}{(1-n)\sinh^{n-1} x}$$

For $n = 1$ and $n = 3$:

26. $$\int \frac{\cosh 3x}{\sinh x}\, dx = 2\sinh^2 x + \ln\sinh x$$

27. $$\int \frac{\cosh 3x}{\sinh^3 x}\, dx = -\frac{1}{2\sinh^2 x} + 4\ln\sinh x$$

28. $$\int \frac{\cosh 3x}{\cosh x}\, dx = \sinh 2x - x$$

29. $$\int \frac{\cosh 3x}{\cosh^2 x}\, dx = 4\sinh x - 3\arcsin(\tanh x)$$

30. $$\int \frac{\cosh 3x}{\cosh^3 x}\, dx = 4x - 3\tanh x$$

## 2.44–2.45  Rational functions of hyperbolic functions

**2.441**

1. $$\int \frac{A + B\sinh x}{(a + b\sinh x)^n}\, dx = \frac{aB - bA}{(n-1)(a^2 + b^2)} \cdot \frac{\cosh x}{(a + b\sinh x)^{n-1}}$$
$$+ \frac{1}{(n-1)(a^2 + b^2)} \int \frac{(n-1)(aA + bB) + (n-2)(aB - bA)\sinh x}{(a + b\sinh x)^{n-1}}\, dx$$

For $n = 1$:

2. $$\int \frac{A + B\sinh x}{a + b\sinh x}\, dx = \frac{B}{b} x - \frac{aB - bA}{b} \int \frac{dx}{a + b\sinh x} \qquad \text{(see \textbf{2.441} 3)}$$

3. $$\int \frac{dx}{a + b\sinh x} = \frac{1}{\sqrt{a^2 + b^2}} \ln \frac{a\tanh\frac{x}{2} - b + \sqrt{a^2 + b^2}}{a\tanh\frac{x}{2} - b - \sqrt{a^2 + b^2}}$$
$$= \frac{2}{\sqrt{a^2 + b^2}} \operatorname{arctanh} \frac{a\tanh\frac{x}{2} - b}{\sqrt{a^2 + b^2}}$$

**2.442**

1. $$\int \frac{A + B\cosh x}{(a + b\sinh x)^n}\, dx = -\frac{B}{(n-1)b\,(a + b\sinh x)^{n-1}} + A \int \frac{dx}{(a + b\sinh x)^n}$$

For $n = 1$:

2. $$\int \frac{A + B\cosh x}{a + b\sinh x}\, dx = \frac{B}{b} \ln(a + b\sinh x) + A \int \frac{dx}{a + b\sinh x}$$

$$\text{(see \textbf{2.441} 3)}$$

**2.443**

1. $$\int \frac{A+B\cosh x}{(a+b\cosh x)^n}\,dx = \frac{aB-bA}{(n-1)(a^2-b^2)}\cdot\frac{\sinh x}{(a+b\cosh x)^{n-1}}$$
$$+\frac{1}{(n-1)(a^2-b^2)}\int \frac{(n-1)(aA-bB)+(n-2)(aB-bA)\cosh x}{(a+b\cosh x)^{n-1}}\,dx$$

For $n=1$:

2. $$\int \frac{A+B\cosh x}{a+b\cosh x}\,dx = \frac{B}{b}x - \frac{aB-bA}{b}\int \frac{dx}{a+b\cosh x} \qquad \text{(see 2.443 3)}$$

3. $$\int \frac{dx}{a+b\cosh x} = \frac{1}{\sqrt{b^2-a^2}}\arcsin\frac{b+a\cosh x}{a+b\cosh x} \qquad [b^2>a^2, \quad x<0]$$
$$= -\frac{1}{\sqrt{b^2-a^2}}\arcsin\frac{b+a\cosh x}{a+b\cosh x} \qquad [b^2>a^2, \quad x>0]$$
$$= \frac{1}{\sqrt{a^2-b^2}}\ln\frac{a+b+\sqrt{a^2-b^2}\tanh\frac{x}{2}}{a+b-\sqrt{a^2-b^2}\tanh\frac{x}{2}} \qquad [a^2>b^2]$$

**2.444**

1. $$\int \frac{dx}{\cosh a+\cosh x} = \operatorname{cosech} a\left[\ln\cosh\frac{x+a}{2} - \ln\cosh\frac{x-a}{2}\right]$$
$$= 2\operatorname{cosech} a \operatorname{arctanh}\left(\tanh\frac{x}{2}\tanh\frac{a}{2}\right)$$

2.[12] $$\int \frac{dx}{\operatorname{cosec} a+\cosh x} = 2\operatorname{cosec} a \arctan\left(\tanh\frac{x}{2}\tan\frac{a}{2}\right)$$

**2.445**

1.[12] $$\int \frac{\sinh x}{(a+b\cosh x)^n}\,dx = -\frac{1}{(n-1)b(a+b\cosh x)^{n-1}} \qquad [n\neq 1]$$

For $n=1$:

2.[12] $$\int \frac{\sinh x}{a+b\cosh x}\,dx = \frac{1}{b}\ln(a+b\cosh x) \qquad \text{(see 2.443 3)}$$

In evaluating definite integrals by use of formulas **2.441–2.443** and **2.445**, one may not take the integral over points at which the integrand becomes infinite, that is, over the points

$$x = \operatorname{arcsinh}\left(-\frac{a}{b}\right)$$

in formulas **2.441** or **2.442** or over the points

$$x = \operatorname{arccosh}\left(-\frac{a}{b}\right)$$

in formulas **2.443** or **2.445**. Formulas **2.443** are not applicable for $a^2=b^2$. Instead, we may use the following formulas in these cases:

**2.446**

1.
$$\int \frac{A + B\cosh x}{(\varepsilon + \cosh x)^n}\,dx$$

$$= \frac{B\sinh x}{(1-n)(\varepsilon + \cosh x)^n} + \left(\varepsilon A + \frac{n}{n-1}B\right)\frac{(n-1)!}{(2n-1)!!}\sinh x \sum_{k=0}^{n-1}\frac{(2n-2k-3)!!}{(n-k-1)!}$$

$$\times \frac{\varepsilon^k}{(\varepsilon + \cosh x)^{n-k}}$$

$$[\varepsilon = \pm 1, \quad n > 1]$$

For $n = 1$:

2.
$$\int \frac{A + B\cosh x}{\varepsilon + \cosh x}\,dx = Bx + (\varepsilon A - B)\frac{\cosh x - \varepsilon}{\sinh x} \qquad [\varepsilon = \pm 1]$$

**2.447**

1.[12]
$$\int \frac{\sinh x\,dx}{a\cosh x + b\sinh x} = \frac{a\ln\cosh\left(x + \operatorname{arctanh}\frac{b}{a}\right) - bx}{a^2 - b^2} \qquad [a > |b|]$$

$$= \frac{bx - a\ln\sinh\left(x + \operatorname{arctanh}\frac{a}{b}\right)}{b^2 - a^2} \qquad [b > |a|] \qquad\qquad \text{MZ 215}$$

For $a = b = 1$:

2.
$$\int \frac{\sinh x\,dx}{\cosh x + \sinh x} = \frac{x}{2} + \frac{1}{4}e^{-2x}$$

For $a = -b = 1$:

3.
$$\int \frac{\sinh x\,dx}{\cosh x - \sinh x} = -\frac{x}{2} + \frac{1}{4}e^{2x} \qquad\qquad\qquad\qquad\qquad \text{MZ 215}$$

**2.448**

1.
$$\int \frac{\cosh x\,dx}{a\cosh x + b\sinh x} = \frac{ax - b\ln\cosh\left(x + \operatorname{arctanh}\frac{b}{a}\right)}{a^2 - b^2} \qquad [a > |b|]$$

$$= \frac{-ax + b\ln\sinh\left(x + \operatorname{arctanh}\frac{a}{b}\right)}{b^2 - a^2} \qquad [b > |a|]$$

For $a = b = 1$:

2.
$$\int \frac{\cosh x\,dx}{\cosh x + \sinh x} = \frac{x}{2} - \frac{1}{4}e^{-2x}$$

For $a = -b = 1$:

3.
$$\int \frac{\cosh x\,dx}{\cosh x - \sinh x} = \frac{x}{2} + \frac{1}{4}e^{2x} \qquad\qquad\qquad\qquad\qquad \text{MZ 214, 215}$$

**2.449**

1.[6]
$$\int \frac{dx}{(a\cosh x + b\sinh x)^n} = \frac{1}{\sqrt{(a^2 - b^2)^n}}\int \frac{dx}{\sinh^n\left(x + \operatorname{arctanh}\frac{b}{a}\right)} \qquad [a > |b|]$$

$$= \frac{1}{\sqrt{(b^2 - a^2)^n}}\int \frac{dx}{\cosh^n\left(x + \operatorname{arctanh}\frac{a}{b}\right)} \qquad [b > |a|]$$

For $n = 1$:

2.    $\displaystyle\int \frac{dx}{a\cosh x + b\sinh x} = \frac{1}{\sqrt{a^2 - b^2}} \arctan \left|\sinh\left(x + \operatorname{arctanh}\frac{b}{a}\right)\right|$     $[a > |b|]$

$\displaystyle = \frac{1}{\sqrt{b^2 - a^2}} \ln\left|\tanh\frac{x + \operatorname{arctanh}\frac{a}{b}}{2}\right|$     $[b > |a|]$

For $a = b = 1$:

3.[12]    $\displaystyle\int \frac{dx}{\cosh x + \sinh x} = -e^{-x} = \sinh x - \cosh x$

For $a = -b = 1$:

4.    $\displaystyle\int \frac{dx}{\cosh x - \sinh x} = e^{x} = \sinh x + \cosh x$     MZ 214

**2.451**

1.    $\displaystyle\int \frac{A + B\cosh x + C\sinh x}{(a + b\cosh x + c\sinh x)^n}\,dx$

$\displaystyle = \frac{Bc - Cb + (Ac - Ca)\cosh x + (Ab - Ba)\sinh x}{(1 - n)(a^2 - b^2 + c^2)(a + b\cosh x + c\sinh x)^{n-1}} + \frac{1}{(n-1)(a^2 - b^2 + c^2)}$

$\displaystyle \times \int \frac{(n-1)(Aa - Bb + Cc) - (n-2)(Ab - Ba)\cosh x - (n-2)(Ac - Ca)\sinh x}{(a + b\cosh x + c\sinh x)^{n-1}}\,dx$

$[a^2 + c^2 \neq b^2]$

$\displaystyle = \frac{Bc - Cb - Ca\cosh x - Ba\sinh x}{(n-1)a(a + b\cosh x + c\sinh x)^n} + \left[\frac{A}{a} + \frac{n(Bb - Cc)}{(n-1)a^2}\right](c\cosh x + b\sinh x)\frac{(n-1)!}{(2n-1)!!}$

$\displaystyle \times \sum_{k=0}^{n-1} \frac{(2n - 2k - 3)!!}{(n - k - 1)!a^k} \frac{1}{(a + b\cosh x + c\sinh x)^{n-k}}$

$[a^2 + c^2 = b^2]$

2.    $\displaystyle\int \frac{A + B\cosh x + C\sinh x}{a + b\cosh x + c\sinh x}\,dx = \frac{Cb - Bc}{b^2 - c^2} \ln\left(a + b\cosh x + c\sinh x\right)$

$\displaystyle + \frac{Bb - Cc}{b^2 - c^2}x + \left(A - a\frac{Bb - Cc}{b^2 - c^2}\right) \int \frac{dx}{a + b\cosh x + c\sinh x}$

$[b^2 \neq c^2]$     (see **2.451** 4)

3.    $\displaystyle\int \frac{A + B\cosh x + C\sinh x}{a + b\cosh x \pm b\sinh x}\,dx = \frac{C \mp B}{2a}(\cosh x \mp \sinh x) + \left[\frac{A}{a} - \frac{(B \mp C)b}{2a^2}\right]x$

$\displaystyle + \left[\frac{C \pm B}{2b} \pm \frac{A}{a} - \frac{(C \mp B)b}{2a^2}\right]\ln\left(a + b\cosh x \pm b\sinh x\right)$

$[ab \neq 0]$

4. 
$$\int \frac{dx}{a + b\cosh x + c\sinh x}$$

$$= \frac{2}{\sqrt{b^2 - a^2 - c^2}} \arctan \frac{(b - a)\tanh \frac{x}{2} + c}{\sqrt{b^2 - a^2 - c^2}} \qquad [b^2 > a^2 + c^2, \quad a \neq b]$$

$$= \frac{1}{\sqrt{a^2 - b^2 + c^2}} \ln \frac{(a - b)\tanh \frac{x}{2} - c + \sqrt{a^2 - b^2 + c^2}}{(a - b)\tanh \frac{x}{2} - c - \sqrt{a^2 - b^2 + c^2}} \qquad [b^2 < a^2 + c^2, \quad a \neq b]$$

$$= \frac{1}{c} \ln \left( a + c\tanh \frac{x}{2} \right) \qquad [a = b, \quad c \neq 0]$$

$$= \frac{2}{(a - b)\tanh \frac{x}{2} + c} \qquad [b^2 = a^2 + c^2]$$

GU (351)(18)

**2.452**

1. 
$$\int \frac{A + B\cosh x + C\sinh x}{(a_1 + b_1\cosh x + c_1\sinh x)(a_2 + b_2\cosh x + c_2\sinh x)} dx$$
$$= A_0 \ln \frac{a_1 + b_1\cosh x + c_1\sinh x}{a_2 + b_2\cosh x + c_2\sinh x} + A_1 \int \frac{dx}{a_1 + b_1\cosh x + c_1\sinh x} + A_2 \int \frac{dx}{a_2 + b_2\cosh x + c_2\sinh x}$$

where

GU (351)(19)

$$A_0 = \frac{\begin{vmatrix} a_1 & b_1 & c_1 \\ A & B & C \\ a_2 & b_2 & c_2 \end{vmatrix}}{\begin{vmatrix} a_1 & b_1 \\ a_2 & b_2 \end{vmatrix}^2 + \begin{vmatrix} b_1 & c_1 \\ b_2 & c_2 \end{vmatrix}^2 - \begin{vmatrix} c_1 & a_1 \\ c_2 & a_2 \end{vmatrix}^2}$$

$$A_1 = \frac{\begin{vmatrix} a_1 & b_1 & c_1 \\ \begin{vmatrix} b_1 & c_1 \\ B & C \end{vmatrix} & \begin{vmatrix} c_1 & a_1 \\ C & A \end{vmatrix} & \begin{vmatrix} a_1 & b_1 \\ A & B \end{vmatrix} \\ a_2 & b_2 & c_2 \end{vmatrix}}{\begin{vmatrix} a_1 & b_1 \\ a_2 & b_2 \end{vmatrix}^2 + \begin{vmatrix} b_1 & c_1 \\ b_2 & c_2 \end{vmatrix}^2 - \begin{vmatrix} c_1 & a_1 \\ c_2 & a_2 \end{vmatrix}^2},$$

$$A_2 = \frac{\begin{vmatrix} a_1 & b_1 & c_1 \\ \begin{vmatrix} C & B \\ c_2 & b_2 \end{vmatrix} & \begin{vmatrix} C & A \\ c_2 & a_2 \end{vmatrix} & \begin{vmatrix} B & A \\ b_2 & a_2 \end{vmatrix} \\ a_2 & b_2 & c_2 \end{vmatrix}}{\begin{vmatrix} a_1 & b_1 \\ a_2 & b_2 \end{vmatrix}^2 + \begin{vmatrix} b_1 & c_1 \\ b_2 & c_2 \end{vmatrix}^2 - \begin{vmatrix} c_1 & a_1 \\ c_2 & a_2 \end{vmatrix}^2},$$

$$\left[ \begin{vmatrix} a_1 & b_1 \\ a_2 & b_2 \end{vmatrix}^2 + \begin{vmatrix} b_1 & c_1 \\ b_2 & c_2 \end{vmatrix}^2 \neq \begin{vmatrix} c_1 & a_1 \\ c_2 & a_2 \end{vmatrix}^2 \right].$$

2. 
$$\int \frac{A\cosh^2 x + 2B\sinh x \cosh x + C\sinh^2 x}{a\cosh^2 x + 2b\sinh x \cosh x + c\sinh^2 x} dx$$

$$= \frac{1}{4b^2 - (a + c)^2} \left\{ [4Bb - (A + C)(a + c)]x \right.$$

$$+ [(A + C)b - B(a + c)] \ln \left( a\cosh^2 x + 2b\sinh x \cosh x + c\sinh^2 x \right)$$

$$\left. + [2(A - C)b^2 - 2Bb(a - c) + (Ca - Ac)(a + c)] f(x) \right\}$$

where                           GU (351)(24)

$$f(x) = \frac{1}{2\sqrt{b^2 - ac}} \ln \frac{c \tanh x + b - \sqrt{b^2 - ac}}{c \tanh x + b + \sqrt{b^2 - ac}} \qquad [b^2 > ac]$$

$$= \frac{1}{\sqrt{ac - b^2}} \arctan \frac{c \tanh x + b}{\sqrt{ac - b^2}} \qquad [b^2 < ac]$$

$$= -\frac{1}{c \tanh x + b} \qquad [b^2 = ac]$$

**2.453**

1.    $$\int \frac{(A + B \sinh x)\, dx}{\sinh x \,(a + b \sinh x)} = \frac{1}{a} \left[ A \ln \left| \tanh \frac{x}{2} \right| + (aB - bA) \int \frac{dx}{a + b \sinh x} \right]$$

(see **2.441** 3)

2.    $$\int \frac{(A + B \sinh x)\, dx}{\sinh x \,(a + b \cosh x)} = \frac{A}{a^2 - b^2} \left( a \ln \left| \tanh \frac{x}{2} \right| + b \ln \left| \frac{a + b \cosh x}{\sinh x} \right| \right) + B \int \frac{dx}{a + b \cosh x}$$

(see **2.443** 3)

For $a^2 = b^2 = 1$:

3.    $$\int \frac{(A + B \sinh x)\, dx}{\sinh x \,(1 + \cosh x)} = \frac{A}{2} \left( \ln \left| \tanh \frac{x}{2} \right| - \frac{1}{2} \tanh^2 \frac{x}{2} \right) + B \tanh \frac{x}{2}$$

4.    $$\int \frac{(A + B \sinh x)\, dx}{\sinh x \,(1 - \cosh x)} = \frac{A}{2} \left( -\ln \left| \coth \frac{x}{2} \right| + \frac{1}{2} \coth^2 \frac{x}{2} \right) + B \coth \frac{x}{2}$$

**2.454**

1.    $$\int \frac{(A + B \sinh x)\, dx}{\cosh x \,(a + b \sinh x)} = \frac{1}{a^2 + b^2} \left[ (Aa + Bb) \arctan (\sinh x) + (Ab - Ba) \ln \left| \frac{a + b \sinh x}{\cosh x} \right| \right]$$

2.    $$\int \frac{(A + B \cosh x)\, dx}{\sinh x \,(a + b \sinh x)} = \frac{1}{a} \left( A \ln \left| \tanh \frac{x}{2} \right| + B \ln \left| \frac{\sinh x}{a + b \sinh x} \right| - Ab \int \frac{dx}{a + b \sinh x} \right)$$

(see **2.441** 3)

**2.455**

1.    $$\int \frac{(A + B \cosh x)\, dx}{\sinh x \,(a + b \cosh x)} = \frac{1}{a^2 - b^2} \left[ (Aa + Bb) \ln \left| \tanh \frac{x}{2} \right| + (Ab - Ba) \ln \left| \frac{a + b \cosh x}{\sinh x} \right| \right]$$

For $a^2 = b^2 = 1$:

2.    $$\int \frac{(A + B \cosh x)\, dx}{\sinh x \,(1 + \cosh x)} = \frac{A + B}{2} \ln \left| \tanh \frac{x}{2} \right| - \frac{A - B}{4} \tanh^2 \frac{x}{2}$$

3.    $$\int \frac{(A + B \cosh x)\, dx}{\sinh x \,(1 - \cosh x)} = \frac{A + B}{4} \coth^2 \frac{x}{2} - \frac{A - B}{2} \ln \coth \frac{x}{2}$$

**2.456**   $$\int \frac{(A + B \cosh x)\, dx}{\cosh x \,(a + b \sinh x)} = \frac{A}{a^2 + b^2} \left[ a \arctan (\sinh x) + b \ln \left| \frac{a + b \sinh x}{\cosh x} \right| \right] + B \int \frac{dx}{a + b \sinh x}$$

(see **2.441** 3)

**2.457**

1.    $\displaystyle \int \frac{(A + B\cosh x)\,dx}{\cosh x\,(a + b\cosh x)} = \frac{1}{a}\left[ A\arctan\sinh x - (Ab - Ba)\int \frac{dx}{a + b\cosh x} \right]$

$$(\text{see } \mathbf{2.443}\ 3)$$

**2.458**

1.    $\displaystyle \int \frac{dx}{a + b\sinh^2 x}$

$$= \frac{1}{\sqrt{a(b-a)}}\arctan\left( \sqrt{\frac{b}{a} - 1}\,\tanh x \right) \qquad \left[ \frac{b}{a} > 1 \right]$$

$$= \frac{1}{\sqrt{a(a-b)}}\operatorname{arctanh}\left( \sqrt{1 - \frac{b}{a}}\,\tanh x \right) \qquad \left[ 0 < \frac{b}{a} < 1 \text{ or } \frac{b}{a} < 0 \text{ and } \sinh^2 x < -\frac{a}{b} \right]$$

$$= \frac{1}{\sqrt{a(a-b)}}\operatorname{arccoth}\left( \sqrt{1 - \frac{b}{a}}\,\tanh x \right) \qquad \left[ \frac{b}{a} < 0 \text{ and } \sinh^2 x > -\frac{a}{b} \right]$$

$$\text{MZ 195}$$

2.    $\displaystyle \int \frac{dx}{a + b\cosh^2 x}$

$$= \frac{1}{\sqrt{-a(a+b)}}\arctan\left( \sqrt{-\left(1 + \frac{b}{a}\right)}\,\coth x \right) \qquad \left[ \frac{b}{a} < -1 \right]$$

$$= \frac{1}{\sqrt{a(a+b)}}\operatorname{arctanh}\left( \sqrt{1 + \frac{b}{a}}\,\coth x \right) \qquad \left[ -1 < \frac{b}{a} < 0 \text{ and } \cosh^2 x > -\frac{a}{b} \right]$$

$$= \frac{1}{\sqrt{a(a+b)}}\operatorname{arccoth}\left( \sqrt{1 + \frac{b}{a}}\,\coth x \right) \qquad \left[ \frac{b}{a} > 0 \text{ or } -1 < \frac{b}{a} < 0 \text{ and } \cosh^2 x < -\frac{a}{b} \right]$$

$$\text{MZ 202}$$

For $a^2 = b^2 = 1$:

3.    $\displaystyle \int \frac{dx}{1 + \sinh^2 x} = \tanh x$

4.    $\displaystyle \int \frac{dx}{1 - \sinh^2 x} = \frac{1}{\sqrt{2}}\operatorname{arctanh}\left( \sqrt{2}\,\tanh x \right) \qquad\qquad [\sinh^2 x < 1]$

$$= \frac{1}{\sqrt{2}}\operatorname{arccoth}\left( \sqrt{2}\,\tanh x \right) \qquad\qquad [\sinh^2 x > 1]$$

5.    $\displaystyle \int \frac{dx}{1 + \cosh^2 x} = \frac{1}{\sqrt{2}}\operatorname{arccoth}\left( \sqrt{2}\,\coth x \right)$

6.    $\displaystyle \int \frac{dx}{1 - \cosh^2 x} = \coth x$

**2.459**

1.    $\displaystyle \int \frac{dx}{\left(a + b\sinh^2 x\right)^2} = \frac{1}{2a(b-a)}\left[ \frac{b\sinh x\cosh x}{a + b\sinh^2 x} + (b - 2a)\int \frac{dx}{a + b\sinh^2 x} \right]$

$$(\text{see } \mathbf{2.458}\ 1) \qquad\qquad \text{MZ 196}$$

2. 
$$\int \frac{dx}{\left(a + b\cosh^2 x\right)^2} = \frac{1}{2a(a+b)}\left[-\frac{b\sinh x\cosh x}{a + b\cosh^2 x} + (2a+b)\int \frac{dx}{a + b\cosh^2 x}\right]$$

(see **2.458** 2)                        MZ 203

3. 
$$\int \frac{dx}{\left(a + b\sinh^2 x\right)^3} = \frac{1}{8pa^3}\left[\left(3 - \frac{2}{p^2} + \frac{3}{p^4}\right)\arctan\left(p\tanh x\right) + \left(3 - \frac{2}{p^2} - \frac{3}{p^4}\right)\frac{p\tanh x}{1 + p^2\tanh^2 x}\right.$$
$$\left. + \left(1 + \frac{2}{p^2} - \frac{1}{p^2}\tanh^2 x\right)\frac{2p\tanh x}{\left(1 + p^2\tanh^2 x\right)^2}\right]$$

$$\left[p^2 = \frac{b}{a} - 1 > 0\right]$$

$$= \frac{1}{8qa^3}\left[\left(3 + \frac{2}{q^2} + \frac{3}{q^4}\right)\operatorname{arctanh}\left(q\tanh x\right) + \left(3 + \frac{2}{q^2} - \frac{3}{q^4}\right)\frac{q\tanh x}{1 - q^2\tanh^2 x}\right.$$
$$\left. + \left(1 - \frac{2}{q^2} + \frac{1}{q^2}\tanh^2 x\right)\frac{2q\tanh x}{\left(1 - q^2\tanh^2 x\right)^2}\right]$$

$$\left[q^2 = 1 - \frac{b}{a} > 0\right]$$

MZ 196

4. 
$$\int \frac{dx}{\left(a + b\cosh^2 x\right)^3} = \frac{1}{8pa^3}\left[\left(3 - \frac{2}{p^2} + \frac{3}{p^4}\right)\arctan\left(p\coth x\right) + \left(3 - \frac{2}{p^2} - \frac{3}{p^4}\right)\frac{p\coth x}{1 + p^2\coth^2 x}\right.$$
$$\left. + \left(1 + \frac{2}{p^2} - \frac{1}{p^2}\coth^2 x\right)\frac{2p\coth x}{\left(1 + p^2\coth^2 x\right)^2}\right]$$

$$\left[p^2 = -1 - \frac{b}{a} > 0\right]$$

$$= \frac{1}{8qa^3}\left[\left(3 + \frac{2}{q^2} + \frac{3}{q^4}\right)\varphi(x)^* + \left(3 + \frac{2}{q^2} - \frac{3}{q^4}\right)\frac{q\coth x}{1 - q^2\coth^2 x}\right.$$
$$\left. + \left(1 - \frac{2}{q^2} + \frac{1}{q^2}\coth^2 x\right)\frac{2q\coth x}{\left(1 - q^2\coth^2 x\right)^2}\right]$$

$$\left[q^2 = 1 + \frac{b}{a} > 0\right]$$

## 2.46 Algebraic functions of hyperbolic functions

**2.461**

1. 
$$\int \sqrt{\tanh x}\,dx = \operatorname{arctanh}\sqrt{\tanh x} - \arctan\sqrt{\tanh x}$$

MZ 221

---

*In 2.459.4, if $\frac{b}{a} < 0$ and $\cosh^2 x > -\frac{a}{b}$, then $\varphi(x) = \operatorname{arctanh}\left(q\coth x\right)$. If $\frac{b}{a} < 0$, but $\cosh^2 x < -\frac{a}{b}$, or if $\frac{b}{a} > 0$, then $\varphi(x) = \operatorname{arccoth}\left(q\coth x\right)$

2.    $\int \sqrt{\coth x}\, dx = \operatorname{arccoth} \sqrt{\coth x} - \arctan \sqrt{\coth x}$        MZ 222

**2.462**

1.    $\displaystyle \int \frac{\sinh x\, dx}{\sqrt{a^2 + \sinh^2 x}} = \operatorname{arcsinh} \frac{\cosh x}{\sqrt{a^2 - 1}} = \ln\left(\cosh x + \sqrt{a^2 + \sinh^2 x}\right)$    $[a^2 > 1]$

$\displaystyle = \operatorname{arccosh} \frac{\cosh x}{\sqrt{1 - a^2}} = \ln\left(\cosh x + \sqrt{a^2 + \sinh^2 x}\right)$    $[a^2 < 1]$

$\displaystyle = \ln \cosh x$    $[a^2 = 1]$

2.    $\displaystyle \int \frac{\sinh x\, dx}{\sqrt{a^2 - \sinh^2 x}} = \arcsin \frac{\cosh x}{\sqrt{a^2 + 1}}$      $\left[\sinh^2 x < a^2\right]$

3.    $\displaystyle \int \frac{\sinh x\, dx}{\sqrt{\sinh^2 x - a^2}} = \operatorname{arccosh} \frac{\cosh x}{\sqrt{a^2 + 1}} = \ln\left(\cosh x + \sqrt{\sinh^2 x - a^2}\right)$

$\left[\sinh^2 x > a^2\right]$      MZ 199

4.[12]    $\displaystyle \int \frac{\cosh x\, dx}{\sqrt{a^2 + \sinh^2 x}} = \operatorname{arcsinh}\left(\frac{\sinh x}{|a|}\right) = \ln\left(\sinh x + \sqrt{a^2 + \sinh^2 x}\right)$

5.[12]    $\displaystyle \int \frac{\cosh x\, dx}{\sqrt{a^2 - \sinh^2 x}} = \arcsin\left(\frac{\sinh x}{|a|}\right)$      $\left[\sinh^2 x < a^2\right]$

6.[12]    $\displaystyle \int \frac{\cosh x\, dx}{\sqrt{\sinh^2 x - a^2}} = \operatorname{arccosh}\left(\frac{\sinh x}{|a|}\right) = \ln\left(\sinh x + \sqrt{\sinh^2 x - a^2}\right)$      $[\sinh x > |a|]$

$\displaystyle = -\operatorname{arccosh}\left(-\frac{\sinh x}{|a|}\right) = -\ln\left(-\sinh x + \sqrt{\sinh^2 x - a^2}\right)$    $[\sinh x < -|a|]$

7.[12]    $\displaystyle \int \frac{\sinh x\, dx}{\sqrt{a^2 + \cosh^2 x}} = \operatorname{arcsinh}\left(\frac{\cosh x}{|a|}\right) = \ln\left(\cosh x + \sqrt{a^2 + \cosh^2 x}\right)$

8.[12]    $\displaystyle \int \frac{\sinh x\, dx}{\sqrt{a^2 - \cosh^2 x}} = \arcsin\left(\frac{\cosh x}{|a|}\right)$      $\left[\cosh^2 x < a^2\right]$

9.[12]    $\displaystyle \int \frac{\sinh x\, dx}{\sqrt{\cosh^2 x - a^2}} = \operatorname{arccosh}\left(\frac{\cosh x}{|a|}\right) = \ln\left(\cosh x + \sqrt{\cosh^2 x - a^2}\right)$

$\left[\cosh^2 x > a^2\right]$      MZ 215, 216

10.    $\displaystyle \int \frac{\cosh x\, dx}{\sqrt{a^2 + \cosh^2 x}} = \operatorname{arcsinh} \frac{\sinh x}{\sqrt{a^2 + 1}} = \ln\left(\sinh x + \sqrt{a^2 + \cosh^2 x}\right)$

11.    $\displaystyle \int \frac{\cosh x\, dx}{\sqrt{a^2 - \cosh^2 x}} = \arcsin \frac{\sinh x}{\sqrt{a^2 - 1}}$      $\left[\cosh^2 x < a^2\right]$

$12.^{12}$ $\displaystyle\int \frac{\cosh x \, dx}{\sqrt{\cosh^2 x - a^2}} = \text{arccosh}\left(\frac{\sinh x}{\sqrt{a^2 - 1}}\right)$ $\qquad \left[a^2 > 1, \quad \sinh x > \sqrt{a^2 - 1}\right]$

$\qquad\qquad\qquad = -\text{arccosh}\left(-\frac{\sinh x}{\sqrt{a^2 - 1}}\right)$ $\qquad \left[a^2 > 1, \quad \sinh x < -\sqrt{a^2 - 1}\right]$

$\qquad\qquad\qquad = \text{arcsinh}\left(\frac{\sinh x}{\sqrt{1 - a^2}}\right)$ $\qquad \left[a^2 < 1\right]$

$\qquad\qquad\qquad = \ln\left(\sinh x + \sqrt{\cosh^2 x - a^2}\right)$ $\qquad \left[a^2 > 1, \quad \sinh x > \sqrt{a^2 - 1}\right]$

$\qquad\qquad\qquad = -\ln\left(-\sinh x + \sqrt{\cosh^2 x - a^2}\right)$ $\qquad \left[a^2 > 1, \quad \sinh x < -\sqrt{a^2 - 1}\right]$

$\qquad\qquad\qquad = \ln\left(\sinh x\right)$ $\qquad \left[a^2 = 1, \quad x > 0\right]$

$\qquad\qquad\qquad = -\ln\left(-\sinh x\right)$ $\qquad \left[a^2 = 1, \quad x < 0\right]$

MZ 206

13. $\displaystyle\int \frac{\coth x \, dx}{\sqrt{a + b \sinh x}} = 2\sqrt{a}\,\text{arccoth}\sqrt{1 + \frac{b}{a}\sinh x}$ $\qquad \left[b \sinh x > 0, \quad a > 0\right]$

$\qquad\qquad\qquad = 2\sqrt{a}\,\text{arctanh}\sqrt{1 + \frac{b}{a}\sinh x}$ $\qquad \left[b \sinh x < 0, \quad a > 0\right]$

$\qquad\qquad\qquad = 2\sqrt{-a}\,\text{arctanh}\sqrt{-\left(1 + \frac{b}{a}\sinh x\right)}$ $\qquad a < 0$

14. $\displaystyle\int \frac{\tanh x \, dx}{\sqrt{a + b \cosh x}} = 2\sqrt{a}\,\text{arccoth}\sqrt{1 + \frac{b}{a}\cosh x}$ $\qquad \left[b \cosh x > 0, \quad a > 0\right]$

$\qquad\qquad\qquad = 2\sqrt{a}\,\text{arctanh}\sqrt{1 + \frac{b}{a}\cosh x}$ $\qquad \left[b \cosh x < 0, \quad a > 0\right]$

$\qquad\qquad\qquad = 2\sqrt{-a}\,\text{arctanh}\sqrt{-\left(1 + \frac{b}{a}\cosh x\right)}$ $\qquad \left[a < 0\right]$

MZ 220, 221

**2.463**

1. $\displaystyle\int \frac{\sinh x \sqrt{a + b \cosh x}}{p + q \cosh x} \, dx$

$\qquad\qquad = 2\sqrt{\frac{aq - bp}{q}}\,\text{arccoth}\sqrt{\frac{q(a + b\cosh x)}{aq - bp}}$ $\qquad \left[b \cosh x > 0, \quad \frac{aq - bp}{q} > 0\right]$

$\qquad\qquad = 2\sqrt{\frac{aq - bp}{q}}\,\text{arctanh}\sqrt{\frac{q(a + b\cosh x)}{aq - bp}}$ $\qquad \left[b \cosh x < 0, \quad \frac{aq - bp}{q} > 0\right]$

$\qquad\qquad = 2\sqrt{\frac{bp - aq}{q}}\,\text{arctanh}\sqrt{\frac{q(a + b\cosh x)}{bp - aq}}$ $\qquad \left[\frac{aq - bp}{q} < 0\right]$

MZ 220

2. 
$$\int \frac{\cosh x \sqrt{a + b \sinh x}}{p + q \sinh x} \, dx$$

$$= 2\sqrt{\frac{aq - bp}{q}} \operatorname{arccoth} \sqrt{\frac{q(a + b \sinh x)}{aq - bp}} \qquad \left[ b \sinh x > 0, \ \frac{aq - bp}{q} > 0 \right]$$

$$= 2\sqrt{\frac{aq - bp}{q}} \operatorname{arctanh} \sqrt{\frac{q(a + b \sinh x)}{aq - bp}} \qquad \left[ b \sinh x < 0, \ \frac{aq - bp}{q} > 0 \right]$$

$$= 2\sqrt{\frac{bp - aq}{q}} \operatorname{arctanh} \sqrt{\frac{q(a + b \sinh x)}{bp - aq}} \qquad \left[ \frac{aq - bp}{q} < 0 \right]$$

MZ 221

**2.464**

1. 
$$\int \frac{dx}{\sqrt{k^2 + k'^2 \cosh^2 x}} = \int \frac{dx}{\sqrt{1 + k'^2 \sinh^2 x}} = F(\arcsin(\tanh x), k)$$

$$[x > 0] \qquad\qquad \text{BY (295.00)(295.10)}$$

2. 
$$\int \frac{dx}{\sqrt{\cosh^2 x - k^2}} = \int \frac{dx}{\sqrt{\sinh^2 x + k'^2}} = F\left(\arcsin\left(\frac{1}{\cosh x}\right), k\right)$$

$$[x > 0] \qquad\qquad \text{BY (295.40)(295.30)}$$

3. 
$$\int \frac{dx}{\sqrt{1 - k'^2 \cosh^2 x}} = F\left(\arcsin\left(\frac{\tanh x}{k}\right), k\right) \qquad \left[ 0 < x < \operatorname{arccosh} \frac{1}{k'} \right] \qquad \text{BY (295.20)}$$

**Notation:** In **2.464** 4–**2.464** 8, we set $\alpha = \arccos \dfrac{1 - \sinh 2ax}{1 + \sinh 2ax}, \ r = \dfrac{1}{\sqrt{2}} \quad [ax > 0]$

4. 
$$\int \frac{dx}{\sqrt{\sinh 2ax}} = \frac{1}{2a} F(\alpha, r) \qquad\qquad \text{BY (296.50)}$$

5. 
$$\int \sqrt{\sinh 2ax} \, dx = \frac{1}{2a} [F(\alpha, r) - 2E(\alpha, r)] + \frac{1}{a} \frac{\sqrt{\sinh 2ax (1 + \sinh^2 2ax)}}{1 + \sinh 2ax} \qquad \text{BY (296.53)}$$

6. 
$$\int \frac{\cosh^2 2ax \, dx}{(1 + \sinh 2ax)^2 \sqrt{\sinh 2ax}} = \frac{1}{2a} E(\alpha, r) \qquad\qquad \text{BY (296.51)}$$

7. 
$$\int \frac{(1 - \sinh 2ax)^2 \, dx}{(1 + \sinh 2ax)^2 \sqrt{\sinh 2ax}} = \frac{1}{2a} [2E(\alpha, r) - F(\alpha, r)] \qquad\qquad \text{BY (296.55)}$$

8. 
$$\int \frac{\sqrt{\sinh 2ax} \, dx}{(1 + \sinh 2ax)^2} = \frac{1}{4a} [F(\alpha, r) - E(\alpha, r)] \qquad\qquad \text{BY (296.54)}$$

**Notation:** In **2.464** 9–**2.464** 15, we set $\alpha = \arcsin \sqrt{\dfrac{\cosh 2ax - 1}{\cosh 2ax}}, \ r = \dfrac{1}{\sqrt{2}} \quad [x \neq 0]$:

9. 
$$\int \frac{dx}{\sqrt{\cosh 2ax}} = \frac{1}{a\sqrt{2}} F(\alpha, r) \qquad\qquad \text{BY (296.00)}$$

10. 
$$\int \sqrt{\cosh 2ax} \, dx = \frac{1}{a\sqrt{2}} [F(\alpha, r) - 2E(\alpha, r)] + \frac{\sinh 2ax}{a\sqrt{\cosh 2ax}} \qquad \text{BY (296.03)}$$

11. 
$$\int \frac{dx}{\sqrt{\cosh^3 2ax}} = \frac{1}{a\sqrt{2}} [2E(\alpha, r) - F(\alpha, r)] \qquad\qquad \text{BY (296.04)}$$

12. $\displaystyle\int \frac{\mathrm{d}x}{\sqrt{\cosh^5 2ax}} = \frac{1}{3\sqrt{2}a} F(\alpha, r) + \frac{\tanh 2ax}{3a\sqrt{\cosh 2ax}}$        BY (296.04)

13. $\displaystyle\int \frac{\sinh^2 2ax \, \mathrm{d}x}{\sqrt{\cosh 2ax}} = -\frac{\sqrt{2}}{3a} F(\alpha, r) + \frac{1}{3a} \sinh 2ax \sqrt{\cosh 2ax}$        BY (296.07)

14. $\displaystyle\int \frac{\tanh^2 2ax \, \mathrm{d}x}{\sqrt{\cosh 2ax}} = \frac{\sqrt{2}}{3a} F(\alpha, r) - \frac{\tanh 2ax}{3a\sqrt{\cosh 2ax}}$        BY (296.05)

15. $\displaystyle\int \frac{\sqrt{\cosh 2ax} \, \mathrm{d}x}{p^2 + (1 - p^2) \cosh 2ax} = \frac{1}{a\sqrt{2}} \Pi\left(\alpha, p^2, r\right)$        BY (296.02)

**Notation**: In **2.464** 16–**2.464** 20, we set:

$$\alpha = \arccos \frac{\sqrt{a^2 + b^2} - a - b\sinh x}{\sqrt{a^2 + b^2} + a + b\sinh x},$$

$$r = \sqrt{\frac{a + \sqrt{a^2 + b^2}}{2\sqrt{a^2 + b^2}}} \qquad \left[a > 0, \quad b > 0, \quad x > -\operatorname{arcsinh}\frac{a}{b}\right]$$

16. $\displaystyle\int \frac{\mathrm{d}x}{\sqrt{a + b\sinh x}} = \frac{1}{\sqrt[4]{a^2 + b^2}} F(\alpha, r)$        BY (298.00)

17. $\displaystyle\int \sqrt{a + b\sinh x} \, \mathrm{d}x = \sqrt[4]{a^2 + b^2}\left[F(\alpha, r) - 2\,E(\alpha, r)\right] + \frac{2b\cosh x\sqrt{a + b\sinh x}}{\sqrt{a^2 + b^2} + a + b\sinh x}$        BY (298.02)

18. $\displaystyle\int \frac{\sqrt{a + b\sinh x}}{\cosh^2 x} \, \mathrm{d}x = \sqrt[4]{a^2 + b^2}\,E(\alpha, r) - \frac{\sqrt{a^2 + b^2} - a}{2\sqrt[4]{a^2 + b^2}} F(\alpha, r)$

$$-\frac{a + \sqrt{a^2 + b^2}}{b} \cdot \frac{\sqrt{a^2 + b^2} - a - b\sinh x}{\sqrt{a^2 + b^2} + a + b\sinh x} \cdot \frac{\sqrt{a + b\sinh x}}{\cosh x}$$

       BY (298.03)

19. $\displaystyle\int \frac{\cosh^2 x \, \mathrm{d}x}{\left[\sqrt{a^2 + b^2} + a + b\sinh x\right]^2 \sqrt{a + b\sinh x}} = \frac{1}{b^2 \sqrt[4]{a^2 + b^2}} E(\alpha, r)$        BY (298.01)

20. $\displaystyle\int \frac{\sqrt{a + b\sinh x} \, \mathrm{d}x}{\left[\sqrt{a^2 + b^2} - a - b\sinh x\right]^2} = -\frac{1}{\sqrt[4]{a^2 + b^2}\left(\sqrt{a^2 + b^2} - a\right)} E(\alpha, r)$

$$+\frac{b}{\sqrt{a^2 + b^2} - a} \cdot \frac{\cosh x\sqrt{a + b\sinh x}}{a^2 + b^2 - (a + b\sinh x)^2}$$

       BY (298.04)

**Notation**: In **2.464** 21–**2.464** 31, we set $\alpha = \arcsin\left(\tanh\frac{x}{2}\right)$, $r = \sqrt{\dfrac{a - b}{a + b}}$    $[0 < b < a, x > 0]$:

21. $\displaystyle\int \frac{\mathrm{d}x}{\sqrt{a + b\cosh x}} = \frac{2}{\sqrt{a + b}} F(\alpha, r)$        BY (297.25)

22. $\displaystyle\int \sqrt{a + b\cosh x} \, \mathrm{d}x = 2\sqrt{a + b}\left[F(\alpha, r) - E(\alpha, r)\right] + 2\tanh\frac{x}{2}\sqrt{a + b\cosh x}$        BY (297.29)

23. $\displaystyle\int \frac{\cosh x \, \mathrm{d}x}{\sqrt{a + b\cosh x}} = \frac{2}{\sqrt{a + b}} F(\alpha, r) - \frac{2\sqrt{a + b}}{b} E(\alpha, r) + \frac{2}{b}\tanh\frac{x}{2}\sqrt{a + b\cosh x}$        BY (297.33)

24. $\displaystyle\int \frac{\tanh^2 \frac{x}{2}}{\sqrt{a + b \cosh x}}\, \mathrm{d}x = \frac{2\sqrt{a+b}}{a-b}\left[F(\alpha,r) - E(\alpha,r)\right]$ 　　　　BY (297.28)

25.[11] $\displaystyle\int \frac{\tanh^4 \frac{x}{2}}{\sqrt{a + b \cosh x}}\, \mathrm{d}x = \frac{2\sqrt{a+b}}{3(a-b)^2}\left[(3a+b)\,F(\alpha,r) - 4a\,E(\alpha,r)\right] + \frac{2}{3(a-b)}\frac{\sinh \frac{x}{2}\sqrt{a+b\cosh x}}{\cosh^3 \frac{x}{2}}$

　　　　BY (297.28)

26. $\displaystyle\int \frac{\cosh x - 1}{\sqrt{a + b \cosh x}}\, \mathrm{d}x = \frac{2}{b}\left[\left(\tanh \frac{x}{2}\right)\sqrt{a+b\cosh x} - \sqrt{a+b}\,E(\alpha,r)\right]$ 　　　　BY (297.31)

27. $\displaystyle\int \frac{(\cosh x - 1)^2}{\sqrt{a + b \cosh x}}\, \mathrm{d}x = \frac{4\sqrt{a+b}}{3b^2}\left[(a+3b)\,E(\alpha,r) - b\,F(\alpha,r)\right]$

　　　　$\displaystyle + \frac{4}{3b^2}\left[b\cosh^2\frac{x}{2} - (a+3b)\right]\tanh\frac{x}{2}\sqrt{a+b\cosh x}$

　　　　BY (297.31)

28. $\displaystyle\int \frac{\sqrt{a + b \cosh x}}{\cosh x + 1}\, \mathrm{d}x = \sqrt{a+b}\,E(\alpha,r)$ 　　　　BY (297.26)

29. $\displaystyle\int \frac{\mathrm{d}x}{(\cosh x + 1)\sqrt{a + b \cosh x}} = \frac{\sqrt{a+b}}{a-b}\,E(\alpha,r) - \frac{2b}{(a-b)\sqrt{a+b}}\,F(\alpha,r)$ 　　　　BY (297.30)

30. $\displaystyle\int \frac{\mathrm{d}x}{(\cosh x + 1)^2\sqrt{a + b \cosh x}} = \frac{1}{3(a-b)^2\sqrt{a+b}}\left[b(5b-a)\,F(\alpha,r)\right.$

　　　　$\displaystyle\left. + (a-3b)(a+b)\,E(\alpha,r)\right] + \frac{1}{6(a-b)}\cdot\frac{\sinh \frac{x}{2}}{\cosh^3 \frac{x}{2}}\sqrt{a+b\cosh x}$

　　　　297.30)

31. $\displaystyle\int \frac{(1 + \cosh x)\,\mathrm{d}x}{[1 + p^2 + (1 - p^2)\cosh x]\sqrt{a + b \cosh x}} = \frac{2}{\sqrt{a+b}}\,\Pi\left(\alpha, p^2, r\right)$ 　　　　BY (297.27)

**Notation**: In **2.464** 32–**2.464** 40, we set:

$$\alpha = \arcsin \sqrt{\frac{a - b\cosh x}{a - b}}$$

$$r = \sqrt{\frac{a-b}{a+b}} \qquad \left[0 < b < a, \quad 0 < x < \operatorname{arccosh}\frac{a}{b}\right]$$

32. $\displaystyle\int \frac{\mathrm{d}x}{\sqrt{a - b \cosh x}} = \frac{2}{\sqrt{a+b}}\,F(\alpha,r)$ 　　　　BY (297.50)

33. $\displaystyle\int \sqrt{a - b \cosh x}\, \mathrm{d}x = 2\sqrt{a+b}\left[F(\alpha,r) - E(\alpha,r)\right]$ 　　　　BY (297.54)

34. $\displaystyle\int \frac{\cosh x\, \mathrm{d}x}{\sqrt{a - b \cosh x}} = \frac{2\sqrt{a+b}}{b}\,E(\alpha,r) - \frac{2}{\sqrt{a+b}}\,F(\alpha,r)$ 　　　　BY (297.56)

35. $\displaystyle\int \frac{\cosh^2 x\, \mathrm{d}x}{\sqrt{a - b \cosh x}} = \frac{2(b-2a)}{3b\sqrt{a+b}}\,F(\alpha,r) + \frac{4a\sqrt{a+b}}{3b^2}\,E(\alpha,r) + \frac{2}{3b}\sinh x\sqrt{a-b\cosh x}$ 　　BY (297.56)

36. $\displaystyle\int \frac{(1 + \cosh x)\, \mathrm{d}x}{\sqrt{a - b \cosh x}} = \frac{2\sqrt{a+b}}{b}\,E(\alpha,r)$ 　　　　BY (297.51)

37. $\displaystyle\int \frac{dx}{\cosh x \sqrt{a - b\cosh x}} = \frac{2b}{a\sqrt{a+b}} \Pi\left(\alpha, \frac{a-b}{a}, r\right)$  BY (297.57)

38. $\displaystyle\int \frac{dx}{(1 + \cosh x)\sqrt{a - b\cosh x}} = \frac{1}{\sqrt{a+b}} E(\alpha, r) - \frac{1}{a+b} \tanh\frac{x}{2}\sqrt{a - b\cosh x}$  BY (297.58)

39. $\displaystyle\int \frac{dx}{(1 + \cosh x)^2 \sqrt{a - b\cosh x}} = \frac{1}{3\sqrt{(a+b)^3}} \left[(a + 3b) E(\alpha, r) - b F(\alpha, r)\right]$

$\displaystyle\qquad\qquad -\frac{1}{3(a+b)^2} \frac{\tanh\frac{x}{2}\sqrt{a - b\cosh x}}{\cosh x + 1} \left[2a + 4b + (a + 3b)\cosh x\right]$

BY (297.58)

40. $\displaystyle\int \frac{dx}{(a - b - ap^2 + bp^2\cosh x)\sqrt{a - b\cosh x}} = \frac{2}{(a-b)\sqrt{a+b}} \Pi\left(\alpha, p^2, r\right)$  BY (297.52)

**Notation:** In **2.464** 41 –**2.464** 47, we set:

$$\alpha = \arcsin\sqrt{\frac{b(\cosh x - 1)}{b\cosh x - a}},$$

$$r = \sqrt{\frac{a+b}{2b}} \qquad [0 < a < b, x > 0]$$

41. $\displaystyle\int \frac{dx}{\sqrt{b\cosh -a}} = \sqrt{\frac{2}{b}} F(\alpha, r)$  BY (297.00)

42. $\displaystyle\int \sqrt{b\cosh x - a}\, dx = (b - a)\sqrt{\frac{2}{b}} F(\alpha, r) - 2\sqrt{2b}\, E(\alpha, r) + \frac{2b\sinh x}{\sqrt{b\cosh x - a}}$  BY (297.05)

43. $\displaystyle\int \frac{dx}{\sqrt{(b\cosh x - a)^3}} = \frac{1}{b^2 - a^2} \cdot \sqrt{\frac{2}{b}} \left[2b\, E(\alpha, r) - (b - a) F(\alpha, r)\right]$  BY (297.06)

44. $\displaystyle\int \frac{dx}{\sqrt{(b\cosh x - a)^5}} = \frac{1}{3(b^2 - a^2)^2} \sqrt{\frac{2}{b}} \left[(b - 3a)(b - a) F(\alpha, r) + 8ab\, E(\alpha, r)\right]$

$\displaystyle\qquad\qquad + \frac{2b}{3(b^2 - a^2)} \frac{\sinh x}{\sqrt{(b\cosh x - a)^3}}$

BY (297.06)

45. $\displaystyle\int \frac{\cosh x\, dx}{\sqrt{b\cosh x - a}} = \sqrt{\frac{2}{b}} \left[F(\alpha, r) - 2\, E(\alpha, r)\right] + \frac{2\sinh x}{\sqrt{b\cosh x - a}}$  BY (297.03)

46. $\displaystyle\int \frac{(\cosh x + 1)\, dx}{\sqrt{(b\cosh x - a)^3}} = \frac{2}{b - a}\sqrt{\frac{2}{b}} E(\alpha, r)$  BY (297.01)

47. $\displaystyle\int \frac{\sqrt{b\cosh x - a}\, dx}{p^2 b - a + b(1 - p^2)\cosh x} = \sqrt{\frac{2}{b}} \Pi\left(\alpha, p^2, r\right)$  BY (297.02)

**Notation:** In **2.464** 48–**2.464** 55, we set $\alpha = \arcsin\sqrt{\dfrac{b\cosh x - a}{b(\cosh x - 1)}}$ and $r = \sqrt{\dfrac{2b}{a+b}}$ for

$\left[0 < b < a, x > \operatorname{arccosh}\dfrac{a}{b}\right]$:

48. $\displaystyle\int \frac{\mathrm{d}x}{\sqrt{b\cosh x - a}} = \frac{2}{\sqrt{a+b}}\, F(\alpha, r)$         BY (297.75)

49. $\displaystyle\int \sqrt{b\cosh x - a}\, \mathrm{d}x = -2\sqrt{a+b}\, E(\alpha, r) + 2\coth\frac{x}{2}\sqrt{b\cosh x - a}$     BY (297.79)

50. $\displaystyle\int \frac{\coth^2 \frac{x}{2}\, \mathrm{d}x}{\sqrt{b\cosh x - a}} = \frac{2\sqrt{a+b}}{a-b}\, E(\alpha, r)$         BY (297.76)

51. $\displaystyle\int \frac{\sqrt{b\cosh x - a}}{\cosh x - 1}\, \mathrm{d}x = \sqrt{a+b}\,[F(\alpha, r) - E(\alpha, r)]$         BY (297.77)

52. $\displaystyle\int \frac{\mathrm{d}x}{(\cosh x - 1)\sqrt{b\cosh x - a}} = \frac{\sqrt{a+b}}{a-b}\, E(\alpha, r) - \frac{1}{\sqrt{a+b}}\, F(\alpha, r)$     BY (297.78)

53. $\displaystyle\int \frac{\mathrm{d}x}{(\cosh x - 1)^2 \sqrt{b\cosh x - a}} = \frac{1}{3(a-b)^2\sqrt{a+b}}\left[(a-2b)(a-b)\,F(\alpha, r)\right.$

$$\left. + (3a-b)(a+b)\,E(\alpha, r)\right] + \frac{a+b}{6b(a-b)}\cdot\frac{\cosh\frac{x}{2}}{\sinh^3\frac{x}{2}}\sqrt{b\cosh x - a}$$

                                                                 BY (297.78)

54. $\displaystyle\int \frac{\mathrm{d}x}{(\cosh x + 1)\sqrt{b\cosh x - a}} = \frac{1}{\sqrt{a+b}}\,[F(\alpha, r) - E(\alpha, r)] + \frac{2\sqrt{b\cosh x - a}}{(a+b)\sinh x}$     BY (297.80)

55.[12] $\displaystyle\int \frac{\mathrm{d}x}{(\cosh x + 1)^2 \sqrt{b\cosh x - a}} = \frac{1}{3\sqrt{(a+b)^3}}\left[(a+2b)\,F(\alpha, r)\right.$

$$\left. - (a+3b)\,E(\alpha, r)\right] + \frac{\sqrt{b\cosh x - a}}{3(a+b)\sinh x}\left(2\frac{a+3b}{a+b} - \tanh^2\frac{x}{2}\right)$$

                                                                 BY (297.80)

**Notation:** In **2.464** 56–**2.464** 60, we set

$$\alpha = \arccos \frac{\sqrt[4]{b^2 - a^2}}{\sqrt{a\sinh x + b\cosh x}},$$

$$r = \frac{1}{\sqrt{2}} \qquad \left[0 < a < b, \quad -\operatorname{arcsinh}\frac{a}{\sqrt{b^2 - a^2}} < x\right]$$

56. $\displaystyle\int \frac{\mathrm{d}x}{\sqrt{a\sinh x + b\cosh x}} = \sqrt[4]{\frac{4}{b^2 - a^2}}\, F(\alpha, r)$         BY (299.00)

57. $\displaystyle\int \sqrt{a\sinh x + b\cosh x}\, \mathrm{d}x = \sqrt[4]{4(b^2 - a^2)}\,[F(\alpha, r) - 2\,E(\alpha, r)] + \frac{2\,(a\cosh x + b\sinh x)}{\sqrt{a\sinh x + b\cosh x}}$

                                                                 BY (299.02)

58. $\displaystyle\int \frac{\mathrm{d}x}{\sqrt{(a\sinh x + b\cosh x)^3}} = \sqrt[4]{\frac{4}{(b^2 - a^2)^3}}\,[2\,E(\alpha, r) - F(\alpha, r)]$     BY (299.03)

59. $\displaystyle\int \frac{\mathrm{d}x}{\sqrt{(a\sinh x + b\cosh x)^5}} = \frac{1}{3}\sqrt[4]{\frac{4}{(b^2 - a^2)^5}}\, F(\alpha, r) + \frac{2}{3\,(b^2 - a^2)}\frac{a\cosh x + b\sinh x}{\sqrt{(a\sinh x + b\cosh x)^3}}$

                                                                 BY (299.03)

60. $\displaystyle\int \frac{\left(\sqrt{b^2 - a^2} + a\sinh x + b\cosh x\right)\,dx}{\sqrt{(a\sinh x + b\cosh x)^3}} = 2\sqrt[4]{\frac{4}{b^2 - a^2}}\,E(\alpha, r)$          BY (299.01)

## 2.47 Combinations of hyperbolic functions and powers

**2.471**

1.[12] $\displaystyle\int x^r \sinh^p x \cosh^q x\,dx$

$$= \frac{1}{(p+q)^2}\left[(p+q)x^r \sinh^{p+1} x \cosh^{q-1} x\right.$$
$$-rx^{r-1}\sinh^p x \cosh^q x + r(r+1)\int x^{r-2}\sinh^p x \cosh^q x\,dx$$
$$\left.+ rp\int x^{r-1}\sinh^{p-1} x \cosh^{q-1} x\,dx + (q-1)(p+q)\int x^r \sinh^p x \cosh^{q-2} x\,dx\right]$$
$$= \frac{1}{(p+q)^2}\left[(p+q)x^r \sinh^{p-1} x \cosh^{q+1} x\right.$$
$$-rx^{r-1}\sinh^p x \cosh^q x + r(r-1)\int x^{r-2}\sinh^p x \cosh^q x\,dx$$
$$\left.- rq\int x^{r-1}\sinh^{p-1} x \cosh^{q-1} x\,dx - (p-1)(p+q)\int x^r \sinh^{p-2} x \cosh^q x\,dx\right]$$

GU (353)(1)

2. $\displaystyle\int x^n \sinh^{2m} x\,dx = (-1)^m \binom{2m}{m}\frac{x^{n+1}}{2^{2m}(n+1)} + \frac{1}{2^{2m-1}}\sum_{k=0}^{m-1}(-1)^k\binom{2m}{k}\int x^n \cosh(2m-2k)x\,dx$

3. $\displaystyle\int x^n \sinh^{2m+1} x\,dx = \frac{1}{2^{2m}}\sum_{k=0}^{m}(-1)^k\binom{2m+1}{k}\int x^n \sinh(2m-2k+1)x\,dx$

4. $\displaystyle\int x^n \cosh^{2m} x\,dx = \binom{2m}{m}\frac{x^{n+1}}{2^{2m}(n+1)} + \frac{1}{2^{2m-1}}\sum_{k=0}^{m-1}\binom{2m}{k}\int x^n \cosh(2m-2k)x\,dx$

5. $\displaystyle\int x^n \cosh^{2m+1} x\,dx = \frac{1}{2^{2m}}\sum_{k=0}^{m}\binom{2m+1}{k}\int x^n \cosh(2m-2k+1)x\,dx$

**2.472**

1. $\displaystyle\int x^n \sinh x\,dx = x^n \cosh x - n\int x^{n-1}\cosh x\,dx$
$$= x^n \cosh x - nx^{n-1}\sinh x + n(n-1)\int x^{n-2}\sinh x\,dx$$

2. $\displaystyle\int x^n \cosh x\,dx = x^n \sinh x - n\int x^{n-1}\sinh x\,dx$
$$= x^n \sinh x - nx^{n-1}\cosh x + n(n-1)\int x^{n-2}\cosh x\,dx$$

3. $\displaystyle\int x^{2n}\sinh x\,dx = (2n)!\left\{\sum_{k=0}^{n}\frac{x^{2k}}{(2k)!}\cosh x - \sum_{k=1}^{n}\frac{x^{2k-1}}{(2k-1)!}\sinh x\right\}$

4. $\displaystyle\int x^{2n+1}\sinh x\,dx=(2n+1)!\sum_{k=0}^{n}\left\{\frac{x^{2k+1}}{(2k+1)!}\cosh x-\frac{x^{2k}}{(2k)!}\sinh x\right\}$

5.[11] $\displaystyle\int x^{2n}\cosh x\,dx=(2n)!\left\{\sum_{k=0}^{n}\frac{x^{2k}}{(2k)!}\sinh x-\sum_{k=1}^{n}\frac{x^{2k-1}}{(2k-1)!}\cosh x\right\}$

6. $\displaystyle\int x^{2n+1}\cosh x\,dx=(2n+1)!\sum_{k=0}^{n}\left\{\frac{x^{2k+1}}{(2k+1)!}\sinh x-\frac{x^{2k}}{(2k)!}\cosh x\right\}$

7. $\displaystyle\int x\sinh x\,dx=x\cosh x-\sinh x$

8. $\displaystyle\int x^2\sinh x\,dx=\left(x^2+2\right)\cosh x-2x\sinh x$

9. $\displaystyle\int x\cosh x\,dx=x\sinh x-\cosh x$

10. $\displaystyle\int x^2\cosh x\,dx=\left(x^2+2\right)\sinh x-2x\cosh x$

**2.473**   **Notation:** $z_1=a+bx$

1. $\displaystyle\int z_1\sinh kx\,dx=\frac{1}{k}z_1\cosh kx-\frac{b}{k^2}\sinh kx$

2. $\displaystyle\int z_1\cosh kx\,dx=\frac{1}{k}z_1\sinh kx-\frac{b}{k^2}\cosh kx$

3. $\displaystyle\int z_1^2\sinh kx\,dx=\frac{1}{k}\left(z_1^2+\frac{2b^2}{k^2}\right)\cosh kx-\frac{2bz_1}{k^2}\sinh kx$

4. $\displaystyle\int z_1^2\cosh kx\,dx=\frac{1}{k}\left(z_1^2+\frac{2b^2}{k^2}\right)\sinh kx-\frac{2bz_1}{k^2}\cosh kx$

5. $\displaystyle\int z_1^3\sinh kx\,dx=\frac{z_1}{k}\left(z_1^2+\frac{6b^2}{k^2}\right)\cosh kx-\frac{3b}{k^2}\left(z_1^2+\frac{2b^2}{k^2}\right)\sinh kx$

6. $\displaystyle\int z_1^3\cosh kx\,dx=\frac{z_1}{k}\left(z_1^2+\frac{6b^2}{k^2}\right)\sinh kx-\frac{3b}{k^2}\left(z_1^3+\frac{2b^2}{k^2}\right)\cosh kx$

7. $\displaystyle\int z_1^4\sinh kx\,dx=\frac{1}{k}\left(z_1^4+\frac{12b^2}{k^2}z_1^2+\frac{24b^4}{k^4}\right)\cosh kx-\frac{4bz_1}{k^2}\left(z_1^2+\frac{6b^2}{k^2}\right)\sinh kx$

8. $\displaystyle\int z_1^4\cosh kx\,dx=\frac{1}{k}\left(z_1^4+\frac{12b^2}{k^2}z_1^2+\frac{24b^4}{k^4}\right)\sinh kx-\frac{4bz_1}{k^2}\left(z_1^2+\frac{6b^2}{k^2}\right)\cosh kx$

9. $\displaystyle\int z_1^5\sinh kx\,dx=\frac{z_1}{k}\left(z_1^4+\frac{20b^2}{k^2}z_1^2+120\frac{b^4}{k^4}\right)\cosh kx-\frac{5b}{k^2}\left(z_1^4+12\frac{b^2}{k^2}z_1^2+24\frac{b^4}{k^4}\right)\sinh kx$

10. $\displaystyle\int z_1^5\cosh kx\,dx=\frac{z_1}{k}\left(z_1^4+20\frac{b^2}{k^2}z_1^2+120\frac{b^4}{k^4}\right)\sinh kx-\frac{5b}{k^2}\left(z_1^4+12\frac{b^2}{k^2}z_1^2+24\frac{b^4}{k^4}\right)\cosh kx$

11. $\displaystyle\int z_1^6\sinh kx\,dx=\frac{1}{k}\left(z_1^6+30\frac{b^2}{k^2}z_1^4+360\frac{b^4}{k^4}z_1^2+720\frac{b^6}{k^6}\right)\cosh kx$
$\displaystyle\qquad\qquad\qquad-\frac{6bz_1}{k^2}\left(z_1^4+20\frac{b^2}{k^2}z_1^2+120\frac{b^4}{k^4}\right)\sinh kx$

12. $\int z_1^6 \cosh kx \, dx = \frac{1}{k}\left(z_1^6 + 30\frac{b^2}{k^2}z_1^4 + 360\frac{b^4}{k^4}z_1^2 + 720\frac{b^6}{k^6}\right)\sinh kx$

$\qquad\qquad - \frac{6bz_1}{k^2}\left(z_1^4 + 20\frac{b^2}{k^2}z_1 + 120\frac{b^4}{k^4}\right)\cosh kx$

## 2.474

1. $\int x^n \sinh^2 x \, dx = -\frac{x^{n+1}}{2(n+1)} + \frac{n!}{4}\sum_{k=0}^{\lfloor n/2 \rfloor}\left\{\frac{x^{n-2k}}{2^{2k}(n-2k)!}\sinh 2x - \frac{x^{n-2k-1}}{2^{2k+1}(n-2k-1)!}\cosh 2x\right\}$

$\hfill\text{GU (353)(2b)}$

2. $\int x^n \cosh^2 x \, dx = \frac{x^{n+1}}{2(n+1)} + \frac{n!}{4}\sum_{k=0}^{\lfloor n/2 \rfloor}\left\{\frac{x^{n-2k}}{2^{2k}(n-2k)!}\sinh 2x - \frac{x^{n-2k-1}}{2^{2k+1}(n-2k-1)!}\cosh 2x\right\}$

$\hfill\text{GU (353)(3e)}$

3. $\int x \sinh^2 x \, dx = \frac{1}{4}x\sinh 2x - \frac{1}{8}\cosh 2x - \frac{x^2}{4}$

4. $\int x^2 \sinh^2 x \, dx = \frac{1}{4}\left(x^2 + \frac{1}{2}\right)\sinh 2x - \frac{x}{4}\cosh 2x - \frac{x^3}{6}$    MZ 257

5. $\int x \cosh^2 x \, dx = \frac{x}{4}\sinh 2x - \frac{1}{8}\cosh 2x + \frac{x^2}{4}$

6. $\int x^2 \cosh^2 x \, dx = \frac{1}{4}\left(x^2 + \frac{1}{2}\right)\sinh 2x - \frac{x}{4}\cosh 2x + \frac{x^3}{6}$    MZ 261

7. $\int x^n \sinh^3 x \, dx$

$\qquad = \frac{n!}{4}\sum_{k=0}^{\lfloor n/2 \rfloor}\left\{\frac{x^{n-2k}}{(n-2k)!}\left(\frac{\cosh 3x}{3^{2k+1}} - 3\cosh x\right) - \frac{x^{n-2k-1}}{(n-2k-1)!}\left(\frac{\sinh 3x}{3^{2k+2}} - 3\sinh x\right)\right\}$

$\hfill\text{GU (353)(2f)}$

8. $\int x^n \cosh^3 x \, dx$

$\qquad = \frac{n!}{4}\sum_{k=0}^{\lfloor n/2 \rfloor}\left\{\frac{x^{n-2k}}{(n-2k)!}\left(\frac{\sinh 3x}{3^{2k+1}} + 3\sinh x\right) - \frac{x^{n-2k-1}}{(n-2k-1)!}\left(\frac{\cosh 3x}{3^{2k+2}} + 3\cosh x\right)\right\}$

$\hfill\text{GU (353)(3f)}$

9. $\int x \sinh^3 x \, dx = \frac{3}{4}\sinh x - \frac{1}{36}\sinh 3x - \frac{3}{4}x\cosh x - \frac{x}{12}\cosh 3x$

10. $\int x^2 \sinh^3 x \, dx = -\left(\frac{3x^2}{4} + \frac{3}{2}\right)\cosh x + \left(\frac{x^2}{12} + \frac{1}{54}\right)\cosh 3x + \frac{3x}{2}\sinh x - \frac{x}{18}\sinh 3x$    MZ 257

11. $\int x \cosh^3 x \, dx = -\frac{3}{4}\cosh x - \frac{1}{36}\cosh 3x + \frac{3}{4}x\sinh x + \frac{x}{12}\sinh 3x$

12. $\int x^2 \cosh^3 x \, dx = \left(\frac{3}{4}x^2 + \frac{3}{2}\right)\sinh x + \left(\frac{x^2}{12} + \frac{1}{54}\right)\sinh 3x - \frac{3}{2}x\cosh x - \frac{x}{18}\cosh 3x$    MZ 262

**2.475**

1.  $$\int \frac{\sinh^q x}{x^p}\,dx = -\frac{(p-2)\sinh^q x + qx\sinh^{q-1} x\cosh x}{(p-1)(p-2)x^{p-1}}$$
    $$+\frac{q(q-1)}{(p-1)(p-2)}\int \frac{\sinh^{q-2} x}{x^{p-2}}\,dx + \frac{q^2}{(p-1)(p-2)}\int \frac{\sinh^q x}{x^{p-2}}\,dx \quad [p>2]$$

    GU (353)(6a)

2.  $$\int \frac{\cosh^q x}{x^p}\,dx = -\frac{(p-2)\cosh^q x + qx\cosh^{q-1} x\sinh x}{(p-1)(p-2)x^{p-1}}$$
    $$-\frac{q(q-1)}{(p-1)(p-2)}\int \frac{\cosh^{q-2} x}{x^{p-2}}\,dx + \frac{q^2}{(p-1)(p-2)}\int \frac{\cosh^q x}{x^{p-2}}\,dx \quad [p>2]$$

    GU (353)(7a)

3.  $$\int \frac{\sinh x}{x^{2n}}\,dx = -\frac{1}{x(2n-1)!}\left\{\sum_{k=0}^{n-2}\frac{(2k+1)!}{x^{2k+1}}\cosh x + \sum_{k=0}^{n-1}\frac{(2k)!}{x^{2k}}\sinh x\right\} + \frac{1}{(2n-1)!}\operatorname{chi}(x)$$

    GU (353)(6b)

4.  $$\int \frac{\sinh x}{x^{2n+1}}\,dx = -\frac{1}{x(2n)!}\left\{\sum_{k=0}^{n-1}\frac{(2k)!}{x^{2k}}\cosh x + \sum_{k=0}^{n-1}\frac{(2k+1)!}{x^{2k+1}}\sinh x\right\} + \frac{1}{(2n)!}\operatorname{shi}(x) \qquad \text{GU (353)(6b)}$$

5.  $$\int \frac{\cosh x}{x^{2n}}\,dx = -\frac{1}{x(2n-1)!}\left\{\sum_{k=0}^{n-2}\frac{(2k+1)!}{x^{2k+1}}\sinh x + \sum_{k=0}^{n-1}\frac{(2k)!}{x^{2k}}\cosh x\right\} + \frac{1}{(2n-1)!}\operatorname{shi}(x)$$

    GU (353)(7b)

6.  $$\int \frac{\cosh x}{x^{2n+1}}\,dx = -\frac{1}{(2n)!x}\left\{\sum_{k=0}^{n-1}\frac{(2k)!}{x^{2k}}\sinh x + \sum_{k=0}^{n-1}\frac{(2k+1)!}{x^{2k+1}}\cosh x\right\} + \frac{1}{(2n)!}\operatorname{chi}(x) \qquad \text{GU (353)(7b)}$$

7.  $$\int \frac{\sinh^{2m} x}{x}\,dx = \frac{1}{2^{2m-1}}\sum_{k=0}^{m-1}(-1)^k\binom{2m}{k}\operatorname{chi}(2m-2k)x + \frac{(-1)^m}{2^{2m}}\binom{2m}{m}\ln x \qquad \text{GU (353)(6c)}$$

8.  $$\int \frac{\sinh^{2m+1} x}{x}\,dx = \frac{1}{2^{2m}}\sum_{k=0}^{m}(-1)^k\binom{2m+1}{k}\operatorname{shi}(2m-2k+1)x \qquad \text{GU (353)(6d)}$$

9.  $$\int \frac{\cosh^{2m} x}{x}\,dx = \frac{1}{2^{2m-1}}\sum_{k=0}^{m-1}\binom{2m}{k}\operatorname{chi}(2m-2k)x + \frac{1}{2^{2m}}\binom{2m}{m}\ln x \qquad \text{GU (353)(7c)}$$

10. $$\int \frac{\cosh^{2m+1} x}{x}\,dx = \frac{1}{2^{2m}}\sum_{k=0}^{m}\binom{2m+1}{k}\operatorname{chi}(2m-2k+1)x \qquad \text{GU (353)(7c)}$$

11. $$\int \frac{\sinh^{2m} x}{x^2}\,dx = \frac{(-1)^{m-1}}{2^{2m}x}\binom{2m}{m}$$
    $$+\frac{1}{2^{2m-1}}\sum_{k=0}^{m-1}(-1)^{k+1}\binom{2m}{k}\left\{\frac{\cosh(2m-2k)x}{x} - (2m-2k)\operatorname{shi}(2m-2k)x\right\}$$

12. $\displaystyle\int \frac{\sinh^{2m+1} x}{x^2}\,dx = \frac{1}{2^{2m}} \sum_{k=0}^{m} (-1)^{k+1} \binom{2m+1}{k}$

$$\times \left\{ \frac{\sinh(2m-2k+1)x}{x} - (2m-2k+1)\,\mathrm{chi}(2m-2k+1)x \right\}$$

13. $\displaystyle\int \frac{\cosh^{2m} x}{x^2}\,dx$

$$= -\frac{1}{2^{2m}x}\binom{2m}{m} - \frac{1}{2^{2m-1}}\sum_{k=0}^{m-1}\binom{2m}{k}\left\{ \frac{\cosh(2m-2k)x}{x} - (2m-2k)\,\mathrm{shi}(2m-2k)x \right\}$$

14. $\displaystyle\int \frac{\cosh^{2m+1} x}{x^2}\,dx$

$$= -\frac{1}{2^{2m}}\sum_{k=0}^{m}\binom{2m+1}{k}\left\{ \frac{\cosh(2m-2k+1)x}{x} - (2m-2k+1)\,\mathrm{shi}(2m-2k+1)x \right\}$$

**2.476**

1. $\displaystyle\int \frac{\sinh kx}{a+bx}\,dx = \frac{1}{b}\left[ \cosh\frac{ka}{b}\,\mathrm{shi}(u) - \sinh\frac{ka}{b}\,\mathrm{chi}(u) \right]$

$$= \frac{1}{2b}\left[ \exp\left(-\frac{ka}{b}\right)\mathrm{Ei}(u) - \exp\left(\frac{ka}{b}\right)\mathrm{Ei}(-u) \right] \qquad \left[ u = \frac{k}{b}(a+bx) \right]$$

2. $\displaystyle\int \frac{\cosh kx}{a+bx}\,dx = \frac{1}{b}\left[ \cosh\frac{ka}{b}\,\mathrm{chi}(u) - \sinh\frac{ka}{b}\,\mathrm{shi}(u) \right]$

$$= \frac{1}{2b}\left[ \exp\left(-\frac{ka}{b}\right)\mathrm{Ei}(u) + \exp\left(\frac{ka}{b}\right)\mathrm{Ei}(-u) \right] \qquad \left[ u = \frac{k}{b}(a+bx) \right]$$

3. $\displaystyle\int \frac{\sinh kx}{(a+bx)^2}\,dx = -\frac{1}{b}\frac{\sinh kx}{a+bx} + \frac{k}{b}\int \frac{\cosh kx}{a+bx}\,dx \qquad$ (see **2.476** 2)

4. $\displaystyle\int \frac{\cosh kx}{(a+bx)^2}\,dx = -\frac{1}{b}\frac{\cosh kx}{a+bx} + \frac{k}{b}\int \frac{\sinh kx}{a+bx}\,dx \qquad$ (see **2.476** 1)

5. $\displaystyle\int \frac{\sinh kx}{(a+bx)^3}\,dx = -\frac{\sinh kx}{2b(a+bx)^2} - \frac{k\cosh kx}{2b^2(a+bx)} + \frac{k^2}{2b^2}\int \frac{\sinh kx}{a+bx}\,dx$

(see **2.476** 1)

6. $\displaystyle\int \frac{\cosh kx}{(a+bx)^3}\,dx = -\frac{\cosh kx}{2b(a+bx)^2} - \frac{k\sinh kx}{2b^2(a+bx)} + \frac{k^2}{2b^2}\int \frac{\cosh kx}{a+bx}\,dx$

(see **2.476** 2)

7. $\displaystyle\int \frac{\sinh kx}{(a+bx)^4}\,dx = -\frac{\sinh kx}{3b(a+bx)^3} - \frac{k\cosh kx}{6b^2(a+bx)^2} - \frac{k^2\sinh kx}{6b^3(a+bx)} + \frac{k^3}{6b^3}\int \frac{\cosh kx}{a+bx}\,dx$

(see **2.476** 2)

8. $\displaystyle\int \frac{\cosh kx}{(a+bx)^4}\,dx = -\frac{\cosh kx}{3b(a+bx)^3} - \frac{k\sinh kx}{6b^2(a+bx)^2} - \frac{k^2\cosh kx}{6b^3(a+bx)} + \frac{k^3}{6b^3}\int \frac{\sinh kx}{a+bx}\,dx$

(see **2.476** 1)

9.  $\displaystyle\int \frac{\sinh kx}{(a+bx)^5}\,dx = -\frac{\sinh kx}{4b(a+bx)^4} - \frac{k\cosh kx}{12b^2(a+bx)^3} - \frac{k^2\sinh kx}{24b^3(a+bx)^2}$
    $\displaystyle\qquad\qquad -\frac{k^3\cosh kx}{24b^4(a+bx)} + \frac{k^4}{24b^4}\int\frac{\sinh kx}{a+bx}\,dx$

    (see **2.476** 1)

10. $\displaystyle\int \frac{\cosh kx}{(a+bx)^5}\,dx = -\frac{\cosh kx}{4b(a+bx)^4} - \frac{k\sinh kx}{12b^2(a+bx)^3} - \frac{k^2\cosh kx}{24b^3(a+bx)^2}$
    $\displaystyle\qquad\qquad -\frac{k^3\sinh kx}{24b^4(a+bx)} + \frac{k^4}{24b^4}\int\frac{\cosh kx}{a+bx}\,dx$

    (see **2.476** 2)

11. $\displaystyle\int \frac{\sinh kx}{(a+bx)^6}\,dx = -\frac{\sinh kx}{5b(a+bx)^5} - \frac{k\cosh kx}{20b^2(a+bx)^4} - \frac{k^2\sinh kx}{60b^3(a+bx)^3} - \frac{k^3\cosh kx}{120b^4(a+bx)^2}$
    $\displaystyle\qquad\qquad -\frac{k^4\sinh kx}{120b^5(a+bx)} + \frac{k^5}{120b^5}\int\frac{\cosh kx}{a+bx}\,dx$

    (see **2.476** 2)

12. $\displaystyle\int \frac{\cosh kx}{(a+bx)^6}\,dx = -\frac{\cosh kx}{5b(a+bx)^5} - \frac{k\sinh kx}{20b^2(a+bx)^4} - \frac{k^2\cosh kx}{60b^3(a+bx)^3} - \frac{k^3\sinh kx}{120b^4(a+bx)^2}$
    $\displaystyle\qquad\qquad -\frac{k^4\cosh kx}{120b^5(a+bx)} + \frac{k^5}{120b^5}\int\frac{\sinh kx}{a+bx}\,dx$

    (see **2.476** 1)

## 2.477

1.  $\displaystyle\int \frac{x^p\,dx}{\sinh^q x} = \frac{-px^{p-1}\sinh x - (q-2)x^p\cosh x}{(q-1)(q-2)\sinh^{q-1} x} + \frac{p(p-1)}{(q-1)(q-2)}\int\frac{x^{p-2}}{\sinh^{q-2} x}\,dx$
    $\displaystyle\qquad\qquad -\frac{q-2}{q-1}\int\frac{x^p\,dx}{\sinh^{q-2} x}$

    $[q > 2]$ \qquad GU (353)(8a)

2.  $\displaystyle\int \frac{x^p\,dx}{\cosh^q x} = \frac{px^{p-1}\cosh x + (q-2)x^p\sinh x}{(q-1)(q-2)\cosh^{q-1} x} - \frac{p(p-1)}{(q-1)(q-2)}\int\frac{x^{p-2}\,dx}{\cosh^{q-2} x}$
    $\displaystyle\qquad\qquad +\frac{q-2}{q-1}\int\frac{x^p\,dx}{\cosh^{q-2} x}$

    $[q > 2]$ \qquad GU (353)(10a)

3.  $\displaystyle\int \frac{x^n}{\sinh x}\,dx = \sum_{k=0}^{\infty}\frac{\left(2 - 2^{2k}\right)B_{2k}}{(n+2k)(2k)!}x^{n+2k}$ \qquad $[|x| < \pi, \quad n > 0]$ \qquad GU(353)(8b)

4.  $\displaystyle\int \frac{x^n}{\cosh x}\,dx = \sum_{k=0}^{\infty}\frac{E_{2k}x^{n+2k+1}}{(n+2k+1)(2k)!}$ \qquad $\left[|x| < \dfrac{\pi}{2}, \quad n \geq 0\right]$ \qquad GU (353)(10b)

5.  $\displaystyle\int \frac{dx}{x^n\sinh x} = -\left[1 + (-1)^n\right]\frac{2^{n-1}-1}{n!}B_n\ln x$
    $\displaystyle\qquad\qquad + \sum_{\substack{k=0 \\ k \neq \frac{n}{2}}}^{\infty}\frac{2 - 2^{2k}}{(2k-n)(2k)!}B_{2k}x^{2k-n}$

    $[|x| < \pi, \quad n \geq 1]$ \qquad GU (353)(9b)

6.$^{12}$ $\displaystyle\int \frac{\mathrm{d}x}{x^n \cosh x} = \sum_{\substack{k=0 \\ k \neq \frac{n-1}{2}}}^{\infty} \frac{E_{2k}}{(2k-n+1)(2k)!} x^{2k-n+1} + \frac{1}{2}\left[1 + (-1)^n\right]\frac{E_{n-1}}{(n-1)!}\ln x$

$$\left[|x| < \frac{\pi}{2}\right]$$                                    GU (353)(11b)

7.    $\displaystyle\int \frac{x^n}{\sinh^2 x}\,\mathrm{d}x = -x^n \coth x + n\sum_{k=0}^{\infty} \frac{2^{2k}B_{2k}}{(n+2k-1)(2k)!}x^{n+2k-1}$

$$[n > 1, \quad |x| < \pi]$$                                    GU (353)(8c)

8.    $\displaystyle\int \frac{x^n}{\cosh^2 x}\,\mathrm{d}x = x^n \tanh x - n\sum_{k=1}^{\infty} \frac{2^{2k}\left(2^{2k}-1\right)B_{2k}}{(n+2k-1)(2k)!}x^{n+2k-1}$

$$\left[n > 1, \quad |x| < \frac{\pi}{2}\right]$$                                    GU (353)(10c)

9.$^{12}$ $\displaystyle\int \frac{\mathrm{d}x}{x^n \sinh^2 x} = -\frac{\coth x}{x^n} - \left[1-(-1)^n\right]\frac{2^n n}{(n+1)!}B_{n+1}\ln x$

$$-\frac{n}{x^{n+1}}\sum_{\substack{k=0 \\ k \neq \frac{n+1}{2}}}^{\infty} \frac{B_{2k}}{(2k-n-1)(2k)!}(2x)^{2k}$$

$$[|x| < \pi]$$                                    GU (353)(9c)

10.$^{12}$ $\displaystyle\int \frac{\mathrm{d}x}{x^n \cosh^2 x} = \frac{\tanh x}{x^n} + \left[1-(-1)^n\right]\frac{2^n\left(2^{n+1}-1\right)n}{(n+1)!}B_{n+1}\ln x$

$$+\frac{n}{x^{n+1}}\sum_{\substack{k=1 \\ k \neq \frac{n+1}{2}}}^{\infty} \frac{\left(2^{2k}-1\right)B_{2k}}{(2k-n-1)(2k)!}(2x)^{2k}$$

$$\left[|x| < \frac{\pi}{2}\right]$$                                    GU (353)(11c)

11.    $\displaystyle\int \frac{x}{\sinh^{2n} x}\,\mathrm{d}x = \sum_{k=1}^{n-1}(-1)^k\frac{(2n-2)(2n-4)\dots(2n-2k+2)}{(2n-1)(2n-3)\dots(2n-2k+1)}$

$$\times \left[\frac{x\cosh x}{\sinh^{2n-2k+1} x} + \frac{1}{(2n-2k)\sinh^{2n-2k} x}\right] + (-1)^{n-1}\frac{(2n-2)!!}{(2n-1)!!}\int \frac{x\,\mathrm{d}x}{\sinh^2 x}$$

(see **2.477** 17)                                    GU (353)(8e)

12.    $\displaystyle\int \frac{x}{\sinh^{2n-1} x}\,\mathrm{d}x = \sum_{k=1}^{n-1}(-1)^k\frac{(2n-3)(2n-5)\dots(2n-2k+1)}{(2n-2)(2n-4)\dots(2n-2k)}$

$$\times \left[\frac{x\cosh x}{\sinh^{2n-2k} x} + \frac{1}{(2n-2k-1)\sinh^{2n-2k-1} x}\right] + (-1)^{n-1}\frac{(2n-3)!!}{(2n-2)!!}\int \frac{x\,\mathrm{d}x}{\sinh x}$$

(see **2.477** 15)                                    GU (353)(8e)

13.    $\displaystyle\int \frac{x}{\cosh^{2n} x}\,\mathrm{d}x = \sum_{k=1}^{n-1}\frac{(2n-2)(2n-4)\dots(2n-2k+2)}{(2n-1)(2n-3)\dots(2n-2k+1)}$

$$\times \left[\frac{x\sinh x}{\cosh^{2n-2k+1} x} + \frac{1}{(2n-2k)\cosh^{2n-2k} x}\right] + \frac{(2n-2)!!}{(2n-1)!!}\int \frac{x\,\mathrm{d}x}{\cosh^2 x}$$

(see **2.477** 18)                                    GU (353)(10e)

14. $\displaystyle \int \frac{x}{\cosh^{2n-1} x}\, dx = \sum_{k=1}^{n-1} \frac{(2n-3)(2n-5)\ldots(2n-2k+1)}{(2n-2)(2n-4)\ldots(2n-2k)}$

$$\times \left[ \frac{x\sinh x}{\cosh^{2n-2k} x} + \frac{1}{(2n-2k-1)\cosh^{2n-2k-1} x} \right] + \frac{(2n-3)!!}{(2n-2)!!} \int \frac{x\, dx}{\cosh x}$$

$$\text{(see } \mathbf{2.477}\ 16) \qquad\qquad \text{GU (353)(10e)}$$

15. $\displaystyle \int \frac{x\, dx}{\sinh x} = \sum_{k=0}^{\infty} \frac{2-2^{2k}}{(2k+1)(2k)!} B_{2k} x^{2k+1}$      $[|x| < \pi]$      GU (353)(8b)a

16. $\displaystyle \int \frac{x\, dx}{\cosh x} = \sum_{k=0}^{\infty} \frac{E_{2k} x^{2k+2}}{(2k+2)(2k)!}$      $\left[|x| < \dfrac{\pi}{2}\right]$      GU (353)(10b)a

17. $\displaystyle \int \frac{x\, dx}{\sinh^2 x} = -x\coth x + \ln\sinh x$      MZ 257

18. $\displaystyle \int \frac{x\, dx}{\cosh^2 x} = x\tanh x - \ln\cosh x$      MZ 262

19. $\displaystyle \int \frac{x\, dx}{\sinh^3 x} = -\frac{x\cosh x}{2\sinh^2 x} - \frac{1}{2\sinh x} - \frac{1}{2}\int \frac{x\, dx}{\sinh x}$    (see $\mathbf{2.477}$ 15)      MZ 257

20. $\displaystyle \int \frac{x\, dx}{\cosh^3 x} = \frac{x\sinh x}{2\cosh^2 x} + \frac{1}{2\cosh x} + \frac{1}{2}\int \frac{x\, dx}{\cosh x}$    (see $\mathbf{2.477}$ 16)      MZ 262

21. $\displaystyle \int \frac{x\, dx}{\sinh^4 x} = -\frac{x\cosh x}{3\sinh^3 x} - \frac{1}{6\sinh^2 x} + \frac{2}{3}x\coth x - \frac{2}{3}\ln\sinh x$      MZ 258

22. $\displaystyle \int \frac{x\, dx}{\cosh^4 x} = \frac{x\sinh x}{3\cosh^3 x} + \frac{1}{6\cosh^2 x} + \frac{2}{3}x\tanh x - \frac{2}{3}\ln\cosh x$      MZ 262

23. $\displaystyle \int \frac{x\, dx}{\sinh^5 x} = -\frac{x\cosh x}{4\sinh^4 x} - \frac{1}{12\sinh^3 x} + \frac{3x\cosh x}{8\sinh^2 x} + \frac{3}{8\sinh x} + \frac{3}{8}\int \frac{x\, dx}{\sinh x}$

$$\text{(see } \mathbf{2.477}\ 15) \qquad\qquad \text{MZ 258}$$

24. $\displaystyle \int \frac{x\, dx}{\cosh^5 x} = \frac{x\sinh x}{4\cosh^4 x} + \frac{1}{12\cosh^3 x} + \frac{3x\sinh x}{8\cosh^2 x} + \frac{3}{8\cosh x} + \frac{3}{8}\int \frac{x\, dx}{\cosh x}$

$$\text{(see } \mathbf{2.477}\ 16) \qquad\qquad \text{MZ 262}$$

**2.478**

1. $\displaystyle \int \frac{x^n \cosh x\, dx}{(a+b\sinh x)^m} = -\frac{x^n}{(m-1)b\,(a+b\sinh x)^{m-1}} + \frac{n}{(m-1)b}\int \frac{x^{n-1}\, dx}{(a+b\sinh x)^{m-1}}$

$$[m \neq 1] \qquad\qquad \text{MZ 263}$$

2. $\displaystyle \int \frac{x^n \sinh x\, dx}{(a+b\cosh x)^m} = -\frac{x^n}{(m-1)b\,(a+b\cosh x)^{m-1}} + \frac{n}{(m-1)b}\int \frac{x^{n-1}\, dx}{(a+b\cosh x)^{m-1}}$

$$[m \neq 1] \qquad\qquad \text{MZ 263}$$

3. $\displaystyle \int \frac{x\, dx}{1+\cosh x} = x\tanh\frac{x}{2} - 2\ln\cosh\frac{x}{2}$

4. $\displaystyle \int \frac{x\, dx}{1-\cosh x} = x\coth\frac{x}{2} - 2\ln\sinh\frac{x}{2}$

5.  $$\int \frac{x \sinh x \, dx}{(1 + \cosh x)^2} = -\frac{x}{1 + \cosh x} + \tanh \frac{x}{2}$$

6.  $$\int \frac{x \sinh x \, dx}{(1 - \cosh x)^2} = \frac{x}{1 - \cosh x} - \coth \frac{x}{2}$$                MZ 262-264

7.  $$\int \frac{x \, dx}{\cosh 2x - \cos 2t} = \frac{1}{2 \sin 2t} \left[ L(u + t) - L(u - t) - 2 \, L(t) \right]$$

$$\left[ u = \arctan (\tanh x \cot t), \quad t \neq \pm n\pi \right]$$
LO III 402

8.  $$\int \frac{x \cosh x \, dx}{\cosh 2x - \cos 2t} = \frac{1}{2 \sin t} \left[ L \left( \frac{u + t}{2} \right) - L \left( \frac{u - t}{2} \right) + L \left( \pi - \frac{v + t}{2} \right) \right.$$
$$\left. + L \left( \frac{v - t}{2} \right) - 2 L \left( \frac{t}{2} \right) - 2 L \left( \frac{\pi - t}{2} \right) \right]$$

$$\left[ u = 2 \arctan \left( \tanh \frac{x}{2} \cdot \cot \frac{t}{2} \right), \quad v = 2 \arctan \left( \coth \frac{x}{2} \cdot \cot \frac{t}{2} \right); \quad t \neq \pm n\pi \right]$$    LO III 403

**2.479**

1.  $$\int x^p \frac{\sinh^{2m} x}{\cosh^n x} \, dx = \sum_{k=0}^{m} (-1)^{m+k} \binom{m}{k} \int \frac{x^p \, dx}{\cosh^{n-2k} x} \qquad \text{(see **4.477** 2)}$$

2.  $$\int x^p \frac{\sinh^{2m+1} x}{\cosh^n x} \, dx = \sum_{k=0}^{m} (-1)^{m+k} \binom{m}{k} \int x^p \frac{\sinh x}{\cosh^{n-2k} x} \, dx$$

$$[n > 1] \qquad \text{(see **2.479** 3)}$$

3.  $$\int x^p \frac{\sinh x}{\cosh^n x} \, dx = -\frac{x^p}{(n - 1) \cosh^{n-1} x} + \frac{p}{n - 1} \int \frac{x^{p-1} \, dx}{\cosh^{n-1} x}$$

$$[n > 1] \qquad \text{(see **2.477** 2)} \qquad \text{GU (353)(12)}$$

4.[12]  $$\int x^p \frac{\cosh^{2m} x}{\sinh^n x} \, dx = \sum_{k=0}^{m} \binom{m}{k} \int \frac{x^p}{\sinh^{n-2k} x} \, dx \qquad \text{(see **2.477** 1)}$$

5.  $$\int x^p \frac{\cosh^{2m+1} x}{\sinh^n x} \, dx = \sum_{k=0}^{m} \binom{m}{k} \int \frac{x^p \cosh x}{\sinh^{n-2k} x} \, dx \qquad \text{(see **2.479** 6)}$$

6.  $$\int x^p \frac{\cosh x}{\sinh^n x} \, dx = -\frac{x^p}{(n - 1) \sinh^{n-1} x} + \frac{p}{n - 1} \int \frac{x^{p-1} \, dx}{\sinh^{n-1} x}$$

$$[n > 1] \qquad \text{(see **2.477** 1)}$$
GU (353)(13c)

7.  $$\int x^p \tanh x \, dx = \sum_{k=1}^{\infty} \frac{2^{2k} (2^{2k} - 1) B_{2k}}{(2k + p)(2k)!} x^{p+2k} \qquad \left[ p \geq 1, \quad |x| < \frac{\pi}{2} \right] \qquad \text{GU (353)(12d)}$$

8.  $$\int x^p \coth x \, dx = \sum_{k=0}^{\infty} \frac{2^{2k} B_{2k}}{(p + 2k)(2k)!} x^{p+2k} \qquad [p \geq +1, \quad |x| < \pi] \qquad \text{GU (353)(13d)}$$

9.  $$\int \frac{x \cosh x}{\sinh^2 x} \, dx = \ln \tanh \frac{x}{2} - \frac{x}{\sinh x}$$

10.    $\int \dfrac{x \sinh x}{\cosh^2 x}\, \mathrm{d}x = -\dfrac{x}{\cosh x} + \arctan(\sinh x)$            MZ 263

## 2.48 Combinations of hyperbolic functions, exponentials, and powers

**2.481**

1.    $\int e^{ax} \sinh(bx + c)\, \mathrm{d}x = \dfrac{e^{ax}}{a^2 - b^2} \left[a \sinh(bx + c) - b \cosh(bx + c)\right]$

$$\left[a^2 \neq b^2\right]$$

2.    $\int e^{ax} \cosh(bx + c)\, \mathrm{d}x = \dfrac{e^{ax}}{a^2 - b^2} \left[a \cosh(bx + c) - b \sinh(bx + c)\right]$

$$\left[a^2 \neq b^2\right]$$

For $a^2 = b^2$:

3.    $\int e^{ax} \sinh(ax + c)\, \mathrm{d}x = -\dfrac{1}{2}xe^{-c} + \dfrac{1}{4a}e^{2ax+c}$

4.    $\int e^{-ax} \sinh(ax + c)\, \mathrm{d}x = \dfrac{1}{2}xe^{c} + \dfrac{1}{4a}e^{-(2ax+c)}$

5.    $\int e^{ax} \cosh(ax + c)\, \mathrm{d}x = \dfrac{1}{2}xe^{-c} + \dfrac{1}{4a}e^{2ax+c}$

6.    $\int e^{-ax} \cosh(ax + c)\, \mathrm{d}x = \dfrac{1}{2}xe^{c} - \dfrac{1}{4a}e^{-(2ax+c)}$           MZ 275-277

**2.482**

1.    $\int x^p e^{ax} \sinh bx\, \mathrm{d}x = \dfrac{1}{2} \left\{ \int x^p e^{(a+b)x}\, \mathrm{d}x - \int x^p e^{(a-b)x}\, \mathrm{d}x \right\}$

$$\left[a^2 \neq b^2\right]$$

2.    $\int x^p e^{ax} \cosh bx\, \mathrm{d}x = \dfrac{1}{2} \left\{ \int x^p e^{(a+b)x}\, \mathrm{d}x + \int x^p e^{(a-b)x}\, \mathrm{d}x \right\}$

$$\left[a^2 \neq b^2\right]$$

For $a^2 = b^2$:

3.    $\int x^p e^{ax} \sinh ax\, \mathrm{d}x = \dfrac{1}{2} \int x^p e^{2ax}\, \mathrm{d}x - \dfrac{x^{p+1}}{2(p+1)}$       (see **2.321**)

4.    $\int x^p e^{-ax} \sinh ax\, \mathrm{d}x = \dfrac{x^{p+1}}{2(p+1)} - \dfrac{1}{2} \int x^p e^{-2ax}\, \mathrm{d}x$       (see **2.321**)

5.    $\int x^p e^{ax} \cosh ax\, \mathrm{d}x = \dfrac{x^{p+1}}{2(p+1)} + \dfrac{1}{2} \int x^p e^{2ax}\, \mathrm{d}x$       (see **2.321**)      MZ 276, 278

**2.483**

1.    $\int xe^{ax} \sinh bx\, \mathrm{d}x = \dfrac{e^{ax}}{a^2 - b^2} \left[ \left(ax - \dfrac{a^2 + b^2}{a^2 - b^2}\right) \sinh bx - \left(bx - \dfrac{2ab}{a^2 - b^2}\right) \cosh bx \right]$

$$\left[a^2 \neq b^2\right]$$

2. $\int xe^{ax}\cosh bx\,dx = \dfrac{e^{ax}}{a^2-b^2}\left[\left(ax-\dfrac{a^2+b^2}{a^2-b^2}\right)\cosh bx - \left(bx-\dfrac{2ab}{a^2-b^2}\right)\sinh bx\right]$

$$\left[a^2\neq b^2\right]$$

3. $\int x^2 e^{ax}\sinh bx\,dx = \dfrac{e^{ax}}{a^2-b^2}\left\{\left[ax^2 - \dfrac{2\left(a^2+b^2\right)}{a^2-b^2}x + \dfrac{2a\left(a^2+3b^2\right)}{\left(a^2-b^2\right)^2}\right]\sinh bx\right.$

$$\left.-\left[bx^2 - \dfrac{4ab}{a^2-b^2}x + \dfrac{2b\left(3a^2+b^2\right)}{\left(a^2-b^2\right)^2}\right]\cosh x\right\} \qquad \left[a^2\neq b^2\right]$$

4. $\int x^2 e^{ax}\cosh bx\,dx = \dfrac{e^{ax}}{a^2-b^2}\left\{\left[ax^2 - \dfrac{2\left(a^2+b^2\right)}{a^2-b^2}x + \dfrac{2a\left(a^2+3b^2\right)}{\left(a^2-b^2\right)^2}\right]\cosh bx\right.$

$$\left.-\left[bx^2 - \dfrac{4ab}{a^2-b^2}x + \dfrac{2b\left(3a^2+b^2\right)}{\left(a^2-b^2\right)^2}\right]\sinh x\right\} \qquad \left[a^2\neq b^2\right]$$

For $a^2 = b^2$:

5. $\int xe^{ax}\sinh ax\,dx = \dfrac{e^{2ax}}{4a}\left(x-\dfrac{1}{2a}\right)-\dfrac{x^2}{4}$

6. $\int xe^{-ax}\sinh ax\,dx = \dfrac{e^{-2ax}}{4a}\left(x+\dfrac{1}{2a}\right)+\dfrac{x^2}{4}$          MZ 276, 278

7. $\int xe^{ax}\cosh ax\,dx = \dfrac{x^2}{4}+\dfrac{e^{2ax}}{4a}\left(x-\dfrac{1}{2a}\right)$

8. $\int xe^{-ax}\cosh ax\,dx = \dfrac{x^2}{4}-\dfrac{e^{-2ax}}{4a}\left(x+\dfrac{1}{2a}\right)$

9. $\int x^2 e^{ax}\sinh ax\,dx = \dfrac{e^{2ax}}{4a}\left(x^2-\dfrac{x}{a}+\dfrac{1}{2a^2}\right)-\dfrac{x^3}{6}$

10. $\int x^2 e^{-ax}\sinh ax\,dx = \dfrac{e^{-2ax}}{4a}\left(x^2+\dfrac{x}{a}+\dfrac{1}{2a^2}\right)+\dfrac{x^3}{6}$

11. $\int x^2 e^{ax}\cosh ax\,dx = \dfrac{x^3}{6}+\dfrac{e^{2ax}}{4a}\left(x^2-\dfrac{x}{a}+\dfrac{1}{2a^2}\right)$

**2.484**

1. $\int e^{ax}\sinh bx\,\dfrac{dx}{x} = \dfrac{1}{2}\left\{\operatorname{Ei}[(a+b)x]-\operatorname{Ei}[(a-b)x]\right\} \qquad \left[a^2\neq b^2\right]$

2. $\int e^{ax}\cosh bx\,\dfrac{dx}{x} = \dfrac{1}{2}\left\{\operatorname{Ei}[(a+b)x]+\operatorname{Ei}[(a-b)x]\right\} \qquad \left[a^2\neq b^2\right]$

3. $\int e^{ax}\sinh bx\,\dfrac{dx}{x^2} = -\dfrac{e^{ax}\sinh bx}{2x}+\dfrac{1}{2}\left\{(a+b)\operatorname{Ei}[(a+b)x]-(a-b)\operatorname{Ei}[(a-b)x]\right\}$

$$\left[a^2\neq b^2\right]$$

4. $\int e^{ax}\cosh bx\,\dfrac{dx}{x^2} = -\dfrac{e^{ax}\cosh bx}{2x}+\dfrac{1}{2}\left\{(a+b)\operatorname{Ei}[(a+b)x]+(a-b)\operatorname{Ei}[(a-b)x]\right\}$

$$\left[a^2 \neq b^2\right]$$

For $a^2 = b^2$:

5.  $\displaystyle\int e^{ax} \sinh ax \frac{\mathrm{d}x}{x} = \frac{1}{2} \left[\mathrm{Ei}(2ax) - \ln x\right]$

6.  $\displaystyle\int e^{-ax} \sinh ax \frac{\mathrm{d}x}{x} = \frac{1}{2} \left[\ln x - \mathrm{Ei}(-2ax)\right]$

7.  $\displaystyle\int e^{ax} \cosh ax \frac{\mathrm{d}x}{x} = \frac{1}{2} \left[\ln x + \mathrm{Ei}(2ax)\right]$

8.  $\displaystyle\int e^{ax} \sinh ax \frac{\mathrm{d}x}{x^2} = -\frac{1}{2x} \left(e^{2ax} - 1\right) + a \,\mathrm{Ei}(2ax)$

9.  $\displaystyle\int e^{-ax} \sinh ax \frac{\mathrm{d}x}{x^2} = -\frac{1}{2x} \left(1 - e^{-2ax}\right) + a \,\mathrm{Ei}(-2ax)$

10. $\displaystyle\int e^{ax} \cosh ax \frac{\mathrm{d}x}{x^2} = -\frac{1}{2x} \left(e^{2ax} + 1\right) + a \,\mathrm{Ei}(2ax)$          MZ 276, 278

# 2.5–2.6 Trigonometric Functions

## 2.50  Introduction

**2.501**  Integrals of the form $\displaystyle\int R\left(\sin x, \cos x\right) \mathrm{d}x$ can always be reduced to integrals of rational functions by means of the substitution $t = \tan\dfrac{x}{2}$.

**2.502**  If $R\left(\sin x, \cos x\right)$ satisfies the relation

$$R\left(\sin x, \cos x\right) = -R\left(-\sin x, \cos x\right),$$

it is convenient to make the substitution $t = \cos x$.

**2.503**  If this function satisfies the relation

$$R\left(\sin x, \cos x\right) = -R\left(\sin x, -\cos x\right),$$

it is convenient to make the substitution $t = \sin x$.

**2.504**  If this function satisfies the relation

$$R\left(\sin x, \cos x\right) = R\left(-\sin x, -\cos x\right),$$

it is convenient to make the substitution $t = \tan x$.

## 2.51–2.52 Powers of trigonometric functions

**2.510**

$$\int \sin^p x \cos^q x \, dx = -\frac{\sin^{p-1} x \cos^{q+1} x}{q+1} + \frac{p-1}{q+1} \int \sin^{p-2} x \cos^{q+2} x \, dx$$

$$= -\frac{\sin^{p-1} x \cos^{q+1} x}{p+q} + \frac{p-1}{p+q} \int \sin^{p-2} x \cos^q x \, dx$$

$$= \frac{\sin^{p+1} x \cos^{q+1} x}{p+1} + \frac{p+q+2}{p+1} \int \sin^{p+2} x \cos^q x \, dx$$

$$= \frac{\sin^{p+1} x \cos^{q-1} x}{p+1} + \frac{q-1}{p+1} \int \sin^{p+2} x \cos^{q-2} x \, dx$$

$$= \frac{\sin^{p+1} x \cos^{q-1} x}{p+q} + \frac{q-1}{p+q} \int \sin^p x \cos^{q-2} x \, dx$$

$$= -\frac{\sin^{p+1} x \cos^{q+1} x}{q+1} + \frac{p+q+2}{q+1} \int \sin^p x \cos^{q+2} x \, dx$$

$$= \frac{\sin^{p-1} x \cos^{q-1} x}{p+q} \left\{ \sin^2 x - \frac{q-1}{p+q-2} \right\}$$

$$+ \frac{(p-1)(q-1)}{(p+q)(p+q-2)} \int \sin^{p-2} x \cos^{q-2} x \, dx$$

FI II 89, TI 214

**2.511**

1.

$$\int \sin^p x \cos^{2n} x \, dx$$

$$= \frac{\sin^{p+1} x}{2n+p} \left\{ \cos^{2n-1} x + \sum_{k=1}^{n-1} \frac{(2n-1)(2n-3)\ldots(2n-2k+1) \cos^{2n-2k-1} x}{(2n+p-2)(2n+p-4)\ldots(2n+p-2k)} \right\}$$

$$+ \frac{(2n-1)!!}{(2n+p)(2n+p-2)\ldots(p+2)} \int \sin^p x \, dx$$

This formula is applicable for arbitrary real $p$ except for the following negative even integers: $-2$, $-4$, $\ldots$, $-2n$. If $p$ is a natural number and $n = 0$, we have:

2.

$$\int \sin^{2l} x \, dx$$

$$= -\frac{\cos x}{2l} \left\{ \sin^{2l-1} x + \sum_{k=1}^{l-1} \frac{(2l-1)(2l-3)\ldots(2l-2k+1)}{2^k (l-1)(l-2)\ldots(l-k)} \sin^{2l-2k-1} x \right\}$$

$$+ \frac{(2l-1)!!}{2^l l!} x \hspace{4cm} \text{(see also } \textbf{2.513 } 1\text{)}$$

TI (232)

3.    $$\int \sin^{2l+1} x \, dx = -\frac{\cos x}{2l+1} \left\{ \sin^{2l} x + \sum_{k=0}^{l-1} \frac{2^{k+1} l(l-1)\ldots(l-k)}{(2l-1)(2l-3)\ldots(2l-2k-1)} \sin^{2l-2k-2} x \right\}$$

(see also 2.513 2)          TI (233)

4.    $$\int \sin^p x \cos^{2n+1} x \, dx = \frac{\sin^{p+1} x}{2n+p+1} \left\{ \cos^{2n} x + \sum_{k=1}^{n} \frac{2^k n(n-1)\ldots(n-k+1) \cos^{2n-2k} x}{(2n+p-1)(2n+p-3)\ldots(2n+p-2k+1)} \right\}$$

This formula is applicable for arbitrary real $p$ except for the negative odd integers: $-1$, $-3$, $\ldots$, $-(2n+1)$.

**2.512**

1.　$\displaystyle\int \cos^p x \sin^{2n} x\,dx$

$$= -\frac{\cos^{p+1} x}{2n+p}\left\{\sin^{2n-1} x + \sum_{k=1}^{n-1}\frac{(2n-1)(2n-3)\ldots(2n-2k+1)\sin^{2n-2k-1} x}{(2n+p-2)(2n+p-4)\ldots(2n+p-2k)}\right\}$$
$$+\frac{(2n-1)!!}{(2n+p)(2n+p-2)\ldots(p+2)}\int \cos^p x\,dx$$

This formula is applicable for arbitrary real $p$ except for the following negative even integers: $-2$, $-4$, $\ldots$, $-2n$. If $p$ is a natural number and $n=0$, we have

2.　$\displaystyle\int \cos^{2l} x\,dx = \frac{\sin x}{2l}\left\{\cos^{2l-1} x + \sum_{k=1}^{l-1}\frac{(2l-1)(2l-3)\ldots(2l-2k+1)}{2^k(l-1)(l-2)\ldots(l-k)}\cos^{2l-2k-1} x\right\}$
$$+\frac{(2l-1)!!}{2^l l!}x$$

$\qquad\qquad\qquad\qquad\qquad\qquad\qquad$ (see also **2.513** 3)$\qquad\qquad$ TI (230)

3.　$\displaystyle\int \cos^{2l+1} x\,dx = \frac{\sin x}{2l+1}\left\{\cos^{2l} x + \sum_{k=0}^{l-1}\frac{2^{k+1}l(l-1)\ldots(l-k)}{(2l-1)(2l-3)\ldots(2l-2k-1)}\cos^{2l-2k-2} x\right\}$

$\qquad\qquad\qquad\qquad\qquad\qquad\qquad$ (see also **2.513** 4)$\qquad\qquad$ TI (231)

4.　$\displaystyle\int \cos^p x \sin^{2n+1} x\,dx$

$$= -\frac{\cos^{p+1} x}{2n+p+1}\left\{\sin^{2n} x + \sum_{k=1}^{n}\frac{2^k n(n-1)\ldots(n-k+1)\sin^{2n-2k} x}{(2n+p-1)(2n+p-3)\ldots(2n+p-2k+1)}\right\}$$

This formula is applicable for arbitrary real $p$ except for the following negative odd integers: $-1$, $-3$, $\ldots$, $-(2n+1)$.

**2.513**

1.　$\displaystyle\int \sin^{2n} x\,dx = \frac{1}{2^{2n}}\binom{2n}{n}x + \frac{(-1)^n}{2^{2n-1}}\sum_{k=0}^{n-1}(-1)^k\binom{2n}{k}\frac{\sin(2n-2k)x}{2n-2k}$　　(see also **2.511** 2)

$\qquad\qquad\qquad\qquad\qquad\qquad\qquad\qquad\qquad\qquad\qquad\qquad\qquad\qquad\qquad\qquad\qquad$ TI (226)

2.　$\displaystyle\int \sin^{2n+1} x\,dx = \frac{1}{2^{2n}}(-1)^{n+1}\sum_{k=0}^{n}(-1)^k\binom{2n+1}{k}\frac{\cos(2n+1-2k)x}{2n+1-2k}$　　(see also **2.511** 3)

$\qquad\qquad\qquad\qquad\qquad\qquad\qquad\qquad\qquad\qquad\qquad\qquad\qquad\qquad\qquad\qquad\qquad$ TI (227)

3.　$\displaystyle\int \cos^{2n} x\,dx = \frac{1}{2^{2n}}\binom{2n}{n}x + \frac{1}{2^{2n-1}}\sum_{k=0}^{n-1}\binom{2n}{k}\frac{\sin(2n-2k)x}{2n-2k}$

$\qquad\qquad\qquad\qquad\qquad\qquad\qquad$ (see also **2.512** 2)$\qquad\qquad$ TI (224)

4.　$\displaystyle\int \cos^{2n+1} x\,dx = \frac{1}{2^{2n}}\sum_{k=0}^{n}\binom{2n+1}{k}\frac{\sin(2n-2k+1)x}{2n-2k+1}$

$\qquad\qquad\qquad\qquad\qquad\qquad\qquad$ (see also **2.512** 3)$\qquad\qquad$ TI (225)

5.    $\int \sin^2 x \, dx = -\frac{1}{4} \sin 2x + \frac{1}{2}x = -\frac{1}{2} \sin x \cos x + \frac{1}{2}x$

6.    $\int \sin^3 x \, dx = \frac{1}{12} \cos 3x - \frac{3}{4} \cos x = \frac{1}{3} \cos^3 x - \cos x$

7.    $\int \sin^4 x \, dx = \frac{3}{8}x - \dfrac{\sin 2x}{4} + \dfrac{\sin 4x}{32}$

$\qquad\qquad\qquad = -\frac{3}{8} \sin x \cos x - \frac{1}{4} \sin^3 x \cos x + \frac{3}{8}x$

8.    $\int \sin^5 x \, dx = -\frac{5}{8} \cos x + \frac{5}{48} \cos 3x - \frac{1}{80} \cos 5x$

$\qquad\qquad\qquad = -\frac{1}{5} \sin^4 x \cos x + \frac{4}{15} \cos^3 x - \frac{4}{5} \cos x$

9.    $\int \sin^6 x \, dx = \frac{5}{16}x - \frac{15}{64} \sin 2x + \frac{3}{64} \sin 4x - \frac{1}{192} \sin 6x$

$\qquad\qquad\qquad = -\frac{1}{6} \sin^5 x \cos x - \frac{5}{24} \sin^3 x \cos x - \frac{5}{16} \sin x \cos x + \frac{5}{16}x$

10.   $\int \sin^7 x \, dx = -\frac{35}{64} \cos x + \frac{7}{64} \cos 3x - \frac{7}{320} \cos 5x + \frac{1}{448} \cos 7x$

$\qquad\qquad\qquad = -\frac{1}{7} \sin^6 x \cos x - \frac{6}{35} \sin^4 x \cos x + \frac{8}{35} \cos^3 x - \frac{24}{35} \cos x$

11.   $\int \cos^2 x \, dx = \frac{1}{4} \sin 2x + \frac{1}{2}x = \frac{1}{2} \sin x \cos x + \frac{1}{2}x$

12.   $\int \cos^3 x \, dx = \frac{1}{12} \sin 3x + \frac{3}{4} \sin x = \sin x - \frac{1}{3} \sin^3 x$

13.   $\int \cos^4 x \, dx = \frac{3}{8}x + \frac{1}{4} \sin 2x + \frac{1}{32} \sin 4x = \frac{3}{8}x + \frac{3}{8} \sin x \cos x + \frac{1}{4} \sin x \cos^3 x$

14.   $\int \cos^5 x \, dx = \frac{5}{8} \sin x + \frac{5}{48} \sin 3x + \frac{1}{80} \sin 5x = \frac{4}{5} \sin x - \frac{4}{15} \sin^3 x + \frac{1}{5} \cos^4 x \sin x$

15.   $\int \cos^6 x \, dx = \frac{5}{16}x + \frac{15}{64} \sin 2x + \frac{3}{64} \sin 4x + \frac{1}{192} \sin 6x$

$\qquad\qquad\qquad = \frac{5}{16}x + \frac{5}{16} \sin x \cos x + \frac{5}{24} \sin x \cos^3 x + \frac{1}{6} \sin x \cos^5 x$

16.   $\int \cos^7 x \, dx = \frac{35}{64} \sin x + \frac{7}{64} \sin 3x + \frac{7}{320} \sin 5x + \frac{1}{448} \sin 7x$

$\qquad\qquad\qquad = \frac{24}{35} \sin x - \frac{8}{35} \sin^3 x + \frac{6}{35} \sin x \cos^4 x + \frac{1}{7} \sin x \cos^6 x$

17.   $\int \sin x \cos^2 x \, dx = -\frac{1}{4} \left( \frac{1}{3} \cos 3x + \cos x \right) = -\dfrac{\cos^3 x}{3}$

18.   $\int \sin x \cos^3 x \, dx = -\dfrac{\cos^4 x}{4}$

19.   $\int \sin x \cos^4 x \, dx = -\dfrac{\cos^5 x}{5}$

20.   $\int \sin^2 x \cos x \, dx = -\frac{1}{4} \left( \frac{1}{3} \sin 3x - \sin x \right) = \dfrac{\sin^3 x}{3}$

21.    $\int \sin^2 x \cos^2 x \, dx = -\frac{1}{8} \left( \frac{1}{4} \sin 4x - x \right)$

22.    $\int \sin^2 x \cos^3 x \, dx = -\frac{1}{16} \left( \frac{1}{5} \sin 5x + \frac{1}{3} \sin 3x - 2 \sin x \right)$

$$= \frac{\sin^3 x}{5} \left( \cos^2 x + \frac{2}{3} \right) = \frac{\sin^3 x}{5} \left( \frac{5}{3} - \sin^2 x \right)$$

23.    $\int \sin^2 x \cos^4 x \, dx = \frac{1}{16} x + \frac{1}{64} \sin 2x - \frac{1}{64} \sin 4x - \frac{1}{192} \sin 6x$

24.    $\int \sin^3 x \cos x \, dx = \frac{1}{8} \left( \frac{1}{4} \cos 4x - \cos 2x \right) = \frac{\sin^4 x}{4}$

25.    $\int \sin^3 x \cos^2 x \, dx = \frac{1}{16} \left( \frac{1}{5} \cos 5x - \frac{1}{3} \cos 3x - 2 \cos x \right)$

$$= \frac{1}{5} \cos^5 x - \frac{1}{3} \cos^3 x$$

26.    $\int \sin^3 x \cos^3 x \, dx = \frac{1}{32} \left( \frac{1}{6} \cos 6x - \frac{3}{2} \cos 2x \right)$

27.    $\int \sin^3 x \cos^4 x \, dx = \frac{1}{7} \cos^3 x \left( -\frac{2}{5} - \frac{3}{5} \sin^2 x + \sin^4 x \right)$

28.    $\int \sin^4 x \cos x \, dx = \frac{\sin^5 x}{5}$

29.    $\int \sin^4 x \cos^2 x \, dx = \frac{1}{16} x - \frac{1}{64} \sin 2x - \frac{1}{64} \sin 4x + \frac{1}{192} \sin 6x$

30.    $\int \sin^4 x \cos^3 x \, dx = \frac{1}{7} \sin^3 x \left( \frac{2}{5} + \frac{3}{5} \cos^2 x - \cos^4 x \right)$

31.    $\int \sin^4 x \cos^4 x \, dx = \frac{3}{128} x - \frac{1}{128} \sin 4x + \frac{1}{1024} \sin 8x$

**2.514**    $\int \dfrac{\sin^p x}{\cos^{2n} x} \, dx$

$$= \frac{\sin^{p+1} x}{2n-1} \left\{ \sec^{2n-1} x + \sum_{k=1}^{n-1} \frac{(2n-p-2)(2n-p-4)\ldots(2n-p-2k)}{(2n-3)(2n-5)\ldots(2n-2k-1)} \sec^{2n-2k-1} x \right\}$$

$$+ \frac{(2n-p-2)(2n-p-4)\ldots(-p+2)(-p)}{(2n-1)!!} \int \sin^p x \, dx$$

This formula is applicable for arbitrary real $p$. For $\int \sin^p x \, dx$, where $p$ is a natural number, see **2.511** 2, 3 and **2.513** 1, 2. If $n = 0$ and $p$ is a negative integer, we have for this integral:

**2.515**

1.    $\int \dfrac{dx}{\sin^{2l} x} = -\dfrac{\cos x}{2l-1} \left\{ \operatorname{cosec}^{2l-1} x + \sum_{k=1}^{l-1} \dfrac{2^k (l-1)(l-2)\ldots(l-k)}{(2l-3)(2l-5)\ldots(2l-2k-1)} \operatorname{cosec}^{2l-2k-1} x \right\}$

TI (242)

2.12    $\int \dfrac{dx}{\sin^{2l+1} x} = -\dfrac{\cos x}{2l} \left\{ \operatorname{cosec}^{2l} x + \sum_{k=1}^{l-1} \dfrac{(2l-1)(2l-3)\ldots(2l-2k+1)}{2^k (l-1)(l-2)\ldots(l-k)} \operatorname{cosec}^{2l-2k} x \right\}$

$$+ \frac{(2l-1)!!}{2^l l!} \ln \tan \frac{x}{2}$$

TI (243)

**2.516**

1. $\displaystyle\int \frac{\sin^p x \, dx}{\cos^{2n+1} x}$

$$= \frac{\sin^{p+1} x}{2n} \left\{ \sec^{2n} x + \sum_{k=1}^{n-1} \frac{(2n-p-1)(2n-p-3)\cdots(2n-p-2k+1)}{2^k(n-1)(n-2)\cdots(n-k)} \sec^{2n-2k} x \right\}$$
$$+ \frac{(2n-p-1)(2n-p-3)\cdots(3-p)(1-p)}{2^n n!} \int \frac{\sin^p x}{\cos x} \, dx$$

This formula is applicable for arbitrary real $p$. For $n = 0$ and $p$ a natural number, we have

2. $\displaystyle\int \frac{\sin^{2l+1} x \, dx}{\cos x} = -\sum_{k=1}^{l} \frac{\sin^{2k} x}{2k} - \ln \cos x$

3. $\displaystyle\int \frac{\sin^{2l} x \, dx}{\cos x} = -\sum_{k=1}^{l} \frac{\sin^{2k-1} x}{2k-1} + \ln \tan \left( \frac{\pi}{4} + \frac{x}{2} \right)$

**2.517**

1. $\displaystyle\int \frac{dx}{\sin^{2m+1} x \cos x} = -\sum_{k=1}^{m} \frac{1}{(2m-2k+2)\sin^{2m-2k+2} x} + \ln \tan x$

2. $\displaystyle\int \frac{dx}{\sin^{2m} x \cos x} = -\sum_{k=1}^{m} \frac{1}{(2m-2k+1)\sin^{2m-2k+1} x} + \ln \tan \left( \frac{\pi}{4} - \frac{x}{2} \right)$

**2.518**

1. $\displaystyle\int \frac{\sin^p x}{\cos^2 x} \, dx = \frac{\sin^{p-1} x}{\cos x} - (p-1) \int \sin^{p-2} x \, dx$

2.[12] $\displaystyle\int \frac{\cos^p x \, dx}{\sin^{2n} x} = -\frac{\cos^{p+1} x}{2n-1} \left\{ \operatorname{cosec}^{2n-1} x \right.$

$$+ \sum_{k=1}^{n-1} \frac{(2n-p-2)(2n-p-4)\ldots(2n-p-2k)}{(2n-3)(2n-5)\ldots(2n-2k-1)} \operatorname{cosec}^{2n-2k-1} x \left.\right\}$$
$$+ \frac{(2n-p-2)(2n-p-4)\ldots(2-p)(-p)}{(2n-1)!!} \int \cos^p x \, dx$$

This formula is applicable for arbitrary real $p$. For $\int \cos^p x \, dx$ where $p$ is a natural number, see **2.512** 2, 3 and **2.513** 3, 4. If $n = 0$ and $p$ is a negative integer, we have for this integral:

**2.519**

1. $\displaystyle\int \frac{dx}{\cos^{2l} x} = \frac{\sin x}{2l-1} \left\{ \sec^{2l-1} x + \sum_{k=1}^{l-1} \frac{2^k(l-1)(l-2)\ldots(l-k)}{(2l-3)(2l-5)\ldots(2l-2k-1)} \sec^{2l-2k-1} x \right\}$     TI (240)

2. $\displaystyle\int \frac{dx}{\cos^{2l+1} x} = \frac{\sin x}{2l} \left\{ \sec^{2l} x + \sum_{k=1}^{l-1} \frac{(2l-1)(2l-3)\ldots(2l-2k+1)}{2^k(l-1)(l-2)\ldots(l-k)} \sec^{2l-2k} x \right\}$
$$+ \frac{(2l-1)!!}{2^l l!} \ln \tan \left( \frac{\pi}{4} + \frac{x}{2} \right)$$

TI (241)

**2.521**

1. $$\int \frac{\cos^p x \, dx}{\sin^{2n+1} x} = -\frac{\cos^{p+1} x}{2n} \left\{ \cosec^{2n} x \right.$$

$$\left. + \sum_{k=1}^{n-1} \frac{(2n-p-1)(2n-p-3)\ldots(2n-p-2k+1)}{2^k(n-1)(n-2)\ldots(n-k)} \cosec^{2n-2k} x \right\}$$

$$+ \frac{(2n-p-1)(2n-p-3)\ldots(3-p)(1-p)}{2^n \cdot n!} \int \frac{\cos^p x}{\sin x} \, dx$$

This formula is applicable for arbitrary real $p$. For $n=0$ and $p$ a natural number, we have

2. $$\int \frac{\cos^{2l+1} x \, dx}{\sin x} = \sum_{k=1}^{l} \frac{\cos^{2k} x}{2k} + \ln \sin x$$

3. $$\int \frac{\cos^{2l} x \, dx}{\sin x} = \sum_{k=1}^{l} \frac{\cos^{2k-1} x}{2k-1} + \ln \tan \frac{x}{2}$$

**2.522**

1. $$\int \frac{dx}{\sin x \cos^{2m+1} x} = \sum_{k=1}^{m} \frac{1}{(2m-2k+2)\cos^{2m-2k+2} x} + \ln \tan x$$

2. $$\int \frac{dx}{\sin x \cos^{2m} x} = \sum_{k=1}^{m} \frac{1}{(2m-2k+1)\cos^{2m-2k+1} x} + \ln \tan \frac{x}{2} \qquad \text{GW (331)(15)}$$

**2.523** $$\int \frac{\cos^m x}{\sin^2 x} \, dx = -\frac{\cos^{m-1} x}{\sin x} - (m-1) \int \cos^{m-2} x \, dx$$

**2.524** In formulas **2.524** 1 and **2.524** 2, $s=1$ for $m$ odd and $m < 2n+1$; in other cases, $s=0$.

1. $$\int \frac{\sin^{2n+1} x}{\cos^m x} \, dx = \sum_{\substack{k=0 \\ k \neq \frac{m-1}{2}}}^{n} (-1)^{k+1} \binom{n}{k} \frac{\cos^{2k-m+1} x}{2k-m+1} + s(-1)^{\frac{m+1}{2}} \binom{n}{\frac{m-1}{2}} \ln \cos x$$

$$\text{GU (331)(11d)}$$

2. $$\int \frac{\cos^{2n+1} x}{\sin^m x} \, dx = \sum_{\substack{k=0 \\ k \neq \frac{m-1}{2}}}^{n} (-1)^{k} \binom{n}{k} \frac{\sin^{2k-m+1} x}{2k-m+1} + s(-1)^{\frac{m-1}{2}} \binom{n}{\frac{m-1}{2}} \ln \sin x$$

**2.525**

1. $$\int \frac{dx}{\sin^{2m} x \cos^{2n} x} = \sum_{k=0}^{m+n-1} \binom{m+n-1}{k} \frac{\tan^{2k-2m+1} x}{2k-2m+1} \qquad \text{TI (267)}$$

2. $$\int \frac{dx}{\sin^{2m+1} x \cos^{2n+1} x} = \sum_{k=0}^{m+n} \binom{m+n}{k} \frac{\tan^{2k-2m} x}{2k-2m} + \binom{m+n}{m} \ln \tan x$$

$$\text{TI (268), GU (331)(15f)}$$

**2.526**

1. $\displaystyle\int \frac{dx}{\sin x} = \ln \tan \frac{x}{2}$

2. $\displaystyle\int \frac{dx}{\sin^2 x} = -\cot x$

3. $\displaystyle\int \frac{dx}{\sin^3 x} = -\frac{1}{2}\frac{\cos x}{\sin^2 x} + \frac{1}{2}\ln \tan \frac{x}{2}$

4. $\displaystyle\int \frac{dx}{\sin^4 x} = -\frac{\cos x}{3\sin^3 x} - \frac{2}{3}\cot x = -\frac{1}{3}\cot^3 x - \cot x$

5. $\displaystyle\int \frac{dx}{\sin^5 x} = -\frac{\cos x}{4\sin^4 x} - \frac{3}{8}\frac{\cos x}{\sin^2 x} + \frac{3}{8}\ln \tan \frac{x}{2}$

6. $\displaystyle\int \frac{dx}{\sin^6 x} = -\frac{\cos x}{5\sin^5 x} - \frac{4}{15}\cot^3 x - \frac{4}{5}\cot x$
   $\displaystyle \qquad = -\frac{1}{5}\cot^5 x - \frac{2}{3}\cot^3 x - \cot x$

7. $\displaystyle\int \frac{dx}{\sin^7 x} = -\frac{\cos x}{6\sin^2 x}\left(\frac{1}{\sin^4 x} + \frac{5}{4\sin^2 x} + \frac{15}{8}\right) + \frac{5}{16}\ln \tan \frac{x}{2}$

8. $\displaystyle\int \frac{dx}{\sin^8 x} = -\left(\frac{1}{7}\cot^7 x + \frac{3}{5}\cot^5 x + \cot^3 x + \cot x\right)$

9. $\displaystyle\int \frac{dx}{\cos x} = \ln \tan \left(\frac{\pi}{4} + \frac{x}{2}\right) = \ln \cot \left(\frac{\pi}{4} - \frac{x}{2}\right) = \ln \sqrt{\frac{1 + \sin x}{1 - \sin x}}$

10. $\displaystyle\int \frac{dx}{\cos^2 x} = \tan x$

11. $\displaystyle\int \frac{dx}{\cos^3 x} = \frac{1}{2}\frac{\sin x}{\cos^2 x} + \frac{1}{2}\ln \tan \left(\frac{\pi}{4} + \frac{x}{2}\right)$

12. $\displaystyle\int \frac{dx}{\cos^4 x} = \frac{\sin x}{3\cos^3 x} + \frac{2}{3}\tan x = \frac{1}{3}\tan^3 x + \tan x$

13. $\displaystyle\int \frac{dx}{\cos^5 x} = \frac{\sin x}{4\cos^4 x} + \frac{3}{8}\frac{\sin x}{\cos^2 x} + \frac{3}{8}\ln \tan \left(\frac{x}{2} + \frac{\pi}{4}\right)$

14. $\displaystyle\int \frac{dx}{\cos^6 x} = \frac{\sin x}{5\cos^5 x} + \frac{4}{15}\tan^3 x + \frac{4}{5}\tan x = \frac{1}{5}\tan^5 x + \frac{2}{3}\tan^3 x + \tan x$

15. $\displaystyle\int \frac{dx}{\cos^7 x} = \frac{\sin x}{6\cos^6 x} + \frac{5\sin x}{24\cos^4 x} + \frac{5\sin x}{16\cos^2 x} + \frac{5}{16}\ln \tan \left(\frac{x}{2} + \frac{\pi}{4}\right)$

16. $\displaystyle\int \frac{dx}{\cos^8 x} = \frac{1}{7}\tan^7 x + \frac{3}{5}\tan^5 x + \tan^3 x + \tan x$

17. $\displaystyle\int \frac{\sin x}{\cos x}\,dx = -\ln \cos x$

18.[12] $\displaystyle\int \frac{\sin^2 x}{\cos x}\,dx = -\sin x + \ln \tan \left(\frac{\pi}{4} + \frac{x}{2}\right)$

19. $\displaystyle\int \frac{\sin^3 x}{\cos x}\,dx = -\frac{\sin^2 x}{2} - \ln \cos x = \frac{1}{2}\cos^2 x - \ln \cos x$

20. $\displaystyle\int \frac{\sin^4 x}{\cos x}\, dx = -\frac{1}{3}\sin^3 x - \sin x + \ln\tan\left(\frac{x}{2} + \frac{\pi}{4}\right)$

21.[12] $\displaystyle\int \frac{\sin x\, dx}{\cos^2 x} = \frac{1}{\cos x}$

22. $\displaystyle\int \frac{\sin^2 x\, dx}{\cos^2 x} = \tan x - x$

23. $\displaystyle\int \frac{\sin^3 x\, dx}{\cos^2 x} = \cos x + \frac{1}{\cos x}$

24. $\displaystyle\int \frac{\sin^4 x\, dx}{\cos^2 x} = \tan x + \frac{1}{2}\sin x\cos x - \frac{3}{2}x$

25. $\displaystyle\int \frac{\sin x\, dx}{\cos^3 x} = \frac{1}{2\cos^2 x} = \frac{1}{2}\tan^2 x$

26. $\displaystyle\int \frac{\sin^2 x\, dx}{\cos^3 x} = \frac{\sin x}{2\cos^2 x} - \frac{1}{2}\ln\tan\left(\frac{\pi}{4} + \frac{x}{2}\right)$

27.[12] $\displaystyle\int \frac{\sin^3 x\, dx}{\cos^3 x} = \frac{1}{2\cos^2 x} + \ln\cos x$

28. $\displaystyle\int \frac{\sin^4 x\, dx}{\cos^3 x} = \frac{1}{2}\frac{\sin x}{\cos^2 x} + \sin x - \frac{3}{2}\ln\tan\left(\frac{x}{2} + \frac{\pi}{4}\right)$

29. $\displaystyle\int \frac{\sin x\, dx}{\cos^4 x} = \frac{1}{3\cos^3 x}$

30. $\displaystyle\int \frac{\sin^2 x\, dx}{\cos^4 x} = \frac{1}{3}\tan^3 x$

31. $\displaystyle\int \frac{\sin^3 x\, dx}{\cos^4 x} = -\frac{1}{\cos x} + \frac{1}{3\cos^3 x}$

32. $\displaystyle\int \frac{\sin^4 x\, dx}{\cos^4 x} = \frac{1}{3}\tan^3 x - \tan x + x$

33. $\displaystyle\int \frac{\cos x\, dx}{\sin x} = \ln\sin x$

34. $\displaystyle\int \frac{\cos^2 x\, dx}{\sin x} = \cos x + \ln\tan\frac{x}{2}$

35. $\displaystyle\int \frac{\cos^3 x\, dx}{\sin x} = \frac{\cos^2 x}{2} + \ln\sin x$

36. $\displaystyle\int \frac{\cos^4 x\, dx}{\sin x} = \frac{1}{3}\cos^3 x + \cos x + \ln\tan\left(\frac{x}{2}\right)$

37. $\displaystyle\int \frac{\cos x}{\sin^2 x}\, dx = -\frac{1}{\sin x}$

38. $\displaystyle\int \frac{\cos^2 x}{\sin^2 x}\, dx = -\cot x - x$

39. $\displaystyle\int \frac{\cos^3 x}{\sin^2 x}\, dx = -\sin x - \frac{1}{\sin x}$

40. $\int \dfrac{\cos^4 x}{\sin^2 x}\, dx = -\cot x - \dfrac{1}{2}\sin x \cos x - \dfrac{3}{2}x$

41. $\int \dfrac{\cos x}{\sin^3 x}\, dx = -\dfrac{1}{2\sin^2 x}$

42. $\int \dfrac{\cos^2 x}{\sin^3 x}\, dx = -\dfrac{\cos x}{2\sin^2 x} - \dfrac{1}{2}\ln \tan \dfrac{x}{2}$

43. $\int \dfrac{\cos^3 x}{\sin^3 x}\, dx = -\dfrac{1}{2\sin^2 x} - \ln \sin x$

44. $\int \dfrac{\cos^4 x}{\sin^3 x}\, dx = -\dfrac{1}{2}\dfrac{\cos x}{\sin^2 x} - \cos x - \dfrac{3}{2}\ln \tan \dfrac{x}{2}$

45. $\int \dfrac{\cos x}{\sin^4 x}\, dx = -\dfrac{1}{3\sin^3 x}$

46. $\int \dfrac{\cos^2 x}{\sin^4 x}\, dx = -\dfrac{1}{3}\cot^3 x$

47. $\int \dfrac{\cos^3 x}{\sin^4 x}\, dx = \dfrac{1}{\sin x} - \dfrac{1}{3\sin^3 x}$

48. $\int \dfrac{\cos^4 x}{\sin^4 x}\, dx = -\dfrac{1}{3}\cot^3 x + \cot x + x$

49. $\int \dfrac{dx}{\sin x \cos x} = \ln \tan x$

50. $\int \dfrac{dx}{\sin x \cos^2 x} = \dfrac{1}{\cos x} + \ln \tan \dfrac{x}{2}$

51. $\int \dfrac{dx}{\sin x \cos^3 x} = \dfrac{1}{2\cos^2 x} + \ln \tan x$

52. $\int \dfrac{dx}{\sin x \cos^4 x} = \dfrac{1}{\cos x} + \dfrac{1}{3\cos^3 x} + \ln \tan \dfrac{x}{2}$

53. $\int \dfrac{dx}{\sin^2 x \cos x} = \ln \tan \left(\dfrac{\pi}{4} + \dfrac{x}{2}\right) - \operatorname{cosec} x$

54. $\int \dfrac{dx}{\sin^2 x \cos^2 x} = -2\cot 2x$

55. $\int \dfrac{dx}{\sin^2 x \cos^3 x} = \left(\dfrac{1}{2\cos^2 x} - \dfrac{3}{2}\right)\dfrac{1}{\sin x} + \dfrac{3}{2}\ln \tan \left(\dfrac{\pi}{4} + \dfrac{x}{2}\right)$

56. $\int \dfrac{dx}{\sin^2 x \cos^4 x} = \dfrac{1}{3\sin x \cos^3 x} - \dfrac{8}{3}\cot 2x$

57. $\int \dfrac{dx}{\sin^3 x \cos x} = -\dfrac{1}{2\sin^2 x} + \ln \tan x$

58. $\int \dfrac{dx}{\sin^3 x \cos^2 x} = -\dfrac{1}{\cos x}\left(\dfrac{1}{2\sin^2 x} - \dfrac{3}{2}\right) + \dfrac{3}{2}\ln \tan \dfrac{x}{2}$

59. $\int \dfrac{dx}{\sin^3 x \cos^3 x} = -\dfrac{2\cos 2x}{\sin^2 2x} + 2\ln \tan x$

60.　$\displaystyle \int \frac{dx}{\sin^3 x \cos^4 x} = \frac{2}{\cos x} + \frac{1}{3\cos^3 x} - \frac{\cos x}{2\sin^2 x} + \frac{5}{2}\ln\tan\frac{x}{2}$

61.　$\displaystyle \int \frac{dx}{\sin^4 x \cos x} = -\frac{1}{\sin x} - \frac{1}{3\sin^3 x} + \ln\tan\left(\frac{x}{2} + \frac{\pi}{4}\right)$

62.　$\displaystyle \int \frac{dx}{\sin^4 x \cos^2 x} = -\frac{1}{3\cos x \sin^3 x} - \frac{8}{3}\cot 2x$

63.　$\displaystyle \int \frac{dx}{\sin^4 x \cos^3 x} = -\frac{2}{\sin x} - \frac{1}{3\sin^3 x} + \frac{\sin x}{2\cos^2 x} + \frac{5}{2}\ln\tan\left(\frac{x}{2} + \frac{\pi}{4}\right)$

64.　$\displaystyle \int \frac{dx}{\sin^4 x \cos^4 x} = -8\cot 2x - \frac{8}{3}\cot^3 2x$

**2.527**

1.　$\displaystyle \int \tan^p x\, dx = \frac{\tan^{p-1} x}{p-1} - \int \tan^{p-2} x\, dx$　　　　　　$[p \neq 1]$

2.　$\displaystyle \int \tan^{2n+1} x\, dx = \sum_{k=1}^{n}(-1)^{n+k}\binom{n}{k}\frac{1}{2k\cos^{2k} x} - (-1)^n \ln\cos x$

$\displaystyle \qquad\qquad\qquad = \sum_{k=1}^{n}\frac{(-1)^{k-1}\tan^{2n-2k+2} x}{2n-2k+2} - (-1)^n \ln\cos x$

3.　$\displaystyle \int \tan^{2n} x\, dx = \sum_{k=1}^{n}(-1)^{k-1}\frac{\tan^{2n-2k+1} x}{2n-2k+1} + (-1)^n x$　　　　　　GU (331)(12)

4.　$\displaystyle \int \cot^p x\, dx = -\frac{\cot^{p-1} x}{p-1} - \int \cot^{p-2} x\, dx$　　　　　　$[p \neq 1]$

5.　$\displaystyle \int \cot^{2n+1} x\, dx = \sum_{k=1}^{n}(-1)^{n+k+1}\binom{n}{k}\frac{1}{2k\sin^{2k} x} + (-1)^n \ln\sin x$

$\displaystyle \qquad\qquad\qquad = \sum_{k=1}^{n}(-1)^k\frac{\cot^{2n-2k+2} x}{2n-2k+2} + (-1)^n \ln\sin x$

6.　$\displaystyle \int \cot^{2n} x\, dx = \sum_{k=1}^{n}(-1)^k\frac{\cot^{2n-2k+1} x}{2n-2k+1} + (-1)^n x$　　　　　　GU (331)(14)

For special formulas for $p = 1, 2, 3, 4$, see **2.526** 17, **2.526** 33, **2.526** 22, **2.526** 38, **2.526** 27, **2.526** 43, **2.526** 32, and **2.526** 48.

## 2.53–2.54 Sines and cosines of multiple angles and of linear and more complicated functions of the argument

**2.531**

1.　$\displaystyle \int \sin(ax + b)\, dx = -\frac{1}{a}\cos(ax + b)$

2.[12]　$\displaystyle \int \cos(ax + b)\, dx = \frac{1}{a}\sin(ax + b)$

**2.532**

1. $$\int \sin(ax+b)\sin(cx+d)\,dx = \frac{\sin[(a-c)x+b-d]}{2(a-c)} - \frac{\sin[(a+c)x+b+d]}{2(a+c)}$$

$$[a^2 \neq c^2]$$

2.8 $$\int \sin(ax+b)\cos(cx+d)\,dx = -\frac{\cos[(a-c)x+b-d]}{2(a-c)} - \frac{\cos[(a+c)x+b+d]}{2(a+c)}$$

$$[a^2 \neq c^2]$$

3. $$\int \cos(ax+b)\cos(cx+d)\,dx = \frac{\sin[(a-c)x+b-d]}{2(a-c)} + \frac{\sin[(a+c)x+b+d]}{2(a+c)}$$

$$[a^2 \neq c^2]$$

For $c = a$:

4. $$\int \sin(ax+b)\sin(ax+d)\,dx = \frac{x}{2}\cos(b-d) - \frac{\sin(2ax+b+d)}{4a}$$

5. $$\int \sin(ax+b)\cos(ax+d)\,dx = \frac{x}{2}\sin(b-d) - \frac{\cos(2ax+b+d)}{4a}$$

6. $$\int \cos(ax+b)\cos(ax+d)\,dx = \frac{x}{2}\cos(b-d) + \frac{\sin(2ax+b+d)}{4a}$$    GU (332)(3)

**2.533**

1.8 $$\int \sin ax \cos bx\,dx = -\frac{\cos(a+b)x}{2(a+b)} - \frac{\cos(a-b)x}{2(a-b)} \qquad [a^2 \neq b^2]$$

2.8 $$\int \sin ax \sin bx \sin cx\,dx = -\frac{1}{4}\left\{ \frac{\cos(a-b+c)x}{a-b+c} + \frac{\cos(b+c-a)x}{b+c-a} \right.$$

$$\left. + \frac{\cos(a+b-c)x}{a+b-c} - \frac{\cos(a+b+c)x}{a+b+c} \right\}$$

PE (376)

3. $$\int \sin ax \cos bx \cos cx\,dx = -\frac{1}{4}\left\{ \frac{\cos(a+b+c)x}{a+b+c} - \frac{\cos(b+c-a)x}{b+c-a} \right.$$

$$\left. + \frac{\cos(a+b-c)x}{a+b-c} + \frac{\cos(a+c-b)x}{a+c-b} \right\}$$

PE (378)

4. $$\int \cos ax \sin bx \sin cx\,dx = \frac{1}{4}\left\{ \frac{\sin(a+b-c)x}{a+b-c} + \frac{\sin(a+c-b)x}{a+c-b} \right.$$

$$\left. - \frac{\sin(a+b+c)x}{a+b+c} - \frac{\sin(b+c-a)x}{b+c-a} \right\}$$

PE (379)

5.    $\displaystyle \int \cos ax \cos bx \cos cx \, dx = \frac{1}{4} \left\{ \frac{\sin(a+b+c)x}{a+b+c} + \frac{\sin(b+c-a)x}{b+c-a} \right.$

$$\left. + \frac{\sin(a+c-b)x}{a+c-b} + \frac{\sin(a+b-c)x}{a+b-c} \right\}$$

PE (377)

**2.534**

1.    $\displaystyle \int \frac{\cos px + i \sin px}{\sin nx} \, dx = -2 \int \frac{z^{p+n-1}}{1-z^{2n}} \, dz$      $[z = \cos x + i \sin x]$      Pe (374)

2.    $\displaystyle \int \frac{\cos px + i \sin px}{\cos nx} \, dx = -2i \int \frac{z^{p+n-1}}{1-z^{2n}} \, dz$      $[z = \cos x + i \sin x]$      Pe (373)

**2.535**

1.    $\displaystyle \int \sin^p x \sin ax \, dx = \frac{1}{p+a} \left\{ -\sin^p x \cos ax + p \int \sin^{p-1} x \cos(a-1)x \, dx \right\}$      GU (332)(5a)

2.    $\displaystyle \int \sin^p x \sin(2n+1)x \, dx$

$$= (2n+1) \left\{ \int \sin^{p+1} x \, dx + \sum_{k=1}^{n} (-1)^k \frac{\left[(2n+1)^2 - 1^2\right]\left[(2n+1)^2 - 3^2\right] \cdots \cdots \left[(2n+1)^2 - (2k-1)^2\right]}{(2k+1)!} \right.$$

$$\left. \times \int \sin^{2k+p+1} x \, dx \right\}$$

TI (299)

$$= \frac{\Gamma(p+1)}{\Gamma\left(\frac{p+3}{2}+n\right)} \left\{ \sum_{k=0}^{n-1} \left[ \frac{(-1)^{k-1}\,\Gamma\left(\frac{p+1}{2}+n-2k\right)}{2^{2k+1}\,\Gamma(p-2k+1)} \sin^{p-2k} x \cos(2n-2k+1)x \right. \right.$$

$$\left. + (-1)^k \frac{\Gamma\left(\frac{p-1}{2}+n-2k\right)}{2^{2k+2}\,\Gamma(p-2k)} \sin^{p-2k-1} x \sin(2n-2k)x \right]$$

$$\left. + \frac{(-1)^n\,\Gamma\left(\frac{p+3}{2}-n\right)}{2^{2n}\,\Gamma(p-2n+1)} \int \sin^{p-2n+1} x \, dx \right\}$$

GU (332)(5c)

3.     $\displaystyle\int \sin^p x \sin 2nx \, dx = 2n \left\{ \frac{\sin^{p+2} x}{p+2} \right.$

$$+ \sum_{k=1}^{n-1} (-1)^k \frac{(4n^2 - 2^2)(4n^2 - 4^2) \dots [4n^2 - (2k)^2]}{(2k+1)!(2k+p+2)} \sin^{2k+p+2} x \left.\right\}$$

TI (303)

$$= \frac{\Gamma(p+1)}{\Gamma\left(\frac{p}{2}+n+1\right)} \left\{ \sum_{k=0}^{n-1} \frac{(-1)^{k-1} \Gamma\left(\frac{p}{2}+n-2k\right)}{2^{2k+1} \Gamma(p-2k+1)} \sin^{p-2k} x \cos(2n-2k)x \right.$$

$$\left. - \frac{(-1)^k \Gamma\left(\frac{p}{2}+n-2k-1\right)}{2^{2k+2} \Gamma(p-2k)} \sin^{p-2k-1} x \sin(2n-2k-1)x \right\}$$

[$p$ is not equal to $-2, -4, \dots, -2n$]<br>GU (332)(5c)

## 2.536

1.[12]     $\displaystyle\int \sin^p x \cos ax \, dx = \frac{1}{p+a} \left\{ \sin^p x \sin ax - p \int \sin^{p-1} x \sin(a-1)x \, dx \right\}$     GU (332)(6a)

2.     $\displaystyle\int \sin^p x \cos(2n+1)x \, dx$

$$= \frac{\sin^{p+1} x}{p+1} + \sum_{k=1}^{n} (-1)^k \frac{\left[(2n+1)^2 - 1^2\right]\left[(2n+1)^2 - 3^2\right] \dots \left[(2n+1)^2 - (2k-1)^2\right]}{(2k)!(2k+p+1)}$$

$$\times \sin^{2k+p+1} x$$

TI (301)

$$= \frac{\Gamma(p+1)}{\Gamma\left(\frac{p+3}{2}+n\right)} \left\{ \sum_{k=0}^{n-1} \left[ \frac{(-1)^k \Gamma\left(\frac{p+1}{2}+n-2k\right)}{2^{2k+1} \Gamma(p-2k+1)} \sin^{p-2k} x \sin(2n-2k+1)x \right.\right.$$

$$\left. + \frac{(-1)^k \Gamma\left(\frac{p-1}{2}+n-2k\right)}{2^{2k+2} \Gamma(p-2k)} \sin^{p-2k-1} x \cos(2n-2k)x \right]$$

$$\left. + \frac{(-1)^n \Gamma\left(\frac{p+3}{2}-n\right)}{2^{2n} \Gamma(p-2n+1)} \int \sin^{p-2n} x \cos x \, dx \right\}$$

[$p$ is not equal to $-3, -5, \dots, -(2n+1)$]<br>GU (332)(6c)

3. $\displaystyle\int \sin^p x \cos 2nx \, \mathrm{d}x$

$$= \int \sin^p x \, \mathrm{d}x + \sum_{k=1}^{n} (-1)^k \frac{4n^2 \cdot \left(4n^2 - 2^2\right) \dots \left[4n^2 - (2k-2)^2\right]}{(2k)!} \int \sin^{2k+p} x \, \mathrm{d}x$$

TI (300)

$$= \frac{\Gamma(p+1)}{\Gamma\left(\frac{p}{2}+n+1\right)} \left\{ \sum_{k=0}^{n-1} \left[ \frac{(-1)^k \Gamma\left(\frac{p}{2}+n-2k\right)}{2^{2k+1} \Gamma(p-2k+1)} \sin^{p-2k} x \sin(2n-2k)x \right. \right.$$

$$\left. + \frac{(-1)^k \Gamma\left(\frac{p}{2}+n-2k-1\right)}{2^{2k+2} \Gamma(p-2k)} \sin^{p-2k-1} x \cos(2n-2k-1)x \right]$$

$$\left. + \frac{(-1)^n \Gamma\left(\frac{p}{2}-n+1\right)}{2^{2n} \Gamma(p-2n+1)} \int \sin^{p-2n} x \, \mathrm{d}x \right\}$$

GU (332)(6c)

**2.537**

1. $\displaystyle\int \cos^p x \sin ax \, \mathrm{d}x = \frac{1}{p+a} \left\{ -\cos^p x \cos ax + p \int \cos^{p-1} x \sin(a-1)x \, \mathrm{d}x \right\}$     GU (332)(7a)

2. $\displaystyle\int \cos^p x \sin(2n+1)x \, \mathrm{d}x$

$$= (-1)^{n+1} \left\{ \frac{\cos^{p+1} x}{p+1} \right.$$

$$+ \sum_{k=1}^{n} (-1)^k \frac{\left[(2n+1)^2 - 1^2\right] \left[(2n+1)^2 - 3^2\right] \dots \left[(2n+1)^2 - (2k-1)^2\right]}{(2k)!(2k+p+1)} \cos^{2k+p+1} x \left. \right\}$$

TI (295)

$$= \frac{\Gamma(p+1)}{\Gamma\left(\frac{p+3}{2}+n\right)} \left\{ -\sum_{k=0}^{n-1} \frac{\Gamma\left(\frac{p+1}{2}+n-k\right)}{2^{2k+1} \Gamma(p-2k+1)} \cos^{p-k} x \cos(2n-k+1)x \right.$$

$$\left. + \frac{\Gamma\left(\frac{p+3}{2}\right)}{2^n \Gamma(p-n+1)} \int \cos^{p-n} x \sin(n+1)x \, \mathrm{d}x \right\}$$

$[p$ is not equal to $-3, -5, \dots, -(2n+1)]$

GU (332)(7b)a

3. $\displaystyle \int \cos^p x \sin 2nx \, dx = (-1)^n \left\{ \frac{\cos^{p+2} x}{p+2} \right.$

$$+ \sum_{k=1}^{n-1} (-1)^k \frac{(4n^2 - 2^2)(4n^2 - 4^2)\ldots[4n^2 - (2k)^2]}{(2k+1)!(2k+p+2)} \cos^{2k+p+2} x \left. \right\}$$

TI (297)

$$= \frac{\Gamma(p+1)}{\Gamma\left(\frac{p}{2} + n + 1\right)} \left\{ -\sum_{k=0}^{n-1} \frac{\Gamma\left(\frac{p}{2} + n - k\right)}{2^{k+1}\Gamma(p-k+1)} \cos^{p-k} x \cos(2n-k)x \right.$$

$$\left. + \frac{\Gamma\left(\frac{p}{2} + 1\right)}{2^n \Gamma(p-n+1)} \int \cos^{p-n} x \sin nx \, dx \right\}$$

$$[p \text{ is not equal to } -2, -4, \ldots, -2n]$$

GU (332)(7b)a

**2.538**

1. $\displaystyle \int \cos^p x \cos ax \, dx = \frac{1}{p+a} \left\{ \cos^p x \sin ax + p \int \cos^{p-1} x \cos(a-1)x \, dx \right\}$     GU (332)(8a)

2. $\displaystyle \int \cos^p x \cos(2n+1)x \, dx$

$$= (-1)^n (2n+1) \left\{ \int \cos^{p+1} x \, dx \right.$$

$$+ \sum_{k=1}^{n} (-1)^k \frac{\left[(2n+1)^2 - 1^2\right]\left[(2n+1)^2 - 3^2\right]\ldots\left[(2n+1)^2 - (2k-1)^2\right]}{(2k+1)!}$$

$$\left. \times \int \cos^{2k+p+1} x \, dx \right\}$$

TI (293)

$$= \frac{\Gamma(p+1)}{\Gamma\left(\frac{p+3}{2} + n\right)} \left\{ \sum_{k=0}^{n-1} \frac{\Gamma\left(\frac{p+1}{2} + n - k\right)}{2^{k+1}\Gamma(p-k+1)} \cos^{p-k} x \sin(2n-k+1)x \right.$$

$$\left. + \frac{\Gamma\left(\frac{p+3}{2}\right)}{2^n \Gamma(p-n+1)} \int \cos^{p-n} x \cos(n+1)x \, dx \right\}$$

GU (332)(8b)a

3. $\displaystyle\int \cos^p x \cos 2nx \, dx$

$$= (-1)^n \left\{ \int \cos^p x \, dx + \sum_{k=1}^{n} (-1)^k \frac{4n^2 \left[4n^2 - 2^2\right] \ldots \left[4n^2 - (2k-2)^2\right]}{(2k)!} \int \cos^{2k+p} x \, dx \right\}$$

TI (294)

$$= \frac{\Gamma(p+1)}{\Gamma\left(\frac{p}{2}+n+1\right)} \left\{ \sum_{k=0}^{n-1} \frac{\Gamma\left(\frac{p}{2}+n-k\right)}{2^{k+1}\,\Gamma(p-k+1)} \cos^{p-k} x \sin(2n-k)x \right.$$

$$\left. + \frac{\Gamma\left(\frac{p}{2}+1\right)}{2^n\,\Gamma(p-n+1)} \int \cos^{p-n} x \cos nx \, dx \right\}$$

GU (332)(8b)a

**2.539**

1. $\displaystyle\int \frac{\sin(2n+1)x}{\sin x} \, dx = 2 \sum_{k=1}^{n} \frac{\sin 2kx}{2k} + x$

2. $\displaystyle\int \frac{\sin 2nx}{\sin x} \, dx = 2 \sum_{k=1}^{n} \frac{\sin(2k-1)x}{2k-1}$   GU (332)(5e)

3. $\displaystyle\int \frac{\cos(2n+1)x}{\sin x} \, dx = 2 \sum_{k=1}^{n} \frac{\cos 2kx}{2k} + \ln \sin x$

4. $\displaystyle\int \frac{\cos 2nx}{\sin x} \, dx = 2 \sum_{k=1}^{n} \frac{\cos(2k-1)x}{2k-1} + \ln \tan \frac{x}{2}$   GI (332)(6e)

5. $\displaystyle\int \frac{\sin(2n+1)x}{\cos x} \, dx = 2 \sum_{k=1}^{n} (-1)^{n-k+1} \frac{\cos 2kx}{2k} + (-1)^{n+1} \ln \cos x$

6. $\displaystyle\int \frac{\sin 2nx}{\cos x} \, dx = 2 \sum_{k=1}^{n} (-1)^{n-k+1} \frac{\cos(2k-1)x}{2k-1}$   GU (332)(7d)

7. $\displaystyle\int \frac{\cos(2n+1)x}{\cos x} \, dx = 2 \sum_{k=1}^{n} (-1)^{n-k} \frac{\sin 2kx}{2k} + (-1)^n x$

8. $\displaystyle\int \frac{\cos 2nx}{\cos x} \, dx = 2 \sum_{k=1}^{n} (-1)^{n-k} \frac{\sin(2k-1)x}{2k-1} + (-1)^n \ln \tan \left(\frac{\pi}{4} + \frac{x}{2}\right).$   GU (332)(8d)

**2.541**

1. $\displaystyle\int \sin(n+1)x \sin^{n-1} x \, dx = \frac{1}{n} \sin^n x \sin nx$   BI (71)(1)a

2. $\displaystyle\int \sin(n+1)x \cos^{n-1} x \, dx = -\frac{1}{n} \cos^n x \cos nx$   BI (71)(2)a

3.    $\int \cos(n+1)x \sin^{n-1} x \, dx = \dfrac{1}{n} \sin^n x \cos nx$       BI (71)(3)a

4.    $\int \cos(n+1)x \cos^{n-1} x \, dx = \dfrac{1}{n} \cos^n x \sin nx$       BI (71)(4)a

5.    $\int \sin\left[(n+1)\left(\dfrac{\pi}{2}-x\right)\right] \sin^{n-1} x \, dx = \dfrac{1}{n} \sin^n x \cos n\left(\dfrac{\pi}{2}-x\right)$       BI (71)(5)a

6.    $\int \cos\left[(n+1)\left(\dfrac{\pi}{2}-x\right)\right] \sin^{n-1} x \, dx = -\dfrac{1}{n} \sin^n x \sin n\left(\dfrac{\pi}{2}-x\right)$       BI (71)(6)a

**2.542**

1.    $\int \dfrac{\sin 2x}{\sin^n x} \, dx = -\dfrac{2}{(n-2)\sin^{n-2} x}$

For $n = 2$:

2.    $\int \dfrac{\sin 2x}{\sin^2 x} \, dx = 2 \ln \sin x$

**2.543**

1.    $\int \dfrac{\sin 2x \, dx}{\cos^n x} = \dfrac{2}{(n-2)\cos^{n-2} x}$

For $n = 2$:

2.    $\int \dfrac{\sin 2x}{\cos^2 x} \, dx = -2 \ln \cos x$

**2.544**

1.    $\int \dfrac{\cos 2x \, dx}{\sin x} = 2 \cos x + \ln \tan \dfrac{x}{2}$

2.    $\int \dfrac{\cos 2x \, dx}{\sin^2 x} = -\cot x - 2x$

3.    $\int \dfrac{\cos 2x \, dx}{\sin^3 x} = -\dfrac{\cos x}{2\sin^2 x} - \dfrac{3}{2} \ln \tan \dfrac{x}{2}$

4.    $\int \dfrac{\cos 2x \, dx}{\cos x} = 2 \sin x - \ln \tan \left(\dfrac{\pi}{4}+\dfrac{x}{2}\right)$

5.    $\int \dfrac{\cos 2x \, dx}{\cos^2 x} = 2x - \tan x$

6.    $\int \dfrac{\cos 2x \, dx}{\cos^3 x} = -\dfrac{\sin x}{2\cos^2 x} + \dfrac{3}{2} \ln \tan \left(\dfrac{\pi}{4}+\dfrac{x}{2}\right)$

7.    $\int \dfrac{\sin 3x \, dx}{\sin x} = x + \sin 2x$

8.    $\int \dfrac{\sin 3x}{\sin^2 x} \, dx = 3 \ln \tan \dfrac{x}{2} + 4 \cos x$

9.    $\int \dfrac{\sin 3x}{\sin^3 x} \, dx = -3 \cot x - 4x$

**2.545**

1. $\displaystyle\int \frac{\sin 3x}{\cos^n x}\,dx = \frac{4}{(n-3)\cos^{n-3} x} - \frac{1}{(n-1)\cos^{n-1} x}$

For $n = 1$ and $n = 3$:

2. $\displaystyle\int \frac{\sin 3x}{\cos x}\,dx = 2\sin^2 x + \ln\cos x$

3. $\displaystyle\int \frac{\sin 3x}{\cos^3 x}\,dx = -\frac{1}{2\cos^2 x} - 4\ln\cos x$

**2.546**

1. $\displaystyle\int \frac{\cos 3x}{\sin^n x}\,dx = \frac{4}{(n-3)\sin^{n-3} x} - \frac{1}{(n-1)\sin^{n-1} x}$

For $n = 1$ and $n = 3$:

2. $\displaystyle\int \frac{\cos 3x}{\sin x}\,dx = -2\sin^2 x + \ln\sin x$

3. $\displaystyle\int \frac{\cos 3x}{\sin^3 x}\,dx = -\frac{1}{2\sin^2 x} - 4\ln\sin x$

**2.547**

1. $\displaystyle\int \frac{\sin nx}{\cos^p x}\,dx = 2\int \frac{\sin(n-1)x\,dx}{\cos^{p-1} x} - \int \frac{\sin(n-2)x\,dx}{\cos^p x}$

2. $\displaystyle\int \frac{\cos 3x}{\cos x}\,dx = \sin 2x - x$

3. $\displaystyle\int \frac{\cos 3x}{\cos^2 x}\,dx = 4\sin x - 3\ln\tan\left(\frac{\pi}{4} + \frac{x}{2}\right)$

4. $\displaystyle\int \frac{\cos 3x}{\cos^3 x}\,dx = 4x - 3\tan x$

**2.548**

1. $\displaystyle\int \frac{\sin^m x\,dx}{\sin(2n+1)x} = \frac{1}{2n+1}\sum_{k=0}^{2n}(-1)^{n+k}\cos^m\left[\frac{2k+1}{2(2n+1)}\pi\right]\ln\frac{\sin\left[\dfrac{(k-n)\pi}{2(2n+1)} + \dfrac{x}{2}\right]}{\sin\left[\dfrac{k+n+1}{(2n+1)}\pi - \dfrac{x}{2}\right]}$

$[m$ a natural number $\le 2n]$      TI (378)

2. $\displaystyle\int \frac{\sin^{2m} x\,dx}{\sin 2nx} = \frac{(-1)^n}{2n}\left\{\ln\cos x + \sum_{k=1}^{n-1}(-1)^k\cos^{2m}\frac{k\pi}{2n}\ln\left(\cos^2 x - \sin^2\frac{k\pi}{2n}\right)\right\}$

$[m$ a natural number $\le n]$      TI (379)

3. $\displaystyle\int \frac{\sin^{2m+1} x}{\sin 2nx}\,dx = \frac{(-1)^n}{2n}\left\{\ln\tan\left(\frac{\pi}{4} - \frac{x}{2}\right)\right.$

$\left.+ \sum_{k=1}^{n-1}(-1)^k\cos^{2m+1}\frac{k\pi}{2n}\ln\left[\tan\left(\frac{n+k}{4n}\pi - \frac{x}{2}\right)\tan\left(\frac{n-k}{4n}\pi - \frac{x}{2}\right)\right]\right\}$

$[m$ a natural number $< n]$

4.   $$\int \frac{\sin^{2m} x \, dx}{\cos(2n+1)x} = \frac{(-1)^{n+1}}{2n+1} \left\{ \ln \tan\left(\frac{\pi}{4} - \frac{x}{2}\right) + \sum_{k=1}^{n} (-1)^k \right.$$

$$\left. \times \cos^{2m} \frac{k\pi}{2n+1} \ln\left[ \tan\left(\frac{2n+2k+1}{4(2n+1)}\pi - \frac{x}{2}\right) \tan\left(\frac{2n-2k+1}{4(2n+1)}\pi - \frac{x}{2}\right) \right] \right\}$$

$$[m \text{ a natural number} \le n] \qquad \text{TI (381)}$$

5.   $$\int \frac{\sin^{2m+1} x \, dx}{\cos(2n+1)x} = \frac{(-1)^{n+1}}{2n+1} \left\{ \ln \cos x + \sum_{k=1}^{n} (-1)^k \cos^{2m+1} \frac{k\pi}{2n+1} \ln\left( \cos^2 x - \sin^2 \frac{k\pi}{2n+1} \right) \right\}$$

$$[m \text{ a natural number} \le n] \qquad \text{TI (382)a}$$

6.   $$\int \frac{\sin^m x \, dx}{\cos 2nx} = \frac{1}{2n} \sum_{k=0}^{2n-1} (-1)^{n+k} \cos^m\left[ \frac{2k+1}{4n}\pi \right] \ln \frac{\sin\left[ \frac{2k-2n+1}{8n}\pi + \frac{x}{2} \right]}{\sin\left[ \frac{2k+2n+1}{8n}\pi - \frac{x}{2} \right]}$$

$$[m \text{ a natural number} < 2n] \qquad \text{TI (377)}$$

7.   $$\int \frac{\cos^{2m+1} x \, dx}{\sin(2n+1)x} = \frac{1}{2n+1} \left\{ \ln \sin x + \sum_{k=1}^{n} (-1)^k \cos^{2m+1} \frac{k\pi}{2n+1} \ln\left( \sin^2 x - \sin^2 \frac{k\pi}{2n+1} \right) \right\}$$

$$[m \text{ a natural number} \le n] \qquad \text{TI (376)}$$

8.   $$\int \frac{\cos^{2m} x \, dx}{\sin(2n+1)x} = \frac{1}{2n+1} \left\{ \ln \tan \frac{x}{2} \right.$$

$$\left. + \sum_{k=1}^{n} (-1)^k \cos^{2m} \frac{k\pi}{2n+1} \ln\left[ \tan\left(\frac{x}{2} + \frac{k\pi}{4n+2}\right) \tan\left(\frac{x}{2} - \frac{k\pi}{4n+2}\right) \right] \right\}$$

$$[m \text{ a natural number} \le n] \qquad \text{TI (375)}$$

9.   $$\int \frac{\cos^{2m+1} x}{\sin 2nx} \, dx = \frac{1}{2n} \left\{ \ln \tan \frac{x}{2} + \sum_{k=1}^{n-1} (-1)^k \cos^{2m+1} \frac{k\pi}{2n} \ln\left[ \tan\left(\frac{x}{2} + \frac{k\pi}{4n}\right) \tan\left(\frac{x}{2} - \frac{k\pi}{4n}\right) \right] \right\}$$

$$[m \text{ a natural number} < n] \qquad \text{TI (374)}$$

10.  $$\int \frac{\cos^{2m} x}{\sin 2nx} \, dx = \frac{1}{2n} \left\{ \ln \sin x + \sum_{k=1}^{n-1} (-1)^k \cos^{2m} \frac{k\pi}{2n} \ln\left( \sin^2 x - \sin^2 \frac{k\pi}{2n} \right) \right\}$$

$$[m \text{ a natural number} \le n] \qquad \text{TI (373)}$$

11.  $$\int \frac{\cos^m x}{\cos nx} \, dx = \frac{1}{n} \sum_{k=0}^{n-1} (-1)^k \cos^m \frac{2k+1}{2n}\pi \ln \frac{\sin\left[ \frac{2k+1}{4n}\pi + \frac{x}{2} \right]}{\sin\left[ \frac{2k+1}{4n}\pi - \frac{x}{2} \right]}$$

$$[m \text{ is a natural number} \le n] \qquad \text{TI (372)}$$

**2.549**

1.   $$\int \sin x^2 \, dx = \sqrt{\frac{\pi}{2}} \, S(x)$$

2.    $\displaystyle \int \cos x^2 \, dx = \sqrt{\frac{\pi}{2}} \, C(x)$

3.[11]    $\displaystyle \int \sin\left(ax^2 + 2bx + c\right) dx = \sqrt{\frac{\pi}{2a}} \left\{ \cos\left(\frac{ac - b^2}{a}\right) S\left(\frac{ax + b}{\sqrt{a}}\right) + \sin\left(\frac{ac - b^2}{a}\right) C\left(\frac{ax + b}{\sqrt{a}}\right) \right\}$

$$[a > 0]$$

4.[11]    $\displaystyle \int \cos\left(ax^2 + 2bx + c\right) dx = \sqrt{\frac{\pi}{2a}} \left\{ \cos\left(\frac{ac - b^2}{a}\right) C\left(\frac{ax + b}{\sqrt{a}}\right) - \sin\left(\frac{ac - b^2}{a}\right) S\left(\frac{ax + b}{\sqrt{a}}\right) \right\}$

$$[a > 0]$$

5.    $\displaystyle \int \sin \ln x \, dx = \frac{x}{2} \left( \sin \ln x - \cos \ln x \right)$             PE (444)

6.    $\displaystyle \int \cos \ln x \, dx = \frac{x}{2} \left( \sin \ln x + \cos \ln x \right)$             PE (445)

## 2.55–2.56 Rational functions of the sine and cosine

**2.551**

1.    $\displaystyle \int \frac{A + B\sin x}{(a + b\sin x)^n} \, dx = \frac{1}{(n-1)(a^2 - b^2)} \left[ \frac{(Ab - aB)\cos x}{(a + b\sin x)^{n-1}} \right.$

$$\left. + \int \frac{(Aa - Bb)(n-1) + (aB - bA)(n-2)\sin x}{(a + b\sin x)^{n-1}} \, dx \right]$$

                                                                    TI (358)a

For $n = 1$:

2.    $\displaystyle \int \frac{A + B\sin x}{a + b\sin x} \, dx = \frac{B}{b}x + \frac{Ab - aB}{b} \int \frac{dx}{a + b\sin x}$     (see **2.551** 3)     TI (342)

3.    $\displaystyle \int \frac{dx}{a + b\sin x} = \frac{2}{\sqrt{a^2 - b^2}} \arctan \frac{a \tan \frac{x}{2} + b}{\sqrt{a^2 - b^2}}$     $\left[a^2 > b^2\right]$

              $\displaystyle = \frac{1}{\sqrt{b^2 - a^2}} \ln \frac{a \tan \frac{x}{2} + b - \sqrt{b^2 - a^2}}{a \tan \frac{x}{2} + b + \sqrt{b^2 - a^2}}$     $\left[a^2 < b^2\right]$

**2.552**

1.    $\displaystyle \int \frac{A + B\cos x}{(a + b\sin x)^n} \, dx = -\frac{B}{(n-1)b\,(a + b\sin x)^{n-1}} + A \int \frac{dx}{(a + b\sin x)^n}$

                                             (see **2.552** 3)     TI (361)

For $n = 1$:

2.    $\displaystyle \int \frac{A + B\cos x}{a + b\sin x} \, dx = \frac{B}{b} \ln\left(a + b\sin x\right) + A \int \frac{dx}{a + b\sin x}$

                                             (see **2.551** 3)     TI (344)

3. $\int \dfrac{dx}{(a+b\sin x)^n} = \dfrac{1}{(n-1)(a^2-b^2)} \left[ \dfrac{b\cos x}{(a+b\sin x)^{n-1}} \right.$

$\left. \qquad + \int \dfrac{(n-1)a - (n-2)b\sin x}{(a+b\sin x)^{n-1}} \, dx \right] \qquad$ (see **2.551** 1)

TI (359)

**2.553**

1. $\int \dfrac{A+B\sin x}{(a+b\cos x)^n} \, dx = \dfrac{B}{(n-1)b\,(a+b\cos x)^{n-1}} + A \int \dfrac{dx}{(a+b\cos x)^n}$

(see **2.554** 3)           TI (355)

For $n = 1$:

2. $\int \dfrac{A+B\sin x}{a+b\cos x} \, dx = -\dfrac{B}{b} \ln(a+b\cos x) + A \int \dfrac{dx}{a+b\cos x}$

(see **2.553** 3)           TI (343)

3. $\int \dfrac{dx}{a+b\cos x} = \dfrac{2}{\sqrt{a^2-b^2}} \arctan \dfrac{\sqrt{a^2-b^2} \tan \frac{x}{2}}{a+b} \qquad [a^2 > b^2]$

$\qquad\qquad = \dfrac{1}{\sqrt{b^2-a^2}} \ln \dfrac{\sqrt{b^2-a^2}\tan\frac{x}{2}+a+b}{\sqrt{b^2-a^2}\tan\frac{x}{2}-a-b} \qquad [a^2 < b^2]$

TI II 93, 94, TI (305)

4.[12] $\int \dfrac{dx}{a+b\cos x}$

$\qquad = \dfrac{2}{\sqrt{a^2-b^2}} \arctan \left( \dfrac{(a-b)\tan\left(\frac{x}{2}\right)}{\sqrt{a^2-b^2}} \right) \qquad [b^2 < a^2]$

$\qquad = \dfrac{2}{\sqrt{b^2-a^2}} \ln \left| \dfrac{(b-a)\tan\left(\frac{x}{2}\right)+\sqrt{b^2-a^2}}{(b-a)\tan\left(\frac{x}{2}\right)-\sqrt{b^2-a^2}} \right| \qquad [b^2 > a^2]$

$\qquad = \dfrac{2}{\sqrt{b^2-a^2}} \operatorname{arctanh} \left( \dfrac{(a-b)\tan\left(\frac{x}{2}\right)}{\sqrt{b^2-a^2}} \right) \qquad \left[ b^2 > a^2, \;\; \left|(b-a)\tan\left(\frac{x}{2}\right)\right| < \sqrt{b^2-a^2} \right]$

$\qquad = \dfrac{2}{\sqrt{b^2-a^2}} \operatorname{arccoth} \left( \dfrac{(a-b)\tan\left(\frac{x}{2}\right)}{\sqrt{b^2-a^2}} \right) \qquad \left[ b^2 > a^2, \;\; \left|(b-a)\tan\left(\frac{x}{2}\right)\right| > \sqrt{b^2-a^2} \right]$

(compare with **2.551** 3)

**2.554**

1. $\int \dfrac{A+B\cos x}{(a+b\cos x)^n} \, dx = \dfrac{1}{(n-1)(a^2-b^2)} \left[ \dfrac{(aB-Ab)\sin x}{(a+b\cos x)^{n-1}} \right.$

$\left. \qquad + \int \dfrac{(Aa-bB)(n-1)+(n-2)(aB-bA)\cos x}{(a+b\cos x)^{n-1}} \, dx \right]$

TI (353)

For $n = 1$:

2. 
$$\int \frac{A + B\cos x}{a + b\cos x}\, dx = \frac{B}{b}x + \frac{Ab - aB}{b} \int \frac{dx}{a + b\cos x} \qquad \text{(see \textbf{2.553} 3)} \qquad \text{TI (341)}$$

3. 
$$\int \frac{dx}{(a + b\cos x)^n} = -\frac{1}{(n-1)(a^2 - b^2)} \left\{ \frac{b\sin x}{(a + b\cos x)^{n-1}} \right.$$
$$\left. - \int \frac{(n-1)a - (n-2)b\cos x}{(a + b\cos x)^{n-1}}\, dx \right\} \qquad \text{(see \textbf{2.554} 1)}$$

TI (354)

In integrating the functions in formulas **2.551** 3 and **2.553** 3, we may not take the integration over points at which the integrand becomes infinite, that is, over the points $x = \arcsin\left(-\dfrac{a}{b}\right)$ in formula **2.551** 3 or over the points $x = \arccos\left(-\dfrac{a}{b}\right)$ in formula **2.553** 3.

**2.555**    Formulas **2.551** 3 and **2.553** 3 are not applicable for $a^2 = b^2$. Instead, we may use the following formulas in these cases:

1. 
$$\int \frac{A + B\sin x}{(1 \pm \sin x)^n}\, dx = -\frac{1}{2^{n-1}} \left\{ 2B \sum_{k=0}^{n-2} \binom{n-2}{k} \frac{\tan^{2k+1}\left(\frac{\pi}{4} \mp \frac{x}{2}\right)}{2k+1} \right.$$
$$\left. \pm (A \mp B) \sum_{k=0}^{n-1} \binom{n-1}{k} \frac{\tan^{2k+1}\left(\frac{\pi}{4} \mp \frac{x}{2}\right)}{2k+1} \right.$$

TI (361)a

2. 
$$\int \frac{A + B\cos x}{(1 \pm \cos x)^n}\, dx = \frac{1}{2^{n-1}} \left\{ 2B \sum_{k=0}^{n-2} \binom{n-2}{k} \frac{\tan^{2k+1}\left[\frac{\pi}{4} \mp \left(\frac{\pi}{4} - \frac{x}{2}\right)\right]}{2k+1} \right.$$
$$\left. \pm (A \mp B) \sum_{k=0}^{n-1} \binom{n-1}{k} \frac{\tan^{2k+1}\left[\frac{\pi}{4} \mp \left(\frac{\pi}{4} - \frac{x}{2}\right)\right]}{2k+1} \right\}$$

TI (356)

For $n = 1$:

3.[12] 
$$\int \frac{A + B\sin x}{1 \pm \sin x}\, dx = \pm Bx + (A \mp B)\tan\left(\frac{\pi}{4} \mp \frac{x}{2}\right) \qquad \text{TI (250)}$$

4. 
$$\int \frac{A + B\cos x}{1 \pm \cos x}\, dx = \pm Bx \pm (A \mp B)\tan\left[\frac{\pi}{4} \mp \left(\frac{\pi}{4} - \frac{x}{2}\right)\right] \qquad \text{TI (248)}$$

**2.556**

1. 
$$\int \frac{(1 - a^2)\, dx}{1 - 2a\cos x + a^2} = 2\arctan\left(\frac{1 + a}{1 - a}\tan\frac{x}{2}\right) \qquad [0 < a < 1, \quad |x| < \pi] \qquad \text{FI II 93}$$

2. 
$$\int \frac{(1 - a\cos x)\, dx}{1 - 2a\cos x + a^2} = \frac{x}{2} + \arctan\left(\frac{1 + a}{1 - a}\tan\frac{x}{2}\right) \qquad [0 < a < 1, \quad |x| < \pi] \qquad \text{FI II 93}$$

**2.557**

1. 
$$\int \frac{dx}{(a\cos x + b\sin x)^n} = \frac{1}{\sqrt{(a^2 + b^2)^n}} \int \frac{dx}{\sin^n\left(x + \arctan\frac{a}{b}\right)}$$

(see **2.515**)

MZ 173a

2.$^{12}$
$$\int \frac{\sin x \, dx}{a\cos x + b\sin x} = \frac{ax - b\ln\sin\left(x + \arctan\frac{a}{b}\right)}{a^2 + b^2}$$

3.
$$\int \frac{\cos x \, dx}{a\cos x + b\sin x} = \frac{ax + b\ln\sin\left(x + \arctan\frac{a}{b}\right)}{a^2 + b^2}$$

MZ 174a

4.
$$\int \frac{dx}{a\cos x + b\sin x} = \frac{\ln\tan\left[\frac{1}{2}\left(x + \arctan\frac{a}{b}\right)\right]}{\sqrt{a^2 + b^2}}$$

5.$^{12}$
$$\int \frac{dx}{(a\cos x + b\sin x)^2} = -\frac{\cot\left(x + \arctan\frac{a}{b}\right)}{a^2 + b^2} = -\frac{1}{a^2 + b^2}\left(\frac{a\cos x - b\sin x}{a\sin x + b\cos x}\right)$$

MZ 174a

**2.558**

1.
$$\int \frac{A + B\cos x + C\sin x}{(a + b\cos x + c\sin x)^n} \, dx$$
$$= \frac{(Bc - Cb) + (Ac - Ca)\cos x - (Ab - Ba)\sin x}{(n-1)(a^2 - b^2 - c^2)(a + b\cos x + c\sin x)^{n-1}} + \frac{1}{(n-1)(a^2 - b^2 - c^2)}$$
$$\times \int \frac{(n-1)(Aa - Bb - Cc) - (n-2)[(Ab - Ba)\cos x - (Ac - Ca)\sin x]}{(a + b\cos x + c\sin x)^{n-1}} \, dx$$
$$\left[n \neq 1, \quad a^2 \neq b^2 + c^2\right]$$
$$= \frac{Cb - Bc + Ca\cos x - Ba\sin x}{(n-1)a(a + b\cos x + c\sin x)^n} + \left(\frac{A}{a} + \frac{n(Bb + Cc)}{(n-1)a^2}\right)(-c\cos x + b\sin x)$$
$$\times \frac{(n-1)!}{(2n-1)!!} \sum_{k=0}^{n-1} \frac{(2n-2k-3)!!}{(n-k-1)!a^k} \cdot \frac{1}{(a + b\cos x + c\sin x)^{n-k}}$$
$$\left[n \neq 1, \quad a^2 = b^2 + c^2\right]$$

For $n = 1$ :

2.$^{11}$
$$\int \frac{A + B\cos x + C\sin x}{a + b\cos x + c\sin x} \, dx = \frac{Bc - Cb}{b^2 + c^2}\ln(a + b\cos x + c\sin x) + \frac{Bb + Cc}{b^2 + c^2}x$$
$$+ \left(A - \frac{Bb + Cc}{b^2 + c^2}a\right)\int \frac{dx}{a + b\cos x + c\sin x} \qquad \text{(see 2.558 4)}$$

GU (331)(18)

3.
$$\int \frac{dx}{(a + b\cos x + c\sin x)^n} = \int \frac{d(x - \alpha)}{[a + r\cos(x - \alpha)]^n},$$
where $b = r\cos\alpha$, $c = r\sin\alpha$ (see **2.554** 3)

4.
$$\int \frac{dx}{a + b\cos x + c\sin x}$$

$$= \frac{2}{\sqrt{a^2 - b^2 - c^2}}\arctan\frac{(a-b)\tan\frac{x}{2} + c}{\sqrt{a^2 - b^2 - c^2}} \qquad \left[a^2 > b^2 + c^2\right] \text{ TI (253), FI II 94}$$

$$= \frac{1}{\sqrt{b^2 + c^2 - a^2}}\ln\frac{(a-b)\tan\frac{x}{2} + c - \sqrt{b^2 + c^2 - a^2}}{(a-b)\tan\frac{x}{2} + c + \sqrt{b^2 + c^2 - a^2}} \qquad \left[a^2 < b^2 + c^2\right] \qquad \text{TI (253)a}$$

$$= \frac{1}{c}\ln\left(a + c\tan\frac{x}{2}\right) \qquad [a = b]$$

$$= \frac{-2}{c + (a-b)\tan\frac{x}{2}} \qquad \left[a^2 = b^2 + c^2\right] \qquad \text{TI (253)a}$$

**2.559**

1.
$$\int \frac{dx}{[a(1+\cos x) + c\sin x]^2} = \frac{1}{c^3}\left[\frac{c(a\sin x - c\cos x)}{a(1+\cos x) + c\sin x} - a\ln\left(a + c\tan\frac{x}{2}\right)\right]$$

2.[12]
$$\int \frac{A + B\cos x + C\sin x}{(a_1 + b_1\cos x + c_1\sin x)(a_2 + b_2\cos x + c_2\sin x)}dx$$
$$= A_0\ln\frac{a_1 + b_1\cos x + c_1\sin x}{a_2 + b_2\cos x + c_2\sin x} + A_1\int\frac{dx}{a_1 + b_1\cos x + c_1\sin x} + A_2\int\frac{dx}{a_2 + b_2\cos x + c_2\sin x}$$

$$\text{(see } \mathbf{2.558\ 4)} \qquad\qquad\qquad \text{GU (331)(19)}$$

where

$$A_0 = \frac{\begin{vmatrix} A & B & C \\ a_1 & b_1 & c_1 \\ a_2 & b_2 & c_2 \end{vmatrix}}{\begin{vmatrix} a_1 & b_1 \\ a_2 & b_2 \end{vmatrix}^2 - \begin{vmatrix} b_1 & c_1 \\ b_2 & c_2 \end{vmatrix}^2 + \begin{vmatrix} c_1 & a_1 \\ c_2 & a_2 \end{vmatrix}^2},$$

$$A_1 = \frac{\begin{vmatrix} \begin{vmatrix} B & C \\ b_1 & c_1 \end{vmatrix} & \begin{vmatrix} A & C \\ a_1 & c_1 \end{vmatrix} & \begin{vmatrix} B & A \\ b_1 & a_1 \end{vmatrix} \\ a_1 & b_1 & c_1 \\ a_2 & b_2 & c_2 \end{vmatrix}}{\begin{vmatrix} a_1 & b_1 \\ a_2 & b_2 \end{vmatrix}^2 - \begin{vmatrix} b_1 & c_1 \\ b_2 & c_2 \end{vmatrix}^2 + \begin{vmatrix} c_1 & a_1 \\ c_2 & a_2 \end{vmatrix}^2},$$

$$A_2 = \frac{\begin{vmatrix} \begin{vmatrix} C & B \\ c_2 & b_2 \end{vmatrix} & \begin{vmatrix} C & A \\ c_2 & a_2 \end{vmatrix} & \begin{vmatrix} A & B \\ a_2 & b_2 \end{vmatrix} \\ a_1 & b_1 & c_1 \\ a_2 & b_2 & c_2 \end{vmatrix}}{\begin{vmatrix} a_1 & b_1 \\ a_2 & b_2 \end{vmatrix}^2 - \begin{vmatrix} b_1 & c_1 \\ b_2 & c_2 \end{vmatrix}^2 + \begin{vmatrix} c_1 & a_1 \\ c_2 & a_2 \end{vmatrix}^2},$$

$$\left[\begin{vmatrix} a_1 & b_1 \\ a_2 & b_2 \end{vmatrix}^2 + \begin{vmatrix} c_1 & a_1 \\ c_2 & a_2 \end{vmatrix}^2 \neq \begin{vmatrix} b_1 & c_1 \\ b_2 & c_2 \end{vmatrix}^2\right]$$

3.
$$\int \frac{A\cos^2 x + 2B\sin x\cos x + C\sin^2 x}{a\cos^2 x + 2b\sin x\cos x + c\sin^2 x}dx$$

$$= \frac{1}{4b^2 + (a-c)^2}\left\{[4Bb + (A-C)(a-c)]x + [(A-C)b - B(a-c)]\right.$$

$$\times \ln\left(a\cos^2 x + 2b\sin x\cos x + c\sin^2 x\right)$$

$$\left. + \left[2(A+C)b^2 - 2Bb(a+c) + (aC - Ac)(a-c)\right]f(x)\right\}$$

$$\text{GU (331)(24)}$$

where

$$f(x) = \frac{1}{2\sqrt{b^2 - ac}}\ln\frac{c\tan x + b - \sqrt{b^2 - ac}}{c\tan x + b + \sqrt{b^2 - ac}} \qquad [b^2 > ac]$$

$$= \frac{1}{\sqrt{ac - b^2}}\arctan\frac{c\tan x + b}{\sqrt{ac - b^2}} \qquad [b^2 < ac]$$

$$= -\frac{1}{c\tan x + b} \qquad [b^2 = ac]$$

**2.561**

1.
$$\int \frac{(A + B\sin x)\,dx}{\sin x\,(a + b\sin x)} = \frac{A}{a}\ln\tan\frac{x}{2} + \frac{Ba - Ab}{a}\int\frac{dx}{a + b\sin x}$$

$$\text{(see } \mathbf{2.551\ 3)} \qquad\qquad\qquad \text{TI (348)}$$

2.    $\displaystyle\int \frac{(A + B\sin x)\,dx}{\sin x\,(a + b\cos x)} = \frac{A}{a^2 - b^2}\left(a\ln\tan\frac{x}{2} + b\ln\frac{a + b\cos x}{\sin x}\right)$

$\displaystyle\qquad\qquad\qquad + B\int \frac{dx}{a + b\cos x}$         (see **2.553** 3)

                                                                            TI (349)

For $a^2 = b^2 (= 1)$:

3.[12]    $\displaystyle\int \frac{(A + B\sin x)\,dx}{\sin x\,(1 + \cos x)} = \frac{A}{2}\left(\ln\tan\frac{x}{2} + \frac{1}{1 + \cos x}\right) + B\tan\frac{x}{2}$

4.    $\displaystyle\int \frac{(A + B\sin x)\,dx}{\sin x\,(1 - \cos x)} = \frac{A}{2}\left(\ln\tan\frac{x}{2} - \frac{1}{1 - \cos x}\right) - B\cot\frac{x}{2}$

5.    $\displaystyle\int \frac{(A + B\sin x)\,dx}{\cos x\,(a + b\sin x)} = \frac{1}{a^2 - b^2}\left((Aa - Bb)\ln\tan\left(\frac{\pi}{4} + \frac{x}{2}\right) - (Ab - aB)\ln\frac{a + b\sin x}{\cos x}\right)$

                                                                            TI (346)

For $a^2 = b^2 (= 1)$:

6.    $\displaystyle\int \frac{(A + B\sin x)\,dx}{\cos x\,(1 \pm \sin x)} = \frac{A \pm B}{2}\ln\tan\left(\frac{\pi}{4} + \frac{x}{2}\right) \mp \frac{A \mp B}{2(1 \pm \sin x)}$

7.    $\displaystyle\int \frac{(A + B\sin x)\,dx}{\cos x\,(a + b\cos x)} = \frac{A}{a}\ln\tan\left(\frac{\pi}{4} + \frac{x}{2}\right) + \frac{B}{a}\ln\frac{a + b\cos x}{\cos x}$

$\displaystyle\qquad\qquad\qquad - \frac{Ab}{a}\int \frac{dx}{a + b\cos x}$        (see **2.553** 3)

                                                                          TI (351)a

8.    $\displaystyle\int \frac{(A + B\cos x)\,dx}{\sin x\,(a + b\sin x)} = \frac{A}{a}\ln\tan\frac{x}{2} - \frac{B}{a}\ln\frac{a + b\sin x}{\sin x} - \frac{Ab}{a}\int \frac{dx}{a + b\sin x}$

                                                (see **2.551** 3)          TI (352)

9.    $\displaystyle\int \frac{(A + B\cos x)\,dx}{\sin x\,(a + b\cos x)} = \frac{1}{a^2 - b^2}\left((Aa - Bb)\ln\tan\frac{x}{2} + (Ab - Ba)\ln\frac{a + b\cos x}{\sin x}\right)$     TI (345)

For $a^2 = b^2 (= 1)$:

10.    $\displaystyle\int \frac{(A + B\cos x)\,dx}{\sin x\,(1 \pm \cos x)} = \pm\frac{A \mp B}{2(1 \pm \cos x)} + \frac{A \pm B}{2}\ln\tan\frac{x}{2}$

11.    $\displaystyle\int \frac{(A + B\cos x)\,dx}{\cos x\,(a + b\sin x)} = \frac{A}{a^2 - b^2}\left(a\ln\tan\left(\frac{\pi}{4} + \frac{x}{2}\right) - b\ln\frac{a + b\sin x}{\cos x}\right) + B\int \frac{dx}{a + b\sin x}$

                                                (see **2.551** 3)          TI (350)

For $a^2 = b^2 (= 1)$:

12.    $\displaystyle\int \frac{(A + B\sin x)\,dx}{\cos x\,(1 \pm \sin x)} = \frac{A \pm B}{2}\ln\tan\left(\frac{\pi}{4} + \frac{x}{2}\right) \mp \frac{A \mp B}{2(1 \pm \sin x)}$

13.    $\displaystyle\int \frac{(A + B\cos x)\,dx}{\cos x\,(a + b\cos x)} = \frac{A}{a}\ln\tan\left(\frac{\pi}{4} + \frac{x}{2}\right) + \frac{Ba - Ab}{a}\int \frac{dx}{a + b\cos x}$

                                                (see **2.553** 3)          TI (347)

**2.562**

1. $\displaystyle\int \frac{\mathrm{d}x}{a + b\sin^2 x} = \frac{\operatorname{sign} a}{\sqrt{a(a+b)}} \arctan\left(\sqrt{\frac{a+b}{a}}\tan x\right) \qquad \left[\dfrac{b}{a} > -1\right]$

$\displaystyle = \frac{\operatorname{sign} a}{\sqrt{-a(a+b)}} \operatorname{arctanh}\left(\sqrt{-\frac{a+b}{a}}\tan x\right) \qquad \left[\dfrac{b}{a} < -1, \quad \sin^2 x < -\dfrac{a}{b}\right]$

$\displaystyle = \frac{\operatorname{sign} a}{\sqrt{-a(a+b)}} \operatorname{arccoth}\left(\sqrt{-\frac{a+b}{a}}\tan x\right) \qquad \left[\dfrac{b}{a} < -1, \quad \sin^2 x > -\dfrac{a}{b}\right]$

MZ 155

2. $\displaystyle\int \frac{\mathrm{d}x}{a + b\cos^2 x} = \frac{-\operatorname{sign} a}{\sqrt{a(a+b)}} \arctan\left(\sqrt{\frac{a+b}{a}}\cot x\right) \qquad \left[\dfrac{b}{a} > -1\right]$

$\displaystyle = \frac{-\operatorname{sign} a}{\sqrt{-a(a+b)}} \operatorname{arctanh}\left(\sqrt{-\frac{a+b}{a}}\cot x\right) \qquad \left[\dfrac{b}{a} < -1, \quad \cos^2 x < -\dfrac{a}{b}\right]$

$\displaystyle = \frac{-\operatorname{sign} a}{\sqrt{-a(a+b)}} \operatorname{arccoth}\left(\sqrt{\frac{a+b}{a}}\cot x\right) \qquad \left[\dfrac{b}{a} < -1, \quad \cos^2 x > -\dfrac{a}{b}\right]$

MZ 162

3. $\displaystyle\int \frac{\mathrm{d}x}{1 + \sin^2 x} = \frac{1}{\sqrt{2}} \arctan\left(\sqrt{2}\tan x\right)$

4. $\displaystyle\int \frac{\mathrm{d}x}{1 - \sin^2 x} = \tan x$

5. $\displaystyle\int \frac{\mathrm{d}x}{1 + \cos^2 x} = -\frac{1}{\sqrt{2}} \arctan\left(\sqrt{2}\cot x\right)$

6. $\displaystyle\int \frac{\mathrm{d}x}{1 - \cos^2 x} = -\cot x$

**2.563**

1. $\displaystyle\int \frac{\mathrm{d}x}{\left(a + b\sin^2 x\right)^2} = \frac{1}{2a(a+b)} \left[(2a+b)\int \frac{\mathrm{d}x}{a + b\sin^2 x} + \frac{b\sin x\cos x}{a + b\sin^2 x}\right]$

(see **2.562** 1)                MZ 155

2. $\displaystyle\int \frac{\mathrm{d}x}{\left(a + b\cos^2 x\right)^2} = \frac{1}{2a(a+b)} \left[(2a+b)\int \frac{\mathrm{d}x}{a + b\cos^2 x} - \frac{b\sin x\cos x}{a + b\cos^2 x}\right]$

(see **2.562** 2)                MZ 163

3. 
$$\int \frac{dx}{\left(a + b\sin^2 x\right)^3} = \frac{1}{8pa^3}\left[\left(3 + \frac{2}{p^2} + \frac{3}{p^4}\right)\arctan\left(p\tan x\right)\right.$$

$$\left. + \left(3 + \frac{2}{p^2} - \frac{3}{p^4}\right)\frac{p\tan x}{1 + p^2\tan^2 x} + \left(1 - \frac{2}{p^2} - \frac{1}{p^2}\tan^2 x\right)\frac{2p\tan x}{\left(1 + p^2\tan^2 x\right)^2}\right]$$

$$\left[p^2 = 1 + \frac{b}{a} > 0\right]$$

$$= \frac{1}{8qa^3}\left[\left(3 - \frac{2}{q^2} + \frac{3}{q^4}\right)\operatorname{arctanh}\left(q\tan x\right)\right.$$

$$\left. + \left(3 - \frac{2}{q^2} - \frac{3}{q^4}\right)\frac{q\tan x}{1 - q^2\tan^2 x} + \left(1 + \frac{2}{q^2} + \frac{1}{q^2}\tan^2 x\right)\frac{2q\tan x}{\left(1 - q^2\tan^2 x\right)^2}\right]$$

$$\left[q^2 = -1 - \frac{b}{a} > 0, \quad \sin^2 x < -\frac{a}{b}; \quad \text{for } \sin^2 x > -\frac{a}{b}, \text{ change } \operatorname{arctanh}\left(q\tan x\right) \text{ to } \operatorname{arccoth}\left(q\tan x\right)\right]$$

MZ 156

4. 
$$\int \frac{dx}{\left(a + b\cos^2 x\right)^3} = -\frac{1}{8pa^3}\left[\left(3 + \frac{2}{p^2} + \frac{3}{p^4}\right)\arctan\left(p\cot x\right)\right.$$

$$\left. + \left(3 + \frac{2}{p^2} - \frac{3}{p^4}\right)\frac{p\cot x}{1 + p^2\cot^2 x} + \left(1 - \frac{2}{p^2} - \frac{1}{p^2}\cot^2 x\right)\frac{2p\cot x}{\left(1 + p^2\cot^2 x\right)^2}\right]$$

$$\left[p^2 = 1 + \frac{b}{a} > 0\right]$$

$$= -\frac{1}{8qa^3}\left[\left(3 - \frac{2}{q^2} + \frac{3}{q^4}\right)\operatorname{arctanh}\left(q\cot x\right)\right.$$

$$\left. + \left(3 - \frac{2}{q^2} - \frac{3}{q^4}\right)\frac{q\cot x}{1 - q^2\cot^2 x} + \left(1 + \frac{2}{q^2} + \frac{1}{q^2}\cot^2 x\right)\frac{2p\cot x}{\left(1 - q^2\cot^2 x\right)^2}\right]$$

$$\left[q^2 = -1 - \frac{b}{a} > 0, \quad \cos^2 x < -\frac{a}{b}; \quad \text{for } \cos^2 x > -\frac{a}{b}, \text{ change } \operatorname{arctanh}\left(q\cot x\right) \text{ to } \operatorname{arccoth}\left(q\cot x\right)\right]$$

MZ 163a

**2.564**

1. 
$$\int \frac{\tan x\, dx}{1 + m^2\tan^2 x} = \frac{\ln\left(\cos^2 x + m^2\sin^2 x\right)}{2\left(m^2 - 1\right)}$$
   LA 210 (10)

2. 
$$\int \frac{\tan\alpha - \tan x}{\tan\alpha + \tan x}\, dx = \sin 2\alpha \ln\sin(x + \alpha) - x\cos 2\alpha$$
   LA 210 (11)a

3. 
$$\int \frac{\tan x\, dx}{a + b\tan x} = \frac{1}{a^2 + b^2}\left\{bx - a\ln\left(a\cos x + b\sin x\right)\right\}$$
   PE (335)

4. 
$$\int \frac{dx}{a + b\tan^2 x} = \frac{1}{a - b}\left[x - \sqrt{\frac{b}{a}}\arctan\left(\sqrt{\frac{b}{a}}\tan x\right)\right]$$
   PE (334)

## 2.57 Integrals containing $\sqrt{a \pm b \sin x}$ or $\sqrt{a \pm b \cos x}$

**Notation:**

$$\alpha = \arcsin \sqrt{\frac{1 - \sin x}{2}}, \qquad \beta = \arcsin \sqrt{\frac{b\,(1 - \sin x)}{a + b}},$$

$$\gamma = \arcsin \sqrt{\frac{b\,(1 - \cos x)}{a + b}}, \qquad \delta = \arcsin \sqrt{\frac{(a + b)\,(1 - \cos x)}{2\,(a - b \cos x)}}, \qquad r = \sqrt{\frac{2b}{a + b}}$$

**2.571**

1. $$\int \frac{dx}{\sqrt{a + b \sin x}} = \frac{-2}{\sqrt{a + b}}\, F(\alpha, r) \qquad\qquad\qquad \left[ a > b > 0, \quad -\frac{\pi}{2} \le x < \frac{\pi}{2} \right]$$

$$= -\sqrt{\frac{2}{b}}\, F\left( \beta, \frac{1}{r} \right) \qquad\qquad\qquad \left[ 0 < |a| < b, \quad -\arcsin \frac{a}{b} < x < \frac{\pi}{2} \right]$$

$$\text{BY (288.00, 288.50)}$$

2. $$\int \frac{\sin x\, dx}{\sqrt{a + b \sin x}}$$

$$= \frac{2a}{b\sqrt{a + b}}\, F(\alpha, r) - \frac{2\sqrt{a + b}}{b}\, E(\alpha, r) \qquad \left[ a > b > 0, \quad -\frac{\pi}{2} \le x < \frac{\pi}{2} \right] \qquad \text{BY (288.03)}$$

$$= \sqrt{\frac{2}{b}} \left\{ F\left( \beta, \frac{1}{r} \right) - 2\, E\left( \beta, \frac{1}{r} \right) \right\} \qquad \left[ 0 < |a| < b, \quad -\arcsin \frac{a}{b} < x < \frac{\pi}{2} \right] \quad \text{BY (288.54)}$$

3. $$\int \frac{\sin^2 x\, dx}{\sqrt{a + b \sin x}} = \frac{4a\sqrt{a + b}}{3b^2}\, E(\alpha, r) - \frac{2\,(2a^2 + b^2)}{3b^2\sqrt{a + b}}\, F(\alpha, r) - \frac{2}{3b} \cos x \sqrt{a + b \sin x}$$

$$\left[ a > b > 0, \quad -\frac{\pi}{2} \le x < \frac{\pi}{2} \right]$$

$$= \sqrt{\frac{2}{b}} \left\{ \frac{4a}{3b}\, E\left( \beta, \frac{1}{r} \right) - \frac{2a + b}{3b}\, F\left( \beta, \frac{1}{r} \right) \right\} - \frac{2}{3b} \cos x \sqrt{a + b \sin x}$$

$$\left[ 0 < |a| < b, \quad -\arcsin \frac{a}{b} < x < \frac{\pi}{2} \right]$$

$$\text{BY (288.03, 288.54)}$$

4. $$\int \frac{dx}{\sqrt{a + b \cos x}} = \frac{2}{\sqrt{a + b}}\, F\left( \frac{x}{2}, r \right) \qquad\qquad \left[ a > b > 0, \quad 0 \le x \le \pi \right]$$

$$= \sqrt{\frac{2}{b}}\, F\left( \gamma, \frac{1}{r} \right) \qquad\qquad\qquad \left[ b \ge |a| > 0, \quad 0 \le x < \arccos\left( -\frac{a}{b} \right) \right]$$

$$\text{BY (289.00)}$$

5. $$\int \frac{dx}{\sqrt{a - b \cos x}} = \frac{2}{\sqrt{a + b}}\, F(\delta, r) \qquad\qquad \left[ a > b > 0, \quad 0 \le x \le \pi \right] \qquad \text{BY (291.00)}$$

6.    $\displaystyle\int \frac{\cos x \, dx}{\sqrt{a + b\cos x}} = \frac{2}{b\sqrt{a+b}}\left\{(a+b)\,E\left(\frac{x}{2},r\right) - a\,F\left(\frac{x}{2},r\right)\right\}$

$$[a > b > 0, \quad 0 \le x \le \pi]$$

BY (289.03)

$$= \sqrt{\frac{2}{b}}\left\{2\,E\left(\gamma,\frac{1}{r}\right) - F\left(\gamma,\frac{1}{r}\right)\right\}$$

$$\left[b > |a| > 0, \quad 0 \le x < \arccos\left(-\frac{a}{b}\right)\right]$$

BY (290.04)

7.[6]    $\displaystyle\int \frac{\cos x \, dx}{\sqrt{a - b\cos x}} = \frac{2}{b\sqrt{a+b}}\left\{(b-a)\,\Pi\left(\delta, r^2, r\right) + a\,F(\delta, r)\right\}$

$$[a > b > 0, \quad 0 \le x \le \pi] \qquad \text{BY (291.03)}$$

8.    $\displaystyle\int \frac{\cos^2 x \, dx}{\sqrt{a + b\cos x}} = \frac{2}{3b^2\sqrt{a+b}}\left\{(2a^2+b^2)\,F\left(\frac{x}{2},r\right) - 2a(a+b)\,E\left(\frac{x}{2},r\right)\right\} + \frac{2}{3b}\sin x\sqrt{a+b\cos x}$

$$[a > b > 0, \quad 0 \le x \le \pi]$$

BY (289.03)

$$= \frac{1}{3b}\sqrt{\frac{2}{b}}\left\{(2a+b)\,F\left(\gamma,\frac{1}{r}\right) - 4a\,E\left(\gamma,\frac{1}{r}\right)\right\} + \frac{2}{3b}\sin x\sqrt{a+b\cos x}$$

$$\left[b \ge |a| > 0, \quad 0 \le x < \arccos\left(-\frac{a}{b}\right)\right]$$

BY (290.04)

9.[12]    $\displaystyle\int \frac{\cos^2 x \, dx}{\sqrt{a - b\cos x}} = \frac{2}{3b^2\sqrt{a+b}}\left\{(2a^2+b^2)\,F(\delta, r) - 2a(a+b)\,E(\delta, r)\right\}$

$$+ \frac{2}{3b}\sin x\frac{a+b\cos x}{\sqrt{a-b\cos x}} \qquad [a > b > 0, \quad 0 \le x < \pi]$$

BY (291.04)a

**2.572**    $\displaystyle\int \frac{\tan^2 x \, dx}{\sqrt{a + b\sin x}} = \frac{1}{\sqrt{a+b}}\,F(\alpha, r) + \frac{a}{(a-b)\sqrt{a+b}}\,E(\alpha, r)$

$$- \frac{b - a\sin x}{(a^2-b^2)\cos x}\sqrt{a+b\sin x} \qquad \left[0 < b < a, \; -\frac{\pi}{2} < x < \frac{\pi}{2}\right]$$

$$= \sqrt{\frac{2}{b}}\left\{\frac{2a+b}{2(a+b)}\,F\left(\beta,\frac{1}{r}\right) + \frac{ab}{power a 2 - b^2}\,E\left(\beta,\frac{1}{r}\right)\right\}$$

$$- \frac{b - a\sin x}{(a^2-b^2)\cos x}\sqrt{a+b\sin x} \qquad \left[0 < |a| < b, \; -\arcsin\frac{a}{b} < x < \frac{\pi}{2}\right]$$

BY(288.08, 288.58)

**2.573**

1.    $\displaystyle\int \frac{1 - \sin x}{1 + \sin x}\frac{dx}{\sqrt{a+b\sin x}} = \frac{2}{a-b}\left\{\sqrt{a+b}\,E(\alpha, r)\right\} - \tan\left(\frac{\pi}{4} - \frac{x}{2}\right)\sqrt{a+b\sin x}$

$$\left[0 < b < a, \quad -\frac{\pi}{2} \le x < \frac{\pi}{2}\right] \qquad \text{BY (288.07)}$$

2.    $\displaystyle\int \frac{1 - \cos x}{1 + \cos x}\frac{dx}{\sqrt{a+b\cos x}} = \frac{2}{a-b}\tan\frac{x}{2}\sqrt{a+b\cos x} - \frac{2\sqrt{a+b}}{a-b}\,E\left(\frac{x}{2},r\right)$

$$[a > b > 0, \quad 0 \le x < \pi] \qquad \text{BY (289.07)}$$

**2.574**

1. $$\int \frac{dx}{(2 - p^2 + p^2 \sin x)\sqrt{a + b \sin x}} = -\frac{1}{a + b} \Pi\left(\alpha, p^2, r\right)$$

$$\left[0 < b < a, \quad -\frac{\pi}{2} \le x < \frac{\pi}{2}\right]$$

BY (288.02)

2. $$\int \frac{dx}{(a + b - p^2 b + p^2 b \sin x)\sqrt{a + b \sin x}} = -\frac{1}{a + b}\sqrt{\frac{2}{b}} \Pi\left(\beta, p^2, \frac{1}{r}\right)$$

$$\left[0 < |a| < b, \quad -\arcsin\frac{a}{b} < x < \frac{\pi}{2}\right]$$

BY (288.52)

3. $$\int \frac{dx}{(2 - p^2 + p^2 \cos x)\sqrt{a + b \cos x}} = \frac{1}{\sqrt{a + b}} \Pi\left(\frac{x}{2}, p^2, r\right)$$

$$[a > b > 0, \quad 0 \le x < \pi] \qquad \text{BY (289.02)}$$

4. $$\int \frac{dx}{(a + b - p^2 b + p^2 b \cos x)\sqrt{a + b \cos x}} = \frac{\sqrt{2}}{(a + b)\sqrt{b}} \Pi\left(\gamma, p^2, \frac{1}{r}\right)$$

$$\left[b \ge |a| > 0, \quad 0 \le x < \arccos\left(-\frac{a}{b}\right)\right]$$

BY (290.02)

**2.575**

1. $$\int \frac{dx}{\sqrt{(a + b \sin x)^3}} = \frac{2b \cos x}{(a^2 - b^2)\sqrt{a + b \sin x}} - \frac{2}{(a - b)\sqrt{a + b}} E(\alpha, r)$$

$$\left[0 < b < a, \quad -\frac{\pi}{2} \le x < \frac{\pi}{2}\right]$$

BY (288.05)

$$= \sqrt{\frac{2}{b}} \left\{ \frac{2b}{b^2 - a^2} E\left(\beta, \frac{1}{r}\right) - \frac{1}{a + b} F\left(\beta, \frac{1}{r}\right) \right\} + \frac{2b}{b^2 - a^2} \cdot \frac{\cos x}{\sqrt{a + b \sin x}}$$

$$\left[0 < |a| < b, \quad -\arcsin\frac{a}{b} < x < \frac{\pi}{2}\right]$$

BY (288.56)

2.

$$\int \frac{dx}{\sqrt{(a+b\sin x)^5}} = \frac{2}{3\left(a^2-b^2\right)^2 \sqrt{a+b}} \left\{ \left(a^2-b^2\right) F(\alpha, r) - 4a(a+b) E(\alpha, r) \right\}$$

$$+ \frac{2b\left(5a^2-b^2+4ab\sin x\right)}{3\left(a^2-b^2\right)^2 \sqrt{(a+b\sin x)^3}} \cos x$$

$$\left[ 0 < b < a, \quad -\frac{\pi}{2} \le x < \frac{\pi}{2} \right]$$
BY (288.05)

$$= -\frac{1}{3\left(a^2-b^2\right)^2} \sqrt{\frac{2}{b}} \left\{ (3a-b)(a-b) F\left(\beta, \frac{1}{r}\right) + 8ab\, E\left(\beta, \frac{1}{r}\right) \right\}$$

$$+ \frac{2b\left[a^2-b^2+4a\left(a+b\sin x\right)\right]}{3\left(a^2-b^2\right)^2 \sqrt{(a+b\sin x)^3}} \cos x$$

$$\left[ 0 < |a| < b, \quad -\arcsin\frac{a}{b} < x < \frac{\pi}{2} \right]$$
BY (288.56)

3.

$$\int \frac{dx}{\sqrt{(a+b\cos x)^3}} = \frac{2}{(a-b)\sqrt{a+b}} E\left(\frac{x}{2}, r\right) - \frac{2b}{a^2-b^2} \cdot \frac{\sin x}{\sqrt{a+b\cos x}}$$

$$[a > b > 0, \quad 0 \le x \le \pi]$$
BY (289.05)

$$= \frac{1}{a^2-b^2} \sqrt{\frac{2}{b}} \left\{ (a-b) F\left(\gamma, \frac{1}{r}\right) + 2b\, E\left(\gamma, \frac{1}{r}\right) \right\} + \frac{2b}{b^2-a^2} \cdot \frac{\sin x}{\sqrt{a+b\cos x}}$$

$$\left[ b \ge |a| > 0, \quad 0 \le x < \arccos\left(-\frac{a}{b}\right) \right]$$
BY (290.06)

4.

$$\int \frac{dx}{\sqrt{(a-b\cos x)^3}} = \frac{2}{(a-b)\sqrt{a+b}} E(\delta, r) \qquad [a > b > 0, \quad 0 \le x \le \pi] \qquad (291.01)$$

5.  $\displaystyle\int \frac{dx}{\sqrt{(a+b\cos x)^5}} = \frac{2\sqrt{a+b}}{3\left(a^2-b^2\right)^2} \left\{ 4a\, E\left(\frac{x}{2},r\right) - (a-b)\, F\left(\frac{x}{2},r\right) \right\}$

$$-\frac{2b}{3\left(a^2-b^2\right)^2} \cdot \frac{5a^2-b^2+4ab\cos x}{\sqrt{(a+b\cos x)^3}}\sin x$$

$$[a>b>0, \quad 0 \le x \le \pi]$$
BY (289.05)

$$= \frac{1}{3\left(a^2-b^2\right)^2}\sqrt{\frac{2}{b}} \left\{ (a-b)(3a-b)\,F\left(\gamma,\frac{1}{r}\right) + 8ab\,E\left(\gamma,\frac{1}{r}\right) \right\}$$

$$+\frac{2b\left(5a^2-b^2+4ab\cos x\right)\sin x}{3\left(a^2-b^2\right)^2\sqrt{(a+b\cos x)^3}}$$

$$\left[b \ge |a| > 0, \quad 0 \le x < \arccos\left(-\frac{a}{b}\right)\right]$$
BY (290.06)

**2.576**

1.  $\displaystyle\int \sqrt{a+b\cos x}\,dx = 2\sqrt{a+b}\,E\left(\frac{x}{2},r\right)$

$$[a>b>0, \quad 0 \le x \le \pi]$$
BY (289.01)

$$= \sqrt{\frac{2}{b}}\left\{ (a-b)\,F\left(\gamma,\frac{1}{r}\right) + 2b\,E\left(\gamma,\frac{1}{r}\right) \right\}$$

$$\left[b \ge |a| > 0, \quad 0 \le x < \arccos\left(-\frac{a}{b}\right)\right]$$
BY (290.03)

2.  $\displaystyle\int \sqrt{a-b\cos x}\,dx = 2\sqrt{a+b}\,E(\delta,r) - \frac{2b\sin x}{\sqrt{a-b\cos x}}$    $[a>b>0, \quad 0 \le x \le \pi]$    BY (291.05)

**2.577**

1.[3]  $\displaystyle\int \frac{\sqrt{a-b\cos x}}{1+p\cos x}\,dx = \frac{2(a-b)}{(1+p)\sqrt{a+b}}\,\Pi\left(\delta,\frac{2ap}{(a+b)(1+p)},r\right)$

$$[a>b>0, \quad 0 \le x \le \pi, \quad p \ne -1]$$
BY (291.02)

2.[12]  $\displaystyle\int \sqrt{\frac{a-b\cos x}{1+p\cos x}}\,dx = \frac{2(a-b)}{\sqrt{(1+p)(a+b)}}\,\Pi\left(\delta,r^2,\sqrt{\frac{2(ap+b)}{(1+p)(a+b)}}\right)$

$$[a>b>0, \quad 0 \le x \le \pi, \quad p \ne -1]$$

**2.578**  $\displaystyle\int \frac{\tan x\,dx}{\sqrt{a+b\tan^2 x}} = \frac{1}{\sqrt{b-a}}\arccos\left(\frac{\sqrt{b-a}}{\sqrt{b}}\cos x\right)$    $[b>a, \quad b>0]$    PE (333)

## 2.58–2.62 Integrals reducible to elliptic and pseudo-elliptic integrals

**2.580**

1. $$\int \frac{d\varphi}{\sqrt{a + b\cos\varphi + c\sin\varphi}} = 2\int \frac{d\psi}{\sqrt{a - p + 2p\cos^2\psi}} \qquad \left[\varphi = 2\psi + \alpha, \tan\alpha = \frac{c}{b}, p = \sqrt{b^2 + c^2}\right]$$

2.[12] $$\int \frac{d\varphi}{\sqrt{a + b\cos\varphi + c\sin\varphi + d\cos^2\varphi + e\sin\varphi\cos\varphi + f\sin^2\varphi}} = 2\int \frac{dx}{\sqrt{A + Bx + Cx^2 + Dx^3 + Ex^4}}$$
$$\left[\tan\frac{\varphi}{2} = x, A = a + b + d, B = 2c + 2e, C = 2a - 2d + 4f, D = 2c - 2e, E = a - b + d\right]$$

**Forms containing $\sqrt{1 - k^2\sin^2 x}$**

ndexsquare roots

**Notation:** $\Delta = \sqrt{1 - k^2\sin^2 x}$, $k' = \sqrt{1 - k^2}$

**2.581**

1.[12] $$\int \sin^m x \cos^n x \Delta^r \, dx$$

$$= \frac{1}{(m + n + r)k^2}\left\{ \sin^{m-3} x \cos^{n+1} x \Delta^{r+2} + \left[m + n - 2(m + r - 1)k^2\right] \right.$$

$$\left. \times \int \sin^{m-2} x \cos^n x \Delta^r \, dx - (m - 3)\int \sin^{m-4} x \cos^n x \Delta^r \, dx \right\}$$

$$= \frac{1}{(m + n + r)k^2}\left\{ \sin^{m+1} x \cos^{n-3} x \Delta^{r+2} + \left[(n + r - 1)k^2 - (m + n - 2)k'^2\right] \right.$$

$$\left. \times \int \sin^m x \cos^{n-2} x \Delta^r \, dx + (n - 3)k'^2 \int \sin^m x \cos^{n-4} x \Delta^r \, dx \right\}$$

$$[m + n + r \neq 0]$$

For $r = -3$ and $r = -5$:

2. $$\int \frac{\sin^m x \cos^n x}{\Delta^3} \, dx = \frac{\sin^{m-1} x \cos^{n-1} x}{k^2 \Delta}$$
$$- \frac{m - 1}{k^2}\int \frac{\sin^{m-2} x \cos^n x}{\Delta} \, dx + \frac{n - 1}{k^2}\int \frac{\sin^m x \cos^{n-2} x}{\Delta} \, dx$$

3. $$\int \frac{\sin^m x \cos^n x}{\Delta^5} \, dx = \frac{\sin^{m-1} x \cos^{n-1} x}{3k^2 \Delta^3}$$
$$- \frac{m - 1}{3k^2}\int \frac{\sin^{m-2} x \cos^n x}{\Delta^3} \, dx + \frac{n - 1}{3k^2}\int \frac{\sin^m x \cos^{n-2} x}{\Delta^3} \, dx$$

For $m = 1$ or $n = 1$:

4. $$\int \sin x \cos^n x \Delta^r \, dx = -\frac{\cos^{n-1} x \Delta^{r+2}}{(n + r + 1)k^2} - \frac{(n - 1)k'^2}{(n + r + 1)k^2}\int \cos^{n-2} x \sin x \Delta^r \, dx$$

5. $$\int \sin^m x \cos x \Delta^r \, dx = -\frac{\sin^{m-1} x \Delta^{r+2}}{(m + r + 1)k^2} + \frac{m - 1}{(m + r + 1)k^2}\int \sin^{m-2} x \cos x \Delta^r \, dx$$

For $m = 3$ or $n = 3$:

6. $\int \sin^3 x \cos^n x \Delta^r \, dx = \dfrac{(n+r+1)k^2 \cos^2 x - \left[(r+2)k^2 + n + 1\right]}{(n+r+1)(n+r+3)k^4} \cos^{n-1} x \Delta^{r+2}$

$\qquad\qquad\qquad -\dfrac{\left[(r+2)k^2 + n + 1\right](n-1)k'^2}{(n+r+1)(n+r+3)k^4} \int \cos^{n-2} x \sin x \Delta^r \, dx$

7.[12] $\int \sin^m x \cos^3 x \Delta^r \, dx$

$\qquad\qquad = \dfrac{(m+r+1)k^2 \sin^2 x - \left[(r+2)k^2 - (m+1)k'^2\right]}{(m+r+1)(m+r+3)k^4}$

$\qquad\qquad \times \sin^{m-1} x \Delta^{r+2} + \dfrac{\left[(r+2)k^2 - (m+1)k'^2\right](m-1)}{(m+r+1)(m+r+3)k^4} \int \sin^{m-2} x \cos x \Delta^r \, dx$

**2.582**

1. $\int \Delta^n \, dx = \dfrac{n-1}{n}\left(2 - k^2\right) \int \Delta^{n-2} \, dx - \dfrac{n-2}{n}\left(1 - k^2\right) \int \Delta^{n-4} \, dx$

$\qquad\qquad + \dfrac{k^2}{n} \sin x \cos x \Delta^{n-2}$

$\qquad\qquad\qquad\qquad\qquad\qquad\qquad\qquad\qquad\qquad\qquad\qquad\qquad$ LA (316)(1)a

2. $\int \dfrac{dx}{\Delta^{n+1}} = -\dfrac{k^2 \sin x \cos x}{(n-1)k'^2 \Delta^{n-1}} + \dfrac{n-2}{n-1}\dfrac{2-k^2}{k'^2} \int \dfrac{dx}{\Delta^{n-1}} - \dfrac{n-3}{n-1}\dfrac{1}{k'^2} \int \dfrac{dx}{\Delta^{n-3}}$ 　　LA 317(8)a

3. $\int \dfrac{\sin^n x}{\Delta} \, dx = \dfrac{\sin^{n-3} x}{(n-1)k^2} \cos x \Delta + \dfrac{n-2}{n-1}\dfrac{1+k^2}{k^2} \int \dfrac{\sin^{n-2} x}{\Delta} \, dx$

$\qquad\qquad -\dfrac{n-3}{(n-1)k^2} \int \dfrac{\sin^{n-4} x}{\Delta} \, dx$

$\qquad\qquad\qquad\qquad\qquad\qquad\qquad\qquad\qquad\qquad\qquad\qquad\qquad$ LA 316(1)a

4. $\int \dfrac{\cos^n x}{\Delta} \, dx = \dfrac{\cos^{n-3} x}{(n-1)k^2} \sin x \Delta + \dfrac{n-2}{n-1}\dfrac{2k^2-1}{k^2} \int \dfrac{\cos^{n-2} x}{\Delta} \, dx$

$\qquad\qquad + \dfrac{n-3}{n-1}\dfrac{k'^2}{k^2} \int \dfrac{\cos^{n-4} x}{\Delta} \, dx$

$\qquad\qquad\qquad\qquad\qquad\qquad\qquad\qquad\qquad\qquad\qquad\qquad\qquad$ LA 316(2)a

5. $\int \dfrac{\tan^n x}{\Delta} \, dx = \dfrac{\tan^{n-3} x}{(n-1)k'^2} \dfrac{\Delta}{\cos^2 x} - \dfrac{(n-2)\left(2-k^2\right)}{(n-1)k'^2} \int \dfrac{\tan^{n-2} x}{\Delta} \, dx$

$\qquad\qquad - \dfrac{n-3}{(n-1)k'^2} \int \dfrac{\tan^{n-4} x}{\Delta} \, dx$

$\qquad\qquad\qquad\qquad\qquad\qquad\qquad\qquad\qquad\qquad\qquad\qquad\qquad$ LA 317(3)

6. $\int \dfrac{\cot^n x}{\Delta} \, dx = -\dfrac{\cot^{n-1} x}{n-1} \dfrac{\Delta}{\cos^2 x} - \dfrac{n-2}{n-1}\left(2 - k^2\right) \int \dfrac{\cot^{n-2} x}{\Delta} \, dx$

$\qquad\qquad - \dfrac{n-3}{n-1}k'^2 \int \dfrac{\cot^{n-4} x}{\Delta} \, dx$

$\qquad\qquad\qquad\qquad\qquad\qquad\qquad\qquad\qquad\qquad\qquad\qquad\qquad$ LA 317(6)

**2.583**

1. $\int \Delta \, dx = E(x, k)$

2. $\displaystyle\int \Delta \sin x \, dx = -\frac{\Delta \cos x}{2} - \frac{k'^2}{2k} \ln \left(k \cos x + \Delta\right)$

3. $\displaystyle\int \Delta \cos x \, dx = \frac{\Delta \sin x}{2} + \frac{1}{2k} \arcsin \left(k \sin x\right)$

4. $\displaystyle\int \Delta \sin^2 x \, dx = -\frac{\Delta}{3} \sin x \cos x + \frac{k'^2}{3k^2} F(x, k) + \frac{2k^2 - 1}{3k^2} E(x, k)$

5. $\displaystyle\int \Delta \sin x \cos x \, dx = -\frac{\Delta^3}{3k^2}$

6. $\displaystyle\int \Delta \cos^2 x \, dx = \frac{\Delta}{3} \sin x \cos x - \frac{k'^2}{3k^2} F(x, k) + \frac{k^2 + 1}{3k^2} E(x, k)$

7. $\displaystyle\int \Delta \sin^3 x \, dx = -\frac{2k^2 \sin^2 x + 3k^2 - 1}{8k^2} \Delta \cos x + \frac{3k^4 - 2k^2 - 1}{8k^3} \ln \left(k \cos x + \Delta\right)$

8. $\displaystyle\int \Delta \sin^2 x \cos x \, dx = \frac{2k^2 \sin^2 x - 1}{8k^2} \Delta \sin x + \frac{1}{8k^3} \arcsin \left(k \sin x\right)$

9. $\displaystyle\int \Delta \sin x \cos^2 x \, dx = -\frac{2k^2 \cos^2 x + k'^2}{8k^2} \Delta \cos x + \frac{k'^4}{8k^3} \ln \left(k \cos x + \Delta\right)$

10. $\displaystyle\int \Delta \cos^3 x \, dx = \frac{2k^2 \cos^2 x + 2k^2 + 1}{8k^2} \Delta \sin x + \frac{4k^2 - 1}{8k^3} \arcsin \left(k \sin x\right)$

11. $\displaystyle\int \Delta \sin^4 x \, dx = -\frac{3k^2 \sin^2 x + 4k^2 - 1}{15k^2} \Delta \sin x \cos x$
$\displaystyle\qquad\qquad - \frac{2 \left(2k^4 - k^2 - 1\right)}{15k^4} F(x, k) + \frac{8k^4 - 3k^2 - 2}{15k^4} E(x, k)$

12. $\displaystyle\int \Delta \sin^3 x \cos x \, dx = \frac{3k^4 \sin^4 x - k^2 \sin^2 x - 2}{15k^4} \Delta$

13. $\displaystyle\int \Delta \sin^2 x \cos^2 x \, dx = -\frac{3k^2 \cos^2 x - 2k^2 + 1}{15k^2} \Delta \sin x \cos x$
$\displaystyle\qquad\qquad - \frac{k'^2 \left(1 + k'^2\right)}{15k^4} F(x, k) + \frac{2 \left(k^4 - k^2 + 1\right)}{15k^4} E(x, k)$

14. $\displaystyle\int \Delta \sin x \cos^3 x \, dx = -\frac{3k^4 \sin^4 x - k^2 \left(5k^2 + 1\right) \sin^2 x + 5k^2 - 2}{15k^4} \Delta$

15. $\displaystyle\int \Delta \cos^4 x \, dx = \frac{3k^2 \cos^2 x + 3k^2 + 1}{15k^2} \Delta \sin x \cos x$
$\displaystyle\qquad\qquad + \frac{2k'^2 \left(k'^2 - 2k^2\right)}{15k^4} F(x, k) + \frac{3k^4 + 7k^2 - 2}{15k^4} E(x, k)$

16. $\displaystyle\int \Delta \sin^5 x \, dx = \frac{-8k^4 \sin^4 x - 2k^2 \left(5k^2 - 1\right) \sin^2 x - 15k^4 + 4k^2 + 3}{48k^4} \Delta \cos x$
$\displaystyle\qquad\qquad + \frac{5k^6 - 3k^4 - k^2 - 1}{16k^5} \ln \left(k \cos x + \Delta\right)$

17. $\displaystyle\int \Delta \sin^4 x \cos x \, dx = \frac{8k^4 \sin^4 x - 2k^2 \sin^2 x - 3}{48k^4} \Delta \sin x + \frac{1}{16k^5} \arcsin \left(k \sin x\right)$

18. $$\int \Delta \sin^3 x \cos^2 x \, dx = \frac{8k^4 \sin^4 x - 2k^2 \left(k^2 + 1\right) \sin^2 x - 3k^4 + 2k^2 - 3}{48k^4} \Delta \cos x$$
$$+ \frac{k'^4 \left(k^2 + 1\right)}{16k^5} \ln \left(k \cos x + \Delta\right)$$

19. $$\int \Delta \sin^2 x \cos^3 x \, dx = \frac{-8k^4 \sin^4 x + 2k^2 \left(6k^2 + 1\right) \sin^2 x - 6k^2 + 3}{48k^4} \Delta \sin x$$
$$+ \frac{2k^2 - 1}{16k^5} \arcsin \left(k \sin x\right)$$

20. $$\int \Delta \sin x \cos^4 x \, dx = \frac{-8k^4 \sin^4 x + 2k^2 \left(7k^2 + 1\right) \sin^2 x - 3k^4 - 8k^2 + 3}{48k^4} \Delta \cos x$$
$$- \frac{k'^6}{16k^5} \ln \left(k \cos x + \Delta\right)$$

21. $$\int \Delta \cos^5 x \, dx = \frac{8k^4 \sin^4 x - 2k^2 \left(12k^2 + 1\right) \sin^2 x + 24k^4 + 12k^2 - 3}{48k^4} \Delta \sin x$$
$$+ \frac{8k^4 - 4k^2 + 1}{16k^5} \arcsin \left(k \sin x\right)$$

22. $$\int \Delta^3 \, dx = \frac{2}{3} \left(1 + k'^2\right) E(x, k) - \frac{k'^2}{3} F(x, k) + \frac{k^2}{3} \Delta \sin x \cos x$$

23. $$\int \Delta^3 \sin x \, dx = \frac{2k^2 \sin^2 x + 3k^2 - 5}{8} \Delta \cos x - \frac{3k'^4}{8k} \ln \left(k \cos x + \Delta\right)$$

24. $$\int \Delta^3 \cos x \, dx = \frac{-2k^2 \sin^2 x + 5}{8} \Delta \sin x + \frac{3}{8k} \arcsin \left(k \sin x\right)$$

25. $$\int \Delta^3 \sin^2 x \, dx = \frac{3k^2 \sin^2 x + 4k^2 - 6}{15} \Delta \sin x \cos x + \frac{k'^2 \left(3 - 4k^2\right)}{15k^2} F(x, k)$$
$$- \frac{8k^4 - 13k^2 + 3}{15k^2} E(x, k)$$

26. $$\int \Delta^3 \sin x \cos x \, dx = -\frac{\Delta^5}{5k^2}$$

27.[12] $$\int \Delta^3 \cos^2 x \, dx = \frac{-3k^2 \sin^2 x + k^2 + 6}{15} \Delta \sin x \cos x - \frac{k'^2 \left(k^2 + 3\right)}{15k^2} F(x, k)$$
$$- \frac{2k^4 - 7k^2 - 3}{15k^2} E(x, k)$$

28. $$\int \Delta^3 \sin^3 x \, dx = \frac{8k^4 \sin^4 x + 2k^2 \left(5k^2 - 7\right) \sin^2 x + 15k^4 - 22k^2 + 3}{48k^2} \Delta \cos x$$
$$- \frac{5k^6 - 9k^4 + 3k^2 + 1}{16k^3} \ln \left(k \cos x + \Delta\right)$$

29. $$\int \Delta^3 \sin^2 x \cos x \, dx = \frac{-8k^4 \sin^4 x + 14k^2 \sin^2 x - 3}{48k^2} \Delta \sin x$$
$$+ \frac{1}{16k^3} \arcsin \left(k \sin x\right)$$

30. $$\int \Delta^3 \sin x \cos^2 x \, dx = \frac{-8k^4 \sin^4 x + 2k^2 \left(k^2 + 7\right) \sin^2 x + 3k^4 - 8k^2 - 3}{48k^2}$$
$$\times \Delta \cos x + \frac{k'^6}{16k^3} \ln \left(k \cos x + \Delta\right)$$

31. $$\int \Delta^3 \cos^3 x \, dx = \frac{8k^4 \sin^4 x - 2k^2 \left(6k^2 + 7\right) \sin^2 x + 30k^2 + 3}{48k^2} \Delta \sin x$$
$$+ \frac{6k^2 - 1}{16k^3} \arcsin \left(k \sin x\right)$$

32.[12] $$\int \frac{\Delta \, dx}{\sin x} = -\frac{1}{2} \ln \frac{\Delta + \cos x}{\Delta - \cos x} + k \ln \left(k \cos x + \Delta\right)$$

33. $$\int \frac{\Delta \, dx}{\cos x} = \frac{k'}{2} \ln \frac{\Delta + k' \sin x}{\Delta - k' \sin x} + k \arcsin \left(k \sin x\right)$$

34. $$\int \frac{\Delta \, dx}{\sin^2 x} = k'^2 \, F(x, k) - E(x, k) - \Delta \cot x$$

35. $$\int \frac{\Delta \, dx}{\sin x \cos x} = \frac{1}{2} \ln \frac{1 - \Delta}{1 + \Delta} + \frac{k'}{2} \ln \frac{\Delta + k'}{\Delta - k'}$$

36. $$\int \frac{\Delta \, dx}{\cos^2 x} = F(x, k) - E(x, k) + \Delta \tan x$$

37. $$\int \frac{\sin x}{\cos x} \Delta \, dx = \int \Delta \tan x \, dx = -\Delta + \frac{k'}{2} \ln \frac{\Delta + k'}{\Delta - k'}$$

38. $$\int \frac{\cos x}{\sin x} \Delta \, dx = \int \Delta \cot x \, dx = \Delta + \frac{1}{2} \ln \frac{1 - \Delta}{1 + \Delta}$$

39. $$\int \frac{\Delta \, dx}{\sin^3 x} = -\frac{\Delta \cos x}{2 \sin^2 x} + \frac{k'^2}{4} \ln \frac{\Delta + \cos x}{\Delta - \cos x}$$

40. $$\int \frac{\Delta \, dx}{\sin^2 x \cos x} = \frac{-\Delta}{\sin x} - \frac{1 + k^2}{2k'} \ln \frac{\Delta - k' \sin x}{\Delta + k' \sin x}$$

41. $$\int \frac{\Delta \, dx}{\sin x \cos^2 x} = \frac{\Delta}{\cos x} + \frac{1}{2} \ln \frac{\Delta + \cos x}{\Delta - \cos x}$$

42. $$\int \frac{\Delta \, dx}{\cos^3 x} = \frac{\Delta \sin x}{2 \cos^2 x} + \frac{1}{4k'} \ln \frac{\Delta + k' \sin x}{\Delta - k' \sin x}$$

43. $$\int \frac{\Delta \sin x \, dx}{\cos^2 x} = \frac{\Delta}{\cos x} - k \ln \left(k \cos x + \Delta\right)$$

44. $$\int \frac{\Delta \cos x \, dx}{\sin^2 x} = -\frac{\Delta}{\sin x} - k \arcsin \left(k \sin x\right)$$

45. $$\int \frac{\Delta \sin^2 x \, dx}{\cos x} = -\frac{\Delta \sin x}{2} + \frac{2k^2 - 1}{2k} \arcsin \left(k \sin x\right) + \frac{k'}{2} \ln \frac{\Delta + k' \sin x}{\Delta - k' \sin x}$$

46. $$\int \frac{\Delta \cos^2 x \, dx}{\sin x} = \frac{\Delta \cos x}{2} + \frac{k^2 + 1}{2k} \ln \left(k \cos x + \Delta\right) + \frac{1}{2} \ln \frac{\Delta + \cos x}{\Delta - \cos x}$$

47. $$\int \frac{\Delta \, dx}{\sin^4 x} = \frac{1}{3} \left\{ -\Delta \cot^3 x + \left(k^2 - 3\right) \Delta \cot x + 2k'^2 \, F(x, k) + \left(k^2 - 2\right) E(x, k) \right\}$$

48. $$\int \frac{\Delta\,dx}{\sin^3 x \cos x} = -\frac{\Delta}{2\sin^2 x} + \frac{k'}{2}\ln\frac{\Delta+k'}{\Delta-k'} + \frac{k^2-2}{4}\ln\frac{1+\Delta}{1-\Delta}$$

49. $$\int \frac{\Delta\,dx}{\sin^2 x \cos^2 x} = \left(\frac{1}{k'^2}\tan x - \cot x\right)\Delta + 2\,F(x,k) - \frac{1+k'^2}{k'^2}\,E(x,k)$$

50. $$\int \frac{\Delta\,dx}{\sin x \cos^3 x} = \frac{\Delta}{2\cos^2 x} - \frac{1}{2}\ln\frac{1+\Delta}{1-\Delta} + \frac{2-k^2}{4k'}\ln\frac{\Delta+k'}{\Delta-k'}$$

51.[12] $$\int \frac{\Delta\,dx}{\cos^4 x} = \frac{1}{3k'^2}\left\{\left[k'^2\tan^3 x - \left(2k^2-3\right)\tan x\right]\Delta + 2k'^2\,F(x,k) + \left(k^2-2\right)E(x,k)\right\}$$

52. $$\int \frac{\sin x}{\cos^3 x}\Delta\,dx = \frac{\Delta}{2\cos^2 x} + \frac{k^2}{4k'}\ln\frac{\Delta+k'}{\Delta-k'}$$

53. $$\int \frac{\cos x}{\sin^3 x}\Delta\,dx = -\frac{\Delta}{2\sin^2 x} + \frac{k^2}{4}\ln\frac{1+\Delta}{1-\Delta}$$

54. $$\int \frac{\sin^2 x}{\cos^2 x}\Delta\,dx = \int \tan^2 x\Delta\,dx = \Delta\tan x + F(x,k) - 2\,E(x,k)$$

55. $$\int \frac{\cos^2 x}{\sin^2 x}\Delta\,dx = \int \cot^2 x\Delta\,dx = -\Delta\cot x + k'^2\,F(x,k) - 2\,E(x,k)$$

56. $$\int \frac{\sin^3 x}{\cos x}\Delta\,dx = -\frac{k^2\sin^2 x + 3k^2 - 1}{3k^2}\Delta + \frac{k'}{2}\ln\frac{\Delta+k'}{\Delta-k'}$$

57. $$\int \frac{\cos^3 x}{\sin x}\Delta\,dx = -\frac{k^2\sin^2 x - 3k^2 - 1}{3k^2}\Delta + \frac{1}{2}\ln\frac{1-\Delta}{1+\Delta}$$

58.[12] $$\int \frac{\Delta\,dx}{\sin^5 x} = \frac{\left(k^2-3\right)\sin^2 x + 2}{8\sin^4 x}\Delta\cos x + \frac{k'^2\left(k^2+3\right)}{16}\ln\frac{\Delta+\cos x}{\Delta-\cos x}$$

59. $$\int \frac{\Delta\,dx}{\sin^4 x \cos x} = -\frac{\left(3-k^2\right)\sin^2 x + 1}{3\sin^3 x}\Delta - \frac{k'}{2}\ln\frac{\Delta-k'\sin x}{\Delta+k'\sin x}$$

60. $$\int \frac{\Delta\,dx}{\sin^3 x \cos^2 x} = \frac{3\sin^2 x - 1}{2\sin^2 x \cos x}\Delta + \frac{k^2-3}{4}\ln\frac{\Delta-\cos x}{\Delta+\cos x}$$

61. $$\int \frac{\Delta\,dx}{\sin^2 x \cos^3 x} = \frac{3\sin^2 x - 2}{2\sin x \cos^2 x}\Delta - \frac{2k^2-3}{4k'}\ln\frac{\Delta+k'\sin x}{\Delta-k'\sin x}$$

62. $$\int \frac{\Delta\,dx}{\sin x \cos^4 x} = \frac{\left(2k^2-3\right)\sin^2 x - 3k^2 + 4}{3k'^2\cos^3 x}\Delta + \frac{1}{2}\ln\frac{\Delta+\cos x}{\Delta-\cos x}$$

63. $$\int \frac{\Delta\,dx}{\cos^5 x} = \frac{\left(2k^2-3\right)\sin^2 x - 4k^2 + 5}{8k'^2\cos^4 x}\sin x\Delta - \frac{4k^2-3}{16k'^3}\ln\frac{\Delta+k'\sin x}{\Delta-k'\sin x}$$

64. $$\int \frac{\sin x}{\cos^4 x}\Delta\,dx = \frac{-\left(2k^2+1\right)k^2\sin^2 x + 3k^4 - k^2 + 1}{3k'^2\cos^3 x}\Delta$$

65. $$\int \frac{\cos x}{\sin^4 x}\Delta\,dx = -\frac{\Delta^3}{3\sin^3 x}$$

66. $$\int \frac{\sin^2 x}{\cos^3 x}\Delta\,dx = \frac{\sin x}{2\cos^2 x}\Delta + \frac{2k^2-1}{4k'}\ln\frac{\Delta+k'\sin x}{\Delta-k'\sin x} - k\arcsin\left(k\sin x\right)$$

67. $$\int \frac{\cos^2 x}{\sin^3 x} \Delta \, dx = -\frac{\cos x}{2 \sin^2 x} \Delta - \frac{k^2+1}{4} \ln \frac{\Delta + \cos x}{\Delta - \cos x} - k \ln (k \cos x + \Delta)$$

68. $$\int \frac{\sin^3 x}{\cos^2 x} \Delta \, dx = -\frac{\sin^2 x - 3}{2 \cos x} \Delta - \frac{3k^2-1}{2k} \ln (k \cos x + \Delta)$$

69. $$\int \frac{\cos^3 x}{\sin^2 x} \Delta \, dx = -\frac{\sin^2 x + 2}{2 \sin x} \Delta - \frac{2k^2+1}{2k} \arcsin (k \sin x)$$

70. $$\int \frac{\sin^4 x}{\cos x} \Delta \, dx = -\frac{2k^2 \sin^2 x + 4k^2 - 1}{8k^2} \sin x \Delta$$
$$+ \frac{8k^4 - 4k^2 - 1}{8k^3} \arcsin (k \sin x) + \frac{k'}{2} \ln \frac{\Delta + k' \sin x}{\Delta - k' \sin x}$$

71. $$\int \frac{\cos^4 x}{\sin x} \Delta \, dx = \frac{-2k^2 \sin^2 x + 5k^2 + 1}{8k^2} \cos x \Delta$$
$$+ \frac{1}{2} \ln \frac{\Delta + \cos x}{\Delta - \cos x} + \frac{3k^4 + 6k^2 - 1}{8k^3} \ln (k \cos x + \Delta)$$

**2.584**

1. $$\int \frac{dx}{\Delta} = F(x, k)$$

2. $$\int \frac{\sin x \, dx}{\Delta} = \frac{1}{2k} \ln \frac{\Delta - k \cos x}{\Delta + k \cos x} = -\frac{1}{k} \ln (k \cos x + \Delta)$$

3. $$\int \frac{\cos x \, dx}{\Delta} = \frac{1}{k} \arcsin (k \sin x) = \frac{1}{k} \arctan \frac{k \sin x}{\Delta}$$

4. $$\int \frac{\sin^2 x \, dx}{\Delta} = \frac{1}{k^2} F(x, k) - \frac{1}{k^2} E(x, k)$$

5. $$\int \frac{\sin x \cos x \, dx}{\Delta} = -\frac{\Delta}{k^2}$$

6. $$\int \frac{\cos^2 x \, dx}{\Delta} = \frac{1}{k^2} E(x, k) - \frac{k'^2}{k^2} F(x, k)$$

7. $$\int \frac{\sin^3 x \, dx}{\Delta} = \frac{\cos x \Delta}{2k^2} - \frac{1+k^2}{2k^3} \ln (k \cos x + \Delta)$$

8. $$\int \frac{\sin^2 x \cos x \, dx}{\Delta} = -\frac{\sin x \Delta}{2k^2} + \frac{\arcsin (k \sin x)}{2k^3}$$

9. $$\int \frac{\sin x \cos^2 x \, dx}{\Delta} = -\frac{\cos x \Delta}{2k^2} + \frac{k'^2}{2k^3} \ln (k \cos x + \Delta)$$

10. $$\int \frac{\cos^3 x \, dx}{\Delta} = \frac{\sin x \Delta}{2k^2} + \frac{2k^2 - 1}{2k^3} \arcsin (k \sin x)$$

11. $$\int \frac{\sin^4 x \, dx}{\Delta} = \frac{\sin x \cos x \Delta}{3k^2} + \frac{2+k^2}{3k^4} F(x, k) - \frac{2(1+k^2)}{3k^4} E(x, k)$$

12. $$\int \frac{\sin^3 x \cos x \, dx}{\Delta} = -\frac{1}{3k^4} (2 + k^2 \sin^2 x) \Delta$$

13. $\displaystyle\int \frac{\sin^2 x \cos^2 x \, dx}{\Delta} = -\frac{\sin x \cos x \Delta}{3k^2} + \frac{2-k^2}{3k^4} E(x,k) + \frac{2k^2 - 2}{3k^4} F(x,k)$

14. $\displaystyle\int \frac{\sin x \cos^3 x \, dx}{\Delta} = -\frac{1}{3k^4}\left(k^2 \cos^2 x - 2k'^2\right)\Delta$

15. $\displaystyle\int \frac{\cos^4 x \, dx}{\Delta} = \frac{\sin x \cos x \Delta}{3k^2} + \frac{4k^2 - 2}{3k^4} E(x,k) + \frac{3k^4 - 5k^2 + 2}{3k^4} F(x,k)$

16. $\displaystyle\int \frac{\sin^5 x \, dx}{\Delta} = \frac{2k^2 \sin^2 x + 3k^2 + 3}{8k^4} \cos x \Delta - \frac{3 + 2k^2 + 3k^4}{8k^5} \ln(k \cos x + \Delta)$

17. $\displaystyle\int \frac{\sin^4 x \cos x \, dx}{\Delta} = -\frac{2k^2 \sin^2 x + 3}{8k^4} \sin x \Delta + \frac{3}{8k^5} \arcsin(k \sin x)$

18.[12] $\displaystyle\int \frac{\sin^3 x \cos^2 x \, dx}{\Delta} = \frac{2k^2 \cos^2 x - k^2 - 3}{8k^4} \cos x \Delta - \frac{k^4 + 2k^2 - 3}{8k^5} \ln(k \cos x + \Delta)$

19. $\displaystyle\int \frac{\sin^2 x \cos^3 x \, dx}{\Delta} = -\frac{2k^2 \cos^2 x + 2k^2 - 3}{8k^4} \sin x \Delta + \frac{4k^2 - 3}{8k^5} \arcsin(k \sin x)$

20. $\displaystyle\int \frac{\sin x \cos^4 x \, dx}{\Delta} = \frac{3 - 5k^2 + 2k^2 \sin^2 x}{8k^4} \cos x \Delta - \frac{3k^4 - 6k^2 + 3}{8k^5} \ln(k \cos x + \Delta)$

21. $\displaystyle\int \frac{\cos^5 x \, dx}{\Delta} = \frac{2k^2 \cos^2 x + 6k^2 - 3}{8k^4} \sin x \Delta + \frac{8k^4 - 8k^2 + 3}{8k^5} \arcsin(k \sin x)$

22. $\displaystyle\int \frac{\sin^6 x \, dx}{\Delta} = \frac{3k^2 \sin^2 x + 4k^2 + 4}{15k^4} \sin x \cos x \Delta$
$\displaystyle\qquad\qquad + \frac{4k^4 + 3k^2 + 8}{15k^6} F(x,k) - \frac{8k^4 + 7k^2 + 8}{15k^6} E(x,k)$

23. $\displaystyle\int \frac{\sin^5 x \cos x \, dx}{\Delta} = -\frac{3k^4 \sin^4 x + 4k^2 \sin^2 x + 8}{15k^6}\Delta$

24.[12] $\displaystyle\int \frac{\sin^4 x \cos x^2 \, dx}{\Delta} = \frac{3k^2 \cos^2 x - 2k^2 - 4}{15k^4} \sin x \cos x \Delta$
$\displaystyle\qquad\qquad + \frac{k^4 + 7k^2 - 8}{15k^6} F(x,k) - \frac{2k^4 + 3k^2 - 8}{15k^6} E(x,k)$

25. $\displaystyle\int \frac{\sin^3 x \cos^3 x \, dx}{\Delta} = \frac{3k^4 \sin^4 x - \left(5k^4 - 4k^2\right)\sin^2 x - 10k^2 + 8}{15k^6}\Delta$

26. $\displaystyle\int \frac{\sin^2 x \cos^4 x \, dx}{\Delta} = -\frac{3k^2 \cos^2 x + 3k^2 - 4}{15k^4} \sin x \cos x \Delta$
$\displaystyle\qquad\qquad + \frac{9k^4 - 17k^2 + 8}{15k^6} F(x,k) - \frac{3k^4 - 13k^2 + 8}{15k^6} E(x,k)$

27. $\displaystyle\int \frac{\sin x \cos^5 x \, dx}{\Delta} = \frac{-3k^4 \cos^4 x + 4k^2 k'^2 \cos^2 x - 8k^4 + 16k^2 - 8}{15k^6}\Delta$

28. $\displaystyle\int \frac{\cos^6 x \, dx}{\Delta} = \frac{3k^2 \cos^2 x + 8k^2 - 4}{15k^4} \sin x \cos x \Delta$
$\displaystyle\qquad\qquad + \frac{15k^6 - 34k^4 + 27k^2 - 8}{15k^6} F(x,k) + \frac{23k^4 - 23k^2 + 8}{15k^6} E(x,k)$

29. $\displaystyle\int \frac{\sin^7 x\,dx}{\Delta} = \frac{8k^4\sin^4 x + 10k^2\left(k^2+1\right)\sin^2 x + 15k^4 + 14k^2 + 15}{48k^6}\cos x\Delta$

$\displaystyle\qquad\qquad - \frac{\left(5k^4 - 2k^2 + 5\right)\left(k^2+1\right)}{16k^7}\ln\left(k\cos x + \Delta\right)$

30. $\displaystyle\int \frac{\sin^6 x\cos x\,dx}{\Delta} = -\frac{8k^4\sin^4 x + 10k^2\sin^2 x + 15}{48k^6}\sin x\Delta + \frac{5}{16k^7}\arcsin\left(k\sin x\right)$

31. $\displaystyle\int \frac{\sin^5 x\cos^2 x\,dx}{\Delta} = \frac{-8k^4\sin^4 x + 2k^2\left(k^2-5\right)\sin^2 x + 3k^4 + 4k^2 - 15}{48k^6}\cos x\Delta$

$\displaystyle\qquad\qquad - \frac{k^6 + k^4 + 3k^2 - 5}{16k^7}\ln\left(k\cos x + \Delta\right)$

32. $\displaystyle\int \frac{\sin^4 x\cos^3 x\,dx}{\Delta} = \frac{8k^4\sin^4 x - 2k^2\left(6k^2-5\right)\sin^2 x - 18k^2 + 15}{48k^6}\sin x\Delta$

$\displaystyle\qquad\qquad + \frac{6k^2-5}{16k^7}\arcsin\left(k\sin x\right)$

33.[12] $\displaystyle\int \frac{\sin^3 x\cos^4 x\,dx}{\Delta} = \frac{8k^4\sin^4 x - 2k^2\left(7k^2-5\right)\sin^2 x + 3k^4 - 22k^2 + 15}{48k^6}\cos x\Delta$

$\displaystyle\qquad\qquad - \frac{k^6 + 3k^4 - 9k^2 + 5}{16k^7}\ln\left(k\cos x + \Delta\right)$

34. $\displaystyle\int \frac{\sin^2 x\cos^5 x\,dx}{\Delta} = \frac{-8k^4\sin^4 x + 2k^2\left(12k^2-5\right)\sin^2 x - 24k^4 + 36k^2 - 15}{48k^6}\sin x\Delta$

$\displaystyle\qquad\qquad + \frac{8k^4 - 12k^2 + 5}{16k^7}\arcsin\left(k\sin x\right)$

35. $\displaystyle\int \frac{\sin x\cos^6 x\,dx}{\Delta} = \frac{-8k^4\sin^4 x + 2k^2\left(13k^2-5\right)\sin^2 x - 33k^4 + 40k^2 - 15}{48k^6}\cos x\Delta$

$\displaystyle\qquad\qquad + \frac{5k'^6}{16k^7}\ln\left(k\cos x + \Delta\right)$

36. $\displaystyle\int \frac{\cos^7 x\,dx}{\Delta} = \frac{8k^4\sin^4 x - 2k^2\left(18k^2-5\right)\sin^2 x + 72k^4 - 54k^2 + 15}{48k^6}\sin x\Delta$

$\displaystyle\qquad\qquad + \frac{16k^6 - 24k^4 + 18k^2 - 5}{16k^7}\arcsin\left(k\sin x\right)$

37. $\displaystyle\int \frac{dx}{\Delta^3} = \frac{1}{k'^2}E(x,k) - \frac{k^2}{k'^2}\frac{\sin x\cos x}{\Delta}$

38. $\displaystyle\int \frac{\sin x\,dx}{\Delta^3} = -\frac{\cos x}{k'^2\Delta}$

39. $\displaystyle\int \frac{\cos x\,dx}{\Delta^3} = \frac{\sin x}{\Delta}$

40.[11] $\displaystyle\int \frac{\sin^2 x\,dx}{\Delta^3} = \frac{1}{k'^2 k^2}E(x,k) - \frac{1}{k^2}F(x,k) - \frac{1}{k'^2}\frac{\sin x\cos x}{\Delta}$

41. $\displaystyle\int \frac{\sin x\cos x\,dx}{\Delta^3} = \frac{1}{k^2\Delta}$

42. $\displaystyle\int \frac{\cos^2 x\,dx}{\Delta^3} = \frac{1}{k^2}F(x,k) - \frac{1}{k^2}E(x,k) + \frac{\sin x\cos x}{\Delta}$

43. $$\int \frac{\sin^3 x \, dx}{\Delta^3} = -\frac{\cos x}{k^2 k'^2 \Delta} + \frac{1}{k^3} \ln\left(k \cos x + \Delta\right)$$

44. $$\int \frac{\sin^2 x \cos x \, dx}{\Delta^3} = \frac{\sin x}{k^2 \Delta} - \frac{1}{k^3} \arcsin\left(k \sin x\right)$$

45. $$\int \frac{\sin x \cos^2 x \, dx}{\Delta^3} = \frac{\cos x}{k^2 \Delta} - \frac{1}{k^3} \ln\left(k \cos x + \Delta\right)$$

46. $$\int \frac{\cos^3 x \, dx}{\Delta^3} = -\frac{k'^2 \sin x}{k^2 \Delta} + \frac{1}{k^3} \arcsin\left(k \sin x\right)$$

47. $$\int \frac{\sin^4 x \, dx}{\Delta^3} = \frac{k'^2 + 1}{k'^2 k^4} E(x,k) - \frac{2}{k^4} F(x,k) - \frac{\sin x \cos x}{k^2 k'^2 \Delta}$$

48. $$\int \frac{\sin^3 x \cos x \, dx}{\Delta^3} = \frac{2 - k^2 \sin^2 x}{k^4 \Delta}$$

49. $$\int \frac{\sin^2 x \cos^2 x \, dx}{\Delta^3} = \frac{2 - k^2}{k^4} F(x,k) - \frac{2}{k^4} E(x,k) + \frac{\sin x \cos x}{k^2 \Delta}$$

50. $$\int \frac{\sin x \cos^3 x \, dx}{\Delta^3} = \frac{k^2 \sin^2 x + k^2 - 2}{k^4 \Delta}$$

51. $$\int \frac{\cos^4 x \, dx}{\Delta^3} = \frac{k'^2 + 1}{k^4} E(x,k) - \frac{2k'^2}{k^4} F(x,k) - \frac{k'^2 \sin x \cos x}{k^2 \Delta}$$

52.[9] $$\int \frac{\sin^5 x \, dx}{\Delta^3} = \frac{k^2 k'^2 \sin^2 x + k^2 - 3}{2 k^4 k'^2 \Delta} \cos x + \frac{k^2 + 3}{2 k^5} \ln\left(k \cos x + \Delta\right)$$

53. $$\int \frac{\sin^4 x \cos x \, dx}{\Delta^3} = \frac{-k^2 \sin^2 x + 3}{2 k^4 \Delta} \sin x - \frac{3}{2 k^5} \arcsin\left(k \sin x\right)$$

54. $$\int \frac{\sin^3 x \cos^2 x \, dx}{\Delta} = \frac{-k^2 \sin^2 x + 3}{2 k^4 \Delta} \cos x + \frac{k^2 - 3}{2 k^5} \ln\left(k \cos x + \Delta\right)$$

55. $$\int \frac{\sin^2 x \cos^3 x \, dx}{\Delta^3} = \frac{k^2 \sin^2 x + 2k^2 - 3}{2 k^4 \Delta} \sin x - \frac{2k^2 - 3}{2 k^5} \arcsin\left(k \sin x\right)$$

56. $$\int \frac{\sin x \cos^4 x \, dx}{\Delta^3} = \frac{k^2 \sin^2 x + 2k^2 - 3}{2 k^4 \Delta} \cos x + \frac{3 k'^2}{2 k^5} \ln\left(k \cos x + \Delta\right)$$

57. $$\int \frac{\cos^5 x \, dx}{\Delta^3} = \frac{-k^2 \sin^2 x + 2k^4 - 4k^2 + 3}{2 k^4 \Delta} \sin x + \frac{4k^2 - 3}{2 k^5} \arcsin\left(k \sin x\right)$$

58. $$\int \frac{dx}{\Delta^5} = \frac{-k^2 \sin x \cos x}{3 k'^2 \Delta^3} - \frac{2 k^2 \left(k'^2 + 1\right) \sin x \cos x}{3 k'^4 \Delta} - \frac{1}{3 k'^2} F(x,k)$$
$$+ \frac{2 \left(k'^2 + 1\right)}{3 k'^4} E(x,k)$$

59. $$\int \frac{\sin x \, dx}{\Delta^5} = \frac{2k^2 \sin^2 x + k^2 - 3}{3 k'^4 \Delta^3} \cos x$$

60. $$\int \frac{\cos x \, dx}{\Delta^5} = \frac{-2k^2 \sin^2 x + 3}{3 \Delta^3} \sin x$$

61. $\displaystyle\int \frac{\sin^2 x \, dx}{\Delta^5} = \frac{k^2+1}{3k'^4 k^2} E(x,k) - \frac{1}{3k'^2 k^2} F(x,k)$
$$+ \frac{k^2\left(k^2+1\right)\sin^2 x - 2}{3k'^4 \Delta^3} \sin x \cos x$$

62. $\displaystyle\int \frac{\sin x \cos x \, dx}{\Delta^5} = \frac{1}{3k^2\Delta^3}$

63.[12] $\displaystyle\int \frac{\cos^2 x \, dx}{\Delta^5} = \frac{1}{3k^2} F(x,k) + \frac{2k^2-1}{3k^2 k'^2} E(x,k) + \frac{k^2\left(2k^2-1\right)\sin^2 x - 3k^2 + 2}{3k'^2 \Delta} \sin x \cos x$

64. $\displaystyle\int \frac{\sin^3 x}{\Delta^5} \, dx = \frac{\left(3k^2-1\right)\sin^2 x - 2}{3k'^4 \Delta^3} \cos x$

65. $\displaystyle\int \frac{\sin^2 x \cos x}{\Delta^5} \, dx = \frac{\sin^3 x}{3\Delta^3}$

66. $\displaystyle\int \frac{\sin x \cos^2 x}{\Delta^5} \, dx = -\frac{\cos^3 x}{3k'^2 \Delta^3}$

67. $\displaystyle\int \frac{\cos^3 x \, dx}{\Delta^5} = \frac{-\left(2k^2+1\right)\sin^2 x + 3}{3\Delta^3} \sin x$

68. $\displaystyle\int \frac{dx}{\Delta \sin x} = -\frac{1}{2} \ln \frac{\Delta + \cos x}{\Delta - \cos x}$

69. $\displaystyle\int \frac{dx}{\Delta \cos x} = -\frac{1}{2k'} \ln \frac{\Delta - k'\sin x}{\Delta + k'\sin x}$

70. $\displaystyle\int \frac{dx}{\Delta \sin^2 x} = \int \frac{1 + \cot^2 x}{\Delta} \, dx = F(x,k) - E(x,k) - \Delta \cot x$

71. $\displaystyle\int \frac{dx}{\Delta \sin x \cos x} = \int (\tan x + \cot x) \frac{dx}{\Delta} = \frac{1}{2} \ln \frac{1-\Delta}{1+\Delta} + \frac{1}{2k'} \ln \frac{\Delta + k'}{\Delta - k'}$

72. $\displaystyle\int \frac{dx}{\Delta \cos^2 x} = \int \left(1 + \tan^2 x\right) \frac{dx}{\Delta} = F(x,k) - \frac{1}{k'^2} E(x,k) + \frac{1}{k'^2} \Delta \tan x$

73. $\displaystyle\int \frac{\sin x \, dx}{\cos x \, \Delta} = \int \tan x \frac{dx}{\Delta} = \frac{1}{2k'} \ln \frac{\Delta + k'}{\Delta - k'}$

74. $\displaystyle\int \frac{\cos x \, dx}{\sin x \, \Delta} = \int \cot x \frac{dx}{\Delta} = \frac{1}{2} \ln \frac{1-\Delta}{1+\Delta}$

75. $\displaystyle\int \frac{dx}{\Delta \sin^3 x} = -\frac{\Delta \cos x}{2\sin^2 x} - \frac{1+k^2}{4} \ln \frac{\Delta + \cos x}{\Delta - \cos x}$

76. $\displaystyle\int \frac{dx}{\Delta \sin^2 x \cos x} = -\frac{\Delta}{\sin x} - \frac{1}{2k'} \ln \frac{\Delta - k'\sin x}{\Delta + k'\sin x}$

77. $\displaystyle\int \frac{dx}{\Delta \sin x \cos^2 x} = \frac{\Delta}{k'^2 \cos x} + \frac{1}{2} \ln \frac{\Delta - \cos x}{\Delta + \cos x}$

78. $\displaystyle\int \frac{dx}{\Delta \cos^3 x} = \frac{\Delta \sin x}{2k'^2 \cos^2 x} + \frac{2k^2-1}{4k'^3} \ln \frac{\Delta - k'\sin x}{\Delta + k'\sin x}$

79. $\displaystyle\int \frac{\sin x \, dx}{\cos^2 x \, \Delta} = \frac{\Delta}{k'^2 \cos x}$

80. $$\int \frac{\cos x}{\sin^2 x} \frac{dx}{\Delta} = -\frac{\Delta}{\sin x}$$

81. $$\int \frac{\sin^2 x}{\cos x} \frac{dx}{\Delta} = \frac{1}{2k'} \ln \frac{\Delta + k' \sin x}{\Delta - k' \sin x} - \frac{1}{k} \arcsin (k \sin x)$$

82. $$\int \frac{\cos^2 x}{\sin x} \frac{dx}{\Delta} = \frac{1}{2} \ln \frac{\Delta + \cos x}{\Delta - \cos x} + \frac{1}{k} \ln (k \cos x + \Delta)$$

83. $$\int \frac{dx}{\Delta \sin^4 x} = \frac{1}{3} \left\{ -\Delta \cot^3 x - \Delta \left(2k^2 + 3\right) \cot x + \left(k^2 + 2\right) F(x, k) - 2 \left(k^2 + 1\right) E(x, k) \right\}$$

84. $$\int \frac{dx}{\Delta \sin^3 x \cos x} = \int \left(\tan x + 2 \cot x + \cot^3 x\right) \frac{dx}{\Delta}$$
$$= -\frac{\Delta}{2 \sin^2 x} + \frac{1}{2k'} \ln \frac{\Delta + k'}{\Delta - k'} - \frac{k^2 + 2}{4} \ln \frac{1 + \Delta}{1 - \Delta}$$

85. $$\int \frac{dx}{\Delta \sin^2 x \cos^2 x} = \int \left(\tan^2 x + 2 + \cot^2 x\right) \frac{dx}{\Delta}$$
$$= \left(\frac{\tan x}{k'^2} - \cot x\right) \Delta + \frac{k^2 - 2}{k'^2} E(x, k) + 2 F(x, k)$$

86.[12] $$\int \frac{dx}{\Delta \sin x \cos^3 x} = \int \left(\cot x + 2 \tan x + \tan^3 x\right) \frac{dx}{\Delta}$$
$$= \frac{\Delta}{2k'^2 \cos^2 x} - \frac{1}{2} \ln \frac{1 + \Delta}{1 - \Delta} + \frac{2 - 3k^2}{4k'^3} \ln \frac{\Delta + k'}{\Delta - k'}$$

87. $$\int \frac{dx}{\Delta \cos^4 x} = \frac{1}{3k'^2} \left\{ \Delta \tan^3 x - \frac{5k^2 - 3}{k'^2} \Delta \tan x - \left(3k^2 - 2\right) F(x, k) \right.$$
$$\left. + \frac{2 \left(2k^2 - 1\right)}{k'^2} E(x, k) \right\}$$

88. $$\int \frac{\sin x}{\cos^3 x} \frac{dx}{\Delta} = \int \tan x \left(1 + \tan^2 x\right) \frac{dx}{\Delta} = \frac{\Delta}{2k'^2 \cos^2 x} - \frac{k^2}{4k'^3} \ln \frac{\Delta + k'}{\Delta - k'}$$

89. $$\int \frac{\cos x}{\sin^3 x} \frac{dx}{\Delta} = -\frac{\Delta}{2 \sin^2 x} - \frac{k^2}{4} \ln \frac{1 + \Delta}{1 - \Delta}$$

90. $$\int \frac{\sin^2 x}{\cos^2 x} \frac{dx}{\Delta} = \int \frac{\tan^2 x}{\Delta} dx = \frac{\Delta}{k'^2} \tan x - \frac{1}{k'^2} E(x, k)$$

91. $$\int \frac{\cos^2 x}{\sin^2 x} \frac{dx}{\Delta} = \int \frac{\cot^2 x}{\Delta} dx = -\Delta \cot x - E(x, k)$$

92. $$\int \frac{\sin^3 x}{\cos x} \frac{dx}{\Delta} = \frac{\Delta}{k^2} + \frac{1}{2k'} \ln \frac{\Delta + k'}{\Delta - k'}$$

93. $$\int \frac{\cos^3 x}{\sin x} \frac{dx}{\Delta} = \frac{\Delta}{k^2} - \frac{1}{2} \ln \frac{1 + \Delta}{1 - \Delta}$$

94.[12] $$\int \frac{dx}{\Delta \sin^5 x} = -\frac{\left[3 \left(1 + k^2\right) \sin^2 x + 2\right]}{8 \sin^4 x} \Delta \cos x + \frac{3k^4 + 2k^2 + 3}{16} \ln \frac{\Delta + \cos x}{\Delta - \cos x}$$

**95.** $\int \dfrac{\mathrm{d}x}{\Delta \sin^4 x \cos x} = -\dfrac{\left(3 + 2k^2\right)\sin^2 x + 1}{3\sin^3 x}\Delta - \dfrac{1}{2k'}\ln\dfrac{\Delta - k'\sin x}{\Delta + k'\sin x}$

**96.** $\int \dfrac{\mathrm{d}x}{\Delta \sin^3 x \cos^2 x} = \dfrac{\left(3 - k^2\right)\sin^2 x - k'^2}{2k'^2 \sin^2 x \cos x}\Delta + \dfrac{k^2 + 3}{4}\ln\dfrac{\Delta - \cos x}{\Delta + \cos x}$

**97.** $\int \dfrac{\mathrm{d}x}{\Delta \sin^2 x \cos^3 x} = \dfrac{\left(3 - 2k^2\right)\sin^2 x - 2k'^2}{2k'^2 \sin x \cos^2 x}\Delta - \dfrac{4k^2 - 3}{4k'^3}\ln\dfrac{\Delta + k'\sin x}{\Delta - k'\sin x}$

**98.** $\int \dfrac{\mathrm{d}x}{\Delta \sin x \cos^4 x} = \dfrac{\left(5k^2 - 3\right)\sin^2 x - 6k^2 + 4}{3k'^4 \cos^3 x}\Delta - \dfrac{1}{2}\ln\dfrac{\Delta + \cos x}{\Delta - \cos x}$

**99.** $\int \dfrac{\mathrm{d}x}{\Delta \cos^5 x} = \dfrac{3\left(2k^2 - 1\right)\sin^2 x - 8k^2 + 5}{8k'^4 \cos^4 x}\Delta \sin x + \dfrac{8k^4 - 8k^2 + 3}{16k'^5}\ln\dfrac{\Delta + k'\sin x}{\Delta - k'\sin x}$

**100.[12]** $\int \dfrac{\sin x}{\cos^4 x}\dfrac{\mathrm{d}x}{\Delta} = -\dfrac{2k^2 \cos^2 x - k'^2}{3k'^4 \cos^3 x}\Delta$

**101.** $\int \dfrac{\cos x}{\sin^4 x}\dfrac{\mathrm{d}x}{\Delta} = -\dfrac{2k^2 \sin^2 x + 1}{3\sin^3 x}\Delta$

**102.** $\int \dfrac{\sin^2 x}{\cos^3 x}\dfrac{\mathrm{d}x}{\Delta} = \dfrac{\Delta \sin x}{2k'^2 \cos^2 x} - \dfrac{1}{4k'^3}\ln\dfrac{\Delta + k'\sin x}{\Delta - k'\sin x}$

**103.[12]** $\int \dfrac{\cos^2 x}{\sin^3 x}\dfrac{\mathrm{d}x}{\Delta} = -\dfrac{\Delta \cos x}{2\sin^2 x} + \dfrac{k'^2}{4}\ln\dfrac{\Delta + \cos x}{\Delta - \cos x}$

**104.** $\int \dfrac{\sin^3 x}{\cos^2 x}\dfrac{\mathrm{d}x}{\Delta} = \dfrac{\Delta}{k'^2 \cos x} + \dfrac{1}{k}\ln\left(k\cos x + \Delta\right)$

**105.** $\int \dfrac{\cos^3 x}{\sin^2 x}\dfrac{\mathrm{d}x}{\Delta} = \dfrac{-\Delta}{\sin x} - \dfrac{1}{k}\arcsin\left(k\sin x\right)$

**106.** $\int \dfrac{\sin^4 x}{\cos x}\dfrac{\mathrm{d}x}{\Delta} = \dfrac{\Delta \sin x}{2k^2} + \dfrac{1}{2k'}\ln\dfrac{\Delta + k'\sin x}{\Delta - k'\sin x} - \dfrac{2k^2 + 1}{2k^3}\arcsin\left(k\sin x\right)$

**107.** $\int \dfrac{\cos^4 x}{\sin x}\dfrac{\mathrm{d}x}{\Delta} = \dfrac{\Delta \cos x}{2k^2} + \dfrac{1}{2}\ln\dfrac{\Delta + \cos x}{\Delta - \cos x} + \dfrac{3k^2 - 1}{2k^3}\ln\left(k\cos x + \Delta\right)$

**2.585**

**1.[12]** $\int \dfrac{\left(a + \sin x\right)^{p+3}\,\mathrm{d}x}{\Delta}$

$$= \dfrac{1}{(p + 2)k^2}\left[\left(a + \sin x\right)^p \cos x \Delta\right.$$

$$+ 2(2p + 3)ak^2 \int \dfrac{\left(a + \sin x\right)^{p+2}\,\mathrm{d}x}{\Delta} + (p + 1)\left(1 + k^2 - 6a^2k^2\right)\int \dfrac{\left(a + \sin x\right)^{p+1}\,\mathrm{d}x}{\Delta}$$

$$- a(2p + 1)\left(1 + k^2 - 2a^2k^2\right)\int \dfrac{\left(a + \sin x\right)^p\,\mathrm{d}x}{\Delta}$$

$$\left. - p\left(1 - a^2\right)\left(1 - a^2k^2\right)\int \dfrac{\left(a + \sin x\right)^{p-1}\,\mathrm{d}x}{\Delta}\right]$$

$$\left[p \ne -2, \quad a \ne \pm 1, \quad a \ne \pm\dfrac{1}{k}\right]$$

For $p = n$ a natural number, this integral can be reduced to the following three integrals:

2.
$$\int \frac{a + \sin x}{\Delta}\, dx = a\, F(x, k) + \frac{1}{2k}\ln\frac{\Delta - k\cos x}{\Delta + k\cos x}$$

3.
$$\int \frac{(a + \sin x)^2}{\Delta}\, dx = \frac{1 + k^2 a^2}{k^2}\, F(x, k) - \frac{1}{k^2}\, E(x, k) + \frac{a}{k}\ln\frac{\Delta - k\cos x}{\Delta + k\cos x}$$

4.[12]
$$\int \frac{dx}{(a + \sin x)\,\Delta} = \frac{1}{a}\, \Pi\left(x, -\frac{1}{a^2}, k\right) - \int \frac{\sin x\, dx}{(a^2 - \sin^2 x)\,\Delta},$$

where

5.
$$\int \frac{\sin x\, dx}{(a^2 - \sin^2 x)\,\Delta} = \frac{-1}{2\sqrt{(1 - a^2)(1 - a^2 k^2)}}\ln\frac{\sqrt{1 - a^2}\,\Delta - \sqrt{1 - k^2 a^2}\,\cos x}{\sqrt{1 - a^2}\,\Delta + \sqrt{1 - k^2 a^2}\,\cos x}$$

**2.586**

1.
$$\int \frac{dx}{(a + \sin x)^n\,\Delta} = \frac{1}{(n-1)(1 - a^2)(1 - a^2 k^2)}\left[-\frac{\cos x\,\Delta}{(a + \sin x)^{n-1}}\right.$$
$$-(2n - 3)\left(1 + k^2 - 2a^2 k^2\right) a\int \frac{dx}{(a + \sin x)^{n-1}\,\Delta}$$
$$-(n - 2)\left(6a^2 k^2 - k^2 - 1\right)\int \frac{dx}{(a + \sin x)^{n-2}\,\Delta}$$
$$\left.-(10 - 4n)ak^2\int \frac{dx}{(a + \sin x)^{n-3}\,\Delta} - (n - 3)k^2\int \frac{dx}{(a + \sin x)^{n-4}\,\Delta}\right]$$
$$\left[n \neq 1, \quad a \neq \pm 1, \quad a \neq \pm\frac{1}{k}\right]$$

This integral can be reduced to the integrals:

2.
$$\int \frac{dx}{(a + \sin x)^2\,\Delta} = \frac{1}{(1 - a^2)(1 - a^2 k^2)}\left[-\frac{\cos x\,\Delta}{a + \sin x} - a\left(1 + k^2 - 2a^2 k^2\right)\int \frac{dx}{(a + \sin x)\,\Delta}\right.$$
$$\left.- 2ak^2\int \frac{(a + \sin x)\, dx}{\Delta} + k^2\int \frac{(a + \sin x)^2\, dx}{\Delta}\right]$$
$$(\text{see } \mathbf{2.585}\ 2, 3, 4)$$

3.
$$\int \frac{dx}{(a + \sin x)^3\,\Delta} = \frac{1}{2(1 - a^2)(1 - a^2 k^2)}\left[-\frac{\cos x\,\Delta}{(a + \sin x)^2} - 3a\left(1 + k^2 - 2a^2 k^2\right)\int \frac{dx}{(a + \sin x)^2\,\Delta}\right.$$
$$\left.-\left(6a^2 k^2 - k^2 - 1\right)\int \frac{dx}{(a + \sin x)\,\Delta} + 2ak^2\, F(x, k)\right]$$
$$(\text{see } \mathbf{2.585}\ 4 \text{ and } \mathbf{2.586}\ 2)$$

For $a = \pm 1$, we have:

4.
$$\int \frac{dx}{(1 \pm \sin x)^n\,\Delta} = \frac{1}{(2n - 1)k'^2}\left[\mp\frac{\cos x\,\Delta}{(1 \pm \sin x)^n} + (n - 1)\left(1 - 5k^2\right)\int \frac{dx}{(1 \pm \sin x)^{n-1}\,\Delta}\right.$$
$$\left.+ 2(2n - 3)k^2\int \frac{dx}{(1 \pm \sin x)^{n-2}\,\Delta} - (n - 2)k^2\int \frac{dx}{(1 \pm \sin x)^{n-3}\,\Delta}\right]$$
$$\text{GU (241)(6a)}$$

This integral can be reduced to the integrals

5. $\displaystyle\int \frac{dx}{(1 \pm \sin x)\,\Delta} = \frac{\mp \cos x \Delta}{k'^2\,(1 \pm \sin x)} + F(x,k) - \frac{1}{k'^2}\,E(x,k)$        GU (241)(6c)

6. $\displaystyle\int \frac{dx}{(1 \pm \sin x)^2\,\Delta} = \frac{1}{3k'^4}\left\{ \mp \frac{k'^2 \cos x \Delta}{(1 \pm \sin x)^2} \mp \frac{(1 - 5k^2)\cos x \Delta}{1 \pm \sin x} \right.$

$\left. + (1 - 3k^2)\,k'^2\,F(x,k) - (1 - 5k^2)\,E(x,k) \right\}$

GU (241)(6b)

For $a = \pm\dfrac{1}{k}$, we have

7. $\displaystyle\int \frac{dx}{(1 \pm k \sin x)^n\,\Delta} = \frac{1}{(2n-1)k'^2}\left[ \pm \frac{k \cos x \Delta}{(1 \pm k \sin x)^n} + (n-1)\,(5 - k^2)\int \frac{dx}{(1 \pm k \sin x)^{n-1}\,\Delta} \right.$

$\left. - 2(2n-3)\int \frac{dx}{(1 \pm k \sin x)^{n-2}\,\Delta} + (n-2)\int \frac{dx}{(1 \pm k \sin x)^{n-3}\,\Delta} \right]$

GU (241)(7a)

This integral can be reduced to the integrals

8. $\displaystyle\int \frac{dx}{(1 \pm k \sin x)\,\Delta} = \pm \frac{k \cos x \Delta}{k'^2\,(1 \pm k \sin x)} + \frac{1}{k'^2}\,E(x,k)$        GU (241)(7b)

9. $\displaystyle\int \frac{dx}{(1 \pm k \sin x)^2\,\Delta} = \frac{1}{3k'^4}\left[ \pm \frac{kk'^2 \cos x \Delta}{(1 \pm k \sin x)^2} \pm \frac{k\,(5 - k^2)\cos x \Delta}{1 \pm k \sin x} \right.$

$\left. - 2k'^2\,F(x,k) + (5 - k^2)\,E(x,k) \right]$

GU(241)(7c)

**2.587**

1.[12] $\displaystyle\int \frac{(b + \cos x)^{p+3}}{\Delta}\,dx = \frac{1}{(p+2)k^2}\left[ (b + \cos x)^p \sin x \Delta + 2(2p+3)bk^2 \int \frac{(b + \cos x)^{p+2}\,dx}{\Delta} \right.$

$- (p+1)\left(k'^2 - k^2 + 6b^2k^2\right)\int \frac{(b + \cos x)^{p+1}\,dx}{\Delta}$

$+ (2p+1)b\left(k'^2 - k^2 + b^2k^2\right)\int \frac{(b + \cos x)^p\,dx}{\Delta}$

$\left. + p\,(1 - b^2)\left(k'^2 + k^2b^2\right)\int \frac{(b + \cos x)^{p-1}\,dx}{\Delta} \right]$

$\left[ p \neq -2, \quad b \neq \pm 1, \quad b \neq \dfrac{ik'}{k} \right]$

For $p = n$ a natural number, this integral can be reduced to the following three integrals:

2. $\displaystyle\int \frac{b + \cos x}{\Delta}\,dx = b\,F(x,k) + \frac{1}{k}\arcsin(k \sin x)$

3. $\displaystyle\int \frac{(b + \cos x)^2}{\Delta}\,dx = \frac{b^2k^2 - k'^2}{k^2}\,F(x,k) + \frac{1}{k^2}\,E(x,k) + \frac{2b}{k}\arcsin(k \sin x)$

4. $\displaystyle\int \frac{dx}{(b + \cos x)\,\Delta} = \frac{b}{b^2 - 1}\,\Pi\left(x, \frac{1}{b^2 - 1}, k\right) + \int \frac{\cos x\,dx}{(1 - b^2 - \sin^2 x)\,\Delta},$

where

**5.**
$$\int \frac{\cos x \, dx}{\left(1 - b^2 - \sin^2 x\right) \Delta} = \frac{1}{2\sqrt{\left(1 - b^2\right)\left(k'^2 + k^2 b^2\right)}} \ln \frac{\sqrt{1 - b^2}\Delta + k\sqrt{k'^2 + k^2 b^2} \sin x}{\sqrt{1 - b^2}\Delta - k\sqrt{k'^2 + k^2 b^2} \sin x}$$

**2.588**

**1.**
$$\int \frac{dx}{\left(b + \cos x\right)^n \Delta} = \frac{1}{\left(n-1\right)\left(1 - b^2\right)\left(k'^2 + b^2 k^2\right)} \left[ \frac{-k'^2 \sin x \Delta}{\left(b + \cos x\right)^{-1}} \right.$$
$$- \left(2n - 3\right)\left(1 - 2k^2 + 2b^2 k^2\right) b \int \frac{dx}{\left(b + \cos x\right)^{n-1} \Delta}$$
$$- \left(n - 2\right)\left(2k^2 - 1 - 6b^2 k^2\right) \int \frac{dx}{\left(b + \cos x\right)^{n-2} \Delta}$$
$$\left. - \left(4n - 10\right)bk^2 \int \frac{dx}{\left(b + \cos x\right)^{n-3} \Delta} + \left(n - 3\right)k^2 \int \frac{dx}{\left(b + \cos x\right)^{n-4} \Delta} \right]$$
$$\left[ n \neq 1, \quad b \neq \pm 1, \quad b \neq \pm\frac{ik'}{k} \right]$$

This integral can be reduced to the following integrals:

**2.**
$$\int \frac{dx}{\left(b + \cos x\right)^2 \Delta} = \frac{1}{\left(1 - b^2\right)\left(k'^2 + b^2 k^2\right)} \left[ \frac{-k'^2 \sin x \Delta}{b + \cos x} - \left(1 - 2k^2 + 2b^2 k^2\right) b \int \frac{dx}{\left(b + \cos x\right) \Delta} \right.$$
$$\left. + 2bk^2 \int \frac{b + \cos x}{\Delta} dx - k^2 \int \frac{\left(b + \cos x\right)^2}{\Delta} dx \right]$$
$$\text{(see **2.587** 2, 3, 4)}$$

**3.**
$$\int \frac{dx}{\left(b + \cos x\right)^3 \Delta} = \frac{1}{2\left(1 - b^2\right)\left(k'^2 + b^2 k^2\right)} \left[ \frac{-k'^2 \sin x \Delta}{\left(b + \cos x\right)^2} \right.$$
$$- 3b\left(1 - 2k^2 + 2k^2 b^2\right) \int \frac{dx}{\left(b + \cos x\right)^2 \Delta}$$
$$\left. - \left(2k^2 - 1 - 6b^2 k^2\right) \int \frac{dx}{\left(b + \cos x\right) \Delta} - 2bk^2 F(x, k) \right]$$
$$\text{(see **2.588** 2 and **2.587** 4)}$$

**2.589**

**1.**
$$\int \frac{\left(c + \tan x\right)^{p+3} dx}{\Delta} = \frac{1}{\left(p + 2\right)k'^2} \left[ \frac{\left(c + \tan x\right)^p \Delta}{\cos^2 x} + 2(2n + 3)ck'^2 \int \frac{\left(c + \tan x\right)^{p+2} dx}{\Delta} \right.$$
$$- \left(p + 1\right)\left(1 + k'^2 + 6c^2 k'^2\right) \int \frac{\left(c + \tan x\right)^{p+1} dx}{\Delta}$$
$$+ \left(2p + 1\right)c\left(1 + k'^2 + 2c^2 k'^2\right) \int \frac{\left(c + \tan x\right)^p dx}{\Delta}$$
$$\left. - p\left(1 + c^2\right)\left(1 + k'^2 c^2\right) \int \frac{\left(c + \tan x\right)^{p-1} dx}{\Delta} \right]$$
$$\left[ p \neq -2 \right]$$

For $p = n$ a natural number, this integral can be reduced to the following three integrals:

2.  $\displaystyle\int \frac{c + \tan x}{\Delta}\, dx = c\, F(x, k) + \frac{1}{2k'} \ln \frac{\Delta + k'}{\Delta - k'}$

3.  $\displaystyle\int \frac{(c + \tan x)^2}{\Delta}\, dx = \frac{1}{k'^2} \tan x \Delta + c^2\, F(x, k) - \frac{1}{k'^2}\, E(x, k) + \frac{c}{k'} \ln \frac{\Delta + k'}{\Delta - k'}$

4.  $\displaystyle\int \frac{dx}{(c + \tan x)\,\Delta} = \frac{c}{1 + c^2}\, F(x, k) + \frac{1}{c\,(1 + c^2)}\, \Pi\left(x, -\frac{1 + c^2}{c^2}, k\right)$
    $\displaystyle\qquad\qquad - \int \frac{\sin x \cos x\, dx}{\left[c^2 - (1 + c^2)\sin^2 x\right]\Delta},$

where

5.  $\displaystyle\int \frac{\sin x \cos x\, dx}{\left[c^2 - (1 + c^2)\sin^2 x\right]\Delta} = \frac{1}{2\sqrt{(1 + c^2)\left(1 + c^2 k'^2\right)}} \ln \frac{\sqrt{1 + c^2 k'^2} + \sqrt{1 + c^2}\,\Delta}{\sqrt{1 + c^2 k'^2} - \sqrt{1 + c^2}\,\Delta}$

**2.591**

1.  $\displaystyle\int \frac{dx}{(c + \tan x)^n\,\Delta} = \frac{1}{(n - 1)\left(1 + c^2\right)\left(1 + k'^2 c^2\right)} \left[ -\frac{\Delta}{(c + \tan x)^{n-1} \cos^2 x} \right.$
    $\displaystyle\qquad\qquad + (2n - 3)c\left(1 + k'^2 + 2c^2 k'^2\right) \int \frac{dx}{(c + \tan x)^{n-1}\,\Delta}$
    $\displaystyle\qquad\qquad - (n - 2)\left(1 + k'^2 + 6c^2 k'^2\right) \int \frac{dx}{(c + \tan x)^{n-2}\,\Delta}$
    $\displaystyle\qquad\qquad \left. + (4n - 10)c k'^2 \int \frac{dx}{(c + \tan x)^{n-3}\,\Delta} - (n - 3)k'^2 \int \frac{dx}{(c + \tan x)^{n-4}\,\Delta} \right]$

This integral can be reduced to the integrals:

2.  $\displaystyle\int \frac{dx}{(c + \tan x)^2\,\Delta} = \frac{1}{\left(1 + c^2\right)\left(1 + k'^2 c^2\right)} \left[ \frac{-\Delta}{(c + \tan x)\cos^2 x} \right.$
    $\displaystyle\qquad\qquad + c\left(1 + k'^2 + 2c^2 k'^2\right) \int \frac{dx}{(c + \tan x)\,\Delta}$
    $\displaystyle\qquad\qquad \left. - 2c k'^2 \int \frac{c + \tan x}{\Delta}\, dx + k'^2 \int \frac{(c + \tan x)^2}{\Delta}\, dx \right]$
    $\qquad\qquad\qquad\qquad\qquad\qquad\qquad\qquad\qquad\qquad\text{(see } \mathbf{2.589}\ 2, 3, 4)$

3.  $\displaystyle\int \frac{dx}{(c + \tan x)^3\,\Delta} = \frac{1}{2\left(1 + c^2\right)\left(1 + k'^2 c^2\right)} \left[ \frac{-\Delta}{(c + \tan x)^2 \cos^2 x} \right.$
    $\displaystyle\qquad\qquad + 3c\left(1 + k'^2 + 2c^2 k'^2\right) \int \frac{dx}{(c + \tan x)^2\,\Delta}$
    $\displaystyle\qquad\qquad \left. - \left(1 + k'^2 + 6c^2 k'^2\right) \int \frac{dx}{(c + \tan x)\,\Delta} + 2c k'^2\, F(x, k) \right]$
    $\qquad\qquad\qquad\qquad\qquad\qquad\qquad\qquad\qquad\text{(see } \mathbf{2.591}\ 2\ \text{and}\ \mathbf{2.589}\ 4)$

**2.592**

1.      $$P_n = \int \frac{\left(a + \sin^2 x\right)^n}{\Delta}\, dx$$

The recursion formula

$$P_{n+2} = \frac{1}{(2n+3)k^2}\left\{ \left(a + \sin^2 x\right)^n \sin x \cos x \Delta + (2n+2)\left(1 + k^2 + 3ak^2\right) P_{n+1} \right.$$

$$\left. - (2n+1)\left[1 + 2a\left(1 + k^2\right) + 3a^2 k^2\right] P_n + 2na(1+a)\left(1 + k^2 a\right) P_{n-1}\right\}$$

reduces this integral (for $n$ an integer) to the integrals

2.      $P_1$                                                                                         (see **2.584** 1 and **2.584** 4)

3.      $P_0$                                                                                         (see **2.584** 1)

4.      $$P_{-1} = \int \frac{dx}{\left(a + \sin^2 x\right)\Delta} = \frac{1}{a}\Pi\left(x, \frac{1}{a}, k\right)$$

For $a = 0$

5.      $$\int \frac{dx}{\sin^2 x \Delta}$$                                        (see **2.584** 70)                                        H (124)a

6.      $$T_n = \int \frac{dx}{\left(h + g\sin^2 x\right)^n \Delta}$$

can be calculated by means of the recursion formula:

$$T_{n-3} = \frac{1}{(2n-5)k^2}\left\{ \frac{-g^2 \sin x \cos x \Delta}{\left(h + g\sin^2 x\right)^{n-1}} + 2(n-2)\left[g\left(1 + k^2\right) + 3hk^2\right] T_{n-2} \right.$$

$$\left. - (2n-3)\left[g^2 + 2hg\left(1 + k^2\right) + 3h^2 k^2\right] T_{n-1} + 2(n-1)h(g+h)\left(g + hk^2\right) T_n\right\}$$

**2.593**

1.[12]      $$Q_n = \int \frac{\left(b + \cos^2 x\right)^n}{\Delta}\, dx$$

The recursion formula

$$Q_{n+2} = \frac{1}{(2n+3)k^2}\left\{ \left(b + \cos^2 x\right)^n \Delta \sin x \cos x - (2n+2)\left(1 - 2k^2 - 3bk^2\right) Q_{n+1} \right.$$

$$\left. + (2n+1)\left[k'^2 + 2b\left(k'^2 - k^2\right) - 3b^2 k^2\right] Q_n - 2nb(1-b)\left(k'^2 - k^2 b\right) Q_{n-1}\right\}$$

reduces this integral (for $n$ an integer) to the integrals:

2.      $Q_1$                                                                                         (see **2.584** 1 and **2.584** 6)

3.      $Q_0$                                                                                         (see **2.584** 1)

4.  $\quad Q_{-1} = \displaystyle\int \frac{dx}{(b + \cos^2 x)\,\Delta} = \frac{1}{b+1}\,\Pi\left(x, -\frac{1}{b+1}, k\right)$

For $b = 0$

5.  $\quad \displaystyle\int \frac{dx}{\cos^2 x \Delta}$                                    (see **2.584** 72)                                    H (123)

**2.594**

1.[12]  $\quad R_n = \displaystyle\int \frac{\left(c + \tan^2 x\right)^n\,dx}{\Delta}$

The recursion formula

$$R_{n+2} = \frac{1}{(2n+3)k'^2}\left\{\frac{\left(c + \tan^2 x\right)^n \tan x \Delta}{\cos^2 x} - (2n+2)\left(1 + k'^2 - 3ck'^2\right)R_{n+1}\right.$$

$$\left. + (2n+1)\left[1 - 2c\left(1 + k'^2\right) + 3c^2 k'^2\right]R_n + 2nc(1-c)\left(1 - k'^2 c\right)R_{n-1}\right\}$$

reduces this integral (for $n$ an integer) to the integrals:

2.  $\quad R_1$                                    (see **2.584** 1 and **2.584** 90)

3.  $\quad R_0$                                    (see **2.584** 1)

4.  $\quad R_{-1} = \displaystyle\int \frac{dx}{(c + \tan^2 x)\,\Delta} = \frac{1}{c-1}\,F(x, k) + \frac{1}{c(1-c)}\,\Pi\left(x, \frac{1-c}{c}, k\right)$

For $c = 0$ see **2.582** 5.

**2.595**  Integrals of the type $\displaystyle\int R\left(\sin x, \cos x, \sqrt{1 - p^2 \sin^2 x}\right)dx$ for $p^2 > 1$.

**Notation:** $\alpha = \arcsin\left(p \sin x\right)$.

Basic formulas

1.  $\quad \displaystyle\int \frac{dx}{\sqrt{1 - p^2 \sin^2 x}} = \frac{1}{p}\,F\left(\alpha, \frac{1}{p}\right)$                                    $\left[p^2 > 1\right]$                                    BY (283.00)

2.  $\quad \displaystyle\int \sqrt{1 - p^2 \sin^2 x}\,dx = p\,E\left(\alpha, \frac{1}{p}\right) - \frac{p^2 - 1}{p}\,F\left(\alpha, \frac{1}{p}\right)$

$\left[p^2 > 1\right]$                                    BY (283.03)

3.  $\quad \displaystyle\int \frac{dx}{\left(1 - r^2 \sin^2 x\right)\sqrt{1 - p^2 \sin^2 x}} = \frac{1}{p}\,\Pi\left(\alpha, \frac{r^2}{p^2}, \frac{1}{p}\right)$    $\left[p^2 > 1\right]$    BY (283.02)

To evaluate integrals of the form $\displaystyle\int R\left(\sin x, \cos x, \sqrt{1 - p^2 \sin^2 x}\right)dx$ for $p^2 > 1$, we may use formulas **2.583** and **2.584**, making the following modifications in them. We replace

(1)  $\quad k$ with $p$;

(2)  $\quad k'^2$ with $1 - p^2$;

(3)     $F(x, k)$ with $\dfrac{1}{p} F\left(\alpha, \dfrac{1}{p}\right)$;

(4)     $E(x, k)$ with $p E\left(\alpha, \dfrac{1}{p}\right) - \dfrac{p^2 - 1}{p} F\left(\alpha, \dfrac{1}{p}\right)$.

For example (see **2.584** 15):

**2.596**

1.[10] 
$$\int \frac{\cos^4 x \, dx}{\sqrt{1 - p^2 \sin^2 x}} = \frac{\sin x \cos x \sqrt{1 - p^2 \sin^2 x}}{3p^2} + \frac{4p^2 - 2}{3p^4}\left[p E\left(\alpha, \frac{1}{p}\right)\right.$$

$$\left. - \frac{p^2 - 1}{p} F\left(\alpha, \frac{1}{p}\right)\right] + \frac{2 - 5p^2 + 3p^4}{3p^4} \cdot \frac{1}{p} F\left(\alpha, \frac{1}{p}\right)$$

$$= \frac{\sin x \cos x \sqrt{1 - p^2 \sin^2 x}}{3p^2} - \frac{p^2 - 1}{3p^3} F\left(\alpha, \frac{1}{p}\right) + \frac{4p^2 - 2}{3p^3} E\left(\alpha, \frac{1}{p}\right) \quad [p^2 > 1]$$

For example (see **2.583** 36):

2. 
$$\int \frac{\sqrt{1 - p^2 \sin^2 x}}{\cos^2 x} \, dx = \tan x \sqrt{1 - p^2 \sin^2 x} + \frac{1}{p} F\left(\alpha, \frac{1}{p}\right)$$

$$- \left[p E\left(\alpha, \frac{1}{p}\right) - \frac{p^2 - 1}{p} F\left(\alpha, \frac{1}{p}\right)\right]$$

$$= p\left[F\left(\alpha, \frac{1}{p}\right) - E\left(\alpha, \frac{1}{p}\right)\right] + \tan x \sqrt{1 - p^2 \sin^2 x}$$

$$[p^2 > 1]$$

For example (see **2.584** 37):

3. 
$$\int \frac{dx}{\sqrt{(1 - p^2 \sin^2 x)^3}} = \frac{-1}{p^2 - 1}\left[p E\left(\alpha, \frac{1}{p}\right) - \frac{p^2 - 1}{p} F\left(\alpha, \frac{1}{p}\right)\right]$$

$$- \frac{p^2}{1 - p^2} \cdot \frac{\sin x \cos x}{\sqrt{1 - p^2 \sin^2 x}} =$$

$$\frac{p^2}{p^2 - 1} \cdot \frac{\sin x \cos x}{\sqrt{1 - p^2 \sin^2 x}} + \frac{1}{p} F\left(\alpha, \frac{1}{p}\right) - \frac{p}{p^2 - 1} E\left(\alpha, \frac{1}{p}\right) \quad [p^2 > 1]$$

**2.597**    Integrals of the form $\displaystyle\int R\left(\sin x, \cos x, \sqrt{1 + p^2 \sin^2 x}\right) dx$

**Notation:** $\alpha = \arcsin\left(\dfrac{\sqrt{1 + p^2} \sin x}{\sqrt{1 + p^2 \sin^2 x}}\right)$

Basic formulas

1.    $\displaystyle\int \frac{dx}{\sqrt{1 + p^2 \sin^2 x}} = \frac{1}{\sqrt{1 + p^2}} F\left(\alpha, \frac{p}{\sqrt{1 + p^2}}\right)$        BY (282.00)

2.    $\displaystyle\int \sqrt{1 + p^2 \sin^2 x} \, dx = \sqrt{1 + p^2} \, E\left(\alpha, \frac{p}{\sqrt{1 + p^2}}\right) - p^2 \frac{\sin x \cos x}{\sqrt{1 + p^2 \sin^2 x}}$        BY (282.03)

3.  $\dfrac{\sqrt{1+p^2 \sin^2 x}\, dx}{1+(p^2-r^2p^2-r^2)\sin^2 x} = \dfrac{1}{\sqrt{1+p^2}}\,\Pi\left(\alpha, r^2, \dfrac{p}{\sqrt{1+p^2}}\right)$    BY (282.02)

4.  $\displaystyle\int \dfrac{\sin x\, dx}{\sqrt{1+p^2 \sin^2 x}} = -\dfrac{1}{p}\arcsin\left(\dfrac{p\cos x}{\sqrt{1+p^2}}\right)$

5.  $\displaystyle\int \dfrac{\cos x\, dx}{\sqrt{1+p^2 \sin^2 x}} = \dfrac{1}{p}\ln\left(p\sin x + \sqrt{1+p^2 \sin^2 x}\right)$

6.  $\displaystyle\int \dfrac{dx}{\sin x\sqrt{1+p^2 \sin^2 x}} = \dfrac{1}{2}\ln\dfrac{\sqrt{1+p^2 \sin^2 x}-\cos x}{\sqrt{1+p^2 \sin^2 x}+\cos x}$

7.  $\displaystyle\int \dfrac{dx}{\cos x\sqrt{1+p^2 \sin^2 x}} = \dfrac{1}{2\sqrt{1+p^2}}\ln\dfrac{\sqrt{1+p^2 \sin^2 x}+\sqrt{1+p^2}\sin x}{\sqrt{1+p^2 \sin^2 x}-\sqrt{1+p^2}\sin x}$

8.  $\displaystyle\int \dfrac{\tan x\, dx}{\sqrt{1+p^2 \sin^2 x}} = \dfrac{1}{2\sqrt{1+p^2}}\ln\dfrac{\sqrt{1+p^2 \sin^2 x}+\sqrt{1+p^2}}{\sqrt{1+p^2 \sin^2 x}-\sqrt{1+p^2}}$

9.  $\displaystyle\int \dfrac{\cot x\, dx}{\sqrt{1+p^2 \sin^2 x}} = \dfrac{1}{2}\ln\dfrac{1-\sqrt{1+p^2 \sin^2 x}}{1+\sqrt{1+p^2 \sin^2 x}}$

**2.598**  To calculate integrals of the form $\int R\left(\sin x, \cos x, \sqrt{1+p^2 \sin^2 x}\right)\, dx$, we may use formulas **2.583** and **2.584**, making the following modifications in them. We replace

(1)  $k^2$ with $-p^2$;

(2)  $k'^2$ with $1+p^2$;

(3)  $F(x,k)$ with $\dfrac{1}{\sqrt{1+p^2}}\, F\left(\alpha, \dfrac{p}{\sqrt{1+p^2}}\right)$;

(4)  $E(x,k)$ with $\sqrt{1+p^2}\, E\left(\alpha, \dfrac{p}{\sqrt{1+p^2}}\right) - p^2\dfrac{\sin x \cos x}{\sqrt{1+p^2 \sin^2 x}}$;

(5)  $\dfrac{1}{k}\ln(k\cos x + \Delta)$ with $\dfrac{1}{p}\arcsin\dfrac{p\cos x}{\sqrt{1+p^2}}$;

(6)  $\dfrac{1}{k}\arcsin(k\sin x)$ with $\dfrac{1}{p}\ln\left(p\sin x + \sqrt{1+p^2 \sin^2 x}\right)$.

For example (see **2.584** 90):

1.  $\displaystyle\int \dfrac{\tan^2 x\, dx}{\sqrt{1+p^2 \sin^2 x}} = \dfrac{1}{(1+p^2)}\left[\tan x\sqrt{1+p^2 \sin^2 x}\right.$

$\left. -\sqrt{1+p^2}\, E\left(\alpha, \dfrac{p}{\sqrt{1+p^2}}\right) + p^2\dfrac{\sin x \cos x}{\sqrt{1+p^2 \sin^2 x}}\right]$

$= -\dfrac{1}{\sqrt{1+p^2}}\, E\left(\alpha, \dfrac{p}{\sqrt{1+p^2}}\right) + \dfrac{\tan x}{\sqrt{1+p^2 \sin^2 x}}$

For example (see **2.584** 37):

2. 
$$\int \frac{dx}{\sqrt{\left(1 + p^2 \sin^2 x\right)^3}} = \frac{1}{\sqrt{1 + p^2}} E\left(\alpha, \frac{p}{\sqrt{1 + p^2}}\right)$$

**2.599**    Integrals of the form $\int R\left(\sin x, \cos x, \sqrt{a^2 \sin^2 x - 1}\right) dx \qquad [a^2 > 1]$

**Notation:** $\alpha = \arcsin\left(\dfrac{a \cos x}{\sqrt{a^2 - 1}}\right).$

**Basic formulas:**

1. 
$$\int \frac{dx}{\sqrt{a^2 \sin^2 x - 1}} = -\frac{1}{a} F\left(\alpha, \frac{\sqrt{a^2 - 1}}{a}\right) \qquad\qquad [a^2 > 1] \qquad\qquad \text{BY (285.00)a}$$

2. 
$$\int \sqrt{a^2 \sin^2 x - 1}\, dx = \frac{1}{a} F\left(\alpha, \frac{\sqrt{a^2 - 1}}{a}\right) - a E\left(\alpha, \frac{\sqrt{a^2 - 1}}{a}\right)$$
$$[a^2 > 1] \qquad\qquad \text{BY (285.06)a}$$

3. 
$$\int \frac{dx}{\left(1 - r^2 \sin^2 x\right) \sqrt{a^2 \sin^2 x - 1}} = \frac{1}{a\left(r^2 - 1\right)} \Pi\left(\alpha, \frac{r^2\left(a^2 - 1\right)}{a^2\left(r^2 - 1\right)}, \frac{\sqrt{a^2 - 1}}{a}\right)$$
$$[a^2 > 1, \quad r^2 > 1] \qquad\qquad \text{BY (285.02)a}$$

4. 
$$\int \frac{\sin x\, dx}{\sqrt{a^2 \sin^2 x - 1}} = -\frac{\alpha}{a} \qquad\qquad [a^2 > 1]$$

5. 
$$\int \frac{\cos x\, dx}{\sqrt{a^2 \sin^2 x - 1}} = \frac{1}{a} \ln\left(a \sin x + \sqrt{a^2 \sin^2 x - 1}\right) \qquad [a^2 > 1]$$

6. 
$$\int \frac{dx}{\sin x \sqrt{a^2 \sin^2 x - 1}} = -\arctan \frac{\cos x}{\sqrt{a^2 \sin^2 x - 1}} \qquad [a^2 > 1]$$

7. 
$$\int \frac{dx}{\cos x \sqrt{a^2 \sin^2 x - 1}} = \frac{1}{2\sqrt{a^2 - 2}} \ln \frac{\sqrt{a^2 - 1} \sin x + \sqrt{a^2 \sin^2 x - 1}}{\sqrt{a^2 - 1} \sin x - \sqrt{a^2 \sin^2 x - 1}}$$
$$[a^2 > 1]$$

8. 
$$\int \frac{\tan x\, dx}{\sqrt{a^2 \sin^2 x - 1}} = \frac{1}{2\sqrt{a^2 - 1}} \ln \frac{\sqrt{a^2 - 1} + \sqrt{a^2 \sin^2 x - 1}}{\sqrt{a^2 - 1} - \sqrt{a^2 \sin^2 x - 1}}$$
$$[a^2 > 1]$$

9. 
$$\int \frac{\cot x\, dx}{\sqrt{a^2 \sin^2 x - 1}} = -\arcsin\left(\frac{1}{a \sin x}\right) \qquad\qquad [a^2 > 1]$$

**2.611**    To calculate integrals of the type $\int R\left(\sin x, \cos x, \sqrt{a^2 \sin^2 x - 1}\right) dx$ for $a^2 > 1$, we may use formulas **2.583** and **2.584**. In doing so, we should follow the procedure outlined below:

(1)      In the right members of these formulas, the following functions should be replaced with integrals equal to them:

$$F(x,k) \quad \text{should be replaced with} \quad \int \frac{dx}{\Delta}$$

$$E(x,k) \quad \text{should be replaced with} \quad \int \Delta\, dx$$

$$-\frac{1}{k}\ln(k\cos x + \Delta) \quad \text{should be replaced with} \quad \int \frac{\sin x\, dx}{\Delta}$$

$$\frac{1}{k}\arcsin(k\sin x) \quad \text{should be replaced with} \quad \int \frac{\cos x\, dx}{\Delta}$$

$$\frac{1}{2}\ln\frac{\Delta - \cos x}{\Delta + \cos x} \quad \text{should be replaced with} \quad \int \frac{dx}{\Delta \sin x}$$

$$\frac{1}{2k'}\ln\frac{\Delta + k'\sin x}{\Delta - k'\sin x} \quad \text{should be replaced with} \quad \int \frac{dx}{\Delta \cos x}$$

$$\frac{1}{2k'}\ln\frac{\Delta + k'}{\Delta - k'} \quad \text{should be replaced with} \quad \int \frac{\tan x}{\Delta}\, dx$$

$$\frac{1}{2}\ln\frac{1 - \Delta}{1 + \Delta} \quad \text{should be replaced with} \quad \int \frac{\cot x}{\Delta}\, dx$$

(2)     Then, on both sides of the equations, we should replace $\Delta$ with $i\sqrt{a^2\sin^2 x - 1}$, $k$ with $a$ and $k'^2$ with $1 - a^2$.

(3)     Both sides of the resulting equations should be multiplied by $i$, as a result of which only real functions $(a^2 > 1)$ should appear on both sides of the equations.

(4)     The integrals on the right sides of the equations should be replaced with their values found from formulas **2.599**.

Examples:

1.     We rewrite equation **2.584** 4 in the form

$$\int \frac{\sin^2 x}{i\sqrt{a^2\sin^2 x - 1}}\, dx = \frac{1}{a^2}\int \frac{dx}{i\sqrt{a^2\sin^2 x - 1}} - \frac{1}{a^2}\int i\sqrt{a^2\sin^2 x - 1}\, dx,$$

from which we get

$$\int \frac{\sin^2 x\, dx}{\sqrt{a^2\sin^2 x - 1}} = \frac{1}{a^2}\left\{\int \frac{dx}{\sqrt{a^2\sin^2 x - 1}} + \int \sqrt{a^2\sin^2 x - 1}\, dx\right\} = -\frac{1}{a}E\left(\alpha, \frac{\sqrt{a^2 - 1}}{a}\right)$$

$$[a^2 > 1]$$

2.     We rewrite equation **2.584** 58 as follows:

$$\int \frac{dx}{i^5\sqrt{(a^2\sin^2 x - 1)^5}} = -\frac{2a^4(a^2 - 2)\sin^2 x - (3a^2 - 5)a^2}{3(1 - a^2)^2 i^3\sqrt{(a^2\sin^2 x - 1)^3}}\sin x \cos x$$

$$-\frac{1}{3(1 - a^2)}\int \frac{dx}{i\sqrt{a^2\sin^2 x - 1}} - \frac{2a^2 - 4}{3(1 - a^2)^2}\int i\sqrt{a^2\sin^2 x - 1}\, dx$$

from which we obtain

$$\int \frac{dx}{\sqrt{\left(a^2 \sin^2 x - 1\right)^5}} = \frac{2a^4\left(a^2 - 2\right)\sin^2 x - \left(3a^2 - 5\right)a^2}{3\left(1 - a^2\right)^2 \sqrt{\left(a^2\sin^2 x - 1\right)^3}} \sin x \cos x + \frac{1}{3\left(1 - a^2\right)^2 a}$$
$$\times \left\{\left(a^2 - 3\right) F\left(\alpha, \frac{\sqrt{a^2 - 1}}{a}\right) - 2a^2\left(a^2 - 2\right) E\left(\alpha, \frac{\sqrt{a^2 - 1}}{a}\right)\right\}$$
$$\left[a^2 > 1\right]$$

3.      We rewrite equation **2.584** 71 in the form

$$\int \frac{dx}{\sin x \cos x i \sqrt{a^2\sin^2 x - 1}} = \int \frac{\cot x \, dx}{i\sqrt{a^2\sin^2 x - 1}} + \int \frac{\tan x \, dx}{i\sqrt{a^2\sin^2 x - 1}},$$

from which we obtain

$$\int \frac{dx}{\sin x \cos x \sqrt{a^2\sin^2 x - 1}} = \frac{1}{2\sqrt{a^2 - 1}} \ln \frac{\sqrt{a^2 - 1} + \sqrt{a^2\sin^2 x - 1}}{\sqrt{a^2 - 1} - \sqrt{a^2\sin^2 x - 1}} - \arcsin\left(\frac{1}{a\sin x}\right)$$
$$\left[a^2 > 1\right]$$

**2.612**    Integrals of the form $\int R\left(\sin x, \cos x, \sqrt{1 - k^2\cos^2 x}\right) dx$.

To find integrals of the form $\int R\left(\sin x, \cos x, \sqrt{1 - k^2\cos^2 x}\right) dx$ we make the substitution $x = \frac{\pi}{2} - y$, which yields

$$\int R\left(\sin x, \cos x, \sqrt{1 - k^2\cos^2 x}\right) dx = -\int R\left(\cos y, \sin y, \sqrt{1 - k^2\sin^2 y}\right) dy$$

The integrals $\int R\left(\cos y, \sin y, \sqrt{1 - k^2\sin^2 y}\right) dy$ are found from formulas **2.583** and **2.584**. As a result of the use of these formulas (where it is assumed that the original integral can be reduced only to integrals of the first and second Legendre forms), when we replace the functions $F(x,k)$ and $E(x,k)$ with the corresponding integrals, we obtain an expression of the form

$$-g\left(\cos y, \sin y\right) - A\int \frac{dy}{\sqrt{1 - k^2\sin^2 y}} - B\int \sqrt{1 - k^2\sin^2 y}\, dy$$

Returning now to the original variable $x$, we obtain

$$\int R\left(\sin x, \cos x, \sqrt{1 - k^2\cos^2 x}\right) dx = -g\left(\sin x, \cos x\right) - A\int \frac{dx}{\sqrt{1 - k^2\cos^2 x}} - B\int \sqrt{1 - k^2\cos^2 x}\, dx$$

The integrals appearing in this expression are found from the formulas

1.    $\int \dfrac{dx}{\sqrt{1 - k^2\cos^2 x}} = F\left(\arcsin\left(\dfrac{\sin x}{\sqrt{1 - k^2\cos^2 x}}\right), k\right)$

2.    $\int \sqrt{1 - k^2\cos^2 x}\, dx = E\left(\arcsin\left(\dfrac{\sin x}{\sqrt{1 - k^2\cos^2 x}}\right), k\right) - \dfrac{k^2\sin x\cos x}{\sqrt{1 - k^2\cos^2 x}}$

**2.613**    Integrals of the form $\int R\left(\sin x, \cos x, \sqrt{1 - p^2\cos^2 x}\right) dx$      $[p > 1]$.

To find integrals of the type $\int R\left(\sin x, \cos x, \sqrt{1 - p^2\cos^2 x}\right) dx$, where $[p > 1]$, we proceed as in section **2.612**. Here, we use the formulas

1.     $\displaystyle\int \frac{\mathrm{d}x}{\sqrt{1-p^2\cos^2 x}} = -\frac{1}{p}\,F\left(\arcsin\left(p\cos x\right),\frac{1}{p}\right)$          $[p>1]$

2.     $\displaystyle\int \sqrt{1-p^2\cos^2 x}\,\mathrm{d}x = \frac{p^2-1}{p}\,F\left(\arcsin\left(p\cos x\right),\frac{1}{p}\right) - p\,E\left(\arcsin\left(p\cos x\right),\frac{1}{p}\right)$

**2.614**   Integrals of the form $\displaystyle\int R\left(\sin x,\cos x,\sqrt{1+p^2\cos^2 x}\right)\,\mathrm{d}x$.

   To find integrals of the type $\displaystyle\int R\left(\sin x,\cos x,\sqrt{1+p^2\cos^2 x}\right)\,\mathrm{d}x$, we need to make the substitution $x=\dfrac{\pi}{2}-y$. This yields

$$\int R\left(\sin x,\cos x,\sqrt{1+p^2\cos^2 x}\right)\,\mathrm{d}x = -\int R\left(\cos y,\sin y,\sqrt{1+p^2\sin^2 y}\right)\,\mathrm{d}y$$

   To calculate the integrals $-\displaystyle\int R\left(\cos y,\sin y,\sqrt{1+p^2\sin^2 y}\right)\,\mathrm{d}y$, we need to use first what was said in **2.598** and **2.612** and then, after returning to the variable $x$, the formulas

1.     $\displaystyle\int \frac{\mathrm{d}x}{\sqrt{1+p^2\cos^2 x}} = \frac{1}{\sqrt{1+p^2}}\,F\left(x,\frac{p}{\sqrt{1+p^2}}\right)$

2.     $\displaystyle\int \sqrt{1+p^2\cos^2 x}\,\mathrm{d}x = \sqrt{1+p^2}\,E\left(x,\frac{p}{\sqrt{1+p^2}}\right)$

**2.615**   Integrals of the form $\displaystyle\int R\left(\sin x,\cos x,\sqrt{a^2\cos^2 x-1}\right)\,\mathrm{d}x$      $[a>1]$.

   To find integrals of the type $\displaystyle\int R\left(\sin x,\cos x,\sqrt{a^2\cos^2 x-1}\right)\,\mathrm{d}x$, we need to make the substitution $x=\dfrac{\pi}{2}-y$. This yields

$$\int R\left(\sin x,\cos x,\sqrt{a^2\cos^2 x-1}\right)\,\mathrm{d}x = -\int R\left(\cos y,\sin y,\sqrt{a^2\sin^2 y-1}\right)\,\mathrm{d}y$$

   To calculate the integrals $-\int R\left(\cos y,\sin y,\sqrt{a^2\sin^2 y-1}\right)\,\mathrm{d}y$, we use what was said in **2.611** and then, after returning to the variable $x$, we use the formulas

1.     $\displaystyle\int \frac{\mathrm{d}x}{\sqrt{a^2\cos^2 x-1}} = \frac{1}{a}\,F\left(\arcsin\left(\frac{a\sin x}{\sqrt{a^2-1}}\right),\frac{\sqrt{a^2-1}}{a}\right)$

$$[a>1]$$

2.     $\displaystyle\int \sqrt{a^2\cos^2 x-1}\,\mathrm{d}x = a\,E\left(\arcsin\left(\frac{a\sin x}{\sqrt{a^2-1}}\right),\frac{\sqrt{a^2-1}}{a}\right)$

$$-\frac{1}{a}\,F\left(\arcsin\left(\frac{a\sin x}{\sqrt{a^2-1}}\right),\frac{\sqrt{a^2-1}}{a}\right)\qquad [a>1]$$

**2.616**[11]    Integrals of the form $\displaystyle\int R\left(\sin x, \cos x, \sqrt{1 - p^2 \sin^2 x}, \sqrt{1 - q^2 \sin^2 x}\right) dx.$

**Notation:** $\alpha = \arcsin\left(\dfrac{\sqrt{1 - p^2}\,\sin x}{\sqrt{1 - p^2 \sin^2 x}}\right).$

1.    $\displaystyle\int \frac{dx}{\sqrt{\left(1 - p^2 \sin^2 x\right)\left(1 - q^2 \sin^2 x\right)}} = \frac{1}{\sqrt{1 - p^2}}\, F\left(\alpha, \sqrt{\frac{q^2 - p^2}{1 - p^2}}\right)$

$$\left[0 < p^2 < q^2 < 1, \quad 0 < x \leq \frac{\pi}{2}\right]$$

BY (284.00)

2.    $\displaystyle\int \frac{\tan^2 x \, dx}{\sqrt{\left(1 - p^2 \sin^2 x\right)\left(1 - q^2 \sin^2 x\right)}} = \frac{\tan x \sqrt{1 - q^2 \sin^2 x}}{\left(1 - q^2\right)\sqrt{1 - p^2 \sin^2 x}}$

$$-\frac{1}{\left(1 - q^2\right)\sqrt{1 - p^2}}\, E\left(\alpha, \sqrt{\frac{q^2 - p^2}{1 - p^2}}\right)$$

$$\left[0 < p^2 < q^2 < 1, \quad 0 < x \leq \frac{\pi}{2}\right] \quad \text{BY (284.07)}$$

3.[12]    $\displaystyle\int \frac{\tan^4 x \, dx}{\sqrt{\left(1 - p^2 \sin^2 x\right)\left(1 - q^2 \sin^2 x\right)}}$

$$= \frac{1}{3\left(1 - q^2\right)^2 \left(1 - p^2\right)^{\frac{3}{2}}} \times \left[2\left(2 - p^2 - q^2\right) E\left(\alpha, \sqrt{\frac{q^2 - p^2}{1 - p^2}}\right) - \left(1 - q^2\right) F\left(\alpha, \sqrt{\frac{q^2 - p^2}{1 - p^2}}\right)\right]$$

$$+ \frac{2p^2 + q^2 - 3 + \sin^2 x \left(4 - 3p^2 - 2q^2 + p^2 q^2\right)}{3\left(1 - p^2\right)\left(1 - q^2\right)^2}\, \frac{\sin x}{\cos^3 x}\sqrt{\frac{1 - q^2 \sin^2 x}{1 - p^2 \sin^2 x}}$$

$$\left[0 < p^2 < q^2 < 1, \quad 0 < x \leq \frac{\pi}{2}\right] \quad \text{BY (284.07)}$$

4.    $\displaystyle\int \frac{\sin^2 x \, dx}{\sqrt{\left(1 - p^2 \sin^2 x\right)\left(1 - q^2 \sin^2 x\right)^3}}$

$$= \frac{\sqrt{1 - p^2}}{\left(1 - q^2\right)\left(q^2 - p^2\right)}\, E\left(\alpha, \sqrt{\frac{q^2 - p^2}{1 - p^2}}\right) - \frac{1}{\left(q^2 - p^2\right)\sqrt{1 - p^2}}\, F\left(\alpha, \sqrt{\frac{q^2 - p^2}{1 - p^2}}\right)$$

$$- \frac{\sin x \cos x}{\left(1 - q^2\right)\sqrt{\left(1 - p^2 \sin^2 x\right)\left(1 - q^2 \sin^2 x\right)}}$$

$$\left[0 < p^2 < q^2 < 1, \quad 0 < x \leq \frac{\pi}{2}\right] \quad \text{BY (284.06)}$$

5.    $\displaystyle\int \frac{\cos^2 x \, dx}{\sqrt{\left(1 - p^2 \sin^2 x\right)^3 \left(1 - q^2 \sin^2 x\right)}}$

$$= \frac{\sqrt{1 - p^2}}{q^2 - p^2}\, E\left(\alpha, \sqrt{\frac{q^2 - p^2}{1 - p^2}}\right) - \frac{1 - q^2}{\left(q^2 - p^2\right)\sqrt{1 - p^2}}\, F\left(\alpha, \sqrt{\frac{q^2 - p^2}{1 - p^2}}\right)$$

$$\left[0 < p^2 < q^2 < 1, \quad 0 < x \leq \frac{\pi}{2}\right] \quad \text{BY (284.05)}$$

6.  $\displaystyle\int \frac{\cos^4 x \, \mathrm{d}x}{\sqrt{\left(1 - p^2 \sin^2 x\right)^5 \left(1 - q^2 \sin^2 x\right)}}$

$$= \frac{\left(1-p^2\right)^{\frac{3}{2}}}{3\left(q^2-p^2\right)^2} \left[ \frac{\left(2+p^2-3q^2\right)\left(1-q^2\right)}{\left(1-p^2\right)^2} F\left(\alpha, \sqrt{\frac{q^2-p^2}{1-p^2}}\right)\right.$$

$$\left. + 2\frac{2q^2-p^2-1}{1-p^2} E\left(\alpha, \sqrt{\frac{q^2-p^2}{1-p^2}}\right)\right] + \frac{\left(1-p^2\right)\sin x \cos x\sqrt{1-q^2\sin^2 x}}{3\left(q^2-p^2\right)\sqrt{\left(1-p^2\sin^2 x\right)^3}}$$

$$\left[0 < p^2 < q^2 < 1, \quad 0 < x \le \frac{\pi}{2}\right] \quad \text{BY (284.05)}$$

7.  $\displaystyle\int \frac{\mathrm{d}x}{1 - p^2 \sin^2 x}\sqrt{\frac{1 - q^2 \sin^2 x}{1 - p^2 \sin^2 x}} = \frac{1}{\sqrt{1-p^2}} E\left(\alpha, \sqrt{\frac{q^2-p^2}{1-p^2}}\right)$

$$\left[0 < p^2 < q^2 < 1, \quad 0 < x \le \frac{\pi}{2}\right]$$
$$\text{BY (284.01)}$$

8.  $\displaystyle\int \sqrt{\frac{1 - p^2 \sin^2 x}{\left(1 - q^2 \sin^2 x\right)^3}}\, \mathrm{d}x = \frac{\sqrt{1-p^2}}{1-q^2} E\left(\alpha, \sqrt{\frac{q^2-p^2}{1-p^2}}\right) - \frac{q^2-p^2}{1-q^2} \frac{\sin x \cos x}{\sqrt{\left(1 - p^2 \sin^2 x\right)\left(1 - q^2 \sin^2 x\right)}}$

$$\left[0 < p^2 < q^2 < 1, \quad 0 < x \le \frac{\pi}{2}\right].$$
$$\text{BY (284.04)}$$

9.  $\displaystyle\int \frac{\mathrm{d}x}{1 + \left(p^2 r^2 - p^2 - r^2\right) \sin^2 x}\sqrt{\frac{1 - p^2 \sin^2 x}{1 - q^2 \sin^2 x}} = \frac{1}{\sqrt{1-p^2}} \Pi\left(\alpha, r^2, \sqrt{\frac{q^2-p^2}{1-p^2}}\right)$

$$\left[0 < p^2 < q^2 < 1, \quad 0 < x \le \frac{\pi}{2}\right].$$
$$\text{BY (284.02)}$$

**2.617**  **Notation:** $\alpha = \arcsin\sqrt{\dfrac{\sqrt{b^2+c^2} - b\sin x - c\cos x}{2\sqrt{b^2+c^2}}}, \quad r = \sqrt{\dfrac{2\sqrt{b^2+c^2}}{a + \sqrt{b^2+c^2}}}.$

1.  $\displaystyle\int \frac{\mathrm{d}x}{\sqrt{a + b\sin x + c\cos x}}$

$$= -\frac{2}{\sqrt{a + \sqrt{b^2+c^2}}} F(\alpha, r)$$

$$\left[0 < \sqrt{b^2+c^2} < a, \quad \arcsin\frac{b}{\sqrt{b^2+c^2}} - \pi \le x < \arcsin\frac{b}{\sqrt{b^2+c^2}}\right]$$
$$\text{BY (294.00)}$$

$$= -\frac{\sqrt{2}}{\sqrt[4]{b^2+c^2}} F(\alpha, r)$$

$$\left[0 < |a| < \sqrt{b^2+c^2}, \quad \arcsin\frac{b}{\sqrt{b^2+c^2}} - \arccos\left(-\frac{a}{\sqrt{b^2+c^2}}\right) \le x < \arcsin\frac{b}{\sqrt{b^2+c^2}}\right]$$
$$\text{BY (293.00)}$$

2.
$$\int \frac{\sin x \, dx}{\sqrt{a + b \sin x + c \cos x}} = -\frac{\sqrt{2b}}{\sqrt[4]{(b^2 + c^2)^3}} \left\{2 E(\alpha, r) - F(\alpha, r)\right\} + \frac{2c}{b^2 + c^2} \sqrt{a + b \sin x + c \cos x}$$
$$\left[0 < |a| < \sqrt{b^2 + c^2}, \quad \arcsin \frac{b}{\sqrt{b^2 + c^2}} - \arccos\left(-\frac{a}{\sqrt{b^2 + c^2}}\right) \le x < \arcsin \frac{b}{\sqrt{b^2 + c^2}}\right]$$

BY (293.05)

3.
$$\int \frac{(b \cos x - c \sin x) \, dx}{\sqrt{a + b \sin x + c \cos x}} = 2\sqrt{a + b \sin x + c \cos x}$$

4.
$$\int \frac{\sqrt{b^2 + c^2} + b \sin x + c \cos x}{\sqrt{a + b \sin x + c \cos x}} \, dx$$
$$= -2\sqrt{a + \sqrt{b^2 + c^2}} \, E(\alpha, r) + \frac{2\left(a - \sqrt{b^2 + c^2}\right)}{\sqrt{a + \sqrt{b^2 + c^2}}} \, F(\alpha, r)$$
$$\left[0 < \sqrt{b^2 + c^2} < a, \quad \arcsin \frac{b}{\sqrt{b^2 + c^2}} - \pi \le x < \arcsin \frac{b}{\sqrt{b^2 + c^2}}\right]$$

BY (294.04)

$$= -2\sqrt{2} \sqrt[4]{b^2 + c^2} \, E(\alpha, r)$$
$$\left[0 < |a| < \sqrt{b^2 + c^2}, \quad \arcsin \frac{b}{\sqrt{b^2 + c^2}} - \arccos\left(-\frac{a}{\sqrt{b^2 + c^2}}\right) \le x < \arcsin \frac{b}{\sqrt{b^2 + c^2}}\right]$$

BY (293.01)

5.
$$\int \sqrt{a + b \sin x + c \cos x} \, dx$$
$$= -2\sqrt{a + \sqrt{b^2 + c^2}} \, E(\alpha, r)$$
$$\left[0 < \sqrt{b^2 + c^2} < a, \quad \arcsin \frac{b}{\sqrt{b^2 + c^2}} - \pi \le x < \arcsin \frac{b}{\sqrt{b^2 + c^2}}\right]$$

BY (294.01)

$$= -2\sqrt{2} \sqrt[4]{b^2 + c^2} \, E(\alpha, r) + \frac{\sqrt{2}\left(\sqrt{b^2 + c^2} - a\right)}{\sqrt[4]{b^2 + c^2}} \, F(\alpha, r)$$
$$\left[0 < |a| < \sqrt{b^2 + c^2}, \quad \arcsin \frac{b}{\sqrt{b^2 + c^2}} - \arccos\left(\frac{-a}{\sqrt{b^2 + c^2}}\right) \le x < \arcsin \frac{b}{\sqrt{b^2 + c^2}}\right]$$

BY (293.03)

**2.618**  Integrals of the form $\int R\left(\sin ax, \cos ax, \sqrt{\cos 2ax}\right) dx = \frac{1}{a}\int R\left(\sin t, \cos t, \sqrt{1 - 2\sin^2 t}\right) dt$
where the substitution $t = ax$ has been used.
**Notation:** $\alpha = \arcsin\left(\sqrt{2}\sin ax\right)$

The integrals $\int R\left(\sin ax, \cos ax, \sqrt{\cos 2ax}\right) dx$ are special cases of the integrals **2.595.** for ($p = 2$).
We give some formulas:

1.
$$\int \frac{dx}{\sqrt{\cos 2ax}} = \frac{1}{a\sqrt{2}} F\left(\alpha, \frac{1}{\sqrt{2}}\right) \qquad\qquad \left[0 < ax \le \frac{\pi}{4}\right]$$

2.
$$\int \frac{\cos^2 ax}{\sqrt{\cos 2ax}} \, dx = \frac{1}{a\sqrt{2}} E\left(\alpha, \frac{1}{\sqrt{2}}\right) \qquad\qquad \left[0 < ax \le \frac{\pi}{4}\right]$$

3. $\displaystyle\int \frac{dx}{\cos^2 ax\sqrt{\cos 2ax}} = \frac{\sqrt{2}}{a} E\left(\alpha, \frac{1}{\sqrt{2}}\right) - \frac{\tan x}{a}\sqrt{\cos 2ax}$

$$\left[0 < ax \leq \frac{\pi}{4}\right]$$

4. $\displaystyle\int \frac{dx}{\cos^4 ax\sqrt{\cos 2ax}} = \frac{2\sqrt{2}}{a} E\left(\alpha, \frac{1}{\sqrt{2}}\right) - \frac{\sqrt{2}}{3a} F\left(\alpha, \frac{1}{\sqrt{2}}\right) - \frac{(6\cos^2 ax + 1)\sin ax}{3a\cos^3 ax}\sqrt{\cos 2ax}$

$$\left[0 < x \leq \frac{\pi}{4}\right]$$

5. $\displaystyle\int \frac{\tan^2 ax\,dx}{\sqrt{\cos 2ax}} = \frac{\sqrt{2}}{a} E\left(\alpha, \frac{1}{\sqrt{2}}\right) - \frac{1}{a\sqrt{2}} F\left(\alpha, \frac{1}{\sqrt{2}}\right) - \frac{1}{a}\tan ax\sqrt{\cos 2ax}$

$$\left[0 < x \leq \frac{\pi}{2}\right]$$

6. $\displaystyle\int \frac{\tan^4 ax\,dx}{\sqrt{\cos 2ax}} = \frac{1}{3a\sqrt{2}} F\left(\alpha, \frac{1}{\sqrt{2}}\right) - \frac{\sin ax}{3a\cos^3 ax}\sqrt{\cos 2ax}$

$$\left[0 < ax \leq \frac{\pi}{4}\right]$$

7. $\displaystyle\int \frac{dx}{(1 - 2r^2\sin^2 ax)\sqrt{\cos 2ax}} = \frac{1}{a\sqrt{2}} \Pi\left(\alpha, r^2, \frac{1}{\sqrt{2}}\right) \quad \left[0 < ax \leq \frac{\pi}{4}\right]$

8. $\displaystyle\int \frac{dx}{\sqrt{\cos^3 2ax}} = \frac{1}{a\sqrt{2}} F\left(\alpha, \frac{1}{\sqrt{2}}\right) - \frac{\sqrt{2}}{a} E\left(\alpha, \frac{1}{\sqrt{2}}\right) + \frac{\sin 2ax}{a\sqrt{\cos 2ax}}$

$$\left[0 < ax \leq \frac{\pi}{4}\right]$$

9. $\displaystyle\int \frac{\sin^2 ax\,dx}{\sqrt{\cos^3 2ax}} = \frac{\sin 2ax}{2a\sqrt{\cos 2ax}} - \frac{1}{a\sqrt{2}} E\left(\alpha, \frac{1}{\sqrt{2}}\right) \quad \left[0 < ax \leq \frac{\pi}{4}\right]$

10. $\displaystyle\int \frac{dx}{\sqrt{\cos^5 2ax}} = \frac{1}{3a\sqrt{2}} F\left(\alpha, \frac{1}{\sqrt{2}}\right) + \frac{\sin 2ax}{3a\sqrt{\cos^3 2ax}} \quad \left[0 < ax \leq \frac{\pi}{4}\right]$

11. $\displaystyle\int \sqrt{\cos 2ax}\,dx = \frac{\sqrt{2}}{a} E\left(\alpha, \frac{1}{\sqrt{2}}\right) - \frac{1}{a\sqrt{2}} F\left(\alpha, \frac{1}{\sqrt{2}}\right)$

$$\left[0 < ax \leq \frac{\pi}{4}\right]$$

12. $\displaystyle\int \frac{\sqrt{\cos 2ax}}{\cos^2 ax}\,dx = \frac{\sqrt{2}}{a}\left\{F\left(\alpha, \frac{1}{\sqrt{2}}\right) - E\left(\alpha, \frac{1}{\sqrt{2}}\right)\right\} + \frac{1}{a}\tan ax\sqrt{\cos 2ax}$

$$\left[0 < x \leq \frac{\pi}{4}\right]$$

**2.619**   Integrals of the form $\displaystyle\int R\left(\sin ax, \cos ax, \sqrt{-\cos 2ax}\right) dx = \frac{1}{a}\int R\left(\sin x, \cos x, \sqrt{2\sin^2 x - 1}\right) dx$

**Notation:** $\alpha = \arcsin\left(\sqrt{2}\cos ax\right)$

The integrals $\displaystyle\int R\left(\sin x, \cos x, \sqrt{2\sin^2 x - 1}\right) dx$ are special cases of the integrals **2.599** and **2.611** for $(a = 2)$. We give some formulas:

1. $$\int \frac{dx}{\sqrt{-\cos 2ax}} = -\frac{1}{a\sqrt{2}} F\left(\alpha, \frac{1}{\sqrt{2}}\right)$$

2. $$\int \frac{\cos^2 ax \, dx}{\sqrt{-\cos 2ax}} = \frac{1}{a\sqrt{2}} \left[ E\left(\alpha, \frac{1}{\sqrt{2}}\right) - F\left(\alpha, \frac{1}{\sqrt{2}}\right) \right]$$

3. $$\int \frac{\cos^4 ax \, dx}{\sqrt{-\cos 2ax}} = \frac{1}{3a\sqrt{2}} \left[ 3 F\left(\alpha, \frac{1}{\sqrt{2}}\right) - \frac{5}{2} E\left(\alpha, \frac{1}{\sqrt{2}}\right) \right] - \frac{1}{12a} \sin 2ax \sqrt{-\cos 2ax}$$

4. $$\int \frac{dx}{\sin^2 ax \sqrt{-\cos 2ax}} = \frac{1}{a} \cot ax \sqrt{-\cos 2ax} - \frac{\sqrt{2}}{a} E\left(\alpha, \frac{1}{\sqrt{2}}\right)$$

5. $$\int \frac{dx}{\sin^4 ax \sqrt{-\cos 2ax}} = \frac{2}{3a\sqrt{2}} \left[ F\left(\alpha, \frac{1}{\sqrt{2}}\right) - 6 E\left(\alpha, \frac{1}{\sqrt{2}}\right) \right]$$
$$+ \frac{1}{3a} \frac{\cos ax}{\sin^3 ax} \left(6 \sin^2 ax + 1\right) \sqrt{-\cos 2ax}$$

6. $$\int \frac{\cot^2 ax \, dx}{\sqrt{-\cos 2ax}} = \frac{1}{a\sqrt{2}} \left[ F\left(\alpha, \frac{1}{\sqrt{2}}\right) - 2 E\left(\alpha, \frac{1}{\sqrt{2}}\right) \right] + \frac{1}{a} \cot ax \sqrt{-\cos 2ax}$$

7. $$\int \frac{dx}{\left(1 - 2r^2 \cos^2 ax\right) \sqrt{-\cos 2ax}} = -\frac{1}{a\sqrt{2}} \Pi\left(\alpha, r^2, \frac{1}{\sqrt{2}}\right)$$

8. $$\int \frac{dx}{\sqrt{-\cos^3 2ax}} = \frac{1}{a\sqrt{2}} \left[ F\left(\alpha, \frac{1}{\sqrt{2}}\right) - 2 E\left(\alpha, \frac{1}{\sqrt{2}}\right) \right] + \frac{\sin 2ax}{a\sqrt{-\cos 2ax}}$$

9. $$\int \frac{\cos^2 ax \, dx}{\sqrt{-\cos^3 2ax}} = \frac{\sin 2ax}{2a\sqrt{-\cos 2ax}} - \frac{1}{a\sqrt{2}} E\left(\alpha, \frac{1}{\sqrt{2}}\right)$$

10. $$\int \frac{dx}{\sqrt{-\cos^5 2ax}} = -\frac{1}{3a\sqrt{2}} F\left(\alpha, \frac{1}{\sqrt{2}}\right) - \frac{\sin 2ax}{3a\sqrt{-\cos^3 2ax}}$$

11. $$\int \sqrt{-\cos 2ax} \, dx = \frac{1}{a\sqrt{2}} \left[ F\left(\alpha, \frac{1}{\sqrt{2}}\right) - 2 E\left(\alpha, \frac{1}{\sqrt{2}}\right) \right]$$

**2.621**   Integrals of the form $\int R\left(\sin ax, \cos ax, \sqrt{\sin 2ax}\right) dx.$

**Notation:** $\alpha = \arcsin \sqrt{\dfrac{2 \sin ax}{1 + \sin ax + \cos ax}}.$

1. $$\int \frac{dx}{\sqrt{\sin 2ax}} = \frac{\sqrt{2}}{a} F\left(\alpha, \frac{1}{\sqrt{2}}\right)$$ 
BY (287.50)

2. $$\int \frac{\sin ax \, dx}{\sqrt{\sin 2ax}} = \frac{\sqrt{2}}{a} \left[ \frac{1+i}{2} \Pi\left(\alpha, \frac{1+i}{2}, \frac{1}{\sqrt{2}}\right) \right.$$
$$\left. + \frac{1-i}{2} \Pi\left(\alpha, \frac{1-i}{2}, \frac{1}{\sqrt{2}}\right) + F\left(\alpha, \frac{1}{\sqrt{2}}\right) - 2 E\left(\alpha, \frac{1}{\sqrt{2}}\right) \right]$$
BY (287.57)

3. $$\int \frac{\sin ax \, dx}{\left(1 + \sin ax + \cos ax\right) \sqrt{\sin 2ax}} = \frac{\sqrt{2}}{a} \left[ F\left(\alpha, \frac{1}{\sqrt{2}}\right) - E\left(\alpha, \frac{1}{\sqrt{2}}\right) \right]$$
BY (287.54)

4.    $\displaystyle \int \frac{\sin ax \, dx}{(1 - \sin ax + \cos ax)\sqrt{\sin 2ax}} = \frac{\sqrt{2}}{a}\left\{\sqrt{\tan ax} - E\left(\alpha, \frac{1}{\sqrt{2}}\right)\right\}$

$$\left[ax \neq \frac{\pi}{2}\right]$$      BY (287.55)

5.    $\displaystyle \int \frac{(1 + \cos ax) \, dx}{(1 + \sin ax + \cos ax)\sqrt{\sin 2ax}} = \frac{\sqrt{2}}{a} E\left(\alpha, \frac{1}{\sqrt{2}}\right)$      BY (287.51)

6.    $\displaystyle \int \frac{(1 + \cos ax) \, dx}{(1 - \sin ax + \cos ax)\sqrt{\sin 2ax}} = \frac{\sqrt{2}}{a}\left\{F\left(\alpha, \frac{1}{\sqrt{2}}\right) - E\left(\alpha, \frac{1}{\sqrt{2}}\right) + \sqrt{\tan ax}\right\}$

$$\left[ax \neq \frac{\pi}{2}\right]$$      BY (287.56)

7.    $\displaystyle \int \frac{(1 - \sin ax + \cos ax) \, dx}{(1 + \sin ax + \cos ax)\sqrt{\sin 2ax}} = \frac{\sqrt{2}}{a}\left\{2 E\left(\alpha, \frac{1}{\sqrt{2}}\right) - F\left(\alpha, \frac{1}{\sqrt{2}}\right)\right\}$      BY (287.53)

8.    $\displaystyle \int \frac{(1 + \sin ax + \cos ax) \, dx}{[1 + \cos ax + (1 - 2r^2)\sin ax]\sqrt{\sin 2ax}} = \frac{\sqrt{2}}{a}\Pi\left(\alpha, r^2, \frac{1}{\sqrt{2}}\right).$      BY (287.52)

## 2.63–2.65 Products of trigonometric functions and powers

**2.631**

1.    $\displaystyle \int x^r \sin^p x \cos^q x \, dx = \frac{1}{(p+q)^2}\left[(p+q)x^r \sin^{p+1} x \cos^{q-1} x\right.$

$$+ rx^{r-1}\sin^p x \cos^q x - r(r-1)\int x^{r-2}\sin^p x \cos^q x \, dx$$

$$\left. - rp\int x^{r-1}\sin^{p-1} x \cos^{q-1} x \, dx + (q-1)(p+q)\int x^r \sin^p x \cos^{q-2} x \, dx\right]$$

$$= \frac{1}{(p+q)^2}\left[-(p+q)x^r \sin^{p-1} x \cos^{q+1} x\right.$$

$$+ rx^{r-1}\sin^p x \cos^q x - r(r-1)\int x^{r-2}\sin^p x \cos^q x \, dx$$

$$\left. + rq\int x^{r-1}\sin^{p-1} x \cos^{q-1} x \, dx + (p-1)(p+q)\int x^r \sin^{p-2} x \cos^q x \, dx\right]$$

GU (331)(1)

2.    $\displaystyle \int x^m \sin^n x \, dx = \frac{x^{m-1}\sin^{n-1} x}{n^2}\{m\sin x - nx\cos x\}$

$$+ \frac{n-1}{n}\int x^m \sin^{n-2} x \, dx - \frac{m(m-1)}{n^2}\int x^{m-2}\sin^n x \, dx$$

3.    $\displaystyle \int x^m \cos^n x \, dx = \frac{x^{m-1}\cos^{n-1} x}{n^2}\{m\cos x + nx\sin x\}$

$$+ \frac{n-1}{n}\int x^m \cos^{n-2} x \, dx - \frac{m(m-1)}{n^2}\int x^{m-2}\cos^n x \, dx$$

4. $$\int x^n \sin^{2m} x \, dx = \binom{2m}{m} \frac{x^{n+1}}{2^{2m}(n+1)}$$
$$+ \frac{(-1)^m}{2^{2m-1}} \sum_{k=0}^{m-1} (-1)^k \binom{2m}{k} \int x^n \cos(2m-2k)x \, dx$$

                                                   (see **2.633** 2)           TI 333

5. $$\int x^n \sin^{2m+1} x \, dx = \frac{(-1)^m}{2^{2m}} \sum_{k=0}^{m} (-1)^k \binom{2m+1}{k} \int x^n \sin(2m-2k+1)x \, dx$$

                                                   (see **2.633** 1)           TI 333

6. $$\int x^n \cos^{2m} x \, dx = \binom{2m}{m} \frac{x^{n+1}}{2^{2m}(n+1)}$$
$$+ \frac{1}{2^{2m-1}} \sum_{k=0}^{m-1} \binom{2m}{k} \int x^n \cos(2m-2k)x \, dx$$

                                                   (see **2.633** 2)           TI 333

7. $$\int x^n \cos^{2m+1} x \, dx = \frac{1}{2^{2m}} \sum_{k=0}^{m} \binom{2m+1}{k} \int x^n \cos(2m-2k+1)x \, dx$$

                                                   (see **2.633** 2)           TI 333

**2.632**

1. $$\int x^{\mu-1} \sin \beta x \, dx = \frac{i}{2} (i\beta)^{-\mu} \gamma(\mu, i\beta x) - \frac{i}{2} (-i\beta)^{-\mu} \gamma(\mu, -i\beta x)$$

                                         $[\operatorname{Re}\mu > -1, \quad x > 0]$      ET I 317(2)

2. $$\int x^{\mu-1} \sin ax \, dx = -\frac{1}{2a^\mu} \left\{ \exp\left[\frac{\pi i}{2}(\mu-1)\right] \Gamma(\mu, -iax) + \exp\left[\frac{\pi i}{2}(1-\mu)\right] \Gamma(\mu, iax) \right\}$$

                                     $[\operatorname{Re}\mu < 1, \quad a > 0, \quad x > 0]$    ET I 317(3)

3. $$\int x^{\mu-1} \cos \beta x \, dx = \frac{1}{2} \left\{ (i\beta)^{-\mu} \gamma(\mu, i\beta x) + (-i\beta)^{-\mu} \gamma(\mu, -i\beta x) \right\}$$

                                         $[\operatorname{Re}\mu > 0, \quad x > 0]$      ET I 319(22)

4. $$\int x^{\mu-1} \cos ax \, dx = -\frac{1}{2a^\mu} \left\{ \exp\left(i\mu\frac{\pi}{2}\right) \Gamma(\mu, -iax) + \exp\left(-i\mu\frac{\pi}{2}\right) \Gamma(\mu, iax) \right\}$$

                                                          ET I 319(23)

**2.633**

1. $$\int x^n \sin ax \, dx = -\sum_{k=0}^{n} k! \binom{n}{k} \frac{x^{n-k}}{a^{k+1}} \cos\left(ax + \frac{1}{2}k\pi\right)$$

                                                                 TI (487)

2.[8] $$\int x^n \cos ax \, dx = \sum_{k=0}^{n} k! \binom{n}{k} \frac{x^{n-k}}{a^{k+1}} \sin\left(ax + \frac{1}{2}k\pi\right)$$

                                                                  TI (486)

3. $$\int x^{2n} \sin x \, dx = (2n)! \left\{ \sum_{k=0}^{n} (-1)^{k+1} \frac{x^{2n-2k}}{(2n-2k)!} \cos x + \sum_{k=0}^{n-1} (-1)^k \frac{x^{2n-2k-1}}{(2n-2k-1)!} \sin x \right\}$$

4.    $\displaystyle \int x^{2n+1} \sin x \, dx = (2n+1)! \left\{ \sum_{k=0}^{n} (-1)^{k+1} \frac{x^{2n-2k+1}}{(2n-2k+1)!} \cos x + \sum_{k=0}^{n} (-1)^k \frac{x^{2n-2k}}{(2n-2k)!} \sin x \right\}$

5.    $\displaystyle \int x^{2n} \cos x \, dx = (2n)! \left\{ \sum_{k=0}^{n} (-1)^k \frac{x^{2n-2k}}{(2n-2k)!} \sin x + \sum_{k=0}^{n-1} (-1)^k \frac{x^{2n-2k-1}}{(2n-2k-1)!} \cos x \right\}$

6.    $\displaystyle \int x^{2n+1} \cos x \, dx = (2n+1)! \left\{ \sum_{k=0}^{n} (-1)^k \frac{x^{2n-2k+1}}{(2n-2k+1)!} \sin x + \sum_{k=0}^{n} (-1)^k \frac{x^{2n-2k}}{(2n-2k)!} \cos x \right\}$

**2.634**

1.    $\displaystyle \int P_n(x) \sin mx \, dx = -\frac{\cos mx}{m} \sum_{k=0}^{\lfloor n/2 \rfloor} (-1)^k \frac{P_n^{(2k)}(x)}{m^{2k}} + \frac{\sin mx}{m} \sum_{k=1}^{\lfloor (n+1)/2 \rfloor} (-1)^{k-1} \frac{P_n^{(2k-1)}(x)}{m^{2k-1}}$

2.    $\displaystyle \int P_n(x) \cos mx \, dx = \frac{\sin mx}{m} \sum_{k=0}^{\lfloor n/2 \rfloor} (-1)^k \frac{P_n^{(2k)}(x)}{m^{2k}} + \frac{\cos mx}{m} \sum_{k=1}^{\lfloor (n+1)/2 \rfloor} (-1)^{k-1} \frac{P_n^{(2k-1)}(x)}{m^{2k-1}}$

In formulas **2.634**, $P_n(x)$ is any $n^{\text{th}}$-degree polynomial and $P_n^{(k)}(x)$ is its $k^{\text{th}}$ derivative with respect to $x$.

**2.635**    **Notation:** $z_1 = a + bx$.

1.    $\displaystyle \int z_1 \sin kx \, dx = -\frac{1}{k} z_1 \cos kx + \frac{b}{k^2} \sin kx$

2.    $\displaystyle \int z_1 \cos kx \, dx = \frac{1}{k} z_1 \sin kx + \frac{b}{k^2} \cos kx$

3.    $\displaystyle \int z_1^2 \sin kx \, dx = \frac{1}{k} \left( \frac{2b^2}{k^2} - z_1^2 \right) \cos kx + \frac{2bz_1}{k^2} \sin kx$

4.    $\displaystyle \int z_1^2 \cos kx \, dx = \frac{1}{k} \left( z_1^2 - \frac{2b^2}{k^2} \right) \sin kx + \frac{2bz_1}{k^2} \cos kx$

5.    $\displaystyle \int z_1^3 \sin kx \, dx = \frac{z_1}{k} \left( \frac{6b^2}{k^2} - z_1^2 \right) \cos kx + \frac{3b}{k^2} \left( z_1^2 - \frac{2b^2}{k^2} \right) \sin kx$

6.    $\displaystyle \int z_1^3 \cos kx \, dx = \frac{z_1}{k} \left( z_1^2 - \frac{6b^2}{k^2} \right) \sin kx + \frac{3b}{k^2} \left( z_1^2 - \frac{2b^2}{k^2} \right) \cos kx$

7.    $\displaystyle \int z_1^4 \sin kx \, dx = -\frac{1}{k} \left( z_1^4 - \frac{12b^2}{k^2} z_1^2 + \frac{24b^4}{k^4} \right) \cos kx + \frac{4bz_1}{k^2} \left( z_1^2 - \frac{6b^2}{k^2} \right) \sin kx$

8.    $\displaystyle \int z_1^4 \cos kx \, dx = \frac{1}{k} \left( z_1^4 - \frac{12b^2}{k^2} z_1^2 + \frac{24b^4}{k^4} \right) \sin kx + \frac{4bz_1}{k^2} \left( z_1^2 - \frac{6b^2}{k^2} \right) \cos kx$

9.    $\displaystyle \int z_1^5 \sin kx \, dx = \frac{5b}{k^2} \left( z_1^4 - \frac{12b^2}{k^2} z_1^2 + \frac{24b^4}{k^4} \right) \sin kx - \frac{z_1}{k} \left( z_1^4 - \frac{20b^2}{k^2} z_1^2 + \frac{120b^4}{k^4} \right) \cos kx$

10.    $\displaystyle \int z_1^5 \cos kx \, dx = \frac{5b}{k^2} \left( z_1^4 - \frac{12b^2}{k^2} z_1^2 + \frac{24b^4}{k^4} \right) \cos kx + \frac{z_1}{k} \left( z_1^4 - \frac{20b^2}{k^2} z_1^2 + \frac{120b^4}{k^4} \right) \sin kx$

11. $\int z_1^6 \sin kx \, dx = \dfrac{6bz_1}{k^2} \left( z_1^4 - \dfrac{20b^2}{k^2} z_1^2 + \dfrac{120b^4}{k^4} \right) \sin kx$

$\qquad\qquad\qquad - \dfrac{1}{k} \left( z_1^6 - \dfrac{30b^2}{k^2} z_1^4 + \dfrac{360b^4}{k^4} z_1^2 - \dfrac{720b^6}{k^6} \right) \cos kx$

12. $\int z_1^6 \cos kx \, dx = \dfrac{6bz_1}{k^2} \left( z_1^4 - \dfrac{20b^2}{k^2} z_1^2 + \dfrac{120b^4}{k^4} \right) \cos kx$

$\qquad\qquad\qquad + \dfrac{1}{k} \left( z_1^6 - \dfrac{30b^2}{k^2} z_1^4 + \dfrac{360b^4}{k^4} z_1^2 - \dfrac{720b^6}{k^6} \right) \sin kx$

**2.636**

1. $\int x^n \sin^2 x \, dx = \dfrac{x^{n+1}}{2(n+1)}$

$\qquad + \dfrac{n!}{4} \left\{ \displaystyle\sum_{k=0}^{\lfloor n/2 \rfloor} \dfrac{(-1)^{k+1} x^{n-2k}}{2^{2k}(n-2k)!} \sin 2x + \displaystyle\sum_{k=0}^{\lfloor (n-1)/2 \rfloor} \dfrac{(-1)^{k+1} x^{n-2k-1}}{2^{2k+1}(n-2k-1)!} \cos 2x \right\}$

$\qquad\qquad\qquad\qquad\qquad\qquad\qquad\qquad\qquad\qquad\qquad\qquad\qquad\qquad\qquad$ GU (333)(2e)

2. $\int x^n \cos^2 x \, dx = \dfrac{x^{n+1}}{2(n+1)}$

$\qquad - \dfrac{n!}{4} \left\{ \displaystyle\sum_{k=0}^{\lfloor n/2 \rfloor} \dfrac{(-1)^{k+1} x^{n-2k}}{2^{2k}(n-2k)!} \sin 2x + \displaystyle\sum_{k=0}^{\lfloor (n-1)/2 \rfloor} \dfrac{(-1)^{k+1} x^{n-2k-1}}{2^{2k+1}(n-2k-1)!} \cos 2x \right\}$

$\qquad\qquad\qquad\qquad\qquad\qquad\qquad\qquad\qquad\qquad\qquad\qquad\qquad\qquad\qquad$ GU (333)(3e)

3. $\int x \sin^2 x \, dx = \dfrac{x^2}{4} - \dfrac{x}{4} \sin 2x - \dfrac{1}{8} \cos 2x$

4. $\int x^2 \sin^2 x \, dx = \dfrac{x^3}{6} - \dfrac{x}{4} \cos 2x - \dfrac{1}{4} \left( x^2 - \dfrac{1}{2} \right) \sin 2x$          MZ 241

5. $\int x \cos^2 x \, dx = \dfrac{x^2}{4} + \dfrac{x}{4} \sin 2x + \dfrac{1}{8} \cos 2x$

6. $\int x^2 \cos^2 x \, dx = \dfrac{x^3}{6} + \dfrac{x}{4} \cos 2x + \dfrac{1}{4} \left( x^2 - \dfrac{1}{2} \right) \sin 2x$          MZ 245

**2.637**

1.[11] $\int x^n \sin^3 x \, dx = \dfrac{n!}{4} \left\{ \displaystyle\sum_{k=0}^{\lfloor n/2 \rfloor} \dfrac{(-1)^k x^{n-2k}}{(n-2k)!} \left( \dfrac{\cos 3x}{3^{2k+1}} - 3\cos x \right) \right.$

$\qquad\qquad\qquad \left. - \displaystyle\sum_{k=0}^{\lfloor (n-1)/2 \rfloor} (-1)^k \dfrac{x^{n-2k-1}}{(n-2k-1)!} \left( \dfrac{\sin 3x}{3^{2k+2}} - 3\sin x \right) \right\}$

$\qquad\qquad\qquad\qquad\qquad\qquad\qquad\qquad\qquad\qquad\qquad\qquad\qquad\qquad\qquad$ GU(333)(2f)

2. $\int x^n \cos^3 x \, dx = \dfrac{n!}{4} \left\{ \displaystyle\sum_{k=0}^{\lfloor n/2 \rfloor} \dfrac{(-1)^k x^{n-2k}}{(n-2k)!} \left( \dfrac{\sin 3x}{3^{2k+1}} + 3 \sin x \right) \right.$

$\left. + \displaystyle\sum_{k=0}^{[(n-1)/2]} (-1)^k \dfrac{x^{n-2k-1}}{(n-2k-1)!} \left( \dfrac{\cos 3x}{3^{2k+2}} + 3 \cos x \right) \right\}$

GU(333)(3f)

3. $\int x \sin^3 x \, dx = \dfrac{3}{4} \sin x - \dfrac{1}{36} \sin 3x - \dfrac{3}{4} x \cos x + \dfrac{x}{12} \cos 3x$

4. $\int x^2 \sin^3 x \, dx = - \left( \dfrac{3}{4} x^2 + \dfrac{3}{2} \right) \cos x + \left( \dfrac{x^2}{12} + \dfrac{1}{54} \right) \cos 3x + \dfrac{3}{2} x \sin x - \dfrac{x}{18} \sin 3x$  MZ 241

5. $\int x \cos^3 x \, dx = \dfrac{3}{4} \cos x + \dfrac{1}{36} \cos 3x + \dfrac{3}{4} x \sin x + \dfrac{x}{12} \sin 3x$

6. $\int x^2 \cos^3 x \, dx = \left( \dfrac{3}{4} x^2 - \dfrac{3}{2} \right) \sin x + \left( \dfrac{x^2}{12} - \dfrac{1}{54} \right) \sin 3x + \dfrac{3}{2} x \cos x + \dfrac{x}{18} \cos 3x$  MZ 245, 246

**2.638**

1. $\int \dfrac{\sin^q x}{x^p} \, dx = - \dfrac{\sin^{q-1} x \left[ (p-2) \sin x + qx \cos x \right]}{(p-1)(p-2)x^{p-1}}$

$- \dfrac{q^2}{(p-1)(p-2)} \int \dfrac{\sin^q x \, dx}{x^{p-2}} + \dfrac{q(q-1)}{(p-1)(p-2)} \int \dfrac{\sin^{q-2} x \, dx}{x^{p-2}}$

$[p \neq 1, \quad p \neq 2]$  TI (496)

2. $\int \dfrac{\cos^q x}{x^p} \, dx = - \dfrac{\cos^{q-1} x \left[ (p-2) \cos x - qx \sin x \right]}{(p-1)(p-2)x^{p-1}}$

$- \dfrac{q^2}{(p-1)(p-2)} \int \dfrac{\cos^q x \, dx}{x^{p-2}} + \dfrac{q(q-1)}{(p-1)(p-2)} \int \dfrac{\cos^{q-2} x \, dx}{x^{p-2}}$

$[p \neq 1, \quad p \neq 2]$  TI (495)

3.[6] $\int \dfrac{\sin x \, dx}{x^p} = - \dfrac{\sin x}{(p-1)x^{p-1}} + \dfrac{1}{p-1} \int \dfrac{\cos x \, dx}{x^{p-1}}$

$= - \dfrac{\sin x}{(p-1)x^{p-1}} - \dfrac{\cos x}{(p-1)(p-2)x^{p-2}} - \dfrac{1}{(p-1)(p-2)} \int \dfrac{\sin x \, dx}{x^{p-2}}$

$(p > 2)$  TI (492)

4.[6] $\int \dfrac{\cos x \, dx}{x^p} = - \dfrac{\cos x}{(p-1)x^{p-1}} - \dfrac{1}{p-1} \int \dfrac{\sin x \, dx}{x^{p-1}}$

$= - \dfrac{\cos x}{(p-1)x^{p-1}} + \dfrac{\sin x}{(p-1)(p-2)x^{p-2}} - \dfrac{1}{(p-1)(p-2)} \int \dfrac{\cos x \, dx}{x^{p-2}}$

$(p > 2)$  TI (491)

**2.639**

1.  $$\int \frac{\sin x \, dx}{x^{2n}} = \frac{(-1)^{n+1}}{x(2n-1)!} \left\{ \sum_{k=0}^{n-2} \frac{(-1)^k (2k+1)!}{x^{2k+1}} \cos x \right.$$
$$\left. + \sum_{k=0}^{n-1} \frac{(-1)^{k+1}(2k)!}{x^{2k}} \sin x \right\} + \frac{(-1)^{n+1}}{(2n-1)!} \operatorname{Ci}(x)$$

GU (333)(6b)a

2.  $$\int \frac{\sin x}{x^{2n+1}} \, dx = \frac{(-1)^{n+1}}{x(2n)!} \left\{ \sum_{k=0}^{n-1} \frac{(-1)^{k+1}(2k)!}{x^{2k}} \cos x \right.$$
$$\left. + \sum_{k=0}^{n-1} \frac{(-1)^{k+1}(2k+1)!}{x^{2k+1}} \sin x \right\} + \frac{(-1)^n}{(2n)!} \operatorname{Si}(x)$$

GU (333)(6b)a

3.  $$\int \frac{\cos x \, dx}{x^{2n}} \, dx = \frac{(-1)^{n+1}}{x(2n-1)!} \left\{ \sum_{k=0}^{n-1} \frac{(-1)^{k+1}(2k)!}{x^{2k}} \cos x \right.$$
$$\left. - \sum_{k=0}^{n-2} \frac{(-1)^k (2k+1)!}{x^{2k+1}} \sin x \right\} + \frac{(-1)^n}{(2n-1)!} \operatorname{Si}(x)$$

GU (333)(7b)

4.  $$\int \frac{\cos x \, dx}{x^{2n+1}} = \frac{(-1)^{n+1}}{x(2n)!} \left\{ \sum_{k=0}^{n-1} \frac{(-1)^{k+1}(2k+1)!}{x^{2k+1}} \cos x \right.$$
$$\left. - \sum_{k=0}^{n-1} \frac{(-1)^{k+1}(2k)!}{x^{2k}} \sin x \right\} + \frac{(-1)^n}{(2n)!} \operatorname{Ci}(x)$$

GU (333)(7b)

**2.641**

1.  $$\int \frac{\sin kx}{a+bx} \, dx = \frac{1}{b} \left[ \cos \frac{ka}{b} \operatorname{si}(u) - \sin \frac{ka}{b} \operatorname{ci}(u) \right] \qquad \left[ u = \frac{k}{b}(a+bx) \right]$$

2.  $$\int \frac{\cos kx}{a+bx} \, dx = \frac{1}{b} \left[ \cos \frac{ka}{b} \operatorname{ci}(u) + \sin \frac{ka}{b} \operatorname{si}(u) \right] \qquad \left[ u = \frac{k}{b}(a+bx) \right]$$

3.  $$\int \frac{\sin kx}{(a+bx)^2} \, dx = -\frac{1}{b}\frac{\sin kx}{a+bx} + \frac{k}{b} \int \frac{\cos kx}{a+bx} \, dx \qquad (\text{see } \mathbf{2.641} \ 2)$$

4.  $$\int \frac{\cos kx}{(a+bx)^2} \, dx = -\frac{1}{b}\frac{\cos kx}{a+bx} - \frac{k}{b} \int \frac{\sin kx}{a+bx} \, dx \qquad (\text{see } \mathbf{2.641} \ 1)$$

5.  $$\int \frac{\sin kx}{(a+bx)^3} \, dx = -\frac{\sin kx}{2b(a+bx)^2} - \frac{k \cos kx}{2b^2(a+bx)} - \frac{k^2}{2b^2} \int \frac{\sin kx}{a+bx} \, dx$$

(see 2.641 1)

6. $\int \dfrac{\cos kx}{(a+bx)^3}\,\mathrm{d}x = -\dfrac{\cos kx}{2b(a+bx)^2} + \dfrac{k\sin kx}{2b^2(a+bx)} - \dfrac{k^2}{2b^2}\int \dfrac{\cos kx}{a+bx}\,\mathrm{d}x$

(see **2.641** 2)

7.[12] $\int \dfrac{\sin kx}{(a+bx)^4}\,\mathrm{d}x = -\dfrac{\sin kx}{3b(a+bx)^3} - \dfrac{k\cos kx}{6b^2(a+bx)^2}$
$+\dfrac{k^2\sin kx}{6b^3(a+bx)} - \dfrac{k^3}{6b^3}\int \dfrac{\cos kx}{a+bx}\,\mathrm{d}x$

(see **2.641** 2)

8. $\int \dfrac{\cos kx}{(a+bx)^4}\,\mathrm{d}x = -\dfrac{\cos kx}{3b(a+bx)^3} + \dfrac{k\sin kx}{6b^2(a+bx)^2} + \dfrac{k^2\cos kx}{6b^3(a+bx)} + \dfrac{k^3}{6b^3}\int \dfrac{\sin kx}{a+bx}\,\mathrm{d}x$

(see **2.641** 1)

9.[12] $\int \dfrac{\sin kx}{(a+bx)^5}\,\mathrm{d}x = -\dfrac{\sin kx}{4b(a+bx)^4} - \dfrac{k\cos kx}{12b^2(a+bx)^3}$
$+\dfrac{k^2\sin kx}{24b^3(a+bx)^2} + \dfrac{k^3\cos kx}{24b^4(a+bx)} + \dfrac{k^4}{24b^4}\int \dfrac{\sin kx}{a+bx}\,\mathrm{d}x$

(see **2.641** 1)

10. $\int \dfrac{\cos kx}{(a+bx)^5}\,\mathrm{d}x = -\dfrac{\cos kx}{4b(a+bx)^4} + \dfrac{k\sin kx}{12b^2(a+bx)^3}$
$+\dfrac{k^2\cos kx}{24b^3(a+bx)^2} - \dfrac{k^3\sin kx}{24b^4(a+bx)} + \dfrac{k^4}{24b^4}\int \dfrac{\cos kx}{a+bx}\,\mathrm{d}x$

(see **2.641** 2)

11. $\int \dfrac{\sin kx}{(a+bx)^6}\,\mathrm{d}x = -\dfrac{\sin kx}{5b(a+bx)^5} - \dfrac{k\cos kx}{20b^2(a+bx)^4} + \dfrac{k^2\sin kx}{60b^3(a+bx)^3} + \dfrac{k^3\cos kx}{120b^4(a+bx)^2}$
$-\dfrac{k^4\sin kx}{120b^5(a+bx)} + \dfrac{k^5}{120b^5}\int \dfrac{\cos kx}{a+bx}\,\mathrm{d}x$

(see **2.641** 2)

12. $\int \dfrac{\cos kx}{(a+bx)^6}\,\mathrm{d}x = -\dfrac{\cos kx}{5b(a+bx)^5} + \dfrac{k\sin kx}{20b^2(a+bx)^4} + \dfrac{k^2\cos kx}{60b^3(a+bx)^3}$
$-\dfrac{k^3\sin kx}{120b^4(a+bx)^2} - \dfrac{k^4\cos kx}{120b^5(a+bx)} - \dfrac{k^5}{120b^5}\int \dfrac{\sin kx}{a+bx}\,\mathrm{d}x$

(see **2.641** 1)

**2.642**

1. $\int \dfrac{\sin^{2m}x}{x}\,\mathrm{d}x = \dbinom{2m}{m}\dfrac{\ln x}{2^{2m}} + \dfrac{(-1)^m}{2^{2m-1}}\sum_{k=0}^{m-1}(-1)^k\dbinom{2m}{k}\operatorname{ci}[(2m-2k)x]$

2. $\int \dfrac{\sin^{2m+1}x}{x}\,\mathrm{d}x = \dfrac{(-1)^m}{2^{2m}}\sum_{k=0}^{m}(-1)^k\dbinom{2m+1}{k}\operatorname{si}[(2m-2k+1)x]$

3. $\int \dfrac{\cos^{2m}x}{x}\,\mathrm{d}x = \dbinom{2m}{m}\dfrac{\ln x}{2^{2m}} + \dfrac{1}{2^{2m-1}}\sum_{k=0}^{m-1}\dbinom{2m}{k}\operatorname{ci}[(2m-2k)x]$

4. $\int \dfrac{\cos^{2m+1}x}{x}\,\mathrm{d}x = \dfrac{1}{2^{2m}}\sum_{k=0}^{m}\dbinom{2m+1}{k}\operatorname{ci}[(2m-2k+1)x]$

5.
$$\int \frac{\sin^{2m} x}{x^2} \, dx = -\binom{2m}{m} \frac{1}{2^{2m} x}$$
$$+ \frac{(-1)^m}{2^{2m-1}} \sum_{k=0}^{m-1} (-1)^{k+1} \binom{2m}{k} \left\{ \frac{\cos(2m-2k)x}{x} + (2m-2k)\operatorname{si}[(2m-2k)x] \right\}$$

6.
$$\int \frac{\sin^{2m+1} x}{x^2} \, dx = \frac{(-1)^m}{2^{2m}} \sum_{k=0}^{m} (-1)^{k+1} \binom{2m+1}{k}$$
$$\times \left\{ \frac{\sin(2m-2k+1)x}{x} - (2m-2k+1)\operatorname{ci}[(2m-2k+1)x] \right\}$$

7.
$$\int \frac{\cos^{2m} x}{x^2} \, dx = -\binom{2m}{m} \frac{1}{2^{2m} x}$$
$$- \frac{1}{2^{2m-1}} \sum_{k=0}^{m-1} \binom{2m}{k} \left\{ \frac{\cos(2m-2k)x}{x} + (2m-2k)\operatorname{si}[(2m-2k)x] \right\}$$

8.
$$\int \frac{\cos^{2m+1} x}{x^2} = -\frac{1}{2^{2m}} \sum_{k=0}^{m} \binom{2m+1}{k} \left\{ \frac{\cos(2m-2k+1)x}{x} \right.$$
$$\left. + (2m-2k+1)\operatorname{si}[(2m-2k+1)x] \right\}$$

**2.643**

1.
$$\int \frac{x^p \, dx}{\sin^q x} = -\frac{x^{p-1}[p \sin x + (q-2)x \cos x]}{(q-1)(q-2)\sin^{q-1} x} + \frac{q-2}{q-1} \int \frac{x^p \, dx}{\sin^{q-2} x} + \frac{p(p-1)}{(q-1)(q-2)} \int \frac{x^{p-2} \, dx}{\sin^{q-2} x}$$

2.
$$\int \frac{x^p \, dx}{\cos^q x} = -\frac{x^{p-1}[p \cos x - (q-2)x \sin x]}{(q-1)(q-2)\cos^{q-1} x}$$
$$+ \frac{q-2}{q-1} \int \frac{x^p \, dx}{\cos^{q-2} x} + \frac{p(p-1)}{(q-1)(q-2)} \int \frac{x^{p-2} \, dx}{\cos^{q-2} x}$$

3.[4]
$$\int \frac{x^n}{\sin x} \, dx = \frac{x^n}{n} + \sum_{k=1}^{\infty} (-1)^{k+1} \frac{2(2^{2k-1}-1)}{(n+2k)(2k)!} B_{2k} x^{n+2k}$$
$$[|x| < \pi, \quad n > 0] \qquad \text{TU (333)(8b)}$$

4.[12]
$$\int \frac{dx}{x^n \sin x} = -\frac{1}{nx^n} - [1+(-1)^n](-1)^{\frac{n}{2}} \frac{2^{n-1}-1}{n!} B_n \ln x - \sum_{\substack{k=1 \\ k \neq \frac{n}{2}}}^{\infty} (-1)^k \frac{2(2^{2n}-1)}{(2k-n) \cdot (2k)!} B_{2k} x^{2k-n}$$
$$[n > 1, \quad |x| < \pi] \qquad \text{GU (333)(9b)}$$

5.[8]
$$\int \frac{x^n \, dx}{\cos x} = \sum_{k=0}^{\infty} \frac{|E_{2k}| x^{n+2k+1}}{(n+2k+1)(2k)!} \qquad \left[|x| < \frac{\pi}{2}, \quad n > 0\right] \qquad \text{GU (333)(10b)}$$

6.
$$\int \frac{dx}{x^n \cos x} = \frac{1}{2}[1-(-1)^n] \frac{|E_{n-1}|}{(n-1)!} \ln x + \sum_{\substack{k=0 \\ k \neq \frac{n-1}{2}}}^{\infty} \frac{|E_{2k}| x^{2k-n+1}}{(2k-n+1) \cdot (2k)!}$$
$$\left[|x| < \frac{\pi}{2}\right] \qquad \text{GU (333)(11b)}$$

7. $\quad \displaystyle\int \frac{x^n \, dx}{\sin^2 x} = -x^n \cot x + \frac{n}{n-1} x^{n-1} + n \sum_{k=1}^{\infty} (-1)^k \frac{2^{2k} x^{n+2k-1}}{(n+2k-1)(2k)!} B_{2k}$

$$[|x| < \pi, \quad n > 1] \qquad \text{GU (333)(8c)}$$

8. $\quad \displaystyle\int \frac{dx}{x^n \sin^2 x} = -\frac{\cot x}{x^n} + \frac{n}{(n+1)x^{n+1}} - [1-(-1)^n](-1)^{\frac{n+1}{2}} \frac{2^n n}{(n+1)!} B_{n+1} \ln x$

$$-\frac{n}{2^{n+1}} \sum_{\substack{k=1 \\ k \neq \frac{n+1}{2}}}^{\infty} \frac{(-1)^k (2x)^{2k}}{(2k-n-1)(2k)!} B_{2k}$$

$$[|x| < \pi] \qquad \text{GU (333)(9c)}$$

9. $\quad \displaystyle\int \frac{x^n \, dx}{\cos^2 x} = x^n \tan x + n \sum_{k=1}^{\infty} (-1)^k \frac{2^{2k} \left(2^{2k} - 1\right) x^{n+2k-1}}{(n+2k-1)\cdot(2k)!} B_{2k}$

$$\left[n > 1, \quad |x| < \frac{\pi}{2}\right] \qquad \text{GU (333)(10c)}$$

10. $\quad \displaystyle\int \frac{dx}{x^n \cos^2 x} = \frac{\tan x}{x^n} - [1-(-1)^n](-1)^{\frac{n+1}{2}} \frac{2^n n}{(n+1)!} \left(2^{n+1} - 1\right) B_{n+1} \ln x$

$$-\frac{n}{x^{n+1}} \sum_{\substack{k=1 \\ k \neq \frac{n+1}{2}}}^{\infty} \frac{(-1)^k \left(2^{2k} - 1\right) (2x)^{2k}}{(2k-n-1)(2k)!} B_{2k}$$

$$\left[|x| < \frac{\pi}{2}\right] \qquad \text{GU (333)(11c)}$$

**2.644**

1. $\quad \displaystyle\int \frac{x \, dx}{\sin^{2n} x} = -\sum_{k=0}^{n-1} \frac{(2n-2)(2n-4)\ldots(2n-2k+2)}{(2n-1)(2n-3)\ldots(2n-2k+3)} \frac{\sin x + (2n-2k)x \cos x}{(2n-2k+1)(2n-2k)\sin^{2n-2k+1} x}$

$$+\frac{2^{n-1}(n-1)!}{(2n-1)!!} \left(\ln \sin x - x \cot x\right)$$

2. $\quad \displaystyle\int \frac{x \, dx}{\sin^{2n+1} x} = -\sum_{k=0}^{n-1} \frac{(2n-1)(2n-3)\ldots(2n-2k+1)}{2n(2n-2)\ldots(2n-2k+2)} \frac{\sin x + (2n-2k-1)x \cos x}{(2n-2k)(2n-2k-1)\sin^{2n-2k} x}$

$$+\frac{(2n-1)!!}{2^n n!} \int \frac{x \, dx}{\sin x}$$

$$\text{(see 2.644 5)}$$

3. $\quad \displaystyle\int \frac{x \, dx}{\cos^{2n} x} = \sum_{k=0}^{n-1} \frac{(2n-2)(2n-4)\ldots(2n-2k+2)}{(2n-1)(2n-3)\ldots(2n-2k+3)} \frac{(2n-2k)x \sin x - \cos x}{(2n-2k+1)(2n-2k)\cos^{2n-2k+1} x}$

$$+\frac{2^{n-1}(n-1)!}{(2n-1)!!} \left(x \tan x + \ln \cos x\right)$$

4. $\quad \displaystyle\int \frac{x \, dx}{\cos^{2n+1} x} = \sum_{k=0}^{n-1} \frac{(2n-1)(2n-3)\ldots(2n-2k+1)}{2n(2n-2)\ldots(2n-2k+2)} \frac{(2n-2k+1)x \sin x - \cos x}{(2n-2k)(2n-2k-1)\cos^{2n-2k} x}$

$$+\frac{(2n-1)!!}{2^n n!} \int \frac{x \, dx}{\cos x}$$

$$\text{(see 2.644 6)}$$

5. $$\int \frac{x\,dx}{\sin x} = x + \sum_{k=1}^{\infty}(-1)^{k+1}\frac{2\left(2^{2k-1}-1\right)}{(2k+1)!}B_{2k}x^{2k+1}$$

6. $$\int \frac{x\,dx}{\cos x} = \sum_{k=0}^{\infty}\frac{|E_{2k}|x^{2k+2}}{(2k+2)(2k)!}$$

7. $$\int \frac{x\,dx}{\sin^2 x} = -x\cot x + \ln\sin x$$

8. $$\int \frac{x\,dx}{\cos^2 x} = x\tan x + \ln\cos x$$

9. $$\int \frac{x\,dx}{\sin^3 x} = -\frac{\sin x + x\cos x}{2\sin^2 x} + \frac{1}{2}\int \frac{x}{\sin x}\,dx \qquad\qquad \text{(see } \mathbf{2.644\ 5})$$

10. $$\int \frac{x\,dx}{\cos^3 x} = \frac{x\sin x - \cos x}{2\cos^2 x} + \frac{1}{2}\int \frac{x\,dx}{\cos x} \qquad\qquad \text{(see } \mathbf{2.644\ 6})$$

11. $$\int \frac{x\,dx}{\sin^4 x} = -\frac{x\cos x}{3\sin^3 x} - \frac{1}{6\sin^2 x} - \frac{2}{3}x\cot x + \frac{2}{3}\ln\left(\sin x\right)$$

12. $$\int \frac{x\,dx}{\cos^4 x} = \frac{x\sin x}{3\cos^3 x} - \frac{1}{6\cos^2 x} + \frac{2}{3}x\tan x - \frac{2}{3}\ln\left(\cos x\right)$$

13. $$\int \frac{x\,dx}{\sin^5 x} = -\frac{x\cos x}{4\sin^4 x} - \frac{1}{12\sin^3 x} - \frac{3x\cos x}{8\sin^2 x} - \frac{3}{8\sin x} + \frac{3}{8}\int \frac{x\,dx}{\sin x}$$

$$\text{(see } \mathbf{2.644\ 5})$$

14. $$\int \frac{x\,dx}{\cos^5 x} = \frac{x\sin x}{4\cos^4 x} - \frac{1}{12\cos^3 x} + \frac{3x\sin x}{8\cos^2 x} - \frac{3}{8\cos x} + \frac{3}{8}\int \frac{x\,dx}{\cos x}$$

$$\text{(see } \mathbf{2.644\ 6})$$

## 2.645

1. $$\int x^p \frac{\sin^{2m} x}{\cos^n x}\,dx = \sum_{k=0}^{m}(-1)^k \binom{m}{k}\int \frac{x^p\,dx}{\cos^{n-2k} x} \qquad\qquad \text{(see } \mathbf{2.643\ 2})$$

2. $$\int x^p \frac{\sin^{2m+1} x}{\cos^n x}\,dx = \sum_{k=0}^{m}(-1)^k \binom{m}{k}\int \frac{x^p\sin x}{\cos^{n-2k} x}\,dx \qquad\qquad \text{(see } \mathbf{2.645\ 3})$$

3. $$\int x^p \frac{\sin x\,dx}{\cos^n x} = \frac{x^p}{(n-1)\cos^{n-1} x} - \frac{p}{n-1}\int \frac{x^{p-1}}{\cos^{n-1} x}\,dx$$

$$[n>1] \qquad \text{(see } \mathbf{2.643\ 2}) \qquad \text{GU (333)(12)}$$

4. $$\int x^p \frac{\cos^{2m} x}{\sin^n x}\,dx = \sum_{k=0}^{m}(-1)^k \binom{m}{k}\int \frac{x^p\,dx}{\sin^{n-2k} x} \qquad\qquad \text{(see } \mathbf{2.643\ 1})$$

5. $$\int x^p \frac{\cos^{2m+1} x}{\sin^n x}\,dx = \sum_{k=0}^{m}(-1)^k \binom{m}{k}\int \frac{x^p\cos x}{\sin^{n-2k} x}\,dx \qquad\qquad \text{(see } \mathbf{2.645\ 6})$$

6. $$\int x^p \frac{\cos x}{\sin^n x} = -\frac{x^p}{(n-1)\sin^{n-1} x} + \frac{p}{n-1}\int \frac{x^{p-1}\,dx}{\sin^{n-1} x} \qquad [n>1] \qquad \text{(see } \mathbf{2.643\ 1}) \qquad \text{GU (333)(13)}$$

7. $\quad \int \dfrac{x \cos x}{\sin^2 x}\, dx = -\dfrac{x}{\sin x} + \ln \tan \dfrac{x}{2}$

8. $\quad \int \dfrac{x \sin x}{\cos^2 x}\, dx = \dfrac{x}{\cos x} - \ln \tan \left( \dfrac{x}{2} + \dfrac{\pi}{4} \right)$

**2.646**

1. $\quad \int x^p \tan x\, dx = \displaystyle\sum_{k=1}^{\infty} (-1)^{k+1} \dfrac{2^{2k} \left( 2^{2k-1} - 1 \right)}{(p + 2k) \cdot (2k)!} B_{2k} x^{p+2k}$

$$\left[ p \geq -1, \quad |x| < \dfrac{\pi}{2} \right] \qquad \text{GU (333)(12d)}$$

2. $\quad \int x^p \cot x\, dx = \displaystyle\sum_{k=0}^{\infty} (-1)^k \dfrac{2^{2k} B_{2k}}{(p + 2k)(2k)!} x^{p+2k}$ $\qquad [p \geq 1, \quad |x| < \pi] \qquad$ GU (333)(13d)

3.[12] $\quad \int x \tan^2 x\, dx = x \tan x + \ln \cos x - \dfrac{x^2}{2}$

4. $\quad \int x \cot^2 x\, dx = -x \cot x + \ln \sin x - \dfrac{x^2}{2}$

**2.647**

1. $\quad \int \dfrac{x^n \cos x\, dx}{(a + b \sin x)^m} = -\dfrac{x^n}{(m-1)b\,(a + b \sin x)^{m-1}} + \dfrac{n}{(m-1)b} \int \dfrac{x^{n-1}\, dx}{(a + b \sin x)^{m-1}}$

$$[m \neq 1] \qquad \text{MZ 247}$$

2. $\quad \int \dfrac{x^n \sin x\, dx}{(a + b \cos x)^m} = \dfrac{x^n}{(m-1)b\,(a + b \cos x)^{m-1}} - \dfrac{n}{(m-1)b} \int \dfrac{x^{n-1}\, dx}{(a + b \cos x)^{m-1}}$

$$[m \neq 1] \qquad \text{MZ 247}$$

3. $\quad \int \dfrac{x\, dx}{1 + \sin x} = -x \tan \left( \dfrac{\pi}{4} - \dfrac{x}{2} \right) + 2 \ln \cos \left( \dfrac{\pi}{4} - \dfrac{x}{2} \right) \qquad$ PE (329)

4. $\quad \int \dfrac{x\, dx}{1 - \sin x} = x \cot \left( \dfrac{\pi}{4} - \dfrac{x}{2} \right) + 2 \ln \sin \left( \dfrac{\pi}{4} - \dfrac{x}{2} \right) \qquad$ PE (330)

5. $\quad \int \dfrac{x\, dx}{1 + \cos x} = x \tan \dfrac{x}{2} + 2 \ln \cos \dfrac{x}{2} \qquad$ PE (331)

6.[12] $\quad \int \dfrac{x\, dx}{1 - \cos x} = -x \cot \dfrac{x}{2} + 2 \ln \sin \dfrac{\pi}{2} \qquad$ PE (332)

7. $\quad \int \dfrac{x \cos x}{(1 + \sin x)^2}\, dx = -\dfrac{x}{1 + \sin x} + \tan \left( \dfrac{x}{2} - \dfrac{\pi}{4} \right)$

8. $\quad \int \dfrac{x \cos x}{(1 - \sin x)^2}\, dx = \dfrac{x}{1 - \sin x} + \tan \left( \dfrac{x}{2} + \dfrac{\pi}{4} \right)$

9. $\quad \int \dfrac{x \sin x}{(1 + \cos x)^2}\, dx = \dfrac{x}{1 + \cos x} - \tan \dfrac{x}{2}$

10. $\quad \int \dfrac{x \sin x}{(1 - \cos x)^2}\, dx = -\dfrac{x}{1 - \cos x} - \cot \dfrac{x}{2} \qquad$ MZ 247a

**2.648**

1. $\int \dfrac{x + \sin x}{1 + \cos x}\, dx = x \tan \dfrac{x}{2}$

2. $\int \dfrac{x - \sin x}{1 - \cos x}\, dx = -x \cot \dfrac{x}{2}$        GU (333)(16)

**2.649** $\int \dfrac{x^2\, dx}{\left[(ax - b)\sin x + (a + bx)\cos x\right]^2} = \dfrac{x \sin x + \cos x}{b\left[(ax - b)\sin x + (a + bx)\cos x\right]}$     GU (333)(17)

**2.651** $\int \dfrac{dx}{\left[a + (ax + b)\tan x\right]^2} = \dfrac{\tan x}{a\left[a + (ax + b)\tan x\right]}$     GU (333)(18)

**2.652** $\int \dfrac{x\, dx}{\cos(x + t)\cos(x - t)} = \operatorname{cosec} 2t \left\{ x \ln \dfrac{\cos(x - t)}{\cos(x + t)} - L(x + t) + L(x - t) \right\}$

$$\left[ t \neq n\pi, \quad |x| < \left| \dfrac{\pi}{2} - |t_0| \right| \right],$$

where $t_0$ is the value of the argument $t$, which is reduced by multiples of the argument $\pi$ to lie in the interval $\left( -\frac{\pi}{2}, \frac{\pi}{2} \right)$.        LO III 288

**2.653**

1. $\int \dfrac{\sin x}{\sqrt{x}}\, dx = \sqrt{2\pi}\, S\left( \sqrt{x} \right)$        (cf. **8.251** 21)

2. $\int \dfrac{\cos x}{\sqrt{x}}\, dx = \sqrt{2\pi}\, C\left( \sqrt{x} \right)$        (cf. **8.251** 3)

**2.654**    **Notation:** $\Delta = \sqrt{1 - k^2 \sin^2 x}, \quad k' = \sqrt{1 - k^2}$:

1. $\int \dfrac{x \sin x \cos x}{\Delta}\, dx = -\dfrac{x\Delta}{k^2} + \dfrac{1}{k^2}\, E(x, k)$

2. $\int \dfrac{x \sin^3 x \cos x}{\Delta}\, dx = -\dfrac{k'^2}{9k^4}\, F(x, k) + \dfrac{2k^2 + 5}{9k^4}\, E(x, k) - \dfrac{1}{9k^4}\left[ 3\left(3 - \Delta^2\right) x + k^2 \sin x \cos x \right] \Delta$

3. $\int \dfrac{x \sin x \cos^3 x}{\Delta}\, dx = -\dfrac{k'^2}{9k^4}\, F(x, k) + \dfrac{7k^2 - 5}{9k^4}\, E(x, k) - \dfrac{1}{9k^4}\left[ 3\left(\Delta^2 - 3k'^2\right) x - k^2 \sin x \cos x \right] \Delta$

4. $\int \dfrac{x \sin x\, dx}{\Delta^3}\, dx = -\dfrac{x \cos x}{k'^2 \Delta} + \dfrac{1}{kk'^2}\, \arcsin\left( k \sin x \right)$

5. $\int \dfrac{x \cos x\, dx}{\Delta^3} = \dfrac{x \sin x}{\Delta} + \dfrac{1}{k}\, \ln\left( k \cos x + \Delta \right)$

6. $\int \dfrac{x \sin x \cos x\, dx}{\Delta^3} = \dfrac{x}{k^2 \Delta} - \dfrac{1}{k^2}\, F(x, k)$

7. $\int \dfrac{x \sin^3 x \cos x\, dx}{\Delta^3} = x\dfrac{2 - k^2 \sin^2 x}{k^4 \Delta} - \dfrac{1}{k^4}\left[ E(x, k) + F(x, k) \right]$

8. $\int \dfrac{x \sin x \cos^3 x\, dx}{\Delta^3} = x\dfrac{k^2 \sin^2 x + k^2 - 2}{k^4 \Delta} + \dfrac{k'^2}{k^4}\, F(x, k) + \dfrac{1}{k^4}\, E(x, k)$

**2.655**   Integrals containing $\sin x^2$ and $\cos x^2$

In integrals containing $\sin x^2$ and $\cos x^2$ it is expedient to make the substitution $x^2 = u$.

1.   $\displaystyle \int x^p \sin x^2 \, dx = -\frac{x^{p-1}}{2} \cos x^2 + \frac{p-1}{2} \int x^{p-2} \cos x^2 \, dx$

2.   $\displaystyle \int x^p \cos x^2 \, dx = \frac{x^{p-1}}{2} \sin x^2 - \frac{p-1}{2} \int x^{p-2} \sin x^2 \, dx$

3.   $\displaystyle \int x^n \sin x^2 \, dx = (n-1)!! \left\{ \sum_{k=1}^{r} (-1)^k \left[ \frac{x^{n-4k+3} \cos x^2}{2^{2k-1}(n-4k+3)!!} - \frac{x^{n-4k+1} \sin x^2}{2^{2k}(n-4k+1)!!} \right] \right.$

$\displaystyle \left. + \frac{(-1)^r}{2^{2r}(n-4r-1)!!} \int x^{n-4r} \sin x^2 \, dx \right\}$

$$\left[ r = \left\lfloor \frac{n}{4} \right\rfloor \right] \qquad\qquad \text{GU (336)(4a)}$$

4.   $\displaystyle \int x^n \cos x^2 \, dx = (n-1)!! \left\{ \sum_{k=1}^{r} (-1)^{k-1} \left[ \frac{x^{n-4k+3} \sin x^2}{2^{2k-1}(n-4k+3)!!} + \frac{x^{n-4k+1} \cos x^2}{2^{2k}(n-4k+1)!!} \right] \right.$

$\displaystyle \left. + \frac{(-1)^r}{2^{2r}(n-4r-1)!!} \int x^{n-4r} \cos x^2 \, dx \right\}$

$$\left[ r = \left\lfloor \frac{n}{4} \right\rfloor \right] \qquad\qquad \text{GU (336)(5a)}$$

5.[12]   $\displaystyle \int x \sin x^2 \, dx = -\frac{\cos x^2}{2}$

6.[12]   $\displaystyle \int x \cos x^2 \, dx = \frac{\sin x^2}{2}$

7.   $\displaystyle \int x^2 \sin x^2 \, dx = -\frac{x}{2} \cos x^2 + \frac{1}{2} \sqrt{\frac{\pi}{2}} \, C(x)$

8.   $\displaystyle \int x^2 \cos x^2 \, dx = \frac{x}{2} \sin x^2 - \frac{1}{2} \sqrt{\frac{\pi}{2}} \, S(x)$

9.   $\displaystyle \int x^3 \sin x^2 \, dx = -\frac{x^2}{2} \cos x^2 + \frac{1}{2} \sin x^2$

10.   $\displaystyle \int x^3 \cos x^2 \, dx = \frac{x^2}{2} \sin x^2 + \frac{1}{2} \cos x^2$

## 2.66 Combinations of trigonometric functions and exponentials

**2.661** $\displaystyle\int e^{ax} \sin^p x \cos^q x \, dx = \frac{1}{a^2 + (p+q)^2} \left\{ e^{ax} \sin^p x \cos^{q-1} x \left[ a \cos x + (p+q) \sin x \right] \right.$

$$- pa \int e^{ax} \sin^{p-1} x \cos^{q-1} x \, dx + (q-1)(p+q) \int e^{ax} \sin^p x \cos^{q-2} x \, dx \Bigg\}$$

TI (523)

$$= \frac{1}{a^2 + (p+q)^2} \left\{ e^{ax} \sin^{p-1} x \cos^q x \left[ a \sin x - (p+q) \cos x \right] \right.$$

$$+ qa \int e^{ax} \sin^{p-1} x \cos^{q-1} x \, dx + (p-1)(p+q) \int e^{ax} \sin^{p-2} x \cos^q x \, dx \Bigg\}$$

TI (524)

$$= \frac{1}{a^2 + (p+q)^2} \left\{ e^{ax} \sin^{p-1} x \cos^{q-1} x \left[ a \sin x \cos x + q \sin^2 x - p \cos^2 x \right] \right.$$

$$+ q(q-1) \int e^{ax} \sin^p x \cos^{q-2} x \, dx + p(p-1) \int e^{ax} \sin^{p-2} x \cos^q x \, dx \Bigg\}$$

TI (525)

$$= \frac{1}{a^2 + (p+q)^2} \left\{ e^{ax} \sin^{p-1} x \cos^{q-1} x \left( a \sin x \cos x + q \sin^2 x - p \cos^2 x \right) \right.$$

$$+ q(q-1) \int e^{ax} \sin^{p-2} x \cos^{q-2} x \, dx$$

$$- (q-p)(p+q-1) \int e^{ax} \sin^{p-2} x \cos^q x \, dx \Bigg\}$$

TI (526)

$$= \frac{1}{a^2 + (p+q)^2} \left[ e^{ax} \sin^{p-1} x \cos^{q-1} x \left( a \sin x \cos x + q \sin^2 x - p \cos^2 x \right) \right.$$

$$+ p(p-1) \int e^{ax} \sin^{p-2} x \cos^{q-2} x \, dx$$

$$+ (q-p)(p+q-1) \int e^{ax} \sin^p x \cos^{q-2} x \, dx \Bigg]$$

GU (334)(1a)

For $p = m$ and $q = n$ even integers, the integral $\displaystyle\int e^{ax} \sin^m x \cos^n x \, dx$ can be reduced by means of these formulas to the integral $\displaystyle\int e^{ax} \, dx$. However, when only $m$ or only $n$ is even, they can be reduced to integrals of the form $\int e^{ax} \cos^n x \, dx$ or $\int e^{ax} \sin^m x \, dx$ respectively.

**2.662**

1.  $\displaystyle \int e^{ax} \sin^n bx \, dx = \frac{1}{a^2 + n^2 b^2} \left[ (a \sin bx - nb \cos bx) e^{ax} \sin^{n-1} bx \right.$

$$\left. + n(n-1)b^2 \int e^{ax} \sin^{n-2} bx \, dx \right]$$

2.  $\displaystyle \int e^{ax} \cos^n bx \, dx = \frac{1}{a^2 + n^2 b^2} \left[ (a \cos bx + nb \sin bx) e^{ax} \cos^{n-1} bx \right.$

$$\left. + n(n-1)b^2 \int e^{ax} \cos^{n-2} bx \, dx \right]$$

3.  $\displaystyle \int e^{ax} \sin^{2m} bx \, dx$

$$= \sum_{k=0}^{m-1} \frac{(2m)! b^{2k} e^{ax} \sin^{2m-2k-1} bx}{(2m-2k)! \left[ a^2 + (2m)^2 b^2 \right] \left[ a^2 + (2m-2)^2 b^2 \right] \cdots \left[ a^2 + (2m-2k)^2 b^2 \right]}$$

$$\times \left[ a \sin bx - (2m-2k)b \cos bx \right] + \frac{(2m)! b^{2m} e^{ax}}{\left[ a^2 + (2m)^2 b^2 \right] \left[ a^2 + (2m-2)^2 b^2 \right] \cdots \left[ a^2 + 4b^2 \right] a}$$

$$= \binom{2m}{m} \frac{e^{ax}}{2^{2m} a} + \frac{e^{ax}}{2^{2m-1}} \sum_{k=1}^{m} (-1)^k \binom{2m}{m-k} \frac{1}{a^2 + 4b^2 k^2} \left( a \cos 2bkx + 2bk \sin 2bkx \right)$$

4.  $\displaystyle \int e^{ax} \sin^{2m+1} bx \, dx$

$$= \sum_{k=0}^{m} \frac{(2m+1)! b^{2k} e^{ax} \sin^{2m-2k} bx \left[ a \sin bx - (2m-2k+1)b \cos bx \right]}{(2m-2k+1)! \left[ a^2 + (2m+1)^2 b^2 \right] \left[ a^2 + (2m-1)^2 b^2 \right] \cdots \left[ a^2 + (2m-2k+1)^2 b^2 \right]}$$

$$= \frac{e^{ax}}{2^{2m}} \sum_{k=0}^{m} \frac{(-1)^k}{a^2 + (2k+1)^2 b^2} \binom{2m+1}{m-k} \left[ a \sin(2k+1)bx - (2k+1)b \cos(2k+1)bx \right]$$

5.[8]  $\displaystyle \int e^{ax} \cos^{2m} bx \, dx = \sum_{k=0}^{m-1} \frac{(2m)! b^{2k} e^{ax} \cos^{2m-2k-1} bx \left[ a \cos bx + (2m-2k)b \sin bx \right]}{(2m-2k)! \left[ a^2 + (2m)^2 b^2 \right] \left[ a^2 + (2m-2)^2 b^2 \right] \cdots \left[ a^2 + (2m-2k)^2 b^2 \right]}$

$$+ \frac{(2m)! b^{2m} e^{ax}}{\left[ a^2 + (2m)^2 b^2 \right] \left[ a^2 + (2m-2)^2 b^2 \right] \cdots \left[ a^2 + 4b^2 \right] a}$$

$$= \binom{2m}{m} \frac{e^{ax}}{2^{2m} a} + \frac{e^{ax}}{2^{2m-1}} \sum_{k=1}^{m} \binom{2m}{m-k} \frac{1}{a^2 + 4b^2 k^2} \left[ a \cos 2kbx + 2kb \sin 2kbx \right]$$

6.[12]  $\displaystyle \int e^{ax} \cos^{2m+1} bx \, dx$

$$= \sum_{k=0}^{m} \frac{(2m+1)! b^{2k} e^{ax} \cos^{2m-2k} bx \left[ a \cos bx + (2m-2k+1)b \sin bx \right]}{(2m-2k+1)! \left[ a^2 + (2m+1)^2 b^2 \right] \left[ a^2 + (2m-1)^2 b^2 \right] \cdots \left[ a^2 + (2m-2k+1)^2 b^2 \right]}$$

$$= \frac{e^{ax}}{2^{2m}} \sum_{k=0}^{m} \binom{2m+1}{m-k} \frac{1}{a^2 + (2k+1)^2 b^2} \left[ a \cos(2k+1)bx + (2k+1)b \sin(2k+1)bx \right]$$

**2.663**

1.  $\displaystyle \int e^{ax} \sin bx \, dx = \frac{e^{ax} (a \sin bx - b \cos bx)}{a^2 + b^2}$

2.
$$\int e^{ax} \sin^2 bx \, dx = \frac{e^{ax} \sin bx \, (a \sin bx - 2b \cos bx)}{4b^2 + a^2} + \frac{2b^2 e^{ax}}{(4b^2 + a^2) \, a}$$
$$= \frac{e^{ax}}{2a} - \frac{e^{ax}}{a^2 + 4b^2} \left( \frac{a}{2} \cos 2bx + b \sin 2bx \right)$$

3.
$$\int e^{ax} \cos bx \, dx = \frac{e^{ax} \, (a \cos bx + b \sin bx)}{a^2 + b^2}$$

4.
$$\int e^{ax} \cos^2 bx \, dx = \frac{e^{ax} \cos bx \, (a \cos bx + 2b \sin bx)}{4b^2 + a^2} + \frac{2b^2 e^{ax}}{(4b^2 + a^2) \, a}$$
$$= \frac{e^{ax}}{2a} + \frac{e^{ax}}{a^2 + 4b^2} \left( \frac{a}{2} \cos 2bx + b \sin 2bx \right)$$

**2.664**

1.
$$\int e^{ax} \sin bx \cos cx \, dx = \frac{e^{ax}}{2} \left[ \frac{a \sin(b + c)x - (b + c) \cos(b + c)x}{a^2 + (b + c)^2} \right.$$
$$\left. + \frac{a \sin(b - c)x - (b - c) \cos(b - c)x}{a^2 + (b - c)^2} \right]$$
GU (334)(6b)

2.
$$\int e^{ax} \sin^2 bx \cos cx \, dx = \frac{e^{ax}}{4} \left[ 2 \frac{a \cos cx + c \sin cx}{a^2 + c^2} - \frac{a \cos(2b + c)x + (2b + c) \sin(2b + c)x}{a^2 + (2b + c)^2} \right.$$
$$\left. - \frac{a \cos(2b - c)x + (2b - c) \sin(2b - c)x}{a^2 + (2b - c)^2} \right]$$
GU (334)(6c)

3.
$$\int e^{ax} \sin bx \cos^2 cx \, dx = \frac{e^{ax}}{4} \left[ 2 \frac{a \sin bx - b \cos bx}{a^2 + b^2} + \frac{a \sin(b + 2c)x - (b + 2c) \cos(b + 2c)x}{a^2 + (b + 2c)^2} \right.$$
$$\left. + \frac{a \sin(b - 2c)x - (b - 2c) \cos(b - 2c)x}{a^2 + (b - 2c)^2} \right]$$
GU (334)(6d)

4.*
$$\int e^{ax} \cos(bx) \cos(cx) \, dx = \frac{e^{ax}[(c + b) \sin((c + b)x) + a \cos((c + b)x)]}{2[a^2 + (c + b)^2]}$$
$$+ \frac{e^{ax}[(c - b) \sin((c - b)x) + a \cos((c - b)x)]}{2[a^2 + (c - b)^2]}$$

5.*
$$\int e^{ax} \sin(bx) \sin(cx) \, dx = \frac{e^{ax}[(c - b) \sin((c - b)x) + a \cos((c - b)x)]}{2[a^2 + (c - b)^2]}$$
$$- \frac{e^{ax}[(c + b) \sin((c + b)x) + a \cos((c + b)x)]}{2[a^2 + (c + b)^2]}$$

**2.665**

1.
$$\int \frac{e^{ax} \, dx}{\sin^p bx} = -\frac{e^{ax} \left[ a \sin bx + (p - 2)b \cos bx \right]}{(p - 1)(p - 2)b^2 \sin^{p-1} bx} + \frac{a^2 + (p - 2)^2 b^2}{(p - 1)(p - 2)b^2} \int \frac{e^{ax} \, dx}{\sin^{p-2} bx} \qquad \text{TI (530)a}$$

2.
$$\int \frac{e^{ax} \, dx}{\cos^p bx} = -\frac{e^{ax} \left[ a \cos bx - (p - 2)b \sin bx \right]}{(p - 1)(p - 2)b^2 \cos^{p-1} bx} + \frac{a^2 + (p - 2)^2 b^2}{(p - 1)(p - 2)b^2} \int \frac{e^{ax} \, dx}{\cos^{p-2} bx} \qquad \text{TI (529)a}$$

By successive applications of formulas **2.665** for $p$ a natural number, we obtain integrals of the form $\int \dfrac{e^{ax} \, dx}{\sin bx}$, $\int \dfrac{e^{ax} \, dx}{\sin^2 bx}$, $\int \dfrac{e^{ax} \, dx}{\cos bx}$, $\int \dfrac{e^{ax} \, dx}{\cos^2 bx}$, which are not expressible in terms of a finite combination of elementary functions.

**2.666**

1. $\displaystyle\int e^{ax} \tan^p x \, dx = \frac{e^{ax}}{p-1} \tan^{p-1} x - \frac{a}{p-1} \int e^{ax} \tan^{p-1} x \, dx - \int e^{ax} \tan^{p-2} x \, dx$     TI (527)

2. $\displaystyle\int e^{ax} \cot^p x \, dx = -\frac{e^{ax} \cot^{p-1} x}{p-1} + \frac{a}{p-1} \int e^{ax} \cot^{p-1} x \, dx - \int e^{ax} \cot^{p-2} x \, dx$     TI (528)

3. $\displaystyle\int e^{ax} \tan x \, dx = \frac{e^{ax} \tan x}{a} - \frac{1}{a} \int \frac{e^{ax} \, dx}{\cos^2 x}$     (see remark following **2.665**)

4. $\displaystyle\int e^{ax} \tan^2 x \, dx = \frac{e^{ax}}{a} (a \tan x - 1) - a \int e^{ax} \tan x \, dx$     (see **2.666** 3)     TI 355

5. $\displaystyle\int e^{ax} \cot x \, dx = \frac{e^{ax} \cot x}{a} + \frac{1}{a} \int \frac{e^{ax} \, dx}{\sin^2 x}$     (see remark following **2.665**)

6. $\displaystyle\int e^{ax} \cot^2 x \, dx = -\frac{e^{ax}}{a} (a \cot x + 1) + a \int e^{ax} \cot x \, dx$

$\qquad\qquad\qquad\qquad\qquad\qquad\qquad\qquad\qquad\qquad$ (see **2.666** 5)

**Integrals of type** $\displaystyle\int R\left(x, e^{ax}, \sin bx, \cos cx\right) \, dx$

**Notation:** $\sin t = -\dfrac{b}{\sqrt{a^2 + b^2}}; \quad \cos t = \dfrac{a}{\sqrt{a^2 + b^2}}.$

**2.667**

1. $\displaystyle\int x^p e^{ax} \sin bx \, dx = \frac{x^p e^{ax}}{a^2 + b^2} (a \sin bx - b \cos bx) - \frac{p}{a^2 + b^2} \int x^{p-1} e^{ax} (a \sin bx - b \cos bx) \, dx$

$\displaystyle\qquad\qquad\qquad = \frac{x^p e^{ax}}{\sqrt{a^2 + b^2}} \sin(bx + t) - \frac{p}{\sqrt{a^2 + b^2}} \int x^{p-1} e^{ax} \sin(bx + t) \, dx$

2. $\displaystyle\int x^p e^{ax} \cos bx \, dx = \frac{x^p e^{ax}}{a^2 + b^2} (a \cos bx + b \sin bx) - \frac{p}{a^2 + b^2} \int x^{p-1} e^{ax} (a \cos bx + b \sin bx) \, dx$

$\displaystyle\qquad\qquad\qquad = \frac{x^p e^{ax}}{\sqrt{a^2 + b^2}} \cos(bx + t) - \frac{p}{\sqrt{a^2 + b^2}} \int x^{p-1} e^{ax} \cos(bx + t) \, dx$

3. $\displaystyle\int x^n e^{ax} \sin bx \, dx = e^{ax} \sum_{k=1}^{n+1} \frac{(-1)^{k+1} n! x^{n-k+1}}{(n-k+1)! \left(a^2 + b^2\right)^{k/2}} \sin(bx + kt)$

4. $\displaystyle\int x^n e^{ax} \cos bx \, dx = e^{ax} \sum_{k=1}^{n+1} \frac{(-1)^{k+1} n! x^{n-k+1}}{(n-k+1)! \left(a^2 + b^2\right)^{k/2}} \cos(bx + kt)$

5. $\displaystyle\int x e^{ax} \sin bx \, dx = \frac{e^{ax}}{a^2 + b^2} \left[ \left( ax - \frac{a^2 - b^2}{a^2 + b^2} \right) \sin bx - \left( bx - \frac{2ab}{a^2 + b^2} \right) \cos bx \right]$

6. $\displaystyle\int x e^{ax} \cos bx \, dx = \frac{e^{ax}}{a^2 + b^2} \left[ \left( ax - \frac{a^2 - b^2}{a^2 + b^2} \right) \cos bx + \left( bx - \frac{2ab}{a^2 + b^2} \right) \sin bx \right]$

7. $$\int x^2 e^{ax} \sin bx \, dx = \frac{e^{ax}}{a^2 + b^2} \left\{ \left[ ax^2 - \frac{2\left(a^2 - b^2\right)}{a^2 + b^2} x + \frac{2a\left(a^2 - 3b^2\right)}{\left(a^2 + b^2\right)^2} \right] \sin bx \right.$$
$$\left. - \left[ bx^2 - \frac{4ab}{a^2 + b^2} x + \frac{2b\left(3a^2 - b^2\right)}{\left(a^2 + b^2\right)^2} \right] \cos bx \right\}$$

8. $$\int x^2 e^{ax} \cos bx \, dx = \frac{e^{ax}}{a^2 + b^2} \left\{ \left[ ax^2 - \frac{2\left(a^2 - b^2\right)}{a^2 + b^2} x + \frac{2a\left(a^2 - 3b^2\right)}{\left(a^2 + b^2\right)^2} \right] \cos bx \right.$$
$$\left. + \left[ bx^2 - \frac{4ab}{a^2 + b^2} x + \frac{2b\left(3a^2 - b^2\right)}{\left(a^2 + b^2\right)^2} \right] \sin bx \right\}$$

GU (335), MZ 274-275

## 2.67 Combinations of trigonometric and hyperbolic functions

**2.671**

1. $$\int \sinh(ax + b) \sin(cx + d) \, dx = \frac{a}{a^2 + c^2} \cosh(ax + b) \sin(cx + d)$$
$$- \frac{c}{a^2 + c^2} \sinh(ax + b) \cos(cx + d)$$

2. $$\int \sinh(ax + b) \cos(cx + d) \, dx = \frac{a}{a^2 + c^2} \cosh(ax + b) \cos(cx + d)$$
$$+ \frac{c}{a^2 + c^2} \sinh(ax + b) \sin(cx + d)$$

3. $$\int \cosh(ax + b) \sin(cx + d) \, dx = \frac{a}{a^2 + c^2} \sinh(ax + b) \sin(cx + d)$$
$$- \frac{c}{a^2 + c^2} \cosh(ax + b) \cos(cx + d)$$

4. $$\int \cosh(ax + b) \cos(cx + d) \, dx = \frac{a}{a^2 + c^2} \sinh(ax + b) \cos(cx + d)$$
$$+ \frac{c}{a^2 + c^2} \cosh(ax + b) \sin(cx + d)$$

GU (354)(1)

**2.672**

1. $$\int \sinh x \sin x \, dx = \frac{1}{2} \left( \cosh x \sin x - \sinh x \cos x \right)$$

2. $$\int \sinh x \cos x \, dx = \frac{1}{2} \left( \cosh x \cos x + \sinh x \sin x \right)$$

3. $$\int \cosh x \sin x \, dx = \frac{1}{2} \left( \sinh x \sin x - \cosh x \cos x \right)$$

4. $$\int \cosh x \cos x \, dx = \frac{1}{2} \left( \sinh x \cos x + \cosh x \sin x \right)$$

**2.673**

$1.^{12}$ $\int \sinh^{2m}(ax+b)\sin^{2n}(cx+d)\,dx$

$$= \frac{(-1)^m}{2^{2m+2n}}\binom{2m}{m}\binom{2n}{n}x + \frac{(-1)^{m+n}}{2^{2m+2n-1}}\binom{2m}{m}\sum_{k=0}^{n-1}\frac{(-1)^k}{(2n-2k)c}\binom{2n}{k}\sin[(2n-2k)(cx+d)]$$

$$+\frac{(-1)^n}{2^{2m+2n-2}}\sum_{j=0}^{m-1}\sum_{k=0}^{n-1}\frac{(-1)^{j+k}\binom{2m}{j}\binom{2n}{k}}{(2m-2j)^2a^2+(2n-2k)^2c^2}$$

$$\times\{(2m-2j)a\sinh[(2m-2j)(ax+b)]\cos[(2n-2k)(cx+d)]$$

$$+(2n-2k)c\cosh[(2m-2j)(ax+b)]\sin[(2n-2k)(cx+d)]\}$$

GU (354)(3a)

2. $\int \sinh^{2m}(ax+b)\sin^{2n-1}(cx+d)\,dx$

$$= \frac{(-1)^{m+n}}{2^{2m+2n-2}}\binom{2m}{m}\sum_{k=0}^{n-1}\frac{(-1)^k}{(2n-2k-1)c}\binom{2n-1}{k}\cos[(2n-2k-1)(cx+d)]$$

$$+\frac{(-1)^{n-1}}{2^{2m+2n-3}}\sum_{j=0}^{m-1}\sum_{k=0}^{n-1}\frac{(-1)^{j+k}\binom{2m}{j}\binom{2n-1}{k}}{(2m-2j)^2a^2+(2n-2k-1)^2c^2}$$

$$\times\{(2m-2j)a\sinh[(2m-2j)(ax+b)]\sin[(2n-2k-1)(cx+d)]$$

$$-(2n-2k-1)c\cosh[(2m-2j)(ax+b)]\cos[(2n-2k-1)(cx+d)]\}$$

GU (354)(3b)

3. $\int \sinh^{2m-1}(ax+b)\sin^{2n}(cx+d)\,dx$

$$= \frac{\binom{2n}{n}}{2^{2m+2n-2}}\sum_{j=0}^{m-1}\frac{(-1)^j\binom{2m-1}{j}}{(2m-2j-1)a}\cosh[(2m-2j-1)(ax+b)]$$

$$+\frac{(-1)^n}{2^{2m+2n-3}}\sum_{j=0}^{m-1}\sum_{k=0}^{n-1}\frac{(-1)^{j+k}\binom{2m-1}{j}\binom{2n}{k}}{(2m-2j-1)^2a^2+(2n-2k)^2c^2}$$

$$\times\{(2m-2j-1)a\cosh[(2m-2j-1)(ax+b)]\cos[(2n-2k)(cx+d)]$$

$$+(2n-2k)c\sinh[(2m-2j-1)(ax+b)]\sin[(2n-2k)(cx+d)]\}$$

GU (354)(3c)

4.    $\displaystyle\int \sinh^{2m-1}(ax+b)\sin^{2n-1}(cx+d)\,\mathrm{d}x$

$$= \frac{(-1)^{n-1}}{2^{2m-2n-4}} \sum_{j=0}^{m-1}\sum_{k=0}^{n-1} \frac{(-1)^{j+k}\binom{2m-1}{j}\binom{2n-1}{k}}{(2m-2j-1)^2 a^2 + (2n-2k-1)^2 c^2}$$

$$\times \{(2m-2j-1)a\cosh[(2m-2j-1)(ax+b)]\sin[(2n-2k-1)(cx+d)]$$

$$-(2n-2k-1)c\sinh[(2m-2j-1)(ax+b)]\cos[(2n-2k-1)(cx+d)]\}$$

GU (354)(3d)

5.[12]    $\displaystyle\int \sinh^{2m}(ax+b)\cos^{2n}(cx+d)\,\mathrm{d}x$

$$= \frac{(-1)^m}{2^{2m+2n}}\binom{2m}{m}\binom{2n}{n} x + \frac{\binom{2n}{n}}{2^{2m+2n-1}}\sum_{j=0}^{m-1} \frac{(-1)^j\binom{2m}{j}}{(2m-2j)a}\sinh[(2m-2j)(ax+b)]$$

$$+ \frac{(-1)^m\binom{2m}{m}}{2^{2m+2n-1}}\sum_{k=0}^{n-1} \frac{\binom{2n}{k}}{(2n-2k)c}\sin[(2n-2k)(cx+d)]$$

$$+ \frac{1}{2^{2m+2n-2}}\sum_{j=0}^{m-1}\sum_{k=0}^{n-1} \frac{(-1)^j\binom{2m}{j}\binom{2n}{k}}{(2m-2j)^2 a^2 + (2n-2k)^2 c^2}$$

$$\times \{(2m-2j)a\sinh[(2m-2j)(ax+b)]\cos[(2n-2k)(cx+d)]$$

$$+(2n-2k)c\cosh[(2m-2j)(ax+b)]\sin[(2n-2k)(cx+d)]\}$$

GU (354)(4a)

6.[12]    $\displaystyle\int \sinh^{2m}(ax+b)\cos^{2n-1}(cx+d)\,\mathrm{d}x$

$$= \frac{(-1)^m\binom{2m}{m}}{2^{2m+2n-2}}\sum_{k=0}^{n-1} \frac{\binom{2n-1}{k}}{(2n-2k-1)c}\sin[(2n-2k-1)(cx+d)]$$

$$+ \frac{1}{2^{2m+2n-3}}\sum_{j=0}^{m-1}\sum_{k=0}^{n-1} \frac{(-1)^j\binom{2m}{j}\binom{2n-1}{k}}{(2m-2j)^2 a^2 + (2n-2k-1)^2 c^2}$$

$$\times \{(2m-2j)a\sinh[(2m-2j)(ax+b)]\cos[(2n-2k-1)(cx+d)]$$

$$+(2n-2k-1)c\cosh[(2m-2j)(ax+b)]\sin[(2n-2k-1)(cx+d)]\}$$

GU (354)(4a)

**7.**[12]   $\displaystyle\int \sinh^{2m-1}(ax+b)\cos^{2n}(cx+d)\,\mathrm{d}x$

$$= \frac{\binom{2n}{n}}{2^{2m+2n-2}}\sum_{j=0}^{m-1}\frac{(-1)^j\binom{2m-1}{j}}{(2m-2j-1)a}\cosh[(2m-2j-1)(ax+b)]$$

$$+\frac{1}{2^{2m-2n-3}}\sum_{j=0}^{m-1}\sum_{k=0}^{n-1}\frac{(-1)^j\binom{2m-1}{j}\binom{2n}{k}}{(2m-2j-1)^2a^2+(2n-2k)^2c^2}$$

$$\times\{(2m-2j-1)a\cosh[(2m-2j-1)(ax+b)]\cos[(2n-2k)(cx+d)]$$

$$+(2n-2k)c\sinh[(2m-2j-1)(ax+b)]\sin[(2n-2k)(cx+d)]\}$$

GU (354)(4b)

**8.**   $\displaystyle\int \sinh^{2m-1}(ax+b)\cos^{2n-1}(cx+d)\,\mathrm{d}x$

$$= \frac{1}{2^{2m+2n-4}}\sum_{j=0}^{m-1}\sum_{k=0}^{n-1}\frac{(-1)^j\binom{2m-1}{j}\binom{2n-1}{k}}{(2m-2j-1)^2a^2+(2n-2k-1)^2c^2}$$

$$\times\{(2m-2j-1)a\cosh[(2m-2j-1)(ax+b)]\cos[(2n-2k-1)(cx+d)]$$

$$+(2n-2k-1)c\sinh[(2m-2j-1)(ax+b)]\sin[(2n-2k-1)(cx+d)]\}$$

GU (354)(4b)

**9.**   $\displaystyle\int \cosh^{2m}(ax+b)\sin^{2n}(cx+d)\,\mathrm{d}x$

$$= \frac{\binom{2m}{m}\binom{2n}{n}}{2^{2m+2n}}x+\frac{(-1)^n\binom{2m}{m}}{2^{2m+2n-1}}\sum_{k=0}^{m-1}\frac{(-1)^k\binom{2n}{k}}{(2n-2k)c}\sin[(2n-2k)(cx+d)]$$

$$+\frac{\binom{2n}{n}}{2^{2m+2n-1}}\sum_{j=0}^{m-1}\frac{\binom{2m}{j}}{(2m-2j)a}\sinh[(2m-2j)(ax+b)]$$

$$+\frac{(-1)^n}{2^{2m+2n-2}}\sum_{j=0}^{m-1}\sum_{k=0}^{n-1}\frac{(-1)^k\binom{2m}{j}\binom{2n}{k}}{(2m-2j)^2a^2+(2n-2k)^2c^2}$$

$$\times\{(2m-2j)a\sinh[(2m-2j)(ax+b)]\cos[(2n-2k)(cx+d)]$$

$$+(2n-2k)c\cosh[(2m-2j)(ax+b)]\sin[(2n-2k)(cx+d)]\}$$

GU (354)(5a)

10. $\displaystyle\int \cosh^{2m-1}(ax+b)\sin^{2n}(cx+d)\,\mathrm{d}x$

$$= \frac{\binom{2n}{n}}{2^{2m+2n-2}}\sum_{j=0}^{m-1}\frac{\binom{2m-1}{j}}{(2m-2j-1)a}\sinh[(2m-2j-1)(ax+b)]$$

$$+\frac{(-1)^n}{2^{2m+2n-3}}\sum_{j=0}^{m-1}\sum_{k=0}^{n-1}\frac{(-1)^k\binom{2m-1}{j}\binom{2n}{k}}{(2m-2j-1)^2a^2+(2n-2k)^2c^2}$$

$$\times\{(2m-2j-1)a\sinh[(2m-2j-1)(ax+b)]\cos[(2n-2k)(cx+d)]$$

$$+(2n-2k)c\cosh[(2m-2j-1)(ax+b)]\sin[(2n-2k)(cx+d)]\}$$

GU (354)(5a)

11. $\displaystyle\int \cosh^{2m}(ax+b)\sin^{2n-1}(cx+d)\,\mathrm{d}x$

$$= \frac{(-1)^{n-1}\binom{2m}{m}}{2^{2m+2n-2}}\sum_{k=0}^{n-1}\frac{(-1)^{k+1}\binom{2n-1}{k}}{(2n-2k-1)c}\cos[(2n-2k-1)(cx+d)]$$

$$+\frac{(-1)^{n-1}}{2^{2m+2n-3}}\sum_{j=0}^{m-1}\sum_{k=0}^{n-1}\frac{(-1)^k\binom{2m}{j}\binom{2n-1}{k}}{(2m-2j)^2a^2+(2n-2k-1)^2c^2}$$

$$\times\{(2m-2j)a\sinh[(2m-2j)(ax+b)]\sin[(2n-2k-1)(cx+d)]$$

$$-(2n-2k-1)c\cosh[(2m-2j)(ax+b)]\cos[(2n-2k-1)(cx+d)]\}$$

GU (354)(5b)

12. $\displaystyle\int \cosh^{2m-1}(ax+b)\sin^{2n-1}(cx+d)\,\mathrm{d}x$

$$= \frac{(-1)^{n-1}}{2^{2m+2n-4}}\sum_{j=0}^{m-1}\sum_{k=0}^{n-1}\frac{(-1)^k\binom{2m-1}{j}\binom{2n-1}{k}}{(2m-2j-1)^2a^2+(2n-2k-1)^2c^2}$$

$$\times\{(2m-2j-1)a\sinh[(2m-2j-1)(ax+b)]\sin[(2n-2k-1)(cx+d)]$$

$$-(2n-2k-1)c\cosh[(2m-2j-1)(ax+b)]\cos[(2n-2k-1)(cx+d)]\}$$

GU (354)(5b)

13. $\displaystyle\int \cosh^{2m}(ax+b)\cos^{2n}(cx+d)\,\mathrm{d}x$

$$= \frac{\binom{2m}{m}\binom{2n}{n}}{2^{2m+2n}}x + \frac{\binom{2m}{m}}{2^{2m+2n-1}}\sum_{k=0}^{n-1}\frac{\binom{2n}{k}}{(2n-2k)c}\sin[(2n-2k)(cx+d)]$$

$$+\frac{\binom{2n}{n}}{2^{2m+2n-1}}\sum_{j=0}^{m-1}\frac{\binom{2m}{j}}{(2m-2j)a}\sinh[(2m-2j)(ax+b)]$$

$$+\frac{1}{2^{2m+2n-2}}\sum_{j=0}^{m-1}\sum_{k=0}^{n-1}\frac{\binom{2m}{j}\binom{2n}{k}}{(2m-2j)^2a^2+(2n-2k)^2c^2}$$

$$\times\{(2m-2j)a\sinh[(2m-2j)(ax+b)]\cos[(2n-2k)(cx+d)]$$

$$+(2n-2k)c\cosh[(2m-2j)(ax+b)]\sin[(2n-2k)(cx+d)]\}$$

GU (354)(6)

14. $\displaystyle\int \cosh^{2m-1}(ax+b)\cos^{2n}(cx+d)\,\mathrm{d}x$

$$= \frac{\binom{2n}{n}}{2^{2m+2n-2}}\sum_{j=0}^{m-1}\frac{\binom{2m-1}{j}}{(2m-2j-1)a}\sinh[(2m-2j-1)(ax+b)]$$

$$+\frac{1}{2^{2m+2n-3}}\sum_{j=0}^{m-1}\sum_{k=0}^{n-1}\frac{\binom{2m-1}{j}\binom{2n}{k}}{(2m-2j-1)^2a^2+(2n-2k)^2c^2}$$

$$\times\{(2m-2j-1)a\sinh[(2m-2j-1)(ax+b)]\cos[(2n-2k)(cx+d)]$$

$$+(2n-2k)c\cosh[(2m-2j-1)(ax+b)]\sin[(2n-2k)(cx+d)]\}$$

GU (354)(6)

15. $\displaystyle\int \cosh^{2m}(ax+b)\cos^{2n-1}(cx+d)\,\mathrm{d}x$

$$= \frac{\binom{2m}{m}}{2^{2m+2n-2}}\sum_{k=0}^{n-1}\frac{\binom{2n-1}{k}}{(2n-2k-1)c}\sin[(2n-2k-1)(cx+d)]$$

$$+\frac{1}{2^{2m+2n-3}}\sum_{j=0}^{m-1}\sum_{k=0}^{n-1}\frac{\binom{2m}{j}\binom{2n-1}{k}}{(2m-2j)^2a^2+(2n-2k-1)^2c^2}$$

$$\times\{(2m-2j)a\sinh[(2m-2j)(ax+b)]\cos[(2n-2k-1)(cx+d)]$$

$$+(2n-2k-1)c\cosh[(2m-2j)(ax+b)]\sin[(2n-2k-1)(cx+d)]\}$$

GU (354)(6)

16. $\displaystyle\int \cosh^{2m-1}(ax+b)\cos^{2n-1}(cx+d)\,dx$

$$= \frac{1}{2^{2m+2n-4}} \sum_{j=0}^{m-1}\sum_{k=0}^{n-1} \frac{\binom{2m-1}{j}\binom{2n-1}{k}}{(2m-2j-1)^2 a^2 + (2n-2k-1)^2 c^2}$$

$$\times \{(2m-2j-1)a\sinh[(2m-2j-1)(ax+b)]\cos[(2n-2k-1)(cx+d)]$$

$$+(2n-2k-1)c\cosh[(2m-2j-1)(ax+b)]\sin[(2n-2k-1)(cx+d)]\}$$

GU (354)(6)

**2.674**

1. $\displaystyle\int e^{ax}\sinh bx \sin cx\,dx = \frac{e^{(a+b)x}}{2\left[(a+b)^2+c^2\right]}\left[(a+b)\sin cx - c\cos cx\right]$

$$-\frac{e^{(a-b)x}}{2\left[(a-b)^2+c^2\right]}\left[(a-b)\sin cx - c\cos cx\right]$$

2. $\displaystyle\int e^{ax}\sinh bx \cos cx\,dx = \frac{e^{(a+b)x}}{2\left[(a+b)^2+c^2\right]}\left[(a+b)\cos cx + c\sin cx\right]$

$$-\frac{e^{(a-b)x}}{2\left[(a-b)^2+c^2\right]}\left[(a-b)\cos cx + c\sin cx\right]$$

3. $\displaystyle\int e^{ax}\cosh bx \sin cx\,dx = \frac{e^{(a+b)x}}{2\left[(a+b)^2+c^2\right]}\left[(a+b)\sin cx - c\cos cx\right]$

$$+\frac{e^{(a-b)x}}{2\left[(a-b)^2+c^2\right]}\left[(a-b)\sin cx - c\cos cx\right]$$

4. $\displaystyle\int e^{ax}\cosh bx \cos cx\,dx = \frac{e^{(a+b)x}}{2\left[(a+b)^2+c^2\right]}\left[(a+b)\cos cx + c\sin cx\right]$

$$+\frac{e^{(a-b)x}}{2\left[(a-b)^2+c^2\right]}\left[(a-b)\cos cx + c\sin cx\right]$$

MZ 379

# 2.7 Logarithms and Inverse-Hyperbolic Functions

## 2.71 The logarithm

**2.711** $\displaystyle\int \ln^m x\,dx = x\ln^m x - m\int \ln^{m-1} x\,dx$

$$= \frac{x}{m+1}\sum_{k=0}^{m}(-1)^k(m+1)m(m-1)\cdots(m-k+1)\ln^{m-k}x$$

$$[m>0]$$

TI (603)

## 2.72–2.73  Combinations of logarithms and algebraic functions

**2.721**

1.[12]   $\displaystyle\int x^n \ln^m x \, dx = \frac{x^{n+1} \ln^m x}{n+1} - \frac{m}{n+1} \int x^n \ln^{m-1} x \, dx$     (see **2.722**)

     For $n = -1$ and $m \neq -1$

2.   $\displaystyle\int \frac{\ln^m x \, dx}{x} = \frac{\ln^{m+1} x}{m+1}$

     For $n = -1$ and $m = -1$

3.   $\displaystyle\int \frac{dx}{x \ln x} = \ln(\ln x)$

4.*   $\displaystyle\int \frac{dx}{\ln x} = \mathrm{Ei}(\ln x)$

**2.722**   $\displaystyle\int x^n \ln^m x \, dx = \frac{x^{n+1}}{m+1} \sum_{k=0}^{m} (-1)^k (m+1) m (m-1) \cdots (m-k+1) \frac{\ln^{m-k} x}{(n+1)^{k+1}}$      TI (604)

**2.723**

1.   $\displaystyle\int x^n \ln x \, dx = x^{n+1} \left[ \frac{\ln x}{n+1} - \frac{1}{(n+1)^2} \right]$      TI 375

2.   $\displaystyle\int x^n \ln^2 x \, dx = x^{n+1} \left[ \frac{\ln^2 x}{n+1} - \frac{2 \ln x}{(n+1)^2} + \frac{2}{(n+1)^3} \right]$      TI 375

3.   $\displaystyle\int x^n \ln^3 x \, dx = x^{n+1} \left[ \frac{\ln^3 x}{n+1} - \frac{3 \ln^2 x}{(n+1)^2} + \frac{6 \ln x}{(n+1)^3} - \frac{6}{(n+1)^4} \right]$

**2.724**

1.   $\displaystyle\int \frac{x^n \, dx}{(\ln x)^m} = -\frac{x^{n+1}}{(m-1)(\ln x)^{m-1}} + \frac{n+1}{m-1} \int \frac{x^n \, dx}{(\ln x)^{m-1}}$

     For $m = 1$

2.   $\displaystyle\int \frac{x^n \, dx}{\ln x} = \mathrm{li}\left(x^{n+1}\right)$

**2.725**

1.   $\displaystyle\int (a+bx)^m \ln x \, dx = \frac{1}{(m+1)b} \left[ (a+bx)^{m+1} \ln x - \int \frac{(a+bx)^{m+1} \, dx}{x} \right]$      TI 374

2.   $\displaystyle\int (a+bx)^m \ln x \, dx = \frac{1}{(m+1)b} \left[ (a+bx)^{m+1} - a^{m+1} \right] \ln x - \sum_{k=0}^{m} \frac{\binom{m}{k} a^{m-k} b^k x^{k+1}}{(k+1)^2}$

     For $m = -1$ see **2.727** 2.

**2.726**

1.   $\displaystyle\int (a+bx) \ln x \, dx = \left[ \frac{(a+bx)^2}{2b} - \frac{a^2}{2b} \right] \ln x - \left( ax + \frac{1}{4}bx^2 \right)$

2.   $\displaystyle\int (a+bx)^2 \ln x \, dx = \frac{1}{3b} \left[ (a+bx)^3 - a^3 \right] \ln x - \left( a^2 x + \frac{abx^2}{2} + \frac{b^2 x^3}{9} \right)$

3.    $\displaystyle\int (a+bx)^3 \ln x \, dx = \frac{1}{4b}\left[(a+bx)^4 - a^4\right]\ln x - \left(a^3 x + \frac{3}{4}a^2 bx^2 + \frac{1}{3}ab^2 x^3 + \frac{1}{16}b^3 x^4\right)$

**2.727**

$1.^8$    $\displaystyle\int \frac{\ln x \, dx}{(a+bx)^m} = \frac{1}{b(m-1)}\left[-\frac{\ln x}{(a+bx)^{m-1}} + \int \frac{dx}{x(a+bx)^{m-1}}\right]$           TI 376

     For $m = 1$

$2.^8$    $\displaystyle\int \frac{\ln x \, dx}{a+bx} = \frac{1}{b}\ln x \ln(a+bx) - \frac{1}{b}\int \frac{\ln(a+bx)\,dx}{x}$      (see **2.728** 2)

3.    $\displaystyle\int \frac{\ln x \, dx}{(a+bx)^2} = -\frac{\ln x}{b(a+bx)} + \frac{1}{ab}\ln\frac{x}{a+bx}$

4.    $\displaystyle\int \frac{\ln x \, dx}{(a+bx)^3} = -\frac{\ln x}{2b(a+bx)^2} + \frac{1}{2ab(a+bx)} + \frac{1}{2a^2 b}\ln\frac{x}{a+bx}$

5.    $\displaystyle\int \frac{\ln x \, dx}{\sqrt{a+bx}} = \frac{2}{b}\left\{(\ln x - 2)\sqrt{a+bx} - 2\sqrt{a}\ln\left[\frac{(a+bx)^{1/2} - a^{1/2}}{x^{1/2}}\right]\right\}$    $[a > 0]$

                    $= \frac{2}{b}\left\{(\ln x - 2)\sqrt{a+bx} + 2\sqrt{-a}\arctan\sqrt{\frac{a+bx}{-a}}\right\}$    $[a < 0]$

$6.^*$    $\displaystyle\int \frac{\ln|x-b|}{\sqrt{x^2+a^2}} x \, dx =$

         $\sqrt{x^2+a^2}(\ln|x-b| - 1) - b\,\operatorname{arcsinh}\left(\frac{x}{|a|}\right) + \sqrt{a^2+b^2}\,\operatorname{arctanh}\left(\frac{a^2+bx}{\sqrt{a^2+b^2}\sqrt{x^2+a^2}}\right)$

$7.^*$    $\displaystyle\int \frac{\ln|x-b|}{\sqrt{a^2-x^2}} x \, dx$

         $= \sqrt{a^2-x^2}(1 - \ln|x-b|) - b\,\operatorname{arcsinh}\left(\frac{x}{|a|}\right)$

         $\quad - \sqrt{a^2-b^2}\ln\left(\frac{\sqrt{a^2-b^2}\sqrt{a^2-x^2} + a^2 - bx}{|a(x-b)|}\right)$    $[|b| \le |a|]$

         $= \sqrt{a^2-x^2}(1 - \ln|x-b|) - b\,\arcsin\left(\frac{x}{|a|}\right)$

         $\quad - \sqrt{b^2-a^2}\arctan\left(\frac{a^2-bx}{\sqrt{b^2-a^2}\sqrt{a^2-x^2}}\right)$    $[|b| \ge |a|]$

$8.^*$    $\displaystyle\int \frac{\ln|x-b|}{\sqrt{x^2-a^2}} x \, dx = \sqrt{x^2-a^2}(\ln|x-b| - 1) + I_1 + I_2$

     where

         $I_1 = -b\ln\left(\frac{\sqrt{x^2-a^2} + x}{|a|}\right)$             $[x \ge |a|]$

         $\quad = b\ln\left(\frac{\sqrt{x^2-a^2} - x}{|a|}\right)$             $[x \le -|a|]$

$$I_2 = \sqrt{a^2 - b^2} \operatorname{arccot}\left(\frac{a^2 - bx}{\sqrt{a^2 - b^2}\sqrt{x^2 - a^2}}\right) \qquad [|b| \le |a|]$$

$$= \sqrt{b^2 - a^2} \ln\left(\frac{\sqrt{b^2 - a^2}\sqrt{x^2 - a^2} - a^2 + bx}{|a(x - b)|}\right)$$

$$[\text{either } b \ge |a|, \ x \ge |a| \quad \text{or} \quad b \le -|a|, \ x \le -|a|]$$

$$= -\sqrt{b^2 - a^2} \ln\left(\frac{\sqrt{b^2 - a^2}\sqrt{x^2 - a^2} + a^2 - bx}{|a(x - b)|}\right)$$

$$[\text{either } b \ge |a|, \ x \le -|a| \quad \text{or} \quad b \le -|a|, \ x \ge -|a|]$$

**2.728**

1.  $$\int x^m \ln(a + bx)\, dx = \frac{1}{m+1}\left[x^{m+1}\ln(a+bx) - b\int \frac{x^{m+1}\, dx}{a+bx}\right]$$

2.[9]  $$\int \frac{\ln(a+bx)}{x} = \ln a \ln x + \frac{bx}{a}\, \Phi\left(-\frac{bx}{a}, 2, 1\right) \qquad [a > 0]$$

**2.729**

1.  $$\int x^m \ln(a+bx)\, dx = \frac{1}{m+1}\left[x^{m+1} - \frac{(-a)^{m+1}}{b^{m+1}}\right]\ln(a+bx) + \frac{1}{m+1}\sum_{k=1}^{m+1}\frac{(-1)^k x^{m-k+2} a^{k-1}}{(m-k+2)b^{k-1}}$$

2.  $$\int x \ln(a+bx)\, dx = \frac{1}{2}\left[x^2 - \frac{a^2}{b^2}\right]\ln(a+bx) - \frac{1}{2}\left[\frac{x^2}{2} - \frac{ax}{b}\right]$$

3.[12]  $$\int x^2 \ln(a+bx)\, dx = \frac{1}{3}\left[x^3 - \frac{a^3}{b^3}\right]\ln(a+bx) - \frac{1}{3}\left[\frac{x^3}{3} - \frac{ax^2}{2b} + \frac{a^2 x}{b^2}\right]$$

4.  $$\int x^3 \ln(a+bx)\, dx = \frac{1}{4}\left[x^4 - \frac{a^4}{b^4}\right]\ln(a+bx) - \frac{1}{4}\left[\frac{x^4}{4} - \frac{ax^3}{3b} + \frac{a^2 x^2}{2b^2} - \frac{a^3 x}{b^3}\right]$$

**2.731**  $$\int x^{2n}\ln\left(x^2 + a^2\right)\, dx = \frac{1}{2n+1}\left[x^{2n+1}\ln\left(x^2+a^2\right) + (-1)^n 2a^{2n+1}\arctan\frac{x}{a}\right.$$

$$\left. - 2\sum_{k=0}^{n}\frac{(-1)^{n-k}}{2k+1}a^{2n-2k}x^{2k+1}\right]$$

**2.732**[7]  $$\int x^{2n+1}\ln\left(x^2+a^2\right)\, dx = \frac{1}{2n+2}\left[\left(x^{2n+2} + (-1)^n a^{2n+2}\right)\ln\left(x^2+a^2\right)\right.$$

$$\left. + \sum_{k=1}^{n+1}\frac{(-1)^{n-k}}{k}a^{2n-2k+2}x^{2k}\right]$$

**2.733**

1.  $$\int \ln\left(x^2 + a^2\right)\, dx = x\ln\left(x^2+a^2\right) - 2x + 2a\arctan\frac{x}{a} \qquad\qquad \text{DW}$$

2.  $$\int x \ln\left(x^2 + a^2\right)\, dx = \frac{1}{2}\left[\left(x^2+a^2\right)\ln\left(x^2+a^2\right) - x^2\right] \qquad\qquad \text{DW}$$

3. $\int x^2 \ln \left(x^2 + a^2\right) \, dx = \frac{1}{3}\left[x^3 \ln \left(x^2 + a^2\right) - \frac{2}{3}x^3 + 2a^2 x - 2a^3 \arctan \frac{x}{a}\right]$     DW

4. $\int x^3 \ln \left(x^2 + a^2\right) \, dx = \frac{1}{4}\left[\left(x^4 - a^4\right) \ln \left(x^2 + a^2\right) - \frac{x^4}{2} + a^2 x^2\right]$     DW

5. $\int x^4 \ln \left(x^2 + a^2\right) \, dx = \frac{1}{5}\left[x^5 \ln \left(x^2 + a^2\right) - \frac{2}{5}x^5 + \frac{2}{3}a^2 x^3 - 2a^4 x + 2a^5 \arctan \frac{x}{a}\right]$     DW

**2.734** $\int x^{2n} \ln \left|x^2 - a^2\right| \, dx$

$$= \frac{1}{2n+1}\left\{x^{2n+1} \ln \left|x^2 - a^2\right| + a^{2n+1} \ln \left|\frac{x+a}{x-a}\right| - 2\sum_{k=0}^{n} \frac{1}{2k+1}a^{2n-2k}x^{2k+1}\right\}$$

**2.735** $\int x^{2n+1} \ln \left|x^2 - a^2\right| \, dx = \frac{1}{2n+2}\left\{\left(x^{2n+2} - a^{2n+2}\right) \ln \left|x^2 - a^2\right| - \sum_{k=1}^{n+1} \frac{1}{k}a^{2n-2k+2}x^{2k}\right\}$

**2.736**

1. $\int \ln \left|x^2 - a^2\right| \, dx = x \ln \left|x^2 - a^2\right| - 2x + a \ln \left|\frac{x+a}{x-a}\right|$     DW

2. $\int x \ln \left|x^2 - a^2\right| \, dx = \frac{1}{2}\left\{\left(x^2 - a^2\right) \ln \left|x^2 - a^2\right| - x^2\right\}$     DW

3. $\int x^2 \ln \left|x^2 - a^2\right| \, dx = \frac{1}{3}\left[x^3 \ln \left|x^2 - a^2\right| - \frac{2}{3}x^3 - 2a^2 x + a^3 \ln \left|\frac{x+a}{x-a}\right|\right]$     DW

4. $\int x^3 \ln \left|x^2 - a^2\right| \, dx = \frac{1}{4}\left[\left(x^4 - a^4\right) \ln \left|x^2 - a^2\right| - \frac{x^4}{2} - a^2 x^2\right]$     DW

5. $\int x^4 \ln \left|x^2 - a^2\right| \, dx = \frac{1}{5}\left[x^5 \ln \left|x^2 - a^2\right| - \frac{2}{5}x^5 - \frac{2}{3}a^2 x^3 - 2a^4 x + a^5 \ln \left|\frac{x+a}{x-a}\right|\right]$     DW

**2.737***

1.* $\int_0^1 \frac{t^x - 1}{\ln t}\left(\sum_{k=0}^{n-1} t^{k/n}\right) dt = \ln \prod_{k=0}^{n-1}\left[\frac{k+n(x+1)}{k+n}\right]$     $[n = 1, 2, \ldots, \quad x > -1]$

2.* $\int_0^1 \frac{t^x - 1}{\ln t}\left(\sum_{k=0}^{n-1} (-1)^k t^{k/n}\right) dt = \ln \prod_{k=0}^{n-1}\left[\frac{k+n(x+1)}{k+n}\right]^{(-1)^k}$

$[n = 2, 4, 6, \ldots, \quad x > -1]$

3.* $\int_0^1 t^x (\ln t)^m \left(\sum_{k=0}^{n-1} t^{k/n}\right) dt = (-1)^m n^{m+1} m! \sum_{k=0}^{n-1} \frac{1}{[k+n(x+1)]^{m+1}}$

$[n = 1, 3, 5, \ldots, \quad m = 1, 2, \ldots, \quad x > -1]$

4.* $\int_0^1 t^x (\ln t)^m \left(\sum_{k=0}^{n-1} (-1)^k t^{k/n}\right) dt = (-1)^m n^{m+1} m! \sum_{k=0}^{n-1} \frac{1}{[k+n(x+1)]^{m+1}}$

$[n = 2, 4, 6, \ldots, \quad m = 1, 2, \ldots, \quad x > -1]$

## 2.74 Inverse hyperbolic functions

Assume $a > 0$

**2.741**

1.  $$\int \operatorname{arcsinh} \frac{x}{a} \, dx = x \operatorname{arcsinh} \frac{x}{a} - \sqrt{x^2 + a^2}$$                                                  DW

2.  $$\int \operatorname{arccosh} \frac{x}{a} \, dx = x \operatorname{arccosh} \frac{x}{a} - \sqrt{x^2 - a^2} \qquad \left[ \operatorname{arccosh} \frac{x}{a} > 0 \right]$$    DW

    $$= x \operatorname{arccosh} \frac{x}{a} + \sqrt{x^2 - a^2} \qquad \left[ \operatorname{arccosh} \frac{x}{a} < 0 \right]$$    DW

3.  $$\int \operatorname{arctanh} \frac{x}{a} \, dx = x \operatorname{arctanh} \frac{x}{a} + \frac{a}{2} \ln \left( a^2 - x^2 \right)$$                              DW

4.  $$\int \operatorname{arccoth} \frac{x}{a} \, dx = x \operatorname{arccoth} \frac{x}{a} + \frac{a}{2} \ln \left( x^2 - a^2 \right)$$                              DW

**2.742**

1.  $$\int x \operatorname{arcsinh} \frac{x}{a} \, dx = \left( \frac{x^2}{2} + \frac{a^2}{4} \right) \operatorname{arcsinh} \frac{x}{a} - \frac{x}{4} \sqrt{x^2 + a^2}$$    DW

2.  $$\int x \operatorname{arccosh} \frac{x}{a} \, dx = \left( \frac{x^2}{2} - \frac{a^2}{4} \right) \operatorname{arccosh} \frac{x}{a} - \frac{x}{4} \sqrt{x^2 - a^2} \qquad \left[ \operatorname{arccosh} \frac{x}{a} > 0 \right]$$

    $$= \left( \frac{x^2}{2} - \frac{a^2}{4} \right) \operatorname{arccosh} \frac{x}{a} + \frac{x}{4} \sqrt{x^2 - a^2} \qquad \left[ \operatorname{arccosh} \frac{x}{a} < 0 \right]$$

                                                                                                                                      DW

## 2.75 Logarithms and exponential functions

**2.751\***

1.  $$\int e^{ax} \ln^p x \, dx = \frac{1}{a} \left[ e^{ax} \ln^p x - p \int \frac{e^{ax} \ln^{p-1} x}{x} dx \right] \qquad [a \neq 0]$$

2.  $$\int e^{ax} \ln x \, dx = \frac{1}{a} \left[ e^{ax} \ln x - \operatorname{Ei}(ax) \right] \qquad [a \neq 0]$$

## 2.8 Inverse Trigonometric Functions

## 2.81 Arcsines and arccosines

Assume $a > 0$

**2.811** $$\int \left( \arcsin \frac{x}{a} \right)^n dx = x \sum_{k=0}^{\lfloor n/2 \rfloor} (-1)^k \binom{n}{2k} \cdot (2k)! \left( \arcsin \frac{x}{a} \right)^{n-2k}$$

$$+ \sqrt{a^2 - x^2} \sum_{k=1}^{\lfloor (n+1)/2 \rfloor} (-1)^{k-1} \binom{n}{2k-1} \cdot (2k-1)! \left( \arcsin \frac{x}{a} \right)^{n-2k+1}$$

**2.812** $\displaystyle\int \left(\arccos \frac{x}{a}\right)^n dx = x \sum_{k=0}^{\lfloor n/2 \rfloor} (-1)^k \binom{n}{2k} \cdot (2k)! \left(\arccos \frac{x}{a}\right)^{n-2k}$

$$+ \sqrt{a^2 - x^2} \sum_{k=1}^{\lfloor (n+1)/2 \rfloor} (-1)^k \binom{n}{2k-1} \cdot (2k-1)! \left(\arccos \frac{x}{a}\right)^{n-2k+1}$$

**2.813**

1.[11]    $\displaystyle\int \arcsin \frac{x}{a}\, dx = \text{sign}(a) \left[x \arcsin \frac{x}{|a|} + \sqrt{a^2 - x^2}\right]$

2.[9]    $\displaystyle\int \left(\arcsin \frac{x}{a}\right)^2 dx = x \left(\arcsin \frac{x}{|a|}\right)^2 + 2\sqrt{a^2 - x^2} \arcsin \frac{x}{|a|} - 2x$

3.    $\displaystyle\int \left(\arcsin \frac{x}{a}\right)^3 dx = \text{sign}(a) \left[x \left(\arcsin \frac{x}{|a|}\right)^3 + 3\sqrt{a^2 - x^2} \left(\arcsin \frac{x}{|a|}\right)^2\right.$

$$\left. - 6x \arcsin \frac{x}{|a|} - 6\sqrt{a^2 - x^2}\right]$$

**2.814**

1.    $\displaystyle\int \arccos \frac{x}{a}\, dx = x \arccos \frac{x}{a} - \sqrt{a^2 - x^2}$

2.    $\displaystyle\int \left(\arccos \frac{x}{a}\right)^2 dx = x \left(\arccos \frac{x}{a}\right)^2 - 2\sqrt{a^2 - x^2} \arccos \frac{x}{a} - 2x$

3.    $\displaystyle\int \left(\arccos \frac{x}{a}\right)^3 dx = x \left(\arccos \frac{x}{a}\right)^3 - 3\sqrt{a^2 - x^2} \left(\arccos \frac{x}{a}\right)^2 - 6x \arccos \frac{x}{a} + 6\sqrt{a^2 - x^2}$

## 2.82 The arcsecant, the arccosecant, the arctangent and the arccotangent

Assume $a > 0$

**2.821**

1.[12]    $\displaystyle\int \text{arccosec} \frac{x}{a}\, dx = \int \arcsin \frac{a}{x}\, dx = x \arcsin \frac{a}{x} + a \ln \left(x + \sqrt{x^2 - a^2}\right)$    $\left[0 < \arcsin \frac{a}{x} < \frac{\pi}{2}\right]$

$$= x \arcsin \frac{a}{x} - a \ln \left(-x + \sqrt{x^2 - a^2}\right) \hspace{2cm} \left[-\frac{\pi}{2} < \arcsin \frac{a}{x} < 0\right]$$

                                                             DW

2.[12]    $\displaystyle\int \text{arcsec} \frac{x}{a}\, dx = \int \arccos \frac{a}{x}\, dx = x \arccos \frac{a}{x} + a \ln \left(x + \sqrt{x^2 - a^2}\right)$    $\left[0 < \arccos \frac{a}{x} < \frac{\pi}{2}\right]$

$$= x \arccos \frac{a}{x} - a \ln \left(-x + \sqrt{x^2 - a^2}\right) \hspace{2cm} \left[\frac{\pi}{2} < \arccos \frac{a}{x} < \pi\right]$$

                                                             DW

**2.822**

1.[8]    $\displaystyle\int \arctan \frac{x}{a}\, dx = x \arctan \frac{x}{a} - \frac{a}{2} \ln \left(a^2 + x^2\right)$                                      DW

$2.^{12}$    $\displaystyle\int \operatorname{arccot}\frac{x}{a}\,\mathrm{d}x = x\operatorname{arccot}\frac{x}{a} + \frac{a}{2}\ln\left(a^2+x^2\right)$          DW

$3.^{9}$    $\displaystyle\int x\arctan\frac{x}{a}\,\mathrm{d}x = \frac{1}{2}\left(x^2+a^2\right)\arctan\frac{x}{a} - \frac{ax}{2}$

$4.^{9}$    $\displaystyle\int x\operatorname{arccot}\frac{x}{a}\,\mathrm{d}x = \frac{ax}{2} + \frac{\pi x^2}{4} - \frac{1}{2}\left(x^2+a^2\right)\arctan\frac{x}{a}$

$5.^{9}$    $\displaystyle\int x^2\arctan\frac{x}{a}\,\mathrm{d}x = \frac{1}{3}x^3\arctan\frac{x}{a} + \frac{1}{6}a^3\ln\left(x^2+a^2\right) - \frac{ax^2}{6}$

$6.^{9}$    $\displaystyle\int x^2\operatorname{arccot}\frac{x}{a}\,\mathrm{d}x = -\frac{1}{3}x^3\arctan\frac{x}{a} - \frac{1}{6}a^3\ln\left(x^2+a^2\right) + \frac{\pi x^3}{6} + \frac{ax^2}{6}$

## 2.83 Combinations of arcsine or arccosine and algebraic functions

Assume $a > 0$

**2.831**   $\displaystyle\int x^n\arcsin\frac{x}{a}\,\mathrm{d}x = \frac{x^{n+1}}{n+1}\arcsin\frac{x}{a} - \frac{1}{n+1}\int\frac{x^{n+1}\,\mathrm{d}x}{\sqrt{a^2-x^2}}$       (see **2.263 1**, **2.264**, **2.27**)

**2.832**   $\displaystyle\int x^n\arccos\frac{x}{a}\,\mathrm{d}x = \frac{x^{n+1}}{n+1}\arccos\frac{x}{a} + \frac{1}{n+1}\int\frac{x^{n+1}\,\mathrm{d}x}{\sqrt{a^2-x^2}}$       (see **2.263 1**, **2.264**, **2.27**)

1.    For $n = -1$, these integrals (that is, $\displaystyle\int\frac{\arcsin x}{x}\,\mathrm{d}x$ and $\displaystyle\int\frac{\arccos x}{x}\,\mathrm{d}x$) cannot be expressed as a finite combination of elementary functions.

2.    $\displaystyle\int\frac{\arccos x}{x}\,\mathrm{d}x = -\frac{\pi}{2}\ln\frac{1}{x} - \int\frac{\arcsin x}{x}\,\mathrm{d}x$

**2.833**[9]

1.    $\displaystyle\int x\arcsin\frac{x}{a}\,\mathrm{d}x = \operatorname{sign}(a)\left[\left(\frac{x^2}{2}-\frac{a^2}{4}\right)\arcsin\frac{x}{|a|} + \frac{x}{4}\sqrt{a^2-x^2}\right]$

2.    $\displaystyle\int x\arccos\frac{x}{a}\,\mathrm{d}x = \frac{\pi x^2}{4} - \operatorname{sign}(a)\left[\frac{1}{4}\left(2x^2-a^2\right)\arcsin\frac{x}{|a|} + \frac{x}{4}\sqrt{a^2-x^2}\right]$

3.    $\displaystyle\int x^2\arcsin\frac{x}{a}\,\mathrm{d}x = \operatorname{sign}(a)\left[\frac{x^3}{3}\arcsin\frac{x}{|a|} + \frac{1}{9}\left(x^2+2a^2\right)\sqrt{a^2-x^2}\right]$

4.    $\displaystyle\int x^2\arccos\frac{x}{a}\,\mathrm{d}x = \frac{\pi x^3}{6} - \operatorname{sign}(a)\left[\frac{x^3}{3}\arcsin\frac{x}{|a|} + \frac{1}{9}\left(x^2+2a^2\right)\sqrt{a^2-x^2}\right]$

5.    $\displaystyle\int x^3\arcsin\frac{x}{a}\,\mathrm{d}x = \operatorname{sign}(a)\left[\left(\frac{x^4}{4}-\frac{3a^4}{32}\right)\arcsin\frac{x}{|a|} + \frac{1}{32}x\left(2x^2+3a^2\right)\sqrt{a^2-x^2}\right]$

6.    $\displaystyle\int x^3\arccos\frac{x}{a}\,\mathrm{d}x = \frac{\pi x^4}{8} - \operatorname{sign}(a)\left[\frac{\left(8x^4-3a^4\right)}{32}\arcsin\frac{x}{|a|} + \frac{1}{32}x\left(2x^2+3a^2\right)\sqrt{a^2-x^2}\right]$

**2.834**

$1.^{12}$   $\displaystyle\int\frac{1}{x^2}\arcsin\frac{x}{a}\,\mathrm{d}x = -\frac{1}{x}\arcsin\frac{x}{a} - \frac{1}{a}\ln\left(\frac{a+\sqrt{a^2-x^2}}{|x|}\right)$

$2.^{12}$   $\int \dfrac{1}{x^2} \arccos \dfrac{x}{a}\, dx = -\dfrac{1}{x} \arccos \dfrac{x}{a} + \dfrac{1}{a} \ln \left( \dfrac{a + \sqrt{a^2 - x^2}}{|x|} \right)$

**2.835**$^{12}$   $\displaystyle\int \dfrac{\arcsin x}{(a + bx)^2}\, dx = -\dfrac{\arcsin x}{b(a + bx)} + \dfrac{1}{b\sqrt{a^2 - b^2}} \arctan \left( \dfrac{b + ax}{\sqrt{a^2 - b^2}\sqrt{1 - x^2}} \right)$     $[a^2 > b^2]$

$\qquad\qquad = -\dfrac{\arcsin x}{b(a + bx)} + \dfrac{1}{2b\sqrt{b^2 - a^2}} \ln \left( \dfrac{b + ax - \sqrt{b^2 - a^2}\sqrt{1 - x^2}}{b + ax + \sqrt{b^2 - a^2}\sqrt{1 - x^2}} \right)$     $[a^2 < b^2]$

$\qquad\qquad = -\dfrac{1}{b^2} \left( \sqrt{\dfrac{1 + x}{1 - x}} + \dfrac{\arcsin x}{x - 1} \right)$     $[a = -b]$

$\qquad\qquad = -\dfrac{1}{b^2} \left( \sqrt{\dfrac{1 - x}{1 + x}} + \dfrac{\arcsin x}{x + 1} \right)$     $[a = b]$

**2.836**$^{12}$   $\displaystyle\int \dfrac{x \arcsin x}{(1 + cx^2)^2}\, dx = \dfrac{\arcsin x}{2c(1 + cx^2)} + \dfrac{1}{2c\sqrt{c + 1}} \arctan \dfrac{\sqrt{c + 1}\,x}{\sqrt{1 - x^2}}$     $[c > -1]$

$\qquad\qquad = -\dfrac{\arcsin x}{2c(1 + cx^2)} + \dfrac{1}{4c\sqrt{-(c + 1)}} \ln \dfrac{\sqrt{1 - x^2} + x\sqrt{-(c + 1)}}{\sqrt{1 - x^2} - x\sqrt{-(c + 1)}}$     $[c < -1]$

**2.837**

1.   $\displaystyle\int \dfrac{x \arcsin x}{\sqrt{1 - x^2}}\, dx = x - \sqrt{1 - x^2}\, \arcsin x$

$2.^{12}$   $\displaystyle\int \dfrac{x^2 \arcsin x}{\sqrt{1 - x^2}}\, dx = \dfrac{x^2}{4} - \dfrac{x}{2} \sqrt{1 - x^2}\, \arcsin x + \dfrac{1}{4} (\arcsin x)^2$

3.   $\displaystyle\int \dfrac{x^3 \arcsin x}{\sqrt{1 - x^2}}\, dx = \dfrac{x^3}{9} + \dfrac{2x}{3} - \dfrac{1}{3} (x^2 + 2) \sqrt{1 - x^2}\, \arcsin x$

**2.838**

1.   $\displaystyle\int \dfrac{\arcsin x}{\sqrt{(1 - x^2)^3}}\, dx = \dfrac{x \arcsin x}{\sqrt{1 - x^2}} + \dfrac{1}{2} \ln (1 - x^2)$

2.   $\displaystyle\int \dfrac{x \arcsin x}{\sqrt{(1 - x^2)^3}}\, dx = \dfrac{\arcsin x}{\sqrt{1 - x^2}} + \dfrac{1}{2} \ln \dfrac{1 - x}{1 + x}$

## 2.84 Combinations of the arcsecant and arccosecant with powers of $x$

Assume $a > 0$

**2.841**

$1.^{12}$   $\displaystyle\int x \operatorname{arcsec} \dfrac{x}{a}\, dx = \int x \arccos \dfrac{a}{x}\, dx = \dfrac{1}{2} \left[ x^2 \arccos \dfrac{a}{x} - a\sqrt{x^2 - a^2} \right]$     $\left[ 0 < \arccos \dfrac{a}{x} < \dfrac{\pi}{2} \right]$

$\qquad\qquad = \dfrac{1}{2} \left[ x^2 \arccos \dfrac{a}{x} + a\sqrt{x^2 - a^2} \right]$     $\left[ \dfrac{\pi}{2} < \arccos \dfrac{a}{x} < \pi \right]$

$2.^{12}$  $\int x^2 \operatorname{arcsec} \dfrac{x}{a} \, dx = \int x^2 \arccos \dfrac{a}{x} \, dx = \dfrac{1}{3} \left[ x^3 \arccos \dfrac{a}{x} - \dfrac{a}{2} x \sqrt{x^2 - a^2} - \dfrac{a^3}{2} \ln \left( x + \sqrt{x^2 - a^2} \right) \right]$

$$\left[ 0 < \arccos \dfrac{a}{x} < \dfrac{\pi}{2} \right]$$

$$= \dfrac{1}{3} \left[ x^3 \arccos \dfrac{a}{x} + \dfrac{a}{2} x \sqrt{x^2 - a^2} - \dfrac{a^3}{2} \ln \left( -x + \sqrt{x^2 - a^2} \right) \right]$$

$$\left[ \dfrac{\pi}{2} < \arccos \dfrac{a}{x} < \pi \right]$$

DW

$3.^{12}$  $\int x \operatorname{arccosec} \dfrac{x}{a} \, dx = \int x \arcsin \dfrac{a}{x} \, dx = \dfrac{1}{2} \left[ x^2 \arcsin \dfrac{a}{x} + a \sqrt{x^2 - a^2} \right] \qquad \left[ 0 < \arcsin \dfrac{a}{x} < \dfrac{\pi}{2} \right]$

$$= \dfrac{1}{2} \left[ x^2 \arcsin \dfrac{a}{x} - a \sqrt{x^2 - a^2} \right] \qquad \left[ -\dfrac{\pi}{2} < \arcsin \dfrac{a}{x} < 0 \right]$$

DW

## 2.85 Combinations of the arctangent and arccotangent with algebraic functions

Assume $a > 0$

**2.851**  $\int x^n \arctan \dfrac{x}{a} \, dx = \dfrac{x^{n+1}}{n+1} \arctan \dfrac{x}{a} - \dfrac{a}{n+1} \int \dfrac{x^{n+1} \, dx}{a^2 + x^2}$

**2.852**

1.    $\int x^n \operatorname{arccot} \dfrac{x}{a} \, dx = \dfrac{x^{n+1}}{n+1} \operatorname{arccot} \dfrac{x}{a} + \dfrac{a}{n+1} \int \dfrac{x^{n+1} \, dx}{a^2 + x^2}$

For $n = -1$

2.    $\int \dfrac{\arctan x}{x} \, dx$ cannot be expressed as a finite combination of elementary functions.

3.    $\int \dfrac{\operatorname{arccot} x}{x} \, dx = \dfrac{\pi}{2} \ln x - \int \dfrac{\arctan x}{x} \, dx$

**2.853**

1.    $\int x \arctan \dfrac{x}{a} \, dx = \dfrac{1}{2} \left( x^2 + a^2 \right) \arctan \dfrac{x}{a} - \dfrac{ax}{2}$

2.    $\int x \operatorname{arccot} \dfrac{x}{a} \, dx = \dfrac{1}{2} \left( x^2 + a^2 \right) \operatorname{arccot} \dfrac{x}{a} + \dfrac{ax}{2}$

$3.^9$    $\int x^2 \arctan \dfrac{x}{a} \, dx = \dfrac{x^3}{3} \arctan \dfrac{x}{a} + \dfrac{a^3}{6} \ln \left( x^2 + a^2 \right) - \dfrac{ax^2}{6}$

$4.^9$    $\int x^2 \operatorname{arccot} \dfrac{x}{a} \, dx = -\dfrac{x^3}{3} \arctan \dfrac{x}{a} - \dfrac{a^3}{6} \ln \left( x^2 + a^2 \right) + \dfrac{\pi x^3}{6} + \dfrac{ax^2}{6}$

**2.854**  $\int \dfrac{1}{x^2} \arctan \dfrac{x}{a} \, dx = -\dfrac{1}{x} \arctan \dfrac{x}{a} - \dfrac{1}{2a} \ln \dfrac{a^2 + x^2}{x^2}$

**2.855**  $\int \dfrac{\arctan x}{(a + bx)^2} \, dx = \dfrac{1}{a^2 + b^2} \left\{ \ln \dfrac{a + bx}{\sqrt{1 + x^2}} - \dfrac{b - ax}{a + bx} \arctan x \right\}$

**2.856**

1.    $\int \dfrac{x \arctan x}{1 + x^2} \, dx = \dfrac{1}{2} \arctan x \ln \left( 1 + x^2 \right) - \dfrac{1}{2} \int \dfrac{\ln \left( 1 + x^2 \right) \, dx}{1 + x^2}$

TI (689)

2.    $\int \dfrac{x^2 \arctan x}{1 + x^2}\, dx = x \arctan x - \dfrac{1}{2} \ln\left(1 + x^2\right) - \dfrac{1}{2}\left(\arctan x\right)^2$        TI (405)

3.    $\int \dfrac{x^3 \arctan x}{1 + x^2}\, dx = -\dfrac{1}{2}x + \dfrac{1}{2}\left(1 + x^2\right) \arctan x - \int \dfrac{x \arctan x}{1 + x^2}\, dx$

(see 2.8511)

4.    $\int \dfrac{x^4 \arctan x}{1 + x^2}\, dx = -\dfrac{1}{6}x^2 + \dfrac{2}{3} \ln\left(1 + x^2\right) + \left(\dfrac{x^3}{3} - x\right) \arctan x + \dfrac{1}{2}\left(\arctan x\right)^2$

**2.857**   $\displaystyle\int \dfrac{\arctan x\, dx}{\left(1 + x^2\right)^{n+1}} = \left[\sum_{k=1}^{n} \dfrac{(2n - 2k)!!(2n - 1)!!}{(2n)!!(2n - 2k + 1)!!} \dfrac{x}{\left(1 + x^2\right)^{n-k+1}} + \dfrac{1}{2}\dfrac{(2n - 1)!!}{(2)!!} \arctan x\right] \arctan x$

$$+ \dfrac{1}{2}\sum_{k=1}^{n} \dfrac{(2n - 1)!!(2n - 2k)!!}{(2n)!!(2n - 2k + 1)!!(n - k + 1)} \dfrac{1}{\left(1 + x^2\right)^{n-k+1}}$$

**2.858**   $\displaystyle\int \dfrac{x \arctan x}{\sqrt{1 - x^2}}\, dx = -\sqrt{1 - x^2}\, \arctan x + \sqrt{2}\,\arctan \dfrac{x\sqrt{2}}{\sqrt{1 - x^2}} - \arcsin x$

**2.859**   $\displaystyle\int \dfrac{\arctan x}{\sqrt{\left(a + bx^2\right)^3}}\, dx = \dfrac{x \arctan x}{a\sqrt{a + bx^2}} - \dfrac{1}{a\sqrt{b - a}} \arctan \sqrt{\dfrac{a + bx^2}{b - a}}$     $[a < b]$

$$= \dfrac{x \arctan x}{a\sqrt{a + bx^2}} + \dfrac{1}{2a\sqrt{a - b}} \ln \dfrac{\sqrt{a + bx^2} - \sqrt{a - b}}{\sqrt{a + bx^2} + \sqrt{a - b}} \quad [a > b]$$

Table of Integrals, Series, and Products. http://dx.doi.org/10.1016/B978-0-12-384933-5.00003-5

# 3–4 Definite Integrals of Elementary Functions

## 3.0 Introduction

### 3.01 Theorems of a general nature

**3.011** Suppose that $f(x)$ is integrable[†] over the largest of the intervals $(p, q), (p, r), (r, q)$. Then (depending on the relative positions of the points $p$, $q$, and $r$) it is also integrable over the other two intervals and we have

$$\int_p^q f(x)\,dx = \int_p^r f(x)\,dx + \int_r^q f(x)\,dx$$

FI II 126

**3.012** *The first mean-value theorem.* Suppose (1) that $f(x)$ is continuous and that $g(x)$ is integrable over the interval $(p, q)$, (2) that $m \leq f(x) \leq M$ and (3) that $g(x)$ does not change sign anywhere in the interval $(p, q)$. Then, there exists at least one point $\xi$ (with $p \leq \xi \leq q$) such that

$$\int_p^q f(x)g(x)\,dx = f(\xi)\int_p^q g(x)\,dx$$

FI II 132

**3.013** *The second mean-value theorem.* If $f(x)$ is monotonic and non-negative throughout the interval $(p, q)$, where $p < q$, and if $g(x)$ is integrable over that interval, then there exists at least one point $\xi$ (with $p \leq \xi \leq q$) such that

1.  $$\int_p^q f(x)g(x)\,dx = f(p)\int_p^\xi g(x)\,dx$$

Under the conditions of Theorem **3.013** 1, if $f(x)$ is nondecreasing, then

2.  $$\int_p^q f(x)g(x)\,dx = f(q)\int_\xi^q g(x)\,dx \qquad\qquad [p \leq \xi \leq q]$$

If $f(x)$ is monotonic in the interval $(p, q)$, where $p < q$, and if $g(x)$ is integrable over that interval, then

---

*We omit the definition of definite and multiple integrals since they are widely known and can easily be found in any textbook on the subject. Here we give only certain theorems of a general nature which provide estimates, or which reduce the given integral to a simpler one.

†A function $f(x)$ is said to be integrable over the interval $(p, q)$, if the integral $\int_p^q f(x)\,dx$ exists. Here, we usually mean the existence of the integral in the sense of Riemann. When it is a matter of the existence of the integral in the sense of Stieltjes or Lebesgue, etc., we shall speak of integrability in the sense of Stieltjes or Lebesgue.

3.
$$\int_p^q f(x)g(x)\,dx = f(p)\int_p^\xi g(x)\,dx + f(q)\int_\xi^q g(x)\,dx \qquad [p \le \xi \le q]\,,$$

or

4.
$$\int_p^q f(x)g(x)\,dx = A\int_p^\xi g(x)\,dx + B\int_\xi^q g(x)\,dx \qquad [p \le \xi \le q]\,,$$

where $A$ and $B$ are any two numbers satisfying the conditions

$$A \ge f(p+0) \quad \text{and} \quad B \le f(q-0) \qquad [\text{if } f \text{ decreases}],$$
$$A \le f(p+0) \quad \text{and} \quad B \ge f(q-0) \qquad [\text{if } f \text{ increases}].$$

In particular,

5.
$$\int_p^q f(x)g(x)\,dx = f(p+0)\int_p^\xi g(x)\,dx + f(q-0)\int_\xi^q g(x)\,dx \qquad \text{FI II 138}$$

## 3.02  Change of variable in a definite integral

**3.020**
$$\int_\alpha^\beta f(x)\,dx = \int_\varphi^\psi f[g(t)]g'(t)\,dt; \qquad x = g(t).$$
This formula is valid under the following conditions:

1.   $f(x)$ is continuous on some interval $A \le x \le B$ containing the original limits of integration $\alpha$ and $\beta$.

2.   The equalities $\alpha = g(\varphi)$ and $\beta = g(\psi)$ hold.

3.   $g(t)$ and its derivative $g'(t)$ are continuous on the interval $\varphi \le t \le \psi$.

4.   As $t$ varies from $\varphi$ to $\psi$, the function $g(t)$ always varies in the same direction from $g(\varphi) = \alpha$ to $g(\psi) = \beta$.*

**3.021**   The integral $\int_\alpha^\beta f(x)\,dx$ can be transformed into another integral with given limits $\varphi$ and $\psi$ by means of the linear substitution

$$x = \frac{\beta - \alpha}{\psi - \varphi}t + \frac{\alpha\psi - \beta\varphi}{\psi - \varphi}:$$

1.
$$\int_\alpha^\beta f(x)\,dx = \frac{\beta - \alpha}{\psi - \varphi}\int_\varphi^\psi f\left(\frac{\beta - \alpha}{\psi - \varphi}t + \frac{\alpha\psi - \beta\varphi}{\psi - \varphi}\right)dt$$

In particular, for $\varphi = 0$ and $\psi = 1$,

---

*If this last condition is not satisfied, the interval $\varphi \le t \le \psi$ should be partitioned into subintervals throughout each of which the condition is satisfied:

$$\int_\alpha^\beta f(x)\,dx = \int_\varphi^{\varphi_1} f[g(t)]g'(t)\,dt + \int_{\varphi_1}^{\varphi_2} f[g(t)]g'(t)\,dt + \cdots + \int_{\varphi_{n-1}}^\psi f[g(t)]g'(t)\,dt.$$

2. $\displaystyle\int_\alpha^\beta f(x)\,\mathrm{d}x = (\beta-\alpha)\int_0^1 f((\beta-\alpha)t+\alpha)\,\mathrm{d}t$

For $\varphi = 0$ and $\psi = \infty$,

3. $\displaystyle\int_\alpha^\beta f(x)\,\mathrm{d}x = (\beta-\alpha)\int_0^\infty f\left(\frac{\alpha+\beta t}{1+t}\right)\frac{\mathrm{d}t}{(1+t)^2}$

**3.022**  The following formulas also hold:

1. $\displaystyle\int_\alpha^\beta f(x)\,\mathrm{d}x = \int_\alpha^\beta f(\alpha+\beta-x)\,\mathrm{d}x$

2. $\displaystyle\int_0^\beta f(x)\,\mathrm{d}x = \int_0^\beta f(\beta-x)\,\mathrm{d}x$

3. $\displaystyle\int_{-\alpha}^\alpha f(x)\,\mathrm{d}x = \int_{-\alpha}^\alpha f(-x)\,\mathrm{d}x$

## 3.03  General formulas

**3.031**

1. Suppose that a function $f(x)$ is integrable over the interval $(-p, p)$ and satisfies the relation $f(-x) = f(x)$ on that interval. (A function satisfying the latter condition is called an *even* function.) Then,

$$\int_{-p}^p f(x)\,\mathrm{d}x = 2\int_0^p f(x)\,\mathrm{d}x \qquad \text{FI II 159}$$

2. Suppose that $f(x)$ is a function that is integrable on the interval $(-p, p)$ and satisfies the relation $f(-x) = -f(x)$ on that interval. (A function satisfying the latter condition is called an *odd* function). Then,

$$\int_{-p}^p f(x)\,\mathrm{d}x = 0. \qquad \text{FI II 159}$$

**3.032**

1. $\displaystyle\int_0^{\frac{\pi}{2}} f(\sin x)\,\mathrm{d}x = \int_0^{\frac{\pi}{2}} f(\cos x)\,\mathrm{d}x,$

where $f(x)$ is a function that is integrable on the interval $(0,1)$.  \qquad FI II 159

2. $\displaystyle\int_0^{2\pi} f(p\cos x + q\sin x)\,\mathrm{d}x = 2\int_0^\pi f\left(\sqrt{p^2+q^2}\,\cos x\right)\mathrm{d}x,$

where $f(x)$ is integrable on the interval $\left(-\sqrt{p^2+q^2},\sqrt{p^2+q^2}\right)$.  \qquad FI II 160

3. $\displaystyle\int_0^{\frac{\pi}{2}} f(\sin 2x)\cos x\,\mathrm{d}x = \int_0^{\frac{\pi}{2}} f\left(\cos^2 x\right)\cos x\,\mathrm{d}x,$

where $f(x)$ is integrable on the interval $(0,1)$.  \qquad FI II 161

**3.033**

1. If $f(x+\pi) = f(x)$ and $f(-x) = f(x)$, then

$$\int_0^\infty f(x)\frac{\sin x}{x}\,dx = \int_0^{\frac{\pi}{2}} f(x)\,dx \qquad\qquad\qquad \text{LO V 277(3)}$$

2.    If $f(x+\pi) = -f(x)$ and $f(-x) = f(x)$, then

$$\int_0^\infty f(x)\frac{\sin x}{x}\,dx = \int_0^{\frac{\pi}{2}} f(x)\cos x\,dx \qquad\qquad \text{LO V 279(4)}$$

In formulas **3.033**, it is assumed that the integrals in the left members of the formulas exist.

**3.034** $\int_0^\infty \dfrac{f(px) - f(qx)}{x}\,dx = [f(0) - f(+\infty)]\ln\dfrac{q}{p}$,

if $f(x)$ is continuous for $x \geq 0$ and if there exists a finite limit $f(+\infty) = \lim\limits_{x \to +\infty} f(x)$.     FI II 633

**3.035**

1.[12]  $\int_0^\pi \dfrac{f\left(\alpha + e^{xi}\right) + f\left(\alpha + e^{-xi}\right)}{1 - 2p\cos x + p^2}\,dx = \dfrac{2\pi}{1 - p^2}f(\alpha + p) \qquad [|p| < 1]$     LA 230(16)

2.  $\int_0^\pi \dfrac{1 - p\cos x}{1 - 2p\cos x + p^2}\left\{f\left(\alpha + e^{xi}\right) + f\left(\alpha + e^{-xi}\right)\right\}\,dx = \pi\left\{f(\alpha + p) + f(\alpha)\right\}$

$$[|p| < 1] \qquad\qquad \text{BE 169}$$

3.[12]  $\int_0^\pi \dfrac{f\left(\alpha + e^{-xi}\right) - f\left(\alpha + e^{xi}\right)}{1 - 2p\cos x + p^2}\sin x\,dx = \dfrac{\pi}{ip}\left\{f(\alpha + p) - f(\alpha)\right\}$

$$[|p| < 1] \qquad\qquad \text{BE 169}$$

In formulas **3.035**, it is assumed that the function $f$ is analytic in the closed unit circle with its center at the point $\alpha$.

**3.036**

1.[12]  $\int_0^\pi f\left(\dfrac{\sin^2 x}{1 + 2p\cos x + p^2}\right)\,dx = \int_0^\pi f\left(\sin^2 x\right)\,dx \qquad [p^2 \geq 1]$

$$= \int_0^\pi f\left(\dfrac{\sin^2 x}{p^2}\right)\,dx \qquad [p^2 < 1]$$

$$\text{LA 228(6)}$$

2.  $\int_0^\pi F^{(n)}\left(\cos x\right)\sin^{2n} x\,dx = (2n - 1)!!\int_0^\pi F\left(\cos x\right)\cos nx\,dx$     B 174

**3.037**  If $f$ is analytic in the circle of radius $r$ and if

$$f\left[r\left(\cos x + i\sin x\right)\right] = f_1(r, x) + if_2(r, x),$$

then

1.  $\int_0^\infty \dfrac{f_1(r, x)}{p^2 + x^2}\,dx = \dfrac{\pi}{2p}f\left(re^{-p}\right)$     LA 230(19)

2.  $\int_0^\infty f_2(r, x)\dfrac{x\,dx}{p^2 + x^2} = \dfrac{\pi}{2}\left[f\left(re^{-p}\right) - f(0)\right]$     LA 230(20)

3.  $\int_0^\infty \dfrac{f_2(r, x)}{x}\,dx = \dfrac{\pi}{2}[f(r) - f(0)]$.     LA 230(21)

4.    $\displaystyle \int_0^\infty \frac{f_2(r,x)}{x\,(p^2+x^2)}\,\mathrm{d}x = \frac{\pi}{2p^2}\left[f(r) - f\left(re^{-p}\right)\right]$                 LA 230(22)

**3.038**    $\displaystyle \int_{-\infty}^\infty \frac{x\,\mathrm{d}x}{\sqrt{1+x^2}} F\left(qx + p\sqrt{1+x^2}\right) = \int_{-\infty}^\infty F\left(p\cosh x + q\sinh x\right)\sinh x\,\mathrm{d}x$

$$= 2q\int_0^\infty F'\left(\operatorname{sign} p \cdot \sqrt{p^2 - q^2}\cosh x\right)\sinh^2 x\,\mathrm{d}x$$

[If $F$ is a function with a continuous derivative in the interval $(-\infty, \infty)$; all these integrals converge.]

## 3.04   Improper integrals

**3.041**    Suppose that a function $f(x)$ is defined on an interval $(p, +\infty)$ and that it is integrable over an arbitrary finite subinterval of the form $(p, P)$. Then, by definition

$$\int_p^{+\infty} f(x)\,\mathrm{d}x = \lim_{P\to+\infty}\int_p^P f(x)\,\mathrm{d}x,$$

if this limit exists. If it does exist, we say that the integral $\displaystyle\int_p^{+\infty} f(x)\,\mathrm{d}x$ exists or that it converges. Otherwise, we say that the integral diverges.

**3.042**    Suppose that a function $f(x)$ is bounded and integrable in an arbitrary interval $(p, q - \eta)$ (for $0 < \eta < q - p$) but is unbounded in every interval $(q - \eta, q)$ to the left of the point $q$. The point $q$ is then called a *singular point*. Then, by definition,

$$\int_p^q f(x)\,\mathrm{d}x = \lim_{\eta\to 0}\int_p^{q-\eta} f(x)\,\mathrm{d}x,$$

if this limit exists. In this case, we say that the integral $\displaystyle\int_p^q f(x)\,\mathrm{d}x$ *exists* or that it *converges*.

**3.043**    If not only the integral of $f(x)$ but also the integral of $|f(x)|$ exists, we say that the integral of $f(x)$ converges *absolutely*.

**3.044**    The integral $\displaystyle\int_p^{+\infty} f(x)\,\mathrm{d}x$ converges absolutely if there exists a number $\alpha > 1$ such that the limit

$$\lim_{x\to+\infty}\{x^\alpha|f(x)|\}$$

exists. On the other hand, if

$$\lim_{x\to+\infty}\{x|f(x)|\} = L > 0,$$

the integral $\displaystyle\int_p^{+\infty}|f(x)|\,\mathrm{d}x$ diverges.

**3.045**    Suppose that the upper limit $q$ of the integral $\displaystyle\int_p^q f(x)\,\mathrm{d}x$ is a singular point. Then, this integral converges absolutely if there exists a number $\alpha < 1$ such that the limit

$$\lim_{x\to q}[(q-x)^\alpha|f(x)|]$$

exists. On the other hand, if

$$\lim_{x\to q}[(q-x)|f(x)|] = L > 0,$$

the integral $\displaystyle\int_p^q f(x)\,\mathrm{d}x$ diverges.

**3.046**   Suppose that the functions $f(x)$ and $g(x)$ are defined on the interval $(p, +\infty)$, that $f(x)$ is integrable over every finite interval of the form $(p, P)$, that the integral

$$\int_p^P f(x)\,\mathrm{d}x$$

is a bounded function of $P$, that $g(x)$ is monotonic, and that $g(x) \to 0$ as $x \to +\infty$. Then, the integral

$$\int_p^{+\infty} f(x)g(x)\,\mathrm{d}x$$

converges.                                                                            FI II 577

## 3.05   The principal values of improper integrals

**3.051**   Suppose that a function $f(x)$ has a singular point $r$ somewhere inside the interval $(p, q)$, that $f(x)$ is defined at $r$, and that $f(x)$ is integrable over every portion of this interval that does not contain the point $r$. Then, by definition

$$\int_p^q f(x)\,\mathrm{d}x = \lim_{\substack{\eta \to 0 \\ \eta' \to 0}} \left\{ \int_p^{r-\eta} f(x)\,\mathrm{d}x + \int_{r+\eta'}^q f(x)\,\mathrm{d}x \right\},$$

Here, the limit must exist for *independent* modes of approach of $\eta$ and $\eta'$ to zero. If this limit does not exist but the limit

$$\lim_{\eta \to 0} \left\{ \int_p^{r-\eta} f(x)\,\mathrm{d}x + \int_{r+\eta}^q f(x)\,\mathrm{d}x \right\},$$

does exist, we say that this latter limit is the *principal value* of the improper integral $\int_p^q f(x)\,\mathrm{d}x$ and we say that the integral $\int_p^q f(x)\,\mathrm{d}x$ exists in the sense of principal values.                       FI II 603

**3.052**   Suppose that the function $f(x)$ is continuous over the interval $(p, q)$ and vanishes at only one point $r$ inside this interval. Suppose that the first derivative $f'(x)$ exists in a neighborhood of the point $r$. Suppose that $f'(r) \neq 0$ and that the second derivative $f''(r)$ exists at the point $r$ itself. Then,

$$\int_p^q \frac{\mathrm{d}x}{f(x)}$$                                                 FI II 605

diverges, but exists in the sense of principal values.

**3.053**   A divergent integral of a positive function cannot exist in the sense of principal values.

**3.054**   Suppose that the function $f(x)$ has no singular points in the interval $(-\infty, +\infty)$. Then, by definition

$$\int_{-\infty}^{+\infty} f(x)\,\mathrm{d}x = \lim_{\substack{P \to -\infty \\ Q \to +\infty}} \int_P^Q f(x)\,\mathrm{d}x,$$

Here, the limit must exist for independent approach of $P$ and $Q$ to $\pm\infty$. If this limit does not exist but the limit

$$\lim_{P \to +\infty} \int_{-P}^{+P} f(x)\,\mathrm{d}x,$$

does exist, this last limit is called the principal value of the improper integral

$$\int_{-\infty}^{+\infty} f(x)\,\mathrm{d}x$$                                        FI II 607

**3.055**   The principal value of an improper integral of an even function exists only when this integral converges (in the ordinary sense).                                     FI II 607

# 3.1–3.2 Power and Algebraic Functions

## 3.11  Rational functions

1.   $\displaystyle\int_{-\infty}^{\infty} \frac{p+qx}{r^2 + 2rx\cos\lambda + x^2}\,dx = \frac{\pi}{r\sin\lambda}\,(p - qr\cos\lambda)$      (principal value)

                            (see also **3.194** 8 and **3.252** 1 and 2)   BI (22)(14)

**3.112**   Integrals of the form $\displaystyle\int_{-\infty}^{\infty} \frac{g_n(x)\,dx}{h_n(x)h_n(-x)}$, where

$$g_n(x) = b_0 x^{2n-2} b_1 x^{2n-4} + \cdots + b_{n-1},$$
$$h_n(x) = a_0 x^n a_1 x^{n-1} + \cdots + a_n$$

[All roots of $h_n(x)$ lie in the upper half-plane.]

1.   $\displaystyle\int_{-\infty}^{\infty} \frac{g_n(x)\,dx}{h_n(x)h_n(-x)} = \frac{\pi i}{a_0}\frac{M_n}{\Delta_n},$                             JE

    where

$$\Delta_n = \begin{vmatrix} a_1 & a_3 & a_5 & & 0 \\ a_0 & a_2 & a_4 & & 0 \\ 0 & a_1 & a_3 & & 0 \\ \vdots & & & \ddots & \\ 0 & 0 & 0 & & a_n \end{vmatrix}, \qquad M_n = \begin{vmatrix} b_0 & b_1 & b_2 & \cdots & b_{n-1} \\ a_0 & a_2 & a_4 & & 0 \\ 0 & a_1 & a_3 & & 0 \\ \vdots & & & \ddots & \\ 0 & 0 & 0 & & a_n \end{vmatrix}.$$

2.   $\displaystyle\int_{-\infty}^{\infty} \frac{g_1(x)\,dx}{h_1(x)h_1(-x)} = \frac{\pi i b_0}{a_0 a_1}$                             JE

3.[8]   $\displaystyle\int_{-\infty}^{\infty} \frac{g_2(x)\,dx}{h_2(x)h_2(-x)} = \pi i\,\frac{-b_0 + \frac{a_0 b_1}{a_2}}{a_0 a_1}$

4.   $\displaystyle\int_{-\infty}^{\infty} \frac{g_3(x)\,dx}{h_3(x)h_3(-x)} = \pi i\,\frac{-a_2 b_0 a_0 b_1 - \dfrac{a_0 a_1 b_2}{a_3}}{a_0\,(a_0 a_3 a_1 a_2)}$             JE

5.[12]   $\displaystyle\int_{-\infty}^{\infty} \frac{g_4(x)\,dx}{h_4(x)h_4(-x)} = -\pi i\,\frac{b_0\,(-a_1 a_4 + a_2 a_3) - a_0 a_3 b_1 + a_0 a_1 b_2 + \dfrac{a_0 b_3}{a_4}\,(a_0 a_3 - a_1 a_2)}{a_0\,(a_0 a_3^2 + a_1^2 a_4 - a_1 a_2 a_3)}$    JE

6.   $\displaystyle\int_{-\infty}^{\infty} \frac{g_5(x)\,dx}{h_5(x)h_5(-x)} = \pi i\,\frac{M_5}{a_0 \Delta_5},$

where

$$M_5 = b_0\left(-a_0 a_4 a_5 + a_1 a_4^2 + a_2^2 a_5 - a_2 a_3 a_4\right) + a_0 b_1\left(-a_2 a_5 + a_3 a_4\right)$$
$$+\, a_0 b_2\,(a_0 a_5 - a_1 a_4) + a_0 b_3\,(-a_0 a_3 + a_1 a_2) + \frac{a_0 b_4}{a_5}\left(-a_0 a_1 a_5 + a_0 a_3^2 + a_1^2 a_4 - a_1 a_2 a_3\right),$$

$$\Delta_5 = a_0^2 a_5^2 - 2a_0 a_1 a_4 a_5 - a_0 a_2 a_3 a_5 + a_0 a_3^2 a_4 + a_1^2 a_4^2 + a_1 a_2^2 a_5 - a_1 a_2 a_3 a_4$$     JE

## 3.12 Products of rational functions and expressions that can be reduced to square roots of first-and second-degree polynomials

**3.121**

1.  $$\int_0^1 \frac{1}{1 - 2x\cos\lambda + x^2} \frac{dx}{\sqrt{x}} = 2\operatorname{cosec}\lambda \sum_{k=1}^{\infty} \frac{\sin k\lambda}{2k - 1}$$

    BI (10)(17)

2.[12]  $$\int_0^1 \frac{1}{q - px} \frac{dx}{\sqrt{x(1 - x)}} = \frac{\pi}{p\sqrt{\frac{q}{p}\left(\frac{q}{p} - 1\right)}}$$

    $$\left[\frac{q}{p} > 1\right]$$

    $$= -\frac{\pi}{p\sqrt{\frac{q}{p}\left(\frac{q}{p} - 1\right)}}$$

    $$\left[\frac{q}{p} < 0\right]$$

    $$= 0$$

    $$\left[0 < \frac{q}{p} < 1\right]$$

    BI (10)(9)

3.  $$\int_0^1 \frac{dx}{1 - 2rx + r^2} \sqrt{\frac{1 \mp x}{1 \pm x}} = \pm\frac{\pi}{4r} \mp \frac{1}{r}\frac{1 \mp r}{1 \pm r} \arctan\frac{1 + r}{1 - r}$$

    LI (14)(5, 16)

## 3.13–3.17 Expressions that can be reduced to square roots of third-and fourth-degree polynomials and their products with rational functions

**Notation:** In **3.131–3.137** we set: $\alpha = \arcsin\sqrt{\dfrac{a - c}{a - u}}$, $\beta = \arcsin\sqrt{\dfrac{c - u}{b - u}}$,

$$\gamma = \arcsin\sqrt{\frac{u - c}{b - c}}, \qquad \delta = \arcsin\sqrt{\frac{(a - c)(b - u)}{(b - c)(a - u)}},$$

$$\kappa = \arcsin\sqrt{\frac{(a - c)(u - b)}{(a - b)(u - c)}}, \qquad \lambda = \arcsin\sqrt{\frac{a - u}{a - b}},$$

$$\mu = \arcsin\sqrt{\frac{u - a}{u - b}}, \qquad \nu = \arcsin\sqrt{\frac{a - c}{u - c}}, \qquad p = \sqrt{\frac{a - b}{a - c}}, \qquad q = \sqrt{\frac{b - c}{a - c}}.$$

**3.131**

1.  $$\int_{-\infty}^{u} \frac{dx}{\sqrt{(a - x)(b - x)(c - x)}} = \frac{2}{\sqrt{a - c}} F(\alpha, p) \qquad [a > b > c \geq u]$$

    BY (231.00)

2.  $$\int_{u}^{c} \frac{dx}{\sqrt{(a - x)(b - x)(c - x)}} = \frac{2}{\sqrt{a - c}} F(\beta, p) \qquad [a > b > c > u]$$

    BY (232.00)

3.  $$\int_{c}^{u} \frac{dx}{\sqrt{(a - x)(b - x)(x - c)}} = \frac{2}{\sqrt{a - c}} F(\gamma, q) \qquad [a > b \geq u > c]$$

    BY (233.00)

4. $\displaystyle\int_u^b \frac{dx}{\sqrt{(a-x)(b-x)(x-c)}} = \frac{2}{\sqrt{a-c}} F(\delta, q)$      $[a > b > u \geq c]$      BY (234.00)

5. $\displaystyle\int_b^u \frac{dx}{\sqrt{(a-x)(x-b)(x-c)}} = \frac{2}{\sqrt{a-c}} F(\kappa, p)$      $[a \geq u > b > c]$      BY (235.00)

6. $\displaystyle\int_u^a \frac{dx}{\sqrt{(a-x)(x-b)(x-c)}} = \frac{2}{\sqrt{a-c}} F(\lambda, p)$      $[a > u \geq b > c]$      BY (236.00)

7. $\displaystyle\int_a^u \frac{dx}{\sqrt{(x-a)(x-b)(x-c)}} = \frac{2}{\sqrt{a-c}} F(\mu, q)$      $[u > a > b > c]$      BY (237.00)

8. $\displaystyle\int_u^\infty \frac{dx}{\sqrt{(x-a)(x-b)(x-c)}} = \frac{2}{\sqrt{a-c}} F(\nu, q)$      $[u \geq a > b > c]$      BY (238.00)

**3.132**

1. $\displaystyle\int_u^c \frac{x\,dx}{\sqrt{(a-x)(b-x)(c-x)}} = \frac{2}{\sqrt{a-c}}\left[c\,F(\beta, p) + (a-c)\,E(\beta, p)\right] - 2\sqrt{\frac{(a-u)(c-u)}{b-u}}$

     $[a > b > c > u]$      BY (232.19)

2. $\displaystyle\int_c^u \frac{x\,dx}{\sqrt{(a-x)(b-x)(x-c)}} = \frac{2a}{\sqrt{a-c}} F(\gamma, q) - 2\sqrt{a-c}\,E(\gamma, q)$

     $[a > b \geq u > c]$      BY (233.17)

3. $\displaystyle\int_u^b \frac{x\,dx}{\sqrt{(a-x)(b-x)(x-c)}} = \frac{2}{\sqrt{a-c}}\left[(b-a)\,\Pi\left(\delta, q^2, q\right) + a\,F(\delta, q)\right]$

     $[a > b > u \geq c]$      BY (234.16)

4. $\displaystyle\int_b^u \frac{x\,dx}{\sqrt{(a-x)(x-b)(x-c)}} = \frac{2}{\sqrt{a-c}}\left[(b-c)\,\Pi\left(\kappa, p^2, p\right) + c\,F\left(\kappa, p\right)\right]$

     $[a \geq u > b > c]$      BY (235.16)

5.[12] $\displaystyle\int_u^a \frac{x\,dx}{\sqrt{(a-x)(x-b)(x-c)}} = \frac{2c}{\sqrt{a-c}} F(\lambda, p) + 2\frac{a}{b}\sqrt{a-c}\,E(\lambda, p)$

     $[a > u \geq b > c]$      BY (236.16)

6.[12] $\displaystyle\int_a^u \frac{x\,dx}{\sqrt{(x-a)(x-b)(x-c)}} = \frac{2}{\sqrt{a-c}}\left[(a-b)\,\Pi(\mu, 1, q) + b\,F(\mu, q)\right]$

     $[u > a > b > c]$      BY (237.16)

**3.133**

1. $\displaystyle\int_{-\infty}^u \frac{dx}{\sqrt{(a-x)^3(b-x)(c-x)}} = \frac{2}{(a-b)\sqrt{a-c}}\left[F(\alpha, p) - E(\alpha, p)\right]$

     $[a > b > c \geq u]$      BY (231.08)

2. $\displaystyle\int_u^c \frac{dx}{\sqrt{(a-x)^3(b-x)(c-x)}} = \frac{2}{(a-b)\sqrt{a-c}}\left[F(\beta, p) - E(\beta, p)\right] + \frac{2}{a-c}\sqrt{\frac{c-u}{(a-u)(b-u)}}$

     $[a > b > c > u]$      BY (232.13)

3.   $\displaystyle\int_c^u \frac{dx}{\sqrt{(a-x)^3(b-x)(x-c)}} = \frac{2}{(a-b)\sqrt{a-c}}\,E(\gamma,q) - \frac{2}{(a-b)(a-c)}\sqrt{\frac{(b-u)(u-c)}{a-u}}$

$$[a > b \geq u > c] \qquad\qquad \text{BY (233.09)}$$

4.[12]   $\displaystyle\int_u^b \frac{dx}{\sqrt{(a-x)^3(b-x)(x-c)}} = \frac{2}{(a-b)\sqrt{a-c}}\,E(\delta,q) \quad [a > b > u \geq c] \qquad \text{BY (234.05)}$

5.   $\displaystyle\int_b^u \frac{dx}{\sqrt{(a-x)^3(x-b)(x-c)}} = \frac{2}{(a-b)\sqrt{a-c}}\,[F(\kappa,p) - E(\kappa,p)] + \frac{2}{a-b}\sqrt{\frac{u-b}{(a-u)(u-c)}}$

$$[a > u > b > c] \qquad\qquad \text{BY (235.04)}$$

6.   $\displaystyle\int_u^\infty \frac{dx}{\sqrt{(x-a)^3(x-b)(x-c)}} = \frac{2}{(b-a)\sqrt{a-c}}\,E(\nu,q) + \frac{2}{a-b}\sqrt{\frac{u-b}{(u-a)(u-c)}}$

$$[u > a > b > c] \qquad\qquad \text{BY (238.05)}$$

7.   $\displaystyle\int_{-\infty}^u \frac{dx}{\sqrt{(a-x)(b-x)^3(c-x)}} = \frac{2\sqrt{a-c}}{(a-b)(b-c)}\,E(\alpha,p) - \frac{2}{(a-b)\sqrt{a-c}}\,F(\alpha,p)$

$$-\frac{2}{b-c}\sqrt{\frac{c-u}{(a-u)(b-u)}}$$

$$[a > b > c \geq u] \qquad\qquad \text{BY (231.09)}$$

8.   $\displaystyle\int_u^c \frac{dx}{\sqrt{(a-x)(b-x)^3(c-x)}} = \frac{2\sqrt{a-c}}{(a-b)(b-c)}\,E(\beta,p) - \frac{2}{(a-b)\sqrt{a-c}}\,F(\beta,p)$

$$[a > b > c > u] \qquad\qquad \text{BY (232.14)}$$

9.   $\displaystyle\int_c^u \frac{dx}{\sqrt{(a-x)(b-x)^3(x-c)}} = \frac{2}{(b-c)\sqrt{a-c}}\,F(\gamma,q) - \frac{2\sqrt{a-c}}{(a-b)(b-c)}\,E(\gamma,q)$

$$+\frac{2}{(a-b)(b-c)}\sqrt{\frac{(a-u)(u-c)}{b-u}}$$

$$[a > b > u > c] \qquad\qquad \text{BY (233.10)}$$

10.   $\displaystyle\int_u^a \frac{dx}{\sqrt{(a-x)(x-b)^3(x-c)}} = \frac{2}{(a-b)\sqrt{a-c}}\,F(\lambda,p) - \frac{2\sqrt{a-c}}{(a-b)(b-c)}\,E(\lambda,p)$

$$+\frac{2}{(a-b)(b-c)}\sqrt{\frac{(a-u)(u-c)}{u-b}}$$

$$[a > u > b > c] \qquad\qquad \text{BY (236.09)}$$

11.   $\displaystyle\int_a^u \frac{dx}{\sqrt{(x-a)(x-b)^3(x-c)}} = \frac{2\sqrt{a-c}}{(a-b)(b-c)}\,E(\mu,q) - \frac{2}{(b-c)\sqrt{a-c}}\,F(\mu,q)$

$$[u > a > b > c] \qquad\qquad \text{BY (237.12)}$$

12.   $\displaystyle\int_u^\infty \frac{dx}{\sqrt{(x-a)(x-b)^3(x-c)}} = \frac{2\sqrt{a-c}}{(a-b)(b-c)}\,E(\nu,q) - \frac{2}{(b-c)\sqrt{a-c}}\,F(\nu,q)$

$$-\frac{2}{a-b}\sqrt{\frac{u-a}{(u-b)(u-c)}}$$

$$[u \geq a > b > c] \qquad\qquad \text{BY (238.04)}$$

13. $$\int_{-\infty}^{u} \frac{dx}{\sqrt{(a-x)(b-x)(c-x)^3}} = \frac{2}{(c-b)\sqrt{a-c}} E(\alpha, p) + \frac{2}{b-c}\sqrt{\frac{b-u}{(a-u)(c-u)}}$$
$$[a > b > c > u] \qquad\qquad \text{BY (231.10)}$$

14. $$\int_{u}^{b} \frac{dx}{\sqrt{(a-x)(b-x)(x-c)^3}} = \frac{2}{(b-c)\sqrt{a-c}} [F(\delta, q) - E(\delta, q)] + \frac{2}{b-c}\sqrt{\frac{b-u}{(a-u)(u-c)}}$$
$$[a > b > u > c] \qquad\qquad \text{BY (234.04)}$$

15. $$\int_{b}^{u} \frac{dx}{\sqrt{(a-x)(x-b)(x-c)^3}} = \frac{2}{(b-c)\sqrt{a-c}} E(\kappa, p)$$
$$[a \ge u > b > c] \qquad\qquad \text{BY (235.01)}$$

16. $$\int_{u}^{a} \frac{dx}{\sqrt{(a-x)(x-b)(x-c)^3}} = \frac{2}{(b-c)\sqrt{a-c}} E(\lambda, p) - \frac{2}{(b-c)(a-c)}\sqrt{\frac{(a-u)(u-b)}{u-c}}$$
$$[a > u \ge b > c] \qquad\qquad \text{BY (236.10)}$$

17.[12] $$\int_{a}^{u} \frac{dx}{\sqrt{(x-a)(x-b)(x-c)^3}} = \frac{2}{(b-c)\sqrt{a-c}} [F(\mu, q) - E(\mu, q)] + \frac{2}{a-c}\sqrt{\frac{u-a}{(u-b)(u-c)}}$$
$$[u > a > b > c] \qquad\qquad \text{BY (237.13)}$$

18. $$\int_{u}^{\infty} \frac{dx}{\sqrt{(x-a)(x-b)(x-c)^3}} = \frac{2}{(b-c)\sqrt{a-c}} [F(\nu, q) - E(\nu, q)]$$
$$[u \ge a > b > c] \qquad\qquad \text{BY (238.03)}$$

**3.134**

1. $$\int_{-\infty}^{u} \frac{dx}{\sqrt{(a-x)^5(b-x)(c-x)}}$$
$$= \frac{2}{3(a-b)^2\sqrt{(a-c)^3}} [(3a - b - 2c) F(\alpha, p) - 2(2a - b - c) E(\alpha, p)]$$
$$+ \frac{2}{3(a-c)(a-b)}\sqrt{\frac{(c-u)(b-u)}{(a-u)^3}}$$
$$[a > b > c \ge u] \qquad\qquad \text{BY (231.08)}$$

2. $$\int_{u}^{c} \frac{dx}{\sqrt{(a-x)^5(b-x)(c-x)}} = \frac{2}{3(a-b)^2\sqrt{(a-c)^3}} [(3a - b - 2c) F(\beta, p) - 2(2a - b - c) E(\beta, p)]$$
$$+ \frac{2[4a^2 - 3ab - 2ac + bc - u(3a - 2b - c)]}{3(a-b)(a-c)^2}\sqrt{\frac{c-u}{(a-u)^3(b-u)}}$$
$$[a > b > c > u] \qquad\qquad \text{BY (232.13)}$$

3.[12] $$\int_{c}^{u} \frac{dx}{\sqrt{(a-x)^5(b-x)(x-c)}} = \frac{2}{3(a-b)^2\sqrt{(a-c)^3}} [2(2a - b - c) E(\gamma, q) - (a - b) F(\gamma, q)]$$
$$- \frac{2[5a^2 - 3ab - 3ac + bc - 2u(2a - b - c)]}{3(a-b)^2(a-c)^2}\sqrt{\frac{(b-u)(u-c)}{(a-u)^3}}$$
$$[a > b \ge u > c] \qquad\qquad \text{BY (233.09)}$$

4.[12] $\displaystyle\int_u^b \frac{dx}{\sqrt{(a-x)^5(b-x)(x-c)}} = \frac{2}{3(a-b)^2\sqrt{(a-c)^3}}[2(2a-b-c)E(\delta,q)-(a-b)F(\delta,q)]$

$\displaystyle\qquad\qquad +\frac{2}{3(a-b)(a-c)}\sqrt{\frac{(b-u)(u-c)}{(a-u)^3}}$

$$[a>b>u\geq c] \qquad\qquad \text{BY (234.05)}$$

5.[12] $\displaystyle\int_b^u \frac{dx}{\sqrt{(a-x)^5(x-b)(x-c)}}$

$\displaystyle\qquad = \frac{2}{3(a-b)^2\sqrt{(a-c)^3}}[(3a-b-2c)F(\kappa,p)-2(2a-b-c)E(\kappa,p)]$

$\displaystyle\qquad +\frac{2\left[4a^2-2ab-3ac+bc-u(3a-b-2c)\right]}{3(a-b)^2(a-c)}\sqrt{\frac{u-b}{(a-u)^3(u-c)}}$

$$[a>u>b>c] \qquad\qquad \text{BY (235.04)}$$

6. $\displaystyle\int_u^\infty \frac{dx}{\sqrt{(x-a)^5(x-b)(x-c)}} = \frac{2}{3(a-b)^2\sqrt{(a-c)^3}}[2(2a-b-c)E(\nu,q)-(a-b)F(\nu,q)]$

$\displaystyle\qquad +\frac{2\left[4a^2-2ab-3ac+bc+u(b+2c-3a)\right]}{3(a-b)^2(a-c)}\sqrt{\frac{u-b}{(u-a)^3(u-c)}}$

$$[u>a>b>c] \qquad\qquad \text{BY (238.05)}$$

7. $\displaystyle\int_{-\infty}^u \frac{dx}{\sqrt{(a-x)(b-x)^5(c-x)}} = \frac{2}{3(a-b)^2(b-c)^2\sqrt{a-c}}$

$\displaystyle\qquad \times [2(a-c)(a+c-2b)E(\alpha,p)+(b-c)(3b-a-2c)F(\alpha,p)]$

$\displaystyle\qquad -\frac{2\left[3ab-ac+2bc-4b^2-u(2a-3b+c)\right]}{3(a-b)(b-c)^2}\sqrt{\frac{c-u}{(a-u)(b-u)^3}}$

$$[a>b>c\geq u] \qquad\qquad \text{BY (231.09)}$$

8. $\displaystyle\int_u^c \frac{dx}{\sqrt{(a-x)(b-x)^5(c-x)}} = \frac{2}{3(a-b)^2(b-c)^2\sqrt{a-c}}$

$\displaystyle\qquad \times [(b-c)(3b-a-2c)F(\beta,p)+2(a-c)(a-2b+c)E(\beta,p)]$

$\displaystyle\qquad +\frac{2}{3(a-b)(b-c)}\sqrt{\frac{(a-u)(c-u)}{(b-u)^3}}$

$$[a>b>c>u] \qquad\qquad \text{BY (232.14)}$$

9. $\displaystyle\int_c^u \frac{dx}{\sqrt{(a-x)(b-x)^5(x-c)}} = \frac{2}{3(a-b)^2(b-c)^2\sqrt{a-c}}$

$\displaystyle\qquad \times [(a-b)(2a-3b+c)F(\gamma,q)+2(a-c)(2b-a-c)E(\gamma,q)]$

$\displaystyle\qquad +\frac{2\left[3ab+3bc-ac-5b^2-2u(a-2b+c)\right]}{3(a-b)^2(b-c)^2}\sqrt{\frac{(a-u)(u-c)}{(b-u)^3}}$

$$[a>b>u>c] \qquad\qquad \text{BY (233.10)}$$

10. $\displaystyle\int_u^a \frac{dx}{\sqrt{(a-x)(x-b)^5(x-c)}} = \frac{2}{3(a-b)^2(b-c)^2\sqrt{a-c}}$

$$\times\,[(b-c)(3b-2c-a)\,F(\lambda,p) + 2(a-c)(a+c-2b)\,E(\lambda,p)]$$

$$+\frac{2\left[3ab + 3bc - ac - 5b^2 + 2u(2b-a-c)\right]}{3(a-b)^2(b-c)^2}\sqrt{\frac{(a-u)(u-c)}{(u-b)^3}}$$

$$[a > u > b > c] \qquad\qquad \text{BY (236.09)}$$

11. $\displaystyle\int_a^u \frac{dx}{\sqrt{(x-a)(x-b)^5(x-c)}} = \frac{2}{3(a-b)^2(b-c)^2\sqrt{a-c}}$

$$\times\,[(a-b)(2a+c-3b)\,F(\mu,q) + 2(a-c)(2b-a-c)\,E(\mu,q)]$$

$$+\frac{2}{3(a-b)(b-c)}\sqrt{\frac{(u-a)(u-c)}{(u-b)^3}}$$

$$[u > a > b > c] \qquad\qquad \text{BY (237.12)}$$

12. $\displaystyle\int_u^\infty \frac{dx}{\sqrt{(x-a)(x-b)^5(x-c)}} = \frac{2}{3(a-b)^2(b-c)^2\sqrt{a-c}}$

$$\times\,[(a-b)(2a+c-3b)\,F(\nu,q) + 2(a-c)(2b-c-a)\,E(\nu,q)]$$

$$-\frac{2\left[3bc + 2ab - ac - 4b^2 + u(3b-a-2c)\right]}{3(a-b)^2(b-c)}\sqrt{\frac{u-a}{(u-b)^3(u-c)}}$$

$$[u \geq a > b > c] \qquad\qquad \text{BY (238.04)}$$

13. $\displaystyle\int_{-\infty}^u \frac{dx}{\sqrt{(a-x)(b-x)(c-x)^5}} = \frac{2}{3(b-c)^2\sqrt{(a-c)^3}}\,[2(a+b-2c)\,E(\alpha,p) - (b-c)\,F(\alpha,p)]$

$$+\frac{2\left[ab - 3ac - 2bc + 4c^2 + u(2a+b-3c)\right]}{3(a-c)(b-c)^2}\sqrt{\frac{b-u}{(a-u)(c-u)^3}}$$

$$[a > b > c > u] \qquad\qquad \text{By (231.10)}$$

14. $\displaystyle\int_u^b \frac{dx}{\sqrt{(a-x)(b-x)(x-c)^5}} = \frac{2}{3(b-c)^2\sqrt{(a-c)^3}}\,[(2a+b-3c)\,F(\delta,q) - 2(a+b-2c)\,E(\delta,q)]$

$$+\frac{2\left[ab - 3ac - 2bc + 4c^2 + u(2a+b-3c)\right]}{3(b-c)^2(a-c)}\sqrt{\frac{b-u}{(a-u)(u-c)^3}}$$

$$[a > b > u > c] \qquad\qquad \text{BY (234.04)}$$

15. $\displaystyle\int_b^u \frac{dx}{\sqrt{(a-x)(x-b)(x-c)^5}} = \frac{2}{3(b-c)^2\sqrt{(a-c)^3}}\,[2(a+b-2c)\,E(\kappa,p) - (b-c)\,F(\kappa,p)]$

$$+\frac{2}{3(a-c)(b-c)}\sqrt{\frac{(a-u)(u-b)}{(u-c)^3}}$$

$$[a \geq u > b > c] \qquad\qquad \text{BY (235.20)}$$

16. $\displaystyle\int_u^a \frac{dx}{\sqrt{(a-x)(x-b)(x-c)^5}} = \frac{2}{3(b-c)^2\sqrt{(a-c)^3}}\,[2(a+b-2c)\,E(\lambda,p) - (b-c)\,F(\lambda,p)]$

$$-\frac{2\left[ab - 3ac - 3bc + 5c^2 + 2u(a+b-2c)\right]}{3(b-c)^2(a-c)^2}\sqrt{\frac{(a-u)(u-b)}{(u-c)^3}}$$

$$[a > u \geq b > c] \qquad\qquad \text{BY (236.10)}$$

17.[12] $\displaystyle\int_a^u \frac{dx}{\sqrt{(x-a)(x-b)(x-c)^5}} = \frac{2}{3(b-c)^2\sqrt{(a-c)^3}}\left[(2a+b-3c)\,F(\mu,q) - 2(a+b-2c)\,E(\mu,q)\right]$

$\displaystyle\qquad\qquad +\frac{2\left[ab-2ac-3bc+4c^2+u(a+2b-3c)\right]}{3(b-c)(a-c)^2}\sqrt{\frac{u-a}{(u-b)(u-c)^3}}$

$[u > a > b > c]$       BY (237.13)

18. $\displaystyle\int_u^\infty \frac{dx}{\sqrt{(x-a)(x-b)(x-c)^5}} = \frac{2}{3(b-c)^2\sqrt{(a-c)^3}}\left[(2a+b-3c)\,F(\nu,q) - 2(a+b-2c)\,E(\nu,q)\right]$

$\displaystyle\qquad\qquad +\frac{2}{3(a-c)(b-c)}\sqrt{\frac{(u-a)(u-b)}{(u-c)^3}}$

$[u \geq a > b > c]$       BY (238.03)

**3.135**

1.[12] $\displaystyle\int_{-\infty}^u \frac{dx}{\sqrt{(a-x)(b-x)^3(c-x)^3}} = \frac{2}{(a-b)(b-c)^2\sqrt{a-c}}\left[(b-c)\,F(\alpha,p) - (2a+b-c)\,E(\alpha,p)\right]$

$\displaystyle\qquad\qquad +\frac{2(b+c-2u)}{(b-c)^2\sqrt{(a-u)(b-u)(c-u)}}$

$[a > b > c > u]$       BY (231.13)

2. $\displaystyle\int_u^a \frac{dx}{\sqrt{(a-x)(x-b)^3(x-c)^3}} = \frac{2}{(a-b)(b-c)^2\sqrt{a-c}}\left[(b-c)\,F(\lambda,p) - 2(2a-b-c)\,E(\lambda,p)\right]$

$\displaystyle\qquad\qquad +\frac{2(a-b-c+u)}{(a-b)(b-c)(a-c)}\sqrt{\frac{a-u}{(u-b)(u-c)}}$

$[a > u > b > c]$       BY (236.15)

3. $\displaystyle\int_a^u \frac{dx}{\sqrt{(x-a)(x-b)^3(x-c)^3}} = \frac{2}{(a-b)(b-c)^2\sqrt{a-c}}\left[(2a-b-c)\,E(\mu,q) - 2(a-b)\,F(\mu,q)\right]$

$\displaystyle\qquad\qquad +\frac{2}{(a-c)(b-c)}\sqrt{\frac{u-a}{(u-b)(u-c)}}$

$[u > a > b > c]$       BY (236.14)

4. $\displaystyle\int_u^\infty \frac{dx}{\sqrt{(x-a)(x-b)^3(x-c)^3}} = \frac{2}{(a-b)(b-c)^2\sqrt{a-c}}\left[(2a-b-c)\,E(\nu,q) - 2(a-b)\,F(\nu,q)\right]$

$\displaystyle\qquad\qquad -\frac{2}{(a-b)(b-c)}\sqrt{\frac{u-a}{(u-b)(u-c)}}$

$[u \geq a > b > c]$       BY (238.13)

5. $\displaystyle\int_{-\infty}^u \frac{dx}{\sqrt{(a-x)^3(b-x)(c-x)^3}} = \frac{2}{(a-b)(b-c)\sqrt{(a-c)^3}}\left[(2b-a-c)\,E(\alpha,p) - (b-c)\,F(\alpha,p)\right]$

$\displaystyle\qquad\qquad +\frac{2}{(b-c)(a-c)}\sqrt{\frac{b-u}{(a-u)(c-u)}}$

$[a > b > c > u]$       BY(231.12)

6. $\displaystyle\int_u^b \frac{dx}{\sqrt{(a-x)^3(b-x)(x-c)^3}} = \frac{2}{(b-c)(a-b)\sqrt{(a-c)^3}}\left[(a-b)\,F(\delta,q) + (2b-a-c)\,E(\delta,q)\right]$

$\displaystyle\qquad\qquad +\frac{2}{(b-c)(a-c)}\sqrt{\frac{b-u}{(a-u)(u-c)}}$

$[a > b > u > c]$       BY (234.03)

7. $\displaystyle\int_b^u \frac{dx}{\sqrt{(a-x)^3(x-b)(x-c)^3}} = \frac{2}{(a-b)(b-c)\sqrt{(a-c)^3}}\left[(b-c)\,F\left(\kappa,p\right)-(2b-a-c)\,E\left(\kappa,p\right)\right]$

$$+\frac{2}{(a-b)(a-c)}\sqrt{\frac{u-b}{(a-u)(u-c)}}$$

$$[a>u>b>c] \qquad\qquad \text{BY (235.15)}$$

8. $\displaystyle\int_u^\infty \frac{dx}{\sqrt{(x-a)^3(x-b)(x-c)^3}} = \frac{2}{(a-b)(b-c)\sqrt{(a-c)^3}}\left[(a+c-2b)\,E(\nu,q)-(a-b)\,F(\nu,q)\right]$

$$+\frac{2}{(a-b)(a-c)}\sqrt{\frac{u-b}{(u-a)(u-c)}}$$

$$[u>a>b>c] \qquad\qquad \text{BY (238.14)}$$

9. $\displaystyle\int_{-\infty}^u \frac{dx}{\sqrt{(a-x)^3(b-x)^3(c-x)}} = \frac{2}{(b-c)(a-b)^2\sqrt{a-c}}\left[(a+b-2c)\,E(\alpha,p)-2(b-c)\,F(\alpha,p)\right]$

$$-\frac{2}{(a-b)(b-c)}\sqrt{\frac{c-u}{(a-u)(b-u)}}$$

$$[a>b>c\geq u] \qquad\qquad \text{BY (231.11)}$$

10. $\displaystyle\int_u^c \frac{dx}{\sqrt{(a-x)^3(b-x)^3(c-x)}} = \frac{2}{(a-b)^2(b-c)\sqrt{a-c}}\left[(a+b-2c)\,E(\beta,p)-2(b-c)\,F(\beta,p)\right]$

$$+\frac{2}{(a-b)(a-c)}\sqrt{\frac{c-u}{(a-u)(b-u)}}$$

$$[a>b>c>u] \qquad\qquad \text{BY (232.15)}$$

11. $\displaystyle\int_c^u \frac{dx}{\sqrt{(a-x)^3(b-x)^3(x-c)}} = \frac{2}{(a-b)^2(b-c)\sqrt{a-c}}\left[(a-b)\,F(\gamma,q)-(a+b-2c)\,E(\gamma,q)\right]$

$$+\frac{2\left[a^2+b^2-ac-bc-u(a+b-2c)\right]}{(a-b)^2(b-c)(a-c)}\sqrt{\frac{u-c}{(a-u)(b-u)}}$$

$$[a>b>u>c] \qquad\qquad \text{BY (233.11)}$$

12.[12] $\displaystyle\int_u^\infty \frac{dx}{\sqrt{(x-a)^3(x-b)^3(x-c)}} = \frac{2}{(a-b)^2(b-c)\sqrt{a-c}}\left[(a-b)\,F(\nu,q)-(a+b-2c)\,E(\nu,q)\right]$

$$+\frac{2(2u-a-b)}{(a-b)^2\sqrt{(u-a)(u-b)(u-c)}}$$

$$[u>a>b>c] \qquad\qquad \text{BY (238.15)}$$

## 3.136

1. $\displaystyle\int_{-\infty}^u \frac{dx}{\sqrt{(a-x)^3(b-x)^3(c-x)^3}}$

$$= \frac{2}{(a-b)^2(b-c)^2\sqrt{(a-c)^3}}$$

$$\times\left[(b-c)(a+b-2c)\,F(\alpha,p)-2\left(c^2+a^2+b^2-ab-ac-bc\right)E(\alpha,p)\right]$$

$$+\frac{2[c(a-c)+b(a-b)-u(2a-c-b)]}{(a-b)(a-c)(b-c)^2\sqrt{(a-u)(b-u)(c-u)}}$$

$$[a>b>c>u] \qquad\qquad \text{BY (231.14)}$$

**2.**
$$\int_u^\infty \frac{dx}{\sqrt{(x-a)^3(x-b)^3(x-c)^3}}$$
$$= \frac{2}{(a-b)^2(b-c)^2\sqrt{(a-c)^3}}$$
$$\times \left[(a-b)(2a-b-c)\,F(\nu,q) - 2\left(a^2+b^2+c^2-ab-ac-bc\right)E(\nu,q)\right]$$
$$+ \frac{2[u(a+b-2c)-a(a-c)-b(b-c)]}{(a-b)^2(a-c)(b-c)\sqrt{(u-a)(u-b)(u-c)}}$$
$$[u > a > b > c] \qquad \text{BY (238.16)}$$

**3.137**

**1.**[6]
$$\int_{-\infty}^u \frac{dx}{(r-x)\sqrt{(a-x)(b-x)(c-x)}} = \frac{2}{(a-r)\sqrt{a-c}}\left[\Pi\left(\alpha, \frac{a-r}{a-c}, p\right) - F(\alpha, p)\right]$$
$$[a > b > c \ge u] \qquad \text{BY (231.15)}$$

**2.**
$$\int_u^c \frac{dx}{(r-x)\sqrt{(a-x)(b-x)(c-x)}} = \frac{2(c-b)}{(r-b)(r-c)\sqrt{a-c}}$$
$$\times \Pi\left(\beta, \frac{r-b}{r-c}, p\right) + \frac{2}{(r-b)\sqrt{a-c}}\,F(\beta, p)$$
$$[a > b > c > u, \quad r \ne 0] \qquad \text{BY (232.17)}$$

**3.**
$$\int_c^u \frac{dx}{(r-x)\sqrt{(a-x)(b-x)(x-c)}} = \frac{2}{(r-c)\sqrt{a-c}}\,\Pi\left(\gamma, \frac{b-c}{r-c}, q\right)$$
$$[a > b \ge u > c, \quad r \ne c] \qquad \text{BY (233.02)}$$

**4.**
$$\int_u^b \frac{dx}{(r-x)\sqrt{(a-x)(b-x)(x-c)}} = \frac{2}{(r-a)(r-b)\sqrt{a-c}}$$
$$\times \left[(b-a)\,\Pi\left(\delta, q^2\frac{r-a}{r-b}, q\right) + (r-b)\,F(\delta, q)\right]$$
$$[a > b > u \ge c, \quad r \ne b] \qquad \text{BY (234.18)}$$

**5.**
$$\int_b^u \frac{dx}{(x-r)\sqrt{(a-x)(x-b)(x-c)}} = \frac{2}{(c-r)(b-r)\sqrt{a-c}}$$
$$\times \left[(c-b)\,\Pi\left(\kappa, p^2\frac{c-r}{b-r}, p\right) + (b-r)\,F(\kappa, p)\right]$$
$$[a \ge u > b > c, \quad r \ne b] \qquad \text{BY (235.17)}$$

**6.**[8]
$$\int_u^a \frac{dx}{(x-r)\sqrt{(a-x)(x-b)(x-c)}} = \frac{2}{(a-r)\sqrt{a-c}}\,\Pi\left(\lambda, \frac{a-b}{a-r}, p\right)$$
$$[a > u \ge b > c, \quad r \ne a] \qquad \text{BY (236.02)}$$

**7.**
$$\int_a^u \frac{dx}{(x-r)\sqrt{(x-a)(x-b)(x-c)}} = \frac{2}{(b-r)(a-r)\sqrt{a-c}}$$
$$\times \left[(b-a)\,\Pi\left(\mu, \frac{b-r}{a-b}, q\right) + (a-p)\,F(\mu, q)\right]$$
$$[u > a > b > c, \quad r \ne a] \qquad \text{BY (237.17)}$$

8. $\displaystyle\int_u^\infty \frac{\mathrm{d}x}{(x-r)\sqrt{(x-a)(x-b)(x-c)}} = \frac{2}{(r-c)\sqrt{a-c}}\left[\Pi\left(\nu, \frac{r-c}{a-c}, q\right) - F(\nu, q)\right]$

$$[u \geq a > b > c] \qquad \text{BY (238.06)}$$

## 3.138

1. $\displaystyle\int_0^u \frac{\mathrm{d}x}{\sqrt{x(1-x)(1-k^2 x)}} = 2F\left(\arcsin\sqrt{u}, k\right)$    $[0 < u < 1]$      PE (532), JA

2. $\displaystyle\int_u^1 \frac{\mathrm{d}x}{\sqrt{x(1-x)\left(k'^2 + k^2 x\right)}} = 2F\left(\arccos\sqrt{u}, k\right)$    $[0 < u < 1]$      PE(533)

3. $\displaystyle\int_u^1 \frac{\mathrm{d}x}{\sqrt{x(1-x)\left(x - k'^2\right)}} = 2F\left(\arcsin\frac{\sqrt{1-u}}{k}, k\right)$    $[0 < u < 1]$      PE (534)

4. $\displaystyle\int_0^u \frac{\mathrm{d}x}{\sqrt{x(1+x)\left(1 + k'^2 x\right)}} = 2F\left(\arctan\sqrt{u}, k\right)$    $[0 < u < 1]$      PE (535)

5. $\displaystyle\int_0^u \frac{\mathrm{d}x}{\sqrt{x\left[1 + x^2 + 2\left(k'^2 - k^2\right)x\right]}} = F\left(2\arctan\sqrt{u}, k\right)$

$$[0 < u < 1] \qquad \text{JA}$$

6. $\displaystyle\int_u^1 \frac{\mathrm{d}x}{\sqrt{x\left[k'^2\left(1 + x^2\right) + 2\left(1 + k^2\right)x\right]}} = F\left(\frac{\pi}{2} - 2\arctan\sqrt{u}, k\right)$

$$[0 < u < 1] \qquad \text{JA}$$

7. $\displaystyle\int_a^u \frac{\mathrm{d}x}{\sqrt{(x-\alpha)\left[(x-m)^2 + n^2\right]}} = \frac{1}{\sqrt{p}} F\left(2\arctan\sqrt{\frac{u-\alpha}{p}}, \sqrt{\frac{p+m-\alpha}{2p}}\right)$

$$[\alpha < u]$$

8. $\displaystyle\int_u^a \frac{\mathrm{d}x}{\sqrt{(\alpha-x)\left[(x-m)^2 + n^2\right]}} = \frac{1}{\sqrt{p}} F\left(2\,\text{arccot}\sqrt{\frac{\alpha-u}{p}}, \sqrt{\frac{p-m+\alpha}{2p}}\right)$

$$[u < \alpha]$$

where $p = \sqrt{(m-\alpha)^2 + n^2}$.

**3.139**    **Notation** $\alpha = \arccos\dfrac{1 - \sqrt{3} - u}{1 + \sqrt{3} - u}$,      $\beta = \arccos\dfrac{\sqrt{3} - 1 + u}{\sqrt{3} + 1 - u}$,

$\gamma = \arccos\dfrac{\sqrt{3} + 1 - u}{\sqrt{3} - 1 + u}$,      $\delta = \arccos\dfrac{u - 1 - \sqrt{3}}{u - 1 + \sqrt{3}}$.

1. $\displaystyle\int_{-\infty}^u \frac{\mathrm{d}x}{\sqrt{1 - x^3}} = \frac{1}{\sqrt[4]{3}} F\left(\alpha, \sin 75°\right)$      H 66 (285)

2. $\displaystyle\int_u^1 \frac{\mathrm{d}x}{\sqrt{1 - x^3}} = \frac{1}{\sqrt[4]{3}} F\left(\beta, \sin 75°\right)$      H 65 (284)

3. $\displaystyle\int_1^u \frac{dx}{\sqrt{x^3-1}} = \frac{1}{\sqrt[4]{3}}\,F\left(\gamma, \sin 15°\right)$ $\hfill$ H 65 (283)

4. $\displaystyle\int_u^\infty \frac{dx}{\sqrt{x^3-1}} = \frac{1}{\sqrt[4]{3}}\,F\left(\delta, \sin 15°\right)$ $\hfill$ H 65 (282)

5. $\displaystyle\int_0^1 \frac{dx}{\sqrt{1-x^3}} = \frac{1}{2\pi\sqrt{3}\sqrt[3]{2}}\left\{\Gamma\left(\frac{1}{3}\right)\right\}^3$ $\hfill$ MO 9

6.[12] $\displaystyle\int_0^1 \frac{x\,dx}{\sqrt{1-x^3}} = \frac{1}{\pi}\frac{\sqrt{3}}{\sqrt[3]{4}}\left\{\Gamma\left(\frac{2}{3}\right)\right\}^2$ $\hfill$ MO 9

7. $\displaystyle\int_u^1 \sqrt{1-x^3}\,dx = \frac{1}{5}\left\{\sqrt[4]{27}\,F\left(\beta, \sin 75°\right) - 2u\sqrt{1-u^3}\right\}$ $\hfill$ BY (244.01)

8. $\displaystyle\int_u^1 \frac{x\,dx}{\sqrt{1-x^3}} = \left(3^{-\frac{1}{4}}-3^{\frac{1}{4}}\right)F\left(\beta, \sin 75°\right) + 2\sqrt[4]{3}\,E\left(\beta, \sin 75°\right) - \frac{2\sqrt{1-u^3}}{\sqrt{3}+1-u}$ $\hfill$ BY (244.05)

9. $\displaystyle\int_u^1 \frac{x^m\,dx}{\sqrt{1-x^3}} = \frac{2u^{m-2}\sqrt{1-u^3}}{2m-1} + \frac{2(m-2)}{2m-1}\int_u^1 \frac{x^{m-3}\,dx}{\sqrt{1-x^3}}$ $\hfill$ BY (244.07)

10. $\displaystyle\int_1^u \frac{x\,dx}{\sqrt{x^3-1}} = \left(3^{-\frac{1}{4}}+3^{\frac{1}{4}}\right)F\left(\gamma, \sin 15°\right) - 2\sqrt[4]{3}\,E\left(\gamma, \sin 15°\right) + \frac{2\sqrt{u^3-1}}{\sqrt{3}-1+u}$ $\hfill$ BY (240.05)

11. $\displaystyle\int_{-\infty}^u \frac{dx}{(1-x)\sqrt{1-x^3}} = \frac{1}{\sqrt[4]{27}}\left[F\left(\alpha, \sin 75°\right) - 2\,E\left(\alpha, \sin 75°\right)\right] + \frac{2}{\sqrt{3}}\frac{\sqrt{1+u+u^2}}{\left(1+\sqrt{3}-u\right)\sqrt{1-u}}$

$\hfill [u \neq 1]$

$\hfill$ BY (246.06)

12. $\displaystyle\int_u^\infty \frac{dx}{(x-1)\sqrt{x^3-1}} = \frac{1}{\sqrt[4]{27}}\left[F\left(\delta, \sin 15°\right) - 2\,E\left(\delta, \sin 15°\right)\right] + \frac{2}{\sqrt{3}}\frac{\sqrt{1+u+u^2}}{\left(u-1+\sqrt{3}\right)\sqrt{u-1}}$

$\hfill [u \neq 1]$

$\hfill$ BY (242.03)

13. $\displaystyle\int_{-\infty}^u \frac{(1-x)\,dx}{\left(1+\sqrt{3}-x\right)^2\sqrt{1-x^3}} = \frac{2-\sqrt{3}}{\sqrt[4]{27}}\left[F\left(\alpha, \sin 75°\right) - E\left(\alpha, \sin 75°\right)\right]$ $\hfill$ BY (246.07)

14. $\displaystyle\int_u^1 \frac{(1-x)\,dx}{\left(1+\sqrt{3}-x\right)^2\sqrt{1-x^3}} = \frac{2-\sqrt{3}}{\sqrt[4]{27}}\left[F\left(\beta, \sin 75°\right) - E\left(\beta, \sin 75°\right)\right]$ $\hfill$ BY (244.04)

15. $\displaystyle\int_1^u \frac{(x-1)\,dx}{\left(1+\sqrt{3}-x\right)^2\sqrt{x^3-1}} = \frac{2\left(\sqrt{3}-2\right)}{\sqrt{3}}\frac{\sqrt{u^3-1}}{u^2-2u-2} - \frac{2-\sqrt{3}}{\sqrt[4]{27}}\,E\left(\gamma, \sin 15°\right)$ $\hfill$ BY (240.08)

16. $\displaystyle\int_u^\infty \frac{(x-1)\,dx}{\left(1+\sqrt{3}-x\right)^2\sqrt{x^3-1}} = \frac{2\left(2-\sqrt{3}\right)}{\sqrt{3}}\frac{\sqrt{u^3-1}}{u^2-2u-2} - \frac{2-\sqrt{3}}{\sqrt[4]{27}}\,E\left(\delta, \sin 15°\right)$ $\hfill$ BY (242.07)

17. $\displaystyle\int_{-\infty}^u \frac{(1-x)\,dx}{\left(1-\sqrt{3}-x\right)^2\sqrt{1-x^3}} = \frac{2+\sqrt{3}}{\sqrt[4]{27}}\left[\frac{2\sqrt[4]{3}\sqrt{1-u^3}}{u^2-2u-2} - E\left(\alpha, \sin 75°\right)\right]$ $\hfill$ BY (246.08)

18. $\displaystyle\int_1^u \frac{(x-1)\,dx}{\left(1-\sqrt{3}-x\right)^2\sqrt{x^3-1}} = \frac{2+\sqrt{3}}{\sqrt[4]{27}}\left[F\left(\gamma, \sin 15°\right) - E\left(\gamma, \sin 15°\right)\right]$ $\hfill$ BY (240.04)

19. $\displaystyle\int_u^\infty \frac{(x-1)\,dx}{\left(1-\sqrt{3}-x\right)^2 \sqrt{x^3-1}} = \frac{2+\sqrt{3}}{\sqrt[4]{27}} \left[F\left(\delta, \sin 15°\right) - E\left(\delta, \sin 15°\right)\right]$      BY (242.05)

20. $\displaystyle\int_{-\infty}^u \frac{\left(x^2+x+1\right)\,dx}{\left(1+\sqrt{3}-x\right)^2 \sqrt{1-x^3}} = \frac{1}{\sqrt[4]{3}} E\left(\alpha, \sin 75°\right)$      BY (246.01)

21. $\displaystyle\int_u^1 \frac{\left(x^2+x+1\right)\,dx}{\left(x-1+\sqrt{3}\right)^2 \sqrt{1-x^3}} = \frac{1}{\sqrt[4]{3}} E\left(\beta, \sin 75°\right)$      BY (244.02)

22. $\displaystyle\int_1^u \frac{\left(x^2+x+1\right)\,dx}{\left(\sqrt{3}+x-1\right)^2 \sqrt{x^3-1}} = \frac{1}{\sqrt[4]{3}} E\left(\gamma, \sin 15°\right)$      BY (240.01)

23. $\displaystyle\int_u^\infty \frac{\left(x^2+x+1\right)\,dx}{\left(x-1+\sqrt{3}\right)^2 \sqrt{x^3-1}} = \frac{1}{\sqrt[4]{3}} E\left(\delta, \sin 15°\right)$      BY (242.01)

24. $\displaystyle\int_1^u \frac{(x-1)\,dx}{\left(x^2+x+1\right) \sqrt{x^3-1}} = \frac{4}{\sqrt[4]{27}} E\left(\gamma, \sin 15°\right) - \frac{2+\sqrt{3}}{\sqrt[4]{27}} F\left(\gamma, \sin 15°\right)$
$\qquad\qquad\qquad\qquad\qquad\qquad - \frac{2-\sqrt{3}}{\sqrt{3}} \frac{2(u-1)\left(\sqrt{3}+1-u\right)}{\left(\sqrt{3}-1+u\right)\sqrt{u^3-1}}$

     BY (240.09)

25. $\displaystyle\int_{-\infty}^u \frac{\left(1+\sqrt{3}-x\right)^2 \,dx}{\left[\left(1+\sqrt{3}-x\right)^2 - 4\sqrt{3}p^2(1-x)\right]\sqrt{1-x^3}} = \frac{1}{\sqrt[4]{3}} \Pi\left(\alpha, p^2, \sin 75°\right)$      BY (246.02)

26. $\displaystyle\int_u^1 \frac{\left(1+\sqrt{3}-x\right)^2 \,dx}{\left[\left(1+\sqrt{3}-x\right)^2 - 4\sqrt{3}p^2(1-x)\right]\sqrt{1-x^3}} = \frac{1}{\sqrt[4]{3}} \Pi\left(\beta, p^2, \sin 75°\right)$      BY (244.03)

27. $\displaystyle\int_1^u \frac{\left(1-\sqrt{3}-x\right)^2 \,dx}{\left[\left(1-\sqrt{3}-x\right)^2 - 4\sqrt{3}p^2(x-1)\right]\sqrt{x^3-1}} = \frac{1}{\sqrt[4]{3}} \Pi\left(\gamma, p^2, \sin 15°\right)$      BY (240.02)

28. $\displaystyle\int_u^\infty \frac{\left(1-\sqrt{3}-x\right)^2 \,dx}{\left[\left(1-\sqrt{3}-x\right)^2 - 4\sqrt{3}p^2(x-1)\right]\sqrt{x^3-1}} = \frac{1}{\sqrt[4]{3}} \Pi\left(\delta, p^2, \sin 15°\right)$      BY (242.02)

**3.141**    **Notation:** In **3.141** and **3.142** we set:

$$\alpha = \arcsin\sqrt{\frac{a-c}{a-u}}, \qquad \beta = \arcsin\sqrt{\frac{c-u}{b-u}}, \qquad \gamma = \arcsin\sqrt{\frac{u-c}{b-c}},$$

$$\delta = \arcsin\sqrt{\frac{(a-c)(b-u)}{(b-c)(a-u)}}, \qquad \kappa = \arcsin\sqrt{\frac{(a-c)(u-b)}{(a-b)(u-c)}}, \qquad \lambda = \arcsin\sqrt{\frac{a-u}{a-b}},$$

$$\mu = \arcsin\sqrt{\frac{u-a}{u-b}}, \qquad \nu = \arcsin\sqrt{\frac{a-c}{u-c}}, \qquad p = \sqrt{\frac{a-b}{a-c}}, \qquad q = \sqrt{\frac{b-c}{a-c}}.$$

1.    $\int_u^c \sqrt{\dfrac{a-x}{(b-x)(c-x)}}\,\mathrm{d}x = 2\sqrt{a-c}\,[F(\beta,p) - E(\beta,p)] + 2\sqrt{\dfrac{(a-u)(c-u)}{b-u}}$

            $[a > b > c > u]$        BY (232.06)

2.    $\int_c^u \sqrt{\dfrac{a-x}{(b-x)(x-c)}}\,\mathrm{d}x = 2\sqrt{a-c}\,E(\gamma,q)$      $[a > b \geq u > c]$        BY (233.01)

3.    $\int_u^b \sqrt{\dfrac{a-x}{(b-x)(x-c)}}\,\mathrm{d}x = 2\sqrt{a-c}\,E(\delta,q) - 2\sqrt{\dfrac{(b-u)(u-c)}{a-u}}$

            $[a > b > u \geq c]$        BY (234.06)

4.    $\int_b^u \sqrt{\dfrac{a-x}{(x-b)(x-c)}}\,\mathrm{d}x = 2\sqrt{a-c}\,[F(\kappa,p) - E(\kappa,p)] + 2\sqrt{\dfrac{(a-u)(u-b)}{u-c}}$

            $[a \geq u > b > c]$        BY (235.07)

5.    $\int_u^a \sqrt{\dfrac{a-x}{(x-b)(x-c)}}\,\mathrm{d}x = 2\sqrt{a-c}\,[F(\lambda,p) - E(\lambda,p)]$

            $[a > u \geq b > c]$        BY (236.04)

6.    $\int_a^u \sqrt{\dfrac{x-a}{(x-b)(x-c)}}\,\mathrm{d}x = -2\sqrt{a-c}\,E(\mu,q) + 2\sqrt{\dfrac{(u-a)(u-c)}{u-b}}$

            $[u > a > b > c]$        BY (237.03)

7.    $\int_u^c \sqrt{\dfrac{b-x}{(a-x)(c-x)}}\,\mathrm{d}x = \dfrac{2(b-c)}{\sqrt{a-c}}\,F(\beta,p) - 2\sqrt{a-c}\,E(\beta,p) + 2\sqrt{\dfrac{(a-u)(c-u)}{b-u}}$

            $[a > b > c > u]$        BY (232.07)

8.    $\int_c^u \sqrt{\dfrac{b-x}{(a-x)(x-c)}}\,\mathrm{d}x = 2\sqrt{a-c}\,E(\gamma,q) - \dfrac{2(a-b)}{\sqrt{a-c}}\,F(\gamma,q)$

            $[a > b \geq u > c]$        BY (233.04)

9.    $\int_u^b \sqrt{\dfrac{b-x}{(a-x)(x-c)}}\,\mathrm{d}x = 2\sqrt{a-c}\,E(\delta,q) - \dfrac{2(a-b)}{\sqrt{a-c}}\,F(\delta,q) - 2\sqrt{\dfrac{(b-u)(u-c)}{a-u}}$

            $[a > b > u \geq c]$        BY (234.07)

10.    $\int_b^u \sqrt{\dfrac{x-b}{(a-x)(x-c)}}\,\mathrm{d}x = 2\sqrt{a-c}\,E(\kappa,p) - \dfrac{2(b-c)}{\sqrt{a-c}}\,F(\kappa,p) - 2\sqrt{\dfrac{(a-u)(u-b)}{u-c}}$

            $[a \geq u > b > c]$        BY (235.06)

11.    $\int_u^a \sqrt{\dfrac{x-b}{(a-x)(x-c)}}\,\mathrm{d}x = 2\sqrt{a-c}\,E(\lambda,p) - \dfrac{2(b-c)}{\sqrt{a-c}}\,F(\lambda,p)$

            $[a > u \geq b > c]$        BY (236.03)

12.    $\int_a^u \sqrt{\dfrac{x-b}{(x-a)(x-c)}}\,\mathrm{d}x = \dfrac{2(a-b)}{\sqrt{a-c}}\,F(\mu,q) - 2\sqrt{a-c}\,E(\mu,q) + 2\sqrt{\dfrac{(u-a)(u-c)}{u-b}}$

            $[u > a > b > c]$        BY (237.04)

13.   $\displaystyle\int_u^c \sqrt{\frac{c-x}{(a-x)(b-x)}}\,dx = -2\sqrt{a-c}\,E(\beta,p) + 2\sqrt{\frac{(a-u)(c-u)}{b-u}}$

$$[a > b > c > u] \qquad\qquad \text{BY (232.08)}$$

14.   $\displaystyle\int_c^u \sqrt{\frac{x-c}{(a-x)(b-x)}}\,dx = 2\sqrt{a-c}\,[F(\gamma,q) - E(\gamma,q)]$

$$[a > b \ge u > c] \qquad\qquad \text{BY (233.03)}$$

15.   $\displaystyle\int_u^b \sqrt{\frac{x-c}{(a-x)(b-x)}}\,dx = 2\sqrt{a-c}\,[F(\delta,q) - E(\delta,q)] + 2\sqrt{\frac{(b-u)(u-c)}{a-u}}$

$$[a > b > u \ge c] \qquad\qquad \text{BY (234.08)}$$

16.   $\displaystyle\int_b^u \sqrt{\frac{x-c}{(a-x)(x-b)}}\,dx = 2\sqrt{a-c}\,E(\kappa,p) - 2\sqrt{\frac{(a-u)(u-b)}{u-c}}$

$$[a \ge u > b > c] \qquad\qquad \text{BY (235.07)}$$

17.   $\displaystyle\int_u^a \sqrt{\frac{x-c}{(a-x)(x-b)}}\,dx = 2\sqrt{a-c}\,E(\lambda,p) \qquad\qquad [a > u \ge b > c] \qquad\qquad \text{BY (236.01)}$

18.   $\displaystyle\int_a^u \sqrt{\frac{x-c}{(x-a)(x-b)}}\,dx = 2\sqrt{a-c}\,[F(\mu,q) - E(\mu,q)] + 2\sqrt{\frac{(u-a)(u-c)}{u-b}}$

$$[u > a > b > c] \qquad\qquad \text{BY (237.05)}$$

19.   $\displaystyle\int_u^c \sqrt{\frac{(b-x)(c-x)}{a-x}}\,dx = \frac{2}{3}\sqrt{a-c}\,[(2a-b-c)\,E(\beta,p) - (b-c)\,F(\beta,p)]$

$$+ \frac{2}{3}(2b - 2a + c - u)\sqrt{\frac{(a-u)(c-u)}{b-u}}$$
$$[a > b > c > u] \qquad\qquad \text{BY (232.11)}$$

20.   $\displaystyle\int_c^u \sqrt{\frac{(x-c)(b-x)}{a-x}}\,dx = \frac{2}{3}\sqrt{a-c}\,[(2a-b-c)\,E(\gamma,q) - 2(a-b)\,F(\gamma,q)]$

$$- \frac{2}{3}\sqrt{(a-u)(b-u)(u-c)}$$
$$[a > b \ge u > c] \qquad\qquad \text{BY (233.06)}$$

21.[11]   $\displaystyle\int_u^b \sqrt{\frac{(x-c)(b-x)}{a-x}}\,dx = \frac{2}{3}\sqrt{a-c}\,[2(b-a)\,F(\delta,q) + (2a-b-c)\,E(\delta,q)]$

$$+ \frac{2}{3}(b+c-a-u)\sqrt{\frac{(b-u)(u-c)}{a-u}}$$
$$[a > b > u \ge c] \qquad\qquad \text{BY (234.11)}$$

22.   $\displaystyle\int_b^u \sqrt{\frac{(x-b)(x-c)}{a-x}}\,dx = \frac{2}{3}\sqrt{a-c}\,[(2a-b-c)\,E(\kappa,p) - (b-c)\,F(\kappa,p)]$

$$+ \frac{2}{3}(b+2c-2a-u)\sqrt{\frac{(a-u)(u-b)}{u-c}}$$
$$[a \ge u > b > c] \qquad\qquad \text{BY (235.10)}$$

23.[11] $\displaystyle\int_u^a \sqrt{\frac{(x-b)(x-c)}{a-x}}\,\mathrm{d}x = \frac{2}{3}\sqrt{a-c}\left[(2a-b-c)\,E(\lambda,p)-(b-c)\,F(\lambda,p)\right]$

$\displaystyle\qquad\qquad +\frac{2}{3}\sqrt{(a-u)(u-b)(u-c)}$

$\qquad\qquad\qquad\qquad\qquad\qquad\qquad [a>u\geq b>c]$ BY (236.07)

24. $\displaystyle\int_a^u \sqrt{\frac{(x-b)(x-c)}{x-a}}\,\mathrm{d}x = \frac{2}{3}\sqrt{a-c}\left[2(a-b)\,F(\mu,q)+(b+c-2a)\,E(\mu,q)\right]$

$\displaystyle\qquad\qquad +\frac{2}{3}(u+2a-2b-c)\sqrt{\frac{(u-a)(u-b)}{u-c}}$

$\qquad\qquad\qquad\qquad\qquad\qquad\qquad [u>a>b>c]$ BY (237.08)

25. $\displaystyle\int_u^c \sqrt{\frac{(a-x)(c-x)}{b-x}}\,\mathrm{d}x = \frac{2}{3}\sqrt{a-c}\left[(2b-a-c)\,E(\beta,p)-(b-c)\,F(\beta,p)\right]$

$\displaystyle\qquad\qquad +\frac{2}{3}(a+c-b-u)\sqrt{\frac{(a-u)(c-u)}{b-u}}$

$\qquad\qquad\qquad\qquad\qquad\qquad\qquad [a>b>c>u]$ BY (232.10)

26. $\displaystyle\int_c^u \sqrt{\frac{(a-x)(x-c)}{b-x}}\,\mathrm{d}x = \frac{2}{3}\sqrt{a-c}\left[(2b-a-c)\,E(\gamma,q)+(a-b)\,F(\gamma,q)\right]$

$\displaystyle\qquad\qquad -\frac{2}{3}\sqrt{(a-u)(b-u)(u-c)}$

$\qquad\qquad\qquad\qquad\qquad\qquad\qquad [a>b\geq u>c]$ BY (233.05)

27. $\displaystyle\int_u^b \sqrt{\frac{(a-x)(x-c)}{b-x}}\,\mathrm{d}x = \frac{2}{3}\sqrt{a-c}\left[(a-b)\,F(\delta,q)+(2b-a-c)\,E(\delta,q)\right]$

$\displaystyle\qquad\qquad +\frac{2}{3}(2a+c-2b-u)\sqrt{\frac{(b-u)(u-c)}{a-u}}$

$\qquad\qquad\qquad\qquad\qquad\qquad\qquad [a>b>u\geq c]$ BY (234.10)

28. $\displaystyle\int_b^u \sqrt{\frac{(a-x)(x-c)}{x-b}}\,\mathrm{d}x = \frac{2}{3}\sqrt{a-c}\left[(b-c)\,F(\kappa,p)+(a+c-2b)\,E(\kappa,p)\right]$

$\displaystyle\qquad\qquad +\frac{2}{3}(2b-a-2c+u)\sqrt{\frac{(a-u)(u-b)}{u-c}}$

$\qquad\qquad\qquad\qquad\qquad\qquad\qquad [a\geq u>b>c]$ BY (235.11)

29. $\displaystyle\int_u^a \sqrt{\frac{(a-x)(x-c)}{x-b}}\,\mathrm{d}x = \frac{2}{3}\sqrt{a-c}\left[(a+c-2b)\,E(\lambda,p)+(b-c)\,F(\lambda,p)\right]$

$\displaystyle\qquad\qquad -\frac{2}{3}\sqrt{(a-u)(u-b)(u-c)}$

$\qquad\qquad\qquad\qquad\qquad\qquad\qquad [a>u\geq b>c]$ BY (236.06)

30.[11] $\displaystyle\int_a^u \sqrt{\frac{(x-a)(x-c)}{x-b}}\,\mathrm{d}x = \frac{2}{3}\sqrt{a-c}\left[(a+c-2b)\,E(\mu,q)-(a-b)\,F(\mu,q)\right]$

$\displaystyle\qquad\qquad +\frac{2}{3}(u+b-a-c)\sqrt{\frac{(u-a)(u-c)}{u-b}}$

$\qquad\qquad\qquad\qquad\qquad\qquad\qquad [u>a>b>c]$ BY (237.06)

31. $\displaystyle\int_u^c \sqrt{\frac{(a-x)(b-x)}{c-x}}\,\mathrm{d}x = \frac{2}{3}\sqrt{a-c}\,[2(b-c)\,F(\beta,p)+(2c-a-b)\,E(\beta,p)]$

$$+\frac{2}{3}(a+2b-2c-u)\sqrt{\frac{(a-u)(c-u)}{b-u}}$$

$$[a>b>c>u] \qquad\qquad \text{BY (232.09)}$$

32. $\displaystyle\int_c^u \sqrt{\frac{(a-x)(b-x)}{x-c}}\,\mathrm{d}x = \frac{2}{3}\sqrt{a-c}\,[(a+b-2c)\,E(\gamma,q)-(a-b)\,F(\gamma,q)]$

$$+\frac{2}{3}\sqrt{(a-u)(b-u)(u-c)}$$

$$[a>b\geq u>c] \qquad\qquad \text{BY (233.07)}$$

33. $\displaystyle\int_u^b \sqrt{\frac{(a-x)(b-x)}{x-c}}\,\mathrm{d}x = \frac{2}{3}\sqrt{a-c}\,[(a+b-2c)\,E(\delta,q)-(a-b)\,F(\delta,q)]$

$$+\frac{2}{3}(2c-2a-b+u)\sqrt{\frac{(b-u)(u-c)}{a-u}}$$

$$[a>b>u\geq c] \qquad\qquad \text{BY (234.09)}$$

34. $\displaystyle\int_b^u \sqrt{\frac{(a-x)(x-b)}{x-c}}\,\mathrm{d}x = \frac{2}{3}\sqrt{a-c}\,[(a+b-2c)\,E\,(\kappa,p)-2(b-c)\,F\,(\kappa,p)]$

$$+\frac{2}{3}(u+c-a-b)\sqrt{\frac{(a-u)(u-b)}{u-c}}$$

$$[a\geq u>b>c] \qquad\qquad \text{BY (235.09)}$$

35. $\displaystyle\int_u^a \sqrt{\frac{(a-x)(x-b)}{x-c}}\,\mathrm{d}x = \frac{2}{3}\sqrt{a-c}\,[(a+b-2c)\,E(\lambda,p)-2(b-c)\,F(\lambda,p)]$

$$-\frac{2}{3}\sqrt{(a-u)(u-b)(u-c)}$$

$$[a>u\geq b>c] \qquad\qquad \text{BY (236.05)}$$

36. $\displaystyle\int_a^u \sqrt{\frac{(x-a)(x-b)}{x-c}}\,\mathrm{d}x = \frac{2}{3}\sqrt{a-c}\,[(a+b-2c)\,E(\mu,q)-(a-b)\,F(\mu,q)]$

$$+\frac{2}{3}(u+2c-a-2b)\sqrt{\frac{(u-a)(u-c)}{u-b}}$$

$$[u>a>b>c] \qquad\qquad \text{BY (237.07)}$$

**3.142**

1. $\displaystyle\int_{-\infty}^u \sqrt{\frac{a-x}{(b-x)(c-x)^3}}\,\mathrm{d}x = \frac{2}{\sqrt{a-c}}\,F(\alpha,p)-\frac{2\sqrt{a-c}}{b-c}\,E(\alpha,p)+\frac{2(a-c)}{b-c}\sqrt{\frac{b-u}{(a-u)(c-u)}}$

$$[a>b>c>u] \qquad\qquad \text{BY (231.05)}$$

2. $\displaystyle\int_u^b \sqrt{\frac{a-x}{(b-x)(x-c)^3}}\,\mathrm{d}x = 2\frac{a-b}{(b-c)\sqrt{a-c}}\,F(\delta,q)-\frac{2\sqrt{a-c}}{b-c}\,E(\delta,q)$

$$+2\frac{a-c}{b-c}\sqrt{\frac{b-u}{(a-u)(u-c)}}$$

$$[a>b>u>c] \qquad\qquad \text{BY (234.13)}$$

3. $\displaystyle\int_b^u \sqrt{\frac{a-x}{(x-b)(x-c)^3}}\,\mathrm{d}x = \frac{2\sqrt{a-c}}{b-c}\,E\,(\kappa,p)-\frac{2}{\sqrt{a-c}}\,F\,(\kappa,p)$

$$[a\geq u>b>c] \qquad\qquad \text{BY (235.12)}$$

4. $\displaystyle\int_u^a \sqrt{\frac{a-x}{(x-b)(x-c)^3}}\,dx = \frac{2\sqrt{a-c}}{b-c}E(\lambda,p) - \frac{2}{\sqrt{a-c}}F(\lambda,p) - \frac{2}{b-c}\sqrt{\frac{(a-u)(u-b)}{u-c}}$

$$[a > u \geq b > c] \qquad \text{BY (236.12)}$$

5. $\displaystyle\int_a^u \sqrt{\frac{x-a}{(x-b)(x-c)^3}}\,dx = \frac{2\sqrt{a-c}}{b-c}E(\mu,q) - \frac{2(a-b)}{(b-c)\sqrt{a-c}}F(\mu,q) - 2\sqrt{\frac{u-a}{(u-b)(u-c)}}$

$$[u > a > b > c] \qquad \text{BY (237.10)}$$

6. $\displaystyle\int_u^\infty \sqrt{\frac{x-a}{(x-b)(x-c)^3}}\,dx = \frac{2\sqrt{a-c}}{b-c}E(\nu,q) - \frac{2(a-b)}{(b-c)\sqrt{a-c}}F(\nu,q)$

$$[u \geq a > b > c] \qquad \text{BY (238.09)}$$

7. $\displaystyle\int_{-\infty}^u \sqrt{\frac{a-x}{(b-x)^3(c-x)}}\,dx = \frac{2\sqrt{a-c}}{b-c}E(\alpha,p) - 2\frac{a-b}{b-c}\sqrt{\frac{c-u}{(a-u)(b-u)}}$

$$[a > b > c \geq u] \qquad \text{BY (231.03)}$$

8. $\displaystyle\int_u^c \sqrt{\frac{a-x}{(b-x)^3(c-x)}}\,dx = \frac{2\sqrt{a-c}}{b-c}E(\beta,p) \qquad [a > b > c > u] \qquad \text{BY (232.01)}$

9. $\displaystyle\int_c^u \sqrt{\frac{a-x}{(b-x)^3(x-c)}}\,dx = \frac{2\sqrt{a-c}}{b-c}[F(\gamma,q) - E(\gamma,q)] + \frac{2}{b-c}\sqrt{\frac{(a-u)(u-c)}{b-u}}$

$$[a > b > u > c] \qquad \text{BY (233.15)}$$

10. $\displaystyle\int_u^a \sqrt{\frac{a-x}{(x-b)^3(x-c)}}\,dx = \frac{2\sqrt{a-c}}{c-b}E(\lambda,p) + \frac{2}{b-c}\sqrt{\frac{(a-u)(u-c)}{u-b}}$

$$[a > u > b > c] \qquad \text{BY (236.11)}$$

11. $\displaystyle\int_a^u \sqrt{\frac{x-a}{(x-b)^3(x-c)}}\,dx = \frac{2\sqrt{a-c}}{b-c}[F(\mu,q) - E(\mu,q)]$

$$[u > a > b > c] \qquad \text{BY (237.09)}$$

12. $\displaystyle\int_u^\infty \sqrt{\frac{x-a}{(x-b)^3(x-c)}}\,dx = \frac{2\sqrt{a-c}}{b-c}[F(\nu,q) - E(\nu,q)] + 2\sqrt{\frac{u-a}{(u-b)(u-c)}}$

$$[u \geq a > b > c] \qquad \text{BY (238.10)}$$

13. $\displaystyle\int_{-\infty}^u \sqrt{\frac{b-x}{(a-x)^3(c-x)}}\,dx = \frac{2}{\sqrt{a-c}}E(\alpha,p) \qquad [a > b > c \geq u] \qquad \text{BY (231.01)}$

14. $\displaystyle\int_u^c \sqrt{\frac{b-x}{(a-x)^3(c-x)}}\,dx = \frac{2}{\sqrt{a-c}}E(\beta,p) - \frac{2(a-b)}{a-c}\sqrt{\frac{c-u}{(a-u)(b-u)}}$

$$[a > b > c > u] \qquad \text{BY (232.05)}$$

15. $\displaystyle\int_c^u \sqrt{\frac{b-x}{(a-x)^3(x-c)}}\,dx = \frac{2}{\sqrt{a-c}}[F(\gamma,q) - E(\gamma,q)] + \frac{2}{a-c}\sqrt{\frac{(b-u)(u-c)}{a-u}}$

$$[a > b \geq u > c] \qquad \text{BY (233.13)}$$

16.    $\displaystyle\int_u^b \sqrt{\frac{b-x}{(a-x)^3(x-c)}}\,\mathrm{d}x = \frac{2}{\sqrt{a-c}}\left[F(\delta,q) - E(\delta,q)\right]$

$[a > b > u \geq c]$    BY (234.15)

17.    $\displaystyle\int_b^u \sqrt{\frac{x-b}{(a-x)^3(x-c)}}\,\mathrm{d}x = -\frac{2}{\sqrt{a-c}}\,E(\kappa,p) + 2\sqrt{\frac{u-b}{(a-u)(u-c)}}$

$[a > u > b > c]$    BY (235.08)

18.    $\displaystyle\int_u^\infty \sqrt{\frac{x-b}{(x-a)^3(x-c)}}\,\mathrm{d}x = \frac{2}{\sqrt{a-c}}\left[F(\nu,q) - E(\nu,q)\right] + 2\sqrt{\frac{u-b}{(u-a)(u-c)}}$

$[u > a > b > c]$    BY (238.07)

19.    $\displaystyle\int_{-\infty}^u \sqrt{\frac{b-x}{(a-x)(c-x)^3}}\,\mathrm{d}x = \frac{2}{\sqrt{a-c}}\left[F(\alpha,p) - E(\alpha,p)\right] + 2\sqrt{\frac{b-u}{(a-u)(c-u)}}$

$[a > b > c > u]$    BY (231.04)

20.    $\displaystyle\int_u^b \sqrt{\frac{b-x}{(a-x)(x-c)^3}}\,\mathrm{d}x = -\frac{2}{\sqrt{a-c}}\,E(\delta,q) + 2\sqrt{\frac{b-u}{(a-u)(u-c)}}$

$[a > b > u > c]$    BY (234.14)

21.    $\displaystyle\int_b^u \sqrt{\frac{x-b}{(a-x)(x-c)^3}}\,\mathrm{d}x = \frac{2}{\sqrt{a-c}}\left[F(\kappa,p) - E(\kappa,p)\right]$

$[a \geq u > b > c]$    BY (235.03)

22.    $\displaystyle\int_u^a \sqrt{\frac{x-b}{(a-x)(x-c)^3}}\,\mathrm{d}x = \frac{2}{\sqrt{a-c}}\left[F(\lambda,p) - E(\lambda,p)\right] + \frac{2}{a-c}\sqrt{\frac{(a-u)(u-b)}{u-c}}$

$[a > u \geq b > c]$    BY (236.14)

23.    $\displaystyle\int_a^u \sqrt{\frac{x-b}{(x-a)(x-c)^3}}\,\mathrm{d}x = \frac{2}{\sqrt{a-c}}\,E(\mu,q) - 2\frac{b-c}{a-c}\sqrt{\frac{u-a}{(u-b)(u-c)}}$

$[u > a > b > c]$    BY (237.11)

24.    $\displaystyle\int_u^\infty \sqrt{\frac{x-b}{(x-a)(x-c)^3}}\,\mathrm{d}x = \frac{2}{\sqrt{a-c}}\,E(\nu,q)$        $[u \geq a > b > c]$    BY (238.01)

25.    $\displaystyle\int_{-\infty}^u \sqrt{\frac{c-x}{(a-x)^3(b-x)}}\,\mathrm{d}x = \frac{2\sqrt{a-c}}{a-b}\,E(\alpha,p) - \frac{2(b-c)}{(a-b)\sqrt{a-c}}\,F(\alpha,p)$

$[a > b > c \geq u]$    BY (231.07)

26.    $\displaystyle\int_u^c \sqrt{\frac{c-x}{(a-x)^3(b-x)}}\,\mathrm{d}x = \frac{2\sqrt{a-c}}{a-b}\,E(\beta,p) - \frac{2(b-c)}{(a-b)\sqrt{a-c}}\,F(\beta,p) - 2\sqrt{\frac{c-u}{(a-u)(b-u)}}$

$[a > b > c > u]$    BY (232.03)

27. $\displaystyle\int_c^u \sqrt{\frac{x-c}{(a-x)^3(b-x)}}\, \mathrm{d}x = \frac{2\sqrt{a-c}}{a-b}\, E(\gamma,q) - \frac{2}{\sqrt{a-c}}\, F(\gamma,q) - \frac{2}{a-b}\sqrt{\frac{(b-u)(u-c)}{a-u}}$

$$[a > b \geq u > c] \qquad\qquad \text{BY (233.14)}$$

28. $\displaystyle\int_u^b \sqrt{\frac{x-c}{(a-x)^3(b-x)}}\, \mathrm{d}x = \frac{2\sqrt{a-c}}{a-b}\, E(\delta,q) - \frac{2}{\sqrt{a-c}}\, F(\delta,q)$

$$[a > b > u \geq c] \qquad\qquad \text{BY (234.20)}$$

29. $\displaystyle\int_b^u \sqrt{\frac{x-c}{(a-x)^3(x-b)}}\, \mathrm{d}x = \frac{2(b-c)}{(a-b)\sqrt{a-c}}\, F(\kappa,p) - \frac{2\sqrt{a-c}}{a-b}\, E(\kappa,p)$

$$+2\frac{a-c}{a-b}\sqrt{\frac{u-b}{(a-u)(u-c)}}$$

$$[a > u > b > c] \qquad\qquad \text{BY (235.13)}$$

30. $\displaystyle\int_u^\infty \sqrt{\frac{x-c}{(x-a)^3(x-b)}}\, \mathrm{d}x = \frac{2}{\sqrt{a-c}}\, F(\nu,q) - \frac{2\sqrt{a-c}}{a-b}\, E(\nu,q) + \frac{2(a-c)}{a-b}\sqrt{\frac{u-b}{(u-a)(u-c)}}$

$$[u > a > b > c] \qquad\qquad \text{BY (238.08)}$$

31. $\displaystyle\int_{-\infty}^u \sqrt{\frac{c-x}{(a-x)(b-x)^3}}\, \mathrm{d}x = \frac{2\sqrt{a-c}}{a-b}\, [F(\alpha,p) - E(\alpha,p)] + 2\sqrt{\frac{c-u}{(a-u)(b-u)}}$

$$[a > b > c \geq u] \qquad\qquad \text{BY (231.06)}$$

32. $\displaystyle\int_u^c \sqrt{\frac{c-x}{(a-x)(b-x)^3}}\, \mathrm{d}x = \frac{2\sqrt{a-c}}{a-b}\, [F(\beta,p) - E(\beta,p)]$

$$[a > b > c > u] \qquad\qquad \text{BY (232.04)}$$

33. $\displaystyle\int_c^u \sqrt{\frac{x-c}{(a-x)(b-x)^3}}\, \mathrm{d}x = -\frac{2\sqrt{a-c}}{a-b}\, E(\gamma,q) + \frac{2}{a-b}\sqrt{\frac{(a-u)(u-c)}{b-u}}$

$$[a > b > u > c] \qquad\qquad \text{BY (233.16)}$$

34. $\displaystyle\int_u^a \sqrt{\frac{x-c}{(a-x)(x-b)^3}}\, \mathrm{d}x = \frac{2\sqrt{a-c}}{a-b}\, [F(\lambda,p) - E(\lambda,p)] + \frac{2}{a-b}\sqrt{\frac{(a-u)(u-c)}{u-b}}$

$$[a > u > b > c] \qquad\qquad \text{BY (236.13)}$$

35. $\displaystyle\int_a^u \sqrt{\frac{x-c}{(x-a)(x-b)^3}}\, \mathrm{d}x = \frac{2\sqrt{a-c}}{a-b}\, E(\mu,q) \qquad\qquad [u > a > b > c] \qquad \text{BY (237.01)}$

36. $\displaystyle\int_u^\infty \sqrt{\frac{x-c}{(x-a)(x-b)^3}}\, \mathrm{d}x = \frac{2\sqrt{a-c}}{a-b}\, E(\nu,q) - 2\frac{b-c}{a-b}\sqrt{\frac{u-a}{(u-b)(u-c)}}$

$$[u \geq a > b > c] \qquad\qquad \text{BY (238.11)}$$

## 3.143

1.[6] $\displaystyle\int_u^1 \frac{\mathrm{d}x}{\sqrt{1+x^4}} = \frac{1}{2}\, F\left(\arctan \frac{(1+\sqrt{2})(1-u)}{(1+u)}, 2\sqrt[4]{2}\left(\sqrt{2}-1\right)\right)$      H 66 (286)

2.    $\displaystyle\int_u^\infty \frac{dx}{\sqrt{1+x^4}} = \frac{1}{2} F\left(\arccos\frac{u^2-1}{u^2+1}, \frac{\sqrt{2}}{2}\right)$       H 66 (287)

**3.144**    **Notation:** $\alpha = \arcsin\dfrac{1}{\sqrt{u^2-u+1}}$.

1.    $\displaystyle\int_u^\infty \frac{dx}{\sqrt{x(x-1)(x^2-x+1)}} = F\left(\alpha, \frac{\sqrt{3}}{2}\right)$      $[u \geq 1]$       BY (261.50)

2.    $\displaystyle\int_u^\infty \frac{dx}{\sqrt{x^3(x-1)^3(x^2-x+1)}} = \frac{2(2u-1)}{\sqrt{u(u-1)(u^2-u+1)}} - 4E\left(\alpha, \frac{\sqrt{3}}{2}\right)$

                             $[u > 1]$       BY (261.54)

3.[12]    $\displaystyle\int_u^\infty \frac{(2x-1)^2\,dx}{\sqrt{x^3(x-1)^3(x^2-x+1)}} = 4\left[F\left(\alpha, \frac{\sqrt{3}}{2}\right) - E\left(\alpha, \frac{\sqrt{3}}{2}\right) + \frac{2u-1}{2\sqrt{u(u-1)(u^2-u+1)}}\right]$

                             $[u > 1]$       BY (261.56)

4.    $\displaystyle\int_u^\infty \frac{dx}{\sqrt{x(x-1)(x^2-x+1)^3}} = \frac{4}{3}\left[F\left(\alpha, \frac{\sqrt{3}}{2}\right) - E\left(\alpha, \frac{\sqrt{3}}{2}\right)\right]$

                             $[u \geq 1]$       BY (261.52)

5.    $\displaystyle\int_u^\infty \frac{(2x-1)^2\,dx}{\sqrt{x(x-1)(x^2-x+1)^3}} = 4E\left(\alpha, \frac{\sqrt{3}}{2}\right)$     $[u > 1]$       BY (261.51)

6.    $\displaystyle\int_u^\infty \sqrt{\frac{x(x-1)}{(x^2-x+1)^3}}\,dx = \frac{4}{3}E\left(\alpha, \frac{\sqrt{3}}{2}\right) - \frac{1}{3}F\left(\alpha, \frac{\sqrt{3}}{2}\right)$

                             $[u > 1]$       BY (261.53)

7.    $\displaystyle\int_u^\infty \frac{dx}{(2x-1)^2}\sqrt{\frac{x(x-1)}{x^2-x+1}} = \frac{1}{3}\left[F\left(\alpha, \frac{\sqrt{3}}{2}\right) - E\left(\alpha, \frac{\sqrt{3}}{2}\right)\right] + \frac{1}{2(2u-1)}\sqrt{\frac{u(u-1)}{u^2-u+1}}$

                             $[u > 1]$       BY (261.57)

8.    $\displaystyle\int_u^\infty \frac{dx}{(2x-1)^2}\sqrt{\frac{x^2-x+1}{x(x-1)}} = E\left(\alpha, \frac{\sqrt{3}}{2}\right) - \frac{3}{2(2u-1)}\sqrt{\frac{u(u-1)}{u^2-u+1}}$

                             $[u > 1]$       BY (261.58)

9.    $\displaystyle\int_u^\infty \frac{dx}{(2x-1)^2\sqrt{x(x-1)(x^2-x+1)}} = \frac{4}{3}E\left(\alpha, \frac{\sqrt{3}}{2}\right) - \frac{1}{3}F\left(\alpha, \frac{\sqrt{3}}{2}\right) - \frac{2}{2u-1}\sqrt{\frac{u(u-1)}{u^2-u+1}}$

                             $[u > 1]$       BY (261.55)

10. $\int_u^\infty \dfrac{dx}{\sqrt{x^5(x-1)^5(x^2-x+1)}} = \dfrac{40}{3} E\left(\alpha, \dfrac{\sqrt{3}}{2}\right) - \dfrac{4}{3} F\left(\alpha, \dfrac{\sqrt{3}}{2}\right) - \dfrac{2(2u-1)\left(9u^2-9u-1\right)}{3\sqrt{u^3(u-1)^3\left(u^2-u+1\right)}}$

$$[u > 1] \qquad\qquad \text{BY (261.54)}$$

11. $\int_u^\infty \dfrac{dx}{\sqrt{x(x-1)\left(x^2-x+1\right)^5}} = \dfrac{44}{27} F\left(\alpha, \dfrac{\sqrt{3}}{2}\right) - \dfrac{56}{27} E\left(\alpha, \dfrac{\sqrt{3}}{2}\right) + \dfrac{2(2u-1)\sqrt{u(u-1)}}{9\sqrt{\left(u^2-u+1\right)^3}}$

$$[u > 1] \qquad\qquad \text{BY (261.52)}$$

12. $\int_u^\infty \dfrac{dx}{(2x-1)^4\sqrt{x(x-1)\left(x^2-x+1\right)}} = \dfrac{16}{27} E\left(\alpha, \dfrac{\sqrt{3}}{2}\right) - \dfrac{1}{27} F\left(\alpha, \dfrac{\sqrt{3}}{2}\right)$

$$-\dfrac{8\left(5u^2-5u+2\right)}{9(2u-1)^3}\sqrt{\dfrac{u(u-1)}{u^2-u+1}}$$

$$[u > 1] \qquad\qquad \text{BY (261.55)}$$

## 3.145

For the following, define $(m-\alpha)^2+n^2=p^2$, and $(m-\beta)^2+n^2=q^2$.[*]

1.[12] $\int_\alpha^u \dfrac{dx}{\sqrt{(x-\alpha)(x-\beta)\left[(x-m)^2+n^2\right]}}$

$$= \dfrac{1}{\sqrt{pq}} \int_0^\phi \dfrac{d\theta}{\sqrt{1-k^2\sin^2\theta}}$$

$$= \dfrac{1}{\sqrt{pq}} F(\phi, k) \qquad\qquad [2m > \alpha+\beta, \quad u < u^*]\,\text{or}\,[2m < \alpha+\beta]$$

$$= \dfrac{1}{\sqrt{pq}}\left[2K(k) - F(\pi-\phi, k)\right] \qquad\qquad [2m > \alpha+\beta, \quad u > u^*]$$

for $\beta < \alpha < u$, $\quad \phi = 2\arctan\sqrt{\dfrac{q(u-\alpha)}{q(u-\beta)}}, \qquad k = \dfrac{1}{2}\sqrt{\dfrac{(p+q)^2-(\alpha-\beta)^2}{pq}}, \qquad u^* = \dfrac{p\beta-q\alpha}{p-q}$

2.[12] $\int_\beta^u \dfrac{dx}{\sqrt{(\alpha-x)(x-\beta)\left[(x-m)^2+n^2\right]}}$

$$= \dfrac{1}{\sqrt{pq}} \int_0^\psi \dfrac{d\theta}{\sqrt{1-\kappa^2\sin^2\theta}}$$

$$= \dfrac{1}{\sqrt{pq}} F(\psi, \kappa) \qquad\qquad [\beta < u < \bar{u} < \alpha]$$

$$= \dfrac{1}{\sqrt{pq}}\left[2K(\kappa) - F(\pi-\psi, \kappa)\right] \qquad\qquad [\beta < \bar{u} < u < \alpha]$$

for $\psi = 2\arctan\sqrt{\dfrac{p(u-\beta)}{q(\alpha-u)}}, \quad \kappa = \dfrac{1}{2}\sqrt{\dfrac{(\alpha-\beta)^2-(p-q)^2}{pq}}, \quad \bar{u} = \dfrac{p\beta-q\alpha}{p+q}$

---

[*]Formulas **3.145** are not valid for $\alpha+\beta=2m$. In this case, we make the substitution $x-m=z$, which leads to one of the formulas **3.152**.

$3.^{12}$

$$\int_u^\beta \frac{dx}{\sqrt{(x-\alpha)(x-\beta)\left[(x-m)^2+n^2\right]}}$$

$$= \frac{1}{\sqrt{pq}} \int_0^\omega \frac{d\theta}{\sqrt{1-\lambda^2\sin^2\theta}}$$

$$= \frac{1}{\sqrt{pq}} F(\omega,\lambda) \qquad\qquad\qquad [2m < \alpha+\beta, \quad \hat{u} < u] \quad \text{or} \quad [2m > \alpha+\beta]$$

$$= \frac{1}{\sqrt{pq}} \left[2\boldsymbol{K}(\lambda) - F(\pi-\omega,\lambda)\right] \qquad [2m < \alpha+\beta, \quad u < \hat{u}]$$

$$\text{for } u < \beta < \alpha, \quad \omega = 2\arctan\sqrt{\frac{q(\alpha-u)}{p(\beta-u)}}, \quad \lambda = \frac{1}{2}\sqrt{\frac{(p+q)^2-(\alpha-\beta)^2}{pq}}, \quad \hat{u} = \frac{p\beta - q\alpha}{p-q}$$

$3(2).^*$

$$\int_u^a \frac{dx}{\sqrt{(\alpha-x)(x-\beta)\left[(x-m)^2+n^2\right]}}$$

$$= \frac{1}{\sqrt{pq}} \int_0^\Omega \frac{d\theta}{\sqrt{1-\Lambda^2\sin^2\theta}}$$

$$= \frac{1}{\sqrt{pq}} F(\Omega,\Lambda) \qquad\qquad\qquad [\beta < \tilde{u} < u < \alpha]$$

$$= \frac{1}{\sqrt{pq}} \left[2\boldsymbol{K}(\Lambda) - F(\pi-\Omega,\Lambda)\right] \qquad [\beta < u < \tilde{u} < \alpha]$$

$$\text{for } \beta < u < \alpha, \quad \Omega = 2\arctan\sqrt{\frac{q(\alpha-u)}{p(u-\beta)}}, \quad \Lambda = \frac{1}{2}\sqrt{\frac{(\alpha-\beta)^2-(p-q)^2}{pq}}, \quad \tilde{u} = \frac{p\beta + q\alpha}{p+q}$$

$4.^{12}$    Set    $(m_1-m)^2+(n_1+n)^2=p^2$, $(m_1-m)^2+(n_1-n)^2=p_1^2$, $\cot\alpha = \sqrt{\dfrac{(p+p_1)^2-4n^2}{4n^2-(p-p_1)^2}}$;

(Note that $m - n\tan\alpha$ and $m + n\cot\alpha$ are the points of intersection in the complex plane of the real line and the circle passing through the 4 points $\{m \pm in, m_1 \pm in_1\}$) then

$$\int_{m-n\tan\alpha}^u \frac{dx}{\sqrt{\left[(x-m)^2+n^2\right]\left[(x-m_1)^2+n_1^2\right]}} = \frac{2}{p+p_1} F\left(\alpha + \arctan\frac{u-m}{n}, \frac{2\sqrt{pp_1}}{p+p_1}\right)$$

$$[m - n\tan\alpha < u < m + n\cot\alpha, \quad m_1 > 1, \quad n > 0]$$

**3.146**

1.    $\displaystyle\int_0^1 \frac{1}{1+x^4}\frac{dx}{\sqrt{1-x^4}} = \frac{\pi}{8} + \frac{1}{4}\sqrt{2}\,\boldsymbol{K}\left(\frac{\sqrt{2}}{2}\right)$          BI (13)(6)

2.    $\displaystyle\int_0^1 \frac{x^2}{1+x^4}\frac{dx}{\sqrt{1-x^4}} = \frac{\pi}{8}$          BI (13)(7)

3.    $\displaystyle\int_0^1 \frac{x^4}{1+x^4}\frac{dx}{\sqrt{1-x^4}} = -\frac{\pi}{8} + \frac{1}{4}\sqrt{2}\,\boldsymbol{K}\left(\frac{\sqrt{2}}{2}\right)$          BI (13)(8)

**3.147**    **Notation:** In **3.147–3.151** we set: $\alpha = \arcsin\sqrt{\dfrac{(a-c)(d-u)}{(a-d)(c-u)}}$,

$$\beta = \arcsin\sqrt{\frac{(a-c)(u-d)}{(c-d)(a-u)}}, \qquad \gamma = \arcsin\sqrt{\frac{(b-d)(c-u)}{(c-d)(b-u)}},$$

$$\delta = \arcsin\sqrt{\frac{(b-d)(u-c)}{(b-c)(u-d)}}, \qquad \kappa = \arcsin\sqrt{\frac{(a-c)(b-u)}{(b-c)(a-u)}},$$

$$\lambda = \arcsin\sqrt{\frac{(a-c)(u-b)}{(a-b)(u-c)}}, \qquad \mu = \arcsin\sqrt{\frac{(b-d)(a-u)}{(a-b)(u-d)}},$$

$$\nu = \arcsin\sqrt{\frac{(b-d)(u-a)}{(a-d)(u-b)}}, \qquad q = \sqrt{\frac{(b-c)(a-d)}{(a-c)(b-d)}}, \qquad r = \sqrt{\frac{(a-b)(c-d)}{(a-c)(b-d)}}.$$

1.    $\displaystyle\int_u^d \frac{\mathrm{d}x}{\sqrt{(a-x)(b-x)(c-x)(d-x)}} = \frac{2}{\sqrt{(a-c)(b-d)}} F(\alpha, q)$

                                             $[a > b > c > d > u]$        BY (251.00)

2.    $\displaystyle\int_d^u \frac{\mathrm{d}x}{\sqrt{(a-x)(b-x)(c-x)(x-d)}} = \frac{2}{\sqrt{(a-c)(b-d)}} F(\beta, r)$

                                             $[a > b > c \geq u > d]$        BY (254.00)

3.    $\displaystyle\int_u^c \frac{\mathrm{d}x}{\sqrt{(a-x)(b-x)(c-x)(x-d)}} = \frac{2}{\sqrt{(a-c)(b-d)}} F(\gamma, r)$

                                             $[a > b > c > u \geq d]$        BY (253.00)

4.    $\displaystyle\int_c^u \frac{\mathrm{d}x}{\sqrt{(a-x)(b-x)(x-c)(x-d)}} = \frac{2}{\sqrt{(a-c)(b-d)}} F(\delta, q)$

                                             $[a > b \geq u > c > d]$        BY (254.00)

5.    $\displaystyle\int_u^b \frac{\mathrm{d}x}{\sqrt{(a-x)(b-x)(x-c)(x-d)}} = \frac{2}{\sqrt{(a-c)(b-d)}} F(\kappa, q)$

                                             $[a > b > u \geq c > d]$        BY (255.00)

6.    $\displaystyle\int_b^u \frac{\mathrm{d}x}{\sqrt{(a-x)(x-b)(x-c)(x-d)}} = \frac{2}{\sqrt{(a-c)(b-d)}} F(\lambda, r)$

                                             $[a \geq u > b > c > d]$        BY (256.00)

7.[11]    $\displaystyle\int_u^a \frac{\mathrm{d}x}{\sqrt{(a-x)(x-b)(x-c)(x-d)}} = \frac{2}{\sqrt{(a-c)(b-d)}} F(\mu, r)$

                                             $[a > u \geq b > c > d]$        BY (257.00)

8.    $\displaystyle\int_a^u \frac{\mathrm{d}x}{\sqrt{(x-a)(x-b)(x-c)(x-d)}} = \frac{2}{\sqrt{(a-c)(b-d)}} F(\nu, q)$

                                             $[u > a > b > c > d]$        BY (258.00)

**3.148**

1.[8] $$\int_u^d \frac{x\,dx}{\sqrt{(a-x)(b-x)(c-x)(d-x)}} = \frac{2}{\sqrt{(a-c)(b-d)}}\left\{(d-c)\,\Pi\left(\alpha,\frac{a-d}{a-c},q\right)+c\,F(\alpha,q)\right\}$$

$$[a>b>c>d>u] \hspace{2cm} \text{BY (251.03)}$$

2. $$\int_d^u \frac{x\,dx}{\sqrt{(a-x)(b-x)(c-x)(x-d)}} = \frac{2}{\sqrt{(a-c)(b-d)}}\left\{(d-a)\,\Pi\left(\beta,\frac{d-c}{a-c},r\right)+a\,F(\beta,r)\right\}$$

$$[a>b>c\geq u>d] \hspace{2cm} \text{BY (252.11)}$$

3. $$\int_u^c \frac{x\,dx}{\sqrt{(a-x)(b-x)(c-x)(x-d)}} = \frac{2}{\sqrt{(a-c)(b-d)}}\left\{(c-b)\,\Pi\left(\gamma,\frac{c-d}{b-d},r\right)+b\,F(\gamma,r)\right\}$$

$$[a>b>c>u\geq d] \hspace{2cm} \text{BY (253.11)}$$

4. $$\int_c^u \frac{x\,dx}{\sqrt{(a-x)(b-x)(x-c)(x-d)}} = \frac{2}{\sqrt{(a-c)(b-d)}}\left\{(c-d)\,\Pi\left(\delta,\frac{b-c}{b-d},q\right)+d\,F(\delta,q)\right\}$$

$$[a>b\geq u>c>d] \hspace{2cm} \text{BY (254.10)}$$

5. $$\int_u^b \frac{x\,dx}{\sqrt{(a-x)(b-x)(x-c)(x-d)}} = \frac{2}{\sqrt{(a-c)(b-d)}}\left\{(b-a)\,\Pi\left(\kappa,\frac{b-c}{a-c},q\right)+a\,F(\kappa,q)\right\}$$

$$[a>b>u\geq c>d] \hspace{2cm} \text{BY (255.17)}$$

6.[8] $$\int_b^u \frac{x\,dx}{\sqrt{(a-x)(x-b)(x-c)(x-d)}} = \frac{2}{\sqrt{(a-c)(b-d)}}\left\{(b-c)\,\Pi\left(\lambda,\frac{a-b}{a-c},r\right)+c\,F(\lambda,r)\right\}$$

$$[a\geq u>b>c>d] \hspace{2cm} \text{BY (256.11)}$$

7. $$\int_u^a \frac{x\,dx}{\sqrt{(a-x)(x-b)(x-c)(x-d)}} = \frac{2}{\sqrt{(a-c)(b-d)}}\left\{(a-d)\,\Pi\left(\mu,\frac{b-a}{b-d},r\right)+d\,F(\mu,r)\right\}$$

$$[a>u\geq b>c>d] \hspace{2cm} \text{BY (257.11)}$$

8. $$\int_a^u \frac{x\,dx}{\sqrt{(x-a)(x-b)(x-c)(x-d)}} = \frac{2}{\sqrt{(a-c)(b-d)}}\left\{(a-b)\,\Pi\left(\nu,\frac{a-d}{b-d},q\right)+b\,F(\nu,q)\right\}$$

$$[u>a>b>c>d] \hspace{2cm} \text{BY (258.11)}$$

**3.149**

1. $$\int_u^d \frac{dx}{x\sqrt{(a-x)(b-x)(c-x)(d-x)}}$$

$$= \frac{2}{cd\sqrt{(a-c)(b-d)}}\left\{(c-d)\,\Pi\left(\alpha,\frac{c(a-d)}{d(a-c)},q\right)+d\,F(\alpha,q)\right\}$$

$$[a>b>c>d>u] \hspace{2cm} \text{BY (251.04)}$$

2. $$\int_d^u \frac{dx}{x\sqrt{(a-x)(b-x)(c-x)(x-d)}}$$

$$= \frac{2}{ad\sqrt{(a-c)(b-d)}}\left\{(a-d)\,\Pi\left(\beta,\frac{a(d-c)}{d(a-c)},r\right)+d\,F(\beta,r)\right\}$$

$$[a>b>c\geq u>d] \hspace{2cm} \text{BY (252.12)}$$

**3.**
$$\int_u^c \frac{dx}{x\sqrt{(a-x)(b-x)(c-x)(x-d)}}$$
$$= \frac{2}{bc\sqrt{(a-c)(b-d)}}\left\{(b-c)\,\Pi\left(\gamma,\frac{b(c-d)}{c(b-d)},r\right)+c\,F(\gamma,r)\right\}$$
$$[a>b>c>u\geq d] \qquad \text{BY (253.12)}$$

**4.**
$$\int_c^u \frac{dx}{x\sqrt{(a-x)(b-x)(x-c)(x-d)}}$$
$$= \frac{2}{cd\sqrt{(a-c)(b-d)}}\left\{(d-c)\,\Pi\left(\delta,\frac{d(b-c)}{c(b-d)},q\right)+c\,F(\delta,q)\right\}$$
$$[a>b\geq u>c>d] \qquad \text{BY (254.11)}$$

**5.**
$$\int_u^b \frac{dx}{x\sqrt{(a-x)(b-x)(x-c)(x-d)}}$$
$$= \frac{2}{ab\sqrt{(a-c)(b-d)}}\times\left\{(a-b)\,\Pi\left(\kappa,\frac{a(b-c)}{b(a-c)},q\right)+b\,F(\kappa,q)\right\}$$
$$[a>b>u\geq c>d] \qquad \text{BY (255.18)}$$

**6.**
$$\int_b^u \frac{dx}{x\sqrt{(a-x)(x-b)(x-c)(x-d)}}$$
$$= \frac{2}{bc\sqrt{(a-c)(b-d)}}\times\left\{(c-b)\,\Pi\left(\lambda,\frac{c(a-b)}{b(a-c)},r\right)+b\,F(\lambda,r)\right\}$$
$$[a\geq u>b>c>d] \qquad \text{BY (256.12)}$$

**7.**
$$\int_u^a \frac{dx}{x\sqrt{(a-x)(x-b)(x-c)(x-d)}}$$
$$= \frac{2}{ad\sqrt{(a-c)(b-d)}}\times\left\{(d-a)\,\Pi\left(\mu,\frac{d(b-a)}{a(b-d)},r\right)+a\,F(\mu,r)\right\}$$
$$[a>u\geq b>c>d] \qquad \text{BY (257.12)}$$

**8.**
$$\int_a^u \frac{dx}{x\sqrt{(x-a)(x-b)(x-c)(x-d)}}$$
$$= \frac{2}{ab\sqrt{(a-c)(b-d)}}\left\{(b-a)\,\Pi\left(\nu,\frac{b(a-d)}{a(b-d)},q\right)+a\,F(\nu,q)\right\}$$
$$[u>a>b>c>d] \qquad \text{BY (258.12)}$$

**3.151**

**1.**
$$\int_u^d \frac{dx}{(p-x)\sqrt{(a-x)(b-x)(c-x)(d-x)}}$$
$$= \frac{2}{(p-c)(p-d)\sqrt{(a-c)(b-d)}}$$
$$\times\left[(d-c)\,\Pi\left(\alpha,\frac{(a-d)(p-c)}{(a-c)(p-d)},q\right)+(p-d)\,F(\alpha,q)\right]$$
$$[a>b>c>d>u,\quad p\neq d] \quad \text{BY (251.39)}$$

2. $$\int_d^u \frac{dx}{(p-x)\sqrt{(a-x)(b-x)(c-x)(x-d)}}$$
$$= \frac{2}{(p-a)(p-d)\sqrt{(a-c)(b-d)}}$$
$$\times \left[ (d-a)\,\Pi\left(\beta, \frac{(d-c)(p-a)}{(a-c)(p-d)}, r\right) + (p-d)\,F(\beta,r) \right]$$
$$[a>b>c\geq u>d, \quad p\neq d] \quad \text{BY (252.39)}$$

3. $$\int_u^c \frac{dx}{(p-x)\sqrt{(a-x)(b-x)(c-x)(x-d)}} = \frac{2}{(p-b)(p-c)\sqrt{(a-c)(b-d)}}$$
$$\times \left[ (c-b)\,\Pi\left(\gamma, \frac{(c-d)(p-b)}{(b-d)(p-c)}, r\right) + (p-c)\,F(\gamma,r) \right]$$
$$[a>b>c>u\geq d, \quad p\neq c] \quad \text{BY (253.39)}$$

4. $$\int_c^u \frac{dx}{(p-x)\sqrt{(a-x)(b-x)(x-c)(x-d)}} = \frac{2}{(p-c)(p-d)\sqrt{(a-c)(b-d)}}$$
$$\times \left[ (c-d)\,\Pi\left(\delta, \frac{(b-c)(p-d)}{(b-d)(p-c)}, q\right) + (p-c)\,F(\delta,q) \right]$$
$$[a>b\geq u>c>d, \quad p\neq c] \quad \text{BY (254.39)}$$

5. $$\int_u^b \frac{dx}{(p-x)\sqrt{(a-x)(b-x)(x-c)(x-d)}}$$
$$= \frac{2}{(p-a)(p-b)\sqrt{(a-c)(b-d)}}$$
$$\times \left[ (b-a)\,\Pi\left(\kappa, \frac{(b-c)(p-a)}{(a-c)(p-b)}, q\right) + (p-b)\,F(\kappa,q) \right]$$
$$[a>b>u\geq c>d, \quad p\neq b] \quad \text{BY (255.38)}$$

6. $$\int_b^u \frac{dx}{(x-p)\sqrt{(a-x)(x-b)(x-c)(x-d)}}$$
$$= \frac{2}{(b-p)(p-c)\sqrt{(a-c)(b-d)}}$$
$$\times \left[ (b-c)\,\Pi\left(\lambda, \frac{(a-b)(p-c)}{(a-c)(p-b)}, r\right) + (p-b)\,F(\lambda,r) \right]$$
$$[a\geq u>b>c>d, \quad p\neq b] \quad \text{BY (256.39)}$$

7. $$\int_u^a \frac{dx}{(p-x)\sqrt{(a-x)(x-b)(x-c)(x-d)}}$$
$$= \frac{2}{(p-a)(p-d)\sqrt{(a-c)(b-d)}}$$
$$\times \left[ (a-d)\,\Pi\left(\mu, \frac{(b-a)(p-d)}{(b-d)(p-a)}, r\right) + (p-a)\,F(\mu,r) \right]$$
$$[a>u\geq b>c>d, \quad p\neq a] \quad \text{BY (257.39)}$$

8.  $$\int_a^u \frac{dx}{(p-x)\sqrt{(x-a)(x-b)(x-c)(x-d)}}$$

$$= \frac{2}{(p-a)(p-b)\sqrt{(a-c)(b-d)}}$$
$$\times \left[ (a-b)\,\Pi\left(\nu, \frac{(a-d)(p-b)}{(b-d)(p-a)}, q\right) + (p-a)\,F(\nu,q) \right]$$
$$[u > a > b > c > d, \quad p \neq a] \quad \text{BY (258.39)}$$

**3.152**  **Notation**: In **3.152–3.163** we set: $\quad \alpha = \arctan\dfrac{u}{b}, \qquad \beta = \operatorname{arccot}\dfrac{u}{a},$

$$\gamma = \arcsin\frac{u}{b}\sqrt{\frac{a^2+b^2}{a^2+u^2}}, \qquad \delta = \arccos\frac{u}{b}, \qquad \varepsilon = \arccos\frac{b}{u}, \qquad \xi = \arcsin\sqrt{\frac{a^2+b^2}{a^2+u^2}},$$

$$\eta = \arcsin\frac{u}{b}, \qquad \zeta = \arcsin\frac{a}{b}\sqrt{\frac{b^2-u^2}{a^2-u^2}}, \qquad \kappa = \arcsin\frac{a}{u}\sqrt{\frac{u^2-b^2}{a^2-b^2}},$$

$$\lambda = \arcsin\sqrt{\frac{a^2-u^2}{a^2-b^2}}, \qquad \mu = \arcsin\sqrt{\frac{u^2-a^2}{u^2-b^2}}, \qquad \nu = \arcsin\frac{a}{u}, \qquad q = \frac{\sqrt{a^2-b^2}}{a},$$

$$r = \frac{b}{\sqrt{a^2+b^2}}, \qquad s = \frac{a}{\sqrt{a^2+b^2}}, \qquad t = \frac{b}{a}.$$

1.  $$\int_0^u \frac{dx}{\sqrt{(x^2+a^2)(x^2+b^2)}} = \frac{1}{a}F(\alpha,q) \qquad\qquad [a>b>0] \qquad \text{H 62(258), BY (221.00)}$$

2.  $$\int_u^\infty \frac{dx}{\sqrt{(x^2+a^2)(x^2+b^2)}} = \frac{1}{a}F(\beta,q) \qquad\qquad [a>b>0] \qquad \text{H 63 (259), BY (222.00)}$$

3.  $$\int_0^u \frac{dx}{\sqrt{(x^2+a^2)(b^2-x^2)}} = \frac{1}{\sqrt{a^2+b^2}}F(\gamma,r) \qquad [b \geq u > 0] \qquad\qquad\qquad \text{H 63 (260)}$$

4.  $$\int_u^b \frac{dx}{\sqrt{(x^2+a^2)(b^2-x^2)}} = \frac{1}{\sqrt{a^2+b^2}}F(\delta,r) \qquad [b > u \geq 0] \qquad \text{H 63 (261), BY (213.00)}$$

5.  $$\int_b^u \frac{dx}{\sqrt{(x^2+a^2)(x^2-b^2)}} = \frac{1}{\sqrt{a^2+b^2}}F(\varepsilon,s) \qquad [u > b > 0] \qquad \text{H 63 (262), BY (211.00)}$$

6.  $$\int_u^\infty \frac{dx}{\sqrt{(x^2+a^2)(x^2-b^2)}} = \frac{1}{\sqrt{a^2+b^2}}F(\xi,s) \qquad [u > b > 0] \qquad \text{H 63 (263), BY (212.00)}$$

7.  $$\int_0^u \frac{dx}{\sqrt{(a^2-x^2)(b^2-x^2)}} = \frac{1}{a}F(\eta,t) \qquad\qquad [a > b \geq u > 0] \qquad \text{H 63 (264), BY (219.00)}$$

8.  $$\int_u^b \frac{dx}{\sqrt{(a^2-x^2)(b^2-x^2)}} = \frac{1}{a}F(\zeta,t) \qquad\qquad [a > b > u \geq 0] \qquad \text{H 63 (265), BY (220.00)}$$

9.  $$\int_b^u \frac{dx}{\sqrt{(a^2-x^2)(x^2-b^2)}} = \frac{1}{a}F(\kappa,q) \qquad\qquad [a \geq u > b > 0] \qquad \text{H 63 (266), BY (217.00)}$$

10.    $\displaystyle\int_u^a \frac{\mathrm{d}x}{\sqrt{(a^2 - x^2)(x^2 - b^2)}} = \frac{1}{a} F(\lambda, q)$      $[a > u \geq b > 0]$    H 63 (257), BY (218.00)

11.    $\displaystyle\int_a^u \frac{\mathrm{d}x}{\sqrt{(x^2 - a^2)(x^2 - b^2)}} = \frac{1}{a} F(\mu, t)$      $[u > a > b > 0]$    H 63 (268), BY (216.00)

12.    $\displaystyle\int_u^\infty \frac{\mathrm{d}x}{\sqrt{(x^2 - a^2)(x^2 - b^2)}} = \frac{1}{a} F(\nu, t)$      $[u \geq a > b > 0]$    H 64(269), BY (215.00)

**3.153**

1.    $\displaystyle\int_0^u \frac{x^2\,\mathrm{d}x}{\sqrt{(x^2 + a^2)(x^2 + b^2)}} = u\sqrt{\frac{a^2 + u^2}{b^2 + u^2}} - a\,E(\alpha, q)$      $[u > 0, \quad a > b]$      BY (221.09)

2.    $\displaystyle\int_0^u \frac{x^2\,\mathrm{d}x}{\sqrt{(a^2 + x^2)(b^2 - x^2)}} = \sqrt{a^2 + b^2}\,E(\gamma, r) - \frac{a^2}{\sqrt{a^2 + b^2}}\,F(\gamma, r) - u\sqrt{\frac{b^2 - u^2}{a^2 + u^2}}$

                                                            $[b \geq u > 0]$      BY (214.05)

3.    $\displaystyle\int_u^b \frac{x^2\,\mathrm{d}x}{\sqrt{(a^2 + x^2)(b^2 - x^2)}} = \sqrt{a^2 + b^2}\,E(\delta, r) - \frac{a^2}{\sqrt{a^2 + b^2}}\,F(\delta, r)$

                                                            $[b > u \geq 0]$      BY (213.06)

4.    $\displaystyle\int_b^u \frac{x^2\,\mathrm{d}x}{\sqrt{(a^2 + x^2)(x^2 - b^2)}} = \frac{b^2}{\sqrt{a^2 + b^2}}\,F(\varepsilon, s) - \sqrt{a^2 + b^2}\,E(\varepsilon, s) + \frac{1}{u}\sqrt{(u^2 + a^2)(u^2 - b^2)}$

                                                            $[u > b > 0]$      BY (211.09)

5.    $\displaystyle\int_0^u \frac{x^2\,\mathrm{d}x}{\sqrt{(a^2 - x^2)(b^2 - x^2)}} = a\,\{F(\eta, t) - E(\eta, t)\}$      $[a > b \geq u > 0]$      BY (219.05)

6.    $\displaystyle\int_u^b \frac{x^2\,\mathrm{d}x}{\sqrt{(a^2 - x^2)(b^2 - x^2)}} = a\,\{F(\zeta, t) - E(\zeta, t)\} + u\sqrt{\frac{b^2 - u^2}{a^2 - u^2}}$

                                                            $[a > b > u \geq 0]$      BY (220.06)

7.    $\displaystyle\int_b^u \frac{x^2\,\mathrm{d}x}{\sqrt{(a^2 - x^2)(x^2 - b^2)}} = a\,E(\kappa, q) - \frac{1}{u}\sqrt{(a^2 - u^2)(u^2 - b^2)}$

                                                            $[a \geq u > b > 0]$      BY (217.05)

8.    $\displaystyle\int_u^a \frac{x^2\,\mathrm{d}x}{\sqrt{(a^2 - x^2)(x^2 - b^2)}} = a\,E(\lambda, q)$      $[a > u \geq b > 0]$      BY (218.06)

9.[6]    $\displaystyle\int_a^u \frac{x^2\,\mathrm{d}x}{\sqrt{(x^2 - a^2)(x^2 - b^2)}} = a\,\{F(\mu, t) - E(\mu, t)\} + u\sqrt{\frac{u^2 - a^2}{u^2 - b^2}}$

                                                            $[u > a > b > 0]$      BY (216.06)

10.    $\displaystyle\int_0^1 \frac{x^2\,\mathrm{d}x}{\sqrt{(1 + x^2)(1 + k^2 x^2)}} = \frac{1}{k^2}\left\{\sqrt{\frac{1 + k^2}{2}} - E\left(\frac{\pi}{4}, \sqrt{1 - k^2}\right)\right\}$      BI (14)(9)

**3.154**

1. $$\int_0^u \frac{x^4\,dx}{\sqrt{(x^2+a^2)(x^2+b^2)}} = \frac{a}{3}\left\{2\left(a^2+b^2\right)E(\alpha,q) - b^2\,F(\alpha,q)\right\} + \frac{u}{3}\left(u^2-2a^2-b^2\right)\sqrt{\frac{a^2+u^2}{b^2+u^2}}$$

$$[a>b,\quad u>0] \qquad \text{BY (221.09)}$$

2. $$\int_0^u \frac{x^4\,dx}{\sqrt{(a^2+x^2)(b^2-x^2)}} = \frac{1}{3\sqrt{a^2+b^2}}\left\{\left(2a^2-b^2\right)a^2\,F(\gamma,r) - 2\left(a^4-b^4\right)E(\gamma,r)\right\}$$

$$-\frac{u}{3}\left(2b^2-a^2+u^2\right)\sqrt{\frac{b^2-u^2}{a^2+u^2}}$$

$$[a \ge u > 0] \qquad \text{BY (214.05)}$$

3. $$\int_u^b \frac{x^4\,dx}{\sqrt{(a^2+x^2)(b^2-x^2)}} = \frac{1}{3\sqrt{a^2+b^2}}\left\{\left(2a^2-b^2\right)a^2\,F(\delta,r) - 2\left(a^4-b^4\right)E(\delta,r)\right\}$$

$$+\frac{u}{3}\sqrt{(a^2+u^2)(b^2-u^2)}$$

$$[b > u \ge 0] \qquad \text{BY (213.06)}$$

4. $$\int_b^u \frac{x^4\,dx}{\sqrt{(a^2+x^2)(x^2-b^2)}} = \frac{1}{3\sqrt{a^2+b^2}}\left\{\left(2b^2-a^2\right)b^2\,F(\varepsilon,s) + 2\left(a^4-b^4\right)E(\varepsilon,s)\right\}$$

$$+\frac{2b^2-2a^2+u^2}{3u}\sqrt{(u^2+a^2)(u^2-b^2)}$$

$$[u > b > 0] \qquad \text{BY (211.09)}$$

5. $$\int_0^u \frac{x^4\,dx}{\sqrt{(a^2-x^2)(b^2-x^2)}} = \frac{a}{3}\left\{\left(2a^2+b^2\right)F(\eta,t) - 2\left(a^2+b^2\right)E(\eta,t)\right\} + \frac{u}{3}\sqrt{(a^2-u^2)(b^2-u^2)}$$

$$[a > b \ge u > 0] \qquad \text{BY (219.05)}$$

6. $$\int_u^b \frac{x^4\,dx}{\sqrt{(a^2-x^2)(b^2-x^2)}} = \frac{a}{3}\left\{\left(2a^2+b^2\right)F(\zeta,t) - 2\left(a^2+b^2\right)E(\zeta,t)\right\}$$

$$+\frac{u}{3}\left(u^2+a^2+2b^2\right)\sqrt{\frac{b^2-u^2}{a^2-u^2}}$$

$$[a > b > u \ge 0] \qquad \text{BY (220.06)}$$

7. $$\int_b^u \frac{x^4\,dx}{\sqrt{(a^2-x^2)(x^2-b^2)}} = \frac{a}{3}\left\{2\left(a^2+b^2\right)E(\kappa,q) - b^2\,F(\kappa,q)\right\}$$

$$-\frac{u^2+2a^2+2b^2}{3u}\sqrt{(a^2-u^2)(u^2-b^2)}$$

$$[a \ge u > b > 0] \qquad \text{BY (217.05)}$$

8. $$\int_u^a \frac{x^4\,dx}{\sqrt{(a^2-x^2)(x^2-b^2)}} = \frac{a}{3}\left\{2\left(a^2+b^2\right)E(\lambda,q) - b^2\,F(\lambda,q)\right\} + \frac{u}{3}\sqrt{(a^2-u^2)(u^2-b^2)}$$

$$[a > u \ge b > 0] \qquad \text{BY (218.06)}$$

9. $$\int_a^u \frac{x^4\,dx}{\sqrt{(x^2-a^2)(x^2-b^2)}} = \frac{a}{3}\left\{\left(2a^2+b^2\right)F(\mu,t) - 2\left(a^2+b^2\right)E(\mu,t)\right\}$$

$$+\frac{u}{3}\left(u^2+2a^2+b^2\right)\sqrt{\frac{u^2-a^2}{u^2-b^2}}$$

$$[u > a > b > 0] \qquad \text{BY (216.06)}$$

**3.155**

1.    $\int_u^a \sqrt{(a^2 - x^2)(x^2 - b^2)}\,dx = \frac{a}{3}\left\{(a^2 + b^2)\,E(\lambda, q) - 2b^2\,F(\lambda, q)\right\} - \frac{u}{3}\sqrt{(a^2 - u^2)(u^2 - b^2)}$

$$[a > u \geq b > 0] \qquad\qquad \text{BY (218.11)}$$

2.    $\int_a^u \sqrt{(x^2 - a^2)(x^2 - b^2)}\,dx = \frac{a}{3}\left\{(a^2 + b^2)\,E(\mu, t) - (a^2 - b^2)\,F(\mu, t)\right\}$

$$+ \frac{u}{3}\,(u^2 - a^2 - 2b^2)\sqrt{\frac{u^2 - a^2}{u^2 - b^2}}$$

$$[u > a > b > 0] \qquad\qquad \text{BY (216.10)}$$

3.    $\int_0^u \sqrt{(x^2 + a^2)(x^2 + b^2)}\,dx = \frac{a}{3}\left\{2b^2\,F(\alpha, q) - (a^2 + b^2)\,E(\alpha, q)\right\}$

$$+ \frac{u}{3}\,(u^2 + a^2 + 2b^2)\sqrt{\frac{a^2 + u^2}{b^2 + u^2}}$$

$$[a > b, \quad u > 0] \qquad\qquad \text{BY (221.08)}$$

4.    $\int_0^u \sqrt{(a^2 + x^2)(b^2 - x^2)}\,dx = \frac{1}{3}\sqrt{a^2 + b^2}\left\{a^2\,F(\gamma, r) - (a^2 - b^2)\,E(\gamma, r)\right\}$

$$+ \frac{u}{3}\,(u^2 + 2a^2 - b^2)\sqrt{\frac{b^2 - u^2}{a^2 + u^2}}$$

$$[a \geq u > 0] \qquad\qquad \text{BY (214.12)}$$

5.[9]    $\int_u^b \sqrt{(a^2 + x^2)(b^2 - x^2)}\,dx = \frac{1}{3}\sqrt{a^2 + b^2}\left\{a^2\,F(\delta, r) + (b^2 - a^2)\,E(\delta, r)\right\}$

$$+ \frac{u}{3}\sqrt{(a^2 + u^2)(b^2 - u^2)}$$

$$[b > u \geq 0] \qquad\qquad \text{BY (213.13)}$$

6.    $\int_b^u \sqrt{(a^2 + x^2)(x^2 - b^2)}\,dx = \frac{1}{3}\sqrt{a^2 + b^2}\left\{(b^2 - a^2)\,E(\varepsilon, s) - b^2\,F(\varepsilon, s)\right\}$

$$+ \frac{u^2 + a^2 - b^2}{3u}\sqrt{(a^2 + u^2)(u^2 - b^2)}$$

$$[u > b > 0] \qquad\qquad \text{BY (211.08)}$$

7.    $\int_0^u \sqrt{(a^2 - x^2)(b^2 - x^2)}\,dx = \frac{a}{3}\left\{(a^2 + b^2)\,E(\eta, t) - (a^2 - b^2)\,F(\eta, t)\right\}$

$$+ \frac{u}{3}\sqrt{(a^2 - u^2)(b^2 - u^2)}$$

$$[a > b \geq u > 0] \qquad\qquad \text{BY (219.11)}$$

8.    $\int_u^b \sqrt{(a^2 - x^2)(b^2 - x^2)}\,dx = \frac{a}{3}\left\{(a^2 + b^2)\,E(\zeta, t) - (a^2 - b^2)\,F(\zeta, t)\right\}$

$$+ \frac{u}{3}\,(u^2 - 2a^2 - b^2)\sqrt{\frac{b^2 - u^2}{a^2 - u^2}}$$

$$[a > b > u \geq 0] \qquad\qquad \text{BY (220.05)}$$

9.    $\int_b^u \sqrt{(a^2 - x^2)(x^2 - b^2)}\,dx = \frac{a}{3}\left\{(a^2 + b^2)\,E(\kappa, q) - 2b^2\,F(\kappa, q)\right\}$

$$+ \frac{u^2 - a^2 - b^2}{3u}\sqrt{(a^2 - u^2)(u^2 - b^2)}$$

$$[a \geq u > b > 0] \qquad\qquad \text{BY (217.09)}$$

**3.156**

1.[6]
$$\int_u^\infty \frac{dx}{x^2 \sqrt{(x^2 + a^2)(x^2 + b^2)}} = \frac{1}{ub^2}\sqrt{\frac{b^2 + u^2}{a^2 + u^2}} - \frac{1}{ab^2} E(\beta, q)$$

$$[a \geq b, \quad u > 0] \qquad \text{BY (222.04)}$$

2.
$$\int_u^b \frac{dx}{x^2 \sqrt{(x^2 + a^2)(b^2 - x^2)}} = \frac{1}{a^2 b^2 \sqrt{a^2 + b^2}} \left\{ a^2 F(\delta, r) - (a^2 + b^2) E(\delta, r) \right\}$$
$$+ \frac{1}{a^2 b^2 u}\sqrt{(a^2 + u^2)(b^2 - u^2)}$$

$$[b > u > 0] \qquad \text{BY (213.09)}$$

3.
$$\int_b^u \frac{dx}{x^2 \sqrt{(x^2 + a^2)(x^2 - b^2)}} = \frac{1}{a^2 b^2 \sqrt{a^2 + b^2}} \left\{ (a^2 + b^2) E(\varepsilon, s) - b^2 F(\varepsilon, s) \right\}$$

$$[u > b > 0] \qquad \text{BY (211.11)}$$

4.
$$\int_u^\infty \frac{dx}{x^2 \sqrt{(x^2 + a^2)(x^2 - b^2)}} = \frac{1}{a^2 b^2 \sqrt{a^2 + b^2}} \left\{ (a^2 + b^2) E(\gamma, s) - b^2 F(\gamma, s) \right\}$$
$$- \frac{1}{b^2 u}\sqrt{\frac{u^2 - b^2}{a^2 + u^2}}$$

$$[u \geq b > 0] \qquad \text{BY (212.06)}$$

5.
$$\int_u^b \frac{dx}{x^2 \sqrt{(a^2 - x^2)(b^2 - x^2)}} = \frac{1}{ab^2} \left\{ F(\zeta, t) - E(\zeta, t) \right\} + \frac{1}{b^2 u}\sqrt{\frac{b^2 - u^2}{a^2 - u^2}}$$

$$[a > b > u > 0] \qquad \text{BY (220.09)}$$

6.
$$\int_b^u \frac{dx}{x^2 \sqrt{(a^2 - x^2)(x^2 - b^2)}} = \frac{1}{ab^2} E(\kappa, q) \qquad [a \geq u > b > 0] \qquad \text{BY (217.01)}$$

7.
$$\int_u^a \frac{dx}{x^2 \sqrt{(a^2 - x^2)(x^2 - b^2)}} = \frac{1}{ab^2} E(\lambda, q) - \frac{1}{a^2 b^2 u}\sqrt{(a^2 - u^2)(u^2 - b^2)}$$

$$[a > u \geq b > 0] \qquad \text{BY (218.12)}$$

8.
$$\int_a^u \frac{dx}{x^2 \sqrt{(x^2 - a^2)(x^2 - b^2)}} = \frac{1}{ab^2} \left\{ F(\mu, t) - E(\mu, t) \right\} + \frac{1}{a^2 u}\sqrt{\frac{u^2 - a^2}{u^2 - b^2}}$$

$$[u > a > b > 0] \qquad \text{BY (216.09)}$$

9.
$$\int_u^\infty \frac{dx}{x^2 \sqrt{(x^2 - a^2)(x^2 - b^2)}} = \frac{1}{ab^2} \left\{ F(\nu, t) - E(\nu, t) \right\}$$

$$[u \geq a > b > 0] \qquad \text{BY (215.07)}$$

**3.157**

1.
$$\int_0^u \frac{dx}{(p - x^2)\sqrt{(x^2 + a^2)(x^2 + b^2)}} = \frac{1}{a(p + b^2)} \left\{ \frac{b^2}{p} \Pi\left(\alpha, \frac{p + b^2}{p}, q\right) + F(\alpha, q) \right\}$$

$$[p \neq 0] \qquad \text{BY (221.13)}$$

2.
$$\int_u^\infty \frac{dx}{(p - x^2)\sqrt{(x^2 + a^2)(x^2 + b^2)}} = -\frac{1}{a(a^2 + p)} \left\{ \Pi\left(\beta, \frac{a^2 + p}{a^2}, q\right) - F(\beta, q) \right\} \qquad \text{BY (222.11)}$$

3.

$$\int_0^u \frac{dx}{(p-x^2)\sqrt{(a^2+x^2)(b^2-x^2)}} = \frac{1}{p(p+a^2)\sqrt{a^2+b^2}} \left\{ a^2 \Pi\left(\gamma, \frac{b^2(p+a^2)}{p(a^2+b^2)}, r\right) + p F(\gamma, r) \right\}$$

$$[b \geq u > 0, \quad p \neq 0] \qquad \text{BY (214.13)a}$$

4.

$$\int_u^b \frac{dx}{(p-x^2)\sqrt{(a^2+x^2)(b^2-x^2)}} = \frac{1}{(p-b^2)\sqrt{a^2+b^2}} \Pi\left(\delta, \frac{b^2}{b^2-p}, r\right)$$

$$[b > u \geq 0, \quad p \neq b^2] \qquad \text{BY (213.02)}$$

5.

$$\int_b^u \frac{dx}{(p-x^2)\sqrt{(a^2+x^2)(x^2-b^2)}} = \frac{1}{p(p-b^2)\sqrt{a^2+b^2}} \left\{ b^2 \Pi\left(\varepsilon, \frac{p}{p-b^2}, s\right) + (p-b^2) F(\varepsilon, s) \right\}$$

$$[u > b > 0, \quad p \neq b^2] \qquad \text{BY (211.14)}$$

6.

$$\int_u^\infty \frac{dx}{(x^2-p)\sqrt{(a^2+x^2)(x^2-b^2)}} = \frac{1}{(a^2+p)\sqrt{a^2+b^2}} \left\{ \Pi\left(\xi, \frac{a^2+p}{a^2+b^2}, s\right) - F(\xi, s) \right\}$$

$$[u \geq b > 0] \qquad \text{BY (212.12)}$$

7.

$$\int_0^u \frac{dx}{(p-x^2)\sqrt{(a^2-x^2)(b^2-x^2)}} = \frac{1}{ap} \Pi\left(\eta, \frac{b^2}{p}, t\right) \qquad [a > b \geq u > 0, \quad p \neq b] \qquad \text{BY (219.02)}$$

8.

$$\int_u^b \frac{dx}{(p-x^2)\sqrt{(a^2-x^2)(b^2-x^2)}} = \frac{1}{a(p-a^2)(p-b^2)}$$

$$\times \left\{ (b^2-a^2) \Pi\left(\zeta, \frac{b^2(p-a^2)}{a^2(p-b^2)}, t\right) + (p-b^2) F(\zeta, t) \right\}$$

$$[a > b > u \geq 0, \quad p \neq b^2] \qquad \text{BY (220.13)}$$

9.

$$\int_b^u \frac{dx}{(p-x^2)\sqrt{(a^2-x^2)(x^2-b^2)}} = \frac{1}{ap(p-b^2)} \left\{ b^2 \Pi\left(\kappa, \frac{p(a^2-b^2)}{a^2(p-b^2)}, q\right) + (p-b^2) F(\kappa, q) \right\}$$

$$[a \geq u > b > 0, \quad p \neq b^2] \qquad \text{BY (217.12)}$$

10.

$$\int_u^a \frac{dx}{(x^2-p)\sqrt{(a^2-x^2)(x^2-b^2)}} = \frac{1}{a(a^2-p)} \Pi\left(\lambda, \frac{a^2-b^2}{a^2-p}, q\right)$$

$$[a > u \geq b > 0, \quad p \neq a^2] \qquad \text{BY (218.02)}$$

11.

$$\int_a^u \frac{dx}{(p-x^2)\sqrt{(x^2-a^2)(x^2-b^2)}}$$

$$= \frac{1}{a(p-a^2)(p-b^2)} \left\{ (a^2-b^2) \Pi\left(\mu, \frac{p-b^2}{p-a^2}, t\right) + (p-a^2) F(\mu, t) \right\}$$

$$[u > a > b > 0, \quad p \neq a^2, \quad p \neq b^2] \quad \text{BY (216.12)}$$

12.

$$\int_u^\infty \frac{dx}{(x^2-p)\sqrt{(x^2-a^2)(x^2-b^2)}} = \frac{1}{ap} \left\{ \Pi\left(\nu, \frac{p}{a^2}, t\right) - F(\nu, t) \right\}$$

$$[u \geq a > b > 0, \quad p \neq 0] \qquad \text{BY (215.12)}$$

**3.158**

1. $\displaystyle\int_0^u \frac{dx}{\sqrt{(x^2+a^2)(x^2+b^2)^3}} = \frac{1}{ab^2(a^2-b^2)}\left\{a^2\,E(\alpha,q) - b^2\,F(\alpha,q)\right\}$

$$[a>b, \quad u>0] \qquad \text{BY (221.05)}$$

2. $\displaystyle\int_u^\infty \frac{dx}{\sqrt{(x^2+a^2)(x^2+b^2)^3}} = \frac{1}{ab^2(a^2-b^2)}\left\{a^2\,E(\beta,q) - b^2\,F(\beta,q)\right\} - \frac{u}{b^2\sqrt{(a^2+u^2)(b^2+u^2)}}$

$$[a>b, \quad u\ge 0] \qquad \text{BY (222.05)}$$

3. $\displaystyle\int_0^u \frac{dx}{\sqrt{(x^2+a^2)^3(x^2+b^2)}} = \frac{1}{a(a^2-b^2)}\left\{F(\alpha,q) - E(\alpha,q)\right\} + \frac{u}{a^2\sqrt{(u^2+a^2)(u^2+b^2)}}$

$$[a>b, \quad u>0] \qquad \text{BY (221.06)}$$

4. $\displaystyle\int_u^\infty \frac{dx}{\sqrt{(a^2+x^2)^3(x^2+b^2)}} = \frac{1}{a(a^2-b^2)}\left\{F(\beta,q) - E(\beta,q)\right\}$

$$[a>b, \quad u\ge 0] \qquad \text{BY (222.03)}$$

5. $\displaystyle\int_0^u \frac{dx}{\sqrt{(a^2+x^2)^3(b^2-x^2)}} = \frac{1}{a^2\sqrt{a^2+b^2}}\,E(\gamma,r) \qquad [b\ge u>0] \qquad \text{BY (214.01)a}$

6. $\displaystyle\int_u^b \frac{dx}{\sqrt{(a^2+x^2)^3(b^2-x^2)}} = \frac{1}{a^2\sqrt{a^2+b^2}}\,E(\delta,r) - \frac{u}{a^2(a^2+b^2)}\sqrt{\frac{b^2-u^2}{a^2+u^2}}$

$$[b>u\ge 0] \qquad \text{BY (213.08)}$$

7. $\displaystyle\int_b^u \frac{dx}{\sqrt{(a^2+x^2)^3(x^2-b^2)}} = \frac{1}{a^2\sqrt{a^2+b^2}}\left\{F(\varepsilon,s) - E(\varepsilon,s)\right\} + \frac{1}{(a^2+b^2)u}\sqrt{\frac{u^2-b^2}{u^2+a^2}}$

$$[u>b>0] \qquad \text{BY (211.05)}$$

8. $\displaystyle\int_u^\infty \frac{dx}{\sqrt{(a^2+x^2)^3(x^2-b^2)}} = \frac{1}{a^2\sqrt{a^2+b^2}}\left\{F(\xi,s) - E(\xi,s)\right\}$

$$[u\ge b>0] \qquad \text{BY (212.03)}$$

9. $\displaystyle\int_0^u \frac{dx}{\sqrt{(a^2+x^2)(b^2-x^2)^3}} = \frac{1}{b^2\sqrt{a^2+b^2}}\left\{F(\gamma,r) - E(\gamma,r)\right\} + \frac{u}{b^2\sqrt{(a^2+u^2)(b^2-u^2)}}$

$$[b>u>0] \qquad \text{BY (214.10)}$$

10. $\displaystyle\int_u^\infty \frac{dx}{\sqrt{(a^2+x^2)(x^2-b^2)^3}} = \frac{u}{b^2\sqrt{(a^2+u^2)(u^2-b^2)}} - \frac{1}{b^2\sqrt{a^2+b^2}}\,E(\xi,s)$

$$[u\ge b>0] \qquad \text{BY (212.04)}$$

11. $\displaystyle\int_0^u \frac{dx}{\sqrt{(a^2-x^2)^3(b^2-x^2)}} = \frac{1}{a^2(a^2-b^2)}\left\{a\,E(\eta,t) - u\sqrt{\frac{b^2-u^2}{a^2-u^2}}\right\}$

$$[a>b\ge u>0] \qquad \text{BY (219.07)}$$

12. $\displaystyle\int_u^b \frac{dx}{\sqrt{(a^2-x^2)^3(b^2-x^2)}} = \frac{1}{a(a^2-b^2)} E(\zeta,t)$    $[a>b>u\geq 0]$    BY (220.10)

13. $\displaystyle\int_b^u \frac{dx}{\sqrt{(a^2-x^2)^3(x^2-b^2)}} = \frac{1}{a(a^2-b^2)}\left\{ F(\kappa,q) - E(\kappa,q) + \frac{a}{u}\sqrt{\frac{u^2-b^2}{a^2-u^2}}\right\}$

$[a>u>b>0]$    BY (217.10)

14. $\displaystyle\int_u^\infty \frac{dx}{\sqrt{(x^2-a^2)^3(x^2-b^2)}} = \frac{1}{a(b^2-a^2)}\left\{ E(\nu,t) - \frac{a}{u}\sqrt{\frac{u^2-b^2}{u^2-a^2}}\right\}$

$[u>a>b>0]$    BY (215.04)

15. $\displaystyle\int_0^u \frac{dx}{\sqrt{(a^2-x^2)(b^2-x^2)^3}} = \frac{1}{ab^2} F(\eta,t) - \frac{1}{b^2(a^2-b^2)}\left\{ a E(\eta,t) - u\sqrt{\frac{a^2-u^2}{b^2-u^2}}\right\}$

$[a>b>u>0]$    BY (219.06)

16. $\displaystyle\int_u^a \frac{dx}{\sqrt{(a^2-x^2)(x^2-b^2)^3}} = \frac{1}{ab^2(a^2-b^2)}\left\{ b^2 F(\lambda,q) - a^2 E(\lambda,q) + au\sqrt{\frac{a^2-u^2}{u^2-b^2}}\right\}$

$[a>u>b>0]$    BY (218.04)

17. $\displaystyle\int_a^u \frac{dx}{\sqrt{(x^2-a^2)(x^2-b^2)^3}} = \frac{a}{b^2(a^2-b^2)} E(\mu,t) - \frac{1}{ab^2} F(\mu,t)$

$[u>a>b>0]$    BY (216.11)

18. $\displaystyle\int_u^\infty \frac{dx}{\sqrt{(x^2-a^2)(x^2-b^2)^3}} = \frac{1}{b^2(a^2-b^2)}\left\{ a E(\nu,t) - \frac{b^2}{u}\sqrt{\frac{u^2-a^2}{u^2-b^2}}\right\} - \frac{1}{ab^2} F(\nu,t)$

$[u\geq a>b>0]$    BY (215.06)

## 3.159

1. $\displaystyle\int_0^u \frac{x^2\,dx}{\sqrt{(x^2+a^2)(x^2+b^2)^3}} = \frac{a}{a^2-b^2}\left\{ F(\alpha,q) - E(\alpha,q)\right\}$

$[a>b, \quad u>0]$    BY (221.12)

2. $\displaystyle\int_u^\infty \frac{x^2\,dx}{\sqrt{(x^2+a^2)(x^2+b^2)^3}} = \frac{a}{a^2-b^2}\left\{ F(\beta,q) - E(\beta,q)\right\} + \frac{u}{\sqrt{(a^2+u^2)(b^2+u^2)}}$

$[a>b, \quad u\geq 0]$    BY (222.10)

3. $\displaystyle\int_0^u \frac{x^2\,dx}{\sqrt{(x^2+a^2)^3(x^2+b^2)}} = \frac{1}{a(a^2-b^2)}\left\{ a^2 E(\alpha,q) - b^2 F(\alpha,q)\right\} - \frac{u}{\sqrt{(a^2+u^2)(b^2+u^2)}}$

$[a>b, \quad u>0]$    BY (221.11)

4.    $\displaystyle\int_u^\infty \frac{x^2\,dx}{\sqrt{(x^2+a^2)^3\,(x^2+b^2)}} = \frac{1}{a\,(a^2-b^2)}\left\{a^2\,E(\beta,q) - b^2\,F(\beta,q)\right\}$

$$[a > b, \quad u \geq 0] \qquad\qquad \text{BY (222.07)}$$

5.    $\displaystyle\int_0^u \frac{x^2\,dx}{\sqrt{(a^2+x^2)^3\,(b^2-x^2)}} = \frac{1}{\sqrt{a^2+b^2}}\left\{F(\gamma,r) - E(\gamma,r)\right\}$

$$[b \geq u > 0] \qquad\qquad \text{BY (214.04)}$$

6.    $\displaystyle\int_u^b \frac{x^2\,dx}{\sqrt{(a^2+x^2)^3\,(b^2-x^2)}} = \frac{1}{\sqrt{a^2+b^2}}\left\{F(\delta,r) - E(\delta,r)\right\} + \frac{u}{a^2+b^2}\sqrt{\frac{b^2-u^2}{a^2+u^2}}$

$$[b > u \geq 0] \qquad\qquad \text{BY (213.07)}$$

7.    $\displaystyle\int_b^u \frac{x^2\,dx}{\sqrt{(a^2+x^2)^3\,(x^2-b^2)}} = \frac{1}{\sqrt{a^2+b^2}}\,E(\varepsilon,s) - \frac{a^2}{u\,(a^2+b^2)}\sqrt{\frac{u^2-b^2}{u^2+a^2}}$

$$[u > b > 0] \qquad\qquad \text{BY (211.13)}$$

8.    $\displaystyle\int_u^\infty \frac{x^2\,dx}{\sqrt{(a^2+x^2)^3\,(x^2-b^2)}} = \frac{1}{\sqrt{a^2+b^2}}\,E(\xi,s) \qquad [u \geq b > 0] \qquad\qquad \text{BY (212.01)}$

9.    $\displaystyle\int_0^u \frac{x^2\,dx}{\sqrt{(a^2+x^2)\,(b^2-x^2)^3}} = \frac{u}{\sqrt{(a^2+u^2)\,(b^2-u^2)}} - \frac{1}{\sqrt{a^2+b^2}}\,E(\gamma,r)$

$$[b > u > 0] \qquad\qquad \text{BY (214.07)}$$

10.    $\displaystyle\int_u^\infty \frac{x^2\,dx}{\sqrt{(a^2+x^2)\,(x^2-b^2)^3}} = \frac{1}{\sqrt{a^2+b^2}}\left\{F(\xi,s) - E(\xi,s)\right\} + \frac{u}{\sqrt{(a^2+u^2)\,(u^2-b^2)}}$

$$[u > b > 0] \qquad\qquad \text{BY (212.10)}$$

11.    $\displaystyle\int_0^u \frac{x^2\,dx}{\sqrt{(a^2-x^2)^3\,(b^2-x^2)}} = \frac{1}{a^2-b^2}\left\{a\,E(\eta,t) - u\sqrt{\frac{b^2-u^2}{a^2-u^2}}\right\} - \frac{1}{a}\,F(\eta,t)$

$$[a > b \geq u > 0] \qquad\qquad \text{BY (219.04)}$$

12.    $\displaystyle\int_u^b \frac{x^2\,dx}{\sqrt{(a^2-x^2)^3\,(b^2-x^2)}} = \frac{a}{a^2-b^2}\,E(\zeta,t) - \frac{1}{a}\,F(\zeta,t)$

$$[a > b > u \geq 0] \qquad\qquad \text{BY (220.08)}$$

13.    $\displaystyle\int_b^u \frac{x^2\,dx}{\sqrt{(a^2-x^2)^3\,(x^2-b^2)}} = \frac{1}{a\,(a^2-b^2)}\left\{b^2\,F(\kappa,q) - a^2\,E(\kappa,q) + \frac{a^3}{u}\sqrt{\frac{u^2-b^2}{a^2-u^2}}\right\}$

$$[a > u > b > 0] \qquad\qquad \text{BY (217.06)}$$

14.    $\displaystyle\int_u^\infty \frac{x^2\,dx}{\sqrt{(x^2-a^2)^3\,(x^2-b^2)}} = \frac{a}{a^2-b^2}\left\{\frac{a}{u}\sqrt{\frac{u^2-b^2}{u^2-a^2}} - E(\nu,t)\right\} + \frac{1}{a}\,F(\nu,t)$

$$[u > a > b > 0] \qquad\qquad \text{BY (215.09)}$$

15. $\int_0^u \dfrac{x^2\,dx}{\sqrt{(a^2-x^2)(b^2-x^2)^3}} = \dfrac{1}{a^2-b^2}\left\{ u\sqrt{\dfrac{a^2-u^2}{b^2-u^2}} - a\,E(\eta,t) \right\}$

$$[a>b>u>0] \qquad \text{BY (219.12)}$$

16. $\int_u^a \dfrac{x^2\,dx}{\sqrt{(a^2-x^2)(x^2-b^2)^3}} = \dfrac{1}{a^2-b^2}\left\{ a\,F(\lambda,q) - a\,E(\lambda,q) + u\sqrt{\dfrac{a^2-u^2}{u^2-b^2}} \right\}$

$$[a>u>b>0] \qquad \text{BY (218.07)}$$

17. $\int_a^u \dfrac{x^2\,dx}{\sqrt{(x^2-a^2)(x^2-b^2)^3}} = \dfrac{a}{a^2-b^2}\,E(\mu,t) \qquad\qquad [u>a>b>0] \qquad \text{BY (216.01)}$

18. $\int_u^\infty \dfrac{x^2\,dx}{\sqrt{(x^2-a^2)(x^2-b^2)^3}} = \dfrac{1}{a^2-b^2}\left\{ a\,E(\nu,t) - \dfrac{b^2}{u}\sqrt{\dfrac{u^2-a^2}{u^2-b^2}} \right\}$

$$[u\geq a>b>0] \qquad \text{BY (215.11)}$$

### 3.161

1. $\int_u^\infty \dfrac{dx}{x^4\sqrt{(x^2+a^2)(x^2+b^2)}} = \dfrac{1}{3a^3b^4}\left\{ 2\left(a^2+b^2\right)E(\beta,q) - b^2\,F(\beta,q) \right\} + \dfrac{a^2b^2 - u^2\left(2a^2+b^2\right)}{3a^2b^4u^3}$

$$[a>b, \quad u>0] \qquad \text{BY (222.04)}$$

2. $\int_u^b \dfrac{dx}{x^4\sqrt{(x^2+a^2)(b^2-x^2)}} = \dfrac{1}{3a^4b^4\sqrt{a^2+b^2}}\left\{ a^2\left(2a^2-b^2\right)F(\delta,r) - 2\left(a^4-b^4\right)E(\delta,r) \right\}$
$\qquad\qquad + \dfrac{a^2b^2 + 2u^2\left(a^2-b^2\right)}{3a^4b^4u^3}\sqrt{(b^2-u^2)(a^2+u^2)}$

$$[b>u>0] \qquad \text{BY (213.09)}$$

3. $\int_b^u \dfrac{dx}{x^4\sqrt{(x^2+a^2)(x^2-b^2)}} = \dfrac{2b^2-a^2}{3a^4b^2\sqrt{a^2+b^2}}\,F(\varepsilon,s) + \dfrac{2}{3}\dfrac{\left(a^2-b^2\right)\sqrt{a^2+b^2}}{a^4b^4}\,E(\varepsilon,s)$
$\qquad\qquad + \dfrac{1}{3a^2b^2u^3}\sqrt{(u^2+a^2)(u^2-b^2)}$

$$[u>b>0] \qquad \text{BY (211.11)}$$

4. $\int_u^\infty \dfrac{dx}{x^4\sqrt{(x^2+a^2)(x^2-b^2)}} = \dfrac{1}{3a^4b^4\sqrt{a^2+b^2}}\left\{ 2\left(a^4-b^4\right)E(\xi,s) + b^2\left(2b^2-a^2\right)F(\xi,s) \right\}$
$\qquad\qquad - \dfrac{a^2b^2 + u^2\left(2a^2-b^2\right)}{3a^2b^4u^3}\sqrt{\dfrac{u^2-b^2}{u^2+a^2}}$

$$[u\geq b>0] \qquad \text{BY (212.06)}$$

5. $\int_u^b \dfrac{dx}{x^4\sqrt{(a^2-x^2)(b^2-x^2)}} = \dfrac{1}{3a^3b^4}\left\{ \left(2a^2+b^2\right)F(\zeta,t) - 2\left(a^2+b^2\right)E(\zeta,t) \right.$

$\qquad\qquad \left. + \dfrac{\left[\left(2a^2+b^2\right)u^2 + a^2b^2\right]a}{u^3}\sqrt{\dfrac{b^2-u^2}{a^2-u^2}} \right\}$

$$[a>b>u>0] \qquad \text{BY (220.09)}$$

6.
$$\int_b^u \frac{dx}{x^4 \sqrt{(a^2 - x^2)(x^2 - b^2)}} = \frac{1}{3a^3b^4} \left\{ 2\left(a^2 + b^2\right) E\left(\kappa, q\right) - b^2 F\left(\kappa, q\right) \right\}$$
$$+ \frac{1}{3a^2b^2u^3} \sqrt{(a^2 - u^2)(u^2 - b^2)}$$

$$[a \geq u > b > 0] \qquad \text{BY (217.14)}$$

7.
$$\int_u^a \frac{dx}{x^4 \sqrt{(a^2 - x^2)(x^2 - b^2)}} = \frac{1}{3a^3b^4} \left\{ 2\left(a^2 + b^2\right) E(\lambda, q) - b^2 F(\lambda, q) \right.$$
$$\left. - \frac{2\left(a^2 + b^2\right) u^2 + a^2b^2}{au^3} \sqrt{(a^2 - u^2)(u^2 - b^2)} \right\}$$

$$[a > u \geq b > 0] \qquad \text{BY (218.12)}$$

8.
$$\int_a^u \frac{dx}{x^4 \sqrt{(x^2 - a^2)(x^2 - b^2)}} = \frac{1}{3a^3b^4} \left\{ \left(2a^2 + b^2\right) F(\mu, t) - 2\left(a^2 + b^2\right) E(\mu, t) \quad [u > a > b > 0] \right.$$
$$\left. + \frac{\left[\left(a^2 + 2b^2\right) u^2 + a^2b^2\right] b^2}{au^3} \sqrt{\frac{u^2 - a^2}{u^2 - b^2}} \right\}$$

$$\text{BY (216.09)}$$

9.
$$\int_u^\infty \frac{dx}{x^4 \sqrt{(x^2 - a^2)(x^2 - b^2)}} = \frac{1}{3a^3b^4} \left\{ \left(2a^2 + b^2\right) F(\nu, t) - 2\left(a^2 + b^2\right) E(\nu, t) \right.$$
$$\left. + \frac{ab^2}{u^3} \sqrt{(u^2 - a^2)(u^2 - b^2)} \right\}$$

$$[u \geq a > b > 0] \qquad \text{BY (215.07)}$$

**3.162**

1.
$$\int_0^u \frac{dx}{\sqrt{(x^2 + a^2)^5 (x^2 + b^2)}} = \frac{1}{3a^3 (a^2 - b^2)^2} \left\{ \left(3a^2 - b^2\right) F(\alpha, q) - 2\left(2a^2 - b^2\right) E(\alpha, q) \right\}$$
$$+ \frac{u \left[a^2 \left(4a^2 - 3b^2\right) + u^2 \left(3a^2 - 2b^2\right)\right]}{3a^4 (a^2 - b^2) \sqrt{(u^2 + a^2)^3 (u^2 + b^2)}}$$

$$[a > b, \quad u > 0] \qquad \text{BY (221.06)}$$

2.
$$\int_u^\infty \frac{dx}{\sqrt{(x^2 + a^2)^5 (x^2 + b^2)}} = \frac{1}{3a^3 (a^2 - b^2)^2} \left\{ \left(3a^2 - b^2\right) F(\beta, q) - 2\left(2a^2 - b^2\right) E(\beta, q) \right\}$$
$$+ \frac{u}{3a^2 (a^2 - b^2)} \sqrt{\frac{u^2 + b^2}{(a^2 + u^2)^3}}$$

$$[a > b, \quad u \geq 0] \qquad \text{BY (222.03)}$$

3.
$$\int_0^u \frac{dx}{\sqrt{(x^2 + a^2)(x^2 + b^2)^5}} = \frac{3b^2 - a^2}{3ab^2 (a^2 - b^2)^2} F(\alpha, q) + \frac{a \left(2a^2 - 4b^2\right)}{3b^4 (a^2 - b^2)^2} E(\alpha, q)$$
$$+ \frac{u}{3b^2 (a^2 - b^2)} \sqrt{\frac{u^2 + a^2}{(u^2 + b^2)^3}}$$

$$[a > b, \quad u > 0] \qquad \text{BY (221.05)}$$

4. $$\int_u^\infty \frac{dx}{\sqrt{(x^2+a^2)(x^2+b^2)^5}} = \frac{1}{3ab^4(a^2-b^2)^2}\left\{2a^2(a^2-2b^2)E(\beta,q)+b^2(3b^2-a^2)F(\beta,q)\right\}$$
$$-\frac{u\left[b^2(3a^2-4b^2)+u^2(2a^2-3b^2)\right]}{3b^4(a^2-b^2)\sqrt{(u^2+a^2)(u^2+b^2)^3}}$$
$$[a>b, \quad u\geq 0] \qquad\qquad \text{BY (222.05)}$$

5. $$\int_0^u \frac{dx}{\sqrt{(a^2+x^2)^5(b^2-x^2)}} = \frac{1}{3a^4\sqrt{(a^2+b^2)^3}}\left\{2(b^2+2a^2)E(\gamma,r)-a^2F(\gamma,r)\right\}$$
$$+\frac{u}{3a^2(a^2+b^2)}\sqrt{\frac{b^2-u^2}{(a^2+u^2)^3}}$$
$$[b\geq u>0] \qquad\qquad \text{BY (214.15)}$$

6. $$\int_u^b \frac{dx}{\sqrt{(a^2+x^2)^5(b^2-x^2)}} = \frac{1}{3a^4\sqrt{(a^2+b^2)^3}}\left\{(4a^2+2b^2)E(\delta,r)-a^2F(\delta,r)\right\}$$
$$-\frac{u\left[a^2(5a^2+3b^2)+u^2(4a^2+2b^2)\right]}{3a^4(a^2+b^2)^2}\sqrt{\frac{b^2-u^2}{(a^2+u^2)^3}}$$
$$[b>u>0] \qquad\qquad \text{BY (213.08)}$$

7. $$\int_b^u \frac{dx}{\sqrt{(a^2+x^3)^5(x^2-b^2)}} = \frac{1}{3a^4\sqrt{(a^2+b^2)^3}}\left\{(3a^2+2b^2)F(\varepsilon,s)-(4a^2+2b^2)E(\varepsilon,s)\right\}$$
$$+\frac{(3a^2+b^2)u^2+2(2a^2+b^2)a^2}{3a^2(a^2+b^2)^2u}\sqrt{\frac{u^2-b^2}{(u^2+a^2)^3}}$$
$$[u>b>0] \qquad\qquad \text{BY (211.05)}$$

8. $$\int_u^\infty \frac{dx}{\sqrt{(a^2+x^2)^5(x^2-b^2)}} = \frac{1}{3a^4\sqrt{(a^2+b^2)^3}}\left\{(3a^2+2b^2)F(\xi,s)-(4a^2+2b^2)E(\xi,s)\right\}$$
$$+\frac{u}{3a^2(a^2+b^2)}\sqrt{\frac{u^2-b^2}{(a^2+u^2)^3}}$$
$$[u>b>0] \qquad\qquad \text{BY (212.03)}$$

9.[12] $$\int_0^u \frac{dx}{\sqrt{(a^2+x^2)(b^2-x^2)^5}} = \frac{1}{3b^4\sqrt{(a^2+b^2)^3}}\left\{(2a^2+3b^2)F(\gamma,r)-(2a^2+4b^2)E(\gamma,r)\right\}$$
$$+\frac{u\left[(3a^2+4b^2)b^2-(2a^2+3b^2)u^2\right]}{3b^4(a^2+b^2)\sqrt{(a^2+u^2)(b^2-u^2)^3}}$$
$$[b>u>0] \qquad\qquad \text{BY (214.10)}$$

10. $$\int_u^\infty \frac{dx}{\sqrt{(a^2+x^2)(x^2-b^2)^5}} = \frac{1}{3b^4\sqrt{(a^2+b^2)^3}}\left\{(2a^2+4b^2)E(\xi,s)-b^2F(\xi,s)\right\}$$
$$+\frac{u\left[(3a^2+4b^2)b^2-(2a^2+3b^2)u^2\right]}{3b^4(a^2+b^2)\sqrt{(a^2+u^2)(u^2-b^2)^3}}$$
$$[u>b>0] \qquad\qquad \text{BY (212.04)}$$

11. $$\int_0^u \frac{dx}{\sqrt{(a^2-x^2)(b^2-x^2)^5}} = \frac{2a^2-3b^2}{3ab^4(a^2-b^2)} F(\eta,t) + \frac{2a(2b^2-a^2)}{3b^4(a^2-b^2)^2} E(\eta,t)$$
$$+ \frac{u\left[(3a^2-5b^2)b^2 - 2(a^2-2b^2)u^2\right]}{3b^4(a^2-b^2)^2(b^2-u^2)}\sqrt{\frac{a^2-u^2}{b^2-u^2}}$$
$$[a > b, \quad a > 0] \qquad \text{BY (219.06)}$$

12. $$\int_u^a \frac{dx}{\sqrt{(a^2-x^2)(x^2-b^2)^5}} = \frac{3b^2-a^2}{3ab^2(a^2-b^2)^2} F(\lambda,q) + \frac{2a(a^2-2b^2)}{3b^4(a^2-b^2)^2} E(\lambda,q)$$
$$+ \frac{u\left[2(2b^2-a^2)u^2 + (3a^2-5b^2)b^2\right]}{3b^4(a^2-b^2)^2(u^2-b^2)}\sqrt{\frac{a^2-u^2}{u^2-b^2}}$$
$$[a > u > b > 0] \qquad \text{BY (218.04)}$$

13. $$\int_a^u \frac{dx}{\sqrt{(x^2-a^2)(x^2-b^2)^5}} = \frac{2a^2-3b^2}{3ab^4(a^2-b^2)} F(\mu,t) + \frac{2a(2b^2-a^2)}{3b^4(a^2-b^2)^2} E(\mu,t)$$
$$+ \frac{u}{3b^2(a^2-b^2)(u^2-b^2)}\sqrt{\frac{u^2-a^2}{u^2-b^2}}$$
$$[u > a > b > 0] \qquad \text{BY (216.11)}$$

14. $$\int_u^\infty \frac{dx}{\sqrt{(x^2-a^2)(x^2-b^2)^5}} = \frac{(4b^2-2a^2)a}{3b^4(a^2-b^2)^2} E(\nu,t) + \frac{2a^2-3b^2}{3ab^4(a^2-b^2)} F(\nu,t)$$
$$- \frac{(3b^2-a^2)u^2 - (4b^2-2a^2)b^2}{3b^2u(a^2-b^2)^2(u^2-b^2)}\sqrt{\frac{u^2-a^2}{u^2-b^2}}$$
$$[u \geq a > b > 0] \qquad \text{BY (215.06)}$$

15. $$\int_0^u \frac{dx}{\sqrt{(a^2-x^2)^5(b^2-x^2)}} = \frac{1}{3a^3(a^2-b^2)^2}\left\{(4a^2-2b^2)E(\eta,t) - (a^2-b^2)F(\eta,t)\right.$$
$$\left. - \frac{u\left[(5a^2-3b^2)a^2 - (4a^2-2b^2)u^2\right]}{a(a^2-u^2)}\sqrt{\frac{b^2-u^2}{a^2-u^2}}\right\}$$
$$[a > b \geq u > 0] \qquad \text{BY (219.07)}$$

16. $$\int_u^b \frac{dx}{\sqrt{(a^2-x^2)^5(b^2-x^2)}} = \frac{2(2a^2-b^2)}{3a^3(a^2-b^2)^2} E(\zeta,r) - \frac{1}{3a^3(a^2-b^2)} F(\zeta,t)$$
$$+ \frac{u}{3a^2(a^2-b^2)(a^2-u^2)}\sqrt{\frac{b^2-u^2}{a^2-u^2}}$$
$$[a > b > u \geq 0] \qquad \text{BY (220.10)}$$

17. $$\int_b^u \frac{dx}{\sqrt{(a^2-x^2)^5(x^2-b^2)}} = \frac{1}{3a^3(a^2-b^2)^2}\left\{(3a^2-b^2)F(\kappa,q) - (4a^2-2b^2)E(\kappa,q)\right\}$$
$$+ \frac{2(2a^2-b^2)a^2 + (b^2-3a^2)u^2}{3a^2u(a^2-b^2)^2(a^2-u^2)}\sqrt{\frac{u^2-b^2}{a^2-u^2}},$$
$$[a > u > b > 0] \qquad \text{BY (217.10)}$$

18. $$\int_u^\infty \frac{dx}{\sqrt{(x^2-a^2)^5(x^2-b^2)}} = \frac{1}{3a^3(a^2-b^2)^2}\left\{(4a^2-2b^2)E(\nu,t) - (a^2-b^2)F(\nu,t)\right\}$$
$$+ \frac{(4a^2-2b^2)a^2 + (b^2-3a^2)u^2}{3a^2u(a^2-b^2)^2(u^2-a^2)}\sqrt{\frac{u^2-b^2}{u^2-a^2}}$$
$$[u > a > b > 0] \qquad \text{BY (215.04)}$$

**3.163**

1.
$$\int_0^u \frac{dx}{\sqrt{(x^2+a^2)^3(x^2+b^2)^3}} = \frac{1}{ab^2(a^2-b^2)^2}\left\{(a^2+b^2)\,E(\alpha,q) - 2b^2\,F(\alpha,q)\right\}$$
$$- \frac{u}{a^2(a^2-b^2)\sqrt{(a^2+u^2)(b^2+u^2)}}$$
$$[a > b, \quad u > 0] \qquad\qquad \text{BY (221.07)}$$

2.
$$\int_u^\infty \frac{dx}{\sqrt{(x^2+a^2)^3(x^2+b^2)^3}} = \frac{1}{ab^2(a^2-b^2)^2}\left\{(a^2+b^2)\,E(\beta,q) - 2b^2\,F(\beta,q)\right\}$$
$$- \frac{u}{b^2(a^2-b^2)\sqrt{(a^2+u^2)(b^2+u^2)}}$$
$$[a > b, \quad u \geq 0] \qquad\qquad \text{BY (222.12)}$$

3.[12]
$$\int_0^u \frac{dx}{\sqrt{(x^2+a^2)^3(b^2-x^2)^3}} = \frac{1}{a^2b^2\sqrt{(a^2+b^2)^3}}\left\{a^2\,F(\gamma,r) - (a^2-b^2)\,E(\gamma,r)\right\}$$
$$+ \frac{u}{b^2(a^2+b^2)\sqrt{(a^2+u^2)(b^2-u^2)}}$$
$$[b > u > 0] \qquad\qquad \text{BY (214.15)}$$

4.
$$\int_u^\infty \frac{dx}{\sqrt{(x^2+a^2)^3(x^2-b^2)^3}} = \frac{b^2-a^2}{a^2b^2\sqrt{(a^2+b^2)^3}}\,E(\xi,s) - \frac{1}{a^2\sqrt{(a^2+b^2)^3}}\,F(\xi,s)$$
$$+ \frac{u}{b^2(a^2+b^2)\sqrt{(u^2+a^2)(u^2-b^2)}}$$
$$[u > b > 0] \qquad\qquad \text{BY (212.05)}$$

5.
$$\int_0^u \frac{dx}{\sqrt{(a^2-x^2)^3(b^2-x^2)^3}} = \frac{1}{ab^2(a^2-b^2)}\,F(\eta,t) - \frac{a^2+b^2}{ab^2(a^2-b^2)^2}\,E(\eta,t)$$
$$+ \frac{\left[a^4+b^4-(a^2+b^2)u^2\right]u}{a^2b^2(a^2-b^2)^2\sqrt{(a^2-u^2)(b^2-u^2)}}$$
$$[a > b > u > 0] \qquad\qquad \text{BY (279.08)}$$

6.
$$\int_u^\infty \frac{dx}{\sqrt{(x^2-a^2)^3(x^2-b^2)^3}} = \frac{1}{ab^2(a^2-b^2)}\,F(\nu,t) - \frac{a^2+b^2}{ab^2(a^2-b^2)^2}\,E(\nu,t)$$
$$+ \frac{1}{u(a^2-b^2)\sqrt{(u^2-a^2)(u^2-b^2)}}$$
$$[u > a > b > 0] \qquad\qquad \text{BY (215.10)}$$

**3.164**    **Notation:** $\alpha = \arccos \dfrac{u^2-\rho\bar\rho}{u^2+\rho\bar\rho}, \qquad r = \dfrac{1}{2}\sqrt{-\dfrac{(\rho-\bar\rho)^2}{\rho\bar\rho}}.$

1.
$$\int_u^\infty \frac{dx}{\sqrt{(x^2+\rho^2)(x^2+\bar\rho^2)}} = \frac{1}{\sqrt{\rho\bar\rho}}\,F(\alpha,r) \qquad\qquad \text{BY (225.00)}$$

2.
$$\int_u^\infty \frac{x^2\,dx}{(x^2-\rho\bar\rho)^2\sqrt{(x^2+\rho^2)(x^2+\bar\rho^2)}} = \frac{2u\sqrt{(u^2+\rho^2)(u^2+\bar\rho^2)}}{(\rho+\bar\rho)^2(u^4-\rho^2\bar\rho^2)} - \frac{1}{(\rho+\bar\rho)^2\sqrt{\rho\bar\rho}}\,E(\alpha,r)$$
$$\text{BY (225.03)}$$

3.     $\displaystyle\int_u^\infty \frac{x^2\,\mathrm{d}x}{\left(x^2+\rho\bar\rho\right)^2\sqrt{\left(x^2+\rho^2\right)\left(x^2+\bar\rho^2\right)}} = -\frac{1}{\left(\rho-\bar\rho\right)^2\sqrt{\rho\bar\rho}}\left[F(\alpha,r)-E(\alpha,r)\right]$     BY (225.07)

4.     $\displaystyle\int_u^\infty \frac{x^2\,\mathrm{d}x}{\sqrt{\left(x^2+\rho^2\right)^3\left(x^2+\bar\rho^2\right)^3}} = -\frac{4\sqrt{\rho\bar\rho}}{\left(\rho^2-\bar\rho^2\right)^2}\,E(\alpha,r)+\frac{1}{\left(\rho-\bar\rho\right)^2\sqrt{\rho\bar\rho}}\,F(\alpha,r)$

$$-\frac{2u\left(u^2-\rho\bar\rho\right)}{\left(\rho+\bar\rho\right)^2\left(u^2+\rho\bar\rho\right)\sqrt{\left(u^2+\rho^2\right)\left(u^2+\bar\rho^2\right)}}$$

BY (225.05)

5.     $\displaystyle\int_u^\infty \frac{\left(x^2-\rho\bar\rho\right)^2\,\mathrm{d}x}{\sqrt{\left(x^2+\rho^2\right)^3\left(x^2+\bar\rho^2\right)^3}} = -\frac{4\sqrt{\rho\bar\rho}}{\left(\rho-\bar\rho\right)^2}\left[F(\alpha,r)-E(\alpha,r)\right]$

$$+\frac{2u\left(u^2-\rho\bar\rho\right)}{\left(u^2+\rho\bar\rho\right)\sqrt{\left(u^2+\rho^2\right)\left(u^2+\bar\rho^2\right)}}$$

BY (225.06)

6.     $\displaystyle\int_u^\infty \frac{\sqrt{\left(x^2+\rho^2\right)\left(x^2+\bar\rho^2\right)}}{\left(x^2+\rho\bar\rho\right)^2}\,\mathrm{d}x = \frac{1}{\sqrt{\rho\bar\rho}}\,E(\alpha,r)$     BY(225.01)

7.     $\displaystyle\int_u^\infty \frac{\left(x^2-\varrho\bar\varrho\right)^2\,\mathrm{d}x}{\left(x^2+\varrho\bar\varrho\right)^2\sqrt{\left(x^2+\varrho^2\right)\left(x^2+\bar\varrho^2\right)}} = -\frac{4\sqrt{\varrho\bar\varrho}}{\left(\varrho-\bar\varrho\right)^2}\,E(\alpha,r)+\frac{\left(\varrho+\bar\varrho\right)^2}{\left(\varrho-\bar\varrho\right)^2\sqrt{\varrho\bar\varrho}}\,F(\alpha,r)$     BY (225.08)

8.     $\displaystyle\int_u^\infty \frac{\left(x^2+\varrho\bar\varrho\right)^2\,\mathrm{d}x}{\left[\left(x^2+\varrho\bar\varrho\right)^2-4p^2\varrho\bar\varrho x^2\right]\sqrt{\left(x^2+\varrho^2\right)\left(x^2+\bar\varrho^2\right)}} = \frac{1}{\sqrt{\varrho\bar\varrho}}\,\Pi\left(\alpha,p^2,r\right)$     BY (225.02)

**3.165**    **Notation:** $\alpha = \arccos\dfrac{u^2-a^2}{u^2+a^2},\qquad r = \dfrac{\sqrt{a^2-b^2}}{a\sqrt{2}}.$

1.     $\displaystyle\int_u^a \frac{\mathrm{d}x}{\sqrt{x^4+2b^2x^2+a^4}} = \frac{\sqrt{2}}{a\sqrt{2}+\sqrt{a^2+b^2}}$

$$\times F\left[\arctan\left(\frac{a\sqrt{2}+\sqrt{a^2-b^2}}{\sqrt{a^2+b^2}}\frac{a-u}{a+u}\right),\ \frac{2\sqrt{a\sqrt{2\left(a^2-b^2\right)}}}{a\sqrt{2}+\sqrt{a^2-b^2}}\right]$$

$[a>b,\quad a>u\ge 0]$     BY (264.00)

2.[12]    $\displaystyle\int_u^\infty \frac{\mathrm{d}x}{\sqrt{x^4+2b^2x^2+a^4}} = \frac{1}{2a}\,F(\alpha,r)$       $\left[a^2>b^2,\quad a^2>0,\quad u\ge 0\right]$

BY (263.00, 266.00)

3.     $\displaystyle\int_u^\infty \frac{\mathrm{d}x}{x^2\sqrt{x^4+2b^2x^2+a^4}} = \frac{1}{2a^3}\left[F(\alpha,r)-2\,E(\alpha,r)\right]+\frac{\sqrt{u^4+2b^2u^2+a^4}}{a^2u\left(u^2+a^2\right)}$

$[a>b>0,\quad u>0]$     BY (263.06)

4.[12]    $\displaystyle\int_u^\infty \frac{x^2\,\mathrm{d}x}{\left(x^2+a^2\right)^2\sqrt{x^4+2b^2x^2+a^4}} = \frac{1}{4a\left(a^2-b^2\right)}\left[F(\alpha,r)-E(\alpha,r)\right]$

$\left[a^2>b^2,\quad a^2>0,\quad u\ge 0\right]$
BY (263.03, 266.05)

5. $$\int_u^\infty \frac{x^2\,dx}{(x^2-a^2)^2\sqrt{x^4+2b^2x^2+a^4}} = \frac{u\sqrt{u^4+2b^2u^2+a^4}}{2(a^2+b^2)(u^4-a^4)} - \frac{1}{4a(a^2+b^2)}E(\alpha,r)$$
$$\left[a^2 > b^2 > -\infty, \quad u^2 > a^2 > 0\right]$$
BY (263.05, 266.02)

6.[12] $$\int_u^\infty \frac{x^2\,dx}{\sqrt{(x^4+2b^2x^2+a^4)^3}} = \frac{a}{2(a^4-b^4)}E(\alpha,r) - \frac{1}{4a(a^2-b^2)}F(\alpha,r)$$
$$- \frac{u(u^2-a^2)}{2(a^2+b^2)(u^2+a^2)\sqrt{u^4+2b^2u^2+a^4}}$$
$$\left[a^2 > b^2, \quad a^2 > 0, \quad u \geq 0\right] \quad \text{BY (263.08, 266.03)}$$

7. $$\int_u^\infty \frac{(x^2-a^2)^2\,dx}{\sqrt{(x^4+2b^2x^2+a^4)^3}} = \frac{a}{a^2-b^2}[F(\alpha,r)-E(\alpha,r)] + \frac{u^2-a^2}{u^2+a^2}\frac{u}{\sqrt{u^4+2b^2u^2+a^4}}$$
$$\left[|b^2| < a^2, \quad u \geq 0\right] \qquad \text{BY (266.08)}$$

8. $$\int_u^\infty \frac{(x^2+a^2)^2\,dx}{\sqrt{(x^2+2b^2x^2+a^4)^3}} = \frac{a}{a^2+b^2}E(\alpha,r) - \frac{a^2-b^2}{a^2+b^2}\cdot\frac{u^2-a^2}{u^2+a^2}\cdot\frac{u}{\sqrt{u^4+2b^2u^2+a^4}}$$
$$\left[|b^2| < a^2, \quad u \geq 0\right] \qquad \text{BY (266.06)a}$$

9.[12] $$\int_u^\infty \frac{(x^2-a^2)^2\,dx}{(x^2+a^2)^2\sqrt{x^4+2b^2x^2+a^4}} = \frac{a}{a^2-b^2}E(\alpha,r) - \frac{a^2+b^2}{2a(a^2-b^2)}F(\alpha,r)$$
$$\left[a^2 > b^2, \quad a^2 > 0, \quad u \geq 0\right]$$
BY (263.04, 266.07)

10. $$\int_u^\infty \frac{\sqrt{x^4+2b^2x^2+a^4}}{(x^2+a^2)^2}\,dx = \frac{1}{2a}E(\alpha,r) \qquad \left[a^2 > b^2, \quad a^2 > 0, \quad u \geq 0\right]$$
BY (263.01, 266.01)

11. $$\int_u^\infty \frac{\sqrt{x^4+2b^2x^2+a^4}}{(x^2-a^2)^2}\,dx = \frac{1}{2a}[F(\alpha,r)-E(\alpha,r)] + \frac{u}{u^4-a^4}\sqrt{u^4+2b^2u^2+a^4}$$
$$\left[a > b > 0, \quad u > a\right] \qquad \text{BY (263)}$$

12. $$\int_u^\infty \frac{(x^2+a^2)^2\,dx}{\left[(x^2+a^2)^2-4a^2p^2x^2\right]\sqrt{x^4+2b^2x^2+a^4}} = \frac{1}{2a}\Pi\left(\alpha,p^2,r\right)$$
$$\left[a > b > 0, \quad u \geq 0\right] \qquad \text{BY (263.02)}$$

**3.166**   **Notation:**   $\alpha = \arccos\dfrac{u^2-1}{u^2+1}, \quad \beta = \arctan\left\{(1+\sqrt{2})\dfrac{1-u}{1+u}\right\},$

$$\gamma = \arccos u, \quad \delta = \arccos\frac{1}{u}, \quad \varepsilon = \arccos\frac{1-u^2}{1+u^2},$$

$$r = \frac{\sqrt{2}}{2}, \quad q = 2\sqrt{3\sqrt{2}-4} = 2\sqrt[4]{2}\left(\sqrt{2}-1\right) \approx 0.985171$$

1. $\displaystyle\int_u^\infty \frac{dx}{\sqrt{x^4+1}} = \frac{1}{2}\,F(\alpha,r)$ $[u \geq 0]$ H (287), BY (263.50)

2. $\displaystyle\int_u^\infty \frac{dx}{x^2\sqrt{x^4+1}} = \frac{1}{2}\left[F(\alpha,r) - 2\,E(\alpha,r)\right] + \frac{\sqrt{u^4+1}}{u\,(u^2+1)}$

$[u > 0]$ BY (263.57)

3. $\displaystyle\int_u^\infty \frac{x^2\,dx}{(x^4+1)\,\sqrt{x^4+1}} = \frac{1}{2}\,E(\alpha,r) - \frac{1}{4}\,F(\alpha,r) - \frac{u\,(u^2-1)}{2\,(u^2+1)\,\sqrt{u^4+1}}$

$[u \geq 0]$ BY (263.59)

4. $\displaystyle\int_u^\infty \frac{x^2\,dx}{(x^2+1)^2\,\sqrt{x^4+1}} = \frac{1}{4}\left[F(\alpha,r) - E(\alpha,r)\right]$ $[u \geq 0]$ BY (263.53)

5. $\displaystyle\int_u^\infty \frac{x^2\,dx}{(x^2-1)^2\,\sqrt{x^4+1}} = \frac{u\sqrt{u^4+1}}{2\,(u^4-1)} - \frac{1}{4}\,E(\alpha,r)$ $[u > 1]$ BY (263.55)

6. $\displaystyle\int_u^\infty \frac{\sqrt{x^4+1}}{(x^2-1)^2}\,dx = \frac{1}{2}\left[F(\alpha,r) - E(\alpha,r)\right] + \frac{u\sqrt{u^4+1}}{u^4-1}$

$[u > 1]$ BY (263.58)

7. $\displaystyle\int_u^\infty \frac{\left(x^2-1\right)^2\,dx}{(x^2+1)^2\,\sqrt{x^4+1}} = E(\alpha,r) - \frac{1}{2}\,F(\alpha,r)$ $[u \geq 0]$ BY (263.54)

8. $\displaystyle\int_u^\infty \frac{\sqrt{x^4+1}\,dx}{(x^2+1)^2} = \frac{1}{2}\,E(\alpha,r)$ $[u \geq 0]$ BY (263.51)

9. $\displaystyle\int_u^\infty \frac{\left(x^2+1\right)^2\,dx}{\left[(x^2+1)^2 - 4p^2x^2\right]\sqrt{x^4+1}} = \frac{1}{2}\,\Pi\left(\alpha,p^2,r\right)$ $[u \geq 0]$ BY (263.52)

10. $\displaystyle\int_0^u \frac{dx}{\sqrt{x^4+1}} = \frac{1}{2}\,F(\varepsilon,r)$ H 66(288)

11. $\displaystyle\int_u^1 \frac{dx}{\sqrt{x^4+1}} = \left(2-\sqrt{2}\right)F(\beta,q)$ $[0 \leq u < 1]$ BY (264.50)

12. $\displaystyle\int_u^1 \frac{\left(x^2 + x\sqrt{2}+1\right)\,dx}{\left(x^2 - x\sqrt{2}+1\right)\sqrt{x^4+1}} = \left(2+\sqrt{2}\right)E(\beta,q)$ $[0 \leq u < 1]$ BY (264.51)

13. $\displaystyle\int_u^1 \frac{(1-x)^2\,dx}{\left(x^2 - x\sqrt{2}+1\right)\sqrt{x^4+1}} = \frac{1}{\sqrt{2}}\left[F(\beta,q) - E(\beta,q)\right]$

$[0 \leq u < 1]$ BY (264.55)

14. $\displaystyle\int_u^1 \frac{(1+x)^2\,dx}{\left(x^2 - x\sqrt{2}+1\right)\sqrt{x^4+1}} = \frac{3\sqrt{2}+4}{2}\,E(\beta,q) - \frac{3\sqrt{2}-4}{2}\,F(\beta,q)$

$[0 \leq u < 1]$ BY (264.56)

15.[12] $\displaystyle\int_u^1 \frac{dx}{\sqrt{1-x^4}} = \frac{1}{\sqrt{2}}\,F(\gamma,r)$ $[0 < u < 1]$ H 66 (290), BY (259.75)

16.    $\displaystyle\int_0^1 \frac{dx}{\sqrt{1-x^4}} = \frac{1}{4\sqrt{2\pi}}\left(\Gamma\left(\frac{1}{4}\right)\right)^2$

17.    $\displaystyle\int_1^u \frac{dx}{\sqrt{x^4-1}} = \frac{1}{\sqrt{2}}F(\delta,r)$          $[u>1]$          H 66 (289), BY (260.75)

18.[12]   $\displaystyle\int_u^1 \frac{x^2\,dx}{\sqrt{1-x^4}} = \sqrt{2}\,E(\gamma,r) - \frac{1}{\sqrt{2}}F(\gamma,r)$      $[0<u<1]$

$$= \frac{1}{\sqrt{2\pi}}\left(\Gamma\left(\frac{3}{4}\right)\right)^2 \qquad\qquad [u=0]$$

                                                                     BY (259.76)

19.    $\displaystyle\int_1^u \frac{x^2\,dx}{\sqrt{x^4-1}} = \frac{1}{\sqrt{2}}F(\delta,r) - \sqrt{2}\,E(\delta,r) + \frac{1}{u}\sqrt{u^4-1}$    $[u>1]$    BY (260.77)

20.[12]   $\displaystyle\int_u^1 \frac{x^4\,dx}{\sqrt{1-x^4}} = \frac{1}{3\sqrt{2}}F(\gamma,r) + \frac{u}{3}\sqrt{1-u^4}$      $[0<u<1]$    BY (259.76)

21.[3]   $\displaystyle\int_1^u \frac{x^4\,dx}{\sqrt{x^4-1}} = \frac{1}{3\sqrt{2}}F(\delta,r) + \frac{1}{3}u\sqrt{u^4-1}$      $[u>1]$    BY (260.77)

22.    $\displaystyle\int_0^u \frac{dx}{\sqrt{x(1+x^3)}} = \frac{1}{\sqrt[4]{3}}F\left(\arccos\frac{1+(1-\sqrt{3})u}{1+(1+\sqrt{3})u},\frac{\sqrt{2+\sqrt{3}}}{2}\right)$

                                                     $[u>0]$          BY (260.50)

23.[12]   $\displaystyle\int_0^u \frac{dx}{\sqrt{x(1-x^3)}} = \frac{1}{\sqrt[4]{3}}F\left(\arccos\frac{1-(1+\sqrt{3})u}{1+(\sqrt{3}-1)u},\frac{\sqrt{2-\sqrt{3}}}{2}\right)$

                                                   $[0<u\le 1]$        BY (259.50)

**3.167**   **Notation: In 3.167 and 3.168 we set:**    $\alpha = \arcsin\sqrt{\dfrac{(a-c)(d-u)}{(a-d)(c-u)}},$

$$\beta = \arcsin\sqrt{\frac{(a-c)(u-d)}{(c-d)(a-u)}}, \qquad \gamma = \arcsin\sqrt{\frac{(b-d)(c-u)}{(c-d)(b-u)}},$$

$$\delta = \arcsin\sqrt{\frac{(b-d)(u-c)}{(b-c)(u-d)}}, \qquad \kappa = \arcsin\sqrt{\frac{(a-c)(b-u)}{(b-c)(a-u)}},$$

$$\lambda = \arcsin\sqrt{\frac{(a-c)(u-b)}{(a-b)(u-c)}}, \qquad \mu = \arcsin\sqrt{\frac{(b-d)(a-u)}{(a-b)(u-d)}},$$

$$\nu = \arcsin\sqrt{\frac{(b-d)(u-a)}{(a-d)(u-b)}}, \qquad q = \sqrt{\frac{(b-c)(a-d)}{(a-c)(b-d)}}, \qquad r = \sqrt{\frac{(a-b)(c-d)}{(a-c)(b-d)}}.$$

1.    $\displaystyle\int_u^d \sqrt{\frac{d-x}{(a-x)(b-x)(c-x)}}\,dx = \frac{2(c-d)}{\sqrt{(a-c)(b-d)}}\left\{\Pi\left(\alpha,\frac{a-d}{a-c},q\right) - F(\alpha,q)\right\}$

                                                       $[a>b>c>d>u]$      BY (251.05)

2.  $\displaystyle\int_d^u \sqrt{\frac{x-d}{(a-x)(b-x)(c-x)}}\,dx = \frac{2(d-a)}{\sqrt{(a-c)(b-d)}}\left\{\Pi\left(\beta,\frac{d-c}{a-c},r\right) - F(\beta,r)\right\}$

$$[a > b > c \geq u > d] \qquad\qquad \text{BY (252.14)}$$

3.  $\displaystyle\int_u^c \sqrt{\frac{x-d}{(a-x)(b-x)(c-x)}}\,dx = \frac{2}{\sqrt{(a-c)(b-d)}}\left\{(c-b)\Pi\left(\gamma,\frac{c-d}{b-d},r\right) + (b-d)F(\gamma,r)\right\}$

$$[a > b > c > u \geq d] \qquad\qquad \text{BY (253.14)}$$

4.  $\displaystyle\int_c^u \sqrt{\frac{x-d}{(a-x)(b-x)(x-c)}}\,dx = \frac{2(c-d)}{\sqrt{(a-c)(b-d)}}\,\Pi\left(\delta,\frac{b-c}{b-d},q\right)$

$$[a > b \geq u > c > d] \qquad\qquad \text{BY (254.02)}$$

5.  $\displaystyle\int_u^b \sqrt{\frac{x-d}{(a-x)(b-x)(x-c)}}\,dx = \frac{2}{\sqrt{(a-c)(b-d)}}\left\{(b-a)\Pi\left(\kappa,\frac{b-c}{a-c},q\right) + (a-d)F(\kappa,q)\right\}$

$$[a > b > u \geq c > d] \qquad\qquad \text{BY (255.20)}$$

6.  $\displaystyle\int_b^u \sqrt{\frac{x-d}{(a-x)(x-b)(x-c)}}\,dx = \frac{2}{\sqrt{(a-c)(b-d)}}\left\{(b-c)\Pi\left(\lambda,\frac{a-b}{a-c},r\right) + (c-d)F(\lambda,r)\right\}$

$$[a \geq u > b > c > d] \qquad\qquad \text{BY (256.13)}$$

7.  $\displaystyle\int_u^a \sqrt{\frac{x-d}{(a-x)(x-b)(x-c)}}\,dx = \frac{2(a-d)}{\sqrt{(a-c)(b-d)}}\,\Pi\left(\mu,\frac{b-a}{b-d},r\right)$

$$[a > u \geq b > c > d] \qquad\qquad \text{BY (257.02)}$$

8.  $\displaystyle\int_a^u \sqrt{\frac{x-d}{(x-a)(x-b)(x-c)}}\,dx = \frac{2}{\sqrt{(a-c)(b-d)}}\left\{(a-b)\Pi\left(\nu,\frac{a-d}{b-d},q\right) + (b-d)F(\nu,q)\right\}$

$$[u > a > b > c > d] \qquad\qquad \text{BY (258.14)}$$

9.  $\displaystyle\int_u^d \sqrt{\frac{c-x}{(a-x)(b-x)(d-x)}}\,dx = \frac{2(c-d)}{\sqrt{(a-c)(b-d)}}\,\Pi\left(\alpha,\frac{a-d}{a-c},q\right)$

$$[a > b > c > d > u] \qquad\qquad \text{BY (251.02)}$$

10. $\displaystyle\int_d^u \sqrt{\frac{c-x}{(a-x)(b-x)(x-d)}}\,dx = \frac{2}{\sqrt{(a-c)(b-d)}}\left[(a-d)\Pi\left(\beta,\frac{d-c}{a-c},r\right) - (a-c)F(\beta,r)\right]$

$$[a > b > c \geq u > d] \qquad\qquad \text{BY (252.13)}$$

11. $\displaystyle\int_u^c \sqrt{\frac{c-x}{(a-x)(b-x)(x-d)}}\,dx = \frac{2(b-c)}{\sqrt{(a-c)(b-d)}}\left[\Pi\left(\gamma,\frac{c-d}{b-d},r\right) - F(\gamma,r)\right]$

$$[a > b > c > u \geq d] \qquad\qquad \text{BY (253.13)}$$

12. $\displaystyle\int_c^u \sqrt{\frac{x-c}{(a-x)(b-x)(x-d)}}\,dx = \frac{2(c-d)}{\sqrt{(a-c)(b-d)}}\left[\Pi\left(\delta,\frac{b-c}{b-d},q\right) - F(\delta,q)\right]$

$$[a > b \geq u > c > d] \qquad\qquad \text{BY (254.12)}$$

13. $$\int_u^b \sqrt{\frac{x-c}{(a-x)(b-x)(x-d)}}\, dx = \frac{2}{\sqrt{(a-c)(b-d)}}\left[(b-a)\,\Pi\left(\kappa, \frac{b-c}{a-c}, q\right) + (a-c)\,F(\kappa, q)\right]$$
$$[a > b > u \geq c > d] \qquad \text{BY (259.19)}$$

14. $$\int_b^u \sqrt{\frac{x-c}{(a-x)(x-b)(x-d)}}\, dx = \frac{2(b-c)}{\sqrt{(a-c)(b-d)}}\,\Pi\left(\lambda, \frac{a-b}{a-c}, r\right)$$
$$[a \geq u > b > c > d] \qquad \text{BY (256.02)}$$

15. $$\int_u^a \sqrt{\frac{x-c}{(a-x)(x-b)(x-d)}}\, dx = \frac{2}{\sqrt{(a-c)(b-d)}}\left[(a-d)\,\Pi\left(\mu, \frac{b-a}{b-d}, r\right) + (d-c)\,F(\mu, r)\right]$$
$$[a > u \geq b > c > d] \qquad \text{BY (257.13)}$$

16. $$\int_a^u \sqrt{\frac{x-c}{(x-a)(x-b)(x-d)}}\, dx = \frac{2}{\sqrt{(a-c)(b-d)}}\left[(a-b)\,\Pi\left(\nu, \frac{a-d}{b-d}, q\right) + (b-c)\,F(\nu, q)\right]$$
$$[u > a > b > c > d] \qquad \text{BY (258.13)}$$

17. $$\int_u^d \sqrt{\frac{b-x}{(a-x)(c-x)(d-x)}}\, dx = \frac{2}{\sqrt{(a-c)(b-d)}}\left[(c-d)\,\Pi\left(\alpha, \frac{a-d}{a-c}, q\right) + (b-c)\,F(\alpha, q)\right]$$
$$[a > b > c > d > u] \qquad \text{BY (251.07)}$$

18. $$\int_d^u \sqrt{\frac{b-x}{(a-x)(c-x)(x-d)}}\, dx = \frac{2}{\sqrt{(a-c)(b-d)}}\left[(a-d)\,\Pi\left(\beta, \frac{d-c}{a-c}, r\right) - (a-b)\,F(\beta, r)\right]$$
$$[a > b > c \geq u > d] \qquad \text{BY (252.15)}$$

19. $$\int_u^c \sqrt{\frac{b-x}{(a-x)(c-x)(x-d)}}\, dx = \frac{2(b-c)}{\sqrt{(a-c)(b-d)}}\,\Pi\left(\gamma, \frac{c-d}{b-d}, r\right)$$
$$[a > b > c > u \geq d] \qquad \text{BY (253.02)}$$

20. $$\int_c^u \sqrt{\frac{b-x}{(a-x)(x-c)(x-d)}}\, dx = \frac{2}{\sqrt{(a-c)(b-d)}}\left[(d-c)\,\Pi\left(\delta, \frac{b-c}{b-d}, q\right) + (b-d)\,F(\delta, q)\right]$$
$$[a > b \geq u > c > d] \qquad \text{BY (254.14)}$$

21. $$\int_u^b \sqrt{\frac{b-x}{(a-x)(x-c)(x-d)}}\, dx = \frac{2(a-b)}{\sqrt{(a-c)(b-d)}}\left[\Pi\left(\kappa, \frac{b-c}{a-c}, q\right) - F(\kappa, q)\right]$$
$$[a > b > u \geq c > d] \qquad \text{BY (255.21)}$$

22. $$\int_b^u \sqrt{\frac{x-b}{(a-x)(x-c)(x-d)}}\, dx = \frac{2(b-c)}{\sqrt{(a-c)(b-d)}}\left[\Pi\left(\lambda, \frac{a-b}{a-c}, r\right) - F(\lambda, r)\right]$$
$$[a \geq u > b > c > d] \qquad \text{BY (256.15)}$$

23.[8] $$\int_u^a \sqrt{\frac{x-b}{(a-x)(x-c)(x-d)}}\, dx = \frac{2}{\sqrt{(a-c)(b-d)}}\left[(a-d)\,\Pi\left(\mu, \frac{b-a}{b-d}, r\right) - (b-d)\,F(\mu, r)\right]$$
$$[a > u \geq b > c > d] \qquad \text{BY (257.15)}$$

24.    $\displaystyle\int_a^u \sqrt{\frac{x-b}{(x-a)(x-c)(x-d)}}\,\mathrm{d}x = \frac{2(a-b)}{\sqrt{(a-c)(b-d)}}\,\Pi\left(\nu,\frac{a-d}{b-d},q\right)$

$$[u > a > b > c > d] \qquad \text{BY (258.02)}$$

25.    $\displaystyle\int_u^d \sqrt{\frac{a-x}{(b-x)(c-x)(d-x)}}\,\mathrm{d}x = \frac{2}{\sqrt{(a-c)(b-d)}}\left[(c-d)\,\Pi\left(\alpha,\frac{a-d}{a-c},q\right) + (a-c)\,F(\alpha,q)\right]$

$$[a > b > c > d > u] \qquad \text{BY (251.06)}$$

26.    $\displaystyle\int_d^u \sqrt{\frac{a-x}{(b-x)(c-x)(x-d)}}\,\mathrm{d}x = \frac{2(a-d)}{\sqrt{(a-c)(b-d)}}\,\Pi\left(\beta,\frac{d-c}{a-c},r\right)$

$$[a > b > c \geq u > d] \qquad \text{BY (252.02)}$$

27.    $\displaystyle\int_u^c \sqrt{\frac{a-x}{(b-x)(c-x)(x-d)}}\,\mathrm{d}x = \frac{2}{\sqrt{(a-c)(b-d)}}\left[(b-c)\,\Pi\left(\gamma,\frac{c-d}{b-d},r\right) + (a-b)\,F(\gamma,r)\right]$

$$[a > b > c > u \geq d] \qquad \text{BY (253.15)}$$

28.    $\displaystyle\int_c^u \sqrt{\frac{a-x}{(b-x)(x-c)(x-d)}}\,\mathrm{d}x = \frac{2}{\sqrt{(a-c)(b-d)}}\left[(d-c)\,\Pi\left(\delta,\frac{b-c}{b-d},q\right) + (a-d)\,F(\delta,q)\right]$

$$[a > b \geq u > c > d] \qquad \text{BY (254.13)}$$

29.    $\displaystyle\int_u^b \sqrt{\frac{a-x}{(b-x)(x-c)(x-d)}}\,\mathrm{d}x = \frac{2(a-b)}{\sqrt{(a-c)(b-d)}}\,\Pi\left(\kappa,\frac{b-c}{a-c},q\right)$

$$[a > b > u \geq c > d] \qquad \text{BY (255.02)}$$

30.    $\displaystyle\int_b^u \sqrt{\frac{a-x}{(x-b)(x-c)(x-d)}}\,\mathrm{d}x = \frac{2}{\sqrt{(a-c)(b-d)}}\left[(c-b)\,\Pi\left(\lambda,\frac{a-b}{a-c},r\right) + (a-c)\,F(\lambda,r)\right]$

$$[a \geq u > b > c > d] \qquad \text{BY (256.14)}$$

31.    $\displaystyle\int_u^a \sqrt{\frac{a-x}{(x-b)(x-c)(x-d)}}\,\mathrm{d}x = \frac{2(d-a)}{\sqrt{(a-c)(b-d)}}\left[\Pi\left(\mu,\frac{b-a}{b-d},r\right) - F(\mu,r)\right]$

$$[a > u \geq b > c > d] \qquad \text{BY (257.14)}$$

32.    $\displaystyle\int_a^u \sqrt{\frac{x-a}{(x-b)(x-c)(x-d)}}\,\mathrm{d}x = \frac{2(a-b)}{\sqrt{(a-c)(b-d)}}\left[\Pi\left(\nu,\frac{a-d}{b-d},q\right) - F(\nu,q)\right]$

$$[u > a > b > c > d] \qquad \text{BY (258.15)}$$

**3.168**

1.    $\displaystyle\int_u^c \sqrt{\frac{c-x}{(a-x)(b-x)(x-d)^3}}\,\mathrm{d}x = \frac{2}{d-a}\left[\sqrt{\frac{a-c}{b-d}}\,E(\gamma,r) - \sqrt{\frac{(a-u)(c-u)}{(b-u)(u-d)}}\right]$

$$[a > b > c > u > d] \qquad \text{BY (253.06)}$$

2.    $\displaystyle\int_c^u \sqrt{\frac{x-c}{(a-x)(b-x)(x-d)^3}}\,\mathrm{d}x = \frac{2}{a-d}\sqrt{\frac{a-c}{b-d}}\,[F(\delta,q) - E(\delta,q)]$

$$[a > b \geq u > c > d] \qquad \text{BY (254.04)}$$

3. $$\int_u^b \sqrt{\frac{x-c}{(a-x)(b-x)(x-d)^3}}\,\mathrm{d}x = \frac{2}{a-d}\sqrt{\frac{a-c}{b-d}}\left[F\left(\kappa,q\right)-E\left(\kappa,q\right)\right]+\frac{2}{b-d}\sqrt{\frac{(b-u)(u-c)}{(a-u)(u-d)}}$$

$$[a>b>u\geq c>d] \qquad \text{BY (255.09)}$$

4. $$\int_b^u \sqrt{\frac{x-c}{(a-x)(x-b)(x-d)^3}}\,\mathrm{d}x = \frac{2}{a-d}\left[\sqrt{\frac{a-c}{b-d}}\,E(\lambda,r)-\frac{c-d}{b-d}\sqrt{\frac{(a-u)(u-b)}{(u-c)(u-d)}}\right]$$

$$[a\geq u>b>c>d] \qquad \text{BY (256.06)}$$

5. $$\int_u^a \sqrt{\frac{x-c}{(a-x)(x-b)(x-d)^3}}\,\mathrm{d}x = \frac{2}{a-d}\sqrt{\frac{a-c}{b-d}}\,E(\mu,r)$$

$$[a>u\geq b>c>d] \qquad \text{BY (257.01)}$$

6. $$\int_a^u \sqrt{\frac{x-c}{(x-a)(x-b)(x-d)^3}}\,\mathrm{d}x = \frac{2}{a-d}\sqrt{\frac{a-c}{b-d}}\left[F\left(\nu,q\right)-E\left(\nu,q\right)\right]$$

$$+\frac{2}{a-d}\sqrt{\frac{(u-a)(u-c)}{(u-b)(u-d)}}$$

$$[u>a>b>c>d] \qquad \text{BY (258.10)}$$

7. $$\int_u^c \sqrt{\frac{b-x}{(a-x)(c-x)(x-d)^3}}\,\mathrm{d}x = \frac{2}{(a-d)(c-d)\sqrt{(a-c)(b-d)}}$$

$$\times\left[(b-c)(a-d)\,F(\gamma,r)-(a-c)(b-d)\,E\left(\gamma,r\right)\right]$$

$$+\frac{2(b-d)}{(a-d)(c-d)}\sqrt{\frac{(a-u)(c-u)}{(b-u)(u-d)}}$$

$$[a>b>c>u>d] \qquad \text{BY (253.03)}$$

8. $$\int_c^u \sqrt{\frac{b-x}{(a-x)(x-c)(x-d)^3}}\,\mathrm{d}x = \frac{2}{(a-d)(c-d)\sqrt{(a-c)(b-d)}}$$

$$\times\left[(a-c)(b-d)\,E(\delta,q)-(a-b)(c-d)\,F\left(\delta,q\right)\right]$$

$$[a>b\geq u>c>d] \qquad \text{BY (254.15)}$$

9. $$\int_u^b \sqrt{\frac{b-x}{(a-x)(x-c)(x-d)^3}}\,\mathrm{d}x = \frac{2}{(a-d)(c-d)\sqrt{(a-c)(b-d)}}$$

$$\times\left[(a-c)(b-d)\,E\left(\kappa,q\right)-(a-b)(c-d)\,F\left(\kappa,q\right)\right]$$

$$-\frac{2}{c-d}\sqrt{\frac{(b-u)(u-c)}{(a-u)(u-d)}}$$

$$[a>b>u\geq c>d] \qquad \text{BY (255.06)}$$

10. $\int_b^u \sqrt{\dfrac{x-b}{(a-x)(x-c)(x-d)^3}}\, \mathrm{d}x = \dfrac{2}{(a-d)(c-d)\sqrt{(a-c)(b-d)}}$

$$\times [(a-c)(b-d)\, E(\lambda,r) - (a-d)(b-c)\, F(\lambda,r)]$$

$$-\dfrac{2}{a-d}\sqrt{\dfrac{(a-u)(u-b)}{(u-c)(u-d)}}$$

$$[a \geq u > b > c > d] \qquad \text{BY (256.03)}$$

11. $\int_u^a \sqrt{\dfrac{x-b}{(a-x)(x-c)(x-d)^3}}\, \mathrm{d}x = 2\dfrac{\sqrt{(a-c)(b-d)}}{(a-d)(c-d)}\, E(\mu,r)$

$$-\dfrac{2(b-c)}{(c-d)\sqrt{(a-c)(b-d)}}\, F(\mu,r)$$

$$[a > u \geq b > c > d] \qquad \text{BY (257.09)}$$

12. $\int_a^u \sqrt{\dfrac{x-b}{(x-a)(x-c)(x-d)^3}}\, \mathrm{d}x$

$$= \dfrac{2(b-d)}{(a-d)(c-d)}\sqrt{\dfrac{(u-a)(u-c)}{(u-b)(u-d)}} + \dfrac{2(a-b)}{(a-d)\sqrt{(a-c)(b-d)}}\, F(\nu,q)$$

$$+2\dfrac{\sqrt{(a-c)(b-d)}}{(a-d)(c-d)}\, E(\nu,q)$$

$$[u > a > b > c > d] \qquad \text{BY (258.09)}$$

13. $\int_u^c \sqrt{\dfrac{a-x}{(b-x)(c-x)(x-d)^3}}\, \mathrm{d}x = \dfrac{2}{c-d}\sqrt{\dfrac{a-c}{b-d}}\, [F(\gamma,r) - E(\gamma,r)] + \dfrac{2}{c-d}\sqrt{\dfrac{(a-u)(c-u)}{(b-u)(u-d)}}$

$$[a > b > c > u > d] \qquad \text{BY (253.04)}$$

14. $\int_c^u \sqrt{\dfrac{a-x}{(b-x)(x-c)(x-d)^3}}\, \mathrm{d}x = \dfrac{2}{c-d}\sqrt{\dfrac{a-c}{b-d}}\, E(\delta,q)$

$$[a > b \geq u > c > d] \qquad \text{BY (254.01)}$$

15. $\int_u^b \sqrt{\dfrac{a-x}{(b-x)(x-c)(x-d)^3}}\, \mathrm{d}x = \dfrac{2}{c-d}\sqrt{\dfrac{a-c}{b-d}}\, E(\kappa,q) - \dfrac{2(a-d)}{(b-d)(c-d)}\sqrt{\dfrac{(b-u)(u-c)}{(a-u)(u-d)}}$

$$[a > b > u \geq c > d] \qquad \text{BY (255.08)}$$

16. $\int_b^u \sqrt{\dfrac{a-x}{(x-b)(x-c)(x-d)^3}}\, \mathrm{d}x = \dfrac{2}{c-d}\sqrt{\dfrac{a-c}{b-d}}\, [F(\lambda,r) - E(\lambda,r)] + \dfrac{2}{b-d}\sqrt{\dfrac{(a-u)(u-b)}{(u-c)(u-d)}}$

$$[a \geq u > b > c > d] \qquad \text{BY (256.05)}$$

17. $\int_u^a \sqrt{\dfrac{a-x}{(x-b)(x-c)(x-d)^3}}\, \mathrm{d}x = \dfrac{2}{c-d}\sqrt{\dfrac{a-c}{b-d}}\, [F(\mu,r) - E(\mu,r)]$

$$[a > u \geq b > c > d] \qquad \text{BY (257.06)}$$

18. $\int_a^u \sqrt{\dfrac{x-a}{(x-b)(x-c)(x-d)^3}}\, \mathrm{d}x = \dfrac{-2}{c-d}\sqrt{\dfrac{a-c}{b-d}}\, E(\nu,q) + \dfrac{2}{c-d}\sqrt{\dfrac{(u-a)(u-c)}{(u-b)(u-d)}}$

$$[u > a > b > c > d] \qquad \text{BY (258.05)}$$

19. $\int_u^d \sqrt{\dfrac{d-x}{(a-x)(b-x)(c-x)^3}}\, dx = \dfrac{2}{b-c}\sqrt{\dfrac{b-d}{a-c}}\,[F(\alpha,q)-E(\alpha,q)]$

$$[a>b>c>d>u] \qquad \text{BY (251.01)}$$

20. $\int_d^u \sqrt{\dfrac{x-d}{(a-x)(b-x)(c-x)^3}}\, dx = \dfrac{-2}{b-c}\sqrt{\dfrac{b-d}{a-c}}\,E(\beta,r)+\dfrac{2}{b-c}\sqrt{\dfrac{(b-u)(u-d)}{(a-u)(c-u)}}$

$$[a>b>c\geq u>d] \qquad \text{BY (252.06)}$$

21. $\int_u^b \sqrt{\dfrac{x-d}{(a-x)(b-x)(x-c)^3}}\, dx = \dfrac{2}{b-c}\sqrt{\dfrac{b-d}{a-c}}\,[F(\kappa,q)-E(\kappa,q)]+\dfrac{2}{b-c}\sqrt{\dfrac{(b-u)(u-d)}{(a-u)(u-c)}}$

$$[a>b>u>c>d] \qquad \text{BY (255.05)}$$

22. $\int_b^u \sqrt{\dfrac{x-d}{(a-x)(x-b)(x-c)^3}}\, dx = \dfrac{2}{b-c}\sqrt{\dfrac{b-d}{a-c}}\,E(\lambda,r)$

$$[a\geq u>b>c>d] \qquad \text{BY (256.01)}$$

23. $\int_u^a \sqrt{\dfrac{x-d}{(a-x)(x-b)(x-c)^3}}\, dx = \dfrac{2}{b-c}\sqrt{\dfrac{b-d}{a-c}}\,E(\mu,r)-\dfrac{2(c-d)}{(a-c)(b-c)}\sqrt{\dfrac{(a-u)(u-b)}{(u-c)(u-d)}}$

$$[a>u\geq b>c>d] \qquad \text{BY (257.06)}$$

24. $\int_a^u \sqrt{\dfrac{x-d}{(x-a)(x-b)(x-c)^3}}\, dx = \dfrac{2}{b-c}\sqrt{\dfrac{b-d}{a-c}}\,[F(\nu,q)-E(\nu,q)]+\dfrac{2}{a-c}\sqrt{\dfrac{(u-a)(u-d)}{(u-b)(u-c)}}$

$$[u>a>b>c>d] \qquad \text{BY (258.06)}$$

25. $\int_u^a \sqrt{\dfrac{b-x}{(a-x)(c-x)^3(d-x)}}\, dx = \dfrac{2}{c-d}\sqrt{\dfrac{b-d}{a-c}}\,E(\alpha,q)$

$$[a>b>c>d>u] \qquad \text{BY (251.01)}$$

26. $\int_d^u \sqrt{\dfrac{b-x}{(a-x)(c-x)^3(x-d)}}\, dx = \dfrac{2}{c-d}\sqrt{\dfrac{b-d}{a-c}}\,[F(\beta,r)-E(\beta,r)]+\dfrac{2}{c-d}\sqrt{\dfrac{(b-u)(u-d)}{(a-u)(c-u)}}$

$$[a>b>c>u>d] \qquad \text{BY (252.03)}$$

27. $\int_u^b \sqrt{\dfrac{b-x}{(a-x)(x-c)^3(x-d)}}\, dx = \dfrac{2}{d-c}\sqrt{\dfrac{b-d}{a-c}}\,E(\kappa,q)+\dfrac{2}{c-d}\sqrt{\dfrac{(b-u)(u-d)}{(a-u)(u-c)}}$

$$[a>b>u>c>d] \qquad \text{BY (255.03)}$$

28. $\int_b^u \sqrt{\dfrac{x-b}{(a-x)(x-c)^3(x-d)}}\, dx = \dfrac{2}{c-d}\sqrt{\dfrac{b-d}{a-c}}\,[F(\lambda,r)-E(\lambda,r)]$

$$[a\geq u>b>c>d] \qquad \text{BY (256.08)}$$

29. $\int_u^a \sqrt{\dfrac{x-b}{(a-x)(x-c)^3(x-d)}}\, dx = \dfrac{2}{c-d}\sqrt{\dfrac{b-d}{a-c}}\,[F(\mu,r)-E(\mu,r)]+\dfrac{2}{a-c}\sqrt{\dfrac{(a-u)(u-b)}{(u-c)(u-d)}}$

$$[a>u\geq b>c>d] \qquad \text{BY (257.03)}$$

30. $\displaystyle\int_a^u \sqrt{\frac{x-b}{(x-a)(x-c)^3(x-d)}}\,\mathrm{d}x = \frac{2}{c-d}\sqrt{\frac{b-d}{a-c}}\,E(\nu,q) - \frac{2(b-c)}{(a-c)(c-d)}\sqrt{\frac{(u-a)(u-d)}{(u-b)(u-c)}}$

$$[u > a > b > c > d] \qquad \text{BY (258.03)}$$

31. $\displaystyle\int_u^d \sqrt{\frac{a-x}{(b-x)(c-x)^3(d-x)}}\,\mathrm{d}x = \frac{2\sqrt{(a-c)(b-d)}}{(b-c)(c-d)}\,E(\alpha,q) - \frac{a-b}{b-c}\frac{2}{\sqrt{(a-c)(b-d)}}\,F(\alpha,q)$

$$[a > b > c > d > u] \qquad \text{BY (251.08)}$$

32. $\displaystyle\int_d^u \sqrt{\frac{a-x}{(b-x)(c-x)^3(x-d)}}\,\mathrm{d}x = \frac{2(a-d)}{(c-d)\sqrt{(a-c)(b-d)}}\,F(\beta,r) - 2\frac{\sqrt{(a-c)(b-d)}}{(b-c)(c-d)}\,E(\beta,r)$

$$+2\frac{a-c}{(b-c)(c-d)}\sqrt{\frac{(b-u)(u-d)}{(a-u)(c-u)}}$$

$$[a > b > c > u > d] \qquad \text{BY (252.04)}$$

33. $\displaystyle\int_u^b \sqrt{\frac{a-x}{(b-x)(x-c)^3(x-d)}}\,\mathrm{d}x = \frac{2(a-b)}{(b-c)\sqrt{(a-c)(b-d)}}\,F(\kappa,q) - 2\sqrt{\frac{(a-c)(b-d)}{(b-c)(c-d)}}\,E(\kappa,q)$

$$+\frac{2(a-c)}{(b-c)(c-d)}\sqrt{\frac{(b-u)(u-d)}{(a-u)(u-c)}}$$

$$[a > b > u > c > d] \qquad \text{BY (255.04)}$$

34. $\displaystyle\int_b^u \sqrt{\frac{a-x}{(x-b)(x-c)^3(x-d)}}\,\mathrm{d}x = \frac{2\sqrt{(a-c)(b-d)}}{(b-c)(c-d)}\,E(\lambda,r) - \frac{2(a-d)}{(c-d)\sqrt{(a-c)(b-d)}}\,F(\lambda,r)$

$$[a \geq u > b > c > d] \qquad \text{BY (256.09)}$$

35. $\displaystyle\int_u^a \sqrt{\frac{a-x}{(x-b)(x-c)^3(x-d)}}\,\mathrm{d}x = \frac{2\sqrt{(a-c)(b-d)}}{(b-c)(c-d)}\,E(\mu,r) - \frac{2(a-d)}{(c-d)\sqrt{(a-c)(b-d)}}\,F(\mu,r)$

$$-\frac{2}{b-c}\sqrt{\frac{(a-u)(u-b)}{(u-c)(u-d)}}$$

$$[a > u \geq b > c > d] \qquad \text{BY (257.04)}$$

36. $\displaystyle\int_a^u \sqrt{\frac{x-a}{(x-b)(x-c)^3(x-d)}}\,\mathrm{d}x = \frac{2\sqrt{(a-c)(b-d)}}{(b-c)(c-d)}\,E(\nu,q) - \frac{2(a-b)}{(b-c)\sqrt{(a-c)(b-d)}}\,F(\nu,q)$

$$-\frac{2}{c-d}\sqrt{\frac{(u-a)(u-d)}{(u-b)(u-c)}}$$

$$[u > a > b > c > d] \qquad \text{BY (258.04)}$$

37. $\displaystyle\int_u^d \sqrt{\frac{d-x}{(a-x)(b-x)^3(c-x)}}\,\mathrm{d}x = \frac{2\sqrt{(a-c)(b-d)}}{(a-b)(b-c)}\,E(\alpha,q) - \frac{2(c-d)}{(b-c)\sqrt{(a-c)(b-d)}}\,F(\alpha,q)$

$$-\frac{2}{a-b}\sqrt{\frac{(a-u)(d-u)}{(b-u)(c-u)}}$$

$$[a > b > c > d > u] \qquad \text{BY (251.11)}$$

38. $$\int_d^u \sqrt{\frac{x-d}{(a-x)(b-x)^3(c-x)}}\, dx = \frac{2\sqrt{(a-c)(b-d)}}{(a-b)(b-c)} E(\beta,r) - \frac{2(a-d)}{(a-b)\sqrt{(a-c)(b-d)}} F(\beta,r)$$
$$+ \frac{2}{b-c}\sqrt{\frac{(c-u)(u-d)}{(a-u)(b-u)}}$$
$$[a>b>c\geq u>d] \qquad\qquad \text{BY (252.07)}$$

39. $$\int_u^c \sqrt{\frac{x-d}{(a-x)(b-x)^3(c-x)}}\, dx = \frac{2\sqrt{(a-c)(b-d)}}{(a-b)(b-c)} E(\gamma,r) - \frac{2(a-d)}{(a-b)\sqrt{(a-c)(b-d)}} F(\gamma,r)$$
$$[a>b>c>u\geq d] \qquad\qquad \text{BY (253.07)}$$

40. $$\int_c^u \sqrt{\frac{x-d}{(a-x)(b-x)^3(x-c)}}\, dx = \frac{2(c-d)}{(b-c)\sqrt{(a-c)(b-d)}} F(\delta,q) - \frac{2\sqrt{(a-c)(b-d)}}{(a-b)(b-c)} E(\delta,q)$$
$$+ \frac{2(b-d)}{(a-b)(b-c)}\sqrt{\frac{(a-u)(u-c)}{(b-u)(u-d)}}$$
$$[a>b>u>c>d] \qquad\qquad \text{BY (254.05)}$$

41. $$\int_u^a \sqrt{\frac{x-d}{(a-x)(x-b)^3(x-c)}}\, dx = \frac{2(a-d)}{(a-b)\sqrt{(a-c)(b-d)}} F(\mu,r) - \frac{2\sqrt{(a-c)(b-d)}}{(a-b)(b-c)} E(\mu,r)$$
$$+ \frac{2(b-d)}{(a-b)(b-c)}\sqrt{\frac{(a-u)(u-c)}{(u-b)(u-d)}}$$
$$[a>u>b>c>d] \qquad\qquad \text{BY (257.07)}$$

42. $$\int_a^u \sqrt{\frac{x-d}{(x-a)(x-b)^3(x-c)}}\, dx = \frac{2\sqrt{(a-c)(b-d)}}{(a-b)(b-c)} E(\nu,q) - \frac{2(c-d)}{(b-c)\sqrt{(a-c)(b-d)}} F(\nu,q)$$
$$[u>a>b>c>d] \qquad\qquad \text{BY (258.07)}$$

43. $$\int_u^d \sqrt{\frac{c-x}{(a-x)(b-x)^3(d-x)}}\, dx = \frac{2}{a-b}\sqrt{\frac{a-c}{b-d}} E(\alpha,q) - \frac{2(b-c)}{(a-b)(b-d)}\sqrt{\frac{(a-u)(d-u)}{(b-u)(c-u)}}$$
$$[a>b>c>d>u] \qquad\qquad \text{BY (...)}$$

44. $$\int_d^u \sqrt{\frac{c-x}{(a-x)(b-x)^3(x-d)}}\, dx = \frac{2}{a-b}\sqrt{\frac{a-c}{b-d}} [F(\beta,r) - E(\beta,r)] + \frac{2}{b-d}\sqrt{\frac{(c-u)(u-d)}{(a-u)(b-u)}}$$
$$[a>b>c\geq u>d] \qquad\qquad \text{BY (252.10)}$$

45. $$\int_u^c \sqrt{\frac{c-x}{(a-x)(b-x)^3(x-d)}}\, dx = \frac{2}{a-b}\sqrt{\frac{a-c}{b-d}} [F(\gamma,r) - E(\gamma,r)]$$
$$[a>b>c>u\geq d] \qquad\qquad \text{BY (254.08)}$$

46. $$\int_c^u \sqrt{\frac{x-c}{(a-x)(b-x)^3(x-d)}}\, dx = \frac{2}{b-a}\sqrt{\frac{a-c}{b-d}} E(\delta,q) + \frac{2}{a-b}\sqrt{\frac{(a-u)(u-c)}{(b-u)(u-d)}}$$
$$[a>b\geq u>c>d] \qquad\qquad \text{BY (254.08)}$$

47. $$\int_u^a \sqrt{\frac{x-c}{(a-x)(x-b)^3(x-d)}}\, dx = \frac{2}{a-b}\sqrt{\frac{a-c}{b-d}} [F(\mu,r) - E(\mu,r)] + \frac{2}{a-b}\sqrt{\frac{(a-u)(u-c)}{(u-b)(u-d)}}$$
$$[a>u\geq b>c>d] \qquad\qquad \text{BY (257.10)}$$

48.
$$\int_a^u \sqrt{\frac{x-c}{(x-a)(x-b)^3(x-d)}}\,dx = \frac{2}{a-b}\sqrt{\frac{a-c}{b-d}}\,E(\nu,q)$$

$$[u > a > b > c > d] \qquad \text{BY (258.01)}$$

49.
$$\int_u^d \sqrt{\frac{a-x}{(b-x)^3(c-x)(d-x)}}\,dx = \frac{2}{b-c}\sqrt{\frac{a-c}{b-d}}\,[F(\alpha,q) - E(\alpha,q)] + \frac{2}{b-d}\sqrt{\frac{(a-u)(d-u)}{(b-u)(c-u)}}$$

$$[a > b > c > d > u] \qquad \text{BY (251.12)}$$

50.
$$\int_d^u \sqrt{\frac{a-x}{(b-x)^3(c-x)(x-d)}}\,dx = \frac{2}{b-c}\sqrt{\frac{a-c}{b-d}}\,E(\beta,r) - \frac{2(a-b)}{(b-c)(b-d)}\sqrt{\frac{(u-d)(c-u)}{(a-u)(b-u)}}$$

$$[a > b > c \geq u > d] \qquad \text{BY (252.09)}$$

51.
$$\int_u^c \sqrt{\frac{a-x}{(b-x)^3(c-x)(x-d)}}\,dx = \frac{2}{b-c}\sqrt{\frac{a-c}{b-d}}\,E(\gamma,r)$$

$$[a > b > c > u \geq d] \qquad \text{BY (253.01)}$$

52.
$$\int_c^u \sqrt{\frac{a-x}{(b-x)^3(x-c)(x-d)}}\,dx = \frac{2}{b-c}\sqrt{\frac{a-c}{b-d}}\,[F(\delta,q) - E(\delta,q)] + \frac{2}{b-c}\sqrt{\frac{(a-u)(u-c)}{(b-u)(u-d)}}$$

$$[a > b > u > c > d] \qquad \text{BY (254.06)}$$

53.
$$\int_u^a \sqrt{\frac{a-x}{(x-b)^3(x-c)(x-d)}}\,dx = \frac{2}{c-b}\sqrt{\frac{a-c}{b-d}}\,E(\mu,r) + \frac{2}{b-c}\sqrt{\frac{(a-u)(u-c)}{(u-b)(u-d)}}$$

$$[a > u > b > c > d] \qquad \text{BY (257.08)}$$

54.
$$\int_a^u \sqrt{\frac{x-a}{(x-b)^3(x-c)(x-d)}}\,dx = \frac{2}{b-c}\sqrt{\frac{a-c}{b-d}}\,[F(\nu,q) - E(\nu,q)]$$

$$[u > a > b > c > d] \qquad \text{BY (258.08)}$$

55.
$$\int_u^d \sqrt{\frac{d-x}{(a-x)^3(b-x)(c-x)}}\,dx = \frac{2}{b-a}\sqrt{\frac{b-d}{a-c}}\,E(\alpha,q) + \frac{2}{a-b}\sqrt{\frac{(b-u)(d-u)}{(a-u)(c-u)}}$$

$$[a > b > c > d > u] \qquad \text{BY (251.09)}$$

56.
$$\int_d^u \sqrt{\frac{x-d}{(a-x)^3(b-x)(c-x)}}\,dx = \frac{2}{a-b}\sqrt{\frac{b-d}{a-c}}\,[F(\beta,q) - E(\beta,q)]$$

$$[a > b > c \geq u > d] \qquad \text{BY (252.05)}$$

57.
$$\int_u^c \sqrt{\frac{x-d}{(a-x)^3(b-x)(c-x)}}\,dx = \frac{2}{a-b}\sqrt{\frac{b-d}{a-c}}\,[F(\gamma,r) - E(\gamma,r)] + \frac{2}{a-c}\sqrt{\frac{(c-u)(u-d)}{(a-u)(b-u)}}$$

$$[a > b > c > u \geq d] \qquad \text{BY (253.05)}$$

58.
$$\int_c^u \sqrt{\frac{x-d}{(a-x)^3(b-x)(x-c)}}\,dx = \frac{2}{a-b}\sqrt{\frac{b-d}{a-c}}\,E(\delta,q) - \frac{2(a-d)}{(a-b)(a-c)}\sqrt{\frac{(b-u)(u-c)}{(a-u)(u-d)}}$$

$$[a > b \geq u > c > d] \qquad \text{BY (254.03)}$$

59. 
$$\int_u^b \sqrt{\frac{x-d}{(a-x)^3(b-x)(x-c)}}\,dx = \frac{2}{a-b}\sqrt{\frac{b-d}{a-c}}\,E\left(\kappa,q\right)$$

$$[a > b > u \ge c > d] \qquad \text{BY (255.01)}$$

60. 
$$\int_b^u \sqrt{\frac{x-d}{(a-x)^3(x-b)(x-c)}}\,dx = \frac{2}{a-b}\sqrt{\frac{b-d}{a-c}}\left[F(\lambda,r) - E(\lambda,r)\right] + \frac{2}{a-b}\sqrt{\frac{(u-b)(u-d)}{(a-u)(u-c)}}$$

$$[a > u > b > c > d] \qquad \text{BY (256.10)}$$

61. 
$$\int_u^d \sqrt{\frac{c-x}{(a-x)^3(b-x)(d-x)}}\,dx = \frac{2(c-d)}{(a-d)\sqrt{(a-c)(b-d)}}\,F(\alpha,q) - \frac{2\sqrt{(a-c)(b-d)}}{(a-b)(a-d)}\,E(\alpha,q)$$
$$+ \frac{2(a-c)}{(a-b)(a-d)}\sqrt{\frac{(b-u)(d-u)}{(a-u)(c-u)}}$$

$$[a > b > c > d > u] \qquad \text{BY (251.15)}$$

62. 
$$\int_d^u \sqrt{\frac{c-x}{(a-x)^3(b-x)(x-d)}}\,dx = \frac{2\sqrt{(a-c)(b-d)}}{(a-b)(a-d)}\,E(\beta,r) - \frac{2(b-c)}{(a-b)\sqrt{(a-c)(b-d)}}\,F(\beta,r)$$

$$[a > b > c \ge u > d] \qquad \text{BY (252.08)}$$

63. 
$$\int_u^c \sqrt{\frac{c-x}{(a-x)^3(b-x)(x-d)}}\,dx = \frac{2\sqrt{(a-c)(b-d)}}{(a-b)(a-d)}\,E(\gamma,r) - \frac{2(b-c)}{(a-b)\sqrt{(a-c)(b-d)}}\,F(\gamma,r)$$
$$- \frac{2}{a-d}\sqrt{\frac{(c-u)(u-d)}{(a-u)(b-u)}}$$

$$[a > b > c > u \ge d] \qquad \text{BY (253.10)}$$

64. 
$$\int_c^u \sqrt{\frac{x-c}{(a-x)^3(b-x)(x-d)}}\,dx = \frac{2\sqrt{(a-c)(b-d)}}{(a-b)(a-d)}\,E(\delta,q) - \frac{2(c-d)}{(a-d)\sqrt{(a-c)(b-d)}}\,F(\delta,q)$$
$$- \frac{2}{a-b}\sqrt{\frac{(b-u)(u-c)}{(a-u)(u-d)}}$$

$$[a > b \ge u > c > d] \qquad \text{BY (254.09)}$$

65. 
$$\int_u^b \sqrt{\frac{x-c}{(a-x)^3(b-x)(x-d)}}\,dx = \frac{2\sqrt{(a-c)(b-d)}}{(a-b)(a-d)}\,E\left(\kappa,q\right) - \frac{2(c-d)}{(a-d)\sqrt{(a-c)(b-d)}}\,F\left(\kappa,q\right)$$

$$[a > b > u \ge c > d] \qquad \text{BY (255.10)}$$

66. 
$$\int_b^u \sqrt{\frac{x-c}{(a-x)^3(x-b)(x-d)}}\,dx = \frac{2(b-c)}{(a-b)\sqrt{(a-c)(b-d)}}\,F(\lambda,r) - \frac{2\sqrt{(a-c)(b-d)}}{(a-b)(a-d)}\,E(\lambda,r)$$
$$+ \frac{2(a-c)}{(a-b)(a-d)}\sqrt{\frac{(u-b)(u-d)}{(a-u)(u-c)}}$$

$$[a > u > b > c > d] \qquad \text{BY (256.07)}$$

67. 
$$\int_u^d \sqrt{\frac{b-x}{(a-x)^3(c-x)(d-x)}}\,dx = \frac{2}{a-d}\sqrt{\frac{b-d}{a-c}}\left[F(\alpha,q) - E(\alpha,q)\right] + \frac{2}{a-d}\sqrt{\frac{(b-u)(d-u)}{(a-u)(c-u)}}$$

$$[a > b > c > d > u] \qquad \text{BY (251.13)}$$

68.   $\displaystyle\int_d^u \sqrt{\frac{b-x}{(a-x)^3(c-x)(x-d)}}\, dx = \frac{2}{a-d}\sqrt{\frac{b-d}{a-c}}\, E(\beta, r)$

$$[a > b > c \geq u > d] \qquad \text{BY (252.01)}$$

69.   $\displaystyle\int_u^c \sqrt{\frac{b-x}{(a-x)^3(c-x)(x-d)}}\, dx = \frac{2}{a-d}\sqrt{\frac{b-d}{a-c}}\, E(\gamma, r) - \frac{2(a-b)}{(a-c)(a-d)}\sqrt{\frac{(c-u)(u-d)}{(a-u)(b-u)}}$

$$[a > b > c > u \geq d] \qquad \text{BY (253.08)}$$

70.   $\displaystyle\int_c^u \sqrt{\frac{b-x}{(a-x)^3(x-c)(x-d)}}\, dx = \frac{2}{a-d}\sqrt{\frac{b-d}{a-c}}\, [F(\delta, q) - E(\delta, q)] + \frac{2}{a-c}\sqrt{\frac{(b-u)(u-c)}{(a-u)(u-d)}}$

$$[a > b \geq u > c > d] \qquad \text{BY (254.07)}$$

71.   $\displaystyle\int_u^b \sqrt{\frac{b-x}{(a-x)^3(x-c)(x-d)}}\, dx = \frac{2}{a-d}\sqrt{\frac{b-d}{a-c}}\, [F(\kappa, q) - E(\kappa, q)]$

$$[a > b > u \geq c > d] \qquad \text{BY (255.07)}$$

72.   $\displaystyle\int_b^u \sqrt{\frac{x-b}{(a-x)^3(x-c)(x-d)}}\, dx = \frac{-2}{a-d}\sqrt{\frac{b-d}{a-c}}\, E(\lambda, r) + \frac{2}{a-d}\sqrt{\frac{(u-b)(u-d)}{(a-u)(u-c)}}$

$$[a \geq u > b > c > d] \qquad \text{BY (256.04)}$$

**3.169**   **Notation:** In **3.169–3.172**, we set:   $\alpha = \arctan\dfrac{u}{b}, \quad \beta = \arctan\dfrac{a}{u},$

$$\gamma = \arcsin\frac{u}{b}\sqrt{\frac{a^2+b^2}{a^2+u^2}}, \qquad \delta = \arccos\frac{u}{b}, \qquad \varepsilon = \arccos\frac{b}{u}, \qquad \xi = \arcsin\sqrt{\frac{a^2+b^2}{a^2+u^2}},$$

$$\eta = \arcsin\frac{u}{b}, \qquad \zeta = \arcsin\frac{a}{b}\sqrt{\frac{b^2-u^2}{a^2-u^2}}, \qquad \kappa = \arcsin\frac{a}{u}\sqrt{\frac{u^2-b^2}{a^2-b^2}},$$

$$\lambda = \arcsin\sqrt{\frac{a^2-u^2}{a^2-b^2}}, \qquad \mu = \arcsin\sqrt{\frac{u^2-a^2}{u^2-b^2}}, \qquad \nu = \arcsin\frac{a}{u}, \qquad q = \frac{\sqrt{a^2-b^2}}{a},$$

$$r = \frac{b}{\sqrt{a^2+b^2}}, \qquad s = \frac{a}{\sqrt{a^2+b^2}} \qquad t = \frac{b}{a}.$$

1.   $\displaystyle\int_0^u \sqrt{\frac{x^2+a^2}{x^2+b^2}}\, dx = a\{F(\alpha, q) - E(\alpha, q)\} + u\sqrt{\frac{a^2+u^2}{b^2+u^2}}$

$$[a > b, \quad u > 0] \qquad \text{BY (221.03)}$$

2.[6]   $\displaystyle\int_0^u \sqrt{\frac{x^2+b^2}{x^2+a^2}}\, dx = \frac{b^2}{a}\, F(\alpha, q) - a\, E(\alpha, q) + u\sqrt{\frac{a^2+u^2}{b^2+u^2}}$

$$[a > b, \quad u > 0] \qquad \text{BY (221.04)}$$

3.   $\displaystyle\int_0^u \sqrt{\frac{x^2+a^2}{b^2-x^2}}\, dx = \sqrt{a^2+b^2}\, E(\gamma, r) - u\sqrt{\frac{b^2-u^2}{a^2+u^2}} \qquad [b \geq u > 0] \qquad \text{BY (214.11)}$

4.   $\displaystyle\int_u^b \sqrt{\frac{a^2+x^2}{b^2-x^2}}\, dx = \sqrt{a^2+b^2}\, E(\delta, r) \qquad [b > u \geq 0] \qquad \text{BY (213.01), ZH 64 (273)}$

5. $\int_b^u \sqrt{\dfrac{a^2 + x^2}{x^2 - b^2}}\, dx = \sqrt{a^2 + b^2}\, \{F(\varepsilon, s) - E(\varepsilon, s)\} + \dfrac{1}{u}\sqrt{(u^2 + a^2)(u^2 - b^2)}$

$[u > b > 0]$     BY (211.03)

6. $\int_0^u \sqrt{\dfrac{b^2 - x^2}{a^2 + x^2}}\, dx = \sqrt{a^2 + b^2}\, \{F(\gamma, r) - E(\gamma, r)\} + u\sqrt{\dfrac{b^2 - u^2}{a^2 + u^2}}$

$[b \geq u > 0]$     BY (214.03)

7. $\int_u^b \sqrt{\dfrac{b^2 - x^2}{a^2 + x^2}}\, dx = \sqrt{a^2 + b^2}\, \{F(\delta, r) - E(\delta, r)\}$     $[b > u \geq 0]$     BY (213.03)

8. $\int_b^u \sqrt{\dfrac{x^2 - b^2}{a^2 + x^2}}\, dx = \dfrac{1}{u}\sqrt{(a^2 + u^2)(u^2 - b^2)} - \sqrt{a^2 + b^2}\, E(\varepsilon, s)$

$[u > b > 0]$     BY (211.04)

9. $\int_0^u \sqrt{\dfrac{b^2 - x^2}{a^2 - x^2}}\, dx = a\, E(\eta, t) - \dfrac{a^2 - b^2}{a}\, F(\eta, t)$     $[a > b \geq u > 0]$     BY (219.03)

10. $\int_u^b \sqrt{\dfrac{b^2 - x^2}{a^2 - x^2}}\, dx = a\, E(\zeta, t) - \dfrac{a^2 - b^2}{a}\, F(\zeta, t) - u\sqrt{\dfrac{b^2 - u^2}{a^2 - u^2}}$

$[a > b > u \geq 0]$     BY (220.04)

11. $\int_b^u \sqrt{\dfrac{x^2 - b^2}{a^2 - x^2}}\, dx = a\, E(\kappa, q) - \dfrac{b^2}{a}\, F(\kappa, q) - \dfrac{1}{u}\sqrt{(a^2 - u^2)(u^2 - b^2)}$

$[a \geq u > b > 0]$     BY (217.04)

12. $\int_u^a \sqrt{\dfrac{x^2 - b^2}{a^2 - x^2}}\, dx = a\, E(\lambda, q) - \dfrac{b^2}{a}\, F(\lambda, q)$     $[a > u \geq b > 0]$     BY (218.03)

13. $\int_a^u \sqrt{\dfrac{x^2 - b^2}{x^2 - a^2}}\, dx = \dfrac{a^2 - b^2}{a}\, F(\mu, t) - a\, E(\mu, t) + u\sqrt{\dfrac{u^2 - a^2}{u^2 - b^2}}$

$[u > a > b > 0]$     BY (216.03)

14. $\int_0^u \sqrt{\dfrac{a^2 - x^2}{b^2 - x^2}}\, dx = a\, E(\eta, t)$     $[a > b \geq u > 0]$     H 64 (276), BY (219.01)

15. $\int_u^b \sqrt{\dfrac{a^2 - x^2}{b^2 - x^2}}\, dx = a\left\{ E(\zeta, t) - \dfrac{u}{a}\sqrt{\dfrac{b^2 - u^2}{a^2 - u^2}} \right\}$     $[a > b > u \geq 0]$     BY (220.03)

16. $\int_b^u \sqrt{\dfrac{a^2 - x^2}{x^2 - b^2}}\, dx = a\, \{F(\kappa, q) - E(\kappa, q)\} + \dfrac{1}{u}\sqrt{(a^2 - u^2)(u^2 - b^2)}$

$[a \geq u > b > 0]$     BY (217.03)

17. $\int_u^a \sqrt{\dfrac{a^2 - x^2}{x^2 - b^2}}\, dx = a\, \{F(\lambda, q) - E(\lambda, q)\}$     $[a > u \geq b > 0]$     BY (218.09)

18. $\int_a^u \sqrt{\dfrac{x^2 - a^2}{x^2 - b^2}}\, dx = u\sqrt{\dfrac{u^2 - a^2}{u^2 - b^2}} - a\, E(\mu, t)$     $[u > a > b > 0]$     BY (216.04)

**3.171**

1.  $\displaystyle\int_b^u \frac{dx}{x^2}\sqrt{\frac{a^2+x^2}{x^2-b^2}} = \frac{\sqrt{a^2+b^2}}{b^2}E(\varepsilon,s)$ $\qquad\qquad [u>b>0]$ $\qquad$ BY (211.01), ZH 64 (274)

2.  $\displaystyle\int_u^\infty \frac{dx}{x^2}\sqrt{\frac{a^2+x^2}{x^2-b^2}} = \frac{\sqrt{a^2+b^2}}{b^2}E(\xi,s) - \frac{a^2}{b^2u}\sqrt{\frac{u^2-b^2}{a^2+u^2}}$

    $\qquad\qquad\qquad\qquad\qquad\qquad\qquad\qquad\qquad [u\geq b>0]$ $\qquad$ BY (212.09)

3.  $\displaystyle\int_u^b \frac{dx}{x^2}\sqrt{\frac{a^2-x^2}{b^2-x^2}} = \frac{a^2-b^2}{ab^2}F(\zeta,t) - \frac{a}{b^2}E(\zeta,t) + \frac{a^2}{b^2u}\sqrt{\frac{b^2-u^2}{a^2-u^2}}$

    $\qquad\qquad\qquad\qquad\qquad\qquad\qquad\qquad\qquad [a>b>u>0]$ $\qquad$ BY (220.12)

4.  $\displaystyle\int_b^u \frac{dx}{x^2}\sqrt{\frac{a^2-x^2}{x^2-b^2}} = \frac{a}{b^2}E(\kappa,q) - \frac{1}{a}F(\kappa,q)$ $\qquad [a\geq u>b>0]$ $\qquad$ BY (217.11)

5.  $\displaystyle\int_u^a \frac{dx}{x^2}\sqrt{\frac{a^2-x^2}{x^2-b^2}} = \frac{a}{b^2}E(\lambda,q) - \frac{1}{a}F(\lambda,q) - \frac{\sqrt{(a^2-u^2)(u^2-b^2)}}{b^2u}$

    $\qquad\qquad\qquad\qquad\qquad\qquad\qquad\qquad\qquad [a>u\geq b>0]$ $\qquad$ BY (218.10)

6.  $\displaystyle\int_a^u \frac{dx}{x^2}\sqrt{\frac{x^2-a^2}{x^2-b^2}} = \frac{a}{b^2}E(\mu,t) - \frac{a^2-b^2}{ab^2}F(\mu,t) - \frac{1}{u}\sqrt{\frac{u^2-a^2}{u^2-b^2}}$

    $\qquad\qquad\qquad\qquad\qquad\qquad\qquad\qquad\qquad [u>a>b>0]$ $\qquad$ BY (216.08)

7.  $\displaystyle\int_u^\infty \frac{dx}{x^2}\sqrt{\frac{x^2+a^2}{x^2+b^2}} = \frac{1}{a}F(\beta,q) - \frac{a}{b^2}E(\beta,q) + \frac{a^2}{b^2u}\sqrt{\frac{b^2+u^2}{a^2+u^2}}$

    $\qquad\qquad\qquad\qquad\qquad\qquad\qquad\qquad\qquad [a>b,\quad u>0]$ $\qquad$ BY (222.08)

8.  $\displaystyle\int_u^\infty \frac{dx}{x^2}\sqrt{\frac{x^2+b^2}{x^2+a^2}} = \frac{1}{a}\{F(\beta,q)-E(\beta,q)\} + \frac{1}{u}\sqrt{\frac{b^2+u^2}{a^2+u^2}}$

    $\qquad\qquad\qquad\qquad\qquad\qquad\qquad\qquad\qquad [a>b,\quad u>0]$ $\qquad$ BY (222.09)

9.  $\displaystyle\int_u^b \frac{dx}{x^2}\sqrt{\frac{b^2-x^2}{a^2+x^2}} = \frac{\sqrt{(b^2-u^2)(a^2+u^2)}}{a^2u} - \frac{\sqrt{a^2+b^2}}{a^2}E(\delta,r)$

    $\qquad\qquad\qquad\qquad\qquad\qquad\qquad\qquad\qquad [b>u>0]$ $\qquad$ BY (213.10)

10. $\displaystyle\int_b^u \frac{dx}{x^2}\sqrt{\frac{x^2-b^2}{a^2+x^2}} = \frac{\sqrt{a^2+b^2}}{a^2}\{F(\varepsilon,s)-E(\varepsilon,s)\}$ $\qquad [u>b>0]$ $\qquad$ BY (211.07)

11. $\displaystyle\int_u^\infty \frac{dx}{x^2}\sqrt{\frac{x^2-b^2}{a^2+x^2}} = \frac{\sqrt{a^2+b^2}}{a^2}\{F(\xi,s)-E(\xi,s)\} + \frac{1}{u}\sqrt{\frac{u^2-b^2}{a^2+u^2}}$

    $\qquad\qquad\qquad\qquad\qquad\qquad\qquad\qquad\qquad [u\geq b>0]$ $\qquad$ BY (212.11)

12. $\displaystyle\int_u^b \frac{dx}{x^2}\sqrt{\frac{a^2+x^2}{b^2-x^2}} = \frac{\sqrt{a^2+b^2}}{b^2}\{F(\delta,r)-E(\delta,r)\} + \frac{\sqrt{(b^2-u^2)(a^2+u^2)}}{b^2u}$

    $\qquad\qquad\qquad\qquad\qquad\qquad\qquad\qquad\qquad [b>u>0]$ $\qquad$ BY (213.05)

13. $\displaystyle\int_u^\infty \frac{dx}{x^2}\sqrt{\frac{x^2-a^2}{x^2-b^2}} = \frac{a}{b^2}E(\nu,t) - \frac{a^2-b^2}{ab^2}F(\nu,t)$ $\qquad [u\geq a>b>0]$ $\qquad$ BY (215.08)

14.   $\displaystyle\int_u^b \frac{dx}{x^2}\sqrt{\frac{b^2-x^2}{a^2-x^2}} = \frac{1}{u}\sqrt{\frac{b^2-u^2}{a^2-u^2}} - \frac{1}{a}E(\zeta,t)$      $[a>b>u>0]$      BY (220.11)

15.   $\displaystyle\int_b^u \frac{dx}{x^2}\sqrt{\frac{x^2-b^2}{a^2-x^2}} = \frac{1}{a}\{F(\kappa,q)-E(\kappa,q)\}$      $[a\geq u>b>0]$      BY (217.08)

16.   $\displaystyle\int_u^a \frac{dx}{x^2}\sqrt{\frac{x^2-b^2}{u^2-x^2}} = \frac{1}{a}\{F(\lambda,q)-E(\lambda,q)\} + \frac{\sqrt{(a^2-u^2)(u^2-b^2)}}{a^2u}$

                                                               $[a>u\geq b>0]$      BY (218.08)

17.   $\displaystyle\int_a^u \frac{dx}{x^2}\sqrt{\frac{x^2-b^2}{x^2-a^2}} = \frac{1}{a}E(\mu,t) - \frac{1}{u}\sqrt{\frac{u^2-a^2}{u^2-b^2}}$      $[u>a>b>0]$      BY (216.07)

18.   $\displaystyle\int_u^\infty \frac{dx}{x^2}\sqrt{\frac{x^2-b^2}{x^2-a^2}} = \frac{1}{a}E(\nu,t)$      $[u\geq a>b>0]$

                                                         BY (215.01), ZH 65 (281)

**3.172**

1.   $\displaystyle\int_0^u \sqrt{\frac{x^2+b^2}{(x^2+a^2)^3}}\,dx = \frac{1}{a}E(\alpha,q) - \frac{a^2-b^2}{a^2}\frac{u}{\sqrt{(a^2+u^2)(b^2+u^2)}}$

                                                     $[a>b,\quad u>0]$      BY (221.10)

2.   $\displaystyle\int_u^\infty \sqrt{\frac{x^2+b^2}{(x^2+a^2)^3}}\,dx = \frac{1}{a}E(\beta,q)$      $[a>b,\quad u\geq 0]$      H 64 (271)

3.   $\displaystyle\int_0^u \sqrt{\frac{x^2+a^2}{(x^2+b^2)^3}}\,dx = \frac{a}{b^2}E(\alpha,q)$      $[a>b,\quad u>0]$      H 64 (270)

4.   $\displaystyle\int_u^\infty \sqrt{\frac{x^2+a^2}{(x^2+b^2)^3}}\,dx = \frac{a}{b^2}E(\beta,q) - \frac{a^2-b^2}{b^2}\frac{u}{\sqrt{(a^2+u^2)(b^2+u^2)}}$

                                                     $[a>b,\quad u\geq 0]$      BY (222.06)

5.   $\displaystyle\int_0^u \sqrt{\frac{b^2-x^2}{(a^2+x^2)^3}}\,dx = \frac{\sqrt{a^2+b^2}}{a^2}E(\gamma,r) - \frac{1}{\sqrt{a^2+b^2}}F(\gamma,r)$

                                                     $[b\geq u>0]$      BY (214.08)

6.   $\displaystyle\int_u^b \sqrt{\frac{b^2-x^2}{(a^2+x^2)^3}}\,dx = \frac{\sqrt{a^2+b^2}}{a^2}E(\delta,r) - \frac{1}{\sqrt{a^2+b^2}}F(\delta,r) - \frac{u}{a^2}\sqrt{\frac{b^2-u^2}{a^2+u^2}}$

                                                     $[b>u\geq 0]$      BY (213.04)

7.   $\displaystyle\int_b^u \sqrt{\frac{x^2-b^2}{(a^2+x^2)^3}}\,dx = \frac{\sqrt{a^2+b^2}}{a^2}E(\varepsilon,s) - \frac{b^2}{a^2\sqrt{a^2+b^2}}F(\varepsilon,s) - \frac{1}{u}\sqrt{\frac{u^2-b^2}{u^2+a^2}}$

                                                     $[u>b>0]$      BY (211.06)

8.   $\displaystyle\int_u^\infty \sqrt{\frac{x^2-b^2}{(a^2+x^2)^3}}\,dx = \frac{\sqrt{a^2+b^2}}{a^2}E(\xi,s) - \frac{b^2}{a^2\sqrt{a^2+b^2}}F(\xi,s)$

                                                     $[u\geq b>0]$      BY (212.08)

9. $\int_0^u \sqrt{\dfrac{x^2+a^2}{(b^2-x^2)^3}}\,dx = \dfrac{a^2}{b^2\sqrt{a^2+b^2}}\,F(\gamma,r) - \dfrac{\sqrt{a^2+b^2}}{b^2}\,E(\gamma,r) + \dfrac{\left(a^2+b^2\right)u}{b^2\sqrt{(a^2+u^2)\left(b^2-u^2\right)}}$

$$[b > u > 0] \qquad\qquad \text{BY (214.09)}$$

10. $\int_u^\infty \sqrt{\dfrac{x^2+a^2}{(x^2-b^2)^3}}\,dx = \dfrac{1}{\sqrt{a^2+b^2}}\,F(\xi,s) - \dfrac{\sqrt{a^2+b^2}}{b^2}\,E(\xi,s) + \dfrac{\left(a^2+b^2\right)u}{b^2\sqrt{(a^2+u^2)\left(u^2-b^2\right)}}$

$$[u > b > 0] \qquad\qquad \text{BY (212.07)}$$

11. $\int_0^u \sqrt{\dfrac{b^2-x^2}{(a^2-x^2)^3}}\,dx = \dfrac{1}{a}\left\{ F(\eta,t) - E(\eta,t) + \dfrac{u}{a}\sqrt{\dfrac{b^2-u^2}{a^2-u^2}}\right\}$

$$[a > b \ge u > 0] \qquad\qquad \text{BY (219.09)}$$

12. $\int_u^b \sqrt{\dfrac{b^2-x^2}{(a^2-x^2)^3}}\,dx = \dfrac{1}{a}\left\{ F(\zeta,t) - E(\zeta,t)\right\} \qquad [a > b > u \ge 0] \qquad\qquad \text{BY (220.07)}$

13. $\int_b^u \sqrt{\dfrac{x^2-b^2}{(a^2-x^2)^3}}\,dx = \dfrac{1}{u}\sqrt{\dfrac{u^2-b^2}{a^2-u^2}} - \dfrac{1}{a}E(\kappa,q) \qquad [a > u > b > 0] \qquad\qquad \text{BY (217.07)}$

14. $\int_u^\infty \sqrt{\dfrac{x^2-b^2}{(x^2-a^2)^3}}\,dx = \dfrac{1}{a}\left[ F(\nu,t) - E(\nu,t)\right] + \dfrac{1}{u}\sqrt{\dfrac{u^2-b^2}{u^2-a^2}}$

$$[u > a > b > 0] \qquad\qquad \text{BY (215.05)}$$

15. $\int_0^u \sqrt{\dfrac{a^2-x^2}{(b^2-x^2)^3}}\,dx = \dfrac{a}{b^2}\left[ F(\eta,t) - E(\eta,t)\right] + \dfrac{u}{b^2}\sqrt{\dfrac{a^2-u^2}{b^2-u^2}}$

$$[a > b > u > 0] \qquad\qquad \text{BY (219.10)}$$

16. $\int_u^a \sqrt{\dfrac{a^2-x^2}{(x^2-b^2)^3}}\,dx = \dfrac{u}{b^2}\sqrt{\dfrac{a^2-u^2}{u^2-b^2}} - \dfrac{a}{b^2}E(\lambda,q) \qquad [a > u > b > 0] \qquad\qquad \text{BY (218.05)}$

17. $\int_a^u \sqrt{\dfrac{x^2-a^2}{(x^2-b^2)^3}}\,dx = \dfrac{a}{b^2}\left[ F(\mu,t) - E(\mu,t)\right] \qquad [u > a > b > 0] \qquad\qquad \text{BY (216.05)}$

18. $\int_u^\infty \sqrt{\dfrac{x^2-a^2}{(x^2-b^2)^3}}\,dx = \dfrac{a}{b^2}\left[ F(\nu,t) - E(\nu,t)\right] + \dfrac{1}{u}\sqrt{\dfrac{u^2-a^2}{u^2-b^2}}$

$$[u \ge a > b > 0] \qquad\qquad \text{BY (215.03)}$$

**3.173**

1. $\int_u^1 \dfrac{dx}{x^2}\sqrt{\dfrac{x^2+1}{1-x^2}} = \sqrt{2}\left[ F\left(\arccos u, \dfrac{\sqrt{2}}{2}\right) - E\left(\arccos u, \dfrac{\sqrt{2}}{2}\right)\right] + \dfrac{\sqrt{1-u^4}}{u}$

$$[u < 1] \qquad\qquad \text{BY (259.77)}$$

2. $\int_1^u \dfrac{dx}{x^2}\sqrt{\dfrac{x^2+1}{x^2-1}} = \sqrt{2}\,E\left(\arccos\dfrac{1}{u}, \dfrac{\sqrt{2}}{2}\right) \qquad [u > 1] \qquad\qquad \text{BY (260.76)}$

**3.174** **Notation:** In **3.174** and **3.175**, we set:　$\alpha = \arccos \dfrac{1 + \left(1 - \sqrt{3}\right) u}{1 + \left(1 + \sqrt{3}\right) u}$,

$$\beta = \arccos \frac{1 - \left(1 + \sqrt{3}\right) u}{1 + \left(\sqrt{3} - 1\right) u}, \qquad p = \frac{\sqrt{2 + \sqrt{3}}}{2}, \qquad q = \frac{\sqrt{2 - \sqrt{3}}}{2}.$$

1.　$\displaystyle \int_0^u \frac{dx}{\left[1 + \left(1 + \sqrt{3}\right) x\right]^2} \sqrt{\frac{1 - x + x^2}{x(1 + x)}} = \frac{1}{\sqrt[4]{3}} E(\alpha, p)$　　　$[u > 0]$　　　BY (260.51)

2.　$\displaystyle \int_0^u \frac{dx}{\left[1 + \left(\sqrt{3} - 1\right) x\right]^2} \sqrt{\frac{1 + x + x^2}{x(1 - x)}} = \frac{1}{\sqrt[4]{3}} E(\beta, q)$　　　$[1 \geq u > 0]$　　　BY (259.51)

3.　$\displaystyle \int_0^u \frac{dx}{1 - x + x^2} \sqrt{\frac{x(1 + x)}{1 - x + x^2}} = \frac{1}{\sqrt[4]{27}} E(\alpha, p) - \frac{2 - \sqrt{3}}{\sqrt[4]{27}} F(\alpha, p) - \frac{2\left(2 + \sqrt{3}\right)}{\sqrt{3}} \frac{1 + \left(1 - \sqrt{3}\right) u}{1 + \left(1 + \sqrt{3}\right) u}$

$$\times \sqrt{\frac{u(1 + u)}{1 - u + u^2}}$$

$[u > 0]$　　　BY (260.54)

4.　$\displaystyle \int_0^u \frac{dx}{1 + x + x^2} \sqrt{\frac{x(1 - x)}{1 + x + x^2}} \quad \frac{4}{\sqrt[4]{27}} E(\beta, q) - \frac{2 + \sqrt{3}}{\sqrt[4]{27}} F(\beta, q) - \frac{2\left(2 - \sqrt{3}\right)}{\sqrt{3}} \frac{1 - \left(1 + \sqrt{3}\right) u}{1 + \left(\sqrt{3} - 1\right) u}$

$$\times \sqrt{\frac{u(1 - u)}{1 + u + u^2}}$$

$[1 \geq u > 0]$　　　BY (259.55)

**3.175**

1.　$\displaystyle \int_0^u \frac{dx}{1 + x} \sqrt{\frac{x}{1 + x^3}} = \frac{1}{\sqrt[4]{27}} \left[ F(\alpha, p) - 2 E(\alpha, p) \right] + \frac{2}{\sqrt{3}} \frac{\sqrt{u \left(1 - u + u^2\right)}}{\sqrt{1 + u} \left[1 + \left(1 + \sqrt{3}\right) u\right]}$

$[u > 0]$　　　BY (260.55)

2.　$\displaystyle \int_0^u \frac{dx}{1 - x} \sqrt{\frac{x}{1 - x^3}} = \frac{1}{\sqrt[4]{27}} \left[ F(\beta, q) - 2 E(\beta, q) \right] + \frac{2}{\sqrt{3}} \frac{\sqrt{u \left(1 + u + u^2\right)}}{\sqrt{1 - u} \left[1 + \left(\sqrt{3} - 1\right) u\right]}$

$[0 < u < 1]$　　　BY (259.52)

## 3.18 Expressions that can be reduced to fourth roots of second-degree polynomials and their products with rational functions

**3.181**

1.　$\displaystyle \int_b^u \frac{dx}{\sqrt[4]{(a - x)(x - b)}} = \sqrt{a - b} \left\{ 2 \left[ E\left(\frac{1}{\sqrt{2}}\right) + E\left(\arccos \sqrt[4]{\frac{4(a - u)(u - b)}{(a - b)^2}}, \frac{1}{\sqrt{2}}\right) \right] \right.$

$$\left. - \left[ K\left(\frac{1}{\sqrt{2}}\right) + F\left(\arccos \sqrt[4]{\frac{4(a - u)(u - b)}{(a - b)^2}}, \frac{1}{\sqrt{2}}\right) \right] \right\}$$

$[a \geq u > b]$　　　BY (271.05)

2.    $\displaystyle \int_a^u \frac{dx}{\sqrt[4]{(x-a)(x-b)}} \sqrt{\frac{a-b}{2}} F\left[\left(\arccos \frac{a-b-2\sqrt{(u-a)(u-b)}}{a-b+2\sqrt{(u-a)(u-b)}}, \frac{1}{\sqrt{2}}\right)\right.$

$$\left. - 2E\left(\arccos \frac{a-b-2\sqrt{(u-a)(u-b)}}{a-b+2\sqrt{(u-a)(u-b)}}, \frac{1}{\sqrt{2}}\right)\right]$$

$$+ \frac{2(2u-a-b)\sqrt[4]{(u-a)(u-b)}}{a-b+2\sqrt{(u-a)(u-b)}}$$

$$[u > a > b] \qquad\qquad \text{BY (272.05)}$$

## 3.182

1.    $\displaystyle \int_b^u \frac{dx}{\sqrt[4]{[(a-x)(x-b)]^3}} = \frac{2}{\sqrt{a-b}}\left[K\left(\frac{1}{\sqrt{2}}\right) + F\left(\arccos\sqrt{\frac{4(a-u)(u-b)}{(a-b)^2}}, \frac{1}{\sqrt{2}}\right)\right]$

$$[a \geq u > b] \qquad\qquad \text{BY (271.01)}$$

2.    $\displaystyle \int_a^u \frac{dx}{\sqrt[4]{[(x-a)(x-b)]^3}} = \frac{\sqrt{2}}{\sqrt{a-b}} F\left(\arccos \frac{a-b-2\sqrt{(u-a)(u-b)}}{a-b+2\sqrt{(u-a)(u-b)}}, \frac{1}{\sqrt{2}}\right)$

$$[u > a > b] \qquad\qquad \text{BY (272.00)}$$

## 3.183    Notation: In **3.183–3.186** we set:

$$\alpha = \arccos \frac{1}{\sqrt[4]{u^2+1}}, \qquad \beta = \arccos \sqrt[4]{1-u^2}, \qquad \gamma = \arccos \frac{1-\sqrt{u^2-1}}{1+\sqrt{u^2-1}}.$$

1.    $\displaystyle \int_0^u \frac{dx}{\sqrt[4]{x^2+1}} = \sqrt{2}\left[F\left(\alpha, \frac{1}{\sqrt{2}}\right) - 2E\left(\alpha, \frac{1}{\sqrt{2}}\right)\right] + \frac{2u}{\sqrt[4]{u^2+1}}$

$$[u > 0] \qquad\qquad \text{BY (273.55)}$$

2.    $\displaystyle \int_0^u \frac{dx}{\sqrt[4]{1-x^2}} = \sqrt{2}\left[2E\left(\beta, \frac{1}{\sqrt{2}}\right) - F\left(\beta, \frac{1}{\sqrt{2}}\right)\right] \qquad [0 < u \leq 1]$    BY (271.55)

3.    $\displaystyle \int_1^u \frac{dx}{\sqrt[4]{x^2-1}} = F\left(\gamma, \frac{1}{\sqrt{2}}\right) - 2E\left(\gamma, \frac{1}{\sqrt{2}}\right) + \frac{2u\sqrt[4]{u^2-1}}{1+\sqrt{u^2-1}}$

$$[u > 1] \qquad\qquad \text{BY (272.55)}$$

## 3.184

1.    $\displaystyle \int_0^u \frac{x^2\,dx}{\sqrt[4]{1-x^2}} = \frac{2\sqrt{2}}{5}\left[2E\left(\beta, \frac{1}{\sqrt{2}}\right) - F\left(\beta, \frac{1}{\sqrt{2}}\right)\right] - \frac{2u}{5}\sqrt[4]{(1-u^2)^3}$

$$[0 < u \leq 1] \qquad\qquad \text{BY (271.59)}$$

2.    $\displaystyle \int_1^u \frac{dx}{x^2\sqrt[4]{x^2-1}} = E\left(\gamma, \frac{1}{\sqrt{2}}\right) - \frac{1}{2}F\left(\gamma, \frac{1}{\sqrt{2}}\right) - \frac{1-\sqrt{u^2-1}}{1+\sqrt{u^2-1}} \cdot \frac{\sqrt{u^2-1}}{u}$

$$[u > 1] \qquad\qquad \text{BY (272.54)}$$

**3.185**

1. $\displaystyle\int_0^u \frac{\mathrm{d}x}{\sqrt[4]{(x^2+1)^3}} = \sqrt{2}\,F\left(\alpha,\frac{1}{\sqrt{2}}\right)$ $\qquad [u>0]$ BY (273.50)

2. $\displaystyle\int_0^u \frac{\mathrm{d}x}{\sqrt[4]{(1-x^2)^3}} = \sqrt{2}\,F\left(\beta,\frac{1}{\sqrt{2}}\right)$ $\qquad [0<u\le 1]$ BY (271.51)

3. $\displaystyle\int_1^u \frac{\mathrm{d}x}{\sqrt[4]{(x^2-1)^3}} = F\left(\gamma,\frac{1}{\sqrt{2}}\right)$ $\qquad [u>1]$ BY (272.50)

4. $\displaystyle\int_0^u \frac{x^2\,\mathrm{d}x}{\sqrt[4]{(1-x^2)^3}} = \frac{2\sqrt{2}}{3}\,F\left(\beta,\frac{1}{\sqrt{2}}\right) - \frac{2}{3}u\,\sqrt[4]{1-u^2}$ $\quad [0<u\le 1]$ BY (271.54)

5. $\displaystyle\int_0^u \frac{\mathrm{d}x}{\sqrt[4]{(x^2+1)^5}} = 2\sqrt{2}\,E\left(\alpha,\frac{1}{\sqrt{2}}\right) - \sqrt{2}\,F\left(\alpha,\frac{1}{\sqrt{2}}\right)$

$\qquad\qquad\qquad\qquad\qquad\qquad\qquad\qquad\qquad [u>0]$ BY (273.54)

6. $\displaystyle\int_0^u \frac{x^2\,\mathrm{d}x}{\sqrt[4]{(x^2+1)^5}} = 2\sqrt{2}\left[F\left(\alpha,\frac{1}{\sqrt{2}}\right) - 2E\left(\alpha,\frac{1}{\sqrt{2}}\right)\right] + \frac{2u}{\sqrt[4]{u^2+1}}$

$\qquad\qquad\qquad\qquad\qquad\qquad\qquad\qquad\qquad [u>0]$ BY (273.56)

7. $\displaystyle\int_0^u \frac{x^2\,\mathrm{d}x}{\sqrt[4]{(x^2+1)^7}} = \frac{1}{3\sqrt{2}}\,F\left(\alpha,\frac{1}{\sqrt{2}}\right) - \frac{u}{6\sqrt[4]{(u^2+1)^3}}$ $\quad [u>0]$ BY (273.53)

**3.186**

1. $\displaystyle\int_0^u \frac{1+\sqrt{x^2+1}}{(x^2+1)\sqrt[4]{x^2+1}}\,\mathrm{d}x = 2\sqrt{2}\,E\left(\alpha,\frac{1}{\sqrt{2}}\right)$ $\qquad [u>0]$ BY (273.51)

2. $\displaystyle\int_0^u \frac{\mathrm{d}x}{(1+\sqrt{1-x^2})\sqrt[4]{1-x^2}} = \sqrt{2}\left[F\left(\beta,\frac{1}{\sqrt{2}}\right) - E\left(\beta,\frac{1}{\sqrt{2}}\right)\right] + \frac{u\sqrt[4]{1-u^2}}{1+\sqrt{1-u^2}}$

$\qquad\qquad\qquad\qquad\qquad\qquad\qquad\qquad\qquad [0<u\le 1]$ BY (271.58)

3. $\displaystyle\int_1^u \frac{\mathrm{d}x}{(x^2+2\sqrt{x^2-1})\sqrt[4]{x^2-1}} = \frac{1}{2}\left[F\left(\gamma,\frac{1}{\sqrt{2}}\right) - E\left(\gamma,\frac{1}{\sqrt{2}}\right)\right]$

$\qquad\qquad\qquad\qquad\qquad\qquad\qquad\qquad\qquad [u>1]$ BY (272.53)

4. $\displaystyle\int_0^u \frac{1-\sqrt{1-x^2}}{1+\sqrt{1-x^2}}\cdot\frac{\mathrm{d}x}{\sqrt[4]{(1-x^2)^3}} = \sqrt{2}\left[2E\left(\beta,\frac{1}{\sqrt{2}}\right) - F\left(\beta,\frac{1}{\sqrt{2}}\right)\right] - \frac{2u\sqrt[4]{1-u^2}}{1+\sqrt{1-u^2}}$

$\qquad\qquad\qquad\qquad\qquad\qquad\qquad\qquad\qquad [0<u\le 1]$ BY (271.57)

5. $\displaystyle\int_1^u \frac{x^2\,\mathrm{d}x}{(x^2+2\sqrt{x^2-1})\sqrt[4]{(x^2-1)^3}} = E\left(\gamma,\frac{1}{\sqrt{2}}\right)$ $\qquad [u>1]$ BY (272.51)

## 3.19–3.23 Combinations of powers of $x$ and powers of binomials of the form $(\alpha+\beta x)$

**3.191**

1. $\quad \displaystyle\int_0^a x^{\nu-1}(a-x)^{\mu-1}\,dx = a^{\mu+\nu-1}\,\mathrm{B}(\mu,\nu)$ $\qquad [\mathrm{Re}\,\mu > 0, \quad \mathrm{Re}\,\nu > 0]$ $\qquad$ ET II 185(7)

2. $\quad \displaystyle\int_u^\infty x^{-\nu}(x-u)^{\mu-1}\,dx = u^{\mu-\nu}\,\mathrm{B}(\nu-\mu,\mu)$ $\qquad [\mathrm{Re}\,\nu > \mathrm{Re}\,\mu > 0]$ $\qquad$ ET II 201(6)

3. $\quad \displaystyle\int_0^1 x^{\nu-1}(1-x)^{\mu-1}\,dx = \int_0^1 x^{\mu-1}(1-x)^{\nu-1}\,dx = \mathrm{B}(\mu,\nu)$

$\qquad\qquad\qquad\qquad\qquad\qquad\qquad\qquad\qquad\qquad [\mathrm{Re}\,\mu > 0, \quad \mathrm{Re}\,\nu > 0]$ $\qquad$ FI II 774(1)

**3.192**

1. $\quad \displaystyle\int_0^1 \frac{x^a\,dx}{(1-x)^a} = a\pi\,\mathrm{cosec}\,a\pi$ $\qquad \left[a^2 < 1\right]$ $\qquad$ BI (3)(4)

2. $\quad \displaystyle\int_0^1 \frac{x^a\,dx}{(1-x)^{a+1}} = -\pi\,\mathrm{cosec}\,a\pi$ $\qquad [-1 < a < 0]$ $\qquad$ BI (3)(5)

3. $\quad \displaystyle\int_0^1 \frac{(1-x)^a}{x^{a+1}}\,dx = -\pi\,\mathrm{cosec}\,a\pi$ $\qquad [-1 < a < 0]$ $\qquad$ BI (4)(6)

4. $\quad \displaystyle\int_1^\infty (x-1)^{a-\frac12}\frac{dx}{x} = \pi\,\sec a\pi$ $\qquad \left[-\frac12 < a < \frac12\right]$ $\qquad$ BI (23)(7)

**3.193**[12] $\displaystyle\int_0^n x^{\nu-1}(n-x)^n\,dx = \frac{n^{\nu+n}n!}{(\nu)_{n+1}} = \frac{n!\,n^{\nu+n}}{\nu(\nu+1)(\nu+2)\ldots(\nu+n)}$ $\quad [\mathrm{Re}\,\nu > 0]$ $\qquad$ EH I 2

**3.194**

1. $\quad \displaystyle\int_0^u \frac{x^{a-1}\,dx}{(1+bx)^\nu} = \frac{u^a}{a}\,{}_2F_1(\nu,a;1+a;-bu)$ $\qquad [|\arg(1+bu)| < \pi, \quad \mathrm{Re}\,a > 0]$

$\qquad\qquad\qquad\qquad\qquad\qquad\qquad\qquad\qquad\qquad\qquad\qquad\qquad$ ET I 310(20)

2.[6] $\quad \displaystyle\int_u^\infty \frac{x^{a-1}\,dx}{(1+bx)^\nu} = \frac{u^{a-\nu}}{b^\nu(\nu-a)}\,{}_2F_1\left(\nu,\nu-a;\;\nu-a+1;\;-\frac{1}{bu}\right)$

$\qquad\qquad\qquad\qquad\qquad\qquad\qquad\qquad\qquad [\mathrm{Re}\,\nu > \mathrm{Re}\,a]$ $\qquad$ ET I 310(21)

3. $\quad \displaystyle\int_0^\infty \frac{x^{a-1}\,dx}{(1+bx)^\nu} = b^{-a}\,\mathrm{B}(a,\nu-a)$ $\qquad [|\arg b| < \pi, \quad \mathrm{Re}\,\nu > \mathrm{Re}\,a > 0]$

$\qquad\qquad\qquad\qquad\qquad\qquad\qquad\qquad\qquad\qquad\qquad$ FI II 775a, ET I 310(19)

4.[11] $\quad \displaystyle\int_0^\infty \frac{x^{a-1}\,dx}{(1+bx)^{n+1}} = (-1)^n\frac{\pi}{b^a}\binom{a-1}{n}\mathrm{cosec}(a\pi)$ $\qquad [|\arg b| < \pi, \quad 0 < \mathrm{Re}\,a < n+1]$

$\qquad\qquad\qquad\qquad\qquad\qquad\qquad\qquad\qquad\qquad\qquad$ ET I 308(6)

5. $\quad \displaystyle\int_0^u \frac{x^{a-1}\,dx}{1+bx} = \frac{u^a}{a}\,{}_2F_1(1,a;1+a;-bu)$ $\qquad [|\arg(1+ub)| < \pi, \quad \mathrm{Re}\,a > 0]$

$\qquad\qquad\qquad\qquad\qquad\qquad\qquad\qquad\qquad\qquad\qquad$ ET I 308(5)

6.[12] $\quad \displaystyle\int_u^\infty \frac{x^{a-1}\,dx}{(1+x)^2} = \frac{(1-a)\pi}{\sin \pi a}$ $\qquad [0 < \mathrm{Re}\,a < 2]$ $\qquad$ BI (16)(4)

7.  $\displaystyle\int_0^\infty \frac{x^m \, dx}{(a+bx)^{n+\frac{1}{2}}} = 2^{m+1} m! \frac{(2n-2m-3)!!}{(2n-1)!!} \frac{a^{m-n+\frac{1}{2}}}{b^{m+1}}$

$$\left[ m < n - \tfrac{1}{2}, \quad a > 0, \quad b > 0 \right]$$

BI (21)(2)

8.  $\displaystyle\int_0^1 \frac{x^{n-1} \, dx}{(1+x)^m} = 2^{-n} \sum_{k=0}^\infty \binom{m-n-1}{k} \frac{(-2)^{-k}}{n+k}$

BI (3)(1)

**3.195**[11] $\displaystyle\int_0^\infty \frac{(1+x)^{p-1}}{(a+x)^{p+1}} \, dx = \frac{1-a^{-p}}{p(a-1)}$ $\qquad [p \neq 0, \quad a > 0, \quad a \neq 1]$

$\qquad\qquad\qquad\qquad\qquad\qquad\qquad = \dfrac{\ln a}{a-1}$ $\qquad\qquad [p = 0, \quad a > 0, \quad a \neq 1]$

$\qquad\qquad\qquad\qquad\qquad\qquad\qquad = 1$ $\qquad\qquad\qquad [a = 1]$

LI (19)(6)

**3.196**

1.  $\displaystyle\int_0^a (x+b)^\nu (a-x)^{\mu-1} \, dx = \frac{b^\nu a^\mu}{\mu} \, {}_2F_1 \left( 1, -\nu; 1+\mu; -\frac{a}{b} \right)$

$$\left[ \left| \arg \frac{a}{b} \right| < \pi \right]$$

ET II 185(8)

2.  $\displaystyle\int_a^\infty (x+b)^{-\nu} (x-a)^{\mu-1} \, dx = (a+b)^{\mu-\nu} \, \mathrm{B}(\nu-\mu, \mu)$

$$\left[ \left| \arg \frac{a}{b} \right| < \pi, \quad \operatorname{Re}\nu > \operatorname{Re}\mu > 0 \right]$$

ET II 201(7)

3.  $\displaystyle\int_a^b (x-a)^{a-1}(b-x)^{\nu-1} \, dx = (b-a)^{a+\nu-1} \, \mathrm{B}(a,\nu)$ $\quad [b > a, \quad \operatorname{Re}a > 0, \quad \operatorname{Re}\nu > 0]$

EH I 10(13)

4.[12]  $\displaystyle\int_1^\infty \frac{dx}{(a-bx)(x-1)^\nu} = -\frac{\pi}{b \sin \pi\nu} \left( 1 - \frac{a}{b} \right)^{-\nu}$ $\quad [a < b, \quad b > 0, \quad 0 < \nu < 1] \quad$ LI (23)(5)

5.[12]  $\displaystyle\int_{-\infty}^1 \frac{dx}{(a-bx)(1-x)^\nu} = \frac{\pi}{b \sin \pi\nu} \left( \frac{a}{b} - 1 \right)^{-\nu}$ $\quad [a > b > 0, \quad 0 < \nu < 1] \quad$ LI (24)(10)

**3.197**

1.  $\displaystyle\int_0^\infty x^{\nu-1}(b+x)^{-\mu}(x+\gamma)^{-\varrho} \, dx = b^{-\mu}\gamma^{\nu-\varrho} \, \mathrm{B}(\nu, \mu-\nu+\varrho) \, {}_2F_1\left( \mu, \nu; \mu+\varrho; 1-\frac{\gamma}{b} \right)$

$[|\arg b| < \pi, \quad |\arg \gamma| < \pi, \quad \operatorname{Re}\nu > 0, \quad \operatorname{Re}\mu > \operatorname{Re}(\nu - \varrho)] \quad$ ET II 233(9)

2.[12]  $\displaystyle\int_a^\infty x^{-\lambda}(x+b)^\nu (x-a)^{\mu-1} \, dx = a^{-\lambda}(b+a)^{\mu+\nu} \, \mathrm{B}(\lambda - \mu - \nu, \mu) \, {}_2F_1\left( \lambda, \mu; \lambda - \nu; -\frac{b}{a} \right)$

$$\left[ \left| \arg \frac{b}{a} \right| < \pi \ \text{ or } \ \left| \frac{b}{a} \right| < 1, \quad 0 < \operatorname{Re}\mu < \operatorname{Re}(\lambda - \nu) \right] \quad \text{ET II 201(8)}$$

3.  $\displaystyle\int_0^1 x^{\lambda-1}(1-x)^{\mu-1}(1-bx)^{-\nu} \, dx = \mathrm{B}(\lambda, \mu) \, {}_2F_1(\nu, \lambda; \lambda+\mu; b)$

$[\operatorname{Re}\lambda > 0, \quad \operatorname{Re}\mu > 0, \quad |b| < 1] \quad$ WH

4.  $\int_0^1 x^{\mu-1}(1-x)^{\nu-1}(1+ax)^{-\mu-\nu}\,\mathrm{d}x = (1+a)^{-\mu}\,\mathrm{B}(\mu,\nu)$

$$[\operatorname{Re}\mu > 0, \quad \operatorname{Re}\nu > 0, \quad a > -1]$$
BI(5)4, EH I 10(11)

5.  $\int_0^\infty x^{\lambda-1}(1+x)^{\nu}(1+ax)^{\mu}\,\mathrm{d}x = \mathrm{B}(\lambda,-\mu-\nu-\lambda)\,{}_2F_1(-\mu,\lambda;-\mu-\nu;1-a)$

$$[|\arg a| < \pi, \quad -\operatorname{Re}(\mu+\nu) > \operatorname{Re}\lambda > 0]$$
EH I 60(12), ET I 310(23)

6.  $\int_1^\infty x^{\lambda-\nu}(x-1)^{\nu-\mu-1}(ax-1)^{-\lambda}\,\mathrm{d}x = a^{-\lambda}\,\mathrm{B}(\mu,\nu-\mu)\,{}_2F_1\left(\nu,\mu;\lambda;a^{-1}\right)$

$$[1+\operatorname{Re}\nu > \operatorname{Re}\lambda > \operatorname{Re}\mu, \quad |\arg(a-1)| < \pi]\quad\text{EH I 115(6)}$$

7.  $\int_0^\infty x^{\mu-\frac{1}{2}}(x+a)^{-\mu}(x+b)^{-\mu}\,\mathrm{d}x = \sqrt{\pi}\left(\sqrt{a}+\sqrt{b}\right)^{1-2\mu}\dfrac{\Gamma\left(\mu-\frac{1}{2}\right)}{\Gamma(\mu)}$

$$[\operatorname{Re}\mu > 0]$$
BI 19(5)

8.  $\int_0^u x^{\nu-1}(x+a)^{\lambda}(u-x)^{\mu-1}\,\mathrm{d}x = a^{\lambda}u^{\mu+\nu-1}\,\mathrm{B}(\mu,\nu)\,{}_2F_1\left(-\lambda,\nu;\mu+\nu;-\dfrac{u}{a}\right)$

$$\left[\left|\arg\left(\dfrac{u}{a}\right)\right| < \pi, \quad \operatorname{Re}\mu > 0, \quad \operatorname{Re}\nu > 0\right]$$
ET II 186(9)

9.  $\int_0^\infty x^{\lambda-1}(1+x)^{-\mu+\nu}(x+\beta)^{-\nu}\,\mathrm{d}x = \mathrm{B}(\mu-\lambda,\lambda)\,{}_2F_1(\nu,\mu-\lambda;\mu;1-\beta)$

$$[\operatorname{Re}\mu > \operatorname{Re}\lambda > 0]$$
EH I 205

10.  $\int_0^1 \dfrac{x^{q-1}\,\mathrm{d}x}{(1-x)^q(1+px)} = \dfrac{\pi}{(1+p)^q}\operatorname{cosec}q\pi$        $[0 < q < 1, \quad p > -1]$        BI (5)(1)

11.  $\int_0^1 \dfrac{x^{p-\frac{1}{2}}\,\mathrm{d}x}{(1-x)^p(1+qx)^p} = \dfrac{2\,\Gamma\left(p+\frac{1}{2}\right)\Gamma(1-p)}{\sqrt{\pi}}\cos^{2p}\left(\arctan\sqrt{q}\right)\dfrac{\sin\left[(2p-1)\arctan\left(\sqrt{q}\right)\right]}{(2p-1)\sin\left[\arctan\left(\sqrt{q}\right)\right]}$

$$\left[-\tfrac{1}{2} < p < 1, \quad q > 0\right]$$
BI (11)(1)

12.  $\int_0^1 \dfrac{x^{p-\frac{1}{2}}\,\mathrm{d}x}{(1-x)^p(1-qx)^p} = \dfrac{\Gamma\left(p+\frac{1}{2}\right)\Gamma(1-p)}{\sqrt{\pi}}\dfrac{\left(1-\sqrt{q}\right)^{1-2p}-\left(1+\sqrt{q}\right)^{1-2p}}{(2p-1)\sqrt{q}}$

$$\left[-\tfrac{1}{2} < p < 1, \quad 0 < q < 1\right]$$
BI (11)(2)

**3.198**  $\int_0^1 x^{\mu-1}(1-x)^{\nu-1}[ax+b(1-x)+c]^{-(\mu+\nu)}\,\mathrm{d}x = (a+c)^{-\mu}(b+c)^{-\nu}\,\mathrm{B}(\mu,\nu)$

$$[a \geq 0, \quad b \geq 0, \quad c > 0, \quad \operatorname{Re}\mu > 0, \quad \operatorname{Re}\nu > 0]\quad\text{FI II 787}$$

**3.199**  $\int_a^b (x-a)^{\mu-1}(b-x)^{\nu-1}(x-c)^{-\mu-\nu}\,\mathrm{d}x = (b-a)^{\mu+\nu-1}(b-c)^{-\mu}(a-c)^{-\nu}\,\mathrm{B}(\mu,\nu)$

$$[\operatorname{Re}\mu > 0, \quad \operatorname{Re}\nu > 0, \quad c < a < b]$$
EH I 10(14)

**3.211**  $\int_0^1 x^{\lambda-1}(1-x)^{\mu-1}(1-ux)^{-\varrho}(1-vx)^{-\sigma}\,\mathrm{d}x = \mathrm{B}(\mu,\lambda)\,F_1\left((\lambda,\varrho,\sigma,\lambda+\mu;u,v)\right)$

$$[\operatorname{Re}\lambda > 0, \quad \operatorname{Re}\mu > 0]$$
EH I 231(5)

**3.212** $\displaystyle\int_0^\infty \left[(1+ax)^{-p} + (1+bx)^{-p}\right] x^{q-1}\,\mathrm{d}x = 2(ab)^{-\frac{q}{2}}\,\mathrm{B}(q,p-q)\cos\left\{q\arccos\left[\dfrac{a+b}{2\sqrt{ab}}\right]\right\}$

$$[p > q > 0]\qquad\qquad\text{BI (19)(9)}$$

**3.213** $\displaystyle\int_0^\infty \left[(1+ax)^{-p} - (1+bx)^{-p}\right] x^{q-1}\,\mathrm{d}x = -2i(ab)^{-\frac{q}{2}}\,\mathrm{B}(q,p-q)\sin\left\{q\arccos\left[\dfrac{a+b}{2\sqrt{ab}}\right]\right\}$

$$[p > q > 0]\qquad\qquad\text{BI (19)(10)}$$

**3.214** $\displaystyle\int_0^1 \left[(1+x)^{\mu-1}(1-x)^{\nu-1} + (1+x)^{\nu-1}(1-x)^{\mu-1}\right]\,\mathrm{d}x = 2^{\mu+\nu-1}\,\mathrm{B}(\mu,\nu)$

$$[\operatorname{Re}\mu > 0,\quad \operatorname{Re}\nu > 0]$$
$$\text{LI(1)(15), EH I 10(10)}$$

**3.215** $\displaystyle\int_0^1 \left\{a^\mu x^{\mu-1}(1-ax)^{\nu-1} + (1-a)^\nu x^{\nu-1}[1-(1-a)x]^{\mu-1}\right\}\,\mathrm{d}x = \mathrm{B}(\mu,\nu)$

$$[\operatorname{Re}\mu > 0,\quad \operatorname{Re}\nu > 0,\quad |a| < 1]$$
$$\text{BI (1)(16)}$$

**3.216**

1. $\displaystyle\int_0^1 \frac{x^{\mu-1} + x^{\nu-1}}{(1+x)^{\mu+\nu}}\,\mathrm{d}x = \mathrm{B}(\mu,\nu)$ $\qquad[\operatorname{Re}\mu > 0,\quad \operatorname{Re}\nu > 0]\qquad$ FI II 775

2. $\displaystyle\int_1^\infty \frac{x^{\mu-1} + x^{\nu-1}}{(1+x)^{\mu+\nu}}\,\mathrm{d}x = \mathrm{B}(\mu,\nu)$ $\qquad[\operatorname{Re}\mu > 0,\quad \operatorname{Re}\nu > 0]\qquad$ FI II 775

**3.217** $\displaystyle\int_0^\infty \left\{\frac{b^p x^{p-1}}{(1+bx)^p} - \frac{(1+bx)^{p-1}}{b^{p-1}x^p}\right\}\,\mathrm{d}x = \pi\cot p\pi$ $\qquad[0 < p < 1,\quad b > 0]\qquad$ BI(18)(13)

**3.218** $\displaystyle\int_0^\infty \frac{x^{2p-1} - (a+x)^{2p-1}}{(a+x)^p x^p}\,\mathrm{d}x = \pi\cot p\pi$ $\qquad[p < 1]\quad(\text{cf. }\mathbf{3.217})\qquad$ BI (18)(7)

**3.219** $\displaystyle\int_0^\infty \left\{\frac{x^\nu}{(x+1)^{\nu+1}} - \frac{x^\mu}{(x+1)^{\mu+1}}\right\}\,\mathrm{d}x = \psi(\mu+1) - \psi(\nu+1)$

$$[\operatorname{Re}\mu > -1,\quad \operatorname{Re}\nu > -1]\qquad\text{BI (19)(13)}$$

**3.221**

1. $\displaystyle\int_a^\infty \frac{(x-a)^{p-1}}{x-b}\,\mathrm{d}x = \pi(a-b)^{p-1}\operatorname{cosec} p\pi$ $\qquad[a > b,\quad 0 < p < 1]\qquad$ LI (24)(8)

2. $\displaystyle\int_{-\infty}^a \frac{(a-x)^{p-1}}{x-b}\,\mathrm{d}x = -\pi(b-a)^{p-1}\operatorname{cosec} p\pi$ $\qquad[a < b,\quad 0 < p < 1]\qquad$ LI (24)(8)

**3.222**

1. $\displaystyle\int_0^1 \frac{x^{\mu-1}\,\mathrm{d}x}{1+x} = \beta(\mu)$ $\qquad\qquad[\operatorname{Re}\mu > 0]\qquad\qquad$ WH

2. $\displaystyle\int_0^\infty \frac{x^{\mu-1}\,\mathrm{d}x}{x+a} = \pi\operatorname{cosec}(\mu\pi)a^{\mu-1}$ $\qquad$ for $a > 0$ $\qquad\qquad$ FI II 718, FI II 737

$\qquad\qquad\qquad\qquad = -\pi\cot(\mu\pi)(-a)^{\mu-1}$ $\qquad$ for $a < 0$ $\qquad$ BI(18)(2), ET II 249(28)

$$[0 < \operatorname{Re}\mu < 1]$$

**3.223**

1. $\displaystyle\int_0^\infty \frac{x^{\mu-1}\,\mathrm{d}x}{(b+x)(c+x)} = \frac{\pi}{c-b}\left(b^{\mu-1} - c^{\mu-1}\right)\operatorname{cosec}(\mu\pi)$

$$[|\arg b| < \pi,\quad |\arg c| < \pi,\quad 0 < \operatorname{Re}\mu < 2]\quad\text{ET I 309(7)}$$

2.    $\int_0^\infty \dfrac{x^{\mu-1}\,dx}{(b+x)(a-x)} = \dfrac{\pi}{a+b}\left[b^{\mu-1}\cosec(\mu\pi) + a^{\mu-1}\cot(\mu\pi)\right]$

$$[|\arg b| < \pi, \quad a > 0, \quad 0 < \operatorname{Re}\mu < 2]$$

ET I 309(8)

3.    $\int_0^\infty \dfrac{x^{\mu-1}\,dx}{(a-x)(b-x)} = \pi\cot(\mu\pi)\dfrac{a^{\mu-1}-b^{\mu-1}}{b-a}$      $[a > b > 0, \quad 0 < \operatorname{Re}\mu < 2]$    ET I 309(9)

**3.224**   $\int_0^\infty \dfrac{(x+b)x^{\mu-1}\,dx}{(x+c)(x+d)} = \pi\cosec(\mu\pi)\left\{\dfrac{c-b}{c-d}c^{\mu-1} + \dfrac{d-b}{d-c}d^{\mu-1}\right\}$

$$[|\arg c| < \pi, \quad |\arg d| < \pi, \quad 0 < \operatorname{Re}\mu < 1]$$    ET I 309(10)

**3.225**

1.    $\int_1^\infty \dfrac{(x-1)^{p-1}}{x^2}\,dx = (1-p)\pi\cosec p\pi$       $[-1 < p < 1]$       BI (23)(8)

2.    $\int_1^\infty \dfrac{(x-1)^{1-p}}{x^3}\,dx = \dfrac{1}{2}p(1-p)\pi\cosec p\pi$       $[0 < p < 1]$       BI (23)(1)

3.    $\int_0^\infty \dfrac{x^p\,dx}{(1+x)^3} = \dfrac{\pi}{2}p(1-p)\cosec p\pi$       $[-1 < p < 2]$       BI (16)(5)

**3.226**

1.    $\int_0^1 \dfrac{x^n\,dx}{\sqrt{1-x}} = 2\dfrac{(2n)!!}{(2n+1)!!}$       BI (8)(1)

2.    $\int_0^1 \dfrac{x^{n-\frac{1}{2}}\,dx}{\sqrt{1-x}} = \dfrac{(2n-1)!!}{(2n)!!}\pi.$       BI (8)(2)

**3.227**

1.    $\int_0^\infty \dfrac{x^{\nu-1}(b+x)^{1-\mu}}{c+x}\,dx = b^{1-\mu}c^{\nu-1}\,\mathrm{B}(\nu,\mu-\nu)\,{}_2F_1\left(\mu-1,\nu;\mu;1-\dfrac{c}{b}\right)$

$$[|\arg b| < \pi, \quad |\arg c| < \pi, \quad 0 < \operatorname{Re}\nu < \operatorname{Re}\mu]$$    ET II 217(9)

2.    $\int_0^\infty \dfrac{x^{-\varrho}(b-x)^{-\sigma}}{c+x}\,dx = \pi c^{-\varrho}(b-c)^{-\sigma}\cosec(\varrho\pi)\,I_{1-c/b}(\sigma,\varrho)$

$$[|\arg b| < \pi, \quad |\arg c| < \pi, \quad -\operatorname{Re}\sigma < \operatorname{Re}\varrho < 1]$$    ET II 217(10)

**3.228**

1.    $\int_a^b \dfrac{(x-a)^\nu(b-x)^{-\nu}}{x-c}\,dx = \pi\cosec(\nu\pi)\left[1 - \left(\dfrac{a-c}{b-c}\right)^\nu\right]$       for $c < a$

$$= \pi\cosec(\nu\pi)\left[1 - \cos(\nu\pi)\left(\dfrac{c-a}{b-c}\right)^\nu\right]$$       for $a < c < b$

$$= \pi\cosec(\nu\pi)\left[1 - \left(\dfrac{c-a}{c-b}\right)^\nu\right]$$       for $c > b$

$$[|\operatorname{Re}\nu| < 1]$$    ET II 250(31)

2.  $\displaystyle\int_a^b \frac{(x-a)^{\nu-1}(b-x)^{-\nu}}{x-c}\,dx = \frac{\pi\operatorname{cosec}(\nu\pi)}{b-c}\left|\frac{a-c}{b-c}\right|^{\nu-1}$  for $c < a$ or $c > b$;

$$= -\frac{\pi(c-a)^{\nu-1}}{(b-c)^\nu}\cot(\nu\pi) \qquad \text{for } a < c < b$$

$$[0 < \operatorname{Re}\nu < 1] \qquad\qquad \text{ET II 250(32)}$$

3.$^{12}$  $\displaystyle\int_a^b \frac{(x-a)^{\nu-1}(b-x)^{\mu-1}}{x-c}\,dx$

$$= \frac{(b-a)^{\mu+\nu-1}}{b-c}\,\mathrm{B}(\mu,\nu)\;{}_2F_1\left(1,\mu;\mu+\nu;\frac{b-a}{b-c}\right) \qquad \text{for } c < a \text{ or } c > b;$$

$$= (c-a)^{\nu-1}(b-c)^{\mu-1}\cot\mu\pi - (b-a)^{\mu+\nu-2}\,\mathrm{B}(\mu-1,\nu)$$

$$\times\;{}_2F_1\left(2-\mu-\nu,1;2-\mu;\frac{b-c}{b-a}\right) \qquad \text{for } a < c < b$$

$$[\operatorname{Re}\mu > 0, \quad \operatorname{Re}\nu > 0, \quad \mu+\nu \neq 1, \quad \mu \neq 1,2,\ldots] \quad \text{ET II 250(33)}$$

4.  $\displaystyle\int_0^1 \frac{(1-x)^{\nu-1}x^{-\nu}}{a-bx}\,dx = \frac{\pi(a-b)^{\nu-1}}{a^\nu}\operatorname{cosec}(\nu\pi)$ $\qquad [0 < \operatorname{Re}\nu < 1, \quad 0 < b < a]$ $\qquad$ BI (5)(8)

5.$^{12}$  $\displaystyle\int_0^\infty \frac{x^{\nu-1}(x+a)^{1-\mu}}{x-c}\,dx = a^{1-\mu}(-c)^{\nu-1}\,\mathrm{B}(\mu-\nu,\nu)\;{}_2F_1\left(\mu-1,\nu;\mu;1+\frac{c}{a}\right)$ $\qquad$ for $c < 0$;

$$= c^{\nu-1}(a+c)^{1-\mu}\cot[(\mu-\nu)\pi] - \frac{a^{1-\mu-\nu}}{\pi(a+c)}\,\mathrm{B}(\mu-\nu-1,\nu)$$

$$\times\;{}_2F_1\left(2-\mu,1;2-\mu+\nu;\frac{a}{a+c}\right) \qquad \text{for } c > 0$$

$$[a > 0, \quad 0 < \operatorname{Re}\nu < \operatorname{Re}\mu] \quad \text{ET II 251(34)}$$

6.$^{12}$  $\displaystyle\int_0^\infty x^{\nu-1}\frac{(\gamma+x)^{-n}}{x+\beta}\,dx = \frac{\pi}{\sin\pi\nu}\frac{\beta^{\nu-1}}{(\gamma-\beta)^n}\left[1-\left(\frac{\gamma}{\beta}\right)^{\nu-1}\sum_{j=0}^{n-1}\frac{(1-\nu)^j}{j!}\left(\frac{\gamma-\beta}{\gamma}\right)^j\right]$

$$[|\arg\beta| < \pi, \quad |\arg\gamma| < \pi, \quad 0 < \operatorname{Re}\nu < n] \quad \text{AS 256 (6.1.22)}$$

**3.229**  $\displaystyle\int_0^1 \frac{x^{\mu-1}\,dx}{(1-x)^\mu(1+ax)(1+bx)} = \frac{\pi\operatorname{cosec}\mu\pi}{a-b}\left[\frac{a}{(1+a)^\mu}-\frac{b}{(1+b)^\mu}\right]$

$$[0 < \operatorname{Re}\mu < 1] \qquad\qquad \text{BI (5)(7)}$$

**3.231**

1.  $\displaystyle\int_0^1 \frac{x^{p-1}-x^{-p}}{1-x}\,dx = \pi\cot p\pi$ $\qquad\qquad [p^2 < 1]$ $\qquad\qquad$ BI (4)(4)

2.$^{11}$  $\displaystyle\int_0^1 \frac{x^{p-1}+x^{-p}}{1+x}\,dx = \pi\operatorname{cosec} p\pi$ $\qquad\qquad [p^2 < 1]$ $\qquad\qquad$ BI (4)(1)

3.  $\displaystyle\int_0^1 \frac{x^p - x^{-p}}{x-1}\,dx = \frac{1}{p} - \pi\cot p\pi$ $\qquad\qquad [p^2 < 1]$ $\qquad\qquad$ BI (4)(3)

4.  $\displaystyle\int_0^1 \frac{x^p - x^{-p}}{1+x}\,dx = \frac{1}{p} - \pi\operatorname{cosec} p\pi$ $\qquad\qquad [p^2 < 1]$ $\qquad\qquad$ BI (4)(2)

5. $\displaystyle\int_0^1 \frac{x^{\mu-1} - x^{\nu-1}}{1-x}\, dx = \psi(\nu) - \psi(\mu)$ $\qquad\qquad$ $[\operatorname{Re}\mu > 0, \quad \operatorname{Re}\nu > 0]$

$\qquad\qquad\qquad\qquad\qquad\qquad\qquad\qquad\qquad\qquad\qquad\qquad\qquad$ FI II 815, BI(4)(5)

6. $\displaystyle\int_0^\infty \frac{x^{p-1} - x^{q-1}}{1-x}\, dx = \pi\left(\cot p\pi - \cot q\pi\right)$ $\qquad$ $[p > 0, \quad q > 0]$ $\qquad$ FI II 718

**3.232** $\displaystyle\int_0^\infty \frac{(c+ax)^{-\mu} - (c+bx)^{-\mu}}{x}\, dx = c^{-\mu}\ln\frac{b}{a}$ $\qquad$ $[\operatorname{Re}\mu > -1, \quad a > 0, \quad b > 0, \quad c > 0]$

$\qquad\qquad\qquad\qquad\qquad\qquad\qquad\qquad\qquad\qquad\qquad\qquad\qquad$ BI (18)(14)

**3.233** $\displaystyle\int_0^\infty \left\{\frac{1}{1+x} - (1+x)^{-\nu}\right\}\frac{dx}{x} = \psi(\nu) + \boldsymbol{C}$ $\qquad$ $[\operatorname{Re}\nu > 0]$ $\qquad$ EH I 17, WH

**3.234**

1.[11] $\displaystyle\int_0^1 \left(\frac{x^{q-1}}{1-ax} - \frac{x^{-q}}{a-x}\right)dx = \pi a^{-q}\cot q\pi$ $\qquad$ $[0 < q < 1, \quad a > 0]$ $\qquad$ BI (5)(11)

2. $\displaystyle\int_0^1 \left(\frac{x^{q-1}}{1+ax} + \frac{x^{-q}}{a+x}\right)dx = \pi a^{-q}\operatorname{cosec} q\pi$ $\qquad$ $[0 < q < 1, \quad a > 0]$ $\qquad$ BI (5)(10)

**3.235** $\displaystyle\int_0^\infty \frac{(1+x)^\mu - 1}{(1+x)^\nu}\frac{dx}{x} = \psi(\nu) - \psi(\nu - \mu)$ $\qquad$ $[\operatorname{Re}\nu > \operatorname{Re}\mu > 0]$ $\qquad$ BI (18)(5)

**3.236**[12] $\displaystyle\int_0^1 \frac{x^{\mu/2}\, dx}{[(1-x)(1-a^2x)]^{(\mu+1)/2}} = \frac{(1-a)^{-\mu} - (1+a)^{-\mu}}{2a\mu\sqrt{\pi}}\,\Gamma\left(1 + \frac{\mu}{2}\right)\Gamma\left(\frac{1-\mu}{2}\right)$

$\qquad\qquad\qquad\qquad\qquad\qquad\qquad\qquad\qquad$ $[-2 < \mu < 1, \quad |a| < 1]$ $\qquad$ BI (12)(32)

**3.237** $\displaystyle\sum_{n=0}^\infty (-1)^{n+1}\int_n^{n+1} \frac{dx}{x+u} = \ln\frac{u\left[\Gamma\left(\frac{u}{2}\right)\right]^2}{2\left[\Gamma\left(\frac{u+1}{2}\right)\right]^2}$ $\qquad$ $[|\arg u| < \pi]$ $\qquad$ ET II 216(1)

**3.238**

1. $\displaystyle\int_{-\infty}^\infty \frac{|x|^{\nu-1}}{x-u}\, dx = -\pi\cot\frac{\nu\pi}{2}|u|^{\nu-1}\operatorname{sign} u$ $\qquad$ $[0 < \operatorname{Re}\nu < 1, \quad u \text{ real}, \quad u \neq 0]$

$\qquad\qquad\qquad\qquad\qquad\qquad\qquad\qquad\qquad\qquad\qquad\qquad\qquad$ ET II 249(29)

2. $\displaystyle\int_{-\infty}^\infty \frac{|x|^{\nu-1}}{x-u}\operatorname{sign} x\, dx = \pi\tan\frac{\nu\pi}{2}|u|^{\nu-1}$ $\qquad$ $[0 < \operatorname{Re}\nu < 1, \quad u \text{ real}, \quad u \neq 0]$

$\qquad\qquad\qquad\qquad\qquad\qquad\qquad\qquad\qquad\qquad\qquad\qquad\qquad$ ET II 249(30)

3. $\displaystyle\int_a^b \frac{(b-x)^{\mu-1}(x-a)^{\nu-1}}{|x-u|^{\mu+\nu}}\, dx = \frac{(b-a)^{\mu+\nu-1}}{|a-u|^\mu|b-u|^\nu}\frac{\Gamma(\mu)\Gamma(\nu)}{\Gamma(\mu+\nu)}$

$\qquad\qquad\qquad\qquad$ $[\operatorname{Re}\mu > 0, \quad \operatorname{Re}\nu > 0, \quad 0 < u < a < b \text{ or } 0 < a < b < u]$ $\qquad$ MO 7

## 3.24–3.27 Powers of $x$, of binomials of the form $\alpha + \beta x^p$ and of polynomials in $x$

**3.241**

1. $\displaystyle\int_0^1 \frac{x^{\mu-1}\, dx}{1+x^p} = \frac{1}{p}\beta\left(\frac{\mu}{p}\right)$ $\qquad\qquad$ $[\operatorname{Re}\mu > 0, \quad p > 0]$ $\qquad$ WH, BI (2)(13)

2.    $\displaystyle\int_0^\infty \frac{x^{\mu-1}\,dx}{1+x^\nu} = \frac{\pi}{\nu}\operatorname{cosec}\frac{\mu\pi}{\nu} = \frac{1}{\nu}\,B\left(\frac{\mu}{\nu},\frac{\nu-\mu}{\nu}\right)$      $[\operatorname{Re}\nu \geq \operatorname{Re}\mu \geq 0]$

ET I 309(15)a, BI (17)(10)

3.[11]    $\displaystyle PV\int_0^\infty \frac{x^{p-1}\,dx}{1-x^q} = \frac{\pi}{q}\cot\frac{p\pi}{q}$      $[p<q]$      BI (17)(11)

4.[12]    $\displaystyle\int_0^\infty \frac{x^{\mu-1}\,dx}{(p+qx^\nu)^{n+1}} = \frac{1}{\nu p^{n+1}}\left(\frac{p}{q}\right)^{\mu/\nu} B\left(1+n-\frac{\mu}{\nu},\frac{\mu}{\nu}\right)$

$\left[0<\dfrac{\mu}{\nu}<n+1,\quad p\neq 0,\quad q\neq 0\right]$<br>BI (17)(22)a

5.    $\displaystyle\int_0^\infty \frac{x^{p-1}\,dx}{(1+x^q)^2} = \frac{(p-q)\pi}{q^2}\operatorname{cosec}\frac{(p-q)\pi}{q}$      $[p<2q]$      BI (17)(18)

6.[10]    $\displaystyle G(x) = \int_a^b \operatorname{sign}\left[\frac{x}{c} - \left(\frac{b-u}{b-a}\right)^p\right]du = (b-a)F\left[\left(\frac{x}{c}\right)^{1/p}\right]$

where

$$F(x) = \int_0^1 \operatorname{sign}(x-t)\,dt = \begin{cases} -1 & x\leq 0 \\ 2x-1 & 0<x<1 \\ 1 & x\geq 1 \end{cases}$$

**3.242**

1.    $\displaystyle\int_{-\infty}^\infty \frac{x^{2m}\,dx}{x^{4n}+2x^{2n}\cos t+1} = \frac{\pi}{n}\sin\left[\frac{(2n-2m-1)}{2n}t\right]\operatorname{cosec} t\,\operatorname{cosec}\frac{(2m+1)\pi}{2n}$

$[m<n,\quad t^2<\pi^2]$      FI II 642

2.[11]    $\displaystyle\int_0^\infty \left[\frac{x^2}{x^4+2ax^2+1}\right]^c \left(\frac{x^2+1}{x^b+1}\right)\frac{dx}{x^2} = 2^{-1/2-c}(1+a)^{1/2-c}\,B\left(c-\frac{1}{2},\frac{1}{2}\right)$

**3.243**[11]    $\displaystyle\int_0^\infty \frac{x^{\mu-1}\,dx}{(1+x^{2\nu})(1+x^{3\nu})}$

$$= \frac{\pi}{48\nu}\left[8\operatorname{cosec}(2\rho)+12\operatorname{cosec}(3\rho)-8\operatorname{cosec}\left(2\rho-\frac{4\pi}{3}\right)+8\operatorname{cosec}\left(2\rho-\frac{2\pi}{3}\right)\right.$$
$$\left.-3\operatorname{cosec}\left(\rho-\frac{\pi}{6}\right)\operatorname{cosec}\left(\rho+\frac{\pi}{6}\right)\sec(\rho)\right]$$

where $\rho = \dfrac{\mu\pi}{6\nu}$,    $[0<\operatorname{Re}\mu<5\operatorname{Re}\nu]$    ET I 312(34)

**3.244**

1.    $\displaystyle\int_0^1 \frac{x^{p-1}+x^{q-p-1}}{1+x^q}\,dx = \frac{\pi}{q}\operatorname{cosec}\frac{p\pi}{q}$      $[q>p>0]$      BI (2)(14)

2.    $\displaystyle\int_0^1 \frac{x^{p-1}-x^{q-p-1}}{1-x^q}\,dx = \frac{\pi}{q}\cot\frac{p\pi}{q}$      $[q>p>0]$      BI (2)(16)

3.    $\displaystyle\int_0^1 \frac{x^{\nu-1}-x^{\mu-1}}{1-x^\nu}\,dx = \frac{1}{\nu}\left[C+\psi\left(\frac{\mu}{\nu}\right)\right]$      $[\operatorname{Re}\mu>\operatorname{Re}\nu>0]$      BI (2)(17)

4.    $\int_{-\infty}^{\infty} \dfrac{x^{2m} - x^{2n}}{1 - x^{2l}} \, dx = \dfrac{\pi}{l} \left[ \cot \left( \dfrac{2m+1}{2l} \pi \right) - \cot \left( \dfrac{2n+1}{2l} \pi \right) \right]$

$$[m < l, \quad n < l]$$      FI II 640

**3.245**    $\int_{0}^{\infty} \left[ x^{\nu - \mu} - x^{\nu} (1+x)^{-\mu} \right] dx = \dfrac{\nu}{\nu - \mu + 1} B(\nu, \mu - \nu)$

$$[\operatorname{Re} \mu > \operatorname{Re} \nu > 0]$$      BI (16)(13)

**3.246**    $\int_{0}^{\infty} \dfrac{1 - x^q}{1 - x^r} x^{p-1} \, dx = \dfrac{\pi}{r} \sin \dfrac{q\pi}{r} \operatorname{cosec} \dfrac{p\pi}{r} \operatorname{cosec} \dfrac{(p+q)\pi}{r}$

$$[p + q < r, \quad p > 0]$$

ET I 331(33), BI (17)(12)

Integrals of the form $\int f \left( x^p \pm x^{-p}, x^q \pm x^{-q}, \ldots \right) \dfrac{dx}{x}$ can be transformed by the substitution $x = e^t$ or $x = e^{-t}$. For example, instead of $\int_{0}^{1} \left( x^{1+p} + x^{1-p} \right)^{-1} dx$, we should seek to evaluate $\int_{0}^{\infty} \operatorname{sech} px \, dx$ and, instead of $\int_{0}^{1} \dfrac{x^{n-m-1} + x^{n+m-1}}{1 + 2x^n \cos a + x^{2n}} \, dx$, we should seek to evaluate $\int_{0}^{\infty} \cosh mx \left( \cosh nx - \cos a \right)^{-1} dx$ (see **3.514** 2).

**3.247**

1.[11]    $\int_{0}^{1} \dfrac{x^{\alpha - 1} (1 - x)^{n-1}}{1 - \xi x^b} \, dx = (n-1)! \sum_{k=0}^{\infty} \dfrac{\xi^k}{(\alpha + kb)(\alpha + kb + 1) \ldots (\alpha + kb + n - 1)}$

$$[b > 0, \quad |\xi| < 1]$$      AD (6704)

2.    $\int_{0}^{\infty} \dfrac{(1 - x^p) x^{\nu - 1}}{1 - x^{np}} \, dx = \dfrac{\pi}{np} \sin \left( \dfrac{\pi}{n} \right) \operatorname{cosec} \dfrac{(p+\nu)\pi}{np} \operatorname{cosec} \dfrac{\pi \nu}{np}$

$$[0 < \operatorname{Re} \nu < (n-1)p]$$      ET I 311(33)

**3.248**

1.    $\int_{0}^{\infty} \dfrac{x^{\mu - 1} \, dx}{\sqrt{1 + x^{\nu}}} = \dfrac{1}{\nu} B \left( \dfrac{\mu}{\nu}, \dfrac{1}{2} - \dfrac{\mu}{\nu} \right)$      $[\operatorname{Re} \nu > \operatorname{Re} 2\mu > 0]$      BI (21)(9)

2.[12]    $\int_{0}^{1} \dfrac{x^{2n+1} \, dx}{\sqrt{1 - x^2}} = \dfrac{\sqrt{\pi} \, n!}{2\Gamma \left( n + \frac{3}{2} \right)} = \dfrac{(2n)!!}{(2n+1)!!}$      BI (8)(14)

3.[12]    $\int_{0}^{1} \dfrac{x^{2n} \, dx}{\sqrt{1 - x^2}} = \dfrac{\sqrt{\pi}}{2n!} \Gamma \left( n + \dfrac{1}{2} \right) = \dfrac{(2n-1)!!}{(2n)!!} \dfrac{\pi}{2}$      BI (8)(13)

4.[3]    $\int_{-\infty}^{\infty} \dfrac{dx}{(1 + x^2) \sqrt{4 + 3x^2}} = \dfrac{\pi}{3}$

6.[12]    $\int_{-\infty}^{\infty} \dfrac{dx}{(1 + x^2) \sqrt{b + ax^2}} = \begin{cases} \dfrac{2}{\sqrt{b - a}} \arctan \left( \sqrt{\dfrac{b}{a} - 1} \right) & \text{if } a < b \\[3mm] \dfrac{2}{\sqrt{a}} & \text{if } a = b \\[3mm] \dfrac{1}{\sqrt{a - b}} \ln \left( \dfrac{\sqrt{a} + \sqrt{a - b}}{\sqrt{a} - \sqrt{a - b}} \right) & \text{if } a > b \end{cases}$

**3.249**

1.$^0$ $\displaystyle\int_0^\infty \frac{\mathrm{d}x}{(x^2+a^2)^n} = \frac{(2n-3)!!}{2\cdot(2n-2)!!}\frac{\pi}{a^{2n-1}}$                          FI II 743

2.$^9$ $\displaystyle\int_0^a \left(a^2-x^2\right)^{n-\frac{1}{2}}\,\mathrm{d}x = a^{2n}\frac{(2n-1)!!}{2(2n)!!}\pi.$                FI II 156

3. $\displaystyle\int_{-1}^1 \frac{\left(1-x^2\right)^n \mathrm{d}x}{(a-x)^{n+1}} = 2^{n+1}\,Q_n(a)$                        EH II 181(31)

4. $\displaystyle\int_0^1 \frac{x^\mu\,\mathrm{d}x}{1+x^2} = \frac{1}{2}\,\beta\left(\frac{\mu+1}{2}\right)$        $[\operatorname{Re}\mu > -1]$          BI (2)(7)

5. $\displaystyle\int_0^1 \left(1-x^2\right)^{\mu-1}\,\mathrm{d}x = 2^{2\mu-2}\,\mathrm{B}(\mu,\mu) = \tfrac{1}{2}\,\mathrm{B}\left(\tfrac{1}{2},\mu\right)$      $[\operatorname{Re}\mu > 0]$      FI II 784

6. $\displaystyle\int_0^1 \left(1-\sqrt{x}\right)^{p-1}\,\mathrm{d}x = \frac{2}{p(p+1)}$          $[p>0]$           BI (7)(7)

7. $\displaystyle\int_0^1 \left(1-x^\mu\right)^{-\frac{1}{\nu}}\,\mathrm{d}x = \frac{1}{\mu}\,\mathrm{B}\left(\frac{1}{\mu}, 1-\frac{1}{\nu}\right)$       $[\operatorname{Re}\mu > 0,\quad |\nu| > 1]$

8.$^{11}$ $\displaystyle\int_{-\infty}^\infty \left(1+\frac{x^2}{n-1}\right)^{-n/2}\,\mathrm{d}x = \frac{\sqrt{\pi(n-1)}}{\Gamma\left(\frac{n}{2}\right)}\Gamma\left(\frac{n-1}{2}\right)$     $[n>1]$

**3.251**

1. $\displaystyle\int_0^1 x^{\mu-1}\left(1-x^\lambda\right)^{\nu-1}\,\mathrm{d}x = \frac{1}{\lambda}\,\mathrm{B}\left(\frac{\mu}{\lambda},\nu\right)$      $[\operatorname{Re}\mu > 0,\quad \operatorname{Re}\nu > 0,\quad \lambda > 0]$

                                                                 FI II 787

2. $\displaystyle\int_0^\infty x^{\mu-1}\left(1+x^2\right)^{\nu-1}\,\mathrm{d}x = \frac{1}{2}\,\mathrm{B}\left(\frac{\mu}{2}, 1-\nu-\frac{\mu}{2}\right)$      $\left[\operatorname{Re}\mu > 0,\quad \operatorname{Re}\left(\nu+\tfrac{1}{2}\mu\right) < 1\right]$

3. $\displaystyle\int_1^\infty x^{\mu-1}\left(x^p-1\right)^{\nu-1}\,\mathrm{d}x = \frac{1}{p}\,\mathrm{B}\left(1-\nu-\frac{\mu}{p},\nu\right)$     $[p>0,\quad \operatorname{Re}\nu > 0,\quad \operatorname{Re}\mu < p-p\operatorname{Re}\nu]$

                                                                  ET I 311(32)

4. $\displaystyle\int_0^\infty \frac{x^{2m}\,\mathrm{d}x}{(ax^2+c)^n} = \frac{(2m-1)!!(2n-2m-3)!!\pi}{2(2n-2)!!a^m c^{n-m-1}\sqrt{ac}}$     $[a>0,\quad c>0,\quad n>m+1]$

                                                                  GU (141)(8a)

5. $\displaystyle\int_0^\infty \frac{x^{2m+1}\,\mathrm{d}x}{(ax^2+c)^n} = \frac{m!(n-m-2)!}{2(n-1)!a^{m+1}c^{n-m-1}}$     $[ac>0,\quad n>m+1\geq 1]$    GU (141)(8b)

6.$^{12}$ $\displaystyle\int_0^\infty \frac{x^{2a+1}\,\mathrm{d}x}{(1+x^2)^2} = \frac{\pi a}{2\sin\pi a}$        $[|\operatorname{Re}a| < 1]$           WH

7.$^{12}$ $\displaystyle\int_0^1 \frac{x^{2a+1}\,\mathrm{d}x}{(1+x^2)^2} = \frac{2a\beta(a)-1}{4}$        $[\operatorname{Re}a > 0]$          LI (3)(11)

8. $\displaystyle\int_0^1 x^{q+p-1}\left(1-x^q\right)^{-\frac{p}{q}}\,\mathrm{d}x = \frac{p\pi}{q^2}\cosec\frac{p\pi}{q}$        $[q>p]$          BI (9)(22)

9.  $\int_0^1 x^{\frac{q}{p}-1}\left(1-x^q\right)^{-\frac{1}{p}}\,\mathrm{d}x = \dfrac{\pi}{q}\operatorname{cosec}\dfrac{\pi}{p}$   $[p>1,\quad q>0]$   BI (9)(23)a

10.  $\int_0^1 x^{p-1}\left(1-x^q\right)^{-\frac{p}{q}}\,\mathrm{d}x = \dfrac{\pi}{q}\operatorname{cosec}\dfrac{p\pi}{q}$   $[q>p>0]$   BI (9)(20)

11.  $\int_0^\infty x^{\mu-1}\left(1+bx^p\right)^{-\nu}\,\mathrm{d}x = \dfrac{1}{p}b^{-\frac{\mu}{p}}\,\mathrm{B}\left(\dfrac{\mu}{p},\nu-\dfrac{\mu}{p}\right)$

$[|\arg b|<\pi,\quad p>0,\quad 0<\operatorname{Re}\mu<p\operatorname{Re}\nu]$   BI (17)(20), EH I 10(16)

**3.252**

1.  $\int_0^\infty \dfrac{\mathrm{d}x}{\left(ax^2+2bx+c\right)^n} = \dfrac{(-1)^{n-1}}{(n-1)!}\dfrac{\partial^{n-1}}{\partial c^{n-1}}\left[\dfrac{1}{\sqrt{ac-b^2}}\operatorname{arccot}\dfrac{b}{\sqrt{ac-b^2}}\right]$

$\left[a>0,\quad ac>b^2\right]$   GW (131)(4)

2.  $\int_{-\infty}^\infty \dfrac{\mathrm{d}x}{\left(ax^2+2bx+c\right)^n} = \dfrac{(2n-3)!!\pi a^{n-1}}{(2n-2)!!\left(ac-b^2\right)^{n-\frac{1}{2}}}$   $\left[a>0,\quad ac>b^2\right]$   GW (131)(5)

3.  $\int_0^\infty \dfrac{\mathrm{d}x}{\left(ax^2+2bx+c\right)^{n+\frac{3}{2}}} = \dfrac{(-2)^n}{(2n+1)!!}\dfrac{\partial^n}{\partial c^n}\left\{\dfrac{1}{\sqrt{c}\left(\sqrt{ac}+b\right)}\right\}$

$\left[a\ge0,\quad c>0,\quad b>-\sqrt{ac}\right]$
GW (213)(4)

4.  $\int_0^\infty \dfrac{x\,\mathrm{d}x}{\left(ax^2+2bx+c\right)^n}$

$= \dfrac{(-1)^n}{(n-1)!}\dfrac{\partial^{n-2}}{\partial c^{n-2}}\left\{\dfrac{1}{2\left(ac-b^2\right)}-\dfrac{b}{2\left(ac-b^2\right)^{\frac{3}{2}}}\operatorname{arccot}\dfrac{b}{\sqrt{ac-b^2}}\right\}$   for $ac>b^2$;

$= \dfrac{(-1)^n}{(n-1)!}\dfrac{\partial^{n-2}}{\partial c^{n-2}}\left\{\dfrac{1}{2\left(ac-b^2\right)}+\dfrac{b}{4\left(b^2-ac\right)^{\frac{3}{2}}}\ln\dfrac{b+\sqrt{b^2-ac}}{b-\sqrt{b^2-ac}}\right\}$   for $b^2>ac>0$;

$= \dfrac{a^{n-2}}{2(n-1)(2n-1)b^{2n-2}}$   for $ac=b^2$

$[a>0,\quad b>0,\quad n\ge2]$   GW (141)(5)

5.  $\int_{-\infty}^\infty \dfrac{x\,\mathrm{d}x}{\left(ax^2+2bx+c\right)^n} = -\dfrac{(2n-3)!!\pi ba^{n-2}}{(2n-2)!!\left(ac-b^2\right)^{\frac{(2n-1)}{2}}}$   $\left[ac>b^2,\quad a>0,\quad n\ge2\right]$

GW (141)(6)

6.  $\int_{-\infty}^\infty \dfrac{x^m\,\mathrm{d}x}{\left(ax^2+2bx+c\right)^n} = \dfrac{(-1)^m\pi a^{n-m-1}b^m}{(2n-2)!!\left(ac-b^2\right)^{n-\frac{1}{2}}}$

$\times \displaystyle\sum_{k=0}^{\lfloor m/2\rfloor}\binom{m}{2k}(2k-1)!!(2n-2k-3)!!\left(\dfrac{ac-b^2}{b^2}\right)^k$

$\left[ac>b^2,\quad 0\le m\le 2n-2\right]$   GW (141)(17)

7.  $\int_0^\infty \dfrac{x^n\,\mathrm{d}x}{\left(ax^2+2bx+c\right)^{n+\frac{3}{2}}} = \dfrac{n!}{(2n+1)!!\sqrt{c}\left(\sqrt{ac}+b\right)^{n+1}}$

$\left[a\ge0,\quad c>0,\quad b>-\sqrt{ac}\right]$
GW (213)(5a)

8. $$\int_0^\infty \frac{x^{n+1}\,dx}{(ax^2+2bx+c)^{n+\frac{3}{2}}} = \frac{n!}{(2n+1)!!\sqrt{a}\left(\sqrt{ac}+b\right)^{n+1}}$$

$$\left[a>0,\quad c\geq 0,\quad b>-\sqrt{ac}\right]$$

GW (213)(5b)

9. $$\int_0^\infty \frac{x^{n+\frac{1}{2}}\,dx}{(ax^2+2bx+c)^{n+1}} = \frac{(2n-1)!!\pi}{2^{2n+\frac{1}{2}}\left(b+\sqrt{ac}\right)^{n+\frac{1}{2}}n!\sqrt{a}}$$

$$\left[a>0,\quad c>0,\quad b+\sqrt{ac}>0\right]$$

LI (21)(19)

10.[12] $$\int_0^\infty \frac{x^{\mu-1}\,dx}{(1+2x\cos t+x^2)^\nu} = 2^{\nu-\frac{1}{2}}(\sin t)^{\frac{1}{2}-\nu}\,\Gamma\left(\nu+\frac{1}{2}\right)\mathrm{B}(\mu,2\nu-\mu)\,P_{\mu-\nu-\frac{1}{2}}^{\frac{1}{2}-\nu}(\cos t)$$

$$\left[0<t<\pi,\quad 0<\mathrm{Re}\,\mu<\mathrm{Re}\,2\nu\right]$$

ET I 310(22)

11. $$\int_0^\infty \left(1+2\beta x+x^2\right)^{\mu-\frac{1}{2}}x^{-\nu-1}\,dx = 2^{-\mu}\left(\beta^2-1\right)^{\frac{\mu}{2}}\Gamma(1-\mu)\,\mathrm{B}(\nu-2\mu+1,-\nu)\,P_{\nu-\mu}^\mu(\beta)$$

$$\left[\mathrm{Re}\,\nu<0,\quad \mathrm{Re}(2\mu-\nu)<1,\quad |\arg(\beta\pm 1)|<\pi\right]$$

EH I 160(33)

$$= -\pi\,\mathrm{cosec}\,\nu\pi\,C_\nu^{\frac{1}{2}-\mu}(\beta)$$

$$\left[-2<\mathrm{Re}\left(\tfrac{1}{2}-\mu\right)<\mathrm{Re}\,\nu<0,\quad |\arg(\beta\pm 1)|<\pi\right]$$

EH I 178(24)

12. $$\int_0^\infty \frac{x^{\mu-1}\,dx}{x^2+2ax\cos t+a^2} = -\pi a^{\mu-2}\,\mathrm{cosec}\,t\,\mathrm{cosec}(\mu\pi)\sin[(\mu-1)t]$$

$$\left[a>0,\quad 0<|t|<\pi,\quad 0<\mathrm{Re}\,\mu<2\right]$$

FI II 738, BI(20)(3)

13. $$\int_0^\infty \frac{x^{\mu-1}\,dx}{(x^2+2ax\cos t+a^2)^2} = \frac{\pi a^{\mu-4}}{2}\,\mathrm{cosec}\,\mu\pi\,\mathrm{cosec}^3 t$$

$$\times\{(\mu-1)\sin t\cos[(\mu-2)t]-\sin[(\mu-1)t]\}$$

$$\left[a>0,\quad 0<|t|<\pi,\quad 0<\mathrm{Re}\,\mu<4\right]\quad \text{LI(20)(8)a, ET I 309(13)}$$

14. $$\int_0^\infty \frac{x^{\mu-1}\,dx}{\sqrt{1+2x\cos t+x^2}} = \pi\,\mathrm{cosec}(\mu\pi)\,P_{\mu-1}(\cos t)\qquad \left[-\pi<t<\pi,\quad 0<\mathrm{Re}\,\mu<1\right]$$

ET I 310(17)

**3.253** $$\int_{-1}^1 \frac{(1+x)^{2\mu-1}(1-x)^{2\nu-1}}{(1+x^2)^{\mu+\nu}}\,dx = 2^{\mu+\nu-2}\,\mathrm{B}(\mu,\nu)\qquad \left[\mathrm{Re}\,\mu>0,\quad \mathrm{Re}\,\nu>0\right]\qquad \text{FI II 787}$$

**3.254**

1.[12] $$\int_0^u x^{\lambda-1}(u-x)^{\mu-1}\left(x^2+\beta^2\right)^\nu\,dx$$

$$= \beta^{2\nu}u^{\lambda+\mu-1}\,\mathrm{B}(\lambda,\mu)\,{}_3F_2\left(-\nu,\frac{\lambda}{2},\frac{\lambda+1}{2};\frac{\lambda+\mu}{2},\frac{\lambda+\mu+1}{2};\frac{-u^2}{\beta^2}\right)$$

$$\left[\mathrm{Re}\left(\frac{u}{\beta}\right)>0,\quad \mathrm{Re}\,\lambda>0,\quad \mathrm{Re}\,\mu>0\right]\quad \text{ET II 186(10)}$$

$2.^{12}$ $\displaystyle\int_u^\infty x^{-\lambda}(x-u)^{\mu-1}\left(x^2+\beta^2\right)^\nu\,dx$

$$= u^{\mu-\lambda+2\nu}B\left(\lambda-2\nu-\mu,\mu\right)$$

$$\times\,{}_3F_2\left(-\nu,\frac{\lambda-\mu}{2}-\nu,\frac{1+\lambda-\mu}{2}-\nu;\frac{\lambda}{2}-\nu,\frac{1+\lambda}{2}-\nu;-\frac{\beta^2}{u^2}\right)$$

$$\left[|u|>|\beta|\text{ and }\operatorname{Re}\left(\frac{\beta}{u}\right)>0,\quad 0<\operatorname{Re}\mu<\operatorname{Re}(\lambda-2\nu)\right]\quad\text{ET II 202(9)}$$

**3.255** $\displaystyle\int_0^1\frac{x^{\mu+\frac12}(1-x)^{\mu-\frac12}}{(c+2bx-ax^2)^{\mu+1}}\,dx=\frac{\sqrt\pi}{\left\{a+\left(\sqrt{c+2b-a}+\sqrt c\right)^2\right\}^{\mu+\frac12}\sqrt{c+2b-a}}\frac{\Gamma\left(\mu+\frac12\right)}{\Gamma(\mu+1)}$

$$\left[a+\left(\sqrt{c+2b-a}+\sqrt c\right)^2>0,\quad c+2b-a>0,\quad\operatorname{Re}\mu>-\frac12\right]$$

$$\text{BI (14)(2)}$$

**3.256**

1. $\displaystyle\int_0^1\frac{x^{p-1}+x^{q-1}}{\left(1-x^2\right)^{\frac{p+q}{2}}}\,dx=\frac12\cos\left(\frac{q-p}{4}\pi\right)\sec\left(\frac{q+p}{4}\pi\right)B\left(\frac p2,\frac q2\right)$

$$[p>0,\quad q>0,\quad p+q<2]\quad\text{BI (8)(25)}$$

2. $\displaystyle\int_0^1\frac{x^{p-1}-x^{q-1}}{\left(1-x^2\right)^{\frac{p+q}{2}}}\,dx=\frac12\sin\left(\frac{q-p}{4}\pi\right)\operatorname{cosec}\left(\frac{q+p}{4}\pi\right)B\left(\frac p2,\frac q2\right)$

$$[p>0,\quad q>0,\quad p+q<2]\quad\text{BI (8)(26)}$$

**3.257**[9] $\displaystyle\int_0^\infty\left[\left(ax+\frac bx\right)^2+c\right]^{-p-1}\,dx$

$$=\frac{\sqrt\pi\,\Gamma\left(p+\frac12\right)}{2ac^{p+\frac12}\,\Gamma(p+1)}\qquad\left[a>0,\quad b<0,\quad c>0,\quad p>-\frac12\right]\quad\text{BI (20)(4)}$$

$$=\frac12\frac{B\left(p+\frac12,\frac12\right)}{a\left(4ab+c\right)^{p+\frac12}}\qquad\left[a>0,\quad b>0,\quad c>-4ab,\quad p>-\frac12\right]$$

**3.258**

1. $\displaystyle\int_b^\infty\left(x-\sqrt{x^2-a^2}\right)^n\,dx=\frac{a^2}{2(n-1)}\left(b-\sqrt{b^2-a^2}\right)^{n-1}-\frac{1}{2(n+1)}\left(b-\sqrt{b^2-a^2}\right)^{n+1}$

$$[0<a\le b,\quad n\ge2]\quad\text{GW (215)(5)}$$

2. $\displaystyle\int_b^\infty\left(\sqrt{x^2+1}-x\right)^n\,dx=\frac{\left(\sqrt{b^2+1}-b\right)^{n-1}}{2(n-1)}+\frac{\left(\sqrt{b^2+1}-b\right)^{n+1}}{2(n+1)}$

$$[n\ge2]\quad\text{GW (214)(7)}$$

3. $\displaystyle\int_0^\infty\left(\sqrt{x^2+a^2}-x\right)^n\,dx=\frac{na^{n+1}}{n^2-1}$

$$[n\ge2]\quad\text{GW (214)(6a)}$$

4.[12] $\displaystyle\int_0^\infty\frac{dx}{\left(x+\sqrt{x^2+a^2}\right)^n}=\frac{n}{a^{n-1}\left(n^2-1\right)}$

$$[n\ge2]\quad\text{GW (214)(5a)}$$

5.    $$\int_0^\infty x^m \left(\sqrt{x^2 + a^2} - x\right)^n dx = \frac{n \cdot m! a^{m+n+1}}{(n-m-1)(n-m+1)\dots(m+n+1)}$$

$$[a > 0, \quad 0 \le m \le n-2] \qquad \text{GW (214)(6)}$$

6.    $$\int_0^\infty \frac{x^m\, dx}{\left(x + \sqrt{x^2 + a^2}\right)^n} = \frac{n \cdot m!}{(n-m-1)(n-m+1)\dots(m+n+1)a^{n-m-1}}$$

$$[a > 0, \quad 0 \le m \le n-2] \qquad \text{GW (214)(5)}$$

7.    $$\int_a^\infty (x-a)^m \left(x - \sqrt{x^2 - a^2}\right)^n dx = \frac{n \cdot (n-m-2)!(2m+1)! a^{m+n+1}}{2^m(n+m+1)!}$$

$$[a > 0, \quad n \ge m+2] \qquad \text{GH (215)(6)}$$

**3.259**

1.⁶    $$\int_0^1 x^{p-1}(1-x)^{n-1}(1+bx^m)^l\, dx = (n-1)! \sum_{k=0}^\infty \binom{l}{k} \frac{b^k\, \Gamma(p+km)}{\Gamma(p+n+km)}$$

$$[|b| < 1 \text{ unless } l = 0, 1, 2, \dots; \quad p, n, p+ml > 0] \quad \text{BI (1)(14)}$$

2.¹¹    $$\int_0^u x^{\nu-1}(u-x)^{\mu-1}(x^m + b^m)^\lambda\, dx$$

$$= b^{m\lambda} u^{\mu+\nu-1} B(\mu, \nu)$$

$$\times\ {}_{m+1}F_m\left(-\lambda, \frac{\nu}{m}, \frac{\nu+1}{m}, \dots, \frac{\nu+m-1}{m}; \frac{\mu+\nu}{m}, \frac{\mu+\nu+1}{m}, \dots, \frac{\mu+\nu+m-1}{m}; \frac{-u^m}{b^m}\right)$$

$$\left[\operatorname{Re}\mu > 0, \quad \operatorname{Re}\nu > 0, \quad \left|\arg\left(\frac{u}{b}\right)\right| < \frac{\pi}{m}\right] \quad \text{ET II 186(11)}$$

3.¹¹    $$\int_0^\infty x^{\lambda-1}(1+ax^p)^{-\mu}(1+bx^p)^{-\nu}\, dx = \frac{1}{p} a^{-\lambda/p} B\left(\frac{\lambda}{p}, \mu+\nu-\frac{\lambda}{p}\right) {}_2F_1\left(\nu, \frac{\lambda}{p}; \mu+\nu; 1-\frac{b}{a}\right)$$

$$[|\arg a| < \pi, \quad |\arg b| < \pi, \quad p > 0, \quad 0 < \operatorname{Re}\lambda < 2\operatorname{Re}(\mu+\nu)] \quad \text{ET I 312(35)}$$

**3.261**

1.¹¹    $$\mathrm{PV}\int_0^1 \frac{(1 - x\cos t) x^{\mu-1}\, dx}{1 - 2x\cos t + x^2} = \sum_{k=0}^\infty \frac{\cos kt}{\mu+k} \qquad [\operatorname{Re}\mu > 0, \quad t \ne 2n\pi] \qquad \text{BI (6)(9)}$$

2.    $$\int_0^1 \frac{(x^\nu + x^{-\nu})\, dx}{1 + 2x\cos t + x^2} = \frac{\pi \sin \nu t}{\sin t \sin \nu\pi} \qquad [\nu^2 < 1, \quad t \ne (2n+1)\pi] \qquad \text{BI (6)(8)}$$

3.    $$\int_0^1 \frac{(x^{1+p} + x^{1-p})\, dx}{(1 + 2x\cos t + x^2)^2} = \frac{\pi(p\sin t \cos pt - \cos t \sin pt)}{2\sin^3 t \sin p\pi}$$

$$[p^2 < 1, \quad t \ne (2n+1)\pi] \qquad \text{BI (6)(18)}$$

4.¹²    $$\int_0^1 \frac{x^{\mu-1}}{1 + 2ax\cos t + a^2 x^2} \cdot \frac{dx}{(1-x)^\mu} = \frac{\pi\operatorname{cosec} t \operatorname{cosec}\mu\pi}{(1 + 2a\cos t + a^2)^{\mu/2}} \sin\left(t - \mu\arctan\frac{a\sin t}{1 + a\cos t}\right)$$

$$[a > 0, \quad 0 < \operatorname{Re}\mu < 1] \qquad \text{BI (6)(21)}$$

**3.262**  $\displaystyle\int_0^\infty \frac{x^{-p}\,dx}{1+x^3} = \frac{\pi}{3}\operatorname{cosec}\frac{(1-p)\pi}{3}$  $\qquad\qquad [-2 < p < 1]$  LI (18)(3)

**3.263**  $\displaystyle\int_0^\infty \frac{x^\nu\,dx}{(x+c)(x^2+b^2)} = \frac{\pi}{2\,(b^2+c^2)}\left[cb^{\nu-1}\sec\frac{\nu\pi}{2} + b^\nu\operatorname{cosec}\frac{\nu\pi}{2} - 2c^\nu\operatorname{cosec}(\nu\pi)\right]$

$\qquad\qquad [\operatorname{Re} b > 0, \quad |\arg c| < \pi, \quad -1 < \operatorname{Re}\nu < 2, \quad \nu \neq 0]$  ET II 216(7)

**3.264**

1.  $\displaystyle\int_0^\infty \frac{x^{p-1}\,dx}{(a^2+x^2)(b^2-x^2)} = \frac{\pi}{2}\,\frac{a^{p-2}+b^{p-2}\cos\frac{p\pi}{2}}{a^2+b^2}\operatorname{cosec}\frac{p\pi}{2}$

$\qquad\qquad [0 < p < 4, \quad a > 0, \quad b > 0]$

$\qquad\qquad$ BI (19)(14)

2.  $\displaystyle\int_0^\infty \frac{x^{\mu-1}\,dx}{(b+x^2)(c+x^2)} = \frac{\pi}{2}\,\frac{c^{\frac{\mu}{2}-1}-b^{\frac{\mu}{2}-1}}{b-c}\operatorname{cosec}\frac{\mu\pi}{2}$

$\qquad\qquad\qquad = \dfrac{\pi}{2(c-b)}\left(\dfrac{1}{\sqrt{b}}-\dfrac{1}{\sqrt{c}}\right)$  $\qquad \left[\mu = \tfrac{1}{2}\right]$

$\qquad\qquad [|\arg b| < \pi, \quad |\arg c| < \pi, \quad 0 < \operatorname{Re}\mu < 4]$  ET I 309(4)

3.  $\displaystyle\int_0^\infty \frac{dx}{(b+x^2)(a+b+x^2)^2} = \frac{\pi}{2}\left(\frac{1}{a^2 b^{1/2}} - \frac{1}{2a\,(a+b)^{3/2}} - \frac{1}{a^2\,(a+b)^{1/2}}\right)$

$\qquad\qquad$ MC

4.  $\displaystyle\int_0^\infty \frac{dx}{(b+x^2)(a+b+x^2)^3} = \frac{\pi}{4}\left(\frac{2}{a^3 b^{1/2}} - \frac{3}{4a\,(a+b)^{5/2}} - \frac{1}{a^2\,(a+b)^{3/2}} - \frac{2}{a^3\,(a+b)^{1/2}}\right)$

5.  $\displaystyle\int_0^\infty \frac{dx}{(b+x^2)(a+b+x^2)^4} = \frac{\pi}{4}\left(\frac{2}{a^4 b^{1/2}} - \frac{5}{8a\,(a+b)^{7/2}} - \frac{3}{4a^2\,(a+b)^{5/2}}\right.$

$\qquad\qquad\qquad\qquad\qquad\left. - \frac{1}{a^3\,(a+b)^{3/2}} - \frac{2}{a^4\,(a+b)^{1/2}}\right)$

6.[12]  $\displaystyle\int_0^\infty \frac{dx}{(b+x^2)(a+b+x^2)^n}$

$\qquad\qquad = \frac{\pi}{2}\,\frac{1}{a^n b^{1/2}} - \frac{1}{2a\,(a+b)^{n-1/2}}\,\mathrm{B}\left(n-\frac{1}{2},\frac{1}{2}\right)\,{}_2F_1\left(1-n,1;\frac{3}{2}-n;\frac{a+b}{a}\right)$

$\qquad\qquad$ AS 263 (6.6.3.2)

$\qquad\qquad = \frac{\pi}{2}\,\frac{1}{a^n b^{1/2}} - \frac{\pi}{2a^n\,(a+b)^{n-1/2}}\sum_{j=0}^{n-1}\frac{\left(\frac{1}{2}\right)^j}{j!}\left(\frac{a}{a+b}\right)^j$

$\qquad\qquad\qquad\qquad\qquad\qquad [n > 0, \quad a+b > 0]$

7.  $\displaystyle\int_0^\infty \frac{x^2\,dx}{(x^2+a^2)(x^2+b^2)(x^2+c^2)} = \frac{\pi}{2a\,(b^2-c^2)}\left[\frac{b}{b+a} - \frac{c}{c+a}\right] = \frac{\pi}{2(a+b)(a+c)(b+c)}$

**3.265**   $\displaystyle\int_0^1 \frac{1 - x^{\mu-1}}{1 - x}\, dx = \psi(\mu) + C$        $[\operatorname{Re}\mu > 0]$        FI II 796, WH, ET I 16(13)

$\displaystyle = \psi(1 - \mu) + C - \pi\cot(\mu\pi)$     $[\operatorname{Re}\mu > 0]$        EH I 16(15)a

**3.266**   $\displaystyle\int_0^\infty \frac{(x^\nu - a^\nu)\, dx}{(x - a)(b + x)} = \frac{\pi}{a + b}\left\{ b^\nu \operatorname{cosec}(\nu\pi) - a^\nu \cot(\nu\pi) - \frac{a^\nu}{\pi}\ln\frac{b}{a} \right\}$

$[|\arg b| < \pi, \quad |\operatorname{Re}\nu| < 1, \quad \nu \neq 0]$

ET II 216(8)

**3.267**

1.   $\displaystyle\int_0^1 \frac{x^{3n}\, dx}{\sqrt[3]{1 - x^3}} = \frac{2\pi}{3\sqrt{3}}\frac{\Gamma\left(n + \frac{1}{3}\right)}{\Gamma\left(\frac{1}{3}\right)\Gamma(n + 1)}$        BI (9)(6)

2.   $\displaystyle\int_0^1 \frac{x^{3n-1}\, dx}{\sqrt[3]{1 - x^3}} = \frac{(n - 1)!\,\Gamma\left(\frac{2}{3}\right)}{3\,\Gamma\left(n + \frac{2}{3}\right)}$        BI (9)(7)

3.   $\displaystyle\int_0^1 \frac{x^{3n-2}\, dx}{\sqrt[3]{1 - x^3}} = \frac{\Gamma\left(n - \frac{1}{3}\right)\Gamma\left(\frac{2}{3}\right)}{3\,\Gamma\left(n + \frac{1}{3}\right)}$

**3.268**

1.   $\displaystyle\int_0^1 \left( \frac{1}{1 - x} - \frac{px^{p-1}}{1 - x^p} \right) dx = \ln p$        BI (5)(14)

2.   $\displaystyle\int_0^1 \frac{1 - x^\mu}{1 - x}x^{\nu-1}\, dx = \psi(\mu + \nu) - \psi(\nu)$     $[\operatorname{Re}\nu > 0, \quad \operatorname{Re}\mu > 0]$      BI (2)(3)

3.   $\displaystyle\int_0^1 \left[ \frac{n}{1 - x} - \frac{x^{\mu-1}}{1 - \sqrt[n]{x}} \right] dx = nC + \sum_{k=1}^n \psi\left( \mu + \frac{n - k}{n} \right)$

$[\operatorname{Re}\mu > 0]$        BI (13)(10)

**3.269**

1.   $\displaystyle\int_0^1 \frac{x^p - x^{-p}}{1 - x^2}x\, dx = \frac{\pi}{2}\cot\frac{p\pi}{2} - \frac{1}{p}$     $[p^2 < 1]$        BI (4)(12)

2.   $\displaystyle\int_0^1 \frac{x^p - x^{-p}}{1 + x^2}x\, dx = \frac{1}{p} - \frac{\pi}{2}\operatorname{cosec}\frac{p\pi}{2}$     $[p^2 < 1]$        BI (4)(8)

3.   $\displaystyle\int_0^1 \frac{x^\mu - x^\nu}{1 - x^2}\, dx = \frac{1}{2}\psi\left( \frac{\nu + 1}{2} \right) - \frac{1}{2}\psi\left( \frac{\mu + 1}{2} \right)$     $[\operatorname{Re}\mu > -1, \quad \operatorname{Re}\nu > -1]$     BI (2)(9)

**3.271**

1.   $\displaystyle\int_0^\infty \frac{x^p - x^q}{x - 1}\frac{dx}{x + a} = \frac{\pi}{1 + a}\left( \frac{a^p - \cos p\pi}{\sin p\pi} - \frac{a^q - \cos q\pi}{\sin q\pi} \right)$

$[p^2 < 1, \quad q^2 < 1, \quad a > 0]$        BI (19)(2)

2.   $\displaystyle\int_0^\infty \frac{x^p - a^p}{x - a}\frac{x^p - 1}{x - 1}\, dx = \frac{\pi}{a - 1}\left\{ \frac{a^{2p} - 1}{\sin(2p\pi)} - \frac{1}{\pi}a^p \ln a \right\}$

$[p^2 < \tfrac{1}{4}]$        BI (19)(3)

3.
$$\int_0^\infty \frac{x^p - a^p}{x-a} \frac{x^{-p}-1}{x-1}\, dx = \frac{\pi}{a-1}\left\{2\left(a^p-1\right)\cot p\pi - \frac{1}{\pi}\left(a^p+1\right)\ln a\right\}$$
$$\left[p^2 < 1\right] \qquad\qquad\qquad \text{BI (18)(9)}$$

4.
$$\int_0^\infty \frac{x^p-a^p}{x-a}\frac{1-x^{-p}}{1-x}x^q\, dx = \frac{\pi}{a-1}\left\{\frac{a^{p+q}-1}{\sin[(p+q)\pi]} + \frac{a^p-a^q}{\sin[(q-p)\pi]}\right\}\frac{\sin p\pi}{\sin q\pi}$$
$$\left[(p+q)^2 < 1, \quad (p-q)^2 < 1\right]$$
$$\text{BI (19)(4)}$$

5.
$$\int_0^\infty \left(\frac{x^p - x^{-p}}{1-x}\right)^2 dx = 2\left(1 - 2p\pi\cot 2p\pi\right) \qquad \left[0 < p^2 < \tfrac{1}{4}\right] \qquad \text{BI (16)(3)}$$

**3.272**

1.
$$\int_0^1 \frac{x^{n-1} + x^{n-\frac12} - 2x^{2n-1}}{1-x}\, dx = 2\ln 2 \qquad\qquad\qquad \text{BI (8)(8)}$$

2.
$$\int_0^1 \frac{x^{n-1} + x^{n-\frac23} + x^{n-\frac13} - 3x^{3n-1}}{1-x}\, dx = 3\ln 3 \qquad\qquad\qquad \text{BI (8)(9)}$$

**3.273**

1.
$$\int_0^1 \frac{\sin t - a^n x^n \sin[(n+1)t] + a^{n+1}x^{n+1}\sin nt}{1 - 2ax\cos t + a^2 x^2}(1-x)^{p-1}\, dx = \Gamma(p)\sum_{k=1}^n \frac{(k-1)!a^{k-1}\sin kt}{\Gamma(p+k)}$$
$$[p > 0] \qquad\qquad \text{BI (6)(13)}$$

2.
$$\int_0^1 \frac{\cos t - ax - a^n x^n \cos[(n+1)t] + a^{n+1}x^{n+1}\cos nt}{1 - 2ax\cos t + a^2 x^2}(1-x)^{p-1}\, dx = \Gamma(p)\sum_{k=1}^n \frac{(k-1)!a^{k-1}\cos kt}{\Gamma(p+k)}$$
$$[p > 0] \qquad\qquad \text{BI (6)(14)}$$

3.
$$\int_0^1 x\frac{\sin t - x^n \sin[(n+1)t] + x^{n+1}\sin nt}{1 - 2x\cos t + x^2}\, dx = \sum_{k=1}^n \frac{\sin kt}{k+1} \qquad\qquad \text{BI (6)(12)}$$

4.
$$\int_0^1 \frac{1 - x\cos t - x^{n+1}\cos[(n+1)t] + x^{n+2}\cos nt}{1 - 2x\cos t + x^2}\, dx = \sum_{k=0}^n \frac{\cos kt}{k+1} \qquad\qquad \text{BI (6)(11)}$$

**3.274**

1.
$$\int_0^\infty \frac{x^{\mu-1}(1-x)}{1-x^n}\, dx = \frac{\pi}{n}\sin\frac{\pi}{n}\,\operatorname{cosec}\frac{\mu\pi}{n}\,\operatorname{cosec}\frac{(\mu+1)\pi}{n}$$
$$\left[0 < \operatorname{Re}\mu < n-1\right] \qquad\qquad \text{BI (20)(13)}$$

2.
$$\int_0^1 \frac{1-x^n}{(1+x)^{n+1}}\frac{dx}{1-x} = \frac{1}{2^{n+1}}\sum_{k=1}^n \frac{2^k}{k} \qquad\qquad\qquad \text{BI (5)(3)}$$

3.
$$\int_0^\infty \frac{x^q-1}{x^p - x^{-p}}\frac{dx}{x} = \frac{\pi}{2p}\tan\frac{q\pi}{2p} \qquad [p > q] \qquad\qquad \text{BI (18)(6)}$$

**3.275**

1.
$$\int_0^1 \left(\frac{x^{n-1}}{1-x^{1/p}} - \frac{px^{np-1}}{1-x}\right) dx = p\ln p \qquad [p > 0] \qquad\qquad \text{BI (13)(9)}$$

2.    $\displaystyle \int_0^1 \left( \frac{nx^{n-1}}{1-x^n} - \frac{x^{mn-1}}{1-x} \right) \, dx = C + \frac{1}{n} \sum_{k=1}^n \psi \left( m + \frac{n-k}{n} \right)$       BI (5)(13)

3.    $\displaystyle \int_0^1 \left( \frac{x^{p-1}}{1-x} - \frac{qx^{pq-1}}{1-x^q} \right) \, dx = \ln q$       $[q > 0]$       BI (5)(12)

4.    $\displaystyle \int_0^\infty \left( \frac{1}{1+x^{2^n}} - \frac{1}{1+x^{2^m}} \right) \frac{dx}{x} = 0.$       BI (18)(17)

**3.276**

$1.^{10}$    $\displaystyle \int_0^\infty \frac{\left[ \left( ax + \dfrac{b}{x} \right)^2 + c \right]^{-p-1}}{x^2} \, dx = \frac{1}{2|b|} \frac{\mathrm{B}\left(p + \frac{1}{2}, \frac{1}{2}\right)}{\left(2a\left(b + |b|\right) + c\right)^{p+\frac{1}{2}}}$

$$\left[ a > 0, \quad c > -4ac, \quad p > -\tfrac{1}{2} \right]$$

$2.^{10}$    $\displaystyle \int_0^\infty \left( a + \frac{b}{x^2} \right) \left[ \left( ax + \frac{b}{x} \right)^2 + c \right]^{-p-1} \, dx = \frac{\mathrm{B}\left(p + \frac{1}{2}, \frac{1}{2}\right)}{\left(4ab + c\right)^{p+\frac{1}{2}}}$

$$\left[ a > 0, \quad b > 0, \quad c > -4ac, \quad p > -\tfrac{1}{2} \right]$$

**3.277**

$1.^{12}$    $\displaystyle \int_0^\infty \frac{x^{\mu-1} \left[ \sqrt{1+x^2} + b \right]^\nu}{\sqrt{1+x^2}} \, dx = 2^{\frac{\mu}{2}-1} \left(b^2 - 1\right)^{\frac{\nu}{2}+\frac{\mu}{4}} \Gamma\left(\frac{\mu}{2}\right) \Gamma(1 - \mu - \nu) \, P_{\frac{\mu}{2}-1}^{\frac{\nu+\mu}{2}}(b)$

$$\left[ \operatorname{Re} b > -1, \quad 0 < \operatorname{Re} \mu < 1 - \operatorname{Re} \nu \right]$$
$$\text{ET I 310(25)}$$

2.    $\displaystyle \int_0^\infty \frac{x^{\mu-1} \left[ \sqrt{b^2 + x^2} + x \right]^\nu}{\sqrt{b^2 + x^2}} \, dx = \frac{b^{\mu+\nu-1}}{2^\mu} \, \mathrm{B}\left( \mu, \frac{1-\mu-\nu}{2} \right)$

$$\left[ \operatorname{Re} b > 0, \quad 0 < \operatorname{Re} \mu < 1 - \operatorname{Re} \nu \right]$$
$$\text{ET I 311(28)}$$

$3.^{12}$    $\displaystyle \int_0^\infty \frac{x^{\mu-1} \left[ \cos t \pm i \sin t \sqrt{1+x^2} \right]^\nu}{\sqrt{1+x^2}} \, dx = 2^{\frac{\mu-1}{2}} \sin^{\frac{1-\mu}{2}} t \frac{\Gamma\left(\frac{\mu}{2}\right) \Gamma(1-\mu-\nu)}{\Gamma(-\nu)}$

$$\times \left[ \pi^{-\frac{1}{2}} \, Q_{-\frac{\mu+1}{2}-\nu}^{\frac{\mu-1}{2}}(\cos t) \mp \frac{i}{2} \pi^{\frac{1}{2}} \, P_{-\frac{\mu+1}{2}-\nu}^{\frac{\mu-1}{2}}(\cos t) \right]$$
$$\left[ \operatorname{Re} \mu > 0 \right] \qquad \text{ET I 311 (27)}$$

$4.^{12}$    $\displaystyle \int_0^\infty \frac{x^{\mu-1} \left[ \sqrt{(b^2-1)(x^2+1)} + b \right]^\nu}{\sqrt{x^2+1}} \, dx$

$$= \frac{2^{(\mu-1)/2}}{\sqrt{\pi}} e^{-\frac{1}{2} i \pi (\mu-1)} \frac{\Gamma\left(\frac{\mu}{2}\right) \Gamma(1-\mu-\nu)}{\Gamma(-\nu)} \left(b^2-1\right)^{(1-\mu)/4} Q_{-\frac{\mu+1}{2}-\nu}^{\frac{\mu-1}{2}}(b)$$
$$\left[ \operatorname{Re} b > 1, \quad \operatorname{Re} \nu < 0, \quad \operatorname{Re} \mu < 1 - \operatorname{Re} \nu \right] \quad \text{ET I 311(26)}$$

5.    $\displaystyle \int_u^\infty \frac{(x-u)^{\mu-1} \left( \sqrt{x+1} - \sqrt{x-1} \right)^{2\nu}}{\sqrt{x^2-1}} \, dx = \frac{2^{\nu+\frac{1}{2}}}{\sqrt{\pi}} e^{\left(\mu-\frac{1}{2}\right)\pi i} \left(u^2-1\right)^{\frac{2\mu-1}{4}} Q_{\nu-\frac{1}{2}}^{\frac{1}{2}-\mu}(u)$

$$\left[ |\arg(u-1)| < \pi, \quad 0 < \operatorname{Re} \mu < 1 + \operatorname{Re} \nu \right] \quad \text{ET II 202(10)}$$

6. $$\int_{1}^{\infty} \frac{x^{\mu-1}\left[\left(x-\sqrt{x^2-1}\right)^{\nu}+\left(x-\sqrt{x^2-1}\right)^{-\nu}\right]}{\sqrt{x^2-1}}\,\mathrm{d}x = 2^{-\mu}\,\mathrm{B}\left(\frac{1-\mu+\nu}{2},\frac{1-\mu-\nu}{2}\right)$$

$$[\operatorname{Re}\mu < 1+\operatorname{Re}\nu] \qquad \text{ET I 311(29)}$$

7. $$\int_{0}^{u} \frac{(u-x)^{\mu-1}\left[\left(\sqrt{x+2}+\sqrt{x}\right)^{2\nu}+\left(\sqrt{x+2}-\sqrt{x}\right)^{2\nu}\right]}{\sqrt{x(x+2)}}\,\mathrm{d}x$$

$$= 2^{(2\mu+1)/2}\sqrt{\pi[u(u+2)]^{\mu-\frac{1}{2}}}\,P_{\nu-\frac{1}{2}}^{\frac{1}{2}-\mu}(u+1)$$

$$[|\arg u| < \pi, \quad \operatorname{Re}\mu > 0] \qquad \text{ET II 186(12)}$$

**3.278$^8$**

1. $$\int_{0}^{\infty}\left(\frac{x^p}{1+x^{2p}}\right)^{q}\frac{\mathrm{d}x}{1-x^2} = 0 \qquad\qquad [pq > 1]$$

# 3.3–3.4 Exponential Functions

## 3.31 Exponential functions

**3.310$^{11}$** $\displaystyle\int_{0}^{\infty} e^{-px}\,\mathrm{d}x = \frac{1}{p}$ $\qquad\qquad\qquad\qquad [\operatorname{Re}p > 0]$

**3.311**

1. $$\int_{0}^{\infty}\frac{\mathrm{d}x}{1+e^{px}} = \frac{\ln 2}{p} \qquad\qquad\qquad\qquad \text{LO III 284a}$$

2. $$\int_{0}^{\infty}\frac{e^{-\mu x}}{1+e^{-x}}\,\mathrm{d}x = \beta(\mu) \qquad\qquad [\operatorname{Re}\mu > 0] \qquad \text{EH I 20(3), ET I 144(7)}$$

3.$^{12}$ $$\int_{-\infty}^{\infty}\frac{e^{-ax}\,\mathrm{d}x}{1+e^{-x}} = \frac{\pi}{\sin\pi a} \qquad [\operatorname{Re}a > 0] \qquad (\text{cf. } \mathbf{3.241}\ 2) \qquad \text{BI (28)(7)}$$

4. $$\int_{0}^{\infty}\frac{e^{-qx}\,\mathrm{d}x}{1-ae^{-px}} = \sum_{k=0}^{\infty}\frac{a^k}{q+kp} \qquad [0 < a < 1] \qquad \text{BI (27)(7)}$$

5. $$\int_{0}^{\infty}\frac{1-e^{\nu x}}{e^x-1}\,\mathrm{d}x = \psi(\nu)+\mathbf{C}+\pi\cot(\pi\nu) \qquad [\operatorname{Re}\nu < 1] \qquad (\text{cf. } \mathbf{3.265}) \qquad \text{EH I 16(16)}$$

6. $$\int_{0}^{\infty}\frac{e^{-x}-e^{-\nu x}}{1-e^{-x}}\,\mathrm{d}x = \psi(\nu)+\mathbf{C} \qquad [\operatorname{Re}\nu > 0] \qquad \text{WH, EH I 16(14)}$$

7. $$\int_{0}^{\infty}\frac{e^{-\mu x}-e^{-\nu x}}{1-e^{-x}}\,\mathrm{d}x = \psi(\nu)-\psi(\mu) \qquad [\operatorname{Re}\mu > 0, \quad \operatorname{Re}\nu > 0] \qquad (\text{cf. } \mathbf{3.231}\ 5)$$

$$\text{BI (27)(8)}$$

8.$^{12}$ $$\mathrm{PV}\int_{-\infty}^{\infty}\frac{e^{-\mu x}\,\mathrm{d}x}{b-e^{-x}} = \pi b^{\mu-1}\cot(\mu\pi) \qquad [b > 0, \quad 0 < \operatorname{Re}\mu < 1] \qquad \text{ET I 120(14)a}$$

9. $$\int_{-\infty}^{\infty}\frac{e^{-\mu x}\,\mathrm{d}x}{b+e^{-x}} = \pi b^{\mu-1}\operatorname{cosec}(\mu\pi) \qquad [|\arg b| < \pi, \quad 0 < \operatorname{Re}\mu < 1]$$

$$\text{ET I 120(15)a}$$

$10.^{11}$    $\displaystyle\int_0^\infty \frac{e^{-px} - e^{-qx}}{1 - e^{-(p+q)x}}\,\mathrm{d}x = \frac{\pi}{p+q}\cot\frac{p\pi}{p+q}$      $[p > 0, \quad q > 0]$      GW (311)(16c)

11.    $\displaystyle\int_0^\infty \frac{e^{px} - e^{qx}}{e^{rx} - e^{sx}}\,\mathrm{d}x = \frac{1}{r-s}\left[\psi\left(\frac{r-q}{r-s}\right) - \psi\left(\frac{r-p}{r-s}\right)\right]$

                                                  $[r > s, r > p, r > q]$      GW (311)(16)

12.    $\displaystyle\int_0^\infty \frac{a^x - b^x}{c^x - d^x}\,\mathrm{d}x = \frac{1}{\ln\frac{c}{d}}\left[\psi\left(\frac{\ln\frac{c}{b}}{\ln\frac{c}{d}}\right) - \psi\left(\frac{\ln\frac{c}{a}}{\ln\frac{c}{d}}\right)\right]$      $[c > a > 0, \quad b > 0, \quad d > 0]$

                                                           GW (311)(16a)

13.    $\displaystyle\int_0^\infty \frac{e^{-px} + e^{-qx}}{1 + e^{-(p+q)x}}\,\mathrm{d}x = \frac{\pi}{p+q}\operatorname{cosec}\left(\frac{\pi p}{p+q}\right)$

**3.312**

1.    $\displaystyle\int_0^\infty \left(1 - e^{-\frac{x}{\beta}}\right)^{\nu-1} e^{-\mu x}\,\mathrm{d}x = \beta\,\mathrm{B}(\beta\mu, \nu)$      $[\operatorname{Re}\beta > 0, \quad \operatorname{Re}\nu > 0, \quad \operatorname{Re}\mu > 0]$

                                                             LI(25)(13), EH I 11(24)

2.    $\displaystyle\int_0^\infty \left(1 - e^{-x}\right)^{-1}\left(1 - e^{-\alpha x}\right)\left(1 - e^{-\beta x}\right)e^{-px}\,\mathrm{d}x = \psi(p+\alpha) + \psi(p+\beta) - \psi(p+\alpha+\beta) - \psi(p)$

            $[\operatorname{Re}p > 0, \quad \operatorname{Re}p > -\operatorname{Re}\alpha, \quad \operatorname{Re}p > -\operatorname{Re}\beta, \quad \operatorname{Re}p > -\operatorname{Re}(\alpha+\beta)]$    ET I 145(15)

$3.^{11}$    $\displaystyle\int_0^\infty \left(1 - e^{-x}\right)^{\nu-1}\left(1 - \beta e^{-x}\right)^{-\varrho}e^{-\mu x}\,\mathrm{d}x = \mathrm{B}(\mu, \nu)\,{}_2F_1(\varrho, \mu; \mu+\nu; \beta)$

                                  $[\operatorname{Re}\mu > 0, \quad \operatorname{Re}\nu > 0, \quad |\arg(1-\beta)| < \pi]$    EH I 116(15)

**3.313**

$1.^7$    $\displaystyle\mathrm{PV}\int_{-\infty}^\infty \frac{e^{-\mu x}\,\mathrm{d}x}{1 - e^{-x}} = \pi\cot\pi\mu$             $[0 < \operatorname{Re}\mu < 1]$

$2.^7$    $\displaystyle\int_{-\infty}^\infty \frac{e^{-\mu x}\,\mathrm{d}x}{(1 + e^{-x})^\nu} = \mathrm{B}(\mu, \nu - \mu)$             $[0 < \operatorname{Re}\mu < \operatorname{Re}\nu]$

**3.314**    $\displaystyle\int_{-\infty}^\infty \frac{e^{-\mu x}\,\mathrm{d}x}{\left(e^{\beta/\gamma} + e^{-x/\gamma}\right)^\nu} = \gamma\exp\left[\beta\left(\mu - \frac{\nu}{\gamma}\right)\right]\mathrm{B}(\gamma\mu, \nu - \gamma\mu)$

                                         $\left[\operatorname{Re}\left(\frac{\nu}{\gamma}\right) > \operatorname{Re}\mu > 0, \quad |\operatorname{Im}\beta| < \pi\operatorname{Re}\gamma\right]$    ET I 120(21)

**3.315**

1.    $\displaystyle\int_{-\infty}^\infty \frac{e^{-\mu x}\,\mathrm{d}x}{(e^\beta + e^{-x})^\nu(e^\gamma + e^{-x})^\varrho} = \exp[\gamma(\mu - \varrho) - \beta\nu]\,\mathrm{B}(\mu, \nu + \varrho - \mu)\,{}_2F_1\left(\nu, \mu; \nu + \varrho; 1 - e^{\nu-\beta}\right)$

                             $[|\operatorname{Im}\beta| < \pi, \quad |\operatorname{Im}\gamma| < \pi, \quad 0 < \operatorname{Re}\mu < \operatorname{Re}(\nu + \varrho)]$    ET I 121(22)

2.    $\displaystyle\int_{-\infty}^\infty \frac{e^{-\mu x}\,\mathrm{d}x}{(\beta + e^{-x})(\gamma + e^{-x})} = \frac{\pi\left(\beta^{\mu-1} - \gamma^{\mu-1}\right)}{\gamma - \beta}\operatorname{cosec}(\mu\pi)$

                               $[|\arg\beta| < \pi, \quad |\arg\gamma| < \pi, \quad \beta \neq \gamma, \quad 0 < \operatorname{Re}\mu < 2]$    ET I 120(18)

**3.316** $\quad \int_{-\infty}^{\infty} \dfrac{(1+e^{-x})^{\nu}-1}{(1+e^{-x})^{\mu}}\,dx = \psi(\mu) - \psi(\mu-\nu) \qquad [\operatorname{Re}\mu > \operatorname{Re}\nu > 0] \qquad \text{(cf. } \mathbf{3.235}\text{)}$

$$\text{BI } (28)(8)$$

**3.317**

1. $\quad \int_{-\infty}^{\infty} \left( \dfrac{1}{1+e^{-x}} - \dfrac{1}{(1+e^{-x})^{\mu}} \right) dx = \mathbf{C} + \psi(\mu) \qquad [\operatorname{Re}\mu > 0] \qquad \text{(cf. } \mathbf{3.233}\text{)} \qquad \text{BI } (28)(10)$

2. $\quad \int_{-\infty}^{\infty} \left( \dfrac{1}{(1+e^{-x})^{\nu}} - \dfrac{1}{(1+e^{-x})^{\mu}} \right) dx = \psi(\mu) - \psi(\nu) \qquad [\operatorname{Re}\mu > 0, \quad \operatorname{Re}\nu > 0] \qquad \text{(cf. } \mathbf{3.219}\text{)}$

$$\text{BI } (28)(11)$$

**3.318**

1. $\quad \displaystyle\int_0^{\infty} \dfrac{\left[b+\sqrt{1-e^{-x}}\right]^{-\nu} + \left[b-\sqrt{1-e^{-x}}\right]^{-\nu}}{\sqrt{1-e^{-x}}} e^{-\mu x}\,dx$

$$= \dfrac{2^{\mu+1} e^{(\mu-\nu)\pi i} \left(b^2-1\right)^{(\mu-\nu)/2} \Gamma(\mu)\, Q_{\mu-1}^{\nu-\mu}(b)}{\Gamma(\nu)}$$

$$[\operatorname{Re}\mu > 0] \qquad\qquad \text{ET I } 145(18)$$

2.[7] $\quad \displaystyle\int_u^{\infty} \dfrac{1}{\sqrt{1-e^{-2x}}} \left( e^{-u}\sqrt{1-e^{-2x}} - e^{-x}\sqrt{1-e^{-2u}} \right)^{\nu} e^{-\mu x}\,dx$

$$= \dfrac{2^{-\frac{1}{2}(\mu+\nu)}\sqrt{\pi}\, e^{-\frac{u}{2}(\mu+\nu)} \Gamma(\mu)\,\Gamma(\nu+1)\, P_{-\frac{1}{2}(\mu-\nu)}^{-\frac{1}{2}(\mu+\nu)}\left(\sqrt{1-e^{-2u}}\right)}{\Gamma[(\mu+\nu+1)/2]}$$

$$[u > 0, \quad \operatorname{Re}\mu > 0, \quad \operatorname{Re}\nu > -1] \quad \text{ET I } 145(19)$$

## 3.32–3.34 Exponentials of more complicated arguments

**3.321**

1.[11] $\quad \dfrac{\sqrt{\pi}}{2}\,\Phi(u) = \dfrac{\sqrt{\pi}}{2}\,\operatorname{erf}(u) = \displaystyle\int_0^u e^{-x^2}\,dx = \sum_{k=0}^{\infty} \dfrac{(-1)^k u^{2k+1}}{k!(2k+1)}$

$$= e^{-u^2} \sum_{k=0}^{\infty} \dfrac{2^k u^{2k+1}}{(2k+1)!!} \qquad\qquad [u > 0]$$

$$\text{(see } \mathbf{8.25}\text{)} \qquad\qquad \text{AD } 6.700$$

2. $\quad \displaystyle\int_0^u e^{-q^2 x^2}\,dx = \dfrac{\sqrt{\pi}}{2q}\,\Phi(qu) \qquad\qquad [q > 0]$

3. $\quad \displaystyle\int_0^{\infty} e^{-q^2 x^2}\,dx = \dfrac{\sqrt{\pi}}{2q} \qquad\qquad [q > 0] \qquad\qquad \text{FI II } 624$

4. $\quad \displaystyle\int_0^u x e^{-q^2 x^2}\,dx = \dfrac{1}{2q^2}\left[1 - e^{-q^2 u^2}\right]$

5. $\quad \displaystyle\int_0^u x^2 e^{-q^2 x^2}\,dx = \dfrac{1}{2q^3}\left[\dfrac{\sqrt{\pi}}{2}\,\Phi(qu) - qu e^{-q^2 u^2}\right]$

6. $\quad \displaystyle\int_0^u x^3 e^{-q^2 x^2}\,dx = \dfrac{1}{2q^4}\left[1 - \left(1 + q^2 u^2\right) e^{-q^2 u^2}\right]$

7. $\displaystyle\int_0^u x^4 e^{-q^2 x^2}\, dx = \frac{1}{2q^5}\left[\frac{3\sqrt{\pi}}{4}\,\Phi(qu) - \left(\frac{3}{2}+q^2 u^2\right) que^{-q^2 u^2}\right]$

**3.322**

1.[12] $\displaystyle\int_u^\infty e^{-x^2-2ax}\,dx = \frac{\sqrt{\pi}}{2}e^{a^2}\left[1-\operatorname{erf}(a+u)\right]$           $[\operatorname{Re}\beta>0,\quad u>0]$          ET I 146(21)

2.[12] $\displaystyle\int_0^\infty e^{-x^2-2ax}\,dx = \frac{\sqrt{\pi}}{2}e^{a^2}\left[1-\operatorname{erf}(a)\right]$                                            NT 27(1)a

3.[11] $\displaystyle\operatorname{PV}\int_0^\infty e^{\pm i\lambda x^2}\,dx = \frac{1}{2}\sqrt{\frac{\pi}{\lambda}}e^{\pm\pi i/4}$              $[\lambda>0]$         PBM 343 (2.3.15(2))

**3.323**

1.[12] $\displaystyle\int_1^\infty e^{-x^2-2ax}\,dx = \frac{\sqrt{\pi}}{2}e^{a^2}\left[1-\operatorname{erf}(a+1)\right]$                                     BI (29)(4)

2.[10] $\displaystyle\int_{-\infty}^\infty \exp\left(-p^2 x^2 \pm qx\right)\,dx = \exp\left(\frac{q^2}{4p^2}\right)\frac{\sqrt{\pi}}{p}$     $[\operatorname{Re}p^2>0]$        BI (28)(1)

3.[12] $\displaystyle\int_0^\infty \exp\left(-x^4-4ax^2\right)\,dx = \frac{\sqrt{a}}{2}e^{2a^2}K_{1/4}(2a^2)$                      ET I 147(34)a

4.* $\displaystyle\int_0^\infty \exp\left(-\beta^2 x^4 \pm 2\gamma^2 x^2\right)\,dx = \frac{\pi}{4}\frac{\gamma}{\beta}\exp\left(\frac{\gamma^4}{2\beta^2}\right)\left[I_{-1/4}\left(\frac{\gamma^4}{2\beta^2}\right)\pm I_{1/4}\left(\frac{\gamma^4}{2\beta^2}\right)\right]$

**3.324**

1. $\displaystyle\int_0^\infty \exp\left(-\frac{\beta}{4x}-\gamma x\right)\,dx = \sqrt{\frac{\beta}{\gamma}}\,K_1\left(\sqrt{\beta\gamma}\right)$     $[\operatorname{Re}\beta\geq 0,\quad \operatorname{Re}\gamma>0]$     ET I 146(25)

2.[11] $\displaystyle\int_{-\infty}^\infty \exp\left[-\left(x-\frac{b}{x}\right)^{2n}\right]\,dx = \frac{1}{n}\Gamma\left(\frac{1}{2n}\right)$          $[b\geq 0]$

**3.325**    $\displaystyle\int_0^\infty \exp\left(-ax^2-\frac{b}{x^2}\right)\,dx = \frac{1}{2}\sqrt{\frac{\pi}{a}}\exp\left(-2\sqrt{ab}\right)$     $[a>0,\quad b>0]$      FI II 644

**3.326**

1.[8] $\displaystyle\int_0^\infty \exp\left(-x^\mu\right)\,dx = \frac{1}{\mu}\Gamma\left(\frac{1}{\mu}\right)$              $[\operatorname{Re}\mu>0]$          BI (26)(4)

2.[10] $\displaystyle\int_0^\infty x^m \exp\left(-\beta x^n\right)\,dx = \frac{\Gamma(\gamma)}{n\beta^\gamma}$     $\gamma = \dfrac{m+1}{n}$     $[\operatorname{Re}\beta>0,\quad \operatorname{Re}m>0,\quad \operatorname{Re}n>0]$

3. $\displaystyle\int_0^\infty (x-a)\exp\left(-\beta(x-b)^n\right)\,dx = \frac{\Gamma\left(\frac{2}{n},\beta(-b)^n\right)}{n\beta^{2/n}} - (a-b)\frac{\Gamma\left(\frac{1}{n},\beta(-b)^n\right)}{n\beta^{1/n}}$

$$[\operatorname{Re}n>0,\quad \operatorname{Re}\beta>0,\quad |\arg b|<\pi]$$

4. $\displaystyle\int_0^u (x-a)\exp\left(-\beta(x-b)^n\right)\,dx = \frac{\Gamma\left(\frac{2}{n},\beta(-b)^n\right)-\Gamma\left(\frac{2}{n},\beta(u-b)^n\right)}{n\beta^{2/n}}$

$$-(a-b)\frac{\Gamma\left(\frac{1}{n},\beta(-b)^n\right)-\Gamma\left(\frac{1}{n},\beta(u-b)^n\right)}{n\beta^{1/n}}$$

$$[\operatorname{Re}n>0,\quad \operatorname{Re}\beta>0,\quad |\arg b|<\pi,\quad |\arg(u-b)|<\pi]$$

5. $\displaystyle\int_u^\infty (x-a)\exp\left(-\beta(x-b)^n\right)\,dx = \frac{\Gamma\left(\frac{2}{n},\beta(-b)^n\right)}{n\beta^{2/n}} - (a-b)\frac{\Gamma\left(\frac{1}{n},\beta(u-b)^n\right)}{n\beta^{1/n}}$

$$[\operatorname{Re}n > 0, \quad \operatorname{Re}\beta > 0, \quad |\arg(u-b)| < \pi]$$

## Exponentials of exponentials

**3.327** $\displaystyle\int_0^\infty \exp\left(-ae^{nx}\right)\,dx = -\frac{1}{n}\operatorname{Ei}(-a)$      $[n \geq 1, \quad \operatorname{Re}a \geq 0, \quad a \neq 0]$    LI (26)(5)

**3.328** $\displaystyle\int_{-\infty}^\infty \exp\left(-e^x\right)e^{\mu x}\,dx = \Gamma(\mu)$      $[\operatorname{Re}\mu > 0]$    NH 145(14)

**3.329** $\displaystyle\int_0^\infty \left[\frac{a\exp\left(-ce^{ax}\right)}{1-e^{-ax}} - \frac{b\exp\left(-ce^{bx}\right)}{1-e^{-bx}}\right]\,dx = e^{-c}\ln\frac{b}{a}$   $[a > 0, \quad b > 0, \quad c > 0]$    BI (27)(12)

**3.331**

1. $\displaystyle\int_0^\infty \exp\left(-\beta e^{-x} - \mu x\right)\,dx = \beta^{-\mu}\,\gamma(\mu,\beta)$      $[\operatorname{Re}\mu > 0]$    ET I 147(36)

2. $\displaystyle\int_0^\infty \exp\left(-\beta e^x - \mu x\right)\,dx = \beta^\mu\,\Gamma(-\mu,\beta)$      $[\operatorname{Re}\beta > 0]$    ET I 147(37)

3.[11] $\displaystyle\int_0^\infty \left(1-e^{-x}\right)^{\nu-1}\exp\left(\beta e^{-x} - \mu x\right)\,dx = \mathrm{B}(\mu,\nu)\beta^{-\frac{\mu+\nu}{2}}e^{\frac{\beta}{2}}M_{\frac{\nu-\mu}{2},\frac{\nu+\mu-1}{2}}(\beta)$

$$[\operatorname{Re}\mu > 0, \quad \operatorname{Re}\nu > 0] \qquad \text{ET I 147(38)}$$

4. $\displaystyle\int_0^\infty \left(1-e^{-x}\right)^{\nu-1}\exp\left(-\beta e^x - \mu x\right)\,dx = \Gamma(\nu)\beta^{\frac{\mu-1}{2}}e^{-\frac{\beta}{2}}\,W_{\frac{1-\mu-2\nu}{2},\frac{-\mu}{2}}(\beta)$

$$[\operatorname{Re}\beta > 0, \quad \operatorname{Re}\nu > 0] \qquad \text{ET I 147(39)}$$

**3.332** $\displaystyle\int_0^\infty \left(1-e^{-x}\right)^{\nu-1}\left(1-\lambda e^{-x}\right)^{-\varrho}\exp\left(\beta e^{-x} - \mu x\right)\,dx = \mathrm{B}(\mu,\nu)\,\Phi_1(\mu,\varrho,\nu,\lambda,\beta)$

$$[\operatorname{Re}\mu > 0, \quad \operatorname{Re}\nu > 0, \quad |\arg(1-\lambda)| < \pi] \quad \text{ET I 147(40)}$$

**3.333**

1.[12] $\displaystyle\int_{-\infty}^\infty \frac{e^{-\mu x}\,dx}{\exp\left(e^{-x}\right)-1} = \Gamma(\mu)\,\zeta(\mu)$      $[\operatorname{Re}\mu < 1]$    ET I 121(24)

2.[3] $\displaystyle\int_{-\infty}^\infty \frac{e^{-\mu x}\,dx}{\exp\left(e^{-x}\right)+1} = \left(1-2^{1-\mu}\right)\Gamma(\mu)\,\zeta(\mu)$      $[\operatorname{Re}\mu > 0, \quad \mu \neq 1]$

$$= \ln 2 \qquad\qquad\qquad [\mu = 1]$$

$$\text{ET I 121(25)}$$

3. $\displaystyle\int_0^\infty \left(\frac{\tanh(x)}{x^3} - \frac{1}{x^2\cosh^2(x)}\right)\,dx = \frac{7\,\zeta(3)}{\pi^2}$

**3.334**[11] $\displaystyle\int_0^\infty \left(e^x-1\right)^{\nu-1}\exp\left[-\frac{\beta}{e^x-1} - \mu x\right]\,dx = \Gamma(\mu-\nu+1)e^{\frac{\beta}{2}}\beta^{\frac{\nu-1}{2}}\,W_{\frac{\nu-2\mu-1}{2},-\frac{\nu}{2}}(\beta)$

$$[\operatorname{Re}\beta > 0, \quad \operatorname{Re}\mu > \operatorname{Re}\nu - 1]$$

$$\text{ET I 137(41)}$$

## Exponentials of hyperbolic functions

**3.335** $\displaystyle\int_0^\infty \left(e^{\nu x} + e^{-\nu x}\cos\nu\pi\right)\exp\left(-\beta\sinh x\right)\,\mathrm{d}x = -\pi\left[\mathbf{E}_\nu(\beta) + Y_\nu(\beta)\right]$

$$[\operatorname{Re}\beta > 0]$$      EH II 35(34)

**3.336**

1.    $\displaystyle\int_0^\infty \exp\left(-\nu x - \beta\sinh x\right)\,\mathrm{d}x = \pi\operatorname{cosec}\nu\pi\left[\mathbf{J}_\nu(\beta) - J_\nu(\beta)\right]$

$$\left[|\arg\beta| < \frac{\pi}{2}\ \text{and}\ |\arg\beta| = \frac{\pi}{2}\ \text{for}\ \operatorname{Re}\nu > 0;\quad \nu\ \text{is not an integer}\right]$$    WA 341(2)

2.    $\displaystyle\int_0^\infty \exp\left(nx - \beta\sinh x\right)\,\mathrm{d}x = \frac{1}{2}\left[S_n(\beta) - \pi\,\mathbf{E}_n(\beta) - \pi\,Y_n(\beta)\right]$

$$[\operatorname{Re}\beta > 0,\quad n = 0, 1, 2, \ldots]$$    WA 342(6)

3.    $\displaystyle\int_0^\infty \exp\left(-nx - \beta\sinh x\right)\,\mathrm{d}x = \frac{1}{2}(-1)^{n+1}\left[S_n(\beta) + \pi\,\mathbf{E}_n(\beta) + \pi\,Y_n(\beta)\right]$

$$[\operatorname{Re}\beta > 0,\quad n = 0, 1, 2, \ldots]$$

   EH II 84(47)

**3.337**

1.    $\displaystyle\int_{-\infty}^\infty \exp\left(-\alpha x - \beta\cosh x\right)\,\mathrm{d}x = 2\,K_\alpha(\beta)$      $\left[|\arg\beta| < \dfrac{\pi}{2}\right]$    WA 201(7)

2.    $\displaystyle\int_{-\infty}^\infty \exp\left(-\nu x + i\beta\cosh x\right)\,\mathrm{d}x = i\pi e^{\frac{i\nu\pi}{2}}\,H_\nu^{(1)}(\beta)$      $[0 < \arg\beta < \pi]$    EH II 21(27)

3.    $\displaystyle\int_{-\infty}^\infty \exp\left(-\nu x - i\beta\cosh x\right)\,\mathrm{d}x = -i\pi e^{-\frac{i\nu\pi}{2}}\,H_\nu^{(2)}(\beta)$      $[-\pi < \arg\beta < 0]$    EH II 21(30)

## Exponentials of trigonometric functions and logarithms

**3.338**

1.    $\displaystyle\int_0^\pi \left\{\exp i\left[(\nu - 1)x - \beta\sin x\right] - \exp i\left[(\nu + 1)x - \beta\sin x\right]\right\}\,\mathrm{d}x = 2\pi\left[\mathbf{J}_\nu'(\beta) + i\,\mathbf{E}_\nu'(\beta)\right]$

$$[\operatorname{Re}\beta > 0]$$    EH II 36

2.    $\displaystyle\int_0^\pi \exp\left[\pm i\left(\nu x - \beta\sin x\right)\right]\,\mathrm{d}x = \pi\left[\mathbf{J}_\nu(\beta) \pm i\,\mathbf{E}_\nu(\beta)\right]$      $[\operatorname{Re}\beta > 0]$    EH II 35(32)

3.[10]    $\displaystyle\int_0^\infty \exp\left[-\gamma\left(x - \beta\sin x\right)\right]\,\mathrm{d}x = \frac{1}{\gamma} + 2\sum_{k=1}^\infty \frac{\gamma\,J_k(k\beta)}{\gamma^2 + k^2}$      $[\operatorname{Re}\gamma > 0]$    WA 619(4)

4.[6]    $\displaystyle\int_{-\pi}^\pi \frac{\exp\left[\dfrac{a + b\sin x + c\cos x}{1 + p\sin x + q\cos x}\right]}{1 + p\sin x + q\cos x}\,\mathrm{d}x = \frac{2\pi}{\sqrt{1 - p^2 - q^2}}e^{-\alpha}\,I_0(\beta),$

$$\text{with } \alpha = \frac{bp + cq - a}{1 - p^2 - q^2};\quad \beta = \sqrt{\alpha^2 - \frac{a^2 - b^2 - c^2}{1 - p^2 - q^2}};\qquad [p^2 + q^2 < 1]$$

5. $\displaystyle\int_0^{\pi/4} \exp\left[-\sum_{n=1}^{\infty} \frac{\tan^{2n} x}{n + \frac{1}{2}}\right] dx = \ln\sqrt{2}$

**3.339**[6] $\displaystyle\int_0^{\pi} \exp\left(z\cos x\right) dx = \pi I_0(z)$ 　　　　　　　　　　　BI (277)(2)a

**3.341** $\displaystyle\int_0^{\frac{\pi}{2}} \exp\left(-p\tan x\right) dx = \text{ci}(p)\sin p - \text{si}(p)\cos(p)$ 　　　$[p > 0]$ 　　　BI (271)(2)a

**3.342**[11] $\displaystyle\int_0^1 \exp\left(-px\ln x\right) dx = \int_0^1 x^{-px}\, dx = \sum_{k=1}^{\infty} \frac{p^{k-1}}{k^k}$ 　　　　　　BI (29)(1)

## 3.35 Combinations of exponentials and rational functions

**3.351**

1.[8] $\displaystyle\int_0^u x^n e^{-\mu x}\, dx = \frac{n!}{\mu^{n+1}} - e^{-u\mu}\sum_{k=0}^{n} \frac{n!}{k!}\frac{u^k}{\mu^{n-k+1}} = \mu^{-n-1}\gamma(n+1, \mu u)$

$$[u > 0, \quad \text{Re}\,\mu > 0, n = 0, 1, 2, \ldots]$$
　　　　　　　　　　　　　　　　　　　　　　　　　　　　　ET I 134(5)

2.[11] $\displaystyle\int_u^{\infty} x^n e^{-\mu x}\, dx = e^{-u\mu}\sum_{k=0}^{n} \frac{n!}{k!}\frac{u^k}{\mu^{n-k+1}} = \mu^{-n-1}\Gamma(n+1, \mu u)$

$$[u > 0, \quad \text{Re}\,\mu > 0, n = 0, 1, 2, \ldots]$$
　　　　　　　　　　　　　　　　　　　　　　　　　　　　　ET I 33(4)

3. $\displaystyle\int_0^{\infty} x^n e^{-\mu x}\, dx = n!\,\mu^{-n-1}$ 　　　　　　　$[\text{Re}\,\mu > 0]$ 　　　ET I 133(3)

4. $\displaystyle\int_u^{\infty} \frac{e^{-px}\, dx}{x^{n+1}} = (-1)^{n+1}\frac{p^n\,\text{Ei}(-pu)}{n!} + \frac{e^{-pu}}{u^n}\sum_{k=0}^{n-1} \frac{(-1)^k p^k u^k}{n(n-1)\ldots(n-k)}$

$$[p > 0]$$ 　　　　　　　　　　　　　　　　　　　　　　　NT 21(3)

5. $\displaystyle\int_1^{\infty} \frac{e^{-\mu x}\, dx}{x} = -\text{Ei}(-\mu)$ 　　　　　　$[\text{Re}\,\mu > 0]$ 　　　BI (104)(10)

6. $\displaystyle\int_{-\infty}^u \frac{e^x}{x}\, dx = \text{li}\left(e^u\right) = \text{Ei}(u)$ 　　　　　　$[u < 0]$

7.[9] $\displaystyle\int_0^u xe^{-\mu x}\, dx = \frac{1}{\mu^2} - \frac{1}{\mu^2}e^{-\mu u}(1 + \mu u)$ 　　　$[u > 0]$

8.[11] $\displaystyle\int_0^u x^2 e^{-\mu x}\, dx = \frac{2}{\mu^3} - \frac{1}{\mu^3}e^{-\mu u}\left(2 + 2\mu u + \mu^2 u^2\right)$ 　　　$[u > 0]$

9.[7] $\displaystyle\int_0^u x^3 e^{-\mu x}\, dx = \frac{6}{\mu^4} - \frac{1}{\mu^4}e^{-\mu u}\left(6 + 6\mu u + 3\mu^2 u^2 + \mu^3 u^3\right)$

$$[u > 0]$$

**3.352**

1. $\displaystyle\int_0^u \frac{e^{-\mu x}\, dx}{x + b} = e^{\mu b}\left[\text{Ei}(-\mu u - \mu b) - \text{Ei}(-\mu b)\right]$ 　　　$[u \geq 0, \quad |\arg b| < \pi]$ 　　　ET II 217(12)

2. $\displaystyle\int_u^\infty \frac{e^{-\mu x}\,dx}{x+b} = -e^{b\mu}\,\mathrm{Ei}(-\mu u - \mu b)$ $\qquad [u \geq 0, \quad |\arg(u+b)| < \pi, \quad \mathrm{Re}\,\mu > 0]$

$\qquad\qquad\qquad$ ET I 134(6), JA

3.12 $\displaystyle\int_u^v \frac{e^{-\mu x}\,dx}{x+a} = e^{a\mu}\left\{\mathrm{Ei}[-(a+v)\mu] - \mathrm{Ei}[-(a+u)\mu]\right\}$ $\qquad [-a < u \text{ or } -a > v, \quad \mathrm{Re}\,\mu > 0]$

$\qquad\qquad\qquad$ ET I 134 (7)

4. $\displaystyle\int_0^\infty \frac{e^{-\mu x}\,dx}{x+b} = -e^{b\mu}\,\mathrm{Ei}(-\mu b)$ $\qquad [|\arg b| < \pi, \quad \mathrm{Re}\,\mu > 0]$ $\qquad$ ET II 217(11)

5.7 $\displaystyle\int_u^\infty \frac{e^{-px}\,dx}{a-x} = e^{-pa}\,\mathrm{Ei}(pa - pu)$

$\quad \left[p > 0, \quad a < u; \text{ for } a > u, \text{ one should replace } \mathrm{Ei}(pa - pu) \text{ in this formula with } \overline{\mathrm{Ei}}(pa - pu)\right]$

$\qquad\qquad\qquad$ ET II 251(37)

6.8 $\displaystyle\int_0^\infty \frac{e^{-\mu x}\,dx}{a-x} = e^{-\mu a}\,\mathrm{Ei}(a\mu)$

$\qquad\qquad\qquad [a < 0, \quad \mathrm{Re}\,\mu > 0]$ $\qquad$ BI (91)(4)

7. $\displaystyle\int_{-\infty}^\infty \frac{e^{ipx}\,dx}{x-a} = i\pi e^{iap}$ $\qquad [p > 0]$ $\qquad$ ET II 251(38)

**3.353**

1.12 $\displaystyle\int_u^\infty \frac{e^{-\mu x}\,dx}{(x+b)^n} = e^{-u\mu}\sum_{k=1}^{n-1}\frac{(k-1)!(-\mu)^{n-k-1}}{(n-1)!(u+b)^k} - \frac{(-\mu)^{n-1}}{(n-1)!}e^{b\mu}\,\mathrm{Ei}[-(u+b)\mu]$

$\qquad\qquad\qquad [n > 2, \quad |\arg(u+b)| < \pi, \quad \mathrm{Re}\,\mu > 0]$
$\qquad\qquad\qquad$ ET I 134(10)

2.7 $\displaystyle\int_0^\infty \frac{e^{-\mu x}\,dx}{(x+b)^n} = \frac{1}{(n-1)!}\sum_{k=1}^{n-1}(k-1)!(-\mu)^{n-k-1}b^{-k} - \frac{(-\mu)^{n-1}}{(n-1)!}e^{b\mu}\,\mathrm{Ei}(-b\mu)$

$\qquad\qquad\qquad [n > 2, \quad |\arg b| < \pi, \quad \mathrm{Re}\,\mu > 0]$
$\qquad\qquad\qquad$ ET I 134(9), BI (92)(2)

3. $\displaystyle\int_0^\infty \frac{e^{-px}\,dx}{(a+x)^2} = pe^{ap}\,\mathrm{Ei}(-ap) + \frac{1}{a}$ $\qquad [p > 0, \quad a > 0]$

$\qquad\qquad\qquad$ LI (281)(28), LI (281)(29)

4. $\displaystyle\int_0^1 \frac{xe^x}{(1+x)^2}\,dx = \frac{e}{2} - 1.$ $\qquad$ BI (80)(6)

5.7 $\displaystyle\int_0^\infty \frac{x^n e^{-\mu x}}{x+b}\,dx = (-1)^{n-1}b^n e^{b\mu}\,\mathrm{Ei}(-b\mu) + \sum_{k=1}^n (k-1)!(-b)^{n-k}\mu^{-k}$

$\qquad\qquad\qquad [|\arg b| < \pi, \quad \mathrm{Re}\,\mu > 0]$
$\qquad\qquad\qquad$ BI (91)(3)a, LET I 135(11)

**3.354**

1. $\displaystyle\int_0^\infty \frac{e^{-\mu x}\,dx}{b^2 + x^2} = \frac{1}{b}\left[\mathrm{ci}(b\mu)\sin b\mu - \mathrm{si}(b\mu)\cos b\mu\right]$ $\qquad [\mathrm{Re}\,b > 0, \quad \mathrm{Re}\,\mu > 0]$ $\qquad$ BI (91)(7)

2.  $\int_0^\infty \dfrac{xe^{-\mu x}\,\mathrm{d}x}{b^2+x^2} = -\operatorname{ci}(b\mu)\cos b\mu - \operatorname{si}(b\mu)\sin b\mu$ $\qquad$ [$\operatorname{Re} b > 0, \quad \operatorname{Re}\mu > 0$] $\qquad$ BI (91)(8)

3.[12] $\operatorname{PV}\int_0^\infty \dfrac{e^{-\mu x}\,\mathrm{d}x}{b^2-x^2} = \dfrac{1}{2b}\left[e^{-b\mu}\operatorname{Ei}(b\mu) - e^{b\mu}\operatorname{Ei}(-b\mu)\right]$ $\qquad$ [$|\arg(\pm b)| < \pi, \quad \operatorname{Re}\mu > 0$] $\qquad$ BI (91)(14)

4.  $\int_0^\infty \dfrac{xe^{-\mu x}\,\mathrm{d}x}{b^2-x^2} = \dfrac{1}{2}\left[e^{-b\mu}\operatorname{Ei}(b\mu) + e^{b\mu}\operatorname{Ei}(-b\mu)\right]$

$\left[|\arg(\pm b)| < \pi, \quad \operatorname{Re}\mu > 0;\ \text{for } b > 0 \text{ one should replace } \operatorname{Ei}(b\mu) \text{ in this formula with } \overline{\operatorname{Ei}}(b\mu)\right]$

$\qquad$ BI (91)(15)

5.[8] $\int_{-\infty}^\infty \dfrac{e^{-ipx}\,\mathrm{d}x}{a^2+x^2} = \dfrac{\pi}{a}e^{-|ap|}$ $\qquad$ [$a \neq 0, \quad p$ real] $\qquad$ ET I 118(1)a

**3.355**

1.  $\int_0^\infty \dfrac{e^{-\mu x}\,\mathrm{d}x}{(b^2+x^2)^2} = \dfrac{1}{2b^3}\left\{\operatorname{ci}(b\mu)\sin(b\mu) - \operatorname{si}(b\mu)\cos(b\mu) - b\mu\left[\operatorname{ci}(b\mu)\cos(b\mu) + \operatorname{si}(b\mu)\sin(b\mu)\right]\right\}$

$\qquad$ LI (92)(6)

2.[12] $\int_0^\infty \dfrac{xe^{-\mu x}\,\mathrm{d}x}{(b^2+x^2)^2} = \dfrac{1}{2b^2}\left\{1 - b\mu\left[\operatorname{ci}(b\mu)\sin b\mu - \operatorname{si}(b\mu)\cos b\mu\right]\right\}$

$\qquad$ [$\operatorname{Re} b > 0, \quad \operatorname{Re}\mu > 0$] $\qquad$ BI (92)(7)

3.[3] $\int_0^\infty \dfrac{e^{-px}\,\mathrm{d}x}{(a^2-x^2)^2} = \dfrac{1}{4a^3}\left[(ap-1)e^{ap}\operatorname{Ei}(-ap) + (1+ap)e^{-ap}\operatorname{Ei}(ap)\right]$

$\qquad$ $\left[\operatorname{Im}(a^2) > 0, \quad p > 0\right]$ $\qquad$ BI (92)(8)

4.[3] $\int_0^\infty \dfrac{xe^{-px}\,\mathrm{d}x}{(a^2-x^2)^2} = \dfrac{1}{4a^2}\left\{-2 + ap\left[e^{-ap}\operatorname{Ei}(ap) - e^{ap}\operatorname{Ei}(-ap)\right]\right\}$

$\qquad$ $\left[\operatorname{Im}(a^2) > 0, \quad p > 0\right]$ $\qquad$ LI (92)(9)

**3.356**

1.  $\int_0^\infty \dfrac{x^{2n+1}e^{-px}}{a^2+x^2}\,\mathrm{d}x = (-1)^{n-1}a^{2n}\left[\operatorname{ci}(ap)\cos ap + \operatorname{si}(ap)\sin ap\right]$

$\qquad + \dfrac{1}{p^{2n}}\sum_{k=1}^n (2n-2k+1)!\left(-a^2p^2\right)^{k-1}$

$\qquad$ [$p > 0$] $\qquad$ BI (91)(12)

2.  $\int_0^\infty \dfrac{x^{2n}e^{-px}}{a^2+x^2}\,\mathrm{d}x = (-1)^n a^{2n-1}\left[\operatorname{ci}(ap)\sin ap - \operatorname{si}(ap)\cos ap\right] + \dfrac{1}{p^{2n-1}}\sum_{k=1}^n (2n-2k)!\left(-a^2p^2\right)^{k-1}$

$\qquad$ [$p > 0$] $\qquad$ BI (91)(11)

3.  $\int_0^\infty \dfrac{x^{2n+1}e^{-px}}{a^2-x^2}\,\mathrm{d}x = \dfrac{1}{2}a^{2n}\left[e^{ap}\operatorname{Ei}(-ap) + e^{-ap}\operatorname{Ei}(ap)\right] - \dfrac{1}{p^{2n}}\sum_{k=1}^n (2n-2k+1)!\left(a^2p^2\right)^{k-1}$

$\qquad$ [$p > 0$] $\qquad$ BI (91)(17)

4.  $\int_0^\infty \dfrac{x^{2n}e^{-px}}{a^2-x^2}\,\mathrm{d}x = \dfrac{1}{2}a^{2n-1}\left[e^{-ap}\operatorname{Ei}(ap) - e^{ap}\operatorname{Ei}(-ap)\right] - \dfrac{1}{p^{2n-1}}\sum_{k=1}^n (2n-2k)!\left(a^2p^2\right)^{k-1}$

$\qquad$ [$p > 0$] $\qquad$ BI (91)(16)

**3.357**

1. $$\int_0^\infty \frac{e^{-\mu x}\, dx}{a^3 + a^2 x + ax^2 + x^3} = \frac{1}{2a^2} \{ \mathrm{ci}(a\mu)\,(\sin a\mu + \cos a\mu)$$
$$+ \mathrm{si}(a\mu)\,(\sin a\mu - \cos a\mu) - e^{a\mu}\,\mathrm{Ei}(-a\mu)\}$$
$$[\mathrm{Re}\,\mu > 0, \quad a > 0] \qquad \text{BI (92)(18)}$$

2.$^{12}$ $$\int_0^\infty \frac{xe^{-\mu x}\, dx}{a^3 + a^2 x + ax^2 + x^3} = \frac{1}{2a} \{ \mathrm{ci}(a\mu)\,(\sin a\mu - \cos a\mu)$$
$$- \mathrm{si}(a\mu)\,(\sin a\mu + \cos a\mu) + e^{a\mu}\,\mathrm{Ei}(-a\mu)\}$$
$$[\mathrm{Re}\,\mu > 0, \quad a > 0] \qquad \text{BI (92)(19)}$$

3. $$\int_0^\infty \frac{x^2 e^{-\mu x}\, dx}{a^3 + a^2 x + ax^2 + x^3} = \frac{1}{2} \{ - \mathrm{ci}(a\mu)\,(\sin a\mu + \cos a\mu)$$
$$- \mathrm{si}(a\mu)\,(\sin a\mu - \cos a\mu) - e^{a\mu}\,\mathrm{Ei}(-a\mu)\}$$
$$[\mathrm{Re}\,\mu > 0, \quad a > 0] \qquad \text{BI (92)(20)}$$

4. $$\int_0^\infty \frac{e^{-\mu x}\, dx}{a^3 - a^2 x + ax^2 - x^3} = \frac{1}{2a^2} \{ \mathrm{ci}(a\mu)\,(\sin a\mu - \cos a\mu)$$
$$- \mathrm{si}(a\mu)\,(\sin a\mu + \cos a\mu) + e^{-a\mu}\,\mathrm{Ei}(a\mu)\}$$
$$[\mathrm{Re}\,\mu > 0, \quad a > 0] \qquad \text{BI (92)(21)}$$

5. $$\int_0^\infty \frac{xe^{-\mu x}\, dx}{a^3 - a^2 x + ax^2 - x^3} = \frac{1}{2a} \{ - \mathrm{ci}(a\mu)\,(\sin a\mu + \cos a\mu)$$
$$- \mathrm{si}(a\mu)\,(\sin a\mu - \cos a\mu) + e^{-a\mu}\,\mathrm{Ei}(a\mu)\}$$
$$[\mathrm{Re}\,\mu > 0, \quad a > 0] \qquad \text{BI (92)(22)}$$

6. $$\int_0^\infty \frac{x^2 e^{-\mu x}\, dx}{a^3 - a^2 x + ax^2 - x^3} = \frac{1}{2} \{ \mathrm{ci}(a\mu)\,(\cos a\mu - \sin a\mu)$$
$$+ \mathrm{si}(a\mu)\,(\cos a\mu + \sin a\mu) + e^{-a\mu}\,\mathrm{Ei}(a\mu)\}$$
$$[\mathrm{Re}\,\mu > 0, \quad a > 0] \qquad \text{BI (92)(23)}$$

**3.358**

1. $$\int_0^\infty \frac{e^{-px}}{a^4 - x^4}\, dx = \frac{1}{4a^3} \{ e^{-ap}\,\mathrm{Ei}(ap) - e^{ap}\,\mathrm{Ei}(-ap) + 2\,\mathrm{ci}(ap)\sin ap - 2\,\mathrm{si}(ap)\cos ap \}$$
$$[p > 0, \quad a > 0] \qquad \text{BI (91)(18)}$$

2. $$\int_0^\infty \frac{xe^{-px}\, dx}{a^4 - x^4} = \frac{1}{4a^2} \{ e^{ap}\,\mathrm{Ei}(-ap) + e^{-ap}\,\mathrm{Ei}(ap) - 2\,\mathrm{ci}(ap)\cos ap - 2\,\mathrm{si}(ap)\sin ap \}$$
$$[p > 0, \quad a > 0] \qquad \text{BI (91)(19)}$$

3. $$\int_0^\infty \frac{x^2 e^{-px}\, dx}{a^4 - x^4} = \frac{1}{4a} \{ e^{-ap}\,\mathrm{Ei}(ap) - e^{ap}\,\mathrm{Ei}(-ap) - 2\,\mathrm{ci}(ap)\sin ap + 2\,\mathrm{si}(ap)\cos ap \}$$
$$[p > 0, \quad a > 0] \qquad \text{BI (91)(20)}$$

4. $$\int_0^\infty \frac{x^3 e^{-px}\, dx}{a^4 - x^4} = \frac{1}{4} \{ e^{ap}\,\mathrm{Ei}(-ap) + e^{-ap}\,\mathrm{Ei}(ap) + 2\,\mathrm{ci}(ap)\cos ap + 2\,\mathrm{si}(ap)\sin ap \}$$
$$[p > 0, \quad a > 0] \qquad \text{BI (91)(21)}$$

5.  $\displaystyle\int_0^\infty \frac{x^{4n}e^{-px}}{a^4-x^4}\,dx = \frac{1}{4}a^{4n-3}\left[e^{-ap}\,\mathrm{Ei}(ap) - e^{ap}\,\mathrm{Ei}(-ap) + 2\,\mathrm{ci}(ap)\sin ap - 2\,\mathrm{si}(ap)\cos ap\right]$

$$-\frac{1}{p^{4n-3}}\sum_{k=1}^n (4n-4k)!\left(a^4p^4\right)^{k-1}$$

$$[p>0,\quad a>0] \qquad\qquad \text{BI (91)(22)}$$

6.  $\displaystyle\int_0^\infty \frac{x^{4n+1}e^{-px}}{a^4-x^4}\,dx = \frac{1}{4}a^{4n-2}\left[e^{ap}\,\mathrm{Ei}(-ap) + e^{-ap}\,\mathrm{Ei}(ap) - 2\,\mathrm{ci}(ap)\cos ap - 2\,\mathrm{si}(ap)\sin ap\right]$

$$-\frac{1}{p^{4n-2}}\sum_{k=1}^n (4n-4k+1)!\left(a^4p^4\right)^{k-1}$$

$$[p>0,\quad a>0] \qquad\qquad \text{BI (91)(23)}$$

7.  $\displaystyle\int_0^\infty \frac{x^{4n+2}e^{-px}}{a^4-x^4}\,dx = \frac{1}{4}a^{4n-1}\left[e^{-ap}\,\mathrm{Ei}(ap) - e^{ap}\,\mathrm{Ei}(-ap) - 2\,\mathrm{ci}(ap)\sin ap + 2\,\mathrm{si}(ap)\cos ap\right]$

$$-\frac{1}{p^{4n-1}}\sum_{k=1}^n (4n-4k+2)!\left(a^4p^4\right)^{k-1}$$

$$[p>0,\quad a>0] \qquad\qquad \text{BI (91)(24)}$$

8.  $\displaystyle\int_0^\infty \frac{x^{4n+3}e^{-px}}{a^4-x^4}\,dx = \frac{1}{4}a^{4n}\left[e^{ap}\,\mathrm{Ei}(-ap) + e^{-ap}\,\mathrm{Ei}(ap) + 2\,\mathrm{ci}(ap)\cos ap + 2\,\mathrm{si}(ap)\sin ap\right]$

$$-\frac{1}{p^{4n}}\sum_{k=1}^n (4n-4k+3)!\left(a^4p^4\right)^{k-1}$$

$$[p>0,\quad a>0] \qquad\qquad \text{BI (91)(25)}$$

**3.359**  $\displaystyle\int_{-\infty}^\infty \frac{(i-x)^n}{(i+x)^n}\frac{e^{-ipx}}{i+x^2}\,dx = (-1)^{n-1}2\pi pe^{-p}\,L_{n-1}(2p) \qquad\text{for } p>0;$

$$= 0 \qquad\qquad\qquad\qquad\qquad\text{for } p<0.$$

$$\text{ET I 118(2)}$$

## 3.36–3.37 Combinations of exponentials and algebraic functions

**3.361**

1.[8]  $\displaystyle\int_0^u \frac{e^{-qx}}{\sqrt{x}}\,dx = \sqrt{\frac{\pi}{q}}\,\Phi\left(\sqrt{qu}\right) \qquad\qquad [q>0]$

2.[8]  $\displaystyle\int_0^\infty \frac{e^{-qx}}{\sqrt{x}}\,dx = \sqrt{\frac{\pi}{q}} \qquad\qquad [q>0] \qquad\qquad \text{BI(98)(10)}$

3.[8]  $\displaystyle\int_{-1}^\infty \frac{e^{-qx}}{\sqrt{1+x}}\,dx = e^q\sqrt{\frac{\pi}{q}} \qquad\qquad [q>0] \qquad\qquad \text{BI (104)(16)}$

**3.362**

1.  $\displaystyle\int_1^\infty \frac{e^{-\mu x}\,dx}{\sqrt{x-1}} = \sqrt{\frac{\pi}{\mu}}e^{-\mu} \qquad\qquad [\mathrm{Re}\,\mu>0] \qquad\qquad \text{BI (104)(11)a}$

2.  $\displaystyle\int_0^\infty \frac{e^{-\mu x}\,dx}{\sqrt{x+b}} = \sqrt{\frac{\pi}{\mu}}e^{b\mu}\left[1-\Phi\left(\sqrt{b\mu}\right)\right] \qquad [\mathrm{Re}\,\mu>0,\quad |\arg b|<\pi] \qquad \text{ET I 135(18)}$

**3.363**

1.
$$\int_u^\infty \frac{\sqrt{x-u}}{x} e^{-\mu x}\, dx = \sqrt{\frac{\pi}{\mu}} e^{-u\mu} - \pi\sqrt{u}\left[1 - \Phi\left(\sqrt{u\mu}\right)\right]$$

$$[u > 0, \quad \operatorname{Re}\mu > 0] \qquad \text{ET I 136(23)}$$

2.
$$\int_u^\infty \frac{e^{-\mu x}\, dx}{x\sqrt{x-u}} = \frac{\pi}{\sqrt{u}}\left[1 - \Phi\left(\sqrt{u\mu}\right)\right] \qquad [u > 0, \quad \operatorname{Re}\mu \geq 0] \qquad \text{ET I 136(26)}$$

**3.364**

1.
$$\int_0^2 \frac{e^{-px}\, dx}{\sqrt{x(2-x)}} = \pi e^{-p} I_0(p) \qquad [p > 0] \qquad \text{GW (312)(7a)}$$

2.
$$\int_{-1}^1 \frac{e^{2x}\, dx}{\sqrt{1-x^2}} = \pi I_0(2) \qquad \text{BI (277)(2)a}$$

3.
$$\int_0^\infty \frac{e^{-px}\, dx}{\sqrt{x(x+a)}} = e^{\frac{ap}{2}} K_0\left(\frac{ap}{2}\right) \qquad [a > 0, \quad p > 0] \qquad \text{GW (312)(8a)}$$

**3.365**

1.
$$\int_0^u \frac{x e^{-\mu x}\, dx}{\sqrt{u^2 - x^2}} = \frac{\pi u}{2}\left[\mathbf{L}_1(\mu u) - I_1(\mu u)\right] + u \qquad [u > 0, \quad \operatorname{Re}\mu > 0] \qquad \text{ET I 136(28)}$$

2.
$$\int_u^\infty \frac{x e^{-\mu x}\, dx}{\sqrt{x^2 - u^2}} = u K_1(u\mu) \qquad [u > 0, \quad \operatorname{Re}\mu > 0] \qquad \text{ET I 136(29)}$$

**3.366**

1.
$$\int_0^{2u} \frac{(u-x)e^{-\mu x}\, dx}{\sqrt{2ux - x^2}} = \pi u e^{-u\mu} I_1(u\mu) \qquad [\operatorname{Re}\mu > 0] \qquad \text{ET I 136(31)}$$

2.
$$\int_0^\infty \frac{(x+b)e^{-\mu x}\, dx}{\sqrt{x^2 + 2bx}} = b e^{b\mu} K_1(b\mu) \qquad [\operatorname{Re}\mu > 0, \quad |\arg b| < \pi] \qquad \text{ET I 136(30)}$$

3.
$$\int_0^\infty \frac{x e^{-\mu x}\, dx}{\sqrt{x^2 + b^2}} = \frac{b\pi}{2}\left[\mathbf{H}_1(b\mu) - Y_1(b\mu)\right] - b \qquad \left[|\arg b| < \frac{\pi}{2}, \quad \operatorname{Re}\mu > 0\right] \qquad \text{ET I 136(27)}$$

**3.367**
$$\int_0^\infty \frac{e^{-\mu x}\, dx}{(1 + \cos t + x)\sqrt{x^2 + 2x}} = \frac{\exp\left(2\mu\cos^2\frac{t}{2}\right)}{\sin t}\left(t - \sin t\int_0^u K_0(v)e^{-v\cos t}\, dv\right)$$

$$[\operatorname{Re}\mu > 0] \qquad \text{ET I 136(33)}$$

**3.368**
$$\int_0^\infty \frac{e^{-\mu x}\, dx}{x + \sqrt{x^2 + b^2}} = \frac{\pi}{2b\mu}\left[\mathbf{H}_1(b\mu) - Y_1(b\mu)\right] - \frac{1}{b^2\mu^2} \qquad \left[|\arg b| < \frac{\pi}{2}, \quad \operatorname{Re}\mu > 0\right] \qquad \text{ET I 136(32)}$$

**3.369**[11]
$$\int_0^\infty \frac{e^{-\mu x}\, dx}{\sqrt{(x+a)^3}} = \frac{2}{\sqrt{a}} - 2\sqrt{\pi\mu}\,e^{a\mu}\left(1 - \Phi\left(\sqrt{a\mu}\right)\right) \qquad [|\arg a| < \pi, \quad \operatorname{Re}\mu > 0] \qquad \text{ET I 135(20)}$$

**3.371**[11]
$$\int_0^\infty x^{n-\frac{1}{2}} e^{-\mu x}\, dx = \sqrt{\pi}\cdot\frac{1}{2}\cdot\frac{3}{2}\cdots\frac{2n-1}{2}\mu^{-n-\frac{1}{2}}$$

$$= \sqrt{\pi}\,2^{-n}\mu^{-n-1/2}(2n-1)!! \qquad [n \geq 0]$$

$$[\operatorname{Re}\mu > 0] \qquad \text{ET I 135(17)}$$

**3.372**
$$\int_0^\infty x^{n-\frac{1}{2}}(2+x)^{n-\frac{1}{2}} e^{-px}\, dx = \frac{(2n-1)!!}{p^n} e^p K_n(p) \qquad [p > 0, \quad n = 0, 1, 2, \ldots] \qquad \text{GW (312)(8)}$$

**3.373**
$$\int_0^\infty \left[\left(x + \sqrt{x^2 + b^2}\right)^n + \left(x - \sqrt{x^2 + b^2}\right)^n\right] e^{-\mu x}\, dx = 2b^{n+1} O_n(b\mu)$$

$$[\operatorname{Re}\mu > 0] \qquad \text{WA 05(1)}$$

**3.374**

1. $\displaystyle\int_0^\infty \frac{\left(x+\sqrt{1+x^2}\right)^n}{\sqrt{1+x^2}} e^{-\mu x}\,dx = \frac{1}{2}\left[S_n(\mu) - \pi\, \mathbf{E}_n(\mu) - \pi\, Y_n(\mu)\right]$

$\qquad\qquad\qquad\qquad\qquad\qquad\qquad\qquad [\operatorname{Re}\mu > 0]$ ET I 37(35)

2. $\displaystyle\int_0^\infty \frac{\left(x-\sqrt{1+x^2}\right)^n}{\sqrt{1+x^2}} e^{-\mu x}\,dx = -\frac{1}{2}\left[S_n(\mu) + \pi\, \mathbf{E}_n(\mu) + \pi\, Y_n(\mu)\right]$

$\qquad\qquad\qquad\qquad\qquad\qquad\qquad\qquad [\operatorname{Re}\mu > 0]$ ET I 137(36)

## 3.38–3.39 Combinations of exponentials and arbitrary powers

**3.381**

1. $\displaystyle\int_0^u x^{\nu-1} e^{-\mu x}\,dx = \mu^{-\nu}\,\gamma(\nu,\mu u)$ $\qquad [\operatorname{Re}\nu > 0]$ $\qquad$ EH I 266(22), EH II 133(1)

2. $\displaystyle\int_0^u x^{p-1} e^{-x}\,dx = \sum_{k=0}^\infty (-1)^k \frac{u^{p+k}}{k!(p+k)}$

$\qquad\qquad\qquad = e^{-u}\sum_{k=0}^\infty \frac{u^{p+k}}{p(p+1)\dots(p+k)}$

$\qquad\qquad\qquad\qquad\qquad\qquad\qquad\qquad\qquad\qquad$ AD 6.705

3.[8] $\displaystyle\int_u^\infty x^{\nu-1} e^{-\mu x}\,dx = \mu^{-\nu}\,\Gamma(\nu,\mu u)$ $\qquad [u > 0,\quad \operatorname{Re}\mu > 0]$

$\qquad\qquad\qquad\qquad\qquad\qquad\qquad\qquad$ EH I 256(21), EH II 133(2)

4. $\displaystyle\int_0^\infty x^{\nu-1} e^{-\mu x}\,dx = \frac{1}{\mu^\nu}\,\Gamma(\nu)$ $\qquad [\operatorname{Re}\mu > 0,\quad \operatorname{Re}\nu > 0]$ FI II 779

5. $\displaystyle\int_0^\infty x^{\nu-1} e^{-(p+iq)x}\,dx = \Gamma(\nu)\left(p^2+q^2\right)^{-\frac{\nu}{2}}\exp\left(-i\nu\arctan\frac{q}{p}\right)$

$\qquad\qquad\qquad [p > 0,\quad \operatorname{Re}\nu > 0 \text{ and } p = 0,\quad 0 < \operatorname{Re}\nu < 1]$ EH I 12(32)

6. $\displaystyle\int_u^\infty \frac{e^{-x}}{x^\nu}\,dx = u^{-\frac{\nu}{2}} e^{-\frac{u}{2}}\, W_{-\frac{\nu}{2},\frac{(1-\nu)}{2}}(u)$ $\qquad [u > 0]$ $\qquad\qquad$ WH

7. $\displaystyle\int_0^\infty x^{k-1} e^{i\mu x}\,dx = \frac{\Gamma(k)}{(-i\mu)^k}$ $\qquad [0 < \operatorname{Re}(k) < 1,\quad \mu \neq 0]$

$\qquad\qquad\qquad\qquad\qquad\qquad\qquad\qquad$ GH2 62 (313.14)

8. $\displaystyle\int_0^u x^m e^{-bx^n}\,dx = \frac{\gamma(v,bu^n)}{nb^v}$ $\qquad v = \frac{m+1}{n}$ $\quad [u > 0,\quad \operatorname{Re}v > 0,\quad \operatorname{Re}n > 0,\quad \operatorname{Re}\beta > 0]$

9. $\displaystyle\int_u^\infty x^m e^{-\beta x^n}\,dx = \frac{\Gamma(v,\beta u^n)}{n\beta^v}$ $\qquad v = \frac{m+1}{n}$ $\quad [\operatorname{Re}n > 0,\quad \operatorname{Re}\beta > 0]$

10. $\displaystyle\int_u^\infty x^m e^{-\beta x^n}\,dx = \frac{\Gamma(v,\beta u^n)}{n\beta^v}$

$\qquad\qquad\qquad v = \frac{m+1}{n}$ $\quad [u > 0,\quad \operatorname{Re}n > 0,\quad \operatorname{Re}\beta > 0]$ $\quad$ (See also **3.326** 2)

11. 
$$\int_{-\infty}^{\infty} x^m e^{-\beta x^n}\, dx = 2\int_{0}^{\infty} x^m e^{-\beta x^n}\, dx = \frac{2\,\Gamma(v)}{n\beta^v}$$

$$v = \frac{m+1}{n} \qquad \left[\operatorname{Re} m > -\tfrac{1}{2}, \quad \operatorname{Re} n > 0, \quad \operatorname{Re}\beta > 0\right]$$

**3.382**

1.[6] 
$$\int_{0}^{u} (u-x)^{\nu} e^{-\mu x}\, dx = (-\mu)^{-\nu-1} e^{-u\mu}\,\gamma(\nu+1, -u\mu) \qquad [\operatorname{Re}\nu > -1, \quad u > 0] \qquad \text{ET I 137(6)}$$

2. 
$$\int_{u}^{\infty} (x-u)^{\nu} e^{-\mu x}\, dx = \mu^{-\nu-1} e^{-u\mu}\,\Gamma(\nu+1) \qquad [u > 0, \quad \operatorname{Re}\nu > -1, \quad \operatorname{Re}\mu > 0]$$

$$\text{ET I 137(5), ET II 202(11)}$$

3. 
$$\int_{0}^{\infty} (1+x)^{-\nu} e^{-\mu x}\, dx = \mu^{\frac{\nu}{2}-1} e^{\frac{\mu}{2}}\, W_{-\frac{\nu}{2}, \frac{(1-\nu)}{2}}(\mu) \qquad [\operatorname{Re}\mu > 0] \qquad \text{WH}$$

4. 
$$\int_{0}^{\infty} (x+b)^{\nu} e^{-\mu x}\, dx = \mu^{-\nu-1} e^{b\mu}\,\Gamma(\nu+1, b\mu) \qquad [|\arg b| < \pi, \quad \operatorname{Re}\mu > 0]$$

$$\text{ET I 137(4), ET II 233(10)}$$

5. 
$$\int_{0}^{u} (a+x)^{\mu-1} e^{-x}\, dx = e^{a}[\gamma(\mu, a+u) - \gamma(\mu, a)] \qquad [\operatorname{Re}\mu > 0] \qquad \text{EH II 139}$$

6. 
$$\int_{-\infty}^{\infty} (b+ix)^{-\nu} e^{-ipx}\, dx = 0 \qquad [\text{for } p > 0]$$

$$= \frac{2\pi(-p)^{\nu-1} e^{bp}}{\Gamma(\nu)} \qquad [\text{for } p < 0]$$

$$[\operatorname{Re}\nu > 0, \quad \operatorname{Re} b > 0] \qquad \text{ET I 118(4)}$$

7. 
$$\int_{-\infty}^{\infty} (b-ix)^{-\nu} e^{-ipx}\, dx = \frac{2\pi p^{\nu-1} e^{-bp}}{\Gamma(\nu)} \qquad [\text{for } p > 0]$$

$$= 0 \qquad [\text{for } p < 0]$$

$$[\operatorname{Re}\nu > 0, \quad \operatorname{Re} b > 0] \qquad \text{ET I 118(3)}$$

**3.383**

1.[11] 
$$\int_{0}^{u} x^{\nu-1}(u-x)^{\mu-1} e^{\beta x}\, dx = \mathrm{B}(\mu, \nu) u^{\mu+\nu-1}\, {}_1F_1(\nu; \mu+\nu; \beta u)$$

$$[\operatorname{Re}\mu > 0, \quad \operatorname{Re}\nu > 0] \qquad \text{ET II 187(14)}$$

2.[11] 
$$\int_{0}^{u} x^{\mu-1}(u-x)^{\mu-1} e^{\beta x}\, dx = \sqrt{\pi}\left(\frac{u}{\beta}\right)^{\mu-\frac{1}{2}} \exp\left(\frac{\beta u}{2}\right) \Gamma(\mu)\, I_{\mu-\frac{1}{2}}\left(\frac{\beta u}{2}\right)$$

$$[\operatorname{Re}\mu > 0] \qquad \text{ET II 187(13)}$$

3. 
$$\int_{u}^{\infty} x^{\mu-1}(x-u)^{\mu-1} e^{-\beta x}\, dx = \frac{1}{\sqrt{\pi}}\left(\frac{u}{\beta}\right)^{\mu-\frac{1}{2}} \Gamma(\mu) \exp\left(-\frac{\beta u}{2}\right) K_{\mu-\frac{1}{2}}\left(\frac{\beta u}{2}\right)$$

$$[\operatorname{Re}\mu > 0, \quad \operatorname{Re}\beta u > 0] \qquad \text{ET II 202(12)}$$

4.[11] 
$$\int_{u}^{\infty} x^{\nu-1}(x-u)^{\mu-1} e^{-\beta x}\, dx = \beta^{-\frac{\mu+\nu}{2}} u^{\frac{\mu+\nu-2}{2}} \Gamma(\mu) \exp\left(-\frac{\beta u}{2}\right) W_{\frac{\nu-\mu}{2}, \frac{1-\mu-\nu}{2}}(\beta u)$$

$$[\operatorname{Re}\mu > 0, \quad \operatorname{Re}\beta u > 0] \qquad \text{ET II 202(13)}$$

$5.^{12}$ $\displaystyle\int_0^\infty e^{-px} x^{q-1} (1+ax)^{-\nu}\,dx$

$$= \frac{\pi^2}{p^q \Gamma(\nu)} \left[ \left(\frac{p}{a}\right)^\nu \frac{L_{-\nu}^{\nu-q}\left(\frac{p}{a}\right)}{\sin(\pi\nu)\sin(\pi(q-\nu))\Gamma(1-q)} - \left(\frac{p}{a}\right)^q \frac{L_{-q}^{q-\nu} - \left(\frac{p}{a}\right)}{\sin(\pi q)\sin(\pi(p-\nu))\Gamma(1-\nu)} \right]$$

$$[\nu \neq \pm 1, \pm 2, \ldots]$$

$$= \frac{\Gamma(q)}{p^q} \qquad\qquad [\nu = 0]$$

$$[\operatorname{Re} q > 0, \quad \operatorname{Re} p > 0, \quad \operatorname{Re} a > 0]$$

6. $\displaystyle\int_0^\infty x^{\nu-1}(x+b)^{-\nu+\frac{1}{2}} e^{-\mu x}\,dx = 2^{\nu-\frac{1}{2}} \Gamma(\nu)\mu^{-\frac{1}{2}} e^{\frac{b\mu}{2}} D_{1-2\nu}\left(\sqrt{2b\mu}\right)$

$$[|\arg b| < \pi, \quad \operatorname{Re}\nu > 0, \quad \operatorname{Re}\mu \geq 0, \quad \mu \neq 0] \quad \text{ET I 39(20), EH II 119(2)a}$$

7. $\displaystyle\int_0^\infty x^{\nu-1}(x+b)^{-\nu-\frac{1}{2}} e^{-\mu x}\,dx = 2^\nu \Gamma(\nu) b^{-\frac{1}{2}} e^{\frac{b\mu}{2}} D_{-2\nu}\left(\sqrt{2b\mu}\right)$

$$[|\arg b| < \pi, \quad \operatorname{Re}\nu > 0, \quad \operatorname{Re}\mu \geq 0]$$
$$\text{ET I 139(21), EH II 119(1)a}$$

8. $\displaystyle\int_0^\infty x^{\nu-1}(x+b)^{\nu-1} e^{-\mu x}\,dx = \frac{1}{\sqrt{\pi}} \left(\frac{b}{\mu}\right)^{\nu-\frac{1}{2}} e^{\frac{b\mu}{2}} \Gamma(\nu) K_{\frac{1}{2}-\nu}\left(\frac{b\mu}{2}\right)$

$$[|\arg b| < \pi, \quad \operatorname{Re}\mu > 0, \quad \operatorname{Re}\nu > 0]$$
$$\text{ET II 233(11), EH II 19(16)a, EH II 82(22)a}$$

9. $\displaystyle\int_u^\infty \frac{(x-u)^\nu e^{-\mu x}}{x}\,dx = u^\nu \Gamma(\nu+1) \Gamma(-\nu, u\mu) \qquad [u > 0, \quad \operatorname{Re}\nu > -1, \quad \operatorname{Re}\mu > 0]$

$$\text{ET I 138(8)}$$

10. $\displaystyle\int_0^\infty \frac{x^{\nu-1} e^{-\mu x}}{x+b}\,dx = b^{\nu-1} e^{b\mu} \Gamma(\nu) \Gamma(1-\nu, b\mu) \qquad [|\arg b| < \pi, \quad \operatorname{Re}\mu > 0, \quad \operatorname{Re}\nu > 0]$

$$\text{EH II 137(3)}$$

**3.384**

1. $\displaystyle\int_{-1}^1 (1-x)^{\nu-1}(1+x)^{\mu-1} e^{-ipx}\,dx = 2^{\mu+\nu-1} \mathrm{B}(\mu,\nu) e^{ip} \, {}_1F_1(\mu; \nu+\mu; -2ip)$

$$[\operatorname{Re}\nu > 0, \quad \operatorname{Re}\mu > 0] \qquad \text{ET I 119(13)}$$

2. $\displaystyle\int_u^v (x-u)^{2\mu-1}(v-x)^{2\nu-1} e^{-px}\,dx$

$$= \mathrm{B}(2\mu, 2\nu)(v-u)^{\mu+\nu-1} p^{-\mu-\nu} \exp\left(-p\frac{u+v}{2}\right) M_{\mu-\nu,\mu+\nu-\frac{1}{2}}(vp-up)$$

$$[v > u > 0, \quad \operatorname{Re}\mu > 0, \quad \operatorname{Re}\nu > 0] \quad \text{ET I 139(23)}$$

3. $\displaystyle\int_u^\infty (x+b)^{2\nu-1}(x-u)^{2\varrho-1} e^{-\mu x}\,dx = \frac{(u+b)^{\nu+\varrho-1}}{\mu^{\nu+\varrho}} \exp\left[\frac{(b-u)\mu}{2}\right] \Gamma(2\varrho) \, W_{\nu-\varrho, \nu+\varrho-\frac{1}{2}}(u\mu+b\mu)$

$$[u > 0, \quad |\arg(b+u)| < \pi, \quad \operatorname{Re}\mu > 0, \quad \operatorname{Re}\varrho > 0] \quad \text{ET I 139(22)}$$

4. $\int_u^\infty (x+b)^\nu (x-u)^{-\nu} e^{-\mu x}\, dx = \frac{1}{\mu}\nu\pi \cosec(\nu\pi) e^{-\frac{(b+u)\mu}{2}} \mathbf{k}_{2\nu}\left[\frac{(b+u)\mu}{2}\right]$

$[\nu \neq 0, \quad u > 0, \quad |\arg(u+b)| < \pi, \quad \operatorname{Re}\mu > 0, \quad \operatorname{Re}\nu < 1]$   **ET I 139(17)**

5. $\int_u^\infty (x-u)^{\nu-1}(x+u)^{-\nu+\frac{1}{2}} e^{-\mu x}\, dx = \frac{1}{\sqrt{\mu}} 2^{\nu-\frac{1}{2}} \Gamma(\nu)\, D_{1-2\nu}\left(2\sqrt{u\mu}\right)$

$[u > 0, \quad \operatorname{Re}\mu > 0, \quad \operatorname{Re}\nu > 0]$
**ET I 139(18)**

6. $\int_u^\infty (x-u)^{\nu-1}(x+u)^{-\nu-\frac{1}{2}} e^{-\mu x}\, dx = \frac{1}{\sqrt{u}} 2^{\nu-\frac{1}{2}} \Gamma(\nu)\, D_{-2\nu}\left(2\sqrt{u\mu}\right)$

$[u > 0, \quad \operatorname{Re}\mu \geq 0, \quad \operatorname{Re}\nu > 0]$
**ET I 139(19)**

7.[6] $\int_{-\infty}^\infty (b-ix)^{-\mu}(c-ix)^{-\nu} e^{-ipx}\, dx = \frac{2\pi e^{-bp} p^{\mu+\nu-1}}{\Gamma(\mu+\nu)}\, {}_1F_1(\nu; \mu+\nu; (b-c)p)$   [for $p > 0$]

$= 0$   [for $p < 0$]

$[\operatorname{Re}b > 0, \quad \operatorname{Re}c > 0, \quad \operatorname{Re}(\mu+\nu) > 1]$   **ET I 119(10)**

8.[6] $\int_{-\infty}^\infty (b+ix)^{-\mu}(c+ix)^{-\nu} e^{-ipx}\, dx = 0$   [for $p > 0$]

$= \frac{2\pi e^{cp}(-p)^{\mu+\nu-1}}{\Gamma(\mu+\nu)}\, {}_1F_1[\mu; \mu+\nu; (b-c)p]$   [for $p < 0$]

$[\operatorname{Re}b > 0, \quad \operatorname{Re}c > 0, \quad \operatorname{Re}(\mu+\nu) > 1]$   **ET I 19(11)**

9.[6] $\int_{-\infty}^\infty (b+ix)^{-2\mu}(c-ix)^{-2\nu} e^{-ipx}\, dx$

$= 2\pi(b+c)^{-\mu-\nu} \frac{p^{\mu+\nu-1}}{\Gamma(2\nu)} \exp\left(\frac{b-c}{2}p\right) W_{\nu-\mu, \frac{1}{2}-\nu-\mu}(bp+cp)$   [for $p > 0$]

$= 2\pi(b+c)^{-\mu-\nu} \frac{(-p)^{\mu+\nu-1}}{\Gamma(2\mu)} \exp\left(\frac{b-c}{2}p\right) W_{\mu-\nu, \frac{1}{2}-\nu-\mu}(-bp-cp)$   [for $p < 0$]

$\left[\operatorname{Re}b > 0, \quad \operatorname{Re}c > 0, \quad \operatorname{Re}(\mu+\nu) > \frac{1}{2}\right]$   **ET I 19(12)**

**3.385**[11] $\int_0^1 x^{\nu-1}(1-x)^{\lambda-1}(1-bx)^{-\varrho} e^{-\mu x}\, dx = \mathrm{B}(\nu, \lambda)\Phi_1(\nu, \varrho, \lambda+\nu, -\mu, b)$

$[\operatorname{Re}\lambda > 0, \quad \operatorname{Re}\nu > 0, \quad |\arg(1-b)| < \pi]$   **ET I 39(24)**

**3.386**

1. $\int_{-\infty}^\infty \frac{(ix)^{\nu_0} \displaystyle\prod_{k=1}^n (\beta_k+ix)^{\nu_k}\, e^{-ipx}\, dx}{\beta_0 - ix} = 2\pi e^{-\beta_0 p}\beta_0^{\nu_0} \prod_{k=1}^n (\beta_0 + \beta_k)^{\nu_k}$

$\left[\operatorname{Re}\nu_0 > -1, \quad \operatorname{Re}\beta_k > 0, \quad \sum_{k=0}^n \operatorname{Re}\nu_k < 1, \quad \arg ix = \frac{\pi}{2}\operatorname{sign} x, \quad p > 0\right]$   **ET I 118(8)**

**2.**

$$\int_{-\infty}^{\infty} \frac{(ix)^{\nu_0} \prod_{k=1}^{n} (\beta_k + ix)^{\nu_k} e^{-ipx} \, \mathrm{d}x}{\beta_0 + ix} = 0$$

$$\left[ \operatorname{Re} \nu_0 > -1, \quad \operatorname{Re} \beta_k > 0, \quad \sum_{k=0}^{n} \operatorname{Re} \nu_k < 1, \quad \arg ix = \frac{\pi}{2} \operatorname{sign} x, \quad p > 0 \right] \quad \text{ET I 119(9)}$$

**3.387**

**1.**[6]

$$\int_{-1}^{1} \left(1 - x^2\right)^{\nu-1} e^{-\mu x} \, \mathrm{d}x = \sqrt{\pi} \left(\frac{2}{\mu}\right)^{\nu-\frac{1}{2}} \Gamma(\nu) \, I_{\nu-\frac{1}{2}}(\mu)$$

$$\left[ \operatorname{Re} \nu > 0, \quad |\arg \mu| < \frac{\pi}{2} \right] \quad \text{WA 172(2)a}$$

**2.**[6]

$$\int_{-1}^{1} \left(1 - x^2\right)^{\nu-1} e^{i\mu x} \, \mathrm{d}x = \sqrt{\pi} \left(\frac{2}{\mu}\right)^{\nu-\frac{1}{2}} \Gamma(\nu) \, J_{\nu-\frac{1}{2}}(\mu)$$

$$[\operatorname{Re} \nu > 0] \quad \text{WA 25(3), WA 48(4)a}$$

**3.**

$$\int_{1}^{\infty} \left(x^2 - 1\right)^{\nu-1} e^{-\mu x} \, \mathrm{d}x = \frac{1}{\sqrt{\pi}} \left(\frac{2}{\mu}\right)^{\nu-\frac{1}{2}} \Gamma(\nu) \, K_{\nu-\frac{1}{2}}(\mu)$$

$$\left[ |\arg \mu| < \frac{\pi}{2}, \quad \operatorname{Re} \nu > 0 \right] \quad \text{WA 190(4)a}$$

**4.**

$$\int_{1}^{\infty} \left(x^2 - 1\right)^{\nu-1} e^{i\mu x} \, \mathrm{d}x$$

$$= i \frac{\sqrt{\pi}}{2} \left(\frac{2}{\mu}\right)^{\nu-\frac{1}{2}} \Gamma(\nu) \, H_{\frac{1}{2}-\nu}^{(1)}(\mu) \quad [\operatorname{Im} \mu > 0, \quad \operatorname{Re} \nu > 0] \quad \text{EH II 83(28)a}$$

$$= -i \frac{\sqrt{\pi}}{2} \left(-\frac{2}{\mu}\right)^{\nu-\frac{1}{2}} \Gamma(\nu) \, H_{\frac{1}{2}-\nu}^{(2)}(-\mu) \quad [\operatorname{Im} \mu < 0, \quad \operatorname{Re} \nu > 0] \quad \text{EH II 83(29)a}$$

**5.**

$$\int_{0}^{u} \left(u^2 - x^2\right)^{\nu-1} e^{\mu x} \, \mathrm{d}x = \frac{\sqrt{\pi}}{2} \left(\frac{2u}{\mu}\right)^{\nu-\frac{1}{2}} \Gamma(\nu) \left[ I_{\nu-\frac{1}{2}}(u\mu) + \mathbf{L}_{\nu-\frac{1}{2}}(u\mu) \right]$$

$$[u > 0, \quad \operatorname{Re} \nu > 0] \quad \text{ET II 188(20)a}$$

**6.**

$$\int_{u}^{\infty} \left(x^2 - u^2\right)^{\nu-1} e^{-\mu x} \, \mathrm{d}x = \frac{1}{\sqrt{\pi}} \left(\frac{2u}{\mu}\right)^{\nu-\frac{1}{2}} \Gamma(\nu) \, K_{\nu-\frac{1}{2}}(u\mu)$$

$$[u > 0, \quad \operatorname{Re} \mu > 0, \quad \operatorname{Re} \nu > 0]$$
$$\text{ET II 203(17)a}$$

**7.**[11]

$$\int_{0}^{\infty} \left(x^2 + u^2\right)^{\nu-1} e^{-\mu x} \, \mathrm{d}x = \frac{\sqrt{\pi}}{2} \left(\frac{2u}{\mu}\right)^{\nu-\frac{1}{2}} \Gamma(\nu) \left[ \mathbf{H}_{\nu-\frac{1}{2}}(u\mu) - Y_{\nu-\frac{1}{2}}(u\mu) \right]$$

$$[|\arg u| < \pi, \quad \operatorname{Re} \mu > 0] \quad \text{ET I 138(10)}$$

**3.388**

**1.**

$$\int_{0}^{2u} \left(2ux - x^2\right)^{\nu-1} e^{-\mu x} \, \mathrm{d}x = \sqrt{\pi} \left(\frac{2u}{\mu}\right)^{\nu-\frac{1}{2}} e^{-u\mu} \Gamma(\nu) \, I_{\nu-\frac{1}{2}}(u\mu)$$

$$[u > 0, \quad \operatorname{Re} \nu > 0] \quad \text{ET I 138(14)}$$

2.  $$\int_0^\infty \left(2\beta x + x^2\right)^{\nu-1} e^{-\mu x}\,dx = \frac{1}{\sqrt{\pi}} \left(\frac{2\beta}{\mu}\right)^{\nu-\frac{1}{2}} e^{\beta\mu}\,\Gamma(\nu)\,K_{\nu-\frac{1}{2}}(\beta\mu)$$

$$[|\arg\beta| < \pi, \quad \operatorname{Re}\nu > 0, \quad \operatorname{Re}\mu > 0]$$
ET I 138(13)

3.  $$\int_0^\infty \left(x^2 + ix\right)^{\nu-1} e^{-\mu x}\,dx = -\frac{i\sqrt{\pi}\,e^{\frac{i\mu}{2}}}{2\mu^{\nu-\frac{1}{2}}}\,\Gamma(\nu)\,H^{(2)}_{\nu-\frac{1}{2}}\left(\frac{\mu}{2}\right)$$

$$[\operatorname{Re}\mu > 0, \quad \operatorname{Re}\nu > 0]$$
ET I 138(15)

4.  $$\int_0^\infty \left(x^2 - ix\right)^{\nu-1} e^{-\mu x}\,dx = \frac{i\sqrt{\pi}\,e^{-\frac{i\mu}{2}}}{2\mu^{\nu-\frac{1}{2}}}\,\Gamma(\nu)\,H^{(1)}_{\nu-\frac{1}{2}}\left(\frac{\mu}{2}\right)$$

$$[\operatorname{Re}\mu > 0, \quad \operatorname{Re}\nu > 0]$$
ET I 138(16)

**3.389**

1.  $$\int_0^u x^{2\nu-1}\left(u^2 - x^2\right)^{\varrho-1} e^{\mu x}\,dx = \frac{1}{2}\,\mathrm{B}(\nu,\varrho)u^{2\nu+2\varrho-2}\ {}_1F_2\left(\nu;\frac{1}{2},\nu+\varrho;\frac{\mu^2 u^2}{4}\right)$$

$$+\frac{\mu}{2}\,\mathrm{B}\left(\nu+\frac{1}{2},\varrho\right)u^{2\nu+2\varrho-1}\ {}_1F_2\left(\nu+\frac{1}{2};\frac{3}{2},\nu+\varrho+\frac{1}{2};\frac{\mu^2 u^2}{4}\right)$$

$$[\operatorname{Re}\varrho > 0, \quad \operatorname{Re}\nu > 0]$$
ET II 188(21)

2.7  $$\int_0^\infty x^{2\nu-1}\left(u^2 + x^2\right)^{\varrho-1} e^{-\mu x}\,dx = \frac{u^{2\nu+2\varrho-2}}{2\sqrt{\pi}\,\Gamma(1-\varrho)}\,G^{31}_{13}\left(\frac{\mu^2 u^2}{4}\,\middle|\,\begin{matrix}1-\nu\\ 1-\varrho-\nu, 0, \frac{1}{2}\end{matrix}\right)$$

$$\left[|\arg u| < \frac{\pi}{2}, \quad \operatorname{Re}\mu > 0, \quad \operatorname{Re}\nu > 0\right]$$
ET II 234(15)a

3.7  $$\int_0^u x\left(u^2 - x^2\right)^{\nu-1} e^{\mu x}\,dx = \frac{u^{2\nu}}{2\nu} + \frac{\sqrt{\pi}}{2}\left(\frac{\mu}{2}\right)^{\frac{1}{2}-\nu} u^{\nu+\frac{1}{2}}\,\Gamma(\nu)\left[I_{\nu+\frac{1}{2}}(\mu u) + \mathbf{L}_{\nu+\frac{1}{2}}(\mu u)\right]$$

$$[\operatorname{Re}\nu > 0]$$
ET II 188(19)a

4.  $$\int_u^\infty x\left(x^2 - u^2\right)^{\nu-1} e^{-\mu x}\,dx = 2^{\nu-\frac{1}{2}}\left(\sqrt{\pi}\right)^{-1}\mu^{\frac{1}{2}-\nu} u^{\nu+\frac{1}{2}}\,\Gamma(\nu)\,K_{\nu+\frac{1}{2}}(u\mu)$$

$$[\operatorname{Re}(u\mu) > 0]$$
ET II 203(16)a

5.  $$\int_{-\infty}^\infty \frac{(ix)^{-\nu}e^{-ipx}\,dx}{b^2 + x^2} = \pi b^{-\nu-1}e^{-|p|b}$$

$$\left[|\nu| < 1, \quad \operatorname{Re}b > 0, \quad \arg ix = \frac{\pi}{2}\operatorname{sign}x\right]$$
ET I 118(5)

6.  $$\int_0^\infty \frac{x^\nu e^{-\mu x}}{b^2 + x^2}\,dx = \frac{1}{2}\,\Gamma(\nu)b^{\nu-1}\left[\exp\left(i\mu b + i\frac{(\nu-1)\pi}{2}\right)\right.$$

$$\left.\times\,\Gamma(1-\nu, ib\mu) + \exp\left(-ib\mu - i\frac{(\nu-1)\pi}{2}\right)\Gamma(1-\nu, -ib\mu)\right]$$

$$[\operatorname{Re}b > 0, \quad \operatorname{Re}\mu > 0, \quad \operatorname{Re}\nu > -1]$$
ET II 218(22)

7.  $$\int_0^\infty \frac{x^{\nu-1}e^{-\mu x}\,dx}{1 + x^2} = \pi\operatorname{cosec}(\nu\pi)V_\nu(2\mu, 0)$$

$$[\operatorname{Re}\mu > 0, \quad \operatorname{Re}\nu > 0]$$
ET I 138(9)

8.    $\displaystyle\int_{-\infty}^{\infty} \frac{(b+ix)^{-\nu}e^{-ipx}}{c^2+x^2}\,dx = \frac{\pi}{c}(b+c)^{-\nu}e^{-pc}$

$$[\operatorname{Re}\nu > -1, \quad p > 0, \quad \operatorname{Re}b > 0, \quad \operatorname{Re}c > 0] \quad \text{ET I 118(6)}$$

9.[6]    $\displaystyle\int_{-\infty}^{\infty} \frac{(b-ix)^{-\nu}e^{-ipx}}{c^2+x^2}\,dx = \frac{\pi}{c}(b+c)^{-\nu}e^{cp}$

$$[p < 0, \quad \operatorname{Re}b > 0, \quad \operatorname{Re}c > 0, \quad \operatorname{Re}\nu > -1] \quad \text{ET I 118(7)}$$

**3.391**    $\displaystyle\int_{0}^{\infty}\left[\left(\sqrt{x+2\beta}+\sqrt{x}\right)^{2\nu} - \left(\sqrt{x+2\beta}-\sqrt{x}\right)^{2\nu}\right]e^{-\mu x}\,dx = 2^{\nu+1}\frac{\nu}{\mu}\beta^{\nu}e^{\beta\mu}\,K_{\nu}(\beta\mu)$

$$[|\arg\beta| < \pi, \quad \operatorname{Re}\mu > 0] \quad \text{ET I 140(30)}$$

**3.392**

1.    $\displaystyle\int_{0}^{\infty}\left(x+\sqrt{1+x^2}\right)^{\nu}e^{-\mu x}\,dx = \frac{1}{\mu}S_{1,\nu}(\mu) + \frac{\nu}{\mu}S_{0,\nu}(\mu)$

$$[\operatorname{Re}\mu > 0] \quad \text{ET I 140(25)}$$

2.    $\displaystyle\int_{0}^{\infty}\left(\sqrt{1+x^2}-x\right)^{\nu}e^{-\mu x}\,dx = \frac{1}{\mu}S_{1,\nu}(\mu) - \frac{\nu}{\mu}S_{0,\nu}(\mu)$

$$[\operatorname{Re}\mu > 0] \quad \text{ET I 140(26)}$$

3.    $\displaystyle\int_{0}^{\infty}\frac{\left(x+\sqrt{1+x^2}\right)^{\nu}}{\sqrt{1+x^2}}e^{-\mu x}\,dx = \pi\operatorname{cosec}\nu\pi\left[\mathbf{J}_{-\nu}(\mu) - J_{-\nu}(\mu)\right]$

$$[\operatorname{Re}\mu > 0] \quad \text{ET I 140(27), EH II 35(33)}$$

4.    $\displaystyle\int_{0}^{\infty}\frac{\left(\sqrt{1+x^2}-x\right)^{\nu}}{\sqrt{1+x^2}}e^{-\mu x}\,dx = S_{0,\nu}(\mu) - \nu\,S_{-1,\nu}(\mu) \quad [\operatorname{Re}\mu > 0] \quad \text{ET I 140(28)}$

**3.393**    $\displaystyle\int_{0}^{\infty}\frac{\left(x+\sqrt{x^2+4b^2}\right)^{2\nu}}{\sqrt{x^3+4b^2x}}e^{-\mu x}\,dx$

$$= \frac{\sqrt{\mu\pi^3}}{2^{2\nu+3/2}b^{2\nu}}\left[J_{\nu+1/4}(b\mu)\,Y_{\nu-1/4}(b\mu) - J_{\nu-1/4}(b\mu)\,Y_{\nu+1/4}(b\mu)\right]$$
$$[\operatorname{Re}b > 0, \quad \operatorname{Re}\mu > 0] \quad \text{ET I 140(33)}$$

**3.394**    $\displaystyle\int_{0}^{\infty}\frac{\left(1+\sqrt{1+x^2}\right)^{\nu+1/2}}{x^{\nu+1}\sqrt{1+x^2}}e^{-\mu x}\,dx = \sqrt{2}\,\Gamma(-\nu)\,D_{\nu}\left(\sqrt{2i\mu}\right)D_{\nu}\left(\sqrt{-2i\mu}\right)$

$$[\operatorname{Re}\mu \geq 0, \quad \operatorname{Re}\nu < 0] \quad \text{ET I 140(32)}$$

**3.395**

1.    $\displaystyle\int_{1}^{\infty}\frac{\left(\sqrt{x^2-1}+x\right)^{\nu} + \left(\sqrt{x^2-1}+x\right)^{-\nu}}{\sqrt{x^2-1}}e^{-\mu x}\,dx = 2\,K_{\nu}(\mu)$

$$[\operatorname{Re}\mu > 0] \quad \text{ET I 140(29)}$$

2.    $\displaystyle\int_{1}^{\infty}\frac{\left(x+\sqrt{x^2-1}\right)^{2\nu} + \left(x-\sqrt{x^2-1}\right)^{2\nu}}{\sqrt{x(x^2-1)}}e^{-\mu x}\,dx = \sqrt{\frac{2\mu}{\pi}}\,K_{\nu+1/4}\left(\frac{\mu}{2}\right)K_{\nu-1/4}\left(\frac{\mu}{2}\right)$

$$[\operatorname{Re}\mu > 0] \quad \text{ET I 140(34)}$$

3.

$$\int_0^\infty \frac{\left(x + \sqrt{x^2 + 1}\right)^\nu + \cos \nu \pi \left(x + \sqrt{x^2 + 1}\right)^{-\nu}}{\sqrt{x^2 + 1}} e^{-\mu x}\, dx = -\pi \left[\mathbf{E}_\nu(\mu) + Y_\nu(\mu)\right]$$

$$[\operatorname{Re}\mu > 0] \hspace{3cm} \text{EH II 35(34)}$$

## 3.41–3.44 Combinations of rational functions of powers and exponentials

### 3.411

1. $$\int_0^\infty \frac{x^{\nu-1}\, dx}{e^{\mu x} - 1} = \frac{1}{\mu^\nu} \Gamma(\nu)\, \zeta(\nu)$$      $[\operatorname{Re}\mu > 0, \quad \operatorname{Re}\nu > 1]$      FI II 792a

2.[12] $$\int_0^\infty \frac{x^{2n-1}}{e^x - 1}\, dx = \frac{(2\pi)^{2n}\, |B_{2n}|}{4n}$$      $[n = 1, 2, \ldots]$      FI II 721a

3.[12] $$\int_0^\infty \frac{x^{a-1}}{e^x + 1}\, dx = \left(1 - 2^{1-a}\right) \Gamma(a)\, \zeta(a)$$      $[\operatorname{Re}a > 0]$      FI II 792a, WH

4.[12] $$\int_0^\infty \frac{x^{2n-1}}{e^x + 1}\, dx = \left(1 - 2^{1-2n}\right)(2\pi)^{2n}\, \frac{|B_{2n}|}{4n}$$      $[n = 1, 2, \ldots]$      BI(83)(2), EH I 39(25)

5. $$\int_0^{\ln 2} \frac{x\, dx}{1 - e^{-x}} = \frac{\pi^2}{12}$$      BI (104)(5)

6.[8] $$\int_0^\infty \frac{x^{\nu-1} e^{-\mu x}}{1 - b e^{-x}}\, dx = \Gamma(\nu) \sum_{n=0}^\infty (\mu + n)^{-\nu} b^n = \Gamma(\nu)\, \Phi(b, \nu; \mu)$$

$$[\operatorname{Re}\mu > 0 \text{ and either } |b| \leq 1, \quad b \neq 1, \quad \operatorname{Re}\nu > 0; \text{ or } b = 1, \quad \operatorname{Re}\nu > 1] \quad \text{EH I 27(3)}$$

7.[11] $$\int_0^\infty \frac{x^{\nu-1} e^{-\mu x}}{1 - e^{-bx}}\, dx = \frac{1}{b^\nu} \Gamma(\nu)\, \zeta\left(\nu, \frac{\mu}{b}\right)$$      $[\operatorname{Re}b > 0, \quad \operatorname{Re}\mu > 0, \quad \operatorname{Re}\nu > 1]$

$$\text{ET I 144(10)}$$

8. $$\int_0^\infty \frac{x^{n-1} e^{-px}}{1 + e^x}\, dx = (n-1)! \sum_{k=1}^\infty \frac{(-1)^{k-1}}{(p + k)^n}$$      $[p > -1, \quad n = 1, 2, \ldots]$      BI (83)(9)

9. $$\int_0^\infty \frac{x e^{-x}\, dx}{e^x - 1} = \frac{\pi^2}{6} - 1$$      (cf. **4.231** 3)      BI (82)(1)

10. $$\int_0^\infty \frac{x e^{-2x}\, dx}{e^{-x} + 1} = 1 - \frac{\pi^2}{12}$$      (cf. **4.251** 6)      BI (82)(2)

11. $$\int_0^\infty \frac{x e^{-3x}}{e^{-x} + 1}\, dx = \frac{\pi^2}{12} - \frac{3}{4}$$      (cf. **4.251** 5)      BI (82)(3)

12.[11] $$\int_0^\infty \frac{x e^{-(2n-1)x}}{1 + e^x}\, dx = -\frac{\pi^2}{12} + \sum_{k=1}^{2n-1} \frac{(-1)^{k-1}}{k^2}$$      (cf. **4.251** 6)      BI (82)(5)

13.[11] $$\int_0^\infty \frac{x e^{-2nx}}{1 + e^x}\, dx = \frac{\pi^2}{12} + \sum_{k=1}^{2n} \frac{(-1)^k}{k^2}$$      (cf. **4.251** 5)      BI (82)(4)

14.[7] $\displaystyle\int_0^\infty \frac{x^2 e^{-nx}}{1-e^{-x}}\,dx = 2\sum_{k=n}^\infty \frac{1}{k^3} = 2\left(\zeta(3) - \sum_{k=1}^{n-1}\frac{1}{k^3}\right)$ $\qquad [n=1,2,\ldots]$ (cf. **4.261** 12)

BI (82)(9)

15.[7] $\displaystyle\int_0^\infty \frac{x^2 e^{-nx}}{1+e^{-x}}\,dx = 2\sum_{k=n}^\infty \frac{(-1)^{n+k}}{k^3} = (-1)^{n+1}\left(\frac{3}{2}\zeta(3) + 2\sum_{k=1}^{n-1}\frac{(-1)^k}{k^3}\right)$

$[n=1,2,\ldots]$ (cf. **4.261** 11)

LI (82)(10)

16. $\displaystyle\int_{-\infty}^\infty \frac{x^2 e^{-\mu x}}{1+e^{-x}}\,dx = \pi^3\cos^3\mu\pi\left(2-\sin^2\mu\pi\right)$ $\qquad [0 < \operatorname{Re}\mu < 1]$ ET I 120(17)a

17. $\displaystyle\int_0^\infty \frac{x^3 e^{-nx}}{1-e^{-x}}\,dx = \frac{\pi^4}{15} - 6\sum_{k=1}^{n-1}\frac{1}{k^4}$ $\qquad$ (cf. **4.262** 5) BI (82)(12)

18.[11] $\displaystyle\int_0^\infty \frac{x^3 e^{-nx}}{1+e^{-x}}\,dx = 6\sum_{k=n}^\infty \frac{(-1)^{n+k}}{k^4} = (-1)^{n+1}\left(\frac{7}{120}\pi^4 + 6\sum_{k=1}^{n-1}\frac{(-1)^k}{k^4}\right)$

(cf. **4.262** 4)

LI (82)(13)

19.[9] $\displaystyle\int_0^\infty e^{-px}\left(e^{-x}-1\right)^n \frac{dx}{x} = -\sum_{k=0}^n (-1)^k \binom{n}{k}\ln(p+n-k)$ LI (89)(10)

20.[9] $\displaystyle\int_0^\infty e^{-px}\left(e^{-x}-1\right)^n \frac{dx}{x^2} = \sum_{k=0}^n (-1)^k \binom{n}{k}(p+n-k)\ln(p+n-k)$ LI (89)(15)

21.[12] $\displaystyle\int_0^\infty x^{n-1}\frac{1-e^{-mx}}{1-e^x}\,dx = -(n-1)!\sum_{k=1}^m \frac{1}{k^n}$ $\qquad$ (cf. **4.272** 11) LI (83)(8)

22.[7] $\displaystyle\int_0^\infty \frac{x^{p-1}}{e^{rx}-q}\,dx = \frac{1}{qr^p}\,\Gamma(p)\sum_{k=1}^\infty \frac{q^k}{k^p} = \Gamma(p)r^{-p}\,\Phi(q,p,1)$

$[p>0, \quad r>0, \quad -1<q<1]$

BI (83)(5)

23. $\displaystyle\int_{-\infty}^\infty \frac{x e^{\mu x}\,dx}{b+e^x} = \pi b^{\mu-1}\operatorname{cosec}(\mu\pi)\left[\ln b - \pi\cot(\mu\pi)\right]$ $\qquad [|\arg b| < \pi, \quad 0 < \operatorname{Re}\mu < 1]$

BI (101)(5), ET I 120(16)a

24. $\displaystyle\int_{-\infty}^\infty \frac{x e^{\mu x}}{e^{\nu x}-1}\,dx = \left(\frac{\pi}{\nu}\operatorname{cosec}\frac{\mu\pi}{\nu}\right)^2$ $\qquad [\operatorname{Re}\nu > \operatorname{Re}\mu > 0]$ (cf. **4.254** 2)

LI (101)(3)

25. $\displaystyle\int_0^\infty x\frac{1+e^{-x}}{e^x-1}\,dx = \frac{\pi^2}{3} - 1$ $\qquad$ (cf. **4.231** 4) BI (82)(6)

26. $\displaystyle\int_0^\infty x\frac{1-e^{-x}}{1+e^{-3x}}e^{-x}\,dx = \frac{2\pi^2}{27}$ $\qquad$ LI (82)(7)a

27. $\displaystyle\int_0^\infty \frac{1-e^{-\mu x}}{1+e^x}\frac{dx}{x} = \ln\left[\frac{\Gamma\left(\frac{\mu}{2}+1\right)}{\Gamma\left(\frac{\mu+1}{2}\right)}\sqrt{\pi}\right]$ $\qquad [\operatorname{Re}\mu > -1]$ BI (93)(4)

28. $$\int_0^\infty \frac{e^{-\nu x} - e^{-\mu x}}{e^{-x}+1}\frac{dx}{x} = \ln\frac{\Gamma\left(\frac{\nu}{2}\right)\Gamma\left(\frac{\mu+1}{2}\right)}{\Gamma\left(\frac{\mu}{2}\right)\Gamma\left(\frac{\nu+1}{2}\right)}$$  $[\operatorname{Re}\mu > 0, \quad \operatorname{Re}\nu > 0]$  BI (93)(6)

29. $$\int_{-\infty}^\infty \frac{e^{px} - e^{qx}}{1+e^{rx}}\frac{dx}{x} = \ln\left[\tan\frac{p\pi}{2r}\cot\frac{q\pi}{2r}\right]$$  $[|r| > |p|, \quad |r| > |q|, \quad rp > 0, \quad rq > 0]$

BI (103)(3)

30. $$\int_{-\infty}^\infty \frac{e^{px} - e^{qx}}{1-e^{rx}}\frac{dx}{x} = \ln\left[\sin\frac{p\pi}{r}\operatorname{cosec}\frac{q\pi}{r}\right]$$  $[|r| > |p|, \quad |r| > |q|, \quad rp > 0, \quad rq > 0]$

BI (103)(4)

31. $$\int_0^\infty \frac{e^{-qx} + e^{(q-p)x}}{1-e^{-px}}x\,dx = \left(\frac{\pi}{p}\operatorname{cosec}\frac{q\pi}{p}\right)^2$$  $[0 < q < p]$  BI (82)(8)

32. $$\int_0^\infty \frac{e^{-px} - e^{(p-q)x}}{e^{-qx}+1}\frac{dx}{x} = \ln\cot\frac{p\pi}{2q}$$  $[0 < p < q]$  BI (93)(7)

**3.412** $$\int_0^\infty \left\{\frac{a+be^{-px}}{ce^{px}+g+he^{-px}} - \frac{a+be^{-qx}}{ce^{qx}+g+he^{-qx}}\right\}\frac{dx}{x} = \frac{a+b}{c+g+h}\ln\frac{p}{q}$$
$$[p > 0, \quad q > 0]$$  BI (96)(7)

**3.413**

1. $$\int_0^\infty \frac{\left(1-e^{-\beta x}\right)\left(1-e^{-\gamma x}\right)e^{-\mu x}}{1-e^{-x}}\frac{dx}{x} = \ln\frac{\Gamma(\mu)\,\Gamma(\beta+\gamma+\mu)}{\Gamma(\mu+\beta)\,\Gamma(\mu+\gamma)}$$
$$[\operatorname{Re}\mu > 0, \quad \operatorname{Re}\mu > -\operatorname{Re}\beta, \quad \operatorname{Re}\mu > -\operatorname{Re}\gamma, \quad \operatorname{Re}\mu > -\operatorname{Re}(\beta+\gamma)] \quad (\text{cf. } \mathbf{4.267}\ 25)$$

BI (93)(13)

2. $$\int_0^\infty \frac{\left\{1-e^{(q-p)x}\right\}^2}{e^{qx} - e^{(q-2p)x}}\frac{dx}{x} = \ln\operatorname{cosec}\frac{q\pi}{2p}$$  $[0 < q < p]$  BI (95)(6)

3. $$\int_0^\infty \frac{e^{-px} - e^{-qx}}{1+e^{-x}}\frac{1+e^{-(2n+1)x}}{x}\,dx$$
$$= \ln\left\{\frac{q(q+2)(q+4)\cdots(q+2n)(p+1)(p+3)\cdots(p+2n-1)}{p(p+2)(p+4)\cdots(p+2n)(q+1)(q+3)\cdots(q+2n-1)}\right\}$$
$$[\operatorname{Re}p > -2n, \quad \operatorname{Re}q > -2n] \quad (\text{cf. } \mathbf{4.267}\ 14) \quad \text{BI (93)(11)}$$

**3.414** $$\int_0^\infty \frac{\left(1-e^{-\beta x}\right)\left(1-e^{-\gamma x}\right)\left(1-e^{-\delta x}\right)e^{-\mu x}}{1-e^{-x}}\frac{dx}{x} = \ln\frac{\Gamma(\mu)\,\Gamma(\mu+\beta+\gamma)\,\Gamma(\mu+\beta+\delta)\,\Gamma(\mu+\gamma+\delta)}{\Gamma(\mu+\beta)\,\Gamma(\mu+\gamma)\,\Gamma(\mu+\delta)\,\Gamma(\mu+\beta+\gamma+\delta)}$$
$$[2\operatorname{Re}\mu > |\operatorname{Re}\beta| + |\operatorname{Re}\gamma| + |\operatorname{Re}\delta|] \quad (\text{cf. } \mathbf{4.267}\ 31) \quad \text{BI (93)(14), ET I 145(17)}$$

**3.415**

1. $$\int_0^\infty \frac{x\,dx}{(x^2+b^2)(e^{\mu x}-1)} = \frac{1}{2}\left[\ln\left(\frac{b\mu}{2\pi}\right) - \frac{\pi}{b\mu} - \psi\left(\frac{b\mu}{2\pi}\right)\right]$$
$$[\operatorname{Re}b > 0, \quad \operatorname{Re}\mu > 0]$$

BI (97)(20), EH I 18(27)

2.[11] $\displaystyle\int_0^\infty \frac{x\,dx}{(x^2+b^2)^2\,(e^{2\pi x}-1)} = -\frac{1}{8b^3} - \frac{1}{4b^2} + \frac{1}{4b}\,\psi'(b)$

$$\sim \frac{1}{4b^4}\sum_{k=0}^\infty \frac{|B_{2k+2}|}{b^{2k}}$$

[asymptotic expansion for $\mathrm{Re}\,b > 0$]     BI(97)(22), EH I 22(12)

3.[11] $\displaystyle\int_0^\infty \frac{x\,dx}{(x^2+b^2)\,(e^{\mu x}+1)} = \frac{1}{2}\left[\psi\left(\frac{b\mu}{2\pi}+\frac{1}{2}\right) - \ln\left(\frac{b\mu}{2\pi}\right)\right]$

[$\mathrm{Re}\,b > 0, \quad \mathrm{Re}\,\mu > 0$]

4.[8] $\displaystyle\int_0^\infty \frac{x\,dx}{(x^2+b^2)^2\,(e^{2\pi x}+1)} = \frac{1}{4b^2} - \frac{1}{4b}\,\psi'\left(b+\frac{1}{2}\right)$     [$\mathrm{Re}\,b > 0, \quad \mathrm{Re}\,\mu > 0$]

5.* $\displaystyle\int_0^\infty \frac{x^3\,dx}{(x^2+b^2)^2(e^{2\pi x}-1)} = \frac{1}{2}\ln b - \frac{1}{8b} + \frac{1}{4} - \frac{1}{2}\psi(b) - \frac{b}{2}\psi'(b)$

[$b > 0$]

6.* $\displaystyle\int_0^\infty \frac{x\,dx}{(x^2+b^2)^3(e^{2\pi x}-1)} = \frac{1}{16}\left(-\frac{2}{b^4} - \frac{3}{2b^5} + \frac{1}{b^3}\psi'(b) - \frac{1}{b^2}\psi''(b)\right)$

[$b > 0$]

7.* $\displaystyle\int_0^\infty \frac{x^3\,dx}{(x^2+b^2)^4(e^{2\pi x}-1)} = \frac{1}{16}\left(-\frac{2}{3b^4} - \frac{1}{4b^5} + \frac{1}{2b^3}\psi'(b) - \frac{1}{2b^2}\psi''(b) - \frac{1}{6b}\psi'''(b)\right)$

[$b > 0$]

**3.416**

1.  $\displaystyle\int_0^\infty \frac{(1+ix)^{2n}-(1-ix)^{2n}}{i}\,\frac{dx}{e^{2\pi x}-1} = \frac{1}{2}\frac{2n-1}{2n+1}$     [$n = 1, 2, \ldots$]     BI (88)(4)

2.  $\displaystyle\int_0^\infty \frac{(1+ix)^{2n}-(1-ix)^{2n}}{i}\,\frac{dx}{e^{\pi x}+1} = \frac{1}{2n+1}$     [$n = 1, 2, \ldots$]     BI (87)(1)

3.[8] $\displaystyle\int_0^\infty \frac{(1+ix)^{2n-1}-(1-ix)^{2n-1}}{i}\,\frac{dx}{e^{\pi x}+1} = \frac{1}{2n}\left[1 - 2^{2^n}B_{2n}\right]$

[$n = 1, 2, \ldots$]     BI (87)(2)

**3.417**

1.  $\displaystyle\int_{-\infty}^\infty \frac{x\,dx}{a^2 e^x + b^2 e^{-x}} = \frac{\pi}{2ab}\ln\frac{b}{a}$     [$ab > 0$]     (cf. **4.231** 8)     BI (101)(1)

2.  $\displaystyle\int_{-\infty}^\infty \frac{x\,dx}{a^2 e^x - b^2 e^{-x}} = \frac{\pi^2}{4ab}$     (cf. **4.231** 10)     LI (101)(2)

**3.418**

1.[6] $\displaystyle\int_0^\infty \frac{x\,dx}{e^x + e^{-x} - 1} = \frac{1}{3}\left[\psi'\left(\frac{1}{3}\right) - \frac{2}{3}\pi^2\right] = 1.1719536193\ldots$     LI (88)(1)

$2.^6 \quad \displaystyle\int_0^\infty \frac{x e^{-x}\,dx}{e^x + e^{-x} - 1} = \frac{1}{6}\left[\psi'\left(\frac{1}{3}\right) - \frac{5}{6}\pi^2\right] = 0.3118211319\ldots$        LI (88)(2)

$3. \quad \displaystyle\int_0^{\ln 2} \frac{x\,dx}{e^x + 2e^{-x} - 2} = \frac{\pi}{8}\ln 2$        BI (104)(7)

**3.419**

$1. \quad \displaystyle\int_{-\infty}^\infty \frac{x\,dx}{(b + e^x)(1 + e^{-x})} = \frac{(\ln b)^2}{2(b-1)}$     $[|\arg b| < \pi]$     (cf. **4.232** 2)

                                                                 BI (101)(16)

$2. \quad \displaystyle\int_{-\infty}^\infty \frac{x\,dx}{(b + e^x)(1 - e^{-x})} = \frac{\pi^2 + (\ln b)^2}{2(b+1)}$     $[|\arg b| < \pi]$     (cf. **4.232** 3)

                                                                 BI (101)(17)

$3. \quad \displaystyle\int_{-\infty}^\infty \frac{x^2\,dx}{(b + e^x)(1 - e^{-x})} = \frac{\left[\pi^2 + (\ln b)^2\right]\ln b}{3(b+1)}$     $[|\arg b| < \pi]$     (cf. **4.261** 4)

                                                                 BI (102)(6)

$4. \quad \displaystyle\int_{-\infty}^\infty \frac{x^3\,dx}{(b + e^x)(1 - e^{-x})} = \frac{\left[\pi^2 + (\ln b)^2\right]^2}{4(b+1)}$     $[|\arg b| < \pi]$     (cf. **4.262** 3)

                                                                 BI (102)(9)

$5.^{12} \quad \displaystyle\int_{-\infty}^\infty \frac{x^4\,dx}{(b + e^x)(1 - e^{-x})} = \frac{\pi^2 + (\ln b)^2}{15(b+1)}\left[7\pi^2 + 3(\ln b)^2\right]\ln b$

                                          (cf. **4.263** 1)            BI (102)(10)

$6.^{11} \quad \displaystyle\int_{-\infty}^\infty \frac{x^5\,dx}{(b + e^x)(1 - e^{-x})} = \frac{\left[\pi^2 + (\ln b)^2\right]^2}{6(b+1)}\left[3\pi^2 + (\ln b)^2\right]$

                                          (cf. **4.264** 3)            BI (102)(11)

$7. \quad \displaystyle\int_{-\infty}^\infty \frac{(x - \ln b)x\,dx}{(b - e^x)(1 - e^{-x})} = \frac{-\left[4\pi^2 + (\ln b)^2\right]\ln b}{6(b-1)}$     $[|\arg b| < \pi]$     (cf. **4.257** 4)

                                                                 BI (102)(7)

**3.421**

$1. \quad \displaystyle\int_0^\infty \left(e^{-\nu x} - 1\right)^n \left(e^{-\rho x} - 1\right)^m e^{-\mu x}\frac{dx}{x^2}$

$$= \sum_{k=0}^n (-1)^k \binom{n}{k} \sum_{l=0}^m (-1)^l \binom{m}{l}$$

$$\times \{(m-l)\rho + (n-k)\nu + \mu\} \ln\left[(m-l)\rho + (n-k)\nu + \mu\right]$$

$$[\operatorname{Re}\nu > 0, \quad \operatorname{Re}\mu > 0, \quad \operatorname{Re}\rho > 0] \quad \text{BI (89)(17)}$$

**2.**
$$\int_0^\infty \left(1 - e^{-\nu x}\right)^n \left(1 - e^{-\rho x}\right) e^{-x} \frac{\mathrm{d}x}{x^3} = \frac{1}{2} \sum_{k=0}^n (-1)^k \binom{n}{k} (\rho + k\nu + 1)^2$$
$$\times \ln(\rho + k\nu + 1) + \frac{1}{2} \sum_{k=1}^n (-1)^{k-1} \binom{n}{k} (k\nu + 1)^2 \ln(k\nu + 1)$$
$$[n \geq 2, \quad \operatorname{Re}\nu > 0, \quad \operatorname{Re}\rho > 0] \quad \text{BI (89)(31)}$$

**3.**
$$\int_{-\infty}^\infty \frac{x e^{-\mu x}\,\mathrm{d}x}{(b + e^{-x})(c + e^{-x})} = \frac{\pi \left(b^{\mu-1}\ln b - c^{\mu-1}\ln c\right)}{(b - c)\sin\mu\pi} + \frac{\pi^2 \left(b^{\mu-1} - c^{\mu-1}\right)\cos\mu\pi}{(c - b)\sin^2\mu\pi}$$
$$[|\arg b| < \pi, \quad |\arg c| < \pi, \quad b \neq c. \quad 0 < \operatorname{Re}\mu < 2] \quad \text{ET I 120(19)}$$

**4.**
$$\int_0^\infty \left(e^{-px} - e^{-qx}\right)\left(e^{-rx} - e^{-sx}\right) e^{-x} \frac{\mathrm{d}x}{x} = \ln \frac{(p + s + 1)(q + r + 1)}{(p + r + 1)(q + s + 1)}$$
$$[p + s > -1, \quad p + r > -1, \quad q > p] \quad \text{(cf. } \textbf{4.267 } 24\text{)} \quad \text{BI (89)(11)}$$

**5.**[12]
$$\int_0^\infty \left(1 - e^{-px}\right)\left(1 - e^{-qx}\right)\left(1 - e^{-rx}\right) e^{-x} \frac{\mathrm{d}x}{x^2}$$
$$= (p + q + 1)\ln(p + q + 1)$$
$$+ (p + r + 1)\ln(p + r + 1) + (q + r + 1)\ln(q + r + 1)$$
$$- (p + 1)\ln(p + 1) - (q + 1)\ln(q + 1) - (r + 1)\ln(r + 1)$$
$$- (p + q + r)\ln(p + q + r)$$
$$[p > 0, \quad q > 0, \quad r > 0] \quad \text{(cf. } \textbf{4.268 } 3\text{)} \quad \text{BI (89)(14)}$$

**3.422**
$$\int_{-\infty}^\infty \frac{x(x - a)e^{\mu x}\,\mathrm{d}x}{(b - e^x)(1 - e^{-x})} = \frac{-\pi^2}{e^a - 1} \operatorname{cosec}^2\mu\pi \left[(e^{a\mu} + 1)\ln\mu - 2\pi\cot\mu\pi\,(e^{a\mu} - 1)\right]$$
$$[a > 0, \quad |\arg b| < \pi, \quad |\operatorname{Re}\mu| < 1] \quad \text{(cf. } \textbf{4.257 } 5\text{)} \quad \text{BI (102)(8)a}$$

**3.423**

**1.**
$$\int_0^\infty \frac{x^{\nu-1}}{(e^x - 1)^2}\,\mathrm{d}x = \Gamma(\nu)\left[\zeta(\nu - 1) - \zeta(\nu)\right] \qquad [\operatorname{Re}\nu > 2] \qquad \text{ET I 313(10)}$$

**2.**[6]
$$\int_0^\infty \frac{x^{\nu-1} e^{-\mu x}}{(e^x - 1)^2}\,\mathrm{d}x = \Gamma(\nu)\left[\zeta(\nu - 1, \mu + 2) - (\mu + 1)\zeta(\nu, \mu + 2)\right]$$
$$[\operatorname{Re}\mu > -2, \quad \operatorname{Re}\nu > 2] \qquad \text{ET I 313(11)}$$

**3.**[8]
$$\int_0^\infty \frac{x^q e^{-px}\,\mathrm{d}x}{(1 - ae^{-px})^2} = \frac{\Gamma(q + 1)}{ap^{q+1}} \sum_{k=1}^\infty \frac{a^k}{k^q} \qquad [a < 1, \quad q > -1, \quad p > 0] \qquad \text{BI (85)(13)}$$

**4.**[7]
$$\int_0^\infty \frac{x^{\nu-1} e^{-\mu x}}{(1 - be^{-x})^2}\,\mathrm{d}x = \Gamma(\nu)\left[\Phi(b; \nu - 1; \mu) - (\mu - 1)\Phi(b; \nu; \mu)\right]$$
$$[\operatorname{Re}\nu > 0, \quad \operatorname{Re}\mu > 0, \quad |\arg(1 - b)| < \pi] \qquad \text{(cf. } \textbf{9.550}\text{)} \quad \text{ET I 313(12)}$$

5. $\int_{-\infty}^{\infty} \dfrac{xe^x\,\mathrm{d}x}{(b+e^x)^2} = \dfrac{1}{b}\ln b$ $\qquad\qquad\qquad$ $[|\arg b|<\pi]$ $\qquad$ (cf. **4.231** 5)

BI (101)(10)

6. $\int_0^t x^5\,\dfrac{e^{-x}}{(1-e^{-x})^2}\,\mathrm{d}x = 120\,\zeta(5) - \sum\limits_{k=1}^{\infty}\dfrac{e^{-kt}}{k^5}\left(y^5 + 5y^4 + 20y^3 + 60y^2 + 120y + 120\right)$

$$= 120\,\zeta(5) - \dfrac{t^5 e^{-t/2}}{2\sinh(t/2)} - 5\sum\limits_{k=1}^{\infty}\dfrac{e^{-kt}}{k^5}\left(y^4 + 4y^3 + 12y^2 + 24y + 24\right)$$

$$[y=kt]$$

### 3.424

1.[7] $\int_0^{\infty}\dfrac{(1+a)e^x - a}{(1-e^x)^2}e^{-ax}x^n\,\mathrm{d}x = n!\,\zeta(n,a)$ $\qquad$ $[a>-1,\quad n=1,2,\ldots]$ $\qquad$ BI (85)(15)

2. $\int_0^{\infty}\dfrac{(1+a)e^x + a}{(1+e^x)^2}e^{-ax}x^n\,\mathrm{d}x = n!\sum\limits_{k=1}^{\infty}\dfrac{(-1)^k}{(a+k)^n}$ $\qquad$ $[a>-1,\quad n=1,2,\ldots]$ $\qquad$ BI (85)(14)

3. $\int_{-\infty}^{\infty}\dfrac{a^2 e^x + b^2 e^{-x}}{(a^2 e^x - b^2 e^{-x})^2}x^2\,\mathrm{d}x = \dfrac{\pi^2}{2ab}$ $\qquad$ $[ab>0]$ $\qquad$ BI (102)(3)a

4. $\int_{-\infty}^{\infty}\dfrac{a^2 e^x - b^2 e^{-x}}{(a^2 e^x + b^2 e^{-x})^2}x^2\,\mathrm{d}x = \dfrac{\pi}{ab}\ln\dfrac{b}{a}$ $\qquad$ $[ab>0]$ $\qquad$ BI (102)(1)

5. $\int_0^{\infty}\dfrac{e^x - e^{-x} + 2}{(e^x - 1)^2}x^2\,\mathrm{d}x = \dfrac{2}{3}\pi^2 - 2$ $\qquad\qquad$ BI (85)(7)

### 3.425

1.[7] $\int_{-\infty}^{\infty}\dfrac{xe^x\,\mathrm{d}x}{(a^2 + b^2 e^{2x})^n} = \dfrac{\sqrt{\pi}\,\Gamma\left(n-\frac{1}{2}\right)}{4a^{2n-1}b\,\Gamma(n)}\left[2\ln\dfrac{a}{2b} - \boldsymbol{C} - \psi\left(n-\dfrac{1}{2}\right)\right]$

$$[ab>0,\quad n>0]$$

BI(101)(13), LI(101)(13)

2.[7] $\int_{-\infty}^{\infty}\dfrac{\left(a^2 e^x - e^{-x}\right)x^2\,\mathrm{d}x}{(a^2 e^x + e^{-x})^{p+1}} = -\dfrac{1}{a^{p+1}}\,\mathrm{B}\left(\dfrac{p}{2},\dfrac{p}{2}\right)\ln a$ $\qquad$ $[a>0,\quad p>0]$ $\qquad$ BI (102)(5)

### 3.426

1. $\int_{-\infty}^{\infty}\dfrac{(e^x - ae^{-x})x^2\,\mathrm{d}x}{(a+e^x)^2(1+e^{-x})^2} = \dfrac{(\ln a)^2}{a-1}$ $\qquad\qquad$ BI (102)(12)

2. $\int_{-\infty}^{\infty}\dfrac{(e^x - ae^{-x})x^2\,\mathrm{d}x}{(a+e^x)^2(1-e^{-x})^2} = \dfrac{\pi^2 + (\ln a)^2}{a+1}$ $\qquad\qquad$ BI (102)(13)

### 3.427

1. $\int_0^{\infty}\left(\dfrac{e^{-x}}{x} + \dfrac{e^{-\mu x}}{e^{-x} - 1}\right)\mathrm{d}x = \psi(\mu)$ $\qquad$ $[\operatorname{Re}\mu>0]$ $\qquad$ (cf. **4.281** 4) $\qquad$ WH

2.[7] $\int_0^{\infty}\left(\dfrac{1}{1-e^{-x}} - \dfrac{1}{x}\right)e^{-x}\,\mathrm{d}x = \boldsymbol{C}$ $\qquad$ (cf. **4.281** 1) $\qquad$ BI (94)(1)

3. $\int_0^\infty \left( \frac{1}{2} - \frac{1}{1 + e^{-x}} \right) \frac{e^{-2x}}{x} \, dx = \frac{1}{2} \ln \frac{\pi}{4}$      BI (94)(5)

4. $\int_0^\infty \left( \frac{1}{2} - \frac{1}{x} + \frac{1}{e^x - 1} \right) \frac{e^{-\mu x}}{x} \, dx = \ln \Gamma(\mu) - \left( \mu - \frac{1}{2} \right) \ln \mu + \mu - \frac{1}{2} \ln(2\pi)$

$\qquad\qquad\qquad\qquad\qquad\qquad\qquad\qquad [\operatorname{Re} \mu > 0]$      WH

5. $\int_0^\infty \left( \frac{1}{2} e^{-2x} - \frac{1}{e^x + 1} \right) \frac{dx}{x} = -\frac{1}{2} \ln \pi$      BI (94)(6)

6. $\int_0^\infty \left( \frac{e^{\mu x} - 1}{1 - e^{-x}} - \mu \right) \frac{e^{-x}}{x} \, dx = -\ln \Gamma(\mu) - \ln \sin(\pi\mu) + \ln \pi$

$\qquad\qquad\qquad\qquad\qquad\qquad\qquad\qquad [\operatorname{Re} \mu < 1]$      EH I 21(6)

7. $\int_0^\infty \left( \frac{e^{-\nu x}}{1 - e^{-x}} - \frac{e^{-\mu x}}{x} \right) dx = \ln \mu - \psi(\nu)$      (cf. **4.281** 5)      BI (94)(3)

8. $\int_0^\infty \left( \frac{n}{x} - \frac{e^{-\mu x}}{1 - e^{-x/n}} \right) e^{-x} \, dx = n \, \psi(n\mu + n) - n \ln n$

$\qquad\qquad\qquad\qquad\qquad\qquad [\operatorname{Re} \mu > 0, \quad n = 1, 2, \ldots]$      BI (94)(4)

9. $\int_0^\infty \left( \mu - \frac{1 - e^{-\mu x}}{1 - e^{-x}} \right) \frac{e^{-x}}{x} \, dx = \ln \Gamma(\mu + 1)$      $[\operatorname{Re} \mu > -1]$      WH

10. $\int_0^\infty \left( \nu e^{-x} - \frac{e^{-\mu x} - e^{-(\mu+\nu)x}}{e^x - 1} \right) \frac{dx}{x} = \ln \frac{\Gamma(\mu + \nu + 1)}{\Gamma(\mu + 1)}$

$\qquad\qquad\qquad\qquad\qquad\qquad [\operatorname{Re} \mu > -1, \quad \operatorname{Re} \nu > 0]$      BI (94)(8)

11. $\int_0^\infty \left[ (1 - e^x)^{-1} + x^{-1} - 1 \right] e^{-xz} \, dx = \psi(z) - \ln z$      $[\operatorname{Re} z > 0]$      EH I 18(24)

**3.428**

1.[12] $\int_0^\infty \left( \nu e^{-\mu x} - \frac{1}{\mu} e^{-x} - \frac{1}{\mu} \frac{e^{-x} - e^{-\mu\nu x}}{1 - e^{-x}} \right) \frac{dx}{x} = \frac{1}{\mu} \ln \Gamma(\mu\nu) - \nu \ln \mu$

$\qquad\qquad\qquad\qquad\qquad\qquad [\operatorname{Re} \mu > 0, \quad \operatorname{Re} \nu > 0]$      BI (94)(18)

2. $\int_0^\infty \left( \frac{n-1}{2} + \frac{n-1}{1 - e^{-x}} + \frac{e^{(1-\mu)x}}{1 - e^{x/n}} + \frac{e^{-n\mu x}}{1 - e^{-x}} \right) e^{-x} \frac{dx}{x} = \frac{n-1}{2} \ln 2\pi - \left( n\mu + \frac{1}{2} \right) \ln n$

$\qquad\qquad\qquad\qquad\qquad\qquad [\operatorname{Re} \mu > 0, \quad n = 1, 2, \ldots]$      BI (94)(14)

3. $\int_0^\infty \left( n\mu - \frac{n-1}{2} - \frac{n}{1 - e^{-x}} - \frac{e^{(1-\mu)x}}{1 - e^{x/n}} \right) \frac{e^{-x}}{x} \, dx = \sum_{k=0}^{n-1} \ln \Gamma \left( \mu - \frac{k}{n} + 1 \right)$

$\qquad\qquad\qquad\qquad\qquad\qquad [\operatorname{Re} \mu > 0, \quad n = 1, 2, \ldots]$      BI (94)(13)

4. $\int_0^\infty \left( \frac{e^{-\nu x}}{1 - e^x} - \frac{e^{-\mu\nu x}}{1 - e^{\mu x}} - \frac{e^x}{1 - e^x} + \frac{e^{\mu x}}{1 - e^{\mu x}} \right) \frac{dx}{x} = \nu \ln \mu$

$\qquad\qquad\qquad\qquad\qquad\qquad [\operatorname{Re} \mu > 0, \quad \operatorname{Re} \nu > 0]$      LI (94)(15)

5. $\displaystyle\int_0^\infty \left[ \frac{1}{e^x - 1} - \frac{\mu e^{-\mu x}}{1 - e^{-\mu x}} + \left( a\mu - \frac{\mu + 1}{2} \right) e^{-\mu x} + (1 - a\mu) e^{-x} \right] \frac{dx}{x}$

$$= \frac{\mu - 1}{2} \ln(2\pi) + \left( \frac{1}{2} - a\mu \right) \ln \mu$$

$$[\operatorname{Re}\mu > 0] \qquad\qquad \text{BI (94)(16)}$$

6.[12] $\displaystyle\int_0^\infty \left[ \frac{e^{-\nu x}}{1 - e^{-x}} - \frac{e^{-\mu\nu x}}{1 - e^{-\mu x}} - \frac{(\mu - 1)e^{-\mu x}}{1 - e^{-\mu x}} - \frac{\mu - 1}{2} e^{-\mu x} \right] \frac{dx}{x} = \frac{\mu - 1}{2} \ln(2\pi) + \left( \frac{1}{2} - \mu\nu \right) \ln \mu$

$$[\operatorname{Re}\mu > 0, \quad \operatorname{Re}\nu > 0] \qquad \text{(cf. } \mathbf{4.267}\ 37) \quad \text{BI (94)(17)}$$

7. $\displaystyle\int_0^\infty \left[ 1 - e^{-x} - \frac{(1 - e^{-\nu x})(1 - e^{-\mu x})}{1 - e^{-x}} \right] \frac{dx}{x} = \ln \mathrm{B}(\mu, \nu)$

$$[\operatorname{Re}\mu > 0, \quad \operatorname{Re}\nu > 0] \qquad\qquad \text{BI (94)(12)}$$

**3.429** $\displaystyle\int_0^\infty \left[ e^{-x} - (1 + x)^{-\mu} \right] \frac{dx}{x} = \psi(\mu) \qquad [\operatorname{Re}\mu > 0] \qquad\qquad \text{NH 184(7)}$

**3.431**

1. $\displaystyle\int_0^\infty \left( e^{-\mu x} - 1 + \mu x - \frac{1}{2}\mu^2 x^2 \right) x^{\nu-1}\, dx = \frac{-1}{\nu(\nu + 1)(\nu + 2)\mu^\nu} \Gamma(\nu + 3)$

$$[\operatorname{Re}\mu > 0, \quad -2 > \operatorname{Re}\nu > -3]$$

$$\text{LI (90)(5)}$$

2. $\displaystyle\int_0^\infty \left[ x^{-1} - \frac{1}{2} x^{-2}(x + 2)\left(1 - e^{-x}\right) \right] e^{-px}\, dx = -1 + \left( p + \frac{1}{2} \right) \ln \left( 1 + \frac{1}{p} \right)$

$$[\operatorname{Re}p > 0] \qquad\qquad \text{ET I 144(6)}$$

**3.432**

1. $\displaystyle\int_0^\infty x^{\nu-1} e^{-mx} \left( e^{-x} - 1 \right)^n dx = \Gamma(\nu) \sum_{k=0}^n (-1)^k \binom{n}{k} \frac{1}{(n + m - k)^\nu}$

$$[n = 0, 1, \ldots, \operatorname{Re}\nu > 0] \qquad \text{LI (90)(10)}$$

2.[12] $\displaystyle\int_0^\infty \left[ x^{\nu-1} e^{-x} - e^{-\mu x} \left( 1 - e^{-x} \right)^{\nu-1} \right] dx = \Gamma(\nu) - \frac{\Gamma(\mu)\Gamma(\nu)}{\Gamma(\mu + \nu)} = \Gamma(\nu) - \mathrm{B}(\mu, \nu)$

$$[\operatorname{Re}\mu > 0, \quad \operatorname{Re}\nu > 0] \qquad \text{LI (81)(14)}$$

**3.433** $\displaystyle\int_0^\infty x^{p-1} \left[ e^{-x} + \sum_{k=1}^n (-1)^k \frac{x^{k-1}}{(k-1)!} \right] dx = \Gamma(p) \qquad [-n < p < -n+1, \quad n = 0, 1, \ldots]$

$$\text{FI II 805}$$

**3.434**

1. $\displaystyle\int_0^\infty \frac{e^{-\nu x} - e^{-\mu x}}{x^{\rho+1}}\, dx = \frac{\mu^\rho - \nu^\rho}{\rho} \Gamma(1 - \rho) \qquad [\operatorname{Re}\mu > 0, \quad \operatorname{Re}\nu > 0, \quad \operatorname{Re}\rho < 1]$

$$\text{BI (90)(6)}$$

2. $\displaystyle\int_0^\infty \frac{e^{-\mu x} - e^{-\nu x}}{x}\, dx = \ln \frac{\nu}{\mu} \qquad [\operatorname{Re}\mu > 0, \quad \operatorname{Re}\nu > 0] \qquad\qquad \text{FI II 634}$

**3.435**

1.    $\int_0^\infty \left\{ (x+1)e^{-x} - e^{-\frac{x}{2}} \right\} \frac{\mathrm{d}x}{x} = 1 - \ln 2$                                    LI (89)(19)

2.[11]    $\int_0^\infty \frac{1 - e^{-\mu x}}{x(x+\beta)}\,\mathrm{d}x = \frac{1}{\beta} \left[ \ln(\beta\mu) + \mathbf{C} - e^{\beta\mu}\,\mathrm{Ei}(-\beta\mu) \right]$        $[|\arg\beta| < \pi, \quad \mathrm{Re}\,\mu > 0]$        ET II 217 (18)

3.    $\int_0^\infty \left( \frac{1}{1+x} - e^{-x} \right) \frac{\mathrm{d}x}{x} = \mathbf{C}$                                    FI II 7 95, 802

4.    $\int_0^\infty \left( e^{-\mu x} - \frac{1}{1+ax} \right) \frac{\mathrm{d}x}{x} = \ln\frac{a}{\mu} - \mathbf{C}$        $[a > 0, \quad \mathrm{Re}\,\mu > 0]$        BI (92)(10)

**3.436**    $\int_0^\infty \left( \frac{e^{-npx} - e^{-nqx}}{n} - \frac{e^{-mpx} - e^{-mqx}}{m} \right) \frac{\mathrm{d}x}{x^2} = (q-p)\ln\frac{m}{n}$        $[p > 0, \quad q > 0]$        BI (89)(28)

**3.437**    $\int_0^\infty \left( pe^{-x} - \frac{1 - e^{-px}}{x} \right) \frac{\mathrm{d}x}{x} = p\ln p - p$        $[p > 0]$        BI (89)(24)

**3.438**

1.    $\int_0^\infty \left\{ \left( \frac{1}{2} + \frac{1}{x} \right) e^{-x} - \frac{1}{x} e^{-\frac{x}{2}} \right\} \frac{\mathrm{d}x}{x} = \frac{\ln 2 - 1}{2}$                                    BI (89)(19)

2.[7]    $\int_0^\infty \left( \frac{p^2}{6} e^{-x} - \frac{p^2}{2x} - \frac{p}{x^2} - \frac{1 - e^{-px}}{x^3} \right) \frac{\mathrm{d}x}{x} = \frac{p^2}{6}\ln p - \frac{11}{36}p^3$

$[p > 0]$        BI (89)(33)

3.    $\int_0^\infty \left( e^{-x} - e^{-2x} - \frac{1}{x} e^{-2x} \right) \frac{\mathrm{d}x}{x} = 1 - \ln 2$                                    BI (89)(25)

4.    $\int_0^\infty \left\{ \left( p - \frac{1}{2} \right) e^{-x} + \frac{x+2}{2x} \left( e^{-px} - e^{-\frac{x}{2}} \right) \right\} \frac{\mathrm{d}x}{x} = \left( p - \frac{1}{2} \right)(\ln p - 1)$

$[p > 0]$        BI (89)(22)

**3.439**    $\int_0^\infty \left( (p-q)e^{-rx} + \frac{1}{mx} \left( e^{-mpx} - e^{-mqx} \right) \right) \frac{\mathrm{d}x}{x} = p\ln p - q\ln q - (p-q)\left( 1 + \ln\frac{r}{m} \right)$

$[p > 0, \quad q > 0, \quad r > 0]$    LI(89)(26), LI(89)(27)

**3.441**    $\int_0^\infty \left\{ (p-r)e^{-qx} + (r-q)e^{-px} + (q-p)e^{-rx} \right\} \frac{\mathrm{d}x}{x^2} = (r-q)p\ln p + (p-r)q\ln q + (q-p)r\ln r$

$[p > 0, \quad q > 0, \quad r > 0]$        (cf. **4.268** 6)    BI (89)(18)

**3.442**

1.    $\int_0^\infty \left( 1 - \frac{x+2}{2x}(1 - e^{-x}) \right) e^{-qx} \frac{\mathrm{d}x}{x} = -1 + \left( q + \frac{1}{2} \right) \ln\frac{q+1}{q}$

$[q > 0]$        BI (89)(23)

2.    $\int_0^\infty \left( \frac{e^{-x} - 1}{x} + \frac{1}{1+x} \right) \frac{\mathrm{d}x}{x} = \mathbf{C} - 1$                                    BI (92)(16)

3.    $\int_0^\infty \left( e^{-px} - \frac{1}{1 + a^2 x^2} \right) \frac{\mathrm{d}x}{x} = -\mathbf{C} + \ln\frac{a}{p}$        $[p > 0]$        BI (92)(11)

**3.443**

1.    $\displaystyle\int_0^\infty \left(\frac{e^{-x}p^2}{2} - \frac{p}{x} + \frac{1 - e^{-px}}{x^2}\right)\frac{dx}{x} = \frac{p^2}{2}\ln p - \frac{3}{4}p^2 \qquad [p > 0]$      BI (89)(32)

2.    $\displaystyle\int_0^\infty \frac{(1 - e^{-px})^n e^{-qx}}{x^3}\,dx = \frac{1}{2}\sum_{k=2}^n (-1)^{k-1}\binom{n}{k}(q + kp)^2 \ln(q + kp)$

$$[n > 2, \quad q > 0, \quad pn + q > 0] \qquad \text{(cf. } \mathbf{4.268}\ 4)\quad \text{BI (89)(30)}$$

3.    $\displaystyle\int_0^\infty \left(1 - e^{-px}\right)^2 e^{-qx}\frac{dx}{x^2} = (2p + q)\ln(2p + q) - 2(p + q)\ln(p + q) + q\ln q$

$$[q > 0, \quad 2p > -q] \qquad \text{(cf. } \mathbf{4.268}\ 2)$$
$$\text{BI (89)(13)}$$

## 3.45 Combinations of powers and algebraic functions of exponentials

**3.451**

1.    $\displaystyle\int_0^\infty xe^{-x}\sqrt{1 - e^{-x}}\,dx = \frac{4}{3}\left(\frac{4}{3} - \ln 2\right)$      BI (99)(1)

2.    $\displaystyle\int_0^\infty xe^{-x}\sqrt{1 - e^{-2x}}\,dx = \frac{\pi}{4}\left(\frac{1}{2} + \ln 2\right)$     (cf. **4.241** 9)     BI (99)(2)

**3.452**

1.    $\displaystyle\int_0^\infty \frac{x\,dx}{\sqrt{e^x - 1}} = 2\pi\ln 2$      FI II 643a,BI(99)(4)

2.    $\displaystyle\int_0^\infty \frac{x^2\,dx}{\sqrt{e^x - 1}} = 4\pi\left((\ln 2)^2 + \frac{\pi^2}{12}\right)$      BI (99)(5)

3.    $\displaystyle\int_0^\infty \frac{xe^{-x}\,dx}{\sqrt{e^x - 1}} = \frac{\pi}{2}\left[2\ln 2 - 1\right]$      BI (99)(6)

4.    $\displaystyle\int_0^\infty \frac{xe^{-x}\,dx}{\sqrt{e^{2x} - 1}} = 1 - \ln 2$      BI (99)(8)

5.    $\displaystyle\int_0^\infty \frac{xe^{-2x}\,dx}{\sqrt{e^x - 1}} = \frac{3}{4}\pi\left(\ln 2 - \frac{7}{12}\right)$      BI (99)(7)

**3.453**

1.    $\displaystyle\int_0^\infty \frac{xe^x}{a^2e^x - (a^2 - b^2)}\frac{dx}{\sqrt{e^x - 1}} = \frac{2\pi}{ab}\ln\left(1 + \frac{b}{a}\right)$    $[ab > 0]$    (cf. **4.298** 17)    BI (99)(16)

2.    $\displaystyle\int_0^\infty \frac{xe^x\,dx}{[a^2e^x - (a^2 + b^2)]\sqrt{e^x - 1}} = \frac{2\pi}{ab}\arctan\frac{b}{a}$    $[ab > 0]$    (cf. **4.298** 18)    BI (99)(17)

**3.454**

1.[11]   $\displaystyle\int_0^\infty \frac{xe^{-2nx}\,dx}{\sqrt{e^{2x} - 1}} = \frac{(2n - 1)!!}{(2n)!!}\frac{\pi}{2}\left(\ln 2 + \sum_{k=1}^{2n}\frac{(-1)^k}{k}\right)$      LI (99)(10)

2. $\quad \int_0^\infty \dfrac{x e^{-(2n-1)x}\,dx}{\sqrt{e^{2x}-1}} = -\dfrac{(2n-2)!!}{(2n-1)!!}\left(\ln 2 + \sum_{k=1}^{2n-1}\dfrac{(-1)^k}{k}\right)$     LI (99)(9)

**3.455**

1. $\quad \int_0^\infty \dfrac{x^2 e^x\,dx}{\sqrt{(e^x-1)^3}} = 8\pi\ln 2$     BI (99)(11)

2. $\quad \int_0^\infty \dfrac{x^3 e^x\,dx}{\sqrt{(e^x-1)^3}} = 24\pi\left[(\ln 2)^2 + \dfrac{\pi^2}{12}\right]$     BI (99)(12)

**3.456**

1. $\quad \int_0^\infty \dfrac{x\,dx}{\sqrt[3]{e^{3x}-1}} = \dfrac{\pi}{3\sqrt{3}}\left[\ln 3 + \dfrac{\pi}{3\sqrt{3}}\right]$     BI (99)(13)

2. $\quad \int_0^\infty \dfrac{x\,dx}{\sqrt[3]{(e^{3x}-1)^2}} = \dfrac{\pi}{3\sqrt{3}}\left[\ln 3 - \dfrac{\pi}{3\sqrt{3}}\right]$    (cf. **4.244** 3)     BI (99)(14)

**3.457**

1. $\quad \int_0^\infty x e^{-x}\left(1-e^{-2x}\right)^{n-1/2}\,dx = \dfrac{(2n-1)!!}{4\cdot(2n)!!}\pi\left[C + \psi(n+1) + 2\ln 2\right]$

                                            (cf. **4.241** 5)     BI (99)(3)

2.[12] $\quad \int_{-\infty}^\infty \dfrac{x e^x\,dx}{(a+e^x)^{n+3/2}} = \dfrac{2}{(2n+1)a^{n+1/2}}\left[\ln(4a) - C - 2\,\psi(2n) + \psi(n)\right]$     BI (101)(12)

3. $\quad \int_{-\infty}^\infty \dfrac{x\,dx}{(a^2 e^x + e^{-x})^\mu} = -\dfrac{1}{2a^\mu}\,B\left(\dfrac{\mu}{2},\dfrac{\mu}{2}\right)\ln a$    $[a>0, \quad \operatorname{Re}\mu > 0]$     BI (101)(14)

**3.458**

1.[7] $\quad \int_0^{\ln 2} x e^x\left(e^x-1\right)^{p-1}\,dx = \dfrac{1}{p}\left[\ln 2 + \sum_{k=0}^\infty \dfrac{(-1)^{k-1}}{p+k+1}\right]$     BI (104)(4)

2. $\quad \int_{-\infty}^\infty \dfrac{x e^x\,dx}{(a+e^x)^{\nu+1}} = \dfrac{1}{\nu a^\nu}\left[\ln a - C - \psi(\nu)\right]$    $[a>0]$

                      $= \dfrac{1}{\nu a^\nu}\left[\ln a - \sum_{k=1}^{\nu-1}\dfrac{1}{k}\right]$    $[a>0,\quad \nu = 1,2,\ldots]$

                                                      BI (101)(11)

## 3.46–3.48 Combinations of exponentials of more complicated arguments and powers

**3.461**

1. $\quad \int_u^\infty \dfrac{e^{-p^2 x^2}}{x^{2n}}\,dx = \dfrac{(-1)^n 2^{n-1} p^{2n-1}\sqrt{\pi}}{(2n-1)!!}\left[1-\Phi(pu)\right]$

                     $+ \dfrac{e^{-p^2 u^2}}{2u^{2n-1}}\sum_{k=0}^{n-1}\dfrac{(-1)^k 2^{k+1}(pu)^{2k}}{(2n-1)(2n-3)\cdots(2n-2k-1)}$

                                                   $[p>0]$     NT 21(4)

$2.^{12}$   $\int_0^\infty x^{2n} e^{-px^2}\, dx = \frac{1}{2} p^{-\left(n+\frac{1}{2}\right)} \Gamma\left(n+\frac{1}{2}\right) = \frac{(2n-1)!!}{2(2p)^n} \sqrt{\frac{\pi}{p}}$

$$[p>0, \quad n=0,1,\ldots] \qquad \text{FI II 743}$$

$2b.^*$   $\int_0^\infty x^m e^{-px^2}\, dx = \frac{1}{2p^{(m+1)/2}} \Gamma\left(\frac{m+1}{2}\right)$

$3.^{12}$   $\int_0^\infty x^{2n+1} e^{-px^2}\, dx = \frac{1}{2} p^{-(n+1)} \Gamma(n+1) = \frac{n!}{2p^{n+1}} \qquad [p>0]$      BI (81)(7)

4.   $\int_{-\infty}^\infty (x+ai)^{2n} e^{-x^2}\, dx = \frac{(2n-1)!!}{2^n} \sqrt{\pi} \sum_{k=0}^n (-1)^k \frac{(2a)^{2k} n!}{(2k)!(n-k)!}$      BI (100)(12)

$5.^{11}$   $\int_u^\infty e^{-\mu x^2} \frac{dx}{x^2} = \frac{1}{u} e^{-\mu u^2} - \sqrt{\mu \pi}\left[1 - \Phi(u\sqrt{\mu})\right]$    $\left[|\arg\mu| < \frac{\pi}{2}, \quad u>0\right]$    ET I 135(19)a

6.   $\int_0^\infty \exp\left(-a\sqrt{x^2+b^2}\right) dx = b\, K_1(ab)$      $[\operatorname{Re} a>0, \quad \operatorname{Re} b>0]$

7.   $\int_0^\infty \exp\left(-a\sqrt{x^2+b^2}\right) dx = \frac{2b}{a^2} K_1(ab) + \frac{b^2}{a} K_0(ab)$

$$[\operatorname{Re} a>0, \quad \operatorname{Re} b>0]$$

8.   $\int_0^\infty x^4 \exp\left(-a\sqrt{x^2+b^2}\right) dx = \frac{12b^2}{a^3} K_2(ab) + \frac{3b^3}{a^2} K_1(ab)$

$$[\operatorname{Re} a>0, \quad \operatorname{Re} b>0]$$

9.   $\int_0^\infty x^6 \exp\left(-a\sqrt{x^2+b^2}\right) dx = \frac{90b^3}{a^4} K_3(ab) + \frac{15b^4}{a^3} K_2(ab)$

$$[\operatorname{Re} a>0, \quad \operatorname{Re} b>0]$$

**3.462**

1.   $\int_0^\infty x^{\nu-1} e^{-\beta x^2 - \gamma x}\, dx = (2\beta)^{-\nu/2} \Gamma(\nu) \exp\left(\frac{\gamma^2}{8\beta}\right) D_{-\nu}\left(\frac{\gamma}{\sqrt{2\beta}}\right)$

$$[\operatorname{Re}\beta>0, \quad \operatorname{Re}\nu>0]$$
$$\text{EH II 119(3)a, ET I 313(13)}$$

$2.^8$   $\int_{-\infty}^\infty x^n e^{-px^2+2qx}\, dx = \frac{1}{2^{n-1}p} \sqrt{\frac{\pi}{p}} \frac{d^{n-1}}{dq^{n-1}}\left(qe^{q^2/p}\right) \qquad [p>0]$      BI (100)(8)

$$= n!e^{q^2/p} \sqrt{\frac{\pi}{p}} \left(\frac{q}{p}\right)^n \sum_{k=0}^{\lfloor n/2 \rfloor} \frac{1}{(n-2k)!(k)!} \left(\frac{p}{4q^2}\right)^k \qquad [p>0] \quad \text{LI (100)(8)}$$

$3.^{12}$   $\int_{-\infty}^\infty (ix)^\nu e^{-\beta^2 x^2 + iqx}\, dx = 2^{-\frac{\nu}{2}} \sqrt{\pi}\, \beta^{-\nu-1} \exp\left(-\frac{q^2}{8\beta^2}\right) D_\nu\left(\frac{q}{\beta\sqrt{2}}\right)$

$$\left[\operatorname{Re}\beta^2>0, \quad \operatorname{Re}\nu>-1, \quad \arg ix = \frac{\pi}{2}\operatorname{sign} x\right] \quad \text{ET I 121(23)}$$

4.   $\int_{-\infty}^\infty x^n \exp\left[-(x-\beta)^2\right] dx = (2i)^{-n} \sqrt{\pi}\, H_n(i\beta)$      EH II 195(31)

5.[11] $\quad \int_0^\infty x e^{-\mu x^2 - 2\nu x} \, \mathrm{d}x = \dfrac{1}{2\mu} - \dfrac{\nu}{2\mu} \sqrt{\dfrac{\pi}{\mu}} e^{\frac{\nu^2}{\mu}} \left[ 1 - \mathrm{erf}\left( \dfrac{\nu}{\sqrt{\mu}} \right) \right]$

$\left[ |\arg \nu| < \frac{\pi}{2}, \quad \mathrm{Re}\,\mu > 0 \right]$ ET I 146(31)a

6. $\quad \int_{-\infty}^\infty x e^{-px^2 + 2qx} \, \mathrm{d}x = \dfrac{q}{p} \sqrt{\dfrac{\pi}{p}} \exp\left( \dfrac{q^2}{p} \right)$ $\qquad [\mathrm{Re}\,p > 0]$ BI (100)(7)

7.[11] $\quad \int_0^\infty x^2 e^{-\mu x^2 - 2\nu x} \, \mathrm{d}x = -\dfrac{\nu}{2\mu^2} + \sqrt{\dfrac{\pi}{\mu^5}} \dfrac{2\nu^2 + \mu}{4} e^{\frac{\nu^2}{\mu}} \left[ 1 - \mathrm{erf}\left( \dfrac{\nu}{\sqrt{\mu}} \right) \right]$

$\left[ |\arg \nu| < \frac{\pi}{2}, \quad \mathrm{Re}\,\mu > 0 \right]$ ET I 146(32)

8. $\quad \int_{-\infty}^\infty x^2 e^{-\mu x^2 + 2\nu x} \, \mathrm{d}x = \dfrac{1}{2\mu} \sqrt{\dfrac{\pi}{\mu}} \left( 1 + 2\dfrac{\nu^2}{\mu} \right) e^{\frac{\nu^2}{\mu}}$ $\qquad [|\arg \nu| < \pi, \quad \mathrm{Re}\,\mu > 0]$ BI (100)(8)a

9. $\quad \int_0^\infty e^{-\beta x^n \pm a} \, \mathrm{d}x = \dfrac{e^{\pm a}}{n \beta^{1/n}} \Gamma\left( \dfrac{1}{n} \right)$ $\qquad [\mathrm{Re}\,\beta > 0, \quad \mathrm{Re}\,n > 0]$

10. $\quad \int_0^\infty (x - a) e^{-\beta(x-a)} \, \mathrm{d}x = e^{a\beta} \dfrac{(1 - a\beta)}{\beta^2}$ $\qquad [\mathrm{Re}\,\beta > 0]$

11. $\quad \int_0^\infty (x - a) e^{-\beta(x+a)} \, \mathrm{d}x = e^{-a\beta} \dfrac{(1 - a\beta)}{\beta^2}$ $\qquad [\mathrm{Re}\,\beta > 0]$

12. $\quad \int_0^\infty (ax \pm b)^m e^{-px} \, \mathrm{d}x = \dfrac{a^m e^{\pm pb/a}}{p^{m+1}} \Gamma\left( m + 1, \pm \dfrac{pb}{a} \right)$ $\quad \left[ p > 0, \quad \left| \arg\left( \frac{b}{a} \right) \right| < \pi \right]$

13. $\quad \int_u^\infty (ax \pm b)^m e^{-px} \, \mathrm{d}x = \dfrac{a^m e^{\pm pb/a}}{p^{m+1}} \Gamma\left( m + 1, pu \pm \dfrac{pb}{a} \right)$

$\left[ p > 0, \quad \left| \arg\left( \frac{b}{a} \pm u \right) \right| < \pi \right]$

14. $\quad \int_0^u (ax \pm b)^m e^{-px} \, \mathrm{d}x = \dfrac{a^m e^{\pm pb/a}}{p^{m+1}} \left[ \Gamma\left( m + 1, \pm \dfrac{pb}{a} \right) - \Gamma\left( m + 1, pu \pm \dfrac{pb}{a} \right) \right]$

$\left[ u > 0, \quad p > 0, \quad \left| \arg\left( \frac{b}{a} \pm u \right) \right| < \pi \right]$

15. $\quad \int_0^\infty \dfrac{e^{-px}}{(ax \pm b)^n} \, \mathrm{d}x = \dfrac{p^{n-1} e^{\pm pb/a}}{a^n} \Gamma\left( -n + 1, \pm \dfrac{pb}{a} \right)$ $\quad \left[ p > 0, \quad \left| \arg\left( \frac{b}{a} \right) \right| < \pi \right]$

16. $\quad \int_u^\infty \dfrac{e^{-px}}{(ax \pm b)^n} \, \mathrm{d}x = \dfrac{p^{n-1} e^{\pm pb/a}}{a^n} \Gamma\left( -n + 1, pu \pm \dfrac{pb}{a} \right)$

$\left[ u > 0, \quad p > 0, \quad \left| \arg\left( \frac{b}{a} \pm u \right) \right| < \pi \right]$

17. $\quad \int_0^u \dfrac{e^{-px}}{(ax \pm b)^n} \, \mathrm{d}x = \dfrac{p^{n-1} e^{\pm pb/a}}{a^n} \left[ \Gamma\left( -n + 1, \pm \dfrac{pb}{a} \right) - \Gamma\left( -n + 1, pu \pm \dfrac{pb}{a} \right) \right]$

$\left[ u > 0, \quad p > 0, \quad \left| \arg\left( \frac{b}{a} \pm u \right) \right| < \pi \right]$

18. $\quad \int_0^\infty \left( \dfrac{x - a}{b} \right)^j \exp\left( -\beta \left( \dfrac{x - a}{b} \right)^k \right) \, \mathrm{d}x = \dfrac{b \Gamma\left( \frac{j+1}{k}, \beta \left( -\frac{a}{b} \right)^k \right)}{k \beta^{(j+1)/k}}$

$\left[ \arg\left( -\frac{a}{b} \right) > 0, \quad \mathrm{Re}\,b > 0, \quad \mathrm{Re}\,\beta > 0, \quad \mathrm{Re}\,k > 0 \right]$

$19.^{12}$ $\displaystyle\int_u^\infty \frac{e^{-\beta x^n}}{x^m}\,dx = \frac{\Gamma(z,\beta u^n)}{n\beta^z}$      $\left[\operatorname{Re}\beta > 0, \quad \operatorname{Re}n > 0, \quad z = \dfrac{1-m}{n}\right]$

$20.^{12}$ $\displaystyle\int_0^\infty \frac{\exp\left(-a\sqrt{x^2+b^2}\right)}{\sqrt{x^2+b^2}}\,dx = K_0(ab)$      $[\operatorname{Re}a > 0, \quad \operatorname{Re}b > 0]$

$21.^{12}$ $\displaystyle\int_0^\infty \frac{x^2 \exp\left(-a\sqrt{x^2+b^2}\right)}{\sqrt{x^2+b^2}}\,dx = \frac{b}{a}K_1(ab)$      $[\operatorname{Re}a > 0, \quad \operatorname{Re}b > 0]$

$22.^{12}$ $\displaystyle\int_0^\infty \frac{x^4 \exp\left(-a\sqrt{x^2+b^2}\right)}{\sqrt{x^2+b^2}}\,dx = \frac{3b^2}{a^2}K_1(ab)$      $[\operatorname{Re}a > 0, \quad \operatorname{Re}b > 0]$

$23.^{12}$ $\displaystyle\int_0^\infty \frac{x^6 \exp\left(-a\sqrt{x^2+b^2}\right)}{\sqrt{x^2+b^2}}\,dx = \frac{15b^3}{a^3}K_3(ab)$      $[\operatorname{Re}a > 0, \quad \operatorname{Re}b > 0]$

$24.^{12}$ $\displaystyle\int_0^\infty \frac{x^{2n} \exp\left(-a\sqrt{x^2+b^2}\right)}{\sqrt{x^2+b^2}}\,dx = (2n-1)!!\left(\frac{b}{a}\right)^n K_n(ab)$

$[\operatorname{Re}a > 0, \quad \operatorname{Re}b > 0]$

$25.$ $\displaystyle\int_0^\infty \frac{\exp\left(-px^2\right)}{\sqrt{a^2+x^2}}\,dx = \frac{1}{2}\exp\left(\frac{a^2 p}{2}\right)K_0\left(\frac{a^2 p}{2}\right)$      $[\operatorname{Re}a > 0, \quad \operatorname{Re}b > 0]$

**3.463** $\displaystyle\int_0^\infty \left(e^{-x^2} - e^{-x}\right)\frac{dx}{x} = \frac{1}{2}C$      BI (89)(5)

**3.464** $\displaystyle\int_0^\infty \left(e^{-\mu x^2} - e^{-\nu x^2}\right)\frac{dx}{x^2} = \sqrt{\pi}\left(\sqrt{\nu} - \sqrt{\mu}\right)$      $[\operatorname{Re}\mu > 0, \quad \operatorname{Re}\nu > 0]$      FI II 645

**3.465** $\displaystyle\int_0^\infty \left(1+2bx^2\right)e^{-\mu x^2}\,dx = \frac{\mu+b}{2}\sqrt{\frac{\pi}{\mu^3}}$      $[\operatorname{Re}\mu > 0]$      ET I 136(24)a

**3.466**

1. $\displaystyle\int_0^\infty \frac{e^{-\mu^2 x^2}}{x^2+b^2}\,dx = [1-\Phi(b\mu)]\frac{\pi}{2b}e^{b^2\mu^2}$      $\left[\operatorname{Re}b > 0, \quad |\arg\mu| < \frac{\pi}{4}\right]$      NT 19(13)

2. $\displaystyle\int_0^\infty \frac{x^2 e^{-\mu^2 x^2}}{x^2+b^2}\,dx = \frac{\sqrt{\pi}}{2\mu} - \frac{\pi b}{2}e^{\mu^2 b^2}[1-\Phi(b\mu)]$      $\left[\operatorname{Re}b > 0, \quad |\arg\mu| < \frac{\pi}{4}\right]$      ET II 217(16)

3. $\displaystyle\int_0^1 \frac{e^{x^2}-1}{x^2}\,dx = \sum_{k=1}^\infty \frac{1}{k!(2k-1)}$      FI II 683

**3.467** $\displaystyle\int_0^\infty \left(e^{-x^2} - \frac{1}{1+x^2}\right)\frac{dx}{x} = -\frac{1}{2}C$      BI (92)(12)

**3.468**

1. $\displaystyle\int_{u\sqrt{2}}^\infty \frac{e^{-x^2}}{\sqrt{x^2-u^2}}\frac{dx}{x} = \frac{\pi}{4u}[1-\Phi(u)]^2$      $[u > 0]$      NT 33(17)

2. $\displaystyle\int_0^\infty \frac{xe^{-\mu x^2}\,dx}{\sqrt{a^2+x^2}} = \frac{1}{2}\sqrt{\frac{\pi}{\mu}}e^{a^2\mu}\left[1-\Phi\left(a\sqrt{u}\right)\right]$      $[\operatorname{Re}\mu > 0, \quad a > 0]$      NT 19(11)

**3.469**

$1.^{12}$ $\displaystyle\int_0^\infty e^{-\mu x^4 - 2\nu x^2}\,dx = \frac{1}{4}\sqrt{\frac{2\nu}{\mu}}\exp\left(\frac{\nu^2}{2\mu}\right)K_{\frac{1}{4}}\left(\frac{\nu^2}{2\mu}\right)$      $[\operatorname{Re}\mu \geq 0, \quad \nu > 0]$      ET I 146(23)

2.    $\displaystyle\int_0^\infty \left(e^{-x^4} - e^{-x}\right) \frac{dx}{x} = \frac{3}{4}C$        (see **3.476** 2)        BI (89)(7)

3.    $\displaystyle\int_0^\infty \left(e^{-x^4} - e^{-x^2}\right) \frac{dx}{x} = \frac{1}{4}C$        (see **3.476** 2)        BI (89)(6)

**3.471**

1.[12]    $\displaystyle\int_0^u \exp\left(-\frac{\beta}{x}\right) \frac{dx}{x^2} = \frac{1}{\beta} \exp\left(-\frac{\beta}{u}\right)$        $[\beta > 0]$        ET II 188(22)

2.    $\displaystyle\int_0^u x^{\nu-1}(u-x)^{\mu-1}e^{-\frac{\beta}{x}}\,dx = \beta^{\frac{\nu-1}{2}}u^{\frac{2\mu+\nu-1}{2}}\exp\left(-\frac{\beta}{2u}\right)\Gamma(\mu)\,W_{\frac{1-2\mu-\nu}{2},\frac{\nu}{2}}\left(\frac{\beta}{u}\right)$

$$[\operatorname{Re}\mu > 0, \quad \operatorname{Re}\beta > 0, \quad u > 0]$$
ET II 187(18)

3.    $\displaystyle\int_0^u x^{-\mu-1}(u-x)^{\mu-1}e^{-\frac{\beta}{x}}\,dx = \beta^{-\mu}u^{\mu-1}\,\Gamma(\mu)\exp\left(-\frac{\beta}{u}\right)$

$$[\operatorname{Re}\mu > 0, \quad u > 0] \qquad \text{ET II 187(16)}$$

4.    $\displaystyle\int_0^u x^{-2\mu}(u-x)^{\mu-1}e^{-\frac{\beta}{x}}\,dx = \frac{1}{\sqrt{\pi u}}\beta^{\frac{1}{2}-\mu}e^{-\frac{\beta}{2u}}\,\Gamma(\mu)\,K_{\mu-\frac{1}{2}}\left(\frac{\beta}{2u}\right)$

$$[u > 0, \quad \operatorname{Re}\beta > 0, \quad \operatorname{Re}\mu > 0]$$
ET II 187(17)

5.    $\displaystyle\int_u^\infty x^{\nu-1}(x-u)^{\mu-1}e^{\frac{\beta}{x}}\,dx = \mathrm{B}(1-\mu-\nu,\mu)u^{\mu+\nu-1}\,{}_1F_1\left(1-\mu-\nu;1-\nu;\frac{\beta}{u}\right)$

$$[0 < \operatorname{Re}\mu < \operatorname{Re}(1-\nu), \quad u > 0]$$
ET II 203(15)

6.    $\displaystyle\int_u^\infty x^{-2\mu}(x-u)^{\mu-1}e^{\frac{\beta}{x}}\,dx = \sqrt{\frac{\pi}{u}}\beta^{\frac{1}{2}-\mu}\,\Gamma(\mu)\exp\left(\frac{\beta}{2u}\right)I_{\mu-\frac{1}{2}}\left(\frac{\beta}{2u}\right)$

$$[\operatorname{Re}\mu > 0, \quad u > 0] \qquad \text{ET II 202(14)}$$

7.    $\displaystyle\int_0^\infty x^{\nu-1}(x+\gamma)^{\mu-1}e^{-\frac{\beta}{x}}\,dx = \beta^{\frac{\nu-1}{2}}\gamma^{\frac{\nu-1}{2}+\mu}\,\Gamma(1-\mu-\nu)e^{\frac{\beta}{2\gamma}}\,W_{\frac{\nu-1}{2}+\mu,-\frac{\nu}{2}}\left(\frac{\beta}{\gamma}\right)$

$$[|\arg\gamma| < \pi, \quad \operatorname{Re}(1-\mu) > \operatorname{Re}\nu > 0]$$
ET II 234(13)a

8.    $\displaystyle\int_0^u x^{-2\mu}\left(u^2-x^2\right)^{\mu-1}e^{-\frac{\beta}{x}}\,dx = \frac{1}{\sqrt{\pi}}\left(\frac{2}{\beta}\right)^{\mu-\frac{1}{2}}u^{\mu-\frac{3}{2}}\,\Gamma(\mu)\,K_{\mu-\frac{1}{2}}\left(\frac{\beta}{u}\right)$

$$[\operatorname{Re}\beta > 0, \quad u > 0, \quad \operatorname{Re}\mu > 0]$$
ET II 188(23)a

9.    $\displaystyle\int_0^\infty x^{\nu-1}e^{-\frac{\beta}{x}-\gamma x}\,dx = 2\left(\frac{\beta}{\gamma}\right)^{\frac{\nu}{2}}K_\nu\left(2\sqrt{\beta\gamma}\right)$        $[\operatorname{Re}\beta > 0, \quad \operatorname{Re}\gamma > 0]$

ET II 82(23)a, LET I 146(29)

10.[12]    $\displaystyle\int_0^\infty x^{\nu-1}\exp\left[\frac{i\mu}{2}\left(x - \frac{\beta^2}{x}\right)\right]dx = 2\beta^\nu e^{\frac{i\nu\pi}{2}}K_{-\nu}(\beta\mu)$

$$[\operatorname{Im}\mu > 0, \quad \operatorname{Im}\left(\beta^2\mu\right) > 0; \text{ note that } K_{-\nu} \equiv K_\nu] \qquad \text{EH II 82(24)}$$

11. $$\int_0^\infty x^{\nu-1} \exp\left[\frac{i\mu}{2}\left(x + \frac{\beta^2}{x}\right)\right] dx = i\pi\beta^\nu e^{-\frac{i\nu\pi}{2}} H^{(1)}_{-\nu}(\beta\mu)$$

$$\left[\operatorname{Im}\mu > 0, \quad \operatorname{Im}\left(\beta^2\mu\right) > 0\right]$$

EH II 21(33)

12. $$\int_0^\infty x^{\nu-1} \exp\left(-x - \frac{\mu^2}{4x}\right) dx = 2\left(\frac{\mu}{2}\right)^\nu K_{-\nu}(\mu)$$

$$\left[|\arg\mu| < \frac{\pi}{2}, \operatorname{Re}\mu^2 > 0; \text{ note that } K_{-\nu} \equiv K_\nu\right] \quad \text{WA 203(15)}$$

13. $$\int_0^\infty \frac{x^{\nu-1} e^{-\frac{\beta}{x}}}{x + \gamma} dx = \gamma^{\nu-1} e^{\frac{\beta}{\gamma}} \Gamma(1-\nu) \Gamma\left(\nu, \frac{\beta}{\gamma}\right)$$    $$[|\arg\gamma| < \pi, \quad \operatorname{Re}\beta > 0, \quad \operatorname{Re}\nu < 1]$$

ET II 218(19)

14. $$\int_0^1 \frac{\exp\left(1 - \frac{1}{x}\right) - x^\nu}{x(1-x)} dx = \psi(\nu)$$    $$[\operatorname{Re}\nu > 0]$$    BI (80)(7)

15. $$\int_0^\infty x^{-\frac{1}{2}} e^{-\gamma x - \beta/x} dx = \sqrt{\frac{\pi}{\gamma}} e^{-2\sqrt{\beta\gamma}}$$    $$[\operatorname{Re}\beta \geq 0, \quad \operatorname{Re}\gamma > 0] \quad \text{ET 245 (5.6.1)}$$

16. $$\int_0^\infty x^{n-\frac{1}{2}} e^{-px - q/x} dx = (-1)^n \sqrt{\pi} \frac{\partial^n}{\partial p^n}\left(p^{-1/2} e^{-2\sqrt{pq}}\right)$$

$$[\operatorname{Re}p > 0, \quad \operatorname{Re}q > 0]$$

PBM 344 (2.3.16(2))

**3.472**

1. $$\int_0^\infty \left(\exp\left(-\frac{a}{x^2}\right) - 1\right) e^{-\mu x^2} dx = \frac{1}{2}\sqrt{\frac{\pi}{\mu}}\left[\exp\left(-2\sqrt{a\mu}\right) - 1\right]$$

$$[\operatorname{Re}\mu > 0, \quad \operatorname{Re}a > 0] \quad \text{ET I 146(30)}$$

2. $$\int_0^\infty x^2 \exp\left(-\frac{a}{x^2} - \mu x^2\right) dx = \frac{1}{4}\sqrt{\frac{\pi}{\mu^3}}\left(1 + 2\sqrt{a\mu}\right)\exp\left(-2\sqrt{a\mu}\right)$$

$$[\operatorname{Re}\mu > 0, \quad \operatorname{Re}a > 0] \quad \text{ET I 146(26)}$$

3. $$\int_0^\infty \exp\left(-\frac{a}{x^2} - \mu x^2\right) \frac{dx}{x^2} = \frac{1}{2}\sqrt{\frac{\pi}{a}} \exp\left(-2\sqrt{a\mu}\right)$$    $$[\operatorname{Re}\mu > 0, \quad a > 0]$$    ET I 146(28)a

4. $$\int_0^\infty \exp\left[-\frac{1}{2a}\left(x^2 + \frac{1}{x^2}\right)\right] \frac{dx}{x^4} = \sqrt{\frac{a\pi}{2}}(1+a)e^{-1/a}$$    $$[a > 0]$$    BI (98)(14)

5. $$\int_0^\infty x^{-n-1/2} e^{-px - q/x} dx = (-1)^n \sqrt{\frac{\pi}{p}} \frac{\partial^n}{\partial q^n} e^{-2\sqrt{pq}}$$    $$[\operatorname{Re}p > 0, \quad \operatorname{Re}q > 0]$$

PBM 344 (2.3.16(3))

**3.473**[12] $$\int_0^\infty \exp(-x^n) x^{(m+1/2)n-1} dx = \frac{(2m-1)!!}{2^m n}\sqrt{\pi}$$    $$[n > 0, \quad m = 0, 1, 2, \ldots]$$    BI (98)(6)

**3.474**

1. $$\int_0^1 \left\{\frac{n\exp\left(1 - x^{-n}\right)}{1 - x^n} - \frac{x^{np}}{1 - x}\right\} \frac{dx}{x} = \frac{1}{n}\sum_{k=1}^n \psi\left(p + \frac{k-1}{n}\right)$$

$$[p > 0]$$

BI (80)(8)

2. $\int_0^1 \left\{ \dfrac{n \exp\left(1 - x^{-n}\right)}{1 - x^n} - \dfrac{\exp\left(1 - \frac{1}{x}\right)}{1 - x} \right\} \dfrac{dx}{x} = -\ln n$         BI (80)(9)

**3.475**

1.[12] $\int_0^\infty \left\{ \exp\left(-x^2\right) - \dfrac{1}{1 + x^{2n+1}} \right\} \dfrac{dx}{x} = -\dfrac{C}{2}$         $[n \in \mathbb{Z}]$         BI (92)(14)

2. $\int_0^\infty \left\{ \exp\left(-x^{2^n}\right) - \dfrac{1}{1 + x^2} \right\} \dfrac{dx}{x} = -2^{-n} C$         BI (92)(13)

3. $\int_0^\infty \left\{ \exp\left(-x^{2^n}\right) - e^{-x} \right\} \dfrac{dx}{x} = \left(1 - 2^{-n}\right) C$         BI (89)(8)

**3.476**

1. $\int_0^\infty \left[ \exp\left(-\nu x^p\right) - \exp\left(-\mu x^p\right) \right] \dfrac{dx}{x} = \dfrac{1}{p} \ln \dfrac{\mu}{\nu}$         $[\operatorname{Re}\mu > 0, \quad \operatorname{Re}\nu > 0]$         BI (89)(3)

2. $\int_0^\infty \left[ \exp\left(-x^p\right) - \exp\left(-x^q\right) \right] \dfrac{dx}{x} = \dfrac{p - q}{pq} C$         $[p > 0, \quad q > 0]$         BI (89)(9)

**3.477**

1.[12] $\int_{-\infty}^\infty \dfrac{e^{-a|x|}}{x - u} dx = e^{au} \operatorname{Ei}(-au) - e^{-au} \operatorname{Ei}(au)$         $[\operatorname{Re}a > 0, \quad \operatorname{Im}u \neq 0, \quad \arg u \neq 0]$         MC

2.[12] $\int_{-\infty}^\infty \dfrac{\operatorname{sign} x \exp\left(-a|x|\right)}{x - u} dx = -e^{au} \operatorname{Ei}(-au) - e^{-au} \operatorname{Ei}(au)$

$[a > 0]$         ET II 251(36)

**3.478**

1. $\int_0^\infty x^{\nu-1} \exp\left(-\mu x^p\right) dx = \dfrac{1}{p} \mu^{-\frac{\nu}{p}} \Gamma\left(\dfrac{\nu}{p}\right)$         $[\operatorname{Re}\mu > 0, \quad \operatorname{Re}\nu > 0, \quad p > 0]$

BI(81)(8)a, ET I 313(15, 16)

2. $\int_0^\infty x^{\nu-1} \left[1 - \exp\left(-\mu x^p\right)\right] dx = -\dfrac{1}{|p|} \mu^{-\frac{\nu}{p}} \Gamma\left(\dfrac{\nu}{p}\right)$
$[\operatorname{Re}\mu > 0 \text{ and } -p < \operatorname{Re}\nu < 0 \text{ for } p > 0, \quad 0 < \operatorname{Re}\nu < -p \text{ for } p < 0]$         ET I 313(18, 19)

3.[11] $\int_0^u x^{\nu-1}(u - x)^{\mu-1} \exp\left(\beta x^n\right) dx = \mathrm{B}(\mu, \nu) u^{\mu+\nu-1} {}_nF_n\left(\dfrac{\nu}{n}, \dfrac{\nu+1}{n}, \dots, \dfrac{\nu+n-1}{n}; \right.$

$\left. \dfrac{\mu+\nu}{n}, \dfrac{\mu+\nu+1}{n}, \dots, \dfrac{\mu+\nu+n-1}{n}; \beta u^n\right)$

$[\operatorname{Re}\mu > 0, \quad \operatorname{Re}\nu > 0, \quad n = 2, 3, \dots]$         ET II 187(15)

4. $\int_0^\infty x^{\nu-1} \exp\left(-\beta x^p - \gamma x^{-p}\right) dx = \dfrac{2}{p} \left(\dfrac{\gamma}{\beta}\right)^{\frac{\nu}{2p}} K_{\frac{\nu}{p}}\left(2\sqrt{\beta\gamma}\right)$

$[\operatorname{Re}\beta > 0, \quad \operatorname{Re}\gamma > 0]$         ET I 313(17)

**3.479**

1.    $\displaystyle \int_0^\infty \frac{x^{\nu-1} \exp\left(-\beta\sqrt{1+x}\right)}{\sqrt{1+x}}\, dx = \frac{2}{\sqrt{\pi}} \left(\frac{\beta}{2}\right)^{\frac{1}{2}-\nu} \Gamma(\nu)\, K_{\frac{1}{2}-\nu}(\beta)$

$$[\operatorname{Re}\beta > 0, \quad \operatorname{Re}\nu > 0] \qquad \text{ET I 313(14)}$$

2.[11]   $\displaystyle \int_0^\infty \frac{x^{\nu-1} \exp\left(i\mu\sqrt{1+x^2}\right)}{\sqrt{1+x^2}}\, dx = i\frac{\sqrt{\pi}}{2}\left(\frac{\mu}{2}\right)^{\frac{1-\nu}{2}} \Gamma\left(\frac{\nu}{2}\right) H^{(1)}_{\frac{1-\nu}{2}}(\mu)$

$$[\operatorname{Im}\mu > 0, \quad \operatorname{Re}\nu > 0] \qquad \text{EH II 83(30)}$$

**3.481**

1.    $\displaystyle \int_{-\infty}^\infty xe^x \exp\left(-\mu e^x\right) dx = -\frac{1}{\mu}\left(\boldsymbol{C} + \ln\mu\right) \qquad [\operatorname{Re}\mu > 0] \qquad \text{BI (100)(13)}$

2.    $\displaystyle \int_{-\infty}^\infty xe^x \exp\left(-\mu e^{2x}\right) dx = -\frac{1}{4}\left[\boldsymbol{C} + \ln(4\mu)\right]\sqrt{\frac{\pi}{\mu}} \qquad [\operatorname{Re}\mu > 0] \qquad \text{BI (100)(14)}$

**3.482**

1.[3]   $\displaystyle \int_0^\infty \exp\left(nx - \beta\sinh x\right) dx = \frac{1}{2}\left[S_n(\beta) - \pi\,\mathbf{E}_n(\beta) - \pi\,Y_n(\beta)\right]$

$$[\operatorname{Re}\beta > 0] \qquad \text{ET I 168(11)}$$

2.    $\displaystyle \int_0^\infty \exp\left(-nx - \beta\sinh x\right) dx = (-1)^{n+1}\frac{1}{2}\left[S_n(\beta) + \pi\,\mathbf{E}_n(\beta) + \pi\,Y_n(\beta)\right]$

$$[\operatorname{Re}\beta > 0] \qquad \text{ET I 168(12)}$$

3.    $\displaystyle \int_0^\infty \exp\left(-\nu x - \beta\sinh x\right) dx = \frac{\pi}{\sin\nu\pi}\left[\mathbf{J}_\nu(\beta) - J_\nu(\beta)\right]$

$$[\operatorname{Re}\beta > 0] \qquad \text{ET I 168(13)}$$

**3.483**   $\displaystyle \int_{-\infty}^\infty \frac{\exp\left(\nu\operatorname{arcsinh} x - iax\right)}{\sqrt{1+x^2}}\, dx = \begin{cases} 2\exp\left(-\dfrac{i\nu\pi}{2}\right) K_\nu(a) & \text{for } a > 0, \\[2mm] 2\exp\left(\dfrac{i\nu\pi}{2}\right) K_\nu(-a) & \text{for } a < 0 \end{cases} \qquad [|\operatorname{Re}\nu| < 1]$

$$\text{ET I 122(32)}$$

**3.484**   $\displaystyle \int_0^\infty \left[\left(1 + \frac{a}{qx}\right)^{qx} - \left(1 + \frac{a}{px}\right)^{px}\right]\frac{dx}{x} = (e^a - 1)\ln\frac{q}{p} \qquad [p > 0, \quad q > 0] \qquad \text{BI (89)(34)}$

**3.485**   $\displaystyle \int_0^{\pi/2} \exp\left(-\tan^2 x\right) dx = \frac{\pi e}{2}\left[1 - \Phi(1)\right]$

**3.486[6]**   $\displaystyle \int_0^1 x^{-x}\, dx = \int_0^1 e^{-x\ln x}\, dx = \sum_{k=1}^\infty k^{-k} = 1.2912859970627\ldots \qquad \text{FI II 483}$

**3.487**

1.    $\displaystyle \int_0^{\pi/4} \exp\left[-\sum_{k=0}^\infty \left(\frac{\tan^{2k+1} x}{k + \frac{1}{2}}\right)\right] dx = \ln 2$

# 3.5 Hyperbolic Functions

## 3.51 Hyperbolic functions

**3.511**

1.[12] $\displaystyle\int_0^\infty \frac{dx}{\cosh x} = \frac{\pi}{2}$

2.[12] $\displaystyle\int_0^\infty \frac{\sinh ax}{\sinh x}\, dx = \frac{\pi}{2}\tan\left(\frac{\pi a}{2}\right)$        $[|a| < 1]$        BI (27)(10)a

3.[12] $\displaystyle\int_0^\infty \frac{\sinh ax}{\cosh x}\, dx = \frac{\pi}{2}\sec\left(\frac{\pi a}{2}\right) - \beta\left(\frac{a+1}{2}\right)$        $[|a| < 1]$        GW (351)(3b)

4.[12] $\displaystyle\int_0^\infty \frac{\cosh ax}{\cosh x}\, dx = \frac{\pi}{2}\sec\left(\frac{\pi a}{2}\right)$        $[|a| < 1]$        BI (4)(14)a

5.[12] $\displaystyle\int_0^\infty \frac{\sinh ax \cosh bx}{\sinh x}\, dx = \frac{\pi}{2}\frac{\sin \pi a}{\cos \pi a + \cos \pi b}$        $[|a| + |b| < 1]$        BI (27)(11)

6.[12] $\displaystyle\int_0^\infty \frac{\cosh ax \cosh bx}{\cosh x}\, dx = \pi\frac{\sin \frac{\pi a}{2} \cos \frac{\pi b}{2}}{\cos \pi a + \cos \pi b}$        $[|a| + |b| < 1]$        BI (27)(5)a

7.[12] $\displaystyle\int_0^\infty \frac{\sinh ax \sinh bx}{\cosh x}\, dx = \pi\frac{\sin \frac{\pi a}{2} \sin \frac{\pi b}{2}}{\cos \pi a + \cos \pi b}$        $[|a| + |b| < 1]$        BI (27)(6)a

8.[11] $\displaystyle\int_0^\infty \frac{dx}{\cosh^2 x} = 1$        BI (98)(25)

9. $\displaystyle\int_{-\infty}^\infty \frac{\sinh^2 ax}{\sinh^2 x}\, dx = 1 - a\pi \cot a\pi$        $[a^2 < 1]$        BI (16)(3)a

10.[12] $\displaystyle\int_0^\infty \frac{\sinh ax \sinh x}{\cosh^2 x}\, dx = \frac{\pi a}{2}\sec\frac{\pi a}{2}$        $[|a| < 1]$        BI (27)(16)a

**3.512**

1.[12] $\displaystyle\int_0^\infty \frac{\cosh 2ax}{(\cosh x)^{2\nu}}\, dx = 4^{\nu-1} B(\nu + a, \nu - a)$        $[\mathrm{Re}\,(\nu \pm a) > 0, \quad a > 0]$

       LI(27)(17)a, EH I 11(26)

2.[12] $\displaystyle\int_0^\infty \frac{\sinh^\mu x}{\cosh^\nu x}\, dx = \frac{1}{2} B\left(\frac{\mu+1}{2}, \frac{\nu-1}{2}\right)$        $[\mathrm{Re}\,\mu > 0, \quad \mathrm{Re}(\mu - \nu) > 0]$    EH I 11(23)

**3.513**

1. $\displaystyle\int_0^\infty \frac{dx}{a + b\sinh x} = \frac{1}{\sqrt{a^2 + b^2}}\ln\frac{a + b + \sqrt{a^2 + b^2}}{a + b - \sqrt{a^2 + b^2}}$        $[ab \neq 0]$        GW (351)(8)

2. $\displaystyle\int_0^\infty \frac{dx}{a + b\cosh x} = \frac{2}{\sqrt{b^2 - a^2}}\arctan\frac{\sqrt{b^2 - a^2}}{a + b}$        $[b^2 > a^2]$

$\displaystyle\qquad\qquad\qquad\qquad = \frac{1}{\sqrt{a^2 - b^2}}\ln\frac{a + b + \sqrt{a^2 - b^2}}{a + b - \sqrt{a^2 - b^2}}$        $[b^2 < a^2]$

       GW (351)(7)

3. 
$$\int_0^\infty \frac{dx}{a\sinh x + b\cosh x} = \frac{2}{\sqrt{b^2 - a^2}} \arctan \frac{\sqrt{b^2 - a^2}}{a + b} \qquad [b^2 > a^2]$$

$$= \frac{1}{\sqrt{a^2 - b^2}} \ln \frac{a + b + \sqrt{a^2 - b^2}}{a + b - \sqrt{a^2 - b^2}} \qquad [a^2 > b^2]$$

GW (351)(9)

4. 
$$\int_0^\infty \frac{dx}{a + b\cosh x + c\sinh x} = \frac{2}{\sqrt{b^2 - a^2 - c^2}} \left[ \arctan \frac{\sqrt{b^2 - a^2 - c^2}}{a + b + c} + \epsilon\pi \right]$$

$$\left[ \text{when } b^2 > a^2 + c^2; \text{ and } \begin{cases} \epsilon = 0 & \text{for } (b - a)(a + b + c) > 0 \\ |\epsilon| = 1 & \text{for } (b - a)(a + b + c) < 0 \\ \epsilon = 1 & \text{for } a < b + c \\ \epsilon = -1 & \text{for } a > b + c \end{cases} \right]$$

$$= \frac{1}{\sqrt{a^2 - b^2 + c^2}} \ln \frac{a + b + c + \sqrt{a^2 - b^2 + c^2}}{a + b + c - \sqrt{a^2 - b^2 + c^2}}$$

$$[b^2 < a^2 + c^2, \quad a^2 \neq b^2]$$

$$= \frac{1}{c} \ln \frac{a + c}{a}$$

$$[a = b \neq 0, \quad c \neq 0]$$

$$= \frac{2(a - b)}{c(a - b - c)}$$

$$[b^2 = a^2 + c^2, \quad c(a - b - c) < 0]$$

GW (351)(6)

**3.514**

1. 
$$\int_0^\infty \frac{dx}{\cosh ax + \cos t} = \frac{t}{a} \operatorname{cosec} t \qquad\qquad [0 < t < \pi, \quad a > 0] \qquad\qquad \text{BI (27)(22)a}$$

2.[12]
$$\int_0^\infty \frac{\cosh ax - \cos t_1}{\cosh bx - \cos t_2} dx = \frac{\pi}{b} \frac{\sin \frac{a(\pi - t_2)}{b}}{\sin t_2 \sin \frac{a}{b}\pi} - \frac{\pi - t_2}{b \sin t_2} \cos t_1$$

$$[0 < |a| < b, \quad 0 < t_2 < \pi] \qquad\qquad \text{BI (6)(20)a}$$

3. 
$$\int_0^\infty \frac{\cosh ax \, dx}{(\cosh x + \cos t)^2} = \frac{\pi(-\cos t \sin at + a\sin t \cos at)}{\sin^3 t \sin a\pi}$$

$$[0 < a^2 < 1, \quad 0 < t < \pi] \qquad\qquad \text{BI (6)(18)a}$$

4. 
$$\int_0^\infty \frac{\sinh ax \sinh bx}{(\cosh ax + \cos t)^2} dx = \frac{b\pi}{a^2} \operatorname{cosec} t \operatorname{cosec} \frac{b\pi}{a} \sin \frac{bt}{a} \qquad [0 < |b| < a, \quad 0 < t < \pi] \qquad \text{BI (27)(27)a}$$

**3.515** 
$$\int_{-\infty}^\infty \left( 1 - \frac{\sqrt{2}\cosh x}{\sqrt{\cosh 2x}} \right) dx = -\ln 2 \qquad\qquad\qquad \text{BI (21)(12)a}$$

**3.516**

1. 
$$\int_0^\infty \frac{dx}{\left(z + \sqrt{z^2 - 1}\cosh x\right)^\mu} = \frac{1}{2} \int_{-\infty}^\infty \frac{dx}{\left(z + \sqrt{z^2 - 1}\cosh x\right)^\mu} = Q_{\mu-1}(z)$$

$$[\operatorname{Re}\mu > -1]$$

For a suitable choice of a single-valued branch of the integrand, this formula is valid for arbitrary values of $z$ in the $z$-plane cut from $-1$ to $+1$ provided $\mu < 0$. If $\mu > 0$, this formula ceases to be valid for points at which the denominator vanishes.                                     CO, WH

2.      $$\int_0^\infty \frac{dx}{\left(b + \sqrt{b^2 - 1}\cosh x\right)^{n+1}} = Q_n(b)$$                                     EH II 181(32)

3.      $$\int_0^\infty \frac{\cosh\gamma x \, dx}{\left(b + \sqrt{b^2 - 1}\cosh x\right)^{\nu+1}} = \frac{e^{-i\gamma\pi}\,\Gamma(\nu - \gamma + 1)\,Q_\nu^\gamma(b)}{\Gamma(\nu + 1)}$$

$$[\operatorname{Re}(\nu \pm \gamma) > -1, \quad \nu \neq -1, -2, -3, \ldots]$$
EH I 157(12)

4.      $$\int_0^\infty \frac{\sinh^{2\mu} x \, dx}{\left(b + \sqrt{b^2 - 1}\cosh x\right)^{\nu+1}} = \frac{2^\mu e^{-i\mu\pi}\,\Gamma(\nu - 2\mu + 1)\,\Gamma\left(\mu + \frac{1}{2}\right)}{\sqrt{\pi}\,(b^2 - 1)^{\frac{\mu}{2}}\,\Gamma(\nu + 1)}\,Q_{\nu-\mu}^\mu(b)$$

$$[\operatorname{Re}(\nu - 2\mu + 1) > 0, \quad \operatorname{Re}(\nu + 1) > 0]$$
EH I 155(2)

**3.517**

1.      $$\int_0^\infty \frac{\cosh\left(\gamma + \frac{1}{2}\right)x \, dx}{\left(b + \cosh x\right)^{\nu+\frac{1}{2}}} = \sqrt{\frac{\pi}{2}}\,(b^2 - 1)^{-\frac{\nu}{2}}\,\frac{\Gamma(\nu + \gamma + 1)\,\Gamma(\nu - \gamma)\,P_\gamma^{-\nu}(b)}{\Gamma\left(\nu + \frac{1}{2}\right)}$$

$$[\operatorname{Re}(\nu - \gamma) > 0, \quad \operatorname{Re}(\nu + \gamma + 1) > 0]$$
EH I 156(11)

2.      $$\int_0^a \frac{\cosh\left(\gamma + \frac{1}{2}\right)x \, dx}{\left(\cosh a - \cosh x\right)^{\nu+\frac{1}{2}}} = \sqrt{\frac{\pi}{2}}\,\frac{\Gamma\left(\frac{1}{2} - \nu\right)}{\sinh^\nu a}\,P_\gamma^\nu(\cosh a)$$

$$\left[\operatorname{Re}\nu < \frac{1}{2}, \quad a > 0\right]$$                                     EH I 156(8)

**3.518**

1.      $$\int_0^\infty \frac{\sinh^{2\mu} x \, dx}{\left(\cosh a + \sinh a \cosh x\right)^{\nu+1}} = \frac{2^\mu e^{-i\mu\pi}}{\sqrt{\pi}\,\sinh^\mu a}\,\frac{\Gamma(\nu - 2\mu + 1)\,\Gamma\left(\mu + \frac{1}{2}\right)}{\Gamma(\nu + 1)}\,Q_{\nu-\mu}^\mu(\cosh a)$$

$$[\operatorname{Re}(\nu + 1) > 0, \quad \operatorname{Re}(\nu - 2\mu + 1) > 0, \quad a > 0] \quad \text{EH I 155(3)a}$$

2.[10]   $$\int_0^\infty \frac{\sinh^{2\mu+1} x \, dx}{\left(b + \cosh x\right)^{\nu+1}} = 2^\mu\,(b^2 - 1)^{\frac{\mu-\nu}{2}}\,\Gamma(\nu - 2\mu)\,\Gamma(\mu + 1)\,P_\mu^{\mu-\nu}(b)$$

$$[\operatorname{Re}(\nu - \mu) > \operatorname{Re}\mu > -1, \quad b \text{ does not lie on the ray } (-\infty, +1) \text{ of the real axis}] \quad \text{EH I 155(1)}$$

3.      $$\int_0^\infty \frac{\sinh^{2\mu-1} x \cosh x \, dx}{\left(1 + a\sinh^2 x\right)^\nu} = \frac{1}{2}a^{-\mu}\,\mathrm{B}(\mu, \nu - \mu) \qquad [\operatorname{Re}\nu > \operatorname{Re}\mu > 0, \quad a > 0] \quad \text{EH I 11(22)}$$

4.[7]    $$\int_0^\infty \frac{\sinh^{\mu-1} x\,(\cosh x + 1)^{\nu-1}\,dx}{\left(b + \cosh x\right)^\varrho} = 2^{\mu+\nu-\varrho}\,\mathrm{B}\left(\frac{1}{2}\mu, \varrho + 2 - \mu - \nu\right)$$

$$\times {}_2F_1\left(\varrho, \varrho + 2 - \mu - \nu; 2 - \frac{1}{2}\mu - \nu; \frac{1}{2} - \frac{1}{2}b\right)$$

$$[\operatorname{Re}\mu > 0, \quad \operatorname{Re}(\varrho - \mu - \nu) > -2, \quad |\arg(1 + b)| < \pi] \quad \text{EH I 115(11)}$$

$$5.^6 \quad \int_0^\infty \frac{\sinh^{\mu-1} x \, (\cosh x - 1)^{\nu-1} \, dx}{(b + \cosh x)^\varrho} = 2^{-(2-\mu-\nu+\varrho)} \, {}_2F_1\left(\varrho, 2 - \mu - \nu + \varrho; 1 + \varrho - \frac{\mu}{2}; \frac{1-b}{2}\right)$$

$$\times B\left(2 - \mu - \nu + \varrho, -1 + \nu + \frac{\mu}{2}\right)$$

$$[b \notin (-\infty, -1), \quad \operatorname{Re}(2 + \varrho)\operatorname{Re}(\mu + \nu), \quad \operatorname{Re}(2\nu + \mu) > 2] \quad \text{EH I 115(10)}$$

$$6.^7 \quad \int_0^\infty \frac{\sinh^{\mu-1} x \cosh^{\nu-1} x}{(\cosh^2 x - b)^\varrho} \, dx = {}_2F_1\left(\varrho, 1 + \varrho - \frac{\mu+\nu}{2}; 1 + \varrho - \frac{\nu}{2}; b\right) 2B\left(\frac{\mu}{2}, 1 + \varrho - \frac{\mu+\nu}{2}\right)$$

$$[b \notin (1, \infty), \quad \operatorname{Re}\mu > 0, \quad 2\operatorname{Re}(1 + \varrho) > \operatorname{Re}(\mu + \nu)] \quad \text{EH I 115(9)}$$

$$\mathbf{3.519}^{12} \quad \int_0^{\pi/2} \frac{\sinh\left[(r - p)\tan x\right]}{\sinh\left(r\tan x\right)} \, dx = \pi \sum_{k=1}^\infty \frac{1}{k\pi + r} \sin \frac{pk\pi}{r} \qquad [p^2 < r^2] \qquad \text{BI (274)(13)}$$

## 3.52–3.53 Combinations of hyperbolic functions and algebraic functions

**3.521**

$$1.^{12} \quad \int_0^\infty \frac{x}{\sinh x} \, dx = \frac{\pi^2}{4} \qquad\qquad\qquad\qquad \text{GW (352)(2b)}$$

$$2. \quad \int_0^\infty \frac{x \, dx}{\cosh x} = 2\mathbf{G} = \pi \ln 2 - 4L\left(\frac{\pi}{4}\right) \qquad\qquad \text{LI III 225(103a), BI(84)(1)a}$$

$$3. \quad \int_1^\infty \frac{dx}{x \sinh ax} = -2\sum_{k=0}^\infty \operatorname{Ei}[-(2k+1)a] \qquad [a > 0] \qquad \text{LI (104)(14)}$$

$$4. \quad \int_1^\infty \frac{dx}{x \cosh ax} = 2\sum_{k=0}^\infty (-1)^{k+1} \operatorname{Ei}[-(2k+1)a] \qquad [a > 0] \qquad \text{LI (104)(13)}$$

**3.522**

$$1. \quad \int_0^\infty \frac{x \, dx}{(b^2 + x^2)\sinh ax} = \frac{\pi}{2ab} + \pi \sum_{k=1}^\infty \frac{(-1)^k}{ab + k\pi} \qquad [a > 0, \quad b > 0]$$

$$2. \quad \int_0^\infty \frac{x \, dx}{(b^2 + x^2)\sinh \pi x} = \frac{1}{2b} - \beta(b + 1) \qquad [b > 0] \qquad \text{BI(97)(16), GW(352)(8)}$$

$$3. \quad \int_0^\infty \frac{dx}{(b^2 + x^2)\cosh ax} = \frac{2\pi}{b} \sum_{k=1}^\infty \frac{(-1)^{k-1}}{2ab + (2k-1)\pi} \qquad [a > 0, \quad b > 0] \qquad \text{BI (97)(5)}$$

$$4. \quad \int_0^\infty \frac{dx}{(b^2 + x^2)\cosh \pi x} = \frac{1}{b}\beta\left(b + \frac{1}{2}\right) \qquad [b > 0] \qquad \text{BI (97)(4)}$$

$$5. \quad \int_0^\infty \frac{x \, dx}{(1 + x^2)\sinh \pi x} = \ln 2 - \frac{1}{2} \qquad\qquad\qquad \text{BI (97)(7)}$$

$$6. \quad \int_0^\infty \frac{dx}{(1 + x^2)\cosh \pi x} = 2 - \frac{\pi}{2} \qquad\qquad\qquad \text{BI (97)(1)}$$

$$7. \quad \int_0^\infty \frac{x \, dx}{(1 + x^2)\sinh \frac{\pi x}{2}} = \frac{\pi}{2} - 1 \qquad\qquad\qquad \text{BI (97)(8)}$$

8.  $\displaystyle\int_0^\infty \frac{\mathrm{d}x}{(1+x^2)\cosh\frac{\pi x}{2}} = \ln 2$  $\qquad$ BI (97)(2)

9.  $\displaystyle\int_0^\infty \frac{x\,\mathrm{d}x}{(1+x^2)\sinh\frac{\pi x}{4}} = \frac{1}{\sqrt{2}}\left[\pi + 2\ln\left(\sqrt{2}+1\right)\right] - 2$  $\qquad$ BI (97)(9)

10.  $\displaystyle\int_0^\infty \frac{\mathrm{d}x}{(1+x^2)\cosh\frac{\pi x}{4}} = \frac{1}{\sqrt{2}}\left[\pi - 2\ln\left(\sqrt{2}+1\right)\right]$  $\qquad$ BI (97)(3)

**3.523**

1.[12]  $\displaystyle\int_0^\infty \frac{x^{b-1}\,\mathrm{d}x}{\sinh x} = \frac{2^b-1}{2^{b-1}}\,\Gamma(b)\,\zeta(b)$  $\qquad$ $[\operatorname{Re} b > 1]$  $\qquad$ WH

2.[12]  $\displaystyle\int_0^\infty \frac{x^{2n-1}}{\sinh x}\,\mathrm{d}x = \frac{2^{2n}-1}{2n}\pi^{2n}|B_{2n}|$  $\qquad$ $[n=1,2,\ldots]$  $\qquad$ WH, GW(352)(2a)

3.[12]  $\displaystyle\int_0^\infty \frac{x^{\beta-1}}{\cosh ax}\,\mathrm{d}x = \frac{2}{(2a)^\beta}\,\Gamma(\beta)\,\Phi\left(-1,\beta,\frac{1}{2}\right)$
$$= \frac{2}{a^\beta}\,\Gamma(\beta)\sum_{k=0}^\infty \frac{(-1)^k}{(2k+1)^\beta}$$

$\qquad$ $[\operatorname{Re}\beta > 0, \quad a > 0]$  $\qquad$ EH I 35, ET I 322(1)

4.[12]  $\displaystyle\int_0^\infty \frac{x^{2n}}{\cosh x}\,\mathrm{d}x = \left(\frac{\pi}{2}\right)^{2n+1}|E_{2n}|$  $\qquad$ $[n=0,1,2,\ldots]$

$\qquad$ BI(84)(12)a, GW(352)(1a)

5.  $\displaystyle\int_0^\infty \frac{x^2\,\mathrm{d}x}{\cosh x} = \frac{\pi^3}{8}$  $\qquad$ (cf. **4.261** 6)  $\qquad$ BI (84)(3)

6.  $\displaystyle\int_0^\infty \frac{x^3\,\mathrm{d}x}{\sinh x} = \frac{\pi^4}{8}$  $\qquad$ (cf. **4.262** 1 and 2)  $\qquad$ BI (84)(5)

7.  $\displaystyle\int_0^\infty \frac{x^4\,\mathrm{d}x}{\cosh x} = \frac{5}{32}\pi^5$  $\qquad$ BI (84)(7)

8.  $\displaystyle\int_0^\infty \frac{x^5}{\sinh x}\,\mathrm{d}x = \frac{\pi^6}{4}$  $\qquad$ BI (84)(8)

9.  $\displaystyle\int_0^\infty \frac{x^6}{\cosh x}\,\mathrm{d}x = \frac{61}{128}\pi^7$  $\qquad$ BI (84)(9)

10.  $\displaystyle\int_0^\infty \frac{x^7}{\sinh x}\,\mathrm{d}x = \frac{17}{16}\pi^8$  $\qquad$ BI (84)(10)

11.  $\displaystyle\int_0^\infty \frac{x^{1/2}\,\mathrm{d}x}{\cosh x} = \sqrt{\pi}\sum_{k=0}^\infty (-1)^k \frac{1}{(2k+1)^{3/2}}$  $\qquad$ BI (98)(7)a

12.  $\displaystyle\int_0^\infty \frac{\mathrm{d}x}{x^{1/2}\cosh x} = 2\sqrt{\pi}\sum_{k=0}^\infty \frac{(-1)^k}{(2k+1)^{1/2}}$  $\qquad$ BI (98)(25)a

**3.524**

1. $$\int_0^\infty x^{\mu-1}\frac{\sinh\beta x}{\sinh\gamma x}\,dx = \frac{\Gamma(\mu)}{(2\gamma)^\mu}\left\{\zeta\left[\mu,\frac{1}{2}\left(1-\frac{\beta}{\gamma}\right)\right]-\zeta\left[\mu,\frac{1}{2}\left(1+\frac{\beta}{\gamma}\right)\right]\right\}$$
$$[\text{Re}\,\gamma > |\text{Re}\,\beta|, \quad \text{Re}\,\mu > -1]$$
ET I 323(10)

2.[11] $$\int_0^\infty x^{2m}\frac{\sinh ax}{\sinh bx}\,dx = \frac{\pi}{2b}\frac{d^{2m}}{da^{2m}}\left(\tan\frac{a\pi}{2b}\right) \qquad [b > |a|]$$
BI (112)(20)a

3. $$\int_0^\infty \frac{\sinh ax}{\sinh bx}\frac{dx}{x^p} = \Gamma(1-p)\sum_{k=0}^\infty\left\{\frac{1}{[b(2k+1)-a]^{1-p}} - \frac{1}{[b(2k+1)+a]^{1-p}}\right\}$$
$$[b > |a|, \quad p < 1]$$
BI (131)(2)a

4.[11] $$\int_0^\infty x^{2m+1}\frac{\sinh ax}{\cosh bx}\,dx = \frac{\pi}{2b}\frac{d^{2m+1}}{da^{2m+1}}\left(\sec\frac{a\pi}{2b}\right) \qquad [b > |a|]$$
BI (112)(18)a

5. $$\int_0^\infty x^{\mu-1}\frac{\cosh\beta x}{\sinh\gamma x}\,dx = \frac{\Gamma(\mu)}{(2\gamma)^\mu}\left\{\zeta\left[\mu,\frac{1}{2}\left(1-\frac{\beta}{\gamma}\right)\right]+\zeta\left[\mu,\frac{1}{2}\left(1+\frac{\beta}{\gamma}\right)\right]\right\}$$
$$[\text{Re}\,\gamma > |\text{Re}\,\beta|, \quad \text{Re}\,\mu > 1]$$
ET I 323(12)

6. $$\int_0^\infty x^{2m}\frac{\cosh ax}{\cosh bx}\,dx = \frac{\pi}{2b}\frac{d^{2m}}{da^{2m}}\left(\sec\frac{a\pi}{2b}\right) \qquad [b > |a|]$$
BI(112)(17)

7. $$\int_0^\infty \frac{\cosh ax}{\cosh bx}\cdot\frac{dx}{x^p} = \Gamma(1-p)\sum_{k=0}^\infty(-1)^k\left\{\frac{1}{[b(2k+1)-a]^{1-p}} + \frac{1}{[b(2k+1)+a]^{1-p}}\right\}$$
$$[b > |a|, \quad p < 1]$$
BI(131)(1)a

8. $$\int_0^\infty x^{2m+1}\frac{\cosh ax}{\sinh bx}\,dx = \frac{\pi}{2b}\frac{d^{2m+1}}{da^{2m+1}}\left(\tan\frac{a\pi}{2b}\right) \qquad [b > |a|]$$
BI (112)(19)a

9.[8] $$\int_0^\infty x^2\frac{\sinh ax}{\sinh bx}\,dx = \frac{\pi^3}{4b^3}\sin\frac{a\pi}{2b}\sec^3\frac{a\pi}{2b} \qquad [b > |a|]$$
BI (84)(18)

10. $$\int_0^\infty x^4\frac{\sinh ax}{\sinh bx}\,dx = 8\left(\frac{\pi}{2b}\sec\frac{a\pi}{2b}\right)^5\sin\frac{a\pi}{2b}\left(2+\sin^2\frac{a\pi}{2b}\right)$$
$$[b > |a|]$$
BI (82)(17)a

11. $$\int_0^\infty x^6\frac{\sinh ax}{\sinh bx}\,dx = 16\left(\frac{\pi}{2b}\sec\frac{a\pi}{2b}\right)^7\sin\frac{a\pi}{2b}\left(45-30\cos^2\frac{a\pi}{2b}+2\cos^4\frac{a\pi}{2b}\right)$$
$$[b > |a|]$$
BI (82)(21)a

12. $$\int_0^\infty x\frac{\sinh ax}{\cosh bx}\,dx = \frac{\pi^2}{4b^2}\sin\frac{a\pi}{2b}\sec^2\frac{a\pi}{2b} \qquad [b > |a|]$$
BI (84)(15)a

13. $$\int_0^\infty x^3\frac{\sinh ax}{\cosh bx}\,dx = \left(\frac{\pi}{2b}\sec\frac{a\pi}{2b}\right)^4\sin\frac{a\pi}{2b}\left(6-\cos^2\frac{a\pi}{2b}\right)$$
$$[b > |a|]$$
BI (82)(14)a

14. $$\int_0^\infty x^5\frac{\sinh ax}{\cosh bx}\,dx = \left(\frac{\pi}{2b}\sec\frac{a\pi}{2b}\right)^6\sin\frac{a\pi}{2b}\left(120-60\cos^2\frac{a\pi}{2b}+\cos^4\frac{a\pi}{2b}\right)$$
$$[b > |a|]$$
BI (82)(18)a

15. $\displaystyle\int_0^\infty x^7 \frac{\sinh ax}{\cosh bx}\,dx = \left(\frac{\pi}{2b}\sec\frac{a\pi}{2b}\right)^8 \sin\frac{a\pi}{2b}\left(5040 - 4200\cos^2\frac{a\pi}{2b} + 546\cos^4\frac{a\pi}{2b} - \cos^6\frac{a\pi}{2b}\right)$

$[b > |a|]$      BI (82)(22)a

16. $\displaystyle\int_0^\infty x\frac{\cosh ax}{\sinh bx}\,dx = \left(\frac{\pi}{2b}\sec\frac{a\pi}{2b}\right)^2$      $[b > |a|]$      BI (84)(16)a

17. $\displaystyle\int_0^\infty x^3 \frac{\cosh ax}{\sinh bx}\,dx = 2\left(\frac{\pi}{2b}\sec\frac{a\pi}{2b}\right)^4\left(1 + 2\sin^2\frac{a\pi}{2b}\right)$      $[b > |a|]$      BI (82)(15)a

18. $\displaystyle\int_0^\infty x^5 \frac{\cosh ax}{\sinh bx}\,dx = 8\left(\frac{\pi}{2b}\sec\frac{a\pi}{2b}\right)^6\left(15 - 15\cos^2\frac{a\pi}{2b} + 2\cos^4\frac{a\pi}{2b}\right)$

$[b > |a|]$      BI (82)(19)a

19. $\displaystyle\int_0^\infty x^7 \frac{\cosh ax}{\sinh bx}\,dx = 16\left(\frac{\pi}{2b}\sec\frac{a\pi}{2b}\right)^8\left(315 - 420\cos^2\frac{a\pi}{2b} + 126\cos^4\frac{a\pi}{2b} - 4\cos^6\frac{a\pi}{2b}\right)$

$[b > |a|]$      BI(82)(23)a

20. $\displaystyle\int_0^\infty x^2 \frac{\cosh ax}{\cosh bx}\,dx = \frac{\pi^3}{8b^3}\left(2\sec^3\frac{a\pi}{2b} - \sec\frac{a\pi}{2b}\right)$      $[b > |a|]$      BI (84)(17)a

21. $\displaystyle\int_0^\infty x^4 \frac{\cosh ax}{\cosh bx}\,dx = \left(\frac{\pi}{2b}\sec\frac{a\pi}{2b}\right)^5\left(24 - 20\cos^2\frac{a\pi}{2b} + \cos^4\frac{a\pi}{2b}\right)$

$[b > |a|]$      BI (82)(16)a

22. $\displaystyle\int_0^\infty x^6 \frac{\cosh ax}{\cosh bx}\,dx = \left(\frac{\pi}{2b}\sec\frac{a\pi}{2b}\right)^7\left(720 - 840\cos^2\frac{a\pi}{2b} + 182\cos^4\frac{a\pi}{2b} - \cos^6\frac{a\pi}{2b}\right)$

$[b > |a|]$      BI (82)(20)a

23. $\displaystyle\int_0^\infty \frac{\sinh ax}{\cosh bx}\frac{dx}{x} = \ln\tan\left(\frac{a\pi}{4b} + \frac{\pi}{4}\right)$      $[b > |a|]$      BI (95)(3)a

**3.525**

1. $\displaystyle\int_0^\infty \frac{\sinh ax}{\sinh \pi x}\frac{dx}{1 + x^2} = -\frac{a}{2}\cos a + \frac{1}{2}\sin a \ln\left[2\left(1 + \cos a\right)\right]$

$[\pi \geq |a|]$      BI (97)(10)a

2. $\displaystyle\int_0^\infty \frac{\sinh ax}{\sinh \frac{\pi}{2}x}\frac{dx}{1 + x^2} = \frac{\pi}{2}\sin a + \frac{1}{2}\cos a \ln\frac{1 - \sin a}{1 + \sin a}$      $[\pi \geq 2|a|]$      BI (97)(11)a

3. $\displaystyle\int_0^\infty \frac{\cosh ax}{\sinh \pi x}\frac{x\,dx}{1 + x^2} = \frac{1}{2}\left(a\sin a - 1\right) + \frac{1}{2}\cos a \ln\left[2\left(1 + \cos a\right)\right]$

$[\pi > |a|]$      BI (97)(12)a

4. $\displaystyle\int_0^\infty \frac{\cosh ax}{\sinh \frac{\pi}{2}x}\frac{x\,dx}{1 + x^2} = \frac{\pi}{2}\cos a - 1 + \frac{1}{2}\sin a \ln\frac{1 + \sin a}{1 - \sin a}$

$\left[\frac{\pi}{2} > |a|\right]$      BI (97)(13)a

5. $\displaystyle\int_0^\infty \frac{\sinh ax}{\cosh \pi x}\frac{x\,dx}{1 + x^2} = -2\sin\frac{a}{2} + \frac{\pi}{2}\sin a - \cos a \ln\tan\frac{a + \pi}{4}$

$[\pi > |a|]$      GW (352)(12)

6. $\displaystyle\int_0^\infty \frac{\cosh ax}{\cosh \pi x}\frac{\mathrm{d}x}{1+x^2} = 2\cos\frac{a}{2} - \frac{\pi}{2}\cos a - \sin a\ln\tan\frac{a+\pi}{4}$

$$[\pi > |a|]$$                    GW (352)(11)

7. $\displaystyle\int_0^\infty \frac{\sinh ax}{\sinh bx}\frac{\mathrm{d}x}{c^2+x^2} = \frac{\pi}{c}\sum_{k=1}^\infty \frac{\sin\frac{k(b-a)}{b}\pi}{bc+k\pi}$                    $[b \geq |a|]$                    BI (97)(18)

8. $\displaystyle\int_0^\infty \frac{\cosh ax}{\sinh bx}\frac{x\,\mathrm{d}x}{c^2+x^2} = \frac{\pi}{2bc} + \pi\sum_{k=1}^\infty \frac{\cos\frac{k(b-a)}{b}\pi}{bc+k\pi}$                    $[b > |a|]$                    BI (97)(19)

**3.526**

1. $\displaystyle\int_0^\infty \frac{\sinh ax\cosh bx}{\cosh cx}\frac{\mathrm{d}x}{x} = \frac{1}{2}\ln\left\{\tan\frac{(a+b+c)\pi}{4c}\cot\frac{(b+c-a)\pi}{4c}\right\}$

$$[c > |a|+|b|]$$                    BI (93)(10)a

2. $\displaystyle\int_0^\infty \frac{\sinh^2 ax}{\sinh bx}\frac{\mathrm{d}x}{x} = \frac{1}{2}\ln\sec\frac{a}{b}\pi$                    $[b > |2a|]$                    BI (95)(5)a

3. $\displaystyle\int_0^\infty \frac{x^{\mu-1}}{\sinh \beta x\cosh \gamma x}\,\mathrm{d}x = \frac{\Gamma(\mu)}{(2\gamma)^\mu}\left\{\Phi\left[-1,\mu,\frac{1}{2}\left(1+\frac{\beta}{\gamma}\right)\right] + \Phi\left[-1,\mu,\frac{1}{2}\left(1-\frac{\beta}{\gamma}\right)\right]\right\}$

$$[\operatorname{Re}\gamma > |\operatorname{Re}\beta|, \quad \operatorname{Re}\mu > 0]$$

ET I 323(11)

**3.527**

1.[12] $\displaystyle\int_0^\infty \frac{x^{b-1}}{\sinh^2 x}\,\mathrm{d}x = 2^{2-b}\,\Gamma(b)\,\zeta(b-1)$                    $[\operatorname{Re}b > 2]$                    BI (86)(7)a

2.[12] $\displaystyle\int_0^\infty \frac{x^{2n}}{\sinh^2 x}\,\mathrm{d}x = \pi^{2n}|B_{2n}|$                    $[n = 1,2,\ldots]$                    BI(86)(5)a

3.[12] $\displaystyle\int_0^\infty \frac{x^{b-1}}{\cosh^2 x}\,\mathrm{d}x = 2^{2-b}\left(1-2^{2-b}\right)\Gamma(b)\,\zeta(b-1)$                    $[\operatorname{Re}b > 0, \quad b \neq 2]$                    BI (86)(6)a

4. $\displaystyle\int_0^\infty \frac{x\,\mathrm{d}x}{\cosh^2 ax} = \frac{\ln 2}{a^2}$                    $[a \neq 0]$                    LO III 396

5. $\displaystyle\int_0^\infty \frac{x^{2m}}{\cosh^2 ax}\,\mathrm{d}x = \frac{\left(2^{2m}-2\right)\pi^{2m}}{(2a)^{2m}a}|B_{2m}|$                    $[a > 0, \quad m = 1,2,\ldots]$                    BI(86)(2)a

6. $\displaystyle\int_0^\infty x^{\mu-1}\frac{\sinh ax}{\cosh^2 ax}\,\mathrm{d}x = \frac{2\,\Gamma(\mu)}{a^\mu}\sum_{k=0}^\infty \frac{(-1)^k}{(2k+1)^{\mu-1}}$                    $[\operatorname{Re}\mu > 1, \quad a > 0]$                    BI (86)(15)a

7. $\displaystyle\int_0^\infty \frac{x\sinh ax}{\cosh^2 ax}\,\mathrm{d}x = \frac{\pi}{2a^2}$                    $[a > 0]$                    BI (86)(8)a

8. $\displaystyle\int_0^\infty x^{2m+1}\frac{\sinh ax}{\cosh^2 ax}\,\mathrm{d}x = \frac{2m+1}{a}\left(\frac{\pi}{2a}\right)^{2m+1}|E_{2m}|$                    $[a > 0, \quad m = 0,1,\ldots]$                    BI (86)(12)a

9. $\displaystyle\int_0^\infty x^{2m+1}\frac{\cosh ax}{\sinh^2 ax}\,\mathrm{d}x = \frac{2^{2m+1}-1}{a^2(2a)^{2m}}(2m+1)!\,\zeta(2m+1)$

$$[a \neq 0, \quad m = 1,2,\ldots]$$                    BI (86)(13)a

$10.^{12}$   $\displaystyle\int_0^\infty x^{2n} \frac{\cosh x}{\sinh^2 x}\, dx = \left(2^{2n}-1\right)\pi^{2n}|B_{2n}|$       $[n=1,2,\ldots]$       BI (86)(14)a

$11.^{12}$   $\displaystyle\int_0^\infty \frac{x\sinh x}{(\cosh x)^{2\mu+1}}\, dx = \frac{\sqrt{\pi}}{4\mu}\frac{\Gamma(\mu)}{\Gamma\left(\mu+\frac12\right)}$       $[\operatorname{Re} u > 0]$       LI (86)(9)

$12.^{12}$   $\displaystyle\int_0^\infty \frac{x^2\, dx}{\sinh^2 x} = \frac{\pi^2}{6}$       BI (102)(2)a

$13.^{12}$   $\displaystyle\int_0^\infty x^2 \frac{\cosh x}{\cosh^2 x}\, dx = \frac{\pi^2}{2}$       BI (86)(11)a

$14.^{11}$   $\displaystyle\int_0^\infty x^2 \frac{\sinh x}{\cosh^2 x}\, dx = 4\boldsymbol{G}$       BI (86)(10)a

$15.^{10}$   $\displaystyle\int_0^\infty \frac{\tanh \frac{x}{2}\, dx}{\cosh x} = \ln 2$       BI (93)(17)a

$16.^{12}$   $\displaystyle\int_0^\infty x^{\mu-1} \frac{\cosh ax}{\sinh^2 ax}\, dx = \frac{2\,\Gamma(\mu)\,\zeta(\mu-1)}{a^\mu}\left(1-2^{1-\mu}\right)$

**3.528**

1.   $\displaystyle\int_0^\infty \frac{(1+xi)^{2n-1}-(1-xi)^{2n-1}}{i\sinh\frac{\pi x}{2}}\, dx = 2$       BI (87)(8)

2.   $\displaystyle\int_0^\infty \frac{(1+xi)^{2n}-(1-xi)^{2n}}{i\sinh\frac{\pi x}{2}}\, dx = (-1)^{n+1}2|E_{2n}|+2$    $[n=0,1,\ldots]$       BI (87)(7)

**3.529**

1.   $\displaystyle\int_0^\infty \left(\frac{1}{\sinh x}-\frac{1}{x}\right)\frac{dx}{x} = -\ln 2$       BI (94)(10)a

2.   $\displaystyle\int_0^\infty \frac{\cosh ax - 1}{\sinh bx}\frac{dx}{x} = -\ln\cos\frac{a\pi}{2b}$       $[b>|a|]$       GW (352)(66)

3.   $\displaystyle\int_0^\infty \left(\frac{a}{\sinh ax}-\frac{b}{\sinh bx}\right)\frac{dx}{x} = (b-a)\ln 2$       BI (94)(11)a

**3.531**

$1.^7$   $\displaystyle\int_0^\infty \frac{x\, dx}{2\cosh x - 1} = \frac{4}{\sqrt{3}}\left[\frac{\pi}{3}\ln 2 - L\left(\frac{\pi}{3}\right)\right] = 1.1719536193\ldots$

                                                $[\text{see } \mathbf{8.26} \text{ for } L(x)]$       LI (88)(1)

$2.^{10}$   $\displaystyle\int_0^\infty \frac{x\, dx}{\cosh 2x + \cos 2t} = \frac{t\ln 2 - L(t)}{\sin 2t}$       LO III 402

3.   $\displaystyle\int_0^\infty \frac{x^2\, dx}{\cosh x + \cos t} = \frac{t}{3}\frac{\pi^2-t^2}{\sin t}$       $[0<t<\pi]$       BI (88)(3)a

4.   $\displaystyle\int_0^\infty \frac{x^4\, dx}{\cosh x + \cos t} = \frac{t}{15}\frac{\left(\pi^2-t^2\right)\left(7\pi^2-3t^2\right)}{\sin t}$       $[0<t<\pi]$       BI (88)(4)a

5.3   $\displaystyle\int_0^\infty \frac{x^{2m}\,dx}{\cosh x - \cos 2a\pi} = 2(2m)!\operatorname{cosec}2a\pi \sum_{k=1}^\infty \frac{\sin 2ka\pi}{k^{2m+1}}$   $\left[0 < a < 1,\quad a \ne \frac{1}{2}\right]$

$$= 2\left(2^{2m-1}-1\right)\pi^{2m}|B_{2m}|\qquad \left[a = \tfrac{1}{2}\right]$$

BI (88)(5)a

6.3   $\displaystyle\int_0^\infty \frac{x^{\mu-1}\,dx}{\cosh x - \cos t}$

$$= \frac{i\,\Gamma(\mu)}{\sin t}\left[e^{-it}\,\Phi\left(e^{-it},\mu,1\right) - e^{it}\,\Phi\left(e^{it},\mu,1\right)\right]\qquad [\operatorname{Re}\mu > 0,\quad 0 < t < 2\pi,\quad t \ne \pi]\ \ \text{ET I 323(5)}$$

$$= \left(2 - 2^{3-\mu}\right)\Gamma(\mu)\,\zeta(\mu-1)\qquad\qquad [\mu \ne 2,\quad t = \pi]$$

$$= 2\ln 2 \qquad\qquad\qquad\qquad\qquad\qquad\ \ [\mu = 2,\quad t = \pi]$$

7.   $\displaystyle\int_0^\infty \frac{x^\mu\,dx}{\cosh x + \cos t} = \frac{2\,\Gamma(\mu+1)}{\sin t}\sum_{k=1}^\infty (-1)^{k-1}\frac{\sin kt}{k^{\mu+1}}$   $[\mu > -1,\quad 0 < t < \pi]$   BII (96)(14)a

8.   $\displaystyle\int_0^u \frac{x\,dx}{\cosh 2x - \cos 2t} = \frac{1}{2}\operatorname{cosec}2t\,[L(\theta+t) - L(\theta-t) - 2\,L(t)]$

$$[\theta = \arctan(\tanh u\cot t),\quad t \ne n\pi]$$

LO III 402

**3.532**

1.[11]   $\displaystyle\int_0^\infty \frac{x^n\,dx}{a\cosh x + b\sinh x} = \frac{2n!}{a+b}\sum_{k=0}^\infty \frac{1}{(2k+1)^{n+1}}\left(\frac{b-a}{b+a}\right)^k$

$$[a > 0,\quad b > 0,\quad n > -1]\quad \text{GW (352)(5)}$$

2.   $\displaystyle\int_0^u \frac{x\cosh x\,dx}{\cosh 2x - \cos 2t} = \frac{1}{2}\operatorname{cosec}t\left\{L\left(\frac{\theta+t}{2}\right) - L\left(\frac{\theta-t}{2}\right) + L\left(\pi - \frac{\psi+t}{2}\right)\right.$

$$\left. + L\left(\frac{\psi-t}{2}\right) - 2L\left(\frac{t}{2}\right) - 2L\left(\frac{\pi-t}{2}\right)\right\}$$

$$\left[\tan\frac{\theta}{2} = \tanh\frac{u}{2}\cot\frac{t}{2},\quad \tan\frac{\psi}{2} = \coth\frac{u}{2}\cot\frac{t}{2};\quad t \ne n\pi\right]\quad \text{LO III 288a}$$

**3.533**

1.   $\displaystyle\int_0^\infty \frac{x\cosh x\,dx}{\cosh 2x - \cos 2t} = \operatorname{cosec}t\left[\frac{\pi}{2}\ln 2 - L\left(\frac{t}{2}\right) - L\left(\frac{(\pi-t)}{2}\right)\right]$

$$[t \ne m\pi]$$

LO III 403

2.[6]   $\displaystyle\int_0^\infty x\frac{\sinh ax\,dx}{(\cosh ax - \cos t)^2} = \frac{\pi-t}{a^2}\operatorname{cosec}t$   $[a > 0,\quad 0 < t < \pi]$   (cf. **3.5141**)

BI (88)(11)a

3.   $\displaystyle\int_0^\infty x^3\frac{\sinh x\,dx}{(\cosh x + \cos t)^2} = \frac{t\left(\pi^2 - t^2\right)}{\sin t}$   $[0 < t < \pi]$   (cf. **3.531** 3)

BI (88)(13)

$4.^{11}$ $\displaystyle\int_0^\infty x^{2m+1}\frac{\sinh x\, dx}{(\cosh x-\cos 2a\pi)^2}=2(2m+1)!\csc 2a\pi\sum_{k=1}^\infty\frac{\sin 2ka\pi}{k^{2m+1}}$ $\quad\left[0<a<1,\quad a\neq\frac12\right]$

$$=2(2m+1)\left(2^{2m-1}-1\right)\pi^{2m}|B_{2m}|\quad\left[a=\frac12\right]$$

BI (88)(14)

**3.534**

1. $\displaystyle\int_0^1\sqrt{1-x^2}\cosh ax\, dx=\frac{\pi}{2a}I_1(a)$ 

WA 94(9)

2. $\displaystyle\int_0^1\frac{\cosh ax}{\sqrt{1-x^2}}\, dx=\frac{\pi}{2}I_0(a)$ 

WA 94(9)

**3.535** $\displaystyle\int_0^1\frac{x}{\sqrt{\cosh 2a-\cosh 2ax}\,\sinh ax}\frac{dx}{}=\frac{\pi}{2\sqrt{2}a^2}\frac{\arcsin\left(\tanh a\right)}{\sinh a}$ $\quad[a>0]$

BI (80)(11)

**3.536**

$1.^{11}$ $\displaystyle\int_0^\infty\frac{x^2}{\cosh^2 x}\, dx=\frac{\pi^2}{12}$ 

BI (98)(7)

2. $\displaystyle\int_0^\infty\frac{x^2\tanh x^2\, dx}{\cosh^2 x}=\frac{\sqrt{\pi}}{2}\sum_{k=0}^\infty\frac{(-1)^k}{\sqrt{2k+1}}$ 

BI (98)(8)

3. $\displaystyle\int_0^\infty\sinh\left(\nu\operatorname{arcsinh}x\right)\frac{x^{\mu-1}}{\sqrt{1+x^2}}\, dx=\frac{\sin\frac{\mu\pi}{2}\sin\frac{\nu\pi}{2}}{2^\mu\pi}\Gamma(\mu)\Gamma\left(\frac{1-\mu-\nu}{2}\right)\times\Gamma\left(\frac{1-\mu+\nu}{2}\right)$

$\quad\left[-1<\operatorname{Re}\mu<1-|\operatorname{Re}\nu|\right]$ ET I 324(14)

4. $\displaystyle\int_0^\infty\cosh\left(\nu\operatorname{arccosh}x\right)\frac{x^{\mu-1}}{\sqrt{1+x^2}}\, dx=\frac{\cos\frac{\mu\pi}{2}\cos\frac{\nu\pi}{2}}{2^\mu\pi}\Gamma(\mu)\Gamma\left(\frac{1-\mu-\nu}{2}\right)\times\Gamma\left(\frac{1-\mu+\nu}{2}\right)$

$\quad\left[0<\operatorname{Re}\mu<1-|\operatorname{Re}\nu|\right]$ ET I 324(15)

## 3.54 Combinations of hyperbolic functions and exponentials

**3.541**

1. $\displaystyle\int_0^\infty e^{-\mu x}\sinh^\nu\beta x\, dx=\frac{1}{2^{\nu+1}\beta}B\left(\frac{\mu}{2\beta}-\frac{\nu}{2},\nu+1\right)$ $\quad[\operatorname{Re}\beta>0,\quad\operatorname{Re}\nu>-1,\operatorname{Re}\mu>\operatorname{Re}\beta\nu]$

EH I 11(25), ET I 163(5)

2. $\displaystyle\int_0^\infty e^{-\mu x}\frac{\sinh\beta x}{\sinh bx}\, dx=\frac{1}{2b}\left[\psi\left(\frac12+\frac{\mu+\beta}{2b}\right)-\psi\left(\frac12+\frac{\mu-\beta}{2b}\right)\right]$

$\quad[\operatorname{Re}(\mu+b\pm\beta)>0]$ EH I 16(14)a

3. $\displaystyle\int_{-\infty}^\infty e^{-\mu x}\frac{\sinh\mu x}{\sinh\beta x}\, dx=\frac{\pi}{2\beta}\tan\frac{\mu\pi}{\beta}$ $\quad[\operatorname{Re}\beta>2|\operatorname{Re}\mu|]$

BI (18)(6)

4. $\displaystyle\int_0^\infty e^{-x}\frac{\sinh ax}{\sinh x}\, dx=\frac{1}{a}-\frac{\pi}{2}\cot\frac{a\pi}{2}$ $\quad[0<a<2]$

BI (4)(3)

5. $\displaystyle\int_0^\infty\frac{e^{-px}\, dx}{(\cosh px)^{2q+1}}=\frac{2^{2q-2}}{p}B(q,q)-\frac{1}{2qp}$ $\quad[p>0,\quad q>0]$

LI (27)(19)

6. $\displaystyle\int_0^\infty e^{-\mu x}\frac{dx}{\cosh x}=\beta\left(\frac{\mu+1}{2}\right)$ 　　　　　　　$[\operatorname{Re}\mu>-1]$　　　　　　ET I 163(7)

7. $\displaystyle\int_0^\infty e^{-\mu x}\tanh x\,dx=\beta\left(\frac{\mu}{2}\right)-\frac{1}{\mu}$ 　　　　　$[\operatorname{Re}\mu>0]$　　　　　ET I 163(9)

8. $\displaystyle\int_0^\infty \frac{e^{-\mu x}}{\cosh^2 x}\,dx=\beta\left(\frac{\mu}{2}\right)-1$ 　　　　　$[\operatorname{Re}\mu>0]$　　　　　ET I 163(8)

9. $\displaystyle\int_0^\infty e^{-\mu x}\frac{\sinh\mu x}{\cosh^2\mu x}\,dx=\frac{1}{\mu}\left(1-\ln 2\right)$ 　　　　$[\operatorname{Re}\mu>0]$　　　　LI (27)(15)

10. $\displaystyle\int_0^\infty e^{-qx}\frac{\sinh px}{\sinh qx}\,dx=\frac{1}{p}-\frac{\pi}{2q}\cot\frac{p\pi}{2q}$ 　　　$[0<p<2q]$　　　BI (27)(9)a

**3.542**

1. $\displaystyle\int_0^\infty e^{-\mu x}\left(\cosh\beta x-1\right)^\nu\,dx=\frac{1}{2^\nu\beta}\,\mathrm{B}\left(\frac{\mu}{\beta}-\nu,2\nu+1\right)$

$$\left[\operatorname{Re}\beta>0,\quad \operatorname{Re}\nu>-\frac{1}{2},\quad \operatorname{Re}\mu>\operatorname{Re}\beta\nu\right]\quad \text{ET I 163(6)}$$

2. $\displaystyle\int_0^\infty e^{-\mu x}\left(\cosh x-\cosh u\right)^{\nu-1}\,dx=-i\sqrt{\frac{2}{\pi}}e^{i\pi\nu}\,\Gamma(\nu)\sinh^{\nu-\frac{1}{2}}u\,Q_{\mu-\frac{1}{2}}^{\frac{1}{2}-\nu}(\cosh u)$

$$[\operatorname{Re}\nu>0,\quad \operatorname{Re}\mu>\operatorname{Re}\nu-1]$$
EH I 155(4), ET I 164(23)

**3.543**

1. $\displaystyle\int_{-\infty}^\infty \frac{e^{-ibx}\,dx}{\sinh x+\sinh t}=-\frac{i\pi e^{itb}}{\sinh\pi b\cosh t}\left(\cosh\pi b-e^{-2itb}\right)$

$$[t>0]\qquad\text{ET I 121(30)}$$

2. $\displaystyle\int_0^\infty \frac{e^{-\mu x}}{\cosh x-\cos t}\,dx=2\operatorname{cosec}t\sum_{k=1}^\infty\frac{\sin kt}{\mu+k}$ 　　$[\operatorname{Re}\mu>-1,\quad t\neq 2n\pi]$　　BI (6)(10)a

3. $\displaystyle\int_0^\infty \frac{1-e^{-x}\cos t}{\cosh x-\cos t}e^{-(\mu-1)x}\,dx=2\sum_{k=0}^\infty\frac{\cos kt}{\mu+k}$ 　　$[\operatorname{Re}\mu>0,\quad t\neq 2n\pi]$　　BI (6)(9)a

4.[12] $\displaystyle\int_0^\infty \frac{e^{px}+\cos t}{\left(\cosh px+\cos t\right)^2}\,dx=\frac{1}{p}\left(t\operatorname{cosec}t+\frac{1}{1+\cos t}\right)$

$$[p>0]\qquad\text{BI (27)(26)a}$$

**3.544** $\displaystyle\int_u^\infty \frac{\exp\left[-\left(n+\frac{1}{2}\right)x\right]}{\sqrt{2\left(\cosh x-\cosh u\right)}}\,dx=Q_n(\cosh u),\qquad[u>0]$ 　　EH II 181(33)

**3.545**

1. $\displaystyle\int_0^\infty \frac{\sinh ax}{e^{px}+1}\,dx=\frac{\pi}{2p}\operatorname{cosec}\frac{a\pi}{p}-\frac{1}{2a}$ 　　　$[p>a,\quad p>0]$　　　BI (27)(3)

2. $\displaystyle\int_0^\infty \frac{\sinh ax}{e^{px}-1}\,dx=\frac{1}{2a}-\frac{\pi}{2p}\cot\frac{a\pi}{p}$ 　　　$[p>a,\quad p>0]$　　　BI (27)(9)

**3.546**

1. $\displaystyle\int_0^\infty e^{-\beta x^2}\sinh ax\,dx = \frac{1}{2}\sqrt{\frac{\pi}{\beta}}\exp\frac{a^2}{4\beta}\,\Phi\left(\frac{a}{2\sqrt{\beta}}\right)$     $[\operatorname{Re}\beta > 0]$     ET I166(38)a

2. $\displaystyle\int_0^\infty e^{-\beta x^2}\cosh ax\,dx = \frac{1}{2}\sqrt{\frac{\pi}{\beta}}\exp\frac{a^2}{4\beta}$     $[\operatorname{Re}\beta > 0]$     FI II 720a

3. $\displaystyle\int_0^\infty e^{-\beta x^2}\sinh^2 ax\,dx = \frac{1}{4}\sqrt{\frac{\pi}{\beta}}\left(\exp\frac{a^2}{\beta}-1\right)$     $[\operatorname{Re}\beta > 0]$     ET I 166(40)

4. $\displaystyle\int_0^\infty e^{-\beta x^2}\cosh^2 ax\,dx = \frac{1}{4}\sqrt{\frac{\pi}{\beta}}\left(\exp\frac{a^2}{\beta}+1\right)$     $[\operatorname{Re}\beta > 0]$     ET I 166(41)

**3.547**

1. $\displaystyle\int_0^\infty \exp(-\beta\sinh x)\sinh\gamma x\,dx = \frac{\pi}{2}\cot\frac{\gamma\pi}{2}\left[J_\gamma(\beta)-\mathbf{J}_\gamma(\beta)\right]-\frac{\pi}{2}\left[\mathbf{E}_\gamma(\beta)+Y_\gamma(\beta)\right]=\gamma\,S_{-1,\gamma}(\beta)$

   $[\operatorname{Re}\beta > 0]$     WA 341(5), ET I 168(14)a

2. $\displaystyle\int_0^\infty \exp(-\beta\cosh x)\sinh\gamma x\sinh x\,dx = \frac{\gamma}{\beta}K_\gamma(\beta)$

3. $\displaystyle\int_0^\infty \exp(-\beta\sinh x)\cosh\gamma x\,dx = \frac{\pi}{2}\tan\frac{\pi\gamma}{2}\left[\mathbf{J}_\gamma(\beta)-J_\gamma(\beta)\right]-\frac{\pi}{2}\left[\mathbf{E}_\gamma(\beta)+Y_\gamma(\beta)\right]=S_{0,\gamma}(\beta)$

   $[\operatorname{Re}\beta > 0,\quad\gamma\text{ not an integer}]$
   ET I 168(16)a, WA 341(4), EH II 84(50)

4. $\displaystyle\int_0^\infty \exp(-\beta\cosh x)\cosh\gamma x\,dx = K_\gamma(\beta)$     $[\operatorname{Re}\beta > 0]$     ET I 168(16)a, WA 201(5)

5. $\displaystyle\int_0^\infty \exp(-\beta\sinh x)\sinh\gamma x\cosh x\,dx = \frac{\gamma}{\beta}S_{0,\gamma}(\beta)$     $[\operatorname{Re}\beta > 0]$     ET I 168(7), EH II 85(51)

6. $\displaystyle\int_0^\infty \exp(-\beta\sinh x)\sinh[(2n+1)x]\cosh x\,dx = O_{2n+1}(\beta)$

   $[\operatorname{Re}\beta > 0]$     ET I 167(5)

7. $\displaystyle\int_0^\infty \exp(-\beta\sinh x)\cosh\gamma x\cosh x\,dx = \frac{1}{\beta}S_{1,\gamma}(\beta)$     $[\operatorname{Re}\beta > 0]$

8. $\displaystyle\int_0^\infty \exp(-\beta\sinh x)\cosh 2nx\cosh x\,dx = O_{2n}(\beta)$     $[\operatorname{Re}\beta > 0]$     ET I 168(6)

9. $\displaystyle\int_0^\infty \exp(-\beta\cosh x)\sinh^{2\nu}x\,dx = \frac{1}{\sqrt{\pi}}\left(\frac{2}{\beta}\right)^\nu\Gamma\left(\nu+\frac{1}{2}\right)K_\nu(\beta)$

   $[\operatorname{Re}\beta > 0,\quad\operatorname{Re}\nu > -\tfrac{1}{2}]$     EH II 82(20)

10.[11] $\displaystyle\int_0^\infty \exp[-2(\beta\coth x+\mu x)]\sinh^{2\nu}x\,dx = \frac{1}{2}\beta^\nu\,\Gamma(\mu-\nu)\,W_{-\mu,\nu-\frac{1}{2}}(4\beta)$

   $[\operatorname{Re}\beta > 0,\quad\operatorname{Re}\mu > \operatorname{Re}\nu]$

11. $\displaystyle\int_0^\infty \exp\left(-\frac{\beta^2}{2}\sinh x\right)\sinh^{\nu-1}x\cosh^\nu x\,dx = -\pi\,D_\nu\left(\beta e^{i\pi/4}\right)D_\nu\left(\beta e^{-i\pi/4}\right)$

   $[\operatorname{Re}\nu > 0,\quad|\arg\beta|\le\tfrac{\pi}{4}]$     EH II 120(10)

12. $\int_0^\infty \dfrac{\exp\left(2\nu x - 2\beta \sinh x\right)}{\sqrt{\sinh x}}\, dx = \dfrac{1}{2}\sqrt{\pi^3\beta}\left[J_{\nu+\frac{1}{4}}(\beta)\,J_{\nu-\frac{1}{4}}(\beta) + Y_{\nu+\frac{1}{4}}(\beta)\,Y_{\nu-\frac{1}{4}}(\beta)\right]$

$[\mathrm{Re}\,\beta > 0]$      EH I 169(20)

13. $\int_0^\infty \dfrac{\exp\left(-2\nu x - 2\beta \sinh x\right)}{\sqrt{\sinh x}}\, dx = \dfrac{1}{2}\sqrt{\pi^3\beta}\left[J_{\nu+\frac{1}{4}}(\beta)\,Y_{\nu-\frac{1}{4}}(\beta) - J_{\nu-\frac{1}{4}}(\beta)\,Y_{\nu+\frac{1}{4}}(\beta)\right]$

$[\mathrm{Re}\,\beta > 0]$      ET I 169(21)

14. $\int_0^\infty \dfrac{\exp\left(-2\beta \sinh x\right)\sinh 2\nu x}{\sqrt{\sinh x}}\, dx = \dfrac{1}{4i}\sqrt{\dfrac{\pi^3\beta}{2}}\left\{e^{\nu\pi i}\,H^{(1)}_{\frac{1}{2}+\nu}(\beta)\,H^{(2)}_{\frac{1}{2}-\nu}(\beta)\right.$

$\left. -\,e^{-\nu\pi i}\,H^{(1)}_{\frac{1}{2}-\nu}(\beta)\,H^{(2)}_{\frac{1}{2}+\nu}(\beta)\right\}$

$[\mathrm{Re}\,\beta > 0]$      ET I 170(24)

15. $\int_0^\infty \dfrac{\exp\left(-2\beta \sinh x\right)\cosh 2\nu x}{\sqrt{\sinh x}}\, dx = \dfrac{1}{4}\sqrt{\dfrac{\pi^3\beta}{2}}\left\{e^{\nu\pi i}\,H^{(1)}_{\frac{1}{2}+\nu}(\beta)\,H^{(2)}_{\frac{1}{2}-\nu}(\beta)\right.$

$\left. +\,e^{-\nu\pi i}\,H^{(1)}_{\frac{1}{2}-\nu}(\beta)\,H^{(2)}_{\frac{1}{2}+\nu}(\beta)\right\}$

$[\mathrm{Re}\,\beta > 0]$      ET I 170(25)

16. $\int_0^\infty \dfrac{\exp\left(-2\beta \cosh x\right)\cosh 2\nu x}{\sqrt{\cosh x}}\, dx = \sqrt{\dfrac{\beta}{\pi}}\,K_{\nu+\frac{1}{4}}(\beta)\,K_{\nu-\frac{1}{4}}(\beta)$

$[\mathrm{Re}\,\beta > 0]$      ET I 170(26)

17.[8] $\int_0^\infty \dfrac{\exp\left[-2\beta\left(\cosh x - 1\right)\right]\cosh 2\nu x}{\sqrt{\cosh x}}\, dx = \sqrt{\dfrac{\beta}{\pi}}\,e^{2\beta}\,K_{\nu+\frac{1}{4}}(\beta)\,K_{\nu-\frac{1}{4}}(\beta)$

$[\mathrm{Re}\,\beta > 0]$      ET I 170(27)

18. $\int_0^\infty \dfrac{\cos\left[\left(\nu+\frac{1}{4}\right)\pi\right]\exp\left(-2\nu x - 2\beta \sinh x\right) + \sin\left[\left(\nu+\frac{1}{4}\right)\pi\right]\exp\left(2\nu x - 2\beta \sinh x\right)}{\sqrt{\sinh x}}\, dx$

$= \dfrac{1}{2}\sqrt{\pi^3\beta}\left[J_{\frac{1}{4}+\nu}(\beta)\,J_{\frac{1}{4}-\nu}(\beta) + Y_{\frac{1}{4}+\nu}(\beta)\,Y_{\frac{1}{4}-\nu}(\beta)\right]$

$[\mathrm{Re}\,\beta > 0]$      ET I 169(22)

19. $\int_0^\infty \dfrac{\sin\left[\left(\nu+\frac{1}{4}\right)\pi\right]\exp\left(-2\nu x - 2\beta \sinh x\right) - \cos\left[\left(\nu+\frac{1}{4}\right)\pi\right]\exp\left(2\nu x - 2\beta \sinh x\right)}{\sqrt{\sinh x}}\, dx$

$= \dfrac{1}{2}\sqrt{\pi^3\beta}\left[J_{\frac{1}{4}+\nu}(\beta)\,Y_{\frac{1}{4}-\nu}(\beta) - J_{\frac{1}{4}-\nu}(\beta)\,Y_{\frac{1}{4}+\nu}(\beta)\right]$

$[\mathrm{Re}\,\beta > 0]$      ET I 169(23)

**3.548**

1. $\int_0^\infty e^{-\mu x^4}\sinh a x^2\, dx = \dfrac{\pi}{4}\sqrt{\dfrac{a}{2\mu}}\exp\left(\dfrac{a^2}{8\mu}\right)I_{\frac{1}{4}}\left(\dfrac{a^2}{8\mu}\right)$      $[\mathrm{Re}\,\mu > 0,\quad a \geq 0]$      ET I 166(42)

2. $\int_0^\infty e^{-\mu x^4}\cosh a x^2\, dx = \dfrac{\pi}{4}\sqrt{\dfrac{a}{2\mu}}\exp\left(\dfrac{a^2}{8\mu}\right)I_{-\frac{1}{4}}\left(\dfrac{a^2}{8\mu}\right)$

$[\mathrm{Re}\,\mu > 0,\quad a > 0]$      ET I 166(43)

**3.549**

1. $\displaystyle\int_0^\infty e^{-\beta x}\sinh\left[(2n+1)\operatorname{arcsinh}x\right]\mathrm{d}x = O_{2n+1}(\beta)$ $\quad[\operatorname{Re}\beta > 0]\quad$ (cf. **3.547 6**)

ET I 167(5)

2. $\displaystyle\int_0^\infty e^{-\beta x}\cosh\left(2n\operatorname{arcsinh}x\right)\mathrm{d}x = O_{2n}(\beta)$ $\quad[\operatorname{Re}\beta > 0]\quad$ (cf. **3.547 8**)

ET I 168(6)

3. $\displaystyle\int_0^\infty e^{-\beta x}\sinh\left(\nu\operatorname{arcsinh}x\right)\mathrm{d}x = \frac{\nu}{\beta}S_{0,\nu}(\beta)$ $\quad[\operatorname{Re}\beta > 0]\quad$ (cf. **3.5475**) ET I 168(7)

4. $\displaystyle\int_0^\infty e^{-\beta x}\cosh\left(\nu\operatorname{arcsinh}x\right)\mathrm{d}x = \frac{1}{\beta}S_{1,\nu}(\beta)$ $\quad[\operatorname{Re}\beta > 0]\quad$ (cf. **3.547 7**)

A number of other integrals containing hyperbolic functions and exponentials, depending on $\operatorname{arcsinh}x$ or $\operatorname{arccosh}x$ can be found by first making the substitution $x = \sinh t$ or $x = \cosh t$.

## 3.55–3.56 Combinations of hyperbolic functions, exponentials, and powers

**3.551**

1.[12] $\displaystyle\int_0^\infty x^{\mu-1}e^{-\beta x}\sinh\gamma x\,\mathrm{d}x = \frac{1}{2}\Gamma(\mu)\left[(\beta-\gamma)^{-\mu} - (\beta+\gamma)^{-\mu}\right]$

$\quad[\operatorname{Re}\mu > -1, \quad \operatorname{Re}\beta > |\operatorname{Re}\gamma|]$

ET I 164(18)

2. $\displaystyle\int_0^\infty x^{\mu-1}e^{-\beta x}\cosh\gamma x\,\mathrm{d}x = \frac{1}{2}\Gamma(\mu)\left[(\beta-\gamma)^{-\mu} + (\beta+\gamma)^{-\mu}\right]$

$\quad[\operatorname{Re}\mu > 0, \quad \operatorname{Re}\beta > |\operatorname{Re}\gamma|]$

ET I 164(19)

3. $\displaystyle\int_0^\infty x^{\mu-1}e^{-\beta x}\coth x\,\mathrm{d}x = \Gamma(\mu)\left[2^{1-\mu}\zeta\left(\mu, \frac{\beta}{2}\right) - \beta^{-\mu}\right]$

$\quad[\operatorname{Re}\mu > 1, \quad \operatorname{Re}\beta > 0]\quad$ ET I 164(21)

4. $\displaystyle\int_0^\infty x^n e^{-(p+mq)x}\sinh^m qx\,\mathrm{d}x = 2^{-m}n!\sum_{k=0}^{m}\binom{m}{k}\frac{(-1)^k}{(p+2kq)^{n+1}}$

$\quad[p > 0, \quad q > 0, \quad m < p+qm]$

LI (81)(4)

5.[11] $\displaystyle\int_0^1 \frac{e^{-\beta x}}{x}\sinh\gamma x\,\mathrm{d}x = \frac{1}{2}\left[\ln\frac{\beta+\gamma}{\beta-\gamma} + \operatorname{Ei}(\gamma-\beta) - \operatorname{Ei}(-\gamma-\beta)\right]$

$\quad[\beta > \gamma]\quad$ BI (80)(4)

6. $\displaystyle\int_0^\infty \frac{e^{-\beta x}}{x}\sinh\gamma x\,\mathrm{d}x = \frac{1}{2}\ln\frac{\beta+\gamma}{\beta-\gamma}$ $\quad[\operatorname{Re}\beta > |\operatorname{Re}\gamma|]\quad$ ET I 163(12)

7. $\displaystyle\int_1^\infty \frac{e^{-\beta x}}{x}\cosh\gamma x\,\mathrm{d}x = \frac{1}{2}\left[-\operatorname{Ei}(\gamma-\beta) - \operatorname{Ei}(-\gamma-\beta)\right]$

$\quad[\operatorname{Re}\beta > |\operatorname{Re}\gamma|]\quad$ ET I 164(15)

8.[6] $\displaystyle\int_0^\infty x e^{-x}\coth x\,dx = \frac{\pi^2}{4} - 1$          BI (82)(6)

9. $\displaystyle\int_0^\infty e^{-\beta x}\tanh x\,\frac{dx}{x} = \ln\frac{\beta}{4} + 2\ln\frac{\Gamma\left(\frac{\beta}{4}\right)}{\Gamma\left(\frac{\beta}{4}+\frac{1}{2}\right)}$      $[\operatorname{Re}\beta > 0]$      ET I 164(16)

10.[6] $\displaystyle\int_0^\infty x e^{-x}\coth(x/2)\,dx = \frac{\pi^2}{3} - 1$

**3.552**

1. $\displaystyle\int_0^\infty \frac{x^{\mu-1}e^{-\beta x}}{\sinh x}\,dx = 2^{1-\mu}\Gamma(\mu)\,\zeta\left[\mu,\frac{1}{2}(\beta+1)\right]$    $[\operatorname{Re}\mu > 1,\quad \operatorname{Re}\beta > -1]$    ET I 164(20)

2. $\displaystyle\int_0^\infty \frac{x^{2m-1}e^{-ax}}{\sinh ax}\,dx = \frac{1}{2m}|B_{2m}|\left(\frac{\pi}{a}\right)^{2m}$    $[a > 0,\quad m = 1,2,\ldots]$    EH I 38(24)a

3. $\displaystyle\int_0^\infty \frac{x^{\mu-1}e^{-x}}{\cosh x}\,dx = 2^{1-\mu}\left(1-2^{1-\mu}\right)\Gamma(\mu)\,\zeta(\mu)$    $[\operatorname{Re}\mu > 0,\quad \mu \neq 1]$

            $= \ln 2$                          $[\text{if } \mu = 1]$

                                                                 EH I 32(5)

4. $\displaystyle\int_0^\infty \frac{x^{2m-1}e^{-ax}}{\cosh ax}\,dx = \frac{1-2^{1-2m}}{2m}|B_{2m}|\left(\frac{\pi}{a}\right)^{2m}$    $[a > 0,\quad m = 1,2,\ldots]$    EH I 39(25)a

5. $\displaystyle\int_0^\infty \frac{x^2 e^{-2nx}}{\sinh x}\,dx = 4\sum_{k=n}^\infty \frac{1}{(2k+1)^3}$    $[n = 0,1,2,\ldots]$    (cf. **4.261** 13)

                                                                        BI(84)(4)

6.[11] $\displaystyle\int_0^\infty \frac{x^3 e^{-2nx}}{\sinh x}\,dx = \frac{\pi^4}{8} - 12\sum_{k=1}^n \frac{1}{(2k-1)^4}$    $[n = 0,1,\ldots]$    (cf. **4.262** 6)

                                                                        BI (84)(6)

**3.553**

1. $\displaystyle\int_0^\infty \frac{\sinh^2 ax}{\sinh x}\,\frac{e^{-x}\,dx}{x} = \frac{1}{2}\ln\left(a\pi\operatorname{cosec} a\pi\right)$    $[a < 1]$    BI (95)(7)

2.[11] $\displaystyle\int_0^\infty \frac{\sinh^2 \frac{x}{2}}{\cosh x}\,\frac{e^{-x}\,dx}{x} = \frac{1}{2}\ln\frac{4}{\pi}$    (cf. **4.267** 2)    BI (95)(4)

**3.554**

1.[11] $\displaystyle\int_0^\infty e^{-\beta x}\left(1-\operatorname{sech} x\right)\frac{dx}{x} = 2\ln\frac{\Gamma\left(\frac{\beta+3}{4}\right)}{\Gamma\left(\frac{\beta+1}{4}\right)} - \ln\frac{\beta}{4}$    $[\operatorname{Re}\beta > 0]$    ET I 164(17)

2. $\displaystyle\int_0^\infty e^{-\beta x}\left(\frac{1}{x} - \operatorname{cosech} x\right)dx = \psi\left(\frac{\beta+1}{2}\right) - \ln\frac{\beta}{2}$    $[\operatorname{Re}\beta > 0]$    ET I 163(10)

3. $\displaystyle\int_0^\infty \left[ \frac{\sinh\left(\frac{1}{2} - \beta\right) x}{\sinh \frac{x}{2}} - (1 - 2\beta)e^{-x} \right] \frac{dx}{x} = 2\ln\Gamma(\beta) - \ln\pi + \ln\left(\sin\pi\beta\right)$

$$[0 < \operatorname{Re}\beta < 1] \qquad\qquad \text{EH I 21(7)}$$

4. $\displaystyle\int_0^\infty e^{-\beta x}\left(\frac{1}{x} - \coth x\right) dx = \psi\left(\frac{\beta}{2}\right) - \ln\frac{\beta}{2} + \frac{1}{\beta} \qquad [\operatorname{Re}\beta > 0] \qquad\qquad \text{ET I 163(11)}$

5. $\displaystyle\int_0^\infty \left\{ -\frac{\sinh qx}{\sinh\frac{x}{2}} + 2qe^{-x} \right\} \frac{dx}{x} = 2\ln\Gamma\left(q + \frac{1}{2}\right) + \ln\cos\pi q - \ln\pi$

$$\left[q^2 < \tfrac{1}{2}\right] \qquad\qquad \text{WH}$$

6. $\displaystyle\int_0^\infty x^{\mu-1} e^{-\beta x} \left(\coth x - 1\right) dx = 2^{1-\mu}\Gamma(\mu)\,\zeta\left(\mu, \frac{\beta}{2} + 1\right)$

$$[\operatorname{Re}\beta > 0, \quad \operatorname{Re}\mu > 1] \qquad \text{ET I 164(22)}$$

**3.555**

1.[12] $\displaystyle\int_0^\infty \frac{\sinh^2(ax)}{e^x - 1}\,\frac{dx}{x} = \frac{1}{4}\ln\left(\frac{2\pi a}{\sin 2\pi a}\right) \qquad \left[0 < |a| < \tfrac{1}{2}\right] \qquad (\text{cf. } \mathbf{3.545}\ 2)$

$$\text{BI (93)(15)}$$

2. $\displaystyle\int_0^\infty \frac{\sinh^2 ax}{e^x + 1}\,\frac{dx}{x} = -\frac{1}{4}\ln\left(a\pi\cot a\pi\right) \qquad \left[0 < |a| < \tfrac{1}{2}\right] \qquad (\text{cf. } \mathbf{3.545}\ 1)$

$$\text{BI (93)(9)}$$

**3.556**

1. $\displaystyle\int_{-\infty}^\infty x\frac{1 - e^{px}}{\sinh x}\,dx = -\frac{\pi^2}{2}\tan^2\frac{p\pi}{2} \qquad [p < 1] \qquad (\text{cf. } \mathbf{4.255}\ 3) \qquad \text{BI (101)(4)}$

2. $\displaystyle\int_0^\infty \frac{1 - e^{-px}}{\sinh x}\frac{1 - e^{-(p+1)x}}{x}\,dx = 2p\ln 2 \qquad [p > -1] \qquad\qquad \text{BI (95)(8)}$

**3.557**

1. $\displaystyle\int_0^\infty \frac{e^{-px} - e^{-qx}}{\cosh x - \cos\frac{m}{n}\pi}\frac{dx}{x}$

$$= 2\operatorname{cosec}\left(\frac{m}{n}\pi\right) \sum_{k=1}^{n-1} (-1)^{k-1}\sin\left(\frac{km}{n}\pi\right)\ln\frac{\Gamma\left(\frac{n+q+k}{2n}\right)\Gamma\left(\frac{p+k}{2n}\right)}{\Gamma\left(\frac{n+p+k}{2n}\right)\Gamma\left(\frac{q+k}{2n}\right)} \qquad [m + n \text{ odd}]$$

$$= 2\operatorname{cosec}\left(\frac{m}{n}\pi\right) \sum_{k=1}^{\frac{n-1}{2}} (-1)^{k-1}\sin\left(\frac{km}{n}\pi\right)\ln\frac{\Gamma\left(\frac{n+q-k}{n}\right)\Gamma\left(\frac{p+k}{n}\right)}{\Gamma\left(\frac{n+p-k}{n}\right)\Gamma\left(\frac{q+k}{n}\right)} \qquad [m + n \text{ even}]$$

$$[p > -1, \quad q > -1] \qquad\qquad \text{BI (96)(1)}$$

2. $$\int_0^\infty \frac{(1-e^{-x})^2}{\cosh x + \cos \frac{m}{n}\pi} \frac{\mathrm{d}x}{x}$$

$$= 2 \operatorname{cosec}\left(\frac{m}{n}\pi\right) \sum_{k=1}^{n-1} (-1)^{k-1} \sin\left(\frac{km}{n}\pi\right) \times \ln \frac{\left[\Gamma\left(\frac{n+k+1}{2n}\right)\right]^2 \Gamma\left(\frac{k+2}{2n}\right) \Gamma\left(\frac{k}{2n}\right)}{\left[\Gamma\left(\frac{k+1}{2n}\right)\right]^2 \Gamma\left(\frac{n+k}{2n}\right) \Gamma\left(\frac{n+k+2}{2n}\right)} \qquad [m+n \text{ odd}]$$

$$= 2 \operatorname{cosec}\left(\frac{m}{n}\pi\right) \sum_{k=1}^{\frac{n-1}{2}} (-1)^{k-1} \sin\left(\frac{km}{n}\pi\right) \times \ln \frac{\left[\Gamma\left(\frac{n-k+1}{n}\right)\right]^2 \Gamma\left(\frac{k+2}{n}\right) \Gamma\left(\frac{k}{n}\right)}{\left[\Gamma\left(\frac{k+1}{n}\right)\right]^2 \Gamma\left(\frac{n-k}{n}\right) \Gamma\left(\frac{n-k+2}{n}\right)} \qquad [m+n \text{ even}]$$

BI (96)(2)

3. $$\int_0^\infty \left[ e^{-x} \tan\frac{m}{2n}\pi - \frac{e^{-px}\sin\frac{m}{n}\pi}{\cosh x + \cos\frac{m}{n}\pi} \right] \frac{\mathrm{d}x}{x}$$

$$= \tan\left(\frac{m}{2n}\pi\right) \ln(2n) + 2 \sum_{k=1}^{n-1} (-1)^{k-1} \sin\left(\frac{km}{n}\pi\right) \ln \frac{\Gamma\left(\frac{p+n+k}{2n}\right)}{\Gamma\left(\frac{p+k}{2n}\right)} \qquad [m+n \text{ odd}]$$

$$= \tan\left(\frac{m}{2n}\pi\right) \ln n + 2 \sum_{k=1}^{\frac{n-1}{2}} (-1)^{k-1} \sin\left(\frac{km}{n}\pi\right) \ln \frac{\Gamma\left(\frac{p+n-k}{n}\right)}{\Gamma\left(\frac{p+k}{n}\right)} \qquad [m+n \text{ even}]$$

BI (96)(3)

4. $$\int_0^\infty \frac{1+e^{-x}}{\cosh x + \cos a} \frac{\mathrm{d}x}{x^{1-p}} = 2 \sec\frac{a}{2} \Gamma(p) \sum_{k=1}^\infty (-1)^{k-1} \frac{\cos\left(k-\frac{1}{2}\right)a}{k^p}$$

$$[p>0]$$

LI (96)(5)

5. $$\int_0^\infty \frac{x^q e^{-\frac{x}{2}} \cosh\frac{x}{2}}{\cosh x + \cos\lambda} \,\mathrm{d}x = \frac{\Gamma(q+1)}{\cos\frac{\lambda}{2}} \sum_{k=1}^\infty (-1)^{k-1} \frac{\cos\left(k-\frac{1}{2}\right)\lambda}{k^{q+1}}$$

$$[q>-1]$$

LI (96)(5)a

6. $$\int_0^\infty x \frac{e^{-x} - \cos a}{\cosh x - \cos a} \,\mathrm{d}x = |a|\pi - \frac{a^2}{2} - \frac{\pi^2}{3}$$

BI (88)(8)

7. $$\int_0^\infty x^{2m+1} \frac{e^{-x} - \cos a\pi}{\cosh x - \cos a\pi} \,\mathrm{d}x = 2 \cdot (2m+1)! \sum_{k=1}^\infty \frac{\cos ka\pi}{k^{2m+2}}$$

BI (88)(6)

**3.558**

1. $$\int_0^\infty x \frac{1-e^{-nx}}{\sinh^2\frac{x}{2}} \,\mathrm{d}x = \frac{2n\pi^2}{3} - 4 \sum_{k=1}^{n-1} \frac{n-k}{k^2}$$

BI (85)(3)

2. $$\int_0^\infty x \frac{1-(-1)^n e^{-nx}}{\cosh^2\frac{x}{2}} \,\mathrm{d}x = \frac{n\pi^2}{3} + 4 \sum_{k=1}^{n-1} (-1)^k \frac{n-k}{k^2}$$

LI (85)(1)

3. $$\int_0^\infty x^2 \frac{1-e^{-nx}}{\sinh^2\frac{x}{2}} \,\mathrm{d}x = 8n\,\zeta(3) - 8 \sum_{k=1}^{n-1} \frac{n-k}{k^3}$$

BI (85)(5)

4. $$\int_0^\infty x^2 e^x \frac{1-e^{-2nx}}{\sinh^2 x} \,\mathrm{d}x = 8n \sum_{k=1}^\infty \frac{1}{(2k-1)^3} - 8 \sum_{k=1}^{n-1} \frac{n-k}{(2k-1)^3}$$

LI (85)(6)

5. $\displaystyle\int_0^\infty x^2 \frac{1+(-1)^n e^{-nx}}{\cosh^2 \frac{x}{2}}\,dx = 6n\,\zeta(3) - 8\sum_{k=1}^{n-1}\frac{n-k}{k^3}$ 　　　LI (85)(4)

6. $\displaystyle\int_0^\infty x^3 \frac{1-e^{-nx}}{\sinh^2 \frac{x}{2}}\,dx = \frac{4}{15}n\pi^4 - 24\sum_{k=1}^{n-1}\frac{n-k}{k^4}$ 　　　BI (85)(9)

7. $\displaystyle\int_0^\infty x^3 \frac{1+(-1)^n e^{-nx}}{\cosh^2 \frac{x}{2}}\,dx = \frac{7}{30}n\pi^4 + 24\sum_{k=1}^{n-1}(-1)^k\frac{n-k}{k^4}$ 　　　BI (85)(8)

**3.559** $\displaystyle\int_0^\infty e^{-x}\left[a - \frac{1}{2} + \frac{(1-e^{-x})(1-ax)-xe^{-x}}{4\sinh^2 \frac{x}{2}}e^{(2-a)x}\right]\frac{dx}{x} = a - \frac{1}{2} + \ln\Gamma(a) - \frac{1}{2}\ln(2\pi)$ 　　$[a>0]$

BI (96)(6)

**3.561** $\displaystyle\int_0^\infty \frac{e^{-2x}\tanh \frac{x}{2}}{x\cosh x}\,dx = 2\ln\frac{\pi}{2\sqrt{2}}$ 　　　BI (93)(18)

**3.562**

1. $\displaystyle\int_0^\infty x^{2\mu-1}e^{-\beta x^2}\sinh\gamma x\,dx = \frac{1}{2}\Gamma(2\mu)(2\beta)^{-\mu}\exp\left(\frac{\gamma^2}{8\beta}\right)\left[D_{-2\mu}\left(-\frac{\gamma}{\sqrt{2\beta}}\right) - D_{-2\mu}\left(\frac{\gamma}{\sqrt{2\beta}}\right)\right]$

　　　$\left[\operatorname{Re}\mu > -\frac{1}{2},\quad \operatorname{Re}\beta > 0\right]$ 　　ET I 166(44)

2. $\displaystyle\int_0^\infty x^{2\mu-1}e^{-\beta x^2}\cosh\gamma x\,dx = \frac{1}{2}\Gamma(2\mu)(2\beta)^{-\mu}\exp\left(\frac{\gamma^2}{8\beta}\right)\left[D_{-2\mu}\left(-\frac{\gamma}{\sqrt{2\beta}}\right) + D_{-2\mu}\left(\frac{\gamma}{\sqrt{2\beta}}\right)\right]$

　　　$\left[\operatorname{Re}\mu > 0,\quad \operatorname{Re}\beta > 0\right]$ 　　ET I 166(45)

3. $\displaystyle\int_0^\infty xe^{-\beta x^2}\sinh\gamma x\,dx = \frac{\gamma}{4\beta}\sqrt{\frac{\pi}{\beta}}\exp\left(\frac{\gamma^2}{4\beta}\right)$ 　　$[\operatorname{Re}\beta > 0]$ 　　BI(81)(12)a,ET I 165(34)

4. $\displaystyle\int_0^\infty xe^{-\beta x^2}\cosh\gamma x\,dx = \frac{\gamma}{4\beta}\sqrt{\frac{\pi}{\beta}}\exp\left(\frac{\gamma^2}{4\beta}\right)\Phi\left(\frac{\gamma}{2\sqrt{\beta}}\right) + \frac{1}{2\beta}$

　　　$[\operatorname{Re}\beta > 0]$ 　　ET I 166(35)

5.[12] $\displaystyle\int_0^\infty x^2 e^{-\beta x^2}\sinh\gamma x\,dx = \frac{\sqrt{\pi}\,(2\beta+\gamma^2)}{8\beta^2\sqrt{\beta}}\exp\left(\frac{\gamma^2}{4\beta}\right)\Phi\left(\frac{\gamma}{2\sqrt{\beta}}\right) - \frac{\gamma}{4\beta^2}$

　　　$[\operatorname{Re}\beta > 0]$ 　　ET I 166(36)

6. $\displaystyle\int_0^\infty x^2 e^{-\beta x^2}\cosh\gamma x\,dx = \frac{\sqrt{\pi}\,(2\beta+\gamma^2)}{8\beta^2\sqrt{\beta}}\exp\left(\frac{\gamma^2}{4\beta}\right)$ 　　$[\operatorname{Re}\beta > 0]$ 　　ET I 166(37)

# 3.6–4.1 Trigonometric Functions

## 3.61 Rational functions of sines and cosines and trigonometric functions of multiple angles

**3.611**

1. $\displaystyle\int_0^{2\pi}(1-\cos x)^n \sin nx\,dx = 0$ 　　　BI (68)(10)

2.    $\displaystyle\int_0^{2\pi} (1-\cos x)^n \cos nx \, dx = (-1)^n \frac{\pi}{2^{n-1}}$          BI (68)(11)

3.    $\displaystyle\int_0^{\pi} (\cos t + i \sin t \cos x)^n \, dx = \int_0^{\pi} (\cos t + i \sin t \cos x)^{-n-1} \, dx = \pi \, P_n(\cos t)$          EH I 158(23)a

### 3.612

1.[12]    $\displaystyle\int_0^{\pi} \frac{\sin nx \cos mx}{\sin x} \, dx = 0$    for $0 < n \le m$;

                 $= \pi$    for $n > m > 0$,    if $m+n$ is odd and positive

                 $= 0$    for $n > m > 0$,    if $m+n$ is even

                                                   LI (64)(3)

2.[12]    $\displaystyle\int_0^{\pi} \frac{\sin nx}{\sin x} \, dx = 0$                 $[n > 0 \text{ and even}]$

            $= \pi$                       $[n > 0 \text{ and odd}]$

                                                   BI (64)(1, 2)

3.[12]    $\displaystyle\int_0^{\pi/2} \frac{\sin(2n-1)x}{\sin x} \, dx = \frac{\pi}{2}$             $[n > 0]$          FI II 145

4.[12]    $\displaystyle\int_0^{\pi/2} \frac{\sin 2nx}{\sin x} \, dx = 2 \left( 1 - \frac{1}{3} + \frac{1}{5} - \cdots + \frac{(-1)^{n-1}}{2n-1} \right)$    $[n > 0]$      GW (332)(21b)

5.[12]    $\displaystyle\int_0^{\pi} \frac{\sin 2nx}{\cos x} \, dx = 2 \int_0^{\pi/2} \frac{\sin 2nx}{\cos x} \, dx = (-1)^{n-1} 4 \left( 1 - \frac{1}{3} + \frac{1}{5} - \cdots + \frac{(-1)^{n-1}}{2n-1} \right)$

                                          $[n > 0]$          GW (332)(22a)

6.[12]    $\displaystyle\int_0^{\pi} \frac{\cos(2n+1)x}{\cos x} \, dx = 2 \int_0^{\frac{\pi}{2}} \frac{\cos(2n+1)x}{\cos x} \, dx = (-1)^n \pi$   $[n \ge 0]$      GW (332)(22b)

7.[12]    $\displaystyle\int_0^{\pi/2} \frac{\sin 2nx \cos x}{\sin x} \, dx = \frac{\pi}{2}$            $[n > 0]$          LI (45)(17)

### 3.613

1.[6]    $\displaystyle\int_0^{\pi} \frac{\cos nx \, dx}{1 + a \cos x} = \frac{\pi}{\sqrt{1-a^2}} \left( \frac{\sqrt{1-a^2}-1}{a} \right)^n$      $[a^2 < 1, \quad n \ge 0]$     BI (64)(12)

2.[6]    $\displaystyle\int_0^{\pi} \frac{\cos nx \, dx}{1 - 2a \cos x + a^2} = \frac{\pi a^n}{1 - a^2}$      $[a^2 < 1, \quad n \ge 0]$

                           $= \frac{\pi}{(a^2 - 1) a^n}$      $[a^2 > 1, \quad n \ge 0]$

                                                         BI (65)(3)

3.    $\displaystyle\int_0^{\pi} \frac{\sin nx \sin x \, dx}{1 - 2a \cos x + a^2} = \frac{\pi}{2} a^{n-1}$      $[a^2 < 1, \quad n \ge 1]$

                           $= \frac{\pi}{2a^{n+1}}$      $[a^2 > 1, \quad n \ge 1]$

                                             BI(65)(4), GW(332)(34a)

4.$^{10}$  $\displaystyle\int_0^\pi \frac{\cos nx \cos x\, dx}{1 - 2a\cos x + a^2} = \frac{\pi}{2}\frac{1+a^2}{1-a^2}a^{n-1}$  $\qquad [a^2 < 1, \quad n \geq 1]$

$\qquad\qquad\qquad\qquad = \dfrac{\pi}{2a^{n+1}}\dfrac{a^2+1}{a^2-1}$  $\qquad [a^2 > 1, \quad n \geq 1]$

$\qquad\qquad\qquad\qquad = \dfrac{\pi a}{1-a^2}$  $\qquad [n = 0, \quad a^2 < 1]$

$\qquad\qquad\qquad\qquad = \dfrac{\pi}{a\left(a^2-1\right)}$  $\qquad [n = 0, \quad a^2 > 1]$

$\qquad\qquad\qquad\qquad\qquad\qquad\qquad\qquad\qquad\qquad\qquad$ BI(65)(5), GW(332)(34b)

5.  $\displaystyle\int_0^\pi \frac{\cos(2n-1)x\, dx}{1 - 2a\cos 2x + a^2} = \int_0^\pi \frac{\cos 2nx \cos x\, dx}{1 - 2a\cos 2x + a^2} = 0$  $\quad [a^2 \neq 1]$  $\qquad$ BI (65)(9, 10)

6.  $\displaystyle\int_0^\pi \frac{\cos(2n-1)x \cos 2x\, dx}{1 - 2a\cos 2x + a^2} = 0$  $\qquad [a^2 \neq 1]$  $\qquad$ BI (65)(12)

7.  $\displaystyle\int_0^\pi \frac{\sin 2nx \sin x\, dx}{1 - 2a\cos 2x + a^2} = \int_0^\pi \frac{\sin(2n-1)x \sin 2x\, dx}{1 - 2a\cos 2x + a^2} = 0$

$\qquad\qquad\qquad\qquad\qquad\qquad\qquad\qquad [a^2 \neq 1]$  $\qquad$ BI (65)(6, 7)

8.  $\displaystyle\int_0^\pi \frac{\sin(2n-1)x \sin x\, dx}{1 - 2a\cos 2x + a^2} = \frac{\pi}{2}\cdot\frac{a^{n-1}}{1+a}$  $\qquad [a^2 < 1]$

$\qquad\qquad\qquad\qquad\qquad\quad = \dfrac{\pi}{2}\cdot\dfrac{1}{(1+a)a^n}$  $\qquad [a^2 > 1]$

$\qquad\qquad\qquad\qquad\qquad\qquad\qquad\qquad\qquad\qquad$ BI (65)(8)

9.  $\displaystyle\int_0^\pi \frac{\cos(2n-1)x \cos x\, dx}{1 - 2a\cos 2x + a^2} = \frac{\pi}{2}\cdot\frac{a^{n-1}}{1-a}$  $\qquad [a^2 < 1]$

$\qquad\qquad\qquad\qquad\qquad\quad = \dfrac{\pi}{2}\cdot\dfrac{1}{(a-1)a^n}$  $\qquad [a^2 > 1]$

$\qquad\qquad\qquad\qquad\qquad\qquad\qquad\qquad\qquad\qquad$ BI (65)(11)

10.  $\displaystyle\int_0^\pi \frac{\sin nx - a\sin(n-1)x}{1 - 2a\cos x + a^2}\sin mx\, dx = 0$  $\qquad$ for $m < n$

$\qquad\qquad\qquad\qquad\qquad\qquad\qquad\quad = \dfrac{\pi}{2}a^{m-n}$  $\qquad$ for $m \geq n$

$\qquad\qquad\qquad\qquad\qquad\qquad\qquad\qquad [a^2 < 1]$  $\qquad$ LI (65)(13)

11.$^6$  $\displaystyle\int_0^\pi \frac{\cos nx - a\cos(n-1)x}{1 - 2a\cos x + a^2}\cos mx\, dx = \frac{\pi}{2}\left(a^{|m|-n} - 1\right)$

$\qquad\qquad\qquad\qquad\qquad\qquad\qquad\qquad [a^2 < 1]$  $\qquad$ BI (65)(14)

12.  $\displaystyle\int_0^\pi \frac{\sin nx - a\sin[(n+1)x]}{1 - 2a\cos x + a^2}\, dx = 0$  $\qquad [a^2 < 1]$  $\qquad$ BI (68)(13)

13.  $\displaystyle\int_0^\pi \frac{\cos nx - a\cos[(n+1)x]}{1 - 2a\cos x + a^2}\, dx = 2\pi a^n$  $\qquad [a^2 < 1]$  $\qquad$ BI (68)(14)

**3.614**$^7$  $\displaystyle\int_0^\pi \frac{\sin x}{a^2 - 2ab\cos x + b^2}\cdot\frac{\sin px \cdot dx}{1 - 2a^p\cos px + a^{2p}}$

$\qquad\qquad\qquad\qquad = \dfrac{\pi b^{p-1}}{2a^{p+1}\left(1 - b^p\right)}$  $\qquad [0 < b \leq a \leq 1, \quad p = 1, 2, 3, \ldots]$

$\qquad\qquad\qquad\qquad = \dfrac{\pi a^{p-1}}{2b\left(b^p - a^{2p}\right)}$  $\qquad [0 < a \leq 1, \quad a^2 < b, \quad p = 1, 2, 3, \ldots]$

$\qquad\qquad\qquad\qquad\qquad\qquad\qquad\qquad\qquad\qquad$ BI (66)(9)

**3.615**

1. 
$$\int_0^{\pi/2} \frac{\cos 2nx \, dx}{1 - a^2 \sin^2 x} = \frac{(-1)^n \pi}{2\sqrt{1-a^2}} \left( \frac{1 - \sqrt{1-a^2}}{a} \right)^{2n} \qquad [a^2 < 1] \qquad \text{BI (47)(27)}$$

2. 
$$\int_0^{\pi} \frac{\cos x \sin 2nx \, dx}{1 + (a + b\sin x)^2} = -\frac{\pi}{b} \sin \left( 2n \arctan \sqrt{\frac{s}{2}} \right) \tan^{2n} \left( \frac{1}{2} \arccos \sqrt{\frac{s}{2a^2}} \right)$$

3. 
$$\int_0^{\pi} \frac{\cos x \cos(2n+1)x \, dx}{1 + (a + b\sin x)^2} = \frac{\pi}{b} \cos \left( (2n+1) \arctan \sqrt{\frac{s}{2}} \right) \tan^{2n+1} \left( \frac{1}{2} \arccos \sqrt{\frac{s}{2a^2}} \right)$$

$$\text{where } s = -\left(1 + b^2 - a^2\right) + \sqrt{\left(1 + b^2 - a^2\right)^2 + 4a^2} \quad \text{BI (65)(21, 22)}$$

**3.616**

1. 
$$\int_0^{\pi} \left( 1 - 2a \cos x + a^2 \right)^n dx = \pi \sum_{k=0}^{n} \binom{n}{k}^2 a^{2k} \qquad\qquad\qquad\qquad \text{BI (63)(1)}$$

2.[10] 
$$\int_0^{\pi} \frac{dx}{(1 - 2a\cos x + a^2)^n} = \frac{1}{2} \int_0^{2\pi} \frac{dx}{(1 - 2a\cos x + a^2)^n}$$

$$= \frac{\pi}{(1-a^2)^n} \sum_{k=0}^{n-1} \frac{(n+k-1)!}{(k!)^2 (n-k-1)!} \left( \frac{a^2}{1-a^2} \right)^k \qquad [a^2 < 1]$$

$$= \frac{\pi}{(a^2-1)^n} \sum_{k=0}^{n-1} \frac{(n+k-1)!}{(k!)^2 (n-k-1)!} \frac{1}{(a^2-1)^k} \qquad [a^2 > 1]$$

$$\text{BI (331)(63)}$$

3. 
$$\int_0^{\pi} \left( 1 - 2a\cos x + a^2 \right)^n \cos nx \, dx = (-1)^n \pi a^n \qquad\qquad\qquad \text{BI (63)(2)}$$

4. 
$$\int_0^{\pi} \left( 1 - 2a\cos x + a^2 \right)^n \cos mx \, dx$$

$$= \frac{1}{2} \int_0^{2\pi} \left( 1 - 2a\cos x + a^2 \right)^n \cos mx \, dx$$

$$= 0 \qquad\qquad\qquad\qquad\qquad\qquad\qquad\qquad\qquad [n < m]$$

$$= \pi(-a)^m \left(1 + a^2\right)^{n-m} \sum_{k=0}^{\lfloor (n-m)/2 \rfloor} \binom{n}{k} \binom{n-k}{m+k} \left( \frac{a}{1+a^2} \right)^{2k} \qquad [n \geq m]$$

$$\text{GW (332)(35a)}$$

5. 
$$\int_0^{2\pi} \frac{\sin nx \, dx}{(1 - 2a\cos 2x + a^2)^m} = 0 \qquad\qquad\qquad\qquad\qquad \text{GW (332)(32a)}$$

6. 
$$\int_0^{\pi} \frac{\sin x \, dx}{(1 - 2a\cos 2x + a^2)^m} = \frac{1}{2(m-1)a} \left[ \frac{1}{(1-a)^{2m-2}} - \frac{1}{(1+a)^{2m-2}} \right] \qquad [a \neq 0, \ \pm 1]$$

$$\text{GW (332)(32c)}$$

7. $\displaystyle\int_0^\pi \frac{\cos nx\,dx}{(1-2a\cos x+a^2)^m} = \frac{1}{2}\int_0^{2\pi}\frac{\cos nx\,dx}{(1-2a\cos x+a^2)^m}$

$$= \frac{a^{2m+n-2}\pi}{(1-a^2)^{2m-1}}\sum_{k=0}^{m-1}\binom{m+n-1}{k}\binom{2m-k-2}{m-1}\left(\frac{1-a^2}{a^2}\right)^k \qquad [a^2<1]$$

$$= \frac{\pi}{a^n(a^2-1)^{2m-1}}\sum_{k=0}^{m-1}\binom{m+n-1}{k}\binom{2m-k-2}{m-1}(a^2-1)^k \qquad [a^2>1]$$

GW (332)(31)

8. $\displaystyle\int_0^{\pi/2}\frac{\cos 2nx\,dx}{(a^2\cos^2 x+b^2\sin^2 x)^{n+1}} = \binom{2n}{n}\frac{(b^2-a^2)^n}{(2ab)^{2n+1}}\pi$

$[a>0, \quad b>0]$ \qquad GW (332)(30b)

9.* $\displaystyle\int_0^{2\pi}\frac{\cos nx}{(\cosh a-\cos x)^2}dx = \frac{8\pi(n+1)e^{-(n+2)a}}{(1-e^{-2a})^3}\left(1-\frac{n-1}{n+1}e^{-2a}\right)$

$[n=0,1,\dots, \quad a>0]$

**3.617**[10] $\displaystyle\int_0^\pi\frac{dx}{(1-2a\cos x+a^2)^{n+1/2}} = \frac{2}{|1+a|^{2n+1}}F_n\left(\frac{2\sqrt{|a|}}{|1+a|}\right), \qquad |a|\neq 1$

with

$$F_n(k) = \int_0^{\pi/2}\frac{dx}{\left(1-k^2\sin^2 x\right)^{n+1/2}}$$

where the $F_n(k)$ satisfy the recurrence relation

$$F_{n+1}(k) = F_n(k) + \frac{k}{2n+1}\frac{dF_n(k)}{dk}, \qquad n=0,1,2,\dots$$

and

$$F_0(k) = \boldsymbol{K}(k) \equiv \int_0^{\pi/2}\frac{dx}{\left(1-k^2\sin^2 x\right)^{1/2}}$$

is the complete elliptic integral of the first kind.
Introducing the complete elliptic integral of the second kind

$$\boldsymbol{E}(k) = \int_0^{\pi/2}\left(1-k^2\sin^2 x\right)^{1/2}dx$$

the derivatives

$$\frac{d\,\boldsymbol{K}(k)}{dk} = \frac{\boldsymbol{E}(k)}{k(1-k^2)} - \frac{\boldsymbol{K}(k)}{k}, \qquad \frac{d\,\boldsymbol{E}(k)}{dk} = \frac{\boldsymbol{E}(k)-\boldsymbol{K}(k)}{k}$$

combined with the recurrence relation lead to

$$F_1(k) = F_0(k) + k\frac{d\,F_0(k)}{dk}$$

$$= \boldsymbol{K}(k) + \frac{\boldsymbol{E}(k)}{1-k^2} - \boldsymbol{K}(k) = \frac{\boldsymbol{E}(k)}{1-k^2},$$

$$F_2(k) = \frac{\boldsymbol{E}(k)}{1-k^2} + \frac{k}{3}\frac{d}{dk}\left(\frac{\boldsymbol{E}(k)}{1-k^2}\right)$$

$$= \frac{1}{3(1-k^2)}\left[\left(\frac{4-2k^2}{1-k^2}\right)\boldsymbol{E}(k) - \boldsymbol{K}(k)\right]$$

## 3.62  Powers of trigonometric functions

**3.621**

1.   $\displaystyle\int_0^{\pi/2} \sin^{\mu-1} x \, dx = \int_0^{\pi/2} \cos^{\mu-1} x \, dx = 2^{\mu-2} \, \mathrm{B}\left(\frac{\mu}{2}, \frac{\mu}{2}\right)$              FI II 789

2.[12]  $\displaystyle\int_0^{\pi/2} \sin^{3/2} x \, dx = \int_0^{\pi/2} \cos^{3/2} x \, dx = \frac{1}{6\sqrt{2\pi}} \Gamma\left(\tfrac{1}{4}\right)$

3.   $\displaystyle\int_0^{\pi/2} \sin^{2m} x \, dx = \int_0^{\pi/2} \cos^{2m} x \, dx = \frac{(2m-1)!!}{(2m)!!} \frac{\pi}{2}$          FI II 151

4.   $\displaystyle\int_0^{\pi/2} \sin^{2m+1} x \, dx = \int_0^{\pi/2} \cos^{2m+1} x \, dx = \frac{(2m)!!}{(2m+1)!!}$      FI II 151

5.   $\displaystyle\int_0^{\pi/2} \sin^{\mu-1} x \cos^{\nu-1} x \, dx = \frac{1}{2} \, \mathrm{B}\left(\frac{\mu}{2}, \frac{\nu}{2}\right)$         $[\operatorname{Re}\mu > 0, \quad \operatorname{Re}\nu > 0]$

                                              LO V 113(50), LO V 122, FI II 788

6.   $\displaystyle\int_0^{\pi/2} \sqrt{\sin x} \, dx = \sqrt{\frac{2}{\pi}} \left(\Gamma\left(\frac{3}{4}\right)\right)^2$

7.   $\displaystyle\int_0^{\pi/2} \frac{dx}{\sqrt{\sin x}} = \frac{\left(\Gamma\left(\frac{1}{4}\right)\right)^2}{2\sqrt{2\pi}}$

**3.622**

1.   $\displaystyle\int_0^{\pi/2} \tan^{\pm\mu} x \, dx = \frac{\pi}{2} \sec \frac{\mu\pi}{2}$           $[|\operatorname{Re}\mu| < 1]$         BI (42)(1)

2.   $\displaystyle\int_0^{\pi/4} \tan^{\mu} x \, dx = \frac{1}{2} \, \beta\left(\frac{\mu+1}{2}\right)$          $[\operatorname{Re}\mu > -1]$       BI (34)(1)

3.   $\displaystyle\int_0^{\pi/4} \tan^{2n} x \, dx = (-1)^n \frac{\pi}{4} + \sum_{k=0}^{n-1} \frac{(-1)^k}{2n-2k-1}$         BI (34)(2)

4.[11]  $\displaystyle\int_0^{\pi/4} \tan^{2n+1} x \, dx = (-1)^n \frac{\ln 2}{2} + \sum_{k=0}^{n-1} \frac{(-1)^k}{2n-2k}$       BI (34)(3)

**3.623**

1.   $\displaystyle\int_0^{\pi/2} \tan^{\mu-1} x \cos^{2\nu-2} x \, dx = \int_0^{\pi/2} \cot^{\mu-1} x \sin^{2\nu-2} x \, dx = \frac{1}{2} \, \mathrm{B}\left(\frac{\mu}{2}, \nu - \frac{\mu}{2}\right)$

                                    $[0 < \operatorname{Re}\mu < 2\operatorname{Re}\nu]$    BI(42)(6), BI(45)(22)

2.[6]  $\displaystyle\int_0^{\pi/4} \tan^{\mu} x \sin^2 x \, dx = \frac{1+\mu}{4} \, \beta\left(\frac{\mu+1}{2}\right) - \frac{1}{4}$       $[\operatorname{Re}\mu > -1]$     BI (34)(4)

3.[6]  $\displaystyle\int_0^{\pi/4} \tan^{\mu} x \cos^2 x \, dx = \frac{1-\mu}{4} \, \beta\left(\frac{\mu+1}{2}\right) + \frac{1}{4}$       $[\operatorname{Re}\mu > -1]$     BI (34)(5)

**3.624**

1.     $\displaystyle\int_0^{\pi/4} \frac{\sin^p x}{\cos^{p+2} x}\, dx = \frac{1}{p+1}$             $[p > -1]$          GW (331)(34b)

2.[12]    $\displaystyle\int_0^{\pi/2} \frac{\sin^{\mu-\frac12} x}{\cos^{2\mu-1} x}\, dx = \int_0^{\pi/2} \frac{\cos^{\mu-\frac12} x}{\sin^{2\mu-1} x}\, dx = -\frac{2^{1/2-\mu}}{\sqrt\pi} \cos\left(\frac{\pi}{4}(1+2\mu)\right)\Gamma(1-\mu)\Gamma(\mu-\tfrac12)$

                                                 $\left[-\tfrac12 < \operatorname{Re}\mu < 1\right]$         LI (55)(12)

3.[12]    $\displaystyle\int_0^{\pi/4} \frac{\cos^{n-\frac12}(2x)}{\cos^{2n+1}(x)}\, dx = \pi \frac{(2n)!}{2^{2n+1}\,(n!)^2}$                          BI (38)(3)

4.[8]    $\displaystyle\int_0^{\pi/4} \frac{\cos^{\mu} 2x}{\cos^{2(\mu+1)} x}\, dx = 2^{2\mu}\, B(\mu+1,\mu+1)$        $[\operatorname{Re}\mu > -1]$          BI (35)(1)

5.     $\displaystyle\int_0^{\pi/4} \frac{\sin^{2\mu-2} x}{\cos^{\mu} 2x}\, dx = 2^{1-2\mu}\, B(2\mu-1, 1-\mu) = \frac{\Gamma\left(\mu-\frac12\right)\Gamma(1-\mu)}{2\sqrt\pi}$

                                              $\left[\tfrac12 < \operatorname{Re}\mu < 1\right]$         BI (35)(4)

6.[6]    $\displaystyle\int_0^{\pi/2} \left(\frac{\sin ax}{\sin x}\right)^2 dx = \frac{a\pi}{2} - \frac12 \sin\pi a\,[2a\,\beta(a) - 1]$        $[a > 0]$

**3.625**

1.     $\displaystyle\int_0^{\pi/4} \frac{\sin^{2n-1} x \cos^p 2x}{\cos^{2p+2n+1} x}\, dx = \frac{(n-1)!}{2}\,\frac{\Gamma(p+1)}{\Gamma(p+n+1)}$

                        $\displaystyle = \frac{(n-1)!}{2(p+n)(p+n-1)\cdots(p+1)} = \frac12\, B(n, p+1)$

                                       $[p > -1]$     (cf. **3.251** 1)      BI (35)(2)

2.     $\displaystyle\int_0^{\pi/4} \frac{\sin^{2n} x \cos^p 2x}{\cos^{2p+2n+2} x}\, dx = \tfrac12\, B\left(n+\tfrac12, p+1\right)$      $[p > -1]$     (cf. **3.251** 1)      BI (35)(3)

3.     $\displaystyle\int_0^{\pi/4} \frac{\sin^{2n-1} x \cos^{m-\frac12} 2x}{\cos^{2n+2m} x}\, dx = \frac{(2n-2)!!(2m-1)!!}{(2n+2m-1)!!}$                    BI (38)(6)

4.[8]    $\displaystyle\int_0^{\pi/4} \frac{\sin^{2n} x \cos^{m-\frac12} 2x}{\cos^{2n+2m+1} x}\, dx = \frac{(2n-1)!!(2m-1)!!}{(2n+2m)!!}\cdot\frac{\pi}{2}$              BI (38)(7)

**3.626**

1.     $\displaystyle\int_0^{\pi/4} \frac{\sin^{2n-1} x}{\cos^{2n+2} x}\sqrt{\cos 2x}\, dx = \frac{(2n-2)!!}{(2n+1)!!}$        (cf. **3.251** 1)       BI (38)(4)

2.     $\displaystyle\int_0^{\pi/4} \frac{\sin^{2n} x}{\cos^{2n+3} x}\sqrt{\cos 2x}\, dx = \frac{(2n-1)!!}{(2n+2)!!}\frac{\pi}{2}$        (cf. **3.251** 1)       BI (38)(5)

**3.627**    $\displaystyle\int_0^{\pi/2} \frac{\tan^{\mu} x}{\cos^{\mu} x}\, dx = \int_0^{\pi/2} \frac{\cot^{\mu} x}{\sin^{\mu} x}\, dx = \frac{\Gamma(\mu)\,\Gamma\left(\frac12 - \mu\right)}{2^{\mu}\sqrt\pi}\,\sin\frac{\mu\pi}{2}$

                                                $\left[-1 < \operatorname{Re}\mu < \tfrac12\right]$       BI (55)(12)a

**3.628**[11]    $\displaystyle\int_0^{\frac{\pi}{2}} \sec^{2p} x \sin^{2p-1} x\, dx = \frac{1}{2\sqrt\pi}\,\Gamma(p)\,\Gamma\left(\tfrac12 - p\right)$        $\left[0 < p < \tfrac12\right]$         WA 691

## 3.63 Powers of trigonometric functions and trigonometric functions of linear functions

**3.631**

1. $$\int_0^\pi \sin^{\nu-1} x \sin ax \, dx = \frac{\pi \sin \frac{a\pi}{2}}{2^{\nu-1} \nu \, B\left(\frac{\nu+a+1}{2}, \frac{\nu-a+1}{2}\right)}$$

$$[\operatorname{Re} \nu > 0] \qquad \text{LO V 121(67a), WA 337a}$$

2.[12] $$\int_0^{\pi/2} \sin^{\nu-2} x \sin \nu x \, dx = \frac{1}{1-\nu} \cos \frac{\nu\pi}{2} \qquad [\operatorname{Re} \nu > 1] \qquad \text{GW(332)(16d), FI I 152}$$

3.[6] $$\int_0^\pi \sin^\nu x \sin \nu x \, dx = 2^{-\nu} \pi \sin \frac{\nu\pi}{2} \qquad [\operatorname{Re} \nu > -1] \qquad \text{LO V 121(69)}$$

4. $$\int_0^\pi \sin^n x \sin 2mx \, dx = 0 \qquad\qquad\qquad \text{GW (332)(11a)}$$

5. $$\int_0^\pi \sin^{2n} x \sin(2m+1)x \, dx = \int_0^{\pi/2} \sin^{2n} x \sin(2m+1)x \, dx$$

$$= \frac{(-1)^m 2^{n+1} n!(2n-1)!!}{(2n-2m-1)!!(2m+2n+1)!!} \qquad [m \le n]^*$$

$$= \frac{(-1)^n 2^{n+1} n!(2m-2n-1)!!(2n-1)!!}{(2m+2n+1)!!} \qquad [m \ge n]^*$$

$$\text{GW (332)(11b)}$$

6. $$\int_0^\pi \sin^{2n+1} x \sin(2m+1)x \, dx = 2\int_0^{\pi/2} \sin^{2n+1} x \sin(2m+1)x \, dx$$

$$= \frac{(-1)^m \pi}{2^{2n+1}} \binom{2n+1}{n-m} \qquad\qquad [n \ge m]$$

$$= 0 \qquad\qquad [n < m]$$

$$\text{BI(40)(12), GW(332)(11c)}$$

7. $$\int_0^\pi \sin^n x \cos(2m+1)x \, dx = 0 \qquad\qquad\qquad \text{GW (332)(12a)}$$

8. $$\int_0^\pi \sin^{\nu-1} x \cos ax \, dx = \frac{\pi \cos \frac{a\pi}{2}}{2^{\nu-1} \nu \, B\left(\frac{\nu+a+1}{2}, \frac{\nu-a+1}{2}\right)}$$

$$[\operatorname{Re} \nu > 0] \qquad \text{LO V 121(68)a, WA 337a}$$

9. $$\int_0^{\pi/2} \cos^{\nu-1} x \cos ax \, dx = \frac{\pi}{2^\nu \nu \, B\left(\frac{\nu+a+1}{2}, \frac{\nu-a+1}{2}\right)}$$

$$[\operatorname{Re} \nu > 0] \qquad \text{GW (332)(9c)}$$

10. $$\int_0^{\pi/2} \sin^{\nu-2} x \cos \nu x \, dx = \frac{1}{\nu-1} \sin \frac{\nu\pi}{2} \qquad [\operatorname{Re} \nu > 1] \qquad \text{GW(332)(16b), FI II 15 2}$$

---

*In 3.631.5, for $m = n$ we should set $(2n - 2m - 1)!! = 1$

11.    $\int_0^\pi \sin^\nu x \cos \nu x \, dx = \dfrac{\pi}{2^\nu} \cos \dfrac{\nu\pi}{2}$                    $[\operatorname{Re}\nu > -1]$                    LO V 121(70)a

12.[12]    $\int_0^\pi \sin^{2n} x \cos 2mx \, dx = 2\int_0^{\pi/2} \sin^{2n} x \cos 2mx \, dx = \dfrac{(-1)^m}{2^{2n}} \begin{pmatrix} 2n \\ n-m \end{pmatrix}$    $[n \ge m]$

$= 0$                    $[n < m]$

BI(40)(16), GW(332)(12b)

13.[12]    $\int_0^\pi \sin^{2n+1} x \cos 2mx \, dx$

$= 2\int_0^{\pi/2} \sin^{2n+1} x \cos 2mx \, dx = \dfrac{(-1)^m 2^{n+1} n!(2n+1)!!}{(2n-2m+1)!!(2m+2n+1)!!}$    $[n \ge m-1]$

$= \dfrac{(-1)^{n+1} 2^{n+1} n!(2m-2n-3)!!(2n+1)!!}{(2m+2n+1)!!}$    $[n < m-1]$

GW (332)(12c)

14.    $\int_0^{\pi/2} \cos^{\nu-2} x \sin \nu x \, dx = \dfrac{1}{\nu-1}$                    $[\operatorname{Re}\nu > 1]$                    GW(332)(16c), FI II 152

15.    $\int_0^\pi \cos^m x \sin nx \, dx = \left[1 - (-1)^{m+n}\right] \int_0^{\pi/2} \cos^m x \sin nx \, dx$

$= \left[1 - (-1)^{m+n}\right] \left\{ \displaystyle\sum_{k=0}^{r-1} \dfrac{m!}{(m-k)!} \dfrac{(m+n-2k-2)!!}{(m+n)!!} + s\dfrac{m!(n-m-2)!!}{(m+n)!!} \right\}$

$\left[ r = \begin{cases} m & \text{if } m \le n \\ n & \text{if } m \ge n \end{cases} \qquad s = \begin{cases} 2 & \text{if } n-m = 4l+2 > 0 \\ 1 & \text{if } n-m = 2l+1 > 0 \\ 0 & \text{if } n-m = 4l \text{ or } n-m < 0 \end{cases} \right]$    GW (332)(13a)

16.    $\int_0^{\pi/2} \cos^n x \sin nx \, dx = \dfrac{1}{2^{n+1}} \displaystyle\sum_{k=1}^n \dfrac{2^k}{k}$                    FI II 153

17.[11]    $\int_0^\pi \cos^n x \sin mx \, dx = \begin{cases} \left[1 + (-1)^{m+n}\right] \dfrac{\pi}{2^{n+1}} \begin{pmatrix} n \\ k \end{pmatrix} & \text{if } m \le n \text{ and } n-m = 2k \\ 0 & \text{otherwise} \end{cases}$

GW (332)(15a)

18.[6]    $\int_0^\pi \cos^m x \cos ax \, dx = \dfrac{(-1)^m \sin a\pi}{2^m(m+a)} \, {}_2F_1\left(-m, -\dfrac{a+m}{2}; 1 - \dfrac{a+m}{2}; -1\right)$

$[a \ne 0, \pm 1, \pm 2, \ldots]$                    WA 313

19.    $\int_0^{\pi/2} \cos^{\nu-2} x \cos \nu x \, dx = 0$                    $[\operatorname{Re}\nu > 1]$                    GW(332)(16a), FI II 152

20.[10]    $\int_0^{\pi/2} \cos^n x \cos nx \, dx = \dfrac{\pi}{2^{n+1}}$                    $[\operatorname{Re}n > -1]$                    LO V 122(78), FI II 153

**3.632**

1.    $\int_0^\pi \sin^{p-1} x \cos\left[a\left(\dfrac{\pi}{2} - x\right)\right] dx = 2^{p-1} \dfrac{\Gamma\left(\frac{p-a}{2}\right) \Gamma\left(\frac{p+a}{2}\right)}{\Gamma(p-a)\,\Gamma(p+a)} \Gamma(p)$

$[p^2 < a^2]$                    BI (62)(11)

2. $\displaystyle\int_{-\frac{\pi}{2}}^{\frac{\pi}{2}} \cos^{p-1} x \sin\left[a\left(x + \frac{\pi}{2}\right)\right] dx = \frac{\pi \sin\frac{a\pi}{2}}{2^{p-1} p\, \mathrm{B}\left(\dfrac{p+a+1}{2}, \dfrac{p-a+1}{2}\right)}$

$$[\operatorname{Re} p > 0] \qquad\qquad \text{WA 337a}$$

3.[10] $\displaystyle\int_{0}^{\pi/2} \cos^{p} x \sin[(p+2n)x]\, dx = (-1)^{n-1} \sum_{k=0}^{n-1} \frac{(-1)^k 2^k}{p+k+1} \binom{n-1}{k}$

$$[n > 0] \qquad\qquad \text{LI (41)(12)}$$

4. $\displaystyle\int_{-\pi}^{\pi} \cos^{n-1} x \cos[m(x-a)]\, dx = \left[1 - (-1)^{n+m}\right] = \int_{-\frac{\pi}{2}}^{\frac{\pi}{2}} \cos^{n-1} x \cos[m(x-a)]\, dx$

$$= \frac{\left[1 - (-1)^{n+m}\right] \pi \cos ma}{2^{n-1} n\, \mathrm{B}\left(\dfrac{n+m+1}{2}, \dfrac{n-m+1}{2}\right)}$$

$$[n \geq m] \qquad \text{LO V 123(80), LO V 139(94a)}$$

5. $\displaystyle\int_{0}^{\pi/2} \cos^{p+q-2} x \cos[(p-q)x]\, dx = \frac{\pi}{2^{p+q-1}(p+q-1)\,\mathrm{B}(p,q)}$

$$[p + q > 1] \qquad\qquad \text{WH}$$

**3.633**

1. $\displaystyle\int_{0}^{\pi/2} \cos^{p-1} x \sin ax \sin x\, dx = \frac{a\pi}{2^{p+1} p(p+1)\, \mathrm{B}\left(\dfrac{p+a}{2}+1, \dfrac{p-a}{2}+1\right)}$      LO V 150(110)

2. $\displaystyle\int_{0}^{\pi/2} \cos^{n} x \sin nx \sin 2mx\, dx = \int_{0}^{\pi/2} \cos^{n} x \cos nx \cos 2mx\, dx = \frac{\pi}{2^{n+2}} \binom{n}{m}$    BI (42)(19, 20)

3. $\displaystyle\int_{0}^{\pi/2} \cos^{n-1} x \cos[(n+1)x] \cos 2mx\, dx = \frac{\pi}{2^{n+1}} \binom{n-1}{m-1}$

$$[n > m - 1] \qquad\qquad \text{BI (42)(21)}$$

4. $\displaystyle\int_{0}^{\pi/2} \cos^{p+q} x \cos px \cos qx\, dx = \frac{\pi}{2^{p+q+2}}\left[1 + \frac{1}{(p+q+1)\,\mathrm{B}(p+1, q+1)}\right]$

$$[p + q > -1] \qquad\qquad \text{GW (332)(10c)}$$

5.[6] $\displaystyle\int_{0}^{\pi/2} \cos^{p+q} x \sin px \sin qx\, dx = \frac{\pi}{2^{p+q+2}} \sum_{k=1}^{\infty} \binom{p}{k}\binom{q}{k} = \frac{\pi}{2^{p+q+2}}\left[\frac{\Gamma(p+q+1)}{\Gamma(p+1)\Gamma(q+1)} - 1\right]$

$$[p + q > -1] \qquad\qquad \text{BI (42)(16)}$$

**3.634**

1. $\displaystyle\int_{0}^{\pi/2} \sin^{\mu-1} x \cos^{\nu-1} x \sin(\mu+\nu)x\, dx = \sin\frac{\mu\pi}{2}\, \mathrm{B}(\mu, \nu)$

$$[\operatorname{Re}\mu > 0, \quad \operatorname{Re}\nu > 0]$$

$$\text{BI(42)(23), FI II 814a}$$

2.    $\displaystyle\int_0^{\pi/2} \sin^{\mu-1} x \cos^{\nu-1} x \cos(\mu+\nu)x\,dx = \cos\frac{\mu\pi}{2}\,B(\mu,\nu)$

$$[\operatorname{Re}\mu > 0, \quad \operatorname{Re}\nu > 0]$$

BI(42)(24), FI II 814a

3.    $\displaystyle\int_0^{\pi/2} \cos^{p+n-1} x \sin px \cos[(n+1)x]\sin x\,dx = \frac{\pi}{2^{p+n+1}}\frac{\Gamma(p+n)}{n!\,\Gamma(p)}$

$$[p > -n]$$

BI (42)(15)

**3.635**

1.    $\displaystyle\int_0^{\pi/4} \cos^{\mu-1} 2x \tan x\,dx = \frac{1}{4}\left[\psi\left(\frac{\mu+1}{2}\right) - \psi\left(\frac{\mu}{2}\right)\right] \quad [\operatorname{Re}\mu > 0]$    BI (34)(7)

2.[12]   $\displaystyle\int_0^{\pi/2} \cos^{p+2n} x \sin px \tan x\,dx = \frac{\pi}{2^{p+2n+1}\,\Gamma(p)}\sum_{k=0}^{\infty}\binom{n}{k}\frac{\Gamma(p+n-k)}{(n-k)!}$

$$= \frac{p\pi}{2^{p+2n+1}}\frac{\Gamma(p+2n)}{\Gamma(n+1)\,\Gamma(p+n+1)}$$

$$[p > -2n]$$

BI (42)(22)

3.    $\displaystyle\int_0^{\pi/2} \cos^{n-1} x \sin[(n+1)x]\cot x\,dx = \frac{\pi}{2}$    BI (45)(18)

**3.636**

1.    $\displaystyle\int_0^{\pi/2} \tan^{\pm\mu} x \sin 2x\,dx = \frac{\mu\pi}{2}\operatorname{cosec}\frac{\mu\pi}{2}$    $[0 < \operatorname{Re}\mu < 2]$    BI (45)(20)a

2.    $\displaystyle\int_0^{\pi/2} \tan^{\pm\mu} x \cos 2x\,dx = \mp\frac{\mu\pi}{2}\sec\frac{\mu\pi}{2}$    $[|\operatorname{Re}\mu| < 1]$    BI (45)(21)

3.[11]   $\displaystyle\int_0^{\pi/2}\frac{\tan^{2\mu} x}{\cos x}\,dx = \int_0^{\pi/2}\frac{\cot^{2\mu} x}{\sin x}\,dx = \frac{\Gamma\left(\mu+\frac{1}{2}\right)\Gamma(-\mu)}{2\sqrt{\pi}}$

$$\left[-\tfrac{1}{2} < \operatorname{Re}\mu < 1\right] \qquad (\text{cf. } \mathbf{3.251}\ 1)$$

BI (45)(13, 14)

**3.637**

1.    $\displaystyle\int_0^{\pi/2} \tan^p x \sin^{q-2} x \sin qx\,dx = -\cos\frac{(p+q)\pi}{2}\,B(p+q-1, 1-p)$

$$[p+q > 1 > p]$$

GW (332)(15d)

2.    $\displaystyle\int_0^{\pi/2} \tan^p x \sin^{q-2} x \cos qx\,dx = \sin\frac{(p+q)\pi}{2}\,B(p+q-1, 1-p)$

$$[p+q > 1 > p]$$

GW (332)(15b)

3.    $\displaystyle\int_0^{\pi/2} \cot^p x \cos^{q-2} x \sin qx\,dx = \cos\frac{p\pi}{2}\,B(p+q-1, 1-p)$

$$[p+q > 1 > p]$$

GW (332)(15c)

4.   $\displaystyle\int_0^{\pi/2} \cot^p x \cos^{q-2} x \cos qx \, dx = \sin\frac{p\pi}{2}\, \mathrm{B}(p+q-1, 1-p)$

$$[p+q > 1 > p] \qquad\qquad \text{GW (332)(15a)}$$

### 3.638

1.   $\displaystyle\int_0^{\pi/4} \frac{\sin^{2\mu} x \, dx}{\cos^{\mu+\frac{1}{2}} 2x \cos x} = \frac{\pi}{2}\sec\mu\pi$      $\left[|\mathrm{Re}\,\mu| < \tfrac{1}{2}\right]$     (cf. **3.192** 2)

$$\text{BI (38)(8)}$$

2.   $\displaystyle\int_0^{\pi/4} \frac{\sin^{\mu-\frac{1}{2}} 2x \, dx}{\cos^{\mu} 2x \cos x} = \frac{2}{2\mu-1}\frac{\Gamma\left(\mu+\frac{1}{2}\right)\Gamma(1-\mu)}{\sqrt{\pi}}\sin\left(\frac{2\mu-1}{4}\pi\right)$

$$\left[-\tfrac{1}{2} < \mathrm{Re}\,\mu < 1\right] \qquad\qquad \text{BI (38)(17)}$$

3.   $\displaystyle\int_0^{\pi/2} \frac{\cos^{p-1} x \sin px}{\sin x}\, dx = \frac{\pi}{2}$      $[p > 0]$     GW(332)(17), BI(45)(5)

## 3.64–3.65 Powers and rational functions of trigonometric functions

### 3.641

1.   $\displaystyle\int_0^{\pi/2} \frac{\sin^{p-1} x \cos^{-p} x}{a\cos x + b\sin x}\, dx = \int_0^{\pi/2}\frac{\sin^{-p} x \cos^{p-1} x}{a\sin x + b\cos x}\,dx = \frac{\pi\cosec\, p\pi}{a^{1-p}b^{p}}$

$$[ab > 0, \quad 0 < p < 1] \qquad\qquad \text{GW (331)(62)}$$

2.   $\displaystyle\int_0^{\pi/2} \frac{\sin^{1-p} x \cos^{p} x}{(\sin x + \cos x)^3}\, dx = \int_0^{\pi/2}\frac{\sin^{p} x \cos^{1-p} x}{(\sin x + \cos x)^3}\,dx = \frac{(1-p)p}{2}\pi\cosec\, p\pi$

$$[-1 < p < 2] \qquad\qquad \text{BI(48)(5)}$$

### 3.642

1.   $\displaystyle\int_0^{\pi/2} \frac{\sin^{2\mu-1} x \cos^{2\nu-1} x \, dx}{\left(a^2\sin^2 x + b^2\cos^2 x\right)^{\mu+\nu}} = \frac{1}{2a^{2\mu}b^{2\nu}}\,\mathrm{B}(\mu,\nu)$    $[\mathrm{Re}\,\mu > 0, \quad \mathrm{Re}\,\nu > 0]$    BI (48)(28)

2.   $\displaystyle\int_0^{\pi/2} \frac{\sin^{n-1} x \cos^{n-1} x \, dx}{\left(a^2\cos^2 x + b^2\sin^2 x\right)^{n}} = \frac{\mathrm{B}\left(\frac{n}{2}, \frac{n}{2}\right)}{2(ab)^n}$      $[ab > 0]$    GW (331)(59a)

3.   $\displaystyle\int_0^{\pi/2} \frac{\sin^{2n} x \, dx}{\left(a^2\cos^2 x + b^2\sin^2 x\right)^{n+1}} = \frac{1}{2}\int_0^{\pi}\frac{\sin^{2n} x \, dx}{\left(a^2\cos^2 x + b^2\sin^2 x\right)^{n+1}}$

$$= \int_0^{\pi/2}\frac{\cos^{2n} x \, dx}{\left(a^2\sin^2 x + b^2\cos^2 x\right)^{n+1}} = \frac{1}{2}\int_0^{\pi}\frac{\cos^{2n} x \, dx}{\left(a^2\sin^2 x + b^2\cos^2 x\right)^{n+1}} = \frac{(2n-1)!!\,\pi}{2^{n+1}n!ab^{2n+1}}$$

$$[ab > 0] \qquad\qquad\qquad\qquad \text{GW (331)(58)}$$

4.   $\displaystyle\int_0^{\pi/2} \frac{\cos^{p+2n} x \cos px \, dx}{\left(a^2\cos^2 x + b^2\sin^2 x\right)^{n+1}} = \pi\sum_{k=0}^{n}\binom{2n-k}{n}\binom{p+k-1}{k}\frac{b^{p-1}}{(2a)^{2n-k+1}(a+b)^{p+k}}$

$$[a > 0, \quad b > 0, \quad p > -2n-1]$$
$$\text{GW (332)(30)}$$

**3.643**

1. $$\int_0^{\pi/2} \frac{\cos^p x \cos px \, dx}{1 - 2a \cos 2x + a^2} = \frac{\pi}{2^{p+1}} \frac{(1+a)^{p-1}}{1-a} \qquad [a^2 < 1, \quad p > -1] \qquad \text{GW (332)(33c)}$$

2.[12] $$\int_0^{\pi/2} \frac{\sin^{2n} x \cos^\mu x \cos \beta x}{(1 - 2a \cos 2x + a^2)^m} \, dx = \frac{(-1)^n \pi (1-a)^{2n-2m+1}}{2^{2m-\beta-1}(1+a)^{2m+\beta+1}} \sum_{k=0}^{m-1} \sum_{l=0}^{m-k-1} \binom{\beta}{k}\binom{2n}{l}$$
$$\times \binom{2m-k-l-2}{m-1}(-2)^l (a-1)^k$$
$$[a^2 < 1, \quad \beta = 2m - 2n - \mu - 2, \quad \mu > -1] \quad \text{GW (332)(33)}$$

**3.644**

1. $$\int_0^\pi \frac{\sin^m x}{p + q \cos x} \, dx = 2^{m-2} \frac{p}{q^2} \sum_{\nu=1}^k \left(\frac{p^2 - q^2}{-4q^2}\right)^{\nu-1} B\left(\frac{m+1-2\nu}{2}, \frac{m+1-2\nu}{2}\right) + \left(\frac{p^2 - q^2}{-q^2}\right)^k A$$
$$\text{where } A = \begin{cases} \dfrac{\pi p}{q^2}\left(1 - \sqrt{1 - \dfrac{q^2}{p^2}}\right) & \text{if } m = 2k + 2 \\[4mm] \dfrac{1}{q} \ln \dfrac{p+q}{p-q} & \text{if } m = 2k + 1 \end{cases} \qquad [k \geq 1, \quad q \neq 0, \quad p^2 - q^2 \geq 0]$$

2. $$\int_0^\pi \frac{\sin^m x}{1 + \cos x} \, dx = 2^{m-1} B\left(\frac{m-1}{2}, \frac{m+1}{2}\right) \qquad [m \geq 2]$$

3. $$\int_0^\pi \frac{\sin^m x}{1 - \cos x} \, dx = 2^{m-1} B\left(\frac{m-1}{2}, \frac{m+1}{2}\right) \qquad [m \geq 2]$$

4. $$\int_0^\pi \frac{\sin^2 x}{p + q \cos x} \, dx = \frac{p\pi}{q^2}\left(1 - \sqrt{1 - \frac{q^2}{p^2}}\right)$$

5. $$\int_0^\pi \frac{\sin^3 x}{p + q \cos x} \, dx = 2\frac{p}{q^2} + \frac{1}{q}\left(1 - \frac{p^2}{q^2}\right) \ln \frac{p+q}{p-q}$$

**3.645** $$\int_0^\pi \frac{\cos^n x \, dx}{(a + b \cos x)^{n+1}} = \frac{\pi}{2^n (a+b)^n \sqrt{a^2 - b^2}} \sum_{k=0}^n (-1)^k \frac{(2n-2k-1)!!(2k-1)!!}{(n-k)!k!} \left(\frac{a+b}{a-b}\right)^k$$
$$[a^2 > b^2] \qquad \text{LI (64)(16)}$$

**3.646**

1. $$\int_0^{\pi/2} \frac{\cos^n x \sin nx \sin 2x}{1 - 2a \cos 2x + a^2} \, dx = \frac{\pi}{4a}\left[\left(\frac{1+a}{2}\right)^n - \frac{1}{2^n}\right] \qquad [a^2 < 1] \qquad \text{BI (50)(6)}$$

2. $$\int_0^{\pi/2} \frac{1 - a \cos 2nx}{1 - 2a \cos 2nx + a^2} \cos^m x \cos mx \, dx = \frac{\pi}{2^{m+2}} \sum_{k=1}^\infty \binom{m}{kn} a^k + \frac{\pi}{2^{m+1}}$$
$$[a^2 < 1] \qquad \text{LI (50)(7)}$$

**3.647** $$\int_0^{\pi/2} \frac{\cos^p x \cos px \, dx}{a^2 \sin^2 x + b^2 \cos^2 x} = \frac{\pi}{2b} \frac{a^{p-1}}{(a+b)^p} \qquad [p > -1, \quad a > 0, \quad b > 0] \qquad \text{BI (47)(20)}$$

**3.648**

1.  $\displaystyle\int_0^{\pi/4} \frac{\tan^l x\,\mathrm{d}x}{1+\cos\frac{m}{n}\pi\sin 2x}$

$$= \frac{1}{2n}\operatorname{cosec}\frac{m}{n}\pi\sum_{k=0}^{n-1}(-1)^{k-1}\sin\frac{km}{n}\pi\left[\psi\left(\frac{n+l+k}{2n}\right)-\psi\left(\frac{l+k}{2n}\right)\right]\qquad [m+n\text{ is odd}]$$

$$= \frac{1}{n}\operatorname{cosec}\frac{m}{n}\pi\sum_{k=0}^{\frac{n-1}{2}}(-1)^{k-1}\sin\frac{km}{n}\pi\left[\psi\left(\frac{n+l-k}{n}\right)-\psi\left(\frac{l+k}{n}\right)\right]\qquad [m+n\text{ is even}]$$

$$[l\text{ is a natural number}]\qquad\text{BI (36)(5)}$$

2.  $\displaystyle\int_0^{\pi/2}\frac{\tan^{\pm\mu}x\,\mathrm{d}x}{1+\cos t\sin 2x}=\pi\operatorname{cosec}t\sin\mu t\operatorname{cosec}(\mu\pi)\qquad\left[|\operatorname{Re}\mu|<1,\quad t^2<\pi^2\right]\qquad\text{BI (47)(4)}$

**3.649**

1.  $\displaystyle\int_0^{\pi/2}\frac{\tan^{\pm\mu}x\sin 2x\,\mathrm{d}x}{1\mp 2a\cos 2x+a^2}=\frac{\pi}{4a}\operatorname{cosec}\frac{\mu\pi}{2}\left[1-\left(\frac{1-a}{1+a}\right)^\mu\right]\qquad\left[a^2<1\right]$

$$=\frac{\pi}{4a}\operatorname{cosec}\frac{\mu\pi}{2}\left[1+\left(\frac{a-1}{a+1}\right)^\mu\right]\qquad\left[a^2>1\right]$$

$$[-2<\operatorname{Re}\mu<1]\qquad\text{BI (50)(3)}$$

2.  $\displaystyle\int_0^{\pi/2}\frac{\tan^{\pm\mu}x\left(1\mp a\cos 2x\right)}{1\mp 2a\cos 2x+a^2}\,\mathrm{d}x=\frac{\pi}{4}\sec\frac{\mu\pi}{2}\left[1+\left(\frac{1-a}{1+a}\right)^\mu\right]\qquad\left[a^2<1\right]$

$$=\frac{\pi}{4}\sec\frac{\mu\pi}{2}\left[1-\left(\frac{a-1}{a+1}\right)^\mu\right]\qquad\left[a^2>1\right]$$

$$[|\operatorname{Re}\mu|<1]\qquad\text{BI (50)(4)}$$

**3.651**

1.  $\displaystyle\int_0^{\pi/4}\frac{\tan^\mu x\,\mathrm{d}x}{1+\sin x\cos x}=\frac{1}{3}\left[\psi\left(\frac{\mu+2}{3}\right)-\psi\left(\frac{\mu+1}{3}\right)\right]\qquad[\operatorname{Re}\mu>-1]\qquad\text{BI (36)(3)}$

2.  $\displaystyle\int_0^{\pi/4}\frac{\tan^\mu x\,\mathrm{d}x}{1-\sin x\cos x}=\frac{1}{3}\left[\beta\left(\frac{\mu+2}{3}\right)+\beta\left(\frac{\mu+1}{3}\right)\right]\qquad[\operatorname{Re}\mu>-1]\qquad\text{BI (36)(4)a}$

**3.652**

1.  $\displaystyle\int_0^{\pi/2}\frac{\tan^\mu x\,\mathrm{d}x}{(\sin x+\cos x)\sin x}=\int_0^{\pi/2}\frac{\cot^\mu x\,\mathrm{d}x}{(\sin x+\cos x)\cos x}=\pi\operatorname{cosec}\mu\pi$

$$[0<\operatorname{Re}\mu<1]\qquad\text{BI (49)(1)}$$

2.  $\displaystyle\int_0^{\pi/2}\frac{\tan^\mu x\,\mathrm{d}x}{(\sin x-\cos x)\sin x}=\int_0^{\pi/2}\frac{\cot^\mu x\,\mathrm{d}x}{(\cos x-\sin x)\cos x}=-\pi\cot\mu\pi$

$$[0<\operatorname{Re}\mu<1]\qquad\text{BI (49)(2)}$$

3.  $\displaystyle\int_0^{\pi/2}\frac{\cot^{\mu+\frac{1}{2}}x\,\mathrm{d}x}{(\sin x+\cos x)\cos x}=\int_0^{\pi/2}\frac{\tan^{\mu-\frac{1}{2}}x\,\mathrm{d}x}{(\sin x+\cos x)\cos x}=\pi\sec\mu\pi$

$$\left[|\operatorname{Re}\mu|<\tfrac{1}{2}\right]\qquad\text{BI (61)(1, 2)}$$

## 3.653

1.
$$\int_0^{\pi/2} \frac{\tan^{1-2\mu} x \, dx}{a^2 \cos^2 x + b^2 \sin^2 x} = \int_0^{\pi/2} \frac{\cot^{1-2\mu} x \, dx}{a^2 \sin^2 x + b^2 \cos^2 x} = \frac{\pi}{2a^{2\mu}b^{2-2\mu} \sin \mu\pi}$$
$$[0 < \operatorname{Re}\mu < 1] \qquad \text{GW (331)(59b)}$$

2.[11]
$$\int_0^{\pi/2} \frac{\tan^\mu x \, dx}{1 - a \sin^2 x} = \int_0^{\pi/2} \frac{\cot^\mu x \, dx}{1 - a \cos^2 x} = \frac{\pi \sec \frac{\mu\pi}{2}}{2\sqrt{(1-a)^{\mu+1}}}$$
$$[|\operatorname{Re}\mu| < 1, \quad a < 1] \qquad \text{BI (49)(6)}$$

3.
$$\int_0^{\pi/2} \frac{\tan^{\pm\mu} x \, dx}{1 - \cos^2 t \sin^2 2x} = \frac{\pi}{2} \operatorname{cosec} t \sec \frac{\mu\pi}{2} \cos\left[\left(\frac{\pi}{2} - t\right)\mu\right]$$
$$[|\operatorname{Re}\mu| < 1, \quad t^2 < \pi^2]$$
$$\text{BI(49)(7), BI(47)(21)}$$

4.
$$\int_0^{\pi/2} \frac{\tan^{\pm\mu} x \sin 2x}{1 - \cos^2 t \sin^2 2x} \, dx = \pi \operatorname{cosec} 2t \operatorname{cosec} \frac{\mu\pi}{2} \sin\left[\left(\frac{\pi}{2} - t\right)\mu\right]$$
$$[|\operatorname{Re}\mu| < 1, \quad t^2 < \pi^2] \qquad \text{BI (47)(22)a}$$

5.
$$\int_0^{\pi/2} \frac{\tan^\mu x \sin^2 x \, dx}{1 - \cos^2 t \sin^2 2x} = \int_0^{\pi/2} \frac{\cot^\mu x \cos^2 x \, dx}{1 - \cos^2 t \sin^2 2x} = \frac{\pi}{2} \operatorname{cosec} 2t \sec \frac{\mu\pi}{2} \cos\left[\frac{\mu\pi}{2} - (\mu+1)t\right]$$
$$[|\operatorname{Re}\mu| < 1, \quad t^2 < \pi^2] \quad \text{BI(47)(23)a, BI(49)(10)}$$

6.
$$\int_0^{\pi/2} \frac{\tan^\mu x \cos^2 x \, dx}{1 - \cos^2 t \sin^2 2x} = \int_0^{\pi/2} \frac{\cot^\mu x \sin^2 x \, dx}{1 - \cos^2 t \sin^2 2x} = \frac{\pi}{2} \operatorname{cosec} 2t \sec \frac{\mu\pi}{2} \cos\left[\frac{\mu\pi}{2} - (\mu-1)t\right]$$
$$[|\operatorname{Re}\mu| < 1, \quad t^2 < \pi^2] \quad \text{BI(47)(24)a, BI(49)(9)}$$

## 3.654

1.
$$\int_0^{\pi/2} \frac{\tan^{\mu+1} x \cos^2 x \, dx}{(1 + \cos t \sin 2x)^2} = \int_0^{\pi/2} \frac{\cot^{\mu+1} x \sin^2 x \, dx}{(1 + \cos t \sin 2x)^2} = \frac{\pi (\mu \sin t \cos \mu t - \cos t \sin \mu t)}{2 \sin \mu\pi \sin^3 t}$$
$$[|\operatorname{Re}\mu| < 1, \quad t^2 < \pi^2]$$
$$\text{BI(48)(3), BI(49)(22)}$$

2.
$$\int_0^{\pi/2} \frac{\tan^{\pm\mu} x \, dx}{(\sin x + \cos x)^2} = \frac{\mu\pi}{\sin \mu\pi}$$
$$[0 < \operatorname{Re}\mu < 1] \qquad \text{BI (56)(9)a}$$

3.
$$\int_0^{\pi/2} \frac{\tan^{\pm(\mu-1)} x \, dx}{\cos^2 x - \sin^2 x} = \pm \frac{\pi}{2} \cot \frac{\mu\pi}{2}$$
$$[0 < \operatorname{Re}\mu < 2] \qquad \text{BI (45)(27, 29)}$$

**3.655**[12]
$$\int_0^{\pi/2} \frac{\tan^{2\mu-1} x \, dx}{1 - 2a\left(\cos t_1 \sin^2 x + \cos t_2 \cos^2 x\right) + a^2} = \int_0^{\pi/2} \frac{\cot^{2\mu-1} x \, dx}{1 - 2a\left(\cos t_1 \cos^2 x + \cos t_2 \sin^2 x\right) + a^2}$$
$$= \frac{\pi \operatorname{cosec} \mu\pi}{\left(1 - 2a \cos t_2 + a^2\right)^\mu \left(1 - 2a \cos t_1 + a^2\right)^{1-\mu}}$$
$$\left[0 < \operatorname{Re}\mu < 1, \quad t_1^2 < \pi^2, \quad t_2^2 < \pi^2\right] \quad \text{BI (50)(18)}$$

**3.656**

1. $$\int_0^{\pi/4} \frac{\tan^\mu x \, dx}{1 - \sin^2 x \cos^2 x} = \frac{1}{12} \left\{ -\psi\left(\frac{\mu+1}{6}\right) - \psi\left(\frac{\mu+2}{6}\right) \right.$$
$$\left. + \psi\left(\frac{\mu+4}{6}\right) + \psi\left(\frac{\mu+5}{6}\right) + 2\psi\left(\frac{\mu+2}{3}\right) - 2\psi\left(\frac{\mu+1}{3}\right) \right\}$$
$$[\operatorname{Re}\mu > -1] \qquad (\text{cf. } \mathbf{3.651} \text{ 1 and 2}) \quad \text{LI (36)(10)}$$

2. $$\int_0^{\pi/2} \frac{\tan^{\mu-1} x \cos^2 x \, dx}{1 - \sin^2 x \cos^2 x} = \int_0^{\pi/2} \frac{\cot^{\mu-1} x \sin^2 x \, dx}{1 - \sin^2 x \cos^2 x} = \frac{\pi}{4\sqrt{3}} \operatorname{cosec} \frac{\mu\pi}{6} \operatorname{cosec}\left(\frac{2+\mu}{6}\pi\right)$$
$$[0 < \operatorname{Re}\mu < 4] \qquad\qquad\qquad \text{LI (47)(26)}$$

## 3.66 Forms containing powers of linear functions of trigonometric functions

**3.661**

1. $$\int_0^{2\pi} (a\sin x + b\cos x)^{2n+1} \, dx = 0 \qquad\qquad\qquad\qquad\qquad\qquad\qquad \text{BI (68)(9)}$$

2. $$\int_0^{2\pi} (a\sin x + b\cos x)^{2n} \, dx = \frac{(2n-1)!!}{(2n)!!} 2\pi (a^2+b^2)^n \qquad\qquad\qquad \text{BI (68)(8)}$$

3. $$\int_0^{\pi} (a + b\cos x)^n \, dx = \frac{1}{2}\int_0^{2\pi} (a + b\cos x)^n \, dx = \pi(a^2-b^2)^{\frac{n}{2}} P_n\left(\frac{a}{\sqrt{a^2-b^2}}\right)$$
$$= \frac{\pi}{2^n} \sum_{k=0}^{\lfloor n/2 \rfloor} \frac{(-1)^k (2n-2k)!}{k!(n-k)!(n-2k)!} a^{n-2k} (a^2-b^2)^k$$
$$[a^2 > b^2] \qquad\qquad \text{GW (332)(37a)}$$

4. $$\int_0^{\pi} \frac{dx}{(a+b\cos x)^{n+1}} = \frac{1}{2}\int_0^{2\pi} \frac{dx}{(a+b\cos x)^{n+1}} = \frac{\pi}{(a^2-b^2)^{\frac{n+1}{2}}} P_n\left(\frac{a}{\sqrt{a^2-b^2}}\right)$$
$$= \frac{\pi}{2^n(a+b)^n\sqrt{a^2-b^2}} \sum_{k=0}^{n} \frac{(2n-2k-1)!!(2k-1)!!}{(n-k)!k!} \left(\frac{a+b}{a-b}\right)^k$$
$$[a > |b|] \qquad\qquad \text{GW(332)(38), LI(64)(14)}$$

**3.662**

1. $$\int_0^{\pi/2} (\sec x - 1)^\mu \sin x \, dx = \int_0^{\pi/2} (\operatorname{cosec} x - 1)^\mu \cos x \, dx = \mu\pi \operatorname{cosec} \mu\pi$$
$$[|\operatorname{Re}\mu| < 1] \qquad\qquad\qquad \text{BI (55)(13)}$$

2. $$\int_0^{\pi/2} (\operatorname{cosec} x - 1)^\mu \sin 2x \, dx = (1-\mu)\mu\pi \operatorname{cosec} \mu\pi \qquad [-1 < \operatorname{Re}\mu < 2] \qquad \text{BI (48)(7)}$$

3. $$\int_0^{\pi/2} (\sec x - 1)^\mu \tan x \, dx = \int_0^{\pi/2} (\operatorname{cosec} x - 1)^\mu \cot x \, dx = -\pi \operatorname{cosec} \mu\pi$$
$$[-1 < \operatorname{Re}\mu < 0] \qquad\qquad\qquad \text{BI (46)(4,6)}$$

4. $$\int_0^{\pi/4} (\cot x - 1)^\mu \frac{dx}{\sin 2x} = -\frac{\pi}{2} \operatorname{cosec} \mu\pi \qquad\qquad [-1 < \operatorname{Re}\mu < 0] \qquad \text{BI (38)(22)a}$$

5. $\displaystyle\int_0^{\pi/4} (\cot x - 1)^\mu \frac{dx}{\cos^2 x} = \mu\pi \csc \mu\pi$  $\qquad [|\mathrm{Re}\,\mu| < 1]$  $\qquad$ BI (38)(11)a

**3.663**

1. $\displaystyle\int_0^u (\cos x - \cos u)^{\nu - \frac{1}{2}} \cos ax \, dx = \sqrt{\frac{\pi}{2}} \sin^\nu u \, \Gamma\left(\nu + \frac{1}{2}\right) P_{a - \frac{1}{2}}^{-\nu}(\cos u)$

$$\left[\mathrm{Re}\,\nu > -\tfrac{1}{2}, \quad a > 0, \quad 0 < u < \pi\right]$$
$$\text{EH I 159(27), ET I 22(28)}$$

2. $\displaystyle\int_0^u (\cos x - \cos u)^{\nu - 1} \cos[(\nu + \beta)x] \, dx = \frac{\sqrt{\pi}\,\Gamma(\beta + 1)\,\Gamma(\nu)\,\Gamma(2\nu)\sin^{2\nu - 1} u}{2^\nu\,\Gamma(\beta + 2\nu)\,\Gamma\left(\nu + \frac{1}{2}\right)} C_\beta^\nu(\cos u)$

$$\left[\mathrm{Re}\,\nu > 0, \quad \mathrm{Re}\,\beta > -1, \quad 0 < u < \pi\right]$$
$$\text{EH I 178(23)}$$

**3.664**

1. $\displaystyle\int_0^\pi \left(z + \sqrt{z^2 - 1}\cos x\right)^q dx = \pi\, P_q(z)$

$$\left[\mathrm{Re}\,z > 0, \quad \arg\left(z + \sqrt{z^2 - 1}\cos x\right) = \arg z \text{ for } x = \frac{\pi}{2}\right] \quad \text{SM 482}$$

2. $\displaystyle\int_0^\pi \frac{dx}{\left(z + \sqrt{z^2 - 1}\cos x\right)^q} = \pi\, P_{q-1}(z)$

$$\left[\mathrm{Re}\,z > 0, \quad \arg\left(z + \sqrt{z^2 - 1}\cos x\right) = \arg z \text{ for } x = \frac{\pi}{2}\right] \quad \text{WH}$$

3. $\displaystyle\int_0^\pi \left(z + \sqrt{z^2 - 1}\cos x\right)^q \cos nx \, dx = \frac{\pi}{(q+1)(q+2)\cdots(q+n)} P_q^n(z)$

$$\left[\mathrm{Re}\,z > 0, \quad \arg\left(z + \sqrt{z^2 - 1}\cos x\right) = \arg z \text{ for } x = \frac{\pi}{2},\right.$$

$$\left. z \text{ lies outside the interval } (-1, 1) \text{ of the real axis}\right]$$

$$\text{WH, SM 483(15)}$$

4. $\displaystyle\int_0^\pi \left(z + \sqrt{z^2 - 1}\cos x\right)^\mu \sin^{2\nu - 1} x \, dx$

$$= \frac{2^{2\nu - 1}\,\Gamma(\mu + 1)\,[\Gamma(\nu)]^2}{\Gamma(2\nu + \mu)} C_\mu^\nu(z)$$

$$= \frac{\sqrt{\pi}\,\Gamma(\nu)\,\Gamma(2\nu)\,\Gamma(\mu + 1)}{\Gamma(2\nu + \mu)\,\Gamma\left(\nu + \frac{1}{2}\right)} C_\mu^\nu(z) = 2^\nu \sqrt{\frac{\pi}{2}}\,(z^2 - 1)^{\frac{1}{4} - \frac{\nu}{2}}\,\Gamma(\nu)\,P_{\mu + \nu - \frac{1}{2}}^{\frac{1}{2} - \nu}(z)$$

$$[\mathrm{Re}\,\nu > 0] \qquad \text{EH I 155(6)a, EH I 178(22)}$$

5. $\displaystyle\int_0^{2\pi} \left[\beta + \sqrt{\beta^2 - 1}\cos(a - x)\right]^\nu \left(\gamma + \sqrt{\gamma^2 - 1}\cos x\right)^{\nu - 1} dx$

$$= 2\pi\, P_\nu\left(\beta\gamma - \sqrt{\beta^2 - 1}\sqrt{\gamma^2 - 1}\cos a\right)$$
$$[\mathrm{Re}\,\beta > 0, \quad \mathrm{Re}\,\gamma > 0] \qquad \text{EH I 157(18)}$$

**3.665**

1.  $$\int_0^\pi \frac{\sin^{\mu-1} x \, dx}{(a + b \cos x)^\mu} = \frac{2^{\mu-1}}{\sqrt{(a^2 - b^2)^\mu}} B\left(\frac{\mu}{2}, \frac{\mu}{2}\right) \qquad [\operatorname{Re}\mu > 0, \quad 0 < b < a] \qquad \text{FI II 790a}$$

2.  $$\int_0^\pi \frac{\sin^{2\mu-1} x \, dx}{(1 + 2a \cos x + a^2)^\nu} = B\left(\mu, \tfrac{1}{2}\right) F\left(\nu, \nu - \mu + \tfrac{1}{2}; \mu + \tfrac{1}{2}; a^2\right)$$
    $$[\operatorname{Re}\mu > 0, \quad |a| < 1] \qquad \text{EH I 81(9)}$$

**3.666**

1.  $$\int_0^\pi (\beta + \cos x)^{\mu-\nu-\frac{1}{2}} \sin^{2\nu} x \, dx = \frac{2^{\nu+\frac{1}{2}} e^{-i\mu\pi} \left(\beta^2 - 1\right)^{\frac{\mu}{2}} \Gamma\left(\nu + \frac{1}{2}\right) Q_{\nu-\frac{1}{2}}^\mu(\beta)}{\Gamma\left(\nu + \mu + \frac{1}{2}\right)}$$
    $$\left[\operatorname{Re}\left(\nu + \mu + \tfrac{1}{2}\right) > 0, \quad \operatorname{Re}\nu > -\tfrac{1}{2}\right]$$
    EH I 155(5)a

2.[6] $$\int_0^\pi (\cosh\beta + \sinh\beta\cos x)^{\mu+\nu} \sin^{-2\nu} x \, dx = \frac{\sqrt{\pi}}{2^\nu} \sinh^\nu(\beta) \Gamma\left(\tfrac{1}{2} - \nu\right) P_\mu^\nu(\cosh\beta)$$
    $$\left[\operatorname{Re}\nu < \tfrac{1}{2}\right] \qquad \text{EH I 156(7)}$$

3.  $$\int_0^\pi (\cos t + i \sin t \cos x)^\mu \sin^{2\nu-1} x \, dx = 2^{\nu-\frac{1}{2}} \sqrt{\pi} \sin^{\frac{1}{2}-\nu} t \, \Gamma(\nu) P_{\mu+\nu-\frac{1}{2}}^{\frac{1}{2}-\nu}(\cos t)$$
    $$[\operatorname{Re}\nu > 0, \quad t^2 < \pi^2] \qquad \text{EH I 158(23)}$$

4.  $$\int_0^{2\pi} [\cos t + i \sin t \cos(a - x)]^\nu \cos mx \, dx = \frac{i^{3m} 2\pi \, \Gamma(\nu + 1)}{\Gamma(\nu + m + 1)} \cos ma \, P_\nu^m(\cos t)$$
    $$\left[0 < t < \tfrac{\pi}{2}\right] \qquad \text{EH I 159(25)}$$

5.[10] $$\int_0^{2\pi} [\cos t + i \sin t \cos(a - x)]^\nu \sin mx \, dx = \frac{i^{3m} 2\pi \, \Gamma(\nu + 1)}{\Gamma(\nu + m + 1)} \sin ma \, P_\nu^m(\cos t)$$
    $$\left[0 < t < \tfrac{\pi}{2}\right] \qquad \text{EH I 159(26)}$$

**3.667**

1.  $$\int_0^{\pi/4} \frac{\sin^{\mu-1} 2x \, dx}{(\cos x + \sin x)^{2\mu}} = \frac{\sqrt{\pi}}{2^{\mu+1}} \frac{\Gamma(\mu)}{\Gamma\left(\mu + \frac{1}{2}\right)} \qquad [\operatorname{Re}\mu > 0] \qquad \text{BI (37)(1)}$$

2.  $$\int_0^{\pi/4} \frac{\sin^\mu x \, dx}{(\cos x - \sin x)^{\mu+1} \cos x} = -\pi \operatorname{cosec}\mu\pi \qquad [-1 < \operatorname{Re}\mu < 0] \qquad (\text{cf. } \mathbf{3.192}\ 2)$$
    BI (37)(16)

3.  $$\int_0^{\pi/4} \frac{(\cos x - \sin x)^\mu}{\sin^\mu x \sin 2x} \, dx = -\frac{\pi}{2} \operatorname{cosec}\mu\pi \qquad [-1 < \operatorname{Re}\mu < 0] \qquad \text{BI (35)(27)}$$

4.  $$\int_0^{\pi/4} \frac{\sin^\mu x \, dx}{(\cos x - \sin x)^\mu \sin 2x} = \frac{\pi}{2} \operatorname{cosec}\mu\pi \qquad [0 < \operatorname{Re}\mu < 1] \qquad \text{LI (37)(20)a}$$

5.  $$\int_0^{\pi/4} \frac{\sin^\mu x \, dx}{(\cos x - \sin x)^\mu \cos^2 x} = \mu\pi \operatorname{cosec}\mu\pi \qquad [|\operatorname{Re}\mu| < 1] \qquad \text{BI (37)(17)}$$

6.  $$\int_0^{\pi/4} \frac{\sin^\mu x \, dx}{(\cos x - \sin x)^{\mu-1} \cos^3 x} = \frac{1 - \mu}{2} \mu\pi \operatorname{cosec}\mu\pi \qquad [|\operatorname{Re}\mu| < 1] \qquad \text{BI(35)(24), BI(37)(18)}$$

7.    $\int_0^{\pi/2} \dfrac{\sin^{\mu-1} x \cos^{\nu-1} x}{(\sin x + \cos x)^{\mu+\nu}} \, dx = B(\mu, \nu)$        $[\operatorname{Re}\mu > 0, \quad \operatorname{Re}\nu > 0]$        BI (48)(8)

**3.668**

1.    $\int_{-\frac{\pi}{4}}^{\frac{\pi}{4}} \left( \dfrac{\cos x + \sin x}{\cos x - \sin x} \right)^{\cos 2t} dx = \dfrac{\pi}{2 \sin(\pi \cos^2 t)}$        FI II 788

2.    $\int_u^v \dfrac{(\cos u - \cos x)^{\mu-1}}{(\cos x - \cos v)^\mu} \dfrac{\sin x \, dx}{1 - 2a \cos x + a^2} = \dfrac{(1 - 2a \cos u + a^2)^{\mu-1}}{(1 - 2a \cos v + a^2)^\mu} \dfrac{\pi}{\sin \mu\pi}$

$\left[ 0 < \operatorname{Re}\mu < 1, \quad a^2 < 1 \right]$        BI (73)(2)

**3.669**    $\int_0^{\pi/2} \dfrac{\sin^{p-1} x \cos^{q-p-1} x \, dx}{(a \cos x + b \sin x)^q} = \int_0^{\pi/2} \dfrac{\sin^{q-p-1} x \cos^{p-1} x}{(a \sin x + b \cos x)^q} \, dx = \dfrac{B(p, q-p)}{a^{q-p} b^p}$

$[q > p > 0, \quad ab > 0]$        BI (331)(9)

## 3.67 Square roots of expressions containing trigonometric functions

**3.671**

1.    $\int_0^{\pi/2} \sin^\alpha x \cos^\beta x \sqrt{1 - k^2 \sin^2 x} \, dx = \dfrac{1}{2} B\left( \dfrac{\alpha+1}{2}, \dfrac{\beta+1}{2} \right) F\left( \dfrac{\alpha+1}{2}, -\dfrac{1}{2}; \dfrac{\alpha+\beta+2}{2}; k^2 \right)$

$[\alpha > -1, \quad \beta > -1, \quad |k| < 1]$
GW (331)(93)

2.    $\int_0^{\pi/2} \dfrac{\sin^\alpha x \cos^\beta x}{\sqrt{1 - k^2 \sin^2 x}} \, dx = \dfrac{1}{2} B\left( \dfrac{\alpha+1}{2}, \dfrac{\beta+1}{2} \right) F\left( \dfrac{\alpha+1}{2}, \dfrac{1}{2}; \dfrac{\alpha+\beta+2}{2}; k^2 \right)$

$[\alpha > -1, \quad \beta > -1, \quad |k| < 1]$
GW (331)(92)

3.    $\int_0^\pi \dfrac{\sin^{2n} x \, dx}{\sqrt{1 - k^2 \sin^2 x}} = \dfrac{\pi}{2^n} \sum_{j=0}^\infty \dfrac{(2j-1)!! \, (2n+2j-1)!!}{2^{2j} j! (n+j)!} k^{2j}$        $[k^2 < 1]$

$= \dfrac{(2n-1)!! \pi}{2^n \sqrt{1-k^2}} \sum_{j=0}^\infty \dfrac{[(2j-1)!!]^2}{2^{2j} j! (n+j)!} \left( \dfrac{k^2}{k^2 - 1} \right)^j$        $[k^2 < \tfrac{1}{2}]$

LI (67)(2)

4.[12]    $\int_0^\pi \sqrt{a \pm b \cos x} \, dx = \int_{-\pi/2}^{\pi/2} \sqrt{a \pm b \sin x} \, dx = 2\sqrt{a+b} \, \boldsymbol{E}\left( \sqrt{\dfrac{2b}{a+b}} \right)$

$[a > b > 0]$

5.[12]    $\int_0^\pi \dfrac{dx}{\sqrt{a \pm b \cos x}} = \int_{-\pi/2}^{\pi/2} \dfrac{dx}{\sqrt{a \pm b \sin x}} = \dfrac{2}{\sqrt{a+b}} \, \boldsymbol{K}\left( \sqrt{\dfrac{2b}{a+b}} \right)$

$[a > b > 0]$

6.*    $\int_0^\pi \dfrac{dx}{\sqrt{1 + 2p \cos x + p^2}} = \dfrac{2}{1+p} \, \boldsymbol{K}\left( \dfrac{2\sqrt{p}}{1+p} \right)$

**3.672**

1. $$\int_0^{\pi/4} \frac{\sin^n x}{\cos^{n+1} x} \frac{dx}{\sqrt{\cos x \,(\cos x - \sin x)}} = 2\frac{(2n)!!}{(2n+1)!!}$$      BI (39)(5)

2. $$\int_0^{\pi/4} \frac{\sin^n x}{\cos^{n+1} x} \frac{dx}{\sqrt{\sin x \,(\cos x - \sin x)}} = \frac{(2n-1)!!}{(2n)!!}\pi$$      BI (39)(6)

**3.673** $$\int_u^{\frac{\pi}{2}} \frac{dx}{\sqrt{\sin x - \sin u}} = \sqrt{2}\,\boldsymbol{K}\left(\sin\frac{\pi - 2u}{4}\right)$$      BI (74)(11)

**3.674**

1.[12] $$\int_0^{2\pi} \frac{dx}{\sqrt{1 \pm 2p\cos x + p^2}} = \frac{2}{1+p}\,\boldsymbol{K}\left(\frac{2\sqrt{p}}{1+p}\right)$$    $[p^2 < 1]$      BI (67)(5)

2. $$\int_0^{\pi} \frac{\sin x \, dx}{\sqrt{1 - 2p\cos x + p^2}} = 2$$      $[p^2 \le 1]$

$$= \frac{2}{p}$$      $[p^2 \ge 1]$

     BI (67)(6)

3.[12] $$\int_0^{\pi} \frac{\cos x \, dx}{\sqrt{1 - 2p\cos x + p^2}} = \frac{2}{p}\left[\boldsymbol{K}(p) - \boldsymbol{E}(p)\right] = \frac{1}{p}\left[\frac{1+p^2}{1+p}\,\boldsymbol{K}\left(\frac{2\sqrt{p}}{1+p}\right) - (1+p)\boldsymbol{E}\left(\frac{2\sqrt{p}}{1+p}\right)\right]$$

     $[p^2 < 1]$      BI (67)(7)

**3.675**

1. $$\int_u^{\pi} \frac{\sin\left(n + \frac{1}{2}\right)x \, dx}{\sqrt{2\,(\cos u - \cos x)}} = \frac{\pi}{2}\,P_n(\cos u)$$      WH

2. $$\int_0^{u} \frac{\cos\left(n + \frac{1}{2}\right)x \, dx}{\sqrt{2\,(\cos x - \cos u)}} = \frac{\pi}{2}\,P_n(\cos u)$$      FI II 684, WH

**3.676**

1. $$\int_0^{\pi/2} \frac{\sin x \, dx}{\sqrt{1 + p^2 \sin^2 x}} = \frac{1}{p}\arctan p$$      BI (60)(5)

2. $$\int_0^{\pi/2} \tan^2 x \sqrt{1 - p^2 \sin^2 x}\, dx = \infty$$      BI (53)(8)

3. $$\int_0^{\pi/2} \frac{dx}{\sqrt{p^2 \cos^2 x + q^2 \sin^2 x}} = \frac{1}{p}\boldsymbol{K}\left(\frac{\sqrt{p^2 - q^2}}{p}\right)$$    $[0 < q < p]$      FI II 165

**3.677**

1. $$\int_0^{\pi/2} \frac{\sin^2 x \, dx}{\sqrt{1 + \sin^2 x}} = \sqrt{2}\boldsymbol{E}\left(\frac{\sqrt{2}}{2}\right) - \frac{1}{\sqrt{2}}\boldsymbol{K}\left(\frac{\sqrt{2}}{2}\right)$$      BI (60)(2)

2. $$\int_0^{\pi/2} \frac{\cos^2 x \, dx}{\sqrt{1 + \sin^2 x}} = \sqrt{2}\left[\boldsymbol{K}\left(\frac{\sqrt{2}}{2}\right) - \boldsymbol{E}\left(\frac{\sqrt{2}}{2}\right)\right]$$      BI (60)(3)

**3.678**

1.  $\displaystyle\int_0^{\pi/4} \left(\sec^{1/2} 2x - 1\right) \frac{\mathrm{d}x}{\tan x} = \ln 2$                    BI (38)(23)

2.  $\displaystyle\int_0^{\pi/4} \frac{\tan^2 x \,\mathrm{d}x}{\sqrt{1 - k^2 \sin^2 2x}} = \sqrt{1 - k^2} - \boldsymbol{E}(k) + \frac{1}{2}\,\boldsymbol{K}(k)$                    BI (39)(2)

3.  $\displaystyle\int_0^u \sqrt{\frac{\cos 2x - \cos 2u}{\cos 2x + 1}}\,\mathrm{d}x = \frac{\pi}{2}\,(1 - \cos u)$ $\qquad\left[u^2 < \dfrac{\pi^2}{4}\right]$                    LI (74)(6)

4.  $\displaystyle\int_0^{\pi/4} \frac{(\cos x - \sin x)^{n-\frac{1}{2}}}{\cos^{n+1} x}\,\sqrt{\operatorname{cosec} x}\,\mathrm{d}x = \frac{(2n-1)!!}{(2n)!!}\,\pi$                    BI (38)(24)

5.  $\displaystyle\int_0^{\pi/4} \frac{(\cos x - \sin x)^{n-\frac{1}{2}}}{\cos^{n+1} x}\,\tan^m x \sqrt{\operatorname{cosec} x}\,\mathrm{d}x = \frac{(2n-1)!!(2m-1)!!}{(2n+2m)!!}\,\pi$                    BI (38)(25)

**3.679**  Set $k' = \sqrt{1 - k^2}$

1.[12]  $\displaystyle\int_0^{\pi/2} \frac{\cos^2 x}{1 - \cos^2 \beta \cos^2 x} \frac{\mathrm{d}x}{\sqrt{1 - k^2 \sin^2 x}}$

$\displaystyle = \frac{1}{\sin^2 \beta \cos^2 \beta}\left[\Pi\left(\frac{\pi}{2}, -\cot^2 \beta, k\right) - \sin^2 \beta\,\boldsymbol{K}(k)\right]$

$\displaystyle = \frac{1}{\sin \beta \cos \beta \sqrt{1 - k'^2 \sin^2 \beta}}\left[\frac{\pi}{2} - \boldsymbol{K} E\left(\beta, k'\right) - \boldsymbol{E} F\left(\beta, k'\right) + \boldsymbol{K} F\left(\beta, k'\right)\right] k' = \sqrt{1 - k^2}$

$[0 < \beta < \pi]$                    MO 138

2.[12]  $\displaystyle\int_0^{\pi/2} \frac{\sin^2 x}{1 - \left(1 - k'^2 \sin^2 \beta\right) \sin^2 x} \frac{\mathrm{d}x}{\sqrt{1 - k^2 \sin^2 x}}$

$\displaystyle = \frac{1}{1 - k'^2 \sin^2 \beta}\left[\Pi\left(\frac{\pi}{2}, 1 - k'^2 \sin^2 \beta, k\right) - \boldsymbol{K}(k)\right]$

$\displaystyle = \frac{1}{k'^2 \sin \beta \cos \beta \sqrt{1 - k'^2 \sin^2 \beta}}\left[\frac{\pi}{2} - \boldsymbol{K} E\left(\beta, k'\right) - \boldsymbol{E} F\left(\beta, k'\right) + \boldsymbol{K} F\left(\beta, k'\right)\right] k' = \sqrt{1 - k^2}$

$[0 \le \beta \le \pi]$                    MO 138

3.[12]  $\displaystyle\int_0^{\pi/2} \frac{\sin^2 x}{1 - k^2 \sin^2 \beta \sin^2 x} \frac{\mathrm{d}x}{\sqrt{1 - k^2 \sin^2 x}} = \frac{1}{k^2 \sin^2 \beta}\left[\Pi\left(\frac{\pi}{2}, k^2 \sin^2 \beta, k\right) - \boldsymbol{K}(k)\right]$

$\displaystyle = \frac{\boldsymbol{K} E(\beta, k) - \boldsymbol{E} F(\beta, k)}{k^2 \sin \beta \cos \beta \sqrt{1 - k^2 \sin^2 \beta}}$

MO 138

**3.679(1)***

Define $\Delta = \sqrt{(a + b\sin x)^2 + (d\cos x)^2 + h^2}$, $s = a^2 - b^2 + h^2 + 2d^2$, $r = \sqrt{(a^2 - b^2 + h^2)^2 + 4b^2h^2}$,

$$I_1 = \boldsymbol{K}\left(\sqrt{\frac{s-r}{s+r}}\right) \qquad\qquad [s \geq r > 0]$$

$$= \sqrt{\frac{s+r}{2r}}\boldsymbol{K}\left(\sqrt{\frac{r-s}{2r}}\right) \qquad\qquad [-r < s < r]$$

$$I_2 = \boldsymbol{E}\left(\sqrt{\frac{s-r}{s+r}}\right) \qquad\qquad [s \geq r > 0]$$

$$= \sqrt{\frac{2r}{s+r}}\boldsymbol{E}\left(\sqrt{\frac{r-s}{2r}}\right) \qquad\qquad [-r < s < r]$$

1. $\displaystyle\int_0^{2\pi} \frac{\mathrm{d}x}{\Delta} = \frac{4\sqrt{2}}{\sqrt{s+r}}I_1$

2. $\displaystyle\int_0^{2\pi} \frac{\mathrm{d}x}{\Delta^3} = \frac{8\sqrt{2}}{(s-r)\sqrt{s+r}}$

$$\times \left[\frac{a^2(b^2 + d^2) - (b^2 - d^2)(b^2 + h^2)}{r}\left(\frac{I_2}{r} - \frac{I_1}{s+r}\right) - (b^2 - d^2)\frac{I_1}{s+r}\right]$$

3. $\displaystyle\int_0^{2\pi} \frac{\sin x}{\Delta^3}dx = -\frac{8\sqrt{2}ab}{(s-r)\sqrt{s+r}}\left(\frac{s}{r}\right)\left(\frac{I_2}{r} - \frac{I_1}{s}\right)$

4. $\displaystyle\int_0^{2\pi} \frac{\cos x}{\Delta^3}dx = 0$

5. $\displaystyle\int_0^{2\pi} \frac{\sin^2 x}{\Delta^3}dx = -\frac{8\sqrt{2}}{(s-r)\sqrt{s+r}}\left[-(a^2 + d^2 + h^2)\frac{I_1}{s+r}\right.$
$$\left. + \frac{(a^2 + d^2 + h^2)(a^2 + h^2) - (a^2 - d^2 - h^2)b^2}{r}\left(\frac{I_2}{r} - \frac{I-1}{s+r}\right)\right]$$

6. $\displaystyle\int_0^{2\pi} \frac{\cos^2 x}{\Delta^3}dx = -\frac{8\sqrt{2}}{(s-r)\sqrt{s+r}}(I_2 - I_1)$

7. $\displaystyle\int_0^{2\pi} \frac{\cos x \sin x}{\Delta^3}dx = 0$

**3.679(2)***

Define $\Delta = \sqrt{(\sinh\alpha\sin\gamma - \sinh\beta\sin x)^2 + (\cosh\alpha\cos\gamma - \cosh\beta\cos x)^2}$,

$p = \sqrt{\sinh^2\alpha + sin^2\gamma} = \sqrt{\cosh^2\alpha - \cos^2\gamma}$, $q = \sqrt{\sinh^2\beta + \sin^2\gamma} = \sqrt{\cosh^2\beta - \cos^2\gamma}$,

$$I_1 = \frac{1}{p-q}\boldsymbol{E}\left(\frac{2\sqrt{pq}}{p+q}\right) + \frac{1}{p+q}\boldsymbol{K}\left(\frac{2\sqrt{pq}}{p+q}\right) \qquad [|\alpha| \neq |\beta|]$$

$$= \frac{2q}{p^2-q^2}\boldsymbol{E}\left(\frac{p}{q}\right) + \frac{2}{q}\boldsymbol{K}\left(\frac{p}{q}\right) \qquad [|\alpha| < |\beta|]$$

$$= \frac{2p}{p^2-q^2}\boldsymbol{E}\left(\frac{q}{p}\right) \qquad [|\alpha| > |\beta|]$$

$$I_2 = \frac{1}{p-q}\boldsymbol{E}\left(\frac{2\sqrt{pq}}{p+q}\right) - \frac{1}{p+q}\boldsymbol{K}\left(\frac{2\sqrt{pq}}{p+q}\right) \qquad [|\alpha| \neq |\beta|]$$

$$= \frac{2q}{p^2-q^2}\boldsymbol{E}\left(\frac{p}{q}\right) \qquad [|\alpha| < |\beta|]$$

$$= \frac{2p}{p^2-q^2}\boldsymbol{E}\left(\frac{p}{q}\right) - \frac{2}{p}\boldsymbol{K}\left(\frac{q}{p}\right) \qquad [|\alpha| > |\beta|]$$

1.     $\displaystyle\int_0^{2\pi}\frac{dx}{\Delta} = 2(I_1 - I_2)$

2.     $\displaystyle\int_0^{2\pi}\frac{dx}{\Delta^3} = \frac{2}{p^2-q^2}\left(\sinh^2\alpha\cosh^2\alpha\frac{I_1}{p^4} + \sinh^2\beta\cosh^2\beta\frac{I_2}{q^4}\right) + \frac{2\sin^2\gamma\cos^2\gamma}{p^2q^2}\left(\frac{I_1}{p^2} - \frac{I_2}{q^2}\right)$

3.     $\displaystyle\int_0^{2\pi}\frac{\sin x}{\Delta^3}dx = \sinh\alpha\sinh\beta\sin\gamma\left[\frac{2}{p^2-q^2}\left(\cosh^2\alpha\frac{I_1}{p^4} + \cosh^2\beta\frac{I_2}{q^4}\right) - \frac{2\cos^2\gamma}{p^2q^2}\left(\frac{I_1}{p^2} - \frac{I_2}{q^2}\right)\right]$

4.     $\displaystyle\int_0^{2\pi}\frac{\cos x}{\Delta^3}dx = \cosh\alpha\cosh\beta\cos\gamma\left[\frac{2}{p^2-q^2}\left(\sinh^2\alpha\frac{I_1}{p^4} + \sinh^2\beta\frac{I_2}{q^4}\right) + \frac{2\sin^2\gamma}{p^2q^2}\left(\frac{I_1}{p^2} - \frac{I_2}{q^2}\right)\right]$

5.     $\displaystyle\int_0^{2\pi}\frac{\sin^2 x}{\Delta^3}dx = \frac{2\sin^2\gamma}{p^2-q^2}\left[\cosh^2\alpha\sinh^2\beta\frac{I_1}{p^4} + \sinh^2\alpha\cosh^2\beta\frac{I_2}{q^4}\right]$

          $+ \dfrac{2\sinh^2\alpha\sinh^2\beta\cos^2\gamma}{p^2q^2}\left(\dfrac{I_1}{p^2} - \dfrac{I_2}{q^2}\right)$

6.     $\displaystyle\int_0^{2\pi}\frac{\cos^2 x}{\Delta^3}dx = \frac{2\cos^2\gamma}{p^2-q^2}\left[\sinh^2\alpha\cosh^2\beta\frac{I_1}{p^4} + \cosh^2\alpha\sinh^2\beta\frac{I_2}{q^4}\right]$

          $+ \dfrac{2\cosh^2\alpha\cosh^2\beta\sin^2\gamma}{p^2q^2}\left(\dfrac{I_1}{p^2} - \dfrac{I_2}{q^2}\right)$

7.     $\displaystyle\int_0^{2\pi}\frac{\cos x\sin x}{\Delta^3}dx = \sinh\alpha\cosh\alpha\sinh\beta\cosh\beta\sin\gamma\cos\gamma$

          $\times\left[\dfrac{2}{p^2-q^2}\left(\dfrac{I_1}{p^4} + \dfrac{I_2}{q^4}\right) - \dfrac{2}{p^2q^2}\left(\dfrac{I_1}{p^2} - \dfrac{I_2}{q^2}\right)\right]$

**3.679(3)***

1.     $\displaystyle\int_0^{\pi/2}\frac{dx}{\sqrt{a\cos^2 x + b\sin^2 x}} = R_F(0, a, b)$          $[\operatorname{Re} a > 0, \quad \operatorname{Re} b > 0]$

DLMF    (19.23.1)

2. $\displaystyle\int_0^{\pi/2} \sqrt{a\cos^2 x + b\sin^2 x}\ dx = 2R_G(0, a, b)$      $[\operatorname{Re} a > 0, \quad \operatorname{Re} b > 0]$      DLMF  (19.23.2)

3. $\displaystyle\int_0^{\pi/2} \frac{\sin^2 x}{(a\cos^2 x + b\sin^2 x)^{3/2}}\ dx = \frac{1}{3}R_D(0, a, b)$      $[\operatorname{Re} a > 0, \quad \operatorname{Re} b > 0]$      DLMF  (19.23.3)

4. $\displaystyle\int_0^\infty \frac{x}{\sqrt{x+a}\sqrt{x+b}\sqrt{x+c}}\left(\frac{a}{x+a} + \frac{b}{x+b} + \frac{c}{x+c}\right) dx = 4R_G(a, b, c)$

$$[a, b, c \in C\backslash(-\infty, 0]]$$      DLMF  (19.23.7)

# 3.68 Various forms of powers of trigonometric functions

**3.681**

1. $\displaystyle\int_0^{\pi/2} \frac{\sin^{2\mu-1} x \cos^{2\nu-1} x\ dx}{\left(1 - k^2\sin^2 x\right)^\varrho} = \frac{1}{2}\,\mathrm{B}(\mu, \nu)\, F\left(\varrho, \mu; \mu+\nu; k^2\right)$

$$[\operatorname{Re}\mu > 0, \quad \operatorname{Re}\nu > 0]$$      EH I 115(7)

2. $\displaystyle\int_0^{\pi/2} \frac{\sin^{2\mu-1} x \cos^{2\nu-1} x\ dx}{\left(1 - k^2\sin^2 x\right)^{\mu+\nu}} = \frac{\mathrm{B}(\mu, \nu)}{2\left(1 - k^2\right)^\mu}$      $[\operatorname{Re}\mu > 0, \quad \operatorname{Re}\nu > 0]$      EH I 10(20)

3.$^{12}$ $\displaystyle\int_0^{\pi/2} \frac{\sin^\mu x\ dx}{\cos^{\mu-3} x \left(1 - k^2\sin^2 x\right)^{\frac{\mu}{2}-1}}$

$$= \frac{\Gamma\left(\frac{\mu+1}{2}\right)\Gamma\left(2 - \frac{\mu}{2}\right)}{k^3(\mu-1)(\mu-3)(\mu-5)\sqrt{\pi}}\left[\frac{1 + (\mu-3)k + k^2}{(1+k)^{\mu-3}} - \frac{1 - (\mu-3)k + k^2}{(1-k)^{\mu-3}}\right]$$

$$[-1 < \operatorname{Re}\mu < 4]$$      BI (54)(10)

4.$^8$ $\displaystyle\int_0^{\pi/2} \frac{\sin^{\mu+1} x\ dx}{\cos^\mu x \left(1 - k^2\sin^2 x\right)^{\frac{\mu+1}{2}}} = \frac{(1-k)^{-\mu} - (1+k)^{-\mu}}{2k\mu\sqrt{\pi}}\Gamma\left(1 + \frac{\mu}{2}\right)\Gamma\left(\frac{1-\mu}{2}\right)$

$$[-2 < \operatorname{Re}\mu < 1]$$      BI (61)(5)

**3.682**    $\displaystyle\int_0^{\pi/2} \frac{\sin^\mu x \cos^\nu x}{(a - b\cos^2 x)^\varrho}\ dx = \frac{1}{2a^\varrho}\,\mathrm{B}\left(\frac{\mu+1}{2}, \frac{\nu+1}{2}\right) F\left(\frac{\nu+1}{2}, \varrho; \frac{\mu+\nu}{2} + 1; \frac{b}{a}\right)$

$$[\operatorname{Re}\mu > -1, \quad \operatorname{Re}\nu > -1, \quad a > |b| \geq 0]$$
$$\text{GW (331)(64)}$$

**3.683**

1. $\displaystyle\int_0^{\pi/4} (\sin^n 2x - 1)\tan\left(\frac{\pi}{4} + x\right) dx = \int_0^{\pi/4} (\cos^n 2x - 1)\cot x\ dx = -\frac{1}{2}\sum_{k=1}^n \frac{1}{k}$

$$= -\frac{1}{2}\left[\boldsymbol{C} + \psi(n+1)\right]$$
$$[n \geq 0]$$      BI(34)(8), BI(35)(11)

2. $\displaystyle\int_0^{\pi/4} (\sin^\mu 2x - 1)\operatorname{cosec}^\mu 2x \tan\left(\frac{\pi}{4} + x\right) dx = \int_0^{\pi/4} (\cos^\mu 2x - 1)\sec^\mu 2x \cot x\ dx$

$$= \frac{1}{2}\left[\boldsymbol{C} + \psi(1 - \mu)\right]$$
$$[\operatorname{Re}\mu < 1]$$      BI (35)(20)

3. 
$$\int_0^{\frac{\pi}{4}} \left(\sin^{2\mu} 2x - 1\right) \operatorname{cosec}^\mu 2x \tan\left(\frac{\pi}{4} + x\right) dx = \int_0^{\pi/4} \left(\cos^{2\mu} 2x - 1\right) \sec^\mu 2x \cot x \, dx$$
$$= -\frac{1}{2\mu} + \frac{\pi}{2} \cot \mu\pi$$

BI (35)(21)

4. 
$$\int_0^{\pi/4} \left(1 - \sec^\mu 2x\right) \cot x \, dx = \int_0^{\pi/4} \left(1 - \operatorname{cosec}^\mu 2x\right) \tan\left(\frac{\pi}{4} + x\right) dx = \frac{1}{2} \left[ \boldsymbol{C} + \psi(1 - \mu) \right]$$
$$[\operatorname{Re} \mu < 1]$$

BI (35)(13)

**3.684** 
$$\int_0^{\pi/4} \frac{(\cot^\mu x - 1) \, dx}{(\cos x - \sin x) \sin x} = \int_0^{\pi/2} \frac{(\tan^\mu x - 1) \, dx}{(\sin x - \cos x) \cos x} = -\boldsymbol{C} - \psi(1 - \mu) \qquad [\operatorname{Re} \mu < 1]$$

BI (37)(9)

**3.685**

1. 
$$\int_0^{\pi/4} \left(\sin^{\mu-1} 2x - \sin^{\nu-1} 2x\right) \tan\left(\frac{\pi}{4} + x\right) dx = \int_0^{\pi/4} \left(\cos^{\mu-1} 2x - \cos^{\nu-1} 2x\right) \cot x \, dx$$
$$= \frac{1}{2} \left[\psi(\nu) - \psi(\mu)\right]$$
$$[\operatorname{Re} \mu > 0, \operatorname{Re} \nu > 0] \quad \text{BI}(34)(9), \text{BI}(35)(12)$$

2. 
$$\int_0^{\pi/2} \left(\sin^{\mu-1} x - \sin^{\nu-1} x\right) \frac{dx}{\cos x} = \int_0^{\pi/2} \left(\cos^{\mu-1} x - \cos^{\nu-1} x\right) \frac{dx}{\sin x} = \frac{1}{2} \left[ \psi\left(\frac{\nu}{2}\right) - \psi\left(\frac{\mu}{2}\right) \right]$$
$$[\operatorname{Re} \mu > 0, \quad \operatorname{Re} \nu > 0] \qquad \text{BI (46)(2)}$$

3. 
$$\int_0^{\pi/2} \left(\sin^\mu x - \operatorname{cosec}^\mu x\right) \frac{dx}{\cos x} = \int_0^{\pi/2} \left(\cos^\mu x - \sec^\mu x\right) \frac{dx}{\sin x} = -\frac{\pi}{2} \tan \frac{\mu\pi}{2}$$
$$[|\operatorname{Re} \mu| < 1] \qquad \text{BI (46)(1, 3)}$$

4. 
$$\int_0^{\pi/4} \left(\sin^\mu 2x - \operatorname{cosec}^\mu 2x\right) \cot\left(\frac{\pi}{4} + x\right) dx = \int_0^{\pi/4} \left(\cos^\mu 2x - \sec^\mu 2x\right) \tan x \, dx$$
$$= \frac{1}{2\mu} - \frac{\pi}{2} \operatorname{cosec} \mu\pi$$
$$[|\operatorname{Re} \mu| < 1] \qquad \text{BI (35)(19, 22)}$$

5. 
$$\int_0^{\pi/4} \left(\sin^\mu 2x - \operatorname{cosec}^\mu 2x\right) \tan\left(\frac{\pi}{4} + x\right) dx = \int_0^{\pi/4} \left(\cos^\mu 2x - \sec^\mu 2x\right) \cot x \, dx$$
$$= -\frac{1}{2\mu} + \frac{\pi}{2} \cot \mu\pi$$
$$[|\operatorname{Re} \mu| < 1] \qquad \text{BI (35)(14)}$$

6. 
$$\int_0^{\pi/4} \left(\sin^{\mu-1} 2x + \operatorname{cosec}^\mu 2x\right) \cot\left(\frac{\pi}{4} + x\right) dx$$
$$= \int_0^{\pi/4} \left(\cos^{\mu-1} 2x + \sec^\mu 2x\right) \tan x \, dx = \frac{\pi}{4} \operatorname{cosec} \mu\pi$$
$$[0 < \operatorname{Re} \mu < 1] \qquad \text{BI (35)(18, 8)}$$

7. 
$$\int_0^{\pi/4} \left(\sin^{\mu-1} 2x - \operatorname{cosec}^\mu 2x\right) \tan\left(\frac{\pi}{4} + x\right) dx = \int_0^{\pi/4} \left(\cos^{\mu-1} 2x - \sec^\mu 2x\right) \cot x \, dx = \frac{\pi}{2} \cot \mu\pi$$
$$[0 < \operatorname{Re} \mu < 1] \qquad \text{BI}(35)(7), \text{LI}(34)(10)$$

**3.686**
$$\int_0^{\pi/2} \frac{\tan x\, dx}{\cos^\mu x + \sec^\mu x} = \int_0^{\pi/2} \frac{\cot x\, dx}{\sin^\mu x + \csc^\mu x} = \frac{\pi}{4\mu}$$
     BI(47)(28), BI(49)(14)

**3.687**

1.
$$\int_0^{\pi/2} \frac{\sin^{\mu-1} x + \sin^{\nu-1} x}{\cos^{\mu+\nu-1} x}\, dx = \int_0^{\pi/2} \frac{\cos^{\mu-1} x + \cos^{\nu-1} x}{\sin^{\mu+\nu-1} x}\, dx = \frac{\cos\left(\frac{\nu-\mu}{4}\pi\right)}{2\cos\left(\frac{\nu+\mu}{4}\pi\right)} B\left(\frac{\mu}{2}, \frac{\nu}{2}\right)$$
$$[\operatorname{Re}\mu > 0, \quad \operatorname{Re}\nu > 0, \quad \operatorname{Re}(\mu+\nu) < 2]$$
     BI (46)(7)

2.
$$\int_0^{\pi/2} \frac{\sin^{\mu-1} x - \sin^{\nu-1} x}{\cos^{\mu+\nu-1} x}\, dx = \int_0^{\pi/2} \frac{\cos^{\mu-1} x - \cos^{\nu-1} x}{\sin^{\mu+\nu-1} x}\, dx = \frac{\sin\left(\frac{\nu-\mu}{4}\pi\right)}{2\sin\left(\frac{\nu+\mu}{4}\pi\right)} B\left(\frac{\mu}{2}, \frac{\nu}{2}\right)$$
$$[\operatorname{Re}\mu > 0, \quad \operatorname{Re}\nu > 0, \quad \operatorname{Re}(\mu+\nu) < 4]$$
     BI(46)(8)

3.
$$\int_0^{\pi/2} \frac{\sin^\mu x + \sin^\nu x}{\sin^{\mu+\nu} x + 1} \cot x\, dx = \int_0^{\frac{\pi}{2}} \frac{\cos^\mu x + \cos^\nu x}{\cos^{\mu+\nu} x + 1} \tan x\, dx = \frac{\pi}{\mu+\nu} \sec\left(\frac{\mu-\nu}{\mu+\nu}\frac{\pi}{2}\right)$$
$$[\operatorname{Re}\mu > 0, \quad \operatorname{Re}\nu > 0]$$
     BI (49)(15)a, BI (47)(29)

4.
$$\int_0^{\pi/2} \frac{\sin^\mu x - \sin^\nu x}{\sin^{\mu+\nu} x - 1} \cot x\, dx = \int_0^{\frac{\pi}{2}} \frac{\cos^\mu x - \cos^\nu x}{\cos^{\mu+\nu} x - 1} \tan x\, dx = \frac{\pi}{\mu+\nu} \tan\left(\frac{\mu-\nu}{\mu+\nu}\frac{\pi}{2}\right)$$
$$[\operatorname{Re}\mu > 0, \quad \operatorname{Re}\nu > 0]$$
     BI(149)(16)a, BI(47)(30)

5.
$$\int_0^{\pi/2} \frac{\cos^\mu x + \sec^\mu x}{\cos^\nu x + \sec^\nu x} \tan x\, dx = \frac{\pi}{2\nu} \sec\left(\frac{\mu}{\nu}\frac{\pi}{2}\right)$$
$$[|\operatorname{Re}\nu| > |\operatorname{Re}\mu|]$$
     BI (49)(12)

6.
$$\int_0^{\pi/2} \frac{\cos^\mu x - \sec^\mu x}{\cos^\nu x - \sec^\nu x} \tan x\, dx = \frac{\pi}{2\nu} \tan\left(\frac{\mu}{\nu}\frac{\pi}{2}\right)$$
$$[|\operatorname{Re}\nu| > |\operatorname{Re}\mu|]$$
     BI (49)(13)

**3.688**

1.
$$\int_0^{\pi/4} \frac{\tan^\nu x - \tan^\mu x}{\cos x - \sin x}\frac{dx}{\sin x} = \psi(\mu) - \psi(\nu)$$
$$[\operatorname{Re}\mu > 0, \quad \operatorname{Re}\nu > 0]$$
     BI (37)(10)

2.
$$\int_0^{\pi/4} \frac{\tan^\mu x - \tan^{1-\mu} x}{\cos x - \sin x}\frac{dx}{\sin x} = \pi \cot\mu\pi$$
$$[0 < \operatorname{Re}\mu < 1]$$
     BI (37)(11)

3.
$$\int_0^{\pi/4} (\tan^\mu x + \cot^\mu x)\, dx = \frac{\pi}{2} \sec\frac{\mu\pi}{2}$$
$$[|\operatorname{Re}\mu| < 1]$$
     BI (35)(9)

4.
$$\int_0^{\pi/4} (\tan^\mu x - \cot^\mu x) \tan x\, dx = \frac{1}{\mu} - \frac{\pi}{2} \csc\frac{\mu\pi}{2}$$
$$[0 < \operatorname{Re}\mu < 2]$$
     BI (35)(15)

5.
$$\int_0^{\pi/4} \frac{\tan^{\mu-1} x - \cot^{\mu-1} x}{\cos 2x}\, dx = \frac{\pi}{2} \cot\frac{\mu\pi}{2}$$
$$[|\operatorname{Re}\mu| < 2]$$
     BI (35)(10)

6.
$$\int_0^{\pi/4} \frac{\tan^\mu x - \cot^\mu x}{\cos 2x} \tan x\, dx = -\frac{1}{\mu} + \frac{\pi}{2} \cot\frac{\mu\pi}{2}$$
$$[-2 < \operatorname{Re}\mu < 0]$$
     BI (35)(23)

7.
$$\int_0^{\pi/4} \frac{\tan^\mu x + \cot^\mu x}{1 + \cos t \sin 2x}\, dx = \pi \csc t \csc \mu\pi \sin\mu t$$
$$[t \neq n\pi, \quad |\operatorname{Re}\mu| < 1]$$
     BI (36)(6)

8.  $\int_0^{\pi/4} \dfrac{\tan^{\mu-1} x + \cot^\mu x}{(\sin x + \cos x)\cos x}\, dx = \pi \operatorname{cosec} \mu\pi$  $\qquad [0 < \operatorname{Re}\mu < 1]$  $\qquad$ BI (37)(3)

9.  $\int_0^{\pi/4} \dfrac{\tan^\mu x - \cot^\mu x}{(\sin x + \cos x)\cos x}\, dx = -\pi \operatorname{cosec} \mu\pi + \dfrac{1}{\mu}$  $\qquad [0 < \operatorname{Re}\mu < 1]$  $\qquad$ BI (37)(4)

10.  $\int_0^{\pi/4} \dfrac{\tan^\nu x - \cot^\mu x}{(\cos x - \sin x)\cos x}\, dx = \psi(1-\mu) - \psi(1+\nu)$  $\qquad [\operatorname{Re}\mu < 1, \quad \operatorname{Re}\nu > -1]$  $\qquad$ BI (37)(5)

11.  $\int_0^{\pi/4} \dfrac{\tan^{\mu-1} x - \cot^\mu x}{(\cos x - \sin x)\cos x}\, dx = \pi \cot \mu\pi$  $\qquad [0 < \operatorname{Re}\mu < 1]$  $\qquad$ BI (37)(7)

12.  $\int_0^{\pi/4} \dfrac{\tan^\mu x - \cot^\mu x}{(\cos x - \sin x)\cos x}\, dx = \pi \cot \mu\pi - \dfrac{1}{\mu}$  $\qquad [0 < \operatorname{Re}\mu < 1]$  $\qquad$ BI (37)(8)

13.  $\int_0^{\pi/4} \dfrac{1}{\tan^\mu x + \cot^\mu x} \dfrac{dx}{\sin 2x} = \dfrac{\pi}{8\mu}$  $\qquad [\operatorname{Re}\mu \neq 0]$  $\qquad$ BI (37)(12)

14.  $\int_0^{\pi/2} \dfrac{1}{(\tan^\mu x + \cot^\mu x)^\nu} \dfrac{dx}{\tan x} = \int_0^{\pi/2} \dfrac{1}{(\tan^\mu x + \cot^\mu x)^\nu} \dfrac{dx}{\sin 2x} = \dfrac{\sqrt{\pi}}{2^{2\nu+1}\mu} \dfrac{\Gamma(\nu)}{\Gamma\left(\nu + \frac{1}{2}\right)}$

   $\qquad\qquad [\nu > 0]$  $\qquad$ BI(49)(25), BI(49)(26)

15.  $\int_0^{\pi/4} (\tan^\mu x - \cot^\mu x)(\tan^\nu x - \cot^\nu x)\, dx = \dfrac{2\pi \sin \frac{\mu\pi}{2} \sin \frac{\nu\pi}{2}}{\cos \mu\pi + \cos \nu\pi}$

   $\qquad\qquad [|\operatorname{Re}\mu| < 1, \quad |\operatorname{Re}\nu| < 1]$  $\qquad$ BI (35)(17)

16.  $\int_0^{\pi/4} (\tan^\mu x + \cot^\mu x)(\tan^\nu x + \cot^\nu x)\, dx = \dfrac{2\pi \cos \frac{\mu\pi}{2} \cos \frac{\nu\pi}{2}}{\cos \mu\pi + \cos \nu\pi}$

   $\qquad\qquad [|\operatorname{Re}\mu| < 1, \quad |\operatorname{Re}\nu| < 1]$  $\qquad$ BI (35)(16)

17.  $\int_0^{\pi/4} \dfrac{(\tan^\mu x - \cot^\mu x)(\tan^\nu x + \cot^\nu x)}{\cos 2x}\, dx = -\pi \dfrac{\sin \mu\pi}{\cos \mu\pi + \cos \nu\pi}$

   $\qquad\qquad [|\operatorname{Re}\mu| < 1, \quad |\operatorname{Re}\nu| < 1]$  $\qquad$ BI (35)(25)

18.  $\int_0^{\pi/4} \dfrac{\tan^\nu x - \cot^\nu x}{\tan^\mu x - \cot^\mu x} \dfrac{dx}{\sin 2x} = \dfrac{\pi}{4\mu} \tan \dfrac{\nu\pi}{2\mu}$  $\qquad [0 < \operatorname{Re}\nu < 1]$  $\qquad$ BI (37)(14)

19.  $\int_0^{\pi/4} \dfrac{\tan^\nu x + \cot^\nu x}{\tan^\mu x + \cot^\mu x} \dfrac{dx}{\sin 2x} = \dfrac{\pi}{4\mu} \sec \dfrac{\nu\pi}{2\mu}$  $\qquad [0 < \operatorname{Re}\nu < 1]$  $\qquad$ BI (37)(13)

20.  $\int_0^{\pi/2} \dfrac{(1 + \tan x)^\nu - 1}{(1 + \tan x)^{\mu+\nu}} \dfrac{dx}{\sin x \cos x} = \psi(\mu + \nu) - \psi(\mu)$  $\qquad [\mu > 0, \quad \nu > 0]$  $\qquad$ BI (49)(29)

**3.689**

1.  $\int_0^{\pi/2} \dfrac{(\sin^\mu x + \operatorname{cosec}^\mu x) \cot x\, dx}{\sin^\nu x - 2\cos t + \operatorname{cosec}^\nu x} = \dfrac{\pi}{\nu} \operatorname{cosec} t \operatorname{cosec} \dfrac{\mu\pi}{\nu} \sin \dfrac{\mu t}{\nu}$

   $\qquad\qquad [\mu < \nu]$  $\qquad$ LI (50)(14)

2.  $\int_0^{\pi/2} \dfrac{\sin^\mu x - 2\cos t_1 + \operatorname{cosec}^\mu x}{\sin^\nu x + 2\cos t_2 + \operatorname{cosec}^\nu x} \cot x\, dx = \dfrac{\pi}{\nu} \operatorname{cosec} t_2 \operatorname{cosec} \dfrac{\mu\pi}{\nu} \sin \dfrac{\mu t_2}{\nu} - \dfrac{t_2}{\nu} \operatorname{cosec} t_2 \cos t_1$

   $\qquad [(\nu > \mu > 0) \text{ or } (\nu < \mu < 0) \text{ or } (\mu > 0, \ \nu < 0, \text{ and } \mu + \nu < 0) \text{ or } (\mu < 0, \ \nu > 0, \text{ and } \mu + \nu > 0)]$

   $\qquad\qquad$ BI (50)(15)

## 3.69–3.71 Trigonometric functions of more complicated arguments

### 3.691

1.[12]   $\displaystyle\int_0^\infty \sin(x^2)\,\mathrm{d}x = \int_0^\infty \cos(x^2)\,\mathrm{d}x = \frac{1}{2}\sqrt{\frac{\pi}{2}}$           FI II 743a, ET I 64(7)a

2.   $\displaystyle\int_0^1 \sin\left(ax^2\right)\,\mathrm{d}x = \sqrt{\frac{\pi}{2a}}\,S\left(\sqrt{a}\right)$         $[a>0]$

3.   $\displaystyle\int_0^1 \cos\left(ax^2\right)\,\mathrm{d}x = \sqrt{\frac{\pi}{2a}}\,C\left(\sqrt{a}\right)$         $[a>0]$        ET I 8(5)a

4.   $\displaystyle\int_0^\infty \sin\left(ax^2\right)\sin 2bx\,\mathrm{d}x = \sqrt{\frac{\pi}{2a}}\left[\cos\frac{b^2}{a}\,C\left(\frac{b}{\sqrt{a}}\right) + \sin\frac{b^2}{a}\,S\left(\frac{b}{\sqrt{a}}\right)\right]$

                                       $[a>0,\quad b>0]$        ET I 82(1)a

5.   $\displaystyle\int_0^\infty \sin\left(ax^2\right)\cos 2bx\,\mathrm{d}x = \frac{1}{2}\sqrt{\frac{\pi}{2a}}\left[\cos\frac{b^2}{a} - \sin\frac{b^2}{a}\right] = \frac{1}{2}\sqrt{\frac{\pi}{a}}\cos\left(\frac{b^2}{a} + \frac{\pi}{4}\right)$

                         $[a>0,\quad b>0]$

                          ET I 82(18), BI(70)(13) GW(334)(5a)

6.   $\displaystyle\int_0^\infty \cos ax^2 \sin 2bx\,\mathrm{d}x = \sqrt{\frac{\pi}{2a}}\left[\sin\frac{b^2}{a}\,C\left(\frac{b}{\sqrt{a}}\right) - \cos\frac{b^2}{a}\,S\left(\frac{b}{\sqrt{a}}\right)\right]$

                                         $[a>0,\quad b>0]$        ET I 83(3)a

7.   $\displaystyle\int_0^\infty \cos ax^2 \cos 2bx\,\mathrm{d}x = \frac{1}{2}\sqrt{\frac{\pi}{2a}}\left[\cos\frac{b^2}{a} + \sin\frac{b^2}{a}\right]$      $[a>0,\quad b>0]$

                          GW(334)(5a), BI(70)(14), ET I 24(7)

8.   $\displaystyle\int_0^\infty (\cos ax + \sin ax)\sin\left(b^2x^2\right)\,\mathrm{d}x$

           $= \dfrac{1}{2b}\sqrt{\dfrac{\pi}{2}}\left[\left(1 + 2\,C\left(\dfrac{a}{2b}\right)\right)\cos\left(\dfrac{a^2}{4b^2}\right) - \left(1 - 2\,S\left(\dfrac{a}{2b}\right)\right)\sin\left(\dfrac{a^2}{4b^2}\right)\right]$

                                $[a>0,\quad b>0]$        ET I 85(22)

9.   $\displaystyle\int_0^\infty (\cos ax + \sin ax)\cos\left(b^2x^2\right)\,\mathrm{d}x$

           $= \dfrac{1}{2b}\sqrt{\dfrac{\pi}{2}}\left[\left(1 + 2\,C\left(\dfrac{a}{2b}\right)\right)\sin\left(\dfrac{a^2}{4b^2}\right) + \left(1 - 2\,S\left(\dfrac{a}{2b}\right)\right)\cos\left(\dfrac{a^2}{4b^2}\right)\right]$

                                $[a>0,\quad b>0]$        ET I 25(21)

10.   $\displaystyle\int_0^\infty \sin\left(a^2x^2\right)\sin 2bx \sin 2cx\,\mathrm{d}x = \frac{\sqrt{\pi}}{2a}\sin\frac{2bc}{a^2}\cos\left(\frac{b^2+c^2}{a^2} - \frac{\pi}{4}\right)$

                               $[a>0,\quad b>0,\quad c>0]$        ET I 84(15)

11.   $\displaystyle\int_0^\infty \sin\left(a^2x^2\right)\cos 2bx \cos 2cx\,\mathrm{d}x = \frac{\sqrt{\pi}}{2a}\cos\frac{2bc}{a^2}\cos\left(\frac{b^2+c^2}{a^2} + \frac{\pi}{4}\right)$

                               $[a>0,\quad b>0,\quad c>0]$        ET I 84(21)

12.   $\displaystyle\int_0^\infty \cos\left(a^2 x^2\right) \sin 2bx \sin 2cx \, dx = \frac{\sqrt{\pi}}{2a} \sin\frac{2bc}{a^2} \sin\left(\frac{b^2+c^2}{a^2} - \frac{\pi}{4}\right)$

$$[a > 0, \quad b > 0, \quad c > 0] \qquad \text{ET I 25(19)}$$

13.   $\displaystyle\int_0^\infty \sin\left(ax^2\right) \cos\left(bx^2\right) \, dx = \frac{1}{4}\sqrt{\frac{\pi}{2}} \left(\frac{1}{\sqrt{a+b}} + \frac{1}{\sqrt{a-b}}\right) \qquad [a > b > 0]$

$$= \frac{1}{4}\sqrt{\frac{\pi}{2}} \left(\frac{1}{\sqrt{b+a}} - \frac{1}{\sqrt{b-a}}\right) \qquad [b > a > 0]$$

$$\text{BI (177)(21)}$$

14.   $\displaystyle\int_0^\infty \left(\sin^2 ax^2 - \sin^2 bx^2\right) \, dx = \frac{1}{8}\left(\sqrt{\frac{\pi}{b}} - \sqrt{\frac{\pi}{a}}\right) \qquad [a > 0, \quad b > 0] \qquad \text{BI (178)(1)}$

15.   $\displaystyle\int_0^\infty \left(\cos^2 ax^2 - \sin^2 bx^2\right) \, dx = \frac{1}{8}\left(\sqrt{\frac{\pi}{b}} + \sqrt{\frac{\pi}{a}}\right) \qquad [a > 0, \quad b > 0] \qquad \text{BI (178)(3)}$

16.   $\displaystyle\int_0^\infty \left(\cos^2 ax^2 - \cos^2 bx^2\right) \, dx = \frac{1}{8}\left(\sqrt{\frac{\pi}{a}} - \sqrt{\frac{\pi}{b}}\right) \qquad [a > 0, \quad b > 0] \qquad \text{BI (178)(5)}$

17.[12]   $\displaystyle\int_0^\infty \left(\sin^4 ax^2 - \sin^4 bx^2\right) \, dx = \frac{1}{64}\left(8 - \sqrt{2}\right)\left(\sqrt{\frac{\pi}{b}} - \sqrt{\frac{\pi}{a}}\right)$

$$[a > 0, \quad b > 0] \qquad \text{BI (178)(2)}$$

18.   $\displaystyle\int_0^\infty \left(\cos^4 ax^2 - \sin^4 bx^2\right) \, dx = \frac{1}{8}\left(\sqrt{\frac{\pi}{a}} + \sqrt{\frac{\pi}{b}}\right) + \frac{1}{32}\left(\sqrt{\frac{\pi}{2a}} - \sqrt{\frac{\pi}{2b}}\right)$

$$[a > 0, \quad b > 0] \qquad \text{BI (178)(4)}$$

19.   $\displaystyle\int_0^\infty \left(\cos^4 ax^2 - \cos^4 bx^2\right) \, dx = \frac{1}{64}\left(8 + \sqrt{2}\right)\left(\sqrt{\frac{\pi}{a}} - \sqrt{\frac{\pi}{b}}\right)$

$$[a > 0, \quad b > 0] \qquad \text{BI (178)(6)}$$

20.   $\displaystyle\int_0^\infty \sin^{2n} ax^2 \, dx = \int_0^\infty \cos^{2n} ax^2 \, dx = \infty \qquad\qquad \text{BI (177)(5, 6)}$

21.   $\displaystyle\int_0^\infty \sin^{2n+1}\left(ax^2\right) \, dx = \frac{1}{2^{2n+1}} \sum_{k=0}^n (-1)^{n+k}\binom{2n+1}{k}\sqrt{\frac{\pi}{2(2n-2k+1)a}}$

$$[a > 0] \qquad \text{BI (70)(9)}$$

22.   $\displaystyle\int_0^\infty \cos^{2n+1}\left(ax^2\right) \, dx = \frac{1}{2^{2n+1}} \sum_{k=0}^n \binom{2n+1}{k}\sqrt{\frac{\pi}{2(2n-2k+1)a}}$

$$[a > 0] \qquad \text{BI(177)(7)a, BI(70)(10)}$$

**3.692**

1.[12]   $\displaystyle\int_0^\infty \left[\sin\left(a - x^2\right) + \cos\left(a - x^2\right)\right] \, dx = \sqrt{\frac{\pi}{2}} \sin a \qquad\qquad \text{GW(333)(30c), BI(178)(7)a}$

2.   $\displaystyle\int_0^\infty \cos\left(\frac{x^2}{2} - \frac{\pi}{8}\right) \cos ax \, dx = \sqrt{\frac{\pi}{2}} \cos\left(\frac{a^2}{2} - \frac{\pi}{8}\right) \qquad [a > 0] \qquad \text{ET I 24(8)}$

3.     $\int_0^\infty \sin\left[a\left(1-x^2\right)\right]\cos bx\, dx = -\dfrac{1}{2}\sqrt{\dfrac{\pi}{a}}\cos\left(a+\dfrac{b^2}{4a}+\dfrac{\pi}{4}\right)$

$$[a > 0]$$       ET I 23(2)

4.     $\int_0^\infty \cos\left[a\left(1-x^2\right)\right]\cos bx\, dx = \dfrac{1}{2}\sqrt{\dfrac{\pi}{a}}\sin\left(a+\dfrac{b^2}{4a}+\dfrac{\pi}{4}\right)$

$$[a > 0]$$       ET I 24(10)

5.     $\int_0^\infty \sin\left(ax^2+\dfrac{b^2}{a}\right)\cos 2bx\, dx = \int_0^\infty \cos\left(ax^2+\dfrac{b^2}{a}\right)\cos 2bx\, dx = \dfrac{1}{2}\sqrt{\dfrac{\pi}{2a}}$

$$[a > 0]$$       BI (70)(19, 20)

6.[8]     $\int_{-\infty}^\infty \left[\cos\sqrt{x^2-1}-\cos\sqrt{x^2+1}\right] dx = \sum_{n=0}^\infty \dfrac{\pi}{2^{4n+1}\left[(2n)!\right]^2\left(n+\frac{1}{2}\right)}$

**3.693**

1.     $\int_0^\infty \sin\left(ax^2+2bx\right) dx = \sqrt{\dfrac{\pi}{2a}}\left[\cos\dfrac{b^2}{a}\left(\frac{1}{2}-S_2\left(\frac{b^2}{a}\right)\right)-\sin\dfrac{b^2}{a}\left(\frac{1}{2}-C_2\left(\frac{b^2}{a}\right)\right)\right]$

$$[a > 0]$$       BI (70)(3)

2.     $\int_0^\infty \cos\left(ax^2+2bx\right) dx = \sqrt{\dfrac{\pi}{2a}}\left[\cos\dfrac{b^2}{a}\left(\frac{1}{2}-C_2\left(\frac{b^2}{a}\right)\right)+\sin\dfrac{b^2}{a}\left(\frac{1}{2}-S_2\left(\frac{b^2}{a}\right)\right)\right]$

$$[a > 0]$$       BI (70)(4)

**3.694**

1.     $\int_0^\infty \sin\left(ax^2+2bx+c\right) dx = \sqrt{\dfrac{\pi}{2a}}\cos\dfrac{b^2}{a}\left[\left(\frac{1}{2}-C_2\left(\frac{b^2}{a}\right)\right)\sin c + \left(\frac{1}{2}-S_2\left(\frac{b^2}{a}\right)\right)\cos c\right]$
$$+\sqrt{\dfrac{\pi}{2a}}\sin\dfrac{b^2}{a}\left[\left(\frac{1}{2}-S_2\left(\frac{b^2}{a}\right)\right)\sin c - \left(\frac{1}{2}-C_2\left(\frac{b^2}{a}\right)\right)\cos c\right]$$

$$[a > 0]$$       GW (334)(4a)

2.     $\int_0^\infty \cos\left(ax^2+2bx+c\right) dx = \sqrt{\dfrac{\pi}{2a}}\cos\dfrac{b^2}{a}\left[\left(\frac{1}{2}-C_2\left(\frac{b^2}{a}\right)\right)\cos c - \left(\frac{1}{2}-S_2\left(\frac{b^2}{a}\right)\right)\sin c\right]$
$$+\sqrt{\dfrac{\pi}{2a}}\sin\dfrac{b^2}{a}\left[\left(\frac{1}{2}-S_2\left(\frac{b^2}{a}\right)\right)\cos c + \left(\frac{1}{2}-C_2\left(\frac{b^2}{a}\right)\right)\sin c\right]$$

$$[a > 0]$$       GW (334)(4b)

**3.695**

1.     $\int_0^\infty \sin\left(a^3x^3\right)\sin(bx)\, dx = \dfrac{\pi}{6a}\sqrt{\dfrac{b}{3a}}\left[J_{\frac{1}{3}}\left(\dfrac{2b}{3a}\sqrt{\dfrac{b}{3a}}\right)+J_{-\frac{1}{3}}\left(\dfrac{2b}{3a}\sqrt{\dfrac{b}{3a}}\right)-\dfrac{\sqrt{3}}{\pi}K_{\frac{1}{3}}\left(\dfrac{2b}{3a}\sqrt{\dfrac{b}{3a}}\right)\right]$

$$[a > 0, \quad b > 0]$$       ET I 83(5)

2.     $\int_0^\infty \cos\left(a^3x^3\right)\cos(bx)\, dx = \dfrac{\pi}{6a}\sqrt{\dfrac{b}{3a}}\left[J_{\frac{1}{3}}\left(\dfrac{2b}{3a}\sqrt{\dfrac{b}{3a}}\right)+J_{-\frac{1}{3}}\left(\dfrac{2b}{3a}\sqrt{\dfrac{b}{3a}}\right)+\dfrac{\sqrt{3}}{\pi}K_{\frac{1}{3}}\left(\dfrac{2b}{3a}\sqrt{\dfrac{b}{3a}}\right)\right]$

$$[a > 0, \quad b > 0]$$       ET I 24(11)

**3.696**

1.  $$\int_0^\infty \sin\left(ax^4\right)\sin\left(bx^2\right)\,\mathrm{d}x = -\frac{\pi}{4}\sqrt{\frac{b}{2a}}\sin\left(\frac{b^2}{8a}-\frac{3}{8}\pi\right)J_{\frac{1}{4}}\left(\frac{b^2}{8a}\right)$$
    $$[a>0,\quad b>0]$$          ET I 83(2)

2.  $$\int_0^\infty \sin\left(ax^4\right)\cos\left(bx^2\right)\,\mathrm{d}x = -\frac{\pi}{4}\sqrt{\frac{b}{2a}}\sin\left(\frac{b^2}{8a}-\frac{\pi}{8}\right)J_{-\frac{1}{4}}\left(\frac{b^2}{8a}\right)$$
    $$[a>0,\quad b>0]$$          ET I 84(19)

3.  $$\int_0^\infty \cos\left(ax^4\right)\sin\left(bx^2\right)\,\mathrm{d}x = \frac{\pi}{4}\sqrt{\frac{b}{2a}}\cos\left(\frac{b^2}{8a}-\frac{3}{8}\pi\right)J_{\frac{1}{4}}\left(\frac{b^2}{8a}\right)$$
    $$[a>0,\quad b>0]$$     ET I 83(4), ET I 25(24)

4.  $$\int_0^\infty \cos\left(ax^4\right)\cos\left(bx^2\right)\,\mathrm{d}x = \frac{\pi}{4}\sqrt{\frac{b}{2a}}\cos\left(\frac{b^2}{8a}-\frac{\pi}{8}\right)J_{-\frac{1}{4}}\left(\frac{b^2}{8a}\right)$$
    $$[a>0,\quad b>0]$$          ET I 25(25)

**3.697**  $$\int_0^\infty \sin\left(\frac{a^2}{x}\right)\sin(bx)\,\mathrm{d}x = \frac{a\pi}{2\sqrt{b}}J_1\left(2a\sqrt{b}\right)\qquad [a>0,\quad b>0]$$          ET I 83(6)

**3.698**

1.  $$\int_0^\infty \sin\left(\frac{a^2}{x^2}\right)\sin\left(b^2x^2\right)\,\mathrm{d}x = \frac{1}{4b}\sqrt{\frac{\pi}{2}}\left[\sin 2ab-\cos 2ab+e^{-2ab}\right]$$
    $$[a>0,\quad b>0]$$          ET I 83(9)

2.[12]  $$\int_0^\infty \sin\left(\frac{a^2}{x^2}\right)\cos\left(b^2x^2\right)\,\mathrm{d}x = \frac{1}{4b}\sqrt{\frac{\pi}{2}}\left[\sin 2ab+\cos 2ab-e^{-2ab}\right]$$
    $$[a>0,\quad b>0]$$          ET I 24(13)

3.  $$\int_0^\infty \cos\left(\frac{a^2}{x^2}\right)\sin\left(b^2x^2\right)\,\mathrm{d}x = \frac{1}{4b}\sqrt{\frac{\pi}{2}}\left[\sin 2ab+\cos 2ab+e^{-2ab}\right]$$
    $$[a>0,\quad b>0]$$          ET I 84(12)

4.  $$\int_0^\infty \cos\left(\frac{a^2}{x^2}\right)\cos\left(b^2x^2\right)\,\mathrm{d}x = \frac{1}{4b}\sqrt{\frac{\pi}{2}}\left[\cos 2ab-\sin 2ab+e^{-2ab}\right]$$
    $$[a>0,\quad b>0]$$          ET I 24(14)

**3.699**

1.  $$\int_0^\infty \sin\left(a^2x^2+\frac{b^2}{x^2}\right)\,\mathrm{d}x = \frac{\sqrt{2\pi}}{4a}\left(\cos 2ab+\sin 2ab\right)\qquad [a>0,\quad b>0]$$          BI (70)(27)

2.  $$\int_0^\infty \cos\left(a^2x^2+\frac{b^2}{x^2}\right)\,\mathrm{d}x = \frac{\sqrt{2\pi}}{4a}\left(\cos 2ab-\sin 2ab\right)\qquad [a>0,\quad b>0]$$          BI (70)(28)

3.  $$\int_0^\infty \sin\left(a^2x^2-2ab+\frac{b^2}{x^2}\right)\,\mathrm{d}x = \int_0^\infty \cos\left(a^2x^2-2ab+\frac{b^2}{x^2}\right)\,\mathrm{d}x = \frac{\sqrt{2\pi}}{4a}$$
    $$[a>0,\quad b>0]$$
    BI(179)(11, 12)a, ET I 83(6)

4.    $\displaystyle\int_0^\infty \sin\left(a^2 x^2 - \frac{b^2}{x^2}\right)\,\mathrm{d}x = \frac{\sqrt{2\pi}}{4a}e^{-2ab}$          $[a > 0, \quad b > 0]$          GW (334)(9b)a

5.    $\displaystyle\int_0^\infty \cos\left(a^2 x^2 - \frac{b^2}{x^2}\right)\,\mathrm{d}x = \frac{\sqrt{2\pi}}{4a}e^{-2ab}$          $[a > 0, \quad b > 0]$          GW (334)(9b)a

**3.711**    $\displaystyle\int_0^u \sin\left(a\sqrt{u^2 - x^2}\right)\cos bx\,\mathrm{d}x = \frac{\pi a u}{2\sqrt{a^2 + b^2}}J_1\left(u\sqrt{a^2 + b^2}\right)$      $[a > 0, \quad b > 0, \quad u > 0]$

                                                     ET I 27(37)

**3.712**

1.    $\displaystyle\int_0^\infty \sin(ax^p)\,\mathrm{d}x = \frac{\Gamma\left(\frac{1}{p}\right)\sin\frac{\pi}{2p}}{pa^{\frac{1}{p}}}$          $[a > 0, \quad p > 1]$          EH I 13(40)

2.    $\displaystyle\int_0^\infty \cos(ax^p)\,\mathrm{d}x = \frac{\Gamma\left(\frac{1}{p}\right)\cos\frac{\pi}{2p}}{pa^{\frac{1}{p}}}$          $[a > 0, \quad p > 1]$          EH I 13(39)

**3.713**

1.    $\displaystyle\int_0^\infty \sin(ax^p + bx^q)\,\mathrm{d}x = \frac{1}{p}\sum_{k=0}^\infty \frac{(-b)^k}{k!}a^{-\frac{kq+1}{p}}\Gamma\left(\frac{kq+1}{p}\right)\sin\left[\frac{k(q-p)+1}{2p}\pi\right]$

                                             $[a > 0, \quad b > 0, \quad p > 0, \quad q > 0]$

                                                 BI (70)(7)

2.    $\displaystyle\int_0^\infty \cos(ax^p + bx^q)\,\mathrm{d}x = \frac{1}{p}\sum_{k=0}^\infty \frac{(-b)^k}{k!}a^{-(kq+1)/p}\Gamma\left(\frac{kq+1}{p}\right)\cos\left[\frac{k(q-p)+1}{2p}\pi\right]$

                                             $[a > 0, \quad b > 0, \quad p > 0, \quad q > 0]$

                                                 BI (70)(0)

**3.714**

1.    $\displaystyle\int_0^\infty \cos(z\sinh x)\,\mathrm{d}x = K_0(z)$          $[\operatorname{Re} z > 0]$          WA 202(14)

2.    $\displaystyle\int_0^\infty \sin(z\cosh x)\,\mathrm{d}x = \frac{\pi}{2}J_0(z)$          $[\operatorname{Re} z > 0]$          MO 36

3.    $\displaystyle\int_0^\infty \cos(z\cosh x)\,\mathrm{d}x = -\frac{\pi}{2}Y_0(z)$          $[\operatorname{Re} z > 0]$          MO 37

4.    $\displaystyle\int_0^\infty \cos(z\sinh x)\cosh \mu x\,\mathrm{d}x = \cos\frac{\mu\pi}{2}K_\mu(z)$          $[\operatorname{Re} z > 0, \quad |\operatorname{Re}\mu| < 1]$          WA 202(13)

5.    $\displaystyle\int_0^\pi \cos(z\cosh x)\sin^{2\mu} x\,\mathrm{d}x = \sqrt{\pi}\left(\frac{2}{z}\right)^\mu\Gamma\left(\mu + \frac{1}{2}\right)I_\mu(z)$

                                     $\left[\operatorname{Re} z > 0, \quad \operatorname{Re}\mu > -\frac{1}{2}\right]$          WH

**3.715**

1.    $\displaystyle\int_0^\pi \sin(z\sin x)\sin ax\,\mathrm{d}x = \sin a\pi\, s_{0,a}(z) = \sin a\pi\sum_{k=1}^\infty \frac{(-1)^{k-1}z^{2k-1}}{(1^2 - a^2)(3^2 - a^2)\ldots[(2k-1)^2 - a^2]}$

                                                   $[a > 0]$          WA 338(13)

2.  $\displaystyle\int_0^\pi \sin\left(z\sin x\right)\sin nx\,dx = \frac{1}{2}\int_{-\pi}^\pi \sin\left(z\sin x\right)\sin nx\,dx$

$$= [1-(-1)^n]\int_0^{\pi/2}\sin\left(z\sin x\right)\sin nx\,dx = [1-(-1)^n]\frac{\pi}{2}J_n(z)$$

$$[n=0,\pm1,\pm2,\ldots]\quad\text{WA 30(6), GW(334)(153a)}$$

3.  $\displaystyle\int_0^{\pi/2}\sin\left(z\sin x\right)\sin 2x\,dx = \frac{2}{z^2}\left(\sin z - z\cos z\right)$            LI (43)(14)

4.  $\displaystyle\int_0^\pi \sin\left(z\sin x\right)\cos ax\,dx = (1+\cos a\pi)\,s_{0,a}(z)$

$$= (1+\cos a\pi)\sum_{k=1}^\infty \frac{(-1)^{k-1}z^{2k-1}}{(1^2-a^2)(3^2-a^2)\ldots[(2k-1)^2-a^2]}$$

$$[a>0]\qquad\qquad\text{WA 338(14)}$$

5.  $\displaystyle\int_0^\pi \sin\left(z\sin x\right)\cos[(2n+1)x]\,dx = 0$            GW (334)(53b)

6.  $\displaystyle\int_0^\pi \cos\left(z\sin x\right)\sin ax\,dx = -a\left(1-\cos a\pi\right)s_{-1,a}(z)$

$$= -a\left(1-\cos a\pi\right)\left\{-\frac{1}{a^2}+\sum_{k=1}^\infty \frac{(-1)^{k-1}z^{2k}}{a^2\left(2^2-a^2\right)\left(4^2-a^2\right)\ldots[(2k)^2-a^2]}\right\}$$

$$[a>0]\qquad\qquad\text{WA 338(12)}$$

7.  $\displaystyle\int_0^\pi \cos\left(z\sin x\right)\sin 2nx\,dx = 0$            GW (334)(54a)

8.  $\displaystyle\int_0^\pi \cos\left(z\sin x\right)\cos ax\,dx = -a\sin a\pi\, s_{-1,a}(z)$

$$= -a\sin a\pi\left\{-\frac{1}{a^2}+\sum_{k=1}^\infty \frac{(-1)^{k-1}z^{2k}}{a^2\left(2^2-a^2\right)\left(4^2-a^2\right)\ldots[(2k)^2-a^2]}\right\}$$

$$[a>0]\qquad\qquad\text{WA 338(11)}$$

9.  $\displaystyle\int_0^\pi \cos\left(z\sin x\right)\cos nx\,dx = \frac{1}{2}\int_{-\pi}^\pi \cos\left(z\sin x\right)\cos nx\,dx$

$$= [1+(-1)^n]\int_0^{\pi/2}\cos\left(z\sin x\right)\cos nx\,dx = [1+(-1)^n]\frac{\pi}{2}J_n(z)$$

$$\text{GW (334)(54b)}$$

10.[8]  $\displaystyle\int_0^{\pi/2}\cos\left(z\sin x\right)\cos^{2n}x\,dx = \frac{\pi}{2}\frac{(2n-1)!!}{z^n}J_n(z)\qquad[n=0,1,2,\ldots]\qquad\text{FI II 486, WA 35a}$

11.  $\displaystyle\int_0^{\pi/2}\sin\left(z\cos x\right)\sin 2x\,dx = \frac{2}{z^2}\left(\sin z - z\cos z\right)$            LI (43)(15)

$12.^8 \quad \displaystyle\int_0^{\pi/2} \sin{(z\cos x)}\cos ax\,\mathrm{d}x = \cos\frac{a\pi}{2}\,s_{0,a}(z) = \frac{\pi}{4}\operatorname{cosec}\frac{a\pi}{2}\left[\mathbf{J}_a(z) - \mathbf{J}_{-a}(z)\right]$

$$= -\frac{\pi}{4}\sec\frac{a\pi}{4}\left[\mathbf{E}_a(z) + \mathbf{E}_{-a}(z)\right]$$

$$= \cos\frac{a\pi}{2}\sum_{k=1}^{\infty}\frac{(-1)^{k-1}z^{2k-1}}{(1^2-a^2)(3^2-a^2)\ldots[(2k-1)^2-a^2]}$$

$$[a>0] \qquad\qquad \text{WA 339}$$

$13. \quad \displaystyle\int_0^{\pi}\sin{(z\cos x)}\cos nx\,\mathrm{d}x = \frac{1}{2}\int_{-\pi}^{\pi}\sin{(z\cos x)}\cos nx\,\mathrm{d}x = \pi\sin\frac{n\pi}{2}\,J_n(z)$     GW (334)(55b)

$14. \quad \displaystyle\int_0^{\pi/2}\sin{(z\cos x)}\cos[(2n+1)x]\,\mathrm{d}x = (-1)^n\frac{\pi}{2}\,J_{2n+1}(z)$     WA 30(8)

$15.^{11} \quad \displaystyle\int_0^{\pi/2}\sin{(a\cos x)}\tan x\,\mathrm{d}x = \operatorname{si}(a) + \frac{\pi}{2}$     $[a>0]$     BI (43)(17)

$16. \quad \displaystyle\int_0^{\pi/2}\sin{(z\cos x)}\sin^{2\nu}x\,\mathrm{d}x = \frac{\sqrt{\pi}}{2}\left(\frac{2}{z}\right)^{\nu}\Gamma\left(\nu+\frac{1}{2}\right)\mathbf{H}_{\nu}(z)$

$$\left[\operatorname{Re}\nu > -\tfrac{1}{2}\right] \qquad\qquad \text{WA 358(1)}$$

$17.^7 \quad \displaystyle\int_0^{\pi/2}\cos{(z\cos x)}\cos ax\,\mathrm{d}x = -a\sin\frac{a\pi}{2}\,s_{-1,a}(z)$

$$= \frac{\pi}{4}\sec\frac{a\pi}{2}\left[\mathbf{J}_a(z) + \mathbf{J}_{-a}(z)\right] = \frac{\pi}{4}\operatorname{cosec}\frac{a\pi}{2}\left[\mathbf{E}_a(z) - \mathbf{E}_{-a}(z)\right]$$

$$= -a\sin\frac{a\pi}{2}\left[-\frac{1}{a^2} + \sum_{k=1}^{\infty}\frac{(-1)^{k-1}z^{2k}}{a^2(2^2-a^2)(4^2-a^2)\ldots[(2k)^2-a^2]}\right]$$

$$[a>0] \qquad\qquad \text{WA 339}$$

$18. \quad \displaystyle\int_0^{\pi}\cos{(z\cos x)}\cos nx\,\mathrm{d}x = \frac{1}{2}\int_{-\pi}^{\pi}\cos{(z\cos x)}\cos nx\,\mathrm{d}x = \pi\cos\frac{n\pi}{2}\,J_n(z)$     GW (334)(56b)

$19. \quad \displaystyle\int_0^{\pi/2}\cos{(z\cos x)}\cos 2nx\,\mathrm{d}x = (-1)^n\frac{\pi}{2}\,J_{2n}(z)$     WA 30(9)

$20. \quad \displaystyle\int_0^{\pi/2}\cos{(z\cos x)}\sin^{2\nu}x\,\mathrm{d}x = \frac{\sqrt{\pi}}{2}\left(\frac{2}{z}\right)^{\nu}\Gamma\left(\nu+\frac{1}{2}\right)J_{\nu}(z)$

$$\left[\operatorname{Re}\nu > -\tfrac{1}{2}\right] \qquad\qquad \text{WA 35, WH}$$

$21. \quad \displaystyle\int_0^{\pi}\cos{(z\cos x)}\sin^{2\mu}x\,\mathrm{d}x = \sqrt{\pi}\left(\frac{2}{z}\right)^{\mu}\Gamma\left(\mu+\frac{1}{2}\right)J_{\mu}(z)$

$$\left[\operatorname{Re}\mu > -\tfrac{1}{2}\right] \qquad\qquad \text{WH}$$

**3.716**

$1. \quad \displaystyle\int_0^{\pi/2}\sin{(a\tan x)}\,\mathrm{d}x = \frac{1}{2}\left[e^{-a}\overline{\operatorname{Ei}}(a) - e^a\operatorname{Ei}(-a)\right]$     $[a>0]$     (cf. **3.723** 1)     BI (43)(1)

$2. \quad \displaystyle\int_0^{\pi/2}\cos{(a\tan x)}\,\mathrm{d}x = \frac{\pi}{2}e^{-a}$     $[a\geq 0]$     BI (43)(2)

3.  $\displaystyle\int_0^{\pi/2} \sin\left(a\tan x\right)\sin 2x\,\mathrm{d}x = \frac{a\pi}{2}e^{-a}$ $\qquad$ $[a \geq 0]$ $\qquad$ BI (43)(7)

4.  $\displaystyle\int_0^{\pi/2} \cos\left(a\tan x\right)\sin^2 x\,\mathrm{d}x = \frac{1-a}{4}\pi e^{-a}$ $\qquad$ $[a \geq 0]$ $\qquad$ BI (43)(8)

5.  $\displaystyle\int_0^{\pi/2} \cos\left(a\tan x\right)\cos^2 x\,\mathrm{d}x = \frac{1+a}{4}\pi e^{-a}$ $\qquad$ $[a \geq 0]$ $\qquad$ BI (43)(9)

6.  $\displaystyle\int_0^{\pi/2} \sin\left(a\tan x\right)\tan x\,\mathrm{d}x = \frac{\pi}{2}e^{-a}$ $\qquad$ $[a > 0]$ $\qquad$ BI (43)(5)

7.  $\displaystyle\int_0^{\pi/2} \cos\left(a\tan x\right)\tan x\,\mathrm{d}x = -\frac{1}{2}\left[e^{-a}\overline{\mathrm{Ei}}(a) + e^a\,\mathrm{Ei}(-a)\right]$

    $\qquad\qquad\qquad\qquad\qquad\qquad\qquad\qquad\qquad\quad [a > 0] \qquad (\text{cf. } \mathbf{3.723}\ 5) \qquad$ BI (43)(6)

8.  $\displaystyle\int_0^{\pi/2} \sin\left(a\tan x\right)\sin^2 x\tan x\,\mathrm{d}x = \frac{2-a}{4}\pi e^{-a}$ $\qquad$ $[a > 0]$ $\qquad$ BI (43)(11)

9.  $\displaystyle\int_0^{\pi/2} \sin^2\left(a\tan x\right)\,\mathrm{d}x = \frac{\pi}{4}\left(1 - e^{-2a}\right)$ $\qquad$ $[a \geq 0]$ $\qquad (\text{cf. } \mathbf{3.742}\ 1) \qquad$ BI (43)(3)

10. $\displaystyle\int_0^{\pi/2} \cos^2\left(a\tan x\right)\,\mathrm{d}x = \frac{\pi}{4}\left(1 + e^{-2a}\right)$ $\qquad$ $[a \geq 0]$ $\qquad (\text{cf. } \mathbf{3.742}\ 3) \qquad$ BI (43)(4)

11. $\displaystyle\int_0^{\pi/2} \sin^2\left(a\tan x\right)\cot^2 x\,\mathrm{d}x = \frac{\pi}{4}\left(e^{-2a} + 2a - 1\right)$ $\qquad$ $[a \geq 0]$ $\qquad$ BI (43)(19)

12. $\displaystyle\int_0^{\pi/2} \left[1 - \sec^2 x\cos\left(\tan x\right)\right]\frac{\mathrm{d}x}{\tan x} = \boldsymbol{C}$ $\qquad$ BI (51)(14)

13. $\displaystyle\int_0^{\pi/2} \sin\left(a\cot x\right)\sin 2x\,\mathrm{d}x = \frac{a\pi}{2}e^{-a}$ $\qquad$ $[a \geq 0]$ $\qquad (\text{cf. } \mathbf{3.716}\ 3.)$

and in general, formulas **3.716** remain valid if we replace $\tan x$ in the argument of the sine or cosine with $\cot x$ if we also replace $\sin x$ with $\cos x$, $\cos x$ with $\sin x$, hence $\tan x$ with $\cot x$, $\cot x$ with $\tan x$, $\sec x$ with $\operatorname{cosec} x$, and $\operatorname{cosec} x$ with $\sec x$ in the factors. Analogously,

**3.717** $\displaystyle\int_0^{\pi/2} \sin\left(a\operatorname{cosec} x\right)\sin\left(a\cot x\right)\frac{\mathrm{d}x}{\cos x} = \int_0^{\pi/2} \sin\left(a\sec x\right)\sin\left(a\tan x\right)\frac{\mathrm{d}x}{\sin x} = \frac{\pi}{2}\sin a$ $\qquad$ $[a \geq 0]$

$\qquad\qquad\qquad\qquad\qquad\qquad\qquad\qquad\qquad\qquad\qquad\qquad\qquad\qquad\qquad\qquad\qquad\qquad\qquad$ BI (52)(11, 12)

**3.718**

1.  $\displaystyle\int_0^{\pi/2} \sin\left(\frac{\pi}{2}p - a\tan x\right)\tan^{p-1} x\,\mathrm{d}x = \int_0^{\pi/2} \cos\left(\frac{\pi}{2}p - a\tan x\right)\tan^p x\,\mathrm{d}x = \frac{\pi}{2}e^{-a}$

    $\qquad\qquad\qquad\qquad\qquad\qquad\qquad\qquad\qquad\qquad\qquad [p^2 < 1, \quad p \neq 0, \quad a \geq 0] \qquad$ BI (44)(5, 6)

2.  $\displaystyle\int_0^{\pi/2} \sin\left(a\tan x - \nu x\right)\sin^{\nu-2} x\,\mathrm{d}x = 0$ $\qquad$ $[\operatorname{Re}\nu > 0, \quad a > 0]$ $\qquad$ NH 157(15)

3.  $\displaystyle\int_0^{\pi/2} \sin\left(n\tan x + \nu x\right)\frac{\cos^{\nu-1} x}{\sin x}\,\mathrm{d}x = \frac{\pi}{2}$ $\qquad$ $[\operatorname{Re}\nu > 0]$ $\qquad$ BI (51)(15)

4. $\displaystyle\int_0^{\pi/2} \cos\left(a\tan x - \nu x\right)\cos^{\nu-2}x\,\mathrm{d}x = \frac{\pi e^{-a}a^{\nu-1}}{\Gamma(\nu)}$       $[\operatorname{Re}\nu > 1, \quad a > 0]$

LO V 153(112), NT 157(14)

5. $\displaystyle\int_0^{\pi/2} \cos\left(a\tan x + \nu x\right)\cos^{\nu}x\,\mathrm{d}x = 2^{-\nu-1}\pi e^{-a}$       $[\operatorname{Re}\nu > -1, \quad a \geq 0]$     BI (44)(4)

6. $\displaystyle\int_0^{\pi/2} \cos\left(a\tan x - \gamma x\right)\cos^{\nu}x\,\mathrm{d}x = \frac{\pi a^{\frac{\nu}{2}}}{2^{\frac{\nu}{2}+1}}\frac{W_{\frac{\gamma}{2},-\frac{\nu+1}{2}}(2a)}{\Gamma\left(1+\frac{\gamma+\nu}{2}\right)}$

$$\left[a > 0, \quad \operatorname{Re}\nu > -1, \quad \frac{\nu+\gamma}{2} \neq -1, -2, \ldots\right] \quad \text{EH I 274(13)a}$$

7. $\displaystyle\int_0^{\pi/2} \frac{\sin nx - \sin\left(nx - a\tan x\right)}{\sin x}\cos^{n-1}x\,\mathrm{d}x = \begin{cases} \pi/2 & [n = 0, \quad a > 0], \\ \pi\left(1 - e^{-a}\right) & [n = 1, \quad a \geq 0] \end{cases}$

LO V 153(114)

**3.719**

1.[6] $\displaystyle\int_0^{\pi} \sin\left(\nu x - z\sin x\right)\,\mathrm{d}x = \pi\,\mathbf{E}_\nu(z)$                                  WA 336(2)

2. $\displaystyle\int_0^{\pi} \cos\left(nx - z\sin x\right)\,\mathrm{d}x = \pi J_n(z)$                                         WH

3. $\displaystyle\int_0^{\pi} \cos\left(\nu x - z\sin x\right)\,\mathrm{d}x = \pi\,\mathbf{J}_\nu(z)$                                        WA 336(1)

# 3.72–3.74 Combinations of trigonometric and rational functions

**3.721**

1. $\displaystyle\int_0^{\infty} \frac{\sin(ax)}{x}\,\mathrm{d}x = \frac{\pi}{2}\operatorname{sign}a$                                        FI II 645

2. $\displaystyle\int_1^{\infty} \frac{\sin(ax)}{x}\,\mathrm{d}x = -\operatorname{si}(a)$                                        BI 203(1)

3.[8] $\displaystyle\int_1^{\infty} \frac{\cos(ax)}{x}\,\mathrm{d}x = -\operatorname{ci}(a)$                                        BI 203(5)

**3.722**

1. $\displaystyle\int_0^{\infty} \frac{\sin(ax)}{x+b}\,\mathrm{d}x = \operatorname{ci}(ab)\sin(ab) - \cos(ab)\operatorname{si}(ab)$       $[|\arg b| < \pi, \quad a > 0]$

BI(16)(1), FI II 646a

2.[11] $\displaystyle\int_{-\infty}^{\infty} \frac{\sin(ax)}{x+b}\,\mathrm{d}x = \pi e^{iab}$       $[a > 0, \quad \operatorname{Im}b > 0]$

3. $\displaystyle\int_0^{\infty} \frac{\cos(ax)}{x+b}\,\mathrm{d}x = -\sin(ab)\operatorname{si}(ab) - \cos(ab)\operatorname{ci}(ab)$       $[|\arg b| < \pi, \quad a > 0]$

ET I 8(7), BI(160)(2)

**4.**[8] $\displaystyle\int_{-\infty}^{\infty}\frac{\cos(ax)}{x+b}\,dx=-i\pi e^{iab}$ $\qquad[a>0,\quad\operatorname{Im}b>0]$

**5.**[10] $\displaystyle\int_{0}^{\infty}\frac{\sin(ax)}{b-x}\,dx=\sin(ba)\operatorname{ci}(ba)-\cos(ba)[\operatorname{si}(ba)+\pi]$ $\qquad[a>0,\quad b\text{ not real and positive}]$

$\qquad\qquad$ FI II 646, BI(161)(1)

**6.**[8] $\displaystyle\int_{-\infty}^{\infty}\frac{\sin(ax)}{b-x}\,dx=-\pi e^{iab}$ $\qquad[a>0,\quad\operatorname{Im}b>0]$

**7.**[12] $\displaystyle\int_{0}^{\infty}\frac{\cos(ax)}{b-x}\,dx=\cos(ab)\operatorname{ci}(ab)+\sin(ab)[\operatorname{si}(ab)+\pi]$ $\qquad[a>0,\quad b\text{ not real and positive}]$

$\qquad\qquad$ ET I 8(8), BI(161)(2)a

**8.**[11] $\displaystyle\int_{-\infty}^{\infty}\frac{\cos(ax)}{b-x}\,dx=i\pi e^{-iab}$ $\qquad[a>0,\quad\operatorname{Im}b>0]$

**3.723**

**1.**[12] $\displaystyle\int_{0}^{\infty}\frac{\sin(ax)}{b^2+x^2}\,dx=\frac{1}{2b}\left[e^{-ab}\overline{\operatorname{Ei}}(ab)-e^{ab}\operatorname{Ei}(-ab)\right]$ $\qquad[a>0,\quad\operatorname{Re}b>0]$

$\qquad\qquad$ ET I 65(14), BI(160)(3)

**2.** $\displaystyle\int_{0}^{\infty}\frac{\cos(ax)}{b^2+x^2}\,dx=\frac{\pi}{2b}e^{-ab}$ $\qquad[a\geq0,\quad\operatorname{Re}b>0]$

$\qquad\qquad$ FI II 741, 750, ET I 8(11), WH

**3.** $\displaystyle\int_{0}^{\infty}\frac{x\sin(ax)}{b^2+x^2}\,dx=\frac{\pi}{2}e^{-ab}$ $\qquad[a>0,\quad\operatorname{Re}b>0]$

$\qquad\qquad$ FI II 741, 750, ET I 65(15), WH

**4.** $\displaystyle\int_{-\infty}^{\infty}\frac{x\sin(ax)}{b^2+x^2}\,dx=\pi e^{-ab}$ $\qquad[a>0,\quad\operatorname{Re}b>0]$ $\quad$ BI (202)(10)

**5.**[12] $\displaystyle\int_{0}^{\infty}\frac{x\cos(ax)}{b^2+x^2}\,dx=-\frac{1}{2}\left[e^{-ab}\overline{\operatorname{Ei}}(ab)+e^{ab}\operatorname{Ei}(-ab)\right]$ $\qquad[a>0,\quad\operatorname{Re}b>0]$ $\quad$ BI (160)(6)

**6.** $\displaystyle\int_{-\infty}^{\infty}\frac{\sin[a(b-x)]}{c^2+x^2}\,dx=\frac{\pi}{c}e^{-ac}\sin(ab)$ $\qquad[a>0,\quad b>0,\quad c>0]$ $\quad$ LI (202)(9)

**7.** $\displaystyle\int_{-\infty}^{\infty}\frac{\cos[a(b-x)]}{c^2+x^2}\,dx=\frac{\pi}{c}e^{-ac}\cos(ab)$ $\qquad[a>0,\quad b>0,\quad c>0]$ $\quad$ LI (202)(11)a

**8.** $\displaystyle\int_{0}^{\infty}\frac{\sin(ax)}{b^2-x^2}\,dx=\frac{1}{b}\left[\sin(ab)\operatorname{ci}(ab)-\cos(ab)\left(\operatorname{si}(ab)+\frac{\pi}{2}\right)\right]$

$\qquad[|\arg b|<\pi,\quad a>0]$ $\quad$ BI (161)(3)

**9.** $\displaystyle\int_{0}^{\infty}\frac{\cos(ax)}{b^2-x^2}\,dx=\frac{\pi}{2b}\sin(ab)$ $\qquad[a>0,\quad b>0]$ $\quad$ BI(161)(5), ET I 9(15)

**10.** $\displaystyle\int_{0}^{\infty}\frac{x\sin(ax)}{b^2-x^2}\,dx=-\frac{\pi}{2}\cos(ab)$ $\qquad[a>0]$ $\quad$ FI II 647, ET II 252(45)

**11.**[12] $\displaystyle\int_{0}^{\infty}\frac{x\cos(ax)}{b^2-x^2}\,dx=\cos(ab)\operatorname{ci}(ab)+\sin(ab)\left[\operatorname{si}(ab)+\frac{\pi}{2}\right]$

$\qquad[|\arg b|<\pi,\quad a>0]$ $\quad$ BI (161)(6)

$12.^{12}$   $\text{PV} \int_{-\infty}^{\infty} \dfrac{\sin(ax)}{x(x-b)}\,dx = \pi \dfrac{\cos(ab)-1}{b}$        $[a>0, \quad b>0]$        ET II 252(44)

**3.724**

1.   $\int_{-\infty}^{\infty} \dfrac{b+cx}{p+2qx+x^2} \sin(ax)\,dx = \left( \dfrac{cq-b}{\sqrt{p-q^2}} \sin(aq) + c\cos(aq) \right) \pi e^{-a\sqrt{p-q^2}}$

$$[a>0, \quad p>q^2]$$      BI (202)(12)

2.   $\int_{-\infty}^{\infty} \dfrac{b+cx}{p+2qx+x^2} \cos(ax)\,dx = \left( \dfrac{b-cq}{\sqrt{p-q^2}} \cos(aq) + c\sin(aq) \right) \pi e^{-a\sqrt{p-q^2}}$

$$[a>0, \quad p>q^2]$$      BI (202)(13)

3.   $\int_{-\infty}^{\infty} \dfrac{\cos[(b-1)t] - x\cos(bt)}{1 - 2x\cos t + x^2} \cos(ax)\,dx = \pi e^{-a\sin t} \sin(bt + a\cos t)$

$$[a>0, \quad t^2 < \pi^2]$$      BI (202)(14)

**3.725**

1.   $\int_0^{\infty} \dfrac{\sin(ax)\,dx}{x(\beta^2+x^2)} = \dfrac{\pi}{2\beta^2}\left(1 - e^{-a\beta}\right)$      $[\operatorname{Re}\beta>0, \quad a>0]$      BI (172)(1)

$2.^{12}$   $\text{PV} \int_0^{\infty} \dfrac{\sin(ax)\,dx}{x(b^2-x^2)} = \dfrac{\pi}{2b^2}\left(1 - \cos(ab)\right)$      $[a>0]$      BI (172)(4)

3.   $\int_0^{\infty} \dfrac{\sin(ax)\cos(bx)}{x(x^2+\beta^2)}\,dx = \dfrac{\pi}{2\beta^2} e^{-\beta b} \sinh(a\beta)$      $[0<a<b]$

$$= -\dfrac{\pi}{2\beta^2} e^{-a\beta} \cosh(b\beta) + \dfrac{\pi}{2\beta^2} \quad [a>b>0]$$

ET I 19(4)

**3.726**

In each of the following integrals, take both lower $(-)$ signs or both upper $(+)$ signs.

$1.^{11}$   $\int_0^{\infty} \dfrac{x\sin(ax)\,dx}{b^3 \pm b^2x + bx^2 \pm x^3}$

$$= \pm\dfrac{1}{4b}\left[ e^{-ab}\overline{\text{Ei}}(ab) - e^{ab}\,\text{Ei}(-ab) - 2\,\text{ci}(ab)\sin(ab) + 2\cos(ab)\left(\text{si}(ab) + \dfrac{\pi}{2}\right) \right]$$
$$+ \dfrac{\pi e^{-ab} - \pi\cos(ab)}{4b}$$

$[a>0, \quad b>0;$   if the lower sign is taken, then the integral is a principal value integral$]$

ET I 65(21)a, BI(176)(10, 13)

$2.^{7}$   $\int_0^{\infty} \dfrac{x^2\sin(ax)\,dx}{b^3 \pm b^2x + bx^2 \pm x^3}$

$$= \dfrac{1}{4}\left[ e^{ab}\,\text{Ei}(-ab) - e^{-ab}\overline{\text{Ei}}(ab) + 2\,\text{ci}(ab)\sin(ab) - 2\cos(ab)\left(\text{si}(ab) + \dfrac{\pi}{2}\right) \right]$$
$$\pm\pi\left(e^{-ab} + \cos(ab)\right)$$

$[a>0, \quad b>0;$   if the lower sign is taken, then the integral is a principal value integral$]$

ET I 66(22), BI(176)(11, 14)

**3.727**

1. $\int_0^\infty \dfrac{\cos(ax)}{b^4 + x^4}\,\mathrm{d}x = \dfrac{\pi\sqrt{2}}{4b^3}\exp\left(-\dfrac{ab}{\sqrt{2}}\right)\left(\cos\dfrac{ab}{\sqrt{2}} + \sin\dfrac{ab}{\sqrt{2}}\right)$

$[a > 0, \quad b > 0]$    BI(160)(25)a, ET I 9(19)

2.[12] $\mathrm{PV}\displaystyle\int_0^\infty \dfrac{\sin(ax)}{b^4 - x^4}\,\mathrm{d}x = \dfrac{1}{4b^3}\left[2\sin(ab)\operatorname{ci}(ab) - 2\cos(ab)\left(\operatorname{si}(ab) + \dfrac{\pi}{2}\right)\right.$

$\left. + e^{-ab}\overline{\operatorname{Ei}}(ab) - e^{ab}\operatorname{Ei}(-ab)\right]$

$[a > 0, \quad b > 0]$    BI (161)(12)

3.[12] $\mathrm{PV}\displaystyle\int_0^\infty \dfrac{\cos(ax)}{b^4 - x^4}\,\mathrm{d}x = \dfrac{\pi}{4b^3}\left[e^{-ab} + \sin(ab)\right]$

$[a > 0, \quad b > 0]$    (cf. **3.723** 2 and **3.723** 9)   BI (161)(16)

4. $\int_0^\infty \dfrac{x\sin(ax)}{b^4 + x^4}\,\mathrm{d}x = \dfrac{\pi}{2b^2}\exp\left(-\dfrac{ab}{\sqrt{2}}\right)\sin\dfrac{ab}{\sqrt{2}}$     $[a > 0, \quad b > 0]$    BI (160)(23)a

5.[12] $\mathrm{PV}\displaystyle\int_0^\infty \dfrac{x\sin(ax)}{b^4 - x^4}\,\mathrm{d}x = \dfrac{\pi}{4b^2}\left[e^{-ab} - \cos(ab)\right]$     $[a > 0, \quad b > 0]$    BI (161)(13)

6.[12] $\mathrm{PV}\displaystyle\int_0^\infty \dfrac{x\cos(ax)}{b^4 - x^4}\,\mathrm{d}x = \dfrac{1}{4b^2}\left[2\cos(ab)\operatorname{ci}(ab) + 2\sin(ab)\left(\operatorname{si}(ab) + \dfrac{\pi}{2}\right)\right.$

$\left. - e^{-ab}\overline{\operatorname{Ei}}(ab) - e^{ab}\operatorname{Ei}(-ab)\right]$

$[a > 0, \quad b > 0]$    (cf. **3.723** 5 and **3.723** 11)   BI (161)(17)

7. $\int_0^\infty \dfrac{x^2\cos(ax)}{b^4 + x^4}\,\mathrm{d}x = \dfrac{\pi\sqrt{2}}{4b}\exp\left(-\dfrac{ab}{\sqrt{2}}\right)\left(\cos\dfrac{ab}{\sqrt{2}} - \sin\dfrac{ab}{\sqrt{2}}\right)$

$[a > 0, \quad b > 0]$    BI (160)(26)a

8.[12] $\mathrm{PV}\displaystyle\int_0^\infty \dfrac{x^2\sin(ax)\,\mathrm{d}x}{b^4 - x^4} = \dfrac{1}{4b}\left[2\sin(ab)\operatorname{ci}(ab)\right.$

$\left. - 2\cos(ab)\left(\operatorname{si}(ab) + \dfrac{\pi}{2}\right) - e^{-ab}\overline{\operatorname{Ei}}(ab) + e^{ab}\operatorname{Ei}(-ab)\right]$

$[a > 0, \quad b > 0]$    BI (161)(14)

9.[12] $\mathrm{PV}\displaystyle\int_0^\infty \dfrac{x^2\cos(ax)}{b^4 - x^4}\,\mathrm{d}x = \dfrac{\pi}{4b}\left(\sin(ab) - e^{-ab}\right)$     $[a > 0, \quad b > 0]$    BI (161)(18)

10. $\int_0^\infty \dfrac{x^3\sin(ax)}{b^4 + x^4}\,\mathrm{d}x = \dfrac{\pi}{2}\exp\left(-\dfrac{ab}{\sqrt{2}}\right)\cos\dfrac{ab}{\sqrt{2}}$     $[a > 0, \quad b > 0]$    BI (160)(24)

11.[12] $\mathrm{PV}\displaystyle\int_0^\infty \dfrac{x^3\sin(ax)}{b^4 - x^4}\,\mathrm{d}x = \dfrac{-\pi}{4}\left[e^{-ab} - \cos(ab)\right]$     $[a > 0, \quad b > 0]$    BI (161)(15)

12.[12] $\mathrm{PV}\displaystyle\int_0^\infty \dfrac{x^3\cos(ax)\,\mathrm{d}x}{b^4 - x^4} = \dfrac{1}{4}\left[2\cos(ab)\operatorname{ci}(ab) + 2\sin(ab)\left(\operatorname{si}(ab) + \dfrac{\pi}{2}\right)\right.$

$\left. + e^{-ab}\overline{\operatorname{Ei}}(ab) + e^{ab}\operatorname{Ei}(-ab)\right]$

$[a > 0, \quad b > 0]$    BI(161)(19)

13. $\displaystyle\int_0^\infty \frac{x^3 \sin ax}{(x^2 + b^2)^3}\, dx = \frac{\pi e^{-ab}}{16b}\left(3a - ba^2\right)$

14. $\displaystyle\int_0^\infty \frac{x^3 \sin ax}{(x^2 + b^2)^4}\, dx = \frac{\pi e^{-ab}a}{96b^3}\left(3 + 3ab - a^2b^2\right)$

**3.728**

1. $\displaystyle\int_0^\infty \frac{\cos(ax)\, dx}{(\beta^2 + x^2)(\gamma^2 + x^2)} = \frac{\pi\left(\beta e^{-a\gamma} - \gamma e^{-a\beta}\right)}{2\beta\gamma\left(\beta^2 - \gamma^2\right)}$        $[a > 0, \quad \operatorname{Re}\beta > 0, \quad \operatorname{Re}\gamma > 0]$

                                                                  BI (175)(1)

2. $\displaystyle\int_0^\infty \frac{x \sin(ax)\, dx}{(\beta^2 + x^2)(\gamma^2 + x^2)} = \frac{\pi\left(e^{-a\beta} - e^{-a\gamma}\right)}{2\left(\gamma^2 - \beta^2\right)}$        $[a > 0, \quad \operatorname{Re}\beta > 0, \quad \operatorname{Re}\gamma > 0]$

                                                                  BI (174)(1)

3. $\displaystyle\int_0^\infty \frac{x^2 \cos(ax)\, dx}{(\beta^2 + x^2)(\gamma^2 + x^2)} = \frac{\pi\left(\beta e^{-a\beta} - \gamma e^{-a\gamma}\right)}{2\left(\beta^2 - \gamma^2\right)}$        $[a > 0, \quad \operatorname{Re}\beta > 0, \quad \operatorname{Re}\gamma > 0]$

                                                                  BI (175)(2)

4. $\displaystyle\int_0^\infty \frac{x^3 \sin(ax)\, dx}{(\beta^2 + x^2)(\gamma^2 + x^2)} = \frac{\pi\left(\beta^2 e^{-a\beta} - \gamma^2 e^{-a\gamma}\right)}{2\left(\beta^2 - \gamma^2\right)}$        $[a > 0, \quad \operatorname{Re}\beta > 0, \quad \operatorname{Re}\gamma > 0]$

                                                                  BI (174)(2)

5.[12] $\displaystyle \mathrm{PV}\int_0^\infty \frac{\cos(ax)\, dx}{(b^2 - x^2)(c^2 - x^2)} = \frac{\pi\left(b\sin(ac) - c\sin(ab)\right)}{2bc\left(b^2 - c^2\right)}$

                                             $[a > 0, \quad b > 0, \quad c > 0]$      BI (175)(3)

6.[12] $\displaystyle \mathrm{PV}\int_0^\infty \frac{x \sin(ax)\, dx}{(b^2 - x^2)(c^2 - x^2)} = \frac{\pi\left(\cos(ab) - \cos(ac)\right)}{2\left(b^2 - c^2\right)}$      $[a > 0]$      BI (174)(3)

7.[12] $\displaystyle \mathrm{PV}\int_0^\infty \frac{x^2 \cos(ax)\, dx}{(b^2 - x^2)(c^2 - x^2)} = \frac{\pi\left(c\sin(ac) - b\sin(ab)\right)}{2\left(b^2 - c^2\right)}$

                                                               $[a > 0, \quad b > 0, \quad c > 0]$      BI (175)(4)

8.[12] $\displaystyle \mathrm{PV}\int_0^\infty \frac{x^3 \sin(ax)\, dx}{(b^2 - x^2)(c^2 - x^2)} = \frac{\pi\left(b^2 \cos(ab) - c^2 \cos(ac)\right)}{2\left(b^2 - c^2\right)}$

                                                               $[a > 0, \quad b > 0, \quad c > 0]$      BI (174)(4)

9.[12] $\displaystyle \mathrm{PV}\int_0^\infty \frac{x \sin ax}{(b^2 - x^2)(c^2 + x^2)}\, dx = \frac{\pi}{2}\frac{e^{-ac} - \cos ba}{a^2 + c^2}$      $[a > 0, \quad c > 0, \quad b \text{ real}]$

**3.729**

1. $\displaystyle\int_0^\infty \frac{\cos(ax)\, dx}{(b^2 + x^2)^2} = \frac{\pi}{4b^3}(1 + ab)e^{-ab}$         $[a > 0, \quad b > 0]$      BI (170)(7)

2. $\displaystyle\int_0^\infty \frac{x \sin(ax)\, dx}{(b^2 + x^2)^2} = \frac{\pi}{4b}ae^{-ab}$         $[a > 0, \quad b > 0]$      BI (170)(3)

3. $\displaystyle\int_0^\infty \cos(px)\frac{1 - x^2}{(1 + x^2)^2}\, dx = \frac{\pi p}{2}e^{-p}$                                BI (43)(10)a

4. $\displaystyle\int_0^\infty \frac{x^3 \sin(ax)\, dx}{(b^2 + x^2)^2} = \frac{\pi}{4}(2 - ab)e^{-ab}$         $[a > 0, \quad b > 0]$      BI (170)(4)

**3.731**   **Notation:**   $2A^2 = \sqrt{b^4 + c^2} + b^2, \quad 2B^2 = \sqrt{b^4 + c^2} - b^2,$

1.   $\displaystyle\int_0^\infty \frac{\cos(ax)\,\mathrm{d}x}{(x^2 + b^2)^2 + c^2} = \frac{\pi}{2c}\,\frac{e^{-aA}\left(B\cos(aB) + A\sin(aB)\right)}{\sqrt{b^4 + c^2}}$

$$[a > 0, \quad b > 0, \quad c > 0] \qquad \text{BI (176)(3)}$$

2.   $\displaystyle\int_0^\infty \frac{x\sin(ax)\,\mathrm{d}x}{(x^2 + b^2)^2 + c^2} = \frac{\pi}{2c}e^{-aA}\sin(aB) \qquad [a > 0, \quad b > 0, \quad c > 0] \qquad \text{BI (176)(1)}$

3.   $\displaystyle\int_0^\infty \frac{\left(x^2 + b^2\right)\cos(ax)\,\mathrm{d}x}{(x^2 + b^2)^2 + c^2} = \frac{\pi}{2}\,\frac{e^{-aA}\left(A\cos(aB) - B\sin(aB)\right)}{\sqrt{b^4 + c^2}}$

$$[a > 0, \quad b > 0, \quad c > 0] \qquad \text{BI (176)(4)}$$

4.   $\displaystyle\int_0^\infty \frac{x\left(x^2 + b^2\right)\sin(ax)\,\mathrm{d}x}{(x^2 + b^2)^2 + c^2} = \frac{\pi}{2}e^{-aA}\cos(aB) \qquad [a > 0, \quad b > 0, \quad c > 0] \qquad \text{BI (176)(2)}$

**3.732**

1.   $\displaystyle\int_0^\infty \left[\frac{1}{\beta^2 + (\gamma - x)^2} - \frac{1}{\beta^2 + (\gamma + x)^2}\right]\sin(ax)\,\mathrm{d}x = \frac{\pi}{\beta}e^{-a\beta}\sin(a\gamma)$

$$[a > 0, \quad \operatorname{Re}\beta > 0, \quad \gamma + i\beta \text{ is not real}]$$
$$\text{ET I 65(16)}$$

2.   $\displaystyle\int_0^\infty \left[\frac{1}{\beta^2 + (\gamma - x)^2} + \frac{1}{\beta^2 + (\gamma + x)^2}\right]\cos(ax)\,\mathrm{d}x = \frac{\pi}{\beta}e^{-a\beta}\cos(a\gamma)$

$$[a > 0, \quad |\operatorname{Im}\gamma| < \operatorname{Re}\beta] \qquad \text{ET I 8(13)}$$

3.   $\displaystyle\int_0^\infty \left[\frac{\gamma + x}{\beta^2 + (\gamma + x)^2} - \frac{\gamma - x}{\beta^2 + (\gamma - x)^2}\right]\sin(ax)\,\mathrm{d}x = \pi e^{-a\beta}\cos(a\gamma)$

$$[a > 0, \quad \operatorname{Re}\beta > 0, \quad \gamma + i\beta \text{ is not real}]$$
$$\text{LI (175)(17)}$$

4.   $\displaystyle\int_0^\infty \left[\frac{\gamma + x}{\beta^2 + (\gamma + x)^2} + \frac{\gamma - x}{\beta^2 + (\gamma - x)^2}\right]\cos(ax)\,\mathrm{d}x = \pi e^{-a\beta}\sin(a\gamma)$

$$[a > 0, \quad |\operatorname{Im}a| < \operatorname{Re}\beta] \qquad \text{LI (176)(21)}$$

**3.733**

1.   $\displaystyle\int_0^\infty \frac{\cos(ax)\,\mathrm{d}x}{x^4 + 2b^2x^2\cos 2t + b^4} = \frac{\pi}{2b^3}\exp\left(-ab\cos t\right)\frac{\sin\left(t + ab\sin t\right)}{\sin 2t}$

$$\left[a > 0, \quad b > 0, \quad |t| < \tfrac{\pi}{2}\right] \qquad \text{BI (176)(7)}$$

2.   $\displaystyle\int_0^\infty \frac{x\sin(ax)\,\mathrm{d}x}{x^4 + 2b^2x^2\cos 2t + b^4} = \frac{\pi}{2b^2}\exp\left(-ab\cos t\right)\frac{\sin\left(ab\sin t\right)}{\sin 2t}$

$$\left[a > 0, \quad b > 0, \quad |t| < \tfrac{\pi}{2}\right]$$
$$\text{BI(176)(5), ET I 66(23)}$$

3.   $\displaystyle\int_0^\infty \frac{x^2\cos(ax)\,\mathrm{d}x}{x^4 + 2b^2x^2\cos 2t + b^4} = \frac{\pi}{2b}\exp\left(-ab\cos t\right)\frac{\sin\left(t - ab\sin t\right)}{\sin 2t}$

$$\left[a > 0, \quad b > 0, \quad |t| < \tfrac{\pi}{2}\right] \qquad \text{BI (176)(8)}$$

4.    $\displaystyle\int_0^\infty \frac{x^3 \sin(ax)\,dx}{x^4 + 2b^2 x^2 \cos 2t + b^4} = \frac{\pi}{2} \exp\left(-ab\cos t\right) \frac{\sin\left(2t - ab\sin t\right)}{\sin 2t}$

$$\left[a > 0, \quad b > 0, \quad |t| < \tfrac{\pi}{2}\right] \qquad \text{BI (176)(6)}$$

5.    $\displaystyle\int_0^\infty \frac{\sin(ax)\,dx}{x\left(x^4 + 2b^2 x^2 \cos 2t + b^4\right)} = \frac{\pi}{2b^4}\left[1 - \exp\left(-ab\cos t\right) \frac{\sin\left(2t + ab\sin t\right)}{\sin 2t}\right]$

$$\left[a > 0, \quad b > 0, \quad |t| < \tfrac{\pi}{2}\right] \qquad \text{BI (176)(22)}$$

**3.734**

1.[12]   $\displaystyle\text{PV}\int_0^\infty \frac{\sin(ax)\,dx}{x\left(b^4 + x^4\right)} = \frac{\pi}{2b^4}\left[1 - \exp\left(-\frac{ab}{\sqrt{2}}\right)\cos\frac{ab}{\sqrt{2}}\right] \quad [a > 0, \quad b > 0]$    BI (172)(7)

2.[12]   $\displaystyle\text{PV}\int_0^\infty \frac{\sin(ax)\,dx}{x\left(b^4 - x^4\right)} = \frac{\pi}{4b^4}\left[2 - e^{-ab} - \cos(ab)\right] \qquad [a > 0, \quad b > 0]$    BI (172)(10)

**3.735**[12]   $\displaystyle\int_0^\infty \frac{\sin x}{x\left(a^2 + x^2\right)^2}\,dx = \frac{\pi}{4a^4}\left[2 - e^{-a}(2 + a)\right] \qquad [a > 0]$    WH, BI (172)(22)

**3.736**

1.[12]   $\displaystyle\text{PV}\int_0^\infty \frac{\cos(ax)\,dx}{\left(b^2 + x^2\right)\left(b^4 - x^4\right)} = \frac{\pi}{8b^5}\left[\sin(ab) + (2 + ab)e^{-ab}\right]$

$$[a > 0, \quad b > 0] \qquad \text{BI (176)(5)}$$

2.[12]   $\displaystyle\text{PV}\int_0^\infty \frac{x\sin(ax)\,dx}{\left(b^2 + x^2\right)\left(b^4 - x^4\right)} = \frac{\pi}{8b^4}\left[(1 + ab)e^{-ab} - \cos(ab)\right]$

$$[a > 0, \quad b > 0] \qquad \text{BI (174)(5)}$$

3.[12]   $\displaystyle\text{PV}\int_0^\infty \frac{x^2\cos(ax)\,dx}{\left(b^2 + x^2\right)\left(b^4 - x^4\right)} = \frac{\pi}{8b^3}\left[\sin(ab) - abe^{-ab}\right] \quad [a > 0, \quad b > 0]$    BI (175)(6)

4.[12]   $\displaystyle\text{PV}\int_0^\infty \frac{x^3\sin(ax)\,dx}{\left(b^2 + x^2\right)\left(b^4 - x^4\right)} = \frac{\pi}{8b^2}\left[(1 - ab)e^{-ab} - \cos(ab)\right]$

$$[a > 0, \quad b > 0] \qquad \text{BI (174)(6)}$$

5.[12]   $\displaystyle\text{PV}\int_0^\infty \frac{x^4\cos(ax)\,dx}{\left(b^2 + x^2\right)\left(b^4 - x^4\right)} = \frac{\pi}{8b}\left[\sin(ab) + (ab - 2)e^{-ab}\right]$

$$[a > 0, \quad b > 0] \qquad \text{BI (175)(7)}$$

6.[12]   $\displaystyle\text{PV}\int_0^\infty \frac{x^5\sin(ax)\,dx}{\left(b^2 + x^2\right)\left(b^4 - x^4\right)} = \frac{\pi}{8}\left[(ab - 3)e^{-ab} - \cos(ab)\right]$

$$[a > 0, \quad b > 0] \qquad \text{BI (174)(7)}$$

**3.737**

1.[12]   $\displaystyle\int_0^\infty \frac{\cos(ax)\,dx}{\left(b^2 + x^2\right)^n} = \frac{\pi e^{-ab}}{(2b)^{2n-1}(n-1)!}\sum_{k=0}^{n-1} \frac{(2n - k - 2)!(2ab)^k}{k!(n - k - 1)!}$

$$= \frac{(-1)^{n-1}\pi}{b^{2n-1}(n-1)!}\left[\frac{d^{n-1}}{dp^{n-1}}\left(\frac{e^{-ab\sqrt{p}}}{\sqrt{p}}\right)\right]_{p=1}$$

$$= \frac{(-1)^{n-1}\pi}{2b^{2n-1}(n-1)!}\left[\frac{d^{n-1}}{dp^{n-1}}\left(\frac{e^{-abp}}{(1 + p)^n}\right)\right]_{p=1}$$

$$[a > 0, \quad b > 0] \quad \text{GW(333)(67b), WA 209, WA 192}$$

2.$^{12}$   $\displaystyle\int_0^\infty \frac{x\sin(ax)\,dx}{(x^2+b^2)^{n+1}} = \frac{\pi a e^{-ab}}{2^{2n}n!b^{2n-1}} \sum_{k=0}^{n-1} \frac{(2n-k-2)!(2ab)^k}{k!(n-k-1)!}$    $[a>0, \quad \mathrm{Re}\, b>0]$

$\displaystyle = \frac{\pi}{2}e^{-ab}$          $[n=0, \quad a>0, \quad b\geq 0]$

GW (333)(66c)

3.   $\displaystyle\int_0^\infty \frac{\sin(ax)\,dx}{x\,(b^2+x^2)^{n+1}} = \frac{\pi}{2b^{2n+2}}\left[1 - \frac{e^{-ab}}{2^n n!}F_n(ab)\right]$

$\left[a>0, \quad \mathrm{Re}\, b>0, \quad F_0(z)=1, \quad F_1(z)=z+2, \ldots, F_n(z)=(z+2n)F_{n-1}(z)-zF_{n-1}'(z)\right]$

GW (333)(66e)

4.   $\displaystyle\int_0^\infty \frac{x\sin(ax)\,dx}{(b^2+x^2)^3} = \frac{\pi a}{16b^3}(1+ab)e^{-ab}$      $[a>0, \quad b>0]$    BI(170)(5), ET I 67(35)a

5.   $\displaystyle\int_0^\infty \frac{x\sin(ax)\,dx}{(b^2+x^2)^4} = \frac{\pi a}{96b^5}\left(3+3ab+a^2b^2\right)e^{-ab}$      $[a>0, \quad b>0]$    BI(170)(6), ET I 67(35)a

6.   $\displaystyle\int_0^\infty \frac{x^3\sin ax}{(x^2+b^2)^{n+1}}\,dx = \frac{\pi e^{-ab}}{2^{2n}n!b^{2n-2}}\left[2^{n-1}(2n-3)!!(2-ba)\right.$

$\displaystyle \left. - \sum_{k=1}^{n-1} \frac{(2n-k-2)!2^k(ba)^{k-1}}{k!(n-k-1)!}\left[k(k+1)-2(k+1)ba+b^2a^2\right]\right]$

## 3.738

1.$^{12}$   $\displaystyle\int_0^\infty \frac{x^{m-1}\sin(ax)}{x^{2n}+b^{2n}}\,dx = -\frac{\pi b^{m-2n}}{2n}\sum_{k=1}^n \exp\left[-ab\sin\frac{(2k-1)\pi}{2n}\right]$

$\displaystyle \times\left[\cos\frac{(2k-1)m\pi}{2n} + ab\cos\frac{(2k-1)\pi}{2n}\right]$

$[m \text{ is even}], \qquad \left[a>0, \quad |\arg b| < \frac{\pi}{2n}, \quad 0\leq m\leq 2n\right]$

$= 0$       $[m \text{ is odd}]$

ET I 67(38)

2.$^{12}$   $\displaystyle\int_0^\infty \frac{x^{m-1}\cos(ax)}{x^{2n}+b^{2n}}\,dx = \frac{\pi b^{m-2n}}{2n}\sum_{k=1}^n \exp\left[-ab\sin\frac{(2k-1)\pi}{2n}\right]$

$\displaystyle \times\left[\sin\frac{(2k-1)m\pi}{2n} + ab\cos\frac{(2k-1)\pi}{2n}\right]$

$[m \text{ is odd}], \qquad \left[a>0, \quad |\arg b| < \frac{\pi}{2n}, \quad 0<m<2n+1\right]$

$= 0$       $[m \text{ is even}]$

BI(160)(29)a, ET I 10(29)

**3.739**

1.
$$\int_0^\infty \frac{\sin(ax)\,dx}{x\left(x^2+2^2\right)\left(x^2+4^2\right)\ldots\left(x^2+4n^2\right)}$$

$$= \frac{\pi(-1)^n}{(2n)!2^{2n+1}}\left[2\sum_{k=0}^{n-1}(-1)^k\binom{2n}{k}e^{2(k-n)a}+(-1)^n\binom{2n}{n}\right]$$

$$[a>0,\quad n\geq 0] \qquad\qquad \text{LI(174)(8)}$$

2.
$$\int_0^\infty \frac{\cos(ax)\,dx}{\left(x^2+1^2\right)\left(x^2+3^2\right)\ldots\left[x^2+(2n+1)^2\right]}$$

$$= \frac{(-1)^n}{(2n+1)!}\frac{\pi}{2^{2n+1}}\sum_{k=0}^{n}(-1)^k\binom{2n+1}{k}e^{(2k-2n-1)a} \qquad [a\geq 0,\quad n\geq 0]$$

$$= \frac{\pi 2^{-2n-1}}{(2n+1)(n!)^2} \qquad\qquad\qquad [a=0,\quad n\geq 0]$$

$$\text{BI(175)(8)}$$

3.
$$\int_0^\infty \frac{x\sin(ax)\,dx}{\left(x^2+1^2\right)\left(x^2+3^2\right)\ldots\left[x^2+(2n+1)^2\right]}$$

$$= \frac{\pi(-1)^n}{(2n+1)!2^{2n+1}}\sum_{k=0}^{n}(-1)^k\binom{2n+1}{k}(2n-2k+1)e^{(2k-2n-1)a}$$

$$[a>0,\quad n\geq 0] \qquad\qquad \text{LI (174)(9)}$$

4.
$$\int_0^\infty \frac{\cos ax\,dx}{\left(x^2+2^2\right)\left(x^2+4^2\right)\ldots\left(x^2+4n^2\right)} = \frac{\pi 2^{1-2n}}{(2n)!}\sum_{k=1}^{n}(-1)^k k\binom{2n}{n-k}e^{-2ak}$$

$$[n\geq 1,\quad a\geq 0]$$

**3.741**

1.
$$\int_0^\infty \frac{\sin(ax)\sin(bx)}{x}\,dx = \frac{1}{4}\ln\left(\frac{a+b}{a-b}\right)^2 \qquad [a>0,\quad b>0,\quad a\neq b] \qquad \text{FI II 647}$$

2.
$$\int_0^\infty \frac{\sin(ax)\cos(bx)}{x}\,dx = \frac{\pi}{2} \qquad\qquad [a>b\geq 0]$$

$$= \frac{\pi}{4} \qquad\qquad [a=b>0]$$

$$= 0 \qquad\qquad [b>a\geq 0]$$

$$\text{FI II 645}$$

3.
$$\int_0^\infty \frac{\sin(ax)\sin(bx)}{x^2}\,dx = \frac{a\pi}{2} \qquad\qquad [0<a\leq b]$$

$$= \frac{b\pi}{2} \qquad\qquad [0<b\leq a]$$

$$\text{BI (157)(1)}$$

**3.742**

1.  $$\int_0^\infty \frac{\sin(ax)\sin(bx)}{c^2+x^2}\,\mathrm{d}x = \frac{\pi}{4c}\left(e^{-|a-b|c} - e^{-(a+b)c}\right) \qquad [a>0, \quad b>0, \quad \mathrm{Re}\,c>0]$$
    $$= \frac{\pi}{2c}e^{-ac}\sinh bc \qquad [c>0, \quad a\geq b\geq 0]$$
    $$= \frac{\pi}{2c}e^{-bc}\sinh ac \qquad [c>0, \quad b\geq a\geq 0]$$

    BI(162)(1)a, GW(333)(71a)

2.  $$\int_0^\infty \frac{\sin(ax)\cos(bx)}{c^2+x^2}\,\mathrm{d}x = \frac{1}{4c}e^{-ac}\left\{e^{bc}\,\mathrm{Ei}\,[c(a-b)] + e^{-bc}\,\mathrm{Ei}\,[c(a+b)]\right\}$$
    $$-\frac{1}{4c}e^{ac}\left\{e^{bc}\,\mathrm{Ei}\,[-c(a+b)] + e^{-bc}\,\mathrm{Ei}\,[c(b-a)]\right\}$$

    BI (162)(3)

3.  $$\int_0^\infty \frac{\cos(ax)\cos(bx)}{c^2+x^2}\,\mathrm{d}x = \frac{\pi}{4c}\left[e^{-|a-b|c} + e^{-(a+b)c}\right] \qquad [a>0, \quad b>0, \quad \mathrm{Re}\,c>0]$$
    $$= \frac{\pi}{2c}e^{-ac}\cosh bc \qquad [c>0, \quad a\geq b\geq 0]$$
    $$= \frac{\pi}{2c}e^{-bc}\cosh ac \qquad [c>0, \quad b\geq a\geq 0]$$

    BI(163)(1)a, GW(333)(71c)

4.  $$\int_0^\infty \frac{x\cos(ax)\cos(bx)}{c^2+x^2}\,\mathrm{d}x = -\frac{1}{4}e^{ac}\left\{e^{bc}\,\mathrm{Ei}\,[-c(a+b)] + e^{-bc}\,\mathrm{Ei}\,[c(b-a)]\right\}$$
    $$-\frac{1}{4}e^{-ac}\left\{e^{bc}\,\mathrm{Ei}\,[c(a-b)] + e^{-bc}\,\mathrm{Ei}\,[c(a+b)]\right\} \qquad [a\neq b]$$
    $$= \infty \qquad [a=b]$$

    BI (163)(2)

5.  $$\int_0^\infty \frac{x\sin(ax)\cos(bx)}{x^2+c^2}\,\mathrm{d}x = \frac{\pi}{2}e^{-ac}\cosh(bc) \qquad [0<b<a]$$
    $$= \frac{\pi}{4}e^{-2ac} \qquad [0<b=a]$$
    $$= -\frac{\pi}{2}e^{-bc}\sinh(ac) \qquad [0<a<b]$$

    BI (162)(4)

6.[12]  $$\mathrm{PV}\int_0^\infty \frac{\sin(ax)\sin(bx)}{p^2-x^2}\,\mathrm{d}x = -\frac{\pi}{2p}\cos(ap)\sin(bp) \qquad [a>b>0]$$
    $$= -\frac{\pi}{4p}\sin(2ap) \qquad [a=b>0]$$
    $$= -\frac{\pi}{2p}\sin(ap)\cos(bp) \qquad [b>a>0]$$

    BI (166)(1)

7.[12]  $$\mathrm{PV}\int_0^\infty \frac{\sin(ax)\cos(bx)}{p^2-x^2}x\,\mathrm{d}x = -\frac{\pi}{2}\cos(ap)\cos(bp) \qquad [a>b>0]$$
    $$= -\frac{\pi}{4}\cos(2ap) \qquad [a=b>0]$$
    $$= \frac{\pi}{2}\sin(ap)\sin(bp) \qquad [b>a>0]$$

    BI (166)(2)

$8.^{12}$   $\text{PV}\displaystyle\int_0^\infty \frac{\cos(ax)\cos(bx)}{p^2-x^2}\,dx = \frac{\pi}{2p}\sin(ap)\cos(bp)$      $[a>b>0]$

$$= \frac{\pi}{4p}\sin(2ap) \qquad\qquad [a=b>0]$$

$$= \frac{\pi}{2p}\cos(ap)\sin(bp) \qquad\qquad [b>a>0]$$

                                                                 BI (166)(3)

**3.743**

$1.^{12}$   $\text{PV}\displaystyle\int_0^\infty \frac{\sin(ax)}{\sin(bx)}\frac{dx}{x^2+c^2} = \frac{\pi}{2c}\frac{\sinh(ac)}{\sinh(bc)}$      $[0<a<b,\quad \operatorname{Re}c>0]$     ET I 80(21)

$2.^{12}$   $\text{PV}\displaystyle\int_0^\infty \frac{\sin(ax)}{\cos(bx)}\frac{x\,dx}{x^2+c^2} = -\frac{\pi}{2}\frac{\sinh(ac)}{\cosh(bc)}$      $[0<a<b,\quad \operatorname{Re}c>0]$     ET I 81(30)

$3.^{12}$   $\text{PV}\displaystyle\int_0^\infty \frac{\cos(ax)}{\sin(bx)}\frac{x\,dx}{x^2+c^2} = \frac{\pi}{2}\frac{\cosh(ac)}{\sinh(bc)}$      $[0<a<b,\quad \operatorname{Re}c>0]$     ET I 23(37)

$4.^{12}$   $\text{PV}\displaystyle\int_0^\infty \frac{\cos(ax)}{\cos(bx)}\frac{dx}{x^2+c^2} = \frac{\pi}{2c}\frac{\cosh(ac)}{\cosh(bc)}$      $[0<a<b,\quad \operatorname{Re}c>0]$     ET I 23(36)

$5.^{6}$   $\text{PV}\displaystyle\int_0^\infty \frac{\sin(ax)}{\sin x}\frac{dx}{b^2-x^2} = 0$      if $0\le a\le 1$

$$= \frac{\pi}{b}\sin(a-1)b \qquad\qquad \text{if } 1\le a\le 2$$

                                                        $[b \text{ real}, b/\pi \notin \mathbb{Z}]$

**3.744**$^{12}$   $\text{PV}\displaystyle\int_0^\infty \frac{\sin(ax)}{\cos(bx)}\frac{dx}{x(x^2+\beta^2)} = \frac{\pi}{2\beta^2}\frac{\sinh(a\beta)}{\cosh(b\beta)}$      $[0<a<b,\quad \operatorname{Re}\beta>0]$     ET I 82(32)

**3.745**$^{12}$   $\text{PV}\displaystyle\int_0^\infty \frac{\sin(ax)}{\cos(bx)}\frac{dx}{x(c^2-x^2)} = 0$      $[0<a<b,\quad c>0]$     ET I 82(31)

**3.746**

1.   $\displaystyle\int_0^\infty \frac{dx}{x^{n+1}}\prod_{k=0}^n \sin(a_k x) = \frac{\pi}{2}\prod_{k=1}^n a_k$      $\left[a_0 > \displaystyle\sum_{k=1}^n a_k,\quad a_k>0\right]$     FI II 646

2.   $\displaystyle\int_0^\infty \frac{\sin(ax)}{x^{n+1}}\,dx\prod_{k=1}^n \sin(a_k x)\prod_{j=1}^m \cos(b_j x) = \frac{\pi}{2}\prod_{k=1}^n a_k$      $\left[a > \displaystyle\sum_{k=1}^n |a_k| + \sum_{j=1}^m |b_j|\right]$     WH

**3.747**

$1.^{7}$   $\displaystyle\int_0^{\pi/2} \frac{x^m}{\sin x}\,dx = \left(\frac{\pi}{2}\right)^m\left[\frac{1}{m} + \sum_{k=1}^\infty \frac{2^{2k-1}-1}{4^{2k-1}(m+2k)}\zeta(2k)\right] = 2\pi G - \frac{7}{2}\zeta(3)$

                                                              $[m=2]$

                                                              LI (206)(2)

2.   $\displaystyle\int_0^{\pi/2} \frac{x\,dx}{\sin x} = \int_0^{\pi/2} \frac{\left(\frac{\pi}{2}-x\right)\,dx}{\cos x} = 2G$      BI(204)(18), BI(206)(1), GW(333)(32)

$3.^{12}$   $\displaystyle\int_0^\infty \frac{x}{(x^2+a^2)\sin x}\,dx = \frac{\pi}{2\sinh a}$      $[a>0]$     GW (333)(79c)

4.   $\displaystyle\int_0^\pi x\tan x\,dx = -\pi\ln 2$      BI (218)(4)

5. $\displaystyle\int_0^{\pi/2} x\tan x\,\mathrm{d}x = \infty$ 
BI (205)(2)

6. $\displaystyle\int_0^{\pi/4} x\tan x\,\mathrm{d}x = -\frac{\pi}{8}\ln 2 + \frac{1}{2}\boldsymbol{G}$ 
BI (204)(1)

7. $\displaystyle\int_0^{\pi/2} x\cot x\,\mathrm{d}x = \frac{\pi}{2}\ln 2$ 
FI II 623

8. $\displaystyle\int_0^{\pi/4} x\cot x\,\mathrm{d}x = \frac{\pi}{8}\ln 2 + \frac{1}{2}\boldsymbol{G}$ 
BI (204)(2)

9. $\displaystyle\int_0^{\pi/2}\left(\frac{\pi}{2}-x\right)\tan x\,\mathrm{d}x = \frac{1}{2}\int_0^{\pi}\left(\frac{\pi}{2}-x\right)\tan x\,\mathrm{d}x = \frac{\pi}{2}\ln 2$ 
GW(333)(33b), BI(218)(12)

10. $\displaystyle\int_0^{\infty}\tan ax\,\frac{\mathrm{d}x}{x} = \frac{\pi}{2}$ $\quad[a>0]$ 
LO V 279(5)

11. $\displaystyle\int_0^{\pi/2}\frac{x\cot x}{\cos 2x}\,\mathrm{d}x = \frac{\pi}{4}\ln 2$ 
BI (206)(12)

**3.748**

1. $\displaystyle\int_0^{\pi/4} x^m\tan x\,\mathrm{d}x = \frac{1}{2}\left(\frac{\pi}{4}\right)^m\sum_{k=1}^{\infty}\frac{\left(4^k-1\right)\zeta(2k)}{4^{2k-1}(m+2k)}$ 
LI (204)(5)

2. $\displaystyle\int_0^{\pi/2} x^p\cot x\,\mathrm{d}x = \left(\frac{\pi}{2}\right)^p\left(\frac{1}{p} - 2\sum_{k=1}^{\infty}\frac{1}{4^k(p+2k)}\,\zeta(2k)\right)$ 
LI (205)(7)

3. $\displaystyle\int_0^{\pi/4} x^m\cot x\,\mathrm{d}x = \frac{1}{2}\left(\frac{\pi}{4}\right)^m\left(\frac{2}{m} - \sum_{k=1}^{\infty}\frac{\zeta(2k)}{4^{2k-1}(m+2k)}\right)$ 
LI (204)(6)

**3.749**

1. $\displaystyle\int_0^{\infty}\frac{x\tan(ax)\,\mathrm{d}x}{x^2+b^2} = \frac{\pi}{e^{2ab}+1}$ $\quad[a>0,\quad b>0]$ 
GW (333)(79a)

2. $\displaystyle\int_0^{\infty}\frac{x\cot(ax)\,\mathrm{d}x}{x^2+b^2} = \frac{\pi}{e^{2ab}-1}$ $\quad[a>0,\quad b>0]$ 
GW (333)(79b)

3. $\displaystyle\int_0^{\infty}\frac{x\tan(ax)\,\mathrm{d}x}{b^2-x^2} = \int_0^{\infty}\frac{x\cot(ax)\,\mathrm{d}x}{b^2-x^2} = \int_0^{\infty}\frac{x\,\mathrm{cosec}(ax)\,\mathrm{d}x}{b^2-x^2} = \infty$ 
BI (161)(7, 8, 9)

## 3.75 Combinations of trigonometric and algebraic functions

**3.751**

1. $\displaystyle\int_0^{\infty}\frac{\sin(ax)\,\mathrm{d}x}{\sqrt{x+b}} = \sqrt{\frac{\pi}{2a}}\left[\cos(ab) - \sin(ab) + 2\,C\left(\sqrt{ab}\right)\sin(ab) - 2\,S\left(\sqrt{ab}\right)\cos(ab)\right]$ 
$\quad[a>0,\quad |\arg b|<\pi]$ ET I 65(12)a

2.[9] $\displaystyle\int_0^{\infty}\frac{\cos(ax)\,\mathrm{d}x}{\sqrt{x+b}} = \sqrt{\frac{\pi}{2a}}\left[\cos(ab) + \sin(ab) - 2\,C\left(\sqrt{ab}\right)\cos(ab) - 2\,S\left(\sqrt{ab}\right)\sin(ab)\right]$ 
$\quad[a>0,\quad |\arg b|<\pi]$ ET I 8(9)a

3.    $\displaystyle\int_u^\infty \frac{\sin(ax)}{\sqrt{x-u}}\,dx = \sqrt{\frac{\pi}{2a}}\,[\sin(au)+\cos(au)]$      $[a>0, \quad u>0]$      ET I 65(13)

4.    $\displaystyle\int_u^\infty \frac{\cos(ax)}{\sqrt{x-u}}\,dx = \sqrt{\frac{\pi}{2a}}\,[\cos(au)-\sin(au)]$      $[a>0, \quad u>0]$      ET I 8(10)

**3.752**

$1.^8$    $\displaystyle\int_0^1 \sin(ax)\sqrt{1-x^2}\,dx = \sum_{k=0}^\infty \frac{(-1)^k a^{2k+1}}{(2k-1)!!(2k+3)!!} = \frac{\pi}{2a}\,\mathbf{H}_1(a)$

                                             $[a>0]$      BI (149)(6)

2.    $\displaystyle\int_0^1 \cos(ax)\sqrt{1-x^2}\,dx = \frac{\pi}{2a}\,J_1(a)$      KU 65(6)a

**3.753**

$1.^8$    $\displaystyle\int_0^1 \frac{\sin(ax)\,dx}{\sqrt{1-x^2}} = \sum_{k=0}^\infty \frac{(-1)^k a^{2k+1}}{[(2k+1)!!]^2} = \frac{\pi}{2}\,\mathbf{H}_0(a)$      BI (149)(9)

2.    $\displaystyle\int_0^1 \frac{\cos(ax)\,dx}{\sqrt{1-x^2}} = \frac{\pi}{2}\,J_0(a)$      WA 30(7)a

3.    $\displaystyle\int_1^\infty \frac{\sin(ax)\,dx}{\sqrt{x^2-1}} = \frac{\pi}{2}\,J_0(a)$      $[a>0]$      WA 200(14)

$4.^{12}$    $\displaystyle\int_1^\infty \frac{\cos(ax)}{\sqrt{x^2-1}}\,dx = -\frac{\pi}{2}\,Y_0(a)$      $[a>0]$      WA 200(15)

5.    $\displaystyle\int_0^1 \frac{x\sin(ax)}{\sqrt{1-x^2}}\,dx = \frac{\pi}{2}\,J_1(a)$      WA 30(6)

**3.754**

1.    $\displaystyle\int_0^\infty \frac{\sin(ax)\,dx}{\sqrt{b^2+x^2}} = \frac{\pi}{2}\,[I_0(ab)-\mathbf{L}_0(ab)]$      $[a>0, \quad \operatorname{Re}b>0]$      ET I 66(26)

2.    $\displaystyle\int_0^\infty \frac{\cos(ax)\,dx}{\sqrt{b^2+x^2}} = K_0(ab)$      $[a>0, \quad \operatorname{Re}b>0]$

                                             WA 191(1), GW(333)(78a)

3.    $\displaystyle\int_0^\infty \frac{x\sin(ax)}{\sqrt{(b^2+x^2)^3}}\,dx = a\,K_0(ab)$      $[a>0, \quad \operatorname{Re}b>0]$      ET I 66(27)

**3.755**

1.    $\displaystyle\int_0^\infty \frac{\sqrt{\sqrt{x^2+b^2}-b}\,\sin(ax)\,dx}{\sqrt{x^2+b^2}} = \sqrt{\frac{\pi}{2a}}\,e^{-ab}$      $[a>0]$      ET I 66(31)

2.    $\displaystyle\int_0^\infty \frac{\sqrt{\sqrt{x^2+b^2}+b}\,\cos(ax)\,dx}{\sqrt{x^2+b^2}} = \sqrt{\frac{\pi}{2a}}\,e^{-ab}$      $[a>0, \quad \operatorname{Re}b>0]$      ET I 10(25)

**3.756**

1.  $$\int_0^\infty \frac{\sin(ax)}{x^{\frac{n}{2}-1}} \prod_{k=2}^n \sin(a_k x)\, dx = 0 \qquad \left[ a_k > 0, \quad a > \sum_{k=2}^n a_k \right] \qquad \text{ET I 80(22)}$$

2.  $$\int_0^\infty x^{\frac{n}{2}-1} \cos(ax) \prod_{k=1}^n \cos(a_k x)\, dx = 0 \qquad \left[ a_k > 0, \quad a > \sum_{k=1}^n a_k \right] \qquad \text{ET I 22(26)}$$

**3.757**

1.[12] $$\int_0^\infty \frac{\sin x}{\sqrt{x}}\, dx = \sqrt{\frac{\pi}{2}} \qquad \text{BI (177)(1)}$$

2.[12] $$\int_0^\infty \frac{\cos x}{\sqrt{x}}\, dx = \sqrt{\frac{\pi}{2}} \qquad \text{BI (177)(2)}$$

## 3.76–3.77 Combinations of trigonometric functions and powers

**3.761**

1.  $$\int_0^1 x^{\mu-1} \sin(ax)\, dx = \frac{-i}{2\mu} \left[ \,_1F_1(\mu; \mu+1; ia) - \,_1F_1(\mu; \mu+1; -ia) \right]$$
    $$[a > 0, \quad \operatorname{Re}\mu > -1, \quad \mu \neq 0]$$
    ET I 68(2)a

2.[8] $$\int_u^\infty x^{\mu-1} \sin x\, dx = \frac{i}{2} \left[ e^{-\frac{\pi}{2}i\mu} \Gamma(\mu, iu) - e^{\frac{\pi}{2}i\mu} \Gamma(\mu, -iu) \right]$$
    $$[\operatorname{Re}\mu < 1] \qquad \text{EH II 149(2)}$$

3.  $$\int_1^\infty \frac{\sin(ax)}{x^{2n}}\, dx = \frac{a^{2n-1}}{(2n-1)!} \left[ \sum_{k=1}^{2n-1} \frac{(2n-k-1)!}{a^{2n-k}} \sin\left( a + (k-1)\frac{\pi}{2} \right) + (-1)^n \operatorname{ci}(a) \right]$$
    $$[a > 0] \qquad \text{LI (203)(15)}$$

4.  $$\int_0^\infty x^{\mu-1} \sin(ax)\, dx = \frac{\Gamma(\mu)}{a^\mu} \sin\frac{\mu\pi}{2} = \frac{\pi \sec\frac{\mu\pi}{2}}{2a^\mu \Gamma(1-\mu)} \qquad [a > 0, \quad 0 < |\operatorname{Re}\mu| < 1]$$
    FI II 809a, BI(150)(1)

5.[10] $$\int_0^\pi x^m \sin(nx)\, dx = \frac{(-1)^{n+1}}{n^{m+1}} \sum_{k=0}^{\lfloor m/2 \rfloor} (-1)^k \frac{m!}{(m-2k)!} (n\pi)^{m-2k}$$
    $$- (-1)^{\lfloor m/2 \rfloor} \frac{m! \left\lfloor m - 2\lfloor \frac{m}{2} \rfloor - 1 \right\rfloor}{n^{m+1}}$$
    GW(333)(6)

6.[8] $$\int_0^1 x^{\mu-1} \cos(ax)\, dx = \frac{1}{2\mu} \left[ \,_1F_1(\mu; \mu+1; ia) + \,_1F_1(\mu; \mu+1; -ia) \right]$$
    $$[a > 0, \quad \operatorname{Re}\mu > 0] \qquad \text{ET I 11(2)}$$

7.  $$\int_u^\infty x^{\mu-1} \cos x\, dx = \frac{1}{2} \left[ e^{-\frac{\pi}{2}i\mu} \Gamma(\mu, iu)s + e^{\frac{\pi}{2}i\mu} \Gamma(\mu, -iu) \right]$$
    $$[\operatorname{Re}\mu < 1] \qquad \text{EH II 149(1)}$$

8.    $$\int_1^\infty \frac{\cos(ax)}{x^{2n+1}}\, dx = \frac{a^{2n}}{(2n)!}\left[\sum_{k=1}^{2n} \frac{(2n-k)!}{a^{2n-k+1}}\cos\left(a+(k-1)\frac{\pi}{2}\right) + (-1)^{n+1}\operatorname{ci}(a)\right]$$

$$[a>0] \qquad\qquad\qquad\qquad \text{LI (203)(16)}$$

9.8    $$\int_0^\infty x^{\mu-1}\cos(ax)\, dx = \frac{\Gamma(\mu)}{a^\mu}\cos\frac{\mu\pi}{2} = \frac{\pi\operatorname{cosec}\frac{\mu\pi}{2}}{2a^\mu\,\Gamma(1-\mu)} \qquad [a>0, \quad 0<\operatorname{Re}\mu<1]$$

$$\text{FI II 809a, BI(150)(2)}$$

10.    $$\int_0^\pi x^m\cos(nx)\, dx = \frac{(-1)^n}{n^{m+1}}\sum_{k=0}^{\lfloor(m-1)/2\rfloor}(-1)^k\frac{m!}{(m-2k-1)!}(n\pi)^{m-2k-1}$$

$$+(-1)^{\lfloor(m+1)/2\rfloor}\frac{2\lfloor(m+1)/2\rfloor-m}{n^{m+1}}m!$$

$$\text{GW (333)(7)}$$

11.    $$\int_0^{\pi/2} x^m\cos x\, dx = \sum_{k=0}^{\lfloor m/2\rfloor}(-1)^k\frac{m!}{(m-2k)!}\left(\frac{\pi}{2}\right)^{m-2k} + (-1)^{\lfloor m/2\rfloor}\left(2\left\lfloor\frac{m}{2}\right\rfloor - m\right)m!$$

$$\text{GW (333)(9c)}$$

12.    $$\int_0^{2n\pi} x^m\cos kx\, dx = -\sum_{j=0}^{m-1}\frac{j!}{k^{j+1}}\binom{m}{j}(2n\pi)^{m-j}\cos\frac{j+1}{2}\pi \qquad\qquad \text{BI (226)(2)}$$

**3.762**

1.    $$\int_0^\infty x^{\mu-1}\sin(ax)\sin(bx)\, dx = \frac{1}{2}\cos\frac{\mu\pi}{2}\Gamma(\mu)\left[|b-a|^{-\mu} - (b+a)^{-\mu}\right]$$

$$[a>0, \quad b>0, \quad a\neq b, \quad -2<\operatorname{Re}\mu<1]$$

$$(\text{for } \mu=0, \text{ see } \mathbf{3.741}\ 1, \text{ for } \mu=-1, \text{ see } \mathbf{3.741}\ 3)$$

$$\text{BI(149)(7), ET I 321(40)}$$

2.    $$\int_0^\infty x^{\mu-1}\sin(ax)\cos(bx)\, dx = \frac{1}{2}\sin\frac{\mu\pi}{2}\Gamma(\mu)\left[(a+b)^{-\mu} + |a-b|^{-\mu}\operatorname{sign}(a-b)\right]$$

$$[a>0, \quad b>0, \quad |\operatorname{Re}\mu|<1] \qquad (\text{for } \mu=0 \text{ see } \mathbf{3.741}\ 2) \quad \text{BI(159)(8)a, ET I 321(41)}$$

3.    $$\int_0^\infty x^{\mu-1}\cos(ax)\cos(bx)\, dx = \frac{1}{2}\cos\frac{\mu\pi}{2}\Gamma(\mu)\left[(a+b)^{-\mu} + |a-b|^{-\mu}\right]$$

$$[a>0, \quad b>0, \quad 0<\operatorname{Re}\mu<1]$$

$$\text{ET I 20(17)}$$

**3.763**

1.    $$\int_0^\infty \frac{\sin(ax)\sin(bx)\sin(cx)}{x^\nu}\, dx = \frac{1}{4}\cos\frac{\nu\pi}{2}\Gamma(1-\nu)\Big\{(c+a-b)^{\nu-1} - (c+a+b)^{\nu-1}$$

$$- |c-a+b|^{\nu-1}\operatorname{sign}(a-b-c) + |c-a-b|^{\nu-1}\operatorname{sign}(a+b-c)\Big\}$$

$$[c>0, \quad 0<\operatorname{Re}\nu<4, \quad \nu\neq 1,2,3,\ldots, \quad a\geq b>0] \quad \text{GW(333)(26a)a, ET I 79(13)}$$

2.    $\displaystyle\int_0^\infty \frac{\sin(ax)\sin(bx)\sin(cx)}{x}\,dx = 0$ $\qquad\qquad$ $[c < a - b \text{ and } c > a + b]$

$\qquad\qquad\qquad\qquad\qquad\qquad\quad = \dfrac{\pi}{8}$ $\qquad\qquad$ $[c = a - b \text{ and } c = a + b]$

$\qquad\qquad\qquad\qquad\qquad\qquad\quad = \dfrac{\pi}{4}$ $\qquad\qquad$ $[a - b < c < a + b]$

$\qquad\qquad\qquad\qquad\qquad\qquad\qquad\qquad\qquad\quad [a \geq b > 0, \quad c > 0]$ $\qquad$ FI II 645

3.    $\displaystyle\int_0^\infty \frac{\sin(ax)\sin(bx)\sin(cx)}{x^2}\,dx = \frac{1}{4}(c + a + b)\ln(c + a + b)$

$\qquad\qquad\qquad\qquad\qquad\qquad\qquad - \dfrac{1}{4}(c + a - b)\ln(c + a - b) - \dfrac{1}{4}|c - a - b|\ln|c - a - b|$

$\qquad\qquad\qquad\qquad\qquad\qquad\qquad \times \operatorname{sign}(a + b - c) + \dfrac{1}{4}|c - a + b|\ln|c - a + b|\operatorname{sign}(a - b - c)$

$\qquad\qquad\qquad\qquad\qquad\qquad\qquad\qquad\qquad [a \geq b > 0, \quad c > 0]$ $\quad$ BI(157)(8)a, ET I 79(11)

4.    $\displaystyle\int_0^\infty \frac{\sin(ax)\sin(bx)\sin(cx)}{x^3}\,dx = \frac{\pi bc}{2}$ $\qquad$ $[0 < c < a - b \text{ and } c > a + b]$

$\qquad\qquad\qquad\qquad\qquad\qquad\qquad\qquad = \dfrac{\pi bc}{2} - \dfrac{\pi(a - b - c)^2}{8}$ $\qquad$ $[a - b < c < a + b]$

$\qquad\qquad\qquad\qquad\qquad\qquad\qquad\qquad\qquad [a \geq b > 0, \quad c > 0]$ $\quad$ BI(157)(20), ET I 79(12)

**3.764**

1.    $\displaystyle\int_0^\infty x^p \sin(ax + b)\,dx = \frac{1}{a^{p+1}}\Gamma(1 + p)\cos\left(b + \frac{p\pi}{2}\right)$ $\qquad$ $[a > 0, \quad -1 < p < 0]$ $\qquad$ GW (333)(30a)

2.    $\displaystyle\int_0^\infty x^p \cos(ax + b)\,dx = -\frac{1}{a^{p+1}}\Gamma(1 + p)\sin\left(b + \frac{\pi p}{2}\right)$

$\qquad\qquad\qquad\qquad\qquad\qquad\qquad\qquad\qquad [a > 0, \quad -1 < p < 0]$ $\qquad$ GW (333)(30b)

**3.765**

1.[12]    $\displaystyle\int_0^\infty \frac{\sin ax}{x^\nu(x + b)}\,dx = \frac{i\,\Gamma(1 - \nu)}{2b^\nu}\left[e^{iab}\,\Gamma(\nu, iab) - e^{-iab}\,\Gamma(\nu, -iab)\right]$

$\qquad\qquad\qquad\qquad\qquad\qquad\qquad [\operatorname{Im} a = 0, \quad -1 < \operatorname{Re} b < 2, \quad \arg b \neq \pi]$ $\quad$ MC

2.    $\displaystyle\int_0^\infty \frac{\cos(ax)}{x^\nu(x + b)}\,dx = \frac{\Gamma(1 - \nu)}{2b^\nu}\left[e^{iab}\,\Gamma(\nu, iab) + e^{-iab}\,\Gamma(\nu, -iab)\right]$

$\qquad\qquad\qquad\qquad\qquad\qquad\qquad [a > 0, \quad |\operatorname{Re}\nu| < 1, \quad |\arg b| < \pi]$

$\qquad\qquad\qquad\qquad\qquad\qquad\qquad\qquad\qquad\qquad\qquad\qquad$ ET II 221(52)

**3.766**

1.[12]    $\displaystyle\int_0^\infty \frac{x^{\mu-1}\sin ax}{1 + x^2}\,dx = \frac{\pi\sinh a}{2\cos\left(\frac{\mu\pi}{2}\right)} + \frac{1}{2}\sin\left(\frac{\mu\pi}{2}\right)\Gamma(mu)\left[e^{-a-i\pi(1-\mu)}\gamma(1 - \mu, -a) - e^a\gamma(1 - \mu, a)\right]$

$\qquad\qquad\qquad\qquad\qquad\qquad\qquad [\operatorname{Im} a = 0, \quad -1 < \operatorname{Re}\mu < 3]$ $\qquad$ MC

2.[12]    $\displaystyle\int_0^\infty \frac{x^{\mu-1}\cos(ax)}{1 + x^2}\,dx = \frac{\pi}{2}\operatorname{cosec}\frac{\mu\pi}{2}\cosh a$

$\qquad\qquad\qquad\qquad\qquad\qquad\qquad + \dfrac{1}{2}\cos\dfrac{\mu\pi}{2}\Gamma(\mu)\left\{e^{-a-i\pi(1-\mu)}\,\gamma(1 - \mu, -a) - e^a\,\gamma(1 - \mu, a)\right\}$

$\qquad\qquad\qquad\qquad\qquad\qquad\qquad [a > 0, \quad 0 < \operatorname{Re}\mu < 3]$ $\qquad$ ET I 319(24)

$3.^9 \quad \int_0^\infty \frac{x^{2\mu+1}\sin(ax)\,dx}{x^2+b^2} = -\frac{\pi}{2}b^{2\mu}\sec(\mu\pi)\sinh(ab)$

$$+\frac{\sin(\mu\pi)}{2a^{2\mu}}\Gamma(2\mu)\left[\,{}_1F_1(1;1-2\mu;ab)+{}_1F_1(1;1-2\mu;-ab)\right]$$

$$\left[a>0,\quad -\tfrac{3}{2}<\operatorname{Re}\mu<\tfrac{1}{2}\right]\quad \text{ET II 220(39)}$$

$4.^{12} \quad \int_0^\infty \frac{x^{2\mu+1}\cos(ax)\,dx}{x^2+b^2}$

$$=-\frac{\pi}{2}b^{2\mu}\frac{\cosh(ab)}{\sin(\mu\pi)}+\frac{\cos(\mu\pi)}{2a^{2\mu}}\Gamma(2u)\left[{}_1F_1(1;1-2\mu;ab)+{}_1F_1(1;1-2\mu;-ab)\right]$$

$$\left[a>0,\;-1<\operatorname{Re}\mu<\tfrac{1}{2}\right]\quad \text{ET II 221(56)}$$

**3.767**

1. $\int_0^\infty \frac{x^{\beta-1}\sin\left(ax-\frac{\beta\pi}{2}\right)}{\gamma^2+x^2}\,dx = -\frac{\pi}{2}\gamma^{\beta-2}e^{-a\gamma}$     $[a>0,\quad \operatorname{Re}\gamma>0,\quad 0<\operatorname{Re}\beta<2]$

         BI (160)(20)

2. $\int_0^\infty \frac{x^\beta \cos\left(ax-\frac{\beta\pi}{2}\right)}{\gamma^2+x^2}\,dx = \frac{\pi}{2}\gamma^{\beta-1}e^{-a\gamma}$     $[a>0,\quad \operatorname{Re}\gamma>0,\quad |\operatorname{Re}\beta|<1]$

         BI (160)(21)

3. $\int_0^\infty \frac{x^{\beta-1}\sin\left(ax-\frac{\beta\pi}{2}\right)}{x^2-b^2}\,dx = \frac{\pi}{2}b^{\beta-2}\cos\left(ab-\frac{\pi\beta}{2}\right)$     $[a>0,\quad b>0,\quad 0<\operatorname{Re}\beta<2]$

         BI (161)(11)

4. $\int_0^\infty \frac{x^\beta \cos\left(ax-\frac{\beta\pi}{2}\right)}{x^2-b^2}\,dx = -\frac{\pi}{2}b^{\beta-1}\sin\left(ab-\frac{\pi\beta}{2}\right)$     $[a>0,\quad b>0,\quad |\beta|<1]$

         GW (333)(82)

**3.768**

1. $\int_u^\infty (x-u)^{\mu-1}\sin(ax)\,dx = \frac{\Gamma(\mu)}{a^\mu}\sin\left(au+\frac{\mu\pi}{2}\right)$     $[a>0,\quad 0<\operatorname{Re}\mu<1]$    ET II 203(19)

2. $\int_u^\infty (x-u)^{\mu-1}\cos(ax)\,dx = \frac{\Gamma(\mu)}{a^\mu}\cos\left(au+\frac{\mu\pi}{2}\right)$     $[a>0,\quad 0<\operatorname{Re}\mu<1]$    ET II 204(24)

3.$^{11}$ $\int_0^1 (1-x)^\nu \sin(ax)\,dx = \frac{1}{a}-\frac{\Gamma(\nu+1)}{a^{\nu+1}}C_\nu(a) = a^{-\nu-1/2}s_{\nu+1/2,1/2}(a)$

$$[a>0,\quad \operatorname{Re}\nu>-1]\quad \text{ET I 11(3)a}$$

Here $C_\nu(a)$ is the Young's function given by:

$$C_\nu(a) = \frac{\frac{1}{2}a^\nu}{\Gamma(\nu+1)}\left[\,{}_1F_1(1;\nu+1;ia)+{}_1F_1(1;\nu+1;-ia)\right] = \sum_{n=0}^\infty \frac{(-1)^n a^{\nu+2n}}{\Gamma(\nu+2n+1)}$$

**4.[3]**
$$\int_0^1 (1-x)^\nu \cos(ax)\,dx = \frac{i}{2}a^{-\nu-1}\left\{\exp\left[\frac{i}{2}(\nu\pi - 2a)\right]\gamma(\nu+1,-ia)\right.$$
$$\left. - \exp\left[-\frac{i}{2}(\nu\pi - 2a)\right]\gamma(\nu+1,ia)\right\}$$
$$= \Gamma(\nu+1)\sum_{n=0}^\infty \frac{(-a^2)^n}{\Gamma(\nu+2+2n)}$$
$$[a > 0, \quad \operatorname{Re}\nu > -1] \qquad \text{ET I 11(3)a}$$

**5.**
$$\int_0^u x^{\nu-1}(u-x)^{\mu-1}\sin(ax)\,dx = \frac{u^{\mu+\nu-1}}{2i}\,B(\mu,\nu)\left[\,{}_1F_1(\nu;\mu+\nu;iau) - {}_1F_1(\nu;\mu+\nu;-iau)\right]$$
$$[a > 0, \quad \operatorname{Re}\mu > 0, \quad \operatorname{Re}\nu > -1, \quad \nu \neq 0] \quad \text{ET II 189(26)}$$

**6.**
$$\int_0^u x^{\nu-1}(u-x)^{\mu-1}\cos(ax)\,dx = \frac{u^{\mu+\nu-1}}{2}\,B(\mu,\nu)\left[\,{}_1F_1(\nu;\mu+\nu;iau) + {}_1F_1(\nu;\mu+\nu;-iau)\right]$$
$$[a > 0, \quad \operatorname{Re}\mu > 0, \quad \operatorname{Re}\nu > 0]$$
$$\text{ET II 189(32)}$$

**7.**
$$\int_0^u x^{\mu-1}(u-x)^{\mu-1}\sin(ax)\,dx = \sqrt{\pi}\left(\frac{u}{a}\right)^{\mu-1/2}\sin\frac{au}{2}\,\Gamma(\mu)\,J_{\mu-1/2}\left(\frac{au}{2}\right)$$
$$[\operatorname{Re}\mu > 0] \qquad \text{ET II 189(25)}$$

**8.**
$$\int_u^\infty x^{\mu-1}(x-u)^{\mu-1}\sin(ax)\,dx$$
$$= \frac{\sqrt{\pi}}{2}\left(\frac{u}{a}\right)^{\mu-1/2}\Gamma(\mu)\left[\cos\frac{au}{2}\,J_{1/2-\mu}\left(\frac{au}{2}\right) - \sin\frac{au}{2}\,Y_{1/2-\mu}\left(\frac{au}{2}\right)\right]$$
$$[a > 0, \quad 0 < \operatorname{Re}\mu < \tfrac{1}{2}] \qquad \text{ET II 203(20)}$$

**9.**
$$\int_0^u x^{\mu-1}(u-x)^{\mu-1}\cos(ax)\,dx = \sqrt{\pi}\left(\frac{u}{a}\right)^{\mu-\frac{1}{2}}\cos\frac{au}{2}\,\Gamma(\mu)\,J_{\mu-\frac{1}{2}}\left(\frac{au}{2}\right)$$
$$[\operatorname{Re}\mu > 0] \qquad \text{ET II 189(31)}$$

**10.**
$$\int_u^\infty x^{\mu-1}(x-u)^{\mu-1}\cos(ax)\,dx = -\frac{\sqrt{\pi}}{2}\left(\frac{u}{a}\right)^{\mu-\frac{1}{2}}\Gamma(\mu)\left[\sin\frac{au}{2}\,J_{\frac{1}{2}-\mu}\left(\frac{au}{2}\right) - \cos\frac{au}{2}\,Y_{\frac{1}{2}-\mu}\left(\frac{au}{2}\right)\right]$$
$$[a > 0, \quad 0 < \operatorname{Re}\mu < \tfrac{1}{2}] \qquad \text{ET II 204(25)}$$

**11.[12]**
$$\int_0^1 x^{\nu-1}(1-x)^{\mu-1}\sin(ax)\,dx = -\frac{i}{2}\,B(\mu,\nu)\left[\,{}_1F_1(\nu;\nu+\mu;ia) - {}_1F_1(\nu;\nu+\mu;-ia)\right]$$
$$[a > 0, \quad \operatorname{Re}\mu > 0, \quad \operatorname{Re}\nu > -1, \quad \nu \neq 0] \quad \text{ET I 68 (5)a, ET I 317(5)}$$

**12.[12]**
$$\int_0^1 x^{\nu-1}(1-x)^{\mu-1}\cos(ax)\,dx = \frac{1}{2}\,B(\mu,\nu)\left[\,{}_1F_1(\nu;\nu+\mu;ia) + {}_1F_1(\nu;\nu+\mu;-ia)\right]$$
$$[a > 0, \quad \operatorname{Re}\mu > 0, \quad \operatorname{Re}\nu > 0]$$
$$\text{ET I 11(5)}$$

**13.**
$$\int_0^1 x^\mu(1-x)^\mu\sin(2ax)\,dx = \frac{\sqrt{\pi}}{(2a)^{\mu+\frac{1}{2}}}\,\Gamma(\mu+1)\,J_{\mu+\frac{1}{2}}(a)\sin a$$
$$[a > 0, \quad \operatorname{Re}\mu > -1] \qquad \text{ET I 68(4)}$$

14.
$$\int_0^1 x^\mu (1-x)^\mu \cos(2ax)\,dx = \frac{\sqrt{\pi}}{(2a)^{\mu+\frac{1}{2}}}\,\Gamma(\mu+1)\,J_{\mu+\frac{1}{2}}(a)\cos a$$

$$[a > 0, \quad \operatorname{Re}\mu > -1] \qquad \text{ET I 11(4)}$$

**3.769**

1.
$$\int_0^\infty \left[(b+ix)^{-\nu} - (b-ix)^{-\nu}\right]\sin(ax)\,dx = -\frac{\pi i a^{\nu-1} e^{-ab}}{\Gamma(\nu)}$$

$$[a > 0, \quad \operatorname{Re}b > 0, \quad \operatorname{Re}\nu > 0]$$
$$\text{ET I 70(15)}$$

2.
$$\int_0^\infty \left[(b+ix)^{-\nu} + (b-ix)^{-\nu}\right]\cos(ax)\,dx = \frac{\pi a^{\nu-1} e^{-ab}}{\Gamma(\nu)}$$

$$[a > 0, \quad \operatorname{Re}b > 0, \quad \operatorname{Re}\nu > 0]$$
$$\text{ET I 13(19)}$$

3.[12]
$$\int_0^\infty x\left[(b+ix)^{-\nu} + (b-ix)^{-\nu}\right]\sin(ax)\,dx = -\frac{\pi a^{\nu-2}(1-ab)}{\Gamma(\nu)}e^{-ab}$$

$$[a > 0, \quad \operatorname{Re}b > 0, \quad \operatorname{Re}\nu > 0]$$
$$\text{ET I 70(16)}$$

4.[12]
$$\int_0^\infty x^{2n}\left[(b-ix)^{-\nu} - (b+ix)^{-\nu}\right]\sin(ax)\,dx = \frac{(-1)^{n+1}i}{\Gamma(\nu)}(2n)!\pi a^{\nu-2n-1}e^{-ab}\,L_{2n}^{\nu-2n-1}(ab)$$

$$[a > 0, \quad \operatorname{Re}b > 0, \quad 0 \le 2n \le \operatorname{Re}\nu]$$
$$\text{ET I 70(17)}$$

5.
$$\int_0^\infty x^{2n}\left[(b+ix)^{-\nu} + (b-ix)^{-\nu}\right]\cos(ax)\,dx = \frac{(-1)^n}{\Gamma(\nu)}(2n)!\pi a^{\nu-2n-1}e^{-ab}\,L_{2n}^{\nu-2n-1}(ab)$$

$$[a > 0, \quad \operatorname{Re}b > 0, \quad 0 \le 2n < \operatorname{Re}\nu]$$
$$\text{ET I 13(20)}$$

6.
$$\int_0^\infty x^{2n+1}\left[(b+ix)^{-\nu} + (b-ix)^{-\nu}\right]\sin(ax)\,dx = \frac{(-1)^{n+1}}{\Gamma(\nu)}(2n+1)!\pi a^{\nu-2n-2}e^{-ab}\,L_{2n+1}^{\nu-2n-2}(ab)$$

$$[a > 0, \quad \operatorname{Re}b > 0, \quad -1 \le 2n+1 < \operatorname{Re}\nu] \qquad \text{ET I 70(18)}$$

7.
$$\int_0^\infty x^{2n+1}\left[(b+ix)^{-\nu} - (b-ix)^{-\nu}\right]\cos(ax)\,dx = \frac{(-1)^{n+1}i}{\Gamma(\nu)}(2n+1)!\pi a^{\nu-2n-2}e^{-ab}\,L_{2n+1}^{\nu-2n-2}(ab)$$

$$[a > 0, \quad \operatorname{Re}b > 0, \quad 0 \le 2n < \operatorname{Re}\nu - 1] \qquad \text{ET I 13(21)}$$

**3.771**

1.
$$\int_0^\infty \left(b^2+x^2\right)^{\nu-\frac{1}{2}}\sin(ax)\,dx = \frac{\sqrt{\pi}}{2}\left(\frac{2b}{a}\right)^\nu\Gamma\left(\nu+\tfrac{1}{2}\right)\left[I_{-\nu}(ab) - \mathbf{L}_\nu(ab)\right]$$

$$[a > 0, \quad \operatorname{Re}b > 0, \quad \operatorname{Re}\nu < \tfrac{1}{2}, \quad \nu \ne -\tfrac{1}{2}, -\tfrac{3}{2}, -\tfrac{5}{2}, \ldots] \qquad \text{EH II 38a, ET I 68(6)}$$

2.
$$\int_0^\infty \left(b^2+x^2\right)^{\nu-\frac{1}{2}}\cos(ax)\,dx = \frac{1}{\sqrt{\pi}}\left(\frac{2b}{a}\right)^\nu\cos(\pi\nu)\,\Gamma\left(\nu+\tfrac{1}{2}\right)K_{-\nu}(ab)$$

$$[a > 0, \quad \operatorname{Re}b > 0, \quad \operatorname{Re}\nu < \tfrac{1}{2}]$$
$$\text{WA 191(1)a, GW(333)(78)a}$$

3.      $\int_0^u x^{2\nu-1} \left(u^2 - x^2\right)^{\mu-1} \sin(ax)\,\mathrm{d}x$

$$= \frac{a}{2} u^{2\mu+2\nu-1} \, \mathrm{B}\left(\mu, \nu + \frac{1}{2}\right) {}_1F_2\left(\nu + \frac{1}{2}; \frac{3}{2}, \mu + \nu + \frac{1}{2}; -\frac{a^2u^2}{4}\right)$$

$$\left[\operatorname{Re}\mu > 0, \quad \operatorname{Re}\nu > -\tfrac{1}{2}\right] \qquad \text{ET II 189(29)}$$

4.      $\int_0^u x^{2\nu-1} \left(u^2 - x^2\right)^{\mu-1} \cos(ax)\,\mathrm{d}x = \frac{1}{2} u^{2\mu+2\nu-2} \, \mathrm{B}(\mu,\nu) \, {}_1F_2\left(\nu; \frac{1}{2}, \mu + \nu; -\frac{a^2u^2}{4}\right)$

$$\left[\operatorname{Re}\mu > 0, \quad \operatorname{Re}\nu > 0\right] \qquad \text{ET II 190(35)}$$

5.[7]    $\int_0^\infty x \left(x^2 + b^2\right)^{\nu-\frac{1}{2}} \sin(ax)\,\mathrm{d}x = \frac{1}{\sqrt{\pi}} b \left(\frac{2b}{a}\right)^\nu \cos\nu\pi\, \Gamma\left(\nu + \frac{1}{2}\right) K_{\nu+1}(ab)$

$$= \sqrt{\pi} b \left(\frac{2b}{a}\right)^\nu \frac{1}{\Gamma\left(\frac{1}{2} - \nu\right)} K_{\nu+1}(ab)$$

$$\left[a > 0, \quad \operatorname{Re}b > 0, \quad \operatorname{Re}\nu < 0\right] \quad \text{ET I 69(11)}$$

6.      $\int_0^u \left(u^2 - x^2\right)^{\nu-\frac{1}{2}} \sin(ax)\,\mathrm{d}x = \frac{\sqrt{\pi}}{2} \left(\frac{2u}{a}\right)^\nu \Gamma\left(\nu + \frac{1}{2}\right) \mathbf{H}_\nu(au)$

$$\left[a > 0, \quad u > 0, \quad \operatorname{Re}\nu > -\tfrac{1}{2}\right]$$
$$\text{ET I 69(7), WA 358(1)a}$$

7.      $\int_u^\infty \left(x^2 - u^2\right)^{\nu-\frac{1}{2}} \sin(ax)\,\mathrm{d}x = \frac{\sqrt{\pi}}{2} \left(\frac{2u}{a}\right)^\nu \Gamma\left(\nu + \frac{1}{2}\right) J_{-\nu}(au)$

$$\left[a > 0, \quad u > 0, \quad |\operatorname{Re}\nu| < \tfrac{1}{2}\right]$$
$$\text{EH II 81(12)a, ET I 69(8), WA 187(3)a}$$

8.      $\int_0^u \left(u^2 - x^2\right)^{\nu-\frac{1}{2}} \cos(ax)\,\mathrm{d}x = \frac{\sqrt{\pi}}{2} \left(\frac{2u}{a}\right)^\nu \Gamma\left(\nu + \frac{1}{2}\right) J_\nu(au)$

$$\left[a > 0, \quad u > 0, \quad \operatorname{Re}\nu > -\tfrac{1}{2}\right]$$
$$\text{ET I 11(8)}$$

9.      $\int_u^\infty \left(x^2 - u^2\right)^{\nu-\frac{1}{2}} \cos(ax)\,\mathrm{d}x = -\frac{\sqrt{\pi}}{2} \left(\frac{2u}{a}\right)^\nu \Gamma\left(\nu + \frac{1}{2}\right) Y_{-\nu}(au)$

$$\left[a > 0, \quad u > 0, \quad |\operatorname{Re}\nu| < \tfrac{1}{2}\right]$$
$$\text{WA 187(4)a, EH II 82(13)a, ET I 11(9)}$$

10.     $\int_0^u x \left(u^2 - x^2\right)^{\nu-\frac{1}{2}} \sin(ax)\,\mathrm{d}x = \frac{\sqrt{\pi}}{2} u \left(\frac{2u}{a}\right)^\nu \Gamma\left(\nu + \frac{1}{2}\right) J_{\nu+1}(au)$

$$\left[a > 0, \quad u > 0, \quad \operatorname{Re}\nu > -\tfrac{1}{2}\right]$$
$$\text{ET I 69(9)}$$

11.     $\int_u^\infty x \left(x^2 - u^2\right)^{\nu-\frac{1}{2}} \sin(ax)\,\mathrm{d}x = \frac{\sqrt{\pi}}{2} u \left(\frac{2u}{a}\right)^\nu \Gamma\left(\nu + \frac{1}{2}\right) Y_{-\nu-1}(au)$

$$\left[a > 0, \quad u > 0, \quad -\tfrac{1}{2} < \operatorname{Re}\nu < 0\right]$$
$$\text{ET I 69(10)}$$

12.[12]    $\displaystyle\int_0^u x\left(u^2-x^2\right)^{\nu-\frac{1}{2}}\cos(ax)\,\mathrm{d}x = -\frac{u^{\nu+1}}{a^\nu}\,s_{\nu,\nu+1}(au)$

$$= \frac{1}{2}\left(\nu+\frac{1}{2}\right)^{-1}u^{2\nu+1} - \frac{\sqrt{\pi}}{2}u\left(\frac{2u}{a}\right)^\nu\Gamma\left(\nu+\tfrac{1}{2}\right)\mathbf{H}_{\nu+1}(au)$$

$$\left[a>0,\quad u>0,\quad \mathrm{Re}\,\nu>-\tfrac{1}{2}\right]\quad\text{ET I 12(10)}$$

13.    $\displaystyle\int_u^\infty x\left(x^2-u^2\right)^{\nu-1/2}\cos(ax)\,\mathrm{d}x\ \frac{\sqrt{\pi}u}{2}\left(\frac{2u}{a}\right)^\nu\Gamma\left(\nu+\tfrac{1}{2}\right)J_{-\nu-1}(au)$

$$\left[a>0,\quad u>0,\quad 0<\mathrm{Re}\,\nu<\tfrac{1}{2}\right]\quad\text{ET I 12(11)}$$

## 3.772

1.    $\displaystyle\int_0^\infty \left(x^2+2bx\right)^{\nu-1/2}\sin(ax)\,\mathrm{d}x = \frac{\sqrt{\pi}}{2}\left(\frac{2b}{a}\right)^\nu\Gamma\left(\nu+\tfrac{1}{2}\right)\left[J_{-\nu}(ab)\cos(ab)+Y_{-\nu}(ab)\sin(ab)\right]$

$$\left[a>0,\quad |\arg b|<\pi,\quad \tfrac{1}{2}>\mathrm{Re}\,\nu>-\tfrac{3}{2}\right]\quad\text{ET I 69(12)}$$

2.    $\displaystyle\int_0^\infty \left(x^2+2bx\right)^{\nu-1/2}\cos(ax)\,\mathrm{d}x$

$$= -\frac{\sqrt{\pi}}{2}\left(\frac{2b}{a}\right)^\nu\Gamma\left(\nu+\tfrac{1}{2}\right)\left[Y_{-\nu}(ab)\cos(ab)-J_{-\nu}(ab)\sin(ab)\right]$$

$$\left[a>0,\quad |\mathrm{Re}\,\nu|<\tfrac{1}{2}\right]\quad\text{ET I 12(13)}$$

3.    $\displaystyle\int_0^{2u}\left(2ux-x^2\right)^{\nu-1/2}\sin(ax)\,\mathrm{d}x = \sqrt{\pi}\left(\frac{2u}{a}\right)^\nu\Gamma\left(\nu+\tfrac{1}{2}\right)\sin(au)\,J_\nu(au)$

$$\left[a>0,\quad u>0,\quad \mathrm{Re}\,\nu>-\tfrac{1}{2}\right]$$
$$\text{ET I 69(13)a}$$

4.    $\displaystyle\int_{2u}^\infty \left(x^2-2ux\right)^{\nu-1/2}\sin(ax)\,\mathrm{d}x = \frac{\sqrt{\pi}}{2}\left(\frac{2u}{a}\right)^\nu\Gamma\left(\nu+\tfrac{1}{2}\right)\left[J_{-\nu}(au)\cos(au)-Y_{-\nu}(au)\sin(au)\right]$

$$\left[a>0,\quad u>0,\quad |\mathrm{Re}\,\nu|<\tfrac{1}{2}\right]\quad\text{ET I 70(14)}$$

5.    $\displaystyle\int_0^{2u}\left(2ux-x^2\right)^{\nu-1/2}\cos(ax)\,\mathrm{d}x = \sqrt{\pi}\left(\frac{2u}{a}\right)^\nu\Gamma\left(\nu+\tfrac{1}{2}\right)J_\nu(au)\cos(au)$

$$\left[a>0,\quad u>0,\quad \mathrm{Re}\,\nu>-\tfrac{1}{2}\right]$$
$$\text{ET I 12(4)}$$

6.    $\displaystyle\int_{2u}^\infty \left(x^2-2ux\right)^{\nu-1/2}\cos(ax)\,\mathrm{d}x = -\frac{\sqrt{\pi}}{2}\left(\frac{2u}{a}\right)^\nu\Gamma\left(\nu+\tfrac{1}{2}\right)\left[J_{-\nu}(au)\sin(au)+Y_{-\nu}(au)\cos(au)\right]$

$$\left[a>0,\quad u>0,\quad |\mathrm{Re}\,\nu|<\tfrac{1}{2}\right]\quad\text{ET I 12(12)}$$

**3.773**

1.[12] $\displaystyle\int_0^\infty \frac{x^{2\nu}}{(x^2+b^2)^{\mu+1}}\sin(ax)\,dx$

$$= \frac{1}{2}b^{2\nu-2\mu}a\,\mathrm{B}\left(1+\nu,\mu-\nu\right)\,_1F_2\left(\nu+1;\ \nu+1-\mu,\frac{3}{2};\ \frac{b^2a^2}{4}\right)$$

$$+ \frac{\sqrt{\pi}a^{2\mu-2\nu+1}}{4^{\mu-\nu+1}} + \frac{\Gamma(\nu-\mu)}{\Gamma\left(\mu-\nu+\frac{3}{2}\right)}\,_1F_2\left(\mu+1;\ \mu-\nu+\frac{3}{2},\mu-\nu+1;\ \frac{b^2a^2}{4}\right)$$

$$= \frac{\sqrt{\pi}}{2\,\Gamma(\mu+1)}b^{2\nu-2\mu-1}\,G^{21}_{13}\left(\frac{a^2b^2}{4}\left|\begin{array}{c}-\nu+\frac{1}{2}\\ \mu-\nu+\frac{1}{2},\frac{1}{2},0\end{array}\right.\right)$$

$$\left[a>0,\quad \operatorname{Re}b>0,\quad -1<\operatorname{Re}\nu<\operatorname{Re}\mu+1\right] \quad \text{ET I 71(28)a, ET II 234(17)}$$

2.[8] $\displaystyle\int_0^\infty \frac{x^{2m+1}\sin(ax)}{(z+x^2)^{n+1}}\,dx = \frac{(-1)^{n+m}}{n!}\frac{\pi}{2}\frac{d^n}{dz^n}\left(z^m e^{-a\sqrt{z}}\right)\quad \left[a>0,\quad 0\le m\le n,\quad |\arg z|<\pi\right]$

$$\text{ET I 68(39)}$$

3.[12] $\displaystyle\int_0^\infty \frac{x^{2m+1}\sin(ax)\,dx}{(b^2+x^2)^{n+\frac{1}{2}}} = \frac{(-1)^{m+1}\sqrt{\pi}}{2^n b^n\,\Gamma\left(n+\frac{1}{2}\right)}\frac{d^{2m+1}}{da^{2m+1}}\left[a^n\,K_n(ab)\right]$

$$\left[a>0,\quad \operatorname{Re}b>0,\quad -1\le m<n\right]$$
$$\text{ET I 67(37)}$$

4. $\displaystyle\int_0^\infty \frac{x^{2\nu}\cos(ax)\,dx}{(x^2+b^2)^{\mu+1}} = \frac{1}{2}b^{2\nu-2\mu-1}\mathrm{B}\left(\nu+\frac{1}{2},\mu-\nu+\frac{1}{2}\right)\,_1F_2\left(\nu+\frac{1}{2};\nu-\mu+\frac{1}{2},\frac{1}{2};\frac{b^2a^2}{4}\right)$

$$+ \frac{\sqrt{\pi}a^{2\mu-2\nu+1}}{4^{\mu-\nu+1}}\frac{\Gamma\left(\nu-\mu-\frac{1}{2}\right)}{\Gamma(\mu-\nu+1)}\,_1F_2\left(\mu+1;\mu-\nu+1,\mu-\nu+\frac{3}{2};\frac{b^2a^2}{4}\right)$$

$$= \frac{\sqrt{\pi}}{2\,\Gamma(\mu+1)}b^{2\nu-2\mu-1}\,G^{21}_{13}\left(\frac{a^2b^2}{4}\left|\begin{array}{c}-\nu+\frac{1}{2}\\ \mu-\nu+\frac{1}{2},0,\frac{1}{2}\end{array}\right.\right)$$

$$\left[a>0,\quad \operatorname{Re}b>0,\quad -\frac{1}{2}<\operatorname{Re}\nu<\operatorname{Re}\mu+1\right] \quad \text{ET I 14(29)a, ET II 235(19)}$$

5. $\displaystyle\int_0^\infty \frac{x^{2m}\cos(ax)\,dx}{(z+x^2)^{n+1}} = (-1)^{m+n}\frac{\pi}{2n!}\frac{d^n}{dz^n}\left(z^{m-\frac{1}{2}}e^{-a\sqrt{z}}\right)$

$$\left[a>0,\quad n+1>m\ge 0,\quad |\arg z|<\pi\right]$$
$$\text{ET I 10(28)}$$

6.[7] $\displaystyle\int_0^\infty \frac{x^{2m}\cos(ax)\,dx}{(b^2+x^2)^{n+\frac{1}{2}}} = \frac{(-1)^m\sqrt{\pi}}{2^n b^n\,\Gamma\left(n+\frac{1}{2}\right)}\frac{d^{2m}}{da^{2m}}\left[a^n\,K_n(ab)\right]$

$$\left[a>0,\quad \operatorname{Re}b>0,\quad 0\le m<n+\frac{1}{2}\right]$$
$$\text{ET I 14(28)}$$

**3.774**

1. $\displaystyle\int_0^\infty \frac{\sin(ax)\,dx}{\sqrt{x^2+b^2}\left(x+\sqrt{x^2+b^2}\right)^\nu} = \frac{\pi}{b^\nu\sin(\nu\pi)}\left[\sin\frac{\nu\pi}{2}\,I_\nu(ab)+\frac{i}{2}\,\mathbf{J}_\nu(iab)-\frac{i}{2}\,\mathbf{J}_\nu(-iab)\right]$

$$\left[a>0,\quad b>0,\quad \operatorname{Re}\nu>-1\right]$$
$$\text{ET I 70(19)}$$

2.    $\displaystyle\int_0^\infty \frac{\cos(ax)\,\mathrm{d}x}{\sqrt{x^2+b^2}\left(x+\sqrt{x^2+b^2}\right)^\nu} = \frac{\pi}{b^\nu \sin(\nu\pi)}\left[\frac{1}{2}\,\mathbf{J}_\nu(iab) + \frac{1}{2}\,\mathbf{J}_\nu(-iab) - \cos\frac{\nu\pi}{2}\,I_\nu(ab)\right]$

$$[a>0, \quad b>0, \quad \operatorname{Re}\nu > -1]$$

ET I 12(15)

3.    $\displaystyle\int_0^\infty \frac{\left(x+\sqrt{x^2+b^2}\right)^\nu}{\sqrt{x\,(x^2+b^2)}}\,\sin(ax)\,\mathrm{d}x = \sqrt{\frac{a\pi}{2}}\,b^\nu\,I_{\frac{1}{4}-\frac{\nu}{2}}\left(\frac{ab}{2}\right)K_{\frac{1}{4}+\frac{\nu}{2}}\left(\frac{ab}{2}\right)$

$$\left[a>0, \quad \operatorname{Re}b > 0, \quad \operatorname{Re}\nu < \tfrac{3}{2}\right]$$

ET I 71(23)

4.    $\displaystyle\int_0^\infty \frac{\left(\sqrt{x^2+b^2}-x\right)^\nu}{\sqrt{x\,(x^2+b^2)}}\,\cos(ax)\,\mathrm{d}x = \sqrt{\frac{a\pi}{2}}\,b^\nu\,I_{-\frac{1}{4}+\frac{\nu}{2}}\left(\frac{ab}{2}\right)K_{-\frac{1}{4}-\frac{\nu}{2}}\left(\frac{ab}{2}\right)$

$$\left[a>0, \quad \operatorname{Re}b > 0, \quad \operatorname{Re}\nu > -\tfrac{3}{2}\right]$$

ET I 12(17)

5.    $\displaystyle\int_0^\infty \frac{\left(b+\sqrt{x^2+b^2}\right)^\nu}{x^{\nu+\frac{1}{2}}\sqrt{x^2+b^2}}\,\sin(ax)\,\mathrm{d}x = \frac{1}{b}\sqrt{\frac{2}{a}}\,\Gamma\left(\frac{3}{4}-\frac{\nu}{2}\right)W_{\frac{\nu}{2},\frac{1}{4}}(ab)\,M_{-\frac{\nu}{2},\frac{1}{4}}(ab)$

$$\left[a>0, \quad \operatorname{Re}b > 0, \quad \operatorname{Re}\nu < \tfrac{3}{2}\right]$$

ET I 71(27)

6.    $\displaystyle\int_0^\infty \frac{\left(b+\sqrt{x^2+b^2}\right)^\nu}{x^{\nu+\frac{1}{2}}\sqrt{b^2+x^2}}\,\cos(ax)\,\mathrm{d}x = \frac{1}{b\sqrt{2a}}\,\Gamma\left(\frac{1}{4}-\frac{\nu}{2}\right)W_{\frac{\nu}{2},-\frac{1}{4}}(ab)\,M_{-\frac{\nu}{2},-\frac{1}{4}}(ab)$

$$\left[a>0, \quad \operatorname{Re}b > 0, \quad \operatorname{Re}\nu < \tfrac{1}{2}\right]$$

ET I 12(18)

**3.775**

1.[12]    $\displaystyle\int_0^\infty \frac{\left(\sqrt{x^2+b^2}+x\right)^\nu - \left(\sqrt{x^2+b^2}-x\right)^\nu}{\sqrt{x^2+h^2}}\,\sin(ax)\,\mathrm{d}x = 2b^\nu\sin\frac{\nu\pi}{2}\,K_\nu(ab)$

$$[a>0, \quad \operatorname{Re}b > 0, \quad \operatorname{Re}\nu < 1]$$

ET I 70(20)

2.[12]    $\displaystyle\int_0^\infty \frac{\left(\sqrt{x^2+b^2}+x\right)^\nu + \left(\sqrt{x^2+b^2}-x\right)^\nu}{\sqrt{x^2+b^2}}\,\cos(ax)\,\mathrm{d}x = 2b^\nu\cos\frac{\nu\pi}{2}\,K_\nu(ab)$

$$[a>0, \quad \operatorname{Re}b > 0, \quad \operatorname{Re}\nu < 1]$$

ET I 13(22)

3.[12]    $\displaystyle\int_u^\infty \frac{\left(x+\sqrt{x^2-u^2}\right)^\nu + \left(x-\sqrt{x^2-u^2}\right)^\nu}{\sqrt{x^2-u^2}}\,\sin(ax)\,\mathrm{d}x = \pi u^\nu\left[J_\nu(au)\cos\frac{\nu\pi}{2} - Y_\nu(au)\sin\frac{\nu\pi}{2}\right]$

$$[a>0, \quad u>0, \quad \operatorname{Re}\nu < 1] \quad \text{ET I 70(22)}$$

4.[12]    $\displaystyle\int_u^\infty \frac{\left(x+\sqrt{x^2-u^2}\right)^\nu + \left(x-\sqrt{x^2-u^2}\right)^\nu}{\sqrt{x^2-u^2}}\,\cos(ax)\,\mathrm{d}x = -\pi u^\nu\left[Y_\nu(au)\cos\frac{\nu\pi}{2} + J_\nu(au)\sin\frac{\nu\pi}{2}\right]$

$$[a>0, \quad u>0, \quad \operatorname{Re}\nu < 1] \quad \text{ET I 13(25)}$$

5.    $\displaystyle\int_0^u \frac{\left(x+i\sqrt{u^2-x^2}\right)^\nu + \left(x-i\sqrt{u^2-x^2}\right)^\nu}{\sqrt{u^2-x^2}}\,\sin(ax)\,\mathrm{d}x = \frac{\pi}{2}u^\nu\operatorname{cosec}\frac{\nu\pi}{2}\left[\mathbf{J}_\nu(au) - \mathbf{J}_{-\nu}(au)\right]$

$$[a>0, \quad u>0]$$

ET I 70(21)

6. $\int_0^u \dfrac{\left(x + i\sqrt{u^2 - x^2}\right)^\nu + \left(x - i\sqrt{u^2 - x^2}\right)^\nu}{\sqrt{u^2 - x^2}} \cos(ax)\,\mathrm{d}x = \dfrac{\pi}{2} u^\nu \sec\dfrac{\nu\pi}{2}\left[\mathbf{J}_\nu(au) + \mathbf{J}_{-\nu}(au)\right]$

$[a > 0, \quad u > 0, \quad |\operatorname{Re}\nu| < 1]$

ET I 13(24)

7.[6] $\int_u^\infty \dfrac{\left(x + \sqrt{x^2 - u^2}\right)^\nu + \left(x - \sqrt{x^2 - u^2}\right)^\nu}{\sqrt{x(x^2 - u^2)}} \sin(ax)\,\mathrm{d}x$

$= -\sqrt{\left(\dfrac{\pi}{2}\right)^3} au^\nu \left[J_{1/4+\nu/2}\left(\dfrac{au}{2}\right) Y_{1/4-\nu/2}\left(\dfrac{au}{2}\right) + J_{1/4-\nu/2}\left(\dfrac{au}{2}\right) Y_{1/4+\nu/2}\left(\dfrac{au}{2}\right)\right]$

$[a > 0, \quad u > 0, \quad \operatorname{Re}\nu < \tfrac{3}{2}]$ ET I 71(25)

8.[6] $\int_u^\infty \dfrac{\left(x + \sqrt{x^2 - u^2}\right)^\nu + \left(x - \sqrt{x^2 - u^2}\right)^\nu}{\sqrt{x(x^2 - u^2)}} \cos(ax)\,\mathrm{d}x$

$= -\sqrt{\left(\dfrac{\pi}{2}\right)^3} au^\nu \left[J_{-1/4+\nu/2}\left(\dfrac{au}{2}\right) Y_{-1/4-\nu/2}\left(\dfrac{au}{2}\right) + J_{-1/4-\nu/2}\left(\dfrac{au}{2}\right) Y_{-1/4+\nu/2}\left(\dfrac{au}{2}\right)\right]$

$[a > 0, \quad u > 0, \quad \operatorname{Re}\nu < \tfrac{3}{2}]$ ET I 13(26)

9. $\int_0^\infty \dfrac{\left(x + b + \sqrt{x^2 + 2bx}\right)^\nu + \left(x + b - \sqrt{x^2 + 2bx}\right)^\nu}{\sqrt{x^2 + 2bx}} \sin(ax)\,\mathrm{d}x$

$= \pi b^\nu \left[Y_\nu(ba) \sin\left(ba - \dfrac{\nu\pi}{2}\right) + J_\nu(ba) \cos\left(ba - \dfrac{\nu\pi}{2}\right)\right]$

$[a > 0, \quad |\arg b| < \pi, \quad \operatorname{Re}\nu < 1]$ ET I 71(26)

10. $\int_0^\infty \dfrac{\left(x + b + \sqrt{x^2 + 2bx}\right)^\nu + \left(x + b - \sqrt{x^2 + 2bx}\right)^\nu}{\sqrt{x^2 + 2bx}} \cos(ax)\,\mathrm{d}x$

$= \pi b^\nu \left[J_\nu(ba) \sin\left(ba - \dfrac{\nu\pi}{2}\right) - Y_\nu(ba) \cos\left(ba - \dfrac{\nu\pi}{2}\right)\right]$

$[a > 0, \quad |\arg b| < \pi, \quad \operatorname{Re}\nu < 1]$ ET I 13(23)

11. $\int_0^{2u} \dfrac{\left(\sqrt{2u + x} + i\sqrt{2u - x}\right)^{4\nu} + \left(\sqrt{2u + x} - i\sqrt{2u - x}\right)^{4\nu}}{\sqrt{4u^2 x - x^3}} \cos(ax)\,\mathrm{d}x$

$= (4u)^{2\nu} \pi^{3/2} \sqrt{\dfrac{a}{2}}\, J_{\nu-1/4}(au)\, J_{-\nu-1/4}(au)$

$[a > 0, \quad u > 0]$ ET I 14(27)

**3.776**

1. $\int_0^\infty \dfrac{a^2(b + x)^2 + p(p + 1)}{(b + x)^{p+2}} \sin(ax)\,\mathrm{d}x = \dfrac{a}{b^p}$ $[a > 0, \quad b > 0, \quad p > 0]$ BI (170)(1)

2. $\int_0^\infty \dfrac{a^2(b + x)^2 + p(p + 1)}{(b + x)^{p+2}} \cos(ax)\,\mathrm{d}x = \dfrac{p}{b^{p+1}}$ $[a > 0, \quad b > 0, \quad p > 0]$ BI (170)(2)

## 3.78–3.81 Rational functions of $x$ and of trigonometric functions

**3.781**

1.    $\displaystyle\int_0^\infty \left(\frac{\sin x}{x} - \frac{1}{1+x}\right)\frac{dx}{x} = 1 - C$           (cf. **3.784** 4 and **3.781** 2)     BI (173)(7)

2.    $\displaystyle\int_0^\infty \left(\cos x - \frac{1}{1+x}\right)\frac{dx}{x} = -C$                                 BI (173)(8)

**3.782**

1.    $\displaystyle\int_0^u \frac{1-\cos x}{x}\,dx - \int_u^\infty \frac{\cos x}{x}\,dx = C + \ln u$     $[u > 0]$         GW (333)(31)

2.    $\displaystyle\int_0^\infty \frac{1-\cos ax}{x^2}\,dx = \frac{a\pi}{2}$              $[a \geq 0]$         BI (158)(1)

3.[12]   $\displaystyle\int_{-\infty}^\infty \frac{1-\cos x}{x(x-b)}\,dx = -\frac{\pi i}{b}\left(e^{ib} - 1\right)$     $\left[\operatorname{Im} b \neq 0, \quad \operatorname{Re} b^2 \leq 0\right]$    ET II 253(48)

**3.783**

1.    $\displaystyle\int_0^\infty \left[\frac{\cos x - 1}{x^2} + \frac{1}{2(1+x)}\right]\frac{dx}{x} = \frac{1}{2}C - \frac{3}{4}$                   BI (173)(19)

2.    $\displaystyle\int_0^\infty \left(\cos x - \frac{1}{1+x^2}\right)\frac{dx}{x} = -C$                   EH I 17, BI(273)(21)

**3.784**

1.    $\displaystyle\int_0^\infty \frac{\cos ax - \cos bx}{x}\,dx = \ln\frac{b}{a}$       $[a > 0, \quad b > 0]$    FI II 635, GW(333)(20)

2.    $\displaystyle\int_0^\infty \frac{a\sin bx - b\sin ax}{x^2}\,dx = ab\ln\frac{a}{b}$       $[a > 0, \quad b > 0]$         FI II 647

3.    $\displaystyle\int_0^\infty \frac{\cos ax - \cos bx}{x^2}\,dx = \frac{(b-a)\pi}{2}$       $[a \geq 0, \quad b \geq 0]$    BI(158)(12), FI II 645

4.    $\displaystyle\int_0^\infty \frac{\sin x - x\cos x}{x^2}\,dx = 1$                                 BI (158)(3)

5.    $\displaystyle\int_0^\infty \frac{\cos ax - \cos bx}{x(x+\beta)}\,dx = \frac{1}{\beta}\left[\operatorname{ci}(a\beta)\cos a\beta + \operatorname{si}(a\beta)\sin a\beta - \operatorname{ci}(b\beta)\cos b\beta - \operatorname{si}(b\beta)\sin b\beta + \ln\frac{b}{a}\right]$

                                             $[a > 0, \quad b > 0, \quad |\arg\beta| < \pi]$    ET II 221(49)

6.    $\displaystyle\int_0^\infty \frac{\cos ax + x\sin ax}{1+x^2}\,dx = \pi e^{-a}$           $[a > 0]$           GW (333)(73)

7.    $\displaystyle\int_0^\infty \frac{\sin ax - ax\cos ax}{x^3}\,dx = \frac{\pi}{4}a^2\operatorname{sign} a$                       LI (158)(5)

8.    $\displaystyle\int_0^\infty \frac{\cos ax - \cos bx}{x^2\left(x^2 + \beta^2\right)}\,dx = \frac{\pi\left[(b-a)\beta + e^{-b\beta} - e^{-a\beta}\right]}{2\beta^3}$

                                             $[a > 0, \quad b > 0, \quad |\arg\beta| < \pi]$

                                             BI(173)(20)a, ET II 222(59)

$9.^{10}$ $\displaystyle\int_0^\infty \frac{\cos mx}{1 + a^2\, T_n(x)} = \frac{\pi}{2n\sqrt{1+a^2}} \sum_{k=1}^n e^{-m\sin u \sinh\phi} \left(\cos\beta \sin u \cosh\phi + \sin\beta \cos u \sinh\phi\right)$

$$[u = (2k-1)\,\pi/\,(2n), \quad \phi = \operatorname{arcsinh}(1/a), \quad \beta = m\cos u \cosh\phi, \quad 0 < |a| < 1]$$

**3.785** $\displaystyle\int_0^\infty \frac{1}{x} \sum_{k=1}^n a_k \cos b_k x\, dx = -\sum_{k=1}^n a_k \ln b_k$ $\qquad \left[ b_k > 0, \quad \sum_{k=1}^n a_k = 0 \right]$ $\qquad$ FI II 649

**3.786**

1. $\displaystyle\int_0^\infty \frac{(1 - \cos ax)\sin bx}{x^2}\, dx = \frac{b}{2}\ln\frac{b^2 - a^2}{b^2} + \frac{a}{2}\ln\frac{a+b}{a-b}$

$$[a > 0, \quad b > 0] \qquad \text{ET I 81(29)}$$

$2.^{11}$ $\displaystyle\int_0^\infty \frac{(1 - \cos ax)\cos bx}{x}\, dx = \ln\frac{\sqrt{|a^2 - b^2|}}{b}$ $\qquad [a > 0, \quad b > 0, \quad a \neq b]$ $\qquad$ FI II 647

$3.^{11}$ $\displaystyle\int_0^\infty \frac{(1 - \cos ax)\cos bx}{x^2}\, dx = \frac{\pi}{2}(a - b)$ $\qquad [a < b \leq 0]$

$$= 0 \qquad [0 < a \leq b]$$

$$\text{ET I 20(16)}$$

**3.787**

1. $\displaystyle\int_0^\infty \frac{(\cos a - \cos nax)\sin mx}{x}\, dx = \frac{\pi}{2}(\cos a - 1)$ $\qquad [m > na > 0]$

$$= \frac{\pi}{2}\cos a \qquad [na > m]$$

$$\text{BI(155)(7)}$$

2. $\displaystyle\int_0^\infty \frac{\sin^2 ax - \sin^2 bx}{x}\, dx = \frac{1}{2}\ln\frac{a}{b}$ $\qquad [a > 0, \quad b > 0]$ $\qquad$ GW (333)(20b)

3. $\displaystyle\int_0^\infty \frac{x^3 - \sin^3 x}{x^5}\, dx = \frac{13}{32}\pi$ $\qquad$ BI (158)(6)

4. $\displaystyle\int_0^\infty \frac{(3 - 4\sin^2 ax)\sin^2 ax}{x}\, dx = \frac{1}{2}\ln 2$ $\qquad [a \text{ real}, a \neq 0]$ $\qquad$ HBI (155)(6)

**3.788** $\displaystyle\int_0^{\pi/2} \left(\frac{1}{x} - \cot x\right) dx = \ln\frac{\pi}{2}$ $\qquad$ GW (333)(61)a

**3.789** $\displaystyle\int_0^{\pi/2} \frac{4x^2 \cos x + (\pi - x)x}{\sin x}\, dx = \pi^2 \ln 2$ $\qquad$ LI (206)(10)

**3.791**

1. $\displaystyle\int_0^{\pi/2} \frac{x\, dx}{1 + \sin x} = \ln 2$ $\qquad$ GW (333)(55a)

2. $\displaystyle\int_0^\pi \frac{x\cos x}{1 + \sin x}\, dx = \pi \ln 2 - 4\mathbf{G}$ $\qquad$ GW (333)(55c)

3. $\displaystyle\int_0^{\pi/2} \frac{x\cos x}{1 + \sin x}\, dx = \pi \ln 2 - 2\mathbf{G}$ $\qquad$ GW (333)(55b)

4. $\int_0^\pi \frac{\left(\frac{\pi}{2} - x\right)\cos x}{1 - \sin x}\, dx = 2\int_0^{\pi/2} \frac{\left(\frac{\pi}{2} - x\right)\cos x}{1 - \sin x}\, dx = \pi \ln 2 + 4\boldsymbol{G}$　　　　　BI(207)(3), GW(333)(56c)

5. $\int_0^{\pi/2} \frac{x^2\, dx}{1 - \cos x} = -\frac{\pi^2}{4} + \pi \ln 2 + 4\boldsymbol{G}$　　　　　BI (207)(3)

6. $\int_0^\pi \frac{x^2\, dx}{1 - \cos x} = 4\pi \ln 2$　　　　　BI (219)(1)

7. $\int_0^{\pi/2} \frac{x^{p+1}\, dx}{1 - \cos x} = -\left(\frac{\pi}{2}\right)^{p+1} + \left(\frac{\pi}{2}\right)^p (p+1)\left[\frac{2}{p} - \sum_{k=1}^\infty \frac{1}{4^{2k-1}(p+2k)}\,\zeta(2k)\right]$

$[p > 0]$　　　　　LI (207)(4)

8. $\int_0^{\pi/2} \frac{x\, dx}{1 + \cos x} = \frac{\pi}{2} - \ln 2$　　　　　GW (333)(55a)

9. $\int_0^{\pi/2} \frac{x \sin x\, dx}{1 - \cos x} = \frac{\pi}{2} \ln 2 + 2\boldsymbol{G}$　　　　　GW (333)(56a)

10. $\int_0^\pi \frac{x \sin x\, dx}{1 - \cos x} = 2\pi \ln 2$　　　　　GW (333)(56b)

11. $\int_0^\pi \frac{x - \sin x}{1 - \cos x}\, dx = \frac{\pi}{2} + \int_0^{\pi/2} \frac{x - \sin x}{1 - \cos x}\, dx = 2$　　　　　GW (333)(57a)

12. $\int_0^{\pi/2} \frac{x \sin x}{1 + \cos x}\, dx = -\frac{\pi}{2} \ln 2 + 2\boldsymbol{G}$　　　　　GW (333)(55b)

**3.792**

1. $\int_{-\pi}^\pi \frac{dx}{1 - 2a \cos x + a^2} = \frac{2\pi}{1 - a^2}$　　　　$[a^2 < 1]$　　　　ГІ ІІ 485

2. $\int_0^{\pi/2} \frac{x \cos x\, dx}{1 + 2a \sin x + a^2} = \frac{\pi}{2a} \ln(1 + a) - \sum_{k=0}^\infty (-1)^k \frac{a^{2k}}{(2k+1)^2}$

$[a^2 < 1]$　　　　LI (241)(2)

3.[12] $\int_0^\pi \frac{x \sin x\, dx}{1 - 2a \cos x + a^2} = \frac{\pi}{a} \ln(1 + a)$　　　　$[a^2 < 1, \quad a \neq 0]$

$\qquad\qquad\qquad\qquad\qquad = \frac{\pi}{a} \ln\left(1 + \frac{1}{a}\right)$　　　　$[a^2 > 1]$

BI (221)(2)

4. $\int_0^{2\pi} \frac{x \sin x\, dx}{1 - 2a \cos x + a^2} = \frac{2\pi}{a} \ln(1 - a)$　　　　$[a^2 < 1, \quad a \neq 0]$

$\qquad\qquad\qquad\qquad\qquad = \frac{2\pi}{a} \ln\left(1 - \frac{1}{a}\right)$　　　　$[a^2 > 1]$

BI (223)(4)

5. $\int_0^{2\pi} \frac{x \sin nx\, dx}{1 - 2a \cos x + a^2} = \frac{2\pi}{1 - a^2}\left[(a^{-n} - a^n)\ln(1 - a) + \sum_{k=1}^{n-1} \frac{a^{-k} - a^k}{n - k}\right]$

$[a^2 < 1, \quad a \neq 0]$　　　　BI (223)(5)

6. $\int_0^\infty \dfrac{\sin x}{1 - 2a\cos x + a^2} \dfrac{dx}{x} = \dfrac{\pi}{4a}\left[\left|\dfrac{1+a}{1-a}\right| - 1\right]$ $\qquad [a \text{ real}, \quad a \neq 0, \quad a \neq 1]$

GW (333)(62b)

7.[8] $\int_0^\infty \dfrac{\sin bx}{1 - 2a\cos x + a^2} \dfrac{dx}{x} = \dfrac{\pi}{2} \dfrac{1 + a - 2a^{\lfloor b\rfloor+1}}{(1-a^2)(1-a)}$ $\qquad [b \neq 0,1,2,\ldots]$

$\qquad\qquad = \dfrac{\pi}{2} \dfrac{1 + a - a^b - a^{b+1}}{(1-a^2)(1-a)}$ $\qquad [b = 1,2,\ldots]; \qquad [0 < a < 1]$

ET I 81(26)

8. $\int_0^\infty \dfrac{\sin x \cos bx}{1 - 2a\cos x + a^2} \dfrac{dx}{x} = \dfrac{\pi}{2(1-a)} a^{\lfloor b\rfloor}$ $\qquad [b \neq 0,1,2,\ldots]$

$\qquad\qquad = \dfrac{\pi}{2(1-a)} a^b + \dfrac{\pi}{4} a^{b-1}$ $\qquad [b = 1,2,3,\ldots];$

$[0 < a < 1, \quad b > 0]; \text{(for } b = 0, \text{ see } \mathbf{3.792}\ 6)$ ET I 19(5)

9. $\int_0^\infty \dfrac{(1 - a\cos x)\sin bx}{1 - 2a\cos x + a^2} \dfrac{dx}{x} = \dfrac{\pi}{2} \dfrac{1 - a^{\lfloor b\rfloor+1}}{1-a}$ $\qquad [b \neq 1,2,3,\ldots]$

$\qquad\qquad = \dfrac{\pi}{2} \dfrac{1 - a^b}{1-a} + \dfrac{\pi a^b}{4}$ $\qquad [b = 1,2,3,\ldots]$

$[0 < a < 1, \quad b > 0]$ ET I 82(33)

10.[3] $\int_0^\infty \dfrac{1}{1 - 2a\cos bx + a^2} \dfrac{dx}{\beta^2 + x^2} = \dfrac{\pi}{2\beta(1-a^2)} \dfrac{1 + ae^{-b\beta}}{1 - ae^{-b\beta}}$

$[a^2 < 1, \quad b \geq 0]$ BI (192)(1)

11. $\int_0^\infty \dfrac{1}{1 - 2a\cos bx + a^2} \dfrac{dx}{\beta^2 - x^2} = \dfrac{a\pi}{\beta(1-a^2)} \dfrac{\sin b\beta}{1 - 2a\cos b\beta + a^2}$

$[a^2 < 1, \quad b > 0]$ BI (193)(1)

12. $\int_0^\infty \dfrac{\sin bcx}{1 - 2a\cos bx + a^2} \dfrac{x\,dx}{\beta^2 + x^2} = \dfrac{\pi}{2} \dfrac{e^{-\beta bc} - a^c}{(1 - ae^{-b\beta})(1 - ae^{b\beta})}$

$[a^2 < 1, \quad b > 0, \quad c > 0]$ BI (192)(8)

13. $\int_0^\infty \dfrac{\sin bx}{1 - 2a\cos bx + a^2} \dfrac{x\,dx}{\beta^2 + x^2} = \dfrac{\pi}{2} \dfrac{1}{e^{b\beta} - a}$ $\qquad [a^2 < 1, \quad b > 0]$

$\qquad\qquad = \dfrac{\pi}{2a} \dfrac{1}{ae^{b\beta} - 1}$ $\qquad [a^2 > 1, \quad b > 0]$

BI (192)(2)

14. $\int_0^\infty \dfrac{\sin bcx}{1 - 2a\cos bx + a^2} \dfrac{x\,dx}{\beta^2 - x^2} = \dfrac{\pi}{2} \dfrac{a^c - \cos\beta bc}{1 - 2a\cos\beta b + a^2}$

$[a^2 < 1, \quad b > 0, \quad c > 0]$ BI (193)(5)

15. $\int_0^\infty \dfrac{\cos bcx}{1 - 2a\cos bx + a^2} \dfrac{dx}{\beta^2 - x^2} = \dfrac{\pi}{2\beta(1-a^2)} \dfrac{(1-a^2)\sin\beta bc + 2a^{c+1}\sin\beta b}{1 - 2a\cos\beta b + a^2}$

$[a^2 < 1, \quad b > 0, \quad c > 0]$ BI (193)(9)

16. $\int_0^\infty \dfrac{1 - a\cos bx}{1 - 2a\cos bx + a^2} \dfrac{dx}{1 + x^2} = \dfrac{\pi}{2} \dfrac{e^b}{e^b - a}$ $\qquad [a^2 < 1, \quad b > 0]$ FI II 719

17.    $\displaystyle\int_0^\infty \frac{\cos bx}{1 - 2a\cos x + a^2}\frac{\mathrm{d}x}{x^2 + \beta^2} = \frac{\pi\left(e^{\beta - \beta b} + ae^{\beta b}\right)}{2\beta\left(1 - a^2\right)\left(e^\beta - a\right)}$

$$[0 \le b < 1, \quad |a| < 1, \quad \operatorname{Re}\beta > 0]$$
$$\text{ET I 21(21)}$$

18.    $\displaystyle\int_0^\infty \frac{\sin bx \sin x}{1 - 2a\cos x + a^2}\frac{\mathrm{d}x}{x^2 + \beta^2} = \frac{\pi}{2\beta}\frac{\sinh b\beta}{e^\beta - a}$       $[0 \le b < 1]$

$$= \frac{\pi}{4\beta\left(ae^\beta - 1\right)}\left[a^m e^{\beta(m+1-b)} - e^{(1-b)\beta}\right]$$
$$-\frac{\pi}{4\beta\left(ae^{-\beta} - 1\right)}\left[a^m e^{-(m+1-b)\beta} - e^{-(1-b)\beta}\right] \quad [m \le b \le m+1]$$
$$[0 < a < 1, \quad \operatorname{Re}\beta > 0] \qquad \text{ET I 81(27)}$$

19.    $\displaystyle\int_0^\infty \frac{(\cos x - a)\cos bx}{1 - 2a\cos x + a^2}\frac{\mathrm{d}x}{x^2 + \beta^2} = \frac{\pi\cosh\beta b}{2\beta\left(e^\beta - a\right)}$     $[0 \le b < 1, \quad |a| < 1, \quad \operatorname{Re}\beta > 0]$

$$\text{ET I 21(23)}$$

20.    $\displaystyle\int_0^\infty \frac{\sin x}{\left(1 - 2a\cos 2x + a^2\right)^{n+1}}\frac{\mathrm{d}x}{x}$

$$= \int_0^\infty \frac{\tan x}{\left(1 - 2a\cos 2x + a^2\right)^{n+1}}\frac{\mathrm{d}x}{x}$$
$$= \int_0^\infty \frac{\tan x}{\left(1 - 2a\cos 4x + a^2\right)^{n+1}}\frac{\mathrm{d}x}{x} = \frac{\pi}{2\left(1 - a^2\right)^{2n+1}}\sum_{k=0}^n \binom{n}{k}^2 a^{2k}$$
$$\text{BI (187)(14)}$$

### 3.793

1.3    $\displaystyle\int_0^{2\pi} \frac{\sin nx - a\sin[(n+1)x]}{1 - 2a\cos x + a^2}x\,\mathrm{d}x = -2\pi a^n\left[\ln(1-a) + \sum_{k=1}^n \frac{1}{ka^k}\right]$

$$[|a| < 1] \qquad\qquad \text{BI (223)(9)}$$

2.12    $\displaystyle\int_0^{2\pi} \frac{\cos nx - a\cos[(n+1)x]}{1 - 2a\cos x + a^2}x\,\mathrm{d}x = 2\pi^2 a^n$      $\left[a^2 < 1\right]$      $\text{BI (223)(13)}$

### 3.794

1.3    $\displaystyle\int_0^\pi \frac{x\,\mathrm{d}x}{1 + a^2 + 2a\cos x} = \frac{\pi^2}{2\left(1 - a^2\right)} + \frac{4}{\left(1 - a^2\right)}\sum_{k=0}^\infty \frac{a^{2k+1}}{(2k+1)^2}$

$$\left[a^2 < 1\right]$$

2.12    $\displaystyle\int_0^{2\pi} \frac{x\sin nx}{1 \pm a\cos x}\,\mathrm{d}x = \frac{2\pi}{\sqrt{1 - a^2}}\left[(\mp 1)^n \frac{\left(1 + \sqrt{1 - a^2}\right)^n - \left(1 - \sqrt{1 - a^2}\right)^n}{a^n}\right.$

$$\left.\times\ln\frac{2\sqrt{1 \pm a}}{\sqrt{1 + a} + \sqrt{1 - a}} + \sum_{k=1}^{n-1}\frac{(\mp 1)^k}{n - k}\frac{\left(1 + \sqrt{1 - a^2}\right)^k - \left(1 - \sqrt{1 - a^2}\right)^k}{a^k}\right]$$
$$\left[a^2 < 1\right] \qquad\qquad \text{BI (223)(2)}$$

3.3    $\displaystyle\int_0^{2\pi} \frac{x\cos nx}{1 \pm a\cos x}\,\mathrm{d}x = \frac{2\pi^2}{\sqrt{1 - a^2}}\left(\frac{1 - \sqrt{1 - a^2}}{\pm a}\right)^n$      $\left[a^2 < 1\right]$      $\text{BI (223)(3)}$

4. $\displaystyle\int_0^\pi \frac{x \sin x \, dx}{a + b \cos x} = \frac{\pi}{b} \ln \frac{a + \sqrt{a^2 - b^2}}{2(a - b)}$ $\qquad [a > |b| > 0]$ $\qquad$ GW (333)(53a)

5. $\displaystyle\int_0^{2\pi} \frac{x \sin x \, dx}{a + b \cos x} = \frac{2\pi}{b} \ln \frac{a + \sqrt{a^2 - b^2}}{2(a + b)}$ $\qquad [a > |b| > 0]$ $\qquad$ GW (333)(53b)

6. $\displaystyle\int_0^\infty \frac{\sin x}{a \pm b \cos 2x} \frac{dx}{x} = \frac{\pi}{2\sqrt{a^2 - b^2}}$ $\qquad \left[a^2 > b^2\right]$

$\displaystyle = 0$ $\qquad \left[a^2 < b^2\right]$

$\qquad$ BI (181)(1)

**3.795** $\displaystyle\int_{-\infty}^\infty \frac{\left(b^2 + c^2 + x^2\right) x \sin ax - \left(b^2 - c^2 - x^2\right) c \sinh ac}{\left[x^2 + (b - c)^2\right]\left[x^2 + (b + c)^2\right](\cos ax + \cosh ac)} \, dx = \pi$ $\qquad [c > b > 0]$

$\displaystyle = \frac{2\pi}{e^{ab} + 1}$ $\qquad [b > c > 0]$

$\qquad [a > 0]$ $\qquad$ BI (202)(18)

**3.796**

1. $\displaystyle\int_0^{\pi/2} \frac{\cos x \pm \sin x}{\cos x \mp \sin x} x \, dx = \mp \frac{\pi}{4} \ln 2 - \boldsymbol{G}$ $\qquad$ BI (207)(8, 9)

2. $\displaystyle\int_0^{\pi/4} \frac{\cos x - \sin x}{\cos x + \sin x} x \, dx = \frac{\pi}{4} \ln 2 - \frac{1}{2} \boldsymbol{G}$ $\qquad$ BI (204)(23)

**3.797**

1. $\displaystyle\int_0^{\pi/4} \left(\frac{\pi}{4} - x \tan x\right) \tan x \, dx = \frac{1}{2} \ln 2 + \frac{\pi^2}{32} - \frac{\pi}{4} + \frac{\pi}{8} \ln 2$ $\qquad$ BI (204)(8)

2. $\displaystyle\int_0^{\pi/4} \frac{\left(\frac{\pi}{4} - x\right) \tan x \, dx}{\cos 2x} = -\frac{\pi}{8} \ln 2 + \frac{1}{2} \boldsymbol{G}$ $\qquad$ BI (204)(19)

3. $\displaystyle\int_0^{\pi/4} \frac{\frac{\pi}{4} - x \tan x}{\cos 2x} \, dx = \frac{\pi}{8} \ln 2 + \frac{1}{2} \boldsymbol{G}$ $\qquad$ BI (204)(20)

**3.798**

1.[8] $\displaystyle\int_0^\infty \frac{\tan x}{a + b \cos 2x} \frac{dx}{x} = \frac{\pi}{2\sqrt{a^2 - b^2}}$ $\qquad [0 < b < a]$

$\displaystyle = 0$ $\qquad [0 < a < b]$

$\qquad$ BI (181)(2)

2.[8] $\displaystyle\int_0^\infty \frac{\tan x}{a + b \cos 4x} \frac{dx}{x} = \frac{\pi}{2\sqrt{a^2 - b^2}}$ $\qquad [0 < b < a]$

$\displaystyle = 0$ $\qquad [0 < a < b]$

$\qquad$ BI (181)(3)

**3.799**

1. $\displaystyle\int_0^{\pi/2} \frac{x \, dx}{(\sin x + a \cos x)^2} = \frac{a}{1 + a^2} \frac{\pi}{2} - \frac{\ln a}{1 + a^2}$ $\qquad [a > 0]$ $\qquad$ BI (208)(5)

2. $\int_0^{\pi/4} \dfrac{x \, dx}{(\cos x + a \sin x)^2} = \dfrac{1}{1+a^2} \ln \dfrac{1+a}{\sqrt{2}} + \dfrac{\pi}{4} \dfrac{1-a}{(1+a)(1+a^2)}$

$[a > 0]$  BI (204)(24)

3. $\int_0^{\pi} \dfrac{a \cos x + b}{(a + b \cos x)^2} x^2 \, dx = \dfrac{2\pi}{b} \ln \dfrac{2(a-b)}{a + \sqrt{a^2 - b^2}}$  $[a > |b| > 0]$  GW (333)(58a)

**3.811**

1. $\int_0^{\pi} \dfrac{\sin x}{1 - \cos t_1 \cos x} \dfrac{x \, dx}{1 - \cos t_2 \cos x} = \pi \operatorname{cosec} \dfrac{t_1 + t_2}{2} \operatorname{cosec} \dfrac{t_1 - t_2}{2} \ln \dfrac{1 + \tan \dfrac{t_1}{2}}{1 + \tan \dfrac{t_2}{2}}$

(cf. **3.794** 4)  BI (222)(5)

2. $\int_0^{\pi/2} \dfrac{x \, dx}{(\cos x \pm \sin x)\sin x} = \dfrac{\pi}{4} \ln 2 + \boldsymbol{G}$  BI (208))(16, 17)

3. $\int_0^{\pi/4} \dfrac{x \, dx}{(\cos x + \sin x)\sin x} = -\dfrac{\pi}{8} \ln 2 + \boldsymbol{G}$  BI (204)(29)

4. $\int_0^{\pi/4} \dfrac{x \, dx}{(\cos x + \sin x)\cos x} = \dfrac{\pi}{8} \ln 2$  BI (204)(28)

5. $\int_0^{\pi/4} \dfrac{\sin x}{\sin x + \cos x} \dfrac{x \, dx}{\cos^2 x} = -\dfrac{\pi}{8} \ln 2 + \dfrac{\pi}{4} - \dfrac{1}{2} \ln 2$  BI (204)(30)

**3.812**

1. $\int_0^{\pi} \dfrac{x \sin x \, dx}{a + b \cos^2 x} = \dfrac{\pi}{\sqrt{ab}} \arctan \sqrt{\dfrac{b}{a}}$  $[a > 0, \quad b > 0]$

$\qquad\qquad\qquad - \dfrac{\pi}{2\sqrt{-ab}} \ln \dfrac{\sqrt{a} + \sqrt{-b}}{\sqrt{a} - \sqrt{-b}}$  $[a > \quad b > 0]$

GW (333)(60a)

2. $\int_0^{\pi/2} \dfrac{x \sin 2x \, dx}{1 + a \cos^2 x} = \dfrac{\pi}{a} \ln \dfrac{1 + \sqrt{1+a}}{2}$  $[a > -1, \quad a \neq 0]$  BI (207)(10)

3. $\int_0^{\pi/2} \dfrac{x \sin 2x \, dx}{1 + a \sin^2 x} = \dfrac{\pi}{a} \ln \dfrac{2(1 + a - \sqrt{1+a})}{2}$  $[a > -1, \quad a \neq 0]$  BI (207)(2)

4.[11] $\int_0^{\pi} \dfrac{x \, dx}{a^2 - \cos^2 x} = \dfrac{\pi^2}{2a\sqrt{a^2 - 1}}$  $[a^2 > 1]$

$\qquad\qquad\qquad = 0$  $[\text{principal value for } 0 < a^2 < 1]$

$\qquad\qquad\qquad = \text{divergent}$  $[a = 0]$

BI (219)(10)

5.[7] $\int_0^{\pi} \dfrac{x \sin x \, dx}{a^2 - \cos^2 x} = \dfrac{\pi}{2a} \ln \left( \dfrac{1+a}{1-a} \right)$  $[0 < a < 1]$  divergent if $a = 0$

BI (219)(13)

6.[11]   $\displaystyle\int_0^\pi \frac{x \sin 2x \, dx}{a^2 - \cos^2 x} = \pi \ln\left[4\left(1 - a^2\right)\right]$      [principal value for $0 \le a^2 < 1$]

$\displaystyle = 2\pi \ln\left[2\left(1 - a^2 + a\sqrt{a^2 - 1}\right)\right]$     $[a^2 > 1]$

$= \text{divergent}$      $[|a| = 1]$

BI (219)(19)

7.   $\displaystyle\int_0^{\pi/2} \frac{x \sin x \, dx}{\cos^2 t - \sin^2 x} = -2 \operatorname{cosec} t \sum_{k=0}^\infty \frac{\sin(2k+1)t}{(2k+1)^2}$      BI (207)(1)

8.   $\displaystyle\int_0^\pi \frac{x \sin x \, dx}{1 - \cos^2 t \sin^2 x} = \pi(\pi - 2t) \operatorname{cosec} 2t$      BI (219)(12)

9.   $\displaystyle\int_0^\pi \frac{x \cos x \, dx}{\cos^2 t - \cos^2 x} = 4 \operatorname{cosec} t \sum_{k=0}^\infty \frac{\sin(2k+1)t}{(2k+1)^2}$      BI (219)(17)

10.   $\displaystyle\int_0^\pi \frac{x \sin x \, dx}{\tan^2 t + \cos^2 x} = \frac{\pi}{2}(\pi - 2t) \cot t$      BI (219)(14)

11.   $\displaystyle\int_0^\infty \frac{x\left(a \cos x + b\right) \sin x \, dx}{\cot^2 t + \cos^2 x} = 2a\pi \ln \cos \frac{t}{2} + \pi b t \tan t$      BI (219)(18)

12.   $\displaystyle\int_0^\pi \frac{x \sin x \cos x}{a - \sin^2 x} \, dx = -\pi \ln 2 + \ln\left[1 + \sqrt{\frac{a-1}{a}}\right]$      $[a > 1]$

13.   $\displaystyle\int_0^{\pi/2} \ln\left(a - \sin^2 x\right) \, dx = -\pi \ln 2 + i\pi \ln \arccos \sqrt{a}$     $[0 < a < 1]$

14.   $\displaystyle\text{PV} \int_0^{\pi/2} \ln\left(\left|a - \sin^2 x\right|\right) \, dx = -\pi \ln 2$     $[0 < a < 1]$

15.   $\displaystyle\text{PV} \int_0^{\pi/2} \ln\left(\left|a - \cos^2 x\right|\right) \, dx = -\pi \ln 2$     $[0 < a < 1]$

**3.813**

1.   $\displaystyle\int_0^\pi \frac{x \, dx}{a^2 \cos^2 x + b^2 \sin^2 x} = \frac{1}{4} \int_0^{2\pi} \frac{x \, dx}{a^2 \cos^2 x + b^2 \sin^2 x} = \frac{\pi^2}{2ab}$

$[a > 0, \quad b > 0]$      GW (333)(36)

2.   $\displaystyle\int_0^\infty \frac{1}{\beta^2 \sin^2 ax + \gamma^2 \cos^2 ax} \frac{dx}{x^2 + \delta^2} = \frac{\pi \sinh(2a\delta)}{4\delta\left(\beta^2 \sinh^2(a\delta) - \gamma^2 \cosh^2(a\delta)\right)}\left[\frac{\beta}{\gamma} - \frac{\gamma}{\beta} - \frac{2}{\sinh(2a\delta)}\right]$

$\left[\left|\arg\dfrac{\beta}{\gamma}\right| < \pi, \quad \operatorname{Re}\delta > 0, \quad a > 0\right]$

GW(333)(81), ET II 222(63)

3.   $\displaystyle\int_0^\infty \frac{\sin x \, dx}{x\left(a^2 \sin^2 x + b^2 \cos^2 x\right)} = \frac{\pi}{2ab}$     $[ab > 0]$      BI (181)(8)

4.   $\displaystyle\int_0^\infty \frac{\sin^2 x \, dx}{x\left(a^2 \cos^2 x + b^2 \sin^2 x\right)} = \frac{\pi}{2b(a+b)}$     $[a > 0, \quad b > 0]$      BI (181)(11)

5.    $\displaystyle\int_0^{\pi/2} \frac{x \sin 2x \, dx}{a^2 \cos^2 x + b^2 \sin^2 x} = \frac{\pi}{a^2 - b^2} \ln \frac{a+b}{2b}$      $[a > 0, \quad b > 0, \quad a \neq b]$    GW (333)(52a)

6.    $\displaystyle\int_0^{\pi} \frac{x \sin 2x \, dx}{a^2 \cos^2 x + b^2 \sin^2 x} = \frac{2\pi}{a^2 - b^2} \ln \frac{a+b}{2a}$      $[a > 0, \quad b > 0, \quad a \neq b]$    GW (333)(52b)

7.    $\displaystyle\int_0^{\infty} \frac{\sin 2x}{a^2 \cos^2 x + b^2 \sin^2 x} \frac{dx}{x} = \frac{\pi}{a(a+b)}$      $[a > 0, \quad b > 0]$    BI (182)(3)

8.    $\displaystyle\int_0^{\infty} \frac{\sin 2ax}{\beta^2 \sin^2 ax + \gamma^2 \cos^2 ax} \frac{x \, dx}{x^2 + \delta^2} = \frac{\pi}{2\left(\beta^2 \sinh^2(a\delta) - \gamma^2 \cosh^2(a\delta)\right)} \left[\frac{\beta - \gamma}{\beta + \gamma} - e^{-2a\delta}\right]$

$$\left[a > 0, \quad \left|\arg \frac{\beta}{\gamma}\right| < \pi, \quad \operatorname{Re} \delta > 0\right]$$

ET II 222(64), GW(333)(80)

9.    $\displaystyle\int_0^{\infty} \frac{(1 - \cos x) \sin x}{a^2 \cos^2 x + b^2 \sin^2 x} \frac{dx}{x} = \frac{\pi}{2b(a+b)}$      $[a > 0, \quad b > 0]$    BI (182)(7)a

10.    $\displaystyle\int_0^{\infty} \frac{\sin x \cos^2 x}{a^2 \cos^2 x + b^2 \sin^2 x} \frac{dx}{x} = \frac{\pi}{2a(a+b)}$      $[a > 0, \quad b > 0]$    BI (182)(4)

11.[12]    $\displaystyle\int_0^{\infty} \frac{\sin^3 x}{a^2 \cos^2 x + b^2 \sin^2 x} \frac{dx}{x} = \frac{\pi}{2b} \frac{1}{a+b}$      $[a > 0, \quad b > 0]$    BI (182)(1)

## 3.814

1.    $\displaystyle\int_0^{\pi/2} \frac{(1 - x \cot x) \, dx}{\sin^2 x} = \frac{\pi}{4}$      BI (206)(9)

2.    $\displaystyle\int_0^{\pi/4} \frac{x \tan x \, dx}{(\sin x + \cos x) \cos x} = -\frac{\pi}{8} \ln 2 + \frac{\pi}{4} - \frac{1}{2} \ln 2$      BI (204)(30)

3.    $\displaystyle\int_0^{\infty} \frac{\tan x}{a^2 \cos^2 x + b^2 \sin^2 x} \frac{dx}{x} = \frac{\pi}{2ab}$      $[a > 0, \quad b > 0]$    DI (101)(9)

4.    $\displaystyle\int_0^{\pi/2} \frac{x \cot x \, dx}{a^2 \cos^2 x + b^2 \sin^2 x} = \frac{\pi}{2a^2} \ln \frac{a+b}{b}$      $[a > 0, \quad b > 0]$    LI (208)(20)

5.    $\displaystyle\int_0^{\pi/2} \frac{\left(\frac{\pi}{2} - x\right) \tan x \, dx}{a^2 \cos^2 x + b^2 \sin^2 x} = \frac{1}{2} \int_0^{\pi} \frac{\left(\frac{\pi}{2} - x\right) \tan x \, dx}{a^2 \cos^2 x + b^2 \sin^2 x}$

$$= \frac{\pi}{2b^2} \ln \frac{a+b}{a}$$

     $[a > 0, \quad b > 0]$    GW (333)(59)

6.    $\displaystyle\int_0^{\infty} \frac{\sin^2 x \tan x}{a^2 \cos^2 x + b^2 \sin^2 x} \frac{dx}{x} = \frac{\pi}{2b(a+b)}$      $[a > 0, \quad b > 0]$    BI (182)(6)

7.    $\displaystyle\int_0^{\infty} \frac{\tan x}{a^2 \cos^2 2x + b^2 \sin^2 2x} \frac{dx}{x} = \frac{\pi}{2ab}$      $[a > 0, \quad b > 0]$    BI (181)(10)a

8.    $\displaystyle\int_0^{\infty} \frac{\sin^2 2x \tan x}{a^2 \cos^2 2x + b^2 \sin^2 2x} \frac{dx}{x} = \frac{\pi}{2b} \frac{1}{a+b}$      $[a > 0, \quad b > 0]$    BI (182)(2)a

9.    $\displaystyle\int_0^{\infty} \frac{\cos^2 2x \tan x}{a^2 \cos^2 2x + b^2 \sin^2 2x} \frac{dx}{x} = \frac{\pi}{2a} \frac{1}{a+b}$      $[a > 0, \quad b > 0]$    BI (182)(5)a

10. $\displaystyle\int_0^\infty \frac{\sin^2 x \cos x}{a^2 \cos^2 2x + b^2 \sin^2 2x} \cdot \frac{dx}{x \cos 4x} = -\frac{\pi}{8b}\frac{a}{a^2+b^2}$  $[a>0, \quad b>0]$  BI (186)(12)a

11. $\displaystyle\int_0^\infty \frac{\sin x}{a^2 \cos^2 x + b^2 \sin^2 x}\frac{dx}{x \cos 2x} = \frac{\pi}{2ab}\frac{b^2-a^2}{b^2+a^2}$  $[a>0, \quad b>0]$  BI (186)(4)a

12. $\displaystyle\int_0^\infty \frac{\sin x \cos x}{a^2 \cos^2 x + b^2 \sin^2 x}\frac{dx}{x \cos 2x} = \frac{\pi}{2a}\frac{b}{a^2+b^2}$  $[a>0, \quad b>0]$  BI (186)(7)a

13. $\displaystyle\int_0^\infty \frac{\sin x \cos^2 x}{a^2 \cos^2 x + b^2 \sin^2 x}\frac{dx}{x \cos 2x} = \frac{\pi}{2ab}\frac{b^2}{a^2+b^2}$  $[a>0, \quad b>0]$  BI (186)(8)a

14. $\displaystyle\int_0^\infty \frac{\sin^3 x}{a^2 \cos^2 x + b^2 \sin^2 x}\frac{dx}{x \cos 2x} = -\frac{\pi}{2b}\frac{a}{a^2+b^2}$  $[a>0, \quad b>0]$  BI (186)(10)

15. $\displaystyle\int_0^\infty \frac{1 - \cos x}{a^2 \cos^2 x + b^2 \sin^2 x}\cdot \frac{dx}{x \sin x} = \frac{\pi}{2ab}$  $[a>0, \quad b>0]$  BI (186)(3)a

**3.815**

1. $\displaystyle\int_0^{\pi/2} \frac{x \sin 2x\, dx}{\left(1 + a\sin^2 x\right)\left(1 + b\sin^2 x\right)} = \frac{\pi}{a-b}\ln\left\{\frac{1+\sqrt{1+b}}{1+\sqrt{1+a}}\cdot\frac{\sqrt{1+a}}{\sqrt{1+b}}\right\}$

$[a>0, \quad b>0]$  (cf. **3.812** 3)

BI (208)(22)

2. $\displaystyle\int_0^{\pi/2} \frac{x \sin 2x\, dx}{\left(1 + a\sin^2 x\right)\left(1 + b\cos^2 x\right)} = \frac{\pi}{a+ab+b}\ln\frac{\left(1+\sqrt{1+b}\right)\sqrt{1+a}}{1+\sqrt{1+a}}$

$[a>0, \quad b>0]$  (cf. **3.812** 2 and 3)

BI (208)(24)

3. $\displaystyle\int_0^{\pi/2} \frac{x \sin 2x\, dx}{\left(1 + a\cos^2 x\right)\left(1 + b\cos^2 x\right)} = \frac{\pi}{a-b}\ln\frac{1+\sqrt{1+a}}{1+\sqrt{1+b}}$

$[a>0, \quad b>0]$  (cf. **3.812** 2)

BI (208)(23)

4. $\displaystyle\int_0^{\pi/2} \frac{x \sin 2x\, dx}{\left(1 - \sin^2 t_1\cos^2 x\right)\left(1 - \sin^2 t_2\cos^2 x\right)} = \frac{2\pi}{\cos^2 t_1 - \cos^2 t_2}\ln\frac{\cos\dfrac{t_1}{2}}{\cos\dfrac{t_2}{2}}$

$[-\pi < t_1 < \pi, \quad -\pi < t_2 < \pi]$

BI (208)(21)

**3.816**

1. $\displaystyle\int_0^{\pi} \frac{x^2 \sin 2x}{\left(a^2 - \cos^2 x\right)^2}\, dx = \pi^2\frac{\sqrt{a^2-1}-a}{a\left(a^2-1\right)}$  $[a>1]$  LI (220)(9)

2.[12] $\displaystyle\int_0^{\pi} \frac{\left(a^2 - 1 - \sin^2 x\right)\cos x}{\left(a^2 - \cos^2 x\right)^2}x^2\, dx = \frac{\pi}{a}\ln\left|\frac{1-a}{1+a}\right|$  $[a^2>1]$  (cf. **3.812** 5)  BI (220)(12)

3.[11] $\displaystyle\int_0^{\pi} \frac{a\cos 2x - \sin^2 x}{\left(a + \sin^2 x\right)^2}x^2\, dx = -2\pi\ln\left[2\left(-a + \sqrt{a}\sqrt{a+1}\right)\right]$

$\left[a < -1 \text{ and } a > 0. \text{ When } a > 0, \text{ can write } \sqrt{a}\sqrt{a+1} \text{ as } \sqrt{a(a+1)}.\right]$  LI (220)(10)

4.[11]  $\displaystyle\int_0^\pi \frac{a\cos 2x + \sin^2 x}{\left(a - \sin^2 x\right)^2} x^2\,dx = 2\pi \ln\left[2\left(a - \sqrt{a}\sqrt{a+1}\right)\right]$

$\left[a < 0 \text{ and } a > 1. \text{ When } a > 1, \text{ can write } \sqrt{a}\sqrt{a+1} \text{ as } \sqrt{a(a+1)}.\right]$    (cf. **3.812** 6)

LI (220)(11)

## 3.817

1.  $\displaystyle\int_0^\infty \frac{\sin x}{\left(a^2\cos^2 x + b^2\sin^2 x\right)^2}\frac{dx}{x} = \frac{\pi}{4}\frac{a^2 + b^2}{a^3 b^3}$    $[ab > 0]$    BI (181)(12)

2.  $\displaystyle\int_0^\infty \frac{\sin x \cos x}{\left(a^2\cos^2 x + b^2\sin^2 x\right)^2}\frac{dx}{x} = \frac{\pi}{4a^3 b}$    $[ab > 0]$    BI (182)(8)

3.  $\displaystyle\int_0^\infty \frac{\sin^3 x}{\left(a^2\cos^2 x + b^2\sin^2 x\right)^2}\frac{dx}{x} = \frac{\pi}{4ab^3}$    $[ab > 0]$    BI (181)(15)

4.  $\displaystyle\int_0^\infty \frac{\sin x \cos^2 x}{\left(a^2\cos^2 x + b^2\sin^2 x\right)^2}\frac{dx}{x} = \frac{\pi}{4a^3 b}$    $[ab > 0]$    BI (182)(9)

5.  $\displaystyle\int_0^\infty \frac{\tan x}{\left(a^2\cos^2 x + b^2\sin^2 x\right)^2}\frac{dx}{x} = \frac{\pi}{4}\frac{a^2 + b^2}{a^3 b^3}$    $[ab > 0]$    BI (181)(13)

6.  $\displaystyle\int_0^\infty \frac{\tan x}{\left(a^2\cos^2 2x + b^2\sin^2 2x\right)^2}\frac{dx}{x} = \frac{\pi}{4}\frac{a^2 + b^2}{a^3 b^3}$    $[ab > 0]$    BI (181)(14)

7.  $\displaystyle\int_0^\infty \frac{\sin^2 x \tan x}{\left(a^2\cos^2 x + b^2\sin^2 x\right)^2}\frac{dx}{x} = \frac{\pi}{4ab^3}$    $[ab > 0]$    BI (182)(11)

8.  $\displaystyle\int_0^\infty \frac{\tan x \cos^2 2x}{\left(a^2\cos^2 2x + b^2\sin^2 2x\right)^2}\frac{dx}{x} = \frac{\pi}{4a^3 b}$    $[ab > 0]$    BI (182)(10)

## 3.818

1.  $\displaystyle\int_0^\infty \frac{\sin x}{\left(a^2\cos^2 x + b^2\sin^2 x\right)^3}\frac{dx}{x} = \frac{\pi}{16}\frac{3a^4 + 2a^2 b^2 + 3b^4}{a^5 b^5}$

    $[ab > 0]$    BI (181)(16)

2.  $\displaystyle\int_0^\infty \frac{\sin x \cos x}{\left(a^2\cos^2 x + b^2\sin^2 x\right)^3}\frac{dx}{x} = \frac{\pi}{16}\frac{a^2 + 3b^2}{a^5 b^3}$    $[ab > 0]$    BI (182)(13)

3.  $\displaystyle\int_0^\infty \frac{\sin x \cos^2 x}{\left(a^2\cos^2 x + b^2\sin^2 x\right)^3}\frac{dx}{x} = \frac{\pi}{16}\frac{a^2 + 3b^2}{a^5 b^3}$    $[ab > 0]$    BI (182)(14)

4.  $\displaystyle\int_0^\infty \frac{\sin^3 x}{\left(a^2\cos^2 x + b^2\sin^2 x\right)^3}\frac{dx}{x} = \frac{\pi}{16}\frac{3a^2 + b^2}{a^3 b^5}$    $[ab > 0]$    LI (181)(19)

5.  $\displaystyle\int_0^\infty \frac{\sin^3 x \cos x}{\left(a^2\cos^2 2x + b^2\sin^2 2x\right)^3}\frac{dx}{x} = \frac{\pi}{64}\frac{3a^2 + b^2}{a^3 b^5}$    $[ab > 0]$    BI (182)(17)

6.    $\displaystyle\int_0^\infty \frac{\tan x}{\left(a^2\cos^2 x + b^2\sin^2 x\right)^3}\frac{dx}{x} = \frac{\pi}{16}\frac{3a^4 + 2a^2b^2 + 3b^4}{a^5b^5}$

$$[ab > 0] \qquad\qquad \text{BI (181)(17)}$$

7.    $\displaystyle\int_0^\infty \frac{\sin^2 x \tan x}{\left(a^2\cos^2 x + b^2\sin^2 x\right)^3}\frac{dx}{x} = \frac{\pi}{16}\frac{3a^2 + b^2}{a^3b^5} \qquad [ab > 0] \qquad \text{BI (182)(16)}$

8.    $\displaystyle\int_0^\infty \frac{\tan x}{\left(a^2\cos^2 2x + b^2\sin^2 2x\right)^3}\frac{dx}{x} = \frac{\pi}{16}\frac{3a^4 + 2a^2b^2 + 3b^4}{a^5b^5}$

$$[ab > 0] \qquad\qquad \text{BI (181)(18)}$$

9.    $\displaystyle\int_0^\infty \frac{\tan x \cos^2 2x}{\left(a^2\cos^2 2x + b^2\sin^2 2x\right)^3}\frac{dx}{x} = \frac{\pi}{16}\frac{a^2 + 3b^2}{a^5b^3} \qquad [ab > 0] \qquad \text{BI (182)(15)}$

**3.819**

1.    $\displaystyle\int_0^\infty \frac{\sin x}{\left(a^2\cos^2 x + b^2\sin^2 x\right)^4}\frac{dx}{x} = \frac{\pi}{32}\frac{5a^6 + 3a^4b^2 + 3a^2b^4 + 5b^6}{a^7b^7}$

$$[ab > 0] \qquad\qquad \text{BI (181)(20)}$$

2.    $\displaystyle\int_0^\infty \frac{\sin x \cos x}{\left(a^2\cos^2 x + b^2\sin^2 x\right)^4}\frac{dx}{x} = \frac{\pi}{32}\frac{a^4 + 2a^2b^2 + 5b^4}{a^7b^5}$

$$[ab > 0] \qquad\qquad \text{BI (182)(18)}$$

3.    $\displaystyle\int_0^\infty \frac{\sin x \cos^2 x}{\left(a^2\cos^2 x + b^2\sin^2 x\right)^4}\frac{dx}{x} = \frac{\pi}{32}\frac{a^4 + 2a^2b^2 + 5b^4}{a^7b^5}$

$$[ab > 0] \qquad\qquad \text{BI (182)(19)}$$

4.    $\displaystyle\int_0^\infty \frac{\sin^3 x}{\left(a^2\cos^2 x + b^2\sin^2 x\right)^4}\frac{dx}{x} = \frac{\pi}{32}\frac{5a^4 + a^2b^2 + b^4}{a^5b^7} \qquad [ab > 0] \qquad \text{BI (181)(23)}$

5.    $\displaystyle\int_0^\infty \frac{\sin^3 x \cos x}{\left(a^2\cos^2 x + b^2\sin^2 x\right)^4}\frac{dx}{x} = \frac{\pi}{32}\frac{a^2 + b^2}{a^5b^5} \qquad [ab > 0] \qquad \text{BI (182)(26)}$

6.    $\displaystyle\int_0^\infty \frac{\sin x \cos^3 x}{\left(a^2\cos^2 x + b^2\sin^2 x\right)^4}\frac{dx}{x} = \frac{\pi}{32}\frac{a^2 + 5b^2}{a^7b^3} \qquad [ab > 0] \qquad \text{BI (182)(23)}$

7.    $\displaystyle\int_0^\infty \frac{\sin^3 x \cos^2 x}{\left(a^2\cos^2 x + b^2\sin^2 x\right)^4}\frac{dx}{x} = \frac{\pi}{32}\frac{a^2 + b^2}{a^5b^5} \qquad [ab > 0] \qquad \text{BI (182)(27)}$

8.    $\displaystyle\int_0^\infty \frac{\sin x \cos^4 x}{\left(a^2\cos^2 x + b^2\sin^2 x\right)^4}\frac{dx}{x} = \frac{\pi}{32}\frac{a^2 + 5b^2}{a^7b^3} \qquad [ab > 0] \qquad \text{BI (182)(24)}$

9.    $\displaystyle\int_0^\infty \frac{\sin^5 x}{\left(a^2\cos^2 x + b^2\sin^2 x\right)^4}\frac{dx}{x} = \frac{\pi}{32}\frac{5a^2 + b^2}{a^3b^7} \qquad [ab > 0] \qquad \text{BI (181)(24)}$

10.    $\displaystyle\int_0^\infty \frac{\sin^3 x \cos x}{\left(a^2\cos^2 2x + b^2\sin^2 2x\right)^4}\frac{dx}{x} = \frac{\pi}{128}\frac{5a^4 + 2a^2b^2 + b^4}{a^5b^7}$

$$[ab > 0] \qquad\qquad \text{BI (182)(22)}$$

11. $\displaystyle\int_0^\infty \frac{\sin^5 x \cos^3 x}{\left(a^2\cos^2 2x + b^2\sin^2 2x\right)^4}\frac{dx}{x} = \frac{\pi}{512}\frac{5a^2+b^2}{a^3 b^7}$      $[ab>0]$          BI (182)(30)

12. $\displaystyle\int_0^\infty \frac{\sin^2 x \tan x}{\left(a^2\cos^2 x + b^2\sin^2 x\right)^4}\frac{dx}{x} = \frac{\pi}{32}\frac{5a^4+2a^2b^2+b^4}{a^5 b^7}$

                                            $[ab>0]$          BI (182)(21)

13. $\displaystyle\int_0^\infty \frac{\sin^4 x \tan x}{\left(a^2\cos^2 x + b^2\sin^2 x\right)^4}\frac{dx}{x} = \frac{\pi}{32}\frac{5a^2+b^2}{a^3 b^7}$      $[ab>0]$          BI (182)(29)

14. $\displaystyle\int_0^\infty \frac{\cos^2 2x \tan x}{\left(a^2\cos^2 2x + b^2\sin^2 2x\right)^4}\frac{dx}{x} = \frac{\pi}{32}\frac{a^4+2a^2b^2+5b^4}{a^7 b^5}$

                                            $[ab>0]$          BI (182)(29)

15. $\displaystyle\int_0^\infty \frac{\sin^3 4x \tan x}{\left(a^2\cos^2 2x + b^2\sin^2 2x\right)^4}\frac{dx}{x} = \frac{\pi}{8}\frac{a^2+b^2}{a^5 b^5}$      $[ab>0]$          BI (182)(28)

16. $\displaystyle\int_0^\infty \frac{\cos^4 2x \tan x}{\left(a^2\cos^2 2x + b^2\sin^2 2x\right)^4}\frac{dx}{x} = \frac{\pi}{32}\frac{a^2+5b^2}{a^7 b^3}$      $[ab>0]$          BI (182)(25)

## 3.82–3.83 Powers of trigonometric functions combined with other powers

### 3.821

1. $\displaystyle\int_0^\pi x\sin^p x\, dx = \frac{\pi^2}{2^{p+1}}\frac{\Gamma(p+1)}{\left[\Gamma\left(\frac{p}{2}+1\right)\right]^2}$      $[p>-1]$        BI(218)(7), LO V 121(71)

2. $\displaystyle\int_0^{r\pi} x\sin^n x\, dx = \frac{\pi^2}{2}\frac{(2m-1)!!}{(2m)!!}r^2$               $[n=2m]$

                   $= (-1)^{r+1}\pi\frac{(2m)!!}{(2m+1)!!}r$        $[n=2m+1]$

                                    $[r \text{ is a natural number}]$      GW (333)(8c)

3.[11] $\displaystyle\int_0^{\pi/2} x\cos^n x\, dx = \frac{\pi^2}{8}\frac{(n-1)!!}{(n)!!} - \frac{1}{2^{n-2}}\sum_{\substack{k=0,m-k\text{ odd}}}^{m-1}\binom{n}{k}\frac{1}{(n-2k)^2}$    $[n=2m]$

                   $= \frac{\pi}{2}\frac{(n-1)!!}{(n)!!} - \frac{1}{2^{n-1}}\sum_{k=0}^{m-1}\binom{n}{k}\frac{1}{(n-2k)^2}$    $[n=2m-1]$

                                                   GW (333)(9b)

4. $\displaystyle\int_0^\pi x\cos^{2m} x\, dx = \frac{\pi^2}{2}\frac{(2m-1)!!}{(2m)!!}$                          BI (218)(10)

5. $\displaystyle\int_{r\pi}^{s\pi} x\cos^{2m} x\, dx = \frac{\pi^2}{2}\left(s^2-r^2\right)\frac{(2m-1)!!}{(2m)!!}$               BI (226)(3)

6. $\displaystyle\int_0^\infty \frac{\sin^p x}{x}\, dx = \frac{\sqrt{\pi}}{2}\frac{\Gamma\left(\frac{p}{2}\right)}{\Gamma\left(\frac{p+1}{2}\right)} = 2^{p-2}\,\mathrm{B}\left(\frac{p}{2},\frac{p}{2}\right)$

       $[p \text{ is a fraction with odd numerator and denominator}]$    LO V 278, FI II 808

7. $\displaystyle\int_0^\infty \frac{\sin^{2n+1} x}{x}\,dx = \frac{(2n-1)!!}{(2n)!!}\frac{\pi}{2}$      BI (151)(4)

8. $\displaystyle\int_0^\infty \frac{\sin^{2n} x}{x}\,dx = \infty$      BI (151)(3)

9. $\displaystyle\int_0^\infty \frac{\sin^2 ax}{x^2}\,dx = \frac{a\pi}{2}$     $[a>0]$      LO V 307, 312, FI II 632

10. $\displaystyle\int_0^\infty \frac{\sin^{2m} ax}{x^2}\,dx = \frac{(2m-3)!!}{(2m-2)!!}\frac{a\pi}{2}$     $[a>0]$      GW (333)(14b)

11. $\displaystyle\int_0^\infty \frac{\sin^{2m+1} ax}{x^3}\,dx = \frac{(2m-3)!!}{(2m)!!}(2m+1)\frac{a^2\pi}{4}$     $[a>0]$      GW (333)(14d)

12. $\displaystyle\int_0^\infty \frac{\sin^p x}{x^m}\,dx$

$$= \frac{p}{m-1}\int_0^\infty \frac{\sin^{p-1} x}{x^{m-1}}\cos x\,dx \qquad [p>m-1>0]$$

$$= \frac{p(p-1)}{(m-1)(m-2)}\int_0^\infty \frac{\sin^{p-2} x}{x^{m-2}}\,dx - \frac{p^2}{(m-1)(m-2)}\int_0^\infty \frac{\sin^p x}{x^{m-2}}\,dx \qquad [p>m-1>1]$$

         GW (333)(17)

13. $\displaystyle\int_0^\infty \frac{\sin^{2n} px}{\sqrt{x}}\,dx = \infty$      BI (177)(5)

14. $\displaystyle\int_0^\infty \sin^{2n+1} px\,\frac{dx}{\sqrt{x}} = \frac{1}{2^{2n}}\sqrt{\frac{\pi}{2p}}\sum_{k=0}^n (-1)^k \binom{2n+1}{n+k+1}\frac{1}{\sqrt{2k+1}}$      BI (177)(7)

**3.822**

1. $\displaystyle\int_0^{\pi/2} x^p \cos^m x\,dx = -\frac{p(p-1)}{m^2}\int_0^{\pi/2} x^{p-2}\cos^m x\,dx + \frac{m-1}{m}\int_0^{\pi/2} x^p \cos^{m-2} x\,dx$

         $[m>1, \quad p>1]$      GW (333)(9a)

2. $\displaystyle\int_0^\infty x^{-1/2}\cos^{2n+1}(px)\,dx = \frac{1}{2^{2n}}\sqrt{\frac{\pi}{2p}}\sum_{k=0}^n \binom{2n+1}{n+k+1}\frac{1}{\sqrt{2k+1}}$      BI (177)(8)

**3.823**[12] $\displaystyle\int_0^\infty x^{b-1}\sin^2 x\,dx = -\frac{\Gamma(b)}{2^{b+1}}\cos\frac{\pi b}{2}$     $[-2<\mathrm{Re}\,b<0]$

         ET I 319(15), GW(333)(19c)a

**3.824**

1.[12] $\displaystyle\int_0^\infty \frac{\sin^2 x}{x^2+a^2}\,dx = \frac{\pi}{4a}\left(1-e^{-2a}\right)$     $[\mathrm{Re}\,a>0]$      BI (160)(10)

2.[12] $\displaystyle\int_0^\infty \frac{\cos^2 x}{x^2+a^2}\,dx = \frac{\pi}{4a}\left(1+e^{-2a}\right)$     $[\mathrm{Re}\,a>0]$      BI (160)(11)

3.[7] $\displaystyle\int_0^\infty \sin^{2m} x\,\frac{dx}{a^2+x^2} = \frac{(-1)^m}{2^{2m+1}}\cdot\frac{\pi}{2}\left[2^{2m}\sinh^{2m} a - 2\sum_{k=0}^m (-1)^k \binom{2m}{k}\sinh[2(m-k)a]\right]$

         $[a>0]$      BI (160)(12)

4.[12] $\displaystyle \int_0^\infty \sin^{2m+1} x \frac{\mathrm{d}x}{a^2+x^2} = \frac{(-1)^{m-1}}{2^{2m+2}a} \left[ e^{(2m+1)a} \sum_{k=0}^{2m+1} (-1)^k \binom{2m+1}{k} e^{-2ka} \operatorname{Ei}[(2k-2m-1)a] \right.$

$$\left. + e^{-(2m+1)a} \sum_{k=1}^{2m+1} (-1)^{k-1} \binom{2m+1}{k} e^{2ka} \operatorname{Ei}[(2m+1-2k)a] \right]$$

$$[a>0] \qquad\qquad \text{BI (160)(14)}$$

5.[7] $\displaystyle \int_0^\infty \sin^{2m+1} x \frac{x\,\mathrm{d}x}{a^2+x^2} = \frac{\pi}{2^{2m+1}} e^{-(2m+1)a} \sum_{k=0}^{m} (-1)^{m+k} \binom{2m+1}{k} e^{2ka}$

$$\left[ |\arg a| < \frac{\pi}{2}, \quad m=0,1,2,\dots \right]$$

6.[7] $\displaystyle \int_0^\infty \cos^{2m} x \frac{\mathrm{d}x}{a^2+x^2} = \frac{\pi}{2^{2m+1}a} \binom{2m}{m} + \frac{\pi}{2^{2m}} \sum_{k=1}^{m} \binom{2m}{m+k} e^{-2ka}$

$$[a>0] \qquad\qquad \text{BI (160)(16)}$$

7. $\displaystyle \int_0^\infty \cos^{2m+1} x \frac{\mathrm{d}x}{a^2+x^2} = \frac{\pi}{2^{2m+1}a} \sum_{k=1}^{m} \binom{2m+1}{m+k+1} e^{-(2k+1)a}$

$$[a>0] \qquad\qquad \text{BI (160)(17)}$$

8. $\displaystyle \int_0^\infty \cos^{2m+1} x \frac{x\,\mathrm{d}x}{a^2+x^2} = -\frac{e^{-(2m+1)a}}{2^{2m+2}} \sum_{k=0}^{2m+1} \binom{2m+1}{k} e^{2ka} \operatorname{Ei}[(2m-2k+1)a]$

$$-\frac{e^{(2m+1)a}}{2^{2m+2}} \sum_{k=0}^{2m+1} \binom{2m+1}{k} e^{-2ka} \operatorname{Ei}[(2k-2m-1)a]$$

$$\text{BI (160)(18)}$$

9.[12] $\displaystyle \operatorname{PV}\int_0^\infty \frac{\cos^2 ax}{b^2-x^2}\,\mathrm{d}x = \frac{\pi}{4b}\sin 2ab \qquad [a>0,\quad b>0] \qquad \text{BI (161)(10)}$

10. $\displaystyle \int_0^\infty \frac{\sin^2 ax \cos^2 bx}{\beta^2+x^2}\,\mathrm{d}x = \frac{\pi}{8\beta}\left[1 - \frac{1}{2}e^{-2(a+b)\beta} + e^{-2b\beta} - \frac{1}{2}e^{2(b-a)\beta} - e^{-2a\beta}\right] \qquad [a>b]$

$$= \frac{\pi}{16\beta}\left[1 - e^{-4a\beta}\right] \qquad [a=b]$$

$$= \frac{\pi}{8\beta}\left[1 - \frac{1}{2}e^{-2(a+b)\beta} + e^{-2b\beta} - \frac{1}{2}e^{2(a-b)\beta} - e^{-2a\beta}\right] \qquad [a<b]$$

$$[a>0,\quad b>0], \qquad (\text{cf. } \mathbf{3.824}\ 1 \text{ and } 3) \quad \text{BI (162)(6)}$$

11. $\displaystyle \int_0^\infty \frac{x\sin 2ax \cos^2 bx}{\beta^2+x^2}\,\mathrm{d}x = \frac{\pi}{8}\left[2e^{-2a\beta} + e^{-2(a+b)\beta} + e^{2(b-a)\beta}\right] \qquad [a>0]$

$$= \frac{\pi}{8}\left[e^{-4a\beta} + 2e^{-2a\beta}\right] \qquad [a=b]$$

$$= \frac{\pi}{8}\left[2e^{-2a\beta} + e^{-2(a+b)\beta} - e^{2(a-b)\beta}\right] \qquad [a<b]$$

$$\text{LI (162)(5)}$$

**3.825**

1.    $\displaystyle\int_0^\infty \frac{\sin^2 ax \, dx}{(b^2 + x^2)(c^2 + x^2)} = \frac{\pi \left(b - c + ce^{-2ab} - be^{-2ac}\right)}{4bc\,(b^2 - c^2)}$

$$[a > 0, \quad b > 0, \quad c > 0] \qquad \text{BI (174)(15)}$$

2.    $\displaystyle\int_0^\infty \frac{\cos^2 ax \, dx}{(b^2 + x^2)(c^2 + x^2)} = \frac{\pi \left(b - c + be^{-2ac} - ce^{-2ab}\right)}{4bc\,(b^2 - c^2)}$

$$[a > 0, \quad b > 0, \quad c > 0] \qquad \text{BI (175)(14)}$$

3.[3]    $\displaystyle\int_0^\infty \frac{\sin^2 ax \, dx}{(b^2 - x^2)(c^2 - x^2)} = \frac{\pi \left(c \sin 2ab - b \sin 2ac\right)}{4bc\,(b^2 - c^2)}$    $[a > 0, \quad b > 0, \quad c > 0, \quad b \neq c]$

$$\text{LI (174)(16)}$$

4.[3]    $\displaystyle\int_0^\infty \frac{\cos^2 ax \, dx}{(b^2 - x^2)(c^2 - x^2)} = \frac{\pi \left(b \sin 2ac - c \sin 2ab\right)}{4bc\,(b^2 - c^2)}$    $[a > 0, \quad b > 0, \quad c > 0, \quad b \neq c]$

$$\text{LI (175)(15)}$$

**3.826**

1.[12]    $\displaystyle\int_0^\infty \frac{\sin^2(ax)}{x^2(1 + x^2)} dx = \frac{\pi}{4}\left[2a - (1 - e^{-2a})\right]$      $[a > 0]$      BI (172)(13)

2.    $\displaystyle\int_0^\infty \frac{\sin^2 ax \, dx}{x^2(1 - x^2)} = \frac{\pi}{4}\left(2a - \sin 2a\right)$      $[a > 0]$      BII (172)(14)

**3.827**

1.[12]    $\displaystyle\int_0^\infty \frac{\sin^3 x}{x^\nu} dx = \frac{3 - 3^{\nu-1}}{4}\Gamma(1 - \nu)\cos\frac{\pi\nu}{2}$      $[1 < \operatorname{Re}\nu < 4, \quad \nu \neq 2, 3]$

$$\text{GW (333)(19f)}$$

2.[8]    $\displaystyle\int_0^\infty \frac{\sin^3 ax}{x} dx = \frac{\pi}{4}$      $[a > 0]$      LO V 277

3.    $\displaystyle\int_0^\infty \frac{\sin^3 ax}{x^2} dx = \frac{3}{4}a \ln 3$      BI (156)(2)

4.[12]    $\displaystyle\int_0^\infty \frac{\sin^3 ax}{x^3} dx = \frac{3}{8}a^2\pi$      $[a > 0]$      BI(156)(7)a, LO V 312

5.    $\displaystyle\int_0^\infty \frac{\sin^4 ax}{x^2} dx = \frac{a\pi}{4}$      $[a > 0]$      BI (156)(3)

6.    $\displaystyle\int_0^\infty \frac{\sin^4 ax}{x^3} dx = a^2 \ln 2$      BI (156)(8)

7.    $\displaystyle\int_0^\infty \frac{\sin^4 ax}{x^4} dx = \frac{a^3\pi}{3}$      $[a > 0]$      BI(156)(11), LO V 312

8.    $\displaystyle\int_0^\infty \frac{\sin^5 ax}{x^2} dx = \frac{5}{16}a\,(3\ln 3 - \ln 5)$      BI (156)(4)

9.    $\displaystyle\int_0^\infty \frac{\sin^5 ax}{x^3} dx = \frac{5}{32}a^2\pi$      $[a > 0]$      BI (156)(9)

10. $\int_0^\infty \dfrac{\sin^5 ax}{x^4}\, dx = \dfrac{5}{96} a^3 \left(25 \ln 5 - 27 \ln 3\right)$                    BI (156)(12)

11. $\int_0^\infty \dfrac{\sin^5 ax}{x^5}\, dx = \dfrac{115}{384} a^4 \pi$                    $[a > 0]$                    BI(156)(13), LO V 312

12. $\int_0^\infty \dfrac{\sin^6 ax}{x^2}\, dx = \dfrac{3}{16} a\pi$                    $[a > 0]$                    BI (156)(5)

13. $\int_0^\infty \dfrac{\sin^6 ax}{x^3}\, dx = \dfrac{3}{16} a^2 \left(8 \ln 2 - 3 \ln 3\right)$                    BI (156)(10)

14. $\int_0^\infty \dfrac{\sin^6 ax}{x^5}\, dx = \dfrac{1}{16} a^4 \left(27 \ln 3 - 32 \ln 2\right)$                    BI (156)(14)

15. $\int_0^\infty \dfrac{\sin^6 ax}{x^6}\, dx = \dfrac{11}{40} a^5 \pi$                    $[a > 0]$                    LO V 312

**3.828**    In **3.828** 1–21 the restrictions $a > 0$, $b > 0$, $c > 0$ apply.

1.[8] $\int_0^\infty \dfrac{\sin ax \sin bx}{x}\, dx = \dfrac{1}{2} \ln \left| \dfrac{a+b}{a-b} \right|$                    $[a \neq b]$                    FI II 647

2.[8] $\int_0^\infty \sin ax \sin bx \dfrac{dx}{x^2} = \dfrac{1}{2} \pi \min(a,b)$                    BI (157)(1)

3.[8] $\int_0^\infty \dfrac{\sin^2 ax \sin bx}{x}\, dx = \dfrac{\pi}{4}$                    $[b < 2a]$

$\qquad\qquad\qquad\qquad\qquad = \dfrac{\pi}{8}$                    $[b = 2a]$

$\qquad\qquad\qquad\qquad\qquad = 0$                    $[b > 2a]$

                                                            BI (151)(10)

4.[8] $\int_0^\infty \dfrac{\sin^2 ax \cos bx}{x}\, dx = \dfrac{1}{4} \ln \dfrac{4a^2 - b^2}{b^2}$                    $[2a \neq b]$                    BI (151)(12)

5.[8] $\int_0^\infty \dfrac{\sin^2 ax \cos 2bx}{x^2}\, dx = \dfrac{1}{2} \pi \max(0, a - b)$

6. $\int_0^\infty \dfrac{\sin 2ax \cos^2 bx}{x}\, dx = \dfrac{\pi}{2}$                    $[a > b]$

$\qquad\qquad\qquad\qquad\qquad = \dfrac{3}{8} \pi$                    $[a = b]$

$\qquad\qquad\qquad\qquad\qquad = \dfrac{\pi}{4}$                    $[a < b]$

                                                            BI (151)(9)

7.[8] $\int_0^\infty \dfrac{\sin^2 ax \sin bx \sin cx}{x^2}\, dx = \dfrac{\pi}{16} \left( |b - 2a - c| - |2a - b - c| + 2c \right)$

$\qquad\qquad\qquad\qquad\qquad\qquad\qquad [a > 0, \quad 0 < c \leq b]$

                                                            BI(157)(9)a, ET I 79(15)

$8.^8$ $\displaystyle\int_0^\infty \frac{\sin^2 ax \sin bx \sin cx}{x}\, dx = \frac{1}{8}\ln\left|\frac{(b+c)^2(2a-b+c)(2a+b-c)}{(b-c)^2(2a+b+c)(2a-b-c)}\right|$

$\qquad\qquad\qquad\qquad [b \neq c, \quad 2a+c \neq b, \quad 2a+b \neq c, \quad 2a \neq b+c]$ \quad LI (152)(2)

9. $\displaystyle\int_0^\infty \frac{\sin^2 ax \sin^2 bx}{x^2}\, dx = \frac{\pi}{4}a$ \qquad\qquad\qquad $[0 \leq a \leq b]$

$\qquad\qquad\qquad\qquad\qquad = \frac{\pi}{4}b$ \qquad\qquad\qquad $[0 \leq b \leq a]$

$\qquad\qquad\qquad\qquad\qquad\qquad$ BI (157)(3)

$10.^8$ $\displaystyle\int_0^\infty \frac{\sin^2 ax \sin^2 bx}{x^4}\, dx = \frac{1}{6}\pi \min\left(a^2, b^2\right)[3\max(a,b) - \min(a,b)]$ \quad BI (157)(27)

$11.^8$ $\displaystyle\int_0^\infty \frac{\sin^2 ax \cos^2 bx}{x^2}\, dx = \frac{1}{4}\pi\,[a + \max(0, a-b)]$ \quad BI (157)(6)

12. $\displaystyle\int_0^\infty \frac{\sin^3 ax \sin 3bx}{x^4}\, dx = \frac{a^3\pi}{2}$ \qquad\qquad $[b > a]$

$\qquad\qquad\qquad\qquad\qquad = \frac{\pi}{16}\left[8a^3 - 9(a-b)^3\right]$ \quad $[a \leq 3b \leq 3a]$ \quad BI (157)(28)

$\qquad\qquad\qquad\qquad\qquad = \frac{9b\pi}{8}\left(a^2 - b^2\right)$ \qquad $[3b \leq a]$ \quad LI (157)(28)

13. $\displaystyle\int_0^\infty \frac{\sin^3 ax \cos bx}{x}\, dx = 0$ \qquad\qquad\qquad $[b > 3a]$

$\qquad\qquad\qquad\qquad\qquad = -\frac{\pi}{16}$ \qquad\qquad\qquad $[b = 3a]$

$\qquad\qquad\qquad\qquad\qquad = -\frac{\pi}{8}$ \qquad\qquad\qquad $[3a > b > a]$

$\qquad\qquad\qquad\qquad\qquad = \frac{\pi}{16}$ \qquad\qquad\qquad $[b = a]$

$\qquad\qquad\qquad\qquad\qquad = \frac{\pi}{4}$ \qquad\qquad\qquad $[a > b]$

$\qquad\qquad\qquad\qquad\qquad\qquad\qquad\qquad [a > 0, \quad b > 0]$ \quad BI (151)(15)

$14.^{12}$ $\displaystyle\int_0^\infty \frac{\sin^3 ax \cos 3bx}{x^2}\, dx$

$\qquad = \frac{3a}{4}\ln 3 + \frac{1}{8}\left[(a+b)\ln|a+b| + (a-b)\ln|a-b| - (a-3b)\ln|a-3b| - (a+3b)\ln|a+3b|\right]$

$\qquad\qquad\qquad\qquad\qquad\qquad [\operatorname{Im} a = 0, \quad \operatorname{Im} b = 0]$ \qquad\qquad MC

15. $\displaystyle\int_0^\infty \frac{\sin^3 ax \cos bx}{x^3}\, dx = \frac{\pi}{8}\left(3a^2 - b^2\right)$ \qquad $[b < a]$

$\qquad\qquad\qquad\qquad\qquad = \frac{\pi b^2}{4}$ \qquad\qquad\qquad $[a = b]$

$\qquad\qquad\qquad\qquad\qquad = \frac{\pi}{16}(3a - b)^2$ \qquad $[a < b < 3a]$

$\qquad\qquad\qquad\qquad\qquad = 0$ \qquad\qquad\qquad $[3a < b]$

$\qquad\qquad\qquad\qquad\qquad\qquad\qquad [a > 0, \quad b > 0]$ \quad BI(157)(19), ET I 19(10)

16. $\displaystyle\int_0^\infty \frac{\sin^3 ax \sin bx}{x^4}\,dx = \frac{b\pi}{24}\left(9a^2 - b^2\right)$ $\qquad [0 < b \le a]$

$\qquad\qquad\qquad = \dfrac{\pi}{48}\left[24a^3 - (3a - b)^3\right]$ $\qquad [0 < a \le b \le 3a]$

$\qquad\qquad\qquad = \dfrac{\pi a^3}{2}$ $\qquad [0 < 3a \le b]$

$\qquad\qquad\qquad\qquad\qquad\qquad\qquad\qquad\qquad\qquad\qquad$ ET I 79(16)

17. $\displaystyle\int_0^\infty \frac{\sin^3 ax \sin^2 bx}{x}\,dx = \frac{\pi}{8}$ $\qquad [2b > 3a]$

$\qquad\qquad\qquad = \dfrac{5\pi}{32}$ $\qquad [2b = 3a]$

$\qquad\qquad\qquad = \dfrac{3\pi}{16}$ $\qquad [3a > 2b > a]$

$\qquad\qquad\qquad = \dfrac{3\pi}{32}$ $\qquad [2b = a]$

$\qquad\qquad\qquad = 0$ $\qquad [a > 2b]$

$\qquad\qquad\qquad\qquad\qquad\qquad [a > 0, \quad b > 0]$ $\qquad$ BI (151)(14)

18.[8] $\displaystyle\int_0^\infty \frac{\sin^2 ax \cos^3 bx}{x}\,dx = \frac{1}{16}\ln\left|\frac{(2a + b)^3(b - 2a)^3(2a + 3b)(3b - 2a)}{9b^8}\right|$

$\qquad\qquad\qquad\qquad\qquad [2a \ne b, \quad 2a \ne 3b]$ $\qquad$ BI (151)(13)

19.[12] $\displaystyle\int_0^\infty \frac{\sin^2 ax \sin^2 bx \sin^2 cx}{x}\,dx$

$\qquad = \dfrac{\pi}{32}\left[\operatorname{sign}(c - a + b) + \operatorname{sign}(c + a - b) - 2\operatorname{sign}(c - a) - 2\operatorname{sign}(c - b)\right]$

$\qquad\qquad\qquad [\operatorname{Im} a = 0, \quad \operatorname{Im} b = 0, \quad \operatorname{Im} c = 0]$ $\qquad$ MC

20. $\displaystyle\int_0^\infty \frac{\sin^2 ax \sin^2 bx \sin 2cx\,dx}{x^2}$

$\qquad = \dfrac{a - b - c}{16}\ln 4(a - b - c)^2 - \dfrac{a + b + c}{16}\ln 4(a + b + c)^2 + \dfrac{a + b - c}{16}\ln 4(a + b - c)^2$

$\qquad\quad - \dfrac{a - b + c}{16}\ln 4(a - b + c)^2 + \dfrac{a + c}{8}\ln 4(a + c)^2 - \dfrac{a - c}{8}\ln 4(a - c)^2$

$\qquad\quad + \dfrac{b + c}{8}\ln 4(b + c)^2 - \dfrac{b - c}{8}\ln 4(b - c)^2 - \dfrac{1}{2}c\ln 2c$

$\qquad\qquad\qquad [a > 0, \quad b > 0, \quad c > 0]$ $\qquad$ BI (157)(10)

21.[8] $\displaystyle\int_0^\infty \frac{\sin^2 ax \sin^3 bx}{x^3}\,dx = \frac{3b^2\pi}{16}$ $\qquad [2a > 3b]$

$\qquad\qquad\qquad = \dfrac{a^2\pi}{12}$ $\qquad [2a = 3b]$

$\qquad\qquad\qquad = \dfrac{6b^2 - (3b - 2a)^2}{32}\pi$ $\qquad [3b > 2a > b]$

$\qquad\qquad\qquad = \dfrac{a^2\pi}{4}$ $\qquad [b \ge 2a]$

$\qquad\qquad\qquad\qquad\qquad\qquad\qquad\qquad\qquad\qquad\qquad$ BI (157)(18)

**3.829**

1. $\displaystyle\int_0^\infty \frac{x^n - \sin^n x}{x^{n+2}}\,dx = \frac{\pi}{2^n (n+1)!}\sum_{k=0}^{\lfloor (n-1)/2 \rfloor} (-1)^k \binom{n}{k}(n - 2k)^{n+1}$ $\qquad$ GW (333)(63)

2. $\int_0^\infty \left(1 - \cos^{2m-1} x\right) \frac{dx}{x^2} = \int_0^\infty \left(1 - \cos^{2m} x\right) \frac{dx}{x^2} = \frac{m\pi}{2^{2m}} \binom{2m}{m}$     BI (158)(7, 8)

**3.831**

1. $\int_0^\infty \frac{\sin^{2n} ax - \sin^{2n} bx}{x} \, dx = \frac{(2n-1)!!}{(2n)!!} \ln \frac{b}{a}$     $[ab > 0, \quad n = 1, 2, \ldots]$     FI II 651

2. $\int_0^\infty \frac{\cos^{2n} ax - \cos^{2n} bx}{x} \, dx = \left[1 - \frac{(2n-1)!!}{(2n)!!}\right] \ln \frac{b}{a}$     $[ab > 0, \quad n = 0, 1, \ldots]$     FI II 651

3. $\int_0^\infty \frac{\cos^{2m+1} ax - \cos^{2m+1} bx}{x} \, dx = \ln \frac{b}{a}$     $[ab > 0, \quad m = 0, 1, \ldots]$     FI II

4. $\int_0^\infty \frac{\cos^m ax \cos max - \cos^m bx \cos mbx}{x} \, dx = \left(1 - \frac{1}{2^m}\right) \ln \frac{b}{a}$

    $[ab > 0, \quad m = 0, 1, \ldots]$     LI (155)(8)

**3.832**

1. $\int_0^{\pi/2} x \cos^{p-1} x \sin ax \, dx = \frac{\pi}{2^{p+1}} \Gamma(p) \frac{\psi\left(\frac{p+a+1}{2}\right) - \psi\left(\frac{p-a+1}{2}\right)}{\Gamma\left(\frac{p+a+1}{2}\right) \Gamma\left(\frac{p-a+1}{2}\right)}$

    $[p > 0, \quad -(p+1) < a < p+1]$

    BI (205)(6)

2.[3] $\int_0^\infty \sin^{2m+1} x \sin 2mx \frac{dx}{a^2 + x^2} = \frac{(-1)^m \pi}{2^{2m+1} a} \left[\left(1 - e^{-2a}\right)^{2m} - 1\right] \sinh a$

    $[a > 0, \quad m = 0, 1, \ldots]$     BI (162)(17)

3. $\int_0^\infty \sin^{2m-1} x \sin[(2m-1)x] \frac{dx}{a^2 + x^2} = \frac{(-1)^{m+1} \pi}{2^{2m} a} \left(1 - e^{-2a}\right)^{2m-1}$

    $[a > 0, \quad m = 1, 2, \ldots]$     BI (162)(11)

4. $\int_0^\infty \sin^{2m-1} x \sin[(2m+1)x] \frac{dx}{a^2 + x^2} = \frac{(-1)^{m-1} \pi}{2^{2m} a} e^{-2a} \left(1 - e^{-2a}\right)^{2m-1}$

    $[a > 0, \quad m = 1, 2, \ldots]$     BI (162)(12)

5. $\int_0^\infty \sin^{2m+1} x \sin[3(2m+1)x] \frac{dx}{a^2 + x^2} = \frac{(-1)^m \pi}{2a} e^{-3(2m+1)a} \sinh^{2m+1} a$

    $[a > 0]$     BI (162)(18)

6.[3] $\int_0^\infty \sin^{2m} x \sin[(2m-1)x] \frac{x \, dx}{a^2 + x^2} = \frac{(-1)^m \pi}{2^{2m+1}} e^a \left[\left(1 - e^{-2a}\right)^{2m} - \left(1 + e^{-2a}\right)\right]$

    $[a \geq 0, \quad m = 0, 1, \ldots]$     BI (162)(13)

7. $\int_0^\infty \sin^{2m} x \sin(2mx) \frac{x \, dx}{a^2 + x^2} = \frac{(-1)^m \pi}{2^{2m+1}} \left[\left(1 - e^{-2a}\right)^{2m} - 1\right]$

    $[a > 0, \quad m = 0, 1, \ldots]$     BI (162)(14)

8. $\int_0^\infty \sin^{2m} x \sin[(2m+2)x] \frac{x \, dx}{a^2 + x^2} = \frac{(-1)^m \pi}{2^{2m+1}} e^{-2a} \left(1 - e^{-2a}\right)^{2m}$

    $[a > 0, \quad m = 0, 1, \ldots]$     BI (162)(15)

9.     $\displaystyle\int_0^\infty \sin^{2m} x \sin 4mx \frac{x\,dx}{a^2+x^2} = \frac{(-1)^m \pi}{2} e^{-4ma} \sinh^{2m} a$

$$[a > 0, \quad m = 1,2,\ldots] \qquad \text{BI (162)(16)}$$

10.    $\displaystyle\int_0^\infty \sin^{2m} x \cos x \frac{dx}{x^2} = \frac{(2m-3)!!}{(2m)!!} \frac{\pi}{2}$         $[m = 1,2,\ldots]$      GW (333)(15a)

11.    $\displaystyle\int_0^\infty \sin^{2m} x \cos[(2m-1)x] \frac{dx}{a^2+x^2} = \frac{(-1)^m \pi}{2^{2m} a}\left[\left(1-e^{-2a}\right)^{2m-1}-1\right]\sinh a$

$$[a > 0, \quad m = 1,2,\ldots] \qquad \text{BI (162)(25)}$$

12.    $\displaystyle\int_0^\infty \sin^{2m} x \cos(2mx) \frac{dx}{a^2+x^2} = \frac{(-1)^m \pi}{2^{2m+1} a}\left(1-e^{-2a}\right)^{2m}$

$$[a > 0, \quad m = 0,1,\ldots] \qquad \text{BI (162)(26)}$$

13.    $\displaystyle\int_0^\infty \sin^{2m} x \cos[(2m+2)x] \frac{dx}{a^2+x^2} = \frac{(-1)^m \pi}{2^{2m+1} a} e^{-2a}\left(1-e^{-2a}\right)^{2m}$

$$[a > 0, \quad m = 0,1,\ldots] \qquad \text{BI (162)(27)}$$

14.    $\displaystyle\int_0^\infty \sin^{2m} x \cos 4mx \frac{dx}{a^2+x^2} = \frac{(-1)^m \pi}{2a} e^{-4ma} \sinh^{2m} a$

$$[a > 0, \quad m = 0,1,\ldots] \qquad \text{BI (162)(28)}$$

15.    $\displaystyle\int_0^\infty \sin^{2m+1} x \cos x \frac{dx}{x} = \frac{(2m-1)!!}{(2m+2)!!} \frac{\pi}{2}$       $[m = 0,1,\ldots]$      GW (333)(15)

16.[3]   $\displaystyle\int_0^\infty \sin^{2m+1} x \cos x \frac{dx}{x^3} = \frac{(2m-3)!!}{(2m)!!} \frac{\pi}{2}$       $[m = 1,2,\ldots]$      GW (333)(15b)

17.    $\displaystyle\int_0^\infty \sin^{2m-1} x \cos[(2m-1)x] \frac{x\,dx}{a^2+x^2} = \frac{(-1)^m \pi}{2^{2m}}\left[\left(1-e^{-2a}\right)^{2m-1}-1\right]$

$$[m = 1,2,\ldots, \quad a > 0] \qquad \text{BI (162)(23)}$$

18.[3]   $\displaystyle\int_0^\infty \sin^{2m+1} x \cos 2mx \frac{x\,dx}{a^2+x^2} = \frac{(-1)^{m-1} \pi}{2^{2m+2}}\left\{e^a\left[\left(1-e^{-2a}\right)^{2m+1}-1\right]-e^{-a}\right\}$

$$[m = 0,1,\ldots, \quad a \geq 0] \qquad \text{BI (162)(29)}$$

19.    $\displaystyle\int_0^\infty \sin^{2m-1} x \cos[(2m+1)x] \frac{x\,dx}{a^2+x^2} = \frac{(-1)^m \pi}{2^{2m}} e^{-2a}\left(1-e^{-2a}\right)^{2m-1}$

$$[m = 1,2,\ldots, \quad a > 0] \qquad \text{BI (162)(24)}$$

20.    $\displaystyle\int_0^\infty \sin^{2m+1} x \cos[2(2m+1)x] \frac{x\,dx}{a^2+x^2} = \frac{(-1)^{m-1} \pi}{2} e^{-2(2m+1)a} \sinh^{2m+1} a$

$$[m = 0,1,\ldots, \quad a > 0] \qquad \text{BI (162)(30)}$$

21.[12]   $\displaystyle\int_0^\infty \cos^m x \sin mx \frac{dx}{a^2+x^2} = \frac{1}{2^{m+1} a} \sum_{k=1}^m \binom{m}{k}\left[e^{-2ka}\,\mathrm{Ei}(2ka)-e^{2ka}\,\mathrm{Ei}(-2ka)\right]$

$$[a > 0] \qquad \text{BI (162)(8)}$$

22. $\displaystyle\int_0^\infty \cos^n sx \sin nsx \frac{x\,dx}{a^2+x^2} = \frac{\pi}{2^{n+1}}\left[\left(1+e^{-2as}\right)^n - 1\right]$

$$[s>0, \quad \operatorname{Re}a>0, \quad n\geq 0] \qquad \text{BI (163)(9)}$$

23. $\displaystyle\int_0^\infty \cos^n sx \sin nsx \frac{x\,dx}{a^2-x^2} = \frac{\pi}{2}\left(2^{-n} - \cos^n as \cos nas\right)$

$$[n=0,1,\ldots] \qquad \text{BI (166)(10)}$$

24. $\displaystyle\int_0^\infty \cos^{m-1} x \sin[(m+1)x]\frac{x\,dx}{a^2+x^2} = \frac{\pi}{2^m}e^{-2a}\left(1+e^{-2a}\right)^{m-1}$

$$[a>0, \quad m=1,2,\ldots] \qquad \text{BI (163)(6)}$$

25. $\displaystyle\int_0^\infty \cos^m x \sin[(m+1)x]\frac{x\,dx}{a^2+x^2} = \frac{\pi}{2^{m+1}}e^{-a}\left(1+e^{-2a}\right)^m$

$$[m=0,1,\ldots, \quad a>0] \qquad \text{BI (163)(10)}$$

26.[3] $\displaystyle\int_0^\infty \cos^m x \sin[(m-1)x]\frac{x\,dx}{a^2+x^2} = \frac{\pi}{2^m}\cosh a\left[\left(1+e^{-2a}\right)^{m-1} - 1\right]$

$$[m=0,1,\ldots, \quad a\geq 0] \qquad \text{BI (163)(7)}$$

27.[12] $\displaystyle\int_0^\infty \cos^m x \sin(3mx)\frac{x\,dx}{a^2+x^2} = \frac{\pi}{2}e^{-3a}\cosh^m a \qquad [a>0, \quad m=1,2,\ldots] \qquad \text{BI (163)(11)}$

28. $\displaystyle\int_0^\infty \cos^n sx \cos nsx \frac{dx}{a^2+x^2} = \frac{\pi}{2^{n+1}a}\left(1+e^{-2as}\right)^n \qquad [n=0,1,\ldots] \qquad \text{BI (163)(16)}$

29. $\displaystyle\int_0^\infty \cos^n sx \cos nsx \frac{dx}{a^2-x^2} = \frac{\pi}{2a}\cos^n as \sin nas \qquad [n=0,1,\ldots]$

30. $\displaystyle\int_0^\infty \cos^{m-1} x \cos[(m+1)x]\frac{dx}{a^2+x^2} = \frac{\pi}{2^m a}e^{-2a}\left(1+e^{-2a}\right)^{m-1}$

$$[m=1,2,\ldots, \quad a>0] \qquad \text{BI (163)(14)}$$

31. $\displaystyle\int_0^\infty \cos^m x \cos[(m-1)x]\frac{dx}{a^2+x^2} = \frac{\pi}{2^{m+1}a}e^a\left[\left(1+e^{-2a}\right)^m - \left(1-e^{-2a}\right)\right]$

$$[m=0,1,\ldots, \quad a>0] \qquad \text{BI (163)(15)}$$

32. $\displaystyle\int_0^\infty \cos^m x \cos[(m+1)x]\frac{dx}{a^2+x^2} = \frac{\pi}{2^{m+1}a}e^{-a}\left(1+e^{-2a}\right)^m$

$$[m=0,1,\ldots, \quad a>0] \qquad \text{BI (163)(17)}$$

33. $\displaystyle\int_0^\infty \sin^p x \cos x \frac{dx}{x^q} = \frac{p}{q-1}\int_0^\infty \frac{\sin^{p-1}x}{x^{q-1}}dx - \frac{p+1}{q-1}\int_0^\infty \frac{\sin^{p+1}x}{x^{q-1}}dx \qquad [p>q-1>0]$

$$= \frac{p(p-1)}{(q-1)(q-2)}\int_0^\infty \sin^{p-2}x \cos x \frac{dx}{x^{q-2}}$$

$$- \frac{(p+1)^2}{(q-1)(q-2)}\int_0^\infty \sin^p x \cos x \frac{dx}{x^{q-2}} \qquad [p>q-1>1]$$

$$\text{GW (333)(18)}$$

34. $\displaystyle\int_0^\infty \cos^{2m} x \cos 2nx \sin x \frac{dx}{x} = \int_0^\infty \cos^{2m-1} x \cos 2nx \sin \frac{dx}{x} = \frac{\pi}{2^{2m+1}}\binom{2m}{m+n}$

$$\text{BI (152)(5, 6)}$$

35. $\displaystyle\int_0^\infty \cos^p ax \sin bx \cos x \frac{\mathrm{d}x}{x} = \frac{\pi}{2}$          $[b > ap, \quad p > -1]$          BI (153)(12)

36. $\displaystyle\int_0^\infty \cos^p ax \sin pax \cos x \frac{\mathrm{d}x}{x} = \frac{\pi}{2^{p+1}}\left(2^p - 1\right)$          $[p > -1]$          BI (153)(2)

37. $\displaystyle\int_0^\infty \frac{\mathrm{d}x}{x^2}\left(\prod_{k=1}^n \cos^{p_k} a_k x\right)\sin bx \sin x = \frac{\pi}{2}$      $\left[b > \displaystyle\sum_{k=1}^n a_k p_k, \quad a_k > 0, \quad p_k > 0\right]$

                                                                     BI (157)(15)

**3.833**

1.[10] $\displaystyle\int_0^\infty \sin^{2m+1} x \cos^{2n} x \frac{\mathrm{d}x}{x} = \int_0^\infty \sin^{2m+1} x \cos^{2n-1} x \frac{\mathrm{d}x}{x} = \frac{(2m-1)!!(2n-1)!!}{2^{m+n+1}(m+n)!}\pi$

                                                                BI (151)(24, 25)

$$= \frac{1}{2}\,\mathrm{B}\left(m + \frac{1}{2}, n + \frac{1}{2}\right)$$

                                                                GW (333)(24)

2. $\displaystyle\int_0^\infty \sin^{2m+1} 2x \cos^{2n-1} 2x \cos^2 x \frac{\mathrm{d}x}{x} = \frac{\pi}{2}\frac{(2m-1)!!(2n-1)!!}{(2m+2n)!!}$          LI (152)(4)

**3.834**

1. $\displaystyle\int_0^\infty \frac{\sin^{2m+1} x}{1 - 2a\cos x + a^2}\cdot\frac{\mathrm{d}x}{x} = \frac{(-1)^m \pi(1+a)^{4m}}{2^{2m+2}a^{2m+1}}\left[\left|\frac{1-a}{1+a}\right|^{2m-1}\right.$

$$\left. - \sum_{k=0}^{2m}(-1)^k\binom{m-\frac{1}{2}}{k}\left(\frac{4a}{(1+a)^2}\right)^k\right]$$

                                                        $[|a| \neq 1]$          GW (333)(62a)

2. $\displaystyle\int_0^\infty \frac{\sin^{2m+1} x \cos^n x}{(1 - 2a\cos x + a^2)^p}\frac{\mathrm{d}x}{x}$

$$= \frac{n!\pi}{2^{n+1}(2m+n+1)!(1+a)^{2p}}\sum_{k=0}^n \frac{(-1)^k(2m+2n-2k+1)!!(2m+2k-1)!!}{k!(n-k)!}$$

$$\times F\left(m+n-k+\frac{3}{2}, p; 2m+n+2; \frac{4a}{(1+a)^2}\right)$$

                                                        $[a \neq \pm 1]$          GW (333)(62)

**3.835**

1. $\displaystyle\int_0^\infty \frac{\cos^{2m} x \cos 2mx \sin x}{a^2 \cos^2 x + b^2 \sin^2 x}\frac{\mathrm{d}x}{x} = \frac{\pi}{2}\frac{b^{2m-1}}{a(a+b)^{2m}}$          $[ab > 0]$          BI (182)(31)a

2. $\displaystyle\int_0^\infty \frac{\cos^{2m-1} x \cos 2mx \sin x}{a^2 \cos^2 x + b^2 \sin^2 x}\frac{\mathrm{d}x}{x} = \frac{\pi}{2a}\frac{b^{2m-1}}{(a+b)^{2m}}$          $[ab > 0]$          LI (182)(32)a

**3.836**

1. $\displaystyle\int_0^\infty \left(\frac{\sin x}{x}\right)^n \frac{\sin mx}{x}\,\mathrm{d}x = \frac{\pi}{2}$          $[m \geq n]$          LI (159)(12)

2.[11] $\displaystyle\int_0^\infty \left(\frac{\sin x}{x}\right)^n \cos mx\, dx = \frac{n\pi}{2^n} \sum_{k=0}^{\left\lfloor \frac{1}{2}(m+n)\right\rfloor} \frac{(-1)^k (n+m-2k)^{n-1}}{k!(n-k)!}$  $\quad [0 \le m < n]$

$\qquad\qquad\qquad\qquad\qquad = 0 \qquad\qquad\qquad\qquad\qquad\qquad [m \ge n \ge 2]$

$\qquad\qquad\qquad\qquad\qquad = \dfrac{\pi}{4} \qquad\qquad\qquad\qquad\qquad\qquad [m = n = 1]$

$\qquad\qquad\qquad\qquad\qquad\qquad\qquad\qquad\qquad\qquad$ GI(159)(14), ET I 20(11)

3. $\displaystyle\int_0^\infty \left(\frac{\sin x}{x}\right)^{n-1} \sin nx \cos x \frac{dx}{x} = \frac{\pi}{2}$ $\qquad\qquad [n \ge 1]$ $\qquad$ BI (159)(20)

4.[8] $\displaystyle\int_0^\infty \left(\frac{\sin x}{x}\right)^n \frac{\sin(anx)}{x}\, dx = \frac{\pi}{2}\left[1 - \frac{1}{2^{n-1}n!} \sum_{k=0}^{\left\lfloor \frac{1}{2}n(1+a)\right\rfloor} (-1)^k \binom{n}{k}(n+an-2k)^n\right]$

$\qquad\qquad\qquad\qquad\qquad\qquad\qquad$ [all real $a$, $n \ge 1$] $\qquad$ ET I 20(11)

5.[10] $I_n(b) = \dfrac{2}{\pi}\displaystyle\int_0^\infty \left(\frac{\sin x}{x}\right)^n \cos bx\, dx = n\left(2^{n-1}n!\right)^{-1} \sum_{k=0}^{\lfloor r\rfloor}(-1)^k \binom{n}{k}(n-b-2k)^{n-1}$

$\qquad\qquad\qquad [0 \le b < n, \quad n \ge 1, \quad r = (n-b)/2]$ $\quad$ LO V 340(14)

6.[11] $\displaystyle\int_0^\infty \left(\frac{\sin x}{x}\right)^n \cos anx\, dx = 0$ $\quad [a \le -1 \text{ or } a \ge 1, \quad n \ge 2; \quad \text{for } n = 1 \text{ see } \mathbf{3.741}\ 2]$

**3.837**

1. $\displaystyle\int_0^{\pi/2} \frac{x^2\, dx}{\sin^2 x} = \pi \ln 2$ $\qquad\qquad\qquad\qquad\qquad\qquad\qquad$ BI (206)(9)

2. $\displaystyle\int_0^{\pi/4} \frac{x^2\, dx}{\sin^2 x} = -\frac{\pi^2}{16} + \frac{\pi}{4}\ln 2 + \mathbf{G}$ $\qquad\qquad\qquad\qquad$ BI (204)(10)

3. $\displaystyle\int_0^{\pi/4} \frac{x^2\, dx}{\cos^2 x} = \frac{\pi^2}{16} + \frac{\pi}{4}\ln 2 - \mathbf{G}$ $\qquad\qquad\qquad\qquad$ GW (333)(35a)

4. $\displaystyle\int_0^{\pi/4} \frac{x^{p+1}}{\sin^2 x}\, dx = -\left(\frac{\pi}{4}\right)^{p+1} + (p+1)\left(\frac{\pi}{4}\right)^p \left[\frac{1}{p} - \frac{1}{2}\sum_{k=1}^\infty \frac{1}{4^{2k-1}(p+2k)}\zeta(2k)\right]$

$\qquad\qquad\qquad\qquad\qquad\qquad\qquad\qquad [p > 0]$ $\qquad\qquad$ LI (204)(14)

5. $\displaystyle\int_0^{\pi/2} \frac{x^2 \cos x}{\sin^2 x}\, dx = -\frac{\pi^2}{4} + 4\mathbf{G}$ $\qquad\qquad\qquad\qquad\qquad$ BI (206)(7)

6. $\displaystyle\int_0^{\pi/2} \frac{x^3 \cos x}{\sin^3 x}\, dx = -\frac{\pi^3}{16} + \frac{3}{2}\pi \ln 2$ $\qquad\qquad\qquad\qquad$ BI (206)(8)

7. $\displaystyle\int_0^\infty \frac{\cos 2nx}{\cos x}\sin^{2n} x \frac{dx}{x^m} = 0$ $\qquad\qquad\left[n > \dfrac{m-1}{2}, \quad m > 0\right]$ $\quad$ BI (180)(16)

8. $\displaystyle\int_0^\infty \frac{\cos 2nx}{\cos x}\sin^{2n+1} x \frac{dx}{x^m} = 0$ $\qquad\qquad\left[n > \dfrac{m-2}{2}, \quad m > 0\right]$ $\quad$ BI (180)(17)

9.    $\displaystyle\int_0^1 \frac{x\,dx}{\cos ax\cos[a(1-x)]} = \frac{1}{a}\operatorname{cosec} a\cdot\ln\sec a$     $\left[a < \dfrac{\pi}{2}\right]$     BI (149)(20)

10.[3]    $\displaystyle\int_0^\pi \frac{x\sin(2n+1)x}{\sin x}\,dx = \frac{1}{2}\pi^2$     $[n = 0,1,2,\dots]$

11.[3]    $\displaystyle\int_0^\pi \frac{x\sin 2nx}{\sin x}\,dx = -4\sum_{k=1}^n (2k-1)^{-2}$     $[n = 1,2,3,\dots]$

**3.838**

1.    $\displaystyle\int_0^{\pi/2} \frac{x\cos^{p-1}x}{\sin^{p+1}x}\,dx = \frac{\pi}{2p}\sec\frac{\pi p}{2}$     $[p < 1]$     BI (206)(13)a

2.    $\displaystyle\int_0^{\pi/4} \frac{x\sin^{p-1}x}{\cos^{p+1}x}\,dx = \frac{\pi}{4p} - \frac{1}{2p}\beta\left(\frac{p+1}{2}\right)$     $[p > -1]$     LI (204)(15)

3.    $\displaystyle\int_0^{\pi/4} \frac{x\sin^{2m-1}x}{\cos^{2m+1}x}\,dx = \frac{\pi}{8m}(1-\cos m\pi) + \frac{1}{2m}\sum_{k=0}^{m-1}\frac{(-1)^{k-1}}{2m-2k-1}$     BI (204)(17)

4.    $\displaystyle\int_0^{\pi/4} \frac{x\sin^{2m}x}{\cos^{2m+2}x}\,dx = \frac{1}{2(2m+1)}\left[\frac{\pi}{2} + (-1)^{m-1}\ln 2 + \sum_{k=0}^{m-1}\frac{(-1)^{k-1}}{m-k}\right]$     BI (204)(16)

**3.839**

1.[11]    $\displaystyle\int_0^{\pi/4} x\tan^2 x\,dx = \frac{\pi}{4} - \frac{\pi^2}{32} - \frac{1}{2}\ln 2$     BI (204)(3)

2.    $\displaystyle\int_0^{\pi/4} x\tan^3 x\,dx = \frac{\pi}{4} - \frac{1}{2} + \frac{\pi}{8}\ln 2 - \frac{1}{2}G$     BI (204)(7)

3.    $\displaystyle\int_0^{\pi/4} \frac{x^2\tan x}{\cos^2 x}\,dx = \frac{1}{2}\ln 2 - \frac{\pi}{4} + \frac{\pi^2}{16}$     (cf. **3.839** 1)     BI (204)(13)

4.    $\displaystyle\int_0^{\pi/4} \frac{x^2\tan^2 x}{\cos^2 x}\,dx = \frac{1}{3}\left(1 - \frac{\pi}{4}\ln 2 - \frac{\pi}{2} + \frac{\pi^2}{16} + G\right)$     (cf. **3.839** 2)     BI (204)(12)

5.    $\displaystyle\int_0^{\pi/2} x\cos^p x\tan x\,dx = \frac{\pi}{2^{p+1}p}\frac{\Gamma(p+1)}{\left[\Gamma\left(\frac{p}{2}+1\right)\right]^2}$     $[p > -1]$     BI (205)(3)

6.    $\displaystyle\int_0^{\pi/2} x\sin^p x\cot x\,dx = \frac{\pi}{2p} - \frac{2^{p-1}}{p}B\left(\frac{p+1}{2}, \frac{p+1}{2}\right)$

                                                                             $[p > -1]$     BI (206)(11)

7.    $\displaystyle\int_0^\infty \sin^{2n} x\tan x\frac{dx}{x} = \frac{\pi}{2}\frac{(2n-1)!!}{(2n)!!}$     GW (333)(16)

8.    $\displaystyle\int_0^\infty \cos^s rx\tan qx\frac{dx}{x} = \frac{\pi}{2}$     $[s > -1]$     BI (151)(26)

9.    $\displaystyle\int_0^\infty \frac{\cos[(2n-1)x]}{\cos x}\left(\frac{\sin x}{x}\right)^{2n} dx = (-1)^{n-1}\frac{2^{2n}-1}{(2n)!}2^{2n-1}\pi|B_{2n}|$     BI (180)(15)

10.    $\displaystyle\int_0^\infty \tan^r px\frac{dx}{q^2+x^2} = \frac{\pi}{2q}\sec\frac{r\pi}{2}\tanh^r pq$     $[r^2 < 1]$     BI (160)(19)

## 3.84 Integrals containing $\sqrt{1 - k^2 \sin^2 x}$, $\sqrt{1 - k^2 \cos^2 x}$, and similar expressions

**Notation:**    $k' = \sqrt{1 - k^2}$

**3.841**

1.    $\displaystyle\int_0^\infty \sin x \sqrt{1 - k^2 \sin^2 x} \, \frac{\mathrm{d}x}{x} = \boldsymbol{E}(k)$            BI (154)(8)

2.    $\displaystyle\int_0^\infty \sin x \sqrt{1 - k^2 \cos^2 x} \, \frac{\mathrm{d}x}{x} = \boldsymbol{E}(k)$            BI (154)(20)

3.    $\displaystyle\int_0^\infty \tan x \sqrt{1 - k^2 \sin^2 x} \, \frac{\mathrm{d}x}{x} = \boldsymbol{E}(k)$            BI (154)(9)

4.    $\displaystyle\int_0^\infty \tan x \sqrt{1 - k^2 \cos^2 x} \, \frac{\mathrm{d}x}{x} = \boldsymbol{E}(k)$            BI (154)(21)

**3.842**

1.[11]    $\displaystyle\int_0^\infty \frac{\sin x}{\sqrt{1 + \sin^2 x}} \frac{\mathrm{d}x}{x}$

$$= \int_0^\infty \frac{\tan x}{\sqrt{1 + \sin^2 x}} \frac{\mathrm{d}x}{x}$$

$$= \int_0^\infty \frac{\sin x}{\sqrt{1 + \cos^2 x}} \frac{\mathrm{d}x}{x} = \int_0^\infty \frac{\tan x}{\sqrt{1 + \cos^2 x}} \frac{\mathrm{d}x}{x} = \frac{1}{\sqrt{2}} \boldsymbol{K}\left(\frac{1}{\sqrt{2}}\right) \approx 1.3110287771$$

                                              BI (183)(4, 5, 9, 10)

2.    $\displaystyle\int_u^{\frac{\pi}{2}} \frac{x \cos x \, \mathrm{d}x}{\sqrt{\sin^2 x - \sin^2 u}} = \frac{\pi}{2} \ln (1 + \cos u)$            BI (226)(4)

3.    $\displaystyle\int_0^\infty \frac{\sin x}{\sqrt{1 - k^2 \sin^2 x}} \frac{\mathrm{d}x}{x} = \int_0^\infty \frac{\tan x}{\sqrt{1 - k^2 \sin^2 x}} \frac{\mathrm{d}x}{x}$

$$= \int_0^\infty \frac{\sin x}{\sqrt{1 - k^2 \cos^2 x}} \frac{\mathrm{d}x}{x} = \int_0^\infty \frac{\tan x}{\sqrt{1 - k^2 \cos^2 x}} \frac{\mathrm{d}x}{x} = \boldsymbol{K}(k)$$

                                              BI (183)(12, 13, 21, 22)

4.    $\displaystyle\int_0^{\pi/2} \frac{x \sin x \cos x}{\sqrt{1 - k^2 \sin^2 x}} \, \mathrm{d}x = \frac{1}{2k^2} \left[ -\pi k' + 2 \boldsymbol{E}(k) \right]$        BI (211)(1)

5.    $\displaystyle\int_0^{\pi/2} \frac{x \sin x \cos x}{\sqrt{1 - k^2 \cos^2 x}} \, \mathrm{d}x = \frac{1}{2k^2} \left[ \pi - 2 \boldsymbol{E}(k) \right]$        BI (214)(1)

6.    $\displaystyle\int_0^\alpha \frac{x \sin x \, \mathrm{d}x}{\cos^2 x \sqrt{\sin^2 \alpha - \sin^2 x}} = \frac{\pi \sin^2 \frac{\alpha}{2}}{\cos^2 \alpha}$        LO III 284

7.    $\displaystyle\int_0^\beta \frac{x \sin x \, \mathrm{d}x}{\left(1 - \sin^2 \alpha \sin^2 x\right) \sqrt{\sin^2 \beta - \sin^2 x}} = \frac{\pi \ln \dfrac{\cos \alpha + \sqrt{1 - \sin^2 \alpha \sin^2 \beta}}{2 \cos \beta \cos^2 \frac{\alpha}{2}}}{2 \cos \alpha \sqrt{1 - \sin^2 \alpha \sin^2 \beta}}$        LO III 284

**3.843**

1.    $\displaystyle\int_0^\infty \tan x \sqrt{1 - k^2 \sin^2 2x} \, \frac{\mathrm{d}x}{x} = \boldsymbol{E}(k)$            BI (154)(10)

2.    $\displaystyle\int_0^\infty \tan x\sqrt{1-k^2\cos^2 2x}\,\frac{dx}{x} = E(k)$        BI (154)(22)

3.[11]    $\displaystyle\int_0^\infty \frac{\tan x}{\sqrt{1+\sin^2 2x}}\,\frac{dx}{x} = \int_0^\infty \frac{\tan x}{\sqrt{1+\cos^2 2x}}\,\frac{dx}{x} = \frac{1}{\sqrt{2}}\,K\left(\frac{1}{\sqrt{2}}\right) \approx 1.3110287771$

         BI (183)(6, 11)

4.    $\displaystyle\int_0^\infty \frac{\tan x}{\sqrt{1-k^2\sin^2 2x}}\,\frac{dx}{x} = \int_0^\infty \frac{\tan x}{\sqrt{1-k^2\cos^2 2x}}\,\frac{dx}{x} = K(k)$        BI (183)(14, 23)

**3.844**

1.    $\displaystyle\int_0^\infty \frac{\sin x\cos x}{\sqrt{1-k^2\cos^2 x}}\,\frac{dx}{x} = \frac{1}{k^2}\left[K(k) - E(k)\right]$        BI (185)(20)

2.    $\displaystyle\int_0^\infty \frac{\sin x\cos^2 x}{\sqrt{1-k^2\cos^2 x}}\,\frac{dx}{x} = \frac{1}{k^2}\left[K(k) - E(k)\right]$        BI (185)(21)

3.    $\displaystyle\int_0^\infty \frac{\sin x\cos^3 x}{\sqrt{1-k^2\cos^2 x}}\,\frac{dx}{x} = \frac{1}{3k^4}\left[(2+k^2)\,K(k) - 2\left(1+k^2\right)E(k)\right]$        BI (185)(22)

4.    $\displaystyle\int_0^\infty \frac{\sin x\cos^4 x}{\sqrt{1-k^2\cos^2 x}}\,\frac{dx}{x} = \frac{1}{3k^4}\left[(2+k^2)\,K(k) - 2\left(1+k^2\right)E(k)\right]$        BI (185)(23)

5.    $\displaystyle\int_0^\infty \frac{\sin^3 x\cos x}{\sqrt{1-k^2\cos^2 x}}\,\frac{dx}{x} = \frac{1}{3k^4}\left[(1+k'^2)\,E(k) - 2k'^2\,K(k)\right]$        BI (185)(24)

6.    $\displaystyle\int_0^\infty \frac{\sin^3 x\cos^2 x}{\sqrt{1-k^2\cos^2 x}}\,\frac{dx}{x} = \frac{1}{3k^4}\left[(1+k'^2)\,E(k) - 2k'^2\,K(k)\right]$        BI (185)(25)

7.    $\displaystyle\int_0^\infty \frac{\sin^2 x\tan x}{\sqrt{1-k^2\cos^2 x}}\,\frac{dx}{x} = \frac{1}{k^2}\left[E(k) - k'^2\,K(k)\right]$        BI (184)(16)

8.    $\displaystyle\int_0^\infty \frac{\sin^4 x\tan x}{\sqrt{1-k^2\cos^2 x}}\,\frac{dx}{x} = \frac{1}{3k^4}\left[(2+3k^2)\,k'^2\,K(k) - 2\left(k'^2 - k^2\right)E(k)\right]$        BI (184)(18)

**3.845**

1.[11]    $\displaystyle\int_0^\infty \frac{\sin x\cos x}{\sqrt{1+\cos^2 x}}\,\frac{dx}{x} = \sqrt{2}\left[E\left(\frac{\sqrt{2}}{2}\right) - \frac{1}{2}K\left(\frac{\sqrt{2}}{2}\right)\right] \approx 0.5990701174$        BI (185)(6)

2.[11]    $\displaystyle\int_0^\infty \frac{\sin x\cos^2 x}{\sqrt{1+\cos^2 x}}\,\frac{dx}{x} = \sqrt{2}\left[E\left(\frac{\sqrt{2}}{2}\right) - \frac{1}{2}K\left(\frac{\sqrt{2}}{2}\right)\right] \approx 0.5990701174$        BI (185)(7)

3.[11]    $\displaystyle\int_0^\infty \frac{\sin^2 x\tan x}{\sqrt{1+\cos^2 x}}\,\frac{dx}{x} = \sqrt{2}\left[K\left(\frac{\sqrt{2}}{2}\right) - E\left(\frac{\sqrt{2}}{2}\right)\right] \approx 0.7119586598$        BU (184)(8)

**3.846**

1.    $\displaystyle\int_0^\infty \frac{\sin x\cos x}{\sqrt{1-k^2\sin^2 x}}\,\frac{dx}{x} = \frac{1}{k^2}\left[E(k) - k'^2\,K(k)\right]$        BI (185)(9)

2.    $\displaystyle\int_0^\infty \frac{\sin x\cos^2 x}{\sqrt{1-k^2\sin^2 x}}\,\frac{dx}{x} = \frac{1}{k^2}\left[E(k) - k'^2\,K(k)\right]$        BI (185)(10)

3. $\displaystyle\int_0^\infty \frac{\sin x \cos^3 x}{\sqrt{1 - k^2 \sin^2 x}} \frac{dx}{x} = \frac{1}{3k^4}\left[\left(2 - 3k^2\right) k'^2 \, \mathbf{K}(k) - 2\left(k'^2 - k^2\right) \mathbf{E}(k)\right]$          BI (185)(11)

4. $\displaystyle\int_0^\infty \frac{\sin x \cos^4 x}{\sqrt{1 - k^2 \sin^2 x}} \frac{dx}{x} = \frac{1}{3k^4}\left[\left(2 - 3k^2\right) k'^2 \, \mathbf{K}(k) - 2\left(k'^2 - k^2\right) \mathbf{E}(k)\right]$          BI (185)(12)

5. $\displaystyle\int_0^\infty \frac{\sin^3 x \cos x}{\sqrt{1 - k^2 \sin^2 x}} \frac{dx}{x} = \frac{1}{3k^4}\left[\left(1 + k'^2\right) \mathbf{E}(k) - 2k'^2 \, \mathbf{K}(k)\right]$          BI (185)(13)

6. $\displaystyle\int_0^\infty \frac{\sin^3 x \cos^2 x}{\sqrt{1 - k^2 \sin^2 x}} \frac{dx}{x} = \frac{1}{3k^4}\left[\left(1 + k'^2\right) \mathbf{E}(k) - 2k'^2 \, \mathbf{K}(k)\right]$          BI (185)(14)

7. $\displaystyle\int_0^\infty \frac{\sin^2 x \tan x}{\sqrt{1 - k^2 \sin^2 x}} \frac{dx}{x} = \frac{1}{k^2}\left[\mathbf{K}(k) - \mathbf{E}(k)\right]$          BI (184)(9)

8. $\displaystyle\int_0^\infty \frac{\sin^4 x \tan x}{\sqrt{1 - k^2 \sin^2 x}} \frac{dx}{x} = \frac{1}{3k^4}\left[\left(2 + k^2\right) \mathbf{K}(k) - 2\left(1 + k^2\right) \mathbf{E}(k)\right]$          BI (184)(11)

**3.847**[11] $\displaystyle\int_0^\infty \frac{\sin x \cos x}{\sqrt{1 + \sin^2 x}} \frac{dx}{x} = \int_0^\infty \frac{\sin x \cos^2 x}{\sqrt{1 + \sin^2 x}} \frac{dx}{x} = \sqrt{2}\left[\mathbf{K}\left(\frac{\sqrt{2}}{2}\right) - \mathbf{E}\left(\frac{\sqrt{2}}{2}\right)\right] \approx 0.7119586598$

BI (185)(3, 4)

**3.848**

1. $\displaystyle\int_0^\infty \frac{\sin^3 x \cos x}{\sqrt{1 - k^2 \sin^2 2x}} \frac{dx}{x} = \frac{1}{4k^2}\left[\mathbf{K}(k) - \mathbf{E}(k)\right]$          BI (185)(15)

2. $\displaystyle\int_0^\infty \frac{\cos^2 2x \tan x}{\sqrt{1 - k^2 \sin^2 2x}} \frac{dx}{x} = \frac{1}{k^2}\left[\mathbf{E}(k) - k'^2 \, \mathbf{K}(k)\right]$          BI (184)(12)

3. $\displaystyle\int_0^\infty \frac{\cos^4 2x \tan x}{\sqrt{1 - k^2 \sin^2 2x}} \frac{dx}{x} = \frac{1}{3k^4}\left[\left(2 - 3k^2\right) k'^2 \, \mathbf{K}(k) - 2\left(k'^2 - k^2\right) \mathbf{E}(k)\right]$          BI (184)(13)

4. $\displaystyle\int_0^\infty \frac{\sin^2 4x \tan x}{\sqrt{1 - k^2 \sin^2 2x}} \frac{dx}{x} = \frac{4}{3k^4}\left[\left(1 + k'^2\right) \mathbf{E}(k) - 2k'^2 \, \mathbf{K}(k)\right]$          BI (184)(17)

5. $\displaystyle\int_0^\infty \frac{\sin^3 x \cos x}{\sqrt{1 - k^2 \cos^2 2x}} \frac{dx}{x} = \frac{1}{4k^2}\left[\mathbf{E}(k) - k'^2 \, \mathbf{K}(k)\right]$          BI (185)(26)

6. $\displaystyle\int_0^\infty \frac{\cos^2 2x \tan x}{\sqrt{1 - k^2 \cos^2 2x}} \frac{dx}{x} = \frac{1}{k^2}\left[\mathbf{K}(k) - \mathbf{E}(k)\right]$          BI (184)(19)

7. $\displaystyle\int_0^\infty \frac{\cos^4 2x \tan x}{\sqrt{1 - k^2 \cos^2 2x}} \frac{dx}{x} = \frac{1}{3k^4}\left[\left(2 + k^2\right) \mathbf{K}(k) - 2\left(1 + k^2\right) \mathbf{E}(k)\right]$          BI (184)(20)

**3.849**

1.[11] $\displaystyle\int_0^\infty \frac{\sin^3 x \cos x}{\sqrt{1 + \cos^2 2x}} \frac{dx}{x} = \frac{1}{2\sqrt{2}}\left[\mathbf{K}\left(\frac{\sqrt{2}}{2}\right) - \mathbf{E}\left(\frac{\sqrt{2}}{2}\right)\right] \approx 0.1779896649$          BI (185)(8)

2.[11] $\displaystyle\int_0^\infty \frac{\sin^3 x \cos x}{\sqrt{1 + \sin^2 2x}} \frac{dx}{x} = \frac{\sqrt{2}}{8}\left[2 \, \mathbf{E}\left(\frac{\sqrt{2}}{2}\right) - \mathbf{K}\left(\frac{\sqrt{2}}{2}\right)\right] \approx 0.1497675293$          BI (185)(5)

3.[11] $\displaystyle\int_0^\infty \frac{\cos^2 2x \tan x}{\sqrt{1 + \sin^2 2x}} \frac{dx}{x} = \sqrt{2}\left[\mathbf{K}\left(\frac{\sqrt{2}}{2}\right) - \mathbf{E}\left(\frac{\sqrt{2}}{2}\right)\right] \approx 0.7119586598$          BI (184)(7)

## 3.85–3.88 Trigonometric functions of more complicated arguments combined with powers

**3.851**

1.* $\displaystyle\int_0^\infty x\sin\left(ax^2\right)\sin\left(2bx\right)\,\mathrm{d}x = \frac{b}{2a}\sqrt{\frac{\pi}{2a}}\left(\cos\frac{b^2}{a}+\sin\frac{b^2}{a}\right)$

$$[a \geq 0, \quad b > 0] \qquad\qquad \text{BI (150)(4)}$$

2.* $\displaystyle\int_0^\infty x\sin\left(ax^2\right)\cos\left(2bx\right)\,\mathrm{d}x = \frac{1}{2a}-\frac{b}{a}\sqrt{\frac{\pi}{2a}}\left[\sin\left(\frac{b^2}{a}\right)C\left(\frac{b}{\sqrt{a}}\right)-\cos\left(\frac{b^2}{a}\right)S\left(\frac{b}{\sqrt{a}}\right)\right]$

$$[a \geq 0] \qquad\qquad \text{BI (150)(5)a}$$

3.* $\displaystyle\int_0^\infty x\cos\left(ax^2\right)\sin\left(2bx\right)\,\mathrm{d}x = \frac{b}{2a}\sqrt{\frac{\pi}{2a}}\left(\sin\frac{b^2}{a}-\cos\frac{b^2}{a}\right)$

$$[a > 0, \quad b > 0] \qquad\qquad \text{BI (150)(7)}$$

4.* $\displaystyle\int_0^\infty x\cos\left(ax^2\right)\cos\left(2bx\right)\,\mathrm{d}x = \frac{b}{a}\sqrt{\frac{\pi}{2a}}\left[\cos\left(\frac{b^2}{a}\right)C\left(\frac{b}{\sqrt{a}}\right)+\sin\left(\frac{b^2}{a}\right)S\left(\frac{b}{\sqrt{a}}\right)\right]$

$$[a \geq 0] \qquad\qquad \text{BI (150)(6)a}$$

5. $\displaystyle\int_0^\infty \sin\left(ax^2\right)\cos(bx)\frac{\mathrm{d}x}{x^2} = \frac{b\pi}{2}\left[S\left(\frac{b}{2\sqrt{a}}\right)-C\left(\frac{b}{2\sqrt{a}}\right)\right]+\sqrt{a\pi}\sin\left(\frac{b^2}{4a}+\frac{\pi}{4}\right)$

$$[a > 0, \quad b > 0], \qquad (\text{cf. } \mathbf{3.691}\ 7)$$
$$\text{ET I 23(3)a}$$

**3.852**

1. $\displaystyle\int_0^\infty \frac{\sin\left(ax^2\right)}{x^2}\,\mathrm{d}x = \sqrt{\frac{a\pi}{2}}$            $[a \geq 0]$            BI (177)(10)a

2. $\displaystyle\int_0^\infty \sin\left(ax^2\right)\cos\left(bx^2\right)\frac{\mathrm{d}x}{x^2} = \frac{1}{2}\sqrt{\frac{\pi}{2}}\left(\sqrt{a+b}+\sqrt{a-b}\right)$ $[a > b > 0]$

$$= \frac{1}{2}\sqrt{\pi a} \qquad\qquad\qquad\qquad [b = a \geq 0]$$

$$= \frac{1}{2}\sqrt{\frac{\pi}{2}}\left(\sqrt{a+b}-\sqrt{b-a}\right)$$

$$[b > a > 0], \quad (\text{cf. } \mathbf{3.852}\ 1) \quad \text{BI (177)(23)}$$

3. $\displaystyle\int_0^\infty \frac{\sin^2\left(a^2x^2\right)}{x^4}\,\mathrm{d}x = \frac{2\sqrt{\pi}}{3}a^3$            $[a \geq 0]$            GW (333)(19e)

4.[10] $\displaystyle\int_0^\infty \frac{\sin^3\left(a^2x^2\right)}{x^2}\,\mathrm{d}x = \frac{a}{4}\sqrt{\frac{\pi}{2}}\left(3-\sqrt{3}\right)$      $\left[\operatorname{Im} a^2 = 0\right]$            MC

5.[12] $\displaystyle\int_0^\infty \frac{\sin\left(x^2\right)-x^2\cos\left(x^2\right)}{x^4}\,\mathrm{d}x = \frac{1}{3}\sqrt{\frac{\pi}{2}}$            BI (178)(8)

6.[12] $\displaystyle\int_0^\infty \left(\cos\left(x^2\right)-\frac{1}{1+x^2}\right)\frac{\mathrm{d}x}{x} = -\frac{1}{2}\mathbf{C}$            BI (173)(22)

7.* $\displaystyle\int_0^\infty x^m \cos(ax^n)\,dx = \frac{\Gamma(\gamma)}{na^\gamma}\cos\left(\frac{\pi\gamma}{2}\right)$ $\qquad \gamma = \dfrac{m+1}{n}$

$$[a>0, \quad \gamma>0]$$

8.* $\displaystyle\int_0^\infty x^m \sin(ax^n)\,dx = \frac{\Gamma(\gamma)}{na^\gamma}\sin\left(\frac{\pi\gamma}{2}\right)$ $\qquad \gamma = \dfrac{m+1}{n}$

$$[a>0, \quad \gamma>0]$$

**3.853**

1.[12] $\displaystyle\int_0^\infty \frac{\sin\left(x^2\right)}{b^2+x^2}\,dx = \frac{\pi}{2b}\left[\sqrt{2}\sin\left(b^2+\frac{\pi}{4}\right)C\left(b\right) - \sqrt{2}\cos\left(b^2+\frac{\pi}{4}\right)S\left(b\right) - \sin\left(b^2\right)\right]$

$$[\operatorname{Re} b > 0] \qquad\qquad \text{ET II 219(33)a}$$

2.[12] $\displaystyle\int_0^\infty \frac{\cos\left(x^2\right)}{b^2+x^2}\,dx = -\frac{\pi}{2b}\left[\sqrt{2}\cos\left(b^2+\frac{\pi}{4}\right)C\left(b\right) - \sqrt{2}\sin\left(b^2+\frac{\pi}{4}\right)S\left(b\right) - \cos\left(b^2\right)\right]$

$$[\operatorname{Re} b > 0] \qquad\qquad \text{ET II 221(51)a}$$

3. $\displaystyle\int_0^\infty \frac{x^2\sin\left(ax^2\right)}{b^2+x^2}\,dx$

$$= \frac{b\pi}{2}\left[\sin\left(ab^2\right) - \sqrt{2}\sin\left(ab^2+\frac{\pi}{4}\right)C\left(\sqrt{ab}\right) + \sqrt{2}\cos\left(ab^2+\frac{\pi}{4}\right)S\left(\sqrt{ab}\right)\right]$$
$$- \frac{1}{2}\sqrt{\frac{\pi}{2a}}$$

$$[a>0, \quad \operatorname{Re} b > 0] \qquad\qquad \text{ET II 219(32)a}$$

4. $\displaystyle\int_0^\infty \frac{x^2\cos\left(ax^2\right)}{b^2+x^2}\,dx = \frac{1}{2}\sqrt{\frac{\pi}{2a}} - \frac{b\pi}{2}\left\{\cos\left(ab^2\right) - \sqrt{2}\cos\left(ab^2+\frac{\pi}{4}\right)C\left(\sqrt{ab}\right)\right.$
$$\left. - \sqrt{2}\sin\left(ab^2+\frac{\pi}{4}\right)S\left(\sqrt{ab}\right)\right\}$$

$$[a>0, \quad \operatorname{Re} b > 0] \qquad\qquad \text{ET II 221(50)a}$$

**3.854**

1.[12] $\displaystyle\int_0^\infty \frac{\cos\left(ax^2\right) - \sin\left(ax^2\right)}{x^4+1}\,dx = \frac{\pi e^{-a}}{2\sqrt{2}}$ $\qquad [a>0] \qquad\qquad \text{LI (178)(11)a, BI (168)(25)}$

2.[12] $\displaystyle\int_0^\infty \frac{\cos\left(ax^2\right) + \sin\left(ax^2\right)}{x^4+1}x^2\,dx = \frac{\pi e^{-a}}{2\sqrt{2}}$ $\qquad [a>0] \qquad\qquad \text{LI (178)(12)}$

3.[12] $\displaystyle\int_0^\infty \frac{\cos\left(ax^2\right) + \sin\left(ax^2\right)}{\left(x^4+1\right)^2}x^2\,dx = \frac{\pi e^{-a}}{4\sqrt{2}}\left(a+\frac{1}{2}\right)$ $\qquad [a>0] \qquad\qquad \text{LI (178)(14)}$

4.[12] $\displaystyle\int_0^\infty \frac{\cos\left(ax^2\right) - \sin\left(ax^2\right)}{\left(x^4+1\right)^2}x^4\,dx = \frac{\pi e^{-a}}{4\sqrt{2}}\left(\frac{1}{2}-a\right)$ $\qquad [a>0] \qquad\qquad \text{BI (178)(15)}$

**3.855**

1.   $\displaystyle\int_0^\infty \frac{\sin\left(ax^2\right)}{\sqrt{b^2+x^4}}\,dx = \frac{1}{2}\sqrt{\frac{a\pi}{2}}\, I_{\frac{1}{4}}\left(\frac{ab}{2}\right) K_{\frac{1}{4}}\left(\frac{ab}{2}\right)$     $[a>0,\quad \mathrm{Re}\,b>0]$     ET I 66(28)

2.   $\displaystyle\int_0^\infty \frac{\cos\left(ax^2\right)}{\sqrt{b^2+x^4}}\,dx = \frac{1}{2}\sqrt{\frac{a\pi}{2}}\, I_{-\frac{1}{4}}\left(\frac{ab}{2}\right) K_{\frac{1}{4}}\left(\frac{ab}{2}\right)$     $[a>0,\quad \mathrm{Re}\,b>0]$     ET I 9(22)

3.[12]   $\displaystyle\int_0^u \frac{\sin\left(a^2x^2\right)}{\sqrt{u^4-x^4}}\,dx = \frac{a}{4}\sqrt{\frac{\pi^3}{2}}\left[J_{\frac{1}{4}}\left(\frac{a^2u^2}{2}\right)\right]^2$     $[a>0]$     ET I 66(29)

4.   $\displaystyle\int_u^\infty \frac{\sin\left(a^2x^2\right)}{\sqrt{x^4-u^4}}\,dx = -\frac{a}{4}\sqrt{\frac{\pi^3}{2}}\, J_{\frac{1}{4}}\left(\frac{a^2u^2}{2}\right) Y_{\frac{1}{4}}\left(\frac{a^2u^2}{2}\right)$

                                                         $[a>0]$     ET I 66(30)

5.   $\displaystyle\int_0^u \frac{\cos\left(a^2x^2\right)}{\sqrt{u^4-x^4}}\,dx = \frac{a}{4}\sqrt{\frac{\pi^3}{2}}\left[J_{-\frac{1}{4}}\left(\frac{a^2u^2}{2}\right)\right]^2$     ET I 9(23)

6.   $\displaystyle\int_u^\infty \frac{\cos\left(a^2x^2\right)}{\sqrt{x^4-u^4}}\,dx = -\frac{a}{4}\sqrt{\frac{\pi^3}{2}}\, J_{-\frac{1}{4}}\left(\frac{a^2u^2}{2}\right) Y_{-\frac{1}{4}}\left(\frac{a^2u^2}{2}\right)$     ET I 10(24)

**3.856**

1.   $\displaystyle\int_0^\infty \frac{\left(\sqrt{\beta^4+x^4}+x^2\right)^\nu}{\sqrt{\beta^4+x^4}}\sin\left(a^2x^2\right)\,dx = \frac{a}{2}\sqrt{\frac{\pi}{2}}\beta^{2\nu}\, I_{\frac{1}{4}-\frac{\nu}{2}}\left(\frac{a^2\beta^2}{2}\right) K_{\frac{1}{4}+\frac{\nu}{2}}\left(\frac{a^2\beta^2}{2}\right)$

                                          $\left[\mathrm{Re}\,\nu < \frac{3}{2},\quad |\arg\beta| < \frac{\pi}{4}\right]$     ET I 71(23)

2.   $\displaystyle\int_0^\infty \frac{\left(\sqrt{\beta^4+x^4}+x^2\right)^\nu}{\sqrt{\beta^4+x^4}}\cos\left(a^2x^2\right)\,dx = \frac{a}{2}\sqrt{\frac{\pi}{2}}\beta^{2\nu}\, I_{-\frac{1}{4}-\frac{\nu}{2}}\left(\frac{a^2\beta^2}{2}\right) K_{-\frac{1}{4}+\frac{\nu}{2}}\left(\frac{a^2\beta^2}{2}\right)$

                                          $\left[\mathrm{Re}\,\nu < \frac{3}{2},\quad |\arg\beta| < \frac{\pi}{4}\right]$     ET I 12(16)

3.   $\displaystyle\int_0^\infty \frac{\left(\sqrt{\beta^4+x^4}-x^2\right)^\nu}{\sqrt{\beta^4+x^4}}\cos\left(a^2x^2\right)\,dx = \frac{a}{2}\sqrt{\frac{\pi}{2}}\beta^{2\nu}\, I_{-\frac{1}{4}+\frac{\nu}{2}}\left(\frac{a^2\beta^2}{2}\right) K_{-\frac{1}{4}-\frac{\nu}{2}}\left(\frac{a^2\beta^2}{2}\right)$

                                          $\left[\mathrm{Re}\,\nu > -\frac{3}{2},\quad |\arg\beta| < \frac{\pi}{4}\right]$     ET I 12(17)

4.   $\displaystyle\int_0^\infty \frac{\sin\left(a^2x^2\right)\,dx}{\sqrt{\beta^4+x^4}\sqrt{x^2+\sqrt{\beta^4+x^4}}} = \frac{\sinh\frac{a^2\beta^2}{2}}{\sqrt{2}\beta^2}\, K_0\left(\frac{a^2\beta^2}{2}\right)$

                                                   $\left[|\arg\beta| < \frac{\pi}{4}\right]$     ET I 66(32)

5.   $\displaystyle\int_0^\infty \frac{\cos\left(a^2x^2\right)\,dx}{\sqrt{\beta^4+x^4}\sqrt{\left(x^2+\sqrt{\beta^4+x^4}\right)^3}} = \frac{\sinh\frac{a^2\beta^2}{2}}{2\sqrt{2}\beta^4}\, K_1\left(\frac{a^2\beta^2}{2}\right)$

                                                   $\left[|\arg\beta| < \frac{\pi}{4}\right]$     ET I 10(27)

6.   $\displaystyle\int_0^\infty \frac{\sqrt{\sqrt{b^4+x^4}+x^2}}{\sqrt{b^4+x^4}}\sin\left(a^2x^2\right)\,dx = \frac{\pi}{2\sqrt{2}}e^{-\frac{a^2b^2}{2}}\, I_0\left(\frac{a^2b^2}{2}\right)$

                                                   $\left[|\arg b| < \frac{\pi}{4}\right]$     ET I 67(33)

**3.857**

1.  $$\int_0^\infty \frac{x^2}{R_1 R_2} \sqrt{\frac{R_2 - R_1}{R_2 + R_1}} \sin\left(ax^2\right)\, dx = \frac{1}{2\sqrt{b}} K_0(ac) \sin ab$$

    $$\left[ R_1 = \sqrt{c^2 + (b - x^2)^2}, \quad R_2 = \sqrt{c^2 + (b + x^2)^2}, \quad a > 0, \quad c > 0 \right] \quad \text{ET I 67(34)}$$

2.  $$\int_0^\infty \frac{x^2}{R_1 R_2} \sqrt{\frac{R_2 + R_1}{R_2 - R_1}} \cos\left(ax^2\right)\, dx = \frac{1}{2\sqrt{b}} K_0(ac) \cos ab$$

    $$\left[ R_1 = \sqrt{c^2 + (b - x^2)^2}, \quad R_2 = \sqrt{c^2 + (b + x^2)^2}, \quad a > 0, \quad c > 0 \right] \quad \text{ET I 10(26)}$$

**3.858**

1.  $$\int_u^\infty \frac{\left(x^2 + \sqrt{x^4 - u^4}\right)^\nu + \left(x^2 - \sqrt{x^4 - u^4}\right)^\nu}{\sqrt{x^4 - u^4}} \sin\left(a^2 x^2\right)\, dx$$

    $$= -\frac{a}{4} \sqrt{\frac{\pi^3}{a}} u^{2\nu} \left[ J_{\frac{1}{4} + \frac{\nu}{2}}\left(\frac{a^2 u^2}{2}\right) Y_{\frac{1}{4} - \frac{\nu}{2}}\left(\frac{a^2 u^2}{2}\right) + J_{\frac{1}{4} - \frac{\nu}{2}}\left(\frac{a^2 u^2}{2}\right) Y_{\frac{1}{4} + \frac{\nu}{2}}\left(\frac{a^2 u^2}{2}\right) \right]$$

    $$\left[ \operatorname{Re} \nu < \tfrac{3}{2} \right] \qquad \text{ET I 71(25)}$$

2.  $$\int_u^\infty \frac{\left(x^2 + \sqrt{x^4 - u^4}\right)^\nu + \left(x^2 - \sqrt{x^4 - u^4}\right)^\nu}{\sqrt{x^4 - u^4}} \cos\left(a^2 x^2\right)\, dx$$

    $$= -\frac{a}{4} \sqrt{\frac{\pi^3}{a}} u^{2\nu} \left[ J_{-\frac{1}{4} + \frac{\nu}{2}}\left(\frac{a^2 u^2}{2}\right) Y_{-\frac{1}{4} - \frac{\nu}{2}}\left(\frac{a^2 u^2}{2}\right) + J_{-\frac{1}{4} - \frac{\nu}{2}}\left(\frac{a^2 u^2}{2}\right) Y_{-\frac{1}{4} + \frac{\nu}{2}}\left(\frac{a^2 u^2}{2}\right) \right]$$

    $$\left[ \operatorname{Re} \nu < \tfrac{3}{2} \right] \qquad \text{ET I 13(26)}$$

**3.859** $\displaystyle\int_0^\infty \left[ \cos\left(x^{2^n}\right) - \frac{1}{1 + x^{2^{n+1}}} \right] \frac{dx}{x} = -\frac{1}{2^n} \mathbf{C}$ $\hfill$ BI (173)(24)

**3.861**

1.  $$\int_0^\infty \sin^{2n+1}\left(ax^2\right) \frac{dx}{x^{2m}} = \pm \frac{\sqrt{\pi} a^{m - \frac{1}{2}}}{2^{2n - m + \frac{1}{2}}(2m - 1)!!} \sum_{k=1}^{n+1} (-1)^{k-1} \binom{2n+1}{n+k} (2k - 1)^{m - \frac{1}{2}}$$

    $$\left[ \begin{matrix} \text{the } + \text{ sign is taken when } m \equiv 0 \pmod 4 \text{ or } m \equiv 1 \pmod 4, \\ \text{the } - \text{ sign is taken when } m \equiv 2 \pmod 4 \text{ or } m \equiv 3 \pmod 4 \end{matrix} \right] \quad \text{BI (177)(19)a}$$

2.  $$\int_0^\infty \sin^{2n}\left(ax^2\right) \frac{dx}{x^{2m}} = \pm \frac{\sqrt{\pi} a^{m - \frac{1}{2}}}{2^{2n - 2m + 1}(2m - 1)!!} \sum_{k=1}^{n} (-1)^k \binom{2n}{n+k} k^{m - \frac{1}{2}}$$

    $$\left[ \begin{matrix} \text{the } + \text{ sign is taken when } m \equiv 0 \pmod 4 \text{ or } m \equiv 3 \pmod 4, \\ \text{the } - \text{ sign is taken when } m \equiv 2 \pmod 4 \text{ or } m \equiv 1 \pmod 4 \end{matrix} \right] \quad \text{BI (177)(18)a, LI (177)(18)}$$

**3.862** $\displaystyle\int_0^\infty \left[ \cos\left(ax^2\sqrt{n}\right) + \sin\left(ax^2\sqrt{n}\right) \right] \left(\frac{\sin^2 x}{x^2}\right)^n dx$

$$= \frac{\sqrt{\pi}}{(2n - 1)!!\sqrt{2}} \sum_{k=0}^{n} (-1)^k \binom{n}{k} \left(n - 2k + a\sqrt{n}\right)^{n - \frac{1}{2}}$$

$$\left[ a > \sqrt{n} > 0 \right] \qquad \text{BI (178)(9)}$$

**3.863**

1.  $$\int_0^\infty x^2 \cos\left(ax^4\right) \sin\left(2bx^2\right)\, dx = -\frac{\pi}{8} \sqrt{\frac{b^3}{a^3}} \left[ \sin\left(\frac{b^2}{2a} - \frac{\pi}{8}\right) J_{-\frac{1}{4}}\left(\frac{b^2}{2a}\right) + \cos\left(\frac{b^2}{2a} - \frac{\pi}{8}\right) J_{\frac{3}{4}}\left(\frac{b^2}{2a}\right) \right]$$

    $$\left[ a > 0, \quad b > 0 \right] \qquad \text{ET I 25(22)}$$

2. $\displaystyle\int_0^\infty x^2 \cos\left(ax^4\right)\cos\left(2bx^2\right)\,dx = -\frac{\pi}{8}\sqrt{\frac{b^3}{a^3}}\left[\sin\left(\frac{b^2}{2a}+\frac{\pi}{8}\right)J_{-\frac{3}{4}}\left(\frac{b^2}{2a}\right)+\cos\left(\frac{b^2}{2a}+\frac{\pi}{8}\right)J_{-\frac{1}{4}}\left(\frac{b^2}{2a}\right)\right]$

$$[a > 0,\quad b > 0] \qquad\qquad \text{ET I 25(23)}$$

**3.864**

1. $\displaystyle\int_0^\infty \sin\frac{b}{x}\sin ax\,\frac{dx}{x} = \frac{\pi}{2}\,Y_0\left(2\sqrt{ab}\right)+K_0\left(2\sqrt{ab}\right)$ $\qquad [a > 0,\quad b > 0]$ $\qquad$ WA 204(3)a

2. $\displaystyle\int_0^\infty \cos\frac{b}{x}\cos ax\,\frac{dx}{x} = -\frac{\pi}{2}\,Y_0\left(2\sqrt{ab}\right)+K_0\left(2\sqrt{ab}\right)$

$$[a > 0,\quad b > 0]$$

$$\text{WA 204(4)a, ET I 24 (12)}$$

**3.865**

1. $\displaystyle\int_0^u \frac{\left(u^2 - x^2\right)^{\mu-1}}{x^{2\mu}}\sin\frac{a}{x}\,dx = \frac{\sqrt{\pi}}{2}\left(\frac{2}{a}\right)^{\mu-\frac{1}{2}}u^{\mu-\frac{3}{2}}\,\Gamma(\mu)\,J_{\frac{1}{2}-\mu}\left(\frac{a}{u}\right)$

$$[a > 0,\quad u > 0,\quad 0 < \operatorname{Re}\mu < 1]$$

$$\text{ET II 189(30)}$$

2. $\displaystyle\int_u^\infty \frac{\left(x - u\right)^{\mu-1}}{x^{2\mu}}\sin\frac{a}{x}\,dx = \sqrt{\frac{\pi}{u}}\,a^{\frac{1}{2}-\mu}\,\Gamma(\mu)\sin\frac{a}{2u}\,J_{\mu-\frac{1}{2}}\left(\frac{a}{2u}\right)$

$$[a > 0,\quad u > 0,\quad \operatorname{Re}\mu > 0]$$

$$\text{ET II 203(21)}$$

3. $\displaystyle\int_0^u \frac{\left(u^2 - x^2\right)^{\mu-1}}{x^{2\mu}}\cos\frac{a}{x}\,dx = -\frac{\sqrt{\pi}}{2}\left(\frac{2}{a}\right)^{\mu-\frac{1}{2}}\Gamma(\mu)u^{\mu-\frac{3}{2}}\,Y_{\frac{1}{2}-\mu}\left(\frac{a}{u}\right)$

$$[a > 0,\quad u > 0,\quad 0 < \operatorname{Re}\mu < 1]$$

$$\text{ET II 190(36)}$$

4. $\displaystyle\int_u^\infty \frac{\left(x - u\right)^{\mu-1}}{x^{2\mu}}\cos\frac{a}{x}\,dx = \sqrt{\frac{\pi}{u}}\,a^{\frac{1}{2}-\mu}\,\Gamma(\mu)\cos\frac{a}{2u}\,J_{\mu-\frac{1}{2}}\left(\frac{a}{2u}\right)$

$$[a > 0,\quad u > 0,\quad \operatorname{Re}\mu > 0]$$

$$\text{ET II 204(26)}$$

**3.866**

1. $\displaystyle\int_0^\infty x^{\mu-1}\sin\frac{b^2}{x}\sin\left(a^2 x\right)\,dx = \frac{\pi}{4}\left(\frac{b}{a}\right)^\mu\operatorname{cosec}\frac{\mu\pi}{2}\left[J_\mu(2ab) - J_{-\mu}(2ab) + I_{-\mu}(2ab) - I_\mu(2ab)\right]$

$$[a > 0,\quad b > 0,\quad |\operatorname{Re}\mu| < 1]$$

$$\text{ET I 322(42)}$$

2. $\displaystyle\int_0^\infty x^{\mu-1}\sin\frac{b^2}{x}\cos\left(a^2 x\right)\,dx = \frac{\pi}{4}\left(\frac{b}{a}\right)^\mu\sec\frac{\mu\pi}{2}\left[J_\mu(2ab) + J_{-\mu}(2ab) + I_\mu(2ab) - I_{-\mu}(2ab)\right]$

$$[a > 0,\quad b > 0,\quad |\operatorname{Re}\mu| < 1]$$

$$\text{ET I 322(43)}$$

3. $\displaystyle\int_0^\infty x^{\mu-1}\cos\frac{b^2}{x}\cos\left(a^2 x\right)\,dx = \frac{\pi}{4}\left(\frac{b}{a}\right)^\mu\operatorname{cosec}\frac{\mu\pi}{2}\left[J_{-\mu}(2ab) - J_\mu(2ab) + I_{-\mu}(2ab) - I_\mu(2ab)\right]$

$$[a > 0,\quad b > 0,\quad |\operatorname{Re}\mu| < 1]$$

$$\text{ET I 322(44)}$$

**3.867**

1.  $$\int_0^1 \frac{\cos ax - \cos \frac{a}{x}}{1 - x^2}\, dx = \frac{1}{2}\int_0^\infty \frac{\cos ax - \cos \frac{a}{x}}{1 - x^2}\, dx = \frac{\pi}{2}\sin a$$

    $$[a > 0] \qquad\qquad \text{GW (334)(7a)}$$

2.  $$\int_0^1 \frac{\cos ax + \cos \frac{a}{x}}{1 + x^2}\, dx = \frac{1}{2}\int_0^\infty \frac{\cos ax + \cos \frac{a}{x}}{1 + x^2}\, dx = \frac{\pi}{2}e^{-a}$$

    $$[a > 0] \qquad\qquad \text{GW (334)(7b)}$$

**3.868**

1.  $$\int_0^\infty \sin\left(a^2 x + \frac{b^2}{x}\right)\frac{dx}{x} = \pi J_0(2ab) \qquad [a > 0, \quad b > 0]$$

    $$\text{GW (334)(11a), WA 200(16)}$$

2.  $$\int_0^\infty \cos\left(a^2 x + \frac{b^2}{x}\right)\frac{dx}{x} = -\pi Y_0(2ab) \qquad [a > 0, \quad b > 0] \qquad \text{GW (334)(11a)}$$

3.  $$\int_0^\infty \sin\left(a^2 x - \frac{b^2}{x}\right)\frac{dx}{x} = 0 \qquad [a > 0, \quad b > 0] \qquad \text{GW (334)(11b)}$$

4.  $$\int_0^\infty \cos\left(a^2 x - \frac{b^2}{x}\right)\frac{dx}{x} = 2K_0(2ab) \qquad [a > 0, \quad b > 0] \qquad \text{GW (334)(11b)}$$

**3.869**

1.  $$\int_0^\infty \sin\left(ax - \frac{b}{x}\right)\frac{x\, dx}{\beta^2 + x^2} = \frac{\pi}{2}\exp\left(-a\beta - \frac{b}{\beta}\right) \qquad [a > 0, \quad b > 0, \quad \operatorname{Re}\beta > 0]$$

    $$\text{ET II 220(42)}$$

2.  $$\int_0^\infty \cos\left(ax - \frac{b}{x}\right)\frac{dx}{\beta^2 + x^2} = \frac{\pi}{2\beta}\exp\left(-a\beta - \frac{b}{\beta}\right) \qquad [a > 0, \quad b > 0, \quad \operatorname{Re}\beta > 0]$$

    $$\text{ET II 222(58)}$$

**3.871**

1.  $$\int_0^\infty x^{\mu-1}\sin\left[a\left(x + \frac{b^2}{x}\right)\right] dx = \pi b^\mu\left[J_\mu(2ab)\cos\frac{\mu\pi}{2} - Y_\mu(2ab)\sin\frac{\mu\pi}{2}\right]$$

    $$[a > 0, \quad b > 0, \quad \operatorname{Re}\mu < 1]$$
    $$\text{ET I 319(17)}$$

2.  $$\int_0^\infty x^{\mu-1}\cos\left[a\left(x + \frac{b^2}{x}\right)\right] dx = -\pi b^\mu\left[J_\mu(2ab)\sin\frac{\mu\pi}{2} + Y_\mu(2ab)\cos\frac{\mu\pi}{2}\right]$$

    $$[a > 0, \quad b > 0, \quad |\operatorname{Re}\mu| < 1]$$
    $$\text{ET I 321(35)}$$

3.  $$\int_0^\infty x^{\mu-1}\sin\left[a\left(x - \frac{b^2}{x}\right)\right] dx = 2b^\mu K_\mu(2ab)\sin\frac{\mu\pi}{2} \qquad [a > 0, \quad b > 0, \quad |\operatorname{Re}\mu| < 1]$$

    $$\text{ET I 319(16)}$$

4.  $$\int_0^\infty x^{\mu-1}\cos\left[a\left(x - \frac{b^2}{x}\right)\right] dx = 2b^\mu K_\mu(2ab)\cos\frac{\mu\pi}{2}$$

    $$[a > 0, \quad b > 0, \quad |\operatorname{Re}\mu| < 1]$$
    $$\text{ET I 321(36)}$$

**3.872**

1. $\displaystyle \int_0^1 \sin\left[a\left(x+\frac{1}{x}\right)\right] \sin\left[a\left(x-\frac{1}{x}\right)\right] \frac{dx}{1-x^2}$
$$= \frac{1}{2}\int_0^\infty \sin\left[a\left(x+\frac{1}{x}\right)\right] \sin\left[a\left(x-\frac{1}{x}\right)\right] \frac{dx}{1-x^2} = -\frac{\pi}{4}\sin 2a$$
$$[a \geq 0] \qquad \text{BI (149)(15), GW (334)(8a)}$$

2. $\displaystyle \int_0^1 \cos\left[a\left(x+\frac{1}{x}\right)\right] \cos\left[a\left(x-\frac{1}{x}\right)\right] \frac{dx}{1+x^2}$
$$= \frac{1}{2}\int_0^\infty \cos\left[a\left(x+\frac{1}{x}\right)\right] \cos\left[a\left(x-\frac{1}{x}\right)\right] \frac{dx}{1+x^2} = \frac{\pi}{4}e^{-2a}$$
$$[a \geq 0] \qquad \text{GW (334)(8b)}$$

**3.873**

1. $\displaystyle \int_0^\infty \sin\frac{a^2}{x^2}\cos b^2 x^2 \frac{dx}{x^2} = \frac{\sqrt{\pi}}{4\sqrt{2a}}\left[\sin(2ab)+\cos(2ab)+e^{-2ab}\right]$
$$[a > 0, \quad b > 0] \qquad \text{ET I 24(15)}$$

2. $\displaystyle \int_0^\infty \cos\frac{a^2}{x^2}\cos b^2 x^2 \frac{dx}{x^2} = \frac{\sqrt{\pi}}{4\sqrt{2a}}\left[\cos(2ab)-\sin(2ab)+e^{-2ab}\right]$
$$[a > 0, \quad b > 0] \qquad \text{ET I 24(16)}$$

**3.874**

1. $\displaystyle \int_0^\infty \sin\left(a^2 x^2 + \frac{b^2}{x^2}\right)\frac{dx}{x^2} = \frac{\sqrt{\pi}}{2b}\sin\left(2ab+\frac{\pi}{4}\right) \qquad [a > 0, \quad b > 0]$
$$\text{BI (179)(6)a, GW(334)(10a)}$$

2. $\displaystyle \int_0^\infty \cos\left(a^2 x^2 + \frac{b^2}{x^2}\right)\frac{dx}{x^2} = \frac{\sqrt{\pi}}{2b}\cos\left(2ab+\frac{\pi}{4}\right) \qquad [a > 0, \quad b > 0]$
$$\text{GI (179)(8)a, GW(334)(10a)}$$

3. $\displaystyle \int_0^\infty \sin\left(a^2 x^2 - \frac{b^2}{x^2}\right)\frac{dx}{x^2} = -\frac{\sqrt{\pi}}{2\sqrt{2b}}e^{-2ab} \qquad [a \geq 0, \quad b > 0] \qquad \text{GW (335)(10b)}$

4. $\displaystyle \int_0^\infty \cos\left(a^2 x^2 - \frac{b^2}{x^2}\right)\frac{dx}{x^2} = \frac{\sqrt{\pi}}{2\sqrt{2b}}e^{-2ab} \qquad [a \geq 0, \quad b > 0] \qquad \text{GW (334)(10b)}$

5. $\displaystyle \int_0^\infty \sin\left(ax - \frac{b}{x}\right)^2 \frac{dx}{x^2} = \frac{\sqrt{2\pi}}{4b} \qquad [a > 0, \quad b > 0] \qquad \text{BI (179)(13)a}$

6. $\displaystyle \int_0^\infty \cos\left(ax - \frac{b}{x}\right)^2 \frac{dx}{x^2} = \frac{\sqrt{2\pi}}{4b} \qquad [a > 0, \quad b > 0] \qquad \text{BI (179)(14)a}$

**3.875**

1. $\displaystyle \int_u^\infty \frac{x\sin\left(p\sqrt{x^2-u^2}\right)}{x^2+a^2}\cos bx\, dx = \frac{\pi}{2}\exp\left(-p\sqrt{a^2+u^2}\right)\cosh ab$
$$[0 < b < p] \qquad \text{ET I 27(39)}$$

2.  $\displaystyle\int_u^\infty \frac{x\sin\left(p\sqrt{x^2-u^2}\right)}{a^2+x^2-u^2}\cos bx\,\mathrm{d}x = \frac{\pi}{2}e^{-ap}\cos\left(b\sqrt{u^2-a^2}\right)$

$$[0<b<p, \quad a>0] \qquad\qquad \text{ET I 27(38)}$$

3.[12]  $\displaystyle\int_0^\infty \frac{\sin\left(p\sqrt{a^2+x^2}\right)}{\left(a^2+x^2\right)^2}\cos bx\,\mathrm{d}x = \frac{\pi p}{2a}e^{-ab}$ $\qquad [0<p<b, \quad b>a>0] \qquad \text{ET I 26(29)}$

4.*  $\displaystyle\int_0^\infty \frac{\sin\left(p\sqrt{x^2+a^2}\right)}{\left(x^2+a^2\right)^2}\cos(bx)\,\mathrm{d}x = \frac{pb}{a}K_1(ab) - \frac{p^3}{6}K_0(ab)$

$$[a>0, \quad b>p>0]$$

5.*  $\displaystyle\int_0^\infty \frac{\cos\left(p\sqrt{x^2+a^2}\right)}{\left(x^2+a^2\right)^{3/2}}\cos(bx)\,\mathrm{d}x = \frac{b}{a}K_1(ab) - \frac{p^2}{2}K_0(ab)$

$$[a>0, \quad b>p>0]$$

6.*  $\displaystyle\int_0^\infty \frac{x\cos\left(p\sqrt{x^2+a^2}\right)}{\left(x^2+a^2\right)^{3/2}}\sin(bx)\,\mathrm{d}x = bK_0(ab) - \frac{ap^2}{2}K_1(ab)$

$$[a>0, \quad b>p>0]$$

**3.876**

1.  $\displaystyle\int_0^\infty \frac{\sin\left(p\sqrt{x^2+a^2}\right)}{\sqrt{x^2+a^2}}\cos bx\,\mathrm{d}x = \frac{\pi}{2}\,J_0\left(a\sqrt{p^2-b^2}\right) \qquad [0<b<p]$

$$= 0 \qquad\qquad\qquad [b>p>0]$$

$$[a>0] \qquad\qquad\qquad \text{ET I 26(30)}$$

2.  $\displaystyle\int_0^\infty \frac{\cos\left(p\sqrt{x^2+a^2}\right)}{\sqrt{x^2+a^2}}\cos bx\,\mathrm{d}x = -\frac{\pi}{2}\,Y_0\left(a\sqrt{p^2-b^2}\right) \qquad [0<b<p]$

$$= K_0\left(a\sqrt{b^2-p^2}\right) \qquad [b>p>0]$$

$$[a>0] \qquad\qquad\qquad \text{ET I 26(34)}$$

3.  $\displaystyle\int_0^\infty \frac{\cos\left(p\sqrt{x^2+a^2}\right)}{x^2+c^2}\cos bx\,\mathrm{d}x = \frac{\pi}{2c}e^{-bc}\cos\left(p\sqrt{a^2-c^2}\right)$

$$[c>0, \quad b>p] \qquad\qquad \text{ET I 26(33)}$$

4.  $\displaystyle\int_0^\infty \frac{\sin\left(p\sqrt{x^2+a^2}\right)}{\left(x^2+c^2\right)\sqrt{x^2+a^2}}\cos bx\,\mathrm{d}x = \frac{\pi}{2c}\frac{e^{-bc}\sin\left(p\sqrt{a^2-c^2}\right)}{\sqrt{a^2-c^2}} \qquad [c\neq a]$

$$= \frac{\pi}{2}e^{-ba}\frac{p}{a} \qquad\qquad [c=a]$$

$$[b>p, \quad c>0] \qquad\qquad \text{ET I 26(31)a}$$

5.[6]  $\displaystyle\int_0^\infty \frac{\cos\left(p\sqrt{x^2+a^2}\right)}{x^2+a^2}\cos bx\,\mathrm{d}x = \frac{\pi}{2a}e^{-ab}$ $\qquad [b>p>0, \quad a>0] \qquad \text{ET I 27(35)a}$

6.[6]  $\displaystyle\int_0^\infty \frac{x\cos\left(p\sqrt{x^2+a^2}\right)}{x^2+a^2}\sin bx\,\mathrm{d}x = \frac{\pi}{2}e^{-ab}$ $\qquad [a>0, \quad b>p>0] \qquad \text{ET I 85(29)a}$

7.  $\displaystyle\int_0^u \frac{\cos\left(p\sqrt{u^2-x^2}\right)}{\sqrt{u^2-x^2}}\cos bx\,\mathrm{d}x = \frac{\pi}{2}\,J_0\left(u\sqrt{b^2+p^2}\right)$ $\qquad\qquad\qquad \text{ET I 28(42)}$

8. $\displaystyle\int_u^\infty \frac{\cos\left(p\sqrt{x^2-u^2}\right)}{\sqrt{x^2-u^2}}\cos bx\,\mathrm{d}x = K_0\left(u\sqrt{p^2-b^2}\right) \qquad [0<b<|p|]$

$\qquad\qquad\qquad\qquad\qquad\qquad\qquad = -\dfrac{\pi}{2}\,Y_0\left(u\sqrt{b^2-p^2}\right) \qquad [b>|p|]$

$\qquad\qquad\qquad\qquad\qquad\qquad\qquad\qquad\qquad\qquad\qquad\qquad$ ET I 28(43)

**3.877**

1. $\displaystyle\int_0^u \frac{\sin\left(p\sqrt{u^2-x^2}\right)}{\sqrt[4]{(u^2-x^2)^3}}\cos bx\,\mathrm{d}x = \sqrt{\frac{\pi^3 p}{8}}\,J_{\frac14}\left[\frac{u}{2}\left(\sqrt{b^2+p^2}-b\right)\right]J_{\frac14}\left[\frac{u}{2}\left(\sqrt{b^2+p^2}+b\right)\right]$

$\qquad\qquad\qquad\qquad\qquad\qquad\qquad\qquad\qquad [b>0,\quad p>0]\qquad\qquad$ ET I 27(40)

2. $\displaystyle\int_u^\infty \frac{\sin\left(p\sqrt{x^2-u^2}\right)}{\sqrt[4]{(x^2-u^2)^3}}\cos bx\,\mathrm{d}x = -\sqrt{\frac{\pi^3 p}{8}}\,J_{\frac14}\left[\frac{u}{2}\left(b-\sqrt{b^2-p^2}\right)\right]Y_{\frac14}\left[\frac{u}{2}\left(b+\sqrt{b^2-p^2}\right)\right]$

$\qquad\qquad\qquad\qquad\qquad\qquad\qquad\qquad\qquad [b>p>0]\qquad\qquad$ ET I 27(41)

3. $\displaystyle\int_0^u \frac{\cos\left(p\sqrt{u^2-x^2}\right)}{\sqrt[4]{(u^2-x^2)^3}}\cos bx\,\mathrm{d}x = \sqrt{\frac{\pi^3 p}{8}}\,J_{-\frac14}\left[\frac{u}{2}\left(\sqrt{p^2+b^2}-b\right)\right]J_{-\frac14}\left[\frac{u}{2}\left(\sqrt{p^2+b^2}+b\right)\right]$

$\qquad\qquad\qquad\qquad\qquad\qquad\qquad\qquad\qquad [u>0,\quad p>0]\qquad\qquad$ ET I 28(44)

4. $\displaystyle\int_u^\infty \frac{\cos\left(p\sqrt{x^2-u^2}\right)}{\sqrt[4]{(x^2-u^2)^3}}\cos bx\,\mathrm{d}x = -\sqrt{\frac{\pi^3 p}{8}}\,J_{-\frac14}\left[\frac{u}{2}\left(b-\sqrt{b^2-p^2}\right)\right]Y_{\frac14}\left[\frac{u}{2}\left(b+\sqrt{b^2-p^2}\right)\right]$

$\qquad\qquad\qquad\qquad\qquad\qquad\qquad\qquad\qquad [b>p>0]\qquad\qquad$ ET I 28(45)

**3.878**

1. $\displaystyle\int_0^\infty \frac{\sin\left(p\sqrt{x^4+a^4}\right)}{\sqrt{x^4+a^4}}\cos bx^2\,\mathrm{d}x = \frac12\sqrt{\left(\frac{\pi}{2}\right)^3}\,b\,J_{-\frac14}\left[\frac{a^2}{2}\left(p-\sqrt{p^2-b^2}\right)\right]J_{\frac14}\left[\frac{a^2}{2}\left(p+\sqrt{p^2-b^2}\right)\right]$

$\qquad\qquad\qquad\qquad\qquad\qquad\qquad [p>b>0]\qquad\qquad$ ET I 26(32)

2. $\displaystyle\int_0^\infty \frac{\cos\left(p\sqrt{x^4+a^4}\right)}{\sqrt{x^4+a^4}}\cos bx^2\,\mathrm{d}x$

$\qquad\qquad\qquad = -\dfrac12\sqrt{\left(\dfrac{\pi}{2}\right)^3}\,b\,J_{-\frac14}\left[\dfrac{a^2}{2}\left(p-\sqrt{p^2-b^2}\right)\right]Y_{\frac14}\left[\dfrac{a^2}{2}\left(p+\sqrt{p^2-b^2}\right)\right]$

$\qquad\qquad\qquad\qquad\qquad\qquad [a>0,\quad p>b>0]\qquad\qquad$ ET I 27(36)

3. $\displaystyle\int_0^u \frac{\cos\left(p\sqrt{u^4-x^4}\right)}{\sqrt{u^4-x^4}}\cos bx^2\,\mathrm{d}x = \frac12\sqrt{\left(\frac{\pi}{2}\right)^3}\,b\,J_{-\frac14}\left[\frac{u^2}{2}\left(\sqrt{p^2+b^2}-p\right)\right]J_{-\frac14}\left[\frac{u^2}{2}\left(\sqrt{p^2+b^2}+p\right)\right]$

$\qquad\qquad\qquad\qquad\qquad\qquad [p>0,\quad b>0]\qquad\qquad$ ET I 28(46)

**3.879** $\displaystyle\int_0^\infty \sin ax^p\,\frac{\mathrm{d}x}{x} = \frac{\pi}{2p} \qquad\qquad [a>0,\quad p>0]\qquad\qquad$ GW (334)(6)

**3.881**

1. $\displaystyle\int_0^{\pi/2} x\sin\left(a\tan x\right)\mathrm{d}x = \frac{\pi}{4}e^{-a}\left[C+\ln 2a - e^{2a}\,\mathrm{Ei}(-2a)\right]$

$\qquad\qquad\qquad\qquad\qquad\qquad\qquad [a>0]\qquad\qquad$ BI (205)(9)

2. $\displaystyle\int_0^\infty \sin\left(a\tan x\right)\frac{\mathrm{d}x}{x} = \frac{\pi}{2}\left(1 - e^{-a}\right)$      $[a > 0]$      BI (151)(6)

3. $\displaystyle\int_0^\infty \sin\left(a\tan x\right)\cos x\frac{\mathrm{d}x}{x} = \frac{\pi}{2}\left(1 - e^{-a}\right)$      $[a > 0]$      BI (151)(19)

4. $\displaystyle\int_0^\infty \cos\left(a\tan x\right)\sin x\frac{\mathrm{d}x}{x} = \frac{\pi}{2}e^{-a}$      $[a > 0]$      BI (151)(20)

5. $\displaystyle\int_0^\infty \sin\left(a\tan x\right)\sin 2x\frac{\mathrm{d}x}{x} = \frac{1+a}{2}\pi e^{-a}$      $[a > 0]$      BI (152)(11)

6. $\displaystyle\int_0^\infty \cos\left(a\tan x\right)\sin^3 x\frac{\mathrm{d}x}{x} = \frac{1-a}{4}\pi e^{-a}$      $[a > 0]$      BI (151)(23)

7. $\displaystyle\int_0^\infty \sin\left(a\tan x\right)\tan\frac{x}{2}\cos^2 x\frac{\mathrm{d}x}{x} = \frac{1+a}{4}\pi e^{-a}$      $[a > 0]$      BI (152)(13)

8. $\displaystyle\int_0^{\pi/2} \cos\left(a\tan x\right)\frac{x\,\mathrm{d}x}{\sin 2x} = -\frac{\pi}{4}\operatorname{Ei}(-a)$      $[a > 0]$      BI (206)(15)

9. $\displaystyle\int_0^{\pi/2} \sin\left(a\cot x\right)\frac{x\,\mathrm{d}x}{\sin^2 x} = \frac{1 - e^{-a}}{2a}\pi$      $[a > 0]$      LI (206)(14)

10. $\displaystyle\int_0^{\pi/2} x\cos\left(a\tan x\right)\tan x\,\mathrm{d}x = -\frac{\pi}{4}e^{-a}\left[\boldsymbol{C} + \ln 2a + e^{2a}\operatorname{Ei}(-2a)\right]$

     $[a > 0]$      BI (205)(10)

11. $\displaystyle\int_0^\infty \cos\left(a\tan x\right)\tan x\frac{\mathrm{d}x}{x} = \frac{\pi}{2}e^{-a}$      $[a > 0]$      BI (151)(21)

12. $\displaystyle\int_0^\infty \cos\left(a\tan x\right)\sin^2 x\tan x\frac{\mathrm{d}x}{x} = \frac{1-a}{16}\pi e^{-a}$      $[a > 0]$      BI (152)(15)

13. $\displaystyle\int_0^\infty \sin\left(a\tan x\right)\tan^2 x\frac{\mathrm{d}x}{x} = \frac{\pi}{2}e^{-a}$      $[a > 0]$      BI (152)(9)

14. $\displaystyle\int_0^\infty \cos\left(a\tan 2x\right)\tan x\frac{\mathrm{d}x}{x} = \frac{\pi}{2}e^{-a}$      $[a > 0]$      BI (151)(22)

15. $\displaystyle\int_0^\infty \sin\left(a\tan 2x\right)\cos^2 2x\tan x\frac{\mathrm{d}x}{x} = \frac{1+a}{4}\pi e^{-a}$      $[a > 0]$      BI (152)(13)

16. $\displaystyle\int_0^\infty \sin\left(a\tan 2x\right)\tan x\tan 2x\frac{\mathrm{d}x}{x} = \frac{\pi}{2}e^{-a}$      $[a > 0]$      BI (152)(10)

17. $\displaystyle\int_0^\infty \sin\left(a\tan 2x\right)\tan x\cot 2x\frac{\mathrm{d}x}{x} = \frac{\pi}{2}\left(1 - e^{-a}\right)$      $[a > 0]$      BI (180)(6)

**3.882**

1. $\displaystyle\int_0^\infty \sin\left(a\tan^2 x\right)\frac{x\,\mathrm{d}x}{b^2 + x^2} = \frac{\pi}{2}\left[\exp\left(-a\tanh b\right) - e^{-a}\right]$

     $[a > 0, \quad b > 0]$      BI (160)(22)

2. $\displaystyle\int_0^\infty \cos\left(a\tan^2 x\right)\cos x\frac{\mathrm{d}x}{b^2 + x^2} = \frac{\pi}{2b}\left[\cosh b\exp\left(-a\tanh b\right) - e^{-a}\sinh b\right]$

     $[a > 0, \quad b > 0]$      BI (163)(3)

3.$$\int_0^\infty \cos\left(a \tan^2 x\right) \operatorname{cosec} 2x \frac{x\,dx}{b^2 + x^2} = \frac{\pi}{2\sinh 2b} \exp\left(-a\tanh b\right)$$

$$[a > 0, \quad b > 0] \qquad\qquad \text{BI (191)(10)}$$

4.$$\int_0^\infty \cos\left(a \tan^2 x\right) \tan x \frac{x\,dx}{b^2 + x^2} = \frac{\pi}{2\cosh b} \left[e^{-a}\cosh b - \exp\left(-a\tanh b\right)\sinh b\right]$$

$$[a > 0, \quad b > 0] \qquad\qquad \text{BI (163)(4)}$$

5.[11]$$\int_0^\infty \cos\left(a \tan^2 x\right) \cot x \frac{x\,dx}{b^2 + x^2} = \frac{\pi}{2} \left[\coth b \exp\left(-a\tanh b\right) - e^{-a}\right]$$

$$[a > 0, \quad b > 0] \qquad\qquad \text{BI (163)(5)}$$

6.$$\int_0^\infty \cos\left(a \tan^2 x\right) \cot 2x \frac{x\,dx}{b^2 + x^2} = \frac{\pi}{2} \left[\coth 2b \exp\left(-a\tanh b\right) - e^{-a}\right]$$

$$[a > 0, \quad b > 0] \qquad\qquad \text{BI (191)(11)}$$

**3.883**

1.$$\int_0^1 \cos\left(a \ln x\right) \frac{dx}{(1+x)^2} = \frac{a\pi}{2\sinh a\pi} \qquad\qquad\qquad \text{BI (404)(4)}$$

2.$$\int_0^1 x^{\mu-1} \sin\left(\beta \ln x\right)\,dx = -\frac{\beta}{\beta^2 + \mu^2} \qquad [\operatorname{Re}\mu > |\operatorname{Im}\beta|] \qquad \text{ET I 319(19)}$$

3.$$\int_0^1 x^{\mu-1} \cos\left(\beta \ln x\right)\,dx = \frac{\mu}{\beta^2 + \mu^2} \qquad [\operatorname{Re}\mu > |\operatorname{Im}\beta|] \qquad \text{ET I 321(38)}$$

**3.884**[11] $$\int_{-\infty}^\infty \frac{\sin a\sqrt{|x|}}{x - b} \operatorname{sign} x\,dx = \pi \left[\exp\left(-a\sqrt{|-b|}\right) + \exp\left(-a\sqrt{|b|}\right)\right]$$

$$[a > 0, \quad \operatorname{Im} b \neq 0] \qquad\qquad \text{ET II 253(46)}$$

## 3.89–3.91 Trigonometric functions and exponentials

**3.891**

1.$$\int_0^{2\pi} e^{imx} \sin nx\,dx = 0 \qquad\qquad\qquad [m \neq n \text{ or } m = n = 0]$$

$$= \pi i \qquad\qquad\qquad [m = n \neq 0]$$

2.$$\int_0^{2\pi} e^{imx} \cos nx\,dx = 0 \qquad\qquad\qquad [m \neq n]$$

$$= \pi \qquad\qquad\qquad [m = n \neq 0]$$

$$= 2\pi \qquad\qquad\qquad [m = n = 0]$$

**3.892**

1.[11] $$\int_0^\pi e^{i\beta x} \sin^{\nu-1} x\,dx = \frac{\pi e^{i\beta\frac{\pi}{2}}}{2^{\nu-1}\nu\,\mathrm{B}\left(\dfrac{\nu + \beta + 1}{2}, \dfrac{\nu - \beta + 1}{2}\right)}$$

$$[\operatorname{Re}\nu > -1] \qquad\qquad \text{NH 158, EH I 12(29)}$$

2. $\displaystyle\int_{-\frac{\pi}{2}}^{\frac{\pi}{2}} e^{i\beta x} \cos^{\nu-1} x \, dx = \frac{\pi}{2^{\nu-1}\nu \, B\left(\dfrac{\nu+\beta+1}{2}, \dfrac{\nu-\beta+1}{2}\right)}$

$\qquad\qquad\qquad\qquad\qquad\qquad [\operatorname{Re}\nu > -1]$ GW (335)(19)

3.6 $\displaystyle\int_0^{\pi/2} e^{i2\beta x} \sin^{2\mu} x \cos^{2\nu} x \, dx = \frac{1}{2^{2\mu+2\nu+1}}\left[ \exp\left[i\pi\left(\beta-\nu-\tfrac{1}{2}\right)\right] B\left(\beta-\mu-\nu, 2\nu+1\right) \right.$

$\qquad\qquad\qquad\qquad \times F\left(-2\mu, \beta-\mu-\nu; 1+\beta-\mu+\nu; -1\right) + \exp\left[i\pi\left(\mu+\tfrac{1}{2}\right)\right]$

$\qquad\qquad\qquad\qquad \left. \times \, B(\beta-\mu-\nu, 2\mu+1) \, F\left(-2\nu, \beta-\mu-\nu; 1+\beta+\mu-\nu; -1\right)\right]$

$\qquad\qquad\qquad\qquad\qquad [\operatorname{Re}\mu > -\tfrac{1}{2}, \quad \operatorname{Re}\nu > -\tfrac{1}{2}]$ EH I 80(6)

4. $\displaystyle\int_0^\pi e^{i2\beta x} \sin^{2\mu} x \cos^{2\nu} x \, dx = \frac{\pi \exp\left[i\pi(\beta-\nu)\right] F(-2\nu, \beta-\mu-\nu; 1+\beta+\mu-\nu; -1)}{4^{\mu+\nu}(2\mu+1) \, B(1-\beta+\mu+\nu, 1+\beta+\mu-\nu)}$

$\qquad\qquad\qquad\qquad\qquad\qquad\qquad\qquad\qquad\qquad\qquad\qquad$ EH I 80(8)

5. $\displaystyle\int_0^{\pi/2} e^{i(\mu+\nu)x} \sin^{\mu-1} x \cos^{\nu-1} x \, dx = e^{i\mu\frac{\pi}{2}} B(\mu, \nu)$

$\qquad\qquad\qquad = \frac{1}{2^{\mu+\nu-1}} e^{i\mu\frac{\pi}{2}}\left[\frac{1}{\mu} F\left(1-\nu, 1; \mu+1; -1\right) + \frac{1}{\nu} F\left(1-\mu, 1; \nu+1; -1\right)\right]$

$\qquad\qquad\qquad\qquad\qquad\qquad [\operatorname{Re}\mu > 0, \quad \operatorname{Re}\nu > 0]$ EH I 80(7)

**3.893**

1.8 $\displaystyle\int_0^\infty e^{-px} \sin(qx+\lambda) \, dx = \frac{1}{p^2+q^2}\left(q\cos\lambda + p\sin\lambda\right)$ $\quad[\operatorname{Re}p > 0]$ BI (261)(3)

2.8 $\displaystyle\int_0^\infty e^{-px} \cos(qx+\lambda) \, dx = \frac{1}{p^2+q^2}\left(p\cos\lambda - q\sin\lambda\right)$ $\quad[\operatorname{Re}p > 0]$ BI (261)(4)

3. $\displaystyle\int_0^\infty e^{-x\cos t} \cos\left(t - x\sin t\right) \, dx = 1$ BI (261)(7)

4.8 $\displaystyle\int_0^\infty \frac{e^{-\beta x} \sin ax}{\sin bx} \, dx = \operatorname{Re}\left\{\frac{1}{2bi}\left[\psi\left(\frac{a+b}{2b} - i\frac{\beta}{2b}\right) - \psi\left(\frac{b-a}{2b} - i\frac{\beta}{2b}\right)\right]\right\}$

$\qquad\qquad\qquad\qquad\qquad [\operatorname{Re}\beta > 0, \quad b \neq 0]$ GW (335)(15)

5.8 $\displaystyle\int_0^\infty \frac{e^{-2px} \sin[(2n+1)x]}{\sin x} \, dx = \frac{1}{2p} + \sum_{k=1}^n \frac{p}{p^2+k^2}$ $\quad[\operatorname{Re}p > 0]$ BI (267)(15)

6.8 $\displaystyle\int_0^\infty \frac{e^{-px} \sin 2nx}{\sin x} \, dx = 2p\sum_{k=0}^{n-1} \frac{1}{p^2+(2k+1)^2}$ $\quad[\operatorname{Re}p > 0]$ GW (335)(15c)

7. $\displaystyle\int_0^\infty e^{-px} \cos[(2n+1)x] \tan x \, dx = \frac{2n+1}{p^2+(2n+1)^2} + (-1)^n 2\sum_{k=0}^{n-1} \frac{(-1)^k(2k+1)}{p^2+(2k+1)^2}$

$\qquad\qquad\qquad\qquad\qquad\qquad [p > 0]$ LI (267)(16)

**3.894** $\displaystyle\int_{-\pi}^\pi \left[\beta + \sqrt{\beta^2-1}\cos x\right]^\nu e^{inx} \, dx = \frac{2\pi\,\Gamma(\nu+1)\,P_\nu^m(\beta)}{\Gamma(\nu+m+1)}$

$\qquad\qquad\qquad\qquad\qquad [\operatorname{Re}\beta > 0]$ ET I 157(15)

**3.895**

1.  $$\int_0^\infty e^{-\beta x} \sin^{2m} x \, dx = \frac{(2m)!}{\beta(\beta^2 + 2^2)(\beta^2 + 4^2)\cdots[\beta^2 + (2m)^2]}$$

    $$[\operatorname{Re}\beta > 0] \qquad\qquad \text{FI II 615, WA 620a}$$

2.[10] $$\int_0^\pi e^{-px} \sin^{2m} x \, dx = \frac{(2m)!\,(1 - e^{-p\pi})}{p\,(p^2 + 2^2)(p^2 + 4^2)\cdots[p^2 + (2m)^2]}$$

    $$\text{GW (335)(4a)}$$

3.[12] $$\int_0^{\pi/2} e^{-px} \sin^{2m} x \, dx$$

    $$= \frac{(2m)!}{p\,(p^2 + 2^2)(p^2 + 4^2)\cdots[p^2 + (2m)^2]}$$
    $$\times \left\{ 1 - e^{-\frac{p\pi}{2}} \left[ 1 + \frac{p^2}{2!} + \frac{p^2\,(p^2 + 2^2)}{4!} + \cdots + \frac{p^2\,(p^2 + 2^2)\cdots[p^2 + (2m-2)^2]}{(2m)!} \right] \right\}$$

    $$[p \ne 0] \qquad\qquad \text{BI (270)(4)}$$

4.  $$\int_0^\infty e^{-\beta x} \sin^{2m+1} x \, dx = \frac{(2m+1)!}{(\beta^2 + 1^2)(\beta^2 + 3^2)\cdots[\beta^2 + (2m+1)^2]}$$

    $$[\operatorname{Re}\beta > 0] \qquad\qquad \text{FI II 615, WA 620a}$$

5.[10] $$\int_0^\pi e^{-px} \sin^{2m+1} x \, dx = \frac{(2m+1)!\,(1 + e^{-p\pi})}{(p^2 + 1^2)(p^2 + 3^2)\cdots[p^2 + (2m+1)^2]}$$

    $$\text{GW (335)(4b)}$$

6.[8] $$\int_0^{\pi/2} e^{-px} \sin^{2m+1} x \, dx$$

    $$= \frac{(2m+1)!}{(p^2 + 1^2)(p^2 + 3^2)\cdots[p^2 + (2m+1)^2]}$$
    $$\times \left\{ 1 - pe^{\frac{-p\pi}{2}} \left[ 1 + \frac{p^2 + 1^2}{3!} + \cdots + \frac{(p^2 + 1^2)(p^2 + 3^2)\cdots[p^2 + (2m-1)^2]}{(2m+1)!} \right] \right\}$$

    $$\text{BI (270)(5)}$$

7.  $$\int_0^\infty e^{-px} \cos^{2m} x \, dx = \frac{(2m)!}{p\,(p^2 + 2^2)\cdots[p^2 + (2m)^2]}$$
    $$\times \left\{ 1 + \frac{p^2}{2!} + \frac{p^2\,(p^2 + 2^2)}{4!} + \cdots + \frac{p^2\,(p^2 + 2^2)\cdots[p^2 + (2m-2)^2]}{(2m)!} \right\}$$

    $$[p > 0] \qquad\qquad \text{BI (262)(3)}$$

8.[12] $$\int_0^{\pi/2} e^{-px} \cos^{2m} x \, dx$$

    $$= \frac{(2m)!}{p\,(p^2 + 2^2)\cdots[p^2 + (2m)^2]}$$
    $$\times \left\{ -e^{-p\frac{\pi}{2}} + 1 + \frac{p^2}{2!} + \frac{p^2\,(p^2 + 2^2)}{4!} + \cdots + \frac{p^2\,(p^2 + 2^2)\cdots[p^2 + (2m-2)^2]}{(2m)!} \right\}$$

    $$[p \ne 0] \qquad\qquad \text{BI (270)(6)}$$

$9.^7$

$$\int_0^\infty e^{-px} \cos^{2m+1} x \, dx$$

$$= \frac{(2m+1)! \, p}{(p^2+1^2)(p^2+3^2)\cdots[p^2+(2m+1)^2]}$$

$$\times \left\{ 1 + \frac{p^2+1^2}{3!} + \frac{(p^2+1^2)(p^2+3^2)}{5!} + \cdots + \frac{(p^2+1^2)(p^2+3^2)\cdots[p^2+(2m-1)^2]}{(2m+1)!} \right\}$$

$$[p>0] \hspace{3cm} \text{BI (262)(4)}$$

$10.^{11}$

$$\int_0^{\pi/2} e^{-px} \cos^{2m+1} x \, dx$$

$$= \frac{(2m+1)!}{(p^2+1^2)(p^2+3^2)\cdots[p^2+(2m+1)^2]}$$

$$\times \left\{ e^{-p\frac{\pi}{2}} + p \left[ 1 + \frac{p^2+1^2}{3!} + \cdots + \frac{(p^2+1)(p^2+3^2)\cdots[p^2+(2m-1)^2]}{(2m+1)!} \right] \right\}$$

$$\text{BI (270)(7)}$$

$11.^8$

$$\int_0^\infty e^{-\beta x} \sin^n ax \begin{Bmatrix} \sin bx \\ \cos bx \end{Bmatrix} dx = \frac{2^{-n-2}}{a(n+1)} e^{\frac{1}{4}(1\mp 1+2n)\pi i}$$

$$\times \left\{ \begin{pmatrix} \frac{b+na+i\beta}{2a} \\ n+1 \end{pmatrix}^{-1} \pm (-1)^n \begin{pmatrix} \frac{b+na-i\beta}{2a} \\ n+1 \end{pmatrix}^{-1} \right\}$$

$$[a>0, \quad b>0, \quad \operatorname{Re}\beta>0]$$

12. $\displaystyle\int_0^\infty e^{-ax} \cos^2 mx \, dx = \frac{a^2+2m^2}{a(a^2+4m^2)}$           DW61 (861.06)

13. $\displaystyle\int_0^\infty e^{-ax} \cos mx \cos nx \, dx = \frac{a(a^2+m^2+n^2)}{(a^2+(m-n)^2)(a^2+(m+n)^2)}$    DW61 (861.15)

14. $\displaystyle\int_0^\infty e^{-ax} \sin mx \cos nx \, dx = \frac{m(a^2+m^2-n^2)}{(a^2+(m-n)^2)(a^2+(m+n)^2)}$    DW61 (861.14)

$15.^{12}$ $\displaystyle\int_0^\infty e^{-ax} \sin^2 mx \, dx = \frac{2m^2}{a(a^2+4m^2)}$     $[a>0]$      DW61 (861.10)

16. $\displaystyle\int_0^\infty e^{-ax} \sin mx \sin nx \, dx = \frac{2amn}{[a^2+(m-n)^2][a^2+(m+n)^2]}$    DW61 (861.13)

**3.896**

1. $\displaystyle\int_{-\infty}^\infty e^{-q^2 x^2} \sin[p(x+\lambda)] \, dx = \frac{\sqrt{\pi}}{q} e^{-\frac{p^2}{4q^2}} \sin p\lambda$      BI (269)(2)

2. $\displaystyle\int_{-\infty}^\infty e^{-q^2 x^2} \cos[p(x+\lambda)] \, dx = \frac{\sqrt{\pi}}{q} e^{-\frac{p^2}{4q^2}} \cos p\lambda$      BI (269)(3)

3. $\displaystyle\int_0^\infty e^{-ax^2} \sin bx \, dx = \frac{b}{2a} \exp\left(-\frac{b^2}{4a}\right) {}_1F_1\left(\frac{1}{2}; \frac{3}{2}; \frac{b^2}{4a}\right)$

$$= \frac{b}{2a} {}_1F_1\left(1; \frac{3}{2}; -\frac{b^2}{4a}\right) \hspace{3cm} \text{ET I 73(18)}$$

$$= \frac{b}{2a} \sum_{k=1}^\infty \frac{1}{(2k-1)!!} \left(-\frac{b^2}{2a}\right)^{k-1} \hspace{1cm} [a>0] \hspace{2cm} \text{FI II 720}$$

4.    $\int_0^\infty e^{-\beta x^2} \cos bx \, dx = \dfrac{1}{2}\sqrt{\dfrac{\pi}{\beta}} \exp\left(-\dfrac{b^2}{4\beta}\right)$        $[\operatorname{Re}\beta > 0]$        BI (263)(2)

**3.897**

1.[12]    $\int_0^\infty e^{-\beta x^2 - \gamma x} \sin bx \, dx = -\dfrac{i}{4}\sqrt{\dfrac{\pi}{\beta}} \left\{ \exp \dfrac{(\gamma - ib)^2}{4\beta} \left[ 1 - \Phi\left(\dfrac{\gamma - ib}{2\sqrt{\beta}}\right) \right] \right.$

$\left. - \exp \dfrac{(\gamma + ib)^2}{4\beta} \left[ 1 - \Phi\left(\dfrac{\gamma + ib}{2\sqrt{\beta}}\right) \right] \right\}$

$[\operatorname{Re}\beta > 0, \quad b > 0]$        ET I 74(27)

2.    $\int_0^\infty e^{-\beta x^2 - \gamma x} \cos bx \, dx = \dfrac{1}{4}\sqrt{\dfrac{\pi}{\beta}} \left\{ \exp \dfrac{(\gamma - ib)^2}{4\beta} \left[ 1 - \Phi\left(\dfrac{\gamma - ib}{2\sqrt{\beta}}\right) \right] \right.$

$\left. + \exp \dfrac{(\gamma + ib)^2}{4\beta} \left[ 1 - \Phi\left(\dfrac{\gamma + ib}{2\sqrt{\beta}}\right) \right] \right\}$

$[\operatorname{Re}\beta > 0, \quad b > 0]$        ET I 15(16)

**3.898**

1.[12]    $\int_0^\infty e^{-\beta x^2} \sin ax \sin bx \, dx = \dfrac{1}{4}\sqrt{\dfrac{\pi}{\beta}} \left[ e^{-\frac{(a-b)^2}{4\beta}} - e^{-\frac{(a+b)^2}{4\beta}} \right]$

$[\operatorname{Re}\beta > 0, \quad a > 0, \quad b > 0]$    BI (263)(4)

2.    $\int_0^\infty e^{-\beta x^2} \cos ax \cos bx \, dx = \dfrac{1}{4}\sqrt{\dfrac{\pi}{\beta}} \left[ e^{-\frac{(a-b)^2}{4\beta}} + e^{-\frac{(a+b)^2}{4\beta}} \right]$

$[\operatorname{Re}\beta > 0]$        BI (263)(5)

3,[8]    $\int_0^\infty e^{-px^2} \sin^2 ax \, dx = \dfrac{1}{4}\sqrt{\dfrac{\pi}{p}} \left(1 - e^{-\frac{a^2}{p}}\right)$        $[\operatorname{Re} p > 0]$        BI (263)(6)

**3.899**

1.[12]    $\int_0^\infty \dfrac{e^{-p^2 x^2} \sin[(2n+1)x]}{\sin x} \, dx = \dfrac{\sqrt{\pi}}{p} \left[ \dfrac{1}{2} + \sum_{k=1}^n e^{-\left(\frac{k}{p}\right)^2} \right]$

$[p > 0]$        BI (267)(17)

2.[12]    $\int_0^\infty \dfrac{e^{-p^2 x^2} \cos[(4n+1)x]}{\cos x} \, dx = \dfrac{\sqrt{\pi}}{p} \left[ \dfrac{1}{2} + \sum_{k=1}^{2n} (-1)^k e^{-\left(\frac{k}{p}\right)^2} \right]$

$[p > 0]$        BI (267)(18)

3.    $\int_0^\infty \dfrac{e^{-px^2} \, dx}{1 - 2a\cos x + a^2} = \dfrac{\sqrt{\frac{\pi}{p}}}{1 - a^2} \left[ \dfrac{1}{2} + \sum_{k=1}^\infty a^k \exp\left(-\dfrac{k^2}{4p}\right) \right]$    $[a^2 < 1, \quad p > 0]$    EI (266)(1)

$= \dfrac{\sqrt{\frac{\pi}{p}}}{a^2 - 1} \left[ \dfrac{1}{2} + \sum_{k=1}^\infty a^{-k} \exp\left(-\dfrac{k^2}{4p}\right) \right]$    $[a^2 > 1, \quad p > 0]$    LI (266)(1)

**3.911**

1.[12]    $\int_0^\infty \dfrac{\sin ax}{e^x + 1} \, dx = \dfrac{1}{2a} - \dfrac{\pi}{2\sinh \pi a}$        $[a > 0]$        BI (264)(1)

2.$^{12}$ $\displaystyle\int_0^\infty \frac{\sin ax}{e^x - 1}\,dx = \frac{\pi}{2}\coth \pi a - \frac{1}{2a}$ $\qquad\qquad [a > 0]$ $\qquad\qquad$ BI (264)(2), WH

3.$^{11}$ $\displaystyle\int_0^\infty \frac{\sin ax}{e^x - 1}e^{x/2}\,dx = \frac{1}{2}\pi \tanh(a\pi)$ $\qquad\qquad [a > 0]$ $\qquad\qquad$ ET I 73(13)

4. $\displaystyle\int_0^\infty \frac{\sin ax}{1 - e^{-x}}e^{-nx}\,dx = \frac{\pi}{2} - \frac{1}{2a} + \frac{\pi}{e^{2\pi a} - 1} - \sum_{k=1}^{n-1}\frac{a}{a^2 + k^2}$

$\qquad\qquad\qquad\qquad\qquad\qquad [a > 0]$ $\qquad\qquad$ BI (264)(8)

5. $\displaystyle\int_0^\infty \frac{\sin ax}{e^{\beta x} - e^{\gamma x}}\,dx = \frac{1}{2i(\beta - \gamma)}\left[\psi\left(\frac{\beta + ia}{\beta - \gamma}\right) - \psi\left(\frac{\beta - ia}{\beta - \gamma}\right)\right]$

$\qquad\qquad\qquad\qquad\qquad [\operatorname{Re}\beta > 0, \quad \operatorname{Re}\gamma > 0]$ $\qquad\qquad$ GW (335)(8)

6. $\displaystyle\int_0^\infty \frac{\sin ax\,dx}{e^{\beta x}(e^{-x} - 1)} = \frac{i}{2}[\psi(\beta + ia) - \psi(\beta - ia)]$ $\qquad [\operatorname{Re}\beta > -1]$ $\qquad\qquad$ ET 73(15)

**3.912**

1. $\displaystyle\int_0^\infty e^{-\beta x}\left(1 - e^{-\gamma x}\right)^{\nu-1}\sin ax\,dx = -\frac{i}{2\gamma}\left[\mathrm{B}\left(\nu, \frac{\beta - ia}{\gamma}\right) - \mathrm{B}\left(\nu, \frac{\beta + ia}{\gamma}\right)\right]$

$\qquad\qquad [\operatorname{Re}\beta > 0, \quad \operatorname{Re}\gamma > 0, \quad \operatorname{Re}\nu > 0, \quad a > 0]$ $\quad$ ET I 73(17)

2. $\displaystyle\int_0^\infty e^{-\beta x}\left(1 - e^{-\gamma x}\right)^{\nu-1}\cos ax\,dx = \frac{1}{2\gamma}\left[\mathrm{B}\left(\nu, \frac{\beta - ia}{\gamma}\right) + \mathrm{B}\left(\nu, \frac{\beta + ia}{\gamma}\right)\right]$

$\qquad\qquad [\operatorname{Re}\beta > 0, \quad \operatorname{Re}\gamma > 0, \quad \operatorname{Re}\nu > 0, \quad a > 0]$ $\quad$ ET I 15(10)

**3.913**

1. $\displaystyle\int_{-\frac{\pi}{2}}^{\frac{\pi}{2}} e^{i\beta x}\cos^\nu x\left(\beta^2 e^{ix} + \nu^2 e^{-ix}\right)^\mu\,dx = \frac{\pi\,{}_2F_1\left(-\mu, \frac{\beta}{2} - \frac{\nu}{2} - \frac{\mu}{2}; 1 + \frac{\beta}{2} + \frac{\nu}{2} - \frac{\mu}{2}; \frac{\beta^2}{\nu^2}\right)}{2^\nu(\nu + 1)\,\mathrm{B}\left(1 + \frac{\beta}{2} + \frac{\nu}{2} - \frac{\mu}{2}, 1 - \frac{\beta}{2} + \frac{\nu}{2} + \frac{\mu}{2}\right)}$

$\qquad\qquad\qquad\qquad\qquad [\operatorname{Re}\nu > -1, \quad |\nu| > |\beta|]$ $\qquad\qquad$ EH I 81(11)a

2.$^{11}$ $\displaystyle\int_{-\frac{\pi}{2}}^{\frac{\pi}{2}} e^{-iux}\cos^\mu x\left(a^2 e^{ix} + b^2 e^{-ix}\right)^\nu\,dx$

$\qquad\qquad = \dfrac{\pi b^{2\nu}\,{}_2F_1\left(-\nu, -\frac{u+\mu+\nu}{2}; 1 + \frac{\mu-\nu-u}{2}; \frac{a^2}{b^2}\right)}{2^\mu(\mu + 1)\,\mathrm{B}\left(1 - \frac{u+\nu-\mu}{2}, 1 + \frac{u+\mu+\nu}{2}\right)}$ $\qquad$ [for $a^2 < b^2$]

$\qquad\qquad = \dfrac{\pi a^{2\nu}\,{}_2F_1\left(-\nu, \frac{u-\mu-\nu}{2}; 1 + \frac{\mu-\nu+u}{2}; \frac{b^2}{a^2}\right)}{2^\mu(\mu + 1)\,\mathrm{B}\left(1 + \frac{u+\mu-\nu}{2}, 1 + \frac{\mu+\nu-u}{2}\right)}$ $\qquad$ [for $b^2 < a^2$]

$\qquad\qquad\qquad\qquad\qquad\qquad [\operatorname{Re}\mu > -1]$ $\qquad\qquad$ ET I 122(31)a

**3.914**

1. $\displaystyle\int_0^\infty e^{-\beta\sqrt{\gamma^2 + x^2}}\cos bx\,dx = \frac{\beta\gamma}{\sqrt{\beta^2 + b^2}}K_1\left(\gamma\sqrt{\beta^2 + b^2}\right)$

$\qquad\qquad\qquad\qquad [\operatorname{Re}\beta > 0, \quad \operatorname{Re}\gamma > 0]$ $\qquad\qquad$ ET I 16(26)

2.    $\displaystyle\int_0^\infty \sqrt{\gamma^2 + x^2}\, e^{-\beta\sqrt{\gamma^2+x^2}} \cos bx\, dx = \frac{\beta^2\gamma^2}{A^2} K_0(\gamma A) + \left(\frac{2\beta^2\gamma}{A^3} - \frac{\gamma}{A}\right) K_1(\gamma A)$

$$\left[A = \sqrt{\beta^2 + b^2}\right]$$

3.    $\displaystyle\int_0^\infty \left(\gamma^2 + x^2\right) e^{-\beta\sqrt{\gamma^2+x^2}} \cos bx\, dx$

$$= \left(-\frac{3\beta\gamma^2}{A^2} + \frac{4\beta^3\gamma^2}{A^4}\right) K_0(\gamma A) + \left(-\frac{6\beta\gamma}{A^3} + \frac{8\beta^3\gamma}{A^5} + \frac{\beta^3\gamma^3}{A^3}\right) K_1(\gamma A)$$

$$\left[A = \sqrt{\beta^2 + b^2}\right]$$

4.    $\displaystyle\int_0^\infty \frac{e^{-\beta\sqrt{\gamma^2+x^2}}}{\sqrt{\gamma^2 + x^2}} \cos bx\, dx = K_0\left(\gamma\sqrt{\beta^2 + b^2}\right)$       $[\operatorname{Re}\beta > 0, \quad \operatorname{Re}\gamma > 0, \quad b > 0]$

ET I 16(27)

5.    $\displaystyle\int_0^\infty \left(\frac{1}{\beta\left(\gamma^2 + x^2\right)^{3/2}} + \frac{1}{\gamma^2 + x^2}\right) e^{-\beta\sqrt{\gamma^2+x^2}} \cos bx\, dx = \frac{1}{\beta\gamma}\sqrt{\beta^2 + b^2}\, K_1\left(\gamma\sqrt{\beta^2 + b^2}\right)$

(6.726(4))

6.    $\displaystyle\int_0^\infty x e^{-\beta\sqrt{\gamma^2+x^2}} \sin bx\, dx = \frac{b\beta\gamma^2}{\beta^2 + b^2} K_2\left(\gamma\sqrt{\beta^2 + b^2}\right)$       ET I 175(35)

7.    $\displaystyle\int_0^\infty x\sqrt{\gamma^2 + x^2}\, e^{-\beta\sqrt{\gamma^2+x^2}} \sin bx\, dx$

$$= \left(-\frac{b\gamma^2}{A^2} + \frac{4b\beta^2\gamma^2}{A^4}\right) K_0(\gamma A) + \left(-\frac{2b\gamma}{A^3} + \frac{8b\beta^2\gamma}{A^5} + \frac{b\beta^2\gamma^3}{A^3}\right) K_1(\gamma A)$$

$$\left[A = \sqrt{\beta^2 + b^2}\right]$$

8.    $\displaystyle\int_0^\infty \left(\gamma^2 + x^2\right) e^{-\beta\sqrt{\gamma^2+x^2}} x\sin bx\, dx = \left(-\frac{12b\beta\gamma^2}{A^4} + \frac{24b\beta^3\gamma^2}{A^6} + \frac{b\beta^3\gamma^4}{A^4}\right) K_0(\gamma A)$

$$+ \left(-\frac{24b\beta\gamma}{A^5} + \frac{48b\beta^3\gamma}{A^7} - \frac{3b\beta\gamma^3}{A^3} + \frac{8b\beta^3\gamma^3}{A^5}\right) K_1(\gamma A)$$

$$\left[A = \sqrt{\beta^2 + b^2}\right]$$

9.    $\displaystyle\int_0^\infty \frac{x e^{-\beta\sqrt{\gamma^2+x^2}}}{\sqrt{\gamma^2 + x^2}} \sin bx\, dx = \frac{\gamma b}{\sqrt{\beta^2 + b^2}} K_1\left(\gamma\sqrt{\beta^2 + b^2}\right)$       ET I 75(36)

10.    $\displaystyle\int_0^\infty \left(\frac{1}{\beta\left(\gamma^2 + x^2\right)^{3/2}} + \frac{1}{\gamma^2 + x^2}\right) e^{-\beta\sqrt{\gamma^2+x^2}} x\sin bx\, dx = \frac{b}{\beta} K_0\left(\gamma\sqrt{\beta^2 + b^2}\right)$       (6.726(3))

### 3.915

1.    $\displaystyle\int_0^\pi e^{a\cos x} \sin x\, dx = \frac{2}{a}\sinh a$       GW (337)(15c)

2.    $\displaystyle\int_0^\pi e^{i\beta\cos x} \cos nx\, dx = i^n \pi J_n(\beta)$       EH II 81(2)

3.[3]   $\displaystyle\int_{-\frac{\pi}{2}}^{\frac{\pi}{2}} e^{i\beta\cos x} \cos^{2\nu} x\, dx = \sqrt{\pi}\left(\frac{2}{\beta}\right)^\nu \Gamma\left(\nu + \frac{1}{2}\right) J_\nu(\beta)$

$$[\operatorname{Re}\nu > -\tfrac{1}{2}]$$       EH II 81(6)

4.  $\displaystyle\int_0^\pi e^{\pm\beta\cos x}\sin^{2\nu}x\,dx = \sqrt{\pi}\left(\frac{2}{\beta}\right)^\nu\Gamma\left(\nu+\frac{1}{2}\right)I_\nu(\beta)$    $\left[\mathrm{Re}\,\nu>-\frac{1}{2}\right]$                GW (337)(15b)

5.  $\displaystyle\int_0^\pi e^{i\beta\cos x}\sin^{2\nu}x\,dx = \sqrt{\pi}\left(\frac{2}{\beta}\right)^\nu\Gamma\left(\nu+\frac{1}{2}\right)J_\nu(\beta)$    $\left[\mathrm{Re}\,\nu>-\frac{1}{2}\right]$                WA 34(2), WA 60(6)

**3.916**

1.  $\displaystyle\int_0^{\pi/2} e^{-p^2\tan x}\frac{\sin\frac{x}{2}\sqrt{\cos x}}{\sin 2x}\,dx = \left[C(p)-\frac{1}{2}\right]^2 + \left[S(p)-\frac{1}{2}\right]^2$                NT 33(18)a

2.  $\displaystyle\int_0^{\pi/2}\frac{\exp(-p\tan x)\,dx}{\sin 2x + a\cos 2x + a} = -\frac{1}{2}e^{ap}\,\mathrm{Ei}(-ap)$    $[p>0]$,    (cf. **3552** 4 and 6)

BI (273)(11)

3.  $\displaystyle\int_0^{\pi/2}\frac{\exp(-p\cot x)\,dx}{\sin 2x + a\cos 2x - a} = -\frac{1}{2}e^{-ap}\,\mathrm{Ei}(ap)$    $[p>0]$,    (cf. **3.552** 4 and 6)

BI (273)(12)

4.  $\displaystyle\int_0^{\pi/2}\frac{\exp(-p\tan x)\sin 2x\,dx}{(1-a^2)-2a^2\cos 2x-(1+a^2)\cos^2 2x} = -\frac{1}{4}\left[e^{-ap}\,\mathrm{Ei}(ap)+e^{ap}\,\mathrm{Ei}(-ap)\right]$

$[p>0]$                BI (273)(13)

5.  $\displaystyle\int_0^{\pi/2}\frac{\exp(-p\cot x)\sin 2x\,dx}{(1-a^2)+2a^2\cos 2x-(1+a^2)\cos^2 2x} = -\frac{1}{4}\left[e^{-ap}\,\mathrm{Ei}(ap)+e^{ap}\,\mathrm{Ei}(-ap)\right]$

$[p>0]$                BI (273)(14)

**3.917**

1.  $\displaystyle\int_0^{\pi/2} e^{-2\beta\cot x}\cos^{\nu-1/2}x\sin^{-(\nu+1)}x\sin\left[\beta-\left(\nu-\frac{1}{2}\right)x\right]\,dx = \frac{\sqrt{\pi}}{2(2\beta)^\nu}\Gamma\left(\nu+\frac{1}{2}\right)J_\nu(\beta)$

$\left[\mathrm{Re}\,\nu>-\frac{1}{2}\right]$                WA 186(7)

2.[12]  $\displaystyle\int_0^{\pi/2} e^{-2\beta\cot x}\cos^{\nu-1/2}x\sin^{-(\nu+1)}x\cos\left[\beta-\left(\nu-\frac{1}{2}\right)x\right]\,dx = \frac{\sqrt{\pi}}{2(2\beta)^\nu}\Gamma\left(\nu+\frac{1}{2}\right)Y_\nu(\beta)$

$\left[\mathrm{Re}\,\nu>-\frac{1}{2}\right]$                WA 186(8)

**3.918**

1.  $\displaystyle\int_0^{\pi/2}\frac{\cos^\mu x}{\sin^{2\mu+2}x}e^{i\gamma(\beta-\mu x)-2\beta\cot x}\,dx = \frac{i\gamma}{2}\sqrt{\frac{\pi}{2\beta}}(2\beta)^{-\mu}\Gamma(\mu+1)H_{\mu+\frac{1}{2}}^{(\varepsilon)}(\beta)$

$\left[\varepsilon=1,2,\quad\gamma=(-1)^{\varepsilon+1},\quad\mathrm{Re}\,\beta>0,\quad\mathrm{Re}\,\mu>-1\right]$    GW (337)(16)

2.  $\displaystyle\int_0^{\pi/2}\frac{\cos^\mu x\sin(\beta-\mu x)}{\sin^{2\mu+2}x}e^{-2\beta\cot x}\,dx = \frac{1}{2}\sqrt{\frac{\pi}{2\beta}}(2\beta)^{-\mu}\Gamma(\mu+1)J_{\mu+\frac{1}{2}}(\beta)$

$\left[\mathrm{Re}\,\beta>0,\quad\mathrm{Re}\,\mu>-1\right]$                WH

3.  $\displaystyle\int_0^{\pi/2}\frac{\cos^\mu x\cos(\beta-\mu x)}{\sin^{2\mu+2}x}e^{-2\beta\cot x}\,dx = -\frac{1}{2}\sqrt{\frac{\pi}{2\beta}}(2\beta)^{-\mu}\Gamma(\mu+1)Y_{\mu+\frac{1}{2}}(\beta)$

$\left[\mathrm{Re}\,\beta>0,\quad\mathrm{Re}\,\mu>-1\right]$    GW (337)(17b)

**3.919**

1.    $\displaystyle\int_0^{\pi/2} \frac{\sin 2nx}{\sin^{2n+2} x} \frac{dx}{\exp\left(2\pi \cot x\right) - 1} = (-1)^{n-1} \frac{2n-1}{4(2n+1)}$      BI (275)(6), LI (275)(6)

2.    $\displaystyle\int_0^{\pi/2} \frac{\sin 2nx}{\sin^{2n+2} x} \frac{dx}{\exp\left(\pi \cot x\right) - 1} = (-1)^{n-1} \frac{n}{2n+1}$      BI (275)(7), LI (275)(7)

## 3.92 Trigonometric functions of more complicated arguments combined with exponentials

**3.921$^6$**

1.$^{12}$    $\displaystyle\int_0^\infty e^{-\gamma x} \cos ax^2 \left(\cos \gamma x - \sin \gamma x\right) dx = \sqrt{\frac{\pi}{8a}} \exp\left(-\frac{\gamma^2}{2a}\right)$

$$[a > 0, \quad \operatorname{Re}\gamma \geq |\operatorname{Im}\gamma|] \qquad \text{ET I 26(28)}$$

2.$^{10}$    $\displaystyle\int_0^{\pi/4} \prod_{n=1}^\infty \exp\left[-\frac{1}{n}\tan^{2n} x\right] = \frac{\pi}{2} - 1$

3.$^{10}$    $\displaystyle\int_0^{\pi/2} \exp\left[-\sum_{n=1}^\infty \frac{1}{n}\sin^{2n} x\right] = \int_0^{\pi/2} \exp\left[-\sum_{n=1}^\infty \frac{1}{n}\cos^{2n} x\right] = \frac{\pi}{4}$

**3.922**

1.    $\displaystyle\int_0^\infty e^{-\beta x^2} \sin ax^2 \, dx = \frac{1}{2}\int_{-\infty}^\infty e^{-\beta x^2} \sin ax^2 \, dx = \sqrt{\frac{\pi}{8}}\sqrt{\frac{\sqrt{\beta^2 + a^2} - \beta}{\beta^2 + a^2}}$

$$= \frac{\sqrt{\pi}}{2\sqrt[4]{\beta^2 + a^2}} \sin\left(\frac{1}{2}\arctan\frac{a}{\beta}\right)$$

$$[\operatorname{Re}\beta > 0, \quad a > 0] \quad \text{FI II 750, BI (263)(8)}$$

2.    $\displaystyle\int_0^\infty e^{-\beta x^2} \cos ax^2 \, dx = \frac{1}{2}\int_{-\infty}^\infty e^{-\beta x^2} \cos ax^2 \, dx = \sqrt{\frac{\pi}{8}}\sqrt{\frac{\sqrt{\beta^2 + a^2} + \beta}{\beta^2 + a^2}}$

$$= \frac{\sqrt{\pi}}{2\sqrt[4]{\beta^2 + a^2}} \cos\left(\frac{1}{2}\arctan\frac{a}{\beta}\right)$$

$$[\operatorname{Re}\beta > 0, \quad a > 0] \quad \text{FI II 750, BI (263)(9)}$$

[In formulas **3.922** 3 and 4, $a > 0$, $b > 0$, $\operatorname{Re}\beta > 0$, and

$$A = \frac{b^2}{4\left(a^2 + \beta^2\right)}, \qquad B = \sqrt{\frac{1}{2}\left(\sqrt{\beta^2 + a^2} + \beta\right)}, \qquad C = \sqrt{\frac{1}{2}\left(\sqrt{\beta^2 + a^2} - \beta\right)}$$

If $a$ is complex, then $\operatorname{Re}\beta > |\operatorname{Im}a|$.]

3.    $\displaystyle\int_0^\infty e^{-\beta x^2} \sin ax^2 \cos bx \, dx = -\frac{1}{2}\sqrt{\frac{\pi}{\beta^2 + a^2}} e^{-A\beta}\left(B\sin Aa - C\cos Aa\right)$

$$= \frac{\sqrt{\pi}}{2\sqrt[4]{\beta^2 + a^2}} \exp\left(-\frac{\beta b^2}{4\left(\beta^2 + a^2\right)}\right)\sin\left[\frac{1}{2}\arctan\frac{a}{\beta} - \frac{ab^2}{4\left(\beta^2 + a^2\right)}\right]$$

$$\text{LI (263)(10), GW (337)(5)}$$

4.     $$\int_0^\infty e^{-\beta x^2} \cos ax^2 \cos bx \, dx = \frac{1}{2}\sqrt{\frac{\pi}{\beta^2 + a^2}} e^{-A\beta} \left( B \cos Aa + C \sin Aa \right)$$

$$= \frac{\sqrt{\pi}}{2\sqrt[4]{\beta^2 + a^2}} \exp\left( -\frac{\beta b^2}{4(\beta^2 + a^2)} \right) \cos\left[ \frac{1}{2} \arctan \frac{a}{\beta} - \frac{ab^2}{4(\beta^2 + a^2)} \right]$$

LI (263)(11), GW (337)(5)

### 3.923

1.     $$\int_{-\infty}^\infty \exp\left[ -\left( ax^2 + 2bx + c \right) \right] \sin\left( px^2 + 2qx + r \right) \, dx$$

$$= \frac{\sqrt{\pi}}{\sqrt[4]{a^2 + p^2}} \exp \frac{a\left( b^2 - ac \right) - \left( aq^2 - 2bpq + cp^2 \right)}{a^2 + p^2}$$

$$\times \sin\left[ \frac{1}{2} \arctan \frac{p}{a} - \frac{p\left( q^2 - pr \right) - \left( b^2 p - 2abq + a^2 r \right)}{a^2 + p^2} \right]$$

[a > 0]          GW (337)(3), BI (296)(6)

2.     $$\int_{-\infty}^\infty \exp\left[ -\left( ax^2 + 2bx + c \right) \right] \cos\left( px^2 + 2qx + r \right) \, dx$$

$$= \frac{\sqrt{\pi}}{\sqrt[4]{a^2 + p^2}} \exp \frac{a\left( b^2 - ac \right) - \left( aq^2 - 2bpq + cp^2 \right)}{a^2 + p^2}$$

$$\times \cos\left[ \frac{1}{2} \arctan \frac{p}{a} - \frac{p\left( q^2 - pr \right) - \left( b^2 p - 2abq + a^2 r \right)}{a^2 + p^2} \right]$$

[a > 0]          GW (337)(3), BI (269)(7)

### 3.924

1.     $$\int_0^\infty e^{-\beta x^4} \sin bx^2 \, dx = \frac{\pi}{4}\sqrt{\frac{b}{2\beta}} \exp\left( -\frac{b^2}{8\beta} \right) I_{\frac{1}{4}}\left( \frac{b^2}{8\beta} \right)$$

[Re β > 0,   b > 0]                    ET 73(22)

2.     $$\int_0^\infty e^{-\beta x^4} \cos bx^2 \, dx = \frac{\pi}{4}\sqrt{\frac{b}{2\beta}} \exp\left( -\frac{b^2}{8\beta} \right) I_{-\frac{1}{4}}\left( \frac{b^2}{8\beta} \right)$$

[Re β > 0,   b > 0]                    ET I 15(12)

### 3.925

1.     $$\int_0^\infty e^{-\frac{p^2}{x^2}} \sin 2a^2 x^2 \, dx = \frac{1}{2}\int_{-\infty}^\infty e^{-\frac{p^2}{x^2}} \sin 2a^2 x^2 \, dx = \frac{\sqrt{\pi}}{4a} e^{-2ap} \left( \cos 2ap + \sin 2ap \right)$$

[a > 0,   b > 0]                    BI (268)(12)

2.     $$\int_0^\infty e^{-\frac{p^2}{x^2}} \cos 2a^2 x^2 \, dx = \frac{1}{2}\int_{-\infty}^\infty e^{-\frac{p^2}{x^2}} \cos 2a^2 x^2 \, dx = \frac{\sqrt{\pi}}{4a} e^{-2ap} \left( \cos 2ap - \sin 2ap \right)$$

[a > 0,   b > 0]                    BI (268)(13)

**3.926**    **Notation:**

$$u = \sqrt{\frac{\sqrt{a^2 + \beta^2} + \beta}{2}}, \qquad v = \sqrt{\frac{\sqrt{a^2 + \beta^2} - \beta}{2}}$$

1.    $\displaystyle \int_0^\infty e^{-\left(\beta x^2 + \frac{\gamma}{x^2}\right)} \sin a x^2 \, dx = \frac{1}{2}\sqrt{\frac{\pi}{a^2 + \beta^2}} e^{-2u\sqrt{\gamma}} \left[ v\cos\left(2v\sqrt{\gamma}\right) + u\sin\left(2v\sqrt{\gamma}\right) \right]$

$$[\operatorname{Re}\beta > 0, \quad \operatorname{Re}\gamma > 0] \qquad \text{BI (268)(14)}$$

2.    $\displaystyle \int_0^\infty e^{-\left(\beta x^2 + \frac{\gamma}{x^2}\right)} \cos a x^2 \, dx = \frac{1}{2}\sqrt{\frac{\pi}{a^2 + \beta^2}} e^{-2u\sqrt{\gamma}} \left[ u\cos\left(2v\sqrt{\gamma}\right) - v\sin\left(2v\sqrt{\gamma}\right) \right]$

$$[\operatorname{Re}\beta > 0, \quad \operatorname{Re}\gamma > 0] \qquad \text{BI (268)(15)}$$

**3.927**    $\displaystyle \int_0^\infty e^{-\frac{p}{x}} \sin^2 \frac{a}{x} \, dx = a\arctan\frac{2a}{p} + \frac{p}{4}\ln\frac{p^2}{p^2 + 4a^2}$      $[a > 0, \quad p > 0]$      LI (268)(4)

**3.928**
**Notation:** $a^2 + p^2 > 0$, $r = \sqrt[4]{a^4 + p^4}$, $s = \sqrt[4]{b^4 + q^4}$, $A = \dfrac{1}{2}\arctan\dfrac{a^2}{p^2}$, and $B = \dfrac{1}{2}\arctan\dfrac{b^2}{q^2}$.

1.    $\displaystyle \int_0^\infty \exp\left[-\left(p^2 x^2 + \frac{q^2}{x^2}\right)\right] \sin\left(a^2 x^2 + \frac{b^2}{x^2}\right) dx = \frac{\sqrt{\pi}}{2r} e^{-2rs\cos(A+B)} \sin\left[A + 2rs\sin(A+B)\right]$

$$\text{BI (268)(22)}$$

2.    $\displaystyle \int_0^\infty \exp\left[-\left(p^2 x^2 + \frac{q^2}{x^2}\right)\right] \cos\left(a^2 x^2 + \frac{b^2}{x^2}\right) dx = \frac{\sqrt{\pi}}{2r} e^{-2rs\cos(A+B)} \cos\left[A + 2rs\sin(A+B)\right]$

$$\text{BI (268)(23)}$$

**3.929**    $\displaystyle \int_0^\infty \left[ e^{-x}\cos\left(p\sqrt{x}\right) + pe^{-x^2}\sin px \right] dx = 1$      LI (268)(3)

## 3.93 Trigonometric and exponential functions of trigonometric functions

**3.931**

1.[12]    $\displaystyle \int_0^\pi e^{-p\cos x} \sin\left(p\sin x\right) dx = \operatorname{Ei}(-p) - \operatorname{ci}(p)$      NT 13(27)

2.    $\displaystyle \int_0^\pi e^{-p\cos x} \sin\left(p\sin x\right) dx = -\int_{-\pi}^0 e^{-p\cos x} \sin\left(p\sin x\right) dx = -2\operatorname{shi}(p)$      GW (337)(11b)

3.    $\displaystyle \int_0^{\pi/2} e^{-p\cos x} \cos\left(p\sin x\right) dx = -\operatorname{si}(p)$      NT 13(26)

4.    $\displaystyle \int_0^\pi e^{-p\cos x} \cos\left(p\sin x\right) dx = \frac{1}{2}\int_0^{2\pi} e^{-p\cos x} \cos\left(p\sin x\right) dx = \pi$      GW (337)(11a)

**3.932**

1.    $\displaystyle \int_0^\pi e^{p\cos x} \sin\left(p\sin x\right)\sin mx \, dx = \frac{1}{2}\int_0^{2\pi} e^{p\cos x} \sin\left(p\sin x\right)\sin mx \, dx = \frac{\pi}{2}\frac{p^m}{m!}$

$$\text{BI (277)(7), GW (337)(13a)}$$

2.    $\int_0^\pi e^{p\cos x}\cos(p\sin x)\cos mx\,dx = \dfrac{1}{2}\int_0^{2\pi} e^{p\cos x}\cos(p\sin x)\cos mx\,dx = \dfrac{\pi}{2}\dfrac{p^m}{m!}$

BI (277)(8), GW (337)(13b)

**3.933**  $\int_0^\pi e^{p\cos x}\sin(p\sin x)\operatorname{cosec} x\,dx = \pi\sinh p$

BI (278)(1)

**3.934**

1.    $\int_0^\pi e^{p\cos x}\sin(p\sin x)\tan\dfrac{x}{2}\,dx = \pi(1-e^p)$

BI (271)(8)

2.    $\int_0^\pi e^{p\cos x}\sin(p\sin x)\cot\dfrac{x}{2}\,dx = \pi(e^p-1)$

BI (272)(5)

**3.935**  $\int_0^\pi e^{p\cos x}\cos(p\sin x)\dfrac{\sin 2nx}{\sin x}\,dx = \pi\sum_{k=0}^{n-1}\dfrac{p^{2k+1}}{(2k+1)!}$   $[p>0]$

LI (278)(3)

**3.936**

1.    $\int_0^{2\pi} e^{p\cos x}\cos(p\sin x - mx)\,dx = 2\int_0^\pi e^{p\cos x}\cos(p\sin x - mx)\,dx = \dfrac{2\pi p^m}{m!}$

BI (277)(9), GW (337)(14a)

2.    $\int_0^{2\pi} e^{p\sin x}\sin(p\cos x + mx)\,dx = \dfrac{2\pi p^m}{m!}\sin\dfrac{m\pi}{2}$   $[p>0]$

GW (337)(14b)

3.    $\int_0^{2\pi} e^{p\sin x}\cos(p\cos x + mx)\,dx = \dfrac{2\pi p^m}{m!}\cos\dfrac{m\pi}{2}$   $[p>0]$

GW (337)(14b)

4.    $\int_0^{2\pi} e^{\cos x}\sin(mx - \sin x)\,dx = 0$

WH

5.    $\int_0^\pi e^{\beta\cos x}\cos(ax + \beta\sin x)\,dx = \beta^{-a}\sin(a\pi)\,\gamma(a,\beta)$

EH II 137(2)

**3.937**  **Notation**: In formulas **3.937** 1 and 2, $(b-p)^2+(a+q)^2>0$, $m=0,1,2,\ldots$, $A=p^2-q^2+a^2-b^2$, $B=2(pq+ab)$, $C=p^2+q^2-a^2-b^2$, and $D=-2(ap+bq)$.

1.[11]   $\int_0^{2\pi}\exp(p\cos x+q\sin x)\sin(a\cos x+b\sin x-mx)\,dx$

$$= i\pi\left[(b-p)^2+(a+q)^2\right]^{-\frac{m}{2}}\left[(A+iB)^{m/2}I_m\left(\sqrt{C-iD}\right) - (A-iB)^{m/2}I_m\left(\sqrt{C+iD}\right)\right]$$

GW (337)(9b)

2.    $\int_0^{2\pi}\exp(p\cos x+q\sin x)\cos(a\cos x+b\sin x-mx)\,dx$

$$= \pi\left[(b-p)^2+(a+q)^2\right]^{-\frac{m}{2}}\left[(A+iB)^{\frac{m}{2}}I_m\left(\sqrt{C-iD}\right) + (A-iB)^{\frac{m}{2}}I_m\left(\sqrt{C+iD}\right)\right]$$

GW (337)(9a)

3.[12]   $\int_0^{2\pi}\exp(p\cos x+q\sin x)\sin(q\cos x - p\sin x+mx)\,dx = \dfrac{2\pi}{m!}\left(p^2+q^2\right)^{\frac{m}{2}}\sin\left(m\arctan\dfrac{q}{p}\right)$

GW (337)(12)

$4.^{12}$ $\displaystyle\int_0^{2\pi} \exp\left(p\cos x + q\sin x\right)\cos\left(q\cos x - p\sin x + mx\right)\,\mathrm{d}x = \frac{2\pi}{m!}\left(p^2 + q^2\right)^{\frac{m}{2}}\cos\left(m\arctan\frac{q}{p}\right)$

GW (337)(12)

**3.938**

1. $\displaystyle\int_0^\pi e^{r(\cos px + \cos qx)}\sin\left(r\sin px\right)\sin\left(r\sin qx\right)\,\mathrm{d}x = \frac{\pi}{2}\sum_{k=1}^\infty \frac{1}{\Gamma(pk+1)\Gamma(qk+1)}r^{(p+q)k}$

BI (277)(14)

2. $\displaystyle\int_0^\pi e^{r(\cos px + \cos qx)}\cos\left(r\sin px\right)\cos\left(r\sin qx\right)\,\mathrm{d}x = \frac{\pi}{2}\left(2 + \sum_{k=1}^\infty \frac{r^{(p+q)k}}{\Gamma(pk+1)\Gamma(qk+1)}\right)$

BI (277)(15)

**3.939**

1. $\displaystyle\int_0^\pi e^{q\cos x}\frac{\sin rx}{1 - 2p^r\cos rx + p^{2r}}\sin\left(q\sin x\right)\,\mathrm{d}x = \frac{\pi}{2pr}\sum_{k=1}^\infty \frac{(pq)^{kr}}{\Gamma(kr+1)}$

$[r > 0, \quad 0 < p < 1]$      BI (278)(15)

$2.^3$ $\displaystyle\int_0^\pi e^{q\cos x}\frac{1 - p^r\cos rx}{1 - 2p^r\cos rx + p^{2r}}\cos\left(q\sin x\right)\,\mathrm{d}x = \frac{\pi}{2}\left[2 + \sum_{k=1}^\infty \frac{(pq)^{kr}}{\Gamma(kr+1)}\right]$

$[r > 0, 0 < p < 1]$      BI (278)(16)

3. $\displaystyle\int_0^{\pi/2}\frac{e^{p\cos 2x}\cos\left(p\sin 2x\right)\,\mathrm{d}x}{\cos^2 x + q^2\sin^2 x} = \frac{\pi}{2q}\exp\left(p\frac{q-1}{q+1}\right)$

BI (273)(8)

## 3.94–3.97 Combinations involving trigonometric functions, exponentials, and powers

**3.941**

1. $\displaystyle\int_0^\infty e^{-px}\sin qx\frac{\mathrm{d}x}{x} = \arctan\frac{q}{p}$          $[p > 0]$      BI (365)(1)

2. $\displaystyle\int_0^\infty e^{-px}\cos qx\frac{\mathrm{d}x}{x} = \infty$                  BI (365)(2)

**3.942**

1. $\displaystyle\int_0^\infty e^{-px}\cos px\frac{x\,\mathrm{d}x}{b^4 + x^4} = \frac{\pi}{4b^2}\exp\left(-bp\sqrt{2}\right)$      $[p > 0, \quad b > 0]$      BI (386)(6)a

2. $\displaystyle\int_0^\infty e^{-px}\cos px\frac{x\,\mathrm{d}x}{b^4 - x^4} = \frac{\pi}{4b^2}e^{-bp}\sin bp$      $[p > 0, \quad b > 0]$      BI (386)(7)a

**3.943** $\displaystyle\int_0^\infty e^{-\beta x}\left(1 - \cos ax\right)\frac{\mathrm{d}x}{x} = \frac{1}{2}\ln\frac{a^2 + \beta^2}{\beta^2}$      $[\operatorname{Re}\beta > 0]$      BI (367)(6)

**3.944**

1. $\displaystyle\int_0^u x^{\mu-1}e^{-\beta x}\sin\delta x\,\mathrm{d}x = \frac{i}{2}\left(\beta + i\delta\right)^{-\mu}\gamma\left[\mu, (\beta + i\delta)\,u\right] - \frac{i}{2}\left(\beta - i\delta\right)^{-\mu}\gamma\left[\mu, (\beta - i\delta)\,u\right]$

$[\operatorname{Re}\mu > -1]$      ET I 318(8)

2. $\displaystyle\int_u^\infty x^{\mu-1}e^{-\beta x}\sin\delta x\,\mathrm{d}x = \frac{i}{2}\left(\beta+i\delta\right)^{-\mu}\Gamma\left[\mu,\left(\beta+i\delta\right)u\right]-\frac{i}{2}\left(\beta-i\delta\right)^{-\mu}\Gamma\left[\mu,\left(\beta-i\delta\right)u\right]$

$$\left[\operatorname{Re}\beta > |\operatorname{Im}\delta|\right] \qquad\qquad \text{ET I 318(9)}$$

3. $\displaystyle\int_0^u x^{\mu-1}e^{-\beta x}\cos\delta x\,\mathrm{d}x = \frac{1}{2}\left(\beta+i\delta\right)^{-\mu}\gamma\left[\mu,\left(\beta+i\delta\right)u\right]+\frac{1}{2}\left(\beta-i\delta\right)^{-\mu}\gamma\left[\mu,\left(\beta-i\delta\right)u\right]$

$$\left[\operatorname{Re}\mu > 0\right] \qquad\qquad \text{ET I 320(28)}$$

4. $\displaystyle\int_u^\infty x^{\mu-1}e^{-\beta x}\cos\delta x\,\mathrm{d}x = \frac{1}{2}\left(\beta+i\delta\right)^{-\mu}\Gamma\left[\mu,\left(\beta+i\delta\right)u\right]+\frac{1}{2}\left(\beta-i\delta\right)^{-\mu}\Gamma\left[\mu,\left(\beta-i\delta\right)u\right]$

$$\left[\operatorname{Re}\beta > |\operatorname{Im}\delta|\right] \qquad\qquad \text{ET I 320(29)}$$

5.[11] $\displaystyle\int_0^\infty x^{\mu-1}e^{-\beta x}\sin\delta x\,\mathrm{d}x = \frac{\Gamma(\mu)}{\left(\beta^2+\delta^2\right)^{\mu/2}}\sin\left(\mu\arctan\frac{\delta}{\beta}\right)$

$$\left[\operatorname{Re}\mu > -1, \quad \operatorname{Re}\beta > |\operatorname{Im}\delta|\right]$$
$$\text{FI II 812, BI (361)(9)}$$

6.[12] $\displaystyle\int_0^\infty x^{\mu-1}e^{-\beta x}\cos\delta x\,\mathrm{d}x = \frac{\Gamma(\mu)}{\left(\delta^2+\beta^2\right)^{\mu/2}}\cos\left(\mu\arctan\frac{\delta}{\beta}\right)$

$$\left[\operatorname{Re}\mu > 0, \quad \operatorname{Re}\beta > |\operatorname{Im}\delta|\right]$$
$$\text{FI II 812, BI (361)(10)}$$

7. $\displaystyle\int_0^\infty x^{\mu-1}\exp\left(-ax\cos t\right)\sin\left(ax\sin t\right)\,\mathrm{d}x = \Gamma(\mu)a^{-\mu}\sin(\mu t)$

$$\left[\operatorname{Re}\mu > -1, \quad a > 0, \quad |t| < \frac{\pi}{2}\right]$$
$$\text{EH I 13(36)}$$

8. $\displaystyle\int_0^\infty x^{\mu-1}\exp\left(-ax\cos t\right)\cos\left(ax\sin t\right)\,\mathrm{d}x = \Gamma(\mu)a^{-\mu}\cos(\mu t)$

$$\left[\operatorname{Re}\mu > -1, \quad a > 0, \quad |t| < \frac{\pi}{2}\right]$$
$$\text{EH I 13(35)}$$

9. $\displaystyle\int_0^\infty x^{p-1}e^{-qx}\sin\left(qx\tan t\right)\,\mathrm{d}x = \frac{1}{q^p}\Gamma(p)\cos^p t\sin pt \qquad \left[|t| < \frac{\pi}{2}, \quad q > 0\right] \qquad \text{LO V 288(16)}$

10. $\displaystyle\int_0^\infty x^{p-1}e^{-qx}\cos\left(qx\tan t\right)\,\mathrm{d}x = \frac{1}{q^p}\Gamma(p)\cos^p(t)\cos pt$

$$\left[|t| < \frac{\pi}{2}, \quad q > 0\right] \qquad \text{LO V 288(15)}$$

11. $\displaystyle\int_0^\infty x^n e^{-\beta x}\sin bx\,\mathrm{d}x = n!\left(\frac{\beta}{\beta^2+b^2}\right)^{n+1}\sum_{0\le 2k\le n}(-1)^k\binom{n+1}{2k+1}\left(\frac{b}{\beta}\right)^{2k+1}$

$$= (-1)^n\frac{\partial^n}{\partial\beta^n}\left(\frac{b}{b^2+\beta^2}\right)$$

$$\left[\operatorname{Re}\beta > 0, \quad b > 0\right] \quad \text{GW (336)(3), ET I 72(3)}$$

12. 
$$\int_0^\infty x^n e^{-\beta x} \cos bx \, dx = n! \left( \frac{\beta}{\beta^2 + b^2} \right)^{n+1} \sum_{0 \le 2k \le n+1} (-1)^k \binom{n+1}{2k} \left( \frac{b}{\beta} \right)^{2k}$$

$$= (-1)^n \frac{\partial^n}{\partial \beta^n} \left( \frac{\beta}{b^2 + \beta^2} \right)$$

$$[\operatorname{Re} \beta > 0, \quad b > 0] \quad \text{GW (336)(4), ET I 14(5)}$$

13. 
$$\int_0^\infty x^{n-1/2} e^{-\beta x} \sin bx \, dx = (-1)^n \sqrt{\frac{\pi}{2}} \frac{d^n}{d\beta^n} \left( \frac{\sqrt{\sqrt{\beta^2 + b^2} - \beta}}{\sqrt{\beta^2 + b^2}} \right)$$

$$[\operatorname{Re} \beta > 0, \quad b > 0] \quad \text{ET I 72(6)}$$

14. 
$$\int_0^\infty x^{n-1/2} e^{-\beta x} \cos bx \, dx = (-1)^n \sqrt{\frac{\pi}{2}} \frac{d^n}{d\beta^n} \left( \frac{\sqrt{\sqrt{\beta^2 + b^2} + \beta}}{\sqrt{\beta^2 + b^2}} \right)$$

$$[\operatorname{Re} \beta > 0, \quad b > 0] \quad \text{ET I 15(6)}$$

**3.945**

1.[12] 
$$\int_0^\infty \left( e^{-\beta x} \sin ax - e^{-\gamma x} \sin bx \right) \frac{dx}{x^r}$$

$$= \Gamma(1-r) \left\{ \left( b^2 + \gamma^2 \right)^{\frac{r-1}{2}} \sin \left[ (r-1) \arctan \frac{b}{\gamma} \right] - \left( a^2 + \beta^2 \right)^{\frac{r-1}{2}} \sin \left[ (r-1) \arctan \frac{a}{\beta} \right] \right\}$$

$$[\operatorname{Re} \beta > 0, \quad \operatorname{Re} \gamma > 0, \quad r < 2, \quad r \ne 1] \quad \text{BI(371)(6)}$$

2.[12] 
$$\int_0^\infty \left( e^{-\beta x} \cos ax - e^{-\gamma x} \cos bx \right) \frac{dx}{x^r}$$

$$= \Gamma(1-r) \left\{ \left( a^2 + \beta^2 \right)^{\frac{r-1}{2}} \cos \left[ (r-1) \arctan \frac{a}{\beta} \right] - \left( b^2 + \gamma^2 \right)^{\frac{r-1}{2}} \cos \left[ (r-1) \arctan \frac{b}{\gamma} \right] \right\}$$

$$[\operatorname{Re} \beta > 0, \quad \operatorname{Re} \gamma > 0, \quad r < 2, \quad r \ne 1] \quad \text{BI (371)(7)}$$

3. 
$$\int_0^\infty \left( a e^{-\beta x} \sin bx - b e^{-\gamma x} \sin ax \right) \frac{dx}{x^2} = ab \left[ \frac{1}{2} \ln \frac{a^2 + \gamma^2}{b^2 + \beta^2} + \frac{\gamma}{a} \operatorname{arccot} \frac{\gamma}{a} - \frac{\beta}{b} \operatorname{arccot} \frac{\beta}{b} \right]$$

$$[\operatorname{Re} \beta > 0, \quad \operatorname{Re} \gamma > 0] \quad \text{BI (368)(22)}$$

**3.946**

1. 
$$\int_0^\infty e^{-px} \sin^{2m+1} ax \frac{dx}{x} = \frac{(-1)^m}{2^{2m}} \sum_{k=0}^m (-1)^k \binom{2m+1}{k} \arctan \frac{(2m - 2k + 1)a}{p}$$

$$[m = 0, 1, \ldots, \quad p > 0] \quad \text{GW (336)(9a)}$$

2. 
$$\int_0^\infty e^{-px} \sin^{2m} ax \frac{dx}{x} = \frac{(-1)^{m+1}}{2^{2m}} \sum_{k=0}^{m-1} (-1)^k \binom{2m}{k} \ln \left[ p^2 + (2m - 2k)^2 a^2 \right] - \frac{1}{2^{2m}} \binom{2m}{m} \ln p$$

$$[m = 1, 2, \ldots, \quad p > 0] \quad \text{GW (336)(9b)}$$

**3.947**

1. 
$$\int_0^\infty e^{-\beta x} \sin \gamma x \sin ax \frac{dx}{x} = \frac{1}{4} \ln \frac{\beta^2 + (a + \gamma)^2}{\beta^2 + (a - \gamma)^2} \qquad [\operatorname{Re} \beta > |\operatorname{Im} \gamma|, \quad a > 0] \quad \text{BI (365)(5)}$$

2.[11] $\displaystyle\int_0^\infty e^{-px}\sin ax\sin bx\,\frac{\mathrm{d}x}{x^2} = \frac{|a+b|}{2}\arctan\left(\frac{|a+b|}{p}\right) - \frac{|a-b|}{2}\arctan\left(\frac{|a-b|}{p}\right)$
$$+\frac{p}{4}\ln\left(\frac{p^2+(a-b)^2}{p^2+(a+b)^2}\right)$$

$\qquad\qquad\qquad\qquad\qquad [p>0,\quad\text{for } p=0 \text{ see } \mathbf{3.741}\ 3]\quad$ BI (368)(1), FI II 744

3.[11] $\displaystyle\int_0^\infty e^{-px}\sin ax\cos bx\,\frac{\mathrm{d}x}{x} = \arctan\frac{a+b}{p} + \arctan\frac{a-b}{p}$

$\qquad\qquad\qquad\qquad\qquad [a\geq 0,\quad p>0]\qquad\qquad$ GW (336)(10b)

**3.948**

1.[11] $\displaystyle\int_0^\infty e^{-px}(\sin ax - \sin bx)\,\frac{\mathrm{d}x}{x} = \arctan\frac{a}{p} - \arctan\frac{b}{p}\quad [\operatorname{Re}p>0],\qquad \text{(cf. } \mathbf{3.951}\ 2)$

$\qquad\qquad\qquad\qquad\qquad\qquad\qquad\qquad\qquad\qquad\qquad$ BI (367)(7)

2. $\displaystyle\int_0^\infty e^{-px}(\cos ax - \cos bx)\,\frac{\mathrm{d}x}{x} = \frac{1}{2}\ln\frac{b^2+p^2}{a^2+p^2}\qquad [\operatorname{Re}p>0],\qquad \text{(cf. } \mathbf{3.951}\ 3)$

$\qquad\qquad\qquad\qquad\qquad\qquad\qquad\qquad\qquad\qquad\qquad$ BI (367)(8), FI II 748a

3. $\displaystyle\int_0^\infty e^{-px}(\cos ax - \cos bx)\,\frac{\mathrm{d}x}{x^2} = \frac{p}{2}\ln\frac{a^2+p^2}{b^2+p^2} + b\arctan\frac{b}{p} - a\arctan\frac{a}{p}$

$\qquad\qquad\qquad\qquad\qquad [\operatorname{Re}p>0]\qquad\qquad$ BI (368)(20)

4. $\displaystyle\int_0^\infty e^{-px}(\sin^2 ax - \sin^2 bx)\,\frac{\mathrm{d}x}{x^2} = a\arctan\frac{2a}{p} - b\arctan\frac{2b}{p} - \frac{p}{4}\ln\frac{p^2+4a^2}{p^2+4b^2}$

$\qquad\qquad\qquad\qquad\qquad [p>0]\qquad\qquad$ BI (368)(25)

5. $\displaystyle\int_0^\infty e^{-px}(\cos^2 ax - \cos^2 bx)\,\frac{\mathrm{d}x}{x^2} = -a\arctan\frac{2a}{p} + b\arctan\frac{2b}{p} + \frac{p}{4}\ln\frac{p^2+4a^2}{p^2+4b^2}$

$\qquad\qquad\qquad\qquad\qquad [p>0]\qquad\qquad$ BI (368)(26)

**3.949**

1. $\displaystyle\int_0^\infty e^{-px}\sin ax\sin bx\sin cx\,\frac{\mathrm{d}x}{x} = -\frac{1}{4}\arctan\frac{a+b+c}{p} + \frac{1}{4}\arctan\frac{a+b-c}{p} + \frac{1}{4}\arctan\frac{a-b+c}{p}$
$$+\frac{1}{4}\arctan\frac{-a+b+c}{p}$$

$\qquad\qquad\qquad\qquad\qquad [p>0]\qquad\qquad$ BI (365)(11)

2.[8] $\displaystyle\int_0^\infty e^{-px}\sin^2 ax\sin bx\,\frac{\mathrm{d}x}{x} = \frac{1}{2}\arctan\frac{b}{p} - \frac{1}{2}\left[\frac{1}{2}\arctan\frac{2pb}{p^2+4a^2-b^2} + s\frac{\pi}{2}\right]$

$$\left[s = \begin{cases} 1 & \text{for } p^2+4a^2-b^2 < 0 \\ 0 & \text{for } p^2+4a^2-b^2 \geq 0 \end{cases}\right]$$

$\qquad\qquad\qquad\qquad\qquad\qquad\qquad\qquad\qquad\qquad\qquad$ BI (365)(8)

3. $\displaystyle\int_0^\infty e^{-px}\sin^2 ax\cos bx\,\frac{\mathrm{d}x}{x} = \frac{1}{8}\ln\frac{\left[p^2+(2a+b)^2\right]\left[p^2+(2a-b)^2\right]}{\left(p^2+b^2\right)^2}$

$\qquad\qquad\qquad\qquad\qquad [p>0]\qquad\qquad$ BI (365)(9)

4.8    $\displaystyle\int_0^\infty e^{-px}\sin ax\cos^2 bx\,\frac{\mathrm{d}x}{x}=\frac{1}{2}\arctan\frac{a}{p}+\frac{1}{2}\left[\frac{1}{2}\arctan\frac{2pa}{p^2+4b^2-a^2}+s\frac{\pi}{2}\right]$

$$\left[s=\begin{cases}1 & \text{for } p^2+4b^2-a^2<0\\ 0 & \text{for } p^2+4b^2-a^2\ge 0\end{cases}\right]$$

BI (365)(10)

5.    $\displaystyle\int_0^\infty e^{-px}\sin^2 ax\sin bx\sin cx\,\frac{\mathrm{d}x}{x}=\frac{1}{8}\ln\frac{p^2+(b+c)^2}{p^2+(b-c)^2}$

$$+\frac{1}{16}\ln\frac{\left[p^2+(2a-b+c)^2\right]\left[p^2+(2a+b-c)^2\right]}{\left[p^2+(2a+b+c)^2\right]\left[p^2+(2a-b-c)^2\right]}$$

$$[p>0]$$

BI (365)(15)

**3.951**

1.    $\displaystyle\int_0^\infty\left(1-e^{-x}\right)\cos x\,\frac{\mathrm{d}x}{x}=\ln\sqrt{2}$                                       FI II 745

2.    $\displaystyle\int_0^\infty\frac{e^{-\gamma x}-e^{-\beta x}}{x}\sin bx\,\mathrm{d}x=\arctan\frac{(\beta-\gamma)b}{b^2+\beta\gamma}$     $[\operatorname{Re}\beta>0,\quad \operatorname{Re}\gamma\ge 0]$     BI (367)(3)

3.    $\displaystyle\int_0^\infty\frac{e^{-\gamma x}-e^{-\beta x}}{x}\cos bx\,\mathrm{d}x=\frac{1}{2}\ln\frac{b^2+\beta^2}{b^2+\gamma^2}$     $[\operatorname{Re}\beta>0,\quad \operatorname{Re}\gamma\ge 0]$     BI (367)(4)

4.11    $\displaystyle\int_0^\infty\frac{e^{-\gamma x}-e^{-\beta x}}{x^2}\sin bx\,\mathrm{d}x=\frac{b}{2}\ln\frac{b^2+\beta^2}{b^2+\gamma^2}+\beta\arctan\frac{b}{\beta}-\gamma\arctan\frac{b}{\gamma}$

$$[\operatorname{Re}\beta>0,\quad \operatorname{Re}\gamma>0]$$

BI (368)(21)a

5.    $\displaystyle\int_0^\infty\frac{x}{e^{\beta x}-1}\cos bx\,\mathrm{d}x=\frac{1}{2b^2}-\frac{\pi^2}{2\beta^2}\operatorname{cosech}^2\frac{b\pi}{\beta}$     $[\operatorname{Re}\beta>0]$     ET I 15(18)

6.    $\displaystyle\int_0^\infty\left(\frac{1}{e^x-1}-\frac{1}{x}\right)\cos bx\,\mathrm{d}x=\ln b-\frac{1}{2}\left[\psi(ib)+\psi(-ib)\right]$

$$[b>0]$$

ET I 15(9)

7.    $\displaystyle\int_0^\infty\frac{1-\cos ax}{e^{2\pi x}-1}\cdot\frac{\mathrm{d}x}{x}=\frac{a}{4}+\frac{1}{2}\ln\frac{1-e^{-a}}{a}$     $[a>0]$     BI (387)(10)

8.    $\displaystyle\int_0^\infty\left(e^{-\beta x}-e^{-\gamma x}\cos ax\right)\frac{\mathrm{d}x}{x}=\frac{1}{2}\ln\frac{a^2+\gamma^2}{\beta^2}$     $[\operatorname{Re}\beta>0,\quad \operatorname{Re}\gamma>0]$     BI (367)(10)

9.    $\displaystyle\int_0^\infty\frac{\cos px-e^{-px}}{b^4+x^4}\frac{\mathrm{d}x}{x}=\frac{\pi}{2b^4}\exp\left(-\frac{1}{2}bp\sqrt{2}\right)\sin\left(\frac{1}{2}bp\sqrt{2}\right)$

$$[p>0]$$

BI (390)(6)

10.    $\displaystyle\int_0^\infty\left(\frac{1}{e^x-1}-\frac{\cos x}{x}\right)\mathrm{d}x=C$                                   NT 65(8)

11.    $\displaystyle\int_0^\infty\left(ae^{-px}-\frac{e^{-qx}}{x}\sin ax\right)\frac{\mathrm{d}x}{x}=\frac{a}{2}\ln\frac{a^2+q^2}{p^2}+q\arctan\frac{a}{q}-a$

$$[p>0,\quad q>0]$$

BI (368)(24)

12. $\displaystyle\int_0^\infty \frac{x^{2m}\sin bx}{e^x-1}\,dx = (-1)^m \frac{\partial^{2m}}{\partial b^{2m}}\left[\frac{\pi}{2}\coth b\pi - \frac{1}{2b}\right]$ $\quad[b>0]$ $\qquad$ GW (336)(15a)

13. $\displaystyle\int_0^\infty \frac{x^{2m+1}\cos bx}{e^x-1}\,dx = (-1)^m \frac{\partial^{2m+1}}{\partial b^{2m+1}}\left[\frac{\pi}{2}\coth b\pi - \frac{1}{2b}\right]$

$\qquad\qquad\qquad\qquad\qquad\qquad\qquad\qquad\qquad\qquad [b>0]$ $\qquad$ GW (336)(15b)

14. $\displaystyle\int_0^\infty \frac{x^{2m}\sin bx\,dx}{e^{(2n+1)cx}-e^{(2n-1)cx}} = (-1)^m \frac{\partial^{2m}}{\partial b^{2m}}\left[\frac{\pi}{4c}\tanh\frac{b\pi}{2c} - \sum_{k=1}^{n}\frac{b}{b^2+(2k-1)^2c^2}\right]$

$\qquad\qquad\qquad\qquad\qquad\qquad\qquad\qquad\qquad\qquad [b>0]$ $\qquad$ GW (336)(14a)

15. $\displaystyle\int_0^\infty \frac{x^{2m+1}\cos bx\,dx}{e^{(2n+1)cx}-e^{(2n-1)cx}} = (-1)^m \frac{\partial^{2m+1}}{\partial b^{2m+1}}\left[\frac{\pi}{4c}\tanh\frac{b\pi}{2c} - \sum_{k=1}^{n}\frac{b}{b^2+(2k-1)^2c^2}\right]$

$\qquad\qquad\qquad\qquad\qquad\qquad\qquad\qquad\qquad\qquad [b>0]$ $\qquad$ GW (336)(14b)

16.[12] $\displaystyle\int_0^\infty \frac{x^{2m}\sin bx\,dx}{e^{2ncx}-e^{(2n-2)cx}} = (-1)^m \frac{\partial^{2m}}{\partial b^{2m}}\left[\frac{\pi}{4c}\coth\frac{b\pi}{2c} - \frac{1}{2b} - \sum_{k=1}^{n-1}\frac{b}{b^2+(2k)^2c^2}\right]$

$\qquad\qquad\qquad\qquad\qquad\qquad\qquad\qquad\qquad [b>0,\quad c>0]$ $\qquad$ GW (336)(14c)

17. $\displaystyle\int_0^\infty \frac{x^{2m+1}\cos bx\,dx}{e^{2ncx}-e^{(2n-2)cx}} = (-1)^m \frac{\partial^{2m+1}}{\partial b^{2m+1}}\left[\frac{\pi}{4c}\coth\frac{b\pi}{2c} - \frac{1}{2b} - \sum_{k=1}^{n-1}\frac{b}{b^2+(2k)^2c^2}\right]$

$\qquad\qquad\qquad\qquad\qquad\qquad\qquad\qquad\qquad [b>0,\quad c>0]$ $\qquad$ GW (336)(14d)

18. $\displaystyle\int_0^\infty \frac{\cos ax - \cos bx}{e^{(2m+1)px}-e^{(2m-1)px}}\cdot\frac{dx}{x} = \frac{1}{2}\ln\frac{\cosh\frac{b\pi}{2p}}{\cosh\frac{a\pi}{2p}} - \frac{1}{2}\sum_{k=1}^{m}\ln\frac{b^2+(2k-1)^2p^2}{a^2+(2k-1)^2p^2}$

$\qquad\qquad\qquad\qquad\qquad\qquad\qquad\qquad\qquad\qquad\qquad [p>0]$ $\qquad$ GW (336)(16a)

19. $\displaystyle\int_0^\infty \frac{\cos ax - \cos bx}{e^{2mpx}-e^{(2m-2)px}}\cdot\frac{dx}{x} = \frac{1}{2}\ln\frac{a\sinh\frac{b\pi}{2p}}{b\sinh\frac{a\pi}{2p}} - \frac{1}{2}\sum_{k=1}^{m-1}\ln\frac{b^2+4k^2p^2}{a^2+4k^2p^2}$

$\qquad\qquad\qquad\qquad\qquad\qquad\qquad\qquad\qquad\qquad\qquad [p>0]$ $\qquad$ GW (336)(16b)

20. $\displaystyle\int_0^\infty \frac{\sin x\sin bx}{1-e^x}\cdot\frac{dx}{x} = \frac{1}{4}\ln\frac{(b+1)\sinh[(b-1)\pi]}{(b-1)\sinh[(b+1)\pi]}$ $\quad[b^2\neq 1]$ $\qquad$ LO V 305

21. $\displaystyle\int_0^\infty \frac{\sin^2 ax}{1-e^x}\cdot\frac{dx}{x} = \frac{1}{4}\ln\frac{2a\pi}{\sinh 2a\pi}$ $\qquad$ LO V 306, BI (387)(5)

**3.952**

1. $\displaystyle\int_0^\infty xe^{-p^2x^2}\sin ax\,dx = \frac{a\sqrt{\pi}}{4p^3}\exp\left(-\frac{a^2}{4p^2}\right)$ $\qquad$ BI (362)(1)

2. $\displaystyle\int_0^\infty xe^{-p^2x^2}\cos ax\,dx = \frac{1}{2p^2} - \frac{a}{4p^3}\sum_{k=0}^{\infty}\frac{(-1)^k k!}{(2k+1)!}\left(\frac{a}{p}\right)^{2k+1}$

$\qquad\qquad\qquad\qquad\qquad\qquad\qquad\qquad\qquad\qquad [a>0]$ $\qquad$ BI (362)(2)

$3.^{12}$    $\displaystyle\int_0^\infty x^2 e^{-p^2 x^2} \sin ax \, dx = \frac{a}{4p^4} + \frac{2p^2 - a^2}{8p^5} \sum_{k=0}^\infty \frac{(-1)^k k!}{(2k+1)!} \left(\frac{a}{p}\right)^{2k+2}$

$$[a > 0] \qquad \text{BI (362)(4)}$$

4.    $\displaystyle\int_0^\infty x^2 e^{-p^2 x^2} \cos ax \, dx = \sqrt{\pi}\,\frac{2p^2 - a^2}{8p^5} \exp\left(-\frac{a^2}{4p^2}\right)$        BI (362)(5)

5.    $\displaystyle\int_0^\infty x^3 e^{-p^2 x^2} \sin ax \, dx = \sqrt{\pi}\,\frac{6ap^2 - a^3}{16p^7} \exp\left(-\frac{a^2}{4p^2}\right)$        BI (362)(6)

$6.^3$    $\displaystyle\int_0^\infty e^{-p^2 x^2} \sin ax \,\frac{dx}{x} = \frac{a\sqrt{\pi}}{2p} \sum_{k=0}^\infty \frac{(-1)^k}{k!(2k+1)} \left(\frac{a}{2p}\right)^{2k} = \frac{\pi}{2}\,\Phi\left(\frac{a}{2p}\right)$        BI (365)(21)

7.    $\displaystyle\int_0^\infty x^{\mu-1} e^{-\beta x^2} \sin \gamma x \, dx = \frac{\gamma e^{-\frac{\gamma^2}{4\beta}}}{2\beta^{\frac{\mu+1}{2}}} \Gamma\left(\frac{1+\mu}{2}\right) {}_1F_1\left(1 - \frac{\mu}{2}; \frac{3}{2}; \frac{\gamma^2}{4\beta}\right)$

$$[\operatorname{Re}\beta > 0, \quad \operatorname{Re}\mu > -1] \qquad \text{ET I 318(10)}$$

$8.^{10}$    $\displaystyle\int_0^\infty x^{\mu-1} e^{-\beta x^2} \cos ax \, dx = \frac{1}{2}\beta^{-\mu/2} \Gamma\left(\frac{\mu}{2}\right) e^{-a^2/4\beta} {}_1F_1\left(-\frac{\mu}{2} + \frac{1}{2}; \frac{1}{2}; \frac{a^2}{4\beta}\right)$

$$[\operatorname{Re}\beta > 0, \quad \operatorname{Re}\mu > 0, \quad a > 0]$$
$$\text{ET I 320(30)}$$

9.    $\displaystyle\int_0^\infty x^{2n} e^{-\beta^2 x^2} \cos ax \, dx = (-1)^n \frac{\sqrt{\pi}}{2^{n+1}\beta^{2n+1}} \exp\left(-\frac{a^2}{8\beta^2}\right) D_{2n}\left(\frac{a}{\beta\sqrt{2}}\right)$

$$= (-1)^n \frac{\sqrt{\pi}}{(2\beta)^{2n+1}} \exp\left(-\frac{a^2}{4\beta^2}\right) H_{2n}\left(\frac{a}{2\beta}\right)$$

$$\left[|\arg\beta| < \frac{\pi}{4}, \quad a > 0\right] \qquad \text{WH, ET I 15(13)}$$

10.    $\displaystyle\int_0^\infty x^{2n+1} e^{-\beta^2 x^2} \sin ax \, dx = (-1)^n \frac{\sqrt{\pi}}{2^{n+\frac{3}{2}}\beta^{2n+2}} \exp\left(-\frac{a^2}{8\beta^2}\right) D_{2n+1}\left(\frac{a}{\beta\sqrt{2}}\right)$

$$= (-1)^n \frac{\sqrt{\pi}}{(2\beta)^{2n+2}} \exp\left(-\frac{a^2}{4\beta^2}\right) H_{2n+1}\left(\frac{a}{2\beta}\right)$$

$$\left[|\arg\beta| < \frac{\pi}{4}, \quad a > 0\right] \qquad \text{WH, ET I 74(23)}$$

**3.953**

1.    $\displaystyle\int_0^\infty x^{\mu-1} e^{-\gamma x - \beta x^2} \sin ax \, dx$

$$= -\frac{i}{2(2\beta)^{\frac{\mu}{2}}} \exp\left(\frac{\gamma^2 - a^2}{8\beta}\right) \Gamma(\mu) \left\{\exp\left(-\frac{ia\gamma}{4\beta}\right) D_{-\mu}\left(\frac{\gamma - ia}{\sqrt{2\beta}}\right) - \exp\left(\frac{ia\gamma}{4\beta}\right) D_{-\mu}\left(\frac{\gamma + ia}{\sqrt{2\beta}}\right)\right\}$$

$$[\operatorname{Re}\mu > -1, \quad \operatorname{Re}\beta > 0, \quad a > 0] \quad \text{ET I 318(11)}$$

2.    $\displaystyle\int_0^\infty x^{\mu-1} e^{-\gamma x - \beta x^2} \cos ax \, dx$

$$= \frac{1}{2(2\beta)^{\frac{\mu}{2}}} \exp\left(\frac{\gamma^2 - a^2}{8\beta}\right) \Gamma(\mu) \left\{\exp\left(-\frac{ia\gamma}{4\beta}\right) D_{-\mu}\left(\frac{\gamma - ia}{\sqrt{2\beta}}\right) + \exp\left(\frac{ia\gamma}{4\beta}\right) D_{-\mu}\left(\frac{\gamma + ia}{\sqrt{2\beta}}\right)\right\}$$

$$[\operatorname{Re}\mu > 0, \quad \operatorname{Re}\beta > 0, \quad a > 0] \quad \text{ET I 16(18)}$$

$3.^{12}$ $\int_0^\infty x e^{-\gamma x - \beta x^2} \sin ax\, dx = \dfrac{i\sqrt{\pi}}{8\sqrt{\beta^3}} \left\{ (\gamma - ia) \exp\left[ \dfrac{(\gamma - ia)^2}{4\beta} \right] \left[ 1 - \Phi\left( \dfrac{\gamma - ia}{2\sqrt{\beta}} \right) \right] \right.$

$$\left. - (\gamma + ia) \exp\left[ \dfrac{(\gamma + ia)^2}{4\beta} \right] \left[ 1 - \Phi\left( \dfrac{\gamma + ia}{2\sqrt{\beta}} \right) \right] \right\}$$

$$[\operatorname{Re} \beta > 0, \quad a > 0] \qquad \text{ET I 74(28)}$$

$4.$ $\int_0^\infty x e^{-\gamma x - \beta x^2} \cos ax\, dx = -\dfrac{\sqrt{\pi}}{8\sqrt{\beta^3}} \left\{ (\gamma - ia) \exp\left[ \dfrac{(\gamma - ia)^2}{4\beta} \right] \left[ 1 - \Phi\left( \dfrac{\gamma - ia}{2\sqrt{\beta}} \right) \right] \right.$

$$\left. + (\gamma + ia) \exp\left[ \dfrac{(\gamma + ia)^2}{4\beta} \right] \left[ 1 - \Phi\left( \dfrac{\gamma + ia}{2\sqrt{\beta}} \right) \right] \right\} + \dfrac{1}{2\beta}$$

$$[\operatorname{Re} \beta > 0, \quad a > 0] \qquad \text{ET I 16(17)}$$

**3.954**

$1.^{11}$ $\int_0^\infty e^{-\beta x^2} \sin ax \dfrac{x\, dx}{\gamma^2 + x^2} = -\dfrac{\pi}{4} e^{\beta \gamma^2} \left[ 2 \sinh a\gamma + e^{-\gamma a} \, \Phi\left( \gamma\sqrt{\beta} - \dfrac{a}{2\sqrt{\beta}} \right) - e^{\gamma a} \, \Phi\left( \gamma\sqrt{\beta} + \dfrac{a}{2\sqrt{\beta}} \right) \right]$

$$[\operatorname{Re} \beta > 0, \quad \operatorname{Re} \gamma > 0, \quad a > 0]$$
$$\text{ET I 74(26)a}$$

$2.^{11}$ $\int_0^\infty e^{-\beta x^2} \cos ax \dfrac{dx}{\gamma^2 + x^2} = \dfrac{\pi}{4\gamma} e^{\beta \gamma^2} \left[ 2 \cosh a\gamma - e^{-\gamma a} \, \Phi\left( \gamma\sqrt{\beta} - \dfrac{a}{2\sqrt{\beta}} \right) - e^{\gamma a} \, \Phi\left( \gamma\sqrt{\beta} + \dfrac{a}{2\sqrt{\beta}} \right) \right]$

$$[\operatorname{Re} \beta > 0, \quad \operatorname{Re} \gamma > 0, \quad a > 0]$$
$$\text{ET I 15(15)}$$

**3.955** $\int_0^\infty x^\nu e^{-\frac{x^2}{2}} \cos\left( \beta x - \nu \dfrac{\pi}{2} \right) dx = \sqrt{\dfrac{\pi}{2}} e^{-\frac{\beta^2}{4}} D_\nu(\beta)$ $\qquad [\operatorname{Re} \nu > -1] \qquad \text{EH II 120(4)}$

**3.956** $\int_0^\infty e^{-x^2} (2x \cos x - \sin x) \sin x \dfrac{dx}{x^2} = \sqrt{\pi} \dfrac{e-1}{2e}$ $\qquad \text{BI (369)(19)}$

**3.957**

$1.$ $\int_0^\infty x^{\mu-1} \exp\left( \dfrac{-\beta^2}{4x} \right) \sin ax\, dx$

$$= \dfrac{i}{2^\mu} \beta^\mu a^{-\frac{\mu}{2}} \left[ \exp\left( -\dfrac{i}{4}\mu\pi \right) K_\mu\left( \beta e^{\frac{\pi i}{4}} \sqrt{a} \right) - \exp\left( \dfrac{i}{4}\mu\pi \right) K_\mu\left( \beta e^{-\pi i/4} \sqrt{a} \right) \right]$$

$$[\operatorname{Re} \beta > 0, \quad \operatorname{Re} \mu < 1, \quad a > 0] \quad \text{ET I 318(12)}$$

$2.$ $\int_0^\infty x^{\mu-1} \exp\left( \dfrac{-\beta^2}{4x} \right) \cos ax\, dx$

$$= \dfrac{1}{2^\mu} \beta^\mu a^{-\frac{\mu}{2}} \left[ \exp\left( -\dfrac{i}{4}\mu\pi \right) K_\mu\left( \beta e^{\pi i/4} \sqrt{a} \right) + \exp\left( \dfrac{i}{4}\mu\pi \right) K_\mu\left( \beta e^{-\pi i/4} \sqrt{a} \right) \right]$$

$$[\operatorname{Re} \beta > 0, \quad \operatorname{Re} \mu < 1, \quad a > 0] \quad \text{ET I 320(32)a}$$

**3.958**

$1.$ $\int_{-\infty}^\infty x^n e^{-(ax^2 + bx + c)} \sin(px + q)\, dx = -\left( \dfrac{-1}{2a} \right)^n \sqrt{\dfrac{\pi}{a}} \exp\left( \dfrac{b^2 - p^2}{4a} - c \right) \sum_{k=0}^{\lfloor n/2 \rfloor} \dfrac{n!}{(n-2k)!k!} a^k$

$$\times \sum_{j=0}^{n-2k} \binom{n-2k}{j} b^{n-2k-j} p^j \sin\left( \dfrac{pb}{2a} - q + \dfrac{\pi}{2}j \right)$$

$$[a > 0] \qquad \text{GW (37)(1b)}$$

$2.^{12}$ $\displaystyle\int_{-\infty}^{\infty} x^n e^{-(ax^2+bx+c)}\cos(px+q)\,dx = \left(\frac{-1}{2a}\right)^n \sqrt{\frac{\pi}{a}}\exp\left(\frac{b^2-p^2}{4a}-c\right)\sum_{k=0}^{\lfloor n/2\rfloor}\frac{n!}{(n-2k)!k!}a^k$

$$\times\sum_{j=0}^{n-2k}\binom{n-2k}{j}b^{n-2k-j}p^j\cos\left(\frac{pb}{2a}-q+\frac{\pi}{2}j\right)$$

$$[a>0] \qquad\qquad \text{GW (337)(1a)}$$

**3.959**    $\displaystyle\int_0^{\infty} xe^{-p^2x^2}\tan ax\,dx = \frac{a\sqrt{\pi}}{p^3}\sum_{k=1}^{\infty}(-1)^k k\exp\left(-\frac{a^2k^2}{p^2}\right)$

$$[p>0] \qquad\qquad \text{BI (362)(15)}$$

**3.961**

1.    $\displaystyle\int_0^{\infty}\exp\left(-\beta\sqrt{\gamma^2+x^2}\right)\sin ax\,\frac{x\,dx}{\sqrt{\gamma^2+x^2}} = \frac{a\gamma}{\sqrt{a^2+\beta^2}}K_1\left(\gamma\sqrt{a^2+\beta^2}\right)$

$$[\operatorname{Re}\beta>0,\quad \operatorname{Re}\gamma>0,\quad a>0]$$
$$\text{ET I 75(36)}$$

2.    $\displaystyle\int_0^{\infty}\exp\left[-\beta\sqrt{\gamma^2+x^2}\right]\cos ax\,\frac{dx}{\sqrt{\gamma^2+x^2}} = K_0\left(\gamma\sqrt{a^2+\beta^2}\right)$

$$[\operatorname{Re}\beta>0,\quad \operatorname{Re}\gamma>0,\quad a>0]$$
$$\text{ET I 17(27)}$$

**3.962**

1.    $\displaystyle\int_0^{\infty}\frac{\sqrt{\sqrt{\gamma^2+x^2}-\gamma}\exp\left(-\beta\sqrt{\gamma^2+x^2}\right)}{\sqrt{\gamma^2+x^2}}\sin ax\,dx = \sqrt{\frac{\pi}{2}}\frac{a\exp\left(-\gamma\sqrt{a^2+\beta^2}\right)}{\sqrt{\beta^2+a^2}\sqrt{\beta+\sqrt{a^2+\beta^2}}}$

$$[\operatorname{Re}\beta>0,\quad \operatorname{Re}\gamma>0,\quad a>0]$$
$$\text{FT I 75(38)}$$

2.    $\displaystyle\int_0^{\infty}\frac{x\exp\left(-\beta\sqrt{\gamma^2+x^2}\right)}{\sqrt{\gamma^2+x^2}\sqrt{\sqrt{\gamma^2+x^2}-\gamma}}\cos ax\,dx = \sqrt{\frac{\pi}{2}}\frac{\sqrt{\beta+\sqrt{a^2+\beta^2}}}{\sqrt{a^2+\beta^2}}\exp\left[-\gamma\sqrt{a^2+\beta^2}\right]$

$$[\operatorname{Re}\beta>0,\quad \operatorname{Re}\gamma>0,\quad a>0]$$
$$\text{ET I 17(29)}$$

**3.963**

1.    $\displaystyle\int_0^{\infty} e^{-\tan^2 x}\frac{\sin x}{\cos^2 x}\frac{dx}{x} = \frac{\sqrt{\pi}}{2}$                           BI (391)(1)

2.    $\displaystyle\int_0^{\pi/2} e^{-p\tan x}\frac{x\,dx}{\cos^2 x} = \frac{1}{p}\left[\operatorname{ci}(p)\sin p-\cos p\operatorname{si}(p)\right]$    $[p>0]$    (cf. **3.339**)    BI (396)(3)

$3.^{8}$    $\displaystyle\int_0^{\pi/2} xe^{-\tan^2 x}\sin 4x\frac{dx}{\cos^8 x} = -\frac{3}{2}\sqrt{\pi}$                      BI (396)(5)

$4.^{8}$    $\displaystyle\int_0^{\pi/2} xe^{-\tan^2 x}\sin^3 2x\frac{dx}{\cos^8 x} = 2\sqrt{\pi}$                      BI (396)(6)

**3.964**

1.    $\displaystyle\int_0^{\pi/2} xe^{-p\tan x}\frac{p\sin x - \cos x}{\cos^3 x}\,\mathrm{d}x = -\sin p\,\mathrm{si}(p) - \mathrm{ci}(p)\cos p$

$[p > 0]$                    LI (396)(4)

2.    $\displaystyle\int_0^{\pi/2} xe^{-p\tan^2 x}\frac{p - \cos^2 x}{\cos^4 x\cot x}\,\mathrm{d}x = \frac{1}{4}\sqrt{\frac{\pi}{p}}$    $[p > 0]$                    BI (396)(7)

3.[8]    $\displaystyle\int_0^{\pi/2} xe^{-p\tan^2 x}\frac{p - 2\cos^2 x}{\cos^6 x\cot x}\,\mathrm{d}x = \frac{1+2p}{8p}\sqrt{\frac{\pi}{p}}$    $[p > 0]$                    BI (396)(8)

**3.965**

1.    $\displaystyle\int_0^{\infty} xe^{-\beta x}\sin ax^2\sin\beta x\,\mathrm{d}x = \frac{\beta}{4}\sqrt{\frac{\pi}{2a^3}}e^{-\frac{\beta^2}{2a}}$    $\left[|\arg\beta| < \dfrac{\pi}{4}, \quad a > 0\right]$                    ET I 84(17)

2.    $\displaystyle\int_0^{\infty} xe^{-\beta x}\cos ax^2\cos\beta x\,\mathrm{d}x = \frac{\beta}{4}\sqrt{\frac{\pi}{2a^3}}e^{-\frac{\beta^2}{2a}}$    $[a > 0, \quad \mathrm{Re}\,\beta > |\mathrm{Im}\,\beta|]$                    ET 26(27)

**3.966**

1.    $\displaystyle\int_0^{\infty} xe^{-px}\cos\left(2x^2 + px\right)\,\mathrm{d}x = 0$    $[p > 0]$                    BI (361)(16)

2.    $\displaystyle\int_0^{\infty} xe^{-px}\cos\left(2x^2 - px\right)\,\mathrm{d}x = \frac{p\sqrt{\pi}}{8}\exp\left(-\frac{1}{4}p^2\right)$    $[p > 0]$                    BI (361)(17)

3.    $\displaystyle\int_0^{\infty} x^2 e^{-px}\left[\sin\left(2x^2 + px\right) + \cos\left(2x^2 + px\right)\right]\,\mathrm{d}x = 0$    $[p > 0]$                    BI (361)(18)

4.    $\displaystyle\int_0^{\infty} x^2 e^{-px}\left[\sin\left(2x^2 - px\right) - \cos\left(2x^2 - px\right)\right]\,\mathrm{d}x = \frac{\sqrt{\pi}}{16}\left(2 - p^2\right)\exp\left(-\frac{1}{4}p^2\right)$                    BI (361)(19)

5.[3]    $\displaystyle\int_0^{\infty} x^{\mu-1}e^{-x}\cos\left(x + ax^2\right)\,\mathrm{d}x = \frac{e^{\frac{1}{4a}}\,\Gamma(\mu)}{(2a)^{\frac{\mu}{2}}}\cos\frac{\mu\pi}{4}\,D_{-\mu}\left(\frac{1}{\sqrt{a}}\right)$

$[\mathrm{Re}\,\mu > 0, \quad a > 0]$                    ET I 321(37)

6.[6]    $\displaystyle\int_0^{\infty} x^{\mu-1}e^{-x}\sin\left(x + ax^2\right)\,\mathrm{d}x = \frac{e^{\frac{1}{4a}}\,\Gamma(\mu)}{(2a)^{\frac{\mu}{2}}}\sin\frac{\mu\pi}{4}\,D_{-\mu}\left(\frac{1}{\sqrt{a}}\right)$

$[\mathrm{Re}\,\mu > -1, \quad a > 0]$                    ET I 319(18)

**3.967**

1.    $\displaystyle\int_0^{\infty} e^{-\frac{\beta^2}{x^2}}\sin a^2 x^2\,\frac{\mathrm{d}x}{x^2} = \frac{\sqrt{\pi}}{2\beta}e^{-\sqrt{2}a\beta}\sin\left(\sqrt{2}a\beta\right)$    $[\mathrm{Re}\,\beta > 0, \quad a > 0]$

ET I 75(30)a, BI(369)(3)a

2.    $\displaystyle\int_0^{\infty} e^{-\frac{\beta^2}{x^2}}\cos a^2 x^2\,\frac{\mathrm{d}x}{x^2} = \frac{\sqrt{\pi}}{2\beta}e^{-\sqrt{2}a\beta}\cos\left(\sqrt{2}a\beta\right)$    $[\mathrm{Re}\,\beta > 0, \quad a > 0]$

BI (369)(4), ET I 16(20)

3.

$$\int_0^\infty x^2 e^{-\beta x^2} \cos ax^2 \, dx = \frac{\sqrt{\pi}}{4\sqrt[4]{(a^2+\beta^2)^3}} \cos\left(\frac{3}{2}\arctan\frac{a}{\beta}\right)$$

$$[\operatorname{Re}\beta > 0] \qquad\qquad \text{ET I 14(3)a}$$

**3.968**

1.[12]
$$\int_0^\infty e^{-\beta x^2}\sin ax^4 \, dx = -\frac{\pi}{8}\sqrt{\frac{\beta}{a}}\left[J_{\frac{1}{4}}\left(\frac{\beta^2}{8a}\right)\cos\left(\frac{\beta^2}{8a}+\frac{\pi}{8}\right) + Y_{\frac{1}{4}}\left(\frac{\beta^2}{8a}\right)\sin\left(\frac{\beta^2}{8a}+\frac{\pi}{8}\right)\right]$$

$$[\operatorname{Re}\beta > 0, \quad a > 0] \qquad \text{ET I 75(34)}$$

2.[12]
$$\int_0^\infty e^{-\beta x^2}\cos ax^4 \, dx = \frac{\pi}{8}\sqrt{\frac{\beta}{a}}\left[J_{\frac{1}{4}}\left(\frac{\beta^2}{8a}\right)\sin\left(\frac{\beta^2}{8a}+\frac{\pi}{8}\right) - Y_{\frac{1}{4}}\left(\frac{\beta^2}{8a}\right)\cos\left(\frac{\beta^2}{8a}+\frac{\pi}{8}\right)\right]$$

$$[\operatorname{Re}\beta > 0, \quad a > 0] \qquad \text{ET I 16(24)}$$

**3.969**

1.
$$\int_0^\infty e^{-p^2 x^4 + q^2 x^2}\left[2px\cos\left(2pqx^3\right) + q\sin\left(2pqx^3\right)\right]\,dx = \frac{\sqrt{\pi}}{2} \qquad\qquad \text{BI (363)(7)}$$

2.
$$\int_0^\infty e^{-p^2 x^4 + q^2 x^2}\left[2px\sin\left(2pqx^3\right) - q\cos\left(2pqx^3\right)\right]\,dx = 0 \qquad\qquad \text{BI (363)(8)}$$

**3.971**   **Notation:** In formulas **3.971** 1 and 2, $p \geq 0$, $q \geq 0$, $r = \sqrt[4]{a^2 + p^2}$, $s = \sqrt[4]{b^2 + q^2}$, $A = \arctan\frac{a}{p}$, and $B = \arctan\frac{b}{q}$.

1.
$$\int_0^\infty \exp\left(-px^2 - \frac{q}{x^2}\right)\sin\left(ax^2 + \frac{b}{x^2}\right)\frac{dx}{x^2} = \frac{1}{2}\int_{-\infty}^\infty \exp\left(-px^2 - \frac{q}{x^2}\right)\sin\left(ax^2 + \frac{b}{x^2}\right)\frac{dx}{x^2}$$
$$= \frac{\sqrt{\pi}}{2s}\exp\left[-2rs\cos(A+B)\right]\sin\left[A + 2rs\sin(A+B)\right]$$

$$\text{BI (369)(16, 17)}$$

2.
$$\int_0^\infty \exp\left(-px^2 - \frac{q}{x^2}\right)\cos\left(ax^2 + \frac{b}{x^2}\right)\frac{dx}{x^2} = \frac{1}{2}\int_{-\infty}^\infty \exp\left(-px^2 - \frac{q}{x^2}\right)\cos\left(ax^2 + \frac{b}{x^2}\right)\frac{dx}{x^2}$$
$$= \frac{\sqrt{\pi}}{2s}\exp\left[-2rs\cos(A+B)\right]\cos\left[A + 2rs\sin(A+B)\right]$$

$$\text{BI (369)(15, 18)}$$

**3.972**

1.
$$\int_0^\infty \exp\left[-\beta\sqrt{\gamma^4 + x^4}\right]\sin ax^2 \frac{dx}{\sqrt{\gamma^4 + x^4}}$$
$$= \sqrt{\frac{a\pi}{8}}\, I_{1/4}\left[\frac{\gamma^2}{2}\left(\sqrt{\beta^2 + a^2} - \beta\right)\right] K_{1/4}\left[\frac{\gamma^2}{4}\left(\sqrt{\beta^2 + a^2} + \beta\right)\right]$$

$$\left[\operatorname{Re}\beta > 0, \quad |\arg\gamma| < \frac{\pi}{4}, \quad a > 0\right] \quad \text{ET I 75(37)}$$

2.
$$\int_0^\infty \exp\left[-\beta\sqrt{\gamma^4 + x^4}\right]\cos ax^2 \frac{dx}{\sqrt{\gamma^4 + x^4}}$$
$$= \sqrt{\frac{a\pi}{8}}\, I_{-1/4}\left[\frac{\gamma^2}{2}\left(\sqrt{\beta^2 + a^2} - \beta\right)\right] K_{1/4}\left[\frac{\gamma^2}{4}\left(\sqrt{\beta^2 + a^2} + \beta\right)\right]$$

$$\left[\operatorname{Re}\beta > 0, \quad |\arg\gamma| < \frac{\pi}{4}, \quad a > 0\right] \quad \text{ET I 17(28)}$$

**3.973**

1. $\displaystyle\int_0^\infty \exp\left(p\cos ax\right)\sin\left(p\sin ax\right)\frac{\mathrm{d}x}{x} = \frac{\pi}{2}\left(e^p - 1\right)$ $\qquad$ $[p > 0, \quad a > 0]$ $\qquad$ WH, FI II 725

2. $\displaystyle\int_0^\infty \exp\left(p\cos ax\right)\sin\left(p\sin ax + bx\right)\frac{x\,\mathrm{d}x}{c^2 + x^2} = \frac{\pi}{2}\exp\left(-cb + pe^{-ac}\right)$

$\qquad\qquad\qquad\qquad\qquad\qquad\qquad [a > 0, \quad b > 0, \quad c > 0, \quad p > 0]$

$\qquad\qquad\qquad\qquad\qquad\qquad\qquad\qquad\qquad\qquad$ BI (372)(3)

3. $\displaystyle\int_0^\infty \exp\left(p\cos ax\right)\cos\left(p\sin ax + bx\right)\frac{\mathrm{d}x}{c^2 + x^2} = \frac{\pi}{2c}\exp\left(-cb + pe^{-ac}\right)$

$\qquad\qquad\qquad\qquad\qquad\qquad\qquad [a > 0, \quad b > 0, \quad c > 0, \quad p > 0]$

$\qquad\qquad\qquad\qquad\qquad\qquad\qquad\qquad\qquad\qquad$ BI (372)(4)

4. $\displaystyle\int_0^\infty \exp\left(p\cos x\right)\sin\left(p\sin x + nx\right)\frac{\mathrm{d}x}{x} = \frac{\pi}{2}e^p$ $\qquad$ $[p > 0]$ $\qquad$ BI (366)(2)

5. $\displaystyle\int_0^\infty \exp\left(p\cos x\right)\sin\left(p\sin x\right)\cos nx\frac{\mathrm{d}x}{x} = \frac{p^n}{n!}\cdot\frac{\pi}{4} + \frac{\pi}{2}\sum_{k=n+1}^\infty \frac{p^k}{k!}$

$\qquad\qquad\qquad\qquad\qquad\qquad\qquad\qquad\qquad [p > 0]$ $\qquad\qquad\qquad$ LI (366)(3)

6. $\displaystyle\int_0^\infty \exp\left(p\cos x\right)\cos\left(p\sin x\right)\sin nx\frac{\mathrm{d}x}{x} = \frac{\pi}{2}\sum_{k=0}^{n-1}\frac{p^k}{k!} + \frac{p^n}{n!}\frac{\pi}{4}$

$\qquad\qquad\qquad\qquad\qquad\qquad\qquad\qquad\qquad [p > 0]$ $\qquad\qquad\qquad$ LI (366)(4)

**3.974**

1. $\displaystyle\int_0^\infty \exp\left(p\cos ax\right)\sin\left(p\sin ax\right)\operatorname{cosec} ax\frac{\mathrm{d}x}{b^2 + x^2} = \frac{\pi\left[e^p - \exp\left(pe^{-ab}\right)\right]}{2b\sinh ab}$

$\qquad\qquad\qquad\qquad\qquad\qquad [a > 0, \quad b > 0, \quad p > 0]$ $\qquad$ BI (391)(4)

2. $\displaystyle\int_0^\infty \left[1 - \exp\left(p\cos ax\right)\cos\left(p\sin ax\right)\right]\operatorname{cosec} ax\frac{x\,\mathrm{d}x}{b^2 + x^2} = \frac{\pi\left[e^p - \exp\left(pe^{-ab}\right)\right]}{2\sinh ab}$

$\qquad\qquad\qquad\qquad\qquad\qquad [a > 0, \quad b > 0, \quad p > 0]$ $\qquad$ BI (391)(5)

3. $\displaystyle\int_0^\infty \exp\left(p\cos ax\right)\sin\left(p\sin ax + ax\right)\operatorname{cosec} ax\frac{\mathrm{d}x}{b^2 + x^2} = \frac{\pi\left[e^p - \exp\left(pe^{-ab} - ab\right)\right]}{2b\sinh ab}$

$\qquad\qquad\qquad\qquad\qquad\qquad [a > 0, \quad b > 0, \quad p > 0]$ $\qquad$ BI (391)(6)

4. $\displaystyle\int_0^\infty \exp\left(p\cos ax\right)\cos\left(p\sin ax + ax\right)\operatorname{cosec} ax\frac{x\,\mathrm{d}x}{b^2 + x^2} = \frac{\pi\left[e^p - \exp\left(pe^{-ab} - ab\right)\right]}{2\sinh ab}$

$\qquad\qquad\qquad\qquad\qquad\qquad [a > 0, \quad b > 0, \quad p > 0]$ $\qquad$ BI (391)(7)

5. $\displaystyle\int_0^\infty \exp\left(p\cos ax\right)\sin\left(p\sin ax\right)\frac{x\,\mathrm{d}x}{b^2 - x^2} = \frac{\pi}{2}\left[1 - \exp\left(p\cos ab\right)\cos\left(p\sin ab\right)\right]$

$\qquad\qquad\qquad\qquad\qquad\qquad\qquad [p > 0, \quad a > 0]$ $\qquad\qquad$ BI (378)(1)

6. $\displaystyle\int_0^\infty \exp\left(p\cos ax\right)\cos\left(p\sin ax\right)\frac{\mathrm{d}x}{b^2 - x^2} = \frac{\pi}{2b}\exp\left(p\cos ab\right)\sin\left(p\sin ab\right)$

$\qquad\qquad\qquad\qquad\qquad\qquad\qquad [a > 0, \quad b > 0, \quad p > 0]$ $\qquad$ BI (378)(2)

7. $\int_0^\infty \exp\left(p\cos ax\right)\sin\left(p\sin ax\right)\tan ax\dfrac{\mathrm{d}x}{b^2+x^2} = \dfrac{\pi}{2b}\cdot\tanh ab\left[\exp\left(pe^{-ab}\right)-e^p\right]$

$$[a>0,\quad b>0,\quad p>0]\qquad \text{BI (372)(14)}$$

8. $\int_0^\infty \exp\left(p\cos ax\right)\sin\left(p\sin ax\right)\cot ax\dfrac{\mathrm{d}x}{b^2+x^2} = \dfrac{\pi}{2b}\coth ab\left[e^p-\exp\left(pe^{-ab}\right)\right]$

$$[a>0,\quad b>0,\quad p>0]\qquad \text{BI (372)(15)}$$

9. $\int_0^\infty \exp\left(p\cos ax\right)\sin\left(p\sin ax\right)\operatorname{cosec} ax\dfrac{\mathrm{d}x}{b^2-x^2} = \dfrac{\pi}{2b}\operatorname{cosec} ab\left[e^p-\exp\left(p\cos ab\right)\cos\left(p\sin ab\right)\right]$

$$[a>0,\quad b>0,\quad p>0]\qquad \text{BI (391)(12)}$$

10. $\int_0^\infty \left[1-\exp\left(p\cos ax\right)\cos\left(p\sin ax\right)\right]\operatorname{cosec} ax\dfrac{x\,\mathrm{d}x}{b^2-x^2} = -\dfrac{\pi}{2}\exp\left(p\cos ab\right)\sin\left(p\sin ab\right)\operatorname{cosec} ab$

$$[a>0,\quad b>0,\quad p>0]\qquad \text{BI (391)(13)}$$

**3.975**

1. $\int_0^\infty \dfrac{\sin\left(b\arctan\frac{x}{\gamma}\right)}{\left(\gamma^2+x^2\right)^{\frac{b}{2}}}\dfrac{\mathrm{d}x}{e^{2\pi x}-1} = \dfrac{1}{2}\zeta(b,\gamma)-\dfrac{1}{4\gamma^b}-\dfrac{\gamma^{1-b}}{2(b-1)}$

$$[\operatorname{Re} b>1,\quad \operatorname{Re}\gamma>0]\qquad \text{WH, ET I 26(7)}$$

2. $\int_0^\infty \dfrac{\sin\left(b\arctan x\right)}{\left(1+x^2\right)^{\frac{b}{2}}}\dfrac{\mathrm{d}x}{e^{2\pi x}+1} = \dfrac{1}{2(b-1)}-\dfrac{\zeta(b)}{2^b}\qquad [\operatorname{Re} b>1]\qquad \text{EH I 33(13)}$

**3.976** $\int_0^\infty \left(1+x^2\right)^{\beta-\frac{1}{2}}e^{-px^2}\cos\left[2px+(2\beta-1)\arctan x\right]\,\mathrm{d}x = \dfrac{e^{-p}}{2p^\beta}\sin\pi\beta\,\Gamma(\beta)$

$$[\operatorname{Re}\beta>0,\quad p>0]\qquad \text{WH}$$

# 3.98–3.99 Combinations of trigonometric and hyperbolic functions

**3.981**

1. $\int_0^\infty \dfrac{\sin ax}{\sinh\beta x}\,\mathrm{d}x = \dfrac{\pi}{2\beta}\tanh\dfrac{a\pi}{2\beta}\qquad [\operatorname{Re}\beta>0,\quad a>0]\qquad \text{BI (264)(16)}$

2. $\int_0^\infty \dfrac{\sin ax}{\cosh\beta x}\,\mathrm{d}x = -\dfrac{\pi}{2\beta}\tanh\dfrac{a\pi}{2\beta}-\dfrac{i}{2\beta}\left[\psi\left(\dfrac{\beta+ai}{4\beta}\right)-\psi\left(\dfrac{\beta-ai}{4\beta}\right)\right]$

$$[\operatorname{Re}\beta>0,\quad a>0]$$
$$\text{GW (335)(12), ET I 88(1)}$$

3. $\int_0^\infty \dfrac{\cos ax}{\cosh\beta x}\,\mathrm{d}x = \dfrac{\pi}{2\beta}\operatorname{sech}\dfrac{a\pi}{2\beta}\qquad [\operatorname{Re}\beta>0,\quad \text{all real } a]\qquad \text{BI (264)(14)}$

4. $\int_0^\infty \sin ax\dfrac{\sinh\beta x}{\sinh\gamma x}\,\mathrm{d}x = \dfrac{\pi}{2\gamma}\dfrac{\sinh\frac{a\pi}{\gamma}}{\cosh\frac{a\pi}{\gamma}+\cos\frac{\beta\pi}{\gamma}}+\dfrac{i}{2\gamma}\left[\psi\left(\dfrac{\beta+\gamma+ia}{2\gamma}\right)-\psi\left(\dfrac{\beta+\gamma-ia}{2\gamma}\right)\right]$

$$[|\operatorname{Re}\beta|<\operatorname{Re}\gamma,\quad a>0]\qquad \text{ET I 88(5)}$$

5. $\int_0^\infty \cos ax\dfrac{\sinh\beta x}{\sinh\gamma x}\,\mathrm{d}x = \dfrac{\pi}{2\gamma}\dfrac{\sin\frac{\pi\beta}{\gamma}}{\cosh\frac{a\pi}{\gamma}+\cos\frac{\beta\pi}{\gamma}}\qquad [|\operatorname{Re}\beta|<\operatorname{Re}\gamma]\qquad \text{BI (265)(7)}$

6. $\displaystyle\int_0^\infty \sin ax \frac{\sinh \beta x}{\cosh \gamma x}\,dx = \frac{\pi}{\gamma}\frac{\sin\frac{\beta\pi}{2\gamma}\sinh\frac{a\pi}{2\gamma}}{\cosh\frac{a\pi}{\gamma}+\cos\frac{\beta\pi}{\gamma}}$ $\qquad [|\mathrm{Re}\,\beta| < \mathrm{Re}\,\gamma, \quad a>0]$ $\qquad$ BI (265)(2)

7. $\displaystyle\int_0^\infty \cos ax \frac{\sinh \beta x}{\cosh \gamma x}\,dx = \frac{1}{4\gamma}\left[\left\{\psi\left(\frac{3\gamma-\beta+ia}{4\gamma}\right)+\psi\left(\frac{3\gamma-\beta-ia}{4\gamma}\right)-\psi\left(\frac{3\gamma+\beta-ia}{4\gamma}\right)\right\}\right.$

$\left.-\psi\left(\frac{3\gamma+\beta+ia}{4\gamma}\right)+\frac{2\pi\sin\frac{\pi\beta}{\gamma}}{\cos\frac{\pi\beta}{\gamma}+\cosh\frac{\pi a}{\gamma}}\right]$

$\qquad\qquad\qquad\qquad [|\mathrm{Re}\,\beta| < \mathrm{Re}\,\gamma, \quad a>0]$ $\qquad$ ET I 31(13)

8. $\displaystyle\int_0^\infty \sin ax \frac{\cosh \beta x}{\sinh \gamma x}\,dx = \frac{\pi}{2\gamma}\cdot\frac{\sinh\frac{\pi a}{\gamma}}{\cosh\frac{\pi a}{\gamma}+\cos\frac{\pi\beta}{\gamma}}$ $\qquad [|\mathrm{Re}\,\beta| < \mathrm{Re}\,\gamma, \quad a>0]$ $\qquad$ BI (265)(4)

9. $\displaystyle\int_0^\infty \sin ax \frac{\cosh \beta x}{\cosh \gamma x}\,dx = \frac{i}{4\gamma}\left[\psi\left(\frac{3\gamma+\beta+ia}{4\gamma}\right)-\psi\left(\frac{3\gamma+\beta-ai}{4\gamma}\right)+\psi\left(\frac{3\gamma-\beta+ia}{4\gamma}\right)\right.$

$\left.-\psi\left(\frac{3\gamma-\beta-ai}{4\gamma}\right)-\frac{2\pi i\sinh\frac{\pi a}{\gamma}}{\cosh\frac{a\pi}{\gamma}+\cos\frac{\beta\pi}{\gamma}}\right]$

$\qquad\qquad\qquad\qquad [|\mathrm{Re}\,\beta| < \mathrm{Re}\,\gamma, \quad a>0]$ $\qquad$ ET I 88(6)

10. $\displaystyle\int_0^\infty \cos ax \frac{\cosh \beta x}{\cosh \gamma x}\,dx = \frac{\pi}{\gamma}\frac{\cos\frac{\beta\pi}{2\gamma}\cosh\frac{a\pi}{2\gamma}}{\cosh\frac{a\pi}{\gamma}+\cos\frac{\beta\pi}{\gamma}}$ $\qquad [|\mathrm{Re}\,\beta| < \mathrm{Re}\,\gamma, \quad \text{all real } a]$ $\qquad$ BI (265)(6)

11.[11] $\displaystyle\int_0^{\pi/2} \cos^{2m} x \cosh \beta x\,dx = \frac{(2m)!\sinh\frac{\pi\beta}{2}}{\beta\left(\beta^2+2^2\right)\ldots\left[\beta^2+(2m)^2\right]}$

$\qquad\qquad\qquad\qquad\qquad\qquad [\beta\neq 0]$ $\qquad$ WA 620a

12.[11] $\displaystyle\int_0^{\pi/2} \cos^{2m+1} x \cosh \beta x\,dx = \frac{(2m+1)!\cosh\frac{\pi\beta}{2}}{\left(\beta^2+1^2\right)\left(\beta^2+3^2\right)\ldots\left[\beta^2+(2m+1)^2\right]}$ $\qquad$ WA 620a

**3.982**

1. $\displaystyle\int_0^\infty \frac{\cos ax}{\cosh^2 \beta x}\,dx = \frac{a\pi}{2\beta^2\sinh\frac{a\pi}{2\beta}}$ $\qquad [\mathrm{Re}\,\beta > 0, \quad a>0]$ $\qquad$ BI (264)(16)

2. $\displaystyle\int_0^\infty \sin ax \frac{\sinh \beta x}{\cosh^2 \gamma x}\,dx = \frac{\pi\left(a\sin\frac{\beta\pi}{2\gamma}\cosh\frac{a\pi}{2\gamma}-\beta\cos\frac{\beta\pi}{2\gamma}\sinh\frac{a\pi}{2\gamma}\right)}{\gamma^2\left(\cosh\frac{a\pi}{\gamma}-\cos\frac{\beta\pi}{\gamma}\right)}$

$\qquad\qquad\qquad\qquad [|\mathrm{Re}\,\beta| < 2\,\mathrm{Re}\,\gamma, \quad a>0]$ $\qquad$ ET I 88(9)

3.[12] $\displaystyle\int_0^\infty \frac{\sin^2 x \cos ax}{\sinh^2 x}\,dx = \frac{\pi}{4}\left\{\frac{a+2}{1-e^{-\pi(a+2)}}-\frac{2a}{1-e^{-\pi a}}+\frac{a-2}{1-e^{-\pi(a-2)}}\right\} = I(a)$

$\qquad\left[I(0) = \frac{1}{2}\left(\pi\coth\pi-1\right), \quad I(\pm 2) = \frac{1}{4}+\frac{\pi}{2}\left(\coth 2\pi-\coth\pi\right)\right]$

## 3.983

1.[6] $\displaystyle\int_0^\infty \frac{\cos ax\, dx}{b\cosh\beta x + c} = \frac{\pi \sin\left(\frac{a}{\beta}\operatorname{arccosh}\frac{c}{b}\right)}{\beta\sqrt{c^2 - b^2}\sinh\frac{a\pi}{\beta}}$ $\qquad [c > b > 0]$

$\displaystyle\qquad\qquad\qquad = \frac{\pi \sinh\left(\frac{a}{\beta}\arccos\frac{c}{b}\right)}{\beta\sqrt{b^2 - c^2}\sinh\frac{a\pi}{\beta}}$ $\qquad [b > |c| > 0]$

$\qquad\qquad\qquad\qquad\qquad\qquad\qquad\qquad [\operatorname{Re}\beta > 0, \quad a > 0]$ $\qquad$ GW (335)(13a)

2. $\displaystyle\int_0^\infty \frac{\cos ax\, dx}{\cosh\beta x + \cos\gamma} = \frac{\pi}{\beta}\frac{\sinh\frac{a\gamma}{\beta}}{\sin\gamma\sinh\frac{a\pi}{\beta}}$ $\qquad \left[\pi\operatorname{Re}\beta < \operatorname{Im}\overline{\beta}\gamma, \quad a > 0\right]$ $\qquad$ BI (267)(3)

3.[12] $\displaystyle\mathrm{PV}\int_0^\infty \frac{\cos ax\, dx}{\cosh x - \cosh b} = -\pi\coth a\pi\frac{\sin ab}{\sinh b}$ $\qquad [a > 0, \quad b > 0]$ $\qquad$ ET I 30(8)

4. $\displaystyle\int_0^\infty \frac{\cos ax\, dx}{1 + 2\cosh\left(\sqrt{\frac{2}{3}}\pi x\right)} = \frac{\sqrt{\frac{\pi}{2}}}{1 + 2\cosh\left(\sqrt{\frac{2}{3}}\pi a\right)}$ $\qquad [a > 0]$ $\qquad$ ET I 30(9)

5. $\displaystyle\int_0^\infty \frac{\sin ax\sinh\beta x}{\cosh\gamma x + \cos\delta}\, dx = \frac{\pi\left\{\sin\left[\frac{\beta}{\gamma}(\pi - \delta)\right]\sinh\left[\frac{a}{\gamma}(\pi + \delta)\right] - \sin\left[\frac{\beta}{\gamma}(\pi + \delta)\right]\sinh\left[\frac{a}{\gamma}(\pi - \delta)\right]\right\}}{\gamma\sin\delta\left(\cosh\frac{2\pi a}{\gamma} - \cos\frac{2\pi\beta}{\gamma}\right)}$

$\qquad\qquad\qquad\qquad\qquad [\pi\operatorname{Re}\gamma > |\operatorname{Re}\overline{\gamma}\delta|, \quad |\operatorname{Re}\beta| < \operatorname{Re}\gamma, \quad a > 0]$ $\quad$ BI (267)(2)

6.[12] $\displaystyle\int_0^\infty \frac{\cos ax\cosh\beta x}{\cosh\gamma x + \cos b}\, dx = \frac{\pi\left\{\cos\left[\frac{\beta}{\gamma}(\pi - b)\right]\cosh\left[\frac{a}{\gamma}(\pi + b)\right] - \cos\left[\frac{\beta}{\gamma}(\pi + b)\right]\cosh\left[\frac{a}{\gamma}(\pi - b)\right]\right\}}{\gamma\sin b\left(\cosh\frac{2\pi a}{\gamma} - \cos\frac{2\pi\beta}{\gamma}\right)}$

$\qquad\qquad\qquad\qquad\qquad\qquad [|\operatorname{Re}\beta| < \operatorname{Re}\gamma, \quad 0 < b < \pi, \quad a \neq 0]$

$\qquad\qquad\qquad\qquad\qquad\qquad\qquad\qquad\qquad\qquad\qquad\qquad\qquad$ BI (267)(6)

7.[12] $\displaystyle\int_0^\infty \frac{\cos ax\, dx}{\left(\beta + \sqrt{\beta^2 - 1}\cosh x\right)^{\nu+1}} = \Gamma(\nu + 1 - ai)e^{a\pi}\frac{Q_\nu^{ai}(\beta)}{\Gamma(\nu + 1)}$

$\qquad\qquad\qquad\qquad\qquad\qquad [\operatorname{Re}\nu > -1, \quad |\arg(\beta \pm 1)| < \pi, \quad a > 0]$

$\qquad\qquad\qquad\qquad\qquad\qquad\qquad\qquad\qquad\qquad\qquad\qquad\qquad$ ET I 30(10)

## 3.984

1.[6] $\displaystyle\lim_{c\uparrow 1}\int_0^\infty \frac{\sin ax\sinh cx}{\cosh x + \cos b}\, dx = \pi\frac{\cosh ab}{\sinh a\pi}$ $\qquad [|b| \leq \pi, \quad a\ \text{real}]$ $\qquad$ BI (267)(1)

2.[6] $\displaystyle\lim_{c\uparrow 1}\int_0^\infty \frac{\cos ax\cosh cx}{\cosh x + \cos b}\, dx = -\pi\cot b\frac{\sinh ab}{\sinh a\pi}$ $\qquad [0 < |b| < \pi, \quad a\ \text{real}]$ $\qquad$ BI (267)(5)

3.[8] $\displaystyle\int_0^\infty \frac{\sin ax\sinh\frac{x}{2}}{\cosh x + \cos\beta}\, dx = \frac{\pi\sinh a\beta}{2\sin\frac{\beta}{2}\cosh a\pi}$ $\qquad [\operatorname{Re}\beta < \pi, \quad a > 0]$ $\qquad$ ET I 80(10)

4. $\displaystyle\int_0^\infty \frac{\cos ax\cosh\frac{\beta}{2}x}{\cosh\beta x + \cosh\gamma}\, dx = \frac{\pi\cos\frac{a\gamma}{\beta}}{2\beta\cosh\frac{\gamma}{2}\cosh\frac{a\pi}{\beta}}$ $\qquad \left[\pi\operatorname{Re}\beta > \left|\operatorname{Im}\left(\overline{\beta}\gamma\right)\right|\right]$ $\qquad$ ET I 31(16)

5. $\displaystyle\int_0^\infty \frac{\sin ax\sinh\beta x}{\cosh 2\beta x + \cos 2ax}\, dx = \frac{a\pi}{4\left(a^2 + \beta^2\right)}$ $\qquad [a > 0, \quad \operatorname{Re}\beta > 0]$ $\qquad$ BI (267)(7)

6.    $\displaystyle\int_0^\infty \frac{\cos ax \cosh \beta x}{\cosh 2\beta x + \cos 2ax}\,dx = \frac{\beta\pi}{4\left(a^2+\beta^2\right)}$      $[\operatorname{Re}\beta > 0, \quad a > 0]$      BI (267)(8)

7.[8]    $\displaystyle\int_0^\infty \frac{\sinh^{2\mu-1} x \cosh^{2\varrho-2\nu+1} x}{\left(\cosh^2 x - \beta \sinh^2 x\right)^\varrho}\,dx = \frac{1}{2}\,\mathrm{B}(\mu,\nu-\mu)\,{}_2F_1(\varrho,\mu;\nu;\beta)$

                                             $[\operatorname{Re}\nu > \operatorname{Re}\mu > 0]$      EH I 115(12)

**3.985**

1.    $\displaystyle\int_0^\infty \frac{\cos ax\,dx}{\cosh^\nu \beta x} = \frac{2^{\nu-2}}{\beta\,\Gamma(\nu)}\,\Gamma\left(\frac{\nu}{2}+\frac{ai}{2\beta}\right)\Gamma\left(\frac{\nu}{2}-\frac{ai}{2\beta}\right)$      $[\operatorname{Re}\beta > 0, \quad \operatorname{Re}\nu > 0, \quad a > 0]$

                                                                              ET I 30(5)

2.    $\displaystyle\int_0^\infty \frac{\cos ax\,dx}{\cosh^{2n}\beta x} = \frac{4^{n-1}\pi a}{2(2n-1)!\beta^2 \sinh\frac{a\pi}{2\beta}} \prod_{k=1}^{n-1}\left(\frac{a^2}{4\beta^2}+k^2\right)$

              $= \dfrac{\pi a\left(a^2+2^2\beta^2\right)\left(a^2+4^2\beta^2\right)\cdots\left[a^2+(2n-2)^2\beta^2\right]}{2(2n-1)!\beta^{2n}\sinh\dfrac{a\pi}{2\beta}}$

                                                         $[n \geq 2, \quad a > 0]$      ET I 30(3)

3.    $\displaystyle\int_0^\infty \frac{\cos ax\,dx}{\cosh^{2n+1}\beta x} = \frac{\pi 2^{2n-1}}{(2n)!\beta\cosh\frac{a\pi}{2\beta}}\prod_{k=1}^{n}\left[\frac{a^2}{4\beta^2}+\left(\frac{2k-1}{2}\right)^2\right]$

              $= \dfrac{\pi\left(a^2+\beta^2\right)\left(a^2+3^2\beta^2\right)\cdots\left[a^2+(2n-1)^2\beta^2\right]}{2(2n)!\beta^{2n+1}\cosh\dfrac{a\pi}{2\beta}}$

                                         $[\operatorname{Re}\beta > 0, \quad n = 0,1,\ldots, \text{ all real } a]$    ET I 30(4)

**3.986**

1.    $\displaystyle\int_0^\infty \frac{\sin\beta x \sin\gamma x}{\cosh\delta x}\,dx = \frac{\pi}{\delta}\cdot\frac{\sinh\frac{\beta\pi}{2\delta}\sinh\frac{\gamma\pi}{2\delta}}{\cosh\frac{\beta}{\delta}\pi + \cosh\frac{\gamma}{\delta}\pi}$      $[|\operatorname{Im}(\beta+\gamma)| < \operatorname{Re}\delta]$      BI (264)(19)

2.    $\displaystyle\int_0^\infty \frac{\sin\alpha x \cos\beta x}{\sinh\gamma x}\,dx = \frac{\pi\sinh\frac{\pi\alpha}{\gamma}}{2\gamma\left(\cosh\frac{\alpha\pi}{\gamma}+\cosh\frac{\beta\pi}{\gamma}\right)}$      $[|\operatorname{Im}(\alpha+\beta)| < \operatorname{Re}\gamma]$      LI (264)(20)

3.    $\displaystyle\int_0^\infty \frac{\cos\beta x \cos\gamma x}{\cosh\delta x}\,dx = \frac{\pi}{\delta}\cdot\frac{\cosh\frac{\beta\pi}{2\delta}\cosh\frac{\gamma\pi}{2\delta}}{\cosh\frac{\beta\pi}{\delta}+\cosh\frac{\gamma\pi}{\delta}}$      $[|\operatorname{Im}(\beta+\gamma)| < \operatorname{Re}\delta]$      BI (264)(21)

4.[3]    $\displaystyle\int_0^\infty \frac{\sin^2\beta x}{\sinh^2\pi x}\,dx = \frac{\beta}{\pi\left(e^{2\beta}-1\right)}+\frac{\beta-1}{2\pi} = \frac{\beta\coth\beta-1}{2\pi}$

                                                             $[|\operatorname{Im}\beta| < \pi]$      EH I 44(3)

**3.987**

1.    $\displaystyle\int_0^\infty \sin ax\,(1-\tanh\beta x)\,dx = \frac{1}{a}-\frac{\pi}{2\beta\sinh\frac{a\pi}{2\beta}}$      $[\operatorname{Re}\beta > 0]$      ET I 88(4)a

2.    $\displaystyle\int_0^\infty \sin ax\,(\coth\beta x - 1)\,dx = \frac{\pi}{2\beta}\coth\frac{a\pi}{2\beta}-\frac{1}{a}$      $[\operatorname{Re}\beta > 0]$      ET I 88(3)

**3.988**

1. $$\int_0^{\pi/2} \frac{\cos ax \sinh\left(2b\cos x\right)}{\sqrt{\cos x}}\, dx = \frac{\pi}{2}\sqrt{\pi b}\, I_{\frac{a}{2}+\frac{1}{4}}(b)\, I_{-\frac{a}{2}+\frac{1}{4}}(b)$$

$$[a > 0] \qquad \text{ET I 37(66)}$$

2. $$\int_0^{\pi/2} \frac{\cos ax \cosh\left(2b\cos x\right)}{\sqrt{\cos x}}\, dx = \frac{\pi}{2}\sqrt{\pi b}\, I_{\frac{a}{2}-\frac{1}{4}}(b)\, I_{-\frac{a}{2}-\frac{1}{4}}(b)$$

$$[a > 0] \qquad \text{ET I 37(67)}$$

3.[12] $$\int_0^\infty \frac{\cos ax\, dx}{\sqrt{\cosh x + \cos b}} = \frac{\pi\, P_{-\frac{1}{2}+ia}\left(\cos b\right)}{\sqrt{2}\cosh a\pi} \qquad [a > 0, \quad b > 0] \qquad \text{ET I 30(7)}$$

**3.989**

1. $$\int_0^\infty \frac{\sin \frac{a^2 x^2}{\pi}\sin bx}{\sinh ax}\, dx = \frac{\pi}{2a}\sin \frac{\pi b^2}{4a^2}\operatorname{cosech}\frac{\pi b}{2a} \qquad [a > 0, \quad b > 0] \qquad \text{ET I 93(44)}$$

2. $$\int_0^\infty \frac{\cos \frac{a^2 x^2}{\pi}\sin bx}{\sinh ax}\, dx = \frac{\pi}{2a}\frac{\cosh \frac{\pi b}{a} - \cos \frac{\pi b^2}{4a^2}}{\sinh \frac{\pi b}{2a}} \qquad [a > 0, \quad b > 0] \qquad \text{ET I 93(45)}$$

3. $$\int_0^\infty \frac{\sin \frac{x^2}{\pi}\cos ax}{\cosh x}\, dx = \frac{\pi}{2}\frac{\cos \frac{a^2\pi}{4} - \frac{1}{\sqrt{2}}}{\cosh \frac{a\pi}{2}} \qquad \text{ET I 36(54)}$$

4. $$\int_0^\infty \frac{\cos \frac{x^2}{\pi}\cos ax}{\cosh x}\, dx = \frac{\pi}{2}\cdot\frac{\sin \frac{a^2\pi}{4} + \frac{1}{\sqrt{2}}}{\cosh \frac{a\pi}{2}} \qquad \text{ET I 36(55)}$$

5. $$\int_0^\infty \frac{\sin\left(\pi a x^2\right)\cos bx}{\cosh \pi x}\, dx = -\sum_{k=0}^\infty \exp\left[-\left(k+\tfrac{1}{2}\right)b\right]\sin\left[\left(k+\tfrac{1}{2}\right)^2\pi a\right]$$
$$+ \frac{1}{\sqrt{a}}\sum_{k=0}^\infty \exp\left[-\frac{b\left(k+\frac{1}{2}\right)}{a}\right]\sin\left[\frac{\pi}{4} - \frac{b^2}{4\pi a} + \frac{\left(k+\frac{1}{2}\right)^2\pi}{a}\right]$$

$$[a > 0, \quad b > 0] \qquad \text{ET I 36(56)}$$

6. $$\int_0^\infty \frac{\cos\left(\pi a x^2\right)\cos bx}{\cosh \pi x}\, dx = \sum_{k=0}^\infty (-1)^k \exp\left[-\left(k+\frac{1}{2}\right)b\right]\cos\left[\left(k+\frac{1}{2}\right)^2\pi a\right]$$
$$+ \frac{1}{\sqrt{a}}\sum_{k=0}^\infty \exp\left[-\frac{b\left(k+\frac{1}{2}\right)}{a}\right]\cos\left[\frac{\pi}{4} - \frac{b^2}{4\pi a} + \frac{\left(k+\frac{1}{2}\right)^2\pi}{a}\right]$$

$$[a > 0, \quad b > 0] \qquad \text{ET I 36(57)}$$

**3.991**

1. $$\int_0^\infty \sin \pi x^2 \sin ax \coth \pi x\, dx = \frac{1}{2}\tanh\frac{a}{2}\sin\left(\frac{\pi}{4} + \frac{a^2}{4\pi}\right) \qquad \text{ET I 93(42)}$$

2.[11] $$\int_0^\infty \cos \pi x^2 \sin ax \coth \pi x\, dx = \frac{1}{2}\tanh\frac{a}{2}\left[1 - \cos\left(\frac{\pi}{4} + \frac{a^2}{4\pi}\right)\right] \qquad \text{ET I 93(43)}$$

**3.992**

1.    $\displaystyle\int_0^\infty \frac{\sin \pi x^2 \cos ax}{1 + 2\cosh\left(\frac{2}{\sqrt{3}}\pi x\right)}\,\mathrm{d}x = -\sqrt{3} + \frac{\cos\left(\frac{\pi}{12} - \frac{a^2}{4\pi}\right)}{4\cosh\frac{a}{\sqrt{3}} - 2}$      ET I 37(60)

2.    $\displaystyle\int_0^\infty \frac{\cos \pi x^2 \cos ax}{1 + 2\cosh\left(\frac{2}{\sqrt{3}}\pi x\right)}\,\mathrm{d}x = 1 - \frac{\sin\left(\frac{\pi}{12} - \frac{a^2}{4\pi}\right)}{4\cosh\frac{a}{\sqrt{3}} - 2}$      ET I 37(61)

**3.993**[12] $\displaystyle\int_0^\infty \frac{\sin\left(x^2\right) + \cos\left(x^2\right)}{\cosh\left(\sqrt{\pi}x\right)}\cos(2ax)\,\mathrm{d}x = \frac{\sqrt{\pi}}{2}\frac{\sin\left(a^2\right) + \cos\left(a^2\right)}{\cosh\left(\sqrt{\pi}a\right)}$      ET I 37(58)

**3.994**

1.    $\displaystyle\int_0^\infty \frac{\sin\left(2a\cosh x\right)\cos bx}{\sqrt{\cosh x}}\,\mathrm{d}x = -\frac{\pi}{4}\sqrt{a\pi}\left[J_{\frac{1}{4}+\frac{ib}{2}}(a)\,Y_{\frac{1}{4}-\frac{ib}{2}}(a) + J_{\frac{1}{4}-\frac{ib}{2}}(a)\,Y_{\frac{1}{4}+\frac{ib}{2}}(a)\right]$

                                   $[a > 0, \quad b > 0]$      ET I 37(62)

2.    $\displaystyle\int_0^\infty \frac{\cos\left(2a\cosh x\right)\cos bx}{\sqrt{\cosh x}}\,\mathrm{d}x = -\frac{\pi}{4}\sqrt{a\pi}\left[J_{-\frac{1}{4}+\frac{ib}{2}}(a)\,Y_{-\frac{1}{4}-\frac{ib}{2}}(a) + J_{-\frac{1}{4}-\frac{ib}{2}}(a)\,Y_{-\frac{1}{4}+\frac{ib}{2}}(a)\right]$

                                   $[a > 0, \quad b > 0]$      ET I 37(63)

3.    $\displaystyle\int_0^\infty \frac{\sin\left(2a\sinh x\right)\sin bx}{\sqrt{\sinh x}}\,\mathrm{d}x = -\frac{i}{2}\sqrt{\pi a}\left[I_{\frac{1}{4}-\frac{ib}{2}}(a)\,K_{-\frac{1}{4}+\frac{ib}{2}}(a) - I_{\frac{1}{4}+\frac{ib}{2}}(a)\,K_{\frac{1}{4}-\frac{ib}{2}}(a)\right]$

                                   $[a > 0, \quad b > 0]$      ET I 93(47)

4.    $\displaystyle\int_0^\infty \frac{\cos\left(2a\sinh x\right)\sin bx}{\sqrt{\sinh x}}\,\mathrm{d}x = -\frac{i}{2}\sqrt{\pi a}\left[I_{-\frac{1}{4}-\frac{ib}{2}}(a)\,K_{-\frac{1}{4}+\frac{ib}{2}}(a) - I_{-\frac{1}{4}+\frac{ib}{2}}(a)\,K_{-\frac{1}{4}-\frac{ib}{2}}(a)\right]$

                                   $[a > 0, \quad b > 0]$      ET I 93(48)

5.    $\displaystyle\int_0^\infty \frac{\sin\left(2a\sinh x\right)\cos bx}{\sqrt{\sinh x}}\,\mathrm{d}x = \frac{\sqrt{\pi a}}{2}\left[I_{\frac{1}{4}-\frac{ib}{2}}(a)\,K_{\frac{1}{4}+\frac{ib}{2}}(a) + I_{\frac{1}{4}+\frac{ib}{2}}(a)\,K_{\frac{1}{4}-\frac{ib}{2}}(a)\right]$

                                   $[a > 0, \quad b > 0]$      ET I 37(64)

6.    $\displaystyle\int_0^\infty \frac{\cos\left(2a\sinh x\right)\cos bx}{\sqrt{\sinh x}}\,\mathrm{d}x = \frac{\sqrt{\pi a}}{2}\left[I_{-\frac{1}{4}-\frac{ib}{2}}(a)\,K_{-\frac{1}{4}+\frac{ib}{2}}(a) + I_{-\frac{1}{4}+\frac{ib}{2}}(a)\,K_{-\frac{1}{4}-\frac{ib}{2}}(a)\right]$

                                   $[a > 0, \quad b > 0]$      ET I 37(65)

7.    $\displaystyle\int_0^\infty \sin\left(a\cosh x\right)\sin\left(a\sinh x\right)\frac{\mathrm{d}x}{\sinh x} = \frac{\pi}{2}\sin a$      $[a > 0]$      BI (264)(22)

**3.995**

1.    $\displaystyle\int_0^{\pi/2} \frac{\sin\left(2a\cos^2 x\right)\cosh\left(a\sin 2x\right)}{b^2\cos^2 x + c^2\sin^2 x}\,\mathrm{d}x = \frac{\pi}{2bc}\sin\frac{2ac}{b+c}$

                                   $[b > 0, \quad c > 0]$      BI (273)(9)

2.    $\displaystyle\int_0^{\pi/2} \frac{\cos\left(2a\cos^2 x\right)\cosh\left(a\sin 2x\right)}{b^2\cos^2 x + c^2\sin^2 x}\,\mathrm{d}x = \frac{\pi}{2bc}\cos\frac{2ac}{b+c}$

                                   $[b > 0, \quad c > 0]$      BI (273)(10)

**3.996**

1.    $\displaystyle\int_0^\infty \sin\left(a\sinh x\right)\sinh bx\,\mathrm{d}x = \sin\frac{b\pi}{2}\,K_b(a)$        $[|\mathrm{Re}\,b| < 1, \quad a > 0]$        EH II 82(26)

2.    $\displaystyle\int_0^\infty \cos\left(a\sinh x\right)\cosh bx\,\mathrm{d}x = \cos\frac{b\pi}{2}\,K_b(a)$        $[|\mathrm{Re}\,b| < 1, \quad a > 0]$        WA 202(13)

3.    $\displaystyle\int_0^{\pi/2} \cos\left(a\sin x\right)\cosh\left(b\cos x\right)\,\mathrm{d}x = \frac{\pi}{2}\,J_0\left(\sqrt{a^2 - b^2}\right)$        MO 40

4.    $\displaystyle\int_0^\infty \sin\left(a\cosh x - \tfrac{1}{2}b\pi\right)\cosh bx\,\mathrm{d}x = \frac{\pi}{2}\,J_b(a)$        $[|\mathrm{Re}\,b| < 1, \quad a > 0]$        WA 199(12)

5.    $\displaystyle\int_0^\infty \cos\left(a\cosh x - \tfrac{1}{2}b\pi\right)\cosh bx\,\mathrm{d}x = -\frac{\pi}{2}\,Y_b(a)$        $[|\mathrm{Re}\,b| < 1, \quad a > 0]$        WA 199(13)

**3.997**

1.    $\displaystyle\int_0^{\pi/2} \sin^\nu x\,\sinh\left(b\cos x\right)\,\mathrm{d}x = \frac{\sqrt{\pi}}{2}\left(\frac{2}{b}\right)^{\frac{\nu}{2}}\Gamma\left(\frac{\nu+1}{2}\right)\mathbf{L}_{\frac{\nu}{2}}(b)$

                                                   $[\mathrm{Re}\,\nu > -1]$        EH II 38(53)

2.    $\displaystyle\int_0^\pi \sin^\nu x\,\cosh\left(b\cos x\right)\,\mathrm{d}x = \sqrt{\pi}\left(\frac{2}{b}\right)^{\frac{\nu}{2}}\Gamma\left(\frac{\nu+1}{2}\right)I_{\frac{\nu}{2}}(b)$

                                                   $[\mathrm{Re}\,\nu > -1]$        WH

3.    $\displaystyle\int_0^{\pi/2} \frac{\mathrm{d}x}{\cosh\left(\tan x\right)\cos x\sqrt{\sin 2x}} = \sqrt{2\pi}\sum_{k=0}^\infty \frac{(-1)^k}{\sqrt{2k+1}}$        BI (276)(13)

4.    $\displaystyle\int_0^{\pi/2} \frac{\tan^q x}{\cosh\left(\tan x\right)\mid\cos\lambda}\frac{\mathrm{d}x}{\sin 2x} = \frac{\Gamma(q)}{\sin\lambda}\sum_{k=1}^\infty (-1)^{k-1}\frac{\sin k\lambda}{k^q}$

                                                   $[q > 0]$        BI (275)(20)

## 4.11–4.12 Combinations involving trigonometric and hyperbolic functions and powers

**4.111**

1. $\displaystyle\int_0^\infty \frac{\sin ax}{\sinh bx} x^{2m}\, dx = (-1)^m \frac{\pi}{2b}\frac{\partial^{2m}}{\partial a^{2m}}\left(\tanh \frac{a\pi}{2b}\right)$      $[\operatorname{Re} b > 0]$     (cf. **3.981** 1)

                                                        GW (336)(17a)

2. $\displaystyle\int_0^\infty \frac{\cos ax}{\sinh bx} x^{2m+1}\, dx = (-1)^m \frac{\pi}{2b}\frac{\partial^{2m+1}}{\partial a^{2m+1}}\left(\tanh \frac{a\pi}{2b}\right)$    $[\operatorname{Re} b > 0]$    (cf. **3.981** 1)

                                                        GW (336)(17b)

3. $\displaystyle\int_0^\infty \frac{\sin ax}{\cosh bx} x^{2m+1}\, dx = (-1)^{m+1} \frac{\pi}{2b}\frac{\partial^{2m+1}}{\partial a^{2m+1}}\left(\frac{1}{\cosh \frac{a\pi}{2b}}\right)$

                                      $[\operatorname{Re} b > 0]$     (cf. **3.981** 3)

                                                        GW (336)(18b)

4. $\displaystyle\int_0^\infty \frac{\cos ax}{\cosh bx} x^{2m}\, dx = (-1)^m \frac{\pi}{2b}\frac{\partial^{2m}}{\partial a^{2m}}\left(\frac{1}{\cosh \frac{a\pi}{2b}}\right)$    $[\operatorname{Re} b > 0]$   (cf. **3.981** 3)

                                                        GW (336)(18a)

5. $\displaystyle\int_0^\infty x\frac{\sin 2ax}{\cosh bx}\, dx = \frac{\pi^2}{4b^2}\cdot\frac{\sinh\frac{a\pi}{b}}{\cosh^2\frac{a\pi}{b}}$      $[\operatorname{Re} b > 0, \quad a > 0]$      BI (364)(6)a

6. $\displaystyle\int_0^\infty x\frac{\cos 2ax}{\sinh bx}\, dx = \frac{\pi^2}{4b^2}\cdot\frac{1}{\cosh^2\frac{a\pi}{b}}$      $[\operatorname{Re} b > 0, \quad a > 0]$      BI (364)(1)a

7. $\displaystyle\int_0^\infty \frac{\sin ax}{\cosh bx}\frac{dx}{x} = 2\arctan\left(\exp\frac{\pi a}{2b}\right) - \frac{\pi}{2}$      $[\operatorname{Re} b > 0, \quad a > 0]$

                                         BI (387)(1), ET I 89(13), LI (298)(17)

**4.112**

1. $\displaystyle\int_0^\infty (x^2 + b^2)\frac{\cos ax}{\cosh\frac{\pi x}{2b}}\, dx = \frac{2b^3}{\cosh^3 ab}$      $[\operatorname{Re} b > 0, \quad a > 0]$      ET I 32(19)

2. $\displaystyle\int_0^\infty x(x^2 + 4b^2)\frac{\cos ax}{\sinh\frac{\pi x}{2b}}\, dx = \frac{6b^4}{\cosh^4 ab}$      $[\operatorname{Re} b > 0, \quad a > 0]$      ET I 32(20)

**4.113**

1. $\displaystyle\int_0^\infty \frac{\sin ax}{\sinh \pi x}\frac{dx}{x^2 + \beta^2} = -\frac{1}{2\beta^2} - \frac{\pi e^{-a\beta}}{\beta\sin\pi\beta}$

$$+\frac{1}{2\beta^2}\left[{}_2F_1\left(1, -\beta; 1-\beta; -e^{-a}\right) + {}_2F_1\left(1, \beta; 1+\beta; -e^{-a}\right)\right]$$

$$= \frac{1}{2\beta^2} - \frac{\pi e^{-a\beta}}{2\beta\sin\pi\beta} - \sum_{k=1}^\infty \frac{(-1)^k e^{-ak}}{k^2 - \beta^2}$$

                                      $[\operatorname{Re}\beta > 0, \quad \beta \neq 0, 1, 2, \ldots, \quad a > 0]$    ET I 90(18)

2. $$\int_0^\infty \frac{\sin ax}{\sinh \pi x}\frac{dx}{x^2+m^2} = \frac{(-1)^m a e^{-ma}}{2m} + \frac{1}{2m}\sum_{k=1}^{m-1}\frac{(-1)^k e^{-ka}}{m-k} + \frac{(-1)^m e^{-ma}}{2m}\ln\left(1+e^{-a}\right)$$
$$+ \frac{1}{2m!}\frac{d^{m-1}}{dz^{m-1}}\left[\frac{(1+z)^{m-1}}{z}\ln(1+z)\right]_{z=e^{-a}}$$
$$[a>0] \qquad\qquad \text{ET I 89(17)}$$

3. $$\int_0^\infty \frac{\sin ax}{\sinh \pi x}\frac{dx}{1+x^2} = \frac{1}{2}\int_{-\infty}^\infty \frac{\sin ax}{\sinh \pi x}\frac{dx}{1+x^2} = -\frac{a}{2}\cosh a + \sinh a \ln\left(2\cosh\frac{a}{2}\right) \qquad \text{GW (336)(21b)}$$

4. $$\int_0^\infty \frac{\sin ax}{\sinh \frac{\pi}{2}x}\frac{dx}{1+x^2} = \frac{1}{2}\int_{-\infty}^\infty \frac{\sin ax}{\sinh \frac{\pi}{2}x}\frac{dx}{1+x^2} = \frac{\pi}{2}\sinh a - \cosh a \arctan\left(\sinh a\right) \qquad \text{GW (336)(21a)}$$

5. $$\int_0^\infty \frac{\sin ax}{\sinh \frac{\pi}{4}x}\frac{dx}{1+x^2} = -\frac{\pi}{\sqrt{2}}e^{-a} + \frac{\sinh a}{\sqrt{2}}\ln\left(\frac{2\cosh a + \sqrt{2}}{2\cosh a - \sqrt{2}}\right) + \sqrt{2}\cosh a \arctan\left(\frac{\sqrt{2}}{2\sinh a}\right)$$
$$[a>0] \qquad\qquad \text{LI (389)(1)}$$

6. $$\int_0^\infty \frac{\sin ax}{\cosh \frac{\pi}{4}x}\frac{x\,dx}{1+x^2} = \frac{\pi}{\sqrt{2}}e^{-a} + \frac{\sinh a}{\sqrt{2}}\ln\left(\frac{2\cosh a + \sqrt{2}}{2\cosh a - \sqrt{2}}\right) - \sqrt{2}\cosh a \arctan\left(\frac{1}{\sqrt{2}\sinh a}\right)$$
$$[a>0] \qquad\qquad \text{BI (388)(1)}$$

7. $$\int_0^\infty \frac{\cos ax}{\sinh \pi x}\frac{x\,dx}{1+x^2} = -\frac{1}{2} + \frac{a}{2}e^{-a} + \cosh a \ln\left(1+e^{-a}\right)$$
$$[a>0] \qquad\qquad \text{BI (389)(14), ET I 32(24)}$$

8. $$\int_0^\infty \frac{\cos ax}{\sinh \frac{\pi}{2}x}\frac{x\,dx}{1+x^2} = 2\sinh a \arctan\left(e^{-a}\right) + \frac{\pi}{2}e^{-a} - 1$$
$$[a>0] \qquad\qquad \text{BI (389)(11)}$$

9.[11] $$\int_0^\infty \frac{\cos ax}{\cosh \pi x}\frac{x\,dx}{x^2+\beta^2} = \frac{\pi e^{-a\beta}}{2\beta \cos(\beta\pi)} - \sum_{k=0}^\infty \frac{(-1)^k e^{-(k+1/2)a}}{\left(k+\frac{1}{2}\right)^2 - \beta^2}$$
$$[\operatorname{Re}\beta>0, \quad a>0] \qquad \text{ET I 32(26)}$$

10.[11] $$\int_0^\infty \frac{\cos ax}{\cosh \pi x}\frac{dx}{x^2+\left(m+\frac{1}{2}\right)^2} = \frac{(-1)^m e^{-a\beta}\left(a\beta+\frac{1}{2}\right)}{2\beta^2} - \sum_{k=0}^\infty \frac{(-1)^k e^{-(k+1/2)a}}{\left(k+\frac{1}{2}\right)^2 - \beta^2}$$
$$[\operatorname{Re}\beta>0, \quad a>0] \qquad \text{ET I 32(25)}$$

11. $$\int_0^\infty \frac{\cos ax}{\cosh \pi x}\frac{dx}{1+x^2} = 2\cosh\frac{a}{2} - \left[e^a \arctan\left(e^{-\frac{a}{2}}\right) + e^{-a}\arctan\left(e^{\frac{a}{2}}\right)\right]$$
$$[a>0] \qquad\qquad \text{ET I 32(21)}$$

12. $$\int_0^\infty \frac{\cos ax}{\cosh \frac{\pi}{2}x}\frac{dx}{1+x^2} = ae^{-a} + \cosh a \ln\left(1+e^{-2a}\right) \qquad [a>0] \qquad\qquad \text{BI (388)(6)}$$

13. $$\int_0^\infty \frac{\cos ax}{\cosh \frac{\pi}{4}x}\frac{dx}{1+x^2} = \frac{\pi}{\sqrt{2}}e^{-a} + \frac{2\sinh a}{\sqrt{2}}\arctan\left(\frac{1}{\sqrt{2}\sinh a}\right) - \frac{\cosh a}{\sqrt{2}}\ln\frac{2\cosh a + \sqrt{2}}{2\cosh a - \sqrt{2}}$$
$$[a>0] \qquad\qquad \text{BI (388)(5)}$$

**4.114**

1. $$\int_0^\infty \frac{\sin ax}{x} \frac{\sinh \beta x}{\sinh \gamma x}\, dx = \arctan\left(\tan\frac{\beta\pi}{2\gamma}\tanh\frac{a\pi}{2\gamma}\right) \qquad [|\mathrm{Re}\,\beta| < \mathrm{Re}\,\gamma, \quad a > 0] \qquad \text{BI (387)(6)a}$$

2. $$\int_0^\infty \frac{\cos ax}{x} \frac{\sinh \beta x}{\cosh \gamma x}\, dx = \frac{1}{2}\ln\frac{\cosh\frac{a\pi}{2\gamma}+\sin\frac{\beta\pi}{2\gamma}}{\cosh\frac{a\pi}{2\gamma}-\sin\frac{\beta\pi}{2\gamma}} \qquad [|\mathrm{Re}\,\beta| < \mathrm{Re}\,\gamma] \qquad \text{ET I 33(34)}$$

**4.115**

1. $$\int_0^\infty \frac{x\sin ax}{x^2+b^2}\frac{\sinh \beta x}{\sinh \pi x}\, dx = \frac{\pi}{2}\frac{e^{-ab}\sin b\beta}{\sin b\pi} + \sum_{k=1}^\infty (-1)^k \frac{ke^{-ak}\sin k\beta}{k^2-b^2}$$
$$[0 < \mathrm{Re}\,\beta < \pi, \quad a > 0, \quad b > 0]$$
$$\text{BI (389)(23)}$$

2. $$\int_0^\infty \frac{x\sin ax}{x^2+1}\frac{\sinh \beta x}{\sinh \pi x}\, dx = \frac{1}{2}e^{-a}(a\sin\beta-\beta\cos\beta) - \frac{1}{2}\sinh a\sin\beta\ln\left[1+2e^{-a}\cos\beta+e^{-2a}\right]$$
$$+ \cosh a\cos\beta\arctan\frac{\sin\beta}{e^a+\cos\beta}$$
$$[|\mathrm{Re}\,\beta| < \pi, \quad a > 0] \qquad \text{LI (389)(10)}$$

3. $$\int_0^\infty \frac{x\sin ax}{x^2+1}\frac{\sinh \beta x}{\sinh \frac{\pi}{2}x}\, dx$$
$$= \frac{\pi}{2}e^{-a}\sin\beta + \frac{1}{2}\cos\beta\sinh a\ln\frac{\cosh a+\sin\beta}{\cosh a-\sin\beta} - \sin\beta\cosh a\arctan\left(\frac{\cos\beta}{\sinh a}\right)$$
$$[|\mathrm{Re}\,\beta| < \frac{\pi}{2}, \quad a > 0] \qquad \text{BI (389)(8)}$$

4. $$\int_0^\infty \frac{\cos ax}{x^2+b^2}\frac{\sinh \beta x}{\sinh \pi x}\, dx = \frac{\pi}{2b}\frac{e^{-ab}\sin b\beta}{\sin b\pi} + \sum_{k=1}^\infty (-1)^k \frac{e^{-ak}\sin k\beta}{k^2-b^2}$$
$$[0 < \mathrm{Re}\,\beta < \pi, \quad a > 0, \quad b > 0]$$
$$\text{BI (389)(22)}$$

5. $$\int_0^\infty \frac{\cos ax}{x^2+1}\frac{\sinh \beta x}{\sinh \pi x}\, dx = \frac{1}{2}e^{-a}(a\sin\beta-\beta\cos\beta) + \frac{1}{2}\cosh a\sin\beta\ln\left(1+2e^{-a}\cos\beta+e^{-2a}\right)$$
$$- \sinh a\cos\beta\arctan\frac{\sin\beta}{e^a+\cos\beta}$$
$$[|\mathrm{Re}\,\beta| < \pi, \quad a > 0, \quad b > 0] \qquad \text{BI (389)(20)a}$$

6. $$\int_0^\infty \frac{\cos ax}{x^2+1}\frac{\sinh \beta x}{\sinh \frac{\pi}{2}x}\, dx = \frac{\pi}{2}e^{-a}\sin\beta - \frac{1}{2}\cosh a\cos\beta\ln\frac{\cosh a+\sin\beta}{\cosh a-\sin\beta} + \sinh a\sin\beta\arctan\frac{\cos\beta}{\sinh a}$$
$$[|\mathrm{Re}\,\beta| < \frac{\pi}{2}, \quad a > 0, \quad b > 0]$$
$$\text{BI (389)(18)}$$

7. $$\int_0^\infty \frac{\sin ax}{x^2+\frac{1}{4}}\frac{\sinh \beta x}{\cosh \pi x}\, dx = e^{-\frac{a}{2}}\left(a\sin\frac{\beta}{2}-\beta\cos\frac{\beta}{2}\right) - \sinh\frac{a}{2}\sin\frac{\beta}{2}\ln\left(1+2e^{-a}\cos\beta+e^{-2a}\right)$$
$$+ \cosh\frac{a}{2}\cos\frac{\beta}{2}\arctan\frac{\sin\beta}{1+e^{-a}\cos\beta}$$
$$[|\mathrm{Re}\,\beta| < \pi, \quad a > 0] \qquad \text{ET I 91(26)}$$

8.    $\displaystyle\int_0^\infty \frac{\sin ax}{x^2+\beta^2}\frac{\cosh\gamma x}{\sinh\pi x}\,dx = \frac{1}{2\beta^2} - \frac{\pi}{2\beta}\frac{e^{-a\beta}\cos\beta\gamma}{\sin\beta\pi} + \sum_{k=1}^\infty (-1)^{k-1}\frac{e^{-ak}\cos k\gamma}{k^2-\beta^2}$

$$[0\le \operatorname{Re}\beta, \quad |\operatorname{Re}\gamma| < \pi, \quad a>0]$$
$$\text{BI (389)(21)}$$

9.    $\displaystyle\int_0^\infty \frac{\sin ax}{x^2+1}\frac{\cosh\beta x}{\sinh\pi x}\,dx = -\frac{1}{2}e^{-a}(a\cos\beta+\beta\sin\beta) + \frac{1}{2}\sinh a\cos\beta \ln\left(1+2e^{-a}\cos\beta+e^{-2a}\right)$

$$+\cosh a\sin\beta\arctan\frac{\sin\beta}{e^a+\cos\beta}$$
$$[|\operatorname{Re}\beta| < \pi, \quad a>0] \quad \text{ET I 91(25), LI (389)(9)}$$

10.    $\displaystyle\int_0^\infty \frac{\sin ax}{x^2+1}\frac{\cosh\beta x}{\sinh\frac{\pi}{2}x}\,dx = -\frac{\pi}{2}e^{-a}\cos\beta + \frac{1}{2}\sinh a\sin\beta\ln\frac{\cosh a+\sin\beta}{\cosh a-\sin\beta} + \cosh a\cos\beta\arctan\frac{\cos\beta}{\sinh a}$

$$\left[|\operatorname{Re}\beta| < \tfrac{\pi}{2}, \quad a>0\right] \qquad \text{BI (389)(7)}$$

11.    $\displaystyle\int_0^\infty \frac{x\cos ax}{x^2+b^2}\frac{\cosh\beta x}{\sinh\pi x}\,dx = \frac{\pi}{2}\frac{e^{-ab}\cos b\beta}{\sin b\pi} + \sum_{k=1}^\infty (-1)^k\frac{ke^{-ak}\cos k\beta}{k^2-b^2}$

$$[|\operatorname{Re}\beta| < \pi, \quad a>0] \qquad \text{BI (389)(24)}$$

12.    $\displaystyle\int_0^\infty \frac{x\cos ax}{x^2+1}\frac{\cosh\beta x}{\sinh\pi x}\,dx = \frac{1}{2}e^{-a}(a\cos\beta+\beta\sin\beta)$

$$-\frac{1}{2}+\frac{1}{2}\cosh a\cos\beta\ln\left[1+2e^{-a}\cos\beta+e^{-2a}\right]$$
$$+\sinh a\sin\beta\arctan\frac{\sin\beta}{e^a+\cos\beta}$$
$$[|\operatorname{Re}\beta| < \pi, \quad a>0] \qquad \text{BI (389)(19)}$$

13.    $\displaystyle\int_0^\infty \frac{x\cos ax}{x^2+1}\frac{\cosh\beta x}{\sinh\frac{\pi}{2}x}\,dx = -1+\frac{\pi}{2}e^{-a}\cos\beta + \frac{1}{2}\cosh a\sin\beta\ln\frac{\cosh a+\sin\beta}{\cosh a-\sin\beta}$

$$+\sinh a\cos\beta\arctan\frac{\cos\beta}{\sinh a}$$
$$\left[|\operatorname{Re}\beta| < \tfrac{\pi}{2}, \quad a>0\right] \qquad \text{BI (389)(17)}$$

14.    $\displaystyle\int_0^\infty \frac{\cos ax}{x^2+1}\frac{\cosh\beta x}{\cosh\frac{\pi}{2}x}\,dx = ae^{-a}\cos\beta + \beta e^{-a}\sin\beta + \sinh a\sin\beta\arctan\frac{e^{-2a}\sin 2\beta}{1+e^{-2a}\cos 2\beta}$

$$+\frac{1}{2}\cosh a\cos\beta\ln\left(1+2e^{-2a}\cos 2\beta+e^{-4a}\right)$$
$$\left[|\operatorname{Re}\beta| < \tfrac{\pi}{2}, \quad a>0\right] \qquad \text{ET I 34(37)}$$

**4.116**

$1.^6$    $\displaystyle\int_0^\infty x\cos 2ax\tanh x\,dx$               the integral is divergent       BI (364)(2)

2.    $\displaystyle\int_0^\infty \cos ax\tanh\beta x\,\frac{dx}{x} = \ln\coth\frac{a\pi}{4\beta}$      $[\operatorname{Re}\beta>0, \quad a>0]$      BI (387)(8)

**4.117**

1.    $\displaystyle\int_0^\infty \frac{\sin ax}{1+x^2}\tanh\frac{\pi x}{2}\,dx = a\cosh a - \sinh a\ln(2\sinh a)$

$$[a>0]$$
$$\text{BI (388)(3)}$$

2. $\displaystyle\int_0^\infty \frac{\sin ax}{1+x^2}\tanh\frac{\pi x}{4}\,\mathrm{d}x = -\frac{\pi}{2}e^a + \sinh a \ln\coth\frac{a}{2} + 2\cosh a \arctan\left(e^a\right)$ BI (388)(4)

3. $\displaystyle\int_0^\infty \frac{\sin ax}{1+x^2}\coth\pi x\,\mathrm{d}x = \frac{a}{2}e^{-a} - \sinh a \ln\left(1-e^{-a}\right)$ $\qquad[a>0]$ BI (389)(5)

4. $\displaystyle\int_0^\infty \frac{\sin ax}{1+x^2}\coth\frac{\pi}{2}x\,\mathrm{d}x = \sinh a \ln\coth\frac{a}{2}$ $\qquad[a>0]$ BI (389)(6)

5.[12] $\displaystyle\int_0^\infty \frac{x\cos ax}{1+x^2}\tanh\frac{\pi}{2}x\,\mathrm{d}x = -ae^{-a} + \sinh a \ln\left(1-e^{-2a}\right)$

$\qquad\qquad\qquad\qquad\qquad\qquad\qquad\qquad\qquad\qquad\qquad[a>0]$ BI (388)(7)

6.[12] $\displaystyle\int_0^\infty \frac{x\cos ax}{1+x^2}\tanh\frac{\pi}{4}x\,\mathrm{d}x = \frac{\pi}{2}e^a + \cosh a \ln\coth\frac{a}{2} + 2\sinh a \arctan\left(e^a\right)$

$\qquad\qquad\qquad\qquad\qquad\qquad\qquad\qquad\qquad\qquad\qquad[a>0]$ BI (388)(8)

7. $\displaystyle\int_0^\infty \frac{x\cos ax}{1+x^2}\coth\pi x\,\mathrm{d}x = -\frac{a}{2}e^{-a} - \frac{1}{2} - \cosh a \ln\left(1-e^{-a}\right)$ BI (389)(15)a, ET I 33(31)a

8. $\displaystyle\int_0^\infty \frac{x\cos ax}{1+x^2}\coth\frac{\pi}{2}x\,\mathrm{d}x = -1 + \cosh a \ln\coth\frac{a}{2}$ $\qquad[a>0]$ BI (389)(12)

9. $\displaystyle\int_0^\infty \frac{x\cos ax}{1+x^2}\coth\frac{\pi}{4}x\,\mathrm{d}x = -2 + \frac{\pi}{2}e^{-a} + \cosh a \ln\coth\frac{a}{2} + 2\sinh a \arctan\left(e^{-a}\right)$

$\qquad\qquad\qquad\qquad\qquad\qquad\qquad\qquad\qquad\qquad\qquad[a>0]$ BI (389)(13)

**4.118**[8] $\displaystyle\int_0^\infty \frac{x\sin ax}{\cosh^2 x}\,\mathrm{d}x = \frac{\pi}{2}\frac{1}{\sinh\frac{1}{2}\pi a}\left(\frac{1}{2}\pi a\coth\frac{1}{2}\pi a - 1\right)$ ET I 89(14)

**4.119** $\displaystyle\int_0^\infty \frac{1-\cos px}{\sinh qx}\frac{\mathrm{d}x}{x} = \ln\left(\cosh\frac{p\pi}{2q}\right)$ BI (387)(2)a

**4.121**

1. $\displaystyle\int_0^\infty \frac{\sin ax - \sin bx}{\cosh\beta x}\frac{\mathrm{d}x}{x} = 2\arctan\frac{\exp\dfrac{a\pi}{2\beta} - \exp\dfrac{b\pi}{2\beta}}{1+\exp\dfrac{(a+b)\pi}{2\beta}}$ $\qquad[\operatorname{Re}\beta>0]$ GW (336)(19b)

2. $\displaystyle\int_0^\infty \frac{\cos ax - \cos bx}{\sinh\beta x}\frac{\mathrm{d}x}{x} = \ln\frac{\cosh\dfrac{b\pi}{2\beta}}{\cosh\dfrac{a\pi}{2\beta}}$ $\qquad[\operatorname{Re}\beta>0]$ GW (336)(19a)

**4.122**

1.[6] $\displaystyle\int_0^\infty \frac{\cos\beta x \sin\gamma x}{\cosh\delta x}\frac{\mathrm{d}x}{x} = \arctan\frac{\sinh\dfrac{\gamma\pi}{2\delta}}{\cosh\dfrac{\beta\pi}{2\delta}}$ $\qquad[\operatorname{Re}\delta>|\operatorname{Im}(\beta+\gamma)|]$ ET I 93(46)a

2. $\displaystyle\int_0^\infty \sin^2 ax\frac{\cosh\beta x}{\sinh x}\frac{\mathrm{d}x}{x} = \frac{1}{4}\ln\frac{\cosh 2a\pi + \cos\beta\pi}{1+\cos\beta\pi}$ $\qquad[|\operatorname{Re}\beta|<1]$ BI (387)(7)

**4.123**

1. $\displaystyle\int_0^\infty \frac{\sin x}{\cosh ax + \cos x}\frac{x\,dx}{x^2 - \pi^2} = \arctan\frac{1}{a} - \frac{1}{a}$      BI (390)(1)

2. $\displaystyle\int_0^\infty \frac{\sin x}{\cosh ax - \cos x}\frac{x\,dx}{x^2 - \pi^2} = \frac{a}{1+a^2} - \arctan\frac{1}{a}$      BI (390)(2)

3. $\displaystyle\int_0^\infty \frac{\sin 2x}{\cosh 2ax - \cos 2x}\frac{x\,dx}{x^2 - \pi^2} = \frac{1}{2a}\frac{1+2a^2}{1+a^2} - \arctan\frac{1}{a}$      BI (390)(4)

4. $\displaystyle\int_0^\infty \frac{\cosh ax \sin x}{\cosh 2ax - \cos 2x}\frac{x\,dx}{x^2 - \pi^2} = \frac{-1}{2a\,(1+a^2)}$      LI (390)(3)

5. $\displaystyle\int_0^\infty \frac{\cos ax}{\cosh \pi x + \cos \pi\beta}\frac{dx}{x^2 + \gamma^2} = \frac{\pi e^{-a\gamma}}{2\gamma\,(\cos\gamma\pi + \cos\beta\pi)}$

$$+\frac{1}{\sinh\beta\pi}\sum_{k=0}^\infty\left\{\frac{e^{-(2k+1-\beta)a}}{\gamma^2 - (2k+1-\beta)^2} - \frac{e^{-(2k+1+\beta)a}}{\gamma^2 - (2k+1+\beta)^2}\right\}$$

$$[0 < \operatorname{Re}\beta < 1, \quad \operatorname{Re}\gamma > 0, \quad a > 0] \quad \text{ET I 33(27)}$$

6. $\displaystyle\int_0^\infty \frac{\sin ax \sinh bx}{\cos 2ax + \cosh 2bx}x^{p-1}\,dx = \frac{\Gamma(p)}{(a^2+b^2)^{\frac{p}{2}}}\sin\left(p\arctan\frac{a}{b}\right)\sum_{k=0}^\infty\frac{(-1)^k}{(2k+1)^p}$

$$[p > 0]$$

     BI (364)(8)

7. $\displaystyle\int_0^\infty \sin ax^2\frac{\sin\frac{\pi x}{2}\sinh\frac{\pi x}{2}}{\cos\pi x + \cosh\pi x}x\,dx = \frac{1}{4}\left[\frac{\partial\,\vartheta_1(z\mid q)}{\partial z}\right]_{z=0,\,q=e^{-2a}}$

$$[a > 0]$$

     ET I 93(49)

**4.124**

1. $\displaystyle\int_0^1 \frac{\cos px\cosh\left(q\sqrt{1-x^2}\right)}{\sqrt{1-x^2}}\,dx = \frac{\pi}{2}\,J_0\left(\sqrt{p^2-q^2}\right)$      MO (40)

2. $\displaystyle\int_u^\infty \cos ax\cosh\sqrt{\beta\,(u^2-x^2)}\frac{dx}{\sqrt{u^2-x^2}} = \frac{\pi}{2}\,J_0\left(\frac{u}{\sqrt{a^2-\beta^2}}\right)$      ET I 34(38)

**4.125**

1. $\displaystyle\int_0^\infty \sinh\left(a\sin x\right)\cos\left(a\cos x\right)\sin x\sin 2nx\frac{dx}{x} = \frac{(-1)^{n-1}a^{2n-1}}{(2n-1)!}\frac{\pi}{8}\left[1 + \frac{a^2}{2n(2n+1)}\right]$

     LI (367)(14)

2. $\displaystyle\int_0^\infty \cosh\left(a\sin x\right)\cos\left(a\cos x\right)\sin x\cos(2n-1)x\frac{dx}{x} = \frac{(-1)^{n-1}a^{2(n-1)}}{[2(n-1)]!}\frac{\pi}{8}\left[1 - \frac{a^2}{2n(2n-1)}\right]$

     LI (367)(15)

3. $\displaystyle\int_0^\infty \sinh\left(a\sin x\right)\cos\left(a\cos x\right)\cos x\cos 2nx\frac{dx}{x} = \frac{\pi}{2}\sum_{k=n+1}^\infty\frac{(-1)^k a^{2k+1}}{(2k+1)!} + \frac{(-1)^n a^{2n+1}}{(2n+1)!}\frac{3\pi}{8}$

$$+\frac{(-1)^{n-1}a^{2n-1}}{(2n-1)!}\frac{\pi}{8}$$

     LI (367)(21)

**4.126**

1.
$$\int_0^\infty \sin\left(a\cos bx\right)\sinh\left(a\sin bx\right)\frac{x\,dx}{c^2-x^2} = \frac{\pi}{2}\left[\cos\left(a\cos bc\right)\cosh\left(a\sin bc\right)-1\right]$$

$$[b>0]$$ BI (381)(2)

2.
$$\int_0^\infty \sin\left(a\cos bx\right)\cosh\left(a\sin bx\right)\frac{dx}{c^2-x^2} = \frac{\pi}{2c}\cos\left(a\cos bc\right)\sinh\left(a\sin bc\right)$$

$$[b>0,\quad c>0]$$ BI (381)(1)

3.
$$\int_0^\infty \cos\left(a\cos bx\right)\sinh\left(a\sin bx\right)\frac{x\,dx}{c^2-x^2} = \frac{\pi}{2}\left[a\cos bc-\sin\left(a\cos bc\right)\cosh\left(a\sin bc\right)\right]$$

$$[b>0]$$ BI (381)(4)

4.
$$\int_0^\infty \cos\left(a\cos bx\right)\cosh\left(a\sin bx\right)\frac{dx}{c^2-x^2} = -\frac{\pi}{2c}\sin\left(a\cos bc\right)\sinh\left(a\sin bc\right)$$

$$[b>0]$$ BI (381)(3)

## 4.13 Combinations of trigonometric and hyperbolic functions and exponentials

**4.131**

1.
$$\int_0^\infty \sin ax\,\sinh^\nu \gamma x\,e^{-\beta x}\,dx = -\frac{i\,\Gamma(\nu+1)}{2^{\nu+2}\gamma}\left\{\frac{\Gamma\left(\frac{\beta-\nu\gamma-ai}{2\gamma}\right)}{\Gamma\left(\frac{\beta+\nu\gamma-ai}{2\gamma}+1\right)} - \frac{\Gamma\left(\frac{\beta-\nu\gamma+ai}{2\gamma}\right)}{\Gamma\left(\frac{\beta+\nu\gamma+ai}{2\gamma}+1\right)}\right\}$$

$$[\operatorname{Re}\nu>-2,\quad \operatorname{Re}\gamma>0,\quad |\operatorname{Re}(\gamma\nu)|<\operatorname{Re}\beta]\quad \text{ET I 91(30)a}$$

2.[12]
$$\int_0^\infty \cos ax\,\sinh^\nu \gamma x\,e^{-\beta x}\,dx = \frac{\Gamma(\nu+1)}{2^{\nu+2}\gamma}\left\{\frac{\Gamma\left(\frac{\beta-\nu\gamma-ai}{2\gamma}\right)}{\Gamma\left(\frac{\beta+\gamma\nu-ai}{2\gamma}+1\right)} + \frac{\Gamma\left(\frac{\beta-\nu\gamma+ai}{2\gamma}\right)}{\Gamma\left(\frac{\beta+\nu\gamma+ai}{2\gamma}+1\right)}\right\}$$

$$[\operatorname{Re}\nu>-1,\quad \operatorname{Re}\gamma>0,\quad |\operatorname{Re}(\gamma\nu)|<\operatorname{Re}\beta]\quad \text{ET I 34(40)a}$$

3.
$$\int_0^\infty e^{-\beta x}\frac{\sin ax}{\sinh \gamma x}\,dx = \sum_{k=1}^\infty \frac{2a}{a^2+[\beta+(2k-1)\gamma]^2}$$ BI (264)(9)a

$$= \frac{1}{2\gamma i}\left[\psi\left(\frac{\beta+\gamma+ia}{2\gamma}\right)-\psi\left(\frac{\beta+\gamma-ia}{2\gamma}\right)\right]\quad [\operatorname{Re}\beta>|\operatorname{Re}\gamma|]\quad \text{ET I 91(28)}$$

4.
$$\int_0^\infty e^{-x}\frac{\sin ax}{\sinh x}\,dx = \frac{\pi}{2}\coth\frac{a\pi}{2}-\frac{1}{a}$$ ET I 91(29)

**4.132**

1.
$$\int_0^\infty \frac{\sin ax\,\sinh bx}{e^{\gamma x}-1}\,dx = -\frac{a}{2\left(a^2+b^2\right)}+\frac{\pi}{2\gamma}\frac{\sinh\frac{2\pi a}{\gamma}}{\cosh\frac{2\pi a}{\gamma}-\cos\frac{2\pi b}{\gamma}}$$

$$+\frac{i}{2\gamma}\left[\psi\left(\frac{b}{\gamma}+i\frac{a}{\gamma}+1\right)-\psi\left(\frac{b}{\gamma}-i\frac{a}{\gamma}+1\right)\right]$$

$$[\operatorname{Re}\gamma>|\operatorname{Re}b|,\quad a>0]\quad \text{ET I 92(33)}$$

**2.**
$$\int_0^\infty \frac{\sin ax \cosh bx}{e^{\gamma x} - 1}\, dx = -\frac{a}{2(a^2 + b^2)} + \frac{\pi}{2\gamma} \frac{\sinh \frac{2\pi a}{\gamma}}{\cosh \frac{2\pi a}{\gamma} - \cos \frac{2\pi b}{\gamma}}$$

$$[\operatorname{Re}\gamma > |\operatorname{Re} b|] \qquad \text{BI (265)(5)a, ET I 92(34)}$$

**3.**
$$\int_0^\infty \frac{\sin ax \cosh bx}{e^{\gamma x} + 1}\, dx = \frac{a}{2(a^2 + b^2)} - \frac{\pi}{\gamma} \frac{\sinh \frac{a\pi}{\gamma} \cos \frac{b\pi}{\gamma}}{\cosh \frac{2a\pi}{\gamma} - \cos \frac{2b\pi}{\gamma}}$$

$$[\operatorname{Re}\gamma > |\operatorname{Re} b|] \qquad \text{ET I 92(35)}$$

**4.**
$$\int_0^\infty \frac{\cos ax \sinh bx}{e^{\gamma x} - 1}\, dx = \frac{b}{2(a^2 + b^2)} - \frac{\pi}{2\gamma} \frac{\sin \frac{2\pi b}{\gamma}}{\cosh \frac{2a\pi}{\gamma} - \cos \frac{2b\pi}{\gamma}}$$

$$[\operatorname{Re}\gamma > |\operatorname{Re} b|] \qquad \text{LI (265)(8)}$$

**5.**
$$\int_0^\infty \frac{\cos ax \sinh bx}{e^{\gamma x} + 1}\, dx = -\frac{b}{2(a^2 + b^2)} + \frac{\pi}{\gamma} \frac{\sin \frac{\pi b}{\gamma} \cosh \frac{\pi a}{\gamma}}{\cosh \frac{2a\pi}{\gamma} - \cos \frac{2b\pi}{\gamma}}$$

$$[\operatorname{Re}\gamma > |\operatorname{Re} b|] \qquad \text{ET I 34(39)}$$

**4.133**

**1.[11]**
$$\int_0^\infty \sin ax \sinh bx \exp\left(-\frac{x^2}{4\gamma}\right) dx = \sqrt{\pi\gamma}\, \exp\left[\gamma\left(b^2 - a^2\right)\right] \sin(2ab\gamma)$$

$$[\operatorname{Re}\gamma > 0] \qquad \text{ET I 92(37)}$$

**2.[11]**
$$\int_0^\infty \cos ax \cosh bx \exp\left(-\frac{x^2}{4\gamma}\right) dx = \sqrt{\pi\gamma}\, \exp\left[\gamma\left(b^2 - a^2\right)\right] \cos(2ab\gamma)$$

$$[\operatorname{Re}\gamma > 0] \qquad \text{ET I 35(41)}$$

**4.134**

**1.[12]**
$$\int_0^\infty e^{-\beta x^2}(\cosh x + \cos x)\, dx = \sqrt{\frac{\pi}{\beta}} \cosh \frac{1}{4\beta} \qquad [\operatorname{Re}\beta > 0] \qquad \text{ME 24}$$

**2.**
$$\int_0^\infty e^{-\beta x^2}(\cosh x - \cos x)\, dx = \sqrt{\frac{\pi}{\beta}} \sinh \frac{1}{4\beta} \qquad [\operatorname{Re}\beta > 0] \qquad \text{ME 24}$$

**4.135**

**1.[12]**
$$\int_0^\infty \sin ax^2 \cosh 2\gamma x e^{-\beta x^2}\, dx = \frac{1}{2} \sqrt[4]{\frac{\pi^2}{a^2 + \beta^2}} \exp\left(\frac{\beta\gamma^2}{a^2 + \beta^2}\right) \sin\left(\frac{a\gamma^2}{a^2 + \beta^2} + \frac{1}{2}\arctan\frac{a}{\beta}\right)$$

$$[\operatorname{Re}\beta > 0] \qquad \text{LI (268)(7)}$$

**2.[12]**
$$\int_0^\infty \cos ax^2 \cosh 2\gamma x e^{-\beta x^2}\, dx = \frac{1}{2} \sqrt[4]{\frac{\pi^2}{a^2 + \beta^2}} \exp\left(\frac{\beta\gamma^2}{a^2 + \beta^2}\right) \cos\left(\frac{a\gamma^2}{a^2 + \beta^2} + \frac{1}{2}\arctan\frac{a}{\beta}\right)$$

$$[\operatorname{Re}\beta > 0] \qquad \text{LI (268)(8)}$$

**4.136**

**1.[12]**
$$\int_0^\infty \left(\sinh x^2 + \sin x^2\right) e^{-\beta x^4}\, dx = \frac{\sqrt{2}\pi}{4\sqrt{\beta}} I_{\frac{1}{4}}\left(\frac{1}{8\beta}\right) \cosh \frac{1}{8\beta}$$

$$[\operatorname{Re}\beta > 0] \qquad \text{ME 24}$$

$2.^{12}$ $\quad \displaystyle\int_0^\infty \left(\sinh x^2 - \sin x^2\right) e^{-\beta x^4}\,\mathrm{d}x = \frac{\sqrt{2}\pi}{4\sqrt{\beta}}\, I_{\frac{1}{4}}\left(\frac{1}{8\beta}\right) \sinh\frac{1}{8\beta}$

$[\operatorname{Re}\beta > 0]$ $\qquad$ ME 24

$3.^{12}$ $\quad \displaystyle\int_0^\infty \left(\cosh x^2 + \cos x^2\right) e^{-\beta x^4}\,\mathrm{d}x = \frac{\sqrt{2}\pi}{4\sqrt{\beta}}\, I_{-\frac{1}{4}}\left(\frac{1}{8\beta}\right) \cosh\frac{1}{8\beta}$

$[\operatorname{Re}\beta > 0]$ $\qquad$ ME 24

$4.^{12}$ $\quad \displaystyle\int_0^\infty \left(\cosh x^2 - \cos x^2\right) e^{-\beta x^4}\,\mathrm{d}x = \frac{\sqrt{2}\pi}{4\sqrt{\beta}}\, I_{-\frac{1}{4}}\left(\frac{1}{8\beta}\right) \sinh\frac{1}{8\beta}$

$[\operatorname{Re}\beta > 0]$ $\qquad$ ME 24

**4.137**

1. $\quad \displaystyle\int_0^\infty \sin 2x^2 \sinh 2x^2 e^{-\beta x^4}\,\mathrm{d}x = \frac{\pi}{\sqrt[4]{128\beta^2}}\, J_{-\frac{1}{4}}\left(\frac{1}{\beta}\right) \cos\left(\frac{1}{\beta} + \frac{\pi}{4}\right)$

$[\operatorname{Re}\beta > 0]$ $\qquad$ MI 32

2. $\quad \displaystyle\int_0^\infty \sin 2x^2 \cosh 2x^2 e^{-\beta x^4}\,\mathrm{d}x = \frac{\pi}{\sqrt[4]{128\beta^2}}\, J_{\frac{1}{4}}\left(\frac{1}{\beta}\right) \cos\left(\frac{1}{\beta} - \frac{\pi}{4}\right)$

$[\operatorname{Re}\beta > 0]$ $\qquad$ MI 32

3. $\quad \displaystyle\int_0^\infty \cos 2x^2 \sinh 2x^2 e^{-\beta x^4}\,\mathrm{d}x = \frac{-\pi}{\sqrt[4]{128\beta^2}}\, J_{\frac{1}{4}}\left(\frac{1}{\beta}\right) \sin\left(\frac{1}{\beta} - \frac{\pi}{4}\right)$

$[\operatorname{Re}\beta > 0]$ $\qquad$ MI 32

4. $\quad \displaystyle\int_0^\infty \cos 2x^2 \cosh 2x^2 e^{-\beta x^4}\,\mathrm{d}x = \frac{\pi}{\sqrt[4]{128\beta^2}}\, J_{-\frac{1}{4}}\left(\frac{1}{\beta}\right) \sin\left(\frac{1}{\beta} + \frac{\pi}{4}\right)$

$[\operatorname{Re}\beta > 0]$ $\qquad$ MI 32

**4.138**

1. $\quad \displaystyle\int_0^\infty \left(\sin 2x^2 \cosh 2x^2 + \cos 2x^2 \sinh 2x^2\right) e^{-\beta x^4}\,\mathrm{d}x = \frac{\pi}{\sqrt[4]{32\beta^2}}\, J_{\frac{1}{4}}\left(\frac{1}{\beta}\right) \cos\left(\frac{1}{\beta}\right)$

$[\operatorname{Re}\beta > 0]$ $\qquad$ MI 32

2. $\quad \displaystyle\int_0^\infty \left(\sin 2x^2 \cosh 2x^2 - \cos 2x^2 \sinh 2x^2\right) e^{-\beta x^4}\,\mathrm{d}x = \frac{\pi}{\sqrt[4]{32\beta^2}}\, J_{\frac{1}{4}}\left(\frac{1}{\beta}\right) \sin\left(\frac{1}{\beta}\right)$

$[\operatorname{Re}\beta > 0]$ $\qquad$ MI 32

3. $\quad \displaystyle\int_0^\infty \left(\cos 2x^2 \cosh 2x^2 + \sin 2x^2 \sinh 2x^2\right) e^{-\beta x^4}\,\mathrm{d}x = \frac{\pi}{\sqrt[4]{32\beta^2}}\, J_{-\frac{1}{4}}\left(\frac{1}{\beta}\right) \cos\left(\frac{1}{\beta}\right)$

$[\operatorname{Re}\beta > 0]$ $\qquad$ MI 32

4. $\quad \displaystyle\int_0^\infty \left(\cos 2x^2 \cosh 2x^2 - \sin 2x^2 \sinh 2x^2\right) e^{-\beta x^4}\,\mathrm{d}x = \frac{\pi}{\sqrt[4]{32\beta^2}}\, J_{-\frac{1}{4}}\left(\frac{1}{\beta}\right) \sin\left(\frac{1}{\beta}\right)$

$[\operatorname{Re}\beta > 0]$ $\qquad$ MI 32

## 4.14 Combinations of trigonometric and hyperbolic functions, exponentials, and powers

**4.141**

1.    $\displaystyle \int_0^\infty x e^{-\beta x^2} \cosh x \sin x \, dx = \frac{1}{4}\sqrt{\frac{\pi}{\beta^3}} \left( \cos \frac{1}{2\beta} + \sin \frac{1}{2\beta} \right)$

$$[\operatorname{Re}\beta > 0] \qquad\qquad \text{MI 32}$$

2.    $\displaystyle \int_0^\infty x e^{-\beta x^2} \sinh x \cos x \, dx = \frac{1}{4}\sqrt{\frac{\pi}{\beta^3}} \left( \cos \frac{1}{2\beta} - \sin \frac{1}{2\beta} \right)$

$$[\operatorname{Re}\beta > 0] \qquad\qquad \text{MI 32}$$

3.    $\displaystyle \int_0^\infty x^2 e^{-\beta x^2} \cosh x \cos x \, dx = \frac{1}{4}\sqrt{\frac{\pi}{\beta^3}} \left( \cos \frac{1}{2\beta} - \frac{1}{\beta} \sin \frac{1}{2\beta} \right)$

$$[\operatorname{Re}\beta > 0] \qquad\qquad \text{MI 32}$$

4.    $\displaystyle \int_0^\infty x^2 e^{-\beta x^2} \sinh x \sin x \, dx = \frac{1}{4}\sqrt{\frac{\pi}{\beta^3}} \left( \sin \frac{1}{2\beta} + \frac{1}{\beta} \cos \frac{1}{2\beta} \right)$

$$[\operatorname{Re}\beta > 0] \qquad\qquad \text{MI 32}$$

**4.142**

1.    $\displaystyle \int_0^\infty x e^{-\beta x^2} \left( \sinh x + \sin x \right) dx = \frac{1}{2}\sqrt{\frac{\pi}{\beta^3}} \cosh \frac{1}{4\beta} \qquad [\operatorname{Re}\beta > 0] \qquad \text{ME 24}$

2.    $\displaystyle \int_0^\infty x e^{-\beta x^2} \left( \sinh x - \sin x \right) dx = \frac{1}{2}\sqrt{\frac{\pi}{\beta^3}} \sinh \frac{1}{4\beta} \qquad [\operatorname{Re}\beta > 0] \qquad \text{ME 24}$

3.    $\displaystyle \int_0^\infty x^2 e^{-\beta x^2} \left( \cosh x + \cos x \right) dx = \frac{1}{2}\sqrt{\frac{\pi}{\beta^3}} \left( \cosh \frac{1}{4\beta} + \frac{1}{2\beta} \sinh \frac{1}{4\beta} \right)$

$$[\operatorname{Re}\beta > 0] \qquad\qquad \text{ME 24}$$

4.    $\displaystyle \int_0^\infty x^2 e^{-\beta x^2} \left( \cosh x - \cos x \right) dx = \frac{1}{2}\sqrt{\frac{\pi}{\beta^3}} \left( \sinh \frac{1}{4\beta} + \frac{1}{2\beta} \cosh \frac{1}{4\beta} \right)$

$$[\operatorname{Re}\beta > 0] \qquad\qquad \text{ME 24}$$

**4.143**

1.    $\displaystyle \int_0^\infty x e^{-\beta x^2} \left( \cosh x \sin x + \sinh x \cos x \right) dx = \frac{1}{2\beta}\sqrt{\frac{\pi}{\beta}} \cos \frac{1}{2\beta}$

$$[\operatorname{Re}\beta > 0] \qquad\qquad \text{MI 32}$$

2.    $\displaystyle \int_0^\infty x e^{-\beta x^2} \left( \cosh x \sin x - \sinh x \cos x \right) dx = \frac{1}{2\beta}\sqrt{\frac{\pi}{\beta}} \sin \frac{1}{2\beta}$

$$[\operatorname{Re}\beta > 0] \qquad\qquad \text{MI 32}$$

**4.144**    $\displaystyle \int_0^\infty e^{-x^2} \sinh x^2 \cos ax \frac{dx}{x^2} = \sqrt{\frac{\pi}{2}} e^{-\frac{a^2}{8}} - \frac{\pi a}{4} \left[ 1 - \Phi\left( \frac{a}{\sqrt{8}} \right) \right]$

$$[a > 0] \qquad\qquad \text{ET I 35(44)}$$

**4.145**

1.    $$\int_0^\infty xe^{-\beta x^2}\cosh\left(2ax\sin t\right)\sin\left(2ax\cos t\right)\,\mathrm{d}x = \frac{a}{2}\sqrt{\frac{\pi}{\beta^3}}\exp\left(-\frac{a^2}{\beta}\cos 2t\right)\cos\left(t-\frac{a^2}{\beta}\sin 2t\right)$$

$$[\operatorname{Re}\beta > 0] \qquad\qquad \text{BI (363)(5)}$$

2.    $$\int_0^\infty xe^{-\beta x^2}\sinh\left(2ax\sin t\right)\cos\left(2ax\cos t\right)\,\mathrm{d}x = \frac{a}{2}\sqrt{\frac{\pi}{\beta^3}}\exp\left(-\frac{a^2}{\beta}\cos 2t\right)\sin\left(t-\frac{a^2}{\beta}\sin 2t\right)$$

$$[\operatorname{Re}\beta > 0] \qquad\qquad \text{BI (363)(6)}$$

**4.146**[10]

1.[8]    $$\int_0^\infty e^{-\beta x^2}\sinh ax\sin bx\,\mathrm{d}x = \frac{1}{2}\sqrt{\frac{\pi}{\beta}}\exp\left(\frac{a^2-b^2}{4\beta}\right)\sin\frac{ab}{2\beta}$$

$$[\operatorname{Re}\beta > 0]$$

2.[8]    $$\int_0^\infty e^{-\beta x^2}\cosh ax\cos bx\,\mathrm{d}x = \frac{1}{2}\sqrt{\frac{\pi}{\beta}}\exp\left(\frac{a^2-b^2}{4\beta}\right)\cos\frac{ab}{2\beta}$$

$$[\operatorname{Re}\beta > 0]$$

3.    $$\int_0^\infty xe^{-\beta x^2}\cosh ax\sin ax\,\mathrm{d}x = \frac{a}{4\beta}\sqrt{\frac{\pi}{\beta}}\left(\cos\frac{a^2}{2\beta}+\sin\frac{a^2}{2\beta}\right)$$

$$[\operatorname{Re}\beta > 0]$$

4.    $$\int_0^\infty xe^{-\beta x^2}\sinh ax\cos ax\,\mathrm{d}x = \frac{a}{4\beta}\sqrt{\frac{\pi}{\beta}}\left(\cos\frac{a^2}{2\beta}-\sin\frac{a^2}{2\beta}\right)$$

$$[\operatorname{Re}\beta > 0]$$

5.[8]    $$\int_0^\infty x^2e^{-\beta x^2}\cosh ax\sin ax\,\mathrm{d}x = \frac{1}{4}\sqrt{\frac{\pi}{\beta^3}}\left(\sin\frac{a^2}{2\beta}+\frac{a^2}{\beta}\cos\frac{a^2}{2\beta}\right)$$

$$[\operatorname{Re}\beta > 0]$$

6.[8]    $$\int_0^\infty x^2e^{-\beta x^2}\cosh ax\cos ax\,\mathrm{d}x = \frac{1}{4}\sqrt{\frac{\pi}{\beta^3}}\left(\cos\frac{a^2}{2\beta}-\frac{a^2}{\beta}\sin\frac{a^2}{2\beta}\right)$$

$$[\operatorname{Re}\beta > 0]$$

# 4.2–4.4 Logarithmic Functions

## 4.21  Logarithmic functions

**4.211**

1.    $$\int_e^\infty \frac{\mathrm{d}x}{\ln\frac{1}{x}} = -\infty$$

$$\text{BI (33)(9)}$$

2.    $$\int_0^u \frac{\mathrm{d}x}{\ln x} = \operatorname{li} u$$

$$\text{FI III 653, FI II 606}$$

**4.212**

1.[7] $\quad \int_0^1 \dfrac{dx}{a + \ln x} = e^{-a}\, \text{Ei}(a)$ $\qquad\qquad [a > 0]$ $\qquad\qquad$ BI (31)(4)

2. $\quad \int_0^1 \dfrac{dx}{a - \ln x} = -e^a\, \text{Ei}(-a)$ $\qquad\qquad [a > 0]$ $\qquad\qquad$ BI (31)(5)

3.[7] $\quad \int_0^1 \dfrac{dx}{(a + \ln x)^2} = -\dfrac{1}{a} + e^{-a}\, \text{Ei}(a)$ $\qquad\qquad [a \geq 0]$ $\qquad\qquad$ BI (31)(14)

4. $\quad \int_0^1 \dfrac{dx}{(a - \ln x)^2} = \dfrac{1}{a} + e^a\, \text{Ei}(-a)$ $\qquad\qquad [a > 0]$ $\qquad\qquad$ BI (31)(16)

5.[8] $\quad \int_0^1 \dfrac{\ln x\, dx}{(a + \ln x)^2} = 1 + (1 - a)e^{-a}\, \text{Ei}(a)$ $\qquad\qquad [a \geq 0]$ $\qquad\qquad$ BI (31)(15)

6. $\quad \int_0^1 \dfrac{\ln x\, dx}{(a - \ln x)^2} = 1 + (1 + a)e^a\, \text{Ei}(-a)$ $\qquad\qquad [a > 0]$ $\qquad\qquad$ BI (31)(17)

7. $\quad \int_1^e \dfrac{\ln x\, dx}{(1 + \ln x)^2} = \dfrac{e}{2} - 1$ $\qquad\qquad$ BI (33)(10)

8.[7] $\quad \int_0^1 \dfrac{dx}{(a + \ln x)^n} = \dfrac{1}{(n-1)!}e^{-a}\, \text{Ei}(a) - \dfrac{1}{(n-1)!}\sum_{k=1}^{n-1}(n-k-1)!a^{k-n}$

$\qquad\qquad\qquad\qquad\qquad\qquad\qquad\qquad\qquad\qquad [a \geq 0]$ $\qquad\qquad$ BI (31))(22)

9. $\quad \int_0^1 \dfrac{dx}{(a - \ln x)^n} = \dfrac{(-1)^n}{(n-1)!}e^a\, \text{Ei}(-a) + \dfrac{(-1)^{n-1}}{(n-1)!}\sum_{k=1}^{n-1}(n-k-1)!(-a)^{k-n}$

$\qquad\qquad\qquad\qquad\qquad\qquad\qquad\qquad\qquad\qquad [a > 0, \quad n \text{ odd}]$ $\qquad\qquad$ DI (31)(23)

In integrals of the form $\int \dfrac{(\ln x)^m}{[a^n + (\ln x)^n]^l}\, dx$ it is convenient to make the substitution $x = e^{-t}$.

Results **4.212** 3, **4.212** 5, and **4.212** 8 [for $n > 1$] and **4.213** 6, **4.213** 8 below are divergent but may be considered to be valid if defined as follows:

$$\int_0^a \frac{f(z)\, dz}{(z - z_0)^n} = \frac{1}{(n-1)!}\left(\frac{d}{dz_0}\right)^{n-1}\left[\text{PV}\int_0^a \frac{f(z)}{z - z_0}\, dz\right]$$

where $a > z_0 > 0, n = 1, 2, 3, \ldots$ and PV indicates the Cauchy principal value.

**4.213**

1. $\quad \int_0^1 \dfrac{dx}{a^2 + (\ln x)^2} = \dfrac{1}{a}\left[\text{ci}(a)\sin a - \text{si}(a)\cos a\right]$ $\qquad\qquad [a > 0]$ $\qquad\qquad$ BI (31)(6)

2.[7] $\quad \int_0^1 \dfrac{dx}{a^2 - (\ln x)^2} = \dfrac{1}{2a}\left[e^{-a}\overline{\text{Ei}}(a) - e^a\, \text{Ei}(-a)\right]$ $\qquad\qquad [a > 0], \quad (\text{cf. } \mathbf{4.212}\ 1 \text{ and } 2)$

$\qquad\qquad\qquad\qquad\qquad\qquad\qquad\qquad\qquad\qquad\qquad\qquad$ BI (31)(8)

3. $\quad \int_0^1 \dfrac{\ln x\, dx}{a^2 + (\ln x)^2} = \text{ci}(a)\cos a + \text{si}(a)\sin a$ $\qquad\qquad [a > 0]$ $\qquad\qquad$ BI (31)(7)

4.7 $\quad \int_0^1 \dfrac{\ln x \, dx}{a^2 - (\ln x)^2} = -\dfrac{1}{2} \left[ e^{-a} \overline{\mathrm{Ei}}(a) + e^a \, \mathrm{Ei}(-a) \right]$ $\qquad [a > 0], \quad$ (cf. **4.212** 1 and 2)

BI (31)(9)

5. $\quad \int_0^1 \dfrac{dx}{\left[ a^2 + (\ln x)^2 \right]^2} = \dfrac{1}{2a^3} \left[ \mathrm{ci}(a) \sin a - \mathrm{si}(a) \cos a \right] - \dfrac{1}{2a^2} \left[ \mathrm{ci}(a) \cos a + \mathrm{si}(a) \sin a \right]$

$[a > 0]$       LI (31)(18)

6.8 $\quad \int_0^1 \dfrac{dx}{\left[ a^2 - (\ln x)^2 \right]^2} \qquad\qquad$ is divergent

7. $\quad \int_0^1 \dfrac{\ln x \, dx}{\left[ a^2 + (\ln x)^2 \right]^2} = \dfrac{1}{2a} \left[ \mathrm{ci}(a) \sin a - \mathrm{si}(a) \cos a \right] - \dfrac{1}{2a^2}$

$[a > 0]$       BI (31)(19)

8.8 $\quad \int_0^1 \dfrac{\ln x \, dx}{\left[ a^2 - (\ln x)^2 \right]^2} \qquad\qquad$ is divergent

**4.214**

1. $\quad \int_0^1 \dfrac{dx}{a^4 - (\ln x)^4} = -\dfrac{1}{4a^3} \left[ e^a \, \mathrm{Ei}(-a) - e^{-a} \overline{\mathrm{Ei}}(a) - 2 \, \mathrm{ci}(a) \sin a + 2 \, \mathrm{si}(a) \cos a \right]$

$[a > 0]$       BI (31)(10)

2. $\quad \int_0^1 \dfrac{\ln x \, dx}{a^4 - (\ln x)^4} = -\dfrac{1}{4a^2} \left[ e^a \, \mathrm{Ei}(-a) + e^{-a} \overline{\mathrm{Ei}}(a) - 2 \, \mathrm{ci}(a) \cos a - 2 \, \mathrm{si}(a) \sin a \right]$

$[a > 0]$       BI (31)(11)

3. $\quad \int_0^1 \dfrac{(\ln x)^2 \, dx}{a^4 - (\ln x)^4} = -\dfrac{1}{4a} \left[ e^a \, \mathrm{Ei}(-a) - e^{-a} \overline{\mathrm{Ei}}(a) + 2 \, \mathrm{ci}(a) \sin a - 2 \, \mathrm{si}(a) \cos a \right]$

$[a > 0]$       BI (31)(12)

4.7 $\quad \int_0^1 \dfrac{(\ln x)^3 \, dx}{a^4 - (\ln x)^4} = -\dfrac{1}{4} \left[ e^a \, \mathrm{Ei}(-a) + e^{-a} \overline{\mathrm{Ei}}(a) + 2 \, \mathrm{ci}(a) \cos a + 2 \, \mathrm{si}(a) \sin a \right]$

$[a > 0]$       BI (31)(13)

**4.215**

1. $\quad \int_0^1 (-\ln x)^{\mu - 1} \, dx = \Gamma(\mu) \qquad\qquad\qquad [\mathrm{Re}\, \mu > 0] \qquad\qquad$ FI II 778

2. $\quad \int_0^1 \dfrac{dx}{(-\ln x)^\mu} = \dfrac{\pi}{\Gamma(\mu)} \, \mathrm{cosec} \, \mu\pi \qquad\qquad [\mathrm{Re}\, \mu < 1] \qquad\qquad$ BI (31)(1)

3. $\quad \int_0^1 \sqrt{-\ln x} \, dx = \dfrac{\sqrt{\pi}}{2} \qquad\qquad\qquad\qquad\qquad\qquad\qquad\qquad$ BI (32)(1)

4.     $\displaystyle \int_0^1 \frac{dx}{\sqrt{-\ln x}} = \sqrt{\pi}$        BI (32)(3)

**4.216**

1.     $\displaystyle \int_0^{1/e} \frac{dx}{\sqrt{(\ln x)^2 - 1}} = K_0(1)$        GW (32)(2)

2.*     $\displaystyle \int_0^{1/e} \frac{dx}{\sqrt{-\ln x - 1}} = \frac{\sqrt{\pi}}{e}$

## 4.22 Logarithms of more complicated arguments

**4.221**

1.     $\displaystyle \int_0^1 \ln x \ln(1 - x) \, dx = 2 - \frac{\pi^2}{6}$        BI (30)(7)

2.     $\displaystyle \int_0^1 \ln x \ln(1 + x) \, dx = 2 - \frac{\pi^2}{12} - 2\ln 2$        BI (30)(8)

3.     $\displaystyle \int_0^1 \ln \frac{1 - ax}{1 - a} \frac{dx}{\ln x} = -\sum_{k=1}^{\infty} a^k \frac{\ln(1 + k)}{k}$     $[a < 1]$        BI (31)(3)

**4.222**

1.     $\displaystyle \int_0^{\infty} \ln \frac{a^2 + x^2}{b^2 + x^2} \, dx = (a - b)\pi$     $[a > 0, \quad b > 0]$        GW (322)(20)

2.     $\displaystyle \int_0^{\infty} \ln x \ln \frac{a^2 + x^2}{b^2 + x^2} \, dx = \pi(b - a) + \pi \ln \frac{a^a}{b^b}$     $[a > 0, \quad b > 0]$        BI (33)(1)

3.     $\displaystyle \int_0^{\infty} \ln x \ln \left(1 + \frac{b^2}{x^2}\right) dx = \pi b \left(\ln b - 1\right)$     $[b > 0]$        BI (33)(2)

4.     $\displaystyle \int_0^{\infty} \ln\left(1 + a^2 x^2\right) \ln\left(1 + \frac{b^2}{x^2}\right) dx = 2\pi \left[\frac{1 + ab}{a} \ln(1 + ab) - b\right]$

                                                      $[a > 0, \quad b > 0]$        BI (33)(3)

5.     $\displaystyle \int_0^{\infty} \ln\left(a^2 + x^2\right) \ln\left(1 + \frac{b^2}{x^2}\right) dx = 2\pi \left[(a + b)\ln(a + b) - a\ln a - b\right]$

                                                       $[a > 0, \quad b > 0]$        BI (33)(4)

6.     $\displaystyle \int_0^{\infty} \ln\left(1 + \frac{a^2}{x^2}\right) \ln\left(1 + \frac{b^2}{x^2}\right) dx = 2\pi \left[(a + b)\ln(a + b) - a\ln a - b\ln b\right]$

                                                       $[a > 0, \quad b > 0]$        BI (33)(5)

7.     $\displaystyle \int_0^{\infty} \ln\left(a^2 + \frac{1}{x^2}\right) \ln\left(1 + \frac{b^2}{x^2}\right) dx = 2\pi \left[\frac{1 + ab}{a} \ln(1 + ab) - b\ln b\right]$

                                                       $[a > 0, \quad b > 0]$        BI (33)(7)

$8.^{12}$ $\displaystyle\int_0^\infty \ln\left(1+ax\right)x^b e^{-x}\,\mathrm{d}x = b!e^{1/a}\mathrm{Ei}\left(-\frac{1}{a}\right) + \sum_{m=0}^{b-1}\frac{b!}{(b-m)!}\left[\frac{(-1)^{b-m-1}}{a^{b-m}}e^{1/a}\,\mathrm{Ei}\left(-\frac{1}{a}\right)\right]$

$\displaystyle\qquad\qquad\qquad\qquad + \sum_{k=1}^{b-m}\frac{(k-1)!}{(-a)^{b-m-k}} \qquad\qquad [b>0, \quad \text{an integer}]$

## 4.223

1. $\displaystyle\int_0^\infty \ln\left(1+e^{-x}\right)\,\mathrm{d}x = \frac{\pi^2}{12}$ BI (256)(10)

2. $\displaystyle\int_0^\infty \ln\left(1-e^{-x}\right)\,\mathrm{d}x = -\frac{\pi^2}{6}$ BI (256)(11)

3. $\displaystyle\int_0^\infty \ln\left(1+2e^{-x}\cos t+e^{-2x}\right)\,\mathrm{d}x = \frac{\pi^2}{6}-\frac{t^2}{2}$ $\qquad [|t|<\pi]$ BI (256)(18)

## 4.224

1. $\displaystyle\int_0^u \ln\sin x\,\mathrm{d}x = L\left(\frac{\pi}{2}-u\right) - L\left(\frac{\pi}{2}\right)$ LO III 186(15)

2. $\displaystyle\int_0^{\pi/4} \ln\sin x\,\mathrm{d}x = -\frac{\pi}{4}\ln 2 - \frac{1}{2}\boldsymbol{G}$ BI (285)(1)

3. $\displaystyle\int_0^{\pi/2} \ln\sin x\,\mathrm{d}x = \frac{1}{2}\int_0^\pi \ln\sin x\,\mathrm{d}x = -\frac{\pi}{2}\ln 2$ FI II 629,643

4. $\displaystyle\int_0^u \ln\cos x\,\mathrm{d}x = -L(u)$ LO III 184(10)

5. $\displaystyle\int_0^{\pi/4} \ln\cos x\,\mathrm{d}x = -\frac{\pi}{4}\ln 2 + \frac{1}{2}\boldsymbol{G}$ BI (286)(1)

6. $\displaystyle\int_0^{\pi/2} \ln\cos x\,\mathrm{d}x = -\frac{\pi}{2}\ln 2$ BI 306(1)

7. $\displaystyle\int_0^{\pi/2} \left(\ln\sin x\right)^2\,\mathrm{d}x = \frac{\pi}{2}\left[(\ln 2)^2 + \frac{\pi^2}{12}\right]$ BI (305)(19)

8. $\displaystyle\int_0^{\pi/2} \left(\ln\cos x\right)^2\,\mathrm{d}x = \frac{\pi}{2}\left[(\ln 2)^2 + \frac{\pi^2}{12}\right]$ BI (306)(14)

$9.^{8}$ $\displaystyle\int_0^\pi \ln\left(a+b\cos x\right)\,\mathrm{d}x = \pi\ln\frac{a+\sqrt{a^2-b^2}}{2}$ $\qquad [a\ge |b|>0]$ GW (322)(15)

10. $\displaystyle\int_0^\pi \ln\left(1\pm\sin x\right)\,\mathrm{d}x = -\pi\ln 2 \pm 4\boldsymbol{G}$ GW (322)(16a)

$11.^{7}$ $\displaystyle\int_0^{\pi/2} \ln\left(1+a\sin x\right)\,\mathrm{d}x = \frac{\pi}{2}\ln\frac{a}{2} + 2\boldsymbol{G} + 2\sum_{k=1}^\infty \frac{b^k}{k}\sum_{n=1}^k\frac{(-1)^{n+1}}{2n-1}$ $\quad [a>0] \qquad b=\frac{1-a}{1+a}$

$\displaystyle\qquad\qquad\qquad\qquad = -\frac{\pi}{2}\ln 2 + 2\boldsymbol{G} \qquad\qquad\qquad [a=1]$

$12.^{12}$   $\displaystyle\int_0^\pi \ln\left(1 + a\cos x\right)\,\mathrm{d}x = \pi\ln\left(\frac{1 + \sqrt{1 - a^2}}{2}\right)$      $\left[a^2 \le 1\right]$

$\displaystyle\qquad\qquad\qquad\qquad\qquad\qquad = 2\pi\ln\frac{|a|}{2}$      $\left[a^2 \ge 1\right]$

                                                            BI (330)(1)

$12\,(1)^{12}$   $\displaystyle\int_0^\pi \ln^2\left(1 + a\cos x\right)\,\mathrm{d}x = \begin{cases} 2\pi\ln\left(\dfrac{1 + \sqrt{1 - a^2}}{2}\right) & \text{for } a^2 \le 1 \\[2mm] \pi\ln\dfrac{a^2}{4} & \text{for } a^2 \ge 1 \end{cases}$

$13.^{12}$   $\displaystyle\int_0^{\pi/2} \ln\left(1 + 2a\sin x + a^2\right)\,\mathrm{d}x = \sum_{k=0}^\infty \frac{2^k\,(k!)^2}{(2k+1)\cdot(2k+1)!!}\left(\frac{2a}{1 + a^2}\right)^{2k+1}$

                                                $\left[a^2 \le 1\right]$         BI (308)(24)

$14.^{11}$   $\displaystyle\int_0^{n\pi} \ln\left(a^2 - 2ab\cos x + b^2\right)\,\mathrm{d}x = 2n\pi\ln\left[\max\left(|a|, |b|\right)\right]$

                                             $\left[ab > 0\right]$       FI II 142, 163, 688

$15.^8$   $\displaystyle\int_0^{n\pi} \ln\left(1 - 2a\cos x + a^2\right)\,\mathrm{d}x = 0$      $\left[a^2 \le 1\right]$

$\displaystyle\qquad\qquad\qquad\qquad\qquad\qquad = n\pi\ln a^2$      $\left[a^2 \ge 1\right]$

**4.225**

1.   $\displaystyle\int_0^{\pi/4} \ln\left(\cos x - \sin x\right)\,\mathrm{d}x = -\frac{\pi}{8}\ln 2 - \frac{1}{2}\boldsymbol{G}$             GW (322)(9b)

2.   $\displaystyle\int_0^{\pi/4} \ln\left(\cos x + \sin x\right)\,\mathrm{d}x = \frac{1}{2}\int_0^{\pi/2} \ln\left(\cos x + \sin x\right)\,\mathrm{d}x = -\frac{\pi}{8}\ln 2 + \frac{1}{2}\boldsymbol{G}$   GW (322)(9a)

3.   $\displaystyle\int_0^{2\pi} \ln\left(1 + a\sin x + b\cos x\right)\,\mathrm{d}x = 2\pi\ln\frac{1 + \sqrt{1 - a^2 - b^2}}{2}$

                                        $\left[a^2 + b^2 < 1\right]$      BI (332)(2)

4.   $\displaystyle\int_0^{2\pi} \ln\left(1 + a^2 + b^2 + 2a\sin x + 2b\cos x\right)\,\mathrm{d}x = 0$     $\left[a^2 + b^2 \le 1\right]$

$\displaystyle\qquad\qquad\qquad\qquad\qquad\qquad\qquad = 2\pi\ln\left(a^2 + b^2\right)$     $\left[a^2 + b^2 \ge 1\right]$

                                                   BI (322)(3)

**4.226**

$1.^{12}$   $\displaystyle\int_0^{\pi/2} \ln^2(a^2 - \sin^2 x)\,\mathrm{d}x = -2\pi\ln 2$          $\left[a^2 \le 1\right]$

$\displaystyle\qquad\qquad = 2\pi\ln\frac{a + \sqrt{a^2 - 1}}{2} = 2\pi\left(\operatorname{arccosh} a - \ln 2\right)$     $\left[a > 1\right]$

                                                   FI II 644, 687

2. $$\int_0^{\pi/2} \ln\left(1 + a\sin^2 x\right)\,dx = \frac{1}{2}\int_0^{\pi} \ln\left(1 + a\sin^2 x\right)\,dx = \int_0^{\pi/2} \ln\left(1 + a\cos^2 x\right)\,dx$$

$$= \frac{1}{2}\int_0^{\pi} \ln\left(1 + a\cos^2 x\right)\,dx = \pi\ln\frac{1 + \sqrt{1+a}}{2}$$

$$[a \geq -1] \qquad \text{BI (308)(15), GW(322)(12)}$$

3. $$\int_0^u \ln\left(1 - \sin^2\alpha \sin^2 x\right)\,dx = (\pi - 2\theta)\ln\cot\frac{\alpha}{2} + 2u\ln\left(\frac{1}{2}\sin\alpha\right) - \frac{\pi}{2}\ln 2$$

$$+ L(\theta + u) - L(\theta - u) + L\left(\frac{\pi}{2} - 2u\right)$$

$$\left[\cot\theta = \cos\alpha\tan u; \quad -\pi \leq \alpha \leq \pi, \quad -\frac{\pi}{2} \leq u \leq \frac{\pi}{2}\right] \quad \text{LO III 287}$$

4. $$\int_0^{\pi/2} \ln\left[1 - \cos^2 x\left(\sin^2\alpha - \sin^2\beta\sin^2 x\right)\right]\,dx = \pi\ln\left[\frac{1}{2}\left(\cos^2\frac{\alpha}{2} + \sqrt{\cos^4\frac{\alpha}{2} + \sin^2\frac{\beta}{2}\cos^2\frac{\beta}{2}}\right)\right]$$

$$[\alpha > \beta > 0] \qquad \text{LO III 283}$$

5. $$\int_0^u \ln\left(1 - \frac{\sin^2 x}{\sin^2\alpha}\right)\,dx = -u\ln\sin^2\alpha - L\left(\frac{\pi}{2} - \alpha + u\right) + L\left(\frac{\pi}{2} - \alpha - u\right)$$

$$\left[-\frac{\pi}{2} \leq u \leq \frac{\pi}{2}, \quad |\sin u| \leq |\sin\alpha|\right]$$

$$\text{LO III 287}$$

6. $$\int_0^{\pi/2} \ln\left(a^2\cos^2 x + b^2\sin^2 x\right)\,dx = \frac{1}{2}\int_0^{\pi} \ln\left(a^2\cos^2 x + b^2\sin^2 x\right)\,dx = \pi\ln\frac{a+b}{2}$$

$$[a > 0, \quad b > 0] \qquad \text{GW (322)(13)}$$

7. $$\int_0^{\pi/2} \ln\frac{1 + \sin t\cos^2 x}{1 - \sin t\cos^2 x}\,dx = \pi\ln\frac{1 + \sin\frac{t}{2}}{\cos\frac{t}{2}} = \pi\ln\cot\frac{\pi - t}{4}$$

$$\left[|t| < \frac{\pi}{2}\right] \qquad \text{LO III 283}$$

**4.227**

1. $$\int_0^u \ln\tan x\,dx = L(u) + L\left(\frac{\pi}{2} - u\right) - L\left(\frac{\pi}{2}\right) \qquad \text{LO III 186(16)}$$

2. $$\int_0^{\pi/4} \ln\tan x\,dx = -\int_{\pi/4}^{\pi/2} \ln\tan x\,dx = -\boldsymbol{G} \qquad \text{BI (286)(11)}$$

3. $$\int_0^{\pi/2} \ln\left(a\tan x\right)\,dx = \frac{\pi}{2}\ln a \qquad [a > 0] \qquad \text{BI (307)(2)}$$

4.[7] $$\int_0^{\pi/4} \left(\ln\tan x\right)^n\,dx = n!(-1)^n\sum_{k=0}^{\infty}\frac{(-1)^k}{(2k+1)^{n+1}}$$

$$= \frac{1}{2}\left(\frac{\pi}{2}\right)^{n+1}|E_n| \qquad [n \text{ even}]$$

$$\text{BI (286)(21)}$$

5.[7] $$\int_0^{\pi/2} \left(\ln\tan x\right)^{2n}\,dx = 2(2n)!\sum_{k=0}^{\infty}\frac{(-1)^k}{(2k+1)^{2n+1}} = \left(\frac{\pi}{2}\right)^{2n+1}|E_{2n}| \qquad \text{BI (307)(15)}$$

6. $\int_0^{\pi/2} (\ln \tan x)^{2n+1} \, dx = 0$ BI (307)(14)

7. $\int_0^{\pi/4} (\ln \tan x)^2 \, dx = \dfrac{\pi^3}{16}$ BI (286)(16)

8. $\int_0^{\pi/4} (\ln \tan x)^4 \, dx = \dfrac{5}{64}\pi^5$ BI (286)(19)

9. $\int_0^{\pi/4} \ln (1 + \tan x) \, dx = \dfrac{\pi}{8}\ln 2$ BI (287)(1)

10. $\int_0^{\pi/2} \ln (1 + \tan x) \, dx = \dfrac{\pi}{4}\ln 2 + \boldsymbol{G}$ BI (308)(9)

11. $\int_0^{\pi/4} \ln (1 - \tan x) \, dx = \dfrac{\pi}{8}\ln 2 - \boldsymbol{G}$ BI (287)(2)

12.[11] $\int_0^{\pi/2} (\ln (1 - \tan x))^2 \, dx = \dfrac{\pi}{2}\ln 2 - 2\boldsymbol{G}$ BI (308)(10)

13. $\int_0^{\pi/4} \ln (1 + \cot x) \, dx = \dfrac{\pi}{8}\ln 2 + \boldsymbol{G}$ BI (287)(3)

14. $\int_0^{\pi/4} \ln (\cot x - 1) \, dx = \dfrac{\pi}{8}\ln 2$ BI (287)(4)

15. $\int_0^{\pi/4} \ln (\tan x + \cot x) \, dx = \dfrac{1}{2}\int_0^{\pi/2} \ln (\tan x + \cot x) \, dx = \dfrac{\pi}{2}\ln 2$ BI (287)(5), BI (308)(11)

16.[11] $\int_0^{\pi/4} (\ln (\cot x - \tan x))^2 \, dx = \dfrac{1}{2}\int_0^{\pi/2} (\ln (\cot x - \tan x))^2 \, dx = \dfrac{\pi}{2}\ln 2$

BI (287)(6), BI (308)(12)

17. $\int_0^{\pi/2} \ln (a^2 + b^2 \tan^2 x) \, dx = \dfrac{1}{2}\int_0^{\pi} \ln (a^2 + b^2 \tan^2 x) \, dx = \pi \ln(a + b)$

$[a > 0, \quad b > 0]$ GW (322)(17)

**4.228**

1. $\int_0^{\pi/2} \ln \left( \sin t \sin x + \sqrt{1 - \cos^2 t \sin^2 x} \right) \, dx = \dfrac{\pi}{2}\ln 2 - 2L\left(\dfrac{t}{2}\right) - 2L\left(\dfrac{\pi - t}{2}\right)$ LO III 290

2. $\int_0^{u} \ln \left( \cos x + \sqrt{\cos^2 x - \cos^2 t} \right) \, dx = -\left(\dfrac{\pi}{2} - t - \varphi\right)\ln \cos t + \dfrac{1}{2}L(u + \varphi) - \dfrac{1}{2}L(u - \varphi) - L(\varphi)$

$\left[\cos \varphi = \dfrac{\sin u}{\sin t}; \quad 0 \le u \le t \le \dfrac{\pi}{2}\right]$

LO III 290

3. $\int_0^{t} \ln \left( \cos x + \sqrt{\cos^2 x - \cos^2 t} \right) \, dx = -\left(\dfrac{\pi}{2} - t\right)\ln \cos t$ LO III 285

4.    $\int_0^u \ln \frac{\sin u + \sin t \cos x \sqrt{\sin^2 u - \sin^2 x}}{\sin u - \sin t \cos x \sqrt{\sin^2 u - \sin^2 x}} \, dx = \pi \ln \left[ \tan \frac{t}{2} \sin u + \sqrt{\tan^2 \frac{t}{2} \sin^2 u + 1} \right]$

$$[t > 0, \quad u > 0]$$      LO III 283

5.    $\int_0^{\pi/4} \sqrt{\ln \cot x} \, dx = \frac{\sqrt{\pi}}{2} \sum_{k=0}^{\infty} \frac{(-1)^k}{\sqrt{(2k+1)^3}}$      BI (297)(9)

6.    $\int_0^{\pi/4} \frac{dx}{\sqrt{\ln \cot x}} = \sqrt{\pi} \sum_{k=0}^{\infty} \frac{(-1)^k}{\sqrt{2k+1}}$      BI (304)(24)

7.    $\int_0^{\pi/4} \ln \left( \sqrt{\tan x} + \sqrt{\cot x} \right) dx = \frac{1}{2} \int_0^{\pi/2} \ln \left( \sqrt{\tan x} + \sqrt{\cot x} \right) dx = \frac{\pi}{8} \ln 2 + \frac{1}{2} G$

     BI (287)(7), BI (308)(22)

8.    $\int_0^{\pi/4} \ln^2 \left( \sqrt{\cot x} - \sqrt{\tan x} \right) dx = \frac{1}{2} \int_0^{\pi/2} \ln^2 \left( \sqrt{\cot x} - \sqrt{\tan x} \right) dx = \frac{\pi}{4} \ln 2 - G$

     BI (287)(8), BI (308)(23)

**4.229**

1.    $\int_0^1 \ln(-\ln x) \, dx = -C$      FI II 807

2.[11]    $\text{PV} \int_0^1 \frac{dx}{\ln(-\ln x)} = \text{PV} \int_0^{\infty} \frac{e^{-u}}{\ln u} \, du \approx -0.154479$      BI (31)(2)

3.    $\int_0^1 \ln(-\ln x) \frac{dx}{\sqrt{-\ln x}} = -(C + 2\ln 2)\sqrt{\pi}$      BI (32)(4)

4.[11]    $\int_0^1 \ln(-\ln x)(-\ln x)^{\mu-1} \, dx = \psi(\mu)\Gamma(\mu)$      $[\text{Re}\,\mu > 0]$      BI (30)(10)

     If the integrand contains $(\ln \ln \frac{1}{x})$, it is convenient to make the substitution $\ln \frac{1}{x} = u$ so that $x = e^{-u}$.

5.[7]    $\int_0^1 \ln(a + \ln x) \, dx = \ln a - e^{-a} \text{Ei}(a)$      $[a > 0]$      BI (30)(5)

6.    $\int_0^1 \ln(a - \ln x) \, dx = \ln a - e^a \text{Ei}(-a)$      $[a > 0]$      BI (30)(6)

7.    $\int_{\pi/4}^{\pi/2} \ln \ln \tan x \, dx = \frac{\pi}{2} \ln \left( \frac{\Gamma\left(\frac{3}{4}\right)}{\Gamma\left(\frac{1}{4}\right)} \sqrt{2\pi} \right)$      BI (308)(28)

## 4.23 Combinations of logarithms and rational functions

**4.231**

1.    $\int_0^1 \frac{\ln x}{1+x} \, dx = -\frac{\pi^2}{12}$      FI II 483a

2.    $\int_0^1 \frac{\ln x}{1-x} \, dx = -\frac{\pi^2}{6}$      FI II 714

3. $\displaystyle\int_0^1 \frac{x \ln x}{1-x}\,dx = 1 - \frac{\pi^2}{6}$ 　　　　　　　　　　　　　　　　　　　　BI (108)(7)

4. $\displaystyle\int_0^1 \frac{1+x}{1-x} \ln x\,dx = 1 - \frac{\pi^2}{3}$ 　　　　　　　　　　　　　　　　　BI (108)(9)

5.[11] $\displaystyle\int_0^\infty \frac{\ln x\,dx}{(x+a)^2} = \frac{\ln a}{a}$ 　　　　　　　　　　$[0 < a]$　　　　　　　BI (139)(1)

6. $\displaystyle\int_0^1 \frac{\ln x}{(1+x)^2}\,dx = -\ln 2$ 　　　　　　　　　　　　　　　　　　BI (111)(1)

7.[7] $\displaystyle\int_0^\infty \ln x \frac{dx}{(a^2+b^2x^2)^n} = \frac{\Gamma\left(n-\frac{1}{2}\right)\sqrt{\pi}}{4(n-1)!a^{2n-1}b}\left[2\ln\frac{a}{2b} - \boldsymbol{C} - \psi\left(n - \frac{1}{2}\right)\right]$

　　　　　　　　　　　　　　　　　　　　　　　　$[a > 0, \quad b > 0]$　　　　　LI (139)(3)

8. $\displaystyle\int_0^\infty \frac{\ln x\,dx}{a^2+b^2x^2} = \frac{\pi}{2ab}\ln\frac{a}{b}$ 　　　　　　　　$[ab > 0]$　　　　　　BI (135)(6)

9. $\displaystyle\int_0^\infty \frac{\ln px}{q^2+x^2}\,dx = \frac{\pi}{2q}\ln pq$ 　　　　　　　　$[p > 0, \quad q > 0]$　　　BI (135)(4)

10. $\displaystyle\int_0^\infty \frac{\ln x\,dx}{a^2-b^2x^2} = -\frac{\pi^2}{4ab}$ 　　　　　　　　$[ab > 0]$

11. $\displaystyle\int_0^a \frac{\ln x\,dx}{x^2+a^2} = \frac{\pi \ln a}{4a} - \frac{\boldsymbol{G}}{a}$ 　　　　　　　　$[a > 0]$　　　　　　GW (324)(7b)

12. $\displaystyle\int_0^1 \frac{\ln x}{1+x^2}\,dx = -\int_1^\infty \frac{\ln x}{1+x^2}\,dx = -\boldsymbol{G}$ 　　　　　　　FI II 482, 614

13. $\displaystyle\int_0^1 \frac{\ln x\,dx}{1-x^2} = -\frac{\pi^2}{8}$ 　　　　　　　　　　　　　　　　　　　　BI (108)(11)

14. $\displaystyle\int_0^1 \frac{x \ln x}{1+x^2}\,dx = -\frac{\pi^2}{48}$ 　　　　　　　　　　　　　　　　　　GW (324)(7b)

15. $\displaystyle\int_0^1 \frac{x \ln x}{1-x^2}\,dx = -\frac{\pi^2}{24}$

16. $\displaystyle\int_0^1 \ln x \frac{1-x^{2n+2}}{(1-x^2)^2}\,dx = -\frac{(n+1)\pi^2}{8} + \sum_{k=1}^n \frac{n-k+1}{(2k-1)^2}$ 　　　　BI (111)(5)

17. $\displaystyle\int_0^1 \ln x \frac{1+(-1)^n x^{n+1}}{(1+x)^2}\,dx = -\frac{(n+1)\pi^2}{12} - \sum_{k=1}^n (-1)^k \frac{n-k+1}{k^2}$ 　　BI (111)(2)

18. $\displaystyle\int_0^1 \ln x \frac{1-x^{n+1}}{(1-x)^2}\,dx = -\frac{(n+1)\pi^2}{6} + \sum_{k=1}^n \frac{n-k+1}{k^2}$ 　　　　BI (111)(3)

19.* $\displaystyle\int_0^1 \frac{x \ln x}{1+x}\,dx = -1 + \frac{\pi^2}{2}$

20.* $\displaystyle\int_0^1 \frac{(1-x)\ln x}{1+x}\,dx = 1 - \frac{\pi^2}{6}$

**4.232**

1.[12] $\displaystyle\int_u^v \frac{\ln x\,dx}{(x+u)(x+v)} = \frac{\log^2 u - \log^2 v}{2(u-v)}$      $[u>0,\quad v>0]$      BI (145)(32)

2. $\displaystyle\int_0^\infty \frac{\ln x\,dx}{(x+\beta)(x+\gamma)} = \frac{(\ln\beta)^2 - (\ln\gamma)^2}{2(\beta-\gamma)}$      $[|\arg\beta|<\pi,\quad |\arg\gamma|<\pi]$

         ET II 218(24)

3. $\displaystyle\int_0^\infty \frac{\ln x}{x+a}\frac{dx}{x-1} = \frac{\pi^2 + (\ln a)^2}{2(a+1)}$      $[a>0]$      BI (140)(10)

**4.233**

1.[3] $\displaystyle\int_0^1 \frac{\ln x\,dx}{1+x+x^2} = \frac{2}{9}\left[\frac{2\pi^2}{3} - \psi'\left(\frac{1}{3}\right)\right] = -0.7813024129\ldots$      LI (113)(1)

2.[3] $\displaystyle\int_0^1 \frac{\ln x\,dx}{1-x+x^2} = \frac{1}{3}\left[\frac{2\pi^2}{3} - \psi'\left(\frac{1}{3}\right)\right] = -1.17195361934\ldots$      LI (113)(2)

3.[11] $\displaystyle\int_0^1 \frac{x\ln x\,dx}{1+x+x^2} = -\frac{1}{9}\left[\frac{7\pi^2}{6} - \psi'\left(\frac{1}{3}\right)\right] = -0.15766014917\ldots$      LI (113)(2)

4.[3] $\displaystyle\int_0^1 \frac{x\ln x\,dx}{1-x+x^2} = \frac{1}{6}\left[\frac{5\pi^2}{6} - \psi'\left(\frac{1}{3}\right)\right] = -0.3118211319\ldots$      LI (113)(4)

5. $\displaystyle\int_0^\infty \frac{\ln x\,dx}{x^2 + 2xa\cos t + a^2} = \frac{t\ln a}{a\sin t}$      $[a>0,\quad 0<t<\pi]$      GW (324)(13c)

**4.234**

1.[11] $\displaystyle\int_1^\infty \frac{\ln x\,dx}{(1+x^2)^2} = \frac{\mathbf{G}}{2} - \frac{\pi}{8}$      BI (144)(18)a

2. $\displaystyle\int_0^1 \frac{x\ln x\,dx}{(1+x^2)^2} = -\frac{1}{4}\ln 2$      BI (111)(4)

3. $\displaystyle\int_0^\infty \frac{1+x^2}{(1-x^2)^2}\ln x\,dx = 0$      BI (142)(2)a

4. $\displaystyle\int_0^\infty \frac{1-x^2}{(1+x^2)^2}\ln x\,dx = -\frac{\pi}{2}$      BI (142)(1)a

5. $\displaystyle\int_0^1 \frac{x^2\ln x\,dx}{(1-x^2)(1+x^4)} = -\frac{\pi^2}{16(2+\sqrt{2})}$      BI (112)(21)

6. $\displaystyle\int_0^\infty \frac{\ln x\,dx}{(a^2+b^2x^2)(1+x^2)} = \frac{b\pi}{2a(b^2-a^2)}\ln\frac{a}{b}$      $[ab>0]$      BI (317)(16)a

7. $\displaystyle\int_0^\infty \frac{\ln x}{x^2+a^2}\cdot\frac{dx}{1+b^2x^2} = \frac{\pi}{2(1-a^2b^2)}\left(\frac{1}{a}\ln a + b\ln b\right)$

         $[a>0,\quad b>0]$      LI (140)(12)

8. $\displaystyle\int_0^\infty \frac{x^2\ln x\,dx}{(a^2+b^2x^2)(1+x^2)} = \frac{a\pi}{2b(b^2-a^2)}\ln\frac{b}{a}$      $[ab>0]$      LI (140)(12), BI (317)(15)a

**4.235**

1.    $\displaystyle\int_0^\infty \ln x \frac{(1-x)x^{n-2}}{1-x^{2n}}\,dx = -\frac{\pi^2}{4n^2}\tan^2\frac{\pi}{2n}$      $[n>1]$      BI (135)(10)

2.    $\displaystyle\int_0^\infty \ln x \frac{(1-x^2)\,x^{m-1}}{1-x^{2n}}\,dx = -\frac{\pi^2\sin\left(\frac{m+1}{n}\right)\pi\sin\left(\frac{\pi}{n}\right)}{4n^2\sin^2\left(\frac{m\pi}{2n}\right)\sin^2\left(\frac{m+2}{2n}\pi\right)}$      LI (135)(12)

3.[11]    $\displaystyle\int_0^\infty \ln x \frac{(1-x^2)\,x^{n-3}}{1-x^{2n}}\,dx = -\frac{\pi^2}{4n^2}\tan^2\left(\frac{\pi}{n}\right)$      $[n>2]$      BI (135)(11)

4.    $\displaystyle\int_0^1 \ln x \frac{x^{m-1}+x^{n-m-1}}{1-x^n}\,dx = -\frac{\pi^2}{n^2\sin^2\left(\frac{m}{n}\pi\right)}$      $[n>m]$      BI (108)(15)

**4.236**

1.    $\displaystyle\int_0^1 \left\{\frac{1+(p-1)\ln x}{1-x} + \frac{x\ln x}{(1-x)^2}\right\} x^{p-1}\,dx = -1+\psi'(p)$

                                              $[p>0]$      BI (111)(6)a, GW (326)(13)

2.    $\displaystyle\int_0^1 \left[\frac{1}{1-x} + \frac{x\ln x}{(1-x)^2}\right]dx = \frac{\pi^2}{6}-1$      GW (326)(13a)

## 4.24 Combinations of logarithms and algebraic functions

**4.241**

1.    $\displaystyle\int_0^1 \frac{x^{2n}\ln x}{\sqrt{1-x^2}}\,dx = \frac{(2n-1)!!}{(2n)!!}\cdot\frac{\pi}{2}\left(\sum_{k=1}^{2n}\frac{(-1)^{k-1}}{k}-\ln 2\right)$      BI (118)(5)a

2.    $\displaystyle\int_0^1 \frac{x^{2n+1}\ln x}{\sqrt{1-x^2}}\,dx = \frac{(2n)!!}{(2n+1)!!}\left(\ln 2 + \sum_{k=1}^{2n+1}\frac{(-1)^k}{k}\right)$      BI (110)(5)a

3.    $\displaystyle\int_0^1 x^{2n}\sqrt{1-x^2}\ln x\,dx = \frac{(2n-1)!!}{(2n+2)!!}\cdot\frac{\pi}{2}\left(\sum_{k=1}^{2n}\frac{(-1)^{k-1}}{k}-\frac{1}{2n+2}-\ln 2\right)$

                                                       LI (117)(4), GW (324)(53a)

4.    $\displaystyle\int_0^1 x^{2n+1}\sqrt{1-x^2}\ln x\,dx = \frac{(2n)!!}{(2n+3)!!}\left(\ln 2 + \sum_{k=1}^{2n+1}\frac{(-1)^k}{k}-\frac{1}{2n+3}\right)$

                                                         BI (117)(5), GW (324)(53b)

5.[12]    $\displaystyle\int_0^1 \ln x\cdot\sqrt{(1-x^2)^{2n-1}}\,dx = -\frac{(2n-1)!!}{4\cdot(2n)!!}\pi\left(2\ln 2 + \sum_{k=1}^n\frac{1}{k}\right)$      BI (117)(3)

6.    $\displaystyle\int_0^{\sqrt{\frac{1}{2}}}\frac{\ln x\,dx}{\sqrt{1-x^2}} = -\frac{\pi}{4}\ln 2 - \frac{1}{2}G$      BI (145)(1)

7.    $\displaystyle\int_0^1 \frac{\ln x\,dx}{\sqrt{1-x^2}} = -\frac{\pi}{2}\ln 2$      FI II 614, 643

8.    $\displaystyle\int_1^\infty \frac{\ln x\,dx}{x^2\sqrt{x^2-1}} = 1-\ln 2$      BI (144)(17)

9. $\int_0^1 \sqrt{1-x^2}\, \ln x\, dx = -\dfrac{\pi}{8} - \dfrac{\pi}{4}\ln 2$  BI (117)(1), GW (324)(53c)

10. $\int_0^1 x\sqrt{1-x^2}\, \ln x\, dx = \frac{1}{3}\ln 2 - \frac{4}{9}$  BI (117)(2)

11. $\int_0^1 \dfrac{\ln x\, dx}{\sqrt{x\,(1-x^2)}} = -\dfrac{\sqrt{2\pi}}{8}\left[\Gamma\left(\dfrac{1}{4}\right)\right]^2$  GW (324)(54a)

**4.242**

1. $\int_0^\infty \dfrac{\ln x\, dx}{\sqrt{(a^2+x^2)\,(x^2+b^2)}} = \dfrac{1}{2a}\, K\left(\dfrac{\sqrt{a^2-b^2}}{a}\right)\ln ab$

$\qquad\qquad\qquad\qquad\qquad\qquad [a > b > 0]$  BY (800.04)

2. $\int_0^b \dfrac{\ln x\, dx}{\sqrt{(a^2+x^2)\,(b^2-x^2)}} = \dfrac{1}{2\sqrt{a^2+b^2}}\left[K\left(\dfrac{b}{\sqrt{a^2+b^2}}\right)\ln ab - \dfrac{\pi}{2}\, K\left(\dfrac{a}{\sqrt{a^2+b^2}}\right)\right]$

$\qquad\qquad\qquad\qquad\qquad\qquad [a > 0,\quad b > 0]$  BY (800.02)

3. $\int_b^\infty \dfrac{\ln x\, dx}{\sqrt{(x^2+a^2)\,(x^2-b^2)}} = \dfrac{1}{2\sqrt{a^2+b^2}}\left[K\left(\dfrac{a}{\sqrt{a^2+b^2}}\right)\ln ab + \dfrac{\pi}{2}\, K\left(\dfrac{b}{\sqrt{a^2+b^2}}\right)\right]$

$\qquad\qquad\qquad\qquad\qquad\qquad [a > 0,\quad b > 0]$  BY (800.06)

4. $\int_0^b \dfrac{\ln x\, dx}{\sqrt{(a^2-x^2)\,(b^2-x^2)}} = \dfrac{1}{2a}\left[K\left(\dfrac{b}{a}\right)\ln ab - \dfrac{\pi}{2}\, K\left(\dfrac{\sqrt{a^2-b^2}}{a}\right)\right]$

$\qquad\qquad\qquad\qquad\qquad\qquad [a > b > 0]$  BY (800.01)

5. $\int_b^a \dfrac{\ln x\, dx}{\sqrt{(a^2-x^2)\,(x^2-b^2)}} = \dfrac{1}{2a}\, K\left(\dfrac{\sqrt{a^2-b^2}}{a}\right)\ln ab$  BY (800.03)

6. $\int_a^\infty \dfrac{\ln x\, dx}{\sqrt{(x^2-a^2)\,(x^2-b^2)}} = \dfrac{1}{2a}\left[K\left(\dfrac{b}{a}\right)\ln ab + \dfrac{\pi}{2}\, K\left(\dfrac{\sqrt{a^2-b^2}}{a}\right)\right]$

$\qquad\qquad\qquad\qquad\qquad\qquad [a > b > 0]$  BY (800.05)

**4.243** $\int_0^1 \dfrac{x\ln x}{\sqrt{1-x^4}}\, dx = -\dfrac{\pi}{8}\ln 2$  GW (324)(56b)

**4.244**

1. $\int_0^1 \dfrac{\ln x\, dx}{\sqrt[3]{x\,(1-x^2)^2}} = -\dfrac{1}{8}\left[\Gamma\left(\dfrac{1}{3}\right)\right]^3$  GW (324)(54b)

2. $\int_0^1 \dfrac{\ln x\, dx}{\sqrt[3]{1-x^3}} = -\dfrac{\pi}{3\sqrt{3}}\left(\ln 3 + \dfrac{\pi}{3\sqrt{3}}\right)$  BI (118)(7)

3. $\int_0^1 \dfrac{x\ln x\, dx}{\sqrt[3]{(1-x^3)^2}} = \dfrac{\pi}{3\sqrt{3}}\left(\dfrac{\pi}{3\sqrt{3}} - \ln 3\right)$  BI (118)(8)

**4.245**

1. 　$\displaystyle\int_0^1 \frac{x^{4n+1}\ln x}{\sqrt{1-x^4}}\,dx = \frac{(2n-1)!!}{(2n)!!}\cdot\frac{\pi}{8}\left(\sum_{k=1}^{2n}\frac{(-1)^{k-1}}{k}-\ln 2\right)$ 　　　　　GW (324)(56a)

2. 　$\displaystyle\int_0^1 \frac{x^{4n+3}\ln x}{\sqrt{1-x^4}}\,dx = \frac{(2n)!!}{4\cdot(2n+1)!!}\left(\ln 2+\sum_{k=1}^{2n+1}\frac{(-1)^k}{k}\right)$ 　　　　　GW (324)(56c)

**4.246** 　$\displaystyle\int_0^1 \left(1-x^2\right)^{n-\frac{1}{2}}\ln x\,dx = -\frac{(2n-1)!!}{(2n)!!}\frac{\pi}{4}\left[2\ln 2+\sum_{k=1}^{n}\frac{1}{k}\right]$ 　　　GW (324)(55)

**4.247**

1.$^6$ 　$\displaystyle\int_0^1 \frac{\ln x}{\sqrt[n]{1-x^{2n}}}\,dx = -\frac{\pi\,\mathrm{B}\left(\dfrac{1}{2n},\dfrac{1}{2n}\right)}{8n^2\sin\dfrac{\pi}{2n}}$ 　　　　$[n>1]$ 　　　GW (324)(54c)a

2.$^6$ 　$\displaystyle\int_0^1 \frac{\ln x\,dx}{\sqrt[n]{x^{n-1}\left(1-x^2\right)}} = -\frac{\pi\,\mathrm{B}\left(\dfrac{1}{2n},\dfrac{1}{2n}\right)}{8\sin\dfrac{\pi}{2n}}$ 　　　　　GW (324)(54)

## 4.25 Combinations of logarithms and powers

**4.251**

1. 　$\displaystyle\int_0^\infty \frac{x^{\mu-1}\ln x}{\beta+x}\,dx = \frac{\pi\beta^{\mu-1}}{\sin\mu\pi}\left(\ln\beta-\pi\cot\mu\pi\right)$ 　　　　$[|\arg\beta|<\pi,\quad 0<\mathrm{Re}\,\mu<1]$

　　　　　　　　　　　　　　　　　　　　　　　　　　　　　　　　BI (135)(1)

2. 　$\displaystyle\int_0^\infty \frac{x^{\mu-1}\ln x}{a-x}\,dx = \pi a^{\mu-1}\left(\cot\mu\pi\ln a-\frac{\pi}{\sin^2\mu\pi}\right)$ 　　$[a>0,\quad 0<\mathrm{Re}\,\mu<1]$ 　　ET I 314(5)

3.$^{10}$ 　$\displaystyle\int_0^1 \frac{x^{\mu-1}\ln x}{x+1}\,dx = \beta'(\mu)$ 　　　　　　　$[\mathrm{Re}\,\mu>0]$ 　　　GW (324)(6), ET I 314(3)

4. 　$\displaystyle\int_0^1 \frac{x^{\mu-1}\ln x}{1-x}\,dx = -\psi'(\mu) = -\zeta(2,\mu)$ 　　　　$[\mathrm{Re}\,\mu>0]$ 　　　BI (108)(8)

5.$^{11}$ 　$\displaystyle\int_0^1 \ln x\frac{x^{2n}}{1+x}\,dx = -\frac{\pi^2}{12}+\sum_{k=1}^{2n}\frac{(-1)^{k-1}}{k^2}$ 　　　　　BI (108)(4)

6.$^{11}$ 　$\displaystyle\int_0^1 \ln x\frac{x^{2n-1}}{1+x}\,dx = \frac{\pi^2}{12}+\sum_{k=1}^{2n-1}\frac{(-1)^k}{k^2}$ 　　　　　BI (108)(5)

**4.252**

1. 　$\displaystyle\int_0^\infty \frac{x^{\mu-1}\ln x}{(x+\beta)(x+\gamma)}\,dx = \frac{\pi}{(\gamma-\beta)\sin\mu\pi}\left[\beta^{\mu-1}\ln\beta-\gamma^{\mu-1}\ln\gamma-\pi\cot\mu\pi\left(\beta^{\mu-1}-\gamma^{\mu-1}\right)\right]$

　　　　　　$[|\arg\beta|<\pi,\quad |\arg\gamma|<\pi,\quad 0<\mathrm{Re}\,\mu<2,\quad \mu\neq 1]$ 　 BI (140)(9)a, ET 314(6)

2.  $$\int_0^\infty \frac{x^{\mu-1}\ln x\,dx}{(x+\beta)(x-1)} = \frac{\pi}{(\beta+1)\sin^2\mu\pi}\left[\pi - \beta^{\mu-1}\left(\sin\mu\pi\ln\beta - \pi\cos\mu\pi\right)\right]$$

$$[|\arg\beta| < \pi, \quad 0 < \operatorname{Re}\mu < 2, \quad \mu \neq 1]$$
BI (140)(11)

3.  $$\int_0^\infty \frac{x^{p-1}\ln x}{1-x^2}\,dx = -\frac{\pi^2}{4}\operatorname{cosec}^2\frac{p\pi}{2}$$   $$[0 < p < 2] \qquad \text{(see also 4.254 2)}$$

4.[6]  $$\int_0^\infty \frac{x^{\mu-1}\ln x}{(x+a)^2}\,dx = \frac{(1-\mu)a^{\mu-2}\pi}{\sin\mu\pi}\left(\ln a - \pi\cot\mu\pi + \frac{1}{\mu-1}\right)$$

$$[|\arg a| < \pi, \quad 0 < \operatorname{Re}\mu < 2, \quad (\mu \neq 1)]$$
GW (324)(13b)

**4.253**

1.[8]  $$\int_0^1 x^{\mu-1}(1-x^r)^{\nu-1}\ln x\,dx = \frac{1}{r^2}\operatorname{B}\left(\frac{\mu}{r},\nu\right)\left[\psi\left(\frac{\mu}{r}\right) - \psi\left(\frac{\mu}{r}+\nu\right)\right]$$

$$[\operatorname{Re}\mu > 0, \quad \operatorname{Re}\nu > 0, \quad r > 0]$$
GW (324)(3b)a, BI (107)(5)a

2.  $$\int_0^1 \frac{x^{p-1}}{(1-x)^{p+1}}\ln x\,dx = -\frac{\pi}{p}\operatorname{cosec}p\pi$$   $$[0 < p < 1] \qquad \text{bi (319)(10)a}$$

3.  $$\int_u^\infty \frac{(x-u)^{\mu-1}\ln x\,dx}{x^\lambda} = u^{\mu-\lambda}\operatorname{B}(\lambda-\mu,\mu)\left[\ln u + \psi(\lambda) - \psi(\lambda-\mu)\right]$$

$$[0 < \operatorname{Re}\mu < \operatorname{Re}\lambda] \qquad \text{ET II 203(18)}$$

4.[11]  $$\int_0^\infty \ln x\left(\frac{x}{a^2+x^2}\right)^p\frac{dx}{x} = \frac{\ln a}{2a^p}\operatorname{B}\left(\frac{p}{2},\frac{p}{2}\right)$$   $$[a > 0, \quad p > 0] \qquad \text{BI (140)(6)}$$

5.  $$\int_1^\infty (x-1)^{p-1}\ln x\,dx = \frac{\pi}{p}\operatorname{cosec}\pi p$$   $$[-1 < p < 0] \qquad \text{BI (289)(12)a}$$

6.[7]  $$\int_0^\infty \ln x\frac{dx}{(a+x)^{\mu+1}} = \frac{1}{\mu a^\mu}\left(\ln a - \boldsymbol{C} - \psi(\mu)\right)$$   $$[\operatorname{Re}\mu > 0, \quad a \neq 0, \quad |\arg a| < \pi]$$
NT 68(7)

7.[7]  $$\int_0^\infty \ln x\frac{dx}{(a+x)^{n+\frac{1}{2}}} = \frac{2}{(2n-1)a^{n-\frac{1}{2}}}\left(\ln a + 2\ln 2 - 2\sum_{k=1}^{n-1}\frac{1}{2k-1}\right)$$

$$[|\arg a| < \pi, \quad n = 1, 2, \ldots] \qquad \text{BI (142)(5)}$$

**4.254**

1.  $$\int_0^1 \frac{x^{p-1}\ln x}{1-x^q}\,dx = -\frac{1}{q^2}\psi'\left(\frac{p}{q}\right)$$   $$[p > 0, \quad q > 0] \qquad \text{GW (324)(5)}$$

2.  $$\int_0^\infty \frac{x^{p-1}\ln x}{1-x^q}\,dx = -\frac{\pi^2}{q^2\sin^2\frac{p\pi}{q}}$$   $$[0 < p < q] \qquad \text{BI (135)(8)}$$

3.  $$\int_0^\infty \frac{\ln x}{x^q-1}\frac{dx}{x^p} = \frac{\pi^2}{q^2\sin^2\frac{p-1}{q}\pi}$$   $$[p < 1, \quad p+q > 1] \qquad \text{BI (140)(2)}$$

4.$^3$ $\quad \displaystyle\int_0^1 \frac{x^{p-1}\ln x}{1+x^q}\,dx = \frac{1}{q^2}\beta'\left(\frac{p}{q}\right)$ $\qquad\qquad [p>0,\quad q>0]$ $\qquad$ GW (324)(7)

5. $\quad \displaystyle\int_0^\infty \frac{x^{p-1}\ln x}{1+x^q}\,dx = -\frac{\pi^2}{q^2}\frac{\cos\dfrac{p\pi}{q}}{\sin^2\dfrac{p\pi}{q}}$ $\qquad [0<p<q]$ $\qquad$ BI (135)(7)

6. $\quad \displaystyle\int_0^1 \frac{x^{q-1}\ln x}{1-x^{2q}}\,dx = -\frac{\pi^2}{8q^2}$ $\qquad\qquad\qquad [q>0]$ $\qquad$ BI (108)(12)

**4.255**

1. $\quad \displaystyle\int_0^1 \ln x\,\frac{(1-x^2)\,x^{p-2}}{1+x^{2p}}\,dx = -\left(\frac{\pi}{2p}\right)^2 \frac{\sin\frac{\pi}{2p}}{\cos^2\left(\frac{\pi}{2p}\right)}$ $\qquad [p>1]$ $\qquad$ BI (108)(13)

2. $\quad \displaystyle\int_0^1 \ln x\,\frac{(1+x^2)\,x^{p-2}}{1-x^{2p}}\,dx = -\left(\frac{\pi}{2p}\right)^2 \sec^2\left(\frac{\pi}{2p}\right)$ $\qquad [p>1]$ $\qquad$ BI (108)(14)

3. $\quad \displaystyle\int_0^\infty \ln x\,\frac{1-x^p}{1-x^2}\,dx = \frac{\pi^2}{4}\tan^2\left(\frac{p\pi}{2}\right)$ $\qquad [p<1]$ $\qquad$ BI (140)(3)

**4.256** $\quad \displaystyle\int_0^1 \ln\frac{1}{x}\,\frac{x^{\mu-1}\,dx}{\sqrt[n]{(1-x^n)^{n-m}}} = \frac{1}{n^2}\,\mathrm{B}\left(\frac{\mu}{n},\frac{m}{n}\right)\left[\psi\left(\frac{\mu+m}{n}\right)-\psi\left(\frac{\mu}{n}\right)\right]$

$\qquad\qquad\qquad\qquad\qquad\qquad\qquad\qquad\qquad [\operatorname{Re}\mu>0]$ $\qquad$ LI (118)(12)

**4.257**

1. $\quad \displaystyle\int_0^\infty \frac{x^\nu \ln\dfrac{x}{\beta}\,dx}{(x+\beta)(x+\gamma)} = \frac{\pi\left[\gamma^\nu\ln\frac{\gamma}{\beta}+\pi\left(\beta^\nu-\gamma^\nu\right)\cot\nu\pi\right]}{\sin\nu\pi\,(\gamma-\beta)}$

$\qquad\qquad\qquad\qquad\qquad\qquad [|\arg\beta|<\pi,\quad |\arg\gamma|<\pi,\quad |\operatorname{Re}\nu|<1]$
$\qquad\qquad\qquad\qquad\qquad\qquad$ ET II 219(30)

2. $\quad \displaystyle\int_0^\infty \ln\frac{x}{q}\left(\frac{x^p}{q^{2p}+x^{2p}}\right)\frac{dx}{x} = 0$ $\qquad [q>0]$ $\qquad$ BI (140)(4)a

3. $\quad \displaystyle\int_0^\infty \ln\frac{x}{q}\left(\frac{x^p}{q^{2p}+x^{2p}}\right)^r \frac{dx}{q^2+x^2} = 0$ $\qquad [q>0]$ $\qquad$ BI (140)(4)a

4. $\quad \displaystyle\int_0^\infty \ln x \ln\frac{x}{a}\,\frac{dx}{(x-1)(x-a)} = \frac{\left[4\pi^2+(\ln a)^2\right]\ln a}{6(a-1)}$ $\qquad [a>0]$ $\quad$ (for $a=1$ see **4.261** 5)

$\qquad\qquad\qquad\qquad\qquad\qquad\qquad\qquad\qquad\qquad$ BI (141)(5)

5. $\quad \displaystyle\int_0^\infty \ln x \ln\frac{x}{a}\,\frac{x^p\,dx}{(x-1)(x-a)} = \frac{\pi^2\left[(a^p+1)\ln a - 2\pi\left(a^p-1\right)\cot p\pi\right]}{(a-1)\sin^2 p\pi}$

$\qquad\qquad\qquad\qquad\qquad\qquad\qquad [p^2<1,\quad a>0]$ $\qquad$ BI (141)(6)

## 4.26–4.27 Combinations involving powers of the logarithm and other powers

**4.261**

1.[7] $\displaystyle\int_0^1 (\ln x)^2 \frac{dx}{1 + 2x\cos t + x^2} = \frac{t\left(\pi^2 - t^2\right)}{6\sin t}$      $[0 \le t \le \pi]$      BI (113)(7)

2. $\displaystyle\int_0^1 \frac{(\ln x)^2\,dx}{x^2 - x + 1} = \frac{1}{2}\int_0^\infty \frac{(\ln x)^2\,dx}{x^2 - x + 1} = \frac{10\pi^3}{81\sqrt{3}}$      GW (324)(16c)

3. $\displaystyle\int_0^1 \frac{(\ln x)^2\,dx}{x^2 + x + 1} = \frac{1}{2}\int_0^\infty \frac{(\ln x)^2\,dx}{x^2 + x + 1} = \frac{8\pi^3}{81\sqrt{3}}$      GW (324)(16b)

4. $\displaystyle\int_0^\infty (\ln x)^2 \frac{dx}{(x-1)(x+a)} = \frac{\left[\pi^2 + (\ln a)^2\right]\ln a}{3(1+a)}$      $[a > 0]$      BI (141)(1)

5. $\displaystyle\int_0^\infty (\ln x)^2 \frac{dx}{(1-x)^2} = \frac{2}{3}\pi^2$      BI (139)(4)

6. $\displaystyle\int_0^1 (\ln x)^2 \frac{dx}{1 + x^2} = \frac{\pi^3}{16}$      BI (109)(3)

7. $\displaystyle\int_0^1 (\ln x)^2 \frac{1 + x^2}{1 + x^4}\,dx = \frac{1}{2}\int_0^\infty (\ln x)^2 \frac{1 + x^2}{1 + x^4}\,dx = \frac{3\sqrt{2}}{64}\pi^3$      BI (109)(5), BI (135)(13)

8.[11] $\displaystyle\int_0^1 (\ln x)^2 \frac{1 - x}{1 - x^6}\,dx = \frac{8\sqrt{3}\pi^3 + 351\,\zeta(3)}{486}$

9. $\displaystyle\int_0^1 (\ln x)^2 \frac{dx}{\sqrt{1 - x^2}} = \frac{\pi}{2}\left[(\ln 2)^2 + \frac{\pi^2}{12}\right]$      BI (118)(13)

10. $\displaystyle\int_0^\infty (\ln x)^2 \frac{x^{\mu-1}}{1 + x}\,dx = \frac{\pi^3\left(2 - \sin^2 \mu\pi\right)}{\sin^3 \mu\pi}$      $[0 < \operatorname{Re}\mu < 1]$      ET I 315(10)

11.[7] $\displaystyle\int_0^1 (\ln x)^2 \frac{x^n\,dx}{1 + x} = 2\sum_{k=n}^\infty \frac{(-1)^{n+k}}{(k+1)^3} = (-1)^n\left(\frac{3}{2}\zeta(3) + 2\sum_{k=1}^n \frac{(-1)^k}{k^3}\right)$

                            $[n = 0, 1, \ldots]$      BI (109)(1)

12.[7] $\displaystyle\int_0^1 (\ln x)^2 \frac{x^n\,dx}{1 - x} = 2\sum_{k=n}^\infty \frac{1}{(k+1)^3} = 2\left(\zeta(3) - \sum_{k=1}^n \frac{1}{k^3}\right)$

                            $[n = 0, 1, \ldots]$      BI (109)(2)

13.[11] $\displaystyle\int_0^1 (\ln x)^2 \frac{x^{2n}\,dx}{1 - x^2} = 2\sum_{k=n}^\infty \frac{1}{(2k+1)^3} = \frac{7}{4}\zeta(3) - 2\sum_{k=1}^n \frac{1}{(2k-1)^3}$

                            $[n = 0, 1, \ldots]$      BI (109)(4)

14. $\displaystyle\int_0^\infty (\ln x)^2 \frac{x^{p-1}\,dx}{x^2 + 2x\cos t + 1} = \frac{\pi\sin(1-p)t}{\sin t\sin p\pi}\left\{\pi^2 - t^2 + 2\pi\cot p\pi\left[\pi\cot p\pi + t\cot(1-p)t\right]\right\}$

                      $[0 < t < \pi, \quad 0 < p < 2, \quad p \ne 1]$

                                          GW (324)(17)

15.    $\displaystyle \int_0^1 (\ln x)^2 \frac{x^{2n}\,\mathrm{d}x}{\sqrt{1-x^2}} = \frac{(2n-1)!!}{2\cdot(2n)!!}\pi\left\{\frac{\pi^2}{12} + \sum_{k=1}^{2n}\frac{(-1)^k}{k^2} + \left[\sum_{k=1}^{2n}\frac{(-1)^k}{k} + \ln 2\right]^2\right\}$      GW (324)(60a)

16.    $\displaystyle \int_0^1 (\ln x)^2 \frac{x^{2n+1}\,\mathrm{d}x}{\sqrt{1-x^2}} = \frac{(2n)!!}{(2n+1)!!}\left\{-\frac{\pi^2}{12} - \sum_{k=1}^{2n+1}\frac{(-1)^k}{k^2} + \left[\sum_{k=1}^{2n+1}\frac{(-1)^k}{k} + \ln 2\right]^2\right\}$

                                                               GW (324)(60b)

17.[7]    $\displaystyle \int_0^1 (\ln x)^2 x^{\mu-1}(1-x)^{\nu-1}\,\mathrm{d}x = \mathrm{B}(\mu,\nu)\left\{[\psi(\mu)-\psi(\nu+\mu)]^2 + \psi'(\mu) - \psi'(\mu+\nu)\right\}$

                                                    $[\operatorname{Re}\mu > 0, \quad \operatorname{Re}\nu > 0]$      ET I 315(11)

18.    $\displaystyle \int_0^1 (\ln x)^2 \frac{1-x^{n+1}}{(1-x)^2}\,\mathrm{d}x = 2(n+1)\zeta(3) - 2\sum_{k=1}^{n}\frac{n-k+1}{k^3}$      LI (111)(8)

19.    $\displaystyle \int_0^1 (\ln x)^2 \frac{1+(-1)^n x^{n+1}}{(1+x)^2}\,\mathrm{d}x = \frac{3}{2}(n+1)\zeta(3) - 2\sum_{k=1}^{n}(-1)^{k-1}\frac{n-k+1}{k^3}$      LI (111)(7)

20.[7]    $\displaystyle \int_0^1 (\ln x)^2 \frac{1-x^{2n+2}}{(1-x^2)^2}\,\mathrm{d}x = \frac{7}{4}(n+1)\zeta(3) - 2\sum_{k=1}^{n}\frac{n-k+1}{(2k-1)^3}$

                                                    $[n = 0, 1, \ldots]$      LI (111)(9)

21.    $\displaystyle \int_0^1 (\ln x)^2 x^{p-1}(1-x^r)^{q-1}\,\mathrm{d}x = \frac{1}{r^3}\mathrm{B}\left(\frac{p}{r},q\right)\left\{\psi'\left(\frac{p}{r}\right) - \psi'\left(\frac{p}{r}+q\right) + \left[\psi\left(\frac{p}{r}\right) - \psi\left(\frac{p}{r}+q\right)\right]^2\right\}$

                                                $[p > 0, \quad q > 0, \quad r > 0]$      GW (324)(8a)

**4.262**

1.    $\displaystyle \int_0^1 (\ln x)^3 \frac{\mathrm{d}x}{1+x} = -\frac{7}{120}\pi^4$      BI (109)(9)

2.    $\displaystyle \int_0^1 (\ln x)^3 \frac{\mathrm{d}x}{1-x} = -\frac{\pi^4}{15}$      BI (109)(11)

3.    $\displaystyle \int_0^\infty (\ln x)^3 \frac{\mathrm{d}x}{(x+a)(x-1)} = \frac{\left[\pi^2 + (\ln a)^2\right]^2}{4(a+1)}$      $[a > 0]$      BI (141)(2)

4.    $\displaystyle \int_0^1 (\ln x)^3 \frac{x^n\,\mathrm{d}x}{1+x} = (-1)^{n+1}\left[\frac{7\pi^4}{120} - 6\sum_{k=0}^{n-1}\frac{(-1)^k}{(k+1)^4}\right]$      $[n = 1, 2, \ldots]$      BI (109)(10)

5.    $\displaystyle \int_0^1 (\ln x)^3 \frac{x^n\,\mathrm{d}x}{1-x} = -\frac{\pi^4}{15} + 6\sum_{k=0}^{n-1}\frac{1}{(k+1)^4}$      $[n = 1, 2, \ldots]$      BI (109)(12)

6.    $\displaystyle \int_0^1 (\ln x)^3 \frac{x^{2n}\,\mathrm{d}x}{1-x^2} = -\frac{\pi^4}{16} + 6\sum_{k=0}^{n-1}\frac{1}{(2k+1)^4}$      $[n = 1, 2, \ldots]$      BI (109)(14)

7.    $\displaystyle \int_0^1 (\ln x)^3 \frac{1-x^{n+1}}{(1-x)^2}\,\mathrm{d}x = -\frac{(n+1)\pi^4}{15} + 6\sum_{k=1}^{n}\frac{n-k+1}{k^4}$      BI (111)(11)

8. $\int_0^1 (\ln x)^3 \dfrac{1 + (-1)^n x^{n+1}}{(1+x)^2}\, dx = -\dfrac{7(n+1)\pi^4}{120} + 6\sum_{k=1}^n (-1)^{k-1} \dfrac{n-k+1}{k^4}$  BI (111)(10)

9. $\int_0^1 (\ln x)^3 \dfrac{1 - x^{2n+2}}{(1-x^2)^2}\, dx = -\dfrac{(n+1)\pi^4}{16} + 6\sum_{k=1}^n \dfrac{n-k+1}{(2k-1)^4}$  BI (111)(12)

**4.263**

1.[8] $\int_0^\infty (\ln x)^4 \dfrac{dx}{(x-1)(x+a)} = \dfrac{\ln a \left[\pi^2 + (\ln a)^2\right]\left[7\pi^2 + 3\,(\ln a)^2\right]}{15(1+a)}$

$[a > 0]$  BI (141)(3)

2. $\int_0^1 (\ln x)^4 \dfrac{dx}{1+x^2} = \dfrac{5\pi^5}{64}$  BI (109)(17)

3. $\int_0^1 (\ln x)^4 \dfrac{dx}{1 + 2x\cos t + x^2} = \dfrac{t\left(\pi^2 - t^2\right)\left(7\pi^2 - 3t^2\right)}{30\sin t}$

$[|t| < \pi]$  BI (113)(8)

**4.264**

1. $\int_0^1 (\ln x)^5 \dfrac{dx}{1+x} = -\dfrac{31\pi^6}{252}$  BI (109)(20)

2. $\int_0^1 (\ln x)^5 \dfrac{dx}{1-x} = -\dfrac{8\pi^6}{63}$  BI (109)(21)

3. $\int_0^\infty (\ln x)^5 \dfrac{dx}{(x-1)(x+a)} = \dfrac{\left[\pi^2 + (\ln a)^2\right]^2 \left[3\pi^2 + (\ln a)^2\right]}{6(1+a)}$

$[a > 0]$  BI (141)(4)

**4.265** $\int_0^1 (\ln x)^6 \dfrac{dx}{1+x^2} = \dfrac{61\pi^7}{256}$  BI (109)(25)

**4.266**

1. $\int_0^1 (\ln x)^7 \dfrac{dx}{1+x} = -\dfrac{127\pi^8}{240}$  BI (109)(28)

2. $\int_0^1 (\ln x)^7 \dfrac{dx}{1-x} = -\dfrac{8\pi^8}{15}$  BI (109)(29)

**4.267**

1. $\int_0^1 \dfrac{1-x}{1+x}\dfrac{dx}{\ln x} = \ln \dfrac{2}{\pi}$  BI (127)(3)

2. $\int_0^1 \dfrac{(1-x)^2}{1+x^2}\dfrac{dx}{\ln x} = \ln \dfrac{\pi}{4}$  BI (128)(2)

3.$^{8}$ $\int_0^1 \dfrac{(1-x)^2}{1+2x\cos\dfrac{mx}{n}+x^2}\cdot\dfrac{\mathrm{d}x}{\ln x}$

$= \dfrac{1}{\sin\left(\frac{m\pi}{n}\right)}\sum_{k=1}^{n-1}(-1)^k\sin\left(\dfrac{km\pi}{n}\right)\ln\dfrac{\left\{\Gamma\left(\frac{n+k+1}{2n}\right)\right\}^2\Gamma\left(\frac{k+2}{2n}\right)\Gamma\left(\frac{k}{2n}\right)}{\left\{\Gamma\left(\frac{k+1}{2n}\right)\right\}^2\Gamma\left(\frac{n+k}{2n}\right)\Gamma\left(\frac{n+k+2}{2n}\right)}$  $\qquad [m+n \text{ is odd}]$

$= \dfrac{1}{\sin\left(\frac{m\pi}{n}\right)}\sum_{k=1}^{\left\lfloor\frac{1}{2}(n-1)\right\rfloor}(-1)^k\sin\left(\dfrac{km\pi}{n}\right)\ln\dfrac{\left\{\Gamma\left(\frac{n-k+1}{n}\right)\right\}^2\Gamma\left(\frac{k+2}{n}\right)\Gamma\left(\frac{k}{n}\right)}{\left\{\Gamma\left(\frac{k+1}{n}\right)\right\}^2\Gamma\left(\frac{n-k}{n}\right)\Gamma\left(\frac{n-k+2}{n}\right)}$  $\qquad [m+n \text{ is even}]$

$\qquad\qquad\qquad\qquad\qquad\qquad\qquad\qquad\qquad\qquad\qquad\qquad [m<n]$  BI (130)(3)

4. $\quad \int_0^1 \dfrac{1-x}{1+x}\cdot\dfrac{1}{1+x^2}\cdot\dfrac{\mathrm{d}x}{\ln x}=-\dfrac{\ln 2}{2}$  $\qquad$ BI (130)(16)

5. $\quad \int_0^1 \dfrac{1-x}{1+x}\cdot\dfrac{x^2}{1+x^2}\cdot\dfrac{\mathrm{d}x}{\ln x}=\ln\dfrac{2\sqrt{2}}{\pi}$  $\qquad$ BI (130)(17)

6.$^{11}$ $\int_0^1 (1-x)^p\dfrac{\mathrm{d}x}{\ln x}=\sum_{k=1}^\infty(-1)^k\binom{p}{k}\ln(1+k)$  $\qquad [p\geq 1]$  BI (123)(2)

7. $\quad \int_0^1 \left(\dfrac{1-x^p}{1-x}-p\right)\dfrac{\mathrm{d}x}{\ln x}=\ln\Gamma(p+1)$  $\qquad$ GW (326)(10)

8. $\quad \int_0^1 \dfrac{x^{p-1}-x^{q-1}}{\ln x}\,\mathrm{d}x=\ln\dfrac{p}{q}$  $\qquad [p>0,\quad q>0]$  FI II 647

9. $\quad \int_0^1 \dfrac{x^{p-1}-x^{q-1}}{\ln x}\cdot\dfrac{\mathrm{d}x}{1+x}=\ln\dfrac{\Gamma\left(\frac{q}{2}\right)\Gamma\left(\frac{p+1}{2}\right)}{\Gamma\left(\frac{p}{2}\right)\Gamma\left(\frac{q+1}{2}\right)}$  $\qquad [p>0,\quad q>0]$  FI II 186

10. $\quad \int_0^1 \dfrac{x^{p-1}-x^{-p}}{(1+x)\ln x}\,\mathrm{d}x-\dfrac{1}{2}\int_0^\infty\dfrac{x^{p-1}-x^{-p}}{(1+x)\ln x}\,\mathrm{d}x-\ln\left(\tan\dfrac{p\pi}{2}\right)$

$\qquad\qquad\qquad\qquad\qquad\qquad\qquad\qquad\qquad [0<p<1]$  FI II 816

11. $\quad \int_0^1 (x^p-x^q)\,x^{r-1}\dfrac{\mathrm{d}x}{\ln x}=\ln\dfrac{p+r}{r+q}$  $\qquad [r>0,\quad p>0,\quad q>0]$  LI (123)(5)

12. $\quad \int_0^1 \dfrac{x^p-x^q}{(1-ax)^n}\dfrac{\mathrm{d}x}{x\ln x}=\sum_{k=0}^\infty\binom{n+k-1}{k}a^k\ln\dfrac{p+k}{q+k}$  $\qquad [p>0,\quad q>0,\quad a^2<1]$  BI (130)(15)

13. $\quad \int_0^1 (x^p-1)(x^q-1)\dfrac{\mathrm{d}x}{\ln x}=\ln\dfrac{p+q+1}{(p+1)(q+1)}$  $\qquad [p>-1,\quad q>-1,\quad p+q>-1]$

$\qquad\qquad\qquad\qquad\qquad\qquad\qquad\qquad\qquad\qquad\qquad\qquad$ GW (324)(19b)

14. $\quad \int_0^1 \dfrac{x^p-x^q}{1+x}\cdot\dfrac{1+x^{2n+1}}{x\ln x}\,\mathrm{d}x=\ln\dfrac{\Gamma\left(\frac{p}{2}+n+1\right)\Gamma\left(\frac{q+1}{2}+n\right)\Gamma\left(\frac{p+1}{2}\right)\Gamma\left(\frac{q}{2}\right)}{\Gamma\left(\frac{q}{2}+n+1\right)\Gamma\left(\frac{p+1}{2}+n\right)\Gamma\left(\frac{q+1}{2}\right)\Gamma\left(\frac{p}{2}\right)}$

$\qquad\qquad\qquad\qquad\qquad\qquad\qquad\qquad\qquad [p>0,\quad q>0]$  BI (127)(7)

15. $\quad \int_0^1 \dfrac{x^p-x^q}{1-x}\cdot\dfrac{1-x^r}{\ln x}\,\mathrm{d}x=\ln\dfrac{\Gamma(q+1)\Gamma(p+r+1)}{\Gamma(p+1)\Gamma(q+r+1)}$

$\qquad\qquad [p>-1,\quad q>-1,\quad p+r>-1,\quad q+r>-1]$  GW (324)(23)

16.    $\int_0^1 \dfrac{x^{p-1} - x^{q-1}}{(1 + x^r) \ln x} \, dx = \ln \dfrac{\Gamma\left(\frac{p+r}{2r}\right) \Gamma\left(\frac{q}{2r}\right)}{\Gamma\left(\frac{q+r}{2r}\right) \Gamma\left(\frac{p}{2r}\right)}$      $[p > 0, \quad q > 0, \quad r > 0]$      GW (324)(21)

17.    $\int_0^1 \dfrac{1 - x^{2p-2q}}{1 + x^{2p}} \dfrac{x^{q-1} \, dx}{\ln x} = \ln \tan \dfrac{q\pi}{4p}$      $[0 < q < p]$      BI (128)(6)

18.    $\int_0^\infty \dfrac{x^{p-1} - x^{q-1}}{(1 + x^r) \ln x} \, dx = \ln \left( \tan \dfrac{p\pi}{2r} \cot \dfrac{q\pi}{2r} \right)$      $[0 < p < r, \quad 0 < q < r]$

                                                             GW (324)(22), BI (143)(2)

19.    $\int_0^\infty \dfrac{x^{p-1} - x^{q-1}}{(1 - x^r) \ln x} \, dx = \ln \left( \dfrac{\sin \frac{p\pi}{r}}{\sin \frac{q\pi}{r}} \right)$      $[0 < p < r, \quad 0 < q < r]$      BI (143)(4)

20.    $\int_0^1 \dfrac{x^{p-1} - x^{q-1}}{1 - x^{2n}} \cdot \dfrac{1 - x^2}{\ln x} \, dx = \ln \dfrac{\Gamma\left(\frac{p+2}{2n}\right) \Gamma\left(\frac{q}{2n}\right)}{\Gamma\left(\frac{q+2}{2n}\right) \Gamma\left(\frac{p}{2n}\right)}$      $[p > 0, \quad q > 0]$      BI (128)(11)

21.    $\int_0^1 \dfrac{x^{p-1} - x^{q-1}}{1 + x^{2(2n+1)}} \dfrac{1 + x^2}{\ln x} \, dx = \ln \dfrac{\Gamma\left(\frac{p+4n+4}{4(2n+1)}\right) \Gamma\left(\frac{q+2}{4(2n+1)}\right) \Gamma\left(\frac{p+4n+2}{4(2n+1)}\right) \Gamma\left(\frac{q}{4(2n+1)}\right)}{\Gamma\left(\frac{q+4n+4}{4(2n+1)}\right) \Gamma\left(\frac{p+2}{4(2n+1)}\right) \Gamma\left(\frac{q+4n+2}{4(2n+1)}\right) \Gamma\left(\frac{p}{4(2n+1)}\right)}$

                                                   $[p > 0, \quad q > 0]$      BI (128)(7)

22.    $\int_0^\infty \dfrac{x^{p-1} - x^{q-1}}{1 + x^{2(2n+1)}} \dfrac{1 + x^2}{\ln x} \, dx = \ln \left\{ \tan \dfrac{p\pi}{4(2n+1)} \tan \dfrac{(p+2)\pi}{4(2n+1)} \cot \dfrac{q\pi}{4(2n+1)} \cot \dfrac{(q+2)\pi}{4(2n+1)} \right\}$

                                         $[0 < p < 4n, \quad 0 < q < 4n]$      BI (143)(5)

23.    $\int_0^\infty \dfrac{x^{p-1} - x^{q-1}}{1 - x^{2n}} \dfrac{1 - x^2}{\ln x} \, dx = \ln \dfrac{\sin \frac{p\pi}{2n} \cdot \sin \frac{(q+2)\pi}{2n}}{\sin \frac{q\pi}{2n} \cdot \sin \frac{(p+2)\pi}{2n}}$

                                         $[0 < p < 2n, \quad 0 < q < 2n]$      BI (143)(6)

24.    $\int_0^1 (1 - x^p)(1 - x^q) \dfrac{x^{r-1} \, dx}{\ln x} = \ln \dfrac{(p + q + r)r}{(p + r)(q + r)}$      $[p > 0, \quad q > 0, \quad r > 0]$      BI (123)(8)

25.    $\int_0^1 (1 - x^p)(1 - x^q) \dfrac{x^{r-1} \, dx}{(1 - x) \ln x} = \ln \dfrac{\Gamma(p + r) \Gamma(q + r)}{\Gamma(p + q + r) \Gamma(r)}$

                               $[r > 0, \quad r + p > 0, \quad r + q > 0, \quad r + p + q > 0]$      FI II 815a

26.    $\int_0^1 (1 - x^p)(1 - x^q)(1 - x^r) \dfrac{dx}{\ln x} = \ln \dfrac{(p + q + 1)(q + r + 1)(r + p + 1)}{(p + q + r + 1)(p + 1)(q + 1)(r + 1)}$

     $[p > -1, \quad q > -1, \quad r > -1, \quad p + q > -1, \quad p + r > -1, \quad q + r > -1, \quad p + q + r > -1]$

                                                       GW (324)(19c)

27.    $\int_0^1 (1 - x^p)(1 - x^q)(1 - x^r) \dfrac{dx}{(1 - x) \ln x} = \ln \dfrac{\Gamma(p + 1) \Gamma(q + 1) \Gamma(r + 1) \Gamma(p + q + r + 1)}{\Gamma(p + q + 1) \Gamma(p + r + 1) \Gamma(q + r + 1)}$

     $[p > -1, \quad q > -1, \quad r > -1, \quad p + q > -1, \quad p + r > -1, \quad q + r > -1, \quad p + q + r > -1]$

                                                                   FI II 815

28.     $\int_0^1 \left(1 - x^p\right)\left(1 - x^q\right)\left(1 - x^r\right) \dfrac{x^{s-1}\, dx}{\ln x} = \ln \dfrac{(p+q+s)(p+r+s)(q+r+s)s}{(p+s)(q+s)(r+s)(p+q+r+s)}$

$$[p > 0, \quad q > 0, \quad r > 0, \quad s > 0]$$
BI (123)(10)

29.     $\int_0^1 \left(1 - x^p\right)\left(1 - x^q\right) \dfrac{x^{s-1}\, dx}{(1 - x^r)\ln x} = \ln \dfrac{\Gamma\left(\frac{p+s}{r}\right)\Gamma\left(\frac{q+s}{r}\right)}{\Gamma\left(\frac{s}{r}\right)\Gamma\left(\frac{p+q+s}{r}\right)}$

$$[p > 0, \quad q > 0, \quad r > 0, \quad s > 0]$$
GW (324)(23a)

30.     $\int_0^\infty \left(1 - x^p\right)\left(1 - x^q\right) \dfrac{x^{s-1}\, dx}{(1 - x^{p+q+2s})\ln x} = 2 \int_0^1 \left(1 - x^p\right)\left(1 - x^q\right) \dfrac{x^{s-1}\, dx}{(1 - x^{p+q+2s})\ln x}$

$$= 2 \ln \left( \sin \frac{s\pi}{p+q+2s}\, \mathrm{cosec}\, \frac{(p+s)\pi}{p+q+2s} \right)$$
$$[s > 0, \quad s + p > 0, \quad s + p + q > 0] \quad \text{GW (324)(23b)a}$$

31.[12]    $\int_0^1 \left(1 - x^p\right)\left(1 - x^q\right)\left(1 - x^r\right) \dfrac{x^{s-1}\, dx}{(1 - x)\ln x} = \ln \dfrac{\Gamma(p+s)\,\Gamma(q+s)\,\Gamma(r+s)\,\Gamma(p+q+r+s)}{\Gamma(p+q+s)\,\Gamma(p+r+s)\,\Gamma(q+r+s)\,\Gamma(s)}$

$$[p > 0, \quad q > 0, \quad r > 0, \quad s > 0]^* \quad \text{BI (127)(11)}$$

32.     $\int_0^1 \left(1 - x^p\right)\left(1 - x^q\right)\left(1 - x^r\right) \dfrac{x^{s-1}\, dx}{(1 - x^t)\ln x} = \ln \dfrac{\Gamma\left(\frac{p+s}{t}\right)\Gamma\left(\frac{q+s}{t}\right)\Gamma\left(\frac{r+s}{t}\right)\Gamma\left(\frac{p+q+r+s}{t}\right)}{\Gamma\left(\frac{p+q+s}{t}\right)\Gamma\left(\frac{q+r+s}{t}\right)\Gamma\left(\frac{p+r+s}{t}\right)\Gamma\left(\frac{s}{t}\right)}$

$$[p > 0, \quad q > 0, \quad r > 0, \quad s > 0, \quad t > 0]^* \quad \text{GW (324)(23b)}$$

33.     $\int_0^1 \left\{ \dfrac{x^p - x^{p+q}}{1 - x} - q \right\} \dfrac{dx}{\ln x} = \ln \dfrac{\Gamma(p+q+1)}{\Gamma(p+1)}$    $\quad [p > -1, \quad p+q > -1] \quad$ BI (127)(19)

34.     $\int_0^1 \left\{ \dfrac{x^\mu - x}{x - 1} - x(\mu - 1) \right\} \dfrac{dx}{x \ln x} = \ln \Gamma(\mu)$    $\quad [\operatorname{Re}\mu > 0] \quad$ WH, BI (127)(18)

35.     $\int_0^1 \left\{ 1 - x - \dfrac{\left(1 - x^p\right)\left(1 - x^q\right)}{1 - x} \right\} \dfrac{dx}{x \ln x} = -\ln \{\mathrm{B}(p, q)\}$

$$[p > 0, \quad q > 0]$$
BI (130)(18)

36.     $\int_0^1 \left\{ \dfrac{x^{p-1}}{1 - x} - \dfrac{x^{pq-1}}{1 - x^q} - \dfrac{1}{x(1 - x)} + \dfrac{1}{x\left(1 - x^q\right)} \right\} \dfrac{dx}{\ln x} = q \ln p$

$$[p > 0]$$
BI (130)(20)

37.     $\int_0^1 \left\{ \dfrac{x^{q-1}}{1 - x} - \dfrac{x^{pq-1}}{1 - x^p} - \dfrac{p-1}{1 - x^p}x^{p-1} - \dfrac{p-1}{2}x^{p-1} \right\} \dfrac{dx}{\ln x} = \dfrac{1-p}{2}\ln(2\pi) + \left(pq - \dfrac{1}{2}\right)\ln p$

$$[p > 0, \quad q > 0]$$
BI (130)(22)

38.     $\int_0^1 \dfrac{\left(1 - x^p\right)\left(1 - x^q\right) - (1 - x)^2}{x(1 - x)\ln x}\, dx = \ln \mathrm{B}(p, q)$    $\quad [p > 0, \quad q > 0] \quad$ GW (324)(24)

---

*In 4.267.31 the restrictions can be somewhat weakened by writing, for example, $s > 0$, $p + s > 0$, $q + s > 0$, $r + s > 0$, $p + q + s > 0$, $p + r + s > 0$, $q + r + s > 0$, $p + q + r + s > 0$, in **4.267** 31 and 32.

$39.^6 \quad \int_0^1 (x^p - 1)^n \frac{dx}{\ln x} = \sum_{k=0}^n \binom{n}{n-k} (-1)^{n-k} \ln(pk+1)$

$$[n > 0, \quad pn > -1]$$
GW (324)(19d), BI (123)(12)a

$40.^6 \quad \int_0^1 \frac{(1-x^p)^n}{1-x} \frac{dx}{\ln x} = \sum_{k=0}^n (-1)^{k-1} \ln \Gamma[(n-k)p+1] \qquad [n > 1, \quad pn > -1]$ BI (127)(12)

$41. \quad \int_0^1 (x^p - 1)^n x^{q-1} \frac{dx}{\ln x} = \sum_{k=0}^n (-1)^k \binom{n}{k} \ln[q + (n-k)p]$

$$[n > 0, \quad q > 0, \quad pn > -q]$$
BI (123)(12)

$42.^6 \quad \int_0^1 (1-x^p)^n x^{q-1} \frac{dx}{(1-x)\ln x} = \sum_{k=0}^n (-1)^{k-1} \ln \Gamma[(n-k)p+q]$

$$[n > 1, \quad q > 0, \quad pn > -q]$$
BI (127)(13)

$43.^{10} \quad \int_0^1 (x^p - 1)^n (x^q - 1)^m \frac{x^{r-1} dx}{\ln x} = \sum_{j=0}^n (-1)^j \binom{n}{j} \sum_{k=0}^m (-1)^k \binom{m}{k} \ln[r + (m-k)q + (n-j)p]$

$$[n \geq 0, \quad m \geq 0, \quad n+m > 0, \quad r > 0, \quad pn + qm + r > 0] \quad \text{BI (123)(16)}$$

**4.268**

$1. \quad \int_0^1 \frac{(x^p - x^q)(1 - x^r)}{(\ln x)^2} dx = (p+1)\ln(p+1) - (q+1)\ln(q+1)$

$$-(p+r+1)\ln(p+r+1) + (q+r+1)\ln(q+r+1)$$
$$[p > -1, \quad q > -1, \quad p+r > -1, \quad q+r > -1] \quad \text{GW (324)(26)}$$

$2. \quad \int_0^1 (x^p - x^q)^2 \frac{dx}{(\ln x)^2} = (2p+1)\ln(2p+1) + (2q+1)\ln(2q+1) - 2(p+q+1)\ln(p+q+1)$

$$\left[p > -\tfrac{1}{2}, \quad q > -\tfrac{1}{2}\right] \qquad \text{GW (324)(26a)}$$

$3. \quad \int_0^1 (1-x^p)(1-x^q)(1-x^r) \frac{dx}{(\ln x)^2}$

$$= (p+q+1)\ln(p+q+1) + (q+r+1)\ln(q+r+1) + (p+r+1)\ln(p+r+1)$$
$$-(p+1)\ln(p+1) - (q+1)\ln(q+1) - (r+1)\ln(r+1) - (p+q+r)\ln(p+q+r)$$
$$[p > -1, \quad q > -1, \quad r > -1, \quad p+q > -1, \quad p+r > -1, \quad q+r > -1, \quad p+q+r > 0]$$
BI (124)(4)

$4. \quad \int_0^1 (1-x^p)^n x^{q-1} \frac{dx}{(\ln x)^2} = \frac{1}{2} \sum_{k=0}^n (-1)^k \binom{n}{k} (pk+q)^2 \ln(pk+q)$

$$\left[q > 0, \quad p > -\frac{q}{n}\right] \qquad \text{BI (124)(14)}$$

5.      $\int_0^1 (1-x^p)^n (1-x^q)^m x^{r-1} \dfrac{dx}{(\ln x)^2} = \left( \displaystyle\sum_{j=0}^n (-1)^j \binom{n}{j} \right) \left( \displaystyle\sum_{k=0}^m (-1)^k \binom{m}{k} \right)$

$\times [(m-k)q + (n-j)p + r] \ln[(m-k)q + (n-j)p + r]$

$[r > 0, \quad mq + r > 0, \quad np + r > 0, \quad mq + np + r > 0]$     BI (124)(8)

6.      $\int_0^1 \left[ (q-r)x^{p-1} + (r-p)x^{q-1} + (p-q)x^{r-1} \right] \dfrac{dx}{(\ln x)^2}$

$= (q-r)p \ln p + (r-p)q \ln q + (p-q)r \ln r$

$[p > 0, \quad q > 0, \quad r > 0]$     BI (124)(9)

7.      $\int_0^1 \left[ \dfrac{x^{p-1}}{(p-q)(p-r)(p-s)} + \dfrac{x^{q-1}}{(q-p)(q-r)(q-s)} + \dfrac{x^{r-1}}{(r-p)(r-q)(r-s)} \right.$

$\left. + \dfrac{x^{s-1}}{(s-p)(s-q)(s-r)} \right] \dfrac{dx}{(\ln x)^2} = \dfrac{1}{2} \left[ \dfrac{p^2 \ln p}{(p-q)(p-r)(p-s)} + \dfrac{q^2 \ln q}{(q-p)(q-r)(q-s)} \right.$

$\left. + \dfrac{r^2 \ln r}{(r-p)(r-q)(r-s)} + \dfrac{s^2 \ln s}{(s-p)(s-q)(s-r)} \right]$

$[p > 0, \quad q > 0, \quad r > 0, \quad s > 0]$     BI (124)(16)

**4.269**

1.      $\int_0^1 \sqrt{-\ln x} \, \dfrac{dx}{1+x^2} = \dfrac{\sqrt{\pi}}{2} \displaystyle\sum_{k=0}^\infty \dfrac{(-1)^k}{\sqrt{(2k+1)^3}}$     BI (115)(33)

2.[11]      $\int_0^1 \dfrac{dx}{\sqrt{-\ln x} \, (1+x^2)} = \sqrt{\pi} \displaystyle\sum_{k=0}^\infty \dfrac{(-1)^k}{\sqrt{2k+1}}$     BI (133)(2)

3.      $\int_0^1 \sqrt{(-\ln x)} \, x^{p-1} \, dx = \dfrac{1}{2} \sqrt{\dfrac{\pi}{p^3}}$     $[p > 0]$     GW (324)(1c)

4.      $\int_0^1 \dfrac{x^{p-1}}{\sqrt{-\ln x}} \, dx = \sqrt{\dfrac{\pi}{p}}$     $[p > 0]$     BI (133)(1)

5.      $\int_0^1 \dfrac{\sin t - x^n \sin[(n+1)t] + x^{n+1} \sin nt}{1 - 2x \cos t + x^2} \cdot \dfrac{dx}{\sqrt{-\ln x}} = \sqrt{\pi} \displaystyle\sum_{k=1}^n \dfrac{\sin kt}{\sqrt{k}}$

$[|t| < \pi]$     BI (133)(5)

6.      $\int_0^1 \dfrac{\cos t - x - x^{n-1} \cos nt + x^n \cos[(n-1)t]}{1 - 2x \cos t + x^2} \cdot \dfrac{dx}{\sqrt{-\ln x}} = \sqrt{\pi} \displaystyle\sum_{k=1}^{n-1} \dfrac{\cos kt}{\sqrt{k}}$

$[|t| < \pi]$     BI (133)(6)

7.      $\int_u^v \dfrac{dx}{x \cdot \sqrt{\ln \dfrac{x}{u} \ln \dfrac{v}{x}}} = \pi$     $[uv > 0]$     BI (145)(37)

**4.271**

1. $\displaystyle\int_0^1 (\ln x)^{2n}\,\frac{dx}{1+x} = \frac{2^{2n}-1}{2^{2n}}\cdot (2n)!\,\zeta(2n+1)$ 

   BI (110)(1)

2. $\displaystyle\int_0^1 (\ln x)^{2n-1}\,\frac{dx}{1+x} = \frac{1-2^{2n-1}}{2n}\pi^{2n}|B_{2n}|$ 

   $[n=1,2,\ldots]$ 

   BI (110)(2)

3. $\displaystyle\int_0^1 (\ln x)^{2n-1}\,\frac{dx}{1-x} = -\frac{1}{n}2^{2n-2}\pi^{2n}|B_{2n}|$ 

   $[n=1,2,\ldots]$ 

   BI (110)(5), GW(324)(9a)

4. $\displaystyle\int_0^1 (\ln x)^{p-1}\,\frac{dx}{1-x} = e^{i(p-1)\pi}\,\Gamma(p)\,\zeta(p)$ 

   $[p>1]$ 

   GW (324)(9b)

5. $\displaystyle\int_0^1 (\ln x)^{n}\,\frac{dx}{1+x^2} = (-1)^n n!\sum_{k=0}^{\infty}\frac{(-1)^k}{(2k+1)^{n+1}}$ 

   BI (110)(11)

6. $\displaystyle\int_0^1 (\ln x)^{2n}\,\frac{dx}{1+x^2} = \frac{1}{2}\int_0^{\infty}(\ln x)^{2n}\,\frac{dx}{1+x^2} = \frac{\pi^{2n+1}}{2^{2n+2}}|E_{2n}|$ 

   GW (324)(10)a

7. $\displaystyle\int_0^{\infty} \frac{(\ln x)^{2n+1}}{1+bx+x^2}\,dx = 0$ 

   $[|b|<2]$ 

   BI (135)(2)

8. $\displaystyle\int_0^1 (\ln x)^{2n}\,\frac{dx}{1-x^2} = \frac{2^{2n+1}-1}{2^{2n+1}}\cdot (2n)!\,\zeta(2n+1)$ 

   $[n=1,2,\ldots]$ 

   BI (110)(12)

9. $\displaystyle\int_0^{\infty} (\ln x)^{2n}\,\frac{dx}{1-x^2} = 0$ 

   BI (312)(7)a

10. $\displaystyle\int_0^1 (\ln x)^{2n-1}\,\frac{dx}{1-x^2} = \frac{1}{2}\int_0^{\infty}(\ln x)^{2n-1}\,\frac{dx}{1-x^2} = \frac{1-2^{2n}}{4n}\pi^{2n}|B_{2n}|$ 

   $[n=1,2,\ldots]$ 

   BI (290)(17)a, BI(312)(6)a

11. $\displaystyle\int_0^1 (\ln x)^{2n-1}\,\frac{x\,dx}{1-x^2} = -\frac{1}{4n}\pi^{2n}|B_{2n}|$ 

   $[n=1,2,\ldots]$ 

   BI (290)(19)a

12. $\displaystyle\int_0^1 (\ln x)^{2n}\,\frac{1+x^2}{(1-x^2)^2}\,dx = \frac{2^{2n}-1}{2}\pi^{2n}|B_{2n}|$ 

   $[n=1,2,\ldots]$ 

   BI (296)(17)a

13. $\displaystyle\int_0^1 (\ln x)^{2n+1}\,\frac{(\cos 2a\pi - x)\,dx}{1-2x\cos 2a\pi + x^2} = -(2n+1)!\sum_{k=1}^{\infty}\frac{\cos 2ak\pi}{k^{2n+2}}$ 

   $[a$ is not an integer$]$ 

   LI (113)(10)

14.[6] $\displaystyle\int_0^{\infty} (\ln x)^{n}\,\frac{x^{\nu-1}\,dx}{a^2 + 2ax\cos t + x^2} = -\pi\csc t\,\frac{d^n}{d\nu^n}\left[a^{\nu-2}\frac{\sin(\nu-1)t}{\sin\nu\pi}\right]$ 

   $[a>0,\quad 0<\operatorname{Re}\nu<2,\quad 0<|t|<\pi]$ 

   ET I 315(12)

15. $\displaystyle\int_0^1 (\ln x)^{n}\,\frac{x^{p-1}}{1-x^q}\,dx = -\frac{1}{q^{n+1}}\psi^{(n)}\left(\frac{p}{q}\right)$ 

   $[p>0,\quad q>0]$ 

   GW (324)(9)

16.[3] $\displaystyle\int_0^1 (\ln x)^{n}\,\frac{x^{p-1}}{1+x^q}\,dx = \frac{1}{q^{n+1}}\beta^{(n)}\left(\frac{p}{q}\right)$ 

   $[p>0,\quad q>0]$ 

   GW (324)(10)

**4.272**

1.   $\displaystyle\int_0^1 \frac{[-\ln x]^{q-1}\,\mathrm{d}x}{1+2x\cos t+x^2} = \operatorname{cosec} t\,\Gamma(q)\sum_{k=1}^{\infty}(-1)^{k-1}\frac{\sin kt}{k^q}$     $[|t|<\pi, q<1]$     LI (130)(1)

2.   $\displaystyle\int_0^1 (-\ln x)^{q-1}\frac{(1+x)\,\mathrm{d}x}{1+2x\cos t+x^2} = \sec\frac{t}{2}\cdot\Gamma(q)\sum_{k=1}^{\infty}(-1)^{k-1}\frac{\cos\left[\left(k-\frac12\right)t\right]}{k^q}$

                                       $\left[|t|<\pi,\quad q<\tfrac12\right]$     LI (130)(5)

3.[9]   $\displaystyle\int_0^1 [-\ln x]^{\mu}\frac{x^{\nu-1}\,\mathrm{d}x}{1-2ax\cos t+x^2a^2} = \frac{\Gamma(\mu+1)}{a\sin t}\sum_{k=1}^{\infty}\frac{a^k\sin kt}{(\nu+k-1)^{\mu+1}}$

                    $[a>0,\quad \operatorname{Re}\mu>0,\quad \operatorname{Re}\nu>0,\quad -\pi<t<\pi]$   BI (140)(14)a

4.   $\displaystyle\int_0^1 (-\ln x)^{r-1}\frac{\cos\lambda - px}{1+p^2x^2-2px\cos\lambda}x^{q-1}\,\mathrm{d}x = \Gamma(r)\sum_{k=1}^{\infty}\frac{p^{k-1}\cos k\lambda}{(q+k-1)^r}$

                                     $[r>0,\quad q>0]$     BI (113)(11)

5.   $\displaystyle\int_1^{\infty}(\ln x)^p\frac{\mathrm{d}x}{x^2} = \Gamma(1+p)$                $[p>-1]$     BI (149)(1)

6.   $\displaystyle\int_0^1 (-\ln x)^{\mu-1}x^{\nu-1}\,\mathrm{d}x = \frac{1}{\nu^{\mu}}\Gamma(\mu)$            $[\operatorname{Re}\mu>0,\quad \operatorname{Re}\nu>0]$     BI (107)(3)

7.   $\displaystyle\int_0^1 (-\ln x)^{n-\frac12}x^{\nu-1}\,\mathrm{d}x = \frac{(2n-1)!!}{(2\nu)^n}\sqrt{\frac{\pi}{\nu}}$         $[\operatorname{Re}\nu>0]$     BI (107)(2)

8.[12]   $\displaystyle\int_0^1 (-\ln x)^{n-1}\frac{x^{\nu-1}}{1+x}\,\mathrm{d}x = (n-1)!\sum_{k=0}^{\infty}\frac{(-1)^k}{(\nu+k)^n}$     $[\operatorname{Re}\nu>0]$     BI (110)(4)

9.   $\displaystyle\int_0^1 (-\ln x)^{n-1}\frac{x^{\nu-1}}{1-x}\,\mathrm{d}x = (n-1)!\,\zeta(n,\nu)$        $[\operatorname{Re}\nu>0]$     BI (110)(7)

10.   $\displaystyle\int_0^1 (-\ln x)^{\mu-1}(x-1)^n\left(a+\frac{nx}{x-1}\right)x^{a-1}\,\mathrm{d}x = \Gamma(\mu)\sum_{k=0}^{n}\frac{(-1)^k n(n-1)\ldots(n-k+1)}{(a+n-k)^{\mu-1}k!}$

                                     $[\operatorname{Re}\mu>0]$     LI (110)(10)

11.   $\displaystyle\int_0^1 (-\ln x)^{n-1}\frac{1-x^m}{1-x}\,\mathrm{d}x = (n-1)!\sum_{k=1}^{m}\frac{1}{k^n}$                  LI (110)(9)

12.   $\displaystyle\int_0^1 (-\ln x)^{\mu-1}\frac{x^{\nu-1}\,\mathrm{d}x}{1-x^2} = \Gamma(\mu)\sum_{k=0}^{\infty}\frac{1}{(\nu+2k)^{\mu}} = \frac{1}{2^{\mu}}\Gamma(\mu)\,\zeta\left(\mu,\frac{\nu}{2}\right)$

                                   $[\operatorname{Re}\mu>0,\quad \operatorname{Re}\nu>0]$     BI (110)(13)

13.   $\displaystyle\int_0^1 \frac{x^q-x^{-q}}{1-x^2}(-\ln x)^p\,\mathrm{d}x = \Gamma(p+1)\sum_{k=1}^{\infty}\left\{\frac{1}{(2k+q-1)^{p+1}}-\frac{1}{(2k-q-1)^{p+1}}\right\}$

                                   $[p>-1,\quad q^2<1]$     LI (326)(12)a

14.
$$\int_0^1 (-\ln x)^{r-1} \frac{x^{p-1}\,dx}{(1+x^q)^s} = \Gamma(r) \sum_{k=0}^\infty \binom{-s}{k} \frac{1}{(p+kq)^r}$$

$$[p > 0, \quad q > 0, \quad r > 0, \quad 0 < s < r+2]$$
GW (324)(11)

15.
$$\int_0^1 (-\ln x)^n (1+x^q)^m x^{p-1}\,dx = n! \sum_{k=0}^m \binom{m}{k} \frac{1}{(p+kq)^{n+1}}$$

$$[p > 0, \quad q > 0]$$
BI (107)(6)

16.
$$\int_0^1 (-\ln x)^n (1-x^q)^m x^{p-1}\,dx = n! \sum_{k=0}^m \binom{m}{k} \frac{(-1)^k}{(p+kq)^{n+1}}$$

$$[p > 0, \quad q > 0]$$
BI (107)(7)

17.
$$\int_0^1 (-\ln x)^{p-1} \frac{x^{q-1}\,dx}{1-ax^q} = \frac{1}{aq^p} \Gamma(p) \sum_{k=1}^\infty \frac{a^k}{k^p} \qquad [p > 0, \quad q > 0, \quad a < 1]$$
LI (110)(8)

18.[12]
$$\int_0^1 (-\ln x)^{2-\frac{1}{n}} \left( x^{p-1} - x^{q-1} \right) dx = \Gamma\left(3 - \frac{1}{n}\right) \left( p^{-3+\frac{1}{n}} - q^{-3+\frac{1}{n}} \right)$$

$$[q > p > 0]$$
BI (133)(4)

19.
$$\int_0^1 (-\ln x)^{2n-1} \frac{x^p - x^{-p}}{1-x^q} x^{q-1}\,dx = \frac{1}{p^{2n}} \sum_{k=n}^\infty \left( \frac{2p\pi}{q} \right)^k \frac{|B_{2k}|}{2k \cdot (2k-2n)!}$$

$$\left[ p < \frac{q}{2} \right]$$
LI (110)(16)

**4.273**
$$\int_u^v \left( \ln \frac{x}{u} \right)^{p-1} \left( \ln \frac{v}{x} \right)^{q-1} \frac{dx}{x} = B(p,q) \left( \ln \frac{v}{u} \right)^{p+q-1} \quad [p > 0, \quad q > 0, \quad uv > 0]$$
BI (145)(36)

**4.274**
$$\int_0^{1/e} \frac{\sqrt[q]{x}\,dx}{x\sqrt{-(1+\ln x)}} = \frac{\sqrt{q\pi}}{\sqrt[q]{e}} \qquad [q > 0]$$
BI (145)(4)

**4.275**

1.
$$\int_0^1 \left[ (-\ln x)^{q-1} - x^{p-1}(1-x)^{q-1} \right] dx = \frac{\Gamma(q)}{\Gamma(p+q)} \left[ \Gamma(p+q) - \Gamma(p) \right]$$

$$[p > 0, \quad q > 0]$$
BI (107)(8)

2.
$$\int_0^1 \left[ x - \left( \frac{1}{1-\ln x} \right)^q \right] \frac{dx}{x \ln x} = -\psi(q) \qquad [q > 0]$$
BI (126)(5)

## 4.28 Combinations of rational functions of $\ln x$ and powers

**4.281**

1.
$$\int_0^1 \left[ \frac{1}{\ln x} + \frac{1}{1-x} \right] dx = C$$
BI (127)(15)

2.
$$\int_1^\infty \frac{dx}{x^2 (\ln p - \ln x)} = \frac{1}{p} \operatorname{li}(p)$$
LA 281(30)

3.    $\displaystyle\int_0^1 \frac{x^{p-1}\,dx}{q \pm \ln x} = \pm e^{\mp pq}\,\mathrm{Ei}\,(\pm pq)$          $[p > 0, \quad q > 0]$          LI (144)(11,12)

4.    $\displaystyle\int_0^1 \left[\frac{1}{\ln x} + \frac{x^{\mu-1}}{1-x}\right] dx = -\,\psi(\mu)$          $[\mathrm{Re}\,\mu > 0]$          WH

5.    $\displaystyle\int_0^1 \left[\frac{x^{p-1}}{\ln x} + \frac{x^{q-1}}{1-x}\right] dx = \ln p - \psi(q)$          $[p > 0, \quad q > 0]$          BI (127)(17)

6.    $\displaystyle\int_0^1 \left[\frac{1}{1-x^2} + \frac{1}{2x\ln x}\right] \frac{dx}{\ln x} = \frac{\ln 2}{2}$          LI (130)(19)

7.    $\displaystyle\int_0^1 \left[q - \frac{1}{2} + \frac{(1-x)(1+q\ln x) + x\ln x}{(1-x)^2}\,x^{q-1}\right] \frac{dx}{\ln x} = \frac{1}{2} - q - \ln\Gamma(q) + \frac{\ln 2\pi}{2}$

                                                             $[q > 0]$          BI (128)(15)

**4.282**

1.[12]    $\displaystyle\int_0^1 \frac{\ln x}{4\pi^2 + \ln x} \cdot \frac{dx}{1-x} = \frac{1}{4} - \frac{1}{2}\,C$          BI (129)(1)

2.    $\displaystyle\int_0^1 \frac{1}{a^2 + (\ln x)^2} \cdot \frac{dx}{1+x^2} = \frac{1}{2a}\,\beta\left(\frac{2a+\pi}{4\pi}\right)$          $\left[a > -\dfrac{\pi}{2}\right]$          BI (129)(9)

3.    $\displaystyle\int_0^1 \frac{1}{\pi^2 + (\ln x)^2} \frac{dx}{1+x^2} = \frac{4-\pi}{4\pi}$          BI (129)(6)

4.    $\displaystyle\int_0^1 \frac{\ln x}{\pi^2 + (\ln x)^2} \cdot \frac{dx}{1-x^2} = \frac{1}{2}\left(\frac{1}{2} - \ln 2\right)$          BI (129)(10)

5.    $\displaystyle\int_0^1 \frac{\ln x}{a^2 + (\ln x)^2} \cdot \frac{x\,dx}{1-x^2} = \frac{1}{2}\left[\frac{\pi}{2a} + \ln\frac{\pi}{a} + \psi\left(\frac{a}{\pi}\right)\right]$    $[a > 0]$          BI (129)(14)

6.    $\displaystyle\int_0^1 \frac{\ln x}{\pi^2 + (\ln x)^2} \cdot \frac{x\,dx}{1-x^2} = \frac{1}{2}\left(\frac{1}{2} - C\right)$          BI (129)(13)

7.    $\displaystyle\int_0^1 \frac{1}{\pi^2 + 4(\ln x)^2} \cdot \frac{dx}{1+x^2} = \frac{\ln 2}{4\pi}$          BI (129)(7)

8.    $\displaystyle\int_0^1 \frac{\ln x}{\pi^2 + 4(\ln x)^2} \cdot \frac{dx}{1-x^2} = \frac{2-\pi}{16}$          BI (129)(11)

9.[10]    $\displaystyle\int_0^1 \frac{1}{\pi^2 + 16(\ln x)^2} \cdot \frac{dx}{1+x^2} = \frac{1}{8\pi\sqrt{2}}\left[\pi + 2\ln\left(\sqrt{2} - 1\right)\right]$          BI (129)(8)

10.    $\displaystyle\int_0^1 \frac{\ln x}{\pi^2 + 16(\ln x)^2} \cdot \frac{dx}{1-x^2} = -\frac{\pi}{32\sqrt{2}} + \frac{1}{16} + \frac{1}{16\sqrt{2}}\ln\left(\sqrt{2} - 1\right)$          BI (129)(12)

11.    $\displaystyle\int_0^1 \frac{\ln x}{\left[a^2 + (\ln x)^2\right]^2} \frac{dx}{1-x} = -\frac{\pi^2}{a^4}\sum_{k=1}^{\infty} |B_{2k}|\left(\frac{2\pi}{a}\right)^{2k-2}$          BI (129)(4)

12.    $\displaystyle\int_0^1 \frac{\ln x}{\left[a^2 + (\ln x)^2\right]^2} \frac{x\,dx}{1-x^2} = -\frac{\pi^2}{4a^4}\sum_{k=1}^{\infty} |B_{2k}|\left(\frac{\pi}{a}\right)^{2k-2}$          BI (129)(16)

13. $$\int_0^1 \frac{x^p - x^{-p}}{x^2 - 1} \frac{dx}{q^2 + (\ln x)^2} = \frac{2\pi}{q} \sum_{k=1}^\infty (-1)^{k-1} \frac{\sin kp\pi}{2q + k\pi} \qquad [p^2 < 1] \qquad \text{BI (132)(13)a}$$

**4.283**

1. $$\int_0^1 \left( \frac{x-1}{\ln x} - x \right) \frac{dx}{\ln x} = \ln 2 - 1 \qquad \text{BI (132)(17)a}$$

2. $$\int_0^1 \left( \frac{1}{\ln x} + \frac{1}{1-x} - \frac{1}{2} \right) \frac{dx}{\ln x} = \frac{\ln 2\pi}{2} - 1 \qquad \text{BI (127)(20)}$$

3. $$\int_0^1 \left( \frac{1}{\ln x} + \frac{x}{1-x} + \frac{x}{2} \right) \frac{dx}{x \ln x} = \frac{\ln 2\pi}{2} \qquad \text{BI (127)(23)}$$

4. $$\int_0^1 \left[ \frac{1}{(\ln x)^2} - \frac{x}{(1-x)^2} \right] dx = \mathbf{C} - \frac{1}{2} \qquad \text{GW (326)(8a)}$$

5. $$\int_0^1 \left( \frac{1}{1-x^2} + \frac{1}{2\ln x} - \frac{1}{2} \right) \frac{dx}{\ln x} = \frac{\ln 2 - 1}{2} \qquad \text{BI (128)(14)}$$

6. $$\int_0^1 \left( \frac{1}{\ln x} + \frac{1}{2} \cdot \frac{1+x}{1-x} - \ln x \right) \frac{dx}{\ln x} = \frac{\ln 2\pi}{2} \qquad \text{BI (127)(22)}$$

7. $$\int_0^1 \left[ \frac{1}{1-\ln x} - x \right] \frac{dx}{x \ln x} = -\mathbf{C} \qquad \text{GW (326)(11a)}$$

8. $$\int_0^1 \left[ \frac{x^q - 1}{x (\ln x)^2} - \frac{q}{\ln x} \right] dx = q \ln q - q \qquad [q > 0] \qquad \text{BI (126)(2)}$$

9. $$\int_0^1 \left[ x + \frac{1}{a \ln x - 1} \right] \frac{dx}{x \ln x} = \ln \frac{a}{q} + \mathbf{C} \qquad [a > 0, \quad q > 0] \qquad \text{BI (126)(8)}$$

10. $$\int_0^1 \left[ \frac{1}{\ln x} + \frac{1+x}{2(1-x)} \right] \frac{x^{p-1}}{\ln x} dx = -\ln \Gamma(p) + \left( p - \frac{1}{2} \right) \ln p - p + \frac{\ln 2\pi}{2}$$
$$[p > 0] \qquad \text{GW (326)(9)}$$

11. $$\int_0^1 \left[ p - 1 - \frac{1}{1-x} + \left( \frac{1}{2} - \frac{1}{\ln x} \right) x^{p-1} \right] \frac{dx}{\ln x} = \left( \frac{1}{2} - p \right) \ln p + p - \frac{\ln 2\pi}{2}$$
$$[p > 0] \qquad \text{BI (127)(25)}$$

12. $$\int_0^1 \left[ -\frac{1}{(\ln x)^2} + \frac{(p-2)x^p - (p-1)x^{p-1}}{(1-x)^2} \right] dx = -\psi(p) + p - \frac{3}{2}$$
$$[p > 0] \qquad \text{GW (326)(8)}$$

13. $$\int_0^1 \left[ \left( p - \frac{1}{2} \right) x^3 + \frac{1}{2} \left( 1 - \frac{1}{\ln x} \right) (x^{2p-1} - 1) \right] \frac{dx}{\ln x} = \left( \frac{1}{2} - p \right) (\ln p - 1)$$
$$[p > 0] \qquad \text{BI (132)(23)a}$$

14. $$\int_0^1 \left[ \left( q - \frac{1}{2} \right) \frac{x^{p-1} - x^{r-1}}{\ln x} + \frac{px^{pq-1}}{1-x^p} - \frac{rx^{rq-1}}{1-x^r} \right] \frac{dx}{\ln x} = (p-r) \left[ \frac{1}{2} - q - \ln \Gamma(q) + \frac{\ln 2\pi}{2} \right]$$
$$[q > 0] \qquad \text{BI (132)(13)}$$

**4.284**

1.   $\displaystyle\int_0^1 \left[\frac{x^q-1}{x\,(\ln x)^3} - \frac{q}{x\,(\ln x)^2} - \frac{q^2}{2\ln x}\right]\,dx = \frac{q^2}{2}\ln q - \frac{3}{4}q^2$

$$[q>0] \qquad\qquad\qquad \text{BI (126)(3)}$$

2.   $\displaystyle\int_0^1 \left[\frac{x^q-1}{x\,(\ln x)^4} - \frac{q}{x\,(\ln x)^3} - \frac{q^2}{2x\,(\ln x)^2} - \frac{q^3}{6\ln x}\right]\,dx = \frac{q^3}{6}\ln q - \frac{11}{36}q^3$

$$[q>0] \qquad\qquad\qquad \text{BI (126)(4)}$$

**4.285**   $\displaystyle\int_0^1 \frac{x^{p-1}\,dx}{(q+\ln x)^n} = \frac{p^{n-1}}{(n-1)!}e^{-pq}\,\mathrm{Ei}(pq) - \frac{1}{(n-1)!q^{n-1}}\sum_{k=1}^{n-1}(n-k-1)!(pq)^{k-1}$

$$[p>0, \quad q<0] \qquad\qquad \text{BI (125)(21)}$$

In integrals of the form $\displaystyle\int \frac{x^a\,(\ln x)^n\,dx}{[b\pm(\ln x)^m]^l}$, we should make the substitution $x=e^t$ or $x=e^{-t}$ and then seek the resulting integrals in **3.351–3.356**.

## 4.29–4.32 Combinations of logarithmic functions of more complicated arguments and powers

**4.291**

1.   $\displaystyle\int_0^1 \frac{\ln(1+x)}{x}\,dx = \frac{\pi^2}{12}$                     FI II 483

2.   $\displaystyle\int_0^1 \frac{\ln(1-x)}{x}\,dx = -\frac{\pi^2}{6}$                     FI II 714

3.   $\displaystyle\int_0^{1/2} \frac{\ln(1-x)}{x}\,dx = \frac{1}{2}(\ln 2)^2 - \frac{\pi^2}{12}$                     BI (145)(2)

4.   $\displaystyle\int_0^1 \ln\left(1-\frac{x}{2}\right)\frac{dx}{x} = \frac{1}{2}(\ln 2)^2 - \frac{\pi^2}{12}$                     BI (114)(18)

5.   $\displaystyle\int_0^1 \frac{\ln\dfrac{1+x}{2}}{1-x}\,dx = \frac{1}{2}(\ln 2)^2 - \frac{\pi^2}{12}$                     BI (115)(1)

6.   $\displaystyle\int_0^1 \frac{\ln(1+x)}{1+x}\,dx = \frac{1}{2}(\ln 2)^2$                     BI (114)(14)a

7.[7]   $\displaystyle\int_0^\infty \frac{\ln(1+ax)}{1+x^2}\,dx = \frac{\pi}{4}\ln\left(1+a^2\right) - \int_0^a \frac{\ln u\,du}{1+u^2}$          $[a>0]$          GI II (2209)

8.   $\displaystyle\int_0^1 \frac{\ln(1+x)}{1+x^2}\,dx = \frac{\pi}{8}\ln 2$                     FI II 157

9.   $\displaystyle\int_0^\infty \frac{\ln(1+x)}{1+x^2}\,dx = \frac{\pi}{4}\ln 2 + \boldsymbol{G}$                     BI (136)(1)

10.   $\displaystyle\int_0^1 \frac{\ln(1-x)}{1+x^2}\,dx = \frac{\pi}{8}\ln 2 - \boldsymbol{G}$                     BI (114)(17)

11.    $\displaystyle\int_1^\infty \frac{\ln(x-1)}{1+x^2}\,dx = \frac{\pi}{8}\ln 2$       BI (144)(4)

12.    $\displaystyle\int_0^1 \frac{\ln(1+x)}{x(1+x)}\,dx = \frac{\pi^2}{12} - \frac{1}{2}\left(\ln 2\right)^2$       BI (144)(4)

13.    $\displaystyle\int_0^\infty \frac{\ln(1+x)}{x(1+x)}\,dx = \frac{\pi^2}{6}$       BI (141)(9)a

14.    $\displaystyle\int_0^1 \frac{\ln(1+x)}{(ax+b)^2}\,dx = \frac{1}{a(a-b)}\ln\frac{a+b}{b} + \frac{2\ln 2}{b^2-a^2}$    $[a \ne b, \quad ab > 0]$

            $= \frac{1}{2a^2}\left(1 - \ln 2\right)$           $[a = b]$

                                                     LI (114)(5)a

15.    $\displaystyle\int_0^\infty \frac{\ln(1+x)}{(ax+b)^2}\,dx = \frac{\ln\frac{a}{b}}{a(a-b)}$       $[ab > 0]$       BI (139)(5)

16.    $\displaystyle\int_0^1 \ln(a+x)\frac{dx}{a+x^2} = \frac{1}{2\sqrt{a}}\operatorname{arccot}\sqrt{a}\ln[(1+a)a]$    $[a > 0]$       BI (114)(20)

17.    $\displaystyle\int_0^\infty \ln(a+x)\frac{dx}{(b+x)^2} = \frac{a\ln a - b\ln b}{b(a-b)}$    $[a > 0, \quad b > 0, \quad a \ne b]$       LI (139)(6)

18.    $\displaystyle\int_0^a \frac{\ln(1+ax)}{1+x^2}\,dx = \frac{1}{2}\arctan a\ln\left(1+a^2\right)$       GI II (2195)

19.    $\displaystyle\int_0^1 \frac{\ln(1+ax)}{1+ax^2}\,dx = \frac{1}{2\sqrt{a}}\arctan\sqrt{a}\ln(1+a)$    $[a > 0]$       BI (114)(21)

20.    $\displaystyle\int_0^1 \frac{\ln(ax+b)}{(1+x)^2}\,dx = \frac{1}{a-b}\left[\frac{1}{2}(a+b)\ln(a+b) - b\ln b - a\ln 2\right]$

                                     $[a > 0, \quad b > 0, \quad a \ne b]$       BI (114)(22)

21.    $\displaystyle\int_0^\infty \frac{\ln(ax+b)}{(1+x)^2}\,dx = \frac{1}{a-b}\left[a\ln a - b\ln b\right]$    $[a > 0, \quad b > 0]$       BI (139)(8)

22.    $\displaystyle\int_0^\infty \ln(a+x)\frac{x\,dx}{(b^2+x^2)^2} = \frac{1}{2\left(a^2+b^2\right)}\left(\ln b + \frac{a\pi}{2b} + \frac{a^2}{b^2}\ln a\right)$

                                     $[a > 0, \quad b > 0]$       BI (139)(9)

23.    $\displaystyle\int_0^1 \ln(1+x)\frac{1+x^2}{(1+x)^4}\,dx = -\frac{1}{3}\ln 2 + \frac{23}{72}$       LI (114)(12)

24.    $\displaystyle\int_0^1 \ln(1+x)\frac{1+x^2}{a^2+x^2}\cdot\frac{dx}{1+a^2x^2} = \frac{1}{2a\left(1+a^2\right)}\left[\frac{\pi}{2}\ln\left(1+a^2\right) - 2\arctan a\cdot\ln a\right]$

                                   $[a > 0]$       LI (114)(11)

25. $\displaystyle\int_0^1 \ln(1+x)\frac{1-x^2}{(ax+b)^2}\frac{\mathrm{d}x}{(bx+a)^2} = \frac{1}{a^2-b^2}\left\{\frac{1}{a-b}\left[\frac{a+b}{ab}\ln(a+b) - \frac{1}{a}\ln b - \frac{1}{b}\ln a\right]\right.$
$$\left. + \frac{4\ln 2}{b^2-a^2}\right\}$$

$$\left[a > 0, \quad b > 0, \quad a^2 \neq b^2\right] \quad \text{LI (114)(13)}$$

26. $\displaystyle\int_0^\infty \ln(1+x)\frac{1-x^2}{(ax+b)^2}\cdot\frac{\mathrm{d}x}{(bx+a)^2} = \frac{1}{ab\,(a^2-b^2)}\ln\frac{b}{a}$

$$[a > 0, \quad b > 0] \qquad\qquad \text{LI (139)(14)}$$

27. $\displaystyle\int_0^1 \ln(1+ax)\frac{1-x^2}{(1+x^2)^2}\,\mathrm{d}x = \frac{1}{2}\frac{(1+a)^2}{1+a^2}\ln(1+a) - \frac{1}{2}\cdot\frac{a}{1+a^2}\ln 2 - \frac{\pi}{4}\cdot\frac{a^2}{1+a^2}$

$$[a > -1] \qquad\qquad \text{BI (114)(23)}$$

28. $\displaystyle\int_0^\infty \ln(a+x)\frac{b^2-x^2}{(b^2+x^2)^2}\,\mathrm{d}x = \frac{1}{a^2+b^2}\left(a\ln\frac{b}{a} - \frac{b\pi}{2}\right)$

$$[a > 0, \quad b > 0] \qquad\qquad \text{BI (139)(11)}$$

29. $\displaystyle\int_0^\infty \ln^2(a-x)\frac{b^2-x^2}{(b^2+x^2)^2}\,\mathrm{d}x = \frac{2}{a^2+b^2}\left(a\ln\frac{a}{b} - \frac{b\pi}{2}\right)$

$$[a > 0, \quad b > 0] \qquad\qquad \text{BI (139)(12)}$$

30. $\displaystyle\int_0^\infty \ln^2(a-x)\frac{x\,\mathrm{d}x}{(b^2+x^2)^2} = \frac{1}{a^2+b^2}\left(\ln b - \frac{a\pi}{2b} + \frac{a^2}{b^2}\ln a\right)$

$$[a > 0, \quad b > 0] \qquad\qquad \text{BI (139)(10)}$$

**4.292**

1. $\displaystyle\int_0^1 \frac{\ln(1\pm x)}{\sqrt{1-x^2}}\,\mathrm{d}x = -\frac{\pi}{2}\ln 2 \pm 2\,\boldsymbol{G}$

$$\text{GW (325)(20)}$$

2. $\displaystyle\int_0^1 \frac{x\ln(1\pm x)}{\sqrt{1-x^2}}\,\mathrm{d}x = -1 \pm \frac{\pi}{2}$

$$\text{GW (325)(22c)}$$

3.[12] $\displaystyle\int_{-a}^a \frac{\ln|1+bx|}{\sqrt{a^2-x^2}}\,\mathrm{d}x = \pi\ln\frac{1+\sqrt{1-a^2b^2}}{2}$ $\qquad\left[|b| \leq \dfrac{1}{|a|}\right]$
$$= \pi\ln\left(\frac{a|b|}{2}\right) \qquad\qquad \left[|b| \geq \dfrac{1}{|a|}\right]$$

$$\text{BI (145)(16, 17)a, GW (325)(21e)}$$

4. $\displaystyle\int_0^1 \frac{x\ln(1+ax)}{\sqrt{1-x^2}}\,\mathrm{d}x = -1 + \frac{\pi}{2}\cdot\frac{1-\sqrt{1-a^2}}{a} + \frac{\sqrt{1-a^2}}{a}\arcsin a \quad [|a| \leq 1]$
$$= -1 + \frac{\pi}{2a} + \frac{\sqrt{a^2-1}}{a}\ln\left(a+\sqrt{a^2-1}\right) \qquad [a \geq 1]$$

$$\text{GW (325)(22)}$$

$5.^{12}$ $\int_0^1 \dfrac{\ln|1+ax|}{x\sqrt{1-x^2}}\,dx = \dfrac{1}{2}\arcsin a\,(\pi - \arcsin a) = \dfrac{\pi^2}{8} - \dfrac{1}{2}(\arccos a)^2$    $[|a| \le 1]$

$$= \frac{\pi^2}{8} + \frac{1}{2}\ln^2\left(a + \sqrt{a^2 - 1}\right) \qquad [a \ge 1]$$

$$= -\frac{3}{8}\pi^2 + \frac{1}{2}\ln^2\left(-a + \sqrt{a^2 - 1}\right) \qquad [a \le -1]$$

BI (120)(4), GW (325)(21a)

**4.293**

1. $\int_0^1 x^{\mu-1}\ln(1+x)\,dx = \dfrac{1}{\mu}[\ln 2 - \beta(\mu+1)]$     $[\operatorname{Re}\mu > -1]$     BI (106)(4)a

$2.^6$ $\int_1^\infty x^{\mu-1}\ln(1+x)\,dx = \dfrac{-1}{\mu}[\beta(-\mu) + \ln 2]$     $[\operatorname{Re}\mu < 0]$     ET I 315(17)

3. $\int_0^\infty x^{\mu-1}\ln(1+x)\,dx = \dfrac{\pi}{\mu\sin\mu\pi}$     $[-1 < \operatorname{Re}\mu < 0]$     GW (325)(3)a

4. $\int_0^1 x^{2n-1}\ln(1+x)\,dx = \dfrac{1}{2n}\sum_{k=1}^{2n}\dfrac{(-1)^{k-1}}{k}$     GW (325)(2b)

5. $\int_0^1 x^{2n}\ln(1+x)\,dx = \dfrac{1}{2n+1}\left[\ln 4 + \sum_{k=1}^{2n+1}\dfrac{(-1)^k}{k}\right]$     GW (325)(2c)

$6.^{11}$ $\int_0^1 x^{n-\frac{1}{2}}\ln(1+x)\,dx = \dfrac{2\ln 2}{2n+1} + \dfrac{(-1)^n \cdot 4}{2n+1}\left[\dfrac{\pi}{4} - \sum_{k=0}^{n}\dfrac{(-1)^k}{2k+1}\right]$     GW (325)(2f)

7. $\int_0^\infty x^{\mu-1}\ln|1-x|\,dx = \dfrac{\pi}{\mu}\cot(\mu\pi)$     $[-1 < \operatorname{Re}\mu < 0]$

BI (134)(4), ET I 315(18)

8. $\int_0^1 x^{\mu-1}\ln(1-x)\,dx = -\dfrac{1}{\mu}[\psi(\mu+1) - \psi(1)] = -\dfrac{1}{\mu}[\psi(\mu+1) + \mathbf{C}]$

$[\operatorname{Re}\mu > -1]$     ET I 316(19)

$9.^7$ $\int_1^\infty x^{\mu-1}\ln(x-1)\,dx = \dfrac{1}{\mu}[\pi\cot(\mu\pi) + \psi(\mu+1) + \mathbf{C}]$

$[\operatorname{Re}\mu < 0]$     ET I 316(20)

10. $\int_0^\infty x^{\mu-1}\ln(1+\gamma x)\,dx = \dfrac{\pi}{\mu\gamma^\mu\sin\mu\pi}$     $[-1 < \operatorname{Re}\mu < 0, \quad |\arg\gamma| < \pi]$

BI (134)(3)

$11.^{11}$ $\int_0^\infty \dfrac{x^{\mu-1}\ln(1+x)}{1+x}\,dx = -\dfrac{\pi}{\sin\mu\pi}[\mathbf{C} + \psi(1-\mu)]$     $[-1 < \operatorname{Re}\mu < 1]$     ET I 316(21)

12. $\int_0^1 \dfrac{\ln(1+x)}{(1+x)^{\mu+1}}\,dx = -\dfrac{\ln 2}{2^\mu\mu} + \dfrac{2^\mu - 1}{2^\mu\mu^2}$     BI (114)(6)

13. $\int_0^1 \dfrac{x^{\mu-1}\ln(1-x)}{(1-x)^{1-\nu}}\,dx = \mathrm{B}(\mu,\nu)[\psi(\nu) - \psi(\mu+\nu)]$     $[\operatorname{Re}\mu > 0, \quad \operatorname{Re}\nu > 0]$     ET I 316(122)

14.    $\displaystyle \int_0^\infty \frac{x^{\mu-1}\ln(\gamma+x)}{(\gamma+x)^\nu}\,dx = \gamma^{\mu-\nu}\,B(\mu,\nu-\mu)\left[\psi(\nu)-\psi(\nu-\mu)+\ln\gamma\right]$

$$[0<\operatorname{Re}\mu<\operatorname{Re}\nu] \qquad\qquad \text{ET I 316(23)}$$

**4.294**

1.    $\displaystyle \int_0^1 \ln(1+x)\frac{(p-1)x^{p-1}-px^{-p}}{x}\,dx = 2\ln 2 - \frac{\pi}{\sin p\pi}$

$$[0<p<1] \qquad\qquad \text{BI (114)(2)}$$

2.    $\displaystyle \int_0^1 \ln(1+x)\frac{1+x^{2n+1}}{1+x}\,dx = 2\ln 2\sum_{k=0}^{n}\frac{1}{2k+1} - \sum_{j=1}^{2n+1}\frac{1}{j}\sum_{k=1}^{j}\frac{(-1)^{k-1}}{k}$    BI (114)(7)

3.    $\displaystyle \int_0^1 \ln(1+x)\frac{1-x^{2n}}{1+x}\,dx = 2\ln 2\cdot\sum_{k=0}^{n-1}\frac{1}{2k+1} - \sum_{j=1}^{2n}\frac{1}{j}\sum_{k=1}^{j}\frac{(-1)^{k-1}}{k}$    BI (114)(8)

4.    $\displaystyle \int_0^1 \ln(1+x)\frac{1-x^{2n}}{1-x}\,dx = 2\ln 2\cdot\sum_{k=0}^{n-1}\frac{1}{2k+1} + \sum_{i=1}^{2n}\frac{(-1)^j}{j}\sum_{k=1}^{j}\frac{(-1)^{k-1}}{k}$    BI (114)(9)

5.    $\displaystyle \int_0^1 \ln(1+x)\frac{1-x^{2n+1}}{1-x}\,dx = 2\ln 2\sum_{k=0}^{n}\frac{1}{2k+1} + \sum_{j=1}^{2n+1}\frac{(-1)^j}{j}\sum_{k=1}^{j}\frac{(-1)^{k-1}}{k}$    BI (114)(10)

6.    $\displaystyle \int_0^1 \ln(1-x)\frac{1-(-1)^n x^n}{1-x}\,dx = \sum_{j=1}^{n}\frac{(-1)^j}{j}\sum_{k=1}^{j}\frac{1}{k}$    BI (114)(15)

7.    $\displaystyle \int_0^1 \ln(1-x)\frac{1-x^n}{1-x}\,dx = -\sum_{j=1}^{n}\frac{1}{j}\sum_{k=1}^{j}\frac{1}{k}$    BI (114)(16)

8.    $\displaystyle \int_0^\infty \ln^2(1-x)x^p\,dx = \frac{2\pi}{p+1}\cot p\pi$      $[-2<p<-1]$    BI (134)(13)a

9.    $\displaystyle \int_0^1 [\ln(1+x)]^n(1+x)^r\,dx = (-1)^{n-1}\frac{n!}{(r+1)^{n+1}} + 2^{r+1}\sum_{k=0}^{n}\frac{(-1)^k n!\,(\ln 2)^{n-k}}{(n-k)!(r+1)^{k+1}}$    LI (106)(34)a

10.    $\displaystyle \int_0^1 [\ln(1-x)]^n(1-x)^r\,dx = (-1)^n\frac{n!}{(r+1)^{n+1}}$      $[r>-1]$    BI (106)(35)a

11.    $\displaystyle \int_0^1 \left(\ln\frac{1}{1-x^2}\right)^n x^{2q-1}\,dx = \frac{n!}{2}\zeta(n+1,q+1)$      $[-1<q<0]$    BI (311)(15)a

12.    $\displaystyle \int_0^1 (\ln x)^{2n}\ln\left(1-x^2\right)\frac{dx}{x} = -\frac{\pi^{2n+2}}{2(n+1)(2n+1)}|B_{2n+2}|$    BI (309)(5)a

13.[6]    $\displaystyle \int_0^1 \left[\ln\frac{1}{x}\right]^m \ln\left(1-x^2\right)\,dx = -\sum_{n=1}^{\infty}\frac{\Gamma(m+1)}{n(2n+1)^{m+1}}$      $[m+1>0,\quad n+1>0]$    BI (309)(5)a

**4.295**

1.[12]    $\displaystyle \int_0^\infty \frac{\ln(ax^2+b)}{1+x^2}\,dx = \pi\ln\left(\sqrt{a}+\sqrt{b}\right)$      $[\operatorname{Re} a>0,\quad \operatorname{Re} b>0]$    ET II 218(27)

2.     $\int_0^1 \ln\left(1+x^2\right) \frac{dx}{x^2} = \frac{\pi}{2} - \ln 2$                                    GW (325)(2g)

3.     $\int_0^\infty \ln\left(1+x^2\right) \frac{dx}{x^2} = \pi$                                    GW (325)(4c)

4.     $\int_0^\infty \ln\left(1+x^2\right) \frac{dx}{(a+x)^2} = \frac{2a}{1+a^2}\left(\frac{\pi}{2a} + \ln a\right)$                $[a>0]$                GW (319)(6)a

5.     $\int_0^1 \ln\left(1+x^2\right) \frac{dx}{1+x^2} = \frac{\pi}{2}\ln 2 - \boldsymbol{G}$                                    BI (114)(24)

6.     $\int_1^\infty \ln\left(1+x^2\right) \frac{dx}{1+x^2} = \frac{\pi}{2}\ln 2 + \boldsymbol{G}$                                    BI (114)(5)

7.     $\int_0^\infty \ln\left(a^2+b^2x^2\right) \frac{dx}{c^2+g^2x^2} = \frac{\pi}{cg}\ln\frac{ag+bc}{g}$                $[a>0, \quad b>0, \quad c>0, \quad g>0]$

                                                                                            BI (136)(11-14)a

8.     $\int_0^\infty \ln\left(a^2+b^2x^2\right) \frac{dx}{c^2-g^2x^2} = -\frac{\pi}{cg}\arctan\frac{bc}{ag}$                $[a>0, \quad b>0, \quad c>0, \quad g>0]$

                                                                                            BI (136)(15)a

9.     $\int_0^\infty \frac{\ln\left(1+p^2x^2\right) - \ln\left(1+q^2x^2\right)}{x^2} dx = \pi(p-q)$                $[p>0, \quad q>0]$                FI II 645

10.    $\int_0^1 \ln\frac{1+a^2x^2}{1+a^2} \frac{dx}{1-x^2} = -\left(\arctan a\right)^2$                                    BI (115)(2)

11.    $\int_0^1 \ln\left(1-x^2\right) \frac{dx}{x} = -\frac{\pi^2}{12}$

12.    $\int_0^\infty \ln^2\left(1-x^2\right) \frac{dx}{x^2} = 0$                                    BI (142)(9)a

13.    $\int_0^1 \ln\left(1-x^2\right) \frac{dx}{1+x^2} = \frac{\pi}{4}\ln 2 - \boldsymbol{G}$                                    GW (325)(17)

14.    $\int_1^\infty \ln\left(x^2-1\right) \frac{dx}{1+x^2} = \frac{\pi}{4}\ln 2 + \boldsymbol{G}$                                    BI (144)(6)

15.    $\int_0^\infty \ln^2\left(a^2-x^2\right) \frac{dx}{b^2+x^2} = \frac{\pi}{b}\ln\left(a^2+b^2\right)$                $[b>0]$                BI (136)(16)

16.    $\int_0^\infty \ln^2\left(a^2-x^2\right) \frac{b^2-x^2}{\left(b^2+x^2\right)^2} dx = -\frac{2b\pi}{a^2+b^2}$                $[b>0]$                BI (136)(20)

17.    $\int_0^1 \ln\left(1+x^2\right) \frac{dx}{x\left(1+x^2\right)} = \frac{1}{2}\left[\frac{\pi^2}{12} - \frac{1}{2}\left(\ln 2\right)^2\right]$                                    BI (114)(25)

18.    $\int_0^\infty \ln\left(1+x^2\right) \frac{dx}{x\left(1+x^2\right)} = \frac{\pi^2}{12}$                                    BI (141)(9)

19.    $\int_0^1 \ln\left(\cos^2 t + x^2\sin^2 t\right) \frac{dx}{1-x^2} = -t^2$                                    BI (114)(27)a

20. $\int_0^\infty \ln\left(a^2 + b^2 x^2\right) \frac{dx}{(c + gx)^2} = \frac{2\ln b}{cg} + \frac{b^2}{a^2 g^2 + b^2 c^2}\left(\frac{a}{b}\pi + 2\frac{c}{g}\ln\frac{c}{g} + 2\frac{a^2 g}{b^2 c}\ln\frac{a}{b}\right)$

$[a > 0, \quad b > 0, \quad c > 0, \quad g > 0]$

BI (139)(16)a

21. $\int_0^1 \ln\left(a^2 + b^2 x^2\right) \frac{dx}{(c + gx)^2}$

$= \frac{2}{c(c+g)}\ln a + \frac{b^2}{a^2 g^2 + b^2 c^2}\left[\frac{2a}{b}\operatorname{arccot}\frac{a}{b} + \frac{cb^2 - ga^2}{b^2(c+g)}\ln\frac{a^2 + b^2}{a^2} - 2\frac{c}{g}\ln\frac{c+g}{c}\right]$

$[a > 0, \quad b > 0, \quad c > 0, \quad g > 0]$   BI (114)(28)a

22.[11] $\int_0^\infty \frac{\ln\left(1 + p^2 x^2\right)}{r^2 + q^2 x^2} dx = \int_0^\infty \frac{\ln\left(p^2 + x^2\right)}{q^2 + r^2 x^2} dx = \frac{\pi}{qr}\ln\frac{q + pr}{r}$

$[qr > 0, \quad p > 0]$

FI II 745a, BI (318)(1)a, BI (318)(4)a

23. $\int_0^\infty \frac{\ln\left(1 + a^2 x^2\right)}{b^2 + c^2 x^2}\frac{dx}{d^2 + g^2 x^2} = \frac{\pi}{b^2 g^2 - c^2 d^2}\left[\frac{g}{d}\ln\left(1 + \frac{ad}{g}\right) - \frac{c}{b}\ln\left(1 + \frac{ab}{c}\right)\right]$

$[a > 0, \quad b > 0, \quad c > 0, \quad d > 0, \quad g > 0, \quad b^2 g^2 \neq c^2 d^2]$   BI (141)(10)

24. $\int_0^\infty \frac{\ln\left(1 + a^2 x^2\right)}{b^2 + c^2 x^2}\frac{x^2\, dx}{d^2 + g^2 x^2} = \frac{\pi}{b^2 g^2 - c^2 d^2}\left[\frac{b}{c}\ln\left(1 + \frac{ab}{c}\right) - \frac{d}{g}\ln\left(1 + \frac{ad}{g}\right)\right]$

$[a > 0, \quad b > 0, \quad c > 0, \quad d > 0, \quad g > 0, \quad b^2 g^2 \neq c^2 d^2]$   BI (141)(11)

25. $\int_0^\infty \ln\left(a^2 + b^2 x^2\right) \frac{dx}{(c^2 + g^2 x^2)^2} = \frac{\pi}{2c^3 q}\left(\ln\frac{ag + bc}{q} - \frac{bc}{ag + bc}\right)$

$[a > 0, \quad b > 0, \quad c > 0, \quad g > 0]$

GW (325)(18a)

26. $\int_0^\infty \ln\left(a^2 + b^2 x^2\right) \frac{x^2\, dx}{(c^2 + g^2 x^2)^2} = \frac{\pi}{2cg^3}\left(\ln\frac{ag + bc}{g} + \frac{bc}{ag + bc}\right)$

$[a > 0, \quad b > 0, \quad c > 0, \quad g > 0]$

GW (325)(18b)

27. $\int_0^1 \ln\left(1 + ax^2\right) \sqrt{1 - x^2}\, dx = \frac{\pi}{2}\left\{\ln\frac{1 + \sqrt{1 + a}}{2} + \frac{1}{2}\frac{1 - \sqrt{1 + a}}{1 + \sqrt{1 + a}}\right\}$

$[a > 0]$

BI (117)(6)

28. $\int_0^1 \ln\left(1 + a - ax^2\right) \sqrt{1 - x^2}\, dx = \frac{\pi}{2}\left\{\ln\frac{1 + \sqrt{1 + a}}{2} - \frac{1}{2}\frac{1 - \sqrt{1 + a}}{1 + \sqrt{1 + a}}\right\}$

$[a > 0]$

BI (117)(7)

29. $\int_0^1 \ln\left(1 - a^2 x^2\right) \frac{dx}{\sqrt{1 - x^2}} = \pi\ln\frac{1 + \sqrt{1 - a^2}}{2}$    $[a^2 < 1]$

BI (119)(1)

30.[12] $\int_0^1 \ln\left(1 - a^2 x^2\right) \frac{dx}{x\sqrt{1 - x^2}} = -\left(\arccos|a| - \frac{\pi}{2}\right)^2$    $[a^2 < 1]$

LI (120)(11)

31. $\displaystyle\int_0^1 \ln\left(1 - x^2\right) \frac{\mathrm{d}x}{\sqrt{\left(1 - x^2\right)\left(1 - k^2 x^2\right)}} = \ln\frac{k'}{k}\,\boldsymbol{K}(k) - \frac{\pi}{2}\,\boldsymbol{K}\left(k'\right)$

BI (120)(12)

32. $\displaystyle\int_0^1 \ln\left(1 \pm kx^2\right) \frac{\mathrm{d}x}{\sqrt{\left(1 - x^2\right)\left(1 - k^2 x^2\right)}} = \frac{1}{2}\ln\frac{2 \pm 2k}{\sqrt{k}}\,\boldsymbol{K}(k) - \frac{\pi}{8}\,\boldsymbol{K}\left(k'\right)$

BI (120)(8), BI (120)(14)

33. $\displaystyle\int_0^1 \frac{\ln\left(1 - k^2 x^2\right)}{\sqrt{\left(1 - x^2\right)\left(1 - k^2 x^2\right)}}\,\mathrm{d}x = \ln k'\,\boldsymbol{K}(k)$

BI (119)(27)

34. $\displaystyle\int_0^1 \ln\left(1 - k^2 x^2\right) \sqrt{\frac{1 - k^2 x^2}{1 - x^2}}\,\mathrm{d}x = \left(2 - k^2\right)\boldsymbol{K}(k) - \left(2 - \ln k'\right)\boldsymbol{E}(k)$

BI (119)(3)

35. $\displaystyle\int_0^1 \sqrt{\frac{1 - x^2}{1 - k^2 x^2}}\,\ln\left(1 - k^2 x^2\right)\,\mathrm{d}x = \frac{1}{k^2}\left(1 + k'^2 - k'^2\ln k'\right)\boldsymbol{K}(k) - \left(2 - \ln k'\right)\boldsymbol{E}(k)$

BI (119)(7)

36. $\displaystyle\int_{-1}^1 \ln\left(1 - x^2\right) \frac{\mathrm{d}x}{(a + bx)\sqrt{1 - x^2}} = \frac{2\pi}{\sqrt{a^2 - b^2}}\ln\frac{\sqrt{a^2 - b^2}}{a + \sqrt{a^2 - b^2}}$

$[a > 0, \quad b > 0, \quad a \neq b]$

BI (145)(15)

37.[12] $\displaystyle\int_0^1 \ln\left(1 - x^2\right)\left(px^{p-1} - qx^{q-1}\right)\mathrm{d}x = \psi\left(\frac{q}{2} + 1\right) - \psi\left(\frac{p}{2} + 1\right)$

$[p > -2, \quad q > -2]$

BI (106)(15)

38. $\displaystyle\int_0^1 \ln\left(1 + ax^2\right) \frac{\mathrm{d}x}{\sqrt{1 - x^2}} = \pi\ln\frac{1 + \sqrt{1 + a}}{2}$ $\qquad [a \geq -1]$

GW (325)(21b)

39. $\displaystyle\int_0^1 \ln\left(1 + x^2\right) x^{\mu-1}\,\mathrm{d}x = \frac{1}{\mu}\left[\ln 2 - \beta\left(\frac{\mu}{2} + 1\right)\right]$ $\qquad [\mathrm{Re}\,\mu > -2]$

BI (106)(12)

40. $\displaystyle\int_0^\infty \ln\left(1 + x^2\right) x^{\mu-1}\,\mathrm{d}x = \frac{\pi}{\mu\sin\dfrac{\mu\pi}{2}}$ $\qquad [-2 < \mathrm{Re}\,\mu < 0]$

BI (311)(4)a, ET I 315(15)

41.[12] $\displaystyle\int_0^\infty \ln\left(1 + x^2\right) \frac{x^{\mu-1}\,\mathrm{d}x}{1 + x} = \frac{\pi}{\sin\mu\pi}\left[\ln 2 + \sin\left(\frac{\mu\pi}{2}\right)\beta\left(\frac{1 - \mu}{2}\right) - \cos\left(\frac{\mu\pi}{2}\right)\beta\left(\frac{2 - \mu}{2}\right)\right]$

ET I 316(25)

## 4.296

1. $\displaystyle\int_0^1 \ln\left(1 + 2x\cos t + x^2\right) \frac{\mathrm{d}x}{x} = \frac{\pi^2}{6} - \frac{t^2}{2}$

BI (114)(34)

2. $\displaystyle\int_{-\infty}^\infty \ln\left(a^2 - 2ax\cos t + x^2\right) \frac{\mathrm{d}x}{1 + x^2} = \pi\ln\left(1 + 2a|\sin t| + a^2\right)$

BI (145)(28)

3. $\displaystyle\int_0^\infty \ln\left(1 + 2x\cos t + x^2\right) x^{\mu-1}\,\mathrm{d}x = \frac{2\pi}{\mu}\frac{\cos\mu t}{\sin\mu\pi}$ $\qquad [|t| < \pi, \quad -1 < \mathrm{Re}\,\mu < 0]$ ET I 316(27)

4. $\displaystyle\int_0^\infty \ln\left(\frac{x^2 + 2ax\cos t + a^2}{x^2 - 2ax\cos t + a^2}\right) \frac{x\,\mathrm{d}x}{x^2 + b^2} = \frac{1}{2}\pi^2 - \pi t + \pi\arctan\frac{\left(a^2 - b^2\right)\cos t}{\left(a^2 + b^2\right)\sin t + 2ab}$

$[a > 0, \quad b > 0, \quad 0 < t < \pi]$

**4.297**

1.    $\int_0^1 \ln \dfrac{ax+b}{bx+a} \dfrac{dx}{(1+x)^2} = \dfrac{1}{a-b}\left[(a+b)\ln\dfrac{a+b}{2} - a\ln a - b\ln b\right]$

                                            $[a > 0, \quad b > 0]$       BI (115)(16)

2.    $\int_0^\infty \ln \dfrac{ax+b}{bx+a} \dfrac{dx}{(1+x)^2} = 0$                   $[ab > 0]$       BI (139)(23)

3.    $\int_0^1 \ln \dfrac{1-x}{x} \dfrac{dx}{1+x^2} = \dfrac{\pi}{8}\ln 2$                              BI (115)(5)

4.    $\int_0^1 \ln \dfrac{1+x}{1-x} \dfrac{dx}{1+x^2} = \boldsymbol{G}$                              BI (115)(17)

5.[11]   $\int_0^\infty \ln\left(\dfrac{1+x}{1-x}\right)^2 \dfrac{dx}{x(1+x^2)} = \dfrac{\pi^2}{2}$                  BI (141)(13)

6.    $\int_u^v \ln \dfrac{v+x}{u+x} \dfrac{dx}{x} = \dfrac{1}{2}\left(\ln\dfrac{v}{u}\right)^2$             $[uv > 0]$       BI (145)(33)

7.    $\int_0^\infty \dfrac{b\ln(1+ax) - a\ln(1+bx)}{x^2}\, dx = ab\ln\dfrac{b}{a}$       $[a > 0, \quad b > 0]$       FI II 647

8.    $\int_0^1 \ln \dfrac{1+ax}{1-ax} \dfrac{dx}{x\sqrt{1-x^2}} = \pi \arcsin a$       $[|a| \le 1]$       GW (325)(21c), BI (122)(2)

9.    $\int_u^v \ln\left(\dfrac{1+ax}{1-ax}\right) \dfrac{dx}{\sqrt{(x^2-u^2)(v^2-x^2)}} = \dfrac{\pi}{v} F\left(\arcsin av, \dfrac{u}{v}\right)$

                                         $[|av| < 1]$       BI (145)(35)

10.[8]   $\mathrm{PV}\int_0^1 \ln\left|\dfrac{a+y}{a-y}\right| \dfrac{dy}{y\sqrt{1-y^2}} = \dfrac{\pi^2}{2}$       $[0 < a \le 1]$

11.*   $\int_a^b \ln\left(\dfrac{c+x}{c-x}\right) \dfrac{x^2\, dx}{\sqrt{(x^2-a^2)(b^2-x^2)}} = \dfrac{\pi}{c}\left(c^2 - \sqrt{(c^2-a^2)(c^2-b^2)}\right) + \pi b\left[F(\phi,k) - E(\phi,k)\right]$

                              $\left[\phi = \arcsin\dfrac{b}{c}, \quad k = \dfrac{a}{b}, \quad 0 < a < b < c\right]$     KM(4.51)

**4.298**

1.[12]   $\int_0^\infty \ln \dfrac{1+x^2}{x} \dfrac{x^{2p-1}}{1+x}\, dx = \dfrac{\pi}{\sin(2p\pi)}\left[\ln 2 + \cos(p\pi)\beta(p) + \sin(p\pi)\beta\left(\dfrac{1}{2}-p\right) - \dfrac{\pi}{\sin(2p\pi)}\right]$

                                        $\left[0 < p < \tfrac{1}{2}\right]$       BI (137)(1)

3.[12]   $\int_0^\infty \ln \dfrac{1+x^2}{x} \dfrac{x^{2p-1}}{1-x}\, dx = \dfrac{\pi}{\sin(2p\pi)}\left[\cos(2p\pi)\ln 2 + \cos(p\pi)\beta(p) - \sin(p\pi)\beta\left(\dfrac{1}{2}-n\right) - \dfrac{\pi}{\tan(2p\pi)}\right]$

                                        $\left[0 < p < \tfrac{1}{2}\right]$       BI (137)(2)

5.[12]   $\int_0^\infty \ln \dfrac{1+x^2}{x} \dfrac{x^{2p-1}}{1+x^2}\, dx = -\dfrac{\pi}{4\sin(p\pi)}\left[2\boldsymbol{C} + 2\psi(p) + \dfrac{\pi}{\tan(p\pi)}\right]$

                                        $[0 < p < 1]$       BI (137)(10)

5b.* $\displaystyle\int_0^\infty \ln\left(\frac{1+x^2}{x}\right)\frac{x^{2p-1}}{1-x^2}\,dx = \frac{\pi}{4\sin p\pi}\left[2\ln 2\cos(p\pi) + 2\beta(p) - \frac{\pi}{\sin(p\pi)}\right]$

$$[0 < p < 1]$$

6. $\displaystyle\int_0^1 \ln\frac{1+x^2}{x}x^{2n}\,dx = \frac{1}{2n+1}\left\{(-1)^n\frac{\pi}{2} + \ln 2 - \frac{1}{2n+1} + 2\sum_{k=0}^{n-1}\frac{(-1)^k}{2n-2k-1}\right\}$    BI (294)(8)

7. $\displaystyle\int_0^1 \ln\frac{1+x^2}{x}x^{2n-1}\,dx = \frac{1}{2n}\left\{(-1)^{n+1}\ln 2 + \ln 2 - \frac{1}{2n} + (-1)^{n+1}\sum_{k=1}^{n-1}\frac{(-1)^k}{k}\right\}$    BI (294)(9)a

8. $\displaystyle\int_0^1 \ln\frac{1+x^2}{x}\frac{dx}{1+x^2} = \frac{\pi}{2}\ln 2$    BI (115)(7)

9. $\displaystyle\int_0^\infty \ln\frac{1+x^2}{x}\frac{dx}{1+x^2} = \pi\ln 2$    BI (137)(8)

10. $\displaystyle\int_0^\infty \ln\frac{1+x^2}{x}\frac{dx}{1-x^2} = 0$    BI (137)(9)

11. $\displaystyle\int_0^1 \ln\frac{1-x^2}{x}\frac{dx}{1+x^2} = \frac{\pi}{4}\ln 2$    BI (115)(9)

12. $\displaystyle\int_1^\infty \ln\frac{1+x^2}{x+1}\frac{dx}{1+x^2} = \frac{3\pi}{8}\ln 2$    BI (144)(8)

13. $\displaystyle\int_0^1 \ln\frac{1+x^2}{x+1}\frac{dx}{1+x^2} = \frac{3\pi}{8}\ln 2 - G$    BI (115)(18)

14. $\displaystyle\int_1^\infty \ln\frac{1+x^2}{x-1}\frac{dx}{1+x^2} = \frac{3\pi}{8}\ln 2 + G$    BI (144)(9)

15. $\displaystyle\int_0^1 \ln\frac{1+x^2}{1-x}\frac{dx}{1+x^2} = \frac{3\pi}{8}\ln 2$    BI (115)(19)

16. $\displaystyle\int_0^\infty \ln\frac{1+x^2}{x^2}\frac{x\,dx}{1+x^2} = \frac{\pi^2}{12}$    BI (138)(3)

17. $\displaystyle\int_0^\infty \ln\frac{a^2+b^2x^2}{x^2}\frac{dx}{c^2+g^2x^2} = \frac{\pi}{cg}\ln\frac{ag+bc}{c}$    $[a > 0, \quad b > 0, \quad c > 0, \quad g > 0]$

    BI (138)(6, 7, 9, 10)a

18. $\displaystyle\int_0^\infty \ln\frac{a^2+b^2x^2}{x^2}\frac{dx}{c^2-g^2x^2} = \frac{1}{cg}\arctan\frac{ag}{bc}$    $[a > 0, b > 0, c > 0, g > 0]$

    BI (138)(8, 11)a

19. $\displaystyle\int_0^\infty \ln\frac{1+x^2}{x^2}\frac{x^2\,dx}{(1+x^2)^2} = \frac{\pi}{4}(\ln 4 - 1)$    BI (139)(21)

20. $\displaystyle\int_0^1 \ln^2\left(\frac{1-x^2}{x^2}\right)\sqrt{1-x^2}\,dx = \pi$    FI II 643a

21. $\displaystyle\int_0^1 \ln\frac{1+2x\cos t + x^2}{(1+x)^2}\frac{dx}{x} = \frac{1}{2}\int_0^\infty \ln\frac{1+2x\cos t + x^2}{(1+x)^2}\frac{dx}{x} = -\frac{t^2}{2}$

$$[|t| < \pi] \qquad \text{BI (115)(23), BI (134)(15)}$$

22. $\displaystyle\int_0^\infty \ln\frac{1+2x\cos t + x^2}{(1+x)^2} x^{p-1}\,dx = -\frac{2\pi(1-\cos pt)}{p\sin p\pi}$     $[0 < |p| < 1, \quad |t| < \pi]$     BI (134)(17)

23. $\displaystyle\int_0^1 \ln\frac{1+x^2\sin t}{1-x^2\sin t}\frac{dx}{\sqrt{1-x^2}} = \pi\ln\cot\left(\frac{\pi-t}{4}\right)$     $[|t| < \pi]$     GW (325)(21d)

**4.299**

1. $\displaystyle\int_0^\infty \ln\frac{(x+1)(x+a^2)}{(x+a)^2}\frac{dx}{x} = (\ln a)^2$     $[a > 0]$     BI (134)(14)

2. $\displaystyle\int_0^1 \ln\frac{(1-ax)(1+ax^2)}{(1-ax^2)^2}\frac{dx}{1+ax^2} = \frac{1}{2\sqrt{a}}\arctan\sqrt{a}\ln(1+a)$

                                                 $[a > 0]$     BI (115)(25)

3. $\displaystyle\int_0^1 \ln\frac{(1-a^2x^2)(1+ax^2)}{(1-ax^2)^2}\frac{dx}{1+ax^2} = \frac{1}{\sqrt{a}}\arctan\sqrt{a}\ln(1+a)$

                                                   $[a > 0]$     BI (115)(26)

4. $\displaystyle\int_0^1 \ln\frac{(x+1)(x+a^2)}{(x+a)^2}x^{\mu-1}\,dx = \frac{\pi(a^\mu-1)^2}{\mu\sin\mu\pi}$     $[a > 0, \quad \operatorname{Re}\mu > 0]$     BI (134)(16)

**4.311**

1.[11] $\displaystyle\int_0^\infty \frac{\ln(1+x^n)}{x^n}\,dx = \frac{\pi\operatorname{cosec}\left(\frac{\pi}{n}\right)}{n-1}$     $[n = 2, 3, \ldots]$

2. $\displaystyle\int_0^\infty \ln(1+x^3)\frac{dx}{1-x+x^2} = \frac{2\pi}{\sqrt{3}}\ln 3$     LI (136)(8)

3. $\displaystyle\int_0^\infty \ln(1+x^3)\frac{dx}{1+x^3} = \frac{\pi}{\sqrt{3}}\ln 3 - \frac{\pi^2}{9}$     LI (136)(6)

4. $\displaystyle\int_0^\infty \ln(1+x^3)\frac{x\,dx}{1+x^3} = \frac{\pi}{\sqrt{3}}\ln 3 + \frac{\pi^2}{9}$     LI (136)(7)

5. $\displaystyle\int_0^\infty \ln(1+x^3)\frac{1-x}{1+x^3}\,dx = -\frac{2}{9}\pi^2$     BI (136)(9)

6.[8] $\displaystyle\int_0^\infty \left|1-\frac{x^3}{a^3}\right|\frac{dx}{x^3} = -\frac{\pi\sqrt{3}}{6a^2}$

**4.312**

1. $\displaystyle\int_0^\infty \ln\frac{1+x^3}{x^3}\frac{dx}{1+x^3} = \frac{\pi}{\sqrt{3}}\ln 3 + \frac{\pi^2}{9}$     BI (138)(12)

2. $\displaystyle\int_0^\infty \ln\frac{1+x^3}{x^3}\frac{x\,dx}{1+x^3} = \frac{\pi}{\sqrt{3}}\ln 3 - \frac{\pi^2}{9}$     BI (138)(13)

**4.313**

1. $\displaystyle\int_0^\infty \ln x\ln(1+a^2x^2)\frac{dx}{x^2} = \pi a(1-\ln a)$     $[a > 0]$     BI (134)(18)

2.　$\displaystyle\int_0^\infty \ln\left(1+c^2x^2\right)\ln\left(a^2+b^2x^2\right)\frac{dx}{x^2} = 2\pi\left[\left(c+\frac{b}{a}\right)\ln(b+ac)-\frac{b}{a}\ln b - c\ln c\right]$

$$[a>0,\quad b>0,\quad c>0]$$

BI (134)(20, 21)a

3.　$\displaystyle\int_0^\infty \ln\left(1+c^2x^2\right)\ln\left(a^2+\frac{b^2}{x^2}\right)\frac{dx}{x^2} = 2\pi\left[\frac{a+bc}{b}\ln(a+bc)-\frac{a}{b}\ln a - c\right]$

$$[a>0,\quad a+bc>0]\qquad \text{BI (134)(22, 23)a}$$

4.　$\displaystyle\int_0^\infty \ln x \ln\frac{1+a^2x^2}{1+b^2x^2}\frac{dx}{x^2} = \pi(a-b)+\pi\ln\frac{b^b}{a^a}$　　$[a>0,\quad b>0]$　　BI (134)(24)

5.　$\displaystyle\int_0^\infty \ln x \ln\frac{a^2+2bx+x^2}{a^2-2bx+x^2}\frac{dx}{x} = 2\pi\ln a\arcsin\frac{b}{a}$　　$[a\geq|b|]$　　BI (134)(25)

6.　$\displaystyle\int_0^\infty \ln(1+x)\frac{x\ln x-x-a}{(x+a)^2}\frac{dx}{x} = \frac{(\ln a)^2}{2(a-1)}$　　$[a>0]$　　BI (141)(7)

7.　$\displaystyle\int_0^\infty \ln^2(1-x)\frac{x\ln x-x-a}{(x+a)^2}\frac{dx}{x} = \frac{\pi^2+(\ln a)^2}{1+a}$　　$[a>0]$　　LI (141)(8)

**4.314**

1.[11]　$\displaystyle\int_0^1 \ln(1+ax)\frac{x^{p-1}-x^{q-1}}{\ln x}\,dx = \sum_{k=1}^\infty (-1)^{k+1}\frac{a^k}{k}\ln\frac{p+k}{q+k}$

$$[|a|<1,\quad p>0,\quad q>0]\qquad \text{BI (123)(18)}$$

2.　$\displaystyle\int_0^\infty \left[\frac{(q-1)x}{(1+x)^2}-\frac{1}{x+1}+\frac{1}{(1+x)^q}\right]\frac{dx}{x\ln(1+x)} = \ln\Gamma(q)$

$$[q>0]\qquad\qquad\qquad\qquad \text{BI (143)(7)}$$

3.　$\displaystyle\int_0^1 \frac{x\ln x+1-x}{x(\ln x)^2}\ln(1+x)\,dx = \ln\frac{4}{\pi}$　　BI (126)(12)

4.　$\displaystyle\int_0^1 \frac{\ln\left(1-x^2\right)dx}{x\left(q^2+(\ln x)^2\right)} = -\frac{\pi}{q}\ln\Gamma\left(\frac{q+\pi}{\pi}\right)+\frac{\pi}{2q}\ln 2q+\ln\frac{q}{\pi}-1$

$$[q>0]\qquad\qquad\qquad \text{LI (327)(12)a}$$

**4.315**

1.　$\displaystyle\int_0^1 \ln(1+x)(\ln x)^{n-1}\frac{dx}{x} = (-1)^{n-1}(n-1)!\left(1-\frac{1}{2^n}\right)\zeta(n+1)$　　BI (116)(3)

2.　$\displaystyle\int_0^1 \ln(1+x)(\ln x)^{2n}\frac{dx}{x} = \frac{2^{2n+1}-1}{(2n+1)(2n+2)}\pi^{2n+2}|B_{2n+2}|$　　BI (116)(1)

3.　$\displaystyle\int_0^1 \ln(1-x)(\ln x)^{n-1}\frac{dx}{x} = (-1)^n(n-1)!\,\zeta(n+1)$　　BI (116)(4)

4.　$\displaystyle\int_0^1 \ln(1-x)(\ln x)^{2n}\frac{dx}{x} = -\frac{2^{2n}}{(n+1)(2n+1)}\pi^{2n+2}\,|B_{2n+2}|$　　BI (116)(2)

**4.316**

1. $\displaystyle \int_0^1 \ln\left(1 - ax^r\right)\left(-\ln x\right)^p \frac{dx}{x} = -\frac{1}{r^{p+1}}\,\Gamma(p+1)\sum_{k=1}^{\infty}\frac{a^k}{k^{p+2}}$

$$[p > -1, \quad a < 1, \quad r > 0]\qquad \text{BI (116)(7)}$$

2. $\displaystyle \int_0^1 \ln\left(1 - 2ax\cos t + a^2x^2\right)\left(-\ln x\right)^p \frac{dx}{x} = -2\,\Gamma(p+1)\sum_{k=1}^{\infty}\frac{a^k \cos kt}{k^{p+2}}$ $\qquad$ LI (116)(8)

**4.317**

1. $\displaystyle \int_0^{\infty} \ln\frac{\sqrt{1+x^2}+a}{\sqrt{1+x^2}-a}\,\frac{dx}{\sqrt{1+x^2}} = \pi\arcsin a$ $\qquad [|a| < 1]$ $\qquad$ BI (142)(11)

2. $\displaystyle \int_0^1 \ln\frac{\sqrt{1-a^2x^2}-x\sqrt{1-a^2}}{1-x}\,\frac{dx}{x} = \frac{1}{2}\left(\arcsin a\right)^2$ $\qquad$ BI (115)(32)

3. $\displaystyle \int_0^1 \ln\frac{1+\cos t\sqrt{1-x^2}}{1-\cos t\sqrt{1-x^2}}\,\frac{dx}{x^2+\tan^2 v} = \pi\cot t\,\frac{\cos\dfrac{v-t}{2}}{\sin\dfrac{v+t}{2}}$ $\qquad$ BI (115)(30)

4. $\displaystyle \int_0^1 \ln^2\left(\frac{x+\sqrt{1-x^2}}{x-\sqrt{1-x^2}}\right)\frac{x\,dx}{1-x^2} = \frac{\pi^2}{2}$ $\qquad$ BI (115)(31)

5. $\displaystyle \int_0^1 \ln\left\{\sqrt{1+kx}+\sqrt{1-kx}\right\}\frac{dx}{\sqrt{\left(1-x^2\right)\left(1-k^2x^2\right)}} = \frac{1}{4}\ln(4k)\,\boldsymbol{K}(k) + \frac{\pi}{8}\,\boldsymbol{K}(k')$ $\qquad$ BI (121)(8)

6. $\displaystyle \int_0^1 \ln\left\{\sqrt{1+kx}-\sqrt{1-kx}\right\}\frac{dx}{\sqrt{\left(1-x^2\right)\left(1-k^2x^2\right)}} = \frac{1}{4}\ln(4k)\,\boldsymbol{K}(k) + \frac{3}{8}\pi\,\boldsymbol{K}(k')$ $\qquad$ BI (121)(9)

7. $\displaystyle \int_0^1 \ln\left\{1+\sqrt{1-k^2x^2}\right\}\frac{dx}{\sqrt{\left(1-x^2\right)\left(1-k^2x^2\right)}} = \frac{1}{2}\ln k\,\boldsymbol{K}(k) + \frac{\pi}{4}\,\boldsymbol{K}(k')$ $\qquad$ BI (121)(6)

8. $\displaystyle \int_0^1 \ln\left\{1-\sqrt{1-k^2x^2}\right\}\frac{dx}{\sqrt{\left(1-x^2\right)\left(1-k^2x^2\right)}} = \frac{1}{2}\ln k\,\boldsymbol{K}(k) - \frac{3}{4}\pi\,\boldsymbol{K}(k')$ $\qquad$ BI (121)(7)

9.$^{12}$ $\displaystyle \int_0^1 \ln\frac{1+p\sqrt{1-x^2}}{1-p\sqrt{1-x^2}}\,\frac{dx}{1-x^2} = \pi\arcsin p$ $\qquad [p^2 < 1]$ $\qquad$ BI (115)(29)

10. $\displaystyle \int_0^1 \ln\frac{1+q\sqrt{1-k^2x^2}}{1-q\sqrt{1-k^2x^2}}\,\frac{dx}{\sqrt{\left(1-x^2\right)\left(1-k^2x^2\right)}} = \pi\,F\left(\arcsin q, k'\right)$

$$[q^2 < 1]\qquad \text{BI (122)(15)}$$

11.$^{10}$ $\displaystyle \int_{-\infty}^{\infty} \ln\left|\frac{1+2\sqrt{1+x^2}}{1-2\sqrt{1+x^2}}\right|\frac{dx}{\sqrt{1+x^2}} = \frac{\pi^2}{3}$

**4.318**

1. $\displaystyle \int_0^1 \frac{\ln\left(1-x^q\right)}{1+(\ln x)^2}\,\frac{dx}{x} = \pi\left[\ln\Gamma\left(\frac{q}{2\pi}+1\right) - \frac{\ln q}{2} + \frac{q}{2\pi}\left(\ln\frac{q}{2\pi}-1\right)\right]$

$$[q > 0]\qquad \text{BI (126)(11)}$$

2. $\displaystyle\int_0^\infty \ln\left(1+x^r\right)\left[\frac{(p-r)x^p-(q-r)x^q}{\ln x}+\frac{x^q-x^p}{(\ln x)^2}\right]\frac{dx}{x^{r+1}}=r\ln\left(\tan\frac{q\pi}{2r}\cot\frac{p\pi}{2r}\right)$

$$[p<r,\quad q<r]\qquad\text{BI (143)(9)}$$

In integrals containing $\ln\left(a+bx^r\right)$, it is useful to make the substitution $x^r=t$ and then to seek the resulting integral in the tables. For example,

$$\int_0^\infty x^{p-1}\ln\left(1+x^r\right)\,dx=\frac{1}{r}\int_0^\infty t^{\frac{p}{r}-1}\ln(1+t)\,dt=\frac{\pi}{p\sin\dfrac{p\pi}{r}}\qquad\text{(see 4.293 3)}$$

**4.319**

1. $\displaystyle\int_0^\infty \ln\left(1-e^{-2a\pi x}\right)\frac{dx}{1+x^2}=-\pi\left[\frac{1}{2}\ln 2a\pi+a\left(\ln a-1\right)-\ln\Gamma(a+1)\right]$

$$[a>0]\qquad\text{BI (354)(6)}$$

2. $\displaystyle\int_0^\infty \ln\left(1+e^{-2a\pi x}\right)\frac{dx}{1+x^2}=\pi\left[\ln\Gamma(2a)-\ln\Gamma(a)+a\left(1-\ln a\right)-\left(2a-\frac{1}{2}\right)\ln 2\right]$

$$[a>0]\qquad\text{BI (354)(7)}$$

3. $\displaystyle\int_0^\infty \ln\frac{a+be^{-px}}{a+be^{-qx}}\frac{dx}{x}=\ln\frac{a}{a+b}\ln\frac{p}{q}$

$$\left[\frac{b}{a}>-1,\quad pq>0\right]$$

$$\text{FI II 635, BI (354)(1)}$$

**4.321**

1. $\displaystyle\int_{-\infty}^\infty x\ln\cosh x\,dx=0$

$$\text{BI (358)(2)a}$$

2. $\displaystyle\int_{-\infty}^\infty \ln\cosh x\frac{dx}{1-x^2}=0$

$$\text{BI (138)(20)a}$$

**4.322**

1.[11] $\displaystyle\int_0^\pi x\ln\sin x\,dx=\frac{1}{2}\int_0^\pi x\ln\cos^2 x\,dx=-\frac{\pi^2}{2}\ln 2$

$$\text{BI (432)(1, 2) FI II 643}$$

2.[12] $\displaystyle\int_0^\infty \frac{\ln\sin^2(ax)}{1+x^2}\,dx=\pi\ln\left(\frac{1-e^{-2a}}{2}\right)$

$$[a>0]\qquad\text{GW (338)(28b)}$$

3.[12] $\displaystyle\int_0^\infty \frac{\ln\cos^2(ax)}{1+x^2}\,dx=\pi\ln\left(\frac{1+e^{-2a}}{2}\right)$

$$[a>0]\qquad\text{GW (338)(28a)}$$

4. $\displaystyle\int_0^\infty \frac{\ln\sin^2 ax}{b^2-x^2}\,dx=-\frac{\pi^2}{2b}+a\pi$

$$[a>0,\quad b>0]\qquad\text{BI (418)(1)}$$

5.[11] $\displaystyle\int_0^\infty \frac{\ln\cos^2 ax}{b^2-x^2}\,dx=\infty$

$$\text{BI (418)(2)}$$

6. $\displaystyle\int_0^\infty \frac{\ln\cos^2 x}{x^2}\,dx=-\pi$

$$\text{FI II 686}$$

7.[7] $\displaystyle\int_0^{\pi/4}\ln\sin x x^{\mu-1}\,dx=-\frac{1}{2\mu}\left(\frac{\pi}{4}\right)^\mu\left[\ln 2+\frac{2}{\mu}-\sum_{k=1}^\infty\frac{\zeta(2k)}{4^{2k-1}(\mu+2k)}\right]$

$$[\operatorname{Re}\mu>0]\qquad\text{LI (425)(1)}$$

8.[7] $\displaystyle\int_0^{\pi/2} \ln \sin x \, x^{\mu-1} \, dx = -\frac{1}{\mu}\left(\frac{\pi}{2}\right)^{\mu}\left[\frac{1}{\mu} - 2\sum_{k=1}^{\infty}\frac{\zeta(2k)}{4^k(\mu+2k)}\right]$

$$[\operatorname{Re}\mu > 0] \qquad \text{LI (430)(1)}$$

9.   $\displaystyle\int_0^{\pi/2} \ln\left(1-\cos x\right) x^{\mu-1} \, dx = -\frac{1}{\mu}\left(\frac{\pi}{2}\right)^{\mu}\left[\frac{2}{\mu} - \sum_{k=1}^{\infty}\frac{\zeta(2k)}{4^{2k-1}(\mu+2k)}\right]$

$$[\operatorname{Re}\mu > 0] \qquad \text{LI (430)(2)}$$

10.  $\displaystyle\int_0^{\infty} \ln\left(1 \pm 2p\cos\beta x + p^2\right)\frac{dx}{q^2+x^2} = \frac{\pi}{q}\ln\left(1 \pm pe^{-\beta q}\right) \qquad [p^2 < 1]$

$$= \frac{\pi}{q}\ln\left(p \pm e^{-\beta q}\right) \qquad [p^2 > 1]$$

$$\text{FI II 718a}$$

**4.323**

1.[11]   $\displaystyle\int_0^{\pi} x \ln \tan^2 x \, dx = 0$                                      BI (432)(3)

2.   $\displaystyle\int_0^{\infty} \frac{\ln\tan^2 ax}{b^2+x^2}\,dx = \frac{\pi}{b}\ln\tanh ab \qquad\qquad [a>0, \quad b>0]$     GW (338)(28c)

3.   $\displaystyle\int_0^{\infty} \ln\left(\frac{1+\tan x}{1-\tan x}\right)^2 \frac{dx}{x} = \frac{\pi^2}{2}$                     GW (338)(26)

**4.324**

1.   $\displaystyle\int_0^{\infty} \ln\left(\frac{1+\sin x}{1-\sin x}\right)^2 \frac{dx}{x} = \pi^2$                          GW (338)(25)

2.   $\displaystyle\int_0^{\infty} \ln\frac{1+2a\cos px + a^2}{1+2a\cos qx + a^2}\frac{dx}{x} = \ln(1+a)\ln\frac{q^2}{p^2} \qquad [-1 < a \leq 1]$

$$= \ln\left(1+\frac{1}{a}\right)\ln\frac{q^2}{p^2} \qquad [a < -1 \text{ or } a \geq 1]$$

$$\text{GW (338)(27)}$$

3.   $\displaystyle\int_0^{\infty} \ln\left(a^2\sin^2 px + b^2\cos^2 px\right)\frac{dx}{c^2+x^2} = \frac{\pi}{c}\left[\ln\left(a\sinh cp + b\cosh cp\right) - cp\right]$

$$[a>0, \quad b>0, \quad c>0, \quad p>0]$$

$$\text{GW (338)(29)}$$

**4.325**

1.[12]   $\displaystyle\int_0^1 \ln\left(-\ln x\right)\frac{dx}{1+x} = -C\ln 2 + \sum_{k=2}^{\infty}(-1)^k\frac{\ln k}{k} = -\frac{1}{2}(\ln 2)^2$     GW (325)(25a)

2.   $\displaystyle\int_0^1 \ln\left(-\ln x\right)\frac{dx}{x+e^{i\lambda}} = \sum_{k=1}^{\infty}\frac{(-1)^k}{k}e^{-ik\lambda}\left(C+\ln k\right)$     GW (325)(26)

3.   $\displaystyle\int_0^1 \ln\left(-\ln x\right)\frac{dx}{(1+x)^2} = \int_1^{\infty}\ln\ln x\frac{dx}{(1+x)^2} = \frac{1}{2}\left[\psi\left(\frac{1}{2}\right) + \ln 2\pi\right] = \frac{1}{2}\left(\ln\frac{\pi}{2} - C\right)$

$$\text{BI (147)(7)}$$

**4.**
$$\int_0^1 \ln(-\ln x) \frac{dx}{1+x^2} = \int_1^\infty \ln\ln x \frac{dx}{1+x^2} = \frac{\pi}{2} \ln \frac{\sqrt{2\pi}\,\Gamma\left(\frac{3}{4}\right)}{\Gamma\left(\frac{1}{4}\right)}$$
BI (148)(1)

**5.**
$$\int_0^1 \ln(-\ln x) \frac{dx}{1+x+x^2} = \int_1^\infty \ln\ln x \frac{dx}{1+x+x^2} = \frac{\pi}{\sqrt{3}} \ln \frac{\sqrt[3]{2\pi}\,\Gamma\left(\frac{2}{3}\right)}{\Gamma\left(\frac{1}{3}\right)}$$
BI (148)(2)

**6.**
$$\int_0^1 \ln(-\ln x) \frac{dx}{1-x+x^2} = \int_1^\infty \ln\ln x \frac{dx}{1-x+x^2} = \frac{2\pi}{\sqrt{3}} \left[\frac{5}{6}\ln 2\pi - \ln\Gamma\left(\frac{1}{6}\right)\right]$$
BI (148)(5)

**7.**
$$\int_0^1 \ln(-\ln x) \frac{dx}{1+2x\cos t+x^2} = \int_1^\infty \ln\ln x \frac{dx}{1+2x\cos t+x^2} = \frac{\pi}{2\sin t} \ln \frac{(2\pi)^{t/\pi}\,\Gamma\left(\frac{1}{2}+\frac{t}{2\pi}\right)}{\Gamma\left(\frac{1}{2}-\frac{t}{2\pi}\right)}$$
BI (147)(9)

**8.**
$$\int_0^1 \ln(-\ln x)\, x^{\mu-1}\, dx = -\frac{1}{\mu}\left(C+\ln\mu\right) \qquad [\operatorname{Re}\mu > 0]$$
BI (147)(1)

**9.**
$$\int_1^\infty \ln\ln x \frac{x^{n-2}\, dx}{1+x^2+x^4+\cdots+x^{2n-2}}$$
$$= \frac{\pi}{2n}\tan\frac{\pi}{2n}\ln 2\pi + \frac{\pi}{n}\sum_{k=1}^{n-1}(-1)^{k-1}\sin\frac{k\pi}{n}\ln\frac{\Gamma\left(\frac{n+k}{2n}\right)}{\Gamma\left(\frac{k}{2n}\right)} \qquad [n \text{ is even}]$$
$$= \frac{\pi}{2n}\tan\frac{\pi}{2n}\ln\pi + \frac{\pi}{n}\sum_{k=1}^{\frac{n-1}{2}}(-1)^{k-1}\sin\frac{k\pi}{n}\ln\frac{\Gamma\left(\frac{n-k}{n}\right)}{\Gamma\left(\frac{k}{n}\right)} \qquad [n \text{ is odd}]$$
BI (148)(4)

**10.**
$$\int_0^1 \ln(-\ln x) \frac{dx}{(1+x^2)\sqrt{-\ln x}} = \int_1^\infty \ln\ln x \frac{dx}{(1+x^2)\sqrt{\ln x}}$$
$$= \sqrt{\pi}\sum_{k=0}^\infty \frac{(-1)^{k+1}}{\sqrt{2k+1}}\left[\ln(2k+1)+2\ln 2 + C\right]$$
BI (147)(4)

**11.**
$$\int_0^1 \ln(-\ln x) \frac{x^{\mu-1}\, dx}{\sqrt{-\ln x}} = -\left(C+\ln 4\mu\right)\sqrt{\frac{\pi}{\mu}} \qquad [\operatorname{Re}\mu > 0]$$
BI (147)(3)

**12.**
$$\int_0^1 \ln(-\ln x)(-\ln x)^{\mu-1}\, x^{\nu-1}\, dx = \frac{1}{\nu^\mu}\Gamma(\mu)\left[\psi(\mu)-\ln(\nu)\right]$$
$$[\operatorname{Re}\mu > 0, \quad \operatorname{Re}\nu > 0] \qquad \text{BI (147)(2)}$$

**4.326**

**1.**
$$\int_0^1 \ln(a-\ln x)\, x^{\mu-1}\, dx = \frac{1}{\mu}\left[\ln a - e^{a\mu}\operatorname{Ei}(-a\mu)\right] \qquad [\operatorname{Re}\mu > 0, \quad a > 0]$$
BI (107)(23)

**2.**
$$\int_0^{1/e} \ln\left(2\ln\frac{1}{x}-1\right) \frac{x^{2\mu-1}}{\ln x}\, dx = -\frac{1}{2}\left[\operatorname{Ei}(-\mu)\right]^2 \qquad [\operatorname{Re}\mu > 0]$$
BI (145)(5)

**4.327**

1.    $\int_0^1 \ln\left[a^2 + (\ln x)^2\right] \dfrac{dx}{1+x^2} = \pi \ln \dfrac{2\,\Gamma\left(\frac{2a+3\pi}{4\pi}\right)}{\Gamma\left(\frac{2a+\pi}{4\pi}\right)} + \dfrac{\pi}{2} \ln \dfrac{\pi}{2}$

$$\left[a > -\tfrac{\pi}{2}\right] \qquad\qquad \text{BI (147)(10)}$$

2.    $\int_0^1 \ln\left[a^2 + 4\,(\ln x)^2\right] \dfrac{dx}{1+x^2} = \pi \ln \dfrac{2\,\Gamma\left(\frac{a+3\pi}{4\pi}\right)}{\Gamma\left(\frac{a+\pi}{4\pi}\right)} + \dfrac{\pi}{2} \ln \pi$

$$\left[a > -\pi\right] \qquad\qquad \text{BI (147)(16)a}$$

3.    $\int_0^\infty \ln\left[a^2 + (\ln x)^2\right] x^{\mu-1}\, dx = \dfrac{2}{\mu} \left[-\cos a\mu\, \mathrm{ci}(a\mu) - \sin a\mu\, \mathrm{si}(a\mu) + \ln a\right]$

$$[a > 0, \quad \mathrm{Re}\,\mu > 0] \qquad\qquad \text{GW (325)(28)}$$

If the integrand contains a logarithm whose argument also contains a logarithm, for example, if the integrand contains $\ln\ln\frac{1}{x}$, it is useful to make the substitution $\ln x = t$ and then seek the transformed integral in the tables.

## 4.33–4.34 Combinations of logarithms and exponentials

**4.331**

1.    $\int_0^\infty e^{-\mu x} \ln x\, dx = -\dfrac{1}{\mu}\left(\boldsymbol{C} + \ln\mu\right)$         $[\mathrm{Re}\,\mu > 0]$         BI (256)(2)

2.    $\int_1^\infty e^{-\mu x} \ln x\, dx = -\dfrac{1}{\mu} \mathrm{Ei}(-\mu)$         $[\mathrm{Re}\,\mu > 0]$         BI (260)(5)

3.    $\int_0^1 e^{\mu x} \ln x\, dx = -\dfrac{1}{\mu}\int_0^1 \dfrac{e^{\mu x} - 1}{x}\, dx$         $[\mu \neq 0]$         GW (324)(81a)

**4.332**

1.    $\int_0^\infty \dfrac{\ln x\, dx}{e^x + e^{-x} - 1} = \dfrac{2\pi}{\sqrt{3}} \left[\dfrac{5}{6} \ln 2\pi - \ln\Gamma\left(\dfrac{1}{6}\right)\right]$         (cf. **4.325** 6)         BI (257)(6)

2.[12]   $\int_0^\infty \dfrac{\ln x\, dx}{e^x + e^{-x} + 1} = \dfrac{\pi}{\sqrt{3}} \ln\left[\dfrac{\Gamma\left(\frac{2}{3}\right)}{\Gamma\left(\frac{1}{3}\right)} \sqrt[3]{2\pi}\right]$         (cf. **4.325** 5)      BI (257)(7)a, LI (260)(3)

**4.333**    $\int_0^\infty e^{-\mu x^2} \ln x\, dx = -\dfrac{1}{4}\left(\boldsymbol{C} + \ln 4\mu\right)\sqrt{\dfrac{\pi}{\mu}}$         $[\mathrm{Re}\,\mu > 0]$         BI (256)(8), FI II 807a

**4.334**    $\int_0^\infty \dfrac{\ln x\, dx}{e^{x^2} + 1 + e^{-x^2}} = \dfrac{1}{2}\sqrt{\dfrac{\pi}{3}} \sum_{k=1}^\infty (-1)^k \dfrac{\boldsymbol{C} + \ln 4k}{\sqrt{k}} \sin\dfrac{k\pi}{3}$         BI (357)(13)

**4.335**

1.    $\int_0^\infty e^{-\mu x} (\ln x)^2\, dx = \dfrac{1}{\mu}\left[\dfrac{\pi^2}{6} + \left(\boldsymbol{C} + \ln\mu\right)^2\right]$         $[\mathrm{Re}\,\mu > 0]$         ET I 149(13)

2.    $\int_0^\infty e^{-x^2} (\ln x)^2\, dx = \dfrac{\sqrt{\pi}}{8}\left[\left(\boldsymbol{C} + 2\ln 2\right)^2 + \dfrac{\pi^2}{2}\right]$         FI II 808

3.[7]   $\int_0^\infty e^{-\mu x} (\ln x)^3\, dx = -\dfrac{1}{\mu}\left[\left(\boldsymbol{C} + \ln\mu\right)^3 + \dfrac{\pi^2}{2}\left(\boldsymbol{C} + \ln\mu\right) - \psi''(1)\right]$         MI 26

**4.336**

1.[7] $\quad \mathrm{PV} \displaystyle\int_0^\infty \frac{e^{-x}}{\ln x}\,\mathrm{d}x \approx -0.154479567$ 

BI (260)(9)

2. $\quad \displaystyle\int_0^\infty \frac{e^{-\mu x}\,\mathrm{d}x}{\pi^2 + (\ln x)^2} = \nu'(\mu) - e^\mu \qquad\qquad [\mathrm{Re}\,\mu > 0]$

MI 26

**4.337**

1. $\quad \displaystyle\int_0^\infty e^{-\mu x}\ln(\beta + x)\mathrm{d}x = \frac{1}{\mu}\left[\ln\beta - e^{\mu\beta}\,\mathrm{Ei}(-\beta\mu)\right] \qquad [|\arg\beta| < \pi, \quad \mathrm{Re}\,\mu > 0]$

BI (256)(3)

2. $\quad \displaystyle\int_0^\infty e^{-\mu x}\ln(1 + \beta x)\mathrm{d}x = -\frac{1}{\mu}e^{\frac{\mu}{\beta}}\,\mathrm{Ei}\left(-\frac{\mu}{\beta}\right) \qquad [|\arg\beta| < \pi, \quad \mathrm{Re}\,\mu > 0]$

ET I 148(4)

3. $\quad \displaystyle\int_0^\infty e^{-\mu x}\ln|a - x|\,\mathrm{d}x = \frac{1}{\mu}\left[\ln a - e^{-a\mu}\,\mathrm{Ei}(a\mu)\right] \qquad [a > 0, \quad \mathrm{Re}\,\mu > 0]$

BI (256)(4)

4.[7] $\quad \displaystyle\int_0^\infty e^{-\mu x}\ln\left|\frac{\beta}{\beta - x}\right|\,\mathrm{d}x = \frac{1}{\mu}\left[e^{-\beta\mu}\,\mathrm{Ei}(\beta\mu)\right] \qquad [\mathrm{Re}\,\mu > 0]$

MI 26

5.[12] $\quad \displaystyle\int_0^\infty \ln(1 + ax)x^n e^{-x}\,\mathrm{d}x$

$$= \sum_{\mu=0}^{n-1} \frac{n!}{(n-\mu)!}\left[\frac{(-1)^{n-\mu-1}}{a^{n-\mu}}e^{1/a}\,\mathrm{Ei}\left(-\frac{1}{a}\right) + \sum_{k=1}^{n-\mu}(k-1)!\left(-\frac{1}{a}\right)^{n-\mu-k}\right]n!e^{1/a}\,\mathrm{Ei}\left(\frac{1}{a}\right)$$

**4.338**

1. $\quad \displaystyle\int_0^\infty e^{-\mu x}\ln\left(\beta^2 + x^2\right)\,\mathrm{d}x = \frac{2}{\mu}\left[\ln\beta - \mathrm{ci}(\beta\mu)\cos(\beta\mu) - \mathrm{si}(\beta\mu)\sin(\beta\mu)\right]$

$$[\mathrm{Re}\,\beta > 0, \quad \mathrm{Re}\,\mu > 0]$$

BI (256)(6)

2. $\quad \displaystyle\int_0^\infty e^{-\mu x}\ln^2\left(x^2 - \beta^2\right)\,\mathrm{d}x = \frac{2}{\mu}\left[\ln^2\beta - e^{\beta\mu}\,\mathrm{Ei}(-\beta\mu) - e^{\beta\mu}\,\mathrm{Ei}(\beta\mu)\right]$

$$[\mathrm{Im}\,\beta > 0, \quad \mathrm{Re}\,\mu > 0]$$

BI (256)(5)

**4.339** $\quad \displaystyle\int_0^\infty e^{-\mu x}\ln\left|\frac{x+1}{x-1}\right|\,\mathrm{d}x = \frac{1}{\mu}\left[e^{-\mu}\left(\ln 2\mu + \gamma\right) - e^\mu\,\mathrm{Ei}(-2\mu)\right]$

$$[\mathrm{Re}\,\mu > 0]$$

MI 27

**4.341** $\quad \displaystyle\int_0^\infty e^{-\mu x}\ln\frac{\sqrt{x+ai} + \sqrt{x-ai}}{\sqrt{2a}}\,\mathrm{d}x = \frac{\pi}{4\mu}\left[\mathbf{H}_0(a\mu) - Y_0(a\mu)\right]$

$$[a > 0, \quad \mathrm{Re}\,\mu > 0]$$

ET I 149(20)

**4.342**

1. $\quad \displaystyle\int_0^\infty e^{-2nx}\ln(\sinh x)\,\mathrm{d}x = \frac{1}{2n}\left[\frac{1}{n} + \ln 2 - 2\,\beta(2n+1)\right]$

BI (256)(17)

2. $\quad \displaystyle\int_0^\infty e^{-\mu x}\ln(\cosh x)\,\mathrm{d}x = \frac{1}{\mu}\left[\beta\left(\frac{\mu}{2}\right) - \frac{1}{\mu}\right] \qquad [\mathrm{Re}\,\mu > 0]$

ET I 165(32)

3.[11] $\quad \displaystyle\int_0^\infty e^{-\mu x}\left[\ln(\sinh x) - \ln x\right]\,\mathrm{d}x = \frac{1}{\mu}\left[\ln\frac{\mu}{2} - \frac{1}{\mu} - \psi\left(\frac{\mu}{2}\right)\right]$

$$[\mathrm{Re}\,\mu > 0]$$

ET I 165(33)

**4.343** $\quad \displaystyle\int_0^\pi e^{\mu\cos x}\left[\ln\left(2\mu\sin^2 x\right) + \mathbf{C}\right]\,\mathrm{d}x = -\pi\,K_0(\mu)$

WA 95(16)

## 4.35–4.36 Combinations of logarithms, exponentials, and powers

**4.351**

1. $$\int_0^1 (1-x)e^{-x} \ln x \, dx = \frac{1-e}{e}$$                   BI (352)(1)

2. $$\int_0^1 e^{\mu x} \left(\mu x^2 + 2x\right) \ln x \, dx = \frac{1}{\mu^2} \left[(1-\mu)e^\mu - 1\right]$$        BI (352)(2)

3. $$\int_1^\infty \frac{e^{-\mu x} \ln x}{1+x} \, dx = \frac{1}{2} e^\mu \left[\text{Ei}(-\mu)\right]^2$$      $[\text{Re}\,\mu > 0]$         NT 32(10)

**4.352**

1. $$\int_0^\infty x^{\nu-1} e^{-\mu x} \ln x \, dx = \frac{1}{\mu^\nu} \Gamma(\nu) \left[\psi(\nu) - \ln \mu\right]$$     $[\text{Re}\,\mu > 0, \quad \text{Re}\,\nu > 0]$

                                                                  BI (353)(3), ET I 315(10)a

2. $$\int_0^\infty x^n e^{-\mu x} \ln x \, dx = \frac{n!}{\mu^{n+1}} \left[1 + \frac{1}{2} + \frac{1}{3} + \cdots + \frac{1}{n} - C - \ln \mu\right]$$

                                                  $[\text{Re}\,\mu > 0]$         ET I 148(7)

3. $$\int_0^\infty x^{n-\frac{1}{2}} e^{-\mu x} \ln x \, dx = \sqrt{\pi} \frac{(2n-1)!!}{2^n \mu^{n+\frac{1}{2}}} \left[2 \left(1 + \frac{1}{3} + \frac{1}{5} + \cdots + \frac{1}{2n-1}\right) - C - \ln 4\mu\right]$$

                                                  $[\text{Re}\,\mu > 0]$       ET I 148(10)

4. $$\int_0^\infty x^{\mu-1} e^{-x} \ln x \, dx = \Gamma'(\mu)$$      $[\text{Re}\,\mu > 0]$     GW (324)(83a)

**4.353**

1. $$\int_0^\infty (x-\nu) x^{\nu-1} e^{-x} \ln x \, dx = \Gamma(\nu)$$      $[\text{Re}\,\nu > 0]$     GW (324)(84)

2. $$\int_0^\infty \left(\mu x - n - \frac{1}{2}\right) x^{n-\frac{1}{2}} e^{-\mu x} \ln x \, dx = \frac{(2n-1)!!}{(2\mu)^n} \sqrt{\frac{\pi}{\mu}}$$

                                                  $[\text{Re}\,\mu > 0]$         BI (357)(2)

3. $$\int_0^1 (\mu x + n + 1) x^n e^{\mu x} \ln x \, dx = e^\mu \sum_{k=0}^n (-1)^{k-1} \frac{n!}{(n-k)! \mu^{k+1}} + (-1)^n \frac{n!}{\mu^{n+1}}$$

                                                            $[\mu \neq 0]$         GW (324)(82)

**4.354**

1.[6] $$\int_0^\infty \frac{x^{\nu-1} \ln x}{e^x + 1} \, dx = \Gamma(\nu) \sum_{k=1}^\infty \frac{(-1)^{k-1}}{k^\nu} \left[\psi(\nu) - \ln k\right]$$     $[\text{Re}\,\nu > 0]$

                      $$= -\frac{1}{2} (\ln 2)^2$$                $[\text{for } \nu = 1]$

                                                         GW (324)(86a)

2.[7] $$\int_0^\infty \frac{x^{\nu-1} \ln x}{(e^x + 1)^2} \, dx = \Gamma(\nu) \sum_{k=2}^\infty \frac{(-1)^k (k-1)}{k^\nu} \left[\psi(\nu) - \ln k\right]$$

                                                  $[\text{Re}\,\nu > 1]$       GW (324)(86b)

3.　　$\displaystyle\int_0^\infty \frac{(x-\nu)e^x - \nu}{(e^x+1)^2}x^{\nu-1}\ln x\,dx = \Gamma(\nu)\sum_{k=1}^\infty \frac{(-1)^{k-1}}{k^\nu}$　　$[\operatorname{Re}\nu > 0]$　　GW (324)(87a)

4.　　$\displaystyle\int_0^\infty \frac{(x-2n)e^x - 2n}{(e^x+1)^2}x^{2n-1}\ln x\,dx = \frac{2^{2n-1}-1}{2n}\pi^{2n}|B_{2n}|$

$[n = 1, 2, \ldots]$　　GW (324)(87b)

5.　　$\displaystyle\int_0^\infty \frac{x^{\nu-1}\ln x}{(e^x+1)^n}\,dx = (-1)^n\frac{\Gamma(\nu)}{(n-1)!}\sum_{k=n}^\infty \frac{(-1)^k(k-1)!}{(k-n)!k^\nu}[\psi(\nu) - \ln k]$

$[\operatorname{Re}\nu > 0]$　　GW (324)(86c)

**4.355**

1.　　$\displaystyle\int_0^\infty x^2 e^{-\mu x^2}\ln x\,dx = \frac{1}{8\mu}\left(2 - \ln 4\mu - \boldsymbol{C}\right)\sqrt{\frac{\pi}{\mu}}$　　$[\operatorname{Re}\mu > 0]$　　BI (357)(1)a

2.　　$\displaystyle\int_0^\infty x\left(\mu x^2 - \nu x - 1\right)e^{-\mu x^2 + 2\nu x}\ln x\,dx = \frac{1}{4\mu} + \frac{\nu}{4\mu}\sqrt{\frac{\pi}{\mu}}\exp\left(\frac{\nu^2}{\mu}\right)\left[1 + \Phi\left(\frac{\nu}{\sqrt{\mu}}\right)\right]$

$[\operatorname{Re}\mu > 0]$　　BI (358)(1)

3.　　$\displaystyle\int_0^\infty \left(\mu x^2 - n\right)x^{2n-1}e^{-\mu x^2}\ln x\,dx = \frac{(n-1)!}{4\mu^n}$　　$[\operatorname{Re}\mu > 0]$　　BI (353)(4)

4.　　$\displaystyle\int_0^\infty \left(2\mu x^2 - 2n - 1\right)x^{2n}e^{-\mu x^2}\ln x\,dx = \frac{(2n-1)!!}{2(2\mu)^n}\sqrt{\frac{\pi}{\mu}}$

$[\operatorname{Re}\mu > 0]$　　BI (353)(5)

**4.356**

1.　　$\displaystyle\int_0^\infty \exp\left[-\mu\left(\frac{x}{a} + \frac{a}{x}\right)\right]\ln x\,\frac{dx}{x} = 2\ln a\,K_0(2\mu)$　　$[a > 0, \quad \operatorname{Re}\mu > 0]$　　GW (324)(91)

2.　　$\displaystyle\int_0^\infty \exp\left(-ax - \frac{b}{x}\right)\ln x\left[2ax^2 - (2n+1)x - 2b\right]x^{n-\frac{1}{2}}\,dx$

$$= 2\left(\frac{b}{a}\right)^{\frac{n}{2}}\sqrt{\frac{\pi}{a}}e^{-2\sqrt{ab}}\sum_{k=0}^\infty \frac{(n+k)!}{(n-k)!(2k)!!\left(2\sqrt{ab}\right)^k}$$

$[a > 0, \quad b > 0]$　　BI (357)(4)

3.　　$\displaystyle\int_0^\infty \exp\left(-ax - \frac{b}{x}\right)\ln x\left[2ax^2 + (2n-1)x - 2b\right]\frac{dx}{x^{n+\frac{3}{2}}}$

$$= 2\left(\frac{a}{b}\right)^{\frac{n}{2}}\sqrt{\frac{\pi}{a}}e^{-2\sqrt{ab}}\sum_{k=0}^\infty \frac{(n+k-1)!}{(n-k-1)!(2k)!!\left(2\sqrt{ab}\right)^k}$$

$[a > 0, \quad b > 0]$　　BI (357)(11)

For $n = \frac{1}{2}$:

4.　　$\displaystyle\int_0^\infty \exp\left(-ax - \frac{a}{x}\right)\ln x\,\frac{ax^2 - b}{x^2}\,dx = 2\,K_0\left(2\sqrt{ab}\right)$　　$[a > 0, \quad b > 0]$　　GW (324)(92c)

For $n = 0$:

5. $\displaystyle\int_0^\infty \exp\left(-ax - \frac{b}{x}\right) \ln x \frac{2ax^2 - x - 2b}{x\sqrt{x}}\, dx = 2\sqrt{\frac{\pi}{a}} e^{-2\sqrt{ab}}$

$$[a > 0, \quad b > 0]$$

BI (357)(7), GW(324)(92a)

For $n = -1$:

6. $\displaystyle\int_0^\infty \exp\left(-ax - \frac{b}{x}\right) \ln x \frac{2ax^2 - 3x - 2b}{\sqrt{x}}\, dx = \frac{1 + 2\sqrt{ab}}{a}\sqrt{\frac{\pi}{a}} e^{-2\sqrt{ab}}$

$$[a > 0, \quad b > 0]$$

LI (357)(6), GW (324)(92b)

7.[9] $\displaystyle\int_0^\infty \exp\left(-ax - \frac{b}{x}\right) \ln x \left(a - \frac{b}{x^2}\right) dx = K_0\left(2\sqrt{ab}\right)$

$$[a > 0, \quad b > 0]$$

8.[9] $\displaystyle\int_0^\infty \exp\left(-ax - \frac{b}{x}\right) \ln x \left[2ax^2 - (2n+1)x - 2b\right] x^{n - \frac{3}{2}}\, dx$

$$= 4\left(\frac{b}{a}\right)^{(2n+1)/4} K_{n+\frac{1}{2}}\left(2\sqrt{ab}\right)$$

$$= 2\left(\frac{b}{a}\right)^{\frac{n}{2}}\sqrt{\frac{\pi}{a}} e^{-2\sqrt{ab}} \sum_{k=0}^n \frac{(n+k)!}{(n-k)!(2k)!!\left(2\sqrt{ab}\right)^k}$$

$$[n = 0, 1, \ldots, \quad a > 0, \quad b > 0]$$

9.[12] $\displaystyle\int_0^\infty \exp\left(-ax - \frac{b}{x}\right) \ln\left[(ax^2 - b)\cos\left(c\ln x\right) + cx\sin\left(c\ln x\right)\right] \frac{dx}{x^2}$

$$= 2\cos\left(c\ln\sqrt{b/a}\right) K_{ic}\left(2\sqrt{ab}\right)$$

$$[a > 0, \quad b > 0]$$

10.[12] $\displaystyle\int_0^\infty \exp\left(-ax - \frac{b}{x}\right) \ln x \left[(ax^2 - b)\sin\left(c\ln x\right) - cx\cos\left(c\ln x\right)\right] \frac{dx}{x^2}$

$$= 2\sin\left(c\ln\sqrt{b/a}\right) K_{ic}\left(2\sqrt{ab}\right)$$

$$[a > 0, \quad b > 0]$$

11.[12] $\displaystyle\int_0^\infty x^c \ln x \left[a - \frac{c}{x} - \frac{b}{x^2}\right] \exp\left(-ax - \frac{b}{x}\right) dx = 2\left(\frac{b}{a}\right)^{c/2} K_c\left(2\sqrt{ab}\right)$

$$[a > 0, \quad b > 0]$$

**4.357**

1. $\displaystyle\int_0^\infty \exp\left(-\frac{1 + x^4}{2ax^2}\right) \ln x \frac{1 + ax^2 - x^4}{x^2}\, dx = -\frac{\sqrt{2a^3\pi}}{2\sqrt[a]{e}}$

$$[a > 0]$$

BI (357)(8)

2. $\displaystyle\int_0^\infty \exp\left(-\frac{1 + x^4}{2ax^2}\right) \ln x \frac{x^4 + ax^2 - 1}{x^4}\, dx = \frac{\sqrt{2a^3\pi}}{2\sqrt[a]{e}} \qquad [a > 0]$

BI (357)(9)

3. $\displaystyle\int_0^\infty \exp\left(-\frac{1+x^4}{2ax^2}\right) \ln x \, \frac{x^4+3ax-1}{x^6} \, dx = \frac{(1+a)\sqrt{2a^3\pi}}{2\sqrt[a]{e}}$

$$[a>0] \qquad\qquad \text{BI (357)(10)}$$

**4.358**

1.[6] $\displaystyle\int_1^\infty x^{\nu-1} e^{-\mu x} (\ln x)^m \, dx = \frac{\partial^m}{\partial \nu^m}\left\{\mu^{-\nu}\,\Gamma(\nu,\mu)\right\}$ $\qquad [m=0,1,\ldots, \quad \operatorname{Re}\mu>0, \quad \operatorname{Re}\nu>0]$

$$\text{MI 26}$$

2.[12] $\displaystyle\int_0^\infty x^{\nu-1} e^{-\mu x} (\ln x)^2 \, dx = \frac{\Gamma(\nu)}{\mu^\nu}\left\{[\psi(\nu)-\ln\mu]^2 + \zeta(2,\nu-1)\right\}$

$$[\operatorname{Re}\mu>0, \quad \operatorname{Re}\nu>0] \qquad\qquad \text{MI 26}$$

3.[9] $\displaystyle\int_0^\infty x^{\nu-1} e^{-\mu x} (\ln x)^3 \, dx = \frac{\Gamma(\nu)}{\mu^\nu}\left\{[\psi(\nu)-\ln\mu]^3 + 3\,\zeta(2,\nu)\,[\psi(\nu)-\ln\mu] - 2\,\zeta(3,\nu)\right\}$

$$[\operatorname{Re}\mu>0, \quad \operatorname{Re}\nu>0] \qquad\qquad \text{MI 26}$$

4.[12] $\displaystyle\int_0^\infty x^{\nu-1} e^{-\mu x} (\ln x)^4 \, dx = \frac{\Gamma(\nu)}{\mu^\nu}\left\{[\psi(\nu)-\ln\mu]^4 + 6\,\zeta(2,\nu)\,[\psi(\nu)-\ln\mu]^2\right.$

$$\left. -8\,\zeta(3,\nu)\,[\psi(\nu)-\ln\mu] + 3\,[\zeta(2,\nu)]^2 + 6\,\zeta(4,\nu)\right\}$$
$$[\operatorname{Re}\mu>0, \quad \operatorname{Re}\nu>0]$$

5.[3] $\displaystyle\int_0^\infty x^{\nu-1} e^{-\mu x} (\ln x)^n \, dx = \frac{\partial^n}{\partial \nu^n}\left\{\mu^{-\nu}\,\Gamma(\nu)\right\}$ $\qquad [n=0,1,2,\ldots]$

**4.359**

1. $\displaystyle\int_0^\infty e^{-\mu x} \frac{x^{p-1}-x^{q-1}}{\ln x} \, dx = \frac{1}{\mu}[\lambda(\mu,p-1) - \lambda(\mu,q-1)]$

$$[\operatorname{Re}\mu>0, \quad p>0, \quad q>0] \qquad\qquad \text{MI 27}$$

2.[11] $\displaystyle\int_0^1 e^{\mu x} \frac{x^{p-1}-x^{q-1}}{\ln x} \, dx = \sum_{k=0}^\infty \frac{\mu^k}{k!} \ln\frac{p+k}{q+k}$ $\qquad [p>0, \quad q>0] \qquad\qquad \text{BI (352)(9)}$

**4.361**

1. $\displaystyle\int_0^\infty \frac{(x+1)e^{-\mu x}}{\pi^2 + (\ln x)^2} \, dx = \nu'(\mu) - \nu''(\mu)$ $\qquad [\operatorname{Re}\mu>0] \qquad\qquad \text{MI 27}$

2. $\displaystyle\int_0^\infty \frac{e^{-\mu x}\,dx}{x\left[\pi^2 + (\ln x)^2\right]} = e^\mu - \nu(\mu)$ $\qquad [\operatorname{Re}\mu>0] \qquad\qquad \text{MI 27}$

**4.362**

1. $\displaystyle\int_0^1 x e^x \ln(1-x) \, dx = 1-e$ $\qquad\qquad \text{BI (352)(5)a}$

2. $\displaystyle\int_1^\infty e^{-\mu x} \ln(2x-1)\frac{dx}{x} = \frac{1}{2}\left[\operatorname{Ei}\left(-\frac{\mu}{2}\right)\right]^2$ $\qquad [\operatorname{Re}\mu>0] \qquad\qquad \text{ET I 148(8)}$

**4.363**

1.  $$\int_0^\infty e^{-\mu x} \ln(a+x) \frac{\mu(x+a)\ln(x+a)-2}{x+a}\, \mathrm{d}x$$

$$= \frac{1}{4}\int_0^\infty e^{-\mu x} \ln^2(a-x) \frac{\mu(x-a)\ln^2(x-a)-4}{x-a}\, \mathrm{d}x = (\ln a)^2$$

$$[\operatorname{Re}\mu > 0, \quad a > 0] \qquad \text{BI (354)(4, 5)}$$

2.  $$\int_0^1 x(1-x)(2-x)e^{-(1-x)^2}\ln(1-x)\, \mathrm{d}x = \frac{1-e}{4e} \qquad\qquad \text{BI (352)(4)}$$

**4.364**

1.  $$\int_0^\infty e^{-\mu x} \ln[(x+a)(x+b)] \frac{\mathrm{d}x}{x+a+b} = e^{(a+b)\mu}\{\operatorname{Ei}(-a\mu)\operatorname{Ei}(-b\mu) - \ln(ab)\operatorname{Ei}[-(a+b)\mu]\}$$

$$[a > 0, \quad b > 0, \quad \operatorname{Re}\mu > 0] \quad \text{BI (354)(11)}$$

2.  $$\int_0^\infty e^{-\mu x} \ln(x+a+b)\left(\frac{1}{x+a} + \frac{1}{x+b}\right)\mathrm{d}x$$

$$= (1 + \ln a \ln b)\ln(a+b) + e^{-(a+b)\mu}\{\operatorname{Ei}(-\alpha\mu)\operatorname{Ei}(-b\mu)$$

$$+ (1 - \ln(ab))\operatorname{Ei}[-(a+b)\mu]\}$$

$$[a > 0, \quad b > 0, \quad \operatorname{Re}\mu > 0] \quad \text{BI (354)(12)}$$

**4.365** $$\int_0^\infty \left[e^{-x} - \frac{x}{(1+x)^{p+1}\ln(1+x)}\right]\frac{\mathrm{d}x}{x} = \ln p \qquad [p > 0] \qquad \text{BI (354)(15)}$$

**4.366**

1.  $$\int_0^\infty e^{-\mu x}\ln\left(1 + \frac{x^2}{a^2}\right)\frac{\mathrm{d}x}{x} = [\operatorname{ci}(a\mu)]^2 + [\operatorname{si}(a\mu)]^2 \qquad [\operatorname{Re}\mu > 0] \qquad \text{NT 32(11)a}$$

2.  $$\int_0^\infty e^{-\mu x}\ln\left|1 - \frac{x^2}{a^2}\right|\frac{\mathrm{d}x}{x} = \operatorname{Ei}(a\mu)\operatorname{Ei}(-a\mu) \qquad [\operatorname{Re}\mu > 0] \qquad \text{ME 18}$$

3.  $$\int_0^\infty xe^{-\mu x^2}\ln\left|\frac{1+x^2}{1-x^2}\right|\mathrm{d}x = \frac{1}{\mu}\left[\cosh\mu\sinh(i\mu) - \sinh\mu\cosh(i\mu)\right]$$

$$[\operatorname{Re}\mu > 0] \qquad (\text{cf. } \mathbf{4.339}) \qquad \text{MI 27}$$

**4.367** $$\int_0^\infty xe^{-\mu x^2}\ln\frac{x+\sqrt{x^2+2\beta}}{\sqrt{2\beta}}\, \mathrm{d}x = \frac{e^{\beta\mu}}{4\mu}K_0(\beta\mu) \qquad [|\arg\beta| < \pi, \quad \operatorname{Re}\mu > 0] \qquad \text{ET I 149(19)}$$

**4.368** $$\int_0^{2u} e^{-\mu x^2}\ln\frac{x^2\left(4u^2-x^2\right)}{u^4}\frac{\mathrm{d}x}{\sqrt{4u^2-x^2}} = \frac{\pi}{2}e^{-2u^2\mu}\left[\frac{\pi}{2}Y_0\left(2iu^2\mu\right) - (\boldsymbol{C} - \ln 2)J_0\left(2iu^2\mu\right)\right]$$

$$[\operatorname{Re}\mu > 0] \qquad \text{ET I 149(21)a}$$

**4.369**

1.  $$\int_0^\infty x^{\nu-1}e^{-\mu x}\left[\psi(\nu) - \ln x\right]\mathrm{d}x = \frac{\Gamma(\nu)\ln\mu}{\mu^\nu} \qquad [\operatorname{Re}\nu > 0] \qquad \text{ET I 149(12)}$$

2. $\displaystyle\int_0^\infty x^n e^{-\mu x}\left\{\left[\ln x - \tfrac{1}{2}\psi(n+1)\right]^2 - \tfrac{1}{2}\psi'(n+1)\right\}dx$

$$= \frac{n!}{\mu^{n+1}}\left\{\left[\ln\mu - \frac{1}{2}\psi(n+1)\right]^2 + \frac{1}{2}\psi'(n+1)\right\}$$

$$[\operatorname{Re}\mu > 0]\qquad\qquad\text{MI 26}$$

## 4.37 Combinations of logarithms and hyperbolic functions

**4.371**

1. $\displaystyle\int_0^\infty \frac{\ln x}{\cosh x}\,dx = \pi\ln\left[\frac{\sqrt{2\pi}\,\Gamma\left(\frac{3}{4}\right)}{\Gamma\left(\frac{1}{4}\right)}\right]$ 　　　　　　　　　　　　LI (260)(1)a

2. $\displaystyle\int_0^\infty \frac{\ln x\,dx}{\cosh x + \cos t} = \frac{\pi}{\sin t}\ln\frac{(2\pi)^{t/\pi}\,\Gamma\left(\dfrac{\pi+t}{2\pi}\right)}{\Gamma\left(\dfrac{\pi-t}{2\pi}\right)}$ 　　$\left[t^2 < \pi^2\right]$ 　　BI (257)(7)a

3. $\displaystyle\int_0^\infty \frac{\ln x\,dx}{\cosh^2 x} = \psi\left(\frac{1}{2}\right) + \ln\pi = \ln\pi - 2\ln 2 - C$ 　　　　　　BI (257)(4)a

4.* $\displaystyle\int_0^\infty \frac{\ln x}{\cosh^4 x}\,dx = -\frac{2}{3}(2\ln 2 + C - \ln\pi) - \frac{7}{3}\frac{\zeta(3)}{\pi^2}$ 　　　　　　GC

5.* $\displaystyle\int_0^\infty \frac{\ln x}{\cosh^6 x}\,dx = -\frac{8}{15}(2\ln 2 + C - \ln\pi) - \frac{1}{15}\left[35\frac{\zeta(3)}{\pi^2} + 93\frac{\zeta(5)}{\pi^4}\right]$ 　　GC

6.* $\displaystyle\int_0^\infty \frac{\ln x}{\cosh^8 x}\,dx = -\frac{16}{35}(2\ln 2 + C - \ln\pi) - \frac{1}{315}\left[686\frac{\zeta(3)}{\pi^2} + 2604\frac{\zeta(5)}{\pi^4} + 5715\frac{\zeta(7)}{\pi^6}\right]$ 　　GC

**4.372**

1. $\displaystyle\int_1^\infty \ln x\,\frac{\sinh mx}{\sinh nx}\,dx = \frac{\pi}{2n}\tan\frac{m\pi}{2n}\ln 2\pi + \frac{\pi}{n}\sum_{k=1}^{n-1}(-1)^{k-1}\sin\frac{km\pi}{n}\ln\frac{\Gamma\left(\frac{n+k}{2n}\right)}{\Gamma\left(\frac{k}{2n}\right)}$ 　　$[m+n\text{ is odd}]$

$$= \frac{\pi}{2n}\tan\frac{m\pi}{2n}\ln\pi + \frac{\pi}{n}\sum_{k=1}^{\lfloor\frac{n-1}{2}\rfloor}(-1)^{k-1}\sin\frac{km\pi}{n}\ln\frac{\Gamma\left(\frac{n-k}{n}\right)}{\Gamma\left(\frac{k}{n}\right)}\qquad [m+n\text{ is even}]$$

$$\text{BI (148)(3)a}$$

2. $\displaystyle\int_1^\infty \ln x\,\frac{\cosh mx}{\cosh nx}\,dx$

$$= \frac{\pi}{2n}\frac{\ln 2\pi}{\cos\frac{m\pi}{2n}} + \frac{\pi}{n}\sum_{k=1}^{n}(-1)^{k-1}\cos\frac{(2k-1)m\pi}{2n}\ln\frac{\Gamma\left(\frac{2n+2k-1}{4n}\right)}{\Gamma\left(\frac{2k-1}{4n}\right)}\qquad [m+n\text{ is odd}]$$

$$= \frac{\pi}{2n}\frac{\ln\pi}{\cos\frac{m\pi}{2n}} + \frac{\pi}{n}\sum_{k=1}^{\lfloor\frac{n-1}{2}\rfloor}(-1)^{k-1}\cos\frac{(2k-1)m\pi}{2n}\ln\frac{\Gamma\left(\frac{2n-2k+1}{2n}\right)}{\Gamma\left(\frac{2k-1}{2n}\right)}\qquad [m+n\text{ is even}]$$

$$\text{BI (148)(6)a}$$

**4.373**

1. $\displaystyle\int_0^\infty \frac{\ln\left(a^2 + x^2\right)}{\cosh bx}\,dx = \frac{\pi}{b}\left[2\ln\frac{2\,\Gamma\left(\frac{2ab+3\pi}{4\pi}\right)}{\Gamma\left(\frac{2ab+\pi}{4\pi}\right)} - \ln\frac{2b}{\pi}\right] \qquad \left[b > 0, \quad a > -\frac{\pi}{2b}\right].$ BI (258)(11)a

2. $\displaystyle\int_0^\infty \ln\left(1 + x^2\right)\frac{dx}{\cosh\frac{\pi x}{2}} = 2\ln\frac{4}{\pi}$ BI (258)(1)a

3. $\displaystyle\int_0^\infty \ln\left(a^2 + x^2\right)\frac{\sinh\left(\frac{2}{3}\pi x\right)}{\sinh\pi x}\,dx = 2\sin\frac{\pi}{3}\ln\frac{6\,\Gamma\left(\frac{a+4}{6}\right)\Gamma\left(\frac{a+5}{6}\right)}{\Gamma\left(\frac{a+1}{6}\right)\Gamma\left(\frac{a+2}{6}\right)}$
$[a > -1].$ BI (258)(12)

4. $\displaystyle\int_0^\infty \ln\left(1 + x^2\right)\frac{dx}{\sinh^2 ax} = \frac{2}{a}\left[\ln\frac{a}{\pi} + \frac{\pi}{2a} - \psi\left(\frac{\pi + a}{\pi}\right)\right]$
$[a > 0]$ BI (258)(5)

5. $\displaystyle\int_0^\infty \ln\left(1 + x^2\right)\frac{\cosh\left(\frac{\pi}{2}x\right)}{\sinh^2\left(\frac{\pi}{2}x\right)}\,dx = \frac{2\pi - 4}{\pi}$ BI (258)(3)

6. $\displaystyle\int_0^\infty \ln\left(1 + x^2\right)\frac{\cosh\left(\frac{\pi}{4}x\right)}{\sinh^2\left(\frac{\pi}{4}x\right)}\,dx = 4\sqrt{2} - \frac{16}{\pi} + \frac{8\sqrt{2}}{\pi}\ln\left(\sqrt{2} + 1\right)$ BI (258)(2)

**4.374**

1. $\displaystyle\int_0^\infty \ln\left(\cos^2 t + e^{-2x}\sin^2 t\right)\frac{dx}{\sinh x} = -2t^2$ BI (259)(10)a

2. $\displaystyle\int_0^\infty \ln\left(a + be^{-2x}\right)\frac{dx}{\cosh^2 x} = \frac{2}{(b - a)}\left[\frac{u + b}{2}\ln(a + b) - a\ln a - b\ln 2\right]$
$[a > 0, \quad a + b > 0]$ LI (259)(14)

**4.375**

1.[11] $\displaystyle\int_0^\infty \ln\cosh\frac{x}{2}\frac{dx}{\cosh x} = \boldsymbol{G} - \frac{\pi}{4}\ln 2$ BI (259)(11)

2. $\displaystyle\int_0^\infty \ln\coth x\,\frac{dx}{\cosh x} = \frac{\pi}{2}\ln 2$ BI (259)(16)

**4.376**

1. $\displaystyle\int_0^\infty \frac{\ln x}{\sqrt{x}\cosh x}\,dx = 2\sqrt{\pi}\sum_{k=0}^\infty \frac{(-1)^{k+1}}{\sqrt{2k+1}}\left\{\ln(2k+1) + 2\ln 2 + \boldsymbol{C}\right\}$ BI (147)(4)

2. $\displaystyle\int_0^\infty \ln x\frac{(\mu + 1)\cosh x - x\sinh x}{\cosh^2 x}x^\mu\,dx = 2\,\Gamma(\mu + 1)\sum_{k=0}^\infty \frac{(-1)^{k+1}}{(2k+1)^{\mu+1}}$
$[\operatorname{Re}\mu > -1]$ BI (356)(10)

3. $\displaystyle\int_0^\infty \ln x\frac{(n + 1)\cosh x - x\sinh x}{\cosh^2 x}x^n\,dx = \frac{(-1)^n}{2^n}\beta^{(n)}\left(\frac{1}{2}\right)$

4.
$$\int_0^\infty \ln 2x \frac{n \sinh 2ax - ax}{\sinh^2 ax} x^{2n-1}\, dx = -\frac{1}{n}\left(\frac{\pi}{a}\right)^{2n}|B_{2n}|$$

$$[n = 1, 2, \ldots]$$  BI (356)(9)a

5.
$$\int_0^\infty \ln x \frac{ax \cosh ax - (2n+1)\sinh ax}{\sinh^2 ax} x^{2n}\, dx = 2\frac{2^{2n+1}-1}{(2a)^{2n+1}}(2n)!\,\zeta(2n+1)$$  BI (356)(14)

6.
$$\int_0^\infty \ln x \frac{ax \cosh ax - 2n \sinh ax}{\sinh^2 ax} x^{2n-1}\, dx = \frac{2^{2n-1}-1}{2n}|B_{2n}|\left(\frac{\pi}{a}\right)^{2n}$$

$$[n = 1, 2, \ldots, \quad a > 0]$$  BI (356)(15)

7.
$$\int_0^\infty \ln \frac{(2n+1)\cosh ax - ax \sinh ax}{\cosh^2 ax} x^{2n}\, dx = -\left(\frac{\pi}{2a}\right)^{2n+1}|E_{2n}|$$

$$[a > 0]$$  BI (356)(11)

8.[6]
$$\int_0^\infty \ln x \frac{2ax \sinh ax - (2n+1)\cosh ax}{\cosh^3 ax} x^{2n}\, dx = \begin{cases} \dfrac{2}{a}(2^{2n-1}-1)\left(\dfrac{\pi}{2a}\right)^{2n}|B_{2n}| & n = 1, 2, \ldots \\ \dfrac{1}{a} & n = 0 \end{cases}$$

$$[a > 0]$$  BI (356)(2)

9.[6]
$$\int_0^\infty \ln x \frac{2ax \cosh ax - (2n+1)\sinh ax}{\sinh^3 ax} x^{2n}\, dx = \frac{1}{a}\left(\frac{\pi}{a}\right)^{2n}|B_{2n}|$$

$$[a > 0, \quad n = 1, 2, \ldots]$$  BI (356)(6)a

10.
$$\int_0^\infty \ln x \frac{x \sinh x - 6 \sinh^2\left(\frac{x}{2}\right) - 6\cos^2\frac{t}{2}}{(\cosh x + \cos t)^2} x^2\, dx = \frac{(\pi - t^2)t}{3\sin t}$$

$$[0 < t < \pi]$$  BI (356)(16)a

11.
$$\int_0^\infty \ln\left(1 + x^2\right) \frac{\cosh \pi x + \pi x \sinh \pi x}{\cosh^2 \pi x} \frac{dx}{x^2} = 4 - \pi$$  BI (356)(12)

12.
$$\int_0^\infty \ln\left(1 + 4x^2\right) \frac{\cosh \pi x + \pi x \sinh \pi x}{\cosh^2 \pi x} \frac{dx}{x^2} = 4\ln 2$$  BI (356)(13)

**4.377**
$$\int_0^\infty \ln 2x \frac{ax - n\left(1 - e^{-2ax}\right)}{\sinh^2 ax} x^{2n-1}\, dx = \frac{1}{2n}\left(\frac{\pi}{a}\right)^{2n}|B_{2n}|$$

$$[n = 1, 2, \ldots]$$  LI (356)(8)a

## 4.38–4.41 Logarithms and trigonometric functions

**4.381**

1.
$$\int_0^1 \ln x \sin ax\, dx = -\frac{1}{a}\left[C + \ln a - \operatorname{ci}(a)\right]$$  $[a > 0]$  GW (338)(2a)

2.
$$\int_0^1 \ln x \cos ax\, dx = -\frac{1}{a}\left[\operatorname{si}(a) + \frac{\pi}{2}\right]$$  $[a > 0]$  BI (284)(2)

3.
$$\int_0^{2\pi} \ln x \sin nx\, dx = -\frac{1}{n}\left[C + \ln(2n\pi) - \operatorname{ci}(2n\pi)\right]$$  GW (338)(1a)

4. $\displaystyle\int_0^{2\pi} \ln x \cos nx\, dx = -\frac{1}{n}\left[\operatorname{si}(2n\pi) + \frac{\pi}{2}\right]$      GW (338)(1b)

**4.382**

1.$^{12}$ $\displaystyle\int_0^\infty \ln\left|\frac{x+a}{x-a}\right| \sin bx\, dx = \frac{\pi}{b}\sin ab$      $[a>0,\quad b>0]$      ET I 77(11)

2.$^{10}$ $\displaystyle\int_0^\infty \ln\left|\frac{x+a}{x-a}\right| \cos bx\, dx = \frac{2}{b}\left[\cos(ab)\left\{\operatorname{si}(ab) + \frac{\pi}{2}\right\} - \sin(ab)\operatorname{ci}(ab)\right]$

$[a>0,\quad b>0]$      ET I 18(9)

3. $\displaystyle\int_0^\infty \ln\frac{a^2+x^2}{b^2+x^2}\cos cx\, dx = \frac{\pi}{c}\left(e^{-bc} - e^{-ac}\right)$      $[a>0,\quad b>0,\quad c>0]$

FI III 648a, BI (337)(5)

4. $\displaystyle\int_0^\infty \ln\frac{x^2+x+a^2}{x^2-x+a^2}\sin bx\, dx = \frac{2\pi}{b}\exp\left(-b\sqrt{a^2-\frac{1}{4}}\right)\sin\frac{b}{2}$

$[b>0]$      ET I 77(12)

5. $\displaystyle\int_0^\infty \ln\frac{(x+\beta)^2+\gamma^2}{(x-\beta)^2+\gamma^2}\sin bx\, dx = \frac{2\pi}{b}e^{-\gamma b}\sin\beta b$      $[\operatorname{Re}\gamma>0,\quad |\operatorname{Im}\beta|\le\operatorname{Re}\gamma,\quad b>0]$

ET I 77(13)

**4.383**

1. $\displaystyle\int_0^\infty \ln\left(1+e^{-\beta x}\right)\cos bx\, dx = \frac{\beta}{2b^2} - \frac{\pi}{2b\sinh\left(\dfrac{\pi b}{\beta}\right)}$      $[\operatorname{Re}\beta>0,\quad b>0]$      ET I 18(13)

2. $\displaystyle\int_0^\infty \ln\left(1-e^{-\beta x}\right)\cos bx\, dx = \frac{\beta}{2b^2} - \frac{\pi}{2b}\coth\left(\frac{\pi b}{\beta}\right)$      $[\operatorname{Re}\beta>0,\quad b>0]$      ET I 18(14)

**4.384**

1. $\displaystyle\int_0^1 \ln(\sin\pi x)\sin 2n\pi x\, dx = 0$      GW (338)(3a)

2.$^7$ $\displaystyle\int_0^1 \ln(\sin\pi x)\sin(2n+1)\pi x\, dx = 2\int_0^{1/2} \ln(\sin\pi x)\sin(2n+1)\pi x\, dx$

$$= \frac{2}{(2n+1)\pi}\left[\ln 2 - \frac{1}{2n+1} - 2\sum_{k=1}^n \frac{1}{2k-1}\right]$$

GW (338)(3b)

3.$^6$ $\displaystyle\int_0^1 \ln(\sin\pi x)\cos 2n\pi x\, dx = 2\int_0^{1/2} \ln(\sin\pi x)\cos 2n\pi x\, dx$

$$= -\ln 2 \qquad [n=0]$$

$$= -\frac{1}{2n} \qquad [n>0]$$

GW (338)(3c)

4.    $\displaystyle\int_0^1 \ln(\sin \pi x)\cos(2n+1)\pi x\, dx = 0$        GW (338)(3d)

5.    $\displaystyle\int_0^{\pi/2} \ln \sin x \sin x\, dx = \ln 2 - 1$        BI (305)(4)

6.    $\displaystyle\int_0^{\pi/2} \ln \sin x \cos x\, dx = -1$        BI (305)(5)

7.    $\displaystyle\int_0^{\pi/2} \ln \sin x \cos 2nx\, dx = \begin{cases} -\dfrac{\pi}{4n}, & \text{for } n > 0 \\ -\dfrac{\pi}{2}\ln 2, & \text{for } n = 0 \end{cases}$        LI (305)(6)

8.    $\displaystyle\int_0^{\pi} \ln \sin x \cos[2m(x-n)]\, dx = -\dfrac{\pi \cos 2mn}{2m}$        LI (330)(8)

9.    $\displaystyle\int_0^{\pi/2} \ln \sin x \sin^2 x\, dx = \dfrac{\pi}{8}(1 - \ln 4)$        BI (305)(7)

10.    $\displaystyle\int_0^{\pi/2} \ln \sin x \cos^2 x\, dx = -\dfrac{\pi}{8}(1 + \ln 4)$        BI (305)(8)

11.    $\displaystyle\int_0^{\pi/2} \ln \sin x \sin x \cos^2 x\, dx = \dfrac{1}{9}(\ln 8 - 4)$        BI (305)(9)

12.    $\displaystyle\int_0^{\pi/2} \ln \sin x \tan x\, dx = -\dfrac{\pi^2}{24}$        BI (305)(11)

13.    $\displaystyle\int_0^{\pi/2} \ln \sin 2x \sin x\, dx = \int_0^{\pi/2} \ln \sin 2x \cos x\, dx = 2(\ln 2 - 1)$        BI (305)(16, 17)

14.    $\displaystyle\int_0^{\pi} \dfrac{\ln(1 + p\cos x)}{\cos x}\, dx = \pi \arcsin p$    $\left[p^2 < 1\right]$        FI II 484

15.    $\displaystyle\int_0^{\pi} \ln \sin x \dfrac{dx}{1 - 2a\cos x + a^2} = \dfrac{\pi}{1 - a^2}\ln\dfrac{1 - a^2}{2}$    $\left[a^2 < 1\right]$

               $= \dfrac{\pi}{a^2 - 1}\ln\dfrac{a^2 - 1}{2a^2}$    $\left[a^2 > 1\right]$

       BI (331)(8)

16.    $\displaystyle\int_0^{\pi} \ln \sin bx \dfrac{dx}{1 - 2a\cos x + a^2} = \dfrac{\pi}{1 - a^2}\ln\dfrac{1 - a^{2b}}{2}$    $\left[a^2 < 1\right]$        BI (331)(10)

17.    $\displaystyle\int_0^{\pi} \ln \cos bx \dfrac{dx}{1 - 2a\cos x + a^2} = \dfrac{\pi}{1 - a^2}\ln\dfrac{1 + a^{2b}}{2}$    $\left[a^2 < 1\right]$        BI (331)(11)

18.    $\displaystyle\int_0^{\pi/2} \ln \sin x \dfrac{dx}{1 - 2a\cos 2x + a^2} = \dfrac{1}{2}\int_0^{\pi} \ln \sin x \dfrac{dx}{1 - 2a\cos 2x + a^2}$

               $= \dfrac{\pi}{2(1 - a^2)}\ln\dfrac{1 - a}{2}$    $\left[a^2 < 1\right]$

               $= \dfrac{\pi}{2(a^2 - 1)}\ln\dfrac{a - 1}{2a}$    $\left[a^2 > 1\right]$

       BI (321)(1), BI (331)(13)

19. $\displaystyle\int_0^\pi \ln\sin bx\, \frac{dx}{1-2a\cos 2x+a^2} = \frac{\pi}{1-a^2}\ln\frac{1-a^b}{2}$ $\qquad [a^2<1]$ $\qquad$ BI (331)(18)

20. $\displaystyle\int_0^\pi \ln\cos bx\, \frac{dx}{1-2a\cos 2x+a^2} = \frac{\pi}{1-a^2}\ln\frac{1+a^b}{2}$ $\qquad [a^2<1]$ $\qquad$ BI (331)(21)

21. $\displaystyle\int_0^{\pi/2} \frac{\ln\cos x\, dx}{1-2p\cos 2x+p^2} = \frac{\pi}{2\,(1-p^2)}\ln\frac{1+p}{2}$ $\qquad [p^2<1]$

$\displaystyle\qquad\qquad\qquad\qquad = \frac{\pi}{2\,(p^2-1)}\ln\frac{p+1}{2p}$ $\qquad [p^2>1]$

$\qquad\qquad\qquad\qquad\qquad\qquad\qquad\qquad\qquad$ BI (321)(8)

22. $\displaystyle\int_0^\pi \ln\sin x\, \frac{\cos x\, dx}{1-2a\cos x+a^2} = \frac{\pi}{2a}\frac{1+a^2}{1-a^2}\ln\left(1-a^2\right) - \frac{a\pi\ln 2}{1-a^2}$ $\qquad [a^2<1]$

$\displaystyle\qquad\qquad\qquad\qquad\qquad = \frac{\pi}{2a}\frac{a^2+1}{a^2-1}\ln\frac{a^2-1}{a^2} - \frac{\pi\ln 2}{a\,(a^2-1)}$ $\qquad [a^2>1]$

$\qquad\qquad\qquad\qquad\qquad\qquad\qquad\qquad\qquad$ LI (331)(9)

23. $\displaystyle\int_0^\pi \ln\sin bx\, \frac{\cos x\, dx}{1-2a\cos 2x+a^2} = \int_0^\pi \ln\cos bx\, \frac{\cos x\, dx}{1-2a\cos 2x+a^2} = 0$

$\qquad\qquad\qquad\qquad\qquad\qquad [0<a<1]$ $\qquad$ BI (331)(19, 22)

24. $\displaystyle\int_0^\pi \ln\sin x\, \frac{\cos^2 x\, dx}{1-2a\cos 2x+a^2} = \frac{\pi}{4a}\frac{1+a}{1-a}\ln(1-a) - \frac{\pi\ln 2}{2(1-a)}$ $\qquad [0<a<1]$

$\displaystyle\qquad\qquad\qquad\qquad\qquad = \frac{\pi}{4a}\frac{a+1}{a-1}\ln\frac{a-1}{a} - \frac{\pi\ln 2}{2a(a-1)}$ $\qquad [a>1]$

$\qquad\qquad\qquad\qquad\qquad\qquad\qquad\qquad\qquad$ BI (331)(16)

25. $\displaystyle\int_0^{\pi/2} \ln\sin x\, \frac{\cos 2x\, dx}{1-2a\cos 2x+a^2} = \frac{1}{2}\int_0^\pi \ln\sin x\, \frac{\cos 2x\, dx}{1-2a\cos 2x+a^2}$

$\displaystyle\qquad\qquad = \frac{\pi}{2a\,(1-a^2)}\left[\frac{1+a^2}{2}\ln(1-a) - a^2\ln 2\right]$ $\qquad [a^2<1]$

$\displaystyle\qquad\qquad = \frac{\pi}{2a\,(a^2-1)}\left[\frac{1+a^2}{2}\ln\frac{a-1}{a} - \ln 2\right]$ $\qquad [a^2>1]$

$\qquad\qquad\qquad\qquad$ BI (321)(2), BI (331)(15), LI (321))(2)

26. $\displaystyle\int_0^{\pi/2} \ln\cos x\, \frac{\cos 2x\, dx}{1-2a\cos 2x+a^2} = \frac{\pi}{2a\,(1-a^2)}\left[\frac{1+a^2}{2}\ln(1+a) - a^2\ln 2\right]$ $\qquad [a^2<1]$

$\displaystyle\qquad\qquad = \frac{\pi}{2a\,(a^2-1)}\left[\frac{1+a^2}{2}\ln\frac{1+a}{a} - \ln 2\right]$ $\qquad [a^2>1]$

$\qquad\qquad\qquad\qquad\qquad\qquad\qquad\qquad\qquad$ BI (321)(9)

**4.385**

1.[12] $\displaystyle\int_0^\pi \ln\sin x\, \frac{dx}{a+b\cos x} = \frac{\pi}{2\sqrt{a^2-b^2}}\ln\frac{\sqrt{a^2-b^2}}{a+\sqrt{a^2-b^2}}$ $\qquad [a>0,\quad a>b]$ $\qquad$ BI (331)(6)

2. $\displaystyle\int_0^{\pi/2} \ln\sin x\, \frac{dx}{(a\sin x\pm b\cos x)^2} = \int_0^{\pi/2} \ln\cos x\, \frac{dx}{(a\cos x\pm b\sin x)^2}$

$\displaystyle\qquad\qquad = \frac{1}{b\,(a^2+b^2)}\left(\mp a\ln\frac{a}{b} - \frac{b\pi}{2}\right)$

$\qquad\qquad\qquad\qquad [a>0,\quad b>0]$ $\qquad$ BI (319)(1,6)a

3.  $$\int_0^{\pi/2} \frac{\ln \sin x \, dx}{a^2 \sin^2 x + b^2 \cos^2 x} = \int_0^{\pi/2} \frac{\ln \cos x \, dx}{b^2 \sin^2 x + a^2 \cos^2 x} = \frac{\pi}{2ab} \ln \frac{b}{a+b}$$
$$[a > 0, \quad b > 0] \qquad \text{BI (317)(4, 10)}$$

4.  $$\int_0^{\pi/2} \ln \sin x \frac{\sin 2x \, dx}{\left(a \sin^2 x + b \cos^2 x\right)^2} = \int_0^{\pi/2} \ln \cos x \frac{\sin 2x \, dx}{\left(b \sin^2 x + a \cos^2 x\right)^2}$$
$$= \frac{1}{2b(b-a)} \ln \frac{a}{b}$$
$$[a > 0, \quad b > 0] \qquad \text{BI (319)(3, 7), LI (319)(3)}$$

5.  $$\int_0^{\pi/2} \ln \sin x \frac{a^2 \sin^2 x - b^2 \cos^2 x}{\left(a^2 \sin^2 x + b^2 \cos^2 x\right)^2} \, dx = \int_0^{\pi/2} \ln \cos x \frac{a^2 \cos^2 x - b^2 \sin^2 x}{\left(a^2 \cos^2 x + b^2 \sin^2 x\right)^2} \, dx$$
$$= \frac{\pi}{2b(a+b)}$$
$$[a > 0, \quad b > 0] \qquad \text{LI (319)(2, 8)}$$

## 4.386

1.  $$\int_0^{\pi/2} \ln \sin x \frac{\sin x}{\sqrt{1 + \sin^2 x}} \, dx = \int_0^{\pi/2} \frac{\cos x \ln \cos x}{\sqrt{1 + \cos^2 x}} \, dx = -\frac{\pi}{8} \ln 2 \qquad \text{BI (322)Zsurround1, 6}$$

2.  $$\int_0^{\pi/2} \frac{\sin^3 x \ln \sin x}{\sqrt{1 + \sin^2 x}} \, dx = \int_0^{\pi/2} \frac{\cos^3 x \ln \cos x}{\sqrt{1 + \cos^2 x}} \, dx = \frac{\ln 2 - 1}{4} \qquad \text{BI (322)(2, 7)}$$

3.  $$\int_0^{\pi/2} \ln \sin x \frac{dx}{\sqrt{1 - k^2 \sin^2 x}} = -\frac{1}{2} \, K(k) \ln k - \frac{\pi}{4} \, K(k') \qquad \text{BI (322)(3)}$$

4.  $$\int_0^{\pi/2} \frac{\ln \cos x \, dx}{\sqrt{1 - k^2 \sin^2 x}} = \frac{1}{2} \, K(k) \ln \frac{k'}{k} - \frac{\pi}{4} \, K(k') \qquad \text{BI (322)(9)}$$

## 4.387

1.  $$\int_0^{\pi/2} \ln \sin x \sin^\mu x \cos^\nu x \, dx = \int_0^{\pi/2} \ln \cos x \cos^\mu x \sin^\nu x \, dx$$
$$= \frac{1}{4} \, \mathrm{B} \left( \frac{\mu+1}{2}, \frac{\nu+1}{2} \right) \left[ \psi \left( \frac{\mu+1}{2} \right) - \psi \left( \frac{\mu+\nu+2}{2} \right) \right]$$
$$[\operatorname{Re} \mu > -1, \quad \operatorname{Re} \nu > -1] \qquad \text{GW (338)(6c)}$$

2.  $$\int_0^{\pi/2} \ln \sin x \sin^{\mu-1} x \, dx = \frac{\sqrt{\pi} \, \Gamma \left( \frac{\mu}{2} \right)}{4 \Gamma \left( \frac{\mu+1}{2} \right)} \left[ \psi \left( \frac{\mu}{2} \right) - \psi \left( \frac{\mu+1}{2} \right) \right]$$
$$[\operatorname{Re} \mu > 0] \qquad \text{GW (338)(6a)}$$

3.  $$\int_0^{\pi/2} \ln \sin x \cos^{\nu-1} x \, dx = \frac{\sqrt{\pi} \, \Gamma \left( \frac{\nu}{2} \right)}{4 \Gamma \left( \frac{\nu+1}{2} \right)} \left[ \psi \left( \frac{\nu}{2} \right) - \psi \left( \frac{\nu+1}{2} \right) \right]$$
$$[\operatorname{Re} \nu > 0] \qquad \text{GW (338)(6b)}$$

4.  $\displaystyle\int_0^{\pi/2} \ln\sin x \sin^{2n} x\, dx = \frac{(2n-1)!!}{(2n)!!}\frac{\pi}{2}\left[\sum_{k=1}^{2n}\frac{(-1)^{k+1}}{k} - \ln 2\right]$

    FI II 811

5.  $\displaystyle\int_0^{\pi/2} \ln\sin x \sin^{2n+1} x\, dx = \frac{(2n)!!}{(2n+1)!!}\left[\sum_{k=1}^{2n+1}\frac{(-1)^k}{k} + \ln 2\right]$

    BI (305)(13)

6.  $\displaystyle\int_0^{\pi/2} \ln\sin x \cos^{2n} x\, dx = -\frac{(2n-1)!!}{(2n)!!}\frac{\pi}{4}\left[\sum_{k=1}^{n}\frac{1}{k} + \ln 4\right]$

    $\displaystyle\qquad\qquad = -\frac{(2n-1)!!}{(2n)!!}\frac{\pi}{4}\left[C + \psi(n+1) + \ln 4\right]$

    BI (305)(14)

7.  $\displaystyle\int_0^{\pi/2} \ln\sin x \cos^{2n+1} x\, dx = -\frac{(2n)!!}{(2n+1)!!}\sum_{k=0}^{n}\frac{1}{2k+1}$

    $\displaystyle\qquad\qquad = -\frac{(2n)!!}{2(2n+1)!!}\left[\psi\left(n+\frac{3}{2}\right) - \psi\left(\frac{1}{2}\right)\right]$

    GW (338)(7b)

8.  $\displaystyle\int_0^{\pi/2} \ln\cos x \sin^{2n} x\, dx = -\frac{(2n-1)!!}{2^{n+1}\cdot n!}\frac{\pi}{2}\left\{C + 2\ln 2 + \psi(n+1)\right\}$

    BI (306)(8)

9.  $\displaystyle\int_0^{\pi/2} \ln\cos x \cos^{2n} x\, dx = -\frac{(2n-1)!!}{2^n n!}\frac{\pi}{2}\left(\ln 2 + \sum_{k=1}^{2n}\frac{(-1)^k}{k}\right)$

    BI (306)(10)

10.[12]  $\displaystyle\int_0^{\pi/2} \ln\cos x \cos^{2n-1} x\, dx = \frac{2^{n-1}(n-1)!}{(2n-1)!!}\left[\ln 2 + \sum_{k=1}^{2n-1}\frac{(-1)^k}{k}\right]$

    BI (306)(9)

**4.388**

1.  $\displaystyle\int_0^{\pi/4} \ln\sin x \frac{\sin^{2n} x}{\cos^{2n+2} x}\, dx = \frac{1}{2n+1}\left[\frac{1}{2}\ln 2 + (-1)^n\frac{\pi}{4} + \sum_{k=0}^{n-1}\frac{(-1)^k}{2n-2k-1}\right]$

    BI (288)(1)

2.  $\displaystyle\int_0^{\pi/4} \ln\sin x \frac{\sin^{2n-1} x}{\cos^{2n+1} x}\, dx = \frac{1}{4n}\left[-\ln 2 + (-1)^n\ln 2 + \sum_{k=1}^{n-1}\frac{(-1)^k}{n-k}\right]$

    LI (288)(2)

3.  $\displaystyle\int_0^{\pi/4} \ln\cos x \frac{\sin^{2n} x}{\cos^{2n+2} x}\, dx = \frac{1}{2n+1}\left[-\frac{1}{2}\ln 2 + (-1)^{n+1}\frac{\pi}{4} + \sum_{k=0}^{n}\frac{(-1)^{k-1}}{2n-2k+1}\right]$

    BI (288)(10)

4.  $\displaystyle\int_0^{\pi/4} \ln\cos x \frac{\sin^{2n-1} x}{\cos^{2n+1} x}\, dx = \frac{1}{4n}\left[-\ln 2 + (-1)^n\ln 2 + \sum_{k=0}^{n-1}\frac{(-1)^k}{n-k}\right]$

    BI (288)(11)

5.  $\displaystyle\int_0^{\pi/2} \ln\sin x \frac{\sin^{p-1} x}{\cos^{p+1} x}\, dx = -\frac{\pi}{2p}\operatorname{cosec}\frac{p\pi}{2}$  $\qquad [0 < p < 2]$

    BI (310)(4)

6. $\int_0^{\pi/2} \ln \sin x \dfrac{dx}{\tan^{p-1} x \sin 2x} = \dfrac{1}{4} \dfrac{\pi}{p-1} \sec \dfrac{p\pi}{2}$      $[p^2 < 1]$      BI (310)(3)

**4.389**

1. $\int_0^{\pi} \ln \sin x \sin^{2n} 2x \cos 2x \, dx = -\dfrac{(2n-1)!!}{(2n)!!} \dfrac{\pi}{4n+2}$      BI (330)(9)

2. $\int_0^{\pi/4} \ln \sin x \cos^n 2x \sin 2x \, dx = -\dfrac{1}{4(n+1)} \left\{ \boldsymbol{C} + \psi(n+2) + \ln 2 \right\}$      BI (285)(2)

3. $\int_0^{\pi/4} \ln \cos x \cos^{\mu-1} 2x \tan 2x \, dx = \dfrac{1}{4(1-\mu)} \beta(\mu)$

                         $[\operatorname{Re} \mu > 0]$      BI (286)(2)

4. $\int_0^{\pi/2} \ln \sin x \sin^{\mu-1} x \cos x \, dx = \int_0^{\pi/2} \ln \cos x \cos^{\mu-1} x \sin x \, dx = -\dfrac{1}{\mu^2}$

                         $[\operatorname{Re} \mu > 0]$      BI (306)(11)

5.[3] $\int_{-\frac{\pi}{2}}^{\frac{\pi}{2}} \ln \cos x \cos^p x \cos px \, dx = \dfrac{\pi}{2^{p+1}} \left[ \boldsymbol{C} + \psi(p+1) - 2 \ln 2 \right]$

                         $[p > -1]$

6. $\int_0^{\pi/2} \ln \cos x \cos^{p-1} x \sin px \sin x \, dx = \dfrac{\pi}{2^{p+2}} \left[ \boldsymbol{C} + \psi(p) - \dfrac{1}{p} - 2 \ln 2 \right]$

                         $[p > 0]$      BI (306)(12)

**4.391**

1. $\int_0^{\pi/4} (\ln \cos 2x)^n \cos^{p-1} 2x \tan x \, dx = \int_0^{\pi/4} (\ln \sin 2x)^n \sin^{p-1} 2x \tan \left( \dfrac{\pi}{4} - x \right) dx = \dfrac{1}{2} \beta^{(n)}(p)$

                         $[p > 0]$      BI (286)(10), BI (285)(18)

2. $\int_0^{\pi/4} (\ln \sin 2x)^n \sin^{p-1} 2x \tan \left( \dfrac{\pi}{4} + x \right) dx = \dfrac{(-1)^n n!}{2} \zeta(n+1, p)$      BI (285)(17)

3. $\int_0^{\pi/4} (\ln \cos 2x)^{2n-1} \tan x \, dx = \dfrac{1 - 2^{2n-1}}{4n} \pi^{2n} |B_{2n}|$      $[n = 1, 2, \ldots]$      BI (286)Zsurround7

4.[12] $\int_0^{\pi/4} (\ln \cos 2x)^{2n} \tan x \, dx = \dfrac{2^{2n-1}}{2^{2n+1}} (2n)! \, \zeta(2n+1)$      BI (286)(8)

**4.392**

1. $\int_0^{\pi/4} \ln(\sin x \cos x) \dfrac{\sin^{2n} x}{\cos^{2n+2} x} dx = \dfrac{1}{2n+1} \left[ (-1)^{n+1} \dfrac{\pi}{2} - \ln 2 + \dfrac{1}{2n+1} + 2 \sum_{k=0}^{n-1} \dfrac{(-1)^{k-1}}{2n-2k-1} \right]$

                         BI (294)(8)

2. $\int_0^{\pi/4} \ln(\sin x \cos x) \dfrac{\sin^{2n-1} x}{\cos^{2n+1} x} dx = \dfrac{1}{2n} \left[ (-1)^n \ln 2 - \ln 2 + \dfrac{1}{2n} + (-1)^n \sum_{k=1}^{n-1} \dfrac{(-1)^k}{k} \right]$

                         BI (294)(9)

**4.393**

1.    $\int_0^{\pi/2} \ln \tan x \sin x \, dx = \ln 2$                  BI (307)(3)

2.    $\int_0^{\pi/2} \ln \tan x \cos x \, dx = -\ln 2$             BI (307)(4)

3.    $\int_0^{\pi/2} \ln \tan x \sin^2 x \, dx = -\int_0^{\pi/2} \ln \tan x \cos^2 x \, dx = \dfrac{\pi}{4}$      BI (307)(5, 6)

4.    $\int_0^{\pi/4} \dfrac{\ln \tan x}{\cos 2x} \, dx = -\dfrac{\pi^2}{8}$            GW (338)(10b)a

5.    $\int_0^{\pi/2} \sin x \ln \cot \dfrac{x}{2} \, dx = \ln 2$           LO III 290

**4.394**

1.    $\int_0^{\pi/2} \dfrac{\ln \tan x \, dx}{1 - 2a \cos 2x + a^2} = \dfrac{\pi}{2(1 - a^2)} \ln \dfrac{1 - a}{1 + a}$      $\left[a^2 < 1\right]$

$\qquad\qquad\qquad\qquad\qquad\qquad = \dfrac{\pi}{2(a^2 - 1)} \ln \dfrac{a - 1}{a + 1}$      $\left[a^2 > 1\right]$

                                                             BI (321)(15)

2.    $\int_0^{\pi/2} \dfrac{\ln \tan x \cos 2x \, dx}{1 - 2a \cos 2x + a^2} = \dfrac{\pi}{4a} \dfrac{1 + a^2}{1 - a^2} \ln \dfrac{1 - a}{1 + a}$      $\left[a^2 < 1\right]$

$\qquad\qquad\qquad\qquad\qquad\qquad\qquad = \dfrac{\pi}{4a} \dfrac{a^2 + 1}{a^2 - 1} \ln \dfrac{a - 1}{a + 1}$      $\left[a^2 > 1\right]$

                                                             BI (321)(16)

3.    $\int_0^{\pi} \dfrac{\ln \tan bx \, dx}{1 - 2a \cos 2x + a^2} = \dfrac{\pi}{1 - a^2} \ln \dfrac{1 - a^b}{1 + a^b}$      $[0 < a < 1, \quad b > 0]$      BI (331)(24)

4.    $\int_0^{\pi} \dfrac{\ln \tan bx \cos x \, dx}{1 - 2a \cos 2x + a^2} = 0$      $[0 < a < 1]$      BI (331)(25)

5.    $\int_0^{\pi/4} \ln \tan x \dfrac{\cos 2x \, dx}{1 - a \sin 2x} = -\dfrac{\arcsin a}{4a} (\pi + \arcsin a)$      $\left[a^2 \leq 1\right]$      BI (291)(2,3)

6.    $\int_0^{\pi/4} \ln \tan x \dfrac{\cos 2x \, dx}{1 - a^2 \sin^2 2x} = -\dfrac{\pi}{4a} \arcsin a$      $\left[a^2 < 1\right]$      BI (291)(9)

7.¹²    $\int_0^{\pi/4} \ln \tan x \dfrac{\cos 2x \, dx}{1 + a^2 \sin^2 x} = -\dfrac{\pi}{4a} \operatorname{arcsinh} a = -\dfrac{\pi}{4a} \ln\left(a + \sqrt{1 + a^2}\right)$

                                                          $\left[a^2 < 1\right]$      BI (291)(10)

8.    $\int_0^{u} \dfrac{\sin x \ln \cot \dfrac{x}{2}}{1 - \cos^2 \alpha \sin^2 x} \, dx = \operatorname{cosec} 2\alpha \left\{ \dfrac{\pi}{2} \ln 2 + L(\varphi - \alpha) - L(\varphi + \alpha) - L\left(\dfrac{\pi}{2} - 2\alpha\right) \right\}$

                                         $[\tan \varphi = \cot \alpha \cos u; \quad 0 < u < \pi]$

                                                             LO III 290

9. $\displaystyle\int_0^{\pi/4} \frac{\ln\tan x \sin 2x\, dx}{1-\cos^2 t \sin^2 2x} = \operatorname{cosec} 2t\left[L\left(\frac{\pi}{2}-t\right)-\left(\frac{\pi}{2}-t\right)\ln 2\right]$      LO III 290a

**4.395**

1.[12] $\displaystyle\int_0^{\pi/2} \frac{\ln\tan x\, dx}{\sqrt{1-k^2\sin^2 x}} = -\frac{\ln k'}{2}\, K(k)$      BI (322)(11)

2. $\displaystyle\int_u^{\pi/4} \frac{\ln\tan x \sin 4x\, dx}{\left(\sin^2 u + \tan^2 v \sin^2 2x\right)\sqrt{\sin^2 2x - \sin^2 u}} = -\frac{\pi}{2}\frac{\cos^2 v}{\sin u \sin v}\ln\frac{\sin v + \sqrt{1-\cos^2 u\cos^2 v}}{\sin u\,(1+\sin v)}$

$$\left[0 < u < \frac{\pi}{2}, \quad 0 < v < \frac{\pi}{2}\right]$$      LO III 285a

**4.396**

1. $\displaystyle\int_0^{\pi/2} \ln\left(a\tan x\right)\sin^{\mu-1} 2x\, dx = 2^{\mu-2}\ln a\,\frac{\left\{\Gamma\left(\frac{a}{2}\right)\right\}^2}{\Gamma(a)}$    $[a>0, \quad \operatorname{Re}\mu>0]$      LI (307)(8)

2. $\displaystyle\int_0^{\pi/2} \ln\tan x \cos^{2(\mu-1)x}\, dx = -\frac{\sqrt{\pi}}{4}\frac{\Gamma\left(u-\frac{1}{2}\right)}{\Gamma(\mu)}\left[C+\psi\left(\frac{2\mu-1}{2}\right)+\ln 4\right]$

$$\left[\operatorname{Re}\mu > \tfrac{1}{2}\right]$$      BI (307)(9)

3. $\displaystyle\int_0^{\pi/2} \ln\tan x \cos^{q-1} x \cot x \sin[(q+1)x]\, dx = -\frac{\pi}{2}\left[C+\psi(q+1)\right]$

$$[q>-1]$$      BI (307)(11)

4. $\displaystyle\int_0^{\pi/2} \ln\tan x \cos^{q-1} x \cos[(q+1)x]\, dx = -\frac{\pi}{2q}$    $[q>0]$      BI (307)(10)

5. $\displaystyle\int_0^{\pi/4} \left(\ln\tan x\right)^n \tan^p x\, dx = \frac{1}{2^{n+1}}\beta^{(n)}\left(\frac{p+1}{2}\right)$    $[p>-1]$      LI (286)(22)

6. $\displaystyle\int_0^{\pi/2} \left(\ln\tan x\right)^{2n-1}\frac{dx}{\cos 2x} = \frac{1-2^{2n}}{2n}\pi^{2n}|B_{2n}|$    $[n=1,2,\ldots]$      BI (312)(6)

7. $\displaystyle\int_0^{\pi/4} \ln\tan x \tan^{2n+1} x\, dx = \frac{(-1)^{n+1}}{4}\left[\frac{\pi^2}{12}+\sum_{k=1}^{n}\frac{(-1)^k}{k^2}\right]$      GW (338)(8a)

**4.397**

1.[12] $\displaystyle\int_0^{\pi/2} \ln\left(1+p\sin x\right)\frac{dx}{\sin x} = \frac{\pi^2}{8}-\frac{1}{2}\left(\arccos p\right)^2$    $[p^2<1]$      BI (313)(1)

2. $\displaystyle\int_0^{\pi/2} \ln\left(1+p\cos x\right)\frac{dx}{\cos x} = \frac{\pi^2}{8}-\frac{1}{2}\left(\arccos p\right)^2$    $[p^2<1]$      BI (313)(8)

3. $\displaystyle\int_0^{\pi} \ln\left(1+p\cos x\right)\frac{dx}{\cos x} = \pi\arcsin p$    $[p^2<1]$      BI (331)(1)

4. $\displaystyle\int_0^{\pi/2} \frac{\cos x \ln\left(1+\cos\alpha\cos x\right)}{1-\cos^2\alpha\cos^2 x}\, dx = \frac{L\left(\frac{\pi}{2}-\alpha\right)-\alpha\ln\sin\alpha}{\sin\alpha\cos\alpha}$

$$\left[0 < \alpha < \frac{\pi}{2}\right]$$      LO III 291

5.    $\displaystyle\int_0^{\pi/2} \frac{\cos x \ln\left(1 - \cos\alpha\cos x\right)}{1 - \cos^2\alpha\cos^2 x}\, dx = \frac{L\left(\frac{\pi}{2} - \alpha\right) + (\pi - \alpha)\ln\sin\alpha}{\sin\alpha\cos\alpha}$

$$\left[0 < \alpha < \tfrac{\pi}{2}\right] \qquad\qquad \text{LO III 291}$$

6.[12]    $\displaystyle\int_0^\pi \ln\left(1 - 2a\cos x + a^2\right)\cos nx\, dx$

$$= \frac{1}{2}\int_0^{2\pi} \ln\left(1 - 2a\cos x + a^2\right)\cos nx\, dx$$

$$= -\frac{\pi}{n}a^n \qquad\qquad \left[a^2 < 1, \quad n = 1,2,3\ldots\right] \quad \text{BI (330)(11), BI (332)(5)}$$

$$= -\frac{\pi}{na^n} \qquad\qquad \left[a^2 > 1, \quad n = 1,2,3\ldots\right] \qquad\qquad \text{GW (338)(13a)}$$

7.[12]    $\displaystyle\int_0^\pi \ln\left(1 - 2a\cos x + a^2\right)\sin nx\sin x\, dx = \frac{1}{2}\int_0^{2\pi} \ln\left(1 - 2a\cos x + a^2\right)\sin nx\sin x\, dx$

$$= \frac{\pi}{2}\left(\frac{a^{n+1}}{n+1} - \frac{a^{n-1}}{n-1}\right)$$

$$\left[a^2 < 1\right] \qquad\qquad \text{BI (330)(10), BI (332)(4)}$$

8.[12]    $\displaystyle\int_0^\pi \ln\left(1 - 2a\cos x + a^2\right)\cos nx\cos x\, dx = \frac{1}{2}\int_0^{2\pi} \ln\left(1 - 2a\cos x + a^2\right)\cos nx\cos x\, dx$

$$= -\frac{\pi}{2}\left(\frac{a^{n+1}}{n+1} + \frac{a^{n-1}}{n-1}\right)$$

$$\text{BI (330)(12), BI (332)(6)}$$

9.    $\displaystyle\int_0^\pi \ln\left(1 - 2a\cos 2x + a^2\right)\cos(2n - 1)x\, dx = 0 \qquad \left[a^2 < 1\right] \qquad\qquad \text{BI (330)(15)}$

10.    $\displaystyle\int_0^\pi \ln\left(1 - 2a\cos 2x + a^2\right)\sin 2nx\sin x\, dx = 0 \qquad \left[a^2 < 1\right] \qquad\qquad \text{BI (330)(13)}$

11.    $\displaystyle\int_0^\pi \ln\left(1 - 2a\cos 2x + a^2\right)\sin(2n - 1)x\sin x\, dx = \frac{\pi}{2}\left(\frac{a^n}{n} - \frac{a^{n-1}}{n-1}\right)$

$$\left[a^2 < 1\right] \qquad\qquad \text{BI (330)(14)}$$

12.    $\displaystyle\int_0^\pi \ln\left(1 - 2a\cos 2x + a^2\right)\cos 2nx\cos x\, dx = 0 \qquad \left[a^2 < 1\right] \qquad\qquad \text{BI (330)(16)}$

13.    $\displaystyle\int_0^\pi \ln\left(1 - 2a\cos 2x + a^2\right)\cos(2n - 1)x\cos x\, dx = -\frac{\pi}{2}\left(\frac{a^n}{n} + \frac{a^{n-1}}{n-1}\right)$

$$\left[a^2 < 1\right] \qquad\qquad \text{BI (330)(17)}$$

14.    $\displaystyle\int_0^{\pi/2} \ln\left(1 + 2a\cos 2x + a^2\right)\sin^2 x\, dx = -\frac{a\pi}{4} \qquad\qquad \left[a^2 < 1\right]$

$$= \frac{\pi(\ln a)^2}{4} - \frac{\pi}{4a} \qquad \left[a^2 > 1\right]$$

$$\text{BI (309)(22), LI (309)(22)}$$

15. $\displaystyle\int_0^{\pi/2} \ln\left(1 + 2a\cos 2x + a^2\right)\cos^2 x\,dx = \frac{a\pi}{4}$ $\qquad\left[a^2 < 1\right]$

$\qquad\qquad\qquad\qquad\qquad\qquad\qquad = \frac{\pi(\ln a)^2}{4} + \frac{\pi}{4a}$ $\quad\left[a^2 > 1\right]$

$\qquad\qquad\qquad\qquad\qquad\qquad\qquad\qquad\qquad\qquad$ BI (309)(23), LI (309)(23)

16. $\displaystyle\int_0^{\pi} \frac{\ln\left(1 - 2a\cos x + a^2\right)}{1 - 2b\cos x + b^2}\,dx = \frac{2\pi\ln(1 - ab)}{1 - b^2}$ $\qquad\left[a^2 \le 1, \quad b^2 < 1\right]$ $\qquad$ BI (331)(26)

**4.398**

1. $\displaystyle\int_0^{\pi} \ln\frac{1 + 2a\cos x + a^2}{1 - 2a\cos x + a^2}\sin(2n+1)x\,dx = (-1)^n\frac{2\pi a^{2n+1}}{2n+1}$

$\qquad\qquad\qquad\qquad\qquad\qquad\qquad\qquad\qquad\left[a^2 < 1\right]$ $\qquad$ BI (330)(18)

2. $\displaystyle\int_0^{2\pi} \ln\frac{1 - 2a\cos x + a^2}{1 - 2a\cos nx + a^2}\cos mx\,dx = 2\pi\left(\frac{n}{m}a^{m/n} - \frac{a^m}{m}\right)$ $\qquad\left[a^2 \le 1\right]$

$\qquad\qquad\qquad\qquad\qquad\qquad\qquad\qquad = 2\pi\left(\frac{n}{m}a^{-m/n} - \frac{a^{-m}}{m}\right)$ $\quad\left[a^2 \ge 1\right]$

$\qquad\qquad\qquad\qquad\qquad\qquad\qquad\qquad\qquad\qquad\qquad$ BI (332)(9)

3. $\displaystyle\int_0^{\pi} \ln\frac{1 + 2a\cos 2x + a^2}{1 + 2a\cos 2nx + a^2}\cot x\,dx = 0$ $\qquad\qquad$ BI (331)(5), LI(331)(5)

**4.399**

1. $\displaystyle\int_0^{\pi/2} \ln\left(1 + a\sin^2 x\right)\sin^2 x\,dx = \frac{\pi}{2}\left(\ln\frac{1 + \sqrt{1+a}}{2} - \frac{1}{2}\frac{1 - \sqrt{1+a}}{1 + \sqrt{1+a}}\right)$

$\qquad\qquad\qquad\qquad\qquad\qquad\qquad\qquad\qquad\left[a > -1\right]$ $\qquad$ BI (309)(14)

2. $\displaystyle\int_0^{\pi/2} \ln\left(1 + a\sin^2 x\right)\cos^2 x\,dx = \frac{\pi}{2}\left(\ln\frac{1 + \sqrt{1+a}}{2} + \frac{1}{2}\frac{1 - \sqrt{1+a}}{1 + \sqrt{1+a}}\right)$

$\qquad\qquad\qquad\qquad\qquad\qquad\qquad\qquad\qquad\left[a > -1\right]$ $\qquad$ BI (309)(15)

3. $\displaystyle\int_0^{\pi/2} \frac{\ln\left(1 - \cos^2\beta\cos^2 x\right)}{1 - \cos^2\alpha\cos^2 x}\,dx = -\frac{\pi}{\sin\alpha}\ln\frac{1 + \sin\alpha}{\sin\alpha + \sin\beta}$

$\qquad\qquad\qquad\qquad\qquad\qquad\left[0 < \beta < \frac{\pi}{2}, \quad 0 < \alpha < \frac{\pi}{2}\right]$ $\qquad$ LO III 285

**4.411**

1. $\displaystyle\int_0^{\pi} \ln\frac{1 + \sin x}{1 + \cos\lambda\sin x}\frac{dx}{\sin x} = \lambda^2$ $\qquad\left[\lambda^2 < \pi^2\right]$ $\qquad$ BI (331)(2)

2. $\displaystyle\int_0^{\pi/2} \ln\frac{p + q\sin ax}{p - q\sin ax}\frac{dx}{\sin ax} = \int_0^{\pi/2} \ln\frac{p + q\cos ax}{p - q\cos ax}\frac{dx}{\cos ax} = \int_0^{\pi/2} \ln\frac{p + q\tan ax}{p - q\tan ax}\frac{dx}{\tan ax} = \pi\arcsin\frac{q}{p}$

$\qquad\qquad\qquad\qquad\qquad\qquad\qquad\qquad\qquad\left[p > q > 0\right]$

$\qquad\qquad\qquad\qquad\qquad\qquad\qquad\qquad\qquad$ FI II 695a, BI (315)(5, 13,17)a

3. $\displaystyle\int_0^{\pi/2} \frac{\cos x}{1 - \cos^2\alpha\cos^2 x}\ln\frac{1 + \cos\beta\cos x}{1 - \cos\beta\cos x}\,dx = \frac{2\pi}{\sin 2\alpha}\ln\frac{\cos\dfrac{\alpha - \beta}{2}}{\sin\dfrac{\alpha + \beta}{2}}$

$\qquad\qquad\qquad\qquad\qquad\qquad\qquad\qquad\left[0 < \alpha \le \beta < \frac{\pi}{2}\right]$ $\qquad$ LO III 284

**4.412**

1. $\displaystyle\int_0^{\pi/4} \ln\tan\left(\frac{\pi}{4}\pm x\right)\frac{\mathrm{d}x}{\sin 2x} = \pm\frac{\pi^2}{8}$         BI (293)(1)

2. $\displaystyle\int_0^{\pi/4} \ln\tan\left(\frac{\pi}{4}\pm x\right)\frac{\mathrm{d}x}{\tan 2x} = \pm\frac{\pi^2}{16}$         BI (293)(2)

3. $\displaystyle\int_0^{\pi/4} \ln\tan\left(\frac{\pi}{4}\pm x\right)(\ln\tan x)^{2n}\frac{\mathrm{d}x}{\sin 2x} = \pm\frac{2^{2n+2}-1}{4(n+1)(2n+1)}\pi^{2n+2}|B_{2n+2}|$         BI (294)(24)

4. $\displaystyle\int_0^{\pi/4} \ln\tan\left(\frac{\pi}{4}\pm x\right)(\ln\tan x)^{2n-1}\frac{\mathrm{d}x}{\sin 2x} = \pm\frac{1-2^{2n+1}}{2^{2n+2}n}(2n)!\,\zeta(2n+1)$         BI (294)(25)

5. $\displaystyle\int_0^{\pi/4} \ln\tan\left(\frac{\pi}{4}\pm x\right)(\ln\sin 2x)^{n-1}\frac{\mathrm{d}x}{\tan 2x} = \frac{(-1)^{n-1}}{2}(n-1)!\,\zeta(n+1)$         LI (294)(20)

**4.413**

1. $\displaystyle\int_0^{\pi/2} \ln\left(p^2+q^2\tan^2 x\right)\frac{\mathrm{d}x}{a^2\sin^2 x + b^2\cos^2 x} = \frac{\pi}{ab}\ln\frac{ap+bq}{a}$

$$[a>0,\quad b>0,\quad p>0,\quad q>0]$$
$$\text{BI (318)(1–4)a}$$

2. $\displaystyle\int_0^{\pi/2} \ln\left(1+q^2\tan^2 x\right)\frac{1}{p^2\sin^2 x + r^2\cos^2 x}\frac{\mathrm{d}x}{s^2\sin^2 x + t^2\cos^2 x}$

$$= \frac{\pi}{p^2 t^2 - s^2 r^2}\left\{\frac{p^2-r^2}{pr}\ln\left(1+\frac{qr}{p}\right) + \frac{t^2-s^2}{st}\ln\left(1+\frac{qt}{s}\right)\right\}$$
$$[q>0,\quad p>0,\quad r>0,\quad s>0,\quad t>0]\quad \text{BI (320)(18)}$$

3. $\displaystyle\int_0^{\pi/2} \ln\left(1+q^2\tan^2 x\right)\frac{\sin^2 x}{p^2\sin^2 x + r^2\cos^2 x}\frac{\mathrm{d}x}{s^2\sin^2 x + t^2\cos^2 x}$

$$= \frac{\pi}{p^2 t^2 - s^2 r^2}\left\{\frac{t}{s}\ln\left(1+\frac{qr}{p}\right) - \frac{r}{p}\ln\left(1+\frac{qt}{s}\right)\right\}$$
$$[q>0,\quad p>0,\quad r>0,\quad s>0,\quad t>0]\quad \text{BI (320)(20)}$$

4. $\displaystyle\int_0^{\pi/2} \ln\left(1+q^2\tan^2 x\right)\frac{\cos^2 x}{p^2\sin^2 x + r^2\cos^2 x}\frac{\mathrm{d}x}{s^2\sin^2 x + t^2\cos^2 x}$

$$= \frac{\pi}{p^2 t^2 - s^2 r^2}\left\{\frac{p}{r}\ln\left(1+\frac{qr}{p}\right) - \frac{s}{t}\ln\left(1+\frac{qt}{s}\right)\right\}$$
$$[q>0,\quad p>0,\quad r>0,\quad s>0,\quad t>0]\quad \text{BI (320)(21)}$$

5. $\displaystyle\int_0^{\pi} \frac{\ln\tan rx\,\mathrm{d}x}{1-2p\cos x + p^2} = \frac{\pi}{1-p^2}\ln\frac{1-p^{2r}}{1+p^{2r}}$       $[p^2<1]$         BI (331)(12)

**4.414**

1. $\displaystyle\int_0^{\pi/2} \ln\left(1-k^2\sin^2 x\right)\frac{\mathrm{d}x}{\sqrt{1-k^2\sin^2 x}} = \ln k'\,\boldsymbol{K}(k)$         BI (323)(1)

2. $\displaystyle\int_0^{\pi/2} \ln\left(1 - k^2 \sin^2 x\right) \frac{\sin^2 x\,\mathrm{d}x}{\sqrt{1 - k^2 \sin^2 x}} = \frac{1}{k^2}\left\{\left(k^2 - 2 + \ln k'\right)\boldsymbol{K}(k) + (2 - \ln k')\,\boldsymbol{E}(k)\right\}$

BI (323)(3)

3.[12] $\displaystyle\int_0^{\pi/2} \ln\left(1 - k^2 \sin^2 x\right)\cos^2 x\frac{\mathrm{d}x}{\sqrt{1 - k^2 \sin^2 x}} = \frac{1}{k^2}\left[\left(1 + k'^2 - k'^2 \ln k'\right)\boldsymbol{K}(k) - (2 - \ln k')\,\boldsymbol{E}(k)\right]$

BI (323)(6)

4. $\displaystyle\int_0^{\pi/2} \ln\left(1 - k^2 \sin^2 x\right) \frac{\mathrm{d}x}{\sqrt{\left(1 - k^2 \sin^2 x\right)^3}} = \frac{1}{k'^2}\left[\left(k^2 - 2\right)\boldsymbol{K}(k) + (2 + \ln k')\,\boldsymbol{E}(k)\right]$

BI (323)(9)

5.[12] $\displaystyle\int_0^{\pi/2} \ln\left(1 - k^2 \sin^2 x\right)\frac{\sin^2 x\,\mathrm{d}x}{\sqrt{\left(1 - k^2 \sin^2 x\right)^3}} = \frac{1}{k^2 k'^2}\left[(2 + \ln k')\,\boldsymbol{E}(k) - \left(1 + k'^2 + k'^2 \ln k'\right)\boldsymbol{K}(k)\right]$

BI (323)(10)

6. $\displaystyle\int_0^{\pi/2} \ln\left(1 - k^2 \sin^2 x\right)\frac{\cos^2 x\,\mathrm{d}x}{\sqrt{\left(1 - k^2 \sin^2 x\right)^3}} = \frac{1}{k^2}\left[\left(1 + k'^2 + \ln k'\right)\boldsymbol{K}(k) - (2 + \ln k')\,\boldsymbol{E}(k)\right]$

BI (323)(16)

7. $\displaystyle\int_0^{\pi/2} \ln\left(1 - k^2 \sin^2 x\right)\sqrt{1 - k^2 \sin^2 x}\,\mathrm{d}x = \left(1 + k'^2\right)\boldsymbol{K}(k) - (2 - \ln k')\,\boldsymbol{E}(k)$     BI (324)(18)

8. $\displaystyle\int_0^{\pi/2} \ln\left(1 - k^2 \sin^2 x\right)\sin^2 x\sqrt{1 - k^2 \sin^2 x}\,\mathrm{d}x = \frac{1}{9k^2}\left\{\left(-2 + 11k^2 - 6k^4 + 3k'^2 \ln k'\right)\boldsymbol{K}(k)\right.$

$$\left. + \left[2 - 10k^2 - 3\left(1 - 2k^2\right)\ln k'\right]\boldsymbol{E}(k)\right\}$$

BI (324)(20)

9. $\displaystyle\int_0^{\pi/2} \ln\left(1 - k^2 \sin^2 x\right)\cos^2 x\sqrt{1 - k^2 \sin^2 x}\,\mathrm{d}x = \frac{1}{9k^2}\left\{\left(2 + 7k^2 - 3k^4 - 3k'^2 \ln k'\right)\boldsymbol{K}(k)\right.$

$$\left. - \left[2 + 8k^2 - 3\left(1 + k^2\right)\ln k'\right]\boldsymbol{E}(k)\right\}$$

BI (324)(21), LI (324)(21)

10. $\displaystyle\int_0^{\pi/2} \ln\left(1 - k^2 \sin^2 x\right)\frac{\sin x \cos x\,\mathrm{d}x}{\sqrt{\left(1 - k^2 \sin^2 x\right)^{2n+1}}} = \frac{2}{(2n - 1)^2 k^2}\left\{\left[1 + (2n - 1)\ln k'\right]k'^{1-2n} - 1\right\}$

BI (324)(17)

**4.415**

1. $\displaystyle\int_0^\infty \ln x \sin ax^2\,\mathrm{d}x = -\frac{1}{4}\sqrt{\frac{\pi}{2a}}\left(\ln 4a + \boldsymbol{C} - \frac{\pi}{2}\right)$      $[a > 0]$      GW (338)(19)

2.[12] $\displaystyle\int_0^\infty \ln x \cos ax^2\,\mathrm{d}x = -\frac{1}{4}\sqrt{\frac{\pi}{2a}}\left(\ln 4a + \boldsymbol{C} + \frac{\pi}{2}\right)$      $[a > 0]$      GW (338)(19)

**4.416**

1.
$$\int_0^{\pi/2} \frac{\cos x \ln\left(1 + \sqrt{\sin^2\beta - \cos^2\beta \tan^2\alpha \sin^2 x}\right)}{1 - \sin^2\alpha \cos^2 x}\, dx$$
$$= \operatorname{cosec} 2\alpha \left\{(2\alpha + 2\gamma - \pi)\ln\cos\beta + 2\,L(\alpha) - 2\,L(\gamma) + L(\alpha + \gamma) - L(\alpha - \gamma)\right\}$$
$$\left[\cos\gamma = \frac{\sin\alpha}{\sin\beta};\quad 0 < \alpha < \beta < \frac{\pi}{2}\right] \quad \text{LO III 291}$$

2.
$$\int_0^{\pi/2} \frac{\cos x \ln\left(1 - \sqrt{\sin^2\beta - \cos^2\beta \tan^2\alpha \sin^2 x}\right)}{1 - \sin^2\alpha \cos^2 x}\, dx$$
$$= \operatorname{cosec} 2\alpha \left\{(\pi + 2\alpha - 2\gamma)\ln\cos\beta + 2\,L(\alpha) + 2\,L(\gamma) - L(\alpha + \gamma) + L(\alpha - \gamma)\right\}$$
$$\left[\cos\gamma = \frac{\sin\alpha}{\sin\beta};\quad 0 < \alpha < \beta < \frac{\pi}{2}\right] \quad \text{LO III 291}$$

3.
$$\int_\beta^{\pi/2} \frac{\ln\left(\sin x + \sqrt{\sin^2 x - \sin^2\beta}\right)}{1 - \cos^2\alpha \cos^2 x}\, dx$$
$$= -\operatorname{cosec}\alpha \left\{\arctan\left(\frac{\tan\beta}{\sin\alpha}\right)\ln\sin\beta + \frac{\pi}{2}\ln\frac{1 + \sin\alpha}{\sin\alpha + \sqrt{1 - \cos^2\alpha \cos^2\beta}}\right\}$$
$$\left[0 < \alpha < \pi,\quad 0 < \beta < \frac{\pi}{2}\right] \quad \text{LO III 285}$$

4.[7]
$$\int_0^{\pi/4} \ln\tan x\,(\ln\cos 2x)^{n-1}\tan 2x\, dx = \tfrac{1}{2}(-1)^n(n-1)!\left(1 - 2^{-(n+1)}\right)\zeta(n+1)$$
$$\text{BI (287)(20)}$$

## 4.42–4.43 Combinations of logarithms, trigonometric functions, and powers

**4.421**

1.
$$\int_0^\infty \ln x \sin ax\,\frac{dx}{x} = -\frac{\pi}{2}\,(C + \ln a) \qquad\qquad [a > 0] \qquad\qquad \text{FI II 810a}$$

2.
$$\int_0^\infty \ln ax \sin bx\,\frac{x\,dx}{\beta^2 + x^2} = \frac{\pi}{2}e^{-b\beta'}\ln(a\beta') - \frac{\pi}{4}\left[e^{b\beta'}\operatorname{Ei}(-b\beta') + e^{-b\beta'}\operatorname{Ei}(b\beta')\right]$$
$$[\beta' = \beta \operatorname{sign}\beta;\quad a > 0,\quad b > 0]$$
$$\text{ET I 76(5), NT 27(10)a}$$

3.
$$\int_0^\infty \ln ax \cos bx\,\frac{\beta'\,dx}{\beta^2 + x^2} = \frac{\pi}{2}e^{-b\beta'}\ln(a\beta') + \frac{\pi}{4}\left[e^{b\beta'}\operatorname{Ei}(-b\beta') - e^{-b\beta'}\operatorname{Ei}(b\beta')\right]$$
$$[\beta' = \beta \operatorname{sign}\beta;\quad a > 0,\quad b > 0]$$
$$\text{ET I 17(3), NT 27(11)a}$$

4.
$$\int_0^\infty \ln ax \sin bx\,\frac{x\,dx}{x^2 - c^2} = \frac{\pi}{2}\left\{-\operatorname{si}(bc)\sin bc + \cos bc\left[\ln ac - \operatorname{ci}(bc)\right]\right\}$$
$$[a > 0,\quad b > 0,\quad c > 0] \qquad \text{BI (422)(5)}$$

5.
$$\int_0^\infty \ln ax \cos bx\,\frac{dx}{x^2 - c^2} = \frac{\pi}{2c}\left\{\sin bc\left[\operatorname{ci}(bc) - \ln ac\right] - \cos bc\,\operatorname{si}(bc)\right\}$$
$$[a > 0,\quad b > 0,\quad c > 0] \qquad \text{BI (422)(6)}$$

**4.422**

1.     $$\int_0^\infty \ln(\sin ax)x^{\mu-1}\,\mathrm{d}x = \frac{\Gamma(\mu)}{a^\mu}\sin\frac{\mu\pi}{2}\left[\psi(\mu)-\ln a+\frac{\pi}{2}\cot\frac{\mu\pi}{2}\right]$$

$$[a>0,\quad |\operatorname{Re}\mu|<1]\qquad\qquad \text{BI (411)(5)}$$

2.     $$\int_0^\infty \ln(\cos ax)x^{\mu-1}\,\mathrm{d}x = \frac{\Gamma(\mu)}{a^\mu}\cos\frac{\mu\pi}{2}\left[\psi(\mu)-\ln a-\frac{\pi}{2}\tan\frac{\mu\pi}{2}\right]$$

$$[a>0,\quad 0<\operatorname{Re}\mu<1]\qquad\qquad \text{BI (411)(6)}$$

**4.423**

1.     $$\int_0^\infty \ln x\frac{\cos ax-\cos bx}{x}\,\mathrm{d}x = \ln\frac{a}{b}\left(\boldsymbol{C}+\frac{1}{2}\ln ab\right)\qquad [a>0,\quad b>0]\qquad\qquad \text{GW (338)(21a)}$$

2.     $$\int_0^\infty \ln x\frac{\cos ax-\cos bx}{x^2}\,\mathrm{d}x = \frac{\pi}{2}\left[(a-b)\left(\boldsymbol{C}-1\right)+a\ln a-b\ln b\right]$$

$$[a>0,\quad b>0]\qquad\qquad \text{GW (338)(21b)}$$

3.     $$\int_0^\infty \ln x\frac{\sin^2 ax}{x^2}\,\mathrm{d}x = -\frac{a\pi}{2}\left(\boldsymbol{C}+\ln 2a-1\right)\qquad [a>0]\qquad\qquad \text{GW (338)(20b)}$$

**4.424**

1.     $$\int_0^\infty (\ln x)^2\sin ax\frac{\mathrm{d}x}{x} = \frac{\pi}{2}\boldsymbol{C}^2 + \frac{\pi^3}{24} + \pi\boldsymbol{C}\ln a + \frac{\pi}{2}\left(\ln a\right)^2$$

$$[a>0]\qquad\qquad \text{ET I 77(9), FI II 810a}$$

2.[12]   $$\int_0^\infty (\ln x)^2\left(\sin ax\right)x^{\mu-1}\,\mathrm{d}x = \frac{\Gamma(\mu)}{a^\mu}\sin\frac{\mu\pi}{2}\left[\psi'(\mu)+\psi^2(\mu)+\pi\,\psi(\mu)\cot\frac{\mu\pi}{2}-2\,\psi(\mu)\ln a\right.$$
$$\left.-\pi\ln a\cot\frac{\mu\pi}{2}+(\ln a)^2-\pi^2\right]$$

$$[a>0,\quad 0<\operatorname{Re}\mu<1]\qquad\qquad \text{ET I 77(10)}$$

**4.425**

1.     $$\int_0^\infty \ln(1+x)\cos ax\frac{\mathrm{d}x}{x} = \frac{1}{2}\left\{[\operatorname{si}(a)]^2+[\operatorname{ci}(a)]^2\right\}\qquad [a>0]\qquad\qquad \text{ET I 18(8)}$$

2.[12]   $$\int_0^\infty \ln^2\left(\frac{b+x}{b-x}\right)\cos ax\frac{\mathrm{d}x}{x} = -2\pi\operatorname{si}(ab)\qquad [a>0,\quad b>0]\qquad\qquad \text{ET I 18(11)}$$

3.     $$\int_0^\infty \ln\left(1+b^2x^2\right)\sin ax\frac{\mathrm{d}x}{x} = -\pi\operatorname{Ei}\left(-\frac{a}{b}\right)\qquad [a>0,\quad b>0]$$

$$\text{GW (338)(24), ET I 77(14)}$$

4.     $$\int_0^1 \ln\left(1-x^2\right)\cos(p\ln x)\frac{\mathrm{d}x}{x} = \frac{1}{2p^2}+\frac{\pi}{2p}\coth\frac{p\pi}{2}\qquad\qquad \text{LI (309)(1)a}$$

**4.426**

1.[11]   $$\int_0^\infty x\ln\frac{b^2+x^2}{c^2+x^2}\sin ax\,\mathrm{d}x = \frac{\pi}{a^2}\left[(1+ac)e^{-ac}-(1+ab)e^{-ab}\right]$$

$$[b\geq 0,\quad c\geq 0,\quad a>0]\qquad\qquad \text{GW (338)(23)}$$

2.    $\int_0^\infty \ln \dfrac{b^2 x^2 + p^2}{c^2 x^2 + p^2} \sin ax \dfrac{dx}{x} = \pi \left[ \operatorname{Ei}\left( -\dfrac{ap}{c} \right) - \operatorname{Ei}\left( -\dfrac{ap}{b} \right) \right]$

$$[b > 0, \quad c > 0, \quad p > 0, \quad a > 0]$$

ET I 77(15)

**4.427**    $\int_0^\infty \ln \left( x + \sqrt{\beta^2 + x^2} \right) \dfrac{\sin ax}{\sqrt{\beta^2 + x^2}}\, dx = \dfrac{\pi}{2} K_0(a\beta) + \dfrac{\pi}{2} \ln(\beta) \left[ I_0(a\beta) - \mathbf{L}(a\beta) \right]$

$$[\operatorname{Re}\beta > 0, \quad a > 0]$$

ET I 77(16)

**4.428**

1.[12]    $\int_0^\infty \ln \cos^2 ax \dfrac{\cos bx}{x^2}\, dx = \pi \left( b \ln 2 - a + \displaystyle\sum_{n=1}^{\lfloor \frac{b}{2a} \rfloor} (-1)^n \dfrac{(b - 2an)}{n} \right)$

$$[a > 0, \quad b > 0]$$

ET I 22(29)

2.    $\int_0^\infty \ln \left( 4 \cos^2 ax \right) \dfrac{\cos bx}{x^2 + c^2}\, dx = \dfrac{\pi}{c} \cosh(bc) \ln \left( 1 + e^{-2ac} \right)$

$$\left[ a < b < 2a < \dfrac{\pi}{c} \right]$$

ET I 22(30)

3.    $\int_0^\infty \ln \cos^2 ax \dfrac{\sin bx}{x(1 + x^2)}\, dx = \pi \ln \left( 1 + e^{-2a} \right) \sinh b - \pi \ln 2 \left( 1 - e^{-b} \right)$

$$[a > 0, \quad b > 0]$$

ET I 82(36)

4.    $\int_0^\infty \ln \cos^2 ax \dfrac{\cos bx}{x^2 (1 + x^2)}\, dx = -\pi \ln \left( 1 + e^{-2a} \right) \cosh b + \left( b + e^{-b} \right) \pi \ln 2 - a\pi$

$$[a > 0, \quad b > 0]$$

ET I 22(31)

**4.429**    $\int_0^1 \dfrac{(1 + x)x}{\ln x} \sin (\ln x)\, dx = \dfrac{\pi}{4}$

BI (326)(2)a

**4.431**

1.    $\int_0^\infty \ln \left( 2 \pm 2 \cos x \right) \dfrac{\sin bx}{x^2 + c^2} x\, dx = - \pi \sinh(bc) \ln \left( 1 \pm e^{-c} \right)$

$$[b > 0, \quad c > 0]$$

ET I 22(32)

2.    $\int_0^\infty \ln \left( 2 \pm 2 \cos x \right) \dfrac{\cos bx}{x^2 + c^2}\, dx = \dfrac{\pi}{c} \cosh(bc) \ln \left( 1 \pm e^{-c} \right)$

$$[b > 0, \quad c > 0]$$

ET I 22(32)

3.    $\int_0^\infty \ln \left( 1 + 2a \cos x + a^2 \right) \dfrac{\sin bx}{x}\, dx = -\dfrac{\pi}{2} \displaystyle\sum_{k=1}^{\lfloor b \rfloor} \dfrac{(-a)^k}{k} \left[ 1 + \operatorname{sign}(b - k) \right]$

$$[0 < a < 1, \quad b > 0]$$

ET I 82(25)

4.    $\int_0^\infty \ln \left( 1 - 2a \cos x + a^2 \right) \dfrac{\cos bx}{x^2 + c^2}\, dx = \dfrac{\pi}{c} \ln \left( 1 - ae^{-c} \right) \cosh(bc) + \dfrac{\pi}{c} \displaystyle\sum_{k=1}^{\lfloor b \rfloor} \dfrac{a^k}{k} \sinh[c(b - k)]$

$$[|a| < 1, \quad b > 0, \quad c > 0]$$

ET I 22(33)

**4.432**

1. $$\int_0^\infty \ln\left(1 - k^2 \sin^2 x\right) \frac{\sin x}{\sqrt{1 - k^2 \sin^2 x}} \frac{\mathrm{d}x}{x} = \int_0^\infty \ln\left(1 - k^2 \cos^2 x\right) \frac{\sin x}{\sqrt{1 - k^2 \cos^2 x}} \frac{\mathrm{d}x}{x} = \ln k' \, \boldsymbol{K}(k)$$

BI ((412, 414))(4)

2. $$\int_0^{\pi/2} \ln\left(1 - k^2 \sin^2 x\right) \frac{\sin x \cos x}{\sqrt{1 - k^2 \sin^2 x}} x \, \mathrm{d}x$$
$$= \frac{1}{k^2}\left\{\pi k'\left(1 - \ln k'\right) + \left(2 - k^2\right)\boldsymbol{K}(k) - \left(4 - \ln k'\right)\boldsymbol{E}(k)\right\}$$

BI (426)(3)

3. $$\int_0^{\pi/2} \ln\left(1 - k^2 \cos^2 x\right) \frac{\sin x \cos x}{\sqrt{1 - k^2 \cos^2 x}} x \, \mathrm{d}x = \frac{1}{k^2}\left\{-\pi - \left(2 - k^2\right)\boldsymbol{K}(k) + \left(4 - \ln k'\right)\boldsymbol{E}(k)\right\}$$

BI (426)(6)

4. $$\int_0^\infty \ln\left(1 - k^2 \sin^2 x\right) \frac{\sin x \cos x}{\sqrt{1 - k^2 \sin^2 x}} \frac{\mathrm{d}x}{x} = \frac{1}{k^2}\left\{\left(2 - k^2 - k'^2 \ln k'\right)\boldsymbol{K}(k) - \left(2 - \ln k'\right)\boldsymbol{E}(k)\right\}$$

BI (412)(5)

5. $$\int_0^\infty \ln\left(1 - k^2 \cos^2 x\right) \frac{\sin x \cos x}{\sqrt{1 - k^2 \cos^2 x}} \frac{\mathrm{d}x}{x} = \frac{1}{k^2}\left\{\left(k^2 - 2 + \ln k'\right)\boldsymbol{K}(k) + \left(2 - \ln k'\right)\boldsymbol{E}(k)\right\}$$

BI (414)(5)

6. $$\int_0^\infty \ln\left(1 \pm k \sin^2 x\right) \frac{\sin x}{\sqrt{1 - k^2 \sin^2 x}} \frac{\mathrm{d}x}{x} = \int_0^\infty \ln\left(1 \pm k \cos^2 x\right) \frac{\sin x}{\sqrt{1 - k^2 \cos^2 x}} \frac{\mathrm{d}x}{x}$$
$$= \int_0^\infty \ln\left(1 \pm k \sin^2 x\right) \frac{\tan x}{\sqrt{1 - k^2 \sin^2 x}} \frac{\mathrm{d}x}{x}$$
$$= \int_0^\infty \ln\left(1 \pm k \cos^2 x\right) \frac{\tan x}{\sqrt{1 - k^2 \cos^2 x}} \frac{\mathrm{d}x}{x}$$
$$= \int_0^\infty \ln\left(1 \pm k \sin^2 2x\right) \frac{\tan x}{\sqrt{1 - k^2 \sin^2 2x}} \frac{\mathrm{d}x}{x}$$
$$= \int_0^\infty \ln\left(1 \pm k^2 \cos^2 2x\right) \frac{\tan x}{\sqrt{1 - k^2 \cos^2 2x}} \frac{\mathrm{d}x}{x}$$
$$= \frac{1}{2}\ln\frac{2\left(1 \pm k\right)}{\sqrt{k}}\boldsymbol{K}(k) - \frac{\pi}{8}\boldsymbol{K}\left(k'\right)$$

BI (413)(1–6), BI (415)(1–6)

7. $$\int_0^\infty \ln\left(1 - k^2 \sin^2 x\right) \frac{\sin^3 x}{\sqrt{1 - k^2 \sin^2 x}} \frac{\mathrm{d}x}{x} = \frac{1}{k^2}\left\{\left(k^2 - 2 + \ln k'\right)\boldsymbol{K}(k) + \left(2 - \ln k'\right)\boldsymbol{E}(k)\right\}$$

BI (412)(6)

8. $$\int_0^\infty \ln\left(1 - k^2 \cos^2 x\right) \frac{\sin^3 x}{\sqrt{1 - k^2 \cos^2 x}} \frac{\mathrm{d}x}{x} = \frac{1}{k^2}\left\{\left(2 - k^2 - k'^2 \ln k'\right)\boldsymbol{K}(k) - \left(2 - \ln k'\right)\boldsymbol{E}(k)\right\}$$

BI (414)(6)a

9. $$\int_0^\infty \ln\left(1 - k^2 \sin^2 x\right) \frac{\sin x \cos^2 x}{\sqrt{1 - k^2 \sin^2 x}} \frac{\mathrm{d}x}{x} = \frac{1}{k^2}\left\{\left(2 - k^2 - k'^2 \ln k'\right)\boldsymbol{K}(k) - \left(2 - \ln k'\right)\boldsymbol{E}(k)\right\}$$

BI (412)(7)

10. $\displaystyle\int_0^\infty \ln\left(1 - k^2\cos^2 x\right) \frac{\sin x \cos^2 x}{\sqrt{1 - k^2\cos^2 x}}\frac{\mathrm{d}x}{x} = \frac{1}{k^2}\left\{\left(k^2 - 2 + \ln k'\right)\boldsymbol{K}(k) + (2 - \ln k')\boldsymbol{E}(k)\right\}$

BI (414)(7)

11. $\displaystyle\int_0^\infty \ln\left(1 - k^2\sin^2 x\right)\frac{\tan x}{\sqrt{1 - k^2\sin^2 x}}\frac{\mathrm{d}x}{x} = \int_0^\infty \ln\left(1 - k^2\cos^2 x\right)\frac{\tan x}{\sqrt{1 - k^2\cos^2 x}}\frac{\mathrm{d}x}{x} = \ln k'\,\boldsymbol{K}(k)$

BI ((412, 414))(9)

12. $\displaystyle\int_0^\infty \ln\left(1 - k^2\sin^2 x\right)\frac{\sin^2 x \tan x}{\sqrt{1 - k^2\sin^2 x}}\frac{\mathrm{d}x}{x} = \frac{1}{k^2}\left\{\left(k^2 - 2 + \ln k'\right)\boldsymbol{K}(k) + (2 - \ln k')\boldsymbol{E}(k)\right\}$

BI (412)(8)

13. $\displaystyle\int_0^\infty \ln\left(1 - k^2\cos^2 x\right)\frac{\sin^2 x \tan x}{\sqrt{1 - k^2\cos^2 x}}\frac{\mathrm{d}x}{x} = \frac{1}{k^2}\left\{\left(2 - k^2 - k'^2\ln k'\right)\boldsymbol{K}(k) - (2 - \ln k')\boldsymbol{E}(k)\right\}$

BI (414)(8)

14.[12] $\displaystyle\int_0^\infty \ln\left(1 - k^2\sin^2 x\right)\frac{\sin x}{\sqrt{\left(1 - k^2\sin^2 x\right)^3}}\frac{\mathrm{d}x}{x} = \int_0^\infty \ln\left(1 - k^2\cos^2 x\right)\frac{\sin x}{\sqrt{\left(1 - k^2\cos^2 x\right)^3}}\frac{\mathrm{d}x}{x}$

$$= \frac{1}{k'^2}\left\{\left(k^2 - 2\right)\boldsymbol{K}(k) + (2 + \ln k')\boldsymbol{E}(k)\right\}$$

BI ((412, 414))(13)

15. $\displaystyle\int_0^{\pi/2} \ln\left(1 - k^2\sin^2 x\right)\frac{\sin x \cos x}{\sqrt{\left(1 - k^2\sin^2 x\right)^3}}x\,\mathrm{d}x = \frac{1}{k^2}\left\{(1 + \ln k')\frac{\pi}{k'} - (2 + \ln k')\boldsymbol{K}(k)\right\}$

BI (426)(9)

16. $\displaystyle\int_0^{\pi/2} \ln\left(1 - k^2\cos^2 x\right)\frac{\sin x \cos x}{\sqrt{\left(1 - k^2\cos^2 x\right)^3}}x\,\mathrm{d}x - \frac{1}{k^2}\left\{-\pi + (2 + \ln k')\boldsymbol{K}(k)\right\}$      BI (426)(15)

17. $\displaystyle\int_0^\infty \ln\left(1 - k^2\sin^2 x\right)\frac{\sin x \cos x}{\sqrt{\left(1 - k^2\sin^2 x\right)^3}}\frac{\mathrm{d}x}{x} = \int_0^\infty \ln\left(1 - k^2\cos^2 x\right)\frac{\sin^3 x}{\sqrt{\left(1 - k^2\cos^2 x\right)^3}}\frac{\mathrm{d}x}{x}$

$$= \frac{1}{k^2}\left\{\left(2 - k^2 + \ln k'\right)\boldsymbol{K}(k) - (2 + \ln k')\boldsymbol{E}(k)\right\}$$

BI (412)(14), BI(414)(15)

18. $\displaystyle\int_0^\infty \ln\left(1 - k^2\sin^2 x\right)\frac{\sin^3 x}{\sqrt{\left(1 - k^2\sin^2 x\right)^3}}\frac{\mathrm{d}x}{x}$

$$= \int_0^\infty \ln\left(1 - k^2\cos^2 x\right)\frac{\sin x \cos x}{\sqrt{\left(1 - k^2\cos^2 x\right)^3}}\frac{\mathrm{d}x}{x}$$

$$= \frac{1}{k^2 k'^2}\left\{(2 + \ln k')\boldsymbol{E}(k) - \left(2 - k^2 + k'^2\ln k'\right)\boldsymbol{K}(k)\right\}$$

BI (412)(15), BI(414)(14)

19. $\displaystyle\int_0^\infty \ln\left(1 - k^2\sin^2 x\right)\frac{\sin x\cos^2 x}{\sqrt{\left(1 - k^2\sin^2 x\right)^3}}\frac{\mathrm{d}x}{x} = \int_0^\infty \ln\left(1 - k^2\cos^2 x\right)\frac{\sin^2 x\tan x}{\sqrt{\left(1 - k^2\cos^2 x\right)^3}}\frac{\mathrm{d}x}{x}$

$$= \frac{1}{k^2}\left\{\left(2 - k^2 + \ln k'\right)\boldsymbol{K}(k) - \left(2 + \ln k'\right)\boldsymbol{E}(k)\right\}$$

BI (412)(16), BI(414)(17)

20. $\displaystyle\int_0^\infty \ln\left(1 - k^2\sin^2 x\right)\frac{\sin^2 x\tan x}{\sqrt{\left(1 - k^2\sin^2 x\right)^3}}\frac{\mathrm{d}x}{x}$

$$= \int_0^\infty \ln\left(1 - k^2\cos^2 x\right)\frac{\sin x\cos^2 x}{\sqrt{\left(1 - k^2\cos^2 x\right)^3}}\frac{\mathrm{d}x}{x}$$

$$= \frac{1}{k^2 k'^2}\left\{\left(2 + \ln k'\right)\boldsymbol{E}(k) - \left(2 - k^2 + k'^2\ln k'\right)\boldsymbol{K}(k)\right\}$$

BI (412)(17), BI(414)(16)

21. $\displaystyle\int_0^\infty \ln\left(1 - k^2\sin^2 x\right)\frac{\tan x}{\sqrt{\left(1 - k^2\sin^2 x\right)^3}}\frac{\mathrm{d}x}{x} = \int_0^\infty \ln\left(1 - k^2\cos^2 x\right)\frac{\tan x}{\sqrt{\left(1 - k^2\cos^2 x\right)^3}}\frac{\mathrm{d}x}{x}$

$$= \frac{1}{k'^2}\left\{\left(k^2 - 2\right)\boldsymbol{K}(k) + \left(2 + \ln k'\right)\boldsymbol{E}(k)\right\}$$

BI ((412, 414))(18)

22. $\displaystyle\int_0^\infty \ln\left(1 - k^2\sin^2 x\right)\sqrt{1 - k^2\sin^2 x}\,\sin x\frac{\mathrm{d}x}{x} = \int_0^\infty \ln\left(1 - k^2\cos^2 x\right)\sqrt{1 - k^2\cos^2 x}\,\sin x\frac{\mathrm{d}x}{x}$

$$= \left(2 - k^2\right)\boldsymbol{K}(k) - \left(2 - \ln k'\right)\boldsymbol{E}(k)$$

BI ((412, 414))(1)

23. $\displaystyle\int_0^{\pi/2} \ln\left(1 - k^2\sin^2 x\right)\sqrt{1 - k^2\sin^2 x}\,\sin x\cos x\cdot x\,\mathrm{d}x$

$$= \frac{1}{27k^2}\left\{3\pi k'^3(1 - 3\ln k') + \left(22k'^2 + 6k^4 - 3k'^2\ln k'\right)\boldsymbol{K}(k) - (2 - k^2)(14 - 6\ln k')\boldsymbol{E}(k)\right\}$$

BI (426)(1)

24. $\displaystyle\int_0^{\pi/2} \ln\left(1 - k^2\cos^2 x\right)\sqrt{1 - k^2\cos^2 x}\,\sin x\cos x\cdot x\,\mathrm{d}x$

$$= \frac{1}{27k^2}\left\{-3\pi - \left(22k'^2 + 6k^4 - 3k'^2\ln k'\right)\boldsymbol{K}(k) + \left(2 - k^2\right)(14 - 6\ln k')\boldsymbol{E}(k)\right\}$$

BI (426)(2)

25. $\displaystyle\int_0^\infty \ln\left(1 - k^2\sin^2 x\right)\sqrt{1 - k^2\sin^2 x}\,\tan x\frac{\mathrm{d}x}{x} = \int_0^\infty \ln\left(1 - k^2\cos^2 x\right)\sqrt{1 - k^2\cos^2 x}\,\tan x\frac{\mathrm{d}x}{x}$

$$= \left(2 - k^2\right)\boldsymbol{K}(k) - \left(2 - \ln k'\right)\boldsymbol{E}(k)$$

((412,414))(2)

**26.**
$$\int_0^\infty \ln\left(\sin^2 x + k'\cos^2 x\right)\frac{\sin x}{\sqrt{1-k^2\cos^2 x}}\frac{dx}{x} = \int_0^\infty \ln\left(\sin^2 x + k'\cos^2 x\right)\frac{\tan x}{\sqrt{1-k^2\cos^2 x}}\frac{dx}{x}$$

$$= \int_0^\infty \ln\left(\sin^2 2x + k'\cos^2 2x\right)\frac{\tan x}{\sqrt{1-k^2\cos^2 2x}}\frac{dx}{x}$$

$$= \frac{1}{2}\ln\left[\frac{2\left(\sqrt{k'}\right)^3}{1+k'}\right]K(k)$$

BI (415)(19–21)

## 4.44 Combinations of logarithms, trigonometric functions, and exponentials

**4.441**

**1.**[12]
$$\int_0^\infty e^{-qx}\sin px\ln x\,dx = \frac{1}{p^2+q^2}\left[q\arctan\frac{p}{q} - p\boldsymbol{C} - \frac{p}{2}\ln\left(p^2+q^2\right)\right]$$

$[q>0,\quad p>0]$        BI (467)(1)

**2.**
$$\int_0^\infty e^{-qx}\cos px\ln x\,dx = -\frac{1}{p^2+q^2}\left[\frac{q}{2}\ln\left(p^2+q^2\right) + p\arctan\frac{p}{q} + q\boldsymbol{C}\right]$$

$[q>0]$        BI (467)(2)

**4.442**    $\displaystyle\int_0^{\pi/2}\frac{e^{-p\tan x}\ln\cos x\,dx}{\sin x\cos x} = -\frac{1}{2}\left[\mathrm{ci}(p)\right]^2 + \frac{1}{2}\left[\mathrm{si}(p)\right]^2$    $[\mathrm{Re}\,p>0]$        NT 32(11)

# 4.5 Inverse Trigonometric Functions

## 4.51   Inverse trigonometric functions

**4.511**    $\displaystyle\int_0^\infty \mathrm{arccot}\,px\,\mathrm{arccot}\,qx\,dx = \frac{\pi}{2}\left\{\frac{1}{p}\ln\left(1+\frac{p}{q}\right) + \frac{1}{q}\ln\left(1+\frac{q}{p}\right)\right\}$

$[p>0,\quad q>0]$        BI (77)(8)

**4.512**    $\displaystyle\int_0^\pi \arctan\left(\cos x\right)\,dx = 0$                                      BI (345)(1)

## 4.52 Combinations of arcsines, arccosines, and powers

**4.521**

**1.**    $\displaystyle\int_0^1 \frac{\arcsin x}{x}\,dx = \frac{\pi}{2}\ln 2$                                           FI II 614, 623

**2.**    $\displaystyle\int_0^1 \frac{\arccos x}{1\pm x}\,dx = \mp\frac{\pi}{2}\ln 2 + 2\boldsymbol{G}$                          BI (231)(7, 8)

**3.**    $\displaystyle\int_0^1 \arcsin x\frac{x}{1+qx^2}\,dx = \frac{\pi}{2q}\ln\frac{2\sqrt{1+q}}{1+\sqrt{1+q}}$    $[q>-1]$        BI (231)(1)

**4.**    $\displaystyle\int_0^1 \arcsin x\frac{x}{1-p^2x^2}\,dx = \frac{\pi}{2p^2}\ln\frac{1+\sqrt{1-p^2}}{2\sqrt{1-p^2}}$    $[p^2<1]$        LI (231)(3)

5.    $\displaystyle\int_0^1 \arccos x \frac{\mathrm{d}x}{\sin^2 \lambda - x^2} = 2\operatorname{cosec}\lambda \sum_{k=0}^{\infty} \frac{\sin[(2k+1)\lambda]}{(2k+1)^2}$ 

     BI (231)(10)

6.    $\displaystyle\int_0^1 \arcsin x \frac{\mathrm{d}x}{x\left(1+qx^2\right)} = \frac{\pi}{2}\ln\frac{1+\sqrt{1+q}}{\sqrt{1+q}}$         $[q > -1]$     BI (235)(10)

7.    $\displaystyle\int_0^1 \arcsin x \frac{x}{\left(1+qx^2\right)^2}\,\mathrm{d}x = \frac{\pi}{4q}\frac{\sqrt{1+q}-1}{1+q}$         $[q > -1]$     BI (234)(2)

8.    $\displaystyle\int_0^1 \arccos x \frac{x}{\left(1+qx^2\right)^2}\,\mathrm{d}x = \frac{\pi}{4q}\frac{\sqrt{1+q}-1}{1+q}$         $[q > -1]$     BI (234)(4)

**4.522**

1.    $\displaystyle\int_0^1 x\sqrt{1-k^2x^2}\,\arccos x\,\mathrm{d}x = \frac{1}{9k^2}\left[\frac{3}{2}\pi + k'^2\,\boldsymbol{K}(k) - 2\left(1+k'^2\right)\boldsymbol{E}(k)\right]$     BI (236)(9)

2.    $\displaystyle\int_0^1 x\sqrt{1-k^2x^2}\,\arcsin x\,\mathrm{d}x = \frac{1}{9k^2}\left[-\frac{3}{2}\pi k'^3 - k'^2\,\boldsymbol{K}(k) + 2\left(1+k'^2\right)\boldsymbol{E}(k)\right]$     BI (236)(1)

3.    $\displaystyle\int_0^1 x\sqrt{k'^2+k^2x^2}\,\arcsin x\,\mathrm{d}x = \frac{1}{9k^2}\left[\frac{3}{2}\pi + k'^2\,\boldsymbol{K}(k) - 2\left(1+k'^2\right)\boldsymbol{E}(k)\right]$     BI(236)(5)

4.    $\displaystyle\int_0^1 \frac{x\arcsin x}{\sqrt{1-k^2x^2}}\,\mathrm{d}x = \frac{1}{k^2}\left[-\frac{\pi}{2}k' + \boldsymbol{E}(k)\right]$     BI (237)(1)

5.    $\displaystyle\int_0^1 \frac{x\arccos x}{\sqrt{1-k^2x^2}}\,\mathrm{d}x = \frac{1}{k^2}\left[\frac{\pi}{2} - \boldsymbol{E}(k)\right]$     BI (240)(1)

6.    $\displaystyle\int_0^1 \frac{x\arcsin x}{\sqrt{k'^2+k^2x^2}}\,\mathrm{d}x = \frac{1}{k^2}\left[\frac{\pi}{2} - \boldsymbol{E}(k)\right]$     BI (238)(1)

7.    $\displaystyle\int_0^1 \frac{x\arccos x}{\sqrt{k'^2+k^2x^2}}\,\mathrm{d}x = \frac{1}{k^2}\left[-\frac{\pi}{2}k' + \boldsymbol{E}(k)\right]$     BI (241)(1)

8.    $\displaystyle\int_0^1 \frac{x\arcsin x\,\mathrm{d}x}{\left(x^2-\cos^2\lambda\right)\sqrt{1-x^2}} = \frac{2}{\sin\lambda}\sum_{k=0}^{\infty}\frac{\sin[(2k+1)\lambda]}{(2k+1)^2}$     BI (243)(11)

9.    $\displaystyle\int_0^1 \frac{x\arcsin kx}{\sqrt{\left(1-x^2\right)\left(1-k^2x^2\right)}}\,\mathrm{d}x = -\frac{\pi}{2k}\ln k'$     BI (239)(1)

10.   $\displaystyle\int_0^1 \frac{x\arccos kx}{\sqrt{\left(1-x^2\right)\left(1-k^2x^2\right)}}\,\mathrm{d}x = \frac{\pi}{2k}\ln(1+k)$     BI (242)(1)

**4.523**

1.    $\displaystyle\int_0^1 x^{2n}\arcsin x\,\mathrm{d}x = \frac{1}{2n+1}\left[\frac{\pi}{2} - \frac{2^n n!}{(2n+1)!!}\right]$     BI (229)(1)

2.    $\displaystyle\int_0^1 x^{2n-1}\arcsin x\,\mathrm{d}x = \frac{\pi}{4n}\left[1 - \frac{(2n-1)!!}{2^n n!}\right]$     BI (229)(2)

3.    $\displaystyle\int_0^1 x^{2n}\arccos x\,\mathrm{d}x = \frac{2^n n!}{(2n+1)(2n+1)!!}$     BI (229)(4)

4.    $\displaystyle\int_0^1 x^{2n-1}\arccos x\,dx = \frac{\pi}{4n}\frac{(2n-1)!!}{2^n n!}$        BI (229)(5)

5.    $\displaystyle\int_{-1}^1 \left(1-x^2\right)^n \arccos x\,dx = \pi\frac{2^n n!}{(2n+1)!!}$        BI (254)(2)

6.    $\displaystyle\int_{-1}^1 \left(1-x^2\right)^{n-\frac{1}{2}} \arccos x\,dx = \frac{\pi^2}{2}\frac{(2n-1)!!}{2^n n!}$        BI (254)(3)

**4.524**

1.    $\displaystyle\int_0^1 (\arcsin x)^2\,\frac{dx}{x^2\sqrt{1-x^2}} = \pi\ln 2$        BI (243)(13)

2.    $\displaystyle\int_0^1 (\arccos x)^2\,\frac{dx}{\left(\sqrt{1-x^2}\right)^3} = \pi\ln 2$        BI (244)(9)

## 4.53–4.54 Combinations of arctangents, arccotangents, and powers

**4.531**

1.    $\displaystyle\int_0^1 \frac{\arctan x}{x}\,dx = \int_1^\infty \frac{\operatorname{arccot} x}{x}\,dx = \boldsymbol{G}$        FI II 482, BI (253)(8)

2.    $\displaystyle\int_0^\infty \frac{\operatorname{arccot} x}{1\pm x}\,dx = \pm\frac{\pi}{4}\ln 2 + \boldsymbol{G}$        BI (248)(6, 7)

3.[12]    $\displaystyle\int_0^1 \frac{\arctan x}{x(1+x)}\,dx = -\frac{\pi}{8}\ln 2 + \boldsymbol{G}$        BI (235)(11)

4.    $\displaystyle\int_0^\infty \frac{\arctan x}{1-x^2}\,dx = -\boldsymbol{G}.$        BI (248)(2)

5.    $\displaystyle\int_0^1 \arctan qx\,\frac{dx}{(1+px)^2} = \frac{1}{2}\frac{q}{p^2+q^2}\ln\frac{(1+p)^2}{1+q^2} + \frac{q^2-p}{(1+p)\left(p^2+q^2\right)}\arctan q$

                                            $[p>-1]$        BI (243)(7)

6.    $\displaystyle\int_0^1 \operatorname{arccot} qx\,\frac{dx}{(1+px)^2} = \frac{1}{2}\frac{q}{p^2+q^2}\ln\frac{1+q^2}{(1+p)^2} + \frac{p}{p^2+q^2}\arctan q + \frac{1}{1+p}\operatorname{arccot} q$

                                            $[p>-1]$        BI (234)(10)

7.    $\displaystyle\int_0^1 \frac{\arctan x}{x\left(1+x^2\right)}\,dx = \frac{\pi}{8}\ln 2 + \frac{1}{2}\boldsymbol{G}$        BI (235)(12)

8.    $\displaystyle\int_0^\infty \frac{x\arctan x}{1+x^4}\,dx = \frac{\pi^2}{16}$        BI (248)(3)

9.    $\displaystyle\int_0^\infty \frac{x\arctan x}{1-x^4}\,dx = -\frac{\pi}{8}\ln 2$        BI (248)(4)

10.    $\displaystyle\int_0^\infty \frac{x\operatorname{arccot} x}{1-x^4}\,dx = \frac{\pi}{8}\ln 2$        BI (248)(12)

11.[12]    $\displaystyle\int_0^\infty \frac{\arctan x}{x\sqrt{1+x^2}}\,dx = \int_0^\infty \frac{\operatorname{arccot} x}{\sqrt{1+x^2}}\,dx = 2\boldsymbol{G}$        BI (251)(3, 10)

12.    $\displaystyle\int_0^1 \frac{\arctan x}{x\sqrt{1-x^2}}\,dx = \frac{\pi}{2}\ln\left(1+\sqrt{2}\right)$       FI II 694

13.    $\displaystyle\int_0^1 \frac{x\arctan x\,dx}{\sqrt{\left(1+x^2\right)\left(1+k'^2 x^2\right)}} = \frac{1}{k^2}\left[F\left(\frac{\pi}{4},k\right) - \frac{\pi}{2\sqrt{2\left(1+k'^2\right)}}\right]$       BI (294)(14)

**4.532**

1.    $\displaystyle\int_0^1 x^p \arctan x\,dx = \frac{1}{2(p+1)}\left[\frac{\pi}{2} - \beta\left(\frac{p}{2}+1\right)\right]$    $[p>-2]$       BI (229)(7)

2.    $\displaystyle\int_0^\infty x^p \arctan x\,dx = \frac{\pi}{2(p+1)}\,\operatorname{cosec}\frac{p\pi}{2}$    $[-1>p>-2]$       BI (246)(1)

3.    $\displaystyle\int_0^1 x^p \operatorname{arccot} x\,dx = \frac{1}{2(p+1)}\left[\frac{\pi}{2} + \beta\left(\frac{p}{2}+1\right)\right]$    $[p>-1]$       BI (229)(8)

4.    $\displaystyle\int_0^\infty x^p \operatorname{arccot} x\,dx = -\frac{\pi}{2(p+1)}\,\operatorname{cosec}\frac{p\pi}{2}$    $[-1<p<0]$       BI (246)(2)

5.    $\displaystyle\int_0^\infty \left(\frac{x^p}{1+x^{2p}}\right)^{2q}\arctan x\,\frac{dx}{x} = \frac{\sqrt{\pi^3}}{2^{2q+2}p}\frac{\Gamma(q)}{\Gamma\left(q+\frac{1}{2}\right)}$    $[q>0]$       BI (250)(10)

**4.533**

1.    $\displaystyle\int_0^\infty \left(1 - x\operatorname{arccot} x\right)\,dx = \frac{\pi}{4}$       BI (246)(3)

2.    $\displaystyle\int_0^1 \left(\frac{\pi}{4} - \arctan x\right)\frac{dx}{1-x} = -\frac{\pi}{8}\ln 2 + \boldsymbol{G}$       BI (232)(2)

3.    $\displaystyle\int_0^1 \left(\frac{\pi}{4} - \arctan x\right)\frac{1+x}{1-x}\frac{dx}{1+x^2} = \frac{\pi}{8}\ln 2 + \frac{1}{2}\boldsymbol{G}$       BI (235)(25)

4.    $\displaystyle\int_0^1 \left(x\operatorname{arccot} x - \frac{1}{x}\arctan x\right)\frac{dx}{1-x^2} = -\frac{\pi}{4}\ln 2$       BI (232)(1)

**4.534**   $\displaystyle\int_0^\infty (\arctan x)^2 \frac{dx}{x^2\sqrt{1+x^2}} = \int_0^\infty (\operatorname{arccot} x)^2 \frac{x\,dx}{\sqrt{1+x^2}} = -\frac{\pi^2}{4} + 4\boldsymbol{G}$       BI (251)(9, 17)

**4.535**

1.    $\displaystyle\int_0^1 \frac{\arctan px}{1+p^2 x}\,dx = \frac{1}{2p^2}\arctan p\ln\left(1+p^2\right)$       BI (231)(19)

2.    $\displaystyle\int_0^1 \frac{\operatorname{arccot} px}{1+p^2 x}\,dx = \frac{1}{p^2}\left[\frac{\pi}{4} + \frac{1}{2}\operatorname{arccot} p\right]\ln\left(1+p^2\right)$    $[p>0]$       BI (231)(24)

3.    $\displaystyle\int_0^\infty \frac{\arctan qx}{(p+x)^2}\,dx = -\frac{q}{1+p^2 q^2}\left(\ln pq - \frac{\pi}{2}pq\right)$    $[p>0,\quad q>0]$       BI (249)(1)

4.    $\displaystyle\int_0^\infty \frac{\operatorname{arccot} qx}{(p+x)^2}\,dx = \frac{q}{1+p^2 q^2}\left(\ln pq + \frac{\pi}{2pq}\right)$    $[p>0,\quad q>0]$       BI (249)(8)

5.    $\displaystyle\int_0^\infty \frac{x\operatorname{arccot} px}{q^2+x^2}\,dx = \frac{\pi}{2}\ln\frac{1+pq}{pq}$    $[p>0,\quad q>0]$       BI (248)(9)

6.    $\displaystyle\int_0^\infty \frac{x \operatorname{arccot} px \, dx}{x^2 - q^2} = \frac{\pi}{4} \ln \frac{1 + p^2 q^2}{p^2 q^2}$        $[p > 0, \quad q > 0]$        BI (248)(10)

7.    $\displaystyle\int_0^\infty \frac{\arctan px}{x\,(1 + x^2)} \, dx = \frac{\pi}{2} \ln(1 + p)$        $[p \geq 0]$        FI II 745

8.    $\displaystyle\int_0^\infty \frac{\arctan px}{x\,(1 - x^2)} \, dx = \frac{\pi}{4} \ln \left(1 + p^2\right)$        $[p \geq 0]$        BI (250)(6)

9.    $\displaystyle\int_0^\infty \arctan qx \frac{dx}{x\,(p^2 + x^2)} = \frac{\pi}{2p^2} \ln(1 + pq)$        $[p > 0, \quad q \geq 0]$        BI (250)(3)

10.[12]    $\displaystyle\int_0^\infty \arctan qx \frac{dx}{x\,(1 - p^2 x^2)} = \frac{\pi}{4} \ln \frac{p^2 + q^2}{p^2}$        $[q \geq 0]$        BI (250)(6)

11.    $\displaystyle\int_0^\infty \frac{x \arctan qx}{(p^2 + x^2)^2} \, dx = \frac{\pi q}{4p(1 + pq)}$        $[p > 0, \quad q \geq 0]$        BI (252)(12)a

12.    $\displaystyle\int_0^\infty \frac{x \operatorname{arccot} qx}{(p^2 + x^2)^2} \, dx = \frac{\pi}{4p^2(1 + pq)}$        $[p > 0, \quad q \geq 0]$        BI (252)(20)a

13.    $\displaystyle\int_0^1 \frac{\arctan qx}{x\sqrt{1 - x^2}} \, dx = \frac{\pi}{2} \ln \left(q + \sqrt{1 + q^2}\right)$        BI (244)(11)

14.[9]    $\displaystyle\int_{-\infty}^\infty \frac{x \arctan(\alpha x)\,dx}{(x^2 + \beta^2)\,(x^2 + \gamma^2)} = \begin{cases} \dfrac{\pi}{\beta^2 - \gamma^2} \ln\left(\dfrac{1 + |\alpha\beta|}{1 + |\alpha\gamma|}\right) \operatorname{sign}(\alpha) & \text{for } \beta \neq \gamma \\[4mm] \dfrac{\pi\alpha}{2|\beta|\,(1 + |\alpha\beta|)} & \text{for } \beta = \gamma \end{cases}$

$$\text{for } \alpha, \beta, \gamma \text{ real}$$

15.[9]    $\displaystyle\int_{-\infty}^\infty \frac{x \arctan (\alpha/x)\,dx}{(x^2 + \beta^2)\,(x^2 + \gamma^2)} = \begin{cases} \dfrac{\pi}{\beta^2 - \gamma^2} \ln\left(\dfrac{1 + |\alpha/\gamma|}{1 + |\alpha/\beta|}\right) \operatorname{sign}(\alpha) & |\alpha, \beta, \gamma \text{ real}; \quad \beta \neq \gamma| \\[4mm] \dfrac{\pi\alpha}{2\beta^2\,(|\beta| + |\alpha|)} & [\beta = \gamma] \end{cases}$

## 4.536

1.    $\displaystyle\int_0^\infty \arctan qx \arcsin x \frac{dx}{x^2} = \frac{1}{2} q\pi \ln \frac{1 + \sqrt{1 + q^2}}{\sqrt{1 + q^2}} + \frac{\pi}{2} \ln \left(q + \sqrt{1 + q^2}\right) - \frac{\pi}{2} - \arctan q$

            BI (230)(7)

2.    $\displaystyle\int_0^\infty \frac{\arctan px - \arctan qx}{x} \, dx = \frac{\pi}{2} \ln \frac{p}{q}$        $[p > 0, \quad q > 0]$        FI II 635

3.    $\displaystyle\int_0^\infty \frac{\arctan px \arctan qx}{x^2} \, dx = \frac{\pi}{2} \ln \frac{(p + q)^{p+q}}{p^p q^q}$        $[p > 0, \quad q > 0]$        FI II 745

## 4.537

1.[8]    $\displaystyle\int_0^1 \arctan \left(\sqrt{1 - x^2}\right) \frac{dx}{1 - x^2 \cos^2 \lambda} = \frac{\pi}{2 \cos \lambda} \ln \left[\cos \left(\frac{\pi - 4\lambda}{8}\right) \operatorname{cosec} \left(\frac{\pi + 4\lambda}{8}\right)\right]$        BI (245)(9)

2.    $\displaystyle\int_0^1 \arctan \left(p\sqrt{1 - x^2}\right) \frac{dx}{1 - x^2} = \frac{1}{2} \pi \ln \left(p + \sqrt{1 + p^2}\right)$

            $[p > 0]$        BI (245)(10)

3.$^{12}$ $\int_0^1 \arctan\left(\tan\lambda\sqrt{1-k^2x^2}\right)\sqrt{\dfrac{1-x^2}{1-k^2x^2}}\,dx = \dfrac{\pi}{2k^2}\left[E(\lambda,k)-k'^2\,F(\lambda,k)\right]$
$$-\dfrac{\pi}{2k^2}\cot\lambda\left(1-\sqrt{1-k^2\sin^2\lambda}\right)$$

BI (245)(12)

4. $\int_0^1 \arctan\left(\tan\lambda\sqrt{1-k^2x^2}\right)\sqrt{\dfrac{1-k^2x^2}{1-x^2}}\,dx = \dfrac{\pi}{2}\,E(\lambda,k)-\dfrac{\pi}{2}\cot\lambda\left(1-\sqrt{1-k^2\sin^2\lambda}\right)$

BI (245)(11)

5. $\int_0^1 \dfrac{\arctan\left(\tan\lambda\sqrt{1-k^2x^2}\right)}{\sqrt{(1-x^2)(1-k^2x^2)}}\,dx = \dfrac{\pi}{2}\,F(\lambda,k)$

BI (245)(13)

**4.538**

1. $\int_0^\infty \arctan x^2\,\dfrac{dx}{1+x^2} = \int_0^\infty \arctan x^3\,\dfrac{dx}{1+x^2}$

BI (252)(10, 11)

$\quad = \int_0^\infty \operatorname{arccot} x^2\,\dfrac{dx}{1+x^2} = \int_0^\infty \operatorname{arccot} x^3\,\dfrac{dx}{1+x^2} = \dfrac{\pi^2}{8}$

BI (252)(18, 19)

2.$^{12}$ $\int_0^1 \dfrac{1-x^2}{x^2}\arctan x^2\,dx = \dfrac{\pi}{2}\left(\sqrt{2}-1\right)$

BI (244)(10)a

3.* $\int_0^\infty \arctan x^n\,\dfrac{dx}{1+x^2} = \int_0^\infty \operatorname{arccot} x^n\,\dfrac{dx}{1+x^2} = \dfrac{\pi^2}{8}$    [$n$ is an integer]

**4.539** $\int_0^\infty x^{s-1}\arctan\left(ae^{-x}\right)\,dx = 2^{-s-1}\,\Gamma(s)a\,\Phi\left(-a^2,s+1,\tfrac{1}{2}\right)$

ET I 222(47)

**4.541** $\int_0^\infty \arctan\left(\dfrac{p\sin qx}{1+p\cos qx}\right)\dfrac{x\,dx}{1+x^2} = \dfrac{\pi}{2}\ln\left(1+pe^{-q}\right)$   [$p>-e^q$]

BI (341)(14)a

## 4.55 Combinations of inverse trigonometric functions and exponentials

**4.551**

1.$^9$ $\int_0^1 (\arcsin x)\,e^{-bx}\,dx = \dfrac{\pi}{2b}\left[I_0(b)-\mathbf{L}_0(b)\right]-\dfrac{\pi e^{-b}}{2b}$

ET I 160(1)

2. $\int_0^1 x\,(\arcsin x)\,e^{-bx}\,dx = \dfrac{\pi}{2b^2}\left[\mathbf{L}_0(b)-I_0(b)+b\,\mathbf{L}_1(b)-b\,I_1(b)\right]+\dfrac{1}{b}$

ET I 161(2)

3.$^9$ $\int_0^\infty \left(\arctan\dfrac{x}{a}\right)e^{-bx}\,dx = \dfrac{1}{b}\left[\operatorname{ci}(ab)\sin(ab)-\operatorname{si}(ab)\cos(ab)\right]$

[$\operatorname{Re}b>0$]
ET I 161(3)

4.$^9$ $\int_0^\infty \left(\operatorname{arccot}\dfrac{x}{a}\right)e^{-bx}\,dx = \dfrac{1}{b}\left[\dfrac{\pi}{2}-\operatorname{ci}(ab)\sin(ab)+\operatorname{si}(ab)\cos(ab)\right]$

[$\operatorname{Re}b>0$]
ET I 161(4)

**4.552** $\int_0^\infty \dfrac{\arctan\dfrac{x}{q}}{e^{2\pi x}-1}\,dx = \dfrac{1}{2}\left[\ln\Gamma(q)-\left(q-\dfrac{1}{2}\right)\ln q+q-\dfrac{1}{2}\ln 2\pi\right]$

[$q>0$]
WH

**4.553** $\displaystyle\int_0^\infty \left(\frac{2}{\pi}\operatorname{arccot} x - e^{-px}\right)\frac{\mathrm{d}x}{x} = C + \ln p$        $[p > 0]$        NT 66(12)

## 4.56   A combination of the arctangent and a hyperbolic function

**4.561** $\displaystyle\int_{-\infty}^\infty \frac{\arctan e^{-x}}{\cosh^{2q} px}\,\mathrm{d}x = \frac{1}{2}\int_{-\infty}^\infty \frac{\Pi(x)}{\cosh^{2q} px}\,\mathrm{d}x = \frac{\sqrt{\pi^3}}{4p}\frac{\Gamma(q)}{\Gamma\left(q + \frac{1}{2}\right)}$

                                                 $[q > 0]$        LI (282)(10)

## 4.57 Combinations of inverse and direct trigonometric functions

**4.571** $\displaystyle\int_0^{\pi/2} \arcsin\left(k\sin x\right)\frac{\sin x\,\mathrm{d}x}{\sqrt{1 - k^2\sin^2 x}} = -\frac{\pi}{2k}\ln k'$        BI (344)(2)

**4.572** $\displaystyle\int_0^\infty \left(\frac{2}{\pi}\operatorname{arccot} x - \cos px\right)\mathrm{d}x = C + \ln p$        $[p > 0]$        NT 66(12)

**4.573**

1.    $\displaystyle\int_0^\infty \operatorname{arccot} qx\,\sin px\,\mathrm{d}x = \frac{\pi}{2p}\left(1 - e^{-\frac{p}{q}}\right)$        $[p > 0,\quad q > 0]$        BI (347)(1)a

2.    $\displaystyle\int_0^\infty \operatorname{arccot} qx\,\cos px\,\mathrm{d}x = \frac{1}{2p}\left[e^{-\frac{p}{q}}\operatorname{Ei}\left(\frac{p}{q}\right) - e^{\frac{p}{q}}\operatorname{Ei}\left(-\frac{p}{q}\right)\right]$

                                                 $[p > 0,\quad q > 0]$        BI (347)(2)a

3.[12]   $\displaystyle\int_0^\infty \operatorname{arccot} rx\,\frac{\sin px\,\mathrm{d}x}{1 \pm 2q\cos px + q^2} = \pm\frac{\pi}{2pq}\ln\frac{1 \pm q}{1 \pm qe^{-\frac{p}{r}}}$    $[q^2 < 1,\quad r > 0,\quad p > 0]$

                                         $= \pm\frac{\pi}{2pq}\ln\frac{q \pm 1}{q \pm e^{-\frac{p}{r}}}$    $[q^2 > 1,\quad r > 0,\quad p > 0]$

                                                     BI (347)(10)

4.    $\displaystyle\int_0^\infty \operatorname{arccot} px\,\frac{\tan x\,\mathrm{d}x}{q^2\cos^2 x + r^2\sin^2 x} = \frac{\pi}{2r^2}\ln\left(1 + \frac{r}{q}\tanh\frac{1}{p}\right)$

                                       $[p > 0,\quad q > 0,\quad r > 0]$    BI (347)(9)

**4.574**

1.    $\displaystyle\int_0^\infty \arctan\left(\frac{2a}{x}\right)\sin(bx)\,\mathrm{d}x = \frac{\pi}{b}e^{-ab}\sinh(ab)$        $[\operatorname{Re} a > 0,\quad b > 0]$        ET I 87(8)

2.[7]   $\displaystyle\int_0^\infty \arctan\frac{a}{x}\cos(bx)\,\mathrm{d}x = \frac{1}{2b}\left[e^{-ab}\operatorname{Ei}(ab) - e^{ab}\operatorname{Ei}(-ab)\right]$

                                               $[a > 0,\quad b > 0]$        ET I 29(7)

3.    $\displaystyle\int_0^\infty \arctan\left[\frac{2ax}{x^2 + c^2}\right]\sin(bx)\,\mathrm{d}x = \frac{\pi}{b}e^{-b\sqrt{a^2 + c^2}}\sinh(ab)$

                                               $[b > 0]$        ET I 87(9)

4.    $\displaystyle\int_0^\infty \arctan\left(\frac{2}{x^2}\right)\cos(bx)\,\mathrm{d}x = \frac{\pi}{b}e^{-b}\sin b$        $[b > 0]$        ET I 29(8)

**4.575**

1. $\int_0^\pi \arctan \dfrac{p\sin x}{1-p\cos x} \sin nx\, dx = \dfrac{\pi}{2n}p^n$ $[p^2<1]$ BI (345)(4)

2. $\int_0^\pi \arctan \dfrac{p\sin x}{1-p\cos x} \sin nx \cos x\, dx = \dfrac{\pi}{4}\left(\dfrac{p^{n+1}}{n+1}+\dfrac{p^{n-1}}{n-1}\right)$

$[p^2<1]$ BI (345)(5)

3. $\int_0^\pi \arctan \dfrac{p\sin x}{1-p\cos x} \cos nx \sin x\, dx = \dfrac{\pi}{4}\left(\dfrac{p^{n+1}}{n+1}-\dfrac{p^{n-1}}{n-1}\right)$

$[p^2<1]$ BI (345)(6)

**4.576**

1. $\int_0^\pi \arctan \dfrac{p\sin x}{1-p\cos x} \dfrac{dx}{\sin x} = \dfrac{\pi}{2}\ln\dfrac{1+p}{1-p}$ $[p^2<1]$ BI(346)(1)

2. $\int_0^\pi \arctan \dfrac{p\sin x}{1-p\cos x} \dfrac{dx}{\tan x} = -\dfrac{\pi}{2}\ln(1-p^2)$ $[p^2<1]$ BI(346)(3)

**4.577**

1. $\int_0^{\pi/2} \arctan\left(\tan\lambda\sqrt{1-k^2\sin^2 x}\right) \dfrac{\sin^2 x\, dx}{\sqrt{1-k^2\sin^2 x}}$

$\qquad = \dfrac{\pi}{2k^2}\left[F(\lambda,k)-E(\lambda,k)+\cot\lambda\left(1-\sqrt{1-k^2\sin^2\lambda}\right)\right]$

BI (344)(4)

2. $\int_0^{\pi/2} \arctan\left(\tan\lambda\sqrt{1-k^2\sin^2 x}\right) \dfrac{\cos^2 x\, dx}{\sqrt{1-k^2\sin^2 x}}$

$\qquad = \dfrac{\pi}{2k^2}\left[E(\lambda,k)-k'^2\,F(\lambda,k)+\cot\lambda\left(\sqrt{1-k^2\sin^2\lambda}-1\right)\right]$

BI (344)(5)

## 4.58 A combination involving an inverse and a direct trigonometric function and a power

**4.581**[10] $\int_0^\infty \arctan x \cos px\, \dfrac{dx}{x} = \int_0^\infty \arctan\dfrac{x}{p}\cos x\, \dfrac{dx}{x} = -\dfrac{\pi}{2}\,\mathrm{Ei}(-p)$

$[\mathrm{Re}(p)>0]$ ET I 29(3), NT 25(13)

## 4.59 Combinations of inverse trigonometric functions and logarithms

**4.591**

1. $\int_0^1 \arcsin x \ln x\, dx = 2-\ln 2-\dfrac{1}{2}\pi$ BI (339)(1)

2.    $\displaystyle\int_0^1 \arccos x \ln x \, dx = \ln 2 - 2$        BI (339)(2)

**4.592**   $\displaystyle\int_0^1 \arccos x \frac{dx}{\ln x} = -\sum_{k=0}^{\infty} \frac{(2k-1)!!}{2^k k!} \frac{\ln(2k+2)}{2k+1}$        BI (339)(8)

**4.593**

1.    $\displaystyle\int_0^1 \arctan x \ln x \, dx = \frac{1}{48}\pi^2 - \frac{\pi}{4} + \frac{1}{2}\ln 2$        BI (339)(3)

2.    $\displaystyle\int_0^1 \text{arccot}\, x \ln x \, dx = -\frac{1}{48}\pi^2 - \frac{\pi}{4} - \frac{1}{2}\ln 2$        BI (339)(4)

**4.594**   $\displaystyle\int_0^1 \arctan x \, (\ln x)^{n-1} (\ln x + n) \, dx = \frac{n!}{(-2)^{n+1}} \left(2^{-n} - 1\right) \zeta(n+1)$        BI (339)(7)

# 4.6 Multiple Integrals

## 4.60 Change of variables in multiple integrals

**4.601**

1.    $\displaystyle\iint\limits_{(\sigma)} f(x,y) \, dx \, dy = \iint\limits_{(\sigma')} f\left[\varphi(u,v), \psi(u,v)\right] |\Delta| \, du \, dv$

     where $x = \varphi(u,v), y = \psi(u,v)$, and $\Delta = \dfrac{\partial\varphi}{\partial u}\dfrac{\partial\psi}{\partial v} - \dfrac{\partial\psi}{\partial u}\dfrac{\partial\varphi}{\partial v} \equiv \dfrac{D(\varphi, \psi)}{D(u,v)}$ is the Jacobian determinant of the functions $\varphi$ and $\psi$.

2.    $\displaystyle\iiint\limits_{(V)} f(x,y,z) \, dx \, dy \, dz = \iiint\limits_{(V')} f\left[\varphi(u,v,w), \psi(u,v,w), \chi(u,v,w)|\Delta| \, du \, dv \, dw\right]$

     where $x = \varphi(u,v,w)$, $y = \psi(u,v,w)$, and $z = \chi(u,v,w)$ and where

$$\Delta = \begin{vmatrix} \frac{\partial\varphi}{\partial u} & \frac{\partial\varphi}{\partial v} & \frac{\partial\varphi}{\partial w} \\ \frac{\partial\psi}{\partial u} & \frac{\partial\psi}{\partial v} & \frac{\partial\psi}{\partial w} \\ \frac{\partial\chi}{\partial u} & \frac{\partial\chi}{\partial v} & \frac{\partial\chi}{\partial w} \end{vmatrix} \equiv \frac{D(\varphi, \psi, \chi)}{D(u,v,w)}$$

     is the Jacobian determinant of the functions $\varphi$, $\psi$, and $\chi$.

     Here, we assume, both in (**4.601** 1) and in (**4.601** 2) that

     (a)     the functions $\varphi, \psi$, and $\chi$ and also their first partial derivatives are continuous in the region of integration;

     (b)     the Jacobian does not change sign in this region;

     (c)     there exists a one-to-one correspondence between the old variables $x, y, z$ and the new ones $u, v, w$ in the region of integration;

     (d)     when we change from the variables $x, y, z$ to the variables $u, v, w$, the region $V$ (resp. $\sigma$) is mapped into the region $V'$ (resp. $\sigma'$).

**4.602**    Transformation to polar coordinates:

$$x = r\cos\varphi, \quad y = r\sin\varphi; \quad \frac{D(x,y)}{D(r,\varphi)} = r$$

**4.603**    Transformation to spherical coordinates:

$$x = r\sin\theta\cos\varphi, \quad y = r\sin\theta\sin\varphi, \quad z = r\cos\theta, \quad \frac{D(x,y,z)}{D(r,\theta,\varphi)} = r^2\sin\theta$$

## 4.61  Change of the order of integration and change of variables

**4.611**

1.
$$\int_0^\alpha dx \int_0^x f(x,y)\,dy = \int_0^\alpha dy \int_y^\alpha f(x,y)\,dx$$

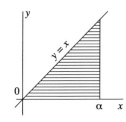

2.
$$\int_0^\alpha dx \int_0^{\frac{\beta}{\alpha}x} f(x,y)\,dy = \int_0^\beta dy \int_{\frac{\alpha}{\beta}y}^\alpha f(x,y)\,dx$$

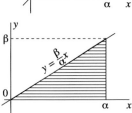

**4.612**

1.
$$\int_0^R dx \int_0^{\sqrt{R^2-x^2}} f(x,y)\,dy = \int_0^R dy \int_0^{\sqrt{R^2-y^2}} f(x,y)\,dx$$

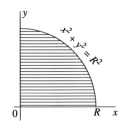

2.
$$\int_0^{2p} dx \int_0^{(q/p)\sqrt{2px-x^2}} f(x,y)\,dy = \int_0^q dy \int_{p\left[1-\sqrt{1-(y/q)^2}\right]}^{p\left[1+\sqrt{1-(y/q)^2}\right]} f(x,y)\,dx$$

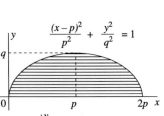

**4.613**

1.
$$\int_0^\alpha dx \int_0^{\beta/(\beta+x)} f(x,y)\,dy = \int_0^{\beta/(\beta+\alpha)} dy \int_0^\alpha f(x,y)\,dx$$

$$+ \int_{\beta/(\beta+\alpha)}^1 dy \int_0^{\beta(1-y)/y} f(x,y)\,dx$$

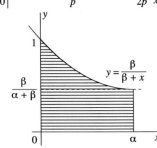

2.
$$\int_0^\alpha dx \int_{\beta x}^{\delta - \nu x} f(x, y)\, dy = \int_0^{\alpha\beta} dy \int_0^{y/\beta} f(x, y)\, dx$$
$$+ \int_{\alpha\beta}^\delta dy \int_0^{(\delta - y)/\gamma} f(x, y)\, dx$$
$$\left[ \alpha = \frac{\delta}{\beta + \gamma}, \quad \alpha > 0, \quad \beta > 0, \quad \gamma > 0 \right]$$

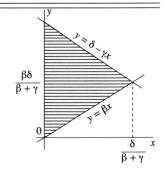

3.
$$\int_0^{2\alpha} dx \int_{x^2/4\alpha}^{3\alpha - x} f(x, y)\, dy = \int_0^\alpha dy \int_0^{2\sqrt{\alpha y}} f(x, y)\, dx +$$
$$+ \int_\alpha^{3\alpha} dy \int_0^{3\alpha - y} f(x, y)\, dx$$

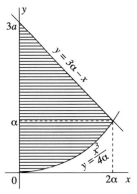

4.
$$\int_0^R dx \int_{\sqrt{R^2 - x^2}}^{x + 2R} f(x, y)\, dy = \int_0^R dy \int_{\sqrt{R^2 - y^2}}^R f(x, y)\, dx$$
$$+ \int_R^{2R} dy \int_0^R f(x, y)\, dx$$
$$+ \int_{2R}^{3R} dy \int_{y - 2R}^R f(x, y)\, dx$$

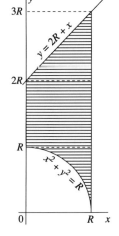

**4.614**
$$\int_0^{\pi/2} d\varphi \int_0^{2R \cos \varphi} f(r, \varphi)\, dr = \int_0^{2R} dr \int_0^{\arccos \frac{r}{2R}} f(r, \varphi)\, d\varphi$$

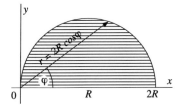
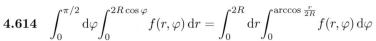
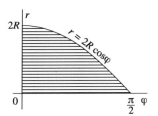

**4.615**   $\displaystyle\int_0^R \mathrm{d}x \int_0^{\sqrt{R^2-x^2}} f(x,y)\,\mathrm{d}y = \int_0^{\pi/2}\mathrm{d}\varphi \int_0^R f\left(r\cos\varphi, r\sin\varphi\right)r\,\mathrm{d}r$

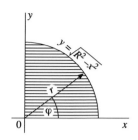

**4.616**[12]   $\displaystyle\int_0^{2R}\mathrm{d}x \int_0^{\sqrt{2Rx-x^2}} f(x,y)\,\mathrm{d}y = \int_0^{\pi/2}\mathrm{d}\varphi \int_0^{2R\cos\varphi} f\left(r\cos\varphi, r\sin\varphi\right)r\,\mathrm{d}r$

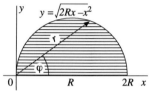

**4.617**   $\displaystyle\int_\alpha^\beta \mathrm{d}x \int_{\varphi_1(x)}^{\varphi_2(x)} f(x,y)\mathrm{d}y = \int_0^\beta \mathrm{d}x \int_0^{\varphi_2(x)} f(x,y)\mathrm{d}y - \int_0^\beta \mathrm{d}x \int_0^{\varphi_1(x)} f(x,y)\mathrm{d}y - \int_0^\alpha \mathrm{d}x \int_0^{\varphi_2(x)} f(x,y)\mathrm{d}y$

$$+ \int_0^\alpha \mathrm{d}x \int_0^{\varphi_1(x)} f(x,y)\,\mathrm{d}y$$

$$[\varphi_1(x) \le \varphi_2(x) \text{ for } \alpha \le x \le \beta]$$

**4.618**   $\displaystyle\int_0^\gamma \mathrm{d}x \int_0^{\varphi(x)} f(x,y)\,\mathrm{d}y = \int_0^\gamma \mathrm{d}x \int_0^1 f\left[x, z\varphi(x)\right]\varphi(x)\mathrm{d}z \quad [y = z\varphi(x)]$

$$= \gamma \int_0^1 \mathrm{d}z \int_0^{\varphi(\gamma z)} f(\gamma z, y)\,\mathrm{d}y \qquad [x = \gamma z]$$

**4.619**   $\displaystyle\int_{x_0}^{x_1} \mathrm{d}x \int_{y_0}^{y_1} f(x,y)\,\mathrm{d}y = \int_{x_0}^{x_1} \mathrm{d}x \int_0^1 (y_1 - y_0) f\left[x, y_0 + (y_1 - y_0)t\right]\mathrm{d}t$

$$[y = y_0 + (y_1 - y_0)\,t]$$

## 4.62  Double and triple integrals with constant limits

**4.620**   General formulas

1.   $\displaystyle\int_0^\pi \mathrm{d}\omega \int_0^\infty f'\left(p\cosh x + q\cos\omega \sinh x\right)\sinh x\,\mathrm{d}x = -\frac{\pi\,\mathrm{sign}\,p}{\sqrt{p^2-q^2}} f\left(\mathrm{sign}\,p\sqrt{p^2-q^2}\right)$

$$\left[p^2 > q^2, \quad \lim_{x\to+\infty} f(x) = 0\right] \qquad \text{LO III 389}$$

2.   $\displaystyle\int_0^{2\pi} \mathrm{d}\omega \int_0^\infty f'\left[p\cosh x + (q\cos\omega + r\sin\omega)\sinh x\right]\sinh x\,\mathrm{d}x$

$$= -\frac{2\pi\,\mathrm{sign}\,p}{\sqrt{p^2-q^2-r^2}} f\left(\mathrm{sign}\,p\sqrt{p^2-q^2-r^2}\right)$$

$$\left[p^2 > q^2 + r^2, \quad \lim_{x\to+\infty} f(x) = 0\right] \qquad \text{LO III 390}$$

3.    $\displaystyle \int_0^\pi \int_0^\pi \frac{dx\,dy}{\sin x \sin^2 y} f'\left[\frac{p - q\cos x}{\sin x \sin y} + r\cot y\right] = -\frac{2\pi\,\mathrm{sign}\,p}{\sqrt{p^2 - q^2 - r^2}} f\left(\mathrm{sign}\,p\sqrt{p^2 - q^2 - r^2}\right)$

$$\left[p^2 > q^2 + r^2, \quad \lim_{x \to +\infty} f(x) = 0\right]$$

LO III 280

4.    $\displaystyle \int_{-\infty}^\infty dx \int_{-\infty}^\infty f'\left(p\cosh x\cosh y + q\sinh x\cosh y + r\sinh y\right)\cosh y\,dy$

$$= -\frac{2\pi\,\mathrm{sign}\,p}{\sqrt{p^2 - q^2 - r^2}} f\left(\mathrm{sign}\,p\sqrt{p^2 - q^2 - r^2}\right)$$

$$\left[p^2 > q^2 + r^2, \quad \lim_{x \to +\infty} f(x) = 0\right] \quad \text{LO III 390}$$

5.    $\displaystyle \int_0^\infty dx \int_0^\pi f\left(p\cosh x + q\cos\omega\sinh x\right)\sinh^2 x\sin\omega\,d\omega = 2\int_0^\infty f\left(\mathrm{sign}\,p\sqrt{p^2 - q^2}\cosh x\right)\sinh^2 x\,dx$

$$\left[\lim_{x \to +\infty} f(x) = 0\right] \quad\quad\quad \text{LO III 391}$$

6.    $\displaystyle \int_0^\infty dx \int_0^{2\pi} d\omega \int_0^\pi f\left[p\cosh x + (q\cos\omega + r\sin\omega)\sin\theta\sinh x\right]\sinh^2 x\sin\theta\,d\theta$

$$= 4\int_0^\infty f\left(\mathrm{sign}\,p\sqrt{p^2 - q^2 - r^2}\cosh x\right)\sinh^2 x\,dx$$

$$\left[p^2 > q^2 + r^2, \quad \lim_{x \to +\infty} f(x) = 0\right] \quad \text{LO III 390}$$

7.    $\displaystyle \int_0^\infty dx \int_0^{2\pi} d\omega \int_0^\pi f\left\{p\cosh x + \left[(q\cos\omega + r\sin\omega)\sin\theta + s\cosh\theta\right]\sinh x\right\}\sinh^2 x\sin\theta\,d\theta$

$$= 4\pi\int_0^\infty f\left(\mathrm{sign}\,p\sqrt{p^2 - q^2 - r^2 - s^2}\cosh x\right)\sinh^2 x\,dx$$

$$\left[p^2 > q^2 + r^2 + s^2, \quad \lim_{x \to +\infty} f(x) = 0\right] \quad \text{LO III 391}$$

**4.621**

1.    $\displaystyle \int_0^{\pi/2} \int_0^{\pi/2} \frac{\sin y\sqrt{1 - k^2\sin^2 x\sin^2 y}}{1 - k^2\sin^2 y}\,dx\,dy = \frac{\pi}{2\sqrt{1 - k^2}}$      LO I 252(90)

2.    $\displaystyle \int_0^{\pi/2} \int_0^{\pi/2} \frac{\cos y\sqrt{1 - k^2\sin^2 x\sin^2 y}}{1 - k^2\sin^2 y}\,dx\,dy = \boldsymbol{K}(k)$      LO I 252(91)

3.    $\displaystyle \int_0^{\pi/2} \int_0^{\pi/2} \frac{\sin\alpha\sin y\,dx\,dy}{\sqrt{1 - \sin^2\alpha\sin^2 x\sin^2 y}} = \frac{\pi\alpha}{2}$      LO I 253

**4.622**

1.    $\displaystyle \int_0^\pi \int_0^\pi \int_0^\pi \frac{dx\,dy\,dz}{1 - \cos x\cos y\cos z} = 4\pi\,\boldsymbol{K}^2\left(\frac{\sqrt{2}}{2}\right)$      MO 137

2.    $\displaystyle \int_0^\pi \int_0^\pi \int_0^\pi \frac{dx\,dy\,dz}{3 - \cos y\cos z - \cos x\cos z - \cos x\cos y} = \sqrt{3}\pi\,\boldsymbol{K}^2\left(\sin\frac{\pi}{12}\right)$      MO 137

3.  $$\int_0^\pi \int_0^\pi \int_0^\pi \frac{dx\,dy\,dz}{3-\cos x - \cos y - \cos z} = 4\pi \left[18 + 12\sqrt{2} - 10\sqrt{3} - 7\sqrt{6}\right] \boldsymbol{K}^2\left[\left(2-\sqrt{3}\right)\left(\sqrt{3}-\sqrt{2}\right)\right]$$

MO 137

**4.623³**  $$\int_0^\infty \int_0^\infty \varphi\left(a^2 x^2 + b^2 y^2\right)\,dx\,dy = \frac{\pi}{2ab}\int_0^\infty \varphi\left(x^2\right) x\,dx$$

**4.624**  $$\int_0^\pi \int_0^{2\pi} f\left(\alpha\cos\theta + \beta\sin\theta\cos\psi + \gamma\sin\theta\sin\psi\right)\sin\theta\,d\theta\,d\psi$$

$$= 2\pi\int_0^\pi f\left(R\cos p\right)\sin p\,dp = 2\pi\int_{-1}^1 f(Rt)\,dt$$
$$\left[R = \sqrt{\alpha^2 + \beta^2 + \gamma^2}\right]$$

**4.625⁸**  $$p_l(a,b) = \int_0^a dx \int_0^b dy\,\left(x^2 + y^2 + 1\right)^{-3/2} P_l\left(1/\sqrt{x^2 + y^2 + 1}\right)$$

Then, for even and odd subscripts:

- $$p_{2l}(a,b) = \frac{1}{l(2l+1)2^{2l}}\frac{ab}{\sqrt{a^2+b^2+1}}\sum_{k=0}^{l-1}\frac{(-1)^{l-k-1}2^{2k}\binom{2l+2k}{l+k}\binom{l+k}{l-k-1}}{\binom{2k}{k}(2k+1)}$$
$$\times (2l+2k+1)\sum_{j=0}^k \frac{\binom{2j}{j}}{2^{2j}}\frac{1}{(a^2+b^2+1)^j}\left(\frac{1}{(a^2+1)^{k-j+1}} + \frac{1}{(b^2+1)^{k-j+1}}\right)$$

- $$p_{2l+1}(a,b) = \frac{1}{2^{2l+1}(2l+1)}\sum_{k=0}^l \frac{(-1)^{l+k}}{2^{2k}}\binom{l}{k}\binom{l+k+1}{k}\binom{2l+2k+1}{l+k}$$
$$\times\left\{\frac{1}{(b^2+1)^k}\frac{b}{\sqrt{b^2+1}}\arctan^{-1}\frac{a}{\sqrt{b^2+1}} + \frac{1}{(a^2+1)^k}\frac{a}{\sqrt{a^2+1}}\arctan^{-1}\frac{b}{\sqrt{a^2+1}}\right.$$
$$\left. +ab\sum_{j=1}^k \frac{2^{2j-1}}{j\binom{2j}{j}}\cdot\frac{1}{(a^2+b^2+1)^j}\left(\frac{1}{(a^2+1)^{k-j+1}} + \frac{1}{(b^2+1)^{k-j+1}}\right)\right\}$$

**4.626***

1.  $$\int_0^1 dx \int_0^1 dy\,\frac{x^{u-1}y^{v-1}}{1-xyz}(-\ln xy)^s = \Gamma(s+1)\frac{\Phi(z,s+1,v) - \Phi(z,s+1,u)}{u-v}$$
$$[z \in C - [1,\infty], \operatorname{Re}(s) > -2] \quad \text{or} \quad [z = 1, \operatorname{Re}(s) > -1] \quad \text{GS}$$

2.  $$\int_0^1 dx \int_0^1 dy\,\frac{(xy)^{u-1}}{1-xyz}(-\ln xy)^s = \Gamma(s+2)\Phi(z,s+2,u)$$
$$[z \in C - [1,\infty], \operatorname{Re}(s) > -2] \quad \text{or} \quad [z = 1, \operatorname{Re}(s) > -1] \quad \text{GS}$$

3.  $$\int_0^1 dx \int_0^1 dy\,\frac{-x\ln xy}{1+x^2y^2} = \boldsymbol{G} - \frac{\pi^2}{48} \qquad\qquad\qquad \text{GS}$$

4.  $$\int_0^1 dx \int_0^1 dy\,\frac{-x\ln xy}{1-x^2y^2} = \frac{\pi^2}{12} \qquad\qquad\qquad \text{GS}$$

5.  $$\int_0^1 dx \int_0^1 dy \frac{1}{1+x^2y^2} = \boldsymbol{G}$$                                                                GS

6.  $$\int_0^1 dx \int_0^1 dy \frac{-1}{(1-xyz)\ln xy} = \frac{-\ln(1-z)}{z} \qquad\qquad [z \neq 1]$$                                GS

7.  $$\int_0^1 dx \int_0^1 dy \frac{1}{1-xy} = \zeta(2)$$                                                                          GS

8.  $$\int_0^1 dx \int_0^1 dy \frac{-\ln xy}{1-xy} = 2\zeta(3)$$                                                                   GS

9.  $$\int_0^1 dx \int_0^1 dy \frac{x^{u-1}y^{v-1}}{-\ln xy} = \frac{1}{u-v}\ln\frac{u}{v}$$                                       GS

10. $$\int_0^1 dx \int_0^1 dy \frac{(xy)^{u-1}}{-\ln xy} = \frac{1}{u}$$                                                           GS

11. $$\int_0^1 dx \int_0^1 dy \frac{-x}{(1+xy)\ln xy} = \ln\frac{\pi}{2}$$                                                         GS

12. $$\int_0^1 dx \int_0^1 dy \frac{-x}{(1+x^2y^2)\ln xy} = \ln\left(\frac{\sqrt{2\pi}}{\Gamma\left(\frac{3}{4}\right)^2}\right)$$  GS

13. $$\int_0^1 dx \int_0^1 dy \frac{-1}{(1+x^2y^2)\ln xy} = \frac{\pi}{4}$$                                                        GS

14. $$\int_0^1 dx \int_0^1 dy \frac{y}{1-x^3y^3} = \frac{\pi}{3\sqrt{3}}$$                                                         GS

15. $$\int_0^1 dx \int_0^1 dy \frac{1}{1-x^2y^2} = \frac{\pi^2}{8}$$                                                               GS

## 4.63–4.64  Multiple integrals

**4.631**  $$\int_p^x dt_{n-1}\int_p^{t_{n-1}} dt_{n-2}\dots\int_p^{t_1} f(t)\,dt = \frac{1}{(n-1)!}\int_p^x (x-t)^{n-1}f(t)\,dt,$$
where $f(t)$ is continuous on the interval $[p,q]$ and $p \leq x \leq q$.                                                         FI II 692

**4.632**

1.  $$\underset{\substack{x_1\geq 0, x_2\geq 0,\dots,x_n\geq 0 \\ x_1+x_2+\dots+x_n\leq h}}{\int\int\cdots\int} dx_1\,dx_2\cdots dx_n = \frac{h^n}{n!}$$

    [the volume of an $n$-dimensional simplex]    FI III 472

2.  $$\underset{x_1^2+x_2^2+\dots+x_n^2\leq R^2}{\int\int\cdots\int} dx_1\,dx_2\cdots dx_n = \frac{\sqrt{\pi^n}}{\Gamma\left(\dfrac{n}{2}+1\right)}R^n \qquad \text{[the volume of an } n\text{-dimensional sphere]}$$

    FI III 473

**4.633**
$$\iint \cdots \int_{x_1^2+x_2^2+\cdots+x_n^2 \le 1} \frac{dx_1\, dx_2 \cdots dx_n}{\sqrt{1 - x_1^2 - x_2^2 - \cdots - x_n^2}} = \frac{\pi^{(n+1)/2}}{\Gamma\left(\dfrac{n+1}{2}\right)} \quad [n > 1]$$

[Half-area of the surface of an $(n+1)$-dimensional sphere $x_1^2 + x_2^2 + \cdots + x_{n+1}^2 = 1$]    FI III 474

**4.634**[8]
$$\iint \cdots \int_{\substack{x_1 \ge 0, x_2 \ge 0, \ldots, x_n \ge 0 \\ \left(\frac{x_1}{q_1}\right)^{\alpha_1} + \left(\frac{x_2}{q_2}\right)^{\alpha_2} + \cdots + \left(\frac{x_n}{q_n}\right)^{\alpha_n} \le 1}} x_1^{p_1-1} x_2^{p_2-1} \cdots x_n^{p_n-1}\, dx_1\, dx_2 \ldots dx_n$$

$$= \frac{q_1^{p_1} q_2^{p_2} \cdots q_n^{p_n}}{\alpha_1 \alpha_2 \ldots \alpha_n} \frac{\Gamma\left(\dfrac{p_1}{\alpha_1}\right) \Gamma\left(\dfrac{p_2}{\alpha_2}\right) \ldots \Gamma\left(\dfrac{p_n}{\alpha_n}\right)}{\Gamma\left(\dfrac{p_1}{\alpha_1} + \dfrac{p_2}{\alpha_2} + \cdots + \dfrac{p_n}{\alpha_n} + 1\right)}$$

$$[\alpha_i > 0, \quad p_i > 0, \quad q_i > 0, \quad i = 1, 2, \ldots, n] \quad \text{FI III 477}$$

**4.635**

**1.**[8]
$$\iint \cdots \int_{\substack{x_1 \ge 0, x_2 \ge 0, \ldots, x_n \ge 0 \\ \left(\frac{x_1}{q_1}\right)^{\alpha_1} + \left(\frac{x_2}{q_2}\right)^{\alpha_2} + \cdots + \left(\frac{x_n}{q_n}\right)^{\alpha_n} \ge 1}} f\left[\left(\frac{x_1}{q_1}\right)^{\alpha_1} + \left(\frac{x_2}{q_2}\right)^{\alpha_2} + \cdots + \left(\frac{x_n}{q_n}\right)^{\alpha_n}\right]$$

$$\times x_1^{p_1-1} x_2^{p_2-1} \cdots x_n^{p_n-1}\, dx_1\, dx_2 \cdots dx_n$$

$$= \frac{q_1^{p_1} q_2^{p_2} \cdots q_n^{p_n}}{\alpha_1 \alpha_2 \cdots \alpha_n} \frac{\Gamma\left(\dfrac{p_1}{\alpha_1}\right) \Gamma\left(\dfrac{p_2}{\alpha_2}\right) \ldots \Gamma\left(\dfrac{p_n}{\alpha_n}\right)}{\Gamma\left(\dfrac{p_1}{\alpha_1} + \dfrac{p_2}{\alpha_2} + \cdots + \dfrac{p_n}{\alpha_n}\right)} \cdot \int_1^\infty f(x) x^{\frac{p_1}{\alpha_1} + \frac{p_2}{\alpha_2} + \cdots + \frac{p_n}{\alpha_n} - 1}\, dx$$

under the assumption that the integral on the right converges absolutely.          FI III 487

**2.**[8]
$$\iint \cdots \int_{\substack{x_1 \ge 0, x_2 \ge 0, \cdots, x_n \ge 0 \\ \left(\frac{x_1}{q_1}\right)^{\alpha_1} + \left(\frac{x_2}{q_2}\right)^{\alpha_2} + \cdots + \left(\frac{x_n}{q_n}\right)^{\alpha_n} \le 1}} f\left[\left(\frac{x_1}{q_1}\right)^{\alpha_1} + \left(\frac{x_2}{q_2}\right)^{\alpha_2} + \cdots + \left(\frac{x_n}{q_n}\right)^{\alpha_n}\right]$$

$$\times x_1^{p_1-1} x_2^{p_2-1} \cdots x_n^{p_n-1}\, dx_1\, dx_2 \cdots dx_n$$

$$= \frac{q_1^{p_1} q_2^{p_2} \cdots q_n^{p_n}}{\alpha_1 \alpha_2 \ldots \alpha_n} \frac{\Gamma\left(\dfrac{p_1}{\alpha_1}\right) \Gamma\left(\dfrac{p_2}{\alpha_2}\right) \ldots \Gamma\left(\dfrac{p_n}{\alpha_n}\right)}{\Gamma\left(\dfrac{p_1}{\alpha_1} + \dfrac{p_2}{\alpha_2} + \cdots + \dfrac{p_n}{\alpha_n}\right)} \int_0^1 f(x) x^{\frac{p_1}{\alpha_1} + \frac{p_2}{\alpha_2} + \cdots + \frac{p_n}{\alpha_n} - 1}\, dx$$

under the assumptions that the one-dimensional integral on the right converges absolutely and that the numbers $q_i, \alpha_i$, and $p_i$ are positive.          FI III 479

In particular,

3. 
$$\iint \cdots \int_{\substack{x_1 \geq 0, x_2 \geq 0, \ldots, x_n \geq 0 \\ x_1 + x_2 + \cdots + x_n \leq 1}} x_1^{p_1 - 1} x_2^{p_2 - 1} \ldots x_n^{p_n - 1} e^{-q(x_1 + x_2 + \cdots + x_n)} \, dx_1 \, dx_2 \ldots dx_n$$

$$= \frac{\Gamma(p_1) \Gamma(p_2) \ldots \Gamma(p_n)}{\Gamma(p_1 + p_2 + \cdots + p_n)} \int_0^1 x^{p_1 + p_2 + \cdots + p_n - 1} e^{-qx} \, dx$$
$$[n > 0, \quad p_i > 0]$$

4.[8]
$$\iint \cdots \int_{\substack{x_1 \geq 0, x_2 \geq 0, \cdots, x_n \geq 0 \\ x_1^{\alpha_1} + x_2^{\alpha_2} + \cdots + x_n^{\alpha_n} \leq 1}} \frac{x_1^{p_1 - 1} x_2^{p_2 - 1} \ldots x_n^{p_n - 1}}{(1 - x_1^{\alpha_1} - x_2^{\alpha_2} - \cdots - x_n^{\alpha_n})^\mu} \, dx_1 \, dx_2 \ldots dx_n$$

$$= \frac{\Gamma(1 - \mu)}{\alpha_1 \alpha_2 \ldots \alpha_n} \frac{\Gamma\left(\frac{p_1}{\alpha_1}\right) \Gamma\left(\frac{p_2}{\alpha_2}\right) \ldots \Gamma\left(\frac{p_n}{\alpha_n}\right)}{\Gamma\left(1 - \mu + \frac{p_1}{\alpha_1} + \frac{p_2}{\alpha_2} + \cdots + \frac{p_n}{\alpha_n}\right)}$$
$$[p_i > 0, \quad \mu < 1] \qquad \text{FI III 480}$$

**4.636**

1.[8]
$$\iint \cdots \int_{\substack{x_1 \geq 0, x_2 \geq 0, \ldots, x_n \geq 0 \\ x_1^{\alpha_1} + x_2^{\alpha_2} + \cdots + x_n^{\alpha_n} \geq 1}} \frac{x_1^{p_1 - 1} x_2^{p_2 - 1} \ldots x_n^{p_n - 1}}{(x_1^{\alpha_1} + x_2^{\alpha_2} + \cdots + x_n^{\alpha_n})^\mu} \, dx_1 \, dx_2 \ldots dx_n$$

$$= \frac{1}{\alpha_1 \alpha_2 \ldots \alpha_n \left(\mu - \frac{p_1}{\alpha_1} - \frac{p_2}{\alpha_2} - \cdots - \frac{p_n}{\alpha_n}\right)} \frac{\Gamma\left(\frac{p_1}{\alpha_1}\right) \Gamma\left(\frac{p_2}{\alpha_2}\right) \ldots \Gamma\left(\frac{p_n}{\alpha_n}\right)}{\Gamma\left(\frac{p_1}{\alpha_1} + \frac{p_2}{\alpha_2} + \cdots + \frac{p_n}{\alpha_n}\right)}$$
$$\left[p_i > 0, \quad \mu > \frac{p_1}{\alpha_1} + \frac{p_2}{\alpha_2} + \cdots + \frac{p_n}{\alpha_n}\right] \qquad \text{FI III 488}$$

2.[8]
$$\iint \cdots \int_{\substack{x_1 \geq 0, x_2 \geq 0, \cdots, x_n \geq 0 \\ x_1^{\alpha_1} + x_2^{\alpha_2} + \cdots + x_n^{\alpha_n} \leq 1}} \frac{x_1^{p_1 - 1} x_2^{p_2 - 1} \ldots x_n^{p_n - 1}}{(x_1^{\alpha_1} + x_2^{\alpha_2} + \cdots + x_n^{\alpha_n})^\mu} \, dx_1 \, dx_2 \ldots dx_n$$

$$= \frac{1}{\alpha_1 \alpha_2 \ldots \alpha_n \left(\frac{p_1}{\alpha_1} + \frac{p_2}{\alpha_2} + \cdots + \frac{p_n}{\alpha_n} - \mu\right)} \frac{\Gamma\left(\frac{p_1}{\alpha_1}\right) \Gamma\left(\frac{p_2}{\alpha_2}\right) \ldots \Gamma\left(\frac{p_n}{\alpha_n}\right)}{\Gamma\left(\frac{p_1}{\alpha_1} + \frac{p_2}{\alpha_2} + \cdots + \frac{p_n}{\alpha_n}\right)}$$
$$\left[\mu < \frac{p_1}{\alpha_1} + \frac{p_2}{\alpha_2} + \cdots + \frac{p_n}{\alpha_n}\right] \qquad \text{FI III 480}$$

3.[12]
$$\iint \cdots \int_{\substack{x_1 \geq 0, x_2 \geq 0, \ldots, x_n \geq 0 \\ x_1^{\alpha_1} + x_2^{\alpha_2} + \cdots + x_n^{\alpha_n} \leq 1}} x_1^{p_1 - 1} x_2^{p_2 - 1} \ldots x_n^{p_n - 1} \sqrt{\frac{1 - x_1^{\alpha_1} - x_2^{\alpha_2} - \cdots - x_n^{\alpha_n}}{1 + x_1^{\alpha_1} + x_2^{\alpha_2} + \cdots + x_n^{\alpha_n}}} \, dx_1 \, dx_2 \ldots dx_n$$

$$= \frac{\sqrt{\pi}}{2} \frac{\Gamma\left(\frac{p_1}{\alpha_1}\right) \Gamma\left(\frac{p_2}{\alpha_2}\right) \ldots \Gamma\left(\frac{p_n}{\alpha_n}\right)}{\alpha_1 \alpha_2 \ldots \alpha_n} \frac{1}{\Gamma(m)} \left[\frac{\Gamma\left(\frac{m}{2}\right)}{\Gamma\left(\frac{m+1}{2}\right)} - \frac{\Gamma\left(\frac{m+1}{2}\right)}{\Gamma\left(\frac{m+2}{2}\right)}\right],$$

where $m = \dfrac{p_1}{\alpha_1} + \dfrac{p_2}{\alpha_2} + \cdots + \dfrac{p_n}{\alpha_n}$.

FI III 480

**4.637**[12] $\displaystyle\iint \cdots \int_{\substack{x_1 \geq 0,\ x_2 \geq 0, \ldots, x_n \geq 0 \\ x_1+x_2+\cdots+x_n \leq 1}} f(x_1 + x_2 + \cdots + x_n) \frac{x_1^{p_1-1} x_2^{p_2-1} \ldots x_n^{p_n-1}\, dx_1\, dx_2 \ldots dx_n}{(q_1 x_1 + q_2 x_2 + \cdots + q_n x_n + r)^{p_1+p_2+\cdots+p_n}}$

$$= \frac{\Gamma(p_1)\Gamma(p_2)\ldots\Gamma(p_n)}{\Gamma(p_1+p_2+\ldots p_n)} \int_0^\infty f(x) \frac{x^{p_1+p_2+\cdots+p_n-1}}{(q_1 x + r)^{p_1}(q_2 x + r)^{p_2}\ldots(q_n x + r)^{p_n}}\, dx,$$
$$[q_i \geq 0, \quad r > 0]$$

where $f(x)$ is continuous on the interval $(0,1)$.

**4.638**

1.[12] $\displaystyle\int_0^\infty \int_0^\infty \cdots \int_0^\infty \frac{x_1^{p_1-1} x_2^{p_2-1} \ldots x_n^{p_n-1} e^{-(q_1 x_1 + q_2 x_2 + \cdots + q_n x_n)}}{(r_0 + r_1 x_1 + r_2 x_2 + \cdots + r_n x_n)^s}\, dx_1\, dx_2 \ldots dx_n$

$$= \frac{\Gamma(p_1)\Gamma(p_2)\ldots\Gamma(p_n)}{\Gamma(s)} \int_0^\infty \frac{e^{-r_0 x} x^{s-1}\, dx}{(q_1 + r_1 x)^{p_1}(q_2 + r_2 x)^{p_2}\ldots(q_n + r_n x)^{p_n}}$$

where $p_i$, $q_i$, $r_i$, and $s$ are positive. This result is also valid for $r_0 = 0$ provided $p_1 + p_2 + \cdots + p_n > s$.

2. $\displaystyle\int_0^\infty \int_0^\infty \cdots \int_0^\infty \frac{x_1^{p_1-1} x_2^{p_2-1} \ldots x_n^{p_n-1}}{(r_0 + r_1 x_1 + r_2 x_2 + \cdots + r_n x_n)^s}\, dx_1\, dx_2 \ldots dx_n$

$$= \frac{\Gamma(p_1)\Gamma(p_2)\ldots\Gamma(p_n)\Gamma(s p_1 p_2 - \cdots - p_n)}{r_1^{p_1} r_2^{p_2} \cdots r_n^{p_n} r_0^{s-p_1-p_2-\cdots-p_n}\Gamma(s)}$$
$$[p_i > 0, \quad r_i > 0, \quad s > 0]$$

3.[12] $\displaystyle\int_0^\infty \int_0^\infty \cdots \int_0^\infty \frac{x_1^{p_1-1} x_2^{p_2-1} \ldots x_n^{p_n-1}}{[1 + (r_1 x_1)^{q_1} + (r_2 x_2)^{q_2} + \cdots + (r_n x_n)^{q_n}]^s}\, dx_1\, dx_2 \ldots dx_n$

$$= \frac{\Gamma\left(\dfrac{p_1}{q_1}\right)\Gamma\left(\dfrac{p_2}{q_2}\right)\ldots\Gamma\left(\dfrac{p_n}{q_n}\right)}{q_1 q_2 \ldots q_n r_1^{p_1} r_2^{p_2} \ldots r_n^{p_n}} \frac{\Gamma\left(s - \dfrac{p_1}{q_1} - \dfrac{p_2}{q_2} - \cdots - \dfrac{p_n}{q_n}\right)}{\Gamma(s)}$$
$$[p_i > 0, \quad q_i > 0, \quad r_i > 0, \quad s > 0]$$

**4.639**

1. $\displaystyle\iint \cdots \int_{x_1^2+x_2^2+\cdots+x_n^2 \leq 1} (p_1 x_1 + p_2 x_2 + \cdots + p_n x_n)^{2m}\, dx_1\, dx_2 \ldots dx_n$

$$= \frac{(2m-1)!!}{2^m} \frac{\sqrt{\pi^n}}{\Gamma\left(\dfrac{n}{2}+m+1\right)} (p_1^2 + p_2^2 + \cdots + p_n^2)^m$$

FI III 482

2. $\displaystyle\iint \cdots \int_{x_1^2+x_2^2+\cdots+x_n^2 \leq 1} (p_1 x_1 + p_2 x_2 + \cdots + p_n x_n)^{2m+1}\, dx_1\, dx_2 \ldots dx_n = 0$
FI III 483

**4.641**

1.[11] $\displaystyle\iint \cdots \int_{x_1^2+x_2^2+\cdots+x_n^2 \leq 1} e^{p_1 x_1 + p_2 x_2 + \cdots + p_n x_n}\, dx_1\, dx_2 \ldots dx_n$

$$= \sqrt{\pi^n} \sum_{k=0}^\infty \frac{1}{k!\,\Gamma\left(\dfrac{n}{2}+k+1\right)} \left(\frac{p_1^2 + p_2^2 + \cdots + p_n^2}{4}\right)^k$$

FI III 483

**2.**[12]
$$\underset{x_1^2+x_2^2+\cdots+x_{2n}^2\leq 1}{\int\int\cdots\int} e^{p_1x_1+p_2x_2+\cdots+p_{2n}x_{2n}}\,dx_1\,dx_2\ldots dx_{2n} = \frac{(2\pi)^n\, I_n\left(\sqrt{p_1^2+p_2^2+\cdots+p_{2n}^2}\right)}{\left(p_1^2+p_2^2+\cdots+p_{2n}^2\right)^n}$$

FI III 483a

**4.642**
$$\underset{x_1^2+x_2^2+\cdots+x_n^2\leq R^2}{\int\int\cdots\int} f\left(\sqrt{x_1^2+x_2^2+\cdots+x_n^2}\right)\,dx_1\,dx_2\ldots dx_n = \frac{2\sqrt{\pi^n}}{\Gamma\left(\dfrac{n}{2}\right)}\int_0^R x^{n-1}f(x)\,dx,$$

where $f(x)$ is a function that is continuous on the interval $(0,R)$.          FI III 485

**4.643** $\displaystyle\int_0^1\int_0^1\cdots\int_0^1 f\,(x_1,x_2,\cdots,x_n)\,(1-x_1)^{p_1-1}\,(1-x_2)^{p_2-1}\ldots(1-x_n)^{p_n-1}$

$$\times x_2^{p_1}\,x_3^{p_1+p_2}\cdots x_n^{p_1+p_2+\cdots+p_{n-1}}\,dx_1\,dx_2\ldots dx_n$$

$$= \frac{\Gamma\,(p_1)\,\Gamma\,(p_2)\ldots\Gamma\,(p_n)}{\Gamma\,(p_1+p_2+\cdots+p_n)}\int_0^1 f(x)(1-x)^{p_1+p_2+\cdots+p_n-1}\,dx$$

under the assumption that the integral on the right converges absolutely.       FI III 488

**4.644** $\displaystyle\overset{n-1}{\overbrace{\underset{x_1^2+x_2^2+\cdots+x_n^2=1}{\int\int\cdots\int}}} f\,(p_1x_1+p_2x_2+\cdots+p_nx_n)\,\frac{dx_1\,dx_2\ldots dx_{n-1}}{|x_n|}$

$$= 2\underset{x_1^2+x_2^2+\cdots+x_{n-1}^2\leq 1}{\int\int\cdots\int} f\,(p_1x_1+p_2x_2+\cdots+p_nx_n)\,\frac{dx_1\,dx_2\cdots dx_{n-1}}{\sqrt{1-x_1^2-x_2^2-\cdots-x_{n-1}^2}}$$

$$= \frac{2\sqrt{\pi^{n-1}}}{\Gamma\left(\dfrac{n-1}{2}\right)}\int_0^\pi f\left(\sqrt{p_1^2+p_2^2+\cdots+p_n^2}\cos x\right)\sin^{n-2}x\,dx \qquad [n\geq 3]$$

where $f(x)$ is continuous on the interval $\left\{-\sqrt{p_1^2+p_2^2+\cdots+p_n^2},\sqrt{p_1^2+p_2^2+\cdots+p_n^2}\right\}$    FI III 489

**4.645**   Suppose that two functions $f\,(x_1,x_2,\ldots,x_n)$ and $g\,(x_1,x_2,\ldots,x_n)$ are continuous in a closed bounded region $D$ and that the smallest and greatest values of the function $g$ in $D$ are $m$ and $M$ respectively. Let $\varphi(u)$ denote a function that is continuous for $m\leq u\leq M$. We denote by $\psi(u)$ the integral

**1.**     $\displaystyle\psi(u) = \underset{m\leq g(x_1,x_2,\ldots,x_n)\leq u}{\int\int\cdots\int} f\,(x_1,x_2,\ldots,x_n)\,dx_1\,dx_2\ldots dx_n,$

over that portion of the region $D$ on which the inequality $m\leq g\,(x_1,x_2,\ldots,x_n)\leq u$ is satisfied. Then

**2.**     $\displaystyle\underset{m\leq g(x_1,x_2,\ldots,x_n)\leq M}{\int\int\cdots\int} f\,(x_1,x_2,\ldots,x_n)\,\varphi\,[g\,(x_1,x_2,\ldots,x_n)]\,dx_1\,dx_2\ldots dx_n$

$$= (S)\int_m^M \varphi(u)\,d\,\psi(u) = (R)\int_m^M \varphi(u)\frac{d\,\psi(u)}{du}\,du$$

where the middle integral must be understood in the sense of Stieltjes. If the derivative $\frac{d\psi}{du}$ exists and is continuous, the Riemann integral on the right exists.

$M$ may be $+\infty$ in formulas **4.645** 2, in which case $\int_m^{+\infty}$ should be understood to mean $\displaystyle\lim_{M\to+\infty}\int_m^M$.

**4.646**[8] $\displaystyle\iint \cdots \int_{\substack{x_1 \geq 0, x_2 \geq 0, \ldots, x_n \geq 0 \\ x_1 + x_2 + \cdots + x_n \leq 1}} \frac{x_1^{p_1-1} x_2^{p_2-1} \ldots x_n^{p_n-1}}{(q_1 x_1 + q_2 x_2 + \cdots + q_n x_n)^r}\, dx_1\, dx_2 \ldots dx_n$

$$= \frac{\Gamma(p_1)\,\Gamma(p_2)\ldots\Gamma(p_n)}{\Gamma(p_1 + p_2 + \cdots + p_n - r + 1)\,\Gamma(r)} \int_0^\infty \frac{x^{r-1}\, dx}{(1 + q_1 x)^{p_1}(1 + q_2 x)^{p_2}\cdots(1 + q_n x)^{p_n}}$$

$$[p_i > 0, \quad q_i > 0, \quad p_1 + p_2 + \cdots + p_n > r > 0] \quad \text{FI III 493}$$

**4.647** $\displaystyle\iint \cdots \int_{0 \leq x_1^2 + x_2^2 + \cdots + x_n^2 \leq 1} \exp\left\{\frac{p_1 x_1 + p_2 x_2 + \cdots + p_n x_n}{\sqrt{x_1^2 + x_2^2 + \cdots + x_n^2}}\right\}\, dx_1\, dx_2 \ldots dx_n$

$$= \frac{2\sqrt{\pi^n}}{n\,(p_1^2 + p_2^2 + \cdots + p_n^2)^{\frac{n}{4}-\frac{1}{2}}}\, I_{\frac{n}{2}-1}\left(\sqrt{p_1^2 + p_2^2 + \cdots + p_n^2}\right)$$

$$\text{FI III 495}$$

**4.648**[12] $\displaystyle\int_0^\infty \int_0^\infty \cdots \int_0^\infty \exp\left[-\left(x_1 + x_2 + \cdots + x_n + \frac{\lambda^{n+1}}{x_1 x_2 \ldots x_n}\right)\right]$

$$\times x_1^{\frac{1}{n+1}-1} x_2^{\frac{2}{n+1}-1} \ldots x_n^{\frac{n}{n+1}-1}\, dx_1\, dx_2 \cdots dx_n$$

$$= \frac{1}{\sqrt{n+1}}\,(2\pi)^{\frac{n}{2}}\, e^{-(n+1)\lambda}$$

$$\text{FI III 496}$$

Table of Integrals, Series, and Products. http://dx.doi.org/10.1016/B978-0-12-384933-5.00005-9

# 5 Indefinite Integrals of Special Functions

## 5.1 Elliptic Integrals and Functions

**Notation:** $k' = \sqrt{1-k^2}$ (cf. 8.1).

### 5.11 Complete elliptic integrals

**5.111**

1.  $$\int K(k)k^{2p+3}\,\mathrm{d}k = \frac{1}{(2p+3)^2}\left\{4(p+1)^2\int K(k)k^{2p+1}\,\mathrm{d}k + k^{2p+2}\left[E(k) - (2p+3)\,K(k)k'^2\right]\right\}$$

    BY (610.04)

2.  $$\int E(k)k^{2p+3}\,\mathrm{d}k = \frac{1}{4p^2+16p+15}\left\{4(p+1)^2\int E(k)k^{2p+1}\,\mathrm{d}k\right.$$
    $$\left. - E(k)k^{2p+2}\left[(2p+3)k'^2 - 2\right] - k^{2p+2}k'^2\,K(k)\right\}$$

    BY (611.04)

**5.112**

1.  $$\int K(k)\,\mathrm{d}k = \frac{\pi k}{2}\left[1 + \sum_{j=1}^{\infty}\frac{[(2j)!]^2\,k^{2j}}{(2j+1)2^{4j}\,(j!)^4}\right]$$

    BY (610.00)

2.[6]  $$\int E(k)\,\mathrm{d}k = \frac{\pi k}{2}\left[1 - \sum_{j=1}^{\infty}\frac{[(2j)!]^2\,k^{2j}}{(4j^2-1)\,2^{4j}\,(j!)^4}\right]$$

    BY (611.00)

3.  $$\int K(k)k\,\mathrm{d}k = E(k) - k'^2\,K(k)$$

    BY (610.01)

4.  $$\int E(k)k\,\mathrm{d}k = \frac{1}{3}\left[(1+k^2)\,E(k) - k'^2\,K(k)\right]$$

    BY (611.01)

5.  $$\int K(k)k^3\,\mathrm{d}k = \frac{1}{9}\left[(4+k^2)\,E(k) - k'^2\,(4+3k^2)\,K(k)\right]$$

    BY (610.02)

6.    $\displaystyle\int E(k)k^3\,\mathrm{d}k = \frac{1}{45}\left[\left(4+k^2+9k^4\right)E(k) - k'^2\left(4+3k^2\right)K(k)\right]$      BY 611.02)

7.    $\displaystyle\int K(k)k^5\,\mathrm{d}k = \frac{1}{225}\left[\left(64+16k^2+9k^4\right)E(k) - k'^2\left(64+48k^2+45k^4\right)K(k)\right]$      BY (610.03)

8.    $\displaystyle\int E(k)k^5\,\mathrm{d}k = \frac{1}{1575}\left[\left(64+16k^2+9k^4+225k^6\right)E(k) - k'^2\left(64+48k^2+45k^4\right)K(k)\right]$

                                                                                       BY (611.03)

9.    $\displaystyle\int \frac{K(k)}{k^2}\,\mathrm{d}k = -\frac{E(k)}{k}$      BY (612.05)

10.    $\displaystyle\int \frac{E(k)}{k^2}\,\mathrm{d}k = \frac{1}{k}\left[k'^2\,K(k) - 2\,E(k)\right]$      BY (612.02)

11.    $\displaystyle\int \frac{E(k)}{k'^2}\,\mathrm{d}k = k\,K(k)$      BY (612.01)

12.    $\displaystyle\int \frac{E(k)}{k^4}\,\mathrm{d}k = \frac{1}{9k^3}\left[2\left(k^2-2\right)E(k) + k'^2\,K(k)\right]$      BY (612.03)

13.    $\displaystyle\int \frac{k\,E(k)}{k'^2}\,\mathrm{d}k = K(k) - E(k)$      BY (612.04)

14.*    $\displaystyle\int \frac{K(k)}{k^4}\,\mathrm{d}k = -\frac{1}{9k^3}\left[\left(1+4k^2\right)E(k) + 2k'^2\,K(k)\right]$

15.*    $\displaystyle\int \frac{(1\pm k')^3}{k^3 k'}\,K(k)\,\mathrm{d}k = \frac{E(k) \mp k'\,K(k)}{1\pm k'}$

**5.113**

1.    $\displaystyle\int \left[K(k) - E(k)\right]\frac{\mathrm{d}k}{k} = -E(k)$      BY (612.06)

2.    $\displaystyle\int \left[E(k) - k'^2\,K(k)\right]\frac{\mathrm{d}k}{k} = 2\,E(k) - k'^2\,K(k)$      BY (612.09)

3.    $\displaystyle\int \left[\left(1+k^2\right)K(k) - E(k)\right]\frac{\mathrm{d}k}{k} = -k'^2\,K(k)$      BY (612.12)

4.    $\displaystyle\int \left[K(k) - E(k)\right]\frac{\mathrm{d}k}{k^2} = \frac{1}{k}\left[E(k) - k'^2\,K(k)\right]$      BY (612.07)

5.    $\displaystyle\int \left[E(k) - k'^2\,K(k)\right]\frac{\mathrm{d}k}{k^2 k'^2} = \frac{1}{k}\left[K(k) - E(k)\right]$

6.    $\displaystyle\int \left[\left(1+k^2\right)E(k) - k'^2\,K(k)\right]\frac{\mathrm{d}k}{k k'^4} = \frac{E(k)}{k'^2}$      BY (612.13)

**5.114**    $\displaystyle\int \frac{k\,K(k)\,\mathrm{d}k}{\left[E(k) - k'^2\,K(k)\right]^2} = \frac{1}{k'^2\,K(k) - E(k)}$      BY (612.11)

**5.115**

1.    $\displaystyle\int \Pi\left(\frac{\pi}{2}, r^2, k\right)k\,\mathrm{d}k = \left(k^2 - r^2\right)\Pi\left(\frac{\pi}{2}, r^2, k\right) - K(k) + E(k)$      BY (612.14)

2.    $\displaystyle\int \left[K(k) - \Pi\left(\frac{\pi}{2}, r^2, k\right)\right]k\,\mathrm{d}k = k^2\,K(k) - \left(k^2 - r^2\right)\Pi\left(\frac{\pi}{2}, r^2, k\right)$      BY (612.15)

3.　　$\int \left[ \dfrac{\boldsymbol{E}(k)}{k'^2} + \Pi \left( \dfrac{\pi}{2}, r^2, k \right) \right] k\, dk = (k^2 - r^2)\, \Pi \left( \dfrac{\pi}{2}, r^2, k \right)$　　　　　　BY (612.16)

## 5.12  Elliptic integrals

**5.121**　$\displaystyle\int_0^x \dfrac{F(x,k)\, dx}{\sqrt{1 - k^2 \sin^2 x}} = \dfrac{[F(x,k)]^2}{2}$　　$\left[ 0 < x \leq \dfrac{\pi}{2} \right]$　　　　BY (630.01)

**5.122**[11]　$\displaystyle\int_0^x E(x,k)\sqrt{1 - k^2 \sin^2 x}\, dx = \dfrac{[E(x,k)]^2}{2}$　　　　　　BY (630.32)

**5.123**

1.　$\displaystyle\int_0^x F(x,k)\sin x\, dx = -\cos x\, F(x,k) + \dfrac{1}{k}\arcsin(k\sin x)$　　　　　BY (630.11)

2.　$\displaystyle\int_0^x F(x,k)\cos x\, dx = \sin x\, F(x,k) + \dfrac{1}{k}\operatorname{arccosh}\sqrt{\dfrac{1 - k^2 \sin^2 x}{k'^2}} - \dfrac{1}{k}\operatorname{arccosh}\left( \dfrac{1}{k'} \right)$　　BY (630.21)

**5.124**

1.　$\displaystyle\int_0^x E(x,k)\sin x\, dx = -\cos x\, E(x,k) + \dfrac{1}{2k}\left[ k\sin x\sqrt{1 - k^2 \sin^2 x} + \arcsin(k\sin x) \right]$　　BY (630.12)

2.　$\displaystyle\int_0^x E(x,k)\cos x\, dx = \sin x\, E(x,k) + \dfrac{1}{2k}\left[ k\cos x\sqrt{1 - k^2 \sin^2 x} \right.$

$$\left. - k'^2 \operatorname{arccosh}\sqrt{\dfrac{1 - k^2 \sin^2 x}{k'^2}} - k + k'^2 \operatorname{arccosh}\left( \dfrac{1}{k'} \right) \right]$$

　　　　　　　　　　　　　　　　　　　　　　　　　　　　BY (630.22)

3.　$\displaystyle\int_0^a \dfrac{x\,\boldsymbol{E}(x)\, dx}{(k'^2 + k^2 x^2)^2 \sqrt{a^2 - x^2}} = \dfrac{\pi}{4}\left( \dfrac{a\sqrt{1 - a^2}}{(k'^2 + k^2 a^2)^2} + \dfrac{a^2\, E(\lambda, k)}{k'^2\,(k'^2 + k^2 a^2)^{3/2}} + \dfrac{(1 - a^2)\, F(\lambda, k)}{(k'^2 + k^2 a^2)^{3/2}} \right)$

$$\lambda = \arcsin\left( \dfrac{a}{\sqrt{k'^2 + k^2 a^2}} \right) \quad k' = \sqrt{1 - k^2} \quad [0 < a < 1, \quad 0 < k < 1]$$

4.　$\displaystyle\int_0^a \dfrac{x\,\boldsymbol{E}(x)\, dx}{(k^2 - x^2)^2 \sqrt{a^2 - x^2}} = \dfrac{\pi}{4}\left( \dfrac{a\sqrt{1 - a^2}}{k^2\,(k^2 - a^2)} + \dfrac{F(\phi, k)}{k^2 \sqrt{k^2 - a^2}} + \dfrac{a^2\, E(\phi, k)}{k^2\,(k^2 - a^2)^{3/2}} \right)$

$$\phi = \arcsin\left( \dfrac{a}{k} \right) \quad [0 < a < k < 1]$$

5.　$\displaystyle\int_0^{\pi/2} \dfrac{E(x, k')\sin x \cos x\, dx}{(1 - k'^2 \cosh^2 v \sin^2 x)\sqrt{1 - k'^2 \sin^2 x}}$

$$= \dfrac{1}{k'^2 \sinh v \cosh v}\left\{ \boldsymbol{E}(k')\operatorname{arctanh}\left( \dfrac{\tanh v}{k} \right) - \dfrac{\pi \tanh v}{2} - \dfrac{\pi}{2}[F(\phi, k) - E(\phi, k)] \right\}$$

$$\phi = \arcsin\left( \dfrac{\tanh v}{k} \right) \quad k' = \sqrt{1 - k^2} \quad [0 < \tanh v < k < 1]$$

6.
$$\int_0^{\pi/2} \frac{E(x,k)\sin x \cos x \, dx}{\left(1 - k^2 \cos^2\psi \sin^2 x\right)\sqrt{1 - k^2 \sin^2 x}}$$
$$= \frac{1}{k^2 \sin\psi \cos\psi} \left\{ E(k) \arctan\left(\frac{\tan\psi}{k'}\right) - \frac{\pi}{2} E(\beta,k) + \frac{\pi}{2} \frac{\tan\psi}{\sqrt{1 - k^2 \cos^2\psi}} \left(1 - \sqrt{1 - k^2 \cos^2\psi}\right) \right\}$$
$$\beta = \arctan\left(\frac{\tan\psi}{k}\right) \quad k' = \sqrt{1 - k^2} \qquad \left[0 < k < 1, \quad 0 < \psi < \frac{\pi}{2}\right]$$

7.
$$\int_0^{\pi/2} \frac{E(x,k')\sin x \cos x \, dx}{\left(1 + k'^2 \sinh^2\mu \sin^2 x\right)\sqrt{1 - k'^2 \sin^2 x}}$$
$$= \frac{-1}{k'^2 \sinh\mu \cosh\mu} \left\{ E(k') \operatorname{arctanh}(k \tanh\mu) - \frac{\pi}{2} \left[ F(\phi,k) - E(\phi,k) + \tanh\mu\sqrt{1 + k'^2 \sinh^2\mu} \right] \right.$$
$$\left. - \frac{\pi}{2} \coth\mu \left(1 - \sqrt{1 + k'^2 \sinh^2\mu}\right) \right\}$$
$$\phi = \arcsin(\tanh\mu) \quad k' = \sqrt{1 - k^2} \qquad [0 < k < 1, \quad 0 < \tanh\mu < 1]$$

8.
$$\int_0^{\pi/2} \frac{F(x,k')\sin x \cos x \, dx}{\left(1 + k'^2 \sinh^2\mu \sin^2 x\right)\sqrt{1 - k'^2 \sin^2 x}}$$
$$= \frac{-1}{k'^2 \sinh\mu \cosh\mu} \left[ K(k') \operatorname{arctanh}(k \tanh\mu) - \frac{\pi}{2} F(\phi,k) \right]$$
$$\phi = \arcsin(\tanh\mu) \quad k' = \sqrt{1 - k^2} \qquad [0 < k < 1, \quad 0 < \tanh\mu < 1]$$

9.
$$\int_0^{\pi/2} \frac{F(x,k')\sin x \cos x \, dx}{\left(1 - k'^2 \cosh^2\nu \sin^2 x\right)\sqrt{1 - k'^2 \sin^2 x}}$$
$$= \frac{1}{k'^2 \sinh\nu \cosh\nu} \left[ K(k') \operatorname{arctanh}\left(\frac{\tanh\nu}{k}\right) - \frac{\pi}{2} F(\phi,k) \right]$$
$$\phi = \arcsin\left(\frac{\tanh\nu}{k}\right) \quad k' = \sqrt{1 - k^2} \qquad [0 < k < 1, \quad 0 < \tanh\nu < 1]$$

10.
$$\int_0^{\pi/2} \frac{F(x,k)\sin x \cos x \, dx}{\left(1 - k^2 \cos^2\psi \sin^2 x\right)\sqrt{1 - k^2 \sin^2 x}}$$
$$= \frac{1}{k^2 \sin\psi \cos\psi} \left[ K(k') \operatorname{arctanh}\left(\frac{\tan\psi}{k'}\right) - \frac{\pi}{2} F(\beta,k) \right]$$
$$\beta = \arctan\left(\frac{\tan\psi}{k'}\right) \quad k' = \sqrt{1 - k^2} \qquad [0 < k < 1, \quad 0 < \psi < 1]$$

11.
$$\int_a^b \ln\left(\frac{\epsilon + x}{\epsilon - x}\right) \frac{x^2 \, dx}{\sqrt{(x^2 - a^2)(b^2 - x^2)}} = \frac{\pi}{\epsilon}\left(\epsilon^2 - \sqrt{(\epsilon^2 - a^2)(\epsilon^2 - b^2)}\right) + \pi\beta \left[F(\phi,k) - E(\phi,k)\right]$$
$$\phi = \arcsin\left(\frac{\beta}{\epsilon}\right) \quad k = \frac{a}{b} \qquad [0 < a < b < \epsilon]$$

**5.125**

1.    $\displaystyle\int_0^x \Pi\left(x, \alpha^2, k\right) \sin x \, dx$

$$= -\cos x \, \Pi\left(x, \alpha^2, k\right) + \frac{1}{\sqrt{k^2 - \alpha^2}} \arctan\left[\sqrt{\frac{k^2 - \alpha^2}{1 - k^2 \sin^2 x}} \sin x\right] \qquad \left[\alpha^2 < k^2\right]$$

$$= -\cos x \, \Pi\left(x, \alpha^2, k\right) + \frac{1}{\sqrt{\alpha^2 - k^2}} \operatorname{arctanh}\left[\sqrt{\frac{\alpha^2 - k^2}{1 - k^2 \sin^2 x}} \sin x\right] \qquad \left[\alpha^2 > k^2\right]$$

$$\text{BY (630.13)}$$

2.    $\displaystyle\int_0^x \Pi\left(x, \alpha^2, k\right) \cos x \, dx = \sin x \, \Pi\left(x, \alpha^2, k\right) - f - f_0$

where

$$f = \frac{1}{2\sqrt{(1 - \alpha^2)(\alpha^2 - k^2)}} \arctan\left[\frac{2(1 - \alpha^2)(\alpha^2 - k^2) + (1 - \alpha^2 \sin^2 x)(2k^2 - \alpha^2 - \alpha^2 k^2)}{2\alpha^2 \sqrt{(1 - \alpha^2)(\alpha^2 - k^2)} \cos x \sqrt{1 - k^2 \sin^2 x}}\right]$$
$$\text{for } (1 - \alpha^2)(\alpha^2 - k^2) > 0;$$

$$= \frac{1}{2\sqrt{(\alpha^2 - 1)(\alpha^2 - k^2)}} \ln\left[\frac{2(\alpha^2 - 1)(\alpha^2 - k^2) + (1 - \alpha^2 \sin^2 x)(\alpha^2 + \alpha^2 k^2 - 2k^2)}{1 - \alpha^2 \sin^2 x}\right.$$
$$\left. + \frac{2\alpha^2 \sqrt{(\alpha^2 - 1)(\alpha^2 - k^2)} \cos x \sqrt{1 - k^2 \sin^2 x}}{1 - \alpha^2 \sin^2 x}\right]$$

$$\text{for } (1 - \alpha^2)(\alpha^2 - k^2) < 0,$$
$$f_0 \text{ is the value of } f \text{ at } x = 0 \quad \text{BY (630.23)}$$

**Integration with respect to the modulus**

**5.126**   $\displaystyle\int F(x, k) k \, dk = E(x, k) - k'^2 F(x, k) + \left(\sqrt{1 - k^2 \sin^2 x} - 1\right) \cot x$        BY (613.01)

**5.127**   $\displaystyle\int E(x, k) k \, dk = \frac{1}{3}\left[(1 + k^2) E(x, k) - k'^2 F(x, k) + \left(\sqrt{1 - k^2 \sin^2 x} - 1\right) \cot x\right]$       BY (613.02)

**5.128**   $\displaystyle\int \Pi\left(x, r^2, k\right) k \, dk = \left(k^2 - r^2\right) \Pi\left(x, r^2, k\right) - F(x, k) + E(x, k) + \left(\sqrt{1 - k^2 \sin^2 x} - 1\right) \cot x$

$$\text{BY (613.03)}$$

## 5.13   Jacobian elliptic functions

**5.131**

1.    $\displaystyle\int \operatorname{sn}^m u \, du = \frac{1}{m+1}\left[\operatorname{sn}^{m+1} u \operatorname{cn} u \operatorname{dn} u + (m+2)(1 + k^2) \int \operatorname{sn}^{m+2} u \, du\right.$

$$\left. - (m+3)k^2 \int \operatorname{sn}^{m+4} u \, du\right]$$

SI 259, PE(567)

2.    $\displaystyle \int \mathrm{cn}^{\,m} u \, du = \frac{1}{(m+1)k'^2} \left[ -\,\mathrm{cn}^{\,m+1} u \, \mathrm{sn}\, u \, \mathrm{dn}\, u \right.$

$$\left. +\, (m+2)\left(1-2k^2\right) \int \mathrm{cn}^{\,m+2} u \, du + (m+3)k^2 \int \mathrm{cn}^{\,m+4} u \, du \right]$$

PE (568)

3.    $\displaystyle \int \mathrm{dn}^{\,m} u \, du = \frac{1}{(m+1)k'^2} \left[ k^2 \, \mathrm{dn}^{\,m+1} u \, \mathrm{sn}\, u \, \mathrm{cn}\, u \right.$

$$\left. +\, (m+2)\left(2-k^2\right) \int \mathrm{dn}^{\,m+2} u \, du - (m+3) \int \mathrm{dn}^{\,m+4} u \, du \right]$$

PE (569)

     By using formulas **5.131**, we can reduce the integrals (for $m \neq 1$) $\int \mathrm{sn}^m u \, du$, $\int \mathrm{cn}^{\,m} u \, du$, and $\int \mathrm{dn}^{\,m} u \, du$ to the integrals **5.132**, **5.133** and **5.134**.

**5.132**

1.    $\displaystyle \int \frac{du}{\mathrm{sn}\, u} = \ln \frac{\mathrm{sn}\, u}{\mathrm{cn}\, u + \mathrm{dn}\, u}$                 H 87(164)

        $\displaystyle = \ln \frac{\mathrm{dn}\, u - \mathrm{cn}\, u}{\mathrm{sn}\, u}$                 SI 266(4)

2.    $\displaystyle \int \frac{du}{\mathrm{cn}\, u} = \frac{1}{k'} \ln \frac{k' \, \mathrm{sn}\, u + \mathrm{dn}\, u}{\mathrm{cn}\, u}$             SI 266(5)

3.    $\displaystyle \int \frac{du}{\mathrm{dn}\, u} = \frac{1}{k'} \arctan \frac{k' \, \mathrm{sn}\, u - \mathrm{cn}\, u}{k' \, \mathrm{sn}\, u + \mathrm{cn}\, u}$       H 88(166)

        $\displaystyle = \frac{1}{k'} \arccos \frac{\mathrm{cn}\, u}{\mathrm{dn}\, u}$                 JA

        $\displaystyle = \frac{1}{ik'} \ln \frac{\mathrm{cn}\, u + ik' \, \mathrm{sn}\, u}{\mathrm{dn}\, u}$           SI 266(6)

        $\displaystyle = \frac{1}{k'} \arcsin \frac{k' \, \mathrm{sn}\, u}{\mathrm{dn}\, u}$               JA

**5.133**

1.    $\displaystyle \int \mathrm{sn}\, u \, du = \frac{1}{k} \ln \left( \mathrm{dn}\, u - k \, \mathrm{cn}\, u \right)$        H 87(161)

        $\displaystyle = \frac{1}{k} \mathrm{arccosh} \frac{\mathrm{dn}\, u - k^2 \, \mathrm{cn}\, u}{1 - k^2}$         JA

        $\displaystyle = \frac{1}{k} \mathrm{arcsinh} \left( k \frac{\mathrm{dn}\, u - \mathrm{cn}\, u}{1 - k^2} \right)$     JA

        $\displaystyle = -\frac{1}{k} \ln \left( \mathrm{dn}\, u + k \, \mathrm{cn}\, u \right)$       SI 365(1)

2.    $\displaystyle \int \mathrm{cn}\, u \, du = \frac{1}{k} \arccos \left( \mathrm{dn}\, u \right)$          H 87(162)

        $\displaystyle = \frac{i}{k} \ln \left( \mathrm{dn}\, u - ik \, \mathrm{sn}\, u \right)$     SI 265(2)a, ZH 87(162)

        $\displaystyle = \frac{1}{k} \arcsin \left( k \, \mathrm{sn}\, u \right)$           JA

3. $\displaystyle\int \mathrm{dn}\,u\,du = \arcsin\left(\mathrm{sn}\,u\right)$ H 87(163)

$\qquad\qquad = \mathrm{am}\,u = i\ln\left(\mathrm{cn}\,u - i\,\mathrm{sn}\,u\right)$ SI 266(3), ZH 87(163)

**5.134**

1. $\displaystyle\int \mathrm{sn}^2 u\,du = \frac{1}{k^2}\left[u - E\left(\mathrm{am}\,u, k\right)\right]$ PE (564)

2. $\displaystyle\int \mathrm{cn}^2 u\,du = \frac{1}{k^2}\left[E\left(\mathrm{am}\,u, k\right) - k'^2 u\right]$ PE (565)

3. $\displaystyle\int \mathrm{dn}^2 u\,du = E\left(\mathrm{am}\,u, k\right)$ PE (566)

**5.135**

1. $\displaystyle\int \frac{\mathrm{sn}\,u}{\mathrm{cn}\,u}\,du = \frac{1}{k'}\ln\frac{\mathrm{dn}\,u + k'}{\mathrm{cn}\,u}$ SI 266(7)

$\qquad\qquad = \dfrac{1}{2k'}\ln\dfrac{\mathrm{dn}\,u + k'}{\mathrm{dn}\,u - k'}$ H 88(167)

2. $\displaystyle\int \frac{\mathrm{sn}\,u}{\mathrm{dn}\,u}\,du = \frac{i}{kk'}\ln\frac{ik' - k\,\mathrm{cn}\,u}{\mathrm{dn}\,u}$ SI 266(8)

$\qquad\qquad = \dfrac{1}{kk'}\operatorname{arccot}\dfrac{k\,\mathrm{cn}\,u}{k'}$

3. $\displaystyle\int \frac{\mathrm{cn}\,u}{\mathrm{sn}\,u}\,du = \ln\frac{1 - \mathrm{dn}\,u}{\mathrm{sn}\,u}$ SI 266(10)

$\qquad\qquad = \dfrac{1}{2}\ln\dfrac{1 - \mathrm{dn}\,u}{1 + \mathrm{dn}\,u}$ H 88(168)

4. $\displaystyle\int \frac{\mathrm{cn}\,u}{\mathrm{dn}\,u}\,du = -\frac{1}{k}\ln\frac{1 - k\,\mathrm{sn}\,u}{\mathrm{dn}\,u}$ SI 266(9)

$\qquad\qquad = \dfrac{1}{2k}\ln\dfrac{1 + k\,\mathrm{sn}\,u}{1 - k\,\mathrm{sn}\,u}$

5. $\displaystyle\int \frac{\mathrm{dn}\,u}{\mathrm{cn}\,u}\,du = \frac{1}{2}\ln\frac{1 + \mathrm{sn}\,u}{1 - \mathrm{sn}\,u}$ H 88(172)

$\qquad\qquad = \ln\dfrac{1 + \mathrm{sn}\,u}{\mathrm{cn}\,u}$ JA

6. $\displaystyle\int \frac{\mathrm{dn}\,u}{\mathrm{sn}\,u}\,du = \frac{1}{2}\ln\frac{1 - \mathrm{cn}\,u}{1 + \mathrm{cn}\,u}$ H 87(170)

**5.136**

1. $\displaystyle\int \mathrm{sn}\,u\,\mathrm{cn}\,u\,du = -\frac{1}{k^2}\,\mathrm{dn}\,u$

2. $\displaystyle\int \mathrm{sn}\,u\,\mathrm{dn}\,u\,du = -\,\mathrm{cn}\,u$

3. $\displaystyle\int \mathrm{cn}\,u\,\mathrm{dn}\,u\,du = \mathrm{sn}\,u$

**5.137**

1. $\displaystyle\int \frac{\mathrm{sn}\,u}{\mathrm{cn}^2 u}\,du = \frac{1}{k'^2}\frac{\mathrm{dn}\,u}{\mathrm{cn}\,u}$ H 88(173)

2.    $\displaystyle\int \frac{\operatorname{sn} u}{\operatorname{dn}^2 u}\, du = -\frac{1}{k'^2}\frac{\operatorname{cn} u}{\operatorname{dn} u}$        H 88(175)

3.    $\displaystyle\int \frac{\operatorname{cn} u}{\operatorname{sn}^2 u}\, du = -\frac{\operatorname{dn} u}{\operatorname{sn} u}$        H 88(174)

4.    $\displaystyle\int \frac{\operatorname{cn} u}{\operatorname{dn}^2 u}\, du = \frac{\operatorname{sn} u}{\operatorname{dn} u}$        H 88(177)

5.    $\displaystyle\int \frac{\operatorname{dn} u}{\operatorname{sn}^2 u}\, du = -\frac{\operatorname{cn} u}{\operatorname{sn} u}$        H 88(176)

6.    $\displaystyle\int \frac{\operatorname{dn} u}{\operatorname{cn}^2 u}\, du = \frac{\operatorname{sn} u}{\operatorname{cn} u}$        H 88(178)

**5.138**

1.    $\displaystyle\int \frac{\operatorname{cn} u}{\operatorname{sn} u \operatorname{dn} u}\, du = \ln \frac{\operatorname{sn} u}{\operatorname{dn} u}$        H 88(183)

2.    $\displaystyle\int \frac{\operatorname{sn} u}{\operatorname{cn} u \operatorname{dn} u}\, du = \frac{1}{k'^2}\ln \frac{\operatorname{dn} u}{\operatorname{cn} u}$        H 88(182)

3.    $\displaystyle\int \frac{\operatorname{dn} u}{\operatorname{sn} u \operatorname{cn} u}\, du = \ln \frac{\operatorname{sn} u}{\operatorname{cn} u}$        H 88(184)

**5.139**

1.[11]    $\displaystyle\int \frac{\operatorname{cn} u \operatorname{dn} u}{\operatorname{sn} u}\, du = \ln \operatorname{sn} u$        H 88(179)

2.    $\displaystyle\int \frac{\operatorname{sn} u \operatorname{dn} u}{\operatorname{cn} u}\, du = \ln \frac{1}{\operatorname{cn} u}$        H 88(180)

3.    $\displaystyle\int \frac{\operatorname{sn} u \operatorname{cn} u}{\operatorname{dn} u}\, du = -\frac{1}{k^2}\ln \operatorname{dn} u$        H 88(181)

## 5.14 Weierstrass elliptic functions

The invariants $g_1$ and $g_2$ used below are defined in 8.161.

**5.141**

1.    $\displaystyle\int \wp(u)\, du = -\zeta(u)$

2.    $\displaystyle\int \wp^2(u)\, du = \frac{1}{6}\wp'(u) + \frac{1}{12}g_2 u$        H 120(192)

3.    $\displaystyle\int \wp^3(u)\, du = \frac{1}{120}\wp'''(u) - \frac{3}{20}g_2\,\zeta(u) + \frac{1}{10}g_3 u$        H 120(193)

4.[8]    $\displaystyle\int \frac{du}{\wp(u) - \wp(v)} = \frac{1}{\wp'(v)}\left[ 2u\,\zeta(v) + \ln \frac{\sigma(u - v)}{\sigma(u + v)} \right]$     $[\wp(v) \neq e_1, e_2, e_3]$     (see **8.162**)

                                                    H 120(194)

5.[12]    $\displaystyle\int \frac{\alpha\wp(u) + \beta}{\gamma\wp(u) + \delta}\, du = \frac{\alpha u}{\gamma} - \frac{\alpha\delta - \beta\gamma}{\gamma^2\wp'(v)}\left[ \ln \frac{\sigma(u + v)}{\sigma(u - v)} - 2u\,\zeta(v) \right]$

                                where $v = (\wp')^{-1}\left( \dfrac{-\delta}{\gamma} \right)$        H 120(195)

# 5.2 The Exponential Integral Function

## 5.21 The exponential integral function

**5.211** $\displaystyle\int_x^\infty \mathrm{Ei}(-\beta x)\,\mathrm{Ei}(-\gamma x)\,\mathrm{d}x = \left(\frac{1}{\beta} + \frac{1}{\gamma}\right)\mathrm{Ei}[-(\beta + \gamma)x]$

$$-x\,\mathrm{Ei}(-\beta x)\,\mathrm{Ei}(-\gamma x) - \frac{e^{-\beta x}}{\beta}\,\mathrm{Ei}(-\gamma x) - \frac{e^{-\gamma x}}{\gamma}\,\mathrm{Ei}(-\beta x)$$

$$[\mathrm{Re}(\beta + \gamma) > 0] \qquad\qquad \text{NT 53(2)}$$

## 5.22 Combinations of the exponential integral function and powers

**5.221**

1. $\displaystyle\int_x^\infty \frac{\mathrm{Ei}[-a(x+b)]}{x^{n+1}}\,\mathrm{d}x = \left[\frac{1}{x^n} - \frac{(-1)^n}{b^n}\right]\frac{\mathrm{Ei}[-a(x+b)]}{n} + \frac{e^{-ab}}{n}\sum_{k=0}^{n-1}\frac{(-1)^{n-k-1}}{b^{n-k}}\int_x^\infty \frac{e^{-ax}}{x^{k+1}}\,\mathrm{d}x$

$$[a > 0, \quad b > 0] \qquad\qquad \text{NT 52(3)}$$

2. $\displaystyle\int_x^\infty \frac{\mathrm{Ei}[-a(x+b)]}{x^2}\,\mathrm{d}x = \left(\frac{1}{x} + \frac{1}{b}\right)\mathrm{Ei}[-a(x+b)] - \frac{e^{-ab}\,\mathrm{Ei}(-ax)}{b}$

$$[a > 0, \quad b > 0] \qquad\qquad \text{NT 52(4)}$$

3. $\displaystyle\int x\,\mathrm{Ei}(-ax)\,\mathrm{d}x = \frac{x^2}{2}\,\mathrm{Ei}(-ax) + \frac{1}{2a^2}e^{-ax} + \frac{xe^{-ax}}{2a} \qquad [a > 0]$

4. $\displaystyle\int x^n\,\mathrm{Ei}(-ax)\,\mathrm{d}x = \frac{x^{n+1}}{n+1}\,\mathrm{Ei}(-ax) + \frac{n!e^{-ax}}{(n+1)a^{n+1}}\sum_{k=0}^{\infty}\frac{(ax)^k}{k!}$

$$[a > 0]$$

5. $\displaystyle\int x\,\mathrm{Ei}(-ax)e^{-bx}\,\mathrm{d}x = \frac{1}{b^2}\,\mathrm{Ei}\left[-(a+b)x\right] - \frac{1}{b^2}\,\mathrm{Ei}(-ax)e^{-bx} - \frac{x}{b}\,\mathrm{Ei}(-ax)e^{-bx} - \frac{1}{b(a+b)}e^{-(a+b)x}$

$$[a > 0, \quad b > 0]$$

6. $\displaystyle\int \mathrm{Ei}^2(-ax)\,\mathrm{d}x = x\,\mathrm{Ei}^2(-ax) + \frac{2}{a}\left[\mathrm{Ei}(-ax)e^{-ax} - \mathrm{Ei}(-2ax)\right]$

$$[a > 0]$$

7. $\displaystyle\int x\,\mathrm{Ei}^2(-ax)\,\mathrm{d}x = \frac{x^2}{2}\,\mathrm{Ei}^2(-ax) + \left(\frac{1}{a^2} + \frac{x}{a}\right)\mathrm{Ei}(-ax)e^{-ax} - \frac{1}{a^2}\,\mathrm{Ei}(-2ax) + \frac{1}{a^2}e^{-2ax}$

$$[a > 0]$$

8. $\displaystyle\int_0^u \mathrm{Ei}(-ax)\,\mathrm{d}x = u\,\mathrm{Ei}(-au) + \frac{e^{-au} - 1}{a} \qquad [a > 0]$

9. $\displaystyle\int_0^\infty x\,\mathrm{Ei}\left(-\frac{x}{a}\right)\mathrm{Ei}\left(-\frac{x}{b}\right)\mathrm{d}x = \left(\frac{a^2 + b^2}{2}\right)\ln(a+b) - \frac{a^2}{2}\ln a - \frac{b^2}{2}\ln b - \frac{ab}{2}$

$$[a > 0, \quad b > 0]$$

10.  $\displaystyle\int_0^\infty x^2 \operatorname{Ei}\left(-\frac{x}{a}\right) \operatorname{Ei}\left(-\frac{x}{b}\right) \mathrm{d}x = \frac{2}{3}\left[(a^3 + b^3)\ln(a + b) - a^3 \ln a - b^3 \ln b - \frac{ab}{a + b}\left(a^2 - ab + b^2\right)\right]$

$$[a > 0, \quad b > 0]$$

## 5.23  Combinations of the exponential integral and the exponential

### 5.231

1.  $\displaystyle\int_0^x e^x \operatorname{Ei}(-x)\,\mathrm{d}x = -\ln x - C + e^x \operatorname{Ei}(-x)$                    ET II 308(11)

2.  $\displaystyle\int_0^x e^{-\beta x} \operatorname{Ei}(-\alpha x)\,\mathrm{d}x = -\frac{1}{\beta}\left\{e^{-\beta x}\operatorname{Ei}(-\alpha x) + \ln\left(1 + \frac{\beta}{\alpha}\right) - \operatorname{Ei}[-(\alpha + \beta)x]\right\}$                    ET II 308(12)

## 5.3  The Sine Integral and the Cosine Integral

### 5.31

1.  $\displaystyle\int \cos\alpha x\,\operatorname{ci}(\beta x)\,\mathrm{d}x = \frac{\sin\alpha x\,\operatorname{ci}(\beta x)}{\alpha} - \frac{\operatorname{si}(\alpha x + \beta x) + \operatorname{si}(\alpha x - \beta x)}{2\alpha}$                    NT 49(1)

2.  $\displaystyle\int \sin\alpha x\,\operatorname{ci}(\beta x)\,\mathrm{d}x = -\frac{\cos\alpha x\,\operatorname{ci}(\beta x)}{\alpha} + \frac{\operatorname{ci}(\alpha x + \beta x) + \operatorname{ci}(\alpha x - \beta x)}{2\alpha}$                    NT 49(2)

### 5.32

1.  $\displaystyle\int \cos\alpha x\,\operatorname{si}(\beta x)\,\mathrm{d}x = \frac{\sin\alpha x\,\operatorname{si}(\beta x)}{\alpha} + \frac{\operatorname{ci}(\alpha x + \beta x) - \operatorname{ci}(\alpha x - \beta x)}{2\alpha}$                    NT 49(3)

2.  $\displaystyle\int \sin\alpha x\,\operatorname{si}(\beta x)\,\mathrm{d}x = -\frac{\cos\alpha x\,\operatorname{si}(\beta x)}{\alpha} + \frac{\operatorname{si}(\alpha x + \beta x) - \operatorname{si}(\alpha x - \beta x)}{2\alpha}$                    NT 49(4)

### 5.33

1.  $\displaystyle\int \operatorname{ci}(\alpha x)\,\operatorname{ci}(\beta x)\,\mathrm{d}x = x\,\operatorname{ci}(\alpha x)\,\operatorname{ci}(\beta x) + \frac{1}{2\alpha}\left[\operatorname{si}(\alpha x + \beta x) + \operatorname{si}(\alpha x - \beta x)\right]$

$\displaystyle\qquad\qquad + \frac{1}{2\beta}\left[\operatorname{si}(\alpha x + \beta x) + \operatorname{si}(\beta x - \alpha x)\right] - \frac{1}{\alpha}\sin\alpha x\,\operatorname{ci}(\beta x) - \frac{1}{\beta}\sin\beta x\,\operatorname{ci}(\alpha x)$

NT 53(5)

2.  $\displaystyle\int \operatorname{si}(\alpha x)\,\operatorname{si}(\beta x)\,\mathrm{d}x = x\,\operatorname{si}(\alpha x)\,\operatorname{si}(\beta x) - \frac{1}{2\beta}\left[\operatorname{si}(\alpha x + \beta x) + \operatorname{si}(\alpha x - \beta x)\right]$

$\displaystyle\qquad\qquad - \frac{1}{2\alpha}\left[\operatorname{si}(\alpha x + \beta x) + \operatorname{si}(\beta x + \alpha x)\right] + \frac{1}{\alpha}\cos\alpha x\,\operatorname{si}(\beta x) + \frac{1}{\beta}\cos\beta x\,\operatorname{si}(\alpha x)$

NT 54(6)

3.  $\displaystyle\int \operatorname{si}(\alpha x)\,\operatorname{ci}(\beta x)\,\mathrm{d}x = x\,\operatorname{si}(\alpha x)\,\operatorname{ci}(\beta x) + \frac{1}{\alpha}\cos\alpha x\,\operatorname{ci}(\beta x)$

$\displaystyle\qquad\qquad - \frac{1}{\beta}\sin\beta x\,\operatorname{si}(\alpha x) - \left(\frac{1}{2\alpha} + \frac{1}{2\beta}\right)\operatorname{ci}(\alpha x + \beta x) - \left(\frac{1}{2\alpha} - \frac{1}{2\beta}\right)\operatorname{ci}(\alpha x - \beta x)$

NT 54(10)

**5.34**

1. $\displaystyle\int_x^\infty \mathrm{si}[a(x+b)]\frac{\mathrm{d}x}{x^2} = \left(\frac{1}{x}+\frac{1}{b}\right)\mathrm{si}[a(x+b)] - \frac{\cos ab\,\mathrm{si}(ax)+\sin ab\,\mathrm{ci}(ax)}{b}$

$$[a>0, \quad b>0] \qquad\qquad \text{NT 52(6)}$$

2. $\displaystyle\int_x^\infty \mathrm{ci}[a(x+b)]\frac{\mathrm{d}x}{x^2} = \left(\frac{1}{x}+\frac{1}{b}\right)\mathrm{ci}[a(x+b)] + \frac{\sin ab\,\mathrm{si}(ax)-\cos ab\,\mathrm{ci}(ax)}{b}$

$$[a>0, \quad b>0] \qquad\qquad \text{NT 52(5)}$$

## 5.4 The Probability Integral and Fresnel Integrals

**5.41[11]** $\displaystyle\int \Phi(\alpha x)\,\mathrm{d}x = x\,\Phi(\alpha x) + \frac{e^{-\alpha^2 x^2}}{\alpha\sqrt{\pi}}$ $\qquad\qquad$ NT 12(20)a

**5.42** $\displaystyle\int S(\alpha x)\,\mathrm{d}x = x\,S(\alpha x) + \frac{\cos^2 \alpha x^2}{\alpha\sqrt{2\pi}}$ $\qquad\qquad$ NT 12(22)a

**5.43** $\displaystyle\int C(\alpha x)\,\mathrm{d}x = x\,C(\alpha x) - \frac{\sin^2 \alpha x^2}{\alpha\sqrt{2\pi}}$ $\qquad\qquad$ NT 12(21)a

## 5.5 Bessel Functions

**Notation**: $Z$ and $\mathfrak{Z}$ denote any of $J$, $N$, $H^{(1)}$, $H^{(2)}$. In formulae 5.52–5.56, $Z_p(x)$ and $\mathfrak{Z}_p(x)$ are arbitrary Bessel functions of the first, second, or third kinds.

**5.51** $\displaystyle\int J_p(x)\,\mathrm{d}x = 2\sum_{k=0}^\infty J_{p+2k+1}(x)$ $\qquad\qquad$ JA, MO 30

**5.52**

1. $\displaystyle\int x^{p+1}\, Z_p(x)\,\mathrm{d}x = x^{p+1}\, Z_{p+1}(x)$ $\qquad\qquad$ WA 132(1)

2.[12] $\displaystyle\int x^{-p+1}\, Z_p(x)\,\mathrm{d}x = -x^{-p+1}\, Z_{p-1}(x)$ $\qquad\qquad$ WA 132(2)

**5.53[10]** $\displaystyle\int\left[(\alpha^2-\beta^2)\,x - \frac{p^2-q^2}{x}\right] Z_p(\alpha x)\,\mathfrak{Z}_q(\beta x)\,\mathrm{d}x$

$$= \alpha x\, Z_{p+1}(\alpha x)\,\mathfrak{Z}_q(\beta x) - \beta x\, Z_p(\alpha x)\,\mathfrak{Z}_{q+1}(\beta x) - (p-q)\,Z_p(\alpha x)\,\mathfrak{Z}_q(\beta x)$$

$$= \beta x\, Z_p(\alpha x)\,\mathfrak{Z}_{q-1}(\beta x) - \alpha x\, Z_{p-1}(\alpha x)\,\mathfrak{Z}_q(\beta x) + (p-q)\,Z_p(\alpha x)\,\mathfrak{Z}_q(\beta x)$$

$$\text{JA, MO 30, WA 134(7)}$$

**5.54**

1.[10] $\displaystyle\int x\, Z_p(\alpha x)\,\mathfrak{Z}_p(\beta x)\,\mathrm{d}x = \frac{\alpha x\, Z_{p+1}(\alpha x)\,\mathfrak{Z}_p(\beta x) - \beta x\, Z_p(\alpha x)\,\mathfrak{Z}_{p+1}(\beta x)}{\alpha^2-\beta^2}$

$$= \frac{\beta x\, Z_p(\alpha x)\,\mathfrak{Z}_{p-1}(\beta x) - \alpha x\, Z_{p-1}(\alpha x)\,\mathfrak{Z}_p(\beta x)}{\alpha^2-\beta^2}$$

$$\text{WA 134(8)}$$

2. $\displaystyle\int x\,[Z_p(\alpha x)]^2\,\mathrm{d}x = \frac{x^2}{2}\left\{[Z_p(\alpha x)]^2 - Z_{p-1}(\alpha x)\,Z_{p+1}(\alpha x)\right\}$ $\qquad$ WA 135(11)

3. $\quad \displaystyle\int x\,Z_p(ax)\,3_p(ax)\,\mathrm{d}x = \frac{x^4}{4}\left[2\,Z_p(ax)\,3_p(ax) - Z_{p-1}(ax)\,3_{p+1}(ax) - Z_{p+1}(ax)\,3_{p-1}(ax)\right]$

**5.55**[10]  $\quad \displaystyle\int \frac{1}{x}\,Z_p(\alpha x)\,3_q(\alpha x)\,\mathrm{d}x = \alpha x\frac{Z_p(\alpha x)\,3_{q+1}(\alpha x) - Z_{p+1}(\alpha x)\,3_q(\alpha x)}{p^2 - q^2} + \frac{Z_p(\alpha x)\,3_q(\alpha x)}{p + q}$

$$= \alpha x\frac{Z_{p-1}(\alpha x)\,3_q(\alpha x) - Z_p(\alpha x)\,3_{q-1}(\alpha x)}{p^2 - q^2} - \frac{Z_p(\alpha x)\,3_q(\alpha x)}{p + q}$$

WA 135(13)

**5.56**

1. $\quad \displaystyle\int Z_1(x)\,\mathrm{d}x = -Z_0(x)$   JA

2. $\quad \displaystyle\int x\,Z_0(x)\,\mathrm{d}x = x\,Z_1(x)$   JA

# 5.6 Orthogonal Polynomials

1. $\quad \displaystyle\int T_n(x)\,\mathrm{d}x = \frac{1}{2}\left[\frac{T_{n+1}(x)}{n+1} - \frac{T_{n-1}(x)}{n-1}\right]$   $[n \geq 2]$

2. $\quad \displaystyle\int U_n(x)\,\mathrm{d}x = \frac{T_{n+1}(x)}{n+1}$   $[n \geq 1]$

3. $\quad \displaystyle\int \frac{T_n(x)}{\sqrt{1-x^2}}\,\mathrm{d}x = -\sqrt{1-x^2}\frac{U_{n-1}(x)}{n}$   $[n \geq 1]$

4. $\quad \displaystyle\int \sqrt{1-x^2}\,U_n(x)\,\mathrm{d}x = \frac{\sqrt{1-x^2}}{2}\left[\frac{U_{n+1}(x)}{n+2} - \frac{U_{n-1}(x)}{n}\right]$   $[n \geq 1]$

# 5.7 Hypergeometric Functions

1. $\quad \displaystyle\int F(\alpha, \beta; \gamma, x)\,\mathrm{d}x = \frac{\gamma - 1}{(\alpha - 1)(\beta - 1)}F(\alpha - 1, \beta - 1; \gamma - 1, x)$   $[\alpha \neq 1, \quad \beta \neq 1, \quad \gamma \neq 1]$

2. $\quad \displaystyle\int x^{\beta-2}F(\alpha, \beta; \gamma, x)\,\mathrm{d}x = \frac{x^{\beta-1}}{\beta - 1}F(\alpha, \beta - 1; \gamma, x)$   $[\beta \neq 1]$

3. $\quad \displaystyle\int x^{\gamma-1}F(\alpha, \beta; \gamma, x)\,\mathrm{d}x = \frac{x^\gamma}{\gamma}F(\alpha, \beta; \gamma + 1, x)$

4. $\quad \displaystyle\int (1-x)^{\beta-2}F(\alpha, \beta; \gamma, x)\,\mathrm{d}x = \frac{\gamma - 1}{(\alpha - \gamma + 1)(\beta - 1)}(1-x)^{\beta-1}F(\alpha, \beta - 1; \gamma - 1, x)$

$$[\beta \neq 1, \quad \gamma - \alpha \neq 1, \quad \gamma \neq 1]$$

5. $\quad \displaystyle\int x^{\gamma-1}(1-x)^{\beta-\gamma-1}F(\alpha, \beta; \gamma, x)\,\mathrm{d}x = \frac{1}{\gamma}x^\gamma(1-x)^{\beta-\gamma}F(\alpha + 1, \beta; \gamma + 1, x)$

**5.71\***

1. $\displaystyle\int {}_1F_1(\alpha;\gamma;x)\,\mathrm{d}x = \frac{\gamma-1}{\alpha-1}\,{}_1F_1(\alpha-1;\gamma-1;x)$         $[\alpha\neq 1,\quad \gamma\neq 1]$

2. $\displaystyle\int x^{\alpha-2}\,{}_1F_1(\alpha;\gamma;x)\,\mathrm{d}x = \frac{x^{\alpha-1}}{\alpha-1}\,{}_1F_1(\alpha-1;\gamma;x)$         $[\alpha\neq 1]$

3. $\displaystyle\int x^{\gamma-1}\,{}_1F_1(\alpha;\gamma;x)\,\mathrm{d}x = \frac{x^{\gamma}}{\gamma}\,{}_1F_1(\alpha;\gamma+1;x)$

4. $\displaystyle\int \Psi(\alpha,\gamma;x)\,\mathrm{d}x = -\frac{1}{\alpha-1}\,\Psi(\alpha-1,\gamma-1;x)$         $[\alpha\neq 1]$

5. $\displaystyle\int x^{\alpha-2}\Psi(\alpha,\gamma;x)\,\mathrm{d}x = -\frac{x^{\alpha-1}}{(\alpha-1)(\gamma-\alpha)}\,\Psi(\alpha-1,\gamma;x)$         $[\alpha\neq 1]$

6. $\displaystyle\int x^{\gamma-1}\Psi(\alpha,\gamma;x)\,\mathrm{d}x = \frac{x^{\gamma}}{\gamma-\alpha}\,\Psi(\alpha,\gamma+1;x)$

Table of Integrals, Series, and Products. http://dx.doi.org/10.1016/B978-0-12-384933-5.00006-0

# 6–7 Definite Integrals of Special Functions

## 6.1 Elliptic Integrals and Functions

**Notation:** $k' = \sqrt{1 - k^2}$ (cf. 8.1).

### 6.11 Forms containing $F(x, k)$

**6.111** $\displaystyle\int_0^{\pi/2} F(x, k) \cot x \, dx = \frac{\pi}{4} \, \boldsymbol{K}(k') + \frac{1}{2} \ln k \, \boldsymbol{K}(k)$  BI (350)(1)

**6.112**

1. $\displaystyle\int_0^{\pi/2} F(x, k) \frac{\sin x \cos x}{1 + k \sin^2 x} \, dx = \frac{1}{4k} \, \boldsymbol{K}(k) \ln \frac{(1+k)\sqrt{k}}{2} + \frac{\pi}{16k} \, \boldsymbol{K}(k')$  BI (350)(6)

2. $\displaystyle\int_0^{\pi/2} F(x, k) \frac{\sin x \cos x}{1 - k \sin^2 x} \, dx = \frac{1}{4k} \, \boldsymbol{K}(k) \ln \frac{2}{(1-k)\sqrt{k}} - \frac{\pi}{16k} \, \boldsymbol{K}(k')$  BI (350)(7)

3. $\displaystyle\int_0^{\pi/2} F(x, k) \frac{\sin x \cos x}{1 - k^2 \sin^2 x} \, dx = -\frac{1}{2k^2} \ln k' \, \boldsymbol{K}(k)$  BI (350)(2)a, BY(802.12)a

**6.113**

1. $\displaystyle\int_0^{\pi/2} F(x, k') \frac{\sin x \cos x \, dx}{\cos^2 x + k \sin^2 x} = \frac{1}{4(1-k)} \ln \frac{2}{(1+k)\sqrt{k}} \, \boldsymbol{K}(k')$  BI (350)(5)

2. $\displaystyle\int_0^{\pi/2} F(x, k) \frac{\sin x \cos x}{1 - k^2 \sin^2 t \sin^2 x} \cdot \frac{dx}{\sqrt{1 - k^2 \sin^2 x}}$

   $$= -\frac{1}{k^2 \sin t \cos t} \left[ \boldsymbol{K}(k) \arctan\left(k' \tan t\right) - \frac{\pi}{2} F(t, k) \right]$$  BI (350)(12)

**6.114** $\displaystyle\int_u^v F(x, k) \frac{dx}{\sqrt{\left(\sin^2 x - \sin^2 u\right)\left(\sin^2 v - \sin^2 x\right)}} = \frac{1}{2 \cos u \sin v} \, \boldsymbol{K}(k) \, \boldsymbol{K}\left(\sqrt{1 - \tan^2 u \cot^2 v}\right)$

$$\left[k^2 = 1 - \cot^2 u \cdot \cot^2 v\right] \qquad \text{BI (351)(9)}$$

**6.115** $\displaystyle\int_0^1 F(\arcsin x, k) \frac{x \, dx}{1 + kx^2} = \frac{1}{4k} \, \boldsymbol{K}(k) \ln \frac{(1+k)\sqrt{k}}{2} + \frac{\pi}{16k} \, \boldsymbol{K}(k')$

$$\text{(cf. } \mathbf{6.112}\ 2) \qquad \text{BI (466)(1)}$$

This and similar formulas can be obtained from formulas **6.111–6.113** by means of the substitution $x = \arcsin t$.

**6.116\*** $\displaystyle\int_0^{\pi/2} F(x,k)\frac{\sin x \cos x}{(1 + k^2 \sinh^2 \mu \sin^2 x)\sqrt{1 - k^2 \sin^2 x}}\, dx$

$$= \frac{-1}{k^2 \sinh \mu \cosh \mu}\left(\boldsymbol{K}(k)\,\operatorname{arctanh}(k' \tanh \mu) - \frac{\pi}{2}F(\phi, k')\right)$$

$$\left[k' = \sqrt{1 - k^2},\quad \phi = \arcsin(\tanh \mu),\qquad 0 < \tanh \mu < 1,\quad 0 < k < 1\right]\quad \text{KM (4.48)}$$

**6.117\*** $\displaystyle\int_0^{\pi/2} F(x,k)\frac{\sin x \cos x}{(1 - k^2 \cosh^2 \nu \sin^2 x)\sqrt{1 - k^2 \sin^2 x}}\, dx$

$$= \frac{1}{k^2 \sinh \nu \cosh \nu}\left(\boldsymbol{K}(k)\operatorname{arctanh}\left(\frac{\tanh \nu}{k'}\right) - \frac{\pi}{2}F(\phi, k')\right)$$

$$\left[k' = \sqrt{1 - k^2},\quad \phi = \arcsin\left(\frac{\tanh \nu}{k'}\right),\qquad 0 < \tanh \nu < k' < 1\right]\qquad \text{KM (4.49)}$$

**6.118\*** $\displaystyle\int_0^{\pi/2} F(x,k)\frac{\sin x \cos x}{(1 - k^2 \cos^2 \psi \sin^2 x)\sqrt{1 - k^2 \sin^2 x}}\, dx$

$$= \frac{1}{k^2 \sin \psi \cos \psi}\left(\boldsymbol{K}(k)\arctan\left(\frac{\tan \psi}{k'}\right) - \frac{\pi}{2}F(\beta, k)\right)$$

$$\left[k' = \sqrt{1 - k^2},\quad \beta = \arctan\left(\frac{\tan \psi}{k'}\right),\quad 0 < \psi < \frac{\pi}{2},\quad 0 < k < 1\right]\qquad \text{KM (4.50)}$$

## 6.12  Forms containing $E(x, k)$

**6.121** $\displaystyle\int_0^{\pi/2} E(x,k)\frac{\sin x \cos x}{1 - k^2 \sin^2 x}\, dx = \frac{1}{2k^2}\left\{\left(1 + k'^2\right)\boldsymbol{K}(k) - (2 + \ln k')\,\boldsymbol{E}(k)\right\}$    BI (350)(4)

**6.122** $\displaystyle\int_0^{\pi/2} E(x,k)\frac{dx}{\sqrt{1 - k^2 \sin^2 x}} = \frac{1}{2}\left\{\boldsymbol{E}(k)\,\boldsymbol{K}(k) - \ln k'\right\}$    BI (350)(10), BY (630.02)

**6.123** $\displaystyle\int_0^{\pi/2} E(x,k)\frac{\sin x \cos x}{1 - k^2 \sin^2 t \sin^2 x}\cdot\frac{dx}{\sqrt{1 - k^2 \sin^2 x}}$

$$= -\frac{1}{k^2 \sin t \cos t}\left[\boldsymbol{E}(k)\arctan(k' \tan t) - \frac{\pi}{2}E(t, k) + \frac{\pi}{2}\cot t\left(1 - \sqrt{1 - k^2 \sin^2 t}\right)\right]$$
$$\text{BI (350)(13)}$$

**6.124**[12] $\displaystyle\int_u^v E(x,k)\frac{dx}{\sqrt{\left(\sin^2 x - \sin^2 u\right)\left(\sin^2 v - \sin^2 x\right)}} = \frac{1}{2\cos u \sin v}\boldsymbol{E}(k)\,\boldsymbol{K}\left(\sqrt{1 - \frac{\tan^2 u}{\tan^2 v}}\right)$

$$+ \frac{k^2 \sin v}{2\cos u}\,\boldsymbol{K}\left(\sqrt{1 - \frac{\sin^2 2u}{\sin^2 2v}}\right)$$

$$\left[k^2 = 1 - \cot^2 u \cot^2 v\right]\qquad \text{BI (351)(10)}$$

**6.125\*** $\displaystyle\int_0^{\pi/2} E(x,k)\frac{\sin x \cos x}{(1-k^2\cosh^2\nu\sin^2 x)\sqrt{1-k^2\sin^2 x}}\,dx$

$$= \frac{1}{k^2\sinh\nu\cosh\nu}\left(\boldsymbol{E}(k)\operatorname{arctanh}\left(\frac{\tanh\nu}{k'}\right)-\frac{\pi\tanh\nu}{2}-\frac{\pi}{2}[F(\phi,k')-E(\phi,k')]\right)$$

$$\left[k'=\sqrt{1-k^2},\quad \phi=\arcsin\left(\frac{\tanh\nu}{k'}\right),\quad 0<\tanh\nu<k'<1\right] \qquad \text{KM (4.45)}$$

**6.126\*** $\displaystyle\int_0^{\pi/2} E(x,k)\frac{\sin x \cos x}{(1-k^2\cos^2\psi\sin^2 x)\sqrt{1-k^2\sin^2 x}}\,dx$

$$= \frac{1}{k^2\sin\psi\cos\psi}\left(\boldsymbol{E}(k)\arctan\left(\frac{\tan\psi}{k'}\right)-\frac{\pi}{2}E(\beta,k)+\frac{\pi}{2}\frac{\tan\psi}{\sqrt{1-k^2\cos^2\psi}}\left(1-\sqrt{1-k^2\cos}\right)\right)$$

$$\left[k'=\sqrt{1-k^2},\quad \beta=\arctan\left(\frac{\tan\psi}{k'}\right),\quad 0<\psi<\frac{\pi}{2},\quad 0<k<1\right] \qquad \text{KM (4.46)}$$

**6.127\*** $\displaystyle\int_0^{\pi/2} E(x,k)\frac{\sin x \cos x}{(1+k^2\sinh^2\mu\sin^2 x)\sqrt{1-k^2\sin^2 x}}\,dx$

$$= \frac{-1}{k^2\sinh\mu\cosh\mu}\left[\boldsymbol{E}(k)\operatorname{arctanh}(k'\tanh\mu)-\frac{\pi}{2}\left(F(\phi,k')-E(\phi,k')+\tanh\mu\sqrt{1+k^2\sinh^2\mu}\right)\right.$$

$$\left.-\frac{\pi}{2}\coth\mu\left(1-\sqrt{1+k^2\sinh^2\mu}\right)\right]$$

$$\left[k'=\sqrt{1-k^2},\quad \phi=\arcsin\left(\tanh\mu\right),\quad 0<\tanh\mu<1,\quad 0<k<1\right] \qquad \text{KM (4.47)}$$

## 6.13 Integration of elliptic integrals with respect to the modulus

**6.131** $\displaystyle\int_0^1 F(x,k)k\,dk = \frac{1-\cos x}{\sin x} = \tan\frac{x}{2}$ \hfill BY (616.03)

**6.132** $\displaystyle\int_0^1 E(x,k)k\,dk = \frac{\sin^2 x + 1 - \cos x}{3\sin x}$ \hfill BY (616.04)

**6.133** $\displaystyle\int_0^1 \Pi\left(x,r^2,k\right)k\,dk = \tan\frac{x}{2} - r\ln\sqrt{\frac{1+r\sin x}{1-r\sin x}} - r^2\,\Pi\left(x,r^2,0\right)$ \hfill BY (616.05)

## 6.14–6.15 Complete elliptic integrals
**6.141**

1. $\displaystyle\int_0^1 \boldsymbol{K}(k)\,dk = 2\boldsymbol{G}$ \hfill FI II 755

2. $\displaystyle\int_0^1 \boldsymbol{K}(k')\,dk = \frac{\pi^2}{4}$ \hfill BY (615.03)

**6.142**

1. $\displaystyle\int_0^1 \left(\boldsymbol{K}(k)-\frac{\pi}{2}\right)\frac{dk}{k} = \pi\ln 2 - 2\boldsymbol{G}$ \hfill BY (615.05)

2.*  $\displaystyle\int_0^1 \left(\boldsymbol{K}(k) - \frac{\pi}{2}\right)\frac{\mathrm{d}k}{k^2} = \frac{\pi}{2} - 1$

**6.143**[7]  $\displaystyle\int_0^1 \boldsymbol{K}(k)\frac{\mathrm{d}k}{k'} = \boldsymbol{K}^2\left(\frac{\sqrt{2}}{2}\right) = \frac{1}{16\pi}\Gamma^4\left(\frac{1}{4}\right)$        BY (615.08)

**6.144**  $\displaystyle\int_0^1 \boldsymbol{K}(k)\frac{\mathrm{d}k}{1+k} = \frac{\pi^2}{8}$        BY (615.09)

**6.145**  $\displaystyle\int_0^1 \left(\boldsymbol{K}(k') - \ln\frac{4}{k}\right)\frac{\mathrm{d}k}{k} = \frac{1}{12}\left[24\left(\ln 2\right)^2 - \pi^2\right]$        BY (615.13)

**6.146**  $\displaystyle n^2\int_0^1 k^n\,\boldsymbol{K}(k)\,\mathrm{d}k = (n-1)^2\int_0^1 k^{n-2}\,\boldsymbol{K}(k)\,\mathrm{d}k + 1$        BY (615.12)

**6.147**  $\displaystyle n\int_0^1 k^n\,\boldsymbol{K}(k')\,\mathrm{d}k = (n-1)\int_0^1 k^{n-2}\,\boldsymbol{E}(k)\,\mathrm{d}k$        $[n>1]$     (see **6.152**)        BY (615.11)

**6.148**

1.  $\displaystyle\int_0^1 \boldsymbol{E}(k)\,\mathrm{d}k = \frac{1}{2} + \boldsymbol{G}$        BY (615.02)

2.  $\displaystyle\int_0^1 \boldsymbol{E}(k')\,\mathrm{d}k = \frac{\pi^2}{8}$        BY (615.04)

3.  $\displaystyle\int_0^1 \frac{\boldsymbol{E}(k)}{1+k}\,\mathrm{d}k = 1$

**6.149**

1.  $\displaystyle\int_0^1 \left(\boldsymbol{E}(k) - \frac{\pi}{2}\right)\frac{\mathrm{d}k}{k} = \pi\ln 2 - 2\boldsymbol{G} + 1 - \frac{\pi}{2}$        BY (615.06)

2.  $\displaystyle\int_0^1 \left(\boldsymbol{E}(k') - 1\right)\frac{\mathrm{d}k}{k} = 2\ln 2 - 1$        BY (615.07)

3.  $\displaystyle\int_0^1 \frac{\boldsymbol{E}(k)}{1+k}\,\mathrm{d}k = 1$

4.  $\displaystyle\int_0^1 \frac{\mathrm{d}x}{x^3}\left(\sqrt{a-x^2}\,\boldsymbol{K}(x) - \frac{\boldsymbol{E}(x)}{\sqrt{1-x^2}} + \frac{\pi}{4}x^2\right) = -\frac{\pi}{4}\ln\left(\frac{4}{\sqrt{e}}\right)$

**6.151** $\quad \int_0^1 E(k) \dfrac{\mathrm{d}k}{k'} = \dfrac{1}{8} \left[ 4\, K^2\left( \dfrac{\sqrt{2}}{2} \right) + \dfrac{\pi^2}{K^2\left( \dfrac{\sqrt{2}}{2} \right)} \right]$ $\hfill$ BY (615.10)

**6.152** $\quad (n+2) \int_0^1 k^n\, E(k')\, \mathrm{d}k = (n+1) \int_0^1 k^n\, K(k')\, \mathrm{d}k \qquad [n>1] \qquad$ (see **6.147**) $\hfill$ BY (615.14)

**6.153**[12] $\quad \int_0^1 \dfrac{x\, E(ax)}{(1-a^2x^2)\sqrt{1-x^2}}\, \mathrm{d}x = \dfrac{\pi}{2\sqrt{1-a^2}} \qquad \left[ a^2 < 1 \right]$

1.* $\quad \int_0^a \dfrac{K(k) - E(k)}{k\sqrt{a^2 - k^2}}\, \mathrm{d}k = \dfrac{\pi}{2a}\left( 1 - \sqrt{1-a^2} \right) \qquad \left[ a^2 \le 1 \right]$

**6.154**[12] $\quad \int_0^{\pi/2} \dfrac{E\,(p\sin x)}{1 - p^2\sin^2 x}\, \sin x\, \mathrm{d}x = \dfrac{\pi}{2\sqrt{1-p^2}} \qquad \left[ p^2 < 1 \right]$ $\hfill$ FI II 489

**6.155***

1. $\quad \int_0^1 \dfrac{x\, K(ax)}{\sqrt{1-x^2}}\, \mathrm{d}x = \dfrac{\pi}{2a}\arcsin(a) \qquad \left[ a^2 \le 1 \right]$

2. $\quad \int_0^1 \dfrac{x\, E(ax)}{\sqrt{1-x^2}}\, \mathrm{d}x = \dfrac{\pi}{4a}\left( \arcsin(a) + a\sqrt{1-a^2} \right) \qquad \left[ a^2 \le 1 \right]$

**6.156***

1. $\quad \int_0^1 [K(k) - E(k)]^2 \dfrac{\mathrm{d}k}{k} = \dfrac{1}{2}$

2. $\quad \int_0^1 [K(k) - E(k)]^2 \dfrac{\mathrm{d}k}{k^3} - \dfrac{\pi^2}{8} - \dfrac{1}{2}$

3. $\quad \int_0^1 [(2-k^2)\, K(k) - 2\, E(k)]^2 \dfrac{\mathrm{d}k}{k^5} = \dfrac{\pi^2}{16} - \dfrac{1}{2}$

4. $\quad \int_0^1 [(2-k^2)\, K(k) - 2\, E(k)]^2 \dfrac{\mathrm{d}k}{k^7} = \dfrac{\pi^2}{32} - \dfrac{1}{6}$

5. $\quad \int_0^1 [(2-k^2)\, K(k) - 2\, E(k)]^2 \dfrac{2-k^2}{k^5 k'}\, \mathrm{d}k = \dfrac{\pi^2}{8}$

**6.157***

1. $\quad \int_0^1 x^a R_F(0,x,1)\, \mathrm{d}x = \dfrac{1}{2}\left[ B\left( a+1, \dfrac{1}{2} \right) \right]^2 \qquad [\operatorname{Re} a > -1]$ $\hfill$ DLMF (19.28.1)

2. $\quad \int_0^1 x^a R_G(0,x,1)\, \mathrm{d}x = \dfrac{a+1}{4a+6}\left[ B\left( a+1, \dfrac{1}{2} \right) \right]^2 \qquad [\operatorname{Re} a > -1]$ $\hfill$ DLMF (19.28.2)

3. $\quad \int_0^1 x^a(1-x) R_D(0,x,1)\, \mathrm{d}x = \dfrac{3}{4a+6}\left[ B\left( a+1, \dfrac{1}{2} \right) \right]^2$

$\hfill [\operatorname{Re} a > -1] \hfill$ DLMF (19.28.3)

4. $\quad \int_c^{\infty} R_D(a,b,x)\, \mathrm{d}x = 6 R_F(a,b,c) \qquad [a,b,c>0]$ $\hfill$ DLMF (19.28.5)

5.  $\displaystyle\int_0^1 R_D(a, b, x^2 c + (1 - x^2)d)\ \mathrm{d}x = R_J(a, b, c, d)$     $[a, b, c, d > 0]$     DLMF (19.28.6)

6.  $\displaystyle\int_0^\infty R_J(a, b, c, x^2)\ \mathrm{d}x = \frac{3\pi}{2} R_F(ab, ac, bc)$     $[a, b, c > 0]$     DLMF (19.28.7)

7.  $\displaystyle\int_0^\infty R_J(ax, b, c, xd)\ \mathrm{d}x = \frac{6}{\sqrt{d}} R_C(d, a) R_F(0, b, c)$     $[a, b, c, d > 0]$     DLMF (19.28.8)

## 6.16  The theta function

**6.161**

1.  $\displaystyle\int_0^\infty x^{s-1}\, \vartheta_2\left(0 \mid ix^2\right)\ \mathrm{d}x = 2^s\left(1 - 2^{-s}\right)\pi^{-\frac{s}{2}}\,\Gamma\left(\tfrac{1}{2}s\right)\zeta(s)$

$[\operatorname{Re} s > 2]$     ET I 339(20)

2.  $\displaystyle\int_0^\infty x^{s-1}\left[\vartheta_3\left(0 \mid ix^2\right) - 1\right]\ \mathrm{d}x = \pi^{-\frac{s}{2}}\,\Gamma\left(\tfrac{1}{2}s\right)\zeta(s)$     $[\operatorname{Re} s > 2]$     ET I 339(21)

3.  $\displaystyle\int_0^\infty x^{s-1}\left[1 - \vartheta_4\left(0 \mid ix^2\right)\right]\ \mathrm{d}x = \left(1 - 2^{1-s}\right)\pi^{-\frac{1}{2}s}\,\Gamma\left(\tfrac{1}{2}s\right)\zeta(s)$

$[\operatorname{Re} s > 2]$     ET I 339(22)

4.  $\displaystyle\int_0^\infty x^{s-1}\left[\vartheta_4\left(0 \mid ix^2\right) + \vartheta_2\left(0 \mid ix^2\right) - \vartheta_3\left(0 \mid ix^2\right)\right]\ \mathrm{d}x = -\left(2^s - 1\right)\left(2^{1-s} - 1\right)\pi^{-\frac{1}{2}s}\,\Gamma\left(\tfrac{1}{2}s\right)\zeta(s)$

ET I 339(24)

**6.162**

1.[11]  $\displaystyle\int_0^\infty e^{-ax}\, \vartheta_4\left(\frac{b\pi}{2l} \,\middle|\, \frac{i\pi x}{l^2}\right)\ \mathrm{d}x = \frac{l}{\sqrt{a}}\cosh\left(b\sqrt{a}\right)\operatorname{cosech}\left(l\sqrt{a}\right)$

$[\operatorname{Re} a > 0, \quad |b| \le l]$     ET I 224(1)a

2.  $\displaystyle\int_0^\infty e^{-ax}\, \vartheta_1\left(\frac{b\pi}{2l} \,\middle|\, \frac{i\pi x}{l^2}\right)\ \mathrm{d}x = -\frac{l}{\sqrt{a}}\sinh\left(b\sqrt{a}\right)\operatorname{sech}\left(l\sqrt{a}\right)$

$[\operatorname{Re} a > 0, \quad |b| \le l]$     ET I 224(2)a

3.[11]  $\displaystyle\int_0^\infty e^{-ax}\, \vartheta_2\left(\frac{(l + b)\pi}{2l} \,\middle|\, \frac{i\pi x}{l^2}\right)\ \mathrm{d}x = -\frac{l}{\sqrt{a}}\sinh\left(b\sqrt{a}\right)\operatorname{sech}\left(l\sqrt{a}\right)$

$[\operatorname{Re} a > 0, \quad |b| \le l]$     ET I 224(3)a

4.[11]  $\displaystyle\int_0^\infty e^{-ax}\, \vartheta_3\left(\frac{(l + b)\pi}{2l} \,\middle|\, \frac{i\pi x}{l^2}\right)\ \mathrm{d}x = \frac{l}{\sqrt{a}}\cosh\left(b\sqrt{a}\right)\operatorname{cosech}\left(l\sqrt{a}\right)$

$[\operatorname{Re} a > 0, \quad |b| \le l]$     ET I 224(4)a

**6.163**[10]

1.[12] $\displaystyle\int_0^\infty e^{-(a-\mu)x}\,\vartheta_3\left(\pi\sqrt{\mu}x \,|\, i\pi x\right)\,dx = \frac{1}{2\sqrt{a}}\left[\tanh\left(\sqrt{a}+\sqrt{\mu}\right) + \tanh\left(\sqrt{a}-\sqrt{\mu}\right)\right]$

$$[\operatorname{Re} a > 0]\qquad\qquad\qquad \text{ET I 224(7)a}$$

2.[10] $\displaystyle\int_0^\infty \vartheta_3\left(i\pi kx \,|\, i\pi x\right)e^{-\left(k^2+l^2\right)x}\,dx = \frac{\sinh 2l}{l\left(\cosh 2l - \cos 2k\right)}$

**6.164**[11] $\displaystyle\int_0^\infty \left[\vartheta_4\left(0 \,|\, ie^{2x}\right) + \vartheta_2\left(0 \,|\, ie^{2x}\right) - \vartheta_3\left(0 \,|\, ie^{2x}\right)\right]e^{\frac{1}{2}x}\cos(ax)\,dx$

$$= \frac{1}{2}\left(2^{\frac{1}{2}+ia}-1\right)\left(1-2^{\frac{1}{2}-ia}\right)\pi^{-\frac{1}{4}-\frac{1}{2}ia}\,\Gamma\left(\tfrac{1}{4}+\tfrac{1}{2}ia\right)\zeta\left(\tfrac{1}{2}+ia\right)$$

$$[a > 0]\qquad\qquad\qquad \text{ET I 61(11)}$$

**6.165** $\displaystyle\int_0^\infty e^{\frac{1}{2}x}\left[\vartheta_3\left(0 \,|\, ie^{2x}\right) - 1\right]\cos(ax)\,dx$

$$= \frac{2}{1+4a^2}\left\{1 + \left[\left(a^2+\tfrac{1}{4}\right)\pi^{-\frac{1}{2}ia-\frac{1}{4}}\,\Gamma\left(\tfrac{1}{2}ia+\tfrac{1}{4}\right)\zeta\left(ia+\tfrac{1}{2}\right)\right]\right\}$$

$$[a > 0]\qquad\qquad\qquad \text{ET I 61(12)}$$

## 6.17[10]   Generalized elliptic integrals

1.    Set

$$\Omega_j(k) \equiv \int_0^\pi \left[1 - k^2\cos\phi\right]^{-\left(j+\frac{1}{2}\right)}\,d\phi,$$

$$\alpha_m(j) = \frac{\pi}{(64)^m}\,\frac{j!}{(2j)!}\,\frac{(4m+2j)!}{(2m+j)!}\left(\frac{1}{m!}\right)^2, \qquad \lambda = \frac{\pi}{2}\sqrt{\frac{(2j+1)k^2}{1-k^2}},$$

then

$$\Omega_j(k) = \sum_{m=0}^\infty \alpha_m(j)k^{4m} = \sqrt{\frac{\pi}{(2j+1)k^2}}\,\left(1-k^2\right)^{-j}\left[\operatorname{erf}\lambda + \frac{1}{2}(2j+1)^{-1}\left(1+\frac{1}{2k^2}\right)\right.$$

$$\times\left\{\operatorname{erf}\lambda - \left(\frac{2}{\sqrt{\pi}}\right)\left(\lambda e^{-\lambda^2}\right)\left(1+\frac{2}{3}\lambda^2\right)\right\} - \frac{1}{12}(2j+1)^{-2}\left(16+\frac{13}{k^2}+\frac{1}{k^4}\right)$$

$$\left.\times\left\{\operatorname{erf}\lambda - \left(\frac{2}{\sqrt{\pi}}\right)\left(\lambda e^{-\lambda^2}\right)\left(1+\frac{2}{3}\lambda^2+\frac{4}{15}\lambda^4\right)\right\} + \cdots\right]$$

while for large $\lambda$

$$\lim_{j\to\infty}\Omega_j(k) = \sqrt{\frac{\pi}{(2j+1)}}\,k^2\left(1-k^2\right)^{-j}$$

$$\times\left[1+\frac{1}{2}(2j+1)^{-1}\left\{1+\frac{1}{2k^2}\right\} - \frac{4}{3}(2j+1)^{-2}\left\{1+\frac{13}{16k^2}+\frac{1}{16k^4}\right\} + \cdots\right]$$

2. Set

$$R_\mu(k, \alpha, \delta) = \int_0^\pi \frac{\cos^{2\alpha-1}(\theta/2) \sin^{2\delta-2\alpha-1}(\theta/2) \, d\theta}{[1 - k^2 \cos\theta]^{\mu+\frac{1}{2}}},$$

$$0 < k < 1, \quad \operatorname{Re}\delta > \operatorname{Re}\alpha > 0, \quad \operatorname{Re}\mu > -1/2,$$

$$M_\nu(\mu, \alpha, \delta) = \frac{(-1)^\nu 2^\nu \left(\mu + \frac{1}{2}\right)_\nu}{\nu!} \frac{\Gamma(\alpha)\,\Gamma(\delta - \alpha + \nu)}{\Gamma(\delta + \nu)},$$

$$\text{with } (\lambda)_\nu = \Gamma(\lambda + \nu)/\Gamma(\lambda), \text{ and}$$

$$W_\nu(\mu, \alpha, \delta) = \frac{2^\nu \left(\mu + \frac{1}{2}\right)_\nu}{\nu!} \frac{\Gamma(\alpha + \nu)\,\Gamma(\delta - \alpha)}{\Gamma(\delta + \nu)},$$

then:

- for small $k$;

$$R_\mu(k, \alpha, \delta) = \left(1 - k^2\right)^{-\left(\mu+\frac{1}{2}\right)} \sum_{\nu=0}^\infty \left[k^2/\left(1 - k^2\right)\right]^\nu M_\nu(\mu, \alpha, \delta)$$

$$= \left(1 + k^2\right)^{-\left(\mu+\frac{1}{2}\right)} \sum_{\nu=0}^\infty \left[k^2/\left(1 + k^2\right)\right]^\nu W_\nu(\mu, \alpha, \delta),$$

- for $k^2$ close to 1;
$R_\mu(k, \alpha, \delta)$

$$= \left[\Gamma(\delta - \alpha)\,\Gamma\left(\mu + \alpha - \delta + \tfrac{1}{2}\right)\Gamma\left(\mu + \tfrac{1}{2}\right)\right] \left(2k^2\right)^{\alpha-\delta} \left(1 - k^2\right)^{\delta-\alpha-\mu-\frac{1}{2}}$$

$$\times \left\{\Gamma\left(\delta - \alpha - \mu - \tfrac{1}{2}\right)\Gamma(\alpha)\left[\Gamma\left(\delta - \mu - \tfrac{1}{2}\right)\left(2k^2\right)^{\mu+\frac{1}{2}}\right]\right\}$$

$$\left[\operatorname{Re}\left(\mu + \alpha - \delta + \tfrac{1}{2}\right) \text{ not an integer}\right]$$

$$= \left[2^{\mu+\frac{1}{2}} k^{2\mu+1} \,\Gamma\left(\mu + \tfrac{1}{2}\right)\Gamma(1 - \alpha)\right]$$

$$\times \sum_{n=0}^\infty \left[\Gamma\left(\delta - \alpha + n\right)\Gamma(1 - \alpha + n)\,\Gamma\left(\alpha - \delta + \mu - n + \tfrac{1}{2}\right)n!\right] \left[2k^2/\left(1 - k^2\right)\right]^{\alpha-\delta+\mu-n+\frac{1}{2}}$$

$$\left[\alpha - \delta + \mu + \tfrac{1}{2} = m, \text{ with } m \text{ a non-negative integer}\right]$$

# 6.2–6.3 The Exponential Integral Function and Functions Generated by It

## 6.21 The logarithm integral

**6.211** $\displaystyle\int_0^1 \operatorname{li}(x)\,dx = -\ln 2$ 

BI (79)(5)

**6.212**

1. $\displaystyle\int_0^1 \operatorname{li}\left(\frac{1}{x}\right) x\,dx = 0$

BI (255)(1)

2. $\displaystyle\int_0^1 \mathrm{li}(x) x^{p-1}\,\mathrm{d}x = -\frac{1}{p}\ln(p+1)$ $\qquad [p > -1]$ $\qquad$ BI (255)(2)

3. $\displaystyle\int_0^1 \mathrm{li}(x)\frac{\mathrm{d}x}{x^{q+1}} = \frac{1}{q}\ln(1-q)$ $\qquad [q < 1]$ $\qquad$ BI (255)(3)

4. $\displaystyle\int_1^\infty \mathrm{li}(x)\frac{\mathrm{d}x}{x^{q+1}} = -\frac{1}{q}\ln(q-1)$ $\qquad [q > 1]$ $\qquad$ BI (255)(4)

**6.213**

1. $\displaystyle\int_0^1 \mathrm{li}\left(\frac{1}{x}\right)\sin(a\ln x)\,\mathrm{d}x = \frac{1}{1+a^2}\left(a\ln a - \frac{\pi}{2}\right)$ $\qquad [a > 0]$ $\qquad$ BI (475)(1)

2. $\displaystyle\int_1^\infty \mathrm{li}\left(\frac{1}{x}\right)\sin(a\ln x)\,\mathrm{d}x = -\frac{1}{1+a^2}\left(\frac{\pi}{2} + a\ln a\right)$ $\qquad [a > 0]$ $\qquad$ BI (475)(9)

3. $\displaystyle\int_0^1 \mathrm{li}\left(\frac{1}{x}\right)\cos(a\ln x)\,\mathrm{d}x = -\frac{1}{1+a^2}\left(\ln a + \frac{\pi}{2}a\right)$ $\qquad [a > 0]$ $\qquad$ BI (475)(2)

4. $\displaystyle\int_1^\infty \mathrm{li}\left(\frac{1}{x}\right)\cos(a\ln x)\,\mathrm{d}x = \frac{1}{1+a^2}\left(\ln a - \frac{\pi}{2}a\right)$ $\qquad [a > 0]$ $\qquad$ BI (475)(10)

5. $\displaystyle\int_0^1 \mathrm{li}(x)\sin(a\ln x)\frac{\mathrm{d}x}{x} = \frac{\ln(1+a^2)}{2a}$ $\qquad [a > 0]$ $\qquad$ BI(479)(1), ET I 98(20)a

6. $\displaystyle\int_0^1 \mathrm{li}(x)\cos(a\ln x)\frac{\mathrm{d}x}{x} = -\frac{\arctan a}{a}$ $\qquad$ BI (479)(2)

7. $\displaystyle\int_0^1 \mathrm{li}(x)\sin(a\ln x)\frac{\mathrm{d}x}{x^2} = \frac{1}{1+a^2}\left(a\ln a + \frac{\pi}{2}\right)$ $\qquad [a > 0]$ $\qquad$ BI (479)(3)

8. $\displaystyle\int_1^\infty \mathrm{li}(x)\sin(a\ln x)\frac{\mathrm{d}x}{x^2} = \frac{1}{1+a^2}\left(\frac{\pi}{2} - a\ln a\right)$ $\qquad [a > 0]$ $\qquad$ BI (479)(13)

9. $\displaystyle\int_0^1 \mathrm{li}(x)\cos(a\ln x)\frac{\mathrm{d}x}{x^2} = \frac{1}{1+a^2}\left(\ln a - \frac{\pi}{2}a\right)$ $\qquad [a > 0]$ $\qquad$ BI (479)(4)

10. $\displaystyle\int_1^\infty \mathrm{li}(x)\cos(a\ln x)\frac{\mathrm{d}x}{x^2} = -\frac{1}{1+a^2}\left(\ln a + \frac{\pi}{2}a\right)$ $\qquad [a > 0]$ $\qquad$ BI (479)(14)

11. $\displaystyle\int_0^1 \mathrm{li}(x)\sin(a\ln x)x^{p-1}\,\mathrm{d}x = \frac{1}{a^2+p^2}\left\{\frac{a}{2}\ln\left[(1+p)^2 + a^2\right] - p\arctan\frac{a}{1+p}\right\}$

$\qquad\qquad [p > 0]$ $\qquad$ BI (477)(1)

12. $\displaystyle\int_0^1 \mathrm{li}(x)\cos(a\ln x)x^{p-1}\,\mathrm{d}x = -\frac{1}{a^2+p^2}\left\{a\arctan\frac{a}{1+p} + \frac{p}{2}\ln\left[(1+p)^2 + a^2\right]\right\}$

$\qquad\qquad [p > 0]$ $\qquad$ BI (477)(2)

**6.214**

1. $\displaystyle\int_0^1 \mathrm{li}\left(\frac{1}{x}\right)(-\ln x)^{p-1}\,\mathrm{d}x = -\pi\cot p\pi \cdot \Gamma(p)$ $\qquad [0 < p < 1]$ $\qquad$ BI (340)(1)

2.     $\displaystyle\int_1^\infty \mathrm{li}\left(\frac{1}{x}\right)(\ln x)^{p-1}\,dx = -\frac{\pi}{\sin p\pi}\,\Gamma(p)$     $[0 < p < 1]$     BI (340)(9)

**6.215**

1.     $\displaystyle\int_0^1 \mathrm{li}(x)\frac{x^{p-1}}{\sqrt{-\ln x}}\,dx = -2\sqrt{\frac{\pi}{p}}\,\mathrm{arcsinh}\,\sqrt{p} = -2\sqrt{\frac{\pi}{p}}\,\ln\left(\sqrt{p}+\sqrt{p+1}\right)$

$[p > 0]$     BI (444)(3)

2.     $\displaystyle\int_0^1 \mathrm{li}(x)\frac{dx}{x^{p+1}\sqrt{-\ln x}} = -2\sqrt{\frac{\pi}{p}}\,\arcsin\sqrt{p}$     $[1 > p > 0]$     BI (444)(4)

**6.216**

1.$^{12}$     $\displaystyle\int_0^1 \mathrm{li}(x)[-\ln x]^{p-1}\frac{dx}{x} = -\frac{1}{p}\,\Gamma(p)$     $[0 < p < 1]$     BI (444)(1)

2.     $\displaystyle\int_0^1 \mathrm{li}(x)[-\ln x]^{p-1}\frac{dx}{x^2} = -\frac{\pi\,\Gamma(p)}{\sin p\pi}$     $[0 < p < 1]$     BI (444)(2)

## 6.22–6.23 The exponential integral function

**6.221**     $\displaystyle\int_0^p \mathrm{Ei}(\alpha x)\,dx = p\,\mathrm{Ei}(\alpha p) + \frac{1-e^{\alpha p}}{\alpha}$     NT 11(7)

**6.222**     $\displaystyle\int_0^\infty \mathrm{Ei}(-px)\,\mathrm{Ei}(-qx)\,dx = \left(\frac{1}{p}+\frac{1}{q}\right)\ln(p+q) - \frac{\ln q}{p} - \frac{\ln p}{q}$

$[p > 0, \quad q > 0]$     FI II 653, NT 53(3)

**6.223**     $\displaystyle\int_0^\infty \mathrm{Ei}(-\beta x)x^{\mu-1}\,dx = -\frac{\Gamma(\mu)}{\mu\beta^\mu}$     $[\mathrm{Re}\,\beta \geq 0, \quad \mathrm{Re}\,\mu > 0]$

NT 55(7), ET I 325(10)

**6.224**

1.     $\displaystyle\int_0^\infty \mathrm{Ei}(-\beta x)e^{-\mu x}\,dx = -\frac{1}{\mu}\ln\left(1+\frac{\mu}{\beta}\right)$     $[\mathrm{Re}(\beta+\mu) \geq 0, \quad \mu > 0]$

$= -1/\beta$     $[\mu = 0]$

FI II 652, NT 48(8)

2.     $\displaystyle\int_0^\infty \mathrm{Ei}(ax)e^{-\mu x}\,dx = -\frac{1}{\mu}\ln\left(\frac{\mu}{a}-1\right)$     $[a > 0, \quad \mathrm{Re}\,\mu > 0, \quad \mu > a]$

ET I 178(23)a, BI (283)(3)

**6.225**

1.     $\displaystyle\int_0^\infty \mathrm{Ei}\left(-x^2\right)e^{-\mu x^2}\,dx = -\sqrt{\frac{\pi}{\mu}}\,\mathrm{arcsinh}\,\sqrt{\mu} = -\sqrt{\frac{\pi}{\mu}}\,\ln\left(\sqrt{\mu}+\sqrt{1+\mu}\right)$

$[\mathrm{Re}\,\mu > 0]$     BI (283)(5), ET I 178(25)a

2.     $\displaystyle\int_0^\infty \mathrm{Ei}\left(-x^2\right)e^{px^2}\,dx = -\sqrt{\frac{\pi}{p}}\,\arcsin\sqrt{p}$     $[1 > p > 0]$     NT 59(9)a

**6.226**

1.    $\int_0^\infty \mathrm{Ei}\left(-\dfrac{1}{4x}\right) e^{-\mu x}\,dx = -\dfrac{2}{\mu} K_0\left(\sqrt{\mu}\right)$        $[\mathrm{Re}\,\mu > 0]$        MI 34

2.    $\int_0^\infty \mathrm{Ei}\left(\dfrac{a^2}{4x}\right) e^{-\mu x}\,dx = -\dfrac{2}{\mu} K_0\left(a\sqrt{\mu}\right)$      $[a > 0, \quad \mathrm{Re}\,\mu > 0]$      MI 34

3.    $\int_0^\infty \mathrm{Ei}\left(-\dfrac{1}{4x^2}\right) e^{-\mu x^2}\,dx = \sqrt{\dfrac{\pi}{\mu}}\,\mathrm{Ei}\left(-\sqrt{\mu}\right)$      $[\mathrm{Re}\,\mu > 0]$      MI 34

4.    $\int_0^\infty \mathrm{Ei}\left(-\dfrac{1}{4x^2}\right) e^{-\mu x^2 + \frac{1}{4x^2}}\,dx = \sqrt{\dfrac{\pi}{\mu}}\left[\cos\sqrt{\mu}\,\mathrm{ci}\,\sqrt{\mu} - \sin\sqrt{\mu}\,\mathrm{si}\,\sqrt{\mu}\right]$

                                                    $[\mathrm{Re}\,\mu > 0]$      MI 34

**6.227**

1.    $\int_0^\infty \mathrm{Ei}(-x) e^{-\mu x} x\,dx = \dfrac{1}{\mu(\mu+1)} - \dfrac{1}{\mu^2}\ln(1+\mu)$      $[\mathrm{Re}\,\mu > 0]$      MI 34

2.    $\int_0^\infty \left[\dfrac{e^{-ax}\,\mathrm{Ei}(ax)}{x-b} - \dfrac{e^{ax}\,\mathrm{Ei}(-ax)}{x+b}\right]\,dx = 0$      $[a > 0, \quad b < 0]$

                                      $= \pi^2 e^{-ab}$      $[a > 0, \quad b > 0]$

                                                              ET II 253(1)a

**6.228**

1.    $\int_0^\infty \mathrm{Ei}(-x) e^x x^{\nu-1}\,dx = -\dfrac{\pi\,\Gamma(\nu)}{\sin\nu\pi}$      $[0 < \mathrm{Re}\,\nu < 1]$      ET II 308(13)

2.    $\int_0^\infty \mathrm{Ei}(-\beta x) c^{-\mu x} x^{\nu-1}\,dx = -\dfrac{\Gamma(\nu)}{\nu(\beta+\mu)^\nu}\,{}_2F_1\left(1,\nu;\nu+1;\dfrac{\mu}{\beta+\mu}\right)$

                         $[|\arg\beta| < \pi, \quad \mathrm{Re}(\beta+\mu) > 0, \quad \mathrm{Re}\,\nu > 0]$    ET II 308(14)

**6.229**    $\int_0^\infty \mathrm{Ei}\left(-\dfrac{1}{4x^2}\right)\exp\left(-\mu x^2 + \dfrac{1}{4x^2}\right)\dfrac{dx}{x^2} = 2\sqrt{\pi}\left(\cos\sqrt{\mu}\,\mathrm{si}\,\sqrt{\mu} - \sin\sqrt{\mu}\,\mathrm{ci}\,\sqrt{\mu}\right)$

                                                      $[\mathrm{Re}\,\mu > 0]$      MI 34

**6.231**    $\int_{-\ln a}^\infty \left[\mathrm{Ei}(-a) - \mathrm{Ei}\left(-e^{-x}\right)\right] e^{-\mu x}\,dx = \dfrac{1}{\mu}\gamma(\mu, a)$      $[a < 1, \quad \mathrm{Re}\,\mu > 0]$      MI 34

**6.232**

1.    $\int_0^\infty \mathrm{Ei}(-ax)\sin bx\,dx = -\dfrac{\ln\left(1+\dfrac{b^2}{a^2}\right)}{2b}$      $[a > 0, \quad b > 0]$      BI (473)(1)a

2.    $\int_0^\infty \mathrm{Ei}(-ax)\cos bx\,dx = -\dfrac{1}{b}\arctan\dfrac{b}{a}$      $[a > 0, \quad b > 0]$      BI (473)(2)a

**6.233**

1.    $\int_0^\infty \mathrm{Ei}(-x) e^{-\mu x}\sin\beta x\,dx = -\dfrac{1}{\beta^2+\mu^2}\left\{\dfrac{\beta}{2}\ln\left[(1+\mu)^2 + \beta^2\right] - \mu\arctan\dfrac{\beta}{1+\mu}\right\}$

                                               $[\mathrm{Re}\,\mu > |\mathrm{Im}\,\beta|]$      BI (473)(7)a

2.
$$\int_0^\infty \mathrm{Ei}(-x)e^{-\mu x}\cos\beta x\,dx = -\frac{1}{\beta^2+\mu^2}\left\{\frac{\mu}{2}\ln\left[(1+\mu)^2+\beta^2\right]+\beta\arctan\frac{\beta}{1+\mu}\right\}$$

$$[\mathrm{Re}\,\mu > |\mathrm{Im}\,\beta|] \qquad\qquad \text{BI (473)(8)a}$$

**6.234** $\displaystyle\int_0^\infty \mathrm{Ei}(-x)\ln x\,dx = \boldsymbol{C}+1$ $\qquad\qquad\qquad\qquad$ NT 56(10)

## 6.24–6.26 The sine integral and cosine integral functions

**6.241**

1.
$$\int_0^\infty \mathrm{si}(px)\,\mathrm{si}(qx)\,dx = \frac{\pi}{2p} \qquad\qquad [p\ge q] \qquad\qquad \text{BI II 653, NT 54(8)}$$

2.
$$\int_0^\infty \mathrm{ci}(px)\,\mathrm{ci}(qx)\,dx = \frac{\pi}{2p} \qquad\qquad [p\ge q] \qquad\qquad \text{FI II 653, NT 54(7)}$$

3.
$$\int_0^\infty \mathrm{si}(px)\,\mathrm{ci}(qx)\,dx = \frac{1}{4q}\ln\left(\frac{p+q}{p-q}\right)^2 + \frac{1}{4p}\ln\frac{\left(p^2-q^2\right)^2}{q^4} \qquad [p\ne q]$$
$$= \frac{1}{q}\ln 2 \qquad\qquad [p=q]$$

$$\text{FI II 653, NT 54(10, 12)}$$

**6.242** $\displaystyle\int_0^\infty \frac{\mathrm{ci}(ax)}{\beta+x}\,dx = -\frac{1}{2}\left\{[\mathrm{si}(a\beta)]^2 + [\mathrm{ci}(a\beta)]^2\right\}$ $\qquad [a>0, \quad |\arg\beta|<\pi]$ $\qquad$ ET II 224(1)

**6.243**

1.
$$\int_{-\infty}^\infty \frac{\mathrm{si}\,(a|x|)}{x-b}\,\mathrm{sign}\,x\,dx = \pi\,\mathrm{ci}\,(a|b|) \qquad\qquad [a>0, \quad b>0] \qquad\qquad \text{ET II 253(3)}$$

2.
$$\int_{-\infty}^\infty \frac{\mathrm{ci}\,(a|x|)}{x-b}\,dx = -\pi\,\mathrm{sign}\,b\cdot\mathrm{si}\,(a|b|) \qquad\qquad [a>0] \qquad\qquad \text{ET II 253(2)}$$

**6.244**

1.[8]
$$\int_0^\infty \mathrm{si}(px)\frac{x\,dx}{q^2+x^2} = \frac{\pi}{2}\,\mathrm{Ei}(-pq) \qquad\qquad [p>0, \quad q>0] \qquad\qquad \text{BI (255)(6)}$$

2.[8]
$$\int_0^\infty \mathrm{si}(px)\frac{x\,dx}{q^2-x^2} = -\frac{\pi}{2}\,\mathrm{ci}(pq) \qquad\qquad [p>0, \quad q>0] \qquad\qquad \text{BI (255)(6)}$$

**6.245**

1.
$$\int_0^\infty \mathrm{ci}(px)\frac{dx}{q^2+x^2} = \frac{\pi}{2q}\,\mathrm{Ei}(-pq) \qquad\qquad [p>0, \quad q>0] \qquad\qquad \text{BI (255)(7)}$$

2.
$$\int_0^\infty \mathrm{ci}(px)\frac{dx}{q^2-x^2} = \frac{\pi}{2q}\,\mathrm{si}(pq) \qquad\qquad [p>0, \quad q>0] \qquad\qquad \text{BI (255)(8)}$$

**6.246**

1.
$$\int_0^\infty \mathrm{si}(ax)x^{\mu-1}\,dx = -\frac{\Gamma(\mu)}{\mu a^\mu}\sin\frac{\mu\pi}{2} \qquad\qquad [a>0, \quad 0<\mathrm{Re}\,\mu<1]$$

$$\text{NT 56(9), ET I 325(12)a}$$

2. $\quad \displaystyle\int_0^\infty \operatorname{ci}(ax) x^{\mu-1}\, dx = -\frac{\Gamma(\mu)}{\mu a^\mu}\cos\frac{\mu\pi}{2}$ $\qquad\qquad [a>0, \quad 0<\operatorname{Re}\mu<1]$

NT 56(8), ET I 325(13)a

**6.247**

1. $\quad \displaystyle\int_0^\infty \operatorname{si}(\beta x) e^{-\mu x}\, dx = -\frac{1}{\mu}\arctan\frac{\mu}{\beta}$ $\qquad\qquad [\operatorname{Re}\mu>0] \qquad$ NT 49(12), ET I 177(18)

2. $\quad \displaystyle\int_0^\infty \operatorname{ci}(\beta x) e^{-\mu x}\, dx = -\frac{1}{\mu}\ln\sqrt{1+\frac{\mu^2}{\beta^2}}$ $\qquad\qquad [\operatorname{Re}\mu>0] \qquad$ NT 49(11), ET I 178(19)a

**6.248**

1.[8] $\quad \displaystyle\int_0^\infty \operatorname{si}(x) e^{-\mu x^2} x\, dx = \frac{\pi}{4\mu}\left[\Phi\left(\frac{1}{2\sqrt{\mu}}\right)-1\right]$ $\qquad\qquad [\operatorname{Re}\mu>0] \qquad$ MI 34

2. $\quad \displaystyle\int_0^\infty \operatorname{ci}(x) e^{-\mu x^2}\, dx = \frac{1}{4}\sqrt{\frac{\pi}{\mu}}\operatorname{Ei}\left(-\frac{1}{4\mu}\right)$ $\qquad\qquad [\operatorname{Re}\mu>0] \qquad$ MI 34

**6.249** $\quad \displaystyle\int_0^\infty \left[\operatorname{si}\left(x^2\right)+\frac{\pi}{2}\right]e^{-\mu x}\, dx = \frac{\pi}{\mu}\left\{\left[S\left(\frac{\mu^2}{4}\right)-\frac{1}{2}\right]^2+\left[C\left(\frac{\mu^2}{4}\right)-\frac{1}{2}\right]^2\right\}$

$[\operatorname{Re}\mu>0] \qquad\qquad$ ME 26

**6.251**

1. $\quad \displaystyle\int_0^\infty \operatorname{si}\left(\frac{1}{x}\right)e^{-\mu x}\, dx = \frac{2}{\mu}\operatorname{kei}\left(2\sqrt{\mu}\right)$ $\qquad\qquad [\operatorname{Re}\mu>0] \qquad$ MI 34

2. $\quad \displaystyle\int_0^\infty \operatorname{ci}\left(\frac{1}{x}\right)e^{-\mu x}\, dx = -\frac{2}{\mu}\operatorname{ker}\left(2\sqrt{\mu}\right)$ $\qquad\qquad [\operatorname{Re}\mu>0] \qquad$ MI 34

**6.252**

1. $\quad \displaystyle\int_0^\infty \sin px\,\operatorname{si}(qx)\, dx = -\frac{\pi}{2p}$ $\qquad\qquad [p^2>q^2]$

$\qquad\qquad\qquad\qquad\qquad\quad = -\frac{\pi}{4p}$ $\qquad\qquad [p^2=q^2]$

$\qquad\qquad\qquad\qquad\qquad\quad = 0$ $\qquad\qquad [p^2<q^2]$

FI II 652, NT 50(8)

2.[6] $\quad \displaystyle\int_0^\infty \cos px\,\operatorname{si}(qx)\, dx = -\frac{1}{4p}\ln\left(\frac{p+q}{p-q}\right)^2$ $\qquad\qquad [p\neq0, \quad p^2\neq q^2]$

$\qquad\qquad\qquad\qquad\qquad\quad = \frac{1}{q}$ $\qquad\qquad [p=0]$

FI II 652, NT 50(10)

3. $\quad \displaystyle\int_0^\infty \sin px\,\operatorname{ci}(qx)\, dx = -\frac{1}{4p}\ln\left(\frac{p^2}{q^2}-1\right)^2$ $\qquad\qquad [p\neq0, \quad p^2\neq q^2]$

$\qquad\qquad\qquad\qquad\qquad\quad = 0$ $\qquad\qquad [p=0]$

FI II 652, NT 50(9)

4.  $\displaystyle\int_0^\infty \cos px\, \mathrm{ci}(qx)\, \mathrm{d}x = -\frac{\pi}{2p}$                     $[p^2 > q^2]$

$\displaystyle\qquad\qquad\qquad = -\frac{\pi}{4p}$                          $[p^2 = q^2]$

$\displaystyle\qquad\qquad\qquad = 0$                                  $[p^2 < q^2]$

FI II 654, NT 50(7)

**6.253**  $\displaystyle\int_0^\infty \frac{\mathrm{si}(ax)\sin bx}{1 - 2r\cos x + r^2}\, \mathrm{d}x = -\frac{\pi\left(r^m + r^{m+1}\right)}{4b(1-r)\left(1-r^2\right)}$            $[b = a - m]$

$\displaystyle\qquad\qquad\qquad\qquad\qquad\quad = -\frac{\pi\left(2 + 2r - r^m - r^{m+1}\right)}{4b(1-r)\left(1-r^2\right)}$        $[b = a + m]$

$\displaystyle\qquad\qquad\qquad\qquad\qquad\quad = -\frac{\pi r^{m+1}}{2b(1-r)\left(1-r^2\right)}$            $[a - m - 1 < b < a - m]$

$\displaystyle\qquad\qquad\qquad\qquad\qquad\quad = -\frac{\pi\left(1 + r - r^{m+1}\right)}{2b(1-r)\left(1-r^2\right)}$          $[a + m < b < a + m + 1]$

ET I 97(10)

**6.254**

1.  $\displaystyle\int_0^\infty \mathrm{ci}(x)\sin^2 x\,\frac{\mathrm{d}x}{x} = \frac{1}{2}\left[L_2\left(\frac{1}{2}\right) - L_2\left(-\frac{1}{2}\right)\right]$

where $L_2(x)$ is the Euler dilogarithm defined as $L_2(z) = -\displaystyle\int_0^z \frac{\log(1-t)}{t}\,\mathrm{d}t$ and this in turn can be expressed as $L_2(z) = \Phi(z, 2, 1)$ in terms of the Lerch function defined in 9.550, with $z$ real.

2.$^{12}$  $\displaystyle\int_0^\infty \left[\mathrm{si}(ax) + \frac{\pi}{2}\right]\cos bx\,\frac{\mathrm{d}x}{x} = \frac{\pi}{2}\ln\frac{a}{b}$          if $a > b > 0$

$\displaystyle\qquad\qquad\qquad\qquad\qquad\quad = 0$                      if $a^2 \le b^2$            ET I 41(11)

**6.255**

1.  $\displaystyle\int_{-\infty}^\infty \left[\cos ax\,\mathrm{ci}\left(a|x|\right) + \sin\left(a|x|\right)\mathrm{si}\left(a|x|\right)\right]\frac{\mathrm{d}x}{x - b} = -\pi\left[\mathrm{sign}\,b\cos ab\,\mathrm{si}\left(a|b|\right) - \sin ab\,\mathrm{ci}\left(a|b|\right)\right]$

$[a > 0]$            ET II 253(4)

2.  $\displaystyle\int_{-\infty}^\infty \left[\sin ax\,\mathrm{ci}\left(a|x|\right) - \mathrm{sign}\,x\cos ax\,\mathrm{si}\left(a|x|\right)\right]\frac{\mathrm{d}x}{x - b} = -\pi\left[\sin\left(a|b|\right)\mathrm{si}\left(a|b|\right) + \cos ab\,\mathrm{ci}\left(a|b|\right)\right]$

$[a > 0]$            ET II 253(5)

**6.256**

1.  $\displaystyle\int_0^\infty \left[\mathrm{si}^2(x) + \mathrm{ci}^2(x)\right]\cos ax\,\mathrm{d}x = \frac{\pi}{a}\ln(1 + a)$            $[a > 0]$

2.*  $\displaystyle\int_0^\infty \left[\mathrm{si}(x)\cos x - \mathrm{ci}(x)\sin x\right]^2\,\mathrm{d}x = \frac{\pi}{2}$

3.*  $\displaystyle\int_0^\infty \mathrm{si}^2(x)\cos(ax)\,\mathrm{d}x = \frac{\pi}{2a}\log(1 + a)$            $[0 \le a \le 2]$

4.* $\displaystyle\int_0^\infty \mathrm{ci}^2(x)\cos(ax)\,\mathrm{d}x = \frac{\pi}{2a}\log(1+a)$         $[0 \le a \le 2]$

**6.257**    $\displaystyle\int_0^\infty \mathrm{si}\left(\frac{a}{x}\right)\sin bx\,\mathrm{d}x = -\frac{\pi}{2b}J_0\left(2\sqrt{ab}\right)$         $[b>0]$         ET I 42(18)

**6.258**

1.   $\displaystyle\int_0^\infty \left[\mathrm{si}(ax)+\frac{\pi}{2}\right]\sin bx\,\frac{\mathrm{d}x}{x^2+c^2}$

$$= \frac{\pi}{4c}\left\{e^{-bc}\left[\mathrm{Ei}(bc)-\mathrm{Ei}(-ac)\right]+e^{bc}\left[\mathrm{Ei}(-ac)-\mathrm{Ei}(-bc)\right]\right\} \quad [0<b\le a, \quad c>0]$$

$$= \frac{\pi}{4c}e^{-bc}\left[\mathrm{Ei}(ac)-\mathrm{Ei}(-ac)\right] \quad [0<a\le b, \quad c>0]$$

BI (460)(1)

2.   $\displaystyle\int_0^\infty \left[\mathrm{si}(ax)+\frac{\pi}{2}\right]\cos bx\,\frac{x\,\mathrm{d}x}{x^2+c^2}$

$$= -\frac{\pi}{4}\left\{e^{-bc}\left[\mathrm{Ei}(bc)-\mathrm{Ei}(-ac)\right]+e^{bc}\left[\mathrm{Ei}(-bc)-\mathrm{Ei}(-ac)\right]\right\} \quad [0<b\le a, \quad c>0]$$

$$= \frac{\pi}{4}e^{-bc}\left[\mathrm{Ei}(-ac)-\mathrm{Ei}(ac)\right] \quad [0<a\le b, \quad c>0]$$

BI (460)(2, 5)

**6.259**

1.   $\displaystyle\int_0^\infty \mathrm{si}(ax)\sin bx\,\frac{\mathrm{d}x}{x^2+c^2} = \frac{\pi}{2c}\,\mathrm{Ei}(-ac)\sinh(bc)$       $[0<b\le a, \quad c>0]$

$$= \frac{\pi}{4c}e^{-cb}\left[\mathrm{Ei}(-bc)+\mathrm{Ei}(bc)-\mathrm{Ei}(-ac)-\mathrm{Ei}(ac)\right]$$

$$+ \frac{\pi}{2c}\,\mathrm{Ei}(-bc)\sinh(bc) \quad [0<a\le b, \quad c>0]$$

ET I 96(8)

2.   $\displaystyle\int_0^\infty \mathrm{ci}(ax)\sin bx\,\frac{x\,\mathrm{d}x}{x^2+c^2} = -\frac{\pi}{2}\sinh(bc)\,\mathrm{Ei}(-ac)$       $[0<b\le a, \quad c>0]$

$$= -\frac{\pi}{2}\sinh(bc)\,\mathrm{Ei}(-bc)+\frac{\pi}{4}e^{-bc}\left[\mathrm{Ei}(-bc)+\mathrm{Ei}(bc)\right.$$

$$\left. - \mathrm{Ei}(-ac)-\mathrm{Ei}(ac)\right] \quad [0<a\le b, \quad c>0]$$

BI (460)(3)a, ET I 97(15)a

3.   $\displaystyle\int_0^\infty \mathrm{ci}(ax)\cos bx\,\frac{\mathrm{d}x}{x^2+c^2}$

$$= \frac{\pi}{2c}\cosh bc\,\mathrm{Ei}(-ac) \quad [0<b\le a, \quad c>0]$$

$$= \frac{\pi}{4c}\left\{e^{-bc}\left[\mathrm{Ei}(ac)+\mathrm{Ei}(-ac)-\mathrm{Ei}(bc)\right]+e^{bc}\,\mathrm{Ei}(-bc)\right\} \quad [0<a\le b, \quad c>0]$$

BI (460)(4), ET I 41(15)

4.*   $\displaystyle\int_0^\infty \left[\mathrm{ci}(x)\sin x - \mathrm{Si}(x)\cos x\right]\sin x\,\frac{x\,\mathrm{d}x}{a^2+x^2} = \frac{1}{8}\left[\mathrm{Ei}(a)e^{-a}-\mathrm{Ei}(-a)e^{a}\right]^2$

$[a \text{ real}]$

5.* $\int_0^\infty [\text{ci}(x)\sin x - \text{Si}(x)\cos x]^2 \dfrac{x\,dx}{a^2+x^2} = \dfrac{\pi^3 e^{-|a|}}{8a}\sinh(a) - \dfrac{\pi}{8|a|}\left[\text{Ei}(a)e^{-a} - \text{Ei}(-a)e^a\right]^2$

$$[a \text{ real}]$$

## 6.261

1. $\int_0^\infty \text{si}(bx)\cos(ax)\,e^{-px}\,dx = -\dfrac{1}{2(a^2+p^2)}\left[\dfrac{a}{2}\ln\dfrac{p^2+(a+b)^2}{p^2+(a-b)^2} + p\arctan\dfrac{2bp}{b^2-a^2-p^2}\right]$

$$[a>0, \quad b>0, \quad p>0] \qquad \text{ET I 40(8)}$$

2. $\int_0^\infty \text{si}(\beta x)\cos(ax)\,e^{-\mu x}\,dx = -\dfrac{\arctan\dfrac{\mu+ai}{\beta}}{2(\mu+ai)} - \dfrac{\arctan\dfrac{\mu-ai}{\beta}}{2(\mu-ai)}$

$$[a>0, \quad \text{Re}\,\mu > |\text{Im}\,\beta|] \qquad \text{ET I 40(9)}$$

## 6.262

1. $\int_0^\infty \text{ci}(bx)\sin(ax)\,e^{-\mu x}\,dx = \dfrac{1}{2(a^2+\mu^2)}\left[\mu\arctan\dfrac{2a\mu}{\mu^2+b^2-a^2} - \dfrac{a}{2}\ln\dfrac{\left(\mu^2+b^2-a^2\right)^2+4a^2\mu^2}{b^4}\right]$

$$[a>0, \quad b>0, \quad \text{Re}\,\mu > 0]$$
$$\text{ET I 98(16)a}$$

2. $\int_0^\infty \text{ci}(bx)\cos(ax)\,e^{-px}\,dx = \dfrac{-1}{2(a^2+p^2)}\left[\dfrac{p}{2}\ln\dfrac{\left[(b^2+p^2-a^2)^2+4a^2p^2\right]}{b^4} + a\arctan\dfrac{2ap}{b^2+p^2-a^2}\right]$

$$[a>0, \quad b>0, \quad \text{Re}\,p > 0] \qquad \text{ET I 41(16)}$$

3. $\int_0^\infty \text{ci}(\beta x)\cos(ax)\,e^{-\mu x}\,dx = \dfrac{-\ln\left[1+\dfrac{(\mu+ai)^2}{\beta^2}\right]}{4(\mu+ai)} - \dfrac{\ln\left[1+\dfrac{(\mu-ai)^2}{\beta^2}\right]}{4(\mu-ai)}$

$$[a>0, \quad \text{Re}\,\mu > |\text{Im}\,\beta|] \qquad \text{ET I 41(17)}$$

## 6.263

1. $\int_0^\infty [\text{ci}(x)\cos x + \text{si}(x)\sin x]\,e^{-\mu x}\,dx = \dfrac{-\dfrac{\pi}{2} - \mu\ln\mu}{1+\mu^2}$    $[\text{Re}\,\mu > 0]$      ME 26a, ET I 178(21)a

2. $\int_0^\infty [\text{si}(x)\cos x - \text{ci}(x)\sin x]\,e^{-\mu x}\,dx = \dfrac{-\dfrac{\pi}{2}\mu + \ln\mu}{1+\mu^2}$    $[\text{Re}\,\mu > 0]$      ME 26a, ET I 178(20)a

3. $\int_0^\infty [\sin x - x\,\text{ci}(x)]\,e^{-\mu x}\,dx = \dfrac{\ln(1+\mu^2)}{2\mu^2}$    $[\text{Re}\,\mu > 0]$      ME 26

## 6.264

1. $\int_0^\infty \text{si}(x)\ln x\,dx = \boldsymbol{C}+1$      NT 46(10)

2. $\int_0^\infty \text{ci}(x)\ln x\,dx = \dfrac{\pi}{2}$      NT 56(11)

## 6.27 The hyperbolic sine integral and hyperbolic cosine integral functions

**6.271**

1.  $\int_0^\infty \mathrm{shi}(x) e^{-\mu x}\, dx = \dfrac{1}{2\mu} \ln \dfrac{\mu+1}{\mu-1} = \dfrac{1}{\mu} \operatorname{arccoth} \mu$  $\qquad [\operatorname{Re}\mu > 1]$  MI 34

2.[11]  $\int_0^\infty \mathrm{chi}(x) e^{-\mu x}\, dx = -\dfrac{1}{2\mu} \ln\left(\mu^2 - 1\right)$  $\qquad [\operatorname{Re}\mu > 1]$  MI 34

**6.272**[11]  $\int_0^\infty \mathrm{chi}(x) e^{-px^2}\, dx = \dfrac{1}{4}\sqrt{\dfrac{\pi}{p}}\,\mathrm{Ei}\left(\dfrac{1}{4p}\right)$  $\qquad [p > 0]$  MI 35

**6.273**

1.[11]  $\int_0^\infty \left[\cosh x\, \mathrm{shi}(x) - \sinh x\, \mathrm{chi}(x)\right] e^{-\mu x}\, dx = \dfrac{\ln\mu}{\mu^2 - 1}$  $\qquad [\operatorname{Re}\mu > 0]$  MI 35

2.[11]  $\int_0^\infty \left[\cosh x\, \mathrm{chi}(x) + \sinh x\, \mathrm{shi}(x)\right] e^{-\mu x}\, dx = \dfrac{\mu\ln\mu}{1 - \mu^2}$  $\qquad [\operatorname{Re}\mu > 2]$  MI 35

**6.274**[11]  $\int_0^\infty \left[\cosh x\, \mathrm{shi}(x) - \sinh x\, \mathrm{chi}(x)\right] e^{-\mu x^2}\, dx = \dfrac{1}{4}\sqrt{\dfrac{\pi}{\mu}}\, e^{\frac{1}{4\mu}}\,\mathrm{Ei}\left(-\dfrac{1}{4\mu}\right)$

$\qquad\qquad [\operatorname{Re}\mu > 0]$  MI 35

**6.275**  $\int_0^\infty \left[x\,\mathrm{chi}(x) - \sinh x\right] e^{-\mu x}\, dx = -\dfrac{\ln\left(\mu^2 - 1\right)}{2\mu^2}$  $\qquad [\operatorname{Re}\mu > 1]$  MI 35

**6.276**  $\int_0^\infty \left[\cosh x\, \mathrm{chi}(x) + \sinh x\, \mathrm{shi}(x)\right] e^{-\mu x^2} x\, dx = \dfrac{1}{8}\sqrt{\dfrac{\pi}{\mu^3}}\,\exp\left(\dfrac{1}{4\mu}\right)\mathrm{Ei}\left(-\dfrac{1}{4\mu}\right)$

$\qquad\qquad [\operatorname{Re}\mu > 0]$  MI 35

**6.277**

1.  $\int_0^\infty \left[\mathrm{chi}(x) + \mathrm{ci}(x)\right] e^{-\mu x}\, dx - \dfrac{\ln\left(\mu^4 - 1\right)}{2\mu}$  $\qquad [\operatorname{Re}\mu > 1]$  MI 34

2.  $\int_0^\infty \left[\mathrm{chi}(x) - \mathrm{ci}(x)\right] e^{-\mu x}\, dx = \dfrac{1}{2\mu} \ln \dfrac{\mu^2 + 1}{\mu^2 - 1}$  $\qquad [\operatorname{Re}\mu > 1]$  MI 35

## 6.28–6.31 The probability integral

**6.281**

1.[6]  $\int_0^\infty \left[1 - \Phi(px)\right] x^{2q-1}\, dx = \dfrac{\Gamma\left(q + \frac{1}{2}\right)}{2\sqrt{\pi}\, q p^{2q}}$  $\qquad [\operatorname{Re} q > 0, \quad \operatorname{Re} p > 0]$

$\qquad\qquad\qquad$ NT 56(12), ET II 306(1)a

2.[6]  $\int_0^\infty \left[1 - \Phi\left(at^\alpha \pm \dfrac{b}{t^\alpha}\right)\right] dt = \dfrac{2b}{\sqrt{\pi}}\left(\dfrac{b}{a}\right)^{\frac{1-\alpha}{2\alpha}}\left[K_{\frac{1+\alpha}{2\alpha}}(2ab) \pm K_{\frac{1-\alpha}{2\alpha}}(2ab)\right] e^{\pm 2ab}$

$\qquad\qquad\qquad [a > 0, \quad b > 0, \quad \alpha \neq 0]$

**6.282**

1.  $\int_0^\infty \Phi(qt) e^{-pt}\, dt = \dfrac{1}{p}\left[1 - \Phi\left(\dfrac{p}{2q}\right)\right]\exp\left(\dfrac{p^2}{4q^2}\right)$  $\qquad [\operatorname{Re} p > 0, \quad |\arg q| < \frac{\pi}{4}]$

$\qquad\qquad\qquad$ MO 175, EH II 148(11)

$2.^{12}$ $\displaystyle\int_0^\infty \left[ \Phi\left( x + \frac{1}{2} \right) - \Phi\left( \frac{1}{2} \right) \right] e^{-\mu x + \frac{1}{4}}\, dx = \frac{1}{\mu} \exp \frac{(\mu+1)^2}{4} \left[ 1 - \Phi\left( \frac{\mu+1}{2} \right) \right]$     ME 27

**6.283**

1. $\displaystyle\int_0^\infty e^{\beta x} \left[ 1 - \Phi\left( \sqrt{\alpha} x \right) \right] dx = \frac{1}{\beta} \left[ \frac{\sqrt{\alpha}}{\sqrt{\alpha - \beta}} - 1 \right]$     $[\operatorname{Re}\alpha > 0, \quad \operatorname{Re}\beta < \operatorname{Re}\alpha]$     ET II 307(5)

2. $\displaystyle\int_0^\infty \Phi\left( \sqrt{qt} \right) e^{-pt}\, dt = \frac{\sqrt{q}}{p} \frac{1}{\sqrt{p+q}}$     $[\operatorname{Re}p > 0, \quad \operatorname{Re}(q+p) > 0]$

EH II 148(12)

**6.284** $\displaystyle\int_0^\infty \left[ 1 - \Phi\left( \frac{q}{2\sqrt{x}} \right) \right] e^{-px}\, dx = \frac{1}{p} e^{-q\sqrt{p}}$     $[\operatorname{Re}p > 0, \quad |\arg q| < \frac{\pi}{4}]$

EF 147(235), EH II 148(13)

**6.285**

1. $\displaystyle\int_0^\infty \left[ 1 - \Phi(x) \right] e^{-\mu^2 x^2}\, dx = \frac{\arctan \mu}{\sqrt{\pi}\mu}$     $[\operatorname{Re}\mu > 0]$     MI 37

2. $\displaystyle\int_0^\infty \Phi(iat) e^{-a^2 t^2 - st}\, dt = \frac{-1}{2ai\sqrt{\pi}} \exp\left( \frac{s^2}{4a^2} \right) \operatorname{Ei}\left( -\frac{s^2}{4a^2} \right)$

$[\operatorname{Re}s > 0, \quad |\arg a| < \frac{\pi}{4}]$     EH II 148(14)a

**6.286**

1. $\displaystyle\int_0^\infty \left[ 1 - \Phi(\beta x) \right] e^{\mu^2 x^2} x^{\nu-1}\, dx = \frac{\Gamma\left( \dfrac{\nu+1}{2} \right)}{\sqrt{\pi}\nu\beta^\nu}\ {}_2F_1\left( \frac{\nu}{2}, \frac{\nu+1}{2}; \frac{\nu}{2} + 1; \frac{\mu^2}{\beta^2} \right)$

$[\operatorname{Re}\beta^2 > \operatorname{Re}\mu^2, \quad \operatorname{Re}\nu > 0]$     ET II 306(2)

2. $\displaystyle\int_0^\infty \left[ 1 - \Phi\left( \frac{\sqrt{2}x}{2} \right) \right] e^{\frac{x^2}{2}} x^{\nu-1}\, dx = 2^{\frac{\nu}{2}-1} \sec \frac{\nu\pi}{2} \Gamma\left( \frac{\nu}{2} \right)$

$[0 < \operatorname{Re}\nu < 1]$     ET I 325(9)

**6.287**

1. $\displaystyle\int_0^\infty \Phi(\beta x) e^{-\mu x^2} x\, dx = \frac{\beta}{2\mu\sqrt{\mu + \beta^2}}$     $[\operatorname{Re}\mu > -\operatorname{Re}\beta^2, \quad \operatorname{Re}\mu > 0]$

ME 27a, ET I 176(4)

2. $\displaystyle\int_0^\infty \left[ 1 - \Phi(\beta x) \right] e^{-\mu x^2} x\, dx = \frac{1}{2\mu}\left( 1 - \frac{\beta}{\sqrt{\mu + \beta^2}} \right)$     $[\operatorname{Re}\mu > -\operatorname{Re}\beta^2, \quad \operatorname{Re}\mu > 0]$

NT 49(14), ET I 177(9)

$3.^{12}$ $\displaystyle\int_0^\infty \frac{r}{\sigma^2} \exp\left( -\frac{r^2}{2\sigma^2} \right) Q(rA)\, Q(rB)\, dr = \frac{1}{4} - \frac{1}{2\pi}\left[ \alpha \arctan\left( \frac{A}{\alpha B} \right) + \beta \arctan\left( \frac{B}{\beta A} \right) \right]$     $B \neq A$

$= \dfrac{1}{4} - \dfrac{1}{\pi}\alpha \arctan \dfrac{1}{\alpha}$     $B = A$

$Q(x) = \dfrac{1}{\sqrt{2\pi}} \displaystyle\int_x^\infty e^{-t^2/2}\, dt = \dfrac{1}{2}\left[ 1 - \operatorname{erf}\left( \dfrac{x}{\sqrt{2}} \right) \right], \quad \alpha = \sqrt{\dfrac{\sigma^2 A^2}{1 + \sigma^2 A^2}}, \quad \beta = \sqrt{\dfrac{\sigma^2 B^2}{1 + \sigma^2 B^2}}$     BEA

4.* $\displaystyle\int_0^\infty r^{2\nu-1} e^{-pr^2} Q(Ar)Q(Br)\,dr = \frac{2^{\nu-2}AB\Gamma(\nu)}{\pi c(1+2\nu)}\left[F_1\left(1,\nu,1;\nu+\frac{3}{2};\frac{A^2+2p}{A^2+B^2+2p},\frac{A^2}{A^2+B^2}\right)\right.$

$\displaystyle\left.+F_1\left(1,\nu,1;\nu+\frac{3}{2};\frac{B^2+2p}{A^2+B^2+2p},\frac{B^2}{A^2+B^2}\right)\right]\quad\left[c=(A^2+B^2)(A^2+B^2+2p)^\nu\right]$    LEI

**6.288** $\displaystyle\int_0^\infty \Phi(iax)e^{-\mu x^2} x\,dx = \frac{ai}{2\mu\sqrt{\mu-a^2}}$        $\left[a>0,\quad \operatorname{Re}\mu>\operatorname{Re}a^2\right]$    MI 37a

**6.289**

1.   $\displaystyle\int_0^\infty \Phi(\beta x)e^{(\beta^2-\mu^2)x^2} x\,dx = \frac{\beta}{2\mu\,(\mu^2-\beta^2)}$      $\left[\operatorname{Re}\mu^2>\operatorname{Re}\beta^2,\quad |\arg\mu|<\frac{\pi}{4}\right]$

                                                                          ET I 176(5)

2.   $\displaystyle\int_0^\infty \left[1-\Phi(\beta x)\right] e^{(\beta^2-\mu^2)x^2} x\,dx = \frac{1}{2\mu(\mu+\beta)}$      $\left[\operatorname{Re}\mu^2>\operatorname{Re}\beta^2,\quad |\arg\mu|<\frac{\pi}{4}\right]$

                                                                         ET I 177(10)

3.   $\displaystyle\int_0^\infty \Phi\left(\sqrt{b-a}\,x\right) e^{-(a+\mu)x^2} x\,dx = \frac{\sqrt{b-a}}{2(\mu+a)\sqrt{\mu+b}}$      $\left[\operatorname{Re}\mu>-a>0,\quad b>a\right]$    ME 27

**6.291**[12] $\displaystyle\int_0^\infty \Phi(ix)e^{-(\mu x+x^2)} x\,dx = \frac{i}{\sqrt{\pi}}\left[\frac{1}{\mu}+\frac{\mu}{a}e^{\mu^2/4}\operatorname{Ei}\left(-\frac{\mu^2}{4}\right)\right]$

                                                                         $\left[\operatorname{Re}\mu>0\right]$    MI 37

**6.292** $\displaystyle\int_0^\infty \left[1-\Phi(x)\right] e^{-\mu^2 x^2} x^2\,dx = \frac{1}{2\sqrt{\pi}}\left\{\frac{\arctan\mu}{\mu^3}-\frac{1}{\mu^2\,(\mu^2+1)}\right\}$

                                                                     $\left[|\arg\mu|<\frac{\pi}{4}\right]$    MI 37

**6.293** $\displaystyle\int_0^\infty \Phi(x)e^{-\mu x^2}\frac{dx}{x} = \frac{1}{2}\ln\frac{\sqrt{\mu+1}+1}{\sqrt{\mu+1}-1} = \operatorname{arccoth}\sqrt{\mu+1}$

                                                                       $\left[\operatorname{Re}\mu>0\right]$    MI 37a

**6.294**

1.   $\displaystyle\int_0^\infty \left[1-\Phi\left(\frac{\beta}{x}\right)\right] e^{-\mu^2 x^2} x\,dx = \frac{1}{2\mu^2}\exp(-2\beta\mu)$      $\left[|\arg\beta|<\frac{\pi}{4},\quad |\arg\mu|<\frac{\pi}{4}\right]$

                                                                          ET I 177(11)

2.   $\displaystyle\int_0^\infty \left[1-\Phi\left(\frac{1}{x}\right)\right] e^{-\mu^2 x^2}\frac{dx}{x} = -\operatorname{Ei}(-2\mu)$      $\left[|\arg\mu|<\frac{\pi}{4}\right]$    MI 37

**6.295**

1.   $\displaystyle\int_0^\infty \left[1-\Phi\left(\frac{1}{x}\right)\right]\exp\left(-\mu^2 x^2+\frac{1}{x^2}\right)dx = \frac{1}{\sqrt{\pi}\mu}\left[\sin 2\mu\,\operatorname{ci}(2\mu)-\cos 2\mu\,\operatorname{si}(2\mu)\right]$

                                                                         $\left[|\arg\mu|<\frac{\pi}{4}\right]$    MI 37

2.   $\displaystyle\int_0^\infty \left[1-\Phi\left(\frac{1}{x}\right)\right]\exp\left(-\mu^2 x^2+\frac{1}{x^2}\right) x\,dx = \frac{\pi}{2\mu}\left[\mathbf{H}_1(2\mu)-Y_1(2\mu)\right]-\frac{1}{\mu^2}$

                                                                         $\left[|\arg\mu|<\frac{\pi}{4}\right]$    MI 37

3. $\int_0^\infty \left[ 1 - \Phi\left(\frac{1}{x}\right) \right] \exp\left( -\mu^2 x^2 + \frac{1}{x^2} \right) \frac{dx}{x} = \frac{\pi}{2} \left[ \mathbf{H}_0(2\mu) - Y_0(2\mu) \right]$

$$\left[ |\arg \mu| < \tfrac{\pi}{4} \right] \qquad\qquad \text{MI 37}$$

**6.296** $\int_0^\infty \left\{ (x^2 + a^2) \left[ 1 - \Phi\left(\frac{a}{\sqrt{2}x}\right) \right] - \sqrt{\frac{2}{\pi}}\, ax \cdot e^{-\frac{a^2}{2x^2}} \right\} e^{-\mu^2 x^2} x\, dx = \frac{1}{2\mu^4} e^{-a\mu\sqrt{2}}$

$$\left[ |\arg \mu| < \tfrac{\pi}{4}, \quad a > 0 \right] \qquad\qquad \text{MI 38a}$$

**6.297**

1. $\int_0^\infty \left[ 1 - \Phi\left( \gamma x + \frac{\beta}{x} \right) \right] e^{(\gamma^2 - \mu)x^2} x\, dx = \frac{1}{2\sqrt{\mu}\,(\sqrt{\mu} + \gamma)} \exp\left[ -2\left( \beta\gamma + \beta\sqrt{\mu} \right) \right]$

$$\left[ \operatorname{Re}\beta > 0, \quad \operatorname{Re}\mu > 0 \right] \qquad\qquad \text{ET I 177(12)a}$$

2. $\int_0^\infty \left[ 1 - \Phi\left( \frac{b + 2ax^2}{2x} \right) \right] \exp\left[ -\left( \mu^2 - a^2 \right)x^2 + ab \right] x\, dx = \frac{e^{-b\mu}}{2\mu(\mu + a)}$

$$\left[ a > 0, \quad b > 0, \quad \operatorname{Re}\mu > 0 \right] \qquad\qquad \text{MI 38}$$

3. $\int_0^\infty \left\{ \left[ 1 - \Phi\left( \frac{b - 2ax^2}{2x} \right) \right] e^{-ab} + \left[ 1 - \Phi\left( \frac{b + 2ax^2}{2x} \right) \right] e^{ab} \right\} e^{-\mu x^2} x\, dx = \frac{1}{\mu} \exp\left( -b\sqrt{a^2 + \mu} \right)$

$$\left[ a > 0, \quad b > 0, \quad \operatorname{Re}\mu > 0 \right] \qquad\qquad \text{MI 38}$$

**6.298** $\int_0^\infty \left\{ 2\cosh ab - e^{-ab}\, \Phi\left( \frac{b - 2ax^2}{2x} \right) - e^{ab}\, \Phi\left( \frac{b + 2ax^2}{2x} \right) \right\} e^{-(\mu - a^2)x^2} x\, dx = \frac{1}{\mu - a^2} \exp\left( -b\sqrt{\mu} \right)$

$$\left[ a > 0, \quad b > 0, \quad \operatorname{Re}\mu > 0 \right] \qquad\qquad \text{MI 38}$$

**6.299** $\int_0^\infty \cosh(2\nu t) \exp\left[ (a\cosh t)^2 \right] \left[ 1 - \Phi\left( a\cosh t \right) \right]\, dt = \frac{1}{2\cos(\nu\pi)} \exp\left( \tfrac{1}{2}a^2 \right) K_\nu\left( a^2 \right)$

$$\left[ \operatorname{Re}a > 0, \quad -\tfrac{1}{2} < \operatorname{Re}\nu < \tfrac{1}{2} \right]$$
$$\text{ET II 308(10)}$$

**6.311** $\int_0^\infty \left[ 1 - \Phi(ax) \right] \sin bx\, dx = \frac{1}{b} \left( 1 - e^{-\frac{b^2}{4a^2}} \right) \qquad\qquad \left[ a > 0, \quad b > 0 \right] \qquad\qquad \text{ET I 96(4)}$

**6.312** $\int_0^\infty \Phi(ax) \sin bx^2\, dx = \frac{1}{4\sqrt{2\pi b}} \left( \ln \frac{b + a^2 + a\sqrt{2b}}{b + a^2 - a\sqrt{2b}} + 2\arctan \frac{a\sqrt{2b}}{b - a^2} \right)$

$$\left[ a > 0, \quad b > 0 \right] \qquad\qquad \text{ET I 96(3)}$$

**6.313**

1. $\int_0^\infty \sin(\beta x) \left[ 1 - \Phi\left( \sqrt{\alpha x} \right) \right] dx = \frac{1}{\beta} - \left( \frac{\frac{\alpha}{2}}{\alpha^2 + \beta^2} \right)^{\frac{1}{2}} \left[ \left( \alpha^2 + \beta^2 \right)^{\frac{1}{2}} - \alpha \right]^{-\frac{1}{2}}$

$$\left[ \operatorname{Re}\alpha > |\operatorname{Im}\beta| \right] \qquad\qquad \text{ET II 307(6)}$$

2. $\int_0^\infty \cos(\beta x) \left[ 1 - \Phi\left( \sqrt{\alpha x} \right) \right] dx = \left( \frac{\frac{\alpha}{2}}{\alpha^2 + \beta^2} \right)^{\frac{1}{2}} \left[ \left( \alpha^2 + \beta^2 \right)^{\frac{1}{2}} + \alpha \right]^{-\frac{1}{2}}$

$$\left[ \operatorname{Re}\alpha > |\operatorname{Im}\beta| \right] \qquad\qquad \text{ET II 307(7)}$$

**6.314**

1.    $\int_0^\infty \sin(bx) \left[ 1 - \Phi\left( \sqrt{\dfrac{a}{x}} \right) \right] dx = b^{-1} \exp\left[ -(2ab)^{\frac{1}{2}} \right] \cos\left[ (2ab)^{\frac{1}{2}} \right]$

$$[\operatorname{Re} a > 0, \quad b > 0] \qquad \text{ET II 307(8)}$$

2.    $\int_0^\infty \cos(bx) \left[ 1 - \Phi\left( \sqrt{\dfrac{a}{x}} \right) \right] dx = -b^{-1} \exp\left[ -(2ab)^{\frac{1}{2}} \right] \sin\left[ (2ab)^{\frac{1}{2}} \right]$

$$[\operatorname{Re} a > 0, \quad b > 0] \qquad \text{ET II 307(9)}$$

**6.315**

1.    $\int_0^\infty x^{\nu-1} \sin(\beta x) \left[ 1 - \Phi(\alpha x) \right] dx = \dfrac{\Gamma\left( 1 + \frac{1}{2}\nu \right) \beta}{\sqrt{\pi}(\nu+1)\alpha^{\nu+1}}\ {}_2F_2\left( \dfrac{\nu+1}{2}, \dfrac{\nu}{2}+1; \dfrac{3}{2}, \dfrac{\nu+3}{2}; -\dfrac{\beta^2}{4\alpha^2} \right)$

$$[\operatorname{Re} \alpha > 0, \quad \operatorname{Re} \nu > -1] \qquad \text{ET II 307(3)}$$

2.    $\int_0^\infty x^{\nu-1} \cos(\beta x) \left[ 1 - \Phi(\alpha x) \right] dx = \dfrac{\Gamma\left( \frac{1}{2} + \frac{1}{2}\nu \right)}{\sqrt{\pi}\nu\alpha^\nu}\ {}_2F_2\left( \dfrac{\nu}{2}, \dfrac{\nu+1}{2}; \dfrac{1}{2}, \dfrac{\nu}{2}+1; -\dfrac{\beta^2}{4\alpha^2} \right)$

$$[\operatorname{Re} \alpha > 0, \quad \operatorname{Re} \nu > 0] \qquad \text{ET II 307(4)}$$

3.    $\int_0^\infty \left[ 1 - \Phi(ax) \right] \cos bx \cdot x\, dx = \dfrac{1}{2a^2} \exp\left( -\dfrac{b^2}{4a^2} \right) - \dfrac{1}{b^2} \left[ 1 - \exp\left( -\dfrac{b^2}{4a^2} \right) \right]$

$$[a > 0, \quad b > 0] \qquad \text{ET I 40(5)}$$

4.    $\int_0^\infty \left[ \Phi(ax) - \Phi(bx) \right] \cos px\, \dfrac{dx}{x} = \dfrac{1}{2} \left[ \operatorname{Ei}\left( -\dfrac{p^2}{4b^2} \right) - \operatorname{Ei}\left( \dfrac{p^2}{4a^2} \right) \right]$

$$[a > 0, \quad b > 0, \quad p > 0] \qquad \text{ET I 40(6)}$$

5.    $\int_0^\infty x^{-\frac{1}{2}}\, \Phi\left( a\sqrt{x} \right) \sin bx\, dx = \dfrac{1}{2\sqrt{2\pi b}} \left\{ \ln\left[ \dfrac{b + a\sqrt{2b} + a^2}{b - a\sqrt{2b} + a^2} \right] + 2\arctan\left[ \dfrac{a\sqrt{2b}}{b - a^2} \right] \right\}$

$$[a > 0, \quad b > 0] \qquad \text{ET I 96(3)}$$

**6.316**    $\int_0^\infty e^{\frac{1}{2}x^2} \left[ 1 - \Phi\left( \dfrac{x}{\sqrt{2}} \right) \right] \sin bx\, dx = \sqrt{\dfrac{\pi}{2}} e^{\frac{b^2}{2}} \left[ 1 - \Phi\left( \dfrac{b}{\sqrt{2}} \right) \right]$

$$[b > 0] \qquad \text{ET I 96(5)}$$

**6.317**[6]    $\int_0^\infty e^{-a^2x^2}\, \Phi(iax) \sin bx\, dx = \dfrac{i}{a} \dfrac{\sqrt{\pi}}{2} e^{-\frac{b^2}{4a^2}}$      $[b > 0]$      ET I 96(2)

**6.318**    $\int_0^\infty \left[ 1 - \Phi(x) \right] \operatorname{si}(2px)\, dx = \dfrac{2}{\pi p} \left( 1 - e^{-p^2} \right) - \dfrac{2}{\sqrt{\pi}} \left( 1 - \Phi(p) \right)$

$$[p > 0] \qquad \text{NT 61(13)a}$$

## 6.32 Fresnel integrals

**6.321**

1.    $\int_0^\infty \left[ \dfrac{1}{2} - S(px) \right] x^{2q-1}\, dx = \dfrac{\sqrt{2}\,\Gamma\left( q + \frac{1}{2} \right) \sin \dfrac{2q+1}{4}\pi}{4\sqrt{\pi}\, q p^{2q}}$

$$\left[ 0 < \operatorname{Re} q < \dfrac{3}{2}, \quad p > 0 \right] \qquad \text{NT 56(14)a}$$

2.    $\displaystyle\int_0^\infty \left[\frac{1}{2} - C(px)\right] x^{2q-1}\,\mathrm{d}x = \frac{\sqrt{2}\,\Gamma\left(q+\frac{1}{2}\right)\cos\dfrac{2q+1}{4}\pi}{4\sqrt{\pi}\,q p^{2q}}$

$$\left[0 < \operatorname{Re} q < \tfrac{3}{2}, \quad p > 0\right]$$      NT 56(13)a

**6.322**

1.    $\displaystyle\int_0^\infty S(t) e^{-pt}\,\mathrm{d}t = \frac{1}{p}\left\{\cos\frac{p^2}{4}\left[\frac{1}{2} - C\left(\frac{p}{2}\right)\right] + \sin\frac{p^2}{4}\left[\frac{1}{2} - S\left(\frac{p}{2}\right)\right]\right\}$      MO 173a

2.    $\displaystyle\int_0^\infty C(t) e^{-pt}\,\mathrm{d}t = \frac{1}{p}\left\{\cos\frac{p^2}{4}\left[\frac{1}{2} - S\left(\frac{p}{2}\right)\right] - \sin\frac{p^2}{4}\left[\frac{1}{2} - C\left(\frac{p}{2}\right)\right]\right\}$      MO 172a

**6.323**

1.    $\displaystyle\int_0^\infty S\left(\sqrt{t}\right) e^{-pt}\,\mathrm{d}x = \frac{\left(\sqrt{p^2+1} - p\right)^{\frac{1}{2}}}{2p\sqrt{p^2+1}}$      EF 122(58)a

2.    $\displaystyle\int_0^\infty C\left(\sqrt{t}\right) e^{-pt}\,\mathrm{d}t = \frac{\left(\sqrt{p^2+1} + p\right)^{\frac{1}{2}}}{2p\sqrt{p^2+1}}$      EF 122(58)a

**6.324**

1.    $\displaystyle\int_0^\infty \left[\frac{1}{2} - S(x)\right]\sin 2px\,\mathrm{d}x = \frac{1 + \sin p^2 - \cos p^2}{4p}$    $[p > 0]$      NT 61(12)a

2.    $\displaystyle\int_0^\infty \left[\frac{1}{2} - C(x)\right]\sin 2px\,\mathrm{d}x = \frac{1 - \sin p^2 - \cos p^2}{4p}$    $[p > 0]$      NT 61(11)a

**6.325**

1.    $\displaystyle\int_0^\infty S(x)\sin b^2 x^2\,\mathrm{d}x = \frac{\sqrt{\pi}}{b} 2^{-\frac{5}{2}}$    $\left[0 < b^2 < 1\right]$

$$= 0 \qquad \left[b^2 > 1\right]$$

ET I 98(21)a

2.    $\displaystyle\int_0^\infty C(x)\cos b^2 x^2\,\mathrm{d}x = \frac{\sqrt{\pi}}{b} 2^{-\frac{5}{2}}$    $\left[0 < b^2 < 1\right]$

$$= 0 \qquad \left[b^2 > 1\right]$$

ET I 42(22)

**6.326**

1.    $\displaystyle\int_0^\infty \left[\frac{1}{2} - S(x)\right]\operatorname{si}(2px)\,\mathrm{d}x = \left(\frac{\pi}{8}\right)^{1/2}[S(p) + C(p) - 1] - \frac{1 + \sin p^2 - \cos p^2}{4p}$

$$[p > 0]$$      NT 61(15)a

2.    $\displaystyle\int_0^\infty \left[\frac{1}{2} - C(x)\right]\operatorname{si}(2px)\,\mathrm{d}x = \left(\frac{\pi}{8}\right)^{1/2}[S(p) - C(p)] - \frac{1 - \sin p^2 - \cos p^2}{4p}$

$$[p > 0]$$      NT 61(14)a

# 6.4 The Gamma Function and Functions Generated by It

## 6.41 The gamma function

**6.411**[12] $\displaystyle\int_{-\infty}^{\infty} \Gamma(\alpha + x)\,\Gamma(\beta - x)\,\mathrm{d}x$

$$= -i\pi 2^{1-\alpha-\beta}\,\Gamma(\alpha+\beta) \qquad [\mathrm{Re}(\alpha+\beta) < 1, \quad \mathrm{Im}\,\alpha > 0, \quad \mathrm{Im}\,\beta > 0] \quad \text{ET II 297(3)}$$

$$= i\pi 2^{1-\alpha-\beta}\,\Gamma(\alpha+\beta) \qquad [\mathrm{Re}(\alpha+\beta) < 1, \quad \mathrm{Im}\,\alpha < 0, \quad \mathrm{Im}\,\beta < 0] \quad \text{ET II 297(2)}$$

$$= 0 \qquad [\mathrm{Re}(\alpha+\beta) < 1, \quad (\mathrm{Im}\,\alpha)\,(\mathrm{Im}\,\beta) < 0] \qquad \text{ET II 297(1)}$$

**6.412**[12] $\displaystyle\int_{-i\infty}^{i\infty} \Gamma(\alpha+s)\,\Gamma(\beta+s)\,\Gamma(\gamma-s)\,\Gamma(\delta-s)\,\mathrm{d}s = 4\pi i\,\frac{\Gamma(\alpha+\gamma)\,\Gamma(\alpha+\delta)\,\Gamma(\beta+\gamma)\,\Gamma(\beta+\delta)}{\Gamma(\alpha+\beta+\gamma+\delta)}$

$$[\mathrm{Re}\,\alpha, \quad \mathrm{Re}\,\beta, \quad \mathrm{Re}\,\gamma, \quad \mathrm{Re}\,\delta > 0]$$
$$\text{ET II 302(32)}$$

**6.413**

1.    $\displaystyle\int_0^\infty |\Gamma(a+ix)\,\Gamma(b+ix)|^2\,\mathrm{d}x = \frac{\sqrt{\pi}\,\Gamma(a)\,\Gamma\left(a+\tfrac{1}{2}\right)\,\Gamma(b)\,\Gamma\left(b+\tfrac{1}{2}\right)\,\Gamma(a+b)}{2\,\Gamma\left(a+b+\tfrac{1}{2}\right)}$

$$[a > 0, \quad b > 0] \qquad \text{ET II 302(27)}$$

2.    $\displaystyle\int_0^\infty \left|\frac{\Gamma(a+ix)}{\Gamma(b+ix)}\right|^2\,\mathrm{d}x = \frac{\sqrt{\pi}\,\Gamma(a)\,\Gamma\left(a+\tfrac{1}{2}\right)\,\Gamma\left(b-a-\tfrac{1}{2}\right)}{2\,\Gamma(b)\,\Gamma\left(b-\tfrac{1}{2}\right)\,\Gamma(b-a)}$

$$\left[0 < a < b - \tfrac{1}{2}\right] \qquad \text{ET II 302(28)}$$

**6.414**

1.    $\displaystyle\int_{-\infty}^{\infty} \frac{\Gamma(\alpha+x)}{\Gamma(\beta+x)}\,\mathrm{d}x = 0 \qquad\qquad\qquad [\mathrm{Im}\,\alpha \neq 0, \quad \mathrm{Re}(\alpha-\beta) < -1]$

$$\text{ET II 297(4)}$$

2.    $\displaystyle\int_{-\infty}^{\infty} \frac{\mathrm{d}x}{\Gamma(\alpha+x)\,\Gamma(\beta-x)} = \frac{2^{\alpha+\beta-2}}{\Gamma(\alpha+\beta-1)} \qquad [\mathrm{Re}(\alpha+\beta) > 1] \qquad \text{ET II 297(5)}$

3.    $\displaystyle\int_{-\infty}^{\infty} \frac{\Gamma(\gamma+x)\,\Gamma(\delta+x)}{\Gamma(\alpha+x)\,\Gamma(\beta+x)}\,\mathrm{d}x = 0$

$$[\mathrm{Re}(\alpha+\beta-\gamma-\delta) > 1, \quad \mathrm{Im}\,\gamma, \quad \mathrm{Im}\,\delta > 0] \quad \text{ET II 299(18)}$$

4.    $\displaystyle\int_{-\infty}^{\infty} \frac{\Gamma(\gamma+x)\,\Gamma(\delta+x)}{\Gamma(\alpha+x)\,\Gamma(\beta+x)}\,\mathrm{d}x = \frac{\pm 2\pi^2 i\,\Gamma(\alpha+\beta-\gamma-\delta-1)}{\sin[\pi(\gamma-\delta)]\,\Gamma(\alpha-\gamma)\,\Gamma(\alpha-\delta)\,\Gamma(\beta-\gamma)\,\Gamma(\beta-\delta)}$

$[\mathrm{Re}(\alpha+\beta-\gamma-\delta) > 1, \quad \mathrm{Im}\,\gamma < 0, \quad \mathrm{Im}\,\delta < 0.$ In the numerator, we take the plus sign if $\mathrm{Im}\,\gamma > \mathrm{Im}\,\delta$ and the minus sign if $\mathrm{Im}\,\gamma < \mathrm{Im}\,\delta.]$    ET II 300(19)

5.    $\displaystyle\int_{-\infty}^{\infty} \frac{\Gamma(\alpha-\beta-\gamma+x+1)\,\mathrm{d}x}{\Gamma(\alpha+x)\,\Gamma(\beta-x)\,\Gamma(\gamma+x)} = \frac{\pi \exp\left(\pm\tfrac{1}{2}\pi(\delta-\gamma)i\right)}{\Gamma(\beta+\gamma-1)\,\Gamma\left(\tfrac{1}{2}(\alpha+\beta)\right)\,\Gamma\left(\tfrac{1}{2}(\gamma-\delta+1)\right)}$

$[\mathrm{Re}(\beta+\gamma) > 1, \quad \delta = \alpha-\beta-\gamma+1, \quad \mathrm{Im}\,\delta \neq 0.$ The sign is plus in the argument if the exponential for $\mathrm{Im}\,\delta > 0$ and minus for $\mathrm{Im}\,\delta < 0].$    ET II 300(20)

6.

$$\int_{-\infty}^{\infty} \frac{dx}{\Gamma(\alpha+x)\,\Gamma(\beta-x)\,\Gamma(\gamma+x)\,\Gamma(\delta-x)} = \frac{\Gamma(\alpha+\beta+\gamma+\delta-3)}{\Gamma(\alpha+\beta-1)\,\Gamma(\beta+\gamma-1)\,\Gamma(\gamma+\delta-1)\,\Gamma(\delta+\alpha-1)}$$

$$[\operatorname{Re}(\alpha+\beta+\gamma+\delta) > 3] \qquad \text{ET II 300(21)}$$

**6.415**

1.

$$\int_{-\infty}^{-\infty} \frac{R(x)\,dx}{\Gamma(\alpha+x)\,\Gamma(\beta-x)\,\Gamma(\gamma+x)\,\Gamma(\delta-x)}$$

$$= \frac{\Gamma(\alpha+\beta+\gamma+\delta-3)}{\Gamma(\alpha+\beta-1)\,\Gamma(\beta+\gamma-1)\,\Gamma(\gamma+\delta-1)\,\Gamma(\delta+\alpha-1)}\int_0^1 R(t)\,dt$$

$$[\operatorname{Re}(\alpha+\beta+\gamma+\delta) > 3, \quad R(x+1) = R(x)] \quad \text{ET II 301(24)}$$

2.

$$\int_{-\infty}^{\infty} \frac{R(x)\,dx}{\Gamma(\alpha+x)\,\Gamma(\beta-x)\,\Gamma(\gamma+x)\,\Gamma(\delta-x)} = \frac{\int_0^1 R(t)\cos\left[\frac{1}{2}\pi(2t+\alpha-\beta)\right]\,dt}{\Gamma\left(\dfrac{\alpha+\beta}{2}\right)\Gamma\left(\dfrac{\gamma+\delta}{2}\right)\Gamma(\alpha+\delta-1)}$$

$$[\alpha+\delta = \beta+\gamma, \quad \operatorname{Re}(\alpha+\beta+\gamma+\delta) > 2, \quad R(x+1) = -R(x)] \quad \text{ET II 301(25)}$$

## 6.42 Combinations of the gamma function, the exponential, and powers

**6.421**

1.

$$\int_{-\infty}^{\infty} \Gamma(\alpha+x)\,\Gamma(\beta-x)\exp\left[2(\pi n+\theta)xi\right]\,dx = 2\pi i\,\Gamma(\alpha+\beta)(2\cos\theta)^{-\alpha-\beta}\exp[(\beta-\alpha)i\theta]$$

$$\times\left[\eta_n(\beta)\exp(2n\pi\beta i) - \eta_n(-\alpha)\exp(-2n\pi\alpha i)\right]$$

$$\left[\operatorname{Re}(\alpha+\beta) < 1, \quad -\frac{\pi}{2} < \theta < \frac{\pi}{2}, \quad n \text{ an integer}, \quad \eta_n(\xi) = \begin{cases} 0 & \text{if } \left(\frac{1}{2}-n\right)\operatorname{Im}\xi > 0 \\ \operatorname{sign}\left(\frac{1}{2}-n\right) & \text{if } \left(\frac{1}{2}-n\right)\operatorname{Im}\xi < 0 \end{cases}\right]$$

$$\text{ET II 298(7)}$$

2.

$$\int_{-\infty}^{\infty} \frac{e^{\pi i c x}\,dx}{\Gamma(\alpha+x)\,\Gamma(\beta-x)\,\Gamma(\gamma+kx)\,\Gamma(\delta-kx)} = 0$$

$$[\operatorname{Re}(\alpha+\beta+\gamma+\delta) > 2, \quad c \text{ and } k \text{ are real}, \quad |c| > |k|+1] \quad \text{ET II 301(26)}$$

3.

$$\int_{-\infty}^{\infty} \frac{\Gamma(\alpha+x)}{\Gamma(\beta+x)}\exp[(2\pi n+\pi-2\theta)xi]\,dx$$

$$= 2\pi i\,\operatorname{sign}\left(n+\tfrac{1}{2}\right)\frac{(2\cos\theta)^{\beta-\alpha-1}}{\Gamma(\beta-\alpha)}\exp[-(2\pi n+\pi-\theta)\alpha i+\theta i(\beta-1)]$$

$$\left[\operatorname{Re}(\beta-\alpha) > 0, \quad -\frac{\pi}{2} < \theta < \frac{\pi}{2}, \quad n \text{ is an integer}, \quad \left(n+\tfrac{1}{2}\right)\operatorname{Im}\alpha < 0\right] \quad \text{ET II 298(8)}$$

4.

$$\int_{-\infty}^{\infty} \frac{\Gamma(\alpha+x)}{\Gamma(\beta+x)}\exp[(2\pi n+\pi-2\theta)xi]\,dx = 0$$

$$\left[\operatorname{Re}(\beta-\alpha) > 0, \quad -\frac{\pi}{2} < \theta < \frac{\pi}{2}, \quad n \text{ is an integer}, \quad \left(n+\tfrac{1}{2}\right)\operatorname{Im}\alpha > 0\right] \quad \text{ET II 297(6)}$$

**6.422**

1.  $$\int_{-i\infty}^{i\infty} \Gamma(s-k-\lambda)\,\Gamma\left(\lambda+\mu-s+\tfrac{1}{2}\right)\Gamma\left(\lambda-\mu-s+\tfrac{1}{2}\right)z^s\,\mathrm{d}s$$

    $$= 2\pi i\,\Gamma\left(\tfrac{1}{2}-k-\mu\right)\Gamma\left(\tfrac{1}{2}-k+\mu\right)z^\lambda e^{\frac{z}{2}}\,W_{k,\mu}(z)$$

    $$\left[\operatorname{Re}(k+\lambda)<0,\quad \operatorname{Re}\lambda>|\operatorname{Re}\mu|-\tfrac{1}{2},\quad |\arg z|<\tfrac{3}{2}\pi\right]\quad \text{ET II 302(29)}$$

2.  $$\int_{\gamma-i\infty}^{\gamma+i\infty}\Gamma(\alpha+s)\,\Gamma(-s)\,\Gamma(1-c-s)x^s\,\mathrm{d}s = 2\pi i\,\Gamma(\alpha)\,\Gamma(\alpha-c+1)\Phi(\alpha,c;x)$$

    $$\left[-\operatorname{Re}\alpha<\gamma<\min\left(0,1-\operatorname{Re}c\right),\quad -\tfrac{3}{2}\pi<\arg x<\tfrac{3}{2}\pi\right]\quad \text{EH I 256(5)}$$

3.  $$\int_{\gamma-i\infty}^{\gamma+i\infty}\Gamma(-s)\,\Gamma(\beta+s)t^s\,\mathrm{d}s = 2\pi i\,\Gamma(\beta)(1+t)^{-\beta}\qquad \left[0>\gamma>\operatorname{Re}(1-\beta),\quad |\arg t|<\pi\right]$$

    $$\text{EH I 256, BU 75}$$

4.  $$\int_{-\infty i}^{\infty i}\Gamma\left(\frac{t-p}{2}\right)\Gamma(-t)\left(\sqrt{2}\right)^{t-p-2}z^t\,\mathrm{d}t = 2\pi i e^{\frac{1}{4}z^2}\,\Gamma(-p)\,D_p(z)$$

    $$\left[|\arg z|<\tfrac{3}{4}\pi,\quad p\text{ is not a positive integer}\right]\quad \text{WH}$$

5.  $$\int_{-i\infty}^{i\infty}\Gamma(s)\,\Gamma\left(\tfrac{1}{2}\nu+\tfrac{1}{4}-s\right)\Gamma\left(\tfrac{1}{2}\nu-\tfrac{1}{4}-s\right)\left(\frac{z^2}{2}\right)^s\,\mathrm{d}s$$

    $$= 2\pi i 2^{\frac{1}{4}-\frac{1}{2}\nu}z^{-\frac{1}{2}}e^{\frac{3}{4}z^2}\,\Gamma\left(\tfrac{1}{2}\nu+\tfrac{1}{4}\right)\Gamma\left(\tfrac{1}{2}\nu-\tfrac{1}{4}\right)D_\nu(z)$$

    $$\left[|\arg z|<\tfrac{3}{4}\pi,\quad \nu\neq\tfrac{1}{2},\ -\tfrac{1}{2},\ -\tfrac{3}{2},\dots\right]\quad \text{EH II 120}$$

6.[3]  $$\int_{c-i\infty}^{c+i\infty}\left(\tfrac{1}{2}x\right)^{-s}\Gamma\left(\tfrac{1}{2}\nu+\tfrac{1}{2}s\right)\left[\Gamma\left(1+\tfrac{1}{2}\nu-\tfrac{1}{2}s\right)\right]^{-1}\,\mathrm{d}s = 4\pi i\,J_\nu(x)$$

    $$\left[x>0,\ -\operatorname{Re}\nu<c<1\right]\qquad \text{EH II 21(34)}$$

7.  $$\int_{-c-i\infty}^{-c+i\infty}\Gamma(-\nu-s)\,\Gamma(-s)\left(-\tfrac{1}{2}iz\right)^{\nu+2s}\,\mathrm{d}s = -2\pi^2 e^{\frac{1}{2}i\nu\pi}\,H_\nu^{(1)}(z)$$

    $$\left[|\arg(-iz)|<\tfrac{\pi}{2},\quad 0<\operatorname{Re}\nu<c\right]$$

    $$\text{EH II 83(34)}$$

8.  $$\int_{-c-i\infty}^{-c+i\infty}\Gamma(-\nu-s)\,\Gamma(-s)\left(\tfrac{1}{2}iz\right)^{\nu+2s}\,\mathrm{d}s = 2\pi^2 e^{-\frac{1}{2}i\nu\pi}\,H_\nu^{(2)}(z)$$

    $$\left[|\arg(iz)|<\tfrac{\pi}{2},\quad 0<\operatorname{Re}\nu<c\right]$$

    $$\text{EH II 83(35)}$$

9.  $$\int_{-i\infty}^{i\infty}\Gamma(-s)\frac{\left(\tfrac{1}{2}x\right)^{\nu+2s}}{\Gamma(\nu+s+1)}\,\mathrm{d}s = 2\pi i\,J_\nu(x)\qquad \left[x>0,\quad \operatorname{Re}\nu>0\right]\qquad \text{EH II 83(36)}$$

10.  $$\int_{-i\infty}^{i\infty}\Gamma(-s)\,\Gamma(-2\nu-s)\,\Gamma\left(\nu+s+\tfrac{1}{2}\right)(-2iz)^s\,\mathrm{d}s = -\pi^{\frac{5}{2}}e^{-i(z-\nu\pi)}\sec(\nu\pi)(2z)^{-\nu}\,H_\nu^{(1)}(z)$$

    $$\left[|\arg(-iz)|<\tfrac{3}{2}\pi,\quad 2\nu\neq\pm1,\pm3\dots\right]$$

    $$\text{EH II 83(37)}$$

11. $\quad \int_{-i\infty}^{i\infty} \Gamma(-s)\,\Gamma(-2\nu-s)\,\Gamma\left(\nu+s+\tfrac{1}{2}\right)(2iz)^s\,\mathrm{d}s = \pi^{\frac{5}{2}} e^{i(z-\nu\pi)} \sec(\nu\pi)(2z)^{-\nu}\,H_\nu^{(2)}(z)$

$$\left[|\arg(iz)| < \tfrac{3}{2}\pi, \quad 2\nu \neq \pm 1, \pm 3 \ldots\right]$$
EH II 84(38)

12. $\quad \int_{-i\infty}^{i\infty} \Gamma(s)\,\Gamma\left(\tfrac{1}{2}-s-\nu\right)\Gamma\left(\tfrac{1}{2}-s+\nu\right)(2z)^s\,\mathrm{d}s = 2^{\frac{3}{2}} \pi^{\frac{3}{2}} i z^{\frac{1}{2}} e^z \sec(\nu\pi)\,K_\nu(z)$

$$\left[|\arg z| < \tfrac{3}{2}\pi, \quad 2\nu \neq \pm 1, \pm 3, \ldots\right]$$
EH II 84(39)

13. $\quad \int_{-\frac{1}{2}-i\infty}^{-\frac{1}{2}+i\infty} \frac{\Gamma(-s)}{s\,\Gamma(1+s)}\,x^{2s}\,\mathrm{d}s = 4\pi \int_{2x}^{\infty} \frac{J_0(t)}{t}\,\mathrm{d}t \qquad [x>0]$ $\qquad$ MO 41

14. $\quad \int_{-i\infty}^{i\infty} \frac{\Gamma(\alpha+s)\,\Gamma(\beta+s)\,\Gamma(-s)}{\Gamma(\gamma+s)}(-z)^s\,\mathrm{d}s = 2\pi i \frac{\Gamma(\alpha)\,\Gamma(\beta)}{\Gamma(\gamma)}\,F(\alpha,\beta;\gamma;z)$

[For $\arg(-z) < \pi$, the path of integration must separate the poles of the integrand at the points $s = 0, 1, 2, 3, \ldots$ from the poles $s = -\alpha - n$ and $s = -\beta - n$ (for $n = 0, 1, 2, \ldots$)].

15. $\quad \int_{\delta-i\infty}^{\delta+i\infty} \frac{\Gamma(\alpha+s)\,\Gamma(-s)}{\Gamma(\gamma+s)}(-z)^s\,\mathrm{d}s = \frac{2\pi i\,\Gamma(\alpha)}{\Gamma(\gamma)}\,{}_1F_1(\alpha;\gamma;z)$

$$\left[-\tfrac{\pi}{2} < \arg(-z) < \tfrac{\pi}{2}, \quad 0 > \delta > -\operatorname{Re}\alpha, \quad \gamma \neq 0, 1, 2, \ldots\right] \quad \text{EH I 62(15), EH I 256(4)}$$

16. $\quad \int_{-i\infty}^{i\infty} \left[\frac{\Gamma\left(\tfrac{1}{2}-s\right)}{\Gamma(s)}\right]^2 z^s\,\mathrm{d}s = 2\pi i z^{\frac{1}{2}}\left[2\pi^{-1}\,K_0\left(4z^{\frac{1}{4}}\right) - Y_0\left(4z^{\frac{1}{4}}\right)\right]$

$$[z>0]$$
ET II 303(33)

17. $\quad \int_{-i\infty}^{i\infty} \frac{\Gamma\left(\lambda+\mu-s+\tfrac{1}{2}\right)\Gamma\left(\lambda-\mu-s+\tfrac{1}{2}\right)}{\Gamma(\lambda-k-s+1)} z^s\,\mathrm{d}s = 2\pi i z^\lambda e^{-\frac{z}{2}}\,W_{k,\mu}(z)$

$$\left[\operatorname{Re}\lambda > |\operatorname{Re}\mu| - \tfrac{1}{2}, \quad |\arg z| < \tfrac{\pi}{2}\right]$$
ET II 302(30)

18. $\quad \int_{-i\infty}^{i\infty} \frac{\Gamma(k-\lambda+s)\,\Gamma\left(\lambda+\mu-s+\tfrac{1}{2}\right)}{\Gamma\left(\mu-\lambda+s+\tfrac{1}{2}\right)} z^s\,\mathrm{d}s = 2\pi i \frac{\Gamma\left(k+\mu+\tfrac{1}{2}\right)}{\Gamma(2\mu+1)} z^\lambda e^{-\frac{z}{2}}\,M_{k,\mu}(z)$

$$\left[\operatorname{Re}(k-\lambda) > 0, \quad \operatorname{Re}(\lambda+\mu) > -\tfrac{1}{2}, \quad |\arg z| < \tfrac{\pi}{2}\right] \quad \text{ET II 302(31)}$$

19.$^{12}$ $\quad \int_{-i\infty}^{i\infty} \dfrac{\displaystyle\prod_{j=1}^{m}\Gamma(b_j-s)\prod_{j=1}^{n}\Gamma(1-a_j+s)}{\displaystyle\prod_{j=m+1}^{q}\Gamma(1-b_j+s)\prod_{j=n+1}^{p}\Gamma(a_j-s)} z^s\,\mathrm{d}s = 2\pi i\,G_{pq}^{mn}\left(z\,\middle|\,\begin{matrix} a_1, \ldots, a_p \\ b_1, \ldots, b_q \end{matrix}\right)$

$$\left[p+q < 2(m+n); \quad |\arg z| < \left(m+n-\tfrac{1}{2}p-\tfrac{1}{2}q\right)\pi;\right.$$
$$\left.\operatorname{Re} a_k < 1, \quad k = 1, \ldots, n; \quad \operatorname{Re} b_j > 0, \quad j = 1, \ldots, m\right]$$
ET II 303(34)

**6.423**

1.    $\displaystyle\int_0^\infty e^{-\alpha x}\frac{\mathrm{d}x}{\Gamma(1+x)} = \nu\left(e^{-\alpha}\right)$                                       MI 39, EH III 222(16)

2.    $\displaystyle\int_0^\infty e^{-\alpha x}\frac{\mathrm{d}x}{\Gamma(x+\beta+1)} = e^{\beta\alpha}\nu\left(e^{-\alpha},\beta\right)$                         MI 39, EH III 222(16)

3.    $\displaystyle\int_0^\infty e^{-\alpha x}\frac{x^m}{\Gamma(x+1)}\,\mathrm{d}x = \mu\left(e^{-\alpha},m\right)\Gamma(m+1)$     $[\operatorname{Re}m>-1]$      MI 39, EH III 222(17)

4.    $\displaystyle\int_0^\infty e^{-\alpha x}\frac{x^m}{\Gamma(x+n+1)}\,\mathrm{d}x = e^{n\alpha}\mu\left(e^{-\alpha},m,n\right)\Gamma(m+1)$          MI 39, EH III 222(17)

**6.424**    $\displaystyle\int_{-\infty}^\infty \frac{R(x)\exp[(2\pi n+\theta)xi]\,\mathrm{d}x}{\Gamma(\alpha+x)\,\Gamma(\beta-x)} = \frac{\left[2\cos\left(\dfrac{\theta}{2}\right)\right]^{\alpha+\beta-2}}{\Gamma(\alpha+\beta-1)}\exp\left[\frac{1}{2}\theta(\beta-\alpha)i\right]\int_0^1 R(t)\exp(2\pi nti)\,\mathrm{d}t$

$[\operatorname{Re}(\alpha+\beta)>1, \quad -\pi<\theta<\pi, \quad n \text{ is an integer}, \quad R(x+1)=R(x)]$     ET II 299(16)

## 6.43 Combinations of the gamma function and trigonometric functions

**6.431**

1.[12]    $\displaystyle\int_{-\infty}^\infty \frac{\sin rx\,\mathrm{d}x}{\Gamma(p+x)\,\Gamma(q-x)} = \frac{\left(2\cos\dfrac{r}{2}\right)^{p+q-2}\sin\dfrac{r(q-p)}{2}}{\Gamma(p+q-1)}$     $[|r|<\pi]$

                                            $=0$     $[|r|>\pi]$

                       $[r \text{ is real}; \quad \operatorname{Re}(p+q)>1]$    MO 10a, ET II 298(9, 10)

2.    $\displaystyle\int_{-\infty}^\infty \frac{\cos rx\,\mathrm{d}x}{\Gamma(p+x)\,\Gamma(q-x)} = \frac{\left(2\cos\dfrac{r}{2}\right)^{p+q-2}\cos\dfrac{r(q-p)}{2}}{\Gamma(p+q-1)}$     $[|r|<\pi]$

                                              $=0$     $[|r|>\pi]$

                       $[r \text{ is real}; \quad \operatorname{Re}(p+q)>1]$    MO 10a, ET II 299(13, 14)

**6.432**    $\displaystyle\int_{-\infty}^\infty \frac{\sin(m\pi x)}{\sin(\pi x)}\frac{\mathrm{d}x}{\Gamma(\alpha+x)\,\Gamma(\beta-x)} = 0$     $[m \text{ is an even integer}]$

                                             $=\dfrac{2^{\alpha+\beta-2}}{\Gamma(\alpha+\beta-1)}$     $[m \text{ is an odd integer}]$

                                            $[\operatorname{Re}(\alpha+\beta)>1]$      ET II 298(11, 12)

**6.433**

1.    $\displaystyle\int_{-\infty}^\infty \frac{\sin\pi x\,\mathrm{d}x}{\Gamma(\alpha+x)\,\Gamma(\beta-x)\,\Gamma(\gamma+x)\,\Gamma(\delta-x)} = \frac{\sin\left[\frac{\pi}{2}(\beta-\alpha)\right]}{2\,\Gamma\left(\dfrac{\alpha+\beta}{2}\right)\Gamma\left(\dfrac{\gamma+\delta}{2}\right)\Gamma(\alpha+\delta-1)}$

                       $[\alpha+\delta=\beta+\gamma, \quad \operatorname{Re}(\alpha+\beta+\gamma+\delta)>2]$    ET II 300(22)

2.
$$\int_{-\infty}^{\infty} \frac{\cos \pi x \, dx}{\Gamma(\alpha + x)\,\Gamma(\beta - x)\,\Gamma(\gamma + x)\,\Gamma(\delta - x)} = \frac{\cos\left[\frac{\pi}{2}(\beta - \alpha)\right]}{2\,\Gamma\left(\dfrac{\alpha + \beta}{2}\right)\Gamma\left(\dfrac{\gamma + \delta}{2}\right)\Gamma(\alpha + \delta - 1)}$$

$$[\alpha + \delta = \beta + \gamma, \quad \operatorname{Re}(\alpha + \beta + \gamma + \delta) > 2]$$    ET II 301(23)

## 6.44 The logarithm of the gamma function*

**6.441**

1.
$$\int_{p}^{p+1} \ln \Gamma(x) \, dx = \ln \sqrt{2\pi} + p \ln p - p$$    FI II 784

2.
$$\int_{0}^{1} \ln \Gamma(x) \, dx = \int_{0}^{1} \ln \Gamma(1 - x) \, dx = \ln \sqrt{2\pi}$$    FI II 783

3.
$$\int_{0}^{1} \ln \Gamma(x + q) \, dx = \ln \sqrt{2\pi} + q \ln q - q$$    $[q \geq 0]$    NH 89(17), ET II 304(40)

4.
$$\int_{0}^{z} \ln \Gamma(x + 1) \, dx = z \ln \sqrt{2\pi} - \frac{z(z + 1)}{2} + z \ln \Gamma(z + 1) - \ln G(z + 1),$$

$$\text{where } G(z + 1) = (2\pi)^{\frac{z}{2}} \exp\left(-\frac{z(z + 1)}{2} - \frac{Cz^2}{2}\right) \prod_{k=1}^{\infty}\left\{\left(1 + \frac{z}{k}\right)^{k} \exp\left(-z + \frac{z^2}{2k}\right)\right\}$$    WH

5.
$$\int_{0}^{n} \ln \Gamma(a + x) \, dx = \sum_{k=0}^{n-1}(a + k)\ln(a + k) - na + n \ln \sqrt{2\pi} - \frac{1}{2}n(n - 1)$$

$$[a \geq 0, \quad n = 1, 2, \ldots]$$    ET II 304(41)

6.*
$$\int_{0}^{1} \ln^2 \Gamma(x)\,dx = \frac{C^2}{12} + \frac{\pi^2}{48} + \frac{1}{3}C \ln \sqrt{2\pi} + \frac{4}{3}\ln^2 \sqrt{2\pi} - \left(C + 2\ln \sqrt{2\pi}\right)\frac{\zeta'(2)}{\pi^2} + \frac{\zeta''(2)}{2\pi^2}$$

**6.442**
$$\int_{0}^{1} \exp(2\pi nxi) \ln \Gamma(a + x) \, dx = (2\pi ni)^{-1}\left[\ln a - \exp(-2\pi nai)\operatorname{Ei}(2\pi nai)\right]$$

$$[a > 0, \quad n = \pm 1, \pm 2, \ldots]$$    ET II 304(38)

**6.443**

1.
$$\int_{0}^{1} \ln \Gamma(x) \sin 2\pi nx \, dx = \frac{1}{2\pi n}\left[\ln(2\pi n) + C\right]$$    NH 203(5), ET II 304(42)

2.
$$\int_{0}^{1} \ln \Gamma(x) \sin(2n + 1)\pi x \, dx = \frac{1}{(2n + 1)\pi}\left[\ln\left(\frac{\pi}{2}\right) + 2\left(1 + \frac{1}{3} + \cdots + \frac{1}{2n - 1}\right) + \frac{1}{2n + 1}\right]$$
    ET II 305(43)

3.
$$\int_{0}^{1} \ln \Gamma(x) \cos 2\pi nx \, dx = \frac{1}{4n}$$    NH 203(6), ET II 305(44)

4.[8]
$$\int_{0}^{1} \ln \Gamma(x) \cos(2n + 1)\pi x \, dx = \frac{2}{\pi^2}\left[\frac{1}{(2n + 1)^2}(C + \ln 2\pi) + 2\sum_{k=2}^{\infty}\frac{\ln k}{4k^2 - (2n + 1)^2}\right]$$    NH 203(6)

---

*Here, we are violating our usual order of presentation of the formulas in order to make it easier to examine the integrals involving the gamma function

5.[12]    $\displaystyle\int_0^1 \sin(2\pi nx)\ln\Gamma(a+x)\,\mathrm{d}x = -(2\pi n)^{-1}\left[\ln a - \cos(2\pi na)\,\mathrm{ci}(2\pi na) - \sin(2\pi na)\,\mathrm{si}(2\pi na)\right]$

$$[a > 0, \quad n = 1, 2, \ldots]\qquad \text{ET II 304(36)}$$

6.[12]    $\displaystyle\int_0^1 \cos(2\pi nx)\ln\Gamma(a+x)\,\mathrm{d}x = -(2\pi n)^{-1}\left[-\sin(2\pi na)\,\mathrm{ci}(2\pi na) + \cos(2\pi na)\,\mathrm{si}(2\pi na)\right]$

$$[a > 0, \quad n = 1, 2, \ldots]\qquad \text{ET II 304(37)}$$

## 6.45 The incomplete gamma function

**6.451**

1.    $\displaystyle\int_0^\infty e^{-\alpha x}\,\gamma(\beta, x)\,\mathrm{d}x = \frac{1}{\alpha}\Gamma(\beta)(1+\alpha)^{-\beta}$          $[\beta > 0]$          MI 39

2.    $\displaystyle\int_0^\infty e^{-\alpha x}\,\Gamma(\beta, x)\,\mathrm{d}x = \frac{1}{\alpha}\Gamma(\beta)\left[1 - \frac{1}{(\alpha+1)^\beta}\right]$      $[\beta > 0]$          MI 39

**6.452**

1.    $\displaystyle\int_0^\infty e^{-\mu x}\,\gamma\left(\nu, \frac{x^2}{8a^2}\right)\,\mathrm{d}x = \frac{1}{\mu}2^{-\nu-1}\Gamma(2\nu)e^{(a\mu)^2}D_{-2\nu}(2a\mu)$

$$\left[|\arg a| < \tfrac{\pi}{4}, \quad \mathrm{Re}\,\nu > -\tfrac{1}{2}, \quad \mathrm{Re}\,\mu > 0\right]$$
$$\text{ET I 179(36)}$$

2.    $\displaystyle\int_0^\infty e^{-\mu x}\,\gamma\left(\frac{1}{4}, \frac{x^2}{8a^2}\right)\,\mathrm{d}x = \frac{2^{\frac{3}{4}}\sqrt{a}}{\sqrt{\mu}}e^{(a\mu)^2}K_{\frac{1}{4}}\left(a^2\mu^2\right)$    $\left[|\arg a| < \tfrac{\pi}{4}, \quad \mathrm{Re}\,\mu > 0\right]$   ET I 179(35)

**6.453**    $\displaystyle\int_0^\infty e^{-\mu x}\,\Gamma\left(\nu, \frac{a}{x}\right)\,\mathrm{d}x = 2a^{\frac{1}{2}\nu}\mu^{\frac{1}{2}\nu-1}K_\nu\left(2\sqrt{\mu a}\right)$    $\left[|\arg u| < \tfrac{\pi}{2}, \quad \mathrm{Re}\,\mu > 0\right]$   ET I 179(32)

**6.454**    $\displaystyle\int_0^\infty e^{-\beta x}\,\gamma\left(\nu, \alpha\sqrt{x}\right)\,\mathrm{d}x = 2^{-\frac{1}{2}\nu}\alpha^\nu\beta^{-\frac{1}{2}\nu-1}\Gamma(\nu)\exp\left(\frac{\alpha^2}{8\beta}\right)D_{-\nu}\left(\frac{\alpha}{\sqrt{2\beta}}\right)$

$$[\mathrm{Re}\,\beta > 0, \quad \mathrm{Re}\,\nu > 0]$$
$$\text{ET II 309(19), MI 39a}$$

**6.455**

1.    $\displaystyle\int_0^\infty x^{\mu-1}e^{-\beta x}\,\Gamma(\nu, \alpha x)\,\mathrm{d}x = \frac{\alpha^\nu\,\Gamma(\mu+\nu)}{\mu(\alpha+\beta)^{\mu+\nu}}\,{}_2F_1\left(1, \mu+\nu; \mu+1; \frac{\beta}{\alpha+\beta}\right)$

$$[\mathrm{Re}(\alpha+\beta) > 0, \quad \mathrm{Re}\,\mu > 0, \quad \mathrm{Re}(\mu+\nu) > 0]\quad \text{ET II 309(16)}$$

2.    $\displaystyle\int_0^\infty x^{\mu-1}e^{-\beta x}\,\gamma(\nu, \alpha x)\,\mathrm{d}x = \frac{\alpha^\nu\,\Gamma(\mu+\nu)}{\nu(\alpha+\beta)^{\mu+\nu}}\,{}_2F_1\left(1, \mu+\nu; \nu+1; \frac{\alpha}{\alpha+\beta}\right)$

$$[\mathrm{Re}(\alpha+\beta) > 0, \quad \mathrm{Re}\,\beta > 0, \quad \mathrm{Re}(\mu+\nu) > 0]\quad \text{ET II 308(15)}$$

**6.456**

1.    $\displaystyle\int_0^\infty e^{-\alpha x}(4x)^{\nu-\frac{1}{2}}\gamma\left(\nu, \frac{1}{4x}\right)\,\mathrm{d}x = \sqrt{\pi}\frac{\gamma\left(2\nu, \sqrt{\alpha}\right)}{\alpha^{\nu+\frac{1}{2}}}$          MI 39a

2.    $\displaystyle\int_0^\infty e^{-\alpha x}(4x)^{\nu-\frac{1}{2}}\Gamma\left(\nu, \frac{1}{4x}\right)\,\mathrm{d}x = \frac{\sqrt{\pi}\,\Gamma\left(2\nu, \sqrt{\alpha}\right)}{\alpha^{\nu+\frac{1}{2}}}$          MI 39a

**6.457**

1.    $\displaystyle \int_0^\infty e^{-\alpha x}\frac{(4x)^\nu}{\sqrt{x}}\,\gamma\left(\nu+1,\frac{1}{4x}\right)\mathrm{d}x = \sqrt{\pi}\,\frac{\gamma\left(2\nu+1,\sqrt{\alpha}\right)}{\alpha^{\nu+\frac{1}{2}}}$          MI 39

2.    $\displaystyle \int_0^\infty e^{-\alpha x}\frac{(4x)^\nu}{\sqrt{x}}\,\Gamma\left(\nu+1,\frac{1}{4x}\right)\mathrm{d}x = \sqrt{\pi}\,\frac{\Gamma\left(2\nu+1,\sqrt{\alpha}\right)}{\alpha^{\nu+\frac{1}{2}}}$          MI 39

**6.458**    $\displaystyle \int_0^\infty x^{1-2\nu}\exp\left(\alpha x^2\right)\sin(bx)\,\Gamma\left(\nu,\alpha x^2\right)\mathrm{d}x = \pi^{\frac{1}{2}}2^{-\nu}\alpha^{\nu-1}\Gamma\left(\tfrac{3}{2}-\nu\right)\exp\left(\frac{b^2}{8\alpha}\right)D_{2\nu-2}\left(\frac{b}{\sqrt{2\alpha}}\right)$

$$\left[|\arg\alpha|<\tfrac{3\pi}{2},\quad 0<\operatorname{Re}\nu<1\right]$$
ET II 309(18)

## 6.46–6.47   The function $\psi(x)$

**6.461**   $\displaystyle \int_1^x \psi(t)\,\mathrm{d}t = \ln\Gamma(x)$

**6.462**   $\displaystyle \int_0^1 \psi(\alpha+x)\,\mathrm{d}x = \ln\alpha$        $[\alpha>0]$        ET II 305(1)

**6.463**   $\displaystyle \int_0^\infty x^{-\alpha}\left[\boldsymbol{C}+\psi(1+x)\right] = -\pi\operatorname{cosec}(\pi\alpha)\,\zeta(\alpha)$        $[1<\operatorname{Re}\alpha<2]$        ET II 305(6)

**6.464**   $\displaystyle \int_0^1 e^{2\pi n x i}\,\psi(\alpha+x)\,\mathrm{d}x = e^{-2\pi n\alpha i}\operatorname{Ei}(2\pi n\alpha i)$        $[\alpha>0;\quad n=\pm i,\pm 2,\ldots]$        ET II 305(2)

**6.465**

1.[8]   $\displaystyle \int_0^1 \psi(x)\sin\pi x\,\mathrm{d}x = -\frac{2}{\pi}\left[\boldsymbol{C}+\ln 2\pi+2\sum_{k=2}^\infty \frac{\ln k}{4k^2-1}\right]$

                                        (see **6.443** 4)        NH 204

2.   $\displaystyle \int_0^1 \psi(x)\sin(2\pi n x)\,\mathrm{d}x = -\frac{1}{2}\pi$        $[n=1,2,\ldots]$        ET II 305(3)

**6.466**   $\displaystyle \int_0^\infty \left[\psi(\alpha+ix)-\psi(\alpha-ix)\right]\sin xy\,\mathrm{d}x = i\pi e^{-\alpha y}\left(1-e^{-y}\right)^{-1}$

                                     $[\alpha>0,\quad y>0]$        ET I 96(1)

**6.467**

1.[12]   $\displaystyle \int_0^1 \sin(2\pi n x)\,\psi(\alpha+x)\,\mathrm{d}x = -\sin(2\pi n\alpha)\operatorname{ci}(2\pi n\alpha)+\cos(2\pi n\alpha)\operatorname{si}(2\pi n\alpha)$

                                     $[\alpha\geq 0;\quad n=1,2,\ldots]$        ET II 305(4)

2.[12]   $\displaystyle \int_0^1 \cos(2\pi n x)\,\psi(\alpha+x)\,\mathrm{d}x = \sin(2\pi n\alpha)\operatorname{si}(2\pi n\alpha)+\cos(2\pi n\alpha)\operatorname{ci}(2\pi n\alpha)$

                                     $[\alpha>0;\quad n=1,2,\ldots]$        ET II 305(5)

**6.468**   $\displaystyle \int_0^1 \psi(x)\sin^2\pi x\,\mathrm{d}x = -\frac{1}{2}\left[\boldsymbol{C}+\ln(2\pi)\right]$        NH 204

**6.469**

1.   $\displaystyle \int_0^1 \psi(x)\sin\pi x\cos\pi x\,\mathrm{d}x = -\frac{\pi}{4}$        NH 204

2.$^8$  $\displaystyle\int_0^1 \psi(x) \sin \pi x \sin(n\pi x)\, \mathrm{d}x = \frac{n}{1-n^2}$        [$n$ is even]

$\displaystyle = \frac{1}{2} \ln \frac{n-1}{n+1}$        [$n > 1$ is odd]

NH 204(8)a

**6.471**

1.  $\displaystyle\int_0^\infty x^{-\alpha} \left[\ln x - \psi(1+x)\right]\, \mathrm{d}x = \pi \operatorname{cosec}(\pi\alpha)\, \zeta(\alpha)$        [$0 < \operatorname{Re}\alpha < 1$]        ET II 306(7)

2.  $\displaystyle\int_0^\infty x^{-\alpha} \left[\ln(1+x) - \psi(1+x)\right]\, \mathrm{d}x = \pi \operatorname{cosec}(\pi\alpha) \left[\zeta(\alpha) - (\alpha-1)^{-1}\right]$

[$0 < \operatorname{Re}\alpha < 1$]        ET II 306(8)

3.  $\displaystyle\int_0^\infty \left[\psi(x+1) - \ln x\right] \cos(2\pi xy)\, \mathrm{d}x = \frac{1}{2}\left[\psi(y+1) - \ln y\right]$        ET II 306(12)

**6.472**

1.  $\displaystyle\int_0^\infty x^{-\alpha} \left[(1+x)^{-1} - \psi'(1+x)\right]\, \mathrm{d}x = -\pi\alpha \operatorname{cosec}(\pi\alpha) \left[\zeta(1+\alpha) - \alpha^{-1}\right]$

[$|\operatorname{Re}\alpha| < 1$]        ET II 306(9)

2.  $\displaystyle\int_0^\infty x^{-\alpha} \left[x^{-1} - \psi'(1+x)\right]\, \mathrm{d}x = -\pi\alpha \operatorname{cosec}(\pi\alpha)\, \zeta(1+\alpha)$

[$-2 < \operatorname{Re}\alpha < 0$]        ET II 306(10)

**6.473**  $\displaystyle\int_0^\infty x^{-\alpha} \psi^{(n)}(1+x)\, \mathrm{d}x = (-1)^{n-1} \frac{\pi\,\Gamma(\alpha+n)}{\Gamma(\alpha)\sin\pi\alpha}\, \zeta(\alpha+n)$

[$n = 1, 2, \ldots;\quad 0 < \operatorname{Re}\alpha < 1$]
ET II 306(11)

# 6.5–6.7 Bessel Functions

## 6.51 Bessel functions

**6.511**

1.  $\displaystyle\int_0^\infty J_\nu(bx)\, \mathrm{d}x = \frac{1}{b}$        [$\operatorname{Re}\nu > -1,\quad b > 0$]        ET II 22(3)

2.  $\displaystyle\int_0^\infty Y_\nu(bx)\, \mathrm{d}x = -\frac{1}{b} \tan\left(\frac{\nu\pi}{2}\right)$        [$|\operatorname{Re}\nu| < 1,\quad b > 0$]

WA 432(7), ET II 96(1)

3.  $\displaystyle\int_0^a J_\nu(x)\, \mathrm{d}x = 2 \sum_{k=0}^\infty J_{\nu+2k+1}(a)$        [$\operatorname{Re}\nu > -1$]        ET II 333(1)

4.  $\displaystyle\int_0^a J_{\frac{1}{2}}(t)\, \mathrm{d}t = 2\, S\left(\sqrt{a}\right)$        WA 599(4)

5.  $\displaystyle\int_0^a J_{-\frac{1}{2}}(t)\, \mathrm{d}t = 2\, C\left(\sqrt{a}\right)$        WA 599(3)

6.  $\int_0^a J_0(x)\,\mathrm{d}x = a\,J_0(a) + \dfrac{\pi a}{2}\left[J_1(a)\,\mathbf{H}_0(a) - J_0(a)\,\mathbf{H}_1(a)\right]$

$\qquad\qquad\qquad\qquad\qquad\qquad\qquad\qquad [a > 0]$         ET II 7(2)

7.  $\int_0^a J_1(x)\,\mathrm{d}x = 1 - J_0(a)$               $[a > 0]$         ET II 18(1)

8.  $\int_a^\infty J_0(x)\,\mathrm{d}x = 1 - a\,J_0(a) + \dfrac{\pi a}{2}\left[J_0(a)\,\mathbf{H}_1(a) - J_1(a)\,\mathbf{H}_0(a)\right]$

$\qquad\qquad\qquad\qquad\qquad\qquad\qquad\qquad [a > 0]$         ET II 7(3)

9.  $\int_a^\infty J_1(x)\,\mathrm{d}x = J_0(a)$              $[a > 0]$         ET II 18(2)

10. $\int_a^b Y_\nu(x)\,\mathrm{d}x = 2\sum_{n=0}^\infty \left[Y_{\nu+2n+1}(b) - Y_{\nu+2n+1}(a)\right]$         ET II 339(46)

11. $\int_0^a I_\nu(x)\,\mathrm{d}x = 2\sum_{n=0}^\infty (-1)^n\, I_{\nu+2n+1}(a)$      $[\operatorname{Re}\nu > -1]$         ET II 364(1)

12. $\int_0^\infty K_0(ax) = \dfrac{\pi}{2a}$              $[a > 0]$

13. $\int_0^\infty K_0^2(ax) = \dfrac{\pi^2}{4a}$              $[a > 0]$

## 6.512

1.[11] $\int_0^\infty J_\mu(ax)\,J_\nu(bx)\,\mathrm{d}x = b^\nu a^{-\nu-1}\,\dfrac{\Gamma\!\left(\dfrac{\mu+\nu+1}{2}\right)}{\Gamma(\nu+1)\,\Gamma\!\left(\dfrac{\mu-\nu+1}{2}\right)}\,F\!\left(\dfrac{\mu+\nu+1}{2},\dfrac{\nu-\mu+1}{2};\nu+1;\dfrac{b^2}{a^2}\right)$

$\qquad\qquad\qquad\qquad [a > 0, \quad b > 0, \quad \operatorname{Re}(\mu+\nu) > -1, \quad b < a.$

$\qquad\qquad$ For $b > a$, the positions of $\mu$ and $\nu$ should be reversed.]

$\qquad\qquad\qquad\qquad\qquad\qquad\qquad\qquad\qquad\qquad\qquad\qquad$ ET II 48(6)

2.[7] $\int_0^\infty J_{\nu+n}(\alpha t)\,J_{\nu-n-1}(\beta t)\,\mathrm{d}t = \dfrac{\beta^{\nu-n-1}\,\Gamma(\nu)}{\alpha^{\nu-n}\,n!\,\Gamma(\nu-n)}\,F\!\left(\nu,-n;\nu-n;\dfrac{\beta^2}{\alpha^2}\right)$    $[0 < \beta < \alpha]$

$\qquad\qquad\qquad\qquad\qquad\qquad\quad = (-1)^n\,\dfrac{1}{2\alpha}$           $[0 < \beta = \alpha]$

$\qquad\qquad\qquad\qquad\qquad\qquad\quad = 0$                $[0 < \alpha < \beta]$

$\qquad\qquad\qquad\qquad\qquad\qquad\qquad\qquad [\operatorname{Re}(\nu) > 0]$           MO 50

3.[8] $\int_0^\infty J_\nu(\alpha x)\,J_{\nu-1}(\beta x)\,\mathrm{d}x = \dfrac{\beta^{\nu-1}}{\alpha^\nu}$         $[\beta < \alpha]$

$\qquad\qquad\qquad\qquad\qquad\qquad\quad = \dfrac{1}{2\beta}$             $[\beta = \alpha]$

$\qquad\qquad\qquad\qquad\qquad\qquad\quad = 0$              $[\beta > \alpha]$

$\qquad\qquad\qquad\qquad\qquad\qquad\qquad\qquad [\operatorname{Re}\nu > 0]$      WA 444(8), KU (40)a

**4.** $\displaystyle\int_0^\infty J_{\nu+2n+1}(ax)\,J_\nu(bx)\,\mathrm{d}x = b^\nu a^{-\nu-1} P_n^{(\nu,0)}\left(1 - \frac{2b^2}{a^2}\right)$    $[\mathrm{Re}\,\nu > -1-n, \quad 0 < b < a]$

$\qquad\qquad\qquad\qquad\qquad\qquad\quad = 0$            $[\mathrm{Re}\,\nu > -1-n, \quad 0 < a < b]$

$\qquad\qquad\qquad\qquad\qquad\qquad\qquad\qquad\qquad\qquad\qquad\qquad\qquad\qquad$ ET II 47(5)

**5.** $\displaystyle\int_0^\infty J_{\nu+n}(ax)\,Y_{\nu-n}(ax)\,\mathrm{d}x = (-1)^{n+1}\frac{1}{2a}$    $\left[\mathrm{Re}\,\nu > -\tfrac{1}{2}, \quad a > 0, \quad n = 0,1,2,\dots\right]$

$\qquad\qquad\qquad\qquad\qquad\qquad\qquad\qquad\qquad\qquad\qquad\qquad\qquad\qquad\qquad$ ET II 347(57)

**6.** $\displaystyle\int_0^\infty J_1(bx)\,Y_0(ax)\,\mathrm{d}x = -\frac{b^{-1}}{\pi}\ln\left(1 - \frac{b^2}{a^2}\right)$    $[0 < b < a]$              ET II 21(31)

**7.** $\displaystyle\int_0^a J_\nu(x)\,J_{\nu+1}(x)\,\mathrm{d}x = \sum_{n=0}^\infty [J_{\nu+n+1}(a)]^2$    $[\mathrm{Re}\,\nu > -1]$              ET II 338(37)

**8.**[9] $\displaystyle\int_0^\infty k\,J_n(ka)\,J_n(kb)\,\mathrm{d}k = \frac{1}{a}\delta(b-a)$    $[n = 0,1,\dots]$              JAC 110

**9.** $\displaystyle\int_0^\infty K_0(ax)\,J_1(bx) = \frac{1}{2b}\ln\left(1 + \frac{b^2}{a^2}\right)$    $[a > 0, \quad b > 0]$

**10.** $\displaystyle\int_0^\infty K_0(ax)\,I_1(bx) = -\frac{1}{2b}\ln\left(1 - \frac{b^2}{a^2}\right)$    $[a > 0, \quad b > 0]$

**6.513**

**1.** $\displaystyle\int_0^\infty [J_\mu(ax)]^2\,J_\nu(bx)\,\mathrm{d}x = a^{2\mu}b^{-2\mu-1}\frac{\Gamma\left(\dfrac{1+\nu+2\mu}{2}\right)}{[\Gamma(\mu+1)]^2\,\Gamma\left(\dfrac{1+\nu-2\mu}{2}\right)}$

$\qquad\qquad\qquad \times \left[F\left(\dfrac{1-\nu+2\mu}{2}, \dfrac{1+\nu+2\mu}{2}; \mu+1; \dfrac{1-\sqrt{1-\dfrac{4a^2}{b^2}}}{2}\right)\right]^2$

$\qquad\qquad\qquad\qquad\qquad\qquad\qquad [\mathrm{Re}\,\nu + \mathrm{Re}\,2\mu > -1, \quad 0 < 2a < b]$    ET II 52(33)

**2.** $\displaystyle\int_0^\infty [J_\mu(ax)]^2\,K_\nu(bx)\,\mathrm{d}x = \frac{b^{-1}}{2}\Gamma\left(\frac{2\mu+\nu+1}{2}\right)\Gamma\left(\frac{2\mu-\nu+1}{2}\right)\left[P_{\frac{1}{2}\nu-\frac{1}{2}}^{-\mu}\left(\sqrt{1+\frac{4a^2}{b^2}}\right)\right]^2$

$\qquad\qquad\qquad\qquad\qquad\qquad [2\,\mathrm{Re}\,\mu > |\mathrm{Re}\,\nu| - 1, \quad \mathrm{Re}\,b > 2|\mathrm{Im}\,a|]$

$\qquad\qquad\qquad\qquad\qquad\qquad\qquad\qquad\qquad\qquad\qquad\qquad\qquad\qquad$ ET II 138(18)

**3.** $\displaystyle\int_0^\infty I_\mu(ax)\,K_\mu(ax)\,J_\nu(bx)\,\mathrm{d}x = \frac{e^{\mu\pi i}\,\Gamma\left(\dfrac{\nu+2\mu+1}{2}\right)}{b\,\Gamma\left(\dfrac{\nu-2\mu+1}{2}\right)}P_{\frac{1}{2}\nu-\frac{1}{2}}^{-\mu}\left(\sqrt{1+\frac{4a^2}{b^2}}\right)Q_{\frac{1}{2}\nu-\frac{1}{2}}^{-\mu}\left(\sqrt{1+\frac{4a^2}{b^2}}\right)$

$\qquad\qquad [\mathrm{Re}\,a > 0, \quad b > 0, \quad \mathrm{Re}\,\nu > -1, \quad \mathrm{Re}(\nu+2\mu) > -1]$    ET II 65(20)

**4.**
$$\int_0^\infty J_\mu(ax)\, J_{-\mu}(ax)\, K_\nu(bx)\, \mathrm{d}x = \frac{\pi}{2b} \sec\left(\frac{\nu\pi}{2}\right) P_{\frac{1}{2}\nu - \frac{1}{2}}^{\mu}\left(\sqrt{1 + \frac{4a^2}{b^2}}\right) P_{\frac{1}{2}\nu - \frac{1}{2}}^{-\mu}\left(\sqrt{1 + \frac{4a^2}{b^2}}\right)$$
$$[|\mathrm{Re}\,\nu| < 1, \quad \mathrm{Re}\,b > 2|\mathrm{Im}\,a|]$$
ET II 138(21)

**5.**
$$\int_0^\infty [K_\mu(ax)]^2\, J_\nu(bx)\, \mathrm{d}x = \frac{e^{2\mu\pi i}\, \Gamma\left(\dfrac{1 + \nu + 2\mu}{2}\right)}{b\, \Gamma\left(\dfrac{1 + \nu - 2\mu}{2}\right)} \left[Q_{\frac{1}{2}\nu - \frac{1}{2}}^{-\mu}\left(\sqrt{1 + \frac{4a^2}{b^2}}\right)\right]^2$$
$$\left[\mathrm{Re}\,a > 0, \quad b > 0, \quad \mathrm{Re}\left(\tfrac{1}{2}\nu \pm \mu\right) > -\tfrac{1}{2}\right] \quad \text{ET II 66(28)}$$

**6.**
$$\int_0^z J_\mu(x)\, J_\nu(z - x)\, \mathrm{d}x = 2 \sum_{k=0}^\infty (-1)^k\, J_{\mu+\nu+2k+1}(z) \qquad [\mathrm{Re}\,\mu > -1, \quad \mathrm{Re}\,\nu > -1] \qquad \text{WA 414(2)}$$

**7.**
$$\int_0^z J_\mu(x)\, J_{-\mu}(z - x)\, \mathrm{d}x = \sin z \qquad [-1 < \mathrm{Re}\,\mu < 1] \qquad \text{WA 415(4)}$$

**8.**
$$\int_0^z J_\mu(x)\, J_{1-\mu}(z - x)\, \mathrm{d}x = J_0(z) - \cos(z) \qquad [-1 < \mathrm{Re}\,\mu < 2] \qquad \text{WA 415(4)}$$

**9.**
$$\int_0^\infty J_0^2(ax)\, J_1(bx) = \frac{1}{b} \qquad\qquad [b > 2a > 0]$$
$$= \frac{2}{\pi b} \arcsin\left(\frac{b}{2a}\right) \qquad [2a > b > 0]$$

## 6.514

**1.**
$$\int_0^\infty J_\nu\left(\frac{a}{x}\right) J_\nu(bx)\, \mathrm{d}x = b^{-1} J_{2\nu}\left(2\sqrt{ab}\right) \qquad \left[a > 0, \quad b > 0, \quad \mathrm{Re}\,\nu > -\tfrac{1}{2}\right]$$
ET II 57(9)

**2.**
$$\int_0^\infty J_\nu\left(\frac{a}{x}\right) Y_\nu(bx)\, \mathrm{d}x = b^{-1}\left[Y_{2\nu}\left(2\sqrt{ab}\right) + \frac{2}{\pi} K_{2\nu}\left(\sqrt{2ab}\right)\right]$$
$$\left[a > 0, \quad b > 0, \quad -\tfrac{1}{2} < \mathrm{Re}\,\nu < \tfrac{3}{2}\right]$$
ET II 110(12)

**3.**
$$\int_0^\infty J_\nu\left(\frac{a}{x}\right) K_\nu(bx)\, \mathrm{d}x = b^{-1} e^{\frac{1}{2}i(\nu+1)\pi} K_{2\nu}\left[2e^{\frac{1}{4}i\pi}\sqrt{ab}\right] + b^{-1} e^{-\frac{1}{2}i(\nu+1)\pi} K_{2\nu}\left[2e^{-\frac{1}{4}\pi i}\sqrt{ab}\right]$$
$$\left[a > 0, \quad \mathrm{Re}\,b > 0, \quad |\mathrm{Re}\,\nu| < \tfrac{5}{2}\right]$$
ET II 141(31)

**4.**
$$\int_0^\infty Y_\nu\left(\frac{a}{x}\right) J_\nu(bx)\, \mathrm{d}x = -\frac{2b^{-1}}{\pi}\left[K_{2\nu}\left(2\sqrt{ab}\right) - \tfrac{\pi}{2} Y_{2\nu}\left(2\sqrt{ab}\right)\right]$$
$$\left[a > 0, \quad b > 0, \quad |\mathrm{Re}\,\nu| < \tfrac{1}{2}\right]$$
ET II 62(37)a

**5.**
$$\int_0^\infty Y_\nu\left(\frac{a}{x}\right) Y_\nu(bx)\, \mathrm{d}x = -b^{-1} J_{2\nu}\left(2\sqrt{ab}\right) \qquad \left[a > 0, \quad b > 0, \quad |\mathrm{Re}\,\nu| < \tfrac{1}{2}\right]$$
ET II 110(14)

6. $\int_0^\infty Y_\nu\left(\dfrac{a}{x}\right) K_\nu(bx)\,\mathrm{d}x = -b^{-1} e^{\frac{1}{2}\nu\pi i} K_{2\nu}\left(2 e^{\frac{1}{4}\pi i}\sqrt{ab}\right) - b^{-1} e^{-\frac{1}{2}\nu\pi i} K_{2\nu}\left(2 e^{-\frac{1}{4}\pi i}\sqrt{ab}\right)$

$$\left[a > 0, \quad \operatorname{Re} b > 0, \quad |\operatorname{Re}\nu| < \tfrac{5}{2}\right]$$
ET II 143(37)

7. $\int_0^\infty K_\nu\left(\dfrac{a}{x}\right) Y_\nu(bx)\,\mathrm{d}x = -2b^{-1}\left[\sin\left(\dfrac{3\nu\pi}{2}\right)\ker_{2\nu}\left(2\sqrt{ab}\right) + \cos\left(\dfrac{3\nu\pi}{2}\right)\kei_{2\nu}\left(2\sqrt{ab}\right)\right]$

$$\left[\operatorname{Re} a > 0, \quad b > 0, \quad |\operatorname{Re}\nu| < \tfrac{1}{2}\right]$$
ET II 113(28)

8. $\int_0^\infty K_\nu\left(\dfrac{a}{x}\right) K_\nu(bx)\,\mathrm{d}x = \pi b^{-1} K_{2\nu}\left(2\sqrt{ab}\right)$ $\qquad \left[\operatorname{Re} a > 0, \quad \operatorname{Re} b > 0\right]$ $\qquad$ ET II 146(54)

**6.515**

1. $\int_0^\infty J_\mu\left(\dfrac{a}{x}\right) Y_\mu\left(\dfrac{a}{x}\right) K_0(bx)\,\mathrm{d}x = -2b^{-1} J_{2\mu}\left(2\sqrt{ab}\right) K_{2\mu}\left(2\sqrt{ab}\right)$

$$\left[a > 0, \quad \operatorname{Re} b > 0\right]$$
ET II 143(42)

2. $\int_0^\infty \left[K_\mu\left(\dfrac{a}{x}\right)\right]^2 K_0(bx)\,\mathrm{d}x = 2\pi b^{-1} K_{2\mu}\left(2 e^{\frac{1}{4}\pi i}\sqrt{ab}\right) K_{2\mu}\left(2 e^{-\frac{1}{4}\pi i}\sqrt{ab}\right)$

$$\left[\operatorname{Re} a > 0, \quad \operatorname{Re} b > 0\right]$$
ET II 147(59)

3. $\int_0^\infty H_\mu^{(1)}\left(\dfrac{a^2}{x}\right) H_\mu^{(2)}\left(\dfrac{a^2}{x}\right) J_0(bx)\,\mathrm{d}x = 16\pi^{-2} b^{-1} \cos\mu\pi\, K_{2\mu}\left(2 e^{\pi i/4} a\sqrt{b}\right) K_{2\mu}\left(2 e^{-\pi i/4} a\sqrt{b}\right)$

$$\left[|\arg a| < \tfrac{\pi}{4}, \quad b > 0, \quad |\operatorname{Re}\mu| < \tfrac{1}{4}\right]$$
ET II 17(36)

**6.516**

1. $\int_0^\infty J_{2\nu}\left(a\sqrt{x}\right) J_\nu(bx)\,\mathrm{d}x = b^{-1} J_\nu\left(\dfrac{a^2}{4b}\right)$ $\qquad \left[a > 0, \quad b > 0, \quad \operatorname{Re}\nu > -\tfrac{1}{2}\right]$

ET II 58(16)

2. $\int_0^\infty J_{2\nu}\left(a\sqrt{x}\right) Y_\nu(bx)\,\mathrm{d}x = -b^{-1} \mathbf{H}_\nu\left(\dfrac{a^2}{4b}\right)$ $\qquad \left[a > 0, \quad b > 0, \quad \operatorname{Re}\nu > -\tfrac{1}{2}\right]$

ET II 111(18)

3. $\int_0^\infty J_{2\nu}\left(a\sqrt{x}\right) K_\nu(bx)\,\mathrm{d}x = \dfrac{\pi}{2} b^{-1}\left[I_\nu\left(\dfrac{a^2}{4b}\right) - \mathbf{L}_\nu\left(\dfrac{a^2}{4b}\right)\right]$

$$\left[\operatorname{Re} b > 0, \quad \operatorname{Re}\nu > -\tfrac{1}{2}\right] \qquad \text{ET II 144(45)}$$

4.[12] $\int_0^\infty Y_{2\nu}\left(a\sqrt{x}\right) J_\nu(bx)\,\mathrm{d}x = \dfrac{\sec(\pi\nu)}{2b}\left(2\cos(\pi\nu)Y_\nu\left(\dfrac{a^2}{4b}\right) - Y_{-\nu}\left(\dfrac{a^2}{4b}\right) + \mathbf{H}_{-\nu}\left(\dfrac{a^2}{4b}\right)\right)$

$$\left[a > 0, \quad b > 0\right] \qquad \text{MC}$$

5. $\int_0^\infty Y_{2\nu}\left(a\sqrt{x}\right) Y_\nu(bx)\,\mathrm{d}x$

$$= \dfrac{b^{-1}}{2}\left[\sec(\nu\pi) J_{-\nu}\left(\dfrac{a^2}{4b}\right) + \csc(\nu\pi) \mathbf{H}_{-\nu}\left(\dfrac{a^2}{4b}\right) - 2\cot(2\nu\pi) \mathbf{H}_\nu\left(\dfrac{a^2}{4b}\right)\right]$$
$$\left[a > 0, \quad b > 0, \quad |\operatorname{Re}\nu| < \tfrac{1}{2}\right] \quad \text{ET II 111(19)}$$

6.   $\displaystyle\int_0^\infty Y_{2\nu}\left(a\sqrt{x}\right) K_\nu(bx)\,\mathrm{d}x = \frac{\pi b^{-1}}{2}\left[\operatorname{cosec}(2\nu\pi)\,\mathbf{L}_{-\nu}\left(\frac{a^2}{4b}\right) - \cot(2\nu\pi)\,\mathbf{L}_\nu\left(\frac{a^2}{4b}\right)\right.$

$\displaystyle\left. - \tan(\nu\pi)\,I_\nu\left(\frac{a^2}{4b}\right) - \frac{\sec(\nu\pi)}{\pi}\,K_\nu\left(\frac{a^2}{4b}\right)\right]$

$$\left[\operatorname{Re} b > 0, \quad |\operatorname{Re}\nu| < \tfrac{1}{2}\right] \qquad \text{ET II 144(46)}$$

7.   $\displaystyle\int_0^\infty K_{2\nu}\left(a\sqrt{x}\right) J_\nu(bx)\,\mathrm{d}x = \frac{1}{4}\pi b^{-1}\sec(\nu\pi)\left[\mathbf{H}_{-\nu}\left(\frac{a^2}{4b}\right) - Y_{-\nu}\left(\frac{a^2}{4b}\right)\right]$

$$\left[\operatorname{Re} a > 0, \quad b > 0, \quad \operatorname{Re}\nu > -\tfrac{1}{2}\right]$$
$$\text{ET II 70(22)}$$

8.   $\displaystyle\int_0^\infty K_{2\nu}\left(a\sqrt{x}\right) Y_\nu(bx)\,\mathrm{d}x$

$\displaystyle= -\frac{1}{4}\pi b^{-1}\left[\sec(\nu\pi)\,J_{-\nu}\left(\frac{a^2}{4b}\right) - \operatorname{cosec}(\nu\pi)\,\mathbf{H}_{-\nu}\left(\frac{a^2}{4b}\right) + 2\operatorname{cosec}(2\nu\pi)\,\mathbf{H}_\nu\left(\frac{a^2}{4b}\right)\right]$

$$\left[\operatorname{Re} a > 0, \quad b > 0, \quad |\operatorname{Re}\nu| < \tfrac{1}{2}\right] \quad \text{ET II 114(34)}$$

9.   $\displaystyle\int_0^\infty K_{2\nu}\left(a\sqrt{x}\right) K_\nu(bx)\,\mathrm{d}x = \frac{\pi b^{-1}}{4\cos(\nu\pi)}\left\{K_\nu\left(\frac{a^2}{4b}\right) + \frac{\pi}{2\sin(\nu\pi)}\left[\mathbf{L}_{-\nu}\left(\frac{a^2}{4b}\right) - \mathbf{L}_\nu\left(\frac{a^2}{4b}\right)\right]\right\}$

$$\left[\operatorname{Re} b > 0, \quad |\operatorname{Re}\nu| < \tfrac{1}{2}\right] \qquad \text{ET II 147(63)}$$

10.   $\displaystyle\int_0^\infty I_{2\nu}\left(a\sqrt{x}\right) K_\nu(bx)\,\mathrm{d}x = \frac{\pi b^{-1}}{2}\left[I_\nu\left(\frac{a^2}{4b}\right) + \mathbf{L}_\nu\left(\frac{a^2}{4b}\right)\right]$

$$\left[\operatorname{Re} b > 0, \quad \operatorname{Re}\nu > -\tfrac{1}{2}\right] \qquad \text{ET II 147(60)}$$

**6.517**   $\displaystyle\int_0^z J_0\left(\sqrt{z^2 - x^2}\right)\,\mathrm{d}x = \sin z$ $\hfill$ MO 48

**6.518**[12]   $\displaystyle\int_0^\infty K_{2\nu}\left(2z\sinh x\right)\,\mathrm{d}x = \frac{\pi^2}{8\cos\nu\pi}\left(J_\nu^2(z) + Y_\nu^2(z)\right)$ $\quad\left[\operatorname{Re} z > 0, \quad -\tfrac{1}{2} < \operatorname{Re}\nu < \tfrac{1}{2}\right]$ $\hfill$ MO 45

**6.519**

1.   $\displaystyle\int_0^{\pi/2} J_{2\nu}\left(2z\cos x\right)\,\mathrm{d}x = \frac{\pi}{2}J_\nu^2(z)$ $\hfill\left[\operatorname{Re}\nu > -\tfrac{1}{2}\right]\hfill$ WH

2.   $\displaystyle\int_0^{\pi/2} J_{2\nu}\left(2z\sin x\right)\,\mathrm{d}x = \frac{\pi}{2}J_\nu^2(z)$ $\hfill\left[\operatorname{Re}\nu > -\tfrac{1}{2}\right]\hfill$ WA 42(1)a

## 6.52 Bessel functions combined with $x$ and $x^2$

**6.521**

1.   $\displaystyle\int_0^1 x J_\nu(\alpha x) J_\nu(\beta x)\,\mathrm{d}x = \frac{\beta J_{\nu-1}(\beta) J_\nu(\alpha) - \alpha J_{\nu-1}(\alpha) J_\nu(\beta)}{\alpha^2 - \beta^2}$ $\hfill\left[\alpha \neq \beta, \quad \nu > -1\right]$

$\displaystyle= \frac{\alpha J_\nu(\beta) J_\nu'(\alpha) - \beta J_\nu(\alpha) J_\nu'(\beta)}{\beta^2 - \alpha^2}$ $\hfill\left[\alpha \neq \beta, \quad \nu > -1\right]$

$\hfill$ WH

2.$^{10}$ $\displaystyle\int_0^\infty x\, K_\nu(ax)\, J_\nu(bx)\, dx = \frac{b^\nu}{a^\nu\,(b^2+a^2)}$ $\qquad$ [$\operatorname{Re} a > 0, \quad b > 0, \quad \operatorname{Re}\nu > -1$]

$\hfill$ ET II 63(2)

3. $\displaystyle\int_0^\infty x\, K_\nu(ax)\, K_\nu(bx)\, dx = \frac{\pi(ab)^{-\nu}\left(a^{2\nu}-b^{2\nu}\right)}{2\sin(\nu\pi)\left(a^2-b^2\right)}$ $\qquad$ [$|\operatorname{Re}\nu| < 1, \quad \operatorname{Re}(a+b) > 0$]

$\hfill$ ET II 145(48)

4. $\displaystyle\int_0^a x\, J_\nu(\lambda x)\, K_\nu(\mu x)\, dx = \left(\mu^2+\lambda^2\right)^{-1}\left[\left(\frac{\lambda}{\mu}\right)^\nu + \lambda a\, J_{\nu+1}(\lambda a)\, K_\nu(\mu a) - \mu a\, J_\nu(\lambda a)\, K_{\nu+1}(\mu a)\right]$

$\hfill$ [$\operatorname{Re}\nu > -1$] $\qquad$ ET II 367(26)

5. $\displaystyle\int_0^\infty x\, K_1(ax) = \frac{\pi}{2a^2}$ $\qquad$ [$a > 0$]

6. $\displaystyle\int_0^\infty x\, K_0^2(ax) = \frac{1}{2a^2}$ $\qquad$ [$a > 0$]

7. $\displaystyle\int_0^\infty x\, K_1(ax)\, J_1(bx) = \frac{b}{a\left(a^2+b^2\right)}$ $\qquad$ [$a > 0, \quad b > 0$]

8. $\displaystyle\int_0^\infty x\, K_0(ax)\, I_0(bx) = \frac{1}{a^2-b^2}$ $\qquad$ [$a > b > 0$]

9. $\displaystyle\int_0^\infty x\, K_1(ax)\, I_1(bx) = \frac{b}{a\left(a^2-b^2\right)}$ $\qquad$ [$a > b > 0$]

10. $\displaystyle\int_0^\infty x^2\, K_0(ax) = \frac{\pi}{2a^3}$ $\qquad$ [$a > 0$]

11. $\displaystyle\int_0^\infty x^2\, K_1(ax) - \frac{2}{a^3}$ $\qquad$ [$a > 0$]

12. $\displaystyle\int_0^\infty x^2\, K_0(ax)\, J_1(bx) = \frac{2b}{\left(a^2+b^2\right)^2}$ $\qquad$ [$a > 0, \quad b > 0$]

13. $\displaystyle\int_0^\infty x^2\, K_1(ax)\, J_0(bx) = \frac{2a}{\left(a^2+b^2\right)^2}$ $\qquad$ [$a > b > 0$]

14. $\displaystyle\int_0^\infty x^2\, K_0(ax)\, I_1(bx) = \frac{2b}{\left(a^2-b^2\right)^2}$ $\qquad$ [$a > b > 0$]

15. $\displaystyle\int_0^\infty x^2\, K_1(ax)\, I_0(bx) = \frac{2a}{\left(a^2-b^2\right)^2}$ $\qquad$ [$a > b > 0$]

**6.522** **Notation:** $\ell_1 = \dfrac{1}{2}\left[\sqrt{(b+c)^2+a^2} - \sqrt{(b-c)^2+a^2}\right], \ell_2 = \dfrac{1}{2}\left[\sqrt{(b+c)^2+a^2} + \sqrt{(b-c)^2+a^2}\right]$

1.$^{12}$ $\displaystyle\int_0^\infty x\,[J_\mu(ax)]^2\, K_\nu(bx)\, dx = \Gamma\left(\mu+\tfrac{1}{2}\nu+1\right)\Gamma\left(\mu-\tfrac{1}{2}\nu+1\right) b^{-2}$

$\hfill \times \left(1+4a^2b^{-2}\right)^{-\frac{1}{2}} P_{\frac{1}{2}\nu}^{-\mu}\left[\left(1+4a^2b^{-2}\right)^{\frac{1}{2}}\right] P_{\frac{1}{2}\nu-1}^{-\mu}\left[\left(1+4a^2b^{-2}\right)^{\frac{1}{2}}\right]$

$\hfill$ [$\operatorname{Re} b > 2|\operatorname{Im} a|, \quad 2\operatorname{Re}\mu > |\operatorname{Re}\nu| - 2$] $\qquad$ ET II 138(19)

**2.**
$$\int_0^\infty x\left[K_\mu(ax)\right]^2 J_\nu(bx)\,dx = \frac{2e^{2\mu\pi i}\,\Gamma\left(1+\tfrac12\nu+\mu\right)}{b\left(4a^2+b^2\right)^{\frac12}\Gamma\left(\tfrac12\nu-\mu\right)}$$
$$\times Q_{\frac12\nu}^{-\mu}\left(\sqrt{1+4a^2b^{-2}}\right)Q_{\frac12\nu-1}^{-\mu}\left(\sqrt{1+4a^2b^{-2}}\right)$$
$$\left[b>0,\quad \operatorname{Re}a>0,\quad \operatorname{Re}\left(\tfrac12\nu\pm\mu\right)>-1\right]\quad \text{ET II 66(27)a}$$

**3.**[12]
$$\int_0^\infty x\,K_0(ax)J_\nu(bx)J_\nu(cx)\,dx = r_1^{-1}r_2^{-1}(r_2-r_1)^\nu(r_2+r_1)^{-\nu} = \frac{\ell_1^\nu}{\ell_2^\nu\left(\ell_2^2-\ell_1^2\right)},$$
$$\left[r_1=\sqrt{a^2+(b-c)^2},\quad r_2=\sqrt{a^2+(b+c)^2},\quad c>0,\quad \operatorname{Re}\nu>-1,\quad \operatorname{Re}a>|\operatorname{Im}b|\right]$$
$$\text{ET II 63(6)}$$

**4.**[10]
$$\int_0^\infty x\,I_0(ax)K_0(bx)J_0(cx)\,dx = \left(a^4+b^4+c^4-2a^2b^2+2a^2c^2+2b^2c^2\right)^{-\frac12}$$
$$\left[\operatorname{Re}b>\operatorname{Re}a,\quad c>0\right]\quad \text{ET II 16(27)}$$

alternatively, with $a$ and $c$ interchanged
$$\int_0^\infty x\,I_0(cx)K_0(bx)J_0(ax)\,dx = \frac{1}{\ell_2^2-\ell_1^2}\quad \left[\operatorname{Re}b>\operatorname{Re}c,\quad a>0\right]$$

**5.**[10]
$$\int_0^\infty x\,J_0(ax)K_0(bx)J_0(cx)\,dx = \left(a^4+b^4+c^4-2a^2c^2+2a^2b^2+2b^2c^2\right)^{-\frac12}$$
$$\left[\operatorname{Re}b>|\operatorname{Im}a|,\quad c>0\right]\quad \text{ET II 15(25)}$$

alternatively, with $a$ and $b$ interchanged
$$\int_0^\infty x\,J_0(bx)K_0(ax)J_0(cx)\,dx = \frac{1}{\ell_2^2-\ell_1^2}\quad \left[\operatorname{Re}a>|\operatorname{Im}b|,\quad c>0\right]$$

**6.**
$$\int_0^\infty x\,J_0(ax)Y_0(ax)J_0(bx)\,dx = 0\quad \left[0<b<2a\right]$$
$$= -2\pi^{-1}b^{-1}\left[b^2-4a^2\right]^{-\frac12}\quad \left[0<2a<b<\infty\right]$$
$$\text{ET II 15(21)}$$

**7.**[12]
$$\int_0^\infty x\,J_\mu(ax)J_{\mu+1}(ax)K_\nu(bx)\,dx = \Gamma\left(\mu+\frac{3+\nu}{2}\right)\Gamma\left(\mu+\frac{3-\nu}{2}\right)b^{-2}\left(1+4a^2b^{-2}\right)^{-\frac12}$$
$$\times P_{\frac12\nu-\frac12}^{-\mu}\left[\sqrt{1+4a^2b^{-2}}\right]P_{\frac12\nu-\frac12}^{-\mu-1}\left[\sqrt{1+4a^2b^{-2}}\right]$$
$$\left[\operatorname{Re}b>2|\operatorname{Im}a|,\quad 2\operatorname{Re}\mu>|\operatorname{Re}\nu|-3\right]\quad \text{ET II 138(20)}$$

**8.**
$$\int_0^\infty x\,K_{\mu-\frac12}(ax)K_{\mu+\frac12}(ax)J_\nu(bx)\,dx$$
$$= -\frac{2e^{2\mu\pi i}\,\Gamma\left(\tfrac12\nu+\mu+1\right)}{b\,\Gamma\left(\tfrac12\nu-\mu\right)\left(b^2+4a^2\right)^{\frac12}}Q_{\frac12\nu-\frac12}^{-\mu+\frac12}\left[\left(1+4a^2b^{-2}\right)^{\frac12}\right]Q_{\frac12\nu-\frac12}^{-\mu-\frac12}\left[\left(1+4a^2b^{-2}\right)^{\frac12}\right]$$
$$\left[b>0,\quad \operatorname{Re}a>0,\quad \operatorname{Re}\nu>-1,\quad |\operatorname{Re}\mu|<1+\tfrac12\operatorname{Re}\nu\right]\quad \text{ET II 67(29)a}$$

**9.**[8]
$$\int_0^\infty x\,I_{\frac12\nu}(ax)K_{\frac12\nu}(ax)J_\nu(bx)\,dx = b^{-1}\left(b^2+4a^2\right)^{-\frac12}$$
$$\left[b>0,\quad \operatorname{Re}a>0,\quad \operatorname{Re}\nu>-1\right]$$
$$\text{ET II 65(16)}$$

10. $\displaystyle\int_0^\infty x\, J_{\frac{1}{2}\nu}(ax)\, Y_{\frac{1}{2}\nu}(ax)\, J_\nu(bx)\, dx$

$$= 0 \qquad [a > 0, \quad \mathrm{Re}\,\nu > -1, \quad 0 < b < 2a]$$

$$= -2\pi^{-1}b^{-1}\left(b^2 - 4a^2\right)^{-\frac{1}{2}} \qquad [a > 0, \quad \mathrm{Re}\,\nu > -1, \quad 2a < b < \infty]$$

ET II 55(48)

11.[8] $\displaystyle\int_0^\infty x\, J_{\frac{1}{2}(\nu+n)}(ax)\, J_{\frac{1}{2}(\nu-n)}(ax)\, J_\nu(bx)\, dx$

$$= 2\pi^{-1}b^{-1}\left(4a^2 - b^2\right)^{-\frac{1}{2}} T_n\left(\frac{b}{2a}\right) \qquad [a > 0, \quad \mathrm{Re}\,\nu > -1, \quad 0 < b < 2a]$$

$$= 0 \qquad [a > 0, \quad \mathrm{Re}\,\nu > -1, \quad 2a < b]$$

ET II 52(32)

12. $\displaystyle\int_0^\infty x\, I_{\frac{1}{2}(\nu-\mu)}(ax)\, K_{\frac{1}{2}(\nu+\mu)}(ax)\, J_\nu(bx)\, dx = 2^{-\mu}a^{-\mu}b^{-1}\left(b^2 + 4a^2\right)^{-\frac{1}{2}}\left[b + \left(b^2 + 4a^2\right)^{\frac{1}{2}}\right]^\mu$

$$[b > 0, \quad \mathrm{Re}\,a > 0, \quad \mathrm{Re}\,\nu > -1, \quad \mathrm{Re}(\nu - \mu) > -2] \quad \text{ET II 66(23)}$$

13.[8] $\displaystyle\int_0^\infty x\, J_\mu\left(xa\sin\varphi\right) K_{\nu-\mu}\left(ax\cos\varphi\cos\psi\right) J_\nu\left(xa\sin\psi\right) dx = \frac{(\sin\varphi)^\mu\,(\sin\psi)^\nu\,(\cos\varphi)^{\nu-\mu}\,(\cos\psi)^{\mu-\nu}}{a^2\left(1 - \sin^2\varphi\sin^2\psi\right)}$

$$\left[a > 0, \quad 0 < \varphi < \frac{\pi}{2}, \quad 0 < \psi < \frac{\pi}{2}, \quad \mathrm{Re}\,\mu > -1, \quad \mathrm{Re}\,\nu > -1\right] \quad \text{ET II 64(10)}$$

14.[8] $\displaystyle\int_0^\infty x\, J_\mu\left(xa\sin\varphi\cos\psi\right) J_{\nu-\mu}(ax)\, J_\nu\left(xa\cos\varphi\sin\psi\right) dx$

$$= -2\pi^{-1}a^{-2}\sin(\mu\pi)\,(\sin\varphi)^\mu\,(\sin\psi)^\nu\,(\cos\varphi)^{-\nu}\,(\cos\psi)^{-\mu}\left[\cos(\varphi+\psi)\cos(\varphi-\psi)\right]^{-1}$$

$$\left[a > 0, \quad 0 < \varphi < \frac{\pi}{2}, \quad 0 < \psi < \tfrac{1}{2}\pi, \quad \mathrm{Re}\,\nu > -1\right] \quad \text{ET II 54(39)}$$

15.[10] $\displaystyle\int_0^\infty x^{\nu+1}\, J_\nu(bx)\, K_\nu(ax)\, J_\nu(cx)\, dx = \frac{2^{3\nu}(abc)^\nu\,\Gamma\left(\nu + \frac{1}{2}\right)}{\sqrt{\pi}\left(\ell_2^2 - \ell_1^2\right)^{2\nu+1}}$

$$[\mathrm{Re}\,a > |\mathrm{Im}\,b|, \quad c > 0]$$

16.[10] $\displaystyle\int_0^\infty x^{\nu+1}\, I_\nu(cx)\, K_\nu(bx)\, J_\nu(ax)\, dx = \frac{2^{3\nu}(abc)^\nu\,\Gamma\left(\nu + \frac{1}{2}\right)}{\sqrt{\pi}\left(\ell_2^2 - \ell_1^2\right)^{2\nu+1}}$

$$[\mathrm{Re}\,b > |\mathrm{Im}\,a| + |\mathrm{Im}\,c|]$$

17.[11] $\displaystyle\int_0^\infty t^{\nu-\mu-\rho+1}\, J_\mu(ct)\, J_\nu(bt)\, K_\rho(at)\, dt$

$$= \frac{2^{1+\nu-\mu-\rho}}{c^\mu b^\nu a^\rho\,\Gamma\left(\mu - \nu + \rho\right)}\int_0^{\ell_1}\frac{x^{1+2\nu-2\rho}\left[\left(\ell_1^2 - x^2\right)\left(\ell_2^2 - x^2\right)\right]^{\mu-\nu+\rho-1}}{\left(b^2 - x^2\right)^{\mu-\nu}}\, dx$$

$$\ell_1 = \frac{1}{2}\left[\sqrt{(b+c)^2 + a^2} - \sqrt{(b-c)^2 + a^2}\right], \quad \ell_2 = \frac{1}{2}\left[\sqrt{(b+c)^2 + a^2} + \sqrt{(b-c)^2 + a^2}\right]$$

$$[\mathrm{Re}\,a > |\mathrm{Im}\,b|, \quad c > 0]$$

18.$^{11}$ $\displaystyle\int_0^\infty t^{\mu-\nu+\rho+1}\, J_\mu(ct)\, J_\nu(bt)\, K_\rho(at)\,\mathrm{d}t$

$$= \frac{2^{1+\mu-\nu+\rho}a^\rho}{c^\mu b^\nu\, \Gamma\left(\nu-\mu-\rho\right)}\int_0^{\ell_1}\frac{x^{1+2\mu+2\rho}\left[\left(\ell_1^2-x^2\right)\left(\ell_2^2-x^2\right)\right]^{\nu-\mu-\rho-1}}{\left(c^2-x^2\right)^{\nu-\mu}}\,\mathrm{d}x$$

$$\ell_1=\frac{1}{2}\left[\sqrt{(b+c)^2+a^2}-\sqrt{(b-c)^2+a^2}\right],\quad \ell_2=\frac{1}{2}\left[\sqrt{(b+c)^2+a^2}+\sqrt{(b-c)^2+a^2}\right]$$

$$[\operatorname{Re}a>|\operatorname{Im}b|,\quad c>0]$$

**6.523** $\displaystyle\int_0^\infty x\left[2\pi^{-1}K_0(ax)-Y_0(ax)\right]K_0(bx)\,\mathrm{d}x = 2\pi^{-1}\left[\left(a^2+b^2\right)^{-1}+\left(b^2-a^2\right)^{-1}\right]\ln\frac{b}{a}$

$$[\operatorname{Re}b>|\operatorname{Im}a|,\quad \operatorname{Re}(a+b)>0]$$

ET II 145(50)

**6.524**

1.  $\displaystyle\int_0^\infty x\, J_\nu^2(ax)\, J_\nu(bx)\, Y_\nu(bx)\,\mathrm{d}x = 0$ $\qquad\left[0<a<b,\quad \operatorname{Re}\nu>-\tfrac{1}{2}\right]$

$$= -(2\pi ab)^{-1}\qquad \left[0<b<a,\quad \operatorname{Re}\nu>-\tfrac{1}{2}\right]$$

ET II 352(14)

2.  $\displaystyle\int_0^\infty x\left[J_0(ax)\, K_0(bx)\right]^2\,\mathrm{d}x = \frac{\pi}{8ab}-\frac{1}{4ab}\arcsin\left(\frac{b^2-a^2}{b^2+a^2}\right)$

$$[a>0,\quad b>0]\qquad \text{ET II 373(9)}$$

**6.525  Notation:** $\ell_1=\dfrac{1}{2}\left[\sqrt{(b+c)^2+a^2}-\sqrt{(b-c)^2+a^2}\right],\ell_2=\dfrac{1}{2}\left[\sqrt{(b+c)^2+a^2}+\sqrt{(b-c)^2+a^2}\right]$

1.$^{10}$ $\displaystyle\int_0^\infty x^2\, J_1(ax)\, K_0(bx)\, J_0(cx)\,\mathrm{d}x = 2a\left(a^2+b^2-c^2\right)\left[\left(a^2+b^2+c^2\right)^2-4a^2c^2\right]^{-\frac{3}{2}}$

$$[c>0,\quad \operatorname{Re}b\ge|\operatorname{Im}a|,\quad \operatorname{Re}a>0]$$

ET II 15(26)

alternatively, with $a$ and $b$ interchanged

$$\int_0^\infty x^2\, J_1(bx)\, K_0(ax)\, J_0(cx)\,\mathrm{d}x = \frac{2b\left(a^2+b^2-c^2\right)}{\left(\ell_2^2-\ell_1^2\right)^3}\qquad [\operatorname{Re}a>|\operatorname{Im}b|,\quad \operatorname{Re}b>0,\quad c>0]$$

2.$^{10}$ $\displaystyle\int_0^\infty x^2\, I_0(ax)\, K_1(bx)\, J_0(cx)\,\mathrm{d}x = 2b\left(b^2+c^2-a^2\right)\left[\left(a^2+b^2+c^2\right)^2-4a^2b^2\right]^{-\frac{3}{2}}$

$$[\operatorname{Re}b>|\operatorname{Re}a|,\quad c>0]\qquad \text{ET II 16(28)}$$

3.$^{10}$ $\displaystyle\int_0^\infty x^2\, I_0(cx)\, K_0(bx)\, J_0(ax)\,\mathrm{d}x = \frac{2b\left(a^2+b^2-c^2\right)}{\left(\ell_2^2-\ell_1^2\right)^3}\qquad [\operatorname{Re}a>|\operatorname{Im}b|,\quad c>0]$

**6.526**

1.  $\displaystyle\int_0^\infty x\, J_{\frac{1}{2}\nu}\left(ax^2\right)\, J_\nu(bx)\,\mathrm{d}x = (2a)^{-1}\, J_{\frac{1}{2}\nu}\left(\frac{b^2}{4a}\right)$

$$[a>0,\quad b>0,\quad \operatorname{Re}\nu>-1]\qquad \text{ET II 56(1)}$$

2.    $\displaystyle\int_0^\infty x\,J_{\frac{1}{2}\nu}\left(ax^2\right)Y_\nu(bx)\,\mathrm{d}x$

$$= (4a)^{-1}\left[Y_{\frac{1}{2}\nu}\left(\frac{b^2}{4a}\right) - \tan\left(\frac{\nu\pi}{2}\right)J_{\frac{1}{2}\nu}\left(\frac{b^2}{4a}\right) + \sec\left(\frac{\nu\pi}{2}\right)\mathbf{H}_{-\frac{1}{2}\nu}\left(\frac{b^2}{4a}\right)\right]$$
$$[a>0, \quad b>0, \quad \mathrm{Re}\,\nu>-1] \quad \text{ET II 109(9)}$$

3.    $\displaystyle\int_0^\infty x\,J_{\frac{1}{2}\nu}\left(ax^2\right)K_\nu(bx)\,\mathrm{d}x = \frac{\pi}{8a\cos\left(\dfrac{\nu\pi}{2}\right)}\left[\mathbf{H}_{-\frac{1}{2}\nu}\left(\frac{b^2}{4a}\right) - Y_{-\frac{1}{2}\nu}\left(\frac{b^2}{4a}\right)\right]$

$$[a>0, \quad \mathrm{Re}\,b>0, \quad \mathrm{Re}\,\nu>-1]$$
$$\text{ET II 140(27)}$$

4.    $\displaystyle\int_0^\infty x\,Y_{\frac{1}{2}\nu}\left(ax^2\right)J_\nu(bx)\,\mathrm{d}x = -(2a)^{-1}\,\mathbf{H}_{\frac{1}{2}\nu}\left(\frac{b^2}{4a}\right)$      $[a>0, \quad \mathrm{Re}\,b>0, \quad \mathrm{Re}\,\nu>-1]$

$$\text{ET II 61(35)}$$

5.    $\displaystyle\int_0^\infty x\,Y_{\frac{1}{2}\nu}\left(ax^2\right)K_\nu(bx)\,\mathrm{d}x$

$$= \frac{\pi}{4a\sin(\nu\pi)}\left[\cos\left(\frac{\nu\pi}{2}\right)\mathbf{H}_{-\frac{1}{2}\nu}\left(\frac{b^2}{4a}\right) - \sin\left(\frac{\nu\pi}{2}\right)J_{-\frac{1}{2}\nu}\left(\frac{b^2}{4a}\right) - \mathbf{H}_{\frac{1}{2}\nu}\left(\frac{b^2}{4a}\right)\right]$$
$$[a>0, \quad \mathrm{Re}\,b>0, \quad |\mathrm{Re}\,\nu|<1] \quad \text{ET II 141(28)}$$

6.    $\displaystyle\int_0^\infty x\,K_{\frac{1}{2}\nu}\left(ax^2\right)J_\nu(bx)\,\mathrm{d}x = \frac{\pi}{4a}\left[I_{\frac{1}{2}\nu}\left(\frac{b^2}{4a}\right) - \mathbf{L}_{\frac{1}{2}\nu}\left(\frac{b^2}{4a}\right)\right]$

$$[\mathrm{Re}\,a>0, \quad b>0, \quad \mathrm{Re}\,\nu>-1]$$
$$\text{ET II 68(9)}$$

7.    $\displaystyle\int_0^\infty x\,K_{\frac{1}{2}\nu}\left(ax^2\right)Y_\nu(bx)\,\mathrm{d}x = \frac{\pi}{4a}\left[\operatorname{cosec}(\nu\pi)\,\mathbf{L}_{-\frac{1}{2}\nu}\left(\frac{b^2}{4a}\right) - \cot(\nu\pi)\,\mathbf{L}_{\frac{1}{2}\nu}\left(\frac{b^2}{4a}\right)\right.$

$$\left. - \tan\left(\frac{\nu\pi}{2}\right)I_{\frac{1}{2}\nu}\left(\frac{b^2}{4a}\right) - \frac{1}{\pi}\sec\left(\frac{\nu\pi}{2}\right)K_{\frac{1}{2}\nu}\left(\frac{b^2}{4a}\right)\right]$$
$$[\mathrm{Re}\,a>0, \quad b>0, \quad |\mathrm{Re}\,\nu|<1] \quad \text{ET II 112(25)}$$

8.    $\displaystyle\int_0^\infty x\,K_{\frac{1}{2}\nu}\left(ax^2\right)K_\nu(bx)\,\mathrm{d}x$

$$= \frac{\pi}{8a}\left\{\sec\left(\frac{\nu\pi}{2}\right)K_{\frac{1}{2}\nu}\left(\frac{b^2}{4a}\right) + \pi\operatorname{cosec}(\nu\pi)\left[\mathbf{L}_{-\frac{1}{2}\nu}\left(\frac{b^2}{4a}\right) - \mathbf{L}_{\frac{1}{2}\nu}\left(\frac{b^2}{4a}\right)\right]\right\}$$
$$[\mathrm{Re}\,a>0, \quad |\mathrm{Re}\,\nu|<1] \quad\quad \text{ET II 146(52)}$$

**6.527**

1.    $\displaystyle\int_0^\infty x^2\,J_{2\nu}(2ax)\,J_{\nu-\frac{1}{2}}\left(x^2\right)\,\mathrm{d}x = \frac{1}{2}a\,J_{\nu+\frac{1}{2}}\left(a^2\right)$      $\left[a>0, \quad \mathrm{Re}\,\nu>-\frac{1}{2}\right]$      ET II 355(33)

2.    $\displaystyle\int_0^\infty x^2\,J_{2\nu}(2ax)\,J_{\nu+\frac{1}{2}}\left(x^2\right)\,\mathrm{d}x = \frac{1}{2}a\,J_{\nu-\frac{1}{2}}\left(a^2\right)$      $\left[a>0, \quad \mathrm{Re}\,\nu>-2\right]$      ET II 355(35)

3.    $\displaystyle\int_0^\infty x^2\,J_{2\nu}(2ax)\,Y_{\nu+\frac{1}{2}}\left(x^2\right)\,\mathrm{d}x = -\frac{1}{2}a\,\mathbf{H}_{\nu-\frac{1}{2}}\left(a^2\right)$      $\left[a>0, \quad \mathrm{Re}\,\nu>-2\right]$      ET II 355(36)

**6.528**
$$\int_0^\infty x\, K_{\frac{1}{4}\nu}\left(\frac{x^2}{4}\right) I_{\frac{1}{4}\nu}\left(\frac{x^2}{4}\right) J_\nu(bx)\, dx = K_{\frac{1}{4}\nu}\left(\frac{b^2}{4}\right) I_{\frac{1}{4}\nu}\left(\frac{b^2}{4}\right)$$

$$[b > 0, \quad \nu > -1]$$                          MO 183a

**6.529**

1.
$$\int_0^\infty x\, J_\nu\left(2\sqrt{ax}\right) K_\nu\left(2\sqrt{ax}\right) J_\nu(bx)\, dx = \frac{1}{2}b^{-2}e^{-\frac{2a}{b}} \qquad [\operatorname{Re} a > 0, \quad b > 0, \quad \operatorname{Re}\nu > -1]$$

ET II 70(23)

2.
$$\int_0^a x\, J_\lambda(2x)\, I_\lambda(2x)\, J_\mu\left(2\sqrt{a^2-x^2}\right) I_\mu\left(2\sqrt{a^2-x^2}\right) dx$$
$$= \frac{a^{2\lambda+2\mu+2}}{2\,\Gamma(\lambda+1)\,\Gamma(\mu+1)\,\Gamma(\lambda+\mu+2)}$$
$$\times\; {}_1F_4\left(\frac{\lambda+\mu+1}{2};\lambda+1,\mu+1,\lambda+\mu+1,\frac{\lambda+\mu+3}{2};-a^4\right)$$
$$[\operatorname{Re}\lambda > -1, \quad \operatorname{Re}\mu > -1] \quad \text{ET II 376(31)}$$

## 6.53–6.54 Combinations of Bessel functions and rational functions

**6.531**

1.[12]
$$\int_0^\infty \frac{Y_\nu(bx)}{x+a}\, dx = \frac{\pi}{\sin(\pi\nu)}\left[\mathbf{E}_\nu(ab) + Y_\nu(ab) + 2\cot(\pi\nu)(\mathbf{J}_\nu(ab) - J_\nu(ab))\right]$$

$$[\operatorname{Re}\nu < 1, \quad \arg a \neq \pi, \quad b > 0] \qquad \text{MC}$$

2.
$$\int_0^\infty \frac{Y_\nu(bx)}{x-a}\, dx = \pi\left\{\cot(\nu\pi)\left[Y_\nu(ab) + \mathbf{E}_\nu(ab)\right] + \mathbf{J}_\nu(ab) + 2\left[\cot(\nu\pi)\right]^2\left[\mathbf{J}_\nu(ab) - J_\nu(ab)\right]\right\}$$

$$[b > 0, \quad a > 0, \quad |\operatorname{Re}\nu| < 1]$$
ET II 98(9)

3.
$$\int_0^\infty \frac{K_\nu(bx)}{x+a}\, dx = \frac{\pi^2}{2}\left[\operatorname{cosec}(\nu\pi)\right]^2\left[I_\nu(ab) + I_{-\nu}(ab) - e^{-\frac{1}{2}i\nu\pi}\mathbf{J}_\nu(iab) - e^{\frac{1}{2}i\nu\pi}\mathbf{J}_{-\nu}(iab)\right]$$

$$[\operatorname{Re} b > 0, \quad |\arg a| < \pi, \quad |\operatorname{Re}\nu| < 1]$$
ET II 128(5)

4.*
$$\int_0^\infty \frac{J_\nu(bx)}{x+a}\, dx = \frac{\pi}{\sin(\pi\nu)}\left(\mathbf{J}_\nu(ab) - J_\nu(ab)\right) \qquad [b > 0, \quad |\arg(a)| < \pi, \quad \operatorname{Re}(\nu) > -1]$$

5.*
$$\int_0^\infty \frac{J_\nu(bx)}{x-a}\, dx = \frac{\pi}{\tan(\pi\nu)}\left(\mathbf{J}_\nu(ab) - J_\nu(ab)\right) + \mathbf{E}_\nu(ab)$$

$$[b > 0, \quad a > 0, \quad \operatorname{Re}(\nu) > -1]$$

**6.532**

1.[12]
$$\int_0^\infty \frac{J_\nu(x)}{x^2+a^2}\, dx = \frac{\pi I_\nu(a)}{2a\cos\left(\frac{\pi\nu}{2}\right)} + \frac{1}{\nu^2-1}\,{}_1F_2\left(1;\frac{3-\nu}{2},\frac{3+\nu}{2};\frac{a^2}{4}\right)$$

$$[\operatorname{Re} a > 0, \quad \operatorname{Re}\nu > -1]$$

1(b).* $\mathrm{PV} \displaystyle\int_0^\infty \frac{J_\nu(x)}{x^2 - a^2}\, \mathrm{d}x = \frac{\pi}{2a} \tan\left(\frac{\pi\nu}{2}\right) (\mathbf{J}_\nu(a) - J_\nu(a)) + \mathbf{E}_\nu(a)$

$$[\operatorname{Re} a > 0, \quad \operatorname{Re}\nu > -1]$$

2.[12] $\displaystyle\int_0^\infty \frac{Y_\nu(bx)}{x^2 + a^2}\, \mathrm{d}x = \frac{1}{\cos\dfrac{\nu\pi}{2}} \left[ -\frac{\pi}{2a} \tan\left(\frac{\nu\pi}{2}\right) I_\nu(ab) - \frac{1}{a} K_\nu(ab) \right.$

$$\left. + \frac{b\sin\left(\dfrac{\nu\pi}{2}\right)}{1 - \nu^2}\ {}_1F_2\left(1; \frac{3-\nu}{2}, \frac{3+\nu}{2}; \frac{a^2 b^2}{4}\right) \right]$$

$$[b > 0, \quad \operatorname{Re} a > 0, \quad |\operatorname{Re}\nu| < 1] \quad \text{ET II 99(13)}$$

3. $\displaystyle\int_0^\infty \frac{Y_\nu(bx)}{x^2 - a^2}\, \mathrm{d}x = \frac{\pi}{2a}\left\{ J_\nu(ab) + \tan\left(\frac{\nu\pi}{2}\right)\left[ \tan\left(\frac{\nu\pi}{2}\right)[\mathbf{J}_\nu(ab) - J_\nu(ab)] - \mathbf{E}_\nu(ab) - Y_\nu(ab)\right]\right\}$

$$[b > 0, \quad a > 0, \quad |\operatorname{Re}\nu| < 1]$$
$$\text{ET II 101(21)}$$

4. $\displaystyle\int_0^\infty \frac{x\,J_0(ax)}{x^2 + k^2}\, \mathrm{d}x = K_0(ak)$ $\qquad [a > 0, \quad \operatorname{Re} k > 0]$ $\qquad$ WA 466(5)

5. $\displaystyle\int_0^\infty \frac{Y_0(ax)}{x^2 + k^2}\, \mathrm{d}x = -\frac{K_0(ak)}{k}$ $\qquad [a > 0, \quad \operatorname{Re} k > 0]$ $\qquad$ WA 466(6)

6. $\displaystyle\int_0^\infty \frac{J_0(ax)}{x^2 + k^2}\, \mathrm{d}x = \frac{\pi}{2k}\left[I_0(ak) - \mathbf{L}_0(ak)\right]$ $\qquad [a > 0, \quad \operatorname{Re} k > 0]$ $\qquad$ WA 467(7)

**6.533**

1. $\displaystyle\int_0^z J_p(x)\,J_q(z - x)\,\frac{\mathrm{d}x}{x} = \frac{J_{p+q}(z)}{p}$ $\qquad [\operatorname{Re} p > 0, \quad \operatorname{Re} q > -1]$ $\qquad$ WA 415(3)

2. $\displaystyle\int_0^z \frac{J_p(x)}{x}\frac{J_q(z - x)}{z - x}\, \mathrm{d}x = \left(\frac{1}{p} + \frac{1}{q}\right)\frac{J_{p+q}(z)}{z}$ $\qquad [\operatorname{Re} p > 0, \quad \operatorname{Re} q > 0]$ $\qquad$ WA 415(5)

3.[11] $\displaystyle\int_0^\infty [J_0(ax) - 1]\,J_1(bx)\frac{\mathrm{d}x}{x^2} = -\frac{b}{4}\left[1 + 2\ln\frac{a}{b}\right]$ $\qquad [0 < b < a]$

$$= -\frac{a^2}{4b} \qquad\qquad\qquad\quad [0 < a < b]$$
$$\text{ET II 21(28)a}$$

3b.[12] $\displaystyle\int_0^\infty [J_0(ax) - 1]\,J_1(bx)\frac{\mathrm{d}x}{x} = \begin{cases} \dfrac{b}{2a}\ {}_2F_1\left(\dfrac{1}{2}, \dfrac{1}{2}; 2, \dfrac{b^2}{a^2}\right) - 1 & [0 < b < a] \\[2ex] \dfrac{2}{\pi}\,\mathbf{E}\left(\dfrac{a^2}{b^2}\right) - 1 & [0 < a < b] \end{cases}$

4. $\displaystyle\int_0^\infty [1 - J_0(ax)]\,J_0(bx)\frac{\mathrm{d}x}{x} = 0$ $\qquad [0 < a < b]$

$$= \ln\frac{a}{b} \qquad\qquad [0 < b < a]$$
$$\text{ET II 14(16)}$$

**6.534** $\displaystyle\int_0^\infty \frac{x^3\,J_0(x)}{x^4 - a^4}\, \mathrm{d}x = \frac{1}{2} K_0(a) - \frac{1}{4}\pi\, Y_0(a)$ $\qquad [a > 0]$ $\qquad$ ET II 340(5)

**6.535** $\displaystyle\int_0^\infty \frac{x}{x^2 + a^2}\left[J_\nu(x)\right]^2 \mathrm{d}x = I_\nu(a)\,K_\nu(a)$ $\qquad [\operatorname{Re} a > 0, \quad \operatorname{Re}\nu > -1]$ $\qquad$ ET II 342(26)

**6.536** $\int_0^\infty \dfrac{x^3 J_0(bx)}{x^4 + a^4}\,dx = \ker(ab)$ $\qquad\qquad\qquad \left[b > 0, \quad |\arg a| < \tfrac{1}{4}\pi\right]$

ET II 8(9), MO 46a

**6.537** $\int_0^\infty \dfrac{x^2 J_0(bx)}{x^4 + a^4}\,dx = -\dfrac{1}{a^2}\,\mathrm{kei}(ab)$ $\qquad \left[b > 0, \quad |\arg a| < \dfrac{\pi}{4}\right]$

MO 46a

**6.538**

1.¹² $\int_0^\infty J_1(ax) J_1(bx) \dfrac{dx}{x^2} = \dfrac{a+b}{3\pi a b^2}\left[(a^2 + b^2)\boldsymbol{E}\left(\dfrac{2\sqrt{ab}}{a+b}\right) - (a-b)^2 \boldsymbol{K}\left(\dfrac{2\sqrt{ab}}{a+b}\right)\right]$

$[a > 0, \quad b > 0]$        ET II 21(30)

2.⁸ $\int_0^\infty x^{-1} J_{\nu+2n+1}(x) J_{\nu+2m+1}(x)\,dx = 0$ $\qquad\qquad [m \neq n \text{ with } m, n \text{ integers}, \nu > -1]$

$\qquad\qquad\qquad\qquad\qquad\qquad = (4n + 2\nu + 2)^{-1} \quad [m = n, \quad \nu > -1]$

EH II 64

**6.539**

1. $\int_a^b \dfrac{dx}{x\left[J_\nu(x)\right]^2} = \dfrac{\pi}{2}\left[\dfrac{Y_\nu(b)}{J_\nu(b)} - \dfrac{Y_\nu(a)}{J_\nu(a)}\right]$ $\qquad [J_\nu(x) \neq 0 \text{ for } x \in [a, b]]$  ET II 338(41)

2. $\int_a^b \dfrac{dx}{x\left[Y_\nu(x)\right]^2} = \dfrac{\pi}{2}\left[\dfrac{J_\nu(a)}{Y_\nu(a)} - \dfrac{J_\nu(b)}{Y_\nu(b)}\right]$ $\qquad [Y_\nu(x) \neq 0 \text{ for } x \in [a, b]]$

ET II 339(49)

3. $\int_a^b \dfrac{dx}{x\,J_\nu(x)\,Y_\nu(x)} = \dfrac{\pi}{2}\ln\left[\dfrac{J_\nu(a)\,Y_\nu(b)}{J_\nu(b)\,Y_\nu(a)}\right]$

ET II 339(50)

**6.541**

1. $\int_0^\infty x\,J_\nu(ax) J_\nu(bx) \dfrac{dx}{x^2 + c^2} = I_\nu(bc) K_\nu(ac)$ $\qquad [0 < b < a, \quad \operatorname{Re} c > 0, \quad \operatorname{Re}\nu > -1]$

$\qquad\qquad\qquad\qquad\qquad\qquad = I_\nu(ac) K_\nu(bc)$ $\qquad [0 < a < b, \quad \operatorname{Re} c > 0, \quad \operatorname{Re}\nu > -1]$

ET II 49(10)

2.⁸ $\int_0^\infty x^{1-2n} J_\nu(ax) J_\nu(bx) \dfrac{dx}{x^2 + c^2}$

$\qquad = \left(-\dfrac{1}{c^2}\right)^n \left[I_\nu(bc) K_\nu(ac) - \dfrac{1}{2}\left(\dfrac{b}{a}\right)^\nu \dfrac{\pi}{\sin(\pi\nu)} \sum_{p=0}^{n-1} \dfrac{\left(a^2 c^2/4\right)^p}{p!\,\Gamma(1 - \nu + p)} \sum_{k=0}^{n-1-p} \dfrac{\left(b^2 c^2/4\right)^k}{k!\,\Gamma(1 - \nu + k)}\right]$

$[0 < b < a]$

$\qquad = \left(-\dfrac{1}{c^2}\right)^n \left[I_\nu(bc) K_\nu(ac) - \dfrac{1}{2\nu}\left(\dfrac{b}{a}\right)^\nu \sum_{p=0}^{n-1} \dfrac{\left(a^2 c^2/4\right)^p}{p!(1 - \nu)_p} \sum_{k=0}^{n-1-p} \dfrac{\left(b^2 c^2/4\right)^k}{k!(1 + \nu)_k}\right]$

$[n = 1, 2, \ldots, \quad \operatorname{Re}\nu > n - 1, \quad \operatorname{Re} c > 0, \quad 0 < b < a]$

$3.^8$
$$\int_0^\infty \frac{x^{\alpha-1}}{(x^2+z^2)^\rho} \, J_\mu(cx) \, J_\nu(cx) \, dx = \frac{1}{2} \left(\frac{c}{2}\right)^{2\rho-\alpha}$$

$$\times \Gamma \left[ \begin{array}{c} (\mu+\nu+\alpha)/2-\rho, 1+2\rho-\alpha \\ (\mu-\nu-\alpha)/2+\rho+1, (\mu+\nu-\alpha)/2+\rho+1, (\nu-\mu-\alpha)/2+\rho+1 \end{array} \right]$$

$$\times {}_3F_4 \left( \frac{1-\alpha}{2}+\rho, 1-\frac{\alpha}{2}+\rho, \rho; \rho+1-\frac{\mu+\nu+\alpha}{2}, \rho+1+\frac{\mu-\nu-\alpha}{2}, \right.$$

$$\left. \rho+1+\frac{\mu+\nu-\alpha}{2}, \rho+1+\frac{\nu-\mu-\alpha}{2}; c^2z^2 \right) + \frac{z^{\alpha-2\rho}}{2} \left(\frac{cz}{2}\right)^{\mu+\nu},$$

$$\Gamma \left[ \begin{array}{c} \rho-(\alpha+\mu+\nu)/2, (\alpha+\mu+\nu)/2 \\ \rho, \mu+1, \nu+1 \end{array} \right] {}_3F_4 \left( \frac{1+\mu+\nu}{2}, 1+\frac{\mu+\nu}{2} \right.$$

$$\left. \frac{\alpha+\mu+\nu}{2}; 1-\rho+\frac{\alpha+\mu+\nu}{2}, \mu+1, \nu+1, \mu+\nu+1; c^2z^2 \right)$$

$$\left[ \Gamma \left[ \begin{array}{c} a_1,\ldots,a_p \\ b_1,\ldots,b_q \end{array} \right] = \frac{\Gamma(a_1)\ldots\Gamma(a_p)}{\Gamma(b_1)\ldots\Gamma(b_q)}, \quad c>0, \quad \operatorname{Re} z>0, \quad \operatorname{Re}(\alpha+\mu+\nu)>0; \quad \operatorname{Re}(\alpha-2\rho)>1 \right]$$

$4.^*$
$$\int_0^\infty \frac{x^{\rho-1}}{x^2+k^2} J_\mu(bx) \left[ \cos\left(\tfrac{1}{2}(\rho-\mu+\nu)\pi\right) J_\nu(ax) + \sin(ax) + \sin\left(\tfrac{1}{2}(\rho+\mu-\nu)\pi\right) Y_\nu(ax) \right]$$

$$= -k^{\rho-2} I_\mu(kb) K_\nu(ka)$$

$$[|\operatorname{Re}\nu| - \operatorname{Re}\mu < \operatorname{Re}\rho < 4, \quad 0 < b \le a] \quad \text{WA 430(3)}$$

**6.542**    $\displaystyle \int_0^\infty \frac{J_\nu(ax) \, Y_\nu(bx) - J_\nu(bx) \, Y_\nu(ax)}{x \left\{ [J_\nu(bx)]^2 + [Y_\nu(bx)]^2 \right\}} \, dx = -\frac{\pi}{2} \left(\frac{b}{a}\right)^\nu$    $[0 < b < a]$      FT II 352(16)

**6.543**    $\displaystyle \int_0^\infty J_\mu(bx) \left\{ \cos\left[\frac{1}{2}(\nu-\mu)\pi\right] J_\nu(ax) - \sin\left[\frac{1}{2}(\nu-\mu)\pi\right] Y_\nu(ax) \right\} \frac{x \, dx}{x^2+r^2} = I_\mu(br) \, K_\nu(ar)$

$$[\operatorname{Re} r > 0, \quad a \ge b > 0, \quad \operatorname{Re}\mu > |\operatorname{Re}\nu| - 2]$$

**6.544**

1.    $\displaystyle \int_0^\infty J_\nu\left(\frac{a}{x}\right) Y_\nu\left(\frac{x}{b}\right) \frac{dx}{x^2} = -\frac{1}{a} \left[ \frac{2}{\pi} K_{2\nu}\left(\frac{2\sqrt{a}}{\sqrt{b}}\right) - Y_{2\nu}\left(\frac{2\sqrt{a}}{\sqrt{b}}\right) \right]$

$$[a>0, \quad b>0, \quad |\operatorname{Re}\nu| < \tfrac{1}{2}]$$
$$\text{EI II 357(47)}$$

2.    $\displaystyle \int_0^\infty J_\nu\left(\frac{a}{x}\right) J_\nu\left(\frac{x}{b}\right) \frac{dx}{x^2} = \frac{1}{a} J_{2\nu}\left(\frac{2\sqrt{a}}{\sqrt{b}}\right)$      $[a>0, \quad b>0, \quad \operatorname{Re}\nu > -\tfrac{1}{2}]$

$$\text{ET II 57(10)}$$

3.    $\displaystyle \int_0^\infty J_\nu\left(\frac{a}{x}\right) K_\nu\left(\frac{x}{b}\right) \frac{dx}{x^2} = \frac{1}{a} e^{\frac{1}{2}i\nu\pi} K_{2\nu}\left(\frac{2\sqrt{a}}{\sqrt{b}} e^{\frac{1}{4}i\pi}\right) + \frac{1}{a} e^{-\frac{1}{2}i\nu\pi} K_{2\nu}\left(\frac{2\sqrt{a}}{\sqrt{b}} e^{-\frac{1}{4}i\pi}\right)$

$$[\operatorname{Re} b > 0, \quad a > 0, \quad |\operatorname{Re}\nu| < \tfrac{1}{2}]$$
$$\text{ET II 142(32)}$$

4. $$\int_0^\infty Y_\nu\left(\frac{a}{x}\right) J_\nu\left(\frac{x}{b}\right) \frac{dx}{x^2} = \frac{2}{a\pi}\left[K_{2\nu}\left(\frac{2\sqrt{a}}{\sqrt{b}}\right) + \frac{\pi}{2} Y_{2\nu}\left(\frac{2\sqrt{a}}{\sqrt{b}}\right)\right]$$

$$\left[a > 0, \quad b > 0, \quad |\mathrm{Re}\,\nu| < \tfrac{1}{2}\right]$$

ET II 62(38)

5.$^{12}$ $$\int_0^\infty Y_\nu\left(\frac{a}{x}\right) K_\nu\left(\frac{x}{b}\right) \frac{dx}{x^2} = \frac{1}{a}\left[e^{\frac{1}{2}i(\nu+1)\pi} K_{2\nu}\left(\frac{2\sqrt{a}}{\sqrt{b}} e^{\frac{1}{4}i\pi}\right) + e^{-\frac{1}{2}i(\nu+1)\pi} K_{2\nu}\left(\frac{2\sqrt{a}}{\sqrt{b}} e^{-\frac{1}{4}i\pi}\right)\right]$$

$$\left[\mathrm{Re}\,b > 0, \quad a > 0, \quad |\mathrm{Re}\,\nu| < \tfrac{1}{2}\right]$$

ET II 143(38)

6. $$\int_0^\infty K_\nu\left(\frac{a}{x}\right) J_\nu\left(\frac{x}{b}\right) \frac{dx}{x^2} = \frac{i}{a}\left[e^{\frac{1}{2}\nu\pi i} K_{2\nu}\left(e^{\frac{1}{4}\pi i}\frac{2\sqrt{a}}{\sqrt{b}}\right) - e^{-\frac{1}{2}\nu\pi i} K_{2\nu}\left(e^{-\frac{1}{4}\pi i}\frac{2\sqrt{a}}{\sqrt{b}}\right)\right]$$

$$\left[\mathrm{Re}\,a > 0, \quad b > 0, \quad |\mathrm{Re}\,\nu| < \tfrac{5}{2}\right]$$

ET II 70(19)

7. $$\int_0^\infty K_\nu\left(\frac{a}{x}\right) Y_\nu\left(\frac{x}{b}\right) \frac{dx}{x^2} = \frac{2}{a}\left[\sin\left(\frac{3}{2}\pi\nu\right) \mathrm{kei}_{2\nu}\left(\frac{2\sqrt{a}}{\sqrt{b}}\right) - \cos\left(\frac{3}{2}\pi\nu\right) \mathrm{ker}_{2\nu}\left(\frac{2\sqrt{a}}{\sqrt{b}}\right)\right]$$

$$\left[\mathrm{Re}\,a > 0, \quad b > 0, \quad |\mathrm{Re}\,\nu| < \tfrac{5}{2}\right]$$

ET II 113(29)

8. $$\int_0^\infty K_\nu\left(\frac{a}{x}\right) K_\nu\left(\frac{x}{b}\right) \frac{dx}{x^2} = \frac{\pi}{a} K_{2\nu}\left(\frac{2\sqrt{a}}{\sqrt{b}}\right) \qquad\qquad \left[\mathrm{Re}\,a > 0, \quad \mathrm{Re}\,b > 0\right] \qquad \text{ET II 146(55)}$$

## 6.55 Combinations of Bessel functions and algebraic functions

### 6.551$^{10}$

1. $$\int_0^1 x^{1/2} J_\nu(xy)\,dx = \sqrt{2} y^{-3/2}\frac{\Gamma\left(\frac{3}{4} + \frac{1}{2}\nu\right)}{\Gamma\left(\frac{1}{4} + \frac{1}{2}\nu\right)}$$

$$+ y^{-1/2}\left[\left(\nu - \tfrac{1}{2}\right) J_\nu(y) S_{-1/2,\nu-1}(y) - J_{\nu-1}(y) S_{1/2,\nu}(y)\right]$$

$$\left[y > 0, \quad \mathrm{Re}\,\nu > -\tfrac{3}{2}\right] \qquad \text{ET II 21(1)}$$

2. $$\int_1^\infty x^{1/2} J_\nu(xy)\,dx = y^{-1/2}\left[J_{\nu-1}(y) S_{1/2,\nu}(y) + \left(\tfrac{1}{2} - \nu\right) J_\nu(y) S_{-1/2,\nu-1}(y)\right]$$

$$\left[y > 0\right] \qquad \text{ET II 22(2)}$$

### 6.552

1. $$\int_0^\infty J_\nu(xy)\frac{dx}{(x^2 + a^2)^{1/2}} = I_{\nu/2}\left(\tfrac{1}{2}ay\right) K_{\nu/2}\left(\tfrac{1}{2}ay\right) \qquad \left[\mathrm{Re}\,a > 0, \quad y > 0, \quad \mathrm{Re}\,\nu > -1\right]$$

ET II 23(11), WA 477(3), MO 44

2. $$\int_0^\infty Y_\nu(xy)\frac{dx}{(x^2 + a^2)^{1/2}} = -\frac{1}{\pi} \sec\left(\tfrac{1}{2}\nu\pi\right) K_{\nu/2}\left(\tfrac{1}{2}ay\right) \left[K_{\nu/2}\left(\tfrac{1}{2}ay\right) + \pi \sin\left(\tfrac{1}{2}\nu\pi\right) I_{\nu/2}\left(\tfrac{1}{2}ay\right)\right]$$

$$\left[y > 0, \quad \mathrm{Re}\,a > 0, \quad |\mathrm{Re}\,\nu| < 1\right]$$

ET II 100(18)

3. $$\int_0^\infty K_\nu(xy)\frac{dx}{(x^2 + a^2)^{1/2}} = \frac{\pi^2}{8} \sec\left(\tfrac{1}{2}\nu\pi\right) \left\{\left[J_{\nu/2}\left(\tfrac{1}{2}ay\right)\right]^2 + \left[Y_{\nu/2}\left(\tfrac{1}{2}ay\right)\right]^2\right\}$$

$$\left[\mathrm{Re}\,a > 0, \quad \mathrm{Re}\,y > 0, \quad |\mathrm{Re}\,\nu| < 1\right]$$

ET II 128(6)

4.    $\displaystyle\int_0^1 J_\nu(xy)\frac{\mathrm{d}x}{(1-x^2)^{1/2}} = \frac{\pi}{2}\left[J_{\nu/2}\left(\tfrac{1}{2}y\right)\right]^2$       $[y>0, \quad \mathrm{Re}\,\nu > -1]$       ET II 24(22)a

5.    $\displaystyle\int_0^1 Y_0(xy)\frac{\mathrm{d}x}{(1-x^2)^{1/2}} = \frac{\pi}{2}\,J_0\left(\tfrac{1}{2}y\right)Y_0\left(\tfrac{1}{2}y\right)$       $[y>0]$       ET II 102(26)a

6.    $\displaystyle\int_1^\infty J_\nu(xy)\frac{\mathrm{d}x}{(x^2-1)^{1/2}} = -\frac{\pi}{2}\,J_{\nu/2}\left(\tfrac{1}{2}y\right)Y_{\nu/2}\left(\tfrac{1}{2}y\right)$       $[y>0]$       ET II 24(23)a

7.    $\displaystyle\int_1^\infty Y_\nu(xy)\frac{\mathrm{d}x}{(x^2-1)^{1/2}} = \frac{\pi}{4}\left\{\left[J_{\nu/2}\left(\tfrac{1}{2}y\right)\right]^2 - \left[Y_{\nu/2}\left(\tfrac{1}{2}y\right)\right]^2\right\}$

                                                     $[y>0]$       ET II 102(27)

**6.553**    $\displaystyle\int_0^\infty x^{-1/2}\,I_\nu(x)\,K_\nu(x)\,K_\mu(2x)\,\mathrm{d}x = \frac{\Gamma\left(\tfrac14+\tfrac12\mu\right)\Gamma\left(\tfrac14-\tfrac12\mu\right)\Gamma\left(\tfrac14+\nu+\tfrac12\mu\right)\Gamma\left(\tfrac14+\nu-\tfrac12\mu\right)}{4\,\Gamma\left(\tfrac34+\nu+\tfrac12\mu\right)\Gamma\left(\tfrac34+\nu-\tfrac12\mu\right)}$

                                         $\left[|\mathrm{Re}\,\mu| < \tfrac12, \quad 2\,\mathrm{Re}\,\nu > |\mathrm{Re}\,\mu| - \tfrac12\right]$

                                                   ET II 372(2)

**6.554**

1.    $\displaystyle\int_0^\infty x\,J_0(xy)\frac{\mathrm{d}x}{(a^2+x^2)^{1/2}} = y^{-1}e^{-ay}$       $[y>0, \quad \mathrm{Re}\,a>0]$       ET II 7(4)

2.    $\displaystyle\int_0^1 x\,J_0(xy)\frac{\mathrm{d}x}{(1-x^2)^{1/2}} = y^{-1}\sin y$       $[y>0]$       ET II 7(5)a

3.    $\displaystyle\int_1^\infty x\,J_0(xy)\frac{\mathrm{d}x}{(x^2-1)^{1/2}} = y^{-1}\cos y$       $[y>0]$       ET II 7(6)a

4.    $\displaystyle\int_0^\infty x\,J_0(xy)\frac{\mathrm{d}x}{(x^2+a^2)^{3/2}} = a^{-1}e^{-ay}$       $[y>0, \quad \mathrm{Re}\,a>0]$       ET II 7(7)a

5.[11]    $\displaystyle\int_0^\infty \frac{x^{\nu+1}\,J_\nu(ax)}{(x^4+4k^4)^{\nu+1/2}}\,\mathrm{d}x = \frac{\left(\tfrac12 a\right)^\nu\sqrt{\pi}}{(2k)^{2\nu}\,\Gamma\left(\nu+\tfrac12\right)}\,J_\nu(ak)\,K_\nu(ak)$

                                     $\left[a>0, \quad |\arg k| > \tfrac{\pi}{4}, \quad \mathrm{Re}\,\nu > -\tfrac12\right]$

                                                   WA 473(1)

**6.555**    $\displaystyle\int_0^\infty x^{1/2}\,J_{2\nu-1}\left(ax^{1/2}\right)Y_\nu(xy)\,\mathrm{d}x = -\frac{a}{2y^2}\,\mathbf{H}_{\nu-1}\left(\frac{a^2}{4y}\right)$

                                     $\left[a>0, \quad y>0, \quad \mathrm{Re}\,\nu > -\tfrac12\right]$

                                                   ET II 111(17)

**6.556**    $\displaystyle\int_0^\infty J_\nu\left[a\left(x^2+1\right)^{1/2}\right]\frac{\mathrm{d}x}{\sqrt{x^2+1}} = -\frac{\pi}{2}\,J_{\nu/2}\left(\frac{a}{2}\right)Y_{\nu/2}\left(\frac{a}{2}\right)$       $[\mathrm{Re}\,\nu > -1, \quad a>0]$       MO 46

## 6.56–6.58 Combinations of Bessel functions and powers

**6.561**

1.    $\displaystyle\int_0^1 x^\nu\,J_\nu(ax)\,\mathrm{d}x = 2^{\nu-1}a^{-\nu}\pi^{\frac12}\,\Gamma\left(\nu+\tfrac12\right)\left[J_\nu(a)\,\mathbf{H}_{\nu-1}(a) - \mathbf{H}_\nu(a)\,J_{\nu-1}(a)\right]$

                                                 $\left[\mathrm{Re}\,\nu > -\tfrac12\right]$       ET II 333(2)a

2. $\int_0^1 x^\nu Y_\nu(ax)\,dx = 2^{\nu-1} a^{-\nu} \pi^{\frac{1}{2}} \Gamma\left(\nu+\tfrac{1}{2}\right) \left[ Y_\nu(a)\,\mathbf{H}_{\nu-1}(a) - \mathbf{H}_\nu(a)\,Y_{\nu-1}(a) \right]$

$$\left[\operatorname{Re}\nu > -\tfrac{1}{2}\right] \qquad \text{ET II 338(43)a}$$

3. $\int_0^1 x^\nu I_\nu(ax)\,dx = 2^{\nu-1} a^{-\nu} \pi^{\frac{1}{2}} \Gamma\left(\nu+\tfrac{1}{2}\right) \left[ I_\nu(a)\,\mathbf{L}_{\nu-1}(a) - \mathbf{L}_\nu(a)\,I_{\nu-1}(a) \right]$

$$\left[\operatorname{Re}\nu > -\tfrac{1}{2}\right] \qquad \text{ET II 364(2)a}$$

4. $\int_0^1 x^\nu K_\nu(ax)\,dx = 2^{\nu-1} a^{-\nu} \pi^{\frac{1}{2}} \Gamma\left(\nu+\tfrac{1}{2}\right) \left[ K_\nu(a)\,\mathbf{L}_{\nu-1}(a) + \mathbf{L}_\nu(a)\,K_{\nu-1}(a) \right]$

$$\left[\operatorname{Re}\nu > -\tfrac{1}{2}\right] \qquad \text{ET II 367(21)a}$$

5. $\int_0^1 x^{\nu+1} J_\nu(ax)\,dx = a^{-1} J_{\nu+1}(a) \qquad \left[\operatorname{Re}\nu > -1\right] \qquad \text{ET II 333(3)a}$

6. $\int_0^1 x^{\nu+1} Y_\nu(ax)\,dx = a^{-1} Y_{\nu+1}(a) + 2^{\nu+1} a^{-\nu-2} \pi^{-1} \Gamma(\nu+1)$

$$\left[\operatorname{Re}\nu > -1\right] \qquad \text{ET II 339(44)a}$$

7. $\int_0^1 x^{\nu+1} I_\nu(ax)\,dx = a^{-1} I_{\nu+1}(a) \qquad \left[\operatorname{Re}\nu > -1\right] \qquad \text{ET II 365(3)a}$

8. $\int_0^1 x^{\nu+1} K_\nu(ax)\,dx = 2^\nu a^{-\nu-2} \Gamma(\nu+1) - a^{-1} K_{\nu+1}(a)$

$$\left[\operatorname{Re}\nu > -1\right] \qquad \text{ET II 367(22)a}$$

9. $\int_0^1 x^{1-\nu} J_\nu(ax)\,dx = \dfrac{a^{\nu-2}}{2^{\nu-1}\Gamma(\nu)} - a^{-1} J_{\nu-1}(a) \qquad \text{ET II 333(4)a}$

10. $\int_0^1 x^{1-\nu} Y_\nu(ax)\,dx = \dfrac{a^{\nu-2}\cot(\nu\pi)}{2^{\nu-1}\Gamma(\nu)} - a^{-1} Y_{\nu-1}(a) \qquad \left[\operatorname{Re}\nu < 1\right] \qquad \text{ET II 339(45)a}$

11. $\int_0^1 x^{1-\nu} I_\nu(ax)\,dx = a^{-1} I_{\nu-1}(a) - \dfrac{a^{\nu-2}}{2^{\nu-1}\Gamma(\nu)} \qquad \text{ET II 365(4)a}$

12. $\int_0^1 x^{1-\nu} K_\nu(ax)\,dx = 2^{-\nu} a^{\nu-2} \Gamma(1-\nu) - a^{-1} K_{\nu-1}(a)$

$$\left[\operatorname{Re}\nu < 1\right] \qquad \text{ET II 367(23)a}$$

13.[7] $\int_0^1 x^\mu J_\nu(ax)\,dx = \dfrac{2^\mu \Gamma\left(\frac{\nu+\mu+1}{2}\right)}{a^{\mu+1}\Gamma\left(\frac{\nu-\mu+1}{2}\right)} + a^{-\mu} \left\{ (\mu+\nu-1) J_\nu(a) S_{\mu-1,\nu-1}(a) - J_{\nu-1}(a) S_{\mu,\nu}(a) \right\}$

$$\left[a > 0, \quad \operatorname{Re}(\mu+\nu) > -1\right] \qquad \text{ET II 22(8)a}$$

14. $\int_0^\infty x^\mu J_\nu(ax)\,dx = 2^\mu a^{-\mu-1} \dfrac{\Gamma\left(\frac{1}{2}+\frac{1}{2}\nu+\frac{1}{2}\mu\right)}{\Gamma\left(\frac{1}{2}+\frac{1}{2}\nu-\frac{1}{2}\mu\right)} \qquad \left[-\operatorname{Re}\nu - 1 < \operatorname{Re}\mu < \tfrac{1}{2}, \quad a > 0\right]$

$$\text{EH II 49(19)}$$

15. $\int_0^\infty x^\mu Y_\nu(ax)\,dx = 2^\mu \cot\left[\tfrac{1}{2}(\nu+1-\mu)\pi\right] a^{-\mu-1} \dfrac{\Gamma\left(\frac{1}{2}+\frac{1}{2}\nu+\frac{1}{2}\mu\right)}{\Gamma\left(\frac{1}{2}+\frac{1}{2}\nu-\frac{1}{2}\mu\right)}$

$$\left[|\operatorname{Re}\nu| - 1 < \mu < \tfrac{1}{2}, \quad a > 0\right]$$
$$\text{ET II 97(3)a}$$

16.    $\displaystyle\int_0^\infty x^\mu K_\nu(ax)\,dx = 2^{\mu-1}a^{-\mu-1}\,\Gamma\left(\frac{1+\mu+\nu}{2}\right)\Gamma\left(\frac{1+\mu-\nu}{2}\right)$

$$[\operatorname{Re}(\mu+1\pm\nu)>0,\quad \operatorname{Re}a>0]$$

EH II 51(27)

17.    $\displaystyle\int_0^\infty \frac{J_\nu(ax)}{x^{\nu-q}}\,dx = \frac{\Gamma\left(\frac12 q+\frac12\right)}{2^{\nu-q}a^{q-\nu+1}\,\Gamma\left(\nu-\frac12 q+\frac12\right)}$      $\left[-1<\operatorname{Re}q<\operatorname{Re}\nu-\frac12\right]$

WA 428(1), KU 144(5)

18.    $\displaystyle\int_0^\infty \frac{Y_\nu(x)}{x^{\nu-\mu}}\,dx = \frac{\Gamma\left(\frac12+\frac12\mu\right)\Gamma\left(\frac12+\frac12\mu-\nu\right)\sin\left(\frac12\mu-\nu\right)\pi}{2^{\nu-\mu}\pi}$

$$\left[|\operatorname{Re}\nu|<\operatorname{Re}(1+\mu-\nu)<\frac32\right]$$

WA 430(5)

19.    $\displaystyle\int_0^1 x^{2m+n+1/2}K_{n+1/2}(\alpha x)\,dx = \sqrt{\frac{\pi}{2}}\sum_{k=0}^n \frac{(n+k)!}{k!(n-k)!}\frac{\gamma(2m+n-k+1,\alpha)}{\alpha^{2m+n+3/2}2^k}$      STR

**6.562**

1.    $\displaystyle\int_0^\infty x^\mu Y_\nu(bx)\frac{dx}{x+a} = (2a)^\mu \pi^{-1}\left\{\sin\left[\tfrac12\pi(\mu-\nu)\right]\Gamma\left[\tfrac12(\mu+\nu+1)\right]\Gamma\left[\tfrac12(1+\mu-\nu)\right]S_{-\mu,\nu}(ab)\right.$

$$\left.-2\cos\left[\tfrac12\pi(\mu-\nu)\right]\Gamma\left(1+\tfrac12\mu+\tfrac12\nu\right)\Gamma\left(1+\tfrac12\mu-\tfrac12\nu\right)S_{-\mu-1,\nu}(ab)\right\}$$

$$\left[b>0,\quad |\arg a|<\pi,\quad \operatorname{Re}(\mu\pm\nu)>-1,\quad \operatorname{Re}\mu<\tfrac32\right]\quad \text{ET II 98(8)}$$

2.    $\displaystyle\int_0^\infty \frac{x^\nu J_\nu(ax)}{x+k}\,dx = \frac{\pi k^\nu}{2\cos\nu\pi}\left[\mathbf{H}_{-\nu}(ak)-Y_{-\nu}(ak)\right]$      $\left[-\tfrac12<\operatorname{Re}\nu<\tfrac32,\quad a>0,\quad |\arg k|<\pi\right]$

WA 479(7)

3.    $\displaystyle\int_0^\infty x^\mu K_\nu(bx)\frac{dx}{x+a}$

$$= 2^{\mu-2}\,\Gamma\left[\tfrac12(\mu+\nu)\right]\Gamma\left[\tfrac12(\mu-\nu)\right]b^{-\mu}\,{}_1F_2\left(1;1-\frac{\mu+\nu}{2},1-\frac{\mu-\nu}{2};\frac{a^2b^2}{4}\right)$$

$$-2^{\mu-3}\,\Gamma\left[\tfrac12(\mu-\nu-1)\right]\Gamma\left[\tfrac12(\mu+\nu-1)\right]ab^{1-\mu}\,{}_1F_2\left(1;\frac{3-\mu-\nu}{2},\frac{3-\mu+\nu}{2};\frac{a^2b^2}{4}\right)$$

$$-\pi a^\mu\operatorname{cosec}[\pi(\mu-\nu)]\left\{K_\nu(ab)+\pi\cos(\mu\pi)\operatorname{cosec}[\pi(\nu+\mu)]I_\nu(ab)\right\}$$

$$[\operatorname{Re}b>0,\quad |\arg a|<\pi,\quad \operatorname{Re}\mu>|\operatorname{Re}\nu|-1]\quad \text{ET II 127(4)}$$

**6.563**    $\displaystyle\int_0^\infty x^{\varrho-1}J_\nu(bx)\frac{dx}{(x+a)^{1+\mu}} = \frac{\pi a^{\varrho-\mu-1}}{\sin[(\varrho+\nu-\mu)\pi]\,\Gamma(\mu+1)}$

$$\times\left\{\sum_{m=0}^\infty \frac{(-1)^m\left(\tfrac12 ab\right)^{\nu+2m}\Gamma(\varrho+\nu+2m)}{m!\,\Gamma(\nu+m+1)\,\Gamma(\varrho+\nu-\mu+2m)}\right.$$

$$\left.-\sum_{m=0}^\infty \frac{\left(\tfrac12 ab\right)^{\mu+1-\varrho+m}\Gamma(\mu+m+1)\sin\left[\tfrac12(\varrho+\nu-\mu-m)\pi\right]}{m!\,\Gamma\left[\tfrac12(\mu+\nu-\varrho+m+3)\right]\Gamma\left[\tfrac12(\mu-\nu-\varrho+m+3)\right]}\right\}$$

$$\left[b>0,\quad |\arg a|<\pi,\quad \operatorname{Re}(\varrho+\nu)>0,\quad \operatorname{Re}(\varrho-\mu)<\tfrac52\right]\quad \text{ET II 23(10), WA 479}$$

**6.564**

1.      $$\int_0^\infty x^{\nu+1}\, J_\nu(bx)\frac{dx}{\sqrt{x^2+a^2}} = \sqrt{\frac{2}{\pi b}}a^{\nu+\frac{1}{2}}\,K_{\nu+\frac{1}{2}}(ab) \qquad \left[\operatorname{Re}a>0, \quad b>0, \quad -1<\operatorname{Re}\nu<\tfrac{1}{2}\right]$$

ET II 23(15)

2.      $$\int_0^\infty x^{1-\nu}\, J_\nu(bx)\frac{dx}{\sqrt{x^2+a^2}} = \sqrt{\frac{\pi}{2b}}a^{\frac{1}{2}-\nu}\left[I_{\nu-\frac{1}{2}}(ab) - \mathbf{L}_{\nu-\frac{1}{2}}(ab)\right]$$

$$\left[\operatorname{Re}a>0, \quad b>0, \quad \operatorname{Re}\nu>-\tfrac{1}{2}\right]$$

ET II 23(16)

**6.565**

1.      $$\int_0^\infty x^{-\nu}\left(x^2+a^2\right)^{-\nu-\frac{1}{2}} J_\nu(bx)\,dx = 2^\nu a^{-2\nu}b^\nu\frac{\Gamma(\nu+1)}{\Gamma(2\nu+1)}I_\nu\left(\frac{ab}{2}\right)K_\nu\left(\frac{ab}{2}\right)$$

$$\left[\operatorname{Re}a>0, \quad b>0, \quad \operatorname{Re}\nu>-\tfrac{1}{2}\right]$$

WA 477(4), ET II 23(17)

2.      $$\int_0^\infty x^{\nu+1}\left(x^2+a^2\right)^{-\nu-\frac{1}{2}} J_\nu(bx)\,dx = \frac{\sqrt{\pi}\,b^{\nu-1}}{2^\nu e^{ab}\,\Gamma\left(\nu+\frac{1}{2}\right)}$$

$$\left[\operatorname{Re}a>0, \quad b>0, \quad \operatorname{Re}\nu>-\tfrac{1}{2}\right]$$

ET II 24(18)

3.      $$\int_0^\infty x^{\nu+1}\left(x^2+a^2\right)^{-\nu-\frac{3}{2}} J_\nu(bx)\,dx = \frac{b^\nu\sqrt{\pi}}{2^{\nu+1}ae^{ab}\,\Gamma\left(\nu+\frac{3}{2}\right)}$$

$$\left[\operatorname{Re}a>0, \quad b>0, \quad \operatorname{Re}\nu>-1\right]$$

ET II 24(19)

4.      $$\int_0^\infty \frac{J_\nu(bx)x^{\nu+1}}{(x^2+a^2)^{\mu+1}}\,dx = \frac{a^{\nu-\mu}b^\mu}{2^\mu\,\Gamma(\mu+1)}K_{\nu-\mu}(ab)$$

$$\left[-1<\operatorname{Re}\nu<\operatorname{Re}\left(2\mu+\tfrac{3}{2}\right), \quad a>0, \quad b>0\right] \quad \text{MO 43}$$

5.[12]   $$\int_0^\infty x^{\nu+1}\left(x^2+a^2\right)^\mu Y_\nu(bx)\,dx$$

$$= \frac{a^{\mu+1}b^{-\mu-\nu-1}}{2\pi\Gamma(-\mu)}\left[\pi^2 2^{\mu+1}(ab)^\nu\csc(\pi(\mu+\nu))[\cot(\pi\mu)I_{-\mu-\nu-1}(ab) + \cot(\pi\nu))I_{\mu+\nu+1}(ab)]\right.$$

$$\left.-2^\nu(ab)^{\mu+1}\Gamma(-\mu-1)\Gamma(\nu)\,{}_1F_2\left(1;\mu+2;1-\nu;\frac{a^2b^2}{4}\right)\right]$$

$$\left[b>0, \quad \operatorname{Re}a>0, \quad -1<\operatorname{Re}\nu<-2\operatorname{Re}\mu\right] \quad \text{ET II 100(19)}$$

6.[12]   $$\int_0^\infty x^{1-\nu}\left(x^2+a^2\right)^\mu Y_\nu(bx)\,dx$$

$$= \frac{2^\mu a^{\mu-\nu+1}}{b^{\mu+1}}\Gamma(\mu+1)\left[\cot(\nu\pi)I_{\mu-\nu+1}(ab) + \frac{2}{\pi}\cos((\mu-\nu)\pi)K_{\mu-\nu+1}(ab)\right]$$

$$-\frac{a^{2\mu+2}b^\nu\cot(\nu\pi)}{2^{\nu+1}(\mu+1)\Gamma(\nu+1)}\,{}_1F_2\left(1;\nu+1,\mu+2;\frac{a^2b^2}{4}\right)$$

$$\left[\operatorname{Re}\nu<1, \quad \operatorname{Re}(\nu-2\mu)>-3, \quad \arg a^2\neq\pi, \quad b>0\right] \quad \text{MC}$$

7.      $$\int_0^\infty x^{1+\nu}\left(x^2+a^2\right)^\mu K_\nu(bx)\,dx = 2^\nu\Gamma(\nu+1)a^{\nu+\mu+1}b^{-1-\mu}S_{\mu-\nu,\mu+\nu+1}(ab)$$

$$\left[\operatorname{Re}a>0, \quad \operatorname{Re}b>0, \quad \operatorname{Re}\nu>-1\right]$$

ET II 128(8)

8.$^{11}$ $\displaystyle\int_0^\infty \frac{x^{\varrho-1} J_\nu(ax)}{(x^2+k^2)^{\mu+1}}\,\mathrm{d}x = \frac{a^\nu k^{\varrho+\nu-2\mu-2}\,\Gamma\left(\frac{1}{2}\varrho+\frac{1}{2}\nu\right)\Gamma\left(\mu+1-\frac{1}{2}\varrho-\frac{1}{2}\nu\right)}{2^{\nu+1}\,\Gamma(\mu+1)\,\Gamma(\nu+1)}$

$$\times\ {}_1F_2\left(\frac{\varrho+\nu}{2};\frac{\varrho+\nu}{2}-\mu,\nu+1;\frac{a^2k^2}{4}\right)$$

$$+\frac{a^{2\mu+2-\varrho}\,\Gamma\left(\frac{1}{2}\nu+\frac{1}{2}\varrho-\mu-1\right)}{2^{2\mu+3-\varrho}\,\Gamma\left(\mu+2+\frac{1}{2}\nu-\frac{1}{2}\varrho\right)}$$

$$\times\ {}_1F_2\left(\mu+1;\mu+2+\frac{\nu-\varrho}{2},\mu+2-\frac{\nu+\varrho}{2};\frac{a^2k^2}{4}\right)$$

$$\left[a>0,\quad -\operatorname{Re}\nu<\operatorname{Re}\varrho<2\operatorname{Re}\mu+\tfrac{7}{2},\quad \operatorname{Re}k>0\right]\quad \text{WA 477(1)}$$

**6.566**

1.    $\displaystyle\int_0^\infty x^\mu\,Y_\nu(bx)\frac{\mathrm{d}x}{x^2+a^2} = 2^{\mu-2}\pi^{-1}b^{1-\mu}$

$$\times\cos\left[\frac{\pi}{2}(\mu-\nu+1)\right]\Gamma\left(\tfrac{1}{2}\mu+\tfrac{1}{2}\nu-\tfrac{1}{2}\right)\Gamma\left(\tfrac{1}{2}\mu-\tfrac{1}{2}\nu-\tfrac{1}{2}\right)$$

$$\times\ {}_1F_2\left(1;2-\frac{\mu+1+\nu}{2},2-\frac{\mu+1-\nu}{2};\frac{a^2b^2}{4}\right)$$

$$-\frac{1}{2}\pi a^{\mu-1}\operatorname{cosec}\left[\frac{\pi}{2}(\mu+\nu+1)\right]\cot\left[\frac{\pi}{2}(\mu-\nu+1)\right]I_\nu(ab)$$

$$-a^{\mu-1}\operatorname{cosec}\left[\frac{\pi}{2}(\mu-\nu+1)\right]K_\nu(ab)$$

$$\left[b>0,\quad \operatorname{Re}a>0,\quad |\operatorname{Re}\nu|-1<\operatorname{Re}\mu<\tfrac{5}{2}\right]\quad \text{ET II 100(17)}$$

2.    $\displaystyle\int_0^\infty x^{\nu+1} J_\nu(ax)\frac{\mathrm{d}x}{x^2+b^2} = b^\nu K_\nu(ab)$        $\left[a>0,\quad \operatorname{Re}b>0,\quad -1<\operatorname{Re}\nu<\tfrac{3}{2}\right]$

$$\text{EH II 96(58)}$$

3.    $\displaystyle\int_0^\infty x^\nu\,K_\nu(ax)\frac{\mathrm{d}x}{x^2+b^2} = \frac{\pi^2 b^{\nu-1}}{4\cos\nu\pi}\left[\mathbf{H}_{-\nu}(ab)-Y_{-\nu}(ab)\right]$

$$\left[a>0,\quad \operatorname{Re}b>0,\quad \operatorname{Re}\nu>-\tfrac{1}{2}\right]$$
$$\text{WA 468(9)}$$

4.    $\displaystyle\int_0^\infty x^{-\nu}\,K_\nu(ax)\frac{\mathrm{d}x}{x^2+b^2} = \frac{\pi^2}{4b^{\nu+1}\cos\nu\pi}\left[\mathbf{H}_\nu(ab)-Y_\nu(ab)\right]$

$$\left[a>0,\quad \operatorname{Re}b>0,\quad \operatorname{Re}\nu<\tfrac{1}{2}\right]$$
$$\text{WA 468(10)}$$

5.    $\displaystyle\int_0^\infty x^{-\nu}\,J_\nu(ax)\frac{\mathrm{d}x}{x^2+b^2} = \frac{\pi}{2b^{\nu+1}}\left[I_\nu(ab)-\mathbf{L}_\nu(ab)\right]$    $\left[a>0,\quad \operatorname{Re}b>0,\quad \operatorname{Re}\nu>-\tfrac{5}{2}\right]$

$$\text{WA 468(11)}$$

**6.567**

1.    $\displaystyle\int_0^1 x^{\nu+1}\left(1-x^2\right)^\mu J_\nu(bx)\,\mathrm{d}x = 2^\mu\,\Gamma(\mu+1)b^{-(\mu+1)}\,J_{\nu+\mu+1}(b)$

$$\left[b>0,\quad \operatorname{Re}\nu>-1,\quad \operatorname{Re}\mu>-1\right]$$
$$\text{ET II 26(33)a}$$

2.    $\displaystyle\int_0^1 x^{\nu+1}\left(1-x^2\right)^{\mu} Y_{\nu}(bx)\,dx$

$$= b^{-(\mu+1)}\left[2^{\mu}\,\Gamma(\mu+1)\,Y_{\mu+\nu+1}(b) + 2^{\nu+1}\pi^{-1}\,\Gamma(\nu+1)\,S_{\mu-\nu,\mu+\nu+1}(b)\right]$$

$$[b>0,\quad \operatorname{Re}\mu>-1,\quad \operatorname{Re}\nu>-1]\quad\text{ET II 103(35)a}$$

3.    $\displaystyle\int_0^1 x^{1-\nu}\left(1-x^2\right)^{\mu} J_{\nu}(bx)\,dx = \frac{2^{1-\nu}\,S_{\nu+\mu,\mu-\nu+1}(b)}{b^{\mu+1}\,\Gamma(\nu)}\qquad [b>0,\quad \operatorname{Re}\mu>-1]\qquad\text{ET II 25(31)a}$

4.    $\displaystyle\int_0^1 x^{1-\nu}\left(1-x^2\right)^{\mu} Y_{\nu}(bx)\,dx = b^{-(\mu+1)}\left[2^{1-\nu}\pi^{-1}\cos(\nu\pi)\,\Gamma\left(1-\nu\right)\right.$

$$\left.\times\, s_{\mu+\nu,\mu-\nu+1}(b) - 2^{\mu}\operatorname{cosec}(\nu\pi)\,\Gamma(\mu+1)\,J_{\mu-\nu+1}(b)\right]$$

$$[b>0,\quad \operatorname{Re}\mu>-1,\quad \operatorname{Re}\nu<1]\quad\text{ET II 104(37)a}$$

5.    $\displaystyle\int_0^1 x^{1-\nu}\left(1-x^2\right)^{\mu} K_{\nu}(bx)\,dx = 2^{-\nu-2}b^{\nu}(\mu+1)^{-1}\,\Gamma(-\nu)\,{}_1F_2\left(1;\nu+1,\mu+2;\frac{b^2}{4}\right)$

$$+\pi 2^{\mu-1}b^{-(\mu+1)}\operatorname{cosec}(\nu\pi)\,\Gamma(\mu+1)\,I_{\mu-\nu+1}(b)$$

$$[\operatorname{Re}\mu>-1,\quad \operatorname{Re}\nu<1]\qquad\text{ET II 129(12)a}$$

6.    $\displaystyle\int_0^1 x^{1-\nu} J_{\nu}(bx)\frac{dx}{\sqrt{1-x^2}} = \sqrt{\frac{\pi}{2b}}\,\mathbf{H}_{\nu-\frac{1}{2}}(b)\qquad [b>0]\qquad\text{ET II 24(24)a}$

7.    $\displaystyle\int_0^1 x^{1+\nu} Y_{\nu}(bx)\frac{dx}{\sqrt{1-x^2}} = \sqrt{\frac{\pi}{2b}}\operatorname{cosec}(\nu\pi)\left[\cos(\nu\pi)\,J_{\nu+\frac{1}{2}}(b) - \mathbf{H}_{-\nu-\frac{1}{2}}(b)\right]$

$$[b>0,\quad \operatorname{Re}\nu>-1]\qquad\text{ET II 102(28)a}$$

8.    $\displaystyle\int_0^1 x^{1-\nu} Y_{\nu}(bx)\frac{dx}{\sqrt{1-x^2}} = \sqrt{\frac{\pi}{2b}}\left\{\cot(\nu\pi)\left[\mathbf{H}_{\nu-\frac{1}{2}}(b) - Y_{\nu-\frac{1}{2}}(b)\right] - J_{\nu-\frac{1}{2}}(b)\right\}$

$$[b>0,\quad \operatorname{Re}\nu<1]\qquad\text{ET II 102(30)a}$$

9.    $\displaystyle\int_0^1 x^{\nu}\left(1-x^2\right)^{\nu-\frac{1}{2}} J_{\nu}(bx)\,dx = 2^{\nu-1}\sqrt{\pi}\,b^{-\nu}\,\Gamma\left(\nu+\tfrac{1}{2}\right)\left[J_{\nu}\left(\frac{b}{2}\right)\right]^2$

$$[b>0,\quad \operatorname{Re}\nu>-\tfrac{1}{2}]\qquad\text{ET II 24(25)a}$$

10.    $\displaystyle\int_0^1 x^{\nu}\left(1-x^2\right)^{\nu-\frac{1}{2}} Y_{\nu}(bx)\,dx = 2^{\nu-1}\sqrt{\pi}\,b^{-\nu}\,\Gamma\left(\nu+\frac{1}{2}\right) J_{\nu}\left(\frac{b}{2}\right) Y_{\nu}\left(\frac{b}{2}\right)$

$$[b>0,\quad \operatorname{Re}\nu>-\tfrac{1}{2}]\qquad\text{ET II 102(31)a}$$

11.    $\displaystyle\int_0^1 x^{\nu}\left(1-x^2\right)^{\nu-\frac{1}{2}} K_{\nu}(bx)\,dx = 2^{\nu-1}\sqrt{\pi}\,b^{-\nu}\,\Gamma\left(\nu+\frac{1}{2}\right) I_{\nu}\left(\frac{b}{2}\right) K_{\nu}\left(\frac{b}{2}\right)$

$$[\operatorname{Re}\nu>-\tfrac{1}{2}]\qquad\text{ET II 129(10)a}$$

12.    $\displaystyle\int_0^1 x^{\nu}\left(1-x^2\right)^{\nu-\frac{1}{2}} I_{\nu}(bx)\,dx = 2^{-\nu-1}\sqrt{\pi}\,b^{-\nu}\,\Gamma\left(\nu+\frac{1}{2}\right)\left[I_{\nu}\left(\frac{b}{2}\right)\right]^2\qquad\text{ET II 365(5)a}$

13.    $\displaystyle\int_0^1 x^{\nu+1}\left(1-x^2\right)^{-\nu-\frac{1}{2}} J_{\nu}(bx)\,dx = 2^{-\nu}\frac{b^{\nu-1}}{\sqrt{\pi}}\,\Gamma\left(\frac{1}{2}-\nu\right)\sin b$

$$[b>0,\quad |\operatorname{Re}\nu|<\tfrac{1}{2}]\qquad\text{ET II 25(27)a}$$

14.
$$\int_1^\infty x^\nu \left(x^2 - 1\right)^{\nu - \frac{1}{2}} Y_\nu(bx)\,dx = 2^{\nu-2}\sqrt{\pi}\,b^{-\nu}\,\Gamma\left(\nu + \frac{1}{2}\right)\left[J_\nu\left(\frac{b}{2}\right)J_{-\nu}\left(\frac{b}{2}\right) - Y_\nu\left(\frac{b}{2}\right)Y_{-\nu}\left(\frac{b}{2}\right)\right]$$
$$\left[|\operatorname{Re}\nu| < \tfrac{1}{2}, \quad b > 0\right] \qquad \text{ET II 103(32)a}$$

15.
$$\int_1^\infty x^\nu \left(x^2 - 1\right)^{\nu - \frac{1}{2}} K_\nu(bx)\,dx = \frac{2^{\nu-1}}{\sqrt{\pi}}\,b^{-\nu}\,\Gamma\left(\nu + \frac{1}{2}\right)\left[K_\nu\left(\frac{b}{2}\right)\right]^2$$
$$\left[\operatorname{Re}b > 0, \quad \operatorname{Re}\nu > -\tfrac{1}{2}\right] \qquad \text{ET II 129(11)a}$$

16.
$$\int_1^\infty x^{-\nu} \left(x^2 - 1\right)^{-\nu - \frac{1}{2}} J_\nu(bx)\,dx = -2^{-\nu-1}\sqrt{\pi}\,b^\nu\,\Gamma\left(\frac{1}{2} - \nu\right)J_\nu\left(\frac{b}{2}\right)Y_\nu\left(\frac{b}{2}\right)$$
$$\left[b > 0, \quad |\operatorname{Re}\nu| < \tfrac{1}{2}\right] \qquad \text{ET II 25(26)a}$$

17.[12]
$$\int_1^\infty x^{-\nu+1} \left(x^2 - 1\right)^{\nu - \frac{1}{2}} J_\nu(bx)\,dx = \frac{2^{-\nu}}{\sqrt{\pi}}\,b^{-\nu-1}\,\Gamma\left(\frac{1}{2} + \nu\right)\cos b$$
$$\left[b > 0, \quad |\operatorname{Re}\nu| < \tfrac{1}{2}\right] \qquad \text{ET II 25(28)}$$

18.*
$$\int_0^1 x(1 - x^2)^k I_0(ax)\,dx = \frac{2^k k!}{a^{k+1}} I_{k+1}(a) \qquad\qquad \text{PBM 2.15.2.6}$$

## 6.568

1.
$$\int_0^\infty x^\nu\, Y_\nu(bx)\frac{dx}{x^2 - a^2} = \frac{\pi}{2}a^{\nu-1}\, J_\nu(ab) \qquad\qquad \left[a > 0, \quad b > 0, \quad -\tfrac{1}{2} < \operatorname{Re}\nu < \tfrac{5}{2}\right]$$
$$\text{ET II 101(22)}$$

2.
$$\int_0^\infty x^\mu\, Y_\nu(bx)\frac{dx}{x^2 - a^2} = \frac{\pi}{2}a^{\mu-1}\, J_\nu(ab) + 2^\mu \pi^{-1} a^{\mu-1}\cos\left[\frac{\pi}{2}(\mu - \nu + 1)\right]$$
$$\times\, \Gamma\left(\frac{\mu - \nu + 1}{2}\right)\Gamma\left(\frac{\mu + \nu + 1}{2}\right) S_{-\mu,\nu}(ab)$$
$$\left[a > 0, \quad b > 0, \quad |\operatorname{Re}\nu| - 1 < \operatorname{Re}\mu < \tfrac{5}{2}\right] \quad \text{ET II (101)(25)}$$

## 6.569

$$\int_0^1 x^\lambda (1 - x)^{\mu - 1} J_\nu(ax)\,dx$$
$$= \frac{\Gamma(\mu)\,\Gamma(1 + \lambda + \nu)2^{-\nu}a^\nu}{\Gamma(\nu + 1)\,\Gamma(1 + \lambda + \mu + \nu)}$$
$$\times\ {}_2F_3\left(\frac{\lambda + 1 + \nu}{2}, \frac{\lambda + 2 + \nu}{2}; \nu + 1, \frac{\lambda + 1 + \mu + \nu}{2}, \frac{\lambda + 2 + \mu + \nu}{2}; -\frac{a^2}{4}\right)$$
$$\left[\operatorname{Re}\mu > 0, \quad \operatorname{Re}(\lambda + \nu) > -1\right] \quad \text{ET II 193(56)a}$$

## 6.571

1.
$$\int_0^\infty \left[\left(x^2 + a^2\right)^{\frac{1}{2}} \pm x\right]^\mu J_\nu(bx)\frac{dx}{\sqrt{x^2 + a^2}} = a^\mu\, I_{\frac{1}{2}(\nu \mp \mu)}\left(\frac{ab}{2}\right)K_{\frac{1}{2}(\nu \pm \mu)}\left(\frac{ab}{2}\right)$$
$$\left[\operatorname{Re}a > 0, \quad b > 0, \quad \operatorname{Re}\nu > -1, \quad \operatorname{Re}\mu < \tfrac{3}{2}\right] \quad \text{ET II 26(38)}$$

2. $\int_0^\infty \left[ (x^2 + a^2)^{\frac{1}{2}} - x \right]^\mu Y_\nu(bx) \dfrac{\mathrm{d}x}{\sqrt{x^2 + a^2}}$

$$= a^\mu \left[ \cot(\nu\pi) \, I_{\frac{1}{2}(\mu+\nu)} \left( \frac{ab}{2} \right) K_{\frac{1}{2}(\mu-\nu)} \left( \frac{ab}{2} \right) - \mathrm{cosec}(\nu\pi) \, I_{\frac{1}{2}(\mu-\nu)} \left( \frac{ab}{2} \right) K_{\frac{1}{2}(\mu+\nu)} \left( \frac{ab}{2} \right) \right]$$

$$\left[ \mathrm{Re}\, a > 0, \quad b > 0, \quad \mathrm{Re}\, \mu > -\tfrac{3}{2}, \quad |\mathrm{Re}\, \nu| < 1 \right] \quad \text{ET II 104(40)}$$

3. $\int_0^\infty \left[ (x^2 + a^2)^{\frac{1}{2}} + x \right]^\mu K_\nu(bx) \dfrac{\mathrm{d}x}{\sqrt{x^2 + a^2}}$

$$= \frac{\pi^2}{4} a^\mu \, \mathrm{cosec}(\nu\pi) \left[ J_{\frac{1}{2}(\nu-\mu)} \left( \frac{ab}{2} \right) Y_{-\frac{1}{2}(\nu+\mu)} \left( \frac{ab}{2} \right) - Y_{\frac{1}{2}(\nu-\mu)} \left( \frac{ab}{2} \right) J_{-\frac{1}{2}(\nu+\mu)} \left( \frac{ab}{2} \right) \right]$$

$$\left[ \mathrm{Re}\, a > 0, \quad \mathrm{Re}\, b > 0 \right] \qquad \text{ET II 130(15)}$$

**6.572**

1. $\int_0^\infty x^{-\mu} \left[ (x^2 + a^2)^{\frac{1}{2}} + a \right]^\mu J_\nu(bx) \dfrac{\mathrm{d}x}{\sqrt{x^2 + a^2}} = \dfrac{\Gamma\left( \frac{1+\nu-\mu}{2} \right)}{ab\,\Gamma(\nu+1)} \, W_{\frac{1}{2}\mu, \frac{1}{2}\nu}(ab) \, M_{-\frac{1}{2}\mu, \frac{1}{2}\nu}(ab)$

$$\left[ \mathrm{Re}\, a > 0, \quad b > 0, \quad \mathrm{Re}(\nu - \mu) > -1 \right]$$
$$\text{ET II 26(40)}$$

2. $\int_0^\infty x^{-\mu} \left[ (x^2 + a^2)^{\frac{1}{2}} + a \right]^\mu K_\nu(bx) \dfrac{\mathrm{d}x}{\sqrt{x^2 + a^2}}$

$$= \frac{\Gamma\left( \frac{1+\nu-\mu}{2} \right) \Gamma\left( \frac{1-\nu-\mu}{2} \right)}{2ab} \, W_{\frac{1}{2}\mu, \frac{1}{2}\nu}(iab) \, W_{\frac{1}{2}\mu, \frac{1}{2}\nu}(-iab)$$

$$\left[ \mathrm{Re}\, a > 0, \quad \mathrm{Re}\, b > 0, \quad \mathrm{Re}\, \mu + |\mathrm{Re}\, \nu| < 1 \right] \quad \text{ET II 130(18), BU 87(6a)}$$

3. $\int_0^\infty x^{-\mu} \left[ (x^2 + a^2)^{\frac{1}{2}} - a \right]^\mu Y_\nu(bx) \dfrac{\mathrm{d}x}{\sqrt{x^2 + a^2}}$

$$= -\frac{1}{ab} \, W_{-\frac{1}{2}\mu, \frac{1}{2}\nu}(ab) \left\{ \frac{\Gamma\left( \frac{1+\nu+\mu}{2} \right)}{\Gamma(\nu+1)} \tan\left( \frac{\nu-\mu}{2}\pi \right) M_{\frac{1}{2}\mu, \frac{1}{2}\nu}(ab) \right.$$

$$\left. + \sec\left( \frac{\nu-\mu}{2}\pi \right) W_{\frac{1}{2}\mu, \frac{1}{2}\nu}(ab) \right\}$$

$$\left[ \mathrm{Re}\, a > 0, \quad b > 0, \quad |\mathrm{Re}\, \nu| < \tfrac{1}{2} + \tfrac{1}{2}\mathrm{Re}\, \mu \right] \quad \text{ET II 105(42)}$$

**6.573**

1. $\int_0^\infty x^{\nu - M + 1} J_\nu(bx) \prod_{i=1}^k J_{\mu_i}(a_i x) \, \mathrm{d}x = 0 \qquad\qquad M = \sum_{i=1}^k \mu_i$

$$\left[ a_i > 0, \quad \sum_{i=1}^k a_i < b < \infty, \quad -1 < \mathrm{Re}\, \nu < \mathrm{Re}\, M + \tfrac{1}{2}k - \tfrac{1}{2} \right] \quad \text{ET II 54(42)}$$

2. $\int_0^\infty x^{\nu - M - 1} J_\nu(bx) \prod_{i=1}^k J_{\mu_i}(a_i x) \, \mathrm{d}x = 2^{\nu - M - 1} b^{-\nu} \Gamma(\nu) \prod_{i=1}^k \dfrac{a_i^{\mu_i}}{\Gamma(1 + \mu_i)}, \qquad M = \sum_{i=1}^k \mu_i$

$$\left[ a_i > 0, \quad \sum_{i=1}^k a_i < b < \infty, \quad 0 < \mathrm{Re}\, \nu < \mathrm{Re}\, M + \tfrac{1}{2}k + \tfrac{3}{2} \right] \quad \text{WA 460(16)a, ET II 54(43)}$$

**6.574**

1.[8] 
$$\int_0^\infty J_\nu(\alpha t)\, J_\mu(\beta t) t^{-\lambda}\, dt = \frac{\alpha^\nu\, \Gamma\left(\dfrac{\nu+\mu-\lambda+1}{2}\right)}{2^\lambda \beta^{\nu-\lambda+1}\, \Gamma\left(\dfrac{-\nu+\mu+\lambda+1}{2}\right)\Gamma(\nu+1)}$$
$$\times F\left(\frac{\nu+\mu-\lambda+1}{2},\frac{\nu-\mu-\lambda+1}{2};\nu+1;\frac{\alpha^2}{\beta^2}\right)$$
$$[\operatorname{Re}(\nu+\mu-\lambda+1)>0,\quad \operatorname{Re}\lambda>-1,\quad 0<\alpha<\beta]\quad \text{WA 439(2)a, MO 49}$$

If we reverse the positions of $\nu$ and $\mu$ and at the same time reverse the positions of $\alpha$ and $\beta$, the function on the right hand side of this equation will change. Thus, the right hand side represents a function of $\dfrac{\alpha}{\beta}$ that is not analytic at $\dfrac{\alpha}{\beta}=1$.

For $\alpha=\beta$, we have the following equation

2. 
$$\int_0^\infty J_\nu(\alpha t)\, J_\mu(\alpha t) t^{-\lambda}\, dt = \frac{\alpha^{\lambda-1}\,\Gamma(\lambda)\,\Gamma\left(\dfrac{\nu+\mu-\lambda+1}{2}\right)}{2^\lambda\,\Gamma\left(\dfrac{-\nu+\mu+\lambda+1}{2}\right)\Gamma\left(\dfrac{\nu+\mu+\lambda+1}{2}\right)\Gamma\left(\dfrac{\nu-\mu+\lambda+1}{2}\right)}$$
$$[\operatorname{Re}(\nu+\mu+1)>\operatorname{Re}\lambda>0,\quad \alpha>0]$$
$$\text{MO 49, WA 441(2)a}$$

If $\mu-\nu+\lambda+1$ (or $\nu-\mu+\lambda+1$) is a negative integer, the right hand side of equation **6.574** 1 (or **6.574** 3) vanishes.

**6.575**

1.[11] 
$$\int_0^\infty J_{\nu+1}(\alpha t)\, J_\mu(\beta t) t^{\mu-\nu}\, dt = 0 \qquad\qquad [\alpha<\beta]$$
$$= \frac{(\alpha^2-\beta^2)^{\nu-\mu}\beta^\mu}{2^{\nu-\mu}\alpha^{\nu+1}\,\Gamma(\nu-\mu+1)}\quad [\alpha\geq\beta]$$
$$[\operatorname{Re}(\nu+1)>\operatorname{Re}\mu>-1]\qquad\qquad \text{MO 51}$$

2. 
$$\int_0^\infty \frac{J_\nu(x)\, J_\mu(x)}{x^{\nu+\mu}}\, dx = \frac{\sqrt{\pi}\,\Gamma(\nu+\mu)}{2^{\nu+\mu}\,\Gamma\left(\nu+\mu+\frac{1}{2}\right)\Gamma\left(\nu+\frac{1}{2}\right)\Gamma\left(\mu+\frac{1}{2}\right)}$$
$$[\operatorname{Re}(\nu+\mu)>0]\qquad \text{KU 147(17), WA 434(1)}$$

**6.576**

1. 
$$\int_0^\infty x^{\mu-\nu+1}\, J_\mu(x)\, K_\nu(x)\, dx = \tfrac{1}{2}\,\Gamma(\mu-\nu+1) \qquad\qquad [\operatorname{Re}\mu>-1,\quad \operatorname{Re}(\mu-\nu)>-1]$$
$$\text{ET II 370(47)}$$

2.[11] 
$$\int_0^\infty x^{-\lambda}\, J_\nu(ax)\, J_\nu(bx)\, dx = \frac{a^\nu b^\nu\, \Gamma\left(\nu+\dfrac{1-\lambda}{2}\right)}{2^\lambda(a+b)^{2\nu-\lambda+1}\,\Gamma(\nu+1)\,\Gamma\left(\dfrac{1+\lambda}{2}\right)}$$
$$\times F\left(\nu+\frac{1-\lambda}{2},\nu+\frac{1}{2};2\nu+1;\frac{4ab}{(a+b)^2}\right)$$
$$[a>0,\quad b>0,\quad 2\operatorname{Re}\nu+1>\operatorname{Re}\lambda>-1]\quad \text{ET II 47(4)}$$

3.    $\int_0^\infty x^{-\lambda} K_\mu(ax) J_\nu(bx)\, \mathrm{d}x = \dfrac{b^\nu \Gamma\left(\dfrac{\nu-\lambda+\mu+1}{2}\right)\Gamma\left(\dfrac{\nu-\lambda-\mu+1}{2}\right)}{2^{\lambda+1}a^{\nu-\lambda+1}\Gamma(1+\nu)}$

$$\times F\left(\frac{\nu-\lambda+\mu+1}{2}, \frac{\nu-\lambda-\mu+1}{2}; \nu+1; -\frac{b^2}{a^2}\right)$$

$$[\mathrm{Re}\,(a\pm ib) > 0, \quad \mathrm{Re}(\nu-\lambda+1) > |\mathrm{Re}\,\mu|] \quad \text{EH II 52(31), ET II 63(4), WA 449(1)}$$

4.    $\int_0^\infty x^{-\lambda} K_\mu(ax) K_\nu(bx)\, \mathrm{d}x = \dfrac{2^{-2-\lambda}a^{-\nu+\lambda-1}b^\nu}{\Gamma(1-\lambda)}\Gamma\left(\dfrac{1-\lambda+\mu+\nu}{2}\right)\Gamma\left(\dfrac{1-\lambda-\mu+\nu}{2}\right)$

$$\times \Gamma\left(\frac{1-\lambda+\mu-\nu}{2}\right)\Gamma\left(\frac{1-\lambda-\mu-\nu}{2}\right)$$

$$\times F\left(\frac{1-\lambda+\mu+\nu}{2}, \frac{1-\lambda-\mu+\nu}{2}; 1-\lambda; 1-\frac{b^2}{a^2}\right)$$

$$[\mathrm{Re}(a+b) > 0, \quad \mathrm{Re}\,\lambda < 1 - |\mathrm{Re}\,\mu| - |\mathrm{Re}\,\nu|] \quad \text{ET II 145(49), EH II 93(36)}$$

5.    $\int_0^\infty x^{-\lambda} K_\mu(ax) I_\nu(bx)\, \mathrm{d}x = \dfrac{b^\nu \Gamma\left(\frac{1}{2}-\frac{1}{2}\lambda+\frac{1}{2}\mu+\frac{1}{2}\nu\right)\Gamma\left(\frac{1}{2}-\frac{1}{2}\lambda-\frac{1}{2}\mu+\frac{1}{2}\nu\right)}{2^{\lambda+1}\Gamma(\nu+1)a^{-\lambda+\nu+1}}$

$$\times F\left(\frac{1}{2}-\frac{1}{2}\lambda+\frac{1}{2}\mu+\frac{1}{2}\nu, \frac{1}{2}-\frac{1}{2}\lambda-\frac{1}{2}\mu+\frac{1}{2}\nu; \nu+1; \frac{b^2}{a^2}\right)$$

$$[\mathrm{Re}\,(\nu+1-\lambda\pm\mu) > 0, \quad a > b] \quad \text{EH II 93(35)}$$

6.    $\int_0^\infty x^{-\lambda} Y_\mu(ax) J_\nu(bx)\, \mathrm{d}x = \dfrac{2}{\pi}\sin\dfrac{\pi(\nu-\mu-\lambda)}{2}\int_0^\infty x^{-\lambda} K_\mu(ax) I_\nu(bx)\, \mathrm{d}x$

$$[a > b, \quad \mathrm{Re}\,\lambda > -1, \quad \mathrm{Re}\,(\nu-\lambda+1\pm\mu) > 0] \quad (\text{see } \mathbf{6.576}\ 5) \quad \text{EH II 93(37)}$$

7.8    $\int_0^\infty x^{\mu+\nu+1} J_\mu(ax) K_\nu(bx)\, \mathrm{d}x = 2^{\mu+\nu}a^\mu b^\nu \dfrac{\Gamma(\mu+\nu+1)}{(a^2+b^2)^{\mu+\nu+1}}$

$$[\mathrm{Re}\,\mu > |\mathrm{Re}\,\nu| - 1, \quad \mathrm{Re}\,b > |\mathrm{Im}\,a|]$$

$$\text{ET 137(16), EH II 93(36)}$$

## 6.577

1.8    $\int_0^\infty x^{\nu-\mu+1+2n} J_\mu(ax) J_\nu(bx)\dfrac{\mathrm{d}x}{x^2+c^2} = (-1)^n c^{\nu-\mu+2n} I_\mu(ac) K_\nu(bc)$

$$[a > 0, \quad b > a, \quad \mathrm{Re}\,c > 0, \quad 2 + \mathrm{Re}\,\mu - 2n > \mathrm{Re}\,\nu > -1 - n, \quad n \geq 0 \text{ an integer}] \quad \text{ET II 49(13)}$$

2.12    $\int_0^\infty x^{\mu-\nu+1+2n} J_\mu(ax) J_\nu(bx)\dfrac{\mathrm{d}x}{x^2+c^2} = (-1)^n c^{\mu-\nu+2n} I_\nu(bc) K_\mu(ac)$

$$[b > 0, \quad a > b, \quad \mathrm{Re}\,c > 0, \quad \mathrm{Re}\,\nu - 2n + 2 > \mathrm{Re}\,\mu > -n - 1, \quad n \geq 0 \text{ an integer}] \quad \text{ET II 49(15)}$$

**6.578**

1.    $\displaystyle\int_0^\infty x^{\varrho-1} J_\lambda(ax) J_\mu(bx) J_\nu(cx)\,dx = \frac{2^{\varrho-1} a^\lambda b^\mu c^{-\lambda-\mu-\varrho}\, \Gamma\left(\frac{\lambda+\mu+\nu+\varrho}{2}\right)}{\Gamma(\lambda+1)\,\Gamma(\mu+1)\,\Gamma\left(1-\dfrac{\lambda+\mu-\nu+\varrho}{2}\right)}$

$$\times F_4\left(\frac{\lambda+\mu-\nu+\varrho}{2}, \frac{\lambda+\mu+\nu+\varrho}{2}; \lambda+1, \mu+1; \frac{a^2}{c^2}, \frac{b^2}{c^2}\right)$$

$$\left[\operatorname{Re}(\lambda+\mu+\nu+\varrho) > 0, \quad \operatorname{Re}\varrho < \frac{5}{2}, \quad a > 0, \quad b > 0, \quad c > 0, \quad c > a+b\right] \quad \text{ET II 351(9)}$$

2.    $\displaystyle\int_0^\infty x^{\varrho-1} J_\lambda(ax) J_\mu(bx) K_\nu(cx)\,dx$

$$= \frac{2^{\varrho-2} a^\lambda b^\mu c^{-\varrho-\lambda-\mu}}{\Gamma(\lambda+1)\,\Gamma(\mu+1)}\, \Gamma\left(\frac{\varrho+\lambda+\mu-\nu}{2}\right)\Gamma\left(\frac{\varrho+\lambda+\mu+\nu}{2}\right)$$

$$\times F_4\left(\frac{\varrho+\lambda+\mu-\nu}{2}, \frac{\varrho+\lambda+\mu+\nu}{2}; \lambda+1, \mu+1; -\frac{a^2}{c^2}, -\frac{b^2}{c^2}\right)$$

$$\left[\operatorname{Re}(\varrho+\lambda+\mu) > |\operatorname{Re}\nu|, \quad \operatorname{Re}c > |\operatorname{Im}a| + |\operatorname{Im}b|\right] \quad \text{ET II 373(8)}$$

3.    $\displaystyle\int_0^\infty x^{\lambda-\mu-\nu+1} J_\nu(ax) J_\mu(bx) J_\lambda(cx)\,dx = 0$

$$\left[\operatorname{Re}\lambda > -1, \quad \operatorname{Re}(\lambda-\mu-\nu) < \frac{1}{2}, \quad c > b > 0, \quad 0 < a < c-b\right] \quad \text{ET II 53(36)}$$

4.    $\displaystyle\int_0^\infty x^{\lambda-\mu-\nu-1} J_\nu(ax) J_\mu(bx) J_\lambda(cx)\,dx = \frac{2^{\lambda-\mu-\nu-1} a^\nu b^\mu\, \Gamma(\lambda)}{c^\lambda\, \Gamma(\mu+1)\,\Gamma(\nu+1)}$

$$\left[\operatorname{Re}\lambda > 0, \quad \operatorname{Re}(\lambda-\mu-\nu) < \frac{5}{2}, \quad c > b > 0, \quad 0 < a < c-b\right] \quad \text{ET II 53(37)}$$

5.    $\displaystyle\int_0^\infty x^{1+\mu} Y_\mu(ax) J_\nu(bx) J_\nu(cx)\,dx = 0 \qquad\qquad \left[0 < b < c, \quad 0 < a < c-b\right]$

$$\text{ET II 352(13)}$$

6.[11]    $\displaystyle\int_0^\infty x^{\mu+1} K_\mu(ax) J_\nu(bx) J_\nu(cx)\,dx = \frac{1}{\sqrt{2\pi}} a^\mu b^{-\mu-1} c^{-\mu-1} e^{-\left(\mu+\frac{1}{2}\right)\pi i}\left(u^2-1\right)^{-\frac{1}{2}\mu-\frac{1}{4}} Q_{\nu-\frac{1}{2}}^{\mu+\frac{1}{2}}(u)$

$$\left[2bcu = a^2+b^2+c^2, \quad \operatorname{Re}a > |\operatorname{Im}b| + |\operatorname{Im}c|, \quad \operatorname{Re}\nu > -1, \quad \operatorname{Re}(\mu+\nu) > -1\right]$$

$$\text{WA 452(2), ET II 64(12)}$$

7.[11]    $\displaystyle\int_0^\infty x^{\mu+1} I_\nu(ax) K_\mu(bx) J_\nu(cx)\,dx = \frac{1}{\sqrt{2\pi}} a^{-\mu-1} b^\mu c^{-\mu-1} e^{-\left(\mu-\frac{1}{2}\nu+\frac{1}{4}\right)\pi i}\left(v^2+1\right)^{-\frac{1}{2}\mu-\frac{1}{4}} Q_{\nu-\frac{1}{2}}^{\mu+\frac{1}{2}}(iv)$

$$\left[2acv = b^2-a^2+c^2, \quad \operatorname{Re}b > |\operatorname{Re}a| + |\operatorname{Im}c|, \quad \operatorname{Re}\nu > -1, \quad \operatorname{Re}(\mu+\nu) > -1\right] \quad \text{ET II 66(22)}$$

8.[11]    $\displaystyle\int_0^\infty x^{1-\mu} J_\mu(ax) J_\nu(bx) J_\nu(cx)\,dx$

$$= \sqrt{\frac{2}{\pi^3}} a^{-\mu}(bc)^{\mu-1}(\sinh u)^{\mu-\frac{1}{2}} \sin[(\mu-\nu)\pi] e^{\left(\mu-\frac{1}{2}\right)\pi i} Q_{\nu-\frac{1}{2}}^{\frac{1}{2}-\mu}(\cosh u) \qquad [a > b+c]$$

$$= \frac{1}{\sqrt{2\pi}} a^{-\mu}(bc)^{\mu-1}(\sin v)^{\mu-\frac{1}{2}} P_{\nu-\frac{1}{2}}^{\frac{1}{2}-\mu}(\cos v) \qquad\qquad [|b-c| < a < b+c]$$

$$= 0 \qquad\qquad\qquad\qquad\qquad\qquad\qquad\qquad\qquad\qquad\qquad [0 < a < |b-c|]$$

$$\left[2bc\cosh u = a^2-b^2-c^2, \quad 2bc\cos v = b^2+c^2-a^2, \quad b > 0, \quad c > 0; \quad \operatorname{Re}\nu > -1, \operatorname{Re}\mu > -\frac{1}{2}\right]$$

9. $\displaystyle\int_0^\infty J_\nu(ax)\, J_\nu(bx)\, J_\nu(cx)\, x^{1-\nu}\, \mathrm{d}x = 0 \qquad\qquad\qquad [0 < c \le |a-b| \text{ or } c \ge a+b]$

$$= \frac{2^{\nu-1}\Delta^{2\nu-1}}{(abc)^\nu\,\Gamma\left(\nu+\frac12\right)\Gamma\left(\frac12\right)} \qquad [|a-b| < c < a+b]$$

$$\left[\Delta = \tfrac14\sqrt{[c^2-(a-b)^2][(a+b)^2-c^2]},\quad a>0,\quad b>0,\quad c>0,\quad \mathrm{Re}\,\nu > -\tfrac12\right]$$

($\Delta > 0$ is equal to the area of a triangle whose sides are $a$, $b$, and $c$.)

10.[11] $\displaystyle\int_0^\infty x^{\nu+1} K_\mu(ax)\, K_\mu(bx)\, J_\nu(cx)\, \mathrm{d}x = \frac{\sqrt\pi\, c^\nu\,\Gamma(\nu+\mu+1)\,\Gamma(\nu-\mu+1)}{2^{\frac23}(ab)^{\nu+1}(u^2-1)^{\frac12\nu+\frac14}}\, P_{\mu-\frac12}^{-\nu-\frac12}(u)$

$$\left[2abu = a^2+b^2+c^2,\quad \mathrm{Re}(a+b) > |\mathrm{Im}\,c|,\quad \mathrm{Re}(\nu\pm\mu) > -1,\quad \mathrm{Re}\,\nu > -1\right]\quad \text{ET II 67(30)}$$

11.[11] $\displaystyle\int_0^\infty x^{\nu+1} K_\mu(ax)\, I_\mu(bx)\, J_\nu(cx)\, \mathrm{d}x = \frac{(ab)^{-\nu-1}c^\nu e^{-\left(\nu+\frac12\right)\pi i}\, Q_{\mu-\frac12}^{\nu+\frac12}(u)}{\sqrt{2\pi}\,(u^2-1)^{\frac12\nu+\frac14}}$

$$\left[2abu = a^2+b^2+c^2,\quad \mathrm{Re}\,a > |\mathrm{Re}\,b| + |\mathrm{Im}\,c|,\quad \mathrm{Re}\,\nu > -1,\quad \mathrm{Re}(\mu+\nu) > -1\right]\quad \text{ET II 66(24)}$$

12.[8] $\displaystyle\int_0^\infty x^{\nu+1}\left[J_\nu(ax)\right]^2 Y_\nu(bx)\, \mathrm{d}x = 0 \qquad\qquad\qquad \left[0 < b < 2a,\quad |\mathrm{Re}\,\nu| < \tfrac12\right]$

$$= \frac{2^{3\nu+1}a^{2\nu}b^{-\nu-1}}{\sqrt\pi\,\Gamma\left(\frac12-\nu\right)}\left(b^2-4a^2\right)^{-\nu-\frac12} \qquad \left[0 < 2a < b,\quad |\mathrm{Re}\,\nu| < \tfrac12\right]$$

$$\text{ET II 109(3)}$$

13.[12] $\displaystyle\int_0^\infty x^{\nu+1} J_\nu(ax)\, Y_\nu(ax)\, J_\nu(bx)\, \mathrm{d}x$

$$= 0 \qquad\qquad\qquad \left[a>0,\quad |\mathrm{Re}\,\nu| < \tfrac12,\quad 0 < b < 2a\right]$$

$$= -\frac{2^{3\nu+1}a^{2\nu}b^{-\nu-1}}{\sqrt\pi\,\Gamma\left(\frac12-\nu\right)}\left(b^2-4a^2\right)^{-\nu-\frac12} \qquad \left[a>0,\quad 2a<b,\quad |\mathrm{Re}\,\nu| < \tfrac12\right]$$

$$\text{ET II 55(49)}$$

14. $\displaystyle\int_0^\infty x^{\nu+1} J_\mu\left(xa\sin\psi\right) J_\nu\left(xa\sin\varphi\right) K_\mu\left(xa\cos\varphi\cos\psi\right)\, \mathrm{d}x$

$$= \frac{2^\nu\,\Gamma(\mu+\nu+1)\,(\sin\varphi)^\nu\left(\cos\frac\alpha2\right)^{2\nu+1}}{a^{\nu+2}\left(\cos\psi\right)^{2\nu+2}}\, P_\nu^{-\mu}(\cos\alpha)$$

$$\left[\tan\tfrac12\alpha = \tan\psi\cos\varphi,\quad a>0,\quad \frac\pi2 > \varphi > 0,\quad 0 < \psi < \frac\pi2,\quad \mathrm{Re}\,\nu > -1,\quad \mathrm{Re}(\mu+\nu) > -1\right]$$

$$\text{ET II 64(11)}$$

15. $\displaystyle\int_0^\infty x^{\nu+1} J_\nu(ax)\, K_\nu(bx)\, J_\nu(cx)\, \mathrm{d}x = \frac{2^{3\nu}(abc)^\nu\,\Gamma\left(\nu+\frac12\right)}{\sqrt\pi\left[\left(a^2+b^2+c^2\right)^2 - 4a^2c^2\right]^{\nu+\frac12}}$

$$\left[\mathrm{Re}\,b > |\mathrm{Im}\,a|,\quad c>0,\quad \mathrm{Re}\,\nu > -\tfrac12\right]$$

$$\text{ET II 63(8)}$$

$16.8$
$$\int_0^\infty x^{\nu+1} I_\nu(ax) K_\nu(bx) J_\nu(cx)\, dx = \frac{2^{3\nu}(abc)^\nu \Gamma\left(\nu+\frac{1}{2}\right)}{\sqrt{\pi}\left[(b^2-a^2+c^2)^2+4a^2c^2\right]^{\nu+\frac{1}{2}}}$$

$$\left[\operatorname{Re} b > |\operatorname{Re} a| + |\operatorname{Im} c|; \quad \operatorname{Re}\nu > -\tfrac{1}{2}\right]$$
ET II 65(18)

**6.579**

1.
$$\int_0^\infty x^{2\nu+1} J_\nu(ax) Y_\nu(ax) J_\nu(bx) Y_\nu(bx)\, dx$$
$$= \frac{a^{2\nu}\Gamma(3\nu+1)}{2\pi b^{4\nu+2}\Gamma\left(\frac{1}{2}-\nu\right)\Gamma\left(2\nu+\frac{3}{2}\right)} F\left(\nu+\tfrac{1}{2}, 3\nu+1; 2\nu+\tfrac{3}{2}; \frac{a^2}{b^2}\right)$$
$$\left[0 < a < b, \quad -\tfrac{1}{3} < \operatorname{Re}\nu < \tfrac{1}{2}\right] \quad \text{EH II 94(45), ET II 352(15)}$$

2.
$$\int_0^\infty x^{2\nu+1} J_\nu(ax) K_\nu(ax) J_\nu(bx) K_\nu(bx)\, dx$$
$$= \frac{2^{\nu-3}a^{2\nu}\Gamma\left(\frac{\nu+1}{2}\right)\Gamma\left(\nu+\frac{1}{2}\right)\Gamma\left(\frac{3\nu+1}{2}\right)}{\sqrt{\pi}b^{4\nu+2}\Gamma(\nu+1)} F\left(\nu+\tfrac{1}{2}, \frac{3\nu+1}{2}; 2\nu+1; 1-\frac{a^4}{b^4}\right)$$
$$\left[0 < a < b, \quad \operatorname{Re}\nu > -\tfrac{1}{3}\right] \quad \text{ET II 373(10)}$$

3.
$$\int_0^\infty x^{1-2\nu}\left[J_\nu(ax)\right]^4 dx = \frac{\Gamma(\nu)\Gamma(2\nu)}{2\pi\left[\Gamma\left(\nu+\frac{1}{2}\right)\right]^2\Gamma(3\nu)} \qquad [\operatorname{Re}\nu > 0] \qquad \text{ET II 342(25)}$$

4.
$$\int_0^\infty x^{1-2\nu}\left[J_\nu(ax)\right]^2\left[J_\nu(bx)\right]^2 dx = \frac{a^{2\nu-1}\Gamma(\nu)}{2\pi b\,\Gamma\left(\nu+\frac{1}{2}\right)\Gamma\left(2\nu+\frac{1}{2}\right)} F\left(\nu, \tfrac{1}{2}-\nu; 2\nu+\tfrac{1}{2}; \frac{a^2}{b^2}\right)$$
ET II 351(10)

**6.581**

1.
$$\int_0^a x^{\lambda-1} J_\mu(x) J_\nu(a-x)\, dx = 2^\lambda \sum_{m=0}^\infty \frac{(-1)^m \Gamma(\lambda+\mu+m)\Gamma(\lambda+m)}{m!\,\Gamma(\lambda)\Gamma(\mu+m+1)} J_{\lambda+\mu+\nu+2m}(a)$$
$$[\operatorname{Re}(\lambda+\mu) > 0, \quad \operatorname{Re}\nu > -1]$$
ET II 354(25)

$2.8$
$$\int_0^a x^{\lambda-1}(a-x)^{-1} J_\mu(x) J_\nu(a-x)\, dx$$
$$= \frac{2^\lambda}{a\nu} \sum_{m=0}^\infty \frac{(-1)^m \Gamma(\lambda+\mu+m)\Gamma(\lambda+m)}{m!\,\Gamma(\lambda)\Gamma(\mu+m+1)}(\lambda+\mu+\nu+2m) J_{\lambda+\mu+\nu+2m}(a)$$
$$[\operatorname{Re}(\lambda+\mu) > 0, \quad \operatorname{Re}\nu > 0] \quad \text{ET II 354(27)}$$

3.
$$\int_0^a x^\mu (a-x)^\nu J_\mu(x) J_\nu(a-x)\, dx = \frac{\Gamma\left(\mu+\frac{1}{2}\right)\Gamma\left(\nu+\frac{1}{2}\right)}{\sqrt{2\pi}\,\Gamma(\mu+\nu+1)} a^{\mu+\nu+\frac{1}{2}} J_{\mu+\nu+\frac{1}{2}}(a)$$
$$\left[\operatorname{Re}\mu > -\tfrac{1}{2}, \quad \operatorname{Re}\nu > -\tfrac{1}{2}\right]$$
ET II 354(28), EH II 46(6)

4.
$$\int_0^a x^\mu (a-x)^{\nu+1} J_\mu(x) J_\nu(a-x)\, dx = \frac{\Gamma\left(\mu+\frac{1}{2}\right)\Gamma\left(\nu+\frac{3}{2}\right)}{\sqrt{2\pi}\,\Gamma(\mu+\nu+2)} a^{\mu+\nu+\frac{3}{2}} J_{\mu+\nu+\frac{1}{2}}(a)$$
$$\left[\operatorname{Re}\nu > -1, \quad \operatorname{Re}\mu > -\tfrac{1}{2}\right]$$
ET II 354(29)

5. $\displaystyle\int_0^a x^\mu (a-x)^{-\mu-1} J_\mu(x) J_\nu(a-x)\,dx = \frac{2^\mu\,\Gamma\left(\mu+\frac{1}{2}\right)\Gamma(\nu-\mu)}{\sqrt{\pi}\,\Gamma(\mu+\nu+1)} a^\mu\,J_\nu(a)$

$$\left[\operatorname{Re}\nu > \operatorname{Re}\mu > -\tfrac{1}{2}\right] \qquad \text{ET II 355(30)}$$

**6.582** $\displaystyle\int_0^\infty x^{\mu-1}|x-b|^{-\mu} K_\mu\left(|x-b|\right) K_\nu(x)\,dx = \frac{1}{\sqrt{\pi}}(2b)^{-\mu}\,\Gamma\left(\tfrac{1}{2}-\mu\right)\Gamma(\mu+\nu)\,\Gamma(\mu-\nu)\,K_\nu(b)$

$$\left[b>0,\quad \operatorname{Re}\mu < \tfrac{1}{2},\quad \operatorname{Re}\mu > |\operatorname{Re}\nu|\right]$$
$$\text{ET II 374(14)}$$

**6.583** $\displaystyle\int_0^\infty x^{\mu-1}(x+b)^{-\mu} K_\mu(x+b) K_\nu(x)\,dx = \frac{\sqrt{\pi}\,\Gamma(\mu+\nu)\,\Gamma(\mu-\nu)}{2^\mu b^\mu\,\Gamma\left(\mu+\frac{1}{2}\right)}\,K_\nu(b)$

$$\left[|\arg b| < \pi,\quad \operatorname{Re}\mu > |\operatorname{Re}\nu|\right]$$
$$\text{ET II 374(15)}$$

**6.584**

$1.^8$ $\displaystyle\int_0^\infty \frac{x^{\varrho-1}\left[H_\nu^{(1)}(ax) - e^{\varrho\pi i} H_\nu^{(1)}\left(axe^{\pi i}\right)\right]}{(x^2-r^2)^{m+1}}\,dx = \frac{\pi i}{2^m m!}\left(\frac{d}{r\,dr}\right)^m\left[r^{\varrho-2}\,H_\nu^{(1)}(ar)\right]$

$$\left[m=0,1,2,\dots,\quad \operatorname{Im}r>0,\quad a>0,\quad |\operatorname{Re}\nu| < \operatorname{Re}\varrho < 2m+\tfrac{7}{2}\right] \quad \text{WA 465}$$

$2.^8$ $\displaystyle\int_0^\infty \left[\cos\tfrac{1}{2}(\varrho-\nu)\pi\,J_\nu(ax) + \sin\tfrac{1}{2}(\varrho-\nu)\pi\,Y_\nu(ax)\right]\frac{x^{\varrho-1}}{(x^2+k^2)^{m+1}}\,dx$

$$= \frac{(-1)^{m+1}}{2^m\cdot m!}\left(\frac{d}{k\,dk}\right)^m\left[k^{\varrho-2}\,K_\nu(ak)\right]$$

$$\left[m=0,1,2,\dots,\quad \operatorname{Re}k>0,\quad a>0,\quad |\operatorname{Re}\nu| < \operatorname{Re}\varrho < 2m+\tfrac{7}{2}\right] \quad \text{WA 466(2)}$$

3. $\displaystyle\int_0^\infty \left\{\cos\nu\pi\,J_\nu(ax) - \sin\nu\pi\,Y_\nu(ax)\right\}\frac{x^{1-\nu}\,dx}{(x^2+k^2)^{m+1}} = \frac{a^m\,K_{\nu+m}(ak)}{2^m\cdot m!k^{\nu+m}}$

$$\left[m=0,1,2,\dots,\quad \operatorname{Re}k>0,\quad a>0,\quad -2m-\tfrac{3}{2} < \operatorname{Re}\nu < 1\right] \quad \text{WA 466(3)}$$

4. $\displaystyle\int_0^\infty \left\{\cos\left[\left(\tfrac{1}{2}\varrho-\tfrac{1}{2}\nu-\mu\right)\pi\right] J_\nu(ax) + \sin\left[\left(\tfrac{1}{2}\varrho-\tfrac{1}{2}\nu-\mu\right)\pi\right] Y_\nu(ax)\right\}\frac{x^{\varrho-1}}{(x^2+k^2)^{\mu+1}}\,dx$

$$= \frac{\pi k^{\varrho-2\mu-2}}{2\sin\nu\pi\cdot\Gamma(\mu+1)}\left[\frac{\left(\tfrac{1}{2}ak\right)^\nu\Gamma\left(\tfrac{1}{2}\varrho+\tfrac{1}{2}\nu\right)}{\Gamma(\nu+1)\,\Gamma\left(\tfrac{1}{2}\varrho+\tfrac{1}{2}\nu-\mu\right)}\,{}_1F_2\left(\frac{\varrho+\nu}{2};\frac{\varrho+\nu}{2}-\mu,\nu+1;\frac{a^2k^2}{4}\right)\right.$$

$$\left. - \frac{\left(\tfrac{1}{2}ak\right)^{-\nu}\Gamma\left(\tfrac{1}{2}\varrho-\tfrac{1}{2}\nu\right)}{\Gamma(1-\nu)\,\Gamma\left(\tfrac{1}{2}\varrho-\tfrac{1}{2}\nu-\mu\right)}\,{}_1F_2\left(\frac{\varrho-\nu}{2};\frac{\varrho-\nu}{2}-\mu,1-\nu;\frac{a^2k^2}{4}\right)\right]$$

$$\left[a>0,\quad \operatorname{Re}k>0,\quad |\operatorname{Re}\nu| < \operatorname{Re}\varrho < 2\operatorname{Re}\mu+\tfrac{7}{2}\right] \quad \text{WA 407(1)}$$

5.$^{12}$  $$\int_0^\infty \left[\prod_{j=1}^n J_{\mu_j}(b_n x)\right] \left\{\cos\left[\frac{1}{2}\left(\varrho + \sum_{j=1}^n \mu_j - \nu\right)\pi\right] J_\nu(ax)\right.$$

$$\left. + \sin\left[\frac{1}{2}\left(\varrho + \sum_{j=1}^n \mu_j - \nu\right)\pi\right] Y_\nu(ax)\right\} \frac{x^{\varrho-1}}{x^2+k^2}\,\mathrm{d}x$$

$$= -\left[\prod_{j=1}^n I_{\mu_j}(b_n k)\right] K_\nu(ak) k^{\varrho-2}$$

$$\left[\operatorname{Re} k > 0, \quad a > \sum_{j=1}^n |\operatorname{Re} b_j|, \quad \operatorname{Re}\left(\varrho + \sum_{j=1}^n \mu_j\right) > |\operatorname{Re}\nu|\right] \qquad \text{WA 472(9)}$$

## 6.59 Combinations of powers and Bessel functions of more complicated arguments

**6.591**

1.  $$\int_0^\infty x^{2\nu+\frac{1}{2}} J_{\nu+\frac{1}{2}}\left(\frac{a}{x}\right) K_\nu(bx)\,\mathrm{d}x = \sqrt{2\pi}\, b^{-\nu-1} a^{\nu+\frac{1}{2}} J_{1+2\nu}\left(\sqrt{2ab}\right) K_{1+2\nu}\left(\sqrt{2ab}\right)$$
    $$[a > 0, \quad \operatorname{Re} b > 0, \quad \operatorname{Re}\nu > -1]$$
    ET II 142(35)

2.  $$\int_0^\infty x^{2\nu+\frac{1}{2}} Y_{\nu+\frac{1}{2}}\left(\frac{a}{x}\right) K_\nu(bx)\,\mathrm{d}x = \sqrt{2\pi}\, b^{-\nu-1} a^{\nu+\frac{1}{2}} Y_{2\nu+1}\left(\sqrt{2ab}\right) K_{2\nu+1}\left(\sqrt{2ab}\right)$$
    $$[a > 0, \quad \operatorname{Re} b > 0, \quad \operatorname{Re}\nu > -1]$$
    ET II 143(41)

3.  $$\int_0^\infty x^{2\nu+\frac{1}{2}} K_{\nu+\frac{1}{2}}\left(\frac{a}{x}\right) K_\nu(bx)\,\mathrm{d}x = \sqrt{2\pi}\, b^{-\nu-1} a^{\nu+\frac{1}{2}} K_{2\nu+1}\left(e^{\frac{1}{4}i\pi}\sqrt{2ab}\right) K_{2\nu+1}\left(e^{-\frac{1}{4}i\pi}\sqrt{2ab}\right)$$
    $$[\operatorname{Re} a > 0, \quad \operatorname{Re} b > 0] \qquad \text{ET II 146(56)}$$

4.  $$\int_0^\infty x^{-2\nu+\frac{1}{2}} J_{\nu-\frac{1}{2}}\left(\frac{a}{x}\right) K_\nu(bx)\,\mathrm{d}x = \sqrt{2\pi}\, b^{\nu-1} a^{\frac{1}{2}-\nu} K_{2\nu-1}\left(\sqrt{2ab}\right)$$
    $$\times \left[\sin(\nu\pi) J_{2\nu-1}\left(\sqrt{2ab}\right) + \cos(\nu\pi) Y_{2\nu-1}\left(\sqrt{2ab}\right)\right]$$
    $$[a > 0, \quad \operatorname{Re} b > 0, \quad \operatorname{Re}\nu < 1] \quad \text{ET II 142(34)}$$

5.  $$\int_0^\infty x^{-2\nu+\frac{1}{2}} Y_{\nu-\frac{1}{2}}\left(\frac{a}{x}\right) K_\nu(bx)\,\mathrm{d}x = -\sqrt{\frac{\pi}{2}}\, b^{\nu-1} a^{\frac{1}{2}-\nu} \sec(\nu\pi) K_{2\nu-1}\left(\sqrt{2ab}\right)$$
    $$\times \left[J_{2\nu-1}\left(\sqrt{2ab}\right) - J_{1-2\nu}\left(\sqrt{2ab}\right)\right]$$
    $$[a > 0, \quad \operatorname{Re}\nu < 1] \qquad \text{ET II 143(40)}$$

6.$^{12}$  $$\int_0^\infty x^{-2\nu+\frac{1}{2}} J_{\frac{1}{2}-\nu}\left(\frac{a}{x}\right) J_\nu(bx)\,\mathrm{d}x$$
    $$= -\sqrt{\frac{\pi}{2}}\, i\operatorname{cosec}(2\nu\pi) b^{\nu-1} a^{\frac{1}{2}-\nu} \left[e^{2\nu\pi i} J_{1-2\nu}(u) J_{2\nu-1}(v) - e^{-2\nu\pi i} J_{2\nu-1}(u) J_{1-2\nu}(v)\right]$$
    $$\left[u = \sqrt{2ab}\,e^{\pi i/4}, \quad v = \sqrt{2ab}\,e^{-\pi i/4}, \quad a > 0, \quad b > 0, \quad -\tfrac{1}{2} < \operatorname{Re}\nu < 3\right] \quad \text{ET II 58(12)}$$

7.  $\int_0^\infty x^{-2\nu+\frac{1}{2}} K_{\nu-\frac{1}{2}}\left(\frac{a}{x}\right) Y_\nu(bx)\,dx = \sqrt{2\pi} b^{\nu-1} a^{\frac{1}{2}-\nu} Y_{2\nu-1}\left(\sqrt{2ab}\right) K_{2\nu-1}\left(\sqrt{2ab}\right)$

$$\left[b > 0, \quad \operatorname{Re} a > 0, \quad \operatorname{Re}\nu > \tfrac{1}{6}\right]$$

ET II 113(30)

8.  $\int_0^\infty x^{\varrho-1} J_\mu(ax) J_\nu\left(\frac{b}{x}\right) dx = \dfrac{a^{\nu-\varrho} b^\nu \,\Gamma\left(\frac{1}{2}\mu + \frac{1}{2}\varrho - \frac{1}{2}\nu\right)}{2^{2\nu-\varrho+1}\,\Gamma(\nu+1)\,\Gamma\left(\frac{1}{2}\mu + \frac{1}{2}\nu - \frac{1}{2}\varrho + 1\right)}$

$$\times {}_0F_3\left(\nu+1, \frac{\nu-\mu-\varrho}{2}+1, \frac{\nu+\mu-\varrho}{2}+1; \frac{a^2 b^2}{16}\right)$$

$$+ \frac{a^\mu b^{\mu+\varrho}\,\Gamma\left(\frac{1}{2}\nu - \frac{1}{2}\mu - \frac{1}{2}\varrho\right)}{2^{2\mu+\varrho+1}\,\Gamma(\mu+1)\,\Gamma\left(\frac{1}{2}\mu + \frac{1}{2}\nu + \frac{1}{2}\varrho + 1\right)}$$

$$\times {}_0F_3\left(\mu+1, \frac{\mu-\nu+\varrho}{2}+1, \frac{\nu+\mu+\varrho}{2}+1; \frac{a^2 b^2}{16}\right)$$

$$\left[a > 0, \quad b > 0, \quad -\operatorname{Re}\left(\mu + \tfrac{3}{2}\right) < \operatorname{Re}\varrho < \operatorname{Re}\left(\nu + \tfrac{3}{2}\right)\right] \quad \text{WA 480(1)}$$

## 6.592

1.¹²  $\int_0^1 x^\lambda (1-x)^{\mu-1} Y_\nu\left(a\sqrt{x}\right) dx = 2^{-\nu} a^\nu \cot(\nu\pi) \dfrac{\Gamma(\mu)\,\Gamma\left(\lambda+1+\frac{1}{2}\nu\right)}{\Gamma(1+\nu)\,\Gamma\left(\lambda+1+\mu+\frac{1}{2}\nu\right)}$

$$\times {}_1F_2\left(\lambda+1+\tfrac{1}{2}\nu; 1+\nu, \lambda+1+\mu+\tfrac{1}{2}\nu; -\frac{a^2}{4}\right)$$

$$- 2^\nu a^{-\nu} \operatorname{cosec}(\nu\pi) \frac{\Gamma(\mu)\,\Gamma\left(\lambda+1-\frac{1}{2}\nu\right)}{\Gamma(1-\nu)\,\Gamma\left(\lambda+1+\mu-\frac{1}{2}\nu\right)}$$

$$\times {}_1F_2\left(\lambda-\tfrac{1}{2}\nu+1; 1-\nu, \lambda+1+\mu-\tfrac{1}{2}\nu; -\frac{a^2}{4}\right)$$

$$\left[\operatorname{Re}\lambda > -1 + \tfrac{1}{2}|\operatorname{Re}\nu|, \quad \operatorname{Re}\mu > 0\right] \quad \text{ET II 197(76)a}$$

2.¹²  $\int_0^1 x^\lambda (1-x)^{\mu-1} K_\nu\left(a\sqrt{x}\right) dx$

$$= 2^{\nu-1} a^{-\nu} \frac{\Gamma(\nu)\,\Gamma(\mu)\,\Gamma\left(\lambda+1-\frac{1}{2}\nu\right)}{\Gamma\left(\lambda+1+\mu-\frac{1}{2}\nu\right)} \,{}_1F_2\left(\lambda+1-\tfrac{1}{2}\nu; 1-\nu, \lambda+1+\mu-\tfrac{1}{2}\nu; \frac{a^2}{4}\right)$$

$$+ 2^{-1-\nu} a^\nu \frac{\Gamma(-\nu)\,\Gamma\left(\lambda+1+\frac{1}{2}\nu\right)\Gamma(\mu)}{\Gamma\left(\lambda+1+\mu+\frac{1}{2}\nu\right)} \,{}_1F_2\left(\lambda+1+\tfrac{1}{2}\nu; 1+\nu, \lambda+1+\mu+\tfrac{1}{2}\nu; \frac{a^2}{4}\right)$$

$$= \frac{2^{\nu-1}}{a^\nu} \Gamma(\mu)\, G_{13}^{21}\left(\frac{a^2}{4} \left|\begin{matrix} \frac{\nu}{2}-\lambda \\ \nu, 0, \frac{\nu}{2}-\lambda-\mu \end{matrix}\right.\right)$$

OB 159 (3.16)

$$\left[\operatorname{Re}\lambda > -1 + \tfrac{1}{2}|\operatorname{Re}\nu|, \quad \operatorname{Re}\mu > 0\right] \quad \text{ET II 198(87)a}$$

3.[11]   $\displaystyle\int_1^\infty x^\lambda (x-1)^{\mu-1} J_\nu\left(a\sqrt{x}\right)\,dx = 2^{2\lambda} a^{-2\lambda}\, G_{13}^{20}\left(\frac{a^2}{4}\,\bigg|\,{0 \atop -\mu,\,\lambda+\frac{1}{2}\nu,\,\lambda-\frac{1}{2}\nu}\right)\Gamma(\mu)$

$$\left[a>0,\quad 0<\operatorname{Re}\mu<\tfrac{3}{4}-\operatorname{Re}\lambda\right]$$
ET II 205(36)a

4.   $\displaystyle\int_1^\infty x^\lambda (x-1)^{\mu-1} K_\nu\left(a\sqrt{x}\right)\,dx = \Gamma(\mu)2^{2\lambda-1} a^{-2\lambda}\, G_{13}^{30}\left(\frac{a^2}{4}\,\bigg|\,{0 \atop -\mu,\,\frac{1}{2}\nu+\lambda,\,-\frac{1}{2}\nu+\lambda}\right)$

$$\left[\operatorname{Re}a>0,\quad \operatorname{Re}\mu>0\right]$$    ET II 209(60)a

5.   $\displaystyle\int_0^1 x^{-\frac{1}{2}}(1-x)^{-\frac{1}{2}} J_\nu\left(a\sqrt{x}\right)\,dx = \pi\left[J_{\frac{1}{2}\nu}\left(\frac{1}{2}a\right)\right]^2$    $\left[\operatorname{Re}\nu>-1\right]$    ET II 194(59)a

6.   $\displaystyle\int_0^1 x^{-\frac{1}{2}}(1-x)^{-\frac{1}{2}} I_\nu\left(a\sqrt{x}\right)\,dx = \pi\left[I_{\frac{1}{2}\nu}\left(\frac{1}{2}a\right)\right]^2$    $\left[\operatorname{Re}\nu>-1\right]$    ET II 197(79)

7.[12]   $\displaystyle\int_0^1 x^{-\frac{1}{2}}(1-x)^{-\frac{1}{2}} K_\nu\left(a\sqrt{x}\right)\,dx = \frac{1}{2}\sqrt{\pi}\sec(\nu\pi)\left[I_{\frac{\nu}{2}}\left(\frac{a}{2}\right)+I_{-\frac{\nu}{2}}\left(\frac{a}{2}\right)\right]K_{\frac{\nu}{2}}\left(\frac{a}{2}\right)$

$$\left[|\operatorname{Re}\nu|<1\right]$$    ET II 198(85)a

8.   $\displaystyle\int_1^\infty x^{-\frac{1}{2}}(x-1)^{-\frac{1}{2}} K_\nu\left(a\sqrt{x}\right)\,dx = \left[K_{\frac{\nu}{2}}\left(\frac{a}{2}\right)\right]^2$    $\left[\operatorname{Re}a>0\right]$    ET II 208(56)a

9.   $\displaystyle\int_0^1 x^{-\frac{1}{2}}(1-x)^{-\frac{1}{2}} Y_\nu\left(a\sqrt{x}\right)\,dx = \pi\left\{\cot(\nu\pi)\left[J_{\frac{\nu}{2}}\left(\frac{a}{2}\right)\right]^2 - \operatorname{cosec}(\nu\pi)\left[J_{-\frac{\nu}{2}}\left(\frac{a}{2}\right)\right]^2\right\}$

$$\left[|\operatorname{Re}\nu|<1\right]$$    ET II 195(68)a

10.   $\displaystyle\int_1^\infty x^{-\frac{1}{2}\nu}(x-1)^{\mu-1} J_\nu\left(a\sqrt{x}\right)\,dx = \Gamma(\mu)2^\mu a^{-\mu} J_{\nu-\mu}(a)$

$$\left[a>0,\quad 0<\operatorname{Re}\mu<\tfrac{1}{2}\operatorname{Re}\nu+\tfrac{3}{4}\right]$$
ET II 205(34)a

11.   $\displaystyle\int_1^\infty x^{-\frac{1}{2}\nu}(x-1)^{\mu-1} J_{-\nu}\left(a\sqrt{x}\right)\,dx = \Gamma(\mu)2^\mu a^{-\mu}\left[\cos(\nu\pi)\,J_{\nu-\mu}(a)-\sin(\nu\pi)\,Y_{\nu-\mu}(a)\right]$

$$\left[a>0,\quad 0<\operatorname{Re}\mu<\tfrac{1}{2}\operatorname{Re}\nu+\tfrac{3}{4}\right]$$
ET II 205(35)a

12.   $\displaystyle\int_1^\infty x^{-\frac{1}{2}\nu}(x-1)^{\mu-1} K_\nu\left(a\sqrt{x}\right)\,dx = \Gamma(\mu)2^\mu a^{-\mu} K_{\nu-\mu}(a)$

$$\left[\operatorname{Re}a>0,\quad \operatorname{Re}\mu>0\right]$$    ET II 209(59)a

13.   $\displaystyle\int_1^\infty x^{-\frac{1}{2}\nu}(x-1)^{\mu-1} Y_\nu\left(a\sqrt{x}\right)\,dx = 2^\mu a^{-\mu} Y_{\nu-\mu}(a)\,\Gamma(\mu)$

$$\left[a>0,\quad 0<\operatorname{Re}\mu<\tfrac{1}{2}\operatorname{Re}\nu+\tfrac{3}{4}\right]$$
ET II 206(40)a

14.   $\displaystyle\int_1^\infty x^{-\frac{1}{2}\nu}(x-1)^{\mu-1} H_\nu^{(1)}\left(a\sqrt{x}\right)\,dx = 2^\mu a^{-\mu} H_{\nu-\mu}^{(1)}(a)\,\Gamma(\mu)$

$$\left[\operatorname{Re}\mu>0,\quad \operatorname{Im}a>0\right]$$    ET II 206(45)a

15.   $\displaystyle\int_1^\infty x^{-\frac{1}{2}\nu}(x-1)^{\mu-1} H_\nu^{(2)}\left(a\sqrt{x}\right)\,dx = 2^\mu a^{-\mu} H_{\nu-\mu}^{(2)}(a)\,\Gamma(\mu)$

$$\left[\operatorname{Re}\mu>0,\quad \operatorname{Im}a<0\right]$$    ET II 207(48)a

16. $\displaystyle\int_0^1 x^{-\frac{1}{2}\nu}(1-x)^{\mu-1} J_\nu\left(a\sqrt{x}\right)\,\mathrm{d}x = \frac{2^{2-\nu}a^{-\mu}}{\Gamma(\nu)}\, s_{\mu+\nu-1,\mu-\nu}(a)$

$\qquad\qquad\qquad\qquad\qquad\qquad [\operatorname{Re}\mu>0]\qquad\qquad$ ET II 194(64)a

17. $\displaystyle\int_0^1 x^{-\frac{1}{2}\nu}(1-x)^{\mu-1} Y_\nu\left(a\sqrt{x}\right)\,\mathrm{d}x = \frac{2^{2-\nu}a^{-\mu}\cot(\nu\pi)}{\Gamma(\nu)}\, s_{\mu+\nu-1,\mu-\nu}(a)$

$$-2^\mu a^{-\mu}\operatorname{cosec}(\nu\pi)\,J_{\mu-\nu}(a)\,\Gamma(\mu)$$

$\qquad\qquad\qquad\qquad [\operatorname{Re}\mu>0,\quad \operatorname{Re}\nu<1]\qquad$ ET II 196(75)a

**6.593**

1. $\displaystyle\int_0^\infty \sqrt{x}\, J_{2\nu-1}\left(a\sqrt{x}\right) J_\nu(bx)\,\mathrm{d}x = \frac{1}{2}ab^{-2} J_{\nu-1}\left(\frac{a^2}{4b}\right)\quad [b>0,\quad \operatorname{Re}\nu>-\tfrac{1}{2}]\qquad$ ET II 58(15)

2. $\displaystyle\int_0^\infty \sqrt{x}\, J_{2\nu-1}\left(a\sqrt{x}\right) K_\nu(bx)\,\mathrm{d}x = \frac{\pi a}{4b^2}\left[I_{\nu-1}\left(\frac{a^2}{4b}\right)-\mathbf{L}_{\nu-1}\left(\frac{a^2}{4b}\right)\right]$

$\qquad\qquad\qquad\qquad\qquad [\operatorname{Re}b>0,\quad \operatorname{Re}\nu>-\tfrac{1}{2}]\qquad$ ET II 144(44)

**6.594**

1. $\displaystyle\int_0^\infty x^\nu I_{2\nu-1}\left(a\sqrt{x}\right) J_{2\nu-1}\left(a\sqrt{x}\right) K_\nu(bx)\,\mathrm{d}x = \sqrt{\pi}2^{-\nu}a^{2\nu-1}b^{-2\nu-\frac{1}{2}} J_{\nu-\frac{1}{2}}\left(\frac{a^2}{2b}\right)$

$\qquad\qquad\qquad\qquad\qquad [\operatorname{Re}b>0,\quad \operatorname{Re}\nu>0]\qquad$ ET II 148(65)

2. $\displaystyle\int_0^\infty x^\nu I_{2\nu-1}\left(a\sqrt{x}\right) Y_{2\nu-1}\left(a\sqrt{x}\right) K_\nu(bx)\,\mathrm{d}x$

$$=\sqrt{\pi}2^{-\nu-1}a^{2\nu-1}b^{-2\nu-\frac{1}{2}}\operatorname{cosec}(\nu\pi)$$

$$\times\left[\mathbf{H}_{\frac{1}{2}-\nu}\left(\frac{a^2}{2b}\right)+\cos(\nu\pi)J_{\nu-\frac{1}{2}}\left(\frac{a^2}{2b}\right)+\sin(\nu\pi)Y_{\nu-\frac{1}{2}}\left(\frac{a^2}{2b}\right)\right]$$

$\qquad\qquad\qquad\qquad [\operatorname{Re}b>0,\quad \operatorname{Re}\nu>0]\qquad$ ET II 148(66)

3. $\displaystyle\int_0^\infty x^\nu J_{2\nu-1}\left(a\sqrt{x}\right) K_{2\nu-1}\left(a\sqrt{x}\right) K_\nu(bx)\,\mathrm{d}x$

$$=\pi^2 2^{-\nu-2}a^{2\nu-1}b^{-2\nu-\frac{1}{2}}\operatorname{cosec}(\nu\pi)\left[\mathbf{H}_{\frac{1}{2}-\nu}\left(\frac{a^2}{2b}\right)-Y_{\frac{1}{2}-\nu}\left(\frac{a^2}{2b}\right)\right]$$

$\qquad\qquad\qquad\qquad [\operatorname{Re}b>0,\quad \operatorname{Re}\nu>0]\qquad$ ET II 148(67)

**6.595**

1. $\displaystyle\int_0^\infty x^{\nu+1} J_\nu(cx)\prod_{i=1}^n z_i^{-\mu_i} J_{\mu_i}(a_i z_i)\,\mathrm{d}x = 0$

$$\left[z_i=\sqrt{x^2+b_i^2},\quad a_i>0,\quad \operatorname{Re}b_i>0,\quad \sum_{i=1}^n a_i<c;\quad \operatorname{Re}\left(\frac{1}{2}n+\sum_{i=1}^n\mu_i-\frac{1}{2}\right)>\operatorname{Re}\nu>-1\right]$$

$\qquad\qquad\qquad\qquad\qquad\qquad\qquad$ EH II 52(33), ET II 60(26)

2. $\displaystyle\int_0^\infty x^{\nu-1} J_\nu(cx)\prod_{i=1}^n z_i^{-\mu_i} J_{\mu_i}(a_i z_i)\,\mathrm{d}x = 2^{\nu-1}\Gamma(\nu)c^{-\nu}\prod_{i=1}^n\left[b_i^{-\mu_i} J_{\mu_i}(a_i b_i)\right]$

$$\left[z_i=\sqrt{x^2+b_i^2},\quad a_i>0,\quad \operatorname{Re}b_i>0,\quad \sum_{i=1}^n a_i<c,\quad \operatorname{Re}\left(\frac{1}{2}n+\sum_{i=1}^n\mu_i+\frac{3}{2}\right)>\operatorname{Re}\nu>0\right]$$

$\qquad\qquad\qquad\qquad\qquad\qquad\qquad$ EH II 52(34), ET II 60(27)

**6.596**

1.
$$\int_0^\infty J_\nu\left(a\sqrt{x^2+z^2}\right)\frac{x^{2\mu+1}}{\sqrt{(x^2+z^2)^\nu}}\,dx = \frac{2^\mu\,\Gamma(\mu+1)}{a^{\mu+1}z^{\nu-\mu-1}}\,J_{\nu-\mu-1}(az)$$

$$\left[a>0,\quad \mathrm{Re}\left(\frac{1}{2}\nu-\frac{1}{4}\right)>\mathrm{Re}\,\mu>-1\right]$$

WA 457(5)

2.
$$\int_0^\infty \frac{J_\nu\left(a\sqrt{t^2+1}\right)}{\sqrt{t^2+1}}\,dt = -\frac{\pi}{2}\,J_{\frac{\nu}{2}}\left(\frac{a}{2}\right)Y_{\frac{\nu}{2}}\left(\frac{a}{2}\right) \qquad [\mathrm{Re}\,\nu>-1,\quad a>0] \qquad \text{MO 46}$$

3.
$$\int_0^\infty K_\nu\left(a\sqrt{x^2+z^2}\right)\frac{x^{2\mu+1}}{\sqrt{(x^2+z^2)^\nu}}\,dx = \frac{2^\mu\,\Gamma(\mu+1)}{a^{\mu+1}z^{\nu-\mu-1}}\,K_{\nu-\mu-1}(az)$$

$$[a>0,\quad \mathrm{Re}\,\mu>-1] \qquad \text{WA 457(6)}$$

4.[8]
$$\int_0^\infty J_\nu(bx)\frac{J_{\mu-1}\left\{a\sqrt{x^2+z^2}\right\}}{(x^2+z^2)^{\frac{1}{2}\mu+\frac{1}{2}}}x^{\nu+1}\,dx = \frac{a^{\mu-1}z^\nu}{2^{\mu-1}\,\Gamma(\mu)}\,K_\nu(bz)$$

$$[a<b,\quad \mathrm{Re}(\mu+2)>\mathrm{Re}\,\nu>-1]$$

ET II 59(19)

5.[8]
$$\int_0^\infty J_\nu(bx)\frac{J_\mu\left\{a\sqrt{x^2+z^2}\right\}}{\sqrt{(x^2+z^2)^\mu}}x^{\nu-1}\,dx = \frac{2^{\nu-1}\,\Gamma(\nu)}{b^\nu}\frac{J_\mu(az)}{z^\mu}$$

$$[\mathrm{Re}(\mu+2)>\mathrm{Re}\,\nu>0,\quad b>a>0]$$

WA 459(12)

6.[6]
$$\int_0^\infty J_\nu(bx)\frac{J_\mu\left(a\sqrt{x^2+z^2}\right)}{\sqrt{(x^2+z^2)^\mu}}x^{\nu+1}\,dx$$

$$= 0 \qquad\qquad\qquad [0<a<b]$$

$$= \frac{b^\nu}{a^\mu}\left(\frac{\sqrt{a^2-b^2}}{z}\right)^{\mu-\nu-1}J_{\mu-\nu-1}\left\{z\sqrt{a^2-b^2}\right\} \qquad [a>b>0]$$

$$[\mathrm{Re}\,\mu>\mathrm{Re}\,\nu>-1] \qquad \text{WA 415(1)}$$

7.[8]
$$\int_0^\infty J_\nu(bx)\frac{K_\mu\left(a\sqrt{x^2+z^2}\right)}{\sqrt{(x^2+z^2)^\mu}}x^{\nu+1}\,dx = \frac{b^\nu}{a^\mu}\left(\frac{\sqrt{a^2+b^2}}{z}\right)^{\mu-\nu-1}K_{\mu-\nu-1}\left(z\sqrt{a^2+b^2}\right)$$

$$\left[a>0,\quad b>0,\quad \mathrm{Re}\,\nu>-1,\quad |\arg z|<\frac{\pi}{2}\right] \qquad \text{KU 151(31), WA 416(2)}$$

8.[8]
$$\int_0^\infty J_\nu(ux)K_\mu\left(v\sqrt{x^2-y^2}\right)(x^2-y^2)^{-\frac{\mu}{2}}x^{\nu+1}\,dx = \frac{\pi}{2}\exp\left[-i\pi\left(\mu-\nu-\frac{1}{2}\right)\right]\cdot\frac{u^\nu}{v^\mu}$$

$$\cdot\left[\frac{\sqrt{u^2+v^2}}{y}\right]^{\mu-\nu-1}H_{\mu-\nu-1}^{(2)}\left(y\sqrt{u^2+v^2}\right)$$

$$\left[\mathrm{Re}\,\mu<1,\quad \mathrm{Re}\,\nu>-1,u>0,\quad v>0,y>0;\quad (x^2-y^2)^{\frac{1}{2}\alpha}=e^{\frac{1}{2}\alpha\pi i}\left(y^2-n^2\right)^{\frac{1}{2}\alpha}\text{ if }x<y\right]$$

$9.^8 \quad \int_0^\infty J_\nu(ux) H_\mu^{(2)}\left(v\sqrt{x^2+y^2}\right)\left(x^2+y^2\right)^{-\frac{\mu}{2}} x^{\nu+1}\,dx$

$$= \frac{u^\nu}{v^\mu}\left[\frac{\sqrt{v^2-u^2}}{y}\right]^{\mu-\nu-1} H_{\mu-\nu-1}^{(2)}\left(y\sqrt{v^2-u^2}\right)$$

$$[u < v]$$

$$\left[\begin{array}{l} \operatorname{Re}\mu > \operatorname{Re}\nu > -1, \quad u > 0, \quad v > 0; \quad \arg\sqrt{v^2-u^2}=0, \text{ for } v > u; \\[2mm] \arg\left(v^2-u^2\right)^\sigma = -\pi\sigma \text{ for } v < u, \text{ where } \sigma = \frac{1}{2} \text{ or } \sigma = \frac{\mu-\nu-1}{2} \end{array}\right]$$

MO 43

$10.^8 \quad \int_0^\infty J_\nu(bx) J_\mu\left(a\sqrt{x^2+z^2}\right) J_\mu\left(\gamma\sqrt{x^2+z^2}\right)\frac{x^{\nu-1}}{(x^2+z^2)^\mu}\,dx = \frac{2^{\nu-1}\Gamma(\nu)}{b^\nu}\frac{J_\mu(az)}{z^\mu}\frac{J_\mu(\gamma z)}{z^\mu}$

$$\left[a > 0; \quad b > a+\gamma; \quad \gamma > 0, \quad \operatorname{Re}\left(2\mu+\tfrac{5}{2}\right) > \operatorname{Re}\nu > 0\right] \quad \text{WA 459(14)}$$

$11.^8 \quad \int_0^\infty J_\nu(bt)t^{\nu-1}\prod_{k=1}^n J_\mu\left(a_k\sqrt{t^2+x^2}\right)\sqrt{(t^2+x^2)^{-n\mu}}\,dt = 2^{\nu-1}b^{-\nu}\Gamma(\nu)\prod_{k=1}^n\left[x^{-\mu}J_\mu(a_k x)\right]$

$$\left[x > 0, \quad a_i > 0, \quad b > \prod_{k=1}^n a_k; \quad \operatorname{Re}\left(n\mu+\frac{1}{2}n+\frac{1}{2}\right) > \operatorname{Re}\nu > 0\right] \quad \text{MO 43}$$

$12.^8 \quad \int_0^\infty \frac{J_\mu^2\left(\sqrt{a^2+x^2}\right)}{(a^2+x^2)^\nu}x^{2\nu-2}\,dx = \frac{\Gamma\left(\nu-\frac{1}{2}\right)}{2a^{\nu+1}\sqrt{\pi}}\mathbf{H}_\nu(2a) \qquad \left[\operatorname{Re}\nu > \tfrac{1}{2}\right]$ \qquad WA 457(8)

**6.597** $\quad \int_0^\infty t^{\nu+1} J_\mu\left[b\left(t^2+y^2\right)^{\frac{1}{2}}\right]\left(t^2+y^2\right)^{-\frac{1}{2}\mu}\left(t^2+\beta^2\right)^{-1}J_\nu(at)\,dt$

$$= \beta^\nu J_\mu\left[b\left(y^2-\beta^2\right)^{\frac{1}{2}}\right]\left(y^2-\beta^2\right)^{-\frac{1}{2}\mu}K_\nu(a\beta)$$

$$[a \ge b, \quad \operatorname{Re}\beta > 0, \quad -1 < \operatorname{Re}\nu < 2+\operatorname{Re}\mu] \quad \text{EH II 95(56)}$$

**6.598** $\quad \int_0^1 x^{\frac{\mu}{2}}(1-x)^{\frac{\nu}{2}}J_\mu\left(a\sqrt{x}\right)J_\nu\left(b\sqrt{1-x}\right)\,dx = 2a^\mu b^\nu\left(a^2+b^2\right)^{-\frac{1}{2}(\nu+\mu+1)}J_{\nu+\mu+1}\left(\sqrt{a^2+b^2}\right)$

$$[\operatorname{Re}\nu > -1, \quad \operatorname{Re}\mu > -1] \qquad \text{EH II 46a}$$

## 6.61 Combinations of Bessel functions and exponentials

**6.611**

1. $\quad \int_0^\infty e^{-\alpha x}J_\nu(bx)\,dx = \frac{b^{-\nu}\left[\sqrt{\alpha^2+b^2}-\alpha\right]^\nu}{\sqrt{\alpha^2+b^2}} \qquad [\operatorname{Re}\nu > -1, \quad \operatorname{Re}(\alpha\pm ib) > 0]$

$$\text{EH II 49(18), WA 422(8)}$$

2. $\quad \int_0^\infty e^{-\alpha x}Y_\nu(bx)\,dx = \left(\alpha^2+b^2\right)^{-\frac{1}{2}}\operatorname{cosec}(\nu\pi)$

$$\times\left\{b^\nu\left[\left(\alpha^2+b^2\right)^{\frac{1}{2}}+\alpha\right]^{-\nu}\cos(\nu\pi)-b^{-\nu}\left[\left(\alpha^2+b^2\right)^{\frac{1}{2}}+\alpha\right]^\nu\right\}$$

$$[\operatorname{Re}\alpha > 0, \quad b > 0, \quad |\operatorname{Re}\nu| < 1] \quad \text{MO 179, ET II 105(1)}$$

3.    $\displaystyle\int_0^\infty e^{-\alpha x} K_\nu(bx)\,dx = \frac{\pi}{b\sin(\nu\pi)}\frac{\sin(\nu\theta)}{\sin\theta}$

$$\left[\cos\theta = \frac{\alpha}{b};\quad \text{with }\theta\to\frac{\pi}{2}\text{ as }b\to\infty\right]$$

ET II 131(22)

$$= \frac{\pi\,\text{cosec}(\nu\pi)}{2\sqrt{\alpha^2-b^2}}\left[b^{-\nu}\left(\alpha+\sqrt{\alpha^2-b^2}\right)^\nu - b^\nu\left(\sqrt{\alpha^2-b^2}+\alpha\right)^{-\nu}\right]$$

$$[|\text{Re}\,\nu|<1,\quad \text{Re}(\alpha+b)>0]$$

ET I 197(24), MO 180

4.[8]    $\displaystyle\int_0^\infty e^{-\alpha x} I_\nu(bx)\,dx = \frac{b^{-\nu}\left[\alpha-\sqrt{\alpha^2-b^2}\right]^\nu}{\sqrt{\alpha^2-b^2}}$      $[\text{Re}\,\nu>-1,\quad \text{Re}\,\alpha>|\text{Re}\,b|]$

MO 180, ET I 195(1)

5.    $\displaystyle\int_0^\infty e^{-\alpha x} H_\nu^{(1,2)}(bx)\,dx = \frac{\left(\sqrt{\alpha^2+b^2}-\alpha\right)^\nu}{b^\nu\sqrt{\alpha^2+b^2}}\left\{1\pm\frac{i}{\sin(\nu\pi)}\left[\cos(\nu\pi)-\frac{\left(\alpha+\sqrt{\alpha^2+b^2}\right)^{2\nu}}{b^{2\nu}}\right]\right\}$

$[-1<\text{Re}\,\nu<1$; a plus sign corresponds to the function $H_\nu^{(1)}$, a minus sign to the function $H_\nu^{(2)}]$.

MO 180, ET I188(54, 55)

6.    $\displaystyle\int_0^\infty e^{-\alpha x} H_0^{(1)}(bx)\,dx = \frac{1}{\sqrt{\alpha^2+b^2}}\left\{1-\frac{2i}{\pi}\ln\left[\frac{\alpha}{b}+\sqrt{1+\left(\frac{\alpha}{b}\right)^2}\right]\right\}$

$[\text{Re}\,\alpha>|\text{Im}\,b|]$     MO 180, ET I 188(53)

7.    $\displaystyle\int_0^\infty e^{-\alpha x} H_0^{(2)}(bx)\,dx = \frac{1}{\sqrt{\alpha^2+b^2}}\left\{1+\frac{2i}{\pi}\ln\left[\frac{\alpha}{b}+\sqrt{1+\left(\frac{\alpha}{b}\right)^2}\right]\right\}$

$[\text{Re}\,\alpha>|\text{Im}\,b|]$     MO 180, ET I 188(53)

8.    $\displaystyle\int_0^\infty e^{-\alpha x} Y_0(bx)\,dx = \frac{-2}{\pi\sqrt{\alpha^2+b^2}}\ln\frac{\alpha+\sqrt{\alpha^2+b^2}}{b}$    $[\text{Re}\,\alpha>|\text{Im}\,b|]$    MO 47, ET I 187(44)

9.[12]    $\displaystyle\int_0^\infty e^{-\alpha x} K_0(bx)\,dx = \frac{\arccos\frac{\alpha}{b}}{\sqrt{b^2-\alpha^2}}$    $[0<a<b]$    WA 424, ET II 131(22)

$$= \frac{1}{\sqrt{\alpha^2-b^2}}\ln\left(\frac{\alpha}{b}+\sqrt{\frac{\alpha^2}{b^2}-1}\right) \quad [0\le b<a] \qquad \text{MO 48}$$

10.[10]    $\displaystyle\int_a^b \alpha\,d\alpha\int_0^\infty dk\,J_1(k\alpha)e^{-k|\beta|} = \int_a^b\left(1-\frac{|\beta|}{\sqrt{\alpha^2+\beta^2}}\right)d\alpha$

(see **3.241** 6)

**6.612**

1.    $\displaystyle\int_0^\infty e^{-2\alpha x} J_0(x)\,Y_0(x)\,dx = \frac{\boldsymbol{K}\left[\alpha\left(\alpha^2+1\right)^{-\frac{1}{2}}\right]}{\pi\left(\alpha^2+1\right)^{\frac{1}{2}}}$    $[\text{Re}\,\alpha>0]$    ET II 347(58)

2.    $\displaystyle\int_0^\infty e^{-2\alpha x} I_0(x)\,K_0(x)\,dx = \frac{1}{2}\boldsymbol{K}\left[\left(1-\alpha^2\right)^{\frac{1}{2}}\right]$    $[0<\alpha<1]$

$$= \frac{1}{2\alpha}\boldsymbol{K}\left[\left(1-\frac{1}{\alpha^2}\right)^{\frac{1}{2}}\right] \quad [1<\alpha<\infty]$$

ET II 370(48)

3. $$\int_0^\infty e^{-\alpha x} J_\nu(bx) J_\nu(\gamma x)\,dx = \frac{1}{\pi\sqrt{\gamma b}}\, Q_{\nu-\frac{1}{2}}\left(\frac{\alpha^2 + b^2 + \gamma^2}{2b\gamma}\right)$$

$$\left[\operatorname{Re}\left(\alpha \pm ib \pm i\gamma\right) > 0, \quad \gamma > 0, \quad \operatorname{Re}\nu > -\tfrac{1}{2}\right] \quad \text{WA 426(2), ET II 50(17)}$$

4. $$\int_0^\infty e^{-\alpha x}\left[J_0(bx)\right]^2\,dx = \frac{2}{\pi\sqrt{\alpha^2 + 4b^2}}\, K\left(\frac{2b}{\sqrt{\alpha^2 + 4b^2}}\right) \qquad\qquad \text{MO 178}$$

5. $$\int_0^\infty e^{-2\alpha x} J_1^2(bx)\,dx = \frac{\left(2\alpha^2 + b^2\right) K\left(\frac{b}{\sqrt{\alpha^2 + b^2}}\right) - 2\left(\alpha^2 + b^2\right) E\left(\frac{b}{\sqrt{\alpha^2 + b^2}}\right)}{\pi b^2 \sqrt{\alpha^2 + b^2}} \qquad \text{WA 428(3)}$$

6. $$\int_0^\infty e^{-3x} I_l(x)\, I_m(x)\, I_n(x)\,dx = r_1 g + \frac{r_2}{\pi^2 g} + r_3$$

where

$$g = \frac{\sqrt{3}-1}{96\pi^3}\,\Gamma^2\left(\frac{1}{24}\right)\Gamma^2\left(\frac{11}{24}\right)$$

and

| (lmn) | $r_1$ | $r_2$ | $r_3$ | (lmn) | $r_1$ | $r_2$ | $r_3$ |
|---|---|---|---|---|---|---|---|
| 000 | $1$ | $0$ | $0$ | 432 | $525/32$ | $-4617/112$ | $0$ |
| 100 | $1$ | $0$ | $-1/3$ | 433 | $-595/72$ | $8809/420$ | $0$ |
| 110 | $5/12$ | $-1/2$ | $0$ | 440 | $6025/36$ | $-620161/1470$ | $0$ |
| 111 | $-1/8$ | $3/4$ | $0$ | 441 | $-29175/224$ | $131379/400$ | $0$ |
| 200 | $10/3$ | $2$ | $-2$ | 442 | $2975/48$ | $-31231/200$ | $0$ |
| 210 | $3/8$ | $-9/4$ | $1/3$ | 443 | $-539/32$ | $119271/2800$ | $0$ |
| 211 | $-2/3$ | $2$ | $0$ | 444 | $77/8$ | $-186003/7700$ | $0$ |
| 220 | $73/36$ | $-29/6$ | $0$ | 500 | $9287/12$ | $3005/2$ | $-2077/3$ |
| 221 | $-15/16$ | $21/8$ | $0$ | 510 | $-189029/180$ | $-138331/50$ | $348$ |
| 222 | $5/8$ | $-27/20$ | $0$ | 511 | $275/4$ | $5751/10$ | $-150$ |
| 300 | $35/2$ | $21$ | $-13$ | 520 | $2897/16$ | $-15123/20$ | $-229/3$ |
| 310 | $-79/36$ | $-85/6$ | $4$ | 521 | $-937/12$ | $27059/30$ | $24$ |
| 311 | $-11/4$ | $21/2$ | $-2/3$ | 522 | $509/8$ | $-4209/28$ | $0$ |
| 320 | $319/48$ | $-119/8$ | $-1/3$ | 530 | $3589/18$ | $-1993883/3075$ | $0$ |
| 321 | $-125/36$ | $269/30$ | $0$ | 531 | $-1329/8$ | $297981/700$ | $-4/3$ |
| 322 | $35/16$ | $-213/40$ | $0$ | 532 | $2555/36$ | $-187777/1050$ | $0$ |
| 330 | $50/3$ | $-1046/25$ | $0$ | 533 | $-2233/48$ | $164399/1400$ | $0$ |
| 331 | $-35/3$ | $148/5$ | $0$ | 540 | $18471/32$ | $-28493109/19600$ | $-1/3$ |
| 332 | $35/9$ | $-1012/105$ | $0$ | 541 | $-1390/3$ | $286274/245$ | $0$ |
| 333 | $-35/16$ | $1587/280$ | $0$ | 542 | $7777/32$ | $-1715589/2800$ | $0$ |
| 400 | $994/9$ | $542/3$ | $-92$ | 543 | $-5621/72$ | $4550057/23100$ | $0$ |
| 410 | $-515/16$ | $-879/8$ | $115/3$ | 544 | $1155/32$ | $-560001/6160$ | $0$ |
| 411 | $-9/2$ | $357/5$ | $-12$ | 550 | $197045/108$ | $-101441689/22050$ | $0$ |
| 420 | $12907/120$ | $-13903/10$ | $-6$ | 551 | $-12023/8$ | $18569853/4900$ | $0$ |
| 421 | $-229/16$ | $1251/40$ | $1$ | 552 | $1683/2$ | $-5718309/2695$ | $0$ |
| 422 | $35/3$ | $-1024/35$ | $0$ | 553 | $-5159/16$ | $2504541/3080$ | $0$ |
| 430 | $2641/48$ | $-28049/200$ | $1/3$ | 554 | $24563/312$ | $-1527851/77000$ | $0$ |
| 431 | $-1505/36$ | $118051/1050$ | $0$ | 555 | $-9251/208$ | $12099711/107800$ | $0$ |

**6.613**[11] $\int_0^\infty e^{-xz} J_{\nu+\frac{1}{2}}\left(\frac{x^2}{2}\right) \, dx = \frac{\Gamma(\nu+1)}{\sqrt{\pi}} D_{-\nu-1}\left(ze^{\frac{\pi}{4}i}\right) D_{-\nu-1}\left(ze^{-\frac{\pi i}{4}}\right)$     $[\operatorname{Re}\nu > -1]$     MO 122

**6.614**

1.    $\int_0^\infty e^{-\alpha x} J_\nu\left(b\sqrt{x}\right) \, dx = \frac{b}{4}\sqrt{\frac{\pi}{\alpha^3}} \exp\left(-\frac{b^2}{8\alpha}\right) \left[I_{\frac{1}{2}(\nu-1)}\left(\frac{b^2}{8\alpha}\right) - I_{\frac{1}{2}(\nu+1)}\left(\frac{b^2}{8\alpha}\right)\right]$

                  $= \frac{1}{\alpha} e^{-b^2/4\alpha}$                                   $[\nu = 0]$

                                                      MO 178

2.    $\int_0^\infty e^{-\alpha x} Y_{2\nu}\left(2\sqrt{bx}\right) \, dx = \frac{e^{-\frac{1}{2}\frac{b}{\alpha}}}{\sqrt{\alpha b}} \left\{\cot(\nu\pi)\frac{\Gamma(\nu+1)}{\Gamma(2\nu+1)} M_{\frac{1}{2},\nu}\left(\frac{b}{\alpha}\right) - \operatorname{cosec}(\nu\pi) W_{\frac{1}{2},nu}\left(\frac{b}{\alpha}\right)\right\}$

                                      $[\operatorname{Re}\alpha > 0, \quad |\operatorname{Re}\nu| < 1]$     ET I 188(50)a

3.    $\int_0^\infty e^{-\alpha x} I_{2\nu}\left(2\sqrt{bx}\right) \, dx = \frac{e^{\frac{1}{2}\frac{b}{\alpha}}}{\sqrt{\alpha b}} \frac{\Gamma(\nu+1)}{\Gamma(2\nu+1)} M_{-\frac{1}{2},\nu}\left(\frac{b}{\alpha}\right)$

                                      $[\operatorname{Re}\alpha > 0, \quad \operatorname{Re}\nu > -1]$     ET I 197(20)a

4.    $\int_0^\infty e^{-\alpha x} K_{2\nu}\left(2\sqrt{bx}\right) \, dx = \frac{e^{\frac{1}{2}\frac{b}{\alpha}}}{2\sqrt{\alpha b}} \Gamma(\nu+1)\Gamma(1-\nu) W_{-\frac{1}{2},\nu}\left(\frac{b}{\alpha}\right)$

                                      $[\operatorname{Re}\alpha > 0, \quad |\operatorname{Re}\nu| < 1]$     ET I 199(37)a

5.    $\int_0^\infty e^{-\alpha x} K_1\left(b\sqrt{x}\right) \, dx = \frac{b}{8}\sqrt{\frac{\pi}{\alpha^3}} \exp\left(\frac{b^2}{8\alpha}\right) \left[K_1\left(\frac{b^2}{8\alpha}\right) - K_0\left(\frac{b^2}{8\alpha}\right)\right]$              MO 181

**6.615**   $\int_0^\infty e^{-\alpha x} J_\nu\left(2\beta\sqrt{x}\right) J_\nu\left(2\gamma\sqrt{x}\right) \, dx = \frac{1}{\alpha} I_\nu\left(\frac{2\beta\gamma}{\alpha}\right) \exp\left(-\frac{\beta^2+\gamma^2}{\alpha}\right)$     $[\operatorname{Re}\nu > -1]$

                                                               MO 178

**6.616**

1.    $\int_0^\infty e^{-\alpha x} J_0\left(b\sqrt{x^2+2\gamma x}\right) \, dx = \frac{1}{\sqrt{\alpha^2+b^2}} \exp\left[\gamma\left(\alpha - \sqrt{\alpha^2+b^2}\right)\right]$            MO 179

2.    $\int_1^\infty e^{-\alpha x} J_0\left(b\sqrt{x^2-1}\right) \, dx = \frac{1}{\sqrt{\alpha^2+b^2}} \exp\left(-\sqrt{\alpha^2+b^2}\right)$                    MO 179

3.    $\int_{-\infty}^\infty e^{itx} H_0^{(1)}\left(r\sqrt{\alpha^2-t^2}\right) \, dt = -2i\frac{e^{i\alpha\sqrt{r^2+x^2}}}{\sqrt{r^2+x^2}}$

                          $\left[0 \le \arg\sqrt{\alpha^2-t^2} < \pi, \quad 0 \le \arg\alpha < \pi; \quad r \text{ and } x \text{ are real}\right]$    MO 49

4.    $\int_{-\infty}^\infty e^{-itx} H_0^{(2)}\left(r\sqrt{\alpha^2-t^2}\right) \, dt = 2i\frac{e^{-i\alpha\sqrt{r^2+x^2}}}{\sqrt{r^2+x^2}}$

                          $\left[-\pi < \arg\sqrt{\alpha^2-t^2} \le 0, \quad -\pi < \arg\alpha \le 0, \quad r \text{ and } x \text{ are real}\right]$    MO 49

5.[3]    $\int_{-1}^1 e^{-ax} I_0\left(b\sqrt{1-x^2}\right) \, dx = 2\left(a^2+b^2\right)^{-1/2} \sinh\sqrt{a^2+b^2}$

                                         $[a > 0, \quad b > 0]$

6.8    $\displaystyle\int_0^\infty e^{-xy} J_0\left(y\sqrt{1-x^2}\right)/(\alpha+y)\,\mathrm{d}y = \sum_{n=0}^\infty n!\,\frac{P_n(x)}{\alpha^{n+1}}$

## 6.617

1.    $\displaystyle\int_0^\infty K_{q-p}\left(2z\sinh x\right) e^{(p+q)x}\,\mathrm{d}x = \frac{\pi^2}{4\sin[(p-q)\pi]}\left[J_p(z)\,Y_q(z) - J_q(z)\,Y_p(z)\right]$

$$[\operatorname{Re} z > 0, \quad -1 < \operatorname{Re}(p-q) < 1]$$

MO 44

2.    $\displaystyle\int_0^\infty K_0\left(2z\sinh x\right) e^{-2px}\,\mathrm{d}x = -\frac{\pi}{4}\left\{J_p(z)\frac{\partial\,Y_p(z)}{\partial p} - Y_p(z)\frac{\partial\,J_p(z)}{\partial p}\right\}$

$$[\operatorname{Re} z > 0]$$

MO 44

## 6.618

1.    $\displaystyle\int_0^\infty e^{-\alpha x^2} J_\nu(bx)\,\mathrm{d}x = \frac{\sqrt{\pi}}{2\sqrt{\alpha}}\exp\left(-\frac{b^2}{8\alpha}\right) I_{\frac{1}{2}\nu}\left(\frac{b^2}{8\alpha}\right) \qquad [\operatorname{Re}\alpha > 0, \quad b > 0, \quad \operatorname{Re}\nu > -1]$

WA 432(5), ET II 29(8)

2.    $\displaystyle\int_0^\infty e^{-\alpha x^2} Y_\nu(bx)\,\mathrm{d}x = -\frac{\sqrt{\pi}}{2\sqrt{\alpha}}\exp\left(-\frac{b^2}{8\alpha}\right)\left[\tan\frac{\nu\pi}{2} I_{\frac{1}{2}\nu}\left(\frac{b^2}{8\alpha}\right) + \frac{1}{\pi}\sec\left(\frac{\nu\pi}{2}\right) K_{\frac{1}{2}\nu}\left(\frac{b^2}{8\alpha}\right)\right]$

$$[\operatorname{Re}\alpha > 0, \quad b > 0, \quad |\operatorname{Re}\nu| < 1]$$

WA 432(6), ET II 106(3)

3.    $\displaystyle\int_0^\infty e^{-\alpha x^2} K_\nu(bx)\,\mathrm{d}x = \frac{1}{4}\sec\left(\frac{\nu\pi}{2}\right)\frac{\sqrt{\pi}}{\sqrt{\alpha}}\exp\left(\frac{b^2}{8\alpha}\right) K_{\frac{1}{2}\nu}\left(\frac{b^2}{8\alpha}\right)$

$$[\operatorname{Re}\alpha > 0, \quad |\operatorname{Re}\nu| < 1]$$

EH II 51(28), ET II 132(24)

4.    $\displaystyle\int_0^\infty e^{-\alpha x^2} I_\nu(bx)\,\mathrm{d}x = \frac{\sqrt{\pi}}{2\sqrt{\alpha}}\exp\left(\frac{b^2}{8\alpha}\right) I_{\frac{1}{2}\nu}\left(\frac{b^2}{8\alpha}\right) \qquad [\operatorname{Re}\nu > -1, \quad \operatorname{Re}\alpha > 0] \qquad \text{EH II 92(27)}$

5.    $\displaystyle\int_0^\infty e^{-\alpha x^2} J_\mu(bx)\,J_\nu(bx)\,\mathrm{d}x$

$$= 2^{-\nu-\mu-1}\alpha^{-\frac{\nu+\mu+1}{2}} b^{\nu+\mu}\frac{\Gamma\left(\frac{\mu+\nu+1}{2}\right)}{\Gamma(\mu+1)\,\Gamma(\nu+1)}$$

$$\times\ {}_3F_3\left(\frac{\nu+\mu+1}{2}, \frac{\nu+\mu+2}{2}, \frac{\nu+\mu+1}{2}; \mu+1, \nu+1, \nu+\mu+1; -\frac{b^2}{\alpha}\right)$$

$$[\operatorname{Re}(\nu+\mu) > -1, \quad \operatorname{Re}\alpha > 0] \quad \text{EH II 50(21)a}$$

## 6.62–6.63 Combinations of Bessel functions, exponentials, and powers

**6.621    Notation:**

$$\ell_1 = \frac{1}{2}\left[\sqrt{(a+b)^2 + z^2} - \sqrt{(a-b)^2 + z^2}\right], \quad \ell_2 = \frac{1}{2}\left[\sqrt{(a+b)^2 + z^2} + \sqrt{(a-b)^2 + z^2}\right]$$

1.  $\displaystyle \int_0^\infty e^{-\alpha x} J_\nu(bx) x^{\mu-1}\, \mathrm{d}x = \frac{\left(\frac{b}{2\alpha}\right)^\nu \Gamma(\nu+\mu)}{\alpha^\mu \, \Gamma(\nu+1)} \, F\left(\frac{\nu+\mu}{2}, \frac{\nu+\mu+1}{2}; \nu+1; -\frac{b^2}{\alpha^2}\right)$

$$\text{WA 421(2)}$$

$$= \frac{\left(\frac{b}{2\alpha}\right)^\nu \Gamma(\nu+\mu)}{\alpha^\mu \, \Gamma(\nu+1)} \left(1+\frac{b^2}{\alpha^2}\right)^{\frac{1}{2}-\mu} F\left(\frac{\nu-\mu+1}{2}, \frac{\nu-\mu}{2}+1; \nu+1; -\frac{b^2}{\alpha^2}\right)$$

$$\text{WA 421(3)}$$

$$= \frac{\left(\frac{b}{2}\right)^\nu \Gamma(\nu+\mu)}{\sqrt{(\alpha^2+b^2)^{\nu+\mu}}\,\Gamma(\nu+1)} \, F\left(\frac{\nu+\mu}{2}, \frac{1-\mu+\nu}{2}; \nu+1; \frac{b^2}{\alpha^2+b^2}\right)$$

$$[\operatorname{Re}(\nu+\mu) > 0, \quad \operatorname{Re}(\alpha+ib) > 0, \quad \operatorname{Re}(\alpha-ib) > 0]$$
$$\text{WA 421(3)}$$

$$= (\alpha^2+b^2)^{-\frac{1}{2}\mu} \, \Gamma(\nu+\mu) \, P_{\mu-1}^{-\nu}\left[\alpha\left(\alpha^2+b^2\right)^{-\frac{1}{2}}\right]$$

$$[\alpha > 0, \quad b > 0, \quad \operatorname{Re}(\nu+\mu) > 0]$$
$$\text{ET II 29(6)}$$

2.  $\displaystyle \int_0^\infty e^{-\alpha x}\, Y_\nu(bx) x^{\mu-1}\, \mathrm{d}x$

$$= \cot\nu\pi \, \frac{\left(\frac{b}{2}\right)^\nu \Gamma(\nu+\mu)}{\sqrt{(\alpha^2+b^2)^{\nu+\mu}}\,\Gamma(\nu+1)} \, F\left(\frac{\nu+\mu}{2}, \frac{\nu-\mu+1}{2}; \nu+1; \frac{b^2}{\alpha^2+b^2}\right)$$

$$- \operatorname{cosec}\nu\pi \, \frac{\left(\frac{b}{2}\right)^{-\nu} \Gamma(\mu-\nu)}{\sqrt{(\alpha^2+b^2)^{\mu-\nu}}\,\Gamma(1-\nu)} \, F\left(\frac{\mu-\nu}{2}, \frac{1-\nu-\mu}{2}; 1-\nu; \frac{b^2}{\alpha^2+b^2}\right)$$

$$[\operatorname{Re}\mu \geq |\operatorname{Re}\nu|, \quad \operatorname{Re}(\alpha \pm ib) > 0]$$
$$\text{WA 421(4)}$$

$$= -\frac{2}{\pi} \, \Gamma(\nu+\mu)\left(b^2+\alpha^2\right)^{-\frac{1}{2}\mu} Q_{\mu-1}^{-\nu}\left[\alpha\left(\alpha^2+b^2\right)^{-\frac{1}{2}}\right]$$

$$[\alpha > 0, \quad b > 0, \quad \operatorname{Re}\mu > |\operatorname{Re}\nu|]$$
$$\text{ET II 105(2)}$$

3.  $\displaystyle \int_0^\infty x^{\mu-1} e^{-\alpha x} K_\nu(bx)\, \mathrm{d}x = \frac{\sqrt{\pi}(2b)^\nu}{(\alpha+b)^{\mu+\nu}} \, \frac{\Gamma(\mu+\nu)\Gamma(\mu-\nu)}{\Gamma\left(\mu+\frac{1}{2}\right)} \, F\left(\mu+\nu, \nu+\frac{1}{2}; \mu+\frac{1}{2}; \frac{\alpha-b}{\alpha+b}\right)$

$$[\operatorname{Re}\mu > |\operatorname{Re}\nu|, \quad \operatorname{Re}(\alpha+b) > 0]$$
$$\text{ET II 131(23)a, EH II 50(26)}$$

4.  $\displaystyle \int_0^\infty x^{m+1} e^{-\alpha x} J_\nu(bx)\, \mathrm{d}x = (-1)^{m+1} b^{-\nu} \frac{\mathrm{d}^{m+1}}{\mathrm{d}\alpha^{m+1}} \left[\frac{\left(\sqrt{\alpha^2+b^2}-\alpha\right)^\nu}{\sqrt{\alpha^2+b^2}}\right]$

$$[b > 0, \quad \operatorname{Re}\nu > -m-2] \qquad \text{ET II 28(3)}$$

5.[10]  $\displaystyle \int_0^\infty e^{-zx} J_1(ax) J_{1/2}(bx) \, x^{-3/2}\, \mathrm{d}x$

$$= \frac{1}{a}\sqrt{\frac{2}{\pi b}}\left\{\frac{\ell_1}{2}\sqrt{a^2-\ell_1^2} + \frac{a^2}{2}\arcsin\left(\frac{\ell_1}{2}\right) + z\left[\sqrt{b^2-\ell_1^2}-b\right]\right\}$$

$$[\arg a > 0, \quad \arg b > 0, \quad \arg z > 0]$$

6.$^{10}$ $\displaystyle\int_0^\infty e^{-zx} J_1(ax) J_{1/2}(bx) x^{-1/2}\,\mathrm{d}x = \frac{1}{a}\sqrt{\frac{2}{\pi b}}\left[b - \sqrt{b^2 - \ell_1^2}\right]$

$$[\arg a > 0, \quad \arg b > 0, \quad \arg z > 0]$$

7.$^{10}$ $\displaystyle\int_0^\infty e^{-zx} J_1(ax) J_{1/2}(bx) x^{1/2}\,\mathrm{d}x = \frac{1}{a}\sqrt{\frac{2}{\pi b}}\,\frac{\ell_1\sqrt{a^2 - \ell_1^2}}{\ell_2^2 - \ell_1^2}$

$$[\arg a > 0, \quad \arg b > 0, \quad \arg z > 0]$$

8.$^{10}$ $\displaystyle\int_0^\infty e^{-zx} J_1(ax) J_{3/2}(bx) x^{1/2}\,\mathrm{d}x = \sqrt{\frac{2}{\pi}}\,\frac{\ell_1^2\sqrt{b^2 - \ell_1^2}}{b^{3/2}a\,(\ell_2^2 - \ell_1^2)}$

$$[\arg a > 0, \quad \arg b > 0, \quad \arg z > 0]$$

9.$^{10}$ $\displaystyle\int_0^\infty e^{-zx} J_1(ax) J_{3/2}(bx) x^{-1/2}\,\mathrm{d}x = \frac{1}{\sqrt{2\pi}}\,\frac{1}{b^{3/2}a}\left[a^2\arcsin\left(\frac{\ell_1}{a}\right) - \ell_1\sqrt{a^2 - \ell_1^2}\right]$

$$[\arg a > 0, \quad \arg b > 0, \quad \arg z > 0]$$

10.$^{10}$ $\displaystyle\int_0^\infty e^{-zx} J_1(ax) J_{5/2}(bx) x^{-1/2}\,\mathrm{d}x = \frac{1}{\sqrt{2\pi}}\,\frac{z}{b^{5/2}a}\left[\ell_1\sqrt{a^2 - \ell_1^2} + \frac{2a^2\ell_1}{\sqrt{a^2 - \ell_1^2}} - 3a^2\arcsin\left(\frac{\ell_1}{a}\right)\right]$

$$[\arg a > 0, \quad \arg b > 0, \quad \arg z > 0]$$

11.$^{12}$ $\displaystyle\int_0^\infty e^{-zx} J_1(ax) J_{5/2}(bx) x^{-3/2}\,\mathrm{d}x$

$$= \frac{1}{\sqrt{2\pi}}\,\frac{1}{b^{5/2}a}\left[\frac{\ell_1}{\sqrt{a^2 - \ell_1^2}}\left(\frac{7a^4}{8} - a^2 z^2 - \frac{\ell_1^4}{4} - \frac{5a^2\ell_1^2}{8}\right)\right.$$

$$\left. - \frac{1}{2}\left(\ell_1^2 + \ell_2^2\right)\ell_1\sqrt{a^2 - \ell_1^2} + \arcsin\left(\frac{\ell_1}{a}\right)\left(\frac{3}{2}a^2 z^2 + \frac{1}{2}a^2 b^2 - \frac{3a^4}{8}\right)\right]$$

$$[\arg a > 0, \quad \arg b > 0, \quad \arg z > 0]$$

12.$^{10}$ $\displaystyle\int_0^\infty e^{-zx} J_1(ax) J_{5/2}(bx) x^{-5/2}\,\mathrm{d}x$

$$= \frac{1}{\sqrt{2\pi}}\,\frac{1}{b^{5/2}a}\left\{\frac{2\left[b^{5/2} - \left(b^2 - \ell_1^2\right)^{5/2}\right]}{15} + za^2\arcsin\left(\frac{\ell_1}{a}\right)\left[\frac{3a^2}{8} - \frac{b^2}{2} - \frac{z^2}{2}\right]\right.$$

$$\left. + z\ell_1\sqrt{a^2 - \ell_1^2}\left[\frac{b^2}{2} - \frac{3a^2}{8} + \frac{z^2}{6} - \frac{\ell_1^2}{4}\right] + \frac{z^3 a^2\ell_1}{3\sqrt{a^2 - \ell_1^2}}\right\}$$

$$[\arg a > 0, \quad \arg b > 0, \quad \arg z > 0]$$

13.$^{10}$ $\displaystyle\int_0^\infty e^{-zx} J_2(ax) J_{3/2}(bx) x^{1/2}\,\mathrm{d}x = \sqrt{\frac{2}{\pi}}\,a^2 b^{3/2}\,\frac{\sqrt{\ell_2^2 - b^2}}{(\ell_2^2 - \ell_1^2)\,\ell_2^4}$

$$[\arg a > 0, \quad \arg b > 0, \quad \arg z > 0]$$

14.$^{10}$ $\displaystyle\int_0^\infty e^{-zx} J_2(ax) J_{3/2}(bx) x^{-1/2}\,\mathrm{d}x = \sqrt{\frac{2}{\pi}}\,\frac{b^{3/2}}{a^2}\left[\frac{2}{3} - \frac{\sqrt{b^2 - \ell_1^2}}{b} + \frac{\left(b^2 - \ell_1^2\right)^{3/2}}{3b^3}\right]$

$$[\arg a > 0, \quad \arg b > 0, \quad \arg z > 0]$$

$15.^{10}$   $\displaystyle\int_0^\infty e^{-zx} J_3(ax) J_{1/2}(bx) x^{-1/2}\,\mathrm{d}x$

$$= \sqrt{\frac{2}{\pi b}}\,\frac{1}{3a^3}\left\{ b\left[3a^2 - 4b^2 + 12z^2\right] - \sqrt{b^2 - \ell_1^2}\left\{12\ell_2^2 - 16b^2 + 4\ell_1^2 - 3a^2\right\}\right\}$$

$$[\arg a > 0, \quad \arg b > 0, \quad \arg z > 0]$$

$16.^{10}$   $\displaystyle\int_0^\infty e^{-zx} J_3(ax) J_{3/2}(bx) x^{1/2}\,\mathrm{d}x$

$$= \sqrt{\frac{2}{\pi}}\,b^{3/2}\left\{ \frac{4}{a^3}\left[\frac{2}{3} - \frac{\sqrt{b^2 - \ell_1^2}}{b} + \frac{\left(b^2 - \ell_1^2\right)^{3/2}}{3b^2}\right] - \frac{a\sqrt{\ell_2^2 - a^2}}{\left(\ell_2^2 - \ell_1^2\right)\ell_2^3}\right\}$$

$$[\arg a > 0, \quad \arg b > 0, \quad \arg z > 0]$$

$17.^{10}$   $\displaystyle\int_0^\infty e^{-zx} J_3(ax) J_{3/2}(bx) x^{-1/2}\,\mathrm{d}x = \sqrt{\frac{2}{\pi}}\,\frac{b^{3/2}}{3a^3}\left[\sqrt{\ell_2^2 - b^2}\left(\frac{4b^2\left(2b^2 - \ell_1^2\right) - \ell_1^4}{b^4}\right) - 8z\right]$

$$[\arg a > 0, \quad \arg b > 0, \quad \arg z > 0]$$

$18.^{10}$   $\displaystyle\int_0^\infty e^{-zx} J_3(ax) J_{3/2}(bx) x^{-3/2}\,\mathrm{d}x$

$$= \sqrt{\frac{2}{\pi}}\,\frac{b^{3/2}}{3a^3}\left\{ a^2 - \frac{4}{5}b^2 + 4z^2 - \sqrt{b^2 - \ell_1^2}\left[\frac{4\ell_2^2}{b} - \frac{24b}{5} + \frac{8\ell_1^2}{5b} - \frac{a^2}{b} + \frac{\ell_1^4}{5b^3}\right]\right\}$$

$$[\arg a > 0, \quad \arg b > 0, \quad \arg z > 0]$$

$19.^{10}$   $\displaystyle\int_0^\infty e^{-zx} J_3(ax) J_{3/2}(bx) x^{-5/2}\,\mathrm{d}x$

$$= -\sqrt{\frac{2}{\pi}}\,\frac{b^{3/2}}{3a^3}\left\{ \left(a^2 - \frac{4}{5}b^2\right)z + \frac{4z^3}{3}\right.$$
$$+ \sqrt{\ell_2^2 - b^2}\left[u^2 + \frac{32}{15}b^2 - \frac{12}{5}\ell_1^2 - \frac{4}{3}\ell_2^2 + \frac{2\ell_1^4}{5b^2} + \frac{a^4\ell_1^2}{16b^4} + \frac{a^2\ell_1^2}{24b^4} + \frac{\ell_1^6}{30b^4}\right]$$
$$\left. - \frac{a^6}{16b^3}\arcsin\left(\frac{b}{\ell_2}\right)\right\}$$

$$[\arg a > 0, \quad \arg b > 0, \quad \arg z > 0]$$

**6.622**

1.   $\displaystyle\int_0^\infty \left(J_0(x) - e^{-\alpha x}\right)\frac{\mathrm{d}x}{x} = \ln 2\alpha$        $[\alpha > 0]$        NT 66(13)

2.   $\displaystyle\int_0^\infty \frac{e^{i(u+x)}}{u+x}\,J_0(x)\,\mathrm{d}x = \frac{\pi}{2}i\,H_0^{(1)}(u)$        MO 44

$3.^8$   $\displaystyle\int_0^\infty e^{-x\cosh\alpha} I_\nu(x) x^{\mu-1}\,\mathrm{d}x = \sqrt{\frac{2}{\pi}}\,e^{-\left(\mu-\frac{1}{2}\right)\pi i}\,\frac{Q_{\nu-\frac{1}{2}}^{\mu-\frac{1}{2}}(\cosh\alpha)}{\sinh^{\mu-\frac{1}{2}}\alpha}$

$$[\operatorname{Re}(\mu+\nu) > 0, \quad \operatorname{Re}(\cosh\alpha) > 1]$$
$$\text{WA 388(6)a}$$

**6.623**

1.   $\displaystyle\int_0^\infty e^{-\alpha x} J_\nu(bx) x^\nu\,\mathrm{d}x = \frac{(2b)^\nu\,\Gamma\left(\nu+\frac{1}{2}\right)}{\sqrt{\pi}\,(\alpha^2 + b^2)^{\nu+\frac{1}{2}}}$        $\left[\operatorname{Re}\nu > -\frac{1}{2}, \quad \operatorname{Re}\alpha > |\operatorname{Im} b|\right]$

$$\text{WA 422(5)}$$

2. $\int_0^\infty e^{-\alpha x} J_\nu(bx) x^{\nu+1}\, dx = \dfrac{2\alpha(2b)^\nu\, \Gamma\left(\nu + \frac{3}{2}\right)}{\sqrt{\pi}\,(\alpha^2 + b^2)^{\nu + \frac{3}{2}}}$      $[\operatorname{Re}\nu > -1, \quad \operatorname{Re}\alpha > |\operatorname{Im} b|]$    WA 422(6)

3. $\int_0^\infty e^{-\alpha x} J_\nu(bx)\dfrac{dx}{x} = \dfrac{\left(\sqrt{\alpha^2 + b^2} - \alpha\right)^\nu}{\nu b^\nu}$

     $[\operatorname{Re}\nu > 0; \quad \operatorname{Re}\alpha > |\operatorname{Im} b|]$    (cf. **6.611** 1)    WA 422(7)

**6.624**

1. $\int_0^\infty x e^{-\alpha x} K_0(bx)\, dx = \dfrac{1}{\alpha^2 - b^2}\left\{\dfrac{\alpha}{\sqrt{\alpha^2 - b^2}}\ln\left[\dfrac{\alpha}{b} + \sqrt{\left(\dfrac{\alpha}{b}\right)^2 - 1}\,\right] - 1\right\}$     MO 181

2. $\int_0^\infty \sqrt{x}\, e^{-\alpha x} K_{\pm\frac{1}{2}}(bx)\, dx = \sqrt{\dfrac{\pi}{2b}}\,\dfrac{1}{\alpha + b}$      MO 181

3. $\int_0^\infty e^{-tz}\left(z^2 - 1\right)^{-1/2} K_\mu(t) t^\nu\, dt = \dfrac{\Gamma(\nu - \mu + 1)}{(z^2 - 1)^{-\frac{1}{2}(\nu+1)}}\, e^{-i\mu\pi}\, Q_\nu^\mu(z)$

     $[\operatorname{Re}(\nu \pm \mu) > -1]$    EH II 57(7)

4. $\int_0^\infty e^{-tz}\left(z^2 - 1\right)^{-1/2} I_{-\mu}(t) t^\nu\, dt = \dfrac{\Gamma(-\nu - \mu)}{(z^2 - 1)^{\frac{1}{2}\nu}}\, P_\nu^\mu(z)$      $[\operatorname{Re}(\nu + \mu) < 0]$    EH II 57(8)

5. $\int_0^\infty e^{-tz}\left(z^2 - 1\right)^{-\frac{1}{2}} I_\mu(t) t^\nu\, dt = \dfrac{\Gamma(\nu + \mu + 1)}{(z^2 - 1)^{-\frac{1}{2}(\nu+1)}}\, P_\nu^{-\mu}(z)$

     $[\operatorname{Re}(\nu + \mu) > -1]$    EH II 57(9)

6. $\int_0^\infty e^{-t\cos\theta} J_\mu(t\sin\theta) t^\nu\, dt = \Gamma(\nu + \mu + 1)\, P_\nu^{-\mu}(\cos\theta)$

     $\left[\operatorname{Re}(\nu + \mu) > -1, \quad 0 \le \theta < \frac{1}{2}\pi\right]$
   EH II 57(10)

7. $\int_0^\infty \dfrac{J_\nu(bx) x^\nu}{e^{\pi x} - 1}\, dx = \dfrac{(2b)^\nu\, \Gamma\left(\nu + \frac{1}{2}\right)}{\sqrt{\pi}} \displaystyle\sum_{n=1}^\infty \dfrac{1}{(n^2\pi^2 + b^2)^{\nu + \frac{1}{2}}}$

     $[\operatorname{Re}\nu > 0, \quad |\operatorname{Im} b| < \pi]$    WA 423(9)

**6.625**

1. $\int_0^1 x^{\lambda - \nu - 1}(1 - x)^{\mu - 1} e^{\pm i\alpha x} J_\nu(\alpha x)\, dx = \dfrac{2^{-\nu}\alpha^\nu\, \Gamma(\lambda)\, \Gamma(\mu)}{\Gamma(\lambda + \mu)\, \Gamma(\nu + 1)}\, {}_2F_2\left(\lambda, \nu + \dfrac{1}{2}; \lambda + \mu, 2\nu + 1; \pm 2i\alpha\right)$

     $[\operatorname{Re}\lambda > 0, \quad \operatorname{Re}\mu > 0]$    ET II 194(58)a

2. $\int_0^1 x^\nu (1 - x)^{\mu - 1} e^{\pm i\alpha x} J_\nu(\alpha x)\, dx = \dfrac{(2\alpha)^\nu\, \Gamma(\mu)\, \Gamma\left(\nu + \frac{1}{2}\right)}{\sqrt{\pi}\, \Gamma(\mu + 2\nu + 1)}\, {}_1F_1\left(\nu + \dfrac{1}{2}; \mu + 2\nu + 1; \pm 2i\alpha\right)$

     $\left[\operatorname{Re}\mu > 0, \quad \operatorname{Re}\nu > -\frac{1}{2}\right]$    ET II 194(57)a

3. $\int_0^1 x^\nu (1 - x)^{\mu - 1} e^{\pm\alpha x} I_\nu(\alpha x)\, dx = \dfrac{(2\alpha)^\nu\, \Gamma\left(\nu + \frac{1}{2}\right)\Gamma(\mu)}{\sqrt{\pi}\, \Gamma(\mu + 2\nu + 1)}\, {}_1F_1\left(\nu + \dfrac{1}{2}; \mu + 2\nu + 1; \pm 2\alpha\right)$

     $\left[\operatorname{Re}\mu > 0, \quad \operatorname{Re}\nu > -\frac{1}{2}\right]$
   BU 9(16a), ET II 197(77)a

4.　$\displaystyle\int_0^1 x^{\lambda-1}(1-x)^{\mu-1}e^{\pm\alpha x}\,I_\nu(\alpha x)\,\mathrm{d}x = \frac{\left(\frac{1}{2}\alpha\right)^\nu \Gamma(\lambda+\nu)\,\Gamma(\mu)}{\Gamma(\nu+1)\,\Gamma(\lambda+\mu+\nu)}$

$$\times\ {}_2F_2\left(\nu+\frac{1}{2},\lambda+\nu;2\nu+1,\mu+\lambda+\nu;\pm2\alpha\right)$$

$$[\operatorname{Re}\mu>0,\quad \operatorname{Re}(\lambda+\nu)>0]\quad\text{ET II 197(78)a}$$

5.　$\displaystyle\int_0^1 x^{\mu-\kappa}(1-x)^{2\kappa-1}\,I_{\mu-\kappa}\left(\frac{1}{2}xz\right)e^{-\frac{1}{2}xz}\,\mathrm{d}x = \frac{\Gamma(2\kappa)}{\sqrt{\pi}\,\Gamma(1+2\mu)}e^{z/2}z^{-\kappa-\frac{1}{2}}\,M_{\kappa,\mu}(z)$

$$\left[\operatorname{Re}\left(\kappa-\tfrac{1}{2}-\mu\right)<0,\quad \operatorname{Re}\kappa>0\right]$$
$$\text{BU 129(14a)}$$

6.　$\displaystyle\int_1^\infty x^{-\lambda}(x-1)^{\mu-1}e^{-\alpha x}\,I_\nu(\alpha x)\,\mathrm{d}x = \frac{(2\alpha)^\lambda\,\Gamma(\mu)}{\sqrt{\pi}}\,G_{23}^{21}\left(2\alpha\ \middle|\ \begin{matrix}\frac{1}{2}-\lambda,0\\ -\mu,\nu-\lambda,-\nu-\lambda\end{matrix}\right)$

$$\left[0<\operatorname{Re}\mu<\tfrac{1}{2}+\operatorname{Re}\lambda,\quad \operatorname{Re}\alpha>0\right]$$
$$\text{ET II 207(50)a}$$

7.　$\displaystyle\int_1^\infty x^{-\lambda}(x-1)^{\mu-1}e^{-\alpha x}\,K_\nu(\alpha x)\,\mathrm{d}x = \Gamma(\mu)\sqrt{\pi}(2\alpha)^\lambda\,G_{23}^{30}\left(2\alpha\ \middle|\ \begin{matrix}0,\frac{1}{2}-\lambda\\ -\mu,\nu-\lambda,-\nu-\lambda\end{matrix}\right)$

$$[\operatorname{Re}\mu>0,\quad \operatorname{Re}\alpha>0]\qquad\text{ET II 208(55)a}$$

8.　$\displaystyle\int_1^\infty x^{-\nu}(x-1)^{\mu-1}e^{-\alpha x}\,I_\nu(\alpha x)\,\mathrm{d}x = \frac{(2\alpha)^{\nu-\mu}\,\Gamma\left(\frac{1}{2}-\mu+\nu\right)\Gamma(\mu)}{\sqrt{\pi}\,\Gamma(1-\mu+2\nu)}$

$$\times\ {}_1F_1\left(\frac{1}{2}-\mu+\nu;1-\mu+2\nu;-2\alpha\right)$$

$$\left[0<\operatorname{Re}\mu<\tfrac{1}{2}+\operatorname{Re}\nu,\quad \operatorname{Re}\alpha>0\right]\quad\text{ET II 207(49)a}$$

9.　$\displaystyle\int_1^\infty x^{\ \nu}(x-1)^{\mu-1}e^{-\alpha x}\,K_\nu(\alpha x)\,\mathrm{d}x = \sqrt{\pi}\,\Gamma(\mu)(2\alpha)^{-\frac{1}{2}\mu-\frac{1}{2}}e^{-\alpha}\,W_{-\frac{1}{2}\mu,\nu-\frac{1}{2}\mu}(2\alpha)$

$$[\operatorname{Re}\mu>0,\quad \operatorname{Re}\alpha>0]\qquad\text{ET II 208(53)a}$$

10.$^{12}$　$\displaystyle\int_1^\infty x^{-\mu-\frac{1}{2}}(x-1)^{\mu-1}e^{-\alpha x}\,K_\nu(\alpha x)\,\mathrm{d}x = \sqrt{\pi}\,\Gamma(\mu)(2\alpha)^{-\frac{1}{2}}e^{-\alpha}\,W_{-\mu,\nu}(2\alpha)$

$$[\operatorname{Re}\mu>0,\quad \operatorname{Re}\alpha>0]\qquad\text{ET II 207(51)a}$$

11.$^3$　$\displaystyle\int_{-1}^1 (1-x^2)^{-1/2}xe^{-ax}\,I_1\left(b\sqrt{1-x^2}\right)\mathrm{d}x = \frac{2}{b}\left\{\sinh a - a\left(a^2+b^2\right)^{-1/2}\sinh\sqrt{a^2+b^2}\right\}$

$$[a>0,\quad b>0]$$

## 6.626

1.$^{11}$　$\displaystyle\int_0^\infty x^{\lambda-1}e^{-\alpha x}\,J_\mu(bx)\,J_\nu(cx)\,\mathrm{d}x = \frac{b^\mu c^\nu}{\Gamma(\nu+1)}2^{-\nu-\mu}\alpha^{-\lambda-\mu-\nu}\sum_{m=0}^\infty \frac{\Gamma(\lambda+\mu+\nu+2m)}{m!\,\Gamma(\mu+m+1)}$

$$\times\ F\left(-m,-\mu-m;\nu+1;\frac{c^2}{b^2}\right)\left(-\frac{b^2}{4\alpha^2}\right)^m$$

$$[\operatorname{Re}(\lambda+\mu+\nu)>0,\quad \operatorname{Re}(\alpha\pm ib\pm ic)>1]\quad\text{EH II 48(15)}$$

2.  $$\int_0^\infty e^{-2\alpha x} J_\nu(bx) J_\mu(bx) x^{\nu+\mu}\, dx = \frac{\Gamma\left(\nu+\mu+\frac{1}{2}\right) b^{\nu+\mu}}{\sqrt{\pi^3}}$$

$$\times \int_0^{\frac{\pi}{2}} \frac{\cos^{\nu+\mu}\varphi \cos(\nu-\mu)\varphi}{(\alpha^2+b^2\cos^2\varphi)^{\nu+\mu}\sqrt{\alpha^2+b^2\cos^2\varphi}}\, d\varphi$$

$$\left[\operatorname{Re}\alpha > |\operatorname{Im} b|, \quad \operatorname{Re}(\nu+\mu) > -\tfrac{1}{2}\right] \quad \text{WA 427(1)}$$

3.  $$\int_0^\infty e^{-2\alpha x} J_0(bx) J_1(bx) x\, dx = \frac{K\left(\frac{b}{\sqrt{\alpha^2+b^2}}\right) - E\left(\frac{b}{\sqrt{\alpha^2+b^2}}\right)}{2\pi b\sqrt{\alpha^2+b^2}}$$

$$\text{WA 427(2)}$$

4.  $$\int_0^\infty e^{-2\alpha x} I_0(bx) I_1(bx) x\, dx = \frac{1}{2\pi b}\left\{\frac{\alpha}{\alpha^2-b^2} E\left(\frac{b}{\alpha}\right) - \frac{1}{\alpha} K\left(\frac{b}{\alpha}\right)\right\}$$

$$[\operatorname{Re}\alpha > \operatorname{Re} b] \qquad \text{WA 428(5)}$$

5.[10]  $$\int_0^\infty x^{\nu-\mu+2n} e^{-zx} J_\mu(\alpha x) J_\nu(\rho x)\, dx = \frac{1}{\sqrt{\pi}}\left(\frac{a}{2}\right)^{\mu-\nu-2n-1}\left(\frac{\rho}{a}\right)^\nu$$

$$\times \frac{1}{\Gamma\left(\mu-\nu-n+\frac{1}{2}\right)} \sum_{q=0}^\infty \frac{\Gamma\left(\nu+n+q+\frac{1}{2}\right)\left(\nu-\mu+n+\frac{1}{2}\right)_q}{q!\,\Gamma\left(\nu+q+\frac{1}{2}\right)}$$

$$\times a^{-2q} \int_0^{\ell_1/\rho} \frac{dx}{\sqrt{1-x^2}} x^{2\nu+2q}\left(\rho^2+\frac{z^2}{1-x^2}\right)^q$$

where $\ell_1 = \frac{1}{2}\left[\sqrt{(a+\rho)^2+z^2} - \sqrt{(a-\rho)^2+z^2}\right]$ $\qquad \left[\mu > \nu+2n, \quad n = 0, 1, \ldots, \quad \nu > -\tfrac{1}{2}\right]$

**6.627** $\quad \int_0^\infty \frac{x^{-1/2}}{x+a} e^{-x} K_\nu(x)\, dx = \frac{\pi e^a K_\nu(a)}{\sqrt{a}\cos(\nu\pi)}$ $\qquad \left[|\arg a| < \pi, \quad |\operatorname{Re}\nu| < \tfrac{1}{2}\right]$ $\qquad$ ET II 368(29)

**6.628**

1.  $$\int_0^\infty e^{-x\cos\beta} J_{-\nu}(x\sin\beta) x^\mu\, dx = \Gamma(\mu-\nu+1) P_\mu^\nu(\cos\beta)$$

$$\left[0 < \beta < \frac{\pi}{2}, \quad \operatorname{Re}(\mu-\nu) > -1\right]$$

$$\text{WA 424(3), WH}$$

2.  $$\int_0^\infty e^{-x\cos\beta} Y_\nu(x\sin\beta) x^\mu\, dx = -\frac{\sin\mu\pi}{\sin(\mu+\nu)\pi}\frac{\Gamma(\mu-\nu+1)}{\pi}$$

$$\times\left[Q_\mu^\nu(\cos\beta+0\cdot i)\, e^{\frac{1}{2}\nu\pi i} + Q_\mu^\nu(\cos\beta-0\cdot i)\, e^{-\frac{1}{2}\nu\pi i}\right]$$

$$\left[\operatorname{Re}(\mu+\nu) > -1, \quad 0 < \beta < \frac{\pi}{2}\right] \quad \text{WA 424(4)}$$

3.  $$\int_0^1 e^{\frac{xu}{2}}(1-x)^{2\nu-1} x^{\mu-\nu} J_{\mu-\nu}\left(\frac{ixu}{2}\right) dx = 2^{2(\nu-\mu)} e^{\frac{\pi}{2}(\mu-\nu)i}\frac{B(2\nu, 2\mu-2\nu+1)}{\Gamma(\mu-\nu+1)}\frac{e^{\frac{u}{2}}}{u^{\nu+\frac{1}{2}}} M_{\nu,\mu}(u)$$

$$\text{MO 118a}$$

4.[8]  $$\int_0^\infty e^{-x\cosh\alpha} I_\nu(x\sinh\alpha) x^\mu\, dx = \Gamma(\nu+\mu+1) P_\mu^{-\nu}(\cosh\alpha)$$

$$\left[\operatorname{Re}(\mu+\nu) > -1, \quad |\operatorname{Im}\alpha| < \tfrac{1}{2}\pi\right]$$

$$\text{WA 423(1)}$$

5.    $\displaystyle\int_0^\infty e^{-x\cosh\alpha}\,K_\nu\,(x\sinh\alpha)\,x^\mu\,\mathrm{d}x = \frac{\sin\mu\pi}{\sin(\nu+\mu)\pi}\,\Gamma(\mu-\nu+1)\,Q_\mu^\nu\,(\cosh\alpha)$

$$[\mathrm{Re}(\mu+1) > |\mathrm{Re}\,\nu|] \qquad \text{WA 423(2)}$$

6.    $\displaystyle\int_0^\infty e^{-x\cosh\alpha}\,I_\nu(x)x^{\mu-1}\,\mathrm{d}x = \frac{\cos\nu\pi}{\sin(\mu+\nu)\pi}\,\frac{Q_{\nu-\frac{1}{2}}^{\mu-\frac{1}{2}}(\cosh\alpha)}{\sqrt{\frac{\pi}{2}}\,(\sinh\alpha)^{\mu-\frac{1}{2}}}$

$$[\mathrm{Re}(\mu+\nu) > 0, \quad \mathrm{Re}\,(\cosh\alpha) > 1]$$
$$\text{WA 424(6)}$$

7.    $\displaystyle\int_0^\infty e^{-x\cosh\alpha}\,K_\nu(x)x^{\mu-1}\,\mathrm{d}x = \sqrt{\frac{\pi}{2}}\,\Gamma(\mu-\nu)\,\Gamma(\mu+\nu)\frac{P_{\nu-\frac{1}{2}}^{\frac{1}{2}-\mu}(\cosh\alpha)}{(\sinh\alpha)^{\mu-\frac{1}{2}}}$

$$[\mathrm{Re}\,\mu > |\mathrm{Re}\,\nu|, \quad \mathrm{Re}\,(\cosh\alpha) > -1]$$
$$\text{WA 424(7)}$$

**6.629**[8]   $\displaystyle\int_0^\infty x^{-1/2}e^{-x\alpha\cos\varphi\cos\psi}\,J_\mu\,(\alpha x\sin\varphi)\,J_\nu\,(\alpha x\sin\psi)\,\mathrm{d}x$

$$= \Gamma\left(\mu+\nu+\tfrac{1}{2}\right)\alpha^{-\frac{1}{2}}\,P_{\nu-\frac{1}{2}}^{-\mu}\,(\cos\varphi)\,P_{\mu-\frac{1}{2}}^{-\nu}\,(\cos\psi)$$

$$\left[\alpha > 0, \quad 0 < \varphi < \frac{\pi}{2}, \quad 0 < \psi < \frac{\pi}{2}, \quad \mathrm{Re}(\mu+\nu) > -\frac{1}{2}\right] \quad \text{ET II 50(19)}$$

**6.631**

1.    $\displaystyle\int_0^\infty x^\mu e^{-\alpha x^2}\,J_\nu(bx)\,\mathrm{d}x = \frac{b^\nu\,\Gamma\left(\frac{1}{2}\nu+\frac{1}{2}\mu+\frac{1}{2}\right)}{2^{\nu+1}\alpha^{\frac{1}{2}(\mu+\nu+1)}\,\Gamma(\nu+1)}\,{}_1F_1\left(\frac{\nu+\mu+1}{2};\nu+1;-\frac{b^2}{4\alpha}\right)$

$$\text{BU 8(15)}$$

$$= \frac{\Gamma\left(\frac{1}{2}\nu+\frac{1}{2}\mu+\frac{1}{2}\right)}{b\alpha^{\frac{1}{2}\mu}\,\Gamma(\nu+1)}\exp\left(-\frac{b^2}{8\alpha}\right)M_{\frac{1}{2}\mu,\frac{1}{2}\nu}\left(\frac{b^2}{4\alpha}\right)$$

$$[\mathrm{Re}\,\alpha > 0, \quad \mathrm{Re}(\mu+\nu) > -1]$$
$$\text{EH II 50(22), ET II 30(14), BU 14(13b)}$$

2.    $\displaystyle\int_0^\infty x^\mu e^{-\alpha x^2}\,Y_\nu(bx)\,\mathrm{d}x$

$$= -\alpha^{-\frac{1}{2}\mu}b^{-1}\sec\left(\frac{\nu-\mu}{2}\pi\right)\exp\left(-\frac{b^2}{8\alpha}\right)$$

$$\times\left\{\frac{\Gamma\left(\frac{1}{2}+\frac{1}{2}\mu+\frac{1}{2}\nu\right)}{\Gamma(1+\nu)}\sin\left(\frac{\nu-\mu}{2}\pi\right)M_{\frac{1}{2}\mu,\frac{1}{2}\nu}\left(\frac{b^2}{4\alpha}\right)+W_{\frac{1}{2}\mu,\frac{1}{2}\nu}\left(\frac{b^2}{4\alpha}\right)\right\}$$

$$[\mathrm{Re}\,\alpha > 0, \quad \mathrm{Re}\,\mu > |\mathrm{Re}\,\nu| - 1, \quad b > 0] \quad \text{ET II 106(4)}$$

3.[12]   $\displaystyle\int_0^\infty x^\mu e^{-\alpha x^2}\,K_\nu(bx)\,\mathrm{d}x = \frac{1}{2}\alpha^{-\frac{1}{2}\mu}b^{-1}\Gamma\left(\frac{1+\nu+\mu}{2}\right)\Gamma\left(\frac{1-\nu+\mu}{2}\right)\exp\left(\frac{b^2}{8\alpha}\right)W_{-\frac{1}{2}\mu,\frac{1}{2}\nu}\left(\frac{b^2}{4\alpha}\right)$

$$[\mathrm{Re}\,\alpha > 0, \quad \mathrm{Re}\,\mu > |\mathrm{Re}\,\nu| - 1]$$
$$\text{ET II 132(25)}$$

4.[11]   $\displaystyle\int_0^\infty x^{\nu+1}e^{-\alpha x^2}\,J_\nu(bx)\,\mathrm{d}x = \frac{b^\nu}{(2\alpha)^{\nu+1}}\exp\left(-\frac{b^2}{4\alpha}\right)$     $[\mathrm{Re}\,\alpha > 0, \quad \mathrm{Re}\,\nu > -1]$

$$\text{WA 431(4), ET II 29(10)}$$

5.[12] $\displaystyle\int_0^\infty x^{\nu-1} e^{-\alpha x^2} J_\nu(bx)\,dx = 2^{\nu-1} b^{-\nu}\,\gamma\left(\nu,\frac{b^2}{4\alpha}\right)$ $\qquad$ $[\operatorname{Re}\alpha > 0, \quad \operatorname{Re}\nu > 0]$ $\qquad$ ET II 30(11)

6. $\displaystyle\int_0^\infty x^{\nu+1} e^{\pm i\alpha x^2} J_\nu(bx)\,dx = \frac{b^\nu}{(2\alpha)^{\nu+1}}\exp\left[\pm i\left(\frac{\nu+1}{2}\pi - \frac{b^2}{4\alpha}\right)\right]$

$$\left[\alpha > 0, \quad -1 < \operatorname{Re}\nu < \tfrac{1}{2}, \quad b > 0\right]$$
$$\text{ET II 30(12)}$$

7. $\displaystyle\int_0^\infty x e^{-\alpha x^2} J_\nu(bx)\,dx = \frac{\sqrt{\pi}\,b}{8\alpha^{\frac{3}{2}}}\exp\left(-\frac{b^2}{8\alpha}\right)\left[I_{\frac{1}{2}\nu-\frac{1}{2}}\left(\frac{b^2}{8\alpha}\right) - I_{\frac{1}{2}\nu+\frac{1}{2}}\left(\frac{b^2}{8\alpha}\right)\right]$

$$\left[\operatorname{Re}\alpha > 0, \quad \operatorname{Re}\nu > -2\right] \qquad \text{ET II 29(9)}$$

8. $\displaystyle\int_0^1 x^{n+1} e^{-\alpha x^2} I_n(2\alpha x)\,dx = \frac{1}{4\alpha}\left[e^\alpha - e^{-\alpha}\sum_{r=-n}^{n} I_r(2\alpha)\right]$

$$[n = 0, 1, \ldots] \qquad \text{ET II 365(8)a}$$

9. $\displaystyle\int_1^\infty x^{1-n} e^{-\alpha x^2} I_n(2\alpha x)\,dx = \frac{1}{4\alpha}\left[e^\alpha - e^{-\alpha}\sum_{r=1-n}^{n-1} I_r(2\alpha)\right]$

$$[n = 1, 2, \ldots] \qquad \text{ET II 367(20)a}$$

10. $\displaystyle\int_0^\infty e^{-x^2} x^{2n+\mu+1} J_\mu\left(2x\sqrt{z}\right)\,dx = \frac{n!}{2} e^{-z} z^{\frac{1}{2}\mu} L_n^\mu(z)$ $\qquad$ $[n = 0, 1, \ldots;\quad n + \operatorname{Re}\mu > -1]$

$$\text{BU 135(5)}$$

**6.632** $\displaystyle\int_0^\infty x^{-\frac{1}{2}}\exp\left[-\left(x^2 + a^2 - 2ax\cos\varphi\right)^{\frac{1}{2}}\right]\left[x^2 + a^2 - 2ax\cos\varphi\right]^{-\frac{1}{2}} K_\nu(x)\,dx$

$$= \pi a^{-\frac{1}{2}}\sec(\nu\pi) P_{\nu-\frac{1}{2}}\left(-\cos\varphi\right) K_\nu(a)$$
$$\left[|\arg a| + |\operatorname{Re}\varphi| < \pi, \quad |\operatorname{Re}\nu| < \tfrac{1}{2}\right] \quad \text{ET II 368(32)}$$

**6.633**

1. $\displaystyle\int_0^\infty x^{\lambda+1} e^{-\alpha x^2} J_\mu(bx) J_\nu(cx)\,dx = \frac{b^\mu c^\nu \alpha^{-\frac{\mu+\nu+\lambda+2}{2}}}{2^{\nu+\mu+1}\Gamma(\nu+1)}\sum_{m=0}^\infty \frac{\Gamma\left(m + \frac{1}{2}\nu + \frac{1}{2}\mu + \frac{1}{2}\lambda + 1\right)}{m!\,\Gamma(m+\mu+1)}\left(-\frac{b^2}{4\alpha}\right)^m$

$$\times F\left(-m, -\mu - m; \nu + 1; \frac{c^2}{b^2}\right)$$
$$[\operatorname{Re}\alpha > 0, \quad \operatorname{Re}(\mu + \nu + \lambda) > -2, \quad b > 0, \quad c > 0] \quad \text{EH II 49(20)a, ET II 51(24)a}$$

2. $\displaystyle\int_0^\infty e^{-\varrho^2 x^2} J_p(ax) J_p(\beta x) x\,dx = \frac{1}{2\varrho^2}\exp\left(-\frac{a^2+\beta^2}{4\varrho^2}\right) I_p\left(\frac{a\beta}{2\varrho^2}\right)$

$$\left[\operatorname{Re}p > -1, \quad |\arg\varrho| < \frac{\pi}{4}, \quad a > 0, \quad \beta > 0\right] \quad \text{KU 146(16)a, WA 433(1)}$$

3. $\displaystyle\int_0^\infty x^{2\nu+1} e^{-\alpha x^2} J_\nu(x) Y_\nu(x)\,dx = -\frac{1}{2\sqrt{\pi}}\alpha^{-\frac{3}{2}\nu-\frac{1}{2}}\exp\left(-\frac{1}{2\alpha}\right) W_{\frac{1}{2}\nu,\frac{1}{2}\nu}\left(\frac{1}{\alpha}\right)$

$$\left[\operatorname{Re}\alpha > 0, \quad \operatorname{Re}\nu > -\tfrac{1}{2}\right] \qquad \text{ET II 347(59)}$$

4.    $\displaystyle\int_0^\infty x e^{-\alpha x^2} I_\nu(bx)\, J_\nu(cx)\, \mathrm{d}x = \frac{1}{2\alpha}\exp\left(\frac{b^2-c^2}{4\alpha}\right) J_\nu\left(\frac{bc}{2\alpha}\right)$

$$[\operatorname{Re}\alpha > 0, \quad \operatorname{Re}\nu > -1] \qquad \text{ET II 63(1)}$$

5.    $\displaystyle\int_0^\infty x^{\lambda-1} e^{-\alpha x^2} J_\mu(bx)\, J_\nu(bx)\, \mathrm{d}x$

$$= 2^{-\nu-\mu-1}\alpha^{-\frac{1}{2}(\nu+\lambda+\mu)} b^{\nu+\mu} \frac{\Gamma\left(\frac{1}{2}\lambda+\frac{1}{2}\mu+\frac{1}{2}\nu\right)}{\Gamma(\mu+1)\,\Gamma(\nu+1)}$$

$$\times\ _3F_3\left[\frac{\nu}{2}+\frac{\mu}{2}+\frac{1}{2},\frac{\nu}{2}+\frac{\mu}{2}+1,\frac{\nu+\mu+\lambda}{2};\mu+1,\nu+1,\mu+\nu+1;-\frac{b^2}{\alpha}\right]$$

$$[\operatorname{Re}(\nu+\lambda+\mu) > 0, \quad \operatorname{Re}\alpha > 0] \quad \text{WA 434, EH II 50(21)}$$

**6.634**    $\displaystyle\int_0^\infty x e^{-\frac{x^2}{2a}}\left[I_\nu(x)+I_{-\nu}(x)\right] K_\nu(x)\, \mathrm{d}x = a e^a\, K_\nu(a)$     $[\operatorname{Re}a > 0, \quad -1 < \operatorname{Re}\nu < 1]$

$$\text{ET II 371(49)}$$

**6.635**

1.    $\displaystyle\int_0^\infty x^{-1} e^{-\frac{\alpha}{x}} J_\nu(bx)\, \mathrm{d}x = 2\, J_\nu\left(\sqrt{2\alpha b}\right) K_\nu\left(\sqrt{2\alpha b}\right)$     $[\operatorname{Re}\alpha > 0, \quad b > 0]$     ET II 30(15)

2.    $\displaystyle\int_0^\infty x^{-1} e^{-\frac{\alpha}{x}} Y_\nu(bx)\, \mathrm{d}x = 2\, Y_\nu\left(\sqrt{2\alpha b}\right) K_\nu\left(\sqrt{2\alpha b}\right)$     $[\operatorname{Re}\alpha > 0, \quad b > 0]$     ET II 106(5)

3.    $\displaystyle\int_0^\infty x^{-1} e^{-\frac{\alpha}{x}-\beta x} J_\nu(\gamma x)\, \mathrm{d}x = 2\, J_\nu\left\{\sqrt{2\alpha}\left[\sqrt{\beta^2+\gamma^2}-\beta\right]^{\frac{1}{2}}\right\} K_\nu\left\{\sqrt{2\alpha}\left[\sqrt{\beta^2+\gamma^2}+\beta\right]^{\frac{1}{2}}\right\}$

$$[\operatorname{Re}\alpha > 0, \quad \operatorname{Re}\beta > 0, \quad \gamma > 0]$$
$$\text{ET II 30(16)}$$

**6.636**    $\displaystyle\int_0^\infty x^{-\frac{1}{2}} e^{-\alpha\sqrt{x}} J_\nu(bx)\, \mathrm{d}x = \frac{\sqrt{2}}{\sqrt{\pi b}}\,\Gamma\left(\nu+\frac{1}{2}\right) D_{-\nu-\frac{1}{2}}\left(2^{-\frac{1}{2}}\alpha e^{\frac{1}{4}\pi i} b^{-\frac{1}{2}}\right) D_{-\nu-\frac{1}{2}}\left(2^{-\frac{1}{2}}\alpha e^{-\frac{1}{4}\pi i} b^{-\frac{1}{2}}\right)$

$$\left[\operatorname{Re}\alpha > 0, \quad b > 0, \quad \operatorname{Re}\nu > -\tfrac{1}{2}\right]$$
$$\text{ET II 30(17)}$$

**6.637**

1.    $\displaystyle\int_0^\infty \left(\beta^2+x^2\right)^{-\frac{1}{2}} \exp\left[-\alpha\left(\beta^2+x^2\right)^{\frac{1}{2}}\right] J_\nu(\gamma x)\, \mathrm{d}x$

$$= I_{\frac{1}{2}\nu}\left\{\frac{1}{2}\beta\left[\left(\alpha^2+\gamma^2\right)^{\frac{1}{2}}-\alpha\right]\right\} K_{\frac{1}{2}\nu}\left\{\frac{1}{2}\beta\left[\left(\alpha^2+\gamma^2\right)^{\frac{1}{2}}+\alpha\right]\right\}$$

$$[\operatorname{Re}\alpha > 0, \quad \operatorname{Re}\beta > 0, \quad \gamma > 0, \quad \operatorname{Re}\nu > -1] \quad \text{ET II 31(20)}$$

2.    $\displaystyle\int_0^\infty \left(\beta^2+x^2\right)^{-\frac{1}{2}} \exp\left[-\alpha\left(\beta^2+x^2\right)^{\frac{1}{2}}\right] Y_\nu(\gamma x)\, \mathrm{d}x$

$$= -\sec\left(\frac{\nu\pi}{2}\right) K_{\frac{1}{2}\nu}\left\{\frac{1}{2}\beta\left[\left(\alpha^2+\gamma^2\right)^{\frac{1}{2}}+\alpha\right]\right\}$$

$$\times\left(\frac{1}{\pi} K_{\frac{1}{2}\nu}\left\{\frac{1}{2}\beta\left[\left(\alpha^2+\gamma^2\right)^{\frac{1}{2}}+\alpha\right]\right\} + \sin\left(\frac{\nu\pi}{2}\right) I_{\frac{1}{2}\nu}\left\{\frac{1}{2}\beta\left[\left(\alpha^2+\gamma^2\right)^{\frac{1}{2}}-\alpha\right]\right\}\right)$$

$$[\operatorname{Re}\alpha > 0, \quad \operatorname{Re}\beta > 0, \quad \gamma > 0, \quad |\operatorname{Re}\nu| < 1] \quad \text{ET II 106(6)}$$

3. $\int_0^\infty (x^2 + \beta^2)^{-\frac{1}{2}} \exp\left[-\alpha (x^2 + \beta^2)^{\frac{1}{2}}\right] K_\nu(\gamma x) \, dx$

$$= \frac{1}{2} \sec\left(\frac{\nu\pi}{2}\right) K_{\frac{1}{2}\nu}\left(\frac{1}{2}\beta\left[\alpha + (\alpha^2 - \gamma^2)^{\frac{1}{2}}\right]\right) K_{\frac{1}{2}\nu}\left(\frac{1}{2}\beta\left[\alpha - (\alpha^2 - \gamma^2)^{\frac{1}{2}}\right]\right)$$

$$[\operatorname{Re}\alpha > 0, \quad \operatorname{Re}\beta > 0, \quad \operatorname{Re}(\gamma + \beta) > 0, \quad |\operatorname{Re}\nu| < 1] \quad \text{ET II 132(26)}$$

## 6.64 Combinations of Bessel functions of more complicated arguments, exponentials, and powers

**6.641** $\int_0^\infty \sqrt{x}\, e^{-\alpha x} J_{\pm\frac{1}{4}}(x^2) \, dx = \frac{\sqrt{\pi\alpha}}{4}\left[\mathbf{H}_{\mp\frac{1}{4}}\left(\frac{\alpha^2}{4}\right) - Y_{\mp\frac{1}{4}}\left(\frac{\alpha^2}{4}\right)\right]$ ⸻ MI 42

**6.642**

1.[10] $\int_0^\infty x^{-1} e^{-\alpha x} Y_\nu\left(\frac{2}{x}\right) dx = 2 K_\nu(2\sqrt{\alpha})\, Y_\nu(2\sqrt{\alpha})$

$$[\operatorname{Re}\alpha > 0] \qquad\qquad \text{MC}$$

2.[12] $\int_0^\infty x^{-1} e^{-\alpha x} H_\nu^{(1,2)}\left(\frac{2}{x}\right) dx = 2 H_\nu^{(1,2)}(\sqrt{\alpha})\, K_\nu(\sqrt{\alpha})$ ⸻ MI 44, EH II 91(26)

**6.643**

1. $\int_0^\infty x^{\mu-\frac{1}{2}} e^{-\alpha x} J_{2\nu}(2b\sqrt{x}) \, dx = \frac{\Gamma\left(\mu+\nu+\frac{1}{2}\right)}{b\,\Gamma(2\nu+1)} \exp\left(-\frac{b^2}{2\alpha}\right) \alpha^{-\mu} M_{\mu,\nu}\left(\frac{b^2}{\alpha}\right)$

$$\left[\operatorname{Re}\left(\mu+\nu+\tfrac{1}{2}\right) > 0\right]$$

$$\text{BU 14(13a), MI 42a}$$

2. $\int_0^\infty x^{\mu-\frac{1}{2}} e^{-\alpha x} I_{2\nu}(2b\sqrt{x}) \, dx = \frac{\Gamma\left(\mu+\nu+\frac{1}{2}\right)}{\Gamma(2\nu+1)} b^{-1} \exp\left(\frac{b^2}{2\alpha}\right) \alpha^{-\mu} M_{-\mu,\nu}\left(\frac{b^2}{\alpha}\right)$

$$\left[\operatorname{Re}\left(\mu+\nu+\tfrac{1}{2}\right) > 0\right] \qquad \text{MI 45}$$

3. $\int_0^\infty x^{\mu-\frac{1}{2}} e^{-\alpha x} K_{2\nu}(2b\sqrt{x}) \, dx = \frac{\Gamma\left(\mu+\nu+\frac{1}{2}\right)\Gamma\left(\mu-\nu+\frac{1}{2}\right)}{2b} \exp\left(\frac{b^2}{2\alpha}\right) \alpha^{-\mu} W_{-\mu,\nu}\left(\frac{b^2}{\alpha}\right)$

$$\left[\operatorname{Re}\left(\mu+\nu+\tfrac{1}{2}\right) > 0\right], \qquad (\text{cf. } \mathbf{6.631}\ 3)$$

$$\text{MI 47a}$$

4. $\int_0^\infty x^{n+\frac{1}{2}\nu} e^{-\alpha x} J_\nu(2b\sqrt{x}) \, dx = n!\, b^\nu \exp\left(-\frac{b^2}{\alpha}\right) \alpha^{-n-\nu-1} L_n^\nu\left(\frac{b^2}{\alpha}\right)$

$$[n + \nu > -1] \qquad\qquad \text{MO 178a}$$

5. $\int_0^\infty x^{-\frac{1}{2}} e^{-\alpha x} Y_{2\nu}(b\sqrt{x}) \, dx = -\sqrt{\frac{\pi}{\alpha}} \frac{\exp\left(-\frac{b^2}{8\alpha}\right)}{\cos(\nu\pi)}\left[\sin(\nu\pi) I_\nu\left(\frac{b^2}{8\alpha}\right) + \frac{1}{\pi} K_\nu\left(\frac{b^2}{8\alpha}\right)\right]$

$$\left[|\operatorname{Re}\nu| < \tfrac{1}{2}\right] \qquad\qquad \text{MI 44}$$

6. $\int_0^\infty x^{\frac{1}{2}m} e^{-\alpha x} K_m(2\sqrt{x}) \, dx = \frac{\Gamma(m+1)}{2\alpha}\left(\frac{1}{\alpha}\right)^{\frac{1}{2}m-\frac{1}{2}} \exp\left(\frac{1}{2\alpha}\right) W_{-\frac{1}{2}(m+1), -\frac{1}{2}m}\left(\frac{1}{\alpha}\right)$ ⸻ MI 48a

**6.644** $\quad \displaystyle\int_0^\infty e^{-\beta x} J_{2\nu}\left(2a\sqrt{x}\right) J_\nu(bx)\, dx = \exp\left(-\frac{a^2\beta}{\beta^2+b^2}\right) J_\nu\left(\frac{a^2 b}{\beta^2+b^2}\right) \frac{1}{\sqrt{\beta^2+b^2}}$

$$\left[\operatorname{Re}\beta > 0, \quad b > 0, \quad \operatorname{Re}\nu > -\tfrac{1}{2}\right]$$

ET II 58(17)

**6.645**

1. $\quad \displaystyle\int_1^\infty \left(x^2-1\right)^{-\frac{1}{2}} e^{-\alpha x} J_\nu\left(\beta\sqrt{x^2-1}\right) dx = I_{\frac{1}{2}\nu}\left[\frac{1}{2}\left(\sqrt{\alpha^2+\beta^2}-\alpha\right)\right] K_{\frac{1}{2}\nu}\left[\frac{1}{2}\left(\sqrt{\alpha^2+\beta^2}+\alpha\right)\right]$

MO 179a

2. $\quad \displaystyle\int_1^\infty \left(x^2-1\right)^{\frac{1}{2}\nu} e^{-\alpha x} J_\nu\left(\beta\sqrt{x^2-1}\right) dx = \sqrt{\frac{2}{\pi}}\beta^\nu \left(\alpha^2+\beta^2\right)^{-\frac{1}{2}\nu-\frac{1}{4}} K_{\nu+\frac{1}{2}}\left(\sqrt{\alpha^2+\beta^2}\right)$

MO 179a

3.³ $\quad \displaystyle\int_{-1}^1 \left(1-x^2\right)^{-1/2} e^{-ax} I_1\left(b\sqrt{1-x^2}\right) dx = \frac{2}{b}\left(\cosh\sqrt{a^2+b^2}-\cosh a\right)$

$$[a > 0, \quad b > 0]$$

**6.646**

1. $\quad \displaystyle\int_1^\infty \left(\frac{x-1}{x+1}\right)^{\frac{1}{2}\nu} e^{-\alpha x} J_\nu\left(b\sqrt{x^2-1}\right) dx = \frac{\exp\left(-\sqrt{\alpha^2+b^2}\right)}{\sqrt{\alpha^2+b^2}}\left(\frac{b}{\alpha+\sqrt{\alpha^2+b^2}}\right)^\nu$

$$[\operatorname{Re}\nu > -1]$$       EF 89(52), MO 179

2. $\quad \displaystyle\int_1^\infty \left(\frac{x-1}{x+1}\right)^{\frac{1}{2}\nu} e^{-\alpha x} I_\nu\left(b\sqrt{x^2-1}\right) dx = \frac{\exp\left(-\sqrt{\alpha^2-b^2}\right)}{\sqrt{\alpha^2-b^2}}\left(\frac{b}{\alpha+\sqrt{\alpha^2-b^2}}\right)^\nu$

$$[\operatorname{Re}\nu > -1, \quad \alpha > b]$$       MO 180

3.⁷ $\quad \displaystyle\int_b^\infty e^{-px} \left(\frac{x-b}{x+b}\right)^{\frac{1}{2}\nu} K_\nu\left(a\sqrt{x^2-b^2}\right) dx = \frac{\Gamma(\nu+1)}{2sa^\nu}\left[x^\nu e^{-bs}\,\Gamma(-\nu,bx) - y^\nu e^{bs}\,\Gamma(-\nu,by)\right]$

$\qquad$ where $x = p - s, \quad y = p + s, \quad s = \left(p^2-a^2\right)^{1/2}$ $\qquad \left[\operatorname{Re}(p+a) > 0, \quad |\operatorname{Re}(\nu)| < 1\right].$

ME 39a

**6.647**

1. $\quad \displaystyle\int_0^\infty x^{-\lambda-\frac{1}{2}} (b+x)^{\lambda-\frac{1}{2}} e^{-\alpha x} K_{2\mu}\left[\sqrt{x(b+x)}\right] dx$

$$= \frac{1}{b} e^{\frac{1}{2}\alpha b}\,\Gamma\left(\tfrac{1}{2}-\lambda+\mu\right)\Gamma\left(\tfrac{1}{2}-\lambda-\mu\right) W_{\lambda,\mu}(z_1)\, W_{\lambda,\mu}(z_2)$$

$$z_1 = \tfrac{1}{2}b\left(\alpha+\sqrt{\alpha^2-1}\right), \quad z_2 = \tfrac{1}{2}b\left(\alpha-\sqrt{\alpha^2-1}\right)$$

$$\left[|\arg b| < \pi, \quad \operatorname{Re}\alpha > -1, \quad \operatorname{Re}\lambda + |\operatorname{Re}\mu| < \tfrac{1}{2}\right]$$    ET II 377(37)

2. $\quad \displaystyle\int_0^\infty (a+x)^{-\frac{1}{2}} x^{-\frac{1}{2}} e^{-x\cosh t} K_\nu\left[\sqrt{x(a+x)}\right] dx$

$$= \frac{1}{2}\sec\left(\frac{\nu\pi}{2}\right) e^{\frac{1}{2}a\cosh t} K_{\frac{1}{2}\nu}\left(\frac{1}{4}ae^t\right) K_{\frac{1}{2}\nu}\left(\frac{1}{4}ae^{-t}\right)$$

$$[-1 < \operatorname{Re}\nu < 1]$$       ET II 377(36)

3.[11]  $\displaystyle\int_0^a x^{\lambda-\frac{1}{2}}(a-x)^{-\lambda-\frac{1}{2}}e^{-x\sinh t}\,I_{2\mu}\left[\sqrt{x(a-x)}\right]dx$

$$= e^{-(a/2)\sinh t}\frac{2\,\Gamma\left(\frac{1}{2}+\lambda+\mu\right)\Gamma\left(\frac{1}{2}-\lambda+\mu\right)}{a\,[\Gamma(2\mu+1)]^2}\,M_{\lambda,\mu}\left(\frac{1}{2}ae^t\right)M_{-\lambda,\mu}\left(\frac{1}{2}ae^{-t}\right)$$

$$\left[\operatorname{Re}\mu>|\operatorname{Re}\lambda|-\tfrac{1}{2}\right]\qquad\text{ET II 377(32)}$$

**6.648**[12]  $\displaystyle\int_{-\infty}^{\infty}e^{\varrho x}\left(\frac{a+be^x}{ae^x+b}\right)K_{2\nu}\left[(a^2+b^2+2ab\cosh x)^{\frac{1}{2}}\right]dx=2\,K_{\nu+\varrho}(a)\,K_{\nu-\varrho}(b)$

$$[\operatorname{Re}a>0,\quad\operatorname{Re}b>0]\qquad\text{ET II 379(45)}$$

**6.649**

1.  $\displaystyle\int_0^{\infty}K_{\mu-\nu}(2z\sinh x)\,e^{(\nu+\mu)x}\,dx=\frac{\pi^2}{4\sin[(\nu-\mu)\pi]}\,[J_\nu(z)\,Y_\mu(z)-J_\mu(z)\,Y_\nu(z)]$

$$[\operatorname{Re}z>0,\quad-1<\operatorname{Re}(\nu-\mu)<1]$$
$$\text{MO 44}$$

2.  $\displaystyle\int_0^{\infty}J_{\nu+\mu}(2x\sinh t)\,e^{(\nu-\mu)t}\,dt=K_\nu(x)\,I_\mu(x)$

$$\left[\operatorname{Re}(\nu-\mu)<\tfrac{3}{2},\quad\operatorname{Re}(\nu+\mu)>-1,\quad x>0\right]\quad\text{EH II 97(68)}$$

3.  $\displaystyle\int_0^{\infty}Y_{\nu-\mu}(2x\sinh t)\,e^{-(\nu+\mu)t}\,dt=\frac{1}{\sin[\pi(\mu-\nu)]}\{I_\mu(x)\,K_\nu(x)-\cos[(\nu-\mu)\pi]\,I_\nu(x)\,K_\mu(x)\}$

$$\left[|\operatorname{Re}(\nu-\mu)|<1,\quad\operatorname{Re}(\nu+\mu)>-\tfrac{1}{2},\quad x>0\right]\quad\text{EH II 97(73)}$$

4.  $\displaystyle\int_0^{\infty}K_0(2z\sinh x)\,e^{-2\nu x}\,dx=-\frac{\pi}{4}\left\{J_\nu(z)\frac{\partial\,Y_\nu(z)}{\partial\nu}-Y_\nu(z)\frac{\partial\,J_\nu(z)}{\partial\nu}\right\}$

## 6.65 Combinations of Bessel and exponential functions of more complicated arguments and powers

**6.651**

1.  $\displaystyle\int_0^{\infty}x^{\lambda+\frac{1}{2}}e^{-\frac{1}{4}\alpha^2x^2}\,I_\mu\left(\tfrac{1}{4}\alpha^2x^2\right)J_\nu(bx)\,dx$

$$=\frac{1}{\sqrt{2\pi}}2^{\lambda+1}b^{-\lambda-\frac{3}{2}}\,G_{23}^{21}\left(\frac{b^2}{2\alpha^2}\left|\begin{matrix}1-\mu,1+\mu\\h,\frac{1}{2},k\end{matrix}\right.\right)$$

$$h=\tfrac{3}{4}+\tfrac{1}{2}\lambda+\tfrac{1}{2}\nu,\quad k=\tfrac{3}{4}+\tfrac{1}{2}\lambda-\tfrac{1}{2}\nu$$

$$\left[|\arg\alpha|<\frac{\pi}{4},\quad b>0,\quad-\tfrac{3}{2}-\operatorname{Re}(2\mu+\nu)<\operatorname{Re}\lambda<0\right]\quad\text{ET II 68(8)}$$

2.  $\displaystyle\int_0^{\infty}x^{\lambda+\frac{1}{2}}e^{-\frac{1}{4}\alpha^2x^2}\,K_\mu\left(\tfrac{1}{4}\alpha^2x^2\right)J_\nu(bx)\,dx$

$$=\sqrt{\frac{\pi}{2}}2^{\lambda+1}b^{-\lambda-\frac{3}{2}}\,G_{23}^{12}\left(\frac{b^2}{2\alpha^2}\left|\begin{matrix}1-\mu,1+\mu\\h,\frac{1}{2},k\end{matrix}\right.\right)$$

$$h=\tfrac{3}{4}+\tfrac{1}{2}\lambda+\tfrac{1}{2}\nu,\quad k=\tfrac{3}{4}+\tfrac{1}{2}\lambda-\tfrac{1}{2}\nu$$

$$\left[|\arg\alpha|<\frac{\pi}{4},\quad\operatorname{Re}(\lambda+\nu\pm2\mu)>-\tfrac{3}{2}\right]\quad\text{ET II 69(15)}$$

3. $$\int_0^\infty x^{2\mu-\nu+1} e^{-\frac{1}{4}\alpha x^2} I_\mu\left(\tfrac{1}{4}\alpha x^2\right) J_\nu(bx)\, dx$$

$$= 2^{\mu-\nu+\frac{1}{2}} (\pi\alpha)^{-\frac{1}{2}} \Gamma\left(\frac{1}{2}+\mu\right) \frac{b^{\nu-2\mu-1}}{\Gamma\left(\frac{1}{2}-\mu+\nu\right)} \, {}_1F_1\left(\frac{1}{2}+\mu; \frac{1}{2}-\mu+\nu; -\frac{b^2}{2\alpha}\right)$$

$$\left[\operatorname{Re}\alpha > 0, \quad b > 0, \quad \operatorname{Re}\nu > 2\operatorname{Re}\mu + \tfrac{1}{2} > -\tfrac{1}{2}\right] \quad \text{ET II 68(6)}$$

4. $$\int_0^\infty x^{2\mu+\nu+1} e^{-\frac{1}{4}\alpha^2 x^2} K_\mu\left(\tfrac{1}{4}\alpha^2 x^2\right) J_\nu(bx)\, dx$$

$$= \sqrt{\pi}\, 2^\mu \alpha^{-2\mu-2\nu-2} b^\nu \frac{\Gamma(1+2\mu+\nu)}{\Gamma\left(\mu+\nu+\frac{3}{2}\right)} \, {}_1F_1\left(1+2\mu+\nu; \mu+\nu+\frac{3}{2}; -\frac{b^2}{2\alpha^2}\right)$$

$$\left[|\arg\alpha| < \tfrac{1}{4}\pi, \quad \operatorname{Re}\nu > -1, \quad \operatorname{Re}(2\mu+\nu) > -1, \quad b > 0\right] \quad \text{ET II 69(13)}$$

5. $$\int_0^\infty x^{2\mu+\nu+1} e^{-\frac{1}{2}\alpha x^2} I_\mu\left(\tfrac{1}{2}\alpha x^2\right) K_\nu(bx)\, dx$$

$$= \frac{2^{\mu-\frac{1}{2}}}{\sqrt{\pi}} b^{-\mu-\frac{3}{2}} \alpha^{-\frac{1}{2}\mu-\frac{1}{2}\nu-\frac{1}{4}} \Gamma(2\mu+\nu+1)\, \Gamma\left(\mu+\tfrac{1}{2}\right) \exp\left(\frac{b^2}{8\alpha}\right) W_{k,m}\left(\frac{b^2}{4\alpha}\right)$$

$$2k = -3\mu-\nu-\tfrac{1}{2}, \quad 2m = \mu+\nu+\tfrac{1}{2}$$

$$\left[\operatorname{Re}\alpha > 0, \quad \operatorname{Re}\mu > -\tfrac{1}{2}, \quad \operatorname{Re}(2\mu+\nu) > -1\right] \quad \text{ET II 146(53)}$$

6. $$\int_0^\infty x e^{-\frac{1}{4}\alpha x^2} J_{\frac{1}{2}\nu}\left(\tfrac{1}{4}bx^2\right) J_\nu(cx)\, dx = 2\left(\alpha^2+b^2\right)^{-\frac{1}{2}} \exp\left(-\frac{\alpha c^2}{\alpha^2+b^2}\right) J_{\frac{1}{2}\nu}\left(\frac{bc^2}{\alpha^2+b^2}\right)$$

$$\left[c > 0, \quad \operatorname{Re}\alpha > |\operatorname{Im}b|, \quad \operatorname{Re}\nu > -1\right]$$

$$\text{ET II 56(2)}$$

7. $$\int_0^\infty x e^{-\frac{1}{4}\alpha x^2} I_{\frac{1}{2}\nu}\left(\tfrac{1}{4}\alpha x^2\right) J_\nu(bx)\, dx = \left(\frac{1}{2}\pi\alpha\right)^{-\frac{1}{2}} b^{-1} \exp\left(-\frac{b^2}{2\alpha}\right)$$

$$\left[\operatorname{Re}\alpha > 0, \quad b > 0, \quad \operatorname{Re}\nu > -1\right]$$

$$\text{ET II 67(3)}$$

8. $$\int_0^\infty x^{1-\nu} e^{-\frac{1}{4}\alpha^2 x^2} I_\nu\left(\tfrac{1}{4}\alpha^2 x^2\right) J_\nu(bx)\, dx = \sqrt{\frac{2}{\pi}} \frac{b^{\nu-1}}{\alpha} \exp\left(-\frac{b^2}{4\alpha^2}\right) D_{-2\nu}\left(\frac{b}{\alpha}\right)$$

$$\left[|\arg\alpha| < \tfrac{1}{4}\pi, \quad b > 0, \quad \operatorname{Re}\nu > -\tfrac{1}{2}\right]$$

$$\text{ET II 67(1)}$$

9. $$\int_0^\infty x^{-\nu-1} e^{-\frac{1}{4}\alpha^2 x^2} I_{\nu+1}\left(\tfrac{1}{4}\alpha^2 x^2\right) J_\nu(bx)\, dx = \sqrt{\frac{2}{\pi}} b^\nu \exp\left(-\frac{b^2}{4\alpha^2}\right) D_{-2\nu-3}\left(\frac{b}{\alpha}\right)$$

$$\left[|\arg\alpha| < \tfrac{1}{4}\pi, \quad \operatorname{Re}\nu > -1, \quad b > 0\right]$$

$$\text{ET II 67(2)}$$

**6.652** $$\int_0^\infty x^{2\nu} e^{-\left(\frac{x^2}{8}+\alpha x\right)} I_\nu\left(\frac{x^2}{8}\right) dx = \frac{\Gamma(4\nu+1)}{2^{4\nu}\,\Gamma(\nu+1)} \frac{e^{\frac{\alpha^2}{2}}}{\alpha^{\nu+1}} W_{-\frac{3}{2}\nu,\frac{1}{2}\nu}\left(\alpha^2\right)$$

$$\left[\operatorname{Re}\left(\nu+\tfrac{1}{4}\right) > 0\right] \qquad \text{MI 45}$$

**6.653**

1.
$$\int_0^\infty \exp\left[-\frac{1}{2}x - \frac{1}{2x}\left(a^2 + b^2\right)\right] I_\nu\left(\frac{ab}{x}\right)\frac{dx}{x} = 2\,I_\nu(a)\,K_\nu(b) \qquad [0 < a < b]$$
$$= 2\,K_\nu(a)\,I_\nu(b) \qquad [0 < b < a]$$
$$[\mathrm{Re}\,\nu > -1] \quad \text{WA 482(2)a, EH II 53(37), WA 482(3)a}$$

2.
$$\int_0^\infty \exp\left[-\frac{1}{2}x - \frac{1}{2x}\left(z^2 + w^2\right)\right] K_\nu\left(\frac{zw}{x}\right)\frac{dx}{x} = 2\,K_\nu(z)\,K_\nu(w)$$
$$\left[|\arg z| < \pi, \quad |\arg w| < \pi, \quad \arg(z+w) < \tfrac{1}{4}\pi\right] \quad \text{WA 483(1), EH II 53(36)}$$

**6.654**
$$\int_0^\infty x^{-\frac{1}{2}} e^{-\frac{\beta^2}{8x} - \alpha x} K_\nu\left(\frac{\beta^2}{8x}\right) dx = \sqrt{4\pi}\,\alpha^{-\frac{1}{2}} K_{2\nu}\left(\beta\sqrt{\alpha}\right) \qquad\qquad \text{ME 39}$$

**6.655**
$$\int_0^\infty x\left(\beta^2 + x^2\right)^{-\frac{1}{2}} \exp\left(-\frac{\alpha^2\beta}{\beta^2 + x^2}\right) J_\nu\left(\frac{\alpha^2 x}{\beta^2 + x^2}\right) J_\nu(\gamma x)\,dx = \gamma^{-1} e^{-\beta\gamma} J_{2\nu}\left(2\alpha\sqrt{\gamma}\right)$$
$$\left[\mathrm{Re}\,\beta > 0, \quad \gamma > 0, \quad \mathrm{Re}\,\nu > -\tfrac{1}{2}\right]$$
$$\text{ET II 58(14)}$$

**6.656**

1.
$$\int_0^\infty e^{-(\xi-z)\cosh t} J_{2\nu}\left[2(z\xi)^{\frac{1}{2}}\sinh t\right] dt = I_\nu(z)\,K_\nu(\xi)$$
$$\left[\mathrm{Re}\,\nu > -\tfrac{1}{2}, \quad \mathrm{Re}(\xi - z) > 0\right]$$
$$\text{EH II 98(78)}$$

2.
$$\int_0^\infty e^{-(\xi+z)\cosh t} K_{2\nu}\left[2(z\xi)^{\frac{1}{2}}\sinh t\right] dt = \frac{1}{2} K_\nu(z)\,K_\nu(\xi)\sec(\nu\pi)$$
$$\left[|\mathrm{Re}\,\nu| < \tfrac{1}{2}, \quad \mathrm{Re}\left(z^{\frac{1}{2}} + \xi^{\frac{1}{2}}\right)^2 \geq 0\right]$$
$$\text{EH II 98(79)}$$

## 6.66 Combinations of Bessel, hyperbolic, and exponential functions

### Bessel and hyperbolic functions

**6.661**

1.
$$\int_0^\infty \sinh(ax) K_\nu(bx)\,dx = \frac{\pi}{2}\frac{\operatorname{cosec}\left(\frac{\nu\pi}{2}\right)\sin\left[\nu\arcsin\left(\frac{a}{b}\right)\right]}{\sqrt{b^2 - a^2}}$$
$$[\mathrm{Re}\,b > |\mathrm{Re}\,a|, \quad |\mathrm{Re}\,\nu| < 2]$$
$$\text{ET II 133(32)}$$

2.
$$\int_0^\infty \cosh(ax) K_\nu(bx)\,dx = \frac{\pi\cos\left[\nu\arcsin\left(\frac{a}{b}\right)\right]}{2\sqrt{b^2 - a^2}\cos\left(\frac{\nu\pi}{2}\right)} \qquad [\mathrm{Re}\,b > |\mathrm{Re}\,a|, \quad |\mathrm{Re}\,\nu| < 1]$$
$$\text{ET II 134(33)}$$

**6.662** **Notation:**
$$\ell_1 = \frac{1}{2}\left[\sqrt{(b+c)^2 + a^2} - \sqrt{(b-c)^2 + a^2}\right], \qquad \ell_2 = \frac{1}{2}\left[\sqrt{(b+c)^2 + a^2} + \sqrt{(b-c)^2 + a^2}\right]$$

1.[10]    $\displaystyle\int_0^\infty \cosh(\beta x)\, K_0(\alpha x)\, J_0(\gamma x)\,\mathrm{d}x = \frac{\boldsymbol{K}(k)}{\sqrt{u+v}}$

$$u = \frac{1}{2}\left\{\sqrt{\left(\alpha^2+\beta^2+\gamma^2\right)^2 - 4\alpha^2\beta^2} + \alpha^2 - \beta^2 - \gamma^2\right\}$$

$$v = \frac{1}{2}\left\{\sqrt{\left(\alpha^2+\beta^2+\gamma^2\right)^2 - 4\alpha^2\beta^2} - \alpha^2 + \beta^2 + \gamma^2\right\}$$

$$k^2 = v(u+v)^{-1} \qquad [\operatorname{Re}\alpha > |\operatorname{Re}\beta|, \quad \gamma > 0]$$

ET II 15(23)

alternatively, with $a = \gamma$, $b = \beta$, $c = \alpha$,

$$\int_0^\infty \cosh(bx)\, K_0(cx)\, J_0(ax)\,\mathrm{d}x = \frac{\boldsymbol{K}(k)}{\sqrt{\ell_2^2 - \ell_1^2}}$$

$$k^2 = \frac{\ell_2^2 - c^2}{\ell_2^2 - \ell_1^2}, \qquad [\operatorname{Re}c > |\operatorname{Re}b|, \quad a > 0]$$

2.[10]    $\displaystyle\int_0^\infty \sinh(\beta x)\, K_1(\alpha x)\, J_0(\gamma x)\,\mathrm{d}x = a^{-1}\left[u\,\boldsymbol{E}(k) - \boldsymbol{K}(k)\,\boldsymbol{E}(u) + \frac{\boldsymbol{K}(k)\,\operatorname{sn}u\,\operatorname{dn}u}{\operatorname{cn}u}\right]$

$$\operatorname{cn}^2 u = 2\gamma^2\left\{\left[\left(\alpha^2+\beta^2+\gamma^2\right)^2 - 4\alpha^2\beta^2\right]^{\frac{1}{2}} - \alpha^2 + \beta^2 + \gamma^2\right\}^{-1}$$

$$k^2 = \frac{1}{2}\left\{1 - \left(\alpha^2-\beta^2-\gamma^2\right)\left[\left(\alpha^2+\beta^2+\gamma^2\right)^2 - 4\alpha^2\beta^2\right]^{-\frac{1}{2}}\right\}$$

$$[\operatorname{Re}\alpha > |\operatorname{Re}\beta|, \quad \gamma > 0]$$

ET II 15(24)

alternatively, with $a = \gamma$, $b = \beta$, $c = \alpha$,

$$\int_0^\infty \sinh(bx)\, K_1(cx)\, J_0(ax)\,\mathrm{d}x = c^{-1}\left[u\,\boldsymbol{E}(k) - \boldsymbol{K}(k)\,\boldsymbol{E}(u) + \frac{\boldsymbol{K}(k)\,\operatorname{sn}u\,\operatorname{dn}u}{\operatorname{cn}u}\right]$$

$$\operatorname{cn}^2 u = \frac{a^2}{\ell_2^2 - c^2}, \qquad k^2 = \frac{\ell_2^2 - c^2}{\ell_2^2 - \ell_1^2} \qquad [\operatorname{Re}c > |\operatorname{Re}b|, \quad a > 0]$$

## 6.663

1.    $\displaystyle\int_0^\infty K_{\nu\pm\mu}\left(2z\cosh t\right)\cosh\left[(\mu\mp\nu)\,t\right]\,\mathrm{d}t = \frac{1}{2}\,K_\mu(z)\,K_\nu(z)$

$$[\operatorname{Re}z > 0] \qquad\qquad \text{WA 484(1), EH II 54(39)}$$

2.    $\displaystyle\int_0^\infty Y_{\mu+\nu}\left(2z\cosh t\right)\cosh[(\mu-\nu)t]\,\mathrm{d}t = \frac{\pi}{4}\left[J_\mu(z)\,J_\nu(z) - Y_\mu(z)\,Y_\nu(z)\right]$

$$[z > 0] \qquad\qquad \text{EH II 96(64)}$$

3.    $\displaystyle\int_0^\infty J_{\mu+\nu}\left(2z\cosh t\right)\cosh[(\mu-\nu)t]\,\mathrm{d}t = -\frac{\pi}{4}\left[J_\mu(z)\,Y_\nu(z) + J_\nu(z)\,Y_\mu(z)\right]$

$$[z > 0] \qquad\qquad \text{EH II 97(65)}$$

4.    $\displaystyle\int_0^\infty J_{\mu+\nu}\left(2z\sinh t\right)\cosh[(\mu-\nu)t]\,\mathrm{d}t = \frac{1}{2}\left[I_\nu(z)\,K_\mu(z) + I_\mu(z)\,K_\nu(z)\right]$

$$\left[\operatorname{Re}(\nu+\mu) > -1, \quad |\operatorname{Re}(\mu-\nu)| < \tfrac{3}{2}, \quad z > 0\right] \quad \text{EH II 97(71)}$$

5.    $\displaystyle\int_0^\infty J_{\mu+\nu}\left(2z\sinh t\right)\sinh[(\mu-\nu)t]\,dt = \frac{1}{2}\left[I_\nu(z)\,K_\mu(z) - I_\mu(z)\,K_\nu(z)\right]$

$$\left[\operatorname{Re}(\nu+\mu) > -1, \quad |\operatorname{Re}(\mu-\nu)| < \tfrac{3}{2}, \quad z > 0\right] \quad \text{EH II 97(72)}$$

**6.664**

1.    $\displaystyle\int_0^\infty J_0\left(2z\sinh t\right)\sinh(2\nu t)\,dt = \frac{\sin(\nu\pi)}{\pi}\left[K_\nu(z)\right]^2 \qquad \left[|\operatorname{Re}\nu| < \tfrac{3}{4}, \quad z > 0\right] \qquad \text{EH II 97(69)}$

2.    $\displaystyle\int_0^\infty Y_0\left(2z\sinh t\right)\cosh(2\nu t)\,dt = -\frac{\cos(\nu\pi)}{\pi}\left[K_\nu(z)\right]^2 \qquad \left[|\operatorname{Re}\nu| < \tfrac{3}{4}, \quad z > 0\right] \qquad \text{EH II 97(70)}$

3.    $\displaystyle\int_0^\infty Y_0\left(2z\sinh t\right)\sinh(2\nu t)\,dt = \frac{1}{\pi}\left[I_\nu(z)\frac{\partial K_\nu(z)}{\partial\nu} - K_\nu(z)\frac{\partial I_\nu(z)}{\partial\nu}\right] - \frac{1}{\pi}\cos(\nu\pi)\left[K_\nu(z)\right]^2$

$$\left[|\operatorname{Re}\nu| < \tfrac{3}{4}, \quad z > 0\right] \qquad \text{EH II 97(75)}$$

4.    $\displaystyle\int_0^\infty K_0\left(2z\sinh t\right)\cosh 2\nu t\,dt = \frac{\pi^2}{8}\left\{J_\nu^2(z) + N_\nu^2(z)\right\} \qquad [\operatorname{Re}z > 0] \qquad \text{MO 44}$

5.    $\displaystyle\int_0^\infty K_{2\mu}\left(z\sinh 2t\right)\coth^{2\nu} t\,dt = \frac{1}{4z}\,\Gamma\left(\frac{1}{2}+\mu-\nu\right)\Gamma\left(\frac{1}{2}-\mu-\nu\right)W_{\nu,\mu}(iz)\,W_{\nu,\mu}(-iz)$

$$\left[|\arg z| \le \frac{\pi}{2}, \quad |\operatorname{Re}\mu| + \operatorname{Re}\nu < \tfrac{1}{2}\right]$$
$$\text{MO 119}$$

6.    $\displaystyle\int_0^\infty \cosh(2\mu x)\,K_{2\nu}\left(2a\cosh x\right)\,dx = \frac{1}{2}\,K_{\mu+\nu}(a)\,K_{\mu-\nu}(a)$

$$[\operatorname{Re}a > 0] \qquad \text{ET II 378(42)}$$

**6.665**    $\displaystyle\int_0^\infty \operatorname{sech}x\,\cosh(2\lambda x)\,I_{2\mu}\left(a\operatorname{sech}x\right)\,dx = \frac{\Gamma\left(\frac{1}{2}+\lambda+\mu\right)\Gamma\left(\frac{1}{2}-\lambda+\mu\right)}{2a\left[\Gamma(2\mu+1)\right]^2}\,M_{\lambda,\mu}(a)\,M_{-\lambda,\mu}(a)$

$$\left[|\operatorname{Re}\lambda| - \operatorname{Re}\mu < \tfrac{1}{2}\right] \qquad \text{ET II 378(43)}$$

## Bessel, hyperbolic, and algebraic functions

**6.666**    $\displaystyle\int_0^\infty x^{\nu+1}\sinh(\alpha x)\operatorname{cosech}(\pi x)\,J_\nu(\beta x)\,dx = \frac{2}{\pi}\sum_{n=1}^\infty (-1)^{n-1}n^{\nu+1}\sin(n\alpha)\,K_\nu(n\beta)$

$$[|\operatorname{Re}\alpha| < \pi, \quad \operatorname{Re}\nu > -1]$$
$$\text{ET II 41(3), WA 469(12)}$$

**6.667**

1.[3]    $\displaystyle\int_0^a \frac{\cosh\left(\sqrt{a^2-x^2}\right)\sinh t\, I_{2\nu}(x)}{\sqrt{a^2-x^2}}\,dx = \frac{\pi}{2}\,I_\nu\left(\frac{1}{2}ae^t\right)I_\nu\left(\frac{1}{2}ae^{-t}\right)$

$$\left[\operatorname{Re}\nu > -\tfrac{1}{2}\right] \qquad \text{ET II 365(10)}$$

2.    $\displaystyle\int_0^a \frac{\cosh\left(\sqrt{a^2-x^2}\sinh t\right)K_{2\nu}(x)}{\sqrt{a^2-x^2}}\,dx = \frac{\pi^2}{4}\operatorname{cosec}(\nu\pi)\left[I_{-\nu}\left(ae^t\right)I_{-\nu}\left(ae^{-t}\right) - I_\nu\left(ae^t\right)I_\nu\left(ae^{-t}\right)\right]$

$$\left[|\operatorname{Re}\nu| < \tfrac{1}{2}\right] \qquad \text{ET II 367(25)}$$

## Exponential, hyperbolic, and Bessel functions

### 6.668    Notation:

$$\ell_1 = \frac{1}{2}\left[\sqrt{(b+c)^2 + a^2} - \sqrt{(b-c)^2 + a^2}\right], \qquad \ell_2 = \frac{1}{2}\left[\sqrt{(b+c)^2 + a^2} + \sqrt{(b-c)^2 + a^2}\right]$$

1.[10] $\displaystyle\int_0^\infty e^{-\alpha x}\sinh(\beta x)\,J_0(\gamma x)\,\mathrm{d}x = (\alpha\beta)^{\frac{1}{2}} r_1^{-1} r_2^{-1} (r_2 + r_1)^{\frac{1}{2}} (r_2 - r_1)^{-\frac{1}{2}}$

$\qquad r_1 = \sqrt{\gamma^2 + (\beta - \alpha)^2}, \qquad r_2 = \sqrt{\gamma^2 + (\beta + \alpha)^2}, \qquad [\operatorname{Re}\alpha > |\operatorname{Re}\beta|, \quad \gamma > 0] \quad$ ET II 12(52)

alternatively, with $a = \gamma$, $b = \beta$, $c = \alpha$,

$\displaystyle\int_0^\infty e^{-cx}\sinh(bx)\,J_0(ax)\,\mathrm{d}x = \frac{\ell_1}{\ell_2^2 - \ell_1^2}$

$$[\operatorname{Re}c > |\operatorname{Re}b|, \quad a > 0]$$

2.[12] $\displaystyle\int_0^\infty e^{-\alpha x}\cosh(\beta x)\,J_0(\gamma x)\,\mathrm{d}x = (\alpha\beta)^{\frac{1}{2}} r_1^{-1} r_2^{-1} (r_2 + r_1)^{\frac{1}{2}} (r_2 - r_1)^{-\frac{1}{2}}$

$\qquad r_1 = \sqrt{\gamma^2 + (\beta - \alpha)^2}, \qquad r_2 = \sqrt{\gamma^2 + (\beta + \alpha)^2}, \qquad [\operatorname{Re}\alpha > |\operatorname{Re}\beta|, \quad \gamma > 0] \quad$ ET II 12(54)

alternatively, with $a = \gamma$, $b = \beta$, $c = \alpha$,

$\displaystyle\int_0^\infty e^{-cx}\cosh(bx)\,J_0(ax)\,\mathrm{d}x = \frac{\ell_2}{\ell_2^2 - \ell_1^2}$

$$[\operatorname{Re}c > |\operatorname{Re}b|, \quad a > 0]$$

### 6.669

1. $\displaystyle\int_0^\infty \left[\coth\left(\frac{1}{2}x\right)\right]^{2\lambda} e^{-\beta\cosh x}\,J_{2\mu}(\alpha\sinh x)\,\mathrm{d}x = \frac{\Gamma\left(\frac{1}{2} - \lambda + \mu\right)}{\alpha\,\Gamma(2\mu + 1)}\,M_{-\lambda,\mu}\left[(\alpha^2 + \beta^2)^{\frac{1}{2}} - \beta\right]$

$$\times\, W_{\lambda,\mu}\left[(\alpha^2 + \beta^2)^{\frac{1}{2}} + \beta\right]$$

$$\left[\operatorname{Re}\beta > |\operatorname{Re}\alpha|, \quad \operatorname{Re}(\mu - \lambda) > -\tfrac{1}{2}\right] \quad \text{BU 86(5b)a, ET II 363(34)}$$

2. $\displaystyle\int_0^\infty \left[\coth\left(\frac{1}{2}x\right)\right]^{2\lambda} e^{-\beta\cosh x}\,Y_{2\mu}(\alpha\sinh x)\,\mathrm{d}x$

$$= -\frac{\sec[(\mu + \lambda)\pi]}{\alpha}\,W_{\lambda,\mu}\left(\sqrt{\alpha^2 + \beta^2} + \beta\right)W_{-\lambda,\mu}\left(\sqrt{\alpha^2 + \beta^2} - \beta\right)$$

$$-\frac{\tan[(\mu + \lambda)\pi]\,\Gamma\left(\frac{1}{2} - \lambda + \mu\right)}{\alpha\,\Gamma(2\mu + 1)}\,W_{\lambda,\mu}\left(\sqrt{\alpha^2 + \beta^2} + \beta\right)M_{-\lambda,\mu}\left(\sqrt{\alpha^2 + \beta^2} - \beta\right)$$

$$\left[\operatorname{Re}\beta > |\operatorname{Re}\alpha|, \quad \operatorname{Re}\lambda < \tfrac{1}{2} - |\operatorname{Re}\mu|\right] \quad \text{ET II 363(35)}$$

3.[12] $\displaystyle\int_0^\infty e^{-\frac{1}{2}(a_1 + a_2)t\cosh x}\left[\coth\left(\frac{1}{2}x\right)\right]^{2\nu} K_{2\mu}(t\sqrt{a_1 a_2}\sinh x)\,\mathrm{d}x$

$$= \frac{\Gamma\left(\frac{1}{2} + \mu - \nu\right)\Gamma\left(\frac{1}{2} - \mu - \nu\right)}{2t\sqrt{a_1 a_2}}\,W_{\nu,\mu}(a_1 t)\,W_{\nu,\mu}(a_2 t)$$

$$\left[\operatorname{Re}\nu < \operatorname{Re}\frac{1 \pm 2\mu}{2}, \quad \operatorname{Re}\left[t\left(\sqrt{a_1} + \sqrt{a_2}\right)^2\right] > 0\right] \quad \text{BU 85(4a)}$$

4.[12] $\int_0^\infty e^{-\frac{1}{2}(a_1+a_2)t\cosh x}\left[\coth\left(\frac{x}{2}\right)\right]^{2\nu} I_{2\mu}(t\sqrt{a_1 a_2}\sinh x)dx = \dfrac{\Gamma\left(\frac{1}{2}+\mu-\nu\right)}{t\sqrt{a_1 a_2}\,\Gamma(1+2\mu)} W_{\nu,\mu}(a_1 t) M_{\nu,\mu}(a_2 t)$

$$\left[\mathrm{Re}\left(\tfrac{1}{2}+\mu-\nu\right)>0,\quad \mathrm{Re}\,\mu>0,\quad a_1>a_2\right]\quad \text{BU 86(5c)}$$

5.    $\int_{-\infty}^\infty e^{2\nu s-\frac{x-y}{2}\tanh s} I_{2\mu}\left(\dfrac{\sqrt{xy}}{\cosh s}\right)\dfrac{ds}{\cosh s} = \dfrac{\Gamma\left(\frac{1}{2}+\mu+\nu\right)\Gamma\left(\frac{1}{2}+\mu-\nu\right)}{\sqrt{xy}\,[\Gamma(1+2\mu)]^2} M_{\nu,\mu}(x) M_{-\nu,\mu}(y)$

$$\left[\mathrm{Re}\left(\pm\nu+\tfrac{1}{2}+\mu\right)>0\right]\quad \text{BU 83(3a)a}$$

6.    $\int_{-\infty}^\infty e^{2\nu s-\frac{x+y}{2}\tanh s} J_{2\mu}\left(\dfrac{\sqrt{xy}}{\cosh s}\right)\dfrac{ds}{\cosh s} = \dfrac{\Gamma\left(\frac{1}{2}+\mu+\nu\right)\Gamma\left(\frac{1}{2}+\mu-\nu\right)}{\sqrt{xy}\,[\Gamma(1+2\mu)]^2} M_{\nu,\mu}(x) M_{\nu,\mu}(y)$

$$\left[\mathrm{Re}\left(\mp\nu+\tfrac{1}{2}+\mu\right)>0\right]\quad \text{BU 84(3b)a}$$

## 6.67–6.68 Combinations of Bessel and trigonometric functions

### 6.671

1.    $\int_0^\infty J_\nu(ax)\sin bx\,dx = \dfrac{\sin\left(\nu\arcsin\frac{b}{a}\right)}{\sqrt{a^2-b^2}}$            $[b<a]$

$$=\infty\text{ or }0\qquad\qquad [b=a]$$

$$=\dfrac{a^\nu\cos\frac{\nu\pi}{2}}{\sqrt{b^2-a^2}\left(b+\sqrt{b^2-a^2}\right)^\nu}\quad [b>a]$$

$$[\mathrm{Re}\,\nu>-2]\qquad\qquad\qquad \text{WA 444(4)}$$

2.    $\int_0^\infty J_\nu(ax)\cos bx\,dx = \dfrac{\cos\left(\nu\arcsin\frac{b}{a}\right)}{\sqrt{a^2-b^2}}$            $[b<a]$

$$=\infty\text{ or }0\qquad\qquad [b=a]$$

$$=\dfrac{-a^\nu\sin\frac{\nu\pi}{2}}{\sqrt{b^2-a^2}\left(b+\sqrt{b^2-a^2}\right)^\nu}\quad [b>a]$$

$$[\mathrm{Re}\,\nu>-1]\qquad\qquad\qquad \text{WA 444(5)}$$

3.    $\int_0^\infty Y_\nu(ax)\sin(bx)\,dx$

$$=\cot\left(\frac{\nu\pi}{2}\right)\left(a^2-b^2\right)^{-\frac{1}{2}}\sin\left[\nu\arcsin\left(\frac{b}{a}\right)\right]\qquad\qquad [0<b<a,|\mathrm{Re}\,\nu|<2]$$

$$=\frac{1}{2}\operatorname{cosec}\left(\frac{\nu\pi}{2}\right)\left(b^2-a^2\right)^{-\frac{1}{2}}$$

$$\times\left\{a^{-\nu}\cos(\nu\pi)\left[b-\left(b^2-a^2\right)^{\frac{1}{2}}\right]^\nu-a^\nu\left[b-\left(b^2-a^2\right)^{\frac{1}{2}}\right]^{-\nu}\right\}\qquad [0<a<b,\quad |\mathrm{Re}\,\nu|<2]$$

$$\text{ET I 103(33)}$$

4.      $\int_0^\infty Y_\nu(ax)\cos(bx)\,\mathrm{d}x$

$$= \frac{\tan\left(\frac{\nu\pi}{2}\right)}{(a^2-b^2)^{\frac{1}{2}}}\cos\left[\nu\arcsin\left(\frac{b}{a}\right)\right] \qquad\qquad [0 < b < a, \quad |\operatorname{Re}\nu| < 1]$$

$$= -\sin\left(\frac{\nu\pi}{2}\right)(b^2-a^2)^{-\frac{1}{2}}\left\{a^{-\nu}\left[b-(b^2-a^2)^{\frac{1}{2}}\right]^\nu + \cot(\nu\pi)\right.$$

$$\left. +a^\nu\left[b-(b^2-a^2)^{\frac{1}{2}}\right]^{-\nu}\operatorname{cosec}(\nu\pi)\right\} \qquad [0 < a < b, \quad |\operatorname{Re}\nu| < 1]$$

ET I 47(29)

5.      $\int_0^\infty K_\nu(ax)\sin(bx)\,\mathrm{d}x$

$$= \frac{1}{4}\pi a^{-\nu}\operatorname{cosec}\left(\frac{\nu\pi}{2}\right)(a^2+b^2)^{-\frac{1}{2}}\left\{\left[(b^2+a^2)^{\frac{1}{2}}+b\right]^\nu - \left[(b^2+a^2)^{\frac{1}{2}}-b\right]^\nu\right\}$$

$[\operatorname{Re}a > 0, \quad b > 0, \quad |\operatorname{Re}\nu| < 2, \quad \nu \neq 0]$    ET I 105(48)

6.      $\int_0^\infty K_\nu(ax)\cos(bx)\,\mathrm{d}x$

$$= \frac{\pi}{4}(b^2+a^2)^{-\frac{1}{2}}\sec\left(\frac{\nu\pi}{2}\right)\left\{a^{-\nu}\left[b+(b^2+a^2)^{\frac{1}{2}}\right]^\nu + a^\nu\left[b+(b^2+a^2)^{\frac{1}{2}}\right]^{-\nu}\right\}$$

$[\operatorname{Re}a > 0, b > 0, |\operatorname{Re}\nu| < 1]$    ET I 49(40)

7.      $\int_0^\infty J_0(ax)\sin(bx)\,\mathrm{d}x = 0 \qquad\qquad\qquad [0 < b < a]$

$$= \frac{1}{\sqrt{b^2-a^2}} \qquad\qquad [0 < a < b]$$

ET I 99(1)

8.      $\int_0^\infty J_0(ax)\cos(bx)\,\mathrm{d}x = \frac{1}{\sqrt{a^2-b^2}} \qquad\qquad [0 < b < a]$

$$= \infty \qquad\qquad\qquad [a = b]$$

$$= 0 \qquad\qquad\qquad [0 < a < b]$$

ET I 43(1)

9.      $\int_0^\infty J_{2n+1}(ax)\sin(bx)\,\mathrm{d}x = (-1)^n\frac{1}{\sqrt{a^2-b^2}}T_{2n+1}\left(\frac{b}{a}\right) \qquad [0 < b < a]$

$$= 0 \qquad\qquad\qquad [0 < a < b]$$

ET I 99(2)

10.     $\int_0^\infty J_{2n}(ax)\cos(bx)\,\mathrm{d}x = (-1)^n\frac{1}{\sqrt{a^2-b^2}}T_{2n}\left(\frac{b}{a}\right) \qquad [0 < b < a]$

$$= 0 \qquad\qquad\qquad [0 < a < b]$$

ET I 43(2)

11. $\int_0^\infty Y_0(ax) \sin(bx)\,\mathrm{d}x = \dfrac{2 \arcsin\left(\frac{b}{a}\right)}{\pi \sqrt{a^2 - b^2}}$ $\qquad [0 < b < a]$

$\qquad\qquad\qquad\qquad = \dfrac{2}{\pi} \dfrac{1}{\sqrt{b^2 - a^2}} \ln\left[\dfrac{b}{a} - \sqrt{\dfrac{b^2}{a^2} - 1}\right]$ $\quad [0 < a < b]$

ET I 103(31)

12. $\int_0^\infty Y_0(ax) \cos(bx)\,\mathrm{d}x = 0$ $\qquad\qquad\qquad\qquad [0 < b < a]$

$\qquad\qquad\qquad\qquad = -\dfrac{1}{\sqrt{b^2 - a^2}}$ $\qquad\qquad [0 < a < b]$

ET I 47(28)

13. $\int_0^\infty K_0(\beta x) \sin \alpha x\,\mathrm{d}x = \dfrac{1}{\sqrt{\alpha^2 + \beta^2}} \ln\left(\dfrac{\alpha}{\beta} + \sqrt{\dfrac{\alpha^2}{\beta^2} + 1}\right)$

$\qquad\qquad\qquad\qquad\qquad\qquad [\alpha > 0, \quad \beta > 0]$ $\qquad$ WA 425(11)a, MO 48

14.[8] $\int_0^\infty K_0(\beta x) \cos \alpha x\,\mathrm{d}x = \dfrac{\pi}{2\sqrt{\alpha^2 + \beta^2}}$ $\qquad\qquad [\alpha > 0]$ $\qquad$ WA 425(10)a, MO 48

**6.672**

1. $\int_0^\infty J_\nu(ax)\, J_\nu(bx) \sin(cx)\,\mathrm{d}x$

$\qquad = 0$ $\qquad\qquad\qquad\qquad\qquad [\operatorname{Re}\nu > -1, \quad 0 < c < b - a, \quad 0 < a < b]$

$\qquad = \dfrac{1}{2\sqrt{ab}}\, P_{\nu-\frac{1}{2}}\left(\dfrac{b^2 + a^2 - c^2}{2ab}\right)$ $\qquad [\operatorname{Re}\nu > -1, \quad b - a < c < b + a, \quad 0 < a < b]$

$\qquad = -\dfrac{\cos(\nu\pi)}{\pi\sqrt{ab}}\, Q_{\nu-\frac{1}{2}}\left(-\dfrac{b^2 + a^2 - c^2}{2ab}\right)$ $\qquad [\operatorname{Re}\nu > -1, \quad b + a < c, \quad 0 < a < b]$

ET I 102(27)

2. $\int_0^\infty J_\nu(x)\, J_{-\nu}(x) \cos(bx)\,\mathrm{d}x = \dfrac{1}{2}\, P_{\nu-\frac{1}{2}}\left(\dfrac{1}{2}b^2 - 1\right)$ $\qquad [0 < b < 2]$

$\qquad\qquad\qquad\qquad\qquad\qquad\quad = 0$ $\qquad\qquad\qquad\qquad [2 < b]$

ET I 46(21)

3. $\int_0^\infty K_\nu(ax)\, K_\nu(bx) \cos(cx)\,\mathrm{d}x = \dfrac{\pi^2}{4\sqrt{ab}} \sec(\nu\pi)\, P_{\nu-\frac{1}{2}}\left[(a^2 + b^2 + c^2)\,(2ab)^{-1}\right]$

$\qquad\qquad\qquad\qquad\qquad\qquad \left[\operatorname{Re}(a + b) > 0, \quad c > 0, \quad |\operatorname{Re}\nu| < \tfrac{1}{2}\right]$

ET I 50(51)

4. $\int_0^\infty K_\nu(ax)\, I_\nu(bx) \cos(cx)\,\mathrm{d}x = \dfrac{1}{2\sqrt{ab}}\, Q_{\nu-\frac{1}{2}}\left(\dfrac{a^2 + b^2 + c^2}{2ab}\right)$

$\qquad\qquad\qquad\qquad\qquad\qquad \left[\operatorname{Re}a > |\operatorname{Re}b|, \quad c > 0, \quad \operatorname{Re}\nu > -\tfrac{1}{2}\right]$

ET I 49(47)

5. $\int_0^\infty \sin(2ax)\, [J_\nu(x)]^2\,\mathrm{d}x = \dfrac{1}{2}\, P_{\nu-\frac{1}{2}}\left(1 - 2a^2\right)$ $\qquad [0 < a < 1, \quad \operatorname{Re}\nu > -1]$

$\qquad\qquad\qquad\qquad\qquad\quad = \dfrac{1}{\pi} \cos(\nu\pi)\, Q_{\nu-\frac{1}{2}}\left(2a^2 - 1\right)$ $\quad [a > 1, \quad \operatorname{Re}\nu > -1]$

ET II 343(30)

6.　　$\displaystyle\int_0^\infty \cos(2ax)\,[J_\nu(x)]^2\,\mathrm{d}x = \frac{1}{\pi}\,Q_{\nu-\frac{1}{2}}\left(1-2a^2\right)$　　　　　$\left[0 < a < 1,\quad \operatorname{Re}\nu > -\tfrac{1}{2}\right]$

$\displaystyle\qquad\qquad\qquad\qquad = -\frac{1}{\pi}\sin(\nu\pi)\,Q_{\nu-\frac{1}{2}}\left(2a^2-1\right)$　　$\left[a > 1,\quad \operatorname{Re}\nu > -\tfrac{1}{2}\right]$

ET II 344(32)

7.　　$\displaystyle\int_0^\infty \sin(2ax)\,J_0(x)\,Y_0(x)\,\mathrm{d}x = 0$　　　　　　　　$[0 < a < 1]$

$\displaystyle\qquad\qquad\qquad\qquad = -\frac{K\left[\left(1-a^{-2}\right)^{\frac{1}{2}}\right]}{\pi a}$　　　$[a > 1]$

ET II 348(60)

8.　　$\displaystyle\int_0^\infty K_0(ax)\,I_0(bx)\cos(cx)\,\mathrm{d}x = \frac{1}{\sqrt{c^2+(a+b)^2}}\,K\left\{\frac{2\sqrt{ab}}{\sqrt{c^2+(a+b)^2}}\right\}$

$[\operatorname{Re}a > |\operatorname{Re}b|,\quad c > 0]$　　　　ET I 49(46)

9.　　$\displaystyle\int_0^\infty \cos(2ax)\,J_0(x)\,Y_0(x)\,\mathrm{d}x = -\frac{1}{\pi}\,K(a)$　　　　$[0 < a < 1]$

$\displaystyle\qquad\qquad\qquad\qquad = -\frac{1}{\pi a}\,K\left(\frac{1}{a}\right)$　　　$[a > 1]$

ET II 348(61)

10.　　$\displaystyle\int_0^\infty \cos(2ax)\,[Y_0(x)]^2\,\mathrm{d}x = \frac{1}{\pi}\,K\left(\sqrt{1-a^2}\right)$　　　$[0 < a < 1]$

$\displaystyle\qquad\qquad\qquad\qquad = \frac{2}{\pi a}\,K\left(\sqrt{1-\frac{1}{a^2}}\right)$　　　$[a > 1]$

ET II 348(62)

**6.673**

1.　　$\displaystyle\int_0^\infty \left[J_\nu(ax)\cos\left(\frac{\nu\pi}{2}\right) - Y_\nu(ax)\sin\left(\frac{\nu\pi}{2}\right)\right]\sin(bx)\,\mathrm{d}x$

$\displaystyle\qquad = 0$　　　　　　　　　　　　　$[0 < b < a,\quad |\operatorname{Re}\nu| < 2]$

$\displaystyle\qquad = \frac{1}{2a^\nu\sqrt{b^2-a^2}}\left\{\left[b+\left(b^2-a^2\right)^{\frac{1}{2}}\right]^\nu + \left[b-\left(b^2-a^2\right)^{\frac{1}{2}}\right]^\nu\right\}$　　$[0 < a < b,\quad |\operatorname{Re}\nu| < 2]$

ET I 104(39)

2.　　$\displaystyle\int_0^\infty \left[Y_\nu(ax)\cos\left(\frac{\nu\pi}{2}\right) + J_\nu(ax)\sin\left(\frac{\nu\pi}{2}\right)\right]\cos(bx)\,\mathrm{d}x$

$\displaystyle\qquad = 0$　　　　　　　　　　　　　$[0 < b < a,\quad |\operatorname{Re}\nu| < 1]$

$\displaystyle\qquad = -\frac{1}{2a^\nu\sqrt{b^2-a^2}}\left\{\left[b+\left(b^2-a^2\right)^{\frac{1}{2}}\right]^\nu + \left[b-\left(b^2-a^2\right)^{\frac{1}{2}}\right]^\nu\right\}$　　$[0 < a < b,\quad |\operatorname{Re}\nu| < 1]$

ET I 48(32)

3.　　$\displaystyle\int_0^{\pi/2} \left[\cos x\,I_0(a\cos x) + I_1(a\cos x)\right]\mathrm{d}x = \frac{e^a-1}{a}$

**6.674**

1. $$\int_0^a \sin(a-x)\, J_\nu(x)\, \mathrm{d}x = a\, J_{\nu+1}(a) - 2\nu \sum_{n=0}^\infty (-1)^n J_{\nu+2n+2}(a)$$

$$[\mathrm{Re}\,\nu > -1] \qquad\qquad \text{ET II 334(12)}$$

2. $$\int_0^a \cos(a-x)\, J_\nu(x)\, \mathrm{d}x = a\, J_\nu(a) - 2\nu \sum_{n=0}^\infty (-1)^n J_{\nu+2n+1}(a)$$

$$[\mathrm{Re}\,\nu > -1] \qquad\qquad \text{ET II 336(23)}$$

3. $$\int_0^a \sin(a-x)\, J_{2n}(x)\, \mathrm{d}x = a\, J_{2n+1}(a) + (-1)^n 2n \left[ \cos a - J_0(a) - 2\sum_{m=1}^n (-1)^m J_{2m}(a) \right]$$

$$[n = 0, 1, 2, \ldots] \qquad\qquad \text{ET II 334(10)}$$

4. $$\int_0^a \cos(a-x)\, J_{2n}(x)\, \mathrm{d}x = a\, J_{2n}(a) - (-1)^n 2n \left[ \sin a - 2\sum_{m=0}^{n-1} (-1)^m J_{2m+1}(a) \right]$$

$$[n = 0, 1, 2, \ldots] \qquad\qquad \text{ET II 335(21)}$$

5. $$\int_0^a \sin(a-x)\, J_{2n+1}(x)\, \mathrm{d}x = a\, J_{2n+2}(a) + (-1)^n (2n+1) \left[ \sin a - 2\sum_{m=0}^{n} (-1)^m J_{2m+1}(a) \right]$$

$$[n = 0, 1, 2, \ldots] \qquad\qquad \text{ET II 334(11)}$$

6. $$\int_0^a \cos(a-x)\, J_{2n+1}(x)\, \mathrm{d}x = a\, J_{2n+1}(a) + (-1)^n (2n+1) \left[ \cos a - J_0(a) - 2\sum_{m=1}^{n} (-1)^m J_{2m}(a) \right]$$

$$[n = 0, 1, 2, \ldots] \qquad\qquad \text{ET II 336(22)}$$

7. $$\int_0^z \sin(z-x)\, J_0(x)\, \mathrm{d}x = z\, J_1(z) \qquad\qquad\qquad\qquad \text{WA 415(2)}$$

8. $$\int_0^z \cos(z-x)\, J_0(x)\, \mathrm{d}x = z\, J_0(z) \qquad\qquad\qquad\qquad \text{WA 415(1)}$$

**6.675**

1. $$\int_0^\infty J_\nu\left(a\sqrt{x}\right) \sin(bx)\, \mathrm{d}x = \frac{a\sqrt{\pi}}{4b^{\frac{3}{2}}} \left[ \cos\left(\frac{a^2}{8b} - \frac{\nu\pi}{4}\right) J_{\frac{1}{2}\nu-\frac{1}{2}}\left(\frac{a^2}{8b}\right) - \sin\left(\frac{a^2}{8b} - \frac{\nu\pi}{4}\right) J_{\frac{1}{2}\nu+\frac{1}{2}}\left(\frac{a^2}{8b}\right) \right]$$

$$[a > 0, \quad b > 0, \quad \mathrm{Re}\,\nu > -4]$$
$$\text{ET I 110(23)}$$

2. $$\int_0^\infty J_\nu\left(a\sqrt{x}\right) \cos(bx)\, \mathrm{d}x$$
$$= -\frac{a\sqrt{\pi}}{4b^{\frac{3}{2}}} \left[ \sin\left(\frac{a^2}{8b} - \frac{\nu\pi}{4}\right) J_{\frac{1}{2}\nu-\frac{1}{2}}\left(\frac{a^2}{8b}\right) + \cos\left(\frac{a^2}{8b} - \frac{\nu\pi}{4}\right) J_{\frac{1}{2}\nu+\frac{1}{2}}\left(\frac{a^2}{8b}\right) \right]$$

$$[a > 0, \quad b > 0, \quad \mathrm{Re}\,\nu > -2] \quad \text{ET I 53(22)a}$$

3. $$\int_0^\infty J_0\left(a\sqrt{x}\right) \sin(bx)\, \mathrm{d}x = \frac{1}{b} \cos\left(\frac{a^2}{4b}\right) \qquad\qquad [a > 0, \quad b > 0] \qquad\qquad \text{ET I 110(22)}$$

4. $\displaystyle\int_0^\infty J_0\left(a\sqrt{x}\right)\cos(bx)\,\mathrm{d}x = \frac{1}{b}\sin\left(\frac{a^2}{4b}\right)$ $\qquad [a>0, \quad b>0]$ $\qquad$ ET I 53(21)

**6.676**

1. $\displaystyle\int_0^\infty J_\nu\left(a\sqrt{x}\right)J_\nu\left(b\sqrt{x}\right)\sin(cx)\,\mathrm{d}x = \frac{1}{c}J_\nu\left(\frac{ab}{2c}\right)\cos\left(\frac{a^2+b^2}{4c}-\frac{\nu\pi}{2}\right)$

$\qquad\qquad\qquad [a>0, \quad b>0, \quad c>0, \quad \operatorname{Re}\nu>-2]$

$\qquad\qquad\qquad$ ET I 111(29)a

2. $\displaystyle\int_0^\infty J_\nu\left(a\sqrt{x}\right)J_\nu\left(b\sqrt{x}\right)\cos(cx)\,\mathrm{d}x = \frac{1}{c}J_\nu\left(\frac{ab}{2c}\right)\sin\left(\frac{a^2+b^2}{4c}-\frac{\nu\pi}{2}\right)$

$\qquad\qquad\qquad [a>0, \quad b>0, \quad c>0, \quad \operatorname{Re}\nu>-1]$

$\qquad\qquad\qquad$ ET I 54(27)

3. $\displaystyle\int_0^\infty J_0\left(a\sqrt{x}\right)K_0\left(a\sqrt{x}\right)\sin(bx)\,\mathrm{d}x = \frac{1}{2b}K_0\left(\frac{a^2}{2b}\right)$ $\quad [\operatorname{Re}a>0, \quad b>0]$ $\quad$ ET I 111(31)

4. $\displaystyle\int_0^\infty J_0\left(\sqrt{ax}\right)K_0\left(\sqrt{ax}\right)\cos(bx)\,\mathrm{d}x = \frac{\pi}{4b}\left[I_0\left(\frac{a}{2b}\right)-\mathbf{L}_0\left(\frac{a}{2b}\right)\right]$

$\qquad\qquad\qquad [\operatorname{Re}a>0, \quad b>0]$

$\qquad\qquad\qquad$ ET I 54(29)

5. $\displaystyle\int_0^\infty K_0\left(\sqrt{ax}\right)Y_0\left(\sqrt{ax}\right)\cos(bx)\,\mathrm{d}x = -\frac{1}{2b}K_0\left(\frac{a}{2b}\right)$

$\qquad\qquad\qquad [\operatorname{Re}\sqrt{a}>0, \quad b>0]$ $\qquad$ ET I 54(30)

6. $\displaystyle\int_0^\infty K_0\left(\sqrt{ax}e^{\frac{1}{4}\pi i}\right)K_0\left(\sqrt{ax}e^{-\frac{1}{4}\pi i}\right)\cos(bx)\,\mathrm{d}x = \frac{\pi^2}{8b}\left[\mathbf{H}_0\left(\frac{a}{2b}\right)-Y_0\left(\frac{a}{2b}\right)\right]$

$\qquad\qquad\qquad [\operatorname{Re}a>0, b>0]$ $\qquad$ ET I 54(31)

**6.677**

1. $\displaystyle\int_a^\infty J_0\left(b\sqrt{x^2-a^2}\right)\sin(cx)\,\mathrm{d}x = 0$ $\qquad [0<c<b]$

$\qquad\qquad\qquad = \frac{\cos\left(a\sqrt{c^2-b^2}\right)}{\sqrt{c^2-b^2}}$ $\qquad [0<b<c]$

$\qquad\qquad\qquad$ ET I 113(47)

2. $\displaystyle\int_a^\infty J_0\left(b\sqrt{x^2-a^2}\right)\cos(cx)\,\mathrm{d}x = \frac{\exp\left(-a\sqrt{b^2-c^2}\right)}{\sqrt{b^2-c^2}}$ $\qquad [0<c<b]$

$\qquad\qquad\qquad = \frac{-\sin\left(a\sqrt{c^2-b^2}\right)}{\sqrt{c^2-b^2}}$ $\qquad [0<b<c]$

$\qquad\qquad\qquad$ ET I 57(48)a

3.[6] $\displaystyle\int_0^\infty J_0\left(a\sqrt{x^2+z^2}\right)\cos\beta x\,\mathrm{d}x = \frac{\cos z\sqrt{a^2-\beta^2}}{\sqrt{a^2-\beta^2}}$ $\qquad [0<\beta<a, \quad z>0]$

$\qquad\qquad\qquad = 0$ $\qquad [0<a<\beta, \quad z>0]$

$\qquad\qquad\qquad$ MO 47a

4. $$\int_0^\infty Y_0\left(a\sqrt{x^2+z^2}\right)\cos\beta x\,dx = \frac{1}{\sqrt{a^2-\beta^2}}\sin\left(z\sqrt{a^2-\beta^2}\right) \qquad [0<\beta<a, \quad z>0]$$
$$= -\frac{1}{\sqrt{\beta^2-a^2}}\exp\left(-z\sqrt{\beta^2-a^2}\right) \qquad [0<a<\beta, \quad z>0]$$

MO 47a

5. $$\int_0^\infty K_0\left[a\sqrt{x^2+\beta^2}\right]\cos(\gamma x)\,dx = \frac{\pi}{2\sqrt{a^2+\gamma^2}}\exp\left(-\beta\sqrt{a^2+\gamma^2}\right)$$
$$[\operatorname{Re}a>0, \quad \operatorname{Re}\beta>0, \quad \gamma>0]$$

ET I 56(43)

6. $$\int_0^a J_0\left(b\sqrt{a^2-x^2}\right)\cos(cx)\,dx = \frac{\sin\left(a\sqrt{b^2+c^2}\right)}{\sqrt{b^2+c^2}} \qquad [b>0]$$

MO 48a, ET I 57(47)

7. $$\int_0^\infty J_0\left(b\sqrt{x^2-a^2}\right)\cos(cx)\,dx = \frac{\cosh\left(a\sqrt{b^2-c^2}\right)}{\sqrt{b^2-c^2}} \qquad [0<c<b, \quad a>0]$$
$$= 0 \qquad [0<b<c, \quad a>0]$$

ET I 57(49)

8. $$\int_0^\infty H_0^{(1)}\left(\alpha\sqrt{\beta^2-x^2}\right)\cos(\gamma x)\,dx = -i\frac{\exp\left(i\beta\sqrt{\alpha^2+\gamma^2}\right)}{\sqrt{\alpha^2+\gamma^2}}$$
$$\left[\pi>\arg\sqrt{\beta^2-x^2}\geq 0, \quad \alpha>0, \quad \gamma>0\right] \quad \text{ET I 59(59)}$$

9. $$\int_0^\infty H_0^{(2)}\left(\alpha\sqrt{\beta^2-x^2}\right)\cos(\gamma x)\,dx = \frac{i\exp\left(-i\beta\sqrt{\alpha^2+\gamma^2}\right)}{\sqrt{\alpha^2+\gamma^2}}$$
$$\left[-\pi<\arg\sqrt{\beta^2-x^2}\leq 0, \quad \alpha>0, \quad \gamma>0\right] \quad \text{ET I 58(58)}$$

**6.678** $$\int_0^\infty\left[K_0\left(2\sqrt{x}\right)+\frac{\pi}{2}Y_0\left(2\sqrt{x}\right)\right]\sin(bx)\,dx = \frac{\pi}{2b}\sin\left(\frac{1}{b}\right) \qquad [b>0] \qquad \text{ET I 111(34)}$$

**6.679**

1. $$\int_0^\infty J_{2\nu}\left[2b\sinh\left(\frac{x}{2}\right)\right]\sin(bx)\,dx = -i\left[I_{\nu-ib}(a)\,K_{\nu+ib}(a)-I_{\nu+ib}(a)\,K_{\nu-ib}(a)\right]$$
$$[a>0, \quad b>0, \quad \operatorname{Re}\nu>-1]$$

ET I 115(59)

2. $$\int_0^\infty J_{2\nu}\left[2a\sinh\left(\frac{x}{2}\right)\right]\cos(bx)\,dx = I_{\nu-ib}(a)\,K_{\nu+ib}(a)+I_{\nu+ib}(a)\,K_{\nu-ib}(a)$$
$$[a>0, \quad b>0, \quad \operatorname{Re}\nu>-\tfrac{1}{2}]$$

ET I 59(64)

3. $$\int_0^\infty J_{2\nu}\left[2a\cosh\left(\frac{x}{2}\right)\right]\cos(bx)\,dx = -\frac{\pi}{2}\left[J_{\nu+ib}(a)\,Y_{\nu-ib}(a)+J_{\nu-ib}(a)\,Y_{\nu+ib}(a)\right] \qquad \text{ET I 59(63)}$$

4. $$\int_0^\infty J_0\left[2a\sinh\left(\frac{x}{2}\right)\right]\sin(bx)\,dx = \frac{2}{\pi}\sinh(\pi b)\left[K_{ib}(a)\right]^2$$
$$[a>0, \quad b>0] \qquad \text{ET I 115(58)}$$

5.    $\int_0^\infty J_0\left[2a\sinh\left(\dfrac{x}{2}\right)\right]\cos(bx)\,\mathrm{d}x = [I_{ib}(a) + I_{-ib}(a)]\,K_{ib}(a)$

$$[a > 0, \quad b > 0] \qquad\qquad \text{ET I 59(62)}$$

6.    $\int_0^\infty Y_0\left[2a\sinh\left(\dfrac{x}{2}\right)\right]\cos(bx)\,\mathrm{d}x = -\dfrac{2}{\pi}\cosh(\pi b)\,[K_{ib}(a)]^2$

$$[a > 0, \quad b > 0] \qquad\qquad \text{ET I 59(65)}$$

7.    $\int_0^\infty K_0\left[2a\sinh\left(\dfrac{x}{2}\right)\right]\cos(bx)\,\mathrm{d}x = \dfrac{\pi^2}{4}\left\{[J_{ib}(a)]^2 + [Y_{ib}(a)]^2\right\}$

$$[\operatorname{Re} a > 0, \quad b > 0] \qquad\qquad \text{ET I 59(66)}$$

**6.681**

1.    $\int_0^{\frac{\pi}{2}} \cos(2\mu x)\, J_{2\nu}(2a\cos x)\,\mathrm{d}x = \dfrac{\pi}{2} J_{\nu+\mu}(a)\, J_{\nu-\mu}(a) \qquad \left[\operatorname{Re}\nu > -\tfrac{1}{2}\right]$      ET II 361(23)

2.    $\int_0^{\frac{\pi}{2}} \cos(2\mu x)\, Y_{2\nu}(2a\cos x)\,\mathrm{d}x = \dfrac{\pi}{2}\left[\cot(2\nu\pi)\, J_{\nu+\mu}(a)\, J_{\nu-\mu}(a) - \operatorname{cosec}(2\nu\pi)\, J_{\mu-\nu}(a)\, J_{-\mu-\nu}(a)\right]$

$$\left[|\operatorname{Re}\nu| < \tfrac{1}{2}\right] \qquad\qquad \text{ET II 361(24)}$$

3.    $\int_0^{\frac{\pi}{2}} \cos(2\mu x)\, I_{2\nu}(2a\cos x)\,\mathrm{d}x = \dfrac{\pi}{2} I_{\nu-\mu}(a)\, I_{\nu+\mu}(a) \qquad \left[\operatorname{Re}\nu > -\tfrac{1}{2}\right]$      ET I 59(61)

4.    $\int_0^{\frac{\pi}{2}} \cos(\nu x)\, K_\nu(2a\cos x)\,\mathrm{d}x = \dfrac{\pi}{2} I_0(a)\, K_\nu(a) \qquad [\operatorname{Re}\nu < 1]$      WA 484(3)

5.    $\int_0^{\pi} J_0(2z\cos x)\cos 2nx\,\mathrm{d}x = (-1)^n \pi J_n^2(z).$      MO 45

6.    $\int_0^{\pi} J_0(2z\sin x)\cos 2nx\,\mathrm{d}x = \pi J_n^2(z).$      WA 43(3), MO 45

7.    $\int_0^{\frac{\pi}{2}} \cos(2nx)\, Y_0(2a\sin x)\,\mathrm{d}x = \dfrac{\pi}{2} J_n(a)\, Y_n(a) \qquad [n = 0, 1, 2, \ldots]$      ET II 360(16)

8.    $\int_0^{\pi} \sin(2\mu x)\, J_{2\nu}(2a\sin x)\,\mathrm{d}x = \pi\sin(\mu\pi)\, J_{\nu-\mu}(a)\, J_{\nu+\mu}(a)$

$$[\operatorname{Re}\nu > -1] \qquad\qquad \text{ET II 360(13)}$$

9.    $\int_0^{\pi} \cos(2\mu x)\, J_{2\nu}(2a\sin x)\,\mathrm{d}x = \pi\cos(\mu\pi)\, J_{\nu-\mu}(a)\, J_{\nu+\mu}(a)$

$$\left[\operatorname{Re}\nu > -\tfrac{1}{2}\right] \qquad\qquad \text{ET II 360(14)}$$

10.    $\int_0^{\frac{\pi}{2}} J_{\nu+\mu}(2z\cos x)\cos[(\nu - \mu)x]\,\mathrm{d}x = \dfrac{\pi}{2} J_\nu(z)\, J_\mu(z)$

$$[\operatorname{Re}(\nu + \mu) > -1] \qquad\qquad \text{MO 42}$$

11.    $\int_0^{\frac{\pi}{2}} \cos[(\mu - \nu)x]\, I_{\mu+\nu}(2a\cos x)\,\mathrm{d}x = \dfrac{\pi}{2} I_\mu(a)\, I_\nu(a) \qquad [\operatorname{Re}(\mu + \nu) > -1]$

$$\text{WA 484(2), ET II 378(39)}$$

$12.^{12}$ $\displaystyle\int_0^{\frac{\pi}{2}} \cos[(\mu-\nu)x]\, K_{\mu+\nu}\,(2a\cos x)\,\mathrm{d}x = \frac{\pi}{2}\,\mathrm{cosec}[(\mu+\nu)\pi]\,[I_{-\mu}(a)\,I_{-\nu}(a) - I_\mu(a)\,I_\nu(a)]$

$$[|\mathrm{Re}(\mu+\nu)| < 1] \qquad \text{ET II 378(40)}$$

$13.^8$ $\displaystyle\int_0^{\frac{\pi}{2}} K_{\nu-m}\,(2a\cos x)\cos[(m+\nu)x]\,\mathrm{d}x = (-1)^m\frac{\pi}{2}\,I_m(a)\,K_\nu(a)$

$$[|\mathrm{Re}(\nu-m)| < 1] \qquad \text{WA 485(4)}$$

## 6.682

$1.^7$ $\displaystyle\int_0^{\frac{\pi}{2}} J_{\nu-\frac{1}{2}}\,(x\sin t)\sin^{\nu+\frac{1}{2}} t\,\mathrm{d}t = \sqrt{\frac{\pi}{2x}}\,J_\nu(x)$

[$\nu$ may be zero, a natural number, one half, or a natural number plus one half; $x > 0$]   MO 42a

2. $\displaystyle\int_0^{\frac{\pi}{2}} J_\nu\,(z\sin x)\sin^\nu x\cos^{2\nu} x\,\mathrm{d}x = 2^{\nu-1}\sqrt{\pi}\,\Gamma\left(\nu+\frac{1}{2}\right)z^{-\nu}J_\nu^2\left(\frac{z}{2}\right)$

$$\left[\mathrm{Re}\,\nu > -\tfrac{1}{2}\right] \qquad \text{MO 42a}$$

## 6.683

1. $\displaystyle\int_0^{\frac{\pi}{2}} J_\nu\,(z\sin x)\,I_\mu\,(z\cos x)\tan^{\nu+1} x\,\mathrm{d}x = \frac{\left(\dfrac{z}{2}\right)^\nu \Gamma\left(\dfrac{\mu-\nu}{2}\right)}{\Gamma\left(\dfrac{\mu+\nu}{2}+1\right)}\,J_\mu(z)$

$$[\mathrm{Re}\,\nu > \mathrm{Re}\,\mu > -1] \qquad \text{WA 407(4)}$$

2. $\displaystyle\int_0^{\frac{\pi}{2}} J_\nu\,(z_1\sin x)\,J_\mu\,(z_2\cos x)\sin^{\nu+1} x\cos^{\mu+1} x\,\mathrm{d}x = \frac{z_1^\nu z_2^\mu\,J_{\nu+\mu+1}\left(\sqrt{z_1^2+z_2^2}\right)}{\sqrt{(z_1^2+z_2^2)^{\nu+\mu+1}}}$

$$[\mathrm{Re}\,\nu > -1, \quad \mathrm{Re}\,\mu > -1] \qquad \text{WA 410(1)}$$

3. $\displaystyle\int_0^{\frac{\pi}{2}} J_\nu\,(z\cos^2 x)\,J_\mu\,(z\sin^2 x)\sin x\cos x\,\mathrm{d}x = \frac{1}{z}\sum_{k=0}^{\infty}(-1)^k\,J_{\nu+\mu+2k+1}(z)$

$$[\mathrm{Re}\,\nu > -1, \quad \mathrm{Re}\,\mu > -1] \qquad (\text{see also } \mathbf{6.513}\ 6) \qquad \text{WA 414(1)}$$

4. $\displaystyle\int_0^{\frac{\pi}{2}} J_\mu\,(z\sin\theta)\,(\sin\theta)^{1-\mu}\,(\cos\theta)^{2\nu+1}\,\mathrm{d}\theta = \frac{s_{\mu+\nu,\nu-\mu+1}(z)}{2^{\mu-1}z^{\nu+1}\,\Gamma(\mu)}$

$$[\mathrm{Re}\,\nu > -1] \qquad \text{WA 407(2)}$$

5. $\displaystyle\int_0^{\frac{\pi}{2}} J_\mu\,(z\sin\theta)\,(\sin\theta)^{1-\mu}\,\mathrm{d}\theta = \frac{\mathbf{H}_{\mu-\frac{1}{2}}(z)}{\sqrt{\dfrac{2z}{\pi}}}$

$$\text{WA 407(3)}$$

6. $\displaystyle\int_0^{\frac{\pi}{2}} J_\mu\,(a\sin\theta)\,(\sin\theta)^{\mu+1}\,(\cos\theta)^{2\varrho+1}\,\mathrm{d}\theta = 2^\varrho\,\Gamma(\varrho+1)a^{-\varrho-1}\,J_{\varrho+\mu+1}(a)$

$$[\mathrm{Re}\,\varrho > -1, \quad \mathrm{Re}\,\mu > -1]$$
$$\text{WA 406(1)}, \quad \text{EH II 46(5)}$$

7.    $\displaystyle\int_0^{\frac{\pi}{2}} J_\nu\left(2z\sin\theta\right)\left(\sin\theta\right)^\nu\left(\cos\theta\right)^{2\nu}\,\mathrm{d}\theta$

$$= \frac{1}{2}\sum_{m=0}^\infty \frac{(-1)^m z^{\nu+2m}\,\Gamma\left(\nu+m+\frac{1}{2}\right)\Gamma\left(\nu+\frac{1}{2}\right)}{m!\,\Gamma(\nu+m+1)\,\Gamma(2\nu+m+1)}$$

$$= \frac{1}{2}z^{-\nu}\sqrt{\pi}\,\Gamma\left(\nu+\tfrac{1}{2}\right)\left[J_\nu(z)\right]^2 \qquad \left[\operatorname{Re}\nu > -\tfrac{1}{2}\right]$$

EH II 47(10)

8.    $\displaystyle\int_0^{\frac{\pi}{2}} J_\nu\left(z\sin\theta\right)\left(\sin\theta\right)^{\nu+1}\left(\cos\theta\right)^{-2\nu}\,\mathrm{d}\theta = 2^{-\nu}\frac{z^{\nu-1}}{\sqrt{\pi}}\,\Gamma\left(\frac{1}{2}-\nu\right)\sin z$

$$\left[-1 < \operatorname{Re}\nu < \tfrac{1}{2}\right] \qquad \text{EH II 68(39)}$$

9.    $\displaystyle\int_0^{\frac{\pi}{2}} J_\nu\left(z\sin^2\theta\right) J_\nu\left(z\cos^2\theta\right)\left(\sin\theta\right)^{2\nu+1}\left(\cos\theta\right)^{2\nu+1}\,\mathrm{d}\theta = \frac{\Gamma\left(\frac{1}{2}+\nu\right)J_{2\nu+\frac{1}{2}}(z)}{2^{2\nu+\frac{3}{2}}\Gamma(\nu+1)\sqrt{z}}$

$$\left[\operatorname{Re}\nu > -\tfrac{1}{2}\right] \qquad \text{WA 409(1)}$$

10.    $\displaystyle\int_0^{\frac{\pi}{2}} J_\mu\left(z\sin^2\theta\right) J_\nu\left(z\cos^2\theta\right)\sin^{2\mu+1}\theta\cos^{2\nu+1}\theta\,\mathrm{d}\theta = \frac{\Gamma\left(\mu+\frac{1}{2}\right)\Gamma\left(\nu+\frac{1}{2}\right)J_{\mu+\nu+\frac{1}{2}}(z)}{2\sqrt{\pi}\,\Gamma(\mu+\nu+1)\sqrt{2z}}$

$$\left[\operatorname{Re}\mu > -\tfrac{1}{2}, \quad \operatorname{Re}\nu > -\tfrac{1}{2}\right] \qquad \text{WA 417(1)}$$

**6.684**

1.[8]    $\displaystyle\int_0^\pi (\sin x)^{2\nu}\frac{J_\nu\left(\sqrt{\alpha^2+\beta^2-2\alpha\beta\cos x}\right)}{\left(\sqrt{\alpha^2+\beta^2-2\alpha\beta\cos x}\right)^\nu}\,\mathrm{d}x = 2^\nu\sqrt{\pi}\,\Gamma\left(\nu+\frac{1}{2}\right)\frac{J_\nu(\alpha)}{\alpha^\nu}\frac{J_\nu(\beta)}{\beta^\nu}$

$$\left[\operatorname{Re}\nu > -\tfrac{1}{2}\right] \qquad \text{ET II 362(27)}$$

2.    $\displaystyle\int_0^\pi (\sin x)^{2\nu}\frac{Y_\nu\left(\sqrt{\alpha^2+\beta^2-2\alpha\beta\cos x}\right)}{\left(\sqrt{\alpha^2+\beta^2-2\alpha\beta\cos x}\right)^\nu}\,\mathrm{d}x = 2^\nu\sqrt{\pi}\,\Gamma\left(\nu+\frac{1}{2}\right)\frac{J_\nu(\alpha)}{\alpha^\nu}\frac{Y_\nu(\beta)}{\beta^\nu}$

$$\left[|\alpha| < |\beta|, \quad \operatorname{Re}\nu > -\tfrac{1}{2}\right] \qquad \text{ET II 362(28)}$$

**6.685**    $\displaystyle\int_0^{\frac{\pi}{2}} \sec x\cos(2\lambda x)\,K_{2\mu}\left(a\sec x\right)\,\mathrm{d}x = \frac{\pi}{2a}\,W_{\lambda,\mu}(a)\,W_{-\lambda,\mu}(a)$    $\left[\operatorname{Re}a > 0\right]$    ET II 378(41)

**6.686**

1.    $\displaystyle\int_0^\infty \sin\left(ax^2\right) J_\nu(bx)\,\mathrm{d}x = -\frac{\sqrt{\pi}}{2\sqrt{a}}\sin\left(\frac{b^2}{8a}-\frac{\nu+1}{4}\pi\right)J_{\frac{1}{2}\nu}\left(\frac{b^2}{8a}\right)$

$$\left[a > 0, \quad b > 0, \quad \operatorname{Re}\nu > -3\right]$$

ET II 34(13)

2.    $\displaystyle\int_0^\infty \cos\left(ax^2\right) J_\nu(bx)\,\mathrm{d}x = \frac{\sqrt{\pi}}{2\sqrt{a}}\cos\left(\frac{b^2}{8a}-\frac{\nu+1}{4}\pi\right)J_{\frac{1}{2}\nu}\left(\frac{b^2}{8a}\right)$

$$\left[a > 0, \quad b > 0, \quad \operatorname{Re}\nu > -1\right]$$

ET II 38(38)

3. $\displaystyle\int_0^\infty \sin\left(ax^2\right) Y_\nu(bx)\,\mathrm{d}x$

$$= -\frac{\sqrt{\pi}}{4\sqrt{a}}\sec\left(\frac{\nu\pi}{2}\right)$$
$$\times\left[\cos\left(\frac{b^2}{8a}-\frac{3\nu+1}{4}\pi\right)J_{\frac{1}{2}\nu}\left(\frac{b^2}{8a}\right)-\sin\left(\frac{b^2}{8a}+\frac{\nu-1}{4}\pi\right)Y_{\frac{1}{2}\nu}\left(\frac{b^2}{8a}\right)\right]$$
$$[a>0,\quad b>0,\quad -3<\operatorname{Re}\nu<3]\quad\text{ET II 107(7)}$$

4. $\displaystyle\int_0^\infty \cos\left(ax^2\right) Y_\nu(bx)\,\mathrm{d}x$

$$= \frac{\sqrt{\pi}}{4\sqrt{a}}\sec\left(\frac{\nu\pi}{2}\right)$$
$$\times\left[\sin\left(\frac{b^2}{8a}-\frac{3\nu+1}{4}\pi\right)J_{\frac{1}{2}\nu}\left(\frac{b^2}{8a}\right)+\cos\left(\frac{b^2}{8a}+\frac{\nu-1}{4}\pi\right)Y_{\frac{1}{2}\nu}\left(\frac{b^2}{8a}\right)\right]$$
$$[a>0,\quad b>0,\quad -1<\operatorname{Re}\nu<1]\quad\text{ET II 107(8)}$$

5. $\displaystyle\int_0^\infty \sin\left(ax^2\right) J_1(bx)\,\mathrm{d}x = \frac{1}{b}\sin\frac{b^2}{4a}$ $\qquad\qquad$ $[a>0,\quad b>0]$ $\qquad\qquad$ ET II 19(16)

6. $\displaystyle\int_0^\infty \cos\left(ax^2\right) J_1(bx)\,\mathrm{d}x = \frac{2}{b}\sin^2\left(\frac{b^2}{8a}\right)$ $\qquad\qquad$ $[a>0,\quad b>0]$ $\qquad\qquad$ ET II 20(20)

7. $\displaystyle\int_0^\infty \sin^2\left(ax^2\right) J_1(bx)\,\mathrm{d}x = \frac{1}{2b}\cos\left(\frac{b^2}{8a}\right)$ $\qquad\qquad$ $[a>0,\quad b>0]$ $\qquad\qquad$ ET II 19(17)

**6.687** $\displaystyle\int_0^\infty \cos\left(\frac{x^2}{2a}\right) K_{2\nu}\left(xe^{i\frac{\pi}{4}}\right) K_{2\nu}\left(xe^{-i\frac{\pi}{4}}\right)\,\mathrm{d}x$

$$= \frac{\Gamma\left(\frac{1}{4}+\nu\right)\Gamma\left(\frac{1}{4}-\nu\right)\sqrt{\pi}}{8\sqrt{a}}W_{\frac{1}{4},\nu}\left(ae^{i\frac{\pi}{2}}\right)W_{\frac{1}{4},\nu}\left(ae^{-i\frac{\pi}{2}}\right)$$
$$\left[a>0,\quad |\operatorname{Re}\nu|<\tfrac{1}{4}\right]\quad\text{ET II 372(1)}$$

**6.688**

1. $\displaystyle\int_0^{\frac{\pi}{2}} J_\nu\left(\mu z\sin t\right)\cos\left(\mu x\cos t\right)\,\mathrm{d}t = \frac{\pi}{2}J_{\frac{\nu}{2}}\left(\mu\frac{\sqrt{x^2+z^2}+x}{2}\right)J_{\frac{\nu}{2}}\left(\mu\frac{\sqrt{x^2+z^2}-x}{2}\right)$

$$[\operatorname{Re}\nu>-1,\quad \operatorname{Re}z>0]\qquad\qquad\text{MO 46}$$

2. $\displaystyle\int_0^{\frac{\pi}{2}} (\sin x)^{\nu+1}\cos\left(\beta\cos x\right) J_\nu\left(\alpha\sin x\right)\,\mathrm{d}x = 2^{-\frac{1}{2}}\sqrt{\pi}\alpha^\nu\left(\alpha^2+\beta^2\right)^{-\frac{1}{2}\nu-\frac{1}{4}}J_{\nu+\frac{1}{2}}\left[\left(\alpha^2+\beta^2\right)^{\frac{1}{2}}\right]$

$$[\operatorname{Re}\nu>-1]\qquad\qquad\text{ET II 361(19)}$$

3. $\displaystyle\int_0^{\frac{\pi}{2}}\cos\left[(z-\zeta)\cos\theta\right] J_{2\nu}\left[2\sqrt{z\zeta}\sin\theta\right]\,\mathrm{d}\theta = \frac{\pi}{2}J_\nu(z)J_\nu(\zeta)$

$$\left[\operatorname{Re}\nu>-\tfrac{1}{2}\right]\qquad\qquad\text{EH II 47(8)}$$

## 6.69–6.74 Combinations of Bessel and trigonometric functions and powers

**6.691** $\displaystyle\int_0^\infty x\sin(bx) K_0(ax)\,\mathrm{d}x = \frac{\pi b}{2}\left(a^2+b^2\right)^{-\frac{3}{2}}$ $\qquad\qquad$ $[\operatorname{Re}a>0,\quad b>0]$ $\qquad\qquad$ ET I 105(47)

**6.692**

1.
$$\int_0^\infty x\, K_\nu(ax)\, I_\nu(bx) \sin(cx)\, dx = -\frac{1}{2}(ab)^{-\frac{3}{2}} c\left(u^2-1\right)^{-\frac{1}{2}} Q_{\nu-\frac{1}{2}}^1(u), \qquad u = (2ab)^{-1}\left(a^2+b^2+c^2\right)$$
$$\left[\operatorname{Re} a > |\operatorname{Re} b|, \quad c > 0, \quad \operatorname{Re} \nu > -\tfrac{3}{2}\right]$$
ET I 106(54)

2.
$$\int_0^\infty x\, K_\nu(ax)\, K_\nu(bx) \sin(cx)\, dx = \frac{\pi}{4}(ab)^{-\frac{3}{2}} c\left(u^2-1\right)^{-\frac{1}{2}} \Gamma\left(\tfrac{3}{2}+\nu\right)\Gamma\left(\tfrac{3}{2}-\nu\right) P_{\nu-\frac{1}{2}}^{-1}(u)$$
$$u = (2ab)^{-1}\left(a^2+b^2+c^2\right) \qquad \left[\operatorname{Re}(a+b) > 0, \quad c > 0, \quad |\operatorname{Re}\nu| < \tfrac{3}{2}\right] \quad \text{ET I 107(61)}$$

**6.693**

1.
$$\int_0^\infty J_\nu(ax) \sin bx\, \frac{dx}{x} = \frac{1}{\nu}\sin\left(\nu \arcsin \frac{b}{a}\right) \qquad\qquad [b \le a]$$
$$= \frac{a^\nu \sin \frac{\nu\pi}{2}}{\nu\left(b+\sqrt{b^2-a^2}\right)^\nu} \qquad\qquad [b \ge a]$$
$$[\operatorname{Re}\nu > -1] \qquad\qquad \text{WA 443(2)}$$

2.[8]
$$\int_0^\infty J_\nu(ax)\cos bx\, \frac{dx}{x} = \frac{1}{\nu}\cos\left(\nu \arcsin \frac{b}{a}\right) \qquad\qquad [b \le a]$$
$$= \frac{a^\nu \cos \frac{\nu\pi}{2}}{\nu\left(b+\sqrt{b^2-a^2}\right)^\nu} \qquad\qquad [b \ge a] \qquad [\operatorname{Re}\nu > 0]$$
$$\text{WA 443(3)}$$

3.
$$\int_0^\infty Y_\nu(ax)\sin(bx)\, \frac{dx}{x}$$
$$= -\frac{1}{\nu}\tan\left(\frac{\nu\pi}{2}\right)\sin\left[\nu \arcsin\left(\frac{b}{a}\right)\right]$$
$$[0 < b < a, \quad |\operatorname{Re}\nu| < 1]$$
$$= \frac{1}{2\nu}\sec\left(\frac{\nu\pi}{2}\right)\left\{a^{-\nu}\cos(\nu\pi)\left[b-\left(b^2-a^2\right)^{\frac{1}{2}}\right]^\nu - a^\nu\left[b-\left(b^2-a^2\right)^{\frac{1}{2}}\right]^{-\nu}\right\}$$
$$[0 < a < b, \quad |\operatorname{Re}\nu| < 1]$$
$$\text{ET I 103(35)}$$

4.
$$\int_0^\infty J_\nu(ax)\sin(bx)\, \frac{dx}{x^2}$$
$$= \frac{\sqrt{a^2-b^2}\,\sin\left[\nu \arcsin\left(\frac{b}{a}\right)\right]}{\nu^2-1} - \frac{b\cos\left[\nu \arcsin\left(\frac{b}{a}\right)\right]}{\nu\left(\nu^2-1\right)} \qquad [0 < b < a, \quad \operatorname{Re}\nu > 0]$$
$$= \frac{-a^\nu\cos\left(\frac{\nu\pi}{2}\right)\left[b+\nu\sqrt{b^2-a^2}\right]}{\nu\left(\nu^2-1\right)\left[b+\sqrt{b^2-a^2}\right]^\nu} \qquad\qquad [0 < a < b, \quad \operatorname{Re}\nu > 0]$$
$$\text{ET I 99(6)}$$

5.
$$\int_0^\infty J_\nu(ax)\cos(bx)\, \frac{dx}{x^2}$$
$$= \frac{a\cos\left[(\nu-1)\arcsin\left(\frac{b}{a}\right)\right]}{2\nu(\nu-1)} + \frac{a\cos\left[(\nu+1)\arcsin\left(\frac{b}{a}\right)\right]}{2\nu(\nu+1)} \qquad [0 < b < a, \quad \operatorname{Re}\nu > 1]$$
$$= \frac{a^\nu \sin\left(\frac{\nu\pi}{2}\right)}{2\nu(\nu-1)\left[b+\sqrt{b^2-a^2}\right]^{\nu-1}} - \frac{a^{\nu+2}\sin\left(\frac{\nu\pi}{2}\right)}{2\nu(\nu+1)\left[b+\sqrt{b^2-a^2}\right]^{\nu+1}} \qquad [0 < a < b, \quad \operatorname{Re}\nu > 1]$$
$$\text{ET I 44(6)}$$

**6.**$^{12}$ $\displaystyle\int_0^\infty J_0(ax)\sin x\,\frac{\mathrm{d}x}{x} = \frac{\pi}{2}$               $[0 < a < 1]$

$\qquad\qquad\qquad\qquad\;\; = \operatorname{arccosec} a$            $[a > 1]$

WH

**7.** $\displaystyle\int_0^\infty J_0(x)\sin bx\,\frac{\mathrm{d}x}{x} = \frac{\pi}{2}$           $[b > 1]$

$\qquad\qquad\qquad\qquad\; = \arcsin b$           $[b^2 < 1]$

$\qquad\qquad\qquad\qquad\; = -\frac{\pi}{2}$          $[b < -1]$

**8.** $\displaystyle\int_0^\infty [J_0(x) - \cos ax]\,\frac{\mathrm{d}x}{x} = \ln 2a$           NT 66(13)

**9.** $\displaystyle\int_0^z J_\nu(x)\sin(z - x)\,\frac{\mathrm{d}x}{x} = \frac{2}{\nu}\sum_{k=0}^\infty (-1)^k J_{\nu+2k+1}(z)$      $[\operatorname{Re}\nu > 0]$      WA 416(4)

**10.** $\displaystyle\int_0^z J_\nu(x)\cos(z - x)\,\frac{\mathrm{d}x}{x} = \frac{1}{\nu} J_\nu(z) + \frac{2}{\nu}\sum_{k=1}^\infty (-1)^k J_{\nu+2k}(z)$

$\qquad\qquad\qquad\qquad\qquad\qquad\qquad\qquad\qquad\qquad$ $[\operatorname{Re}\nu > 0]$      WA 416(5)

**6.694**$^{12}$ $\displaystyle\int_0^\infty \left[\frac{J_1(ax)}{x}\right]^2 \sin(bx)\,\mathrm{d}x = \frac{b}{2} - \frac{2a + b}{12\pi a^2}\left[(4a^2 + b^2)\boldsymbol{E}\left(\frac{2\sqrt{2ab}}{2a + b}\right) - (2a - b)^2 \boldsymbol{K}\left(\frac{2\sqrt{2ab}}{2a + b}\right)\right]$

$\qquad\qquad\qquad\qquad\qquad\qquad\qquad\qquad\qquad\qquad\qquad\qquad\qquad\quad$ $[a > 0, \quad b > 0]$

$\qquad\qquad\qquad\qquad\qquad\qquad\qquad\qquad\qquad\qquad\qquad\qquad\qquad\quad$ ET I 102(22)

**6.695**

**1.** $\displaystyle\int_0^\infty \frac{\sin ax}{b^2 + x^2} J_0(ux)\,\mathrm{d}x = \frac{\sinh ab}{b} K_0(bu)$      $[a > 0, \quad \operatorname{Re} b > 0, \quad u > a]$      MO 46

**2.** $\displaystyle\int_0^\infty \frac{\cos ax}{b^2 + x^2} J_0(ux)\,\mathrm{d}x = \frac{\pi}{2}\frac{e^{-ab}}{b} I_0(bu)$      $[a > 0, \quad \operatorname{Re} b > 0, \quad -a < u < a]$

$\qquad\qquad\qquad\qquad\qquad\qquad\qquad\qquad\qquad\qquad\qquad\qquad\qquad\qquad$ MO 46

**3.** $\displaystyle\int_0^\infty \frac{x}{x^2 + b^2} \sin(ax) J_0(\gamma x)\,\mathrm{d}x = \frac{\pi}{2} e^{-ab} I_0(\gamma b)$      $[a > 0, \quad \operatorname{Re} b > 0, \quad 0 < \gamma < a]$

$\qquad\qquad\qquad\qquad\qquad\qquad\qquad\qquad\qquad\qquad\qquad\qquad\qquad\qquad$ ET II 10(36)

**4.** $\displaystyle\int_0^\infty \frac{x}{x^2 + \beta^2} \cos(\alpha x) J_0(\gamma x)\,\mathrm{d}x = \cosh(\alpha\beta) K_0(\beta\gamma)$      $[\alpha > 0, \quad \operatorname{Re}\beta > 0, \quad \alpha < \gamma]$

$\qquad\qquad\qquad\qquad\qquad\qquad\qquad\qquad\qquad\qquad\qquad\qquad\qquad\qquad$ ET II 11(45)

**6.696** $\displaystyle\int_0^\infty [1 - \cos(ax)] J_0(bx)\,\frac{\mathrm{d}x}{x} = \operatorname{arccosh}\left(\frac{a}{b}\right)$      $[0 < b < a]$

$\qquad\qquad\qquad\qquad\qquad\qquad\qquad\qquad\;\; = 0$                   $[0 < a < b]$

$\qquad\qquad\qquad\qquad\qquad\qquad\qquad\qquad\qquad\qquad\qquad\qquad\qquad\qquad$ ET II 11(43)

**6.697**

1.    $\displaystyle\int_{-\infty}^{\infty} \frac{\sin[a(x+b)]}{x+b}\, J_0(x)\, dx = 2\int_0^a \frac{\cos bu}{\sqrt{1-u^2}}\, du$      $[0 \le a \le 1]$       WA 463(2)

                          $= \pi\, J_0(b)$         $[1 \le a < \infty]$    WA 463(1), ET II 345(42)

2.    $\displaystyle\int_0^{\infty} \frac{\sin(x+t)}{x+t}\, J_0(t)\, dt = \frac{\pi}{2}\, J_0(x)$          $[x > 0]$           WA 475(4)

3.    $\displaystyle\int_0^{\infty} \frac{\cos(x+t)}{x+t}\, J_0(t)\, dt = -\frac{\pi}{2}\, Y_0(x)$        $[x > 0]$           WA 475(5)

4.    $\displaystyle\int_{-\infty}^{\infty} \frac{|x|}{x+\beta}\, \sin[\alpha(x+\beta)]\, J_0(bx)\, dx = 0$      $[0 \le \alpha < b]$     WA 464(5), ET II 345(43)a

5.[12]   $\displaystyle\int_{-\infty}^{\infty} \frac{\sin[a(x+b)]}{x+b}\, \left[J_{n+\frac{1}{2}}(x)\right]^2 dx = \pi \left[J_{n+\frac{1}{2}}(b)\right]^2$     $[2 \le a, \quad n = 0,1,\ldots]$     ET II 346(45)

6.[12]   $\displaystyle\int_{-\infty}^{\infty} \frac{\sin[a(x+b)]}{x+b}\, J_{n+\frac{1}{2}}(x)\, J_{-n-\frac{1}{2}}(x)\, dx = \pi\, J_{n+\frac{1}{2}}(b)\, J_{-n-\frac{1}{2}}(b)$

                                          $[2 \le a, \quad n = 0,1,\ldots]$     ET II 346(46)

7.    $\displaystyle\int_{-\infty}^{\infty} \frac{J_\mu[a(z+x)]}{(z+x)^\mu}\, \frac{J_\nu[a(\zeta+x)]}{(\zeta+x)^\nu}\, dx = \frac{\Gamma(\mu+\nu)\sqrt{\pi}\sqrt{\frac{2}{a}}}{\Gamma\left(\mu+\frac{1}{2}\right)\Gamma\left(\nu+\frac{1}{2}\right)} \cdot \frac{J_{\mu+\nu-\frac{1}{2}}[a(z-\zeta)]}{(z-\zeta)^{\mu+\nu-\frac{1}{2}}}$

                                          $[\mathrm{Re}(\mu+\nu) > 0]$      WA 463(3)

**6.698**

1.    $\displaystyle\int_0^{\infty} \sqrt{x}\, J_{\nu+\frac{1}{4}}(ax)\, J_{-\nu+\frac{1}{4}}(ax)\, \sin(bx)\, dx = \sqrt{\frac{2}{\pi b}}\, \frac{\cos\left[2\nu \arccos\left(\frac{b}{2a}\right)\right]}{\sqrt{4a^2-b^2}}$    $[0 < b < 2a]$

                                     $= 0$                          $[0 < 2a < b]$

                                                         ET I 102(26)

2.    $\displaystyle\int_0^{\infty} \sqrt{x}\, J_{\nu-\frac{1}{4}}(ax)\, J_{-\nu-\frac{1}{4}}(ax)\, \cos(bx)\, dx = \sqrt{\frac{2}{\pi b}}\, \frac{\cos\left[2\nu \arccos\left(\frac{b}{2a}\right)\right]}{\sqrt{4a^2-b^2}}$    $[0 < b < 2a]$

                                     $= 0$                          $[0 < 2a < b]$

                                                         ET I 46(24)

3.    $\displaystyle\int_0^{\infty} \sqrt{x}\, I_{\frac{1}{4}-\nu}\left(\frac{1}{2}ax\right) K_{\frac{1}{4}+\nu}\left(\frac{1}{2}ax\right) \sin(bx)\, dx = \sqrt{\frac{\pi}{2b}}\, a^{-2\nu}\, \frac{\left(b+\sqrt{a^2+b^2}\right)^{2\nu}}{\sqrt{a^2+b^2}}$

                                   $\left[\mathrm{Re}\, a > 0, \quad b > 0, \quad \mathrm{Re}\, \nu < \frac{5}{4}\right]$

                                                          ET I 106(56)

4.    $\displaystyle\int_0^{\infty} \sqrt{x}\, I_{-\frac{1}{4}-\nu}\left(\frac{1}{2}ax\right) K_{-\frac{1}{4}+\nu}\left(\frac{1}{2}ax\right) \cos(bx)\, dx = \sqrt{\frac{\pi}{2b}}\, a^{-2\nu}\, \frac{\left(b+\sqrt{a^2+b^2}\right)^{2\nu}}{\sqrt{a^2+b^2}}$

                                   $\left[\mathrm{Re}\, a > 0, \quad b > 0, \quad \mathrm{Re}\, \nu < \frac{3}{4}\right]$

                                                          ET I 50(49)

**6.699**

1.     $\displaystyle\int_0^\infty x^\lambda\, J_\nu(ax)\sin(bx)\,\mathrm{d}x = 2^{1+\lambda}a^{-(2+\lambda)}b\frac{\Gamma\left(\frac{2+\lambda+\nu}{2}\right)}{\Gamma\left(\frac{\nu-\lambda}{2}\right)}\,F\left(\frac{2+\lambda+\nu}{2},\frac{2+\lambda-\nu}{2};\frac{3}{2};\frac{b^2}{a^2}\right)$

$$\left[0<b<a,\quad -\operatorname{Re}\nu-1<1+\operatorname{Re}\lambda<\tfrac{3}{2}\right]$$

$$= \left(\frac{1}{2}a\right)^\nu b^{-(\nu+\lambda+1)}\frac{\Gamma(\nu+\lambda+1)}{\Gamma(\nu+1)}\sin\left[\pi\left(\frac{1+\lambda+\nu}{2}\right)\right]$$

$$\times F\left(\frac{2+\lambda+\nu}{2},\frac{1+\lambda+\nu}{2};\nu+1;\frac{a^2}{b^2}\right)$$

$$\left[0<a<b,\quad -\operatorname{Re}\nu-1<1+\operatorname{Re}\lambda<\tfrac{3}{2}\right]$$

ET I 100(11)

2.     $\displaystyle\int_0^\infty x^\lambda\, J_\nu(ax)\cos(bx)\,\mathrm{d}x$

$$= \frac{2^\lambda a^{-(1+\lambda)}\,\Gamma\left(\frac{1+\lambda+\nu}{2}\right)}{\Gamma\left(\frac{\nu-\lambda+1}{2}\right)}\,F\left(\frac{1+\lambda+\nu}{2},\frac{1+\lambda-\nu}{2};\frac{1}{2};\frac{b^2}{a^2}\right)$$

$$\left[0<b<a,\quad -\operatorname{Re}\nu<1+\operatorname{Re}\lambda<\tfrac{3}{2}\right]$$

$$= \frac{\left(\frac{a}{2}\right)^\nu b^{-(\nu+1+\lambda)}\,\Gamma(1+\lambda+\nu)\cos\left[\frac{\pi}{2}(1+\lambda+\nu)\right]}{\Gamma(\nu+1)}\,F\left(\frac{1+\lambda+\nu}{2},\frac{2+\lambda+\nu}{2};\nu+1;\frac{a^2}{b^2}\right)$$

$$\left[0<a<b,\quad -\operatorname{Re}\nu<1+\operatorname{Re}\lambda<\tfrac{3}{2}\right]$$

ET I 45(13)

3.     $\displaystyle\int_0^\infty x^\lambda\, K_\mu(ax)\sin(bx)\,\mathrm{d}x = \frac{2^\lambda b\,\Gamma\left(\frac{2+\mu+\lambda}{2}\right)\Gamma\left(\frac{2+\lambda-\mu}{2}\right)}{a^{2+\lambda}}\,F\left(\frac{2+\mu+\lambda}{2},\frac{2+\lambda-\mu}{2};\frac{3}{2};-\frac{b^2}{a^2}\right)$

$$\left[\operatorname{Re}\left(-\lambda\pm\mu\right)<2,\quad \operatorname{Re}a>0,\quad b>0\right]$$

ET I 106(50)

4.     $\displaystyle\int_0^\infty x^\lambda\, K_\mu(ax)\cos(bx)\,\mathrm{d}x = 2^{\lambda-1}a^{-\lambda-1}\Gamma\left(\frac{\mu+\lambda+1}{2}\right)\Gamma\left(\frac{1+\lambda-\mu}{2}\right)$

$$\times F\left(\frac{\mu+\lambda+1}{2},\frac{1+\lambda-\mu}{2};\frac{1}{2};-\frac{b^2}{a^2}\right)$$

$$\left[\operatorname{Re}\left(-\lambda\pm\mu\right)<1,\quad \operatorname{Re}a>0,\quad b>0\right]\quad \text{ET I 49(42)}$$

5.     $\displaystyle\int_0^\infty x^\nu \sin(ax)\, J_\nu(bx)\,\mathrm{d}x = \frac{\sqrt{\pi}\,2^\nu b^\nu\left(a^2-b^2\right)^{-\nu-\frac{1}{2}}}{\Gamma\left(\frac{1}{2}-\nu\right)}$      $\left[0<b<a,\quad -1<\operatorname{Re}\nu<\tfrac{1}{2}\right]$

$$= 0 \qquad\qquad\qquad \left[0<a<b,\quad -1<\operatorname{Re}\nu<\tfrac{1}{2}\right]$$

ET II 32(4)

6.     $\displaystyle\int_0^\infty x^\nu \cos(ax)\, J_\nu(bx)\,\mathrm{d}x = -2^\nu\frac{\sin(\nu\pi)}{\sqrt{\pi}}\Gamma\left(\frac{1}{2}+\nu\right)b^\nu\left(a^2-b^2\right)^{-\nu-\frac{1}{2}}$    $\left[0<b<a,\quad |\operatorname{Re}\nu|<\tfrac{1}{2}\right]$

$$= 2^\nu\frac{b^\nu}{\sqrt{\pi}}\Gamma\left(\frac{1}{2}+\nu\right)\left(b^2-a^2\right)^{-\nu-\frac{1}{2}} \qquad \left[0<a<b,\quad |\operatorname{Re}\nu|<\tfrac{1}{2}\right]$$

ET II 36(29)

7.    $\displaystyle\int_0^\infty x^{\nu+1} \sin(ax) \, J_\nu(bx) \, dx$

$$= -2^{1+\nu} a \frac{\sin(\nu\pi)}{\sqrt{\pi}} b^\nu \, \Gamma\left(\nu + \frac{3}{2}\right) \left(a^2 - b^2\right)^{-\nu-\frac{3}{2}} \qquad \left[0 < b < a, \quad -\frac{3}{2} < \operatorname{Re}\nu < -\frac{1}{2}\right]$$

$$= -\frac{2^{1+\nu}}{\sqrt{\pi}} ab^\nu \, \Gamma\left(\nu + \frac{3}{2}\right) \left(b^2 - a^2\right)^{-\nu-\frac{3}{2}} \qquad \left[0 < a < b, \quad -\frac{3}{2} < \operatorname{Re}\nu < -\frac{1}{2}\right]$$

ET II 32(3)

8.    $\displaystyle\int_0^\infty x^{\nu+1} \cos(ax) \, J_\nu(bx) \, dx = 2^{1+\nu} \sqrt{\pi} ab^\nu \frac{\left(a^2 - b^2\right)^{-\nu-\frac{3}{2}}}{\Gamma\left(-\frac{1}{2} - \nu\right)} \qquad \left[0 < b < a, \quad -1 < \operatorname{Re}\nu < -\frac{1}{2}\right]$

$$= 0 \qquad \left[0 < a < b, \quad -1 < \operatorname{Re}\nu < -\frac{1}{2}\right]$$

ET II 36(28)

9.    $\displaystyle\int_0^1 x^\nu \sin(ax) \, J_\nu(ax) \, dx = \frac{1}{2\nu + 1} \left[\sin a J_\nu(a) - \cos a \, J_{\nu+1}(a)\right]$

$\left[\operatorname{Re}\nu > -1\right]$        ET II 334(9)a

10.    $\displaystyle\int_0^1 x^\nu \cos(ax) \, J_\nu(ax) \, dx = \frac{1}{2\nu + 1} \left[\cos a J_\nu(a) + \sin a \, J_{\nu+1}(a)\right]$

$\left[\operatorname{Re}\nu > -\frac{1}{2}\right]$        ET II 335(20)

11.    $\displaystyle\int_0^\infty x^{1+\nu} K_\nu(ax) \sin(bx) \, dx = \sqrt{\pi} (2a)^\nu \, \Gamma\left(\frac{3}{2} + \nu\right) b \left(b^2 + a^2\right)^{-\frac{3}{2} - \nu}$

$\left[\operatorname{Re} a > 0, \quad b > 0, \quad \operatorname{Re}\nu > -\frac{3}{2}\right]$<br>ET I 105(49)

12.    $\displaystyle\int_0^\infty x^\mu K_\mu(ax) \cos(bx) \, dx = \frac{1}{2} \sqrt{\pi} (2a)^\mu \, \Gamma\left(\mu + \frac{1}{2}\right) \left(b^2 + a^2\right)^{\mu \frac{1}{2}}$

$\left[\operatorname{Re} a > 0, \quad b > 0, \quad \operatorname{Re}\mu > -\frac{1}{2}\right]$<br>ET I 49(41)

13.    $\displaystyle\int_0^\infty x^\nu Y_{\nu-1}(ax) \sin(bx) \, dx = 0 \qquad \left[0 < b < a, \quad |\operatorname{Re}\nu| < \frac{1}{2}\right]$

$$= \frac{2^\nu \sqrt{\pi} a^{\nu-1} b}{\Gamma\left(\frac{1}{2} - \nu\right)} \left(b^2 - a^2\right)^{-\nu-\frac{1}{2}} \qquad \left[0 < a < b, \quad |\operatorname{Re}\nu| < \frac{1}{2}\right]$$

ET I 104(36)

14.    $\displaystyle\int_0^\infty x^\nu Y_\nu(ax) \cos(bx) \, dx = 0 \qquad \left[0 < b < a, \quad |\operatorname{Re}\nu| < \frac{1}{2}\right]$

$$= -2^\nu \sqrt{\pi} a^\nu \frac{\left(b^2 - a^2\right)^{-\nu-\frac{1}{2}}}{\Gamma\left(\frac{1}{2} - \nu\right)} \qquad \left[0 < a < b, \quad |\operatorname{Re}\nu| < \frac{1}{2}\right]$$

ET I 47(30)

**6.711**

1.    $\displaystyle\int_0^\infty x^{\nu-\mu} J_\mu(ax) \, J_\nu(bx) \sin(cx) \, dx = 0 \qquad \left[0 < c < b - a, \quad -1 < \operatorname{Re}\nu < 1 + \operatorname{Re}\mu\right]$

ET I 103(28)

2. $\int_0^\infty x^{\nu-\mu+1} J_\mu(ax) J_\nu(bx) \cos(cx)\,dx = 0$

$$[0 < c < b - a, \quad a > 0, \quad b > 0, \quad -1 < \operatorname{Re}\nu < \operatorname{Re}\mu] \quad \text{ET I 47(25)}$$

3. $\int_0^\infty x^{\nu-\mu-2} J_\mu(ax) J_\nu(bx) \sin(cx)\,dx = 2^{\nu-\mu-1} a^\mu b^{-\nu} \dfrac{c\,\Gamma(\nu)}{\Gamma(\mu+1)}$

$$[0 < a, \quad 0 < b, \quad 0 < c < b - a, \quad 0 < \operatorname{Re}\nu < \operatorname{Re}\mu + 3] \quad \text{ET I 103(29)}$$

4. $\int_0^\infty x^{\varrho-\mu-1} J_\mu(ax) J_\varrho(bx) \cos(cx)\,dx = 2^{\varrho-\mu-1} b^{-\varrho} a^\mu \dfrac{\Gamma(\varrho)}{\Gamma(\mu+1)}$

$$[b > 0, \quad a > 0, \quad 0 < c < b - a, \quad 0 < \operatorname{Re}\varrho < \operatorname{Re}\mu + 2] \quad \text{ET I 47(26)}$$

5. $\int_0^\infty x^{1-2\nu} \sin(2ax) J_\nu(x) Y_\nu(x)\,dx = -\dfrac{\Gamma\left(\frac{3}{2}-\nu\right)a}{2\,\Gamma\left(2\nu-\frac{1}{2}\right)\Gamma(2-\nu)} F\left(\dfrac{3}{2}-\nu, \dfrac{3}{2}-2\nu; 2-\nu; a^2\right)$

$$\left[0 < \operatorname{Re}\nu < \frac{3}{2}, \quad 0 < a < 1\right]$$
$$\text{ET II 348(63)}$$

6.[10] $\int_0^\infty \arg\sin(zx)x^{\nu-\mu-4} J_\mu(ax) J_\nu(\rho x)\,dx = z\dfrac{\Gamma(\nu)a^\mu \rho^{-\nu}}{2^{\mu-\nu+3}\Gamma(\mu+1)}\left[\dfrac{\rho^2}{\nu-1} - \dfrac{a^2}{\mu+1} - \dfrac{2z^2}{3}\right]$

7.[10] $\int_0^\infty \cos(zx)x^{\nu-\mu-3} J_\mu(ax) J_\nu(\rho x)\,dx = \dfrac{\Gamma(\nu)a^\mu \rho^{-\nu}}{2^{\mu-\nu+3}\Gamma(\mu+1)}\left[\dfrac{\rho^2}{\nu-1} - \dfrac{a^2}{\mu+1} - 2z^2\right]$

**6.712**

1. $\int_0^\infty x^\nu [J_\nu(ax)\cos(ax) + Y_\nu(ax)\sin(ax)]\sin(bx)\,dx = \dfrac{\sqrt{\pi}(2a)^\nu}{\Gamma\left(\frac{1}{2}-\nu\right)}\left(b^2 + 2ab\right)^{-\nu-\frac{1}{2}}$

$$\left[b > 0, \quad -1 < \operatorname{Re}\nu < \frac{1}{2}\right] \quad \text{ET I 104(40)}$$

2. $\int_0^\infty x^\nu [Y_\nu(ax)\cos(ax) - J_\nu(ax)\sin(ax)]\cos(bx)\,dx = -\dfrac{\sqrt{\pi}(2a)^\nu}{\Gamma\left(\frac{1}{2}-\nu\right)}\left(b^2 + 2ab\right)^{-\nu-\frac{1}{2}}$

$$\text{ET I 48(35)}$$

3. $\int_0^\infty x^\nu [J_\nu(ax)\cos(ax) - Y_\nu(ax)\sin(ax)]\sin(bx)\,dx$

$$= 0 \qquad \left[0 < b < 2a, \quad -1 < \operatorname{Re}\nu < \frac{1}{2}\right]$$

$$= \dfrac{2^\nu \sqrt{\pi} b^\nu}{\Gamma\left(\frac{1}{2}-\nu\right)}\left(b^2 - 2ab\right)^{-\nu-\frac{1}{2}} \qquad \left[2a < b, \quad -1 < \operatorname{Re}\nu < \frac{1}{2}\right]$$

$$\text{ET I 104(41)}$$

4. $\int_0^\infty x^\nu [J_\nu(ax)\sin(ax) + Y_\nu(ax)\cos(ax)]\cos(bx)\,dx$

$$= 0 \qquad \left[0 < b < 2a, \quad |\operatorname{Re}\nu| < \frac{1}{2}\right]$$

$$= -\dfrac{\sqrt{\pi}(2a)^\nu}{\Gamma\left(\frac{1}{2}-\nu\right)}\left(b^2 - 2ab\right)^{-\nu-\frac{1}{2}} \qquad \left[0 < 2a < b, \quad |\operatorname{Re}\nu| < \frac{1}{2}\right]$$

$$\text{ET I 48(33)}$$

**6.713**

1.    $\displaystyle\int_0^\infty x^{1-2\nu}\sin(2ax)\left\{[J_\nu(x)]^2-[Y_\nu(x)]^2\right\}dx$

$$=\frac{\sin(2\nu\pi)\,\Gamma\left(\frac{3}{2}-\nu\right)\Gamma\left(\frac{3}{2}-2\nu\right)a}{\pi\,\Gamma(2-\nu)}F\left(\frac{3}{2}-\nu,\frac{3}{2}-2\nu;2-\nu;a^2\right)$$

$$\left[0<\operatorname{Re}\nu<\tfrac{3}{4},\quad 0<a<1\right]\quad\text{ET II 348(64)}$$

2.    $\displaystyle\int_0^\infty x^{2-2\nu}\sin(2ax)\left[J_\nu(x)\,J_{\nu-1}(x)-Y_\nu(x)\,Y_{\nu-1}(x)\right]dx$

$$=-\frac{\sin(2\nu\pi)\,\Gamma\left(\frac{3}{2}-\nu\right)\Gamma\left(\frac{5}{2}-2\nu\right)a}{\pi\,\Gamma(2-\nu)}F\left(\frac{3}{2}-\nu,\frac{5}{2}-2\nu;2-\nu;a^2\right)$$

$$\left[\tfrac{1}{2}<\operatorname{Re}\nu<\tfrac{5}{4},\quad 0<a<1\right]\quad\text{ET II 348(65)}$$

3.    $\displaystyle\int_0^\infty x^{2-2\nu}\sin(2ax)\left[J_\nu(x)\,Y_{\nu-1}(x)+Y_\nu(x)\,J_{\nu-1}(x)\right]dx$

$$=-\frac{\Gamma\left(\frac{3}{2}-\nu\right)a}{\Gamma\left(2\nu-\frac{3}{2}\right)\Gamma(2-\nu)}F\left(\frac{3}{2}-\nu,\frac{5}{2}-2\nu;2-\nu;a^2\right)$$

$$\left[\tfrac{1}{2}<\operatorname{Re}\nu<\tfrac{5}{2},\quad 0<a<1\right]\quad\text{ET II 349(66)}$$

**6.714**

1.    $\displaystyle\int_0^\infty \sin(2ax)\left[x^\nu J_\nu(x)\right]^2 dx$

$$=\frac{a^{-2\nu}\,\Gamma\left(\frac{1}{2}+\nu\right)}{2\sqrt{\pi}\,\Gamma(1-\nu)}F\left(\frac{1}{2}+\nu,\frac{1}{2};1-\nu;a^2\right)\qquad\left[0<a<1,\quad|\operatorname{Re}\nu|<\tfrac{1}{2}\right]$$

$$=\frac{a^{-4\nu-1}\,\Gamma\left(\frac{1}{2}+\nu\right)}{2\,\Gamma(1+\nu)\,\Gamma\left(\frac{1}{2}-2\nu\right)}F\left(\frac{1}{2}+\nu,\frac{1}{2}+2\nu;1+\nu;\frac{1}{a^2}\right)\qquad\left[a>1,\quad|\operatorname{Re}\nu|<\tfrac{1}{2}\right]$$

$$\text{ET II 343(31)}$$

2.    $\displaystyle\int_0^\infty \cos(2ax)\left[x^\nu J_\nu(x)\right]^2 dx$

$$=\frac{a^{-2\nu}\,\Gamma(\nu)}{2\sqrt{\pi}\,\Gamma\left(\frac{1}{2}-\nu\right)}F\left(\nu+\frac{1}{2},\frac{1}{2};1-\nu;a^2\right)$$

$$+\frac{\Gamma(-\nu)\,\Gamma\left(\frac{1}{2}+2\nu\right)}{2\pi\,\Gamma\left(\frac{1}{2}-\nu\right)}F\left(\frac{1}{2}+\nu,\frac{1}{2}+2\nu;1+\nu;a^2\right)\qquad\left[0<a<1,\quad-\tfrac{1}{4}<\operatorname{Re}\nu<\tfrac{1}{2}\right]$$

$$=-\frac{\sin(\nu\pi)a^{-4\nu-1}\,\Gamma\left(\frac{1}{2}+2\nu\right)}{\Gamma(1+\nu)\,\Gamma\left(\frac{1}{2}-\nu\right)}F\left(\frac{1}{2}+\nu,\frac{1}{2}+2\nu;1+\nu;\frac{1}{a^2}\right)\qquad\left[a>1,\quad-\tfrac{1}{4}<\operatorname{Re}\nu<\tfrac{1}{2}\right]$$

$$\text{ET II 344(33)}$$

**6.715**

1.    $\displaystyle\int_0^\infty \frac{x^\nu}{x+b}\sin(x+b)\,J_\nu(x)\,dx=\frac{\pi}{2}\sec(\nu\pi)b^\nu\,J_{-\nu}(b)\qquad\left[|\arg b|<\pi,\quad|\operatorname{Re}\nu|<\tfrac{1}{2}\right]\quad\text{ET II 340(8)}$

2.    $\displaystyle\int_0^\infty \frac{x^\nu}{x+b}\cos(x+b)\,J_\nu(x)\,dx=-\frac{\pi}{2}\sec(\nu\pi)b^\nu\,Y_{-\nu}(b)$

$$\left[|\arg b|<\pi,\quad|\operatorname{Re}\nu|<\tfrac{1}{2}\right]\quad\text{ET II 340(9)}$$

**6.716**

1.    $$\int_0^a x^\lambda \sin(a-x)\, J_\nu(x)\, \mathrm{d}x = 2a^{\lambda+1} \sum_{n=0}^\infty \frac{(-1)^n \Gamma(\nu-\lambda+2n)\,\Gamma(\nu+\lambda+1)}{\Gamma(\nu-\lambda)\,\Gamma(\nu+\lambda+3+2n)}(\nu+2n+1)\, J_{\nu+2n+1}(a)$$

$$[\operatorname{Re}(\lambda+\nu) > -1] \qquad \text{ET II 335(16)}$$

2.    $$\int_0^a x^\lambda \cos(a-x)\, J_\nu(x)\, \mathrm{d}x = \frac{a^{\lambda+1} J_\nu(a)}{\lambda+\nu+1} + 2a^{\lambda+1}$$

$$\times \sum_{n=1}^\infty \frac{(-1)^n \Gamma(\nu-\lambda+2n-1)\,\Gamma(\nu+\lambda+1)}{\Gamma(\nu-\lambda)\,\Gamma(\nu+\lambda+2n+2)}(\nu+2n)\, J_{\nu+2n}(a)$$

$$[\operatorname{Re}(\lambda+\nu) > -1] \qquad \text{ET II 335(26)}$$

**6.717**    $$\int_{-\infty}^\infty \frac{\sin[a(x+b)]}{x^\nu(x+b)}\, J_{\nu+2n}(x)\, \mathrm{d}x = \pi b^{-\nu} J_{\nu+2n}(b)$$

$$\left[1 \le a < \infty, \quad n = 0,1,2,\ldots; \quad \operatorname{Re}\nu > -\tfrac{3}{2}\right] \qquad \text{ET II 345(44)}$$

**6.718**

1.    $$\int_0^\infty \frac{x^\nu}{x^2+b^2} \sin(ax)\, J_\nu(cx)\, \mathrm{d}x = b^{\nu-1} \sinh(ab)\, K_\nu(bc)$$

$$\left[0 < a \le c, \quad \operatorname{Re}b > 0, \quad -1 < \operatorname{Re}\nu < \tfrac{3}{2}\right] \qquad \text{ET II 33(8)}$$

2.    $$\int_0^\infty \frac{x^{\nu+1}}{x^2+b^2} \cos(ax)\, J_\nu(cx)\, \mathrm{d}x = b^\nu \cosh(ab)\, K_\nu(bc)$$

$$\left[0 < a \le c, \quad \operatorname{Re}b > 0, \quad -1 < \operatorname{Re}\nu < \tfrac{1}{2}\right] \qquad \text{ET II 37(33)}$$

3.    $$\int_0^\infty \frac{x^{1-\nu}}{x^2+b^2} \sin(ax)\, J_\nu(cx)\, \mathrm{d}x = \frac{\pi}{2} b^{-\nu} e^{-ab}\, I_\nu(bc) \qquad \left[0 < c \le a, \quad \operatorname{Re}b > 0, \quad \operatorname{Re}\nu > -\tfrac{1}{2}\right]$$

$$\text{ET II 33(9)}$$

4.    $$\int_0^\infty \frac{x^{-\nu}}{x^2+b^2} \cos(ax)\, J_\nu(cx)\, \mathrm{d}x = \frac{\pi}{2} b^{-\nu-1} e^{-ab}\, I_\nu(bc) \qquad \left[0 < c \le a, \quad \operatorname{Re}b > 0, \quad \operatorname{Re}\nu > -\tfrac{3}{2}\right]$$

$$\text{ET II 37(34)}$$

**6.719**

1.[6]    $$\int_0^a \frac{\sin(bx)}{\sqrt{a^2-x^2}}\, J_\nu(x)\, \mathrm{d}x = \pi \sum_{n=0}^\infty (-1)^n\, J_{2n+1}(ab)\, J_{\frac{1}{2}\nu+n+\frac{1}{2}}\left(\tfrac{1}{2}a\right) J_{\frac{1}{2}\nu-n-\frac{1}{2}}\left(\tfrac{1}{2}a\right)$$

$$[\operatorname{Re}\nu > -2] \qquad \text{ET II 335(17)}$$

2.    $$\int_0^a \frac{\cos(bx)}{\sqrt{a^2-x^2}}\, J_\nu(x)\, \mathrm{d}x = \frac{\pi}{2} J_0(ab)\left[J_{\frac{1}{2}\nu}\left(\tfrac{1}{2}a\right)\right]^2 + \pi \sum_{n=1}^\infty (-1)^n\, J_{2n}(ab)\, J_{\frac{1}{2}\nu+n}\left(\tfrac{1}{2}a\right) J_{\frac{1}{2}\nu-n}\left(\tfrac{1}{2}a\right)$$

$$[\operatorname{Re}\nu > -1] \qquad \text{ET II 336(27)}$$

**6.721**

1.    $$\int_0^\infty \sqrt{x}\, J_{\frac{1}{4}}\left(a^2 x^2\right) \sin(bx)\, \mathrm{d}x = 2^{-3/2} a^{-2} \sqrt{\pi b}\, J_{\frac{1}{4}}\left(\frac{b^2}{4a^2}\right)$$

$$[b > 0] \qquad \text{ET I 108(1)}$$

2.    $\displaystyle\int_0^\infty \sqrt{x}\, J_{-\frac{1}{4}}\left(a^2 x^2\right)\cos(bx)\,\mathrm{d}x = 2^{-3/2} a^{-2}\sqrt{\pi b}\, J_{-\frac{1}{4}}\left(\frac{b^2}{4a^2}\right)$

$$[b>0]$$           ET I 51(1)

3.    $\displaystyle\int_0^\infty \sqrt{x}\, Y_{\frac{1}{4}}\left(a^2 x^2\right)\sin(bx)\,\mathrm{d}x = -2^{-3/2}\sqrt{\pi b}\, a^{-2}\, \mathbf{H}_{\frac{1}{4}}\left(\frac{b^2}{4a^2}\right)$        ET I 108(7)

4.    $\displaystyle\int_0^\infty \sqrt{x}\, Y_{-\frac{1}{4}}\left(a^2 x^2\right)\cos(bx)\,\mathrm{d}x = -2^{-3/2}\sqrt{\pi b}\, a^{-2}\, \mathbf{H}_{-\frac{1}{4}}\left(\frac{b^2}{4a^2}\right)$        ET I 52(7)

5.    $\displaystyle\int_0^\infty \sqrt{x}\, K_{\frac{1}{4}}\left(a^2 x^2\right)\sin(bx)\,\mathrm{d}x = 2^{-5/2}\sqrt{\pi^3 b}\, a^{-2}\left[I_{\frac{1}{4}}\left(\frac{b^2}{4a^2}\right) - \mathbf{L}_{\frac{1}{4}}\left(\frac{b^2}{4a^2}\right)\right]$

$$\left[|\arg a| < \frac{\pi}{4}, \quad b>0\right]$$        ET I 109(11)

6.    $\displaystyle\int_0^\infty \sqrt{x}\, K_{-\frac{1}{4}}\left(a^2 x^2\right)\cos(bx)\,\mathrm{d}x = 2^{-5/2}\sqrt{\pi^3 b}\, a^{-2}\left[I_{-\frac{1}{4}}\left(\frac{b^2}{4a^2}\right) - \mathbf{L}_{-\frac{1}{4}}\left(\frac{b^2}{4a^2}\right)\right]$

$$[b>0]$$           ET I 52(10)

**6.722**

1.    $\displaystyle\int_0^\infty \sqrt{x}\, K_{\frac{1}{8}+\nu}\left(a^2 x^2\right) I_{\frac{1}{8}-\nu}\left(a^2 x^2\right)\sin(bx)\,\mathrm{d}x = \sqrt{2\pi}\, b^{-3/2}\frac{\Gamma\left(\frac{5}{8}-\nu\right)}{\Gamma\left(\frac{5}{4}\right)}\, W_{\nu,\frac{1}{8}}\left(\frac{b^2}{8a^2}\right) M_{-\nu,\frac{1}{8}}\left(\frac{b^2}{8a^2}\right)$

$$\left[\operatorname{Re}\nu < \frac{5}{8}, \quad |\arg a| < \frac{\pi}{4}, \quad b>0\right]$$
$$\text{ET I 109(13)}$$

2.[12]    $\displaystyle\int_0^\infty \sqrt{x}\, J_{-\frac{1}{8}-\nu}\left(a^2 x^2\right) J_{-\frac{1}{8}+\nu}\left(a^2 x^2\right)\cos(bx)\,\mathrm{d}x$

$$= \sqrt{\frac{2}{b^3\pi}}\left[e^{i\pi/8} W_{nu,-1/8}\left(\frac{b^2 e^{-\pi i/2}}{8a^2}\right) W_{-nu,-1/8}\left(\frac{b^2 e^{-\pi i/2}}{8a^2}\right)\right.$$
$$\left. + e^{-i\pi/8} W_{nu,-1/8}\left(\frac{b^2 e^{\pi i/2}}{8a^2}\right) W_{-nu,-1/8}\left(\frac{b^2 e^{\pi i/2}}{8a^2}\right)\right]$$
$$[a^2>0, \quad \operatorname{Im} b = 0]$$        MC

3.[12]    $\displaystyle\int_0^\infty \sqrt{x}\, J_{\frac{1}{8}-\nu}\left(a^2 x^2\right) J_{\frac{1}{8}+\nu}\left(a^2 x^2\right)\sin(bx)\,\mathrm{d}x$

$$= \sqrt{\frac{2}{\pi}}\, b^{-3/2}\left[e^{\pi i/8}\, W_{\nu,\frac{1}{8}}\left(\frac{b^2 e^{\pi i/2}}{8a^2}\right) W_{-\nu,\frac{1}{8}}\left(\frac{b^2 e^{\pi i/2}}{8a^2}\right)\right.$$
$$\left. + e^{-i\pi/8}\, W_{\nu,\frac{1}{8}}\left(\frac{b^2 e^{-\pi i/2}}{8a^2}\right) W_{-\nu,\frac{1}{8}}\left(\frac{b^2 e^{-\frac{\pi i}{2}}}{8a^2}\right)\right]$$
$$[a^2>0, \quad b>0]$$        ET I 108(6)

4.[12]    $\displaystyle\int_0^\infty \sqrt{x}\, K_{\frac{1}{8}-\nu}\left(a^2 x^2\right) I_{-\frac{1}{8}-\nu}\left(a^2 x^2\right)\cos(bx)\,\mathrm{d}x$

$$= \sqrt{2\pi}\, b^{-3/2}\frac{\Gamma\left(\frac{3}{8}-\nu\right)}{\Gamma\left(\frac{3}{4}\right)}\, W_{\nu,-\frac{1}{8}}\left(\frac{b^2}{8a^2}\right) M_{-\nu,-\frac{1}{8}}\left(\frac{b^2}{8a^2}\right)$$
$$\left[\operatorname{Re}\nu < \frac{3}{8}, \quad |\arg a| < \frac{\pi}{4}, \quad b>0\right] \text{ ET I 52(12)}$$

**6.723** $\quad \int_0^\infty x J_\nu\left(x^2\right)\left[\sin(\nu\pi) J_\nu\left(x^2\right) - \cos(\nu\pi) Y_\nu\left(x^2\right)\right] J_{4\nu}(4ax)\,dx = \frac{1}{4} J_\nu\left(a^2\right) J_{-\nu}\left(a^2\right)$

$$[a > 0, \quad \operatorname{Re}\nu > -1] \qquad \text{ET II 375(20)}$$

**6.724**

1. $\quad \int_0^\infty x^{2\lambda} J_{2\nu}\left(\frac{a}{x}\right) \sin(bx)\,dx$

$$= \frac{\sqrt{\pi} a^{2\nu}\, \Gamma(\lambda - \nu + 1) b^{2\nu - 2\lambda - 1}}{4^{2\nu - \lambda}\, \Gamma(2\nu + 1)\, \Gamma\left(\nu - \lambda + \frac{1}{2}\right)}\, {}_0F_3\left(2\nu + 1, \nu - \lambda, \nu - \lambda + \frac{1}{2}; \frac{a^2 b^2}{16}\right)$$

$$+ \frac{a^{2\lambda + 2}\, \Gamma(\nu - \lambda - 1) b}{2^{2\lambda + 3}\, \Gamma(\nu + \lambda + 2)}\, {}_0F_3\left(\frac{3}{2}, \lambda - \nu + 2, \lambda + \nu + 2; \frac{a^2 b^2}{16}\right)$$

$$\left[-\tfrac{5}{4} < \operatorname{Re}\lambda < \operatorname{Re}\nu, \quad a > 0, \quad b > 0\right] \qquad \text{ET I 109(15)}$$

2. $\quad \int_0^\infty x^{2\lambda} J_{2\nu}\left(\frac{a}{x}\right) \cos(bx)\,dx$

$$= 4^{\lambda - 2\nu} \sqrt{\pi} a^{2\nu} b^{2\nu - 2\lambda - 1} \frac{\Gamma\left(\lambda - \nu + \frac{1}{2}\right)}{\Gamma(2\nu + 1)\, \Gamma(\nu - \lambda)}\, {}_0F_3\left(2\nu + 1, \nu - \lambda + \frac{1}{2}, \nu - \lambda; \frac{a^2 b^2}{16}\right)$$

$$+ 4^{-\lambda - 1} a^{2\lambda + 1} \frac{\Gamma\left(\nu - \lambda - \frac{1}{2}\right)}{\Gamma\left(\nu + \lambda + \frac{3}{2}\right)}\, {}_0F_3\left(\frac{1}{2}, \lambda - \nu + \frac{3}{2}, \nu + \lambda + \frac{3}{2}; \frac{a^2 b^2}{16}\right)$$

$$\left[-\tfrac{3}{4} < \operatorname{Re}\lambda < \operatorname{Re}\nu - \tfrac{1}{2}, \quad a > 0, \quad b > 0\right] \qquad \text{ET I 53(14)}$$

**6.725**

1. $\quad \int_0^\infty \frac{\sin(bx)}{\sqrt{x}} J_\nu\left(a\sqrt{x}\right) dx = -\sqrt{\frac{\pi}{b}} \sin\left(\frac{a^2}{8b} - \frac{\nu\pi}{4} - \frac{\pi}{4}\right) J_{\frac{\nu}{2}}\left(\frac{a^2}{8b}\right)$

$$[\operatorname{Re}\nu > -3, \quad a > 0, \quad b > 0]$$

$$\text{ET I 110(27)}$$

2. $\quad \int_0^\infty \frac{\cos(bx)}{\sqrt{x}} J_\nu\left(a\sqrt{x}\right) dx = \sqrt{\frac{\pi}{b}} \cos\left(\frac{a^2}{8b} - \frac{\nu\pi}{4} - \frac{\pi}{4}\right) J_{\frac{1}{2}\nu}\left(\frac{a^2}{8b}\right)$

$$[\operatorname{Re}\nu > -1, \quad a > 0, \quad b > 0]$$

$$\text{ET I 54(25)}$$

3. $\quad \int_0^\infty x^{\frac{1}{2}\nu} J_\nu\left(a\sqrt{x}\right) \sin(bx)\,dx = 2^{-\nu} a^\nu b^{-\nu - 1} \cos\left(\frac{a^2}{4b} - \frac{\nu\pi}{2}\right)$

$$\left[-2 < \operatorname{Re}\nu < \tfrac{1}{2}, \quad a > 0, \quad b > 0\right]$$

$$\text{ET I 110(28)}$$

4. $\quad \int_0^\infty x^{\frac{1}{2}\nu} J_\nu\left(a\sqrt{x}\right) \cos(bx)\,dx = 2^{-\nu} b^{-\nu - 1} a^\nu \sin\left(\frac{a^2}{4b} - \frac{\nu\pi}{2}\right)$

$$\left[-1 < \operatorname{Re}\nu < \tfrac{1}{2}, \quad a > 0, \quad b > 0\right]$$

$$\text{ET I 54(26)}$$

**6.726**

1. $$\int_0^\infty x \left(x^2 + b^2\right)^{-\frac{1}{2}\nu} J_\nu \left(a\sqrt{x^2 + b^2}\right) \sin(cx)\, dx$$

$$= \sqrt{\frac{\pi}{2}} a^{-\nu} b^{-\nu+\frac{3}{2}} c \left(a^2 - c^2\right)^{\frac{1}{2}\nu-\frac{3}{4}} J_{\nu-\frac{3}{2}} \left(b\sqrt{a^2 - c^2}\right) \qquad \left[0 < c < a, \quad \operatorname{Re}\nu > \tfrac{1}{2}\right]$$

$$= 0 \qquad \left[0 < a < c, \quad \operatorname{Re}\nu > \tfrac{1}{2}\right]$$

                                                   ET I 111(37)

2. $$\int_0^\infty \left(x^2 + b^2\right)^{-\frac{1}{2}\nu} J_\nu \left(a\sqrt{x^2 + b^2}\right) \cos(cx)\, dx$$

$$= \sqrt{\frac{\pi}{2}} a^{-\nu} b^{-\nu+\frac{1}{2}} \left(a^2 - c^2\right)^{\frac{1}{2}\nu-\frac{1}{4}} J_{\nu-\frac{1}{2}} \left(b\sqrt{a^2 - c^2}\right) \qquad \left[0 < c < a, \quad b > 0, \quad \operatorname{Re}\nu > -\tfrac{1}{2}\right]$$

$$= 0 \qquad \left[0 < a < c, \quad b > 0, \quad \operatorname{Re}\nu > -\tfrac{1}{2}\right]$$

                                                   ET I 55(37)

3. $$\int_0^\infty x \left(x^2 + b^2\right)^{\frac{1}{2}\nu} K_{\pm\nu} \left(a\sqrt{x^2 + b^2}\right) \sin(cx)\, dx$$

$$= \sqrt{\frac{\pi}{2}} a^\nu b^{\nu+\frac{3}{2}} c \left(a^2 + c^2\right)^{-\frac{1}{2}\nu-\frac{3}{4}} K_{-\nu-\frac{3}{2}} \left(b\sqrt{a^2 + c^2}\right)$$

$$\left[\operatorname{Re} a > 0, \quad \operatorname{Re} b > 0, \quad c > 0\right] \quad \text{ET I 113(45)}$$

4.[11] $$\int_0^\infty \left(x^2 + b^2\right)^{\mp\frac{1}{2}\nu} K_\nu \left(a\sqrt{x^2 + b^2}\right) \cos(cx)\, dx$$

$$= \sqrt{\frac{\pi}{2}} a^{\mp\nu} b^{\frac{1}{2}\mp\nu} \left(a^2 + c^2\right)^{\pm\frac{1}{2}\nu-\frac{1}{4}} K_{\pm\nu-\frac{1}{2}} \left(b\sqrt{a^2 + c^2}\right)$$

$$\left[\operatorname{Re} a > 0, \quad \operatorname{Re} b > 0, \quad c \text{ is real}\right] \quad \text{ET I 56(45)}$$

5. $$\int_0^\infty \left(x^2 + a^2\right)^{-\frac{1}{2}\nu} Y_\nu \left(b\sqrt{x^2 + a^2}\right) \cos(cx)\, dx$$

$$= \sqrt{\frac{a\pi}{2}} (ab)^{-\nu} \left(b^2 - c^2\right)^{\frac{1}{2}\nu-\frac{1}{4}} Y_{\nu-\frac{1}{2}} \left(a\sqrt{b^2 - c^2}\right) \qquad \left[0 < c < b, \quad a > 0, \quad \operatorname{Re}\nu > -\tfrac{1}{2}\right]$$

$$= -\sqrt{\frac{2a}{\pi}} (ab)^{-\nu} \left(c^2 - b^2\right)^{\frac{1}{2}\nu-\frac{1}{4}} K_{\nu-\frac{1}{2}} \left(a\sqrt{c^2 - b^2}\right) \qquad \left[0 < b < c, \quad a > 0, \quad \operatorname{Re}\nu > -\tfrac{1}{2}\right]$$

                                                   ET I 56(41)

**6.727**

1.[9] $$\int_0^a \frac{\cos(cx)}{\sqrt{a^2 - x^2}} J_\nu \left(b\sqrt{a^2 - x^2}\right) dx = \frac{\pi}{2} J_{\frac{1}{2}\nu}\left[\frac{a}{2}\left(\sqrt{b^2 + c^2} - c\right)\right] J_{\frac{1}{2}\nu}\left[\frac{a}{2}\left(\sqrt{b^2 + c^2} + c\right)\right]$$

$$\left[\operatorname{Re}\nu > -1, \quad c > 0, \quad a > 0\right]$$

                                                   ET I 113(48)

2.[12] $$\int_a^\infty \frac{\sin(cx)}{\sqrt{x^2 - a^2}} J_\nu \left(b\sqrt{x^2 - a^2}\right) dx = \frac{\pi}{2} J_{\frac{1}{2}\nu}\left[\frac{a}{2}\left(c - \sqrt{c^2 - b^2}\right)\right] J_{-\frac{1}{2}\nu}\left[\frac{a}{2}\left(c + \sqrt{c^2 - b^2}\right)\right]$$

$$\left[0 < b < c, \quad a > 0, \quad \operatorname{Re}\nu > -1\right]$$

                                                   ET I 113(49)

3. $$\int_a^\infty \frac{\cos(cx)}{\sqrt{x^2 - a^2}} J_\nu \left(b\sqrt{x^2 - a^2}\right) dx = -\frac{\pi}{2} J_{\frac{1}{2}\nu}\left[\frac{a}{2}\left(c - \sqrt{c^2 - b^2}\right)\right] Y_{-\frac{1}{2}\nu}\left[\frac{a}{2}\left(c + \sqrt{c^2 - b^2}\right)\right]$$

$$\left[0 < b < c, \quad a > 0, \quad \operatorname{Re}\nu > -1\right]$$

                                                   ET I 58(54)

4.8    $\displaystyle\int_0^a \left(a^2 - x^2\right)^{\frac{1}{2}\nu} \cos x\, I_\nu\left(\sqrt{a^2 - x^2}\right) dx = \frac{\sqrt{\pi}\, a^{2\nu+1}}{2^{\nu+1}\,\Gamma\left(\nu + \frac{3}{2}\right)}$

$$\left[\operatorname{Re}\nu > -\tfrac{1}{2}\right] \qquad \text{WA 409(2)}$$

**6.728**

1.    $\displaystyle\int_0^\infty x \sin\left(ax^2\right) J_\nu(bx)\, dx$

$$= \frac{\sqrt{\pi}\, b}{8a^{3/2}} \left[\cos\left(\frac{b^2}{8a} - \frac{\nu\pi}{4}\right) J_{\frac{1}{2}\nu - \frac{1}{2}}\left(\frac{b^2}{8a}\right) - \sin\left(\frac{b^2}{8a} - \frac{\nu\pi}{4}\right) J_{\frac{1}{2}\nu + \frac{1}{2}}\left(\frac{b^2}{8a}\right)\right]$$

$$\left[a > 0, \quad b > 0, \quad \operatorname{Re}\nu > -4\right] \quad \text{ET II 34(14)}$$

2.    $\displaystyle\int_0^\infty x \cos\left(ax^2\right) J_\nu(bx)\, dx$

$$= \frac{\sqrt{\pi}\, b}{8a^{3/2}} \left[\cos\left(\frac{b^2}{8a} - \frac{\nu\pi}{4}\right) J_{\frac{1}{2}\nu + \frac{1}{2}}\left(\frac{b^2}{8a}\right) + \sin\left(\frac{b^2}{8a} - \frac{\nu\pi}{4}\right) J_{\frac{1}{2}\nu - \frac{1}{2}}\left(\frac{b^2}{8a}\right)\right]$$

$$\left[a > 0, \quad b > 0, \quad \operatorname{Re}\nu > -2\right] \quad \text{ET II 38(39)}$$

3.    $\displaystyle\int_0^\infty J_0(bx) \sin\left(ax^2\right) x\, dx = \frac{1}{2a} \cos\frac{b^2}{4a}$        $\left[a > 0, \quad b > 0\right]$       MO 47

4.    $\displaystyle\int_0^\infty J_0(bx) \cos\left(ax^2\right) x\, dx = \frac{1}{2a} \sin\frac{b^2}{4a}$        $\left[a > 0, \quad b > 0\right]$       MO 47

5.    $\displaystyle\int_0^\infty x^{\nu+1} \sin\left(ax^2\right) J_\nu(bx)\, dx = \frac{b^\nu}{2^{\nu+1} a^{\nu+1}} \cos\left(\frac{b^2}{4a} - \frac{\nu\pi}{2}\right)$

$$\left[a > 0, \quad b > 0, \quad -2 < \operatorname{Re}\nu < \tfrac{1}{2}\right]$$
$$\text{ET II 34(15)}$$

6.    $\displaystyle\int_0^\infty x^{\nu+1} \cos\left(ax^2\right) J_\nu(bx)\, dx = \frac{b^\nu}{2^{\nu+1} a^{\nu+1}} \sin\left(\frac{b^2}{4a} - \frac{\nu\pi}{2}\right)$

$$\left[a > 0, \quad b > 0, \quad -1 < \operatorname{Re}\nu < \tfrac{1}{2}\right]$$
$$\text{ET II 38(40)}$$

**6.729**

1.    $\displaystyle\int_0^\infty x \sin\left(ax^2\right) J_\nu(bx) J_\nu(cx)\, dx = \frac{1}{2a} \cos\left(\frac{b^2 + c^2}{4a} - \frac{\nu\pi}{2}\right) J_\nu\left(\frac{bc}{2a}\right)$

$$\left[a > 0, \quad b > 0, \quad c > 0, \quad \operatorname{Re}\nu > -2\right]$$
$$\text{ET II 51(26)}$$

2.    $\displaystyle\int_0^\infty x \cos\left(ax^2\right) J_\nu(bx) J_\nu(cx)\, dx = \frac{1}{2a} \sin\left(\frac{b^2 + c^2}{4a} - \frac{\nu\pi}{2}\right) J_\nu\left(\frac{bc}{2a}\right)$

$$\left[a > 0, \quad b > 0, \quad c > 0, \quad \operatorname{Re}\nu > -1\right]$$
$$\text{ET II 51(27)}$$

**6.731**

1.[11] $\quad \int_0^\infty x \sin\left(ax^2\right) J_\nu\left(bx^2\right) J_{2\nu}(2cx)\,dx$

$$= \frac{1}{2\sqrt{b^2 - a^2}} \sin\left(\frac{ac^2}{b^2 - a^2}\right) J_\nu\left(\frac{bc^2}{b^2 - a^2}\right) \qquad [0 < a < b, \quad \operatorname{Re}\nu > -1]$$

$$= \frac{1}{2\sqrt{a^2 - b^2}} \cos\left(\frac{ac^2}{a^2 - b^2}\right) J_\nu\left(\frac{bc^2}{a^2 - b^2}\right) \qquad [0 < b < a, \quad \operatorname{Re}\nu > -1]$$

ET II 356(41)a

2.[10] $\quad \int_0^\infty x \cos\left(ax^2\right) J_\nu\left(bx^2\right) J_{2\nu}(2cx)\,dx$

$$= \frac{1}{2\sqrt{b^2 - a^2}} \cos\left(\frac{ac^2}{b^2 - a^2}\right) J_\nu\left(\frac{bc^2}{b^2 - a^2}\right) \qquad \left[0 < a < b, \quad \operatorname{Re}\nu > -\tfrac{1}{2}\right]$$

$$= \frac{1}{2\sqrt{a^2 - b^2}} \sin\left(\frac{ac^2}{a^2 - b^2}\right) J_\nu\left(\frac{bc^2}{a^2 - b^2}\right) \qquad \left[0 < b < a, \quad \operatorname{Re}\nu > -\tfrac{1}{2}\right]$$

ET II 356(42)a

**6.732**[12] $\quad \int_0^\infty x^3 \cos\left(\dfrac{x^2}{2a}\right) Y_1(x) K_1(x)\,dx = -a^3 K_0(a) \qquad [a > 0] \qquad$ ET II 371(52)

**6.733**

1. $\quad \int_0^\infty \sin\left(\dfrac{a}{2x}\right) [\sin x\, J_0(x) + \cos x\, Y_0(x)] \dfrac{dx}{x} = \pi J_0\left(\sqrt{a}\right) Y_0\left(\sqrt{a}\right)$

$[a > 0] \qquad$ ET II 346(51)

2. $\quad \int_0^\infty \cos\left(\dfrac{a}{2x}\right) [\sin x\, Y_0(x) - \cos x\, J_0(x)] \dfrac{dx}{x} = \pi J_0\left(\sqrt{a}\right) Y_0\left(\sqrt{a}\right)$

$[a > 0] \qquad$ ET II 347(52)

3. $\quad \int_0^\infty x \sin\left(\dfrac{a}{2x}\right) K_0(x)\,dx = \dfrac{\pi a}{2} J_1\left(\sqrt{a}\right) K_1\left(\sqrt{a}\right) \qquad [a > 0] \qquad$ ET II 368(34)

4. $\quad \int_0^\infty x \cos\left(\dfrac{a}{2x}\right) K_0(x)\,dx = -\dfrac{\pi a}{2} Y_1\left(\sqrt{a}\right) K_1\left(\sqrt{a}\right) \qquad [a > 0] \qquad$ ET II 369(35)

**6.734** $\quad \int_0^\infty \cos\left(a\sqrt{x}\right) K_\nu(bx) \dfrac{dx}{\sqrt{x}}$

$$= \frac{\pi}{2\sqrt{b}} \sec(\nu\pi) \left[ D_{\nu-\frac{1}{2}}\left(\frac{a}{\sqrt{2b}}\right) D_{-\nu-\frac{1}{2}}\left(-\frac{a}{\sqrt{2b}}\right) + D_{\nu-\frac{1}{2}}\left(-\frac{a}{\sqrt{2b}}\right) D_{-\nu-\frac{1}{2}}\left(\frac{a}{\sqrt{2b}}\right) \right]$$

$\left[\operatorname{Re} b > 0, \quad |\operatorname{Re}\nu| < \tfrac{1}{2}\right] \qquad$ ET II 132(27)

**6.735**

1. $\quad \int_0^\infty x^{1/4} \sin\left(2a\sqrt{x}\right) J_{-\frac{1}{4}}(x)\,dx = \sqrt{\pi} a^{3/2} J_{\frac{3}{4}}\left(a^2\right) \qquad [a > 0] \qquad$ ET II 341(10)

2. $\quad \int_0^\infty x^{1/4} \cos\left(2a\sqrt{x}\right) J_{\frac{1}{4}}(x)\,dx = \sqrt{\pi} a^{3/2} J_{-\frac{3}{4}}\left(a^2\right) \qquad [a > 0] \qquad$ ET II 341(12)

3. $\quad \int_0^\infty x^{1/4} \sin\left(2a\sqrt{x}\right) J_{\frac{3}{4}}(x)\,dx = \sqrt{\pi} a^{3/2} J_{-\frac{1}{4}}\left(a^2\right) \qquad [a > 0] \qquad$ ET II 341(11)

4. $\int_0^\infty x^{1/4} \cos\left(2a\sqrt{x}\right) J_{-\frac{3}{4}}(x)\,dx = \sqrt{\pi}a^{3/2} J_{\frac{1}{4}}\left(a^2\right)$ $\qquad [a>0]$ $\qquad$ ET II 341(13)

**6.736**

1.[11] $\int_0^\infty x^{-1/2} \sin x \cos\left(4a\sqrt{x}\right) J_0(x)\,dx = -2^{-3/2}\sqrt{\pi}\left[\cos\left(a^2-\frac{\pi}{4}\right) J_0\left(a^2\right) - \sin\left(a^2-\frac{\pi}{4}\right) Y_0\left(a^2\right)\right]$

$\qquad\qquad [a>0]$ $\qquad\qquad$ ET II 341(18)

2. $\int_0^\infty x^{-1/2} \cos x \cos\left(4a\sqrt{x}\right) J_0(x)\,dx = -2^{-3/2}\sqrt{\pi}\left[\sin\left(a^2-\frac{\pi}{4}\right) J_0\left(a^2\right) + \cos\left(a^2-\frac{\pi}{4}\right) Y_0\left(a^2\right)\right]$

$\qquad\qquad [a>0]$ $\qquad\qquad$ ET II 342(22)

3. $\int_0^\infty x^{-1/2} \sin x \sin\left(4a\sqrt{x}\right) J_0(x)\,dx = \sqrt{\frac{\pi}{2}} \cos\left(a^2+\frac{\pi}{4}\right) J_0\left(a^2\right)$

$\qquad\qquad [a>0]$ $\qquad\qquad$ ET II 341(16)

4. $\int_0^\infty x^{-1/2} \cos x \sin\left(4a\sqrt{x}\right) J_0(x)\,dx = \sqrt{\frac{\pi}{2}} \cos\left(a^2-\frac{\pi}{4}\right) J_0\left(a^2\right)$

$\qquad\qquad [a>0]$ $\qquad\qquad$ ET II 342(20)

5. $\int_0^\infty x^{-1/2} \sin x \cos\left(4a\sqrt{x}\right) Y_0(x)\,dx = 2^{-3/2}\sqrt{\pi}\left[3\sin\left(a^2-\frac{\pi}{4}\right) J_0\left(a^2\right) - \cos\left(a^2-\frac{\pi}{4}\right) Y_0\left(a^2\right)\right]$

$\qquad\qquad [a>0]$ $\qquad\qquad$ ET II 347(55)

6. $\int_0^\infty x^{-1/2} \cos x \cos\left(4a\sqrt{x}\right) Y_0(x)\,dx$

$\qquad\qquad = -2^{-3/2}\sqrt{\pi}\left[3\cos\left(a^2-\frac{\pi}{4}\right) J_0\left(a^2\right) + \sin\left(a^2-\frac{\pi}{4}\right) Y_0\left(a^2\right)\right]$

$\qquad\qquad [a>0]$ $\qquad\qquad$ ET II 347(56)

**6.737**

1. $\int_0^\infty \frac{\sin\left(a\sqrt{x^2+b^2}\right)}{\sqrt{x^2+b^2}} J_\nu(cx)\,dx = \frac{\pi}{2} J_{\frac{1}{2}\nu}\left[\frac{b}{2}\left(a-\sqrt{a^2-c^2}\right)\right] J_{-\frac{1}{2}\nu}\left[\frac{b}{2}\left(a+\sqrt{a^2-c^2}\right)\right]$

$\qquad [a>0, \quad \mathrm{Re}\,b>0, \quad c>0, \quad a>c, \quad \mathrm{Re}\,\nu>-1]$ $\quad$ ET II 35(19)

2. $\int_0^\infty \frac{\cos\left(a\sqrt{x^2+b^2}\right)}{\sqrt{x^2+b^2}} J_\nu(cx)\,dx = -\frac{\pi}{2} J_{\frac{1}{2}\nu}\left[\frac{b}{2}\left(a-\sqrt{a^2-c^2}\right)\right] Y_{-\frac{1}{2}\nu}\left[\frac{b}{2}\left(a+\sqrt{a^2-c^2}\right)\right]$

$\qquad [a>0, \quad \mathrm{Re}\,b>0, \quad c>0, \quad a>c, \quad \mathrm{Re}\,\nu>-1]$ $\quad$ ET II 39(44)

3. $\int_0^a \frac{\cos\left(b\sqrt{a^2-x^2}\right)}{\sqrt{a^2-x^2}} J_\nu(cx)\,dx = \frac{\pi}{2} J_{\frac{1}{2}\nu}\left[\frac{a}{2}\left(\sqrt{b^2+c^2}-b\right)\right] J_{\frac{1}{2}\nu}\left[\frac{a}{2}\left(\sqrt{b^2+c^2}+b\right)\right]$

$\qquad\qquad [c>0, \quad \mathrm{Re}\,\nu>-1]$ $\qquad\qquad$ ET II 39(47)

4. $\int_0^a x^{\nu+1} \frac{\cos\left(\sqrt{a^2-x^2}\right)}{\sqrt{a^2-x^2}} I_\nu(x)\,dx = \frac{\sqrt{\pi}a^{2\nu+1}}{2^{\nu+1}\Gamma\left(\nu+\frac{3}{2}\right)}$ $\qquad [\mathrm{Re}\,\nu>-1]$ $\qquad$ ET II 365(9)

5. $$\int_0^\infty x^{\nu+1} \frac{\sin\left(a\sqrt{b^2+x^2}\right)}{\sqrt{b^2+x^2}} \, J_\nu(cx)\,dx$$

$$= \sqrt{\frac{\pi}{2}} b^{\frac{1}{2}+\nu} c^\nu \left(a^2-c^2\right)^{-\frac{1}{4}-\frac{1}{2}\nu} J_{-\nu-\frac{1}{2}}\left(b\sqrt{a^2-c^2}\right) \qquad \left[0<c<a, \quad \operatorname{Re} b>0, \quad -1<\operatorname{Re}\nu<\tfrac{1}{2}\right]$$

$$= 0 \qquad\qquad\qquad\qquad\qquad\qquad\qquad\qquad\qquad\qquad \left[0<a<c, \quad \operatorname{Re} b>0, \quad -1<\operatorname{Re}\nu<\tfrac{1}{2}\right]$$

ET II 35(20)

6. $$\int_0^\infty x^{\nu+1} \frac{\cos\left(a\sqrt{x^2+b^2}\right)}{\sqrt{x^2+b^2}} \, J_\nu(cx)\,dx = -\sqrt{\frac{\pi}{2}} b^{\frac{1}{2}+\nu} c^\nu \left(a^2-c^2\right)^{-\frac{1}{4}-\frac{1}{2}\nu} Y_{-\nu-\frac{1}{2}}\left(b\sqrt{a^2-c^2}\right)$$

$$\left[0<c<a, \quad \operatorname{Re} b>0, \quad -1<\operatorname{Re}\nu<\frac{1}{2}\right]$$

$$= \sqrt{\frac{2}{\pi}} b^{\frac{1}{2}+\nu} c^\nu \left(c^2-a^2\right)^{-\frac{1}{4}-\frac{1}{2}\nu} K_{\nu+\frac{1}{2}}\left(b\sqrt{c^2-a^2}\right)$$

$$\left[0<a<c, \quad \operatorname{Re} b>0, \quad -1<\operatorname{Re}\nu<\frac{1}{2}\right]$$

ET II 39(45)

**6.738**

1. $$\int_0^a x^{\nu+1} \sin\left(b\sqrt{a^2-x^2}\right) J_\nu(x)\,dx = \sqrt{\frac{\pi}{2}} a^{\nu+\frac{3}{2}} b \left(1+b^2\right)^{-\frac{1}{2}\nu-\frac{3}{4}} J_{\nu+\frac{3}{2}}\left(a\sqrt{1+b^2}\right)$$

$$[\operatorname{Re}\nu>-1] \qquad\qquad \text{ET II 335(19)}$$

2. $$\int_0^\infty x^{\nu+1} \cos\left(a\sqrt{x^2+b^2}\right) J_\nu(cx)\,dx$$

$$= \sqrt{\frac{\pi}{2}} a b^{\nu+\frac{3}{2}} c^\nu \left(a^2-c^2\right)^{-\frac{1}{2}\nu-\frac{3}{4}} \left[\cos(\pi\nu) J_{\nu+\frac{3}{2}}\left(b\sqrt{a^2-c^2}\right) - \sin(\pi\nu) Y_{\nu+\frac{3}{2}}\left(b\sqrt{a^2-c^2}\right)\right]$$

$$\left[0<c<a, \quad \operatorname{Re} b>0, \quad -1<\operatorname{Re}\nu<-\tfrac{1}{2}\right]$$

$$= 0$$

$$\left[0<a<c, \quad \operatorname{Re} b>0, \quad -1<\operatorname{Re}\nu<-\tfrac{1}{2}\right]$$

ET II 39(43)

**6.739** $$\int_0^t x^{-1/2} \frac{\cos\left(b\sqrt{t-x}\right)}{\sqrt{t-x}} \, J_{2\nu}\left(a\sqrt{x}\right)\,dx = \pi J_\nu\left[\frac{\sqrt{t}}{2}\left(\sqrt{a^2+b^2}+b\right)\right] J_\nu\left[\frac{\sqrt{t}}{2}\left(\sqrt{a^2+b^2}-b\right)\right]$$

$$\left[\operatorname{Re}\nu>-\tfrac{1}{2}\right] \qquad\qquad \text{EH II 47(7)}$$

**6.741**

1. $$\int_0^1 \frac{\cos\left(\mu \arccos x\right)}{\sqrt{1-x^2}} \, J_\nu(ax)\,dx = \frac{\pi}{2} J_{\frac{1}{2}(\mu+\nu)}\left(\frac{a}{2}\right) J_{\frac{1}{2}(\nu-\mu)}\left(\frac{a}{2}\right)$$

$$[\operatorname{Re}(\mu+\nu)>-1, \quad a>0] \qquad \text{ET II 41(54)}$$

2. $$\int_0^1 \frac{\cos\left[(\nu+1)\arccos x\right]}{\sqrt{1-x^2}} \, J_\nu(ax)\,dx = \sqrt{\frac{\pi}{a}} \cos\left(\frac{a}{2}\right) J_{\nu+\frac{1}{2}}\left(\frac{a}{2}\right)$$

$$[\operatorname{Re}\nu>-1, \quad a>0] \qquad \text{ET II 40(53)}$$

3. $$\int_0^1 \frac{\cos\left[(\nu-1)\arccos x\right]}{\sqrt{1-x^2}} \, J_\nu(ax)\,dx = \sqrt{\frac{\pi}{a}} \sin\left(\frac{a}{2}\right) J_{\nu-\frac{1}{2}}\left(\frac{a}{2}\right)$$

$$[\operatorname{Re}\nu>0, \quad a>0] \qquad \text{ET II 40(52)a}$$

## 6.75 Combinations of Bessel, trigonometric, and exponential functions and powers

**6.751**   **Notation:** $\ell_1 = \dfrac{1}{2}\left[\sqrt{(b+c)^2 + a^2} - \sqrt{(b-c)^2 + a^2}\right]$, $\ell_2 = \dfrac{1}{2}\left[\sqrt{(b+c)^2 + a^2} + \sqrt{(b-c)^2 + a^2}\right]$

1.   $\displaystyle\int_0^\infty e^{-\frac{1}{2}ax} \sin(bx) I_0\left(\frac{1}{2}ax\right) dx = \frac{1}{\sqrt{2b}}\frac{1}{\sqrt{b^2+a^2}}\sqrt{b + \sqrt{b^2+a^2}}$

$$[\operatorname{Re} a > 0, \quad b > 0] \qquad \text{ET I 105(44)}$$

2.   $\displaystyle\int_0^\infty e^{-\frac{1}{2}ax} \cos(bx) I_0\left(\frac{1}{2}ax\right) dx = \frac{a}{\sqrt{2b}}\frac{1}{\sqrt{a^2+b^2}\sqrt{b + \sqrt{a^2+b^2}}}$

$$[\operatorname{Re} a > 0, \quad b > 0] \qquad \text{ET I 48(38)}$$

3.¹⁰   $\displaystyle\int_0^\infty e^{-bx} \cos(ax) J_0(cx)\, dx = \frac{\left[\sqrt{(b^2+c^2-a^2)^2 + 4a^2b^2} + b^2 + c^2 - a^2\right]^{1/2}}{\sqrt{2}\sqrt{(b^2+c^2-a^2)^2 + 4a^2b^2}}$

$$[c > 0] \qquad \text{ET II 11(46)}$$

alternatively, with $a$ and $b$ interchanged,

$$\int_0^\infty e^{-ax} \cos(bx) J_0(cx)\, dx = \frac{\sqrt{\ell_2^2 - b^2}}{\ell_2^2 - \ell_1^2} \qquad [c > 0]$$

**6.752**

1.¹⁰   $\displaystyle\int_0^\infty e^{-ax} J_0(bx) \sin(cx)\frac{dx}{x} = \arcsin\left(\frac{2c}{\sqrt{a^2 + (c+b)^2} + \sqrt{a^2 + (c-b)^2}}\right) = \arcsin\left(\frac{c}{\ell_2}\right)$

$$[\operatorname{Re} a > |\operatorname{Im} b|, \quad c > 0] \qquad \text{ET I 101(17)}$$

2.¹⁰   $\displaystyle\int_0^\infty e^{-ax} J_1(cx) \sin(bx)\frac{dx}{x} = \frac{b}{c}(1 - r) = \frac{b - \sqrt{b^2 - \ell_1^2}}{c},$

$$\left[b^2 = \frac{c^2}{1 - r^2} - \frac{a^2}{r^2}, \quad c > 0\right]$$

$$\text{ET II 19(15)}$$

**Notation:** For integrals 6.752 3–6.752 5 we define the auxiliary functions

$$\ell_1(a) \equiv \ell_1(a, \rho, z) = \frac{1}{2}\left[\sqrt{(a+\rho)^2 + z^2} - \sqrt{(a-\rho)^2 + z^2}\right]$$

$$\ell_2(a) \equiv \ell_1(a, \rho, z) = \frac{1}{2}\left[\sqrt{(a+\rho)^2 + z^2} + \sqrt{(a-\rho)^2 + z^2}\right]$$

when $a \geq 0$, $\rho \geq 0$, and $z \geq 0$.

3.¹⁰   $\sqrt{\dfrac{\pi}{2}}\displaystyle\int_0^\infty e^{-zx} J_{\nu+1/2}(ax) J_{\nu+1}(\rho x)\sqrt{x}\, dx$

$$= a^{-\nu - 3/2}\rho^{-\nu - 1}\frac{\ell_1^{2\nu+2}}{\sqrt{\rho^2 - \ell_1^2}}\frac{a\left(\rho^2 - \ell_1^2\right)}{\ell_1\left(\ell_2^2 - \ell_1^2\right)}$$

$$= a^{\nu + 1/2}\frac{\rho^{\nu+1}}{\ell_2^{2\nu+2}}\frac{\sqrt{\ell_2^2 - a^2}}{\ell_2^2 - \ell_1^2} \qquad [\operatorname{Re} z > |\operatorname{Im} a| + |\operatorname{Im} \rho|]$$

$4.^{10}$   $\sqrt{\dfrac{\pi}{2}} \displaystyle\int_0^\infty e^{-zx} J_{\nu+1/2}(ax) J_\nu(\rho x) \dfrac{dx}{\sqrt{x}}$

$$= a^{\nu+1/2} \rho^\nu \int_0^{1/\ell_2} \frac{1}{\ell_2^{2\nu}} \frac{1}{\sqrt{1 - a^2/\ell_2^2}} \, d\left(\frac{1}{\ell_2}\right)$$

$$= a^{-\nu-1/2} \rho^\nu \int_0^{a/\ell_2} x^{2\nu} \frac{dx}{\sqrt{1 - x^2}} \qquad \left[\nu > -\tfrac{1}{2}, \quad \operatorname{Re} z > |\operatorname{Im} a| + |\operatorname{Im} \rho|\right]$$

$5.^{10}$   $\displaystyle\int_0^\infty e^{-zx} \sin(ax) J_1(\rho x) \frac{dx}{x^2} = \frac{\sqrt{\ell_2^2 - a^2}\left(a - \sqrt{a^2 - \ell_1^2}\right)^2}{2a\rho} + \frac{\rho}{2} \arcsin\left(\frac{a}{\ell_2}\right)$

$$[\operatorname{Re} z > |\operatorname{Im} a| + |\operatorname{Im} \rho|]$$

**6.753**

$1.^8$   $\displaystyle\int_0^\infty \frac{\sin(xa \sin \psi)}{x} e^{-xa \cos \varphi \cos \psi} J_\nu(xa \sin \varphi) \, dx = \nu^{-1} \left(\tan \frac{\varphi}{2}\right)^\nu \sin(\nu\psi)$

$$\left[\operatorname{Re}\nu > -1, \quad a > 0, \quad 0 < \varphi < \frac{\pi}{2}, \quad 0 < \psi < \frac{\pi}{2}\right] \quad \text{ET II 33(10)}$$

$2.$   $\displaystyle\int_0^\infty \frac{\cos(xa \sin \psi)}{x} e^{-xa \cos \varphi \cos \psi} J_\nu(xa \sin \varphi) \, dx = \nu^{-1} \left(\tan \frac{\varphi}{2}\right)^\nu \cos(\nu\psi)$

$$\left[\operatorname{Re}\nu > 0, \quad a > 0, \quad 0 < \varphi, \quad \psi < \frac{\pi}{2}\right]$$
$$\text{ET II 38(35)}$$

$3.^8$   $\displaystyle\int_0^\infty x^{\nu+1} e^{-sx} \sin(bx) J_\nu(ax) \, dx = -\frac{2(2a)^\nu}{\sqrt{\pi}} \Gamma(\nu + \tfrac{3}{2}) R^{-2\nu-3} \left[b \cos(\nu + \tfrac{3}{2})\varphi + s \sin(\nu + \tfrac{3}{2})\varphi\right]$

$$\left[\operatorname{Re}\nu > -\tfrac{3}{2}, \quad \operatorname{Re} s > |\operatorname{Im} a| + |\operatorname{Im} b|,\right.$$

$$\left. R^4 = \left(s^2 + a^2 - b^2\right)^2 + 4b^2 s^2, \quad \varphi = \arg\left(s^2 + a^2 - b^2 - 2ibs\right)\right]$$

$4.^8$   $\displaystyle\int_0^\infty x^{\nu+1} e^{-sx} \cos(bx) J_\nu(ax) \, dx = \frac{2(2a)^\nu}{\sqrt{\pi}} \Gamma(\nu + \tfrac{3}{2}) R^{-2\nu-3} \left[s \cos(\nu + \tfrac{3}{2})\varphi - b \sin(\nu + \tfrac{3}{2})\varphi\right],$

$$\left[\operatorname{Re}\nu > -1, \quad \operatorname{Re} s > |\operatorname{Im} a| + |\operatorname{Im} b|,\right.$$

$$\left. R^4 = \left(s^2 + a^2 - b^2\right)^2 + 4b^2 s^2, \quad \varphi = \arg\left(s^2 + a^2 - b^2 - 2ibs\right)\right]$$

$5.^{10}$   $\displaystyle\int_0^\infty x^\nu e^{-ax \cos \varphi \cos \psi} \sin(ax \sin \psi) J_\nu(ax \sin \varphi) \, dx$

$$= 2^\nu \frac{\Gamma\left(\nu + \tfrac{1}{2}\right)}{\sqrt{\pi}} a^{-\nu-1} (\sin \varphi)^\nu \left(\cos^2 \psi + \sin^2 \psi \cos^2 \varphi\right)^{-\nu - \frac{1}{2}} \sin\left[\left(\nu + \tfrac{1}{2}\right)\beta\right]$$

$$\tan \frac{\beta}{2} = \tan \psi \cos \varphi \qquad \left[a > 0, \quad 0 < \varphi < \frac{\pi}{2}, \quad 0 < \psi < \frac{\pi}{2}, \quad \operatorname{Re}\nu > -1\right] \quad \text{ET II 34(12)}$$

$6.$   $\displaystyle\int_0^\infty x^\nu e^{-ax \cos \varphi \cos \psi} \cos(ax \sin \psi) J_\nu(ax \sin \varphi) \, dx$

$$= 2^\nu \frac{\Gamma\left(\nu + \tfrac{1}{2}\right)}{\sqrt{\pi}} a^{-\nu-1} (\sin \varphi)^\nu \left(\cos^2 \psi + \sin^2 \psi \cos^2 \varphi\right)^{-\nu - \frac{1}{2}} \cos\left[\left(\nu + \tfrac{1}{2}\right)\beta\right]$$

$$\tan \frac{\beta}{2} = \tan \psi \cos \varphi \qquad \left[a > 0, \quad 0 < \varphi, \quad \psi < \frac{\pi}{2}, \quad \operatorname{Re}\nu > -\tfrac{1}{2}\right] \quad \text{ET II 38(37)}$$

**6.754**

1.  $$\int_0^\infty e^{-x^2} \sin(bx)\, I_0\left(x^2\right)\, dx = \frac{\sqrt{\pi}}{2^{3/2}} e^{-\frac{b^2}{8}} I_0\left(\frac{b^2}{8}\right) \qquad [b > 0] \qquad\qquad \text{ET I 108(9)}$$

2.  $$\int_0^\infty e^{-ax} \cos\left(x^2\right) J_0\left(x^2\right)\, dx = \frac{1}{4}\sqrt{\frac{\pi}{2}} \left[ J_0\left(\frac{a^2}{16}\right) \cos\left(\frac{a^2}{16} - \frac{\pi}{4}\right) - Y_0\left(\frac{a^2}{16}\right) \cos\left(\frac{a^2}{16} + \frac{\pi}{4}\right) \right]$$
    $$[a > 0] \qquad\qquad \text{MI 42}$$

3.  $$\int_0^\infty e^{-ax} \sin\left(x^2\right) J_0\left(x^2\right)\, dx = \frac{1}{4}\sqrt{\frac{\pi}{2}} \left[ J_0\left(\frac{a^2}{16}\right) \sin\left(\frac{a^2}{16} - \frac{\pi}{4}\right) - Y_0\left(\frac{a^2}{16}\right) \sin\left(\frac{a^2}{16} + \frac{\pi}{4}\right) \right]$$
    $$[a > 0] \qquad\qquad \text{MI 42}$$

**6.755**

1.  $$\int_0^\infty x^{-\nu} e^{-x} \sin\left(4a\sqrt{x}\right) I_\nu(x)\, dx = \left(2^{3/2} a\right)^{\nu-1} e^{-a^2} W_{\frac{1}{2}-\frac{3}{2}\nu, \frac{1}{2}-\frac{1}{2}\nu}\left(2a^2\right)$$
    $$[a > 0, \quad \operatorname{Re}\nu > 0] \qquad \text{ET II 366(14)}$$

2.  $$\int_0^\infty x^{-\nu-\frac{1}{2}} e^{-x} \cos\left(4a\sqrt{x}\right) I_\nu(x)\, dx = 2^{\frac{3}{2}\nu-1} a^{\nu-1} e^{-a^2} W_{-\frac{3}{2}\nu, \frac{1}{2}\nu}\left(2a^2\right)$$
    $$[a > 0, \quad \operatorname{Re}\nu > -\tfrac{1}{2}] \qquad \text{ET II 366(16)}$$

3.  $$\int_0^\infty x^{-\nu} e^{x} \sin\left(4a\sqrt{x}\right) K_\nu(x)\, dx = \left(2^{3/2} a\right)^{\nu-1} \pi \frac{\Gamma\left(\frac{3}{2}-2\nu\right)}{\Gamma\left(\frac{1}{2}+\nu\right)} e^{a^2} W_{\frac{3}{2}\nu-\frac{1}{2}, \frac{1}{2}-\frac{1}{2}\nu}\left(2a^2\right)$$
    $$[a > 0, \quad 0 < \operatorname{Re}\nu < \tfrac{3}{4}] \qquad \text{ET II 369(38)}$$

4.  $$\int_0^\infty x^{-\nu-\frac{1}{2}} e^{x} \cos\left(4a\sqrt{x}\right) K_\nu(x)\, dx = 2^{\frac{3}{2}\nu-1} \pi a^{\nu-1} \frac{\Gamma\left(\frac{1}{2}-2\nu\right)}{\Gamma\left(\frac{1}{2}+\nu\right)} e^{a^2} W_{\frac{3}{2}\nu, -\frac{1}{2}\nu}\left(2a^2\right)$$
    $$[a > 0, \quad -\tfrac{1}{2} < \operatorname{Re}\nu < \tfrac{1}{4}] \qquad \text{ET II 369(42)}$$

5.  $$\int_0^\infty x^{\varrho-\frac{3}{2}} e^{-x} \sin\left(4a\sqrt{x}\right) K_\nu(x)\, dx = \frac{\sqrt{\pi}\, a\, \Gamma(\varrho+\nu)\, \Gamma(\varrho-\nu)}{2^{\varrho-2}\, \Gamma\left(\varrho+\frac{1}{2}\right)}\, {}_2F_2\left(\varrho+\nu, \varrho-\nu; \frac{3}{2}, \varrho+\frac{1}{2}; -2a^2\right)$$
    $$[\operatorname{Re}\varrho > |\operatorname{Re}\nu|] \qquad \text{ET II 369(39)}$$

6.  $$\int_0^\infty x^{\varrho-1} e^{-x} \cos\left(4a\sqrt{x}\right) K_\nu(x)\, dx = \frac{\sqrt{\pi}\, \Gamma(\varrho+\nu)\, \Gamma(\varrho-\nu)}{2^{\varrho}\, \Gamma\left(\varrho+\frac{1}{2}\right)}\, {}_2F_2\left(\varrho+\nu, \varrho-\nu; \frac{1}{2}, \varrho+\frac{1}{2}; -2a^2\right)$$
    $$[\operatorname{Re}\varrho > |\operatorname{Re}\nu|] \qquad \text{ET II 370(43)}$$

7.  $$\int_0^\infty x^{-1/2} e^{-x} \cos\left(4a\sqrt{x}\right) I_0(x)\, dx = \frac{1}{\sqrt{2\pi}} e^{-a^2} K_0\left(a^2\right)$$
    $$[a > 0] \qquad\qquad \text{ET II 366(15)}$$

8.  $$\int_0^\infty x^{-1/2} e^{x} \cos\left(4a\sqrt{x}\right) K_0(x)\, dx = \sqrt{\frac{\pi}{2}} e^{a^2} K_0\left(a^2\right) \qquad [a > 0] \qquad\qquad \text{ET II 369(40)}$$

9.  $$\int_0^\infty x^{-1/2} e^{-x} \cos\left(4a\sqrt{x}\right) K_0(x)\, dx = \frac{1}{\sqrt{2}} \pi^{3/2} e^{-a^2} I_0\left(a^2\right) \qquad\qquad \text{ET II 369(41)}$$

**6.756**

1.    $\displaystyle \int_0^\infty x^{-\frac{1}{2}} e^{-a\sqrt{x}} \sin\left(a\sqrt{x}\right) J_\nu(bx)\, dx$

$$= \frac{i}{\sqrt{2\pi b}}\, \Gamma\left(\nu + \frac{1}{2}\right) D_{-\nu-\frac{1}{2}}\left(\frac{a}{\sqrt{b}}\right) \left[D_{-\nu-\frac{1}{2}}\left(\frac{ia}{\sqrt{b}}\right) - D_{-\nu-\frac{1}{2}}\left(-\frac{ia}{\sqrt{b}}\right)\right]$$

$$[a > 0, \quad b > 0, \quad \operatorname{Re}\nu > -1] \quad \text{ET II 34(17)}$$

2.    $\displaystyle \int_0^\infty x^{-\frac{1}{2}} e^{-a\sqrt{x}} \cos\left(a\sqrt{x}\right) J_\nu(bx)\, dx$

$$= \frac{1}{\sqrt{2\pi b}}\, \Gamma\left(\nu + \frac{1}{2}\right) D_{-\nu-\frac{1}{2}}\left(\frac{a}{\sqrt{b}}\right) \left[D_{-\nu-\frac{1}{2}}\left(\frac{ia}{\sqrt{b}}\right) + D_{-\nu-\frac{1}{2}}\left(-\frac{ia}{\sqrt{b}}\right)\right]$$

$$[a > 0, \quad b > 0, \quad \operatorname{Re}\nu > -\tfrac{1}{2}] \quad \text{ET II 39(42)}$$

3.    $\displaystyle \int_0^\infty x^{-1/2} e^{-a\sqrt{x}} \sin\left(a\sqrt{x}\right) J_0(bx)\, dx = \frac{1}{2b} a\, I_{\frac{1}{4}}\left(\frac{a^2}{4b}\right) K_{\frac{1}{4}}\left(\frac{a^2}{4b}\right)$

$$\left[|\arg a| < \frac{\pi}{4}, \quad b > 0\right] \quad \text{ET II 11(40)}$$

4.    $\displaystyle \int_0^\infty x^{-1/2} e^{-a\sqrt{x}} \cos\left(a\sqrt{x}\right) J_0(bx)\, dx = \frac{a}{2b}\, I_{-\frac{1}{4}}\left(\frac{a^2}{4b}\right) K_{\frac{1}{4}}\left(\frac{a^2}{4b}\right)$

$$\left[|\arg a| < \frac{\pi}{4}, \quad b > 0\right] \quad \text{ET II 12(49)}$$

**6.757**

1.    $\displaystyle \int_0^\infty e^{-bx} \sin\left[a\left(1 - e^{-x}\right)\right] J_\nu\left(ae^{-x}\right) dx$

$$= 2\sum_{n=0}^\infty \frac{(-1)^n\, \Gamma(\nu - b + 2n + 1)\, \Gamma(\nu + b)}{\Gamma(\nu - b + 1)\,\Gamma(\nu + b + 2n + 2)} (\nu + 2n - 1)\, J_{\nu+2n+1}(a)$$

$$[\operatorname{Re} b > -\operatorname{Re}\nu] \quad \text{ET I 193(26)}$$

2.    $\displaystyle \int_0^\infty e^{-bx} \cos\left[a\left(1 - e^{-x}\right)\right] J_\nu\left(ae^{-x}\right) dx$

$$= \frac{J_\nu(a)}{\nu + b} + \sum_{n=0}^\infty 2(-1)^n \frac{\Gamma(\nu - b + 2n)\,\Gamma(\nu + b)}{\Gamma(\nu - b + 1)\,\Gamma(\nu + b + 2n + 1)} (\nu + 2n)\, J_{\nu+2n}(a)$$

$$[\operatorname{Re} b > -\operatorname{Re}\nu] \quad \text{ET I 193(27)}$$

**6.758**    $\displaystyle \int_{-\frac{\pi}{2}}^{\frac{\pi}{2}} e^{i(\mu-\nu)\theta}\, (\cos\theta)^{\nu+\mu}\, (\lambda z)^{-\nu-\mu}\, J_{\nu+\mu}(\lambda z)\, d\theta = \pi(2az)^{-\mu}(2bz)^{-\nu}\, J_\mu(az)\, J_\nu(bz)$

$$\lambda = \sqrt{2\cos\theta\left(a^2 e^{i\theta} + b^2 e^{-i\theta}\right)} \quad [\operatorname{Re}(\nu + \mu) > -1] \quad \text{EH II 48(12)}$$

## 6.76 Combinations of Bessel, trigonometric, and hyperbolic functions

**6.761**    $\displaystyle \int_0^\infty \cosh x \cos\left(2a\sinh x\right) J_\nu\left(be^x\right) J_\nu\left(be^{-x}\right) dx = \frac{J_{2\nu}\left(2\sqrt{b^2 - a^2}\right)}{2\sqrt{b^2 - a^2}}$    $[0 < a < b, \quad \operatorname{Re}\nu > -1]$

$$= 0 \qquad [0 < b < a, \quad \operatorname{Re}\nu > -1]$$

$$\text{ET II 359(10)}$$

**6.762** $\displaystyle\int_0^\infty \cosh x \sin\left(2a\sinh x\right)\left[J_\nu\left(be^x\right)Y_\nu\left(be^{-x}\right) - Y_\nu\left(be^x\right)J_\nu\left(be^{-x}\right)\right]\,dx$

$$= 0 \qquad\qquad\qquad\qquad\qquad \left[0 < a < b, \quad |\operatorname{Re}\nu| < \tfrac{1}{2}\right]$$

$$= -\frac{2}{\pi}\cos(\nu\pi)\left(a^2 - b^2\right)^{-1/2}K_{2\nu}\left[2\left(a^2 - b^2\right)^{1/2}\right] \quad \left[0 < b < a, \quad |\operatorname{Re}\nu| < \tfrac{1}{2}\right]$$

$$\text{ET II 360(12)}$$

**6.763** $\displaystyle\int_0^\infty \cosh x \cos\left(2a\sinh x\right)Y_\nu\left(be^x\right)Y_\nu\left(be^{-x}\right)\,dx$

$$= -\frac{1}{2}\left(b^2 - a^2\right)^{-1/2}J_{2\nu}\left[2\left(b^2 - a^2\right)^{1/2}\right] \qquad \left[0 < a < b, \quad |\operatorname{Re}\nu| < 1\right]$$

$$= \frac{2}{\pi}\cos(\nu\pi)\left(a^2 - b^2\right)^{-1/2}K_{2\nu}\left[2\left(a^2 - b^2\right)^{1/2}\right] \qquad \left[0 < b < a, \quad |\operatorname{Re}\nu| < 1\right]$$

$$\text{ET II 360(11)}$$

## 6.77 Combinations of Bessel functions and the logarithm, or arctangent

**6.771** $\displaystyle\int_0^\infty x^{\mu+\frac{1}{2}}\ln x\, J_\nu(ax)\,dx = \frac{2^{\mu-\frac{1}{2}}\,\Gamma\left(\frac{\mu+\nu}{2} + \frac{3}{4}\right)}{\Gamma\left(\frac{\nu-\mu}{2} + \frac{1}{4}\right)a^{\mu+\frac{3}{2}}}\left[\psi\left(\frac{\mu+\nu}{2} + \frac{3}{4}\right) + \psi\left(\frac{\nu-\mu}{2} + \frac{1}{4}\right) - \ln\frac{a^2}{4}\right]$

$$\left[a > 0, \quad -\operatorname{Re}\nu - \tfrac{3}{2} < \operatorname{Re}\mu < 0\right]$$

$$\text{ET II 32(25)}$$

**6.772**

1. $\displaystyle\int_0^\infty \ln x\, J_0(ax)\,dx = -\frac{1}{a}\left[\ln(2a) + \boldsymbol{C}\right]$ 　　　　　　　 WA 430(4)a, ET II 10(27)

2. $\displaystyle\int_0^\infty \ln x\, J_1(ax)\,dx = -\frac{1}{a}\left[\ln\left(\frac{a}{2}\right) + \boldsymbol{C}\right]$ 　　　　　　　　　　　 ET II 19(11)

3. $\displaystyle\int_0^\infty \ln\left(a^2 + x^2\right)J_1(bx)\,dx = \frac{2}{b}\left[K_0(ab) + \ln a\right]$ 　　　　　　　 ET II 19(12)

4. $\displaystyle\int_0^\infty J_1(tx)\ln\sqrt{1 + t^4}\,dt = \frac{2}{x}\ker x$ 　　　　　　　　　　　　 MO 46

**6.773** $\displaystyle\int_0^\infty \frac{\ln\left(x + \sqrt{x^2 + a^2}\right)}{\sqrt{x^2 + a^2}}J_0(bx)\,dx = \left[\frac{1}{2}K_0^2\left(\frac{ab}{2}\right) + \ln a\, I_0\left(\frac{ab}{2}\right)K_0\left(\frac{ab}{2}\right)\right]$

$$\left[a > 0, \quad b > 0\right] \qquad \text{ET II 10(28)}$$

**6.774** $\displaystyle\int_0^\infty \ln\frac{\sqrt{x^2 + a^2} + x}{\sqrt{x^2 + a^2} - x}J_0(bx)\frac{dx}{\sqrt{x^2 + a^2}} = K_0^2\left(\frac{ab}{2}\right)$ 　　 $\left[\operatorname{Re}a > 0, \quad b > 0\right]$ 　　 ET II 10(29)

**6.775**[12] $\displaystyle\int_0^\infty x\left[\ln\left(a + \sqrt{a^2 + x^2}\right) - \ln x\right]J_0(bx)\,dx = \frac{1}{b^2}\left(1 - e^{-ab}\right)$

$$\left[\operatorname{Re}a > 0, \quad b > 0\right] \qquad \text{ET II 12(55)}$$

**6.776** $\displaystyle\int_0^\infty x\ln\left(1 + \frac{a^2}{x^2}\right)J_0(bx)\,dx = \frac{2}{b}\left[\frac{1}{b} - a\,K_1(ab)\right]$ 　 $\left[\operatorname{Re}a > 0, \quad b > 0\right]$ 　 ET II 10(30)

**6.777** $\displaystyle\int_0^\infty J_1(tx)\arctan t^2\,dt = -\frac{2}{x}\ker x$ 　　　　　　　　　　　　 MO 46

## 6.78   Combinations of Bessel and other special functions

**6.781**    $\displaystyle\int_0^\infty \operatorname{si}(ax)\, J_0(bx)\, \mathrm{d}x = -\frac{1}{b}\arcsin\left(\frac{b}{a}\right)$             $[0 < b < a]$

$\displaystyle = 0$             $[0 < a < b]$

ET II 13(6)

**6.782**

1.    $\displaystyle\int_0^\infty \operatorname{Ei}(-x)\, J_0\left(2\sqrt{zx}\right)\, \mathrm{d}x = \frac{e^{-z}-1}{z}$        NT 60(4)

2.    $\displaystyle\int_0^\infty \operatorname{si}(x)\, J_0\left(2\sqrt{zx}\right)\, \mathrm{d}x = -\frac{\sin z}{z}$        NT 60(6)

3.    $\displaystyle\int_0^\infty \operatorname{ci}(x)\, J_0\left(2\sqrt{zx}\right)\, \mathrm{d}x = \frac{\cos z - 1}{z}$        NT 60(5)

4.    $\displaystyle\int_0^\infty \operatorname{Ei}(-x)\, J_1\left(2\sqrt{zx}\right)\frac{\mathrm{d}x}{\sqrt{x}} = \frac{\operatorname{Ei}(-z)-\boldsymbol{C}-\ln z}{\sqrt{z}}$        NT 60(7)

5.    $\displaystyle\int_0^\infty \operatorname{si}(x)\, J_1\left(2\sqrt{zx}\right)\frac{\mathrm{d}x}{\sqrt{x}} = -\frac{\frac{\pi}{2}-\operatorname{si}(z)}{\sqrt{z}}$        NT 60(9)

6.[12]    $\displaystyle\int_0^\infty \operatorname{ci}(x)\, J_1\left(2\sqrt{zx}\right)\frac{\mathrm{d}x}{\sqrt{x}} = \frac{\operatorname{ci}(x)-\boldsymbol{C}-\ln z}{\sqrt{z}}$        NT 60(8)

7.[12]    $\displaystyle\int_0^\infty \operatorname{Ei}(-x)\, Y_0\left(2\sqrt{zx}\right)\, \mathrm{d}x = \frac{\boldsymbol{C}+\ln z - e^z\operatorname{Ei}(-z)}{\pi z}$        NT 63(5)

**6.783**

1.    $\displaystyle\int_0^\infty x\operatorname{si}\left(a^2 x^2\right) J_0(bx)\, \mathrm{d}x = -\frac{2}{b^2}\sin\left(\frac{b^2}{4a^2}\right)$     $[a>0]$      ET II 13(7)a

2.    $\displaystyle\int_0^\infty x\operatorname{ci}\left(a^2 x^2\right) J_0(bx)\, \mathrm{d}x = \frac{2}{b^2}\left[1-\cos\left(\frac{b^2}{4a^2}\right)\right]$     $[a>0]$      ET II 13(8)a

3.    $\displaystyle\int_0^\infty \operatorname{ci}\left(a^2 x^2\right) J_0(bx)\, \mathrm{d}x = \frac{1}{b}\left[\operatorname{ci}\left(\frac{b^2}{4a^2}\right)+\ln\left(\frac{b^2}{4a^2}\right)+2\boldsymbol{C}\right]$

                                                           $[a>0]$      ET II 13(8)a

4.    $\displaystyle\int_0^\infty \operatorname{si}\left(a^2 x^2\right) J_1(bx)\, \mathrm{d}x = \frac{1}{b}\left[-\operatorname{si}\left(\frac{b^2}{4a^2}\right)-\frac{\pi}{2}\right]$     $[a>0]$      ET II 20(25)a

**6.784**

1.    $\displaystyle\int_0^\infty x^{\nu+1}\left[1-\Phi(ax)\right] J_\nu(bx)\, \mathrm{d}x = a^{-\nu}\frac{\Gamma\left(\nu+\frac{3}{2}\right)}{b^2\,\Gamma(\nu+2)}\exp\left(-\frac{b^2}{8a^2}\right) M_{\frac12\nu+\frac12,\,\frac12\nu+\frac12}\left(\frac{b^2}{4a^2}\right)$

$\left[|\arg a| < \frac{\pi}{4}, \quad b>0, \quad \operatorname{Re}\nu > -1\right]$

ET II 92(22)

2. $\displaystyle\int_0^\infty x^\nu \left[1 - \Phi(ax)\right] J_\nu(bx)\,dx = \sqrt{\frac{2}{\pi}}\,\frac{a^{\frac{1}{2}-\nu}\,\Gamma\left(\nu+\frac{1}{2}\right)}{b^{3/2}\,\Gamma\left(\nu+\frac{3}{2}\right)}\exp\left(-\frac{b^2}{8a^2}\right) M_{\frac{1}{2}\nu-\frac{1}{4},\frac{1}{2}\nu+\frac{1}{4}}\left(\frac{b^2}{4a^2}\right)$

$$\left[|\arg a| < \tfrac{\pi}{4}, \quad \operatorname{Re}\nu > -\tfrac{1}{2}, \quad b > 0\right]$$
ET II 92(23)

**6.785** $\displaystyle\int_0^\infty \frac{\exp\left(\frac{a^2}{2x}-x\right)}{x}\left[1 - \Phi\left(\frac{a}{\sqrt{2x}}\right)\right] K_\nu(x)\,dx = \frac{\pi^{5/2}}{4}\sec(\nu\pi)\left\{[J_\nu(a)]^2 + [Y_\nu(a)]^2\right\}$

$$\left[\operatorname{Re} a > 0, \quad |\operatorname{Re}\nu| < \tfrac{1}{2}\right] \qquad \text{ET II 370(46)}$$

**6.786** $\displaystyle\int_0^\infty x^{\nu-2\mu+2n+2} e^{x^2}\,\Gamma\left(\mu, x^2\right) Y_\nu(bx)\,dx$

$$= (-1)^n \frac{\Gamma\left(\frac{3}{2}-\mu+\nu+n\right)\Gamma\left(\frac{3}{2}-\mu+n\right)}{b\,\Gamma(1-\mu)}\exp\left(\frac{b^2}{8}\right) W_{\mu-\frac{1}{2}\nu-n-1,\frac{1}{2}\nu}\left(\frac{b^2}{4}\right)$$

$$\left[n \text{ is an integer}, \quad b > 0, \quad \operatorname{Re}(\nu-\mu+n) > -\tfrac{3}{2}, \quad \operatorname{Re}(-\mu+n) > -\tfrac{3}{2}, \quad \operatorname{Re}\nu < \tfrac{1}{2}-2n\right]$$
ET II 108(2)

**6.787** $\displaystyle\int_0^\infty \frac{x^{\nu+2n-\frac{1}{2}}}{\mathrm{B}(a+x,a-x)} J_\nu(bx)\,dx = 0$

$$\left[\pi \le b < \infty, \quad -1 < \operatorname{Re}\nu < 2a - 2n - \tfrac{7}{2}\right] \quad \text{ET II 92(21)}$$

## 6.79 Integration of Bessel functions with respect to the order

**6.791**

1. $\displaystyle\int_{-\infty}^\infty K_{ix+iy}(a)\, K_{ix+iz}(b)\,dx = \pi\, K_{iy-iz}(a+b)$　　　　$[|\arg a| + |\arg b| < \pi]$　　　ET II 382(21)

2. $\displaystyle\int_{-\infty}^\infty J_{\nu-x}(a)\, J_{\mu+x}(a)\,dx = J_{\mu+\nu}(2a)$　　　　　　　　$[\operatorname{Re}(\mu+\nu) > 1]$　　　ET II 379(1)

3. $\displaystyle\int_{-\infty}^\infty J_{\kappa+x}(a)\, J_{\lambda-x}(a)\, J_{\mu+x}(a)\, J_{\nu-x}(a)\,dx$

$$= \frac{\Gamma(\kappa+\lambda+\mu+\nu+1)}{\Gamma(\kappa+\lambda+1)\,\Gamma(\lambda+\mu+1)\,\Gamma(\mu+\nu+1)\,\Gamma(\nu+\kappa+1)}$$

$$\times\ _4F_5\left(\frac{\kappa+\lambda+\mu+\nu+1}{2}, \frac{\kappa+\lambda+\mu+\nu+1}{2}, \frac{\kappa+\lambda+\mu+\nu}{2}+1, \frac{\kappa+\lambda+\mu+\nu}{2}+1;\right.$$

$$\left.\kappa+\lambda+\mu+\nu+1, \kappa+\lambda+1, \lambda+\mu+1, \mu+\nu+1, \nu+\kappa+1; -4a^2\right)$$

$$[\operatorname{Re}(\kappa+\lambda+\mu+\nu) > -1] \qquad \text{ET II 379(3)}$$

**6.792**

1. $\displaystyle\int_{-\infty}^\infty e^{\pi x}\, K_{ix+iy}(a)\, K_{ix+iz}(b)\,dx = \pi e^{-\pi z}\, K_{i(y-z)}(a-b)$

$$[a > b > 0] \qquad \text{ET II 382(22)}$$

$2.^{12}$  $\displaystyle\int_{-\infty}^{\infty} e^{i\rho x}\, K_{\nu+ix}(\alpha)\, K_{\nu-ix}(\beta)\, dx = \pi \left(\frac{\alpha + \beta e^{\rho}}{\alpha e^{\rho} + \beta}\right)^{\nu} K_{2\nu}\left(\sqrt{\alpha^2 + \beta^2 + 2\alpha\beta\cosh\rho}\right)$

$$[|\arg\alpha| + |\arg\beta| + |\mathrm{Im}\,\rho| < \pi]$$
ET II 382(23)

3.  $\displaystyle\int_{-\infty}^{\infty} e^{(\pi-\gamma)x}\, K_{ix+iy}(a)\, K_{ix+iz}(b)\, dx = \pi e^{-\beta y - \alpha z}\, K_{iy-iz}(c)$

$$[0 < \gamma < \pi, \quad a > 0, \quad b > 0, \quad c > 0, \quad \alpha,\beta,\gamma\text{—the angles of the triangle with sides } a,\, b,\, c]$$
ET II 382(24), EH II 55(44)a

$4.^{11}$  $\displaystyle\int_{-\infty}^{\infty} e^{-cxi}\, H^{(2)}_{\nu-ix}(a)\, H^{(2)}_{\nu+ix}(b)\, dx = 2i\left(\frac{h}{k}\right)^{2\nu} H^{(2)}_{2\nu}(hk)$

$$h = \sqrt{ae^{\frac{1}{2}c} + be^{-\frac{1}{2}c}}, \quad k = \sqrt{ae^{-\frac{1}{2}c} + be^{\frac{1}{2}c}} \qquad [a, b > 0, \quad c \text{ is real}] \quad \text{ET II 380(11)}$$

5.  $\displaystyle\int_{-\infty}^{\infty} a^{-\mu-x} b^{-\nu+x} e^{cxi}\, J_{\mu+x}(a)\, J_{\nu-x}(b)\, dx$

$$= \left[\frac{2\cos\left(\frac{c}{2}\right)}{a^2 e^{-\frac{1}{2}ci} + b^2 e^{\frac{1}{2}ci}}\right]^{\frac{1}{2}\mu + \frac{1}{2}\nu} \exp\left[\frac{c}{2}(\nu-\mu)i\right] J_{\mu+\nu}\left\{\left[2\cos\left(\frac{c}{2}\right)\left(a^2 e^{-\frac{1}{2}ci} + b^2 e^{\frac{1}{2}ci}\right)\right]^{1/2}\right\}$$

$$[a > 0, \quad b > 0, \quad |c| < \pi, \quad \mathrm{Re}(\mu+\nu) > 1]$$

$$= 0$$

$$[a > 0, \quad b > 0, \quad |c| \geq \pi, \quad \mathrm{Re}(\mu+\nu) > 1]$$
EH II 54(41), ET II 379(2)

## 6.793

1.  $\displaystyle\int_{-\infty}^{\infty} e^{-cxi}\left[J_{\nu-ix}(a)\, Y_{\nu+ix}(b) + Y_{\nu-ix}(a)\, J_{\nu+ix}(b)\right] dx = -2\left(\frac{h}{k}\right)^{2\nu} J_{2\nu}(hk)$

$$h = \sqrt{ae^{\frac{1}{2}c} + be^{-\frac{1}{2}c}}, \qquad k = \sqrt{ae^{-\frac{1}{2}c} + be^{\frac{1}{2}c}} \qquad [a, b > 0, \quad \mathrm{Im}\,c = 0] \quad \text{ET II 380(9)}$$

2.  $\displaystyle\int_{-\infty}^{\infty} e^{-cxi}\left[J_{\nu-ix}(a)\, J_{\nu+ix}(b) - Y_{\nu-ix}(a)\, Y_{\nu+ix}(b)\right] dx = 2\left(\frac{h}{k}\right)^{2\nu} Y_{2\nu}(hk)$

$$h = \sqrt{ae^{\frac{1}{2}c} + be^{-\frac{1}{2}c}}, \qquad k = \sqrt{ae^{-\frac{1}{2}c} + be^{\frac{1}{2}c}} \qquad [a, b > 0, \quad \mathrm{Im}\,c = 0] \quad \text{ET II 380(10)}$$

$3.^{10}$  $\displaystyle\int_{-\infty}^{\infty} e^{i\gamma x}\,\mathrm{sech}(\pi x)\left[J_{-ix}(\alpha)\, J_{ix}(\beta) - J_{ix}(\alpha)\, J_{-ix}(\beta)\right] dx = 2i\, \mathrm{H}(\sigma)\,\mathrm{sign}(\beta - \alpha)\, J_0\left(\sigma^{1/2}\right)$

$$[\alpha, \beta, \gamma \in \mathbb{R}, \quad \alpha, \beta > 0, \quad \sigma = \alpha^2 + \beta^2 - 2\alpha\beta\cosh\gamma]$$

## 6.794

1.  $\displaystyle\int_{0}^{\infty} K_{ix}(a)\, K_{ix}(b)\,\cosh[(\pi-\varphi)x]\, dx = \frac{\pi}{2}\, K_0\left(\sqrt{a^2 + b^2 - 2ab\cos\varphi}\right)$
EH II 55(42)

2.  $\displaystyle\int_{0}^{\infty}\cosh\left(\frac{\pi}{2}x\right) K_{ix}(a)\, dx = \frac{\pi}{2}$  $\qquad [a > 0]$
ET II 382(19)

3. $\displaystyle\int_0^\infty \cosh(\varrho x)\, K_{ix+\nu}(a)\, K_{-ix+\nu}(a)\, \mathrm{d}x = \frac{\pi}{2}\, K_{2\nu}\left[2a\cos\left(\frac{\varrho}{2}\right)\right]$

$\qquad\qquad\qquad\qquad\qquad\qquad\qquad$ $[2|\arg a| + |\operatorname{Re}\varrho| < \pi]$ $\qquad$ ET II 383(28)

4. $\displaystyle\int_{-\infty}^\infty \operatorname{sech}\left(\frac{\pi}{2}x\right) J_{ix}(a)\, \mathrm{d}x = 2\sin a$ $\qquad$ $[a > 0]$ $\qquad$ ET II 380(6)

5. $\displaystyle\int_{-\infty}^\infty \operatorname{cosech}\left(\frac{\pi}{2}x\right) J_{ix}(a)\, \mathrm{d}x = -2i\cos a$ $\qquad$ $[a > 0]$ $\qquad$ ET II 380(7)

6. $\displaystyle\int_0^\infty \operatorname{sech}(\pi x)\left\{[J_{ix}(a)]^2 + [Y_{ix}(a)]^2\right\} \mathrm{d}x = -Y_0(2a) - \mathbf{E}_0(2a)$

$\qquad\qquad\qquad\qquad\qquad\qquad\qquad\qquad$ $[a > 0]$ $\qquad$ ET II 380(12)

7. $\displaystyle\int_0^\infty x\sinh\left(\frac{\pi}{2}x\right) K_{ix}(a)\, \mathrm{d}x = \frac{\pi a}{2}$ $\qquad$ $[a > 0]$ $\qquad$ ET II 382(20)

8. $\displaystyle\int_0^\infty x\tanh(\pi x)\, K_{ix}(b)\, K_{ix}(a)\, \mathrm{d}x = \frac{\pi}{2}\sqrt{ab}\,\frac{\exp(-b-a)}{a+b}$

$\qquad\qquad\qquad\qquad\qquad\qquad$ $[|\arg b| < \pi, \quad |\arg a| < \pi]$ $\qquad$ ET II 175(4)

9. $\displaystyle\int_0^\infty x\sinh(\pi x)\, K_{2ix}(a)\, K_{ix}(b)\, \mathrm{d}x = \frac{\pi^{3/2}a}{2^{5/2}\sqrt{b}}\exp\left(-b - \frac{a^2}{8b}\right)$

$\qquad\qquad\qquad\qquad\qquad\qquad\qquad$ $\left[b > 0, \quad |\arg a| < \dfrac{\pi}{4}\right]$ $\qquad$ ET II 175(5)

10. $\displaystyle\int_0^\infty \frac{x\sinh(\pi x)}{x^2 + n^2}\, K_{ix}(a)\, K_{ix}(b)\, \mathrm{d}x = \frac{\pi^2}{2}\, I_n(b)\, K_n(a)$ $\quad$ $[0 < b < a; \quad n = 0, 1, 2, \ldots]$

$\qquad\qquad\qquad\qquad\qquad\qquad\qquad\qquad$ $= \frac{\pi^2}{2}\, I_n(a)\, K_n(b)$ $\quad$ $[0 < a < b; \quad n = 0, 1, 2, \ldots]$

$\qquad\qquad\qquad\qquad\qquad\qquad\qquad\qquad\qquad\qquad\qquad\qquad\qquad$ ET II 176(8)

11. $\displaystyle\int_0^\infty x\sinh(\pi x)\, K_{ix}(a)\, K_{ix}(b)\, K_{ix}(c)\, \mathrm{d}x = \frac{\pi^2}{4}\exp\left[-\frac{c}{2}\left(\frac{a}{b} + \frac{b}{a} + \frac{ab}{c^2}\right)\right]$

$\qquad\qquad\qquad\qquad\qquad\qquad\qquad$ $\left[|\arg a| + |\arg b| < \dfrac{\pi}{2}, \quad c > 0\right]$

$\qquad\qquad\qquad\qquad\qquad\qquad\qquad\qquad\qquad\qquad\qquad\qquad$ ET II 176(9)

12. $\displaystyle\int_0^\infty x\sinh\left(\frac{\pi}{2}x\right) K_{\frac{1}{2}ix}(a)\, K_{\frac{1}{2}ix}(b)\, K_{ix}(c)\, \mathrm{d}x = \frac{\pi^2 c}{2\sqrt{c^2 + 4ab}}\exp\left[-\frac{(a+b)\sqrt{c^2 + 4ab}}{2\sqrt{ab}}\right]$

$\qquad\qquad\qquad\qquad\qquad\qquad$ $[|\arg a| + |\arg b| < \pi, \quad c > 0]$

$\qquad\qquad\qquad\qquad\qquad\qquad\qquad\qquad\qquad\qquad\qquad\qquad$ ET II 176(10)

13. $\displaystyle\int_0^\infty x\sinh(\pi x)\, K_{\frac{1}{2}ix+\lambda}(\alpha)\, K_{\frac{1}{2}ix-\lambda}(\alpha)\, K_{ix}(\gamma)\, \mathrm{d}x = 0$ $\qquad$ $[0 < \gamma < 2\alpha]$

$\qquad\qquad\qquad\qquad\qquad\qquad\qquad\qquad = \frac{\pi^2\gamma}{2^{2\lambda+1}\alpha^{2\lambda}z}\left[(\gamma + z)^{2\lambda} + (\gamma - z)^{2\lambda}\right]$

$\qquad\qquad\qquad\qquad\qquad\qquad\qquad z = \sqrt{\gamma^2 - 4\alpha^2}$ $\qquad$ $[0 < 2\alpha < \gamma]$ $\quad$ ET II 176(11)

**6.795**

1. $$\int_0^\infty \cos(bx)\, K_{ix}(a)\, \mathrm{d}x = \frac{\pi}{2} e^{-a\cosh b} \qquad\qquad \left[|\mathrm{Im}\, b| < \tfrac{\pi}{2}, \quad a > 0\right]$$

$$\text{EH II 55(46), ET II 175(2)}$$

2. $$\int_0^\infty J_x(ax)\, J_{-x}(ax)\, \cos(\pi x)\, \mathrm{d}x = \frac{1}{4}\left(1-a^2\right)^{-1/2} \qquad [|a| < 1] \qquad\qquad \text{ET II 380(4)}$$

3. $$\int_0^\infty x \sin(ax)\, K_{ix}(b)\, \mathrm{d}x = \frac{\pi b}{2}\sinh a \exp\left(-b\cosh a\right) \qquad \left[|\mathrm{Im}\, a| < \tfrac{\pi}{2}, \quad b > 0\right] \qquad \text{ET II 175(1)}$$

4. $$\int_{-\infty}^{-\infty} \frac{\sin[(\nu+ix)\pi]}{n+\nu+ix}\, K_{\nu+ix}(a)\, K_{\nu-ix}(b)\, \mathrm{d}x = \pi^2\, I_n(a)\, K_{n+2\nu}(b) \qquad [0 < a < b; \quad n = 0, 1, \ldots]$$

$$= \pi^2\, K_{n+2\nu}(a)\, I_n(b) \qquad [0 < b < a; \quad n = 0, 1, \ldots]$$

$$\text{ET II 382(25)}$$

5. $$\int_0^\infty x \sin\left(\frac{1}{2}\pi x\right) K_{\frac{1}{2}ix}(a)\, K_{ix}(b)\, \mathrm{d}x = \frac{\pi^{3/2} b}{\sqrt{2a}}\exp\left(-a-\frac{b^2}{8a}\right)$$

$$\left[|\arg a| < \tfrac{\pi}{2}, \quad b > 0\right] \qquad \text{ET II 175(6)}$$

**6.796**

1. $$\int_{-\infty}^\infty \frac{e^{\frac{1}{2}\pi x}\cos(bx)}{\sinh(\pi x)}\, J_{ix}(a)\, \mathrm{d}x = -i \exp\left(ia\cosh b\right) \qquad [a > 0, \quad b > 0] \qquad \text{ET II 380(8)}$$

2. $$\int_0^\infty \cos(bx)\cosh\left(\frac{1}{2}\pi x\right) K_{ix}(a)\, \mathrm{d}x = \frac{\pi}{2}\cos\left(a\sinh b\right) \qquad\qquad \text{EH II 55(47)}$$

3. $$\int_0^\infty \sin(bx)\sinh\left(\frac{1}{2}\pi x\right) K_{ix}(a)\, \mathrm{d}x - \frac{\pi}{2}\sin\left(a\sinh b\right) \qquad\qquad \text{EH II 55(48)}$$

4. $$\int_0^\infty \cos(bx)\cosh(\pi x)\left[K_{ix}(a)\right]^2 \mathrm{d}x = -\frac{\pi^2}{4}\, Y_0\left[2a\sinh\left(\frac{b}{2}\right)\right]$$

$$[a > 0, \quad b > 0] \qquad \text{ET II 383(27)}$$

5. $$\int_0^\infty \sin(bx)\sinh(\pi x)\left[K_{ix}(a)\right]^2 \mathrm{d}x = \frac{\pi^2}{4}\, J_0\left[2a\sinh\left(\frac{b}{2}\right)\right]$$

$$[a > 0, \quad b > 0] \qquad \text{ET II 382(26)}$$

**6.797**

1. $$\int_0^\infty x e^{\pi x}\sinh(\pi x)\, \Gamma(\nu+ix)\, \Gamma(\nu-ix)\, H_{ix}^{(2)}(a)\, H_{ix}^{(2)}(b)\, \mathrm{d}x$$

$$= i 2^\nu \sqrt{\pi}\, \Gamma\left(\tfrac{1}{2}+\nu\right)(ab)^\nu (a+b)^{-\nu}\, K_\nu(a+b)$$

$$[a > 0, \quad b > 0, \quad \mathrm{Re}\,\nu > 0] \quad \text{ET II 381(14)}$$

2. $$\int_0^\infty x e^{\pi x}\sinh(\pi x)\cosh(\pi x)\, \Gamma(\nu+ix)\, \Gamma(\nu-ix)\, H_{ix}^{(2)}(a)\, H_{ix}^{(2)}(b)\, \mathrm{d}x = \frac{i\pi^{3/2} 2^\nu}{\Gamma\left(\tfrac{1}{2}-\nu\right)}(b-a)^{-\nu}\, H_\nu^{(2)}(b-a)$$

$$\left[0 < a < b, \quad 0 < \mathrm{Re}\,\nu < \tfrac{1}{2}\right]$$

$$\text{ET II 381(15)}$$

3.  $\int_0^\infty xe^{\pi x}\sinh(\pi x)\,\Gamma\left(\dfrac{\nu+ix}{2}\right)\Gamma\left(\dfrac{\nu-ix}{2}\right)H_{ix}^{(2)}(a)\,H_{ix}^{(2)}(b)\,dx$

$$= i\pi 2^{2-\nu}(ab)^\nu\left(a^2+b^2\right)^{-\frac{1}{2}\nu}H_\nu^{(2)}\left(\sqrt{a^2+b^2}\right)$$

$$[a>0,\quad b>0,\quad \operatorname{Re}\nu>0]\quad \text{ET II 381(16)}$$

4.[11]  $\int_0^\infty x\sinh(\pi x)\,\Gamma(\lambda+ix)\,\Gamma(\lambda-ix)\,K_{ix}(a)\,K_{ix}(b)\,dx = 2^{\lambda-1}\pi^{3/2}(ab)^\lambda(a+b)^{-\lambda}\Gamma\left(\lambda+\tfrac{1}{2}\right)K_\lambda(a+b)$

$$[|\arg a|<\pi,\quad \operatorname{Re}\lambda>0,\quad b>0]$$
$$\text{ET II 176(12)}$$

5.  $\int_0^\infty x\sinh(2\pi x)\,\Gamma(\lambda+ix)\,\Gamma(\lambda-ix)\,K_{ix}(a)\,K_{ix}(b)\,dx = \dfrac{2^\lambda\pi^{\frac{5}{2}}}{\Gamma\left(\frac{1}{2}-\lambda\right)}\left(\dfrac{ab}{|b-a|}\right)^\lambda K_\lambda\left(|b-a|\right)$

$$[a>0,\quad 0<\operatorname{Re}\lambda<\tfrac{1}{2},\quad b>0]$$
$$\text{ET II 176(13)}$$

6.  $\int_0^\infty x\sinh(\pi x)\,\Gamma\left(\lambda+\tfrac{1}{2}ix\right)\Gamma\left(\lambda-\tfrac{1}{2}ix\right)K_{ix}(a)\,K_{ix}(b)\,dx = 2\pi^2\left(\dfrac{ab}{2\sqrt{a^2+b^2}}\right)K_{2\lambda}\left(\sqrt{a^2+b^2}\right)$

$$\left[|\arg a|<\tfrac{\pi}{2},\quad \operatorname{Re}\lambda>0,\quad b>0\right]$$
$$\text{ET II 177(14)}$$

7.  $\int_0^\infty \dfrac{x\tanh(\pi x)\,K_{ix}(a)\,K_{ix}(b)}{\Gamma\left(\frac{3}{4}+\frac{1}{2}ix\right)\Gamma\left(\frac{3}{4}-\frac{1}{2}ix\right)}\,dx = \dfrac{1}{2}\sqrt{\dfrac{\pi ab}{a^2+b^2}}\exp\left(-\sqrt{a^2+b^2}\right)$

$$\left[|\arg a|<\tfrac{\pi}{2},\quad b>0\right],\qquad \text{ET II 177(15)}$$

# 6.8 Functions Generated by Bessel Functions

## 6.81 Struve functions

### 6.811

1.  $\int_0^\infty \mathbf{H}_\nu(bx)\,dx = -\dfrac{\cot\left(\frac{\nu\pi}{2}\right)}{b}$  $\qquad [-2<\operatorname{Re}\nu<0,\quad b>0]\qquad \text{ET II 158(1)}$

2.  $\int_0^\infty \mathbf{H}_\nu\left(\dfrac{a^2}{x}\right)\mathbf{H}_\nu(bx)\,dx = -\dfrac{J_{2\nu}\left(2a\sqrt{b}\right)}{b}$  $\qquad \left[a>0,\quad b>0,\quad \operatorname{Re}\nu>-\tfrac{3}{2}\right]$

$$\text{ET II 170(37)}$$

3.  $\int_0^\infty \mathbf{H}_{\nu-1}\left(\dfrac{a^2}{x}\right)\mathbf{H}_\nu(bx)\,\dfrac{dx}{x} = -\dfrac{1}{a\sqrt{b}}\,J_{2\nu-1}\left(2a\sqrt{b}\right)$  $\quad \left[a>0,\quad b>0,\quad \operatorname{Re}\nu>-\tfrac{1}{2}\right]$

$$\text{ET II 170(38)}$$

### 6.812

1.  $\int_0^\infty \dfrac{\mathbf{H}_1(bx)\,dx}{x^2+a^2} = \dfrac{\pi}{2a}\left[I_1(ab)-\mathbf{L}_1(ab)\right]$  $\qquad [\operatorname{Re}a>0,\quad b>0]\qquad \text{ET II 158(6)}$

2.    $\displaystyle\int_0^\infty \frac{\mathbf{H}_\nu(bx)}{x^2+a^2}\,dx = -\frac{\pi}{2a\sin\left(\frac{\nu\pi}{2}\right)}\,\mathbf{L}_\nu(ab) + \frac{b\cot\left(\frac{\nu\pi}{2}\right)}{1-\nu^2}\,{}_1F_2\left(1;\frac{3-\nu}{2},\frac{3+\nu}{2};\frac{a^2b^2}{2}\right)$

$$[\operatorname{Re}a>0,\quad b>0,\quad |\operatorname{Re}\nu|<2]$$
ET II 159(7)

**6.813**

1.    $\displaystyle\int_0^\infty x^{s-1}\,\mathbf{H}_\nu(ax)\,dx = \frac{2^{s-1}\,\Gamma\left(\frac{s+\nu}{2}\right)}{a^s\,\Gamma\left(\frac{1}{2}\nu-\frac{1}{2}s+1\right)}\tan\left(\frac{s+\nu}{2}\pi\right)$

$$\left[a>0,\quad -1-\operatorname{Re}\nu<\operatorname{Re}s<\min\left(\frac{3}{2},1-\operatorname{Re}\nu\right)\right]\quad \text{WA 429(2), ET I 335(52)}$$

2.    $\displaystyle\int_0^\infty x^{-\nu-1}\,\mathbf{H}_\nu(x)\,dx = \frac{2^{-\nu-1}\pi}{\Gamma(\nu+1)}$      $[\operatorname{Re}\nu>-\tfrac{3}{2}]$      ET II 383(2)

3.    $\displaystyle\int_0^\infty x^{-\mu-\nu}\,\mathbf{H}_\mu(x)\,\mathbf{H}_\nu(x)\,dx = \frac{2^{-\mu-\nu}\sqrt{\pi}\,\Gamma(\mu+\nu)}{\Gamma\left(\mu+\frac{1}{2}\right)\Gamma\left(\nu+\frac{1}{2}\right)\Gamma\left(\mu+\nu+\frac{1}{2}\right)}$

$$[\operatorname{Re}(\mu+\nu)>0]\quad \text{WA 435(2), ET II 384(8)}$$

4.    $\displaystyle\int_0^1 x^{\nu+1}\,\mathbf{H}_\nu(ax)\,dx = \frac{1}{a}\,\mathbf{H}_{\nu+1}(a)$      $[a>0,\quad \operatorname{Re}\nu>-\tfrac{3}{2}]$      ET II 158(2)a

5.    $\displaystyle\int_0^1 x^{1-\nu}\,\mathbf{H}_\nu(ax)\,dx = \frac{a^{\nu-1}}{2^{\nu-1}\sqrt{\pi}\,\Gamma\left(\nu+\frac{1}{2}\right)} - \frac{1}{a}\,\mathbf{H}_{\nu-1}(a)$

$$[a>0]$$
ET II 158(3)a

**6.814**

1.    $\displaystyle\int_0^\infty \frac{x^{\nu+1}\,\mathbf{H}_\nu(bx)}{(x^2+a^2)^{1-\mu}}\,dx = \frac{2^{\mu-1}\,a^{\mu+\nu}b^{-\mu}}{\Gamma(1-\mu)\cos[(\mu+\nu)\pi]}\left[I_{-\mu-\nu}(ab)-\mathbf{L}_{\mu+\nu}(ab)\right]$

$$\left[\operatorname{Re}a>0,\quad b>0,\quad \operatorname{Re}\nu>-\tfrac{3}{2},\quad \operatorname{Re}(\mu+\nu)<\tfrac{1}{2},\quad \operatorname{Re}(2\mu+\nu)<\tfrac{3}{2}\right]\quad \text{ET II 159(8)}$$

**6.815**

1.    $\displaystyle\int_0^1 x^{\frac{1}{2}\nu}(1-x)^{\mu-1}\,\mathbf{H}_\nu\left(a\sqrt{x}\right)\,dx = 2^\mu a^{-\mu}\,\Gamma(\mu)\,\mathbf{H}_{\mu+\nu}(a)$

$$\left[\operatorname{Re}\nu>-\tfrac{3}{2},\quad \operatorname{Re}\mu>0\right]\quad \text{ET II 199(88)a}$$

2.    $\displaystyle\int_0^1 x^{\lambda-\frac{1}{2}\nu-\frac{3}{2}}(1-x)^{\mu-1}\,\mathbf{H}_\nu\left(a\sqrt{x}\right)\,dx = \frac{B(\lambda,\mu)a^{\nu+1}}{2^\nu\sqrt{\pi}\,\Gamma\left(\nu+\frac{3}{2}\right)}\,{}_2F_3\left(1,\lambda;\frac{3}{2},\nu+\frac{3}{2},\lambda+\mu;-\frac{a^2}{4}\right)$

$$\left[\operatorname{Re}\lambda>0,\quad \operatorname{Re}\mu>0\right]\quad \text{ET II 199(89)a}$$

## 6.82   Combinations of Struve functions, exponentials, and powers

**6.821**

1.[6]    $\displaystyle\int_0^\infty e^{-\alpha x}\,\mathbf{H}_{-n-\frac{1}{2}}(\beta x)\,dx = (-1)^n\beta^{n+\frac{1}{2}}\left(\alpha+\sqrt{\alpha^2+\beta^2}\right)^{-n-\frac{1}{2}}\frac{1}{\sqrt{\alpha^2+\beta^2}}$

$$[\operatorname{Re}\alpha>|\operatorname{Im}\beta|]$$
ET I 206(6)

$2.^6$ $\qquad \int_0^\infty e^{-\alpha x}\, \mathbf{L}_{-n-\frac{1}{2}}(\beta x)\, dx = \beta^{n+\frac{1}{2}} \left( \alpha + \sqrt{\alpha^2 - \beta^2} \right)^{-n-\frac{1}{2}} \dfrac{1}{\sqrt{\alpha^2 - \beta^2}}$

$$[\operatorname{Re}\alpha > |\operatorname{Re}\beta|] \qquad\qquad \text{ET I 208(26)}$$

$3.$ $\qquad \int_0^\infty e^{-\alpha x}\, \mathbf{H}_0(\beta x)\, dx = \dfrac{2}{\pi}\, \dfrac{\ln\left( \dfrac{\sqrt{\alpha^2+\beta^2}+\beta}{\alpha} \right)}{\sqrt{\alpha^2 + \beta^2}}$ $\qquad [\operatorname{Re}\alpha > |\operatorname{Im}\beta|] \qquad \text{ET II 205(1)}$

$4.$ $\qquad \int_0^\infty e^{-\alpha x}\, \mathbf{L}_0(\beta x)\, dx = \dfrac{2}{\pi}\, \dfrac{\arcsin\left( \dfrac{\beta}{\alpha} \right)}{\sqrt{\alpha^2 + \beta^2}}$ $\qquad [\operatorname{Re}\alpha > |\operatorname{Re}\beta|] \qquad \text{ET II 207(18)}$

**6.822** $\qquad \int_0^\infty e^{(\nu+1)x}\, \mathbf{H}_\nu(a\sinh x)\, dx = \sqrt{\dfrac{\pi}{a}}\, \operatorname{cosec}(\nu\pi) \left[ \sinh\left( \dfrac{a}{2} \right) I_{\nu+\frac{1}{2}}\left( \dfrac{a}{2} \right) - \cosh\left( \dfrac{a}{2} \right) I_{-\nu-\frac{1}{2}}\left( \dfrac{a}{2} \right) \right]$

$$[\operatorname{Re} a > 0, \quad -2 < \operatorname{Re}\nu < 0]$$
$$\text{ET II 385(11)}$$

**6.823**

$1.$ $\qquad \int_0^\infty x^\lambda e^{-\alpha x}\, \mathbf{H}_\nu(bx)\, dx = \dfrac{b^{\nu+1}\,\Gamma(\lambda+\nu+2)}{2^\nu a^{\lambda+\nu+2}\sqrt{\pi}\,\Gamma\left( \nu+\dfrac{3}{2} \right)}\; {}_3F_2\left( 1, \dfrac{\lambda+\nu}{2}+1, \dfrac{\lambda+\nu+3}{2}; \dfrac{3}{2}, \nu+\dfrac{3}{2}; -\dfrac{b^2}{a^2} \right)$

$$[\operatorname{Re} a > 0, \quad b > 0, \quad \operatorname{Re}(\lambda+\nu) > -2]$$
$$\text{ET II 161(19)}$$

$2.$ $\qquad \int_0^\infty x^\nu e^{-\alpha x}\, \mathbf{L}_\nu(\beta x)\, dx = \dfrac{(2\beta)^\nu\,\Gamma\left( \nu+\dfrac{1}{2} \right)}{\sqrt{\pi}\left( \sqrt{\alpha^2-\beta^2} \right)^{2\nu+1}} - \dfrac{\Gamma(2\nu+1)\left( \dfrac{\beta}{\alpha} \right)^\nu}{\sqrt{\dfrac{\pi}{2}}\,\alpha\,(\beta^2-\alpha^2)^{\frac{1}{2}\nu+\frac{1}{4}}}\, P_{-\nu-\frac{1}{2}}^{-\nu-\frac{1}{2}}\left( \dfrac{\beta}{\alpha} \right)$

$$\left[\operatorname{Re}\alpha > |\operatorname{Re}\beta|, \quad \operatorname{Re}\nu > -\dfrac{1}{2}\right]$$
$$\text{ET I 209(35)a}$$

**6.824**

$1.$ $\qquad \int_0^\infty t^\nu e^{-at}\, \mathbf{L}_{2\nu}\left( 2\sqrt{t} \right)\, dt = \dfrac{1}{a^{2\nu+1}}\, e^{\frac{1}{a}}\, \Phi\left( \dfrac{1}{\sqrt{a}} \right)$ $\qquad\qquad\qquad\qquad \text{MI 51}$

$2.$ $\qquad \int_0^\infty t^\nu e^{-at}\, \mathbf{L}_{-2\nu}\left( \sqrt{t} \right)\, dt = \dfrac{1}{\Gamma\left( \dfrac{1}{2}-2\nu \right) a^{2\nu+1}}\, e^{\frac{1}{a}}\, \gamma\left( \dfrac{1}{2}-2\nu, \dfrac{1}{a} \right)$ $\qquad\qquad \text{MI 51}$

**6.825** $\qquad \int_0^\infty x^{s-1} e^{-\alpha^2 x^2}\, \mathbf{H}_\nu(\beta x)\, dx = \dfrac{\beta^{\nu+1}\,\Gamma\left( \dfrac{1}{2}+\dfrac{s}{2}+\dfrac{\nu}{2} \right)}{2^{\nu+1}\sqrt{\pi}\,\alpha^{\nu+s+1}\,\Gamma\left( \nu+\dfrac{3}{2} \right)}\; {}_2F_2\left( 1, \dfrac{\nu+s+1}{2}; \dfrac{3}{2}, \nu+\dfrac{3}{2}; -\dfrac{\beta^2}{4\alpha^2} \right)$

$$\left[\operatorname{Re} s > -\operatorname{Re}\nu - 1, \quad |\arg\alpha| < \dfrac{\pi}{4}\right]$$
$$\text{ET I 335(51)a, ET II 162(20)}$$

## 6.83 Combinations of Struve and trigonometric functions

**6.831** $\qquad \int_0^\infty x^{-\nu} \sin(ax)\, \mathbf{H}_\nu(bx)\, dx = 0$ $\qquad\qquad\qquad \left[ 0 < b < a, \quad \operatorname{Re}\nu > -\dfrac{1}{2} \right]$

$$= \sqrt{\pi}\, 2^{-\nu} b^{-\nu}\, \dfrac{(b^2-a^2)^{\nu-\frac{1}{2}}}{\Gamma\left( \nu+\dfrac{1}{2} \right)} \qquad \left[ 0 < a < b, \quad \operatorname{Re}\nu > -\dfrac{1}{2} \right]$$

$$\text{ET II 162(21)}$$

**6.832** $\quad \displaystyle\int_0^\infty \sqrt{x} \sin(ax) \, \mathbf{H}_{\frac{1}{4}}\left(b^2 x^2\right) dx = -2^{-3/2} \sqrt{\pi} \, \frac{\sqrt{a}}{b^2} \, Y_{\frac{1}{4}}\left(\frac{a^2}{4b^2}\right)$

$$[a > 0]$$

ET I 109(14)

## 6.84–6.85   Combinations of Struve and Bessel functions

**6.841** $\quad \displaystyle\int_0^\infty \mathbf{H}_{\nu-1}(ax) \, Y_\nu(bx) \, dx = -a^{\nu-1} b^{-\nu} \qquad\qquad \left[0 < b < a, \quad |\mathrm{Re}\,\nu| < \tfrac{1}{2}\right]$

$$= 0 \qquad\qquad \left[0 < a < b, \quad |\mathrm{Re}\,\nu| < \tfrac{1}{2}\right]$$

ET II 114(36)

**6.842** $\quad \displaystyle\int_0^\infty \left[\mathbf{H}_0(ax) - Y_0(ax)\right] J_0(bx) \, dx = \frac{4}{\pi(a+b)} \, K\left(\frac{|a-b|}{a+b}\right)$

$$[a > 0, \quad b > 0]$$
ET II 15(22)

**6.843**

1. $\quad \displaystyle\int_0^\infty J_{2\nu}\left(a\sqrt{x}\right) \mathbf{H}_\nu(bx) \, dx = -\frac{1}{b} \, Y_\nu\left(\frac{a^2}{4b}\right) \qquad \left[a > 0, \quad b > 0, \quad -1 < \mathrm{Re}\,\nu < \tfrac{5}{4}\right]$

ET II 164(10)

2. $\quad \displaystyle\int_0^\infty K_{2\nu}\left(2a\sqrt{x}\right) \mathbf{H}_\nu(bx) \, dx = \frac{2^\nu}{\pi b} \, \Gamma(\nu+1) \, S_{-\nu-1,\nu}\left(\frac{a^2}{b}\right)$

$$\left[\mathrm{Re}\,a > 0, \quad b > 0, \quad \mathrm{Re}\,\nu > -1\right]$$
ET II 168(27)

**6.844** $\quad \displaystyle\int_0^\infty \left[\cos\left(\frac{\mu-\nu}{2}\pi\right) J_\mu\left(a\sqrt{x}\right) - \sin\left(\frac{\mu-\nu}{2}\pi\right) Y_\mu\left(a\sqrt{x}\right)\right] K_\mu\left(a\sqrt{x}\right) \mathbf{H}_\nu(bx) \, dx$

$$= \frac{1}{a^2} \, W_{\frac{1}{2}\nu, \frac{1}{2}\mu}\left(\frac{a^2}{2b}\right) W_{-\frac{1}{2}\nu, \frac{1}{2}\mu}\left(\frac{a^2}{2b}\right)$$

$$\left[|\arg a| < \tfrac{\pi}{4}, \quad b > 0, \quad \mathrm{Re}\,\nu > |\mathrm{Re}\,\mu| - 2\right] \quad \text{ET II 169(35)}$$

**6.845**

1. $\quad \displaystyle\int_0^\infty \left[\mathbf{H}_{-\nu}\left(\frac{a}{x}\right) - Y_{-\nu}\left(\frac{a}{x}\right)\right] J_\nu(bx) \, dx = \frac{4}{\pi b} \cos(\nu\pi) \, K_{2\nu}\left(2\sqrt{ab}\right)$

$$\left[|\arg a| < \pi, \quad b > 0, \quad |\mathrm{Re}\,\nu| < \tfrac{1}{2}\right]$$
ET II 73(7)

2. $\quad \displaystyle\int_0^\infty \left[J_{-\nu}\left(\frac{a^2}{x}\right) + \sin(\nu\pi) \mathbf{H}_\nu\left(\frac{a^2}{x}\right)\right] \mathbf{H}_\nu(bx) \, dx = \frac{1}{b}\left[\frac{2}{\pi} K_{2\nu}\left(2a\sqrt{b}\right) - Y_{2\nu}\left(2a\sqrt{b}\right)\right]$

$$\left[a > 0, \quad b > 0, \quad -\tfrac{3}{2} < \mathrm{Re}\,\nu < 0\right]$$
ET II 170(39)

**6.846** $\quad \displaystyle\int_0^\infty \left[\frac{2}{\pi} K_{2\nu}\left(2a\sqrt{x}\right) + Y_{2\nu}\left(2a\sqrt{x}\right)\right] \mathbf{H}_\nu(bx) \, dx = \frac{1}{b} J_\nu\left(\frac{a^2}{b}\right)$

$$\left[a > 0, \quad b > 0, \quad |\mathrm{Re}\,\nu| < \tfrac{1}{2}\right]$$
ET II 169(30)

**6.847** $\quad \displaystyle\int_0^\infty \left[\cos\frac{\nu\pi}{2} J_\nu(ax) + \sin\frac{\nu\pi}{2} \mathbf{H}_\nu(ax)\right] \frac{dx}{x^2 + k^2} = \frac{\pi}{2k}\left[I_\nu(ak) - \mathbf{L}_\nu(ak)\right]$

$$\left[a > 0, \quad \mathrm{Re}\,k > 0, \quad -\tfrac{1}{2} < \mathrm{Re}\,\nu < 2\right]$$
ET II 384(5)a, WA 467(8)

**6.848**

1. 
$$\int_0^\infty x\left[I_\nu(ax) - \mathbf{L}_{-\nu}(ax)\right] J_\nu(bx)\,dx = \frac{2}{\pi}\left(\frac{a}{b}\right)^{\nu-1}\cos(\nu\pi)\frac{1}{a^2+b^2}$$
$$\left[\operatorname{Re} a > 0, \quad b > 0, \quad -1 < \operatorname{Re}\nu < -\tfrac{1}{2}\right]$$
ET II 74(12)

2. 
$$\int_0^\infty x\left[\mathbf{H}_{-\nu}(ax) - Y_{-\nu}(ax)\right] J_\nu(bx)\,dx = 2\frac{\cos(\nu\pi)}{a^\nu\pi}b^{\nu-1}\frac{1}{a+b}$$
$$\left[|\arg a| < \pi, \quad -\tfrac{1}{2} < \operatorname{Re}\nu, \quad b > 0\right]$$
ET II 73(5)

**6.849**

1. 
$$\int_0^\infty x\,K_\nu(ax)\,\mathbf{H}_\nu(bx)\,dx = a^{-\nu-1}b^{\nu+1}\frac{1}{a^2+b^2} \qquad \left[\operatorname{Re} a > 0, \quad b > 0, \quad \operatorname{Re}\nu > -\tfrac{3}{2}\right]$$
ET II 164(12)

2. 
$$\int_0^\infty x\left[K_\mu(ax)\right]^2 \mathbf{H}_0(bx)\,dx = -2^{-\mu-1}\pi a^{-2\mu}\frac{\left[(z+b)^{2\mu} + (z-b)^{2\mu}\right]}{bz}\sec(\mu\pi),$$
$$z = \sqrt{4a^2+b^2} \qquad \left[\operatorname{Re} a > 0, \quad b > 0, \quad |\operatorname{Re}\mu| < \tfrac{3}{2}\right] \quad \text{ET II 166(18)}$$

**6.851**

1. 
$$\int_0^\infty x\left\{\left[J_{\frac{1}{2}\nu}(ax)\right]^2 - \left[Y_{\frac{1}{2}\nu}(ax)\right]^2\right\}\mathbf{H}_\nu(bx)\,dx = 0 \qquad \left[0 < b < 2a, \quad -\tfrac{3}{2} < \operatorname{Re}\nu < 0\right]$$
$$= \frac{4}{\pi b\sqrt{b^2-4a^2}} \qquad \left[0 < 2a < b, \quad -\tfrac{3}{2} < \operatorname{Re}\nu < 0\right]$$
ET II 164(7)

2. 
$$\int_0^\infty x^{\nu+1}\left\{\left[J_\nu(ax)\right]^2 - \left[Y_\nu(ax)\right]^2\right\}\mathbf{H}_\nu(bx)\,dx$$
$$= 0 \qquad \left[0 < b < 2a, \quad -\tfrac{3}{4} < \operatorname{Re}\nu < 0\right]$$
$$= \frac{2^{3\nu+2}a^{2\nu}b^{-\nu-1}}{\sqrt{\pi}\,\Gamma\left(\tfrac{1}{2}-\nu\right)}\left(b^2-4a^2\right)^{-\nu-\frac{1}{2}} \qquad \left[0 < 2a < b, \quad -\tfrac{3}{4} < \operatorname{Re}\nu < 0\right]$$
ET II 163(6)

**6.852**

1. 
$$\int_0^\infty x^{1-\mu-\nu} J_\nu(x)\,\mathbf{H}_\mu(x)\,dx = \frac{(2\nu-1)2^{-\mu-\nu}}{(\mu+\nu-1)\,\Gamma\left(\mu+\tfrac{1}{2}\right)\Gamma\left(\nu+\tfrac{1}{2}\right)}$$
$$\left[\operatorname{Re}\nu > \tfrac{1}{2}, \quad \operatorname{Re}(\mu+\nu) > 1\right]$$
ET II 383(4)

2. 
$$\int_0^\infty x^{\mu-\nu+1} Y_\mu(ax)\,\mathbf{H}_\nu(bx)\,dx$$
$$= 0 \qquad \left[0 < b < a, \quad \operatorname{Re}(\nu-\mu) > 0, \quad -\tfrac{3}{2} < \operatorname{Re}\mu < \tfrac{1}{2}\right]$$
$$= \frac{2^{1+\mu-\nu}a^\mu b^{-\nu}}{\Gamma(\nu-\mu)}\left(b^2-a^2\right)^{\nu-\mu-1} \qquad \left[0 < a < b, \quad \operatorname{Re}(\nu-\mu) > 0, \quad -\tfrac{3}{2} < \operatorname{Re}\mu < \tfrac{1}{2}\right]$$
ET II 163(3)

3.    $\displaystyle\int_0^\infty x^{\mu+\nu+1}\, K_\mu(ax)\,\mathbf{H}_\nu(bx)\,\mathrm{d}x = \frac{2^{\mu+\nu+1}b^{\nu+1}}{\sqrt{\pi}\,a^{\mu+2\nu+3}}\,\Gamma\!\left(\mu+\nu+\frac{3}{2}\right)F\!\left(1,\mu+\nu+\frac{3}{2};\frac{3}{2};-\frac{b^2}{a^2}\right)$

$$\left[\mathrm{Re}\,a>0,\quad b>0,\quad \mathrm{Re}\,\nu>-\tfrac{3}{2},\quad \mathrm{Re}(\mu+\nu)>-\tfrac{3}{2}\right]\quad \text{ET II 165(13)}$$

**6.853**

1.    $\displaystyle\int_0^\infty x^{1-\mu}\left[\sin\left(\mu\pi\right)J_{\mu+\nu}(ax)+\cos(\mu\pi)\,Y_{\mu+\nu}(ax)\right]\mathbf{H}_\nu(bx)\,\mathrm{d}x$

$$= 0 \qquad\qquad \left[0<b<a,\quad 1<\mathrm{Re}\,\mu<\tfrac{3}{2},\quad \mathrm{Re}\,\nu>-\tfrac{3}{2},\quad \mathrm{Re}(\nu-\mu)<\tfrac{1}{2}\right]$$

$$= \frac{b^\nu\left(b^2-a^2\right)^{\mu-1}}{2^{\mu-1}a^{\mu+\nu}\,\Gamma(\mu)}\qquad \left[0<a<b,\quad 1<\mathrm{Re}\,\mu<\tfrac{3}{2},\quad \mathrm{Re}\,\nu>-\tfrac{3}{2},\quad \mathrm{Re}(\nu-\mu)<\tfrac{1}{2}\right]$$

$$\text{ET II 163(4)}$$

2.    $\displaystyle\int_0^\infty x^{\lambda+\frac{1}{2}}\left[I_\mu(ax)-\mathbf{L}_{-\mu}(ax)\right]J_\nu(bx)\,\mathrm{d}x$

$$= 2^{\lambda+\frac{1}{2}}\frac{\cos(\mu\pi)}{\pi}b^{-\lambda-\frac{3}{2}}\,G^{22}_{33}\!\left(\frac{b^2}{a^2}\left|\begin{array}{c}\frac{1+\mu}{2},1-\frac{\mu}{2},1+\frac{\mu}{2}\\[4pt]\frac{3}{4}+\frac{\lambda+\nu}{2},\frac{1+\mu}{2},\frac{3}{4}+\frac{\lambda-\nu}{2}\end{array}\right.\right)$$

$$\left[\mathrm{Re}\,a>0,\quad b>0,\quad \mathrm{Re}(\mu+\nu+\lambda)>-\tfrac{3}{2},\quad -\mathrm{Re}\,\nu-\tfrac{5}{2}<\mathrm{Re}(\lambda-\mu)<1\right]\quad \text{ET II 76(21)}$$

3.    $\displaystyle\int_0^\infty x^{\lambda+\frac{1}{2}}\left[\mathbf{H}_\mu(ax)-Y_\mu(ax)\right]J_\nu(bx)\,\mathrm{d}x$

$$= 2^{\lambda+\frac{1}{2}}\frac{\cos(\mu\pi)}{\pi^2}b^{-\lambda-\frac{3}{2}}\,G^{23}_{33}\!\left(\frac{b^2}{a^2}\left|\begin{array}{c}\frac{1-\mu}{2},1-\frac{\mu}{2},1+\frac{\mu}{2}\\[4pt]\frac{3}{4}+\frac{\lambda+\nu}{2},\frac{1-\mu}{2},\frac{3}{4}+\frac{\lambda-\nu}{?}\end{array}\right.\right)$$

$$\left[b>0,\quad |\arg a|<\pi,\quad \mathrm{Re}(\lambda+\mu)<1,\quad \mathrm{Re}(\lambda+\nu)+\tfrac{3}{2}>|\mathrm{Re}\,\mu|\right]\quad \text{ET II 73(6)}$$

4.    $\displaystyle\int_0^\infty \sqrt{x}\left[I_{\nu-\frac{1}{2}}(ax)-\mathbf{L}_{\nu-\frac{1}{2}}(ax)\right]J_\nu(bx)\,\mathrm{d}x = \sqrt{\frac{2}{\pi}}\,a^{\nu-\frac{1}{2}}b^{-\nu}\frac{1}{\sqrt{a^2+b^2}}$

$$\left[\mathrm{Re}\,a>0,\quad b>0,\quad |\mathrm{Re}\,\nu|<\tfrac{1}{2}\right]$$
$$\text{ET II 74(11)}$$

5.    $\displaystyle\int_0^\infty x^{\mu-\nu+1}\left[I_\mu(ax)-\mathbf{L}_\mu(ax)\right]J_\nu(bx)\,\mathrm{d}x = \frac{2^{\mu-\nu+1}a^{\mu-1}b^{\nu-2\mu-1}}{\sqrt{\pi}\,\Gamma\left(\nu-\mu+\frac{1}{2}\right)}\,F\!\left(1,\frac{1}{2};\nu-\mu+\frac{1}{2};-\frac{b^2}{a^2}\right)$

$$\left[-1<2\,\mathrm{Re}\,\mu+1<\mathrm{Re}\,\nu+\tfrac{1}{2},\quad \mathrm{Re}\,a>0,\quad b>0\right]\quad \text{ET II 74(13)}$$

6.    $\displaystyle\int_0^\infty x^{\mu-\nu+1}\left[I_\mu(ax)-\mathbf{L}_{-\mu}(ax)\right]J_\nu(bx)\,\mathrm{d}x = \frac{2^{\mu-\nu+1}a^{-\mu-1}b^{\nu-1}}{\Gamma\left(\frac{1}{2}-\mu\right)\Gamma\left(\frac{1}{2}+\nu\right)}\,F\!\left(1,\frac{1}{2}+\mu;\frac{1}{2}+\nu;-\frac{b^2}{a^2}\right)$

$$\left[\mathrm{Re}\,a>0,\quad \mathrm{Re}\,\nu>-\tfrac{1}{2},\quad \mathrm{Re}\,\mu>-1,\quad b>0\right]\quad \text{ET II 75(18)}$$

**6.854**

1.    $\displaystyle\int_0^\infty x\,\mathbf{H}_{\frac{1}{2}\nu}\left(ax^2\right)K_\nu(bx)\,\mathrm{d}x = \frac{\Gamma\left(\frac{1}{2}\nu+1\right)}{2^{1-\frac{1}{2}\nu}a\pi}\,S_{-\frac{1}{2}\nu-1,\frac{1}{2}\nu}\!\left(\frac{b^2}{4a}\right)$

$$\left[a>0,\quad \mathrm{Re}\,b>0,\quad \mathrm{Re}\,\nu>-2\right]$$
$$\text{ET II 150(75)}$$

2. 
$$\int_0^\infty x \, \mathbf{H}_{\frac{1}{2}\nu}\left(ax^2\right) J_\nu(bx) \, dx = -\frac{1}{2a} \, Y_{\frac{1}{2}\nu}\left(\frac{b^2}{4a}\right) \qquad \left[a > 0, \quad b > 0, \quad -2 < \operatorname{Re}\nu < \tfrac{3}{2}\right]$$
ET II 73(3)

**6.855**

1. 
$$\int_0^\infty x^{2\nu+\frac{1}{2}} \left[I_{\nu+\frac{1}{2}}\left(\frac{a}{x}\right) - \mathbf{L}_{\nu+\frac{1}{2}}\left(\frac{a}{x}\right)\right] J_\nu(bx) \, dx = 2^{\frac{3}{2}} \frac{a^{\nu+\frac{1}{2}}}{\sqrt{\pi}b^{\nu+1}} J_{2\nu+1}\left(\sqrt{2ab}\right) K_{2\nu+1}\left(\sqrt{2ab}\right)$$
$$\left[\operatorname{Re}a > 0, \quad b > 0, \quad -1 < \operatorname{Re}\nu < \tfrac{1}{2}\right]$$
ET II 76(22)

2. 
$$\int_0^\infty \left[\mathbf{H}_{-\nu-1}\left(\frac{a}{x}\right) - Y_{-\nu-1}\left(\frac{a}{x}\right)\right] J_\nu(bx) \frac{dx}{x} = -\frac{4}{\pi\sqrt{ab}} \cos(\nu\pi) \, K_{-2\nu-1}\left(2\sqrt{ab}\right)$$
$$\left[|\arg a| < \pi, \quad b > 0, \quad |\operatorname{Re}\nu| < \tfrac{1}{2}\right]$$
ET II 74(8)

3. 
$$\int_0^\infty x^{2\nu+\frac{1}{2}} \left[\mathbf{H}_{\nu+\frac{1}{2}}\left(\frac{a}{x}\right) - Y_{\nu+\frac{1}{2}}\left(\frac{a}{x}\right)\right] J_\nu(bx) \, dx$$
$$= -2^{5/2}\pi^{-3/2} a^{\nu+\frac{1}{2}} b^{-\nu-1} \sin(\nu\pi) \, K_{2\nu+1}\left(\sqrt{2ab}e^{\frac{1}{4}\pi i}\right) K_{2\nu+1}\left(\sqrt{2ab}e^{-\frac{1}{4}\pi i}\right)$$
$$\left[|\arg a| < \pi, \quad b > 0, \quad -1 < \operatorname{Re}\nu < -\tfrac{1}{6}\right] \quad \text{ET II 74(9)}$$

**6.856** 
$$\int_0^\infty x \, Y_\nu\left(a\sqrt{x}\right) K_\nu\left(a\sqrt{x}\right) \mathbf{H}_\nu(bx) \, dx = \frac{1}{2b^2} \exp\left(-\frac{a^2}{2b}\right)$$
$$\left[b > 0, \quad |\arg a| < \tfrac{\pi}{4}, \quad \operatorname{Re}\nu > -\tfrac{3}{2}\right]$$
ET II 169(32)

**6.857**

1. 
$$\int_0^\infty x \exp\left(\frac{a^2 x^2}{8}\right) K_{\frac{1}{2}\nu}\left(\frac{a^2 x^2}{8}\right) \mathbf{H}_\nu(bx) \, dx$$
$$= \frac{2}{\sqrt{\pi}} a^{-\frac{\nu}{2}-1} b^{\frac{\nu}{2}-1} \cos\left(\frac{\nu\pi}{2}\right) \Gamma\left(-\frac{1}{2}\nu\right) \exp\left(\frac{b^2}{2a^2}\right) W_{k,m}\left(\frac{b^2}{a^2}\right)$$
$$k = \tfrac{1}{4}\nu, \qquad m = \tfrac{1}{2} + \tfrac{1}{4}\nu \qquad \left[|\arg a| < \tfrac{3}{4}\pi, \quad b > 0, \quad -\tfrac{3}{2} < \operatorname{Re}\nu < 0\right] \quad \text{ET II 167(24)}$$

2. 
$$\int_0^\infty x^{\sigma-2} \exp\left(-\frac{1}{2}a^2 x^2\right) K_\mu\left(\frac{1}{2}a^2 x^2\right) \mathbf{H}_\nu(bx) \, dx$$
$$= \frac{\sqrt{\pi}}{2^{\nu+2}} a^{-\nu-\sigma} b^{\nu+1} \frac{\Gamma\left(\frac{\nu+\sigma}{2} + \mu\right) \Gamma\left(\frac{\nu+\sigma}{2} - \mu\right)}{\Gamma\left(\frac{3}{2}\right) \Gamma\left(\nu + \frac{3}{2}\right) \Gamma\left(\frac{\nu+\sigma}{2}\right)}$$
$$\times \, _3F_3\left(1, \frac{\nu+\sigma}{2} + \mu, \frac{\nu+\sigma}{2} - \mu; \frac{3}{2}, \nu + \frac{3}{2}, \frac{\nu+\sigma}{2}; -\frac{b^2}{4a^2}\right)$$
$$\left[b > 0, \quad |\arg a| < \tfrac{\pi}{4}, \quad \operatorname{Re}(\sigma+\nu) > 2|\operatorname{Re}\mu|\right] \quad \text{ET II 167(23)}$$

## 6.86 Lommel functions

**6.861**

1. $$\int_0^\infty x^{\lambda-1} S_{\mu,\nu}(x)\,dx = \frac{\Gamma\left[\frac{1}{2}(1+\lambda+\mu)\right]\Gamma\left[\frac{1}{2}(1-\lambda-\mu)\right]\Gamma\left[\frac{1}{2}(1+\mu+\nu)\right]\Gamma\left[\frac{1}{2}(1+\mu-\nu)\right]}{2^{2-\lambda-\mu}\,\Gamma\left[\frac{1}{2}(\nu-\lambda)+1\right]\Gamma\left[1-\frac{1}{2}(\lambda+\nu)\right]}$$
$$\left[-\operatorname{Re}\mu < \operatorname{Re}\lambda+1 < \tfrac{5}{2}\right] \qquad \text{ET II 385(17)}$$

**6.862**

1.[12] $$\int_0^u x^{\lambda-\frac{1}{2}\mu-\frac{1}{2}}(u-x)^{\sigma-1} s_{\mu,\nu}\left(a\sqrt{x}\right)dx$$
$$= \Gamma(\sigma)\frac{a^\mu u^{\lambda+\sigma}\,\Gamma(\lambda+1)}{(\mu-\nu+1)(\mu+\nu+1)\,\Gamma(\lambda+\sigma+1)}$$
$$\times\ {}_2F_3\left(1,1+\lambda;\frac{\mu-\nu+3}{2},\frac{\mu+\nu+3}{2},\lambda+\sigma+1;-\frac{a^2 u}{4}\right)$$
$$[\operatorname{Re}\lambda>-1,\quad \operatorname{Re}\sigma>0] \qquad \text{ET II 199(92)}$$

2. $$\int_u^\infty x^{\frac{1}{2}\nu}(x-u)^{\mu-1} S_{\lambda,\nu}\left(a\sqrt{x}\right)dx = \frac{\mathrm{B}\left[\mu,\frac{1}{2}(1-\lambda-\nu)-\mu\right] u^{\frac{1}{2}\mu+\frac{1}{2}\nu}}{a^\mu} S_{\lambda+\mu,\mu+\nu}\left(a\sqrt{u}\right)$$
$$\left[\left|\arg\left(a\sqrt{u}\right)\right|<\pi,\quad 0<2\operatorname{Re}\mu<1-\operatorname{Re}(\lambda+\nu)\right] \qquad \text{ET II 211(71)}$$

**6.863** $$\int_0^\infty \sqrt{x}\,e^{-\alpha x}\, s_{\mu,\frac{1}{4}}\left(\frac{x^2}{2}\right)dx = 2^{-2\mu-1}\sqrt{\alpha}\,\Gamma\left(2\mu+\frac{3}{2}\right) S_{-\mu-1,\frac{1}{4}}\left(\frac{\alpha^2}{2}\right)$$
$$\left[\operatorname{Re}\alpha>0,\quad \operatorname{Re}\mu>-\tfrac{3}{4}\right] \qquad \text{ET I 209(38)}$$

**6.864** $$\int_0^\infty \exp[(\mu+1)x]\, s_{\mu,\nu}(a\sinh x)\,dx = 2^{\mu-2}\pi\operatorname{cosec}(\mu\pi)\,\Gamma(\varrho)\,\Gamma(\sigma)$$
$$\times\left[I_\varrho\left(\frac{a}{2}\right)I_\sigma\left(\frac{a}{2}\right)-I_{-\varrho}\left(\frac{a}{2}\right)I_{-\sigma}\left(\frac{a}{2}\right)\right]$$
$$2\varrho=\mu+\nu+1,\quad 2\sigma=\mu-\nu+1 \qquad [a>0,\quad -2<\operatorname{Re}\mu<0] \qquad \text{ET II 386(22)}$$

**6.865** $$\int_0^\infty \sqrt{\sinh x}\,\cosh(\nu x)\, S_{\mu,\frac{1}{2}}(a\cosh x)\,dx = \frac{\mathrm{B}\left(\frac{1}{4}-\frac{\mu+\nu}{2},\frac{1}{4}-\frac{\mu-\nu}{2}\right)}{\sqrt{a}\,2^{\mu+\frac{3}{2}}} S_{\mu+\frac{1}{2},\nu}(a)$$
$$\left[|\arg a|<\pi,\quad \operatorname{Re}\mu+|\operatorname{Re}\nu|<\tfrac{1}{2}\right]$$
$$\text{ET II 388(31)}$$

**6.866**

1.[12] $$\int_0^\infty x^{-\mu-1}\cos(ax)\, s_{\mu,\nu}(x)\,dx$$
$$= 0 \qquad\qquad\qquad\qquad\qquad\qquad\qquad [a>1]$$
$$= 2^{\mu-\frac{1}{2}}\sqrt{\pi}\,\Gamma\left(\frac{\mu+\nu+1}{2}\right)\Gamma\left(\frac{\mu-\nu+1}{2}\right)\left(1-a^2\right)^{\frac{1}{2}\mu+\frac{1}{4}} P_{\nu-\frac{1}{2}}^{-\mu-\frac{1}{2}}(a) \qquad [0<a<1]$$
$$\text{ET II 386(18)}$$

2. $$\int_0^\infty x^{-\mu}\sin(ax)\, S_{\mu,\nu}(x)\,dx = 2^{-\mu-\frac{1}{2}}\sqrt{\pi}\,\Gamma\left(1-\frac{\mu+\nu}{2}\right)\Gamma\left(1-\frac{\mu-\nu}{2}\right)\left(a^2-1\right)^{\frac{1}{2}\mu-\frac{1}{4}} P_{\nu-\frac{1}{2}}^{\mu-\frac{1}{2}}(a)$$
$$[a>1,\quad \operatorname{Re}\mu<1-|\operatorname{Re}\nu|]$$
$$\text{ET II 387(23)}$$

**6.867**

1.
$$\int_0^{\pi/2} \cos(2\mu x)\, S_{2\mu-1,2\nu}\,(a\cos x)\, \mathrm{d}x$$
$$= \frac{\pi 2^{2\mu-3} a^{2\mu} \operatorname{cosec}(2\nu\pi)}{\Gamma(1-\mu-\nu)\,\Gamma(1-\mu+\nu)} \left[ J_{\mu+\nu}\left(\frac{a}{2}\right) Y_{\mu-\nu}\left(\frac{a}{2}\right) - J_{\mu-\nu}\left(\frac{a}{2}\right) Y_{\mu+\nu}\left(\frac{a}{2}\right) \right]$$
$$[\operatorname{Re}\mu > -2, \quad |\operatorname{Re}\nu| < 1] \quad \text{ET II 388(29)}$$

2.
$$\int_0^{\pi/2} \cos\left[(\mu+1)\,x\right] s_{\mu,\nu}\,(a\cos x)\, \mathrm{d}x = 2^{\mu-2}\pi\,\Gamma(\varrho)\,\Gamma(\sigma)\, J_\varrho\left(\frac{a}{2}\right) J_\sigma\left(\frac{a}{2}\right)$$
$$2\varrho = \mu+\nu+1, \quad 2\sigma = \mu-\nu+1 \qquad [\operatorname{Re}\mu > -2] \quad \text{ET II 386(21)}$$

**6.868**
$$\int_0^{\pi/2} \frac{\cos(2\mu x)}{\cos x}\, S_{2\mu,2\nu}\,(a\sec x)\, \mathrm{d}x = \frac{\pi 2^{2\mu-1}}{a}\, W_{\mu,\nu}\left(ae^{i\frac{\pi}{2}}\right) W_{\mu,\nu}\left(ae^{-i\frac{\pi}{2}}\right)$$
$$[|\arg a| < \pi, \quad \operatorname{Re}\mu < 1] \quad \text{ET II 388(30)}$$

**6.869**

1.
$$\int_0^\infty x^{1-\mu-\nu}\, J_\nu(ax)\, S_{\mu,-\mu-2\nu}(x)\, \mathrm{d}x = \frac{\sqrt{\pi}\, a^{\nu-1}\,\Gamma(1-\mu-\nu)}{2^{\mu+2\nu}\,\Gamma\left(\nu+\frac{1}{2}\right)}\,(a^2-1)^{\frac{1}{2}(\mu+\nu-1)}\, P_{\mu+\nu}^{\mu+\nu-1}(a)$$
$$\left[a > 1, \quad \operatorname{Re}\nu > -\frac{1}{2}, \quad \operatorname{Re}(\mu+\nu) < 1\right]$$
$$\text{ET II 388(28)}$$

2.
$$\int_0^\infty x^{-\mu}\, J_\nu(ax)\, s_{\nu+\mu,-\nu+\mu+1}(x)\, \mathrm{d}x$$
$$= 2^{\nu-1}\,\Gamma(\nu)a^{-\nu}\left(1-a^2\right)^\mu \quad \left[0 < a < 1, \quad \operatorname{Re}\mu > -1, \quad -1 < \operatorname{Re}\nu < \tfrac{3}{2}\right]$$
$$= 0 \quad \left[1 < a, \quad \operatorname{Re}\mu > -1, \quad -1 < \operatorname{Re}\nu < \tfrac{3}{2}\right]$$
$$\text{ET II 388(28)}$$

3.
$$\int_0^\infty x\, K_\nu(bx)\, s_{\mu,\frac{1}{2}\nu}\left(ax^2\right)\, \mathrm{d}x = \frac{1}{4a}\,\Gamma\left(\mu+\frac{1}{2}\nu+1\right)\Gamma\left(\mu-\frac{1}{2}\nu+1\right) S_{-\mu-1,\frac{1}{2}\nu}\left(\frac{b^2}{4a}\right)$$
$$\left[\operatorname{Re}\mu > \tfrac{1}{2}|\operatorname{Re}\nu| - 2, \quad a > 0, \quad \operatorname{Re}b > 0\right] \quad \text{ET II 151(78)}$$

## 6.87 Thomson functions

**6.871**

1.
$$\int_0^\infty e^{-\beta x}\, \operatorname{ber} x\, \mathrm{d}x = \frac{\left(\sqrt{\beta^4+1}+\beta^2\right)^{1/2}}{\sqrt{2\left(\beta^4+1\right)}}$$
$$\text{ME 40}$$

2.
$$\int_0^\infty e^{-\beta x}\, \operatorname{bei} x\, \mathrm{d}x = \frac{\left(\sqrt{\beta^4+1}-\beta^2\right)^{1/2}}{\sqrt{2\left(\beta^4+1\right)}}$$
$$\text{ME 40}$$

**6.872**

1.  $\int_0^\infty e^{-\beta x}\,\mathrm{ber}_\nu\left(2\sqrt{x}\right)\,\mathrm{d}x = \frac{1}{2\beta}\sqrt{\frac{\pi}{\beta}}\left[J_{\frac{1}{2}(\nu-1)}\left(\frac{1}{2\beta}\right)\cos\left(\frac{1}{2\beta}+\frac{3\nu\pi}{4}\right)\right.$

    $\left.- J_{\frac{1}{2}(\nu+1)}\left(\frac{1}{2\beta}\right)\cos\left(\frac{1}{2\beta}+\frac{3\nu+6}{4}\pi\right)\right]$

    MI 49

2.  $\int_0^\infty e^{-\beta x}\,\mathrm{bei}_\nu\left(2\sqrt{x}\right)\,\mathrm{d}x = \frac{1}{2\beta}\sqrt{\frac{\pi}{\beta}}\left[J_{\frac{1}{2}(\nu-1)}\left(\frac{1}{2\beta}\right)\sin\left(\frac{1}{2\beta}+\frac{3\nu}{4}\pi\right)\right.$

    $\left.- J_{\frac{1}{2}(\nu+1)}\left(\frac{1}{2\beta}\right)\sin\left(\frac{1}{2\beta}+\frac{3\nu+6}{4}\pi\right)\right]$

    MI 49

3.  $\int_0^\infty e^{-\beta x}\,\mathrm{ber}\left(2\sqrt{x}\right)\,\mathrm{d}x = \frac{1}{\beta}\cos\frac{1}{\beta}$                                                  ME 40

4.  $\int_0^\infty e^{-\beta x}\,\mathrm{bei}\left(2\sqrt{x}\right)\,\mathrm{d}x = \frac{1}{\beta}\sin\frac{1}{\beta}$                                                  ME 40

5.  $\int_0^\infty e^{-\beta x}\,\mathrm{ker}\left(2\sqrt{x}\right)\,\mathrm{d}x = -\frac{1}{2\beta}\left[\cos\frac{1}{\beta}\,\mathrm{ci}\,\frac{1}{\beta}+\sin\frac{1}{\beta}\,\mathrm{si}\,\frac{1}{\beta}\right]$                                                  MI 50

6.  $\int_0^\infty e^{-\beta x}\,\mathrm{kei}\left(2\sqrt{x}\right)\,\mathrm{d}x = -\frac{1}{2\beta}\left[\sin\frac{1}{\beta}\,\mathrm{ci}\,\frac{1}{\beta}-\cos\frac{1}{\beta}\,\mathrm{si}\,\frac{1}{\beta}\right]$                                                  MI 50

7.  $\int_0^\infty e^{-\beta x}\,\mathrm{ber}_\nu\left(2\sqrt{x}\right)\mathrm{bei}_\nu\left(2\sqrt{x}\right)\,\mathrm{d}x = \frac{1}{2\beta}J_\nu\left(\frac{2}{\beta}\right)\sin\left(\frac{2}{\beta}+\frac{3\nu\pi}{2}\right)$

    $[\mathrm{Re}\,\nu > -1]$                                                  MI 49

**6.873**   $\int_0^\infty \left[\mathrm{ber}_\nu^2\left(2\sqrt{x}\right)+\mathrm{bei}_\nu^2\left(2\sqrt{x}\right)\right]e^{-\beta x}\,\mathrm{d}x = \frac{1}{\beta}I_\nu\left(\frac{2}{\beta}\right)$

$[\mathrm{Re}\,\nu > -1]$                                                  ME 40

**6.874**

1.  $\int_0^\infty \frac{e^{-\beta x}}{\sqrt{x}}\,\mathrm{ber}_{2\nu}\left(2\sqrt{2x}\right)\,\mathrm{d}x = \sqrt{\frac{\pi}{\beta}}J_\nu\left(\frac{1}{\beta}\right)\cos\left(\frac{1}{\beta}-\frac{3\pi}{4}+\frac{3\nu\pi}{2}\right)$

    $[\mathrm{Re}\,\nu > -\frac{1}{2}]$                                                  MI 49

2.  $\int_0^\infty \frac{e^{-\beta x}}{\sqrt{x}}\,\mathrm{bei}_{2\nu}\left(2\sqrt{2x}\right)\,\mathrm{d}x = \sqrt{\frac{\pi}{\beta}}J_\nu\left(\frac{1}{\beta}\right)\sin\left(\frac{1}{\beta}-\frac{3\pi}{4}+\frac{3\nu\pi}{2}\right)$

    $[\mathrm{Re}\,\nu > -\frac{1}{2}]$                                                  MI 49

3.  $\int_0^\infty x^{\frac{\nu}{2}}\,\mathrm{ber}_\nu\left(\sqrt{x}\right)e^{-\beta x}\,\mathrm{d}x = \frac{2^{-\nu}}{\beta^{1+\nu}}\cos\left(\frac{1}{4\beta}+\frac{3\nu\pi}{4}\right)$   $[\mathrm{Re}\,\nu > -1]$                                                  ME 40

4.  $\int_0^\infty x^{\frac{\nu}{2}}\,\mathrm{bei}_\nu\left(\sqrt{x}\right)e^{-\beta x}\,\mathrm{d}x = \frac{2^{-\nu}}{\beta^{1+\nu}}\sin\left(\frac{1}{4\beta}+\frac{3\nu\pi}{4}\right)$   $[\mathrm{Re}\,\nu > -1]$                                                  ME 40

**6.875**

1. $\int_0^\infty e^{-\beta x} \left[ \ker\left(2\sqrt{x}\right) - \frac{1}{2} \ln x \, \text{ber}\left(2\sqrt{x}\right) \right] dx = \frac{1}{\beta} \left[ \ln \beta \cos \frac{1}{\beta} + \frac{\pi}{4} \sin \frac{1}{\beta} \right]$    MI 50

2. $\int_0^\infty e^{-\beta x} \left[ \text{kei}\left(2\sqrt{x}\right) - \frac{1}{2} \ln x \, \text{bei}\left(2\sqrt{x}\right) \right] dx = \frac{1}{\beta} \left[ \ln \beta \sin \frac{1}{\beta} - \frac{\pi}{4} \cos \frac{1}{\beta} \right]$    MI 50

**6.876**

1. $\int_0^\infty x \, \text{kei} \, x \, J_1(ax) \, dx = -\frac{1}{2a} \arctan a^2$          $[a > 0]$          ET II 21(32)

2. $\int_0^\infty x \, \ker x \, J_1(ax) \, dx = \frac{1}{2a} \ln \sqrt{(1 + a^4)}$          $[a > 0]$          ET II 21(33)

# 6.9 Mathieu Functions

**Notation:** $k^2 = q$. For definition of the coefficients $A_p^{(m)}$ and $B_p^{(m)}$ see section 8.6.

## 6.91  Mathieu functions

**6.911**

1. $\int_0^{2\pi} \text{ce}_m(z, q) \, \text{ce}_p(z, q) \, dz = 0$          $[m \neq p]$          MA

2. $\int_0^{2\pi} [\text{ce}_{2n}(z, q)]^2 \, dz = 2\pi \left[ A_0^{(2n)} \right]^2 + \pi \sum_{r=1}^\infty \left[ A_{2r}^{(2n)} \right]^2 = \pi$          MA

3. $\int_0^{2\pi} [\text{ce}_{2n+1}(z, q)]^2 \, dz = \pi \sum_{r=0}^\infty \left[ A_{2r+1}^{(2n+1)} \right]^2 = \pi$          MA

4. $\int_0^{2\pi} \text{se}_m(z, q) \, \text{se}_p(z, q) \, dz = 0$          $[m \neq p]$          MA

5. $\int_0^{2\pi} [\text{se}_{2n+1}(z, q)]^2 \, dz = \pi \sum_{r=0}^\infty \left[ B_{2r+1}^{(2n+1)} \right]^2 = \pi$          MA

6. $\int_0^{2\pi} [\text{se}_{2n+2}(z, q)]^2 \, dz = \pi \sum_{r=0}^\infty \left[ B_{2r+2}^{(2n+2)} \right]^2 = \pi$          MA

7. $\int_0^{2\pi} \text{se}_m(z, q) \, \text{ce}_p(z, q) \, dz = 0$          $[m = 1, 2, \ldots; \quad p = 1, 2, \ldots]$          MA

## 6.92 Combinations of Mathieu, hyperbolic, and trigonometric functions

**6.921**

1. $\int_0^\pi \cosh\left(2k \cos u \sinh z\right) \text{ce}_{2n}(u, q) \, du = \frac{\pi A_0^{(2n)}}{\text{ce}_{2n}\left(\frac{\pi}{2}, q\right)} (-1)^n \, \text{Ce}_{2n}(z, -q)$

$[q > 0]$          MA

2.    $\displaystyle\int_0^\pi \cosh\left(2k\sin u \cosh z\right) \mathrm{ce}_{2n}(u,q)\,\mathrm{d}u = \frac{\pi A_0^{(2n)}}{\mathrm{ce}_{2n}(0,q)}(-1)^n \, \mathrm{Ce}_{2n}(z,-q)$

$$[q>0] \hspace{4cm} \text{MA}$$

3.    $\displaystyle\int_0^\pi \sinh\left(2k\sin u \cosh z\right) \mathrm{se}_{2n+1}(u,q)\,\mathrm{d}u = \frac{\pi k B_1^{(2n+1)}}{\mathrm{se}'_{2n+1}(0,q)}(-1)^n \, \mathrm{Ce}_{2n+1}(z,-q)$

$$[q>0] \hspace{4cm} \text{MA}$$

4.    $\displaystyle\int_0^\pi \sinh\left(2k\cos u \sinh z\right) \mathrm{ce}_{2n+1}(u,q)\,\mathrm{d}u = \frac{\pi k A_1^{(2n+1)}}{\mathrm{ce}'_{2n+1}\left(\frac{\pi}{2},q\right)}(-1)^{n+1} \, \mathrm{Se}_{2n+1}(z,-q)$

$$[q>0] \hspace{4cm} \text{MA}$$

5.    $\displaystyle\int_0^\pi \sinh\left(2k\sin u \sin z\right) \mathrm{se}_{2n+1}(u,q)\,\mathrm{d}u = \frac{\pi k B_1^{(2n+1)}}{\mathrm{se}'_{2n+1}(0,q)}\, \mathrm{se}_{2n+1}(z,q)$

$$[q>0] \hspace{4cm} \text{MA}$$

**6.922**

1.    $\displaystyle\int_0^\pi \cos u \cosh z \cos\left(2k\sin u \sinh z\right) \mathrm{ce}_{2n+1}(u,q)\,\mathrm{d}u = \frac{\pi A_1^{(2n+1)}}{2\,\mathrm{ce}_{2n+1}(0,q)}\, \mathrm{Ce}_{2n+1}(z,q)$

$$[q>0] \hspace{4cm} \text{MA}$$

2.    $\displaystyle\int_0^\pi \sin u \sinh z \cos\left(2k\cos u \cosh z\right) \mathrm{se}_{2n+1}(u,q)\,\mathrm{d}u = \frac{\pi B_1^{(2n+1)}}{2\,\mathrm{se}_{2n+1}\left(\frac{\pi}{2},q\right)}\, \mathrm{Se}_{2n+1}(z,q)$

$$[q>0] \hspace{4cm} \text{MA}$$

3.    $\displaystyle\int_0^\pi \sin u \sinh z \sin\left(2k\cos u \cosh z\right) \mathrm{se}_{2n+2}(u,q)\,\mathrm{d}u = -\frac{\pi k B_2^{(2n+2)}}{2\,\mathrm{se}'_{2n+2}\left(\frac{\pi}{2},q\right)}\, \mathrm{Se}_{2n+2}(z,q)$

$$[q>0] \hspace{4cm} \text{MA}$$

4.    $\displaystyle\int_0^\pi \cos u \cosh z \sin\left(2k\sin u \sinh z\right) \mathrm{se}_{2n+2}(u,q)\,\mathrm{d}u = \frac{\pi k B_2^{(2n+2)}}{2\,\mathrm{se}'_{2n+2}(0,q)}\, \mathrm{Se}_{2n+2}(z,q)$

$$[q>0] \hspace{4cm} \text{MA}$$

5.    $\displaystyle\int_0^\pi \sin u \cosh z \cosh\left(2k\cos u \sinh z\right) \mathrm{se}_{2n+1}(u,q)\,\mathrm{d}u = \frac{\pi B_1^{(2n+1)}}{2\,\mathrm{se}_{2n+1}\left(\frac{\pi}{2},q\right)}(-1)^n \, \mathrm{Ce}_{2n+1}(z,-q)$

$$[q>0] \hspace{4cm} \text{MA}$$

6.    $\displaystyle\int_0^\pi \cos u \sinh z \cosh\left(2k\sin u \cosh z\right) \mathrm{ce}_{2n+1}(u,q)\,\mathrm{d}u = \frac{\pi A_1^{(2n+1)}}{2\,\mathrm{ce}_{2n+1}(0,q)}(-1)^n \, \mathrm{Se}_{2n+1}(z,-q)$

$$[q>0] \hspace{4cm} \text{MA}$$

7.    $\displaystyle\int_0^\pi \sin u \cosh z \sinh\left(2k\cos u \sinh z\right) \mathrm{se}_{2n+2}(u,q)\,\mathrm{d}u = \frac{\pi k B_2^{(2n+2)}}{2\,\mathrm{se}'_{2n+2}\left(\frac{\pi}{2},q\right)}(-1)^{n+1} \, \mathrm{Se}_{2n+2}(z,-q)$

$$[q>0] \hspace{4cm} \text{MA}$$

8.　　$\displaystyle\int_0^\pi \cos u \sinh z \sinh\left(2k \sin u \cosh z\right) \operatorname{se}_{2n+2}(u,q)\,\mathrm{d}u = \frac{\pi k B_2^{(2n+2)}}{2\,\operatorname{se}'_{2n+2}(0,q)}(-1)^n \operatorname{Se}_{2n+2}(z,-q)$

$$[q>0] \qquad\qquad \text{MA}$$

**6.923**

1.　　$\displaystyle\int_0^\infty \sin\left(2k \cosh z \cosh u\right) \sinh z \sinh u\, \operatorname{Se}_{2n+1}(u,q)\,\mathrm{d}u = -\frac{\pi B_1^{(2n+1)}}{4\,\operatorname{se}_{2n+1}\left(\frac{1}{2}\pi,q\right)} \operatorname{Se}_{2n+1}(z,q)$

$$[q>0] \qquad\qquad \text{MA}$$

2.　　$\displaystyle\int_0^\infty \cos\left(2k \cosh z \cosh u\right) \sinh z \sinh u\, \operatorname{Se}_{2n+1}(u,q)\,\mathrm{d}u = -\frac{\pi B_1^{(2n+1)}}{4\,\operatorname{se}_{2n+1}\left(\frac{1}{2}\pi,q\right)} \operatorname{Gey}_{2n+1}(z,q)$

$$[q>0] \qquad\qquad \text{MA}$$

3.　　$\displaystyle\int_0^\infty \sin\left(2k \cosh z \cosh u\right) \sinh z \sinh u\, \operatorname{Se}_{2n+2}(u,q)\,\mathrm{d}u = -\frac{k\pi B_2^{(2n+2)}}{4\,\operatorname{se}'_{2n+2}\left(\frac{1}{2}\pi,q\right)} \operatorname{Gey}_{2n+2}(z,q)$

$$[q>0] \qquad\qquad \text{MA}$$

4.　　$\displaystyle\int_0^\infty \cos\left(2k \cosh z \cosh u\right) \sinh z \sinh u\, \operatorname{Se}_{2n+2}(u,q)\,\mathrm{d}u = -\frac{k\pi B_2^{(2n+2)}}{4\,\operatorname{se}_{2n+2}\left(\frac{1}{2}\pi,q\right)} \operatorname{Se}_{2n+2}(z,q)$

$$[q>0] \qquad\qquad \text{MA}$$

5.　　$\displaystyle\int_0^\infty \sin\left(2k \cosh z \cosh u\right) \operatorname{Ce}_{2n}(u,q)\,\mathrm{d}u = \frac{\pi A_0^{(2n)}}{2\,\operatorname{ce}_{2n}\left(\frac{1}{2}\pi,q\right)} \operatorname{Ce}_{2n}(z,q)$

$$[q>0] \qquad\qquad \text{MA}$$

6.　　$\displaystyle\int_0^\infty \cos\left(2k \cosh z \cosh u\right) \operatorname{Ce}_{2n}(u,q)\,\mathrm{d}u = -\frac{\pi A_0^{(2n)}}{2\,\operatorname{ce}_{2n}\left(\frac{1}{2}\pi,q\right)} \operatorname{Fey}_{2n}(z,q)$

$$[q>0] \qquad\qquad \text{MA}$$

7.　　$\displaystyle\int_0^\infty \sin\left(2k \cosh z \cosh u\right) \operatorname{Ce}_{2n+1}(u,q)\,\mathrm{d}u = \frac{k\pi A_1^{(2n+1)}}{2\,\operatorname{ce}'_{2n+1}\left(\frac{1}{2}\pi,q\right)} \operatorname{Fey}_{2n+1}(z,q)$

$$[q>0] \qquad\qquad \text{MA}$$

8.　　$\displaystyle\int_0^\infty \cos\left(2k \cosh z \cosh u\right) \operatorname{Ce}_{2n+1}(u,q)\,\mathrm{d}u = \frac{k\pi A_1^{(2n+1)}}{2\,\operatorname{ce}'_{2n+1}\left(\frac{1}{2}\pi,q\right)} \operatorname{Ce}_{2n+1}(z,q)$

$$[q>0] \qquad\qquad \text{MA}$$

**6.924**

1.　　$\displaystyle\int_0^\pi \cos\left(2k \cos u \cos z\right) \operatorname{ce}_{2n}(u,q)\,\mathrm{d}u = \frac{\pi A_0^{(2n)}}{\operatorname{ce}_{2n}\left(\frac{1}{2}\pi,q\right)} \operatorname{ce}_{2n}(z,q)$

$$[q>0] \qquad\qquad \text{MA}$$

2.　　$\displaystyle\int_0^\pi \sin\left(2k \cos u \cos z\right) \operatorname{ce}_{2n+1}(u,q)\,\mathrm{d}u = -\frac{\pi k A_1^{(2n+1)}}{\operatorname{ce}'_{2n+1}\left(\frac{1}{2}\pi,q\right)} \operatorname{ce}_{2n+1}(z,q)$

$$[q>0] \qquad\qquad \text{MA}$$

3. $\displaystyle \int_0^\pi \cos\left(2k\cos u\cosh z\right)\mathrm{ce}_{2n}(u,q)\,\mathrm{d}u = \frac{\pi A_0^{(2n)}}{\mathrm{ce}_{2n}\left(\frac{1}{2}\pi,q\right)}\,\mathrm{Ce}_{2n}(z,q)$

$[q>0]$                     MA

4. $\displaystyle \int_0^\pi \cos\left(2k\sin u\sinh z\right)\mathrm{ce}_{2n}(u,q)\,\mathrm{d}u = \frac{\pi A_0^{(2n)}}{\mathrm{ce}_{2n}(0,q)}\,\mathrm{Ce}_{2n}(z,q)$

$[q>0]$                     MA

5. $\displaystyle \int_0^\pi \sin\left(2k\cos u\cosh z\right)\mathrm{ce}_{2n+1}(u,q)\,\mathrm{d}u = -\frac{\pi k A_1^{(2n+1)}}{\mathrm{ce}'_{2n+1}\left(\frac{1}{2}\pi,q\right)}\,\mathrm{Ce}_{2n+1}(z,q)$

$[q>0]$                     MA

6. $\displaystyle \int_0^\pi \sin\left(2k\sin u\sinh z\right)\mathrm{se}_{2n+1}(u,q)\,\mathrm{d}u = \frac{\pi k B_1^{(2n+1)}}{\mathrm{se}'_{2n+1}(0,q)}\,\mathrm{Se}_{2n+1}(z,q)$

$[q>0]$                     MA

**6.925**   **Notation:** $z_1 = 2k\sqrt{\cosh^2\xi - \sin^2\eta}$, and $\tan\alpha = \tanh\xi\tan\eta$

1. $\displaystyle \int_0^{2\pi} \sin\left[z_1\cos(\theta-\alpha)\right]\mathrm{ce}_{2n}(\theta,q)\,\mathrm{d}\theta = 0.$                     MA

2. $\displaystyle \int_0^{2\pi} \cos\left[z_1\cos(\theta-\alpha)\right]\mathrm{ce}_{2n}(\theta,q)\,\mathrm{d}\theta = \frac{2\pi A_0^{(2n)}}{\mathrm{ce}_{2n}(0,q)\,\mathrm{ce}_{2n}\left(\frac{1}{2}\pi,q\right)}\,\mathrm{Ce}_{2n}(\xi,q)\,\mathrm{ce}_{2n}(\eta,q)$                     MA

3. $\displaystyle \int_0^{2\pi} \sin\left[z_1\cos(\theta-\alpha)\right]\mathrm{ce}_{2n+1}(\theta,q)\,\mathrm{d}\theta = -\frac{2\pi k A_1^{(2n+1)}}{\mathrm{ce}_{2n+1}(0,q)\,\mathrm{ce}'_{2n+1}\left(\frac{1}{2}\pi,q\right)}\,\mathrm{Ce}_{2n+1}(\xi,q)\,\mathrm{ce}_{2n+1}(\eta,q)$

MA

4. $\displaystyle \int_0^{2\pi} \cos\left[z_1\cos(\theta-\alpha)\right]\mathrm{ce}_{2n+1}(\theta,q)\,\mathrm{d}\theta = 0$                     MA

5. $\displaystyle \int_0^{2\pi} \sin\left[z_1\cos(\theta-\alpha)\right]\mathrm{se}_{2n+1}(\theta,q)\,\mathrm{d}\theta = \frac{2\pi k B_1^{(2n+1)}}{\mathrm{se}_{2n+1}(0,q)\,\mathrm{se}_{2n+1}\left(\frac{1}{2}\pi,q\right)}\,\mathrm{Se}_{2n+1}(\xi,q)\,\mathrm{se}_{2n+1}(\eta,q)$

MA

6. $\displaystyle \int_0^{2\pi} \cos\left[z_1\cos(\theta-\alpha)\right]\mathrm{se}_{2n+1}(\theta,q)\,\mathrm{d}\theta = 0$                     MA

7. $\displaystyle \int_0^{2\pi} \sin\left[z_1\cos(\theta-\alpha)\right]\mathrm{se}_{2n+2}(\theta,q)\,\mathrm{d}\theta = 0$                     MA

8. $\displaystyle \int_0^{2\pi} \cos\left[z_1\cos(\theta-\alpha)\right]\mathrm{se}_{2n+2}(\theta,q)\,\mathrm{d}\theta = \frac{2\pi k^2 B_2^{(2n+2)}}{\mathrm{se}'_{2n+2}(0,q)\,\mathrm{se}'_{2n+2}\left(\frac{1}{2}\pi,q\right)}\,\mathrm{Se}_{2n+2}(\xi,q)\,\mathrm{se}_{2n+2}(\eta,q)$

MA

**6.926**   $\displaystyle \int_0^\pi \sin u\sin z\sin\left(2k\cos u\cos z\right)\mathrm{se}_{2n+2}(u,q)\,\mathrm{d}u = -\frac{\pi k B_2^{(2n+2)}}{2\,\mathrm{se}'_{2n+2}\left(\frac{\pi}{2},q\right)}\,\mathrm{se}_{2n+2}(z,q)$

$[q>0]$                     MA

## 6.93 Combinations of Mathieu and Bessel functions

**6.931**

1.
$$\int_0^\pi J_0 \left\{ k \left[ 2 \left( \cos 2u + \cos 2z \right) \right]^{1/2} \right\} ce_{2n}(u,q)\, du = \frac{\pi \left[ A_0^{(2n)} \right]^2}{ce_{2n}(0,q)\, ce_{2n}\left( \frac{\pi}{2}, q \right)} ce_{2n}(z,q) \qquad \text{MA}$$

2.
$$\int_0^{2\pi} Y_0 \left\{ k \left[ 2 \left( \cos 2u + \cosh 2z \right) \right]^{1/2} \right\} ce_{2n}(u,q)\, du = \frac{2\pi \left[ A_0^{(2n)} \right]^2}{ce_{2n}(0,q)\, ce_{2n}\left( \frac{\pi}{2}, q \right)} \text{Fey}_{2n}(z,q) \qquad \text{MA}$$

## 6.94 Relationships between eigenfunctions of the Helmholtz equation in different coordinate systems

**Notation**: Particular solutions of the Helmholtz equation in three-dimensional infinite space
$$\nabla^2 \Psi + k^2 \Psi = 0$$
in Cartesian $(x, y, z)$, spherical $(r, \theta, \phi)$, and cylindrical $(\rho, z, \phi)$ coordinates are

$$\Psi_{k_x k_y k_z}(x,y,z) \propto e^{i(k_x x + k_y y + k_z z)} \quad \text{with} \quad k^2 = k_x^2 + k_y^2 + k_z^2$$

$$\Psi_{lm}(r,\theta,\phi) \propto e^{im\phi} \sqrt{\frac{k}{r}}\, Z_{l+1/2}(kr)\, P_l^m(\cos\theta)$$

$$\Psi_{mk_z}(\rho,z,\phi) \propto e^{i(m\phi + k_z z)}\, Z_{l+1/2}\left( \rho\sqrt{k^2 - k_z^2} \right)$$

with $P_l^m(\cos\theta)$ the associated Legendre function, $Z$ is any Bessel function, $m = 0, 1, \ldots, l$; $l \in \mathbb{N}$, $r^2 = \rho^2 + z^2$, $\rho = r\sin\theta$, $z = r\cos\theta$, $\phi = \text{arccot}(x/y)$, and $k_t^2 = k^2 - k_z^2$.

**6.941**

1.
$$\int_{-k}^{k} e^{i\rho z} J_m\left( \rho\sqrt{k^2 - \rho^2} \right) P_l^m\left( \frac{p}{k} \right) dp = i^{l-m} \sqrt{\frac{2\pi k}{r}}\, J_{l+1/2}(kr)\, P_l^m\left( \frac{z}{r} \right)$$

$$[\rho > 0, \quad l \geq m \geq 0]$$

2.
$$\int_{-\infty}^{\infty} e^{-i\rho z} J_{l+1/2}(kr)\, P_l^m\left( \frac{z}{r} \right) dz = i^{m-l} \sqrt{\frac{2\pi r}{k}}\, J_m\left( \rho\sqrt{k^2 - \rho^2} \right) P_l^m\left( \frac{\rho}{k} \right)$$

$$[\rho > 0, \quad l \geq m \geq 0]$$

3.
$$\int_0^{\infty} J_m(\rho k_t) \cos\left[ k_x x + m \arcsin\left( \frac{x}{\rho} \right) \right] dx$$
$$= \frac{(-1)^m}{\sqrt{k_t^2 - k_x^2}} \cos\left[ y\sqrt{k_t^2 - k_x^2} + m \arccos\left( \frac{k_x}{k_t} \right) \right] \qquad [k_x^2 < k_t^2]$$
$$= 0 \qquad\qquad\qquad\qquad\qquad\qquad\qquad\qquad\qquad\qquad\quad [k_x^2 > k_t^2]$$

4.    $\displaystyle\int_0^\infty Y_m\left(\rho k_t\right)\cos\left[k_x x + m\arcsin\left(\frac{x}{\rho}\right)\right]\,\mathrm{d}x$

$$= \frac{(-1)^m}{\sqrt{k_t^2 - k_x^2}}\sin\left[y\sqrt{k_t^2 - k_x^2} + m\arccos\left(\frac{k_x}{k_t}\right)\right] \qquad [k_x^2 < k_t^2]$$

$$= \frac{(-1)^m}{\sqrt{k_x^2 - k_t^2}}\exp\left[-y\sqrt{k_x^2 - k_t^2} - m\,\mathrm{sign}\,(k_x)\,\mathrm{arccosh}\left(\frac{|k_x|}{k_t}\right)\right] \qquad [k_x^2 > k_t^2]$$

5.    $\displaystyle\int_{-\infty}^\infty H^{(j)}_{l+1/2}(kr)\,P_l^m\left(\frac{z}{r}\right)e^{-ik_z z}\,\mathrm{d}z = i^{m-l}\sqrt{\frac{2\pi r}{k}}\,H^{(j)}_m\left(\rho\sqrt{k^2 - k_z^2}\right)P_l^m\left(\frac{k_z}{k}\right)$

$$[\rho > 0]$$

The result is true for $j = 1$ if $\pi > \arg\sqrt{k^2 - k_z^2} \geq 0$, for $j = 2$ if $-\pi < \arg\sqrt{k^2 - k_z^2} \leq 0$.

6.    $\displaystyle\int_{-\infty}^\infty H^{(j)}_m\left(\rho\sqrt{k^2 - k_z^2}\right)P_l^m\left(\frac{k_z}{k}\right)e^{ik_z z}\,\mathrm{d}k_z = i^{l-m}\sqrt{\frac{2\pi k}{r}}\,H^{(j)}_{l+1/2}(kr)\,P_l^m\left(\frac{z}{r}\right)$

The result is true for $j = 1$ if $\pi > \arg\sqrt{k^2 - k_z^2} \geq 0$, for $j = 2$ if $-\pi < \arg\sqrt{k^2 - k_z^2} \leq 0$.

7.    $\displaystyle\int_{-\infty}^\infty J_{l+1/2}(kr)\,P_l^m\left(\frac{z}{r}\right)e^{-ik_z z}\,\mathrm{d}z = i^{m-l}\sqrt{\frac{2\pi r}{k}}\,J_m\left(\rho\sqrt{k^2 - k_z^2}\right)P_l^m\left(\frac{k_z}{k}\right) \qquad [k_z^2 < k^2]$

$$= 0 \qquad [k_z^2 > k^2]$$

8.    $\displaystyle\int_{-k}^k J_m\left(\rho\sqrt{k^2 - k_z^2}\right)P_l^m\left(\frac{k_z}{k}\right)e^{ik_z z}\,\mathrm{d}k_z = i^{l-m}\sqrt{\frac{2\pi k}{r}}\,J_{l+1/2}(kr)\,P_l^m\left(\frac{z}{r}\right)$

9.    $\displaystyle\int_{-\infty}^\infty Y_{l+1/2}(kr)\,P_l^m\left(\frac{z}{r}\right)e^{-ik_z z}\,\mathrm{d}z = i^{m-l}\sqrt{\frac{2\pi r}{k}}\,Y_m\left(\rho\sqrt{k^2 - k_z^2}\right)P_l^m\left(\frac{k_z}{k}\right) \qquad [k_z^2 < k^2]$

$$= -2i^{m-l}\sqrt{\frac{2r}{k\pi}}\,K_m\left(\rho\sqrt{k_z^2 - k^2}\right)P_l^m\left(\frac{k_z}{k}\right) \qquad [k_z^2 > k^2]$$

10.    $\displaystyle i^{l-m}\int_{-k}^k Y_m\left(\rho\sqrt{k^2 - k_z^2}\right)P_l^m\left(\frac{k_z}{k}\right)e^{ik_z z}\,\mathrm{d}k_z$

$$-\frac{4}{\pi}\int_k^\infty \cos\left[k_z z + \tfrac{1}{2}\pi(m-l)\right]P_l^m\left(\frac{k_z}{k}\right)K_m\left(\rho\sqrt{k_z^2 - k^2}\right)e^{ik_z z}\,\mathrm{d}k_z$$

$$= \sqrt{\frac{2\pi k}{r}}\,Y_{l+1/2}(kr)\,P_l^m\left(\frac{z}{r}\right)$$

# 7.1–7.2 Associated Legendre Functions

## 7.11 Associated Legendre functions

**7.111**
$$\int_{\cos\varphi}^{1} P_\nu(x)\,\mathrm{d}x = \sin\varphi\, P_\nu^{-1}(\cos\varphi)$$
MO 90

**7.112**

1.
$$\int_{-1}^{1} P_n^m(x)\, P_k^m(x)\,\mathrm{d}x = 0 \qquad\qquad [n \neq k]$$
$$= \frac{2}{2n+1}\frac{(n+m)!}{(n-m)!} \qquad [n = k]$$
SM III 185, WH

2.
$$\int_{-1}^{1} Q_n^m(x)\, P_k^m(x)\,\mathrm{d}x = (-1)^m \frac{1-(-1)^{n+k}(n+m)!}{(k-n)(k+n+1)(n-m)!}$$
EH I 171(18)

3.
$$\int_{-1}^{1} P_\nu(x)\, P_\sigma(x)\,\mathrm{d}x$$
$$= \frac{2\pi \sin\pi(\sigma-\nu) + 4\sin(\pi\nu)\sin(\pi\sigma)\,[\psi(\nu+1)-\psi(\sigma+1)]}{\pi^2(\sigma-\nu)(\sigma+\nu+1)} \qquad [\sigma+\nu+1 \neq 0] \quad \text{EH I 170(7)}$$
$$= \frac{\pi^2 - 2(\sin\pi\nu)^2\,\psi'(\nu+1)}{\pi^2\left(\nu+\dfrac{1}{2}\right)} \qquad\qquad [\sigma=\nu] \qquad \text{EH I 170(9)a}$$

4.
$$\int_{-1}^{1} Q_\nu(x)\, Q_\sigma(x)\,\mathrm{d}x = \frac{[\psi(\nu+1)-\psi(\sigma+1)]\,[1+\cos(\pi\sigma)\cos(\nu\pi)] - \frac{\pi}{2}\sin\pi(\nu-\sigma)}{(\sigma-\nu)(\sigma+\nu+1)}$$
$$[\sigma+\nu+1 \neq 0;\quad \nu,\sigma \neq -1,-2,-3,\ldots]$$
EH I 170(11)
$$= \frac{\frac{1}{2}\pi^2 - \psi'(\nu+1)\left[1+(\cos\nu\pi)^2\right]}{2\nu+1}$$
$$[\nu=\sigma,\quad \nu \neq -1,-2,-3,\ldots]$$
EH I 170(12)

5.
$$\int_{-1}^{1} P_\nu(x)\, Q_\sigma(x)\,\mathrm{d}x = \frac{1-\cos\pi(\sigma-\nu) - 2\pi^{-1}\sin(\pi\nu)\cos(\pi\sigma)\,[\psi(\nu+1)-\psi(\sigma+1)]}{(\nu-\sigma)(\nu+\sigma+1)}$$
$$[\operatorname{Re}\nu > 0,\quad \operatorname{Re}\sigma > 0,\quad \sigma \neq \nu]$$
EH I 170(13)
$$= -\frac{\sin(2\nu\pi)\,\psi'(\nu+1)}{\pi(2\nu+1)}$$
$$[\operatorname{Re}\nu > 0,\quad \sigma = \nu]$$
EH I 171(14)

**7.113**    **Notation:**    $A = \dfrac{\Gamma\left(\frac{1}{2}+\frac{\nu}{2}\right)\Gamma\left(1+\frac{\sigma}{2}\right)}{\Gamma\left(\frac{1}{2}+\frac{\sigma}{2}\right)\Gamma\left(1+\frac{\nu}{2}\right)}$

1.
$$\int_{0}^{1} P_\nu(x)\, P_\sigma(x)\,\mathrm{d}x = \frac{A\sin\frac{\pi\sigma}{2}\cos\frac{\pi\nu}{2} - A^{-1}\sin\frac{\pi\nu}{2}\cos\frac{\pi\sigma}{2}}{\frac{1}{2}\pi(\sigma-\nu)(\sigma+\nu+1)}$$
EH I 171(15)

$2.^{12}$   $\displaystyle\int_0^1 Q_\nu(x)\,Q_\sigma(x)\,\mathrm{d}x = \frac{\psi(\nu+1)-\psi(\sigma+1)-\frac{\pi}{2}\left[\left(A-A^{-1}\right)\sin\frac{\pi(\sigma+\nu)}{2}-\left(A+A^{-1}\right)\sin\frac{\pi(\sigma-\nu)}{2}\right]}{(\sigma-\nu)(\sigma+\nu+1)}$

$$[\mathrm{Re}\,\nu > 0, \quad \mathrm{Re}\,\sigma > 0] \qquad \text{EH I 171(16)}$$

3.   $\displaystyle\int_0^1 P_\nu(x)\,Q_\sigma(x)\,\mathrm{d}x = \frac{A^{-1}\cos\frac{\pi(\nu-\sigma)}{2}-1}{(\sigma-\nu)(\sigma+\nu+1)}$      $[\mathrm{Re}\,\nu > 0, \quad \mathrm{Re}\,\sigma > 0]$      EH I 171(17)

**7.114**

1.   $\displaystyle\int_1^\infty P_\nu(x)\,Q_\sigma(x)\,\mathrm{d}x = \frac{1}{(\sigma-\nu)(\sigma+\nu+1)}$      $[\mathrm{Re}(\sigma-\nu) > 0, \quad \mathrm{Re}(\sigma+\nu) > -1]$

$$\text{ET II 324(19)}$$

2.   $\displaystyle\int_1^\infty Q_\nu(x)\,Q_\sigma(x)\,\mathrm{d}x = \frac{\psi(\sigma+1)-\psi(\nu+1)}{(\sigma-\nu)(\sigma+\nu+1)}$

$$[\mathrm{Re}(\nu+\sigma) > -1; \quad \sigma, \nu \neq -1, -2, -3, \ldots] \quad \text{EH I 170(5)}$$

3.   $\displaystyle\int_1^\infty [Q_\nu(x)]^2\,\mathrm{d}x = \frac{\psi'(\nu+1)}{2\nu+1}$      $\left[\mathrm{Re}\,\nu > -\frac{1}{2}\right]$      EH I 170(6)

**7.115**   $\displaystyle\int_1^\infty Q_\nu(x)\,\mathrm{d}x = \frac{1}{\nu(\nu+1)}$      $[\mathrm{Re}\,\nu > 0]$      ET II 324(18)

## 7.12–7.13 Combinations of associated Legendre functions and powers

**7.121**$^{12}$ $\displaystyle\int_{\cos\varphi}^1 x\,P_\nu(x)\,\mathrm{d}x = \frac{\sin\varphi}{(\nu-1)(\nu+2)}\left[\sin\varphi P_\nu(\cos\varphi)+\cos\varphi P_\nu^1(\cos\varphi)\right]$      MO 90

**7.122**

1.   $\displaystyle\int_0^1 \frac{[P_n^m(x)]^2}{1-x^2}\,\mathrm{d}x = \frac{1}{2m}\frac{(n+m)!}{(n-m)!}$      $[0 < m \leq n]$      MO 74

2.   $\displaystyle\int_0^1 [P_\nu^\mu(x)]^2\frac{\mathrm{d}x}{1-x^2} = -\frac{\Gamma(1+\mu+\nu)}{2\mu\,\Gamma(1-\mu+\nu)}$      $[\mathrm{Re}\,\mu < 0, \quad \nu+\mu \text{ is a positive integer}]$

$$\text{EH I 172(26)}$$

3.   $\displaystyle\int_0^1 \left[P_\nu^{n-\nu}(x)\right]^2\frac{\mathrm{d}x}{1-x^2} = -\frac{n!}{2(n-\nu)\,\Gamma(1-n+2\nu)}$      $[n = 0, 1, 2, \ldots; \quad \mathrm{Re}\,\nu > n]$

$$\text{ET II 315(9)}$$

**7.123**   $\displaystyle\int_{-1}^1 P_n^m(x)\,P_n^k(x)\frac{\mathrm{d}x}{1-x^2} = 0$      $[0 \leq m \leq n, \quad 0 \leq k \leq n; \quad m \neq k]$

$$\text{MO 74}$$

**7.124**$^{12}$ $\displaystyle\int_{-1}^1 x^k(z-x)^{-1}\left(1-x^2\right)^{\frac{1}{2}m}P_n^m(x)\,\mathrm{d}x = 2\left(z^2-1\right)^{\frac{1}{2}m}Q_n^m(z)z^k$      $[m \leq n; k = 0, 1, \ldots, n-m;$

$$z \text{ is in the complex plane with a cut along the interval } (-1, 1) \text{ on the real axis}]$$
$$\text{ET II 279(26)}$$

**7.125**
$$\int_{-1}^{1} \left(1-x^2\right)^{\frac{1}{2}m} P_k^m(x) P_l^m(x) P_n^m(x)\, dx = (-1)^m \pi^{-3/2} \frac{(k+m)!(l+m)!(n+m)!(s-m)!}{(k-m)!(l-m)!(n-m)!(s-k)!}$$
$$\times \frac{\Gamma\left(m+\tfrac{1}{2}\right)\Gamma\left(t-k+\tfrac{1}{2}\right)\Gamma\left(t-l+\tfrac{1}{2}\right)\Gamma\left(t-n+\tfrac{1}{2}\right)}{(s-l)!(s-n)!\,\Gamma\left(s+\tfrac{3}{2}\right)}$$

$[2s = k+l+n+m$ and $2t = k+l-n-m$ are both even
$l \geq m, \quad m \leq k-l-m \leq n \leq k+l+m]$

ET II 280(32)

**7.126**

1.
$$\int_0^1 P_\nu(x) x^\sigma\, dx = \frac{\sqrt{\pi}\,2^{-\sigma-1}\Gamma(1+\sigma)}{\Gamma\left(1+\tfrac{1}{2}\sigma-\tfrac{1}{2}\nu\right)\Gamma\left(\tfrac{1}{2}\sigma+\tfrac{1}{2}\nu+\tfrac{3}{2}\right)} \qquad [\operatorname{Re}\sigma > -1]$$
EH I 171(23)

2.
$$\int_0^1 x^\sigma P_\nu^m(x)\, dx = \frac{(-1)^m \pi^{1/2} 2^{-2m-1}\Gamma\left(\tfrac{1+\sigma}{2}\right)\Gamma(1+m+\nu)}{\Gamma\left(\tfrac{1}{2}+\tfrac{1}{2}m\right)\Gamma\left(\tfrac{3}{2}+\tfrac{\sigma}{2}+\tfrac{m}{2}\right)\Gamma(1-m+\nu)}$$
$$\times {}_3F_2\left(\frac{m+\nu+1}{2}, \frac{m-\nu}{2}, \frac{m}{2}+1; m+1, \frac{3+\sigma+m}{2}; 1\right)$$
$[\operatorname{Re}\sigma > -1; \quad m = 0,1,2,\dots]$   ET II 313(2)

3.
$$\int_0^1 x^\sigma P_\nu^\mu(x)\, dx = \frac{\pi^{1/2} 2^{2\mu-1}\Gamma\left(\tfrac{1+\sigma}{2}\right)}{\Gamma\left(\tfrac{1-\mu}{2}\right)\Gamma\left(\tfrac{3+\sigma-\mu}{2}\right)} {}_3F_2\left(\frac{\nu-\mu+1}{2}, -\frac{\mu+\nu}{2}, 1-\frac{\mu}{2}; 1-\mu, \frac{3+\sigma-\mu}{2}; 1\right)$$
$[\operatorname{Re}\sigma > -1, \quad \operatorname{Re}\mu < 2]$   ET II 313(3)

4.
$$\int_1^\infty x^{\mu-1} Q_\nu(ax)\, dx = e^{\mu\pi i}\Gamma(\mu) a^{-\mu}\left(a^2-1\right)^{\frac{1}{2}\mu} Q_\nu^{-\mu}(a)$$
$[|\arg(a-1)| < \pi, \quad \operatorname{Re}\mu > 0, \quad \operatorname{Re}(\nu-\mu) > -1]$   ET II 325(26)

**7.127**
$$\int_{-1}^1 (1+x)^\sigma P_\nu(x)\, dx = \frac{2^{\sigma+1}[\Gamma(\sigma+1)]^2}{\Gamma(\sigma+\nu+2)\Gamma(1+\sigma-\nu)} \qquad [\operatorname{Re}\sigma > -1]$$
ET II 316(15)

**7.128**

1.
$$\int_{-1}^1 (1-x)^{-\frac{1}{2}\mu}(1+x)^{\frac{1}{2}\mu-\frac{1}{2}}(z+x)^{\mu-\frac{3}{2}} P_\nu^\mu(x)\, dx$$
$$= -\frac{\Gamma\left(\mu-\tfrac{1}{2}\right)(z-1)^{\mu-\frac{1}{2}}(z+1)^{-1/2}}{\pi^{1/2} e^{2\mu\pi i}\Gamma(\mu+\nu)\Gamma(\mu-\nu-1)}$$
$$\times \left\{ Q_\nu^\mu\left[\left(\frac{1+z}{2}\right)^{1/2}\right] Q_{-\nu-1}^{\mu-1}\left[\left(\frac{1+z}{2}\right)^{1/2}\right] + Q_\nu^{\mu-1}\left[\left(\frac{1+z}{2}\right)^{1/2}\right] Q_{-\nu-1}^\mu\left[\left(\frac{1+z}{2}\right)^{1/2}\right] \right\}$$
$[-\tfrac{1}{2} < \operatorname{Re}\mu < 1,$
$z$ is in the complex plane with a cut along the interval $(-1,1)$ of the real axis]

ET II 317(20)

2.
$$\int_{-1}^1 (1-x)^{-\frac{1}{2}\mu}(1+x)^{\frac{1}{2}\mu-\frac{1}{2}}(z+x)^{\mu-\frac{1}{2}} P_\nu^\mu(x)\, dx$$
$$= \frac{2e^{-2\mu\pi i}\Gamma\left(\tfrac{1}{2}+\mu\right)}{\pi^{1/2}\Gamma(\mu-\nu)\Gamma(\mu+\nu+1)}(z-1)^\mu Q_\nu^\mu\left[\left(\frac{1+z}{2}\right)^{1/2}\right] Q_{-\nu-1}^\mu\left[\left(\frac{1+z}{2}\right)^{1/2}\right]$$
$[-\tfrac{1}{2} < \operatorname{Re}\mu < 1,$
$z$ is in the complex plane with a cut along the interval $(-1,1)$ of the real axis]

ET II 316(18)

**7.129** $\displaystyle\int_{-1}^{1} P_\nu(x)\, P_\lambda(x)(1+x)^{\lambda+\nu}\, dx = \frac{2^{\lambda+\nu+1}\,[\Gamma(\lambda+\nu+1)]^4}{[\Gamma(\lambda+1)\,\Gamma(\nu+1)]^2\,\Gamma(2\lambda+2\nu+2)}$

$$[\operatorname{Re}(\nu+\lambda+1) > 0] \qquad \text{EH I 172(30)}$$

**7.131**

1. $\displaystyle\int_{1}^{\infty} (x-1)^{-\frac{1}{2}\mu}\,(x+1)^{\frac{1}{2}\mu-\frac{1}{2}}\,(z+x)^{\mu-\frac{1}{2}}\,P_\nu^\mu(x)\,dx$

$$= \pi^{1/2}\frac{\Gamma(-\mu-\nu)\,\Gamma(1-\mu+\nu)}{\Gamma\left(\frac{1}{2}-\mu\right)}(z-1)^\mu\left\{P_\nu^\mu\left[\left(\frac{1+z}{2}\right)^{1/2}\right]\right\}^2$$

$$[\operatorname{Re}(\mu+\nu) < 0,\quad \operatorname{Re}(\mu-\nu) < 1,\quad |\arg(z+1)| < \pi]\quad \text{ET II 321(6)}$$

2. $\displaystyle\int_{1}^{\infty} (x-1)^{-\frac{1}{2}\mu}\,(x+1)^{\frac{1}{2}\mu-\frac{1}{2}}\,(z+x)^{\mu-\frac{3}{2}}\,P_\nu^\mu(x)\,dx$

$$= \frac{\pi^{1/2}\,\Gamma(1-\mu-\nu)\,\Gamma(2-\mu+\nu)\,(z-1)^{\mu-\frac{1}{2}}\,(z+1)^{-1/2}}{\Gamma\left(\frac{3}{2}-\mu\right)}P_\nu^\mu\left[\left(\frac{1+z}{2}\right)^{1/2}\right]P_\nu^{\mu-1}\left[\left(\frac{1+z}{2}\right)^{1/2}\right]$$

$$[\operatorname{Re}\mu < 1,\quad \operatorname{Re}(\mu+\nu) < 1,\quad \operatorname{Re}(\mu-\nu) < 2,\quad |\arg(1+z)| < \pi]\quad \text{ET II 321(7)}$$

**7.132**

1.[12] $\displaystyle\int_{-1}^{1} \left(1-x^2\right)^{\lambda-1}P_\nu^\mu(x)\,dx = \frac{\pi 2^\mu\,\Gamma\left(\lambda+\frac{1}{2}\mu\right)\Gamma\left(\lambda-\frac{1}{2}\mu\right)}{\Gamma\left(\lambda+\frac{1}{2}\nu+1\right)\Gamma\left(\lambda-\frac{1}{2}\nu\right)\Gamma\left(-\frac{1}{2}\mu+\frac{1}{2}\nu+1\right)\Gamma\left(-\frac{1}{2}\mu-\frac{1}{2}\nu+\frac{1}{2}\right)}$

$$[2\operatorname{Re}\lambda > |\operatorname{Re}\mu|] \qquad \text{ET II 316(16)}$$

2. $\displaystyle\int_{1}^{\infty} \left(x^2-1\right)^{\lambda-1}P_\nu^\mu(x)\,dx = \frac{2^{\mu-1}\,\Gamma\left(\lambda-\frac{1}{2}\mu\right)\Gamma\left(1-\lambda+\frac{1}{2}\nu\right)\Gamma\left(\frac{1}{2}-\lambda-\frac{1}{2}\nu\right)}{\Gamma\left(1-\frac{1}{2}\mu+\frac{1}{2}\nu\right)\Gamma\left(\frac{1}{2}-\frac{1}{2}\mu-\frac{1}{2}\nu\right)\Gamma\left(1-\lambda-\frac{1}{2}\mu\right)}$

$$[\operatorname{Re}\lambda > \operatorname{Re}\mu,\quad \operatorname{Re}(1-2\lambda-\nu) > 0,\quad \operatorname{Re}(2-2\lambda+\nu) > 0]\quad \text{ET II 320(2)}$$

3.[9] $\displaystyle\int_{1}^{\infty} \left(x^2-1\right)^{\lambda-1}Q_\nu^\mu(x)\,dx = e^{\mu\pi i}\frac{\Gamma\left(\frac{1}{2}+\frac{1}{2}\nu+\frac{1}{2}\mu\right)\Gamma\left(1-\lambda+\frac{1}{2}\nu\right)\Gamma\left(\lambda+\frac{1}{2}\mu\right)\Gamma\left(\lambda-\frac{1}{2}\mu\right)}{2^{2-\mu}\,\Gamma\left(1+\frac{1}{2}\nu-\frac{1}{2}\mu\right)\Gamma\left(\frac{1}{2}+\lambda+\frac{1}{2}\nu\right)}$

$$[|\operatorname{Re}\mu| < 2\operatorname{Re}\lambda < \operatorname{Re}\nu+2]$$

$$\text{ET II 324(23)}$$

4. $\displaystyle\int_{0}^{1} x^\sigma\left(1-x^2\right)^{-\frac{1}{2}\mu}P_\nu^\mu(x)\,dx = \frac{2^{\mu-1}\,\Gamma\left(\frac{1}{2}+\frac{1}{2}\sigma\right)\Gamma\left(1+\frac{1}{2}\sigma\right)}{\Gamma\left(1+\frac{1}{2}\sigma-\frac{1}{2}\nu-\frac{1}{2}\mu\right)\Gamma\left(\frac{1}{2}\sigma+\frac{1}{2}\nu-\frac{1}{2}\mu+\frac{3}{2}\right)}$

$$[\operatorname{Re}\mu < 1,\quad \operatorname{Re}\sigma > -1] \qquad \text{EH I 172(24)}$$

5. $\displaystyle\int_{0}^{1} x^\sigma\left(1-x^2\right)^{\frac{1}{2}m}P_\nu^m(x)\,dx = \frac{(-1)^m 2^{-m-1}\,\Gamma\left(\frac{1}{2}+\frac{1}{2}\sigma\right)\Gamma\left(1+\frac{1}{2}\sigma\right)\Gamma(1+m+\nu)}{\Gamma(1-m+\nu)\,\Gamma\left(1+\frac{1}{2}\sigma+\frac{1}{2}m-\frac{1}{2}\nu\right)\Gamma\left(\frac{3}{2}+\frac{1}{2}\sigma+\frac{1}{2}m+\frac{1}{2}\nu\right)}$

$$[\operatorname{Re}\sigma > -1,\quad m \text{ is a positive integer}] \quad \text{EH I 172(25), ET II 313(4)}$$

6.[12] $\displaystyle\int_{0}^{1} x^\sigma\left(1-x^2\right)^\eta P_\nu^\mu(x)\,dx = \frac{2^{\mu-1}\,\Gamma\left(1+\eta-\frac{1}{2}\mu\right)\Gamma\left(\frac{1}{2}+\frac{1}{2}\sigma\right)}{\Gamma(1-\mu)\,\Gamma\left(\frac{3}{2}+\eta+\frac{1}{2}\sigma-\frac{1}{2}\mu\right)}$

$$\times\ {}_3F_2\left(\frac{\nu-\mu+1}{2},-\frac{\mu+\nu}{2},1+\eta-\frac{\mu}{2};1-\mu,\frac{3+\sigma-\mu}{2}+\eta;1\right)$$

$$\left[\operatorname{Re}\left(\eta-\tfrac{1}{2}\mu\right) > -1, \operatorname{Re}\sigma > -1\right]\quad \text{ET II 314(6)}$$

7.  $$\int_1^\infty x^{-\rho} \left(x^2 - 1\right)^{-\frac{1}{2}\mu} P_\nu^\mu(x)\, \mathrm{d}x = \frac{2^{\rho+\mu-2}\, \Gamma\left(\frac{\rho+\mu+\nu}{2}\right) \Gamma\left(\frac{\rho+\mu-\nu-1}{2}\right)}{\sqrt{\pi}\, \Gamma(\rho)}$$

$$[\mathrm{Re}\,\mu < 1, \quad \mathrm{Re}(\rho + \mu + \nu) > 0, \quad \mathrm{Re}(\rho + \mu - \nu) > 1] \quad \text{ET II 320(3)}$$

**7.133**

1.  $$\int_u^\infty Q_\nu(x)(x-u)^{\mu-1}\, \mathrm{d}x = \Gamma(\mu) e^{\mu\pi i} \left(u^2 - 1\right)^{\frac{1}{2}\mu} Q_\nu^{-\mu}(u)$$

$$[|\arg(u-1)| < \pi, \quad 0 < \mathrm{Re}\,\mu < 1 + \mathrm{Re}\,\nu] \quad \text{MO 90a}$$

2.  $$\int_u^\infty \left(x^2 - 1\right)^{\frac{1}{2}\lambda} Q_\nu^{-\lambda}(x)(x-u)^{\mu-1}\, \mathrm{d}x = \Gamma(\mu) e^{\mu\pi i} \left(u^2 - 1\right)^{\frac{1}{2}\lambda+\frac{1}{2}\mu} Q_\nu^{-\lambda-\mu}(u)$$

$$[|\arg(u-1)| < \pi, \quad 0 < \mathrm{Re}\,\mu < 1 + \mathrm{Re}(\nu - \lambda)] \quad \text{ET II 204(30)}$$

**7.134**

1.  $$\int_1^\infty (x-1)^{\lambda-1} \left(x^2 - 1\right)^{\frac{1}{2}\mu} P_\nu^\mu(x)\, \mathrm{d}x = \frac{2^{\lambda+\mu}\, \Gamma(\lambda)\, \Gamma(-\lambda-\mu-\nu)\, \Gamma(1-\lambda-\mu+\nu)}{\Gamma(1-\mu+\nu)\, \Gamma(-\mu-\nu)\, \Gamma(1-\lambda-\mu)}$$

$$[\mathrm{Re}\,\lambda > 0, \quad \mathrm{Re}(\lambda+\mu+\nu) < 0, \quad \mathrm{Re}(\lambda+\mu-\nu) < 1] \quad \text{ET II 321(4)}$$

2.  $$\int_1^\infty (x-1)^{\lambda-1} \left(x^2 - 1\right)^{-\frac{1}{2}\mu} P_\nu^\mu(x)\, \mathrm{d}x = -\frac{2^{\lambda-\mu} \sin\pi\nu\, \Gamma(\lambda-\mu)\, \Gamma(-\lambda+\mu-\nu)\, \Gamma(1-\lambda+\mu+\nu)}{\pi\, \Gamma(1-\lambda)}$$

$$[\mathrm{Re}(\lambda-\mu) > 0, \quad \mathrm{Re}(\mu-\lambda-\nu) > 0, \quad \mathrm{Re}(\mu-\lambda+\nu) > -1] \quad \text{ET II 321(5)}$$

**7.135**

1.  $$\int_{-1}^1 \left(1 - x^2\right)^{-\frac{1}{2}\mu} (z-x)^{-1} P_{\mu+n}^\mu(x)\, \mathrm{d}x = 2 e^{-i\mu\pi} \left(z^2 - 1\right)^{-\frac{1}{2}\mu} Q_{\mu+n}^\mu(z)$$

$[n = 0, 1, 2, \ldots, \quad \mathrm{Re}\,\mu + n > -1, \; z \text{ is in the complex plane with a cut along the interval } (-1, 1)$
of the real axis.]                                    ET II 316(17)

2.  $$\int_1^\infty (x-1)^{\lambda-1} \left(x^2 - 1\right)^{\mu/2} (x+z)^{-\rho} P_\nu^\mu(x)\, \mathrm{d}x$$

$$= \frac{2^{\lambda+\mu-\rho}\, \Gamma(\lambda-\rho)\, \Gamma(\rho-\lambda-\mu-\nu)\, \Gamma(\rho-\lambda-\mu+\nu+1)}{\Gamma(1-\mu+\nu)\, \Gamma(-\mu-\nu)\, \Gamma(1+\rho-\lambda-\mu)}$$

$$\times\; {}_3F_2\left(\rho, \rho-\lambda-\mu-\nu, \rho-\lambda-\mu+\nu+1; \rho-\lambda+1, \rho-\lambda-\mu+1; \frac{1+z}{2}\right)$$

$$+ \frac{\Gamma(\rho-\lambda)\, \Gamma(\lambda)}{\Gamma(\rho)\, \Gamma(1-\mu)} 2^\mu (z+1)^{\lambda-\rho}\; {}_3F_2\left(\lambda, -\mu-\nu, 1-\mu+\nu; 1-\mu, 1-\rho+\lambda; \frac{1+z}{2}\right)$$

$$[\mathrm{Re}\,\lambda > 0, \quad \mathrm{Re}(\rho-\lambda-\mu-\nu) > 0, \quad \mathrm{Re}(\rho-\lambda-\mu+\nu+1) > 0, \quad |\arg(z+1)| < \pi]$$

$$\text{ET II 322(9)}$$

3.    $\displaystyle\int_1^\infty (x-1)^{\lambda-1}\left(x^2-1\right)^{-\mu/2}(x+z)^{-\rho}P_\nu^\mu(x)\,\mathrm{d}x$

$$= -\frac{\sin(\nu\pi)\,\Gamma(\lambda-\mu-\rho)\,\Gamma(\rho-\lambda+\mu-\nu)\,\Gamma(\rho-\lambda+\mu+\nu+1)}{2^{\rho-\lambda+\mu}\pi\,\Gamma(1+\rho-\lambda)}$$

$$\times\,{}_3F_2\left(\rho,\rho-\lambda+\mu-\nu,\rho-\lambda+\mu+\nu+1;1+\rho-\lambda,1+\rho-\lambda+\mu;\frac{1+z}{2}\right)$$

$$+\frac{\Gamma(\lambda-\mu)\,\Gamma(\rho-\lambda+\mu)}{\Gamma(\rho)\,\Gamma(1-\mu)}(z+1)^{\lambda-\rho-\mu}$$

$$\times\,{}_3F_2\left(\lambda-\mu,-\nu,\nu+1;1+\lambda-\mu-\rho,1-\mu;\frac{1+z}{2}\right)$$

$$[\operatorname{Re}(\lambda-\mu)>0,\quad \operatorname{Re}(\rho-\lambda+\mu-\nu)>0,\quad \operatorname{Re}(\rho-\lambda+\mu+\nu+1)>0,\quad |\arg(z+1)|<\pi]$$

ET II 322(10)

**7.136**

1.    $\displaystyle\int_{-1}^1 \left(1-x^2\right)^{\lambda-1}\left(1-a^2x^2\right)^{\mu/2}P_\nu(ax)\,\mathrm{d}x$

$$=\frac{\pi 2^\mu\,\Gamma(\lambda)}{\Gamma\left(\frac12+\lambda\right)\Gamma\left(\frac12-\frac12\mu-\frac12\nu\right)\Gamma\left(1-\frac12\mu+\frac12\nu\right)}\,{}_2F_1\left(-\frac{\mu+\nu}{2},\frac{1-\mu+\nu}{2};\frac12+\lambda;a^2\right)$$

$$[\operatorname{Re}\lambda>0,\quad -1<a<1]\quad \text{ET II 318(31)}$$

2.    $\displaystyle\int_1^\infty \left(x^2-1\right)^{\lambda-1}\left(a^2x^2-1\right)^{\mu/2}P_\nu^\mu(ax)\,\mathrm{d}x$

$$=\frac{\Gamma(\lambda)\,\Gamma\left(1-\lambda-\frac12\mu+\frac12\nu\right)\Gamma\left(\frac12-\lambda-\frac12\mu-\frac12\nu\right)}{\Gamma\left(1-\frac12\mu+\frac12\nu\right)\Gamma\left(\frac12-\frac12\nu-\frac12\mu\right)\Gamma(1-\lambda-\mu)}$$

$$\times 2^{\mu-1}a^{\mu-\nu-1}\,{}_2F_1\left(\frac{1-\mu+\nu}{2},1-\lambda-\frac{\mu-\nu}{2};1-\lambda-\mu;1-\frac{1}{a^2}\right)$$

$$[\operatorname{Re}a>0,\quad \operatorname{Re}\lambda>0,\quad \operatorname{Re}(\nu-\mu-2\lambda)>-2,\quad \operatorname{Re}(2\lambda+\mu+\nu)<1]\quad \text{ET II 325(25)}$$

3.    $\displaystyle\int_1^\infty \left(x^2-1\right)^{\lambda-1}\left(a^2x^2-1\right)^{-\frac12\mu}Q_\nu^\mu(ax)\,\mathrm{d}x=\frac{\Gamma\left(\frac{\mu+\nu+1}{2}\right)\Gamma(\lambda)\,\Gamma\left(1-\lambda+\frac{\mu+\nu}{2}\right)2^{\mu-2}e^{\mu\pi i}a^{-\mu-\nu-1}}{\Gamma\left(\nu+\frac32\right)}$

$$\times\,{}_2F_1\left(\frac{\mu+\nu+1}{2},1-\lambda+\frac{\mu+\nu}{2};\nu+\frac32;a^{-2}\right)$$

$$[|\arg(a-1)|<\pi,\quad \operatorname{Re}\lambda>0,\quad \operatorname{Re}(2\lambda-\mu-\nu)<2]\quad \text{ET II 325(27)}$$

**7.137**

1.    $\displaystyle\int_1^\infty x^{-\frac12\mu-\frac12}(x-1)^{-\mu-\frac12}(1+ax)^{\frac12\mu}Q_\nu^\mu(1+2ax)\,\mathrm{d}x$

$$=\pi^{-1/2}e^{-\mu\pi i}\,\Gamma\left(\frac12-\mu\right)a^{\frac12\mu}\left\{Q_\nu^\mu\left[(1+a)^{1/2}\right]\right\}^2$$

$$\left[|\arg a|<\pi,\quad \operatorname{Re}\mu<\frac12,\quad \operatorname{Re}(\mu+\nu)>-1\right]\quad \text{ET II 325(28)}$$

2.    $\displaystyle\int_1^\infty x^{-\frac12\mu-\frac12}(x-1)^{-\mu-\frac32}(1+ax)^{\frac12\mu}Q_\nu^\mu(1+2ax)\,\mathrm{d}x$

$$=-\pi^{-1/2}e^{-\mu\pi i}\,\Gamma\left(-\mu-\frac12\right)a^{\frac12\mu+\frac12}\left(1+a^2\right)^{-1/2}Q_\nu^{\mu+1}\left[(1+a)^{1/2}\right]Q_\nu^\mu\left[(1+a)^{1/2}\right]$$

$$\left[|\arg a|<\pi,\quad \operatorname{Re}\mu<-\frac12,\quad \operatorname{Re}(\mu+\nu+2)>0\right]\quad \text{ET II 326(29)}$$

3. $\displaystyle\int_0^1 x^{-\frac{1}{2}\mu-\frac{1}{2}}(1-x)^{-\mu-\frac{1}{2}}(1+ax)^{\frac{1}{2}\mu}P_\nu^\mu(1+2ax)\,\mathrm{d}x = \pi^{1/2}\,\Gamma\left(\tfrac{1}{2}-\mu\right)a^{\frac{1}{2}\mu}\left\{P_\nu^\mu\left[(1+a)^{1/2}\right]\right\}^2$

$$\left[\operatorname{Re}\mu<\tfrac{1}{2},\quad |\arg a|<\pi\right]\quad\text{ET II 319(32)}$$

4. $\displaystyle\int_0^1 x^{-\frac{1}{2}\mu-\frac{1}{2}}(1-x)^{-\mu-\frac{3}{2}}(1+ax)^{\frac{1}{2}\mu}P_\nu^\mu(1+2ax)\,\mathrm{d}x$

$$= \pi^{1/2}\,\Gamma\left(-\tfrac{1}{2}-\mu\right)a^{\frac{1}{2}\mu+\frac{1}{2}}P_\nu^{\mu+1}\left[(1+a)^{1/2}\right]P_\nu^\mu\left[(1+a)^{1/2}\right]$$
$$\left[\operatorname{Re}\mu<-\tfrac{1}{2},\quad |\arg a|<\pi\right]\quad\text{ET II 319(33)}$$

5. $\displaystyle\int_0^1 x^{\frac{1}{2}\mu-\frac{1}{2}}(1-x)^{\mu-\frac{1}{2}}(1+ax)^{-\frac{1}{2}\mu}P_\nu^\mu(1+2ax)\,\mathrm{d}x$

$$= \pi^{1/2}\,\Gamma\left(\tfrac{1}{2}+\mu\right)a^{-\frac{1}{2}\mu}P_\nu^\mu\left[(1+a)^{1/2}\right]P_\nu^{-\mu}\left[(1+a)^{1/2}\right]$$
$$\left[\operatorname{Re}\mu>-\tfrac{1}{2},\quad |\arg a|<\pi\right]\quad\text{ET II 319(34)}$$

6. $\displaystyle\int_0^1 x^{\frac{1}{2}\mu-\frac{1}{2}}(1-x)^{\mu-\frac{3}{2}}(1+ax)^{-\frac{1}{2}\mu}P_\nu^\mu(1+2ax)\,\mathrm{d}x$

$$= \frac{1}{2}\pi^{1/2}\,\Gamma\left(\mu-\tfrac{1}{2}\right)a^{\frac{1}{2}-\frac{1}{2}\mu}(1+a)^{-1/2}\left\{P_\nu^{1-\mu}\left[(1+a)^{1/2}\right]P_\nu^\mu\left[(1+a)^{1/2}\right]\right.$$
$$\left.+(\mu+\nu)(1-\mu+\nu)P_\nu^{-\mu}\left[(1+a)^{1/2}\right]P_\nu^\mu\left[(1+a)^{1/2}\right]\right\}$$
$$\left[\operatorname{Re}\mu>\tfrac{1}{2},\quad |\arg a|<\pi\right]\quad\text{ET II 319(35)}$$

7. $\displaystyle\int_0^1 x^{-\frac{\mu}{2}-\frac{1}{2}}(1-x)^{-\mu-\frac{1}{2}}(1+ax)^{\frac{1}{2}\mu}Q_\nu^\mu(1+2ax)\,\mathrm{d}x$

$$= \pi^{1/2}\,\Gamma\left(\tfrac{1}{2}-\mu\right)a^{\frac{1}{2}\mu}P_\nu^\mu\left[(1+a)^{1/2}\right]Q_\nu^\mu\left[(1+a)^{1/2}\right]$$
$$\left[\operatorname{Re}\mu<\tfrac{1}{2},\quad |\arg a|<\pi\right]\quad\text{ET II 320(38)}$$

8. $\displaystyle\int_0^1 x^{-\frac{\mu}{2}-\frac{1}{2}}(1-x)^{-\mu-\frac{3}{2}}(1+ax)^{\frac{1}{2}\mu}Q_\nu^\mu(1+2ax)\,\mathrm{d}x$

$$= \frac{1}{2}\pi^{1/2}\,\Gamma\left(-\mu-\tfrac{1}{2}\right)(1+a)^{-1/2}a^{\frac{1}{2}\mu+\frac{1}{2}}$$
$$\times\left\{P_\nu^{\mu+1}\left[(1+a)^{1/2}\right]Q_\nu^\mu\left[(1+a)^{1/2}\right]+P_\nu^\mu\left[(1+a)^{1/2}\right]Q_\nu^{\mu+1}\left[(1+a)^{1/2}\right]\right\}$$
$$\left[\operatorname{Re}\mu<-\tfrac{1}{2},\quad |\arg a|<\pi\right]\quad\text{ET II 320(39)}$$

9. $\displaystyle\int_0^y (y-x)^{\mu-1}\left[x\left(1+\tfrac{1}{2}\gamma x\right)\right]^{-\frac{1}{2}\lambda}P_\nu^\lambda(1+\gamma x)\,\mathrm{d}x$

$$= \Gamma(\mu)\left(\frac{2}{\gamma}\right)^{\frac{1}{2}\mu}\left[y\left(1+\frac{1}{2}\gamma y\right)\right]^{\frac{1}{2}\mu-\frac{1}{2}\lambda}P_\nu^{\lambda-\mu}(1+\gamma y)$$
$$\left[\operatorname{Re}\lambda<1,\quad\operatorname{Re}\mu>0,\quad |\arg\gamma y|<\pi\right]\quad\text{ET II 193(52)}$$

10. $\displaystyle\int_0^y (y-x)^{\mu-1}x^{\sigma+\frac{1}{2}\lambda-1}\left(1+\tfrac{1}{2}\gamma x\right)^{-\frac{1}{2}\lambda}P_\nu^\lambda(1+\gamma x)\,\mathrm{d}x$

$$= \frac{\left(\frac{\gamma}{2}\right)^{-\frac{1}{2}\lambda}\Gamma(\sigma)\,\Gamma(\mu)y^{\sigma+\mu-1}}{\Gamma(1-\lambda)\,\Gamma(\sigma+\mu)}\;{}_3F_2\left(-\nu,1+\nu,\sigma;1-\lambda,\sigma+\mu;-\frac{1}{2}\gamma y\right)$$
$$\left[\operatorname{Re}\sigma>0,\quad\operatorname{Re}\mu>0,\quad |\gamma y|<1\right]\quad\text{ET II 193(53)}$$

11.    $\int_0^y (y-x)^{\mu-1}[x(1-x)]^{-\frac{1}{2}\lambda} P_\nu^\lambda(1-2x)\,dx = \Gamma(\mu)[y(1-y)]^{\frac{1}{2}\mu-\frac{1}{2}\lambda} P_\nu^{\lambda-\mu}(1-2y)$

$$[\operatorname{Re}\lambda < 1, \quad \operatorname{Re}\mu > 0, \quad 0 < y < 1]$$
ET II 193(54)

12.    $\int_0^y (y-x)^{\mu-1} x^{\sigma+\frac{1}{2}\lambda-1}(1-x)^{-\frac{1}{2}\lambda} P_\nu^\lambda(1-2x)\,dx$

$$= \frac{\Gamma(\mu)\,\Gamma(\sigma)y^{\sigma+\mu-1}}{\Gamma(\sigma+\mu)\,\Gamma(1-\lambda)} \,{}_3F_2\left(-\nu,1+\nu,\sigma;1-\lambda,\sigma+\mu;y\right)$$
$$[\operatorname{Re}\sigma > 0, \quad \operatorname{Re}\mu > 0, \quad 0 < y < 1] \quad \text{ET II 193(155)}$$

**7.138**    $\int_0^\infty (a+x)^{-\mu-\nu-2} P_\mu\left(\dfrac{a-x}{a+x}\right) P_\nu\left(\dfrac{a-x}{a+x}\right) dx = \dfrac{a^{-\mu-\nu-1}\left[\Gamma(\mu+\nu+1)\right]^4}{\left[\Gamma(\mu+1)\,\Gamma(\nu+1)\right]^2 \Gamma(2\mu+2\nu+2)}$

$$[|\arg a| < \pi, \quad \operatorname{Re}(\mu+\nu) > -1]$$
ET II 326(3)

## 7.14 Combinations of associated Legendre functions, exponentials, and powers

**7.141**

1.    $\int_1^\infty e^{-ax}(x-1)^{\lambda-1}\left(x^2-1\right)^{\frac{1}{2}\mu} P_\nu^\mu(x)\,dx = \dfrac{a^{-\lambda-\mu}e^{-a}}{\Gamma(1-\mu+\nu)\,\Gamma(-\mu-\nu)} G_{23}^{31}\left(2a\,\Bigg|\begin{matrix}1+\mu,1\\ \lambda+\mu,-\nu,1+\nu\end{matrix}\right)$

$$[\operatorname{Re}a > 0, \quad \operatorname{Re}\lambda > 0] \qquad \text{ET II 323(13)}$$

2.    $\int_1^\infty e^{-ax}(x-1)^{\lambda-1}\left(x^2-1\right)^{\frac{1}{2}\mu} Q_\nu^\mu(x)\,dx$

$$= \frac{\Gamma(\nu+\mu+1)e^{\mu\pi i}}{2\,\Gamma(\nu-\mu+1)} a^{-\lambda-\mu}e^{-a} G_{23}^{22}\left(2a\,\Bigg|\begin{matrix}1+\mu,1\\ \lambda+\mu,\nu+1,-\nu\end{matrix}\right)$$
$$[\operatorname{Re}a > 0, \quad \operatorname{Re}\lambda > 0, \quad \operatorname{Re}(\lambda+\mu) > 0] \quad \text{ET II 325(24)}$$

3.    $\int_1^\infty e^{-ax}(x-1)^{\lambda-1}\left(x^2-1\right)^{-\frac{1}{2}\mu} P_\nu^\mu(x)\,dx = -\pi^{-1}\sin(\nu\pi)a^{\mu-\lambda}e^{-a} G_{23}^{31}\left(2a\,\Bigg|\begin{matrix}1,1-\mu\\ \lambda-\mu,1+\nu,-\nu\end{matrix}\right)$

$$[\operatorname{Re}a > 0, \quad \operatorname{Re}(\lambda-\mu) > 0]$$
ET II 323(15)

4.    $\int_1^\infty e^{-ax}(x-1)^{\lambda-1}\left(x^2-1\right)^{-\frac{1}{2}\mu} Q_\nu^\mu(x)\,dx = \dfrac{1}{2}e^{\mu\pi i}a^{\mu-\lambda}e^{-a} G_{23}^{22}\left(2a\,\Bigg|\begin{matrix}1-\mu,1\\ \lambda-\mu,\nu+1,-\nu\end{matrix}\right)$

$$[\operatorname{Re}a > 0, \quad \operatorname{Re}\lambda > 0, \quad \operatorname{Re}(\lambda-\mu) > 0]$$
ET II 323(14)

5.    $\int_1^\infty e^{-ax}\left(x^2-1\right)^{-\frac{1}{2}\mu} P_\nu^\mu(x)\,dx = 2^{1/2}\pi^{-1/2}a^{\mu-\frac{1}{2}} K_{\nu+\frac{1}{2}}(a)$

$$[\operatorname{Re}a > 0, \quad \operatorname{Re}\mu < 1]$$
ET II 323(11), MO 90

**7.142**    $\int_1^\infty e^{-\frac{1}{2}ax}\left(\dfrac{x+1}{x-1}\right)^{\frac{1}{2}\mu} P_{\nu-\frac{1}{2}}^\mu(x)\,dx = \dfrac{2}{a} W_{\mu,\nu}(a)$     $\left[\operatorname{Re}\mu < 1, \quad \nu-\frac{1}{2} \neq 0,\pm1,\pm2,\dots\right]$

BU 79(34), MO 118

**7.143**

1. $$\int_0^\infty [x(1+x)]^{-\frac{1}{2}\mu} e^{-\beta x} P_\nu^\mu(1+2x)\,dx = \frac{\beta^{\mu-\frac{1}{2}}}{\sqrt{\pi}} e^{\frac{1}{2}\beta} K_{\nu+\frac{1}{2}}\left(\frac{\beta}{2}\right)$$

$$[\operatorname{Re}\mu < 1, \quad \operatorname{Re}\beta > 0] \qquad \text{ET I 179(1)}$$

2. $$\int_0^\infty \left(1+\frac{1}{x}\right)^{\frac{1}{2}\mu} e^{-\beta x} P_\nu^\mu(1+2x)\,dx = \frac{e^{\frac{1}{2}\beta}}{\beta} W_{\mu,\nu+\frac{1}{2}}(\beta)$$

$$[\operatorname{Re}\mu < 1, \quad \operatorname{Re}\beta > 0] \qquad \text{ET I 179(2)}$$

**7.144**

1. $$\int_0^\infty e^{-\beta x} x^{\lambda+\frac{1}{2}\mu-1}(x+2)^{\frac{1}{2}\mu} Q_\nu^\mu(1+x)\,dx$$

$$= \frac{\Gamma(\nu+\mu+1)}{\Gamma(\nu-\mu+1)}\left\{\frac{\sin(\nu\pi)}{2\beta^{\lambda+\mu}\sin(\mu\pi)} E\left(-\nu,\nu+1,\lambda+\mu;\mu+1:2\beta\right)\right.$$

$$\left. - \frac{\sin[(\mu+\nu)\pi]}{2^{1-\mu}\beta^\lambda\sin(\mu\pi)} E\left(\nu-\mu+1,-\nu-\mu,\lambda:1-\mu:2\beta\right)\right\}$$

$$[\operatorname{Re}\beta > 0, \quad \operatorname{Re}\lambda > 0, \quad \operatorname{Re}(\lambda+\mu) > 0] \quad \text{ET I 181(16)}$$

2. $$\int_0^\infty e^{-\beta x} x^{\lambda-\frac{1}{2}\mu-1}(x+2)^{\frac{1}{2}\mu} Q_\nu^\mu(1+x)\,dx = -\frac{\sin(\nu\pi)}{2\beta^{\lambda-\mu}\sin(\mu\pi)} E(-\nu,\nu+1,\lambda-\mu:1-\mu:2\beta)$$

$$- \frac{\sin[(\mu-\nu)\pi]}{2^{1+\mu}\beta^\lambda\sin(\mu\pi)} E(\mu+\nu+1,\mu-\nu,\lambda:1+\mu:2\beta)$$

$$[\operatorname{Re}\beta > 0, \quad \operatorname{Re}\lambda > 0, \quad \operatorname{Re}(\lambda-\mu)] > 0 \quad \text{ET I 181(17)}$$

**7.145**

1. $$\int_0^\infty \frac{e^{-\beta x}}{1+x} P_\nu\left[\frac{1}{(1+x)^2}-1\right]dx = \frac{e^\beta}{\beta} W_{\nu+\frac{1}{2},0}(\beta)\, W_{-\nu-\frac{1}{2},0}(\beta)$$

$$[\operatorname{Re}\beta > 0] \qquad \text{ET I 180(6)}$$

2. $$\int_0^\infty x^{-1} e^{-\beta x} Q_{-\frac{1}{2}}\left(1+2x^{-2}\right)dx = \frac{\pi^2}{8}\left\{\left[J_0\left(\frac{1}{2}\beta\right)\right]^2 + \left[Y_0\left(\frac{1}{2}\beta\right)\right]^2\right\}$$

$$[\operatorname{Re}\beta > 0] \qquad \text{ET II 327(5)}$$

3. $$\int_0^\infty x^{-1} e^{-ax} Q_\nu\left(1+2x^{-2}\right)dx = \frac{1}{2}\left[\Gamma(\nu+1)\right]^2 a^{-1} W_{-\nu-\frac{1}{2},0}(ai)\, W_{-\nu-\frac{1}{2},0}(-ai)$$

$$[\operatorname{Re}a > 0, \quad \operatorname{Re}\nu > -1] \qquad \text{ET II 327(6)}$$

**7.146**

1. $$\int_0^\infty x^{-\frac{1}{2}\mu} e^{-\beta x} P_\nu^\mu\left(\sqrt{1+x}\right)dx = 2^\mu \beta^{\frac{1}{2}\mu-\frac{5}{4}} e^{\frac{\beta}{2}} W_{\frac{1}{2}\mu+\frac{1}{4},\frac{1}{2}\nu+\frac{1}{4}}(\beta)$$

$$[\operatorname{Re}\mu < 1, \quad \operatorname{Re}\beta > 0] \qquad \text{ET I 180(7)}$$

2. $$\int_0^\infty x^{-\frac{1}{2}\mu} \frac{e^{-\beta x}}{\sqrt{1+x}} P_\nu^\mu\left(\sqrt{1+x}\right)dx = 2^\mu \beta^{\frac{1}{2}\mu-\frac{3}{4}} e^{\frac{\beta}{2}} W_{\frac{1}{2}\mu+\frac{1}{4},\frac{1}{2}\nu+\frac{1}{4}}(\beta)$$

$$[\operatorname{Re}\mu < 1, \operatorname{Re}\beta > 0] \qquad \text{ET I 180(8)a}$$

3. $\quad \int_0^\infty \sqrt{x}\, e^{-\beta x}\, P_\nu^{1/4}\left(\sqrt{1+x^2}\right) P_\nu^{-1/4}\left(\sqrt{1+x^2}\right)\,\mathrm{d}x = \frac{1}{2}\sqrt{\frac{\pi}{2\beta}}\, H_{\nu+\frac{1}{2}}^{(1)}\left(\frac{1}{2}\beta\right) H_{\nu+\frac{1}{2}}^{(2)}\left(\frac{1}{2}\beta\right)$

$$[\operatorname{Re}\beta > 0]\qquad\qquad \text{ET I 180(9)}$$

**7.147** $\quad \int_0^\infty x^{\lambda-1}\left(x^2+a^2\right)^{\frac{1}{2}\nu} e^{-\beta x}\, P_\nu^\mu\left[\frac{x}{(x^2+a^2)^{1/2}}\right]\mathrm{d}x$

$$= \frac{2^{-\nu-2}a^{\lambda+\nu}}{\pi\,\Gamma(-\mu-\nu)}\, G_{24}^{32}\left(\frac{a^2\beta^2}{4}\,\middle|\,\begin{matrix}1-\frac{\lambda}{2},\ \frac{1-\lambda}{2}\\[2pt]0,\ \frac{1}{2},\ -\frac{\lambda+\mu+\nu}{2},\ -\frac{\lambda-\mu+\nu}{2}\end{matrix}\right)$$

$$[a>0,\quad \operatorname{Re}\beta>0,\quad \operatorname{Re}\lambda>0]\quad \text{ET II 327(7)}$$

**7.148** $\quad \int_{-1}^1 (1-x)^{-\frac{1}{2}\mu}(1+x)^{\frac{1}{2}\mu+\nu-1}\exp\left(-\frac{1-x}{1+x}y\right) P_\nu^\mu(x)\,\mathrm{d}x = 2^\nu y^{\frac{1}{2}\mu+\nu-\frac{1}{2}} e^{\frac{1}{2}y}\, W_{\frac{1}{2}\mu-\nu-\frac{1}{2},\frac{1}{2}\mu}(y)$

$$[\operatorname{Re}y>0]\qquad\qquad \text{ET II 317(21)}$$

**7.149** $\quad \int_1^\infty \left(\alpha^2+\beta^2+2\alpha\beta x\right)^{-1/2}\exp\left[-\left(\alpha^2+\beta^2+2\alpha\beta x\right)^{1/2}\right] P_\nu(x)\,\mathrm{d}x$

$$= 2\pi^{-1}(\alpha\beta)^{-1/2}\, K_{\nu+\frac{1}{2}}(\alpha)\, K_{\nu+\frac{1}{2}}(\beta)$$

$$[\operatorname{Re}\alpha>0,\quad \operatorname{Re}\beta>0]\qquad \text{ET II 323(16)}$$

## 7.15 Combinations of associated Legendre and hyperbolic functions

**7.151**

1. $\quad \int_0^\infty (\sinh x)^{\alpha-1}\, P_\nu^{-\mu}(\cosh x)\,\mathrm{d}x = \frac{2^{-1-\mu}\,\Gamma\left(\frac{1}{2}\alpha+\frac{1}{2}\mu\right)\Gamma\left(\frac{1}{2}\nu-\frac{1}{2}\alpha+1\right)\Gamma\left(\frac{1}{2}-\frac{1}{2}\alpha-\frac{1}{2}\nu\right)}{\Gamma\left(\frac{1}{2}\mu+\frac{1}{2}\nu+1\right)\Gamma\left(\frac{1}{2}+\frac{1}{2}\mu-\frac{1}{2}\nu\right)\Gamma\left(1+\frac{1}{2}\mu-\frac{1}{2}\alpha\right)}$

$$[\operatorname{Re}(\alpha+\mu)>0,\quad \operatorname{Re}(\nu-\alpha+2)>0,\quad \operatorname{Re}(1-\alpha-\nu)>0]\quad \text{EH I 172(28)}$$

2. $\quad \int_0^\infty (\sinh x)^{\alpha-1}\, Q_\nu^\mu(\cosh x)\,\mathrm{d}x\, \frac{e^{i\mu\pi}2^{\mu-\alpha}\,\Gamma\left(\frac{1}{2}+\frac{1}{2}\nu+\frac{1}{2}\mu\right)\Gamma\left(1+\frac{1}{2}\nu-\frac{1}{2}\alpha\right)}{\Gamma\left(1+\frac{1}{2}\nu-\frac{1}{2}\mu\right)\Gamma\left(\frac{1}{2}+\frac{1}{2}\nu+\frac{1}{2}\alpha\right)}$

$$\times\,\Gamma\left(\tfrac{1}{2}\alpha+\tfrac{1}{2}\mu\right)\Gamma\left(\tfrac{1}{2}\alpha-\tfrac{1}{2}\mu\right)$$

$$[\operatorname{Re}(\alpha\pm\mu)>0,\quad \operatorname{Re}(\nu-\alpha+2)>0]\quad \text{EH I 172(29)}$$

**7.152** $\quad \int_0^\infty e^{-\alpha x}\sinh^{2\mu}\left(\tfrac{1}{2}x\right) P_{2n}^{-2\mu}\left[\cosh\left(\tfrac{1}{2}x\right)\right]\mathrm{d}x = \frac{\Gamma\left(2\mu+\frac{1}{2}\right)\Gamma(\alpha-n-\mu)\Gamma\left(\alpha+n-\mu+\frac{1}{2}\right)}{4^\mu\sqrt{\pi}\,\Gamma(\alpha+n+\mu+1)\Gamma\left(\alpha-n+\mu+\frac{1}{2}\right)}$

$$[\operatorname{Re}\alpha>n+\operatorname{Re}\mu,\quad \operatorname{Re}\mu>-\tfrac{1}{4}]$$

$$\text{ET I 181(15)}$$

**7.153\*** $\quad \int_{-1}^1 (\cosh a - x)^{-3/2}\, P_n(x)\,\mathrm{d}x = \frac{2\sqrt{2}}{\sinh|a|}\, e^{-(n+1/2)|a|}$

## 7.16 Combinations of associated Legendre functions, powers, and trigonometric functions

**7.161**

1.  $\displaystyle\int_0^1 x^{\lambda-1}\left(1-x^2\right)^{-\frac{1}{2}\mu}\sin(ax)\,P_\nu^\mu(x)\,\mathrm{d}x$

$$= \frac{\pi^{1/2}2^{\mu-\lambda-1}\,\Gamma\left(\lambda+1\right)a}{\Gamma\left(1+\frac{\lambda-\mu-\nu}{2}\right)\Gamma\left(\frac{3+\lambda-\mu+\nu}{2}\right)}$$

$$\times\,{}_2F_3\left(\frac{1+\lambda}{2},1+\frac{\lambda}{2};\frac{3}{2},1+\frac{\lambda-\mu-\nu}{2},\frac{3+\lambda-\mu+\nu}{2};-\frac{a^2}{4}\right)$$

$$[\operatorname{Re}\lambda>-1,\quad\operatorname{Re}\mu<1]\qquad\text{ET II 314(7)}$$

2.  $\displaystyle\int_0^1 x^{\lambda-1}\left(1-x^2\right)^{-\frac{1}{2}\mu}\cos(ax)\,P_\nu^\mu(x)\,\mathrm{d}x$

$$= \frac{\pi^{1/2}2^{\mu-\lambda}\,\Gamma(\lambda)}{\Gamma\left(1+\dfrac{\lambda-\mu+\nu}{2}\right)\Gamma\left(\dfrac{1+\lambda-\mu-\nu}{2}\right)}$$

$$\times\,{}_2F_3\left(\frac{\lambda}{2},\frac{\lambda+1}{2};\frac{1}{2},\frac{1+\lambda-\mu-\nu}{2},1+\frac{\lambda-\mu+\nu}{2};-\frac{a^2}{4}\right)$$

$$[\operatorname{Re}\lambda>0,\quad\operatorname{Re}\mu<1]\qquad\text{ET II 314(8)}$$

3.  $\displaystyle\int_0^\infty\left(x^2-1\right)^{\frac{1}{2}\mu}\sin(ax)\,P_\nu^\mu(x)\,\mathrm{d}x=\frac{2^\mu\pi^{1/2}a^{-\mu-\frac{1}{2}}}{\Gamma\left(\frac{1}{2}-\frac{1}{2}\mu-\frac{1}{2}\nu\right)\Gamma\left(1-\frac{1}{2}\mu+\frac{1}{2}\nu\right)}\,S_{\mu+\frac{1}{2},\nu+\frac{1}{2}}(a)$

$$\left[a>0,\quad\operatorname{Re}\mu<\tfrac{3}{2},\quad\operatorname{Re}(\mu+\nu)<1\right]$$
$$\text{ET II 320(1)}$$

**7.162**

1.  $\displaystyle\int_a^\infty P_\nu\left(2x^2a^{-2}-1\right)\sin(bx)\,\mathrm{d}x=-\frac{\pi a}{4\cos(\nu\pi)}\left\{\left[J_{\nu+\frac{1}{2}}\left(\frac{ab}{2}\right)\right]^2-\left[J_{-\nu-\frac{1}{2}}\left(\frac{ab}{2}\right)\right]^2\right\}$

$$[a>0,\quad b>0,\quad-1<\operatorname{Re}\nu<0]$$
$$\text{ET II 326(1)}$$

2.  $\displaystyle\int_a^\infty P_\nu\left(2x^2a^{-2}-1\right)\cos(bx)\,\mathrm{d}x$

$$=-\frac{\pi}{4}a\left[J_{\nu+\frac{1}{2}}\left(\frac{ab}{2}\right)J_{-\nu-\frac{1}{2}}\left(\frac{ab}{2}\right)-Y_{\nu+\frac{1}{2}}\left(\frac{ab}{2}\right)Y_{-\nu-\frac{1}{2}}\left(\frac{ab}{2}\right)\right]$$

$$[a>0,\quad b>0,\quad-1<\operatorname{Re}\nu<0]\qquad\text{ET II 326(2)}$$

3.  $\displaystyle\int_0^\infty\left(x^2+2\right)^{-1/2}\sin(ax)\,P_\nu^{-1}\left(x^2+1\right)\,\mathrm{d}x=2^{-1/2}\pi^{-1}a\sin(\nu\pi)\left[K_{\nu+\frac{1}{2}}\left(2^{-1/2}a\right)\right]^2$

$$[a>0,\quad-2<\operatorname{Re}\nu<1]\qquad\text{ET I 98(22)}$$

4.  $\displaystyle\int_0^\infty\left(x^2+2\right)^{-1/2}\sin(ax)\,Q_\nu^1\left(x^2+1\right)\,\mathrm{d}x=-2^{-3/2}\pi a\,K_{\nu+\frac{1}{2}}\left(2^{-1/2}a\right)I_{\nu+\frac{1}{2}}\left(2^{-1/2}a\right)$

$$\left[a>0,\quad\operatorname{Re}\nu>-\tfrac{3}{2}\right]\qquad\text{ET 98(23)}$$

5.    $\displaystyle\int_0^\infty \cos(ax)\, P_\nu\left(1+x^2\right)\,\mathrm{d}x = -\frac{\sqrt{2}}{\pi}\sin(\nu\pi)\left[K_{\nu+\frac{1}{2}}\left(\frac{a}{\sqrt{2}}\right)\right]^2$

$$[a>0,\quad -1<\operatorname{Re}\nu<0]\qquad \text{ET I 42(23)}$$

6.    $\displaystyle\int_0^\infty \cos(ax)\, Q_\nu\left(1+x^2\right)\,\mathrm{d}x = \frac{\pi}{\sqrt{2}}K_{\nu+\frac{1}{2}}\left(\frac{a}{\sqrt{2}}\right)I_{\nu+\frac{1}{2}}\left(\frac{a}{\sqrt{2}}\right)$

$$[a>0,\quad \operatorname{Re}\nu>-1]\qquad \text{ET I 42(24)}$$

7.    $\displaystyle\int_0^1 \cos(ax)\, P_\nu\left(2x^2-1\right)\,\mathrm{d}x = \frac{\pi}{2}J_{\nu+\frac{1}{2}}\left(\frac{a}{2}\right)J_{-\nu-\frac{1}{2}}\left(\frac{a}{2}\right)$

$$[a>0]\qquad \text{ET I 42(25)}$$

**7.163**

1.    $\displaystyle\int_a^\infty \left(x^2-a^2\right)^{\frac{1}{2}\nu-\frac{1}{4}}\sin(bx)\, P_0^{\frac{1}{2}-\nu}\left(ax^{-1}\right)\,\mathrm{d}x = b^{-\nu-\frac{1}{2}}\cos\left(ab-\frac{\nu\pi}{2}+\frac{\pi}{4}\right)$

$$\left[a>0,\quad |\operatorname{Re}\nu|<\tfrac{1}{2}\right]\qquad \text{ET I 98(24)}$$

2.    $\displaystyle\int_0^1 x^{-1}\cos(ax)\, P_\nu\left(2x^{-2}-1\right)\,\mathrm{d}x = -\frac{1}{2}\pi\operatorname{cosec}(\nu\pi)\,{}_1F_1\left(\nu+1;1;ai\right){}_1F_1\left(\nu+1;1;-ai\right)$

$$[a>0,\quad -1<\operatorname{Re}\nu<0]\qquad \text{ET II 327(4)}$$

**7.164**

1.    $\displaystyle\int_0^\infty x^{1/2}\sin(bx)\left[P_\nu^{-1/4}\left(\sqrt{1+a^2x^2}\right)\right]^2\,\mathrm{d}x = \frac{\sqrt{\frac{2}{\pi}}\,a^{-1}b^{-1/2}}{\Gamma\left(\frac{5}{4}+\nu\right)\Gamma\left(\frac{1}{4}-\nu\right)}\left[K_{\nu+\frac{1}{2}}\left(\frac{b}{2a}\right)\right]^2$

$$\left[\operatorname{Re}a>0,\quad b>0,\quad -\tfrac{5}{4}<\operatorname{Re}\nu<\tfrac{1}{4}\right]$$
$$\text{ET II 327(8)}$$

2.$^{12}$    $\displaystyle\int_0^\infty x^{1/2}\sin(bx)\, P_\nu^{-1/4}\left(\sqrt{1+a^2x^2}\right)Q_\nu^{-1/4}\left(\sqrt{1+a^2x^2}\right)\,\mathrm{d}x$

$$= \frac{\sqrt{\frac{\pi}{2}}\,e^{-\frac{1}{4}\pi i}\,\Gamma\left(\nu+\frac{5}{4}\right)}{ab^{\frac{1}{2}}\,\Gamma\left(\nu+\frac{3}{4}\right)}I_{\nu+\frac{1}{2}}\left(\frac{b}{2a}\right)K_{\nu+\frac{1}{2}}\left(\frac{b}{2a}\right)$$
$$\left[\operatorname{Re}a>0,\quad b>0,\quad \operatorname{Re}\nu>-\tfrac{5}{4}\right]\quad \text{ET II 328(9)}$$

3.    $\displaystyle\int_0^\infty x^{1/2}\sin(bx)\, P_\nu^{-1/4}\left(\sqrt{1+a^2x^2}\right)P_{\nu-1}^{-1/4}\left(\sqrt{1+a^2x^2}\right)\frac{\mathrm{d}x}{\sqrt{1+a^2x^2}}$

$$= \frac{a^{-2}b^{1/2}}{\sqrt{2\pi}\,\Gamma\left(\frac{5}{4}+\nu\right)\Gamma\left(\frac{5}{4}-\nu\right)}K_{\nu-\frac{1}{2}}\left(\frac{b}{2a}\right)K_{\nu+\frac{1}{2}}\left(\frac{b}{2a}\right)$$
$$\left[\operatorname{Re}a>0,\quad b>0,\quad -\tfrac{5}{4}<\operatorname{Re}\nu<\tfrac{5}{4}\right]\quad \text{ET II 328(10)}$$

4.    $\displaystyle\int_0^\infty x^{1/2}\sin(bx)\, P_\nu^{1/4}\left(\sqrt{1+a^2x^2}\right)P_\nu^{-3/4}\left(\sqrt{1+a^2x^2}\right)\frac{\mathrm{d}x}{\sqrt{1+a^2x^2}}$

$$= \frac{a^{-2}b^{1/2}}{\sqrt{2\pi}\,\Gamma\left(\frac{7}{4}+\nu\right)\Gamma\left(\frac{3}{4}-\nu\right)}\left[K_{\nu+\frac{1}{2}}\left(\frac{b}{2a}\right)\right]^2$$
$$\left[\operatorname{Re}a>0,\quad b>0,\quad -\tfrac{7}{4}<\operatorname{Re}\nu<\tfrac{3}{4}\right]\quad \text{ET II 328(11)}$$

5. $\displaystyle\int_0^\infty x^{1/2}\cos(bx)\left[P_\nu^{1/4}\left(\sqrt{1+a^2x^2}\right)\right]^2\,dx = \frac{a^{-1}\left(\frac{\pi b}{2}\right)^{-1/2}}{\Gamma\left(\frac{3}{4}+\nu\right)\Gamma\left(-\frac{1}{4}-\nu\right)}\left[K_{\nu+\frac{1}{2}}\left(\frac{b}{2a}\right)\right]^2$

$$\left[\operatorname{Re} a > 0, \quad b > 0, \quad -\tfrac{3}{4} < \operatorname{Re}\nu < -\tfrac{1}{4}\right]$$
ET II 328(12)

6. $\displaystyle\int_0^\infty x^{1/2}\cos(bx)\,P_\nu^{1/4}\left(\sqrt{1+a^2x^2}\right)Q_\nu^{1/4}\left(\sqrt{1+a^2x^2}\right)\,dx$

$$= \frac{\sqrt{\frac{\pi}{2}}e^{\frac{1}{4}\pi i}\Gamma\left(\nu+\frac{3}{4}\right)}{ab^{1/2}\Gamma\left(\nu+\frac{5}{4}\right)}I_{\nu+\frac{1}{2}}\left(\frac{b}{2a}\right)K_{\nu+\frac{1}{2}}\left(\frac{b}{2a}\right)$$
$$\left[\operatorname{Re} a > 0, \quad b > 0, \quad \operatorname{Re}\nu > -\tfrac{3}{4}\right] \quad \text{ET II 328(13)}$$

7. $\displaystyle\int_0^\infty x^{1/2}\cos(bx)\,P_\nu^{-1/4}\left(\sqrt{1+a^2x^2}\right)P_\nu^{3/4}\left(\sqrt{1+a^2x^2}\right)\frac{dx}{\sqrt{1+a^2x^2}}$

$$= \frac{a^{-2}b^{1/2}}{\sqrt{2\pi}\,\Gamma\left(\frac{5}{4}+\nu\right)\Gamma\left(\frac{1}{4}-\nu\right)}\left[K_{\nu+\frac{1}{2}}\left(\frac{b}{2a}\right)\right]^2$$
$$\left[\operatorname{Re} a > 0, \quad b > 0, \quad -\tfrac{5}{4} < \operatorname{Re}\nu < \tfrac{1}{4}\right] \quad \text{ET II 328(14)}$$

8. $\displaystyle\int_0^\infty x^{1/2}\cos(bx)\,P_\nu^{1/4}\left(\sqrt{1+a^2x^2}\right)P_{\nu-1}^{1/4}\left(\sqrt{1+a^2x^2}\right)\frac{dx}{\sqrt{1+a^2x^2}}$

$$= \frac{a^{-2}b^{1/2}}{\sqrt{2\pi}\,\Gamma\left(\frac{3}{4}+\nu\right)\Gamma\left(\frac{3}{4}-\nu\right)}K_{\nu-\frac{1}{2}}\left(\frac{b}{2a}\right)K_{\nu+\frac{1}{2}}\left(\frac{b}{2a}\right)$$
$$\left[\operatorname{Re} a > 0, \quad b > 0, \quad |\operatorname{Re}\nu| < \tfrac{3}{4}\right] \quad \text{ET II 329(15)}$$

**7.165** $\displaystyle\int_0^\infty \cos(ax)\,P_\nu(\cosh x)\,dx$

$$= -\frac{\sin(\nu\pi)}{4\pi^2}\Gamma\left(\frac{1+\nu+i\alpha}{2}\right)\Gamma\left(\frac{1+\nu-i\alpha}{2}\right)\Gamma\left(-\frac{\nu+i\alpha}{2}\right)\Gamma\left(-\frac{\nu-i\alpha}{2}\right)$$
$$[a > 0, \quad -1 < \operatorname{Re}\nu < 0] \quad \text{ET II 329(18)}$$

**7.166** $\displaystyle\int_0^\pi P_\nu^{-\mu}(\cos\varphi)\sin^{\alpha-1}\varphi\,d\varphi = \frac{2^{-\mu}\pi\,\Gamma\left(\frac{1}{2}\alpha+\frac{1}{2}\mu\right)\Gamma\left(\frac{1}{2}\alpha-\frac{1}{2}\mu\right)}{\Gamma\left(\frac{1}{2}+\frac{1}{2}\alpha+\frac{1}{2}\nu\right)\Gamma\left(\frac{1}{2}\alpha-\frac{1}{2}\nu\right)\Gamma\left(\frac{1}{2}\mu+\frac{1}{2}\nu+1\right)\Gamma\left(\frac{1}{2}\mu-\frac{1}{2}\nu+\frac{1}{2}\right)}$

$$[\operatorname{Re}(\alpha\pm\mu) > 0] \quad \text{MO 90, EH I 172(27)}$$

**7.167** $\displaystyle\int_0^a P_\nu^{-\mu}(\cos x)\,P_\nu^{-\eta}[\cos(a-x)]\left[\frac{\sin(a-x)}{\sin x}\right]^\eta\frac{dx}{\sin x} = \frac{2^\eta\,\Gamma(\mu-\eta)\,\Gamma\left(\eta+\frac{1}{2}\right)(\sin a)^\eta}{\sqrt{\pi}\,\Gamma(\eta+\mu+1)}P_\nu^{-\mu}(\cos a)$

$$\left[\operatorname{Re}\mu > \operatorname{Re}\eta > -\tfrac{1}{2}\right] \quad \text{ET II 329(16)}$$

## 7.17 A combination of an associated Legendre function and the probability integral

**7.171** $\displaystyle\int_1^\infty \left(x^2-1\right)^{-\frac{1}{2}\mu}\exp\left(a^2x^2\right)[1-\Phi(ax)]\,P_\nu^\mu(x)\,dx$

$$= \pi^{-1}2^{\mu-1}\Gamma\left(\frac{1+\mu+\nu}{2}\right)\Gamma\left(\frac{\mu-\nu}{2}\right)a^{\mu-\frac{3}{2}}e^{\frac{a^2}{2}}\,W_{\frac{1}{4}-\frac{1}{2}\mu,\,\frac{1}{4}+\frac{1}{2}\nu}\left(a^2\right)$$

$$[\operatorname{Re} a > 0, \quad \operatorname{Re}\mu < 1, \quad \operatorname{Re}(\mu+\nu) > -1, \quad \operatorname{Re}(\mu-\nu) > 0]$$
ET II 324(17)

## 7.18 Combinations of associated Legendre and Bessel functions

**7.181**

1. $\displaystyle\int_1^\infty P_{\nu-\frac{1}{2}}(x)\,x^{1/2}\,Y_\nu(ax)\,\mathrm{d}x = 2^{-1/2}a^{-1}\left[\cos\left(\tfrac{1}{2}a\right)J_\nu\left(\tfrac{1}{2}a\right) - \sin\left(\tfrac{1}{2}a\right)Y_\nu\left(\tfrac{1}{2}a\right)\right]$

$$\left[a>0,\quad \operatorname{Re}\nu<\tfrac{1}{2}\right] \qquad \text{ET II 108(3)a}$$

2. $\displaystyle\int_1^\infty P_{\nu-\frac{1}{2}}(x)\,x^{1/2}\,J_\nu(ax)\,\mathrm{d}x = -\frac{1}{\sqrt{2a}}\left[\cos\left(\tfrac{1}{2}a\right)Y_\nu\left(\tfrac{1}{2}a\right) + \sin\left(\tfrac{1}{2}a\right)J_\nu\left(\tfrac{1}{2}a\right)\right]$

$$\left[\left|\operatorname{Re}\nu\right|<\tfrac{1}{2}\right] \qquad \text{ET II 344(36)a}$$

**7.182**

1. $\displaystyle\int_1^\infty x^\nu\left(x^2-1\right)^{\frac{1}{2}\lambda-\frac{1}{2}}P_\lambda^{\lambda-1}(x)\,J_\nu(ax)\,\mathrm{d}x = \frac{2^{\lambda+\nu}a^{-\lambda}\,\Gamma\left(\tfrac{1}{2}+\nu\right)}{\pi^{1/2}\,\Gamma(1-\lambda)}S_{\lambda-\nu,\lambda+\nu}(a)$

$$\left[a>0,\quad \operatorname{Re}\nu<\tfrac{5}{2},\quad \operatorname{Re}(2\lambda+\nu)<\tfrac{3}{2}\right]$$
$$\text{ET II 345(38)a}$$

2. $\displaystyle\int_1^\infty x^{\frac{1}{2}-\mu}\left(x^2-1\right)^{-\frac{1}{2}\mu}P_{\nu-\frac{1}{2}}^\mu(x)\,J_\nu(ax)\,\mathrm{d}x$

$$= -2^{-3/2}\pi^{1/2}a^{\mu-\frac{1}{2}}\left[J_{\mu-\frac{1}{2}}\left(\frac{a}{2}\right)Y_\nu\left(\frac{a}{2}\right) + Y_{\mu-\frac{1}{2}}\left(\frac{a}{2}\right)J_\nu\left(\frac{a}{2}\right)\right]$$
$$\left[-\tfrac{1}{4}<\operatorname{Re}\mu<1,\quad a>0,\quad \left|\operatorname{Re}\nu\right|<\tfrac{1}{2}+2\operatorname{Re}\mu\right] \quad \text{ET II 344(37)a}$$

3. $\displaystyle\int_1^\infty x^{\frac{1}{2}-\mu}\left(x^2-1\right)^{-\frac{1}{2}\mu}P_{\nu-\frac{1}{2}}^\mu(x)\,Y_\nu(ax)\,\mathrm{d}x$

$$= 2^{-3/2}\pi^{1/2}a^{\mu-\frac{1}{2}}\left[J_\nu\left(\frac{a}{2}\right)J_{\mu-\frac{1}{2}}\left(\frac{a}{2}\right) - Y_\nu\left(\frac{a}{2}\right)Y_{\mu-\frac{1}{2}}\left(\frac{a}{2}\right)\right]$$
$$\left[-\tfrac{1}{4}<\operatorname{Re}\mu<1,\quad a>0,\quad \operatorname{Re}(2\mu-\nu)>-\tfrac{1}{2}\right] \quad \text{ET II 349(67)a}$$

4. $\displaystyle\int_0^1 x^{\frac{1}{2}-\mu}\left(1-x^2\right)^{-\frac{1}{2}\mu}P_\nu^\mu(x)\,J_{\nu+\frac{1}{2}}(ax)\,\mathrm{d}x = \sqrt{\frac{\pi}{2}}a^{\mu-\frac{1}{2}}J_{\frac{1}{2}-\mu}\left(\tfrac{1}{2}a\right)J_{\nu+\frac{1}{2}}\left(\tfrac{1}{2}a\right)$

$$\left[\operatorname{Re}\mu<1,\quad \operatorname{Re}(\mu-\nu)<2\right]$$
$$\text{ET II 337(33)a}$$

5. $\displaystyle\int_1^\infty x^{\frac{1}{2}-\mu}\left(x^2-1\right)^{-\frac{1}{2}\mu}P_{\nu-\frac{1}{2}}^\mu(x)\,K_\nu(ax)\,\mathrm{d}x = (2\pi)^{-1/2}a^{\mu-\frac{1}{2}}K_\nu\left(\tfrac{1}{2}a\right)K_{\mu-\frac{1}{2}}\left(\tfrac{1}{2}a\right)$

$$\left[\operatorname{Re}\mu<1,\quad \operatorname{Re}a>0\right] \qquad \text{ET II 135(5)a}$$

6. $\displaystyle\int_1^\infty x^{\mu+\frac{1}{2}}\left(x^2-1\right)^{-\frac{1}{2}\mu}P_{\nu-\frac{1}{2}}^\mu(x)\,K_\nu(ax)\,\mathrm{d}x = \sqrt{\frac{\pi}{2}}a^{-3/2}e^{-\frac{1}{2}a}W_{\mu,\nu}(a)$

$$\left[\operatorname{Re}\mu<1,\quad \operatorname{Re}a>0\right] \qquad \text{ET II 135(3)a}$$

7. $\displaystyle\int_1^\infty x^{\mu-\frac{3}{2}}\left(x^2-1\right)^{-\frac{1}{2}\mu}P_{\nu-\frac{1}{2}}^\mu(x)\,K_\nu(ax)\,\mathrm{d}x = \sqrt{\frac{\pi}{2}}a^{-1/2}e^{-\frac{1}{2}a}W_{\mu-1,\nu}(a)$

$$\left[\operatorname{Re}\mu<1,\quad \operatorname{Re}a>0\right] \qquad \text{ET II 135(4)a}$$

8. $\displaystyle\int_1^\infty x^{\mu-\frac{1}{2}}\left(x^2-1\right)^{-\frac{1}{2}\mu}P_{\nu-\frac{3}{2}}^\mu(x)\,K_\nu(ax)\,\mathrm{d}x = \sqrt{\frac{\pi}{2}}a^{-1}e^{-\frac{1}{2}a}W_{\mu-\frac{1}{2},\nu-\frac{1}{2}}(a)$

$$\left[\operatorname{Re}\mu<1\right] \qquad \text{ET II 135(6)a}$$

9.  $$\int_1^\infty x^{1/2} \left(x^2 - 1\right)^{\frac{1}{2}\nu - \frac{1}{4}} P_\mu^{\frac{1}{2} - \nu} \left(2x^2 - 1\right) K_\nu(ax) \, dx = \pi^{-1/2} a^{-\nu} 2^{\nu - 1} \left[K_{\mu + \frac{1}{2}} \left(\frac{a}{2}\right)\right]^2$$

$$\left[\operatorname{Re}\nu > -\tfrac{1}{2}, \quad \operatorname{Re}a > 0\right] \quad \text{ET II 136(11)a}$$

10. $$\int_1^\infty x^{1/2} \left(x^2 - 1\right)^{\frac{1}{2}\nu - \frac{1}{4}} P_\mu^{\frac{1}{2} - \nu} \left(2x^2 - 1\right) Y_\nu(ax) \, dx$$

$$= \pi^{1/2} 2^{\nu - 2} a^{-\nu} \left[J_{\mu + \frac{1}{2}} \left(\frac{a}{2}\right) J_{-\mu - \frac{1}{2}} \left(\frac{a}{2}\right) - Y_{\mu + \frac{1}{2}} \left(\frac{a}{2}\right) Y_{-\mu - \frac{1}{2}} \left(\frac{a}{2}\right)\right]$$

$$\left[\operatorname{Re}\nu > -\tfrac{1}{2}, \quad a > 0, \quad \operatorname{Re}\nu + |2\operatorname{Re}\mu + 1| < \tfrac{3}{2}\right] \quad \text{ET II 108(5)a}$$

11. $$\int_1^\infty x^{1/2} \left(x^2 - 1\right)^{\frac{1}{2}\nu - \frac{1}{4}} P_\mu^{\frac{1}{2} - \nu} \left(2x^2 - 1\right) J_\nu(ax) \, dx$$

$$= -2^{\nu - 2} a^{-\nu} \pi^{1/2} \sec(\mu\pi) \left\{\left[J_{\mu + \frac{1}{2}} \left(\frac{a}{2}\right)\right]^2 - \left[J_{-\mu - \frac{1}{2}} \left(\frac{a}{2}\right)\right]^2\right\}$$

$$\left[\operatorname{Re}\nu > -\tfrac{1}{2}, \quad a > 0, \quad \operatorname{Re}\nu - \tfrac{3}{2} < 2\operatorname{Re}\mu < \tfrac{1}{2} - \operatorname{Re}\nu\right] \quad \text{ET II 345(39)a}$$

12. $$\int_1^\infty x \left(x^2 - 1\right)^{-\frac{1}{2}\nu} P_\mu^\nu \left(2x^2 - 1\right) K_\nu(ax) \, dx = 2^{-\nu} a^{\nu - 1} K_{\mu + 1}(a)$$

$$\left[\operatorname{Re}a > 0, \quad \operatorname{Re}\nu < 1\right] \quad \text{ET II 136(10)a}$$

13. $$\int_0^\infty x \left(x^2 + a^2\right)^{\frac{1}{2}\nu} P_\mu^\nu \left(1 + 2x^2 a^{-2}\right) K_\nu(xy) \, dx = 2^{-\nu} a y^{-\nu - 1} S_{2\nu, 2\mu + 1}(ay)$$

$$\left[\operatorname{Re}a > 0, \quad \operatorname{Re}y > 0, \quad \operatorname{Re}\nu < 1\right]$$
$$\text{ET II 135(7)}$$

14. $$\int_0^\infty x \left(x^2 + a^2\right)^{\frac{1}{2}\nu} \left[(\mu - \nu) P_\mu^\nu \left(1 + 2x^2 a^{-2}\right) + (\mu + \nu) P_{-\mu}^\nu \left(1 + 2x^2 a^{-2}\right)\right] K_\nu(xy) \, dx$$

$$= 2^{1 - \nu} \mu y^{-\nu - 2} S_{2\nu + 1, 2\mu}(ay)$$

$$\left[\operatorname{Re}a > 0, \quad \operatorname{Re}y > 0, \quad \operatorname{Re}\nu < 1\right] \quad \text{ET II 136(8)}$$

15. $$\int_0^\infty x \left(x^2 + a^2\right)^{\frac{1}{2}\nu - 1} \left[P_\mu^\nu \left(1 + 2x^2 a^{-2}\right) + P_{-\mu}^\nu \left(1 + 2x^2 a^{-2}\right)\right] K_\nu(xy) \, dx = 2^{1 - \nu} y^{-\nu} S_{2\nu - 1, 2\mu}(ay)$$

$$\left[\operatorname{Re}a > 0, \quad \operatorname{Re}y > 0, \quad \operatorname{Re}\nu < 1\right]$$
$$\text{ET II 136(9)}$$

16. $$\int_0^\infty x^{1/2} \left(x^2 + 2\right)^{-\frac{1}{2}\nu - \frac{1}{4}} P_\mu^{-\nu - \frac{1}{2}} \left(x^2 + 1\right) J_\nu(xy) \, dx = \frac{y^{-1/2} 2^{\frac{1}{2} - \nu} \pi^{-1/2} \left[K_{\mu + \frac{1}{2}} \left(2^{-1/2} y\right)\right]^2}{\Gamma\left(\nu + \mu + \frac{3}{2}\right) \Gamma\left(\nu - \mu + \frac{1}{2}\right)}$$

$$\left[-\tfrac{3}{2} - \operatorname{Re}\nu < \operatorname{Re}\mu < \operatorname{Re}\nu + \tfrac{1}{2}, \quad y > 0\right]$$
$$\text{ET II 44(1)}$$

17. $$\int_0^\infty x^{1/2} \left(x^2 + 2\right)^{-\frac{1}{2}\nu - \frac{1}{4}} Q_\mu^{\nu + \frac{1}{2}} \left(x^2 + 1\right) J_\nu(xy) \, dx$$

$$= 2^{-\nu - \frac{1}{2}} \pi^{1/2} e^{\left(\nu + \frac{1}{2}\right)\pi i} y^\nu K_{\mu + \frac{1}{2}} \left(2^{-1/2} y\right) I_{\mu + \frac{1}{2}} \left(2^{-1/2} y\right)$$

$$\left[\operatorname{Re}\nu > -1, \quad \operatorname{Re}(2\mu + \nu) > -\tfrac{5}{2}, \quad y > 0\right] \quad \text{ET II 46(12)}$$

**7.183** $\displaystyle\int_0^\infty x^{1-\mu}\left(1+a^2x^2\right)^{-\frac{1}{2}\mu-\frac{1}{4}} Q_{\nu-\frac{1}{2}}^{\mu+\frac{1}{2}}(\pm iax)\, J_\nu(xy)\, dx$

$$= i(2\pi)^{1/2} e^{i\pi\left(\mu\mp\frac{1}{2}\nu\mp\frac{1}{4}\right)} a^{-1} y^{\mu-1} I_\nu\left(\tfrac{1}{2}a^{-1}y\right) K_\mu\left(\tfrac{1}{2}a^{-1}y\right)$$

$$\left[-\tfrac{3}{4}-\tfrac{1}{2}\operatorname{Re}\nu < \operatorname{Re}\mu < 1+\operatorname{Re}\nu, \quad y>0, \quad \operatorname{Re}a>0\right] \quad \text{ET II 46(11)}$$

**7.184**

**1.**[12] $\displaystyle\int_1^\infty x^{1/2}\left(x^2-1\right)^{\frac{1}{2}\mu-\frac{1}{4}} P_{-\frac{1}{2}+\nu}^{\frac{1}{2}-\mu}\left(x^{-1}\right) J_\nu(xa)\, dx = 2^{1/2} a^{-1-\mu}\pi^{-1/2}\cos\left[a+\tfrac{1}{2}(\nu-\mu)\pi\right]$

$$\left[|\operatorname{Re}\mu|<\tfrac{1}{2}, \quad \operatorname{Re}\nu>-1, \quad a>0\right]$$
$$\text{ET II 44(2)a}$$

**2.** $\displaystyle\int_1^\infty x^{-\nu}\left(x^2-1\right)^{\frac{1}{4}-\frac{1}{2}\nu} P_\mu^{\nu-\frac{1}{2}}\left(2x^{-2}-1\right) K_\nu(ax)\, dx$

$$= \pi^{1/2} 2^{-\nu} a^{-2+\nu} W_{\mu+\frac{1}{2},\nu-\frac{1}{2}}(a)\, W_{-\mu-\frac{1}{2},\nu-\frac{1}{2}}(a)$$
$$\left[\operatorname{Re}\nu<\tfrac{3}{2}, \quad a>0\right] \qquad \text{ET II 370(45)a}$$

**3.** $\displaystyle\int_0^\infty x^\nu\left(1+x^2\right)^{\frac{1}{4}+\frac{\nu}{2}} Q_\mu^{\nu+\frac{1}{2}}\left(1+\frac{2}{x^2}\right) J_\nu(ax)\, dx$

$$= -ie^{i\pi\nu}\pi^{-\frac{1}{2}} 2^\nu a^{-\nu-2}\left[\Gamma\left(\tfrac{3}{2}+\mu+\nu\right)\right]^2 \Gamma\left(\tfrac{1}{2}+\nu-\mu\right)$$

$$\times\, W_{-\mu-\frac{1}{2},\nu+\frac{1}{2}}(a)\left[\frac{\cos(\mu\pi)}{\Gamma(2+2\nu)} M_{\mu+\frac{1}{2},\nu+\frac{1}{2}}(a) + \frac{\sin(\nu\pi)}{\Gamma\left(\nu+\mu+\frac{3}{2}\right)} W_{\mu+\frac{1}{2},\nu+\frac{1}{2}}(a)\right]$$
$$\left[a>0, \quad \operatorname{Re}(\mu+\nu)>-\tfrac{3}{2}, \quad \operatorname{Re}(\mu-\nu)<\tfrac{1}{2}\right] \quad \text{ET II 46(14)}$$

**4.** $\displaystyle\int_0^1 x^\nu\left(1-x^2\right)^{\frac{1}{2}\nu+\frac{1}{4}} P_\mu^{-\nu-\frac{1}{2}}\left(2x^{-2}-1\right) J_\nu(xy)\, dx$

$$= 2^{\nu+\frac{1}{2}} y^\nu \frac{\Gamma\left(\tfrac{3}{2}+\mu+\nu\right)\Gamma\left(\tfrac{1}{2}+\nu-\mu\right)}{(2\pi)^{1/2}\left[\Gamma\left(\tfrac{3}{2}+\nu\right)\right]^2}$$

$$\times\, {}_1F_1\left(\nu+\mu+\frac{3}{2};2\nu+2;iy\right) {}_1F_1\left(\nu+\mu+\frac{3}{2};2\nu+2;-iy\right)$$
$$\left[y>0, \quad -\tfrac{3}{2}-\operatorname{Re}\nu<\operatorname{Re}\mu<\operatorname{Re}\nu+\tfrac{1}{2}\right] \quad \text{ET II 45(3)}$$

**5.** $\displaystyle\int_0^\infty x^{-\nu}\left(x^2+a^2\right)^{\frac{1}{4}-\frac{1}{2}\nu} Q_\mu^{\frac{1}{2}-\nu}\left(1+2a^2x^{-2}\right) K_\nu(xy)\, dx$

$$= ie^{-i\pi\nu}\pi^{1/2} 2^{-\nu-1} a^{-\nu-\frac{1}{2}} y^{\nu-2}\left[\Gamma\left(\tfrac{3}{2}+\mu-\nu\right)\right]^2 W_{-\mu-\frac{1}{2},\nu-\frac{1}{2}}(iay)\, W_{-\mu-\frac{1}{2},\nu-\frac{1}{2}}(-iay)$$
$$\left[\operatorname{Re}a>0, \quad \operatorname{Re}y>0, \quad \operatorname{Re}\mu>-\tfrac{3}{2}, \quad \operatorname{Re}(\mu-\nu)>-\tfrac{3}{2}\right] \quad \text{ET II 137(13)}$$

**6.** $\displaystyle\int_0^\infty x^{-\nu}\left(x^2+1\right)^{\frac{1}{4}-\frac{1}{2}\nu} Q_\mu^{\frac{1}{2}-\nu}\left(1+2x^{-2}\right) J_\nu(ax)\, dx$

$$= 2^{-\nu} a^{-\nu-2}\frac{ie^{-i\nu\pi}\pi^{1/2}\Gamma\left(\tfrac{3}{2}+\mu-\nu\right)}{\Gamma(2\nu)} M_{\mu+\frac{1}{2},\nu-\frac{1}{2}}(a)\, W_{-\mu-\frac{1}{2},\nu-\frac{1}{2}}(a)$$
$$\left[a>0, \quad 0<\operatorname{Re}\nu<\operatorname{Re}\mu+\tfrac{3}{2}\right] \quad \text{ET II 47(15)a}$$

7.
$$\int_0^\infty x^{-\nu} \left(x^2+a^2\right)^{\frac{1}{4}-\frac{1}{2}\nu} Q_{-\frac{1}{2}}^{\frac{1}{2}-\nu} \left(1+2a^2x^{-2}\right) K_\nu(xy)\,\mathrm{d}x$$
$$= ie^{-i\pi\nu}\pi^{3/2}2^{-\nu-3}a^{\frac{1}{2}-\nu}y^{\nu-1}\left[\Gamma(1-\nu)\right]^2 \times \left\{\left[J_{\nu-\frac{1}{2}}\left(\frac{ay}{2}\right)\right]^2 + \left[Y_{\nu-\frac{1}{2}}\left(\frac{ay}{2}\right)\right]^2\right\}$$
$$\left[\operatorname{Re}a>0,\quad \operatorname{Re}y>0,\quad \operatorname{Re}\nu<1\right]\quad \text{ET II 136(12)}$$

**7.185**
$$\int_0^\infty x^{1/2} Q_{\nu-\frac{1}{2}}\left[\left(a^2+x^2\right)x^{-1}\right] J_\nu(xy)\,\mathrm{d}x = 2^{-1/2}\pi y^{-1}\exp\left[-\left(a^2-\tfrac{1}{4}\right)^{1/2}y\right] J_\nu\left(\tfrac{1}{2}y\right)$$
$$\left[\operatorname{Re}\nu>-\tfrac{1}{2},\quad y>0\right]\qquad \text{ET II 46(10)}$$

**7.186**
$$\int_0^\infty x\left(1+x^2\right)^{-\nu-1} P_\nu\left(\frac{1-x^2}{1+x^2}\right) J_0(xy)\,\mathrm{d}x = y^{2\nu}\left[2^\nu\Gamma(\nu+1)\right]^{-2} K_0(y)$$
$$\left[\operatorname{Re}\nu>0\right]\qquad \text{ET II 13(10)}$$

**7.187**

1.
$$\int_0^\infty x P_\mu^\nu\left(\sqrt{1+x^2}\right) K_\nu(xy)\,\mathrm{d}x = y^{-3/2} S_{\nu+\frac{1}{2},\mu+\frac{1}{2}}(y)$$
$$\left[\operatorname{Re}\nu<1,\quad \operatorname{Re}y>0\right]\qquad \text{ET II 137(14)}$$

2.
$$\int_0^\infty x\left[P_{\lambda-\frac{1}{2}}\left(\sqrt{1+a^2x^2}\right)\right]^2 J_0(xy)\,\mathrm{d}x = 2\pi^{-2}y^{-1}a^{-1}\cos(\lambda\pi)\left[K_\lambda\left(\frac{y}{2a}\right)\right]^2$$
$$\left[\operatorname{Re}a>0,\quad |\operatorname{Re}\lambda|<\tfrac{1}{4},\quad y>0\right]$$
$$\text{ET II 13(11)}$$

3.
$$\int_0^\infty x\left(1+x^2\right)^{-1/2} P_\mu^\nu\left(\sqrt{1+x^2}\right) K_\nu(xy)\,\mathrm{d}x = y^{-1/2} S_{\nu-\frac{1}{2},\mu+\frac{1}{2}}(y)$$
$$\left[\operatorname{Re}\nu<1,\quad \operatorname{Re}y>0\right]\qquad \text{ET II 137(15)}$$

4.
$$\int_0^\infty x P_\mu^{-\frac{1}{2}\nu}\left(\sqrt{1+a^2x^2}\right) Q_\mu^{-\frac{1}{2}\nu}\left(\sqrt{1+a^2x^2}\right) J_\nu(xy)\,\mathrm{d}x$$
$$= \frac{y^{-1}e^{-\frac{1}{2}\nu\pi i}\Gamma\left(1+\mu+\frac{1}{2}\nu\right)}{a\,\Gamma\left(1+\mu-\frac{1}{2}\nu\right)} I_{\mu+\frac{1}{2}}\left(\frac{y}{2a}\right) K_{\mu+\frac{1}{2}}\left(\frac{y}{2a}\right)$$
$$\left[\operatorname{Re}a>0,\quad y>0,\quad \operatorname{Re}\mu>-\tfrac{3}{4},\quad \operatorname{Re}\nu>-1\right]\quad \text{ET II 47(16)}$$

5.
$$\int_0^\infty x P_{\sigma-\frac{1}{2}}^\mu\left(\sqrt{1+a^2x^2}\right) Q_{\sigma-\frac{1}{2}}^\mu\left(\sqrt{1+a^2x^2}\right) J_0(xy)\,\mathrm{d}x$$
$$= y^{-2}e^{\mu\pi i}\frac{\Gamma\left(\frac{1}{2}+\sigma-\mu\right)}{\Gamma(1+2\sigma)} W_{\mu,\sigma}\left(\frac{y}{a}\right) M_{-\mu,\sigma}\left(\frac{y}{a}\right)$$
$$\left[\operatorname{Re}a>0,\quad y>0,\quad \operatorname{Re}\sigma>-\tfrac{1}{4},\quad \operatorname{Re}\mu<1\right]\quad \text{ET II 14(15)}$$

6.
$$\int_0^\infty x P_{\sigma-\frac{1}{2}}^\mu\left(\sqrt{1+a^2x^2}\right) P_{\sigma-\frac{1}{2}}^{-\mu}\left(\sqrt{1+a^2x^2}\right) J_0(xy)\,\mathrm{d}x$$
$$= 2\pi^{-1}y^{-2}\cos(\sigma\pi) W_{\mu,\sigma}\left(\frac{y}{a}\right) W_{-\mu,\sigma}\left(\frac{y}{a}\right)$$
$$\left[\operatorname{Re}a>0,\quad y>0,\quad |\operatorname{Re}\sigma|<\tfrac{1}{4}\right]\quad \text{ET II 14(14)}$$

7.
$$\int_0^\infty x\left\{P_{\sigma-\frac{1}{2}}^\mu\left(\sqrt{1+a^2x^2}\right)\right\}^2 J_0(xy)\,\mathrm{d}x = -i\pi^{-1}y^{-2} W_{\mu,\sigma}\left(\frac{y}{a}\right)\left[W_{\mu,\sigma}\left(e^{\pi i}\frac{y}{a}\right) - W_{\mu,\sigma}\left(e^{-\pi i}\frac{y}{a}\right)\right]$$
$$\left[\operatorname{Re}a>0,\quad y>0,\quad |\operatorname{Re}\sigma|<\tfrac{1}{4},\quad \operatorname{Re}\mu<1\right]\quad \text{ET II 14(13)}$$

8.    $\displaystyle\int_0^\infty x \left(1 + a^2 x^2\right)^{-1/2} P_\mu^{-\frac{1}{2}-\frac{1}{2}\nu}\left(\sqrt{1+a^2x^2}\right) P_\mu^{\frac{1}{2}-\frac{1}{2}\nu}\left(\sqrt{1+a^2x^2}\right) J_\nu(xy)\,\mathrm{d}x$

$$= \frac{\left[K_{\mu+\frac{1}{2}}\left(\frac{y}{2a}\right)\right]^2}{\pi a^2\,\Gamma\left(\frac{\nu}{2}+\mu+\frac{3}{2}\right)\Gamma\left(\frac{\nu}{2}-\mu+\frac{1}{2}\right)}$$

$$\left[\operatorname{Re} a > 0, \quad y > 0, \quad -\tfrac{5}{4} < \operatorname{Re}\mu < \tfrac{1}{4}\right] \quad \text{ET II 46(9)}$$

9.    $\displaystyle\int_0^\infty x\left\{P_\mu^{-\frac{1}{2}\nu}\left(\sqrt{1+a^2x^2}\right)\right\}^2 J_\nu(xy)\,\mathrm{d}x = \frac{2\left[K_{\mu+\frac{1}{2}}\left(\frac{y}{2a}\right)\right]^2 y^{-1}}{\pi a\,\Gamma\left(1+\mu+\frac{1}{2}\nu\right)\Gamma\left(\frac{1}{2}\nu-\mu\right)}$

$$\left[\operatorname{Re} a > 0, \quad y > 0, \quad -\tfrac{3}{4} < \operatorname{Re}\mu < -\tfrac{1}{4}, \quad \operatorname{Re}\nu > -1\right] \quad \text{ET II 45(7)}$$

10.   $\displaystyle\int_0^\infty x\left(1+a^2x^2\right)^{-1/2} P_\mu^{-\frac{1}{2}\nu}\left(\sqrt{1+a^2x^2}\right) P_{\mu+1}^{-\frac{1}{2}\nu}\left(\sqrt{1+a^2x^2}\right) J_\nu(xy)\,\mathrm{d}x$

$$= \frac{K_{\mu+\frac{1}{2}}\left(\frac{y}{2a}\right) K_{\mu+\frac{3}{2}}\left(\frac{y}{2a}\right)}{\pi a^2\,\Gamma\left(2+\frac{1}{2}\nu+\mu\right)\Gamma\left(\frac{1}{2}\nu-\mu\right)}$$

$$\left[\operatorname{Re} a > 0, \quad y > 0, \quad -\tfrac{7}{4} < \operatorname{Re}\mu < -\tfrac{1}{4}\right] \quad \text{ET II 45(8)}$$

**7.188**

1.    $\displaystyle\int_0^\infty x\left(a^2+x^2\right)^{-\frac{1}{2}\mu} P_{\mu-1}^{-\nu}\left[\frac{a}{\sqrt{a^2+x^2}}\right] J_\nu(xy)\,\mathrm{d}x = \frac{y^{\mu-2}e^{-ay}}{\Gamma(\mu+\nu)}$

$$\left[\operatorname{Re} a > 0, \quad y > 0, \quad \operatorname{Re}\nu > -1, \quad \operatorname{Re}\mu > \tfrac{1}{2}\right] \quad \text{ET II 45(4)}$$

2.    $\displaystyle\int_0^\infty x^{\nu+1}\left(x^2+a^2\right)^{\frac{1}{2}\nu} P_\nu\left(\frac{x^2+2a^2}{2a\sqrt{x^2+a^2}}\right) J_\nu(ry)\,\mathrm{d}x - \frac{(2a)^{\nu+1}y^{-\nu-1}}{\pi\,\Gamma(-\nu)}\left[K_{\nu+\frac{1}{2}}\left(\frac{ya}{2}\right)\right]^2$

$$\left[\operatorname{Re} a > 0, \quad -1 < \operatorname{Re}\nu < 0, \quad y > 0\right]$$
$$\text{ET II 45(5)}$$

3.    $\displaystyle\int_0^\infty x^{1-\nu}\left(x^2+a^2\right)^{-\frac{1}{2}\nu} P_{\nu-1}\left(\frac{x^2+2a^2}{2a\sqrt{x^2+a^2}}\right) J_\nu(xy)\,\mathrm{d}x = \frac{(2a)^{1-\nu}y^{\nu-1}}{\Gamma(\nu)} I_{\nu-\frac{1}{2}}\left(\frac{ay}{2}\right) K_{\nu-\frac{1}{2}}\left(\frac{ay}{2}\right)$

$$\left[\operatorname{Re} a > 0, \quad y > 0, \quad 0 < \operatorname{Re}\nu < 1\right]$$
$$\text{ET II 45(6)}$$

**7.189**

1.    $\displaystyle\int_0^\infty (a+x)^\mu e^{-x} P_\nu^{-2\mu}\left(1+\frac{2x}{a}\right) I_\mu(x)\,\mathrm{d}x = 0$

$$\left[-\tfrac{1}{2} < \operatorname{Re}\mu < 0, \quad -\tfrac{1}{2}+\operatorname{Re}\mu < \operatorname{Re}\nu < -\tfrac{1}{2}-\operatorname{Re}\mu\right] \quad \text{ET II 366(18)}$$

2.    $\displaystyle\int_0^\infty (x+a)^{-\mu} e^{-x} P_\nu^{-2\mu}\left(1+\frac{2x}{a}\right) I_\mu(x)\,\mathrm{d}x$

$$= \frac{2^{\mu-1}\Gamma\left(\mu+\nu+\frac{1}{2}\right)\Gamma\left(\mu-\nu-\frac{1}{2}\right)e^a}{\pi^{1/2}\,\Gamma(2\mu+\nu+1)\Gamma(2\mu-\nu)} W_{\frac{1}{2}-\mu,\frac{1}{2}+\nu}(2a)$$

$$\left[|\arg a| < \pi, \quad \operatorname{Re}\mu > \left|\operatorname{Re}\nu+\tfrac{1}{2}\right|\right] \quad \text{ET II 367(19)}$$

3. $\displaystyle\int_0^\infty x^{-\mu} e^x P_\nu^{2\mu}\left(1+\frac{2x}{a}\right) K_\mu(x+a)\,dx$

$$= \pi^{-1/2} 2^{\mu-1} \cos(\mu\pi)\,\Gamma\left(\mu+\nu+\tfrac{1}{2}\right)\Gamma\left(\mu-\nu+\tfrac{1}{2}\right) W_{\frac{1}{2}-\mu,\frac{1}{2}+\nu}(2a)$$

$$\left[|\arg a| < \pi, \quad \operatorname{Re}\mu > \left|\operatorname{Re}\nu+\tfrac{1}{2}\right|\right] \quad \text{ET II 373(11)}$$

4. $\displaystyle\int_0^\infty x^{-\frac{1}{2}\mu}(x+a)^{-1/2} e^{-x} P_{\nu-\frac{1}{2}}^\mu\left(\frac{a-x}{a+x}\right) K_\nu(a+x)\,dx = \sqrt{\frac{\pi}{2}}\,a^{-\frac{1}{2}\mu}\,\Gamma(\mu,2a)$

$$\left[a>0, \quad \operatorname{Re}\mu<1\right] \qquad \text{ET II 374(12)}$$

5. $\displaystyle\int_0^\infty (\sinh x)^{\mu+1}(\cosh x)^{-2\mu-\frac{3}{2}} P_\nu^{-\mu}[\cosh(2x)]\,I_{\mu-\frac{1}{2}}\left(a\operatorname{sech} x\right)dx$

$$= \frac{2^{\mu-\frac{1}{2}}\,\Gamma(\mu-\nu)\,\Gamma(\mu+\nu+1)}{\pi^{1/2} a^{\mu+\frac{3}{2}}\left[\Gamma(\mu+1)\right]^2}\,M_{\nu+\frac{1}{2},\mu}(a)\,M_{-\nu-\frac{1}{2},\mu}(a)$$

$$\left[\operatorname{Re}\mu>\operatorname{Re}\nu, \quad \operatorname{Re}\mu>-\operatorname{Re}\nu-1\right] \quad \text{ET II 378(44)}$$

## 7.19 Combinations of associated Legendre functions and functions generated by Bessel functions

### 7.191

1. $\displaystyle\int_a^\infty x^{1/2}\left(x^2-a^2\right)^{-\frac{1}{4}-\frac{1}{2}\nu} P_\mu^{\nu+\frac{1}{2}}\left(2x^2 a^{-2}-1\right)\left[\mathbf{H}_\nu(x)-Y_\nu(x)\right]dx$

$$= 2^{-\nu-2}\pi^{1/2} a\operatorname{cosec}(\mu\pi)\cos(\nu\pi)\left\{\left[Y_\nu\left(\tfrac{1}{2}a\right)\right]^2-\left[J_\nu\left(\tfrac{1}{2}a\right)\right]^2\right\}$$

$$\left[-1<\operatorname{Re}\mu<0, \quad \operatorname{Re}\nu<\tfrac{1}{2}\right] \quad \text{ET II 384(6)}$$

2. $\displaystyle\int_0^\infty x^{1/2}\left(x^2-a^2\right)^{-1/4-\nu/2} P_\mu^{\nu+1/2}\left(2x^2 a^{-2}-1\right)\left[I_{-\nu}(x)-\mathbf{L}_\nu(x)\right]dx$

$$= 2^{-\nu-1}\pi^{1/2} a\operatorname{cosec}(2\mu\pi)\cos(\nu\pi)\left\{\left[I_\nu\left(\tfrac{1}{2}a\right)\right]^2-\left[I_{-\nu}\left(\tfrac{1}{2}a\right)\right]^2\right\}$$

$$\left[-1<\operatorname{Re}\mu<0, \quad \operatorname{Re}\nu<\tfrac{1}{2}\right] \quad \text{ET II 385(15)}$$

### 7.192

1. $\displaystyle\int_0^1 x^{(\nu-\mu-1)/2}\left(1-x^2\right)^{(\nu-\mu-2)/4} P_{\nu-1/2}^{(\mu-\nu+2)/2}(x)\,S_{\mu,\nu}(ax)\,dx$

$$= 2^{\mu-3/2}\pi^{1/2} a^{-(\nu-\mu-1)/2}\,\Gamma\left(\frac{\mu+\nu+3}{4}\right)\Gamma\left(\frac{\mu-3\nu+3}{4}\right)\cos\left(\frac{\mu-\nu}{2}\pi\right)$$

$$\times\left[J_\nu\left(\tfrac{1}{2}a\right)Y_{-(\mu-\nu+1)/2}\left(\tfrac{1}{2}a\right)-Y_\nu\left(\tfrac{1}{2}a\right)J_{-(\mu-\nu+1)/2}\left(\tfrac{1}{2}a\right)\right]$$

$$\left[\operatorname{Re}(\mu-\nu)<0, \quad a>0, \quad |\operatorname{Re}(\mu+\nu)|<1, \quad \operatorname{Re}(\mu-3\nu)<1\right] \quad \text{ET II 387(24)a}$$

2.    $\displaystyle\int_1^\infty x^{1/2}\left(x^2-1\right)^{-\beta/2} P_\nu^\beta(x)\, S_{\mu,1/2}(ax)\,\mathrm{d}x$

$$= \frac{2^{-3/2+\beta-\mu}a^{\beta-1}\Gamma\left(\frac{\beta-\mu+\nu}{2}+\frac14\right)\Gamma\left(\frac{\beta-\mu-\nu}{2}-\frac14\right)}{\pi^{1/2}\Gamma\left(\frac12-\mu\right)}\, S_{\mu-\beta+1,\nu+1/2}(a)$$

$$\left[\operatorname{Re}\beta<1,\quad a>0,\quad \operatorname{Re}(\mu+\nu-\beta)<-\tfrac12,\quad \operatorname{Re}(\mu-\nu-\beta)<\tfrac12\right] \qquad \text{ET II 387(25)a}$$

**7.193**

1.    $\displaystyle\int_1^\infty x^{-\nu}\left(x^2-1\right)^{1/4-\nu/2} P_{\mu/2-\nu/2}^{\nu-1/2}\left(2x^{-2}-1\right) S_{\mu,\nu}(ax)\,\mathrm{d}x$

$$= \frac{2^{\mu-\nu}a^{\nu-2}\pi^{1/2}\Gamma\left(\frac{3\nu-\mu-1}{2}\right)}{\Gamma\left(\frac{1+\nu-\mu}{2}\right)}\, W_{\rho,\sigma}\left(ae^{i\pi/2}\right) W_{\rho,\sigma}\left(ae^{-i\pi/2}\right)$$

$$\rho=\tfrac12(\mu+1-\nu),\quad \sigma=\nu-\tfrac12,\qquad \left[\operatorname{Re}(\mu-\nu)<0,\quad a>0,\quad \operatorname{Re}\nu<\tfrac32,\quad \operatorname{Re}(3\nu-\mu)>1\right]$$

$$\text{ET II 387(27)a}$$

2.    $\displaystyle\int_1^\infty x\left(x^2-1\right)^{-\nu/2} P_\lambda^\nu\left(2x^2-1\right) S_{\mu,\nu}(ax)\,\mathrm{d}x$

$$= \frac{a^{\nu-1}\Gamma\left(\frac{\nu-\mu+1}{2}+\lambda\right)\Gamma\left(\frac{\nu-\mu-1}{2}-\lambda\right)}{2\,\Gamma\left(\frac{1-\mu-\nu}{2}\right)\Gamma\left(\frac{1-\mu+\nu}{2}\right)}\, S_{\mu-\nu+1,2\lambda+1}(a)$$

$$\left[\operatorname{Re}\nu<1,\quad a>0,\quad \operatorname{Re}(\mu-\nu+\lambda)<-1,\quad \operatorname{Re}(\mu-\nu+\lambda)<0\right] \qquad \text{ET II 387(26)a}$$

## 7.21 Integration of associated Legendre functions with respect to the order

**7.211**

1.    $\displaystyle\int_0^\infty P_{-x-\frac12}(\cos\theta)\,\mathrm{d}x = \frac12\operatorname{cosec}\left(\frac12\theta\right)$       $[0<\theta<\pi]$       ET II 329(19)

2.    $\displaystyle\int_{-\infty}^\infty P_x(\cos\theta)\,\mathrm{d}x = \operatorname{cosec}\left(\frac12\theta\right)$       $[0<\theta<\pi]$       ET II 329(20)

**7.212**    $\displaystyle\int_0^\infty x^{-1}\tanh(\pi x)\, P_{-\frac12+ix}(\cosh a)\,\mathrm{d}x = 2e^{-\frac12 a}\, K\left(e^{-a}\right)$

$$[a>0] \qquad \text{ET II 330(22)}$$

**7.213**    $\displaystyle\int_0^\infty \frac{x\tanh(\pi x)}{a^2+x^2}\, P_{-\frac12+ix}(\cosh b)\,\mathrm{d}x = Q_{a-\frac12}(\cosh b)$    $[\operatorname{Re}a>0]$       ET II 387(23)

**7.214**    $\displaystyle\int_0^\infty \sinh(\pi x)\cos(ax)\, P_{-\frac12+ix}(b)\,\mathrm{d}x = \frac{1}{\sqrt{2(b+\cosh a)}}$

$$[a>0,\quad |b|<1] \qquad \text{ET I 42(27)}$$

**7.215**    $\displaystyle\int_0^\infty \cos(bx)\, P_{-\frac12+ix}^\mu(\cosh a)\,\mathrm{d}x = 0$       $[0<a<b]$

$$= \frac{\sqrt{\frac{\pi}{2}}(\sinh a)^\mu}{\Gamma\left(\frac12-\mu\right)(\cosh a-\cosh b)^{\mu+\frac12}} \qquad [0<b<a]$$

$$\text{ET II 330(21)}$$

**7.216** $\displaystyle \int_0^\infty \cos(bx)\,\Gamma(\mu+ix)\,\Gamma(\mu-ix)\,P_{-\frac{1}{2}+ix}^{\frac{1}{2}-\mu}(\cosh a)\,\mathrm{d}x = \frac{\sqrt{\frac{\pi}{2}}\,\Gamma(\mu)\,(\sinh a)^{\mu-\frac{1}{2}}}{(\cosh a + \cosh b)^\mu}$

$$[a>0, \quad b>0, \quad \operatorname{Re}\mu>0]$$

ET II 330(24)

**7.217**

1. $\displaystyle \int_{-\infty}^\infty \left(\nu-\frac{1}{2}+ix\right)\Gamma\left(\frac{1}{2}-ix\right)\Gamma\left(2\nu-\frac{1}{2}+ix\right)P_{\nu+ix-1}^{\frac{1}{2}-\nu}(\cos\theta)\,I_{\nu-\frac{1}{2}+ix}(a)\,K_{\nu-\frac{1}{2}+ix}(b)\,\mathrm{d}x$

$$= \sqrt{2\pi}\,(\sin\theta)^{\nu-\frac{1}{2}}\left(\frac{ab}{\omega}\right)^\nu K_\nu(\omega)$$

$$\left[\omega=\left(a^2+b^2+2ab\cos\theta\right)^{1/2}\right] \quad \text{ET II 383(29)}$$

2. $\displaystyle \int_0^\infty xe^{\pi x}\tanh(\pi x)\,P_{-\frac{1}{2}+ix}(-\cos\theta)\,H_{ix}^{(2)}(ka)\,H_{ix}^{(2)}(kb)\,\mathrm{d}x = -\frac{2(ab)^{1/2}}{\pi R}e^{-ikR};$

$$R=\left(a^2+b^2-2ab\cos\theta\right)^{1/2} \quad [a>0, \quad b>0, \quad 0<\theta<\pi, \quad \operatorname{Im}k\le 0] \quad \text{ET II 381(17)}$$

3. $\displaystyle \int_0^\infty xe^{\pi x}\sinh(\pi x)\,\Gamma(\nu+ix)\,\Gamma(\nu-ix)\,P_{-\frac{1}{2}+ix}^{\frac{1}{2}-\nu}(-\cos\theta)\,H_{ix}^{(2)}(a)\,H_{ix}^{(2)}(b)\,\mathrm{d}x$

$$= i(2\pi)^{1/2}\,(\sin\theta)^{\nu-\frac{1}{2}}\left(\frac{ab}{R}\right)^\nu H_\nu^{(2)}(R)$$

$$R=\left(a^2+b^2-2ab\cos\theta\right)^{1/2} \quad [a>0, \quad b>0, \quad 0<\theta<\pi, \quad \operatorname{Re}\nu>0] \quad \text{ET II 381 (18)}$$

4. $\displaystyle \int_0^\infty x\sinh(\pi x)\,\Gamma(\lambda+ix)\,\Gamma(\lambda-ix)\,K_{ix}(a)\,K_{ix}(b)\,P_{-\frac{1}{2}+ix}^{\frac{1}{2}-\lambda}(\beta)\,\mathrm{d}x = \frac{\pi^{1/2}}{\sqrt{2}}\left(\frac{ab}{z}\right)^\lambda (\beta^2-1)^{\frac{1}{2}\lambda-\frac{1}{4}}K_\lambda(z)$

$$z=\sqrt{a^2+b^2+2ab\beta} \quad \left[|\arg a|<\frac{\pi}{2}, \quad |\arg(\beta-1)|<\pi, \quad \operatorname{Re}\lambda>0\right] \quad \text{ET II 177(16)}$$

**7.218\***

1. $\displaystyle \int_0^\infty \frac{\cos(ax)}{\cosh(\pi x)}P_{-\frac{1}{2}+ix}(z)\,\mathrm{d}x = \frac{\sqrt{2}}{2}\frac{1}{\sqrt{z+\cosh a}}$

2. $\displaystyle \int_0^\infty \frac{\cos(ax)}{\cosh^2(\pi x)}P_{-\frac{1}{2}+ix}(z)\,\mathrm{d}x = \frac{\sqrt{2}}{\pi}\frac{1}{\sqrt{z-\cosh a}}\arctan\left(\sqrt{\frac{z-\cosh a}{1+\cosh a}}\right)$

3. $\displaystyle \int_0^\infty \frac{\cosh(bx)}{\cosh(\pi x)}P_{-\frac{1}{2}+ix}(z)\,\mathrm{d}x = \frac{\sqrt{2}}{\pi}\frac{1}{\sqrt{z-\cos b}}\operatorname{arccot}\left(\sqrt{\frac{1+\cos b}{z-\cos b}}\right) \qquad [|b|<2\pi]$

## 7.22 Combinations of Legendre polynomials, rational functions, and algebraic functions

**7.221**

1. $\displaystyle \int_{-1}^1 P_n(x)\,P_m(x)\,\mathrm{d}x = 0 \qquad [m\ne n]$

$$= \frac{2}{2n+1} \qquad [m=n]$$

WH, EH I 170(8, 10)

$2.^6$ $\qquad \displaystyle\int_0^1 P_n(x)\,P_m(x)\,\mathrm{d}x = \frac{1}{2n+1}$ $\qquad\qquad\qquad\qquad$ $[m = n]$

$\qquad\qquad\qquad\qquad = 0$ $\qquad\qquad\qquad\qquad\qquad$ $[n - m$ is even, $\quad m \neq n]$

$\qquad\qquad\qquad\qquad = \dfrac{(-1)^{\frac{1}{2}(m+n-1)}\, m!\,n!}{2^{m+n-1}(m-n)\,(n+m+1)\left[\left(\frac{n}{2}\right)!\left(\frac{m-1}{2}\right)!\right]^2}$ $\qquad$ $[n$ is even, $m$ is odd$]$

$\qquad\qquad\qquad\qquad\qquad\qquad\qquad\qquad\qquad\qquad\qquad\qquad\qquad\qquad\qquad\qquad$ WH

3. $\qquad \displaystyle\int_0^{2\pi} P_{2n}\left(\cos\varphi\right)\,\mathrm{d}\varphi = 2\pi\left[\binom{2n}{n}2^{-2n}\right]^2.$ $\qquad\qquad\qquad\qquad$ MO 70, EH II 183(50)

**7.222**

1. $\qquad \displaystyle\int_{-1}^1 x^m\,P_n(x)\,\mathrm{d}x = 0$ $\qquad\qquad\qquad\qquad$ $[m < n]$

2. $\qquad \displaystyle\int_{-1}^1 (1+x)^{m+n}\,P_m(x)\,P_n(x)\,\mathrm{d}x = \dfrac{2^{m+n+1}\left[(m+n)!\right]^4}{\left(m!\,n!\right)^2(2m+2n+1)!}$ $\qquad$ ET II 277(15)

3. $\qquad \displaystyle\int_{-1}^1 (1+x)^{m-n-1}\,P_m(x)\,P_n(x)\,\mathrm{d}x = 0$ $\qquad\qquad$ $[m > n]$ $\qquad\qquad$ ET II 278(16)

4. $\qquad \displaystyle\int_{-1}^1 \left(1-x^2\right)^n\,P_{2m}(x)\,\mathrm{d}x = \dfrac{2n^2}{(n-m)(2m+2n+1)}\int_{-1}^1 \left(1-x^2\right)^{n-1}\,P_{2m}(x)\,\mathrm{d}x$

$\qquad\qquad\qquad\qquad\qquad\qquad\qquad\qquad\qquad\qquad$ $[m < n]$ $\qquad\qquad\qquad\qquad\qquad\qquad$ WH

5. $\qquad \displaystyle\int_0^1 x^2\,P_{n+1}(x)\,P_{n-1}(x)\,\mathrm{d}x = \dfrac{n(n+1)}{(2n-1)(2n+1)(2n+3)}$ $\qquad\qquad\qquad$ WH

**7.223** $\qquad \displaystyle\int_{-1}^1 \dfrac{1}{z-x}\left\{P_n(x)\,P_{n-1}(z) - P_{n-1}(x)\,P_n(z)\right\}\,\mathrm{d}x = -\dfrac{2}{n}$ $\qquad\qquad\qquad\qquad$ WH

**7.224** $\quad [z$ belongs to the complex plane with a discontinuity along the interval from $-1$ to $+1.]$

1. $\qquad \displaystyle\int_{-1}^1 (z-x)^{-1}\,P_n(x)\,\mathrm{d}x = 2\,Q_n(z)$ $\qquad\qquad\qquad\qquad\qquad\qquad$ ET II 277(7)

2. $\qquad \displaystyle\int_{-1}^1 x(z-x)^{-1}\,P_0(x)\,\mathrm{d}x = 2\,Q_1(z)$ $\qquad\qquad\qquad\qquad\qquad\qquad$ ET II 277(8)

3. $\qquad \displaystyle\int_{-1}^1 x^{n+1}(z-x)^{-1}\,P_n(x)\,\mathrm{d}x = 2z^{n+1}\,Q_n(z) - \dfrac{2^{n+1}\,(n!)^2}{(2n+1)!}$ $\qquad\qquad$ ET II 277(9)

4. $\qquad \displaystyle\int_{-1}^1 x^m(z-x)^{-1}\,P_n(x)\,\mathrm{d}x = 2z^m\,Q_n(z)$ $\qquad\qquad$ $[m \leq n]$ $\qquad\qquad$ ET II 277(10)a

5. $\qquad \displaystyle\int_{-1}^1 (z-x)^{-1}\,P_m(x)\,P_n(x)\,\mathrm{d}x = 2\,P_m(z)\,Q_n(z)$ $\qquad\qquad$ $[m \leq n]$ $\qquad\qquad$ ET II 278(18)a

6. $\qquad \displaystyle\int_{-1}^1 (z-x)^{-1}\,P_n(x)\,P_{n+1}(x)\,\mathrm{d}x = 2\,P_{n+1}(z)\,Q_n(z) - \dfrac{2}{n+1}$ $\qquad\qquad$ ET II 278(19)

7. $\qquad \displaystyle\int_{-1}^1 x(z-x)^{-1}\,P_m(x)\,P_n(x)\,\mathrm{d}x = 2z\,P_m(z)\,Q_n(z)$ $\qquad\qquad$ $[m < n]$ $\qquad\qquad$ ET II 278(21)

8. $\displaystyle\int_{-1}^{1} x(z-x)^{-1}\,[P_n(x)]^2\,\mathrm{d}x = 2z\,P_n(z)\,Q_n(z) - \frac{2}{2n+1}$ 　　　　　ET II 278(20)

**7.225**

1. $\displaystyle\int_{-1}^{x}(x-t)^{-1/2}\,P_n(t)\,\mathrm{d}t = \left(n+\frac{1}{2}\right)^{-1}(1+x)^{-1/2}\left[T_n(x)+T_{n+1}(x)\right]$ 　　　EH II 187(43)

2. $\displaystyle\int_{x}^{1}(t-x)^{-1/2}\,P_n(t)\,\mathrm{d}t = \left(n+\frac{1}{2}\right)^{-1}(1-x)^{-1/2}\left[T_n(x)-T_{n+1}(x)\right]$ 　　　EH II 187(44)

3. $\displaystyle\int_{-1}^{1}(1-x)^{-1/2}\,P_n(x)\,\mathrm{d}x = \frac{2^{3/2}}{2n+1}$ 　　　　　EH II 183(49)

4. $\displaystyle\int_{-1}^{1}(\cosh 2p - x)^{-1/2}\,P_n(x)\,\mathrm{d}x = \frac{2\sqrt{2}}{2n+1}\exp[-(2n+1)p]$

　　　　　　　　　　　　　　　　　　$[p>0]$ 　　　　　WH

5.[10] $\displaystyle\frac{1}{2}\int_{-1}^{1}\frac{P_\ell(z)\,\mathrm{d}z}{\sqrt{(xy-z)^2-(x^2-1)(y^2-1)}} = P_\ell(x)\,Q_\ell(y) \qquad (1<x\le y)$

　　　　　　　　　　　　　　　　　　　　　　　$= P_\ell(y)\,Q_\ell(x) \qquad (1<y\le x)$

**7.226**

1. $\displaystyle\int_{-1}^{1}\left(1-x^2\right)^{-1/2}\,P_{2m}(x)\,\mathrm{d}x = \left[\frac{\Gamma\left(\frac{1}{2}+m\right)}{m!}\right]^2$ 　　　　　ET II 276(4)

2. $\displaystyle\int_{-1}^{1}x\left(1-x^2\right)^{-1/2}\,P_{2m+1}(x)\,\mathrm{d}x = \frac{\Gamma\left(\frac{1}{2}+m\right)\Gamma\left(\frac{3}{2}+m\right)}{m!(m+1)!}$ 　　　　　ET II 276(5)

3. $\displaystyle\int_{-1}^{1}\left(1+px^2\right)^{-m-3/2}\,P_{2m}(x)\,\mathrm{d}x = \frac{2}{2m+1}(-p)^m(1+p)^{-m-1/2}$

　　　　　　　　　　　　　　　　　　$[|p|<1]$ 　　　　　MO 71

**7.227** $\displaystyle\int_{0}^{1}x\left(a^2+x^2\right)^{-1/2}\,P_n\left(1-2x^2\right)\,\mathrm{d}x = \frac{\left[a+\left(a^2+1\right)^{1/2}\right]^{-2n-1}}{2n+1}$

　　　　　　　　　　　　　　　　　$[\mathrm{Re}\,a>0]$ 　　　　　ET II 278(23)

**7.228**[6] $\displaystyle\frac{1}{2}\Gamma(1+\mu)\int_{-1}^{1}P_l(x)(z-x)^{-\mu-1}\,\mathrm{d}x = \left(z^2-1\right)^{-\mu/2}e^{-i\pi\mu}\,Q_l^\mu(z)$

　　　　　　　　　　　　　　　　$[l=0,1,2,\ldots,\quad |\arg(z-1)|<\pi]$

## 7.23 Combinations of Legendre polynomials and powers

**7.231**

1. $\displaystyle\int_{0}^{1}x^\lambda\,P_{2m}(x)\,\mathrm{d}x = \frac{(-1)^m\,\Gamma\left(m-\frac{1}{2}\lambda\right)\Gamma\left(\frac{1}{2}+\frac{1}{2}\lambda\right)}{2\,\Gamma\left(-\frac{1}{2}\lambda\right)\Gamma\left(m+\frac{3}{2}+\frac{1}{2}\lambda\right)}$ 　　$[\mathrm{Re}\,\lambda>-1]$ 　　EH II 183(51)

**2.**[6] $\quad \int_0^1 x^\lambda P_{2m+1}(x)\, dx = \dfrac{(-1)^m \, \Gamma\left(m + \frac{1}{2} - \frac{1}{2}\lambda\right) \Gamma\left(1 + \frac{1}{2}\lambda\right)}{2\,\Gamma\left(\frac{1}{2} - \frac{1}{2}\lambda\right)\Gamma\left(m + 2 + \frac{1}{2}\lambda\right)}$

$$[\operatorname{Re}\lambda > -2] \qquad\qquad \text{EH II 183(52)}$$

**3.**[*] $\quad \int_0^1 x^\lambda P_n(x)\, dx = \dfrac{1}{2^n} \dfrac{\Gamma(\lambda+1)}{\Gamma(\lambda - n + 2)} \dfrac{\Gamma\left(\frac{\lambda - n + 3}{2}\right)}{\Gamma\left(\frac{\lambda + n + 3}{2}\right)}$

$$[n \text{ is even}, \operatorname{Re}(\nu) > -1] \quad \text{or} \quad [n \text{ is odd}, \operatorname{Re}(\nu) > -2]$$

**7.232**

**1.** $\quad \displaystyle\int_{-1}^1 (1-x)^{a-1} P_m(x)\, P_n(x)\, dx$

$$= \dfrac{2^a\,\Gamma(a)\,\Gamma(n - a + 1)}{\Gamma(1-a)\,\Gamma(n + a + 1)} \, {}_4F_3\left(-m, m+1, a, a; 1, a+n+1, a-n; 1\right)$$

$$[\operatorname{Re} a > 0] \qquad\qquad \text{ET II 278(17)}$$

**2.** $\quad \displaystyle\int_{-1}^1 (1-x)^{a-1}(1+x)^{b-1} P_n(x)\, dx = \dfrac{2^{a+b-1}\,\Gamma(a)\,\Gamma(b)}{\Gamma(a+b)}\, {}_3F_2(-n, 1+n, a; 1, a+b; 1)$

$$[\operatorname{Re} a > 0, \quad \operatorname{Re} b > 0] \qquad\qquad \text{ET II 276(6)}$$

**3.** $\quad \displaystyle\int_0^1 (1-x)^{\mu-1} P_n(1 - \gamma x)\, dx = \dfrac{\Gamma(\mu)\, n!}{\Gamma(\mu + n + 1)}\, P_n^{(\mu, -\mu)}(1 - \gamma)$

$$[\operatorname{Re}\mu > 0] \qquad\qquad \text{ET II 190(37)a}$$

**4.** $\quad \displaystyle\int_0^1 (1-x)^{\mu-1} x^{\nu-1} P_n(1 - \gamma x)\, dx = \dfrac{\Gamma(\mu)\,\Gamma(\nu)}{\Gamma(\mu + \nu)}\, {}_3F_2\left(-n, n+1, \nu; 1, \mu + \nu; \tfrac{1}{2}\gamma\right)$

$$[\operatorname{Re}\mu > 0, \quad \operatorname{Re}\nu > 0] \qquad\qquad \text{ET II 190(38)}$$

**7.233**[12] $\quad \displaystyle\int_0^1 x^{2\mu-1} P_n\left(1 - 2x^2\right) dx = \dfrac{(-1)^n\,[\Gamma(\mu)]^2}{2\,\Gamma(\mu + n)\,\Gamma(\mu - n)} \qquad [\operatorname{Re}\mu > 0] \qquad \text{ET II 278(22)}$

## 7.24 Combinations of Legendre polynomials and other elementary functions

**7.241** $\quad \displaystyle\int_0^\infty P_n(1 - x)e^{-ax}\, dx = e^{-a}a^n\left(\dfrac{1}{a}\dfrac{d}{da}\right)^n\left(\dfrac{e^a}{a}\right)$

$$= a^n\left(1 + \dfrac{1}{2}\dfrac{d}{da}\right)^n\left(\dfrac{1}{a^{n+1}}\right)$$

$$[\operatorname{Re} a > 0] \qquad\qquad \text{ET I 171(2)}$$

**7.242** $\quad \displaystyle\int_0^\infty P_n\left(e^{-x}\right) e^{-ax}\, dx = \dfrac{(a-1)(a-2)\cdots(a-n+1)}{(a+n)(a+n-2)\cdots(a-n+2)}$

$$[n \geq 2, \quad \operatorname{Re} a > 0] \qquad\qquad \text{ET I 171(3)}$$

**7.243**

**1.** $\quad \displaystyle\int_0^\infty P_{2n}(\cosh x)e^{-ax}\, dx = \dfrac{\left(a^2 - 1^2\right)\left(a^2 - 3^2\right)\cdots\left[a^2 - (2n-1)^2\right]}{a\left(a^2 - 2^2\right)\left(a^2 - 4^2\right)\cdots\left[a^2 - (2n)^2\right]}$

$$[\operatorname{Re} a > 2n] \qquad\qquad \text{ET I 171(6)}$$

**2.** $\quad \displaystyle\int_0^\infty P_{2n+1}(\cosh x)e^{-ax}\, dx = \dfrac{a\left(a^2 - 2^2\right)\left(a^2 - 4^2\right)\cdots\left[a^2 - (2n)^2\right]}{\left(a^2 - 1\right)\left(a^2 - 3^2\right)\cdots\left[a^2 - (2n+1)^2\right]}$

$$[\operatorname{Re} a > 2n + 1] \qquad\qquad \text{ET I 171(7)}$$

3.    $\int_0^\infty P_{2n}\left(\cos x\right) e^{-ax}\,\mathrm{d}x = \dfrac{\left(a^2+1^2\right)\left(a^2+3^2\right)\cdots\left[a^2+(2n-1)^2\right]}{a\left(a^2+2^2\right)\left(a^2+4^2\right)\cdots\left[a^2+(2n)^2\right]}$

$$[\operatorname{Re}a>0] \qquad\qquad \text{ET I 171(4)}$$

4.    $\int_0^\infty P_{2n+1}\left(\cos x\right) e^{-ax}\,\mathrm{d}x = \dfrac{a\left(a^2+2^2\right)\left(a^2+4^2\right)\cdots\left[a^2+(2n)^2\right]}{\left(a^2+1^2\right)\left(a^2+3^2\right)\cdots\left[a^2+(2n+1)^2\right]}$

$$[\operatorname{Re}a>0] \qquad\qquad \text{ET I 171(5)}$$

5.[11]    $\int_{-1}^1 e^{ix\alpha}\,P_n(x)\,\mathrm{d}x = i^n\sqrt{\dfrac{2\pi}{\alpha}}\,J_{n+\frac{1}{2}}(\alpha)$       $[n=0,1,2,\ldots,\quad a>0]$

$$\text{GH2 24 (171.10)}$$

**7.244**

1.    $\int_0^1 P_n\left(1-2x^2\right)\sin ax\,\mathrm{d}x = \dfrac{\pi}{2}\left[J_{n+\frac{1}{2}}\left(\dfrac{a}{2}\right)\right]^2$     $[a>0]$       ET I 94(2)

2.    $\int_0^1 P_n\left(1-2x^2\right)\cos ax\,\mathrm{d}x = \dfrac{\pi}{2}(-1)^n\,J_{n+\frac{1}{2}}\left(\dfrac{a}{2}\right)J_{-n-\frac{1}{2}}\left(\dfrac{a}{2}\right)$

$$[a>0] \qquad\qquad \text{ET I 38(1)}$$

**7.245**

1.    $\int_0^{2\pi} P_{2m+1}\left(\cos\theta\right)\cos\theta\,\mathrm{d}\theta = \dfrac{\pi}{2^{4m+1}}\dbinom{2m}{m}\dbinom{2m+2}{m+1}$       MO 70, EH II 183(5)

2.    $\int_0^{\pi} P_m\left(\cos\theta\right)\sin n\theta\,\mathrm{d}\theta = \dfrac{2(n-m+1)(n-m+3)\cdots(n+m-1)}{(n-m)(n-m+2)\cdots(n+m)}$    $[n>m$ and $n+m$ is odd$]$

$$\qquad\qquad\qquad\qquad\qquad\qquad = 0 \qquad\qquad\qquad [n\le m\text{ or }n+m\text{ is even}]$$

$$\text{MO 71}$$

3.[10]    $\int_0^{2\pi} P_{2n+1}\left(\sin\alpha\sin\phi\right)\sin\phi\,\mathrm{d}\phi = (-1)^{n+1}\dfrac{2\sqrt{\pi}\,\Gamma\left(n+\frac{3}{2}\right)}{(2n+1)\,\Gamma(n+2)}\,P^1_{2n+1}\left(\cos\alpha\right)$

$$\left[\alpha\ne\tfrac{1}{2}(2n+1)\pi,\quad n\text{ an integer}\right]$$

4.    $\int_{-1}^1 \cos(\alpha x)\,P_n(x)\,\mathrm{d}x = 0$       $[n$ is odd$]$

$$\qquad\qquad\qquad = (-1)^v\sqrt{\dfrac{2\pi}{\alpha}}\,J_{2v+\frac{1}{2}}(\alpha) \qquad\qquad [n=2v\text{ is even}]$$

$$\text{GH2 24 (171.10a)}$$

**7.246**    $\int_0^{\pi} P_n\left(1-2\sin^2 x\sin^2\theta\right)\sin x\,\mathrm{d}x = \dfrac{2\sin(2n+1)\theta}{(2n+1)\sin\theta}$       MO 71

**7.247**    $\int_0^1 P_{2n+1}(x)\sin ax\,\dfrac{\mathrm{d}x}{\sqrt{x}} = (-1)^{n+1}\sqrt{\dfrac{\pi}{2a}}\,J_{2n+\frac{3}{2}}(a)$     $[a>0]$       ET I 94(1)

**7.248**

1.　　$\displaystyle\int_{-1}^{1}\left(a^2+b^2-2abx\right)^{-1/2}\sin\left[\lambda\left(a^2+b^2-2abx\right)^{1/2}\right]P_n(x)\,\mathrm{d}x=\pi(ab)^{-1/2}\,J_{n+\frac12}(a\lambda)\,J_{n+\frac12}(b\lambda)$

$$[a>0,\quad b>0]\qquad\qquad\text{ET II 277(11)}$$

2.　　$\displaystyle\int_{-1}^{1}\left(a^2+b^2-2abx\right)^{-1/2}\cos\left[\lambda\left(a^2+b^2-2abx\right)^{1/2}\right]P_n(x)\,\mathrm{d}x=-\pi(ab)^{-1/2}\,J_{n+\frac12}(a\lambda)\,Y_{n+\frac12}(b\lambda)$

$$[0\le a\le b]\qquad\qquad\text{ET II 277(12)}$$

3.*　　$\displaystyle\int_{-1}^{1}\frac{P_n(x)}{(\cosh a-x)^{3/2}}\,\mathrm{d}x=\frac{2\sqrt2}{\sinh a}e^{-(n+\frac12)a}\qquad[n=0,1,\ldots,\qquad a>0]$

**7.249**

1.　　$\displaystyle\int_{-1}^{1}P_n(x)\arcsin x\,\mathrm{d}x=0\qquad\qquad[n\text{ is even}]$

$$=\pi\left\{\frac{(n-2)!!}{2^{\frac12(n+1)}\left(\dfrac{n+1}{2}\right)!}\right\}^2\qquad[n\text{ is odd}]$$

WH

2.　　$\displaystyle P_n(x)=\frac1t\sum_{t=0}^{t-1}\left(x+\sqrt{x^2-1}\cos\frac{2\pi r}{t}\right)^n\qquad[t>n]$

## 7.25 Combinations of Legendre polynomials and Bessel functions

**7.251**

1.　　$\displaystyle\int_{0}^{1}x\,P_n\left(1-2x^2\right)Y_\nu(xy)\,\mathrm{d}x=\pi^{-1}y^{-1}\left[S_{2n+1}(y)+\pi\,Y_{2n+1}(y)\right]$

$$[n=0,1,\ldots;\quad y>0,\quad \nu>0]$$
$$\text{ET II 108(1)}$$

2.　　$\displaystyle\int_{0}^{1}x\,P_n\left(1-2x^2\right)K_0(xy)\,\mathrm{d}x=y^{-1}\left[(-1)^{n+1}K_{2n+1}(y)+\frac{i}{2}S_{2n+1}(iy)\right]$

$$[y>0]\qquad\qquad\text{ET II 134(1)}$$

3.　　$\displaystyle\int_{0}^{1}x\,P_n\left(1-2x^2\right)J_0(xy)\,\mathrm{d}x=y^{-1}J_{2n+1}(y)\qquad[y>0]\qquad\qquad\text{ET II 13(1)}$

4.　　$\displaystyle\int_{0}^{1}x\,P_n\left(1-2x^2\right)[J_0(ax)]^2\,\mathrm{d}x=\frac{1}{2(2n+1)}\left\{[J_n(a)]^2+[J_{n+1}(a)]^2\right\}\qquad\text{ET II 338(39)a}$

5.　　$\displaystyle\int_{0}^{1}x\,P_n\left(1-2x^2\right)J_0(ax)\,Y_0(ax)\,\mathrm{d}x=\frac{1}{2(2n+1)}\left[J_n(a)\,Y_n(a)+J_{n+1}(a)\,Y_{n+1}(a)\right]$

$$\text{ET II 339(48)a}$$

6.　　$\displaystyle\int_{0}^{1}x^2\,P_n\left(1-2x^2\right)J_1(xy)\,\mathrm{d}x=y^{-1}(2n+1)^{-1}\left[(n+1)\,J_{2n+2}(y)-n\,J_{2n}(y)\right]$

$$[y>0]\qquad\qquad\text{ET II 20(23)}$$

7. $\int_0^1 x^{\mu-1} P_n \left(2x^2 - 1\right) J_\nu(ax) \, dx = \dfrac{2^{-\nu-1} a^\nu \left[\Gamma\left(\frac{1}{2}\mu + \frac{1}{2}\nu\right)\right]^2}{\Gamma(\nu+1)\,\Gamma\left(\frac{1}{2}\mu + \frac{1}{2}\nu + n + 1\right)\Gamma\left(\frac{1}{2} + \frac{1}{2}\nu - n\right)}$

$$\times \, {}_2F_3\left(\frac{\mu+\nu}{2}, \frac{\mu+\nu}{2}; \nu+1, \frac{\mu+\nu}{2}+n+1, \frac{\mu+\nu}{2} - n; -\frac{a^2}{4}\right)$$

$$[a > 0, \quad \operatorname{Re}(\mu + \nu) > 0] \qquad \text{ET II 337(32)a}$$

**7.252** $\displaystyle\int_0^1 e^{-ax} P_n(1 - 2x) I_0(ax) \, dx = \frac{e^{-a}}{2n+1}\left[I_n(a) + I_{n+1}(a)\right]$

$$[a > 0] \qquad \text{ET II 366(11)a}$$

**7.253** $\displaystyle\int_0^{\pi/2} \sin(2x)\, P_n\left(\cos 2x\right) J_0\left(a \sin x\right)\, dx = a^{-1} J_{2n+1}(a) \qquad \text{ET II 361(20)}$

**7.254** $\displaystyle\int_0^1 x\, P_n\left(1 - 2x^2\right)\left[I_0(ax) - \mathbf{L}_0(ax)\right]\, dx = (-1)^n\left[I_{2n+1}(a) - \mathbf{L}_{2n+1}(a)\right]$

$$[a > 0] \qquad \text{ET II 385(14)a}$$

## 7.3–7.4 Orthogonal Polynomials

## 7.31 Combinations of Gegenbauer polynomials $C_n^\nu(x)$ and powers

**7.311**

1. $\displaystyle\int_{-1}^1 \left(1 - x^2\right)^{\nu - \frac{1}{2}} C_n^\nu(x) \, dx = 0 \qquad\qquad \left[n > 0, \quad \operatorname{Re}\nu > -\frac{1}{2}\right] \qquad \text{ET II 280(1)}$

2. $\displaystyle\int_0^1 x^{n+2\rho} \left(1 - x^2\right)^{\nu - \frac{1}{2}} C_n^\nu(x)\, dx = \frac{\Gamma(2\nu + n)\,\Gamma(2\rho + n + 1)\,\Gamma\left(\nu + \frac{1}{2}\right)\Gamma\left(\rho + \frac{1}{2}\right)}{2^{n+1}\,\Gamma(2\nu)\,\Gamma(2\rho+1)\, n!\,\Gamma(n + \nu + \rho + 1)}$

$$\left[\operatorname{Re}\rho > -\frac{1}{2}, \quad \operatorname{Re}\nu > -\frac{1}{2}\right] \qquad \text{ET II 280(2)}$$

3. $\displaystyle\int_{-1}^1 (1 - x)^{\nu - \frac{1}{2}} (1 + x)^\beta \, C_n^\nu(x)\, dx = \frac{2^{\beta + \nu + \frac{1}{2}}\,\Gamma(\beta + 1)\,\Gamma\left(\nu + \frac{1}{2}\right)\Gamma(2\nu + n)\,\Gamma\left(\beta - \nu + \frac{3}{2}\right)}{n!\,\Gamma(2\nu)\,\Gamma\left(\beta - \nu - n + \frac{3}{2}\right)\Gamma\left(\beta + \nu + n + \frac{3}{2}\right)}$

$$\left[\operatorname{Re}\beta > -1, \quad \operatorname{Re}\nu > -\frac{1}{2}\right] \qquad \text{ET II 280(3)}$$

4. $\displaystyle\int_{-1}^1 (1 - x)^\alpha \, (1 + x)^\beta \, C_n^\nu(x)\, dx = \frac{2^{\alpha + \beta + 1}\,\Gamma(\alpha + 1)\,\Gamma(\beta + 1)\,\Gamma(n + 2\nu)}{n!\,\Gamma(2\nu)\,\Gamma(\alpha + \beta + 2)}$

$$\times \, {}_3F_2\left(-n, n + 2\nu, \alpha + 1; \nu + \frac{1}{2}, \alpha + \beta + 2; 1\right)$$

$$[\operatorname{Re}\alpha > -1, \quad \operatorname{Re}\beta > -1] \qquad \text{ET II 281(4)}$$

**7.312** In the following integrals, $z$ belongs to the complex plane with a cut along the interval of the real axis from $-1$ to $1$.

1. $\displaystyle\int_{-1}^1 x^m (z - x)^{-1} \left(1 - x^2\right)^{\nu - \frac{1}{2}} C_n^\nu(x)\, dx = \frac{\pi^{1/2} 2^{\frac{3}{2} - \nu}}{\Gamma(\nu)} e^{-\left(\nu - \frac{1}{2}\right)\pi i} z^m \left(z^2 - 1\right)^{\frac{1}{2}\nu - \frac{1}{4}} Q_{n+\nu-\frac{1}{2}}^{\nu - \frac{1}{2}}(z)$

$$\left[m \le n, \quad \operatorname{Re}\nu > -\frac{1}{2}\right] \qquad \text{ET II 281(5)}$$

$2.^{12}$    $\displaystyle\int_{-1}^{1} x^{n+1}(z-x)^{-1}\left(1-x^2\right)^{\nu-\frac{1}{2}} C_n^\nu(x)\,\mathrm{d}x = \frac{\pi^{1/2}2^{\frac{1}{2}-\nu}}{\Gamma(\nu)} e^{-\left(\nu-\frac{1}{2}\right)\pi i} z^{n+1}\left(z^2-1\right)^{\frac{1}{2}\nu-\frac{1}{4}} Q_{n+\nu-\frac{1}{2}}^{\nu-\frac{1}{2}}(z)$

$$-\frac{\pi 2^{1-2\nu-n}n!}{\Gamma(\nu)\,\Gamma(\nu+n+1)}$$

$$\left[\operatorname{Re}\nu > -\tfrac{1}{2}\right] \qquad\qquad \text{ET II 281(6)}$$

$3.^{6}$    $\displaystyle\int_{-1}^{1} (z-x)^{-1}\left(1-x^2\right)^{\nu-\frac{1}{2}} C_m^\nu(x)\,C_n^\nu(x)\,\mathrm{d}x = \frac{\pi^{1/2}2^{\frac{3}{2}-\nu}}{\Gamma(\nu)} e^{-\left(\nu-\frac{1}{2}\right)\pi i}\left(z^2-1\right)^{\frac{1}{2}\nu-\frac{1}{4}} C_m^\nu(z)\,Q_{n+\nu-\frac{1}{2}}^{\nu-\frac{1}{2}}(z)$

$$\left[m \le n, \quad \operatorname{Re}\nu > -\tfrac{1}{2}\right] \qquad \text{ET II 283(17)}$$

**7.313**

1.    $\displaystyle\int_{-1}^{1} \left(1-x^2\right)^{\nu-\frac{1}{2}} C_m^\nu(x)\,C_n^\nu(x)\,\mathrm{d}x = 0$        $\left[m \ne n, \quad \operatorname{Re}\nu > -\tfrac{1}{2}\right]$

$$\text{ET II 282(12), MO 98a, EH I 177(16)}$$

2.    $\displaystyle\int_{-1}^{1} \left(1-x^2\right)^{\nu-\frac{1}{2}} \left[C_n^\nu(x)\right]^2\,\mathrm{d}x = \frac{\pi 2^{1-2\nu}\,\Gamma(2\nu+n)}{n!(n+\nu)\left[\Gamma(\nu)\right]^2}$      $\left[\operatorname{Re}\nu > -\tfrac{1}{2}\right]$

$$\text{ET II 281(8), MO 98a, EH I 177(17)}$$

**7.314**

1.    $\displaystyle\int_{-1}^{1} (1-x)^{\nu-\frac{3}{2}}(1+x)^{\nu-\frac{1}{2}}\left[C_n^\nu(x)\right]^2\,\mathrm{d}x = \frac{\pi^{1/2}\,\Gamma\left(\nu-\frac{1}{2}\right)\Gamma(2\nu+n)}{n!\,\Gamma(\nu)\,\Gamma(2\nu)}$

$$\left[\operatorname{Re}\nu > \tfrac{1}{2}\right] \qquad\qquad \text{ET II 281(9)}$$

2.    $\displaystyle\int_{-1}^{1} (1-x)^{\nu-\frac{1}{2}}(1+x)^{2\nu-1}\left[C_n^\nu(x)\right]^2\,\mathrm{d}x = \frac{2^{3\nu-\frac{1}{2}}\left[\Gamma(2\nu+n)\right]^2\Gamma\left(2n+\nu+\frac{1}{2}\right)}{(n!)^2\,\Gamma(2\nu)\,\Gamma\left(3\nu+2n+\frac{1}{2}\right)}$

$$\left[\operatorname{Re}\nu > 0\right] \qquad\qquad \text{ET II 282(10)}$$

3.    $\displaystyle\int_{-1}^{1} (1-x)^{3\nu+2n-\frac{3}{2}}(1+x)^{\nu-\frac{1}{2}}\left[C_n^\nu(x)\right]^2\,\mathrm{d}x$

$$= \frac{\pi^{1/2}\left[\Gamma\left(\nu+\frac{1}{2}\right)\right]^2\Gamma\left(\nu+2n+\frac{1}{2}\right)\Gamma\left(2\nu+2n\right)\Gamma\left(3\nu+2n-\frac{1}{2}\right)}{2^{2\nu+2n}\left[n!\,\Gamma\left(\nu+n+\frac{1}{2}\right)\Gamma(2\nu)\right]^2\Gamma\left(2\nu+2n+\frac{1}{2}\right)}$$

$$\left[\operatorname{Re}\nu > \tfrac{1}{6}\right] \qquad\qquad \text{ET II 282(11)}$$

4.    $\displaystyle\int_{-1}^{1} (1-x)^{\nu-\frac{1}{2}}(1+x)^{\nu+m-n-\frac{3}{2}} C_m^\nu(x)\,C_n^\nu(x)\,\mathrm{d}x$

$$= (-1)^m \frac{2^{2-2\nu-m+n}\pi^{3/2}\,\Gamma(2\nu+n)}{m!(n-m)!\left[\Gamma(\nu)\right]^2\Gamma\left(\frac{1}{2}+\nu+m\right)} \frac{\Gamma\left(\nu-\frac{1}{2}+m-n\right)\Gamma\left(\frac{1}{2}-\nu+m-n\right)}{\Gamma\left(\frac{1}{2}-\nu-n\right)\Gamma\left(\frac{1}{2}+m-n\right)}$$

$$\left[\operatorname{Re}\nu > -\tfrac{1}{2}; \quad n \ge m\right] \qquad \text{ET II 282(13)a}$$

5.    $\displaystyle\int_{-1}^{1} (1-x)^{2\nu-1}(1+x)^{\nu-\frac{1}{2}} C_m^\nu(x)\,C_n^\nu(x)\,\mathrm{d}x$

$$= \frac{2^{3\nu-\frac{1}{2}}\,\Gamma\left(\nu+\frac{1}{2}\right)\Gamma(2\nu+m)\Gamma(2\nu+n)}{m!n!\,\Gamma(2\nu)\,\Gamma\left(\frac{1}{2}-\nu\right)} \frac{\Gamma\left(\nu+\frac{1}{2}+m+n\right)\Gamma\left(\frac{1}{2}-\nu+n-m\right)}{\Gamma\left(\nu+\frac{1}{2}+n-m\right)\Gamma\left(3\nu+\frac{1}{2}+m+n\right)}$$

$$\left[\operatorname{Re}\nu > 0\right] \qquad\qquad \text{ET II 282(14)}$$

6.　　$\int_{-1}^{1} (1-x)^{\nu-\frac{1}{2}} (1+x)^{3\nu+m+n-\frac{3}{2}} C_m^{\nu}(x) C_n^{\nu}(x) \, dx$

$$= \frac{2^{4\nu+m+n-1} \left[ \Gamma \left( \nu + \frac{1}{2} \right) \Gamma (2\nu + m + n) \right]^2}{\Gamma \left( \nu + m + \frac{1}{2} \right) \Gamma \left( \nu + n + \frac{1}{2} \right) \Gamma (2\nu + m)} \frac{\Gamma \left( \nu + m + n + \frac{1}{2} \right) \Gamma \left( 3\nu + m + n - \frac{1}{2} \right)}{\Gamma (2\nu + n) \Gamma (4\nu + 2m + 2n)}$$

$$\left[ \operatorname{Re} \nu > \frac{1}{6} \right] \qquad \text{ET II 282(15)}$$

7.　　$\int_{-1}^{1} (1-x)^{\alpha} (1+x)^{\nu-\frac{1}{2}} C_m^{\mu}(x) C_n^{\nu}(x) \, dx$

$$= \frac{2^{\alpha+\nu+\frac{1}{2}} \Gamma(\alpha+1) \Gamma \left( \nu + \frac{1}{2} \right) \Gamma \left( \nu - \alpha + n - \frac{1}{2} \right)}{m! n! \Gamma \left( \nu - \alpha - \frac{1}{2} \right) \Gamma \left( \nu - \alpha + n + \frac{3}{2} \right)} \frac{\Gamma (2\mu + m) \Gamma (2\nu + n)}{\Gamma (2\mu) \Gamma (2\nu)}$$

$$\times {}_4F_3 \left( -m, m + 2\mu, \alpha + 1, \alpha - \nu + \frac{3}{2}; \mu + \frac{1}{2}, \nu + \alpha + n + \frac{3}{2}, \alpha - \nu - n + \frac{3}{2}; 1 \right)$$

$$\left[ \operatorname{Re} \alpha > -1, \quad \operatorname{Re} \nu > -\frac{1}{2} \right] \qquad \text{ET II 283(16)}$$

**7.315**　　$\int_{-1}^{1} \left( 1 - x^2 \right)^{\frac{1}{2}\nu-1} C_{2n}^{\nu}(ax) \, dx = \frac{\pi^{1/2} \Gamma \left( \frac{1}{2}\nu \right)}{\Gamma \left( \frac{1}{2}\nu + \frac{1}{2} \right)} C_n^{\frac{1}{2}\nu} \left( 2a^2 - 1 \right)$

$$[\operatorname{Re} \nu > 0] \qquad \text{ET II 283(19)}$$

**7.316**　　$\int_{-1}^{1} \left( 1 - x^2 \right)^{\nu-1} C_n^{\nu} \left( \cos\alpha \cos\beta + x \sin\alpha \sin\beta \right) \, dx = \frac{2^{2\nu-1} n! \left[ \Gamma(\nu) \right]^2}{\Gamma(2\nu + n)} C_n^{\nu} \left( \cos\alpha \right) C_n^{\nu} \left( \cos\beta \right)$

$$[\operatorname{Re} \nu > 0] \qquad \text{ET II 283(20)}$$

**7.317**

1.　　$\int_{0}^{1} (1-x)^{\mu-1} x^{\lambda-\frac{1}{2}} C_n^{\lambda} (1 - \gamma x) \, dx = \frac{\Gamma (2\lambda + n) \Gamma \left( \lambda + \frac{1}{2} \right) \Gamma(\mu)}{\Gamma(2\lambda) \Gamma \left( \lambda + \mu + n + \frac{1}{2} \right)} P_n^{(\alpha,\beta)}(1 - \gamma)$

$$\alpha = \lambda + \mu - \frac{1}{2}, \qquad \beta = \lambda - \mu - \frac{1}{2} \qquad \left[ \operatorname{Re} \lambda > -1, \quad \lambda \neq 0, \quad -\frac{1}{2}, \quad \operatorname{Re} \mu > 0 \right] \qquad \text{ET II 190(39)a}$$

2.　　$\int_{0}^{1} (1-x)^{\mu-1} x^{\nu-1} C_n^{\lambda}(1 - \gamma x) \, dx = \frac{\Gamma (2\lambda + n) \Gamma (\mu) \Gamma (\nu)}{n! \Gamma (2\lambda) \Gamma (\mu + \nu)}$

$$\times {}_3F_2 \left( -n, n + 2\lambda, \nu; \lambda + \frac{1}{2}, \mu + \nu; \frac{\gamma}{2} \right)$$

$$[2\lambda \neq 0, -1, -2, \ldots, \quad \operatorname{Re} \mu > 0, \quad \operatorname{Re} \nu > 0] \qquad \text{ET II 191(40)a}$$

**7.318**　　$\int_{0}^{1} x^{2\nu} \left( 1 - x^2 \right)^{\sigma-1} C_n^{\nu} \left( 1 - x^2 y \right) \, dx = \frac{\Gamma(2\nu + n) \Gamma \left( \nu + \frac{1}{2} \right) \Gamma(\sigma)}{2 \Gamma(2\nu) \Gamma \left( n + \nu + \sigma + \frac{1}{2} \right)} P_n^{(\alpha,\beta)}(1 - y),$

$$\alpha = \nu + \sigma - \frac{1}{2}, \qquad \beta = \nu - \sigma - \frac{1}{2} \qquad \left[ \operatorname{Re} \nu > -\frac{1}{2}, \quad \operatorname{Re} \sigma > 0 \right] \qquad \text{ET II 283(21)}$$

**7.319**

1.　　$\int_{0}^{1} (1-x)^{\mu-1} x^{\nu-1} C_{2n}^{\lambda} \left( \gamma x^{1/2} \right) \, dx = (-1)^n \frac{\Gamma(\lambda + n) \Gamma(\mu) \Gamma(\nu)}{n! \Gamma(\lambda) \Gamma(\mu + \nu)} \, {}_3F_2 \left( -n, n + \lambda, \nu; \frac{1}{2}, \mu + \nu; \gamma^2 \right)$

$$[\operatorname{Re} \mu > 0, \quad \operatorname{Re} \nu > 0] \qquad \text{ET II 191(41)a}$$

2.　　$\int_{0}^{1} (1-x)^{\mu-1} x^{\nu-1} C_{2n+1}^{\lambda} \left( \gamma x^{1/2} \right) \, dx = \frac{(-1)^n 2\gamma \, \Gamma(\mu) \Gamma(\lambda + n + 1) \Gamma \left( \nu + \frac{1}{2} \right)}{n! \Gamma(\lambda) \Gamma \left( \mu + \nu + \frac{1}{2} \right)}$

$$\times {}_3F_2 \left( -n, n + \lambda + 1, \nu + \frac{1}{2}; \frac{3}{2}, \mu + \nu + \frac{1}{2}; \gamma^2 \right)$$

$$\left[ \operatorname{Re} \mu > 0, \quad \operatorname{Re} \nu > -\frac{1}{2} \right] \qquad \text{ET II 191(42)}$$

## 7.32 Combinations of Gegenbauer polynomials $C_n^\nu(x)$ and elementary functions

**7.321** $$\int_{-1}^{1} \left(1 - x^2\right)^{\nu - \frac{1}{2}} e^{iax} \, C_n^\nu(x) \, dx = \frac{\pi 2^{1-\nu} i^n \, \Gamma(2\nu + n)}{n! \, \Gamma(\nu)} a^{-\nu} \, J_{\nu+n}(a)$$

$$\left[\operatorname{Re} \nu > -\tfrac{1}{2}\right] \qquad \text{ET II 281(7), MO 99a}$$

**7.322** $$\int_{0}^{2a} [x(2a - x)]^{\nu - \frac{1}{2}} \, C_n^\nu \left(\frac{x}{a} - 1\right) e^{-bx} \, dx = (-1)^n \frac{\pi \, \Gamma(2\nu + n)}{n! \, \Gamma(\nu)} \left(\frac{a}{2b}\right)^\nu e^{-ab} \, I_{\nu+n}(ab)$$

$$\left[\operatorname{Re} \nu > -\tfrac{1}{2}\right] \qquad \text{ET I 171(9)}$$

**7.323**

1. $$\int_{0}^{\pi} C_n^\nu (\cos\varphi) \, (\sin\varphi)^{2\nu} \, d\varphi = 0 \qquad\qquad [n = 1, 2, 3, \ldots]$$

$$= 2^{-2\nu} \pi \, \Gamma(2\nu + 1) \left[\Gamma(1 + \nu)\right]^{-2} \quad [n = 0]$$

$$\text{EH I 177(18)}$$

2.[11] $$\int_{0}^{\pi} C_n^\nu (\cos\psi \cos\psi' + \sin\psi \sin\psi' \cos\varphi) \, (\sin\varphi)^{2\nu - 1} \, d\varphi$$

$$= 2^{2\nu - 1} n! \left[\Gamma(\nu)\right]^2 C_n^\nu(\cos\psi) \, C_n^\nu(\cos\psi') \left[\Gamma(2\nu + n)\right]^{-1}$$

$$\left[\operatorname{Re} \nu > 0\right] \qquad \text{EH I 177(20)}$$

**7.324**

1. $$\int_{0}^{1} \left(1 - x^2\right)^{\nu - \frac{1}{2}} C_{2n+1}^\nu(x) \sin ax \, dx = (-1)^n \pi \frac{\Gamma(2n + 2\nu + 1) \, J_{2n+\nu+1}(a)}{(2n + 1)! \, \Gamma(\nu) (2a)^\nu}$$

$$\left[\operatorname{Re} \nu > -\tfrac{1}{2}, \quad a > 0\right] \qquad \text{ET I 94(4)}$$

2. $$\int_{0}^{1} \left(1 - x^2\right)^{\nu - \frac{1}{2}} C_{2n}^\nu(x) \cos ax \, dx = \frac{(-1)^n \pi \, \Gamma(2n + 2\nu) \, J_{\nu+2n}(a)}{(2n)! \, \Gamma(\nu) (2a)^\nu}$$

$$\left[\operatorname{Re} \nu > -\tfrac{1}{2}, \quad a > 0\right] \qquad \text{ET I 38(3)a}$$

## 7.325* Complete System of Orthogonal Step Functions

Let $s_j(x) = (-1)^{\lfloor 2jx \rfloor}$ for $j \in \mathbb{N}$ and $c_j(x) = (-1)^{\lfloor 2jx + 1/2 \rfloor}$ for $j \in 0 + \mathbb{N}$ where $\lfloor z \rfloor$ denotes the integer part of $z$. Thus, $c_j(z)$ and $s_j(z)$ have minimal period $j^{-1}$ and manifest even and odd symmetry about $x = 1/2$, respectively, and so are the discrete analogues of $\cos 2\pi jx$ and $\sin 2\pi jx$. Furthermore, for $j \in \mathbb{N}$ let $\underline{j}$ denotes its odd part: the quotient of $j$ by its highest power-of-two factor. Then for all $j$ and $k \in \mathbb{N}$, if $(\overline{j}, k)$ denotes their highest common factor and $[j, k]$ denotes their lowest common multiple:

1. $$\int_{0}^{1} s_j(x) s_k(x) \, dx = \begin{cases} \frac{(j,k)}{[j,k]} & \text{if } \frac{j}{\underline{j}} = \frac{k}{\underline{k}} \\ 0 & \text{otherwise} \end{cases}$$

2. $$\int_{0}^{1} c_j(x) c_k(x) \, dx = \begin{cases} (-1)^{(j+k)/2+1} \frac{(j,k)}{[j,k]} & \text{if } \frac{j}{\underline{j}} = \frac{k}{\underline{k}} \\ 0 & \text{otherwise} \end{cases}$$

## 7.33 Combinations of the polynomials $C_n^\nu(x)$ and Bessel functions. Integration of Gegenbauer functions with respect to the index.

**7.331**

1.
$$\int_1^\infty x^{2n+1-\nu} \left(x^2 - 1\right)^{\nu - 2n - \frac{1}{2}} C_{2n}^{\nu - 2n}\left(\frac{1}{x}\right) J_\nu(xy)\, dx$$
$$= (-1)^n 2^{2n-\nu+1} y^{-\nu+2n-1} [(2n)!]^{-1} \Gamma(2\nu - 2n) [\Gamma(\nu - 2n)]^{-1} \cos y$$
$$\left[y > 0, \quad 2n - \tfrac{1}{2} < \operatorname{Re}\nu < 2n + \tfrac{1}{2}\right] \quad \text{ET II 44(10)a}$$

**7.332**

1.
$$\int_0^\infty x^{\nu+1} \left(x^2 + \beta^2\right)^{-\frac{1}{2}\nu - \frac{3}{4}} C_{2n+1}^{\nu + \frac{1}{2}}\left[\left(x^2 + \beta^2\right)^{-1/2} \beta\right] J_{\nu + \frac{3}{2} + 2n}\left[\left(x^2 + \beta^2\right)^{1/2} a\right] J_\nu(xy)\, dx$$
$$= (-1)^n 2^{1/2} \pi^{-1/2} a^{\frac{1}{2} - \nu} y^\nu \left(a^2 - y^2\right)^{-1/2} \sin\left[\beta \left(a^2 - y^2\right)^{1/2}\right] C_{2n+1}^{\nu + \frac{1}{2}}\left[\left(1 - \frac{y^2}{a^2}\right)^{1/2}\right]$$
$$[0 < y < a]$$
$$= 0$$
$$[a < y < \infty] \qquad [a > 0, \quad \operatorname{Re}\beta > 0, \quad \operatorname{Re}\nu > -1]$$
$$\text{ET II 59(23)}$$

2.
$$\int_0^\infty x^{\nu+1} \left(x^2 + \beta^2\right)^{-\frac{1}{2}\nu - \frac{3}{4}} C_{2n}^{\nu + \frac{1}{2}}\left[\beta \left(x^2 + \beta^2\right)^{-1/2}\right] J_{\nu + \frac{1}{2} + 2n}\left[\left(x^2 + \beta^2\right)^{1/2} a\right] J_\nu(xy)\, dx$$
$$= (-1)^n 2^{1/2} \pi^{-1/2} a^{\frac{1}{2} - \nu} y^\nu \left(a^2 - y^2\right)^{-1/2} \cos\left[\beta \left(a^2 - y^2\right)^{1/2}\right] C_{2n}^{\nu + \frac{1}{2}}\left[\left(1 - \frac{y^2}{a^2}\right)^{1/2}\right]$$
$$[0 < y < a]$$
$$= 0$$
$$[a < y < \infty] \qquad [a > 0, \quad \operatorname{Re}\beta > 0, \quad \operatorname{Re}\nu > -1]$$
$$\text{ET II 59(24)}$$

**7.333**

1.
$$\int_0^\pi (\sin x)^{\nu+1} \cos\left(a \cos\theta \cos x\right) C_n^{\nu + \frac{1}{2}}(\cos x) J_\nu\left(a \sin\theta \sin x\right) dx$$
$$= (-1)^{\frac{n}{2}} \left(\frac{2\pi}{a}\right)^{1/2} (\sin\theta)^\nu C_n^{\nu + \frac{1}{2}}(\cos\theta) J_{\nu + \frac{1}{2} + n}(a) \quad [n = 0, 2, 4, \ldots]$$
$$= 0 \qquad\qquad\qquad [n = 1, 3, 5, \ldots]$$
$$[\operatorname{Re}\nu > -1] \qquad\qquad \text{WA 414(2)a}$$

2.
$$\int_0^\pi (\sin x)^{\nu+1} \sin\left(a \cos\theta \cos x\right) C_n^{\nu + \frac{1}{2}}(\cos x) J_\nu\left(a \sin\theta \sin x\right) dx$$
$$= 0 \qquad\qquad\qquad [n = 0, 2, 4, \ldots]$$
$$= (-1)^{\frac{n-1}{2}} \left(\frac{2\pi}{a}\right)^{1/2} (\sin\theta)^\nu C_n^{\nu + \frac{1}{2}}(\cos\theta) J_{\nu + \frac{1}{2} + n}(a) \quad [n = 1, 3, 5, \ldots]$$
$$[\operatorname{Re}\nu > -1] \qquad\qquad \text{WA 414(3)a}$$

**7.334**

1. 
$$\int_0^\pi (\sin x)^{2\nu} \, C_n^\nu (\cos x) \, \frac{J_\nu(\omega)}{\omega^\nu} \, dx = \frac{\pi \, \Gamma(2\nu + n)}{2^{\nu-1} n! \, \Gamma(\nu)} \, \frac{J_{\nu+n}(\alpha)}{\alpha^\nu} \, \frac{J_{\nu+n}(\beta)}{\beta^\nu},$$

$$\omega = \left(\alpha^2 + \beta^2 - 2\alpha\beta \cos x\right)^{1/2} \qquad \left[n = 0, 1, 2, \ldots; \quad \operatorname{Re}\nu > -\tfrac{1}{2}\right] \quad \text{ET II 362(29)}$$

2.$^{12}$ 
$$\int_0^\pi (\sin x)^{2\nu} \, C_n^\nu (\cos x) \, \frac{Y_\nu(\omega)}{\omega^\nu} \, dx = \frac{\pi \, \Gamma(2\nu + n)}{2^{\nu-1} n! \, \Gamma(\nu)} \, \frac{J_{\nu+n}(\alpha)}{\alpha^\nu} \, \frac{Y_{\nu+n}(\beta)}{\beta^\nu},$$

$$\omega = \left(\alpha^2 + \beta^2 - 2\alpha\beta \cos x\right)^{1/2} \qquad \left[|\alpha| < |\beta|, \quad \operatorname{Re}\nu > -\tfrac{1}{2}\right] \quad \text{ET II 362(30)}$$

**Integration of Gegenbauer functions with respect to the index**

**7.335** 
$$\int_{c-i\infty}^{c+i\infty} [\sin(\alpha\pi)]^{-1} \, t^\alpha \, C_\alpha^\nu(z) \, d\alpha = -2i \left(1 + 2tz + t^2\right)^{-\nu}$$

$$\left[-2 < \operatorname{Re}\nu < c < 0, \quad |\arg(z \pm 1)| < \pi\right]$$
$$\text{EH I 178(25)}$$

**7.336** 
$$\int_{-\infty}^{\infty} \operatorname{sech}(\pi x) \left(\nu - \frac{1}{2} + ix\right) K_{\nu-\frac{1}{2}+ix}(a) \, I_{\nu-\frac{1}{2}+ix}(b) \, C_{-\frac{1}{2}+ix}^\nu (-\cos\varphi) \, dx$$

$$= \frac{2^{-\nu+1}(ab)^\nu}{\Gamma(\nu)} \omega^{-\nu} \, K_\nu(\omega)$$

$$\omega = \sqrt{a^2 + b^2 - 2ab\cos\varphi} \qquad \text{EH II 55(45)}$$

## 7.34 Combinations of Chebyshev polynomials and powers

**7.341**

1. 
$$\int_{-1}^{1} [T_n(x)]^2 \, dx = 1 - \left(4n^2 - 1\right)^{-1} \qquad\qquad\qquad \text{ET II 271(6)}$$

2.* 
$$\int_{-1}^{1} T_m(x) T_n(x) \, dx = \frac{1}{1-(m-n)^2} + \frac{1}{1-(m+n)^2} \qquad [m+n \text{ is even}]$$

$$= 0 \qquad [m+n \text{ is odd}]$$

3.* 
$$\int_{-1}^{1} \left(1 - x^2\right) U_m(x) U_n(x) \, dx = \frac{1}{1-(m-n)^2} - \frac{1}{1-(m+n)^2} \qquad [m+n \text{ is even}]$$

$$= 0 \qquad [m+n \text{ is odd}]$$

**7.342** 
$$\int_{-1}^{1} U_n \left[x \left(1-y^2\right)^{1/2} \left(1-z^2\right)^{1/2} + yz\right] dx = \frac{2}{n+1} \, U_n(y) \, U_n(z)$$

$$[|y| < 1, \quad |z| < 1] \qquad \text{ET II 275(34)}$$

**7.343**

1. 
$$\int_{-1}^{1} T_n(x) \, T_m(x) \frac{dx}{\sqrt{1-x^2}} = 0 \qquad [m \neq n]$$

$$= \frac{\pi}{2} \qquad [m = n \neq 0]$$

$$= \pi \qquad [m = n = 0]$$

2. $\displaystyle\int_{-1}^{1}\sqrt{1-x^2}\,U_n(x)\,U_m(x)\,dx=0 \qquad [m\neq n]$ 　　　　ET II 274(28)

$\displaystyle\qquad\qquad\qquad\qquad = \frac{\pi}{2} \qquad [m=n]$ 　　　　ET II 274(27), MO 105a

**7.344**

1.[12] $\displaystyle\int_{-1}^{1}(y-x)^{-1}\big(1-y^2\big)^{-1/2}T_{n+1}(y)\,dy=\pi U_n(x) \qquad\qquad [|x|<1]$

$\displaystyle\qquad\qquad\qquad = \pi U_n(x)-\pi T_{n+1}(x)\frac{\mathrm{sign}(x)}{\sqrt{x^2-1}} \qquad [|x|\geq 1]$

EH II 187(47)

2.[12] $\displaystyle\int_{-1}^{1}(y-x)^{-1}\big(1-y^2\big)^{1/2}U_n(y)\,dy=-\pi T_{n+1}(x) \qquad\qquad [|x|<1]$

$\displaystyle\qquad\qquad\qquad = -\pi T_{n+1}(x)+\pi U_n(x)\,\mathrm{sign}(x)\sqrt{x^2-1} \quad [|x|\geq 1]$

EH II 187(48)

**7.345**

1. $\displaystyle\int_{-1}^{1}(1-x)^{-1/2}(1+x)^{m-n-\frac{3}{2}}\,T_m(x)\,T_n(x)\,dx=0 \qquad [m>n]$ 　　　ET II 272(10)

2. $\displaystyle\int_{-1}^{1}(1-x)^{-1/2}(1+x)^{m+n-\frac{3}{2}}\,T_m(x)\,T_n(x)\,dx=\frac{\pi(2m+2n-2)!}{2^{m+n}(2m-1)!(2n-1)!}$

$\qquad\qquad\qquad\qquad\qquad\qquad\qquad\qquad [m+n\neq 0]$ 　　　ET II 272(11)

3. $\displaystyle\int_{-1}^{1}(1-x)^{1/2}(1+x)^{m+n+\frac{3}{2}}\,U_m(x)\,U_n(x)\,dx=\frac{\pi(2m+2n+2)!}{2^{m+n+2}(2m+1)!(2n+1)!}$ 　ET II 274(31)

4. $\displaystyle\int_{-1}^{1}(1-x)^{1/2}(1+x)^{m-n-\frac{1}{2}}\,U_m(x)\,U_n(x)\,dx=0 \qquad [m>n]$ 　　　ET II 274(30)

5.[12] $\displaystyle\int_{-1}^{1}(1-x)(1+x)^{1/2}\,U_m(x)\,U_n(x)\,dx=\frac{\sqrt{2}}{1-4(m-n)^2}-\frac{\sqrt{2}}{1-4(m+n+2)^2}$ 　ET II 274(29)

6. $\displaystyle\int_{-1}^{1}(1+x)^{-1/2}(1-x)^{\alpha-1}\,T_m(x)\,T_n(x)\,dx$

$\displaystyle\qquad = \frac{\pi^{1/2}2^{\alpha-\frac{1}{2}}\,\Gamma(\alpha)\,\Gamma\left(n-\alpha+\frac{1}{2}\right)}{\Gamma\left(\frac{1}{2}-\alpha\right)\Gamma\left(\alpha+n+\frac{1}{2}\right)}\;{}_4F_3\left(-m,m,\alpha,\alpha+\frac{1}{2};\frac{1}{2},\alpha+n+\frac{1}{2},\alpha-n+\frac{1}{2};1\right)$

$\qquad\qquad\qquad\qquad\qquad [\mathrm{Re}\,\alpha>0]$ 　　　ET II 272(12)

7. $\displaystyle\int_{-1}^{1}(1+x)^{1/2}(1-x)^{\alpha-1}\,U_m(x)\,U_n(x)\,dx$

$\displaystyle\qquad\qquad = \frac{\pi^{1/2}2^{\alpha-\frac{1}{2}}(m+1)(n+1)\,\Gamma(\alpha)\,\Gamma\left(n-\alpha+\frac{3}{2}\right)}{\Gamma\left(\frac{3}{2}-\alpha\right)\Gamma\left(\frac{3}{2}+\alpha+n\right)}$

$\displaystyle\qquad\qquad\qquad\times\;{}_4F_3\left(-m,m+2,\alpha,\alpha-\frac{1}{2};\frac{3}{2},\alpha+n+\frac{3}{2},\alpha-n-\frac{1}{2};1\right)$

$\qquad\qquad\qquad\qquad [\mathrm{Re}\,\alpha>0]$ 　　　ET II 275(32)

**8.***    $\displaystyle \int_{-1}^{1} (1+x)^{-1/2}\, T_m(x) T_n(x)\, dx = \frac{\sqrt{2}}{1 - 4(m-n)^2} + \frac{\sqrt{2}}{1 - 4(m+n)^2}$

**7.346**    $\displaystyle \int_0^1 x^{s-1}\, T_n(x) \frac{dx}{\sqrt{1-x^2}} = \frac{\pi}{s 2^s\, B\left(\frac{1}{2} + \frac{1}{2}s + \frac{1}{2}n,\ \frac{1}{2} + \frac{1}{2}s - \frac{1}{2}n\right)}$

$$[\operatorname{Re} s > 0]$$      ET II 324(2)

**7.347**

**1.**[12]   $\displaystyle \int_{-1}^{1} (1-x)^\alpha (1+x)^\beta\, T_n(x)\, dx = \frac{2^{\alpha+\beta+1}\,\Gamma(\alpha+1)\,\Gamma(\beta+1)}{\Gamma(\alpha+\beta+2)}$

$$\times\ {}_3F_2\left(-n, n, \alpha+1;\ \frac{1}{2},\ \alpha+\beta+2;\ 1\right)$$

$$[\operatorname{Re}\alpha > -1,\quad \operatorname{Re}\beta > -1]$$    ET II 271(2)

**2.**[12]   $\displaystyle \int_{-1}^{1} (1-x)^\alpha (1+x)^\beta\, U_n(x)\, dx = \frac{2^{\alpha+\beta+1}(n+1)\,\Gamma(\alpha+1)\,\Gamma(\beta+1)}{\Gamma(\alpha+\beta+2)}$

$$\times\ {}_3F_2\left(-n, n+2, \alpha+1;\ \frac{3}{2},\ \alpha+\beta+2;\ 1\right)$$

ET II 273(22)

**7.348**    $\displaystyle \int_{-1}^{1} (1-x^2)^{-1/2}\, U_{2n}(xz)\, dx = \pi\, P_n\left(2z^2 - 1\right)$      $[|z| < 1]$      ET II 275(33)

**7.349**    $\displaystyle \int_{-1}^{1} (1-x^2)^{-1/2}\, T_n\left(1 - x^2 y\right)\, dx = \frac{1}{2}\pi\left[P_n(1-y) + P_{n-1}(1-y)\right]$    ET II 222(14)

## 7.35 Combinations of Chebyshev polynomials and elementary functions

**7.351**    $\displaystyle \int_0^1 x^{-1/2}\left(1-x^2\right)^{-\frac{1}{2}} e^{-\frac{2a}{x}}\, T_n(x)\, dx = \pi^{1/2}\, D_{n-\frac{1}{2}}\left(2a^{1/2}\right) D_{-n-\frac{1}{2}}\left(2a^{1/2}\right)$

$$[\operatorname{Re} a > 0]$$      ET II 272(13)

**7.352**

**1.**   $\displaystyle \int_0^\infty \frac{x\, U_n\left[a\left(a^2 + x^2\right)^{-1/2}\right]}{\left(a^2 + x^2\right)^{\frac{1}{2}n+1}\left(e^{\pi x} + 1\right)}\, dx = \frac{a^{-n}}{2n} - 2^{-n-1}\,\zeta\left(n+1, \frac{a+1}{2}\right)$

$$[\operatorname{Re} a > 0]$$      ET II 275(39)

**2.**   $\displaystyle \int_0^\infty \frac{x\, U_n\left[a\left(a^2 + x^2\right)^{-1/2}\right]}{\left(a^2 + x^2\right)^{\frac{1}{2}n+1}\left(e^{2\pi x} - 1\right)}\, dx = \frac{1}{2}\,\zeta(n+1, a) - \frac{a^{-n-1}}{4} - \frac{a^{-n}}{2n}$

$$[\operatorname{Re} a > 0]$$      ET II 276(40)

**7.353**

**1.**   $\displaystyle \int_0^\infty \left(a^2 + x^2\right)^{-\frac{1}{2}n} \operatorname{sech}\left(\frac{1}{2}\pi x\right) T_n\left[a\left(a^2 + x^2\right)^{-1/2}\right] dx = 2^{1-2n}\left[\zeta\left(n, \frac{a+1}{4}\right) - \zeta\left(n, \frac{a+3}{4}\right)\right]$

$$= 2^{1-n}\,\Phi\left(-1, n, \frac{a+1}{2}\right)$$

$$[\operatorname{Re} a > 0]$$      ET II 273(19)

2. $\int_0^\infty (a^2 + x^2)^{-\frac{1}{2}n} \left[\cosh\left(\frac{1}{2}\pi x\right)\right]^{-2} T_n\left[a\left(a^2 + x^2\right)^{-1/2}\right] dx = \pi^{-1} n 2^{1-n} \zeta\left(n+1, \frac{a+1}{2}\right)$

$$[\operatorname{Re} a > 0]$$
$$\text{ET II 273(20)}$$

**7.354**

1.[12] $\int_{-1}^1 \frac{\sin(xyz)}{\sqrt{1-x^2}} \cos\left[\left(1-x^2\right)^{1/2}\left(1-y^2\right)^{1/2} z\right] T_{2n+1}(x)\, dx = (-1)^n \pi\, T_{2n+1}(y)\, J_{2n+1}(x)$

$$\text{ET II 271(4)}$$

2. $\int_{-1}^1 \sin(xyz) \sin\left[\left(1-x^2\right)^{1/2}\left(1-y^2\right)^{1/2} z\right] U_{2n+1}(x)\, dx = (-1)^n \pi \left(1-y^2\right)^{1/2} U_{2n+1}(y)\, J_{2n+2}(z)$

$$\text{ET II 274(25)}$$

3.[12] $\int_{-1}^1 \frac{\cos(xyz)}{\sqrt{1-x^2}} \cos\left[\left(1-x^2\right)^{1/2}\left(1-y^2\right)^{1/2} z\right] T_{2n}(x)\, dx = (-1)^n \pi\, T_{2n}(y)\, J_{2n}(z)$  $\quad$ ET II 271(5)

4. $\int_{-1}^1 \cos(xyz) \sin\left[\left(1-x^2\right)^{1/2}\left(1-y^2\right)^{1/2} z\right] U_{2n}(x)\, dx = (-1)^n \pi \left(1-y^2\right)^{1/2} U_{2n}(y)\, J_{2n+1}(z)$

$$\text{ET II 274(24)}$$

**7.355**

1. $\int_0^1 T_{2n+1}(x) \sin ax \frac{dx}{\sqrt{1-x^2}} = (-1)^n \frac{\pi}{2} J_{2n+1}(a)$  $\qquad [a>0]$  $\qquad$ ET I 94(3)a

2. $\int_0^1 T_{2n}(x) \cos ax \frac{dx}{\sqrt{1-x^2}} = (-1)^n \frac{\pi}{2} J_{2n}(a)$  $\qquad [a>0]$  $\qquad$ ET I 38(2)a

## 7.36 Combinations of Chebyshev polynomials and Bessel functions

**7.361** $\int_0^1 \left(1-x^2\right)^{-1/2} T_n(x)\, J_\nu(xy)\, dx = \frac{1}{2}\pi\, J_{\frac{1}{2}(\nu+n)}\left(\frac{1}{2}y\right) J_{\frac{1}{2}(\nu-n)}\left(\frac{1}{2}y\right)$

$$[y>0, \quad \operatorname{Re}\nu > -n-1] \qquad \text{ET II 42(1)}$$

**7.362** $\int_1^\infty \left(x^2-1\right)^{-\frac{1}{2}} T_n\left(\frac{1}{x}\right) K_{2\mu}(ax)\, dx = \frac{\pi}{2a} W_{\frac{1}{2}n,\mu}(a)\, W_{-\frac{1}{2}n,\mu}(a)$

$$[\operatorname{Re} a > 0] \qquad \text{ET II 366(17)a}$$

## 7.37–7.38 Hermite polynomials

**7.371** $\int_0^x H_n(y)\, dy = [2(n+1)]^{-1} [H_{n+1}(x) - H_{n+1}(0)]$  $\qquad$ EH II 194(27)

**7.372** $\int_{-1}^1 \left(1-t^2\right)^{\alpha-\frac{1}{2}} H_{2n}\left(\sqrt{x}t\right) dt = \dfrac{(-1)^n \pi^{1/2}(2n)!\, \Gamma\left(\alpha+\frac{1}{2}\right) L_n^\alpha(x)}{\Gamma(n+\alpha+1)}$

$$\left[\operatorname{Re} a > -\tfrac{1}{2}\right] \qquad \text{EH II 195(34)}$$

**7.373**

1. $\int_0^x e^{-y^2} H_n(y)\, dy = H_{n-1}(0) - e^{-x^2} H_{n-1}(x)$  $\qquad$ [see **8.956**]  $\qquad$ EH II 194(26)

2. $\int_{-\infty}^{\infty} e^{-x^2} H_{2m}(xy)\, dx = \sqrt{\pi}\dfrac{(2m)!}{m!}\left(y^2-1\right)^m$ 

EH II 195(28)

**7.374**

1. $\int_{-\infty}^{\infty} e^{-x^2} H_n(x)\, H_m(x)\, dx = 0 \qquad [m \neq n]$ 

SM II 567

$$= 2^n \cdot n!\sqrt{\pi} \qquad [m = n]$$

SM II 568

2.[11] $\int_{-\infty}^{\infty} e^{-2x^2} H_m(x)\, H_n(x)\, dx = (-1)^{\left\lfloor \frac{m}{2}\right\rfloor + \left\lfloor \frac{n}{2}\right\rfloor} 2^{\frac{m+n-1}{2}}\, \Gamma\!\left(\dfrac{m+n+1}{2}\right) \quad [m+n \text{ is even}]$ 

$$= 0 \qquad\qquad\qquad\quad [m+n \text{ is odd}]$$

ET II 289(10)a

3. $\int_{-\infty}^{\infty} e^{-x^2} H_m(ax)\, H_n(x)\, dx = 0 \qquad [m < n]$ 

ET II 290(20)a

4.[12] $\int_{-\infty}^{\infty} e^{-x^2} H_{2m+n}(ax)\, H_n(x)\, dx = \sqrt{\pi}\,2^{-m+\frac{1}{2}}\dfrac{(2m+n)!}{m!}\left(a^2-1\right)^m a^n$ 

ET II 291(21)a

5.[12] $\int_{-\infty}^{\infty} e^{-2\alpha^2 x^2} H_m(x)\, H_n(x)\, dx = 2^{\frac{m+n-1}{2}}\alpha^{-m-n-1}\left(1-2\alpha^2\right)^{\frac{m+n}{2}}\Gamma\!\left(\dfrac{m+n+1}{2}\right)$ 

$$\times\ {}_2F_1\!\left(-m,-n;\dfrac{1-m-n}{2};\dfrac{\alpha^2}{2\alpha^2-1}\right)$$

$$\left[\operatorname{Re}\alpha^2 > 0, \quad \alpha^2 \neq \tfrac{1}{2}, \quad m+n \text{ is even}\right] \quad \text{ET II 289(12)a}$$

6. $\int_{-\infty}^{\infty} e^{-(x-y)^2} H_n(x)\, dx = \pi^{1/2} y^n 2^n$ 

ET II 288(2)a, EH II 195(31)

7. $\int_{-\infty}^{\infty} e^{-(x-y)^2} H_m(x)\, H_n(x)\, dx = 2^n \pi^{1/2} m!\, y^{n-m}\, L_n^{n-m}\left(-2y^2\right)$ 

$$[m \leq n] \qquad \text{BU 148(15), ET II 289(13)a}$$

8. $\int_{-\infty}^{\infty} e^{-(x-y)^2} H_n(\alpha x)\, dx = \pi^{1/2}\left(1-\alpha^2\right)^{\frac{n}{2}} H_n\!\left[\dfrac{\alpha y}{\left(1-\alpha^2\right)^{1/2}}\right]$ 

ET II 290(17)a

9. $\int_{-\infty}^{\infty} e^{-(x-y)^2} H_m(\alpha x)\, H_n(\alpha x)\, dx$ 

$$= \pi^{1/2}\sum_{k=0}^{\min(m,n)} 2^k k!\binom{m}{k}\binom{n}{k}\left(1-\alpha^2\right)^{\frac{m+n}{2}-k} H_{m+n-2k}\!\left[\dfrac{\alpha y}{\left(1-\alpha^2\right)^{1/2}}\right]$$

ET II 291(26)a

10. $\int_{-\infty}^{\infty} e^{-\frac{(x-y)^2}{2u}} H_n(x)\, dx = (2\pi u)^{1/2}(1-2u)^{\frac{n}{2}} H_n\!\left[y(1-2u)^{-1/2}\right]$ 

$$\left[0 \leq u < \tfrac{1}{2}\right] \qquad \text{EH II 195(30)}$$

**7.375**

1.

$$\int_{-\infty}^{\infty} e^{-2x^2} H_k(x) H_m(x) H_n(x)\, dx = \pi^{-1} 2^{\frac{1}{2}(m+n+k-1)} \Gamma(s-k)\Gamma(s-m)\Gamma(s-n)$$

$$2s = k + m + n + 1 \qquad [k+m+n \text{ is even}] \qquad \text{ET II 290(14)a}$$

2.

$$\int_{-\infty}^{\infty} e^{-x^2} H_k(x) H_m(x) H_n(x)\, dx = \frac{2^{\frac{m+n+k}{2}} \pi^{1/2} k! m! n!}{(s-k)!(s-m)!(s-n)!},$$

$$2s = m+n+k \qquad [k+m+n \text{ is even}]$$

$$\text{ET II 290(15)a}$$

**7.376**

1.

$$\int_{-\infty}^{\infty} e^{ixy} e^{-\frac{x^2}{2}} H_n(x)\, dx = (2\pi)^{1/2} e^{-\frac{y^2}{2}} H_n(y) i^n \qquad\qquad\qquad \text{MO 165a}$$

2.

$$\int_{0}^{\infty} e^{-2\alpha x^2} x^{\nu} H_{2n}(x)\, dx = (-1)^n 2^{2n-\frac{3}{2}-\frac{1}{2}\nu} \frac{\Gamma\left(\frac{\nu+1}{2}\right)\Gamma\left(n+\frac{1}{2}\right)}{\sqrt{\pi}\alpha^{\frac{1}{2}(\nu+1)}} F\left(-n, \frac{\nu+1}{2}; \frac{1}{2}; \frac{1}{2\alpha}\right)$$

$$[\operatorname{Re}\alpha > 0, \quad \operatorname{Re}\nu > -1] \qquad \text{BU 150(18a)}$$

3.[12]

$$\int_{0}^{\infty} e^{-2\alpha x^2} x^{\nu} H_{2n+1}(x)\, dx = (-1)^n 2^{2n-\frac{1}{2}\nu} \frac{\Gamma\left(\frac{\nu+1}{2}\right)\Gamma\left(n+\frac{1}{2}\right)}{\sqrt{\pi}\alpha^{\frac{1}{2}\nu+1}} F\left(-n, \frac{\nu}{2}+1; \frac{3}{2}; \frac{1}{2\alpha}\right)$$

$$[\operatorname{Re}\alpha > 0, \quad \operatorname{Re}\nu > -2] \qquad \text{BU 150(18b)}$$

4.[*]

$$\int_{0}^{\infty} e^{-x^2} x^{n-1} H_n(x)\, dx = \Gamma(n) = (n-1)!$$

**7.377**[8]
$$\int_{-\infty}^{\infty} e^{-x^2} H_m(x+y) H_n(x+z)\, dx = 2^n \pi^{1/2} m! y^{n-m} L_m^{n-m}(-2yz)$$

$$[m \le n] \qquad \text{ET II 292(30)a}$$

**7.378**
$$\int_{0}^{\infty} x^{\alpha-1} e^{-\beta x} H_n(x)\, dx = 2^n \sum_{m=0}^{\lfloor\frac{n}{2}\rfloor} \frac{n!\,\Gamma(\alpha+n-2m)}{m!(n-2m)!} (-1)^m 2^{-2m} \beta^{2m-\alpha-n}$$

$$[\operatorname{Re}\alpha > 0, \text{ if } n \text{ is even}; \quad \operatorname{Re}\alpha > -1, \text{ if } n \text{ is odd}; \quad \operatorname{Re}\beta > 0] \qquad \text{ET I 172(11)a}$$

**7.379**

1.

$$\int_{-\infty}^{\infty} x e^{-x^2} H_{2m+1}(xy)\, dx = \pi^{1/2} \frac{(2m+1)!}{m!} y\left(y^2 - 1\right)^m \qquad\qquad \text{EH II 195(28)}$$

2.

$$\int_{-\infty}^{\infty} x^n e^{-x^2} H_n(xy)\, dx = \pi^{1/2} n! P_n(y) \qquad\qquad\qquad\qquad\qquad \text{EH II 195(29)}$$

**7.381**
$$\int_{-\infty}^{\infty} (x \pm ic)^{\nu} e^{-x^2} H_n(x)\, dx = 2^{n-1-\nu} \pi^{1/2} \frac{\Gamma\left(\frac{n-\nu}{2}\right)}{\Gamma(-\nu)} \exp\left[\pm\tfrac{1}{2}\pi(\nu+n)i\right]$$

$$[c > 0] \qquad \text{ET II 288(3)a}$$

**7.382**
$$\int_{0}^{\infty} x^{-1}\left(x^2 + a^2\right)^{-1} e^{-x^2} H_{2n+1}(x)\, dx = (-2)^n \pi^{1/2} a^{-2}\left[2^n n! - (2n+1)! e^{\frac{1}{2}a^2} D_{-2n-2}\left(a\sqrt{2}\right)\right]$$

$$\text{ET II 288(4)a}$$

**7.383**

1.    $\int_0^\infty e^{-xp} H_{2n+1}\left(\sqrt{x}\right)\,dx = (-1)^n 2^n (2n+1)!!\,\pi^{1/2}(p-1)^n p^{-n-\frac{3}{2}}$

$$[\operatorname{Re} p > 0] \qquad \text{EF 151(261)a, ET I 172(12)a}$$

2.    $\int_0^\infty e^{-(b-\beta x)} H_{2n+1}\left(\sqrt{(\alpha-\beta)x}\right)\,dx = (-1)^n \sqrt{\pi}\sqrt{\alpha-\beta}\frac{(2n+1)!}{n!}\frac{(b-\alpha)^n}{(b-\beta)^{n+\frac{3}{2}}}$

$$[\operatorname{Re}(b-\beta) > 0] \qquad \text{ET I 172(15)a}$$

3.    $\int_0^\infty \frac{1}{\sqrt{x}} e^{-(b-\beta)x} H_{2n}\left(\sqrt{(\alpha-\beta)x}\right)\,dx = (-1)^n \sqrt{\pi}\frac{(2n)!}{n!}\frac{(b-\alpha)^n}{(b-\beta)^{n+\frac{1}{2}}}$

$$[\operatorname{Re}(b-\beta) > 0] \qquad \text{ET I 172(16)a}$$

4.[12]    $\int_0^\infty x^{a-\frac{1}{2}n-1} e^{-bx} H_n\left(\sqrt{x}\right)\,dx = 2^{n/2}\,\Gamma(a)b^{-a}\,{}_2F_1\left(-\tfrac{1}{2}n, \tfrac{1}{2}-\tfrac{1}{2}n; 1-a; b\right)$

$$\left[\operatorname{Re} a > \tfrac{1}{2}n,\ \text{if } n \text{ is even}, \quad \operatorname{Re} a > \tfrac{1}{2}n - \tfrac{1}{2},\ \text{if } n \text{ is odd}, \quad \operatorname{Re} b > 0,\right.$$

$$\left.\text{If } a \text{ is even, only the first } 1 + \left\lfloor\frac{n}{2}\right\rfloor \text{ terms are kept in the series for } {}_2F_1\right]$$

$$\text{ET I 172(14)a}$$

5.    $\int_0^\infty x^{-1/2} e^{-px} H_{2n}\left(\sqrt{x}\right)\,dx = (-1)^n 2^n (2n-1)!!\,\pi^{1/2}(p-1)^n p^{-n-\frac{1}{2}}$

$$\text{MO 177a}$$

**7.384**    $\int_0^\infty \frac{1}{\sqrt{x}} e^{-bx}\left[H_n\left(\frac{\alpha+\sqrt{x}}{\lambda}\right) + H_n\left(\frac{a-\sqrt{x}}{\lambda}\right)\right]\,dx = \sqrt{\frac{2\pi}{b}}\left(1-\lambda^{-2}b^{-1}\right)^{\frac{n}{2}} H_n\left(\frac{\alpha}{\sqrt{\lambda^2-\frac{1}{b}}}\right)$

$$[\operatorname{Re} b > 0] \qquad \text{ET I 173(17)a}$$

**7.385**

1.    $\int_0^\infty \frac{e^{-bx}}{\sqrt{e^x-1}} H_{2n}\left[\sqrt{s(1-e^{-x})}\right]\,dx = (-1)^n 2^{2n}\sqrt{\pi}\frac{(2n)!\,\Gamma\left(b+\frac{1}{2}\right)}{\Gamma(n+b+1)} L_n^b(s)$

$$\left[\operatorname{Re} b > -\tfrac{1}{2}\right] \qquad \text{ET I 174(23)a}$$

2.    $\int_0^\infty e^{-bx} H_{2n+1}\left[\sqrt{s}\sqrt{1-e^{-x}}\right]\,dx = (-1)^n 2^{2n}\sqrt{\pi}s\frac{(2n+1)!\,\Gamma(b)}{\Gamma\left(n+b+\frac{3}{2}\right)} L_n^b(s)$

$$[\operatorname{Re} b > 0] \qquad \text{ET I 174(24)a}$$

**7.386**    $\int_0^\infty x^{-\frac{n+1}{2}} e^{-\frac{q^2}{4x}} H_n\left(\frac{q}{2\sqrt{x}}\right) e^{-px}\,dx = 2^n \pi^{1/2} p^{\frac{n-1}{2}} e^{-q\sqrt{p}}$

$$\text{EF 129(117)}$$

**7.387**

1.    $\int_0^\infty e^{-x^2}\sinh\left(\sqrt{2}\beta x\right) H_{2n+1}(x)\,dx = 2^{n-\frac{1}{2}}\pi^{1/2}\beta^{2n+1} e^{\frac{1}{2}\beta^2}$

$$\text{ET II 289(7)a}$$

2.    $\int_0^\infty e^{-x^2}\cosh\left(\sqrt{2}\beta x\right) H_{2n}(x)\,dx = 2^{n-1}\pi^{1/2}\beta^{2n} e^{\frac{1}{2}\beta^2}$

$$\text{ET II 289(8)a}$$

**7.388**

1.    $\int_0^\infty e^{-x^2} \sin\left(\sqrt{2}\beta x\right) H_{2n+1}(x)\,dx = (-1)^n 2^{n-\frac{1}{2}} \pi^{1/2} \beta^{2n+1} e^{-\frac{1}{2}\beta^2}$      ET II 288(5)a

2.    $\int_0^\infty e^{-x^2} \sin\left(\sqrt{2}\beta x\right) H_{2n+1}(ax)\,dx = (-1)^n 2^{-1}\pi^{1/2}\left(a^2-1\right)^{n+\frac{1}{2}} e^{-\frac{1}{2}\beta^2} H_{2n+1}\left(\dfrac{a\beta}{\sqrt{2}\left(a^2-1\right)^{1/2}}\right)$

                                                                 ET II 290(18)a

3.    $\int_0^\infty e^{-x^2} \cos\left(\sqrt{2}\beta x\right) H_{2n}(x)\,dx = (-1)^n 2^{n-1}\pi^{1/2}\beta^{2n} e^{-\frac{1}{2}\beta^2}$      ET II 289(6)a

4.    $\int_0^\infty e^{-x^2} \cos\left(\sqrt{2}\beta x\right) H_{2n}(ax)\,dx = 2^{-1}\pi^{1/2}\left(1-a^2\right)^n e^{-\frac{1}{2}\beta^2} H_{2n}\left[\dfrac{a\beta}{\sqrt{2}\left(a^2-1\right)^{1/2}}\right]$

                                                                 ET II 290(19)a

5.[12]    $\int_0^\infty e^{-y^2} [H_n(y)]^2 \cos\left(\sqrt{2}\beta y\right)\,dy = \pi^{1/2} 2^{n-1} n!\, L_n\left(\beta^2\right)$      EH II 195(33)

6.[11]    $\int_0^\infty e^{-x^2} \sin(bx) H_n(x) H_{n+2m+1}(x)\,dx = 2^{n-1}(-1)^m \sqrt{\pi}\, n!\, b^{2m+1} e^{-\frac{b^2}{4}} L_n^{2m+1}\left(\dfrac{b^2}{2}\right)$

                                             $[b > 0]$      ET I 39(11)a

7.    $\int_0^\infty e^{-x^2} \cos(bx) H_n(x) H_{n+2m}(x)\,dx = 2^{n-\frac{1}{2}}\sqrt{\dfrac{\pi}{2}}\, n!(-1)^m b^{2m} e^{-\frac{b^2}{4}} L_n^{2m}\left(\dfrac{b^2}{2}\right)$

                                               $[b > 0]$      ET I 39(11)a

**7.389**    $\int_0^\pi (\cos x)^n H_{2n}\left[a\left(1-\sec x\right)^{1/2}\right]\,dx = 2^{-n}(-1)^n \pi \dfrac{(2n)!}{(n!)^2} [H_n(a)]^2$      ET II 292(31)

## 7.39   Jacobi polynomials

**7.391**

1.    $\int_{-1}^1 (1-x)^\alpha (1+x)^\beta P_n^{(\alpha,\beta)}(x) P_m^{(\alpha,\beta)}(x)\,dx$

       $= 0$                          $[m \neq n, \quad \operatorname{Re}\alpha > -1, \quad \operatorname{Re}\beta > -1]$

       $= \dfrac{2^{\alpha+\beta+1}\,\Gamma(\alpha+n+1)\,\Gamma(\beta+n+1)}{n!\,(\alpha+\beta+1+2n)\,\Gamma(\alpha+\beta+n+1)}$    $[m = n, \quad \operatorname{Re}\alpha > -1, \quad \operatorname{Re}\beta > -1]$

                                                          ET II 285(5, 9)

2.    $\int_{-1}^1 (1-x)^\rho (1+x)^\sigma P_n^{(\alpha,\beta)}(x)\,dx = \dfrac{2^{\rho+\sigma+1}\,\Gamma(\rho+1)\,\Gamma(\sigma+1)\,\Gamma(n+1+\alpha)}{n!\,\Gamma(\rho+\sigma+2)\,\Gamma(1+\alpha)}$

                     $\times\, {}_3F_2\left(-n, \alpha+\beta+n+1, \rho+1; \alpha+1, \rho+\sigma+2; 1\right)$

                                 $[\operatorname{Re}\rho > -1, \quad \operatorname{Re}\sigma > -1]$      ET II 284(3)

3.[12]    $\int_{-1}^1 (1-x)^\alpha (1+x)^\sigma P_n^{(\alpha,\beta)}(x)\,dx = \dfrac{2^{\alpha+\sigma+1}\Gamma(\sigma+1)\,\Gamma(n+\alpha+1)\,\Gamma(\sigma-\beta+1)}{n!\,\Gamma(\sigma-\beta-n+1)\,\Gamma(\alpha+\sigma+n+2)}$

                                 $[\operatorname{Re}\alpha > -1, \quad \operatorname{Re}\sigma > -1]$      ET II 284(1)

4.    $\displaystyle\int_{-1}^{1}(1-x)^{\rho}(1+x)^{\beta}\,P_{n}^{(\alpha,\beta)}(x)\,\mathrm{d}x = \frac{2^{\beta+\rho+1}\,\Gamma(\rho+1)\,\Gamma(\beta+n+1)\,\Gamma(\alpha-\rho+n)}{n!\,\Gamma(\alpha-\rho)\,\Gamma(\beta+\rho+n+2)}$

$$[\operatorname{Re}\rho>-1,\quad \operatorname{Re}\beta>-1] \qquad \text{ET II 284(2)}$$

5.    $\displaystyle\int_{-1}^{1}(1-x)^{\alpha-1}(1+x)^{\beta}\left[P_{n}^{(\alpha,\beta)}(x)\right]^{2}\,\mathrm{d}x = \frac{2^{\alpha+\beta}\,\Gamma(\alpha+n+1)\,\Gamma(\beta+n+1)}{n!\,\alpha\,\Gamma(\alpha+\beta+n+1)}$

$$[\operatorname{Re}\alpha>0,\quad \operatorname{Re}\beta>-1] \qquad \text{ET II 285(6)}$$

6.    $\displaystyle\int_{-1}^{1}(1-x)^{2\alpha}(1+x)^{\beta}\left[P_{n}^{(\alpha,\beta)}(x)\right]^{2}\,\mathrm{d}x = \frac{2^{4\alpha+\beta+1}\,\Gamma\left(\alpha+\frac{1}{2}\right)\left[\Gamma(\alpha+n+1)\right]^{2}\,\Gamma(\beta+2n+1)}{\sqrt{\pi}\,(n!)^{2}\,\Gamma(\alpha+1)\,\Gamma(2\alpha+\beta+2n+2)}$

$$\left[\operatorname{Re}\alpha>-\tfrac{1}{2},\quad \operatorname{Re}\beta>-1\right] \qquad \text{ET II 285(7)}$$

7.    $\displaystyle\int_{-1}^{1}(1-x)^{\rho}(1+x)^{\beta}\,P_{n}^{(\alpha,\beta)}(x)\,P_{n}^{(\rho,\beta)}(x)\,\mathrm{d}x$

$$= \frac{2^{\rho+\beta+1}\,\Gamma(\rho+n+1)\,\Gamma(\beta+n+1)\,\Gamma(\alpha+\beta+2n+1)}{n!\,\Gamma(\beta+\rho+2n+2)\,\Gamma(\alpha+\beta+n+1)}$$

$$[\operatorname{Re}\rho>-1,\quad \operatorname{Re}\beta>-1] \qquad \text{ET II285(10)}$$

8.    $\displaystyle\int_{-1}^{1}(1-x)^{\rho-1}(1+x)^{\beta}\,P_{n}^{(\alpha,\beta)}(x)\,P_{n}^{(\rho,\beta)}(x)\,\mathrm{d}x = \frac{2^{\rho+\beta}\,\Gamma(\alpha+n+1)\,\Gamma(\beta+n+1)\,\Gamma(\rho)}{n!\,\Gamma(\alpha+1)\,\Gamma(\rho+\beta+n+1)}$

$$[\operatorname{Re}\beta>-1,\quad \operatorname{Re}\rho>0] \qquad \text{ET II 286(11)}$$

9.[7]    $\displaystyle\int_{-1}^{1}(1-x)^{\alpha}(1+x)^{\sigma}\,P_{n}^{(\alpha,\beta)}(x)\,P_{m}^{(\alpha,\sigma)}(x)\,\mathrm{d}x$

$$= \frac{2^{\alpha+\sigma+1}\,\Gamma(\alpha+n+1)\,\Gamma(\alpha+\beta+m+n+1)\,\Gamma(\sigma+m+1)\,\Gamma(\sigma-\beta+1)}{m!\,(n-m)!\,\Gamma(\alpha+\beta+n+1)\,\Gamma(\alpha+\sigma+m+n+2)\,\Gamma(\alpha-\beta+m-n+1)}$$

$$[\operatorname{Re}\alpha>-1,\quad \operatorname{Re}\sigma>-1] \qquad \text{ET II 286(12)}$$

10.[12]    $\displaystyle\int_{-1}^{1}(1-x)^{\rho}(1+x)^{\beta}\,P_{n}^{(\alpha,\beta)}(x)\,P_{m}^{(\rho,\beta)}(x)\,\mathrm{d}x$

$$= \frac{2^{\beta+\rho+1}\,\Gamma(\alpha+\beta+m+n+1)\,\Gamma(\beta+n+1)\,\Gamma(\rho+m+1)}{n!\,(n-m)!\,\Gamma(\alpha+\beta+n+1)\,\Gamma(\beta+\rho+m+n+2)}\frac{\Gamma(\rho-\alpha-m+n)}{\Gamma(\rho-\alpha)}$$

$$[\operatorname{Re}\beta>-1,\quad \operatorname{Re}\rho>-1] \qquad \text{ET II 287(16)}$$

11.    $\displaystyle\int_{0}^{x}(1-y)^{\alpha}(1+y)^{\beta}\,P_{n}^{(\alpha,\beta)}(y)\,\mathrm{d}y = \frac{1}{2n}\left[P_{n-1}^{(\alpha+1,\beta+1)}(0)-(1-x)^{\alpha+1}(1+x)^{\beta+1}\,P_{n-1}^{(\alpha+1,\beta+1)}(x)\right]$

$$\text{EH II 173(38)}$$

**7.392**

1.    $\displaystyle\int_{0}^{1}x^{\lambda-1}(1-x)^{\mu-1}\,P_{n}^{(\alpha,\beta)}(1-\gamma x)\,\mathrm{d}x$

$$= \frac{\Gamma(\alpha+n+1)\,\Gamma(\lambda)\,\Gamma(\mu)}{n!\,\Gamma(\alpha+1)\,\Gamma(\lambda+\mu)}\;{}_{3}F_{2}\left(-n,n+\alpha+\beta+1,\lambda;\alpha+1,\lambda+\mu;\frac{1}{2}\gamma\right)$$

$$[\operatorname{Re}\lambda>0,\quad \operatorname{Re}\mu>0] \qquad \text{ET II 192(46)a}$$

2. 
$$\int_0^1 x^{\lambda-1}(1-x)^{\mu-1} P_n^{(\alpha,\beta)}(\gamma x - 1)\,\mathrm{d}x$$
$$= (-1)^n \frac{\Gamma(\beta+n+1)\,\Gamma(\lambda)\,\Gamma(\mu)}{n!\,\Gamma(\beta+1)\,\Gamma(\lambda+\mu)}\;{}_3F_2\left(-n, n+\alpha+\beta+1, \lambda; \beta+1, \lambda+\mu; \frac{1}{2}\gamma\right) a$$

$$[\operatorname{Re}\lambda > 0, \quad \operatorname{Re}\mu > 0]$$   ET II 192(47)a

3. 
$$\int_0^1 x^{\alpha}(1-x)^{\mu-1} P_n^{(\alpha,\beta)}(1-\gamma x)\,\mathrm{d}x = \frac{\Gamma(\alpha+n+1)\,\Gamma(\mu)}{\Gamma(\alpha+\mu+n+1)}\,P_n^{(\alpha+\mu,\beta-\mu)}(1-\gamma)$$

$$[\operatorname{Re} a > -1, \quad \operatorname{Re}\mu > 0]$$   ET II 191(43)a

4. 
$$\int_0^1 x^{\beta}(1-x)^{\mu-1} P_n^{(\alpha,\beta)}(\gamma x - 1)\,\mathrm{d}x = \frac{\Gamma(\beta+n+1)\,\Gamma(\mu)}{\Gamma(\beta+\mu+n+1)}\,P_n^{(\alpha-\mu,\beta+\mu)}(\gamma-1)$$

$$[\operatorname{Re}\beta > -1, \quad \operatorname{Re}\mu > 0]$$   ET II 191(44)a

**7.393**

1. 
$$\int_0^1 (1-x^2)^{\nu}\sin bx\, P_{2n+1}^{(\nu,\nu)}(x)\,\mathrm{d}x = \frac{(-1)^n\sqrt{\pi}\,\Gamma(2n+\nu+2)\,J_{2n+\nu+\frac{3}{2}}(b)}{2^{\frac{1}{2}-\nu}(2n+1)!\,b^{\nu+\frac{1}{2}}}$$

$$[b > 0, \quad \operatorname{Re}\nu > -1]$$   ET I 94(5)

2. 
$$\int_0^1 (1-x^2)^{\nu}\cos bx\, P_{2n}^{(\nu,\nu)}(x)\,\mathrm{d}x = \frac{(-1)^n 2^{\nu-\frac{1}{2}}\sqrt{\pi}\,\Gamma(2n+\nu+1)\,J_{2n+\nu+\frac{1}{2}}(b)}{(2n)!\,b^{\nu+\frac{1}{2}}}$$

$$[b > 0, \quad \operatorname{Re}\nu > -1]$$   ET I 38(4)

## 7.41–7.42 Laguerre polynomials

**7.411**

1. 
$$\int_0^t L_n(x)\,\mathrm{d}x = L_n(t) - L_{n+1}(t)/(n+1)$$   MO 110

2. 
$$\int_0^t L_n^{\alpha}(x)\,\mathrm{d}x = L_n^{\alpha}(t) - L_{n+1}^{\alpha}(t) - \binom{n+\alpha}{n} + \binom{n+1+\alpha}{n+1}$$   EH II 189(16)a

3. 
$$\int_0^t L_{n-1}^{\alpha+1}(x)\,\mathrm{d}x = -L_n^{\alpha}(t) + \binom{n+\alpha}{n}$$   EH II 189(15)a

4. 
$$\int_0^t L_m(x)\,L_n(t-x)\,\mathrm{d}x = L_{m+n}(t) - L_{m+n+1}(t)$$   EH II 191(31)

5. 
$$\sum_{k=0}^{\infty}\left[\int_0^t \frac{L_k(x)}{k!}\,\mathrm{d}x\right]^2 = e^t - 1$$   $$[t \geq 0]$$   MO 110

**7.412**

1. 
$$\int_0^1 (1-x)^{\mu-1} x^{\alpha} L_n^{\alpha}(ax)\,\mathrm{d}x = \frac{\Gamma(\alpha+n+1)\,\Gamma(\mu)}{\Gamma(\alpha+\mu+n+1)}\,L_n^{\alpha+\mu}(a)$$

$$[\operatorname{Re}\alpha > -1, \quad \operatorname{Re}\mu > 0]$$
EH II 191(30)a, BU 129(14c)

2. $\int_0^1 (1-x)^{\mu-1} x^{\lambda-1} L_n^\alpha(\beta x)\,dx = \dfrac{\Gamma(\alpha+n+1)\,\Gamma(\lambda)\,\Gamma(\mu)}{n!\,\Gamma(\alpha+1)\,\Gamma(\lambda+\mu)}\; {}_2F_2\left(-n,\lambda;\alpha+1,\lambda+\mu:\beta\right)$

$$[\operatorname{Re}\lambda>0,\quad \operatorname{Re}\mu>0]\qquad \text{ET II 192(50)a}$$

**7.413**　　$\int_0^1 x^\alpha(1-x)^\beta L_m^\alpha(xy)\,L_n^\beta[(1-x)y]\,dx = \dfrac{(m+n)!\,\Gamma(\alpha+m+1)\,\Gamma(\beta+n+1)}{m!n!\,\Gamma(\alpha+\beta+m+n+2)}\,L_{m+n}^{\alpha+\beta+1}(y)$

$$[\operatorname{Re}\alpha>-1,\quad \operatorname{Re}\beta>-1]\qquad \text{ET II 293(7)}$$

**7.414**

1.[11]　　$\int_y^\infty e^{-x} L_n^\alpha(x)\,dx = e^{-y}\left[L_n^\alpha(y)-L_{n-1}^\alpha(y)\right]$

$$\text{EH II 191(29)}$$

2.　　$\int_0^\infty e^{-bx} L_n(\lambda x)\,L_n(\mu x)\,dx = \dfrac{(b-\lambda-\mu)^n}{b^{n+1}}\,P_n\left[\dfrac{b^2-(\lambda+\mu)b+2\lambda\mu}{b(b-\lambda-\mu)}\right]$

$$[\operatorname{Re}b>0]\qquad \text{ET I 175(34)}$$

3.[8]　　$\int_0^\infty e^{-x} x^\alpha L_n^\alpha(x)\,L_m^\alpha(x)\,dx = 0 \qquad\qquad [m\neq n,\quad \operatorname{Re}\alpha>-1]\qquad \text{BU 115(8), ET II 293(3)}$

$$= \dfrac{\Gamma(\alpha+n+1)}{n!}\qquad [m=n,\quad \operatorname{Re}\alpha>0]\qquad \text{BU 115(8), ET II 292(2)}$$

4.　　$\int_0^\infty e^{-bx} x^\alpha L_n^\alpha(\lambda x)\,L_m^\alpha(\mu x)\,dx = \dfrac{\Gamma(m+n+\alpha+1)}{m!n!}\,\dfrac{(b-\lambda)^n(b-\mu)^m}{b^{m+n+\alpha+1}}$

$$\times F\left[-m,-n;-m-n-\alpha;\dfrac{b(b-\lambda-\mu)}{(b-\lambda)(b-\mu)}\right]$$

$$[\operatorname{Re}\alpha>-1,\quad \operatorname{Re}b>0]\qquad \text{ET I 175(35)}$$

4(1)[9].　　$\int_0^\infty e^{-x} x^{\alpha+1/2} L_n^\alpha(x)\,L_m^\alpha(x)\,dx = \dfrac{\Gamma(\alpha+n+1)^2\,\Gamma(\alpha+m+1)\,\Gamma\left(\alpha+\tfrac{3}{2}\right)\Gamma\left(m-\tfrac{1}{2}\right)}{n!m!\,\Gamma(\alpha+1)\,\Gamma\left(-\tfrac{1}{2}\right)}$

$$\times\, {}_3F_2\left(-n,\alpha+\tfrac{3}{2},\tfrac{3}{2};\alpha+1,\tfrac{3}{2}-m;1\right)$$

5.　　$\int_0^\infty e^{-bx} L_n^a(x)\,dx = \displaystyle\sum_{m=0}^n \binom{a+m-1}{m}\dfrac{(b-1)^{n-m}}{b^{n-m+1}}\qquad [\operatorname{Re}b>0]\qquad \text{ET I 174(27)}$

6.　　$\int_0^\infty e^{-bx} L_n(x)\,dx = (b-1)^n b^{-n-1}\qquad\qquad [\operatorname{Re}b>0]\qquad \text{ET I 174(25)}$

7.　　$\int_0^\infty e^{-st} t^\beta L_n^\alpha(t)\,dt = \dfrac{\Gamma(\beta+1)\,\Gamma(\alpha+n+1)}{n!\,\Gamma(\alpha+1)}\,s^{-\beta-1}\,F\left(-n,\beta+1;\alpha+1;\dfrac{1}{s}\right)$

$$[\operatorname{Re}\beta>-1,\quad \operatorname{Re}s>0]$$
$$\text{BU 119(4b), EH II 191(133)}$$

8.　　$\int_0^\infty e^{-st} t^\alpha L_n^\alpha(t)\,dt = \dfrac{\Gamma(\alpha+n+1)(s-1)^n}{n!\,s^{\alpha+n+1}}\qquad [\operatorname{Re}\alpha>-1,\quad \operatorname{Re}s>0]$

$$\text{EH II 191(32), MO 176a}$$

9.　　$\int_0^\infty e^{-x} x^{\alpha+\beta} L_m^\alpha(x)\,L_n^\beta(x)\,dx = (-1)^{m+n}(\alpha+\beta)!\,\binom{\alpha+m}{n}\binom{\beta+n}{m}$

$$[\operatorname{Re}(\alpha+\beta)>-1]\qquad \text{ET II 293(4)}$$

**10.**[6] $\int_0^\infty e^{-bx} x^{2a} \left[ L_n^a(x) \right]^2 dx = \dfrac{2^{2a} \Gamma\left(a + \frac{1}{2}\right) \Gamma\left(n + \frac{1}{2}\right)}{\pi \, (n!)^2 \, b^{2a+1}}$

$$\times F\left(-n, a + \tfrac{1}{2}; \tfrac{1}{2} - n; \left(1 - \tfrac{2}{b}\right)^2\right) \Gamma(a + n + 1)$$

$$\left[\operatorname{Re} a > -\tfrac{1}{2}, \quad \operatorname{Re} b > 0\right] \qquad \text{ET I 174(30)}$$

**11.** $\int_0^\infty e^{-x} x^{\gamma-1} L_n^\mu(x) \, dx = \dfrac{\Gamma(\gamma) \Gamma(1 + \mu + n - \gamma)}{n! \, \Gamma(1 + \mu - \gamma)}$ $\qquad [\operatorname{Re}\gamma > 0]$ $\qquad\qquad$ BU 120(4b)

**12.** $\int_0^\infty e^{-x\left(s + \frac{a_1 + a_2}{2}\right)} x^{\mu+\beta} L_k^\mu(a_1 x) L_k^\mu(a_2 x) \, dx$

$$= \dfrac{\Gamma(1 + \mu + \beta)\Gamma(1 + \mu + k)}{(k!)^2 \, \Gamma(1 + \mu)} \left\{ \dfrac{d^k}{dh^k} \left[ \dfrac{F\left(\frac{1+\mu+\beta}{2}, 1 + \frac{\mu+\beta}{2}; 1 + \mu; \frac{A^2}{B^2}\right)}{(1-h)^{1+\mu} B^{1+\mu+\beta}} \right] \right\}_{h=0}$$

$$A^2 = \dfrac{4a_1 a_2 h}{(1-h)^2}; \qquad B = s + \dfrac{a_1 + a_2}{2} \dfrac{1+h}{1-h}$$

$$\left[\operatorname{Re}\left(s + \dfrac{a_1 + a_2}{2}\right) > 0, \quad a_1 > 0, \quad a_2 > 0, \quad \operatorname{Re}(\mu + \beta) > -1\right] \quad \text{BU 142(19)}$$

**13.**[12] $\int_0^\infty \exp\left[-x\left(s + \dfrac{a_1 + a_2}{2}\right)\right] x^\mu L_k^\mu(a_1 x) L_k^\mu(a_2 x) \, dx = \dfrac{\Gamma(1 + \mu + k)}{b_0^{1+\mu+k}} \dfrac{b_2^k}{k!} P_k^{(\mu,0)} \left(\dfrac{b_1^2}{b_0 b_2}\right)$

$$b_0 = s + \dfrac{a_1 + a_2}{2}, \qquad b_1^2 = b_0 b_2 + 2a_1 a_2, \qquad b_2 = s - \dfrac{a_1 + a_2}{2}$$

$$\left[\operatorname{Re}\mu > -1, \quad \operatorname{Re}\left(s + \dfrac{a_1 + a_2}{2}\right) > 0\right]$$

$$\text{BU 144(22)}$$

**7.415** $\int_0^1 (1-x)^{\mu-1} x^{\lambda-1} e^{-\beta x} L_n^\alpha(\beta x) \, dx = \dfrac{\Gamma(\alpha + n + 1)}{n! \, \Gamma(\alpha + 1)} \, \mathrm{B}(\lambda, \mu) \, {}_2F_2\left(\alpha + n + 1, \lambda; \alpha + 1, \lambda + \mu; -\beta\right)$

$$[\operatorname{Re}\lambda > 0, \quad \operatorname{Re}\mu > 0] \qquad \text{ET II 193(51)a}$$

**7.416** $\int_{-\infty}^\infty x^{m-n} \exp\left[-\dfrac{1}{2}(x-y)^2\right] L_n^{m-n}(x^2) \, dx = \dfrac{(2\pi)^{1/2}}{n!} i^{n-m} 2^{-\frac{n+m}{2}} H_n\left(\dfrac{iy}{\sqrt{2}}\right) H_m\left(\dfrac{iy}{\sqrt{2}}\right)$

$$\text{BU 149(15b), ET II 293(8)a}$$

**7.417**

**1.** $\int_0^\infty x^{\nu-2n-1} e^{-ax} \sin(bx) L_{2n}^{\nu-2n-1}(ax) \, dx = (-1)^n i \, \Gamma(\nu) \dfrac{b^{2n}\left[(a-ib)^{-\nu} - (a+ib)^{-\nu}\right]}{2(2n)!}$

$$[b > 0, \quad \operatorname{Re} a > 0, \quad \operatorname{Re}\nu > 2n]$$

$$\text{ET I 95(12)}$$

**2.** $\int_0^\infty x^{\nu-2n-2} e^{-ax} \sin(bx) L_{2n+1}^{\nu-2n-2}(ax) \, dx = (-1)^{n+1} \Gamma(\nu) \dfrac{b^{2n+1}\left[(a+ib)^{-\nu} + (a-ib)^{-\nu}\right]}{2(2n+1)!}$

$$[b > 0, \quad \operatorname{Re} a > 0, \quad \operatorname{Re}\nu > 2n+1]$$

$$\text{ET I 95(13)}$$

**3.**[12] $\int_0^\infty x^{\nu-2n} e^{-ax} \cos(bx) L_{2n-1}^{\nu-2n}(ax) \, dx = i(-1)^{n+1} \Gamma(\nu) \dfrac{b^{2n-1}\left[(a-ib)^{-\nu} - (a+ib)^{-\nu}\right]}{2(2n-1)!}$

$$[b > 0, \quad \operatorname{Re} a > 0, \quad \operatorname{Re}\nu > 2n-1]$$

$$\text{ET I 39(12)}$$

4.    $\int_0^\infty x^{\nu-2n-1} e^{-ax} \cos(bx) L_{2n}^{\nu-2n-1}(ax)\, \mathrm{d}x = (-1)^n \Gamma(\nu) \dfrac{b^{2n}\left[(a+ib)^{-\nu} + (a-ib)^{-\nu}\right]}{2(2n)!}$

$$[b>0, \quad \operatorname{Re}\nu > 2n, \quad \operatorname{Re}a > 0]$$

ET I 39(13)

**7.418**

1.    $\int_0^\infty e^{-\frac{1}{2}x^2} \sin(bx) L_n\left(x^2\right) \mathrm{d}x = (-1)^n \dfrac{i}{2} n! \dfrac{1}{\sqrt{2\pi}} \left\{ \left[D_{-n-1}(ib)\right]^2 - \left[D_{-n-1}(-ib)\right]^2 \right\}$

$$[b>0]$$

ET I 95(14)

2.    $\int_0^\infty e^{-\frac{1}{2}x^2} \cos(bx) L_n\left(x^2\right) \mathrm{d}x = \sqrt{\dfrac{\pi}{2}}\, (n!)^{-1} e^{-\frac{1}{2}b^2} 2^{-n} \left[H_n\left(\dfrac{b}{\sqrt{2}}\right)\right]^2$

$$[b>0]$$

ET I 39(14)

3.    $\int_0^\infty x^{2n+1} e^{-\frac{1}{2}x^2} \sin(bx) L_n^{n+\frac{1}{2}}\left(\dfrac{1}{2}x^2\right) \mathrm{d}x = \sqrt{\dfrac{\pi}{2}}\, b^{2n+1} e^{-\frac{1}{2}b^2} L_n^{n+\frac{1}{2}}\left(\dfrac{b^2}{2}\right)$

$$[b>0]$$

ET I 95(15)

4.    $\int_0^\infty x^{2n} e^{-\frac{1}{2}x^2} \cos(bx) L_n^{n-\frac{1}{2}}\left(\dfrac{1}{2}x^2\right) \mathrm{d}x = \sqrt{\dfrac{\pi}{2}}\, b^{2n} e^{-\frac{1}{2}b^2} L_n^{n+\frac{1}{2}}\left(\dfrac{1}{2}b^2\right)$

$$[b>0]$$

ET I 39(16)

5.    $\int_0^\infty x e^{-\frac{1}{2}x^2} L_n^\alpha\left(\dfrac{1}{2}x^2\right) L_n^{\frac{1}{2}-\alpha}\left(\dfrac{1}{2}x^2\right) \sin(xy)\, \mathrm{d}x = \left(\dfrac{\pi}{2}\right)^{1/2} y e^{-\frac{1}{2}y^2} L_n^\alpha\left(\dfrac{1}{2}y^2\right) L_n^{\frac{1}{2}-\alpha}\left(\dfrac{1}{2}y^2\right)$

ET II 294(11)

6.    $\int_0^\infty e^{-\frac{1}{2}x^2} L_n^\alpha\left(\dfrac{1}{2}x^2\right) L_n^{-\frac{1}{2}-\alpha}\left(\dfrac{1}{2}x^2\right) \cos(xy)\, \mathrm{d}x = \left(\dfrac{\pi}{2}\right)^{1/2} e^{-\frac{1}{2}y^2} L_n^\alpha\left(\dfrac{1}{2}y^2\right) L_n^{-\alpha-\frac{1}{2}}\left(\dfrac{1}{2}y^2\right)$

ET II 294(12)

**7.419**   $\int_0^\infty x^{n+2\nu-\frac{1}{2}} \exp[-(1+a)x] L_n^{2\nu}(ax) K_\nu(x)\, \mathrm{d}x$

$$= \dfrac{\pi^{1/2}\, \Gamma\left(n+\nu+\frac{1}{2}\right) \Gamma\left(n+3\nu+\frac{1}{2}\right)}{2^{n+2\nu+\frac{1}{2}} n!\, \Gamma(2\nu+1)} F\left(n+\nu+\dfrac{1}{2}, n+3\nu+\dfrac{1}{2}; 2\nu+1; -\dfrac{1}{2}a\right)$$

$$\left[\operatorname{Re}a > -2, \quad \operatorname{Re}(n+\nu) > -\tfrac{1}{2}, \quad \operatorname{Re}(n+3\nu) > -\tfrac{1}{2}\right] \quad \text{ET II 370(44)}$$

**7.421**

1.    $\int_0^\infty x e^{-\frac{1}{2}\alpha x^2} L_n\left(\dfrac{1}{2}\beta x^2\right) J_0(xy)\, \mathrm{d}x = \dfrac{(\alpha-\beta)^n}{\alpha^{n+1}} e^{-\frac{1}{2\alpha}y^2} L_n\left[\dfrac{\beta y^2}{2\alpha(\beta-\alpha)}\right]$

$$[y>0, \quad \operatorname{Re}\alpha > 0]$$

ET II 13(4)a

2.    $\int_0^\infty x e^{-x^2} L_n\left(x^2\right) J_0(xy)\, \mathrm{d}x = \dfrac{2^{-2n-1}}{n!} y^{2n} e^{-\frac{1}{4}y^2}$

ET II 13(5)

3.    $\int_0^\infty x^{2n+\nu+1} e^{-\frac{1}{2}x^2} L_n^{\nu+n}\left(\dfrac{1}{2}x^2\right) J_\nu(xy)\, \mathrm{d}x = y^{2n+\nu} e^{-\frac{1}{2}y^2} L_n^{\nu+n}\left(\dfrac{1}{2}y^2\right)$

$$[y>0, \quad \operatorname{Re}\nu > -1]$$

MO 183

4.
$$\int_0^\infty x^{\nu+1} e^{-\beta x^2} L_n^\nu \left(\alpha x^2\right) J_\nu(xy)\,dx = 2^{-\nu-1}\beta^{-\nu-n-1}(\beta-\alpha)^n y^\nu e^{-\frac{y^2}{4\beta}} L_n^\nu \left[\frac{\alpha y^2}{4\beta(\alpha-\beta)}\right]$$

ET II 43(5)

5.
$$\int_0^\infty e^{-\frac{1}{2q}x^2} x^{\nu+1} L_n^\nu \left[\frac{x^2}{2q(1-q)}\right] J_\nu(xy)\,dx = \frac{q^{n+\nu+1}}{(q-1)^n} e^{-\frac{qy^2}{2}} y^\nu L_n^\nu \left(\frac{y^2}{2}\right)$$

$[\nu > 0]$

MO 183

6.
$$\int_0^\infty x^{\nu+1} e^{-x^2} L_n^\nu \left(x^2\right) J_\nu(xy)\,dx = \frac{1}{2n!}\left(\frac{y}{2}\right)^{2n+\nu} e^{-\frac{1}{4}y^2}$$

**7.422**

1.
$$\int_0^\infty x^{\nu+1} e^{-\beta x^2} \left[L_n^{\frac{1}{2}\nu}\left(\alpha x^2\right)\right]^2 J_\nu(xy)\,dx$$

$$= \frac{y^\nu}{\pi n!}\Gamma\left(n+1+\tfrac{1}{2}\nu\right)(2\beta)^{-\nu-1}e^{-\frac{y^2}{4\beta}}$$

$$\times \sum_{l=0}^n \frac{(-1)^l \Gamma\left(n-l+\tfrac{1}{2}\right)\Gamma\left(l+\tfrac{1}{2}\right)}{\Gamma\left(l+1+\tfrac{1}{2}\nu\right)(n-l)!}\left(\frac{2\alpha-\beta}{\beta}\right)^{2l} L_{2l}^\nu \left[\frac{\alpha y^2}{2\beta(2\alpha-\beta)}\right]$$

$[y > 0, \quad \operatorname{Re}\beta > 0, \quad \operatorname{Re}\nu > -1]$    ET II 43(7)

2.[12]
$$\int_0^\infty x^{\nu+1} e^{-\alpha x^2} L_m^{\nu-\sigma}\left(\alpha x^2\right) L_n^\sigma \left(\alpha x^2\right) J_\nu(xy)\,dx$$

$$= (-1)^{m+n}(2\alpha)^{-\nu-1} y^\nu e^{-\frac{y^2}{4\alpha}} L_n^{\sigma-m+n}\left(\frac{y^2}{4\alpha}\right) L_m^{\nu-\sigma+m-n}\left(\frac{y^2}{4\alpha}\right)$$

$[y > 0, \quad \operatorname{Re}\alpha > 0, \quad \operatorname{Re}\nu > -1, \quad n \neq 0, \quad \sigma \neq 0]$    ET II 43(8)

**7.423**

1.
$$\int_0^\infty e^{-\frac{1}{2}x^2} L_n\left(\frac{1}{2}x^2\right) H_{2n+1}\left(\frac{x}{2\sqrt{2}}\right)\sin(xy)\,dx = \left(\frac{\pi}{2}\right)^{1/2} e^{-\frac{1}{2}y^2} L_n\left(\frac{1}{2}y^2\right) H_{2n+1}\left(\frac{y}{2\sqrt{2}}\right)$$

ET II 294(13)a

2.
$$\int_0^\infty e^{-\frac{1}{2}x^2} L_n\left(\frac{1}{2}x^2\right) H_{2n}\left(\frac{x}{2\sqrt{2}}\right)\cos(xy)\,dx = \left(\frac{\pi}{2}\right)^{1/2} e^{-\frac{1}{2}y^2} L_n\left(\frac{1}{2}y^2\right) H_{2n}\left(\frac{y}{2\sqrt{2}}\right)$$

ET II 294(14)a

# 7.5 Hypergeometric Functions

## 7.51 Combinations of hypergeometric functions and powers

**7.511**
$$\int_0^\infty F(a,b;c;-x)x^{-s-1}\,dx = \frac{\Gamma(a+s)\,\Gamma(b+s)\,\Gamma(c)\,\Gamma(-s)}{\Gamma(a)\,\Gamma(b)\,\Gamma(c+s)}$$

$[c \neq 0, -1, -2, \ldots, \quad \operatorname{Re}s < 0, \quad \operatorname{Re}(a+s) > 0, \quad \operatorname{Re}(b+s) > 0]$    EH I 79(4)

**7.512**

1.    $\displaystyle\int_0^1 x^{\alpha-\gamma}(1-x)^{\gamma-\beta-1}\,F(\alpha,\beta;\gamma;x)\,dx = \frac{\Gamma\left(1+\dfrac{\alpha}{2}\right)\Gamma(\gamma)\,\Gamma(\alpha-\gamma+1)\,\Gamma\left(\gamma-\dfrac{\alpha}{2}-\beta\right)}{\Gamma(1+\alpha)\,\Gamma\left(1+\dfrac{\alpha}{2}-\beta\right)\Gamma\left(\gamma-\dfrac{\alpha}{2}\right)}$

$$\left[\operatorname{Re}\alpha+1 > \operatorname{Re}\gamma > \operatorname{Re}\beta, \quad \operatorname{Re}\left(\gamma-\frac{\alpha}{2}-\beta\right) > 0\right] \quad \text{ET II 398(1)}$$

2.    $\displaystyle\int_0^1 x^{\rho-1}(1-x)^{\beta-\gamma-n}\,F(-n,\beta;\gamma;x)\,dx = \frac{\Gamma(\gamma)\,\Gamma(\rho)\,\Gamma(\beta-\gamma+1)\,\Gamma(\gamma-\rho+n)}{\Gamma(\gamma+n)\,\Gamma(\gamma-\rho)\,\Gamma(\beta-\gamma+\rho+1)}$

$$[n=0,1,2\ldots;\quad \operatorname{Re}\rho > 0, \quad \operatorname{Re}(\beta-\gamma) > n-1] \quad \text{ET II 398(2)}$$

3.    $\displaystyle\int_0^1 x^{\rho-1}(1-x)^{\beta-\rho-1}\,F(\alpha,\beta;\gamma;x)\,dx = \frac{\Gamma(\gamma)\,\Gamma(\rho)\,\Gamma(\beta-\rho)\,\Gamma(\gamma-\alpha-\rho)}{\Gamma(\beta)\,\Gamma(\gamma-\alpha)\,\Gamma(\gamma-\rho)}$

$$[\operatorname{Re}\rho > 0, \quad \operatorname{Re}(\beta-\rho) > 0, \quad \operatorname{Re}(\gamma-\alpha-\rho) > 0] \quad \text{ET II 399(3)}$$

4.    $\displaystyle\int_0^1 x^{\gamma-1}(1-x)^{\rho-1}\,F(\alpha,\beta;\gamma;x)\,dx = \frac{\Gamma(\gamma)\,\Gamma(\rho)\,\Gamma(\gamma+\rho-\alpha-\beta)}{\Gamma(\gamma+\rho-\alpha)\,\Gamma(\gamma+\rho-\beta)}$

$$[\operatorname{Re}\gamma > 0, \quad \operatorname{Re}\rho > 0, \quad \operatorname{Re}(\gamma+\rho-\alpha-\beta) > 0] \quad \text{ET II 399(4)}$$

5.    $\displaystyle\int_0^1 x^{\rho-1}(1-x)^{\sigma-1}\,F(\alpha,\beta;\gamma;x)\,dx = \frac{\Gamma(\rho)\,\Gamma(\sigma)}{\Gamma(\rho+\sigma)}\, {}_3F_2(\alpha,\beta,\rho;\gamma,\rho+\sigma;1)$

$$[\operatorname{Re}\rho > 0, \quad \operatorname{Re}\sigma > 0, \quad \operatorname{Re}(\gamma+\sigma-\alpha-\beta) > 0] \quad \text{ET II 399(5)}$$

6.[12]    $\displaystyle\int_0^1 x^{\lambda-1}(1-x)^{\beta-\lambda-1}\,F(\alpha,\beta;\lambda;zx)\,dx = B(\lambda,\beta-\lambda)\,F(\alpha,\lambda;\gamma;z)$

$$[\operatorname{Re}\beta > \operatorname{Re}\lambda, \quad |\arg(1-z)| < \pi] \quad \text{BU 9}$$

7.[11]    $\displaystyle\int_0^1 x^{\gamma-1}(1-x)^{\delta-\gamma-1}\,F(\alpha,\beta;\gamma;xz)\,F(\delta-\alpha,\delta-\beta;\delta-\gamma;(1-x)\zeta)\,dx$

$$= \frac{\Gamma(\gamma)\,\Gamma(\delta-\gamma)}{\Gamma(\delta)}(1-\zeta)^{\alpha+\beta-\delta}\,F(\alpha,\beta;\delta;z+\zeta-z\zeta)$$

$$[0 < \operatorname{Re}\gamma < \operatorname{Re}\delta, \quad |\arg(1-z)| < \pi, \quad |\arg(1-\zeta)| < \pi] \quad \text{ET II 400(11)}$$

8.[12]    $\displaystyle\int_0^1 x^{\gamma-1}(1-x)^{\epsilon-1}(1-xz)^{-\delta}\,F(\alpha,\beta;\gamma;xz)\,F\left[\delta,\beta-\gamma;\epsilon;\frac{(1-x)z}{1-xz}\right]dx$

$$= \frac{\Gamma(\gamma)\,\Gamma(\epsilon)}{\Gamma(\gamma+\epsilon)}\,F(\alpha+\delta,\beta;\gamma+\epsilon;z)$$

$$[\operatorname{Re}\gamma > 0, \quad \operatorname{Re}\epsilon > 0, \quad |\arg(1-z)| < \pi] \quad \text{ET II 400(12), Eh I 78(3)}$$

9.[12]    $\displaystyle\int_0^1 x^{\gamma-1}(1-x)^{\rho-1}(1-zx)^{-\sigma}\,F(\alpha,\beta;\gamma;x)\,dx$

$$= \frac{\Gamma(\gamma)\,\Gamma(\rho)\,\Gamma(\gamma+\rho-\alpha-\beta)}{\Gamma(\gamma+\rho-\alpha)\,\Gamma(\gamma+\rho-\beta)}(1-z)^{\sigma}$$

$$\times\, {}_3F_2\left(\rho,\sigma,\gamma+\rho-\alpha-\beta;\gamma+\rho-\alpha,\gamma+\rho-\beta;\frac{z}{z-1}\right)$$

$$[\operatorname{Re}\gamma > 0, \quad \operatorname{Re}\rho > 0, \quad \operatorname{Re}(\gamma+\rho-\alpha-\beta) > 0, \quad |\arg(1-z)| < \pi] \quad \text{ET II 399(6)}$$

10.    $\displaystyle\int_0^\infty x^{\gamma-1}(x+z)^{-\sigma}\,F(\alpha,\beta;\gamma;-x)\,dx = \frac{\Gamma(\gamma)\,\Gamma(\alpha-\gamma+\sigma)\,\Gamma(\beta-\gamma+\sigma)}{\Gamma(\sigma)\,\Gamma(\alpha+\beta-\gamma+\sigma)}$

$$\times F\left(\alpha-\gamma+\sigma, \beta-\gamma+\sigma; \alpha+\beta-\gamma+\sigma; 1-z\right)$$

$$[\operatorname{Re}\gamma>0, \quad \operatorname{Re}(\alpha-\gamma+\sigma)>0, \quad \operatorname{Re}(\beta-\gamma+\sigma)>0, \quad |\arg z|<\pi] \quad \textbf{ET II 400(10)}$$

11.    $\displaystyle\int_0^1 (1-x)^{\mu-1}x^{\nu-1}\,{}_pF_q\left(a_1,\ldots,a_p;\nu,b_2,\ldots,b_q;ax\right)\,dx$

$$= \frac{\Gamma(\mu)\,\Gamma(\nu)}{\Gamma(\mu+\nu)}\,{}_pF_q\left(a_1,\ldots,a_p;\mu+\nu,b_2,\ldots,b_q;a\right)$$

$$[\operatorname{Re}\mu>0, \quad \operatorname{Re}\nu>0, \quad p\le q+1;\text{ if }p=q+1,\text{ then }|a|<1] \quad \textbf{ET II 200(94)}$$

12.    $\displaystyle\int_0^1 (1-x)^{\mu-1}x^{\nu-1}\,{}_pF_q\left(a_1,\ldots,a_p;b_1,\ldots,b_q;ax\right)\,dx$

$$= \frac{\Gamma(\mu)\,\Gamma(\nu)}{\Gamma(\mu+\nu)}\,{}_{p+1}F_{q+1}\left(\nu,a_1,\ldots,a_p;\mu+\nu,b_1,\ldots,b_q;a\right)$$

$$[\operatorname{Re}\mu>0, \quad \operatorname{Re}\nu>0, \quad p\le q+1,\text{ if }p=q+1,\text{ then }|a|<1] \quad \textbf{ET II 200(95)}$$

13.*    $\displaystyle\int_0^1 x^{\gamma-1}(1-x)^{\sigma-1}F(\alpha,\beta;\gamma;zx)F(-n,n+\gamma+\sigma-1;\gamma;x)\,dx$

$$= \frac{\Gamma(\alpha+n)\Gamma(\beta+n)\Gamma^2(\gamma)\Gamma(\sigma+n)}{\Gamma(\alpha)\Gamma(\beta)\Gamma(\gamma+n)\Gamma(\gamma+\sigma+2n)}\times(-z)^n F(\alpha+n,\beta+n;\gamma+\sigma+2n;z)$$

$$[\operatorname{Re}\gamma>0, \quad \operatorname{Re}\sigma>0, \quad n=0,1,2,\ldots, \quad |\arg(1-z)|<\pi]$$

**7.513**    $\displaystyle\int_0^1 x^{s-1}\left(1-x^2\right)^\nu F\left(-n,a;b;x^2\right)\,dx = \frac{1}{2}\,B\left(\nu+1,\frac{s}{2}\right)\,{}_3F_2\left(-n,a,\frac{s}{2};b,\nu+1+\frac{s}{2};1\right)$

$$[\operatorname{Re}s>0, \quad \operatorname{Re}\nu>-1] \quad\quad \textbf{ET I 336(4)}$$

## 7.52 Combinations of hypergeometric functions and exponentials

**7.521**    $\displaystyle\int_0^\infty e^{-st}\,{}_pF_q\left(a_1,\ldots,a_p;b_1,\ldots,b_q;t\right)\,dt = \frac{1}{s}\,{}_{p+1}F_q\left(1,a_1,\ldots,a_p;b_1,\ldots,b_q;s^{-1}\right)$

$$[p\le q] \quad\quad \textbf{EH I 192}$$

**7.522**

1.[11]    $\displaystyle\int_0^\infty e^{-\lambda x}x^{\gamma-1}\,{}_2F_1(\alpha,\beta;\delta;-x)\,dx = \frac{\Gamma(\delta)\lambda^{-\gamma}}{\Gamma(\alpha)\,\Gamma(\beta)}\,E(\alpha;\beta:\gamma;\delta:\lambda)$

$$[\operatorname{Re}\lambda>0, \quad \operatorname{Re}\gamma>0] \quad\quad \textbf{EH I 205(10)}$$

2.[6]    $\displaystyle\int_0^\infty e^{-bx}x^{a-1}F\left(\frac{1}{2}+\nu,\frac{1}{2}-\nu;a;-\frac{x}{2}\right)\,dx = 2^a e^b \frac{1}{\sqrt{\pi}}\,\Gamma(a)(2b)^{\frac{1}{2}-a}K_\nu(b)$

$$[\operatorname{Re}a>0, \quad \operatorname{Re}b>0] \quad\quad \textbf{ET I 212(1)}$$

3.    $\int_0^\infty e^{-bx} x^{\gamma-1} F(2\alpha, 2\beta; \gamma; -\lambda x)\,\mathrm{d}x = \Gamma(\gamma) b^{-\gamma} \left(\dfrac{b}{\lambda}\right)^{\alpha+\beta-\frac{1}{2}} e^{\frac{b}{2\lambda}}\, W_{\frac{1}{2}-\alpha-\beta,\alpha-\beta}\left(\dfrac{b}{\lambda}\right)$

$[\operatorname{Re} b > 0, \quad \operatorname{Re}\gamma > 0, \quad |\arg\lambda| < \pi]$

BU 78(30), ET I 212(4)

4.[6]    $\int_0^\infty e^{-xt} t^{b-1} F(a, a-c+1; b; -t)\,\mathrm{d}t = x^{a-b}\,\Gamma(b)\Psi(a, c; x)$

$[\operatorname{Re} b > 0, \quad \operatorname{Re} x > 0]$    EH I 273(11)

5.    $\int_0^\infty e^{-x} x^{s-1}\, {}_pF_q(a_1, \ldots, a_p; b_1, \ldots, b_q; ax)\,\mathrm{d}x = \Gamma(s)\, {}_{p+1}F_q(s, a_1, \ldots, a_p; b_1, \ldots, b_q; a)$

$[p < q, \quad \operatorname{Re} s > 0]$    ET I 337(11)

6.    $\int_0^\infty x^{\beta-1} e^{-\mu x}\, {}_2F_2(-n, n+1; 1, \beta; x)\,\mathrm{d}x = \Gamma(\beta)\mu^{-\beta}\, P_n\left(1 - \dfrac{2}{\mu}\right)$

$[\operatorname{Re}\mu > 0, \quad \operatorname{Re}\beta > 0]$    ET I 218(6)

7.    $\int_0^\infty x^{\beta-1} e^{-\mu x}\, {}_2F_2\left(-n, n; \beta, \dfrac{1}{2}; x\right)\mathrm{d}x = \Gamma(\beta)\mu^{-\beta}\cos\left[2n\arcsin\left(\dfrac{1}{\sqrt{\mu}}\right)\right]$

$[\operatorname{Re}\mu > 0, \quad \operatorname{Re}\beta > 0]$    ET I 218(7)

8.    $\int_0^\infty x^{\rho_n-1} e^{-\mu x}\, {}_mF_n(a_1, \ldots, a_m; \rho_1, \ldots, \rho_n; \lambda x)\,\mathrm{d}x$

$= \Gamma(\rho_n)\,\mu^{-\rho_n}\, {}_mF_{n-1}\left(a_1, \ldots, a_m; \rho_1, \ldots, \rho_{n-1}; \dfrac{\lambda}{\mu}\right)$

$[m \le n; \quad \operatorname{Re}\rho_n > 0, \quad \operatorname{Re}\mu > 0,\ \text{if } m < n; \operatorname{Re}\mu > \operatorname{Re}\lambda,\ \text{if } m = n]$    ET I 219(16)a

9.    $\int_0^\infty x^{\sigma-1} e^{-\mu x}\, {}_mF_n(a_1, \ldots, a_m; \rho_1, \ldots, \rho_n; \lambda x)\,\mathrm{d}x$

$= \Gamma(\sigma)\mu^{-\sigma}\, {}_{m+1}F_n\left(a_1, \ldots, a_m, \sigma; \rho_1, \ldots, \rho_n; \dfrac{\lambda}{\mu}\right)$

$[m \le n, \quad \operatorname{Re}\sigma > 0, \quad \operatorname{Re}\mu > 0,\ \text{if } m < n; \operatorname{Re}\mu > \operatorname{Re}\lambda,\ \text{if } m = n]$    ET I 219(17)

**7.523**    $\int_1^\infty (x-1)^{\mu-1} x^{-\mu-\frac{1}{2}} e^{-\frac{1}{2}ax}\, W_{2\mu+\frac{1}{2},\lambda}(ax)\,\mathrm{d}x = \Gamma(\mu) e^{-\frac{1}{2}a}\, W_{\mu+\frac{1}{2},\lambda}(a)$

$[\operatorname{Re}\mu > 0, \quad \operatorname{Re} a > 0]$

**7.524**

1.    $\int_0^\infty e^{-\lambda x} F\left(\alpha, \beta; \dfrac{1}{2}; -x^2\right)\mathrm{d}x = \lambda^{\alpha+\beta-1} S_{1-\alpha-\beta,\alpha-\beta}(\lambda)$

$[\operatorname{Re}\lambda > 0]$    ET II 401(13)

2.    $\int_0^\infty e^{-st}\, {}_pF_q(a_1, \ldots, a_p; b_1, \ldots, b_q; t^2)\,\mathrm{d}x = s^{-1}\, {}_{p+2}F_q\left(a_1, \ldots, a_p, 1, \dfrac{1}{2}; b_1, \ldots, b_q; \dfrac{4}{s^2}\right)$

$[p < q]$    MO 176

3.    $\int_0^\infty e^{-st}\, {}_0F_q\left(\dfrac{1}{q}, \dfrac{2}{q}, \ldots, \dfrac{q-1}{q}, 1; \dfrac{t^q}{q^q}\right)\mathrm{d}t = s^{-1}\exp\left(s^{-q}\right)$    MO 176

**7.525**

1.[12] $\displaystyle\int_0^\infty x^{\sigma-1} e^{-\mu x}\; {}_mF_n\left(a_1,\ldots,a_m;\rho_1,\ldots,\rho_n;(\lambda x)^k\right)\,\mathrm{d}x$

$$= \Gamma(\sigma)\mu^{-\sigma}\; {}_{m+k}F_n\left(a_1,\ldots,a_m,\frac{\sigma}{k},\frac{\sigma+1}{k},\ldots,\frac{\sigma+k-1}{k};\rho_1,\ldots,\rho_n;\left(\frac{k\lambda}{\mu}\right)^k\right)$$

$$\left[m+k\leq n+1,\quad \operatorname{Re}\sigma>0;\quad \operatorname{Re}\mu>0,\text{ if } m+k\leq n;\right.$$

$$\left.\operatorname{Re}\left(\mu+k\lambda e^{2\pi i r/k}\right)>0;\quad r=0,1,\ldots,k-1\text{ for } m+k=n+1\right]$$

ET I 220(19)

2. $\displaystyle\int_0^\infty x e^{-\lambda x}\, F\left(\alpha,\beta;\tfrac{3}{2};-x^2\right)\,\mathrm{d}x = \lambda^{\alpha+\beta-2}\, S_{1-\alpha-\beta,\alpha-\beta}(\lambda)$

$$[\operatorname{Re}\lambda>0] \qquad\qquad \text{ET II 401(14)}$$

**7.526**

1. $\displaystyle\int_{\gamma-i\infty}^{\gamma+i\infty} e^{xt} x^{-b}\, F\left(a,b;a+b-c+1;1-\frac{1}{x}\right)\,\mathrm{d}x = 2\pi i\frac{\Gamma(a+b-c+1)}{\Gamma(b)\,\Gamma(b-c+1)}t^{b-1}\,\Psi(a;c;t)$

$$\left[\operatorname{Re}b>0,\quad \operatorname{Re}(b-c)>-1,\quad \gamma>\tfrac{1}{2}\right]$$

EH I 273(12)

2. $\displaystyle\int_0^\infty e^{-t}t^{\gamma-1}(x+t)^{-\alpha}(y+t)^{-\beta}\, F\left[\alpha,\beta;\gamma;\frac{t(x+y+t)}{(x+t)(y+t)}\right]\,\mathrm{d}t = \Gamma(\gamma)\Psi(a,c;x)\Psi(\beta,c;y),$

$$\gamma = a+\beta-c+1 \qquad [\operatorname{Re}\gamma>0,\quad xy\neq 0] \quad \text{EH I 287(21)}$$

3. $\displaystyle\int_0^\infty x^{\gamma-1}(x+y)^{-\alpha}(x+z)^{-\beta}e^{-x}\, F\left[\alpha,\beta;\gamma;\frac{x(x+y+z)}{(x+y)(x+z)}\right]\,\mathrm{d}x$

$$= \Gamma(\gamma)(zy)^{-\frac{1}{2}-\mu}e^{\frac{y+z}{2}}\, W_{\nu,\mu}(y)\, W_{\lambda,\mu}(z)$$

$$2\nu = 1-\alpha+\beta-\gamma;\quad 2\lambda = 1+\alpha-\beta-\gamma;\quad 2\mu = \alpha+\beta-\gamma$$

$$[\operatorname{Re}\gamma>0,\quad |\arg y|<\pi,\quad |\arg z|<\pi]$$

ET II 401(15)

**7.527**

1. $\displaystyle\int_0^\infty \left(1-e^{-x}\right)^{\lambda-1} e^{-\mu x}\, F\left(\alpha,\beta;\gamma;\delta e^{-x}\right)\,\mathrm{d}x = \mathrm{B}(\mu,\lambda)\, {}_3F_2(\alpha,\beta,\mu;\gamma,\mu+\lambda;\delta)$

$$[\operatorname{Re}\lambda>0,\quad \operatorname{Re}\mu>0,\quad |\arg(1-\delta)|<\pi] \quad \text{ET I 213(9)}$$

2. $\displaystyle\int_0^\infty \left(1-e^{-x}\right)^{\mu} e^{-\alpha x}\, F\left(-n,\mu+\beta+n;\beta;e^{-x}\right)\,\mathrm{d}x = \frac{\mathrm{B}(\alpha,\mu+n+1)\,\mathrm{B}(\alpha,\beta+n-\alpha)}{\mathrm{B}(\alpha,\beta-\alpha)}$

$$[\operatorname{Re}\alpha>0,\quad \operatorname{Re}\mu>-1] \qquad \text{ET I 213(10)}$$

3. $\displaystyle\int_0^\infty \left(1-e^{-x}\right)^{\gamma-1} e^{-\mu x}\, F\left(\alpha,\beta;\gamma;1-e^{-x}\right)\,\mathrm{d}x = \frac{\Gamma(\mu)\,\Gamma(\gamma-\alpha-\beta+\mu)\,\Gamma(\gamma)}{\Gamma(\gamma-\alpha+\mu)\,\Gamma(\gamma-\beta+\mu)}$

$$[\operatorname{Re}\mu>0,\quad \operatorname{Re}\mu>\operatorname{Re}(\alpha+\beta-\gamma),\quad \operatorname{Re}\gamma>0] \quad \text{ET I 213(11)}$$

4. $\displaystyle\int_0^\infty \left(1 - e^{-x}\right)^{\gamma-1} e^{-\mu x} F\left[\alpha, \beta; \gamma; \delta\left(1 - e^{-x}\right)\right]\, dx = \mathrm{B}(\mu, \gamma)\, F(\alpha, \beta; \mu + \gamma; \delta)$

$$\left[\operatorname{Re}\mu > 0, \quad \operatorname{Re}\gamma > 0, \quad |\arg(1 - \delta)| < \pi\right] \quad \text{ET I 213(12)}$$

## 7.53 Hypergeometric and trigonometric functions

**7.531**

1. $\displaystyle\int_0^\infty x \sin \mu x\, F\left(\alpha, \beta; \frac{3}{2}; -c^2 x^2\right)\, dx = 2^{-\alpha-\beta+1}\pi c^{-\alpha-\beta}\mu^{\alpha+\beta-2}\frac{K_{\alpha-\beta}\left(\frac{\mu}{c}\right)}{\Gamma(\alpha)\,\Gamma(\beta)}$

$$\left[\mu > 0, \quad \operatorname{Re}\alpha > \tfrac{1}{2}, \quad \operatorname{Re}\beta > \tfrac{1}{2}\right]$$
$$\text{ET I 115(6)}$$

2. $\displaystyle\int_0^\infty \cos \mu x\, F\left(\alpha, \beta; \frac{1}{2}; -c^2 x^2\right)\, dx = 2^{-\alpha-\beta+1}\pi c^{-\alpha-\beta}\mu^{\alpha+\beta-1}\frac{K_{\alpha-\beta}\left(\frac{\mu}{c}\right)}{\Gamma(\alpha)\,\Gamma(\beta)}$

$$\left[\mu > 0, \quad \operatorname{Re}\alpha > 0, \quad \operatorname{Re}\beta > 0, \quad c > 0\right]$$
$$\text{ET I 61(9)}$$

## 7.54 Combinations of hypergeometric and Bessel functions

**7.541** $\displaystyle\int_0^\infty x^{\alpha+\beta-2\nu-1}(x+1)^{-\nu}e^{xz}\, K_\nu[(x+1)z]\, F\left(\alpha, \beta; \alpha+\beta-2\nu; -x\right)\, dx$

$$= \pi^{-\frac{1}{2}}\cos(\nu\pi)\,\Gamma\left(\tfrac{1}{2} - \alpha + \nu\right)\Gamma\left(\tfrac{1}{2} - \beta + \nu\right)\Gamma(\gamma)(2z)^{-\frac{1}{2}-\frac{1}{2}\gamma}\, W_{\frac{1}{2}\gamma, \frac{1}{2}(\beta-\alpha)}(2z)$$

$\gamma = \alpha + \beta - 2\nu$    $\left[\operatorname{Re}(\alpha+\beta-2\nu) > 0, \quad \operatorname{Re}\left(\tfrac{1}{2} - \alpha + \nu\right) > 0, \quad \operatorname{Re}\left(\tfrac{1}{2} - \beta + \nu\right) > 0, \quad |\arg z| < \tfrac{3}{2}\pi\right]$
$$\text{ET II 401(16)}$$

**7.542**

1.[12] $\displaystyle\int_0^\infty x^{\sigma-1}\, {}_pF_{p-1}\left(a_1, \ldots, a_p; b_1, \ldots, b_{p-1}; -\lambda x^2\right) Y_\nu(xy)\, dx$

$$= \frac{\Gamma(b_1)\ldots\Gamma(b_{p-1})}{2\lambda^{\frac{1}{2}\sigma}\,\Gamma(a_1)\ldots\Gamma(a_p)}\, G_{p+2,p+3}^{p+2,1}\left(\frac{y^2}{4\lambda}\left|\begin{matrix} b_0^*, \ldots, b_{p-1}^*, l \\ h, k, a_1^*, \ldots, a_p^*, l \end{matrix}\right.\right)$$

$$a_j^* = a_j - \frac{\sigma}{2}, \quad j = 1, \ldots, p; \quad b_0^* = 1 - \frac{\sigma}{2}; \quad b_j^* = b_j - \frac{\sigma}{2},$$

$$j = 1, \ldots, p-1; h = \frac{\nu}{2}, \quad k = -\frac{\nu}{2}, \quad l = -\frac{1+\nu}{2}$$

$$\left[|\arg\lambda| < \pi, \quad \operatorname{Re}\sigma > |\operatorname{Re}\nu|, \quad \operatorname{Re}a_j > \tfrac{1}{2}\operatorname{Re}\sigma - \tfrac{3}{4}, \quad y > 0\right]$$
$$\text{ET II 118(53)}$$

2. $\displaystyle\int_0^\infty x^{\sigma-1}\, {}_pF_p\left(a_1, \ldots, a_p; b_1, \ldots, b_p; -\lambda x^2\right) Y_\nu(xy)\, dx$

$$= \frac{\Gamma(b_1)\ldots\Gamma(b_p)}{2\lambda^{\frac{1}{2}\sigma}\,\Gamma(a_1)\ldots\Gamma(a_p)}\, G_{p+2,p+3}^{p+2,1}\left(\frac{y^2}{4\lambda}\left|\begin{matrix} b_0^*, \ldots, b_p^*, l \\ h, k, a_1^*, \ldots, a_p^*, l \end{matrix}\right.\right)$$

$$b_0^* = 1 - \frac{\sigma}{2}; \quad a_j^* = a_j - \frac{\sigma}{2}, \quad b_j* = b_j - \frac{\sigma}{2}; \quad j = 1, \ldots, p; \quad h = \frac{\nu}{2}, \quad k = -\frac{\nu}{2}, \quad l = -\frac{1+\nu}{2}$$

$$\left[\operatorname{Re}\lambda > 0, \quad \operatorname{Re}\sigma > |\operatorname{Re}\nu|, \quad \operatorname{Re}a_j > \tfrac{1}{2}\operatorname{Re}\sigma - \tfrac{3}{4}, \quad y > 0\right]$$
$$\text{ET II 119(54)}$$

3. $\displaystyle\int_0^\infty x^{\sigma-1}\,{}_pF_q\left(a_1,\ldots,a_p;b_1,\ldots,b_q;-\lambda x^2\right)Y_\nu(xy)\,\mathrm{d}x$

$$= -\pi^{-1}2^{\sigma-1}y^{-\sigma}\cos\left[\frac{\pi}{2}(\sigma-\nu)\right]\Gamma\left(\frac{\sigma+\nu}{2}\right)\Gamma\left(\frac{\sigma-\nu}{2}\right)$$

$$\times\ {}_{p+2}F_q\left(a_1,\ldots,a_p,\frac{\sigma+\nu}{2},\frac{\sigma-\nu}{2};b_1,\ldots,b_q;-\frac{4\lambda}{y^2}\right)$$

$$[y>0,\quad p\le q-1,\quad \mathrm{Re}\,\sigma>|\mathrm{Re}\,\nu|]\quad \text{ET II 119(55)}$$

4. $\displaystyle\int_0^\infty x^{\sigma-1}\,{}_pF_q\left(a_1,\ldots,a_p;b_1,\ldots,b_q;-\lambda x^2\right)K_\nu(xy)\,\mathrm{d}x$

$$= 2^{\sigma-2}y^{-\sigma}\Gamma\left(\frac{\sigma+\nu}{2}\right)\Gamma\left(\frac{\sigma-\nu}{2}\right){}_{p+2}F_q\left(a_1,\ldots,a_p,\frac{\sigma+\nu}{2},\frac{\sigma-\nu}{2};b_1,\ldots,b_q;\frac{4\lambda}{y^2}\right)$$

$$[\mathrm{Re}\,y>0,\quad p\le q-1,\quad \mathrm{Re}\,\sigma>|\mathrm{Re}\,\nu|]\quad \text{ET II 153(88)}$$

5. $\displaystyle\int_0^\infty x^{2\rho}\,{}_pF_p\left(a_1,\ldots,a_p;b_1,\ldots,b_p;-\lambda x^2\right)J_\nu(xy)\,\mathrm{d}x$

$$= \frac{2^{2\rho}\,\Gamma(b_1)\ldots\Gamma(b_p)}{y^{2\rho+1}\,\Gamma(a_1)\ldots\Gamma(a_p)}\,G_{p+1,p+2}^{p+1,1}\left(\frac{y^2}{4\lambda}\left|\begin{array}{c}1,\quad b_1,\ldots,b_p\\ h,\quad a_1,\ldots,a_p,\quad k\end{array}\right.\right)$$

$$h=\tfrac12+\rho+\tfrac12\nu,\qquad k=\tfrac12+\rho-\tfrac12\nu$$

$$\left[y>0,\quad \mathrm{Re}\,\lambda>0,\quad -1-\mathrm{Re}\,\nu<2\,\mathrm{Re}\,\rho<\tfrac12+2\,\mathrm{Re}\,a_r,\quad r=1,\ldots,p\right]\quad \text{ET II 91(18)}$$

6. $\displaystyle\int_0^\infty x^{2\rho}\,{}_{m+1}F_m\left(a_1,\ldots,a_{m+1};b_1,\ldots,b_m;-\lambda^2 x^2\right)J_\nu(xy)\,\mathrm{d}x$

$$= \frac{2^{2\rho}\,\Gamma(b_1)\ldots\Gamma(b_m)\,y^{-2\rho-1}}{\Gamma(a_1)\ldots\Gamma(a_{m+1})}\,G_{m+1,m+3}^{m+2,1}\left(\frac{y^2}{4\lambda^2}\left|\begin{array}{c}1,\quad b_1,\ldots,b_m\\ h,\quad a_1,\ldots,a_{m+1},\quad k\end{array}\right.\right)$$

$$h=\tfrac12+\rho+\tfrac12\nu,\qquad k=\tfrac12+\rho-\tfrac12\nu,$$

$$\left[y>0,\quad \mathrm{Re}\,\lambda>0,\quad \mathrm{Re}(2\rho+\nu)>-1,\quad \mathrm{Re}\,(\rho-a_r)<\tfrac14;\quad r=1,\ldots,m+1\right]\quad \text{ET II 91(19)}$$

7. $\displaystyle\int_0^\infty x^\delta\,F\left(\alpha,\beta;\gamma;-\lambda^2 x^2\right)J_\nu(xy)\,\mathrm{d}x$

$$= \frac{2^\delta\,\Gamma(\gamma)}{\Gamma(\alpha)\,\Gamma(\beta)}y^{-\delta-1}\,G_{24}^{22}\left(\frac{y^2}{4\lambda^2}\left|\begin{array}{cccc}1-\alpha,&1-\beta\\ \dfrac{1+\delta+\nu}{2},&0,&1-\gamma,&\dfrac{1+\delta-\nu}{2}\end{array}\right.\right)$$

$$\left[y>0,\quad \mathrm{Re}\,\lambda>0,\quad -1-\mathrm{Re}\,\nu-2\min(\mathrm{Re}\,\alpha,\quad \mathrm{Re}\,\beta)<\mathrm{Re}\,\delta<-\tfrac12\right]\quad \text{ET II 82(9)}$$

8. $\displaystyle\int_0^\infty x^\delta\,F\left(\alpha,\beta;\gamma;-\lambda^2 x^2\right)J_\nu(xy)\,\mathrm{d}x = \frac{2^\delta y^{-\delta-1}\,\Gamma(\gamma)}{\Gamma(\alpha)\,\Gamma(\beta)}\,G_{24}^{31}\left(\frac{y^2}{4\lambda^2}\left|\begin{array}{cccc}1,&\gamma\\ \dfrac{1+\delta+\nu}{2},&\alpha,\beta,&\dfrac{1+\delta-\nu}{2}\end{array}\right.\right)$

$$\left[y>0,\quad \mathrm{Re}\,\lambda>0,\quad -\mathrm{Re}\,\nu-1<\mathrm{Re}\,\delta<2\max(\mathrm{Re}\,\alpha,\quad \mathrm{Re}\,\beta)-\tfrac12\right]\quad \text{ET II 81(6)}$$

9. $\displaystyle\int_0^\infty x^{\nu+1}\,F\left(\alpha,\beta;\gamma;-\lambda^2 x^2\right)J_\nu(xy)\,\mathrm{d}x = \frac{2^{\nu+1}\,\Gamma(\gamma)}{\Gamma(\alpha)\,\Gamma(\beta)}y^{-\nu-2}\,G_{13}^{30}\left(\frac{y^2}{4\lambda^2}\left|\begin{array}{c}\gamma\\ \nu+1,\quad \alpha,\quad \beta\end{array}\right.\right)$

$$\left[y>0,\quad \mathrm{Re}\,\lambda>0,\quad -1<\mathrm{Re}\,\nu<2\max(\mathrm{Re}\,\alpha,\quad \mathrm{Re}\,\beta)-\tfrac32\right]\quad \text{ET II 81(5)}$$

10. $\int_0^\infty x^{\nu+1} F\left(\alpha, \beta; \nu+1; -\lambda^2 x^2\right) J_\nu(xy)\, dx = \frac{2^{\nu-\alpha-\beta+2}\,\Gamma(\nu+1)}{\lambda^{\alpha+\beta}\,\Gamma(\alpha)\,\Gamma(\beta)} y^{\alpha+\beta-\nu-2}\, K_{\alpha-\beta}\left(\frac{y}{\lambda}\right)$

$\left[y > 0, \quad \operatorname{Re}\lambda > 0, \quad -1 < \operatorname{Re}\nu < 2\max\left(\operatorname{Re}\alpha, \ \operatorname{Re}\beta\right) - \frac{3}{2}\right]$ ET II 81(3)

11. $\int_0^\infty x^{\nu+1} F\left(\alpha, \beta; \nu+1; -\lambda^2 x^2\right) K_\nu(xy)\, dx = 2^{\nu+1}\lambda^{-\alpha-\beta} y^{\alpha+\beta-\nu-2}\,\Gamma(\nu+1)\, S_{1-\alpha-\beta,\alpha-\beta}\left(\frac{y}{\lambda}\right)$

$\left[\operatorname{Re} y > 0, \quad \operatorname{Re}\lambda > 0, \quad \operatorname{Re}\nu > -1\right]$
ET II 152(86)

12. $\int_0^\infty x^{\nu+1} F\left(\alpha, \beta; \frac{\beta+\nu}{2}+1; -\lambda^2 x^2\right) J_\nu(xy)\, dx = \frac{\Gamma\left(\frac{\beta+\nu+2}{2}\right) y^{\beta-1}\lambda^{-\nu-\beta-1}}{\pi^{\frac{1}{2}}\,\Gamma(\alpha)\,\Gamma(\beta) 2^{\beta-1}}\left[K_{\frac{1}{2}(\nu-\beta+1)}\left(\frac{y}{2\lambda}\right)\right]^2$

$\left[y > 0, \quad -1 < \operatorname{Re}\nu < \left(2\max\left(\operatorname{Re}\alpha, \operatorname{Re}\beta\right) - \frac{3}{2}\right)\right]$ ET II 81(4)

13. $\int_0^\infty x^{\sigma+\frac{1}{2}} F\left(\alpha, \beta; \gamma; -\lambda^2 x^2\right) Y_\nu(xy)\, dx = \frac{\lambda^{-\sigma-1} y^{-\frac{1}{2}}\,\Gamma(\gamma)}{\sqrt{2}\,\Gamma(\alpha)\,\Gamma(\beta)}\, G_{35}^{41}\left(\frac{y^2}{4\lambda^2}\,\bigg|\, \begin{matrix} 1-p, \gamma-p, l \\ h, \quad k, \alpha-p, \beta-p, l \end{matrix}\right)$

$h = \frac{1}{4}+\frac{1}{2}\nu, \quad k = \frac{1}{4}-\frac{1}{2}\nu, \quad l = -\frac{1}{4}-\frac{1}{2}\nu, \quad p = \frac{1}{2}+\frac{1}{2}\sigma$

$\left[y > 0, \quad \operatorname{Re}\lambda > 0, \quad \operatorname{Re}\sigma > |\operatorname{Re}\nu| - \frac{3}{2}, \quad \operatorname{Re}\sigma < 2\operatorname{Re}\alpha, \quad \operatorname{Re}\sigma < 2\operatorname{Re}\beta\right]$ ET II 118(52)

14. $\int_0^\infty x^{\nu+2} F\left(\frac{1}{2}, \frac{1}{2}-\nu; \frac{3}{2}; -\lambda^2 x^2\right) Y_\nu(xy)\, dx = \frac{2^\nu y^{-\nu-1}}{\pi^{\frac{1}{2}}\lambda^2\,\Gamma\left(\frac{1}{2}-\nu\right)}\, K_\nu\left(\frac{y}{2\lambda}\right) K_{\nu+1}\left(\frac{y}{2\lambda}\right)$

$\left[y > 0, \quad \operatorname{Re}\lambda > 0, \quad -\frac{3}{2} < \operatorname{Re}\nu < -\frac{1}{2}\right]$
ET II 117(49)

15.[12] $\int_0^\infty x^{\nu+2} F\left(1, 2\nu+\frac{3}{2}; \nu+2; -\lambda^2 x^2\right) Y_\nu(xy)\, dx = \pi^{-\frac{1}{2}} 2^{-\nu}\lambda^{-2\nu-3} y^\nu \frac{\Gamma(\nu+2)}{\Gamma\left(2\nu+\frac{3}{2}\right)}\left[K_\nu\left(\frac{y}{2\lambda}\right)\right]^2$

$\left[y > 0, \quad \operatorname{Re}\lambda > 0, \quad -\frac{1}{2} < \operatorname{Re}\nu < \frac{1}{2}\right]$
ET II 117(50)

16. $\int_0^\infty x^{\nu+2} F\left(1, \mu+\nu+\frac{3}{2}; \frac{3}{2}; -\lambda^2 x^2\right) Y_\nu(xy)\, dx = \frac{\pi^{\frac{1}{2}} 2^{-\mu-\nu-1}\lambda^{-\mu-2\nu-3} y^{\mu+\nu}}{\Gamma\left(\mu+\nu+\frac{3}{2}\right)}\, K_\mu\left(\frac{y}{\lambda}\right)$

$\left[y > 0, \quad \operatorname{Re}\lambda > 0, \quad -\frac{3}{2} < \operatorname{Re}\nu < \frac{1}{2}, \quad \operatorname{Re}(2\mu+\nu) > -\frac{3}{2}\right]$ ET II 118(51)

17. $\int_0^\infty x^{2\alpha+\nu} F\left(\alpha-\nu-\frac{1}{2}, \alpha; 2\alpha; -\lambda^2 x^2\right) J_\nu(xy)\, dx$

$= \frac{i\,\Gamma\left(\frac{1}{2}+\alpha\right)\Gamma\left(\frac{1}{2}+\alpha+\nu\right)}{\pi 2^{1-\nu-2\alpha}\lambda^{2\alpha-1} y^{\nu+2}}\, W_{\frac{1}{2}-\alpha, -\frac{1}{2}-\nu}\left(\frac{y}{\lambda}\right)\left[W_{\frac{1}{2}-\alpha, -\frac{1}{2}-\nu}\left(e^{-i\pi}\frac{y}{\lambda}\right) - W_{\frac{1}{2}-\alpha, -\frac{1}{2}-\nu}\left(e^{i\pi}\frac{y}{\lambda}\right)\right]$

$\left[y > 0, \quad \operatorname{Re}\lambda > 0, \quad \operatorname{Re}\nu < -\frac{1}{2}, \quad \operatorname{Re}(\alpha+\nu) > -\frac{1}{2}\right]$ ET II 80(1)

18. $\int_0^\infty x^{2\alpha-\nu} F\left(\nu+\alpha-\frac{1}{2}, \alpha; 2\alpha; -\lambda^2 x^2\right) J_\nu(xy)\, dx$

$= \frac{2^{2\alpha-\nu}\,\Gamma\left(\frac{1}{2}+\alpha\right) y^{\nu-2}}{\lambda^{2\alpha-1}\,\Gamma(2\nu)}\, M_{\alpha-\frac{1}{2}, \nu-\frac{1}{2}}\left(\frac{y}{\lambda}\right) W_{\frac{1}{2}-\alpha, \nu-\frac{1}{2}}\left(\frac{y}{\lambda}\right)$

ET II 80(2)

**7.543**

1.　$\displaystyle\int_0^\infty x^{-2\alpha-1} F\left(\frac{1}{2}+\alpha, 1+\alpha; 1+2\alpha; -\frac{4\lambda^2}{x^2}\right) J_\nu(xy)\,dx = \lambda^{-2\alpha} I_{\frac{1}{2}\nu+\alpha}(\lambda y) K_{\frac{1}{2}\nu-\alpha}(\lambda y)$

$$\left[y>0, \quad \operatorname{Re}\lambda>0, \quad \operatorname{Re}\nu>-1, \quad \operatorname{Re}\alpha>-\tfrac{1}{2}\right] \quad \text{ET II 81(7)}$$

2.　$\displaystyle\int_0^\infty x^{\nu+1-4\alpha} F\left(\alpha, \alpha+\frac{1}{2}; \nu+1; -\frac{\lambda^2}{x^2}\right) J_\nu(xy)\,dx$

$$= \frac{\Gamma(\nu)}{\Gamma(2\alpha)} 2^\nu \lambda^{1-2\alpha} y^{2\alpha-\nu-1} I_\nu\left(\frac{1}{2}\lambda y\right) K_{2\alpha-\nu-1}\left(\frac{1}{2}\lambda y\right)$$

$$\left[y>0, \quad \operatorname{Re}\lambda>0, \quad \operatorname{Re}\alpha-1<\operatorname{Re}\nu<4\operatorname{Re}\alpha-\tfrac{3}{2}\right] \quad \text{ET II 81(8)}$$

**7.544**　$\displaystyle\int_0^\infty x^{\nu+1}(1+x)^{-2\alpha} F\left[\alpha, \nu+\frac{1}{2}; 2\nu+1; \frac{4x}{(1+x)^2}\right] J_\nu(xy)\,dx$

$$= \frac{\Gamma(\nu+1)\,\Gamma(\nu-\alpha+1)}{\Gamma(\alpha)} 2^{2\nu-2\alpha+1} y^{2(\alpha-\nu-1)} J_\nu(y)$$

$$\left[y>0, \quad -1<\operatorname{Re}\nu<2\operatorname{Re}\alpha-\tfrac{3}{2}\right] \quad \text{ET II 82(10)}$$

# 7.6 Confluent Hypergeometric Functions

## 7.61 Combinations of confluent hypergeometric functions and powers

**7.611**

1.　$\displaystyle\int_0^\infty x^{-1} W_{k,\mu}(x)\,dx = \frac{\pi^{\frac{3}{2}} 2^k \sec(\mu\pi)}{\Gamma\left(\frac{3}{4}-\frac{1}{2}k+\frac{1}{2}\mu\right)\Gamma\left(\frac{3}{4}-\frac{1}{2}k-\frac{1}{2}\mu\right)}$

$$\left[|\operatorname{Re}\mu|<\tfrac{1}{2}\right] \qquad \text{ET II 406(22)}$$

2.　$\displaystyle\int_0^\infty x^{-1} M_{k,\mu}(x) W_{\lambda,\mu}(x)\,dx = \frac{\Gamma(2\mu+1)}{(k-\lambda)\Gamma\left(\frac{1}{2}+\mu-\lambda\right)}$

$$\left[\operatorname{Re}\mu>-\tfrac{1}{2}, \quad \operatorname{Re}(k-\lambda)>0\right]$$
$$\text{BU 116(11), ET II 409(39)}$$

3.[12]　$\displaystyle\int_0^\infty x^{-1} W_{k,\mu}(x) W_{\lambda,\mu}(x)\,dx$

$$= \frac{\pi}{(k-\lambda)\sin(2\mu\pi)}\left[\frac{1}{\Gamma\left(\frac{1}{2}-k+\mu\right)\Gamma\left(\frac{1}{2}-\lambda-\mu\right)} - \frac{1}{\Gamma\left(\frac{1}{2}-k-\mu\right)\Gamma\left(\frac{1}{2}-\lambda+\mu\right)}\right]$$
$$\left[|\operatorname{Re}\mu|<\tfrac{1}{2}\right] \qquad \text{BU 116(12), ET II 409(40)}$$

4.　$\displaystyle\int_0^\infty \left\{W_{\kappa,\mu}(z)\right\}^2 \frac{dz}{z} = \frac{\pi}{\sin 2\pi\mu} \frac{\psi\left(\frac{1}{2}+\mu-\kappa\right)-\psi\left(\frac{1}{2}-\mu-\kappa\right)}{\Gamma\left(\frac{1}{2}+\mu-\kappa\right)\Gamma\left(\frac{1}{2}-\mu-\kappa\right)}$

$$\left[|\operatorname{Re}\mu|<\tfrac{1}{2}\right] \qquad \text{BU 117(12a)}$$

5.　$\displaystyle\int_0^\infty \frac{1}{z}\left[W_{\kappa,0}(z)\right]^2 dz = \frac{\psi'\left(\frac{1}{2}-\kappa\right)}{\left[\Gamma\left(\frac{1}{2}-\kappa\right)\right]^2}$

$$\text{BU 117(12b)}$$

**6.** $\displaystyle\int_0^\infty x^{\rho-1}\, W_{k,\mu}(x)\, W_{-k,\mu}(x)\, \mathrm{d}x = \frac{\Gamma(\rho+1)\,\Gamma\left(\frac{1}{2}\rho+\frac{1}{2}+\mu\right)\Gamma\left(\frac{1}{2}\rho+\frac{1}{2}-\mu\right)}{2\,\Gamma\left(1+\frac{1}{2}\rho+k\right)\Gamma\left(1+\frac{1}{2}\rho-k\right)}$

$$[\operatorname{Re}\rho > 2|\operatorname{Re}\mu|-1] \qquad \text{ET II 409(41)}$$

**7.11** $\displaystyle\int_0^\infty x^{\rho-1}\, W_{k,\mu}(x)\, W_{\lambda,\nu}(x)\, \mathrm{d}x$

$$= \frac{\Gamma(1-\mu+\nu+\rho)\,\Gamma(1+\mu+\nu+\rho)\,\Gamma(-2\nu)}{\Gamma\left(\frac{1}{2}-\lambda-\nu\right)\Gamma\left(\frac{3}{2}-k+\nu+\rho\right)}$$

$$\times\; {}_3F_2\left(1-\mu+\nu+\rho, 1+\mu+\nu+\rho, \tfrac{1}{2}-\lambda+\nu; 1+2\nu, \tfrac{3}{2}-k+\nu+\rho; 1\right)$$

$$+\frac{\Gamma(1+\mu-\nu+\rho)\,\Gamma(1-\mu-\nu+\rho)\,\Gamma(2\nu)}{\Gamma\left(\frac{1}{2}-\lambda+\nu\right)\Gamma\left(\frac{3}{2}-k-\nu+\rho\right)}$$

$$\times\; {}_3F_2\left(1+\mu-\nu+\rho, 1-\mu-\nu+\rho, \tfrac{1}{2}-\lambda-\nu; 1-2\nu, \tfrac{3}{2}-k-\nu+\rho; 1\right)$$

$$[|\operatorname{Re}\mu|+|\operatorname{Re}\nu| < \operatorname{Re}\rho+1] \qquad \text{ET II 410(42)}$$

**7.612**

**1.** $\displaystyle\int_0^\infty t^{b-1}\, {}_1F_1(a;c;-t)\, \mathrm{d}t = \frac{\Gamma(b)\,\Gamma(c)\,\Gamma(a-b)}{\Gamma(a)\,\Gamma(c-b)} \qquad [0<\operatorname{Re}b<\operatorname{Re}a] \qquad \text{EH I 285(10)}$

**2.** $\displaystyle\int_0^\infty t^{b-1}\, \Psi(a,c;t)\, \mathrm{d}t = \frac{\Gamma(b)\,\Gamma(a-b)\,\Gamma(b-c+1)}{\Gamma(a)\,\Gamma(a-c+1)} \qquad [0<\operatorname{Re}b<\operatorname{Re}a \quad \operatorname{Re}c<\operatorname{Re}b+1]$

$$\text{EH I 285(11)}$$

**7.613**

**1.** $\displaystyle\int_0^t x^{\gamma-1}(t-x)^{c-\gamma-1}\, {}_1F_1(a;\gamma;x)\, \mathrm{d}x = t^{c-1}\frac{\Gamma(\gamma)\,\Gamma(c-\gamma)}{\Gamma(c)}\, {}_1F_1(a;c;t)$

$$[\operatorname{Re}c>\operatorname{Re}\gamma>0]$$
$$\text{BU 9(16)a, EH I 271(16)}$$

**2.** $\displaystyle\int_0^t x^{\beta-1}(t-x)^{\gamma-1}\, {}_1F_1(t;\beta;x)\, \mathrm{d}x = \frac{\Gamma(\beta)\,\Gamma(\gamma)}{\Gamma(\beta+\gamma)} t^{\beta+\gamma-1}\, {}_1F_1(t;\beta+\gamma;t)$

$$[\operatorname{Re}\beta>0, \quad \operatorname{Re}\gamma>0] \qquad \text{ET II 401(1)}$$

**3.** $\displaystyle\int_0^1 x^{\lambda-1}(1-x)^{2\mu-\lambda}\, {}_1F_1\left(\frac{1}{2}+\mu-\nu;\lambda;xz\right)\, \mathrm{d}x = \mathrm{B}(\lambda,1+2\mu-\lambda)e^{\frac{1}{2}z}z^{-\frac{1}{2}-\mu}M_{\nu,\mu}(z)$

$$[\operatorname{Re}\lambda>0, \quad \operatorname{Re}(2\mu-\lambda)>-1]$$
$$\text{BU 14(14)}$$

**4.** $\displaystyle\int_0^t x^{\beta-1}(t-x)^{\delta-1}\, {}_1F_1(t;\beta;x)\, {}_1F_1(\gamma;\delta;t-x)\, \mathrm{d}x = \frac{\Gamma(\beta)\,\Gamma(\delta)}{\Gamma(\beta+\delta)} t^{\beta+\delta-1}\, {}_1F_1(t+\gamma;\beta+\delta;t)$

$$[\operatorname{Re}\beta>0, \quad \operatorname{Re}\delta>0]$$
$$\text{ET II 402(2), EH I 271(15)}$$

**5.** $\displaystyle\int_0^t x^{\mu-\frac{1}{2}}(t-x)^{\nu-\frac{1}{2}}\, M_{k,\mu}(x)\, M_{\lambda,\nu}(t-x)\, \mathrm{d}x = \frac{\Gamma(2\mu+1)\,\Gamma(2\nu+1)}{\Gamma(2\mu+2\nu+2)} t^{\mu+\nu} M_{k+\lambda,\mu+\nu+\frac{1}{2}}(t)$

$$\left[\operatorname{Re}\mu>-\tfrac{1}{2}, \quad \operatorname{Re}\nu>-\tfrac{1}{2}\right]$$
$$\text{BU 128(14), ET II 402(7)}$$

6.
$$\int_0^1 x^{\beta-1}(1-x)^{\sigma-\beta-1}\,{}_1F_1(\alpha;\beta;\lambda x)\,{}_1F_1[\sigma-\alpha;\sigma-\beta;\mu(1-x)]\,dx$$

$$= \frac{\Gamma(\beta)\,\Gamma(\sigma-\beta)}{\Gamma(\sigma)}e^\lambda\,{}_1F_1\left(\alpha;\sigma;\mu-\lambda\right)$$

$$[0 < \operatorname{Re}\beta < \operatorname{Re}\sigma]$$               ET II 402(3)

## 7.62–7.63 Combinations of confluent hypergeometric functions and exponentials

**7.621**

1.
$$\int_0^\infty e^{-st}t^\alpha\,M_{\mu,\nu}(t)\,dt = \frac{\Gamma\left(\alpha+\nu+\frac{3}{2}\right)}{\left(\frac{1}{2}+s\right)^{\alpha+\nu+\frac{3}{2}}}\,F\left(\alpha+\nu+\frac{3}{2},-\mu+\nu+\frac{1}{2};2\nu+1;\frac{2}{2s+1}\right)$$

$$\left[\operatorname{Re}\left(\alpha+\mu+\tfrac{3}{2}\right) > 0, \quad \operatorname{Re}s > \tfrac{1}{2}\right]$$
BU 118(1), MO 176a, EH I 270(12)a

2.
$$\int_0^\infty e^{-st}t^{\mu-\frac{1}{2}}\,M_{\lambda,\mu}(qt)\,dt = q^{\mu+\frac{1}{2}}\,\Gamma(2\mu+1)\left(s-\tfrac{1}{2}q\right)^{\lambda-\mu-\frac{1}{2}}\left(s+\tfrac{1}{2}q\right)^{-\lambda-\mu-\frac{1}{2}}$$

$$\left[\operatorname{Re}\mu > -\frac{1}{2}, \quad \operatorname{Re}s > \frac{|\operatorname{Re}q|}{2}\right]$$
BU 119(4c), MO 176a, EH I 271(13)a

3.
$$\int_0^\infty e^{-st}t^\alpha\,W_{\lambda,\mu}(qt)\,dt = \frac{\Gamma\left(\alpha+\mu+\frac{3}{2}\right)\Gamma\left(\alpha-\mu+\frac{3}{2}\right)q^{\mu+\frac{1}{2}}}{\Gamma(\alpha-\lambda+2)}\left(s+\tfrac{1}{2}q\right)^{-\alpha-\mu-\frac{3}{2}}$$

$$\times F\left(\alpha+\mu+\frac{3}{2},\mu-\lambda+\frac{1}{2};\alpha-\lambda+2;\frac{2s-q}{2s+q}\right)$$

$$\left[\operatorname{Re}\left(\alpha\pm\mu+\frac{3}{2}\right) > 0, \quad \operatorname{Re}s > -\frac{q}{2}, \quad q > 0\right]$$   EH I 271(14)a, BU 121(6), MO 176

4.
$$\int_0^\infty e^{-st}t^{b-1}\,{}_1F_1(a;c;kt)\,dt = \Gamma(b)s^{-b}\,F\left(a,b;c;ks^{-1}\right)$$               $$[|s| > |k|]$$

$$= \Gamma(b)(s-k)^{-b}\,F\left(c-a,b;c;\frac{k}{k-s}\right)$$   $$[|s-k > |k||]$$

$$[\operatorname{Re}b > 0, \quad \operatorname{Re}s > \max(0,\operatorname{Re}k)]$$   EH I 269(5)

5.
$$\int_0^\infty t^{c-1}\,{}_1F_1(a;c;t)e^{-st}\,dt = \Gamma(c)s^{-c}\left(1-s^{-1}\right)^{-a}$$   $$[\operatorname{Re}c > 0, \quad \operatorname{Re}s > 1]$$   EH I 270(6)

6.
$$\int_0^\infty t^{b-1}\Psi(a,c;t)e^{-st}\,dt = \frac{\Gamma(b)\,\Gamma(b-c+1)}{\Gamma(a+b-c+1)}\,F\left(b,b-c+1;a+b-c+1;1-s\right)$$

$$[\operatorname{Re}b > 0, \quad \operatorname{Re}c < \operatorname{Re}b+1, \quad |1-s| < 1]$$

$$= \frac{\Gamma(b)\,\Gamma(b-c+1)}{\Gamma(a+b-c+1)}\,s^{-b}\,F\left(a,b;a+b-c+1;1-s^{-1}\right)$$

$$\left[\operatorname{Re}s > \tfrac{1}{2}\right]$$
EH I 270(7)

7.
$$\int_0^\infty e^{-\frac{b}{2}x}x^{\nu-1}\,M_{\kappa,\mu}(bx)\,dx = \frac{\Gamma(1+2\mu)\,\Gamma(\kappa-\nu)\,\Gamma\left(\frac{1}{2}+\mu+\nu\right)}{\Gamma\left(\frac{1}{2}+\mu+\kappa\right)\Gamma\left(\frac{1}{2}+\mu-\nu\right)}\,b^\nu$$

$$\left[\operatorname{Re}\left(\nu+\tfrac{1}{2}+\mu\right) > 0, \quad \operatorname{Re}(\kappa-\nu) > 0\right]$$
BU 119(3)a, ET I 215(11)a

8. $$\int_0^\infty e^{-sx} M_{\kappa,\mu}(x) \frac{\mathrm{d}x}{x} = \frac{2\,\Gamma(1+2\mu)e^{-i\pi\kappa}}{\Gamma\left(\frac{1}{2}+\mu+\kappa\right)} \left(\frac{s-\frac{1}{2}}{s+\frac{1}{2}}\right)^{\frac{\kappa}{2}} Q^\kappa_{\mu-\frac{1}{2}}(2s)$$

$$\left[\mathrm{Re}\left(\tfrac{1}{2}+\mu\right)>0, \quad \mathrm{Re}\,s>\tfrac{1}{2}\right]$$

BU 119(4a)

9.[12] $$\int_0^\infty e^{-sx} W_{\kappa,\mu}(x) \frac{\mathrm{d}x}{x} = \frac{\pi}{\cos(\pi\mu)} \left(\frac{s-\frac{1}{2}}{s+\frac{1}{2}}\right)^{\frac{\kappa}{2}} P^\kappa_{\mu-\frac{1}{2}}(2s)$$

$$\left[\mathrm{Re}\left(\tfrac{1}{2}\pm\mu\right)>0, \quad \mathrm{Re}\,s>-\tfrac{1}{2}\right]$$

BU 121(7)

10.[12] $$\int_0^\infty x^{k+2\mu-1} e^{-\frac{3}{2}x} W_{k,\mu}(x)\,\mathrm{d}x = \frac{\Gamma\left(k+\mu+\frac{1}{2}\right)\Gamma\left[\frac{1}{4}(2k+6\mu+5)\right]}{\Gamma\left(k+3\mu+\frac{1}{2}\right)\Gamma\left[\frac{1}{4}(2\mu-2k+3)\right]}$$

$$\left[\mathrm{Re}(k+\mu)>-\tfrac{1}{2}, \quad \mathrm{Re}(k+3\mu)>-\tfrac{1}{2}\right]$$

BU 122(8a), ET II 406(23)

11. $$\int_0^\infty e^{-\frac{1}{2}x} x^{\nu-1} W_{\kappa,\mu}(x)\,\mathrm{d}x = \frac{\Gamma\left(\nu+\frac{1}{2}-\mu\right)\Gamma\left(\nu+\frac{1}{2}+\mu\right)}{\Gamma(\nu-\kappa+1)}$$

$$\left[\mathrm{Re}\left(\nu+\tfrac{1}{2}\pm\mu\right)>0\right] \qquad \text{BU 122(8b)}$$

12. $$\int_0^\infty e^{\frac{1}{2}x} x^{\nu-1} W_{\kappa,\mu}(x)\,\mathrm{d}x = \Gamma(-\kappa-\mu)\frac{\Gamma\left(\frac{1}{2}+\mu+\nu\right)\Gamma\left(\frac{1}{2}-\mu+\nu\right)}{\Gamma\left(\frac{1}{2}-\mu-\kappa\right)\Gamma\left(\frac{1}{2}+\mu-\kappa\right)}$$

$$\left[\mathrm{Re}\left(\nu+\tfrac{1}{2}\pm\mu\right)>0, \quad \mathrm{Re}(\kappa+\nu)<0\right]$$

BU 122(8c)a

**7.622**

1. $$\int_0^\infty e^{-st} t^{c-1}\,{}_1F_1(a;c;t)\,{}_1F_1(\alpha;c;\lambda t)\,\mathrm{d}t$$

$$= \Gamma(c)(s-1)^{-a}(s-\lambda)^{-\alpha} s^{a+\alpha-c} F\left[a,\alpha;c;\lambda(s-1)^{-1}(s-\lambda)^{-1}\right]$$

$$\left[\mathrm{Re}\,c>0, \quad \mathrm{Re}\,s>\mathrm{Re}\,\lambda+1\right] \qquad \text{EH I 287(22)}$$

2. $$\int_0^\infty e^{-t} t^\rho\,{}_1F_1(a;c;t)\Psi(a';c';\lambda t)\,\mathrm{d}t$$

$$= C\frac{\Gamma(c)\,\Gamma(\beta)}{\Gamma(\gamma)}\lambda^\sigma\,F\left(c-a,\beta;\gamma;1-\lambda^{-1}\right),$$

$$\rho=c-1, \quad \sigma=-c, \quad \beta=c-c'+1, \quad \gamma=c-a+a'-c'+1, \quad C=\frac{\Gamma(a'-a)}{\Gamma(a')}, \text{ or}$$

$$\rho=c+c'-2, \quad \sigma=1-c-c', \quad \beta=c+c'-1, \quad \gamma=a'-a+c, \quad C=\frac{\Gamma(a'-a-c'+1)}{\Gamma(a'-c'+1)}$$

EH I 287(24)

$3.^{12}$ $\quad \displaystyle\int_0^\infty x^{\nu-1} e^{-bx}\, M_{\lambda_1,\mu_1-\frac{1}{2}}(a_1 x) \ldots M_{\lambda_n,\mu_n-\frac{1}{2}}(a_n x)\, \mathrm{d}x$

$$= a_1^{\mu_1} \ldots a_n^{\mu_n} (b+A)^{-\nu-M}\, \Gamma(\nu+M)$$

$$\times F_A\left(\nu+M : \mu_1 - \lambda_1, \ldots, \mu_n - \lambda_n; 2\mu_1, \ldots, 2\mu_n : \frac{a_1}{b+A}, \ldots, \frac{a_n}{b+A}\right),$$

$$M = \mu_1 + \cdots + \mu_n, \qquad A = \tfrac{1}{2}(a_1 + \cdots + a_n)$$

$$\left[\mathrm{Re}(\nu+M) > 0, \quad \mathrm{Re}\left(b \pm \tfrac{1}{2}a_1 \pm \cdots \pm \tfrac{1}{2}a_n\right) > 0\right] \quad \textbf{ET I 216(14)}$$

## 7.623

1. $\quad \displaystyle\int_0^\infty e^{-x} x^{c+n-1} (x+y)^{-1}\, {}_1F_1(a; c; x)\, \mathrm{d}x = (-1)^n\, \Gamma(c)\, \Gamma(1-a)\, y^{c+n-1}\, \Psi(c-a, c; y)$

$$\left[-\,\mathrm{Re}\,c < n < 1 - \mathrm{Re}\,a, \quad n = 0, 1, 2, \ldots, \quad |\arg y| < \pi\right] \quad \textbf{EH I 285(16)}$$

2. $\quad \displaystyle\int_0^t x^{-1} (t-x)^{k-1} e^{\frac{1}{2}(t-x)}\, M_{k,\mu}(x)\, \mathrm{d}x = \frac{\Gamma(k)\,\Gamma(2\mu+1)}{\Gamma\left(k+\mu+\frac{1}{2}\right)} \pi^{\frac{1}{2}} t^{k-\frac{1}{2}} I_\mu\left(\frac{1}{2}t\right)$

$$\left[\mathrm{Re}\,k > 0, \quad \mathrm{Re}\,\mu > -\tfrac{1}{2}\right] \quad \textbf{ET II 402(5)}$$

3. $\quad \displaystyle\int_0^t x^{k-1} (t-x)^{\lambda-1} e^{\frac{1}{2}(t-x)}\, M_{k+\lambda,\mu}(x)\, \mathrm{d}x = \frac{\Gamma(\lambda)\,\Gamma\left(k+\mu+\frac{1}{2}\right) t^{k+\lambda-1}}{\Gamma\left(k+\lambda+\mu+\frac{1}{2}\right)}\, M_{k,\mu}(t)$

$$\left[\mathrm{Re}(k+\mu) > -\tfrac{1}{2}, \quad \mathrm{Re}\,\lambda > 0\right]$$
$$\textbf{ET II 402(6)}$$

4. $\quad \displaystyle\int_0^t x^{-k-\lambda-1} (t-x)^{\lambda-1} e^{\frac{1}{2}x}\, W_{k,\mu}(x)\, \mathrm{d}x = \frac{\Gamma(\lambda)\,\Gamma\left(\frac{1}{2} - k - \lambda + \mu\right)\Gamma\left(\frac{1}{2} - k - \lambda - \mu\right)}{t^{k+1}\,\Gamma\left(\frac{1}{2} - k + \mu\right)\Gamma\left(\frac{1}{2} - k - \mu\right)}\, W_{k+\lambda,\mu}(t)$

$$\left[\mathrm{Re}\,\lambda > 0, \quad \mathrm{Re}(k+\lambda) < \tfrac{1}{2} - |\mathrm{Re}\,\mu|\right]$$
$$\textbf{ET II 405(21)}$$

5. $\quad \displaystyle\int_1^\infty (x-1)^{\mu-1} x^{\lambda-\frac{1}{2}} e^{\frac{1}{2}ax}\, W_{k,\lambda}(ax)\, \mathrm{d}x = \frac{\Gamma(\mu)\,\Gamma\left(\frac{1}{2} - k - \lambda - \mu\right)}{\Gamma\left(\frac{1}{2} - k - \lambda\right)}\, a^{-\frac{1}{2}\mu} e^{\frac{1}{2}a}\, W_{k+\frac{1}{2}\mu,\lambda+\frac{1}{2}\mu}(a)$

$$\left[|\arg(a)| < \tfrac{3}{2}\pi, \quad 0 < \mathrm{Re}\,\mu < \tfrac{1}{2} - \mathrm{Re}(k+\lambda)\right] \quad \textbf{ET II 211(72)a}$$

$6.^{11}$ $\quad \displaystyle\int_1^\infty (x-1)^{\mu-1} x^{\lambda-\frac{1}{2}} e^{-\frac{1}{2}ax}\, W_{k,\lambda}(ax)\, \mathrm{d}x = a^{-\frac{1}{2}\mu}\, \Gamma(\mu) e^{-\frac{1}{2}a}\, W_{k-\frac{1}{2}\mu,\lambda-\frac{1}{2}\mu}(a)$

$$\left[\mathrm{Re}\,\mu > 0, \quad \mathrm{Re}\,a > 0\right] \quad \textbf{ET II 211(74)a}$$

7. $\quad \displaystyle\int_1^\infty (x-1)^{\mu-1} x^{k-\mu-1} e^{-\frac{1}{2}ax}\, W_{k,\lambda}(ax)\, \mathrm{d}x = \Gamma(\mu) e^{-\frac{1}{2}a}\, W_{k-\mu,\lambda}(a)$

$$\left[\mathrm{Re}\,\mu > 0, \quad \mathrm{Re}\,a > 0\right] \quad \textbf{ET II 211(73)a}$$

8.    $\displaystyle\int_0^1 (1-x)^{\mu-1} x^{k-\mu-1} e^{-\frac{1}{2}ax}\, W_{k,\lambda}(ax)\, \mathrm{d}x$

$$= \Gamma(\mu) e^{-\frac{1}{2}a} \sec[(k-\mu-\lambda)\pi]$$

$$\times \left\{ \sin(\mu\pi) \frac{\Gamma\left(k-\mu+\lambda+\frac{1}{2}\right)}{\Gamma(2\lambda+1)}\, M_{k-\mu,\lambda}(a) + \cos[(k-\lambda)\pi]\, W_{k-\mu,\lambda}(a) \right\}$$

$$\left[0 < \operatorname{Re}\mu < \operatorname{Re}k - |\operatorname{Re}\lambda| + \tfrac{1}{2}\right] \quad \text{ET II 200(93)a}$$

**7.624**

1.    $\displaystyle\int_0^\infty x^{\rho-1} \left[x^{\frac{1}{2}} + (a+x)^{\frac{1}{2}}\right]^{2\sigma} e^{-\frac{1}{2}x}\, M_{k,\mu}(x)\, \mathrm{d}x$

$$= \frac{-\sigma\,\Gamma(2\mu+1)a^\sigma}{\pi^{\frac{1}{2}}\,\Gamma\left(\frac{1}{2}+k+\mu\right)}\, G_{34}^{23}\left(a \left|\begin{array}{l} \frac{1}{2}, 1, 1-k+\rho \\ \frac{1}{2}+\mu+\rho, -\sigma, \sigma, \frac{1}{2}-\mu+\rho \end{array}\right.\right)$$

$$\left[|\arg a| < \pi, \quad \operatorname{Re}(\mu+\rho) > -\tfrac{1}{2}, \quad \operatorname{Re}(k-\rho-\sigma) > 0\right] \quad \text{ET II 403(8)}$$

2.    $\displaystyle\int_0^\infty x^{\rho-1} \left[x^{\frac{1}{2}} + (a+x)^{\frac{1}{2}}\right]^{2\sigma} e^{-\frac{1}{2}x}\, W_{k,\mu}(x)\, \mathrm{d}x = -\pi^{-\frac{1}{2}}\sigma a^\sigma\, G_{34}^{32}\left(a \left|\begin{array}{l} \frac{1}{2}, 1, 1-k+\rho \\ \frac{1}{2}+\mu+\rho, \frac{1}{2}-\mu+\rho, -\sigma, \sigma \end{array}\right.\right)$

$$\left[|\arg a| < \pi, \quad \operatorname{Re}\rho > |\operatorname{Re}\mu| - \tfrac{1}{2}\right] \quad \text{ET II 406(24)}$$

3.    $\displaystyle\int_0^\infty x^{\rho-1} \left[x^{\frac{1}{2}} + (a+x)^{\frac{1}{2}}\right]^{2\sigma} e^{\frac{1}{2}x}\, W_{k,\mu}(x)\, \mathrm{d}x$

$$= -\frac{\sigma\pi^{-\frac{1}{2}}a^\sigma}{\Gamma\left(\frac{1}{2}-k+\mu\right)\Gamma\left(\frac{1}{2}-k-\mu\right)}\, G_{34}^{33}\left(a \left|\begin{array}{l} \frac{1}{2}, 1, 1+k+\rho \\ \frac{1}{2}+\mu+\rho, \frac{1}{2}-\mu+\rho, -\sigma, \sigma \end{array}\right.\right)$$

$$\left[|\arg a| < \pi, \quad \operatorname{Re}\rho > |\operatorname{Re}\mu| - \tfrac{1}{2}, \quad \operatorname{Re}(k+\rho+\sigma) < 0\right] \quad \text{ET II 406(25)}$$

4.    $\displaystyle\int_0^\infty x^{\rho-1}(a+x)^{-\frac{1}{2}} \left[x^{\frac{1}{2}} + (a+x)^{\frac{1}{2}}\right]^{2\sigma} e^{-\frac{1}{2}x}\, M_{k,\mu}(x)\, \mathrm{d}x$

$$= \frac{\Gamma(2\mu+1)a^\sigma}{\pi^{\frac{1}{2}}\,\Gamma\left(\frac{1}{2}+k+\mu\right)}\, G_{34}^{23}\left(a \left|\begin{array}{l} 0, \frac{1}{2}, \frac{1}{2}-k-\rho \\ -\sigma, \rho+\mu, \rho-\mu, \sigma \end{array}\right.\right)$$

$$\left[|\arg a| < \pi, \quad \operatorname{Re}(\rho+\mu) > -\tfrac{1}{2}, \quad \operatorname{Re}(k-\rho-\sigma) > -\tfrac{1}{2}\right] \quad \text{ET II 403(9)}$$

5.    $\displaystyle\int_0^\infty x^{\rho-1}(a+x)^{-\frac{1}{2}} \left[x^{\frac{1}{2}} + (a+x)^{\frac{1}{2}}\right]^{2\sigma} e^{\frac{1}{2}x}\, W_{k,\mu}(x)\, \mathrm{d}x$

$$= \frac{\pi^{-\frac{1}{2}}a^\sigma}{\Gamma\left(\frac{1}{2}-k+\mu\right)\Gamma\left(\frac{1}{2}-k-\mu\right)}\, G_{34}^{33}\left(a \left|\begin{array}{l} 0, \frac{1}{2}, \frac{1}{2}+k+\rho \\ -\sigma, \rho+\mu, \rho-\mu, \sigma \end{array}\right.\right)$$

$$\left[|\arg a| < \pi, \quad \operatorname{Re}\rho > |\operatorname{Re}\mu| - \tfrac{1}{2}, \quad \operatorname{Re}(k+\rho+\sigma) < \tfrac{1}{2}\right] \quad \text{ET II 406(26)}$$

6.    $\displaystyle\int_0^\infty x^{\rho-1}(a+x)^{-\frac{1}{2}} \left[x^{\frac{1}{2}} + (a+x)^{\frac{1}{2}}\right]^{2\sigma} e^{-\frac{1}{2}x}\, W_{k,\mu}(x)\, \mathrm{d}x = \pi^{-\frac{1}{2}}a^\sigma\, G_{34}^{32}\left(a \left|\begin{array}{l} 0, \frac{1}{2}, \frac{1}{2}-k+\rho \\ -\sigma, \rho+\mu, \rho-\mu, \sigma \end{array}\right.\right)$

$$\left[|\arg a| < \pi, \quad \operatorname{Re}\rho > |\operatorname{Re}\mu| - \tfrac{1}{2}\right] \quad \text{ET II 406(27)}$$

**7.625**

1.
$$\int_0^\infty x^{\rho-1} \exp\left[-\tfrac{1}{2}(\alpha+\beta)x\right] M_{k,\mu}(\alpha x)\, W_{\lambda,\nu}(\beta x)\, dx$$

$$= \frac{\Gamma(1+\mu+\nu+\rho)\,\Gamma(1+\mu-\nu+\rho)}{\Gamma\left(\tfrac{3}{2}-\lambda+\mu+\rho\right)} \alpha^{\mu+\frac{1}{2}} \beta^{-\mu-\rho-\frac{1}{2}}$$

$$\times\ {}_3F_2\left(\frac{1}{2}+k+\mu, 1+\mu+\nu+\rho, 1+\mu-\nu+\rho; 2\mu+1, \frac{3}{2}-\lambda+\mu+\rho; -\frac{\alpha}{\beta}\right)$$

$$[\operatorname{Re}\alpha>0, \quad \operatorname{Re}\beta>0, \quad \operatorname{Re}(\rho+\mu)>|\operatorname{Re}\nu|-1] \quad \textbf{ET II 410(43)}$$

2.[12]
$$\int_0^\infty x^{\rho-1} \exp\left[\frac{1}{2}(\alpha+\beta)x\right] W_{k,\mu}(\alpha x)\, W_{\lambda,\nu}(\beta x)\, dx$$

$$= \beta^{-\rho}\left[\Gamma\left(\tfrac{1}{2}-k+\mu\right)\Gamma\left(\tfrac{1}{2}-k-\mu\right)\Gamma\left(\tfrac{1}{2}-\lambda+\nu\right)\Gamma\left(\tfrac{1}{2}-\lambda-\nu\right)\right]^{-1}$$

$$\times\ G_{33}^{33}\left(\frac{\beta}{\alpha}\ \middle|\ \begin{matrix} \frac{1}{2}+\mu, \frac{1}{2}-\mu, 1+\lambda+\rho \\ \frac{1}{2}+\nu+\rho, \frac{1}{2}-\nu+\rho, -k \end{matrix}\right)$$

$$[|\operatorname{Re}\mu|+|\operatorname{Re}\nu|<\operatorname{Re}\rho+1, \quad \operatorname{Re}(k+\lambda+\rho)<0] \quad \textbf{ET II 410(44)a}$$

3.[12]
$$\int_0^\infty x^{\rho-1} \exp\left[-\frac{1}{2}(\alpha+\beta)x\right] W_{k,\mu}(\alpha x)\, W_{\lambda,\nu}(\beta x)\, dx = \beta^{-\rho}\, G_{33}^{22}\left(\frac{\beta}{\alpha}\ \middle|\ \begin{matrix} \frac{1}{2}+\mu, \frac{1}{2}-\nu, 1-\lambda+\rho \\ \frac{1}{2}+\nu+\rho, \frac{1}{2}-\nu+\rho, k \end{matrix}\right)$$

$$[\operatorname{Re}(\alpha+\beta)>0, \quad |\operatorname{Re}\mu|+|\operatorname{Re}\nu|<\operatorname{Re}\rho+1] \quad \textbf{ET II 411(46)}$$

4.
$$\int_0^\infty x^{\rho-1} \exp\left[-\frac{1}{2}(\alpha-\beta)x\right] W_{k,\mu}(\alpha x)\, W_{\lambda,\nu}(\beta x)\, dx$$

$$= \beta^{-\rho}\left[\Gamma\left(\tfrac{1}{2}-\lambda+\nu\right)\Gamma\left(\tfrac{1}{2}-\lambda-\nu\right)\right]^{-1} G_{33}^{23}\left(\frac{\beta}{\alpha}\ \middle|\ \begin{matrix} \frac{1}{2}+\mu, \frac{1}{2}-\mu, 1+\lambda+\rho \\ \frac{1}{2}+\nu+\rho, \frac{1}{2}-\nu+\rho, k \end{matrix}\right)$$

$$[\operatorname{Re}\alpha>0, \quad |\operatorname{Re}\mu|+|\operatorname{Re}\nu|<\operatorname{Re}\rho+1] \quad \textbf{ET II 411(45)}$$

**7.626**

1.[12]
$$\int_0^1 \left(\frac{k}{x}-\frac{1}{4}(\xi+\eta)\right) \exp\left(-\frac{1}{2}(\xi+\eta)x\right) x^c\ {}_1F_1(a;c;\xi x)\ {}_1F_1(a;c;\eta x)\, dx$$

$$= 0 \qquad\qquad\qquad [\xi\neq\eta, \quad \operatorname{Re}c>0]$$

$$= \frac{a}{\xi} e^{-\xi}\left[\,{}_1F_1(a+1;c;\xi)\right]^2 \qquad [\xi=\eta, \quad \operatorname{Re}c>0]$$

$$[\text{where } \xi \text{ and } \eta \text{ are any two zeros of the function } {}_1F_1(a;c;x)] \quad \textbf{EH I 285}$$

2.
$$\int_1^\infty \left[\frac{k}{x}-\frac{1}{4}(\xi+\eta)\right] e^{-\frac{1}{2}(\xi+\eta)x} x^c\, \Psi(a,c;\xi x)\, \Psi(a,c;\eta x)\, dx = 0 \qquad\qquad [\xi\neq\eta];$$

$$= -\xi^{-1} e^{-\xi}[\Psi(a-1,c;\xi)]^2 \qquad [\xi=\eta]$$

$$[\text{where } \xi \text{ and } \eta \text{ are any two zeros of the function } \Psi(a,c;x)] \quad \textbf{EH I 286}$$

**7.627**

1.     $\displaystyle\int_0^\infty x^{2\lambda-1}(a+x)^{-\mu-\frac{1}{2}}e^{\frac{1}{2}x}\,W_{k,\mu}(a+x)\,\mathrm{d}x = \frac{\Gamma(2\lambda)\,\Gamma\left(\frac{1}{2}-k+\mu-2\lambda\right)}{\Gamma\left(\frac{1}{2}-k+\mu\right)}a^{\lambda-\mu-\frac{1}{2}}\,W_{k+\lambda,\mu-\lambda}(a)$

$$\left[|\arg a| < \pi, \quad 0 < 2\operatorname{Re}\lambda < \tfrac{1}{2}-\operatorname{Re}(k+\mu)\right] \quad \text{ET II 411(50)}$$

2.     $\displaystyle\int_0^\infty x^{2\lambda-1}(a+x)^{-\mu-\frac{1}{2}}e^{-\frac{1}{2}x}\,M_{k,\mu}^{-\frac{1}{2}x}(a+x)\,\mathrm{d}x$

$$= \frac{\Gamma(2\lambda)\,\Gamma(2\mu+1)\,\Gamma\left(k+\mu-2\lambda+\frac{1}{2}\right)}{\Gamma\left(k+\mu+\frac{1}{2}\right)\Gamma(1-2\lambda+2\mu)}a^{\lambda-\mu-\frac{1}{2}}\,M_{k-\lambda,\mu-\lambda}(a)$$

$$\left[\operatorname{Re}\lambda > 0, \quad \operatorname{Re}(k+\mu-2\lambda) > -\tfrac{1}{2}\right] \quad \text{ET II 405(20)}$$

3.     $\displaystyle\int_0^\infty x^{2\lambda-1}(a+x)^{-\mu-\frac{1}{2}}e^{-\frac{1}{2}x}\,W_{k,\mu}(a+x)\,\mathrm{d}x = \Gamma(2\lambda)a^{\lambda-\mu-\frac{1}{2}}\,W_{k-\lambda,\mu-\lambda}(a)$

$$\left[|\arg a| < \pi, \quad \operatorname{Re}\lambda > 0\right] \quad \text{ET II 411(47)}$$

4.     $\displaystyle\int_0^\infty x^{\lambda-1}(a+x)^{k-\lambda-1}e^{-\frac{1}{2}x}\,W_{k,\mu}(a+x)\,\mathrm{d}x = \Gamma(\lambda)a^{k-1}\,W_{k-\lambda,\mu}(a)$

$$\left[|\arg a| < \pi, \quad \operatorname{Re}\lambda > 0\right] \quad \text{ET II 411(48)}$$

5.     $\displaystyle\int_0^\infty x^{\rho-1}(a+x)^{-\sigma}e^{-\frac{1}{2}x}\,W_{k,\mu}(a+x)\,\mathrm{d}x = \Gamma(\rho)a^\rho e^{\frac{1}{2}a}\,G_{23}^{30}\left(a\left|\begin{matrix}0,1-k-\sigma\\-\rho,\frac{1}{2}+\mu-\sigma,\frac{1}{2}-\mu-\sigma\end{matrix}\right.\right)$

$$\left[|\arg a| < \pi, \quad \operatorname{Re}\rho > 0\right] \quad \text{ET II 411(49)}$$

6.     $\displaystyle\int_0^\infty x^{\rho-1}(a+x)^{-\sigma}e^{\frac{1}{2}x}\,W_{k,\mu}(a+x)\,\mathrm{d}x$

$$= \frac{\Gamma(\rho)a^\rho e^{\frac{1}{2}a}}{\Gamma\left(\frac{1}{2}-k+\mu\right)\Gamma\left(\frac{1}{2}-k-\mu\right)}\,G_{23}^{31}\left(a\left|\begin{matrix}k-\sigma+1,0\\-\rho,\frac{1}{2}+\mu-\sigma,\frac{1}{2}-\mu-\sigma\end{matrix}\right.\right)$$

$$\left[|\arg a| < \pi, \quad 0 < \operatorname{Re}\rho < \operatorname{Re}(\sigma-k)\right] \quad \text{ET II 412(51)}$$

7.     $\displaystyle\int_0^\infty e^{-\frac{1}{2}(a+x)}\frac{(a+x)^{2\kappa-1}}{(ax)^\kappa}\,W_{\kappa,\mu}(x)\frac{\mathrm{d}x}{x} = \frac{\Gamma\left(\frac{1}{2}-\mu-\kappa\right)\Gamma\left(\frac{1}{2}+\mu-\kappa\right)}{a\,\Gamma(1-2\kappa)}\,W_{\kappa,\mu}(a)$

$$\left[\operatorname{Re}\left(\tfrac{1}{2}\pm\mu-\kappa\right) > 0\right] \quad \text{BU 126(7a)}$$

8.     $\displaystyle\int_0^\infty e^{-\frac{1}{2}x}x^{\gamma+\alpha-1}\,M_{\kappa,\mu}(x)\frac{\mathrm{d}x}{(x+a)^\alpha}$

$$= \frac{\Gamma(1+2\mu)\,\Gamma\left(\frac{1}{2}+\mu+\gamma\right)\Gamma(\kappa-\gamma)}{\Gamma\left(\frac{1}{2}+\mu-\gamma\right)\Gamma\left(\frac{1}{2}+\mu+\kappa\right)}\,{}_2F_2\left(\alpha,\kappa-\gamma;\tfrac{1}{2}+\mu-\gamma,\tfrac{1}{2}-\mu-\gamma;a\right)$$

$$+ \frac{\Gamma\left(\alpha+\gamma+\frac{1}{2}+\mu\right)\Gamma\left(-\gamma-\frac{1}{2}-\mu\right)}{\Gamma(\alpha)}a^{\gamma+\frac{1}{2}+\mu}$$

$$\times\,{}_2F_2\left(\alpha+\gamma+\mu+\tfrac{1}{2},\kappa+\mu+\tfrac{1}{2};1+2\mu,\tfrac{3}{2}+\mu+\gamma;a\right)$$

$$\left[\operatorname{Re}\left(\gamma+\alpha+\tfrac{1}{2}+\mu\right) > 0, \quad \operatorname{Re}(\gamma-\kappa) < 0\right] \quad \text{BU 126(8)a}$$

9. 
$$\int_0^\infty e^{-\frac{1}{2}x} x^{n+\mu+\frac{1}{2}} M_{\kappa,\mu}(x) \frac{dx}{x+a} = (-1)^{n+1} a^{n+\mu+\frac{1}{2}} e^{\frac{1}{2}a} \Gamma(1+2\mu) \Gamma\left(\frac{1}{2} - \mu + \kappa\right) W_{-\kappa,\mu}(a)$$

$$\left[ n = 0, 1, 2, \ldots, \quad \operatorname{Re}\left(\mu + 1 + \frac{n}{2}\right) > 0, \quad \operatorname{Re}\left(\kappa - \mu - \frac{1}{2}\right) < n, \quad |\arg a| < \pi \right] \quad \text{BU 127(10a)a}$$

**7.628**

1. 
$$\int_0^\infty e^{-st} e^{-t^2} t^{2c-2} \, {}_1F_1\left(a; c; t^2\right) dt = 2^{1-2c} \Gamma(2c-1) \Psi\left(c - \frac{1}{2}, a + \frac{1}{2}; \frac{1}{4} s^2\right)$$

$$\left[\operatorname{Re} c > \frac{1}{2}, \quad \operatorname{Re} s > 0\right] \quad \text{EH I 270(11)}$$

2. 
$$\int_0^\infty t^{2\nu-1} e^{-\frac{1}{2a}t^2} e^{-st} M_{-3\nu,\nu}\left(\frac{t^2}{a}\right) dt = \frac{1}{2\sqrt{\pi}} \Gamma(4\nu+1) a^{-\nu} s^{-4\nu} e^{as^2/8} K_{2\nu}\left(\frac{as^2}{8}\right)$$

$$\left[\operatorname{Re} a > 0, \quad \operatorname{Re} \nu > -\tfrac{1}{4}, \quad \operatorname{Re} s > 0\right]$$
$$\text{ET I 215(12)}$$

3. 
$$\int_0^\infty t^{2\mu-1} e^{-\frac{1}{2a}t^2} e^{-st} M_{\lambda,\mu}\left(\frac{t^2}{a}\right) dt$$
$$= 2^{-3\mu-\lambda} \Gamma(4\mu+1) a^{\frac{1}{2}(\lambda+\mu-1)} s^{\lambda-\mu-1} e^{\frac{as^2}{8}} W_{-\frac{1}{2}(\lambda+3\mu),\frac{1}{2}(\lambda-\mu)}\left(\frac{as^2}{4}\right)$$
$$\left[\operatorname{Re} a > 0, \quad \operatorname{Re} \mu > -\tfrac{1}{4}, \quad \operatorname{Re} s > 0\right] \quad \text{ET I 215(13)}$$

**7.629**

1.[12] 
$$\int_0^\infty t^k \exp\left(\frac{a}{2t}\right) e^{-st} W_{k,\mu}\left(\frac{a}{t}\right) dt = 2^{1-2k} \sqrt{a} s^{-k-\frac{1}{2}} S_{2k,2\mu}\left(2\sqrt{as}\right)$$

$$\left[|\arg a| < \pi, \quad \operatorname{Re}(k \pm \mu) > -\tfrac{1}{2}, \quad \operatorname{Re} s > 0\right] \quad \text{ET I 217(21)}$$

2. 
$$\int_0^\infty t^{-k} \exp\left(-\frac{a}{2t}\right) e^{-st} W_{k,\mu}\left(\frac{a}{t}\right) dt = 2\sqrt{a} s^{k-\frac{1}{2}} K_{2\mu}\left(2\sqrt{as}\right)$$

$$\left[\operatorname{Re} a > 0, \quad \operatorname{Re} s > 0\right] \quad \text{ET I 217(22)}$$

**7.631**

1. 
$$\int_0^\infty x^{\rho-1} \exp\left[\frac{1}{2}\left(\alpha^{-1}x - \beta x^{-1}\right)\right] W_{k,\mu}\left(\alpha^{-1}x\right) W_{\lambda,\nu}\left(\beta x^{-1}\right) dx$$
$$= \beta^\rho \left[\Gamma\left(\tfrac{1}{2} - k + \mu\right) \Gamma\left(\tfrac{1}{2} - k - \mu\right)\right]^{-1}$$
$$\times G_{24}^{41}\left(\frac{\beta}{\alpha} \left|\begin{array}{cc} 1+k, & 1-\lambda-\rho \\ \frac{1}{2}+\mu, & \frac{1}{2}-\mu, \quad \frac{1}{2}+\nu-\rho, \quad \frac{1}{2}-\nu-\rho \end{array}\right.\right)$$
$$\left[|\arg \alpha| < \tfrac{3}{2}\pi, \quad \operatorname{Re} \beta > 0, \quad \operatorname{Re}(k+\rho) < -|\operatorname{Re} \nu| - \tfrac{1}{2}\right] \quad \text{ET II 412(55)}$$

2. 
$$\int_0^\infty x^{\rho-1} \exp\left[\frac{1}{2}\left(\alpha^{-1}x - \beta x^{-1}\right)\right] W_{k,\mu}\left(\alpha^{-1}x\right) W_{\lambda,\nu}\left(\beta x^{-1}\right) dx$$
$$= \beta^\rho \left[\Gamma\left(\tfrac{1}{2} - k + \mu\right) \Gamma\left(\tfrac{1}{2} - k - \mu\right) \Gamma\left(\tfrac{1}{2} - \lambda + \nu\right) \Gamma\left(\tfrac{1}{2} - \lambda - \nu\right)\right]^{-1}$$
$$\times G_{24}^{42}\left(\frac{\beta}{\alpha} \left|\begin{array}{cc} 1+k, & 1+\lambda-\rho \\ \frac{1}{2}+\mu, & \frac{1}{2}-\mu, \quad \frac{1}{2}+\nu-\rho, \quad \frac{1}{2}-\nu-\rho \end{array}\right.\right)$$
$$\left[|\arg \alpha| < \tfrac{3}{2}\pi, \quad |\arg \beta| < \tfrac{3}{2}\pi, \quad \operatorname{Re}(\lambda-\rho) < \tfrac{1}{2} - |\operatorname{Re} \mu|, \quad \operatorname{Re}(k+\rho) < \tfrac{1}{2} - |\operatorname{Re} \nu|\right]$$
$$\text{ET II 412(57)}$$

3.    $\displaystyle \int_0^\infty x^{\rho-1} \exp\left[\tfrac{1}{2}\left(\alpha^{-1}x + \beta x^{-1}\right)\right] W_{k,\mu}\left(\alpha^{-1}x\right) W_{\lambda,\nu}\left(\beta x^{-1}\right) \mathrm{d}x$

$$= \beta^\rho \, G_{24}^{40}\left(\frac{\beta}{\alpha} \left| \begin{array}{ll} 1-k, & 1-\lambda-\rho \\ \tfrac{1}{2}+\mu, \;\; \tfrac{1}{2}-\mu, & \tfrac{1}{2}+\nu-\rho, \;\; \tfrac{1}{2}-\nu-\rho \end{array} \right.\right)$$

$$[\mathrm{Re}\,\alpha > 0, \quad \mathrm{Re}\,\beta > 0] \qquad \text{ET II 412(54)}$$

**7.632**    $\displaystyle \int_0^\infty e^{-st}\left(e^t - 1\right)^{\mu-\frac{1}{2}} \exp\left(-\tfrac{1}{2}\lambda e^t\right) M_{k,\mu}\left(\lambda e^t - \lambda\right) \mathrm{d}t$

$$= \frac{\Gamma(2\mu+1)\,\Gamma\left(\tfrac{1}{2}+k-\mu+s\right)}{\Gamma(s+1)} W_{-k-\frac{1}{2}s,\mu-\frac{1}{2}s}(\lambda)$$

$$\left[\mathrm{Re}\,\mu > -\tfrac{1}{2}, \quad \mathrm{Re}\,s > \mathrm{Re}(\mu-k) - \tfrac{1}{2}\right] \qquad \text{ET I 216(15)}$$

## 7.64 Combinations of confluent hypergeometric and trigonometric functions

**7.641**[12]   $\displaystyle \int_0^\infty \cos(ax)\; {}_1F_1(\nu+1;1;ix)\; {}_1F_1\left(\nu+1;1;-ix\right) \mathrm{d}x$

$$= -a^{-1}\sin(\nu\pi)\, P_\nu\left(2a^{-2} - 1\right) \quad [0 < a < 1];$$

$$= 0 \qquad\qquad\qquad\qquad [1 < a]$$

$$[-1 < \mathrm{Re}\,\nu < 0] \qquad \text{ET II 402(4)}$$

**7.642**[11]   $\displaystyle \int_0^\infty \cos(2xy)\; {}_1F_1\left(a;c;-x^2\right) \mathrm{d}x = \frac{1}{2}\pi^{\frac{1}{2}}\frac{\Gamma(c)}{\Gamma(a)}|y|^{2a-1}e^{-y^2}\Psi\left(c-\tfrac{1}{2},a+\tfrac{1}{2};y^2\right)$    EH I 285(12)

**7.643**

1.    $\displaystyle \int_0^\infty x^{4\nu}e^{-\frac{1}{2}x^2}\sin(bx)\; {}_1F_1\left(\tfrac{1}{2}-2\nu;2\nu+1;\tfrac{1}{2}x^2\right) \mathrm{d}x = \sqrt{\frac{\pi}{2}}b^{4\nu}c^{-\frac{1}{2}b^2}\; {}_1F_1\left(\tfrac{1}{2}-2\nu;1+2\nu;\tfrac{1}{2}b^2\right)$

$$\left[b > 0, \quad \mathrm{Re}\,\nu > -\tfrac{1}{4}\right] \qquad \text{ET I 115(5)}$$

2.    $\displaystyle \int_0^\infty x^{2\nu-1}e^{-\frac{1}{4}x^2}\sin(bx) M_{3\nu,\nu}\left(\tfrac{1}{2}x^2\right) \mathrm{d}x = \sqrt{\frac{\pi}{2}}b^{2\nu-1}e^{-\frac{1}{4}b^2} M_{3\nu,\nu}\left(\tfrac{1}{2}b^2\right)$

$$\left[b > 0, \quad \mathrm{Re}\,\nu > -\tfrac{1}{4}\right] \qquad \text{ET I 116(10)}$$

3.    $\displaystyle \int_0^\infty x^{-2\nu-1}e^{\frac{1}{4}x^2}\cos(bx) W_{3\nu,\nu}\left(\tfrac{1}{2}x^2\right) \mathrm{d}x = \sqrt{\frac{\pi}{2}}b^{-2\nu-1}e^{\frac{1}{4}b^2} W_{3\nu,\nu}\left(\tfrac{1}{2}b^2\right)$

$$\left[\mathrm{Re}\,\nu < \tfrac{1}{4}, \quad b > 0\right] \qquad \text{ET I 61(7)}$$

4.    $\displaystyle \int_0^\infty x^{-2\nu}e^{\frac{1}{4}x^2}\sin(bx) W_{3\nu-1,\nu}\left(\tfrac{1}{2}x^2\right) \mathrm{d}x = \sqrt{\frac{\pi}{2}}b^{-2\nu}e^{\frac{1}{4}b^2} W_{3\nu-1,\nu}\left(\tfrac{1}{2}b^2\right)$

$$\left[\mathrm{Re}\,\nu < \tfrac{1}{2}, \quad b > 0\right] \qquad \text{ET I 116(9)}$$

**7.644**

1.[11]    $\displaystyle \int_0^\infty x^{-\mu-\frac{1}{2}}e^{-\frac{1}{2}x}\sin\left(2ax^{\frac{1}{2}}\right) M_{k,\mu}(x) \mathrm{d}x = \pi^{\frac{1}{2}}a^{k+\mu-1}\frac{\Gamma(3-2\mu)}{\Gamma\left(\tfrac{1}{2}+k+\mu\right)}\exp\left(-\frac{a^2}{2}\right) W_{\rho,\sigma}\left(a^2\right),$

$$2\rho = k - 3\mu + 1, \qquad 2\sigma = k + \mu - 1 \qquad [a > 0, \quad \mathrm{Re}(k+\mu) > 0] \qquad \text{ET II 403(10)}$$

2. $\displaystyle\int_0^\infty x^{\rho-1}\sin\left(cx^{\frac12}\right)e^{-\frac12 x}\,W_{k,\mu}(x)\,dx=\frac{c\,\Gamma(1+\mu+\rho)\,\Gamma(1-\mu+\rho)}{\Gamma\left(\frac32-k+\rho\right)}$

$$\times\ {}_2F_2\left(1+\mu+\rho,1-\mu+\rho;\frac32,\frac32-k+\rho;-\frac{c^2}{4}\right)$$

$$[\operatorname{Re}\rho>|\operatorname{Re}\mu|-1]\qquad\text{ET II 407(28)}$$

3. $\displaystyle\int_0^\infty x^{\rho-1}\sin\left(cx^{\frac12}\right)e^{\frac12 x}\,W_{k,\mu}(x)\,dx$

$$=\frac{\pi^{\frac12}}{\Gamma\left(\frac12-k+\mu\right)\Gamma\left(\frac12-k-\mu\right)}\,G^{22}_{23}\left(\frac{c^2}{4}\left|\begin{array}{l}\frac12+\mu-\rho,\frac12-\mu-\rho\\\frac12,-k-\rho,0\end{array}\right.\right)$$

$$\left[c>0,\quad \operatorname{Re}\rho>|\operatorname{Re}\mu|-1,\quad \operatorname{Re}(k+\rho)<\tfrac12\right]\quad\text{ET II 407(29)}$$

4. $\displaystyle\int_0^\infty x^{\rho-1}\cos\left(cx^{\frac12}\right)e^{-\frac12 x}\,W_{k,\mu}(x)\,dx=\frac{\Gamma\left(\frac12+\mu+\rho\right)\Gamma\left(\frac12-\mu+\rho\right)}{\Gamma(1-k+\rho)}$

$$\times\ {}_2F_2\left(\frac12+\mu+\rho,\frac12-\mu+\rho;\frac12,1-k+\rho;-\frac{c^2}{4}\right)$$

$$\left[\operatorname{Re}\rho>|\operatorname{Re}\mu|-\tfrac12\right]\qquad\text{ET II 407(30)}$$

5. $\displaystyle\int_0^\infty x^{\rho-1}\cos\left(cx^{\frac12}\right)e^{\frac12 x}\,W_{k,\mu}(x)\,dx$

$$=\frac{\pi^{\frac12}}{\Gamma\left(\frac12-k+\mu\right)\Gamma\left(\frac12-k-\mu\right)}\,G^{22}_{23}\left(\frac{c^2}{4}\left|\begin{array}{l}\frac12+\mu-\rho,\frac12-\mu-\rho\\0,-k-\rho,\frac12\end{array}\right.\right)$$

$$\left[c>0,\quad \operatorname{Re}\rho>|\operatorname{Re}\mu|-\tfrac12,\quad \operatorname{Re}(k+\rho)<\tfrac12\right]\quad\text{ET II 407(31)}$$

## 7.65 Combinations of confluent hypergeometric functions and Bessel functions

**7.651**

1. $\displaystyle\int_0^\infty J_\nu(xy)\,M_{-\frac12\mu,\frac12\nu}(ax)\,W_{\frac12\mu,\frac12\nu}(ax)\,dx$

$$=ay^{-\mu-1}\frac{\Gamma(\nu+1)}{\Gamma\left(\frac12-\frac12\mu+\frac12\nu\right)}\left[a+\left(a^2+y^2\right)^{\frac12}\right]^\mu\left(a^2+y^2\right)^{-\frac12}$$

$$\left[y>0,\quad \operatorname{Re}\nu>-1,\quad \operatorname{Re}\mu<\tfrac12,\quad \operatorname{Re}a>0\right]\quad\text{ET II 85(19)}$$

2. $\displaystyle\int_0^\infty M_{k,\frac12\nu}(-iax)\,M_{-k,\frac12\nu}(-iax)\,J_\nu(xy)\,dx$

$$=\frac{ae^{-\frac12(\nu+1)\pi i}\left[\Gamma(1+\nu)\right]^2}{\Gamma\left(\frac12+k+\frac12\nu\right)\Gamma\left(\frac12-k+\frac12\nu\right)}\,y^{-1-2k}$$

$$\times\left(a^2-y^2\right)^{-\frac12}\left\{\left[a+\left(a^2-y^2\right)^{\frac12}\right]^{2k}+\left[a-\left(a^2-y^2\right)^{\frac12}\right]^{2k}\right\}\qquad[0<y<a]\,;$$

$$=0\qquad\qquad\qquad\qquad\qquad\qquad\qquad\qquad[a<y<\infty]$$

$$\left[a>0,\quad \operatorname{Re}\nu>-1,\quad |\operatorname{Re}k|<\tfrac14\right]\quad\text{ET II 85(18)}$$

**7.652** $\quad \displaystyle\int_0^\infty M_{-\mu,\frac{1}{2}\nu}\left\{a\left[\left(b^2+x^2\right)^{\frac{1}{2}}-b\right]\right\} W_{\mu,\frac{1}{2}\nu}\left\{a\left[\left(b^2+x^2\right)^{\frac{1}{2}}+b\right]\right\} J_\nu(xy)\,\mathrm{d}x$

$$= \frac{ay^{-2\mu-1}\,\Gamma(1+\nu)\left[\left(a^2+y^2\right)^{\frac{1}{2}}+a\right]^{2\mu}}{\Gamma\left(\frac{1}{2}+\frac{1}{2}\nu-\mu\right)\left(a^2+y^2\right)^{\frac{1}{2}}}\exp\left[-b\left(a^2+y^2\right)^{\frac{1}{2}}\right]$$

$$\left[y>0,\quad \operatorname{Re}\nu>-1,\quad \operatorname{Re}\mu<\tfrac{1}{4},\quad \operatorname{Re}a>0,\quad \operatorname{Re}b>0\right] \quad \text{ET II 87(29)}$$

## 7.66 Combinations of confluent hypergeometric functions, Bessel functions, and powers

**7.661**

1. $\quad \displaystyle\int_0^\infty x^{-1}\,W_{k,\mu}(ax)\,M_{-k,\mu}(ax)\,J_0(xy)\,\mathrm{d}x$

$$= e^{-ik\pi}\frac{\Gamma(1+2\mu)}{\Gamma\left(\frac{1}{2}+\mu+k\right)}\,P^k_{\mu-\frac{1}{2}}\left[\left(1+\frac{y^2}{a^2}\right)^{\frac{1}{2}}\right]Q^k_{\mu-\frac{1}{2}}\left[\left(1+\frac{y^2}{a^2}\right)^{\frac{1}{2}}\right]$$

$$\left[y>0,\quad \operatorname{Re}a>0,\quad \operatorname{Re}\mu>-\tfrac{1}{2},\quad \operatorname{Re}k<\tfrac{3}{4}\right] \quad \text{ET II 18(44)}$$

2. $\quad \displaystyle\int_0^\infty x^{-1}\,W_{k,\mu}(ax)\,W_{-k,\mu}(ax)\,J_0(xy)\,\mathrm{d}x = \frac{1}{2}\pi\cos(\mu\pi)\,P^k_{\mu-\frac{1}{2}}\left[\left(1+\frac{y^2}{a^2}\right)^{\frac{1}{2}}\right]P^{-k}_{\mu-\frac{1}{2}}\left[\left(1+\frac{y^2}{a^2}\right)^{\frac{1}{2}}\right]$

$$\left[y>0,\quad \operatorname{Re}a>0,\quad |\operatorname{Re}\mu|<\tfrac{1}{2}\right]$$
$$\text{ET II 18(45)}$$

3. $\quad \displaystyle\int_0^\infty x^{2\mu-\nu}\,W_{k,\mu}(ax)\,M_{-k,\mu}(ax)\,J_\nu(xy)\,\mathrm{d}x$

$$= 2^{2\mu-\nu+2k}a^{2k}y^{\nu-2\mu-2k-1}\frac{\Gamma(2\mu+1)}{\Gamma\left(\nu-k-\mu+\frac{1}{2}\right)}$$

$$\times\;_3F_2\left(\frac{1}{2}-k,1-k,\frac{1}{2}-k+\mu;1-2k,\frac{1}{2}-k-\mu+\nu;-\frac{y^2}{a^2}\right)$$

$$\left[y>0,\quad \operatorname{Re}\mu>-\tfrac{1}{2},\quad \operatorname{Re}a>0,\quad \operatorname{Re}(2\mu+2k-\nu)<\tfrac{1}{2}\right] \quad \text{ET II 85(20)}$$

4. $\quad \displaystyle\int_0^\infty x^{2\rho-\nu}\,W_{k,\mu}(iax)\,W_{k,\mu}(-iax)\,J_\nu(xy)\,\mathrm{d}x$

$$= 2^{2\rho-\nu}y^{\nu-2\rho-1}\pi^{-\frac{1}{2}}\left[\Gamma\left(\frac{1}{2}-k+\mu\right)\Gamma\left(\frac{1}{2}-k-\mu\right)\right]^{-1}G^{24}_{44}\left(\frac{y^2}{a^2}\left|\begin{array}{c}\frac{1}{2},0,\frac{1}{2}-\mu,\frac{1}{2}+\mu\\ \rho+\frac{1}{2},-k,k,\rho-\nu+\frac{1}{2}\end{array}\right.\right)$$

$$\left[y>0,\quad \operatorname{Re}a>0,\quad \operatorname{Re}\rho>|\operatorname{Re}\mu|-1,\quad \operatorname{Re}(2\rho+2k-\nu)<\tfrac{1}{2}\right] \quad \text{ET II 86(23)a}$$

5. $\quad \displaystyle\int_0^\infty x^{2\rho-\nu}\,W_{k,\mu}(ax)\,M_{-k,\mu}(ax)\,J_\nu(xy)\,\mathrm{d}x$

$$= \frac{2^{2\rho-\nu}\,\Gamma(2\mu+1)}{\pi^{\frac{1}{2}}\,\Gamma\left(\frac{1}{2}-k+\mu\right)}y^{\nu-2\rho-1}G^{23}_{44}\left(\frac{y^2}{a^2}\left|\begin{array}{c}\frac{1}{2},0,\frac{1}{2}-\mu,\frac{1}{2}+\mu\\ \rho+\frac{1}{2},-k,k,\rho-\nu+\frac{1}{2}\end{array}\right.\right)$$

$$\left[y>0,\quad \operatorname{Re}a>0,\quad \operatorname{Re}\rho>-1,\quad \operatorname{Re}(\rho+\mu)>-1,\quad \operatorname{Re}(2\rho+2k+\nu)<\tfrac{1}{2}\right] \quad \text{ET II 86(21)a}$$

6.   $\displaystyle\int_0^\infty x^{2\rho-\nu}\, W_{k,\mu}(ax)\, W_{-k,\mu}(ax)\, J_\nu(xy)\, dx$

$$= \frac{\Gamma(\rho+1+\mu)\,\Gamma(\rho+1-\mu)\,\Gamma(2\rho+2)}{\Gamma\left(\frac{3}{2}+k+\rho\right)\Gamma\left(\frac{3}{2}-k+\rho\right)\Gamma(1+\nu)}\, y^\nu 2^{-\nu-1} a^{-2\rho-1}$$

$$\times\, {}_4F_3\left(\rho+1,\rho+\frac{3}{2},\rho+1+\mu,\rho+1-\mu;\frac{3}{2}+k+\rho,\frac{3}{2}-k+\rho,1+\nu;-\frac{y^2}{a^2}\right)$$

$$\left[y>0,\quad \operatorname{Re}\rho>|\operatorname{Re}\mu|-1,\quad \operatorname{Re}a>0\right]\quad \text{ET II 86(22)a}$$

**7.662**

1.   $\displaystyle\int_0^\infty x^{-1}\, M_{-\mu,\frac{1}{4}\nu}\left(\tfrac{1}{2}x^2\right) W_{\mu,\frac{1}{4}\nu}\left(\tfrac{1}{2}x^2\right) J_\nu(xy)\, dx = \frac{\Gamma\left(1+\frac{1}{2}\nu\right)}{\Gamma\left(\frac{1}{2}+\frac{1}{4}\nu-\mu\right)}\, I_{\frac{1}{4}\nu-\mu}\left(\tfrac{1}{4}y^2\right) K_{\frac{1}{4}\nu+\mu}\left(\tfrac{1}{4}y^2\right)$

$$\left[y>0,\quad \operatorname{Re}\nu>-1\right]\qquad \text{ET II 86(24)}$$

2.   $\displaystyle\int_0^\infty x^{-1}\, M_{\alpha-\beta,\frac{1}{4}\nu-\gamma}\left(\tfrac{1}{2}x^2\right) W_{\alpha+\beta,\frac{1}{4}\nu+\gamma}\left(\tfrac{1}{2}x^2\right) J_\nu(xy)\, dx$

$$= \frac{\Gamma\left(1+\frac{1}{2}\nu-2\gamma\right)}{\Gamma\left(1+\frac{1}{2}\nu-2\beta\right)}\, y^{-2}\, M_{\alpha-\gamma,\frac{1}{4}\nu-\beta}\left(\tfrac{1}{2}y^2\right) W_{\alpha+\gamma,\frac{1}{4}\nu+\beta}\left(\tfrac{1}{2}y^2\right)$$

$$\left[y>0,\quad \operatorname{Re}\beta<\tfrac{1}{8},\quad \operatorname{Re}\nu>-1,\quad \operatorname{Re}(\nu-4\gamma)>-2\right]\quad \text{ET II 86(25)}$$

3.   $\displaystyle\int_0^\infty x^{-1}\, M_{k,0}\left(iax^2\right) M_{k,0}\left(-iax^2\right) K_0(xy)\, dx = \frac{\pi}{16}\left\{\left[J_k\left(\frac{y^2}{8a}\right)\right]^2 + \left[Y_k\left(\frac{y^2}{8a}\right)\right]^2\right\}$

$$\left[a>0\right]\qquad \text{ET II 152(83)}$$

4.   $\displaystyle\int_0^\infty x^{-1}\, M_{k,\mu}\left(iax^2\right) M_{k,\mu}\left(-iax^2\right) K_0(xy)\, dx = ay^{-2}\left[\Gamma(2\mu+1)\right]^2 W_{-\mu,k}\left(\frac{iy^2}{4a}\right) W_{-\mu,k}\left(-\frac{iy^2}{4a}\right)$

$$\left[a>0,\quad \operatorname{Re}y>0,\quad \operatorname{Re}\mu>-\tfrac{1}{2}\right]$$
$$\text{ET II 152(84)}$$

**7.663**

1.   $\displaystyle\int_0^\infty x^{2\rho}\, {}_1F_1\left(a;b;-\lambda x^2\right) J_\nu(xy)\, dx = \frac{2^{2\rho}\,\Gamma(b)}{\Gamma(a)y^{2\rho+1}}\, G^{21}_{23}\left(\frac{y^2}{4\lambda}\,\middle|\, \begin{matrix} 1,b \\ \frac{1}{2}+\rho+\frac{1}{2}\nu,a,\frac{1}{2}+\rho-\frac{1}{2}\nu \end{matrix}\right)$

$$\left[y>0,\quad -1-\operatorname{Re}\nu<2\operatorname{Re}\rho<\tfrac{1}{2}+2\operatorname{Re}a,\quad \operatorname{Re}\lambda>0\right]\quad \text{ET II 88(6)}$$

2.   $\displaystyle\int_0^\infty x^{\nu+1}\, {}_1F_1\left(2a-\nu;a+1;-\frac{1}{2}x^2\right) J_\nu(xy)\, dx = \frac{2^{\nu-a+\frac{1}{2}}\,\Gamma(a+1)}{\pi^{\frac{1}{2}}\,\Gamma(2a-\nu)}\, y^{2a-\nu-1} e^{-\frac{1}{4}y^2} K_{a-\nu-\frac{1}{2}}\left(\tfrac{1}{4}y^2\right)$

$$\left[y>0,\quad \operatorname{Re}\nu>-1,\quad \operatorname{Re}(4a-3\nu)>\tfrac{1}{2}\right]\quad \text{ET II 87(1)}$$

3.   $\displaystyle\int_0^\infty x^a\, {}_1F_1\left(a;\frac{1+a+\nu}{2};-\frac{1}{2}x^2\right) J_\nu(xy)\, dx = y^{a-1}\, {}_1F_1\left(a;\frac{1+a+\nu}{2};-\frac{y^2}{2}\right)$

$$\left[y>0,\quad \operatorname{Re}a>-\tfrac{1}{2},\quad \operatorname{Re}(a+\nu)>-1\right]$$
$$\text{ET II 87(2)}$$

4.　$\displaystyle\int_0^\infty x^{\nu+1-2a}\ {}_1F_1\left(a;1+\nu-a;-\tfrac{1}{2}x^2\right)J_\nu(xy)\,\mathrm{d}x$

$$=\frac{\pi^{\frac{1}{2}}\,\Gamma(1+\nu-a)}{\Gamma(a)}2^{-2a+\nu+\frac{1}{2}}y^{2a-\nu-1}e^{-\frac{1}{4}y^2}\,I_{a-\frac{1}{2}}\left(\tfrac{1}{4}y^2\right)$$

$$\left[y>0,\quad \operatorname{Re}a-1<\operatorname{Re}\nu<4\operatorname{Re}a-\tfrac{1}{2}\right]\quad\text{ET II 87(3)}$$

5.　$\displaystyle\int_0^\infty x\ {}_1F_1\left(\lambda;1;-x^2\right)J_0(xy)\,\mathrm{d}x=\left[2^{2\lambda-1}\Gamma(\lambda)\right]^{-1}y^{2\lambda-2}e^{-\frac{1}{4}y^2}$

$$\left[y>0,\quad \operatorname{Re}\lambda>0\right]\qquad\qquad\text{ET II 18(46)}$$

6.　$\displaystyle\int_0^\infty x^{\nu+1}\ {}_1F_1\left(a;b;-\lambda x^2\right)J_\nu(xy)\,\mathrm{d}x$

$$=\frac{2^{1-a}\Gamma(b)}{\Gamma(a)\lambda^{\frac{1}{2}a+\frac{1}{2}\nu}}y^{a-2}e^{-\frac{y^2}{8\lambda}}\,W_{k,\mu}\left(\frac{y^2}{4\lambda}\right),\quad 2k=a-2b+\nu+2,\quad 2\mu=a-\nu-1$$

$$\left[y>0,\quad -1<\operatorname{Re}\nu<2\operatorname{Re}a-\tfrac{1}{2},\quad \operatorname{Re}\lambda>0\right]\quad\text{ET II 88(4)}$$

7.　$\displaystyle\int_0^\infty x^{2b-\nu-1}\ {}_1F_1\left(a;b;-\lambda x^2\right)J_\nu(xy)\,\mathrm{d}x=\frac{2^{2b-2a-\nu-1}\Gamma(b)}{\Gamma(a-b+\nu+1)}\lambda^{-a}y^{2a-2b+\nu}$

$$\times\ {}_1F_1\left(a;1+a-b+\nu;-\frac{y^2}{4\lambda}\right)$$

$$\left[y>0,\quad 0<\operatorname{Re}b<\tfrac{3}{4}+\operatorname{Re}\left(a+\tfrac{1}{2}\nu\right),\quad \operatorname{Re}\lambda>0\right]\quad\text{ET II 88(5)}$$

**7.664**

1.　$\displaystyle\int_0^\infty x\,W_{\frac{1}{2}\nu,\mu}\left(\frac{a}{x}\right)W_{\frac{1}{2}\nu,\mu}\left(\frac{a}{x}\right)K_\nu(xy)\,\mathrm{d}x=2ay^{-1}\,K_{2\mu}\left[(2ay)^{\frac{1}{2}}e^{\frac{1}{4}i\pi}\right]K_{2\mu}\left[(2ay)^{\frac{1}{2}}e^{-\frac{1}{4}i\pi}\right]$

$$\left[\operatorname{Re}y>0,\quad \operatorname{Re}a>0\right]\qquad\text{ET II 152(85)}$$

2.　$\displaystyle\int_0^\infty x\,W_{\frac{1}{2}\nu,\mu}\left(\frac{2}{x}\right)W_{-\frac{1}{2}\nu,\mu}\left(\frac{2}{x}\right)J_\nu(xy)\,\mathrm{d}x$

$$=-4y^{-1}\left\{\sin\left[\left(\mu-\tfrac{1}{2}\nu\right)\pi\right]J_{2\mu}\left(2y^{\frac{1}{2}}\right)+\cos\left[\left(\mu-\tfrac{1}{2}\nu\right)\pi\right]Y_{2\mu}\left(2y^{\frac{1}{2}}\right)\right\}K_{2\mu}\left(2y^{\frac{1}{2}}\right)$$

$$\left[y>0,\quad \operatorname{Re}(\nu\pm2\mu)>-1\right]\quad\text{ET II 87(27)}$$

3.　$\displaystyle\int_0^\infty x\,W_{\frac{1}{2}\nu,\mu}\left(\frac{2}{x}\right)W_{-\frac{1}{2}\nu,\mu}\left(\frac{2}{x}\right)Y_\nu(xy)\,\mathrm{d}x$

$$=4y^{-1}\left\{\left\{\cos\left[\left(\mu-\tfrac{1}{2}\nu\right)\pi\right]J_{2\mu}\left(2y^{\frac{1}{2}}\right)-\sin\left[\left(\mu-\tfrac{1}{2}\nu\right)\pi\right]Y_{2\mu}\left(2y^{\frac{1}{2}}\right)\right\}K_{2\mu}\left(2y^{\frac{1}{2}}\right)\right\}$$

$$\left[y>0,\quad |\operatorname{Re}\mu|<\tfrac{1}{4}\right]\qquad\text{ET II 117(48)}$$

4.　$\displaystyle\int_0^\infty x\,W_{-\frac{1}{2}\nu,\mu}\left(\frac{2}{x}\right)M_{\frac{1}{2}\nu,\mu}\left(\frac{2}{x}\right)J_\nu(xy)\,\mathrm{d}x=\frac{4\Gamma(1+2\mu)y^{-1}}{\Gamma\left(\frac{1}{2}+\frac{1}{2}\nu+\mu\right)}J_{2\mu}\left(2y^{\frac{1}{2}}\right)K_{2\mu}\left(2y^{\frac{1}{2}}\right)$

$$\left[y>0,\quad \operatorname{Re}\nu>-1,\quad \operatorname{Re}\mu>-\tfrac{1}{4}\right]$$
$$\text{ET II 86(26)}$$

5.    $\displaystyle\int_0^\infty x\, W_{-\frac{1}{2}\nu,\mu}\left(\frac{ia}{x}\right) W_{-\frac{1}{2}\nu,\mu}\left(-\frac{ia}{x}\right) J_\nu(xy)\,\mathrm{d}x$

$$= 4ay^{-1}\left[\Gamma\left(\tfrac{1}{2}+\mu+\tfrac{1}{2}\nu\right)\Gamma\left(\tfrac{1}{2}-\mu+\tfrac{1}{2}\nu\right)\right]^{-1} K_\mu\left[(2iay)^{\frac{1}{2}}\right] K_\mu\left[(-2iay)^{\frac{1}{2}}\right]$$
$$\left[y>0,\quad \operatorname{Re}a>0,\quad |\operatorname{Re}\mu|<\tfrac{1}{2},\quad \operatorname{Re}\nu>-1\right]\quad \text{ET II 87(28)}$$

**7.665**

1.    $\displaystyle\int_0^\infty x^{-\frac{1}{2}} J_\nu\left(ax^{\frac{1}{2}}\right) K_{\frac{1}{2}\nu-\mu}\left(\tfrac{1}{2}x\right) M_{k,\mu}(x)\,\mathrm{d}x$

$$= \frac{\Gamma(2\mu+1)}{a\,\Gamma\left(k+\frac{1}{2}\nu+1\right)}\, W_{\frac{1}{2}(k-\mu),\frac{1}{2}k-\frac{1}{4}\nu}\left(\frac{a^2}{2}\right) M_{\frac{1}{2}(k+\mu),\frac{1}{2}k+\frac{1}{4}\nu}\left(\frac{a^2}{2}\right)$$
$$\left[a>0,\quad \operatorname{Re}k>-\tfrac{1}{4},\quad \operatorname{Re}\mu>-\tfrac{1}{2},\quad \operatorname{Re}\nu>-1\right]\quad \text{ET II 405(18)}$$

2.    $\displaystyle\int_0^\infty x^{\frac{1}{2}c+\frac{1}{2}c'-1}\Psi(a,c;x)\, {}_1F_1\left(a';c';-x\right) J_{c+c'-2}\left[2(xy)^{\frac{1}{2}}\right]\,\mathrm{d}x$

$$= \frac{\Gamma(c')}{\Gamma(a+a')}y^{\frac{1}{2}c+\frac{1}{2}c'-1}\Psi\left(c'-a',c+c'-a-a';y\right) {}_1F_1\left(a';a+a';-y\right)$$
$$\left[\operatorname{Re}c'>0,\quad 1<\operatorname{Re}(c+c')<2\operatorname{Re}(a+a')+\tfrac{1}{2}\right]\quad \text{EH I 287(23)}$$

**7.666**    $\displaystyle\int_0^\infty x^{\frac{1}{2}c-\frac{1}{2}}\,{}_1F_1\left(a;c;-2x^{\frac{1}{2}}\right) \Psi\left(a,c;2x^{\frac{1}{2}}\right) J_{c-1}\left[2(xy)^{\frac{1}{2}}\right]\,\mathrm{d}x$

$$= 2^{-c}\frac{\Gamma(c)}{\Gamma(a)}y^{a-\frac{1}{2}c-\frac{1}{2}}\left[1+(1+y)^{\frac{1}{2}}\right]^{c-2a}(1+y)^{-\frac{1}{2}}$$
$$\left[\operatorname{Re}c>2,\quad \operatorname{Re}(c-2a)<\tfrac{1}{2}\right]\quad \text{EH I 285(13)}$$

## 7.67 Combinations of confluent hypergeometric functions, Bessel functions, exponentials, and powers

**7.671**

1.    $\displaystyle\int_0^\infty x^{k-\frac{3}{2}}\exp\left[-\tfrac{1}{2}(a+1)x\right] K_\nu\left(\tfrac{1}{2}ax\right) M_{k,\nu}(x)\,\mathrm{d}x$

$$= \frac{\pi^{\frac{1}{2}}\,\Gamma(k)\,\Gamma(k+2\nu)}{a^{k+\nu}\,\Gamma\left(k+\nu+\frac{1}{2}\right)}\, {}_2F_1\left(k,k+2\nu;2\nu+1;-a^{-1}\right)$$
$$\left[\operatorname{Re}a>0,\quad \operatorname{Re}k>0,\quad \operatorname{Re}(k+2\nu)>0\right]\quad \text{ET II 405(17)}$$

2.    $\displaystyle\int_0^\infty x^{-k-\frac{3}{2}}\exp\left[-\tfrac{1}{2}(a-1)x\right] K_\mu\left(\tfrac{1}{2}ax\right) W_{k,\mu}(x)\,\mathrm{d}x$

$$= \frac{\pi\,\Gamma(-k)\,\Gamma(2\mu-k)\,\Gamma(-2\mu-k)}{\Gamma\left(\frac{1}{2}-k\right)\Gamma\left(\frac{1}{2}+\mu-k\right)\Gamma\left(\frac{1}{2}-\mu-k\right)}2^{2k+1}a^{k-\nu}\,{}_2F_1\left(-k,2\mu-k;-2k;1-a^{-1}\right)$$
$$\left[\operatorname{Re}a>0,\quad \operatorname{Re}k<2\operatorname{Re}\mu<-\operatorname{Re}k\right]\quad \text{ET II 408(36)}$$

**7.672**

1.
$$\int_0^\infty x^{2\rho} e^{-\frac{1}{2}ax^2} M_{k,\mu}\left(ax^2\right) J_\nu(xy)\,dx$$

$$= \frac{\Gamma(2\mu+1)}{\Gamma\left(\mu+k+\frac{1}{2}\right)} 2^{2\rho} y^{-2\rho-1} G^{21}_{23}\left(\frac{y^2}{4a}\,\middle|\,\begin{array}{c}\frac{1}{2}-\mu,\frac{1}{2}+\mu\\ \frac{1}{2}+\rho+\frac{1}{2}\nu,k,\frac{1}{2}+\rho-\frac{1}{2}\nu\end{array}\right)$$

$$\left[y>0,\quad -1-\mathrm{Re}\left(\tfrac{1}{2}\nu+\mu\right)<\mathrm{Re}\,\rho<\mathrm{Re}\,k-\tfrac{1}{4},\quad \mathrm{Re}\,a>0\right]\quad \text{ET II 83(10)}$$

2.
$$\int_0^\infty x^{2\rho} e^{-\frac{1}{2}ax^2} W_{k,\mu}\left(ax^2\right) J_\nu(xy)\,dx$$

$$= \frac{\Gamma\left(1+\mu+\frac{1}{2}\nu+\rho\right)\Gamma\left(1-\mu+\frac{1}{2}\nu+\rho\right)2^{-\nu-1}}{\Gamma(\nu+1)\Gamma\left(\frac{3}{2}-k+\frac{1}{2}\nu+\rho\right)} a^{-\frac{1}{2}\nu-\rho-\frac{1}{2}} y^\nu$$

$$\times {}_2F_2\left(\lambda+\mu,\lambda-\mu;\nu+1,\frac{1}{2}-k+\lambda;-\frac{y^2}{4a}\right),$$

$$\lambda=1+\tfrac{1}{2}\nu+\rho \qquad \left[y>0,\quad \mathrm{Re}\,a>0,\quad \mathrm{Re}\left(\rho\pm\mu+\tfrac{1}{2}\nu\right)>-1\right]\quad \text{ET II 85(16)}$$

3.
$$\int_0^\infty x^{2\rho} e^{\frac{1}{2}ax^2} W_{k,\mu}\left(ax^2\right) J_\nu(xy)\,dx = \frac{2^{2\rho} y^{-2\rho-1}}{\Gamma\left(\frac{1}{2}+\mu-k\right)\Gamma\left(\frac{1}{2}-\mu-k\right)}$$

$$\times G^{22}_{23}\left(\frac{y^2}{4a}\,\middle|\,\begin{array}{c}\frac{1}{2}-\mu,\quad \frac{1}{2}+\mu\\ \frac{1}{2}+\rho+\frac{1}{2}\nu,\quad -k,\quad \frac{1}{2}+\rho-\frac{1}{2}\nu\end{array}\right)$$

$$\left[y>0,\quad |\arg a|<\pi,\quad -1-\mathrm{Re}\left(\tfrac{1}{2}\nu\pm\mu\right)<\mathrm{Re}\,\rho<-\tfrac{1}{4}-\mathrm{Re}\,k\right]\quad \text{ET II 85(17)}$$

4.[12]
$$\int_0^\infty x^{2\lambda+\frac{1}{2}} e^{-\frac{1}{4}x^2} M_{k,\mu}\left(\tfrac{1}{2}x^2\right) Y_\nu(xy)\,dx$$

$$= \frac{2^\lambda y^{-1/2}\Gamma(2\mu+1)}{\Gamma\left(\frac{1}{2}+k+\mu\right)} G^{31}_{34}\left(\frac{y^2}{2}\,\middle|\,\begin{array}{c}-\mu-\lambda,\quad \mu-\lambda,\quad l\\ h,\quad \kappa,\quad k-\lambda-\frac{1}{2},\quad l\end{array}\right)$$

$$h=\tfrac{1}{4}+\tfrac{1}{2}\nu,\quad \kappa=\tfrac{1}{4}-\tfrac{1}{2}\nu,\quad l=-\tfrac{1}{4}-\tfrac{1}{2}\nu$$

$$\left[y>0,\quad \mathrm{Re}(k-\lambda)>0,\quad \mathrm{Re}\left(2\lambda+2\mu\pm\nu\right)>-\tfrac{5}{2}\right]\quad \text{ET II 116(45)}$$

5.
$$\int_0^\infty x^{2\lambda+\frac{1}{2}} e^{\frac{1}{4}x^2} W_{k,\mu}\left(\tfrac{1}{2}x^2\right) Y_\nu(xy)\,dx$$

$$= 2^\lambda \left[\Gamma\left(\tfrac{1}{2}-k+\mu\right)\Gamma\left(\tfrac{1}{2}-k-\mu\right)\right]^{-1} G^{32}_{34}\left(\frac{y^2}{2}\,\middle|\,\begin{array}{c}-\mu-\lambda,\quad \mu-\lambda,\quad l\\ h,\quad \kappa,\quad -\frac{1}{2}-k-\lambda,\quad l\end{array}\right) y^{-1/2},$$

$$h=\tfrac{1}{4}+\tfrac{1}{2}\nu,\quad \kappa=\tfrac{1}{4}-\tfrac{1}{2}\nu,\quad l=-\tfrac{1}{4}-\tfrac{1}{2}\nu$$

$$\left[y>0,\quad \mathrm{Re}(k+\lambda)<0,\quad \mathrm{Re}\left(2\lambda\pm2\mu\pm\nu\right)>-\tfrac{5}{2}\right]\quad \text{ET II 117(47)}$$

6.
$$\int_0^\infty x^{-1/2} e^{-\frac{1}{2}x^2} M_{\frac{1}{2}\nu-\frac{1}{4},\frac{1}{2}\nu+\frac{1}{4}}\left(x^2\right) J_\nu(xy)\,dx = (2\nu+1)2^{-\nu} y^{\nu-1}\left[1-\Phi\left(\tfrac{1}{2}y\right)\right]$$

$$\left[y>0,\quad \mathrm{Re}\,\nu>-\tfrac{1}{2}\right]\qquad \text{ET II 82(1)}$$

7.
$$\int_0^\infty x^{-1} e^{-\frac{1}{2}x^2} M_{\frac{1}{2}\nu+\frac{1}{2},\frac{1}{2}\nu+\frac{1}{2}}\left(x^2\right) J_\nu(xy)\,dx = \frac{\Gamma(\nu+2)y^\nu}{\Gamma\left(\nu+\frac{3}{2}\right)2^\nu}\left[1-\Phi\left(\tfrac{1}{2}y\right)\right]$$

$$\left[y>0,\mathrm{Re}\,\nu>-1\right]\qquad \text{ET II 82(2)}$$

8.     $$\int_0^\infty e^{-\frac{1}{4}x^2} M_{k,\frac{1}{2}\nu}\left(\frac{1}{2}x^2\right) J_\nu(xy)\,dx = \frac{2^{-k}\,\Gamma(\nu+1)}{\Gamma\left(k+\frac{1}{2}\nu+\frac{1}{2}\right)} y^{2k-1} e^{-\frac{1}{2}y^2}$$

$$\left[y>0,\quad \operatorname{Re}\nu>-1,\quad \operatorname{Re}k<\tfrac{1}{2}\right]$$

ET II 83(7)

9.     $$\int_0^\infty x^{\nu-2\mu} e^{-\frac{1}{4}x^2} M_{k,\mu}\left(\frac{1}{2}x^2\right) J_\nu(xy)\,dx$$

$$= 2^{\frac{1}{2}\left(\frac{1}{2}-k-3\mu+\nu\right)} \frac{\Gamma(2\mu+1)}{\Gamma\left(\mu+k+\frac{1}{2}\right)} y^{k+\mu-\frac{3}{2}} e^{-\frac{1}{4}y^2}\, W_{\alpha,\beta}\left(\frac{1}{2}y^2\right),$$

$$2\alpha = k-3\mu+\nu+\tfrac{1}{2},\qquad 2\beta = k+\mu-\nu-\tfrac{1}{2}$$

$$\left[y>0,\quad -1<\operatorname{Re}\nu<2\operatorname{Re}(k+\mu)-\tfrac{1}{2}\right]$$

ET II 83(9)

10.    $$\int_0^\infty x^{\nu-2\mu} e^{\frac{1}{4}x^2}\, W_{k,\pm\mu}\left(\tfrac{1}{2}x^2\right) J_\nu(xy)\,dx = \frac{\Gamma(1+\nu-2\mu)}{\Gamma(1+2\beta)} 2^{\beta-\mu} y^{k+\mu-\frac{3}{2}} e^{-\frac{1}{4}y^2} M_{\alpha,\beta}\left(\tfrac{1}{2}y^2\right)$$

$$2\alpha = \tfrac{1}{2}+k+\nu-3\mu,\qquad 2\beta = \tfrac{1}{2}-k+\nu-\mu$$

$$\left[y>0,\quad \operatorname{Re}\nu>-1,\quad \operatorname{Re}(\nu-2\mu)>-1\right]$$

ET II 84(14)

11.    $$\int_0^\infty x^{\nu-2\mu} e^{-\frac{1}{4}x^2}\, W_{k,\pm\mu}\left(\tfrac{1}{2}x^2\right) J_\nu(xy)\,dx$$

$$= \frac{\Gamma(1+\nu-2\mu)}{\Gamma\left(\frac{1}{2}+\mu-k\right)} 2^{\frac{1}{2}\left(\frac{1}{2}+k-3\mu+\nu\right)} y^{\mu-k-\frac{3}{2}} e^{\frac{1}{4}y^2}\, W_{\alpha,\beta}\left(\tfrac{1}{2}y^2\right),$$

$$2\alpha = k+3\mu-\nu-\tfrac{1}{2},\qquad 2\beta = k-\mu+\nu+\tfrac{1}{2}$$

$$\left[y>0,\quad \operatorname{Re}\nu>-1,\quad \operatorname{Re}(\nu-2\mu)>-1,\quad \operatorname{Re}\left(k-\mu+\tfrac{1}{2}\nu\right)<-\tfrac{1}{4}\right]$$ ET II 84(15)

12.    $$\int_0^\infty x^{2\mu-\nu} e^{-\frac{1}{4}x^2} M_{k,\mu}\left(\tfrac{1}{2}x^2\right) J_\nu(xy)\,dx$$

$$= \frac{\Gamma(2\mu+1)}{\Gamma\left(\frac{1}{2}+k-\mu+\nu\right)} 2^{\frac{1}{2}\left(\frac{1}{2}-k+3\mu-\nu\right)} y^{k-\mu-\frac{3}{2}} e^{-\frac{1}{4}y^2} M_{\alpha,\beta}\left(\tfrac{1}{2}y^2\right)$$

$$2\alpha = \tfrac{1}{2}+k+3\mu-\nu,\qquad 2\beta = -\tfrac{1}{2}+k-\mu+\nu$$

$$\left[y>0,\quad -\tfrac{1}{2}<\operatorname{Re}\mu<\operatorname{Re}\left(k+\tfrac{1}{2}\nu\right)-\tfrac{1}{4}\right]$$ ET II 83(8)

13.    $$\int_0^\infty x^{2\mu-\nu} e^{-\frac{1}{4}x^2} M_{k,\mu}\left(\tfrac{1}{2}x^2\right) Y_\nu(xy)\,dx$$

$$= \pi^{-1} 2^{\mu+\beta} y^{k-\mu-\frac{3}{2}} e^{-\frac{1}{4}y^2}\,\Gamma(2\mu+1)$$

$$\times \Gamma\left(\tfrac{1}{2}-k-\mu\right) \left\{ \cos[(\nu-2\mu)\pi] \frac{\Gamma(2\mu-\nu-1)}{\Gamma(2\beta+1)} M_{\alpha,\beta}\left(\tfrac{1}{2}y^2\right)\right.$$

$$\left. - \sin[(\nu+k-\mu)\pi]\, W_{\alpha,\beta}\left(\tfrac{1}{2}y^2\right)\right\}$$

$$2\alpha = 3\mu-\nu+k+\tfrac{1}{2},\qquad 2\beta = \mu-\nu-k+\tfrac{1}{2}$$

$$\left[y>0,\quad -1<2\operatorname{Re}\mu<\operatorname{Re}(2k+\nu)+\tfrac{1}{2},\quad \operatorname{Re}(2\mu-\nu)>-1\right]$$ ET II 116(44)

14.    $\displaystyle\int_0^\infty x^{2\mu+\nu} e^{-\frac{1}{4}x^2} M_{k,\mu}\left(\tfrac{1}{2}x^2\right) Y_\nu(xy)\,\mathrm{d}x$

$$= \pi^{-1} 2^{\mu+\beta} y^{k-\mu-\frac{3}{2}} \Gamma(2\mu+1)$$

$$\times \Gamma\left(\tfrac{1}{2}-\mu-k\right) e^{-\frac{1}{4}y^2} \left\{ \cos(2\mu\pi)\frac{\Gamma(2\mu+\nu+1)}{\Gamma\left(\mu+\nu-k+\frac{3}{2}\right)} M_{\alpha,\beta}\left(\tfrac{1}{2}y^2\right)\right.$$

$$\left. + \sin[(\mu-k)\pi]\, W_{\alpha,\beta}\left(\tfrac{1}{2}y^2\right)\right\}$$

$$2\alpha = 3\mu+\nu+k+\tfrac{1}{2}, \qquad 2\beta = \mu+\nu-k+\tfrac{1}{2}$$

$$\left[y > 0, \quad -1 < 2\operatorname{Re}\mu < \operatorname{Re}(2k-\nu)+\tfrac{1}{2}, \quad \operatorname{Re}(2\mu+\nu) > -1\right] \quad \text{ET II 116(43)}$$

15.    $\displaystyle\int_0^\infty x^{2\mu+\nu} e^{-\frac{1}{2}ax^2} M_{k,\mu}\left(ax^2\right) K_\nu(xy)\,\mathrm{d}x = 2^{\mu-k-\frac{1}{2}} a^{\frac{1}{4}-\frac{1}{2}(\mu+\nu+k)} y^{k-\mu-\frac{3}{2}}$

$$\times \Gamma(2\mu+1)\,\Gamma(2\mu+\nu+1) \exp\left(\frac{y^2}{8a}\right) W_{\kappa,m}\left(\frac{y^2}{4a}\right),$$

$$2\kappa = -3\mu-\nu-k-\tfrac{1}{2}, \qquad 2m = \mu+\nu-k+\tfrac{1}{2}$$

$$\left[\operatorname{Re} y > 0, \quad \operatorname{Re} a > 0, \quad \operatorname{Re}\mu > -\tfrac{1}{2}, \quad \operatorname{Re}(2\mu+\nu) > -1\right] \quad \text{ET II 152(82)}$$

**7.673**

1.[10]    $\displaystyle\int_0^\infty e^{-\frac{1}{2}ax} x^{\frac{1}{2}(\mu-\nu-1)} M_{\kappa,\frac{1}{2}\mu}(ax) J_\nu\left(2\sqrt{bx}\right)\,\mathrm{d}x$

$$= \left(\frac{b}{a}\right)^{\frac{\kappa-1}{2}-\frac{1+\mu}{4}} a^{-\frac{1}{2}(\mu+1-\nu)} \Gamma(1+\mu) e^{-\frac{b}{2a}} \frac{1}{\Gamma\left(1+\dfrac{\kappa+\nu}{2}-\dfrac{1+\mu}{4}\right)}$$

$$\times M_{\frac{1}{2}(\kappa-\nu-1)+\frac{3}{4}(1+\mu),\,\frac{\kappa+\nu}{2}-\frac{1+\mu}{4}}\left(\frac{b}{a}\right)$$

$$\left[\operatorname{Re}(1+\mu) > 0, \quad \operatorname{Re}\left(\kappa+\frac{\nu-\mu}{2}\right) > -\frac{3}{4}, \quad \operatorname{Im} b = 0\right] \quad \text{BU 128(12)a}$$

2.    $\displaystyle\int_0^\infty e^{\frac{1}{2}ax} x^{\frac{1}{2}(\nu-1\mp\mu)} W_{\kappa,\frac{1}{2}\mu}(ax) J_\nu\left(2\sqrt{bx}\right)\,\mathrm{d}x = a^{-\frac{1}{2}(\nu+1\mp\mu)} \frac{\Gamma(\nu+1\mp\mu) e^{\frac{b}{2a}}}{\Gamma\left(\frac{1\pm\mu}{2}-\kappa\right)} \left(\frac{a}{b}\right)^{\frac{1}{2}(\kappa+1)+\frac{1}{4}(1\mp\mu)}$

$$\times W_{\frac{1}{2}(\kappa+1-\nu)-\frac{3}{4}(1\mp\mu),\,\frac{1}{2}(\kappa+\nu)+\frac{1}{4}(1\mp\mu)}\left(\frac{b}{a}\right)$$

$$\left[\operatorname{Re}\left(\frac{\nu\mp\mu}{2}+\kappa\right) < \frac{3}{4}, \quad \operatorname{Re}\nu > -1\right] \quad \text{BU 128(13)}$$

**7.674**

1.  $$\int_0^\infty x^{\rho-1} e^{-\frac{1}{2}\kappa} J_{\lambda+\nu}\left(ax^{1/2}\right) J_{\lambda-\nu}\left(ax^{1/2}\right) W_{k,\mu}(x)\,\mathrm{d}x$$

    $$= \frac{\left(\frac{1}{2}a\right)^{2\lambda} \Gamma\left(\frac{1}{2}+\lambda+\mu+\rho\right) \Gamma\left(\frac{1}{2}+\lambda-\mu+\rho\right)}{\Gamma\left(1+\lambda+\nu\right)\Gamma(1+\lambda-\nu)\Gamma(1+\lambda-k+\rho)}$$

    $$\times\; {}_4F_4\left(1+\lambda, \frac{1}{2}+\lambda, \frac{1}{2}+\lambda+\mu+\rho, \frac{1}{2}+\lambda-\mu+\rho; 1+\lambda+\nu,\right.$$

    $$\left. 1+\lambda-\nu, 1+2\lambda, 1+\lambda-k+\rho; -a^2\right)$$

    $$\left[|\mathrm{Re}\,\mu| < \mathrm{Re}(\lambda+\rho)+\tfrac{1}{2}\right] \qquad \text{ET II 409(37)}$$

2.  $$\int_0^\infty x^{\rho-1} e^{-\frac{1}{2}\kappa} I_{\lambda+\nu}\left(ax^{1/2}\right) K_{\lambda-\nu}\left(ax^{1/2}\right) W_{k,\mu}(x)\,\mathrm{d}x$$

    $$= \frac{\pi^{-1/2}}{2} G_{45}^{24}\left(a^2 \left|\begin{matrix} 0, \frac{1}{2}, \frac{1}{2}+\mu-\rho, \frac{1}{2}-\mu-\rho \\ \lambda, \nu, -\lambda, -\nu, k-\rho \end{matrix}\right.\right)$$

    $$\left[|\mathrm{Re}\,\mu| < \mathrm{Re}(\lambda+\rho)+\tfrac{1}{2}, \quad |\mathrm{Re}\,\mu| < \mathrm{Re}(\nu+\rho)+\tfrac{1}{2}\right] \qquad \text{ET II 409(38)}$$

## Combinations of Struve functions and confluent hypergeometric functions

**7.675**

1.  $$\int_0^\infty x^{2\lambda+\frac{1}{2}} e^{-\frac{1}{4}x^2} M_{k,\mu}\left(\tfrac{1}{2}x^2\right) \mathbf{H}_\nu(xy)\,\mathrm{d}x = \frac{2^{-\lambda}\Gamma(2\mu+1)}{y^{1/2}\Gamma\left(\frac{1}{2}+k+\mu\right)} G_{34}^{22}\left(\frac{y^2}{2}\left|\begin{matrix} l, -\mu-\lambda, \mu-\lambda \\ l, k-\lambda-\frac{1}{2}, h, \kappa \end{matrix}\right.\right)$$

    $$h = \tfrac{1}{4}+\tfrac{1}{2}\nu, \quad \kappa = \tfrac{1}{4}-\tfrac{1}{2}\nu, \quad l = \tfrac{3}{4}+\tfrac{1}{2}\nu$$

    $$\left[\mathrm{Re}(2\lambda+2\mu+\nu) > -\tfrac{7}{2}, \quad \mathrm{Re}(k-\lambda) > 0, \quad y > 0, \quad \mathrm{Re}(2\lambda-2k+\nu) < -\tfrac{1}{2}\right]$$

    $$\text{ET II 171(42)}$$

2.[12]  $$\int_0^\infty x^{2\lambda+\frac{1}{2}} e^{-\frac{1}{4}x^2} W_{k,\mu}\left(\tfrac{1}{2}x^2\right) \mathbf{H}_\nu(xy)\,\mathrm{d}x$$

    $$= 2^{\frac{1}{4}-\lambda-\frac{1}{2}\nu}\pi^{-1/2}y^{\nu+1}\frac{\Gamma\left(\frac{7}{4}+\frac{1}{2}\nu+\lambda+\mu\right)\Gamma\left(\frac{7}{4}+\frac{1}{2}\nu+\lambda-\mu\right)}{\Gamma\left(\nu+\frac{3}{2}\right)\Gamma\left(\frac{9}{4}+\lambda-k-\frac{1}{2}\nu\right)}$$

    $$\times\; {}_3F_3\left(1, \frac{7}{4}+\frac{\nu}{2}+\lambda+\mu, \frac{7}{4}+\frac{\nu}{2}+\lambda-\mu; \frac{3}{2}, \nu+\frac{3}{2}, \frac{9}{4}+\lambda-k+\frac{\nu}{2}; -\frac{y^2}{2}\right)$$

    $$\left[\mathrm{Re}(2\lambda+\nu) > 2|\mathrm{Re}\,\mu| - \tfrac{7}{2}, \quad y > 0\right] \qquad \text{ET II 171(43)}$$

3.  $$\int_0^\infty x^{2\lambda+\frac{1}{2}} e^{\frac{1}{4}x^2} W_{k,\mu}\left(\tfrac{1}{2}x^2\right) \mathbf{H}_\nu(xy)\,\mathrm{d}x$$

    $$= \left[2^\lambda \Gamma\left(\tfrac{1}{2}-k+\mu\right)\Gamma\left(\tfrac{1}{2}-k-\mu\right)\right]^{-1} y^{-1/2} G_{34}^{23}\left(\frac{y^2}{2}\left|\begin{matrix} l, -\mu-\lambda, \mu-\lambda \\ l, -k-\lambda-\frac{1}{2}, h, \kappa \end{matrix}\right.\right)$$

    $$h = \tfrac{1}{4}+\tfrac{1}{2}\nu, \quad \kappa = \tfrac{1}{4}-\tfrac{1}{2}\nu, \quad l = \tfrac{3}{4}+\tfrac{1}{2}\nu$$

    $$\left[y > 0, \quad \mathrm{Re}(2\lambda+\nu) > 2|\mathrm{Re}\,\mu| - \tfrac{7}{2}, \quad \mathrm{Re}(2k+2\lambda+\nu) < -\tfrac{1}{2}, \quad \mathrm{Re}(k+\lambda) < 0\right] \quad \text{ET II 172(46)a}$$

4.  $$\int_0^\infty e^{\frac{1}{2}x^2} W_{-\frac{1}{2}\nu-\frac{1}{2},\frac{1}{2}\nu}\left(x^2\right) \mathbf{H}_\nu(xy)\,\mathrm{d}x = 2^{-\nu-1} y^\nu \pi e^{\frac{1}{4}y^2}\left[1-\Phi\left(\frac{y}{2}\right)\right]$$

    $$\left[y > 0, \quad \mathrm{Re}\,\nu > -1\right] \qquad \text{ET II 171(44)}$$

## 7.68 Combinations of confluent hypergeometric functions and other special functions

### Combinations of confluent hypergeometric functions and associated Legendre functions

**7.681**

1.
$$\int_0^\infty x^{-1/2}(a+x)^\mu e^{-\frac{1}{2}x} P_\nu^{-2\mu}\left(1+2\frac{x}{a}\right) M_{k,\mu}(x)\,dx$$
$$= -\frac{\sin(\nu\pi)}{\pi\,\Gamma(k)}\Gamma(2\mu+1)\,\Gamma\left(k-\mu+\nu+\tfrac{1}{2}\right)\Gamma\left(k-\mu-\nu-\tfrac{1}{2}\right)e^{\frac{1}{2}a}\,W_{\rho,\sigma}(a),$$
$$\rho = \tfrac{1}{2}-k+\mu, \qquad \sigma = \tfrac{1}{2}+\nu$$
$$\left[|\arg a| < \pi, \quad \operatorname{Re}\mu > -\tfrac{1}{2}, \quad \operatorname{Re}(k-\mu) > \left|\operatorname{Re}\nu+\tfrac{1}{2}\right|\right] \quad \text{ET II 403(11)}$$

2.
$$\int_0^\infty x^{-1/2}(a+x)^{-\mu} e^{-\frac{1}{2}x} P_\nu^{-2\mu}\left(1+2\frac{x}{a}\right) M_{k,\mu}(x)\,dx$$
$$= \frac{\Gamma(2\mu+1)\,\Gamma\left(k+\mu+\nu+\tfrac{1}{2}\right)\Gamma\left(k+\mu-\nu-\tfrac{1}{2}\right)e^{\frac{1}{2}a}}{\Gamma\left(k+\mu+\tfrac{1}{2}\right)\Gamma(2\mu+\nu+1)\,\Gamma(2\mu-\nu)}\,W_{\frac{1}{2}-k-\mu,\frac{1}{2}+\nu}(a)$$
$$\left[|\arg a| < \pi, \quad \operatorname{Re}\mu > -\tfrac{1}{2}, \quad \operatorname{Re}(k+\mu) > \left|\operatorname{Re}\nu+\tfrac{1}{2}\right|\right] \quad \text{ET II 403(12)}$$

3.
$$\int_0^\infty x^{-\frac{1}{2}-\frac{1}{2}\mu-\nu}(a+x)^{\frac{1}{2}\mu} e^{-\frac{1}{2}x} P_{k+\nu-\frac{3}{2}}^\mu\left(1+2\frac{x}{a}\right) W_{k,\nu}(x)\,dx$$
$$= \frac{\Gamma(1-\mu-2\nu)}{\Gamma\left(\tfrac{3}{2}-k-\mu-\nu\right)}a^{-\frac{1}{4}+\frac{1}{2}k-\frac{1}{2}\nu}e^{\frac{1}{2}a}\,W_{\rho,\sigma}(a)$$
$$2\rho = \tfrac{1}{2}+2\mu+\nu-k, \quad 2\sigma = k+3\nu-\tfrac{3}{2}$$
$$\left[|\arg a| < \pi, \quad \operatorname{Re}\mu < 1, \quad \operatorname{Re}(\mu+2\nu) < 1\right]$$
$$\text{ET II 407(32)}$$

4.
$$\int_0^\infty x^{-\frac{1}{2}-\frac{1}{2}\mu-\nu}(a+x)^{-\frac{1}{2}\mu} e^{-\frac{1}{2}x} P_{k+\mu+\nu-\frac{3}{2}}^\mu\left(1+2\frac{x}{a}\right) W_{k,\nu}(x)\,dx$$
$$= \frac{\Gamma(1-\mu-2\nu)}{\Gamma\left(\tfrac{3}{2}-k-\mu-\nu\right)}a^{-\frac{1}{2}+\frac{1}{2}k-\frac{1}{2}\nu}e^{\frac{1}{2}a}\,W_{\rho,\sigma}(a)$$
$$2\rho = \tfrac{1}{2}-k+\nu, \qquad 2\sigma = k+2\mu+3\nu-\tfrac{3}{2}$$
$$\left[|\arg a| < \pi, \quad \operatorname{Re}\mu < 1, \quad \operatorname{Re}(\mu+2\nu) < 1\right]$$
$$\text{ET II 408(33)}$$

5.
$$\int_0^\infty x^{\mu-\frac{1}{4}k-\frac{1}{2}\nu-\frac{1}{2}}(a+x)^{\frac{1}{2}\nu} e^{-\frac{1}{2}x} Q_{\mu-k+\frac{3}{2}}^\nu\left(1+2\frac{x}{a}\right) M_{k,\mu}(x)\,dx$$
$$= \frac{e^{\nu\pi i}\,\Gamma(1+2\mu-\nu)\,\Gamma(1+2\mu)\,\Gamma\left(\tfrac{5}{2}-k+\mu+\nu\right)}{2\,\Gamma\left(\tfrac{1}{2}+k+\mu\right)}a^{\frac{1}{4}(k+2\mu-2\nu+5)}e^{\frac{1}{2}a}\,W_{\rho,\sigma}(a)$$
$$2\rho = \tfrac{1}{2}-k-\mu+2\nu, \qquad 2\sigma = k-3\mu-\tfrac{3}{2}$$
$$\left[|\arg a| < \pi, \quad \operatorname{Re}\mu > -\tfrac{1}{2}, \quad \operatorname{Re}(2\mu-\nu) > -1\right]$$
$$\text{ET II 404(14)}$$

**7.682**

1. $$\int_0^\infty x^{-1/2} e^{-\frac{1}{2}x} P_\nu^{-2\mu}\left[\left(1+\frac{x}{a}\right)^{1/2}\right] M_{k,\mu}(x)\,\mathrm{d}x$$

$$= \frac{\Gamma(2\mu+1)\,\Gamma\left(k+\frac{1}{2}\nu\right)\Gamma\left(k-\frac{1}{2}\nu-\frac{1}{2}\right)e^{\frac{1}{2}a}}{2^{2\mu}a^{1/4}\,\Gamma\left(k+\mu+\frac{1}{2}\right)\Gamma\left(\mu+\frac{1}{2}\nu+\frac{1}{2}\right)\Gamma\left(\mu-\frac{1}{2}\nu\right)}\,W_{\frac{3}{4}-k,\frac{1}{4}+\frac{1}{2}\nu}(a)$$

$$\left[\left|\arg a\right|<\pi,\quad \operatorname{Re}k>\tfrac{1}{2}\operatorname{Re}\nu-\tfrac{1}{2},\quad \operatorname{Re}k>-\tfrac{1}{2}\operatorname{Re}\nu\right]\quad \text{ET II 404(13)}$$

2. $$\int_0^\infty x^{\frac{1}{2}(k+\mu+\nu)-1}(a+x)^{-1/2}e^{-\frac{1}{2}x}Q_{k-\mu-\nu-1}^{1-k+\mu-\nu}\left[\left(1+\frac{x}{a}\right)^{1/2}\right]M_{k,\mu}(x)\,\mathrm{d}x$$

$$= e^{(1-k+\mu-\nu)\pi i}2^{\mu-k-\nu}a^{\frac{1}{2}(k+\mu-1)}\frac{\Gamma\left(\frac{1}{2}-\nu\right)\Gamma(1+2\mu)\,\Gamma(k+\mu+\nu)}{\Gamma\left(k+\mu+\frac{1}{2}\right)}e^{\frac{1}{2}a}\,W_{\rho,\sigma}(a),$$

$$\rho=\tfrac{1}{2}-k-\tfrac{1}{2}\nu,\qquad \sigma=\mu+\tfrac{1}{2}\nu$$

$$\left[\left|\arg a\right|<\pi,\quad \operatorname{Re}\mu>-\tfrac{1}{2},\quad \operatorname{Re}(k+\mu+\nu)>0\right]\quad \text{ET II 404(15)}$$

3. $$\int_0^\infty x^{\nu-\frac{1}{2}}e^{-\frac{1}{2}x}Q_{2k-2\nu-3}^{2\mu-2\nu}\left[\left(1+\frac{x}{a}\right)^{1/2}\right]M_{k,\mu}(x)\,\mathrm{d}x$$

$$= e^{2(\mu-\nu)\pi i}2^{2\mu-2\nu-1}a^{\frac{1}{2}(k+\mu-1)}e^{\frac{1}{2}a}\frac{\Gamma(2\mu+1)\,\Gamma(\nu+1)\,\Gamma\left(k+\mu-2\nu-\frac{1}{2}\right)}{\Gamma\left(k+\mu+\frac{1}{2}\right)}\,W_{\rho,\sigma}(a),$$

$$2\rho=1-k+\mu-2\nu,\qquad 2\sigma=k-\mu-2\nu-2$$

$$\left[\left|\arg a\right|<\pi,\quad \operatorname{Re}\mu>-\tfrac{1}{2},\quad \operatorname{Re}\nu>-1,\quad \operatorname{Re}(k+\mu-2\nu)>\tfrac{1}{2}\right]\quad \text{ET II 404(16)}$$

4. $$\int_0^\infty x^{-\frac{1}{2}-\frac{1}{2}\mu-\nu}e^{-\frac{1}{2}x}P_{2k+\mu+2\nu-3}^{\mu}\left[\left(1+\frac{x}{a}\right)^{\frac{1}{2}}\right]W_{k,\nu}(x)\,\mathrm{d}x$$

$$= \frac{2^\mu\,\Gamma(1-\mu-2\nu)}{\Gamma\left(\frac{3}{2}-k-\mu-\nu\right)}a^{-\frac{1}{2}+\frac{1}{2}k-\frac{1}{2}\nu}e^{\frac{1}{2}a}\,W_{\rho,\sigma}(a),$$

$$2\rho=1-k+\mu+\nu,\qquad 2\sigma=k+\mu+3\nu-2$$

$$\left[\left|\arg a\right|<\pi,\quad \operatorname{Re}\mu<1,\quad \operatorname{Re}(\mu+2\nu)<1\right]$$

$$\text{ET II 408(34)}$$

5.[8] $$\int_0^\infty x^{-\frac{1}{2}-\frac{1}{2}\mu-\nu}(a+x)^{-1/2}e^{-\frac{1}{2}x}P_{2k+\mu+2\nu-2}^{\mu}\left[\left(1+\frac{x}{a}\right)^{1/2}\right]W_{k,\nu}(x)\,\mathrm{d}x$$

$$= \frac{2^\mu\,\Gamma(1-\mu-2\nu)}{\Gamma\left(\frac{3}{2}-k-\mu-\nu\right)}a^{-\frac{1}{2}+\frac{1}{2}k-\frac{1}{2}\nu}e^{\frac{1}{2}a}\,W_{\rho,\sigma}(a),\qquad 2\rho=\mu+\nu-k,\qquad 2\sigma=k+\mu+3\nu-1$$

$$\left[\left|\arg a\right|<\pi,\quad \operatorname{Re}\mu>0,\quad \operatorname{Re}\nu>0\right]\quad \text{ET II 408(35)}$$

**A combination of confluent hypergeometric functions and orthogonal polynomials**

**7.683**[8] $$\int_0^1 e^{-\frac{1}{2}ax}x^\alpha(1-x)^{\frac{\mu-\alpha}{2}-1}L_n^\alpha(ax)\,M_{k-\frac{1+\alpha}{2},\frac{\mu-\alpha-1}{2}}\left[a(1-x)\right]\,\mathrm{d}x$$

$$= \frac{\Gamma(\mu-\alpha)\,\Gamma(1+n+\alpha)}{\Gamma(1+\mu)}\frac{}{n!}a^{-\frac{1+\alpha}{2}}\,M_{k+n,\frac{\mu}{2}}(a)$$

$$\left[\operatorname{Re}a>-1,\quad \operatorname{Re}(\mu-\alpha)>0,\quad n=0,1,2,\ldots\right]\quad \text{BU 129(14b)}$$

**A combination of hypergeometric and confluent hypergeometric functions**

**7.684** $\displaystyle\int_0^\infty x^{\rho-1} e^{-\frac{1}{2}x} M_{\gamma+\rho,\beta+\rho+\frac{1}{2}}(x)\; {}_2F_1\left(\alpha,\beta;\gamma;-\frac{\lambda}{x}\right)\mathrm{d}x$

$$= \frac{\Gamma(\alpha+\beta+2\rho)\,\Gamma(2\beta+2\rho)\,\Gamma(\gamma)}{\Gamma(\beta)\,\Gamma(\beta+\gamma+2\rho)}\lambda^{\frac{1}{2}\beta+\rho-\frac{1}{2}}e^{\frac{1}{2}\lambda}\,W_{k,\mu}(\lambda);$$

$$k=\tfrac{1}{2}-\alpha-\tfrac{1}{2}\beta-\rho, \qquad \mu=\tfrac{1}{2}\beta+\rho$$

$$[|\arg\lambda|<\pi,\quad \operatorname{Re}(\beta+\rho)>0,\quad \operatorname{Re}(\alpha+\beta+2\rho)>0,\quad \operatorname{Re}\gamma>0]$$

ET II 405(19)

## 7.69 Integration of confluent hypergeometric functions with respect to the index

**7.691** $\displaystyle\int_{-\infty}^\infty \operatorname{sech}(\pi x)\, W_{ix,0}(\alpha)\, W_{-ix,0}(\beta)\,\mathrm{d}x = 2\frac{(\alpha\beta)^{1/2}}{\alpha+\beta}\exp\left[-\frac{1}{2}(\alpha+\beta)\right]$      ET II 414(61)

**7.692** $\displaystyle\int_{-i\infty}^{i\infty}\Gamma(-a)\,\Gamma(c-a)\Psi(a,c;x)\Psi(c-a,c;y)\,\mathrm{d}a = 2\pi i\,\Gamma(c)\Psi(c,2c;x+y)$      EH I 285(15)

**7.693**

1.   $\displaystyle\int_{-\infty}^\infty \Gamma(ix)\,\Gamma(2k+ix)\,W_{k+ix,k-\frac{1}{2}}(\alpha)\,W_{-k-ix,k-\frac{1}{2}}(\beta)\,\mathrm{d}x$

$$= 2\pi^{1/2}\,\Gamma(2k)(\alpha\beta)^k(\alpha+\beta)^{\frac{1}{2}-2k}K_{2k-\frac{1}{2}}\left(\frac{\alpha+\beta}{2}\right)$$

ET II 414(62)

2.   $\displaystyle\int_{-i\infty}^{i\infty}\Gamma\left(\tfrac{1}{2}+\nu+\mu+x\right)\Gamma\left(\tfrac{1}{2}+\nu+\mu-x\right)\Gamma\left(\tfrac{1}{2}+\nu-\mu+x\right)\Gamma\left(\tfrac{1}{2}+\nu-\mu-x\right)$

$$\times M_{\mu+ix,\nu}(\alpha)\,M_{\mu-ix,\nu}(\beta)\,\mathrm{d}x$$

$$= \frac{2\pi(\alpha\beta)^{\nu+\frac{1}{2}}\,[\Gamma(2\nu+1)]^2\,\Gamma(2\nu+2\mu+1)\,\Gamma(2\nu-2\mu+1)}{(\alpha+\beta)^{2\nu+1}\,\Gamma(4\nu+2)}M_{2\mu,2\nu+\frac{1}{2}}(\alpha+\beta)$$

$$\left[\operatorname{Re}\nu>|\operatorname{Re}\mu|-\tfrac{1}{2}\right]$$

ET II 413(59)

**7.694**[11] $\displaystyle\int_{-\infty}^\infty e^{-2\rho x i}\,\Gamma\left(\tfrac{1}{2}+\nu+ix\right)\Gamma\left(\tfrac{1}{2}+\nu-ix\right)M_{ix,\nu}(\alpha)\,M_{ix,\nu}(\beta)\,\mathrm{d}x$

$$= \pi\sqrt{\alpha\beta}\,[\Gamma(2\nu+1)]^2\,\operatorname{sech}\rho\exp\left[-\frac{1}{2}(\alpha+\beta)\tanh\rho\right]J_{2\nu}\left(\sqrt{\alpha\beta}\,\operatorname{sech}\rho\right)$$

$$\left[|\operatorname{Im}\rho|<\tfrac{1}{2}\pi,\quad \operatorname{Re}\nu>-\tfrac{1}{2}\right]$$

# 7.7 Parabolic Cylinder Functions

## 7.71 Parabolic cylinder functions

**7.711**

1.   $\displaystyle\int_{-\infty}^\infty D_n(x)\,D_m(x)\,\mathrm{d}x = 0$                $[m\neq n]$

$$= n!(2\pi)^{1/2}$$                $[m=n]$

WH

2.  $\int_0^\infty D_\mu\left(\pm t\right) D_\nu(t)\,\mathrm{d}t = \dfrac{\pi 2^{\frac{1}{2}(\mu+\nu+1)}}{\mu-\nu}\left[\dfrac{1}{\Gamma\left(\frac{1}{2}-\frac{1}{2}\mu\right)\Gamma\left(-\frac{1}{2}\nu\right)}\mp\dfrac{1}{\Gamma\left(\frac{1}{2}-\frac{1}{2}\nu\right)\Gamma\left(-\frac{1}{2}\mu\right)}\right]$

[when the lower sign is taken, $\operatorname{Re}\mu > \operatorname{Re}\nu$]    BU 11 117(13a), EH II 122(21)

3.  $\int_0^\infty \left[D_\nu(t)\right]^2\,\mathrm{d}t = \pi^{1/2}2^{-3/2}\dfrac{\psi\left(\frac{1}{2}-\frac{1}{2}\nu\right)-\psi\left(-\frac{1}{2}\nu\right)}{\Gamma(-\nu)}$    BU 117(13b)a, EH II 122(22)a

## 7.72 Combinations of parabolic cylinder functions, powers, and exponentials

### 7.721

1.  $\int_{-\infty}^\infty e^{-\frac{1}{4}x^2}(x-z)^{-1}D_n(x)\,\mathrm{d}x = \pm ie^{\mp n\pi i}(2\pi)^{1/2}n!e^{-\frac{1}{4}z^2}D_{-n-1}\left(\mp iz\right)$

[The upper or lower sign is taken according as the imaginary part of $z$ is positive or negative.]

WH

2.  $\int_1^\infty x^\nu(x-1)^{\frac{1}{2}\mu-\frac{1}{2}\nu-1}\exp\left[-\dfrac{(x-1)^2a^2}{4}\right]D_\mu(ax)\,\mathrm{d}x = 2^{\mu-\nu-2}a^{\frac{\mu}{2}-\frac{\nu}{2}-1}\Gamma\left(\dfrac{\mu-\nu}{2}\right)D_\nu(a)$

[$\operatorname{Re}(\mu-\nu)>0$]    ET II 395(4)a

### 7.722

1.  $\int_0^\infty e^{-\frac{3}{4}x^2}x^\nu D_{\nu+1}(x)\,\mathrm{d}x = 2^{-\frac{1}{2}-\frac{1}{2}\nu}\Gamma(\nu+1)\sin\dfrac{1}{4}(1-\nu)\pi$

[$\operatorname{Re}\nu>-1$]    WH

2.  $\int_0^\infty e^{-\frac{1}{4}x^2}x^{\mu-1}D_{-\nu}(x)\,\mathrm{d}x = \dfrac{\pi^{1/2}2^{-\frac{1}{2}\mu-\frac{1}{2}\nu}\Gamma(\mu)}{\Gamma\left(\frac{1}{2}\mu+\frac{1}{2}\nu+\frac{1}{2}\right)}$    [$\operatorname{Re}\mu>0$]    EH II 122(20)

3.[11]  $\int_0^\infty e^{-\frac{3}{4}x^2}x^\nu D_{\nu-1}(x)\,\mathrm{d}x = 2^{-\frac{1}{2}\nu}\Gamma(\nu)\sin\left(\dfrac{1}{4}\pi\nu\right)$    [$\operatorname{Re}\nu>-1$]    ET II 395(2)

### 7.723

1.  $\int_0^\infty e^{-\frac{1}{4}x^2}x^\nu\left(x^2+y^2\right)^{-1}D_\nu(x)\,\mathrm{d}x = \left(\dfrac{\pi}{2}\right)^{1/2}\Gamma(\nu+1)y^{\nu-1}e^{\frac{1}{4}y^2}D_{-\nu-1}(y)$

[$\operatorname{Re}y>0,\quad \operatorname{Re}\nu>-1$]

EH II 121(18)a, ET II 396(6)a

2.  $\int_0^\infty e^{-\frac{1}{4}x^2}x^{\nu-1}\left(x^2+y^2\right)^{-1/2}D_\nu(x)\,\mathrm{d}x = y^{\nu-1}\Gamma(\nu)e^{\frac{1}{4}y^2}D_{-\nu}(y)$

[$\operatorname{Re}y>0,\quad \operatorname{Re}\nu>0$]    ET II 396(7)

3.  $\int_0^1 x^{2\nu-1}\left(1-x^2\right)^{\lambda-1}e^{\frac{a^2x^2}{4}}D_{-2\lambda-2\nu}(ax)\,\mathrm{d}x = \dfrac{\Gamma(\lambda)\Gamma(2\nu)}{\Gamma(2\lambda+2\nu)}2^{\lambda-1}e^{\frac{a^2}{4}}D_{-2\nu}(a)$

[$\operatorname{Re}\lambda>0,\quad \operatorname{Re}\nu>0$]    ET II 395(3)a

**7.724**  $\int_{-\infty}^\infty e^{-\frac{(x-y)^2}{2\mu}}e^{\frac{1}{4}x^2}D_\nu(x)\,\mathrm{d}x = (2\pi\mu)^{1/2}(1-\mu)^{\frac{1}{2}\nu}e^{\frac{y^2}{4-4\mu}}D_\nu\left[y(1-\mu)^{-1/2}\right]$    [$0<\operatorname{Re}\mu<1$]

EH II 121(15)

**7.725**

1.  $$\int_0^\infty e^{-pt}(2t)^{\frac{\nu-1}{2}}e^{-\frac{t}{2}}D_{-\nu-2}\left(\sqrt{2t}\right)\,dt = \left(\frac{\pi}{2}\right)^{1/2}\frac{\left(\sqrt{p+1}-1\right)^{\nu+1}}{(\nu+1)p^{\nu+1}}$$
    $$[\operatorname{Re}\nu > -1]\qquad\qquad\text{MO 175}$$

2.  $$\int_0^\infty e^{-pt}(2t)^{\frac{\nu-1}{2}}e^{-\frac{t}{2}}D_{-\nu}\left(\sqrt{2t}\right)\,dt = \left(\frac{\pi}{2}\right)^{1/2}\frac{\left(\sqrt{p+1}-1\right)^{\nu}}{p^\nu\sqrt{p+1}}$$
    $$[\operatorname{Re}\nu > -1]\qquad\qquad\text{MO 175}$$

3.  $$\int_0^\infty e^{-bx}D_{2n+1}\left(\sqrt{2x}\right)\,dx = (-2)^n\,\Gamma\left(n+\tfrac{3}{2}\right)\left(b-\tfrac{1}{2}\right)^n\left(b+\tfrac{1}{2}\right)^{-n-\frac{3}{2}}$$
    $$\left[\operatorname{Re}b > -\tfrac{1}{2}\right]\qquad\qquad\text{ET I 210(3)}$$

4.  $$\int_0^\infty \left(\sqrt{x}\right)^{-1}e^{-bx}D_{2n}\left(\sqrt{2x}\right)\,dx = (-2)^n\,\Gamma\left(n+\tfrac{1}{2}\right)\left(b-\tfrac{1}{2}\right)^n\left(b+\tfrac{1}{2}\right)^{-n-\frac{1}{2}}$$
    $$\left[\operatorname{Re}b > -\tfrac{1}{2}\right]\qquad\qquad\text{ET I 210(5)}$$

5.  $$\int_0^\infty x^{-\frac{1}{2}(\nu+1)}e^{-sx}D_\nu\left(\sqrt{x}\right)\,dx = \sqrt{\pi}\left(1+\sqrt{\tfrac{1}{2}+2s}\right)^\nu\frac{1}{\sqrt{\tfrac{1}{4}+s}}$$
    $$\left[\operatorname{Re}s > -\tfrac{1}{4},\quad \operatorname{Re}\nu < 1\right]\qquad\text{ET I 210(7)}$$

6.  $$\int_0^\infty e^{-zt}t^{-1+\frac{\beta}{2}}D_{-\nu}\left[2(kt)^{1/2}\right]\,dt = \frac{2^{1-\beta-\frac{\nu}{2}}\pi^{1/2}\,\Gamma(\beta)}{\Gamma\left(\tfrac{1}{2}\nu+\tfrac{1}{2}\beta+\tfrac{1}{2}\right)}(z+k)^{-\frac{\beta}{2}}F\left(\frac{\nu}{2},\frac{\beta}{2};\frac{\nu+\beta+1}{2};\frac{z-k}{z+k}\right)$$
    $$\left[\operatorname{Re}(z+k) > 0,\quad \operatorname{Re}\frac{z}{k} > 0\right]$$
    $$\text{EH II 121(11)}$$

**7.726** $$\int_{-\infty}^\infty e^{ixy-\frac{(1+\lambda)x^2}{4}}D_\nu\left[x(1-\lambda)^{1/2}\right]\,dx = (2\pi)^{1/2}\lambda^{\frac{1}{2}\nu}e^{-\frac{(1+\lambda)y^2}{4\lambda}}D_\nu\left[i\left(\lambda^{-1}-1\right)^{1/2}y\right]\qquad[\operatorname{Re}\lambda > 0]$$
$$\text{EH II 121(16)}$$

**7.727** $$\int_0^\infty \frac{e^{\frac{1}{2}x}e^{-bx}}{(e^x-1)^{\mu+\frac{1}{2}}}\exp\left(-\frac{a}{1-e^{-x}}\right)D_{2\mu}\left(\frac{2\sqrt{a}}{\sqrt{1-e^{-x}}}\right)\,dx = e^{-a}2^{b+\mu}\,\Gamma(b+\mu)\,D_{-2b}\left(2\sqrt{a}\right)$$
$$[\operatorname{Re}a > 0,\quad \operatorname{Re}b > -\operatorname{Re}\mu]$$
$$\text{ET I 211(13)}$$

**7.728** $$\int_0^\infty (2t)^{-\frac{\nu}{2}}e^{-pt}e^{-\frac{q^2}{8t}}D_{\nu-1}\left(\frac{q}{\sqrt{2t}}\right)\,dt = \left(\frac{\pi}{2}\right)^{\frac{1}{2}}p^{\frac{1}{2}\nu-1}e^{-q\sqrt{p}}\qquad\qquad\text{MO 175}$$

## 7.73 Combinations of parabolic cylinder and hyperbolic functions

**7.731**

1.  $$\int_0^\infty \cosh(2\mu x)\exp\left[-(a\sinh x)^2\right]D_{2k}\left(2a\cosh x\right)\,dx = 2^{k-\frac{3}{2}}\pi^{1/2}a^{-1}W_{k,\mu}\left(2a^2\right)$$
    $$\left[\operatorname{Re}^2 a > 0\right]\qquad\qquad\text{ET II 398(20)}$$

2.  $$\int_0^\infty \cosh(2\mu x)\exp\left[(a\sinh x)^2\right]D_{2k}\left(2a\cosh x\right)\,dx = \frac{\Gamma(\mu-k)\,\Gamma(-\mu-k)}{2^{k+\frac{5}{2}}a\,\Gamma(-2k)}W_{k+\frac{1}{2},\mu}\left(2a^2\right)$$
    $$\left[|\arg a| < \tfrac{3\pi}{4},\quad \operatorname{Re}k+|\operatorname{Re}\mu| < 0\right]$$
    $$\text{ET II 398(21)}$$

## 7.74 Combinations of parabolic cylinder and trigonometric functions

**7.741**

1.  $$\int_0^\infty \sin(bx)\left\{[D_{-n-1}(ix)]^2 - [D_{-n-1}(-ix)]^2\right\}\,\mathrm{d}x = (-1)^{n+1}\frac{i}{n!}\pi\sqrt{2\pi}e^{-\frac{1}{2}b^2}L_n\left(b^2\right)$$

$$[b > 0] \qquad\qquad \text{ET I 115(3)}$$

2.  $$\int_0^\infty e^{-\frac{1}{4}x^2}\sin(bx)\,D_{2n+1}(x)\,\mathrm{d}x = (-1)^n\sqrt{\frac{\pi}{2}}b^{2n+1}e^{-\frac{1}{2}b^2}$$

$$[b > 0] \qquad\qquad \text{ET I 115(1)}$$

3.  $$\int_0^\infty e^{-\frac{1}{4}x^2}\cos(bx)\,D_{2n}(x)\,\mathrm{d}x = (-1)^n\sqrt{\frac{\pi}{2}}b^{2n}e^{-\frac{1}{2}b^2} \qquad [b > 0] \qquad\qquad \text{ET I 60(2)}$$

4.  $$\int_0^\infty e^{-\frac{1}{4}x^2}\sin(bx)\left[D_{2\nu-\frac{1}{2}}(x) - D_{2\nu-\frac{1}{2}}(-x)\right]\,\mathrm{d}x = \sqrt{2\pi}\sin\left[\left(\nu - \tfrac{1}{4}\right)\pi\right]b^{2\nu-\frac{1}{2}}e^{-\frac{1}{2}b^2}$$

$$\left[\operatorname{Re}\nu > \tfrac{1}{4}, \quad b > 0\right] \qquad\qquad \text{ET I 115(2)}$$

5.  $$\int_0^\infty e^{-\frac{1}{2}x^2}\cos(bx)\left[D_{2\nu-\frac{1}{2}}(x) + D_{2\nu-\frac{1}{2}}(-x)\right]\,\mathrm{d}x = \frac{2^{\frac{1}{4}-2\nu}\sqrt{\pi}b^{2\nu-\frac{1}{2}}e^{-\frac{1}{4}b^2}}{\operatorname{cosec}\left[\left(\nu + \tfrac{1}{4}\right)\pi\right]}$$

$$\left[\operatorname{Re}\nu > \tfrac{1}{4}, \quad b > 0\right] \qquad\qquad \text{ET I 61(4)}$$

**7.742**

1.  $$\int_0^\infty x^{2\rho-1}\sin(ax)e^{-\frac{x^2}{4}}D_{2\nu}(x)\,\mathrm{d}x = 2^{\nu-\rho-\frac{1}{2}}\pi^{1/2}a\frac{\Gamma(2\rho+1)}{\Gamma(\rho-\nu+1)}$$
    $$\times\,{}_2F_2\left(\rho+\frac{1}{2},\rho+1;\frac{3}{2},\rho-\nu+1;-\frac{a^2}{2}\right)$$

$$\left[\operatorname{Re}\rho > -\tfrac{1}{2}\right] \qquad\qquad \text{ET II 396(8)}$$

2.  $$\int_0^\infty x^{2\rho-1}\sin(ax)e^{\frac{x^2}{4}}D_{2\nu}(x)\,\mathrm{d}x = \frac{2^{\rho-\nu-2}}{\Gamma(-2\nu)}G_{23}^{22}\left(\frac{a^2}{2}\,\middle|\,\begin{matrix}\frac{1}{2}-\rho, 1-\rho\\ -\rho-\nu, \frac{1}{2}, 0\end{matrix}\right)$$

$$\left[a > 0, \quad \operatorname{Re}\rho > -\tfrac{1}{2}, \quad \operatorname{Re}(\rho+\nu) < \tfrac{1}{2}\right]$$
$$\text{ET II 396(9)}$$

3.  $$\int_0^\infty x^{2\rho-1}\cos(ax)e^{-\frac{x^2}{4}}D_{2\nu}(x)\,\mathrm{d}x = \frac{2^{\nu-\rho}\Gamma(2\rho)\pi^{1/2}}{\Gamma\left(\rho-\nu+\frac{1}{2}\right)}\,{}_2F_2\left(\rho,\rho+\frac{1}{2};\frac{1}{2},\rho-\nu+\frac{1}{2};-\frac{a^2}{2}\right)$$

$$\left[\operatorname{Re}\rho > 0\right] \qquad\qquad \text{ET II 396(10)a}$$

4.  $$\int_0^\infty x^{2\rho-1}\cos(ax)e^{\frac{x^2}{4}}D_{2\nu}(x)\,\mathrm{d}x = \frac{2^{\rho-\nu-2}}{\Gamma(-2\nu)}G_{23}^{22}\left(\frac{a^2}{2}\,\middle|\,\begin{matrix}\frac{1}{2}-\rho, 1-\rho\\ -\rho-\nu, 0, \frac{1}{2}\end{matrix}\right)$$

$$\left[a > 0, \quad \operatorname{Re}\rho > 0, \quad \operatorname{Re}(\rho+\nu) < \tfrac{1}{2}\right]$$
$$\text{ET II 396(11)}$$

**7.743**   $$\int_0^{\pi/2}(\cos x)^{-\mu-2}(\sin x)^{-\nu}D_\nu(a\sin x)D_\mu(a\cos x)\,\mathrm{d}x = -\left(\tfrac{1}{2}\pi\right)^{1/2}(1+\mu)^{-1}D_{\mu+\nu+1}(a)$$

$$\left[\operatorname{Re}\nu < 1, \quad \operatorname{Re}\mu < -1\right] \qquad \text{ET II 397(19)}$$

**7.744**

1. $\int_0^\infty \sin(bx) \left[ D_{-\nu-\frac{1}{2}}\left(\sqrt{2x}\right) - D_{-\nu-\frac{1}{2}}\left(-\sqrt{2x}\right) \right] D_{\nu-\frac{1}{2}}\left(\sqrt{2x}\right) dx$

$$= -\sqrt{2\pi} \sin\left[\left(\tfrac{1}{4} + \tfrac{1}{2}\nu\right)\pi\right] b^{-\nu-\frac{1}{2}} \frac{\left(1 + \sqrt{1+b^2}\right)^\nu}{\sqrt{1+b^2}}$$

$$[b > 0] \qquad \text{ET I 115(4)}$$

2. $\int_0^\infty \cos(bx) \left[ D_{-2\nu-\frac{1}{2}}\left(\sqrt{2x}\right) + D_{-2\nu-\frac{1}{2}}\left(-\sqrt{2x}\right) \right] D_{2\nu-\frac{1}{2}}\left(\sqrt{2x}\right) dx$

$$= -\frac{\sqrt{\pi} \sin\left[\left(\nu - \tfrac{1}{4}\right)\pi\right] \left(1 + \sqrt{1+b^2}\right)^{2\nu}}{\sqrt{1+b^2} b^{2\nu+\frac{1}{2}}}$$

$$[b > 0] \qquad \text{ET I 60(3)}$$

## 7.75 Combinations of parabolic cylinder and Bessel functions

**7.751**

1. $\int_0^\infty [D_n(ax)]^2 \, J_1(xy) \, dx = (-1)^{n-1} y^{-1} \left[ D_n\left(\frac{y}{a}\right) \right]^2 \qquad [y > 0] \qquad \text{ET II 20(24)}$

2. $\int_0^\infty J_0(xy) \, D_n(ax) \, D_{n+1}(ax) \, dx = (-1)^n y^{-1} D_n\left(\frac{y}{a}\right) D_{n+1}\left(\frac{y}{a}\right)$

$$[y > 0, \quad |\arg a| < \tfrac{1}{4}\pi] \qquad \text{ET II 17(42)}$$

3. $\int_0^\infty J_0(xy) \, D_\nu(x) \, D_{\nu+1}(x) \, dx = 2^{-1} y^{-1} \left[ D_\nu(-y) \, D_{\nu+1}(y) - D_{\nu+1}(-y) \, D_\nu(y) \right] \qquad \text{ET II 397(17)a}$

**7.752**

1. $\int_0^\infty x^\nu e^{-\frac{1}{4}x^2} D_{2\nu-1}(x) \, J_\nu(xy) \, dx = -\frac{1}{2} \sec(\nu\pi) y^{\nu-1} e^{-\frac{1}{4}y^2} \left[ D_{2\nu-1}(y) - D_{2\nu-1}(-y) \right]$

$$\left[ y > 0, \quad \operatorname{Re}\nu > -\tfrac{1}{2} \right]$$

$$\text{ET II 76(1), MO 183}$$

2. $\int_0^\infty x^\nu e^{\frac{1}{4}x^2} D_{2\nu-1}(x) \, J_\nu(xy) \, dx = 2^{\frac{1}{2}-\nu} \pi \sin(\nu\pi) y^{-\nu} \Gamma(2\nu) e^{\frac{1}{4}y^2} K_\nu\left(\tfrac{1}{4}y^2\right)$

$$\left[ y > 0, \quad -\tfrac{1}{2} < \operatorname{Re}\nu < \tfrac{1}{2} \right] \qquad \text{ET II 77(4)}$$

3. $\int_0^\infty x^{\nu+1} e^{-\frac{1}{4}x^2} D_{2\nu}(x) \, J_\nu(xy) \, dx = \frac{1}{2} \sec(\nu\pi) y^{\nu-1} e^{-\frac{1}{4}y^2} \left[ D_{2\nu+1}(y) - D_{2\nu+1}(-y) \right]$

$$[y > 0, \quad \operatorname{Re}\nu > -1] \qquad \text{ET II 78(13)}$$

4. $\int_0^\infty x^\nu e^{-\frac{1}{4}x^2} D_{2\nu+1}(x) \, J_\nu(xy) \, dx = \frac{1}{2} \sec(\nu\pi) e^{-\frac{1}{4}y^2} y^\nu \left[ D_{2\nu}(y) + D_{2\nu}(-y) \right]$

$$\left[ y > 0, \quad \operatorname{Re}\nu > -\tfrac{1}{2} \right] \qquad \text{ET II 77(5)}$$

5. $\int_0^\infty x^{\nu+1} e^{-\frac{1}{4}x^2} D_{2\nu+2}(x) \, J_\nu(xy) \, dx = -\tfrac{1}{2} \sec(\nu\pi) y^\nu e^{-\frac{1}{4}y^2} \left[ D_{2\nu+2}(y) + D_{2\nu+2}(-y) \right]$

$$[\operatorname{Re}\nu > -1, \quad y > 0] \qquad \text{ET II 78(16)}$$

6.  $\displaystyle\int_0^\infty x^{\nu+1} e^{\frac{1}{4}x^2} D_{2\nu+2}(x)\, J_\nu(xy)\, \mathrm{d}x = \pi^{-1}\sin(\nu\pi)\,\Gamma(2\nu+3)\, y^{-\nu-2} e^{\frac{1}{4}y^2}\, K_{\nu+1}\left(\tfrac{1}{4}y^2\right)$

$$\left[y>0,\quad -1<\operatorname{Re}\nu<-\tfrac{5}{6}\right]$$

ET II 78(19)

7.  $\displaystyle\int_0^\infty x^\nu e^{-\frac{1}{4}x^2} D_{-2\nu}(x)\, J_\nu(xy)\, \mathrm{d}x = 2^{-1/2}\pi^{1/2} y^{-\nu} e^{-\frac{1}{4}y^2}\, I_\nu\left(\tfrac{1}{4}y^2\right)$

$$\left[y>0,\quad \operatorname{Re}\nu>-\tfrac{1}{2}\right]\qquad \text{ET II 77(8)}$$

8.  $\displaystyle\int_0^\infty x^\nu e^{\frac{1}{4}x^2} D_{-2\nu}(x)\, J_\nu(xy)\, \mathrm{d}x = y^{\nu-1} e^{\frac{1}{4}y^2}\, D_{-2\nu}(y)\qquad \left[\operatorname{Re}\nu>-\tfrac{1}{2},\quad y>0\right]$

ET II 77(9), EH II 121(17)

9.  $\displaystyle\int_0^\infty x^\nu e^{\frac{1}{4}x^2} D_{-2\nu-2}(x)\, J_\nu(xy)\, \mathrm{d}x = (2\nu+1)^{-1} y^\nu e^{\frac{1}{4}y^2}\, D_{-2\nu-1}(y)$

$$\left[y>0,\quad \operatorname{Re}\nu>-\tfrac{1}{2}\right]\qquad \text{ET II 77(10)}$$

10. $\displaystyle\int_0^\infty x^\nu e^{-\frac{1}{4}a^2 x^2} D_{2\mu}(ax)\, J_\nu(xy)\, \mathrm{d}x = \frac{2^{\mu-\frac{1}{2}}\,\Gamma\left(\nu+\frac{1}{2}\right) y^\nu}{\Gamma(\nu-\mu+1)a^{1+2\nu}}\, {}_1F_1\left(\nu+\frac{1}{2}; \nu-\mu+1; -\frac{y^2}{2a^2}\right)$

$$\left[y>0,\quad |\arg a|<\tfrac{1}{4}\pi,\quad \operatorname{Re}\nu>-\tfrac{1}{2}\right]$$

ET II 77(11)

11. $\displaystyle\int_0^\infty x^\nu e^{\frac{1}{4}a^2 x^2} D_{2\mu}(ax)\, J_\nu(xy)\, \mathrm{d}x = \frac{\Gamma\left(\frac{1}{2}+\nu\right) a^{2k} 2^{m+\mu}}{\Gamma\left(\frac{1}{2}-\mu\right) y^{\mu+\frac{3}{2}}}\, e^{\frac{y^2}{4a^2}}\, W_{k,m}\left(\frac{y^2}{4a^2}\right)$

$$2k=\tfrac{1}{2}+\mu-\nu,\quad 2m=\tfrac{1}{2}+\mu+\nu$$

$$\left[y>0,\quad |\arg a|<\tfrac{1}{4}\pi,\quad -\tfrac{1}{2}<\operatorname{Re}\nu<\operatorname{Re}\left(\tfrac{1}{2}-2\mu\right)\right]$$

ET II 78(12)

12. $\displaystyle\int_0^\infty x^{\nu+1} e^{-\frac{1}{4}a^2 x^2} D_{2\mu}(ax)\, J_\nu(xy)\, \mathrm{d}x = \frac{2^\mu\,\Gamma\left(\nu+\frac{3}{2}\right) y^\nu}{\Gamma\left(\nu-\mu+\frac{3}{2}\right) a^{2\nu+2}}\, {}_1F_1\left(\nu+\frac{3}{2}; \nu-\mu+\frac{3}{2}; -\frac{y^2}{2a^2}\right)$

$$\left[y>0,\quad |\arg a|<\tfrac{1}{4}\pi,\quad \operatorname{Re}\nu>-1\right]$$

ET II 79(23)

13. $\displaystyle\int_0^\infty x^{\nu+1} e^{\frac{1}{4}a^2 x^2} D_{2\mu}(ax)\, J_\nu(xy)\, \mathrm{d}x = \frac{\Gamma\left(\frac{3}{2}+\nu\right) 2^{\frac{1}{2}+m+\mu} a^{2k+1}}{\Gamma(-\mu) y^{\mu+2}}\, e^{\frac{y^2}{4a^2}}\, W_{k,m}\left(\frac{y^2}{2a^2}\right)$

$$2k=\mu-\nu-1,\quad 2m=\mu+\nu+1$$

$$\left[y>0,\quad |\arg a|<\tfrac{3}{4}\pi,\quad -1<\operatorname{Re}\nu<-\tfrac{1}{2}-2\operatorname{Re}\mu\right]$$

ET II 79(24)

14. $\displaystyle\int_0^\infty x^{\lambda+\frac{1}{2}} e^{\frac{1}{4}a^2 x^2} D_\mu(ax)\, J_\nu(xy)\, \mathrm{d}x = \frac{2^{\lambda-\frac{1}{2}\mu}\pi^{-\frac{1}{2}}}{\Gamma(-\mu) y^{\lambda+\frac{3}{2}}}\, G_{23}^{22}\left(\frac{y^2}{2a^2}\ \middle|\ \begin{matrix}\frac{1}{2},1\\ \frac{3}{4}+\frac{\lambda+\nu}{2}, -\frac{\mu}{2}, \frac{3}{4}+\frac{\lambda-\nu}{2}\end{matrix}\right)$

$$\left[y>0,\quad |\arg a|<\tfrac{3}{4}\pi,\quad \operatorname{Re}\mu<-\operatorname{Re}\lambda<\operatorname{Re}\nu+\tfrac{3}{2}\right]\qquad \text{ET II 80(26)}$$

15. $\displaystyle\int_0^\infty x^{\nu+1} e^{\frac{1}{4}x^2} D_{-2\nu-1}(x)\, J_\nu(xy)\, \mathrm{d}x = (2\nu+1) y^{\nu-1} e^{\frac{1}{4}y^2}\, D_{-2\nu-2}(y)$

$$\left[y>0,\quad \operatorname{Re}\nu>-\tfrac{1}{2}\right]\qquad \text{ET II 79(20)}$$

16.    $\displaystyle\int_0^\infty x^{\nu+1} e^{-\frac{1}{4}x^2} D_{-2\nu-3}(x) \, J_\nu(xy) \, dx = 2^{-1/2} \pi^{1/2} y^{-\nu-2} e^{-\frac{1}{4}y^2} I_{\nu+1}\left(\tfrac{1}{4}y^2\right)$

$$[y > 0, \quad \operatorname{Re}\nu > -1] \qquad \text{ET II 79(21)}$$

17.    $\displaystyle\int_0^\infty x^{\nu+1} e^{\frac{1}{4}x^2} D_{-2\nu-3}(x) \, J_\nu(xy) \, dx = y^\nu e^{\frac{1}{4}y^2} D_{-2\nu-3}(y)$

$$[y > 0, \quad \operatorname{Re}\nu > -1] \qquad \text{ET II 79(22)}$$

18.    $\displaystyle\int_0^\infty x^\nu e^{\frac{1}{4}a^2 x^2} D_{\frac{1}{2}\nu-\frac{1}{2}}(ax) \, Y_\nu(xy) \, dx = -\pi^{-1} 2^{\frac{3}{4}\nu+\frac{3}{4}} a^{-\nu} y^{-1} \Gamma(\nu+1) e^{\frac{y^2}{4a^2}} W_{-\frac{1}{2}\nu-\frac{1}{2},\frac{1}{2}\nu}\left(\dfrac{y^2}{2a^2}\right)$

$$\left[y > 0, \quad |\arg a| < \tfrac{3}{4}\pi, \quad -\tfrac{1}{2} < \operatorname{Re}\nu < \tfrac{2}{3}\right] \qquad \text{ET II 115(39)}$$

**7.753**

1.    $\displaystyle\int_0^\infty x^{\nu-\frac{1}{2}} e^{-(x+a)^2} I_{\nu-\frac{1}{2}}(2ax) \, D_\nu(2x) \, dx = \frac{1}{2} \pi^{-1/2} \Gamma(\nu) a^{\nu-\frac{1}{2}} D_{-\nu}(2a)$

$$[\operatorname{Re}a > 0, \quad \operatorname{Re}\nu > 0] \qquad \text{ET II 397(12)}$$

2.    $\displaystyle\int_0^\infty x^{\nu-\frac{3}{2}} e^{-(x+a)^2} I_{\nu-\frac{3}{2}}(2ax) \, D_\nu(2x) \, dx = \frac{1}{2} \pi^{-1/2} \Gamma(\nu) a^{\nu-\frac{3}{2}} D_{-\nu}(2a)$

$$[\operatorname{Re}a > 0, \quad \operatorname{Re}\nu > 1] \qquad \text{ET II 397(13)}$$

**7.754**

1.    $\displaystyle\int_0^\infty x^\nu e^{-\frac{1}{4}x^2} \left\{[1 \mp 2\cos(\nu\pi)] D_{2\nu-1}(x) - D_{2\nu-1}(-x)\right\} J_\nu(xy) \, dx$

$$= \pm y^{\nu-1} e^{-\frac{1}{4}y^2} \left\{[1 \mp 2\cos(\nu\pi)] D_{2\nu-1}(y) - D_{2\nu-1}(-y)\right\}$$
$$\left[y > 0, \quad \operatorname{Re}\nu > -\tfrac{1}{2}\right] \qquad \text{ET II 76(2, 3)}$$

2.    $\displaystyle\int_0^\infty x^\nu e^{-\frac{1}{4}x^2} \left\{[1 \mp 2\cos(\nu\pi)] D_{2\nu+1}(x) - D_{2\nu+1}(-x)\right\} J_\nu(xy) \, dx$

$$= \mp y^\nu e^{-\frac{1}{4}y^2} \left\{[1 \mp 2\cos(\nu\pi)] D_{2\nu}(y) + D_{2\nu}(-y)\right\}$$
$$\left[y > 0, \quad \operatorname{Re}\nu > -\tfrac{1}{2}\right] \qquad \text{ET II 77(6, 7)}$$

3.    $\displaystyle\int_0^\infty x^{\nu+1} e^{-\frac{1}{4}x^2} \left\{[1 \pm 2\cos(\nu\pi)] D_{2\nu}(x) + D_{2\nu}(-x)\right\} J_\nu(xy) \, dx$

$$= \pm y^{\nu-1} e^{-\frac{1}{4}y^2} \left\{[1 \pm 2\cos(\nu\pi)] D_{2\nu+1}(y) - D_{2\nu+1}(-y)\right\}$$
$$[y > 0, \quad \operatorname{Re}\nu > -1] \qquad \text{ET II 78(14, 15)}$$

4.    $\displaystyle\int_0^\infty x^{\nu+1} e^{-\frac{1}{4}x^2} \left\{[1 \mp 2\cos(\nu\pi)] D_{2\nu+2}(x) + D_{2\nu+2}(-x)\right\} J_\nu(xy) \, dx$

$$= \pm y^\nu e^{-\frac{1}{4}y^2} \left\{[1 \mp 2\cos(\nu\pi)] D_{2\nu+2}(y) + D_{2\nu+2}(-y)\right\}$$
$$[y > 0, \quad \operatorname{Re}\nu > -1] \qquad \text{ET II 78(17, 18)}$$

**7.755**

1.[12] $\displaystyle\int_0^\infty x^{-1/2}\, D_\nu\left(\sqrt{ax}\right) D_{-\nu-1}\left(\sqrt{ax}\right) J_0(xy)\,dx$

$$= 2^{-3/2}\pi a^{-1/2}\, P_{-\frac14}^{\frac12\nu+\frac14}\left[\left(1+\frac{4y^2}{a^2}\right)^{1/2}\right] P_{-\frac14}^{-\frac12\nu-\frac14}\left[\left(1+\frac{4y^2}{a^2}\right)^{1/2}\right]$$
$$[y>0,\ \mathrm{Re}\,a>0]\qquad\qquad\text{ET II 17(43)}$$

2. $\displaystyle\int_0^\infty x^{1/2}\, D_{-\frac12-\nu}\left(ae^{\frac14\pi i}x^{1/2}\right) D_{-\frac12-\nu}\left(ae^{-\frac14\pi i}x^{1/2}\right) J_\nu(xy)\,dx$

$$= 2^{-\nu}\pi^{1/2}y^{-\nu-1}\left(a^2+2y\right)^{-1/2}\left[\Gamma\left(\nu+\tfrac12\right)\right]^{-1}\left[\left(a^2+2y\right)^{1/2}-a\right]^{2\nu}$$
$$[y>0,\quad \mathrm{Re}\,a>0,\quad \mathrm{Re}\,\nu>-\tfrac12]\quad\text{ET II 80(27)}$$

3.[12] $\displaystyle\int_0^\infty D_{-\frac12-\nu}\left(ae^{\frac14\pi i}x^{-1/2}\right) D_{-\frac12-\nu}\left(ae^{-\frac14\pi i}x^{-1/2}\right) J_\nu(xy)\,dx$

$$= 2^{1/2}\pi^{1/2}y^{-1}\left[\Gamma\left(\nu+\tfrac12\right)\right]^{-1}\exp\left[-a(2y)^{1/2}\right]$$
$$[y>0,\quad \mathrm{Re}\,a>0,\quad \mathrm{Re}\,\nu>-\tfrac12]\quad\text{ET II 80(28)a}$$

4. $\displaystyle\int_0^\infty x^{1/2}\, D_{\nu-\frac12}\left(ax^{-1/2}\right) D_{-\nu-\frac12}\left(ax^{-1/2}\right) Y_\nu(xy)\,dx$

$$= y^{-3/2}\exp\left(-ay^{1/2}\right)\sin\left[ay^{1/2}-\tfrac12\left(\nu-\tfrac12\right)\pi\right]$$
$$[y>0,\quad |\arg a|<\tfrac14\pi]\qquad\text{ET II 115(40)}$$

5. $\displaystyle\int_0^\infty x^{1/2}\, D_{\nu-\frac12}\left(ax^{-1/2}\right) D_{-\nu-\frac12}\left(ax^{-1/2}\right) K_\nu(xy)\,dx = 2^{-1}y^{-3/2}\pi\exp\left[-a(2y)^{1/2}\right]$

$$[\mathrm{Re}\,y>0,\quad |\arg a|<\tfrac14\pi]\quad\text{ET II 151(81)}$$

**Combinations of parabolic cylinder and Struve functions**

**7.756** $\displaystyle\int_0^\infty x^{-\nu}e^{-\frac14x^2}\left[D_\mu(x)-D_\mu(-x)\right]\mathbf{H}_\nu(xy)\,dx$

$$= \frac{2^{3/2}\,\Gamma\left(\tfrac12\mu+\tfrac12\right)}{\Gamma\left(\tfrac12\mu+\nu+1\right)}y^{\mu+\nu}\sin\left(\tfrac12\mu\pi\right)\,{}_1F_1\left(\tfrac12\mu+\tfrac12;\tfrac12\mu+\nu+1;-\tfrac12y^2\right)$$
$$[y>0,\quad \mathrm{Re}(\mu+\nu)>-\tfrac32,\quad \mathrm{Re}\,\mu>-1]\quad\text{ET II 171(41)}$$

## 7.76 Combinations of parabolic cylinder functions and confluent hypergeometric functions

**7.761**

1.    $\displaystyle\int_0^\infty e^{\frac{1}{4}t^2} t^{2c-1} D_{-\nu}(t)\, {}_1F_1\left(a; c; -\frac{1}{2}pt^2\right) dt$

$$= \frac{\pi^{1/2}}{2^{c+\frac{1}{2}\nu}} \frac{\Gamma(2c)\,\Gamma\left(\frac{1}{2}\nu - c + a\right)}{\Gamma\left(\frac{1}{2}\nu\right)\Gamma\left(a + \frac{1}{2} + \frac{1}{2}\nu\right)} F\left(a, c + \frac{1}{2}; a + \frac{1}{2} + \frac{1}{2}\nu; 1 - p\right)$$

$$[|1-p| < 1, \quad \operatorname{Re} c > 0, \quad \operatorname{Re}\nu > 2\operatorname{Re}(c - a)] \quad \text{EH II 121(12)}$$

2.    $\displaystyle\int_0^\infty e^{\frac{1}{4}t^2} t^{2c-2} D_{-\nu}(t)\, {}_1F_1\left(a; c; -\frac{1}{2}pt^2\right) dt$

$$= \frac{\pi^{1/2}}{2^{c+\frac{1}{2}\nu-\frac{1}{2}}} \frac{\Gamma(2c-1)\,\Gamma\left(\frac{1}{2}\nu + \frac{1}{2} - c + a\right)}{\Gamma\left(\frac{1}{2} + \frac{1}{2}\nu\right)\Gamma\left(a + \frac{1}{2}\nu\right)} F\left(a, c - \frac{1}{2}; a + \frac{1}{2}\nu; 1 - p\right)$$

$$\left[|1-p| < 1, \quad \operatorname{Re} c > \frac{1}{2}, \quad \operatorname{Re}\nu > 2\operatorname{Re}(c - a) - 1\right] \quad \text{EH II 121(13)}$$

## 7.77 Integration of a parabolic cylinder function with respect to the index

**7.771**    $\displaystyle\int_0^\infty \cos(ax)\, D_{x-\frac{1}{2}}(\beta)\, D_{-x-\frac{1}{2}}(\beta)\, dx = \frac{1}{2}\left(\frac{\pi}{\cos a}\right)^{1/2} \exp\left(-\frac{\beta^2 \cos a}{2}\right)$    $\left[|a| < \frac{1}{2}\pi\right]$

$$= 0 \qquad\qquad \left[|a| > \tfrac{1}{2}\pi\right]$$

$$\text{ET II 298(22)}$$

**7.772**

1.    $\displaystyle\int_{-\frac{1}{2}-i\infty}^{-\frac{1}{2}+i\infty} \left[\frac{\left(\tan\frac{1}{2}\varphi\right)^\nu}{\cos\frac{1}{2}\varphi} D_\nu\left(-e^{\frac{1}{4}i\pi}\xi\right) D_{-\nu-1}\left(e^{\frac{1}{4}i\pi}\eta\right)\right.$

$$\left. + \frac{\left(\cot\frac{1}{2}\varphi\right)^\nu}{\sin\frac{1}{2}\varphi} D_{-\nu-1}\left(e^{\frac{1}{4}i\pi}\xi\right) D_\nu\left(-e^{\frac{1}{4}i\pi}\eta\right)\right] \frac{d\nu}{\sin\nu\pi}$$

$$= -2i(2\pi)^{1/2} \exp\left[-\tfrac{1}{4}i\left(\xi^2 - \eta^2\right)\cos\varphi - \tfrac{1}{2}i\xi\eta\sin\varphi\right]$$

$$\text{EH II 125(7)}$$

2.    $\displaystyle\int_{-\frac{1}{2}-i\infty}^{-\frac{1}{2}+i\infty} \frac{\left(\tan\frac{1}{2}\varphi\right)^\nu}{\cos\frac{1}{2}\varphi} D_\nu\left(-e^{\frac{1}{4}i\pi}\zeta\right) D_{-\nu-1}\left(e^{\frac{1}{4}i\pi}\eta\right) \frac{d\nu}{\sin\nu\pi}$

$$= -2i\, D_0\left[e^{\frac{1}{4}i\pi}\left(\zeta\cos\tfrac{1}{2}\varphi + \eta\sin\tfrac{1}{2}\varphi\right)\right] D_{-1}\left[e^{\frac{1}{4}i\pi}\left(\eta\cos\tfrac{1}{2}\varphi - \zeta\sin\tfrac{1}{2}\varphi\right)\right]$$

$$\text{EH II 125(8)}$$

**7.773**

1.    $\displaystyle\int_{c-i\infty}^{c+i\infty} D_\nu(z) t^\nu \, \Gamma(-\nu)\, d\nu = 2\pi i e^{-\frac{1}{4}z^2 - zt - \frac{1}{2}t^2}$     $\left[c < 0, \quad |\arg t| < \dfrac{\pi}{4}\right]$    EH II 126(10)

$2.^{12}$ $\displaystyle\int_{c-i\infty}^{c+i\infty} \left[ D_\nu(x)\, D_{-\nu-1}(iy) + D_\nu(-x)\, D_{-\nu-1}(-iy) \right] \frac{t^{-\nu-1}\, d\nu}{\sin(-\nu\pi)}$

$$= \frac{2\pi i}{\left(\dfrac{\pi}{2}\right)^{1/2}} \left( 1 + t^2 \right)^{-\frac{1}{2}} \exp\left[ \frac{1}{4} \frac{1 - t^2}{1 + t^2} \left( x^2 + y^2 \right) + i \frac{txy}{1 + t^2} \right]$$

$$\left[ -1 < c < 0, \quad |\arg t| < \frac{1}{2}\pi \right] \qquad \text{EH II 126(11)}$$

**7.774** $\displaystyle\int_{c-i\infty}^{c+i\infty} D_\nu\left[ k^{\frac{1}{2}}(1+i)\xi \right] D_{-\nu-1}\left[ k^{\frac{1}{2}}(1+i)\eta \right] \Gamma\left( -\tfrac{1}{2}\nu \right) \Gamma\left( \tfrac{1}{2} + \tfrac{1}{2}\nu \right)\, d\nu = 2^{1/2}\pi^2\, H_0^{(2)}\left[ \tfrac{1}{2}k\left( \xi^2 + \eta^2 \right) \right]$

$$\left[ -1 < c < 0, \quad \operatorname{Re} ik \geq 0 \right] \qquad \text{EH II 125(9)}$$

# 7.8 Meijer's and MacRobert's Functions ($G$ and $E$)

## 7.81 Combinations of the functions $G$ and $E$ and the elementary functions

**7.811**

1.    $\displaystyle\int_0^\infty G_{p,q}^{m,n}\left( \eta x \, \middle| \, \begin{matrix} a_1, \ldots, a_p \\ b_1, \ldots, b_q \end{matrix} \right) G_{\sigma,\tau}^{\mu,\nu}\left( \omega x \, \middle| \, \begin{matrix} c_1, \ldots, c_\sigma \\ d_1, \ldots, d_\tau \end{matrix} \right)\, dx$

$$= \frac{1}{\eta} G_{q+\sigma,p+\tau}^{n+\mu,m+\nu}\left( \frac{\omega}{\eta} \, \middle| \, \begin{matrix} -b_1, \ldots, -b_m, c_1, \ldots, c_\sigma, -b_{m+1}, \ldots, -b_q \\ -a_1, \ldots, -a_n, d_1, \ldots, d_\tau, -a_{n+1}, \ldots, -a_p \end{matrix} \right)$$

subject to the following constraints

- $m, n, p, q, \mu, \nu, \sigma, \tau$ are integers;
- $1 \leq n \leq p < q < p + \tau - \sigma$
- $\frac{1}{2}p + \frac{1}{2}q - n < m \leq q, \quad 0 \leq \nu \leq \sigma, \quad \frac{1}{2}\sigma + \frac{1}{2}\tau - \nu < \mu \leq \tau$;
- $\operatorname{Re}(b_j + d_k) > -1 \qquad (j = 1, \ldots, m; k = 1, \ldots, \mu)$,
- $\operatorname{Re}(a_j + c_k) < 1 \qquad (j = 1, \ldots, n; k = 1, \ldots, \tau)$;
- $\omega \neq 0, \quad \eta \neq 0, \quad |\arg \eta| < \left( m + n - \frac{1}{2}p - \frac{1}{2}q \right)\pi, \quad |\arg \omega| < \left( \mu + \nu - \frac{1}{2}\sigma - \frac{1}{2}\tau \right)\pi$
- The following must not be integers:

$$\begin{aligned} b_j - b_k &\quad (j = 1, \ldots, m; k = 1, \ldots, m; j \neq k), \\ a_j - a_k &\quad (j = 1, \ldots, n; k = 1, \ldots, n; j \neq k), \\ d_j - d_k &\quad (j = 1, \ldots, \mu; k = 1, \ldots, \mu; j \neq k), \\ a_j + d_k &\quad (j = 1, \ldots, n; k = 1, \ldots, n); \end{aligned}$$

- The following must not be positive integers:

$$\begin{aligned} a_j - b_k &\quad (j = 1, \ldots, n; k = 1, \ldots, m), \\ c_j - d_k &\quad (j = 1, \ldots, \nu; k = 1, \ldots, \mu); \end{aligned}$$

Formula **7.811** 1 also holds for four sets of restrictions. See C. S. Meijer, Neue Integraldarstellungen für Whittakersche Funktionen, Nederl. Akad. Wetensch. Proc. **44** (1941), 82–92.

ET II 422(14)

Hereafter, $G_{p,q}^{m,n}$ will be written as $G_{pq}^{mn}$, and commas will only be inserted in entries like $G_{p+1,q+1}^{m,n+1}$ where their omission could cause ambiguity.

2.  $$\int_0^1 x^{\rho-1}(1-x)^{\sigma-1}\, G_{pq}^{mn}\left(\alpha x\left|\begin{matrix}a_1,\ldots,a_p\\b_1,\ldots,b_q\end{matrix}\right.\right)\,\mathrm{d}x = \Gamma(\sigma)\, G_{p+1,q+1}^{m,n+1}\left(\alpha\left|\begin{matrix}1-\rho,a_1,\ldots,a_p\\b_1,\ldots,b_q,1-\rho-\sigma\end{matrix}\right.\right)$$

    where

    - $(p+q) < 2(m+n)$;
    - $|\arg a| < \left(m+n-\frac{1}{2}p-\frac{1}{2}q\right)\pi$;
    - $\mathrm{Re}\,(\rho+b_j) > 0,\ j=1,\ldots,m$;
    - $\mathrm{Re}\,\sigma > 0$,
    - either

    $$p+q \le 2(m+n), \quad |\arg\alpha| \le \left(m+n-\tfrac{1}{2}\rho-\tfrac{1}{2}q\right)\pi,$$
    $$\mathrm{Re}\,(\rho+b_j) > 0; \quad j=1,\ldots,m; \quad \mathrm{Re}\,\sigma > 0,$$

    $$\mathrm{Re}\left[\sum_{j=1}^p a_j - \sum_{j=1}^q b_j + (p-q)\left(\rho-\frac{1}{2}\right)\right] > -\frac{1}{2},$$

    or

    $$p < q \quad (\text{or } p \le q \text{ for } |\alpha| < 1), \quad \mathrm{Re}\,(p+b_j) > 0; \quad j=1,\ldots,m; \quad \mathrm{Re}\,\sigma > 0$$

    ET II 417(1)

3.  $$\int_1^\infty x^{-\rho}(x-1)^{\sigma-1}\, G_{pq}^{mn}\left(\alpha x\left|\begin{matrix}a_1,\ldots,a_p\\b_1,\ldots,b_q\end{matrix}\right.\right)\,\mathrm{d}x = \Gamma(\sigma)\, G_{p+1,q+1}^{m+1,n}\left(\alpha\left|\begin{matrix}a_1,\ldots,a_p,\rho\\\rho-\sigma,b_1,\ldots,b_q\end{matrix}\right.\right)$$

    where

    - $p+q < 2(m+n)$
    - $|\arg\alpha| < \left(m+n-\frac{1}{2}p-\frac{1}{2}q\right)\pi$
    - $\mathrm{Re}\,(\rho-\sigma-a_j) > -1; \quad j=1,\ldots,n$
    - $\mathrm{Re}\,\sigma > 0$
    - either

    $$p+q \le 2(m+n), \quad |\arg\alpha| \le \left(m+n-\tfrac{1}{2}p-\tfrac{1}{2}q\right)\pi,$$
    $$\mathrm{Re}\,(\rho-\sigma-a_j) > -1; \quad j=1,\ldots,n; \quad \mathrm{Re}\,\sigma > 0,$$

    $$\mathrm{Re}\left[\sum_{j=1}^p a_j - \sum_{j=1}^q b_j + (q-p)\left(\rho-\sigma+\frac{1}{2}\right)\right] > -\frac{1}{2},$$

    or

    $$q < p \quad (\text{or } q \le p \text{ for } |\alpha| > 1), \quad \mathrm{Re}\,(\rho-\sigma-a_j) > -1; \quad j=1,\ldots,n; \quad \mathrm{Re}\,\sigma > 0$$

    ET II 417(2)

4.  $$\int_0^\infty x^{\rho-1}\, G_{pq}^{mn}\left(\alpha x\left|\begin{matrix}a_1,\ldots,a_p\\b_1,\ldots,b_q\end{matrix}\right.\right)\,\mathrm{d}x = \frac{\prod_{j=1}^m \Gamma(b_j+\rho)\prod_{j=1}^n \Gamma(1-a_j-\rho)}{\prod_{j=m+1}^q \Gamma(1-b_j-\rho)\prod_{j=n+1}^p \Gamma(a_j+\rho)}\alpha^{-\rho}$$
    $$p+q < 2(m+n), \quad |\arg\alpha| < \left(m+n-\tfrac{1}{2}p-\tfrac{1}{2}q\right)\pi, \quad -\min_{1\le j\le m}\mathrm{Re}\,b_j < \mathrm{Re}\,\rho < 1-\max_{1\le j\le n}\mathrm{Re}\,a_j$$

    ET II 418(3)a, ET I 337(14)

$5.^{12}$  $\displaystyle\int_0^\infty x^{\rho-1}(x+\beta)^{-\sigma}\, G_{pq}^{mn}\left(\alpha x\,\middle|\,\begin{matrix}a_1,\ldots,a_p\\b_1,\ldots,b_q\end{matrix}\right)\mathrm{d}x = \frac{\beta^{\rho-\sigma}}{\Gamma(\sigma)}\, G_{p+1,q+1}^{m+1,n+1}\left(\alpha\beta\,\middle|\,\begin{matrix}1-\rho,a_1,\ldots,a_p\\\sigma-\rho,b_1,\ldots,b_q\end{matrix}\right)$

where

- $p+q < 2(m+n)$
- $|\arg\alpha| < \left(m+n-\frac{1}{2}p-\frac{1}{2}q\right)\pi$
- $|\arg\beta| < \pi$
- $\mathrm{Re}\,(\rho+b_j) > 0, \quad j=1,\ldots,m$
- $\mathrm{Re}\,(\rho-\sigma+a_j) < 1, \quad j=1,\ldots,n$
- either

$$p \le q, \quad p+q \le 2(m+n), \quad |\arg\alpha| \le \left(m+n-\tfrac{1}{2}p-\tfrac{1}{2}q\right)\pi, \quad |\arg\beta| < \pi$$
$$\mathrm{Re}\,(\rho+b_j) > 0, \quad j=1,\ldots,m, \quad \mathrm{Re}\,(\rho-\sigma+a_j) < 1, \quad j=1,\ldots,n,$$
$$\mathrm{Re}\left[\sum_{j=1}^p a_j - \sum_{j=1}^q b_j - (q-p)\left(\rho-\sigma-\frac{1}{2}\right)\right] > 1,$$

or

$$p \ge q, \quad p+q \le 2(m+n), \quad |\arg\alpha| \le \left(m+n-\frac{1}{2}p-\frac{1}{2}q\right)\pi, \quad |\arg\beta| < \pi,$$
$$\mathrm{Re}\,(\rho+b_j) > 0, \quad j=1,\ldots,m, \quad \mathrm{Re}\,(\rho-\sigma+a_j) < 1, \quad j=1,\ldots,n,$$
$$\mathrm{Re}\left[\sum_{j=1}^p a_j - \sum_{j=1}^q b_j + (p-q)\left(\rho-\frac{1}{2}\right)\right] > 1$$

ET II 418(4)

**7.812**

1.  $\displaystyle\int_0^1 x^{\beta-1}(1-x)^{\gamma-\beta-1}\, E\left(a_1,\ldots,a_p:\rho_1,\ldots,\rho_q:\frac{z}{x^m}\right)\mathrm{d}x$

$$= \Gamma(\gamma-\beta)m^{\beta-\gamma}\, E\left(a_1,\ldots,a_{p+m}:\rho_1,\ldots,\rho_{q+m}:z\right)$$
$$a_{p+k} = \frac{\beta+k-1}{m}, \quad \rho_{q+k} = \frac{\gamma+k-1}{m}, \quad k=1,\ldots,m$$
$$[\mathrm{Re}\,\gamma > \mathrm{Re}\,\beta > 0, \quad m=1,2,\ldots]\quad \text{ET II 414(2)}$$

2.  $\displaystyle\int_0^\infty x^{\rho-1}(1+x)^{-\sigma}\, E\left[a_1,\ldots,a_p:\rho_1,\ldots,\rho_q:(1+x)z\right]\mathrm{d}x$

$$= \Gamma(\rho)\, E\left(a_1,\ldots,a_p,\sigma-\rho:\rho_1,\ldots,\rho_q,\sigma:z\right)$$
$$[\mathrm{Re}\,\sigma > \mathrm{Re}\,\rho > 0]\qquad \text{ET II 415(3)}$$

3.    $$\int_0^\infty (1+x)^{-\beta} x^{s-1} G_{pq}^{mn} \left( \frac{ax}{1+x} \, \middle| \, \begin{matrix} a_1, \ldots, a_p \\ b_1, \ldots, b_q \end{matrix} \right) dx = \Gamma(\beta - s) \, G_{p+1,q+1}^{m,n+1} \left( a \, \middle| \, \begin{matrix} 1-s, a_1, \ldots, a_p \\ b_1, \ldots, b_q, 1-\beta \end{matrix} \right)$$

$$\left[ -\min \operatorname{Re} b_k < \operatorname{Re} s < \operatorname{Re} \beta, \quad 1 \le k \le m; \quad (p+q) < 2(m+n), \right.$$

$$\left. |\arg a| < \left( m + n - \tfrac{1}{2} p - \tfrac{1}{2} q \right) \pi \right]$$

ET I 338(19)

**7.813**

1.    $$\int_0^\infty x^{-\rho} e^{-\beta x} G_{pq}^{mn} \left( \alpha x \, \middle| \, \begin{matrix} a_1, \ldots, a_p \\ b_1, \ldots, b_q \end{matrix} \right) dx = \beta^{\rho-1} G_{p+1,q}^{m,n+1} \left( \frac{\alpha}{\beta} \, \middle| \, \begin{matrix} \rho, a_1, \ldots, a_p \\ b_1, \ldots, b_q \end{matrix} \right)$$

$$\left[ p + q < 2(m+n), \quad |\arg \alpha| < \left( m + n - \tfrac{1}{2} p - \tfrac{1}{2} q \right) \pi, \right.$$

$$\left. |\arg \beta| < \tfrac{1}{2}\pi, \quad \operatorname{Re}(b_j - \rho) > -1, \quad j = 1, \ldots, m \right]$$

ET II 419(5)

2.    $$\int_0^\infty e^{-\beta x} G_{pq}^{mn} \left( \alpha x^2 \, \middle| \, \begin{matrix} a_1, \ldots, a_p \\ b_1, \ldots, b_q \end{matrix} \right) dx = \pi^{-1/2} \beta^{-1} G_{p+2,q}^{m,n+2} \left( \frac{4\alpha}{\beta^2} \, \middle| \, \begin{matrix} 0, \tfrac{1}{2}, a_1, \ldots, a_p \\ b_1, \ldots, b_q \end{matrix} \right)$$

$$\left[ p + q < 2(m+n), \quad |\arg \alpha| < \left( m + n - \tfrac{1}{2} p - \tfrac{1}{2} q \right) \pi, \right.$$

$$\left. |\arg \beta| < \tfrac{1}{2}\pi, \quad \operatorname{Re} b_j > -\tfrac{1}{2}; \quad j = 1, \ldots, m \right]$$

ET II 419(6)

**7.814**

1.    $$\int_0^\infty x^{\beta-1} e^{-x} E\left( a_1, \ldots, a_p : \rho_1, \ldots, \rho_q : xz \right) dx$$

$$= \pi \operatorname{cosec}(\beta\pi) \left[ E\left( a_1, \ldots, a_p : 1 - \beta, \rho_1, \ldots, \rho_q : e^{\pm i\pi} z \right) \right.$$

$$\left. - z^{-\beta} E\left( a_1 + \beta, \ldots, a_p + \beta : 1 + \beta, \rho_1 + \beta, \ldots, \rho_l + \beta : e^{\pm i\pi} z \right) \right]$$

$[p \ge q + 1, \operatorname{Re}(a_r + \beta) > 0, r = 1, \ldots, p, |\arg z| < \pi.$ The formula holds also for $p < q + 1$, provided the integral converges].
ET II 415(4)

2.    $$\int_0^\infty x^{\beta-1} e^{-x} E\left( a_1, \ldots, a_p : \rho_1, \ldots, \rho_q : x^{-m} z \right) dx$$

$$= (2\pi)^{\frac{1}{2} - \frac{1}{2} m} m^{\beta - \frac{1}{2}} E\left( a_1, \ldots, a_{p+m} : \rho_1, \ldots, \rho_q : m^{-m} z \right)$$

$$\left[ \operatorname{Re} \beta > 0, \quad a_{p+k} = \frac{\beta + k - 1}{m}, \quad k = 1, \ldots, m; \quad m = 1, 2, \ldots \right] \quad \text{ET II 415(5)}$$

**7.815**

1.    $$\int_0^\infty \sin(cx) \, G_{pq}^{mn} \left( \alpha x^2 \, \middle| \, \begin{matrix} a_1, \ldots, a_p \\ b_1, \ldots, b_q \end{matrix} \right) dx = \sqrt{\pi} \, c^{-1} G_{p+2,q}^{m,n+1} \left( \frac{4\alpha}{c^2} \, \middle| \, \begin{matrix} 0, a_1, \ldots, a_p, \tfrac{1}{2} \\ b_1, \ldots, b_q \end{matrix} \right)$$

$$\left[ p + q < 2(m+n), \quad |\arg \alpha| < \left( m + n - \tfrac{1}{2} p - \tfrac{1}{2} q \right) \pi, \right.$$

$$\left. c > 0, \quad \operatorname{Re} b_j > -1, \quad j = 1, 2, \ldots, m, \quad \operatorname{Re} a_j < \tfrac{1}{2}, \quad j = 1, \ldots, n \right]$$

ET II 420(7)

2. 
$$\int_0^\infty \cos(cx)\, G_{pq}^{mn}\left(\alpha x^2 \left|\begin{array}{c} a_1,\dots,a_p \\ b_1,\dots,b_q \end{array}\right.\right) dx = \pi^{1/2} c^{-1}\, G_{p+2,q}^{m,n+1}\left(\frac{4\alpha}{c^2}\left|\begin{array}{c} \frac{1}{2},a_1,\dots,a_p,0 \\ b_1,\dots,b_q \end{array}\right.\right)$$

$$\left[p+q<2(m+n),\quad |\arg\alpha|<\left(m+n-\tfrac{1}{2}p-\tfrac{1}{2}q\right)\pi,\right.$$
$$\left. c>0,\quad \mathrm{Re}\, b_j>-\tfrac{1}{2},\quad j=1,\dots,m,\quad \mathrm{Re}\, a_j<\tfrac{1}{2},\quad j=1,\dots,n\right]$$

ET II 420(8)

## 7.82 Combinations of the functions $G$ and $E$ and Bessel functions

**7.821**

1. 
$$\int_0^\infty x^{-\rho}\, J_\nu\left(2\sqrt{x}\right) G_{pq}^{mn}\left(\alpha x \left|\begin{array}{c} a_1,\dots,a_p \\ b_1,\dots,b_q \end{array}\right.\right) dx = G_{p+2,q}^{m,n+1}\left(\alpha \left|\begin{array}{c} \rho-\frac{1}{2}\nu,a_1,\dots,a_p,\rho+\frac{1}{2}\nu \\ b_1,\dots,b_q \end{array}\right.\right)$$

$$\left[p+q<2(m+n),\quad |\arg\alpha|<\left(m+n-\tfrac{1}{2}p-\tfrac{1}{2}q\right)\pi\right.$$
$$\left. -\tfrac{3}{4}+\max_{1\le j\le n}\mathrm{Re}\, a_j<\mathrm{Re}\,\rho<1+\tfrac{1}{2}\mathrm{Re}\,\nu+\min_{1\le j\le m}\mathrm{Re}\, b_j\right]$$

ET II 420(9)

2. 
$$\int_0^\infty x^{-\rho}\, Y_\nu\left(2\sqrt{x}\right) G_{pq}^{mn}\left(\alpha x \left|\begin{array}{c} a_1,\dots,a_p \\ b_1,\dots,b_q \end{array}\right.\right) dx$$

$$= G_{p+3,q+1}^{m,n+2}\left(\alpha \left|\begin{array}{c} \rho-\frac{1}{2}\nu,\rho+\frac{1}{2}\nu,a_1,\dots,a_p,\rho+\frac{1}{2}+\frac{1}{2}\nu \\ b_1,\dots,b_q,\rho+\frac{1}{2}+\frac{1}{2}\nu \end{array}\right.\right)$$

$$\left[p+q<2(m+n),\quad |\arg\alpha|<\left(m+n-\tfrac{1}{2}p-\tfrac{1}{2}q\right)\pi,\right.$$
$$\left. -\tfrac{3}{4}+\max_{1\le j\le n}\mathrm{Re}\, a_j<\mathrm{Re}\,\rho<\min_{1\le j\le m}\mathrm{Re}\, b_j+\tfrac{1}{2}|\mathrm{Re}\,\nu|+1\right]$$

ET II 420(10)

3. 
$$\int_0^\infty x^{-\rho}\, K_\nu\left(2\sqrt{x}\right) G_{pq}^{mn}\left(\alpha x \left|\begin{array}{c} a_1,\dots,a_p \\ b_1,\dots,b_q \end{array}\right.\right) dx = \frac{1}{2}\, G_{p+2,q}^{m,n+2}\left(\alpha \left|\begin{array}{c} \rho-\frac{1}{2}\nu,\rho+\frac{1}{2}\nu,a_1,\dots,a_p \\ b_1,\dots,b_q \end{array}\right.\right)$$

$$\left[p+q<2(m+n),\quad |\arg\alpha|<\left(m+n-\tfrac{1}{2}p-\tfrac{1}{2}q\right)\pi,\right.$$
$$\left. \mathrm{Re}\,\rho<1-\tfrac{1}{2}|\mathrm{Re}\,\nu|+\min_{1\le j\le m}\mathrm{Re}\, b_j\right]$$

ET II 421(11)

**7.822**

1. 
$$\int_0^\infty x^{2\rho}\, J_\nu(xy)\, G_{pq}^{mn}\left(\lambda x^2 \left|\begin{array}{c} a_1,\dots,a_p \\ b_1,\dots,b_q \end{array}\right.\right) dx = \frac{2^{2\rho}}{y^{2\rho+1}}\, G_{p+2,q}^{m,n+1}\left(\frac{4\lambda}{y^2}\left|\begin{array}{c} h,a_1,\dots,a_p,k \\ b_1,\dots,b_q \end{array}\right.\right)$$
$$h=\tfrac{1}{2}-\rho-\tfrac{1}{2}\nu,\quad k=\tfrac{1}{2}-\rho+\tfrac{1}{2}\nu$$

$$\left[p+q<2(m+n),\quad |\arg\lambda|<\left(m+n-\tfrac{1}{2}p-\tfrac{1}{2}q\right)\pi,\quad \mathrm{Re}\left(b_j+\rho+\tfrac{1}{2}\nu\right)>-\tfrac{1}{2},\right.$$
$$\left. j=1,2,\dots,m,\quad \mathrm{Re}\left(a_j+\rho\right)<\tfrac{3}{4},\quad j=1,\dots,n,\quad y>0\right]$$

ET II 91(20)

2.     $$\int_0^\infty x^{1/2}\, Y_\nu(xy)\, G_{pq}^{\,mn}\left(\lambda x^2 \left|\begin{matrix} a_1,\dots,a_p \\ b_1,\dots,b_q \end{matrix}\right.\right) dx$$

$$= (2\lambda)^{-1/2} y^{-1/2}\, G_{q+1,p+3}^{\,n+2,m}\left(\frac{y^2}{4\lambda}\left|\begin{matrix} \frac{1}{2}-b_1,\dots,\frac{1}{2}-b_q, l \\ h, k, \frac{1}{2}-a_1,\dots,\frac{1}{2}-a_p, l \end{matrix}\right.\right)$$

$$h = \tfrac{1}{4}+\tfrac{1}{2}\nu, \quad k = \tfrac{1}{4}-\tfrac{1}{2}\nu, \quad l = -\tfrac{1}{4}-\tfrac{1}{2}\nu$$

$$\left[ p+q < 2(m+n), \quad |\arg\lambda| < \left(m+n-\tfrac{1}{2}p-\tfrac{1}{2}q\right)\pi, \quad y > 0, \right.$$

$$\left. \operatorname{Re} a_j < 1, \quad j = 1,\dots,n, \quad \operatorname{Re}\left(b_j \pm \tfrac{1}{2}\nu\right) > -\frac{3}{4}, \quad j = 1,\dots,m \right]$$

ET II 119(56)

3.     $$\int_0^\infty x^{1/2}\, K_\nu(xy)\, G_{pq}^{\,mn}\left(\lambda x^2 \left|\begin{matrix} a_1,\dots,a_p \\ b_1,\dots,b_q \end{matrix}\right.\right) dx$$

$$= 2^{-3/2}\lambda^{-1/2} y^{-1/2}\, G_{q,p+2}^{\,n+2,m}\left(\frac{y^2}{4\lambda}\left|\begin{matrix} \frac{1}{2}-b_1,\dots,\frac{1}{2}-b_q \\ h, k, \frac{1}{2}-a_1,\dots,\frac{1}{2}-a_p \end{matrix}\right.\right)$$

$$h = \tfrac{1}{4}+\tfrac{1}{2}\nu, \quad k = \tfrac{1}{4}-\tfrac{1}{2}\nu$$

$$\left[ \operatorname{Re} y > 0, \quad p+q < 2(m+n), \quad |\arg\lambda| < \left(m+n-\tfrac{1}{2}p-\tfrac{1}{2}q\right)\pi, \right.$$

$$\left. \operatorname{Re} b_j > \tfrac{1}{2}|\operatorname{Re}\nu| - \tfrac{3}{4}, \quad j = 1,\dots,m \right]$$

ET II 153(90)

**7.823**

1.     $$\int_0^\infty x^{\beta-1}\, J_\nu(x)\, E\left(a_1,\dots,a_p : \rho_1,\dots,\rho_q : x^{-2m}z\right) dx$$

$$= (2\pi)^{-m}(2m)^{\beta-1}\left\{ \exp\left[\tfrac{1}{2}\pi(\beta-\nu-1)i\right] E\left[a_1,\dots,a_{p+2m} : \rho_1,\dots,\rho_q : (2m)^{-2m} z e^{-m\pi i}\right] \right.$$

$$\left. + \exp\left[-\tfrac{1}{2}\pi(\beta-\nu-1)i\right] E\left[a_1,\dots,a_{p+2m} : \rho_1,\dots,\rho_q : (2m)^{-2m} z e^{m\pi i}\right] \right\},$$

$$a_{p+k} = \frac{\beta+\nu+2k-2}{2m}, \quad a_{p+m+k} = \frac{\beta-\nu+2k-2}{2m}, \quad m = 1,2,\dots; \quad k = 1,\dots,m$$

$$\left[\operatorname{Re}(\beta+\nu) > 0, \quad \operatorname{Re}\left(2a_r m-\beta\right) > -\tfrac{3}{2}, \quad r = 1,\dots,p\right] \quad \text{ET II 415(7)}$$

2.     $$\int_0^\infty x^{\beta-1}\, K_\nu(x)\, E\left(a_1,\dots,a_p : \rho_1,\dots,\rho_q : x^{-2m}z\right) dx$$

$$= (2\pi)^{1-m} 2^{\beta-2} m^{\beta-1}$$

$$\times E\left[a_1,\dots,a_{p+2m} : \rho_1,\dots,\rho_q : (2m)^{-2m} z\right],$$

$$a_{p+k} = \frac{\beta+\nu+2k-2}{2m}, \quad a_{p+m+k} = \frac{\beta-\nu+2k-2}{2m}, \quad k = 1,2,\dots,m$$

$$\left[\operatorname{Re}\beta > |\operatorname{Re}\nu|, \quad m = 1,2,\dots\right]$$

ET II 416(8)

**7.824**

1. $$\int_0^\infty x^{1/2}\, \mathbf{H}_\nu(xy)\, G_{pq}^{mn}\left(\lambda x^2 \left| \begin{matrix} a_1,\ldots,a_p \\ b_1,\ldots,b_q \end{matrix}\right.\right) dx$$

$$= (2\lambda y)^{-1/2} G_{q+1,p+3}^{n+1,m+1}\left(\frac{y^2}{4\lambda} \left| \begin{matrix} l, \frac{1}{2}-b_1,\ldots,\frac{1}{2}-b_q \\ l, \frac{1}{2}-a_1,\ldots,\frac{1}{2}-a_p, h, k \end{matrix}\right.\right)$$

$$h = \frac{1}{4}+\frac{\nu}{2}, \quad k = \frac{1}{4}-\frac{\nu}{2}, \quad l = \frac{3}{4}+\frac{\nu}{2}$$

$$\left[ p+q < 2(m+n), \quad |\arg\lambda| < \left(m+n-\tfrac{1}{2}p-\tfrac{1}{2}q\right)\pi, \quad y>0, \right.$$

$$\left. \operatorname{Re} a_j < \min\left(1, \tfrac{3}{4}-\tfrac{1}{2}\nu\right), \quad j=1,\ldots,n, \quad \operatorname{Re}(2b_j+\nu) > -\tfrac{5}{2}, \quad j=1,\ldots,m \right]$$

ET II 172(47)

2. $$\int_0^\infty x^{-\rho}\, \mathbf{H}_\nu\left(2\sqrt{x}\right) G_{pq}^{mn}\left(\alpha x \left| \begin{matrix} a_1,\ldots,a_p \\ b_1,\ldots,b_q \end{matrix}\right.\right) dx$$

$$= G_{p+3,q+1}^{m+1,n+1}\left(\alpha \left| \begin{matrix} \rho-\frac{1}{2}-\frac{1}{2}\nu, a_1,\ldots,a_p, \rho+\frac{1}{2}\nu, \rho-\frac{1}{2}\nu \\ \rho-\frac{1}{2}-\frac{1}{2}\nu, b_1,\ldots,b_q \end{matrix}\right.\right)$$

$$\left[ p+q < 2(m+n), \quad |\arg\alpha| < \left(m+n-\tfrac{1}{2}p-\tfrac{1}{2}q\right)\pi, \right.$$

$$\left. \max\left(-\frac{3}{4}, \operatorname{Re}\frac{\nu-1}{2}\right) + \max_{1\le j\le n}\operatorname{Re} a_j < \operatorname{Re}\rho < \min_{1\le j\le m}\operatorname{Re} b_j + \frac{1}{2}\operatorname{Re}\nu + \frac{3}{2} \right]$$

ET II 421(12)

## 7.83 Combinations of the functions $G$ and $E$ and other special functions

**7.831** $$\int_1^\infty x^{-\rho}(x-1)^{\sigma-1} F(k+\sigma-\rho, \lambda+\sigma-\rho; \sigma; 1-x)\, G_{pq}^{mn}\left(\alpha x \left| \begin{matrix} a_1,\ldots,a_p \\ b_1,\ldots,b_q \end{matrix}\right.\right) dx$$

$$= \Gamma(\sigma)\, G_{p+2,q+2}^{m+2,n}\left(\alpha \left| \begin{matrix} a_1,\ldots,a_p, k+\lambda+\sigma-\rho, \rho \\ k, \lambda, b_1,\ldots,b_q \end{matrix}\right.\right)$$

where
ET II 421(13)

- $\operatorname{Re}\left[\sum_{j=1}^p a_j - \sum_{j=1}^q b_j + (q-p)\left(k+\frac{1}{2}\right)\right] > -\frac{1}{2}$,

- $\operatorname{Re}\left[\sum_{j=1}^p a_j - \sum_{j=1}^q b_j + (q-p)\left(\lambda+\frac{1}{2}\right)\right] > -\frac{1}{2}$

- either

$$p+q < 2(m+n), \quad |\arg\alpha| < \left(m+n-\tfrac{1}{2}p-\tfrac{1}{2}q\right)\pi,$$
$$\operatorname{Re}\sigma > 0, \quad \operatorname{Re} k \ge \operatorname{Re}\lambda > \operatorname{Re} a_j - 1, \quad j=1,\ldots,n,$$

or

$$p + q \leq 2(m + n), \quad |\arg \alpha| \leq \left(m + n - \tfrac{1}{2}p - \tfrac{1}{2}q\right)\pi,$$
$$\operatorname{Re}\sigma > 0, \quad \operatorname{Re}k \geq \operatorname{Re}\lambda > \operatorname{Re}a_j - 1, \quad j = 1, \ldots, n,$$

**7.832** $\quad \displaystyle\int_0^\infty x^{\beta-1} e^{-\frac{1}{2}x}\, W_{\kappa,\mu}(x)\, E\left(a_1, \ldots, a_p : \rho_1, \ldots, \rho_q : x^{-m}z\right)\, \mathrm{d}x$

$$= (2\pi)^{\frac{1}{2}-\frac{1}{2}m} m^{\beta+\kappa-\frac{1}{2}} E\left(a_1, \ldots, a_{p+2m} : \rho_1, \ldots, \rho_{q+m} : m^{-m}z\right),$$

$$a_{p+k} = \frac{\beta + k + \mu - \frac{1}{2}}{m}, \quad a_{p+m+k} = \frac{\beta - \mu + k - \frac{1}{2}}{m}, \quad \rho_{q+k} = \frac{\beta - \kappa + k}{m}, \quad k = 1, \ldots, m$$

$$\left[\operatorname{Re}\beta > |\operatorname{Re}\mu| - \tfrac{1}{2}, \quad m = 1, 2, \ldots\right] \quad \text{ET II 416(10)}$$

Table of Integrals, Series, and Products. http://dx.doi.org/10.1016/B978-0-12-384933-5.00008-4

# 8–9 Special Functions

## 8.1 Elliptic Integrals and Functions

### 8.11 Elliptic integrals

**8.110**

1. Every integral of the form $\int R\left(x, \sqrt{P(x)}\right)\,\mathrm{d}x$, where $P(x)$ is a third- or fourth-degree polynomial, can be reduced to a linear combination of integrals leading to elementary functions and the following three integrals:

$$\int \frac{\mathrm{d}x}{\sqrt{(1-x^2)(1-k^2x^2)}}, \qquad \int \frac{\sqrt{1-k^2x^2}}{\sqrt{1-x^2}}\,\mathrm{d}x, \qquad \int \frac{\mathrm{d}x}{(1-nx^2)\sqrt{(1-x^2)(1-k^2x^2)}},$$

which are called respectively *elliptic integrals of the first, second, and third kind in the Legendre normal form*. The results of this reduction for the more frequently encountered integrals are given in formulas **3.13–3.17**. The number $k$ is called the *modulus** of these integrals, the number $k' = \sqrt{1-k^2}$ is called the complementary modulus, and the number $n$ is called the parameter of the integral of the third kind.   BY (110.04)

2. By means of the substitution $x = \sin\varphi$, elliptic integrals can be reduced to the normal trigonometric forms

$$\int \frac{\mathrm{d}\varphi}{\sqrt{1-k^2\sin^2\varphi}}, \qquad \int \sqrt{1-k^2\sin^2\varphi}\,\mathrm{d}\varphi, \qquad \int \frac{\mathrm{d}\varphi}{(1-n\sin^2\varphi)\sqrt{1-k^2\sin^2\varphi}}\qquad \text{BY (110.04)}$$

The results of reducing integrals of trigonometric functions to normal form are given in **2.58–2.62**.

3.[11] Elliptic integrals from 0 to 1 in the **8.110 1** formulation (or from 0 to $\dfrac{\pi}{2}$ in the **8.110 2** formulation) are called *complete elliptic integrals*.

4.* Take note that in mathematical software, and elsewhere, the notation for elliptic integrals is often modified by replacing the parameter $k^2$ that is used here by $k$.

---

*The quantity $k$ is sometimes called the *module* of the functions.

### 8.111

Notations:

1.      $\Delta\varphi = \sqrt{1 - k^2 \sin^2 \varphi}; \quad k' = \sqrt{1 - k^2}; \quad k^2 < 1.$

2.      The elliptic integral of the first kind:

$$F(\varphi, k) = \int_0^\varphi \frac{d\alpha}{\sqrt{1 - k^2 \sin^2 \alpha}} = \int_0^{\sin \varphi} \frac{dx}{\sqrt{(1 - x^2)(1 - k^2 x^2)}}.$$

3.      The elliptic integral of the second kind:

$$E(\varphi, k) = \int_0^\varphi \sqrt{1 - k^2 \sin^2 \alpha} \, d\alpha = \int_0^{\sin \varphi} \frac{\sqrt{1 - k^2 x^2}}{\sqrt{1 - x^2}} \, dx \qquad \text{FI II 135}$$

4.[11]    The elliptic integral of the third kind:

$$\Pi(\varphi, n, k) = \int_0^\varphi \frac{d\alpha}{(1 - n \sin^2 \alpha) \sqrt{1 - k^2 \sin^2 \alpha}} = \int_0^{\sin \varphi} \frac{dx}{(1 - nx^2) \sqrt{(1 - x^2)(1 - k^2 x^2)}}$$

BY (110.04)

5.      $D(\varphi, k) = \dfrac{F(\varphi, k) - E(\varphi, k)}{k^2} = \displaystyle\int_0^\varphi \frac{\sin^2 \alpha \, d\alpha}{\sqrt{1 - k^2 \sin^2 \alpha}} = \int_0^{\sin \varphi} \frac{x^2 \, dx}{\sqrt{(1 - x^2)(1 - k^2 x^2)}}$

6.*    $\displaystyle\int_0^{\pi/2} \frac{dx}{\sqrt{a^2 + \sin^2 x}} \arctan\left(\frac{b}{\sqrt{a^2 + \sin^2 x}}\right) = \frac{\pi}{2|a|} F\left(\arcsin\left(\frac{b}{\sqrt{a^2 + b^2 + 1}}\right), \frac{i}{a}\right)$

[a and b are real]

7.*    Carlson has introduced a notation for the elliptic functions that preserves certain symmetries; see http://dlmf.nist.gov/19. The Carlson elliptic integrals are:

7.1     $R_F(x, y, z) = \dfrac{1}{2} \displaystyle\int_0^\infty \frac{dt}{\sqrt{(t + x)(t + y)(t + z)}}$

7.2     $R_J(x, y, z, p) = \dfrac{3}{2} \displaystyle\int_0^\infty \frac{dt}{(t + p)\sqrt{(t + x)(t + y)(t + z)}}$

7.3     $R_C(x, y) = R_F(x, y, y) = \dfrac{1}{2} \displaystyle\int_0^\infty \frac{dt}{(t + y)\sqrt{(t + x)}}$

7.4     $R_D(x, y, z) = R_J(x, y, z, z) = \dfrac{3}{2} \displaystyle\int_0^\infty \frac{dt}{(t + z)\sqrt{(t + x)(t + y)(t + z)}}$

$R_C$ and $R_J$ are interpreted as Cauchy principal values when the last argument is negative. Each function has the value unity when all of its arguments are unity.

Use of Carlson's notation for elliptical integrals reduces the number of separate cases that need to be tabulated.

8.*      The incomplete elliptic integrals can be written in Carlson's notation as follows:

8.1      $F(\phi, k) = \sin \phi R_F(\cos^2 \phi, 1 - k^2 \sin^2 \phi, 1)$

8.2      $E(\phi, k) = \sin \phi R_F(\cos^2 \phi, 1 - k^2 \sin^2 \phi, 1)$

$$- \frac{1}{3} k^2 \sin^3 \phi R_D(\cos^2 \phi, 1 - k^2 \sin^2 \phi, 1)$$

8.3      $\Pi(\phi, n, k) = \sin \phi R_F(\cos^2 \phi, 1 - k^2 \sin^2 \phi, 1)$

$$+ \frac{1}{3} n \sin^3 \phi R_J(\cos^2 \phi, 1 - k^2 \sin^2 \phi, 1, 1 - n \sin^2 \phi)$$

9.*      The complete elliptic integrals (i.e., the incomplete elliptic integrals with $\phi = \frac{\pi}{2}$) can be written in Carlson's notation as follows:

9.1      $\boldsymbol{K}(k) = R_F(0, 1 - k^2, 1)$

9.2      $\boldsymbol{K}E(k) = R_F(0, 1 - k^2, 1) - \frac{1}{3} k^2 R_D(0, 1 - k^2, 1)$

9.3      $\boldsymbol{\Pi}(n, k) = R_F(0, 1 - k^2, 1) + \frac{1}{3} n R_J(0, 1 - k^2, 1, 1 - n)$

**8.112**    Complete elliptic integrals

1.      $\boldsymbol{K}(k) = F\left(\frac{\pi}{2}, k\right) = \boldsymbol{K}'(k')$

2.      $\boldsymbol{E}(k) = E\left(\frac{\pi}{2}, k\right) = \boldsymbol{E}'(k')$

3.      $\boldsymbol{K}'(k) = F\left(\frac{\pi}{2}, k'\right) = \boldsymbol{K}(k')$

4.      $\boldsymbol{E}'(k) = E\left(\frac{\pi}{2}, k'\right) = \boldsymbol{E}(k')$

5.      $\boldsymbol{D} = D\left(\frac{\pi}{2}, k\right) = \dfrac{\boldsymbol{K} - \boldsymbol{E}}{k^2}$

In writing complete elliptic integrals, the modulus $k$, which acts as an independent variable, is often omitted and we write

$$\boldsymbol{K} \left(\equiv \boldsymbol{K}(k)\right), \quad \boldsymbol{K}' \left(\equiv \boldsymbol{K}'(k)\right), \quad \boldsymbol{E} \left(\equiv \boldsymbol{E}(k)\right), \quad \boldsymbol{E}' \left(\equiv \boldsymbol{E}'(k)\right)$$

**Series representations**

**8.113**

1.      $\boldsymbol{K} = \dfrac{\pi}{2} \left\{ 1 + \left(\dfrac{1}{2}\right)^2 k^2 + \left(\dfrac{1 \cdot 3}{2 \cdot 4}\right)^2 k^4 + \cdots + \left(\dfrac{(2n-1)!!}{2^n n!}\right)^2 k^{2n} + \ldots \right\} = \dfrac{\pi}{2} F\left(\dfrac{1}{2}, \dfrac{1}{2}; 1; k^2\right)$

FI II 487, WH 499

2. $\quad K = \dfrac{\pi}{1+k'}\left\{1 + \left(\dfrac{1}{2}\right)^2\left(\dfrac{1-k'}{1+k'}\right)^2 + \left(\dfrac{1\cdot 3}{2\cdot 4}\right)^2\left(\dfrac{1-k'}{1+k'}\right)^4 + \cdots + \left(\dfrac{(2n-1)!!}{2^n n!}\right)^2\left(\dfrac{1-k'}{1+k'}\right)^{2n} + \cdots\right\}$

DW

3. $\quad K = \ln\dfrac{4}{k'} + \left(\dfrac{1}{2}\right)^2\left(\ln\dfrac{4}{k'} - \dfrac{2}{1\cdot 2}\right)k'^2 + \left(\dfrac{1\cdot 3}{2\cdot 4}\right)^2\left(\ln\dfrac{4}{k'} - \dfrac{2}{1\cdot 2} - \dfrac{2}{3\cdot 4}\right)k'^4$

$\qquad + \left(\dfrac{1\cdot 3\cdot 5}{2\cdot 4\cdot 6}\right)^2\left(\ln\dfrac{4}{k'} - \dfrac{2}{1\cdot 2} - \dfrac{2}{3\cdot 4} - \dfrac{2}{5\cdot 6}\right)k'^6 + \cdots$

DW

See also **8.197** 1 and **8.197** 2.

**8.114**

1.[12] $\quad E = \dfrac{\pi}{2}\left\{1 - \dfrac{1}{2^2}k^2 - \dfrac{1^2\cdot 3^2}{2^2\cdot 4^2}k^4 - \cdots - \left(\dfrac{(2n-1)!!}{2^n n!}\right)^2\dfrac{k^{2n}}{2n-1} - \cdots\right\} = \dfrac{\pi}{2}F\left(-\dfrac{1}{2},\dfrac{1}{2};1;k^2\right)$

WH 518, FI II 487

2. $\quad E = \dfrac{(1+k')\pi}{4}\left\{1 + \dfrac{1}{2^2}\left(\dfrac{1-k'}{1+k'}\right)^2 + \dfrac{1^2}{2^2\cdot 4^2}\left(\dfrac{1-k'}{1+k'}\right)^4 + \cdots + \left(\dfrac{(2n-3)!!}{2^n n!}\right)^2\left(\dfrac{1-k'}{1+k'}\right)^{2n} + \cdots\right\}$

DW

3. $\quad E = 1 + \dfrac{1}{2}\left(\ln\dfrac{4}{k'} - \dfrac{1}{1\cdot 2}\right)k'^2 + \dfrac{1^2\cdot 3}{2^2\cdot 4}\left(\ln\dfrac{4}{k'} - \dfrac{2}{1\cdot 2} - \dfrac{1}{3\cdot 4}\right)k'^4$

$\qquad + \dfrac{1^2\cdot 3^2\cdot 5}{2^2\cdot 4^2\cdot 6}\left(\ln\dfrac{4}{k'} - \dfrac{2}{1\cdot 2} - \dfrac{2}{3\cdot 4} - \dfrac{1}{5\cdot 6}\right)k'^6 + \cdots$

DW

**8.115**[12] $D = \pi\left\{\dfrac{1}{1}\left(\dfrac{1}{2}\right)^2 + \dfrac{2}{3}\left(\dfrac{1\cdot 3}{2\cdot 4}\right)^2 k^2 + \cdots + \dfrac{n}{2n-1}\left[\dfrac{(2n-1)!!}{2^n n!}\right]^2 k^{2(n-1)} + \cdots\right\}$

$\qquad = \dfrac{\pi}{4}F\left(\dfrac{1}{2},\dfrac{3}{2};2;k^2\right)$

ZH 43(158)

**8.116** $\displaystyle\int_0^{\frac{\pi}{2}}\dfrac{\sqrt{1-k^2\sin^2\varphi}}{1-n^2\sin^2\varphi}\,d\varphi = \sqrt{n'^2-k'^2}\left(\dfrac{\arccos\frac{1}{n'}}{n'\sqrt{n'^2-1}} + \mathbf{R}\right),\qquad\text{where}$

ZH 44(163)

$\mathbf{R} = \dfrac{k'^2}{2}\left(p + \dfrac{1}{2}\right)\dfrac{1}{n'^3} + \dfrac{k'^4}{16}\left[-1 + \left(p + \dfrac{1}{4}\right)\dfrac{1}{n'^3}\left(1 + \dfrac{6}{n'^2}\right)\right]$

$\qquad + \dfrac{k'^6}{16}\left[-\dfrac{7}{16} - \dfrac{1}{n'^2} + \left(p + \dfrac{1}{6}\right)\dfrac{1}{n'^3}\left(\dfrac{3}{8} + \dfrac{1}{n'^2} + \dfrac{5}{n'^4}\right)\right]$

$\qquad + \dfrac{15k'^8}{256}\left[-\dfrac{37}{144} - \dfrac{21}{40n'^2} - \dfrac{1}{n'^4} + \left(p + \dfrac{1}{8}\right)\dfrac{1}{n'^3}\left(\dfrac{5}{24} + \dfrac{9}{20n'^2} + \dfrac{1}{n'^4} + \dfrac{14}{3n'^6}\right)\right] + \cdots,$

$\qquad\qquad p = \ln\dfrac{4}{k'},\quad k' = 4e^{-p},\quad k'^2 = 1 - k^2,\quad n'^2 = 1 - n^2$   ZH 44(163)

## Trigonometric series

**8.117**  For *small* values of $k$ and $\varphi$, we may use the series

1.  $$F(\varphi, k) = \frac{2}{\pi} \boldsymbol{K} \varphi - \sin \varphi \cos \varphi \left( a_0 + \frac{2}{3} a_1 \sin^2 \varphi + \frac{2 \cdot 4}{3 \cdot 5} a_2 \sin^4 \varphi + \ldots \right), \qquad \text{where}$$

$$a_0 = \frac{2}{\pi} \boldsymbol{K} - 1; \quad a_n = a_{n-1} - \left[ \frac{(2n-1)!!}{2^n n!} \right]^2 k^{2n} \qquad \text{ZH 10(19)}$$

2.[12]  $$E(\varphi, k) = \frac{2}{\pi} \boldsymbol{E} \varphi - \sin \varphi \cos \varphi \left( b_0 + \frac{2}{3} b_1 \sin^2 \varphi + \frac{2 \cdot 4}{3 \cdot 5} b_2 \sin^4 \varphi + \ldots \right), \qquad \text{where}$$

$$b_0 = 1 - \frac{2}{\pi} \boldsymbol{E}, \quad b_n = b_{n-1} - \left[ \frac{(2n-1)!!}{2^n n!} \right]^2 \frac{k^{2n}}{2n-1} \qquad \text{ZH 27(86)}$$

**8.118**  For $k$ close to 1, we may use the series

1.  $$F(\varphi, k) = \frac{2}{\pi} \boldsymbol{K}' \ln \tan \left( \frac{\varphi}{2} + \frac{\pi}{4} \right) - \frac{\tan \varphi}{\cos \varphi} \left( a_0' - \frac{2}{3} a_1' \tan^2 \varphi + \frac{2 \cdot 4}{3 \cdot 5} a_2' \tan^4 \varphi - \ldots \right), \qquad \text{where}$$

$$a_0' = \frac{2}{\pi} \boldsymbol{K}' - 1; \quad a_n' = a_{n-1} - \left[ \frac{(2n-1)!!}{2^n n!} \right]^2 k'^{2n} \qquad \text{ZH 10(23)}$$

2.  $$E(\varphi, k) = \frac{2}{\pi} \left( \boldsymbol{K}' - \boldsymbol{E}' \right) \ln \tan \left( \frac{\varphi}{2} + \frac{\pi}{2} \right)$$
$$+ \frac{\tan \varphi}{\cos \varphi} \left( b_1' - \frac{2}{3} b_2' \tan^2 \varphi + \frac{2 \cdot 4}{3 \cdot 5} b_3' \tan^4 \varphi - \ldots \right) + \frac{1}{\sin \varphi} \left[ 1 - \cos \varphi \sqrt{1 - k^2 \sin \varphi} \right],$$

where

$$b_0' = \frac{2}{\pi} \left( \boldsymbol{K}' - \boldsymbol{E}' \right), b_n' = b_{n-1}' - \left[ \frac{(2n-3)!!}{2^{n-1}(n-1)!} \right]^2 \left( \frac{2n-1}{2n} \right) k'^{2n} \qquad \text{ZH 27(90)}$$

For the expansion of complete elliptic integrals in Legendre polynomials, see **8.928**.

**8.119**  Representation in the form of an infinite product:

1.  $$\boldsymbol{K}(k) = \frac{\pi}{2} \prod_{n=1}^{\infty} \left( 1 + k_n \right), \qquad \text{where}$$

$$k_n = \frac{1 - \sqrt{1 - k_{n-1}^2}}{1 + \sqrt{1 - k_{n-1}^2}}; \qquad k_0 = k \qquad \text{FI II 166}$$

See also **8.197**.

## 8.12  Functional relations between elliptic integrals

**8.121**

1.    $F(-\varphi, k) = -F(\varphi, k)$                                    JA

2.    $E(-\varphi, k) = -E(\varphi, k)$                                    JA

3.    $F(n\pi \pm \varphi, k) = 2n\,\boldsymbol{K}(k) \pm F(\varphi, k)$                  JA

4.    $E(n\pi \pm \varphi, k) = 2n\,\boldsymbol{E}(k) \pm E(\varphi, k)$                  JA

**8.122**  $\boldsymbol{E}(k)\,\boldsymbol{K}'(k) + \boldsymbol{E}'(k)\,\boldsymbol{K}(k) - \boldsymbol{K}(k)\,\boldsymbol{K}'(k) = \dfrac{\pi}{2}$          FI II 691, 791

**8.123**

1.    $\dfrac{\partial F}{\partial k} = \dfrac{1}{k'^2}\left(\dfrac{E - k'^2 F}{k} - \dfrac{k\sin\varphi\cos\varphi}{\sqrt{1 - k^2\sin^2\varphi}}\right)$          MO 138, BY (710.07)

2.    $\dfrac{d\,\boldsymbol{K}(k)}{dk} = \dfrac{\boldsymbol{E}(k)}{kk'^2} - \dfrac{\boldsymbol{K}(k)}{k}$                        FI II 691

3.    $\dfrac{\partial E}{\partial k} = \dfrac{E - F}{k}$                                MO 138

4.    $\dfrac{d\,\boldsymbol{E}(k)}{dk} = \dfrac{\boldsymbol{E}(k) - \boldsymbol{K}(k)}{k}$                          FI II 690

**8.124**

1.    The functions $\boldsymbol{K}$ and $\boldsymbol{K}'$ satisfy the equation

$$\frac{d}{dk}\left\{kk'^2\frac{du}{dk}\right\} - ku = 0.$$          WH 499, WH 502

2.    The functions $\boldsymbol{E}$ and $\boldsymbol{E}' - \boldsymbol{K}'$ satisfy the equation

$$k'^2\frac{d}{dk}\left(k\frac{du}{dk}\right) + ku = 0.$$          WH

**8.125**

1.    $F\left(\psi, \dfrac{1 - k'}{1 + k'}\right) = (1 + k')\,F(\varphi, k)$          $[\tan(\psi - \varphi) = k'\tan\varphi]$          MO 130

2.    $E\left(\psi, \dfrac{1 - k'}{1 + k'}\right) = \dfrac{2}{1 + k'}\left[E(\varphi, k) + k'\,F(\varphi, k)\right] - \dfrac{1 - k'}{1 + k'}\sin\psi$

$[\tan(\psi - \varphi) = k'\tan\varphi]$          MO 131

3.    $F\left(\psi, \dfrac{2\sqrt{k}}{1 + k}\right) = (1 + k)\,F(\varphi, k)$          $\left[\sin\psi = \dfrac{(1 + k)\sin\varphi}{1 + k\sin^2\varphi}\right]$

4.    $E\left(\psi, \dfrac{2\sqrt{k}}{1 + k}\right) = \dfrac{1}{1 + k}\left[2E(\varphi, k) - k'^2\,F(\varphi, k) + 2k\dfrac{\sin\varphi\cos\varphi}{1 + k\sin^2\varphi}\sqrt{1 - k^2\sin^2\varphi}\right]$

$\left[\sin\psi = \dfrac{(1 + k)\sin\varphi}{1 + k\sin^2\varphi}\right]$          MO 131

**8.126** In particular,

1. $\quad K\left(\dfrac{1-k'}{1+k'}\right) = \dfrac{1+k'}{2}\,K(k)$ MO 130

2. $\quad E\left(\dfrac{1-k'}{1+k'}\right) = \dfrac{1}{1+k'}\left[E(k)+k'\,K(k)\right]$ MO 130

3. $\quad K\left(\dfrac{2\sqrt{k}}{1+k}\right) = (1+k)\,K(k)$ MO 130

4. $\quad E\left(\dfrac{2\sqrt{k}}{1+k}\right) = \dfrac{1}{1+k}\left[2\,E(k) - k'^2\,K(k)\right]$ MO 130

**8.127**[12]

| $k_1$ | $\sin\varphi_1$ | $\cos\varphi_1$ | $F(\varphi_1,k_1)$ | $E(\varphi_1,k_1)$ |
|---|---|---|---|---|
| $i\dfrac{k}{k'}$ | $k'\dfrac{\sin\varphi}{\Delta\varphi}$ | $\dfrac{\cos\varphi}{\Delta\varphi}$ | $k'\,F(\varphi,k)$ | $\dfrac{1}{k'}\left[E(\varphi,k)-\dfrac{k^2\sin\varphi\cos\varphi}{\Delta\varphi}\right]$ |
| $k'$ | $-i\tan\varphi$ | $\sec\varphi$ | $-i\,F(\varphi,k)$ | $i\left[E(\varphi,k)-F(\varphi,k)-\Delta\varphi\tan\varphi\right]$ |
| $\dfrac{1}{k}$ | $k\sin\varphi$ | $\Delta\varphi$ | $k\,F(\varphi,k)$ | $\dfrac{1}{k}\left[E(\varphi,k)-k'^2\,F(\varphi,k)\right]$ |
| $\dfrac{1}{k'}$ | $-ik'\tan\varphi$ | $\dfrac{\Delta\varphi}{\cos\varphi}$ | $-ik'\,F(\varphi,k)$ | $\dfrac{i}{k'}\left[E(\varphi,k)-k'^2\,F(\varphi,k)-\Delta\varphi\tan\varphi\right]$ |
| $\dfrac{k'}{ik}$ | $\dfrac{-ik\sin\varphi}{\Delta\varphi}$ | $\dfrac{1}{\Delta\varphi}$ | $-ik\,F(\varphi,k)$ | $\dfrac{i}{k}\left[E(\varphi,k)-F(\varphi,k)-\dfrac{k^2\sin\varphi\cos\varphi}{\Delta\varphi}\right]$ |

(see 8.111 1)    MO 131

**8.128** In particular,

1. $\quad K\left(i\dfrac{k}{k'}\right) = k'\,K(k)$ <div style="float:right">MO 130</div> $\qquad\qquad\qquad\qquad [\mathrm{Im}(k) < 0]$

2.[12] $\quad K'\left(i\dfrac{k}{k'}\right) = k'\left[K(k') - i\,K(k)\right]$ $\qquad\qquad\qquad [\mathrm{Im}(k) < 0]$ MO 130

3. $\quad K\left(\dfrac{1}{k}\right) = k\left[K(k) + i\,K'(k)\right]$ $\qquad\qquad\qquad\qquad [\mathrm{Im}(k) < 0]$ MO 130

For integrals of elliptic integrals, see **6.11–6.15**. For indefinite integrals of complete elliptic integrals, see **5.11**.

**8.129** Special values:

1. $\quad K\left(\sin\dfrac{\pi}{4}\right) = K\left(\dfrac{\sqrt{2}}{2}\right) = K'\left(\dfrac{\sqrt{2}}{2}\right) = \sqrt{2}\displaystyle\int_0^1 \dfrac{dt}{\sqrt{1-t^4}} = \dfrac{1}{4\sqrt{\pi}}\left[\Gamma\left(\dfrac{1}{4}\right)\right]^2$ MO 130

2.[12] $\quad K'\left(\tan\dfrac{\pi}{8}\right) = K'\left(\sqrt{2}-1\right) = \sqrt{2}\,K\left(\sqrt{2}-1\right)$ MO 130

3.[12]    $K'\left(\sin\dfrac{\pi}{12}\right) = \sqrt{3}\,K\left(\sin\dfrac{\pi}{12}\right) = \sqrt{3}K\left(\dfrac{\sqrt{6}-\sqrt{2}}{4}\right) = \dfrac{3^{3/4}}{2^{7/3}\pi}\Gamma^3\left(\dfrac{1}{3}\right)$      MO 130

4.    $K'\left(\tan^2\dfrac{\pi}{8}\right) = K'\left(\dfrac{2-\sqrt{2}}{2+\sqrt{2}}\right) = 2\,K\left(\tan^2\dfrac{\pi}{8}\right)$      MO 130

5.    $K\left(\sin\dfrac{\pi}{12}\right) = \dfrac{\sqrt{3}-1}{2\sqrt{2}}$

6.    $E = \dfrac{\pi\sqrt{3}}{12K} + \sqrt{\dfrac{2}{3}}k'K$

7.    $E' = \dfrac{\pi\sqrt{3}}{4K'} + \sqrt{\dfrac{2}{3}}kK'$

8.*    $E\left(\sin\dfrac{\pi}{4}\right) = E\left(\dfrac{\sqrt{2}}{2}\right) = E'\left(\dfrac{\sqrt{2}}{2}\right) = \dfrac{1}{8\sqrt{\pi}}\Gamma^2\left(\dfrac{1}{4}\right) + \dfrac{1}{2\sqrt{\pi}}\Gamma^2\left(\dfrac{3}{4}\right)$

## 8.13   Elliptic functions

**8.130**   Definition and general properties.

1.    A single-valued function $f(z)$ of a complex variable, which is not a constant, is said to be elliptic if it has two periods $2\omega_1$ and $2\omega_2$, that is

$$f(z + 2m\omega_1 + 2n\omega_2) = f(z) \qquad [m, n \text{ integers}]$$

*The ratio of the periods of an analytic function cannot be a real number.* For an elliptic function $f(z)$, the $z$-plane can be partitioned into parallelograms—the period parallelograms—the vertices of which are the points $z_0 + 2m\omega_1 + 2n\omega_2$. At corresponding points of these parallelograms, the function $f(z)$ has the same value.      ZH 117, SI 299

2.    Suppose that $\alpha$ is the angle between the sides $a$ and $b$ of one of the period parallelograms. Then,

$$\tau = \frac{\omega_1}{\omega_2} = \frac{a}{b}e^{i\alpha}, \quad q = e^{i\pi\tau} = e^{-\frac{a}{b}\pi\sin\alpha}\left[\cos\left(\frac{a}{b}\pi\cos\alpha\right) + i\sin\left(\frac{a}{b}\pi\cos\alpha\right)\right]$$

3.    The *derivative* of an elliptic function is also an elliptic function with the same periods.

     SM III 598

4.[12]    A non-constant elliptic function has a finite number of poles in a period parallelogram: it has at least two simple poles or one second-order pole in such a parallelogram. Suppose that these poles lie at the points $a_1, a_2, \ldots, a_n$ and that their orders are $\alpha_1, \alpha_2, \ldots, \alpha_n$. Suppose that the zeros of an analytic function that occur in a single parallelogram are $b_1, b_2, \ldots, b_m$ and that the orders of the zeros are $\beta_1, \beta_2, \ldots, \beta_m$, respectively. Then,

$$\gamma = \alpha_1 + \alpha_2 + \cdots + \alpha_n = \beta_1 + \beta_2 + \cdots + \beta_m \qquad \text{ZH 118}$$

The number $\gamma$ representing this sum is called the *order* of the elliptic function.

5.    The sum of the residues of an elliptic function with respect to all the poles belonging to a period parallelogram is equal to zero.

6.    The difference between the sum of all the zeros and the sum of all the poles of an elliptic function that are located in a period parallelogram is equal to one of its periods.

7.      Every two elliptic functions with the same periods are related by an algebraic relationship.

GO II 151

8.[7]    A non-constant single-valued function which is not constant cannot have more than two periods.

GO II 147

9.      An elliptic function of order $\gamma$ assumes *an arbitrary value* $\gamma$ times in a period parallelogram.

SM 601, SI 301

## 8.14   Jacobian elliptic functions

**8.141**    Consider the upper limit $\varphi$ of the integral

$$u = \int_0^\varphi \frac{d\alpha}{\sqrt{1 - k^2 \sin^2 \alpha}}$$

as a function of $u$. Using the notation

$$\varphi = \operatorname{am} u$$

we call this upper limit the *amplitude*. The quantity $u$ is called the *argument*, and its dependence on $\varphi$ is written

$$u = \arg \varphi$$

**8.142**    The amplitude is an *infinitely-many-valued* function of $u$ and has a period of $4\boldsymbol{K}i$. The *branch points* of the amplitude correspond to the values of the argument

$$u = 2m\boldsymbol{K} + (2n + 1)\boldsymbol{K}'i, \qquad\qquad \text{ZH 67–69}$$

where $m$ and $n$ are arbitrary integers (see also **8.151**).

**8.143**    The first two of the following functions

$$\operatorname{sn} u = \sin \varphi = \sin \operatorname{am} u, \qquad \operatorname{cn} u = \cos \varphi = \cos \operatorname{am} u,$$

$$\operatorname{dn} u = \Delta \varphi = \sqrt{1 - k^2 \sin^2 \varphi} = \frac{d\varphi}{du}$$

are called, respectively, the *sine-amplitude* and the *cosine-amplitude* while the third may be called the *delta amplitude*. All these elliptic functions were exhibited by Jacobi and they bear his name.    SI 16

     The Jacobian elliptic functions are *doubly-periodic* functions and have *two simple poles* in a period parallelogram.       ZH 69

**8.144**

1.    $u = \displaystyle\int_0^{\operatorname{sn} u} \frac{dt}{\sqrt{(1 - t^2)(1 - k^2 t^2)}}$        SI 21(23)

2.    $u = \displaystyle\int_1^{\operatorname{cn} u} \frac{dt}{\sqrt{(1 - t^2)(k'^2 + k^2 t^2)}}$        SI 21(23)

3.    $u = \displaystyle\int_1^{\operatorname{dn} u} \frac{dt}{\sqrt{(1 - t^2)(t^2 - k'^2)}}$        SI 21(23)

**8.145**    Power series representations:

1.[11]   $\operatorname{sn} u = u - \dfrac{1 + k^2}{3!} u^3 + \dfrac{1 + 14k^2 + k^4}{5!} u^5 - \dfrac{1 + 135k^2 + 135k^4 + k^6}{7!} u^7$

$$+ \frac{1 + 1228k^2 + 5478k^4 + 1228k^6 + k^8}{9!} u^9 - \ldots$$

$[|u| < |\boldsymbol{K}'|]$        ZH 81(97)

2.　　$\operatorname{cn} u = 1 - \dfrac{1}{2!}u^2 + \dfrac{1+4k^2}{4!}u^4 - \dfrac{1+44k^2+16k^4}{6!}u^6 + \dfrac{1+408k^2+912k^4+64k^6}{8!}u^8 - \dots$

$$[|u| < |\boldsymbol{K}'|]$$ 　　　　　　　ZH 81(98)

3.　　$\operatorname{dn} u =$

$$1 - \dfrac{k^2}{2!}u^2 + \dfrac{k^2\left(4+k^2\right)}{4!}u^4 - \dfrac{k^2\left(16+44k^2+k^4\right)}{6!}u^6 + \dfrac{k^2\left(64+912k^2+408k^4+k^6\right)}{8!}u^8 - \dots$$

$$[|u| < |\boldsymbol{K}'|]$$ 　　　　　　　ZH 81(99)

4.　　$\operatorname{am} u$

$$= u - \dfrac{k^2}{3!}u^3 + \dfrac{k^2\left(4+k^2\right)}{5!}u^5 - \dfrac{k^2\left(16+44k^2+k^4\right)}{7!}u^7 + \dfrac{k^2\left(64+912k^2+408k^4+k^6\right)}{9!}u^9 - \dots$$

$$[|u| < |\boldsymbol{K}'|]$$ 　　　　　　　LA 380(4)

**8.146** 　Representation as a trigonometric series or a product $\left(q = e^{-\frac{\pi K'}{K}} = e^{\pi i \tau}\right)^*$

1.[11]　　$\operatorname{sn} u = \dfrac{2\pi}{k\boldsymbol{K}} \sum\limits_{n=1}^{\infty} \dfrac{q^{n-\frac{1}{2}}}{1-q^{2n-1}} \sin(2n-1)\dfrac{\pi u}{2\boldsymbol{K}}$　　　　　WH 511a, ZH 84(108)

2.[11]　　$\operatorname{cn} u = \dfrac{2\pi}{k\boldsymbol{K}} \sum\limits_{n=1}^{\infty} \dfrac{q^{n-\frac{1}{2}}}{1+q^{2n-1}} \cos(2n-1)\dfrac{\pi u}{2\boldsymbol{K}}$　　　　　WH 511a, ZH 84(109)

3.　　$\operatorname{dn} u = \dfrac{\pi}{2\boldsymbol{K}} + \dfrac{2\pi}{\boldsymbol{K}} \sum\limits_{n=1}^{\infty} \dfrac{q^n}{1+q^{2n}} \cos\dfrac{n\pi u}{\boldsymbol{K}}$　　　　　WH 511a, ZH 84(110)

4.[11]　　$\operatorname{am} u = \dfrac{\pi u}{2\boldsymbol{K}} + 2\sum\limits_{n=1}^{\infty} \dfrac{1}{n}\dfrac{q^n}{1+q^{2n}} \sin\dfrac{n\pi u}{\boldsymbol{K}}$　　　　　WH 511a

5.　　$\dfrac{1}{\operatorname{sn} u} = \dfrac{\pi}{2\boldsymbol{K}} \left[ \dfrac{1}{\sin\frac{\pi u}{2\boldsymbol{K}}} + 4\sum\limits_{n=1}^{\infty} \dfrac{q^{2n-1}}{1-q^{2n-1}} \sin(2n-1)\dfrac{\pi u}{2\boldsymbol{K}} \right]$　　　　　LA 369(3)

6.　　$\dfrac{1}{\operatorname{cn} u} = \dfrac{\pi}{2k'\boldsymbol{K}} \left[ \dfrac{1}{\cos\frac{\pi u}{2\boldsymbol{K}}} + 4\sum\limits_{n=1}^{\infty} (-1)^n \dfrac{q^{2n-1}}{1+q^{2n-1}} \cos(2n-1)\dfrac{\pi u}{2\boldsymbol{K}} \right]$　　　　　LA 369(3)

7.　　$\dfrac{1}{\operatorname{dn} u} = \dfrac{\pi}{2k'\boldsymbol{K}} \left[ 1 + 4\sum\limits_{n=1}^{\infty} (-1)^n \dfrac{q^n}{1+q^{2n}} \cos\dfrac{n\pi u}{\boldsymbol{K}} \right]$　　　　　LA 369(3)

8.　　$\dfrac{\operatorname{sn} u}{\operatorname{cn} u} = \dfrac{\pi}{2k'\boldsymbol{K}} \left[ \tan\dfrac{\pi u}{2\boldsymbol{K}} + 4\sum\limits_{n=1}^{\infty} (-1)^n \dfrac{q^{2n}}{1+q^{2n}} \sin\dfrac{n\pi u}{\boldsymbol{K}} \right]$　　　　　LA 369(4)

9.[11]　　$\dfrac{\operatorname{sn} u}{\operatorname{dn} u} = -\dfrac{2\pi}{kk'\boldsymbol{K}} \sum\limits_{n=1}^{\infty} (-1)^n \dfrac{q^{n-\frac{1}{2}}}{1+q^{2n-1}} \sin(2n-1)\dfrac{\pi u}{2\boldsymbol{K}}$　　　　　LA 369(4)

10.　　$\dfrac{\operatorname{cn} u}{\operatorname{sn} u} = \dfrac{\pi}{2\boldsymbol{K}} \left[ \cot\dfrac{\pi u}{2\boldsymbol{K}} - 4\sum\limits_{n=1}^{\infty} \dfrac{q^{2n}}{1+q^{2n}} \sin\dfrac{\pi n u}{\boldsymbol{K}} \right]$　　　　　LA 369(5)

---

*The expansions 1–22 are valid in every strip of the form $\left|\operatorname{Im}\dfrac{\pi u}{2\boldsymbol{K}}\right| < \dfrac{1}{2}\pi \operatorname{Im}\tau$. The expansions 23–25 are valid in an arbitrary bounded portion of $u$.

11. $\dfrac{\operatorname{cn} u}{\operatorname{dn} u} = -\dfrac{2\pi}{kK} \sum_{n=1}^{\infty} (-1)^n \dfrac{q^{n-\frac{1}{2}}}{1-q^{2n-1}} \cos(2n-1)\dfrac{\pi u}{2K}$

LA 369(5)

12. $\dfrac{\operatorname{dn} u}{\operatorname{sn} u} = \dfrac{\pi}{2K}\left[\dfrac{1}{\sin\frac{\pi u}{2K}} - 4\sum_{n=1}^{\infty}\dfrac{q^{2n-1}}{1+q^{2n-1}}\sin(2n-1)\dfrac{\pi u}{2K}\right]$

LA 369(6)

13. $\dfrac{\operatorname{dn} u}{\operatorname{cn} u} = \dfrac{\pi}{2K}\left[\dfrac{1}{\cos\frac{\pi u}{2K}} - 4\sum_{n=1}^{\infty}(-1)^n\dfrac{q^{2n-1}}{1-q^{2n-1}}\cos(2n-1)\dfrac{\pi u}{2K}\right]$

LA 369(6)

14. $\dfrac{\operatorname{cn} u\,\operatorname{dn} u}{\operatorname{sn} u} = \dfrac{\pi}{2K}\left[\cot\dfrac{\pi u}{2K} - 4\sum_{n=1}^{\infty}\dfrac{q^n}{1+q^n}\sin\dfrac{n\pi u}{K}\right]$

LA 369(7)

15. $\dfrac{\operatorname{sn} u\,\operatorname{dn} u}{\operatorname{cn} u} = \dfrac{\pi}{2K}\left\{\tan\dfrac{\pi u}{2K} + 4\sum_{n=1}^{\infty}\dfrac{q^n}{1+(-1)^n q^n}\sin\dfrac{n\pi u}{K}\right\}$

LA 369(7)

16. $\dfrac{\operatorname{sn} u\,\operatorname{cn} u}{\operatorname{dn} u} = \dfrac{4\pi^2}{k^2 K}\sum_{n=1}^{\infty}\dfrac{q^{2n-1}}{1-q^{2(2n-1)}}\sin(2n-1)\dfrac{\pi u}{K}$

LA 369(7)

17. $\dfrac{\operatorname{sn} u}{\operatorname{cn} u\,\operatorname{dn} u} = \dfrac{\pi}{2(1-k^2)K}\left[\tan\dfrac{\pi u}{2K} + 4\sum_{n=1}^{\infty}(-1)^n\dfrac{q^n}{1-q^n}\sin\dfrac{n\pi u}{K}\right]$

LA 369(8)

18. $\dfrac{\operatorname{cn} u}{\operatorname{sn} u\,\operatorname{dn} u} = \dfrac{\pi}{2K}\left[\cot\dfrac{\pi u}{2K} - 4\sum_{n=1}^{\infty}\dfrac{(-1)^n q^n}{1+(-1)^n q^n}\sin\dfrac{n\pi u}{K}\right]$

LA 369(8)

19. $\dfrac{\operatorname{dn} u}{\operatorname{sn} u\,\operatorname{cn} u} = \dfrac{\pi}{K}\left[\dfrac{1}{\sin\frac{\pi u}{K}} + 4\sum_{n=1}^{\infty}\dfrac{q^{2(2n-1)}}{1-q^{2(2n-1)}}\sin(2n-1)\dfrac{\pi u}{K}\right]$

LA 369(8)

20.[11] $\ln\operatorname{sn} u = \ln\dfrac{2K}{\pi} + \ln\sin\dfrac{\pi u}{2K} - 4\sum_{n=1}^{\infty}\dfrac{1}{n}\dfrac{q^n}{1+q^n}\sin^2\dfrac{n\pi u}{2K}$

LA 369(2)

21. $\ln\operatorname{cn} u = \ln\cos\dfrac{\pi u}{2K} - 4\sum_{n=1}^{\infty}\dfrac{1}{n}\dfrac{q^n}{1+(-1)^n q^n}\sin^2\dfrac{n\pi u}{2K}$

LA 369(2)

22. $\ln\operatorname{dn} u = -8\sum_{n=1}^{\infty}\dfrac{1}{2n-1}\dfrac{q^{2n-1}}{1-q^{2(2n-1)}}\sin^2(2n-1)\dfrac{\pi u}{2K}$

LA 369(2)

23.[11] $\operatorname{sn} u = \dfrac{2\sqrt[4]{q}}{\sqrt{k}}\sin\dfrac{\pi u}{2K}\prod_{n=1}^{\infty}\dfrac{1-2q^{2n}\cos\frac{\pi u}{K}+q^{4n}}{1-2q^{2n-1}\cos\frac{\pi u}{K}+q^{4n-2}}$

WH 508a, ZH 86(145)

24. $\operatorname{cn} u = \dfrac{2\sqrt{k'}\sqrt[4]{q}}{\sqrt{k}}\cos\dfrac{\pi u}{2K}\prod_{n=1}^{\infty}\dfrac{1+2q^{2n}\cos\frac{\pi u}{K}+q^{4n}}{1-2q^{2n-1}\cos\frac{\pi u}{K}+q^{4n-2}}$

WH 508a, ZH 86(146)

25. $\operatorname{dn} u = \sqrt{k'}\prod_{n=1}^{\infty}\dfrac{1+2q^{2n-1}\cos\frac{\pi u}{K}+q^{4n-2}}{1-2q^{2n-1}\cos\frac{\pi u}{K}+q^{4n-2}}$

WH 508a, ZH 86(147)

26.[12] $\operatorname{sn}^2 u = \sum_{n=0}^{\infty}\left[\dfrac{1+k^2}{2k^3} - \dfrac{(2n+1)^2}{2k^3}\dfrac{\pi^2}{4K^2}\right]\dfrac{2\pi q^{n+\frac{1}{2}}\sin(2n+1)\frac{\pi u}{2K}}{K(1-q^{2n+1})}$

$\left[\left|\operatorname{Im}\dfrac{u}{2K}\right| < \operatorname{Im}\tau\right]$

MO 147

27. $\qquad \dfrac{1}{\operatorname{sn}^2 u} = \dfrac{\pi^2}{4\boldsymbol{K}^2}\operatorname{cosec}^2 \dfrac{\pi u}{2\boldsymbol{K}} + \dfrac{\boldsymbol{K}-\boldsymbol{E}}{\boldsymbol{K}} - \dfrac{2\pi^2}{\boldsymbol{K}^2}\sum\limits_{n=1}^{\infty}\dfrac{nq^{2n}\cos\frac{n\pi u}{\boldsymbol{K}}}{1-q^{2n}}$

$$\left[\left|\operatorname{Im}\dfrac{u}{2\boldsymbol{K}}\right| < \dfrac{1}{2}\operatorname{Im}\tau\right] \qquad\qquad \text{MO 148}$$

**8.147**

1. $\qquad \operatorname{sn} u = \dfrac{\pi}{2k\boldsymbol{K}}\sum\limits_{n=-\infty}^{\infty}\dfrac{1}{\sin\dfrac{\pi}{2\boldsymbol{K}}[u-(2n-1)i\boldsymbol{K'}]} \qquad\qquad\qquad\qquad \text{MO 149}$

2. $\qquad \operatorname{cn} u = \dfrac{\pi i}{2k\boldsymbol{K}}\sum\limits_{n=-\infty}^{\infty}\dfrac{(-1)^n}{\sin\dfrac{\pi}{2\boldsymbol{K}}[u-(2n-1)i\boldsymbol{K'}]} \qquad\qquad\qquad \text{MO 150}$

3. $\qquad \operatorname{dn} u = \dfrac{\pi i}{2\boldsymbol{K}}\sum\limits_{n=-\infty}^{\infty}\dfrac{(-1)^n}{\tan\dfrac{\pi}{2\boldsymbol{K}}[u-(2n-1)i\boldsymbol{K'}]} \qquad\qquad\qquad\quad \text{MO150}$

**8.148** The Weierstrass expansions of the functions $\operatorname{sn} u$, $\operatorname{cn} u$, $\operatorname{dn} u$:

$$\operatorname{sn} u = \dfrac{B}{A}, \qquad \operatorname{cn} u = \dfrac{C}{A}, \qquad \operatorname{dn} u = \dfrac{D}{A}, \qquad\qquad \text{ZH 82–83(105,106,107)}$$

where

$$A = 1 - \sum_{n=1}^{\infty}(-1)^{n+1}a_{n+1}\dfrac{u^{2n+2}}{(2n+2)!} \qquad\qquad B = \sum_{n=0}^{\infty}(-1)^n b_n\dfrac{u^{2n+1}}{(2n+1)!}$$

$$C = \sum_{n=0}^{\infty}(-1)^n c_n\dfrac{u^{2n}}{(2n)!} \qquad\qquad\qquad D = \sum_{n=0}^{\infty}(-1)^n d_n\dfrac{u^{2n}}{(2n)!}$$

and

$a_2 = 2k^2, \quad a_3 = 8\left(k^2 + k^4\right), \quad a_4 = 32\left(k^2 + k^6\right) + 68k^4, \quad a_5 = 128\left(k^2 + k^8\right) + 480\left(k^4 + k^6\right),$

$a_6 = 512\left(k^2 + k^{10}\right) + 3008\left(k^4 + k^8\right) + 5400k^6, \quad \ldots$

$b_0 = 1, \quad b_1 = 1 + k^2, \quad b_2 = 1 + k^4 + 4k^2, \quad b_3 = 1 + k^6 + 9\left(k^2 + k^4\right),$

$b_4 = 1 + k^8 + 16\left(k^2 + k^6\right) - 6k^4, \quad b_5 = 1 + k^{10} + 25\left(k^2 + k^8\right) - 494\left(k^4 + k^6\right),$

$b_6 = 1 + k^{12} + 36\left(k^2 + k^{10}\right) - 5781\left(k^4 + k^8\right) - 12184k^6, \quad \ldots$

$c_0 = 1, \quad c_1 = 1, \quad c_2 = 1 + 2k^2, \quad c_3 = 1 + 6k^2 + 8k^4, \quad c_4 = 1 + 12k^2 + 60k^4 + 32k^6,$

$c_5 = 1 + 20k^2 + 348k^4 + 448k^6 + 128k^8, \quad c_6 = 1 + 30k^2 + 2372k^4 + 4600k^6 + 2880k^8 + 512k^{10}, \quad \ldots$

$d_0 = 1, \quad d_1 = k^2, \quad d_2 = 2k^2 + k^4, \quad d_3 = 8k^2 + 6k^4 + k^6, \quad d_4 = 32k^2 + 60k^4 + 12k^6 + k^8,$

$d_5 = 128k^2 + 448k^4 + 348k^6 + 20k^8 + k^{10},$

$d_6 = 512k^2 + 2880k^4 + 4600k^6 + 2372k^8 + 30k^{10} + k^{12}, \quad \ldots$

## 8.15 Properties of Jacobian elliptic functions and functional relationships between them

**8.151**    The periods, zeros, poles, and residues of Jacobian elliptic functions:

1.[12]

|  | Periods | Zeros | Poles | Residues |
|---|---|---|---|---|
| $\operatorname{sn} u$ | $4m\boldsymbol{K} + 2n\boldsymbol{K}'i$ | $2m\boldsymbol{K} + 2n\boldsymbol{K}'i$ | $2m\boldsymbol{K} + (2n+1)\boldsymbol{K}'i$ | $(-1)^m \frac{1}{k}$ |
| $\operatorname{cn} u$ | $4m\boldsymbol{K} + 2n\left(\boldsymbol{K} + \boldsymbol{K}'i\right)$ | $(2m+1)\boldsymbol{K} + 2n\boldsymbol{K}'i$ | $2m\boldsymbol{K} + (2n+1)\boldsymbol{K}'i$ | $(-1)^{m+n-1}\frac{i}{k}$ |
| $\operatorname{dn} u$ | $2m\boldsymbol{K} + 4n\boldsymbol{K}'i$ | $(2m+1)\boldsymbol{K} + (2n+1)\boldsymbol{K}'i$ | $2m\boldsymbol{K} + (2n+1)\boldsymbol{K}'i$ | $(-1)^{n-1}i$ |

SM 630, ZH 69–72

2.

| $u^* = u + \boldsymbol{K}$ | $u + i\boldsymbol{K}$ | $u + \boldsymbol{K} + i\boldsymbol{K}'$ | $u + 2\boldsymbol{K}$ | $u + 2i\boldsymbol{K}'$ | $u + 2\boldsymbol{K} + 2i\boldsymbol{K}'$ |
|---|---|---|---|---|---|
| $\operatorname{sn} u^* = \dfrac{\operatorname{cn} u}{\operatorname{dn} u}$ | $\dfrac{1}{k\operatorname{sn} u}$ | $\dfrac{1}{k}\dfrac{\operatorname{dn} u}{\operatorname{cn} u}$ | $-\operatorname{sn} u$ | $\operatorname{sn} u$ | $-\operatorname{sn} u$ |
| $\operatorname{cn} u^* = -k'\dfrac{\operatorname{sn} u}{\operatorname{dn} u}$ | $-\dfrac{i}{k}\dfrac{\operatorname{dn} u}{\operatorname{sn} u}$ | $-\dfrac{ik'}{k\operatorname{cn} u}$ | $-\operatorname{cn} u$ | $-\operatorname{cn} u$ | $\operatorname{cn} u$ |
| $\operatorname{dn} u^* = k'\dfrac{1}{\operatorname{dn} u}$ | $-i\dfrac{\operatorname{cn} u}{\operatorname{sn} u}$ | $ik'\dfrac{\operatorname{sn} u}{\operatorname{cn} u}$ | $\operatorname{dn} u$ | $-\operatorname{dn} u$ | $-\operatorname{dn} u$ |

SM 630

3.

| $u^* = 0$ | $-u$ | $\tfrac{1}{2}\boldsymbol{K}$ | $\tfrac{1}{2}\left(\boldsymbol{K} + i\boldsymbol{K}'\right)$ | $\tfrac{1}{2}i\boldsymbol{K}'$ | $u + 2m\boldsymbol{K} + 2n\boldsymbol{K}'i$ |
|---|---|---|---|---|---|
| $\operatorname{sn} u^* = 0$ | $-\operatorname{sn} u$ | $\dfrac{1}{\sqrt{1 + k'}}$ | $\dfrac{\sqrt{1 + k} + i\sqrt{1 - k}}{\sqrt{2k}}$ | $\dfrac{i}{\sqrt{k}}$ | $(-1)^m \operatorname{sn} u$ |
| $\operatorname{cn} u^* = 1$ | $\operatorname{cn} u$ | $\dfrac{\sqrt{k'}}{\sqrt{1 + k'}}$ | $\dfrac{(1 - i)\sqrt{k'}}{\sqrt{2k}}$ | $\dfrac{\sqrt{1 + k}}{\sqrt{k}}$ | $(-1)^{m+n} \operatorname{cn} u$ |
| $\operatorname{dn} u^* = 1$ | $\operatorname{dn} u$ | $\sqrt{k'}$ | $\dfrac{\sqrt{k'}\left(\sqrt{1 + k'} - i\sqrt{1 - k'}\right)}{\sqrt{2}}$ | $\sqrt{1 + k}$ | $(-1)^n \operatorname{dn} u$ |

SI 19, SI 18(13), WH,                          WH        WH        WH

**8.152**[12]    Transformation formulas

| $u_1$ | $k_1$ | $sn\,(u_1, k_1)$ | $cn\,(u_1, k_1)$ | $dn\,(u_1, k_1)$ |
|---|---|---|---|---|
| $ku$ | $\dfrac{1}{k}$ | $k\,\mathrm{sn}(u,k)$ | $\mathrm{dn}(u,k)$ | $\mathrm{cn}(u,k)$ |
| $iu$ | $k'$ | $i\,\dfrac{\mathrm{sn}(u,k)}{\mathrm{cn}(u,k)}$ | $\dfrac{1}{\mathrm{cn}(u,k)}$ | $\dfrac{\mathrm{dn}(u,k)}{\mathrm{cn}(u,k)}$ |
| $k'u$ | $i\,\dfrac{k}{k'}$ | $k'\,\dfrac{\mathrm{sn}(u,k)}{\mathrm{dn}(u,k)}$ | $\dfrac{\mathrm{cn}(u,k)}{\mathrm{dn}(u,k)}$ | $\dfrac{1}{\mathrm{dn}(u,k)}$ |
| $iku$ | $i\,\dfrac{k'}{k}$ | $ik\,\dfrac{\mathrm{sn}(u,k)}{\mathrm{dn}(u,k)}$ | $\dfrac{1}{\mathrm{dn}(u,k)}$ | $\dfrac{\mathrm{cn}(u,k)}{\mathrm{dn}(u,k)}$ |
| $ik'u$ | $\dfrac{1}{k'}$ | $ik'\,\dfrac{\mathrm{sn}(u,k)}{\mathrm{cn}(u,k)}$ | $\dfrac{\mathrm{dn}(u,k)}{\mathrm{cn}(u,k)}$ | $\dfrac{1}{\mathrm{cn}(u,k)}$ |
| $(1+k)u$ | $\dfrac{2\sqrt{k}}{1+k}$ | $\dfrac{(1+k)\,\mathrm{sn}(u,k)\,\mathrm{cn}(u,k)}{1+k\,\mathrm{sn}^2(u,k)}$ | $\dfrac{\mathrm{cn}(u,k)\,\mathrm{dn}(u,k)}{1+k\,\mathrm{sn}^2(u,k)}$ | $\dfrac{1-k\,\mathrm{sn}^2(u,k)}{1+k\,\mathrm{sn}^2(u,k)}$ |
| $(1+k')\,u$ | $\dfrac{1-k'}{1+k'}$ | $(1+k')\,\dfrac{\mathrm{sn}(u,k)\,\mathrm{cn}(u,k)}{\mathrm{dn}(u,k)}$ | $1-(1+k')\,\dfrac{\mathrm{sn}^2(u,k)}{\mathrm{dn}(u,k)}$ | $1-(1-k')\,\dfrac{\mathrm{sn}^2(u,k)}{\mathrm{dn}(u,k)}$ |
| $\dfrac{\left(1+\sqrt{k'}\right)^2}{2}\,u$ | $\left(\dfrac{1-\sqrt{k'}}{1+\sqrt{k'}}\right)^2$ | $\dfrac{k^2\,\mathrm{sn}(u,k)\,\mathrm{cn}(u,k)}{\sqrt{k_1}\,[1+\mathrm{dn}(u,k)]\,[k'+\mathrm{dn}(u,k)]}$ | $\dfrac{\mathrm{dn}(u,k)-\sqrt{k'}}{1-\sqrt{k'}}$ $\times\sqrt{\dfrac{2(1+k')}{[1+\mathrm{dn}(u,k)]\,[k'+\mathrm{dn}(u,k)]}}$ | $\dfrac{\sqrt{1+k'}\left(\mathrm{dn}(u,k)+\sqrt{k'}\right)}{\sqrt{[1+\mathrm{dn}(u,k)]\,[k'+\mathrm{dn}(u,k)]}}$ |

JA

## 8.153

1.     $\operatorname{sn}(iu, k) = i \dfrac{\operatorname{sn}(u, k')}{\operatorname{cn}(u, k')}$        SI 50(64)

2.     $\operatorname{cn}(iu, k) = \dfrac{1}{\operatorname{cn}(u, k')}$        SI 50(65)

3.     $\operatorname{dn}(iu, k) = \dfrac{\operatorname{dn}(u, k')}{\operatorname{cn}(u, k')}$        SI 50(65)

4.     $\operatorname{sn}(u, k) = k^{-1} \operatorname{sn}(ku, k^{-1})$

5.     $\operatorname{cn}(u, k) = \operatorname{dn}(ku, k^{-1})$

6.     $\operatorname{dn}(u, k) = \operatorname{cn}(ku, k^{-1})$

7.[11]    $\operatorname{sn}(u, ik) = \dfrac{1}{\sqrt{1 + k^2}} \dfrac{\operatorname{sn}\left(u\sqrt{1 + k^2}, k\left(1 + k^2\right)^{-1/2}\right)}{\operatorname{dn}\left(u\sqrt{1 + k^2}, k\left(1 + k^2\right)^{-1/2}\right)}$

8.[12]    $\operatorname{cn}(u, ik) = \dfrac{\operatorname{cn}\left(u\left(1 + k^2\right)^{1/2}, k\left(1 + k^2\right)^{-1/2}\right)}{\operatorname{dn}\left(u\left(1 + k^2\right)^{1/2}, k\left(1 + k^2\right)^{-1/2}\right)}$

9.[11]    $\operatorname{dn}(u, ik) = \dfrac{1}{\operatorname{dn}\left(u\left(1 + k^2\right)^{1/2}, k\left(1 + k^2\right)^{-1/2}\right)}$

## Functional relations

## 8.154

1.     $\operatorname{sn}^2 u = \dfrac{1 - \operatorname{cn} 2u}{1 + \operatorname{dn} 2u}$        MO 146

2.     $\operatorname{cn}^2 u = \dfrac{\operatorname{cn} 2u + \operatorname{dn} 2u}{1 + \operatorname{dn} 2u}$        MO 146

3.     $\operatorname{dn}^2 u = \dfrac{\operatorname{dn} 2u + k^2 \operatorname{cn} 2u + k'^2}{1 + \operatorname{dn} 2u}$        MO 146

4.     $\operatorname{sn}^2 u + \operatorname{cn}^2 u = 1$        SI 16(9)

5.     $\operatorname{dn}^2 u + k^2 \operatorname{sn}^2 u = 1$        SI 16(9)

## 8.155

1.     $\dfrac{1 - \operatorname{dn} 2u}{1 + \operatorname{dn} 2u} = k^2 \dfrac{\operatorname{sn}^2 u \operatorname{cn}^2 u}{\operatorname{dn}^2 u}$        MO 146

2.     $\dfrac{1 - \operatorname{cn} 2u}{1 + \operatorname{cn} 2u} = \dfrac{\operatorname{sn}^2 u \operatorname{dn}^2 u}{\operatorname{cn}^2 u}$        MO 146

## 8.156

1.     $\operatorname{sn}(u \pm v) = \dfrac{\operatorname{sn} u \operatorname{cn} v \operatorname{dn} v \pm \operatorname{sn} v \operatorname{cn} u \operatorname{dn} u}{1 - k^2 \operatorname{sn}^2 u \operatorname{sn}^2 v}$        SI 46(56)

2.      $\operatorname{cn}(u \pm v) = \dfrac{\operatorname{cn} u \operatorname{cn} v \mp \operatorname{sn} u \operatorname{sn} v \operatorname{dn} u \operatorname{dn} v}{1 - k^2 \operatorname{sn}^2 u \operatorname{sn}^2 v}$              SI 46(57)

3.      $\operatorname{dn}(u \pm v) = \dfrac{\operatorname{dn} u \operatorname{dn} v \mp k^2 \operatorname{sn} u \operatorname{sn} v \operatorname{cn} u \operatorname{cn} v}{1 - k^2 \operatorname{sn}^2 u \operatorname{sn}^2 v}$              SI 46(58)

**8.157**

1.      $\operatorname{sn} \dfrac{u}{2} = \pm \dfrac{1}{k} \sqrt{\dfrac{1 - \operatorname{dn} u}{1 + \operatorname{cn} u}} = \pm \sqrt{\dfrac{1 - \operatorname{cn} u}{1 + \operatorname{dn} u}}$            SI 47(61), SU 67(15)

2.      $\operatorname{cn} \dfrac{u}{2} = \pm \sqrt{\dfrac{\operatorname{cn} u + \operatorname{dn} u}{1 + \operatorname{dn} u}} = \pm \dfrac{k'}{k} \sqrt{\dfrac{1 - \operatorname{dn} u}{\operatorname{dn} u - \operatorname{cn} u}}$            SI 48(62), SI 67(16)

3.[12]    $\operatorname{dn} \dfrac{u}{2} = \pm \sqrt{\dfrac{\operatorname{cn} u + \operatorname{dn} u}{1 + \operatorname{cn} u}} = \pm k' \sqrt{\dfrac{1 - \operatorname{cn} u}{\operatorname{dn} u - \operatorname{cn} u}}$            SI 48(63), SI 67(17)

**8.158**

1.      $\dfrac{d}{du} \operatorname{sn} u = \operatorname{cn} u \operatorname{dn} u$                            SI 21(21)

2.      $\dfrac{d}{du} \operatorname{cn} u = -\operatorname{sn} u \operatorname{dn} u$                       SI 21(21)

3.[8]    $\dfrac{d}{du} \operatorname{dn} u = -k^2 \operatorname{dn} u \operatorname{cn} u$                     SI 21(21)

**8.159**    Jacobian elliptic functions are solutions of the following differential equations:

1.      $\dfrac{d}{du} \operatorname{sn} u = \sqrt{(1 - \operatorname{sn}^2 u)(1 - k^2 \operatorname{sn}^2 u)}$           SI 21(22)

2.      $\dfrac{d}{du} \operatorname{cn} u = -\sqrt{(1 - \operatorname{cn}^2 u)(k'^2 + k^2 \operatorname{cn}^2 u)},$      SI 21(22)

3.      $\dfrac{d}{du} \operatorname{dn} u = -\sqrt{(1 - \operatorname{dn}^2 u)(\operatorname{dn}^2 u - k'^2)}$        SI 21(22)

For the indefinite integrals of Jacobi's elliptic functions, see **5.13**.

## 8.16 The Weierstrass function $\wp(u)$

**8.160**    The Weierstrass elliptic function $\wp(u)$ is defined by

1.      $\wp(u) = \dfrac{1}{u^2} + \sum_{m,n}' \left\{ \dfrac{1}{(u - 2m\omega_1 - 2n\omega_2)^2} - \dfrac{1}{(2m\omega_1 + 2n\omega_2)^2} \right\},$      SI 307(6)

where the symbol $\sum'$ means that the summation is made over all combinations of integers $m$ and $n$ except for the combination $m = n = 0$; $2\omega_1$ and $2\omega_2$ are the periods of the function $\wp(u)$. Obviously,

2.      $\wp(u + 2m\omega_1 + 2n\omega_2) = \wp(u)$ and $\operatorname{Im}\left(\dfrac{\omega_1}{\omega_2}\right) \neq 0,$

3. $$\frac{d}{du}\wp(u) = -2\sum_{m,n}\frac{1}{(u - 2m\omega_1 - 2n\omega_2)^3},$$

where the summation is made over all integral values of $m$ and $n$.

The series **8.160** 1 and **8.160** 3 converge everywhere except at the poles, that is, at the points $2m\omega_1 + 2n\omega_2$ (where $m$ and $n$ are integers).

4. The function $\wp(u)$ is a *doubly- periodic function* and has *one second-order pole* in a period parallelogram.                                                                 SI 306

**8.161** The function $\wp(u)$ satisfies the differential equation

1. $$\left[\frac{d\wp(u)}{du}\right]^2 = 4\wp^3(u) - g_2\wp(u) - g_3,$$                SI 142, 310, WH

where

2. $$g_2 = 60\sum_{m,n}{}'(m\omega_1 + n\omega_2)^{-4}; \qquad g_3 = 140\sum_{m,n}{}'(m\omega_1 + n\omega_2)^{-6}$$                WH, SI 310

The functions $g_2$ and $g_3$ are called the *invariants* of the function $\wp(u)$.

**8.162**   $$u = \int_{\wp(u)}^{\infty}\frac{dz}{\sqrt{4z^3 - g_2 z - g_3}} = \int_{\wp(u)}^{\infty}\frac{dz}{\sqrt{4(z - e_1)(z - e_2)(z - e_3)}},$$

where $e_1$, $e_2$, and $e_3$ are the roots of the equation $4z^3 - g_2 z - g_3 = 0$; that is,

$$e_1 + e_2 + e_3 = 0, \quad e_1 e_2 + e_2 e_3 + e_3 e_1 = -\frac{g_2}{4}, \quad e_1 e_2 e_3 = \frac{g_3}{4}$$                SI 142, 143, 144

**8.163**   $\wp(\omega_1) = e_1$, $\wp(\omega_1) + \omega_2 = e_2$, $\wp(\omega_2) = e_3$. Here, it is assumed that if $e_1$, $e_2$, and $e_3$ lie on a straight line in the complex plane, $e_2$ lies between $e_1$ and $e_3$.

**8.164**   The number $\Delta = g_2^3 - 27g_3^2$ is called the *discriminant* of the function $\wp(u)$. If $\Delta > 0$, all roots $e_1$, $e_2$, and $e_3$ of the equation $4z^3 - g_2 z - g_3 = 0$ (where $g_2$ and $g_3$ are real numbers) are *real*. In this case, the roots $e_1$, $e_2$, and $e_3$ are numbered in such a way that $e_1 > e_2 > e_3$.

1. If $\Delta > 0$, then

$$\omega_1 = \int_{e_1}^{\infty}\frac{dz}{\sqrt{4z^3 - g_2 z - g_3}}, \qquad \omega_2 = i\int_{-\infty}^{e_3}\frac{dz}{\sqrt{g_3 + g_2 z - 4z^3}},$$

where $\omega_1$ is real and $\omega_2$ is a purely imaginary number. Here, the values of the radical in the integrand are chosen in such a way that $\omega_1$ and $\dfrac{\omega_2}{i}$ will be positive.

2. If $\Delta < 0$, the root $e_2$ of the equation $4z^3 - g_2 z - g_3 = 0$ is *real* and the remaining two roots ($e_1$ and $e_3$) are *complex conjugates*. Suppose that $e_1 = \alpha + i\beta$, and $e_3 = \alpha - i\beta$. In this case, it is convenient to take

$$\omega' = \int_{e_1}^{\infty}\frac{dz}{\sqrt{4z^3 - g_2 z - g_3}} \quad \text{and} \quad \omega'' = \int_{e_3}^{\infty}\frac{dz}{\sqrt{4z^3 - g_2 z - g_3}}$$

as basic semiperiods.

In the first integral, the integration is taken over a path lying entirely in the upper half-plane and in the second over a path lying entirely in the lower-half plane.                SI 151(21, 22)

**8.165**    Series representation:

1.　　$\wp(u) = \dfrac{1}{u^2} + \dfrac{g_2 u^2}{4 \cdot 5} + \dfrac{g_3 u^4}{4 \cdot 7} + \dfrac{g_2^2 u^6}{2^4 \cdot 3 \cdot 5^2} + \dfrac{3 g_2 g_3 u^8}{2^4 \cdot 5 \cdot 7 \cdot 11} + \ldots$　　　　　　　WH

**8.166**    Functional relations

1.　　$\wp(u) = \wp(-u), \quad \wp'(u) = -\wp'(-u)$

2.　　$\wp(u+v) = -\wp(u) - \wp(v) + \dfrac{1}{4} \left[ \dfrac{\wp'(u) - \wp'(v)}{\wp(u) - \wp(v)} \right]^2$　　　　　SI 163(32)

**8.167**　　$\wp\left(u; g_2, g_3\right) = \mu^2 \, \wp\left(\mu u; \dfrac{g_2}{\mu^4}, \dfrac{g_3}{\mu^6}\right)$　　　　　(the formula for homogeneity)

　　　　　　　　　　　　　　　　　　　　　　　　　　　　　　　　　　SI 149(13)

The special case: $\mu = i$.

1.　　$\wp\left(u; g_2, g_3\right) = -\wp\left(iu; g_2, -g_3\right)$

**8.168**    An arbitrary elliptic function can be expressed in terms of the elliptic function $\wp(u)$ having the same periods as the original function and its derivative $\wp'(u)$. This expression is rational with respect to $\wp(u)$ and linear with respect to $\wp'(u)$.

**8.169**    A connection with the Jacobian elliptic functions. For $\Delta > 0$ (see **8.164** 1).

1.[12]　$\wp\left(\dfrac{u}{\sqrt{e_1 - e_3}}\right) = e_1 + (e_1 - e_3)\dfrac{\mathrm{cn}^2(u, k)}{\mathrm{sn}^2(u, k)}$

$$= e_2 + (e_1 - e_3)\dfrac{\mathrm{dn}^2(u, k)}{\mathrm{sn}^2(u, k)}$$

$$= e_3 + (e_1 - e_3)\dfrac{1}{\mathrm{sn}^2(u, k)}$$

　　　　　　　　　　　　　　　　　　　　　　　SI 145(5), ZH 120(197–199)a

2.　　$\omega_1 = \dfrac{K}{\sqrt{e_1 - e_3}}, \qquad \omega_2 = \dfrac{iK'}{\sqrt{e_1 - e_3}},$　　　　　SI 154(29)

where

3.　　$k = \sqrt{\dfrac{e_2 - e_3}{e_1 - e_3}}, \qquad k' = \sqrt{\dfrac{e_1 - e_2}{e_1 - e_3}}$　　　　　SI 145(7)

For $\Delta < 0$ (see **8.164** 2)

4.　　$\wp\left(\dfrac{u}{\sqrt[4]{9\alpha^2 + \beta^2}}\right) = e_2 + \sqrt{9\alpha^2 + \beta^2}\,\dfrac{1 + \mathrm{cn}(2u, k)}{1 - \mathrm{cn}(2u, k)};$　　　　SI 147(12)

5.　　$\omega' = \dfrac{K - iK'}{2\sqrt{9\alpha^2 + \beta^2}}, \qquad \omega'' = \dfrac{K + iK'}{\sqrt[4]{9\alpha^2 + \beta^2}},$　　　　SI 153(28)

where

6.[11]　$k = \sqrt{\dfrac{1}{2} - \dfrac{3e_2}{4\sqrt{9\alpha^2 + \beta^2}}}; \qquad k' = \sqrt{\dfrac{1}{2} + \dfrac{3e_2}{4\sqrt{9\alpha^2 + \beta^2}}}$　　　SI 147

For $\Delta = 0$, all the roots $e_1$, $e_2$, and $e_3$ are real and if $g_2 g_3 \neq 0$, two of them are equal to each other. If $e_1 = e_2 \neq e_3$, then

7.     $\wp(u) = \dfrac{3g_3}{g_2} - \dfrac{9g_3}{2g_2} \coth^2\left(u\sqrt{-\dfrac{9g_3}{2g_2}}\right)$            SI 148

      If $e_1 \neq e_2 = e_3$, then

8.     $\wp(u) = -\dfrac{3g_3}{2g_2} + \dfrac{9g_3}{2g_2}\dfrac{1}{\sin^2\left(u\sqrt{\frac{9g_3}{2g_2}}\right)}$            SI 149

      If $g_2 = g_3 = 0$, then $e_1 = e_2 = e_3 = 0$, and

9.     $\wp(u) = \dfrac{1}{u^2}$            SI 149

## 8.17 The functions $\zeta(u)$ and $\sigma(u)$

**8.171**   Definitions:

1.     $\zeta(u) = \dfrac{1}{u} - \displaystyle\int_0^u \left(\wp(z) - \dfrac{1}{z^2}\right) dz$            SI 181(45)

2.     $\sigma(u) = u\exp\left\{\displaystyle\int_0^u \left(\wp(z) - \dfrac{1}{z^2}\right) dz\right\}$            SI 181(46)

**8.172**   Series and infinite-product representation

1.     $\zeta(u) = \dfrac{1}{u} + \displaystyle\sum_{m,n}{}' \left(\dfrac{1}{u - 2m\omega_1 - 2n\omega_2} + \dfrac{1}{2m\omega_1 + 2n\omega_2} + \dfrac{u}{(2m\omega_1 - 2n\omega_2)^2}\right)$            SI 307(8)

2.     $\sigma(u) = u\displaystyle\prod_{m,n}{}' \left(1 - \dfrac{u}{2m\omega_1 + 2n\omega_2}\right)\exp\left\{\dfrac{u}{2m\omega_1 + 2n\omega_2} + \dfrac{u^2}{2(2m\omega_1 + 2n\omega_2)^2}\right\}$            SI 308(9)

**8.173**

1.     $\zeta(u) = u - \dfrac{g_2 u^3}{2^2 \cdot 3 \cdot 5} - \dfrac{g_3 u^5}{2^2 \cdot 5 \cdot 7} - \dfrac{g_2^2 u^7}{2^4 \cdot 3 \cdot 5^2 \cdot 7} - \dfrac{3g_2 g_3 u^9}{2^4 \cdot 5 \cdot 7 \cdot 9 \cdot 11} - \cdots$            SI 181(49)

2.[12]     $\sigma(u) = u - \dfrac{g_2 u^5}{2^4 \cdot 3 \cdot 5} - \dfrac{g_3 u^7}{2^3 \cdot 3 \cdot 5 \cdot 7} - \dfrac{g_2^2 u^9}{2^9 \cdot 3^2 \cdot 5 \cdot 7} - \dfrac{g_2 g_3 u^{11}}{2^7 \cdot 3^2 \cdot 5^2 \cdot 7 \cdot 11} - \cdots$            SI 181(49)

**8.174**   $\zeta(u) = \dfrac{\zeta(\omega_1)}{\omega_1}u + \dfrac{\pi}{2\omega_1}\cot\dfrac{\pi u}{2\omega_1} + \dfrac{\pi}{2\omega_1}\displaystyle\sum_{n=1}^{\infty}\left\{\cot\left(\dfrac{\pi u}{2\omega_1} + n\pi\dfrac{\omega_2}{\omega_1}\right)\right.$

              $\left. + \cot\left(\dfrac{\pi u}{2\omega_1} - n\pi\dfrac{\omega_2}{\omega_1}\right)\right\}$            MO 154

              $= \dfrac{\zeta(\omega_1)}{\omega_1}u + \dfrac{\pi}{2\omega_1}\cot\dfrac{\pi u}{2\omega_1} + \dfrac{2\pi}{\omega_1}\displaystyle\sum_{n=1}^{\infty}\dfrac{q^{2n}}{1-q^{2n}}\sin\dfrac{\pi n u}{\omega_1}$            MO 155

### Functional relations and properties

**8.175**   $\zeta(u) = -\zeta(-u), \quad \sigma(u) = -\sigma(-u)$            SI 181

**8.176**

1.     $\zeta(u + 2\omega_1) = \zeta(u) + 2\zeta(\omega_1)$            SI 184(57)

2. $\zeta (u + 2\omega_2) = \zeta(u) + 2\zeta (\omega_2)$ SI 184(57)

3. $\sigma (u + 2\omega_1) = - \sigma(u) \exp \{2 (u + \omega_1) \zeta (\omega_1)\}.$ SI 185(60)

4. $\sigma (u + 2\omega_2) = - \sigma(u) \exp \{2 (u + \omega_2) \zeta (\omega_2)\}.$ SI 185(60)

5. $\omega_2 \zeta (\omega_1) - \omega_1 \zeta (\omega_2) = \dfrac{\pi}{2} i$ SI 186(62)

**8.177**

1. $\zeta(u + v) - \zeta(u) - \zeta(v) = \dfrac{1}{2} \dfrac{\wp'(u) - \wp'(v)}{\wp(u) - \wp(v)}$ SI 182(53)

2. $\wp(u) - \wp(v) = - \dfrac{\sigma(u - v) \sigma (u + v)}{\sigma^2(u) \sigma^2(v)}$ SI 183(54)

3. $\zeta(u - v) + \zeta(u + v) - 2\zeta(u) = \dfrac{\wp'(u)}{\wp(u) - \wp(v)}$ SI 182(51)

**8.178**

1. $\zeta (u; \omega_1, \omega_2) = t \zeta (tu; t\omega_1, t\omega_2)$ MO 154

2.[8] $\sigma (u; \omega_1, \omega_2) = t^{-1} \sigma (tu; t\omega_1, t\omega_2)$ MO 156

For the indefinite integrals of Weierstrass elliptic functions, see **5.14**.

## 8.18–8.19  Theta functions

**8.180**  *Theta functions* are defined as the sums (for $|q| < 1$) of the following series:

1. $\vartheta_4(u) = \displaystyle\sum_{n=-\infty}^{\infty} (-1)^n q^{n^2} e^{2nui} = 1 + 2\sum_{n=1}^{\infty}(-1)^n q^{n^2} \cos 2nu$ WH

2. $\vartheta_1(u) = \dfrac{1}{i} \displaystyle\sum_{n=-\infty}^{\infty} (-1)^n q^{\left(n+\frac{1}{2}\right)^2} e^{(2n+1)ui} = 2\sum_{n=1}^{\infty}(-1)^{n+1} q^{\left(n-\frac{1}{2}\right)^2} \sin(2n - 1)u$ WH

3.[11] $\vartheta_2(u) = \displaystyle\sum_{n=-\infty}^{\infty} q^{\left(n+\frac{1}{2}\right)^2} e^{(2n+1)ui} = 2\sum_{n=1}^{\infty} q^{\left(n-\frac{1}{2}\right)^2} \cos(2n - 1)u$ WH

4. $\vartheta_3(u) = \displaystyle\sum_{n=-\infty}^{\infty} q^{n^2} e^{2nui} = 1 + 2\sum_{n=1}^{\infty} q^{n^2} \cos 2nu$ WH

The notations $\vartheta(u, q)$ and $\vartheta (u \mid \tau)$, where $\tau$ and $q$ are related by $q = e^{i\pi\tau}$, are also used. Here, $q$ is called the *nome* of the theta function and $\tau$ its *parameter*.

**8.181**  Representation of theta functions in terms of infinite products

1. $\vartheta_4(u) = \displaystyle\prod_{n=1}^{\infty} \left(1 - 2q^{2n-1} \cos 2u + q^{2(2n-1)}\right) \left(1 - q^{2n}\right)$ SI 200(9), ZH 90(9)

2. $\vartheta_3(u) = \displaystyle\prod_{n=1}^{\infty} \left(1 + 2q^{2n-1} \cos 2u + q^{2(2n-1)}\right) \left(1 - q^{2n}\right)$ SI 200(9), ZH 90(9)

3.      $\vartheta_1(u) = 2\sqrt[4]{q}\sin u \prod_{n=1}^{\infty}\left(1 - 2q^{2n}\cos 2u + q^{4n}\right)\left(1 - q^{2n}\right)$         SI 200(9), ZH 90(9)

4.[8]     $\vartheta_2(u) = 2\sqrt[4]{q}\cos u \prod_{n=1}^{\infty}\left(1 + 2q^{2n}\cos 2u + q^{4n}\right)\left(1 - q^{2n}\right)$        SI 200(0), ZH 90(9)

## Functional relations and properties

**8.182**   Quasiperiodicity. Suppose that $q = e^{\pi\tau i}$ ($\operatorname{Im}\tau > 0$). Then, theta functions that are periodic functions of $u$ are called *quasiperiodic functions* of $\tau$ and $u$. This property follows from the equations

1.      $\vartheta_4\left(u + \pi\right) = \vartheta_4(u)$                                       SI 200(10)

2.      $\vartheta_4\left(u + \tau\pi\right) = -\dfrac{1}{q}e^{-2iu}\,\vartheta_4(u)$                      SI 200(10)

3.      $\vartheta_1\left(u + \pi\right) = -\,\vartheta_1(u)$                                    SI 200(10)

4.      $\vartheta_1\left(u + \tau\pi\right) = -\dfrac{1}{q}e^{-2iu}\,\vartheta_1(u)$                       SI 200(10)

5.      $\vartheta_2\left(u + \pi\right) = -\,\vartheta_2(u)$                                      SI 200(10)

6.      $\vartheta_2\left(u + \tau\pi\right) = \dfrac{1}{q}e^{-2iu}\,\vartheta_2(u)$                         SI 200(10)

7.      $\vartheta_3\left(u + \pi\right) = \vartheta_3(u)$                                         SI 200(10)

8.      $\vartheta_3\left(u + \tau\pi\right) = \dfrac{1}{q}e^{-2iu}\,\vartheta_3(u)$                         SI 200(10)

**8.183**

1.      $\vartheta_4\left(u + \tfrac{1}{2}\pi\right) - \vartheta_3(u)$                                    WH

2.      $\vartheta_1\left(u + \tfrac{1}{2}\pi\right) = \vartheta_2(u)$                                    WH

3.      $\vartheta_2\left(u + \tfrac{1}{2}\pi\right) = -\,\vartheta_1(u)$                                WH

4.      $\vartheta_3\left(u + \tfrac{1}{2}\pi\right) = \vartheta_4(u)$                                    WH

5.      $\vartheta_4\left(u + \tfrac{1}{2}\pi\tau\right) = iq^{-1/4}e^{-iu}\,\vartheta_1(u)$               WH

6.      $\vartheta_1\left(u + \tfrac{1}{2}\pi\tau\right) = iq^{-1/4}e^{-iu}\,\vartheta_4(u)$               WH

7.      $\vartheta_2\left(u + \tfrac{1}{2}\pi\tau\right) = q^{-1/4}e^{-iu}\,\vartheta_3(u)$                WH

8.      $\vartheta_3\left(u + \tfrac{1}{2}\pi\tau\right) = q^{-1/4}e^{-iu}\,\vartheta_2(u)$                WH

**8.184**   Even and odd theta functions

1.      $\vartheta_1(-u) = -\,\vartheta_1(u)$                                       WH

2.      $\vartheta_2(-u) = \vartheta_2(u)$                                         WH

3.      $\vartheta_3(-u) = \vartheta_3(u)$                                         WH

4.      $\vartheta_4(-u) = \vartheta_4(u)$                                         WH

**8.185**    $\vartheta_4^4(u) + \vartheta_2^4(u) = \vartheta_1^4(u) + \vartheta_3^4(u)$                       WH

**8.186**[7]    Considering the theta functions as functions of two independent variables $u$ and $\tau$, we have

$$\pi i \frac{\partial^2 \vartheta_k(u \mid \tau)}{\partial u^2} + 4 \frac{\partial \vartheta_k(u \mid \tau)}{\partial \tau} = 0 \qquad [k = 1, 2, 3, 4] \qquad \text{WH}$$

**8.187** We denote the partial derivatives of the theta functions with respect to $u$ by a prime and consider them as functions of the single argument $u$. Then,

1.     $\vartheta_1'(0) = \vartheta_2(0)\,\vartheta_3(0)\,\vartheta_4(0)$           WH

2.     $\dfrac{\vartheta_1'''(0)}{\vartheta_1'(0)} = \dfrac{\vartheta_2''(0)}{\vartheta_2(0)} + \dfrac{\vartheta_3''(0)}{\vartheta_3(0)} + \dfrac{\vartheta_4''(0)}{\vartheta_4(0)}$         WH

**8.188**[12] $\vartheta_1(u)\,\vartheta_2(u)\,\vartheta_3(u)\,\vartheta_4(u) = \frac{1}{2}\vartheta_1(2u)\,\vartheta_2(0)\,\vartheta_3(0)\,\vartheta_4(0)$        WH

**8.189** The zeros of the theta functions:

1.[8]    $\vartheta_4(u) = 0$ for $u = 2m\dfrac{\pi}{2} + (2n - 1)\dfrac{\pi\tau}{2}$         SI 201

2.[10]   $\vartheta_1(u) = 0$ for $u = 2m\dfrac{\pi}{2} + 2n\dfrac{\pi\tau}{2}$           SI 201

3.     $\vartheta_2(u) = 0$ for $u = (2m - 1)\dfrac{\pi}{2} + 2n\dfrac{\pi\tau}{2}$       SI 201

4.     $\vartheta_3(u) = 0$ for $u = (2m - 1)\dfrac{\pi}{2} + (2n - 1)\dfrac{\pi\tau}{2}$    [$m$ and $n$ are integers or zero]    SI 201

For integrals of theta functions, see **6.16**.

**8.191** Connections with the Jacobian elliptic functions:

For $\tau = i\dfrac{K'}{K}$, i.e. for $q = \exp\left(-\pi \dfrac{K'}{K}\right)$,

1.     $\operatorname{sn} u = \dfrac{1}{\sqrt{k}} \dfrac{\vartheta_1\left(\dfrac{\pi u}{2K}\right)}{\vartheta_4\left(\dfrac{\pi u}{2K}\right)} = \dfrac{1}{\sqrt{k}} \dfrac{H(u)}{\Theta(u)}$     SI 206(22), SI 209(35)

2.     $\operatorname{cn} u = \sqrt{\dfrac{k'}{k}} \dfrac{\vartheta_2\left(\dfrac{\pi u}{2K}\right)}{\vartheta_4\left(\dfrac{\pi u}{2K}\right)} = \sqrt{\dfrac{k'}{k}} \dfrac{H_1(u)}{\Theta(u)}$     SI 207(23), SI 209(35)

3.     $\operatorname{dn} u = \sqrt{k'} \dfrac{\vartheta_3\left(\dfrac{\pi u}{2K}\right)}{\vartheta_4\left(\dfrac{\pi u}{2K}\right)} = \sqrt{k'} \dfrac{\Theta_1(u)}{\Theta(u)}$     SI 207(24), SI 209(35)

**8.192** Series representation of the functions $H$, $H_1$, $\Theta$, $\Theta_1$.

In these formulas, $q = \exp\left(-\pi \dfrac{K'}{K}\right)$.

1.     $\Theta(u) = \vartheta_4\left(\dfrac{\pi u}{2K}\right) = 1 + 2\displaystyle\sum_{n=1}^{\infty} (-1)^n q^{n^2} \cos\dfrac{n\pi u}{K}$     SI 207(25), SI 212(42)

2.     $H(u) = \vartheta_1\left(\dfrac{\pi u}{2K}\right) = 2\displaystyle\sum_{n=1}^{\infty} (-1)^{n+1} \sqrt[4]{q^{(2n+1)^2}} \sin(2n - 1)\dfrac{\pi u}{2K}$     SI 207(25), SI 212(43)

3.     $\Theta_1(u) = \vartheta_3\left(\dfrac{\pi u}{2K}\right) = 1 + 2\displaystyle\sum_{n=1}^{\infty} q^{n^2} \cos\dfrac{n\pi u}{K}$     SI 207(25), SI 212(45)

4.    $H_1(u) = \vartheta_2\left(\dfrac{\pi u}{2\boldsymbol{K}}\right) = 2\displaystyle\sum_{n=1}^{\infty} \sqrt[4]{q^{(2n-1)^2}}\cos(2n-1)\dfrac{\pi u}{2\boldsymbol{K}}$    SI 207(25), SI 212(44)

**8.193**   Connections with the Weierstrass elliptic functions

1.[12]    $\wp(u) = e_1 + \left[\dfrac{H_1\left(u\sqrt{\lambda}\right)H'(0)}{H_1(0)\,H\left(u\sqrt{\lambda}\right)}\right]^2 \lambda = e_2 + \left[\dfrac{\Theta_1\left(u\sqrt{\lambda}\right)H'(0)}{\Theta_1(0)\,H\left(u\sqrt{\lambda}\right)}\right]^2 \lambda = e_3 + \left[\dfrac{\Theta\left(u\sqrt{\lambda}\right)H'(0)}{\Theta(0)\,H\left(u\sqrt{\lambda}\right)}\right]^2 \lambda$

SI 235(77,78)

2.    $\zeta(u) = \dfrac{\eta_1 u}{\omega_1} + \sqrt{\lambda}\,\dfrac{H'\left(u\sqrt{\lambda}\right)}{H\left(u\sqrt{\lambda}\right)}$    SI 234(73)

3.    $\sigma(u) = \dfrac{1}{\sqrt{\lambda}}\exp\left(\dfrac{\eta_1 u^2}{2\omega_1}\right)\dfrac{H\left(u\sqrt{\lambda}\right)}{H'(0)}$    SI 234(72)

where

$$\lambda = e_1 - e_3; \qquad \eta_1 = \zeta\left(\omega_1\right) = -\dfrac{\omega_1\lambda}{3}\dfrac{H'''(0)}{H'(0)}$$    SI 236

**8.194**   The connection with elliptic integrals:

1.[12]    $E(\operatorname{am}u, k) = u - u\dfrac{\Theta''(0)}{\Theta(0)} + \dfrac{\Theta'(u)}{\Theta(u)}$    SI 228(65)

2.[12]    $\Pi\left(\operatorname{am}u, k^2\operatorname{sn}^2 a, k\right) = \displaystyle\int_0^u \dfrac{d\varphi}{1 - k^2\operatorname{sn}^2 a\operatorname{sn}^2\varphi} = u + \dfrac{\operatorname{sn}a}{\operatorname{cn}a\operatorname{dn}a}\left[\dfrac{\Theta'(a)}{\Theta(a)}u + \dfrac{1}{2}\ln\dfrac{\Theta(u-a)}{\Theta(u\mid a)}\right]$

SI 228(65)

**q-series and products, $q = \exp\left(-\pi\dfrac{\mathsf{K}'}{\mathsf{K}}\right)$**

**8.195**    $\dfrac{\pi}{2}\left[1 + 2\displaystyle\sum_{n=1}^{\infty} q^{n^2}\right]^2 = \boldsymbol{K} = \dfrac{\pi}{2}\Theta^2(\boldsymbol{K})$    (cf. **8.197** 1)    SI 219

**8.196**    $\boldsymbol{E} = \boldsymbol{K} - \boldsymbol{K}\dfrac{\Theta''(0)}{\Theta(0)} = \boldsymbol{K} - \dfrac{2\pi^2}{\boldsymbol{K}}\dfrac{\displaystyle\sum_{n=1}^{\infty}(-1)^{n+1}n^2 q^{n^2}}{1 + 2\displaystyle\sum_{n=1}^{\infty}(-1)^n q^{n^2}}$    SI 230(67)

**8.197**

1.    $1 + 2\displaystyle\sum_{n=1}^{\infty} q^{n^2} = \sqrt{\dfrac{2\boldsymbol{K}}{\pi}} = \vartheta_3(0)$    (cf. **8.195**)    WH

2.    $\displaystyle\sum_{n=1}^{\infty} q^{\left(\frac{2n-1}{2}\right)^2} = \sqrt{\dfrac{k\boldsymbol{K}}{2\pi}} = \dfrac{1}{2}\vartheta_2(0)$    WH

3.    $4\sqrt{q}\prod_{n=1}^{\infty}\left(\dfrac{1+q^{2n}}{1+q^{2n-1}}\right)^4 = k$        SI 206(17, 18)

4.    $\prod_{n=1}^{\infty}\left(\dfrac{1-q^{2n-1}}{1+q^{2n-1}}\right)^4 = k'$        SI 206(19, 20)

5.    $2\sqrt[4]{q}\prod_{n=1}^{\infty}\left(\dfrac{1-q^{2n}}{1-q^{2n-1}}\right)^2 = 2\sqrt{k}\dfrac{K}{\pi}$        WH

6.    $\prod_{n=1}^{\infty}\left(\dfrac{1-q^{2n}}{1+q^{2n}}\right)^2 = 2\sqrt{k'}\dfrac{K}{\pi}$        WH

**8.198**

1.    $\lambda = \dfrac{1}{2}\dfrac{1-\sqrt{k'}}{1+\sqrt{k'}} = \dfrac{\sum_{n=0}^{\infty} q^{(2n+1)^2}}{1+2\sum_{n=1}^{\infty} q^{4n^2}}$        [for $0 < k < 1$, we have $0 < \lambda < \frac{1}{2}$]    WH

The series

2.    $q = \lambda + 2\lambda^5 + 15\lambda^9 + 150\lambda^{13} + 1707\lambda^{17} + \ldots$        WH

is used to determine $q$ from the given modulus $k$.

**8.199**[10]    Identities involving products of theta functions

1.    $\vartheta_1(x,q)\,\vartheta_1(y,q) = \vartheta_3\left(x+y,q^2\right)\vartheta_2\left(x-y,q^2\right) - \vartheta_2\left(x+y,q^2\right)\vartheta_3\left(x-y,q^2\right)$        LW 7(1.4.7)

2.    $\vartheta_1(x,q)\,\vartheta_2(y,q) = \vartheta_1\left(x+y,q^2\right)\vartheta_4\left(x-y,q^2\right) + \vartheta_4\left(x+y,q^2\right)\vartheta_1\left(x-y,q^2\right)$        LW 8(1.4.8)

3.    $\vartheta_2(x,q)\,\vartheta_2(y,q) = \vartheta_2\left(x+y,q^2\right)\vartheta_3\left(x-y,q^2\right) + \vartheta_3\left(x+y,q^2\right)\vartheta_2\left(x-y,q^2\right)$        LW 8(1.4.9)

4.    $\vartheta_3(x,q)\,\vartheta_3(y,q) = \vartheta_3\left(x+y,q^2\right)\vartheta_3\left(x-y,q^2\right) + \vartheta_2\left(x+y,q^2\right)\vartheta_2\left(x-y,q^2\right)$        LW 8(1.4.10)

5.    $\vartheta_3(x,q)\,\vartheta_4(y,q) = \vartheta_4\left(x+y,q^2\right)\vartheta_4\left(x-y,q^2\right) - \vartheta_1\left(x+y,q^2\right)\vartheta_1\left(x-y,q^2\right)$        LW 8(1.4.11)

6.    $\vartheta_4(x,q)\,\vartheta_4(y,q) = \vartheta_3\left(x+y,q^2\right)\vartheta_3\left(x-y,q^2\right) - \vartheta_2\left(x+y,q^2\right)\vartheta_2\left(x-y,q^2\right)$        LW 8(1.4.12)

7.    $\vartheta_1(x+y)\,\vartheta_1(x-y)\,\vartheta_4^2(0) = \vartheta_3^2(x)\,\vartheta_2^2(y) - \vartheta_2^2(x)\,\vartheta_3^2(y) = \vartheta_1^2(x)\,\vartheta_4^2(y) - \vartheta_4^2(x)\,\vartheta_1^2(y)$        LW 8(1.4.16)

8.    $\vartheta_2(x+y)\,\vartheta_2(x-y)\,\vartheta_4^2(0) = \vartheta_4^2(x)\,\vartheta_2^2(y) - \vartheta_1^2(x)\,\vartheta_3^2(y) = \vartheta_2^2(x)\,\vartheta_4^2(y) - \vartheta_3^2(x)\,\vartheta_1^2(y)$        LW 8(1.4.17)

9.    $\vartheta_3(x+y)\,\vartheta_3(x-y)\,\vartheta_4^2(0) = \vartheta_4^2(x)\,\vartheta_3^2(y) - \vartheta_1^2(x)\,\vartheta_2^2(y) = \vartheta_3^2(x)\,\vartheta_4^2(y) - \vartheta_2^2(x)\,\vartheta_1^2(y)$        LW 8(1.4.18)

10.    $\vartheta_4(x+y)\,\vartheta_4(x-y)\,\vartheta_4^2(0) = \vartheta_4^2(x)\,\vartheta_4^2(y) - \vartheta_1^2(x)\,\vartheta_1^2(y)$        LW 8(1.4.15)

11.    $\vartheta_4(x+y)\,\vartheta_4(x-y)\,\vartheta_4^2(0) = \vartheta_3^2(x)\,\vartheta_3^2(y) - \vartheta_2^2(x)\,\vartheta_2^2(y) = \vartheta_4^2(x)\,\vartheta_4^2(y) - \vartheta_1^2(x)\,\vartheta_1^2(y)$        LW 9(1.4.19)

12.    $\vartheta_1(x+y)\,\vartheta_1(x-y)\,\vartheta_3^2(0) = \vartheta_1^2(x)\,\vartheta_3^2(y) - \vartheta_3^2(x)\,\vartheta_1^2(y) = \vartheta_4^2(x)\,\vartheta_2^2(y) - \vartheta_2^2(x)\,\vartheta_4^2(y)$        LW 9(1.4.23)

13.    $\vartheta_2(x+y)\,\vartheta_2(x-y)\,\vartheta_3^2(0) = \vartheta_2^2(x)\,\vartheta_3^2(y) - \vartheta_4^2(x)\,\vartheta_1^2(y) = \vartheta_3^2(x)\,\vartheta_2^2(y) - \vartheta_1^2(x)\,\vartheta_4^2(y)$        LW 9(1.4.24)

14.    $\vartheta_3(x+y)\,\vartheta_3(x-y)\,\vartheta_3^2(0) = \vartheta_1^2(x)\,\vartheta_1^2(y) + \vartheta_3^2(x)\,\vartheta_3^2(y) = \vartheta_2^2(x)\,\vartheta_2^2(y) + \vartheta_4^2(x)\,\vartheta_4^2(y)$        LW 9(1.4.25)

15.    $\vartheta_4(x+y)\,\vartheta_4(x-y)\,\vartheta_3^2(0) = \vartheta_1^2(x)\,\vartheta_2^2(y) + \vartheta_3^2(x)\,\vartheta_4^2(y) = \vartheta_2^2(x)\,\vartheta_1^2(y) + \vartheta_4^2(x)\,\vartheta_3^2(y)$        LW 9(1.4.26)

16.    $\vartheta_1(x+y)\,\vartheta_1(x-y)\,\vartheta_2^2(0) = \vartheta_1^2(x)\,\vartheta_2^2(y) - \vartheta_2^2(x)\,\vartheta_1^2(y) = \vartheta_4^2(x)\,\vartheta_3^2(y) - \vartheta_3^2(x)\,\vartheta_4^2(y)$        LW 9(1.4.30)

17.    $\vartheta_2(x+y)\,\vartheta_2(x-y)\,\vartheta_2^2(0) = \vartheta_2^2(x)\,\vartheta_2^2(y) - \vartheta_1^2(x)\,\vartheta_1^2(y) = \vartheta_3^2(x)\,\vartheta_3^2(y) - \vartheta_4^2(x)\,\vartheta_4^2(y)$     LW 10(1.4.31)

18.    $\vartheta_3(x+y)\,\vartheta_3(x-y)\,\vartheta_2^2(0) = \vartheta_3^2(x)\,\vartheta_2^2(y) + \vartheta_4^2(x)\,\vartheta_1^2(y) = \vartheta_2^2(x)\,\vartheta_3^2(y) + \vartheta_1^2(x)\,\vartheta_4^2(y)$     LW 10(1.4.32)

19.    $\vartheta_4(x+y)\,\vartheta_4(x-y)\,\vartheta_2^2(0) = \vartheta_4^2(x)\,\vartheta_2^2(y) + \vartheta_3^2(x)\,\vartheta_1^2(y) = \vartheta_1^2(x)\,\vartheta_3^2(y) + \vartheta_2^2(x)\,\vartheta_4^2(y)$     LW 10(1.4.33)

20.    $\vartheta_3^2(x)\,\vartheta_3^2(0) = \vartheta_4^2(x)\,\vartheta_4^2(0) + \vartheta_2^2(x)\,\vartheta_2^2(0)$     LW 11(1.4.49)

21.    $\vartheta_4^2(x)\,\vartheta_3^2(0) = \vartheta_1^2(x)\,\vartheta_2^2(0) + \vartheta_3^2(x)\,\vartheta_4^2(0)$     LW 11(1.4.50)

22.    $\vartheta_4^2(x)\,\vartheta_2^2(0) = \vartheta_1^2(x)\,\vartheta_3^2(0) + \vartheta_2^2(x)\,\vartheta_4^2(0)$     LW 11(1.4.51)

23.    $\vartheta_3^2(x)\,\vartheta_2^2(0) = \vartheta_1^2(x)\,\vartheta_4^2(0) + \vartheta_2^2(x)\,\vartheta_3^2(0)$     LW 11(1.4.52)

24.[8]    $\vartheta_3^4(x) = \vartheta_2^4(0) + \vartheta_4^4(0)$     LW 11(1.4.53)

**8.199(2)[10]**    Derivatives of ratios of theta functions

1.    $\dfrac{d}{dx}\,(\vartheta_1 / \vartheta_4) = \vartheta_4^2(0)\,\vartheta_2(x)\,\vartheta_3(x)/\,\vartheta_4^2(x)$     LW 19(1.9.3)

2.    $\dfrac{d}{dx}\,(\vartheta_2 / \vartheta_4) = -\,\vartheta_3^2(0)\,\vartheta_1(x)\,\vartheta_3(x)/\,\vartheta_4^2(x)$     LW 19(1.9.6)

3.    $\dfrac{d}{dx}\,(\vartheta_3 / \vartheta_4) = -\,\vartheta_2^2(0)\,\vartheta_1(x)\,\vartheta_2(x)/\,\vartheta_4^2(x)$     LW 19(1.9.7)

4.    $\dfrac{d}{dx}\,(\vartheta_1 / \vartheta_3) = \vartheta_3^2(0)\,\vartheta_2(x)\,\vartheta_4(x)/\,\vartheta_3^2(x)$     LW 19(1.9.8)

5.    $\dfrac{d}{dx}\,(\vartheta_2 / \vartheta_3) = -\,\vartheta_4^2(0)\,\vartheta_1(x)\,\vartheta_4(x)/\,\vartheta_3^2(x)$     LW 19(1.9.9)

6.    $\dfrac{d}{dx}\,(\vartheta_1 / \vartheta_2) = \vartheta_2^2(0)\,\vartheta_3(x)\,\vartheta_4(x)/\,\vartheta_2^2(x)$     LW 19(1.9.10)

7.    $\dfrac{d}{dx}\,(\vartheta_4 / \vartheta_1) = -\,\vartheta_4^2(0)\,\vartheta_2(x)\,\vartheta_3(x)/\,\vartheta_1^2(x)$     LW 19(1.9.11)

8.    $\dfrac{d}{dx}\,(\vartheta_4 / \vartheta_2) = \vartheta_3^2(0)\,\vartheta_1(x)\,\vartheta_3(x)/\,\vartheta_2^2(x)$     LW 20(1.9.12)

9.    $\dfrac{d}{dx}\,(\vartheta_4 / \vartheta_3) = \vartheta_2^2(0)\,\vartheta_1(x)\,\vartheta_2(x)/\,\vartheta_3^2(x)$     LW 20(1.9.13)

10.    $\dfrac{d}{dx}\,(\vartheta_3 / \vartheta_1) = -\,\vartheta_3^2(0)\,\vartheta_2(x)\,\vartheta_4(x)/\,\vartheta_1^2(x)$     LW 20(1.9.14)

11.    $\dfrac{d}{dx}\,(\vartheta_3 / \vartheta_2) = \vartheta_4^2(0)\,\vartheta_1(x)\,\vartheta_4(x)/\,\vartheta_2^2(x)$     LW 20(1.9.15)

12.    $\dfrac{d}{dx}\,(\vartheta_2 / \vartheta_1) = -\,\vartheta_2^2(0)\,\vartheta_3(x)\,\vartheta_4(x)/\,\vartheta_1^2(x)$     LW 20(1.9.16)

**8.199(3)[10]**    Derivatives of theta functions

1.    $\dfrac{d}{du}\ln\vartheta_1(u) = \cot u + 4\sin 2u \displaystyle\sum_{n=1}^{\infty} \dfrac{q^{2n}}{1 - 2q^{2n}\cos 2u + q^{4n}}$

2.    $\dfrac{d}{du}\ln\vartheta_2(u) = -\tan u - 4\sin 2u \displaystyle\sum_{n=1}^{\infty} \dfrac{q^{2n}}{1 + 2q^{2n}\cos 2u + q^{4n}}$

$3.^{12}$ $\quad \dfrac{d}{du} \ln \vartheta_3(u) = -4 \sin 2u \displaystyle\sum_{n=1}^{\infty} \dfrac{q^{2n-1}}{1 + 2q^{2n-1} \cos 2u + q^{4n-2}}$

$4. \quad \dfrac{d}{du} \ln \vartheta_4(u) = 4 \sin 2u \displaystyle\sum_{n=1}^{\infty} \dfrac{q^{2n-1}}{1 - 2q^{2n} \cos 2u + q^{4n-2}}$

$5. \quad \dfrac{d^2}{du^2} \ln \vartheta_2(u) = -\displaystyle\sum_{n=-\infty}^{\infty} \operatorname{sech}^2 \{i(u + n\pi\tau)\}$

**8.199(4)**　Relations among theta function arguments

$1.^* \quad (-i\tau)^{1/2} \vartheta_1(z|\tau) = -i \exp\left(\dfrac{z^2}{i\pi\tau}\right) \vartheta_1\left(-\dfrac{z}{\tau} \,\middle|\, -\dfrac{1}{\tau}\right)$

$2.^* \quad (-i\tau)^{1/2} \vartheta_2(z|\tau) = \exp\left(\dfrac{z^2}{i\pi\tau}\right) \vartheta_4\left(-\dfrac{z}{\tau} \,\middle|\, -\dfrac{1}{\tau}\right)$

$3.^* \quad (-i\tau)^{1/2} \vartheta_3(z|\tau) = \exp\left(\dfrac{z^2}{i\pi\tau}\right) \vartheta_3\left(-\dfrac{z}{\tau} \,\middle|\, -\dfrac{1}{\tau}\right)$

$4.^* \quad (-i\tau)^{1/2} \vartheta_4(z|\tau) = \exp\left(\dfrac{z^2}{i\pi\tau}\right) \vartheta_2\left(-\dfrac{z}{\tau} \,\middle|\, -\dfrac{1}{\tau}\right)$

# 8.2 The Exponential Integral Function and Functions Generated by It

## 8.21 The exponential integral function Ei(x)

**8.211**

$1. \quad \operatorname{Ei}(x) = -\displaystyle\int_{-x}^{\infty} \dfrac{e^{-t}}{t}\,dt = \int_{-\infty}^{x} \dfrac{e^t}{t}\,dt = \operatorname{li}(e^x) \qquad\qquad [\operatorname{Re} x > 0]$

$2.^{11} \quad \operatorname{Ei}(x) = -\displaystyle\lim_{\varepsilon \to 0+}\left[\int_{-x}^{-\varepsilon} \dfrac{e^{-t}}{t}\,dt + \int_{\varepsilon}^{\infty} \dfrac{e^{-t}}{t}\,dt\right] = \mathrm{PV}\int_{-\infty}^{x} \dfrac{e^t}{t}\,dt$

$\qquad\qquad\qquad\qquad\qquad\qquad\qquad\qquad\qquad\qquad\qquad [x > 0]$

$3.^7 \quad \operatorname{Ei}(x) = \tfrac{1}{2}\{\operatorname{Ei}(x + i0) + \operatorname{Ei}(x - i0)\} \qquad\qquad [x > 0] \qquad\qquad\qquad\text{ET I 386}$

**8.212**

$1.^8 \quad \operatorname{Ei}(-x) = C + \ln x + \displaystyle\int_0^x \dfrac{e^{-t} - 1}{t}\,dt \qquad [x > 0] \qquad\qquad\qquad\qquad\text{NT 11(1)}$

$\qquad\quad = C + e^{-x}\ln x + \displaystyle\int_0^x e^{-t}\ln t\,dt \qquad [x > 0] \qquad\qquad\qquad\text{NT 11(10)}$

$2.^7 \quad \operatorname{Ei}(x) = e^x\left[\dfrac{1}{x} + \displaystyle\int_0^{\infty} \dfrac{e^{-t}\,dt}{(x - t)^2}\right] \qquad\qquad\qquad [x > 0] \qquad (\text{cf. } \mathbf{8.211\ 1})$

$3. \quad \operatorname{Ei}(-x) = e^{-x}\left[-\dfrac{1}{x} + \displaystyle\int_0^{\infty} \dfrac{e^{-t}\,dt}{(x + t)^2}\right] \qquad\qquad [x > 0] \qquad (\text{cf. } \mathbf{8.211\ 1}) \qquad \text{LA 281(28)}$

$4. \quad \operatorname{Ei}(\pm x) = \pm e^{\pm x}\displaystyle\int_0^1 \dfrac{dt}{x \pm \ln t} \qquad\qquad\qquad\qquad [x > 0] \qquad (\text{cf. } \mathbf{8.211\ 1})$

5.     $\mathrm{Ei}\,(\pm xy) = \pm e^{\pm xy} \displaystyle\int_0^\infty \frac{e^{-xt}}{y \mp t}\, dt$        $[\mathrm{Re}\, y > 0, \quad x > 0]$        NT 19(11)

6.     $\mathrm{Ei}\,(\pm x) = -e^{\pm x} \displaystyle\int_0^\infty \frac{e^{-it}}{t \pm ix}\, dt$        $[x > 0]$        NT 23(2, 3)

7.[8]     $\mathrm{Ei}(xy) = e^{xy} \displaystyle\int_0^1 \frac{t^{y-1}}{x + \ln t}\, dt$        LA 282(44)a

8.     $\mathrm{Ei}(-xy) = -e^{-xy} \displaystyle\int_0^1 \frac{t^{y-1}}{x - \ln t}\, dt$        LA 282(45)a

       $= x^{-1} e^{-xy} \left[ \displaystyle\int_0^1 \frac{t^{x-1}}{(y - \ln t)^2}\, dt - y^{-1} \right]$        $[x > 0, \quad y > 0]$        LA 283(47)a

9.     $\mathrm{Ei}(x) = e^x \displaystyle\int_1^\infty \frac{1}{x - \ln t} \frac{dt}{t^2}$        $[x > 0]$        LA 283(48)

10.     $\mathrm{Ei}(-x) = -e^{-x} \displaystyle\int_1^\infty \frac{1}{x + \ln t} \frac{dt}{t^2}$        $[x > 0]$        LA 283(48)

11.     $\mathrm{Ei}(-x) = -e^{-x} \displaystyle\int_0^\infty \frac{t \cos t + x \sin t}{t^2 + x^2}\, dt$        $[x > 0]$        NT 23(6)

12.     $\mathrm{Ei}(-x) = -e^{-x} \displaystyle\int_0^\infty \frac{t \cos t - x \sin t}{t^2 + x^2}\, dt$        $[x < 0]$        NT 23(6)

13.     $\mathrm{Ei}(-x) = \dfrac{2}{\pi} \displaystyle\int_0^\infty \frac{\cos t}{t} \arctan \frac{t}{x}\, dt$        $[\mathrm{Re}\, x > 0]$        NT 25(13)

14.     $\mathrm{Ei}(-x) = \dfrac{2 e^{-x}}{\pi} \displaystyle\int_0^\infty \frac{x \cos t - t \sin t}{t^2 + x^2} \ln t\, dt$        $[x > 0]$        NT 26(7)

15.     $\mathrm{Ei}(x) = 2 \ln x - \dfrac{2 e^x}{\pi} \displaystyle\int_0^\infty \frac{x \cos t + t \sin t}{t^2 + x^2} \ln t\, dt$        $[x > 0]$        NT 27(8)

16.     $\mathrm{Ei}(-x) = -x \displaystyle\int_1^\infty e^{-tx} \ln t\, dt$        $[x > 0]$        NT 32(12)

See also **3.327**, **3.881** 8, **3.916** 2 and 3, **4.326** 1, **4.326** 2, **4.331** 2, **4.351** 3, **4.425** 3, **4.581**. For integrals of the exponential integral function, see **6.22–6.23**, **6.78**.

### Series and asymptotic representations

**8.213**

1.     $\mathrm{li}(x) = \boldsymbol{C} + \ln(-\ln x) + \displaystyle\sum_{k=1}^\infty \frac{(\ln x)^k}{k \cdot k!}$        $[0 < x < 1]$        NT 3(9)

2.     $\mathrm{li}(x) = \boldsymbol{C} + \ln \ln x + \displaystyle\sum_{k=1}^\infty \frac{(\ln x)^k}{k \cdot k!}$        $[x > 1]$        NT 3(10)

**8.214**

1.     $\mathrm{Ei}(x) = \boldsymbol{C} + \ln(-x) + \displaystyle\sum_{k=1}^\infty \frac{x^k}{k \cdot k!}$        $[x < 0]$

2.      $\operatorname{Ei}(x) = \boldsymbol{C} + \ln x + \displaystyle\sum_{k=1}^{\infty} \frac{x^k}{k \cdot k!}$          $[x > 0]$

3.      $\operatorname{Ei}(x) - \operatorname{Ei}(-x) = 2x \displaystyle\sum_{k=0}^{\infty} \frac{x^{2k}}{(2k+1)(2k+1)!}$          $[x > 0]$          NT 39(13)

**8.215[7]**    $\operatorname{Ei}(z) = \dfrac{e^z}{z} \left[ \displaystyle\sum_{k=0}^{n} \frac{k!}{z^k} + R_n(z) \right]$      $|R_n(z)| = O\left( |z|^{-n-1} \right)$

         $[z \to \infty, \quad |\arg(-z)| \leq \pi - \delta; \quad \delta > 0 \text{ small}]$,      $|R_n(z)| \leq (n+1)! |z|^{-n-1}$    $[\operatorname{Re} z \leq 0]$

**8.216[7]**    $\operatorname{Ei}(nx) - \operatorname{Ei}(-nx) = e^{nx'} \left( \dfrac{1}{nx} + \dfrac{1}{n^2 x^2} + \dfrac{k_n}{n^3 x^3} \right),$

                        where $x' = x \operatorname{sign} \operatorname{Re}(x)$,    $k_n = O(1)$, and $n \to \infty$    NT 39(15)

**8.217**    Functional relations:

1.      $e^{x'} \operatorname{Ei}(-x') - e^{-x'} \operatorname{Ei}(x') = -2 \displaystyle\int_0^{\infty} \frac{x' \sin t}{t^2 + x^2} \, dt$          NT 24(11)

             $= \dfrac{4}{\pi} \displaystyle\int_0^{\infty} \frac{x' \cos t}{t^2 + x^2} \ln t \, dt - 2 e^{-x'} \ln x'$      $[x' = x \operatorname{sign} \operatorname{Re} x]$      NT 27(9)

2.      $e^{x'} \operatorname{Ei}(-x') + e^{-x'} \operatorname{Ei}(x') = -2 \displaystyle\int_0^{\infty} \frac{t \cos t}{t^2 + x^2} \, dt = 2 e^{-x'} \ln x' - \dfrac{4}{\pi} \displaystyle\int_0^{\infty} \frac{t \sin t}{t^2 + x^2} \ln t \, dt$

                                      $[x' = x \operatorname{sign} \operatorname{Re} x]$      NT 24(10), NT 27(10)

3.      $\operatorname{Ei}(-x) - \operatorname{Ei}\left( -\dfrac{1}{x} \right) = \dfrac{2}{\pi} \displaystyle\int_0^{\infty} \frac{\cos t}{t} \arctan \frac{t\left(x - \frac{1}{x}\right)}{1 + t^2} \, dt$

                                      $[\operatorname{Re} x > 0]$      NT 25(14)

4.      $\operatorname{Ei}(-\alpha x) \operatorname{Ei}(-\beta x) - \ln(\alpha\beta) \operatorname{Ei}[-(\alpha + \beta)x] = e^{-(\alpha+\beta)x} \displaystyle\int_0^{\infty} \frac{e^{-tx} \ln[(\alpha+t)(\beta+t)]}{t + \alpha + \beta} \, dt$      NT 32(9)

See also **3.723** 1 and 5, **3.742** 2 and 4, **3.824** 4, **4.573** 2.

- For a connection with a confluent hypergeometric function, see **9.237**.
- For integrals of the exponential integral function, see **5.21**, **5.22**, **5.23**, **6.22**, and **6.23**.

**8.218**    Two numerical values:

1.      $\operatorname{Ei}(-1) = -0.219\,383\,934\,395\,520\,273\,665\ldots$          NT 89

2.      $\operatorname{Ei}(1) = 1.895\,117\,816\,355\,936\,755\,478\ldots$          NT 89

**8.219\***    Definite integrals of exponential functions

1.\*      $\displaystyle\int_0^{\infty} \operatorname{Ei}^2(x) e^{-2x} \, dx = \dfrac{\pi^2}{4}$

2.\*      $\displaystyle\int_0^{\infty} \operatorname{Ei}^2(-x) e^{2x} \, dx = \dfrac{\pi^2}{4}$

3.\*      $\displaystyle\int_0^{\infty} \operatorname{Ei}(x) \operatorname{Ei}(-x) \, dx = 0$

## 8.22 The hyperbolic sine integral $\operatorname{shi} x$ and the hyperbolic cosine integral $\operatorname{chi} x$

**8.221**

1. $\quad \operatorname{shi} x = \displaystyle\int_0^x \frac{\sinh t}{t}\, dt = -i\left[\frac{\pi}{2} + \operatorname{si}(ix)\right]$ (see **8.230** 1) EH II 146(17)

2.[11] $\quad \operatorname{chi} x = \boldsymbol{C} + \ln x + \displaystyle\int_0^x \frac{\cosh t - 1}{t}\, dt$ EH II 146(18)

## 8.23 The sine integral and the cosine integral: $\operatorname{si} x$ and $\operatorname{ci} x$

**8.230**

1.[10] $\quad \operatorname{si}(x) = -\displaystyle\int_x^\infty \frac{\sin t}{t}\, dt = -\frac{\pi}{2} + \operatorname{Si}(x), \quad \text{where } \operatorname{Si}(x) = \int_0^x \frac{\sin t}{t}\, dt$ NT 11(3)

2.[10] $\quad \operatorname{ci}(x) = -\displaystyle\int_x^\infty \frac{\cos t}{t}\, dt = \boldsymbol{C} + \ln x + \int_0^x \frac{\cos t - 1}{t}\, dt \qquad [\operatorname{ci}(x) \text{ is also written } \operatorname{Ci}(x)]$ NT 11(2)

**8.231**

1. $\quad \operatorname{si}(xy) = -\displaystyle\int_x^\infty \frac{\sin ty}{t}\, dt$ NT 18(7)

2. $\quad \operatorname{ci}(xy) = -\displaystyle\int_x^\infty \frac{\cos ty}{t}\, dt$ NT 18(6)

3. $\quad \operatorname{si}(x) = -\displaystyle\int_0^{\pi/2} e^{-x\cos t} \cos(x\sin t)\, dt$ NT 13(26)

**8.232**

1. $\quad \operatorname{si}(x) = -\dfrac{\pi}{2} + \displaystyle\sum_{k=1}^\infty \frac{(-1)^{k+1} x^{2k-1}}{(2k-1)(2k-1)!}$ NT 7(4)

2.[7] $\quad \operatorname{ci}(x) = \boldsymbol{C} + \ln(x) + \displaystyle\sum_{k=1}^\infty (-1)^k \frac{x^{2k}}{2k(2k)!}$ NT 7(3)

**8.233**

1. $\quad \operatorname{ci}(x) \pm i\operatorname{si}(x) = \operatorname{Ei}(\pm ix)$ NT 6a

2. $\quad \operatorname{ci}(x) - \operatorname{ci}\left(xe^{\pm\pi i}\right) = \mp\pi i$ NT 7(5)

3. $\quad \operatorname{si}(x) + \operatorname{si}(-x) = -\pi$ NT 7(7)

**8.234**

1.[7] $\quad \operatorname{Ei}(-x) - \operatorname{ci}(x) = \displaystyle\int_0^{\pi/2} e^{-x\cos\varphi} \sin(s\sin\varphi)\, d\varphi$ NT 13(27)

2. $\quad [\operatorname{ci}(x)]^2 + [\operatorname{si}(x)]^2 = -2\displaystyle\int_0^{\pi/2} \frac{\exp(-x\tan\varphi)\ln\cos\varphi}{\sin\varphi\cos\varphi}\, d\varphi$

$$[\operatorname{Re} x > 0] \qquad (\text{see also } \textbf{4.366})$$

NT 32(11)

See also **3.341**, **3.351** 1 and 2, **3.354** 1 and 2, **3.721** 2 and 3, **3.722** 1, 3, 5 and 7, **3.723** 8 and 11, **4.338** 1, **4.366** 1.

**8.235**

1.[12]    $\lim\limits_{x\to+\infty}\left(x^{\rho}\,\mathrm{si}(x)\right)=0,\quad \lim\limits_{x\to+\infty}\left(x^{\rho}\,\mathrm{ci}(x)\right)=0$      $[\rho<1]$        NT 38(5)

2.    $\lim\limits_{x\to-\infty}\mathrm{si}(x)=-\pi,\quad \lim\limits_{x\to-\infty}\mathrm{ci}(x)=\pm\pi i$        NT 38(6)

- For integrals of the sine integral and cosine integral, see **6.24–6.26**, **6.781**, **6.782**, and **6.783**.
- For indefinite integrals of the sine integral and cosine integral, see **5.3**.

## 8.24 The logarithm integral $\mathrm{li}(x)$

**8.240**

1.    $\mathrm{li}(x)=\displaystyle\int_0^x\frac{dt}{\ln t}=\mathrm{Ei}\,(\ln x)$      $[x<1]$        JA

2.    $\mathrm{li}(x)=\displaystyle\lim_{\varepsilon\to0}\left[\int_0^{1-\varepsilon}\frac{dt}{\ln t}+\int_{1+\varepsilon}^x\frac{dt}{\ln t}\right]=\mathrm{Ei}\,(\ln x)$      $[x>1]$        JA

3.    $\mathrm{li}\left\{\exp\left(-xe^{\pm\pi i}\right)\right\}=\mathrm{Ei}\left(-xe^{\pm i\pi}\right)=\mathrm{Ei}\,(x\mp i0)=\mathrm{Ei}(x)\pm i\pi=\mathrm{li}\,(e^x)\pm i\pi$

                                                     $[x>0]$        JA, NT 2(6)

### Integral representations

**8.241**

1.[12]    $\mathrm{li}(x)=\displaystyle\int_{-\infty}^{\ln x}\frac{e^t}{t}\,dt=x\ln\ln\frac{1}{x}-\int_{-\ln x}^{\infty}e^{-t}\ln t\,dt$      $[0<x<1]$        LA 281(33)

2.    $\mathrm{li}(x)=x\displaystyle\int_0^1\frac{dt}{\ln x+\ln t}$                                            LA 280(22)

       $=\dfrac{x}{\ln x}+x\displaystyle\int_0^1\frac{dt}{(\ln x+\ln t)^2}$                          LA 280(29)

       $=x\displaystyle\int_1^{\infty}\frac{1}{\ln x-\ln t}\frac{dt}{t^2}$      $[x<1]$        LA 280(30)

3.[12]    $\mathrm{li}\,(a^x)=\displaystyle\int_{-\infty}^x\frac{a^t}{t}\,dt$      $[x>0]$

       For integrals of the logarithm integral, see **6.21**

## 8.25 The probability integral $\Phi(x)$, the Fresnel integrals $S(x)$, $C(x)$, the error function $\mathrm{erf}(x)$, and the complementary error function $\mathrm{erfc}(x)$

**8.250**    Definition:

1.[11]    $\Phi(x)=\mathrm{erf}(x)=\dfrac{2}{\sqrt{\pi}}\displaystyle\int_0^x e^{-t^2}\,dt$        (called the error function)

2.    $S(x)=\dfrac{2}{\sqrt{2\pi}}\displaystyle\int_0^x\sin t^2\,dt$

**3.**     $C(x) = \dfrac{2}{\sqrt{2\pi}} \displaystyle\int_0^x \cos t^2 \, dt$

**4.**[11]     $\mathrm{erfc}(x) = 1 - \mathrm{erf}(x)$     (called the complementary error function)

**5.**     $\displaystyle\int_0^\infty \dfrac{e^{-(p+x)y}}{\pi(p+x)} \sin\left(a\sqrt{x}\right) \, dx$

$$= -\sinh\left(a\sqrt{p}\right) + \frac{1}{2} e^{-a\sqrt{p}} \, \Phi\left(\frac{a}{2\sqrt{y}} - \sqrt{py}\right) + \frac{1}{2} e^{a\sqrt{p}} \, \Phi\left(\frac{a}{2\sqrt{y}} + \sqrt{py}\right)$$

**6.**     $\displaystyle\int_0^\infty \dfrac{e^{-(p+x)y}}{\pi(p+x)} \cos\left(a\sqrt{x}\right) \, dx = \dfrac{1}{\sqrt{\pi y}} \exp\left(-\dfrac{a^2}{4y} - py\right) - \dfrac{\sqrt{p}}{2} e^{-a\sqrt{p}} \, \Phi\left(\dfrac{a}{a\sqrt{y}} - \sqrt{py}\right)$

$$+ \frac{\sqrt{p}}{2} e^{\sqrt{p}} \, \Phi\left(\frac{a}{2\sqrt{y}} + \sqrt{py}\right) - \sqrt{p} \cosh\left(a\sqrt{p}\right)$$

$$[\operatorname{Re} p > 0, \quad a, b \text{ are real}]$$

**7.**     $\displaystyle\int_0^p \exp\left(-x^2\right) \Phi(p-x) \, dx = \int_0^p \exp\left(-x^2\right) \mathrm{erf}(p-x) \, dx = \dfrac{\sqrt{\pi}}{2} \left[\Phi\left(\dfrac{p}{\sqrt{2}}\right)\right]^2$

**8.**     $\displaystyle\int_0^p x^2 \exp\left(-x^2\right) \Phi(p-x) \, dx = \int_0^p x^2 \exp\left(-x^2\right) \mathrm{erf}(p-x) \, dx$

$$= \frac{\sqrt{\pi}}{4} \left[\Phi\left(\frac{p}{\sqrt{2}}\right)\right]^2 - \frac{p}{2\sqrt{2}} \Phi\left(-\frac{x^2}{2}\right) \mathrm{erf}\left(\frac{p}{\sqrt{2}}\right)$$

**9.**     $\displaystyle\int_{(a+b)/\sqrt{2}}^{(b-a)/\sqrt{2}} \exp\left(-x^2\right) \Phi\left(b\sqrt{2} - x\right) \, dx + \int_{(a+b)/\sqrt{2}}^{(a-b)/\sqrt{2}} \exp\left(-x^2\right) \Phi\left(a\sqrt{2} - x\right) \, dx = \sqrt{\pi} \, \Phi(a) \, \Phi(b)$

## Integral representations

**8.251**

**1.**     $\Phi(x) = \dfrac{1}{\sqrt{\pi}} \displaystyle\int_0^{x^2} \dfrac{e^{-t}}{\sqrt{t}} \, dt$                    (see also **3.361** 1)

**2.**     $S(x) = \dfrac{1}{\sqrt{2\pi}} \displaystyle\int_0^{x^2} \dfrac{\sin t}{\sqrt{t}} \, dt$

**3.**     $C(x) = \dfrac{1}{\sqrt{2\pi}} \displaystyle\int_0^{x^2} \dfrac{\cos t}{\sqrt{t}} \, dt$

**8.252**

**1.**     $\Phi(xy) = \dfrac{2y}{\sqrt{\pi}} \displaystyle\int_0^x e^{-t^2 y^2} \, dt$

**2.**     $S(xy) = \dfrac{2y}{\sqrt{2\pi}} \displaystyle\int_0^x \sin\left(t^2 y^2\right) \, dt$

**3.**     $C(xy) = \dfrac{2y}{\sqrt{2\pi}} \displaystyle\int_0^x \cos\left(t^2 y^2\right) \, dt$

4.    $\Phi(xy) = 1 - \dfrac{2}{\sqrt{\pi}} e^{-x^2 y^2} \displaystyle\int_0^\infty \dfrac{e^{-t^2 y^2} t y\, dt}{\sqrt{t^2 + x^2}}$    $\left[\operatorname{Re}^2 y > 0\right]$      NT 19(11)a

       $= 1 - \dfrac{2x}{\pi} e^{-x^2 y^2} \displaystyle\int_0^\infty \dfrac{e^{-t^2 y^2}\, dt}{t^2 + x^2}$    $\left[\operatorname{Re}^2 y > 0\right]$      NT 19(13)a

5.[12]    $\Phi\left(\dfrac{-y}{2xi}\right) - \Phi\left(\dfrac{y}{2xi}\right) = \dfrac{4xi e^{y^2/4x^2}}{\sqrt{\pi}} \displaystyle\int_0^\infty e^{-t^2 y^2} \sin(ty)\, dt$

                                               $\left[\operatorname{Re} x^2 > 0\right]$      NT 28(3)a

6.[8]    $\Phi\left(\dfrac{y}{2x}\right) = 1 - \dfrac{2}{\sqrt{\pi}} x e^{-y^2/4x^2} \displaystyle\int_0^\infty e^{-t^2 x^2 - ty}\, dt$    $\left[\operatorname{Re} x^2 > 0\right]$      NT 27(1)a

See also **3.322**, **3.362** 2, **3.363**, **3.468**, **3.897**, **6.511** 4 and 5.

**8.253**[8]    Series representations:

1.[11]    $\operatorname{erf}(x) = \dfrac{2}{\sqrt{\pi}} e^{-x^2} x\, F_1\left(1; \dfrac{3}{2}; x^2\right) = \dfrac{2}{\sqrt{\pi}} \displaystyle\sum_{k=1}^\infty (-1)^{k+1} \dfrac{x^{2k-1}}{(2k-1)(k-1)!}$      NT 7(9)a

       $= \dfrac{2}{\sqrt{\pi}} e^{-x^2} \displaystyle\sum_{k=0}^\infty \dfrac{2^k x^{2k+1}}{(2k+1)!!}$      NT 10(11)a

2.[12]    $S(x) = \dfrac{2}{\sqrt{2\pi}} \left( x \sin^2 x\, F\left(1, \dfrac{5}{4}; \dfrac{3}{4}; -\dfrac{1}{4} x^2\right) - \dfrac{2}{3} x^3 \cos^2 x\, F\left(1, \dfrac{7}{4}; \dfrac{5}{4}; -\dfrac{1}{4} x^2\right) \right)$

       $\dfrac{2}{\sqrt{2\pi}} \displaystyle\sum_{k=0}^\infty \dfrac{(-1)^k x^{4k+3}}{(2k+1)!(4k+3)}$      NT 8(14)a

       $= \dfrac{2}{\sqrt{2\pi}} \left\{ \sin^2 x \displaystyle\sum_{k=0}^\infty \dfrac{(-1)^k 2^{2k} x^{4k+1}}{(4k+1)!!} - \cos^2 x \displaystyle\sum_{k=0}^\infty \dfrac{(-1)^k 2^{2k+1} x^{4k+3}}{(4k+3)!!} \right\}$      NT 10(13)a

3.[12]    $C(x) = \dfrac{2}{\sqrt{2\pi}} \left( \dfrac{2}{3} x^3 \sin^2 x\, F\left(1, \dfrac{7}{4}; \dfrac{5}{4}; -\dfrac{1}{4} x^2\right) - x \cos^2 x\, F\left(1, \dfrac{5}{4}; \dfrac{3}{4}; -\dfrac{1}{4} x^2\right) \right)$

       $= \dfrac{2}{\sqrt{2\pi}} \displaystyle\sum_{k=0}^\infty \dfrac{(-1)^k x^{4k+1}}{(2k)!(4k+1)}$      NT 8(13)a

       $= \dfrac{2}{\sqrt{2\pi}} \left\{ \sin^2 x \displaystyle\sum_{k=0}^\infty \dfrac{(-1)^k 2^{2k+1} x^{4k+3}}{(4k+3)!!} + \cos^2 x \displaystyle\sum_{k=0}^\infty \dfrac{(-1)^k 2^{2k} x^{4k+1}}{(4k+1)!!} \right\}$      NT 10(12)a

For the expansions in Bessel functions, see **8.515** 2, **8.515** 3.

## Asymptotic representations

**8.254**[12]    $\Phi(z) = 1 - \dfrac{e^{-z^2}}{\sqrt{\pi} z} \left[ \displaystyle\sum_{k=0}^n (-1)^k \dfrac{(2k-1)!!}{(2z^2)^k} + O\left(|z|^{-2n-2}\right) \right],$

                                     $\left[z \to \infty, \quad |\arg(-z)| \le \pi - \delta; \quad \delta > 0 \text{ small}\right]$

where

$|R_n| < \dfrac{\Gamma\left(n + \frac{1}{2}\right)}{|x|^{n+\frac{1}{2}} \cos \frac{\varphi}{2}}, \quad x = |x| e^{i\varphi} \text{ and } \varphi^2 < \pi^2$      NT 37(10)a

**8.255**

1.[12]    $S(x) = \dfrac{1}{2} - \dfrac{1}{\sqrt{2\pi x}} \cos^2 x + O\left(\dfrac{1}{x^2}\right)$    $[x \to \infty]$      MO 127a

$2.^{12}$    $C(x) = \dfrac{1}{2} + \dfrac{1}{\sqrt{2\pi x}} \sin^2 x + O\left(\dfrac{1}{x^2}\right)$        $[x \to \infty]$        MO 127a

**8.256**    Functional relations:

1.        $C(z) + i\, S(z) = \sqrt{\dfrac{i}{2}}\, \Phi\left(\dfrac{z}{\sqrt{i}}\right) = \dfrac{2}{\sqrt{2\pi}} \displaystyle\int_0^z e^{it^2}\, dt$

2.        $C(z) - i\, S(z) = \dfrac{1}{\sqrt{2i}}\, \Phi\left(z\sqrt{i}\right) = \dfrac{2}{\sqrt{2\pi}} \displaystyle\int_0^z e^{-it^2}\, dt$

$3.^{12}$    $\left[\cos^2 u\, C(u) + \sin^2 u\, S(u)\right] = \dfrac{1}{2}\left[\cos^2 u + \sin^2 u\right] + \sqrt{\dfrac{2}{\pi}} \displaystyle\int_0^\infty e^{-2ut} \sin^2 t\, dt$

            $[\operatorname{Re} u \geq 0]$        NT 28(6)a

$4.^{12}$    $\left[\cos^2 u\, S(u) - \sin^2 u\, C(u)\right] = \dfrac{1}{2}\left[\cos^2 u - \sin^2 u\right] - \sqrt{\dfrac{2}{\pi}} \displaystyle\int_0^\infty e^{-2ut} \cos^2 t\, dt$

            $[\operatorname{Re} u \geq 0]$        NT 28(5)a

$5.^{11}$    $\left[C(x) - \dfrac{1}{2}\right]^2 + \left[S(x) - \dfrac{1}{2}\right]^2 = \dfrac{2}{\pi} \displaystyle\int_0^{\pi/2} \dfrac{\exp\left(-x^2 \tan\varphi\right) \sin\frac{\varphi}{2} \sqrt{\cos\varphi}}{\sin 2\varphi}\, d\varphi$

            (see also **6.322**)        NT 33(18)a

- For a connection with a confluent hypergeometric function, see **9.236**.
- For a connection with a parabolic cylinder function, see **9.254**.

**8.257**

1.        $\displaystyle\lim_{x \to +\infty}\left(x^\rho\left[S(x) - \tfrac{1}{2}\right]\right) = 0$        $[\rho < 1]$        NT 38(11)

2.        $\displaystyle\lim_{x \to +\infty}\left(x^\rho\left[C(x) - \tfrac{1}{2}\right]\right) = 0$        $[\rho < 1]$        NT 38(11)

3.        $\displaystyle\lim_{x \to +\infty} S(x) = \dfrac{1}{2}$        NT 38(12)a

4.        $\displaystyle\lim_{x \to +\infty} C(x) = \dfrac{1}{2}$        NT 38(12)a

- For integrals of the probability integral, see **6.28–6.31**.
- For integrals of Fresnel's sine integral and cosine integral, see **6.32**.

$\mathbf{8.258}^{10}$    Integrals involving the complementary error function

1.        $\displaystyle\int_0^\infty \operatorname{erfc}^2(x) e^{-\beta x^2}\, dx = \dfrac{1}{\sqrt{\beta\pi}}\left(-\arccos\left(\dfrac{1}{1+\beta}\right) + 2\arctan\left(\sqrt{\beta}\right)\right)$

            $[\beta > 0]$

2.        $\displaystyle\int_0^\infty x\, \operatorname{erfc}^2(x) e^{-\beta x^2}\, dx = \dfrac{1}{2\beta}\left(1 - \dfrac{4}{\pi}\dfrac{\arctan\left(\sqrt{1+\beta}\right)}{\sqrt{1+\beta}}\right)$

            $[\beta > 0]$

3.    $\displaystyle\int_0^\infty x^3 \operatorname{erfc}^2(x) e^{-\beta x^2}\, dx = \frac{1}{2\beta^2}\left(1 - \frac{4}{\pi}\frac{\arctan\left(\sqrt{1+\beta}\right)}{\sqrt{1+\beta}}\right)$

$\displaystyle + \frac{1}{\beta\pi}\left(\frac{1}{(1+\beta)(\beta^2+2\beta+2)} - \frac{\arctan\left(\sqrt{1+\beta}\right)}{(1+\beta)^{\frac{3}{2}}}\right)$

$[\beta > 0]$

4.    $\displaystyle\int_0^\infty x \operatorname{erfc}\left(\sqrt{x}\right) e^{-\beta x}\, dx = \frac{1}{\beta^2}\left[1 - \frac{1+\frac{3}{2}\beta}{(1+\beta)^{\frac{3}{2}}}\right]$      $[\beta > 0]$

5.[12]    $\displaystyle\int_0^\infty \sqrt{x}\,\operatorname{erfc}\left(\sqrt{x}\right) e^{-\beta x}\, dx = \frac{1}{\sqrt{\pi}}\left(\frac{1}{\beta^{\frac{3}{2}}}\arctan\left(\sqrt{\beta}\right) - \frac{1}{\beta(1+\beta)}\right)$

$[\beta > 0]$

**8.259\***    Integrals involving the error function and an exponential function

1.    $\displaystyle\int_{-\infty}^\infty e^{-px^2}\Phi(a+bx)\, dx = \sqrt{\frac{\pi}{p}}\,\Phi\left(\frac{a\sqrt{p}}{\sqrt{b^2+p}}\right)$      $[\operatorname{Re} p > 0],\quad a, b \text{ real}$

2.    $\displaystyle\int_{-\infty}^\infty x^2 e^{-px^2}\Phi(a+bx)\, dx = \frac{1}{2p}\sqrt{\frac{\pi}{p}}\,\Phi\left(\frac{a\sqrt{p}}{\sqrt{b^2+p}}\right) - \frac{ab^2}{p(b^2+p)^{3/2}}\exp\left(-\frac{a^2 p}{b^2+p}\right)$

$[\operatorname{Re} p > 0,\quad a, b \text{ are real}]$

3.    $\displaystyle\int_{-\infty}^\infty x^{2n} e^{-px^2}\Phi(a+bx)\, dx = (-1)^n \frac{\partial^n}{\partial p^n}\left[\sqrt{\frac{\pi}{p}}\,\Phi\left(\frac{a\sqrt{p}}{\sqrt{b^2+p}}\right)\right]$

$[n = 0, 1, \ldots,\quad \operatorname{Re} p > 0,\quad a, b \text{ are real}]$

## 8.26 Lobachevskiy's function $L(x)$

**8.260**    Definition:

$$L(x) = -\int_0^x \ln\cos t\, dt$$    LO III 184(10)

For integral representations of the function $L(x)$, see also **3.531** 8, **3.532** 2, **3.533**, and **4.224**.

**8.261**    Representation in the form of a series:

$$L(x) = x\ln 2 - \frac{1}{2}\sum_{k=1}^\infty (-1)^{k-1}\frac{\sin 2kx}{k^2}$$    LO III 185(11)

**8.262**    Functional relationships:

1.    $L(-x) = -L(x)$      $\left[-\frac{\pi}{2} \le x \le \frac{\pi}{2}\right]$      LO III 185(13)

2.    $L(\pi - x) = \pi\ln 2 - L(x)$      LO III 286

3.    $L(\pi + x) = \pi\ln 2 + L(x)$      LO III 286

4.    $L(x) - L\left(\frac{\pi}{2} - x\right) = \left(x - \frac{\pi}{4}\right)\ln 2 - \frac{1}{2}L\left(\frac{\pi}{2} - 2x\right)$      $\left[0 \le x < \frac{\pi}{4}\right]$      LO III 186(14)

# 8.3 Euler's Integrals of the First and Second Kinds and Functions Generated by Them

## 8.31 The gamma function (Euler's integral of the second kind): $\Gamma(z)$

**8.310**    Definition:

1.      $\Gamma(z) = \displaystyle\int_0^\infty e^{-t} t^{z-1}\, dt$                     $[\operatorname{Re} z > 0]$      (Euler)        FI II 777(6)

       Generalization:

2.      $\Gamma(z) = -\dfrac{1}{2i \sin \pi z} \displaystyle\int_C (-t)^{z-1} e^{-t}\, dt$

       for $z$ not an integer. The contour $C$ is shown in the drawing.        WH

       $\Gamma(z)$ is an analytic function $z$ with simple poles at the points $z = -l$ (for $l = 0, 1, 2, \ldots$) to which correspond to residues $\dfrac{(-1)^l}{l!}$. $\Gamma(z)$ satisfies the relation $\Gamma(1) = 1$.        WH, MO 1

**Integral representations**

**8.311**    $\Gamma(z) = \dfrac{1}{e^{2\pi i z} - 1} \displaystyle\int_\infty^{(0+)} e^{-t} t^{z-1}\, dt$        MO 2

**8.312**

1.      $\Gamma(z) = \displaystyle\int_0^1 \left( \ln \dfrac{1}{t} \right)^{z-1} dt$                $[\operatorname{Re} z > 0]$        FI II 778

2.      $\Gamma(z) = x^z \displaystyle\int_0^\infty e^{-xt} t^{z-1}\, dt$            $[\operatorname{Re} z > 0, \quad \operatorname{Re} x > 0]$        FI II 779(8)

3.      $\Gamma(z) = \dfrac{2 a^z e^a}{\sin \pi z} \displaystyle\int_0^\infty e^{-at^2} \left(1 + t^2\right)^{z-\frac{1}{2}} \cos\left[2at + (2z-1)\arctan t\right] dt$

                                                  $[a > 0]$        WH

4.      $\Gamma(z) = \dfrac{1}{2 \sin \pi z} \displaystyle\int_0^\infty e^{-t^2} t^{z-1} \left(1 + t^2\right)^{\frac{z}{2}} \{3\sin[t + z \operatorname{arccot}(-t)] + \sin[t + (z-2)\operatorname{arccot}(-t)]\}\, dt$

                                     [arccot denotes an obtuse angle]        WH

5.      $\Gamma(y) = x^y e^{-i\beta y} \displaystyle\int_0^\infty t^{y-1} \exp\left(-xte^{-i\beta}\right) dt$

                         $\left[x, y, \beta \text{ real}, \quad x > 0, \quad y > 0, \quad |\beta| < \dfrac{\pi}{2}\right]$     MO 8

6.      $\Gamma(z) = \dfrac{b^z}{2 \sin \pi z} \displaystyle\int_{-\infty}^\infty e^{bti} (it)^{z-1}\, dt$           $[b > 0, \quad 0 < \operatorname{Re} z < 1]$        NH 154(3)

$7.^{12}$  $\Gamma(z) = \dfrac{\left(\sqrt{a^2 + b^2}\right)^z}{\cos\left(z \arctan \frac{b}{a}\right)} \displaystyle\int_0^\infty e^{-at} \cos(bt) t^{z-1}\, dt$                          NH 152(1)a

$\qquad = \dfrac{\left(\sqrt{a^2 + b^2}\right)^z}{\sin\left(z \arctan \frac{b}{a}\right)} \displaystyle\int_0^\infty e^{-at} \sin(bt) t^{z-1}\, dt$                          NH 152(2)

$$[a > 0, \quad b \geq 0, \quad \operatorname{Re} z > -1]$$

8.  $\Gamma(z) = \dfrac{b^z}{\cos \frac{\pi z}{2}} \displaystyle\int_0^\infty \cos(bt) t^{z-1}\, dt$

$\qquad = \dfrac{b^z}{\sin \frac{\pi z}{2}} \displaystyle\int_0^\infty \sin(bt) t^{z-1}\, dt$

$$[b > 0, \quad 0 < \operatorname{Re} z < 1] \qquad \text{NH 152(5)}$$

9.  $\Gamma(z) = \displaystyle\int_0^\infty e^{-t}(t - z) t^{z-1} \ln t\, dt$

$$[\operatorname{Re} z > 0] \qquad \text{NH 173(7)}$$

10.  $\Gamma(z) = \displaystyle\int_{-\infty}^\infty \exp\left(zt - e^t\right) dt$

$$[\operatorname{Re} z > 0] \qquad \text{NH 145(14)}$$

$11.^{12}$  $\Gamma(z) \cos \alpha z = \lambda^z \displaystyle\int_0^\infty t^{z-1} e^{-\lambda t \cos \alpha} \cos\left(\lambda t \sin \alpha\right) dt$

$$\left[\lambda > 0, \quad \operatorname{Re} z > 0, \quad -\frac{\pi}{2} < \alpha < \frac{\pi}{2}\right] \quad \text{WH}$$

$12.^{12}$  $\Gamma(z) \sin \alpha z = \lambda^z \displaystyle\int_0^\infty t^{z-1} e^{-\lambda t \cos \alpha} \sin\left(\lambda t \sin \alpha\right) dt$

$$\left[\lambda > 0, \quad \operatorname{Re} z > -1, \quad -\frac{\pi}{2} < \alpha < \frac{\pi}{2}\right] \quad \text{WH}$$

13.  $\Gamma(-z) = \displaystyle\int_0^\infty \left[\dfrac{e^{-t} - \displaystyle\sum_{k=0}^n (-1)^k \dfrac{t^k}{k!}}{t^{z+1}}\right] dt$                $[n = \lfloor \operatorname{Re} z \rfloor]$                          MO 2

**8.313**  $\Gamma\left(\dfrac{z+1}{v}\right) = v u^{\frac{z+1}{v}} \displaystyle\int_0^\infty \exp\left(-u t^v\right) t^z\, dt$                $[\operatorname{Re} u > 0, \quad \operatorname{Re} v > 0, \quad \operatorname{Re} z > -1]$

$$\text{JA, MO 7a}$$

**8.314\***  $\Gamma(z) = \displaystyle\int_1^\infty e^{-t} t^{z-1}\, dt + \displaystyle\sum_{n=0}^\infty \dfrac{(-1)^k}{k!(z+k)}$

**8.315**

$1.^{11}$  $\dfrac{1}{\Gamma(z)} = \dfrac{i}{2\pi} \displaystyle\int_C (-t)^{-z} e^{-t}\, dt$                          [For the contour $C$ see **8.310** 2]

$2.^{12}$  $\displaystyle\int_{-\infty}^\infty \dfrac{e^{bti}}{(a + it)^z}\, dt = \dfrac{2\pi e^{-ab} b^{z-1}}{\Gamma(z)}$

$\qquad \displaystyle\int_{-\infty}^\infty \dfrac{e^{-bti}}{(a + it)^z}\, dt = 0 \qquad \left[\operatorname{Re} a > 0, \quad b < 0, \quad \operatorname{Re} z > 0, \quad |\arg(a + it)| < \tfrac{1}{2}\pi\right]$

3.    $$\frac{1}{\Gamma(z)} = a^{1-z}\frac{e^a}{\pi}\int_0^{\pi/2}\cos\left(a\tan\theta - z\theta\right)\cos^{z-2}\theta\,d\theta \qquad [\operatorname{Re} z > 1] \qquad \text{NH 157(14)}$$

See also **3.324** 2, **3.326**, **3.328**, **3.381** 4, **3.382** 2, **3.389** 2, **3.433**, **3.434**, **3.478** 1, **3.551** 1, 2, **3.827** 1, **4.267** 7, **4.272**, **4.353** 1, **4.369** 1, **6.214**, **6.223**, **6.246**, **6.281**.

## 8.32  Representation of the gamma function as series and products

**8.321**   Representation in the form of a series:

$1.^6$    $$\Gamma(z+1) = \sum_{k=0}^{\infty} c_k z^k$$

$$\left[c_0 = 1, \quad c_{n+1} = \frac{\sum_{k=0}^{n}(-1)^{k+1}s_{k+1}c_{n-k}}{n+1}; \quad s_1 = \boldsymbol{C}, \quad s_n = \zeta(n) \text{ for } n \geq 2, \quad |z| < 1\right]$$

NH 40(1, 3)

$2.^{11}$    $$\frac{1}{\Gamma(z+1)} = \sum_{k=0}^{\infty} d_k z^k$$

$$\left[d_0 = 1, \quad d_{n+1} = \frac{\sum_{k=0}^{n}(-1)^{k}s_{k+1}d_{n-k}}{n+1}; \quad s_1 = \boldsymbol{C}, \quad s_n = \zeta(n) \text{ for } n \geq 2\right] \quad \text{NH 41(4, 6)}$$

### Infinite-product representation

**8.322**$^{11}$ $$\Gamma(z) = e^{-Cz}\frac{1}{z}\prod_{k=1}^{\infty}\frac{e^{z/k}}{1+\frac{z}{k}} \qquad [\operatorname{Re} z > 0] \qquad \text{SM 269}$$

$$= \frac{1}{z}\prod_{k=1}^{\infty}\frac{\left(1+\frac{1}{k}\right)^z}{1+\frac{z}{k}} \qquad [\operatorname{Re} z > 0] \qquad \text{WH}$$

$$= \lim_{n\to\infty}\frac{n^z}{z}\prod_{k=1}^{n}\frac{k}{z+k} \qquad [\operatorname{Re} z > 0] \qquad \text{SM 267(130)}$$

**8.323**$^7$ $$\Gamma(z) = 2z^z e^{-z}\prod_{k=1}^{\infty}\sqrt[2^k]{\mathrm{B}\left(2^{k-1}z, \tfrac{1}{2}\right)} \qquad \text{NH 98(12)}$$

**8.324**$^7$ $$\Gamma(1+z) = 4^z\prod_{k=1}^{\infty}\frac{\Gamma\left(\frac{1}{2}+\frac{z}{2^k}\right)}{\sqrt{\pi}} \qquad \text{MO 3}$$

**8.325**

1.    $$\frac{\Gamma(\alpha)\,\Gamma(\beta)}{\Gamma(\alpha+\gamma)\,\Gamma(\beta-\gamma)} = \prod_{k=0}^{\infty}\left[\left(1+\frac{\gamma}{\alpha+k}\right)\left(1-\frac{\gamma}{\beta+k}\right)\right] \qquad \text{NH 62(2)}$$

$2.^{11}$    $$\frac{e^{Cx}\,\Gamma(z+1)}{\Gamma(z-x+1)} = \prod_{k=1}^{\infty}\left[\left(1-\frac{x}{z+k}\right)e^{x/k}\right] \qquad [z \neq 0, -1, -2, \ldots; \quad \operatorname{Re} z > 0, \quad \operatorname{Re}(z-x) > 0]$$

$3.^7$    $$\frac{\sqrt{\pi}}{\Gamma\left(1+\frac{z}{2}\right)\Gamma\left(\frac{1}{2}-\frac{z}{2}\right)} = \prod_{k=1}^{\infty}\left(1-\frac{z}{2k-1}\right)\left(1+\frac{z}{2k}\right) \qquad \text{MO 2}$$

**8.326**

1.[12]
$$\frac{\dfrac{[\Gamma(x)]^2}{\Gamma(2x)}}{B(x+iy, x-iy)} = \left|\frac{\Gamma(x)}{\Gamma(x+iy)}\right|^2 = \prod_{k=0}^{\infty}\left(1 + \frac{y^2}{(x+k)^2}\right)$$

$$[x, y \text{ are real}, \quad x \neq 0, -1, -2, \ldots]$$

LO V, NH 63(4)

2.[11]
$$\frac{\Gamma(x+iy)}{\Gamma(x)} = \frac{xe^{-iCy}}{x+iy}\prod_{n=1}^{\infty}\frac{\exp\left(\frac{iy}{n}\right)}{1 + \frac{iy}{x+n}}$$

$$[x, y \text{ are real}, \quad x \neq 0, -1, -2, \ldots]$$

MO 2

**8.327**    Asymptotic representation for large arguments:

1.
$$\Gamma(z) \sim z^{z-\frac{1}{2}}e^{-z}\sqrt{2\pi}\left\{1 + \frac{1}{12z} + \frac{1}{288z^2} - \frac{139}{51840z^3} - \frac{571}{2488320z^4} + O\left(z^{-5}\right)\right\}$$

$$[|\arg z| < \pi] \qquad \text{WH}$$

For $z$ real and positive, the remainder of the series is less than the last term that is retained.

2.    $n! \sim \sqrt{2\pi n}\left(\dfrac{n}{e}\right)^n$ or equivalently $\Gamma(n+1) \sim \sqrt{2\pi n}\left(\dfrac{n}{e}\right)^n$

$$[n \to \infty] \qquad \text{AS 6.1.38}$$

3.
$$\ln\Gamma(z) \sim \left(z - \frac{1}{2}\right)\ln z - z + \frac{1}{2}\ln(2\pi) + \frac{1}{12z} - \frac{1}{360z^3} + \frac{1}{1260z^5} - \frac{1}{1680z^7} + \cdots$$

$$[z \to \infty, \quad |\arg z| < \pi] \qquad \text{AS 6.1.38}$$

**8.328**

1.
$$\lim_{|y| \to \infty}|\Gamma(x+iy)|e^{\frac{\pi}{2}|y|}|y|^{\frac{1}{2}-x} = \sqrt{2\pi} \qquad [x \text{ and } y \text{ are real}] \qquad \text{MO 6}$$

2.
$$\lim_{|z| \to \infty}\frac{\Gamma(z+a)}{\Gamma(z)}e^{-a\ln z} = 1 \qquad\qquad \text{MO 6}$$

## 8.33 Functional relations involving the gamma function

**8.331**

1.    $\Gamma(x+1) = x\,\Gamma(x)$

2.    $\Gamma(x+a) = (x+a-1)\,\Gamma(x+a-1)$
$$= \frac{\Gamma(x+a+1)}{(x+a)}$$

3.    $\Gamma(x-a) = (x-a-1)\,\Gamma(x-a-1)$
$$= \frac{\Gamma(x-a+1)}{(x-a)}$$

**8.332**

1.    $\left|\Gamma(iy)\right|^2 = \dfrac{\pi}{y \sinh \pi y}$          [$y$ is real]          MO 3

2.    $\left|\Gamma\left(\tfrac{1}{2}+iy\right)\right|^2 = \dfrac{\pi}{\cosh \pi y}$          [$y$ is real]

3.    $\Gamma(1+ix)\,\Gamma(1-ix) = \dfrac{\pi x}{\sinh x\pi}$          [$x$ is real]          LO V

4.    $\Gamma(1+x+iy)\,\Gamma(1-x+iy)\,\Gamma\left(1+x-iy\right)\Gamma(1-x-iy) = \dfrac{2\pi^2\left(x^2+y^2\right)}{\cosh 2y\pi - \cos 2x\pi}$

                                                    [$x$ and $y$ are real]          LO V

**8.333**    $\left[\Gamma(n+1)\right]^n = G(n+1) \displaystyle\prod_{k=1}^{n} k^k,$

     where $n$ is a natural number and

$$G(z+1) = (2\pi)^{\frac{z}{2}} \exp\left[-\frac{z(z+1)}{2} - \frac{C}{2}z^2\right] \prod_{n=1}^{\infty} \left\{\left(1+\frac{z}{n}\right)^n \exp\left(-z+\frac{z^2}{2n}\right)\right\}$$      WH

**8.334**

1.    $\displaystyle\prod_{k=1}^{n} \frac{1}{\Gamma\left(-z \exp \frac{2\pi k i}{n}\right)} = -z^n \prod_{k=1}^{\infty} \left[1 - \left(\frac{z}{k}\right)^n\right]$          [$n=2,3\ldots$]          MO 2

2.    $\Gamma\left(\tfrac{1}{2}+x\right)\Gamma\left(\tfrac{1}{2}-x\right) = \dfrac{\pi}{\cos \pi x}$

3.    $\Gamma(1-x)\,\Gamma(x) = \dfrac{\pi}{\sin \pi x}$          FI II 430

## Special cases

**8.335**[7]    $\Gamma(nx) = (2\pi)^{\frac{1-n}{2}} n^{nx-\frac{1}{2}} \displaystyle\prod_{k=0}^{n-1} \Gamma\left(x+\frac{k}{n}\right)$          [product theorem]          FI II 782a, WH

1.*    $\Gamma(2x) = \dfrac{2^{2x-1}}{\sqrt{\pi}}\, \Gamma(x)\,\Gamma\left(x+\tfrac{1}{2}\right)$          [doubling formula]

2.    $\Gamma(3x) = \dfrac{3^{3x-\frac{1}{2}}}{2\pi}\, \Gamma(x)\,\Gamma\left(x+\tfrac{1}{3}\right)\Gamma\left(x+\tfrac{2}{3}\right)$

3.    $\displaystyle\prod_{k=1}^{n-1} \Gamma\left(\frac{k}{n}\right)\Gamma\left(1-\frac{k}{n}\right) = \frac{(2\pi)^{n-1}}{n}$          WH

4.[10]    $\displaystyle\sum_{n=0}^{\infty} \frac{\Gamma^2\left(n-\frac{1}{2}\right)}{4\,(n!)^2\,\Gamma^2\left(-\frac{1}{2}\right)} = \frac{1}{4} + \frac{1}{16} + \frac{1}{256} + \frac{1}{1024} + \frac{25}{65536} + \cdots = \frac{1}{\pi}$

**8.336**[12]    $\Gamma\left(-\dfrac{yz+xi}{2y}\right)\Gamma(1+z) = (2i)^{z+1}y\,\Gamma\left(1+\dfrac{yz-xi}{2y}\right)\displaystyle\int_0^\infty e^{-tx}\sin^z(ty)\,dt$

$$[\operatorname{Re}(yi) > 0, \quad \operatorname{Re}(x-yzi) > 0]$$

NH 133(10)

- For a connection with the psi function, **8.361** 1.
- For a connection with the beta function, see **8.384** 1.
- For integrals of the gamma function, see **8.412** 4, **8.414**, **9.223**, **9.242** 3, **9.242** 4.

**8.337**

1.    $\left[\Gamma'(x)\right]^2 < \Gamma(x)\,\Gamma''(x)$ $\qquad\qquad\qquad\qquad\qquad [x > 0]$ MO 1

2.    For $x > 0$, $\min\Gamma(1+x) = 0.88560\ldots$ is attained when $x = 0.46163\ldots$ JA

**Particular values**

**8.338**

1.    $\Gamma(1) = \Gamma(2) = 1$

2.    $\Gamma\left(\tfrac{1}{2}\right) = \sqrt{\pi}$

3.    $\Gamma\left(-\tfrac{1}{2}\right) = -2\sqrt{\pi}$

4.    $\left[\Gamma\left(\dfrac{1}{4}\right)\right]^4 = 16\pi^2 \displaystyle\prod_{k=1}^\infty \frac{(4k-1)^2\left[(4k+1)^2 - 1\right]}{\left[(4k-1)^2 - 1\right](4k+1)^2}$ MO 1a

5.    $\displaystyle\prod_{k=1}^{8}\Gamma\left(\frac{k}{3}\right) = \frac{640}{3^6}\left(\frac{\pi}{\sqrt{3}}\right)^3$ WH

**8.339**    For $n$ a natural number

1.    $\Gamma(n) = (n-1)!$

2.[12]    $\Gamma\left(n+\tfrac{1}{2}\right) = \dfrac{\sqrt{\pi}}{2^n}(2n-1)!! = \dfrac{\sqrt{\pi}}{2^{2n}}\dfrac{(2n)!}{n!}$

3.    $\Gamma\left(\tfrac{1}{2}-n\right) = (-1)^n\dfrac{2^n\sqrt{\pi}}{(2n-1)!!}$

4.    $\dfrac{\Gamma\left(p+n+\tfrac{1}{2}\right)}{\Gamma\left(p-n+\tfrac{1}{2}\right)} = \dfrac{\left(4p^2-1^2\right)\left(4p^2-3^2\right)\ldots\left[4p^2-(2n-1)^2\right]}{2^{2n}}$ WA 221

5.    $\Gamma(n+k) = (n+k-1)!$

$\qquad = \dfrac{\Gamma(n+k+1)}{(n+k)}$ $\qquad\qquad\qquad [n+k \geq 0, 1, \ldots]$

6.    $\Gamma(n-k) = (n-k-1)!$

$\qquad = \dfrac{\Gamma(n-k+1)}{(n-k)}$ $\qquad\qquad\qquad [n-k \geq 0, 1, \ldots]$

## 8.34  The logarithm of the gamma function

**8.341**   Integral representation:

1.    $\ln \Gamma(z) = \left(z - \dfrac{1}{2}\right) \ln z - z + \dfrac{1}{2} \ln 2\pi + \displaystyle\int_0^\infty \left(\dfrac{1}{2} - \dfrac{1}{t} + \dfrac{1}{e^t - 1}\right) \dfrac{e^{-tz}}{t} \, dt$

$\qquad\qquad\qquad\qquad\qquad\qquad\qquad\qquad\qquad$ [$\operatorname{Re} z > 0$] $\qquad\qquad\qquad$ WH

2.[11]   $\ln \Gamma(z) = z \ln z - z - \dfrac{1}{2} \ln z + \ln \sqrt{2\pi} + 2 \displaystyle\int_0^\infty \dfrac{\arctan \frac{t}{z}}{e^{2\pi t} - 1} \, dt$

$\qquad\left[\operatorname{Re} z > 0 \text{ and } \arctan w = \displaystyle\int_0^w \dfrac{du}{1 + u^2} \text{ is taken over a rectangular path in the } w\text{-plane}\right]$   WH

3.    $\ln \Gamma(z) = \displaystyle\int_0^\infty \left\{\dfrac{e^{-zt} - e^{-t}}{1 - e^{-t}} + (z - 1)e^{-t}\right\} \dfrac{dt}{t}$ $\qquad$ [$\operatorname{Re} z > 0$] $\qquad\qquad$ WH

4.    $\ln \Gamma(z) = \displaystyle\int_0^\infty \left\{(z - 1)e^{-t} + \dfrac{(1 + t)^{-z} - (1 + t)^{-1}}{\ln(1 + t)}\right\} \dfrac{dt}{t}$

$\qquad\qquad\qquad\qquad\qquad\qquad\qquad\qquad\qquad$ [$\operatorname{Re} z > 0$] $\qquad\qquad\qquad$ WH

5.    $\ln \Gamma(x) = \dfrac{\ln \pi - \ln \sin \pi x}{2} + \dfrac{1}{2} \displaystyle\int_0^\infty \left\{\dfrac{\sinh \left(\frac{1}{2} - x\right) t}{\sinh \frac{t}{2}} - (1 - 2x)e^{-t}\right\} \dfrac{dt}{t}$

$\qquad\qquad\qquad\qquad\qquad\qquad\qquad\qquad\qquad$ [$0 < x < 1$] $\qquad\qquad\qquad$ WH

6.    $\ln \Gamma(z) = \displaystyle\int_0^1 \left\{\dfrac{t^z - t}{t - 1} - t(z - 1)\right\} \dfrac{dt}{t \ln t}$ $\qquad$ [$\operatorname{Re} z > 0$] $\qquad\qquad$ WH

7.    $\ln \Gamma(z) = \displaystyle\int_0^\infty \left[(z - 1)e^{-t} + \dfrac{e^{-tz} - e^{-t}}{1 - e^{-t}}\right] \dfrac{dt}{t}$ $\qquad$ [$\operatorname{Re} z > 0$] $\qquad$ NH 187(7)

See also **3.427** 9, **3.554** 5.

**8.342**   Series representations:

1.[11]   $\ln \Gamma(z + 1)$

$$= \dfrac{1}{2} \left[\ln\left(\dfrac{\pi z}{\sin \pi z}\right) - \ln \dfrac{1 + z}{1 - z}\right] + (1 - \boldsymbol{C}) z + \sum_{k=1}^\infty \dfrac{1 - \zeta(2k + 1)}{2k + 1} z^{2k+1}$$

$$= -\boldsymbol{C}z + \sum_{k=2}^\infty (-1)^k \dfrac{z^k}{k} \zeta(k) \qquad\qquad\qquad [|z| < 1] \text{ NH 38(16, 12)}$$

2.    $\ln \Gamma(1 + x) = \dfrac{1}{2} \ln \dfrac{\pi x}{\sin \pi x} - \boldsymbol{C}x - \displaystyle\sum_{n=1}^\infty \dfrac{x^{2n+1}}{2n + 1} \zeta(2n + 1)$

$\qquad\qquad\qquad\qquad\qquad\qquad\qquad\qquad\qquad$ [$|x| < 1$] $\qquad\qquad$ NH 38(14)

**8.343**

1.    $\ln \Gamma(x) = \ln \sqrt{2\pi} + \displaystyle\sum_{n=1}^\infty \left\{\dfrac{1}{2n} \cos 2n\pi x + \dfrac{1}{n\pi} \left(\boldsymbol{C} + \ln 2n\pi\right) \sin 2n\pi x\right\}$

$\qquad\qquad\qquad\qquad\qquad\qquad\qquad\qquad\qquad$ [$0 < x < 1$] $\qquad\qquad$ FI III 558

2.      $\ln \Gamma(z) = z \ln z - z - \frac{1}{2} \ln z + \ln \sqrt{2\pi} + \frac{1}{2} \sum_{m=1}^{\infty} \frac{m}{(m+1)(m+2)} \sum_{n=1}^{\infty} \frac{1}{(z+n)^{m+1}}$

$$[|\arg z| < \pi]$$      MO 9

**8.344**[7]    Asymptotic expansion for large values of $|z|$:

$$\ln \Gamma(z) = z \ln z - z - \frac{1}{2} \ln z + \ln \sqrt{2\pi} + \sum_{k=1}^{n-1} \frac{B_{2k}}{2k(2k-1)z^{2k-1}} + R_n(z),$$

where

$$|R_n(z)| < \frac{|B_{2n}|}{2n(2n-1)|z|^{2n-1} \cos^{2n-1}\left(\frac{1}{2}\arg z\right)}$$      MO5

For integrals of $\ln \Gamma(x)$, see **6.44**.

## 8.35 The incomplete gamma function

**8.350**    Definition:

1.      $\gamma(\alpha, x) = \displaystyle\int_0^x e^{-t} t^{\alpha-1}\, dt$      $[\operatorname{Re}\alpha > 0]$      EH II 133(1), NH 1(1)

2.[11]    $\Gamma(\alpha, x) = \displaystyle\int_x^\infty e^{-t} t^{\alpha-1}\, dt$      EH II 133(2), NH 2(2), LE 339

3.      $\Gamma(z, 0) = \Gamma(z)$

4.      $\Gamma(a, \infty) = 0$

5.      $\gamma(a, 0) = 0$

**8.351**

1.      $\gamma^*(\alpha, x) = \dfrac{x^{-\alpha}}{\Gamma(\alpha)} \gamma(\alpha, x)$ is an analytic function with respect to $\alpha$ and $x$      EH II 133(5)

2.      Another definition of $\Gamma(\alpha, x)$, that is also suitable for the case $\operatorname{Re}\alpha \le 0$:

$$\gamma(\alpha, x) = \frac{x^\alpha}{\alpha} e^{-x}\, \Phi(1, 1+\alpha; x) = \frac{x^\alpha}{\alpha}\, \Phi(a, 1+a; -x)$$      EH II 133(3)

3.      For fixed $x$, $\Gamma(\alpha, x)$ is an entire function of $\alpha$. For non-integral $\alpha$, $\Gamma(\alpha, x)$ is a multiple-valued function of $x$ with a branch point at $x = 0$.

4.      A second definition of $\Gamma(\alpha, x)$:

$$\Gamma(\alpha, x) = x^\alpha e^{-x} \Psi(1, 1+\alpha; x) = e^{-x} \Psi(1-\alpha, 1-\alpha; x)$$      EH II 133(4)

**8.352**    Special cases:

1.      $\gamma(1+n, x) = n!\left[1 - e^{-x}\left(\displaystyle\sum_{m=0}^n \frac{x^m}{m!}\right)\right]$      $[n = 0, 1, \ldots]$

     EH II 136(17, 16), NH 6(11)

2.      $\Gamma(1+n, x) = n! e^{-x} \displaystyle\sum_{m=0}^n \frac{x^m}{m!}$      $[n = 0, 1, \ldots]$      EH II 136(16, 18)

3.$^{12}$ $\quad \Gamma(-n, x) = \dfrac{(-1)^{n+1}}{n!} \operatorname{Ei}(-x) - e^{-x} \displaystyle\sum_{k=1}^{m} \dfrac{x^{k-n-1}}{(-n)_k}$ $\qquad$ $[\operatorname{Re} x > 0, \quad n = 1, 2, \ldots]$

4. $\quad \Gamma(n, x) = (n-1)! e^{-x} \displaystyle\sum_{m=0}^{n-1} \dfrac{x^m}{m!}$ $\qquad$ $[n = 1, 2, \ldots]$

5.$^{12}$ $\quad \Gamma(-n+1, x) = \dfrac{(-1)^{n-1}}{(n-1)!} \left[ \Gamma(0, x) - e^{-x} \displaystyle\sum_{m=0}^{n-2} (-1)^m \dfrac{m!}{x^{m+1}} \right]$

$\qquad\qquad\qquad\qquad\qquad\qquad\qquad\qquad\qquad\qquad\qquad\qquad [n = 2, 3, \ldots]$

6. $\quad \gamma(n, x) = (n-1)! \left[ 1 - e^{-x} \displaystyle\sum_{m=0}^{n-1} \dfrac{x^m}{m!} \right]$ $\qquad$ $[n = 1, 2, \ldots]$

7. $\quad \Gamma(-n+k, x) = \dfrac{(-1)^{n-k}}{(n-k)!} \left[ \Gamma(0, x) - e^{-z} \displaystyle\sum_{m=0}^{n-k-1} (-1)^m \dfrac{m!}{x^{m+1}} \right]$

$\qquad\qquad\qquad\qquad\qquad\qquad\qquad\qquad\qquad\qquad\qquad\qquad [n - k \geq 1, \quad k = 0, 1, \ldots]$

**8.353**  Integral representations:

1.$^{12}$ $\quad \gamma(\alpha, x) = x^{\alpha} \operatorname{cosec} \pi\alpha \displaystyle\int_0^{\pi} e^{x \cos\theta} \cos(\alpha\theta + x \sin\theta)\, d\theta$ $\qquad$ $[x \neq 0, \quad \operatorname{Re}\alpha > 0, \quad \alpha \neq 1, 2, \ldots]$

$\qquad\qquad\qquad\qquad\qquad\qquad\qquad\qquad\qquad\qquad\qquad\qquad\qquad\qquad$ EH II 137(2)

2. $\quad \gamma(\alpha, x) = x^{\frac{1}{2}\alpha} \displaystyle\int_0^{\infty} e^{-t} t^{\frac{1}{2}\alpha - 1} J_{\alpha}\left(2\sqrt{xt}\right) dt$ $\qquad$ $[\operatorname{Re}\alpha > 0]$ $\qquad$ EH II 138(4)

3.$^{12}$ $\quad \Gamma(\alpha, x) = \dfrac{e^{-x} x^{\alpha}}{\Gamma(1-\alpha)} \displaystyle\int_0^{\infty} \dfrac{e^{-t} t^{-\alpha}}{x + t}\, dt$ $\qquad$ $[\operatorname{Re}\alpha < 1, \quad x > 0]$

$\qquad\qquad\qquad\qquad\qquad\qquad\qquad\qquad\qquad\qquad\qquad$ EH II 137(3), NH 19(12)

4. $\quad \Gamma(\alpha, x) = \dfrac{2x^{\frac{1}{2}\alpha} e^{-x}}{\Gamma(1-\alpha)} \displaystyle\int_0^{\infty} e^{-t} t^{-\frac{1}{2}\alpha} K_{\alpha}\left[2\sqrt{xt}\right] dt$ $\qquad$ $[\operatorname{Re}\alpha < 1]$ $\qquad$ EH II 138(5)

5. $\quad \Gamma(\alpha, xy) = y^{\alpha} e^{-xy} \displaystyle\int_0^{\infty} e^{-ty} (t + x)^{\alpha - 1}\, dt$

$\qquad\qquad\qquad\qquad [\operatorname{Re} y > 0, \quad x > 0, \quad \operatorname{Re}\alpha > 1]$ $\qquad$ (See also **3.936** 5, **3.944** 1–4) $\quad$ NH 19(10)

For integrals of the gamma function, see **6.45**.

**8.354**  Series representations:

1. $\quad \gamma(\alpha, x) = \displaystyle\sum_{n=0}^{\infty} \dfrac{(-1)^n x^{\alpha+n}}{n!(\alpha + n)}$ $\qquad$ EH II 135(4)

2.    $\Gamma(\alpha, x) = \Gamma(\alpha) - \sum\limits_{n=0}^{\infty} \dfrac{(-1)^n x^{\alpha+n}}{n!(\alpha+n)}$          $[\alpha \neq 0, -1, -2, \dots]$

                               EH II 135(5), LE 340(2)

3.    $\Gamma(\alpha, x) - \Gamma(\alpha, x+y) = \gamma(\alpha, x+y) - \gamma(\alpha, x)$

$$= e^{-x} x^{\alpha-1} \sum_{k=0}^{\infty} \frac{(-1)^k \left[1 - e^{-y} e_k(y)\right] \Gamma(1-\alpha+k)}{x^k \, \Gamma(1-\alpha)}$$

                                     $e_k(x) = \sum\limits_{m=0}^{k} \dfrac{x^m}{m!}$     $[|y| < |x|]$    EH II 139(2)

4.    $\gamma(\alpha, x) = \Gamma(\alpha) e^{-x} x^{\frac{1}{2}\alpha} \sum\limits_{n=0}^{\infty} x^{\frac{1}{2}n} I_{n+\alpha}\left(2\sqrt{x}\right) \sum\limits_{m=0}^{n} \dfrac{(-1)^m}{m!}$     $[x \neq 0, \quad \alpha \neq 0, \quad -1, -2, \dots]$

                                                         EH II 139(3)

5.[12]    $\Gamma(\alpha, x) = e^{-x} x^{\alpha} \sum\limits_{n=0}^{\infty} \dfrac{L_n^{\alpha}(x)}{n+1}$          $[x > 0, \quad \alpha > -1]$        EH II 140(5)

**8.355**    $\Gamma(\alpha, x)\, \gamma(\alpha, y) = e^{-x-y}(xy)^{\alpha} \sum\limits_{n=0}^{\infty} \dfrac{n!\, \Gamma(\alpha)}{(n+1)\, \Gamma(\alpha+n+1)}\, L_n^{\alpha}(x)\, L_n^{\alpha}(y)$

                                 $[y > 0, \quad x \geq y, \quad \alpha \neq 0, -1, \dots]$

                                                          EH II 139(4)

**8.356**   Functional relations:

1.[11]    $\gamma(\alpha+1, x) = \alpha\,\gamma(\alpha, x) - x^{\alpha} e^{-x}$                                      EH II 134(2)

2.    $\Gamma(\alpha+1, x) = \alpha\,\Gamma(\alpha, x) + x^{\alpha} e^{-x}$                                       EH II 134(3)

3.    $\Gamma(\alpha, x) + \gamma(\alpha, x) = \Gamma(\alpha)$                                            EH II 134(1)

4.    $\dfrac{d\,\gamma(\alpha, x)}{dx} = -\dfrac{d\,\Gamma(\alpha, x)}{dx} = x^{\alpha-1} e^{-x}$                                EH II 135(8)

5.    $\dfrac{\Gamma(\alpha+n, x)}{\Gamma(\alpha+n)} = \dfrac{\Gamma(\alpha, x)}{\Gamma(\alpha)} + e^{-x} \sum\limits_{s=0}^{n-1} \dfrac{x^{\alpha+s}}{\Gamma(\alpha+s+1)}$                   NH 4(3)

6.[11]    $\Gamma(\alpha)\,\Gamma(\alpha+n, x) - \Gamma(\alpha+n)\,\Gamma(\alpha, x) = \Gamma(\alpha+n)\,\gamma(\alpha, x) - \Gamma(\alpha)\,\gamma(\alpha+n, x)$      NH 5

7.    $\Gamma(a+k, x) = (a+k-1)\,\Gamma(a+k-1, x) + x^{a+k-1} e^{-x}$

$$= \frac{1}{a+k}\left[\Gamma(a+k+1, x) - x^{a+k} e^{-x}\right]$$

8.    $\Gamma(a-k, x) = (a-k-1)\,\Gamma(a-k-1, x) + x^{a-k-1} e^{-x}$

$$= \frac{1}{a-k}\left[\Gamma(a-k+1, x) - x^{a-k} e^{-x}\right]$$

9.[12]    $\gamma(a+k, x) = (a+k-1)\,\gamma(a+k-1, x) - x^{a+k-1} e^{-x}$

$$= \frac{1}{a+k}\left[\gamma(a+k+1, x) + x^{a+k} e^{-x}\right]$$

10.     $\gamma(a-k,x) = (a-k-1)\,\gamma(a-k-1,x) - x^{a-k-1}e^{-x}$

$$= \frac{1}{a-k}\left[\gamma(a-k+1,x) + x^{a-k}e^{-x}\right]$$

**8.357**    Asymptotic representation for large values of $|x|$:

1.     $\Gamma(\alpha,x) = x^{\alpha-1}e^{-x}\left[\displaystyle\sum_{m=0}^{M-1}\frac{(-1)^m\,\Gamma(1-\alpha+m)}{x^m\,\Gamma(1-\alpha)} + O\left(|x|^{-M}\right)\right]$

$$\left[|x|\to\infty, -\tfrac{3\pi}{2} < \arg x < \tfrac{3\pi}{2}, \quad M = 1,2,\ldots\right] \quad \text{EH II 135(6), NH 37(7), LE 340(3)}$$

**8.358**    Representation as a continued fraction:

$$\Gamma(\alpha,x) = \cfrac{e^{-x}x^\alpha}{x + \cfrac{1-\alpha}{1 + \cfrac{1}{x + \cfrac{2-\alpha}{1 + \cfrac{2}{x + \cfrac{3-\alpha}{1+\ldots}}}}}} \qquad\qquad \text{EH II 136(13), NH 42(9)}$$

**8.359**    Relationships with other functions:

1.     $\Gamma(0,x) = -\operatorname{Ei}(-x)$                                                EH II 143(1)

2.     $\Gamma\left(0,\ln\dfrac{1}{x}\right) = -\operatorname{li}(x)$                                 EH II 143(2)

3.     $\Gamma\left(\tfrac{1}{2},x^2\right) = \sqrt{\pi} - \sqrt{\pi}\,\Phi(x)$                        EH II 147(2)

4.[11]   $\gamma\left(\tfrac{1}{2},x^2\right) = \sqrt{\pi}\,\Phi(x)$                                EH II 147(1)

## 8.36   The psi function $\psi(x)$

**8.360**    Definition:

1.     $\psi(x) = \dfrac{d}{dx}\ln\Gamma(x)$

**8.361**    Integral representations:

1.[8]   $\psi(z) = \dfrac{d\ln\Gamma(z)}{dz} = \displaystyle\int_0^\infty\left(\frac{e^{-t}}{t} - \frac{e^{-zt}}{1-e^{-t}}\right)dt$      $[\operatorname{Re} z > 0]$         NH 183(1), WH

2.     $\psi(z) = \displaystyle\int_0^\infty\left\{e^{-t} - \frac{1}{(1+t)^z}\right\}\frac{dt}{t}$               $[\operatorname{Re} z > 0]$         NH 184(7), WH

3.     $\psi(z) = \ln z - \dfrac{1}{2z} - 2\displaystyle\int_0^\infty\frac{t\,dt}{(t^2+z^2)(e^{2\pi t}-1)}$      $[\operatorname{Re} z > 0]$            WH

4.     $\psi(z) = \displaystyle\int_0^1\left(\frac{1}{-\ln t} - \frac{t^{z-1}}{1-t}\right)dt$                 $[\operatorname{Re} z > 0]$            WH

5.      $\psi(z) = \displaystyle\int_0^\infty \frac{e^{-t} - e^{-zt}}{1 - e^{-t}}\, dt - C,$                        $[\operatorname{Re} z > 0]$                    WH

6.      $\psi(z) = \displaystyle\int_0^\infty \left\{(1+t)^{-1} - (1+t)^{-z}\right\} \frac{dt}{t} - C,$                        $[\operatorname{Re} z > 0]$                    WH

7.      $\psi(z) = \displaystyle\int_0^1 \frac{t^{z-1} - 1}{t - 1}\, dt - C$                        $[\operatorname{Re} z > 0]$                    FI II 796, WH

8.      $\psi(z) = \ln z + \displaystyle\int_0^\infty e^{-tz} \left[\frac{1}{t} - \frac{1}{1 - e^{-t}}\right] dt$                        $[\operatorname{Re} z > 0]$                    MO 4

See also **3.244** 3, **3.311** 6, **3.317** 1, **3.457**, **3.458** 2, **3.471** 14, **4.253** 1 and 6, **4.275** 2, **4.281** 4, **4.482** 5. For integrals of the psi function, see **6.46**, **6.47**.

## Series representation

**8.362**

1.      $\psi(x) = -C - \displaystyle\sum_{k=0}^\infty \left(\frac{1}{x + k} - \frac{1}{k + 1}\right)$                        FI II 799(26), KU 26(1)

        $= -C - \dfrac{1}{x} + x \displaystyle\sum_{k=1}^\infty \frac{1}{k(x + k)}$                        FI II 495

2.      $\psi(x) = \ln x - \displaystyle\sum_{k=0}^\infty \left[\frac{1}{x + k} - \ln\left(1 + \frac{1}{x + k}\right)\right]$                        MO 4

3.      $\psi(x) = -C + \dfrac{\pi^2}{6}(x - 1) - (x - 1)\displaystyle\sum_{k=1}^\infty \left(\frac{1}{k + 1} - \frac{1}{x + k}\right)\sum_{n=0}^{k-1}\frac{1}{x + n}$                        NH 54(12)

**8.363**

1.      $\psi(x + 1) = -C + \displaystyle\sum_{k=2}^\infty (-1)^k \zeta(k) x^{k-1}$                        NH 37(5)

2.      $\psi(x + 1) = \dfrac{1}{2x} - \dfrac{\pi}{2}\cot \pi x - \dfrac{x^2}{1 - x^2} - C + \displaystyle\sum_{k=1}^\infty \left[1 - \zeta(2k + 1)\right] x^{2k}$                        NH 38(10)

3.      $\psi(x) - \psi(y) = \displaystyle\sum_{k=0}^\infty \left(\frac{1}{y + k} - \frac{1}{x + k}\right)$

        (see also **3.219**, **3.231** 5, **3.311** 7, **3.688** 20, **4.253** 1, **4.295** 37)   NH 99(3)

4.      $\psi(x + iy) - \psi(x - iy) = \displaystyle\sum_{k=0}^\infty \frac{2yi}{y^2 + (x + k)^2}$

5.      $\psi\left(\dfrac{p}{q}\right) = -C + \displaystyle\sum_{k=0}^\infty \left(\frac{1}{k + 1} - \frac{q}{p + kq}\right)$                        (see also **3.244** 3)                    NH 29(1)

$6.^{8}$    $\psi\left(\dfrac{p}{q}\right) = -\boldsymbol{C} - \ln(2q) - \dfrac{\pi}{2}\cot\dfrac{p\pi}{q} + 2\displaystyle\sum_{k=1}^{\left\lfloor \frac{q+1}{2}\right\rfloor - 1}\left[\cos\dfrac{2kp\pi}{q}\ln\sin\dfrac{k\pi}{q}\right]$

$$[q = 2, 3, \ldots, p = 1, 2, \ldots, q-1]$$

MO 4, EH I 19(29)

7.    $\psi\left(\dfrac{p}{q}\right) - \psi\left(\dfrac{p-1}{q}\right) = q\displaystyle\sum_{n=2}^{\infty}\sum_{k=0}^{\infty}\dfrac{1}{(p+kq)^n - 1}$          NH 59(3)

8.    $\psi^{(n)}(x) = (-1)^{n+1}n!\displaystyle\sum_{k=0}^{\infty}\dfrac{1}{(x+k)^{n+1}} = (-1)^{n+1}n!\,\zeta(n+1, x)$          NH 37(1)

## Infinite-product representation

**8.364**

1.    $e^{\psi(x)} = x\displaystyle\prod_{k=0}^{\infty}\left(1 + \dfrac{1}{x+k}\right)e^{-\frac{1}{x+k}}$          NH 65(12)

2.    $e^{y\,\psi(x)} = \dfrac{\Gamma(x+y)}{\Gamma(x)}\displaystyle\prod_{k=0}^{\infty}\left(1 + \dfrac{y}{x+k}\right)e^{-\frac{y}{x+k}}$          NH 65(11)

See also **8.37**.

- For a connection with Riemann's zeta function, see **9.533** 2.
- For a connection with the gamma function, see **4.325** 12 and **4.352** 1.
- For a connection with the beta function, see **4.253** 1.
- For series of psi functions, see **8.403** 2, **8.440**, and **8.447** 3 (Bessel functions), **8.761** (derivatives of associated Legendre functions with respect to the degree), **9.153**, **9.154** (hypergeometric function), **9.237** (confluent hypergeometric function).
- For integrals containing psi functions, see **6.46**–**6.47**.

**8.365**    Functional relations:

1.    $\psi(x+1) = \psi(x) + \dfrac{1}{x}$          JA

2.    $\psi\left(\dfrac{x+1}{2}\right) - \psi\left(\dfrac{x}{2}\right) = 2\,\beta(x)$          (cf. **8.37** 0)

3.    $\psi(x+n) = \psi(x) + \displaystyle\sum_{k=0}^{n-1}\dfrac{1}{x+k}$          GA 154(64)a

4.    $\psi(n+1) = -\boldsymbol{C} + \displaystyle\sum_{k=1}^{n}\dfrac{1}{k}$          MO 4

5.    $\displaystyle\lim_{n\to\infty}\left[\psi(z+n) - \ln n\right] = 0$          MO 3

6.    $\psi(nz) = \dfrac{1}{n}\displaystyle\sum_{k=0}^{n-1}\psi\left(z + \dfrac{k}{n}\right) + \ln n$          $[n = 2, 3, 4, \ldots]$          MO 3

7.     $\psi(x-n) = \psi(x) - \sum\limits_{k=1}^{n} \dfrac{1}{x-k}$

8.     $\psi(1-z) = \psi(z) + \pi \cot \pi z$                                            GA 155(68)a

9.     $\psi\left(\frac{1}{2}+z\right) = \psi\left(\frac{1}{2}-z\right) + \pi \tan \pi z$                            JA

10.     $\psi\left(\frac{3}{4}-n\right) = \psi\left(\frac{1}{4}+n\right) + \pi$            $[n = 0, \quad \pm 1, \quad \pm 2, \ldots]$

**8.366**    Particular values

1.     $\psi(1) = -\boldsymbol{C}$                              (cf. **8.367** 1)

2.     $\psi\left(\frac{1}{2}\right) = -\boldsymbol{C} - 2\ln 2 = -1.963\,510\,026\ldots$                GA 155a

3.     $\psi\left(\frac{1}{2}\pm n\right) = -\boldsymbol{C} + 2\left[\sum\limits_{k=1}^{n} \dfrac{1}{2k-1} - \ln 2\right]$              JA

4.     $\psi\left(\frac{1}{4}\right) = -\boldsymbol{C} - \dfrac{\pi}{2} - 3\ln 2$                         GA 157a

5.     $\psi\left(\frac{3}{4}\right) = -\boldsymbol{C} + \dfrac{\pi}{2} - 3\ln 2$                         GA 157a

6.     $\psi\left(\frac{1}{3}\right) = -\boldsymbol{C} - \dfrac{\pi}{2}\sqrt{\frac{1}{3}} - \dfrac{3}{2}\ln 3$                 GA 157a

7.     $\psi\left(\frac{2}{3}\right) = -\boldsymbol{C} + \dfrac{\pi}{2}\sqrt{\frac{1}{3}} - \dfrac{3}{2}\ln 3$                 GA 157a

8.     $\psi'(1) = \dfrac{\pi^2}{6} = 1.644\,934\,066\,848\ldots$                     JA

9.     $\psi'\left(\frac{1}{2}\right) = \dfrac{\pi^2}{2} = 4.934\,802\,200\,5\ldots$                    JA

10.     $\psi'(-n) = \infty$                   $[n$ is a natural number$]$          JA

11.     $\psi'(n) = \dfrac{\pi^2}{6} - \sum\limits_{k=1}^{n-1} \dfrac{1}{k^2}$            $[n$ is a natural number$]$          JA

12.     $\psi'\left(\frac{1}{2}+n\right) = \dfrac{\pi^2}{2} - 4\sum\limits_{k=1}^{n} \dfrac{1}{(2k-1)^2}$       $[n$ is a natural number$]$          JA

13.     $\psi'\left(\frac{1}{2}-n\right) = \dfrac{\pi^2}{2} + 4\sum\limits_{k=1}^{n} \dfrac{1}{(2k-1)^2}$       $[n$ is a natural number$]$          JA

**8.367**    Euler's constant (also denoted by $\gamma$):

1.     $\boldsymbol{C} = -\psi(1) = 0.577\,215\,664\,90\ldots$                   FI II 319, 795

2.     $\boldsymbol{C} = \lim\limits_{n\to\infty}\left[\sum\limits_{k=1}^{n-1} \dfrac{1}{k} - \ln n\right]$                   FI II 801a

3.[12]     $\boldsymbol{C} = \lim\limits_{x\to 0}\left[\zeta(x) - \dfrac{1}{x-1}\right]$                     FI II 804

Integral representations:

4. $\quad C = -\int_0^\infty e^{-t} \ln t \, dt$ FI II 807

5. $\quad C = -\int_0^1 \ln\left(\ln\frac{1}{t}\right) dt$ FI II 807

6. $\quad C = \int_0^1 \left[\frac{1}{\ln t} + \frac{1}{1-t}\right] dt$ DW

7. $\quad C = -\int_0^\infty \left[\cos t - \frac{1}{1+t}\right] \frac{dt}{t}$ MO 10

8. $\quad C = 1 - \int_0^\infty \left[\frac{\sin t}{t} - \frac{1}{1+t}\right] \frac{dt}{t}$ MO 10

9. $\quad C = -\int_0^\infty \left[e^{-t} - \frac{1}{1+t}\right] \frac{dt}{t}$ FI II 795, 802

10. $\quad C = -\int_0^\infty \left[e^{-t} - \frac{1}{1+t^k}\right] \frac{dt}{t}$ $\qquad [K \geq 1]$

11. $\quad C = \int_0^\infty \left[\frac{1}{e^t - 1} - \frac{1}{te^t}\right] dt$ DW

12. $\quad C = \int_0^1 \left(1 - e^{-t}\right) \frac{dt}{t} - \int_1^\infty \frac{e^{-t}}{t} dt$ FI II 802

See also **8.361** 5–**8.361** 7, **3.311** 6, **3.435** 3 and 4, **3.476** 2, **3.481** 1 and 2, **3.951** 10, **4.283** 9, **4.331** 1, **4.421** 1, **4.424** 1, **4.553**, **4.572**, **6.234**, **6.264** 1, **6.468**.

13. Asymptotic expansions

$$C = \sum_{k=1}^{n-1} \frac{1}{k} - \ln n + \frac{1}{2n} + \frac{1}{12n^2} - \frac{1}{120n^4} + \frac{1}{252n^6} - \frac{1}{240n^8} + \cdots$$

$$\cdots + \frac{B_{2r}}{2r} \frac{1}{n^{2r}} + \frac{B_{2r+2}}{2(r+1)} \frac{\theta}{n^{2r+2}}$$

$[0 < \theta < 1]$    FI II 827

14.* $\quad C = -\int_1^\infty \frac{\ln(\ln t)}{t^2} dt$

15.* $\quad C = -\int_{-\infty}^\infty t \exp(t - e^t) \, dt$

## 8.37 The function $\beta(x)$

Definition:

**8.370** $\quad \beta(x) = \frac{1}{2}\left[\psi\left(\frac{x+1}{2}\right) - \psi\left(\frac{x}{2}\right)\right]$ NH 16(13)

**8.371** Integral representations:

1.³ $\quad \beta(x) = \int_0^1 \frac{t^{x-1}}{1+t} dt$ $\qquad [\operatorname{Re} x > 0]$ WH

2.      $\beta(x) = \int_0^\infty \frac{e^{-xt}}{1+e^{-t}}\, dt$ $\qquad\qquad$ $[\operatorname{Re} x > 0]$ $\qquad\qquad$ MO 4

3.      $\beta\left(\frac{x+1}{2}\right) = \int_0^\infty \frac{e^{-xt}}{\cosh t}\, dt$ $\qquad\qquad$ $[\operatorname{Re} x > -1]$

See also **3.241** 1, **3.251** 7, **3.522** 2 and 4, **3.623** 2 and 3, **4.282** 2, **4.389** 3, **4.532** 1 and 3.

## Series representation

**8.372**

$1.^7$      $\beta(x) = \sum_{k=0}^\infty \frac{(-1)^k}{x+k}$ $\qquad\qquad$ $[-x \notin \mathbb{N}]$ $\qquad\qquad$ NH 37, 101(1)

$2.^7$      $\beta(x) = \sum_{k=0}^\infty \frac{1}{(x+2k)(x+2k+1)}$ $\qquad$ $[-x \notin \mathbb{N}]$ $\qquad\qquad$ NH 101(2)

$3.^8$      $\beta(x) = \frac{1}{2}\sum_{k=0}^\infty \frac{k!}{x(x+1)\dots(x+k)}\frac{1}{2^k}$ $\qquad$ $[-x \notin \mathbb{N}]$

$\qquad\qquad\qquad\qquad$ $[\beta$ has simple poles at $x=-n$ with residue $(-1)^n]$ $\quad$ NH 246(7)

**8.373**

$1.^6$      $\beta(x+1) = \ln 2 + \sum_{k=1}^\infty (-1)^k \left(1-2^{-k}\right) \zeta(k+1)x^k$ $\qquad$ $[|x| < 1]$ $\qquad$ NH 37(5)

$2.^6$      $\beta(x+1) = \ln 2 - 1 + \frac{1}{2x} - \frac{\pi}{2\sin \pi x} + \frac{1}{1-x^2} - \sum_{k=1}^\infty \left[1 - \left(1-2^{-2k}\right)\zeta(2k+1)\right]x^{2k}$

$\qquad\qquad\qquad\qquad\qquad\qquad$ $[0 < |x| < 2; \quad x \neq \pm 1]$ $\qquad$ NH 38(11)

**8.374**    $\frac{d^n}{dx^n}\beta(x) = (-1)^n n! \sum_{k=0}^\infty \frac{(-1)^k}{(x+k)^{n+1}}$ $\qquad$ $[-x \in \mathbb{N}]$ $\qquad$ NH 37(2)

**8.375**    Representation in the form of a finite sum:

$1.^6$      $\beta\left(\frac{p}{q}\right) = \frac{\pi}{2\sin\frac{p\pi}{q}} - \sum_{k=0}^{\lfloor \frac{q-1}{2}\rfloor} \cos\frac{p(2k+1)\pi}{q}\ln\sin\frac{(2k+1)\pi}{2q}$

$\qquad\qquad\qquad$ $[q = 2, 3, \dots, p = 1, 2, 3, \dots, q-1]$ $\qquad$ (see also **8.362** 5–7) $\quad$ NH 23(9)

2.      $\beta(n) = (-1)^{n+1}\ln 2 + \sum_{k=1}^{n-1}\frac{(-1)^{k+n+1}}{k}$

## Functional relations

**8.376**  $\sum_{k=0}^{2n}(-1)^k \beta\left(\frac{x+k}{2n+1}\right) = (2n+1)\beta(x)$ $\qquad\qquad\qquad$ NH 19

**8.377**  $\sum_{k=1}^n \beta\left(2^k x\right) = \psi\left(2^n x\right) - \psi(x) - n\ln 2$ $\qquad\qquad\qquad$ NH 20(10)

## 8.38 The beta function (Euler's integral of the first kind): $\mathrm{B}(x, y)$

**Integral representation**

**8.380**

1. $$\mathrm{B}\,(x,y) = \int_0^1 t^{x-1}(1-t)^{y-1}\,\mathrm{d}t^*$$
$$= 2\int_0^1 t^{2x-1}\left(1-t^2\right)^{y-1}\,\mathrm{d}t \qquad [\mathrm{Re}\,x > 0, \quad \mathrm{Re}\,y > 0] \qquad\qquad \text{FI II 774(1)}$$

2. $$\mathrm{B}(x,y) = 2\int_0^{\pi/2} \sin^{2x-1}\varphi \cos^{2y-1}\varphi\,\mathrm{d}\varphi \qquad [\mathrm{Re}\,x > 0, \quad \mathrm{Re}\,y > 0] \qquad\qquad \text{KU 10}$$

3. $$\mathrm{B}(x,y) = \int_0^\infty \frac{t^{x-1}}{(1+t)^{x+y}}\,\mathrm{d}t = 2\int_0^\infty \frac{t^{2x-1}}{(1+t^2)^{x+y}}\,\mathrm{d}t \qquad [\mathrm{Re}\,x > 0, \quad \mathrm{Re}\,y > 0] \qquad\qquad \text{FI II 775}$$

4. $$\mathrm{B}(x,y) = 2^{2-y-x}\int_{-1}^1 \frac{(1+t)^{2x-1}(1-t)^{2y-1}}{(1+t^2)^{x+y}}\,\mathrm{d}t \qquad [\mathrm{Re}\,x > 0, \quad \mathrm{Re}\,y > 0] \qquad\qquad \text{MO 7}$$

5. $$\mathrm{B}(x,y) = \int_0^1 \frac{t^{x-1}+t^{y-1}}{(1+t)^{x+y}}\,\mathrm{d}t = \int_1^\infty \frac{t^{x-1}+t^{y-1}}{(1+t)^{x+y}}\,\mathrm{d}t \qquad [\mathrm{Re}\,x > 0, \quad \mathrm{Re}\,y > 0] \qquad\qquad \text{BI (1)(15)}$$

6. $$\mathrm{B}(x,y) = \frac{1}{2^{x+y-1}}\int_0^1 \left[(1+t)^{x-1}(1-t)^{y-1} + (1+t)^{y-1}(1-t)^{x-1}\right]\mathrm{d}t$$
$$[\mathrm{Re}\,x > 0, \quad \mathrm{Re}\,y > 0] \qquad\qquad \text{BI (1)(15)}$$

7. $$\mathrm{B}\,(x,y) = z^y(1+z)^x \int_0^1 \frac{t^{x-1}(1-t)^{y-1}}{(t+z)^{x+y}}\,\mathrm{d}t$$
$$[\mathrm{Re}\,x > 0, \quad \mathrm{Re}\,y > 0, \quad 0 > z > 1, \quad \mathrm{Re}(x+y) < 1] \qquad \text{NH 163(8)}$$

8. $$\mathrm{B}(x,y) = z^y(1+z)^x \int_0^{\pi/2} \frac{\cos^{2x-1}\varphi \sin^{2y-1}\varphi}{(z+\cos^2\varphi)^{x+y}}\,\mathrm{d}\varphi$$
$$[\mathrm{Re}\,x > 0, \quad \mathrm{Re}\,y > 0, \quad 0 > z > -1, \quad \mathrm{Re}(x+y) < 1] \qquad \text{NH 163(8)}$$

See also **3.196** 3, **3.198**, **3.199**, **3.215**, **3.238** 3, **3.251** 1–3, 11, **3.253**, **3.312** 1, **3.512** 1 and 2, **3.541** 1, **3.542** 1, **3.621** 5, **3.623** 1, **3.631** 1, 8, 9, **3.632** 2, **3.633** 1, 4, **3.634** 1, 2, **3.637**, **3.642** 1, **3.667** 8, **3.681** 2.

9. $$\mathrm{B}(x,x) = \frac{1}{2^{2x-2}}\int_0^1 \left(1-t^2\right)^{x-1}\mathrm{d}t = \frac{1}{2^{2x-1}}\int_0^1 \frac{(1-t)^{x-1}}{\sqrt{t}}\,\mathrm{d}t$$
See **8.384** 4, **8.382** 3, and also **3.621** 1, **3.642** 2, **3.665** 1, **3.821** 6, **3.839** 6.

10. $$\mathrm{B}(x+y, x-y) = 4^{1-x}\int_0^\infty \frac{\cosh 2yt}{\cosh^{2x}t}\,\mathrm{d}t \qquad [\mathrm{Re}\,x > |\mathrm{Re}\,y|, \quad \mathrm{Re}\,x > 0] \qquad\qquad \text{MO 9}$$

11. $$\mathrm{B}\left(x, \frac{y}{z}\right) = z\int_0^1 \left(1-t^z\right)^{x-1}t^{y-1}\,\mathrm{d}t \qquad \left[\mathrm{Re}\,z > 0, \quad \mathrm{Re}\,\frac{y}{z} > 0, \quad \mathrm{Re}\,x > 0\right]$$
$$\text{FI II 787a}$$

---

*This equation is used as the definition of the function $\mathrm{B}(x,y)$.

**8.381**

1.    $\displaystyle\int_{-\infty}^{\infty} \frac{dt}{(a+it)^x(b-it)^y} = \frac{2\pi\,(a+b)^{1-x-y}}{(x+y-1)\,\mathrm{B}(x,y)}$

$$[a>0, \quad b>0; \quad x \text{ and } y \text{ are real}, \quad x+y>1] \quad \text{MO 7}$$

2.    $\displaystyle\int_{-\infty}^{\infty} \frac{dt}{(a-it)^x(b-it)^y} = 0$

$$[a>0, \quad b>0; \quad x \text{ and } y \text{ are real}, \quad x+y>1] \quad \text{MO 7}$$

3.    $\displaystyle\mathrm{B}(x+iy,x-iy) = 2^{1-2x}\alpha e^{-2i\gamma y}\int_{-\infty}^{\infty}\frac{e^{2i\alpha yt}\,dt}{\cosh^{2x}(\alpha t-\gamma)}$

$$[y,\alpha,\gamma \text{ are real}, \quad \alpha>0; \quad \operatorname{Re} x>0]$$
$$\text{MI 8a}$$

For an integral representation of $\ln\mathrm{B}(x,y)$, see **3.428** 7.

4.    $\displaystyle\frac{1}{\mathrm{B}(x,y)} = \frac{2^{x+y-1}(x+y-1)}{\pi}\int_0^{\pi/2}\cos[(x-y)t]\cos^{x+y-2}t\,dt$        NH 158(5)a

$\displaystyle\qquad\quad = \frac{2^{x+y-2}(x+y-1)}{\pi\cos\left[(x-y)\frac{\pi}{2}\right]}\int_0^{\pi}\cos[(x-y)t]\sin^{x+y-2}t\,dt$      NH 159(8)a

$\displaystyle\qquad\quad = \frac{2^{x+y-2}(x+y-1)}{\pi\sin\left[(x-y)\frac{\pi}{2}\right]}\int_0^{\pi}\sin[(x-y)t]\sin^{x+y-2}t\,dt$      NH 159(9)a

## Series representation

**8.382**

1.    $\displaystyle\mathrm{B}(x,y) = \frac{1}{y}\sum_{n=0}^{\infty}(-1)^n y\frac{(y-1)\ldots(y-n)}{n!(x+n)}$       $[y>0]$       WH

2.    $\displaystyle\ln\mathrm{B}\left(\frac{1+x}{2},\frac{1}{2}\right) = \ln\sqrt{2\pi} + \frac{1}{2}\left[\ln\left(\frac{\tan\frac{\pi x}{2}}{x}\right) - \ln\left(\frac{1+x}{1-x}\right)\right] + \sum_{k=0}^{\infty}\frac{1-\left(1-2^{-2k}\right)\zeta(2k+1)}{2k+1}x^{2k+1}$

$$[|x|<2]$$
$$\text{NH 39(17)}$$

3.    $\displaystyle\mathrm{B}\left(z,\frac{1}{2}\right) = \sum_{k=1}^{\infty}\frac{(2k-1)!!}{2^k k!}\frac{1}{z+k} + \frac{1}{z}$      (see also **8.384** and **8.380** 9)      WH

**8.383**   Infinite-product representation:

$$(x+y+1)\,\mathrm{B}(x+1,y+1) = \prod_{k=1}^{\infty}\frac{k(x+y+k)}{(x+k)(y+k)} \qquad [x, \quad y \neq -1, \quad -2,\ldots] \qquad \text{MO 2}$$

**8.384**   Functional relations involving the beta function:

1.    $\displaystyle\mathrm{B}(x,y) = \frac{\Gamma(x)\,\Gamma(y)}{\Gamma(x+y)} = \mathrm{B}(y,x)$            FI II 779

2.    $\mathrm{B}(x,y)\,\mathrm{B}(x+y,z) = \mathrm{B}(y,z)\,\mathrm{B}(y+z,x)$        MO 6

3. $\displaystyle\sum_{k=0}^{\infty} \mathrm{B}(x, y+k) = \mathrm{B}(x-1, y)$ WH

4. $\mathrm{B}(x, x) = 2^{1-2x}\, \mathrm{B}\left(\frac{1}{2}, x\right)$ (see also **8.380** 9 and **8.382** 3)

FI II 784

5. $\mathrm{B}(x, x)\, \mathrm{B}\left(x+\frac{1}{2}, x+\frac{1}{2}\right) = \dfrac{\pi}{2^{4x-1} x}$ WH

6. $\dfrac{1}{\mathrm{B}(n, m)} = m\dbinom{n+m-1}{n-1} = n\dbinom{n+m-1}{m-1}$ [$m$ and $n$ are natural numbers]

For a connection with the psi function, see **4.253** 1.

## 8.39 The incomplete beta function $\mathrm{B}_x(p, q)$

**8.391**[7] $\mathrm{B}_x(p, q) = \displaystyle\int_0^x t^{p-1}(1-t)^{q-1}\, \mathrm{d}t = \dfrac{x^p}{p}\, {}_2F_1\left(p, 1-q; p+1; x\right)$ ET I 373

**8.392** $I_x(p, q) = \dfrac{\mathrm{B}_x(p, q)}{\mathrm{B}(p, q)}$ ET II 429

# 8.4–8.5 Bessel Functions and Functions Associated with Them

## 8.40 Definitions

**8.401** Bessel functions $Z_\nu(z)$ are solutions of the differential equation

$$\frac{d^2 Z_\nu}{dz^2} + \frac{1}{z}\frac{d Z_\nu}{dz} + \left(1 - \frac{\nu^2}{z^2}\right) Z_\nu = 0$$ KU 37(1)

Special types of Bessel functions are what are called Bessel functions of the first kind $J_\nu(z)$, Bessel functions of the second kind $Y_\nu(z)$ (also called Neumann functions and often written $N_\nu(z)$), and Bessel functions of the third kind $H_\nu^{(1)}(z)$ and $H_\nu^{(2)}(z)$ (also called Hankel's functions).

**8.402** $J_\nu(z) = \dfrac{z^\nu}{2^\nu}\displaystyle\sum_{k=0}^{\infty}(-1)^k\dfrac{z^{2k}}{2^{2k} k!\, \Gamma(\nu+k+1)}$ [$|\arg z| < \pi$] KU 55(1)

**8.403**

1. $Y_\nu(z) = \dfrac{1}{\sin\nu\pi}\left[\cos\nu\pi\, J_\nu(z) - J_{-\nu}(z)\right]$ [for non-integer $\nu$, $\quad |\arg z| < \pi$]

KU 41(3)

2. $\quad \pi\, Y_n(z) = 2\, J_n(z)\ln\dfrac{z}{2} - \displaystyle\sum_{k=0}^{n-1}\dfrac{(n-k-1)!}{k!}\left(\dfrac{z}{2}\right)^{2k-n}$

$$-\sum_{k=0}^{\infty}(-1)^k\dfrac{1}{k!\,(k+n)!}\left(\dfrac{z}{2}\right)^{n+2k}[\psi(k+1)+\psi(k+n+1)]$$

KU 43(10)

$$= 2\, J_n(z)\left(\ln\dfrac{z}{2}+C\right) - \sum_{k=0}^{n-1}\dfrac{(n-k-1)!}{k!}\left(\dfrac{z}{2}\right)^{2k-n}$$

$$-\left(\dfrac{z}{2}\right)^n\dfrac{1}{n!}\sum_{k=1}^{n}\dfrac{1}{k} - \sum_{k=1}^{\infty}\dfrac{(-1)^k\left(\dfrac{z}{2}\right)^{n+2k}}{k!\,(k+n)!}\left[\sum_{m=1}^{n+k}\dfrac{1}{m}+\sum_{m=1}^{k}\dfrac{1}{m}\right]$$

$[n+1$ a positive integer, $|\arg z| < \pi]$
KU 44, WA 75(3)a

## 8.404

1. $\quad Y_{-n}(z) = (-1)^n\, Y_n(z)$            $[n$ is a natural number$]$      KU 41(2)

2. $\quad J_{-n}(z) = (-1)^n\, J_n(z)$            $[n$ is a natural number$]$      KU 41(2)

## 8.405[7]

1. $\quad H_\nu^{(1)}(z) = J_\nu(z) + i\, Y_\nu(z)$        KU 44(1)

2. $\quad H_\nu^{(2)}(z) = J_\nu(z) - i\, Y_\nu(z)$        KU 44(1)

In all relationships that hold for an arbitrary Bessel function $Z_\nu(z)$, that is, for the functions $J_\nu(z)$, $Y_\nu(z)$, and linear combinations of them, for example, $H_\nu^{(1)}(z)$ and $H_\nu^{(2)}(z)$, we shall write simply the letter $Z$ instead of the letters $J$, $Y$, $H^{(1)}$, and $H^{(2)}$.

### Modified Bessel functions of imaginary argument $I_\nu(z)$ and $K_\nu(z)$

## 8.406

1. $\quad I_\nu(z) = e^{-\frac{\pi}{2}\nu i}\, J_\nu\left(e^{\frac{\pi}{2}i}z\right)$        $\left[-\pi < \arg z \le \dfrac{\pi}{2}\right]$      WA 92

2. $\quad I_\nu(z) = e^{\frac{3}{2}\pi\nu i}\, J_\nu\left(e^{-\frac{3}{2}\pi i}z\right)$        $\left[\dfrac{\pi}{2} < \arg z \le \pi\right]$      WA 92

For integer $\nu$,

3. $\quad I_n(z) = i^{-n}\, J_n(iz)$        KU 46(1)

## 8.407

1.[8] $\quad K_\nu(z) = \dfrac{\pi i}{2}e^{\frac{\pi}{2}\nu i}\, H_\nu^{(1)}\left(ze^{\frac{1}{2}\pi i}\right)$        $\left[-\pi < \arg z \le \tfrac{1}{2}\pi\right]$

2.[8] $\quad K_\nu(z) = \dfrac{-\pi i}{2}e^{-\frac{\pi}{2}\nu i}\, H_{-\nu}^{(2)}\left(ze^{-\frac{1}{2}\pi i}\right)$        $\left[-\tfrac{1}{2}\pi < \arg z \le \pi\right]$      WA 92(8)

For the differential equation defining these functions, see **8.494**.

## 8.41 Integral representations of the functions $J_\nu(z)$ and $N_\nu(z)$

### 8.411

1.[11] $$J_n(z) = \frac{1}{2\pi} \int_{-\pi}^{\pi} e^{-ni\theta + iz\sin\theta} \, d\theta$$
$$= \frac{1}{\pi} \int_0^{\pi} \cos(n\theta - z\sin\theta) \, d\theta \qquad [n = 0, 1, 2, \ldots] \qquad \text{WH}$$

2. $$J_{2n}(z) = \frac{1}{\pi} \int_0^{\pi} \cos 2n\theta \cos(z\sin\theta) \, d\theta = \frac{2}{\pi} \int_0^{\pi/2} \cos 2n\theta \cos(z\sin\theta) \, d\theta$$
$$[n \text{ an integer}] \qquad \text{WA 30(7)}$$

3.[11] $$J_{2n+1}(z) = \frac{1}{\pi} \int_0^{\pi} \sin(2n+1)\theta \sin(z\sin\theta) \, d\theta$$
$$= \frac{2}{\pi} \int_0^{\pi/2} \sin(2n+1)\theta \sin(z\sin\theta) \, d\theta \qquad [n \text{ an integer}] \qquad \text{WA 30(6)}$$

4. $$J_\nu(z) = 2 \frac{\left(\frac{z}{2}\right)^\nu}{\Gamma\left(\nu + \frac{1}{2}\right)\Gamma\left(\frac{1}{2}\right)} \int_0^{\pi/2} \sin^{2\nu}\theta \cos(z\cos\theta) \, d\theta$$
$$\left[\operatorname{Re}\nu > -\tfrac{1}{2}\right] \qquad \text{WH}$$

5. $$J_\nu(z) = \frac{\left(\frac{z}{2}\right)^\nu}{\Gamma\left(\nu + \frac{1}{2}\right)\Gamma\left(\frac{1}{2}\right)} \int_0^{\pi} \sin^{2\nu}\theta \cos(z\cos\theta) \, d\theta \qquad \left[\operatorname{Re}\nu > -\tfrac{1}{2}\right]$$

6. $$J_\nu(z) = \frac{\left(\frac{z}{2}\right)^\nu}{\Gamma\left(\nu + \frac{1}{2}\right)\Gamma\left(\frac{1}{2}\right)} \int_{-\pi/2}^{\pi/2} \cos(z\sin\theta) \cos^{2\nu}\theta \, d\theta$$
$$\left[\operatorname{Re}\nu > -\tfrac{1}{2}\right] \qquad \text{KU 65(5), WA 35(4)a}$$

7. $$J_\nu(z) = \frac{\left(\frac{z}{2}\right)^\nu}{\Gamma\left(\nu + \frac{1}{2}\right)\Gamma\left(\frac{1}{2}\right)} \int_0^{\pi} e^{\pm iz\cos\varphi} \sin^{2\nu}\varphi \, d\varphi \qquad \left[\operatorname{Re}\left(\nu + \tfrac{1}{2}\right) > 0\right] \qquad \text{WH}$$

8. $$J_\nu(z) = \frac{\left(\frac{z}{2}\right)^\nu}{\Gamma\left(\nu + \frac{1}{2}\right)\Gamma\left(\frac{1}{2}\right)} \int_{-1}^{1} \left(1 - t^2\right)^{\nu - \frac{1}{2}} \cos zt \, dt \qquad \left[\operatorname{Re}\nu > -\tfrac{1}{2}\right] \qquad \text{KU 65(6), WH}$$

9. $$J_\nu(x) = 2 \frac{\left(\frac{x}{2}\right)^{-\nu}}{\Gamma\left(\frac{1}{2} - \nu\right)\Gamma\left(\frac{1}{2}\right)} \int_1^{\infty} \frac{\sin xt}{(t^2 - 1)^{\nu + \frac{1}{2}}} \, dt \qquad \left[-\tfrac{1}{2} < \operatorname{Re}\nu < \tfrac{1}{2}, \quad x > 0\right] \qquad \text{MO 37}$$

10. $$J_\nu(z) = \frac{\left(\frac{z}{2}\right)^\nu}{\Gamma\left(\nu + \frac{1}{2}\right)\Gamma\left(\frac{1}{2}\right)} \int_{-1}^{1} e^{izt} \left(1 - t^2\right)^{\nu - \frac{1}{2}} \, dt \qquad \left[\operatorname{Re}\nu > -\tfrac{1}{2}\right] \qquad \text{WA 34(3)}$$

11. $$J_\nu(x) = \frac{2}{\pi} \int_0^{\infty} \sin\left(x\cosh t - \frac{\nu\pi}{2}\right) \cosh\nu t \, dt \qquad \text{WA 199(12)}$$

12. $$J_\nu(z) = \frac{2^{\nu+1} z^\nu}{\Gamma\left(\nu + \frac{1}{2}\right)\Gamma\left(\frac{1}{2}\right)} \int_0^{\pi/2} \frac{\left(\cos^{\nu-\frac{1}{2}}\theta\right) \sin\left(z - \nu\theta + \frac{1}{2}\theta\right)}{\sin^{2\nu+1}\theta} e^{-2z\cot\theta} \, d\theta$$
$$\left[|\arg z| < \frac{\pi}{2}, \quad \operatorname{Re}\left(\nu + \tfrac{1}{2}\right) > 0\right] \qquad \text{WH}$$

13.[10] $\quad J_\nu(z) = \dfrac{1}{\pi} \displaystyle\int_0^\pi \cos(\nu\theta - z\sin\theta)\,\mathrm{d}\theta - \dfrac{\sin\nu\pi}{\pi} \displaystyle\int_0^\infty e^{-\nu\theta - z\sinh\theta}\,\mathrm{d}\theta$

$$[\operatorname{Re} z > 0] \qquad\qquad \text{WA 195(4)}$$

14. $\quad J_\nu(z) = \dfrac{e^{\pm\nu\pi i}}{\pi} \left[ \displaystyle\int_0^\pi \cos(\nu\theta + z\sin\theta)\,\mathrm{d}\theta - \sin\nu\pi \displaystyle\int_0^\infty e^{-\nu\theta + z\sinh\theta}\,\mathrm{d}\theta \right]$

$$\left[ \text{for } \tfrac{\pi}{2} < |\arg z| < \pi, \text{ with the upper sign taken for } |\arg z| > \tfrac{\pi}{2} \right.$$

$$\left. \text{and the lower sign taken for } |\arg z| < -\tfrac{\pi}{2} \right]$$

$$\text{WH}$$

**8.412**

1. $\quad J_\nu(z) = \dfrac{1}{2\pi i} \displaystyle\int_{-\infty}^{(0+)} t^{-\nu-1} \exp\left[ \dfrac{z}{2}\left( t - \dfrac{1}{t} \right) \right] \mathrm{d}t \qquad\qquad \left[ |\arg z| < \dfrac{\pi}{2} \right] \qquad \text{WH, WA 195(2)}$

2. $\quad J_\nu(z) = \dfrac{z^\nu}{2^{\nu+1}\pi i} \displaystyle\int_{-\infty}^{(0+)} t^{-\nu-1} \exp\left( t - \dfrac{z^2}{4t} \right) \mathrm{d}t \qquad\qquad \text{WA 195(1)}$

3.[12] $\quad J_\nu(z) = \dfrac{z^\nu}{2^{\nu+1}\pi i} \displaystyle\sum_{k=0}^\infty \dfrac{(-1)^k z^{2k}}{2^{2k} k!} \displaystyle\int_{-\infty}^{(0+)} e^t t^{-\nu-k-1}\,\mathrm{d}t \qquad\qquad \text{WA 195(1)}$

4. $\quad J_\nu(x) = \dfrac{1}{2\pi i} \displaystyle\int_{-i\infty}^{i\infty} \dfrac{\Gamma(-t)}{\Gamma(\nu+t+1)} \left( \dfrac{x}{2} \right)^{\nu+2t} \mathrm{d}t \qquad [\operatorname{Re}\nu \geq 0, \quad x > 0] \qquad \text{WA 214(7)}$

5.[7] $\quad J_\nu(z) = \dfrac{\Gamma\left(\frac{1}{2} - \nu\right)\left(\frac{z}{2}\right)^\nu}{2\pi i\,\Gamma\left(\frac{1}{2}\right)} \displaystyle\int_A^{(1+,-1-)} (t^2-1)^{\nu-\frac{1}{2}} \cos(zt)\,\mathrm{d}t$

$$\left[ \nu \neq \tfrac{1}{2}, \tfrac{3}{2}, \dots; \text{ The point } A \text{ falls to the right of the point } t = 1, \right.$$

$$\left. \text{and } \arg(t-1) = \arg(t+1) = 0 \text{ at the point } A \right]$$

$$\text{WH}$$

6.[8] $\quad J_\nu(z) = \dfrac{1}{2\pi} \displaystyle\int_{-\pi+\infty i}^{\pi+\infty i} e^{-iz\sin\theta + i\nu\theta}\,\mathrm{d}\theta \qquad\qquad [\operatorname{Re} z > 0]$

The path of integration being taken around the semi-infinite strip $y \geq 0$, $-\pi \leq x \leq \pi$.

**8.413**[8] $\quad \dfrac{J_\nu\left(\sqrt{z^2+\zeta^2}\right)}{(z^2-\zeta^2)^{\frac{\nu}{2}}} = \dfrac{1}{\pi(z+\zeta)^\nu} \left\{ \displaystyle\int_0^\infty e^{\zeta\cos t} \cos(z\sin t - \nu t)\,\mathrm{d}t \right.$

$$\left. - \sin\nu\pi \displaystyle\int_0^\infty \exp(-z\sinh t - \zeta\cosh t - \nu t)\,\mathrm{d}t \right\}$$

$$[\operatorname{Re}(z+\zeta) > 0] \qquad\qquad \text{MO 40}$$

**8.414** $\quad \displaystyle\int_{2x}^\infty \dfrac{J_0(t)}{t}\,\mathrm{d}t = \dfrac{1}{4\pi} \displaystyle\int_{-\frac{1}{2}-i\infty}^{-\frac{1}{2}+i\infty} \dfrac{\Gamma(-t)}{t\,\Gamma(1+t)} x^{2t}\,\mathrm{d}t \qquad [x > 0] \qquad\qquad \text{MO 41}$

See **3.715** 2, 9, 10, 13, 14, 19–21, **3.865** 1, 2, 4, **3.996** 4.

- For an integral representation of $J_0(z)$ see **3.714** 2, **3.753** 2, 3, and **4.124**.
- For an integral representation of $J_1(z)$ see **3.697**, **3.711**, **3.752** 2, and **3.753** 5.

**8.415**

1.     $Y_0(x) = \dfrac{4}{\pi^2} \displaystyle\int_0^1 \dfrac{\arcsin t}{\sqrt{1-t^2}} \sin(xt)\,dt - \dfrac{4}{\pi^2} \int_1^\infty \dfrac{\ln\left(t + \sqrt{t^2-1}\right)}{\sqrt{t^2-1}} \sin(xt)\,dt$

$$[x > 0] \qquad\qquad \text{MO 37}$$

2.     $Y_\nu(x) = -2\dfrac{\left(\frac{x}{2}\right)^{-\nu}}{\Gamma\left(\frac{1}{2} - \nu\right)\Gamma\left(\frac{1}{2}\right)} \displaystyle\int_1^\infty \dfrac{\cos xt}{(t^2-1)^{\nu + \frac{1}{2}}}\,dt \qquad \left[-\frac{1}{2} < \operatorname{Re}\nu < \frac{1}{2}, \quad x > 0\right]$

$$\text{KU 89(28)a, MO 38}$$

3.     $Y_\nu(x) = -\dfrac{2}{\pi} \displaystyle\int_0^\infty \cos\left(x\cosh t - \dfrac{\nu\pi}{2}\right) \cosh\nu t\,dt \qquad [-1 < \operatorname{Re}\nu < 1, \quad x > 0] \qquad \text{WA 199(13)}$

4.[8]     $Y_\nu(z) = \dfrac{1}{\pi} \displaystyle\int_0^\pi \sin\left(z\sin\theta - \nu\theta\right)\,d\theta - \dfrac{1}{\pi} \int_0^\infty \left(e^{\nu t} + e^{-\nu t}\cos\nu\pi\right) e^{-z\sinh t}\,dt$

$$[\operatorname{Re} z > 0] \qquad\qquad \text{WA 197(1)}$$

5.     $Y_\nu(z) = \dfrac{2\left(\frac{z}{2}\right)^\nu}{\Gamma\left(\nu + \frac{1}{2}\right)\Gamma\left(\frac{1}{2}\right)} \left[\displaystyle\int_0^{\pi/2} \sin\left(z\sin\theta\right)\cos^{2\nu}\theta\,d\theta - \int_0^\infty e^{-z\sinh\theta}\cosh^{2\nu}\theta\,d\theta\right]$

$$\left[\operatorname{Re}\nu > -\frac{1}{2}, \quad \operatorname{Re} z > 0\right] \qquad \text{WA 181(5)a}$$

6.[12]     $Y_\nu(z) = -\dfrac{2^{\nu+1}z^\nu}{\Gamma\left(\nu + \frac{1}{2}\right)\Gamma\left(\frac{1}{2}\right)} \displaystyle\int_0^{\frac{\pi}{2}} \dfrac{\cos^{\nu - \frac{1}{2}}\theta \cos\left(z - \nu\theta + \frac{1}{2}\theta\right)}{\sin^{2\nu+1}\theta} e^{-2z\cot\theta}\,d\theta$

$$\left[|\arg z| < \frac{\pi}{2}, \quad \operatorname{Re}\left(\nu + \frac{1}{2}\right) > 0\right]$$
$$\text{WA 186(8)}$$

For an integral representation of $Y_0(z)$, see **3.714** 3, **3.753** 4, **3.864**. See also **3.865** 3.

## 8.42   Integral representations of the functions $H_\nu^{(1)}(z)$ and $H_\nu^{(2)}(z)$

**8.421**

1.     $H_\nu^{(1)}(x) = \dfrac{e^{-\frac{\nu\pi i}{2}}}{\pi i} \displaystyle\int_{-\infty}^\infty e^{ix\cosh t - \nu t}\,dt$

$\qquad\qquad = \dfrac{2e^{-\frac{\nu\pi i}{2}}}{\pi i} \displaystyle\int_0^\infty e^{ix\cosh t}\cosh\nu t\,dt$

$$[-1 < \operatorname{Re}\nu < 1, \quad x > 0] \qquad \text{WA 199(10)}$$

2.     $H_\nu^{(2)}(x) = -\dfrac{e^{\frac{\nu\pi i}{2}}}{\pi i} \displaystyle\int_{-\infty}^\infty e^{-ix\cosh t - \nu t}\,dt$

$\qquad\qquad = -\dfrac{2e^{\frac{\nu\pi i}{2}}}{\pi i} \displaystyle\int_0^\infty e^{-ix\cosh t}\cosh\nu t\,dt$

$$[-1 < \operatorname{Re}\nu < 1, \quad x > 0] \qquad \text{WA 199(11)}$$

$3.^{12}$   $H_\nu^{(1)}(z) = -\dfrac{2^{\nu+1} i z^\nu}{\Gamma\left(\nu+\frac{1}{2}\right)\Gamma\left(\frac{1}{2}\right)} \displaystyle\int_0^{\pi/2} \dfrac{\cos^{\nu-\frac{1}{2}} t\, e^{i\left(z-\nu t+\frac{t}{2}\right)}}{\sin^{2\nu+1} t}\, \exp\left(-2z \cot t\right) dt$

$$\left[\operatorname{Re}\nu > -\tfrac{1}{2}, \quad \operatorname{Re}z > 0\right] \qquad \text{WA 168(5)}$$

$4.^{12}$   $H_\nu^{(2)}(z) = \dfrac{2^{\nu+1} i z^\nu}{\Gamma\left(\nu+\frac{1}{2}\right)\Gamma\left(\frac{1}{2}\right)} \displaystyle\int_0^{\pi/2} \dfrac{\cos^{\nu-\frac{1}{2}} t\, e^{-i\left(z-\nu t+\frac{t}{2}\right)}}{\sin^{2\nu+1} t}\, \exp\left(-2z \cot t\right) dt$

$$\left[\operatorname{Re}\nu > -\tfrac{1}{2}, \quad \operatorname{Re}z > 0\right] \qquad \text{WA 168(6)}$$

5.   $H_\nu^{(1)}(x) = -\dfrac{2i\left(\frac{x}{2}\right)^{-\nu}}{\sqrt{\pi}\,\Gamma\left(\frac{1}{2}-\nu\right)} \displaystyle\int_1^\infty \dfrac{e^{ixt}}{(t^2-1)^{\nu+\frac{1}{2}}}\, dt \qquad \left[-\tfrac{1}{2} < \operatorname{Re}\nu < \tfrac{1}{2}, \quad x > 0\right] \qquad \text{WA 87(1)}$

6.   $H_\nu^{(2)}(x) = \dfrac{2i\left(\frac{x}{2}\right)^{-\nu}}{\sqrt{\pi}\,\Gamma\left(\frac{1}{2}-\nu\right)} \displaystyle\int_1^\infty \dfrac{e^{-ixt}}{(t^2-1)^{\nu+\frac{1}{2}}}\, dt \qquad \left[-\tfrac{1}{2} < \operatorname{Re}\nu < \tfrac{1}{2}, \quad x > 0\right] \qquad \text{WA 187(2)}$

7.   $H_\nu^{(1)}(z) = -\dfrac{i}{\pi} e^{-\frac{1}{2}i\nu\pi} \displaystyle\int_0^\infty \exp\left[\tfrac{1}{2}iz\left(t+\tfrac{1}{t}\right)\right] t^{-\nu-1}\, dt$

$$\left[0 < \arg z < \pi; \text{ or } \arg z = 0 \text{ and } -1 < \operatorname{Re}\nu < 1\right] \quad \text{MO 38}$$

8.   $H_\nu^{(1)}(xz) = -\dfrac{i}{\pi} e^{-\frac{1}{2}i\nu\pi} z^\nu \displaystyle\int_0^\infty \exp\left[\tfrac{1}{2}ix\left(t+\tfrac{z^2}{t}\right)\right] t^{-\nu-1}\, dt$

$$\left[0 < \arg z < \frac{\pi}{2}, \quad x > 0, \quad \operatorname{Re}\nu > -1; \text{ or } \arg z = \frac{\pi}{2}, \quad x > 0 \text{ and } -1 < \operatorname{Re}\nu < 1\right] \quad \text{MO 38}$$

9.   $H_\nu^{(1)}(xz) = \sqrt{\dfrac{2}{\pi z}}\, \dfrac{x^\nu \exp\left[i\left(xz - \frac{\pi}{2}\nu - \frac{\pi}{4}\right)\right]}{\Gamma\left(\nu+\frac{1}{2}\right)} \displaystyle\int_0^\infty \left(1+\dfrac{it}{2z}\right)^{\nu-\frac{1}{2}} t^{\nu-\frac{1}{2}} e^{-xt}\, dt$

$$\left[\operatorname{Re}\nu > -\tfrac{1}{2}, \quad -\tfrac{1}{2}\pi < \arg z < \tfrac{3}{2}\pi, \quad x > 0\right] \quad \text{MO 39}$$

10.   $H_\nu^{(1)}(z) = \dfrac{-2i e^{-i\nu\pi}\left(\frac{z}{2}\right)^\nu}{\sqrt{\pi}\,\Gamma\left(\nu+\frac{1}{2}\right)} \displaystyle\int_0^\infty e^{iz \cosh t} \sinh^{2\nu} t\, dt$

$$\left[0 < \arg z < \pi, \quad \operatorname{Re}\nu > -\tfrac{1}{2} \text{ or } \arg z = 0 \text{ and } -\tfrac{1}{2} < \operatorname{Re}\nu < \tfrac{1}{2}\right] \quad \text{MO 38}$$

11.   $H_0^{(1)}(x) = -\dfrac{i}{\pi} \displaystyle\int_{-\infty}^\infty \dfrac{\exp\left(i\sqrt{x^2+t^2}\right)}{\sqrt{x^2+t^2}}\, dt \qquad [x > 0] \qquad \text{MO 38}$

**8.422**

1.   $H_\nu^{(1)}(z) = \dfrac{\Gamma\left(\frac{1}{2}-\nu\right)\left(\frac{z}{2}\right)^\nu}{\pi i\,\Gamma\left(\frac{1}{2}\right)} \displaystyle\int_{1+\infty i}^{(1+)} e^{izt}\left(t^2-1\right)^{\nu-\frac{1}{2}}\, dt \qquad [-\pi < \arg z < 2\pi] \qquad \text{WA 183(4)}$

$2.^{12}$   $H_\nu^{(2)}(z) = \dfrac{\Gamma\left(\frac{1}{2}-\nu\right)\left(\frac{z}{2}\right)^\nu}{\pi i\,\Gamma\left(\frac{1}{2}\right)} \displaystyle\int_{-1+\infty i}^{(-1-)} e^{izt}\left(t^2-1\right)^{\nu-\frac{1}{2}}\, dt$

$$[-2\pi < \arg z < \pi]$$

The paths of integration are shown in the drawing.

**8.423**

1.    $H_\nu^{(1)}(z) = -\dfrac{1}{\pi} \displaystyle\int_{-\infty i}^{-\pi+\infty i} e^{-iz\sin\theta + i\nu\theta}\, d\theta$        $[\operatorname{Re} z > 0]$        WA 197(2)a

2.    $H_\nu^{(2)}(z) = -\dfrac{1}{\pi} \displaystyle\int_{\pi+\infty i}^{-\infty i} e^{-iz\sin\theta + i\nu\theta}\, d\theta$        $[\operatorname{Re} z > 0]$        WA 197(3)a

The path of integration for **8.423** 1 is shown in the left hand drawing and for **8.423** 2 in the right hand drawing.

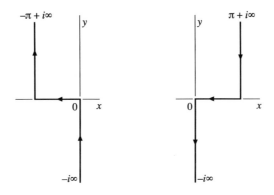

**8.424**

1.    $H_\nu^{(1)}(z) J_\nu(\zeta) = \dfrac{1}{\pi i} \displaystyle\int_0^{\gamma+i\infty} \exp\left[\dfrac{1}{2}\left(t - \dfrac{z^2+\zeta^2}{t}\right)\right] I_\nu\left(\dfrac{z\zeta}{t}\right) \dfrac{dt}{t}$

                   $[\gamma > 0, \quad \operatorname{Re}\nu > -1, \quad |\zeta| < |z|]$    MO 45

2.    $H_\nu^{(2)}(z) J_\nu(\zeta) = \dfrac{i}{\pi} \displaystyle\int_0^{\gamma-i\infty} \exp\left[\dfrac{1}{2}\left(t - \dfrac{z^2+\zeta^2}{t}\right)\right] I_\nu\left(\dfrac{z\zeta}{t}\right) \dfrac{dt}{t}$

                   $[\gamma > 0, \quad \operatorname{Re}\nu > -1, \quad |\zeta| < |z|]$    MO 45

## 8.43 Integral representations of the functions $I_\nu(z)$ and $K_\nu(z)$

### The function $I_\nu(z)$

**8.431**

1.    $I_\nu(z) = \dfrac{\left(\frac{z}{2}\right)^\nu}{\Gamma\left(\nu+\frac{1}{2}\right)\Gamma\left(\frac{1}{2}\right)} \displaystyle\int_{-1}^{1} (1-t^2)^{\nu-\frac{1}{2}} e^{\pm zt}\, dt$        $\left[\operatorname{Re}\left(\nu+\frac{1}{2}\right) > 0\right]$        WA 94(9)

2.    $I_\nu(z) = \dfrac{\left(\frac{z}{2}\right)^\nu}{\Gamma\left(\nu+\frac{1}{2}\right)\Gamma\left(\frac{1}{2}\right)} \displaystyle\int_{-1}^{1} (1-t^2)^{\nu-\frac{1}{2}} \cosh zt\, dt$        $\left[\operatorname{Re}\left(\nu+\frac{1}{2}\right) > 0\right]$        WA 94(9)

3.    $I_\nu(z) = \dfrac{\left(\frac{z}{2}\right)^\nu}{\Gamma\left(\nu+\frac{1}{2}\right)\Gamma\left(\frac{1}{2}\right)} \displaystyle\int_0^\pi e^{\pm z\cos\theta} \sin^{2\nu}\theta\, d\theta$        $\left[\operatorname{Re}\left(\nu+\frac{1}{2}\right) > 0\right]$        WA 94(9)

4.    $I_\nu(z) = \dfrac{\left(\frac{z}{2}\right)^\nu}{\Gamma\left(\nu+\frac{1}{2}\right)\Gamma\left(\frac{1}{2}\right)} \displaystyle\int_0^\pi \cosh(z\cos\theta) \sin^{2\nu}\theta\, d\theta$        $\left[\operatorname{Re}\left(\nu+\frac{1}{2}\right) > 0\right]$        WA 94(9)

5.      $I_\nu(z) = \dfrac{1}{\pi} \displaystyle\int_0^\pi e^{z\cos\theta} \cos\nu\theta \, d\theta - \dfrac{\sin\nu\pi}{\pi} \displaystyle\int_0^\infty e^{-z\cosh t - \nu t} \, dt$

$$\left[ |\arg z| \le \tfrac{\pi}{2}, \quad \operatorname{Re}\nu > 0 \right] \qquad \text{WA 201(4)}$$

See also **3.383** 2, **3.387** 1, **3.471** 6, **3.714** 5.
     For an integral representation of $I_0(z)$ and $I_1(z)$, see **3.366** 1, **3.534 3.856** 6.

## The function $K_\nu(z)$

**8.432**

1.      $K_\nu(z) = \displaystyle\int_0^\infty e^{-z\cosh t} \cosh\nu t \, dt \qquad\qquad \left[ |\arg z| < \tfrac{\pi}{2} \text{ or } \operatorname{Re} z = 0 \text{ and } \nu = 0 \right]$

$$\text{MO 39}$$

2.      $K_\nu(z) = \dfrac{\left(\tfrac{z}{2}\right)^\nu \Gamma\left(\tfrac{1}{2}\right)}{\Gamma\left(\nu + \tfrac{1}{2}\right)} \displaystyle\int_0^\infty e^{-z\cosh t} \sinh^{2\nu} t \, dt$

$$\left[ \operatorname{Re}\nu > -\tfrac{1}{2}, \quad \operatorname{Re} z > 0; \text{ or } \operatorname{Re} z = 0 \text{ and } -\tfrac{1}{2} < \operatorname{Re}\nu < \tfrac{1}{2} \right] \quad \text{WA 190(5), WH}$$

3.      $K_\nu(z) = \dfrac{\left(\tfrac{z}{2}\right)^\nu \Gamma\left(\tfrac{1}{2}\right)}{\Gamma\left(\nu + \tfrac{1}{2}\right)} \displaystyle\int_1^\infty e^{-zt} \left(t^2 - 1\right)^{\nu - \tfrac{1}{2}} \, dt$

$$\left[ \operatorname{Re}\left(\nu + \tfrac{1}{2}\right) > 0, \quad |\arg z| < \tfrac{\pi}{2}; \text{ or } \operatorname{Re} z = 0 \text{ and } \nu = 0 \right] \quad \text{WA 190(4)}$$

4.      $K_\nu(x) = \dfrac{1}{\cos\tfrac{\nu\pi}{2}} \displaystyle\int_0^\infty \cos\left(x\sinh t\right) \cosh\nu t \, dt \qquad\qquad [x > 0, \quad -1 < \operatorname{Re}\nu < 1] \qquad \text{WA 202(13)}$

5.      $K_\nu(xz) = \dfrac{\Gamma\left(\nu + \tfrac{1}{2}\right)(2z)^\nu}{x^\nu \, \Gamma\left(\tfrac{1}{2}\right)} \displaystyle\int_0^\infty \dfrac{\cos xt \, dt}{\left(t^2 + z^2\right)^{\nu + \tfrac{1}{2}}} \qquad\qquad \left[ \operatorname{Re}\left(\nu + \tfrac{1}{2}\right) \ge 0, \quad x > 0, \quad |\arg z| < \dfrac{\pi}{2} \right]$

$$\text{WA 191(1)}$$

6.[12]    $K_\nu(z) = \dfrac{1}{2} \left(\dfrac{z}{2}\right)^\nu \displaystyle\int_0^\infty \dfrac{e^{-t - z^2/4t} \, dt}{t^{\nu+1}} \qquad\qquad \left[ |\arg z| < \dfrac{\pi}{4}, \quad \operatorname{Re} z^2 > 0 \right] \quad \text{WA 203(15)}$

7.[7]    $K_\nu(xz) = \dfrac{z^\nu}{2} \displaystyle\int_0^\infty \exp\left[ -\dfrac{x}{2}\left(t + \dfrac{z^2}{t}\right) \right] t^{-\nu-1} \, dt$

$$\left[ |\arg z| < \tfrac{\pi}{4} \text{ or } |\arg z| = \tfrac{\pi}{4} \text{ and } \operatorname{Re}\nu < 1 \right] \quad \text{MO 39}$$

8.      $K_\nu(xz) = \sqrt{\dfrac{\pi}{2z}} \dfrac{x^\nu e^{-xz}}{\Gamma\left(\nu + \tfrac{1}{2}\right)} \displaystyle\int_0^\infty e^{-xt} t^{\nu - \tfrac{1}{2}} \left(1 + \dfrac{t}{2z}\right)^{\nu - \tfrac{1}{2}} \, dt$

$$\left[ |\arg z| < \pi, \quad \operatorname{Re}\nu > -\tfrac{1}{2}, x > 0 \right]$$
$$\text{MO 39}$$

9.      $K_\nu(xz) = \dfrac{\sqrt{\pi}}{\Gamma\left(\nu + \tfrac{1}{2}\right)} \left(\dfrac{x}{2z}\right)^\nu \displaystyle\int_0^\infty \dfrac{\exp\left(-x\sqrt{t^2 + z^2}\right)}{\sqrt{t^2 + z^2}} t^{2\nu} \, dt$

$$\left[ \operatorname{Re}\nu > -\tfrac{1}{2}, \quad \operatorname{Re} z > 0, \quad \operatorname{Re}\sqrt{t^2 + z^2} > 0, \quad x > 0 \right] \quad \text{MO 39}$$

See also **3.383** 3, **3.387** 3, 6, **3.388** 2, **3.389** 4, **3.391**, **3.395** 1, **3.471** 9, **3.483**, **3.547** 2, **3.856**, **3.871** 3, 4, **7.141** 5.

**8.433**    $K_{\frac{1}{3}}\left(\dfrac{2x\sqrt{x}}{3\sqrt{3}}\right) = \dfrac{3}{\sqrt{x}}\displaystyle\int_0^\infty \cos\left(t^3 + xt\right)\,dt$             KU 98(31), WA 211(2)

For an integral representation of $K_0(z)$, see **3.754** 2, **3.864**, **4.343**, **4.356**, **4.367**.

## 8.44 Series representation

### The function $J_\nu(z)$

**8.440**    $J_\nu(z) = \left(\dfrac{z}{2}\right)^\nu \displaystyle\sum_{k=0}^\infty \dfrac{(-1)^k}{k!\,\Gamma(\nu + k + 1)}\left(\dfrac{z}{2}\right)^{2k}$        $[|\arg z| < \pi]$        WH 358 a

**8.441**    Special cases:

1.      $J_0(z) = \displaystyle\sum_{k=0}^\infty (-1)^k \dfrac{z^{2k}}{2^{2k}\,(k!)^2}$

2.      $J_1(z) = -J_0'(z) = \dfrac{z}{2}\displaystyle\sum_{k=0}^\infty \dfrac{(-1)^k z^{2k}}{2^{2k}k!(k+1)!}$

3.      $J_{\frac{1}{3}}(z) = \dfrac{1}{\Gamma\left(\frac{4}{3}\right)}\sqrt[3]{\dfrac{z}{2}}\displaystyle\sum_{k=0}^\infty (-1)^k \dfrac{\left(z\sqrt{3}\right)^{2k}}{2^{2k}k!\cdot 1\cdot 4\cdot 7\cdots (3k+1)}$

4.      $J_{-\frac{1}{3}}(z) = \dfrac{1}{\Gamma\left(\frac{2}{3}\right)}\sqrt[3]{\dfrac{2}{z}}\left\{1 + \displaystyle\sum_{k=1}^\infty (-1)^k \dfrac{\left(z\sqrt{3}\right)^{2k}}{2^{2k}k!\cdot 2\cdot 5\cdot 8\cdots (3k-1)}\right\}$

For the expansion of $J_\nu(z)$ in Laguerre polynomials, see **8.975** 3.

**8.442**

1.[7]    $J_\nu(z)\,J_\mu(z) = \displaystyle\sum_{m=0}^\infty \dfrac{(-1)^m \left(\frac{1}{2}z\right)^{\mu+\nu+2m}(\mu+\nu+m+1)_m}{m!\,\Gamma(\mu+m+1)\,\Gamma(\nu+m+1)}$

2.[12]    $J_\nu(az)\,J_\mu(bz) = \dfrac{\left(\frac{az}{2}\right)^\nu \left(\frac{bz}{2}\right)^\mu}{\Gamma(\mu+1)}\displaystyle\sum_{k=0}^\infty \dfrac{(-1)^k \left(\frac{az}{2}\right)^{2k} F\left(-k, -\nu-k; \mu+1; \frac{b^2}{a^2}\right)}{k!\,\Gamma(\nu+k+1)}$      MO 28

### The function $Y_\nu(z)$

**8.443**[11]   $Y_\nu(z) = \dfrac{1}{\sin\nu\pi}\left\{\cos\nu\pi \left(\dfrac{z}{2}\right)^\nu \displaystyle\sum_{k=0}^\infty (-1)^k \dfrac{z^{2k}}{2^{2k}k!\,\Gamma(\nu+k+1)}\right.$

$\left. - \left(\dfrac{z}{2}\right)^{-\nu}\displaystyle\sum_{k=0}^\infty (-1)^k \dfrac{z^{2k}}{2^{2k}k!\,\Gamma(k-\nu+1)}\right\}$

                              $[\nu \neq$ an integer$]$      (cf. **8.403** 1)

For $\nu + 1$ a natural number, see **8.403** 2.; for $\nu$ a negative integer see **8.404** 1

**8.444**   Special cases,

1.      $\pi\,Y_0(z) = 2\,J_0(z)\left(\ln\dfrac{z}{2} + C\right) - 2\sum_{k=1}^{\infty}\dfrac{(-1)^k}{(k!)^2}\left(\dfrac{z}{2}\right)^{2k}\sum_{m=1}^{k}\dfrac{1}{m}$          KU 44

2.[11]   $\pi\,Y_1(z) = 2\,J_1(z)\left(\ln\dfrac{z}{2} + C\right) - \dfrac{2}{z} - \dfrac{z}{2} - \sum_{k=2}^{\infty}\dfrac{(-1)^{k+1}\left(\dfrac{z}{2}\right)^{2k-1}}{k!(k-1)!}\left(\dfrac{1}{k} + 2\sum_{m=1}^{k-1}\dfrac{1}{m}\right)$

## The functions $I_\nu(z)$ and $K_n(z)$

**8.445**   $I_\nu(z) = \sum_{k=0}^{\infty}\dfrac{1}{k!\,\Gamma(\nu+k+1)}\left(\dfrac{z}{2}\right)^{\nu+2k}$          WH 372a

**8.446**[8]   $K_n(z) = \dfrac{1}{2}\sum_{k=0}^{n-1}(-1)^k\dfrac{(n-k-1)!}{k!\left(\frac{z}{2}\right)^{n-2k}}$

$$+(-1)^{n+1}\sum_{k=0}^{\infty}\dfrac{\left(\frac{z}{2}\right)^{n+2k}}{k!(n+k)!}\left[\ln\dfrac{z}{2} - \dfrac{1}{2}\,\psi(k+1) - \dfrac{1}{2}\,\psi(n+k+1)\right]$$

WA 95(15)

$$= (-1)^{n+1}\,I_n(z)\left(\ln\dfrac{1}{2}z + C\right) + \dfrac{1}{2}(-1)^n\sum_{l=0}^{\infty}\dfrac{\left(\frac{z}{2}\right)^{n+2l}}{l!(n+l)!}\left(\sum_{k=1}^{l}\dfrac{1}{k} + \sum_{k=1}^{n+l}\dfrac{1}{k}\right)$$

$$+\dfrac{1}{2}\sum_{l=0}^{n-1}\dfrac{(-1)^l(n-l-1)!}{l!}\left(\dfrac{z}{2}\right)^{2l-n}$$

[$n+1$ is a natural number]<br>MO 29

**8.447**   Special cases:

1.      $I_0(z) = \sum_{k=0}^{\infty}\dfrac{\left(\frac{z}{2}\right)^{2k}}{(k!)^2}$

2.      $I_1(z) = I_0'(z) = \sum_{k=0}^{\infty}\dfrac{\left(\frac{z}{2}\right)^{2k+1}}{k!(k+1)!}$

3.      $K_0(z) = -\ln\dfrac{z}{2}\,I_0(z) + \sum_{k=0}^{\infty}\dfrac{z^{2k}}{2^{2k}\,(k!)^2}\,\psi(k+1)$          WA 95(14)

## 8.45 Asymptotic expansions of Bessel functions

**8.451** For large values of $|z|$ *

1. $$J_{\pm\nu}(z) = \sqrt{\frac{2}{\pi z}} \left\{ \cos\left(z \mp \frac{\pi}{2}\nu - \frac{\pi}{4}\right) \left[ \sum_{k=0}^{n-1} \frac{(-1)^k}{(2z)^{2k}} \frac{\Gamma\left(\nu + 2k + \frac{1}{2}\right)}{(2k)!\,\Gamma\left(\nu - 2k + \frac{1}{2}\right)} + R_1 \right] \right.$$
$$\left. - \sin\left(z \mp \frac{\pi}{2}\nu - \frac{\pi}{4}\right) \left[ \sum_{k=0}^{n-1} \frac{(-1)^k}{(2z)^{2k+1}} \frac{\Gamma\left(\nu + 2k + \frac{3}{2}\right)}{(2k+1)!\,\Gamma\left(\nu - 2k - \frac{1}{2}\right)} + R_2 \right] \right\}$$
$$[|\arg z| < \pi] \qquad \text{(see } \mathbf{8.339}\ 4) \quad \text{WA } 222(1,\,3)$$

2.[11] $$Y_{\pm\nu}(z) = \sqrt{\frac{2}{\pi z}} \left\{ \sin\left(z \mp \frac{\pi}{2}\nu - \frac{\pi}{4}\right) \left[ \sum_{k=0}^{n-1} \frac{(-1)^k}{(2z)^{2k}} \frac{\Gamma\left(\nu + 2k + \frac{1}{2}\right)}{(2k)!\,\Gamma\left(\nu - 2k + \frac{1}{2}\right)} + R_1 \right] \right.$$
$$\left. + \cos\left(z \mp \frac{\pi}{2}\nu - \frac{\pi}{4}\right) \left[ \sum_{k=0}^{n-1} \frac{(-1)^k}{(2z)^{2k+1}} \frac{\Gamma\left(\nu + 2k + \frac{3}{2}\right)}{(2k+1)!\,\Gamma\left(\nu - 2k - \frac{1}{2}\right)} + R_2 \right] \right\}$$
$$[|\arg z| < \pi] \qquad \text{(see } \mathbf{8.339}\ 4) \quad \text{WA } 222(2,\,4,\,5)$$

3.[12] $$H_{\nu}^{(1)}(z) = \sqrt{\frac{2}{\pi z}}\, e^{i\left(z - \frac{\pi}{2}\nu - \frac{\pi}{4}\right)} \left[ \sum_{k=0}^{n-1} \frac{(-1)^k}{(2iz)^k} \frac{\Gamma\left(\nu + k + \frac{1}{2}\right)}{k!\,\Gamma\left(\nu - k + \frac{1}{2}\right)} + \theta_1 \frac{(-1)^n}{(2iz)^n} \frac{\Gamma\left(\nu + n + \frac{1}{2}\right)}{n!\,\Gamma\left(\nu - n + \frac{1}{2}\right)} \right]$$
$$\left[\operatorname{Re}\nu > -\tfrac{1}{2}, \quad |\arg z| < \pi\right] \qquad \text{(see } \mathbf{8.339}\ 4) \quad \text{WA } 221(5)$$

4.[12] $$H_{\nu}^{(2)}(z) = \sqrt{\frac{2}{\pi z}}\, e^{-i\left(z - \frac{\pi}{2}\nu - \frac{\pi}{4}\right)} \left[ \sum_{k=0}^{n-1} \frac{1}{(2iz)^k} \frac{\Gamma\left(\nu + k + \frac{1}{2}\right)}{k!\,\Gamma\left(\nu - k + \frac{1}{2}\right)} + \theta_2 \frac{1}{(2iz)^n} \frac{\Gamma\left(\nu + n + \frac{1}{2}\right)}{n!\,\Gamma\left(\nu - n + \frac{1}{2}\right)} \right]$$
$$\left[\operatorname{Re}\nu > -\tfrac{1}{2}, \quad |\arg z| < \pi\right] \qquad \text{(see } \mathbf{8.339}\ 4) \quad \text{WA } 221(6)$$

For indices of the form $\nu = \frac{2n-1}{2}$ (where $n$ is a natural number), the series **8.451** terminate. In this case, the closed formulas **8.46** are valid for all values.

5. $$I_\nu(z) \sim \frac{e^z}{\sqrt{2\pi z}} \sum_{k=0}^{\infty} \frac{(-1)^k}{(2z)^k} \frac{\Gamma\left(\nu + k + \frac{1}{2}\right)}{k!\,\Gamma\left(\nu - k + \frac{1}{2}\right)}$$
$$+ \frac{\exp\left[-z \pm \left(\nu + \frac{1}{2}\right)\pi i\right]}{\sqrt{2\pi z}} \sum_{k=0}^{\infty} \frac{1}{(2z)^k} \frac{\Gamma\left(\nu + k + \frac{1}{2}\right)}{k!\,\Gamma\left(\nu - k + \frac{1}{2}\right)}$$

[The $+$ sign is taken for $-\frac{1}{2}\pi < \arg z < \frac{3}{2}\pi$, the $-$ sign for $-\frac{3}{2}\pi < \arg z < \frac{1}{2}\pi$][†] $\qquad$ (see **8.339** 4)]
$$\text{WA } 226(2,3)$$

6.[11] $$K_\nu(z) = \sqrt{\frac{\pi}{2z}}\, e^{-z} \left[ \sum_{k=0}^{n-1} \frac{1}{(2z)^k} \frac{\Gamma\left(\nu + k + \frac{1}{2}\right)}{k!\,\Gamma\left(\nu - k + \frac{1}{2}\right)} + \theta_3 \frac{\Gamma\left(\nu + n + \frac{1}{2}\right)}{(2z)^n n!\,\Gamma\left(\nu - n + \frac{1}{2}\right)} \right]$$
$$\text{(see } \mathbf{8.339}\ 4) \qquad\qquad \text{WA } 231,\, 245(9)$$

An estimate of the remainders of the asymptotic series in formulas **8.451**:

---

*An estimate of the remainders in formulas **8.451** is given in **8.451** 7 and **8.451** 8.

†The contradiction that this condition contains at first glance is explained by the so-called Stokes phenomenon (see Watson, G.N., *A Treatise on the Theory of Bessel Functions*, 2nd Edition, Cambridge Univ. Press, 1944, page 201).

7. $\quad |R_1| < \left| \dfrac{\Gamma\left(\nu + 2n + \frac{1}{2}\right)}{(2z)^{2n}(2n)!\,\Gamma\left(\nu - 2n + \frac{1}{2}\right)} \right| \qquad \left[ n > \dfrac{\nu}{2} - \dfrac{1}{4} \right] \qquad$ WA 231

8.$^{12}$ $\quad |R_2| < \left| \dfrac{\Gamma\left(\nu + 2n + \frac{3}{2}\right)}{(2z)^{2n+1}(2n+1)!\,\Gamma\left(\nu - 2n - \frac{1}{2}\right)} \right| \qquad \left[ n \geq \dfrac{\nu}{2} - \dfrac{3}{4} \right] \qquad$ WA 231

For $-\frac{\pi}{2} < \arg z < \frac{3}{2}\pi$, $\nu$ real, and $n + \frac{1}{2} > |\nu|$ $\qquad$ WA 245

$$|\theta_1| < \begin{cases} 1, & \text{if } \operatorname{Im} z \geq 0 \\ |\sec(\arg z)|, & \text{if } \operatorname{Im} z \leq 0 \end{cases}$$

For $-\frac{3}{2}\pi < \arg z < \frac{\pi}{2}$, $\nu$ real, and $n + \frac{1}{2} > |\nu|$ $\qquad$ WA 246

$$|\theta_2| < \begin{cases} 1, & \text{if } \operatorname{Im} z \leq 0 \\ |\sec(\arg z)|, & \text{if } \operatorname{Im} z \geq 0 \end{cases}$$

For $\nu$ real, $\qquad$ WA 245

$$|\theta_3| < \begin{cases} 1 \text{ and } \operatorname{Re} \theta_3 \geq 0 & \text{if } \operatorname{Re} z \geq 0 \\ |\operatorname{cosec}(\arg z)| & \text{if } \operatorname{Re} z < 0 \end{cases}$$

$$\operatorname{Re}\theta_3 \geq 0, \qquad \text{if } \operatorname{Re} z \geq 0$$

For $\nu$ and $z$ real and $n \geq \nu - \frac{1}{2}$, $\qquad$ WA 231

$$0 \leq |\theta_3| \leq 1$$

In particular, it follows from **8.451** 7 and **8.451** 8 that for real positive values of $z$ and $\nu$, the errors $|R_1|$ and $|R_2|$ are less than the absolute value of the first discarded term. For values of $|\arg z|$ close to $\pi$, the series **8.451** 1 and **8.451** 2 may not be suitable for calculations. In particular, the error for $|\arg z| > \pi$ can be greater in absolute value than the first discarded term.

### "Approximation by tangents"

**8.452**$^{11}$ $\quad$ For large values of the index (where the argument is less than the index).

Suppose that $x > 0$ and $\nu > 0$. Let us set $\nu/x = \cosh\alpha$. Then, for large values of $\nu$, the following expansions are valid:

1. $\quad J_\nu\left(\dfrac{\nu}{\cosh\alpha}\right) \sim \dfrac{\exp(\nu\tanh\alpha - \nu\alpha)}{\sqrt{2\nu\pi\tanh\alpha}} \left\{ 1 + \dfrac{1}{\nu}\left(\dfrac{1}{8}\coth\alpha - \dfrac{5}{24}\coth^3\alpha\right) \right.$
$\qquad\qquad \left. + \dfrac{1}{\nu^2}\left(\dfrac{9}{128}\coth^2\alpha - \dfrac{231}{576}\coth^4\alpha + \dfrac{1155}{3456}\coth^6\alpha\right) + \ldots \right\}$

$\qquad\qquad\qquad\qquad\qquad\qquad$ WA 269(3)

2.     $Y_\nu \left( \dfrac{\nu}{\cosh \alpha} \right) \sim -\dfrac{\exp \left( \nu\alpha - \nu \tanh \alpha \right)}{\sqrt{\frac{\pi}{2}\nu \tanh \alpha}} \left\{ 1 - \dfrac{1}{\nu} \left( \dfrac{1}{8} \coth \alpha - \dfrac{5}{24} \coth^3 \alpha \right) \right.$

$$\left. + \dfrac{1}{\nu^2} \left( \dfrac{9}{128} \coth^2 \alpha - \dfrac{231}{576} \coth^4 \alpha + \dfrac{1155}{3456} \coth^6 \alpha \right) + \ldots \right\}$$

WA 270(5)

**8.453**   For large values of the index (where the argument is greater than the index).

Suppose that $x > 0$ and $\nu > 0$. Let us set $\nu/x = \cos \beta$. Then, for large values of $\nu$, the following expansions are valid:

1.     $J_\nu \left( \nu \sec \beta \right) \sim \sqrt{\dfrac{2}{\nu\pi \tan \beta}} \left\{ \left[ 1 - \dfrac{1}{\nu^2} \left( \dfrac{9}{128} \cot^2 \beta + \dfrac{231}{576} \cot^4 \beta \right. \right. \right.$

$$\left. \left. + \dfrac{1155}{3456} \cot^6 \beta \right) + \ldots \right] \cos \left( \nu \tan \beta - \nu\beta - \dfrac{\pi}{4} \right)$$

$$\left. + \left[ \dfrac{1}{\nu} \left( \dfrac{1}{8} \cot \beta + \dfrac{5}{24} \cot^3 \beta \right) - \ldots \right] \sin \left( \nu \tan \beta - \nu\beta - \dfrac{\pi}{4} \right) \right\}$$

WA 271(4)

2.     $Y_\nu \left( \nu \sec \beta \right) \sim \sqrt{\dfrac{2}{\nu\pi \tan \beta}} \left\{ \left[ 1 - \dfrac{1}{\nu^2} \left( \dfrac{9}{128} \cot^2 \beta + \dfrac{231}{576} \cot^4 \beta \right. \right. \right.$

$$\left. \left. + \dfrac{1155}{3456} \cot^6 \beta \right) + \ldots \right] \sin \left( \nu \tan \beta - \nu\beta - \dfrac{\pi}{4} \right)$$

$$\left. - \left[ \dfrac{1}{\nu} \left( \dfrac{1}{8} \cot \beta + \dfrac{5}{24} \cot^3 \beta \right) - \ldots \right] \cos \left( \nu \tan \beta - \nu\beta - \dfrac{\pi}{4} \right) \right\}$$

WA 271(5)

3.     $H_\nu^{(1)} \left( \nu \sec \beta \right) \sim \dfrac{\exp \left[ \nu i \left( \tan \beta - \beta \right) - \frac{\pi}{4} i \right]}{\sqrt{\frac{\pi}{2}\nu \tan \beta}} \left\{ 1 - \dfrac{i}{\nu} \left( \dfrac{1}{8} \cot \beta + \dfrac{5}{24} \cot^3 \beta \right) \right.$

$$\left. - \dfrac{1}{\nu^2} \left( \dfrac{9}{128} \cot^2 \beta + \dfrac{231}{576} \cot^4 \beta + \dfrac{1155}{3456} \cot^6 \beta \right) + \ldots \right\}$$

WA 271(1)

4.     $H_\nu^{(2)} \left( \nu \sec \beta \right) \sim \dfrac{\exp \left[ -\nu i \left( \tan \beta - \beta \right) + \frac{\pi}{4} i \right]}{\sqrt{\frac{\pi}{2}\nu \tan \beta}} \left\{ 1 + \dfrac{i}{\nu} \left( \dfrac{1}{8} \cot \beta + \dfrac{5}{24} \cot^3 \beta \right) \right.$

$$\left. - \dfrac{1}{\nu^2} \left( \dfrac{9}{128} \cot^2 \beta + \dfrac{231}{576} \cot^4 \beta + \dfrac{1155}{3456} \cot^6 \beta \right) + \ldots \right\}$$

WA 271(2)

Formulas **8.453** are not valid when $|x - \nu|$ is of a size comparable to $x^{\frac{1}{3}}$. For arbitrary small (and also large) values of $|x - \nu|$, we may use the following formulas:

**8.454**   Suppose that $x > 0$ and $\nu > 0$, we set

$$w = \sqrt{\frac{x^2}{\nu^2} - 1};$$

Then,

1.    $H_\nu^{(1)}(x) = \dfrac{w}{\sqrt{3}} \exp\left\{\left[\dfrac{\pi}{6} + \nu\left(w - \dfrac{w^3}{3} - \arctan w\right)\right]i\right\} H_{\frac{1}{3}}^{(1)}\left(\dfrac{\nu}{3}w^3\right) + O\left(\dfrac{1}{|\nu|}\right)$

2.    $H_\nu^{(2)}(x) = \dfrac{w}{\sqrt{3}} \exp\left\{\left[-\dfrac{\pi}{6} - \nu\left(w - \dfrac{w^3}{3} - \arctan w\right)\right]i\right\} H_{\frac{1}{3}}^{(2)}\left(\dfrac{\nu}{3}w^3\right) + O\left(\dfrac{1}{|\nu|}\right)$      MO 34

The absolute value of the error $O\left(\dfrac{1}{|\nu|}\right)$ is then less than $24\sqrt{2}\left|\dfrac{1}{\nu}\right|$.

**8.455**   For $x$ real and $\nu$ a natural number $(\nu = n)$, if $n \gg 1$, the following approximations are valid:

1.[7]    $J_n(x) \approx \dfrac{1}{\pi}\sqrt{\dfrac{2(n - x)}{3x}} K_{\frac{1}{3}}\left\{\dfrac{[2(n-x)]^{\frac{3}{2}}}{3\sqrt{x}}\right\}$

         $[n > x]$      (see also **8.433**)

         WA 276(1)

$$\approx \dfrac{1}{2}e^{\frac{2}{3}\pi i}\sqrt{\dfrac{2(n-x)}{3x}} H_{\frac{1}{3}}^{(1)}\left\{\dfrac{i}{3}\dfrac{[2(n-x)]^{\frac{3}{2}}}{\sqrt{x}}\right\}$$

         $[n > x]$

         MO 34

$$\approx \dfrac{1}{\sqrt{3}}\sqrt{\dfrac{2(x-n)}{3x}}\left\{J_{\frac{1}{3}}\left[\dfrac{\{2(x-n)\}^{\frac{3}{2}}}{3\sqrt{x}}\right] + J_{-\frac{1}{3}}\left[\dfrac{\{2(x-n)\}^{\frac{3}{2}}}{3\sqrt{x}}\right]\right\}$$

         (see also **8.441** 3, **8.441** 4)

         WA 276(2)

2.[12]    $Y_n(x) \approx -\sqrt{\dfrac{2(x-n)}{3x}}\left\{J_{-\frac{1}{3}}\left[\dfrac{\{2(x-n)\}^{\frac{3}{2}}}{3\sqrt{x}}\right] - J_{\frac{1}{3}}\left[\dfrac{\{2(x-n)\}^{\frac{3}{2}}}{3\sqrt{x}}\right]\right\}$

         $[x > n]$      WA 276(3)

An estimate of the error in formulas **8.455** has not yet been achieved.

**8.456**[11] $J_\nu^2(z) + Y_\nu^2(z) \approx \dfrac{2}{\pi z}\displaystyle\sum_{k=0}^{\infty}\dfrac{(2k-1)!!}{2^k z^{2k}}\dfrac{\Gamma\left(\nu + k + \frac{1}{2}\right)}{k!\,\Gamma\left(\nu - k + \frac{1}{2}\right)}$   $[|\arg z| < \pi]$      (see also **8.479** 1)

         WA 250(5)

**8.457**    $J_\nu^2(x) + J_{\nu+1}^2(x) \approx \dfrac{2}{\pi x}$          $[x \gg |\nu|]$      WA 223

## 8.46   Bessel functions of order equal to an integer plus one-half

**The function $J_\nu(z)$**

**8.461**

1.[11]   $J_{n+\frac{1}{2}}(z) = \sqrt{\dfrac{2}{\pi z}} \left\{ \sin\left(z - \dfrac{\pi}{2}n\right) \displaystyle\sum_{k=0}^{\lfloor \frac{n}{2} \rfloor} \dfrac{(-1)^k (n+2k)!}{(2k)!(n-2k)!}(2z)^{-2k} \right.$

$\left. + \cos\left(z - \dfrac{\pi}{2}n\right) \displaystyle\sum_{k=0}^{\lfloor \frac{n-1}{2} \rfloor} \dfrac{(-1)^k (n+2k+1)!}{(2k+1)!(n-2k-1)!}(2z)^{-(2k+1)} \right\}$

               [$n+1$ is a natural number]       (cf. **8.451** 1)    KU 59(6), WA 66(2)

2.   $J_{-n-\frac{1}{2}}(z) = \sqrt{\dfrac{2}{\pi z}} \left\{ \cos\left(z + \dfrac{\pi}{2}n\right) \displaystyle\sum_{k=0}^{\lfloor \frac{n}{2} \rfloor} \dfrac{(-1)^k (n+2k)!}{(2k)!(n-2k)!(2z)^{2k}} \right.$

$\left. - \sin\left(z + \dfrac{\pi}{2}n\right) \displaystyle\sum_{k=0}^{\lfloor \frac{n-1}{2} \rfloor} \dfrac{(-1)^k (n+2k+1)!}{(2k+1)!(n-2k-1)!(2z)^{2k+1}} \right\}$

               [$n+1$ is a natural number]       (cf. **8.451** 1)    KU 58(7), WA 67(5)

**8.462**

1.   $J_{n+\frac{1}{2}}(z) = \dfrac{1}{\sqrt{2\pi z}} \left\{ e^{iz} \displaystyle\sum_{k=0}^{n} \dfrac{i^{-n+k-1}(n+k)!}{k!(n-k)!(2z)^k} + e^{-iz} \displaystyle\sum_{k=0}^{n} \dfrac{(-i)^{-n+k-1}(n+k)!}{k!(n-k)!(2z)^k} \right\}$

               [$n+1$ is a natural number]

               KU 59(6), WA 66(1)

2.   $J_{-n-\frac{1}{2}}(z) = \dfrac{1}{\sqrt{2\pi z}} \left\{ e^{iz} \displaystyle\sum_{k=0}^{n} \dfrac{i^{n+k}(n+k)!}{k!(n-k)!(2z)^k} + e^{-iz} \displaystyle\sum_{k=0}^{n} \dfrac{(-i)^{n+k}(n+k)!}{k!(n-k)!(2z)^k} \right\}$

               [$n+1$ is a natural number]

               KU 59(7), WA 67(4)

**8.463**

1.   $J_{n+\frac{1}{2}}(z) = (-1)^n z^{n+\frac{1}{2}} \sqrt{\dfrac{2}{\pi}} \dfrac{d^n}{(z\,dz)^n}\left(\dfrac{\sin z}{z}\right)$                               KU 58(4)

2.   $J_{-n-\frac{1}{2}}(z) = z^{n+\frac{1}{2}} \sqrt{\dfrac{2}{\pi}} \dfrac{d^n}{(z\,dz)^n}\left(\dfrac{\cos z}{z}\right)$                                  KU 58(5)

**8.464**   Special cases:

1.   $J_{\frac{1}{2}}(z) = \sqrt{\dfrac{2}{\pi z}} \sin z$                                                   DW

2.   $J_{-\frac{1}{2}}(z) = \sqrt{\dfrac{2}{\pi z}} \cos z$                                                   DW

3.     $J_{\frac{3}{2}}(z) = \sqrt{\dfrac{2}{\pi z}}\left(\dfrac{\sin z}{z} - \cos z\right)$        DW

4.     $J_{-\frac{3}{2}}(z) = \sqrt{\dfrac{2}{\pi z}}\left(-\sin z - \dfrac{\cos z}{z}\right)$        DW

5.[8]     $J_{\frac{5}{2}}(z) = \sqrt{\dfrac{2}{\pi z}}\left\{\left(\dfrac{3}{z^2} - 1\right)\sin z - \dfrac{3}{z}\cos z\right\}$        DW

6.     $J_{-\frac{5}{2}}(z) = \sqrt{\dfrac{2}{\pi z}}\left\{\dfrac{3}{z}\sin z + \left(\dfrac{3}{z^2} - 1\right)\cos z\right\}$        DW

## The function $Y_{n+\frac{1}{2}}(z)$

**8.465**

1.     $Y_{n+\frac{1}{2}}(z) = (-1)^{n-1} J_{-n-\frac{1}{2}}(z)$        JA

2.     $Y_{-n-\frac{1}{2}}(z) = (-1)^{n} J_{n+\frac{1}{2}}(z)$        JA

## The functions $H^{(1,2)}_{n-\frac{1}{2}}(z),\ I_{n+\frac{1}{2}}(z),\ K_{n+\frac{1}{2}}(z)$

**8.466**

1.     $H^{(1)}_{n-\frac{1}{2}}(z) = \sqrt{\dfrac{2}{\pi z}}\, i^{-n} e^{iz} \sum_{k=0}^{n-1} (-1)^k \dfrac{(n+k-1)!}{k!(n-k-1)!}\dfrac{1}{(2iz)^k}$

$$\text{(cf. } \mathbf{8.451}\ 3)$$

2.     $H^{(2)}_{n-\frac{1}{2}}(z) = \sqrt{\dfrac{2}{\pi z}}\, i^{n} e^{-iz} \sum_{k=0}^{n-1} \dfrac{(n+k-1)!}{k!(n-k-1)!}\dfrac{1}{(2iz)^k}$    (cf. $\mathbf{8.451}\ 4$)

**8.467**    $I_{\pm(n+\frac{1}{2})}(z) = \dfrac{1}{\sqrt{2\pi z}}\left[ e^{z}\sum_{k=0}^{n}\dfrac{(-1)^k(n+k)!}{k!(n-k)!(2z)^k} \pm (-1)^{n+1} e^{-z}\sum_{k=0}^{n}\dfrac{(n+k)!}{k!(n-k)!(2z)^k}\right]$

$$\text{(cf. } \mathbf{8.451}\ 5) \qquad\qquad \text{KU 60a}$$

**8.468**    $K_{n+\frac{1}{2}}(z) = \sqrt{\dfrac{\pi}{2z}}\, e^{-z} \sum_{k=0}^{n}\dfrac{(n+k)!}{k!(n-k)!(2z)^k}$    (cf. $\mathbf{8.451}\ 6$)        KU 60

**8.469**    Special cases:

1.     $Y_{\frac{1}{2}}(z) = -\sqrt{\dfrac{2}{\pi z}}\cos z$

2.     $Y_{-\frac{1}{2}}(z) = \sqrt{\dfrac{2}{\pi z}}\sin z$

3.     $K_{\pm\frac{1}{2}}(z) = \sqrt{\dfrac{\pi}{2z}}\, e^{-z}$        WA 95(13)

4.     $H^{(1)}_{\frac{1}{2}}(z) = \sqrt{\dfrac{2}{\pi z}}\dfrac{e^{iz}}{i}$        MO 27

5.       $H^{(2)}_{\frac{1}{2}}(z) = \sqrt{\dfrac{2}{\pi z}} \dfrac{e^{-iz}}{-i}$                                           MO 27

6.       $H^{(1)}_{-\frac{1}{2}}(z) = \sqrt{\dfrac{2}{\pi z}} e^{iz}$                                           MO 27

7.       $H^{(2)}_{-\frac{1}{2}}(z) = \sqrt{\dfrac{2}{\pi z}} e^{-iz}$                                          MO 27

## 8.47–8.48   Functional relations

**8.471**[8]   Recursion formulas:

1.       $z\, Z_{\nu-1}(z) + z\, Z_{\nu+1}(z) = 2\nu\, Z_\nu(z)$            KU 56(13), WA 56(1), WA 79(1), WA 88(3)

2.       $Z_{\nu-1}(z) - Z_{\nu+1}(z) = 2\dfrac{d}{dz} Z_\nu(z)$            KU 56(12), WA 56(2), WA 79(2), We 88(4)

     Sonin and Nielsen, in their construction of the theory of Bessel functions, defined Bessel functions as analytic functions of $z$ that satisfy the recursion relations **8.471**. $Z$ denotes $J$, $N$, $H^{(1)}$, $H^{(2)}$ or any linear combination of these functions, the coefficients in which are independent of $z$ and $\nu$.

**8.472**   Consequences of the recursion formulas for $Z$ defined as above:

1.       $z\dfrac{d}{dz} Z_\nu(z) + \nu\, Z_\nu(z) = z\, Z_{\nu-1}(z)$            KU 56(11), WA 56(3), WA 79(3), WA 88(5)

2.       $z\dfrac{d}{dz} Z_\nu(z) - \nu\, Z_\nu(z) = -z\, Z_{\nu+1}(z)$            KU 56(10), WA 56(4), WA 79(4), WA 88(6)

3.       $\left(\dfrac{d}{z\,dz}\right)^m (z^\nu\, Z_\nu(z)) = z^{\nu-m}\, Z_{\nu-m}(z)$            KU 56(8), WA 57(5), WA 89(9)

4.       $\left(\dfrac{d}{z\,dz}\right)^m (z^{-\nu}\, Z_\nu(z)) = (-1)^m z^{-\nu-m}\, Z_{\nu+m}(z)$            WA 89(10), Ku 55(5), WA 57(6)

5.       $Z_{-n}(z) = (-1)^n Z_n(z)$            [$n$ is a natural number]     (cf. **8.404**)

**8.473**   Special cases:

1.       $J_2(z) = \dfrac{2}{z} J_1(z) - J_0(z)$

2.       $Y_2(z) = \dfrac{2}{z} Y_1(z) - Y_0(z)$

3.       $H^{(1,2)}_2(z) = \dfrac{2}{z} H^{(1,2)}_1(z) - H^{(1,2)}_0(z)$

4.       $\dfrac{d}{dz} J_0(z) = -J_1(z)$

5.       $\dfrac{d}{dz} Y_0(z) = -Y_1(z)$

6.       $\dfrac{d}{dz} H^{(1,2)}_0(z) = -H^{(1,2)}_1(z)$

**8.474**[12]   Each of the pairs of functions $J_\nu(z)$ and $J_{-\nu}(z)$ (for $\nu \neq 0, \pm 1, \pm 2, \dots$), $J_\nu(z)$ and $Y_\nu(z)$, and $H^{(1)}_\nu(z)$ and $H^{(2)}_\nu(z)$, which are solutions of equation **8.401**, and also the pair $I_\nu(z)$ and $K_\nu(z)$ is a pair of linearly independent functions. The Wronskians of these pairs are, respectively,

$$\frac{2}{\pi z}\sin\nu\pi, \quad \frac{2}{\pi z}, \quad -\frac{4i}{\pi z}, \quad -\frac{1}{z} \qquad\qquad \text{KU 52(10, 11, 12), WA 90(1, 4)}$$

**8.475**[6]    The functions $J_\nu(z)$, and $Y_\nu(z)$, $H_\nu^{(1,2)}(z)$, $I_\nu(z)$, $K_\nu(z)$ with the exception of $J_n(z)$ and $I_n(z)$ for $n$ an integer are *non-single-valued*: $z = 0$ is a branch point for these functions. The branches of these functions that lie on opposite sides of the cut $(-\infty, 0)$ are connected by the relations

**8.476**

1.     $J_\nu\left(e^{m\pi i}z\right) = e^{m\nu\pi i}\,J_\nu(z)$                            WA 90(1)

2.     $Y_\nu\left(e^{m\pi i}z\right) = e^{-m\nu\pi i}\,Y_\nu(z) + 2i\sin m\nu\pi\cot\nu\pi\,J_\nu(z)$       WA 90(3)

3.     $Y_{-\nu}\left(e^{m\pi i}z\right) = e^{-m\nu\pi i}\,Y_{-\nu}(z) + 2i\sin m\nu\pi\operatorname{cosec}\nu\pi\,J_\nu(z)$     WA 90(4)

4.     $I_\nu\left(e^{m\pi i}z\right) = e^{m\nu\pi i}\,I_\nu(z)$                           WA 95(17)

5.     $K_\nu\left(e^{m\pi i}z\right) = e^{-m\nu\pi i}\,K_\nu(z) - i\pi\dfrac{\sin m\nu\pi}{\sin\nu\pi}\,I_\nu(z)$     [$\nu$ not an integer]     WA 95(18)

6.     $H_\nu^{(1)}\left(e^{m\pi i}z\right) = e^{-m\nu\pi i}\,H_\nu^{(1)}(z) - 2e^{-\nu\pi i}\dfrac{\sin m\nu\pi}{\sin\nu\pi}\,J_\nu(z)$

$$= \frac{\sin(1-m)\nu\pi}{\sin\nu\pi}\,H_\nu^{(1)}(z) - e^{-\nu\pi i}\frac{\sin m\nu\pi}{\sin\nu\pi}\,H_\nu^{(2)}(z)$$

                                                 WA 95(5)

7.     $H_\nu^{(2)}\left(e^{m\pi i}z\right) = e^{-m\nu\pi i}\,H_\nu^{(2)}(z) + 2e^{\nu\pi i}\dfrac{\sin m\nu\pi}{\sin\nu\pi}\,J_\nu(z)$

$$= \frac{\sin(1+m)\nu\pi}{\sin\nu\pi}\,H_\nu^{(2)}(z) + e^{\nu\pi i}\frac{\sin m\nu\pi}{\sin\nu\pi}\,H_\nu^{(1)}(z)$$

                                    [$m$ an integer]     WA 90(6)

8.     $H_\nu^{(1)}\left(e^{i\pi}z\right) = -H_{-\nu}^{(2)}(z) = -e^{-i\pi\nu}\,H_\nu^{(2)}(z)$            MO 26

9.     $H_\nu^{(2)}\left(e^{-i\pi}z\right) = -H_{-\nu}^{(1)}(z) = -e^{i\pi\nu}\,H_\nu^{(1)}(z)$           MO 26

10.[8]    $\overline{H}_\nu^{(2)}(z) = H_{\overline{\nu}}^{(1)}(\overline{z})$                                    MO 26

**8.477**

1.     $J_\nu(z)\,Y_{\nu+1}(z) - J_{\nu+1}(z)\,Y_\nu(z) = -\dfrac{2}{\pi z}$          WA 91(12)

2.[12]   $I_\nu(z)\,K_{\nu+1}(z) + J_{\nu+1}(z)\,K_\nu(z) = \dfrac{1}{z}$          WA 95(20)

See also **3.863**.

- For a connection with Legendre functions, see **8.722**.
- For a connection with the polynomials $C_n^\lambda(t)$, see **8.936** 4.
- For a connection with a confluent hypergeometric function, see **9.235**.

**8.478**    For $\nu > 0$ and $x > 0$, the product

$$x\left[J_\nu^2(x) + Y_\nu^2(x)\right],$$

considered as a function of $x$, decreases monotonically, if $\nu > \frac{1}{2}$ and increases monotonically if $0 < \nu < \frac{1}{2}$.

                                                               MO 35

## 8.479

1.[11]    $\dfrac{1}{\sqrt{x^2 - \nu^2}} > \dfrac{\pi}{2}\left[J_\nu^2(x) + Y_\nu^2(x)\right] \geq \dfrac{1}{x}$       $\left[x \geq \nu \geq \frac{1}{2}\right]$       MO 35

2.    $|J_n(nz)| \leq 1$

$$\left[\left|\frac{z \exp\sqrt{1 - z^2}}{1 + \sqrt{1 - z^2}}\right| < 1, n \text{ a natural number}\right] \quad \text{MO 35}$$

## Relations between Bessel functions of the first, second, and third kinds

**8.481**   $J_\nu(z) = \dfrac{Y_{-\nu}(z) - Y_\nu(z)\cos\nu\pi}{\sin\nu\pi} = H_\nu^{(1)}(z) - i\,Y_\nu(z)$

$$= H_\nu^{(2)}(z) + i\,Y_\nu(z) = \frac{1}{2}\left(H_\nu^{(1)}(z) + H_\nu^{(2)}(z)\right)$$

                                (cf. **8.403** 1, **8.405**)       WA 89(1), JA

**8.482**   $Y_\nu(z) = \dfrac{J_\nu(z)\cos\nu\pi - J_{-\nu}(z)}{\sin\nu\pi} = i\,J_\nu(z) - i\,H_\nu^{(1)}(z)$

$$= i\,H_\nu^{(2)}(z) - i\,J_\nu(z) = \frac{i}{2}\left(H_\nu^{(2)}(z) - H_\nu^{(1)}(z)\right)$$

                                (cf. **8.403** 1, **8.405**)       WA 89(3), JA

## 8.483

1.    $H_\nu^{(1)}(z) = \dfrac{J_{-\nu}(z) - e^{-\nu\pi i}\,J_\nu(z)}{i\sin\nu\pi} = \dfrac{Y_{-\nu}(z) - e^{-\nu\pi i}\,Y_\nu(z)}{\sin\nu\pi} = J_\nu(z) + i\,Y_\nu(z)$       WA 89(5)

2.    $H_\nu^{(2)}(z) = \dfrac{e^{\nu\pi i}\,J_\nu(z) - J_{-\nu}(z)}{i\sin\nu\pi} = \dfrac{Y_{-\nu}(z) - e^{\nu\pi i}\,Y_\nu(z)}{\sin\nu\pi} = J_\nu(z) - i\,Y_\nu(z)$

                                (cf. **8.405**)       WA 89(6)

## 8.484

1.    $H_{-\nu}^{(1)}(z) = e^{\nu\pi i}\,H_\nu^{(1)}(z)$       WA 89(7)

2.    $H_{-\nu}^{(2)}(z) = e^{-\nu\pi i}\,H_\nu^{(2)}(z)$       WA 89(7)

**8.485**[7]   $K_\nu(z) = \dfrac{\pi}{2}\dfrac{I_{-\nu}(z) - I_\nu(z)}{\sin\nu\pi}$       [$\nu$ not an integer]       (see also **8.407**)

                                                                   WA 92(6)

**8.486**   Recursion formulas for the functions $I_\nu(z)$ and $K_\nu(z)$ and their consequences:

1.    $z\,I_{\nu-1}(z) - z\,I_{\nu+1}(z) = 2\nu\,I_\nu(z)$       WA 93(1)

2.    $I_{\nu-1}(z) + I_{\nu+1}(z) = 2\dfrac{d}{dz}\,I_\nu(z)$       WA 93(2)

3.    $z\dfrac{d}{dz}\,I_\nu(z) + \nu\,I_\nu(z) = z\,I_{\nu-1}(z)$       WA 93(3)

4.    $z\dfrac{d}{dz}\,I_\nu(z) - \nu\,I_\nu(z) = z\,I_{\nu+1}(z)$       WA 93(4)

5.    $\left(\dfrac{d}{z\,dz}\right)^m\{z^\nu\,I_\nu(z)\} = z^{\nu-m}\,I_{\nu-m}(z)$       WA 93(5)

6.    $\left(\dfrac{d}{z\,dz}\right)^m \{z^{-\nu} I_\nu(z)\} = z^{-\nu-m} I_{\nu+m}(z)$     WA 93(6)

7.[12]    $I_{-n}(z) = I_n(z)$     [$n$ a natural number]     WA 93(8)

8.[12]    $I_2(z) = -\dfrac{2}{z} I_1(z) + I_0(z)$

9.    $\dfrac{d}{dz} I_0(z) = I_1(z)$     WA 93(7)

10.    $z\,K_{\nu-1}(z) - z\,K_{\nu+1}(z) = -2\nu\,K_\nu(z)$     WA 93(1)

11.    $K_{\nu-1}(z) + K_{\nu+1}(z) = -2\dfrac{d}{dz} K_\nu(z)$     WA 93(2)

12.    $z\dfrac{d}{dz} K_\nu(z) + \nu\,K_\nu(z) = -z\,K_{\nu-1}(z)$     WA 93(3)

13.    $z\dfrac{d}{dz} K_\nu(z) - \nu\,K_\nu(z) = -z\,K_{\nu+1}(z)$     WA 93(4)

14.    $\left(\dfrac{d}{z\,dz}\right)^m \{z^\nu K_\nu(z)\} = (-1)^m z^{\nu-m} K_{\nu-m}(z)$     WA 93(5)

15.    $\left(\dfrac{d}{z\,dz}\right)^m \{z^{-\nu} K_\nu(z)\} = (-1)^m z^{-\nu-m} K_{\nu+m}(z)$     WA 93(6)

16.    $K_{-\nu}(z) = K_\nu(z)$     WA 93(8)

17.    $K_2(z) = \dfrac{2}{z} K_1(z) + K_0(z)$

18.    $\dfrac{d}{dz} K_0(z) = -K_1(z)$     WA 93(7)

19.    $\dfrac{\partial J_\nu(z)}{\partial \nu} = \left[\ln \dfrac{z}{2} - \psi(\nu+1)\right] J_\nu(z) + \dfrac{(z/2)^{\nu+1}}{\Gamma(\nu+1)} \sum_{n=0}^{\infty} \dfrac{(z/2)^n J_{n+1}(z)}{n!(\nu+n+1)^2}$     LUKE 360

**8.486(1)**[7]    Differentiation with respect to order

1.    $\dfrac{\partial J_\nu(z)}{\partial \nu} = J_\nu(z) \ln\left(\dfrac{1}{2}z\right) - \sum_{k=0}^{\infty} (-1)^k \left(\dfrac{1}{2}z\right)^{\nu+2k} \dfrac{\psi(\nu+k+1)}{k!\,\Gamma(\nu+k+1)}$

$\left[\nu \neq n \text{ or } n+\tfrac{1}{2}, \quad n \text{ integer}\right]$     MS 3.1.3

2.    $\dfrac{\partial J_{-\nu}(z)}{\partial \nu} = -J_{-\nu}(z) \ln\left(\dfrac{1}{2}z\right) + \sum_{k=0}^{\infty} (-1)^k \left(\dfrac{1}{2}z\right)^{-\nu+2k} \dfrac{\psi(-\nu+k+1)}{k!\,\Gamma(-\nu+k+1)}$

$\left[\nu \neq n \text{ or } n+\tfrac{1}{2}, \quad n \text{ integer}\right]$     MS 3.1.3

3.    $\dfrac{\partial Y_\nu(z)}{\partial \nu} = \cot \pi\nu \dfrac{\partial J_\nu(z)}{\partial \nu} - \operatorname{cosec} \pi\nu \dfrac{\partial J_{-\nu}(z)}{\partial \nu} - \pi \operatorname{cosec} \pi\nu\, Y_\nu(z)$

$\left[\nu \neq n \text{ or } n+\tfrac{1}{2}, \quad n \text{ integer}\right]$     MS 3.1.3

4.[12]    $\dfrac{\partial I_\nu(z)}{\partial \nu} = I_\nu(z) \ln\left(\dfrac{1}{2}z\right) - \sum_{k=0}^{\infty} \left(\dfrac{1}{2}z\right)^{\nu+2k} \dfrac{\psi(\nu+k+1)}{k!\,\Gamma(\nu+k+1)}$     MS 3.1.3

5.$^{12}$   $\dfrac{\partial K_\nu(z)}{\partial \nu} = -\pi \cot \pi\nu \, K_\nu(z) + \dfrac{1}{2}\pi \operatorname{cosec} \pi\nu \left[ \dfrac{\partial I_{-\nu}(z)}{\partial \nu} - \dfrac{\partial I_\nu(z)}{\partial \nu} \right]$

$$[\nu \neq n, \quad n \text{ integer}] \qquad \text{MS 3.1.3}$$

6.   $\left[ \dfrac{\partial J_\nu(z)}{\partial \nu} \right]_{\nu = \pm n} = \dfrac{1}{2}\pi \, (\pm 1)^n \, Y_n(z) \pm (\pm 1)^n \dfrac{1}{2} n! \sum_{k=0}^{n-1} \dfrac{\left(\frac{1}{2}z\right)^{k-n} J_k(z)}{k!(n-k)} \qquad [n = 0, 1, \ldots] \qquad \text{MS 3.2.3}$

7.   $\left[ \dfrac{\partial Y_\nu(z)}{\partial \nu} \right]_{\nu = \pm n} = -\dfrac{1}{2}\pi \, (\pm 1)^n \, J_n(z) \pm (\pm 1)^n \dfrac{1}{2} n! \sum_{k=0}^{n-1} \dfrac{\left(\frac{1}{2}z\right)^{k-n} Y_k(z)}{k!(n-k)} \qquad [n = 0, 1, \ldots]$

$$\text{MS 3.2.3}$$

8.   $\left[ \dfrac{\partial I_\nu(z)}{\partial \nu} \right]_{\nu = \pm n} = (-1)^{n+1} K_n(z) \pm (-1)^n \dfrac{1}{2} n! \sum_{k=0}^{n-1} \dfrac{(-1)^k \left(\frac{1}{2}z\right)^{k-n} I_k(z)}{k!(n-k)} \qquad [n = 0, 1, \ldots]$

$$\text{MS 3.2.3}$$

9.   $\left[ \dfrac{\partial K_\nu(z)}{\partial \nu} \right]_{\nu = \pm n} = \pm \dfrac{1}{2} n! \sum_{k=0}^{n-1} \dfrac{\left(\frac{1}{2}z\right)^{k-n} K_k(z)}{k!(n-k)} \qquad [n = 0, 1, \ldots] \qquad \text{MS 3.2.3}$

10.   $(-1)^n \left[ \dfrac{\partial}{\partial \nu} I_\nu(z) \right]_{\nu = n} = - K_n(z) + \dfrac{1}{2} n! \sum_{k=0}^{n-1} \dfrac{(-1)^k \left(\frac{1}{2}z\right)^{k-n} I_k(z)}{k!(n-k)}$

$$[n = 0, 1, \ldots] \qquad \text{AS 9.6.44}$$

11.$^{11}$   $\left[ \dfrac{\partial K_\nu(z)}{\partial \nu} \right]_{\nu = n} = \dfrac{1}{2} n! \sum_{k=0}^{n-1} \dfrac{\left(\frac{1}{2}z\right)^{k-n} K_k(z)}{k!(n-k)} \qquad [n = 0, 1, \ldots] \qquad \text{AS 9.6.45}$

Special cases

12.   $\left[ \dfrac{\partial J_\nu(z)}{\partial \nu} \right]_{\nu = 0} = \frac{1}{2}\pi \, Y_0(z) \qquad\qquad\qquad\qquad \text{MS 3.2.3}$

13.   $\left[ \dfrac{\partial Y_\nu(z)}{\partial \nu} \right]_{\nu = 0} = -\frac{1}{2}\pi \, J_0(z) \qquad\qquad\qquad\qquad \text{MS 3.2.3}$

14.   $\left[ \dfrac{\partial I_\nu(z)}{\partial \nu} \right]_{\nu = 0} = - K_0(z) \qquad\qquad\qquad\qquad\quad \text{MS 3.2.3}$

15.   $\left[ \dfrac{\partial K_\nu(z)}{\partial \nu} \right]_{\nu = 0} = 0 \qquad\qquad\qquad\qquad\qquad\quad \text{MS 3.2.3}$

16.   $\left[ \dfrac{\partial J_\nu(x)}{\partial \nu} \right]_{\nu = \frac{1}{2}} = \left(\frac{1}{2}\pi x\right)^{-1/2} \left[\sin x \operatorname{Ci}(3x) - \cos x \operatorname{Si}(2x)\right] \qquad \text{MS 3.3.3}$

17.   $\left[ \dfrac{\partial J_\nu(x)}{\partial \nu} \right]_{\nu = -\frac{1}{2}} = \left(\frac{1}{2}\pi x\right)^{-1/2} \left[\cos x \operatorname{Ci}(2x) + \sin x \operatorname{Si}(2x)\right] \qquad \text{MS 3.3.3}$

18.   $\left[ \dfrac{\partial Y_\nu(x)}{\partial \nu} \right]_{\nu = \frac{1}{2}} = \left(\frac{1}{2}\pi x\right)^{-1/2} \left\{\cos x \operatorname{Ci}(2x) + \sin x \left[\operatorname{Si}(2x) - \pi\right]\right\} \qquad \text{MS 3.3.3}$

19.   $\left[ \dfrac{\partial Y_\nu(x)}{\partial \nu} \right]_{\nu = -\frac{1}{2}} = - \left(\frac{1}{2}\pi x\right)^{-1/2} \left\{\sin x \operatorname{Ci}(2x) - \cos x \left[\operatorname{Si}(2x) - \pi\right]\right\} \qquad \text{MS 3.3.3}$

$20.^{12}$ $\left[\dfrac{\partial I_\nu(x)}{\partial \nu}\right]_{\nu=\pm\frac{1}{2}}$

$$= (2\pi x)^{-1/2}\left[e^x\,\mathrm{Ei}(-2x) \mp e^{-x}\overline{\mathrm{Ei}}(2x)\right]$$

$$= \left(\frac{x}{2}\right)^{\pm 1/2}\sum_{k=0}^{\infty}\frac{(x^2/4)^k}{k!\,\Gamma(1\pm\frac{1}{2}+k)}\left[\ln\frac{x}{2} - \Psi\left(1\pm\frac{1}{2}+k\right)\right]$$

$$= \frac{\pi}{2}\left(\frac{x}{2}\right)^{-1/2}\sum_{k=0}^{\infty}\frac{(x^2/4)^k}{k!\,\Gamma(\frac{1}{2}+k)}\left[\Psi\left(\frac{1}{2}+k\right) - \ln\frac{x}{2}\right] + \frac{\pi}{2}\left(\frac{x}{2}\right)^{1/2}\sum_{k=0}^{\infty}\frac{(x^2/4)^k}{k!\,\Gamma(\frac{3}{2}+k)}\left[\Psi\left(\frac{3}{2}+k\right) - \ln\frac{x}{2}\right]$$

MS 3.3.3

21.  $\left[\dfrac{\partial K_\nu(x)}{\partial \nu}\right]_{\nu=\pm\frac{1}{2}} = \mp\left(\dfrac{\pi}{2x}\right)^{\frac{1}{2}}e^x\,\mathrm{Ei}(-2x)$

MS 3.3.3

**8.487**  Continuity with respect to the order*:

1.  $\displaystyle\lim_{\nu\to n} Y_\nu(z) = Y_n(z)$            [$n$ an integer]      WA 76

2.  $\displaystyle\lim_{\nu\to n} H_\nu^{(1,2)}(z) = H_n^{(1,2)}(z)$     [$n$ an integer]      WA 183

3.  $\displaystyle\lim_{\nu\to n} K_\nu(z) = K_n(z)$            [$n$ an integer]      WA 92

## 8.49  Differential equations leading to Bessel functions

See also **8.401**

**8.491**

1.  $\dfrac{1}{z}\dfrac{d}{dz}(zu') + \left(\beta^2 - \dfrac{\nu^2}{z^2}\right)u = 0$        $u = Z_\nu(\beta z)$      JA

2.  $\dfrac{1}{z}\dfrac{d}{dz}(zu') + \left[(\beta\gamma z^{\gamma-1})^2 - \left(\dfrac{\nu\gamma}{z}\right)^2\right]u = 0$    $u = Z_\nu(\beta z^\gamma)$      JA

3.  $u'' + \dfrac{1-2\alpha}{z}u' + \left[(\beta\gamma z^{\gamma-1})^2 + \dfrac{\alpha^2 - \nu^2\gamma^2}{z^2}\right]u = 0$    $u = z^\alpha Z_\nu(\beta z^\gamma)$      JA

4.  $u'' + \left[(\beta\gamma z^{\gamma-1})^2 - \dfrac{4\nu^2\gamma^2 - 1}{4z^2}\right]u = 0$    $u = \sqrt{z}\,Z_\nu(\beta z^\gamma)$      JA

5.  $u'' + \left(\beta^2 - \dfrac{4\nu^2 - 1}{4z^2}\right)u = 0$    $u = \sqrt{z}\,Z_\nu(\beta z)$      JA

6.  $u'' + \dfrac{1-2\alpha}{z}u' + \left(\beta^2 + \dfrac{\alpha^2 - \nu^2}{z^2}\right)u = 0$    $u = z^\alpha Z_\nu(\beta z)$      JA

7.  $u'' + bz^m u = 0$    $u = \sqrt{z}\,Z_{\frac{1}{m+2}}\left(\dfrac{2\sqrt{b}}{m+2}z^{\frac{m+2}{2}}\right)$

JA 111(5)

---

*The continuity of the functions $J_\nu(z)$ and $I_\nu(z)$ follows directly from the series representations of these functions.

8.　　$u'' + \dfrac{1}{z}u' + 4\left(z^2 - \dfrac{\nu^2}{z^2}\right)u = 0$　　　　　　　　$u = Z_\nu\left(z^2\right)$　　　　WA 111(6)

9.　　$u'' + \dfrac{1}{z}u' + \dfrac{1}{4z}\left(1 - \dfrac{\nu^2}{z}\right)u = 0$　　　　　　　$u = Z_\nu\left(\sqrt{z}\right)$　　　　WA 111(7)

10.　　$u'' + \dfrac{1-\nu}{z}u' + \dfrac{1}{4}\dfrac{u}{z} = 0$　　　　　　　$u = z^{\frac{\nu}{2}}Z_\nu\left(\sqrt{z}\right)$　　　WA 111(9)a

11.　　$u'' + \beta^2\gamma^2 z^{2\beta-2}u = 0$　　　　　　　　$u = z^{1/2}Z_{\frac{1}{2\beta}}\left(\gamma z^\beta\right)$　　WA 110(3)

12.　　$z^2 u'' + (2\alpha - 2\beta\nu + 1)zu' + \left[\beta^2\gamma^2 z^{2\beta} + \alpha(\alpha - 2\beta\nu)\right]u = 0$

　　　　　　　　　　　　　　　　　　　$u = z^{\beta\nu-\alpha}Z_\nu\left(\gamma z^\beta\right)$　　WA 112(21)

**8.492**

1.　　$u'' + \left(e^{2z} - \nu^2\right)u = 0$　　　　　　　　$u = Z_\nu\left(e^z\right)$　　　　WA 112(22)

2.　　$u'' + \dfrac{e^{2/z} - \nu^2}{z^4}u = 0$　　　　　　　$u = z\,Z_\nu\left(e^{1/z}\right)$　　WA 112(22)

**8.493**

1.　　$u'' + \left(\dfrac{1}{z} - 2\tan z\right)u' - \left(\dfrac{\nu^2}{z^2} + \dfrac{\tan z}{z}\right)u = 0$　　　$u = \sec z\,Z_\nu(z)$　　JA

2.　　$u'' + \left(\dfrac{1}{z} + 2\cot z\right)u' - \left(\dfrac{\nu^2}{z^2} - \dfrac{\cot z}{z}\right)u = 0$　　　$u = \operatorname{cosec} z\,Z_\nu(z)$　　JA

**8.494**

1.　　$u'' + \dfrac{1}{z}u' - \left(1 + \dfrac{\nu^2}{z^2}\right)u = 0$　　　　$u = Z_\nu(iz) = C_1\,I_\nu(z) + C_2\,K_\nu(z)$　　JA

2.　　$u'' + \dfrac{1}{z}u' - \left[\dfrac{1}{z} + \left(\dfrac{\nu}{2z}\right)^2\right]u = 0$　　　　$u = Z_\nu\left(2i\sqrt{z}\right)$　　JA

3.　　$u'' + u' + \dfrac{1}{z^2}\left(\dfrac{1}{4} - \nu^2\right)u = 0$　　　　$u = \sqrt{z}\,e^{-\frac{z}{2}}Z_\nu\left(\dfrac{iz}{2}\right)$　　JA

4.[12]　　$u'' + \left(\dfrac{2\nu+1}{z} - k\right)u' - \dfrac{2\nu+1}{2z}ku = 0$　　　　$u = z^{-\nu}e^{\frac{1}{2}kz}Z_\nu\left(\dfrac{ikz}{2}\right)$　　JA

5.　　$u'' + \dfrac{1-\nu}{z}u' - \dfrac{1}{4}\dfrac{u}{z} = 0$　　　　$u = z^{\frac{\nu}{2}}Z_\nu\left(i\sqrt{z}\right)$　　WA 111(8)

6.　　$u'' \pm \dfrac{u}{\sqrt{z}} = 0$

　　　　　　　　$u = \sqrt{z}\,Z_{\frac{2}{3}}\left(\dfrac{4}{3}z^{\frac{3}{4}}\right),\qquad u = \sqrt{z}\,Z_{\frac{2}{3}}\left(\dfrac{4}{3}iz^{\frac{3}{4}}\right)$　　WA 111(10)

7.　　$u'' \pm zu = 0$

　　　　　　　　$u = \sqrt{z}\,Z_{\frac{1}{3}}\left(\dfrac{2}{3}z^{\frac{3}{2}}\right),\qquad u = \sqrt{z}\,Z_{\frac{1}{3}}\left(\dfrac{2}{3}iz^{\frac{3}{2}}\right)$　　WA 111(10)

8.　　$u'' - \left(c^2 + \dfrac{\nu(\nu+1)}{z^2}\right)u = 0$　　　　　　$u = \sqrt{z}\,Z_{\nu+\frac{1}{2}}(icz)$　　WA 108(1)

9.  $\quad u'' - \dfrac{2\nu}{z}u' - c^2 u = 0$ $\qquad\qquad\qquad\qquad u = z^{\nu+\frac{1}{2}}\, Z_{\nu+\frac{1}{2}}(icz)$ $\qquad$ WA 109(3, 4)

10. $\quad u'' - c^2 z^{2\nu-2} u = 0$ $\qquad\qquad\qquad\qquad u = \sqrt{z}\, Z_{\frac{1}{2\nu}}\left(i\dfrac{c}{\nu}z^\nu\right)$ $\qquad$ WA 109(5, 6)

**8.495**

1.  $\quad u'' + \dfrac{1}{z}u' + \left(i - \dfrac{\nu^2}{z^2}\right)u = 0$ $\qquad\qquad u = Z_\nu\left(z\sqrt{i}\right)$ $\qquad\qquad\qquad$ JA

2.  $\quad u'' + \left(\dfrac{1}{z}\mp 2i\right)u' - \left(\dfrac{\nu^2}{z^2}\pm\dfrac{i}{z}\right)u = 0$ $\qquad u = e^{\pm iz}\, Z_\nu(z)$ $\qquad\qquad\qquad$ JA

3.  $\quad u'' + \dfrac{1}{z}u' + se^{i\alpha}u = 0$ $\qquad\qquad\qquad u = Z_0\left(\sqrt{sz}\,e^{\frac{i}{2}\alpha}\right)$ $\qquad\qquad$ JA

4.  $\quad u'' + \left(se^{i\alpha} + \dfrac{1}{4z^2}\right)u = 0$ $\qquad\qquad u = \sqrt{z}\, Z_0\left(\sqrt{sz}\,e^{\frac{i}{2}\alpha}\right)$ $\qquad$ JA

**8.496**

1.  $\quad \dfrac{d^2}{dz^2}\left(z^4\dfrac{d^2u}{dz^2}\right) - z^2 u = 0$ $\qquad\qquad u = \dfrac{1}{z}\left\{Z_2\left(2\sqrt{z}\right) + \overline{Z_2\left(2i\sqrt{z}\right)}\right\}$

$\qquad\qquad\qquad\qquad\qquad\qquad\qquad\qquad\qquad\qquad\qquad\qquad\qquad$ WA 122(7)

2.  $\quad \dfrac{d^2}{dz^2}\left(z^{\frac{16}{5}}\dfrac{d^2u}{dz^2}\right) - z^{\frac{8}{5}}u = 0$ $\qquad\qquad u = z^{-7/10}\left\{Z_{\frac{5}{6}}\left(\dfrac{5}{3}z^{\frac{3}{5}}\right) + \overline{Z_{\frac{5}{6}}\left(\dfrac{5}{3}iz^{\frac{3}{5}}\right)}\right\}$

$\qquad\qquad\qquad\qquad\qquad\qquad\qquad\qquad\qquad\qquad\qquad\qquad\qquad$ WA 122(8)

3.  $\quad \dfrac{d^2}{dz^2}\left(z^{12}\dfrac{d^2u}{dz^2}\right) - z^6 u = 0$

$\qquad\qquad\qquad u = z^{-4}\left\{Z_{10}\left(2z^{-1/2}\right) + \overline{Z_{10}\left(2iz^{-1/2}\right)}\right\}$ $\quad$ WA 122(9)

4.  $\quad \dfrac{d^4u}{dz^4} + \dfrac{2}{z}\dfrac{d^3u}{dz^3} - \dfrac{2\nu^2+1}{z^2}\dfrac{d^2u}{dz^2} + \dfrac{2\nu^2+1}{z^3}\dfrac{du}{dz} + \left(\dfrac{\nu^4-4\nu^2}{z^4}-1\right)u = 0,$

$\qquad u = A_1\, J_\nu(z) + A_2\, Y_\nu(z) + A_3\, I_\nu(z) + A_4\, K_\nu(z)$, where $A_1, A_2, A_3, A_4$ are constants $\quad$ MO 29

## 8.51–8.52 Series of Bessel functions

**8.511** Generating functions for Bessel functions:

1.  $\quad \exp\dfrac{1}{2}\left(t - \dfrac{1}{t}\right)z = J_0(z) + \displaystyle\sum_{k=1}^{\infty}\left[t^k + (-t)^{-k}\right]J_k(z) = \sum_{k=-\infty}^{\infty} J_k(z)t^k$

$\qquad\qquad\qquad\qquad\qquad\qquad\qquad\qquad\qquad [|z| < |t|]$ $\qquad\qquad$ KU 119(12)

2.  $\quad \exp\left(t - \dfrac{1}{t}\right)z = \left\{\displaystyle\sum_{k=-\infty}^{\infty} t^k J_k(z)\right\}\left\{\sum_{m=-\infty}^{\infty} t^m J_m(z)\right\}$ $\qquad\qquad$ WA 40

3.  $\quad \exp\left(\pm iz\sin\varphi\right) = J_0(z) + 2\displaystyle\sum_{k=1}^{\infty} J_{2k}(z)\cos 2k\varphi \pm 2i\sum_{k=0}^{\infty} J_{2k+1}(z)\sin(2k+1)\varphi$ $\qquad$ KU 120(13)

4.     $\exp\left(iz\cos\varphi\right) = \sqrt{\dfrac{\pi}{2z}}\sum_{k=0}^{\infty}(2k+1)i^{k}\,J_{k+\frac{1}{2}}(z)\,P_{k}\left(\cos\varphi\right)$         WA 401(1)

$$= \sum_{k=-\infty}^{\infty} i^{k}\,J_{k}(z)e^{ik\varphi}$$         MO 27

$$= J_{0}(z) + 2\sum_{k=1}^{\infty} i^{k}\,J_{k}(z)\cos k\varphi$$         MO 27

5.     $\sqrt{\dfrac{i}{\pi}}e^{iz\cos 2\varphi}\displaystyle\int_{-\infty}^{\sqrt{2z}\cos\varphi} e^{-it^{2}}\,\mathrm{d}t = \dfrac{1}{2}\,J_{0}(z) + \sum_{k=1}^{\infty} e^{\frac{1}{4}k\pi i}\,J_{\frac{k}{2}}(z)\cos k\varphi$         MO 28

## The series $\sum J_{k}(z)$

**8.512**

1.     $J_{0}(z) + 2\displaystyle\sum_{k=1}^{\infty} J_{2k}(z) = 1$         WA 44

2.     $\displaystyle\sum_{k=0}^{\infty} \dfrac{(n+2k)(n+k-1)!}{k!}\,J_{n+2k}(z) = \left(\dfrac{z}{2}\right)^{n}$     $[n=1,2,\ldots]$         WA 45

3.     $\displaystyle\sum_{k=0}^{\infty} \dfrac{(4k+1)(2k-1)!!}{2^{k}k!}\,J_{2k+\frac{1}{2}}(z) = \sqrt{\dfrac{2z}{\pi}}$

**8.513**

**Notation**: In formulas **8.513**     $Q_{k}^{(p)} = \displaystyle\sum_{m=0}^{\lfloor\frac{k-1}{2}\rfloor} \dfrac{(-1)^{m}\binom{k}{m}(k-2m)^{p}}{2^{k}k!}$

1.     $\displaystyle\sum_{k=1}^{\infty}(2k)^{2p}\,J_{2k}(z) = \sum_{k=0}^{p} Q_{2k}^{(2p)} z^{2k}$     $[p=1,2,3,\ldots]$         WA 46(1)

2.     $\displaystyle\sum_{k=0}^{\infty}(2k+1)^{2p+1}\,J_{2k+1}(z) = \sum_{k=0}^{p} Q_{2k+1}^{(2p+1)} z^{2k+1}$     $[p=0,1,2,3,\ldots]$         WA 46(2)

In particular:

3.     $\displaystyle\sum_{k=0}^{\infty}(2k+1)^{3}\,J_{2k+1}(z) = \dfrac{1}{2}\left(z+z^{3}\right)$         WA 47(4)

4.     $\displaystyle\sum_{k=1}^{\infty}(2k)^{2}\,J_{2k}(z) = \dfrac{1}{2}z^{2}$         WA 47(4)

5.     $\displaystyle\sum_{k=1}^{\infty} 2k(2k+1)(2k+2)\,J_{2k+1}(z) = \dfrac{1}{2}z^{3}$         WA 47(4)

**8.514**

1.     $\displaystyle\sum_{k=0}^{\infty}(-1)^{k}\,J_{2k+1}(z) = \dfrac{\sin z}{2}$         WH

2.  $J_0(z) + 2 \sum_{k=1}^{\infty} (-1)^k J_{2k}(z) = \cos z$  WH

3.  $\sum_{k=1}^{\infty} (-1)^{k+1} (2k)^2 J_{2k}(z) = \dfrac{z \sin z}{2}$  WA 32(9)

4.  $\sum_{k=0}^{\infty} (-1)^k (2k+1)^2 J_{2k+1}(z) = \dfrac{z \cos z}{2}$  WA 32(10)

5.  $J_0(z) + 2 \sum_{k=1}^{\infty} J_{2k}(z) \cos 2k\theta = \cos(z \sin \theta)$  KU 120(14), WA 32

6.  $\sum_{k=0}^{\infty} J_{2k+1}(z) \sin(2k+1)\theta = \dfrac{\sin(z \sin \theta)}{2}$  KU 120(15), WA 32

7.  $\sum_{k=0}^{\infty} J_{2k+1}(x) = \dfrac{1}{2} \int_0^x J_0(t)\, dt$  [x is real]  WA 638

8.*  $\sum_{k=1}^{\infty} \dfrac{(-1)^k}{2k} J_{2k}(z) = -\dfrac{\pi}{8} N_0(z) + \dfrac{1}{4} \left( \ln \dfrac{z}{2} + \boldsymbol{C} \right) J_0(z)$

9.*  $\sum_{k=1}^{\infty} \dfrac{(-1)^k (2k+1)}{(2k+1)^2 - 1} J_{2k+1}(z) = -\dfrac{\pi}{8} N_1(z) + \dfrac{1}{4} \left( \ln \dfrac{z}{2} + \boldsymbol{C} - 1 \right) J_1(z) - \dfrac{1}{4z} J_0(z)$

**8.515**

1.  $\sum_{k=0}^{\infty} \dfrac{(-1)^k t^k}{k!} \left( \dfrac{2z+t}{2z} \right)^k J_{\nu+k}(z) = \left( \dfrac{z}{z+t} \right)^{\nu} J_{\nu}(z+t)$  AD (9140)

2.  $\sum_{k=1}^{\infty} J_{2k-\frac{1}{2}}(x^2) = S(x)$  MO 127a

3.  $\sum_{k=0}^{\infty} J_{2k+\frac{1}{2}}(x^2) = C(x)$  MO 127a

**8.516**  $\sum_{k=0}^{\infty} \dfrac{(2n+2k)(2n+k-1)!}{k!} J_{2n+2k}(2z \sin \theta) = (z \sin \theta)^{2n}$  WA 47

**The series $\sum a_k J_k(kx)$ and $\sum a_k J'_k(kx)$**

**8.517**

1.  $\sum_{k=1}^{\infty} J_k(kz) = \dfrac{z}{2(1-z)}$  $\left[ \left| \dfrac{z \exp \sqrt{1-z^2}}{1 + \sqrt{1-z^2}} \right| < 1 \right]$  WA 615(1)

2.  $\sum_{k=1}^{\infty} (-1)^k J_k(kz) = -\dfrac{z}{2(1+z)}$  $\left[ \left| \dfrac{z \exp \sqrt{1-z^2}}{1 + \sqrt{1-z^2}} \right| < 1 \right]$  WA 622(1)

3.  $\sum_{k=1}^{\infty} J_{2k}(2kz) = \dfrac{z^2}{2(1-z^2)}$  $\left[ \left| \dfrac{z \exp \sqrt{1-z^2}}{1 + \sqrt{1-z^2}} \right| < 1 \right]$  MO 58

**8.518**

1.[11] $\displaystyle\sum_{k=1}^{\infty} \frac{J_k'(kx)}{k} = \frac{1}{2} + \frac{x}{4}$ $\qquad [0 \le x < 1]$ $\qquad$ MO 58

2.[11] $\displaystyle\sum_{k=1}^{\infty} (-1)^{k-1} \frac{J_k'(kx)}{k} = \frac{1}{2} - \frac{x}{4}$ $\qquad [0 \le x < 1]$ $\qquad$ MO 58

3. $\displaystyle\sum_{k=1}^{\infty} k\, J_k'(kx) = \frac{1}{2(1-x)^2}$ $\qquad [0 \le x < 1]$ $\qquad$ MO 58

4. $\displaystyle\sum_{k=1}^{\infty} (-1)^{k-1} J_k'(kx) k = \frac{1}{2(1+x)^2}$ $\qquad [0 \le x < 1]$ $\qquad$ MO 58

## The series $\sum a_k J_0(kx)$

**8.519** If, on the interval $[0 \le x \le \pi]$, a function $f(x)$ possesses a continuous derivative with respect to $x$ that is of bounded variation, then

1. $\qquad f(x) = \dfrac{a_0}{2} + \displaystyle\sum_{k=1}^{\infty} a_k J_0(kx)$ $\qquad [0 < x < \pi]$

where

2. $\qquad a_0 = 2f(0) + \dfrac{2}{\pi} \displaystyle\int_0^{\pi} du \int_0^{\pi/2} u f'\left(u \sin \varphi\right) d\varphi$

3. $\qquad a_n = \dfrac{2}{\pi} \displaystyle\int_0^{\pi} du \int_0^{\pi/2} u f'\left(u \sin \varphi\right) \cos nu\, d\varphi$ $\qquad$ WH

**8.521** Examples:

1. $\displaystyle\sum_{k=1}^{\infty} J_0(kx) = -\frac{1}{2} + \frac{1}{x} + 2\sum_{m=1}^{n} \frac{1}{\sqrt{x^2 - 4m^2\pi^2}}$ $\qquad [2n\pi < x < 2(n+1)\pi]$ $\qquad$ MO 59

2. $\displaystyle\sum_{k=1}^{\infty} (-1)^{k+1} J_0(kx) = \frac{1}{2}$ $\qquad [0 < x < \pi]$ $\qquad$ KU 124(12)

3. $\displaystyle\sum_{k=1}^{\infty} \frac{1}{(2k-1)^2} J_0\left\{(2k-1)x\right\} = \frac{\pi^2}{8} - \frac{|x|}{2}$ $\qquad [-\pi < x < \pi]$ $\qquad$ KU 124

$\qquad\qquad\qquad = \dfrac{\pi^2}{8} + \sqrt{x^2 - \pi^2} - \dfrac{x}{2} - \pi \arccos \dfrac{\pi}{x}$ $\qquad [\pi < x < 2\pi]$ $\qquad$ MO 59

4.[12] $\displaystyle\sum_{k=1}^{\infty} e^{-kz} J_0\left(k\sqrt{x^2 + y^2}\right)$

$\qquad = \dfrac{1}{r} - \dfrac{1}{2} + \displaystyle\sum_{k=1}^{\infty} \left\{ \frac{1}{\sqrt{(2ki\pi + z)^2 + x^2 + y^2}} + \frac{1}{\sqrt{(2ki\pi - z)^2 + x^2 + y^2}} \right\}$

$\qquad = \dfrac{1}{r} - \dfrac{1}{2} + \displaystyle\sum_{k=1}^{\infty} \frac{1}{(2k)!} B_{2k} r^{2k-1} P_{2k-1}\left(\frac{z}{r}\right)$ $\qquad [0 < r < 2\pi]$ MO 59

where $r = \sqrt{x^2 + y^2 + z^2}$ and where the radical indicates the square root with a positive real part. In formula **8.521** 4, the first equation holds when $x$ and $y$ are real and $\operatorname{Re} z > 0$; the second equation holds when $x$, $y$, and $z$ are all real.

**The series** $\sum a_k Z_0(kx) \sin kx$ **and** $\sum a_k Z_0(kx) \cos kx$

**8.522**

1. $$\sum_{k=1}^{\infty} J_0(kx) \cos kxt = -\frac{1}{2} + \sum_{l=1}^{m} \frac{1}{\sqrt{x^2 - (2\pi l + tx)^2}} + \frac{1}{x\sqrt{1-t^2}} + \sum_{l=1}^{n} \frac{1}{\sqrt{x^2 - (2\pi l - tx)^2}}$$ MO 59

2. $$\sum_{k=1}^{\infty} J_0(kx) \sin kxt = \frac{1}{2\pi}\left\{\sum_{l=1}^{n}\frac{1}{l} - \sum_{l=1}^{m}\frac{1}{l}\right\} + \sum_{l=m+1}^{\infty}\left\{\frac{1}{\sqrt{(2\pi l + tx)^2 - x^2}} - \frac{1}{2\pi l}\right\}$$
$$- \sum_{l=n+1}^{\infty}\left\{\frac{1}{\sqrt{(2\pi l - tx)^2 - x^2}} - \frac{1}{2\pi l}\right\}$$

MO 59

3. $$\sum_{k=1}^{\infty} Y_0(kx) \cos kxt = -\frac{1}{\pi}\left(C + \ln\frac{x}{4\pi}\right) + \frac{1}{2\pi}\left\{\sum_{l=1}^{m}\frac{1}{l} + \sum_{l=1}^{n}\frac{1}{l}\right\}$$
$$- \sum_{l=m+1}^{\infty}\left\{\frac{1}{\sqrt{(2\pi l + tx)^2 - x^2}} - \frac{1}{2\pi l}\right\}$$
$$- \sum_{l=n+1}^{\infty}\left\{\frac{1}{\sqrt{(2\pi l - tx)^2 - x^2}} - \frac{1}{2\pi l}\right\}$$

MO 60

In formulas **8.522**, $x > 0, 0 \le t < 1, 2\pi m < x(1-t) < 2(m+1)\pi, 2n\pi < x(1+t) < 2(n+1)\pi, m+1$ and $n+1$ are natural numbers.

**8.523**

1. $$\sum_{k=1}^{\infty}(-1)^k J_0(kx) \cos kxt = -\frac{1}{2} + \sum_{l=1}^{m}\frac{1}{\sqrt{x^2 - [(2l-1)\pi + tx]^2}} + \sum_{l=1}^{n}\frac{1}{\sqrt{x^2 - [(2l-1)\pi - tx]^2}}$$

MO 60

2. $$\sum_{k=1}^{\infty}(-1)^k J_0(kx) \sin kxt \quad \frac{1}{2\pi}\left\{\sum_{l=1}^{n}\frac{1}{l} - \sum_{l=1}^{m}\frac{1}{l}\right\} + \sum_{l=m+1}^{\infty}\left\{\frac{1}{\sqrt{[(2l-1)\pi + tx]^2 - x^2}} - \frac{1}{2l\pi}\right\}$$
$$- \sum_{l=n+1}^{\infty}\left\{\frac{1}{\sqrt{[(2l-1)\pi - tx]^2 - x^2}} - \frac{1}{2l\pi}\right\}$$

MO 60

3.[12] $$\sum_{k=1}^{\infty}(-1)^k Y_0(kx) \cos kxt = -\frac{1}{\pi}\left(C + \ln\frac{x}{4\pi}\right) + \frac{1}{2\pi}\left\{\sum_{l=1}^{m}\frac{1}{l} + \sum_{l=1}^{n}\frac{1}{l}\right\}$$
$$- \sum_{l=m+1}^{\infty}\left\{\frac{1}{\sqrt{[(2l-1)\pi + tx]^2 - x^2}} - \frac{1}{2l\pi}\right\}$$
$$- \sum_{l=n+1}^{\infty}\left\{\frac{1}{\sqrt{[(2l-1)\pi - tx]^2 - x^2}} - \frac{1}{2l\pi}\right\}$$

MO 60

In formulas **8.523**, $x > 0, 0 \le t < 1, (2m-1)\pi < x(1-t) < (2m+1)\pi, (2n-1)\pi < x(1+t) < (2n+1)\pi,$ $m$ and $n$ are natural numbers.

**8.524**

1. $$\sum_{k=1}^{\infty} J_0(kx)\cos kxt = -\frac{1}{2} + \sum_{l=m+1}^{n} \frac{1}{\sqrt{x^2 - (2l\pi - tx)^2}}$$ MO 60

2. $$\sum_{k=1}^{\infty} J_0(kx)\sin kxt \sum_{l=0}^{m} \frac{1}{\sqrt{(2l\pi - tx)^2 - x^2}} + \sum_{l=1}^{\infty} \left\{ \frac{1}{\sqrt{(2l\pi + tx)^2 - x^2}} - \frac{1}{2l\pi} \right\}$$
$$- \sum_{l=n+1}^{\infty} \left\{ \frac{1}{\sqrt{(2l\pi - tx)^2 - x^2}} - \frac{1}{2l\pi} \right\} + \frac{1}{2\pi}\sum_{l=1}^{n}\frac{1}{l}$$

MO 60

3.[6] $$\sum_{k=1}^{\infty} Y_0(kx)\cos kxt \quad -\frac{1}{\pi}\left(C + \ln\frac{x}{4\pi}\right) - \sum_{l=0}^{m} \frac{1}{\sqrt{(2\pi l - tx)^2 - x^2}} + \frac{1}{2\pi}\sum_{l=1}^{n}\frac{1}{l}$$
$$- \sum_{l=1}^{\infty}\left\{\frac{1}{\sqrt{(2l\pi + tx)^2 - x^2}} - \frac{1}{2l\pi}\right\}$$
$$- \sum_{l=n+1}^{\infty}\left\{\frac{1}{\sqrt{(2l\pi - tx)^2 - x^2}} - \frac{1}{2l\pi}\right\}$$

MO 61

In formulas **8.524**, $x > 0, t > 1, 2m\pi < x(t-1) < 2(m+1)\pi, 2n\pi < x(t+1) < 2(n+1)\pi, m+1$ and $n+1$ are natural numbers.

**8.525**

1. $$\sum_{k=1}^{\infty}{}'(-1)^k J_0(kx)\cos kxt = -\frac{1}{2} + \sum_{l=m+1}^{n} \frac{1}{\sqrt{x^2 - [(2l-1)\pi - tx]^2}}$$ MO 61

2. $$\sum_{k=1}^{\infty}(-1)^k J_0(kx)\sin kxt = \sum_{l=1}^{m} \frac{1}{\sqrt{[(2l-1)\pi - tx]^2 - x^2}} + \frac{1}{2\pi}\sum_{l=1}^{n}\frac{1}{l}$$
$$+ \sum_{l=1}^{\infty}\left\{\frac{1}{\sqrt{[(2l-1)\pi + tx]^2 - x^2}} - \frac{1}{2l\pi}\right\}$$
$$- \sum_{l=n+1}^{\infty}\left\{\frac{1}{\sqrt{[(2l-1)\pi - tx]^2 - x^2}} - \frac{1}{2l\pi}\right\}$$

MO 61

3.  $$\sum_{k=1}^{\infty}(-1)^k\,Y_0(kx)\cos kxt = -\frac{1}{\pi}\left(C+\ln\frac{x}{4\pi}\right)+\frac{1}{2\pi}\sum_{l=1}^{n}\frac{1}{l}$$

$$-\sum_{l=1}^{m}\frac{1}{\sqrt{[(2l-1)\pi-tx]^2-x^2}}$$

$$-\sum_{l=1}^{\infty}\left\{\frac{1}{\sqrt{[(2l-1)\pi+tx]^2-x^2}}-\frac{1}{2l\pi}\right\}$$

$$-\sum_{l=n+1}^{\infty}\left\{\frac{1}{\sqrt{[(2l-1)\pi-tx]^2-x^2}}-\frac{1}{2l\pi}\right\}$$

MO 61

In formulas **8.525**, $x>0, t>1, (2m-1)\pi < x(t-1) < (2m+1)\pi, (2n-1)\pi < x(t+1) < (2n+1)\pi,$ $m$ and $n$ are natural numbers.

**8.526**

1.  $$\sum_{k=1}^{\infty}K_0(kx)\cos kxt = \frac{1}{2}\left(C+\ln\frac{x}{4\pi}\right)+\frac{\pi}{2x\sqrt{1+t^2}}+\frac{\pi}{2}\sum_{l=1}^{\infty}\left\{\frac{1}{\sqrt{x^2+(2l\pi-tx)^2}}-\frac{1}{2l\pi}\right\}$$

$$+\frac{\pi}{2}\sum_{l=1}^{\infty}\left\{\frac{1}{\sqrt{x^2+(2l\pi+tx)^2}}-\frac{1}{2l\pi}\right\}$$

MO 61

2.  $$\sum_{k=1}^{\infty}(-1)^k\,K_0(kx)\cos kxt = \frac{1}{2}\left(C+\ln\frac{x}{4\pi}\right)+\frac{\pi}{2}\sum_{l=1}^{\infty}\left\{\frac{1}{\sqrt{x^2+[(2l-1)\pi-xt]^2}}-\frac{1}{2l\pi}\right\}$$

$$+\frac{\pi}{2}\sum_{l=1}^{\infty}\left\{\frac{1}{\sqrt{x^2+[(2l-1)\pi+xt]^2}}-\frac{1}{2l\pi}\right\}$$

$[x>0,\quad t\ \text{real}]$      (see also **8.66**)    MO 62

## 8.53  Expansion in products of Bessel functions

"Summation theorems"

**8.530**[12]   Suppose that $r>0, \rho>0, \varphi>0,$ and $R=\sqrt{r^2+\rho^2-2r\rho\cos\varphi}$; that is, suppose that $r, \rho,$ and $R$ are the sides of a triangle such that the angle between the sides $r$ and $\rho$ is equal to $\varphi$. Suppose also that $\rho < r$ and that $\psi$ is the angle opposite the side $\rho$, so that

1.[12]    $$0<\psi<\frac{\pi}{2}, \quad e^{2i\psi}=\frac{r-\rho e^{-i\varphi}}{r-\rho e^{i\varphi}}$$

When these conditions are satisfied, we have the "summation theorem" for Bessel functions:

2.[12]    $e^{i\nu\psi} Z_\nu(mR) = \sum\limits_{k=-\infty}^{\infty} J_k(m\rho) Z_{\nu+k}(mr) e^{ik\varphi}$            [$m$ is an arbitrary complex number]

                                                                   WA 394(6)

For $Z_\nu = J_\nu$ and $\nu$ an integer, the restriction $\rho < r$ is superfluous.           MO 31

**8.531**[12]    Special cases:

1.[12]    $J_0(mR) = J_0(m\rho) J_0(mr) + 2 \sum\limits_{k=1}^{\infty} J_k(m\rho) J_k(mr) \cos k\varphi$          WA 391(1)

2.[12]    $H_0^{(1,2)}(mR) = J_0(m\rho) H_0^{(1,2)}(mr) + 2 \sum\limits_{k=1}^{\infty} J_k(m\rho) H_k^{(1,2)}(mr) \cos k\varphi$      MO 31

3.    $J_0(z \sin \alpha) = J_0^2\left(\dfrac{z}{2}\right) + 2 \sum\limits_{k=1}^{\infty} J_k^2\left(\dfrac{z}{2}\right) \cos 2k\alpha$

            $= \sqrt{\dfrac{2\pi}{z}} \sum\limits_{k=0}^{\infty} \left(2k + \dfrac{1}{2}\right) \dfrac{(2k-1)!!}{2^k k!} J_{2k+\frac{1}{2}}(z) P_{2k}(\cos \alpha)$          MO 31

**8.532**    The term "summation theorem" is also applied to the formula

1.[12]    $\dfrac{Z_\nu(mR)}{R^\nu} = 2^\nu m^{-\nu} \Gamma(\nu) \sum\limits_{k=0}^{\infty} (\nu + k) \dfrac{J_{\nu+k}(m\rho)}{\rho^\nu} \dfrac{Z_{\nu+k}(mr)}{r^\nu} C_k^\nu(\cos \varphi)$

[$\nu \neq -1, -2, -3, \ldots$; the conditions on $r, \rho, R, \varphi$, and $m$ are the same as in formula **8.530**; for $Z_\nu = J_\nu$ and $\nu$ an integer, formula **8.532** 1 is valid for arbitrary $r, \rho$, and $\varphi$].          WA 398(4)

**8.533**    Special cases:

1.[12]    $\dfrac{e^{imR}}{R} = \dfrac{\pi i}{2\sqrt{r\rho}} \sum\limits_{k=0}^{\infty} (2k + 1) J_{k+\frac{1}{2}}(m\rho) H_{k+\frac{1}{2}}^{(1)}(mr) P_k(\cos \varphi)$          MO 31

2.[12]    $\dfrac{e^{-imR}}{R} = -\dfrac{\pi i}{2\sqrt{r\rho}} \sum\limits_{k=0}^{\infty} (2k + 1) J_{k+\frac{1}{2}}(m\rho) H_{k+\frac{1}{2}}^{(2)}(mr) P_k(\cos \varphi)$          MO 31

**8.534**[12]    A degenerate addition theorem ($r \to \infty$):

$e^{im\rho \cos \varphi} = \sqrt{\dfrac{\pi}{2m\rho}} \sum\limits_{k=0}^{\infty} i^k (2k + 1) J_{k+\frac{1}{2}}(m\rho) P_k(\cos \varphi)$          WA 401(1)

          $= 2^\nu \Gamma(\nu) \sum\limits_{k=0}^{\infty} (\nu + k) i^k (m\rho)^{-\nu} J_{\nu+k}(m\rho) C_k^\nu(\cos \varphi)$    [$\nu \neq 0, -1, -2, \ldots$]    WA 401(2)

**8.535**   The term "product theorem" is also applied to the formula

$$Z_\nu(\lambda z) = \lambda^\nu \sum_{k=0}^\infty \frac{1}{k!} Z_{\nu+k}(z) \left(\frac{1-\lambda^2}{2}z\right)^k \qquad \left[|1-\lambda|^2 < 1\right]$$

For $Z_\nu = J_\nu$, it is valid for all values of $\lambda$ and $z$.                     MO 32

**8.536**

1.   $$\sum_{k=0}^\infty \frac{(2n+2k)(2n+k-1)!}{k!} J_{n+k}^2(z) = \frac{(2n)!}{(n!)^2} \left(\frac{z}{2}\right)^{2n} \qquad [n>0]$$                     WA 47(1)

2.   $$2\sum_{k=n}^\infty \frac{k\,\Gamma(n+k)}{\Gamma(k-n+1)} J_k^2(z) = \frac{(2n)!}{(n!)^2} \left(\frac{z}{2}\right)^{2n} \qquad [n>0]$$                     WA 47(2)

3.   $$J_0^2(z) + 2\sum_{k=1}^\infty J_k^2(z) = 1$$                     WA 41(3)

**8.537**

1.   $$\sum_{k=-\infty}^\infty Z_{\nu-k}(t)\, J_k(z) = Z_\nu(z+t) \qquad [|z| < |t|]$$                     WA 158(2)

2.   $$\sum_{k=-\infty}^\infty J_k(z)\, J_{n-k}(z) = J_n(2z)$$                     WA 41

**8.538**

1.   $$\sum_{k=-\infty}^\infty (-1)^k J_{-\nu+k}(t)\, J_k(z) = J_{-\nu}(z+t) \qquad [|z| < |t|]$$                     WA 159

2.   $$\sum_{k=-\infty}^\infty Z_{\nu+k}(t)\, J_k(z) = Z_\nu(t-z) \qquad [|z| < |t|]$$                     WA 159(5)

## 8.54   The zeros of Bessel functions

**8.541**   For arbitrary real $\nu$, the function $J_\nu(z)$ has infinitely many real zeros. For $\nu > -1$, all its zeros are real.                     WA 526, 530

A Bessel function $Z_\nu(z)$ has no multiple zeros except possibly the coordinate origin.                     WA 528

**8.542**   All zeros of the function $Y_0(z)$ with positive real parts are real.                     WA 531

**8.543**   If $-(2s+2) < \nu < -(2s+1)$, where $s$ is a natural number or 0, then $J_\nu(z)$ has exactly $4s+2$ complex roots, two of which are purely imaginary. If $-(2s+1) < \nu < -2s$, where $s$ is a natural number, then the function $J_\nu(z)$ has exactly $4s$ complex zeros none of which are purely imaginary.                     WA 532

**8.544**   If $x_\nu$ and $x'_\nu$ are, respectively, the smallest positive zeros of the functions $J_\nu(z)$ and $J'_\nu(z)$ for $\nu > 0$, then $x_\nu > \nu$ and $x'_\nu > \nu$. Suppose also that $y_\nu$ is the smallest positive zero of the function $Y_\nu(z)$. Then, $x_\nu < y_\nu < x'_\nu$.                     WA 534, 536

Suppose that $z_{\nu,m}$ (for $m = 1, 2, 3, \ldots$) are the zeros of the function $z^{-\nu} J_\nu(z)$, numbered in order of the absolute value of their real parts. Here, we assume that $\nu \neq -1, -2, -3, \ldots$. Then, for arbitrary $z$

$$J_\nu(z) = \frac{\left(\frac{z}{2}\right)^\nu}{\Gamma(\nu+1)} \prod_{m=1}^{\infty} \left(1 - \frac{z^2}{z_{\nu,m}^2}\right) \qquad\qquad \text{WA 550}$$

**8.545**[8]    The number of zeros of the function $z^{-\nu} J_\nu(z)$ that occur between the imaginary axis and the line on which

$$\operatorname{Re} z = \left(m + \tfrac{1}{2}\operatorname{Re}\nu + \tfrac{1}{4}\right)\pi, \qquad\qquad \text{WA 497}$$

is exactly $m$.

**8.546**    For $\nu \ge 0$, the number of zeros of the function $K_\nu(z)$ that occur in the region $\operatorname{Re} z < 0$, $|\arg z| < \pi$ is equal to the even number closest to $\nu - \frac{1}{2}$.                 WA 562

**8.547**    Large zeros of the functions $J_\nu(z)\cos\alpha - Y_\nu(z)\sin\alpha$, where $\nu$ and $\alpha$ are real numbers, are given by the asymptotic expansion

$$x_{\nu,m} \sim \left(m + \frac{1}{2}\nu - \frac{1}{4}\right)\pi - \alpha - \frac{4\nu^2 - 1}{8\left[\left(m + \frac{1}{2}\nu - \frac{1}{4}\right)\pi - \alpha\right]}$$
$$- \frac{\left(4\nu^2 - 1\right)\left(28\nu^2 - 31\right)}{384\left[\left(m + \frac{1}{2}\nu - \frac{1}{4}\right)\pi - \alpha\right]^3} - \cdots$$

KU 109(24), WA 558

**8.548**    In particular, large zeros of the function $J_0(z)$ are given by the expansion

$$x_{0,m} \sim \frac{\pi}{4}(4m-1) + \frac{1}{2\pi(4m-1)} - \frac{31}{6\pi^3(4m-1)^3} + \frac{3779}{15\pi^5(4m-1)^5} - \cdots \qquad \text{KU 109(25), WA 556}$$

This series is suitable for calculating all (except the smallest $x_{01}$) zeros of the function $J_0(z)$ correctly to at least five digits.

**8.549**    To calculate the roots $x_{\nu,m}$ of the function $J_\nu(z)$ of smallest absolute value, we may use the identity

$$\sum_{m=1}^{\infty} \frac{1}{x_{\nu,m}^{16}} = \frac{429\nu^5 + 7640\nu^4 + 53752\nu^3 + 185430\nu^2 + 311387\nu + 202738}{2^{16}(\nu+1)^8(\nu+2)^4(\nu+3)^2(\nu+4)^2(\nu+5)(\nu+6)(\nu+7)(\nu+8)} \quad \text{KU 112(27)a, WA 554}$$

## 8.55   Struve functions

**8.550**    Definitions:

1.    $$\mathbf{H}_\nu(z) = \sum_{m=0}^{\infty} (-1)^m \frac{\left(\frac{z}{2}\right)^{2m+\nu+1}}{\Gamma\left(m+\frac{3}{2}\right)\Gamma\left(\nu+m+\frac{3}{2}\right)} \qquad\qquad \text{WA 358(2)}$$

2.    $$\mathbf{L}_\nu(z) = -ie^{-i\nu\frac{\pi}{2}}\,\mathbf{H}_\nu\left(ze^{i\frac{\pi}{2}}\right) = \sum_{m=0}^{\infty} \frac{\left(\frac{z}{2}\right)^{2m+\nu+1}}{\Gamma\left(m+\frac{3}{2}\right)\Gamma\left(\nu+m+\frac{3}{2}\right)} \qquad \text{WA 360(11)}$$

**8.551**    Integral representations:

1.    $$\mathbf{H}_\nu(z) = \frac{2\left(\frac{z}{2}\right)^\nu}{\sqrt{\pi}\,\Gamma\left(\nu+\frac{1}{2}\right)} \int_0^1 \left(1-t^2\right)^{\nu-\frac{1}{2}}\sin zt\,dt = \frac{2\left(\frac{z}{2}\right)^\nu}{\sqrt{\pi}\,\Gamma\left(\nu+\frac{1}{2}\right)} \int_0^{\pi/2} \sin(z\cos\varphi)(\sin\varphi)^{2\nu}\,d\varphi$$
$$\left[\operatorname{Re}\nu > -\tfrac{1}{2}\right] \qquad\qquad \text{WA 358(1)}$$

2.    $$\mathbf{L}_\nu(z) = \frac{2\left(\frac{z}{2}\right)^\nu}{\sqrt{\pi}\,\Gamma\left(\nu+\frac{1}{2}\right)} \int_0^{\pi/2} \sinh(z\cos\varphi)(\sin\varphi)^{2\nu}\,d\varphi$$
$$\left[\operatorname{Re}\nu > -\tfrac{1}{2}\right] \qquad\qquad \text{WA 360(11)}$$

**8.552** Special cases:

1.[6] $\quad \mathbf{H}_n(z) = \dfrac{1}{\pi} \displaystyle\sum_{m=0}^{\lfloor \frac{n-1}{2} \rfloor} \dfrac{\Gamma\left(m+\frac{1}{2}\right)\left(\frac{z}{2}\right)^{n-2m-1}}{\Gamma\left(n+\frac{1}{2}-m\right)} - \mathbf{E}_n(z) \qquad [n=1,2,\ldots] \qquad$ EH II 40(66), WA 337(1)

2.[6] $\quad \mathbf{H}_{-n}(z) = (-1)^{n+1}\dfrac{1}{\pi} \displaystyle\sum_{m=0}^{\lfloor \frac{n-1}{2} \rfloor} \dfrac{\Gamma\left(n-m-\frac{1}{2}\right)\left(\frac{z}{2}\right)^{-n+2m+1}}{\Gamma\left(m+\frac{3}{2}\right)} - \mathbf{E}_{-n}(z)$

$\qquad\qquad\qquad\qquad\qquad\qquad\qquad\qquad\qquad [n=1,2,\ldots] \qquad$ EH II 40(67), WA 337(2)

3. $\quad \mathbf{H}_{n+\frac{1}{2}}(z) = Y_{n+\frac{1}{2}}(z) + \dfrac{1}{\pi} \displaystyle\sum_{m=0}^{n} \dfrac{\Gamma\left(m+\frac{1}{2}\right)\left(\frac{z}{2}\right)^{-2m+n-\frac{1}{2}}}{\Gamma(n+1-m)}$

$\qquad\qquad\qquad\qquad\qquad\qquad\qquad\qquad\qquad [n=0,1,\ldots] \qquad$ EH II 39(64)

4. $\quad \mathbf{H}_{-\left(n+\frac{1}{2}\right)}(z) = (-1)^n J_{n+\frac{1}{2}}(z) \qquad\qquad [n=0,1,\ldots] \qquad$ EH II 39(65)

5. $\quad \mathbf{L}_{-\left(n+\frac{1}{2}\right)}(z) = I_{n+\frac{1}{2}}(z) \qquad\qquad\qquad [n=0,1,\ldots] \qquad$ EH II 39(65)

6. $\quad \mathbf{H}_{\frac{1}{2}}(z) = \dfrac{\sqrt{2}}{\sqrt{\pi z}}(1-\cos z) \qquad\qquad\qquad\qquad$ EH II 39, WA 364(3)

7. $\quad \mathbf{H}_{\frac{3}{2}}(z) = \left(\dfrac{z}{2\pi}\right)^{1/2}\left(1+\dfrac{2}{z^2}\right) - \left(\dfrac{2}{\pi z}\right)^{1/2}\left(\sin z + \dfrac{\cos z}{z}\right) \qquad$ WA 364(3)

**8.553** Functional relations:

1. $\quad \mathbf{H}_\nu\left(ze^{im\pi}\right) = e^{i\pi(\nu+1)m}\,\mathbf{H}_\nu(z) \qquad\qquad [m=1,2,3,\ldots] \qquad$ WA 362(5)

2. $\quad \dfrac{d}{dz}\left[z^\nu\,\mathbf{H}_\nu(z)\right] = z^\nu\,\mathbf{H}_{\nu-1}(z) \qquad\qquad\qquad\qquad$ WA 358

3. $\quad \dfrac{d}{dz}\left[z^{-\nu}\,\mathbf{H}_\nu(z)\right] = 2^{-\nu}\pi^{-1/2}\left[\Gamma\left(\nu+\frac{3}{2}\right)\right]^{-1} - z^{-\nu}\,\mathbf{H}_{\nu+1}(z) \qquad$ WA 359

4. $\quad \mathbf{H}_{\nu-1}(z) + \mathbf{H}_{\nu+1}(z) = 2\nu z^{-1}\,\mathbf{H}_\nu(z) + \pi^{-1/2}\left(\dfrac{z}{2}\right)^\nu\left[\Gamma\left(\nu+\frac{3}{2}\right)\right]^{-1} \qquad$ WA 359(5)

5. $\quad \mathbf{H}_{\nu-1}(z) - \mathbf{H}_{\nu+1}(z) = 2\,\mathbf{H}'_\nu(z) - \pi^{-1/2}\left(\dfrac{z}{2}\right)^\nu\left[\Gamma\left(\nu+\frac{3}{2}\right)\right]^{-1} \qquad$ WA 359(6)

**8.554** Asymptotic representations:

$$\mathbf{H}_\nu(\xi) = Y_\nu(\xi) + \dfrac{1}{\pi}\sum_{m=0}^{p-1} \dfrac{\Gamma\left(m+\frac{1}{2}\right)\left(\frac{\xi}{2}\right)^{-2m+\nu-1}}{\Gamma\left(\nu+\frac{1}{2}-m\right)} + O\left(|\xi|^{\nu-2p-1}\right)$$

$\qquad\qquad\qquad\qquad\qquad\qquad\qquad\qquad [|\arg\xi| < \pi] \qquad$ EH II 39(63), WA 363(2)

For the asymptotic representation of $Y_\nu(\xi)$, see **8.451** 2.

**8.555** The differential equation for Struve functions:

$$z^2 y'' + zy' + \left(z^2 - \nu^2\right)y = \dfrac{1}{\sqrt{\pi}}\dfrac{4\left(\frac{z}{2}\right)^{\nu+1}}{\Gamma\left(\nu+\frac{1}{2}\right)} \qquad\qquad \text{WA 359(10)}$$

## 8.56 Thomson functions and their generalizations

$\mathrm{ber}_\nu(z)$, $\mathrm{bei}_\nu(z)$, $\mathrm{her}_\nu(z)$, $\mathrm{hei}_\nu(z)$, $\mathrm{ker}_\nu(z)$, $\mathrm{kei}_\nu(z)$

**8.561**

1.      $\mathrm{ber}_\nu(z) + i\,\mathrm{bei}_\nu(z) = J_\nu\left(ze^{\frac{3}{4}\pi i}\right)$                                         WA 96(6)

2.      $\mathrm{ber}_\nu(z) - i\,\mathrm{bei}_\nu(z) = J_\nu\left(ze^{-\frac{3}{4}\pi i}\right)$                                     WA 96(6)

**8.562**

1.[12]    $\mathrm{her}_\nu(z) + i\,\mathrm{hei}_\nu(z) = H_\nu^{(1)}\left(ze^{\frac{3}{4}\pi i}\right)$           (see also **8.567**)         WA 96(7)

2.[12]    $\mathrm{her}_\nu(z) - i\,\mathrm{hei}_\nu(z) = H_\nu^{(1)}\left(ze^{-\frac{3}{4}\pi i}\right)$         (see also **8.567**)         WA 96(7)

**8.563**

1.      $\mathrm{ber}_0(z) \equiv \mathrm{ber}(z);\quad \mathrm{bei}_0(z) \equiv \mathrm{bei}(z)$                             WA 96(8)

2.[12]    $\mathrm{ker}(z) \equiv -\dfrac{\pi}{2}\,\mathrm{hei}_0(z);\quad \mathrm{kei}(z) \equiv \dfrac{\pi}{2}\,\mathrm{her}_0(z)$           WA 96(8)

For integral representations, see **6.251**, **6.536**, **6.537**, **6.772** 4, **6.777**.

### Series representation

**8.564**

1.      $\mathrm{ber}(z) = \displaystyle\sum_{k=0}^{\infty} \frac{(-1)^k z^{4k}}{2^{4k}\,[(2k)!]^2}$                                 WA 96(3)

2.      $\mathrm{bei}(z) = \displaystyle\sum_{k=0}^{\infty} \frac{(-1)^k z^{4k+2}}{2^{4k+2}\,[(2k+1)!]^2}$                          WA 96(4)

3.      $\mathrm{ker}(z) = \left(\ln\dfrac{2}{z} - C\right)\mathrm{ber}(z) + \dfrac{\pi}{4}\mathrm{bei}(z) + \displaystyle\sum_{k=1}^{\infty}(-1)^k \frac{z^{4k}}{2^{4k}\,[(2k)!]^2}\sum_{m=1}^{2k}\frac{1}{m}$     WA 96(9)a, DW

4.      $\mathrm{kei}(z) = \left(\ln\dfrac{2}{z} - C\right)\mathrm{bei}(z) - \dfrac{\pi}{4}\mathrm{ber}(z) + \displaystyle\sum_{k=0}^{\infty}(-1)^k \frac{z^{4k+2}}{2^{4k+2}\,[(2k+1)!]^2}\sum_{m=1}^{2k+1}\frac{1}{m}$     WA 96(10)a, DW

**8.565**    $\mathrm{ber}_\nu^2(z) + \mathrm{bei}_\nu^2(z) = \displaystyle\sum_{k=0}^{\infty} \frac{(z/2)^{2\nu+4k}}{k!\,\Gamma(\nu+k+1)\,\Gamma(\nu+2k+1)}$               WA 163(6)

### Asymptotic representation

**8.566**

1.      $\mathrm{ber}(z) = \dfrac{e^{\alpha(z)}}{\sqrt{2\pi z}}\cos\beta(z)$                  $\left[|\arg z| < \frac{\pi}{4}\right]$             WA 227(1)

2.      $\mathrm{bei}(z) = \dfrac{e^{\alpha(z)}}{\sqrt{2\pi z}}\sin\beta(z)$                   $\left[|\arg z| < \frac{\pi}{4}\right]$             WA 227(1)

3.      $\mathrm{ker}(z) = \sqrt{\dfrac{\pi}{2z}}\,e^{\alpha(-z)}\cos\beta(-z)$            $\left[|\arg z| < \frac{5}{4}\pi\right]$          WA 227(2)

4.     $\mathrm{kei}(z) = \sqrt{\dfrac{\pi}{2z}} e^{\alpha(-z)} \sin \beta(-z)$ $\qquad\qquad$ $\left[|\arg z| < \frac{5}{4}\pi\right],$ $\qquad\qquad$ WA 227(2)

where

$$\alpha(z) \sim \frac{z}{\sqrt{2}} + \frac{1}{8z\sqrt{2}} - \frac{25}{384z^3\sqrt{2}} - \frac{13}{128z^4} - \cdots,$$

$$\beta(z) \sim \frac{z}{\sqrt{2}} - \frac{\pi}{8} - \frac{1}{8z\sqrt{2}} - \frac{1}{16z^2} - \frac{25}{384z^3\sqrt{2}} + \cdots$$

**8.567**  **Functional relations**

1.     $\mathrm{ker}(z) + i\,\mathrm{kei}(z) = K_0\left(z\sqrt{i}\right)$ $\qquad\qquad$ (see **8.562**) $\qquad\qquad$ WA 96(5), DW

2.     $\mathrm{ker}(z) - i\,\mathrm{kei}(z) = K_0\left(z\sqrt{-i}\right)$ $\qquad\qquad$ (see **8.562**) $\qquad\qquad$ WA 96(5), DW

For integrals of Thomson's functions, see **6.87**.

## 8.57  Lommel functions

**8.570**  Definitions of the Lommel functions $s_{\mu,\nu}(z)$ and $S_{\mu,\nu}(z)$:

1.[12]     $s_{\mu,\nu}(z) = \displaystyle\sum_{m=0}^{\infty} \frac{(-1)^m z^{\mu+1+2m}}{[(\mu+1)^2 - \nu^2]\,[(\mu+3)^2 - \nu^2]\ldots\left[(\mu+2m+1)^2 - \nu^2\right]}$

$\qquad\qquad\qquad\qquad\qquad\qquad$ [$\mu \pm \nu$ is not a negative odd integer]

$\qquad\quad = z^{\mu-1} \displaystyle\sum_{m=0}^{\infty} \frac{(-1)^m \left(\frac{z}{2}\right)^{2m+2} \Gamma\left(\frac{1}{2}\mu - \frac{1}{2}\nu + \frac{1}{2}\right) \Gamma\left(\frac{1}{2}\mu + \frac{1}{2}\nu + \frac{1}{2}\right)}{\Gamma\left(\frac{1}{2}\mu - \frac{1}{2}\nu + m + \frac{3}{2}\right) \Gamma\left(\frac{1}{2}\mu + \frac{1}{2}\nu + m + \frac{3}{2}\right)}$

$\qquad\qquad$ [$\mu \pm \nu$ is not a negative odd integer]$\quad$ EH II 40(69), WA 377(2)

2.[12]     $S_{\mu,\nu}(z) = s_{\mu,\nu}(z) + 2^{\mu-1} \Gamma\left(\frac{1}{2}\mu - \frac{1}{2}\nu + \frac{1}{2}\right) \Gamma\left(\frac{1}{2}\mu + \frac{1}{2}\nu + \frac{1}{2}\right)$

$\qquad\qquad \times \dfrac{\cos\left[\frac{1}{2}(\mu-\nu)\pi\right] J_{-\nu}(z) - \cos\left[\frac{1}{2}(\mu+\nu)\pi\right] J_{\nu}(z)}{\sin\nu\pi}$

$\qquad\qquad\qquad\qquad$ [$\mu \pm \nu$ is a positive odd integer, $\nu$ is an odd integer]

$\qquad\qquad\qquad\qquad\qquad\qquad$ EH II 40(71), WA 379(2)

$\qquad\quad = s_{\mu,\nu}(z) + 2^{\mu-1} \Gamma\left(\frac{1}{2}\mu - \frac{1}{2}\nu + \frac{1}{2}\right) \Gamma\left(\frac{1}{2}\mu + \frac{1}{2}\nu + \frac{1}{2}\right)$

$\qquad\qquad \times \left\{\sin\left[\frac{1}{2}(\mu-\nu)\pi\right] J_{\nu}(z) - \cos\left[\frac{1}{2}(\mu-\nu)\pi\right] Y_{\nu}(z)\right\}$

$\qquad\qquad\qquad\qquad$ [$\mu \pm \nu$ is a positive odd integer, $\nu$ is an integer]

$\qquad\qquad\qquad\qquad\qquad\qquad$ EH II 41(71), WA 379(3)

## Integral representations

**8.571**   $s_{\mu,\nu}(z) = \dfrac{\pi}{2}\left[Y_\nu(z)\displaystyle\int_0^z z^\mu\, J_\nu(z)\,\mathrm{d}z - J_\nu(z)\int_0^z z^\mu\, Y_\nu(z)\,\mathrm{d}z\right]$       WA 378(9)

**8.572**   $s_{\mu,\nu}(z)$

$$= 2^\mu\left(\frac{z}{2}\right)^{\frac{1}{2}(1+\nu+\mu)}\Gamma\left(\frac{1}{2}+\frac{1}{2}\mu-\frac{1}{2}\nu\right)\int_0^{\pi/2} J_{\frac{1}{2}(1+\mu-\nu)}\left(z\sin\theta\right)(\sin\theta)^{\frac{1}{2}(1+\nu-\mu)}(\cos\theta)^{\nu+\mu}\,\mathrm{d}\theta$$

$$[\mathrm{Re}(\nu+\mu+1)>0] \qquad \text{EH II 42(86)}$$

**8.573**   Special cases:

1.     $S_{1,2n}(z) = zO_{2n}(z)$                                                        WA 382(1)

2.     $S_{0,2n+1}(z) = \dfrac{z}{2n+1}\, O_{2n+1}(z)$                                   WA 382(1)

3.     $S_{-1,2n}(z) = \dfrac{1}{4n}\, S_{2n}(z)$                                        WA 382(2)

4.     $S_{0,2n+1}(z) = \dfrac{1}{2}\, S_{2n+1}(z)$                                       WA 382(2)

5.     $s_{\nu,\nu}(z) = \Gamma\left(\nu+\dfrac{1}{2}\right)\sqrt{\pi}2^{\nu-1}\,\mathbf{H}_\nu(z)$                        EH II 42(84)

6.     $S_{\nu,\nu}(z) = [\mathbf{H}_\nu(z) - Y_\nu(z)]\,2^{\nu-1}\sqrt{\pi}\,\Gamma\left(\nu+\frac{1}{2}\right)$               EH II 42(84)

7.*    $s_{\nu+1,\nu}(z) = z^\nu - 2^\nu\Gamma(\nu+1)J_\nu(z)$

**8.574**   Connections with other special functions:

1.     $\mathbf{J}_\nu(z) = \dfrac{1}{\pi}\sin(\nu\pi)\left[s_{0,\nu}(z) - \nu\, s_{-1,\nu}(z)\right]$                    EH II 41(82)

2.[12]   $\mathbf{E}_\nu(z) = -\dfrac{1}{\pi}\left[(1+\cos\nu\pi)\,s_{0,\nu}(z) + \nu\left(1-\cos\nu\pi\right)s_{-1,\nu}(z)\right]$       EH II 42(83)

### A connection with a generalized hypergeometric function

3.     $s_{\mu,\nu}(z) = \dfrac{z^{\mu+1}}{(\mu-\nu+1)(\mu+\nu+1)}\;{}_1F_2\left(1;\dfrac{\mu-\nu+3}{2},\dfrac{\mu+\nu+3}{2};-\dfrac{z^2}{4}\right)$

$$\text{EH II 40(69), WA 378(10)}$$

**8.575**   Functional relations:

1.     $s_{\mu+2,\nu}(z) = z^{\mu+1} - \left[(\mu+1)^2 - \nu^2\right]s_{\mu,\nu}(z)$               EH II 41(73), WA 380(1)

2.[8]   $s'_{\mu,\nu}(z) + \left(\dfrac{\nu}{z}\right)s_{\mu,\nu}(z) = (\mu+\nu-1)\,s_{\mu-1,\nu-1}(z)$       EH II 41(74), WA 380(2)

3.     $s'_{\mu,\nu}(z) - \left(\dfrac{\nu}{z}\right)s_{\mu,\nu}(z) = (\mu-\nu-1)\,s_{\mu-1,\nu+1}(z)$       EH II 41(75), WA 380(3)

4.     $\left(2\dfrac{\nu}{z}\right)s_{\mu,\nu}(z) = (\mu+\nu-1)\,s_{\mu-1,\nu-1}(z) - (\mu-\nu-1)\,s_{\mu-1,\nu+1}(z)$    EH II 41(76), WA 380(4)

5.[8]   $2\,s'_{\mu,\nu}(z) = (\mu+\nu-1)\,s_{\mu-1,\nu-1}(z) + (\mu-\nu-1)\,s_{\mu-1,\nu+1}(z)$    EH II 41(77), WA 380(5)

In formulas **8.575** 1–5, $s_{\mu,\nu}(z)$ can be replaced with $S_{\mu,\nu}(z)$.

**8.576**   Asymptotic expansion of $S_{\mu,\nu}(z)$.

In the case in which $\mu \pm \nu$ is not a positive odd integer, $S_{\mu,\nu}(z)$ has the following asymptotic expansion:

$$S_{\mu,\nu}(z) \sim z^{\mu-1} \sum_{m=0}^{\infty} (-1)^m \left(\frac{1-\mu+\nu}{2}\right)_m \left(\frac{1-\mu-\nu}{2}\right)_m \left(\frac{z}{2}\right)^{-2m}$$

$$[|z| \to \infty, \quad |\arg z| < \pi] \qquad \text{WA 347, 352}$$

The series terminates and is equal to $S_{\mu,\nu}(z)$ when $\mu \pm \nu$ is a positive odd integer.

**8.577**   Lommel functions satisfy the following differential equation:

$$z^2 w'' + z w' + \left(z^2 - \nu^2\right) w = z^{\mu+1} \qquad \text{WA 377(1), EH II 40(68)}$$

**8.578**   Lommel functions of two variables $U_\nu(w,z)$ and $V_\nu(w,z)$:

## Definition

1.   $$U_\nu(w,z) = \sum_{m=0}^{\infty} (-1)^m \left(\frac{w}{z}\right)^{\nu+2m} J_{\nu+2m}(z) \qquad \text{EH II 42(87), WA 591(5)}$$

2.   $$V_\nu(w,z) = \cos\left[\frac{1}{2}\left(w + \frac{z^2}{w} + \nu\pi\right)\right] + U_{-\nu+2}(w,z) \qquad \text{EH II 42(88), WA 591(6)}$$

Particular values:

3.   $$U_0(z,z) = V_0(z,z) = \tfrac{1}{2}\{J_0(z) + \cos z\} \qquad \text{WA 591(9)}$$

4.   $$U_1(z,z) = -V_1(z,z) = \tfrac{1}{2}\sin z \qquad \text{WA 591(10)}$$

5.   $$U_{2n}(z,z) = \frac{(-1)^n}{2}\left\{\cos z - \sum_{m=0}^{n-1}(-1)^m \varepsilon_{2m} J_{2m}(z)\right\}$$

$$[n \geq 1], \quad \varepsilon_m = \begin{cases} 2, & m > 0, \\ 1, & m = 0 \end{cases} \qquad \text{WA 591(11)}$$

6.   $$U_{2n+1}(z,z) = \frac{(-1)^n}{2}\left\{\sin z - \sum_{m=0}^{n-1}(-1)^m \varepsilon_{2m+1} J_{2m+1}(z)\right\}$$

$$[n \geq 0], \quad \varepsilon_m = \begin{cases} 2, & m > 0, \\ 1, & m = 0 \end{cases} \qquad \text{WA 591(12)}$$

7.   $$V_n(w,z) = (-1)^n U_n\left(\frac{z^2}{w},z\right)$$

8.   $$U_\nu(w,0) = \frac{\left(\frac{w}{2}\right)^{1/2}}{\Gamma(\nu-1)} s_{\nu-\frac{3}{2},\frac{1}{2}}\left(\frac{w}{2}\right) \qquad \text{WA 593(9)}$$

9.   $$V_{-\nu+2}(w,0) = \frac{\left(\frac{w}{2}\right)^{1/2}}{\Gamma(\nu-1)} S_{\nu-\frac{3}{2},\frac{1}{2}}\left(\frac{w}{2}\right) \qquad \text{WA 593(10)}$$

**8.579**   Functional relations:

1.   $$2\frac{\partial}{\partial w} U_\nu(w,z) = U_{\nu-1}(w,z) + \left(\frac{z}{w}\right)^2 U_{\nu+1}(w,z) \qquad \text{WA 593(2)}$$

2.   $$2\frac{\partial}{\partial w} V_\nu(w,z) = V_{\nu+1}(w,z) + \left(\frac{z}{w}\right)^2 V_{\nu-1}(w,z) \qquad \text{WA 593(4)}$$

3.      The function $U_\nu(w,z)$ is a particular solution of the differential equation

$$\frac{\partial^2 U}{\partial z^2} - \frac{1}{z}\frac{\partial U}{\partial z} + \frac{z^2 U}{w^2} = \left(\frac{w}{z}\right)^{\nu-2} J_\nu(z) \qquad\qquad \text{WA 592(2)}$$

4.      The function $V_\nu(w,z)$ is a particular solution of the differential equation

$$\frac{\partial^2 V}{\partial z^2} - \frac{1}{z}\frac{\partial V}{\partial z} + \frac{z^2 V}{w^2} = \left(\frac{w}{z}\right)^{-\nu} J_{-\nu+2}(z) \qquad\qquad \text{WA 592(3)}$$

## 8.58   Anger and Weber functions J$_\nu(z)$ and E$_\nu(z)$

### 8.580   Definitions:

1.      The Anger function $\mathbf{J}_\nu(z)$:

$$\mathbf{J}_\nu(z) = \frac{1}{\pi}\int_0^\pi \cos\left(\nu\theta - z\sin\theta\right)\,\mathrm{d}\theta \qquad\qquad \text{WA 336(1), EH II 35(32)}$$

2.      The Weber function $\mathbf{E}_\nu(z)$:

$$\mathbf{E}_\nu(z) = \frac{1}{\pi}\int_0^\pi \sin\left(\nu\theta - z\sin\theta\right)\,\mathrm{d}\theta \qquad\qquad \text{WA 336(2), EH II 35(32)}$$

### 8.581   Series representations:

1.      $\displaystyle \mathbf{J}_\nu(z) = \cos\frac{\nu\pi}{2}\sum_{n=0}^\infty \frac{(-1)^n \left(\frac{z}{2}\right)^{2n}}{\Gamma\left(n+1+\frac{1}{2}\nu\right)\Gamma\left(n+1-\frac{1}{2}\nu\right)}$

$$+ \sin\frac{\nu\pi}{2}\sum_{n=0}^\infty \frac{(-1)^n \left(\frac{z}{2}\right)^{2n+1}}{\Gamma\left(n+\frac{3}{2}+\frac{1}{2}\nu\right)\Gamma\left(n+\frac{3}{2}-\frac{1}{2}\nu\right)}$$

$$\text{EH II 36(36), WA 337(3)}$$

2.      $\displaystyle \mathbf{E}_\nu(z) = \sin\frac{\nu\pi}{2}\sum_{n=0}^\infty \frac{(-1)^n \left(\frac{z}{2}\right)^{2n}}{\Gamma\left(n+1+\frac{1}{2}\nu\right)\Gamma\left(n+1-\frac{1}{2}\nu\right)}$

$$- \cos\frac{\nu\pi}{2}\sum_{n=0}^\infty \frac{(-1)^n \left(\frac{z}{2}\right)^{2n+1}}{\Gamma\left(n+\frac{3}{2}+\frac{1}{2}\nu\right)\Gamma\left(n+\frac{3}{2}-\frac{1}{2}\nu\right)}$$

$$\text{EH II 36(37), WA 338(4)}$$

### 8.582   Functional relations:

1.[6]    $2\,\mathbf{J}'_\nu(z) = \mathbf{J}_{\nu-1}(z) - \mathbf{J}_{\nu+1}(z)$                      EH II 36(40), WA 340(2)

2.[6]    $2\,\mathbf{E}'_\nu(z) = \mathbf{E}_{\nu-1}(z) - \mathbf{E}_{\nu+1}(z)$                      EH II 36(41), WA 340(6)

3.[6]    $\mathbf{J}_{\nu-1}(z) + \mathbf{J}_{\nu+1}(z) = 2\nu z^{-1}\,\mathbf{J}_\nu(z) - 2(\pi z)^{-1}\sin(\nu\pi)$       EH II 36(42), WA 340(1)

4.[6]    $\mathbf{E}_{\nu-1}(z) + \mathbf{E}_{\nu+1}(z) = 2\nu z^{-1}\,\mathbf{E}_\nu(z) - 2(\pi z)^{-1}(1 - \cos\nu\pi)$       EH II 36(43), WA 340(5)

**8.583**   Asymptotic expansions:

1.[6]   $$\mathbf{J}_\nu(z) = J_\nu(z) + \frac{\sin\nu\pi}{\pi z}\left[\sum_{n=0}^{p-1}(-1)^n 2^{2n}\frac{\Gamma\left(n+\frac{1+\nu}{2}\right)\Gamma\left(n+\frac{1-\nu}{2}\right)}{\Gamma\left(\frac{1+\nu}{2}\right)\Gamma\left(\frac{1-\nu}{2}\right)}z^{-2n}\right.$$

$$\left. + O\left(|z|^{-2p}\right) - \nu\sum_{n=0}^{p-1}(-1)^n 2^{2n}\frac{\Gamma\left(n+1+\frac{1}{2}\nu\right)\Gamma\left(n+1-\frac{1}{2}\nu\right)}{\Gamma\left(1+\frac{1}{2}\nu\right)\Gamma\left(1-\frac{1}{2}\nu\right)}z^{-2n-1} + \nu\, O\left(|z|^{-2p-1}\right)\right]$$

$$[|\arg z| < \pi] \qquad \text{EH II 37(47), WA 344(1)}$$

2.   $$\mathbf{E}_\nu(z) = -Y_\nu(z)$$

$$-\frac{1+\cos(\nu\pi)}{\pi z}\left[\sum_{n=0}^{p-1}(-1)^n 2^{2n}\frac{\Gamma\left(n+\frac{1+\nu}{2}\right)\Gamma\left(n+\frac{1-\nu}{2}\right)}{\Gamma\left(\frac{1+\nu}{2}\right)\Gamma\left(\frac{1-\nu}{2}\right)}z^{-2n} + O\left(|z|^{-2p}\right)\right]$$

$$-\frac{\nu(1-\cos\nu\pi)}{z\pi}\left[\sum_{n=0}^{p-1}(-1)^n 2^{2n}\frac{\Gamma\left(n+1+\frac{1}{2}\nu\right)\Gamma\left(n+1-\frac{1}{2}\nu\right)}{\Gamma\left(1+\frac{1}{2}\nu\right)\Gamma\left(1-\frac{1}{2}\nu\right)}z^{-2n-1} + O\left(|z|^{-2p-1}\right)\right]$$

$$\text{WA344(2), EH II 37(48)}$$

For the asymptotic expansion of $J_\nu(z)$ and $Y_\nu(z)$, see **8.451**.

**8.584**   The Anger and Weber functions satisfy the differential equation

$$y'' + z^{-1}y' + \left(1 - \frac{\nu^2}{z^2}\right)y = f(\nu, z),$$

where $f(\nu, z) = \dfrac{z-\nu}{\pi z^2}\sin\nu\pi$ for $\mathbf{J}_\nu(z)$          WA 341(9), EH II 37(44)

and $f(\nu, z) = -\dfrac{1}{\pi z^2}\left[z + \nu + (z-\nu)\cos\nu\pi\right]$ for $\mathbf{E}_\nu(z)$          EH II 37(45), WA 341(10)

## 8.59 Neumann's and Schläfli's polynomials: $O_n(z)$ and $S_n(z)$

**8.590**   Definition of Neumann's polynomials

1.   $$O_n(z) = \frac{1}{4}\sum_{m=0}^{\lfloor\frac{n}{2}\rfloor}\frac{n(n-m-1)!}{m!}\left(\frac{z}{2}\right)^{2m-n-1} \qquad [n\geq 1] \qquad \text{WA 299(2), EH II 33(6)}$$

2.   $$O_{-n}(z) = (-1)^n O_n(z) \qquad [n\geq 1] \qquad \text{WA 303(8)}$$

3.   $$O_0(z) = \frac{1}{z} \qquad \text{WA 299(3), EH II 33(7)}$$

4.   $$O_1(z) = \frac{1}{z^2} \qquad \text{EH II 33(7)}$$

5.   $$O_2(z) = \frac{1}{z} + \frac{4}{z^3} \qquad \text{EH II 33(7)}$$

In general, $O_n(z)$ is a polynomial in $z^{-1}$ of degree $n+1$.

**8.591**   Functional relations:

1.   $$O_0'(z) = -O_1(z) \qquad \text{EH II 33(9), WA 301(3)}$$

2.   $$2\,O_n'(z) = O_{n-1}(z) - O_{n+1}(z) \qquad [n\geq 1] \qquad \text{EH II 33(10), WA 301(2)}$$

3.      $(n-1)\,O_{n+1}(z) + (n+1)\,O_{n-1}(z) - 2z^{-1}\left(n^2-1\right)O_n(z) = 2nz^{-1}\left(\sin n\dfrac{\pi}{2}\right)^2$

                                           $[n \geq 1]$          EH II 33(11), WA 301(1)

4.[12]    $nz\,O_{n-1}(z) - \left(n^2-1\right)O_n(z) = (n-1)z\,O_n'(z) + n\left(\sin n\dfrac{\pi}{2}\right)^2$       EH II 33(12), WA 303(4)

5.      $nz\,O_{n+1}(z) - \left(n^2-1\right)O_n(z) = -(n+1)z\,O_n'(z) + n\left(\sin n\dfrac{\pi}{2}\right)^2$      EH II 33(13), WA 303(5)a

**8.592**    The generating function:

$$\frac{1}{z-\xi} = J_0(\xi)z^{-1} + 2\sum_{n=1}^{\infty} J_n(\xi)\,O_n(z) \qquad [|\xi| < |z|]$$

                                                                 EH II 32(1), WA 298(1)

**8.593**    The integral representation:

$$O_n(z) = \int_0^{\infty} \frac{\left[u + \sqrt{u^2+z^2}\right]^n + \left[u - \sqrt{u^2+z^2}\right]^n}{2z^{n+1}} e^{-u}\,du$$

See also **3.547** 6, 8, **3.549** 1, 2.                                            EH II 32(3), WA 305(1)

**8.594**    The inequality

$$|O_n(z)| \leq 2^{n-1}n!\,|z|^{-n-1}e^{\frac{1}{4}|z|^2} \qquad [n > 1]$$

                                                                 EH II 33(8), WA 300(8)

**8.595**    Neumann's polynomial $O_n(z)$ satisfies the differential equation

$$z^2\frac{d^2y}{dz^2} + 3z\frac{dy}{dz} + \left(z^2 + 1 - n^2\right)y = z\left(\cos n\frac{\pi}{2}\right)^2 + n\left(\sin n\frac{\pi}{2}\right)^2$$      EH II 33(14), WA 303(1)

**8.596**    Schläfli's polynomials $S_n(z)$. These are the functions that satisfy the formulas

1.      $S_0(z) = 0$                                                            EH II 34(18), WA 312(2)

2.      $S_n(z) - \dfrac{1}{n}\left[2zO_n(z) - 2\left(\cos n\dfrac{\pi}{2}\right)^2\right] \qquad [n \geq 1]$          EH II 34(19), WA 312(3)

         $= \displaystyle\sum_{m=0}^{\lfloor\frac{n}{2}\rfloor} \frac{(n-m-1)!}{m!}\left(\frac{z}{2}\right)^{2m-n} \qquad [n \geq 1]$                           EH II 34(18)

3.      $S_{-n}(z) = (-1)^{n+1}\,S_n(z)$                                                WA 313(6)

**8.597**    Functional relations:

1.      $S_{n-1}(z) + S_{n+1}(z) = 4\,O_n(z)$                                           WA 313(7)

Other functional relations may be obtained from **8.591** by replacing $O_n(z)$ with the expression for $S_n(z)$ given by **8.596** 2.

# 8.6 Mathieu Functions

## 8.60 Mathieu's equation

$$\frac{d^2y}{dz^2} + \left(a - 2k^2\cos 2z\right)y = 0, \quad k^2 = q$$

                                                                         MA

## 8.61 Periodic Mathieu functions

**8.610**   In general, Mathieu's equation **8.60** does not have periodic solutions. If $k$ is a real number, there exist infinitely many *eigenvalues a*, not identically equal to zero, corresponding to the periodic solutions

$$y(z) = y(2\pi + z),$$

If $k$ is nonzero, there are no other linearly independent periodic solutions. Periodic solutions of Mathieu's equations are called *Mathieu's periodic functions* or *Mathieu functions of the first kind*, or, more simply, *Mathieu functions*.

**8.611**   Mathieu's equation has four series of distinct periodic solutions:

1.      $\displaystyle \mathrm{ce}_{2n}(z, q) = \sum_{r=0}^{\infty} A_{2r}^{(2n)} \cos 2rz$                                  MA

2.      $\displaystyle \mathrm{ce}_{2n+1}(z, q) = \sum_{r=0}^{\infty} A_{2r+1}^{(2n+1)} \cos(2r+1)z$                                  MA

3.      $\displaystyle \mathrm{se}_{2n+1}(z, q) = \sum_{r=0}^{\infty} B_{2r+1}^{(2n+1)} \sin(2r+1)z$                                  MA

4.      $\displaystyle \mathrm{se}_{2n+2}(z, q) = \sum_{r=0}^{\infty} B_{2r+2}^{(2n+2)} \sin(2r+2)z$                                  MA

5.      The coefficients $A$ and $B$ depend on $q$. The eigenvalues $a$ of the functions $\mathrm{ce}_{2n}$, $\mathrm{ce}_{2n+1}$, $\mathrm{se}_{2n}$, $\mathrm{se}_{2n+1}$ are denoted by $a_{2n}$, $a_{2n+1}$, $b_{2n}$, $b_{2n+1}$.

**8.612**   The solutions of Mathieu's equation are normalized so that

$$\int_0^{2\pi} y^2 \, \mathrm{d}x = \pi$$                                  MO 65

**8.613**

1.      $\displaystyle \lim_{q \to 0} \mathrm{ce}_0(x) = \frac{1}{\sqrt{2}}$

2.      $\displaystyle \lim_{q \to 0} \mathrm{ce}_n(x) = \cos nx$                                  $[n \neq 0]$

3.      $\displaystyle \lim_{q \to 0} \mathrm{se}_n(x) = \sin nx$                                  MO 65

## 8.62 Recursion relations for the coefficients $A_{2r}^{(2n)}$, $A_{2r+1}^{(2n+1)}$, $B_{2r+1}^{(2n+1)}$, $B_{2r+2}^{(2n+2)}$

**8.621**

1.      $a A_0^{(2n)} - q A_2^{(2n)} = 0$                                  MA

2.      $(a - 4) A_2^{(2n)} - q \left( A_4^{(2n)} + 2 A_0^{(2n)} \right) = 0$                                  MA

3.      $\left( a - 4r^2 \right) A_{2r}^{(2n)} - q \left( A_{2r+2}^{(2n)} + A_{2r-2}^{(2n)} \right) = 0$                                  $[r \geq 2]$                                  MA

**8.622**

1.     $(a - 1 - q)A_1^{(2n+1)} - qA_3^{(2n+1)} = 0$                                          MA

2.     $\left[a - (2r + 1)^2\right] A_{2r+1}^{(2n+1)} - q\left(A_{2r+3}^{(2n+1)} + A_{2r-1}^{(2n+1)}\right) = 0$     $[r \geq 1]$        MA

**8.623**

1.     $(a - 1 + q)B_1^{(2n+1)} - qB_3^{(2n+1)} = 0$                                          MA

2.     $\left[a - (2r + 1)^2\right] B_{2r+1}^{(2n+1)} - q\left(B_{2r+3}^{(2n+1)} + B_{2r-1}^{(2n+1)}\right) = 0$

                                                       $[r \geq 1]$        MA

**8.624**

1.     $(a - 4)B_2^{(2n+2)} - qB_4^{(2n+2)} = 0$                                             MA

2.[11]    $\left(a - 4r^2\right) B_{2r}^{(2n+2)} - q\left(B_{2r+2}^{(2n+2)} + B_{2r-2}^{(2n+2)}\right) = 0$         $[r \geq 2]$        MA

**8.625**    We can determine the coefficients $A$ and $B$ from equations **8.612, 8.613** and **8.621-8.624** provided $a$ is known. Suppose, for example, that we need to determine the coefficients $A_{2r}^{(2n)}$ for the function $\mathrm{ce}_{2n}(z, q)$. From the recursion formulas, we have

1.    
$$\begin{vmatrix} a & -q & 0 & 0 & 0 & \cdots \\ -2q & a-4 & -q & 0 & 0 & \cdots \\ 0 & -q & a-16 & -q & 0 & \cdots \\ 0 & 0 & -q & a-36 & -q & \\ 0 & 0 & 0 & -q & a-64 & \\ \vdots & \vdots & \vdots & & & \ddots \end{vmatrix} = 0$$
       ST

     For given $q$ in equation **8.625** 1, we may determine the eigenvalues

2.[12]    $a = a_0, a_2, a_4, \ldots$                               $[|a_0| \leq |a_2| \leq |a_4| \leq \ldots]$

     If we now set $a = a_{2n}$, we can determine the coefficients $A_{2r}^{(2n)}$ from the recursion formulas **8.621** up to a proportionality coefficient. This coefficient is determined from the formula

3.     $2\left[A_0^{(2n)}\right]^2 + \sum\limits_{r=1}^{\infty}\left[A_{2r}^{(2n)}\right]^2 = 1,$                                  MA

     which follows from the conditions of normalization.

## 8.63   Mathieu functions with a purely imaginary argument

**8.630**    If, in equation **8.60**, we replace $z$ with $iz$, we arrive at the differential equation

1.[11]    $\dfrac{d^2y}{dz^2} + (-a + 2q\cosh 2z)\, y = 0$

We can find the solutions of this equation if we replace the argument $z$ with $iz$ in the functions $\mathrm{ce}_n(z, q)$ and $\mathrm{se}_n(z, q)$. The functions obtained in this way are called *associated Mathieu functions of the first kind* and are denoted as follows:

1.      $\mathrm{Ce}_{2n}(z,q), \quad \mathrm{Ce}_{2n+1}(z,q), \quad \mathrm{Se}_{2n+1}(z,q), \quad \mathrm{Se}_{2n+2}(z,q)$

**8.631**

1.      $\mathrm{Ce}_{2n}(z,q) = \sum_{r=0}^{\infty} A_{2r}^{(2n)} \cosh 2rz$                                    MA

2.      $\mathrm{Ce}_{2n+1}(z,q) = \sum_{r=0}^{\infty} A_{2r+1}^{(2n+1)} \cosh(2r+1)z$                                    MA

3.      $\mathrm{Se}_{2n+1}(z,q) = \sum_{r=0}^{\infty} B_{2r+1}^{(2n+1)} \sinh(2r+1)z$                                    MA

4.      $\mathrm{Se}_{2n+2}(z,q) = \sum_{r=0}^{\infty} B_{2r+2}^{(2n+2)} \sinh(2r+2)z$                                    MA

## 8.64  Non-periodic solutions of Mathieu's equation

Along with each periodic solution of equation **8.60**, there exists a second non-periodic solution that is linearly independent. The non-periodic solutions are denoted as follows:

$$\mathrm{fe}_{2n}(z,q), \quad \mathrm{fe}_{2n+1}(z,q), \quad \mathrm{ge}_{2n+1}(z,q), \quad \mathrm{ge}_{2n+2}(z,q)$$

Analogously, the second solutions of equation **8.630** 1 are denoted by

$$\mathrm{Fe}_{2n}(z,q), \quad \mathrm{Fe}_{2n+1}(z,q), \quad \mathrm{Ge}_{2n+1}(z,q), \quad \mathrm{Ge}_{2n+2}(z,q)$$

## 8.65  Mathieu functions for negative $q$

**8.651**   If we replace the argument $z$ in equation **8.60** with$\pm\left(\dfrac{\pi}{2} \pm z\right)$, we get the equation

$$\frac{d^2 y}{dz^2} + (a + 2q\cos 2z)\, y = 0 \qquad\qquad \text{MA}$$

This equation has the following solutions:

**8.652**

1.      $\mathrm{ce}_{2n}(z,-q) = (-1)^n \, \mathrm{ce}_{2n}\left(\tfrac{1}{2}\pi - z, q\right)$                                    MA

2.      $\mathrm{ce}_{2n+1}(z,-q) = (-1)^n \, \mathrm{se}_{2n+1}\left(\tfrac{1}{2}\pi - z, q\right)$                                    MA

3.      $\mathrm{se}_{2n+1}(z,-q) = (-1)^n \, \mathrm{ce}_{2n+1}\left(\tfrac{1}{2}\pi - z, q\right)$                                    MA

4.      $\mathrm{se}_{2n+2}(z,-q) = (-1)^n \, \mathrm{se}_{2n+2}\left(\tfrac{1}{2}\pi - z, q\right)$                                    MA

5.      $\mathrm{fe}_{2n}(z,-q) = (-1)^{n+1} \, \mathrm{fe}_{2n}\left(\tfrac{1}{2}\pi - z, q\right)$                                    MA

6.      $\mathrm{fe}_{2n+1}(z,-q) = (-1)^n \, \mathrm{ge}_{2n+1}\left(\tfrac{1}{2}\pi - z, q\right)$                                    MA

7.      $\mathrm{ge}_{2n+1}(z,-q) = (-1)^n \, \mathrm{fe}_{2n+1}\left(\tfrac{1}{2}\pi - z, q\right)$                                    MA

8.      $\mathrm{ge}_{2n+2}(z,-q) = (-1)^n \, \mathrm{ge}_{2n+2}\left(\tfrac{1}{2}\pi - z, q\right)$                                    MA

**8.653**  Analogously, if we replace $z$ with $\dfrac{\pi}{2}i + z$ in equation **8.630** 1 we get the equation

$$\frac{d^2 y}{dz^2} - (a + 2q \cosh z)\, y = 0$$

It has the following solutions:

**8.654**

1.  $\quad \mathrm{Ce}_{2n}(z, -q) = (-1)^n\, \mathrm{Ce}_{2n}\left(\dfrac{\pi}{2}i + z, q\right)$                                                        MA

2.  $\quad \mathrm{Ce}_{2n+1}(z, -q) = (-1)^{n+1} i\, \mathrm{Se}_{2n+1}\left(\tfrac{1}{2}\pi i + z, q\right)$                                          MA

3.  $\quad \mathrm{Se}_{2n+1}(z, -q) = (-1)^{n+1} i\, \mathrm{Ce}_{2n+1}\left(\tfrac{1}{2}\pi i + z, q\right)$                                          MA

4.  $\quad \mathrm{Se}_{2n+2}(z, -q) = (-1)^{n+1}\, \mathrm{Se}_{2n+2}\left(\tfrac{1}{2}\pi i + z, q\right)$                                            MA

5.  $\quad \mathrm{Fe}_{2n}(z, -q) = (-1)^n\, \mathrm{Fe}_{2n}\left(\tfrac{1}{2}\pi i + z, q\right)$                                                    MA

6.[11]  $\quad \mathrm{Fe}_{2n+1}(z, -q) = (-1)^{n+1} i\, \mathrm{Ge}_{2n+1}\left(\tfrac{1}{2}\pi i + z, q\right)$                                      MA

7.[11]  $\quad \mathrm{Ge}_{2n+1}(z, -q) = (-1)^{n+1} i\, \mathrm{Fe}_{2n+1}\left(\tfrac{1}{2}\pi i + z, q\right)$                                      MA

8.[11]  $\quad \mathrm{Ge}_{2n+2}(z, -q) = (-1)^{n+1}\, \mathrm{Ge}_{2n+2}\left(\tfrac{1}{2}\pi i + z, q\right)$                                        MA

## 8.66 Representation of Mathieu functions as series of Bessel functions

**8.661**

1.  $\quad \mathrm{ce}_{2n}(z, q) = \dfrac{\mathrm{ce}_{2n}\left(\tfrac{\pi}{2}, q\right)}{A_0^{(2n)}} \displaystyle\sum_{r=0}^{\infty} (-1)^r A_{2r}^{(2n)}\, J_{2r}(2k \cos z)$                                              MA

   $\quad = \dfrac{\mathrm{ce}_{2n}(0, q)}{A_0^{(2n)}} \displaystyle\sum_{r=0}^{\infty} (-1)^r A_{2r}^{(2n)}\, I_{2r}(2k \sin z)$                                              MA

2.  $\quad \mathrm{ce}_{2n+1}(z, q) = -\dfrac{\mathrm{ce}_{2n+1}'\left(\tfrac{\pi}{2}, q\right)}{k A_1^{(2n+1)}} \displaystyle\sum_{r=0}^{\infty} (-1)^r A_{2r+1}^{(2n+1)}\, J_{2r+1}(2k \cos z)$                                              MA

   $\quad = \dfrac{\mathrm{ce}_{2n+1}(0, q)}{k A_1(2n+1)} \cot z \displaystyle\sum_{r=0}^{\infty} (-1)^r (2r+1) A_{2r+1}^{(2n+1)}\, I_{2r+1}(2k \sin z)$                                              MA

3.  $\quad \mathrm{se}_{2n+1}(z, q) = \dfrac{\mathrm{se}_{2n+1}\left(\tfrac{\pi}{2}, q\right)}{k B_1^{(2n+1)}} \tan z \displaystyle\sum_{r=0}^{\infty} (-1)^r (2r+1) B_{2r+1}^{(2n+1)}\, J_{2r+1}(2k \cos z)$                                              MA

   $\quad = \dfrac{\mathrm{se}_{2n+1}'(0, q)}{k B_1^{(2n+1)}} \displaystyle\sum_{r=0}^{\infty} (-1)^r B_{2r+1}^{(2n+1)}\, I_{2r+1}(2k \sin z)$                                              MA

4.  $\quad \mathrm{se}_{2n+2}(z, q) = \dfrac{-\,\mathrm{se}_{2n+2}'\left(\tfrac{\pi}{2}, q\right)}{k^2 B_2^{(2n+2)}} \tan z \displaystyle\sum_{r=0}^{\infty} (-1)^r (2r+2) B_{2r+2}^{(2n+2)}\, J_{2r+2}(2k \cos z)$                                              MA

   $\quad = \dfrac{\mathrm{se}_{2n+2}'(0, q)}{k^2 B_2^{(2n+2)}} \cot z \displaystyle\sum_{r=0}^{\infty} (-1)^r (2r+2) B_{2r+2}^{(2n+2)}\, I_{2r+2}(2k \sin z)$                                              MA

**8.662**

1.  $\quad \mathrm{fe}_{2n}(z, q) = -\dfrac{\pi\, \mathrm{fe}_{2n}'(0, q)}{2\, \mathrm{ce}_{2n}\left(\tfrac{\pi}{2}, q\right)} \displaystyle\sum_{r=0}^{\infty} (-1)^r A_{2r}^{(2n)}\, \mathrm{Im}\left[ J_r\left(k e^{iz}\right) Y_r\left(k e^{-iz}\right) \right]$                                              MA

2. $\quad \mathrm{fe}_{2n+1}(z,q) = \dfrac{\pi k \, \mathrm{fe}_{2n+1}'(0,q)}{2 \, \mathrm{ce}_{2n+1}'\left(\frac{\pi}{2},q\right)}$

$$\times \sum_{r=0}^{\infty} (-1)^r A_{2r+1}^{(2n+1)} \, \mathrm{Im} \left[ J_r\left(ke^{iz}\right) Y_{r+1}\left(ke^{-iz}\right) + J_{r+1}\left(ke^{iz}\right) Y_r\left(ke^{-iz}\right) \right]$$

MA

3. $\quad \mathrm{ge}_{2n+1}(z,q) = -\dfrac{\pi k \, \mathrm{ge}_{2n+1}(0,q)}{2 \, \mathrm{se}_{2n+1}\left(\frac{\pi}{2},q\right)}$

$$\times \sum_{r=0}^{\infty} (-1)^r B_{2r+1}^{(2n+1)} \, \mathrm{Re} \left[ J_r\left(ke^{iz}\right) Y_{r+1}\left(ke^{-iz}\right) - J_{r+1}\left(ke^{iz}\right) Y_r\left(ke^{-iz}\right) \right]$$

MA

4. $\quad \mathrm{ge}_{2n+2}(z,q) = -\dfrac{\pi k^2 \, \mathrm{ge}_{2n+2}(0,q)}{2 \, \mathrm{se}_{2n+2}'\left(\frac{1}{2}\pi,q\right)}$

$$\times \sum_{r=0}^{\infty} (-1)^r \, \mathrm{Re} \left[ J_k\left(ke^{iz}\right) Y_{r+2}\left(ke^{-iz}\right) - J_{r+2}\left(ke^{iz}\right) Y_r\left(ke^{-iz}\right) \right]$$

MA

The expansions of the functions $\mathrm{Fe}_n$ and $\mathrm{Ge}_n$ as series of the functions $Y_\nu$ are denoted, respectively, by $\mathrm{Fey}_n$ and $\mathrm{Gey}_n$ and the expansions of these functions as series of the functions $K_\nu$ are denoted, respectively, by $\mathrm{Fek}_n$ and $\mathrm{Gek}_n$.

**8.663**

1.[12] $\quad \mathrm{Fey}_{2n}(z,q) = \dfrac{\mathrm{ce}_{2n}(0,q)}{A_0^{(2n)}} \sum_{r=0}^{\infty} A_{2r}^{(2n)} \, Y_{2r}(2k \sinh z)$

$$\left[ k^2 = q, \quad |\sinh z| > 1, \quad \mathrm{Re}\, z > 0 \right]$$

MA

$$= \dfrac{\mathrm{ce}_{2n}\left(\frac{\pi}{2},q\right)}{A_0^{(2n)}} \sum_{r=0}^{\infty} (-1)^r A_{2r}^{(2n)} \, Y_{2r}(2k \cosh z)$$

$$\left[ |\cosh z| > 1 \right]$$

MA

$$= \dfrac{\mathrm{ce}_{2n}(0,q)\, \mathrm{ce}_{2n}\left(\frac{\pi}{2},q\right)}{\left[ A_0^{(2n)} \right]^2} \sum_{r=0}^{\infty} (-1)^r A_{2r}^{(2n)} \, J_r\left(ke^{-z}\right) Y_r\left(ke^{z}\right)$$

MA

2.    $\text{Fey}_{2n+1}(z,q) = \dfrac{\text{ce}_{2n+1}(0,q)\coth z}{kA_1(2n+1)}\displaystyle\sum_{r=0}^{\infty}(2r+1)A_{2r+1}^{(2n+1)}\,Y_{2r+1}(2k\sinh z),$

$$k^2 = q, \quad [|\sinh z| > 1, \quad \operatorname{Re} z > 0]$$

MA

$$= -\frac{\text{ce}'_{2n+1}\left(\frac{\pi}{2},q\right)}{kA_1^{(2n+1)}}\sum_{r=0}^{\infty}(-1)^r A_{2r+1}^{(2n+1)}\,Y_{2r+1}(2k\cosh z)$$

$$[|\cosh z| > 1]$$

MA

$$= -\frac{\text{ce}_{2n+1}(0,q)\,\text{ce}'_{2n+1}\left(\frac{\pi}{2},q\right)}{k\left[A_1^{(2n+1)}\right]^2}$$

$$\times\sum_{r=0}^{\infty}(-1)^r A_{2r+1}^{(2n+1)}\left[J_r\left(ke^{-z}\right)Y_{r+1}\left(ke^z\right)+J_{r+1}\left(ke^{-z}\right)Y_r\left(ke^z\right)\right]$$

MA

3.[12]    $\text{Gey}_{2n+1}(z,q) = \dfrac{\text{se}'_{2n+1}(0,q)}{kB_1^{(2n+1)}}\displaystyle\sum_{r=0}^{\infty}B_{2r+1}^{(2n+1)}\,Y_{2r+1}(2k\sinh z)$

$$[|\sinh z| > 1, \quad \operatorname{Re} z > 0]$$

MA

$$= \frac{\text{se}_{2n+1}\left(\frac{\pi}{2},q\right)}{kB_1^{(2n+1)}}\tanh z\sum_{r=0}^{\infty}(-1)^r(2r+1)B_{2r+1}^{(2n+1)}\,Y_{2r+1}(2k\cosh z)$$

$$[|\cosh z| > 1]$$

MA

$$= \frac{\text{se}_{2n+1}(0,q)\,\text{se}_{2n+1}\left(\frac{\pi}{2},q\right)}{k\left[B_1^{(2n+1)}\right]^2}\sum_{r=0}^{\infty}(-1)^r B_{2r+1}^{(2n+1)}$$

$$\times\left[J_r\left(ke^{-z}\right)Y_{r+1}\left(ke^z\right)\right]-J_{r+1}\left(ke^{-z}\right)Y_r\left(ke^z\right)$$

MA

4.  $$\mathrm{Gey}_{2n+2}(z,q) = \frac{\mathrm{se}'_{2n+2}(0,q)}{k^2 B_2^{(2n+2)}} \coth z \sum_{r=0}^{\infty} (2r+2) B_{2r+2}^{(2n+2)} \, Y_{2r+2} \, (2k \sinh z)$$

$$[|\sinh z| > 1, \quad \mathrm{Re}\, z > 0]$$
MA

$$= -\frac{\mathrm{se}'_{2n+2}\left(\frac{\pi}{2},q\right)}{k^2 B_2^{(2n+2)}} \tanh z \sum_{r=0}^{\infty} (-1)^r (2r+2) B_{2r+2}^{(2n+2)} \, Y_{2r+2} \, (2k \cosh z)$$

$$[|\cosh z| > 1]$$
MA

$$= \frac{\mathrm{se}'_{2n+2}(0,q)\, \mathrm{se}'_{2n+2}\left(\frac{\pi}{2},q\right)}{k^2 \left[B_2^{(2n+2)}\right]^2} \sum_{r=0}^{\infty} (-1)^r B_{2r+2}^{(2n+2)}$$

$$\times \left[J_r\left(ke^{-z}\right) Y_{r+2}\left(ke^z\right)\right] - J_{r+2}\left(ke^{-z}\right) Y_r\left(ke^z\right)$$

MA

**8.664**

1.  $$\mathrm{Fek}_{2n}(z,q) = \frac{\mathrm{ce}_{2n}(0,q)}{\pi A_0^{(2n)}} \sum_{r=0}^{\infty} (-1)^r A_{2r}^{(2n)} \, K_{2r}\left(-2ik \sinh z\right)$$

$$k^2 = q, \quad [|\sinh z| > 1, \quad \mathrm{Re}\, z > 0]$$
MA

2.  $$\mathrm{Fek}_{2n+1}(z,q) = \frac{\mathrm{ce}_{2n+1}(0,q)}{\pi k A_1^{(2n+1)}} \coth z \sum_{r=0}^{\infty} (-1)^r (2r+1) A_{2r+1}^{(2n+1)} \, K_{2r+1}\left(-2ik \sinh z\right)$$

$$k^2 = q \quad [|\sinh z| > 1, \quad \mathrm{Re}\, z > 0]$$
MA

3.  $$\mathrm{Gek}_{2n+1}(z,q) = \frac{\mathrm{se}_{2n+1}\left(\frac{\pi}{2},q\right)}{\pi k B_1^{(2n+1)}} \tanh z \sum_{r=0}^{\infty} (2r+1) B_{2r+1}^{(2n+1)} \, K_{2r+1}\left(-2ik \cosh z\right)$$
MA

4.  $$\mathrm{Gek}_{2n+2}(z,q) = \frac{\mathrm{se}'_{2n+2}\left(\frac{\pi}{2},q\right)}{\pi k^2 B_2^{(2n+2)}} \tanh z \sum_{r=0}^{\infty} (2r+2) B_{2r+2}^{(2n+2)} \, K_{2r+2}\left(-2ik \cosh z\right)$$
MA

## 8.67 The general theory

If $i\mu$ is not an integer, the general solution of equation **8.60** can be found in the form

**8.671**

1.  $$y = Ae^{\mu z} \sum_{r=-\infty}^{\infty} c_{2r} e^{2rzi} + Be^{-\mu z} \sum_{r=-\infty}^{\infty} c_{2r} e^{-2rzi}$$
MA

The coefficients $c_{2r}$ can be determined from the homogeneous system of linear algebraic equations

2.[11]  $$c_{2r} + \xi_{2r}\left(c_{2r+2} + c_{2r-2}\right) = 0, \qquad r = \ldots, -2, -1, 0, 1, 2, \ldots,$$
MA

where

$$\xi_{2r} = \frac{q}{(2r - i\mu)^2 - a}$$

The condition that this system be compatible yields an equation that $\mu$ must satisfy:

3.[7]    $\Delta(i\mu) = \begin{vmatrix} \cdot & \cdot & \cdot & \cdot & \cdot & \cdot & \cdot & \cdot & \cdot \\ \cdot & \xi_{-4} & 1 & \xi_{-4} & 0 & 0 & 0 & 0 & \cdot \\ \cdot & 0 & \xi_{-2} & 1 & \xi_{-2} & 0 & 0 & 0 & \cdot \\ \cdot & 0 & 0 & \xi_0 & 1 & \xi_0 & 0 & 0 & \cdot \\ \cdot & 0 & 0 & 0 & \xi_2 & 1 & \xi_2 & 0 & \cdot \\ \cdot & \cdot & \cdot & \cdot & \cdot & \cdot & \cdot & \cdot & \cdot \end{vmatrix} = 0$          MA

This equation can also be written in the form

4.    $\cosh \mu\pi = 1 - 2\Delta(0)\sin^2\left(\dfrac{\pi\sqrt{a}}{2}\right)$, where $\Delta(0)$ is the value that is assumed by the determinant of the preceding article if we set $\mu = 0$ in the expressions for $\xi_{2r}$.

5.    If the pair $(a, q)$ is such that $|\cosh \mu\pi| < 1$, then $\mu = i\beta, \operatorname{Im} \beta = 0$, and the solution **8.671** 1 is bounded on the real axis.

6.    If $|\cosh \mu\pi| > 1, \mu$ may be real or complex and the solution **8.671** 1 will not be bounded on the real axis.

7.    If $\cosh \mu\pi = \pm 1$, then $i\mu$ will be an integer. In this case, one of the solutions will be of period $\pi$ or $2\pi$ (depending on whether $n$ is even or odd). The second solution is non-periodic (see **8.61** and **8.64**).

# 8.7–8.8 Associated Legendre Functions

## 8.70 Introduction

**8.700**   An *associated Legendre function* is a solution of the differential equation

1.    $(1 - z^2)\dfrac{d^2 u}{dz^2} - 2z\dfrac{du}{dz} + \left[\nu(\nu + 1) - \dfrac{\mu^2}{1 - z^2}\right]u = 0,$

in which $\nu$ and $\mu$ are arbitrary complex constants.

This equation is a special case of (Riemann's) hypergeometric equation (see **9.151**). The points

$$+1, -1, \infty$$

are, in general, its *singular points*, specifically, its ordinary branch points.

We are interested, on the one hand, in solutions of the equation that correspond to real values of the independent variable $z$ that lie in the interval $[-1, 1]$ and, on the other hand, in solutions corresponding to an arbitrary complex number $z$ such that $\operatorname{Re} z > 1$. These are multiple-valued in the $z$-plane. To separate these functions into single-valued branches, we make a cut along the real axis from $-\infty$ to $+1$. We are also interested in those solutions of equation **8.700** 1 for which $\nu$ or $\mu$ or both are integers. Of special significance is the case in which $\mu = 0$.

**8.701**   In connection with this, we shall use the following notations:

The letter $z$ will denote *an arbitrary complex variable*; the letter $x$ will denote a *real* variable that varies over the interval $[-1, +1]$. We shall sometimes set $x = \cos \varphi$, where $\varphi$ is a real number.

We shall use the symbols $P_\nu^\mu(z)$, $Q_\nu^\mu(z)$ to denote those solutions of equation **8.700** 1, that are single-valued and regular for $|z| < 1$ and, in particular, uniquely determined for $z = x$.

We shall use the symbols $P_\nu^\mu(z)$, $Q_\nu^\mu(z)$ to denote those solutions of equation **8.700** 1 that are single-valued and regular for Re $z > 1$. When these functions cannot be unrestrictedly extended without violating their single-valuedness we make a cut along the real axis to the left of the point $z = 1$. The values of the functions $P_\nu^\mu(z)$ and $Q_\nu^\mu(z)$ on the upper and lower boundaries of that portion of the cuts lying between the points $-1$ and $+1$ are denoted respectively by

$$P_\nu^\mu(x \pm i0), \quad Q_\nu^\mu(x \pm i0)$$

The letters $n$ and $m$ denote natural numbers or zero. The letters $\nu$ and $\mu$ denote arbitrary complex numbers unless the contrary is stated.

The upper index will be omitted when it is equal to zero. That is, we set

$$P_\nu^0(z) = P_\nu(z), \quad Q_\nu^0(z) = Q_\nu(z)$$

The *linearly independent* functions

**8.702**    $P_\nu^\mu(z) = \dfrac{1}{\Gamma(1 - \mu)} \left( \dfrac{z+1}{z-1} \right)^{\frac{\mu}{2}} F \left( -\nu, \nu + 1; \quad 1 - \mu; \quad \dfrac{1-z}{2} \right)$

$$\left[ \arg \frac{z+1}{z-1} = 0, \text{ if } z \text{ is real and greater than 1 and} \right] \qquad \text{MO 80, WH}$$

**8.703**   $Q_\nu^\mu(z) = \dfrac{e^{\mu \pi i} \Gamma(\nu + \mu + 1) \Gamma\left(\frac{1}{2}\right)}{2^{\nu+1} \Gamma\left(\nu + \frac{3}{2}\right)} \left( z^2 - 1 \right)^{\frac{\mu}{2}} z^{-\nu-\mu-1} F \left( \dfrac{\nu+\mu+2}{2}, \dfrac{\nu+\mu+1}{2}; \nu + \dfrac{3}{2}; \dfrac{1}{z^2} \right)$

$[\arg(z^2 - 1) = 0$ when $z$ is real and greater than 1; $\arg z = 0$ when $z$ is real and greater than zero] which are solutions of the differential equation **8.700** 1, are called *associated Legendre functions* (or *spherical functions*) of *the first* and *second kinds* respectively. They are uniquely defined, respectively, in the intervals $|1 - z| < 2$ and $|z| > 1$ with the portion of the real axis that lies between $-\infty$ and $+1$ excluded. They can be extended by means of hypergeometric series to the entire $z$-plane where the above-mentioned cut was made. These expressions for $P_\nu^\mu(z)$ and $Q_\nu^\mu(z)$ lose their meaning when $1 - \mu$ and $\nu + \frac{3}{2}$ are non-positive integers respectively.      MO 80

When $z$ is a real number lying on the interval $[-1, +1]$, so that $(z = x = \cos \varphi)$, we take the following functions as linearly independent solutions of the equation

**8.704**   $P_\nu^\mu(x) = \frac{1}{2} \left[ e^{\frac{1}{2}\mu\pi i} P_\nu^\mu(\cos \varphi + i0) + e^{-\frac{1}{2}\mu\pi i} P_\nu^\mu(\cos \varphi - i0) \right]$      EH I 143(1)

$$= \frac{1}{\Gamma(1-\mu)} \left( \frac{1+x}{1-x} \right)^{\frac{\mu}{2}} F \left( -\nu, \nu + 1; 1 - \mu; \frac{1-x}{2} \right) \qquad \text{EH I 143(6)}$$

**8.705**   $Q_\nu^\mu(x) = \frac{1}{2} e^{-\mu\pi i} \left[ e^{-\frac{1}{2}\mu\pi i} Q_\nu^\mu(x + i0) + e^{\frac{1}{2}\mu\pi i} Q_\nu^\mu(x - i0) \right]$      EH I 143(2)

$$= \frac{\pi}{2 \sin \mu\pi} \left[ P_\nu^\mu(x) \cos \mu\pi - \frac{\Gamma(\nu + \mu + 1)}{\Gamma(\nu - \mu + 1)} P_\nu^{-\mu}(x) \right] \qquad \text{(cf. 8.732 5)}$$

If $\mu = \pm m$ is an integer, the last equation loses its meaning. In this case, we get the following formulas by passing to the limit:

**8.706**

1.      $Q_\nu^m(x) = (-1)^m \left( 1 - x^2 \right)^{\frac{m}{2}} \dfrac{d^m}{dx^m} Q_\nu(x)$      (cf. **8.752** 1)      EH I 149(7)

2.[11]    $Q_\nu^{-m}(x) = \dfrac{\Gamma(\nu - m + 1)}{\Gamma(\nu + m + 1)} Q_\nu^m(x)$      EH I 144(18)

The functions $Q_\nu^\mu(z)$ are not defined when $\nu + \mu$ is equal to a negative integer. Therefore, we must exclude the cases when $\nu + \mu = -1, -2, -3, \ldots$ for these formulas.

The functions

$$P_\nu^{\pm\mu}(\pm z), \quad Q_\nu^{\pm\mu}(\pm z), \quad P_{-\nu-1}^{\pm\mu}(\pm z), \quad Q_{-\nu-1}^{\pm\mu}(\pm z)$$

are *linearly independent solutions* of the differential equation for $\nu + \mu \neq 0, \pm 1, \pm 2, \ldots$.

**8.707**  Nonetheless, two linearly independent solutions can always be found. Specifically, for $\nu \pm \mu$ not an integer, the differential equation **8.700** 1 has the following solutions:

1.  $P_\nu^{\pm\mu}(\pm z), \quad Q_\nu^{\pm\mu}(\pm z), \quad P_{-\nu-1}^{\pm\mu}(\pm z), \quad Q_{-\nu-1}^{\pm\mu}(\pm z)$

    respectively, for $z = x = \cos\varphi$,

2.  $P_\nu^{\pm\mu}(\pm x), \quad Q_\nu^{\pm\mu}(\pm x), \quad P_{-\nu-1}^{\pm\mu}(\pm x), \quad Q_{-\nu-1}^{\pm\mu}(\pm x)$

    If $\nu \pm \mu$ is not an integer, the solutions

3.  $P_\nu^\mu(z), \quad Q_\nu^\mu(z)$, respectively, and $P_\nu^\mu(x), \quad Q_\nu^\mu(x)$

    are linearly independent. If $\nu \pm \mu$ is an integer but $\mu$ itself is not an integer, the following functions are linearly independent solutions of equation **8.700** 1:

4.  $P_\nu^\mu(z), \quad P_\nu^{-\mu}(z)$, respectively, and $P_\nu^\mu(x), \quad P_\nu^{-\mu}(x)$

    If $\mu = \pm m, \nu = n$, or $\nu = -n-1$, the following functions are linearly independent solutions of equation **8.700** 1 for $n \geq$m:

5.  $P_n^m(z), \quad Q_n^m(z)$, respectively, and $P_n^m(x), \quad Q_n^m(x)$,

    and for $n <$m, the following functions will be linearly independent solutions

6.  $P_n^{-m}(z), \quad Q_n^m(z)$, respectively, and $P_n^{-m}(x), \quad Q_n^m(x)$

## 8.71 Integral representations

**8.711**

1.  $$P_\nu^{-\mu}(z) = \frac{(z^2-1)^{\frac{\mu}{2}}}{2^\mu \sqrt{\pi}\, \Gamma\left(\mu+\frac{1}{2}\right)} \int_{-1}^{1} \frac{(1-t^2)^{\mu-\frac{1}{2}}}{(z+t\sqrt{z^2-1})^{\mu-\nu}}\, dt \qquad \left[\operatorname{Re}\mu > -\frac{1}{2}, \quad |\arg(z\pm 1)| < \pi\right]$$

    MO 88

2.  $$P_\nu^m(z) = \frac{(\nu+1)(\nu+2)\ldots(\nu+m)}{\pi} \int_0^\pi \left[z+\sqrt{z^2-1}\cos\varphi\right]^\nu \cos m\varphi\, d\varphi$$

    $$= (-1)^m \frac{\nu(\nu-1)\ldots(\nu-m+1)}{\pi} \int_0^\pi \frac{\cos m\varphi\, d\varphi}{\left[z+\sqrt{z^2-1}\cos\varphi\right]^{\nu+1}}$$

    $$\left[|\arg z| < \frac{\pi}{2}, \quad \arg\left(z+\sqrt{z^2-1}\cos\varphi\right) = \arg z \text{ for } \varphi = \frac{\pi}{2}\right] \quad \text{(cf. \textbf{8.822} 1)} \quad \text{SM 483(15), WH}$$

3.  $$Q_\nu^\mu(z) = \sqrt{\pi}\, \frac{e^{\mu\pi i}\, \Gamma(\nu+\mu+1)}{2^\mu \Gamma\left(\mu+\frac{1}{2}\right)\Gamma(\nu-\mu+1)} (z^2-1)^{\frac{\mu}{2}} \int_0^\infty \frac{\sinh^{2\mu} t\, dt}{(z+\sqrt{z^2-1}\cosh t)^{\nu+\mu+1}}$$

    $$\left[\operatorname{Re}(\nu\pm\mu) > -1, \quad |\arg(z\pm 1)| < \pi\right] \qquad \text{(cf. \textbf{8.822} 2)} \quad \text{MO 88}$$

4. $\quad Q_\nu^\mu(z) = \dfrac{e^{\mu\pi i}\,\Gamma(\nu+1)}{\Gamma(\nu-\mu+1)} \displaystyle\int_0^\infty \dfrac{\cosh\mu t\,dt}{\left(z+\sqrt{z^2-1}\,\cosh t\right)^{\nu+1}}$

$$[\operatorname{Re}(\nu+\mu) > -1, \nu \neq -1, -2, -3, \ldots, \quad |\arg(z\pm 1)| < \pi] \quad \text{WH, MO 88}$$

5.[12] $\quad \displaystyle\int_{-1}^1 P_l^2(x)\,P_l^0(x)\,dx = -\dfrac{l!}{(l-2)!}\dfrac{1}{2l+1} = -2\dfrac{l(l-1)}{2l+1}$

6.* $\quad \displaystyle\int_{-1}^1 P_l^{2m}(x)\,P_l(x) = \dfrac{(-1)^m}{2l+1}\dfrac{2l!}{(l-2m)!}$

**8.712**[12] $\quad Q_\nu^\mu(z) = \dfrac{e^{\mu\pi i}\,\Gamma(\nu+\mu+1)}{2^{\nu+1}\,\Gamma(\nu+1)}(z^2-1)^{\frac{\mu}{2}}\displaystyle\int_{-1}^1 (1-t^2)^\nu\,(z-t)^{-\nu-\mu-1}\,dt$

$$[\operatorname{Re}(\nu+\mu) > -1, \quad \operatorname{Re}\mu > -1, \quad |\arg(z\pm 1)| < \pi] \qquad (\text{cf. } \mathbf{8.821}\ 2) \quad \text{MO 88a, EH I 155(5)a}$$

**8.713**

1. $\quad Q_\nu^\mu(z) = \dfrac{e^{\mu\pi i}\,\Gamma\left(\mu+\dfrac{1}{2}\right)}{\sqrt{2\pi}}(z^2-1)^{\frac{\mu}{2}}\left\{\displaystyle\int_0^\pi \dfrac{\cos\left(\nu+\frac{1}{2}\right)t\,dt}{(z-\cos t)^{\mu+\frac{1}{2}}} - \cos\nu\pi\displaystyle\int_0^\infty \dfrac{e^{-\left(\nu+\frac{1}{2}\right)t}\,dt}{(z+\cosh t)^{\mu+\frac{1}{2}}}\right\}$

$$\left[\operatorname{Re}\mu > -\tfrac{1}{2}, \quad \operatorname{Re}(\nu+\mu) > -1, \quad |\arg(z\pm 1)| < \pi\right] \quad \text{MO 89}$$

2. $\quad P_\nu^{-\mu}(z) = \dfrac{(z^2-1)^{\frac{\mu}{2}}}{2^\nu\,\Gamma(\mu-\nu)\,\Gamma(\nu+1)}\displaystyle\int_0^\infty \dfrac{\sinh^{2\nu+1} t}{(z+\cosh t)^{\nu+\mu+1}}\,dt$

$$[\operatorname{Re} z > -1, \quad |\arg(z\pm 1)| < \pi, \quad \operatorname{Re}(\nu+1) > 0, \quad \operatorname{Re}(\mu-\nu) > 0] \quad \text{MO 89}$$

3. $\quad P_\nu^{-\mu}(z) = \sqrt{\dfrac{2}{\pi}}\dfrac{\Gamma\left(\mu+\frac{1}{2}\right)(z^2-1)^{\frac{\mu}{2}}}{\Gamma(\nu+\mu+1)\,\Gamma(\mu-\nu)}\displaystyle\int_0^\infty \dfrac{\cosh\left(\nu+\frac{1}{2}\right)t\,dt}{(z+\cosh t)^{\mu+\frac{1}{2}}}$

$$[\operatorname{Re} z > -1, \quad |\arg(z\pm 1)| < \pi, \quad \operatorname{Re}(\nu+\mu) > -1, \quad \operatorname{Re}(\mu-\nu) > 0] \quad \text{MO 89}$$

**8.714**

1. $\quad P_\nu^\mu(\cos\varphi) = \sqrt{\dfrac{2}{\pi}}\dfrac{\sin^\mu\varphi}{\Gamma\left(\frac{1}{2}-\mu\right)}\displaystyle\int_0^\varphi \dfrac{\cos\left(\nu+\frac{1}{2}\right)t\,dt}{(\cos t-\cos\varphi)^{\mu+\frac{1}{2}}} \qquad \left[0<\varphi<\pi, \quad \operatorname{Re}\mu<\tfrac{1}{2}\right]; \quad (\text{cf. } \mathbf{8.823})$

$$\text{MO 87}$$

2. $\quad P_\nu^{-\mu}(\cos\varphi) = \dfrac{\Gamma(2\mu+1)\sin^\mu\varphi}{2^\mu\,\Gamma(\mu+1)\,\Gamma(\nu+\mu+1)\,\Gamma(\mu-\nu)}\displaystyle\int_0^\infty \dfrac{t^{\nu+\mu}\,dt}{(1+2t\cos\varphi+t^2)^{\mu+\frac{1}{2}}}$

$$[\operatorname{Re}(\nu+\mu) > -1, \quad \operatorname{Re}(\mu-\nu) > 0]$$
$$\text{MO 89}$$

3. $\quad Q_\nu^\mu(\cos\varphi) = \dfrac{1}{2^{\mu+1}}\dfrac{\Gamma(\nu+\mu+1)}{\Gamma(\nu-\mu+1)}\dfrac{\sin^\mu\varphi}{\Gamma\left(\mu+\frac{1}{2}\right)}$

$$\times\displaystyle\int_0^\infty\left[\dfrac{\sinh^{2\mu} t}{(\cos\varphi+i\sin\varphi\cosh t)^{\nu+\mu+1}} + \dfrac{\sinh^{2\mu} t}{(\cos\varphi-i\sin\varphi\cosh t)^{\nu+\mu+1}}\right]dt$$

$$\left[\operatorname{Re}(\nu+\mu+1) > 0, \quad \operatorname{Re}(\nu-\mu+1) > 0, \quad \operatorname{Re}\mu > -\tfrac{1}{2}\right] \quad \text{MO 89}$$

4. $\quad P_\nu^\mu (\cos\varphi) = \dfrac{i}{2^\mu} \dfrac{\Gamma(\nu+\mu+1)}{\Gamma(\nu-\mu+1)} \dfrac{\sin^\mu \varphi}{\Gamma\left(\mu+\frac{1}{2}\right)}$

$$\times \int_0^\infty \left[ \frac{\sinh^{2\mu} t}{(\cos\varphi + i\sin\varphi\cosh t)^{\nu+\mu+1}} - \frac{\sinh^{2\mu} t}{(\cos\varphi - i\sin\varphi\cosh t)^{\nu+\mu+1}} \right] dt$$

$$\left[ \mathrm{Re}\,(\nu \pm \mu + 1) > 0, \quad \mathrm{Re}\,\mu > -\tfrac{1}{2} \right] \quad \text{MO 89}$$

**8.715**

1. $\quad P_\nu^\mu (\cosh\alpha) = \dfrac{\sqrt{2}\sinh^\mu \alpha}{\sqrt{\pi}\,\Gamma\left(\frac{1}{2}-\mu\right)} \displaystyle\int_0^\alpha \dfrac{\cosh\left(\nu+\frac{1}{2}\right) t\, dt}{(\cosh\alpha - \cosh t)^{\mu+\frac{1}{2}}}$

$$\left[ \alpha > 0, \quad \mathrm{Re}\,\mu < \tfrac{1}{2} \right] \qquad \text{MO 87}$$

2. $\quad Q_\nu^\mu (\cosh\alpha) = \sqrt{\dfrac{\pi}{2}}\, \dfrac{e^{\mu\pi i}\sinh^\mu \alpha}{\Gamma\left(\frac{1}{2}-\mu\right)} \displaystyle\int_\alpha^\infty \dfrac{e^{-\left(\nu+\frac{1}{2}\right)t}\, dt}{(\cosh t - \cosh\alpha)^{\mu+\frac{1}{2}}}$

$$\left[ \alpha > 0, \quad \mathrm{Re}\,\mu < \tfrac{1}{2}, \quad \mathrm{Re}(\nu+\mu) > -1 \right]$$
$$\text{MO 87}$$

See also **3.277** 1, 4, 5, 7, **3.318**, **3.516** 3, **3.518** 1, 2, **3.542** 2, **3.663** 1, **3.894**, **3.988** 3, **6.622** 3, **6.628** 1, 4–7, and also **8.742**.

## 8.72 Asymptotic series for large values of $|\nu|$

**8.721**[6]    For real values of $\mu, |\nu| \gg 1, |\nu| \gg |\mu|, |\arg\nu| < \pi$, we have:

1.[12] $\quad P_\nu^\mu(\cos\varphi) = \dfrac{2}{\sqrt{\pi}}\Gamma(\nu+\mu+1) \displaystyle\sum_{k=0}^\infty \dfrac{\Gamma\left(\mu+k+\frac{1}{2}\right)}{\Gamma\left(\mu-k+\frac{1}{2}\right)} \dfrac{\cos\left[\left(\nu+k+\frac{1}{2}\right)\varphi + \frac{\pi}{4}(2k-1) + \frac{\mu\pi}{2}\right]}{k!\,\Gamma\left(\nu-k+\frac{3}{2}\right)(2\sin\varphi)^{k+\frac{1}{2}}}$

$$\left[ \nu+\mu \neq -1, -2, -3, \ldots; \quad \nu \neq -\tfrac{3}{2}, -\tfrac{5}{2}, -\tfrac{7}{2} \ldots; \; \text{for } \tfrac{\pi}{6} < \varphi < \tfrac{5\pi}{6} \right.$$

This series also converges for complex values of $\nu$ and $\mu$.

In the remaining cases, it is an asymptotic expansion for

$$\left. |\nu| \gg |\mu|, |\nu| \gg 1, \text{if } \nu > 0, \mu > 0 \text{ and } 0 < \varepsilon \leq \varphi \leq \pi - \varepsilon \right]$$
$$\text{MO 92}$$

2.[6] $\quad Q_\nu^\mu(\cos\varphi) = \sqrt{\pi}\,\Gamma(\nu+\mu+1)$

$$\times \sum_{k=0}^\infty (-1)^k \frac{\Gamma\left(\mu+k+\frac{1}{2}\right)}{\Gamma\left(\mu-k+\frac{1}{2}\right)} \frac{\cos\left[\left(\nu+k+\frac{1}{2}\right)\varphi - \frac{\pi}{4}(2k-1) + \frac{\mu\pi}{2}\right]}{k!\,\Gamma\left(\nu+k+\frac{3}{2}\right)(2\sin\varphi)^{k+\frac{1}{2}}}$$

$$\left[ \nu+\mu \neq -1, -2, -3, \ldots; \quad \nu \neq -\tfrac{3}{2}, -\tfrac{5}{2}, -\tfrac{7}{2}, \ldots; \; \text{for } \tfrac{\pi}{6} < \varphi < \tfrac{5\pi}{6} \right.$$

This series also converges for complex values of $\nu$ and $\mu$.

In the remaining cases, it is an asymptotic expansion for

$$\left. |\nu| \gg |\mu|, \quad |\nu| \gg 1, \text{if } \nu > 0, \quad \mu > 0, \quad 0 < \varepsilon \leq \varphi \leq \pi - \varphi \right]$$

EH I 147(6), MO 92

3.    $P_\nu^\mu (\cos\varphi) = \dfrac{2}{\sqrt{\pi}} \dfrac{\Gamma(\nu+\mu+1)}{\Gamma(\nu+\frac{3}{2})} \dfrac{\cos\left[\left(\nu+\frac{1}{2}\right)\varphi - \frac{\pi}{4} + \frac{\mu\pi}{2}\right]}{\sqrt{2\sin\varphi}} \left[1 + O\left(\dfrac{1}{\nu}\right)\right]$

$$\left[0 < \varepsilon \le \varphi \le \pi - \varepsilon, \quad |\nu| \gg \dfrac{1}{\varepsilon}\right] \qquad \text{MO 92}$$

For $\nu > 0, \mu > 0$ and $\nu > \mu$, it follows from formulas **8.721** 1 and **8.721** 2 that

4.    $\nu^{-\mu} P_\nu^\mu (\cos\varphi) = \sqrt{\dfrac{2}{\nu\pi\sin\varphi}} \cos\left[\left(\nu+\dfrac{1}{2}\right)\varphi - \dfrac{\pi}{4} + \dfrac{\mu\pi}{2}\right] + O\left(\dfrac{1}{\sqrt{\nu^3}}\right)$

5.    $\nu^{-\mu} Q_\nu^\mu (\cos\varphi) = \sqrt{\dfrac{\pi}{2\nu\sin\varphi}} \cos\left[\left(\nu+\dfrac{1}{2}\right)\varphi + \dfrac{\pi}{4} + \dfrac{\mu\pi}{2}\right] O\left(\dfrac{1}{\sqrt{\nu^3}}\right)$

$$\left[0 < \varepsilon \le \varphi \le \pi - \varepsilon; \quad \nu \gg \dfrac{1}{\varepsilon}\right] \qquad \text{MO 92}$$

**8.722**   If $\varphi$ is sufficiently close to 0 or $\pi$ that $\nu\varphi$ or $\nu(\pi-\varphi)$ is small in comparison with 1, the asymptotic formulas **8.721** become unsuitable. In this case, the following asymptotic representation is applicable for $\mu \le 0, \nu \gg 1$, and *small* values of $\varphi$:

1.    $\left[\left(\nu+\dfrac{1}{2}\right)\cos\dfrac{\varphi}{2}\right]^\mu P_\nu^{-\mu} (\cos\varphi) = J_\mu(\eta) + \sin^2\dfrac{\varphi}{2}\left[\dfrac{J_{\mu+1}(\eta)}{2\eta} - J_{\mu+2}(\eta) + \dfrac{\eta}{6} J_{\mu+3}(\eta)\right] + O\left(\sin^4\dfrac{\varphi}{2}\right)$

where $\eta = (2\nu + 1)\sin\frac{\varphi}{2}$. In particular, it follows that

1.    $\displaystyle\lim_{\nu\to\infty} \nu^\mu P_\nu^{-\mu}\left(\cos\dfrac{x}{\nu}\right) = J_\mu(x)$          $[x \ge 0, \mu \ge 0]$                           MO 93

**8.723**   We can see how the functions $P_\nu^\mu(z)$ and $Q_\nu^\mu(z)$ behave for large $|\nu|$ and real values of $z > \frac{3}{2\sqrt{2}}$:

1.[12]  $P_\nu^\mu (\cosh\alpha) = \dfrac{2^\mu}{\sqrt{\pi}}\left\{\dfrac{\Gamma\left(-\nu-\frac{1}{2}\right)}{\Gamma(-\nu-\mu)} \dfrac{e^{(\mu-\nu)\alpha}\sinh^\mu\alpha}{(e^{2\alpha}-1)^{\mu+\frac{1}{2}}} F\left(\mu+\dfrac{1}{2}, -\mu+\dfrac{1}{2}; \nu+\dfrac{3}{2}; \dfrac{1}{1-e^{2\alpha}}\right)\right.$

$$\left. + \dfrac{\Gamma\left(\nu+\frac{1}{2}\right)}{\Gamma(\nu-\mu+1)} \dfrac{e^{(\nu+\mu+1)\alpha}\sinh^\mu\alpha}{(e^{2\alpha}-1)^{\mu+\frac{1}{2}}} F\left(\mu+\dfrac{1}{2}, -\mu+\dfrac{1}{2}; -\nu+\dfrac{1}{2}; \dfrac{1}{1-e^{2\alpha}}\right)\right\}$$

$$\left[\nu \neq \pm\tfrac{1}{2}, \pm\tfrac{3}{2}, \pm\tfrac{5}{2}, \ldots; \quad \alpha > \tfrac{1}{2}\ln 2\right] \qquad \text{MO 94}$$

2.    $Q_\nu^\mu (\cosh\alpha) = e^{\mu\pi i} 2^\mu \sqrt{\pi} \dfrac{\Gamma(\nu+\mu+1)}{\Gamma\left(\nu+\frac{3}{2}\right)} \dfrac{e^{-(\nu+\mu+1)\alpha}}{(1-e^{-2\alpha})^{\mu+\frac{1}{2}}} \sinh^\mu\alpha$

$$\times F\left(\mu+\dfrac{1}{2}, -\mu+\dfrac{1}{2}; \nu+\dfrac{3}{2}; \dfrac{1}{1-e^{2\alpha}}\right)$$

$$\left[\mu+\nu+1 \neq 0, -1, -2, \ldots; \quad \alpha > \tfrac{1}{2}\ln 2\right] \qquad \text{MO 94}$$

See also **8.776**.

**8.724**   For the inequalities in **8.776** 1–4, $\nu$ and $\mu$ are arbitrary real numbers satisfying the inequalities $\nu \ge 1, \nu - \mu + 1 > 0$, and $\mu \ge 0$:

1.    $\left|P_\nu^{\pm\mu} (\cos\varphi)\right| < \sqrt{\dfrac{8}{\nu\pi}} \dfrac{\Gamma(\nu \pm \mu + 1)}{\Gamma(\nu+1)} \dfrac{1}{\sin^{\mu+\frac{1}{2}}\varphi}$          MO 91-92

2. $\quad \left| Q_\nu^{\pm\mu}(\cos\varphi) \right| < \sqrt{\dfrac{2\pi}{\nu}} \dfrac{\Gamma(\nu \pm \mu + 1)}{\Gamma(\nu + 1)} \dfrac{1}{\sin^{\mu+\frac{1}{2}}\varphi}$ \hfill MO 91-92

3. $\quad \left| P_\nu^{\pm\mu}(\cos\varphi) \right| < \dfrac{2}{\sqrt{\nu\pi}} \dfrac{\Gamma(\nu \pm \mu + 1)}{\Gamma(\nu + 1)} \dfrac{1}{\sin^{\mu+\frac{1}{2}}\varphi}$ \hfill MO 91-92

4. $\quad \left| Q_\nu^{\pm\mu}(\cos\varphi) \right| < \sqrt{\dfrac{\pi}{\nu}} \dfrac{\Gamma(\nu \pm \mu + 1)}{\Gamma(\nu + 1)} \dfrac{1}{\sin^{\mu+\frac{1}{2}}\varphi}$ \hfill MO 91-92

5.[8] $\quad \left| \sqrt{\sin\varphi}\, P_n^m(\cos\varphi) \right| < \dfrac{\Gamma\left(n + \frac{1}{2}\right)}{\Gamma(n - m + 1)} 2^{(m+n)^2/n} \sup_{0<t<\infty} \left| \sqrt{t}\, J_m(t) \right|$

$$[\text{uniformly } 0 \le m \le n]$$

**8.725**[10]   For fixed $z$ and $\nu$ and $\operatorname{Re}\mu \to \infty$, with $z$ not on the real axis between $-\infty$ and $-1$ and $+\infty$ and $+1$, the following are asymptotic expansions in which the upper and lower signs are taken according to whether $\operatorname{Im} z$ is greater than or less than 0.

1. $\quad P_\nu^\mu(z) = \dfrac{\Gamma(\nu + \mu + 1)\,\Gamma(\mu - \nu)}{\pi\,\Gamma(\mu + 1)} \left(\dfrac{z+1}{z-1}\right)^{\frac{1}{2}\mu} \sin\mu\pi \left[ F\left(-\nu, \nu+1; 1+\mu; \dfrac{1}{2} + \dfrac{1}{2}z\right) \right.$

$$\left. - \dfrac{\sin\nu\pi}{\sin\mu\pi} e^{\mp i\mu\pi} \left(\dfrac{z-1}{z+1}\right)^\mu F\left(-\nu, \nu+1; 1+\mu; \dfrac{1}{2} - \dfrac{1}{2}z\right) \right]$$

\hfill AS 8.10.1

2. $\quad Q_\nu^\mu(z) = \dfrac{1}{2} e^{i\mu\pi} \dfrac{\Gamma(\nu + \mu + 1)}{\Gamma(\mu + 1)} \left(\dfrac{z+1}{z-1}\right)^{\frac{1}{2}\mu} \Gamma(\mu - \nu) \left[ F\left(-\nu, \nu+1; 1+\mu; \dfrac{1}{2} + \dfrac{1}{2}z\right) \right.$

$$\left. - e^{\mp i\nu\pi} \left(\dfrac{z-1}{z+1}\right)^\mu F\left(-\nu, \nu+1; 1+\mu; \dfrac{1}{2} - \dfrac{1}{2}z\right) \right]$$

\hfill AS 8.10.2

3. $\quad Q_\nu^{-\mu}(z) = \dfrac{e^{-i\mu\pi} \operatorname{cosec}[\pi(\nu - \mu)]}{2\pi\,\Gamma(1 + \mu)} \left[ e^{\mp i\nu\pi} \left(\dfrac{z+1}{z-1}\right)^{-\frac{1}{2}\mu} F\left(-\nu, \nu+1; 1+\mu; \dfrac{1}{2} - \dfrac{1}{2}z\right) \right.$

$$\left. - \left(\dfrac{z-1}{z+1}\right)^{-\frac{1}{2}\mu} F\left(-\nu, \nu+1; 1+\mu; \dfrac{1}{2} + \dfrac{1}{2}z\right) \right]$$

\hfill AS 8.10.3

## 8.73–8.74  Functional relations

**8.731**

1. $\quad (z^2 - 1)\dfrac{d\,P_\nu^\mu(z)}{dz} = (\nu - \mu + 1)\,P_{\nu+1}^\mu(z) - (\nu + 1)z\,P_\nu^\mu(z)$

$$(\text{cf. } \mathbf{8.832}\ 1,\ \mathbf{8.914}\ 2)$$

\hfill EH I 161(10), MO 81

$1(1)^9 \quad (z^2 - 1) \dfrac{d\,P_\nu^\mu(z)}{dz} = \nu z\,P_\nu^\mu(z) - (\nu + \mu)\,P_{\nu-1}^\mu(z)$ 
AS 8.5.4

$1(2)^{12} \quad (z^2 - 1) \dfrac{d\,P_\nu^\mu(z)}{dz} = -(\nu + \mu)(\nu - \mu + 1)\sqrt{1 - z^2}\,P_\nu^{\mu-1}(z) - \mu z\,P_\nu^\mu(z)$
AS 8.5.2

$2. \qquad (2\nu + 1)z\,P_\nu^\mu(z) = (\nu - \mu + 1)\,P_{\nu+1}^\mu(z) + (\nu + \mu)\,P_{\nu-1}^\mu(z)$

(cf. 8.832 2, 8.914 1)
EH I 160(2), MO 81

$3. \qquad P_\nu^{\mu+2}(z) + 2(\mu + 1)\dfrac{z}{\sqrt{z^2 - 1}}\,P_\nu^{\mu+1}(z) = (\nu - \mu)(\nu + \mu + 1)\,P_\nu^\mu(z)$
MO 82, EH I 160(1)

$3(1)^9 \quad P_\nu^{\mu+1}(z) = (z^2 - 1)^{-1/2}\left[(\nu - \mu)z\,P_\nu^\mu(z) - (\nu + \mu)\,P_{\nu-1}^\mu(z)\right]$
AS 8.5.1

$4. \qquad P_{\nu+1}^\mu(z) - P_{\nu-1}^\mu(z) = (2\nu + 1)\sqrt{z^2 - 1}\,P_\nu^{\mu-1}(z)$
EH I 160(3), MO 82

$4(1)^9 \quad (\nu - \mu + 1)\,P_{\nu+1}^\mu(z) = (2\nu + 1)z\,P_\nu^\mu(z) - (\nu + \mu)\,P_{\nu-1}^\mu(z)$
AS 334(8.5.3)

$4(2)^9 \quad P_{\nu+1}^\mu(z) = P_{\nu-1}^\mu(z) + (2\nu + 1)(z^2 - 1)^{1/2}\,P_\nu^{\mu-1}(z)$
AS 334(8.5.5)

$5. \qquad P_{-\nu-1}^\mu(z) = P_\nu^\mu(z) \qquad$ (cf. 8.820, 8.832 4)
EH I 140(1), MO 82

**8.732**

$1. \qquad (z^2 - 1)\dfrac{d\,Q_\nu^\mu(z)}{dz} = (\nu - \mu + 1)\,Q_{\nu+1}^\mu(z) - (\nu + 1)z\,Q_\nu^\mu(z)$

(cf. 8.832 3)
MO 82

$2.^{10} \quad (2\nu + 1)z\,Q_\nu^\mu(z) = (\nu - \mu + 1)\,Q_{\nu+1}^\mu(z) + (\nu + \mu)\,Q_{\nu-1}^\mu(z)$

(cf. 8.832 4)
MO 82

$3. \qquad Q_\nu^{\mu+2}(z) + 2(\mu + 1)\dfrac{z}{\sqrt{z^2 - 1}}\,Q_\nu^{\mu+1}(z) = (\nu - \mu)(\nu + \mu + 1)\,Q_\nu^\mu(z)$
MO 82

$4. \qquad Q_{\nu-1}^\mu(z) - Q_{\nu+1}^\mu(z) = -(2\nu + 1)\sqrt{z^2 - 1}\,Q_\nu^{\mu-1}(z)$
MO 82a

$5. \qquad e^{-\mu\pi i}\,Q_\nu^\mu(x \pm i0) = e^{\pm\frac{1}{2}\mu\pi i}\left[Q_\nu^\mu(x) \mp i\dfrac{\pi}{2}\,P_\nu^\mu(x)\right]$
MO 83

**8.733**

$1.^{12} \quad (1 - x^2)\dfrac{d\,P_\nu^\mu(x)}{dx} = (\nu + 1)x\,P_\nu^\mu(x) - (\nu - \mu + 1)\,P_{\nu+1}^\mu(x) \qquad$ (cf. 8.731 1)

$\qquad\qquad\qquad\qquad = -\nu x\,P_\nu^\mu(x) + (\nu + \mu)\,P_{\nu-1}^\mu(x)$

$\qquad\qquad\qquad\qquad = -\sqrt{1 - x^2}\,P_\nu^{\mu+1}(x) - \mu x\,P_\nu^\mu(x)$

$\qquad\qquad\qquad\qquad = (\nu - \mu + 1)(\nu + \mu)\sqrt{1 - x^2}\,P_\nu^{\mu-1}(x) + \mu x\,P_\nu^\mu(x)$
MO 82

$2. \qquad (2\nu + 1)x\,P_\nu^\mu(x) = (\nu - \mu + 1)\,P_{\nu+1}^\mu(x) + (\nu + \mu)\,P_{\nu-1}^\mu(x)$

(cf. 8.731 2)
MO 82

$3.^{11} \quad P_\nu^{\mu+2}(x) + 2(\mu + 1)\dfrac{x}{\sqrt{1 - x^2}}\,P_\nu^{\mu+1}(x) + (\nu - \mu)(\nu + \mu + 1)\,P_\nu^\mu(x) = 0$

(cf. 8.731 3)
MO 82

4.      $P_{\nu-1}^{\mu}(x) - P_{\nu+1}^{\mu}(x) = (2\nu + 1)\sqrt{1 - x^2}\, P_{\nu}^{\mu-1}(x)$      (cf. **8.731** 4)      MO 82

5.      $P_{-\nu-1}^{\mu}(x) = P_{\nu}^{\mu}(x)$      (cf. **8.731** 5)

**8.734**

1.[12]    $(\nu + \mu + 1)z\, Q_{\nu}^{\mu}(z) + \sqrt{z^2 - 1}\, Q_{\nu}^{\mu+1}(z) = (\nu - \mu + 1)\, Q_{\nu+1}^{\mu}(z)$      MO 82

2.      $(\nu + \mu)\, Q_{\nu-1}^{\mu}(z) + \sqrt{z^2 - 1}\, Q_{\nu}^{\mu+1}(z) = (\nu - \mu)z\, Q_{\nu}^{\mu}(z)$      MO 82

3.      $Q_{\nu-1}^{\mu}(z) - z\, Q_{\nu}^{\mu}(z) = -(\nu - \mu + 1)\sqrt{z^2 - 1}\, Q_{\nu}^{\mu-1}(z)$      MO 82

4.      $z\, Q_{\nu}^{\mu}(z) - Q_{\nu+1}^{\mu}(z) = -(\nu + \mu)\sqrt{z^2 - 1}\, Q_{\nu}^{\mu-1}(z)$      MO 82

5.      $(\nu + \mu)(\nu + \mu + 1)\, Q_{\nu-1}^{\mu}(z) + (2\nu + 1)\sqrt{z^2 - 1}\, Q_{\nu}^{\mu+1}(z) = (\nu - \mu)(\nu - \mu + 1)\, Q_{\nu+1}^{\mu}(z)$      MO 82

**8.735**

1.      $(\nu + \mu + 1)x\, P_{\nu}^{\mu}(x) + \sqrt{1 - x^2}\, P_{\nu}^{\mu+1}(x) = (\nu - \mu + 1)\, P_{\nu+1}^{\mu}(x)$      MO 83

2.      $(\nu - \mu)x\, P_{\nu}^{\mu}(x) - (\nu + \mu)\, P_{\nu-1}^{\mu}(x) = \sqrt{1 - x^2}\, P_{\nu}^{\mu+1}(x)$      MO 83

3.      $P_{\nu-1}^{\mu}(x) - x\, P_{\nu}^{\mu}(x) = (\nu - \mu + 1)\sqrt{1 - x^2}\, P_{\nu}^{\mu-1}(x)$      MO 83

4.      $x\, P_{\nu}^{\mu}(x) - P_{\nu+1}^{\mu}(x) = (\nu + \mu)\sqrt{1 - x^2}\, P_{\nu}^{\mu-1}(x)$      MO 83

5.      $(\nu - \mu)(\nu - \mu + 1)\, P_{\nu+1}^{\mu}(x) = (\nu + \mu)(\nu + \mu + 1)\, P_{\nu-1}^{\mu}(x) + (2\nu + 1)\sqrt{1 - x^2}\, P_{\nu}^{\mu+1}(x)$      MO 83

**8.736**

1.      $P_{\nu}^{-\mu}(z) = \dfrac{\Gamma(\nu - \mu + 1)}{\Gamma(\nu + \mu + 1)}\left[P_{\nu}^{\mu}(z) - \dfrac{2}{\pi}e^{-\mu\pi i}\sin\mu\pi\, Q_{\nu}^{\mu}(z)\right]$      MO 83

2.      $P_{\nu}^{\mu}(-z) = e^{\nu\pi i}\, P_{\nu}^{\mu}(z) - \dfrac{2}{\pi}\sin[(\nu + \mu)\pi]e^{-\mu\pi i}\, Q_{\nu}^{\mu}(z)$      [Im $z < 0$]      (cf. **8.833** 1)      MO 83

3.      $P_{\nu}^{\mu}(-z) = e^{-\nu\pi i}\, P_{\nu}^{\mu}(z) - \dfrac{2}{\pi}\sin[(\nu + \mu)\pi]e^{-\mu\pi i}\, Q_{\nu}^{\mu}(z)$

                                        [Im $z > 0$]      (cf. **8.833** 2)      MO 83

4.      $Q_{\nu}^{-\mu}(z) = e^{-2\mu\pi i}\dfrac{\Gamma(\nu - \mu + 1)}{\Gamma(\nu + \mu + 1)}\, Q_{\nu}^{\mu}(z)$      MO 82

5.      $Q_{\nu}^{\mu}(-z) = -e^{-\nu\pi i}\, Q_{\nu}^{\mu}(z)$      [Im $z < 0$]      MO 82

6.      $Q_{\nu}^{\mu}(-z) = -e^{\nu\pi i}\, Q_{\nu}^{\mu}(z)$      [Im $z > 0$]      MO 82

7.[6]    $Q_{\nu}^{\mu}(z)\sin[(\nu + \mu)\pi] - Q_{-\nu-1}^{\mu}(z)\sin[(\nu - \mu)\pi] = \pi e^{\mu\pi i}\cos\nu\pi\, P_{\nu}^{\mu}(z)$      MO 83

**8.737**

1.      $P_{\nu}^{-\mu}(x) = \dfrac{\Gamma(\nu - \mu + 1)}{\Gamma(\nu + \mu + 1)}\left[\cos\mu\pi\, P_{\nu}^{\mu}(x) - \dfrac{2}{\pi}\sin(\mu\pi)\, Q_{\nu}^{\mu}(x)\right]$      MO 84

2.      $P_{\nu}^{\mu}(-x) = \cos[(\nu + \mu)\pi]\, P_{\nu}^{\mu}(x) - \dfrac{2}{\pi}\sin[(\nu + \mu)\pi]\, Q_{\nu}^{\mu}(x)$      MO 84

3.      $Q_{\nu}^{\mu}(-x) = -\cos[(\nu + \mu)\pi]\, Q_{\nu}^{\mu}(x) - \dfrac{\pi}{2}\sin[(\nu + \mu)\pi]\, P_{\nu}^{\mu}(x)$      MO 83, EH I 144(15)

4.    $Q^{\mu}_{-\nu-1}(x) = \dfrac{\sin[(\nu+\mu)\pi]}{\sin[(\nu-\mu)\pi]}\, Q^{\mu}_{\nu}(x) - \dfrac{\pi\cos\nu\pi\cos\mu\pi}{\sin[(\nu-\mu)\pi]}\, P^{\mu}_{\nu}(x)$    MO 84

**8.738**

1.[11]    $Q^{\mu}_{\nu}(i\cot\varphi) = \exp\left[i\pi\left(\mu - \dfrac{\nu+1}{2}\right)\right]\sqrt{\pi}\,\Gamma(\nu+\mu+1)\sqrt{\dfrac{1}{2}\sin\varphi}\; P^{-\nu-\frac{1}{2}}_{-\mu-\frac{1}{2}}(\cos\varphi)$

$$\left[0 < \varphi < \dfrac{\pi}{2}\right]$$    MO 83

2.[6]    $P^{\mu}_{\nu}(i\cot\varphi) = \sqrt{\dfrac{2}{\pi}}\exp\left[i\pi\left(\nu+\dfrac{1}{4}\right)\right]\dfrac{\sqrt{\sin\varphi}}{\Gamma(-\nu-\mu)}\, Q^{-\nu-\frac{1}{2}}_{-\mu-\frac{1}{2}}(\cos\varphi - i0)$

$$\left[0 < \varphi < \dfrac{\pi}{2}\right]$$    MO 83

**8.739**    $e^{-\mu\pi i}\, Q^{\mu}_{\nu}(\cosh\alpha) = \dfrac{\sqrt{\pi}\,\Gamma(\nu+\mu+1)}{\sqrt{2\sinh\alpha}}\, P^{-\nu-\frac{1}{2}}_{-\mu-\frac{1}{2}}(\coth\alpha)$    $[\mathrm{Re}\,(\cosh\alpha) > 0]$    MO 83

**8.741**

1.    $P^{-\mu}_{\nu}(x)\dfrac{d\,P^{\mu}_{\nu}(x)}{dx} - P^{\mu}_{\nu}(x)\dfrac{d\,P^{-\mu}_{\nu}(x)}{dx} = \dfrac{2\sin\mu\pi}{\pi(1-x^2)}$    MO 83

2.    $P^{\mu}_{\nu}(x)\dfrac{d\,Q^{\mu}_{\nu}(x)}{dx} - Q^{\mu}_{\nu}(x)\dfrac{d\,P^{\mu}_{\nu}(x)}{dx} = \dfrac{2^{2\mu}}{1-x^2}\dfrac{\Gamma\left(\frac{\nu+\mu+1}{2}\right)\Gamma\left(\frac{\nu+\mu}{2}+1\right)}{\Gamma\left(\frac{\nu-\mu+1}{2}\right)\Gamma\left(\frac{\nu-\mu}{2}+1\right)}$    MO 83

**8.742**

1.[12]    $\dfrac{\Gamma(\nu-\mu+1)}{\Gamma(\nu+\mu+1)}\left\{\cos\mu\pi\, P^{\mu}_{\nu}(\cos\varphi) - \dfrac{2}{\pi}\sin\mu\pi\, Q^{\mu}_{\nu}(\cos\varphi)\right\} = \sqrt{\dfrac{2}{\pi}}\dfrac{\operatorname{cosec}^{\mu}\varphi}{\Gamma\left(\mu+\frac{1}{2}\right)}\displaystyle\int_0^{\varphi}\dfrac{\cos\left(\nu+\frac{1}{2}\right)t\,dt}{(\cos t - \cos\varphi)^{\frac{1}{2}-\mu}}$

$$\left[\mathrm{Re}\,\mu > -\tfrac{1}{2}\right]$$    MO 88

2.    $\dfrac{\Gamma(\nu-\mu+1)}{\Gamma(\nu+\mu+1)}\left\{\cos\nu\pi\, P^{\mu}_{\nu}(\cos\varphi) - \dfrac{2}{\pi}\sin\nu\pi\, Q^{\mu}_{\nu}(\cos\varphi)\right\}$

$$= \sqrt{\dfrac{2}{\pi}}\dfrac{\operatorname{cosec}^{\mu}\varphi}{\Gamma\left(\mu+\frac{1}{2}\right)}\int_{\varphi}^{\pi}\dfrac{\cos\left[\left(\nu+\frac{1}{2}\right)(t-\pi)\right]dt}{(\cos\varphi - \cos t)^{\frac{1}{2}-\mu}}$$

$$\left[\mathrm{Re}\,\mu > -\tfrac{1}{2}\right]$$    MO 88

3.    $P^{\mu}_{\nu}(\cos\varphi)\cos(\nu+\mu)\pi - \dfrac{2}{\pi}Q^{\mu}_{\nu}(\cos\varphi)\sin(\nu+\mu)\pi = \sqrt{\dfrac{2}{\pi}}\dfrac{\sin^{\mu}\varphi}{\Gamma\left(\frac{1}{2}-\mu\right)}\displaystyle\int_{\varphi}^{\pi}\dfrac{\cos\left[\left(\nu+\frac{1}{2}\right)(t-\pi)\right]dt}{(\cos\varphi - \cos t)^{\mu+\frac{1}{2}}}$

$$\left[\mathrm{Re}\,\mu < \tfrac{1}{2}\right]$$    MO 88

4.    $\cos\mu\pi\, P^{\mu}_{\nu}(\cos\varphi) - \dfrac{2}{\pi}\sin\mu\pi\, Q^{\mu}_{\nu}(\cos\varphi)$

$$= \dfrac{1}{2^{\mu}\sqrt{\pi}}\dfrac{\Gamma(\nu+\mu+1)}{\Gamma(\nu-\mu+1)}\dfrac{\sin^{\mu}\varphi}{\Gamma\left(\mu+\frac{1}{2}\right)}\int_0^{\pi}\dfrac{\sin^{2\mu}t\,dt}{(\cos\varphi \pm i\sin\varphi\cos t)^{\nu-\mu}}$$

$$\left[\mathrm{Re}\,\mu > -\tfrac{1}{2},\quad 0 < \varphi < \pi\right]$$    MO 38

For integrals of Legendre functions, see **7.11**–**7.21**.

## 8.75   Special cases and particular values

### 8.751

1.    $P_\nu^m(x) = (-1)^m \dfrac{\Gamma(\nu + m + 1)\left(1 - x^2\right)^{\frac{m}{2}}}{2^m \, \Gamma(\nu - m + 1) m!} \, F\left(m - \nu, m + \nu + 1; m + 1; \dfrac{1 - x}{2}\right)$       MO 84

2.    $P_\nu^m(z) = \dfrac{\Gamma(\nu + m + 1)\left(z^2 - 1\right)^{\frac{m}{2}}}{2^m m! \, \Gamma(\nu - m + 1)} \, F\left(m - \nu, m + \nu + 1; m + 1; \dfrac{1 - z}{2}\right)$       MO 84

3.[8]   $Q_{n + \frac{1}{2}}^\mu(z) = \dfrac{e^{\mu \pi i} \, \Gamma\left(\mu + n + \dfrac{3}{2}\right)}{2^{n + \frac{3}{2}} (n + 1)!} \left(z^2 - 1\right)^{\frac{\mu}{2}} \pi^{1/2} z^{-n - \mu - 3/2} \, F\left(\dfrac{\mu + n + \frac{5}{2}}{2}, \dfrac{\mu + n + \frac{3}{2}}{2}; n + 2; \dfrac{1}{z^2}\right)$

                                                                               MO 84

### 8.752

1.    $P_\nu^m(x) = (-1)^m \left(1 - x^2\right)^{\frac{m}{2}} \dfrac{d^m}{dx^m} P_\nu(x)$          WH, MO 84, EH I 148(6)

2.    $P_\nu^{-m}(x) = (-1)^m \dfrac{\Gamma(\nu - m + 1)}{\Gamma(\nu + m + 1)} P_\nu^m(x) = \left(1 - x^2\right)^{-\frac{m}{2}} \displaystyle\int_x^1 \cdots \int_x^1 P_\nu(x) (dx)^m$

                                      $[m \geq 1]$       HO 99a, MO 85, EH I 149(10)a

3.    $P_\nu^{-m}(z) = \left(z^2 - 1\right)^{-\frac{m}{2}} \displaystyle\int_1^z \cdots \int_1^z P_\nu(z) (dz)^m$       $[m \geq 1]$       MO 85, EH I 149(8)

4.    $Q_\nu^m(z) = \left(z^2 - 1\right)^{\frac{m}{2}} \dfrac{d^m}{dz^m} Q_\nu(z)$          WH, MO 85, EH I 148(5)

5.    $Q_\nu^{-m}(z) = (-1)^m \left(z^2 - 1\right)^{-\frac{m}{2}} \displaystyle\int_z^\infty \cdots \int_z^\infty Q_\nu(z) (dz)^m$

                                         $[m \geq 1]$       MO 85, EH I 149(9)

### Special values of the indices

### 8.753

1.    $P_0^\mu(\cos\varphi) = \dfrac{1}{\Gamma(1 - \mu)} \cot^\mu \dfrac{\varphi}{2}$       MO 84

2.    $P_\nu^{-1}(\cos\varphi) = -\dfrac{1}{\nu(\nu + 1)} \dfrac{d\, P_\nu(\cos\varphi)}{d\varphi}$       MO 84

3.    $P_n^m(z) \equiv 0,$                            $P_n^m(x) \equiv 0$ for $m > n$       MO 85

### 8.754

1.    $P_{\nu - \frac{1}{2}}^{1/2}(\cosh\alpha) = \sqrt{\dfrac{2}{\pi \sinh\alpha}} \cosh\nu\alpha$       MO 85

2.    $P_{\nu - \frac{1}{2}}^{1/2}(\cos\varphi) = \sqrt{\dfrac{2}{\pi \sin\varphi}} \cos\nu\varphi$       MO 85

3.
$$P_{\nu-\frac{1}{2}}^{-1/2}(\cos\varphi) = \sqrt{\frac{2}{\pi\sin\varphi}}\,\frac{\sin\nu\varphi}{\nu}$$
MO 85

4.
$$Q_{\nu-\frac{1}{2}}^{1/2}(\cosh\alpha) = i\sqrt{\frac{\pi}{2\sinh\alpha}}\,e^{-\nu\alpha}$$
MO 85

**8.755**

1.
$$P_\nu^{-\nu}(\cos\varphi) = \frac{1}{\Gamma(1+\nu)}\left(\frac{\sin\varphi}{2}\right)^\nu$$
MO 85

2.
$$P_\nu^{-\nu}(\cosh\alpha) = \frac{1}{\Gamma(1+\nu)}\left(\frac{\sinh\alpha}{2}\right)^\nu$$
MO 85

**Special values of Legendre functions**

**8.756**

1.
$$P_\nu^\mu(0) = \frac{2^\mu\sqrt{\pi}}{\Gamma\left(\frac{\nu-\mu}{2}+1\right)\Gamma\left(\frac{-\nu-\mu+1}{2}\right)}$$
MO 84

2.
$$\frac{d\,P_\nu^\mu(0)}{dx} = \frac{2^{\mu+1}\sin\frac{1}{2}(\nu+\mu)\pi\,\Gamma\left(\frac{\nu+\mu}{2}+1\right)}{\sqrt{\pi}\,\Gamma\left(\frac{\nu-\mu+1}{2}\right)}$$
MO 84

3.[12]
$$Q_\nu^\mu(0) = -2^{\mu-1}\sqrt{\pi}\sin\left(\frac{1}{2}(\nu+\mu)\pi\right)\frac{\Gamma\left(\frac{\nu+\mu+1}{2}\right)}{\Gamma\left(\frac{\nu-\mu}{2}+1\right)}$$
MO84

4.[12]
$$\frac{d\,Q_\nu^\mu(0)}{dx} = 2^\mu\sqrt{\pi}\cos\left(\frac{1}{2}(\nu+\mu)\pi\right)\frac{\Gamma\left(\frac{\nu+\mu}{2}+1\right)}{\Gamma\left(\frac{\nu-\mu+1}{2}\right)}$$
MO 84

## 8.76 Derivatives with respect to the order

**8.761**
$$\frac{\partial P_\nu^{-\mu}(x)}{\partial\nu} = \frac{1}{\Gamma(\mu+1)}\left(\frac{1-x}{1+x}\right)^{\frac{\mu}{2}}\sum_{n=1}^\infty \frac{(-\nu)(1-\nu)\ldots(n-1-\nu)(\nu+1)(\nu+2)\ldots(\nu+n)}{(\mu+1)(\mu+2)\ldots(\mu+n)1\cdot2\ldots n}$$
$$\times\,[\psi(\nu+n+1)-\psi(\nu-n+1)]\left(\frac{1-x}{2}\right)^n$$
$$[\nu\neq0,\pm1,\pm2,\ldots;\quad\operatorname{Re}\mu>-1]\quad\text{MO 94}$$

**8.762**

1.
$$\left[\frac{\partial\,P_\nu(\cos\varphi)}{\partial\nu}\right]_{\nu=0} = 2\ln\cos\frac{\varphi}{2}$$
MO 94

2.
$$\left[\frac{\partial P_\nu^{-1}(\cos\varphi)}{\partial\nu}\right]_{\nu=0} = -\tan\frac{\varphi}{2} - 2\cot\frac{\varphi}{2}\ln\cos\frac{\varphi}{2}$$
MO 94

3.
$$\left[\frac{\partial P_\nu^{-1}(\cos\varphi)}{\partial\nu}\right]_{\nu=1} = -\frac{1}{2}\tan\frac{\varphi}{2}\sin^2\frac{\varphi}{2} + \sin\varphi\ln\cos\frac{\varphi}{2}$$
MO 94

- For a connection with the polynomials $C_n^\lambda(x)$, see **8.936**.
- For a connection with a hypergeometric function, see **8.77**.

## 8.77 Series representation

For a representation in the form of a series, see **8.721**. It is also possible to represent associated Legendre functions in the form of a series by expressing them in terms of a hypergeometric function.

**8.771**

1. 
$$P_\nu^\mu(z) = \left(\frac{z+1}{z-1}\right)^{\frac{\mu}{2}} \frac{1}{\Gamma(1-\mu)} F\left(-\nu, \nu+1; 1-\mu; \frac{1-z}{2}\right)$$
MO 15

2.[8] 
$$Q_\nu^\mu(z) = \frac{e^{\mu\pi i}}{2^{\nu+1}} \frac{\Gamma(\nu+\mu+1)}{\Gamma\left(\nu+\frac{3}{2}\right)} \frac{\Gamma\left(\frac{1}{2}\right)(z^2-1)^{\frac{\mu}{2}}}{z^{\nu+\mu+1}} F\left(\frac{\nu+\mu}{2}+1, \frac{\nu+\mu+1}{2}; \nu+\frac{3}{2}; \frac{1}{z^2}\right)$$
MO 15

See also **8.702, 8.703, 8.704, 8.723, 8.751, 8.772**.

### The analytic continuation for $|z| \gg 1$

The formulas are consequences of theorems on the analytic continuation of hypergeometric series (see **9.154** and **9.155**):

**8.772**

1. 
$$P_\nu^\mu(z) = \frac{\sin(\nu+\mu)\pi \, \Gamma(\nu+\mu+1)}{2^{\nu+1}\sqrt{\pi} \cos\nu\pi \, \Gamma\left(\nu+\frac{3}{2}\right)} (z^2-1)^{\frac{\mu}{2}} z^{-\nu-\mu-1} F\left(\frac{\nu+\mu}{2}+1, \frac{\nu+\mu+1}{2}; \nu+\frac{3}{2}; \frac{1}{z^2}\right)$$
$$+ \frac{2^\nu \, \Gamma\left(\nu+\frac{1}{2}\right)}{\sqrt{\pi}\,\Gamma(\nu-\mu+1)} (z^2-1)^{\frac{\mu}{2}} z^{\nu-\mu} F\left(\frac{\mu-\nu+1}{2}, \frac{\mu-\nu}{2}; \frac{1}{2}-\nu; \frac{1}{z^2}\right)$$
$$[2\nu \neq \pm 1, \pm 3, \pm 5, \ldots; \quad |z| > 1; \quad |\arg(z\pm 1)| < \pi] \quad \text{MO 85}$$

2.[12] 
$$P_\nu^\mu(z) = \frac{\Gamma\left(-\nu-\frac{1}{2}\right)(z^2-1)^{-\frac{\nu+1}{2}}}{2^{\nu+1}\sqrt{\pi}\,\Gamma(-\nu-\mu)} F\left(\frac{\nu-\mu+1}{2}, \frac{\nu+\mu+1}{2}; \nu+\frac{3}{2}; \frac{1}{1-z^2}\right)$$
$$+ \frac{2^\nu \, \Gamma\left(\nu+\frac{1}{2}\right)}{\sqrt{\pi}\,\Gamma(\nu-\mu+1)} (z^2-1)^{\frac{\nu}{2}} F\left(\frac{\mu-\nu}{2}, -\frac{\mu+\nu}{2}; \frac{1}{2}-\nu; \frac{1}{1-z^2}\right)$$
$$[2\nu \neq \pm 1, \pm 3, \pm 5, \ldots; \quad |1-z^2| > 1; \quad |\arg(z\pm 1)| < \pi] \quad \text{MO 85}$$

3.[12] 
$$P_\nu^\mu(z) = \frac{1}{\Gamma(1-\mu)} \left(\frac{z-1}{z+1}\right)^{-\frac{\mu}{2}} \left(\frac{z+1}{2}\right)^{-\nu} F\left(-\nu, -\nu-\mu; 1-\mu; \frac{z-1}{z+1}\right)$$
$$\left[\left|\frac{z-1}{z+1}\right| < 1\right] \quad \text{MO 86}$$

**8.773**

1. 
$$Q_\nu^\mu(z) = e^{\mu\pi i} \frac{\sqrt{\pi}\,\Gamma(\nu+\mu+1)}{2^{\nu+1}\Gamma\left(\nu+\frac{3}{2}\right)} (z^2-1)^{-\frac{\nu+1}{2}} F\left(\frac{\nu+\mu+1}{2}, \frac{\nu-\mu+1}{2}; \nu+\frac{3}{2}; \frac{1}{1-z^2}\right)$$
$$[\nu+\mu \neq -1, -2, -3, \ldots; \quad |\arg(z\pm 1)| < \pi; \quad |1-z^2| > 1] \quad \text{MO 86}$$

2. 
$$Q_\nu^\mu(z) = \frac{1}{2} e^{\mu\pi i} \left\{ \Gamma(\mu) \left(\frac{z+1}{z-1}\right)^{\frac{\mu}{2}} F\left(-\nu, \nu+1; 1-\mu; \frac{1-z}{2}\right) \right.$$
$$\left. + \frac{\Gamma(-\mu)\Gamma(\nu+\mu+1)}{\Gamma(\nu-\mu+1)} \left(\frac{z-1}{z+1}\right)^{\frac{\mu}{2}} F\left(-\nu, \nu+1; 1+\mu; \frac{1-z}{2}\right) \right\}$$
$$[|\arg(z\pm 1)| < \pi, \quad |1-z| < 2] \quad \text{MO 86}$$

**8.774**  $P_\nu^\mu \left( i \cot \varphi \right) = \sqrt{\dfrac{\sin \varphi}{2\pi}} \dfrac{\Gamma \left( -\nu - \frac{1}{2} \right)}{\Gamma(-\nu - \mu)} e^{-i(\nu+1)\frac{\pi}{2}} \left( \tan \dfrac{\varphi}{2} \right)^{\nu+\frac{1}{2}} F \left( \dfrac{1}{2} + \mu, \dfrac{1}{2} - \mu; \nu + \dfrac{3}{2}; \sin^2 \dfrac{\varphi}{2} \right)$

$\qquad\qquad + \sqrt{\dfrac{\sin \varphi}{2\pi}} \dfrac{\Gamma \left( \nu + \frac{1}{2} \right)}{\Gamma(\nu - \mu + 1)} e^{i\nu\frac{\pi}{2}} \left( \cot \dfrac{\varphi}{2} \right)^{\nu+\frac{1}{2}} F \left( \dfrac{1}{2} + \mu, \dfrac{1}{2} - \mu; \dfrac{1}{2} - \nu; \sin^2 \dfrac{\varphi}{2} \right)$

$\qquad\qquad\qquad\qquad\qquad\qquad \left[ 2\nu \neq \pm 1, \pm 3, \pm 5, \ldots, \quad 0 < \varphi < \dfrac{\pi}{2} \right]$   MO 86

**8.775**

1.[6]  $P_\nu^\mu(x) = \dfrac{2^\mu \cos \left( \frac{1}{2} (\nu + \mu) \pi \right) \Gamma \left( \frac{\nu+\mu+1}{2} \right)}{\sqrt{\pi} \, \Gamma \left( \frac{\nu-\mu}{2} + 1 \right)} \left( 1 - x^2 \right)^{\frac{\mu}{2}} F \left( \dfrac{\nu + \mu + 1}{2}, \dfrac{\mu - \nu}{2}; \dfrac{1}{2}; x^2 \right)$

$\qquad\qquad + \dfrac{2^{\mu+1} \sin \left( \frac{1}{2} (\nu + \mu) \pi \right) \Gamma \left( \frac{\nu+\mu}{2} + 1 \right)}{\sqrt{\pi}} \dfrac{}{\Gamma \left( \frac{\nu-\mu+1}{2} \right)} x \left( 1 - x^2 \right)^{\frac{\mu}{2}} F \left( \dfrac{\nu+\mu}{2} + 1, \dfrac{-\nu+\mu+1}{2}; \dfrac{3}{2}; x^2 \right)$

$\qquad\qquad\qquad\qquad\qquad\qquad\qquad\qquad\qquad\qquad\qquad\qquad\qquad\qquad\qquad$   MO 87

2.[6]  $Q_\nu^\mu(x) = -\dfrac{\sqrt{\pi}}{2^{1-\mu}} \dfrac{\sin \left( \frac{1}{2} (\nu + \mu) \pi \right) \Gamma \left( \frac{\nu+\mu+1}{2} \right)}{\Gamma \left( \frac{\nu-\mu}{2} + 1 \right)} \left( 1 - x^2 \right)^{\frac{\mu}{2}} F \left( \dfrac{\nu + \mu + 1}{2}, \dfrac{\mu - \nu}{2}; \dfrac{1}{2}; x^2 \right)$

$\qquad\qquad + 2^\mu \sqrt{\pi} \dfrac{\cos \left( \frac{1}{2} (\nu + \mu) \pi \right) \Gamma \left( \frac{\nu+\mu}{2} + 1 \right)}{\Gamma \left( \frac{\nu-\mu+1}{2} \right)} x \left( 1 - x^2 \right)^{\frac{\mu}{2}} F \left( \dfrac{\nu+\mu}{2} + 1, \dfrac{\mu-\nu+1}{2}; \dfrac{3}{2}; x^2 \right)$

$\qquad\qquad\qquad\qquad\qquad\qquad\qquad\qquad\qquad\qquad\qquad\qquad\qquad\qquad\qquad\qquad$   MO 87

**8.776**  For $|z| \gg 1$

1.  $P_\nu^\mu(z) = \left\{ \dfrac{2^\nu \Gamma \left( \nu + \frac{1}{2} \right)}{\sqrt{\pi} \, \Gamma(\nu - \mu + 1)} z^\nu + \dfrac{\Gamma \left( -\nu - \frac{1}{2} \right)}{2^{\nu+1} \sqrt{\pi} \, \Gamma(-\nu - \mu)} z^{-\nu-1} \right\} \left( 1 + O \left( \dfrac{1}{z^2} \right) \right)$

$\qquad\qquad\qquad\qquad\qquad\qquad\qquad [2\nu \neq \pm 1, \pm 3, \pm 5, \ldots, \quad |\arg z| < \pi]$
$\qquad\qquad\qquad\qquad\qquad\qquad\qquad\qquad\qquad\qquad\qquad\qquad\qquad\qquad$   MO 87

2.  $Q_\nu^\mu(z) = \sqrt{\pi} \dfrac{e^{\mu\pi i}}{2^{\nu+1}} \dfrac{\Gamma(\mu + \nu + 1)}{\Gamma \left( \nu + \frac{3}{2} \right)} z^{-\nu-1} \left( 1 + O \left( \dfrac{1}{z^2} \right) \right)$

$\qquad\qquad\qquad\qquad\qquad\qquad\qquad [2\nu \neq -3, -5, -7, \ldots; \quad |\arg z| < \pi]$
$\qquad\qquad\qquad\qquad\qquad\qquad\qquad\qquad\qquad\qquad\qquad\qquad\qquad\qquad$   MO 87

**8.777**  Set $\zeta = z + \sqrt{z^2 - 1}$. The variable $\zeta$ is uniquely defined by this equation on the entire $z$-plane in which a cut is made from $-\infty$ to $+1$. Here, we are considering that branch of the variable $\zeta$ for which values of $\zeta$ exceeding 1 correspond to real values of $z$ exceeding 1. In this case,

1.  $P_\nu^\mu(z) = \dfrac{2^\mu \, \Gamma \left( -\nu - \frac{1}{2} \right)}{\sqrt{\pi} \, \Gamma(-\nu - \mu)} \dfrac{\left( z^2 - 1 \right)^{\frac{\mu}{2}}}{\zeta^{\nu+\mu+1}} F \left( \dfrac{1}{2} + \mu, \nu + \mu + 1; \nu + \dfrac{3}{2}; \dfrac{1}{\zeta^2} \right)$

$\qquad\qquad + \dfrac{2^\mu}{\sqrt{\pi}} \dfrac{\Gamma \left( \nu + \frac{1}{2} \right)}{\Gamma(\nu - \mu + 1)} \dfrac{\left( z^2 - 1 \right)^{\frac{\mu}{2}}}{\zeta^{\mu-\nu}} F \left( \dfrac{1}{2} + \mu, \mu - \nu; \dfrac{1}{2} - \nu; \dfrac{1}{\zeta^2} \right)$

$\qquad\qquad\qquad\qquad\qquad [2\nu \neq \pm 1, \pm 3, \pm 5, \ldots; \quad |\arg(z - 1)| < \pi]$   MO 86

2.  $Q_\nu^\mu(z) = 2^\mu e^{\mu\pi i} \sqrt{\pi} \dfrac{\Gamma(\nu + \mu + 1)}{\Gamma \left( \nu + \frac{3}{2} \right)} \dfrac{\left( z^2 - 1 \right)^{\frac{\mu}{2}}}{\zeta^{\nu+\mu+1}} F \left( \dfrac{1}{2} + \mu, \nu + \mu + 1; \nu + \dfrac{3}{2}; \dfrac{1}{\zeta^2} \right)$

$\qquad\qquad\qquad\qquad\qquad\qquad\qquad\qquad [|\arg(z - 1)| < \pi]$   MO 86

## 8.78 The zeros of associated Legendre functions

**8.781**    The function $P_\nu^{-\mu}(\cos\varphi)$, considered as a function of $\nu$ has infinitely many zeros for $\mu \geq 0$. These are all simple and real. If a number $\nu_0$ is a zero of the function $P_\nu^{-\mu}(\cos\varphi)$, the number $-\nu_0 - 1$ is also a zero of this function.          MO 91

**8.782**    If $\nu$ and $\mu$ are both real and $\mu \leq 0$, or if $\nu$ and $\mu$ are integers, the function $P_\nu^\mu(t)$ has no *real* zeros exceeding 1. If $\nu$ and $\mu$ are both real with $\nu < \mu < 0$, the function $P_\nu^\mu(t)$ has no real zeros exceeding 1 when $\sin\mu\pi\sin(\mu - \nu)\pi > 0$, but does have one such zero when $\sin\mu\pi\sin(\mu - \nu)\pi < 0$. Finally, if $\mu \leq \nu$, the function $P_\nu^\mu(t)$ has no zeros exceeding 1 for $\lfloor\mu\rfloor$ even but does have one zero for $\lfloor\mu\rfloor$ odd.

**8.783**    If $\nu > -\frac{3}{2}$ and $\nu + \mu + 1 > 0$, the function $Q_\nu^\mu(t)$ has no real zeros exceeding 1.       MO 91

**8.784**    The function $P_{-\frac{1}{2}+i\lambda}(z)$ has infinitely many zeros for real $\lambda$. All these zeros are *real* and *greater than unity.*

**8.785**    For $n$ a natural number, the function $P_n(x)$ has exactly $n$ real zeros which lie in the closed interval $-1, +1$.

**8.786**    The function $Q_n(z)$ has no zeros for which $|\arg(z-1)| < \pi$ if $n$ is a natural number. The function $Q_n(\cos\varphi)$ has exactly $n+1$ zeros in the interval $0 \leq \varphi \leq \pi$.      MO 91

**8.787**    The following approximate formula can be used to calculate the values of $\nu$ for which the equation $P_\nu^{-\mu}(\cos\varphi) = 0$ holds for given small values of $\varphi$:

$$\nu + \frac{1}{2} = -\frac{j_\mu}{2\sin\frac{\varphi}{2}}\left\{1 - \frac{\sin^2\frac{\varphi}{2}}{6}\left(1 - \frac{4\mu^2 - 1}{j_\mu^2}\right) + O\left(\sin^4\frac{\varphi}{2}\right)\right\} \qquad \text{MO 93}$$

Here, $j_\mu$ denotes an arbitrary nonzero root of the equation $J_\mu(z) = 0$ (for $\mu \geq 0$). If $\varphi$ is close to $\pi$ then, instead of this formula, we can use the following formulas:

1.    $\nu \approx \mu + k + \dfrac{\Gamma(2\mu + k + 1)}{\Gamma(\mu)\,\Gamma(\mu + 1)\,\Gamma(k + 1)}\left(\dfrac{\pi - \varphi}{3}\right)^{2\mu}$       $[\mu > 0, \quad k = 0, 1, 2, \ldots]$    MO 93

2.    $\nu \approx k + \dfrac{1}{2\ln\left(\dfrac{2}{\pi - \varphi}\right)}$                  $[\mu = 0, \quad k = 0, 1, 2, \ldots]$    MO 93

## 8.79 Series of associated Legendre functions

**8.791**

1.    $\dfrac{1}{z - t} = \sum\limits_{k=0}^{\infty}(2k + 1)\,P_k(t)\,Q_k(z)$       $\left[\left|t + \sqrt{t^2 - 1}\right| < \left|z + \sqrt{z^2 - 1}\right|\right]$

Here, $t$ must lie inside an ellipse passing through the point $z$ with foci at the points $\pm 1$.

2.    $\dfrac{1}{\sqrt{1 - 2tz + t^2}}\ln\dfrac{z - t + \sqrt{1 - 2tz + t^2}}{\sqrt{z^2 - 1}} = \sum\limits_{k=0}^{\infty}t^k\,Q_k(z)$

                                            $[\operatorname{Re} z > 1, \quad |t| < 1]$    MO 78

**8.792**[12]   $P_\nu^{-\alpha}(\cos\varphi)\,P_\nu^{-\beta}(\cos\psi) = \dfrac{\sin\nu\pi}{\pi}\sum\limits_{k=1}^{\infty}(-1)^k\left[\dfrac{1}{\nu - k} - \dfrac{1}{\nu + k + 1}\right]P_k^{-\alpha}(\cos\varphi)\,P_k^{-\beta}(\cos\psi)$

                           $[a \geq 0, \quad \beta \geq 0, \quad \nu \text{ real}, \quad -\pi < \varphi \pm \psi < \pi]$    MO 94

**8.793**  $P_\nu^{-\mu}(\cos\varphi) = \dfrac{\sin\nu\pi}{\pi}\sum_{k=0}^{\infty}(-1)^k\left(\dfrac{1}{\nu-k}-\dfrac{1}{\nu+k+1}\right)P_k^{-\mu}(\cos\varphi)$     $[\mu\geq 0,\quad 0<\varphi<\pi]$

MO 94

## Addition theorems

**8.794**

$1.^{11}$     $P_\nu(\cos\psi_1\cos\psi_2+\sin\psi_1\sin\psi_2\cos\varphi)$

$$= P_\nu(\cos\psi_1)\,P_\nu(\cos\psi_2)+2\sum_{k=1}^{\infty}(-1)^k\,P_\nu^{-k}(\cos\psi_1)\,P_\nu^k(\cos\psi_2)\cos k\varphi$$

$$= P_\nu(\cos\psi_1)\,P_\nu(\cos\psi_2)+2\sum_{k=1}^{\infty}\frac{\Gamma(\nu-k+1)}{\Gamma(\nu+k+1)}\,P_\nu^k(\cos\psi_1)\,P_\nu^k(\cos\psi_2)\cos k\varphi$$

$[0\leq\psi_1<\pi,\quad 0\leq\psi_2<\pi,\quad \psi_1+\psi_2<\pi,\quad \varphi\ \text{real}]$     (cf. **8.814**, **8.844** 1)   MO 90

$2.^{12}$     $Q_\nu(\cos\psi_1\cos\psi_2+\sin\psi_1\sin\psi_2\cos\varphi)$

$$= P_\nu(\cos\psi_1)\,Q_\nu(\cos\psi_2)+2\sum_{k=1}^{\infty}(-1)^k\,P_\nu^{-k}(\cos\psi_1)\,Q_\nu^k(\cos\psi_2)\cos k\varphi$$

$\left[0<\psi_1<\dfrac{\pi}{2},\quad 0<\psi_2<\pi,\quad 0<\psi_1+\psi_2<\pi;\quad \varphi\ \text{real}\right]$     (cf. **8.844** 3)   MO 90

**8.795**

1.     $P_\nu\left(z_1 z_2-\sqrt{z_1^2-1}\sqrt{z_2^2-1}\cos\varphi\right)=P_\nu(z_1)\,P_\nu(z_2)+2\sum_{k=1}^{\infty}(-1)^k\,P_\nu^k(z_1)\,P_\nu^{-k}(z_2)\cos k\varphi$

$[\operatorname{Re} z_1>0,\quad \operatorname{Re} z_2>0,\quad |\arg(z_1-1)|<\pi,\quad |\arg(z_2-1)|<\pi]$   MO 91

2.     $Q_\nu\left(x_1 x_2-\sqrt{x_1^2-1}\sqrt{x_2^2-1}\cos\varphi\right)=P_\nu(x_1)\,Q_\nu(x_2)+2\sum_{k=1}^{\infty}(-1)^k\,P_\nu^{-k}(x_1)\,Q_\nu^k(x_2)\cos k\varphi$

$[1<x_1<x_2,\quad \nu\neq-1,-2,-3,\ldots,\quad \varphi\ \text{real}]$   MO 91

3.     $Q_n\left(x_1 x_2+\sqrt{x_1^2+1}\sqrt{x_2^2+1}\cosh\alpha\right)=\sum_{k=n+1}^{\infty}\frac{1}{(k-n-1)!(k+n)!}\,Q_n^k(ix_1)\,Q_n^k(ix_2)\,e^{-k\alpha}$

$[x_1>0,\quad x_2>0,\quad \alpha>0]$   MO 91

**8.796**  $P_\nu(-\cos\psi_1\cos\psi_2-\sin\psi_1\sin\psi_2\cos\varphi)=P_\nu(-\cos\psi_1)\,P_\nu(\cos\psi_2)+2\sum_{k=1}^{\infty}(-1)^k\dfrac{\Gamma(\nu+k+1)}{\Gamma(\nu-k+1)}$

$$\times P_\nu^{-k}(-\cos\psi_1)\,P_\nu^{-k}(\cos\psi_2)\cos k\varphi$$

$[0<\psi_2<\psi_1<\pi,\quad \varphi\ \text{real}]$     (cf. **8.844** 2)   MO 91

See also **8.934** 3.

## 8.81   Associated Legendre functions with integer indices

**8.810**    For *integer* values of $\nu$ and $\mu$, the differential equation **8.700** 1. (with $|\nu| > |\mu|$) has a simple solution in the real domain, namely:

$$u = P_n^m(x) = (-1)^m \left(1 - x^2\right)^{\frac{m}{2}} \frac{d^m}{dx^m} P_n(x)$$

The functions $P_n^m(x)$ are called *associated Legendre functions* (or *spherical functions*) *of the first kind*. The number $n$ is called the *degree* and the number $m$ is called the *order* of the function $P_n^m(x)$. The functions $\{\cos m\vartheta \, P_n^m(\cos\varphi), \quad \sin m\vartheta \, P_n^m(\cos\varphi)\}$, which depend on the angles $\varphi$ and $\vartheta$, are also called Legendre functions of the first kind, or, more specifically, *tesseral harmonics* for $m < n$ and *sectoral harmonics* for $m = n$. These last functions are periodic with respect to the angles $\varphi$ and $\vartheta$. Their periods are respectively $\pi$ and $2\pi$. They are single-valued and continuous everywhere on the surface of the unit sphere $x_1^2 + x_2^2 + x_3^2 = 1$ (where $x_1 = \sin\varphi\cos\vartheta$, $x_2 = \sin\varphi\sin\vartheta$, $x_3 = \cos\varphi$) and they are solutions of the differential equation

$$\frac{1}{\sin\varphi} \frac{\partial}{\partial\varphi} \left( \sin\varphi \frac{\partial Y}{\partial\varphi} \right) + \frac{1}{\sin^2\varphi} \frac{\partial^2 Y}{\partial\vartheta^2} + n(n+1)Y = 0$$

**8.811**    The integral representation

$$P_n^m(\cos\varphi) = \frac{(-1)^m(n+m)!}{\Gamma\left(m+\frac{1}{2}\right)(n-m)!} \sqrt{\frac{2}{\pi}} \sin^{-m}\varphi \int_0^\varphi (\cos t - \cos\varphi)^{m-\frac{1}{2}} \cos\left(n+\tfrac{1}{2}\right) t \, dt \qquad \text{MO 75}$$

**8.812**    The series representation:

$$P_n^m(x) = \frac{(-1)^m(n+m)!}{2^m m!(n-m)!} \left(1-x^2\right)^{\frac{m}{2}} \left\{ 1 - \frac{(n-m)(m+n+1)}{1!(m+1)} \frac{1-x}{2} \right.$$
$$\left. + \frac{(n-m)(n-m+1)(m+n+1)(m+n+2)}{2!(m+1)(m+2)} \left(\frac{1-x}{2}\right)^2 - \cdots \right\} \qquad \text{MO 73}$$

$$= \frac{(-1)^m(2n-1)!!}{(n-m)!} \left(1-x^2\right)^{\frac{m}{2}} \left\{ x^{n-m} - \frac{(n-m)(n-m-1)}{2(2n-1)} x^{n-m-2} \right.$$
$$\left. + \frac{(n-m)(n-m-1)(n-m-2)(n-m-3)}{2\cdot 4(2n-1)(2n-3)} x^{n-m-4} - \cdots \right\} \qquad \text{MO 73}$$

$$= \frac{(-1)^m(2n-1)!!}{(n-m)!} \left(1-x^2\right)^{\frac{m}{2}} x^{n-m} F\left(\frac{m-n}{2}, \frac{m-n+1}{2}; \frac{1}{2} - n; \frac{1}{x^2}\right) \qquad \text{MO 73}$$

**8.813**   Special cases:

1.    $P_1^1(x) = -\left(1-x^2\right)^{1/2} = -\sin\varphi$                                             MO 73

2.    $P_2^1(x) = -3\left(1-x^2\right)^{1/2} x = -\frac{3}{2}\sin 2\varphi$                      MO 73

3.    $P_2^2(x) = 3\left(1-x^2\right) = \frac{3}{2}(1-\cos 2\varphi)$                       MO 73

4.    $P_3^1(x) = -\frac{3}{2}\left(1-x^2\right)^{1/2}\left(5x^2-1\right) = -\frac{3}{8}(\sin\varphi + 5\sin 3\varphi)$      MO 73

5.    $P_3^2(x) = 15\left(1-x^2\right) x = \frac{15}{4}(\cos\varphi - \cos 3\varphi)$                 MO 73

6.    $P_3^3(x) = -15\left(1-x^2\right)^{3/2} = -\frac{15}{4}(3\sin\varphi - \sin 3\varphi)$         MO 73

## Functional relations

For recursion formulas, see **8.731**.

**8.814**   $P_n \left( \cos \varphi_1 \cos \varphi_2 + \sin \varphi_1 \sin \varphi_2 \cos \Theta \right)$

$$= P_n \left( \cos \varphi_1 \right) P_n \left( \cos \varphi_2 \right) + 2 \sum_{m=1}^{n} \frac{(n-m)!}{(n+m)!} P_n^m \left( \cos \varphi_1 \right) P_n^m \left( \cos \varphi_2 \right) \cos m \Theta$$

$$[0 \leq \varphi_1 \leq \pi, \quad 0 \leq \varphi_2 \leq \pi] \qquad \text{(``addition theorem'')} \quad \text{MO 74}$$

**8.815**[12]   If

$$Y_{n_1}(\varphi, \vartheta) = a_0 P_{n_1} \left( \cos \varphi \right) + \sum_{m=1}^{n_1} \left( a_m \cos m \vartheta + b_m \sin m \vartheta \right) P_{n_1}^m \left( \cos \varphi \right),$$

$$Z_{n_2}(\varphi, \vartheta) = \alpha_0 P_{n_2} \left( \cos \varphi \right) + \sum_{m=1}^{n_2} \left( \alpha_m \cos m \vartheta + \beta_m \sin m \vartheta \right) P_{n_2}^m \left( \cos \varphi \right),$$

then

$$\int_0^{2\pi} d\vartheta \int_0^{\pi} \sin \varphi \, d\varphi \, Y_{n_1}(\varphi, \vartheta) \, Z_{n_2}(\varphi, \vartheta) = 0,$$

$$\int_0^{2\pi} d\vartheta \int_0^{\pi} \sin \varphi \, d\varphi \, Y_n(\varphi, \vartheta) \, P_n \left[ \cos \varphi \cos \psi + \sin \varphi \sin \psi \cos(\vartheta - \theta) \right] = \frac{4\pi}{2n+1} \, Y_n(\psi, \theta) \qquad \text{MO 75}$$

**8.816**   $\left( \cos \varphi + i \sin \varphi \cos \vartheta \right)^n = P_n \left( \cos \varphi \right) + 2 \sum_{m=1}^{n} (-1)^m \frac{n!}{(n+m)!} \cos m \vartheta \, P_n^m \left( \cos \varphi \right)$ \qquad MO 75

For integrals of the functions, $P_n^m(x)$, see **7.112** 1, **7.122** 1.

## 8.82–8.83  Legendre functions

**8.820**   The differential equation

$$\frac{d}{dz} \left[ (1 - z^2) \frac{du}{dz} \right] + \nu(\nu + 1) u = 0 \qquad \text{(cf. **8.700** 1),}$$

where the parameter $\nu$ can be an arbitrary number, has the following two linearly independent solutions:

1.   $P_\nu(z) = F \left( -\nu, \nu + 1; 1; \frac{1 - z}{2} \right)$

2.   $Q_\nu(z) = \frac{\Gamma(\nu + 1) \Gamma \left( \frac{1}{2} \right)}{2^{\nu+1} \Gamma \left( \nu + \frac{3}{2} \right)} z^{-\nu-1} F \left( \frac{\nu + 2}{2}, \frac{\nu + 1}{2}; \frac{2\nu + 3}{2}; \frac{1}{z^2} \right)$ \qquad SM 518(137)

   The functions $P_\nu(z)$ and $Q_\nu(z)$ are called *Legendre functions of the first* and *second kind* respectively. If $\nu$ is not an integer, the function $P_\nu(z)$ has *singularities* at $z = -1$ and $z = \infty$. However, if $\nu = n = 0, 1, 2, \ldots$, the function $P_\nu(z)$ becomes the *Legendre polynomial* $P_n(z)$ (see **8.91**) For $\nu = -n = -1, -2, \ldots$, we have

$$P_{-n-1}(z) = P_n(z)$$

3.   If $\nu \neq 0, 1, 2, \ldots$, the function $Q_\nu(z)$ has singularities at the points $z = \pm 1$ and $z = \infty$. These points are branch points of the function. On the other hand, if $\nu = n = 0, 1, 2, \ldots$, the function $Q_n(z)$ is single-valued for $|z| > 1$ and regular for $z = \infty$.

4. In the right half-plane,

$$P_\nu(z) = \left(\frac{1+z}{2}\right)^\nu F\left(-\nu, -\nu; 1; \frac{z-1}{z+1}\right) \qquad [\text{Re}\, z > 0]$$

5. The function $P_\nu(z)$ is uniquely determined by equations **8.820** 1 and **8.820** 4 within a circle of radius 2 with its center at the point $z = 1$ in the right half-plane.

For $z = x = \cos\varphi$, a solution of equation **8.820** is the function

6. $\quad P_\nu(x) = P_\nu(\cos\varphi) = F\left(-\nu, \nu + 1; 1; \sin^2\frac{\varphi}{2}\right);$

In general,

7. $\quad P_\nu(z) = P_{-\nu-1}(z) = P_\nu(x) = P_{-\nu-1}(x), \text{ for } z = x$

8. The function $Q_\nu(z)$ for $|z| > 1$ is uniquely determined by equation **8.820** 2 everywhere in the $z$-plane in which a cut is made from the point $z = -\infty$ to the point $z = 1$. By means of a hypergeometric series, the function can be continued analytically inside the unit circle. On the cut $(-1 \le x \le +1)$ of the real axis, the function $Q_\nu(x)$ is determined by the equation

9. $\quad Q_\nu(x) = \frac{1}{2}\left[Q_\nu(x+i0) + Q_\nu(x-i0)\right]$ HO 52(53), WH

**Integral representations**

**8.821**

1. $\quad P_\nu(z) = \frac{1}{2\pi i}\int_A^{(1+,z+)} \frac{\left(t^2 - 1\right)^\nu}{2^\nu(t-z)^{\nu+1}}\, dt$

Here, $A$ is a point on the real axis to the right of the point $t = 1$ and to the right of $z$ if $z$ is real. At the point $A$, we set

$$\arg(t-1) = \arg(t+1) = 0 \text{ and } [|\arg(t-z)| < \pi] \qquad \text{WH}$$

2. $\quad Q_\nu(z) = \frac{1}{4i\sin\nu\pi}\int_A^{(1-,1+)} \frac{\left(t^2 - 1\right)^\nu}{2^\nu(z-t)^{\nu+1}}\, dt$

[$\nu$ is not an integer; the point $A$ is at the end of the major axis of an ellipse to the right of $t = 1$ drawn in the $t$-plane with foci at the points $\pm 1$ and with a minor axis sufficiently small that the point $z$ lies outside it. The contour begins at the point $A$, follows the path $(1-,-1+)$ and returns to $A$; $|\arg z| \le \pi$ and $|\arg(z-t)| \to \arg z$ as $t \to 0$ on the contour; $\arg(t+1) = \arg(t-1) = 0$ at the point $A$; $z$ does not lie on the real axis between $-1$ and $1$.]

For $\nu = n$ an integer,

3. $\quad Q_n(z) = \frac{1}{2^{n+1}}\int_{-1}^1 \left(1 - t^2\right)^n (z-t)^{-n-1}\, dt$ SM 517(134), WH

**8.822**

1. $\quad P_\nu(z) = \frac{1}{\pi}\int_0^\pi \frac{d\varphi}{\left(z + \sqrt{z^2 - 1}\cos\varphi\right)^{\nu+1}} = \frac{1}{\pi}\int_0^\pi \left(z + \sqrt{z^2 - 1}\cos\varphi\right)^\nu d\varphi$

$\left[\text{Re}\, z > 0 \text{ and } \arg\left\{z + \sqrt{z^2 - 1}\cos\varphi\right\} = \arg z \text{ for } \varphi = \frac{\pi}{2}\right]$ WH

2.     $Q_\nu(z) = \int_0^\infty \dfrac{d\varphi}{\left(z + \sqrt{z^2 - 1}\cosh\varphi\right)^{\nu+1}}$,

$\left[\operatorname{Re}\nu > -1; \quad \text{if } \nu \text{ is not an integer, } \left\{\left(z + \sqrt{z^2 - 1}\right)\cosh\varphi\right\} \text{ for } \varphi = 0 \text{ has its principal value}\right]$

WH

**8.823**    $P_\nu(\cos\theta) = \dfrac{2}{\pi} \int_0^\theta \dfrac{\cos\left(\nu + \frac{1}{2}\right)\varphi}{\sqrt{2(\cos\varphi - \cos\theta)}} \, d\varphi$

WH

**8.824**    $Q_n(z) = 2^n n! \displaystyle\int_z^\infty \cdots \int_z^\infty \dfrac{(dz)^{n+1}}{(z^2-1)^{n+1}} = 2^n \int_z^\infty \dfrac{(t-z)^n}{(t^2-1)^{n+1}} \, dt$

$\qquad\qquad = \dfrac{(-1)^n}{(2n-1)!!} \dfrac{d^n}{dz^n}\left[(z^2-1)^n \displaystyle\int_z^\infty \dfrac{dt}{(t^2-1)^{n+1}}\right]$     $[\operatorname{Re} z > 1]$

WH, MO 78

**8.825**    $Q_n(z) = \dfrac{1}{2} \displaystyle\int_{-1}^1 \dfrac{P_n(t)}{z-t} \, dt$     $[|\arg(z-1)| < \pi]$

WH, MO 78

See also **6.622** 3, **8.842**.

**8.826**   Fourier series:

1.     $P_n(\cos\varphi) = \dfrac{2^{n+2}}{\pi} \dfrac{n!}{(2n+1)!!} \left[\sin(n+1)\varphi + \dfrac{1}{1}\dfrac{n+1}{2n+3}\sin(n+3)\varphi \right.$

$\qquad\qquad\qquad \left. + \dfrac{1\cdot3(n+1)(n+2)}{1\cdot2(2n+3)(2n+5)}\sin(n+5)\varphi + \ldots\right]$

$\qquad\qquad\qquad\qquad\qquad\qquad\qquad\qquad [0 < \varphi < \pi]$

MO 79

2.     $Q_n(\cos\varphi) = 2^{n+1}\dfrac{n!}{(2n+1)!!}\left[\cos(n+1)\varphi + \dfrac{1}{1}\dfrac{n+1}{2n+3}\cos(n+3)\varphi\right.$

$\qquad\qquad\qquad + \dfrac{1\cdot3}{1\cdot2}\dfrac{(n+1)(n+2)}{(2n+3)(2n+5)}\cos(n+5)\varphi + \ldots\bigg]$

$\qquad\qquad\qquad\qquad\qquad\qquad\qquad\qquad [0 < \varphi < \pi]$

MO 79

The expressions for Legendre functions in terms of a hypergeometric function (see **8.820**) provide other series representations of these functions.

## Special cases and particular values

**8.827**

1.     $Q_0(x) = \dfrac{1}{2}\ln\dfrac{1+x}{1-x} = \operatorname{arctanh} x$

JA

2.     $Q_1(x) = \dfrac{x}{2}\ln\dfrac{1+x}{1-x} - 1$

JA

3.     $Q_2(x) = \dfrac{1}{4}\left(3x^2 - 1\right)\ln\dfrac{1+x}{1-x} - \dfrac{3}{2}x$

JA

4.     $Q_3(x) = \dfrac{1}{4}\left(5x^3 - 3x\right)\ln\dfrac{1+x}{1-x} - \dfrac{5}{2}x^2 + \dfrac{2}{3}$

JA

5.     $Q_4(x) = \dfrac{1}{16}\left(35x^4 - 30x^2 + 3\right)\ln\dfrac{1+x}{1-x} - \dfrac{35}{8}x^3 + \dfrac{55}{24}x$         JA

6.     $Q_5(x) = \dfrac{1}{16}\left(63x^5 - 70x^3 + 15x\right)\ln\dfrac{1+x}{1-x} - \dfrac{63}{8}x^4 + \dfrac{49}{8}x^2 - \dfrac{8}{15}$         JA

**8.828**

1.     $P_\nu(1) = 1$         MO 79

2.     $P_\nu(0) = -\dfrac{1}{2}\dfrac{\sin\nu\pi}{\sqrt{\pi^3}}\,\Gamma\left(\dfrac{\nu+1}{2}\right)\Gamma\left(-\dfrac{\nu}{2}\right)$         MO 79

**8.829**    $Q_\nu(0) = \dfrac{1}{4\sqrt{\pi}}\left(1 - \cos\nu\pi\right)\Gamma\left(\dfrac{\nu+1}{2}\right)\Gamma\left(-\dfrac{\nu}{2}\right)$         MO 79

## Functional relationships

**8.831**

1.     $Q_\nu(x) = \dfrac{\pi}{2\sin\nu\pi}\left[\cos\nu\pi\, P_\nu(x) - P_\nu(-x)\right]$      $[\nu \neq 0, \quad \pm 1, \pm 2, \ldots]$         MO 76

2.     $Q_n(x) = \dfrac{1}{2}\,P_n(x)\ln\dfrac{1+x}{1-x} - W_{n-1}(x)$      $[n = 0, 1, 2, \ldots],$

    where

3.[12]   $W_{n-1}(x) = \displaystyle\sum_{k=0}^{2\left\lfloor\frac{n-1}{2}\right\rfloor} \dfrac{2(n-2k)-1}{(2k+1)(n-k)}\,P_{n-2k-1}(x) = \sum_{k=1}^{n}\dfrac{1}{k}\,P_{k-1}(x)\,P_{n-k}(x)$

    and

4.     $W_{-1}(x) \equiv 0$         (see also **8.839**)      SM 516(131), MO 76

5.     $\displaystyle\sum_{k=0}^{\infty}(-1)^k\left(\dfrac{1}{\nu-k} - \dfrac{1}{\nu+k+1}\right)P_k(\cos\varphi) = \dfrac{\pi}{\sin\nu\pi}\,P_\nu(\cos\varphi)$

                                      $[\nu \text{ not an integer}; \quad 0 \leq \varphi < \pi]$      MO 77

6.     $\displaystyle\sum_{k=0}^{\infty}(-1)^k\left(\dfrac{1}{\nu-k} - \dfrac{1}{\nu+k+1}\right)P_k(\cos\varphi)\,P_k(\cos\psi) = \dfrac{\pi}{\sin\nu\pi}\,P_\nu(\cos\varphi)\,P_\nu(\cos\psi)$

           $[\nu \text{ not an integer}, \quad -\pi < \varphi + \psi < \pi, \quad -\pi < \varphi - \psi < \pi]$      MO 77

See also **8.521** 4.

**8.832**

1.     $\left(z^2 - 1\right)\dfrac{d}{dz}P_\nu(z) = (\nu+1)\left[P_{\nu+1}(z) - z\,P_\nu(z)\right]$         WH

2.     $(2\nu+1)z\,P_\nu(z) = (\nu+1)\,P_{\nu+1}(z) + \nu\,P_{\nu-1}(z)$         WH

3.     $\left(z^2 - 1\right)\dfrac{d}{dz}Q_\nu(z) = (\nu+1)\left[Q_{\nu+1}(z) - z\,Q_\nu(z)\right]$         WH

4.     $(2\nu+1)z\,Q_\nu(z) = (\nu+1)\,Q_{\nu+1}(z) + \nu\,Q_{\nu-1}(z)$         WH

**8.833**

1.     $P_\nu(-z) = e^{\nu\pi i} P_\nu(z) - \dfrac{2}{\pi} \sin \nu\pi \, Q_\nu(z)$          $[\operatorname{Im} z < 0]$          MO 77

2.     $P_\nu(-z) = e^{-\nu\pi i} P_\nu(z) - \dfrac{2}{\pi} \sin \nu\pi \, Q_\nu(z)$          $[\operatorname{Im} z > 0]$          MO 77

3.     $Q_\nu(-z) = -e^{-\nu\pi i} Q_\nu(z)$          $[\operatorname{Im} z < 0]$          MO 77

4.     $Q_\nu(-z) = -e^{\nu\pi i} Q_\nu(z)$          $[\operatorname{Im} z > 0]$          MO 77

**8.834**

1.     $Q_\nu(x \pm i0) = Q_\nu(x) \mp \dfrac{\pi i}{2} P_\nu(x)$          MO 77

2.     $Q_n(z) = \dfrac{1}{2} P_n(z) \ln \dfrac{z+1}{z-1} - W_{n-1}(z)$          (see **8.831** 3)          MO 77

**8.835**

1.     $Q_\nu(z) - Q_{-\nu-1}(z) = \pi \cot \nu\pi \, P_\nu(z)$          $[\sin \nu\pi \neq 0]$          MO 77

2.     $Q_{-\nu-1}(\cos \varphi) = Q_\nu(\cos \varphi) - \pi \cot \nu\pi \, P_\nu(\cos \varphi)$          $[\sin \nu\pi \neq 0]$          MO 77

3.     $Q_\nu(-\cos \varphi) = -\cos \nu\pi \, Q_\nu(\cos \varphi) - \dfrac{\pi}{2} \sin \nu\pi \, P_\nu(\cos \varphi)$          MO 77

**8.836**

1.     $Q_n(z) = \dfrac{1}{2^n n!} \dfrac{d^n}{dz^n} \left[ (z^2 - 1)^n \ln \dfrac{z+1}{z-1} \right] - \dfrac{1}{2} P_n(z) \ln \dfrac{z+1}{z-1}$          MO 79

2.     $Q_n(x) = \dfrac{1}{2^n n!} \dfrac{d^n}{dx^n} \left[ (x^2 - 1)^n \ln \dfrac{1+x}{1-x} \right] - \dfrac{1}{2} P_n(x) \ln \dfrac{1+x}{1-x}$          MO 79

**8.837**

1.     $P_\nu(x) = P_\nu(\cos \varphi) = F\left(-\nu, \nu+1; 1; \sin^2 \dfrac{\varphi}{2}\right)$          (cf. **8.820** 6)          MO 76

2.     $P_\nu(z) = \dfrac{\tan \nu\pi}{2^{\nu+1} \sqrt{\pi}} \dfrac{\Gamma(\nu+1)}{\Gamma\left(\nu + \frac{3}{2}\right)} z^{-\nu-1} F\left(\dfrac{\nu}{2} + 1, \dfrac{\nu+1}{2}; \nu + \dfrac{3}{2}; \dfrac{1}{z^2}\right)$

         $+ \dfrac{2^\nu}{\sqrt{\pi}} \dfrac{\Gamma\left(\nu + \frac{1}{2}\right)}{\Gamma(\nu+1)} z^\nu F\left(\dfrac{1-\nu}{2}, -\dfrac{\nu}{2}; \dfrac{1}{2} - \nu; \dfrac{1}{z^2}\right)$

         MO 78

See also **8.820**.

     For integrals of Legendre functions, see **7.1–7.2**.

**8.838**    Inequalities ($0 \leq \varphi \leq \pi$, $\nu > 1$, and $C_0$ is a number that does not depend on the values of $\nu$ or $\varphi$):

1.     $|P_\nu(\cos \varphi) - P_{\nu+2}(\cos \varphi)| \leq 2C_0 \sqrt{\dfrac{1}{\nu\pi}}$          MO 78

2.     $|Q_\nu(\cos \varphi) - Q_{\nu+2}(\cos \varphi)| < C_0 \sqrt{\dfrac{\pi}{\nu}}$          MO 78

     With regard to the zeros of Legendre functions of the second kind, see **8.784**, **8.785**, and **8.786**. For the expansion of Legendre functions in series of associated Legendre functions, see **8.794**, **8.795**, and **8.796**.

**8.839**   A differential equation leading to the functions $W_{n-1}$ (see **8.831** 3):

$$\left(1 - x^2\right) \frac{d^2 W_{n-1}}{dx^2} - 2x \frac{dW_{n-1}}{dx} + (n+1)n W_{n-1} = 2 \frac{d P_\nu}{dx}$$

MO 76

## 8.84   Conical functions

**8.840**   Let us set

$$\nu = -\tfrac{1}{2} + i\lambda,$$

where $\lambda$ is a real parameter, in the defining differential equation **8.700** 1 for associated Legendre functions. We then obtain the differential equation of the so-called conical functions. A conical function is a special case of the associated Legendre function. However, the Legendre functions

$$P_{-\frac{1}{2}+i\lambda}(x), \quad Q_{-\frac{1}{2}+i\lambda}(x)$$

have certain peculiarities that make us distinguish them as a special class—the class of conical functions. The most important of these peculiarities is the following

**8.841**   The functions

$$P_{-\frac{1}{2}+i\lambda}\left(\cos\varphi\right) = 1 + \frac{4\lambda^2 + 1^2}{2^2} \sin^2 \frac{\varphi}{2} + \frac{\left(4\lambda^2 + 1^2\right)\left(4\lambda^2 + 3^2\right)}{2^2 4^2} \sin^4 \frac{\varphi}{2} + \dots$$

are real for real values of $\varphi$. Also,

$$P_{-\frac{1}{2}+i\lambda}(x) \equiv P_{-\frac{1}{2}-i\lambda}(x)$$

MO 95

**8.842**   Integral representations:

1.      $$P_{-\frac{1}{2}+i\lambda}\left(\cos\varphi\right) = \frac{2}{\pi} \int_0^\varphi \frac{\cosh \lambda u \, du}{\sqrt{2\left(\cos u - \cos\varphi\right)}} = \frac{2}{\pi} \cosh \lambda\pi \int_0^\infty \frac{\cos \lambda u \, du}{\sqrt{2\left(\cos\varphi + \cosh u\right)}}$$

MO 95

2.[6]     $$Q_{-\frac{1}{2}\mp\lambda i}\left(\cos\varphi\right) = \pm i \sinh \lambda\pi \int_0^\infty \frac{\cos \lambda u \, du}{\sqrt{2\left(\cosh u + \cos\varphi\right)}} + \int_0^\infty \frac{\cos \lambda u \, du}{\sqrt{2\left(\cosh u - \cos\varphi\right)}}$$

MO 95

### Functional relations

(See also **8.73**)

**8.843**   $$P_{-\frac{1}{2}+i\lambda}\left(-\cos\varphi\right) = \frac{\cosh \lambda\pi}{\pi} \left[ Q_{-\frac{1}{2}+i\lambda}\left(\cos\varphi\right) + Q_{-\frac{1}{2}-i\lambda}\left(\cos\varphi\right) \right]$$

MO 95

**8.844**

1.      $$P_{-\frac{1}{2}+i\lambda}\left(\cos\psi \cos\vartheta + \sin\psi \sin\vartheta \cos\varphi\right)$$

$$= P_{-\frac{1}{2}+i\lambda}\left(\cos\psi\right) P_{-\frac{1}{2}+i\lambda}\left(\cos\vartheta\right) + 2\sum_{k=1}^\infty \frac{(-1)^k 2^{2k} \, P^k_{-\frac{1}{2}+i\lambda}\left(\cos\psi\right) P^k_{-\frac{1}{2}+i\lambda}\left(\cos\vartheta\right) \cos k\varphi}{\left(4\lambda^2 + 1^2\right)\left(4\lambda^2 + 3^2\right) \cdots \left[4\lambda^2 + (2k-1)^2\right]}$$

$$\left[0 < \vartheta < \tfrac{\pi}{2}, \quad 0 < \psi < \pi, \quad 0 < \psi + \vartheta < \pi\right] \qquad \text{(cf. } \textbf{8.794} \text{ 1)} \quad \text{MO 95}$$

2.      $$P_{-\frac{1}{2}+i\lambda}\left(-\cos\psi \cos\vartheta - \sin\psi \sin\vartheta \cos\varphi\right)$$

$$= P_{-\frac{1}{2}+i\lambda}\left(\cos\psi\right) P_{-\frac{1}{2}+i\lambda}\left(-\cos\vartheta\right) + 2\sum_{k=1}^\infty \frac{(-1)^k 2^{2k} \, P^k_{-\frac{1}{2}+i\lambda}\left(\cos\psi\right) P^k_{-\frac{1}{2}+i\lambda}\left(-\cos\vartheta\right) \cos k\varphi}{\left(4\lambda^2 + 1\right)\left(4\lambda^2 + 3^2\right) \cdots \left[4\lambda^2 + (2k-1)^2\right]}$$

$$\left[0 < \psi < \tfrac{\pi}{2} < \vartheta, \quad \psi + \vartheta < \pi\right] \qquad \text{(cf. } \textbf{8.796}\text{)} \quad \text{MO 95}$$

3.  $Q_{-\frac{1}{2}+i\lambda}\left(\cos\psi\cos\vartheta+\sin\psi\sin\vartheta\cos\varphi\right)$

$$= P_{-\frac{1}{2}+i\lambda}\left(\cos\psi\right)Q_{-\frac{1}{2}+i\lambda}\left(\cos\vartheta\right)+2\sum_{k=1}^{\infty}\frac{(-1)^{k}2^{2k}\,P_{-\frac{1}{2}+i\lambda}^{k}\left(\cos\psi\right)Q_{-\frac{1}{2}+i\lambda}^{k}\left(\cos\vartheta\right)\cos k\varphi}{(4\lambda^{2}+1)\,(4\lambda^{2}+3^{2})\cdots\left[4\lambda^{2}+(2k-1)^{2}\right]}$$

$$\left[0<\psi<\tfrac{\pi}{2}<\vartheta,\quad\psi+\vartheta<\pi\right]\qquad(\text{cf. }\mathbf{8.794}\,2)\quad\text{MO 96}$$

Regarding the zeros of conical functions, see **8.784**.

## 8.85  Toroidal functions*

**8.850**  Solutions of the differential equation

1.  $$\frac{d^{2}u}{d\eta^{2}}+\frac{\cosh\eta}{\sinh\eta}\frac{du}{d\eta}-\left(n^{2}-\frac{1}{4}+\frac{m^{2}}{\sinh^{2}\eta}\right)u=0,$$

are called toroidal functions. They are equivalent (under a coordinate transformation) to associated Legendre functions. In particular, the functions

$$P_{n-\frac{1}{2}}^{m}\left(\cosh\eta\right),\quad Q_{n-\frac{1}{2}}^{m}\left(\sinh\eta\right)\qquad\text{MO 96}$$

are solutions of equation **8.850** 1.

The following formulas, obtained from the formulas obtained earlier for associated Legendre functions, are valid for toroidal functions:

**8.851**  Integral representations:

1.  $$P_{n-\frac{1}{2}}^{m}\left(\cosh\eta\right)=\frac{\Gamma\left(n+m+\frac{1}{2}\right)}{\Gamma\left(n-m+\frac{1}{2}\right)}\frac{(\sinh\eta)^{m}}{2^{m}\sqrt{\pi}\,\Gamma\left(m+\frac{1}{2}\right)}\int_{0}^{\pi}\frac{\sin^{2m}\varphi\,d\varphi}{(\cosh\eta+\sinh\eta\cos\varphi)^{n+m+\frac{1}{2}}}$$

$$=\frac{(-1)^{m}}{2\pi}\frac{\Gamma\left(n+\frac{1}{2}\right)}{\Gamma\left(n-m+\frac{1}{2}\right)}\int_{0}^{2\pi}\frac{\cos m\varphi\,d\varphi}{(\cosh\eta+\sinh\eta\cos\varphi)^{n+\frac{1}{2}}}$$

MO 96

2.  $$Q_{n-\frac{1}{2}}^{m}\left(\cosh\eta\right)=(-1)^{m}\frac{\Gamma\left(n+\frac{1}{2}\right)}{\Gamma\left(n-m+\frac{1}{2}\right)}\int_{0}^{\infty}\frac{\cosh mt\,dt}{(\cosh\eta+\sinh\eta\cosh t)^{n+\frac{1}{2}}}\qquad[n\geq m]$$

$$=(-1)^{m}\frac{\Gamma\left(n+m+\frac{1}{2}\right)}{\Gamma\left(n+\frac{1}{2}\right)}\int_{0}^{\ln\coth\frac{\eta}{2}}(\cosh\eta-\sinh\eta\cosh t)^{n-\frac{1}{2}}\cosh mt\,dt$$

MO 96

**8.852**  Functional relations:

1.[12]  $$Q_{n-\frac{1}{2}}^{m}\left(\cosh\eta\right)=(-1)^{m}\frac{2^{m}\,\Gamma\left(n+m+\frac{1}{2}\right)\sqrt{\pi}}{\Gamma(n+1)}\sinh^{m}\left(\eta\right)e^{-\left(n+m+\frac{1}{2}\right)\eta}$$

$$\times F\left(m+\tfrac{1}{2},n+m+\tfrac{1}{2};n+1;e^{-2\eta}\right)$$

---

*Sometimes called *torus functions*

$2.^{12}$ $P_{n-\frac{1}{2}}^{-m}(\cosh\eta) = \dfrac{2^{-m}}{\Gamma(m+1)}\left(1-e^{-2\eta}\right)^m e^{-\left(n+\frac{1}{2}\right)\eta}F\left(m+\tfrac{1}{2},n+m+\tfrac{1}{2};2m+1;1-e^{-2\eta}\right)$

MO 96

**8.853** An asymptotic representation $P_{n-\frac{1}{2}}(\cosh\eta)$ for large values of $n$:

$$P_{n-\frac{1}{2}}(\cosh\eta) = \frac{\Gamma(n)e^{\left(n-\frac{1}{2}\right)\eta}}{\sqrt{\pi}\,\Gamma\left(n+\frac{1}{2}\right)}$$

$$\times\left[\frac{2\,\Gamma^2\left(n+\frac{1}{2}\right)}{\pi n!\,\Gamma(n)}\ln\left(4e^\eta\right)e^{-2n\eta}F\left(\frac{1}{2},n+\frac{1}{2};n+1;e^{-2\eta}\right)+A+B\right],$$

where

$$A = 1+\frac{1}{2^2}\frac{1\cdot(2n-1)}{1\cdot(n-1)}e^{-2\eta}+\frac{1}{2^4}\frac{1\cdot 3\cdot(2n-1)(2n-3)}{1\cdot 2\cdot(n-1)(n-2)}e^{-4\eta}+\cdots+\frac{1}{2^{2n-2}}\left(\frac{(2n-1)!!}{(n-1)!}\right)^2 e^{-2(n-1)\eta}$$

$$B = \frac{\Gamma\left(n+\frac{1}{2}\right)}{\sqrt{\pi^3}\,\Gamma(n)}\sum_{k=1}^{\infty}\frac{\Gamma\left(k+\frac{1}{2}\right)\Gamma\left(n+k+\frac{1}{2}\right)}{\Gamma(n+k+1)\,\Gamma(k+1)}\left(u_{n+k}+u_k-v_{n+k-\frac{1}{2}}-v_{k-\frac{1}{2}}\right)e^{-2(n+k)\eta}$$

Here,

$$u_r = \sum_{s=1}^{r}\frac{1}{s},\qquad v_{r-\frac{1}{2}} = \sum_{s=1}^{r}\frac{2}{2s-1}\qquad [r\text{ is a natural number}]$$

MO 97

# 8.9 Orthogonal Polynomials

## 8.90 Introduction

**8.901** Suppose that $w(x)$ is a nonnegative real function of a real variable $x$. Let $(a,b)$ be a fixed interval on the $x$-axis. Let us suppose further that, for $n = 0, 1, 2, \ldots$, the integral

$$\int_a^b x^n w(x)\,dx$$

exists and that the integral

$$\int_a^b w(x)\,dx$$

is positive. In this case, there exists a sequence of polynomials $p_0(x), p_1(x), \ldots, p_n(x), \ldots$, that is uniquely determined by the following conditions:

1. $p_n(x)$ is a polynomial of degree $n$ and the coefficient of $x^n$ in this polynomial is positive.

2. The polynomials $p_0(x), p_1(x), \ldots$ are orthonormal; that is,

$$\int_a^b p_n(x)p_m(x)w(x)\,dx = \begin{cases}0 & \text{for } n\neq m,\\ 1 & \text{for } n = m.\end{cases}$$

We say that the polynomials $p_n(x)$ constitute *a system of orthogonal polynomials on the interval* $(a,b)$ *with the weight function* $w(x)$.

**8.902**  If $q_n$ is the coefficient of $x^n$ in the polynomial $p_n(x)$, then

1.  $$\sum_{k=0}^{n} p_k(x)p_k(y) = \frac{q_n}{q_{n+1}} \frac{p_{n+1}(x)p_n(y) - p_n(x)p_{n+1}(y)}{x - y} \qquad \text{(Darboux–Christoffel formula)}$$

EH II 159(10)

2.[11]  $$\sum_{k=0}^{n} [p_k(x)]^2 = \frac{q_n}{q_{n+1}} \left[ p_n(x)p'_{n+1}(x) - p'_n(x)p_{n+1}(x) \right]$$

EH II 159(11)

**8.903**  Between any three consecutive orthogonal polynomials, there is a dependence

$$p_n(x) = (A_n x + B_n) \, p_{n-1}(x) - C_n p_{n-2}(x) \qquad [n = 2, 3, 4, \ldots]$$

In this formula, $A_n, B_n$, and $C_n$ are constants and

$$A_n = \frac{q_n}{q_{n-1}}, \qquad C_n = \frac{q_n q_{n-2}}{q_{n-1}^2}$$

MO 102

**8.904**  Examples of normalized systems of orthogonal polynomials:

| Notation and name | | Interval | Weight |
|---|---|---|---|
| $\left(n + \frac{1}{2}\right)^{1/2} P_n(x)$ | see **8.91** | $(-1, +1)$ | $1$ |
| $2^\lambda \, \Gamma(\lambda) \left[ \dfrac{(n + \lambda)\, n!}{2\pi \, \Gamma(2\lambda + n)} \right]^{1/2} C_n^\lambda(x)$ | see **8.93** | $(-1, +1)$ | $\left(1 - x^2\right)^{\lambda - \frac{1}{2}}$ |
| $\sqrt{\dfrac{\varepsilon_n}{\pi}} \, T_n(x), \quad \varepsilon_0 = 1, \varepsilon_n = 2$ for $n = 1, 2, 3, \ldots$ | see **8.94** | $(-1, +1)$ | $\left(1 - x^2\right)^{-1/2}$ |
| $2^{-\frac{n}{2}} \pi^{-1/4} \, (n!)^{-1/2} H_n(x)$ | see **8.95** | $(-\infty, \infty)$ | $e^{-x^2}$ |
| $\left[ \dfrac{\Gamma(n+1)\,\Gamma(\alpha + \beta + 1 + n)(\alpha + \beta + 1 + 2n)}{\Gamma(\alpha + 1 + n)\,\Gamma(\beta + 1 + n)2^{\alpha + \beta + 1}} \right]^{1/2} P_n^{(\alpha, \beta)}(x)$ | see **8.96** | $(-1, +1)$ | $(1 - x)^\alpha (1 + x)^\beta$ |
| $\left[ \dfrac{\Gamma(n+1)}{\Gamma(\alpha + n + 1)} \right]^{1/2} (-1)^n \, L_n^\alpha(x)$ | see **8.97** | $(0, \infty)$ | $x^\alpha e^{-x}$ |

Cf. **7.221** 1, **7.313**, **7.343**, **7.374** 1, **7.391** 1, **7.414** 3.

## 8.91  Legendre polynomials

**8.910**  Definition.    The Legendre polynomials $P_n(z)$ are polynomials satisfying equation **8.700** 1 with $\mu = 0$ and $\nu = n$: that is, they satisfy the differential equation

1.  $$\left(1 - z^2\right) \frac{d^2 u}{dz^2} - 2z \frac{du}{dz} + n(n+1)u = 0$$

This equation has a polynomial solution if, and only if, $n$ is an integer. Thus, Legendre polynomials constitute a special type of associated Legendre function.

Legendre polynomials of degree $n$ are of the form

2.  $$P_n(z) = \frac{1}{2^n n!} \frac{d^n}{dz^n} \left(z^2 - 1\right)^n$$

**8.911**  Legendre polynomials written in expanded form:

1.  $$P_n(z) = \frac{1}{2^n} \sum_{k=0}^{\lfloor \frac{n}{2} \rfloor} \frac{(-1)^k (2n-2k)!}{k!(n-k)!(n-2k)!} z^{n-2k}$$

    $$= \frac{(2n)!}{2^n (n!)^2} \left( z^n - \frac{n(n-1)}{2(2n-1)} z^{n-2} + \frac{n(n-1)(n-2)(n-3)}{2 \cdot 4(2n-1)(2n-3)} z^{n-4} - \cdots \right)$$

    $$= \frac{(2n-1)!!}{n!} z^n F\left( -\frac{n}{2}, \frac{1-n}{2}; \frac{1}{2} - n; \frac{1}{z^2} \right)$$

    HO 13, AD (9001), MO 69

2.  $$P_{2n}(z) = (-1)^n \frac{(2n-1)!!}{2^n n!} \left( 1 - \frac{2n(2n+1)}{2!} z^2 + \frac{2n(2n-2)(2n+1)(2n+3)}{4!} z^4 - \cdots \right)$$

    $$= (-1)^n \frac{(2n-1)!!}{2^n n!} F\left( -n, n+\frac{1}{2}; \frac{1}{2}; z^2 \right)$$

    AD (9002), MO 69

3.  $$P_{2n+1}(z) = (-1)^n \frac{(2n+1)!!}{2^n n!} \left( z - \frac{2n(2n+3)}{3!} z^3 + \frac{2n(2n-2)(2n+3)(2n+5)}{5!} z^5 - \cdots \right)$$

    $$= (-1)^n \frac{(2n+1)!!}{2^n n!} z F\left( -n, n+\frac{3}{2}; \frac{3}{2}; z^2 \right)$$

    AD (9002), MO 69

4.[12]  $$P_n(\cos\varphi) = 2\frac{(2n-1)!!}{(2n)!!} \left( \cos n\varphi + \frac{1}{1}\frac{n}{2n-1} \cos(n-2)\varphi \right.$$

    $$+ \frac{1 \cdot 3}{1 \cdot 2}\frac{n(n-1)}{(2n-1)(2n-3)} \cos(n-4)\varphi$$

    $$\left. + \frac{1 \cdot 3 \cdot 5}{1 \cdot 2 \cdot 3}\frac{n(n-1)(n-2)}{(2n-1)(2n-3)(2n-5)} \cos(n-6)\varphi - \cdots \right)$$

    [there are $\lceil \frac{n+1}{2} \rceil$ terms; divide last term by 2 when $n$ is even]    WH

5.  $$P_{2n}(\cos\varphi) = (-1)^n \frac{(2n-1)!!}{2^n n!}$$

    $$\times \left\{ \sin^{2n}\varphi - \frac{(2n)^2}{2!} \sin^{2n-2}\varphi \cos^2\varphi + \cdots + (-1)^n \frac{2^n n!}{(2n-1)!!} \cos^{2n}\varphi \right\}$$

    AD (9011)

6.  $$P_{2n+1}(\cos\varphi) = (-1)^n \frac{(2n+1)!!}{2^n n!} \cos\varphi$$

    $$\times \left\{ \sin^{2n}\varphi - \frac{(2n)^2}{3!} \sin^{2n-2}\varphi \cos^2\varphi + \cdots + (-1)^n \frac{2^n n!}{(2n+1)!!} \cos^{2n}\varphi \right\}$$

    AD (9012)

7.  $$P_n(z) = \sum_{k=0}^{n} \frac{(-1)^k (n+k)!}{(n-k)!(k!)^2 2^{k+1}} \left[ (1-z)^k + (-1)^n (1+z)^k \right]$$

    WH

**8.912**    Special cases:

1.      $P_0(x) = 1$                                                                                                JA

2.      $P_1(x) = x = \cos\varphi$                                                                                  JA

3.      $P_2(x) = \dfrac{1}{2}\left(3x^2 - 1\right) = \dfrac{1}{4}\left(3\cos 2\varphi + 1\right)$                    JA

4.      $P_3(x) = \dfrac{1}{2}\left(5x^3 - 3x\right) = \dfrac{1}{8}\left(5\cos 3\varphi + 3\cos\varphi\right)$        JA

5.      $P_4(x) = \dfrac{1}{8}\left(35x^4 - 30x^2 + 3\right) = \dfrac{1}{64}\left(35\cos 4\varphi + 20\cos 2\varphi + 9\right)$    JA

6.      $P_5(x) = \dfrac{1}{8}\left(63x^5 - 70x^3 + 15x\right) = \dfrac{1}{128}\left(63\cos 5\varphi + 35\cos 3\varphi + 30\cos\varphi\right)$    JA

7.[10]  $P_6(x) = \dfrac{1}{16}\left(231x^6 - 315x^4 + 105x^2 - 5\right) = \dfrac{1}{512}\left(231\cos 6\varphi + 126\cos 4\varphi + 105\cos 2\varphi + 50\right)$

8.      $P_7(x) = \dfrac{1}{16}\left(429x^7 - 693x^5 + 315x^3 - 35x\right)$

$$= \dfrac{1}{1024}\left(429\cos 7\varphi + 231\cos 5\varphi + 189\cos 3\varphi + 175\cos\varphi\right)$$

9.      $P_8(x) = \dfrac{1}{128}\left(6435x^8 - 12012x^6 + 6930x^4 - 1260x^2 + 35\right)$

$$= \dfrac{1}{16384}\left(6435\cos 8\varphi - 3432\cos 6\varphi + 2772\cos 4\varphi - 2520\cos 2\varphi + 1225\right)$$

**8.913**    Integral representations:

1.      $P_n\left(\cos\varphi\right) = \dfrac{2}{\pi}\displaystyle\int_\varphi^\pi \dfrac{\sin\left(n + \frac{1}{2}\right)t}{\sqrt{2\left(\cos\varphi - \cos t\right)}}\,dt$          WH

See also **3.611** 3, **3.661** 3, 4.

2.[7]   Schläfli's integral formula:

$$P_n(z) = \dfrac{1}{2\pi i}\int_C \dfrac{\left(t^2 - 1\right)^n}{2^n(t - z)^{n+1}}\,dt,$$

with $C$ a simple contour containing $z$.                                                                          SA 175(9)

3.[10]  Laplace integral formula:

$$P_n(z) = \dfrac{1}{\pi}\int_0^\pi \left[x + \left(x^2 - 1\right)^{1/2}\cos\varphi\right]^n\,d\varphi \qquad [|x| \le 1]$$          SA 180(19)

**Functional relations**

**8.914**    Recurrence formulas:

1.      $(n + 1)\,P_{n+1}(z) - (2n + 1)z\,P_n(z) + n\,P_{n-1}(z) = 0$                                              WH

2.      $\left(z^2 - 1\right)\dfrac{d\,P_n}{dz} = n\left[z\,P_n(z) - P_{n-1}(z)\right] = \dfrac{n(n+1)}{2n+1}\left[P_{n+1}(z) - P_{n-1}(z)\right]$          WH

**8.915**

1.[10]    $\displaystyle\sum_{k=0}^{n}(2k+1)\,P_k(x)\,P_k(y) = (n+1)\frac{P_n(x)\,P_{n+1}(y) - P_n(y)\,P_{n+1}(x)}{y-x}$

(Christoffel summation formula)<br>MO 70

1(1)[10].    $(y-x)\displaystyle\sum_{k=0}^{n}(2k+1)\,P_k(x)\,Q_k(y) = 1 - (n+1)\left[P_{n+1}(x)\,Q_n(y) - P_n(x)\,Q_{n+1}(y)\right]$

AS 335(8.9.2)

2.[7]    $\displaystyle\sum_{k=0}^{\lfloor\frac{n-1}{2}\rfloor}(2n-4k-1)\,P_{n-2k-1}(z) = P'_n(z)$          (summation theorem)      MO 70

3.[7]    $\displaystyle\sum_{k=0}^{\lfloor\frac{n-2}{2}\rfloor}(2n-4k-3)\,P_{n-2k-2}(z) = z\,P'_n(z) - n\,P_n(z)$       SM 491(42), WH

4.[10]    $\displaystyle\sum_{k=1}^{\lfloor\frac{n}{2}\rfloor}(2n-4k+1)[k(2n-2k+1)-2]\,P_{n-2k}(z) = z^2\,P''_n(z) - n(n-1)\,P_n(z)$      WH

5.[11]    $\displaystyle\sum_{k=0}^{m}\frac{a_{m-k}a_k a_{n-k}}{a_{n+m-k}}\left(\frac{2n+2m-4k+1}{2n+2m-2k+1}\right)P_{n+m-2k}(z) = P_n(z)\,P_m(z)$

$\left[a_k = \dfrac{(2k-1)!!}{k!}, \quad m \leq n\right]$     AD (9036)

**8.916**

1.    $P_n(\cos\varphi) = \dfrac{(2n-1)!!}{2^n n!}\,e^{\mp in\varphi}\,F\left(\dfrac{1}{2}, -n; \dfrac{1}{2}-n; e^{\pm 2i\varphi}\right)$      MO 69

2.    $P_n(\cos\varphi) = F\left(n+1, -n; 1; \sin^2\dfrac{\varphi}{2}\right)$      MO 69

3.    $P_n(\cos\varphi) = (-1)^n\,F\left(n+1, -n; 1; \cos^2\dfrac{\varphi}{2}\right)$      WH

4.    $P_n(\cos\varphi) = \cos^n\varphi\,F\left(-\dfrac{1}{2}n, \dfrac{1}{2}-\dfrac{1}{2}n; 1; -\tan^2\varphi\right)$      HO 23

5.    $P_n(\cos\varphi) = \cos^{2n}\dfrac{\varphi}{2}\,F\left(-n, -n; 1; -\tan^2\dfrac{\varphi}{2}\right)$      HO 23, 29, WH

See also **8.911** 1, **8.911** 2, **8.911** 3. For a connection with other functions, see **8.936** 3, **8.836**, **8.962** 2.

- For integrals of Legendre polynomials, see **7.22–7.25**.
- For the zeros of Legendre polynomials, see **8.785**.

**8.917**   Inequalities:

1.    $P_0(x) < P_1(x) < P_2(x) < \cdots < P_n(x) < \ldots$          $[x > 1]$          MO 71

2.    For $x > -1$, $P_0(x) + P_1(x) + \cdots + P_n(x) > 0$.          MO 71

3.[12]   $[P_n(\cos\varphi)]^2 > \dfrac{\sin(2n+1)\varphi}{(2n+1)\sin\varphi}$          $[0 < \varphi < \pi]$        MO 71

4.    $\sqrt{n\sin\varphi}\,|P_n(\cos\varphi)| \leq 1.$                   MO 71

5.    $|P_n(\cos\varphi)| \leq 1.$                          WH

6.[10]   Let $n \geq 2$.    The successive relative maxima of $|P_n(x)|$, when $x$ decreases from 1 to 0, form a decreasing sequence. More precisely, if $\mu_1, \mu_2, \ldots, \mu_{\lfloor n/2 \rfloor}$ denote these maxima corresponding to decreasing values of $x$, we have

$$1 > \mu_1 > \mu_2 > \cdots > \mu_{\lfloor n/2 \rfloor}$$        SZ 162(7.3.1)

7.[10]   Let $n \geq 2$.    The successive relative maxima of $(\sin\theta)^{1/2}\,|P_n(\cos\theta)|$ when $\theta$ increases from 0 to $\pi/2$, form an increasing sequence.        SZ 163(7.3.2)

8.[10]   We have

$$(\sin\theta)^{1/2}\,|P_n(\cos\theta)| < (2/\pi)^{1/2}n^{-1/2} \qquad [0 \leq q\theta \leq q\pi]$$        SZ 163(7.3.8)

Here the constant $(2/\pi)^{1/2}$ cannot be replaced by a smaller one.

9.[10]   $\displaystyle\max_{0 \leq q\theta \leq q\pi} (\sin\theta)^{1/2}\,|P_n(\cos\theta)| \cong (2/\pi)^{1/2}n^{-\frac{1}{2}}$          $[n \to \infty]$        SZ 164(7.3.12)

10.[10]   Stieltjes' first theorem:

$$|P_n(\cos\theta)| \leq \left(\frac{2}{\pi}\right)^{1/2} \frac{4}{\sqrt{n\sin\theta}} \qquad [n = 1, 2, \ldots, 0 < \theta < \pi]$$        SA 197(8)

11.[10]   Stieltjes' second theorem:

$$|P_n(x) - P_{n+2}(x)| < \frac{4}{\sqrt{\pi}\sqrt{n+2}} \qquad [|x| \leq 1]$$        SA 199(15)

12.[10]   $\left|\dfrac{d\,P_n(x)}{dx}\right| < \dfrac{2}{\sqrt{\pi}}\dfrac{\sqrt{n}}{1-x^2}$          $[|x| < 1, \quad n = 1, 2, \ldots]$        SA 201(18)

13.[10]   $|P_{n+1}(x) + P_n(x)| < 6\left(\dfrac{2}{\pi n}\right)^{\frac{1}{2}}(1-x)^{-1/2}$          $[|x| < 1, \quad n = 0, 1, \ldots]$        SA 201(19)

**8.918**[10]    Asymptotic approximations:

1.[12]   $P_n(\cos\theta) = \left(\dfrac{2}{\pi n\sin\theta}\right)^{1/2}\cos\left[\left(n+\dfrac{1}{2}\right)\theta - \dfrac{\pi}{4}\right] + O\left(n^{-3/2}\right)$

$$[\varepsilon \leq \theta \leq \pi - \varepsilon, \quad 0 < \varepsilon < \pi/2m]$$      (Laplace's formula)    SA 208(1)

2.[12]   $P_n(\cos\theta) = \left(\dfrac{2}{\pi n\sin\theta}\right)^{1/2}\left\{\left(1 - \dfrac{1}{4n}\right)\cos\left[\left(n+\dfrac{1}{2}\right)\theta - \dfrac{\pi}{4}\right] + \dfrac{1}{8n}\cot\theta\sin\left[\left(n+\dfrac{1}{2}\right)\theta - \dfrac{\pi}{4}\right]\right\}$

$$+ O\left(n^{-5/2}\right)$$

$$[\varepsilon \leq \theta \leq \pi - \varepsilon, \quad 0 < \varepsilon < \pi/2]$$      (Bonnet–Heine formula)    SA 208(2)

## 8.919[10]   Series of products of Legendre and Chebyshev polynomials

1.    $\displaystyle 2\int_{-1}^{1} T_n(x)\, P_n(x)\, \mathrm{d}x = \sum_{\substack{i,j=0}}^{i+j=n} \int_{-1}^{1} P_i(x)\, P_j(x)\, P_n(x)\, \mathrm{d}x$

## 8.92   Series of Legendre polynomials

**8.921**    The generating function:

$$\frac{1}{\sqrt{1-2tz+t^2}} = \sum_{k=0}^{\infty} t^k\, P_k(z) \qquad \left[|t| < \min\left|z \pm \sqrt{z^2-1}\right|\right] \qquad \text{SM 489(31), WH}$$

$$= \sum_{k=0}^{\infty} \frac{1}{t^{k+1}}\, P_k(z) \qquad \left[|t| > \max\left|z \pm \sqrt{z^2-1}\right|\right] \qquad \text{MO 70}$$

**8.922**

1.    $\displaystyle z^{2n} = \frac{1}{2n+1} P_0(z) + \sum_{k=1}^{\infty} (4k+1) \frac{2n(2n-2)\ldots(2n-2k+2)}{(2n+1)(2n+3)\ldots(2n+2k+1)} P_{2k}(z)$      MO 72

2.    $\displaystyle z^{2n+1} = \frac{3}{2n+3} P_1(z) + \sum_{k=1}^{\infty} (4k+3) \frac{2n(2n-2)\ldots(2n-2k+2)}{(2n+3)(2n+5)\ldots(2n+2k+3)} P_{2k+1}(z)$      MO 72

3.    $\displaystyle \frac{1}{\sqrt{1-x^2}} = \frac{\pi}{2} \sum_{k=0}^{\infty} (4k+1) \left\{\frac{(2k-1)!!}{2^k k!}\right\}^2 P_{2k}(x) \qquad [|x| < 1, \quad (-1)!! \equiv 1]$

                                                                                 MO 72, LA 385(15)

4.    $\displaystyle \frac{x}{\sqrt{1-x^2}} = \frac{\pi}{2} \sum_{k=0}^{\infty} (4k+3) \frac{(2k-1)!!(2k+1)!!}{2^{2k+1} k!(k+1)!} P_{2k+1}(x)$

                                                     $[|x| < 1, \quad (-1)!! \equiv 1]$      LA 385(17)

5.    $\displaystyle \sqrt{1-x^2} = \frac{\pi}{2} \left\{\frac{1}{2} - \sum_{k=1}^{\infty} (4k+1) \frac{(2k-3)!!(2k-1)!!}{2^{2k+1} k!(k+1)!} P_{2k}(x)\right\}$

                                                     $[|x| < 1, \quad (-1)!! \equiv 1]$      LA 385(18)

6.[10]    $\displaystyle \sqrt{\frac{1-x}{2}} = \frac{2}{3} P_0(x) - 2 \sum_{n=1}^{\infty} \frac{1}{(2n-1)(2n+3)} P_n(x) \qquad [-1 \leq x \leq 1]$

7.[12]    $\displaystyle \frac{1-\rho^2}{(1-2\rho x+\rho^2)^{3/2}} = \sum_{n=0}^{\infty} (2n+1)\rho^n\, P_n(x) \qquad [|\rho| < 1, \quad |x| \leq 1] \qquad \text{SA 170(4)}$

**8.923**    $\displaystyle \arcsin x = \frac{\pi}{2} \sum_{k=1}^{\infty} \left\{\frac{(2k-1)!!}{2^k k!}\right\}^2 [P_{2k+1}(x) - P_{2k-1}(x)] + \pi x/2$

                                                     $[|x| < 1, \quad (-1)!! \equiv 1]$      WH

**8.924**

1.
$$-\frac{1+\cos n\pi}{2\left(n^2-1\right)}P_0\left(\cos\theta\right)-\frac{1+\cos n\pi}{2}\sum_{k=0}^{\infty}\frac{(4k+5)n^2\left(n^2-2^2\right)\ldots\left[n^2-(2k)^2\right]}{\left(n^2-1^2\right)\left(n^2-3^2\right)\ldots\left[n^2-(2k+3)^2\right]}P_{2k+2}\left(\cos\theta\right)$$

$$-\frac{3\left(1-\cos n\pi\right)}{2\left(n^2-2^2\right)}P_1\left(\cos\theta\right)$$

$$-\frac{1-\cos n\pi}{2}\sum_{k=1}^{\infty}\frac{(4k+3)\left(n^2-1^2\right)\ldots\left[n^2-(2k-1)^2\right]}{\left(n^2-2^2\right)\left(n^2-4^2\right)\ldots\left[n^2-(2k+2)^2\right]}P_{2k+1}\left(\cos\theta\right)=\cos n\theta$$

AD (9060.1)

2.
$$\frac{-\sin n\pi}{2\left(n^2-1\right)}P_0\left(\cos\theta\right)-\frac{\sin n\pi}{2}\sum_{k=0}^{\infty}\frac{(4k+5)n^2\left(n^2-2^2\right)\ldots\left[n^2-(2k)^2\right]}{\left(n^2-1^2\right)\left(n^2-3^2\right)\ldots\left[n^2-(2k+3)^2\right]}P_{2k+2}\left(\cos\theta\right)$$

$$+\frac{3\sin n\pi}{2\left(n^2-2^2\right)}P_1\left(\cos\theta\right)$$

$$+\frac{\sin n\pi}{2}\sum_{k=1}^{\infty}\frac{(4k+3)\left(n^2-1^2\right)\left(n^2-3^2\right)\ldots\left[n^2-(2k-1)^2\right]}{\left(n^2-2^2\right)\left(n^2-4^2\right)\ldots\left[n^2-(2k+2)^2\right]}P_{2k+1}\left(\cos\theta\right)=\sin n\theta$$

AD (9060.2)

3.[3]
$$\frac{2^{n-1}n!}{(2n-1)!!}P_n\left(\cos\theta\right)-n\sum_{k=1}^{\lfloor n/2\rfloor}(2n-4k+1)\frac{2^{n-2k-1}(n-k-1)!(2k-3)!!}{(2n-2k+1)!!k!}P_{n-2k}\left(\cos\theta\right)$$

$$=\cos n\theta$$

AD (9061.1)

4.
$$\frac{(2n-1)!!P_{n-1}\left(\cos\theta\right)}{2^{n-1}(n-1)!}-\frac{n}{2^{n+1}}\sum_{k=0}^{\infty}\frac{(2n+2k-1)!!(2k-1)!!\left(2n+4k+3\right)}{2^{2k}(n+k+1)!(k+1)!}P_{n+2k+1}\left(\cos\theta\right)$$

$$=\frac{4\sin n\theta}{\pi}$$

AD (9061.2)

**8.925**

1.
$$\sum_{k=1}^{\infty}\frac{4k-1}{2^{2k}(2k-1)^2}\left[\frac{(2k-1)!!}{k!}\right]^2P_{2k-1}\left(\cos\theta\right)=1-\frac{2\theta}{\pi}$$

2.
$$\sum_{k=1}^{\infty}\frac{4k+1}{2^{2k+1}(2k-1)(k+1)}\left[\frac{(2k-1)!!}{k!}\right]^2P_{2k}\left(\cos\theta\right)=\frac{1}{2}-\frac{2\sin\theta}{\pi}$$

AD (9062.2)

3.
$$\sum_{k=1}^{\infty}\frac{k(4k-1)}{2^{2k-1}(2k-1)}\left[\frac{(2k-1)!!}{k!}\right]^2P_{2k-1}\left(\cos\theta\right)=\frac{2\cot\theta}{\pi}$$

AD (9062.3)

4.
$$\sum_{k=1}^{\infty}\frac{4k+1}{2^{2k}}\left[\frac{(2k-1)!!}{k!}\right]^2P_{2k}\left(\cos\theta\right)=\frac{2}{\pi\sin\theta}-1$$

AD (9062.4)

**8.926**

1.
$$\sum_{n=1}^{\infty}\frac{1}{n}P_n\left(\cos\theta\right)=\ln\frac{2\tan\frac{\pi-\theta}{4}}{\sin\theta}=-\ln\sin\frac{\theta}{2}-\ln\left(1+\sin\frac{\theta}{2}\right)$$

AD (9063.2)

2.    $\displaystyle\sum_{n=1}^{\infty} \frac{1}{n+1} P_n\left(\cos\theta\right) = \ln\frac{1+\sin\frac{\theta}{2}}{\sin\frac{\theta}{2}} - 1$          AD (9063.1)

**8.927**   $\displaystyle\sum_{k=0}^{\infty} \cos\left(k+\tfrac{1}{2}\right)\beta\, P_k\left(\cos\varphi\right) = \frac{1}{\sqrt{2\left(\cos\beta - \cos\varphi\right)}}$      $\left[0 \leq \beta < \varphi < \pi\right]$

                                         $= 0$      $\left[0 < \varphi < \beta < \pi\right]$

                                                     MO 72

**8.928**

1.    $\displaystyle\sum_{n=1}^{\infty} \frac{(-1)^n(4k+1)\left[(2n-1)!!\right]^3}{2^{3n}\left(n!\right)^3} P_{2n}\left(\cos\theta\right) = \frac{4\,\boldsymbol{K}\left(\sin\theta\right)}{\pi^2} - 1$      AD (9064.1)

2.    $\displaystyle\sum_{n=1}^{\infty}(-1)^{n+1} \frac{(4n+1)\left[(2n-1)!!\right]^3}{(2n-1)(2n+2)2^{3n}\left(n!\right)^3} P_{2n}\left(\cos\theta\right) = \frac{4\,\boldsymbol{E}\left(\sin\theta\right)}{\pi^2} - \frac{1}{2}$      AD (9064.2)

- For series of products of Bessel functions and Legendre polynomials, see **8.511** 4, **8.531** 3, **8.533** 1, **8.533** 2, and **8.534**.
- For series of products of Legendre and Chebyshev polynomials, see **8.919**.

# 8.93   Gegenbauer polynomials $C_n^\lambda(t)$

**8.930**   Definition.     The polynomials $C_n^\lambda(t)$ of degree $n$ are the coefficients of $\alpha^n$ in the power-series expansion of the function

$$\left(1 - 2t\alpha + \alpha^2\right)^{-\lambda} = \sum_{n=0}^{\infty} C_n^\lambda(t)\alpha^n \qquad\qquad \text{WH}$$

Thus, the polynomials $C_n^\lambda(t)$ are a *generalization of the Legendre polynomials*.

1.[10]    $C_0^\lambda(t) = 1$

2.[10]    $C_1^\lambda(t) = 2\lambda t$

3.[10]    $C_2^\lambda(t) = 2\lambda(\lambda+1)t^2 - \lambda$

4.[10]    $C_3^\lambda(t) = \frac{1}{3}\lambda\left(4\lambda^2 + 12\lambda + 8\right)t^3 - 2\lambda(\lambda+1)t$

5.[11]    $C_4^\lambda(t) = \frac{2}{3}\lambda\left(\lambda^3 + 6\lambda^2 + 11\lambda + 6\right)t^4 - 2\lambda\left(\lambda^2 + 3\lambda + 2\right)t^2 + \frac{1}{2}\lambda(\lambda+1)$

6.[10]    $C_5^\lambda(t) = \frac{1}{15}\lambda\left(4\lambda^4 + 40\lambda^3 + 140\lambda^2 + 200\lambda + 96\right)t^5$

                $- \frac{1}{3}\lambda\left(4\lambda^3 + 24\lambda^2 + 44\lambda + 24\right)t^3 + \lambda\left(\lambda^2 + 3\lambda + 2\right)t$

7.[10]    $C_6^\lambda(t) = \frac{1}{45}\lambda\left(\lambda^5 + 60\lambda^4 + 340\lambda^3 + 900\lambda^2 + 1096\lambda + 480\right)t^6$

                $- \frac{1}{3}\lambda\left(2\lambda^4 + 20\lambda^3 + 70\lambda^2 + 100\lambda + 48\right)t^4$

                $+ \lambda\left(\lambda^3 + 6\lambda^2 + 11\lambda + 6\right)t^2 + \frac{1}{6}\lambda\left(\lambda^2 + 3\lambda + 2\right)$

**8.931**    Integral representation:

$$C_n^\lambda(t) = \frac{1}{\sqrt{\pi}} \frac{\Gamma(2\lambda + n)}{n!\,\Gamma(2\lambda)} \frac{\Gamma\left(\frac{2\lambda+1}{2}\right)}{\Gamma(\lambda)} \int_0^\pi \left(t + \sqrt{t^2 - 1}\cos\varphi\right)^n \sin^{2\lambda-1}\varphi\,d\varphi \qquad \text{MO 99}$$

See also **3.252** 11, **3.663** 2, **3.664** 4.

## Functional relations

**8.932**    Expressions in terms of hypergeometric functions:

1. 
$$C_n^\lambda(t) = \frac{\Gamma(2\lambda + n)}{\Gamma(n+1)\,\Gamma(2\lambda)} F\left(2\lambda + n, -n; \lambda + \frac{1}{2}; \frac{1-t}{2}\right)^* \qquad \text{MO 97}$$

$$= \frac{2^n\,\Gamma(\lambda + n)}{n!\,\Gamma(\lambda)} t^n\, F\left(-\frac{n}{2}, \frac{1-n}{2}; 1 - \lambda - n; \frac{1}{t^2}\right) \qquad \text{MO 99}$$

2. 
$$C_{2n}^\lambda(t) = \frac{(-1)^n}{(\lambda + n)\,\mathrm{B}(\lambda, n+1)} F\left(-n, n + \lambda; \frac{1}{2}; t^2\right) \qquad \text{MO 99}$$

3. 
$$C_{2n+1}^\lambda(t) = \frac{(-1)^n 2t}{\mathrm{B}(\lambda, n+1)} F\left(-n, n + \lambda + 1; \frac{3}{2}; t^2\right) \qquad \text{MO 99}$$

**8.933**    Recursion formulas:

1. 
$$(n + 2)\, C_{n+2}^\lambda(t) = 2(\lambda + n + 1)t\, C_{n+1}^\lambda(t) - (2\lambda + n)\, C_n^\lambda(t) \qquad \text{Mo 98}$$

2. 
$$n\, C_n^\lambda(t) = 2\lambda\left[t\, C_{n-1}^{\lambda+1}(t) - C_{n-2}^{\lambda+1}(t)\right] \qquad \text{WH}$$

3. 
$$(2\lambda + n)\, C_n^\lambda(t) = 2\lambda\left[C_n^{\lambda+1}(t) - t\, C_{n-1}^{\lambda+1}(t)\right] \qquad \text{WH}$$

4. 
$$n\, C_n^\lambda(t) = (2\lambda + n - 1)t\, C_{n-1}^\lambda(t) - 2\lambda\left(1 - t^2\right) C_{n-2}^{\lambda+1}(t) \qquad \text{WH}$$

**8.934**

1. 
$$C_n^\lambda(t) = \frac{(-1)^n}{2^n} \frac{\Gamma(2\lambda + n)\,\Gamma\left(\frac{2\lambda+1}{2}\right)}{\Gamma(2\lambda)\,\Gamma\left(\frac{2\lambda+1}{2} + n\right)} \frac{\left(1 - t^2\right)^{\frac{1}{2} - \lambda}}{n!} \frac{d^n}{dt^n}\left[\left(1 - t^2\right)^{\lambda + n - \frac{1}{2}}\right] \qquad \text{WH}$$

2. 
$$C_n^\lambda(\cos\varphi) = \sum_{\substack{k,l=0 \\ k+l=n}}^n \frac{\Gamma(\lambda + k)\,\Gamma(\lambda + l)}{k!\,l!\,[\Gamma(\lambda)]^2} \cos(k - l)\varphi \qquad \text{MO 99}$$

3. 
$$C_n^\lambda\left(\cos\psi\cos\vartheta + \sin\psi\sin\vartheta\cos\varphi\right)$$

$$= \frac{\Gamma(2\lambda - 1)}{[\Gamma(\lambda)]^2} \sum_{k=0}^n \frac{2^{2k}(n - k)!\,[\Gamma(\lambda + k)]^2}{\Gamma(2\lambda + n + k)}(2\lambda + 2k - 1)\sin^k\psi\sin^k\vartheta$$

$$\times\, C_{n-k}^{\lambda+k}(\cos\psi)\, C_{n-k}^{\lambda+k}(\cos\vartheta)\, C_k^{\lambda-\frac{1}{2}}(\cos\varphi)$$

$$\left[\psi, \vartheta, \varphi \text{ real}; \quad \lambda \neq \tfrac{1}{2}\right] \qquad [\text{"summation theorem"}] \quad (\text{see also } \mathbf{8.794}\text{–}\mathbf{8.796}) \quad \text{WH}$$

4. 
$$\lim_{\lambda \to 0} \Gamma(\lambda)\, C_n^\lambda(\cos\varphi) = \frac{2\cos n\varphi}{n} \qquad \text{MO 98}$$

For orthogonality, see **8.904**, **7.313**.

---

*Equation 8.932.1 defines the generalized functions $C_n^\lambda(t)$, where the subscript $n$ can be an arbitrary number

**8.935**    Derivatives:

1.      $\dfrac{d^k}{dt^k}\, C_n^\lambda(t) = 2^k \dfrac{\Gamma(\lambda + k)}{\Gamma(\lambda)}\, C_{n-k}^{\lambda + k}(t)$           MO 99

        In particular,

2.[11]     $\dfrac{d\, C_n^\lambda(t)}{dt} = 2\lambda\, C_{n-1}^{\lambda + 1}(t)$           WH

For integrals of the polynomials $C_n^\lambda(x)$ see **7.31–7.33**.

**8.936**    Connections with other functions:

1.[12]    $C_n^\lambda(t) = \dfrac{\Gamma(2\lambda + n)\,\Gamma\left(\lambda + \frac{1}{2}\right)}{\Gamma(2\lambda)\,\Gamma(n + 1)} \left\{ \dfrac{1}{4}\left(1 - t^2\right) \right\}^{\frac{1}{4} - \frac{\lambda}{2}} P_{\lambda + n - \frac{1}{2}}^{\frac{1}{2} - \lambda}(t)$           MO 98

2.      $C_{n-m}^{m + \frac{1}{2}}(t) = \dfrac{1}{(2m - 1)!!} \dfrac{d^m\, P_n(t)}{dt^m} = (-1)^m \dfrac{\left(1 - t^2\right)^{-\frac{m}{2}}\, m!\, 2^m}{(2m)!}\, P_n^m(t)$

                                             $[m + 1$ a natural number$]$        MO 98, WH

3.      $C_n^{1/2}(t) = P_n(t)$

4.      $J_{\lambda - \frac{1}{2}}\left(r \sin \vartheta \sin \alpha\right)\left(r \sin \vartheta \sin \alpha\right)^{-\lambda + \frac{1}{2}} e^{-ir \cos \vartheta \cos \alpha}$

$$= \sqrt{2}\, \dfrac{\Gamma(\lambda)}{\Gamma\left(\lambda + \frac{1}{2}\right)} \sum_{k=0}^{\infty} (\lambda + k) i^{-k} \dfrac{J_{\lambda + k}(r)\, C_k^\lambda(\cos \vartheta)\, C_k^\lambda(\cos \alpha)}{r^\lambda\, C_k^\lambda(1)}$$

                                                     MO 99

5.      $\lim\limits_{\lambda \to \infty} \lambda^{-\frac{n}{2}}\, C_n^{\frac{\lambda}{2}}\left(t\sqrt{\dfrac{2}{\lambda}}\right) = \dfrac{2^{-\frac{n}{2}}}{n!}\, H_n(t)$           MO 99a

See also **8.932**.

**8.937**    Special cases and particular values:

1.      $C_n^1(\cos \varphi) = \dfrac{\sin(n + 1)\varphi}{\sin \varphi}$           MO 99

2.      $C_0^0(\cos \varphi) = 1$           MO 98

3.      $C_0^\lambda(t) \equiv 1$           MO 98

4.      $C_n^\lambda(1) \equiv \dbinom{2\lambda + n - 1}{n}$           MO 98

**8.938**    A differential equation leading to the polynomials $C_n^\lambda(t)$:

$$y'' + \dfrac{(2\lambda + 1)t}{t^2 - 1} y' - \dfrac{n(2\lambda + n)}{t^2 - 1} y = 0 \quad \text{(cf. **9.174**)} \qquad \text{WH}$$

For series of products of Bessel functions and the polynomials $C_n^\lambda(x)$, see **8.532, 8.534**.

**8.939**[10]    Differentiation and Rodrigues' formulas and orthogonality relation

1.    $\dfrac{d}{dt} C_n^\lambda(t) = 2\lambda\, C_{n-1}^{\lambda+1}(t)$                          MS 5.3.2

2.    $\dfrac{d^m}{dt^m} C_n^\lambda(t) = 2^m \lambda(\lambda+1)(\lambda+2)\ldots(\lambda+m-1)\, C_{n-m}^{\lambda+m}(t)$      MS 5.3.2

3.    $\dfrac{d}{dt} C_{n-1}^\lambda(t) = t\dfrac{d}{dt} C_n^\lambda(t) - n\, C_n^\lambda(t)$                MS 5.3.2

4.    $\dfrac{d}{dt} C_{n+1}^\lambda(t) = t\dfrac{d}{dt} C_n^\lambda(t) + (2\lambda+n)\, C_n^\lambda(t)$             MS 5.3.2

5.    $\left(1-t^2\right)\dfrac{d}{dt} C_n^\lambda(t) = (n+2\lambda-1)\, C_{n-1}^\lambda(t) - nt\, C_n^\lambda(t) = (n+2\lambda)t\, C_n^\lambda(t) - (n+1)\, C_{n+1}^\lambda(t)$

$$= 2\lambda\left(1-t^2\right) C_{n-1}^{\lambda+1}(t)$$

                                                         MS 5.3.2

6.    $\dfrac{d}{dt}\left[C_{n+1}^\lambda(t) - C_{n-1}^\lambda(t)\right] = 2(n+\lambda)\, C_n^\lambda(t)$          MS 5.3.2

7.    $C_n^\lambda(t) = \dfrac{(-1)^n 2\lambda(2\lambda+1)(2\lambda+2)\ldots(2\lambda+n-1)\left(1-t^2\right)^{\frac12-\lambda}}{2^n n!\left(\lambda+\frac12\right)\left(\lambda+\frac32\right)\ldots\left(\lambda+n-\frac12\right)}\dfrac{d^n}{dt^n}\left[\left(1-t^2\right)^{n+\lambda-\frac12}\right]$

$$= \dfrac{(-1)^n\,\Gamma\left(\lambda+\frac12\right)\Gamma(n+2\lambda)\left(1-t^2\right)^{\frac12-\lambda}}{2^n n!\,\Gamma(2\lambda)\,\Gamma\left(n+\lambda+\frac12\right)}\dfrac{d^n}{dt^n}\left[\left(1-t^2\right)^{n+\lambda-\frac12}\right]$$

                                      [Rodrigues' formula]      MS 5.3.2

8.    $\displaystyle\int_{-1}^{1} C_n^\lambda(t)\, C_m^\lambda(t)\left(1-t^2\right)^{\lambda-\frac12}\, dt = 0$           $n \neq m$

$$= \dfrac{\pi 2^{1-2\lambda}\,\Gamma(n+2\lambda)}{n!(\lambda+n)\,[\Gamma(\lambda)]^2} \qquad n = m$$

                             $[\lambda \neq 0]$    [Orthogonality relation]    MS 5.3.2

## 8.94 The Chebyshev polynomials $T_n(x)$ and $U_n(x)$

**8.940**    Definition

1.    Chebyshev's polynomials of the first kind

$$T_n(x) = \cos\left(n\arccos x\right) = \frac12\left[\left(x+i\sqrt{1-x^2}\right)^n + \left(x-i\sqrt{1-x^2}\right)^n\right]$$

$$= x^n - \binom{n}{2}x^{n-2}\left(1-x^2\right) + \binom{n}{4}x^{n-4}\left(1-x^2\right)^2 - \binom{n}{6}x^{n-6}\left(1-x^2\right)^3 + \ldots$$

                                                          NA 66, 71

2.    Chebyshev's polynomials of the second kind:

$$U_n(x) = \frac{\sin\left[(n+1)\arccos x\right]}{\sin\left[\arccos x\right]} = \frac{1}{2i\sqrt{1-x^2}}\left[\left(x+i\sqrt{1-x^2}\right)^{n+1} - \left(x-i\sqrt{1-x^2}\right)^{n+1}\right]$$

$$= \binom{n+1}{1}x^n - \binom{n+1}{3}x^{n-2}\left(1-x^2\right) + \binom{n+1}{5}x^{n-4}\left(1-x^2\right)^2 - \ldots$$

## Functional relations

**8.941**    Recursion formulas:

1.     $T_{n+1}(x) - 2x\,T_n(x) + T_{n-1}(x) = 0$             NA 358

2.     $U_{n+1}(x) - 2x\,U_n(x) + U_{n-1}(x) = 0$

3.     $T_n(x) = U_n(x) - x\,U_{n-1}(x)$             EH II 184(3)

4.     $\left(1 - x^2\right) U_{n-1}(x) = x\,T_n(x) - T_{n+1}(x)$        EH II 184(4)

5.*    $T_{m\pm n}(x) = T_m(x)T_n(x) \mp (1 - x^2)U_{m-1}(x)U_{n-1}(x)$

6.*    $U_{m\pm n}(x) = U_m(x)T_n(x) \pm T_{m+1}(x)U_{n-1}(x)$

7.*    $T_{mn}(x) = T_m(x)(T_n(x))$

8.*    $U_{mn-1}(x) = U_{m-1}(T_n(x))U_{n-1}(x)$

For the orthogonality, see **7.343** and **8.904**.

**8.942**    Relations with other functions:

1.     $T_n(x) = F\left(n, -n; \dfrac{1}{2}; \dfrac{1-x}{2}\right)$             MO 104

2.     $T_n(x) = (-1)^n \dfrac{\sqrt{1-x^2}}{(2n-1)!!} \dfrac{d^n}{dx^n}\left(1 - x^2\right)^{n-\frac{1}{2}}$        MO 104

3.     $U_n(x) = \dfrac{(-1)^n(n+1)}{\sqrt{1-x^2}(2n+1)!!} \dfrac{d^n}{dx^n}\left(1 - x^2\right)^{n+\frac{1}{2}}$      EH II 185(15)

4 *    $U_n(x) = (n+1)F\left(n+2; -n; \dfrac{3}{2}; \dfrac{1-x}{2}\right)$

See also **8.962** 3.

**8.943**[10]    Special cases

1.     $T_0(x) = 1$                        10.     $U_0(x) = 1$

2.     $T_1(x) = x$                     11.     $U_1(x) = 2x$

3.     $T_2(x) = 2x^2 - 1$            12.     $U_2(x) = 4x^2 - 1$

4.     $T_3(x) = 4x^3 - 3x$          13.     $U_3(x) = 8x^3 - 4x$

5.     $T_4(x) = 8x^4 - 8x^2 + 1$     14.     $U_4(x) = 16x^4 - 12x^2 + 1$

6.     $T_5(x) = 16x^5 - 20x^3 + 5x$    15.     $U_5(x) = 32x^5 - 32x^3 + 6x$

7.     $T_6(x) = 32x^6 - 48x^4 + 18x^2 - 1$    16.     $U_6(x) = 64x^6 - 80x^4 + 24x^2 - 1$

8.     $T_7(x) = 64x^7 - 112x^5 + 56x^3 - 7x$    17.     $U_7(x) = 128x^7 - 192x^5 + 80x^3 - 8x$

9.     $T_8(x) = 128x^8 - 256x^6 + 160x^4 - 32x^2 + 1$    18.     $U_8(x) = 256x^8 - 448x^6 + 240x^4 - 40x^2 + 1$

**8.944**    Particular values:

1.    $T_n(1) = 1$                                          5.    $U_{2n+1}(0) = 0$

2.    $T_n(-1) = (-1)^n$                                    6.    $U_{2n}(0) = (-1)^n$

3.    $T_{2n}(0) = (-1)^n$

4.    $T_{2n+1}(0) = 0$

**8.945**    The generating function:

1.[11]    $\dfrac{1 - t^2}{1 - 2tx + t^2} = T_0(x) + 2\sum\limits_{k=1}^{\infty} T_k(x)t^k$              $[|t| < 1]$                          MO 104

2.[11]    $\dfrac{1}{1 - 2tx + t^2} = \sum\limits_{k=0}^{\infty} U_k(x)t^k$              $[|t| < 1]$                          MO 104a, EH II 186(31)

**8.946**    Zeros.    The polynomials $T_n(x)$ and $U_n(x)$ only have real simple zeros. All these zeros lie in the interval $(-1, +1)$.

**8.947**    The functions $T_n(x)$ and $\sqrt{1 - x^2}\, U_{n-1}(x)$ are two linearly independent solutions of the differential equation

$$\left(1 - x^2\right) \frac{d^2 y}{dx^2} - x \frac{dy}{dx} + n^2 y = 0 \qquad \text{NA 69(58)}$$

**8.948**    Of all polynomials of degree $n$ with leading coefficient equal to 1, the one that deviates the least from zero on the interval $[-1, +1]$ is the polynomial $2^{-n+1}\, T_n(x)$.

**8.949**[10]    Differentiation and Rodrigues' formulas and orthogonality relations

1.    $\dfrac{d}{dx}\, T_n(x) = n\, U_{n-1}(x)$                                                        MS 5.7.2

2.    $\dfrac{d^m}{dx^m}\, T_n(x) = 2^{m-1}\, \Gamma(m) n\, C_{n-m}^m(x)$                                    MS 5.7.2

3.    $\left(1 - x^2\right) \dfrac{d}{dx}\, T_n(x) = n\left[T_{n-1}(x) - x\, T_n(x)\right] = n\left[x\, T_n(x) - T_{n+1}(x)\right]$      MS 5.7.2

4.    $\dfrac{d}{dx}\, U_n(x) = 2\, C_{n-1}^2(x)$                                                    MS 5.7.2

5.    $\dfrac{d^m}{dx^m}\, U_n(x) = 2^m m!\, C_{n-m}^{m+1}(x)$                                          MS 5.7.2

6.    $\left(1 - x^2\right) \dfrac{d}{dx}\, U_n(x) = (n+1)\, U_{n-1}(x) - nx\, U_n(x) = (n+2)x\, U_n(x) - (n+1)\, U_{n+1}(x)$      MS 5.7.2

7.    $T_n(x) = \dfrac{(-1)^n \pi^{1/2} \left(1 - x^2\right)^{c\frac{1}{2}}}{2^{n+1}\, \Gamma\left(n + \frac{1}{2}\right)} \dfrac{d^n}{dx^n}\left[\left(1 - x^2\right)^{n - \frac{1}{2}}\right]$    [Rodrigues' formula]      MS 5.7.2

8.    $U_n(x) = \dfrac{(-1)^n \pi^{1/2}(n + 1)\left(1 - x^2\right)^{-1/2}}{2^{n+1}\, \Gamma\left(n + \frac{3}{2}\right)} \dfrac{d^n}{dx^n}\left[\left(1 - x^2\right)^{n + \frac{1}{2}}\right]$

[Rodrigues' formula]      MS 5.7.2

9.    $\displaystyle \int_{-1}^{1} T_m(x)\, T_n(x)\, \left(1 - x^2\right)^{-1/2}\, \mathrm{d}x = \begin{cases} 0, & m \neq n \\ \pi/2, & m = n \neq 0 \\ \pi, & m = n = 0 \end{cases}$

[Orthogonality relation]      MS 5.7.2

10.    $\displaystyle \int_{-1}^{1} U_m(x)\, U_n(x)\, \left(1 - x^2\right)^{-1/2}\, \mathrm{d}x = \begin{cases} 0, & m \neq n \\ \pi/8, & m = n \end{cases}$

[Orthogonality relation]      MS 5.7.2

## 8.95 The Hermite polynomials $H_n(x)$

**8.950**    Definition

1.    $H_n(x) = (-1)^n e^{x^2} \dfrac{d^n}{dx^n} \left(e^{-x^2}\right)$              SM 567(14)

      or

2.    $H_n(x) = 2^n x^n - 2^{n-1} \dbinom{n}{2} x^{n-2} + 2^{n-2} \cdot 1 \cdot 3 \cdot \dbinom{n}{4} x^{n-4} - 2^{n-3} \cdot 1 \cdot 3 \cdot 5 \cdot \dbinom{n}{6} x^{n-6} + \ldots$    MO 105a

3.[10]    $H_0(x) = 1$

4.[10]    $H_1(x) = 2x$

5.[10]    $H_2(x) = 4x^2 - 2$

6.[10]    $H_3(x) = 8x^3 - 12x$

7.[10]    $H_4(x) = 16x^4 - 48x^2 + 12$

8.[10]    $H_5(x) = 32x^5 - 160x^3 + 120x$

9.[10]    $H_6(x) = 64x^6 - 480x^4 + 720x^2 - 120$

10.[10]    $H_7(x) = 128x^7 - 1344x^5 + 3360x^3 - 1680x$

11.[10]    $H_8(x) = 256x^8 - 3584x^6 + 13440x^4 - 13440x^2 + 1680$

**8.951**    The integral representation:

$$H_n(x) = \frac{2^n}{\sqrt{\pi}} \int_{-\infty}^{\infty} (x + it)^n e^{-t^2}\, \mathrm{d}t$$      MO 106a

**Functional relations**

**8.952**    Recursion formulas:

1.    $\dfrac{d\, H_n(x)}{dx} = 2n\, H_{n-1}(x)$             SM 569(22)

2.    $H_{n+1}(x) = 2x\, H_n(x) - 2n\, H_{n-1}(x)$         SM 570(23)

      For the orthogonality, see **7.374** 1 and **8.904**.

3.[10]    $n\, H_n(x) = -n\, H'_{n-1}(x) + x\, H'_n(x)$        MS 5.6.2

$4.^{10}$     $H_n(x) = 2x\, H_{n-1}(x) - H'_{n-1}(x)$     MS 5.6.2

**8.953**   The connection with other functions:

1.     $H_{2n}(x) = (-1)^n \dfrac{(2n)!}{n!}\, \Phi\left(-n, \tfrac{1}{2}; x^2\right)$     MO 106a

2.     $H_{2n+1}(x) = (-1)^n 2 \dfrac{(2n+1)!}{n!}\, x\, \Phi\left(-n, \tfrac{3}{2}; x^2\right)$     MO 106a

- For a connection with the polynomials $C_n^\lambda(x)$, see **8.936** 5.
- For a connection with the Laguerre polynomials, see **8.972** 2 and **8.972** 3.
- For a connection with functions of a parabolic cylinder, see **9.253**.

**8.954**   Inequalities:

$1.^{10}$     $|H_n(x)| \le 2^{\frac{n}{2} - \lfloor \frac{n}{2} \rfloor} \dfrac{n!}{\lfloor n/2 \rfloor!} e^{2x\sqrt{\lfloor n/2 \rfloor}}$     MO 106a

$2.^{10}$     $|H_n(x)| < k\sqrt{n!}\, 2^{n/2} e^{x^2/2}, \quad k \approx 1.086435$     SA 324

**8.955**   Asymptotic representation:

1.     $H_{2n}(x) = (-1)^n 2^n (2n-1)!!\, e^{x^2/2} \left[\cos\left(\sqrt{4n+1}x\right) + O\left(\dfrac{1}{\sqrt[4]{n}}\right)\right]$     SM 579

2.     $H_{2n+1}(x) = (-1)^n 2^{n+\frac{1}{2}} (2n-1)!!\, \sqrt{2n+1}\, e^{x^2/2} \left[\sin\left(\sqrt{4n+3}x\right) + O\left(\dfrac{1}{\sqrt[4]{n}}\right)\right]$     SM 579

**8.956**   Special cases and particular values:

1.     $H_0(x) = 1$

2.     $H_1(x) = 2x$

3.     $H_2(x) = 4x^2 - 2$

4.     $H_3(x) = 8x^3 - 12x$

5.     $H_4(x) = 16x^4 - 48x^2 + 12$

6.     $H_{2n}(0) = (-1)^n 2^n (2n-1)!!$     SM 570(24)

7.     $H_{2n+1}(0) = 0$

## Series of Hermite polynomials

**8.957**   The generating function:

1.     $\exp\left(-t^2 + 2tx\right) = \displaystyle\sum_{k=0}^{\infty} \dfrac{t^k}{k!}\, H_k(x)$     SM 569(21)

2.     $\dfrac{1}{e}\sinh 2x = \displaystyle\sum_{k=0}^{\infty} \dfrac{1}{(2k+1)!}\, H_{2k+1}(x)$     MO 106a

3.   $\dfrac{1}{e}\cosh 2x = \displaystyle\sum_{k=0}^{\infty} \dfrac{1}{(2k)!}\, H_{2k}(x)$                                        MO 106a

4.   $e\sin 2x = \displaystyle\sum_{k=0}^{\infty} (-1)^k \dfrac{1}{(2k+1)!}\, H_{2k+1}(x)$                            MO 106a

5.   $e\cos 2x = \displaystyle\sum_{k=0}^{\infty} (-1)^k \dfrac{1}{(2k)!}\, H_{2k}(x)$                                   MO 106a

**8.958**   "The summation theorem":

1.[12]   $\dfrac{\left(\displaystyle\sum_{k=1}^{r} a_k^2\right)^{\frac{n}{2}}}{n!}\, H_n\left(\dfrac{\displaystyle\sum_{k=1}^{r} a_k x_k}{\sqrt{\displaystyle\sum_{k=1}^{r} a_k^2}}\right) = \displaystyle\sum_{\substack{m_1+m_2+\cdots+m_r=n \\ m_i \geq 0}} \prod_{k=1}^{r}\left\{\dfrac{a_k^{m_k}}{m_k!}\, H_{m_k}(x_k)\right\}$                MO 106a

2.   A special case:

$$2^{\frac{n}{2}}\, H_n(x+y) = \sum_{k=0}^{n} \binom{n}{k} H_{n-k}\left(x\sqrt{2}\right) H_k\left(y\sqrt{2}\right)$$                MO 107a

**8.959**   Hermite polynomials satisfy the differential equation

1.   $\dfrac{d^2 u_n}{dx^2} - 2x\dfrac{du_n}{dx} + 2n u_n = 0;$                                        SM 566(9)

   A second solution of this differential equation is provided by the functions ($A$ and $B$ are arbitrary constants):

2.   $u_{2n} = Ax\,\Phi\left(\tfrac{1}{2}-n; \tfrac{3}{2}; x^2\right),$

3.   $u_{2n+1} = B\,\Phi\left(-\tfrac{1}{2}-n; \tfrac{1}{2}; x^2\right)$                                        MO 107

**8.959(1)**[10]   Rodrigues' formula and orthogonality relation

1.   $H_n(x) = (-1)^n e^{x^2} \dfrac{d^n}{dx^n}\left[e^{-x^2}\right]$                [Rodrigues' formula]           MS 5.6.2

2.   $\displaystyle\int_{-\infty}^{\infty} e^{-x^2} H_m(x) H_n(x)\,dx = \begin{cases} 0 & \text{for } m \neq n \\ \pi^{1/2} 2^n n! & \text{for } m = n \end{cases}$                MS 5.6.2

## 8.96  Jacobi's polynomials

**8.960**   Definition

1.   $P_n^{(\alpha,\beta)}(x) = \dfrac{(-1)^n}{2^n n!}(1-x)^{-\alpha}(1+x)^{-\beta}\dfrac{d^n}{dx^n}\left[(1-x)^{\alpha+n}(1+x)^{\beta+n}\right]$                EH II 169(10), CO

$\qquad\quad = \dfrac{1}{2^n}\displaystyle\sum_{m=0}^{n}\binom{n+\alpha}{m}\binom{n+\beta}{n-m}(x-1)^{n-m}(x+1)^m$                EH II 169(2)

**8.961**    Functional relations:

1.[11]    $P_n^{(\alpha,\alpha)}(-x) = (-1)^n P_n^{(\alpha,\alpha)}(x)$        EH II 169(13)

2.    $2(n+1)(n+\alpha+\beta+1)(2n+\alpha+\beta) P_{n+1}^{(\alpha,\beta)}(x)$

$$= (2n+\alpha+\beta+1) \left[ (2n+\alpha+\beta)(2n+\alpha+\beta+2)x + \alpha^2 - \beta^2 \right] P_n^{(\alpha,\beta)}(x)$$

$$-2(n+\alpha)(n+\beta)(2n+\alpha+\beta+2) P_{n-1}^{(\alpha,\beta)}(x)$$

EH II 169(11)

3.    $(2n+\alpha+\beta)\left(1-x^2\right)\dfrac{d}{dx} P_n^{(\alpha,\beta)}(x) = n[(\alpha-\beta) - (2n+\alpha+\beta)x] P_n^{(\alpha,\beta)}(x)$

$$+2(n+\alpha)(n+\beta) P_{n-1}^{(\alpha,\beta)}(x)$$

EH II 170(15)

4.[11]    $\dfrac{d^m}{dx^m}\left[ P_n^{(\alpha,\beta)}(x) \right] = \dfrac{1}{2^m} \dfrac{\Gamma(n+m+\alpha+\beta+1)}{\Gamma(n+\alpha+\beta+1)} P_{n-m}^{(\alpha+m,\beta+m)}(x)$

$$[m = 1, 2, \ldots, n]$$      EH II 170(17)

5.    $\left(n+\frac{1}{2}\alpha+\frac{1}{2}\beta+1\right)(1-x) P_n^{(\alpha+1,\beta)}(x) = (n+\alpha+1) P_n^{(\alpha,\beta)}(x) - (n+1) P_{n+1}^{(\alpha,\beta)}(x)$     EH II 173(32)

6.    $\left(n+\frac{1}{2}\alpha+\frac{1}{2}\beta+1\right)(1+x) P_n^{(\alpha,\beta+1)}(x) = (n+\beta+1) P_n^{(\alpha,\beta)}(x) + (n+1) P_{n+1}^{(\alpha,\beta)}(x)$     EH II 173(33)

7.    $(1-x) P_n^{(\alpha+1,\beta)}(x) + (1+x) P_n^{(\alpha,\beta+1)}(x) = 2 P_n^{(\alpha,\beta)}(x)$     EH II 173(34)

8.    $(2n+\alpha+\beta) P_n^{(\alpha-1,\beta)}(x) = (n+\alpha+\beta) P_n^{(\alpha,\beta)}(x) - (n+\beta) P_{n-1}^{(\alpha,\beta)}(x)$     EH II 173(35)

9.    $(2n+\alpha+\beta) P_n^{(\alpha,\beta-1)}(x) = (n+\alpha+\beta) P_n^{(\alpha,\beta)}(x) + (n+\alpha) P_{n-1}^{(\alpha,\beta)}(x)$     EH II 173(36)

10.    $P_n^{(\alpha,\beta-1)}(x) - P_n^{(\alpha-1,\beta)}(x) = P_{n-1}^{(\alpha,\beta)}(x)$     EH II 173(37)

**8.962**    Connections with other functions:

1.    $P_n^{(\alpha,\beta)}(x) = \dfrac{(-1)^n \Gamma(n+1+\beta)}{n!\,\Gamma(1+\beta)} F\left(n+\alpha+\beta+1, -n; 1+\beta; \dfrac{1+x}{2}\right)$     CO, EH II 170(16)

$$= \dfrac{\Gamma(n+1+\alpha)}{n!\,\Gamma(1+\alpha)} F\left(n+\alpha+\beta+1, -n; 1+\alpha; \dfrac{1-x}{2}\right)$$     EH II 170(16)

$$= \dfrac{\Gamma(n+1+\alpha)}{n!\,\Gamma(1+\alpha)} \left(\dfrac{1+x}{2}\right)^n F\left(-n, -n-\beta; \alpha+1; \dfrac{x-1}{x+1}\right)$$     EH II 170(16)

$$= \dfrac{\Gamma(n+1+\beta)}{n!\,\Gamma(1+\beta)} \left(\dfrac{x-1}{2}\right)^n F\left(-n, -n-\alpha; \beta+1; \dfrac{x+1}{x-1}\right)$$     EH II 170(16)

2.    $P_n(x) = P_n^{(0,0)}(x)$     CO, EH II 179(3)

3.    $T_n(x) = \dfrac{2^{2n}(n!)^2}{(2n)!} P_n^{\left(-\frac{1}{2},-\frac{1}{2}\right)}(x)$     CO, EH II 184(5)a

4.    $C_n^\nu(x) = \dfrac{\Gamma(n+2\nu)\,\Gamma\left(\nu+\frac{1}{2}\right)}{\Gamma(2\nu)\,\Gamma\left(n+\nu+\frac{1}{2}\right)} P_n^{(\nu-1/2,\nu-1/2)}(x)$     MO 108a, EH II 174(4)

5.*    $U_n(x) = \dfrac{2^n(n+1)!}{(2n+1)!!} P_n^{\left(\frac{1}{2},\frac{1}{2}\right)}(x)$

**8.963** The generating function:

$$\sum_{n=0}^{\infty} P_n^{(\alpha,\beta)}(x)z^n = 2^{\alpha+\beta} R^{-1}(1 - z + R)^{-\alpha}(1 + z + R)^{-\beta},$$

$$R = \sqrt{1 - 2xz + z^2} \qquad [|z| < 1]$$

EH II 172(29)

**8.964** The Jacobi polynomials constitute the *unique* rational solution of the differential (hypergeometric) equation

$$\left(1 - x^2\right)y'' + [\beta - \alpha - (\alpha + \beta + 2)x]y' + n(n + \alpha + \beta + 1)y = 0 \qquad \text{EH II 169(14)}$$

**8.965** Asymptotic representation

$$P_n(\alpha, \ \beta)(\cos \ \theta) =$$

$$\frac{\cos\left\{\left[n + \frac{1}{2}(\alpha + \beta + 1)\right]\theta - \left(\frac{1}{2}\alpha + \frac{1}{4}\right)\pi\right\}}{\sqrt{\pi n}\left(\sin\frac{1}{2}\theta\right)^{\alpha+\frac{1}{2}}\left(\cos\frac{1}{2}\theta\right)^{\beta+\frac{1}{2}}} + O\left(n^{-3/2}\right) \qquad [\text{Im}\,\alpha = \text{Im}\,\beta = 0, \quad 0 < \theta < \pi]$$

EH II 198(10)

**8.966** A limit relationship:

$$\lim_{n\to\infty}\left[n^{-\alpha}P_n^{(\alpha,\beta)}\left(\cos\frac{z}{n}\right)\right] = \left(\frac{z}{2}\right)^{-\alpha}J_\alpha(z) \qquad \text{EH II 173(41)}$$

**8.967** If $\alpha > -1$ and $\beta > -1$, all the zeros of the polynomial $P_n^{(\alpha,\beta)}(x)$ are simple and they lie in the interval $(-1, 1)$.

# 8.97 The Laguerre polynomials

**8.970** Definition.

1. $$L_n^\alpha(x) = \frac{1}{n!}e^x x^{-\alpha}\frac{d^n}{dx^n}\left(e^{-x}x^{n+\alpha}\right) \qquad \text{[Rodrigues' formula]} \qquad \text{EH II 188(5), MO 108}$$

$$= \sum_{m=0}^{n}(-1)^m\binom{n+\alpha}{n-m}\frac{x^m}{m!} \qquad \text{MO 109, EH II 188(7)}$$

2. $L_n^0(x) = L_n(x)$ \hfill ET I 369

3.[10] $L_0^\alpha(x) = 1$

4.[10] $L_1^\alpha(x) = -x + \alpha + 1$

5.[10] $L_2^\alpha(x) = \frac{1}{2}\left[x^2 - 2(\alpha + 2)x + (\alpha + 1)(\alpha + 2)\right]$

6.[10] $L_3^\alpha(x) = -\frac{1}{6}\left[x^3 - 3(\alpha + 3)x^2 + 3(\alpha + 2)(\alpha + 3)x - (\alpha + 1)(\alpha + 2)(\alpha + 3)\right]$

7.[10] $L_4^\alpha(x) = \frac{1}{24}\Big[x^4 - 4(\alpha + 4)x^3 + 6(\alpha + 3)(\alpha + 4)x^2 - 4(\alpha + 2)(\alpha + 3)(\alpha + 4)x$

$\qquad\qquad + (\alpha + 1)(\alpha + 2)(\alpha + 3)(\alpha + 4)\Big]$

8.[10] $L_5^\alpha(x) = -\frac{1}{120}\Big[x^5 - 5(\alpha + 5)x^4 + 10(\alpha + 4)(\alpha + 5)x^3 - 10(\alpha + 3)(\alpha + 4)(\alpha + 5)x^2$

$\qquad\qquad + 5(\alpha + 2)(\alpha + 3)(\alpha + 4)(\alpha + 5)x - (\alpha + 1)(\alpha + 2)(\alpha + 3)(\alpha + 4)(\alpha + 5)\Big]$

**8.971** Functional relations:

1. $\dfrac{d}{dx}\left[L_n^\alpha(x) - L_{n+1}^\alpha(x)\right] = L_n^\alpha(x)$ EH II 189(16)

2.[11] $\dfrac{d}{dx}\,L_n^\alpha(x) = -L_{n-1}^{\alpha+1}(x) = \dfrac{n\,L_n^\alpha(x) - (n+\alpha)\,L_{n-1}^\alpha(x)}{x}$ EH II 189(15), SM 575(42)a

3. $x\dfrac{d}{dx}\,L_n^\alpha(x) = n\,L_n^\alpha(x) - (n+\alpha)\,L_{n-1}^\alpha(x)$

$\qquad\qquad = (n+1)\,L_{n+1}^\alpha(x) - (n+\alpha+1-x)\,L_n^\alpha(x)$

EH II 189(12), MO 109

4. $x\,L_n^{\alpha+1}(x) = (n+\alpha+1)\ L_n^\alpha(x) - (n+1)\,L_{n+1}^\alpha(x)$

$\qquad\qquad = (n+\alpha)\,L_{n-1}^\alpha(x) - (n-x)\,L_n^\alpha(x)$

SM 575(43)a, EH II 190(23)

5. $L_n^{\alpha-1}(x) = L_n^\alpha(x) - L_{n-1}^\alpha(x)$ SM 575(44)a, EH II 190(24)

6. $(n+1)\,L_{n+1}^\alpha(x) - (2n+\alpha+1-x)\,L_n^\alpha(x) + (n+\alpha)\,L_{n-1}^\alpha(x) = 0$

$\qquad\qquad\qquad\qquad\qquad [n = 1, 2, \ldots]$ MO 109, EH II 190(25, 24)

7.[10] $(n+\alpha)\,L_n^{\alpha-1}(x) = (n+1)\,L_{n+1}^\alpha(x) - (n+1-x)\,L_n^\alpha(x)$ MS 5.5.2

8.[10] $n\,L_n^\alpha(x) = (2n+\alpha-1-x)\,L_{n-1}^\alpha(x) - (n+\alpha-1)\,L_{n-2}^\alpha(x)$

$\qquad\qquad\qquad\qquad\qquad [n = 2, 3, \ldots]$ MS 5.5.2

**8.972** Connections with other functions:

1. $L_n^\alpha(x) = \dbinom{n+\alpha}{n}\Phi(-n, \alpha+1; x)$ MO 109, FI II 189(14)

2. $H_{2n}(x) = (-1)^n 2^{2n} n!\,L_n^{-1/2}\left(x^2\right)$ EH II 193(2), SM 576(47)

3. $H_{2n+1}(x) = (-1)^n 2^{2n+1} n!\,x\,L_n^{1/2}\left(x^2\right)$ EH II 193(3), SM 577(48)

**8.973** Special cases:

1. $L_0^\alpha(x) = 1$ EH II 188(6)

2. $L_1^\alpha(x) = \alpha + 1 - x$ EH II 188(6)

3. $L_n^\alpha(0) = \dbinom{n+\alpha}{n}$ EH II 189(13)

4. $L_n^{-n}(x) = (-1)^n \dfrac{x^n}{n!}$ MO 109

5. $L_1(x) = 1 - x$

6. $L_2(x) = 1 - 2x + \dfrac{x^2}{2}$ MO 109

**8.974**   Finite sums:

1.   $\displaystyle\sum_{m=0}^{n} \frac{m!}{\Gamma(m+\alpha+1)} L_m^\alpha(x)\, L_m^\alpha(y) = \frac{(n+1)!}{\Gamma(n+\alpha+1)(x-y)} \left[ L_n^\alpha(x)\, L_{n+1}^\alpha(y) - L_{n+1}^\alpha(x)\, L_n^\alpha(y) \right]$

                                                                           EH II 188(9)

2.[12]   $\displaystyle\sum_{m=0}^{n} \frac{\Gamma(\alpha-\beta+m)}{\Gamma(\alpha-\beta)m!} L_{n-m}^\beta(x) = L_n^\alpha(x)$                       MO 110, EH II 192(39)

3.   $\displaystyle\sum_{m=0}^{n} L_m^\alpha(x) = L_n^{\alpha+1}(x)$                                       EH II 192(38)

4.[11]   $\displaystyle\sum_{m=0}^{n} L_m^\alpha(x)\, L_{n-m}^\beta(y) = L_n^{\alpha+\beta+1}(x+y)$                       EH II 192(41)

**8.975**   Arbitrary functions:

1.   $(1-z)^{-\alpha-1} \exp\dfrac{xz}{z-1} = \displaystyle\sum_{n=0}^{\infty} L_n^\alpha(x) z^n$        $[|z|<1]$         EH II 189(17), MO 109

2.   $e^{-xz}(1+z)^\alpha = \displaystyle\sum_{n=0}^{\infty} L_n^{\alpha-n}(x) z^n$              $[|z|<1]$         MO 110, EH II 189(19)

3.   $J_\alpha\left(2\sqrt{xz}\right) e^z (xz)^{-\frac{1}{2}\alpha} = \displaystyle\sum_{n=0}^{\infty} \frac{z^n}{\Gamma(n+\alpha+1)} L_n^\alpha(x)$     $[\alpha>-1]$         EH II 189(18), MO 109

**8.976**   Other series of Laguerre polynomials:

1.   $\displaystyle\sum_{n=0}^{\infty} n! \frac{L_n^\alpha(x)\, L_n^\alpha(y) z^n}{\Gamma(n+\alpha+1)} = \frac{(xyz)^{-\frac{1}{2}\alpha}}{1-z} \exp\left(-z\frac{x+y}{1-z}\right) I_\alpha\left(2\frac{\sqrt{xyz}}{1-z}\right)$

                                                                $[|z|<1]$         EH II 189(20)

2.   $\displaystyle\sum_{n=0}^{\infty} \frac{L_n^\alpha(x)}{n+1} = e^x x^{-\alpha} \Gamma(\alpha, x)$          $[\alpha>-1, \quad x>0]$         EH II 215(19)

3.[6]   $L_n^\alpha(x)^2 = \dfrac{\Gamma(n+\alpha+1)}{2^{2n}n!} \displaystyle\sum_{k=0}^{n} \binom{2n-2k}{n-k} \frac{(2k)!}{k!} \frac{1}{\Gamma(\alpha+k+1)} L_{2k}^{2\alpha}(2x)$         MO 110

4.[6]   $L_n^\alpha(x)\, L_n^\alpha(y) = \dfrac{\Gamma(1+\alpha+n)}{n!} \displaystyle\sum_{k=0}^{n} \frac{L_{n-k}^{\alpha+2k}(x+y)}{\Gamma(1+\alpha+k)} \frac{(xy)^k}{k!}$         MO 110, EH II 192(42)

**8.977**   Summation theorems:

1.[12]   $L_n^{\alpha_1+\alpha_2+\cdots+\alpha_k+k-1}(x_1+x_2+\cdots+x_k) = \displaystyle\sum_{i_1+i_2+\cdots+i_k=n} L_{i_1}^{\alpha_1}(x_1)\, L_{i_2}^{\alpha_2}(x_2) \cdots L_{i_k}^{\alpha_k}(x_k)$         MO 110

2.   $L_n^\alpha(x+y) = e^y \displaystyle\sum_{k=0}^{\infty} \frac{(-1)^k}{k!} y^k\, L_n^{\alpha+k}(x)$                             MO 110

**8.978**  Limit relations and asymptotic behavior:

1.    $$L_n^\alpha(x) = \lim_{\beta \to \infty} P_n^{(\alpha,\beta)}\left(1 - \frac{2x}{\beta}\right)$$                                    EH II 191(35)

2.    $$\lim_{n \to \infty}\left[n^{-\alpha} L_n^\alpha\left(\frac{x}{n}\right)\right] = x^{-\frac{1}{2}\alpha} J_\alpha\left(2\sqrt{x}\right)$$                                    EH II 191(36)

3.    $$L_n^\alpha(x) = \frac{1}{\sqrt{\pi}} e^{\frac{1}{2}x} x^{-\frac{1}{2}\alpha - \frac{1}{4}} n^{\frac{1}{2}\alpha - \frac{1}{4}} \cos\left[2\sqrt{nx} - \frac{\alpha\pi}{2} - \frac{\pi}{4}\right] + O\left(n^{\frac{1}{2}\alpha - \frac{3}{4}}\right)$$

$$[\operatorname{Im}\alpha = 0, \quad x > 0]$$                                    EH II 199(1)

**8.979**  Laguerre polynomials satisfy the following differential equation:

$$x\frac{d^2u}{dx^2} + (\alpha - x + 1)\frac{du}{dx} + nu = 0$$                                    EH II 188(10), SM 574(34)

**8.980**[11]  Orthogonality relation

$$\int_0^\infty e^{-x} x^\alpha\, L_n^\alpha(x)\, L_m^\alpha(x)\, dx = \begin{cases} 0, & m \neq n \\ \Gamma(1+\alpha)\binom{n+\alpha}{n}, & m = n \end{cases}$$                                    MS 5.5.2

**8.981**[10]  Behavior of relative maxima of $|L_n^\alpha(x)|$

1.    Let $\alpha$ be arbitrary and real. The sequence formed by the relative maxima of $|L_n^\alpha(x)|$ and by the value of this function at $x = 0$, is decreasing for $x < \alpha + \frac{1}{2}$, and increasing for $x > \alpha + \frac{1}{2}$. The successive relative maxima of $|L_n^\alpha(x)|$ form a decreasing sequence for $x \leq 0$, and an increasing sequence for $x \geq 0$.                                    SZ 174(7.6.1)

2.[12]    Let $\alpha$ be an arbitrary real number. The successive relative maxima of

$$e^{-x/2} x^{(\alpha+1)/2} |L_n^\alpha(x)| \text{ and } e^{-x/2} x^{\alpha/2 + \frac{1}{4}} |L_n^\alpha(x)|$$

form an increasing sequence provided $x > x_0$. In the first case

$$x_0 = \begin{cases} 0 & \text{if } \alpha^2 \leq 1, \\ \dfrac{\alpha^2 - 1}{2n + \alpha + 1} & \text{if } \alpha^2 > 1 \end{cases}$$

In the second case

$$x_0 = \begin{cases} 0 & \text{if } \alpha^2 \leq \frac{1}{4}, \\ \left(\alpha^2 - \frac{1}{4}\right)^{\frac{1}{2}} & \text{if } \alpha^2 > \frac{1}{4} \end{cases}$$                                    SZ 174(7.6.2)

In the first case we take $n$ so large that $2n + \alpha + 1 > 0$.

**8.982**[10]  Asymptotic and limiting behavior of $L_n^\alpha(x)$

1.    Let $\alpha$ be arbitrary and real, $c$ and $w$ fixed positive constants, and let $n \to \infty$. Then

$$L_n^\alpha(x) = \begin{cases} x^{-\alpha/2 - \frac{1}{4}} O\left(n^{\alpha/2 - \frac{1}{4}}\right) & \text{if } cn^{-1} \leq qx \leq qw \\ O(n^\alpha) & \text{if } 0 \leq qx \leq qcn^{-1} \end{cases}$$

These bounds are precise as regards their orders in $n$. For $\alpha \geq q - \frac{1}{2}$, both bounds hold in both intervals, that is,

$$L_n^\alpha(x) = \begin{cases} x^{-\alpha/2-\frac{1}{4}} O\left(n^{\alpha/2-\frac{1}{4}}\right), \\ O\left(n^\alpha\right), \end{cases} \quad 0 < x \leq q\omega, \quad \alpha \geq q - \frac{1}{2} \qquad \text{SZ 175(7.6.4)}$$

$2.^{12}$    Let $\alpha$ be arbitrary and real. Then for an arbitrary complex $z$

$$\lim_{n\to\infty} n^{-\alpha} L_n^\alpha(z) = z^{-\alpha/2} J_\alpha\left(2z^{1/2}\right), \qquad \text{SZ 191(8.1.3)}$$

uniformly if $z$ is bounded.

# 9.1 Hypergeometric Functions

## 9.10 Definition

**9.100**   A *hypergeometric series* is a series of the form

$$F(\alpha, \beta; \gamma; z) = 1 + \frac{\alpha \cdot \beta}{\gamma \cdot 1} z + \frac{\alpha(\alpha+1)\beta(\beta+1)}{\gamma(\gamma+1) \cdot 1 \cdot 2} z^2 + \frac{\alpha(\alpha+1)(\alpha+2)\beta(\beta+1)(\beta+2)}{\gamma(\gamma+1)(\gamma+2) \cdot 1 \cdot 2 \cdot 3} z^3 + \cdots$$

**9.101**   A hypergeometric series terminates if $\alpha$ or $\beta$ is equal to a negative integer or to zero. For $\gamma = -n \, (n = 0, 1, 2, \ldots)$, the hypergeometric series is indeterminate if neither $\alpha$ nor $\beta$ is equal to $-m$ (where $m < n$ and $m$ is a natural number). However,

1.   $$\lim_{\gamma \to -n} \frac{F(\alpha, \beta; \gamma; z)}{\Gamma(\gamma)} = \frac{\alpha(\alpha+1)\ldots(\alpha+n)\beta(\beta+1)\ldots(\beta+n)}{(n+1)!}$$

$$\times z^{n+1} F(\alpha+n+1, \beta+n+1; n+2; z)$$

EH I 62(16)

**9.102**   If we exclude these values of the parameters $\alpha, \beta, \gamma$, a hypergeometric series converges in the unit circle $|z| < 1$. $F$ then has a branch point at $z = 1$. Then we have the following conditions for convergence on the unit circle:

1.   $1 > \operatorname{Re}(\alpha + \beta - \gamma) \geq 0$.   The series converges throughout the entire unit circle except at the point $z = 1$.

2.   $\operatorname{Re}(\alpha + \beta - \gamma) < 0$.   The series converges (absolutely) throughout the entire unit circle.

3.   $\operatorname{Re}(\alpha + \beta - \gamma) \geq 1$.   The series diverges on the entire unit circle.     FI II 410, WH

**9.103***   Derivatives of hypergeometric functions

1.   $$\frac{\mathrm{d}}{\mathrm{d}z} F(\alpha, \beta; \gamma; z) = \frac{\alpha\beta}{\gamma} F(\alpha+1, \beta+1; \gamma+1; z)$$

2.*   $$\frac{\mathrm{d}^k}{\mathrm{d}z^k} F(\alpha, \beta; \gamma; z) = \frac{\Gamma(\alpha+k)\Gamma(\beta+k)\Gamma(\gamma)}{\Gamma(\alpha)\Gamma(\beta)\Gamma(\gamma+k)} F(\alpha+k, \beta+k; \gamma+k; z)$$

3.*   $$\frac{\mathrm{d}}{\mathrm{d}z} (z^{\gamma-1} F(\alpha, \beta; \gamma; z)) = (\gamma-1) z^{\gamma-2} F(\alpha, \beta; \gamma-1; z)$$

4.*   $$\frac{\mathrm{d}^k}{\mathrm{d}z^k} (z^{\gamma-1} F(\alpha, \beta; \gamma; z)) = \frac{\Gamma(\gamma)}{\Gamma(\gamma-k)} z^{\gamma-1-k} F(\alpha, \beta; \gamma-k; z)$$

## 9.11 Integral representations

**9.111**   $$F(\alpha, \beta; \gamma; z) = \frac{1}{\mathrm{B}(\beta, \gamma-\beta)} \int_0^1 t^{\beta-1}(1-t)^{\gamma-\beta-1}(1-tz)^{-\alpha} \, \mathrm{d}t \qquad [\operatorname{Re} \gamma > \operatorname{Re} \beta > 0] \qquad \text{WH}$$

**9.112**[8]   $$F(p, n+p; n+1; z^2) = \frac{z^{-n}}{2\pi} \frac{\Gamma(p)n!}{\Gamma(p+n)} \int_0^{2\pi} \frac{\cos nt \, \mathrm{d}t}{(1 - 2z\cos t + z^2)^p}$$

$$[n = 0, 1, 2, \ldots; \quad p \neq 0, -1, -2, \ldots; \quad |z| < 1] \qquad \text{WH, MO 16}$$

**9.113**   $F(\alpha, \beta; \gamma; z) = \dfrac{\Gamma(\gamma)}{\Gamma(\alpha)\,\Gamma(\beta)}\dfrac{1}{2\pi i}\displaystyle\int_{-\infty i}^{\infty i}\dfrac{\Gamma(\alpha+t)\,\Gamma(\beta+t)\,\Gamma(-t)}{\Gamma(\gamma+t)}(-z)^t\,dt$

Here, $|\arg(-z)| < \pi$ and the path of integration is chosen in such a way that the poles of the functions $\Gamma(\alpha+t)$ and $\Gamma(\beta+t)$ lie to the left of the path of integration and the poles of the function $\Gamma(-t)$ lie to the right of it.

**9.114**     $F\left(-m, -\dfrac{p+m}{2}; 1-\dfrac{p+m}{2}; -1\right) = \dfrac{(-2)^m(p+m)}{\sin p\pi}\displaystyle\int_0^{\pi}\cos^m\varphi\cos p\varphi\,d\varphi$

$[m+1$ is a natural number;    $p \neq 0, \quad \pm 1, \dots ]$    EH I 80(8), MO 16

See also **3.194** 1, 2, 5, **3.196** 1, **3.197** 6, 9, **3.259** 3, **3.312** 3, **3.518** 4–6, **3.665** 2, **3.671** 1, 2, **3.681** 1, **3.984** 7.

## 9.12 Representation of elementary functions in terms of a hypergeometric functions

### 9.121

1.[8]     $F(-n, \beta; \beta; -z) = (1+z)^n$                        EH I 101(4), GA 127 Ia

2.     $F\left(-\dfrac{n}{2}, -\dfrac{n-1}{2}; \dfrac{1}{2}; \dfrac{z^2}{t^2}\right) = \dfrac{(t+z)^n + (t-z)^n}{2t^n}$        GA 127 II

3.     $\displaystyle\lim_{\omega\to\infty} F\left(-n, \omega; 2\omega; -\dfrac{z}{t}\right) = \left(1+\dfrac{z}{2t}\right)^n$        GA 127 IIIa

4.     $F\left(-\dfrac{n-1}{2}, -\dfrac{n-2}{2}; \dfrac{3}{2}; \dfrac{z^2}{t^2}\right) = \dfrac{(t+z)^n - (t-z)^n}{2nzt^{n-1}}$        GA 127 IV

5.     $F\left(1-n, 1; 2; -\dfrac{z}{t}\right) = \dfrac{(t+z)^n - t^n}{nzt^{n-1}}$        GA 127 V

6.     $F(1, 1; 2; -z) = \dfrac{\ln(1+z)}{z}$        GA 127 VI

7.     $F\left(\dfrac{1}{2}, 1; \dfrac{3}{2}; z^2\right) = \dfrac{\ln\frac{1+z}{1-z}}{2z}$        GA 127 VII

8.     $\displaystyle\lim_{k\to\infty} F\left(1, k; 1; \dfrac{z}{k}\right) = 1 + z\lim_{k\to\infty} F\left(1, k; 2; \dfrac{z}{k}\right)$

                     $= 1 + z + \dfrac{z^2}{2}\displaystyle\lim_{k\to\infty} F\left(1, k; 3; \dfrac{z}{k}\right) = \cdots = e^z$

                                                 GA 127 VIII

9.     $\displaystyle\lim_{\substack{k\to\infty \\ k'\to\infty}} F\left(k, k'; \dfrac{1}{2}; \dfrac{z^2}{4kk'}\right) = \dfrac{e^z + e^{-z}}{2} = \cosh z$        GA 127 IX

10.     $\displaystyle\lim_{\substack{k\to\infty \\ k'\to\infty}} F\left(k, k'; \dfrac{3}{2}; \dfrac{z^2}{4kk'}\right) = \dfrac{e^z - e^{-z}}{2z} = \dfrac{\sinh z}{z}$        GA 127 X

11.     $\displaystyle\lim_{\substack{k\to\infty \\ k'\to\infty}} F\left(k, k'; \dfrac{3}{2}; -\dfrac{z^2}{4kk'}\right) = \dfrac{\sin z}{z}$        GA 127 XI

12.     $\displaystyle\lim_{\substack{k\to\infty \\ k'\to\infty}} F\left(k, k'; \dfrac{1}{2}; -\dfrac{z^2}{4kk'}\right) = \cos z$        GA 127 XII

13. $\quad F\left(\dfrac{1}{2}, \dfrac{1}{2}; \dfrac{3}{2}; \sin^2 z\right) = \dfrac{z}{\sin z}$ $\hfill$ GA 127 XIII

14. $\quad F\left(1, 1; \dfrac{3}{2}; \sin^2 z\right) = \dfrac{z}{\sin z \cos z}$ $\hfill$ GA 127 XIV

15. $\quad F\left(\dfrac{1}{2}, 1; \dfrac{3}{2}; -\tan^2 z\right) = \dfrac{z}{\tan z}$ $\hfill$ GA 127 XV

16. $\quad F\left(\dfrac{n+1}{2}, -\dfrac{n-1}{2}; \dfrac{3}{2}; \sin^2 z\right) = \dfrac{\sin nz}{n \sin z}$ $\hfill$ GA 127 XVI

17. $\quad F\left(\dfrac{n+2}{2}, -\dfrac{n-2}{2}; \dfrac{3}{2}; \sin^2 z\right) = \dfrac{\sin nz}{n \sin z \cos z}$ $\hfill$ GA 127 XVII

18. $\quad F\left(-\dfrac{n-2}{2}, -\dfrac{n-1}{2}; \dfrac{3}{2}; -\tan^2 z\right) = \dfrac{\sin nz}{n \sin z \cos^{n-1} z}$ $\hfill$ GA 127 XVIII

19. $\quad F\left(\dfrac{n+2}{2}, \dfrac{n+1}{2}; \dfrac{3}{2}; -\tan^2 z\right) = \dfrac{\sin nz \cos^{n+1} z}{n \sin z}$ $\hfill$ GA 127 XIX

20. $\quad F\left(\dfrac{n}{2}, -\dfrac{n}{2}; \dfrac{1}{2}; \sin^2 z\right) = \cos nz$ $\hfill$ EH I 101(11), GA 127 XX

21. $\quad F\left(\dfrac{n+1}{2}, -\dfrac{n-1}{2}; \dfrac{1}{2}; \sin^2 z\right) = \dfrac{\cos nz}{\cos z}$ $\hfill$ EH I 101(11), GA 127 XXI

22. $\quad F\left(-\dfrac{n}{2}, -\dfrac{n-1}{2}; \dfrac{1}{2}; -\tan^2 z\right) = \dfrac{\cos nz}{\cos^n z}$ $\hfill$ EH I 101(11), GA 127 XXII

23. $\quad F\left(\dfrac{n+1}{2}, \dfrac{n}{2}; \dfrac{1}{2}; -\tan^2 z\right) = \cos nz \cos^n z$ $\hfill$ GA 127 XXIII

24. $\quad F\left(\dfrac{1}{2}, 1; 2; 4z(1-z)\right) = \dfrac{1}{1-z}$ $\hfill$ $\left[|z| \le \tfrac{1}{2}; \quad |z(1-z)| \le \tfrac{1}{4}\right]$

25. $\quad F\left(\dfrac{1}{2}, 1; 1; \sin^2 z\right) = \sec z$

26. $\quad F\left(\dfrac{1}{2}, \dfrac{1}{2}; \dfrac{3}{2}; z^2\right) = \dfrac{\arcsin z}{z}$ $\hfill$ (cf. **9.121** 13)

27. $\quad F\left(\dfrac{1}{2}, 1; \dfrac{3}{2}; -z^2\right) = \dfrac{\arctan z}{z}$ $\hfill$ (cf. **9.121** 15)

28. $\quad F\left(\dfrac{1}{2}, \dfrac{1}{2}; \dfrac{3}{2}; -z^2\right) = \dfrac{\operatorname{arcsinh} z}{z}$ $\hfill$ (cf. **9.121** 26)

29. $\quad F\left(\dfrac{1+n}{2}, \dfrac{1-n}{2}; \dfrac{3}{2}; z^2\right) = \dfrac{\sin (n \arcsin z)}{nz}$ $\hfill$ (cf. **9.121** 16)

30. $\quad F\left(1+\dfrac{n}{2}, 1-\dfrac{n}{2}; \dfrac{3}{2}; z^2\right) = \dfrac{\sin (n \arcsin z)}{nz\sqrt{1-z^2}}$ $\hfill$ (cf. **9.121** 17)

31. $\quad F\left(\dfrac{n}{2}, -\dfrac{n}{2}; \dfrac{1}{2}; z^2\right) = \cos\left(n \arcsin z\right)$ $\qquad$ (cf. **9.121** 20)

32. $\quad F\left(\dfrac{1+n}{2}, \dfrac{1-n}{2}; \dfrac{1}{2}; z^2\right) = \dfrac{\cos\left(n \arcsin z\right)}{\sqrt{1-z^2}}$ $\qquad$ (cf. **9.121** 21)

The representation of special functions in terms of a hypergeometric function:

- for complete elliptic integrals, see **8.113** 1 and **8.114** 1;
- for integrals of Bessel functions, see **6.574** 1, 3, **6.576** 2–5, **6.621** 1–3;
- for Legendre polynomials, see **8.911** and **8.916**. (All these hypergeometric series terminate; that is, these series are finite sums);
- for Legendre functions, see **8.820** and **8.837**;
- for associated Legendre functions, see **8.702**, **8.703**, **8.751**, **8.77**, **8.852**, and **8.853**;
- for Chebyshev polynomials, see **8.942** 1;
- for Jacobi's polynomials, see **8.962**;
- for Gegenbauer polynomials, see **8.932**;
- for integrals of parabolic cylinder functions, see **7.725** 6.
- for Lerch function $\Phi(z, s, v)$; see **9.559** 1.

**9.122** Particular values:

1. $\quad F(\alpha, \beta; \gamma; 1) = \dfrac{\Gamma(\gamma)\,\Gamma(\gamma - \alpha - \beta)}{\Gamma(\gamma - \alpha)\,\Gamma(\gamma - \beta)}$ $\qquad$ $[\operatorname{Re}\gamma > \operatorname{Re}(\alpha + \beta)]$

$\qquad\qquad\qquad\qquad\qquad\qquad\qquad\qquad\qquad\qquad\qquad$ GA 147(48), FI II 793

2. $\quad F(\alpha, \beta; \gamma; 1) = F(-\alpha, -\beta; \gamma - \alpha - \beta; 1)$ $\qquad$ $[\operatorname{Re}\gamma > \operatorname{Re}(\alpha + \beta)]$ $\qquad$ GA 148(49)

$\qquad\quad = \dfrac{1}{F(-\alpha, \beta; \gamma - \alpha; 1)}$ $\qquad$ $[\operatorname{Re}\gamma > \operatorname{Re}(\alpha + \beta)]$ $\qquad$ GA 148(50)

$\qquad\quad = \dfrac{1}{F(\alpha, -\beta; \gamma - \beta; 1)}$ $\qquad$ $[\operatorname{Re}\gamma > \operatorname{Re}(\alpha + \beta)]$ $\qquad$ GA 148(51)

3. $\quad F\left(1, 1; \dfrac{3}{2}; \dfrac{1}{2}\right) = \dfrac{\pi}{2}$ $\qquad$ (cf. **9.121** 14)

## 9.13 Transformation formulas and the analytic continuation of functions defined by hypergeometric series

**9.130** The series $F(\alpha, \beta; \gamma; z)$ defines an analytic function that, speaking generally, has singularities at the points $z = 0$, 1, and $\infty$ (In the general case, there are branch points). We make a cut in the $z$-plane along the real axis from $z = 1$ to $z = \infty$; that is, we require that $|\arg(-z)| < \pi$ for $|z| \geq 1$. Then, the series $f(\alpha, \beta; \gamma; z)$ will, in the cut plane, yield a single-valued analytic continuation which we can obtain by means of the formulas below (provided $\gamma + 1$ is not a natural number and $\alpha - \beta$ and $\gamma - \alpha - \beta$ are not integers). These formulas make it possible to calculate the values of $F$ in the given region even in the case in which $|z| > 1$. There are other closely related transformation formulas that can also be used to get the analytic continuation when the corresponding relationships hold between $\alpha, \beta, \gamma$.

## Transformation formulas

**9.131**

1.[11]    $F(\alpha, \beta; \gamma; z) = (1-z)^{-\alpha} F\left(\alpha, \gamma - \beta; \gamma; \dfrac{z}{z-1}\right)$        GA 218(91)

$$= (1-z)^{-\beta} F\left(\beta, \gamma - \alpha; \gamma; \dfrac{z}{z-1}\right)$$        GA 218(92)

$$= (1-z)^{\gamma-\alpha-\beta} F(\gamma - \alpha, \gamma - \beta; \gamma; z)$$

2.    $F(\alpha, \beta; \gamma; z) = \dfrac{\Gamma(\gamma)\,\Gamma(\gamma - \alpha - \beta)}{\Gamma(\gamma - \alpha)\,\Gamma(\gamma - \beta)} F(\alpha, \beta; \alpha + \beta - \gamma + 1; 1 - z)$

$$+ (1-z)^{\gamma-\alpha-\beta} \dfrac{\Gamma(\gamma)\,\Gamma(\alpha + \beta - \gamma)}{\Gamma(\alpha)\,\Gamma(\beta)} F(\gamma - \alpha, \gamma - \beta; \gamma - \alpha - \beta + 1; 1 - z)$$

EH I 94, MO 13

**9.132**

1.    $F(\alpha, \beta; \gamma; z) = \dfrac{(1-z)^{-\alpha}\,\Gamma(\gamma)\,\Gamma(\beta - \alpha)}{\Gamma(\beta)\,\Gamma(\gamma - \alpha)} F\left(\alpha, \gamma - \beta; \alpha - \beta + 1; \dfrac{1}{1-z}\right)$

$$+ (1-z)^{-\beta} \dfrac{\Gamma(\gamma)\,\Gamma(\alpha - \beta)}{\Gamma(\alpha)\,\Gamma(\gamma - \beta)} F\left(\beta, \gamma - \alpha; \beta - \alpha + 1; \dfrac{1}{1-z}\right)$$

MO 13

2.[12]    $F(\alpha, \beta; \gamma; z) = \dfrac{\Gamma(\gamma)\,\Gamma(\beta - \alpha)}{\Gamma(\beta)\,\Gamma(\gamma - \alpha)} (-z)^{-\alpha} F\left(\alpha, \alpha + 1 - \gamma; \alpha + 1 - \beta; \dfrac{1}{z}\right)$

$$+ \dfrac{\Gamma(\gamma)\,\Gamma(\alpha - \beta)}{\Gamma(\alpha)\,\Gamma(\gamma - \beta)} (-z)^{-\beta} F\left(\beta, \beta + 1 - \gamma; \beta + 1 - \alpha; \dfrac{1}{z}\right)$$

$$[|\arg(-z)| < \pi, \quad \alpha - \beta \neq \pm m, \quad m = 0, 1, 2, \ldots]$$    GA 220(93)

**9.133**    $F\left(2\alpha, 2\beta; \alpha + \beta + \tfrac{1}{2}; z\right) = F\left(\alpha, \beta; \alpha + \beta + \tfrac{1}{2}; 4z(1-z)\right)$

$$\left[|z| \leq \tfrac{1}{2}, \quad |z(1-z)| \leq \tfrac{1}{4}\right]$$    WH

**9.134**

1.    $F(\alpha, \beta; 2\beta; z) = \left(1 - \dfrac{z}{2}\right)^{-\alpha} F\left(\dfrac{\alpha}{2}, \dfrac{\alpha + 1}{2}; \beta + \dfrac{1}{2}; \left(\dfrac{z}{2-z}\right)^2\right)$    MO 13, EH I 111(4)

2.    $F(2\alpha, 2\alpha + 1 - \gamma; \gamma; z) = (1 + z)^{-2\alpha} F\left(\alpha, \alpha + \dfrac{1}{2}; \gamma; \dfrac{4z}{(1+z)^2}\right)$    GA 225(100)

3.    $F\left(\alpha, \alpha + \dfrac{1}{2} - \beta; \beta + \dfrac{1}{2}; z^2\right) = (1 + z)^{-2\alpha} F\left(\alpha, \beta; 2\beta; \dfrac{4z}{(1+z)^2}\right)$    GA 225(101)

**9.135**    $F\left(\alpha, \beta; \alpha + \beta + \dfrac{1}{2}; \sin^2 \varphi\right) = F\left(2\alpha, 2\beta; \alpha + \beta + \dfrac{1}{2}; \sin^2 \dfrac{\varphi}{2}\right)$

$$\left[x = \sin^2 \dfrac{\varphi}{2} \text{ real}; \quad \dfrac{1 - \sqrt{2}}{2} < x < \dfrac{1}{2}\right]$$

MO 13

**9.136**[8]    We set

$$A = \frac{\Gamma\left(\alpha + \beta + \frac{1}{2}\right)\sqrt{\pi}}{\Gamma\left(\alpha + \frac{1}{2}\right)\Gamma\left(\beta + \frac{1}{2}\right)}, \qquad B = \frac{-\Gamma\left(\alpha + \beta + \frac{1}{2}\right)2\sqrt{\pi}}{\Gamma(\alpha)\Gamma(\beta)};$$

then

1.    $$F\left(2\alpha, 2\beta; \alpha + \beta + \frac{1}{2}; \frac{1 - \sqrt{z}}{2}\right) = AF\left(\alpha, \beta; \frac{1}{2}; z\right) + B\sqrt{z}\, F\left(\alpha + \frac{1}{2}, \beta + \frac{1}{2}; \frac{3}{2}; z\right)$$

GA 227(106)

2.    $$F\left(2\alpha, 2\beta; \alpha + \beta + \frac{1}{2}; \frac{1 + \sqrt{z}}{2}\right) = AF\left(\alpha, \beta; \frac{1}{2}; z\right) - B\sqrt{z}\, F\left(\alpha + \frac{1}{2}, \beta + \frac{1}{2}; \frac{3}{2}; z\right)$$

GA 227(107)

3.    $$\frac{\left(\alpha - \frac{1}{2}\right)\left(\beta - \frac{1}{2}\right)}{\alpha + \beta - \frac{1}{2}} A\sqrt{z}\, F\left(\alpha, \beta; \frac{3}{2}; z\right) = F\left(2\alpha - 1, 2\beta - 1; \alpha + \beta - \frac{1}{2}; \frac{1 + \sqrt{z}}{2}\right)$$
$$- F\left(2\alpha - 1, 2\beta - 1; \alpha + \beta - \frac{1}{2}; \frac{1 - \sqrt{z}}{2}\right)$$

GA 229(110)

**9.137**[7]    Gauss' recursion functions:

1.    $\gamma[\gamma - 1 - (2\gamma - \alpha - \beta - 1)z]\, F(\alpha, \beta; \gamma; z) + (\gamma - \alpha)(\gamma - \beta)z\, F(\alpha, \beta; \gamma + 1; z) + \gamma(\gamma - 1)(z - 1)\, F(\alpha, \beta; \gamma - 1; z) = 0$

2.    $(2\alpha - \gamma - \alpha z + \beta z)\, F(\alpha, \beta; \gamma; z) + (\gamma - \alpha)\, F(\alpha - 1, \beta; \gamma; z) + \alpha(z - 1)\, F(\alpha + 1, \beta; \gamma; z) = 0$

3.    $(2\beta - \gamma - \beta z + \alpha z)\, F(\alpha, \beta; \gamma; z) + (\gamma - \beta)\, F(\alpha, \beta - 1; \gamma; z) + \beta(z - 1)\, F(\alpha, \beta + 1; \gamma; z) = 0$

4.    $\gamma\, F(\alpha, \beta - 1; \gamma; z) - \gamma\, F(\alpha - 1, \beta; \gamma; z) + (\alpha - \beta)z\, F(\alpha, \beta; \gamma + 1; z) = 0$

5.[8]    $\gamma(\alpha - \beta)\, F(\alpha, \beta; \gamma; z) - \alpha(\gamma - \beta)\, F(\alpha + 1, \beta; \gamma + 1; z) + \beta(\gamma - \alpha)\, F(\alpha, \beta + 1; \gamma + 1; z) = 0$

6.    $\gamma(\gamma + 1)\, F(\alpha, \beta; \gamma; z) - \gamma(\gamma + 1)\, F(\alpha, \beta; \gamma + 1; z) - \alpha\beta z\, F(\alpha + 1, \beta + 1; \gamma + 2; z) = 0$

7.    $\gamma\, F(\alpha, \beta; \gamma; z) - (\gamma - \alpha)\, F(\alpha, \beta + 1; \gamma + 1; z) - \alpha(1 - z)\, F(\alpha + 1, \beta + 1; \gamma + 1; z) = 0$

8.    $\gamma\, F(\alpha, \beta; \gamma; z) + (\beta - \gamma)\, F(\alpha + 1, \beta; \gamma + 1; z) - \beta(1 - z)\, F(\alpha + 1, \beta + 1; \gamma + 1; z) = 0$

9.    $\gamma(\gamma - \beta z - \alpha)\, F(\alpha, \beta; \gamma; z) - \gamma(\gamma - \alpha)\, F(\alpha - 1, \beta; \gamma; z) + \alpha\beta z(1 - z)\, F(\alpha + 1, \beta + 1; \gamma + 1; z) = 0$

10.    $\gamma(\gamma - \alpha z - \beta)\, F(\alpha, \beta; \gamma; z) - \gamma(\gamma - \beta)\, F(\alpha, \beta - 1; \gamma; z) + \alpha\beta z(1 - z)\, F(\alpha + 1, \beta + 1; \gamma + 1; z) = 0$

11.    $\gamma\, F(\alpha, \beta; \gamma; z) - \gamma\, F(\alpha, \beta + 1; \gamma; z) + \alpha z\, F(\alpha + 1, \beta + 1; \gamma + 1; z) = 0$

12.[8]    $\gamma\, F(\alpha, \beta; \gamma; z) - \gamma\, F(\alpha + 1, \beta; \gamma; z) + \beta z\, F(\alpha + 1, \beta + 1; \gamma + 1; z) = 0$

13.    $\gamma[\alpha - (\gamma - \beta)z]\, F(\alpha, \beta; \gamma; z) - \alpha\gamma(1 - z)\, F(\alpha + 1, \beta; \gamma; z) + (\gamma - \alpha)(\gamma - \beta)z\, F(\alpha, \beta; \gamma + 1; z) = 0$

14.    $\gamma[\beta - (\gamma - \alpha)z]\, F(\alpha, \beta; \gamma; z) - \beta\gamma(1 - z)\, F(\alpha, \beta + 1; \gamma; z) + (\gamma - \alpha)(\gamma - \beta)z\, F(\alpha, \beta; \gamma + 1; z) = 0$

15.[8]    $\gamma(\gamma + 1)\, F(\alpha, \beta; \gamma; z) - \gamma(\gamma + 1)\, F(\alpha, \beta + 1; \gamma + 1; z) + \alpha(\gamma - \beta)z\, F(\alpha + 1, \beta + 1; \gamma + 2; z) = 0$

16.    $\gamma(\gamma + 1)\, F(\alpha, \beta; \gamma; z) - \gamma(\gamma + 1)\, F(\alpha + 1, \beta; \gamma + 1; z) + \beta(\gamma - \alpha)z\, F(\alpha + 1, \beta + 1; \gamma + 2; z) = 0$

17.    $\gamma\, F(\alpha, \beta; \gamma; z) - (\gamma - \beta)\, F(\alpha, \beta; \gamma + 1; z) - \beta\, F(\alpha, \beta + 1; \gamma + 1; z) = 0$

18.[8]    $\gamma\, F(\alpha, \beta; \gamma; z) - (\gamma - \alpha)\, F(\alpha, \beta; \gamma + 1; z) - \alpha\, F(\alpha + 1, \beta; \gamma + 1; z) = 0$    MO 13–14

**9.138***

1.  $$\gamma\,(\gamma+1)\,F\,(\alpha,\beta;\gamma;z) - \gamma\,(\gamma-\alpha+1)\,F\,(\alpha,\beta+1;\gamma+2;z)$$
$$-\,\alpha\,[\gamma-(\gamma-\beta)\,z]\,F\,(\alpha+1,\beta+1;\gamma+2;z) = 0$$

2.  $$\gamma\,(\gamma+1)\,F\,(\alpha,\beta;\gamma;z) - \gamma\,(\gamma-\beta+1)\,F\,(\alpha+1,\beta;\gamma+2;z)$$
$$-\,\beta\,[\gamma-(\gamma-\alpha)\,z]\,F\,(\alpha+1,\beta+1;\gamma+2;z) = 0$$

## 9.14  A generalized hypergeometric series

The series

1.  $$_pF_q\,(\alpha_1,\alpha_2,\ldots,\alpha_p;\ \beta_1,\beta_2,\ldots,\beta_q;\ z) = \sum_{k=0}^{\infty} \frac{(\alpha_1)_k\,(\alpha_2)_k\cdots(\alpha_p)_k}{(\beta_1)_k\,(\beta_2)_k\cdots(\beta_q)_k}\,\frac{z^k}{k!}$$   MO 14

   is called a *generalized hypergeometric series* (see also 9.210).

2.  $$_2F_1(\alpha,\beta;\gamma;z) \equiv F(\alpha,\beta;\gamma;z)$$   MO 15

For integral representations, see **3.254** 2, **3.259** 2, and **3.478** 3.
For a representation in terms of the Meijer G-function, see **9.343** 8.

## 9.15  The hypergeometric differential equation

**9.151**  A hypergeometric series is one of the solutions of the differential equation

$$z(1-z)\frac{d^2u}{dz^2} + [\gamma - (\alpha+\beta+1)z]\frac{du}{dz} - \alpha\beta u = 0,$$   WH

which is called the *hypergeometric equation*.

### The solution of the hypergeometric differential equation

**9.152**  The hypergeometric differential equation **9.151** possesses *two linearly independent solutions*. These solutions have analytic continuations to the entire $z$-plane except possibly for the three points $0$, $1$, and $\infty$. Generally speaking, the points $z = 0, 1, \infty$ are branch points of at least one of the branches of each solution of the hypergeometric differential equation. The ratio $w(z)$ of two linearly independent solutions satisfies the differential equation

$$2\frac{w'''}{w'} - 3\left(\frac{w''}{w'}\right)^2 = \frac{1-a_1^2}{z^2} + \frac{1-a_2^2}{(z-1)^2} + \frac{a_1^2+a_2^2-a_3^2-1}{z(z-1)},$$

where

$$a_1^2 = (1-\gamma)^2, \quad a_2^2 = (\gamma-\alpha-\beta)^2, \quad a_3^2 = (\alpha-\beta)^2.$$

If $\alpha, \beta, \gamma$ are real, the function $w(z)$ maps the upper (Im $z > 0$) or the lower (Im $z < 0$) half-plane onto a curvilinear triangle whose angles are $\pi a_1, \pi a_2, \pi a_3$. The vertices of this triangle are the images of the points $z = 0, z = 1$, and $z = \infty$.

**9.153**  Within the unit circle $|z| < 1$, the linearly independent solutions $u_1(z)$ and $u_2(z)$ of the hypergeometric differential equation are given by the following formulas:

1.  If $\gamma$ is not an integer,

$$u_1 = F(\alpha,\beta;\gamma;z),$$
$$u_2 = z^{1-\gamma}e\,F(\alpha-\gamma+1,\beta-\gamma+1;2-\gamma;z)$$

2.     If $\gamma = 1$, then

$$u_1 = F(\alpha, \beta; 1; z),$$

$$u_2 = F(\alpha, \beta; 1; z) \ln z + \sum_{k=1}^{\infty} z^k \frac{(\alpha)_k (\beta)_k}{(k!)^2}$$

$$\times \left\{ \psi(\alpha + k) - \psi(\alpha) + \psi(\beta + k) - \psi(\beta) - 2\,\psi(k + 1) + 2\,\psi(1) \right\}$$

(see **9.14** 2)

3.     If $\gamma = m + 1$ (where $m$ is a natural number), and if neither $\alpha$ nor $\beta$ is a positive number not exceeding $m$, then

$$u_1 = F(\alpha, \beta; m + 1; z),$$

$$u_2 = F(\alpha, \beta; m + 1; z) \ln z + z^m \sum_{k=1}^{\infty} z^k \frac{(\alpha)_k (\beta)_k}{(1 + m)_k} \left\{ h(k) - h(0) \right\} - \sum_{k=1}^{m} \frac{(k - 1)!(-m)_k}{(1 - \alpha)_k (1 - \beta)_k} z^{-k}$$

(see **9.14** 2)

where

$$h(n) = \psi(\alpha + n) + \psi(\beta + n) - \psi(m + 1 + n) - \psi(n + 1) \qquad [n + 1 \text{ is a natural number}]$$

4.[11]    Suppose that $\gamma = m + 1$ (where $m$ is a natural number) and that $\alpha$ or $\beta$ is equal to $m' + 1$, where $0 \le m' < m$. Then, for example, for $\alpha = m' + 1$, we obtain

$$u_1 = F\left(1 + m', \beta; 1 + m; z\right),$$

$$u_2 = z^{-m} F\left(1 + m' - m, \beta - m; 1 - m; z\right)$$

In this case, $u_2$ is a polynomial in $z^{-1}$.

5.[12]    If $\gamma = 1 - m$ (where $m$ is a natural number) and if $\alpha$ and $\beta$ are both different from the numbers $0, -1, -2, \ldots, 1 - m$, then

$$u_1 = z^m F(\alpha + m, \beta + m; 1 + m; z),$$

$$u_2 = z^m F(\alpha + m, \beta + m; 1 + m; z) \ln z + z^m \sum_{k=1}^{\infty} z^k \frac{(\alpha + m)_k (\beta + m)_k}{(1 + m)_k k!} \left\{ h^*(k) - h^*(0) \right\}$$

$$- \sum_{k=1}^{\infty} \frac{(k - 1)!(-m)_k}{(1 - \alpha - m)_k (1 - \beta - m)_k} z^{m-n}$$

(see **9.14** 2)

where

$$h^*(n) = \psi(\alpha + m + n) + \psi(\beta + m + n) - \psi(1 + m + n) - \psi(1 + n)$$

We note that

$$\psi(\alpha + n) - \psi(\alpha) = \frac{1}{\alpha} + \frac{1}{\alpha + 1} + \cdots + \frac{1}{\alpha + n - 1} \qquad \text{(cf. **8.365** 3)}$$

and that, for $\alpha = -\lambda$, where $\lambda$ is a natural number or zero and $n = \lambda + 1, \lambda + 2, \ldots$ the expression

$$(\alpha)_k \left[ \psi(\alpha + n) - \psi(\alpha) \right]$$

in formulas **9.153** 2–5 should be replaced with the expression

$$(-1)^\lambda \lambda!(n - \lambda - 1)!$$

6.    Suppose that $\gamma = 1 - m$ (where $m$ is a natural number) and that $\alpha$ or $\beta$ is an integer $(-m')$, where $m'$ is one of the following numbers: $0, 1, \ldots, m - 1$. Suppose, for example, that $\alpha = -m'$. Then,

$$u_1 = F(-m', \beta; 1 - m; z),$$
$$u_2 = F(-m' + m, \beta + m; 1 + m; z)$$

MO 18

7.    For $\gamma = \frac{1}{2}(\alpha + \beta + 1)$

$$u_1 = F\left(\alpha, \beta; \tfrac{1}{2}(\alpha + \beta + 1); z\right),$$
$$u_2 = F\left(\alpha, \beta; \tfrac{1}{2}(\alpha + \beta + 1); 1 - z\right)$$

are two linearly independent solutions of the hypergeometric differential equation provided $\alpha, \beta$ and $\gamma$ are not zero or negative integers.

MO 17–19

**The analytic continuation of a solution that is regular at the point $z = 0$**

**9.154**    Formulas **9.153** make possible the analytic continuation, by means of the hypergeometric series, of the function $F(\alpha, \beta; \gamma; z)$ defined inside the circle $|z| < 1$ to the region $|z| > 1$, and $|\arg(-z)| < \pi$. Here, it is assumed that $\alpha - \beta$ is not an integer. In the event that $\alpha - \beta$ is an integer (for example, if $\beta = \alpha + m$, where $m$ is a natural number), then, for $|z| > 1$, and $|\arg(-z)| < \pi$ we have:

1.
$$\frac{\Gamma(\alpha)\Gamma(\alpha + m)}{\Gamma(\gamma)} F(\alpha, \alpha + m; \gamma; z)$$

$$= \frac{\sin \pi(\gamma - \alpha)}{\pi}\left\{ \sum_{k=0}^{m-1} \frac{\Gamma(\alpha + k)\Gamma(1 - \gamma + \alpha + k)\Gamma(m - k)}{k!}(-z)^{-\alpha-k}\right.$$

$$\left. + (-z)^{-\alpha-m} \sum_{k=0}^{\infty} \frac{\Gamma(\alpha + m + k)\Gamma(1 - \gamma + \alpha + m + k)}{k!(k + m)!}g(k)z^{-k}\right\}$$

where

2.    $g(n) = \ln(-z) + \pi \cot \pi(\gamma - \alpha) + \psi(n + 1) + \psi(n + m + 1)$

$$- \psi(\alpha + m + n) - \psi(1 - \gamma + \alpha + m + n)$$

For $m = 0$, we should set $\displaystyle\sum_{k=0}^{m-1} = 0$.

**9.155**    This formula loses its meaning when $\alpha, \gamma$, or $\alpha - \gamma + 1$ is equal to one of the numbers $0, -1, -2, \ldots$. In this last case, we have

1.    If $\alpha$ is a non-positive integer and $\gamma$ is not an integer, $F(\alpha, \alpha + m; \gamma; z)$ is a polynomial in $z$.

2.    Suppose that $\gamma$ is a non-positive integer and that $\alpha$ is not an integer. We then set $\gamma = -\lambda$, where $\lambda = 0, 1, 2, \ldots$. Then,

$$\frac{\Gamma(\alpha + \lambda + 1)\Gamma(\alpha + \lambda + m + 1)}{\Gamma(\lambda + 2)}z^{\lambda+1} F(\alpha + \lambda + 1, \alpha + \lambda + m + 1; \lambda + 2; z)$$

is a solution of the hypergeometric equation that is regular at the point $z = 0$. This solution is equal to the right hand member of formula **9.154** 1 if we replace $\gamma$ with $\lambda$ in this equation and in formula **9.154** 2.

3.      If $\alpha - \gamma + 1$ is a non-positive integer and if $\alpha$ and $\gamma$ are not themselves integers, we may use the formula

$$F(\alpha, \alpha + m; \gamma; z) = (1 - z)^{\gamma - 2\alpha - m} F(\gamma - \alpha - m, \gamma - \alpha; \gamma; z)$$

and apply formula **9.154** 1 to its right hand member provided $\gamma - \alpha - m > 0$. However, if $\alpha - \gamma - m \leq 0$, the right member of this expression is a polynomial taken to the $(1 - z)^{\text{th}}$ power.

4.      If $\alpha, \beta$, and $\gamma$ are integers, the hypergeometric differential equation always has a solution that is regular for $z = 0$ and that is of the form

$$R_1(z) + \ln(1 - z)R_2(z),$$

where $R_1(z)$ and $R_2(z)$ are rational functions of $z$. To get a solution of this form, we need to apply formulas **9.137** 1–**9.137** 3 to the function $F(\alpha, \beta; \gamma; z)$. However, if $\gamma = -\lambda$, where $\lambda + 1$ is a natural number, formulas **9.137** 1 and **9.137** 2 should be applied not to $F(\alpha, \beta; \gamma; z)$ but to the function $z^{\lambda+1} F(\alpha + \lambda + 1, \beta + \lambda + 1; \lambda + 2, z)$.

By successive applications of these formulas, we can reduce the positive values of the parameters to the pair, unity and zero. Furthermore, we can obtain the desired form of the solution from the formulas

$$F(1, 1; 2; z) = -z^{-1} \ln(1 - z),$$
$$F(0, \beta; \gamma; z) = F(\alpha, 0; \gamma; z) = 1$$

MO 19–20

## 9.16   Riemann's differential equation

**9.160**    The hypergeometric differential equation is a particular case of Riemann's differential equation

1.[11]
$$\frac{d^2u}{dz^2} + \left[ \frac{1 - \alpha - \alpha'}{z - a} + \frac{1 - \beta - \beta'}{z - b} + \frac{1 - \gamma - \gamma'}{z - c} \right] \frac{du}{dz}$$
$$+ \left[ \frac{\alpha\alpha'(a - b)(a - c)}{z - a} + \frac{\beta\beta'(b - c)(b - a)}{z - b} + \frac{\gamma\gamma'(c - a)(c - b)}{z - c} \right] \frac{u}{(z - a)(z - b)(z - c)} = 0$$

WH

The coefficients of this equation have poles at the points $a$, $b$, and $c$, and the numbers $\alpha, \alpha'; \beta, \beta'; \gamma, \gamma'$ are called the indices corresponding to these poles. The indices $\alpha, \alpha'; \beta, \beta'; \gamma, \gamma'$ are related by the following equation:

$$\alpha + \alpha' + \beta + \beta' + \gamma + \gamma' - 1 = 0$$

WH

2.      The differential equations **9.160** 1 are written diagramatically as follows:

3.     
$$u = P \left\{ \begin{matrix} a & b & c & \\ \alpha & \beta & \gamma & z \\ \alpha' & \beta' & \gamma' & \end{matrix} \right\}$$

The singular points of the equation appear in the first row in this scheme, the indices corresponding to them appear beneath them, and the independent variable appears in the fourth column.      WH

**9.161**    The two following transformation formulas are valid for Riemann's $P$-equation:

1.    $$\left(\frac{z-a}{z-b}\right)^k \left(\frac{z-c}{z-b}\right)^l P \left\{ \begin{matrix} a & b & c & \\ \alpha & \beta & \gamma & z \\ \alpha' & \beta' & \gamma' & \end{matrix} \right\} = P \left\{ \begin{matrix} a & b & c & \\ \alpha+k & \beta-k-1 & \gamma+l & z \\ \alpha'+k & \beta'-k-l & \gamma'+l & \end{matrix} \right\}$$     WH

2.    $$P \left\{ \begin{matrix} a & b & c & \\ \alpha & \beta & \gamma & z \\ \alpha' & \beta' & \gamma' & \end{matrix} \right\} = P \left\{ \begin{matrix} a_1 & b_1 & c_1 & \\ \alpha & \beta & \gamma & z_1 \\ \alpha' & \beta' & \gamma' & \end{matrix} \right\}$$     WH

The first of these formulas means that if

$$u = P \left\{ \begin{matrix} a & b & c & \\ \alpha & \beta & \gamma & z \\ \alpha' & \beta' & \gamma' & \end{matrix} \right\},$$

then the function

$$u_1 = \left(\frac{z-a}{z-b}\right)^k \left(\frac{z-c}{z-b}\right)^l u$$

satisfies a second-order differential equation having the same singular points as equation **9.161** 2 and indices equal to $\alpha+k, \alpha'+k; \beta-k-l, \beta'-k-l; \gamma+l, \gamma'+l$. The second transformation formula converts a differential equation with singularities at the points $a$,$b$, and $c$, indices $\alpha, \alpha'; \beta, \beta'; \gamma, \gamma'$, and an independent variable $z$ into a differential equation with the same indices, singular points $a_1$, $b_1$, and $c_1$, and independent variable $z_1$. The variable $z_1$ is connected with the variable $z$ by the fractional transformation

$$z = \frac{Az_1 + B}{Cz_1 + D} \qquad [AD - BC \neq 0]$$

The same transformation connects the points $a_1$, $b_1$, and $c_1$ with the points $a$, $b$, and $c$.

WH, MO 20

**9.162**    By the successive application of the two transformation formulas **9.161** 1 and **9.161** 2, we can convert Riemann's differential equation into the hypergeometric differential equation. Thus, the solution of Riemann's differential equation can be expressed in terms of a hypergeometric function.

For $k = -\alpha, l = -\gamma$, and $z_1 = \frac{(z-a)(c-b)}{(z-b)(c-a)}$, we have

1.    $$u = P \left\{ \begin{matrix} a & b & c & \\ \alpha & \beta & \gamma & z \\ \alpha' & \beta' & \gamma' & \end{matrix} \right\} = \left(\frac{z-a}{z-b}\right)^\alpha \left(\frac{z-c}{z-b}\right)^\gamma P \left\{ \begin{matrix} a & b & c & \\ 0 & \beta+\alpha+\gamma & 0 & z \\ \alpha'-\alpha & \beta'+\alpha+\gamma & \gamma'-\gamma & \end{matrix} \right\}$$

$$= \left(\frac{z-a}{z-b}\right)^\alpha \left(\frac{z-c}{z-b}\right)^\gamma P \left\{ \begin{matrix} 0 & \infty & 1 & \\ 0 & \beta+\alpha+\gamma & 0 & \frac{(z-a)(c-b)}{(z-b)(c-a)} \\ \alpha'-\alpha & \beta'+\alpha+\gamma & \gamma'-\gamma & \end{matrix} \right\}$$

MO 23

Thus, this solution can be expressed as a hypergeometric series as follows:

2.    $$u = \left(\frac{z-a}{z-b}\right)^\alpha \left(\frac{z-c}{z-b}\right)^\gamma F \left( \alpha+\beta+\gamma, \alpha+\beta'+\gamma; 1+\alpha-\alpha'; \frac{(z-a)(c-b)}{(z-b)(c-a)} \right)$$

If the constants $a$, $b$, $c$; $\alpha$, $\alpha'$; $\beta$, $\beta'$; $\gamma$, $\gamma'$ are permuted in a suitable manner, Riemann's equation remains unchanged. Thus, we obtain a set of 24 solutions of differential equations having the following form (provided none of the differences $\alpha - \alpha'$, $\beta - \beta'$, $\gamma - \gamma'$ are integers):

WH, MO 23

**9.163**

1. $u_1 = \left(\dfrac{z-a}{z-b}\right)^{\alpha} \left(\dfrac{z-c}{z-b}\right)^{\gamma} F\left\{\alpha+\beta+\gamma, \alpha+\beta'+\gamma; 1+\alpha-\alpha'; \dfrac{(c-b)(z-a)}{(c-a)(z-b)}\right\}$

2. $u_2 = \left(\dfrac{z-a}{z-b}\right)^{\alpha'} \left(\dfrac{z-c}{z-b}\right)^{\gamma} F\left\{\alpha'+\beta+\gamma, \alpha'+\beta'+\gamma; 1+\alpha'-\alpha; \dfrac{(c-b)(z-a)}{(c-a)(z-b)}\right\}$

3. $u_3 = \left(\dfrac{z-a}{z-b}\right)^{\alpha} \left(\dfrac{z-c}{z-b}\right)^{\gamma'} F\left\{\alpha+\beta+\gamma', \alpha+\beta'+\gamma'; 1+\alpha-\alpha'; \dfrac{(c-b)(z-a)}{(c-a)(z-b)}\right\}$

4.[12] $u_4 = \left(\dfrac{z-a}{z-b}\right)^{\alpha'} \left(\dfrac{z-c}{z-b}\right)^{\gamma'} F\left\{\alpha'+\beta+\gamma', \alpha'+\beta'+\gamma'; 1+\alpha'-\alpha; \dfrac{(c-b)(z-a)}{(c-a)(z-b)}\right\}$

**9.164**

1.[10] $u_5 = \left(\dfrac{z-b}{z-c}\right)^{\beta} \left(\dfrac{z-a}{z-c}\right)^{\alpha} F\left\{\beta+\gamma+\alpha, \beta+\gamma'+\alpha; 1+\beta-\beta'; \dfrac{(a-c)(z-b)}{(a-b)(z-c)}\right\}$

2. $u_6 = \left(\dfrac{z-b}{z-c}\right)^{\beta'} \left(\dfrac{z-a}{z-c}\right)^{\alpha} F\left\{\beta'+\gamma+\alpha, \beta'+\gamma'+\alpha; 1+\beta'-\beta; \dfrac{(a-c)(z-b)}{(a-b)(z-c)}\right\}$

3. $u_7 = \left(\dfrac{z-b}{z-c}\right)^{\beta} \left(\dfrac{z-a}{z-c}\right)^{\alpha'} F\left\{\beta+\gamma+\alpha', \beta+\gamma'+\alpha'; 1+\beta-\beta'; \dfrac{(a-c)(z-b)}{(a-b)(z-c)}\right\}$

4. $u_8 = \left(\dfrac{z-b}{z-c}\right)^{\beta'} \left(\dfrac{z-a}{z-c}\right)^{\alpha'} F\left\{\beta'+\gamma+\alpha', \beta'+\alpha'+\gamma'; 1+\beta'-\beta; \dfrac{(a-c)(z-b)}{(a-b)(z-c)}\right\}$

**9.165**

1. $u_9 = \left(\dfrac{z-c}{z-a}\right)^{\gamma} \left(\dfrac{z-b}{z-a}\right)^{\beta} F\left\{\gamma+\alpha+\beta, \gamma+\alpha'+\beta; 1+\gamma-\gamma'; \dfrac{(b-a)(z-c)}{(b-c)(z-a)}\right\}$

2. $u_{10} = \left(\dfrac{z-c}{z-a}\right)^{\gamma'} \left(\dfrac{z-b}{z-a}\right)^{\beta} F\left\{\gamma'+\alpha+\beta, \gamma'+\alpha'+\beta; 1+\gamma'-\gamma; \dfrac{(b-a)(z-c)}{(b-c)(z-a)}\right\}$

3. $u_{11} = \left(\dfrac{z-c}{z-a}\right)^{\gamma} \left(\dfrac{z-b}{z-a}\right)^{\beta'} F\left\{\gamma+\alpha+\beta', \gamma+\alpha'+\beta'; 1+\gamma-\gamma'; \dfrac{(b-a)(z-c)}{(b-c)(z-a)}\right\}$

4. $u_{12} = \left(\dfrac{z-c}{z-a}\right)^{\gamma'} \left(\dfrac{z-b}{z-a}\right)^{\beta'} F\left\{\gamma'+\alpha+\beta', \gamma'+\alpha'+\beta'; 1+\gamma'-\gamma; \dfrac{(b-a)(z-c)}{(b-c)(z-a)}\right\}$

**9.166**

1. $u_{13} = \left(\dfrac{z-a}{z-c}\right)^{\alpha} \left(\dfrac{z-b}{z-c}\right)^{\beta} F\left\{\alpha+\gamma+\beta, \alpha+\gamma'+\beta; 1+\alpha-\alpha'; \dfrac{(b-c)(z-a)}{(b-a)(z-c)}\right\}$

2. $u_{14} = \left(\dfrac{z-a}{z-c}\right)^{\alpha'} \left(\dfrac{z-b}{z-c}\right)^{\beta} F\left\{\alpha'+\gamma+\beta, \alpha'+\gamma'+\beta; 1+\alpha'-\alpha; \dfrac{(b-c)(z-a)}{(b-a)(z-c)}\right\}$

3. $u_{15} = \left(\dfrac{z-a}{z-c}\right)^{\alpha} \left(\dfrac{z-b}{z-c}\right)^{\beta'} F\left\{\alpha+\gamma+\beta', \alpha+\gamma'+\beta'; 1+\alpha-\alpha'; \dfrac{(b-c)(z-a)}{(b-a)(z-c)}\right\}$

4. $\quad u_{16} = \left(\dfrac{z-a}{z-c}\right)^{\alpha'} \left(\dfrac{z-b}{z-c}\right)^{\beta'} F\left\{\alpha' + \gamma + \beta', \alpha' + \gamma' + \beta'; 1 + \alpha' - \alpha; \dfrac{(b-c)(z-a)}{(b-a)(z-c)}\right\}$

**9.167**

1. $\quad u_{17} = \left(\dfrac{z-c}{z-b}\right)^{\gamma} \left(\dfrac{z-a}{z-b}\right)^{\alpha} F\left\{\gamma + \beta + \alpha, \gamma + \beta' + \alpha; 1 + \gamma - \gamma'; \dfrac{(a-b)(z-c)}{(a-c)(z-b)}\right\}$

2. $\quad u_{18} = \left(\dfrac{z-c}{z-b}\right)^{\gamma'} \left(\dfrac{z-a}{z-b}\right)^{\alpha} F\left\{\gamma' + \beta + \alpha, \gamma' + \beta' + \alpha; 1 + \gamma' - \gamma; \dfrac{(a-b)(z-c)}{(a-c)(z-b)}\right\}$

3. $\quad u_{19} = \left(\dfrac{z-c}{z-b}\right)^{\gamma} \left(\dfrac{z-a}{z-b}\right)^{\alpha'} F\left\{\gamma + \beta + \alpha', \gamma + \beta' + \alpha'; 1 + \gamma - \gamma'; \dfrac{(a-b)(z-c)}{(a-c)(z-b)}\right\}$

4. $\quad u_{20} = \left(\dfrac{z-c}{z-b}\right)^{\gamma'} \left(\dfrac{z-a}{z-b}\right)^{\alpha'} F\left\{\gamma' + \beta + \alpha', \gamma' + \beta' + \alpha'; 1 + \gamma' - \gamma; \dfrac{(a-b)(z-c)}{(a-c)(z-b)}\right\}$

**9.168**

1. $\quad u_{21} = \left(\dfrac{z-b}{z-a}\right)^{\beta} \left(\dfrac{z-c}{z-a}\right)^{\gamma} F\left\{\beta + \alpha + \gamma, \beta + \alpha' + \gamma; 1 + \beta - \beta'; \dfrac{(c-a)(z-b)}{(c-b)(z-a)}\right\}$

2. $\quad u_{22} = \left(\dfrac{z-b}{z-a}\right)^{\beta'} \left(\dfrac{z-c}{z-a}\right)^{\gamma} F\left\{\beta' + \alpha + \gamma, \beta' + \alpha' + \gamma; 1 + \beta' - \beta; \dfrac{(c-a)(z-b)}{(c-b)(z-a)}\right\}$

3. $\quad u_{23} = \left(\dfrac{z-b}{z-a}\right)^{\beta} \left(\dfrac{z-c}{z-a}\right)^{\gamma'} F\left\{\beta + \alpha + \gamma', \beta + \alpha' + \gamma'; 1 + \beta - \beta'; \dfrac{(c-a)(z-b)}{(c-b)(z-a)}\right\}$

4. $\quad u_{24} = \left(\dfrac{z-b}{z-a}\right)^{\beta'} \left(\dfrac{z-c}{z-a}\right)^{\gamma'} F\left\{\beta' + \alpha + \gamma', \beta' + \alpha' + \gamma'; 1 + \beta' - \beta; \dfrac{(c-a)(z-b)}{(c-b)(z-a)}\right\}$ WH

## 9.17 Representing the solutions to certain second-order differential equations using a Riemann scheme

**9.171** The hypergeometric equation (see **9.151**):

$$u = P\left\{\begin{array}{ccc} 0 & \infty & 1 \\ 0 & \alpha & 0 \\ 1-\gamma & \beta & \gamma - \alpha - \beta \end{array} z\right\}$$ WH

**9.172** The associated Legendre's equation defining the functions $P_n^m(z)$ for $n$ and $m$ integers (see **8.700 1**):

1. $\quad u = P\left\{\begin{array}{ccc} 0 & \infty & 1 \\ \frac{1}{2}m & n+1 & \frac{1}{2}m \\ -\frac{1}{2}m & -n & -\frac{1}{2}m \end{array} \dfrac{1-z}{2}\right\}$ WH

2. $\quad u = P\left\{\begin{array}{ccc} 0 & \infty & 1 \\ -\frac{1}{2}n & \frac{1}{2}m & 0 \\ \frac{n+1}{2} & -\frac{1}{2}m & \frac{1}{2} \end{array} \dfrac{1}{1-z^2}\right\}$ WH

**9.173** The function $P_n^m \left( 1 - \dfrac{z^2}{2n^2} \right)$ satisfies the equation

$$u = P \left\{ \begin{array}{ccc} 4n^2 & \infty & 0 \\ \frac{1}{2}m & n+1 & \frac{1}{2}m & z^2 \\ -\frac{1}{2}m & -n & -\frac{1}{2}m \end{array} \right\}$$

WH

The function $J_m(z)$ satisfies the limiting form of this equation obtained as $n \to \infty$.

**9.174** The equation defining the Gegenbauer polynomials $C_n^\lambda(z)$, (see **8.938**):

$$u = P \left\{ \begin{array}{ccc} -1 & \infty & 1 \\ \frac{1}{2} - \lambda & n+2\lambda & \frac{1}{2} - \lambda & z \\ 0 & -n & 0 \end{array} \right\}$$

WH

**9.175** Bessel's equation (see **8.401**) is the limiting form of the equations:

1. $$u = P \left\{ \begin{array}{ccc} 0 & \infty & c \\ n & ic & \frac{1}{2} + ic & z \\ -n & -ic & \frac{1}{2} - ic \end{array} \right\}$$

WH

2. $$u = e^{iz} P \left\{ \begin{array}{ccc} 0 & \infty & c \\ n & \frac{1}{2} & 0 & z \\ -n & \frac{3}{2} - 2ic & 2ic - 1 \end{array} \right\}$$

WH

3. $$u = P \left\{ \begin{array}{ccc} 0 & \infty & c^2 \\ \frac{1}{2}n & \frac{1}{2}(c-n) & 0 & z^2 \\ -\frac{1}{2}n & -\frac{1}{2}(c+n) & n+1 \end{array} \right\}$$

WH

as $c \to \infty$.

## 9.18 Hypergeometric functions of two variables

### 9.180

1.  $$F_1 \left( \alpha, \beta, \beta', \gamma; x, y \right) = \sum_{m=0}^{\infty} \sum_{n=0}^{\infty} \frac{(\alpha)_{m+n} (\beta)_m (\beta')_n}{(\gamma)_{m+n} m! n!} x^m y^n$$

$$[|x| < 1, \quad |y| < 1]$$

EH I 224(6), AK 14(11)

2.  $$F_2 \left( \alpha, \beta, \beta', \gamma, \gamma'; x, y \right) = \sum_{m=0}^{\infty} \sum_{n=0}^{\infty} \frac{(\alpha)_{m+n} (\beta)_m (\beta')_n}{(\gamma)_m (\gamma')_n m! n!} x^m y^n$$

$$[|x| + |y| < 1] \qquad \text{EH I 224(7), AK 14(12)}$$

3.  $$F_3 \left( \alpha, \alpha', \beta, \beta', \gamma; x, y \right) = \sum_{m=0}^{\infty} \sum_{n=0}^{\infty} \frac{(\alpha)_m (\alpha')_n (\beta)_m (\beta')_n}{(\gamma)_{m+n} m! n!} x^m y^n$$

$$[|x| < 1, \quad |y| < 1]$$

EH I 224(8), AK 14(13)

4.  $$F_4 \left( \alpha, \beta, \gamma, \gamma'; x, y \right) = \sum_{m=0}^{\infty} \sum_{n=0}^{\infty} \frac{(\alpha)_{m+n} (\beta)_{m+n}}{(\gamma)_m (\gamma')_n m! n!} x^m y^n \qquad [|\sqrt{x}| + |\sqrt{y}| < 1]$$

EH I 224(9), AK 14(14)

**9.181** The functions $F_1$, $F_2$, $F_3$, and $F_4$ satisfy the following systems of partial differential equations for $z$:

1.  System of equations for $z = F_1$:

$$x(1-x)\frac{\partial^2 z}{\partial x^2} + y(1-x)\frac{\partial^2 z}{\partial x\,\partial y} + [\gamma - (\alpha+\beta+1)x]\frac{\partial z}{\partial x} - \beta y\frac{\partial z}{\partial y} - \alpha\beta z = 0, \qquad \text{EH I 233(9)}$$

$$y(1-y)\frac{\partial^2 z}{\partial y^2} + x(1-y)\frac{\partial^2 z}{\partial x\,\partial y} + [\gamma - (\alpha+\beta'+1)\,y]\frac{\partial z}{\partial x} - \beta' x\frac{\partial z}{\partial x} - \alpha\beta' z = 0$$

2.  System of equations for $z = F_2$:

$$x(1-x)\frac{\partial^2 z}{\partial x^2} - xy\frac{\partial^2 z}{\partial x\,\partial y} + [\gamma - (\alpha+\beta+1)x]\frac{\partial z}{\partial x} - \beta y\frac{\partial z}{\partial y} - \alpha\beta z = 0, \qquad \text{EH I 234(10)}$$

$$y(1-y)\frac{\partial^2 z}{\partial y^2} - xy\frac{\partial^2 z}{\partial x\,\partial y} + [\gamma' - (\alpha+\beta'+1)\,y]\frac{\partial z}{\partial y} - \beta' x\frac{\partial z}{\partial x} - \alpha\beta' z = 0$$

3.  System of equations for $z = F_3$:

$$x(1-x)\frac{\partial^2 z}{\partial x^2} + y\frac{\partial^2 z}{\partial x\,\partial y} + [\gamma - (\alpha+\beta+1)x]\frac{\partial z}{\partial x} - \alpha\beta z = 0,$$

$$y(1-y)\frac{\partial^2 z}{\partial y^2} + x\frac{\partial^2 z}{\partial x\,\partial y} + [\gamma - (\alpha'+\beta'+1)\,y]\frac{\partial z}{\partial y} - \alpha'\beta' z = 0$$

$$\text{EH I 234(11)}$$

4.  System of equations for $z = F_4$:

$$x(1-x)\frac{\partial^2 z}{\partial x^2} - y^2\frac{\partial^2 z}{\partial y^2} - 2xy\frac{\partial^2 z}{\partial x\,\partial y} + [\gamma - (\alpha+\beta+1)x]\frac{\partial z}{\partial x} - (\alpha+\beta+1)y\frac{\partial z}{\partial y} - \alpha\beta z = 0,$$

$$\text{EH I 234(12)}$$

$$y(1-y)\frac{\partial^2 z}{\partial y^2} - x^2\frac{\partial^2 z}{\partial x^2} - 2xy\frac{\partial^2 z}{\partial x\,\partial y} + [\gamma' - (\alpha+\beta+1)y]\frac{\partial z}{\partial y} - (\alpha+\beta+1)x\frac{\partial z}{\partial x} - \alpha\beta z = 0$$

$$\text{AK 44}$$

**9.182** For certain relationships between the parameters and the argument, hypergeometric functions of two variables can be expressed in terms of hypergeometric functions of a single variable or in terms of elementary functions:

1.  $F_1\left(\alpha, \beta, \beta', \beta + \beta'; x, y\right) = (1-y)^{-\alpha}\, F\left(\alpha, \beta; \beta + \beta'; \dfrac{x-y}{1-y}\right)$      EH I 238(1), AK 24(28)

2.  $F_2\left(\alpha, \beta, \beta', \beta, \gamma'; x, y\right) = (1-x)^{-\alpha}\, F\left(\alpha, \beta'; \gamma'; \dfrac{y}{1-x}\right)$      EH I 238(2), AK 23

3.  $F_2\left(\alpha, \beta, \beta', \alpha, \alpha; x, y\right) = (1-x)^{-\beta}(1-y)^{-\beta'}\, F\left(\beta, \beta'; \alpha; \dfrac{xy}{(1-x)(1-y)}\right)$      EH I 238(3)

4.  $F_3\left(\alpha, \gamma - \alpha, \beta, \gamma - \beta, \gamma; x, y\right) = (1-y)^{\alpha+\beta-\gamma}\, F(\alpha, \beta; \gamma; x + y - xy)$      EH I 238(4), AK 25(35)

5.  $F_4\left(\alpha, \gamma + \gamma' - \alpha - 1, \gamma, \gamma'; x(1-y), y(1-x)\right)$

$$= F\left(\alpha, \gamma + \gamma' - \alpha - 1; \gamma; x\right) F\left(\alpha, \gamma + \gamma' - \alpha - 1; \gamma'; y\right)$$

$$\text{EH I 238(5)}$$

6.　$F_4\left(\alpha, \beta, \alpha, \beta; -\dfrac{x}{(1-x)(1-y)}, \dfrac{-y}{(1-x)(1-y)}\right) = \dfrac{(1-x)^\beta (1-y)^\alpha}{(1-xy)}$　　　　EH I 238(6)

7.　$F_4\left(\alpha, \beta, \beta, \beta; -\dfrac{x}{(1-x)(1-y)}, -\dfrac{y}{(1-x)(1-y)}\right) = (1-x)^\alpha (1-y)^\alpha F(\alpha, 1+\alpha-\beta; \beta; xy)$

　　　　　　　　　　　　　　　　　　　　　　　　　　　　　　　　　　　　　　　EH I 238(7)

8.　$F_4\left(\alpha, \beta, 1+\alpha-\beta, \beta; -\dfrac{x}{(1-x)(1-y)}, -\dfrac{y}{(1-x)(1-y)}\right)$

　　　　　　　　　　　　　$= (1-y)^\alpha F\left[\alpha, \beta; 1+\alpha-\beta; -\dfrac{x(1-y)}{1-x}\right]$

　　　　　　　　　　　　　　　　　　　　　　　　　　　　　　　　　　　　EH I 238(8)

9.　$F_4\left(\alpha, \alpha+\dfrac{1}{2}, \gamma, \dfrac{1}{2}; x, y\right) = \dfrac{1}{2}(1+\sqrt{y})^{-2\alpha} F\left(\alpha, \alpha+\dfrac{1}{2}; \gamma; \dfrac{x}{(1+\sqrt{y})^2}\right)$

　　　　　　　　　　　　$+ \dfrac{1}{2}(1-\sqrt{y})^{-2\alpha} F\left(\alpha, \alpha+\dfrac{1}{2}; \gamma; \dfrac{x}{(1-\sqrt{y})^2}\right)$

　　　　　　　　　　　　　　　　　　　　　　　　　　　　　　　　　　　　AK 23

10.　$F_1(\alpha, \beta, \beta', \gamma; x, 1) = \dfrac{\Gamma(\gamma)\,\Gamma(\gamma-\alpha-\beta')}{\Gamma(\gamma-\alpha)\,\Gamma(\gamma-\beta')} F(\alpha, \beta; \gamma-\beta'; x)$　　　EH I 239(10), AK 22(23)

11.　$F_1(\alpha, \beta, \beta', \gamma; x, x) = F(\alpha, \beta+\beta'; \gamma; x)$　　　　　　　　　EH I 239(11), AK 23(25)

**9.183**　Functional relations between hypergeometric functions of two variables:

1.[12]　$F_1(\alpha, \beta, \beta', \gamma; x, y) = (1-x)^{-\beta}(1-y)^{-\beta'} F_1\left(\gamma-\alpha, \beta, \beta', \gamma; \dfrac{x}{x-1}, \dfrac{y}{y-1}\right)$

　　　　　　　　　　　　　　　　　　　　　　　　　　　　　　　　　　　EH I 239(1)

　　　　　$= (1-x)^{-\alpha} F_1\left(\alpha, \gamma-\beta-\beta', \beta', \gamma; \dfrac{x}{x-1}, \dfrac{y-x}{1-x}\right)$

　　　　　　　　　　　　　　　　　　　　　　　　　　　　　　　　　　　EH I 239(2)

　　　　　$= (1-y)^{-\alpha} F_1\left(\alpha, \beta, \gamma-\beta-\beta', \gamma; \dfrac{y-x}{y-1}, \dfrac{y}{y-1}\right)$

　　　　　　　　　　　　　　　　　　　　　　　　　　　　　　　　　　　EH I 239(3)

　　　　　$= (1-x)^{\gamma-\alpha-\beta}(1-y)^{-\beta'} F_1\left(\gamma-\alpha, \gamma-\beta-\beta', \beta', \gamma; x, \dfrac{x-y}{1-y}\right)$

　　　　　　　　　　　　　　　　　　　　　　　　　　　　　　　　　　　EH I 240(4)

　　　　　$= (1-x)^{-\beta}(1-y)^{\gamma-\alpha-\beta'} F_1\left(\gamma-\alpha, \beta, \gamma-\beta-\beta', \gamma; \dfrac{x-y}{x-1}, y\right)$

　　　　　　　　　　　　　　　　　　　　　　　　　　　　　　　　　　　EH I 240(5), AK 30(5)

$2.^8$　$F_2\left(\alpha,\beta,\beta',\gamma,\gamma';x,y\right)=(1-x)^{-\alpha}F_2\left(\alpha,\gamma-\beta,\beta',\gamma,\gamma';\dfrac{x}{x-1},\dfrac{y}{1-x}\right)$

EH I 240(6)

$$=(1-y)^{-\alpha}F_2\left(\alpha,\beta,\gamma'-\beta',\gamma,\gamma';\frac{x}{1-y},\frac{y}{y-1}\right)$$

EH I 240(7)

$$=(1-x-y)^{-\alpha}F_2\left(\alpha,\gamma-\beta,\gamma'-\beta',\gamma,\gamma';\frac{x}{x+y-1},\frac{y}{x+y-1}\right)$$

EH I 240(8), AK 32(6)

$3.^7$　$F_4\left(\alpha,\beta,\gamma,\gamma';x,y\right)=\dfrac{\Gamma(\gamma')\,\Gamma(\beta-\alpha)}{\Gamma\left(\gamma'-\alpha\right)\Gamma(\beta)}(-y)^{-\alpha}F_4\left(\alpha,\alpha+1-\gamma',\gamma,\alpha+1-\beta;\dfrac{x}{y},\dfrac{1}{y}\right)$

$$+\frac{\Gamma(\gamma')\,\Gamma(\alpha-\beta)}{\Gamma\left(\gamma'-\beta\right)\Gamma(\alpha)}(-y)^{\beta}F_4\left(\beta+1-\gamma',\beta,\gamma,\beta+1-\alpha;\frac{x}{y},\frac{1}{y}\right)$$

EH I 240(9), AK 26(37)

**9.184**　Integral representations: Double integrals of the Euler type

1.　$F_1\left(\alpha,\beta,\beta',\gamma;x,y\right)=\dfrac{\Gamma(\gamma)}{\Gamma(\beta)\,\Gamma\left(\beta'\right)\Gamma\left(\gamma-\beta-\beta'\right)}$

$$\times\iint\limits_{\substack{u\geq0,v\geq0\\u+v\leq1}}u^{\beta-1}v^{\beta'-1}(1-u-v)^{\gamma-\beta-\beta'-1}\,(1-ux-vy)^{-\alpha}\,du\,dv$$

$[\operatorname{Re}\beta>0,\quad\operatorname{Re}\beta'>0,\quad\operatorname{Re}\left(\gamma-\beta-\beta'\right)>0]$　EH I 230(1), AK 28(1)

2.　$F_2\left(\alpha,\beta,\beta',\gamma,\gamma';x,y\right)=\dfrac{\Gamma(\gamma)\,\Gamma\left(\gamma'\right)}{\Gamma(\beta)\,\Gamma\left(\beta'\right)\Gamma\left(\gamma-\beta\right)\Gamma\left(\gamma'-\beta'\right)}$

$$\times\int_0^1\int_0^1u^{\beta-1}v^{\beta'-1}(1-u)^{\gamma-\beta-1}(1-v)^{\gamma'-\beta'-1}(1-ux-vy)^{-\alpha}\,du\,dv$$

$[\operatorname{Re}\beta>0,\quad\operatorname{Re}\beta'>0,\quad\operatorname{Re}\left(\gamma-\beta\right)>0,\quad\operatorname{Re}\left(\gamma'-\beta'\right)>0]$　EH I 230(2), AK 28(2)

3.　$F_3\left(\alpha,\alpha',\beta,\beta',\gamma;x,y\right)$

$$=\frac{\Gamma(\gamma)}{\Gamma(\beta)\,\Gamma\left(\beta'\right)\Gamma\left(\gamma-\beta-\beta'\right)}$$

$$\times\iint\limits_{\substack{u\geq0,v\geq0\\u+v\leq1}}u^{\beta-1}v^{\beta'-1}(1-u-v)^{-\gamma-\beta-\beta'-1}(1-ux)^{-\alpha}(1-vy)^{-\alpha'}\,du\,dv$$

$[\operatorname{Re}\beta>0,\quad\operatorname{Re}\beta'>0,\quad\operatorname{Re}\left(\gamma-\beta-\beta'\right)>0]$　EH I 230(3), AK 28(3)

4.　$F_4\left(\alpha,\beta,\gamma,\gamma';x(1-y),y(1-x)\right)$

$$=\frac{\Gamma(\gamma)\,\Gamma\left(\gamma'\right)}{\Gamma(\alpha)\,\Gamma(\beta)\,\Gamma(\gamma-\alpha)\,\Gamma\left(\gamma'-\beta\right)}\int_0^1\int_0^1u^{\alpha-1}v^{\beta-1}(1-u)^{\gamma-\alpha-1}(1-v)^{\gamma'-\beta-1}$$

$$\times(1-ux)^{\alpha-\gamma-\gamma'+1}(1-vy)^{\beta-\gamma-\gamma'+1}(1-ux-vy)^{\gamma+\gamma'-\alpha-\beta-1}\,du\,dv$$

$[\operatorname{Re}\alpha>0,\quad\operatorname{Re}\beta>0,\quad\operatorname{Re}\left(\gamma-\alpha\right)>0,\quad\operatorname{Re}\left(\gamma'-\beta\right)>0]$　EH I 230(4)

**9.185**   Integral representations: Integrals of the Mellin–Barnes type

The functions $F_1$, $F_2$, $F_3$ and $F_4$ can be represented by means of double integrals of the following form:

$$F(x,y) = \frac{\Gamma(\gamma)}{\Gamma(\alpha)\,\Gamma(\beta)(2\pi i)^2} \int_{-i\infty}^{i\infty} \int_{-i\infty}^{i\infty} \Psi(s,t)\,\Gamma(-s)\,\Gamma(-t)(-x)^s(-y)^t \,\mathrm{d}s\,\mathrm{d}t$$

| $\Psi(s,t)$ | $F(x,y)$ |
|---|---|
| $\dfrac{\Gamma(\alpha+s+t)\,\Gamma(\beta+s)\,\Gamma(\beta'+t)}{\Gamma(\beta')\,\Gamma(\gamma+s+t)}$ | $F_1\,(\alpha,\beta,\beta',\gamma;x,y)$ |
| $\dfrac{\Gamma(\alpha+s+t)\,\Gamma(\beta+s)\,\Gamma(\beta'+t)\,\Gamma(\gamma')}{\Gamma(\beta')\,\Gamma(\gamma+s)\,\Gamma(\gamma'+t)}$ | $F_2\,(\alpha,\beta,\beta',\gamma,\gamma';x,y)$ |
| $\dfrac{\Gamma(\alpha+s)\,\Gamma(\alpha'+t)\,\Gamma(\beta+s)\,\Gamma(\beta'+t)}{\Gamma(\alpha')\,\Gamma(\beta')\,\Gamma(\gamma+s+t)}$ | $F_3\,(\alpha,\alpha',\beta,\beta',\gamma;x,y)$ |
| $\dfrac{\Gamma(\alpha+s+t)\,\Gamma(\beta+s+t)\,\Gamma(\gamma')}{\Gamma(\gamma+s)\,\Gamma(\gamma'+t)}$ | $F_4\,(\alpha,\beta,\gamma,\gamma';x,y)$ |
| $[\alpha, \alpha', \beta, \beta'$ may not be negative integers$]$ | EH I 232(9–13), AK 41(33) |

## 9.19  A hypergeometric function of several variables

$F_A\,(\alpha;\beta_1,\ldots,\beta_n;\gamma_1,\ldots,\gamma_n;z_1,\ldots,z_n)$

$$= \sum_{m_1=0}^{\infty} \sum_{m_2=0}^{\infty} \cdots \sum_{m_n=0}^{\infty} \frac{(\alpha)_{m_1+\cdots+m_n}\,(\beta_1)_{m_1}\cdots(\beta_n)_{m_n}}{(\gamma_1)_{m_1}\cdots(\gamma_n)_{m_n}\,m_1!\cdots m_n!}\,z_1^{m_1} z_2^{m_2}\cdots z_n^{m_n}$$

FT I 385

# 9.2 Confluent Hypergeometric Functions

## 9.20  Introduction

**9.201**[10]   A *confluent hypergeometric function* is obtained by taking the limit as $c \to \infty$ in the solution of Riemann's differential equation

$$u = P \left\{ \begin{matrix} 0 & \infty & c & \\ \frac{1}{2}+\mu & -c & c-\lambda & z \\ \frac{1}{2}-\mu & 0 & \lambda & \end{matrix} \right\}$$

WH

**9.202**   The equation obtained by means of this limiting process is of the form

1.   $\dfrac{\mathrm{d}^2 u}{\mathrm{d}z^2} + \dfrac{\mathrm{d}u}{\mathrm{d}z} + \left( \dfrac{\lambda}{z} + \dfrac{\frac{1}{4}-\mu^2}{z^2} \right) u = 0$

WH

Equation **9.202** 1 has the following two linearly independent solutions:

2.   $z^{\frac{1}{2}+\mu} e^{-z}\, \Phi\left(\frac{1}{2}+\mu-\lambda, 2\mu+1; z\right)$

3.   $z^{\frac{1}{2}-\mu} e^{-z}\, \Phi\left(\frac{1}{2}-\mu-\lambda, -2\mu+1; z\right)$

which are defined for all values of $\mu \neq \pm\frac{1}{2}, \pm\frac{2}{2}, \pm\frac{3}{2}, \ldots$

MO 111

## 9.21  The functions $\Phi(\alpha, \gamma; z)$ and $\Psi(\alpha, \gamma; z)$

**9.210**[10]   The series

1.    $$\Phi(\alpha, \gamma; z) = 1 + \frac{\alpha}{\gamma} \frac{z}{1!} + \frac{\alpha(\alpha+1)}{\gamma(\gamma+1)} \frac{z^2}{2!} + \frac{\alpha(\alpha+1)(\alpha+2)}{\gamma(\gamma+1)(\gamma+2)} \frac{z^3}{3!} + \cdots$$

is also called a *confluent hypergeometric function*.

A second notation: $\Phi(\alpha, \gamma; z) = {}_1F_1(\alpha; \gamma; z)$.

2.    $$\Psi(\alpha, \gamma; z) = \frac{\Gamma(1-\gamma)}{\Gamma(\alpha-\gamma+1)} \Phi(\alpha, \gamma; z) + \frac{\Gamma(\gamma-1)}{\Gamma(\alpha)} z^{1-\gamma} \Phi(\alpha-\gamma+1, 2-\gamma; z) \qquad \text{EH I 257(7)}$$

3.    Bateman's function $k_\nu(x)$ is defined by

$$k_\nu(x) = \frac{2}{\pi} \int_0^{\pi/2} \cos(x \tan\theta - \nu\theta)\, d\theta \qquad [x, \nu \text{ real}] \qquad \text{EH I 267}$$

**9.211**   Integral representation:

1.    $$\Phi(\alpha, \gamma; z) = \frac{2^{1-\gamma} e^{\frac{1}{2}z}}{B(\alpha, \gamma-\alpha)} \int_{-1}^{1} (1-t)^{\gamma-\alpha-1}(1+t)^{\alpha-1} e^{\frac{1}{2}zt}\, dt$$

$$[0 < \operatorname{Re}\alpha < \operatorname{Re}\gamma] \qquad \text{MO 114}$$

2.    $$\Phi(\alpha, \gamma; z) = \frac{1}{B(\alpha, \gamma-\alpha)} z^{1-\gamma} \int_0^z e^t t^{\alpha-1}(z-t)^{\gamma-\alpha-1}\, dt$$

$$[0 < \operatorname{Re}\alpha < \operatorname{Re}\gamma] \qquad \text{MO 114}$$

3.    $$\Phi(-\nu, \alpha+1; z) = \frac{\Gamma(\alpha+1)}{\Gamma(\alpha+\nu+1)} e^z z^{-\frac{\alpha}{2}} \int_0^\infty e^{-t} t^{\nu+\frac{\alpha}{2}} J_\alpha\left(2\sqrt{zt}\right) dt$$

$$\left[\operatorname{Re}(\alpha+\nu+1) > 0, \quad |\arg z| < \tfrac{\pi}{2}\right]$$
$$\text{MO 115}$$

4.[8]    $$\Psi(\alpha, \gamma; z) = \frac{1}{\Gamma(\alpha)} \int_0^\infty e^{-zt} t^{\alpha-1}(1+t)^{\gamma-\alpha-1}\, dt \qquad [\operatorname{Re}\alpha > 0, \quad \operatorname{Re}z > 0] \qquad \text{EH I 255(2)}$$

### Functional relations

**9.212**

1.    $$\Phi(\alpha, \gamma; z) = e^z \Phi(\gamma-\alpha, \gamma; -z) \qquad\qquad \text{MO 112}$$

2.    $$\frac{z}{\gamma} \Phi(\alpha+1, \gamma+1; z) = \Phi(\alpha+1, \gamma; z) - \Phi(\alpha, \gamma; z) \qquad\qquad \text{MO 112}$$

3.    $$\alpha \Phi(\alpha+1, \gamma+1; z) = (\alpha-\gamma) \Phi(\alpha, \gamma+1; z) + \gamma \Phi(\alpha, \gamma; z) \qquad\qquad \text{MO 112}$$

4.    $$\alpha \Phi(\alpha+1, \gamma; z) = (z+2a-\gamma) \Phi(\alpha, \gamma; z) + (\gamma-\alpha) \Phi(\alpha-1, \gamma; z) \qquad\qquad \text{MO 112}$$

5.*    $$\Psi(\alpha, \gamma; z) = z^{1-\gamma} \Psi(\alpha-\gamma+1, 2-\gamma; z)$$

6.*    $$z\Psi(\alpha+1, \gamma+1; z) = (\gamma-\alpha-1)\Psi(\alpha+1, \gamma; z) + \Psi(\alpha, \gamma; z)$$

7.*    $$\alpha\Psi(\alpha+1, \gamma+1; z) = \Psi(\alpha, \gamma+1; z) - \Psi(\alpha, \gamma; z)$$

**9.213**

1.[12]    $\dfrac{\mathrm{d}\Phi(\alpha,\gamma;z)}{\mathrm{d}z} = \dfrac{\alpha}{\gamma}\,\Phi(\alpha+1,\gamma+1;z)$    MO 112

2.*    $\dfrac{\mathrm{d}\Psi(\alpha,\gamma;z)}{\mathrm{d}z} = -\alpha\Psi(\alpha+1,\gamma+1;z)$

**9.214**    $\displaystyle\lim_{\gamma\to -n}\dfrac{1}{\Gamma(\gamma)}\,\Phi(\alpha,\gamma;z) = z^{n+1}\binom{\alpha+n}{n+1}\Phi(\alpha+n+1,n+2;z)$     $[n=0,1,2,\ldots]$

MO 112

**9.215**[10]

1.    $\Phi(\alpha,\alpha;z) = e^{z}$    MO 15

2.    $\Phi(\alpha,2\alpha;2z) = 2^{\alpha-\frac{1}{2}}\exp\left[\tfrac{1}{4}(1-2\alpha)\pi i\right]\Gamma\left(\alpha+\tfrac{1}{2}\right)e^{z}z^{\frac{1}{2}-\alpha}J_{\alpha-\frac{1}{2}}\left(ze^{\frac{\pi}{2}i}\right)$    MO 112

3.    $\Phi\left(p+\tfrac{1}{2},2p+1;2iz\right) = \Gamma(p+1)\left(\dfrac{z}{2}\right)^{-p}e^{iz}J_{p}(z)$    MO 15

4.*    $\Psi(\alpha,\alpha+1;z) = z^{-\alpha}$

For a representation of special functions in terms of a confluent hypergeometric function $\Phi(\alpha,\gamma;z)$, see:

- for the probability integral, **9.236**;
- for integrals of Bessel functions, **6.631** 1;
- for Hermite polynomials, **8.953** and **8.959**;
- for Laguerre polynomials, **8.972** 1;
- for parabolic cylinder functions, **9.240**;
- for the Whittaker functions $M_{\lambda,\mu}(z)$, **9.220** 2 and **9.220** 3.

**9.216**    The function $\Phi(\alpha,\gamma;z)$ is a solution of the differential equation

1.    $z\dfrac{\mathrm{d}^{2}F}{\mathrm{d}z^{2}} + (\gamma-z)\dfrac{\mathrm{d}F}{\mathrm{d}z} - \alpha F = 0$    MO 111

This equation has two linearly independent solutions:

2.    $\Phi(\alpha,\gamma;z)$

3.    $z^{1-\gamma}\,\Phi(\alpha-\gamma+1,2-\gamma;z)$    MO 112

Or alternatively $\Phi(\alpha,\gamma;z)$ and $\Psi(\alpha,\gamma;z)$.

## 9.22–9.23  The Whittaker functions $M_{\lambda,\mu}(z)$ and $W_{\lambda,\mu}(z)$

**9.220**    If we make the change of variable $u = e^{-\frac{z}{2}}W$ in equation **9.202** 1, we obtain the equation

1.    $\dfrac{\mathrm{d}^{2}W}{\mathrm{d}z^{2}} + \left(-\dfrac{1}{4} + \dfrac{\lambda}{z} + \dfrac{\frac{1}{4}-\mu^{2}}{z^{2}}\right)W = 0$    MO 115

Equation **9.220** 1 has the following two linearly independent solutions:

2.    $M_{\lambda,\mu}(z) = z^{\mu+\frac{1}{2}}e^{-z/2}\,\Phi\left(\mu-\lambda+\tfrac{1}{2},2\mu+1;z\right)$

3.[11]    $M_{\lambda,-\mu}(z) = z^{-\mu+\frac{1}{2}}e^{-z/2}\,\Phi\left(-\mu-\lambda+\tfrac{1}{2},-2\mu+1;z\right)$    MO 115

To obtain solutions that are also suitable for $2\mu = \pm1,\pm2,\ldots$, we introduce Whittaker's function

4.$^{12}$     $W_{\lambda,\mu}(z) = z^{\mu+\frac{1}{2}} e^{-z/2} \Psi\left(\mu - \lambda + \frac{1}{2}, 2\mu + 1; z\right)$

$$= \frac{\Gamma(-2\mu)}{\Gamma\left(\frac{1}{2} - \mu - \lambda\right)} M_{\lambda,\mu}(z) + \frac{\Gamma(2\mu)}{\Gamma\left(\frac{1}{2} + \mu - \lambda\right)} M_{\lambda,-\mu}(z) \qquad \text{WH}$$

which, for $2\mu$ approaching an integer, is also a solution of equation **9.220** 1.

For the functions $M_{\lambda,\mu}(z)$ and $W_{\lambda,\mu}(z)$, $z = 0$ is a branch point and $z = \infty$ is an essential singular point. Therefore, we shall examine these functions only for $|\arg z| < \pi$.

These functions $W_{\lambda,\mu}(z)$ and $W_{-\lambda,\mu}(-z)$ are linearly independent solutions of equation **9.220** 1.

**Integral representations**

**9.221**     $M_{\lambda,\mu}(z) = \dfrac{z^{\mu+\frac{1}{2}}}{2^{2\mu}\, \mathrm{B}\left(\mu + \lambda + \frac{1}{2}, \mu - \lambda + \frac{1}{2}\right)} \displaystyle\int_{-1}^{1} (1+t)^{\mu-\lambda-\frac{1}{2}} (1-t)^{\mu+\lambda-\frac{1}{2}} e^{\frac{1}{2}zt}\, dt,$     WH

if the integral converges. See also **6.631** 1 and **7.623** 3.

**9.222**

1.$^{11}$     $W_{\lambda,\mu}(z) = \dfrac{z^{\mu+\frac{1}{2}} e^{-z/2}}{\Gamma\left(\mu - \lambda + \frac{1}{2}\right)} \displaystyle\int_{0}^{\infty} e^{-zt} t^{\mu-\lambda-\frac{1}{2}} (1+t)^{\mu+\lambda-\frac{1}{2}}\, dt$

$$\left[\mathrm{Re}(\mu - \lambda) > -\tfrac{1}{2}, \quad |\arg z| < \tfrac{\pi}{2}\right]$$

MO 118

2.     $W_{\lambda,\mu}(z) = \dfrac{z^{\lambda} e^{-z/2}}{\Gamma\left(\mu - \lambda + \frac{1}{2}\right)} \displaystyle\int_{0}^{\infty} t^{\mu-\lambda-\frac{1}{2}} e^{-t} \left(1 + \dfrac{t}{z}\right)^{\mu+\lambda-\frac{1}{2}}\, dt$

$$\left[\mathrm{Re}(\mu - \lambda) > -\tfrac{1}{2}, \quad |\arg z| < \pi\right] \qquad \text{WH}$$

**9.223**     $W_{\lambda,\mu}(z) = \dfrac{e^{-\frac{z}{2}}}{2\pi i} \displaystyle\int_{-i\infty}^{i\infty} \dfrac{\Gamma(u - \lambda)\, \Gamma\left(-u - \mu + \frac{1}{2}\right) \Gamma\left(-u + \mu + \frac{1}{2}\right)}{\Gamma\left(-\lambda + \mu + \frac{1}{2}\right) \Gamma\left(-\lambda - \mu + \frac{1}{2}\right)} z^{u}\, du$

[the path of integration is chosen in such a way that the poles of the function $\Gamma(u - \lambda)$ are separated from the poles of the functions $\Gamma\left(-u - \mu + \frac{1}{2}\right)$ and $\Gamma\left(-u + \mu + \frac{1}{2}\right)$]. See also **7.142**.     MO 118

**9.224**     $W_{\mu,\frac{1}{2}+\mu}(z) = z^{\mu+1} e^{-\frac{1}{2}z} \displaystyle\int_{0}^{\infty} (1+t)^{2\mu} e^{-zt}\, dt = z^{-\mu} e^{\frac{1}{2}z} \int_{z}^{\infty} t^{2\mu} e^{-t}\, dt$     [$\mathrm{Re}\, z > 0$]     WH

**9.225**

1.     $W_{\lambda,\mu}(x)\, W_{-\lambda,\mu}(x) = -x \displaystyle\int_{0}^{\infty} \tanh^{2\lambda} \dfrac{t}{2} \left\{ J_{2\mu}\left(x \sinh t\right) \sin(\mu - \lambda)\pi \right.$

$$\left. + \, Y_{2\mu}\left(x \sinh t\right) \cos(\mu - \lambda)\pi \right\}\, dt$$

$$\left[|\mathrm{Re}\,\mu| - \mathrm{Re}\,\lambda < \tfrac{1}{2}; \quad x > 0\right] \qquad \text{MO 119}$$

2.     $W_{\kappa,\mu}(z_1)\, W_{\lambda,\mu}(z_2) = \dfrac{(z_1 z_2)^{\mu+\frac{1}{2}} \exp\left[-\frac{1}{2}(z_1 + z_2)\right]}{\Gamma(1 - \kappa - \lambda)}$

$$\times \int_{0}^{\infty} e^{-t} t^{-\kappa-\lambda} (z_1 + t)^{-\frac{1}{2}+\kappa-\mu} (z_2 + t)^{-\frac{1}{2}+\lambda-\mu}$$

$$\times F\left(\tfrac{1}{2} - \kappa + \mu, \tfrac{1}{2} - \lambda + \mu; 1 - \kappa - \lambda; \Theta\right)\, dt$$

$$\Theta = \dfrac{t(z_1 + z_2 + t)}{(z_1 + t)(z_2 + t)}, \qquad \left[z_1 \neq 0, \quad z_2 \neq 0, \quad |\arg z_1| < \pi, \quad |\arg z_2| < \pi, \quad \mathrm{Re}(\kappa + \lambda) < 1\right]$$

MO 119

See also **3.334**, **3.381** 6, **3.382** 3, **3.383** 4, 8, **3.384** 3, **3.471** 2.

**9.226**   Series representations

$$M_{0,\mu}(z) = z^{\frac{1}{2}+\mu}\left\{1 + \sum_{k=1}^{\infty} \frac{z^{2k}}{2^{4k}k!(\mu+1)(\mu+2)\dots(\mu+k)}\right\} \qquad \text{WH}$$

### Asymptotic representations

**9.227**[7]   For large values of $|z|$

$$W_{\lambda,\mu}(z) \sim e^{-z/2}z^{\lambda}\left(1 + \sum_{k=1}^{\infty}\frac{\left[\mu^2 - \left(\lambda-\frac{1}{2}\right)^2\right]\left[\mu^2 - \left(\lambda-\frac{3}{2}\right)^2\right]\dots\left[\mu^2 - \left(\lambda-k+\frac{1}{2}\right)^2\right]}{k!z^k}\right)$$

$$[|\arg z| \leq \pi - \alpha < \pi] \qquad \text{WH}$$

**9.228**   For large values of $|\lambda|$

$$M_{\lambda,\mu}(z) \sim \frac{1}{\sqrt{\pi}}\Gamma(2\mu+1)\lambda^{-\mu-\frac{1}{4}}z^{1/4}\cos\left(2\sqrt{\lambda z} - \mu\pi - \frac{1}{4}\pi\right) \qquad \text{MO 118}$$

**9.229**

1.    $W_{\lambda,\mu} \sim -\left(\dfrac{4z}{\lambda}\right)^{\frac{1}{4}}e^{-\lambda+\lambda\ln\lambda}\sin\left(2\sqrt{\lambda z} - \lambda\pi - \dfrac{\pi}{4}\right)$         MO 118

2.    $W_{-\lambda,\mu} \sim \left(\dfrac{z}{4\lambda}\right)^{\frac{1}{4}}e^{\lambda-\lambda\ln\lambda-2\sqrt{\lambda z}}$         MO 118

Formulas **9.228** and **9.229** are applicable for

$$|\lambda| \gg 1, \quad |\lambda| \gg |z|, \quad |\lambda| \gg |\mu|, \quad z \neq 0, \quad |\arg\sqrt{z}| < \tfrac{3\pi}{4} \text{ and } |\arg\lambda| < \tfrac{\pi}{2}. \qquad \text{MO 118}$$

### Functional relations

**9.231**

1.    $M_{n+\mu+\frac{1}{2},\mu}(z) = \dfrac{z^{\frac{1}{2}-\mu}e^{\frac{1}{2}z}}{(2\mu+1)(2\mu+2)\dots(2\mu+n)}\dfrac{\mathrm{d}^n}{\mathrm{d}z^n}\left(z^{n+2\mu}e^{-z}\right)$

$$[n = 0, 1, 2, \dots; \quad 2\mu \neq -1, -2, -3, \dots]$$
$$\text{MO 117}$$

2.    $z^{-\frac{1}{2}-\mu}M_{\lambda,\mu}(z) = (-z)^{-\frac{1}{2}-\mu}M_{-\lambda,\mu}(-z)$     $[2\mu \neq -1, -2, -3, \dots]$     WH

**9.232**

1.    $W_{\lambda,\mu}(z) = W_{\lambda,-\mu}(z)$         MO 116

**9.233**

1.    $M_{\lambda,\mu}(z) = \dfrac{\Gamma(2\mu+1)}{\Gamma\left(\mu-\lambda+\frac{1}{2}\right)}e^{i\pi\lambda}\,W_{-\lambda,\mu}\left(e^{i\pi}z\right) + \dfrac{\Gamma(2\mu+1)}{\Gamma\left(\mu+\lambda+\frac{1}{2}\right)}\exp\left[i\pi\left(\lambda-\mu-\frac{1}{2}\right)\right]W_{\lambda,\mu}(z)$

$$\left[-\tfrac{3}{2}\pi < \arg z < \tfrac{1}{2}\pi; \quad 2\mu \neq -1, -2, \dots\right]$$
$$\text{MO 117}$$

2.    $M_{\lambda,\mu}(z) = \dfrac{\Gamma(2\mu+1)}{\Gamma\left(\mu-\lambda+\frac{1}{2}\right)}e^{-i\pi\lambda}\,W_{-\lambda,\mu}\left(e^{-i\pi}z\right) + \dfrac{\Gamma(2\mu+1)}{\Gamma\left(\mu+\lambda+\frac{1}{2}\right)}\exp\left[-i\pi\left(\lambda-\mu-\frac{1}{2}\right)\right]W_{\lambda,\mu}(z)$

$$\left[-\tfrac{1}{2}\pi < \arg z < \tfrac{3}{2}\pi; \quad 2\mu \neq -1, -2, \dots\right]$$
$$\text{MO 117}$$

**9.234** Recursion formulas

1.
$$W_{\lambda,\mu}(z) = \sqrt{z}\, W_{\lambda-\frac{1}{2},\mu-\frac{1}{2}}(z) + \left(\tfrac{1}{2} + \mu - \lambda\right) W_{\lambda-1,\mu}(z) \qquad \text{WH}$$

2.[11]
$$W_{\lambda,\mu}(z) = \sqrt{z}\, W_{\lambda-\frac{1}{2},\mu+\frac{1}{2}}(z) + \left(\tfrac{1}{2} - \mu - \lambda\right) W_{\lambda-1,\mu}(z) \qquad \text{WH}$$

3.
$$z\frac{\mathrm{d}}{\mathrm{d}z} W_{\lambda,\mu}(z) = \left(\lambda - \tfrac{1}{2}z\right) W_{\lambda,\mu}(z) - \left[\mu^2 - \left(\lambda - \tfrac{1}{2}\right)^2\right] W_{\lambda-1,\mu}(z) \qquad \text{WH}$$

4.
$$\left[\left(\mu + \frac{1-z}{2}\right) W_{\lambda,\mu}(z) - z\frac{\mathrm{d}}{\mathrm{d}z} W_{\lambda,\mu}(z)\right]\left(\mu + \tfrac{1}{2} + \lambda\right)$$
$$= \left[\left(\mu + \frac{1+z}{2}\right) W_{\lambda,\mu+1}(z) + z\frac{\mathrm{d}}{\mathrm{d}z} W_{\lambda,\mu+1}(z)\right]\left(\mu + \tfrac{1}{2} - \lambda\right)$$
$$\text{MO 117}$$

5.
$$\left(\tfrac{3}{2} + \lambda + \mu\right)\left(\tfrac{1}{2} + \lambda + \mu\right) z\, W_{\lambda,\mu}(z) = z(z + 2\mu + 1)\frac{\mathrm{d}}{\mathrm{d}z} W_{\lambda+1,\mu+1}(z)$$
$$+ \left[\tfrac{1}{2}z^2 + \left(\mu - \lambda - \tfrac{1}{2}\right)z + 2\mu^2 + 2\mu + \tfrac{1}{2}\right] W_{\lambda+1,\mu+1}(z)$$
$$\text{MO 117}$$

6.*
$$\left(\frac{1}{2} - \lambda - \mu\right) M_{\lambda,\mu}(z) = -2\mu\sqrt{z} M_{\lambda-\frac{1}{2},\mu+\frac{1}{2}}(z) + \left(\frac{1}{2} + \mu - \lambda\right) M_{\lambda-1,\mu}(z)$$

7.*
$$(2\mu + 1) M_{\lambda,\mu}(z) = -\sqrt{z} M_{\lambda-\frac{1}{2},\mu+\frac{1}{2}}(z) + (2\mu + 1) M_{\lambda-1,\mu}(z)$$

8.*
$$z\frac{\mathrm{d}M_{\lambda,\mu}(z)}{\mathrm{d}z} = \left(\lambda - \frac{1}{2}z\right) M_{\lambda,\mu}(z) + \left(\mu - \lambda + \frac{1}{2}\right) M_{\lambda-1,\mu}(z)$$

## Connections with other functions

**9.235**

1.
$$M_{0,\mu}(z) = 2^{2\mu}\,\Gamma(\mu + 1)\sqrt{z}\, I_\mu\left(\frac{z}{2}\right) \qquad \text{MO 125a}$$

2.
$$W_{0,\mu}(z) = \sqrt{\frac{z}{\pi}}\, K_\mu\left(\frac{z}{2}\right) \qquad \text{MO 125}$$

**9.236**

1.[12]
$$\Phi(x) = 1 - \frac{e^{-\frac{x^2}{2}}}{\sqrt{\pi x}} W_{-\frac{1}{4},\frac{1}{4}}\left(x^2\right) = \frac{2x}{\sqrt{\pi}}\, \Phi\left(\tfrac{1}{2}, \tfrac{3}{2}; -x^2\right) \qquad \text{WH, MO 126}$$

2.[12]
$$\mathrm{li}(z) = -\frac{\sqrt{z}}{\sqrt{\ln\frac{1}{z}}} W_{-\frac{1}{2},0}\left(-\ln z\right) \qquad \text{WH}$$

3.
$$\Gamma(\alpha, x) = e^{-x}\, \Psi(1 - \alpha, 1 - \alpha; x) \qquad \text{EH I 266(21)}$$

4.
$$\gamma(\alpha, x) = \frac{x^\alpha}{\alpha}\, \Phi(\alpha, \alpha + 1; -x) \qquad \text{EH I 266(22)}$$

**9.237**

1. $$W_{\lambda,\mu}(z) = \frac{(-1)^{2\mu} z^{\mu+\frac{1}{2}} e^{-\frac{1}{2}z}}{\Gamma\left(\frac{1}{2} - \mu - \lambda\right) \Gamma\left(\frac{1}{2} + \mu - \lambda\right)}$$
$$\times \left\{ \sum_{k=0}^{\infty} \frac{\Gamma\left(\mu + k - \lambda + \frac{1}{2}\right)}{k!(2\mu+k)!} z^k \left[\Psi(k+1) + \Psi(2\mu+k+1) - \Psi\left(\mu+k-\lambda+\frac{1}{2}\right) - \ln z\right] \right.$$
$$\left. + (-z)^{-2\mu} \sum_{k=0}^{2\mu-1} \frac{\Gamma(2\mu-k)\Gamma\left(k-\mu-\lambda+\frac{1}{2}\right)}{k!} (-z)^k \right\}$$
$$\left[|\arg z| < \tfrac{3}{2}\pi; \quad 2\mu+1 \text{ is a natural number}\right] \quad \text{MO 116}$$

2. Set $\lambda - \mu - \frac{1}{2} = l$, where $l+1$ is a natural number. Then

3. $$W_{l+\mu+\frac{1}{2},\mu}(z) = (-1)^l z^{\mu+\frac{1}{2}} e^{-\frac{1}{2}z}(2\mu+1)(2\mu+2)\cdots(2\mu+l)\,\Phi(-l, 2\mu+1; z)$$
$$= (-1)^l z^{\mu+\frac{1}{2}} e^{-\frac{1}{2}z} L_l^{2\mu}(z)$$

MO 116

**9.238**

1. $$J_\nu(x) = \frac{2^{-\nu}}{\Gamma(\nu+1)} x^\nu e^{-ix}\, \Phi\left(\tfrac{1}{2}+\nu, 1+2\nu; 2ix\right)$$ EH I 265(9)

2. $$I_\nu(x) = \frac{2^{-\nu}}{\Gamma(\nu+1)} x^\nu e^{-x}\, \Phi\left(\tfrac{1}{2}+\nu, 1+2\nu; 2x\right)$$ EH I 265(10)

3. $$K_\nu(x) = \sqrt{\pi} e^{-x}(2x)^\nu\, \Psi\left(\tfrac{1}{2}+\nu, 1+2\nu; 2x\right)$$ EH I 265(13)

## 9.24–9.25 Parabolic cylinder functions $D_p(z)$

**9.240**[12] $D_p(z) = 2^{p/2} e^{-z^2/4} \Psi\left(-\dfrac{p}{2}, \dfrac{1}{2}; \dfrac{z^2}{2}\right) = 2^{\frac{1}{4}+\frac{p}{2}} W_{\frac{1}{4}+\frac{p}{2}, -\frac{1}{4}}\left(\dfrac{z^2}{2}\right) z^{-1/2}$

$$= 2^{\frac{p}{2}} e^{-\frac{z^2}{4}} \left\{ \frac{\sqrt{\pi}}{\Gamma\left(\dfrac{1-p}{2}\right)} \Phi\left(-\dfrac{p}{2}, \dfrac{1}{2}; \dfrac{z^2}{2}\right) - \frac{\sqrt{2\pi}\,z}{\Gamma\left(-\dfrac{p}{2}\right)} \Phi\left(\dfrac{1-p}{2}, \dfrac{3}{2}; \dfrac{z^2}{2}\right) \right\}$$

MO 120a

are called *parabolic cylinder functions*.

**Integral representations**

**9.241**

1. $$D_p(z) = \frac{1}{\sqrt{\pi}} 2^{p+\frac{1}{2}} e^{-\frac{\pi}{2}pi} e^{\frac{z^2}{4}} \int_{-\infty}^{\infty} x^p e^{-2x^2+2ixz}\, dx \qquad [\operatorname{Re} p > -1; \quad \text{for } x < 0, \quad \arg x^p = p\pi i]$$

MO 122

2. $$D_p(z) = \frac{e^{-\frac{z^2}{4}}}{\Gamma(-p)} \int_0^{\infty} e^{-xz-\frac{x^2}{2}} x^{-p-1}\, dx \qquad [\operatorname{Re} p < 0] \qquad (\text{cf. } \mathbf{3.462}\ 1) \qquad \text{MO 122}$$

**9.242**

1.¹⁰  $$D_p(z) = -\frac{\Gamma(p+1)}{2\pi i} e^{-\frac{1}{4}z^2} \int_\infty^{(0+)} e^{-zt-\frac{1}{2}t^2}(-t)^{-p-1}\,dt \qquad [|\arg(-t)| \le \pi] \qquad \text{WH}$$

2.  $$D_p(z) = 2^{\frac{1}{2}(p-1)}\frac{\Gamma\left(\frac{p}{2}+1\right)}{i\pi} \int_{-\infty}^{(-1+)} e^{\frac{1}{4}z^2 t}(1+t)^{-\frac{1}{2}p-1}(1-t)^{\frac{1}{2}(p-1)}\,dt$$

$$[|\arg z| < \tfrac{\pi}{4};\quad |\arg(1+t)| \le \pi] \qquad \text{WH}$$

3.  $$D_p(z) = \frac{1}{2\pi i} e^{-\frac{1}{4}z^2} \int_{-\infty i}^{\infty i} \frac{\Gamma\left(\frac{1}{2}t - \frac{1}{2}p\right)\Gamma(-t)}{\Gamma(-p)} \left(\sqrt{2}\right)^{t-p-2} z^t\,dt$$

$$[|\arg z| < \tfrac{3}{4}\pi;\quad p \text{ is not a positive integer}] \qquad \text{WH}$$

4.  $$D_p(z) = \frac{1}{2\pi i} e^{-\frac{1}{4}z^2} \int_\infty^{(0-)} \frac{\Gamma\left(\frac{1}{2}t - \frac{1}{2}p\right)\Gamma(-t)}{\Gamma(-p)} \left(\sqrt{2}\right)^{t-p-2} z^t\,dt$$

[for all values of arg $z$; also, the contours encircle the poles of the function $\Gamma(-t)$ but they do not encircle the poles of the function $\Gamma\left(\frac{1}{2}t - \frac{1}{2}p\right)$].                      WH

**9.243**

1.  $$D_n(z) = (-1)^\mu \left(\frac{\pi}{2}\right)^{-1/2} \left(\sqrt{n}\right)^{n+1} e^{\frac{1}{4}z^2 - \frac{1}{2}n} \left\{ \int_{-\infty}^\infty e^{-n(t-1)^2} \frac{\cos}{\sin}\left(zt\sqrt{n}\right)\,dt \right.$$

$$\left. + \int_0^\infty \left[e^{\frac{1}{2}n\left(1-t^2\right)} t^n - e^{-n(t-1)^2}\right] \frac{\cos}{\sin}\left(zt\sqrt{n}\right)\,dt - \int_{-\infty}^0 e^{-n(t-1)^2} \frac{\cos}{\sin}\left(zt\sqrt{n}\right)\,dt \right\}$$

[$n$ is a natural number]                      WH

2.  $$D_n(z) = (-1)^\mu 2^{n+2}(2\pi)^{-1/2} e^{\frac{1}{4}z^2} \int_0^\infty t^n e^{-2t^2} \frac{\cos}{\sin}(2zt)\,dt$$

[$n$ is a natural number, $\mu = \left\lfloor \frac{n}{2} \right\rfloor$, and the cosine or sine is chosen according as $n$ is even or odd]

WH

**9.244**

1.  $$D_{-p-1}[(1+i)z] = \frac{e^{-\frac{iz^2}{2}}}{2^{\frac{p-1}{2}}\Gamma\left(\frac{p+1}{2}\right)} \int_0^\infty \frac{e^{-ix^2z^2}x^p}{(1+x^2)^{1+\frac{p}{2}}}\,dx \qquad [\operatorname{Re}p > -1,\quad \operatorname{Re}\left(iz^2\right) \ge 0] \qquad \text{MO 122}$$

2.  $$D_p[(1+i)z] = \frac{2^{\frac{p+1}{2}}}{\Gamma\left(-\frac{p}{2}\right)} \int_1^\infty e^{-\frac{i}{2}z^2 x}\frac{(x+1)^{\frac{p-1}{2}}}{(x-1)^{1+\frac{p}{2}}}\,dx \qquad [\operatorname{Re}p < 0;\quad \operatorname{Re}\left(iz^2\right) \ge 0] \qquad \text{MO 122}$$

See also **3.383** 6, 7, **3.384** 2, 6, **3.966** 5, 6.

**9.245**

1.¹⁰  $$D_p(x)\,D_{-p-1}(x) = -\frac{1}{\sqrt{\pi}} \int_0^\infty \coth^{p+\frac{1}{2}}\left(\frac{t}{2}\right) \frac{1}{\sqrt{\sinh t}} \sin\left(\frac{x^2\sinh t + p\pi}{2}\right)\,dt$$

[$x$ is real,   $\operatorname{Re}p < 0$]                      MO 122

2.     $D_p\left(ze^{\frac{\pi}{4}i}\right) D_p\left(ze^{-\frac{\pi}{4}i}\right) = \dfrac{1}{\Gamma(-p)} \displaystyle\int_0^\infty \coth^p t \exp\left(-\dfrac{z^2}{2}\sinh 2t\right) \dfrac{dt}{\sinh t}$

$$\left[|\arg z| < \tfrac{\pi}{4}; \quad \operatorname{Re} p < 0\right] \qquad \text{MO 122}$$

See also **6.613**.

**9.246**    Asymptotic expansions. If $|z| \gg 1$ and $|z| \gg |p|$, then

1.     $D_p(z) \sim e^{-\frac{z^2}{4}} z^p \left(1 - \dfrac{p(p-1)}{2z^2} + \dfrac{p(p-1)(p-2)(p-3)}{2\cdot 4z^4} - \cdots\right)$

$$\left[|\arg z| < \tfrac{3}{4}\pi\right] \qquad \text{MO 121}$$

2.[11]   $D_p(z) \sim e^{-z^2/4} z^p \left(1 - \dfrac{p(p-1)}{2z^2} + \dfrac{p(p-1)(p-2)(p-3)}{2\cdot 4z^4} - \cdots\right)$

$$-\dfrac{\sqrt{2\pi}}{\Gamma(-p)} e^{p\pi i} e^{z^2/4} z^{-p-1} \left(1 + \dfrac{(p+1)(p+2)}{2z^2} + \dfrac{(p+1)(p+2)(p+3)(p+4)}{2\cdot 4z^4} + \cdots\right)$$

$$\left[\tfrac{1}{4}\pi < \arg z < \tfrac{5}{4}\pi\right] \qquad \text{MO 121}$$

3.[11]   $D_p(z) \sim e^{-z^2/4} z^p \left(1 - \dfrac{p(p-1)}{2z^2} + \dfrac{p(p-1)(p-2)(p-3)}{2\cdot 4z^4} - \cdots\right)$

$$-\dfrac{\sqrt{2\pi}}{\Gamma(-p)} e^{-p\pi i} e^{z^2/4} z^{-p-1} \left(1 + \dfrac{(p+1)(p+2)}{2z^2} + \dfrac{(p+1)(p+2)(p+3)(p+4)}{2\cdot 4z^4} + \cdots\right)$$

$$\left[-\tfrac{1}{4}\pi > \arg z > -\tfrac{5}{4}\pi\right] \qquad \text{MO 121}$$

**Functional relations**

**9.247**    Recursion formulas:

1.     $D_{p+1}(z) - z\, D_p(z) + p\, D_{p-1}(z) = 0$         WH

2.     $\dfrac{d}{dz} D_p(z) + \dfrac{1}{2} z\, D_p(z) - p\, D_{p-1}(z) = 0$      WH

3.     $\dfrac{d}{dz} D_p(z) - \dfrac{1}{2} z\, D_p(z) + D_{p+1}(z) = 0$      MO 121

**9.248**    Linear relations:

1.[12]   $D_p(z) = \dfrac{\Gamma(p+1)}{\sqrt{2\pi}} \left[e^{\pi pi/2} D_{-p-1}(iz) + e^{-\pi pi/2} D_{-p-1}(-iz)\right]$

$$= e^{-p\pi i} D_p(-z) + \dfrac{\sqrt{2\pi}}{\Gamma(-p)} e^{-\pi(p+1)i/2} D_{-p-1}(iz)$$

$$= e^{p\pi i} D_p(-z) + \dfrac{\sqrt{2\pi}}{\Gamma(-p)} e^{\pi(p+1)i/2} D_{-p-1}(-iz)$$

$$\text{MO 121}$$

**9.249**[10]  $D_p[(1+i)x] + D_p[-(1+i)x] = \dfrac{2^{1+p/2}}{\Gamma(-p)} \exp\left[-\dfrac{i}{2}\left(x^2 + p\dfrac{\pi}{2}\right)\right] \displaystyle\int_0^\infty \dfrac{\cos xt}{t^{p+1}} e^{-it^2/4}\,\mathrm{d}t$

$[x \text{ real}; \quad -1 < \operatorname{Re} p < 0]$          MO 122

**9.251**[10] $D_n(z) = (-1)^n e^{z^2/4} \dfrac{\mathrm{d}^n}{\mathrm{d}z^n}\left(e^{-z^2/2}\right) \qquad [n = 0,1,2,\ldots]$          WH

**9.252**  $D_p(ax+by) = \exp\dfrac{(bx-ay)^2}{4}\left(\dfrac{a}{\sqrt{a^2+b^2}}\right)^p \displaystyle\sum_{k=0}^\infty \binom{p}{k} D_{p-k}\left(\sqrt{a^2+b^2}\,x\right) D_k\left(\sqrt{a^2+b^2}\,y\right)\left(\dfrac{b}{a}\right)^k$

$[a > b > 0, \quad x > 0, \quad y > 0, \quad \operatorname{Re} p \geq 0]$     "summation theorem"     MO 124

## Connections with other functions

**9.253**[11] $D_n(z) = 2^{-\frac{n}{2}} e^{-\frac{z^2}{4}} H_n\left(\dfrac{z}{\sqrt{2}}\right)$          MO 123a

**9.254**

1.     $D_{-1}(z) = e^{\frac{z^2}{4}}\sqrt{\dfrac{\pi}{2}}\left[1 - \Phi\left(\dfrac{z}{\sqrt{2}}\right)\right]$          MO 123

2.[11]  $D_{-2}(z) = e^{\frac{z^2}{4}}\sqrt{\dfrac{\pi}{2}}\left\{\sqrt{\dfrac{2}{\pi}}\,e^{-\frac{z^2}{2}} - z\left[1 - \Phi\left(\dfrac{z}{\sqrt{2}}\right)\right]\right\}$          MO 123

**9.255**   Differential equations leading to parabolic cylinder functions:

1.     $\dfrac{\mathrm{d}^2 u}{\mathrm{d}z^2} + \left(p + \dfrac{1}{2} - \dfrac{z^2}{4}\right)u = 0$

The solutions are $u = D_p(z),\ D_p(-z),\ D_{-p-1}(iz),$ and $D_{-p-1}(-iz)$

(These four solutions are linearly dependent. See **9.248**)

2.     $\dfrac{\mathrm{d}^2 u}{\mathrm{d}z^2} + (z^2 + \lambda)u = 0,$ $\qquad\qquad\qquad u = D_{-\frac{1+i\lambda}{2}}[\pm(1+i)z]$

EH II 118(12,13)a, MO 123

3.[7]   $\dfrac{\mathrm{d}^2 u}{\mathrm{d}z^2} + z\dfrac{\mathrm{d}u}{\mathrm{d}z} + (p+1)u = 0,$ $\qquad\qquad u = e^{-\frac{z^2}{4}} D_p(z)$          MO 123

## 9.26  Confluent hypergeometric series of two variables

**9.261**

1.[12]  $\Phi_1(\alpha,\beta,\gamma,x,y) = \displaystyle\sum_{m,n=0}^\infty \dfrac{(\alpha)_{m+n}(\beta)_n}{(\gamma)_{m+n} m!\, n!} x^m y^n$ $\qquad [|x| < 1]$          EH I 225(20)

2.[12]  $\Phi_2(\beta,\beta',\gamma,x,y) = \displaystyle\sum_{m,n=0}^\infty \dfrac{(\beta)_m (\beta')_n}{(\gamma)_{m+n} m!\, n!} x^m y^n$          EH I 225(21)a, ET I 385

3.     $\Phi_3(\beta,\gamma,x,y) = \displaystyle\sum_{m,n=0}^\infty \dfrac{(\beta)_m}{(\gamma)_{m+n} m!\, n!} x^m y^n$          EH I 225(22)

The functions $\Phi_1$, $\Phi_2$, $\Phi_3$ satisfy the following systems of partial differential equations:

**9.262**

1. $\quad z = \Phi_1 \left( \alpha, \beta, \gamma, x, y \right)$ EH I 235(23)

$$x(1-x)\frac{\partial^2 z}{\partial x^2} + y(1-x)\frac{\partial^2 z}{\partial x\,\partial y} + [\gamma - (\alpha + \beta + 1)x]\frac{\partial z}{\partial x} - \beta y\frac{\partial z}{\partial y} - \alpha\beta z = 0,$$

$$y\frac{\partial^2 z}{\partial y^2} + x\frac{\partial^2 z}{\partial x\,\partial y} + (\gamma - y)\frac{\partial z}{\partial y} - x\frac{\partial z}{\partial x} - \alpha z = 0$$

2. $\quad z = \Phi_2 \left( \beta, \beta', \gamma, x, y \right)$ EH I 235(24)

$$x\frac{\partial^2 z}{\partial x^2} + y\frac{\partial^2 z}{\partial x\,\partial y} + (\gamma - x)\frac{\partial z}{\partial x} - \beta z = 0,$$

$$y\frac{\partial^2 z}{\partial y^2} + x\frac{\partial^2 z}{\partial x\,\partial y} + (\gamma - y)\frac{\partial z}{\partial y} - \beta' z = 0$$

3. $\quad z = \Phi_3(\beta, \gamma, x, y)$ EH I 235(25)

$$x\frac{\partial^2 z}{\partial x^2} + y\frac{\partial^2 z}{\partial x\,\partial y} + (\gamma - x)\frac{\partial z}{\partial x} - \beta z = 0,$$

$$y\frac{\partial^2 z}{\partial y^2} + x\frac{\partial^2 z}{\partial x\,\partial y} + \gamma\frac{\partial z}{\partial y} - z = 0$$

# 9.3 Meijer's $G$-Function

## 9.30 Definition

**9.301**[12] $\quad G_{p,q}^{m,n} \left( x \left| \begin{matrix} a_1, \ldots, a_p \\ b_1, \ldots, b_q \end{matrix} \right. \right) = \dfrac{1}{2\pi i} \displaystyle\int_L \dfrac{\displaystyle\prod_{j=1}^{m} \Gamma(b_j - s) \prod_{j=1}^{n} \Gamma(1 - a_j + s)}{\displaystyle\prod_{j=m+1}^{q} \Gamma(1 - b_j + s) \prod_{j=n+1}^{p} \Gamma(a_j - s)} x^s \, ds$

$[0 \le m \le q, \quad 0 \le n \le p$, and the poles of $\Gamma(b_j - s)$ must not coincide with the poles of $\Gamma(1 - a_k + s)$ for any $j$ and $k$ (where $j = 1, \ldots, m; \quad k = 1, \ldots, n]$). Besides **9.301**, the following notations are also used:

$$G_{pq}^{mn} \left( x \left| \begin{matrix} a_r \\ b_s \end{matrix} \right. \right), \qquad G_{pq}^{mn}(x), \qquad G(x) \qquad\qquad \text{EH I 207(1)}$$

**9.302** Three types of integration paths $L$ in the right member of **9.301** can be exhibited:

1.[12]   The path $L$ runs from $-i\infty$ to $+i\infty$ along the imaginary axis so that the poles of the functions $\Gamma\left(1 - a_k + s\right)$ lie to the left, and the poles of the functions $\Gamma\left(b_j - s\right)$ lie to the right of $L$ (for $j = 1, 2, \ldots, m$ and $k = 1, 2, \ldots, n$). In this case, the conditions under which the integral **9.301** converges are of the form

$$p + q < 2(m + n), \quad |\arg x| < \left(m + n - \tfrac{1}{2}p - \tfrac{1}{2}q\right)\pi \qquad \text{EH I 207(2)}$$

2.   $L$ is a loop, beginning and ending at $+\infty$, that encircles the poles of the functions $\Gamma\left(b_j - s\right)$ (for $j = 1, 2, \ldots, m$) once in the negative direction. All the poles of the functions $\Gamma\left(1 - a_k + s\right)$ must remain outside this loop. Then, the conditions under which the integral **9.301** converges are:

$$q \geq 1 \text{ and either } p < q \text{ or } p = q \text{ and } |x| < 1 \qquad \text{EH I 207(3)}$$

3.   $L$ is a loop, beginning and ending at $-\infty$, that encircles the poles of the functions $\Gamma\left(1 - a_k + s\right)$ (for $k = 1, 2, \ldots, n$) once in the positive direction. All the poles of the functions $\Gamma\left(b_j - s\right)$ (for $j = 1, 2, \ldots, m$) must remain outside this loop.

The conditions under which the integral in **9.301** converges are

$$p \geq 1 \text{ and either } p > q \text{ or } p = q \text{ and } |x| > 1 \qquad \text{EH I 207(4)}$$

The function $G_{pq}^{mn}\left(x \,\middle|\, \begin{matrix} a_r \\ b_s \end{matrix}\right)$ is analytic with respect to $x$; it is symmetric with respect to the parameters $a_1, \ldots, a_n$ and also with respect to $a_{n+1}, \ldots, a_p$; $b_1, \ldots, b_m$; $b_{m+1}, \ldots, b_q$.

EH I 208

**9.303**[12]   If no two $b_j$ (for $j = 1, 2, \ldots, m$) differ by an integer, then, under the conditions that either $p < q$ or $p = q$ and $|x| < 1$,

$$G_{pq}^{mn}\left(x \,\middle|\, \begin{matrix} a_r \\ b_s \end{matrix}\right) = \sum_{h=1}^{m} \frac{\displaystyle\prod_{j=1}^{m}{}' \Gamma\left(b_j - b_h\right) \prod_{j=1}^{n} \Gamma\left(1 + b_h - a_j\right)}{\displaystyle\prod_{j=m+1}^{q} \Gamma\left(1 + b_h - b_j\right) \prod_{j=n+1}^{p} \Gamma\left(a_j - b_h\right)} x^{b_h}$$

$$\times\ {}_pF_{q-1}\Bigg[1 + b_h - a_1, \ldots, 1 + b_h - a_p;\quad 1 + b_h - b_1, \ldots$$

$$\ldots, *, \ldots, 1 + b_h - b_q;\quad (-1)^{p-m-n} x\Bigg]$$

EH I 208(5)

   The prime by the product symbol denotes the omission of the product when $j = h$. The asterisk in the function ${}_pF_{q-1}$ denotes the omission of the $h^{\text{th}}$ parameter.

**9.304[7]** If no two $a_k$ (for $k = 1, 2, \ldots, n$) differ by an integer then, under the conditions that $q < p$ or $q = p$ and $|x| > 1$,

$$G_{pq}^{mn}\left(x \left| \begin{matrix} a_r \\ b_s \end{matrix} \right.\right) = \sum_{h=1}^{n} \frac{\prod\limits_{j=1}^{n}{}' \Gamma(a_h - a_j) \prod\limits_{j=1}^{m} \Gamma(b_j - a_h + 1)}{\prod\limits_{j=n+1}^{p} \Gamma(a_j - a_h + 1) \prod\limits_{j=m+1}^{q} \Gamma(a_h - b_j)} x^{a_h - 1}$$

$$\times \; {}_qF_{p-1}\left[1 + b_1 - a_h, \ldots, 1 + b_q - a_h; \quad 1 + a_1 - a_h, \ldots \right.$$

$$\left. \ldots, *, \ldots, 1 + a_p - a_h; \quad (-1)^{q-m-n} x^{-1}\right]$$

EH I 208(6)

## 9.31 Functional relations

If one of the parameters $a_j$ (for $j = 1, 2, \ldots, n$) coincides with one of the parameters $b_j$ (for $j = m+1, m+2, \ldots, q$), the order of the $G$-function decreases. For example,

1. $\quad G_{pq}^{mn}\left(x \left| \begin{matrix} a_1, \ldots, a_p \\ b_1, \ldots, b_{q-1}, a_1 \end{matrix} \right.\right) = G_{p-1, q-1}^{m, n-1}\left(x \left| \begin{matrix} a_2, \ldots, a_p \\ b_1, \ldots, b_{q-1} \end{matrix} \right.\right)$

$[n, p, q \geq 1]$

An analogous relationship occurs when one of the parameters $b_j$ (for $j = 1, 2, \ldots, m$) coincides with one of the $a_j$ (for $j = n+1, \ldots, p$). In this case, it is $m$ and not $n$ that decreases by one unit.

The $G$-function with $p > q$ can be transformed into the $G$-function with $p < q$ by means of the relationships.

2. $\quad G_{pq}^{mn}\left(x^{-1} \left| \begin{matrix} a_r \\ b_s \end{matrix} \right.\right) = G_{qp}^{nm}\left(x \left| \begin{matrix} 1 - b_s \\ 1 - a_r \end{matrix} \right.\right)$ EH I 209(9)

3.[12] $\quad x \dfrac{\mathrm{d}}{\mathrm{d}x} G_{pq}^{mn}\left(x \left| \begin{matrix} a_r \\ b_s \end{matrix} \right.\right) = G_{p,q}^{m,n}\left(x \left| \begin{matrix} a_1 - 1, a_2, \ldots, a_p \\ \mathbf{b}_q \end{matrix} \right.\right) + (a_1 - 1) G_{p,q}^{m,n}\left(x \left| \begin{matrix} \mathbf{a}_p \\ \mathbf{b}_q \end{matrix} \right.\right)$ $[n \geq 1]$

$$= -G_{p,q}^{m,n}\left(x \left| \begin{matrix} a_1, \ldots, a_{p-1}, a_p - 1 \\ \mathbf{b}_q \end{matrix} \right.\right) + (a_p - 1) G_{p,q}^{m,n}\left(x \left| \begin{matrix} \mathbf{a}_p \\ \mathbf{b}_q \end{matrix} \right.\right) \qquad [n < p]$$

$$= -G_{p,q}^{m,n}\left(x \left| \begin{matrix} \mathbf{a}_p \\ b_1 + 1, b_2, \ldots, b_q \end{matrix} \right.\right) + b_1 G_{p,q}^{m,n}\left(x \left| \begin{matrix} \mathbf{a}_p \\ \mathbf{b}_q \end{matrix} \right.\right) \qquad [m \geq 1]$$

$$= -G_{p,q}^{m,n}\left(x \left| \begin{matrix} \mathbf{a}_p \\ b_1, b_2, \ldots, b_{q-1}, b_q + 1 \end{matrix} \right.\right) + b_q G_{p,q}^{m,n}\left(x \left| \begin{matrix} \mathbf{a}_p \\ \mathbf{b}_q \end{matrix} \right.\right) \qquad [m < q]$$

EH I 210(13)

4. $\quad G_{p+1, q+1}^{m+1, n}\left(z \left| \begin{matrix} \mathbf{a}_p, 1 - r \\ 0, \mathbf{b}_q \end{matrix} \right.\right) = (-1)^r \, G_{p+1, q+1}^{m, n+1}\left(z \left| \begin{matrix} 1 - r, \mathbf{a}_p \\ \mathbf{b}_q, 1 \end{matrix} \right.\right)$

$[r = 0, 1, 2, \ldots]$    MS2 6 (1.2.2)

5. $\quad z^k \, G_{pq}^{mn}\left(z \left| \begin{matrix} \mathbf{a}_p \\ \mathbf{b}_q \end{matrix} \right.\right) = G_{pq}^{mn}\left(z \left| \begin{matrix} \mathbf{a}_p + k \\ \mathbf{b}_q + k \end{matrix} \right.\right)$

MS2 7 (1.2.7)

6.* $\qquad G^{m,n+1}_{p+1,q+1}\left(z \left|\begin{matrix} \alpha, \mathbf{a}_p \\ \mathbf{b}_q, \beta \end{matrix}\right.\right) = (-1)^{\alpha-\beta} G^{m+1,n}_{p+1,q+1}\left(z \left|\begin{matrix} \mathbf{a}_p, \alpha \\ \beta, \mathbf{b}_q \end{matrix}\right.\right) \qquad [q \geq m, \quad \beta - \alpha = 0, 1, 2, \ldots]$

where $\mathbf{a}_p = a_1, a_2, \ldots, a_p$ and $\mathbf{b}_q = b_1, b_2, \ldots, b_q$.

7.* $\qquad x^k \dfrac{d^k}{dx^k} G^{m,n}_{p,q}\left(x \left|\begin{matrix} \mathbf{a}_p \\ \mathbf{b}_q \end{matrix}\right.\right) = G^{m,n+1}_{p+1,q+1}\left(x \left|\begin{matrix} 0, \mathbf{a}_p \\ \mathbf{b}_q, k \end{matrix}\right.\right) \qquad [n \geq 1]$

## 9.32 A differential equation for the G-function

$G^{mn}_{pq}\left(x \left|\begin{matrix} a_r \\ b_s \end{matrix}\right.\right)$ satisfies the following linear $q^{\text{th}}$-order differential equation

$$\left[(-1)^{p-m-n} x \prod_{j=1}^{p}\left(x\frac{d}{dx} - a_j + 1\right) - \prod_{j=1}^{q}\left(x\frac{d}{dx} - b_j\right)\right] y = 0 \qquad [p \leq q] \qquad \text{EH I 210(1)}$$

## 9.33 Series of G-functions

$$G^{mn}_{pq}\left(\lambda x \left|\begin{matrix} a_1, \ldots, a_p \\ b_1, \ldots, b_q \end{matrix}\right.\right) = \lambda^{b_1} \sum_{r=0}^{\infty} \frac{1}{r!}(1-\lambda)^r G^{mn}_{pq}\left(x \left|\begin{matrix} a_1, \ldots, a_p \\ b_1 + r, b_2, \ldots, b_q \end{matrix}\right.\right)$$
$$[|\lambda - 1| < 1, \quad m \geq 1, \quad \text{if } m = 1 \text{ and } p < q, \lambda \text{ may be arbitrary}]$$
$$\text{EH I 213(1)}$$

$$= \lambda^{b_q} \sum_{r=0}^{\infty} \frac{1}{r!}(\lambda - 1)^r G^{mn}_{pq}\left(x \left|\begin{matrix} a_1, \ldots, a_p \\ b_1, \ldots, b_{q-1}, b_q + r \end{matrix}\right.\right)$$
$$[m < q, \quad |\lambda - 1| < 1]$$
$$\text{EH I 213(2)}$$

$$= \lambda^{a_1-1} \sum_{r=0}^{\infty} \frac{1}{r!}\left(\lambda - \frac{1}{\lambda}\right)^r G^{mn}_{pq}\left(x \left|\begin{matrix} a_1 - r, a_2, \ldots, a_p \\ b_1, \ldots, b_q \end{matrix}\right.\right)$$
$$\left[n \geq 1, \quad \text{Re}\,\lambda > \tfrac{1}{2}, \quad (\text{if } n = 1 \text{ and } p > q, \text{ then } \lambda \text{ may be arbitrary})\right]$$
$$\text{EH I 213(3)}$$

$$= \lambda^{a_p-1} \sum_{r=0}^{\infty} \frac{1}{r!}\left(\frac{1}{\lambda} - 1\right)^r G^{mn}_{pq}\left(x \left|\begin{matrix} a_1, \ldots, a_{p-1}, a_p - r \\ b_1, \ldots, b_q \end{matrix}\right.\right)$$
$$\left[n < p, \quad \text{Re}\,\gamma > \tfrac{1}{2}\right]$$
$$\text{EH I 213(4)}$$

For integrals of the $G$-function, see **7.8**.

## 9.34 Connections with other special functions

1. $\qquad J_\nu(x) x^\mu = 2^\mu\, G^{10}_{02}\left(\frac{1}{4}x^2 \left|\begin{matrix} \\ \frac{1}{2}\nu + \frac{1}{2}\mu, \frac{1}{2}\mu - \frac{1}{2}\nu \end{matrix}\right.\right) \qquad\qquad \text{EH I 219(44)}$

2. $\qquad Y_\nu(x) x^\mu = 2^\mu\, G^{20}_{13}\left(\frac{1}{4}x^2 \left|\begin{matrix} \frac{1}{2}\mu - \frac{1}{2}\nu - \frac{1}{2} \\ \frac{1}{2}\mu - \frac{1}{2}\nu, \frac{1}{2}\mu + \frac{1}{2}\nu, \frac{1}{2}\mu - \frac{1}{2}\nu - \frac{1}{2} \end{matrix}\right.\right) \qquad \text{EH I 219(46)}$

3. $\qquad K_\nu(x) x^\mu = 2^{\mu-1}\, G^{20}_{02}\left(\frac{1}{4}x^2 \left|\begin{matrix} \\ \frac{1}{2}\mu + \frac{1}{2}\nu, \frac{1}{2}\mu - \frac{1}{2}\nu \end{matrix}\right.\right) \qquad\qquad \text{EH I 219(47)}$

4. $\quad K_\nu(x) = e^x \sqrt{\pi}\, G_{12}^{20}\left(2x \,\middle|\, \begin{matrix} \frac{1}{2} \\ \nu, -\nu \end{matrix}\right)$ EH I 219(49)

5. $\quad \mathbf{H}_\nu(x)x^\mu = 2^\mu\, G_{13}^{11}\left(\frac{1}{4}x^2 \,\middle|\, \begin{matrix} \frac{1}{2}+\frac{1}{2}\nu+\frac{1}{2}\mu \\ \frac{1}{2}+\frac{1}{2}\nu+\frac{1}{2}\mu, \frac{1}{2}\mu-\frac{1}{2}\nu, \frac{1}{2}\mu+\frac{1}{2}\nu \end{matrix}\right)$ EH I 220(51)

6. $\quad S_{\mu,\nu}(x) = 2^{\mu-1}\dfrac{1}{\Gamma\left(\frac{1-\mu-\nu}{2}\right)\Gamma\left(\frac{1-\mu+\nu}{2}\right)}\, G_{13}^{31}\left(\frac{1}{4}x^2 \,\middle|\, \begin{matrix} \frac{1}{2}+\frac{1}{2}\mu \\ \frac{1}{2}+\frac{1}{2}\mu, \frac{1}{2}\nu, -\frac{1}{2}\nu \end{matrix}\right)$ EH I 220(55)

7.[7] $\quad {}_2F_1(a,b;c;-x) = \dfrac{\Gamma(c)x}{\Gamma(a)\Gamma(b)}\, G_{22}^{12}\left(x \,\middle|\, \begin{matrix} -a, -b \\ -1, -c \end{matrix}\right)$ EH I 222(74)a

8. $\quad {}_pF_q\left(a_1,\ldots,a_p; b_1,\ldots,b_q; x\right) = \dfrac{\prod_{j=1}^q \Gamma(b_j)}{\prod_{j=1}^p \Gamma(a_j)}\, G_{p,q+1}^{1,p}\left(-x \,\middle|\, \begin{matrix} 1-a_1,\ldots,1-a_p \\ 0, 1-b_1,\ldots,1-b_q \end{matrix}\right)$

$\qquad\qquad\qquad\qquad\qquad\quad = \dfrac{\prod_{j=1}^q \Gamma(b_j)}{\prod_{j=1}^p \Gamma(a_j)}\, G_{q+1,p}^{p,1}\left(-\frac{1}{x} \,\middle|\, \begin{matrix} 1, b_1,\ldots,b_q \\ a_1,\ldots,a_p \end{matrix}\right)$

EH I 215(1)

9. $\quad W_{k,m}(x) = \dfrac{2^k\sqrt{x}\,e^{\frac{1}{2}x}}{\sqrt{2\pi}}\, G_{24}^{40}\left(\frac{x^2}{4} \,\middle|\, \begin{matrix} \frac{1}{4}-\frac{1}{2}k, \frac{3}{4}-\frac{1}{2}k \\ \frac{1}{2}+\frac{1}{2}m, \frac{1}{2}-\frac{1}{2}m, \frac{1}{2}m, -\frac{1}{2}m \end{matrix}\right)$ EH I 221(70)

# 9.4 MacRobert's $E$-Function

## 9.41 Representation by means of multiple integrals

$$E\left(p; \alpha_r : q; \varrho_s : x\right) = \frac{\Gamma\left(\alpha_{q+1}\right)}{\Gamma\left(\varrho_1-\alpha_1\right)\Gamma\left(\varrho_2 \quad \alpha_2\right)\cdots\Gamma\left(\varrho_q-\alpha_q\right)}$$

$$\times \prod_{\mu=1}^q \int_0^\infty \lambda_\mu^{\varrho_\mu-\alpha_\mu-1}\left(1-\lambda_\mu\right)^{-\varrho_\mu}\, \mathrm{d}\lambda_\mu \prod_{\nu=2}^{p-q-1}\int_0^\infty e^{-\lambda_{q+\nu}}\lambda_{q+\nu}^{\alpha_{q+\nu}-1}\, \mathrm{d}\lambda_{q+\nu}$$

$$\times \int_0^\infty e^{-\lambda_p}\lambda_p^{\alpha_p-1}\left[1+\frac{\lambda_{q+2}\lambda_{q+3}\cdots\lambda_p}{(1+\lambda_1)\cdots(1+\lambda_q)x}\right]^{-\alpha_{q+1}}\, \mathrm{d}\lambda_p$$

$[|\arg x| < \pi,\ p \geq q+1,\ \alpha_r$ and $\varrho_s$ are bounded by the condition that the integrals on the right be convergent.] $\qquad$ EH I 204(3)

## 9.42 Functional relations

1. $\quad \alpha_1 x\, E\left(\alpha_1,\ldots,\alpha_p : \varrho_1,\ldots,\varrho_q : x\right) = x\, E\left(\alpha_1+1,\alpha_2,\ldots,\alpha_p : \varrho_1,\ldots,\varrho_q : x\right)$

$\qquad\qquad\qquad\qquad\qquad + E\left(\alpha_1+1,\alpha_2+1,\ldots,\alpha_p+1 : \varrho_1+1,\ldots,\varrho_q+1 : x\right)$

EH I 205(7)

2. $\quad (\varrho_1-1)\, x\, E\left(\alpha_1,\ldots,\alpha_p : \varrho_1,\ldots,\varrho_q : x\right) = x\, E\left(\alpha_1,\ldots,\alpha_p : \varrho_1-1,\varrho_2,\ldots,\varrho_q : x\right)$

$\qquad\qquad\qquad\qquad\qquad + E\left(\alpha_1+1,\ldots,\alpha_p+1 : \varrho_1+1,\ldots,\varrho_q+1 : x\right)$

EH I 205(9)

3. $\quad \dfrac{d}{dx} E\left(\alpha_1,\ldots,\alpha_p:\varrho_1,\ldots,\varrho_q:x\right) = x^{-2} E\left(\alpha_1+1,\ldots,\alpha_p+1:\varrho_1+1,\ldots,\varrho_q+1:x\right)$

EH I 205(8)

# 9.5 Riemann's Zeta Functions $\zeta(z,q)$, and $\zeta(z)$, and the Functions $\Phi(z,s,v)$ and $\xi(s)$

## 9.51 Definition and integral representations

**9.511** $\quad \zeta(z,q) = \dfrac{1}{\Gamma(z)} \displaystyle\int_0^\infty \dfrac{t^{z-1}e^{-qt}}{1-e^{-t}}\,dt;$

WH

$$= \frac{1}{2}q^{-z} + \frac{q^{1-z}}{z-1} + 2\int_0^\infty \left(q^2+t^2\right)^{-\frac{z}{2}} \left[\sin\left(z\arctan\frac{t}{q}\right)\right] \frac{dt}{e^{2\pi t}-1}$$

$[0 < q < 1,\quad \operatorname{Re} z > 1]$<br>WH

**9.512** $\quad \zeta(z,q) = -\dfrac{\Gamma(1-z)}{2\pi i} \displaystyle\int_\infty^{(0+)} \dfrac{(-\theta)^{z-1}e^{-q\theta}}{1-e^{-\theta}}\,d\theta$

This equation is valid for all values of $z$ except for $z = 1,2,3,\ldots$. It is assumed that the path of integration (see drawing below) does not pass through the points $2n\pi i$ (where $n$ is a natural number).

See also **4.251** 4, **4.271** 1, 4, 8, **4.272** 9, 12, **4.294** 11.

**9.513**

1. $\quad \zeta(z) = \dfrac{1}{(1-2^{1-z})\Gamma(z)} \displaystyle\int_0^\infty \dfrac{t^{z-1}}{e^t+1}\,dt \qquad [\operatorname{Re} z > 0]$

WH

2. $\quad \zeta(z) = \dfrac{2^z}{(2^z-1)\Gamma(z)} \displaystyle\int_0^\infty \dfrac{t^{z-1}e^t}{e^{2t}-1}\,dt \qquad [\operatorname{Re} z > 1]$

WH

3.[11] $\quad \zeta(z) = \dfrac{\pi^{\frac{z}{2}}}{\Gamma\left(\frac{z}{2}\right)} \left[\dfrac{1}{z(z-1)} + \displaystyle\int_1^\infty \left(t^{\frac{1-z}{2}} + t^{\frac{z}{2}}\right) t^{-1} \sum_{k=1}^\infty e^{-k^2\pi t}\,dt\right]$

WH

4. $\quad \zeta(z) = \dfrac{2^{z-1}}{z-1} - 2^z \displaystyle\int_0^\infty \left(1+t^2\right)^{-\frac{z}{2}} \sin\left(z\arctan t\right) \dfrac{dt}{e^{\pi t}+1}$

WH

5. $\quad \zeta(z) = \dfrac{2^{z-1}}{2^z-1}\dfrac{z}{z-1} + \dfrac{2}{2^z-1} \displaystyle\int_0^\infty \left(\dfrac{1}{4}+t^2\right)^{-z/2} \sin\left(z\arctan 2t\right) \dfrac{dt}{e^{2\pi t}-1}$

WH

See also **3.411** 1, **3.523** 1, **3.527** 1, 3, **4.271** 8.

## 9.52   Representation as a series or as an infinite product

**9.521**

1.    $\zeta(z, q) = \sum\limits_{n=0}^{\infty} \dfrac{1}{(q+n)^z}$                        $[\operatorname{Re} z > 1, \quad q \neq 0, -1, -2, \ldots]$      WH

2.[12]   $\zeta(z, q) = \dfrac{2\,\Gamma(1-z)}{(2\pi)^{1-z}} \left[ \sin \dfrac{z\pi}{2} \sum\limits_{n=1}^{\infty} \dfrac{\cos 2\pi qn}{n^{1-z}} + \cos \dfrac{z\pi}{2} \sum\limits_{n=1}^{\infty} \dfrac{\sin 2\pi qn}{n^{1-z}} \right]$

                                                 $[\operatorname{Re} z > 0, \quad 0 < q \le 1]$      WH

3.[12]   $\zeta(z, q) = \sum\limits_{n=0}^{N} \dfrac{1}{(q+n)^z} - \dfrac{1}{(1-z)(N+q)^{z-1}} - \sum\limits_{n=N}^{\infty} F_n(z),$

                                         $[\operatorname{Re} z > 1, \quad N \text{ is a natural number}]$

where

$$F_n(z) = \dfrac{1}{1-z}\left( \dfrac{1}{(n+1+q)^{z-1}} - \dfrac{1}{(n+q)^{z-1}} \right) - \dfrac{1}{(n+1+q)^z}$$

$$= z \int_n^{n+1} \dfrac{(t-n)\,dt}{(t+q)^{z+1}}$$

     WH

**9.522**

1.    $\zeta(z) = \sum\limits_{n=1}^{\infty} \dfrac{1}{n^z}$                                     $[\operatorname{Re} z > 1]$      WH

2.    $\zeta(z) = \dfrac{1}{1 - 2^{1-z}} \sum\limits_{n-1}^{\infty} (-1)^{n+1} \dfrac{1}{n^z}$                 $[\operatorname{Re} z > 0]$      WH

**9.523**   The following product and summation are taken over all primes $p$:

1.[7]   $\zeta(z) = \prod\limits_{p} \dfrac{1}{1 - p^{-z}}$                                $[\operatorname{Re} z > 1]$      WH

2.    $\ln \zeta(z) = \sum\limits_{p} \sum\limits_{k=1}^{\infty} \dfrac{1}{k p^{kz}}$                         $[\operatorname{Re} z > 1]$      WH

**9.524**[11]   $\dfrac{\zeta'(z)}{\zeta(z)} = -\sum\limits_{k=1}^{\infty} \dfrac{\Lambda(k)}{k^z},$                         $[\operatorname{Re} z > 1]$

where $\Lambda(k) = 0$ when $k$ is not a power of a prime and $\Lambda(k) = \ln p$ when $k$ is a power of a prime $p$.      WH

## 9.53 Functional relations

**9.531**

1.[12]     $\zeta(-n, q) = -\dfrac{B'_{n+2}(q)}{(n+1)(n+2)} = \dfrac{-B_{n+1}(q)}{n+1}$

$\qquad\qquad\qquad\qquad\qquad$ [$n$ is a nonnegative integer]      see EH I 27 (11)      WH

2.*     $\zeta(s, q) = \zeta(s, q+m) + \displaystyle\sum_{k=0}^{m-1} (q+k)^{-s}$      [$m = 1, 2, 3 \ldots$,      $q \neq 0, -1, -2, \ldots$]

3.*     $\zeta(s, q) = m^{-s} \displaystyle\sum_{k=0}^{m-1} \zeta\left(s, \dfrac{q+k}{m}\right)$      [$m = 1, 2, 3 \ldots$,      $q \neq 0, -1, -2, \ldots$]

**9.532**     $\displaystyle\sum_{k=2}^{\infty} \dfrac{(-1)^{k-1}}{k} z^k \zeta(k, q) = \ln \dfrac{e^{-Cz}\,\Gamma(q)}{\Gamma(z+q)} - \dfrac{z}{q} + \sum_{k=1}^{\infty} \dfrac{qz}{k(q+k)}$      [$|z| < q$]      WH

**9.533**

1.     $\displaystyle\lim_{z \to 1} \dfrac{\zeta(z, q)}{\Gamma(1-z)} = -1$      WH

2.     $\displaystyle\lim_{z \to 1} \left\{ \zeta(z, q) - \dfrac{1}{z-1} \right\} = -\Psi(q)$      WH

3.     $\left\{ \dfrac{\mathrm{d}}{\mathrm{d}z} \zeta(z, q) \right\}_{z=0} = \ln \Gamma(q) - \dfrac{1}{2} \ln 2\pi$      WH

**9.534**     $\zeta(z, 1) = \zeta(z)$

**9.535**

1.     $\zeta(z) = \dfrac{1}{2^z - 1} \zeta\left(z, \tfrac{1}{2}\right)$      [$\operatorname{Re} z > 1$]      WH

2.[11]     $2^z\, \Gamma(1-z)\, \zeta(1-z) \sin\left(\dfrac{z\pi}{2}\right) = \pi^{1-z}\, \zeta(z)$      WH

3.     $2^{1-z}\, \Gamma(z)\, \zeta(z) \cos \dfrac{z\pi}{2} = \pi^z\, \zeta(1-z)$      WH

4.     $\Gamma\left(\dfrac{z}{2}\right) \pi^{-\frac{z}{2}} \zeta(z) = \Gamma\left(\dfrac{1-z}{2}\right) \pi^{\frac{z-1}{2}} \zeta(1-z)$      WH

5.*     $\zeta\left(1-s, \dfrac{m}{n}\right) = \dfrac{2\Gamma(s)}{(2\pi n)^s} \displaystyle\sum_{k=1}^{n} \cos\left(s\dfrac{\pi}{2} - 2km\dfrac{\pi}{n}\right) \zeta\left(s\dfrac{k}{n}\right)$

$\qquad\qquad\qquad\qquad\qquad$ [$n = 1, 2, 3 \ldots$,      $m = 1, 2, \ldots, n$]

**9.536**     $\displaystyle\lim_{z \to 1} \left\{ \zeta(z) - \dfrac{1}{z-1} \right\} = C$

**9.537**     Set $z = \tfrac{1}{2} + it$. Then, $\Xi(t) = \dfrac{(z-1)\,\Gamma\left(\frac{z}{2}+1\right)}{\sqrt{\pi^z}} \zeta(z) = \Xi(-t)$ is an even function of $t$ with real

coefficients in its expansion in powers of $t^2$.

$\qquad\qquad\qquad\qquad\qquad\qquad\qquad\qquad\qquad\qquad\qquad\qquad\qquad\qquad\qquad$ JA

## 9.54 Singular points and zeros

### 9.541[7]

1.    $z = 1$ is the only singular point of the function $\zeta(z)$                                         WH

2.    The function $\zeta(z)$ has simple zeros at the points $-2n$, where $n$ is a natural number. All other zeros of the function $\zeta(z)$ lie in the strip $0 \leq \operatorname{Re} z < 1$.

3.[8]   Riemann's hypothesis: All zeros of the function $\zeta(z)$ lie on the straight line $\operatorname{Re} z = \frac{1}{2}$. It has been shown that a countably infinite set of zeros of the zeta function lie on this line. The first 1,500,000,001 zeros lying in $0 < \operatorname{Im} z < 545,439,823.215$ are known to have $\operatorname{Re} z = \frac{1}{2}$.     WH

### 9.542   Particular values:

1.    $\zeta(2m) = \dfrac{2^{2m-1}\pi^{2m}|B_{2m}|}{(2m)!}$                 [$m$ is a natural number]                    WH

2.    $\zeta(1 - 2m) = -\dfrac{B_{2m}}{2m}$                   [$m$ is a natural number]                    WH

3.    $\zeta(-2m) = 0$                           [$m$ is a natural number]                    WH

4.    $\zeta'(0) = -\frac{1}{2}\ln 2\pi$                                                           WH

5.*   $\zeta(0) = -\dfrac{1}{2}$

## 9.55  The Lerch function $\Phi(z, s, v)$

### 9.550   Definition:

$$\Phi(z, s, v) = \sum_{n=0}^{\infty} (v + n)^{-s} z^n \qquad [|z| < 1, \quad v \neq 0, -1, \ldots]$$                    EH I 27(1)

### Functional relations

### 9.551

1.[12]   $\Phi(z, s, v) = z^m\, \Phi(z, s, m + v) + \displaystyle\sum_{n=0}^{m-1} (v + n)^{-s} z^n$      $[m = 1, 2, 3, \ldots, \quad v \neq 0, -1, -2, \ldots]$

EH I 27(1)

2.*   $\Phi(z, s, mv) = m^{-s} \displaystyle\sum_{n=0}^{m-1} \Phi\left(z^m, s, v + \frac{n}{m}\right) z^n$      $[m = 1, 2, 3, \ldots]$

3.*   $\Phi(z^m, s, v) = m^{s-1} \displaystyle\sum_{n=0}^{m-1} \Phi\left(z e^{2\pi i n/m}, s, mv\right)$      $[m = 1, 2, 3, \ldots]$

**9.552**[12]  $\Phi(z, s, v) = (-z)^{-v}(2\pi)^{s-1}\,\Gamma(1 - s)\left[i^{1-s} e^{-i\pi v}\, \Phi\left(e^{-2\pi i v}, 1 - s, \dfrac{1}{2} + \dfrac{\ln(-z)}{2\pi i}\right)\right.$

$$\left. + i^{s-1} e^{i\pi v}\, \Phi\left(e^{2\pi i v}, 1 - s, \dfrac{1}{2} - \dfrac{\ln(-z)}{2\pi i}\right)\right]$$

$[\operatorname{Im}(v) \leq 0, 0 < \operatorname{Re}(v) \leq 1]$ or $[\operatorname{Im}(v) > 0, 0 \leq \operatorname{Re}(v) < 1]$   EH I 29(7)

## Series representation

**9.553**     $\Phi(z,s,v) = z^{-v}\,\Gamma(1-s)\sum_{n=-\infty}^{\infty}(-\ln z + 2\pi ni)^{s-1}\,e^{2\pi nvi}$

$$[0 < v \le 1, \quad \operatorname{Re} s < 0, \quad |\arg(-\ln z + 2\pi ni)| \le \pi]\quad \text{EH I 28(6)}$$

**9.554**[12]     $\Phi(z,m,v) = z^{-v}\left\{\sum_{n=0}^{\infty}{}' \zeta(m-n,v)\dfrac{(\ln z)^n}{n!} + \dfrac{(\ln z)^{m-1}}{(m-1)!}\left[\Psi(m) - \Psi(v) - \ln(-\ln z)\right]\right\}$

$$[m = 1,2,3,\ldots, \quad |\ln z| < 2\pi, \quad v \ne 0, -1, -2,\ldots]\quad \text{EH I 30(9)}$$

**9.555**

**1.**[12]     $\Phi(z,-m,v) = \dfrac{m!}{z^v}(-\ln z)^{-m-1} - \dfrac{1}{z^v}\sum_{r=0}^{\infty}\dfrac{B_{m+r+1}(v)\,(\ln z)^r}{r!(m+r+1)}$ $\qquad [m = 0,1,2,\ldots, \quad |\ln z| < 2\pi]$

$$\text{EH I 30(11)}$$

**2.***     $\Phi(z,s,v) = z^{-v}\left[\Gamma(1-s)(-\ln z)^{s-1} + \sum_{n=0}^{\infty}\zeta(s-n,v)\dfrac{(\ln z)^n}{n!}\right]$ $\qquad [s \ne 1,2,3,\ldots, \quad |\ln z| < 2\pi]$

## Integral representation

**9.556**

**1.**[12]     $\Phi(z,s,v) = \dfrac{1}{\Gamma(s)}\displaystyle\int_0^{\infty}\dfrac{t^{s-1}e^{-vt}}{1 - ze^{-t}}\,dt = \dfrac{1}{\Gamma(s)}\int_0^{\infty}\dfrac{t^{s-1}e^{-(v-1)t}\,dt}{e^t - z}$

$$[\operatorname{Re} v > 0] \text{ or } [|\arg(1-z)| < \pi, \quad \operatorname{Re} s > 0] \text{ or } [z = 1, \quad \operatorname{Re} s > 1]\quad \text{EH I 27(3)}$$

**2.***     $z\dfrac{\partial}{\partial z}\Phi(z,s,v) = \Phi(z,s-1,v) - v\Phi(z,s,v)$

**3.***     $\dfrac{\partial}{\partial v}\Phi(z,s,v) = -s\Phi(z,s+1,v)$

## Limit relationships

**9.557**     $\lim\limits_{z\to 1}(1-z)^{1-s}\,\Phi(z,s,v) = \Gamma(1-s)$ $\qquad\qquad [\operatorname{Re} s < 1]$ $\qquad\qquad$ EH I 30(12)

**9.558**     $\lim\limits_{z\to 1}\dfrac{\Phi(z,1,v)}{-\ln(1-z)} = 1$ $\qquad\qquad$ EH I 30(13)

## Relations to other functions

**9.559**

**1.**[12]     $\Phi(z,1,v) = v^{-1}\,{}_2F_1(1,v;1+v;z)$ $\qquad\qquad$ EH I 30(10)

**2.***     $\Phi(1,s,q) = \zeta(s,q)$

---

*In 9.554 the prime on the symbol $\sum$ means that the term corresponding to $n = m - 1$ is omitted.

## 9.56  The function $\xi(s)$

**9.561**  $\xi(s) = \dfrac{1}{2}s(s-1)\dfrac{\Gamma\left(\frac{1}{2}s\right)}{\pi^{\frac{1}{2}s}}\zeta(s)$                                     EH III 190(10)

**9.562**  $\xi(1-s) = \xi(s)$                                     EH III 190(11)

# 9.6 Bernoulli Numbers and Polynomials, Euler Numbers, the Functions $\nu(x)$, $\nu(x,\alpha)$, $\mu(x,\beta)$, $\mu(x,\beta,\alpha)$, $\lambda(x,y)$ and Euler Polynomials

## 9.61  Bernoulli numbers

**9.610**   The numbers $B_n$, representing the coefficients of $\dfrac{t^n}{n!}$ in the expansion of the function

$$\frac{t}{e^t - 1} = \sum_{n=0}^{\infty} B_n \frac{t^n}{n!} \qquad [0 < |t| < 2\pi],$$

are called *Bernoulli* numbers. Thus, the function $\dfrac{t}{e^t - 1}$ is a generating function for the Bernoulli numbers.

GE 48(57), FI II 520

**9.611**   Integral representations

1.      $B_{2n} = (-1)^{n-1}4n\displaystyle\int_0^\infty \frac{x^{2n-1}}{e^{2\pi x} - 1}\,\mathrm{d}x$                    $[n = 1, 2, \ldots]$     (cf. **3.411** 2, 4)

FI II 721a

2.[12]   $B_{2n} = (-1)^{n-1}\pi^{2n}\displaystyle\int_0^\infty \frac{x^{2n}}{\sinh^2 x}\,\mathrm{d}x$                    $[n = 1, 2, \ldots]$

3.      $B_{2n} = (-1)^{n-1}\dfrac{2n(1-2n)}{\pi}\displaystyle\int_0^\infty x^{2n-2}\ln\left(1 - e^{-2\pi x}\right)\,\mathrm{d}x$

$[n = 1, 2, \ldots]$

4.      $B_n = \displaystyle\lim_{x\to 0}\frac{\mathrm{d}^n}{\mathrm{d}x^n}\left(\frac{x}{e^x - 1}\right)$

See also **3.523** 2, **4.271** 3.

**Properties and functional relations**

**9.612**[8]   A symbolic notation:

$$(B + \alpha)^{[n]} = \sum_{k=0}^n \binom{n}{k} B_k \alpha^{n-k} \qquad [n \geq 2]$$

in particular

$$B_n = (B + 1)^{[n]} = \sum_{k=0}^n \binom{n}{k} B_k \qquad [n \geq 2]$$

hence by recursion

$$B_n = -n!\sum_{k=0}^{n-1} \frac{B_k}{k!(n+1-k)!} \qquad [n \geq 2]$$

**9.613**　All the Bernoulli numbers are rational numbers.

**9.614**　Every number $B_n$ can be represented in the form

$$B_n = C_n - \sum \frac{1}{k+1},$$

where $C_n$ is an integer and the sum is taken over all $k > 0$ such that $k + 1$ is a prime and $k$ is a divisor of $n$.

GE 64

**9.615**[12]　All the Bernoulli numbers with odd index are equal to zero except that $B_1 = -\frac{1}{2}$; that is, $B_{2n+1} = 0$ for $n$ a natural number.

GE 52, FI II 521

$$B_{2n} = -\frac{1}{2n+1} + \frac{1}{2} - \sum_{k=1}^{n-1} \frac{2n(2n-1)\ldots(2n-2k+2)}{(2k)!} B_{2k} \qquad [n \geq 1]$$

**9.616**　$B_{2n} = \dfrac{(-1)^{n-1}(2n)!}{2^{2n-1}\pi^{2n}} \zeta(2n) \qquad [n \geq 0]$ 　　　　　(cf. **9.542**)　　GE 56(79), FI II 721a

**9.617**[7]　$B_{2n} = (-1)^{n-1} \dfrac{2(2n)!}{(2\pi)^{2n}} \dfrac{1}{\displaystyle\prod_{p=2}^{\infty} \left(1 - \dfrac{1}{p^{2n}}\right)} \qquad [n \geq 1] \qquad$ (cf. **9.523**)

(where the product is taken over all primes $p$).

- For a connection with Riemann's zeta function, see **9.542**.
- For a connection with the Euler numbers, see **9.635**.
- For a table of values of the Bernoulli numbers, see **9.71**

**9.619**　An inequality $\left|(B - \theta)^{[n]}\right| \leq |B_n| \qquad [0 < \theta < 1]$

## 9.62　Bernoulli polynomials

**9.620**　The Bernoulli polynomials $B_n(x)$ are defined by

$$B_n(x) = \sum_{k=0}^{n} \binom{n}{k} B_k x^{n-k}$$

GE 51(62)

or symbolically, $B_n(x) = (B + x)^{[n]}$.

GE 52(68)

**9.621**　The generating function

$$\frac{e^{xt}}{e^t - 1} = \sum_{n=0}^{\infty} B_n(x) \frac{t^{n-1}}{n!} \qquad [0 < |t| < 2\pi] \qquad \text{(cf. 1.213)}$$

GE 65(89)a

**9.622**　Series representation

1.[7]　$B_n(x) = -2 \dfrac{n!}{(2\pi)^n} \displaystyle\sum_{k=1}^{\infty} \frac{\cos\left(2\pi kx - \frac{1}{2}\pi n\right)}{k^n}$

$$[n > 1, \quad 1 \geq x \geq 0; \quad n = 1, \quad 1 > x > 0] \quad \text{AS 805(23.1.16)}$$

$2.^7 \qquad B_{2n-1}(x) = 2\dfrac{(-1)^n 2(2n-1)!}{(2\pi)^{2n-1}} \displaystyle\sum_{k=1}^{\infty} \dfrac{\sin 2k\pi x}{k^{2n-1}}$

$$[n > 1, \quad 1 \geq x \geq 0; \quad n = 1, \quad 1 > x > 0] \quad \text{AS } 805(23.1.17)$$

$3.^{10} \qquad B_{2n}(x) = \dfrac{(-1)^{n-1} 2(2n)!}{(2\pi)^{2n}} \displaystyle\sum_{k=1}^{\infty} \dfrac{\cos 2k\pi x}{k^{2n}} \qquad\qquad [0 \leq x \leq 1, \quad n = 1, 2, \ldots] \qquad\qquad \text{GE } 71$

**9.623**  Functional relations and properties:

1. $\qquad B_{m+1}(n) = B_{m+1} + (m+1) \displaystyle\sum_{k=1}^{n-1} k^m$

$$[\text{n and m are natural numbers}] \qquad (\text{see also } \mathbf{0.121}) \quad \text{GE } 51(65)$$

2. $\qquad B_n(x+1) - B_n(x) = nx^{n-1}$ $\qquad\qquad\qquad\qquad\qquad\qquad\qquad$ GE 65(90)

3. $\qquad B'_n(x) = n\, B_{n-1}(x)$ $\qquad\qquad\qquad\qquad\qquad [n = 1, 2, \ldots]$ $\qquad\qquad$ GE 66

4. $\qquad B_n(1-x) = (-1)^n\, B_n(x)$ $\qquad\qquad\qquad\qquad\qquad\qquad\qquad\qquad$ GE 66

$5.^{10} \qquad (-1)^n\, B_n(-x) = B_n(x) + nx^{n-1}$ $\qquad\qquad\quad [n = 0, 1, \ldots]$ $\qquad$ AS 804(23.1.9)

**9.624**$^7$ $\quad B_n(mx) = m^{n-1} \displaystyle\sum_{k=0}^{m-1} B_n\left(x + \dfrac{k}{m}\right)$

$$[m = 1, 2, \ldots n = 0, 1, \ldots]; \qquad \text{``summation theorem''} \quad \text{GE } 67$$

**9.625**  For $n$ odd, the differences

$$B_n(x) - B_n$$

vanish on the interval $[0, 1]$ only at the points $0, \frac{1}{2}$, and 1. They change sign at the point $x = \frac{1}{2}$. For $n$ even, these differences vanish at the end points of the interval $[0, 1]$. Within this interval, they do not change sign and their greatest absolute value occurs at the point $x = \frac{1}{2}$.

**9.626**  The polynomials

$$B_{2n}(x) - B_{2n} \text{ and } B_{2n+2}(x) - B_{2n+2}$$

have opposite signs in the interval $(0, 1)$. $\qquad\qquad\qquad\qquad\qquad\qquad\qquad\qquad$ GE 87

**9.627**  Special cases:

1. $\qquad B_1(x) = x - \frac{1}{2}$ $\qquad\qquad\qquad\qquad\qquad\qquad\qquad\qquad\qquad\qquad\qquad$ GE 70

2. $\qquad B_2(x) = x^2 - x + \frac{1}{6}$ $\qquad\qquad\qquad\qquad\qquad\qquad\qquad\qquad\qquad\qquad$ GE 70

3. $\qquad B_3(x) = x^3 - \frac{3}{2}x^2 + \frac{1}{2}x$ $\qquad\qquad\qquad\qquad\qquad\qquad\qquad\qquad\qquad$ GE 70

4. $\qquad B_4(x) = x^4 - 2x^3 + x^2 - \frac{1}{30}$ $\qquad\qquad\qquad\qquad\qquad\qquad\qquad\qquad$ GE 70

5. $\qquad B_5(x) = x^5 - \frac{5}{2}x^4 + \frac{5}{3}x^3 - \frac{1}{6}x$ $\qquad\qquad\qquad\qquad\qquad\qquad\qquad$ GE 70

**9.628**  Particular values:

1. $\qquad B_n(0) = B_n$

$2.^{12} \qquad B_1(1) = -B_1 = \frac{1}{2}, \qquad B_n(1) = (-1)^n B_n$ $\qquad\qquad [n \neq 1]$ $\qquad\qquad$ GE 76

## 9.63　Euler numbers

**9.630**　The numbers $E_n$, representing the coefficients of $\frac{t^n}{n!}$ in the expansion of the function

$$\frac{1}{\cosh t} = \sum_{n=0}^{\infty} E_n \frac{t^n}{n!} \qquad \left[|t| < \frac{\pi}{2}\right]$$

are known as the *Euler numbers*. Thus, the function $\frac{1}{\cosh t}$ is a generating function for the Euler numbers.

CE 330

**9.631**　A recursion formula

$$(E+1)^{[n]} + (E-1)^{[n]} = 0 \qquad [n \geq 1], \qquad E_0 = 1$$

CE 329

**Properties of the Euler numbers**

**9.632**　The Euler numbers are integers.

**9.633**　The Euler numbers of odd index are equal to zero; the signs of two adjacent numbers of even indices are opposite; that is,

$$E_{2n+1} = 0, \quad E_{4n} > 0, \quad E_{4n+2} < 0$$

CE 329

**9.634**　If $\alpha, \beta\gamma, \ldots$ are the divisors of the number $n - m$, the difference $E_{2n} - E_{2m}$ is divisible by those of the numbers $2\alpha + 1, 2\beta + 1, 2\gamma + 1, \ldots$, that are primes.

**9.635**　A connection with the Bernoulli numbers (symbolic notation):

1.[11]　$E_{n-1} + 4(-1)^n \left(3^{n-1} - 1\right) B_1 = \dfrac{(4B-1)^{[n]} - (4B-3)^{[n]}}{2n} + 4(-1)^{n+1} \left(3^{n-1} - 1\right) B_1$ 　　　CE 330

2.　$B_n = \dfrac{n(E+1)^{[n-1]}}{2^n (2^n - 1)}$ 　　　　　　　　　$[n \geq 2]$ 　　　　　CE 330

3.[6]　$\left(B + \tfrac{1}{4}\right)^{[2n+1]} = -4^{-2n-1}(2n+1)E_{2n}$ 　　　$[n \geq 0]$ 　　　　CE 341

4.　$E_{n-1} = \dfrac{(4B+3)^{[n]} - (4B+1)^{[n]}}{2n}$ 　　　　　$[n \geq 1]$

For a table of values of the Euler numbers, see **9.72**.

## 9.64　The functions $\nu(x)$, $\nu(x, \alpha)$, $\mu(x, \beta)$, $\mu(x, \beta, \alpha)$, $\lambda(x, y)$

**9.640**

1.　$\nu(x) = \displaystyle\int_0^\infty \frac{x^t \, \mathrm{d}t}{\Gamma(t+1)}$ 　　　　　　　　　　　EH III 217(1)

2.　$\nu(x, \alpha) = \displaystyle\int_0^\infty \frac{x^{\alpha+t} \, \mathrm{d}t}{\Gamma(\alpha + t + 1)}$ 　　　　　　　EH III 217(1)

3.　$\mu(x, \beta) = \displaystyle\int_0^\infty \frac{x^t t^\beta \, \mathrm{d}t}{\Gamma(\beta + 1)\Gamma(t+1)}$ 　　　　　EH III 217(2)

4.　$\mu(x, \beta, \alpha) = \displaystyle\int_0^\infty \frac{x^{\alpha+t} t^\beta \, \mathrm{d}t}{\Gamma(\beta + 1)\Gamma(\alpha + t + 1)}$ 　　EH III 217(2)

5.　$\lambda(x, y) = \displaystyle\int_0^y \frac{\Gamma(u + 1) \, \mathrm{d}u}{x^u}$ 　　　　　　　　MI 9

# 9.65$^{10}$   Euler polynomials

**9.650**   The Euler polynomials are defined by

$$E_n(x) = \sum_{k=0}^{n} \binom{n}{k} \frac{E_k}{2^k} \left( x - \frac{1}{2} \right)^{n-k}$$

AS 804 (23.1.7)

**9.651**   The generating function:

$$\frac{2e^{xt}}{e^t + 1} = \sum_{n=0}^{\infty} E_n(x) \frac{t^n}{n!}$$

AS 804 (23.1.1)

**9.652**   Series representation:

1.      $$E_n(x) = 4 \frac{n!}{\pi^{n+1}} \sum_{k=0}^{\infty} \frac{\sin\left((2k+1)\pi x - \frac{1}{2}\pi n\right)}{(2k+1)^{n+1}}$$

$$[n > 0, \quad 1 \geq x \geq 0, \quad n = 1, \quad 1 > x > 0] \quad \text{AS 804 (23.1.16)}$$

2.$^{10}$    $$E_{2n-1}(x) = \frac{(-1)^n 4(2n-1)!}{\pi^{2n}} \sum_{k=0}^{\infty} \frac{\cos(2k+1)\pi x}{(2k+1)^{2n}} \qquad [n = 1, 2, \ldots, \quad 1 \geq x \geq 0]$$

AS 804 (23.1.17)

3.      $$E_{2n}(x) = \frac{(-1)^n 4(2n)!}{\pi^{2n+1}} \sum_{k=0}^{\infty} \frac{\sin(2k+1)\pi x}{(2k+1)^{2n+1}}$$

$$[n > 0, \quad 1 \geq x \geq 0, \quad n = 0, \quad 1 > x > 0] \quad \text{AS 804 (23.1.18)}$$

**9.653**   Functional relations and properties:

1.      $$E_m(n+1) = 2 \sum_{k=1}^{n} (-1)^{n-k} k^m + (-1)^{n+1} E_m(0), \qquad [m \text{ and } n \text{ are natural numbers}]$$

AS 804 (23.1.4)

2.      $E_n'(x) = n E_{n-1}(x).$                $[n = 1, 2, \ldots]$      AS 804 (23.1.5)

3.      $E_n(x+1) + E_n(x) = 2x^n$        $[n = 0, 1, \ldots]$      AS 804 (23.1.6)

4.$^8$    $$E_n(mx) = m^n \sum_{k=0}^{m-1} (-1)^k E_n\left( x - \frac{k}{m} \right) \qquad [n = 0, 1, \ldots, m = 1, 3, \ldots]$$

AS 804 (23.1.10)

5.      $$E_n(mx) = \frac{-2}{n+1} m^n \sum_{k=0}^{m-1} (-1)^k B_{n+1}\left( x + \frac{k}{m} \right) \qquad [n = 0, 1, \ldots, m = 2, 4, \ldots]$$

AS 804 (23.1.10)

**9.654** Special cases:

1. $E_1(x) = x - \frac{1}{2}$

2. $E_2(x) = x^2 - x$

3. $E_3(x) = x^3 - \frac{3}{2}x^2 + \frac{1}{4}$

4. $E_4(x) = x^4 - 2x^3 + x$

5. $E_5(x) = x^5 - \frac{5}{2}x^4 + \frac{5}{2}x^2 - \frac{1}{2}$

**9.655** Particular values:

1. $E_{2n+1} = 0$ $\qquad [n = 0, 1, \ldots]$ $\qquad$ AS 805 (23.1.19)

2. $E_n(0) = -E_n(1) = -2(n+1)^{-1}\left(2^{n+1} - 1\right)B_{n+1}$ $\quad [n = 1, 2, \ldots]$ $\quad$ AS 805 (23.1.20)

3. $E_n\left(\frac{1}{2}\right) = 2^{-n}E_n$ $\qquad [n = 0, 1, \ldots]$ $\qquad$ AS 805 (23.1.21)

4. $E_{2n-1}\left(\frac{1}{3}\right) = -E_{2n-1}\left(\frac{2}{3}\right) = -(2n)^{-1}\left(1 - 3^{1-2n}\right)\left(2^{2n} - 1\right)B_{2n}$

$\qquad\qquad\qquad\qquad\qquad\qquad\qquad\qquad [n = 1, 2, \ldots]$ $\qquad$ AS 806 (23.1.22)

# 9.7 Constants

## 9.71 Bernoulli numbers

- $B_0 = 1$
- $B_1 = -1/2$
- $B_2 = 1/6$
- $B_4 = -1/30$
- $B_6 = 1/42$
- $B_8 = -1/30$
- $B_{10} = 5/66$
- $B_{12} = -691/2730$
- $B_{14} = 7/6$
- $B_{16} = -3617/510$

- $B_{18} = 43\,867/798$
- $B_{20} = -174\,611/330$
- $B_{22} = 854\,513/138$
- $B_{24} = -236\,364\,091/2730$
- $B_{26} = 8\,553\,103/6$
- $B_{28} = -23\,749\,461\,029/870$
- $B_{30} = 8\,615\,841\,276\,005/14\,322$
- $B_{32} = -7\,709\,321\,041\,217/510$
- $B_{34} = 2\,577\,687\,858\,367/6$

## 9.72 Euler numbers

- $E_0 = 1$
- $E_2 = -1$
- $E_4 = 5$
- $E_6 = -61$
- $E_8 = 1385$
- $E_{10} = -50\,521$

- $E_{12} = 2\,702\,765$
- $E_{14} = -199\,360\,981$
- $E_{16} = 19\,391\,512\,145$
- $E_{18} = -2\,404\,879\,675\,441$
- $E_{20} = 370\,371\,188\,237\,525$

The Bernoulli and Euler numbers of odd index (with the exception of $B_1$) are equal to zero.

## 9.73 Euler's and Catalan's constants

**Euler's constant**

$$C = 0.577\,215\,664\,901\,532\,860\,606\,512\ldots \qquad\qquad (\text{cf. } \mathbf{8.367})$$

**Catalan's constant**

$$G = \sum_{k=0}^{\infty} \frac{(-1)^k}{(2k+1)^2} = 0.915\,965\,594\ldots$$

## 9.74$^{10}$ Stirling numbers

**9.740** The **Stirling number of the first kind** $S_n^{(m)}$ is defined by the requirement that $(-1)^{n-m}S_n^{(m)}$ is the number of permutations of $n$ symbols which have exactly $m$ cycles. AS 824 (23.1.3)

**9.741** Generating functions:

1. $$x(x-1)\cdots(x-n+1) = \sum_{m=0}^{n} S_n^{(m)} x^m \qquad\qquad \text{AS 824 (24.1.3)}$$

2. $$\{\ln(1+x)\}^m = m! \sum_{n=m}^{\infty} S_n^{(m)} \frac{x^n}{n!} \qquad [|x|<1] \qquad \text{AS 824 (24.1.3)}$$

**9.742** Recurrence relations:

1.[8] $$S_{n+1}^{(m)} = S_n^{(m-1)} - nS_n^{(m)}; \quad S_n^{(0)} = \delta_{0n}; \quad S_n^{(1)} = (-1)^{n-1}(n-1)!; \quad S_n^{(n)} = 1$$
$$[n \geq m \geq 1] \qquad \text{AS 824 (24.1.3)}$$

2. $$\binom{m}{r} S_n^{(m)} = \sum_{k=m-r}^{n-r} \binom{n}{k} S_{n-k}^{(r)} S_k^{(m+r)} \qquad [n \geq m \geq r] \qquad \text{AS 824 (24.1.3)}$$

**9.743** Functional relations and properties

1. $$x(x-h)(x-2h)\cdots(x-mh+h) = \frac{h^m\,\Gamma\left(\frac{x}{h}+1\right)}{\Gamma\left(\frac{x}{h}-m+1\right)} = h^m \sum_{k=1}^{m} \left(\frac{x}{h}\right)^k S_k^{(m)}$$

2. $$[(x+1)(x+2)\cdots(x+m)]^{-1} = \left[\binom{x+m}{m} m!\right]^{-1} = \left[\sum_{k=1}^{p} (x+m)^k S_k^{(m)}\right]^{-1}$$

3. $$[(x+h)(x+2h)\cdots(x+mh)]^{-1} = \frac{\Gamma\left(\frac{x}{h}+1\right)}{h^m\,\Gamma\left(\frac{x}{h}+m+1\right)} = \left[h^m \sum_{k=1}^{m} \left(\frac{x}{h}+m\right)^k S_k^{(m)}\right]^{-1}$$

**9.744** The Stirling number of the second kind $\mathfrak{S}_n^{(m)}$ is the number of ways of partitioning a set of $n$ elements into $m$ non-empty subsets.

**9.745** Generating functions:

1. $\quad x^n = \sum_{m=0}^{n} \mathfrak{S}_n^{(m)} x(x-1)\cdots(x-m+1)$ AS 824 (24.1.4)

2. $\quad (e^x - 1)^m = m! \sum_{n=m}^{\infty} \mathfrak{S}_n^{(m)} \dfrac{x^n}{n!}$ AS 824 (24.1.4)

3. $\quad [(1-x)(1-2x)\cdots(1-mx)]^{-1} = \sum_{n=m}^{\infty} \mathfrak{S}_n^{(m)} x^{n-m} \qquad \left[|x| < m^{-1}\right]$ AS 824 (24.1.4)

**9.746** Closed form expression:

1. $\quad \mathfrak{S}_n^{(m)} = \dfrac{1}{m!} \sum_{k=0}^{m} (-1)^{m-k} \dbinom{m}{k} k^n$ AS 824 (24.1.4)

**9.747** Recurrence relations:

1.[8] $\quad \mathfrak{S}_{n+1}^{(m)} = m\mathfrak{S}_n^{(m)} + \mathfrak{S}_n^{(m-1)}, \quad \mathfrak{S}_n^{(0)} = \delta_{0n}, \quad \mathfrak{S}_n^{(1)} = \mathfrak{S}_n^{(n)} = 1$

$\qquad\qquad\qquad\qquad\qquad\qquad\qquad\qquad [n \geq m \geq 1]$ AS 825(24.1.4)

2. $\quad \dbinom{m}{r}\mathfrak{S}_n^{(m)} = \sum_{k=m-r}^{n-r} \dbinom{n}{k}\mathfrak{S}_{n-k}^{(r)}\mathfrak{S}_k^{(m-r)} \qquad [n \geq m \geq r]$ AS 825 (24.1.4)

3. $\quad S_n^{(m)} = \sum_{k=0}^{n-m} (-1)^k \dbinom{n-1+k}{n-m+k}\dbinom{2n-m}{n-m-k}\mathfrak{S}_{n-m+k}^{(k)}$ AS 824 (24.1.3)

**9.748**[7] Particular values:

Stirling numbers of the first kind $S_n^{(m)}$

| $m$ | $S_1^{(m)}$ | $S_2^{(m)}$ | $S_3^{(m)}$ | $S_4^{(m)}$ | $S_5^{(m)}$ | $S_6^{(m)}$ | $S_7^{(m)}$ | $S_8^{(m)}$ | $S_9^{(m)}$ |
|---|---|---|---|---|---|---|---|---|---|
| 1 | 1 | -1 | 2 | -6 | 24 | -120 | 720 | -5040 | 40320 |
| 2 | | 1 | -3 | 11 | -50 | 274 | -1764 | 13068 | -109584 |
| 3 | | | 1 | -6 | 35 | -225 | 1624 | -13132 | 118121 |
| 4 | | | | 1 | -10 | 85 | -735 | 6769 | -67284 |
| 5 | | | | | 1 | -15 | 175 | -1960 | 22449 |
| 6 | | | | | | 1 | -21 | 332 | -4536 |
| 7 | | | | | | | 1 | -28 | 546 |
| 8 | | | | | | | | 1 | -36 |
| 9 | | | | | | | | | 1 |

Stirling numbers of the second kind $\mathfrak{S}_n^{(m)}$

| $m$ | $\mathfrak{S}_1^{(m)}$ | $\mathfrak{S}_2^{(m)}$ | $\mathfrak{S}_3^{(m)}$ | $\mathfrak{S}_4^{(m)}$ | $\mathfrak{S}_5^{(m)}$ | $\mathfrak{S}_6^{(m)}$ | $\mathfrak{S}_7^{(m)}$ | $\mathfrak{S}_8^{(m)}$ | $\mathfrak{S}_9^{(m)}$ |
|---|---|---|---|---|---|---|---|---|---|
| 1 | 1 | 1 | 1 | 1 | 1 | 1 | 1 | 1 | 1 |
| 2 | | 1 | 3 | 7 | 15 | 31 | 63 | 127 | 255 |
| 3 | | | 1 | 6 | 25 | 90 | 301 | 966 | 3025 |
| 4 | | | | 1 | 10 | 65 | 350 | 1701 | 7770 |
| 5 | | | | | 1 | 15 | 140 | 1050 | 6951 |
| 6 | | | | | | 1 | 21 | 266 | 2646 |
| 7 | | | | | | | 1 | 28 | 462 |
| 8 | | | | | | | | 1 | 36 |
| 9 | | | | | | | | | 1 |

**9.749[8]**    Relationship between Stirling numbers of the first kind and derivatives of $(\ln x)^{-m}$:

1. $$\frac{d^n}{dx^n}\left(\frac{1}{\ln^m x}\right) = \frac{1}{\ln^m x}\sum_{k=1}^{n}\frac{(-1)^k (m)_k S_n^{(k)}}{x^n \ln^k x}$$

$$\text{where } (m)_k = \Gamma(m+k)/\Gamma(m), \qquad [m, n \text{ are positive integers}]$$

Table of Integrals, Series, and Products. http://dx.doi.org/10.1016/B978-0-12-384933-5.00010-2

# 10 Vector Field Theory

## 10.1–10.8 Vectors, Vector Operators, and Integral Theorems

### 10.11  Products of vectors

Let $\mathbf{a} = (a_1, a_2, a_3)$, $\mathbf{b} = (b_1, b_2, b_3)$, and $\mathbf{c} = (c_1, c_2, c_2)$ be arbitrary vectors, and $\mathbf{i}$, $\mathbf{j}$, $\mathbf{k}$ be the set of orthogonal unit vectors in terms of which the components of $\mathbf{a}$, $\mathbf{b}$, and $\mathbf{c}$ are expressed. Two different products involving pairs of vectors are defined, namely, the scalar product, written $\mathbf{a} \cdot \mathbf{b}$, and the vector product, written either $\mathbf{a} \times \mathbf{b}$ or $\mathbf{a} \wedge \mathbf{b}$. Their properties are as follows:

1.  $\mathbf{a} \cdot \mathbf{b} = a_1 b_1 + a_2 b_2 + a_3 b_3$        (scalar product)

2.  $\mathbf{a} \times \mathbf{b} = \begin{vmatrix} \mathbf{i} & \mathbf{j} & \mathbf{k} \\ a_1 & a_2 & a_3 \\ b_1 & b_2 & b_3 \end{vmatrix}$        (vector product)

3.  $\mathbf{a} \times \mathbf{b} \cdot \mathbf{c} = \begin{vmatrix} a_1 & a_2 & a_3 \\ b_1 & b_2 & b_3 \\ c_1 & c_2 & c_3 \end{vmatrix}$        (triple scalar product)

4.  $\mathbf{a} \times (\mathbf{b} \times \mathbf{c}) = (\mathbf{a} \cdot \mathbf{c})\, \mathbf{b} - (\mathbf{a} \cdot \mathbf{b})\, \mathbf{c}$        (triple vector product)

### 10.12  Properties of scalar product

1.  $\mathbf{a} \cdot \mathbf{b} = \mathbf{b} \cdot \mathbf{a}$        (commutative)

2.  $\mathbf{a} \times \mathbf{b} \cdot \mathbf{c} = \mathbf{b} \times \mathbf{c} \cdot \mathbf{a} = \mathbf{c} \times \mathbf{a} \cdot \mathbf{b} = -\mathbf{a} \times \mathbf{c} \cdot \mathbf{b} = -\mathbf{b} \times \mathbf{a} \cdot \mathbf{c} = -\mathbf{c} \times \mathbf{b} \cdot \mathbf{a}$.

    *Note:* $\mathbf{a} \times \mathbf{b} \cdot \mathbf{c}$ is also written $[\mathbf{a}, \mathbf{b}, \mathbf{c}]$; thus (2) may also be written

3.  $[\mathbf{a}, \mathbf{b}, \mathbf{c}] = [\mathbf{b}, \mathbf{c}, \mathbf{a}] = [\mathbf{c}, \mathbf{a}, \mathbf{b}] = - [\mathbf{a}, \mathbf{c}, \mathbf{b}] = - [\mathbf{b}, \mathbf{a}, \mathbf{c}] = - [\mathbf{c}, \mathbf{b}, \mathbf{a}]$

### 10.13  Properties of vector product

1.  $\mathbf{a} \times \mathbf{b} = -\mathbf{b} \times \mathbf{a}$        (anticommutative)

2.  $\mathbf{a} \times (\mathbf{b} \times \mathbf{c}) = -\mathbf{a} \times (\mathbf{c} \times \mathbf{b}) = - (\mathbf{b} \times \mathbf{c}) \times \mathbf{a}$

3.  $\mathbf{a} \times (\mathbf{b} \times \mathbf{c}) + \mathbf{b} \times (\mathbf{c} \times \mathbf{a}) + \mathbf{c} \times (\mathbf{a} \times \mathbf{b}) = \mathbf{0}$

## 10.14 Differentiation of vectors

If $\mathbf{a}(t) = (a_1(t), a_2(t), a_3(t))$, $\mathbf{b}(t) = (b_1(t), b_2(t), b_3(t))$, $\mathbf{c}(t) = (c_1(t), c_2(t), c_3(t))$, $\phi(t)$ is a scalar and all functions of $t$ are differentiable, then

1. $\dfrac{d\mathbf{a}}{dt} = \dfrac{da_1}{dt}\mathbf{i} + \dfrac{da_2}{dt}\mathbf{j} + \dfrac{da_3}{dt}\mathbf{k}$

2. $\dfrac{d}{dt}(\mathbf{a} + \mathbf{b}) = \dfrac{d\mathbf{a}}{dt} + \dfrac{d\mathbf{b}}{dt}$

3. $\dfrac{d}{dt}(\phi\mathbf{a}) = \dfrac{d\phi}{dt}\mathbf{a} + \phi\dfrac{d\mathbf{a}}{dt}$

4. $\dfrac{d}{dt}(\mathbf{a} \cdot \mathbf{b}) = \dfrac{d\mathbf{a}}{dt} \cdot \mathbf{b} + \mathbf{a} \cdot \dfrac{d\mathbf{b}}{dt}$

5. $\dfrac{d}{dt}(\mathbf{a} \times \mathbf{b}) = \dfrac{d\mathbf{a}}{dt} \times \mathbf{b} + \mathbf{a} \times \dfrac{d\mathbf{b}}{dt}$

6. $\dfrac{d}{dt}(\mathbf{a} \times \mathbf{b} \cdot \mathbf{c}) = \dfrac{d\mathbf{a}}{dt} \times \mathbf{b} \cdot \mathbf{c} + \mathbf{a} \times \dfrac{d\mathbf{b}}{dt} \cdot \mathbf{c} + \mathbf{a} \times \mathbf{b} \cdot \dfrac{d\mathbf{c}}{dt}$

7. $\dfrac{d}{dt}\{\mathbf{a} \times (\mathbf{b} \times \mathbf{c})\} = \dfrac{d\mathbf{a}}{dt} \times (\mathbf{b} \times \mathbf{c}) + \mathbf{a} \times \left(\dfrac{d\mathbf{b}}{dt} \times \mathbf{c}\right) + \mathbf{a} \times \left(\mathbf{b} \times \dfrac{d\mathbf{c}}{dt}\right)$

## 10.21 Operators grad, div, and curl

In cartesian coordinates $O\{x_1, x_2, x_3\}$, in which system it is convenient to denote the triad of unit vectors by $\mathbf{e}_1, \mathbf{e}_2, \mathbf{e}_3$, the vector operator $\nabla$, called either "del" or "nabla", has the form

1. $\nabla \equiv \mathbf{e}_1\dfrac{\partial}{\partial x_1} + \mathbf{e}_2\dfrac{\partial}{\partial x_2} + \mathbf{e}_3\dfrac{\partial}{\partial x_3}$

   If $\Phi(x, y, z)$ is any differentiable scalar function, the gradient of $\Phi$, written grad $\Phi$, is

2. $\operatorname{grad} \Phi \equiv \nabla\,\Phi = \dfrac{\partial \Phi}{\partial x_1}\mathbf{e}_1 + \dfrac{\partial \Phi}{\partial x_2}\mathbf{e}_2 + \dfrac{\partial \Phi}{\partial x_3}\mathbf{e}_3$

   The divergence of the differentiable vector function $\mathbf{f} = (f_1, f_2, f_3)$, written div $\mathbf{f}$, is

3. $\operatorname{div} f \equiv \nabla \cdot \mathbf{f} = \dfrac{\partial f_1}{\partial x_1} + \dfrac{\partial f_2}{\partial x_2} + \dfrac{\partial f_3}{\partial x_3}$

   The curl, or rotation, of the differentiable vector function $\mathbf{f} = (f_1, f_2, f_3)$, written either curl $\mathbf{f}$ or rot $\mathbf{f}$, is

4. $\operatorname{curl} \mathbf{f} \equiv \operatorname{rot} \mathbf{f} \equiv \nabla \times \mathbf{f} = \left(\dfrac{\partial f_3}{\partial x_2} - \dfrac{\partial f_2}{\partial x_3}\right)\mathbf{e}_1 + \left(\dfrac{\partial f_1}{\partial x_3} - \dfrac{\partial f_3}{\partial x_1}\right)\mathbf{e}_2 + \left(\dfrac{\partial f_2}{\partial x_1} - \dfrac{\partial f_1}{\partial x_2}\right)\mathbf{e}_3,$

   or equivalently,

$$\operatorname{curl} \mathbf{f} = \begin{vmatrix} \mathbf{e}_1 & \mathbf{e}_2 & \mathbf{e}_2 \\ \dfrac{\partial}{\partial x_1} & \dfrac{\partial}{\partial x_2} & \dfrac{\partial}{\partial x_3} \\ f_1 & f_2 & f_3 \end{vmatrix}$$

## 10.31 Properties of the operator $\nabla$

Let $\Phi\,(x_1, x_2, x_3)$, $\Psi\,(x_1, x_2, x_3)$ be any two differentiable scalar functions, $\mathbf{f}\,(x_1, x_2, x_3)$, $\mathbf{g}\,(x_1, x_2, x_3)$ any two differentiable vector functions, and $\mathbf{a}$ an arbitrary vector. Define the scalar operator $\nabla^2$, called the Laplacian, by

$$\nabla^2 \equiv \frac{\partial^2}{\partial x_1^2} + \frac{\partial^2}{\partial x_2^2} + \frac{\partial^2}{\partial x_3^2}$$

Then, in terms of the operator $\nabla$, we have the following:      MF I 114

1.    $\nabla(\Phi + \Psi) = \nabla\,\Phi + \nabla\,\Psi$

2.    $\nabla(\Phi\Psi) = \Phi\,\nabla\,\Psi + \Psi\,\nabla\,\Phi$

3.    $\nabla\,(\mathbf{f} \cdot \mathbf{g}) = (\mathbf{f} \cdot \nabla)\,\mathbf{g} + (\mathbf{g} \cdot \nabla)\,\mathbf{f} + \mathbf{f} \times (\nabla \times \mathbf{g}) + \mathbf{g} \times (\nabla \times \mathbf{f})$

4.    $\nabla \cdot (\Phi\mathbf{f}) = \Phi\,(\nabla \cdot \mathbf{f}) + \mathbf{f} \cdot \nabla\,\Phi$

5.    $\nabla \cdot (\mathbf{f} \times \mathbf{g}) = \mathbf{g} \cdot (\nabla \times \mathbf{f}) - \mathbf{f} \cdot (\nabla \times \mathbf{g})$

6.    $\nabla \times (\Phi\mathbf{f}) = \Phi\,(\nabla \times \mathbf{f}) + (\nabla\,\Phi) \times \mathbf{f}$

7.    $\nabla \times (\mathbf{f} \times \mathbf{g}) = \mathbf{f}\,(\nabla \cdot \mathbf{g}) - \mathbf{g}\,(\nabla \cdot \mathbf{f}) + (\mathbf{g} \cdot \nabla)\,\mathbf{f} - (\mathbf{f} \cdot \nabla)\,\mathbf{g}$

8.    $\nabla \times (\nabla \times \mathbf{f}) = \nabla\,(\nabla \cdot \mathbf{f}) - \nabla^2\,\mathbf{f}$

9.    $\nabla \times (\nabla\,\Phi) \equiv \mathbf{0}$

10.   $\nabla \cdot (\nabla \times \mathbf{f}) \equiv 0$

11.[10]   $\nabla^2(\Phi\Psi) = \Phi\,\nabla^2\,\Psi + 2\,(\nabla\,\Phi) \cdot (\nabla\,\Psi) + \Psi\,\nabla^2\,\Phi$

The equivalent results in terms of grad, div, and curl are as follows:

1.    $\mathrm{grad}(\Phi + \Psi) = \mathrm{grad}\,\Phi + \mathrm{grad}\,\Psi$

2.    $\mathrm{grad}(\Phi\Psi) = \Phi\,\mathrm{grad}\,\Psi + \Psi\,\mathrm{grad}\,\Phi$

3.    $\mathrm{grad}\,(\mathbf{f} \cdot \mathbf{g}) = (\mathbf{f} \cdot \mathrm{grad})\,\mathbf{g} + (\mathbf{g} \cdot \mathrm{grad})\,\mathbf{f} + \mathbf{f} \times \mathrm{curl}\,\mathbf{g} + \mathbf{g} \times \mathrm{curl}\,\mathbf{f}$

4.    $\mathrm{div}\,(\Phi\mathbf{f}) = \Phi\,\mathrm{div}\,\mathbf{f} + \mathbf{f} \cdot \mathrm{grad}\,\Phi$

5.    $\mathrm{div}\,(\mathbf{f} \times \mathbf{g}) = \mathbf{g} \cdot \mathrm{curl}\,\mathbf{f} - \mathbf{f} \cdot \mathrm{curl}\,\mathbf{g}$

6.    $\mathrm{curl}\,(\Phi\mathbf{f}) = \Phi\,\mathrm{curl}\,\mathbf{f} + \mathrm{grad}\,\Phi \times \mathbf{f}$

7.    $\mathrm{curl}\,(\mathbf{f} \times \mathbf{g}) = \mathbf{f}\,\mathrm{div}\,\mathbf{g} - \mathbf{g}\,\mathrm{div}\,\mathbf{f} + (\mathbf{g} \cdot \mathrm{grad})\,\mathbf{f} - (\mathbf{f} \cdot \mathrm{grad})\,\mathbf{g}$

8.    $\mathrm{curl}\,(\mathrm{curl}\,\mathbf{f}) = \mathrm{grad}\,(\mathrm{div}\,\mathbf{f}) - \nabla^2\,\mathbf{f}$

9.    $\mathrm{curl}\,(\mathrm{grad}\,\Phi) \equiv \mathbf{0}$

10.   $\mathrm{div}\,(\mathrm{curl}\,\mathbf{f}) \equiv 0$

11.   $\nabla^2(\Phi\Psi) = \Phi\,\nabla^2\,\Psi + 2\,\mathrm{grad}\,\Phi \cdot \mathrm{grad}\,\Psi + \Psi\,\nabla^2\,\Phi$

The expression $(\mathbf{a} \cdot \nabla)$ or, equivalently $(\mathbf{a} \cdot \mathrm{grad})$, defined by

$$(\mathbf{a} \cdot \nabla) \equiv a_1\frac{\partial}{\partial x_1} + a_2\frac{\partial}{\partial x_2} + a_3\frac{\partial}{\partial x_3},$$

is the directional derivative operator in the direction of vector $\mathbf{a}$.

## 10.41 Solenoidal fields

A vector field $\mathbf{f}$ is said to be solenoidal if div $\mathbf{f} \equiv 0$. We have the following representation.

**10.411**   *Representation theorem for vector Helmholtz equation.* If $u$ is a solution of the scalar Helmholtz equation

$$\nabla^2 u + \lambda^2 u = 0,$$

and $\mathbf{m}$ is a constant unit vector, then the vectors

$$\mathbf{X} = \operatorname{curl}(\mathbf{m}u), \qquad \mathbf{Y} = \frac{1}{\lambda}\operatorname{curl}\mathbf{X}$$

are independent solutions of the vector Helmholtz equation

$$\nabla^2 \mathbf{H} + \lambda^2 \mathbf{H} = \mathbf{0}$$

involving a solenoidal vector $\mathbf{H}$. The general solution of the equation is

$$\mathbf{H} = \operatorname{curl}(\mathbf{m}u) + \frac{1}{\lambda}\operatorname{curl}\operatorname{curl}(\mathbf{m}u)$$

## 10.51–10.61 Orthogonal curvilinear coordinates

Consider a transformation from the cartesian coordinates $O\{x_1, x_2, x_3\}$ to the general orthogonal curvilinear coordinates $O\{u_1, u_2, u_3\}$:

$$x_1 = x_1(u_1, u_2, u_3), \qquad x_2 = x_2(u_1, u_2, u_3), \qquad x_3 = x_3(u_1, u_2, u_3)$$

Then,

1.    $\mathrm{d}x_i = \dfrac{\partial x_i}{\partial u_1}\,\mathrm{d}u_1 + \dfrac{\partial x_i}{\partial u_2}\,\mathrm{d}u_2 + \dfrac{\partial x_i}{\partial u_3}\,\mathrm{d}u_3 \qquad (i = 1, 2, 3),$

     and the length element $\mathrm{d}l$ may be determined from

2.    $\mathrm{d}l^2 = g_{11}\,\mathrm{d}u_1^2 + g_{22}\,\mathrm{d}u_2^2 + g_{33}\,\mathrm{d}u_3^2 + 2g_{23}\,\mathrm{d}u_2\,\mathrm{d}u_3 + 2g_{31}\,\mathrm{d}u_3\,\mathrm{d}u_1 + 2g_{12}\,\mathrm{d}u_1\,\mathrm{d}u_2,$

     where

3.3    $g_{ij} = \dfrac{\partial x_1}{\partial u_i}\dfrac{\partial x_1}{\partial u_j} + \dfrac{\partial x_2}{\partial u_i}\dfrac{\partial x_2}{\partial u_j} + \dfrac{\partial x_3}{\partial u_i}\dfrac{\partial x_3}{\partial u_j} = g_{ji}, \qquad\qquad g_{ij} = 0, \quad i \neq j.$

     provided the Jacobian of the transformation

4.    $J = \begin{vmatrix} \dfrac{\partial x_1}{\partial u_1} & \dfrac{\partial x_2}{\partial u_1} & \dfrac{\partial x_3}{\partial u_1} \\[4pt] \dfrac{\partial x_1}{\partial u_2} & \dfrac{\partial x_2}{\partial u_2} & \dfrac{\partial x_3}{\partial u_2} \\[4pt] \dfrac{\partial x_1}{\partial u_3} & \dfrac{\partial x_2}{\partial u_3} & \dfrac{\partial x_3}{\partial u_3} \end{vmatrix}$

     does not vanish (see **14.313**).

     Define the metrical coefficients

5.    $h_1 = \sqrt{g_{11}}, \quad h_2 = \sqrt{g_{22}}, \quad h_3 = \sqrt{g_{33}};$

     then the volume element $\mathrm{d}V$ in orthogonal curvilinear coordinates is

6.    $\mathrm{d}V = h_1 h_2 h_3\,\mathrm{d}u_1\,\mathrm{d}u_2\,\mathrm{d}u_3,$

     and the surface elements of area $\mathrm{d}s_i$ on the surfaces $u_i = \text{constant}$, for $i = 1, 2, 3$, are

7.    $\mathrm{d}s_1 = h_2 h_3\,\mathrm{d}u_2\,\mathrm{d}u_3, \quad \mathrm{d}s_2 = h_1 h_3\,\mathrm{d}u_1\,\mathrm{d}u_3, \quad \mathrm{d}s_3 = h_1 h_2\,\mathrm{d}u_1\,\mathrm{d}u_2$

     Denote by $\mathbf{e}_1, \mathbf{e}_2,$ and $\mathbf{e}_3$ the triad of orthogonal unit vectors that are tangent to the $u_1, u_2,$ and $u_3$ coordinate lines through any given point $P$, and choose their sense so that they form a right-handed set in this order. Then in terms of this triad of vectors and the components $f_{u_1}, f_{u_2},$ and $f_{u_3}$ of $\mathbf{f}$ along the coordinate line,

8.      $\mathbf{f} = f_{u_1}\mathbf{e}_1 + f_{u_2}\mathbf{e}_2 + f_{u_3}\mathbf{e}_3$                                      MF I 115

**10.611**   $\nabla\,\Phi$, div $\mathbf{f}$, curl $\mathbf{f}$, *and* $\nabla^2$ *in general orthogonal curvilinear coordinates.*

1.      $\operatorname{grad}\Phi = \dfrac{\mathbf{e}_1}{h_1}\dfrac{\partial\Phi}{\partial u_1} + \dfrac{\mathbf{e}_2}{h_2}\dfrac{\partial\Phi}{\partial u_2} + \dfrac{\mathbf{e}_3}{h_3}\dfrac{\partial\Phi}{\partial u_3}$

2.[3]    $\operatorname{div}\mathbf{f} = \dfrac{1}{h_1 h_2 h_3}\left(\dfrac{\partial}{\partial u_1}(h_2 h_3 f_{u_1}) + \dfrac{\partial}{\partial u_2}(h_3 h_1 f_{u_2}) + \dfrac{\partial}{\partial u_3}(h_1 h_2 f_{u_3})\right)$

3.      $\operatorname{curl}\mathbf{f} = \dfrac{1}{h_1 h_2 h_3}\begin{vmatrix} h_1\mathbf{e}_1 & h_2\mathbf{e}_2 & h_3\mathbf{e}_3 \\ \frac{\partial}{\partial u_1} & \frac{\partial}{\partial u_2} & \frac{\partial}{\partial u_3} \\ h_1 f_{u_1} & h_2 f_{u_2} & h_3 f_{u_3} \end{vmatrix}$

4.      $\nabla^2 \equiv \dfrac{1}{h_1 h_2 h_3}\left(\dfrac{\partial}{\partial u_1}\left(\dfrac{h_2 h_3}{h_1}\dfrac{\partial}{\partial u_1}\right) + \dfrac{\partial}{\partial u_2}\left(\dfrac{h_3 h_1}{h_2}\dfrac{\partial}{\partial u_2}\right) + \dfrac{\partial}{\partial u_3}\left(\dfrac{h_1 h_2}{h_3}\dfrac{\partial}{\partial u_3}\right)\right)$      MF I 21-31

**10.612**   *Cylindrical polar coordinates.* In terms of the coordinates $O\{r,\phi,z\}$, that is, $u_1 = r$, $u_2 = \phi$, $u_3 = z$, where $x_1 = r\cos\phi$, $x_2 = r\sin\phi$, $x_3 = z$ for $-\pi < \phi \le \pi$, it follows that

1.      $h_1 = 1, \quad h_2 = r, \quad h_3 = 1,$

      and

2.      $\operatorname{grad}\Phi = \dfrac{\partial\Phi}{\partial r}\mathbf{e}_r + \dfrac{1}{r}\dfrac{\partial\Phi}{\partial\phi}\mathbf{e}_\phi + \dfrac{\partial\Phi}{\partial z}\mathbf{e}_z,$

3.      $\operatorname{div}\mathbf{f} = \dfrac{1}{r}\dfrac{\partial}{\partial r}(rf_r) + \dfrac{1}{r}\dfrac{\partial f_\phi}{\partial\phi} + \dfrac{\partial f_z}{\partial z},$

4.      $\operatorname{curl}\mathbf{f} = \dfrac{1}{r}\begin{vmatrix} \mathbf{e}_r & r\mathbf{e}_\phi & \mathbf{e}_z \\ \frac{\partial}{\partial r} & \frac{\partial}{\partial\phi} & \frac{\partial}{\partial z} \\ f_r & r f_\phi & f_z \end{vmatrix},$

5.      $\nabla^2 \equiv \dfrac{1}{r}\dfrac{\partial}{\partial r}\left(r\dfrac{\partial}{\partial r}\right) + \dfrac{1}{r^2}\dfrac{\partial^2}{\partial\phi^2} + \dfrac{\partial^2}{\partial z^2}$                  MF I 116

**10.613**   *Spherical polar coordinates.* In terms of the coordinates $O\{r,\theta,\phi\}$, that is, $u_1 = r$, $u_2 = \theta$, $u_3 = \phi$, where $x_1 = r\sin\theta\cos\phi$, $x_2 = r\sin\theta\sin\phi$, $x_3 = r\cos\theta$, for $0 \le \theta \le \pi$, $-\pi < \phi \le \pi$, we have

1.      $h_1 = 1, \quad h_2 = r, \quad h_3 = r\sin\theta,$

      and

2.[10]   $\operatorname{grad}\Phi = \dfrac{\partial\Phi}{\partial r}\mathbf{e}_r + \dfrac{1}{r}\dfrac{\partial\Phi}{\partial\theta}\mathbf{e}_\theta + \dfrac{1}{r\sin\theta}\dfrac{\partial\Phi}{\partial\phi}\mathbf{e}_\phi,$

3.      $\operatorname{div}\mathbf{f} = \dfrac{1}{r^2}\dfrac{\partial}{\partial r}(r^2 f_r) + \dfrac{1}{r\sin\theta}\dfrac{\partial}{\partial\theta}(f_\theta\sin\theta) + \dfrac{1}{r\sin\theta}\dfrac{\partial f_\phi}{\partial\phi},$

4.      $\operatorname{curl}\mathbf{f} = \dfrac{1}{r^2\sin\theta}\begin{vmatrix} \mathbf{e}_r & r\mathbf{e}_\theta & r\sin\theta\mathbf{e}_\phi \\ \frac{\partial}{\partial r} & \frac{\partial}{\partial\theta} & \frac{\partial}{\partial\phi} \\ f_r & r f_\theta & r\sin\theta f_\phi \end{vmatrix},$

5.      $\nabla^2 \equiv \dfrac{1}{r^2}\dfrac{\partial}{\partial r}\left(r^2\dfrac{\partial}{\partial r}\right) + \dfrac{1}{r^2\sin\theta}\dfrac{\partial}{\partial\theta}\left(\sin\theta\dfrac{\partial}{\partial\theta}\right) + \dfrac{1}{r^2\sin^2\theta}\dfrac{\partial^2}{\partial\phi^2}$      MF I 116

**Special Orthogonal Curvilinear Coordinates and their Metrical Coefficients $h_1, h_2, h_3$**

**10.614**  *Elliptic cylinder coordinates* $O\{u_1, u_2, u_3\}$.

1. $\quad x_1 = u_1 u_2, \qquad x_2 = \sqrt{(u_1^2 - c^2)(1 - u_2^2)}, \qquad x_3 = u_3$

2. $\quad h_1 = \sqrt{\dfrac{u_1^2 - c^2 u_2^2}{u_1^2 - c^2}}, \qquad h_2 = \sqrt{\dfrac{u_1^2 - c^2 u_2^2}{1 - u_2^2}}, \qquad h_3 = 1$ $\hspace{2cm}$ MF I 657

**10.615**  *Parabolic cylinder coordinates* $O\{u_1, u_2, u_3\}$.

1. $\quad x_1 = \dfrac{1}{2}\left(u_1^2 - u_2^2\right), \qquad x_2 = u_1 u_2, \qquad x_3 = u_3$

2. $\quad h_1 = \sqrt{u_1^2 + u_2^2}, \qquad h_2 = \sqrt{u_1^2 + u_2^2}, \qquad h_3 = 1$ $\hspace{2cm}$ MF I 658

**10.616**  *Conical coordinates* $O\{u_1, u_2, u_3\}$.

1. $\quad x_1 = \dfrac{u_1}{a}\sqrt{(a^2 - u_2^2)(a^2 + u_3^2)}, \qquad x_2 = \dfrac{u_1}{b}\sqrt{(b^2 + u_2^2)(b^2 - u_3^2)}, \qquad x_3 = \dfrac{u_1 u_2 u_3}{ab}$

$$\text{with } a^2 + b^2 = 1$$

2. $\quad h_1 = 1, \qquad h_2 = u_1\sqrt{\dfrac{u_2^2 + u_3^2}{(a^2 - u_2^2)(b^2 + u_2^2)}}, \qquad h_3 = u_1\sqrt{\dfrac{u_2^2 + u_3^2}{(a^2 + u_3^2)(b^2 - u_3^2)}}$ $\hspace{1cm}$ MF I 659

**10.617**  *Rotational parabolic coordinates* $O\{u_1, u_2, u_3\}$.

1. $\quad x_1 = u_1 u_2 u_3, \qquad x_2 = u_1 u_2\sqrt{1 - u_3^2}, \qquad x_3 = \dfrac{1}{2}\left(u_1^2 - u_2^2\right)$

2. $\quad h_1 = \sqrt{u_1^2 + u_2^2}, \qquad h_2 = \sqrt{u_1^2 + u_2^2}, \qquad h_3 = \dfrac{u_1 u_2}{\sqrt{1 - u_3^2}}$ $\hspace{2cm}$ MF I 660

**10.618**  *Rotational prolate spheroidal coordinates* $O\{u_1, u_2, u_3\}$.

1. $\quad x_1 = \sqrt{(u_1^2 - a^2)(1 - u_2^2)}, \qquad x_2 = \sqrt{(u_1^2 - a^2)(1 - u_2^2)(1 - u_3^2)}, \qquad x_3 = u_1 u_2$

2. $\quad h_1 = \sqrt{\dfrac{u_1^2 - a^2 u_2^2}{u_1^2 - a^2}}, \qquad h_2 = \sqrt{\dfrac{u_1^2 - a^2 u_2^2}{1 - u_2^2}}, \qquad h_3 = \sqrt{\dfrac{(u_1^2 - a^2)(1 - u_2^2)}{1 - u_3^2}}$ $\hspace{1cm}$ MF I 661

**10.619**  *Rotational oblate spheroidal coordinates* $O\{u_1, u_2, u_3\}$.

1. $\quad x_1 = u_3\sqrt{(u_1^2 + a^2)(1 - u_2^2)}, \qquad x_2 = \sqrt{(u_1^2 + a^2)(1 - u_2^2)(1 - u_3^2)}, \qquad x_3 = u_1 u_2$

2. $\quad h_1 = \sqrt{\dfrac{u_1^2 + a^2 u_2^2}{u_1^2 + a^2}}, \qquad h_2 = \sqrt{\dfrac{u_1^2 + a^2 u_2^2}{1 - u_2^2}}, \qquad h_3 = \sqrt{\dfrac{(u_1^2 + a^2)(1 - u_2^2)}{1 - u_3^2}}$ $\hspace{1cm}$ MF I 662

**10.620**   *Ellipsoidal coordinates $O\{u_1, u_2, u_3\}$.*

1.      $x_1 = \sqrt{\dfrac{(u_1^2 - a^2)(u_2^2 - a^2)(u_3^2 - a^2)}{a^2(a^2 - b^2)}}, \quad x_2 = \sqrt{\dfrac{(u_1^2 - b^2)(u_2^2 - b^2)(u_3^2 - b^2)}{b^2(b^2 - a^2)}}, \quad x_3 = \dfrac{u_1 u_2 u_3}{ab}$

2.      $h_1 = \sqrt{\dfrac{(u_1^2 - u_2^2)(u_1^2 - u_3^2)}{(u_1^2 - a^2)(u_1^2 - b^2)}}, \quad h_2 = \sqrt{\dfrac{(u_2^2 - u_1^2)(u_2^2 - u_3^2)}{(u_2^2 - a^2)(u_2^2 - b^2)}}, \quad h_3 = \sqrt{\dfrac{(u_3^2 - u_1^2)(u_3^2 - u_2^2)}{(u_3^2 - a^2)(u_3^2 - b^2)}}$

MF I 663

**10.621**   *Paraboloidal coordinates $O\{u_1, u_2, u_3\}$.*

1.      $x_1 = \sqrt{\dfrac{(u_1^2 - a^2)(u_2^2 - a^2)(u_3^2 - a^2)}{a^2 - b^2}}, \quad x_2 = \sqrt{\dfrac{(u_1^2 - b^2)(u_2^2 - b^2)(u_3^2 - b^2)}{b^2 - a^2}},$

        $x_3 = \tfrac{1}{2}\left(u_1^2 + u_2^2 + u_3^2 - a^2 - b^2\right)$

2.      $h_1 = \sqrt{\dfrac{(u_1^2 - u_2^2)(u_1^2 - u_3^2)}{(u_1^2 - a^2)(u_1^2 - b^2)}}, \quad h_2 = u_2\sqrt{\dfrac{(u_3^2 - u_1^2)(u_3^2 - u_2^2)}{(u_2^2 - a^2)(u_2^2 - b^2)}}, \quad h_3 = u_3\sqrt{\dfrac{(u_3^2 - u_1^2)(u_3^2 - u_2^2)}{(u_3^2 - a^2)(u_3^2 - b^2)}}$

MF I 664

**10.622**   *Bispherical coordinates $O\{u_1, u_2, u_3\}$.*

1.      $x_1 = a u_3 \dfrac{\sqrt{1 - u_2^2}}{u_1 - u_2}, \quad x_2 = a\dfrac{\sqrt{(1 - u_2^2)(1 - u_3^2)}}{u_1 - u_2}, \quad x_3 = \dfrac{\sqrt{u_1^2 - 1}}{u_1 - u_2}$

2.      $h_1 = \dfrac{a}{(u_1 - u_2)\sqrt{u_1^2 - 1}},$

                  $h_2 = \dfrac{a}{(u_1 - u_2)\sqrt{1 - u_2^2}}, \quad h_3 = \left(\dfrac{a}{u_1 - u_2}\right)\sqrt{\dfrac{1 - u_2^2}{1 - u_3^2}} \qquad$ MF I 665

## 10.71–10.72  Vector integral theorems

**10.711**   *Gauss's divergence theorem.* Let $V$ be a volume bounded by a simple closed surface $S$ and let $\mathbf{f}$ be a continuously differentiable vector field defined in $V$ and on $S$. Then, if $\mathrm{d}\mathbf{S}$ is the outward drawn vector element of area,

$$\int_S \mathbf{f} \cdot \mathrm{d}\mathbf{S} = \int_V \operatorname{div} \mathbf{f} \, \mathrm{d}V \qquad\qquad \text{KE 39}$$

**10.712**   *Green's theorems.* Let $\Phi$ and $\Psi$ be scalar fields which, together with $\nabla^2 \Phi$ and $\nabla^2 \Psi$, are defined both in a volume $V$ and on its surface $S$, which we assume to be simple and closed. Then, if $\partial/\partial n$ denotes differentiation along the outward drawn normal to $S$, we have

**10.713**   *Green's first theorem*

$$\int_S \Phi \frac{\partial \Psi}{\partial n} \, \mathrm{d}S = \int_V \left(\Phi \nabla^2 \Psi + \operatorname{grad} \Phi \cdot \operatorname{grad} \Psi\right) \mathrm{d}V \qquad\qquad \text{KE 212}$$

**10.714** *Green's second theorem*

$$\int_S \left( \Phi \frac{\partial \Psi}{\partial n} - \Psi \frac{\partial \Phi}{\partial n} \right) dS = \int_V \left( \Phi \nabla^2 \Psi - \Psi \nabla^2 \Phi \right) dV \qquad \text{KE 215}$$

**10.715** *Special cases*

1.  $$\int_S (\Phi \operatorname{grad} \Phi) \cdot d\mathbf{S} = \int_V \left( \Phi \nabla^2 \Phi + (\operatorname{grad} \Phi)^2 \right) dV$$

2.  $$\int_S \frac{\partial \Phi}{\partial n} dS = \int_V \nabla^2 \Phi \, dV \qquad \text{MV 81}$$

**10.716** *Green's reciprocal theorem.* If $\Phi$ and $\Psi$ are harmonic, so that $\nabla^2 \Phi = \nabla^2 \Psi = 0$, then

3.  $$\int_S \Phi \frac{\partial \Psi}{\partial n} dS = \int_S \Psi \frac{\partial \Phi}{\partial n} dS \qquad \text{MM 105}$$

**10.717** *Green's representation theorem.* If $\Phi$ and $\nabla^2 \Phi$ are defined within a volume $V$ bounded by a simple closed surface $S$, and $P$ is an interior point of $V$, then in three dimensions

4.  $$\Phi(P) = -\frac{1}{4\pi} \int_V \frac{1}{r} \nabla^2 \Phi \, dV + \frac{1}{4\pi} \int_S \frac{1}{r} \frac{\partial \Phi}{\partial n} dS - \frac{1}{4\pi} \int_S \Phi \frac{\partial}{\partial n} \left( \frac{1}{r} \right) dS \qquad \text{KE 219}$$

   If $\Phi$ is harmonic within $V$, so that $\nabla^2 \Phi = 0$, then the previous result becomes

5.  $$\Phi(P) = \frac{1}{4\pi} \int_S \frac{1}{r} \frac{\partial \Phi}{\partial n} dS - \frac{1}{4\pi} \int_S \Phi \frac{\partial}{\partial n} \left( \frac{1}{r} \right) dS$$

   In the case of two dimensions, result (4) takes the form

6.  $$\Phi(p) = \frac{1}{2\pi} \int_S \nabla^2 \Phi(q) \ln |p - q| \, dS$$
   $$+ \frac{1}{2\pi} \int_C \Phi(q) \frac{\partial}{\partial n_q} \ln |p - q| \, dq - \frac{1}{2\pi} \int \ln |p - q| \frac{\partial}{\partial n_q} \Phi(q) \, dq$$
   
   $$\text{MM 116}$$

   where $C$ is the boundary of the planar region $S$, and result (5) takes the form

7.  $$\Phi(p) = \frac{1}{2\pi} \int_C \Phi(q) \frac{\partial}{\partial n_q} \ln |p - q| \, dq - \frac{1}{2\pi} \int_C \ln |p - q| \frac{\partial}{\partial n_q} \Phi(q) \, dq \qquad \text{VL 280}$$

**10.718** *Green's representation theorem in $R^n$.* If $\Phi$ is twice differentiable within a region $\Omega$ in $R^n$ bounded by the surface $\Sigma$ with outward drawn unit normal $\mathbf{n}$, then for $p \notin \Sigma$ and $n > 3$

$$\Phi(p) = \frac{-1}{(n-2)\sigma_n} \int_\Omega \frac{\nabla^2 \Phi(q)}{|p-q|^{n-2}} d\Omega_q + \frac{1}{(n-2)\sigma_n} \int_\Sigma \left( \frac{1}{|p-q|^{n-2}} \frac{\partial \Phi(q)}{\partial n_q} - \Phi(q) \frac{\partial}{\partial n_q} \frac{1}{|p-q|^{n-2}} \right) d\Sigma_q,$$

where

$$\sigma_n = \frac{2\pi^{n/2}}{\Gamma(n/2)} \qquad \text{VL 279}$$

is the area of the unit sphere in $R^n$.

**10.719** *Green's theorem of the arithmetic mean.* If $\Phi$ is harmonic in a sphere, then the value of $\Phi$ at the center of the sphere is the arithmetic mean of its value on the surface. $\qquad \text{KE 223}$

**10.720** *Poisson's integral in three dimensions.* If $\Phi$ is harmonic in the interior of a spherical volume $V$ of radius $R$ and is continuous on the surface of the sphere on which, in terms of the spherical polar coordinates $(r, \theta, \phi)$, it satisfies the boundary condition $\Phi(R, \theta, \phi) = f(\theta, \phi)$, then

$$\Phi(r, \theta, \phi) = \frac{R\left(R^2 - r^2\right)}{4\pi} \int_0^\pi \int_{-\pi}^\pi \frac{f\left(\theta', \phi'\right) \sin \theta' \, d\theta' \, d\phi'}{\left(r^2 + R^2 - 2rR\cos\gamma\right)^{3/2}},$$

where

$$\cos\gamma = \cos\theta\cos\theta' + \sin\theta\sin\theta'\cos\left(\phi - \phi'\right) \qquad \text{KE 241}$$

**10.721**  *Poisson's integral in two dimensions.* If $\Phi$ is harmonic in the interior of a circular disk $S$ of radius $R$ and is continuous on the boundary of the disk on which, in terms of the polar coordinates $(r, \theta)$, it satisfies the boundary condition $\Phi(R, \theta) = f(\theta)$, then

$$\Phi(r, \theta) = \frac{\left(R^2 - r^2\right)}{2\pi} \int_{-\pi}^\pi \frac{f(\phi)\, d\phi}{r^2 + R^2 - 2rR\cos(\theta - \phi)}$$

**10.722**  *Stokes' theorem.* Let a simple closed curve $C$ be spanned by a surface $S$. Define the positive normal $\mathbf{n}$ to $S$, and the positive sense of description of the curve $C$ with line element $d\mathbf{r}$, such that the positive sense of the contour $C$ is clockwise when we look through the surface $S$ in the direction of the normal. Then, if $\mathbf{f}$ is continuously differentiable vector field defined on $S$ and $C$ with vector element $\mathbf{S} = \mathbf{n}\, dS$,

$$\oint_C \mathbf{f} \cdot d\mathbf{r} = \int_S \operatorname{curl} \mathbf{f} \cdot d\mathbf{S}, \qquad \text{MM 143}$$

where the line integral around $C$ is taken in the positive sense.

**10.723**  *Planar case of Stokes' theorem.* If a region $R$ in the $(x, y)$-plane is bounded by a simple closed curve $C$, and $f_1(x, y), f_2(x, y)$ are any two functions having continuous first derivatives in $R$ and on $C$, then

$$\oint_C \left(f_1\, dx + f_2\, dy\right) = \int\int_R \left(\frac{\partial f_2}{\partial x} - \frac{\partial f_1}{\partial y}\right) dx\, dy, \qquad \text{MM 143}$$

where the line integral is taken in the anticlockwise sense.

## 10.81  Integral rate of change theorems

**10.811**  *Rate of change of volume integral bounded by a moving closed surface.* Let $f$ be a continuous scalar function of position and time $t$ defined throughout the volume $V(t)$, which is itself bounded by a simple closed surface $S(t)$ moving with velocity $\mathbf{v}$. Then the rate of change of the volume integral of $f$ is given by

$$\frac{D}{Dt} \int_{V(t)} f\, dV = \int_{V(t)} \frac{\partial f}{\partial t}\, dV + \int_{S(t)} f\mathbf{v} \cdot d\mathbf{S},$$

where $d\mathbf{S}$ is the outward drawn vector element of area, and

$$\frac{D}{Dt} \equiv \frac{\partial}{\partial t} + \mathbf{v} \cdot \nabla.$$

By virtue of Gauss's theorem this also takes the form

$$\frac{D}{Dt} \int_{V(t)} f\, dV = \int_{V(t)} \left(\frac{Df}{Dt} + f \operatorname{div} \mathbf{v}\right) dV \qquad \text{MV 88}$$

**10.812**   *Rate of change of flux through a surface.* Let $\mathbf{q}$ be a vector function that may also depend on the time $t$, and $\mathbf{n}$ be the unit outward drawn normal to the surface $S$ that moves with velocity $\mathbf{v}$. Defining the flux of $\mathbf{q}$ through $S$ as

$$m = \int_S \mathbf{q} \cdot \mathbf{n} \, dS,$$

then

$$\frac{Dm}{Dt} = \int_S \left( \frac{\partial \mathbf{q}}{\partial t} + \mathbf{v} \operatorname{div} \mathbf{q} + \operatorname{curl} \left( \mathbf{q} \times \mathbf{v} \right) \right) \cdot \mathbf{n} \, dS \qquad\qquad \text{MV 90}$$

**10.813**   *Rate of change of the circulation around a given moving curve.* Let $C$ be a closed curve, moving with velocity $\mathbf{v}$, on which is defined a vector field $\mathbf{q}$. Defining the circulation $\zeta$ of $\mathbf{q}$ around $C$ by

$$\zeta = \int_C \mathbf{q} \cdot d\mathbf{r},$$

then

$$\frac{D\zeta}{Dt} = \int_C \left( \frac{\partial \mathbf{q}}{\partial t} + (\operatorname{curl} \mathbf{q}) \times \mathbf{v} \right) \cdot d\mathbf{r} \qquad\qquad \text{MV 94}$$

Table of Integrals, Series, and Products. http://dx.doi.org/10.1016/B978-0-12-384933-5.00011-4

# 11 Integral Inequalities

## 11.11 Mean Value Theorems

### 11.111 First mean value theorem

Let $f(x)$ and $g(x)$ be two bounded functions integrable in $[a, b]$ and let $g(x)$ be of one sign in this interval. Then

$$\int_a^b f(x)g(x)\,\mathrm{d}x = f(\xi)\int_a^b g(x)\,\mathrm{d}x,$$

CA 105

with $a \leq \xi \leq b$.

### 11.112 Second mean value theorem

(i)     Let $f(x)$ be a bounded, monotonic decreasing, and nonnegative function in $[a, b]$, and let $g(x)$ be a bounded integrable function. Then,

$$\int_a^b f(x)g(r)\,\mathrm{d}x - f(a)\int_a^\xi g(x)\,\mathrm{d}x,$$

with $a \leq \xi \leq b$.

(ii)    Let $f(x)$ be a bounded, monotonic increasing, and nonnegative function in $[a, b]$, and let $g(x)$ be a bounded integrable function. Then,

$$\int_a^b f(x)g(x)\,\mathrm{d}x = f(b)\int_\eta^b g(x)\,\mathrm{d}x,$$

with $a \leq \eta \leq b$.

(iii)   Let $f(x)$ be bounded and monotonic in $[a, b]$, and let $g(x)$ be a bounded integrable function which experiences only a finite number of sign changes in $[a, b]$. Then,

$$\int_a^b f(x)g(x)\,\mathrm{d}x = f(a+0)\int_a^\xi g(x)\,\mathrm{d}x + f(b-0)\int_\xi^b g(x)\,\mathrm{d}x,$$

CA 107

with $a \leq \xi \leq b$.

### 11.113 First mean value theorem for infinite integrals

Let $f(x)$ be bounded for $x \geq a$, and integrable in the arbitrary interval $[a, b]$, and let $g(x)$ be of one sign in $x \geq a$ and such that $\int_a^\infty g(x)\,\mathrm{d}x$ is finite. Then,

$$\int_a^\infty f(x)g(x)\,\mathrm{d}x = \mu \int_a^\infty g(x)\,\mathrm{d}x, \qquad\qquad \text{CA 123}$$

where $m \le \mu \le M$ and $m, M$ are, respectively, the lower and upper bounds of $f(x)$ for $x \ge a$.

## 11.114 Second mean value theorem for infinite integrals

Let $f(x)$ be bounded and monotonic when $x \ge a$, and $g(x)$ be bounded and integrable in the arbitrary interval $[a, b]$ in which it experiences only a finite number of changes of sign. Then, provided $\int_a^\infty g(x)\,\mathrm{d}x$ is finite,

$$\int_a^\infty f(x)g(x)\,\mathrm{d}x = f(a+0) \int_a^\xi g(x)\,\mathrm{d}x + f(\infty) \int_\xi^\infty g(x)\,\mathrm{d}x, \qquad\qquad \text{CA 123}$$

with $a \le \xi \le \infty$.

# 11.21 Differentiation of Definite Integral Containing a Parameter

## 11.211 Differentiation when limits are finite

Let $\phi(\alpha)$ and $\psi(\alpha)$ be twice differentiable functions in some interval $c \le \alpha \le d$, and let $f(x, \alpha)$ be both integrable with respect to $x$ over the interval $\phi(\alpha) \le x \le \psi(\alpha)$ and differentiable with respect to $\alpha$. Then,

$$\frac{d}{d\alpha} \int_{\phi(\alpha)}^{\psi(\alpha)} f(x, \alpha)\,\mathrm{d}x = \left(\frac{d\psi}{d\alpha}\right) f\left(\psi(\alpha), \alpha\right) - \left(\frac{d\phi}{d\alpha}\right) f\left(\phi(\alpha), \alpha\right) + \int_{\phi(\alpha)}^{\psi(\alpha)} \frac{\partial f}{\partial \alpha}\,\mathrm{d}x \qquad\qquad \text{FI II 680}$$

## 11.212 Differentiation when a limit is infinite

Let $f(x, \alpha)$ and $\partial f / \partial \alpha$ both be integrable with respect to $x$ over the semi-infinite region $x \ge a$, $b \le \alpha < c$. Then, if the integral

$$f(\alpha) = \int_a^\infty f(x, \alpha)\,\mathrm{d}x$$

exists for all $b \le \alpha \le c$, and if $\int_a^\infty \frac{\partial f}{\partial \alpha}\,\mathrm{d}x$ is uniformly convergent for $\alpha$ in $[b, c]$, it follows that

$$\frac{d}{d\alpha} \int_a^\infty f(x, \alpha)\,\mathrm{d}x = \int_a^\infty \frac{\partial f}{\partial \alpha}\,\mathrm{d}x$$

# 11.31 Integral Inequalities

## 11.311 Cauchy–Schwarz–Buniakowsky inequality for integrals

Let $f(x)$ and $g(x)$ be any two real integrable functions on $[a, b]$. Then,

$$\left(\int_a^b f(x)g(x)\,\mathrm{d}x\right)^2 \le \left(\int_a^b f^2(x)\,\mathrm{d}x\right)\left(\int_a^b g^2(x)\,\mathrm{d}x\right),$$

and the equality will hold if, and only if, $f(x) = kg(x)$, with $k$ real.        BB 21

## 11.312 Hölder's inequality for integrals

Let $f(x)$ and $g(x)$ be any two real functions for which $|f(x)|^p$ and $|g(x)|^q$ are integrable on $[a, b]$ with $p > 1$ and $\frac{1}{p} + \frac{1}{q} = 1$; then

$$\int_a^b f(x)g(x)\,\mathrm{d}x \le \left(\int_a^b |f(x)|^p\,\mathrm{d}x\right)^{1/p}\left(\int_a^b |g(x)|^q\,\mathrm{d}x\right)^{1/q}.$$

The equality holds if, and only if, $\alpha|f(x)|^p = \beta|g(x)|^q$, where $\alpha$ and $\beta$ are positive constants. **BB 21**

## 11.313 Minkowski's inequality for integrals

Let $f(x)$ and $g(x)$ be any two real functions for which $|f(x)|^p$ and $|g(x)|^p$ are integrable on $[a,b]$ for $p > 0$; then

$$\left(\int_a^b |f(x)+g(x)|^p\,\mathrm{d}x\right)^{1/p} \le \left(\int_a^b |f(x)|^p\,\mathrm{d}x\right)^{1/p} + \left(\int_a^b |g(x)|^p\,\mathrm{d}x\right)^{1/p}.$$

The equality holds if, and only if, $f(x) = kg(x)$ for some real $k \ge 0$. n **BB 21**

## 11.314 Chebyshev's inequality for integrals

Let $f_1, f_2, \ldots, f_n$ be nonnegative integrable functions on $[a,b]$ which are all either monotonic increasing or monotonic decreasing; then

$$\int_a^b f_1(x)\,\mathrm{d}x \int_a^b f_2(x)\,\mathrm{d}x \ldots \int_a^b f_n(x)\,\mathrm{d}x \le (b-a)^{n-1}\int_a^b f_1(x)f_2(x)\ldots f_n(x)\,\mathrm{d}x \qquad \textbf{MT 39}$$

## 11.315 Young's inequality for integrals

Let $f(x)$ be a real-valued continuous strictly monotonic increasing function on the interval $[0,a]$, with $f(0) = 0$ and $b \le f(a)$. Then

$$ab \le \int_0^a f(x)\,\mathrm{d}x + \int_0^b f^{-1}(y)\,\mathrm{d}y,$$

where $f^{-1}(y)$ denotes the function inverse to $f(x)$. The equality holds if, and only if, $b = f(a)$. **BB 15**

## 11.316 Steffensen's inequality for integrals

Let $f(x)$ be nonnegative and monotonic decreasing in $[a,b]$ and $g(x)$ be such that $0 \le g(x) \le 1$ in $[a,b]$. Then

$$\int_{b-k}^b f(x)\,\mathrm{d}x \le \int_a^b f(x)g(x)\,\mathrm{d}x \le \int_a^{a+k} f(x)\,\mathrm{d}x,$$

where $k = \int_a^b g(x)\,\mathrm{d}x$. **MT 107**

## 11.317 Gram's inequality for integrals

Let $f_1(x), f_2(x), \ldots, f_n(x)$ be real square integrable functions on $[a,b]$; then

$$\begin{vmatrix} \int_a^b f_1^2(x)\,\mathrm{d}x & \int_a^b f_1(x)f_2(x)\,\mathrm{d}x & \cdots & \int_a^b f_1(x)f_n(x)\,\mathrm{d}x \\ \int_a^b f_2(x)f_1(x)\,\mathrm{d}x & \int_a^b f_2^2(x)\,\mathrm{d}x & \cdots & \int_a^b f_2(x)f_n(x)\,\mathrm{d}x \\ \vdots & \vdots & \ddots & \vdots \\ \int_a^b f_n(x)f_1(x)\,\mathrm{d}x & \int_a^b f_n(x)f_2(x)\,\mathrm{d}x & \cdots & \int_a^b f_n^2(x)\,\mathrm{d}x \end{vmatrix} \ge 0. \qquad \textbf{MT 47}$$

## 11.318  Ostrowski's inequality for integrals

Let $f(x)$ be a monotonic function integrable on $[a, b]$, and let $f(a)f(b) \geq 0, |f(a)| \geq |f(b)|$. Then, if $g$ is a real function integrable on $[a, b]$,

$$\left| \int_a^b f(x)g(x)\,dx \right| \leq |f(a)| \max_{a \leq \xi \leq b} \left| \int_a^\xi g(x)\,dx \right|.$$

# 11.41  Convexity and Jensen's Inequality

A function $f(x)$ is said to be **convex** on an interval $[a, b]$ if for any two points $x_1, x_2$ in $[a, b]$

$$f\left( \frac{x_1 + x_2}{2} \right) \leq \frac{f(x_1) + f(x_2)}{2}.$$

A function $f(x)$ is said to be **concave** on an interval $[a, b]$ if for any two points $x_1, x_2$ in $[a, b]$ the function $-f(x)$ is convex in that interval.

If the function $f(x)$ possesses a second derivative in the interval $[a, b]$, then a necessary and sufficient condition for it to be convex on that interval is that $f''(x) \geq 0$ for all $x$ in $[a, b]$.

A function $f(x)$ is said to be **logarithmically convex** on the interval $[a, b]$ if $f > 0$ and $\log f(x)$ is concave on $[a, b]$.

If $f(x)$ and $g(x)$ are logarithmically convex on the interval $[a, b]$, then the functions $f(x) + g(x)$ and $f(x)g(x)$ are also logarithmically convex on $[a, b]$.                          MT 17

## 11.411  Jensen's inequality

Let $f(x), p(x)$ be two functions defined for $a \leq x \leq b$ such that $\alpha \leq f(x) \leq \beta$ and $p(x) \geq 0$, with $p(x) \not\equiv 0$. Let $\phi(u)$ be a convex function defined on the interval $\alpha \leq u \leq \beta$; then

$$\phi\left( \frac{\int_a^b f(x)p(x)\,dx}{\int_a^b p(x)\,dx} \right) \leq \frac{\int_a^b \phi(f)p(x)\,dx}{\int_a^b p(x)\,dx}.$$                          HL 151

## 11.412  Carleman's inequality for integrals

If $f(x) \geq 0$ and the integrals exist, then

$$\int_0^\infty \exp\left( \frac{1}{x} \int_0^x f(t)\,dt \right) dx \leq e \int_0^\infty f(x)\,dx$$

# 11.51  Fourier Series and Related Inequalities

The trigonometric **Fourier series** representation of the function $f(x)$ integrable on $[-\pi, \pi]$ is

$$f(x) \sim \frac{a_0}{2} + \sum_{n=1}^{\infty} (a_n \cos nx + b_n \sin nx),$$

where the **Fourier coefficients** $a_n$ and $b_n$ of $f(x)$ are given by

$$a_n = \frac{1}{2\pi} \int_{-\pi}^{\pi} f(x) \cos nx\,dx, \qquad b_n = \frac{1}{2\pi} \int_{-\pi}^{\pi} f(x) \sin nx\,dx.$$

(See **0.320–0.328** for convergence of Fourier series on $(-l, l)$.)                          TF 1

## 11.511  Riemann–Lebesgue lemma

If $f(x)$ is integrable on $[-\pi, \pi]$, then

$$\lim_{t \to \infty} \int_{-\pi}^{\pi} f(x) \sin tx \, dx \to 0$$

and

$$\lim_{t \to \infty} \int_{-\pi}^{\pi} f(x) \cos tx \, dx \to 0. \qquad\qquad \text{TF 11}$$

## 11.512  Dirichlet lemma

$$\int_{0}^{\pi} \frac{\sin\left(n + \frac{1}{2}\right) x}{2 \sin \frac{1}{2} x} \, dx = \frac{\pi}{2},$$

in which $\sin\left(n + \frac{1}{2}\right) x \; m/2 \sin \frac{1}{2} x$ is called the **Dirichlet kernel**. $\qquad$ ZY 21

## 11.513  Parseval's theorem for trigonometric Fourier series

If $f(x)$ is square integrable on $[-\pi, \pi]$, then

$$\frac{a_0^2}{2} + \sum_{r=1}^{\infty} \left(a_r^2 + b_r^2\right) = \frac{1}{\pi} \int_{-\pi}^{\pi} f^2(x) \, dx. \qquad\qquad \text{Y 10}$$

## 11.514  Integral representation of the $n^{\text{th}}$ partial sum

If $f(x)$ is integrable on $[-\pi, \pi]$, then the $n^{\text{th}}$ partial sum

$$s_n(x) = \frac{a_0}{2} + \sum_{r=1}^{n} \left(a_r \cos rx + b_r \sin rx\right)$$

has the following integral representation in terms of the Dirichlet kernel,

$$s_n(x) = \frac{1}{\pi} \int_{-\pi}^{\pi} f(x - t) \frac{\sin\left(n + \frac{1}{2}\right) t}{2 \sin \frac{1}{2} t} \, dt. \qquad\qquad \text{Y 20}$$

## 11.515  Generalized Fourier series

Let the set of functions $\{\phi_n\}_{n=0}^{\infty}$ form an **orthonormal set** over $[a, b]$, so that

$$\int_{a}^{b} \phi_m(x) \phi_n(x) \, dx = \begin{cases} 1 & \text{for} \quad m = n, \\ 0 & \text{for} \quad m \neq n. \end{cases}$$

Then the **generalized Fourier series** representation of an integrable function $f(x)$ on $[a, b]$ is

$$f(x) \sim \sum_{n=0}^{\infty} c_n \phi_n(x),$$

where the generalized Fourier coefficients of $f(x)$ are given by

$$c_n = \int_{a}^{b} f(x) \phi_n(x) \, dx.$$

## 11.516 Bessel's inequality for generalized Fourier series

For any square integrable function defined on $[a, b]$,

$$\sum_{n=0}^{\infty} c_n^2 \leq \int_a^b f^2(x) \, dx,$$

where the $c_n$ are the generalized Fourier coefficients of $f(x)$.

## 11.517 Parseval's theorem for generalized Fourier series

If $f(x)$ is a square integrable function defined on $[a, b]$ and $\{\phi_n(x)\}_{n=0}^{\infty}$ is a **complete orthonormal** set of continuous functions defined on $[a, b]$, then

$$\sum_{n=0}^{\infty} c_n^2 = \int_a^b f^2(x) \, dx,$$

where the $c_n$ are generalized Fourier coefficients of $f(x)$.

Table of Integrals, Series, and Products. http://dx.doi.org/10.1016/B978-0-12-384933-5.00012-6

# 12 Fourier, Laplace, and Mellin Transforms

## 12.1– 12.4 Integral Transforms

### 12.11  Laplace transform

The **Laplace transform** of the function $f(x)$, denoted by $F(s)$, is defined by the integral

$$F(s) = \int_0^\infty f(x)e^{-sx}\,dx, \qquad \operatorname{Re} s > 0.$$

The functions $f(x)$ and $F(s)$ are called a **Laplace transform pair**, and knowledge of either one enables the other to be recovered.

If $f$ is summable over all finite intervals, and there is a constant $c$ for which

$$\int_0^\infty |f(x)|e^{-c|x|}\,dx,$$

is finite, then the Laplace transform exists when $s = \sigma + i\tau$ is such that $\sigma \geq c$.

Setting

$$F(s) = \mathcal{L}\left[f(x); s\right],$$

to emphasize the nature of the transform, we have the symbolic inverse result

$$f(x) = \mathcal{L}^{-1}\left[F(s); x\right].$$

The inversion of the Laplace transform is accomplished for analytic functions $F(s)$ of order $O\left(s^{-k}\right)$ with $k > 1$ by means of the **inversion integral**

$$f(x) = \frac{1}{2\pi i}\int_{\gamma - i\infty}^{\gamma + i\infty} F(s)e^{sx}\,ds,$$

where $\gamma$ is a real constant that exceeds the real part of all the singularities of $F(s)$.                SN 30

### 12.12  Basic properties of the Laplace transform

1.[8]     For $a$ and $b$ arbitrary constants,

$$\mathcal{L}\left[af(x) + bg(x)\right] = aF(s) + bG(s) \qquad \text{(linearity)}$$

2.     If $n > 0$ is an integer and $\lim\limits_{x \to \infty} f(x)e^{-sx} = 0$, then for $x > 0$,

$$\mathcal{L}\left[f^{(n)}(x); s\right] = s^n F(s) - s^{n-1}f(0) - s^{n-2}f^{(1)}(0) - \cdots - f^{(n-1)}(0) \quad \text{(transform of a derivative)}$$

SN 32

$3.^{11}$  If $\lim_{x\to\infty} \left( e^{-sx} \int_0^x f(\zeta)\,\mathrm{d}\zeta \right) = 0$, then

$$\mathcal{L}\left[ \int_0^x f(\xi)\,\mathrm{d}\xi; s \right] = \frac{1}{s}F(s) \qquad \text{(transform of an integral)} \qquad \text{SN 37}$$

4.    $\mathcal{L}\left[ e^{-ax} f(x); s \right] = F(s+a)$                    (shift theorem)                    SU 143

5.    The **Laplace convolution** $f * g$ of two functions $f(x)$ and $g(x)$ is defined by the integral

$$f * g(x) = \int_0^x f(x-\xi)g(\xi)\,\mathrm{d}\xi,$$

and it has the property that $f * g = g * f$ and $f * (g * h) = (f * g) * h$. In terms of the convolution operation

$$\mathcal{L}\left[ f * g(x); s \right] = F(s)G(s) \qquad \text{(convolution (Faltung) theorem)} \qquad \text{SN 30}$$

## 12.13  Table of Laplace transform pairs

| | $f(x)$ | | $F(s)$ | | |
|---|---|---|---|---|---|
| **1** | $1$ | | $1/s$ | | |
| **2** | $x^n,$ | $n = 0, 1, 2, \ldots$ | $\dfrac{n!}{s^{n+1}},$ | $\operatorname{Re} s > 0$ | ET I 133(3) |
| **3** | $x^\nu,$ | $\nu > -1$ | $\dfrac{\Gamma(\nu+1)}{s^{\nu+1}},$ | $\operatorname{Re} s > 0$ | ET I 137(1) |
| **4** | $x^{n-\frac{1}{2}}$ | | $\dfrac{\Gamma\left(n+\frac{1}{2}\right)}{s^{n+\frac{1}{2}}},$ | $\operatorname{Re} s > 0$ | ET I 135(17) |
| **5** | $x^{-1/2}(x+a)^{-1},$ | $\lvert \arg a \rvert < \pi$ | $\pi a^{-1/2} e^{as} \operatorname{erfc}\left( a^{1/2} s^{1/2} \right),$ $\operatorname{Re} s \geq 0$ | | ET I 136(25) |
| **6** | $\begin{cases} x & \text{for } 0 < x < 1 \\ 1 & \text{for } x > 1 \end{cases}$ | | $\dfrac{1 - e^{-s}}{s^2},$ | $\operatorname{Re} s > 0$ | ET I 142(14) |
| **7** | $e^{-ax}$ | | $\dfrac{1}{s+a},$ | $\operatorname{Re} s > -\operatorname{Re} a$ | ET I 143(1) |
| **8** | $xe^{-ax}$ | | $\dfrac{1}{(s+a)^2},$ | $\operatorname{Re} s > -\operatorname{Re} a$ | ET I 144(2) |
| **9a** | $\dfrac{e^{-ax} - e^{-bx}}{b-a}$ | | $(s+a)^{-1}(s+b)^{-1},$ $\operatorname{Re} s > \{-\operatorname{Re} a, -\operatorname{Re} b\}$ | | AS 1022(29.3.12) |

*continued on next page*

*continued from previous page*

| | $f(x)$ | $F(s)$ |
|---|---|---|
| **9b$^{11}$** | $\dfrac{\alpha e^{-ax} + \beta e^{-bx} + \gamma e^{-cx}}{(a-b)(b-c)(c-a)}$ <br><br> $a, b, c$ distinct , $\quad \alpha = c - b,$ <br> $\beta = a - c, \quad \gamma = b - a$ | $(s+a)^{-1}(s+b)^{-1}(s+c)^{-1},$ <br><br> $\qquad \mathrm{Re}\,s > \{-\mathrm{Re}\,a, -\mathrm{Re}\,b, -\mathrm{Re}\,c\}$ |
| **10$^{11}$** | $\dfrac{ae^{-ax} - be^{-bx}}{b-a}$ | $s(s+a)^{-1}(s+b)^{-1},$ <br><br> $\qquad \mathrm{Re}\,s > \{-\mathrm{Re}\,a, -\mathrm{Re}\,b\} \qquad \text{AS 1022(29.3.13)}$ |
| **11** | $\dfrac{e^{ax} - 1}{a}$ | $s^{-1}(s-a)^{-1}, \qquad\qquad\qquad \mathrm{Re}\,s > \mathrm{Re}\,a$ |
| **12** | $\dfrac{e^{ax} - ax - 1}{a^2}$ | $s^{-2}(s-a)^{-1}, \qquad\qquad\qquad \mathrm{Re}\,s > \mathrm{Re}\,a$ |
| **13** | $\dfrac{\left(e^{ax} - \frac{1}{2}a^2x^2 - ax - 1\right)}{a^3}$ | $s^{-3}(s-a)^{-1}, \qquad\qquad\qquad \mathrm{Re}\,s > \mathrm{Re}\,a$ |
| **14** | $(1 + ax)e^{ax}$ | $\dfrac{s}{(s-a)^2}, \qquad\qquad\qquad \mathrm{Re}\,s > \mathrm{Re}\,a$ |
| **15** | $\dfrac{1 + (ax-1)e^{ax}}{a^2}$ | $s^{-1}(s-a)^{-2}, \qquad\qquad\qquad \mathrm{Re}\,s > \mathrm{Re}\,a$ |
| **16** | $\dfrac{2 + ax + (ax-2)e^{ax}}{a^3}$ | $s^{-2}(s-a)^{-2}, \qquad\qquad\qquad \mathrm{Re}\,s > \mathrm{Re}\,a$ |
| **17** | $x^n e^{ax}, \qquad\qquad n = 0, 1, 2, \ldots$ | $n!(s-a)^{-(n+1)}, \qquad\qquad \mathrm{Re}\,s > \mathrm{Re}\,a$ |
| **18** | $\left(x + \frac{1}{2}ax^2\right)e^{ax}$ | $\dfrac{s}{(s-a)^3}, \qquad\qquad\qquad \mathrm{Re}\,s > \mathrm{Re}\,a$ |
| **19** | $\left(1 + 2ax + \frac{1}{2}a^2x^2\right)e^{ax}$ | $\dfrac{s^2}{(s-a)^3}, \qquad\qquad\qquad \mathrm{Re}\,s > \mathrm{Re}\,a$ |
| **20** | $\frac{1}{6}x^3 e^{ax}$ | $(s-a)^{-4}, \qquad\qquad\qquad \mathrm{Re}\,s > \mathrm{Re}\,a$ |
| **21** | $\left(\frac{1}{2}x^2 + \frac{1}{6}ax^3\right)e^{ax}$ | $\dfrac{s}{(s-a)^4}, \qquad\qquad\qquad \mathrm{Re}\,s > \mathrm{Re}\,a$ |
| **22** | $\left(x + ax^2 + \frac{1}{6}a^2x^3\right)e^{ax}$ | $s^2(s-a)^{-4}, \qquad\qquad\qquad \mathrm{Re}\,s > \mathrm{Re}\,a$ |

*continued on next page*

| | $f(x)$ | | $F(s)$ | |
|---|---|---|---|---|

*continued from previous page*

| | $f(x)$ | | $F(s)$ |
|---|---|---|---|
| **23** | $\left(1 + 3ax + \frac{3}{2}a^2x^2 + \frac{1}{6}a^3x^3\right)e^{ax}$ | | $s^3(s-a)^{-4}$, $\qquad\qquad\qquad\qquad$ $\mathrm{Re}\,s > \mathrm{Re}\,a$ |
| **24** | $\dfrac{ae^{ax} - be^{bx}}{a - b}$ | | $s(s-a)^{-1}(s-b)^{-1}$, $\qquad\quad$ $\mathrm{Re}\,s > \{\mathrm{Re}\,a, \mathrm{Re}\,b\}$ |
| **25** | $\dfrac{\left(\frac{1}{a}e^{ax} - \frac{1}{b}e^{bx} + \frac{1}{b} - \frac{1}{a}\right)}{a - b}$ | | $s^{-1}(s-a)^{-1}(s-b)^{-1}$, $\qquad$ $\mathrm{Re}\,s > \{\mathrm{Re}\,a, \mathrm{Re}\,b\}$ |
| **26** | $x^{\nu-1}e^{-ax}$, | $\mathrm{Re}\,\nu > 0$ | $\Gamma(\nu)(s+a)^{-\nu}$, $\quad$ $\mathrm{Re}\,s > -\mathrm{Re}\,a$ $\qquad$ ET I 144(3) |
| **27** | $xe^{-x^2/(4a)}$, | $\mathrm{Re}\,a > 0$ | $2a - 2\pi^{1/2}a^{3/2}se^{as^2}\,\mathrm{erfc}\left(sa^{1/2}\right)$ $\qquad\qquad\qquad\qquad\qquad\qquad\qquad$ ET I 146(22) |
| **28** | $\exp\left(-ae^x\right)$, | $\mathrm{Re}\,a > 0$ | $a^s\,\Gamma\left(-s, a\right)$ $\qquad\qquad\qquad$ ET I 147(37) |
| **29**[8] | $x^{1/2}e^{-a/(4x)}$, | $\mathrm{Re}\,a \geq 0$ | $\frac{1}{2}\pi^{1/2}s^{-3/2}\left(1 + a^{1/2}s^{1/2}\right)\exp\left[(-as)^{1/2}\right]$, $\qquad\qquad\qquad\qquad\qquad$ $\mathrm{Re}\,s > 0$ $\quad$ ET I 146(26) |
| **30**[12] | $x^{-1/2}e^{-a/(4x)}$, | $\mathrm{Re}\,a \geq 0$ | $\pi^{1/2}s^{-1/2}\exp\left[-(as)^{1/2}\right]$, $\qquad\qquad\qquad\qquad\qquad$ $\mathrm{Re}\,s > 0$ $\quad$ ET I 146(27) |
| **31**[12] | $x^{-3/2}e^{-a/(4x)}$, | $\mathrm{Re}\,a > 0$ | $2\pi^{1/2}a^{-1/2}\exp\left[-(as)^{1/2}\right]$, $\qquad\qquad\qquad\qquad\qquad$ $\mathrm{Re}\,s \geq 0$ $\quad$ ET I 146(28) |
| **32** | $\sin(ax)$ | | $a\left(s^2 + a^2\right)^{-1}$, $\qquad$ $\mathrm{Re}\,s > |\mathrm{Im}\,a|$ $\quad$ ET I 150(1) |
| **33** | $\cos(ax)$ | | $s\left(s^2 + a^2\right)^{-1}$, $\qquad$ $\mathrm{Re}\,s > |\mathrm{Im}\,a|$ $\quad$ ET I 154(3) |
| **34** | $|\sin(ax)|$, | $a > 0$ | $a\left(s^2 + a^2\right)^{-1}\coth\left(\dfrac{\pi s}{2a}\right)$, $\qquad\qquad\qquad\qquad$ $\mathrm{Re}\,s > 0$ $\quad$ ET I 150(2) |

*continued on next page*

| | $f(x)$ | $F(s)$ |
|---|---|---|
| **35**[11] | $|\cos(ax)|,$  $a > 0$ | $\left(s^2 + a^2\right)^{-1}\left[s + a\operatorname{cosech}\left(\dfrac{\pi s}{2a}\right)\right],$   $\operatorname{Re} s > 0$   ET I 155(44) |
| **36** | $\dfrac{1 - \cos(ax)}{a^2}$ | $s^{-1}\left(s^2 + a^2\right)^{-1},$   $\operatorname{Re} s > |\operatorname{Im} a|$   AS 1022(29.3.19) |
| **37** | $\dfrac{ax - \sin(ax)}{a^3}$ | $s^{-2}\left(s^2 + a^2\right)^{-1},$   $\operatorname{Re} s > |\operatorname{Im} a|$   AS 1022(29.3.20) |
| **38** | $\dfrac{\sin(ax) - ax\cos(ax)}{2a^3}$ | $\left(s^2 + a^2\right)^{-2},$   $\operatorname{Re} s > |\operatorname{Im} a|$   AS 1022(29.3.21) |
| **39** | $\dfrac{x\sin(ax)}{2a}$ | $s\left(s^2 + a^2\right)^{-2},$   $\operatorname{Re} s > |\operatorname{Im} a|$   ET I 152(14) |
| **40** | $\dfrac{\sin(ax) + ax\cos(ax)}{2a}$ | $s^2\left(s^2 + a^2\right)^{-2},$   $\operatorname{Re} s > |\operatorname{Im} a|$   AS 1023(29.3.23) |
| **41** | $x\cos(ax)$ | $\left(s^2 - a^2\right)\left(s^2 + a^2\right)^{-2},$   $\operatorname{Re} s > |\operatorname{Im} a|$   ET I 157(57) |
| **42** | $\dfrac{\cos(ax) - \cos(bx)}{b^2 - a^2}$ | $s\left(s^2 + a^2\right)^{-1}\left(s^2 + b^2\right)^{-1},$   $\operatorname{Re} s > \{|\operatorname{Im} a|, |\operatorname{Im} b|\}$   AS 1023(29.3.25) |
| **43** | $\dfrac{\left[\frac{1}{2}a^2x^2 - 1 + \cos(ax)\right]}{a^4}$ | $s^{-3}\left(s^2 + a^2\right)^{-1},$   $\operatorname{Re} s > |\operatorname{Im} a|$ |
| **44** | $\dfrac{\left[1 - \cos(ax) - \frac{1}{2}ax\sin(ax)\right]}{a^4}$ | $s^{-1}\left(s^2 + a^2\right)^{-2},$   $\operatorname{Re} s > |\operatorname{Im} a|$ |
| **45** | $\dfrac{\left[\frac{1}{b}\sin(bx) - \frac{1}{a}\sin(ax)\right]}{a^2 - b^2}$ | $\left(s^2 + a^2\right)^{-1}\left(s^2 + b^2\right)^{-1},$   $\operatorname{Re} s > \{|\operatorname{Im} a|, |\operatorname{Im} b|\}$ |
| **46**[11] | $\dfrac{\left[1 - \cos(ax) + \frac{1}{2}ax\sin(ax)\right]}{a^2}$ | $s^{-1}\left(s^2 + a^2\right)^{-2}\left(2s^2 + a^2\right),$   $\operatorname{Re} s > |\operatorname{Im} a|$ |

*continued from previous page*

*continued on next page*

*continued from previous page*

| | $f(x)$ | $F(s)$ |
|---|---|---|
| **47** | $\dfrac{a\sin(ax) - b\sin(bx)}{a^2 - b^2}$ | $s^2 \left(s^2 + a^2\right)^{-1} \left(s^2 + b^2\right)^{-1},$<br><br>$\operatorname{Re}s > \{|\operatorname{Im}a|, |\operatorname{Im}b|\}$ |
| **48** | $\sin(a + bx)$ | $(s\sin a + b\cos a)\left(s^2 + b^2\right)^{-1}, \qquad \operatorname{Re}s > |\operatorname{Im}b|$ |
| **49** | $\cos(a + bx)$ | $(s\cos a - b\sin a)\left(s^2 + b^2\right)^{-1}, \qquad \operatorname{Re}s > |\operatorname{Im}b|$ |
| **50** | $\dfrac{\left[\frac{1}{a}\sinh(ax) - \frac{1}{b}\sin(bx)\right]}{a^2 + b^2}$ | $\left(s^2 - a^2\right)^{-1} \left(s^2 + b^2\right)^{-1},$<br><br>$\operatorname{Re}s > \{|\operatorname{Re}a|, |\operatorname{Im}b|\}$ |
| **51** | $\dfrac{\cosh(ax) - \cos(bx)}{a^2 + b^2}$ | $s\left(s^2 - a^2\right)^{-1} \left(s^2 + b^2\right)^{-1},$<br><br>$\operatorname{Re}s > \{|\operatorname{Re}a|, |\operatorname{Im}b|\}$ |
| **52** | $\dfrac{a\sinh(ax) + b\sin(bx)}{a^2 + b^2}$ | $s^2 \left(s^2 - a^2\right)^{-1} \left(s^2 + b^2\right)^{-1},$<br><br>$\operatorname{Re}s > \{|\operatorname{Re}a|, |\operatorname{Im}b|\}$ |
| **53** | $\sin(ax)\sin(bx)$ | $2abs\left[s^2 + (a - b)^2\right]^{-1}\left[s^2 + (a + b)^2\right]^{-1},$<br><br>$\operatorname{Re}s > \{|\operatorname{Im}a|, |\operatorname{Im}b|\}$ |
| **54** | $\cos(ax)\cos(bx)$ | $s\left(s^2 + a^2 + b^2\right)\left[s^2 + (a - b)^2\right]^{-1}\left[s^2 + (a + b)^2\right]^{-1},$<br><br>$\operatorname{Re}s > \{|\operatorname{Im}a|, |\operatorname{Im}b|\}$ |
| **55** | $\sin(ax)\cos(bx)$ | $a\left(s^2 + a^2 - b^2\right)\left[s^2 + (a - b)^2\right]^{-1}\left[s^2 + (a + b)^2\right]^{-1},$<br><br>$\operatorname{Re}s > \{|\operatorname{Im}a|, |\operatorname{Im}b|\}$ |
| **56** | $\sin^2(ax)$ | $2a^2 s^{-1}\left(s^2 + 4a^2\right)^{-1}, \qquad \operatorname{Re}s > |\operatorname{Im}a|$ |
| **57** | $\cos^2(ax)$ | $\left(s^2 + 2a^2\right)s^{-1}\left(s^2 + 4a^2\right)^{-1}, \qquad \operatorname{Re}s > |\operatorname{Im}a|$ |
| **58** | $\sin(ax)\cos(ax)$ | $a\left(s^2 + 4a^2\right)^{-1}, \qquad \operatorname{Re}s > |\operatorname{Im}a|$ |
| **59** | $e^{-ax}\sin(bx)$ | $b\left[(s + a)^2 + b^2\right]^{-1}, \qquad \operatorname{Re}s > \{-\operatorname{Re}a, \quad |\operatorname{Im}b|\}$ |

*continued on next page*

*continued from previous page*

| | $f(x)$ | $F(s)$ |
|---|---|---|
| **60** | $e^{-ax}\cos(bx)$ | $(s+a)\left[(s+a)^2 + b^2\right]^{-1}$, <br><br> $\operatorname{Re} s > \{-\operatorname{Re} a, \quad \lvert\operatorname{Im} b\rvert\}$ |
| **61** | $x^{-1}\sin(ax)$ | $\arctan(a/s)$, $\qquad \operatorname{Re} s > \lvert\operatorname{Im} a\rvert$ $\qquad$ ET I 152(16) |
| **62** | $x^{-1}\left[1 - \cos(ax)\right]$ | $\frac{1}{2}\ln\left(1 + a^2/s^2\right)$, <br><br> $\operatorname{Re} s > \lvert\operatorname{Im} a\rvert$ $\qquad$ ET I 157(59) |
| **63** | $\sinh(ax)$ | $a\left(s^2 - a^2\right)^{-1}$, $\qquad \operatorname{Re} s > \lvert\operatorname{Re} a\rvert$ $\qquad$ ET I 162(1) |
| **64** | $\cosh(ax)$ | $s\left(s^2 - a^2\right)^{-1}$, $\qquad \operatorname{Re} s > \lvert\operatorname{Re} a\rvert$ $\qquad$ ET I 162(2) |
| **65** | $x^{\nu-1}\sinh(ax)$, $\qquad \operatorname{Re}\nu > -1$ | $\frac{1}{2}\Gamma(\nu)\left[(s-a)^{-\nu} - (s+a)^{-\nu}\right]$, <br><br> $\operatorname{Re} s > \lvert\operatorname{Re} a\rvert$ $\qquad$ ET I 164(18) |
| **66** | $x^{\nu-1}\cosh(ax)$, $\qquad \operatorname{Re}\nu > 0$ | $\frac{1}{2}\Gamma(\nu)\left[(s-a)^{-\nu} + (s+a)^{-\nu}\right]$, <br><br> $\operatorname{Re} s > \lvert\operatorname{Re} a\rvert$ $\qquad$ ET I 164(19) |
| **67** | $x\sinh(ax)$ | $2as\left(s^2 - a^2\right)^{-2}$, $\qquad \operatorname{Re} s > \lvert\operatorname{Re} a\rvert$ |
| **68** | $x\cosh(ax)$ | $\left(s^2 + a^2\right)\left(s^2 - a^2\right)^{-2}$, $\qquad \operatorname{Re} s > \lvert\operatorname{Re} a\rvert$ |
| **69** | $\sinh(ax) - \sin(ax)$ | $2a^3\left(s^4 - a^4\right)^{-1}$, <br><br> $\operatorname{Re} s > \{\lvert\operatorname{Re} a\rvert, \lvert\operatorname{Im} a\rvert\}$ $\qquad$ AS 1023(29.3.31) |
| **70** | $\cosh(ax) - \cos(ax)$ | $2a^2 s\left(s^4 - a^4\right)^{-1}$, <br><br> $\operatorname{Re} s > \{\lvert\operatorname{Re} a\rvert, \lvert\operatorname{Im} a\rvert\}$ $\qquad$ AS 1023(29.3.32) |
| **71** | $\sinh(ax) + ax\cosh(ax)$ | $2as^2\left(a^2 - s^2\right)^{-2}$, $\qquad \operatorname{Re} s > \lvert\operatorname{Re} a\rvert$ |
| **72** | $ax\cosh(ax) - \sinh(ax)$ | $2a^3\left(a^2 - s^2\right)^{-2}$, $\qquad \operatorname{Re} s > \lvert\operatorname{Re} a\rvert$ |

*continued on next page*

| | $f(x)$ | $F(s)$ |
|---|---|---|
| | *continued from previous page* | |
| **73** | $x\sinh(ax) - \cosh(ax)$ | $s\left(a^2 + 2a - s^2\right)\left(a^2 - s^2\right)^{-2}$, $\qquad$ Re $s > |\text{Re}\,a|$ |
| **74** | $\dfrac{\left[\frac{1}{a}\sinh(ax) - \frac{1}{b}\sinh(bx)\right]}{a^2 - b^2}$ | $\left(a^2 - s^2\right)^{-1}\left(b^2 - s^2\right)^{-1}$, $\qquad\qquad$ Re $s > \{|\text{Re}\,a|, |\text{Re}\,b|\}$ |
| **75** | $\dfrac{\cosh(ax) - \cosh(bx)}{a^2 - b^2}$ | $s\left(s^2 - a^2\right)^{-1}\left(s^2 - b^2\right)^{-1}$, $\qquad\qquad$ Re $s > \{|\text{Re}\,a|, |\text{Re}\,b|\}$ |
| **76** | $\dfrac{a\sinh(ax) - b\sinh(bx)}{a^2 - b^2}$ | $s^2\left(s^2 - a^2\right)^{-1}\left(s^2 - b^2\right)^{-1}$, $\qquad\qquad$ Re $s > \{|\text{Re}\,a|, |\text{Re}\,b|\}$ |
| **77** | $\sinh(a + bx)$ | $(b\cosh a + s\sinh a)\left(s^2 - b^2\right)^{-1}$, $\qquad$ Re $s > |\text{Re}\,b|$ |
| **78** | $\cosh(a + bx)$ | $(s\cosh a + b\sinh a)\left(s^2 - b^2\right)^{-1}$, $\qquad$ Re $s > |\text{Re}\,b|$ |
| **79** | $\sinh(ax)\sinh(bx)$ | $2abs\left[s^2 - (a + b)^2\right]^{-1}\left[s^2 - (a - b)^2\right]^{-1}$, $\qquad\qquad$ Re $s > \{|\text{Re}\,a|, |\text{Re}\,b|\}$ |
| **80**[8] | $\cosh(ax)\cosh(bx)$ | $s\left(s^2 - a^2 - b^2\right)\left[s^2 - (a + b)^2\right]^{-1}\left[s^2 - (a - b)^2\right]^{-1}$, $\qquad\qquad$ Re $s > \{|\text{Re}\,a|, |\text{Re}\,b|\}$ |
| **81** | $\sinh(ax)\cosh(bx)$ | $a\left(s^2 - a^2 + b^2\right)\left[s^2 - (a + b)^2\right]^{-1}\left[s^2 - (a - b)^2\right]^{-1}$, $\qquad\qquad$ Re $s > \{|\text{Re}\,a|, |\text{Re}\,b|\}$ |
| **82** | $\sinh^2(ax)$ | $2a^2 s^{-1}\left(s^2 - 4a^2\right)^{-1}$, $\qquad\qquad$ Re $s > |\text{Re}\,a|$ |
| **83** | $\cosh^2(ax)$ | $\left(s^2 - 2a^2\right)s^{-1}\left(s^2 - 4a^2\right)^{-1}$, $\qquad$ Re $s > |\text{Re}\,a|$ |
| **84** | $\sinh(ax)\cosh(ax)$ | $a\left(s^2 - 4a^2\right)^{-1}$, $\qquad\qquad$ Re $s > |\text{Re}\,a|$ |
| **85** | $\dfrac{\cosh(ax) - 1}{a^2}$ | $s^{-1}\left(s^2 - a^2\right)^{-1}$, $\qquad\qquad$ Re $s > |\text{Re}\,a|$ |
| | | *continued on next page* |

| | $f(x)$ | $F(s)$ |
|---|---|---|
| | *continued from previous page* | |
| **86** | $\dfrac{\sinh(ax) - ax}{a^3}$ | $s^{-2}\left(s^2 - a^2\right)^{-1}$, $\qquad\qquad$ $\operatorname{Re} s > |\operatorname{Re} a|$ |
| **87** | $\dfrac{\left[\cosh(ax) - \frac{1}{2}a^2 x^2 - 1\right]}{a^4}$ | $s^{-3}\left(s^2 - a^2\right)^{-1}$, $\qquad\qquad$ $\operatorname{Re} s > |\operatorname{Re} a|$ |
| **88** | $\dfrac{\left[1 - \cosh(ax) + \frac{1}{2}ax\sinh(ax)\right]}{a^4}$ | $s^{-1}\left(s^2 - a^2\right)^{-2}$, $\qquad\qquad$ $\operatorname{Re} s > |\operatorname{Re} a|$ |
| **89** | $x^{1/2}\sinh(ax)$ | $\left(\pi^{1/2}/4\right)\left[(s-a)^{3/2} - (s+a)^{3/2}\right]$, $\qquad\qquad\qquad\qquad\qquad$ $\operatorname{Re} s > |\operatorname{Re} a|$ |
| **90** | $\ln x$ | $-s^{-1}\ln\left(\mathbf{C}s\right)$, $\qquad$ $\operatorname{Re} s > 0$ $\qquad$ ET I 148(1) |
| **91** | $\ln(1 + ax)$, $\qquad$ $|\arg a| < \pi$ | $s^{-1}e^{s/a}\operatorname{Ei}(-s/a)$, $\qquad$ $\operatorname{Re} s > 0$ $\qquad$ ET I 148(4) |
| **92** | $x^{-1/2}\ln x$ | $-\left(\pi/s\right)^{1/2}\ln\left(4\mathbf{C}s\right)$, $\qquad$ $\operatorname{Re} s > 0$ $\qquad$ ET I 148(9) |
| **93** | $\mathrm{H}(x-a) = \begin{cases} 0 & \text{for } x < a \\ 1 & \text{for } x > a \end{cases}$ <br> (Heaviside step function) | $s^{-1}e^{-as}$, $\qquad\qquad\qquad\qquad$ $a \geq 0$ |
| **94** | $\delta(x)$ $\qquad$ (Dirac delta function) | $1$ |
| **95** | $\delta(x - a)$ | $e^{-as}$, $\qquad\qquad\qquad\qquad$ $a \geq 0$ |
| **96** | $\delta'(x - a)$ | $se^{-as}$, $\qquad\qquad\qquad\qquad$ $a \geq 0$ |
| **97** | $\operatorname{Si}(x) \equiv \displaystyle\int_0^x \frac{\sin\xi}{\xi}\,d\xi \equiv \frac{1}{2}\pi + \operatorname{si}(x)$ | $s^{-1}\operatorname{arccot} s$, $\qquad$ $\operatorname{Re} s > 0$ $\qquad$ ET I 177(17) |
| **98** | $\operatorname{Ci}(x) \equiv \operatorname{ci}(x) \equiv -\displaystyle\int_x^\infty \frac{\cos\xi}{\xi}\,d\xi$ | $-\dfrac{1}{2}s^{-1}\ln\left(1 + s^2\right)$, $\qquad$ $\operatorname{Re} s > 0$ $\qquad$ ET I 178(19) |
| **99**[8] | $\operatorname{erf}\left(\dfrac{x}{2a}\right)$ | $s^{-1}e^{a^2 s^2}\operatorname{erfc}(as)$, <br> $\qquad\qquad$ $\operatorname{Re} s > 0, |\arg a| < \pi/4$ $\qquad$ ET I 176(2) |

*continued on next page*

*continued from previous page*

| | $f(x)$ | | $F(s)$ |
|---|---|---|---|
| **100** | $\operatorname{erf}\left(a\sqrt{x}\right)$ | | $as^{-1}\left(s+a^2\right)^{-1/2}$, <div align="right">$\operatorname{Re}s > \left\{0, -\operatorname{Re}a^2\right\}$     ET I 176(4)</div> |
| **101** | $\operatorname{erfc}\left(a\sqrt{x}\right)$ | | $s^{-1}\left(s+a^2\right)^{-\frac{1}{2}}\left[\left(s+a^2\right)^{1/2}-a\right]$, <div align="right">$\operatorname{Re}s > 0$     ET I 177(9)</div> |
| **102**[8] | $\operatorname{erfc}\left(\dfrac{a}{\sqrt{x}}\right)$ | | $s^{-1}e^{-2a\sqrt{s}}$, <div align="right">$\operatorname{Re}s > 0$,    $\operatorname{Re}a > 0$     ET I 177(11)</div> |
| **103**[8] | $J_\nu(ax)$, | $\operatorname{Re}\nu > -1$ | $a^{-\nu}\left(\sqrt{s^2+a^2}-s\right)^\nu\left(s^2+a^2\right)^{-1/2}$, <div align="right">$\operatorname{Re}s > |\operatorname{Im}a|$     ET I 182(1)</div> |
| **104** | $x\,J_\nu(ax)$, | $\operatorname{Re}\nu > -2$ | $a^\nu\left[s+\nu\left(s^2+a^2\right)^{1/2}\right]\left[s+\left(s^2+a^2\right)^{1/2}\right]^{-\nu}$ $\times\left(s^2+a^2\right)^{-3/2}$, <div align="right">$\operatorname{Re}s > |\operatorname{Im}a|$     ET I 182(2)</div> |
| **105** | $\dfrac{J_\nu(ax)}{x}$ | | $a^\nu\nu^{-1}\left[s+\left(s^2+a^2\right)^{1/2}\right]^{-\nu}$, <div align="right">$\operatorname{Re}s \geq |\operatorname{Im}a|$     ET I 182(5)</div> |
| **106** | $x^n\,J_n(ax)$ | | $1\cdot3\cdot5\cdots(2n-1)a^n\left(s^2+a^2\right)^{-\left(n+\frac{1}{2}\right)}$, <div align="right">$\operatorname{Re}s > |\operatorname{Im}a|$     ET I 182(4)</div> |
| **107** | $x^\nu\,J_\nu(ax)$, | $\operatorname{Re}\nu > -\frac{1}{2}$ | $2^\nu\pi^{-1/2}\Gamma\left(\nu+\frac{1}{2}\right)a^\nu\left(s^2+a^2\right)^{-\left(\nu+\frac{1}{2}\right)}$, <div align="right">$\operatorname{Re}s > |\operatorname{Im}a|$     ET I 182(7)</div> |
| **108** | $x^{\nu+1}\,J_\nu(ax)$, | $\operatorname{Re}\nu > -1$ | $2^{\nu+1}\pi^{-1/2}\Gamma\left(\nu+\frac{3}{2}\right)a^\nu s\left(s^2+a^2\right)^{-\left(\nu+\frac{3}{2}\right)}$, <div align="right">$\operatorname{Re}s > |\operatorname{Im}a|$     ET I 182(8)</div> |
| **109**[8] | $I_\nu(ax)$, | $\operatorname{Re}\nu > -1$ | $a^{-\nu}\left[s-\sqrt{s^2-a^2}\right]^\nu\left(s^2-a^2\right)^{-1/2}$, <div align="right">$\operatorname{Re}s > |\operatorname{Re}a|$     ET I 195(1)</div> |

continued on next page

| | $f(x)$ | | $F(s)$ | |
|---|---|---|---|---|

continued from previous page

| | $f(x)$ | | $F(s)$ | |
|---|---|---|---|---|
| **110** | $x^{\nu} I_{\nu}(ax),$ | $\operatorname{Re}\nu > -\frac{1}{2}$ | $2^{\nu}\pi^{-1/2}\Gamma\left(\nu+\frac{1}{2}\right)a^{\nu}\left(s^2-a^2\right)^{-\left(\nu+\frac{1}{2}\right)},$ $\operatorname{Re}s > \lvert\operatorname{Re}a\rvert$ | ET I 195(6) |
| **111** | $x^{\nu+1} I_{\nu}(ax),$ | $\operatorname{Re}\nu > -1$ | $2^{\nu+1}\pi^{-1/2}\Gamma\left(\nu+\frac{3}{2}\right)a^{\nu}s\left(s^2-a^2\right)^{-\left(\nu+\frac{3}{2}\right)},$ $\operatorname{Re}s > \lvert\operatorname{Re}a\rvert$ | ET I 196(7) |
| **112** | $x^{-1} I_{\nu}(ax),$ | $\operatorname{Re}\nu > 0$ | $\nu^{-1}a^{\nu}\left[s+\left(s^2-a^2\right)^{1/2}\right]^{-\nu},$ $\operatorname{Re}s > \lvert\operatorname{Re}a\rvert$ | ET I 195(4) |
| **113** | $\sin\left(2a^{1/2}x^{1/2}\right)$ | | $(\pi a)^{1/2}s^{-3/2}e^{-a/s},\qquad \operatorname{Re}s > 0$ | ET I 153(32) |
| **114** | $x^{-1/2}\cos\left(2a^{1/2}x^{1/2}\right)$ | | $\pi^{1/2}s^{-1/2}e^{-a/s},\qquad \operatorname{Re}s > 0$ | ET I 158(67) |
| **115** | $x^{-1}e^{-ax} I_1(ax)$ | | $\left[(s+2a)^{1/2}-s^{1/2}\right]\left[(s+2a)^{1/2}+s^{1/2}\right]^{-1},$ $\operatorname{Re}s > \lvert\operatorname{Re}a\rvert$ | AS 1024(29.3.52) |
| **116** | $\dfrac{J_k(ax)}{x}$ | | $k^{-1}a^{-k}\left[\left(s^2+a^2\right)^{1/2}-s\right]^{k},$ $\operatorname{Re}s > \lvert\operatorname{Im}a\rvert, k > -1$ | AS 1025(29.3.58) |
| **117** | $\left(\dfrac{x}{2a}\right)^{k-\frac{1}{2}} J_{k-\frac{1}{2}}(ax)$ | | $\Gamma(k)\pi^{-1/2}\left(s^2+a^2\right)^{k},$ $\operatorname{Re}s > \lvert\operatorname{Im}a\rvert,\quad k > 0$ | AS 1024(29.3.57) |
| **118** | $J_0(ax) - ax\,J_1(ax)$ | | $s^2\left(s^2+a^2\right)^{-3/2},\qquad\qquad \operatorname{Re}s > \lvert\operatorname{Im}a\rvert$ | |
| **119** | $I_0(ax) + ax\,I_1(ax)$ | | $s^2\left(s^2-a^2\right)^{-3/2},\qquad\qquad \operatorname{Re}s > \lvert\operatorname{Im}a\rvert$ | |

## 12.21  Fourier transform

The **Fourier transform**, also called the **exponential** or **complex Fourier transform**, of the function $f(x)$, denoted by $F(\xi)$, is defined by the integral

$$F(\xi) = \frac{1}{\sqrt{2\pi}}\int_{-\infty}^{\infty} f(x)e^{i\xi x}\,\mathrm{d}x.$$

The functions $f(x)$ and $F(\xi)$ are called a **Fourier transform pair**, and knowledge of either one enables the other to be recovered. Setting $F(\xi) = \mathcal{F}\left[f(x);\xi\right]$, to emphasize the nature of the transform, we have the symbolic inverse result $f(x) = \mathcal{F}^{-1}[F(\xi); x]$. The inversion of the Fourier transform is accomplished by means of the **inversion integral**

$$f(x) = \frac{1}{\sqrt{2\pi}} \int_{-\infty}^{\infty} F(\xi) e^{-i\xi x} \, d\xi.$$

## 12.22  Basic properties of the Fourier transform

1.     For $a$ and $b$ arbitrary constants,

$$\mathcal{F}[af(x) + bg(x)] = aF(\xi) + bG(\xi) \qquad \text{(linearity)}$$

2.     If $n > 0$ is an integer, and $\lim_{|x| \to \infty} f^{(r)}(x) = 0$ for $r = 0, 1, \ldots, n-1$ with $f^{(0)}(x) \equiv f(x)$, then

$$\mathcal{F}\left[f^{(n)}(x); \xi\right] = (-i\xi)^n F(\xi) \qquad \text{(transform of a derivative)} \qquad \text{SN 27}$$

3.     The **Fourier convolution** $f * g$ of two functions $f(x)$ and $g(x)$ is defined by the integral

$$f * g(x) = \frac{1}{\sqrt{2\pi}} \int_{-\infty}^{\infty} f(x - \xi) g(\xi) \, d\xi,$$

and it has the property $f * g = g * f$, and $f * (g * h) = (f * g) * h$. In terms of the convolution operation.

$$\mathcal{F}[f * g(x); \xi] = F(\xi) G(\xi) \qquad \text{(convolution (Faltung) theorem).} \qquad \text{SN 24}$$

## 12.23  Table of Fourier transform pairs

| | $f(x)$ | | $F(\xi)$ | |
|---|---|---|---|---|
| **1** | $1$ | | $(2\pi)^{1/2} \delta(\xi)$ | SU 496 |
| **2⁷** | $\dfrac{1}{x}$ | | $(\pi/2)^{1/2} i \operatorname{sign} \xi$ | SU 50 |
| **3** | $\delta(x)$ | | $(2\pi)^{-1/2}$ | SU 496 |
| **4⁸** | $\delta(ax + b),$ | $a, b \in \mathbb{R}, \quad a \neq 0$ | $(2\pi)^{-1/2} e^{ib\xi/a}$ | SU 517 |
| **5** | $\begin{cases} 1 & \lvert x \rvert < a \\ 0 & \lvert x \rvert > a \end{cases},$ | $a > 0$ | $(2/\pi)^{1/2} \xi^{-1} \sin(a\xi)$ | |
| **6⁸** | $\mathrm{H}(x) = \begin{cases} 0 & x < 0 \\ 1 & x > 0 \end{cases}$ | | $-\dfrac{1}{i\xi\sqrt{2\pi}} + \sqrt{\dfrac{\pi}{2}}\, \delta(\xi)$ | SN 523 |
| | | | *continued on next page* | |

| | continued from previous page | | | |
|---|---|---|---|---|
| | $f(x)$ | | $F(\xi)$ | |
| **7** | $\dfrac{1}{\|x\|^a},$ $\qquad 0 < \operatorname{Re} a < 1$ | | $\dfrac{(2/\pi)^{1/2}\,\Gamma(1-a)\sin\left(\frac{1}{2}a\pi\right)}{\|\xi\|^{1-a}}$ | SN 523 |
| **8** | $e^{iax},$ $\qquad a \in \mathbb{R}$ | | $(2\pi)^{1/2}\,\delta(\xi + a)$ | SU 50 |
| **9** | $e^{-a\|x\|},$ $\qquad a > 0$ | | $\dfrac{a(2/\pi)^{1/2}}{a^2 + \xi^2}$ | SU 50 |
| **10**[7] | $xe^{-a\|x\|},$ $\qquad a > 0$ | | $\dfrac{2ai\xi(2/\pi)^{1/2}}{\left(a^2 + \xi^2\right)^2},$ $\qquad \xi > 0$ | SU 50 |
| **11** | $\|x\|e^{-a\|x\|},$ $\qquad a > 0$ | | $\dfrac{(2/\pi)^{1/2}\left(a^2 - \xi^2\right)}{\left(a^2 + \xi^2\right)^2}$ | SU 50 |
| **12** | $\dfrac{e^{-a\|x\|}}{\|x\|^{1/2}},$ $\qquad a > 0$ | | $\dfrac{\left[a + \left(a^2 + \xi^2\right)^{1/2}\right]^{1/2}}{x\left(a^2 + \xi^2\right)^{1/2}}$ | SN 523 |
| **13** | $e^{-a^2x^2},$ $\qquad a > 0$ | | $\left(a\sqrt{2}\right)^{-1}e^{-\xi^2/4a^2}$ | SU 51 |
| **14** | $\dfrac{1}{a^2 + x^2},$ $\qquad \operatorname{Re} a > 0$ | | $\dfrac{(\pi/2)^{1/2}\,e^{-a\|\xi\|}}{a}$ | SU 51 |
| **15**[7] | $\dfrac{x}{a^2 + x^2},$ $\qquad \operatorname{Re} a > 0$ | | $i\operatorname{sign}\xi\,(\pi/2)^{1/2}\,e^{-a\|\xi\|}$ | |
| **16**[9] | $\sin\left(ax^2\right)$ | | $\dfrac{1}{(2a)^{1/2}}\cos\left(\dfrac{\xi^2}{4a} + \dfrac{\pi}{4}\right)$ | SN 523 |
| **17** | $\cos\left(ax^2\right)$ | | $\dfrac{1}{(2a)^{1/2}}\cos\left(\dfrac{\xi^2}{4a} - \dfrac{\pi}{4}\right)$ | SN 523 |
| **18** | $e^{-a\|x\|}\cos(bx),$ $\quad a > 0, \quad b > 0$ | | $a(2\pi)^{-1/2}\left[\dfrac{1}{a^2 + (b+\xi)^2} + \dfrac{1}{a^2 + (b-\xi)^2}\right]$ | |
| **19** | $e^{-\frac{1}{2}ax^2}\sin(bx),$ $\quad a > 0, \quad b > 0$ | | $\dfrac{1}{2}ia^{-1/2}\left\{\exp\left[-\dfrac{1}{2}\dfrac{(\xi-b)^2}{a}\right]\right.$ $\left. - \exp\left[-\dfrac{1}{2}\dfrac{(\xi+b)^2}{a}\right]\right\}$ | |
| **20**[9] | $\dfrac{\sinh(ax)}{\sinh(bx)},$ $\qquad \|a\| < \|b\|$ | | $\dfrac{(\pi/2)^{1/2}\sin(\pi a/b)}{b\left[\cosh(\pi\xi/b) + \cos(\pi a/b)\right]}$ | SU 123 |
| **21**[9] | $\dfrac{\cosh(ax)}{\sinh(bx)},$ $\qquad \|a\| < \|b\|$ | | $\dfrac{i(\pi/2)^{1/2}\sinh(\pi\xi/b)}{b\left[\cosh(\pi\xi/b) + \cos(\pi a/b)\right]}$ | SU 123 |
| | | | continued on next page | |

| | $f(x)$ | $F(\xi)$ | |
|---|---|---|---|
| *continued from previous page* | | | |
| **22**[12] | $\dfrac{\sin(ax)}{x}$ $\qquad a > 0$ | $\begin{cases} (\pi/2)^{1/2} & |\xi| < a, \\ 0 & |\xi| > a \end{cases}$ | SN 523 |
| **23**[11] | $\dfrac{x}{\sinh x}$ | $\dfrac{\left(2Constantpi^3\right)^{1/2} e^{\pi\xi}}{\left(1 + e^{\pi\xi}\right)^2}$ | SU 123 |
| **24**[7] | $x^n \operatorname{sign} x, \qquad n = 1, 2, \ldots$ | $(2/\pi)^{1/2} (-i\xi)^{-(1+n)} n!$ | SU 506 |
| **25**[7] | $|x|^\nu,$ $-1 < \nu < 0, \text{but not integral}$ | $(2/\pi)^{1/2} \Gamma(\nu + 1) |\xi|^{-\nu-1} \cos\left[\pi(\nu + 1)/2\right]$ | SU506 |
| **26**[7] | $|x|^\nu \operatorname{sign} x,$ $-1 < \nu < 0, \text{but not integral}$ | $\dfrac{i \operatorname{sign} \xi (2/\pi)^{1/2} \sin\left[(\pi/2)(\nu + 1)\right] \Gamma(\nu + 1)}{|\xi|^{\nu+1}}$ | SU 506 |
| **27** | $e^{-ax} \ln\left|1 - e^{-x}\right|,$ $-1 < \operatorname{Re} a < 0$ | $\left(\dfrac{\pi}{2}\right)^{1/2} \dfrac{\cot\left(\pi a - i\xi\pi\right)}{a - i\xi}$ | ET I 121(26) |
| **28** | $e^{-ax} \ln\left(1 + e^{-x}\right),$ $-1 < \operatorname{Re} a < 0$ | $\left(\dfrac{\pi}{2}\right)^{1/2} \dfrac{\csc\left(\pi a - i\xi\pi\right)}{a - i\xi}$ | ET I 121 (27) |

In deriving results for the preceding table from ET I, account has been taken of the fact that the normalization factor $1/(2\pi)^{1/2}$ employed in our definition of $F$ has not been used in those tables, and that there is a difference of sign between the exponents used in the definitions of the exponential Fourier transform.

## 12.24 Table of Fourier transform pairs for spherically symmetric functions

| | $f(\|\mathbf{r}\|) = \dfrac{1}{(2\pi)^{3/2}} \iiint E(\|\mathbf{k}\|) e^{i\mathbf{k}\cdot\mathbf{r}} \, d\mathbf{k}$ | $E(\|\mathbf{k}\|) = \dfrac{1}{(2\pi)^{3/2}} \iiint f(\|\mathbf{r}\|) e^{-i\mathbf{k}\cdot\mathbf{r}} \, d\mathbf{r}$ |
|---|---|---|
| 1 | $f(r) = \sqrt{\dfrac{2}{\pi}} \dfrac{1}{r} \displaystyle\int_0^\infty E(k) \sin(kr) k \, dk$ | $E(k) = \sqrt{\dfrac{2}{\pi}} \dfrac{1}{k} \displaystyle\int_0^\infty f(r) \sin(kr) r \, dr$ |
| 2 | $e^{-ar}$ | $\sqrt{\dfrac{2}{\pi}} \dfrac{2a}{\left(a^2 + k^2\right)^2}$ |
| 3[12] | $\dfrac{e^{-ar}}{r}$ | $\sqrt{\dfrac{2}{\pi}} \dfrac{1}{a^2 + k^2}$ |
| 4[11] | $1$ | $(2\pi)^{3/2} \delta(\mathbf{k})$ |

## 12.31 Fourier sine and cosine transforms

The **Fourier sine** and **cosine transforms** of the function $f(x)$, denoted by $F_s(\xi)$ and $F_c(\xi)$, respectively, are defined by the integrals

$$F_s(\xi) = \sqrt{\frac{2}{\pi}} \int_0^\infty f(x) \sin(\xi x)\, dx \quad \text{and} \quad F_c(\xi) = \sqrt{\frac{2}{\pi}} \int_0^\infty f(x) \cos(\xi x)\, dx$$

The functions $f(x)$ and $F_s(\xi)$ are called a **Fourier sine transform pair**, and the functions $f(x)$ and $F_c(\xi)$ a **Fourier cosine transform pair**, and knowledge of either $F_s(\xi)$ or $F_c(\xi)$ enables $f(x)$ to be recovered.

Setting

$$F_s(\xi) = \mathcal{F}_s\left[f(x); \xi\right] \quad \text{and} \quad F_c(\xi) = \mathcal{F}_c\left[f(x); \xi\right],$$

to emphasize the nature of the transforms, we have the symbolic inverses

$$f(x) = \mathcal{F}_s^{-1}\left[F_s(\xi); x\right] \quad \text{and} \quad f(x) = \mathcal{F}_c^{-1}\left[F_c(\xi); x\right]$$

The inversion of the Fourier sine transform is accomplished by means of the **inversion integral**

$$f(x) = \sqrt{\frac{2}{\pi}} \int_0^\infty F_s(\xi) \sin(\xi x)\, d\xi \qquad [x \geq 0]$$

and the inversion of the Fourier cosine transform is accomplished by means of the **inversion integral**

$$f(x) = \sqrt{\frac{2}{\pi}} \int_0^\infty F_c(\xi) \cos(\xi x)\, d\xi \qquad [x \geq 0]$$

SN 17

## 12.32 Basic properties of the Fourier sine and cosine transforms

1.      For $a$ and $b$ arbitrary constants,

$$\mathcal{F}_s\left[af(x) + bg(x)\right] = aF_s(\xi) + bG_s(\xi)$$

and

$$\mathcal{F}_c\left[af(x) + bg(x)\right] = aF_c(\xi) + bG_c(\xi) \qquad \text{(linearity)}$$

2.[12]    If $\lim_{x \to \infty} f^{(r-1)}(x) = 0$ and $\lim_{x \to 0} \sqrt{\frac{2}{\pi}} f^{(r-1)}(x) = a_{r-1}$, then denoting the Fourier sine and cosine transforms of $f^{(r)}(x)$ by $F_s^{(r)}$ and $F_c^{(r)}$, respectively,

(i)     $F_c^{(r)}(\xi) = -a_{r-1} + \xi F_s^{(r-1)}$.

(ii)    $F_s^{(r)}(\xi) = -\xi F_c^{(r-1)}(\xi),$

(iii)   $F_c^{(2r)}(\xi) = -\sum_{n=0}^{r-1}(-1)^n a_{2r-2n-1}\xi^{2n} + (-1)^r \xi^{2n} F_c(\xi),$

(iv)   $F_c^{(2r+1)}(\xi) = -\sum_{n=0}^{r-1}{}'(-1)^n a_{2r-2n}\xi^{2n} + (-1)^r \xi^{2r+1} F_s(\xi),$

(v)    $F_s^{(r)}(\xi) = \xi a_{r-2} - \xi^2 F_s^{(r-2)}(\xi),$

(vi)[6]   $F_s^{(2r)}(\xi) = -\sum_{n=1}^{r}(-1)^n \xi^{2n-1} a_{2r-2n} + (-1)^r \xi^{2r} F_s(\xi),$

(vii)   $F_s^{(2r+1)}(\xi) = -\sum_{n=1}^{r}{}'(-1)^n \xi^{2n-1} a_{2r-2n+1} + (-1)^{r+1} \xi^{2r+1} F_c(\xi).$

SN 28

3.   (i)    $\displaystyle\int_0^\infty F_s(\xi)G_s(\xi)\cos(\xi x)\,\mathrm{d}\xi = \frac{1}{2}\int_0^\infty g(s)\left[f(s+x)+f(s-x)\right]\mathrm{d}s,$

  (ii)    $\displaystyle\int_0^\infty F_c(\xi)G_c(\xi)\cos(\xi x)\,\mathrm{d}\xi = \frac{1}{2}\int_0^\infty g(s)\left[f(s+x)+f\left(|x-s|\right)\right]\mathrm{d}s$

$\hspace{7cm}$ (convolution (Faltung) theorem)   **SN 24**

4.   (i)    If $F_s(\xi)$ is the Fourier sine transform of $f(x)$, then the Fourier sine transform of $F_s(x)$ is $f(\xi)$.

  (ii)    If $F_c(\xi)$ is the Fourier cosine transform of $f(x)$, then the Fourier cosine transform of $F_c(x)$ is $f(\xi)$.

  (iii)    If $f(x)$ is an odd function in $(-\infty,\infty)$, then the Fourier sine transform of $f(x)$ in $(0,\infty)$ is $-iF(\xi)$.

  (iv)    If $f(x)$ is an even function in $(-\infty,\infty)$, then the Fourier cosine transform of $f(x)$ in $(0,\infty)$ is $F(\xi)$.

  (v)    The Fourier sine transform of $f(x/a)$ is $aF_s(a\xi)$.

  (vi)    The Fourier cosine transform of $f(x/a)$ is $aF_c(a\xi)$.

  (vii)    $\mathcal{F}_s\left[f(x);\xi\right] = F_s\left(|\xi|\right)\operatorname{sign}\xi$ $\hspace{4cm}$ **SU 45**

## 12.33   Table of Fourier sine transforms

| | $f(x)$ | $F_s(\xi)$      $(\xi > 0)$ | |
|---|---|---|---|
| **1** | $x^{-1}$ | $(\pi/2)^{1/2},$              $\xi > 0$ | ET I 64(3) |
| **2** | $x^{-\nu},$        $0 < \operatorname{Re}\nu < 2$ | $(2/\pi)^{1/2}\xi^{\nu-1}\,\Gamma(1-\nu)\cos\left(\nu\pi/2\right),$              $\xi > 0$ | ET I 68(1) |
| **3** | $x^{-1/2}$ | $\xi^{-1/2},$              $\xi > 0$ | ET I 64(6) |
| **4** | $x^{-3/2}$ | $2\xi^{1/2},$              $\xi > 0$ | ET I 64(9) |
| **5** | $\begin{cases}1 & 0 < x < a \\ 0 & x > a\end{cases}$ | $(2/\pi)^{1/2}\xi^{-1}\left[1-\cos(a\xi)\right],$    $\xi > 0$ | ET I 63(1) |
| **6** | $\begin{cases}x^{-1} & 0 < x < a \\ 0 & x > a\end{cases}$ | $(2/\pi)^{1/2}\operatorname{Si}(a\xi),$           $\xi > 0$ | ET I 64(4) |
| **7** | $\dfrac{1}{a-x},$        $a > 0$ | $(2/\pi)^{1/2}\left\{\sin(a\xi)\operatorname{Ci}(a\xi)-\cos(a\xi)\left[\tfrac{1}{2}\pi+\operatorname{Si}(a\xi)\right]\right\},$              $\xi > 0$      ET I 64(11) | |
| | | *continued on next page* | |

| | $f(x)$ | | $F_s(\xi)$ $(\xi > 0)$ | | |
|---|---|---|---|---|---|
| **8**[7] | $\dfrac{1}{x^2 + a^2}$, | $a > 0$ | $(2\pi)^{-1/2} a^{-1} \left[ e^{-a\xi} \, \mathrm{Ei}(a\xi) - e^{a\xi} \, \mathrm{Ei}(-a\xi) \right]$, | | |
| | | | | $\xi > 0$ | ET I 65(14) |
| **9** | $x \left( x^2 + a^2 \right)^{-3/2}$, | $\operatorname{Re} a > 0$ | $(2/\pi)^{1/2} \xi \, K_0(a\xi)$, | $\xi > 0$ | ET I 66(27) |
| **10** | $x^{-1/2} \left( x^2 + a^2 \right)^{-1/2}$, | $\operatorname{Re} a > 0$ | $\xi^{1/2} I_{\frac{1}{4}} \left( \tfrac{1}{2} a\xi \right) K_{\frac{1}{4}} \left( \tfrac{1}{2} a\xi \right)$, | $\xi > 0$ | ET I 66(28) |
| **11**[7] | $x \left( x^2 + a^2 \right)^{-\nu - \frac{3}{2}}$, | | $\dfrac{\xi^{\nu+1}}{\sqrt{2}(2a)^\nu \, \Gamma\left( \nu + \frac{3}{2} \right)} K_\nu(a\xi)$, | | |
| | $\operatorname{Re}\nu > -1$, | $\operatorname{Re} a > 0$ | | | |
| **12** | $\dfrac{x}{a^2 + x^2}$, | $\operatorname{Re} a > 0$ | $\left( \dfrac{\pi}{2} \right)^{1/2} e^{-a\xi}$, | $\xi > 0$ | ET I 65(15) |
| **13** | $\dfrac{x}{\left( a^2 + x^2 \right)^2}$ | | $\sqrt{\pi/8} \, a^{-1} \xi e^{-a\xi}$, | $\xi > 0$ | ET I 67(35) |
| **14** | $x^{-1} \left( x^2 + a^2 \right)^{-1}$, | $\operatorname{Re} a > 0$ | $\dfrac{\sqrt{\pi/2}}{a^2} \left( 1 - e^{-a\xi} \right)$, | $\xi > 0$ | ET I 65(20) |
| **15** | $x^{-1} e^{-ax}$, | $\operatorname{Re} a > 0$ | $(2/\pi)^{1/2} \tan^{-1} \left( \dfrac{\xi}{a} \right)$, | $\xi > 0$ | ET I 72(2) |
| **16** | $x^{\nu-1} e^{-ax}$, | | $(2/\pi)^{1/2} \, \Gamma(\nu) \left( a^2 + \xi^2 \right)^{-\nu/2} \sin \left[ \nu \tan^{-1} \left( \dfrac{\xi}{a} \right) \right]$, | | |
| | $\operatorname{Re}\nu > -1$, | $\operatorname{Re} a > 0$ | | $\xi > 0$ | ET I 72(7) |
| **17** | $e^{-ax}$, | $\operatorname{Re} a > 0$ | $\dfrac{\sqrt{2/\pi} \, \xi}{a^2 + \xi^2}$, | $\xi > 0$ | ET I 72(1) |
| **18** | $xe^{-ax}$, | $\operatorname{Re} a > 0$ | $\dfrac{(2/\pi)^{1/2} 2a\xi}{\left( a^2 + \xi^2 \right)^2}$, | $\xi > 0$ | ET I 72(3) |
| **19** | $xe^{-ax^2}$, | $|\arg a| < \pi/2$ | $(2a)^{-3/2} \xi \exp \left( \dfrac{-\xi^2}{4a} \right)$, | $\xi > 0$ | ET I 73(19) |
| **20** | $\dfrac{\sin ax}{x}$, | $a > 0$ | $\dfrac{1}{(2\pi)^{1/2}} \ln \left| \dfrac{\xi + a}{\xi - a} \right|$, | $\xi > 0$ | ET I 78(1) |
| **21** | $\dfrac{\sin ax}{x^2}$, | $a > 0$ | $\begin{cases} \xi \left( \frac{\pi}{2} \right)^{1/2} & 0 < \xi < a \\ a \left( \frac{\pi}{2} \right)^{1/2} & a < \xi < \infty \end{cases}$, | $\xi > 0$ | ET I 78(2) |

*continued from previous page*

*continued on next page*

continued from previous page

| | $f(x)$ | | $F_s(\xi)$ $\quad$ $(\xi > 0)$ |
|---|---|---|---|
| **22** | $\sin\left(\dfrac{a^2}{x}\right)$, | $a > 0$ | $a\left(\dfrac{\pi}{2}\right)^{1/2}\xi^{-1/2}\,J_1\left(2a\xi^{\frac{1}{2}}\right)$, $\qquad\qquad\qquad\qquad\qquad \xi > 0 \qquad$ ET I 83(6) |
| **23** | $x^{-1}\sin\left(\dfrac{a^2}{x}\right)$, | $a > 0$ | $\left(\dfrac{\pi}{2}\right)^{1/2}Y_0\left(2a\xi^{1/2}\right) + \left(\dfrac{2}{\pi}\right)^{1/2}K_0\left(2a\xi^{1/2}\right)$ $\qquad\qquad\qquad\qquad\qquad\qquad\qquad$ ET I 83(7) |
| **24** | $x^{-2}\sin\left(\dfrac{a^2}{x}\right)$, | $a > 0$ | $\left(\dfrac{\pi}{2}\right)^{1/2}a^{-1}\xi^{1/2}\,J_1\left(2a\xi^{1/2}\right)$, $\qquad\qquad\qquad\qquad\qquad \xi > 0 \qquad$ ET I 83(8) |
| **25**[10] | $\operatorname{cosech}(ax)$, | $\operatorname{Re} a > 0$ | $(\pi/2)^{1/2}\,a^{-1}\tanh\left(\tfrac{1}{2}\pi a^{-1}\xi\right)$, $\qquad\qquad\qquad\qquad\qquad \xi > 0 \qquad$ ET I 88(2) |
| **26** | $\coth\left(\dfrac{1}{2}ax\right) - 1$, | $\operatorname{Re} a > 0$ | $(2\pi)^{1/2}a^{-1}\coth\left(\pi a^{-1}\xi\right) - \xi$, $\qquad\qquad\qquad\qquad\qquad \xi > 0 \qquad$ ET I 88(3) |
| **27** | $\left(1 - x^2\right)^{-1}\sin(\pi x)$ | | $\begin{cases}(2/\pi)^{1/2}\sin\xi & 0 \le \xi \le \pi \\ 0 & \pi < \xi\end{cases}$ $\qquad$ ET I 78(4) |
| **28** | $e^{-ax^2}\sin(bx)$, | $\operatorname{Re} a > 0$ | $(2a)^{-1/2}\exp\left[-\left(\xi^2 + b^2\right)/(4a)\right]\sinh\left(b\xi/2a\right)$, $\qquad\qquad\qquad\qquad\qquad \xi > 0 \qquad$ ET I 78(7) |
| **29** | $\dfrac{\sin^2(ax)}{x}$, | $a > 0$ | $\begin{cases}\pi^{1/2}2^{-3/2} & 0 < \xi < 2a \\ \pi^{1/2}2^{-5/2} & \xi = 2a \\ 0 & 2a < \xi\end{cases}$ $\qquad$ ET I 78(8) |
| **30** | $\sin\left(ax^2\right)$, | $a > 0$ | $a^{-1/2}\left\{\cos\left(\xi^2/4a\right)C\left[(2\pi a)^{-1/2}\xi\right]\right\}$ $+\sin\left(\xi^2/4a\right)S\left[(2\pi a)^{-1/2}\xi\right]$, $\qquad\qquad\qquad \xi > 0 \qquad$ ET I 82(1) |
| **31** | $\cos\left(ax^2\right)$, | $a > 0$ | $a^{-1/2}\left\{\sin\left(\xi^2/4a\right)C\left[(2\pi a)^{-1/2}\xi\right]\right\}$ $-\cos\left(\xi^2/4a\right)S\left[(2\pi a)^{-1/2}\xi\right]$, $\qquad\qquad\qquad\qquad\qquad \xi > 0$ |

continued on next page

continued from previous page

| | $f(x)$ | | $F_s(\xi)$ $\quad (\xi > 0)$ | | |
|---|---|---|---|---|---|
| **32** | $\arctan\left(\dfrac{x}{a}\right),$ | $a > 0$ | $(\pi/2)^{1/2}\,\xi^{-1}e^{-a\xi},$ | $\xi > 0$ | ET I 87(3) |
| **33**[7] | $\arctan\left(\dfrac{2a}{x}\right),$ | $\operatorname{Re} a > 0$ | $(2\pi)^{-1/2}e^{-a\xi}\sinh(a\xi),$ | $\xi > 0$ | ET I 87(8) |
| **34** | $\dfrac{\ln x}{x}$ | | $-(\pi/2)^{1/2}\left(\boldsymbol{C}+\ln\xi\right),$ | $\xi > 0$ | ET I 76(2) |
| **35** | $\ln\left|\dfrac{x+a}{x-a}\right|,$ | $a > 0$ | $(2\pi)^{1/2}\xi^{-1}\sin(a\xi),$ | $\xi > 0$ | ET I 77(11) |
| **36**[7] | $\dfrac{\ln\left(1+a^2x^2\right)}{x},$ | $a > 0$ | $-(2\pi)^{1/2}\operatorname{Ei}\left(-\xi/a\right),$ | $\xi > 0$ | ET I 77(14) |
| **37** | $J_0(ax),$ | $a > 0$ | $\begin{cases} 0 & 0 < \xi < a \\ (2/\pi)^{1/2}\left(\xi^2-a^2\right)^{-1/2} & a < \xi < \infty \end{cases}$ | | ET I 99(1) |
| **38** | $J_\nu(ax),$ $\quad \operatorname{Re}\nu > -2,$ | $a > 0$ | $(2/\pi)^{1/2}\left(a^2-\xi^2\right)^{-1/2}\sin\left[\nu\sin^{-1}\left(\dfrac{\xi}{a}\right)\right]$ $\qquad\qquad\qquad\qquad$ for $0 < \xi < a$ $\dfrac{a^\nu\cos\left(\frac{1}{2}\nu\pi\right)}{\left(\xi^2-a^2\right)^{1/2}\left[\xi+\left(\xi^2-a^2\right)^{1/2}\right]^\nu}$ $\quad$ for $a < \xi < \infty$ | | ET I 99(3) |
| **39** | $\dfrac{J_0(ax)}{x},$ | $a > 0$ | $\begin{cases} (2/\pi)^{1/2}\sin^{-1}\left(\frac{\xi}{a}\right) & 0 < \xi < a \\ (\pi/2)^{1/2} & a < \xi < \infty \end{cases}$ | | ET I 99(4) |
| **40**[7] | $\left(x^2+b^2\right)^{-1}J_0(ax),$ $\qquad\qquad a > 0, \quad \operatorname{Re} b > 0$ | | $(2/\pi)^{1/2}\sinh(b\xi)\,K_0(ab)/b,$ $\qquad\qquad\qquad 0 < \xi < a$ | | ET I 100(12) |
| **41** | $x\left(x^2+b^2\right)^{-1}J_0(ax),$ $\qquad\qquad a > 0, \quad \operatorname{Re} b > 0$ | | $(\pi/2)^{1/2}\,e^{-b\xi}I_0(ab),$ $\qquad\qquad\qquad a < \xi < \infty$ | | ET I 100(13) |

In deriving results for the preceding table from ET I, account has been taken of the fact that the normalization factor $\sqrt{2/\pi}$ employed in our definition of $F_s$ has not been used in those tables.

## 12.34  Table of Fourier cosine transforms

| | $f(x)$ | | $F_c(\xi)$ | | |
|---|---|---|---|---|---|
| $\mathbf{1^{12}}$ | $x^{-\nu}$, | $0 < \operatorname{Re}\nu < 1$ | $(2/\pi)^{1/2}\zeta^{\nu-1}\Gamma(1-\nu)\sin(\nu\pi/2)$, $\qquad\qquad\qquad\qquad \xi > 0$ | | ET I 10(1) |
| $\mathbf{2}$ | $\begin{cases} 1 & 0 < x < a \\ 0 & x > a \end{cases}$ | | $(2/\pi)^{1/2}\dfrac{\sin(a\xi)}{\xi}$, | $\xi > 0$ | ET I 7(1) |
| $\mathbf{3}$ | $\begin{cases} 0 & 0 < x < a \\ 1/x & x > a \end{cases}$ | | $-(2/\pi)^{1/2}\operatorname{Ci}(a\xi)$, | $\xi > 0$ | ET I 8(3) |
| $\mathbf{4}$ | $\begin{cases} x^{-1/2} & 0 < x < a \\ 0 & x > a \end{cases}$ | | $2\xi^{-1/2}\,C(a\xi)$, | $\xi > 0$ | ET I 8(5) |
| $\mathbf{5}$ | $\begin{cases} 0 & 0 < x < a \\ x^{-1/2} & x > a \end{cases}$ | | $2\xi^{-1/2}\left[\tfrac{1}{2} - C(a\xi)\right]$, | $\xi > 0$ | ET I 8(6) |
| $\mathbf{6^{9}}$ | $x^{\nu-1}$, | $0 < \nu < 1$ | $(2/\pi)^{1/2}\,\Gamma(\nu)\xi^{-\nu}\cos\left(\tfrac{1}{2}\nu\pi\right)$, $\qquad\qquad\qquad\qquad 0 < \nu < 1$ | | ET I 10(1) |
| $\mathbf{7}$ | $\dfrac{1}{x^2 + a^2}$, | $\operatorname{Re}a > 0$ | $\dfrac{(\pi/2)^{1/2}\,e^{-a\xi}}{a}$, | $\xi > 0$ | ET I 11(7) |
| $\mathbf{8^{11}}$ | $\dfrac{1}{\left(x^2 + a^2\right)^2}$, | $\operatorname{Re}a > 0$ | $\dfrac{(\pi/2)^{\frac{1}{2}}\,(1 + a\xi)e^{-a\xi}}{2a^3}$, | $\xi > 0$ | ET I 11(7) |
| $\mathbf{9}$ | $\left(x^2 + a^2\right)^{-\nu-\frac{1}{2}}$, $\operatorname{Re}a > 0, \operatorname{Re}\nu > -\dfrac{1}{2}$ | | $\sqrt{2}\left(\dfrac{\xi}{2a}\right)^{\nu}\dfrac{K_\nu(a\xi)}{\Gamma\left(\nu + \frac{1}{2}\right)}$, | $\xi > 0$ | ET I 11(7) |
| $\mathbf{10}$ | $\begin{cases} \left(a^2 - x^2\right)^{\nu} & 0 < x < a \\ 0 & x > a \end{cases}$, $\operatorname{Re}\nu > -1$ | | $2^{\nu}\,\Gamma(\nu + 1)(a/\xi)^{\nu+\frac{1}{2}}\,J_{\nu+\frac{1}{2}}(a\xi)$, $\qquad\qquad\qquad\qquad \xi > 0$ | | ET I 11(8) |
| $\mathbf{11}$ | $\begin{cases} 0 & 0 < x < a \\ \left(x^2 - a^2\right)^{-\nu-\frac{1}{2}} & x > a \end{cases}$, $-\dfrac{1}{2} < \operatorname{Re}\nu < \dfrac{1}{2}$ | | $-2^{-\left(\nu+\frac{1}{2}\right)}\Gamma\left(\tfrac{1}{2} - \nu\right)(\xi/a)^{\nu}\,Y_\nu(a\xi)$, $\qquad\qquad\qquad\qquad \xi > 0$ | | ET I 11(9) |
| $\mathbf{12}$ | $e^{-ax}$, | $\operatorname{Re}a > 0$ | $(2/\pi)^{1/2}a\left(a^2 + \xi^2\right)^{-1}$, | $\xi > 0$ | ET I 14(1) |
| | | | | *continued on next page* | |

continued from previous page

| | $f(x)$ | | $F_c(\xi)$ |
|---|---|---|---|
| **13** | $xe^{-ax}$, | $\operatorname{Re} a > 0$ | $(2/\pi)^{1/2} \left(a^2 - \xi^2\right) \left(a^2 + \xi^2\right)^{-2}$, |
| | | | $\xi > 0$ ET I 15(7) |
| **14[7]** | $x^{\nu-1}e^{-ax}$, | | $(2/\pi)^{1/2} \Gamma(\nu) \left(a^2 + \xi^2\right)^{-\nu/2} \cos\left[\nu \tan^{-1}\left(\dfrac{\xi}{a}\right)\right]$, |
| | | $\operatorname{Re} a > 0$, $\operatorname{Re}\nu > a$ | $\xi > 0$ ET I 15(7) |
| **15** | $x^{-1/2}e^{-ax}$, | $\operatorname{Re} a > 0$ | $\left(a^2 + \xi^2\right)^{-1/2} \left[\left(a^2 + \xi^2\right)^{1/2} + a\right]^{1/2}$, |
| | | | $\xi > 0$ ET I 14(4) |
| **16[7]** | $e^{-a^2 x^2}$, | $\operatorname{Re} a > 0$ | $2^{-1/2}|a|^{-1}e^{-\xi^2/4a^2}$, $\xi > 0$ ET I 15(11) |
| **17** | $x^{-1}e^{-x}\sin x$ | | $(2\pi)^{-1/2}\tan^{-1}\left(\dfrac{2}{\xi^2}\right)$, $\xi > 0$ ET I 19(7) |
| **18** | $\sin\left(ax^2\right)$, | $a > 0$ | $\dfrac{1}{2\sqrt{a}}\left[\cos\left(\dfrac{\xi^2}{4a}\right) - \sin\left(\dfrac{\xi^2}{4a}\right)\right]$, |
| | | | $\xi > 0$ ET I 23(1) |
| **19** | $\cos\left(ax^2\right)$, | $a > 0$ | $\dfrac{1}{2\sqrt{a}}\left[\cos\left(\dfrac{\xi^2}{4a}\right) + \sin\left(\dfrac{\xi^2}{4a}\right)\right]$, |
| | | | $\xi > 0$ ET I 24(7) |
| **20** | $\dfrac{\sin(ax)}{x}$, | $a > 0$ | $\begin{cases} (\pi/2)^{1/2} & \xi < a \\ \frac{1}{2}\left(\pi/2\right)^{1/2} & \xi = a \\ 0 & \xi > a \end{cases}$  ET I 18(1) |
| **21[7]** | $\dfrac{\sin^2(ax)}{x^2}$, | $a > 0$ | $\begin{cases} (\pi/2)^{1/2}\left(a - \frac{1}{2}\xi\right) & \xi < 2a \\ 0 & 2a < \xi \end{cases}$  ET I 19(8) |
| **22[7]** | $e^{-bx}\sin(ax)$, $a > 0$, $\operatorname{Re} b > 0$ | | $(2\pi)^{-1/2}\left[\dfrac{a + \xi}{b^2 + (a + \xi)^2} + \dfrac{a - \xi}{b^2 + (a - \xi)^2}\right]$, |
| | | | $\xi > 0$ ET I 19(6) |
| **23** | $\dfrac{\sin\left[b\left(x^2 + a^2\right)^{1/2}\right]}{\left(x^2 + a^2\right)^2}$, | $a > 0$ | $(b/a)\left(\pi/2\right)^{1/2}e^{-a\xi}$, $\xi > 0$ ET I 26(29) |
| **24** | $\left(x^2 + a^2\right)^{-1/2}\sin\left[b\left(x^2 + a^2\right)^{1/2}\right]$, $a > 0$ | | $\begin{cases} (\pi/2)^{1/2} J_0\left[a\left(b^2 - \xi^2\right)^{1/2}\right] & 0 < \xi < b \\ 0 & b < \xi \end{cases}$ |
| | | | ET I 26(30) |

continued on next page

| | $f(x)$ | $F_c(\xi)$ | |
|---|---|---|---|
| **25** | $\dfrac{1-\cos(ax)}{x^2}$, $\qquad a>0$ | $\begin{cases} (\pi/2)^{1/2}\,(a-\xi) & \xi < a \\ 0 & a < \xi \end{cases}$ | ET I 20(16) |
| **26** | $e^{-ax^2}\sin\left(bx^2\right)$, $\quad \operatorname{Re} a > \lvert\operatorname{Im} b\rvert$ | $2^{-1/2}\left(a^2+b^2\right)^{-1/4}\exp\left\{-a\xi^2/\left[4\left(a^2+b^2\right)\right]\right\}$ $\times \sin\left[\tfrac{1}{2}\arctan(b/a)-\tfrac{1}{4}b\xi^2\left(a^2+b^2\right)^{-1}\right],$ $\xi > 0 \qquad$ ET I 23(5) | |
| **27** | $e^{-ax^2}\cos\left(bx^2\right)$, $\quad \operatorname{Re} a > \lvert\operatorname{Im} b\rvert$ | $2^{-1/2}\left(a^2+b^2\right)^{-1/4}\exp\left\{-a\xi^2/\left[4\left(a^2+b^2\right)\right]\right\}$ $\times \cos\left[\tfrac{1}{4}b\xi^2\left(a^2+b^2\right)^{-1}-\tfrac{1}{2}\arctan(b/a)\right],$ $\xi > 0 \qquad$ ET I 24(6) | |
| **28** | $\dfrac{\sinh(ax)}{\sinh(bx)}$ $\qquad \lvert\operatorname{Re} a\rvert < \operatorname{Re} b$ | $\left(\dfrac{\pi}{2}\right)^{1/2}\dfrac{\sin(\pi a/b)}{b\left[\cosh\left(\pi\xi/b\right)+\cos(\pi a/b)\right]},$ $\xi > 0 \qquad$ ET I 31(14) | |
| **29** | $\dfrac{\cosh(ax)}{\cosh(bx)}$, $\qquad \lvert\operatorname{Re} a\rvert < \operatorname{Re} b$ | $\dfrac{(2\pi)^{1/2}\cos(\pi a/2b)\cosh\left(\pi\xi/2b\right)}{b\left[\cosh\left(\pi\xi/b\right)+\cos(\pi a/b)\right]},$ $\xi > 0 \qquad$ ET I 31(12) | |
| **30** | $\operatorname{sech}(ax)$, $\qquad \operatorname{Re} a > 0$ | $a^{-1}\left(\pi/2\right)^{1/2}\operatorname{sech}\left(\pi\xi/2a\right),$ $\xi > 0 \qquad$ ET I 30(1) | |
| **31** | $\left(x^2+a^2\right)\operatorname{sech}\left(\dfrac{\pi x}{2a}\right)$, $\quad \operatorname{Re} a > 0$ | $2(2/\pi)^{1/2}a^3\operatorname{sech}^3(a\xi)$, $\qquad \xi > 0$ | ET I 32(19) |
| **32** | $\ln\left(1+\dfrac{a^2}{x^2}\right)$, $\qquad \operatorname{Re} a > 0$ | $(2\pi)^{1/2}\xi^{-1}\left(1-e^{-a\xi}\right)$, $\qquad \xi > 0$ | ET I 18(10) |
| **33**[7] | $\ln\left(\dfrac{a^2+x^2}{b^2+x^2}\right)$, $\operatorname{Re} a > 0, \operatorname{Re} b > 0$ | $(2\pi)^{1/2}\left(e^{-b\xi}-e^{-a\xi}\right)$, $\qquad \xi > 0$ | ET I 18(12) |
| **34** | $\left(x^2+b^2\right)^{-1}J_0(ax)$, $a > 0, \quad \operatorname{Re} b > 0$ | $(\pi/2)^{1/2}b^{-1}e^{-b\xi}I_0(ab)$, $a < \xi < \infty \qquad$ ET I 45(14) | |

continued on next page

| | $f(x)$ | $F_c(\xi)$ |
|---|---|---|
| | *continued from previous page* | |
| **35** | $x\left(x^2+b^2\right)^{-1}J_0(ax),$ | $(2/\pi)^{1/2}\cosh(b\xi)\,K_0(ab),$ |
| | $a>0, \quad \mathrm{Re}\,b>0$ | $0<\xi<a \qquad$ ET I 45(15) |

In deriving results for the preceding table from ET I, account has been taken of the fact that the normalization factor $\sqrt{2/\pi}$ employed in our definition of $F_c$ has not been used in those tables.

## 12.35  Relationships between transforms

The following relationships exist between transforms and they may be used to derive further transform pairs from among the results given in Sections 12.13–12.34. The appropriate sections of the main body of the tables may also be used to extend the list of transform pairs.

### 12.351

*Fourier cosine transform and Laplace transform relationship*

$$\mathcal{F}_c\left[f(x);\xi\right] = \frac{1}{\sqrt{2\pi}}\mathcal{L}\left[f(x);i\xi\right] + \frac{1}{\sqrt{2\pi}}\mathcal{L}\left[f(x);-i\xi\right].$$

### 12.352

*Fourier sine transform and Laplace transform relationship.*

$$\mathcal{F}_s\left[f(x);\xi\right] = \frac{i}{\sqrt{2\pi}}\mathcal{L}\left[f(x);i\xi\right] - \frac{i}{\sqrt{2\pi}}\mathcal{L}\left[f(x);-i\xi\right].$$

### 12.353

*Exponential Fourier transform and Laplace transform relationship*

$$\mathcal{F}\left[f(x);\xi\right] = \sqrt{2\pi}\mathcal{L}\left[f(x);-i\xi\right] + \sqrt{2\pi}\mathcal{L}\left[f(-x);i\xi\right].$$

## 12.41[10]  Mellin transform

The **Mellin transform** of the function $f(x)$, denoted by $f^*(s)$, is defined by the integral

$$f^*(s) = \int_0^\infty f(x)x^{s-1}\,\mathrm{d}x.$$

The functions $f(x)$ and $f^*(s)$ are called a **Mellin transform pair**, and knowledge of either one enables the other to be recovered.

The transform exists provided the integral

$$\int_0^\infty |f(x)|x^{k-1}\,\mathrm{d}x$$

is bounded for some $k>0$, and then the inversion of the Mellin transform is accomplished by means of the **inversion integral**

$$f(x) = \frac{1}{2\pi i} \int_{c-i\infty}^{c+i\infty} f^*(s) x^{-s} \, ds,$$

where $c > k$.

Setting

$$f^*(s) = \mathcal{M}\left[f(x); s\right]$$

to denote the Mellin transform, we have the symbolic expression for the inverse result

$$f(x) = \mathcal{M}^{-1}\left[f^*(s); x\right].$$                       MS 397(6)

## 12.42  Basic properties of the Mellin transform

1.      For $a$ and $b$ arbitrary constants,

$$\mathcal{M}\left[af(x) + bg(x)\right] = af^*(s) + bg^*(s) \qquad \text{(linearity)}$$

2.      If $\lim_{x \to 0} x^{s-r-1} f^{(r)}(x) = 0, \quad r = 0, 1, \ldots, n-1,$

(i)      $$\mathcal{M}\left[f^{(n)}(x); s\right] = (-1)^n \frac{\Gamma(s)}{\Gamma(s-n)} f^*(s-n)$$

                                   (transform of a derivative)   SU 267 (4.2.3)

(ii)      $$\mathcal{M}\left[x^n f^{(n)}(x); s\right] = (-1)^n \frac{\Gamma(s+n)}{\Gamma(s)} f^*(s)$$

                                   (transform of a derivative)   SU 267 (4.2.5)

3.      Denoting the $n^{\text{th}}$ repeated integral of $f(x)$ by $I_n\left[f(x)\right]$, where

$$I_n\left[f(x)\right] = \int_0^x I_{n-1}\left[f(u)\right] \, du,$$

(i)      $$\mathcal{M}\left[I_n\left[f(x)\right]; s\right] = (-1)^n \frac{\Gamma(s)}{\Gamma(n+s)} f^*(s+n)$$

                                   (transform of an integral)   SU 269 (4.2.15)

(ii)      $$\mathcal{M}\left[I_n^\infty\left[f(x)\right]; s\right] = \frac{\Gamma(s)}{\Gamma(s+n)} f^*(s+n),$$

        where

$$I_n^\infty\left[f(x)\right] = \int_x^\infty I_{n-1}^\infty f(u) \, du \qquad \text{(transform of an integral)} \qquad \text{SU 269 (4.2.18)}$$

4.      $$\mathcal{M}\left[f(x)g(x); s\right] = \frac{1}{2\pi i} \int_{c-i\infty}^{c+i\infty} f^*(u) g^*(s-u) \, du$$

                       (Mellin convolution theorem)   SU 275(4.4.1)

## 12.43  Table of Mellin transforms

| | $f(x)$ | $f^*(s)$ | |
|---|---|---|---|
| **1** | $e^{-x}$ | $\Gamma(s),$ $\qquad\qquad \operatorname{Re} s > 0$ | SU 521(M13) |
| **2** | $e^{-x^2}$ | $\frac{1}{2}\Gamma\left(\frac{1}{2}s\right),$ $\qquad\qquad \operatorname{Re} s > 0$ | SU 521(M14) |
| **3** | $\cos x$ | $\Gamma(s)\cos\left(\frac{1}{2}\pi s\right),$ $\quad 0 < \operatorname{Re} s < 1$ | SU 521(M15) |
| **4** | $\sin x$ | $\Gamma(s)\sin\left(\frac{1}{2}\pi s\right),$ $\quad 0 < \operatorname{Re} s < 1$ | SU 521(M16) |
| **5** | $\dfrac{1}{1-x}$ | $\pi\cot(\pi s),$ $\qquad\quad 0 < \operatorname{Re} s < 1$ | SU 521(M1) |
| **6** | $\dfrac{1}{1+x}$ | $\pi\operatorname{cosec}(\pi s),$ $\qquad 0 < \operatorname{Re} s < 1$ | SU 521(M2) |
| **7** | $(1+x^a)^{-b}$ | $\dfrac{\Gamma(s/a)\,\Gamma(b-s/a)}{a\,\Gamma(b)},\quad 0 < \operatorname{Re} s < ab$ | SU 521(M3) |
| **8** | $\dfrac{T_n(x)\,\mathrm{H}(1-x)}{\sqrt{(1-x^2)}}$ | $\dfrac{2^{-s}\pi\,\Gamma(s)}{\Gamma\left(\frac{1}{2}+\frac{1}{2}s+\frac{1}{2}n\right)\Gamma\left(\frac{1}{2}+\frac{1}{2}s-\frac{1}{2}n\right)},$ $\operatorname{Re} s > 0$ | SU 521(M4) |
| **9** | $\dfrac{T_n\left(x^{-1}\right)\mathrm{H}(1-x)}{\sqrt{(1-x^2)}}$ | $\dfrac{2^{s-2}\Gamma\left(\frac{1}{2}n+\frac{1}{2}s\right)\Gamma\left(\frac{1}{2}s-\frac{1}{2}n\right)}{\Gamma(s)},$ $\operatorname{Re} s > n$ | SU 521(M5) |
| **10** | $P_n(x)\,\mathrm{H}(1-x)$ | $\dfrac{\Gamma\left(\frac{1}{2}s\right)\Gamma\left(\frac{1}{2}s+\frac{1}{2}\right)}{2\,\Gamma\left(\frac{1}{2}s-\frac{1}{2}n+\frac{1}{2}\right)\Gamma\left(\frac{1}{2}s+\frac{1}{2}n+1\right)},$ $\operatorname{Re} s > 0$ | SU 521(M6) |
| **11** | $P_n\left(x^{-1}\right)\mathrm{H}(1-x)$ | $\dfrac{2^{s-1}\Gamma\left(\frac{1}{2}s+\frac{1}{2}n+\frac{1}{2}\right)\Gamma\left(\frac{1}{2}s-\frac{1}{2}n\right)}{\sqrt{\pi}\,\Gamma(s+1)},$ $\operatorname{Re} s > n$ | SU 521(M7) |
| **12** | $\dfrac{1+x\cos\phi}{1-2x\cos\phi+x^2}$ | $\dfrac{\pi\cos(s\phi)}{\sin(s\pi)},\quad 0 < \operatorname{Re} s < 1$ | SU 521(M11) |
| **13** | $\dfrac{x\sin\phi}{1-2x\cos\phi+x^2},\quad -\pi < \phi < \pi$ | $\dfrac{\pi\sin(s\phi)}{\sin(s\pi)},\quad 0 < \operatorname{Re} s < 1$ | SU 521(M12) |

*continued on next page*

| | $f(x)$ | $f^*(s)$ |
|---|---|---|
| | *continued from previous page* | |

*continued from previous page*

| | $f(x)$ | $f^*(s)$ | |
|---|---|---|---|
| **14** | $e^{-x\cos\phi}\cos\left(x\sin\phi\right),$ $\frac{1}{2}\pi<\phi<\frac{1}{2}\pi$ | $\Gamma(s)\cos(s\phi),$ $\qquad\operatorname{Re}s>0$ | SU 522(M17) |
| **15** | $e^{-x\sin\phi}\sin\left(x\sin p\phi\right),$ $-\frac{1}{2}\pi<\phi<\frac{1}{2}\pi$ | $\Gamma(s)\sin(s\phi),$ $\qquad\operatorname{Re}s>-1$ | SU 522(M18) |
| **16** | $x^{-\nu}J_\nu(x),\qquad\nu>-\frac{1}{2}$ | $\dfrac{2^{s-\nu-1}\Gamma\left(\frac{1}{2}s\right)}{\Gamma\left(\nu-\frac{1}{2}s+1\right)},\quad 0<\operatorname{Re}s<1$ | SU 522(M19) |
| **17** | $Y_\nu(x),\qquad\nu\in\mathbb{R}$ | $-2^{s-1}\pi^{-1}\Gamma\left(\frac{1}{2}s+\frac{1}{2}\nu\right)\Gamma\left(\frac{1}{2}s-\frac{1}{2}\nu\right)$ $\times\cos\left(\frac{1}{2}s-\frac{1}{2}\nu\right)\pi,$ $\qquad\qquad\qquad\;\|\nu\|<\operatorname{Re}s<\frac{3}{2}$ | SU 522(M20) |
| **18** | $K_\nu(x),\qquad\nu\in\mathbb{R}$ | $2^{s-2}\Gamma\left(\frac{1}{2}s+\frac{1}{2}\nu\right)\Gamma\left(\frac{1}{2}s-\frac{1}{2}\nu\right),$ $\qquad\qquad\operatorname{Re}s>\nu>0$ | SU 522(M21) |
| **19** | $\mathbf{H}_\nu(x),\qquad\nu\in\mathbb{R}$ | $\dfrac{2^{s-1}\tan\left(\frac{1}{2}\pi s+\frac{1}{2}\pi\nu\right)\Gamma\left(\frac{1}{2}s+\frac{1}{2}\nu\right)}{\Gamma\left(\frac{1}{2}\nu-\frac{1}{2}s+1\right)},$ $-1-\nu<\operatorname{Re}s<\min\left(\frac{3}{2},1-\nu\right)$ | SU 522(M22) |
| **20** | $\dfrac{1}{a+x^n},$ $\|\arg a\|<\pi,\quad n=1,2,3,\ldots,$ | $\pi n^{-1}\operatorname{cosec}\left(\dfrac{\pi s}{n}\right)a^{(s/n)-1},$ $\qquad\qquad 0<\operatorname{Re}s<n$ | MS 453 |
| **21** | $\left(1+ax^h\right)^{-\nu},$ $h>0,\quad\|\arg a\|<\pi$ | $h^{-1}a^{-s/h}\,\mathrm{B}\left(s/h,\nu-(s/h)\right)$ $\qquad\qquad 0<\operatorname{Re}s<h\operatorname{Re}\nu$ | MS 454 |
| **22** | $\begin{cases}\left(1-x^h\right)^{\nu-1} & \text{for } 0<x<1\\ 0 & \text{for } x>1\end{cases},$ $h>0,\quad\operatorname{Re}\nu>0$ | $h^{-1}\,\mathrm{B}\left(\nu,s/h\right)$ | MS 454 |
| **23** | $\ln(1+ax),\qquad\|\arg a\|<\pi$ | $\pi s^{-1}a^{-s}\operatorname{cosec}(\pi s),\quad -1<\operatorname{Re}s<0$ | MS 454 |
| **24** | $\arctan x$ | $-\frac{1}{2}\pi s^{-1}\sec(\pi s/2),\quad -1<\operatorname{Re}s<0$ | MS 454 |
| | | *continued on next page* | |

*continued on next page*

| | $f(x)$ | | $f^*(s)$ | |
|---|---|---|---|---|
| | *continued from previous page* | | | |
| **25** | $\operatorname{arccot} x$ | | $\frac{1}{2}\pi s^{-1}\sec(\pi s/2),\qquad 0<\operatorname{Re}s<1$ | MS 454 |
| **26** | $\operatorname{cosech}(ax)$ | $\operatorname{Re}a>0$ | $a^{-s}2\left(1-2^{-s}\right)\Gamma(s)\zeta(s),\qquad \operatorname{Re}s>1$ | MS 454 |
| **27** | $\operatorname{sech}^2(ax),$ | $\operatorname{Re}a>0$ | $4a^{-s}(1-2^{2-s})\Gamma(s)2^{-s}\zeta(s-1),$ $\operatorname{Re}s>2$ | MS 454 |
| **28** | $\operatorname{cosech}^2(ax),$ | $\operatorname{Re}a>0$ | $4a^{-s}\Gamma(s)2^{-s}\zeta(s-1),\qquad \operatorname{Re}s>2$ | MS 454 |
| **29**[11] | $\left(x^2+b^2\right)^{-\frac{1}{2}\nu}J_\nu\left[a\left(x^2+b^2\right)^{1/2}\right]$ | | $2^{\frac{1}{2}s-1}a^{-\frac{1}{2}s}b^{\frac{1}{2}s-\nu}\Gamma\left(\frac{1}{2}s\right)J_{\nu-s/2}(ab),$ $0<\operatorname{Re}s<\frac{3}{2}+\operatorname{Re}\nu$ | ET I 328 |
| **30** | $\begin{cases}\left(a^2-x^2\right)^{\frac{1}{2}\nu}J_\nu\left[a\left(b^2-x^2\right)^{1/2}\right]\\ \qquad\qquad\text{for }0<x<a\\ 0\qquad\qquad\text{for }x>a\end{cases}$ $\operatorname{Re}\nu>-1$ | | $2^{\frac{1}{2}s-1}\Gamma\left(\frac{1}{2}s\right)b^{-\frac{1}{2}s}a^{\nu+\frac{1}{2}s}J_{\nu+\frac{1}{2}s}(ab),$ $\operatorname{Re}s>0$ | MS 455 |
| **31** | $\begin{cases}\left(a^2-x^2\right)^{-\frac{1}{2}\nu}J_\nu\left[b\left(a^2-x^2\right)^{1/2}\right]\\ \qquad\qquad\text{for }0<x<a\\ 0\qquad\qquad\text{for }x>a\end{cases}$ | | $2^{1-\nu}\left[\Gamma(\nu)\right]^{-1}a^{\frac{1}{2}s-\nu}b^{-\frac{1}{2}\nu}s_{\nu-1+\frac{1}{2}s,\frac{1}{2}s-\nu}(ab),$ $\operatorname{Re}s>0$ | MS 455 |
| **32** | $K_\nu(\alpha x)$ | | $\alpha^{-s}2^{s-2}\Gamma\left(\frac{1}{2}s-\frac{1}{2}\nu\right)\Gamma\left(\frac{1}{2}s+\frac{1}{2}\nu\right),$ $\operatorname{Re}s>\lvert\operatorname{Re}\nu\rvert$ | MS 455 |
| **33** | $\left(\beta a^2+x^2\right)^{-\frac{1}{2}\nu}$ $\times K_\nu\left[\alpha\left(\beta a^2+x^2\right)^{1/2}\right]$ $\operatorname{Re}(\alpha,\beta)>0$ | | $\alpha^{-\frac{1}{2}s}2^{\frac{1}{2}s-1}\beta^{\frac{1}{2}s-\nu}\Gamma,\left(\frac{1}{2}s\right)K_{\nu-\frac{1}{2}s}(\alpha\beta),$ $\operatorname{Re}s>0$ | MS 455 |

# Bibliographic References

(See the introduction for an explanation of the letters preceding each bibliographic reference.)

**AS**    Abramowitz, M. and Stegun, I. A., *Handbook of Mathematical Functions*, Dover Publications, New York, 1972.

**AD**    Adams, E. P. and Hippisley, R. L., *Smithsonian Mathematical Formulae and Tables of Elliptic Functions*, Smithsonian Institute, Washington, D.C., 1922.

**AK**    Appell, P. and Kampé de Fériet, *Fonctions hypergéometriques et hypersphériques, polynomes d'Hermite*, Gauthier Villars, Paris, 1926.

**BB**    Beckenbach, E. F. and Bellman, R., *Inequalities*, 3rd printing. Springer–Verlag, Berlin, 1971.

**BE**    Bertrand, J., *Traite de calcul différentiel et de calcul intégral*, vol. 2, *Calcul intégral, intégrales définies et indéfinies*, Gauthier-Villars, Paris, 1870.

**BEA**    Beaulieu, N. C., A Useful Integral for Wireless Communication Theory and Its Application to Rectangular Signaling Constellation Error Rates, *IEEE Trans. Commun.*, **54**(5), pages 802–805, 2006.

**BI**    Bierens de Haan, D., *Nouvelles tables d'intégrales définies*, Amsterdam, 1867. (Reprint) G. E. Stechert & Co., New York, 1939.

**BL**    Bellman, R., *Introduction to Matrix Analysis*, McGraw Hill, New York, 1960.

**BR**    Bromwich, T. I'A., *An Introduction to the Theory of Infinite Series*, Macmillan, London, 1908, 2nd edition, 1926.*

**BS**    Bellman, R., *Stability Theory of Differential Equations*, McGraw-Hill, New York, 1953.

**BU**    Buchholz, H., *Die konfluente hypergeometrische Funktion mit besonderer Berücksichtigung ihrer Anwendungen*, Springer–Verlag, Berlin, 1953. Also an English edition: *The confluent Hypergeometric Function*, Springer–Verlag, Berlin, 1969.

**BY**    Byrd, P. F. and Friedman, M. D., *Handbook of Elliptic Integrals for Engineers and Physicists*, Springer–Verlag, Berlin, 1954.

**CA**    Carslaw, H. S., *Introduction to the Theory of Fourier's Series and Integrals*, Macmillan, London, 1930.

**CE**    Cesàro, Z., *Elementary Class Book of Algebraic Analysis and the Calculation of Infinite Limits*, 1st ed. ONTI, Moscow and Leningrad, 1936.

**CL**    Coddington, E. A. and Levinson, N., *Theory of Ordinary Differential Equations*, McGraw Hill, New York, 1955.

**CO**    Courant, R. and Hilbert, D., *Methods of Mathematical Physics*, vol. I, Wiley (Interscience), New York, 1953.

---

*The Bibliographic Reference BR* refers to the 1908 edition of Bromwich T. I.'A., *An Introduction to the Theory of Infinite Series*; BR refers to the 1926 edition.

**DLMF**  *Digital Library of Mathematical Functions*, 2011-08-29, National Institute of Standards and Technology, http://dlmf.nist.gov/

**DW**  Dwight, H. B., *Tables of Integrals and Other Mathematical Data*, Macmillan, New York, 1934.

**DW61**  Dwight, H. B., *Tables of Integrals and Other Mathematical Data*, Macmillan, New York, 1961.

**EF**  Efros, A. M. and Danilevskiy, A. M., *Operatsionnoye ischisleniye i konturnyye integraly* (Operational calculus and contour integrals). GNTIU, Khar'kov, 1937.

**EH**  Erdélyi, A., et al., *Higher Transcendental Functions*, vols. I, II, and III. McGraw Hill, New York, 1953–1955.

**ET**  Erdélyi, A. et al., *Tables of Integral Transforms*, vols. I and II. McGraw Hill, New York, 1954.

**EU**  Euler, L., *Introductio in Analysin Infinitorum*, Bousquet, Lausanne, 1748.

**FI**  Fikhtengol'ts, G. M., *Kurs differentsial'nogo i integral'nogo ischisleniya* (Course in differential and integral calculus), vols. I, II, and III. Gostekhizdat, Moscow and Leningrad, 1947–1949. Also a German edition: *Differential-und Integralrechnung I–III*, VEB Deutscher Verlag der Wissenschaften, Berlin, 1986–1987.

**GA**  Gauss, K. F., *Werke*, Bd. III. Göttingen, 1876.

**GC**  Gonczarek and Czerwonko, *Bulletin of the Polish Academy of Sciences*, vol. 48, no. 4, 2000.

**GE**  Gel'fond, A. O., *Ischisleniye konechnykh raznostey* (Calculus of finite differences), part I. ONTI, Moscow and Leningrad, 1936.

**GH2**  Gröbner, W. and Hofreiter, N., *Integraltafel*, vol. 2, *Bestimmte Integrale*, Springer, Wien, 1961.

**GI**  Giunter, N. M. and Kuz'min, R. O. (eds.), *Sbornik zadach po vysshey matematike* (Collection of problems in higher mathematics), vols. I, II, and III. Gostekhizdat, Moscow and Leningrad, 1947.

**GM**  Gantmacher, F. R., *Applications of the Theory of Matrices*, translation by J. L. Brenner. Wiley (Interscience), New York, 1959.

**GO**  Goursat, E. J. B., *Cours d'Analyse*, vol. I, Gauthier–Villars, Paris, 1923.

**GS**  Guillera, J. and Sondow, J., *Double Integrals and Infinite Products for some Classical Constants via Analytic Continuation of Lerch's Transcendent*, arXiv:math/0506319 v2, 2005.

**GU**  Gröbner, W. et al., *Integraltafel*, Teil I, *Unbestimmte Integrale*, Akad. Verlag, Braunschweig, 1944.

**GW**  Gröbner, W. and Hofreiter, N., *Integraltafel*, Teil II, *Bestimmte Integrale*, Springer–Verlag, Wien and Innsbruck, 1958.

**HI**  Hille, E., *Lectures on Ordinary Differential Equations*, Addison- Wesley, Reading, Massachusetts, 1969.

**HL**  Hardy, G. H., Littlewood, J. E., and Polya, G., *Inequalities*, Cambridge University Press, London, 2nd ed., 1952.

**HO**  Hobson, E. W., *The Theory of Spherical and Ellipsoidal Harmonics*, Cambridge University Press, London, 1931.

**HU**  Hurewicz, W., *Lectures on Ordinary Differential Equations*, MIT Press, Cambridge, Massachusetts, 1958.

**IN**  Ince, E. L., *Ordinary Differential Equations*, Dover, New York, 1944.

**JA**  Jahnke, E. and Emde, F., *Tables of Functions with Formulas and Curves*, Dover, New York, 1943.

**JAC**  Jackson, J. D., *Classical Electrodynamics*, Wiley, New York, 1975.

**JE**  James, H. M. et al. (eds.), *Theory of Servomechanisms*, McGraw Hill, New York, 1947.

**JO**  Jolley, L., *Summation of Series*, Chapman and Hall, London, 1925.

**KE**   Kellogg, O. D., *Foundations of Potential Theory*, Dover, New York, 1958.

**KM**   Krajcik, R. A. and McLenitham K. D., *Integrals and Series Related to the Surface Area of Arbitrary Ellipsoids*, Los Alamos LA-UR-04-4398, http://arxiv.org/abs/math/0605216, 2004.

**KR**   Krechmar, V. A., *Zadachnik po algebre* (Problem book in algebra), 2nd ed. Gostekhizdat, Moscow and Leningrad, 1950.

**KU**   Kuzmin, R. O., *Besselevy funktsii* (Bessel functions). ONTI, Moscow and Leningrad, 1935.

**LA**   Laska, W., *Sammlung von Formeln der reinen und angewandten Mathematik*, Friedrich Viewig und Sohn, Braunschweig, 1888–1894.

**LE**   Legendre, A. M., *Exercises calcul intégral*, Paris, 1811.

**LEI**  Lei, X., Fan, P. and Chen, Q., Exact Symbol Error Probability of General Order Rectangular QAM with MRC Diversity Reception over Nakagami-$m$ Fading Channels, *IEEE Commun. Lett.*, **11**(12), pages 958–960, 2007.

**LI**   Lindeman, C. E., *Examen des nouvelles tables d'intégrales définies de M. Bierens de Haan*, Amsterdam, 1867, Norstedt, Stockholm, 1891.

**LO**   Lobachevskiy, N. I., *Poloye sobraniye sochineniy* (Complete works), vols. I, III, and V. Gostekhizdat, Moscow and Leningrad, 1946–1951.

**LUKE** Luke, Y. L., *Mathematical Functions and their Approximations*, Academic Press, New York, 1975.

**LW**   Lawden, D. F., *Elliptic Functions and Applications*, Springer–Verlag, Berlin, 1989.

**MA**   McLachlan, N. W., *Theory and Application of Mathieu Functions*, Oxford University Press, London, 1947.

**MC**   Computation by Mathematica.

**ME**   McLachlan, N. W. and Humbert, P., *Formulaire pour le calcul symbolique*, L'Acad. des Sciences de Paris, Fasc. 100, 1950.

**MF**   Morse, M. P. and Feshbach, H., *Methods of Theoretical Physics*, vol. I, McGraw Hill, New York, 1953.

**MG**   Marden, M., *Geometry of Polynomials*, American Mathematical Society, Mathematical Survey 3, Providence, Rhode Island, 1966.

**MI**   McLachlan, N. W. et al., *Supplément au formulaire pour le calcul symbolique*, L'Acad. des Sciences de Paris, Fasc. 113, 1950.

**ML**   Mirsky L., *An Introduction to Linear Algebra*, Oxford University Press, London, 1963.

**MM**   MacMillan, W. D., *The Theory of the Potential*, Dover, New York, 1958.

**MO**   Magnus, W. and Oberhettinger, F., *Formeln und Sätze für die speziellen Funktionen der mathematischen Physik*, Springer–Verlag, Berlin, 1948.

**MS**   Magnus, W., Oberhettinger, F. and Soni, R. P., *Formulas and Theorems for the Special Functions of Mathematical Physics*, 3rd ed. Springer–Verlag, Berlin, 1966.

**MS2**  Mathai, A. M. and Saxens, R. K., *Generalized Hypergeometrics Functions With Applications in Statistics and Physical Science*, Springer–Verlag, Berlin, 1973.

**MT**   Mitrinović, D. S., *Analytic Inequalities*, Springer–Verlag, Berlin, 1970.

**MV**   Milne, E. A., *Vectorial Mechanics*, Methuen, London, 1948.

**MZ**   Meyer Zur Capellen, W., *Integraltafeln, Sammlung unbestimmer Integrale elementarer Funktionen*, Springer–Verlag, Berlin, 1950.

**NA**   Natanson, I. P., *Konstruktivnaya teoriya funktsiy* (Constructive theory of functions). Gostekhizdat, Moscow and Leningrad, 1949.

**NH**   Nielsen, N., *Handbuch der Theorie der Gammafunktion*, Teubner, Leipzig, 1906.

**NO**      Noble, B., *Applied Linear Algebra*, Prentice Hall, Englewood Cliffs, New Jersey, 1969.

**NT**      Nielsen, N., *Theorie des Integrallogarithmus und verwandter Transcendenten*, Teubner, Leipzig, 1906.

**NV**      Novoselov, S. I., *Obratnyye trigonometricheskiye funktsii, posobive dlya uchiteley* (Inverse trigonometric functions, textbook for students), 3rd ed. Uchpedgiz, Moscow and Leningrad, 1950.

**OB**      Oberhettinger, F., *Tables of Bessel Transforms*, Springer–Verlag, New York: 1972.

**PBM**     Prudnikov, A. P., Brychkov, Yu. A., and Marichev, O. I., *Integrals and Series*, Gordan and Breach, New York, vols. I (1986), II (1986), III (1990).

**PE**      Peirce, B. O., *A Short Table of Integrals*, 3rd ed. Ginn, Boston, 1929.

**SA**      Sansone, G., *Orthogonal Functions* (Revised English Edition), Interscience, New York, 1959.

**SI**      Sikorskiy, Yu. S., *Elementy teorii ellipticheskikh funktsiy s prilozheniyama k mekhanike* (Elements of theory of elliptic functions with applications to mechanics). ONTI, Moscow and Leningrad, 1936.

**SN**      Sneddon, I. N., *Fourier Transforms*, 1st ed. McGraw Hill, New York, 1951.

**SM**      Smirnov, V. I., *Kurs vysshey matematiki* (A course of higher mathematics), vol. III, Part 2, 4th ed. Gostekhizdat, Moscow and Leningrad, 1949.

**ST**      Strutt, M. J. O., *Lamésche, Mathieusche und verwandte Funktionen in Physik and Technik*, Springer–Verlag, Berlin, 1932.

**STR**     Stratton, J. C., *Phys. Rev A*, **43**(3), pages 1381–1388, 1991.

**SU**      Sneddon, I. N., *The Use of Integral Transforms*, McGraw Hill, New York, 1972.

**SZ**      Szegö, G., *Orthogonal Polynomials*, Revised Edition, Colloquium Publications XXIII, American Mathematical Society, New York, 1959.

**TF**      Titchmarsh, E. C., *Introduction to the Theory of Fourier Integrals*, 2nd ed. Oxford University Press, London, 1948.

**TI**      Timofeyev, A. F. *Integrirovaniye funktsiy* (Integration of functions), part I. GTTI, Moscow and Leningrad, 1933.

**VA**      Varga, R. S., *Matrix Iterative Analysis*, Prentice Hall, Englewood Cliffs, New Jersey, 1963.

**VL**      Vladimirov, V. S., *Equations of Mathematical Physics*, Dekker, New York, 1971.

**WA**      Watson, G. N., *A Treatise on the Theory of Bessel Functions*, 2nd ed. Cambridge University Press, London, 1966.

**WH**      Whittaker, E. T. and Watson, G. N., *Modern Analysis*, 4th ed. Cambridge University Press, London, 1927, part II, 1934.

**ZH**      Zhuravskiy, A. M., *Spravochnik po ellipticheskim funktsiyam* (Reference book on elliptic functions). Izd. Akad. Nauk. U.S.S.R., Moscow and Leningrad, 1941.

**ZY**      Zygmund, A., *Trigonometrical Series*, 2nd ed. Chelsea, New York, 1952.

# Supplementary References

(Prepared by Alan Jeffrey for the English language edition.)

## General reference books

1. Bromwich, T. I'A., *An Introduction to the Theory of Infinite Series*, 2nd ed., Macmillan, London, 1926 (Reprinted 1942).
2. Carlitz, L., "Generating Functions", 1969, *Fibonacci Quarterly*, 7 (4): 359–393.
3. Copson, E. T., *An Introduction to the Theory of Functions of a Complex Variable*, Oxford University Press, London, 1935.
4. Courant, R. and Hilbert, D., *Methods of Mathematical Physics*, vol. I, Interscience Publishers, New York, 1953.
5. Davis, H. T., *Summation of Series*, Trinity University Press, San Antonio, Texas, 1962.
6. Erdélyi, A. et al. *Higher Transcendental Functions*, vols. I to III, McGraw Hill, New York 1953–1955.
7. Erdélyi, A. et al., *Tables of Integral Transforms*, vols. I and II. McGraw Hill, New York, 1954.
8. Fletcher, A., Miller, J. C. P., and Rosenhead, L., *An Index of Mathematical Tables*, 2nd ed., Scientific Computing Service, London, 1962.
9. Gröbner, W. and Hofreiter, N., *Integraltafel*, I, II. Springer–Verlag, Wien and Innsbruck, 1949.
10. Hardy, G. H., Littlewood, J. E., and Pólya, G., *Inequalities*, 2nd ed., Cambridge University Press, London, 1952.
11. Hartley, H. O. and Greenwood, J. A., *Guide to Tables in Mathematical Statistics*, Princeton University Press, Princeton, New Jersey, 1962.
12. Jeffreys, H. and Jeffreys, B. S., *Methods of Mathematical Physics*, Cambridge University Press, London, 1956.
13. Jolley, L. B. W., *Summation of Series*, Dover Publications, New York, 1962.
14. Knopp, K., *Theory and Application of Infinite Series*, Blackie, London, 1946, Hafner, New York, 1948.
15. Lebedev, N. N., *Special Functions and their Applications*, Prentice Hall, Englewood Cliffs, New Jersey, 1965.
16. Magnus, W. and Oberhettinger, F., *Formulas and Theorems for the Special Functions of Mathematical Physics*, Chelsea, New York, 1949.
17. McBride, E. B., *Obtaining Generating Functions*, Springer–Verlag, Berlin, 1971.
18. National Bureau of Standards, *Handbook of Mathematical Functions*, U.S. Government Printing Office, Washington, D.C., 1964.
19. Prudnikov, A. P., Brychkov, Yu. A., and Marichev, O. I., *Integrals and Series*, Vols. 1–5, Gordon and Breach, New York, 1986–1992.
20. Truesdell, C. *A Unified Theory of Special Functions*, Princeton University Press, Princeton, New Jersey, 1948.

21. Vein, R. and Dale, P., *Determinants and Their Applications in Mathematical Physics*, Springer–Verlag, New York, 1999.
22. Whittaker, E. T. and Watson, G. N., *A Course of Modern Analysis*, 4th ed., Cambridge University Press, London, 1940.

## Asymptotic expansions

1. De Bruijn, N. G., *Asymptotic Methods in Analysis*, North-Holland Publishing Co., Amsterdam, 1958.
2. Cesari, L., *Asymptotic Behavior and Stability Problems in Ordinary Differential Equations*, 3rd ed., Springer, New York, 1971.
3. Copson, E. T., *Asymptotic Expansions*, Cambridge University Press, London, 1965.
4. Erdélyi, A., *Asymptotic Expansions*, Dover Publications, New York, 1956.
5. Ford, W. B., *Studies on Divergent Series and Summability*, Macmillan, New York, 1916.
6. Hardy, G. H., *Divergent Series*, Clarendon Press, Oxford, 1949.
7. Watson, G. N., *A Treatise on the Theory of Bessel Functions*, 2nd ed., Cambridge University Press, London, 1958.

## Bessel functions

1. Bickley, W. G., *Bessel Functions and Formulae*, Cambridge University Press, London, 1953.
2. Erdélyi, A. et al., *Higher Transcendental Functions*, vols. I and II. McGraw Hill, New York, 1954.
3. Erdélyi, A. et al., *Tables of Integral Transforms*, vols. I and II. McGraw Hill, New York, 1954.
4. Gray, A., Mathews, G. B. and MacRobert, T. M., *A Treatise on Bessel Functions and Their Applications to Physics*, 2nd ed., Macmillan, 1922.
5. McLachlan, N. W., *Bessel Functions for Engineers*, 2nd ed., Oxford University Press, London, 1955.
6. Luke, Y. L., *Integrals of Bessel Functions*, McGraw Hill, New York, 1962.
7. Petiau, G., *La théorie des fonctions de Bessel*, Centre National de la Recherche Scientifique, Paris, 1955.
8. Relton, F. E., *Applied Bessel Functions*, Blackie, London, 1946.
9. Watson, G. N., *A Treatise on the Theory of Bessel Functions*, 2nd ed., Cambridge University Press, London, 1958.
10. Wheelon, A. D., *Tables of Summable Series and Integrals Involving Bessel Functions*, Holden-Day, San Francisco, 1968.

## Complex analysis

1. Ahlfors, L. V., *Complex Analysis*, 3rd ed., McGraw Hill, New York, 1979.
2. Ahlfors, L. V. and Sario, L., *Riemann Surfaces*, Princeton University Press, Princeton, New Jersey, 1971.
3. Bieberbach, L., *Conformal Mapping*, Chelsea, New York, 1964.
4. Henrici, P., *Applied and Computational Complex Analysis*, 3 vols, Wiley, New York, 1988, 1991, 1977.
5. Hille, E., *Analytic Function Theory*, 2 vols. 2nd ed., Chelsea, New York, 1990, 1987.
6. Kober, H., *Dictionary of Conformal Representations*, Dover Publications, New York, 1952.
7. Titchmarsh, E. C., *The Theory of Functions*, 2nd ed., Oxford University Press, London, 1939. (Reprinted 1975).

# Error function and Fresnel integrals

1. Erdélyi, A. et al., *Higher Transcendental Functions*, vol. II, McGraw Hill, New York, 1953.
2. Erdélyi, A. et al., *Tables of Integral Transforms*, vol. I, McGraw Hill, New York, 1954.
3. Slater, L. J., *Confluent Hypergeometric Functions*, Cambridge University Press, London, 1960.
4. Tricomi, F. G., *Funzioni ipergeometriche confluenti*, Edizioni Cremonese, Turan, Italy, 1954.
5. Watson, G. N., *A Treatise on the Theory of Bessel Functions*, 2nd ed., Cambridge University Press, London, 1958.

# Exponential integrals, gamma function and related functions

1. Artin, E., *The Gamma Function*, Holt, Rinehart, and Winston, New York, 1964.
2. Busbridge, I. W., *The Mathematics of Radiative Transfer*, Cambridge University Press, London, 1960.
3. Erdélyi, A. et al., *Higher Transcendental Functions*, vol. II, McGraw Hill, New York, 1953.
4. Erdélyi, A. et al., *Tables of Integral Transforms*, vols. I and II, McGraw Hill, New York, 1954.
5. Hastings, Jr., C., *Approximations for Digital Computers*, Princeton University Press, Princeton, New Jersey, 1955.
6. Kourganoff, V., *Basic Methods in Transfer Problems*, Oxford University Press, London, 1952.
7. Lösch, F. and Schoblik, F., *Die Fakultät (Gammafunktion) und verwandte Funktionen*, Teubner, Leipzig, 1951.
8. Nielsen, N., *Handbuch der Theorie der Gammafunktion*, Teubner, Leipzig, 1906.
9. Oberhettinger, F., *Tabellen zur Fourier Transformation*, Springer–Verlag, Berlin, 1957.

# Hypergeometric and confluent hypergeometric functions

1. Appell, P., *Sur les Fonctions Hypergéometriques de Plusieures Variables*, Gauthier-Villars, Paris, 1926.
2. Bailey, W. N., *Generalized Hypergeometric Functions*, Cambridge University Press, London, 1935.
3. Buchholz, H., *Die konfluente hypergeometrische Funktion*, Springer–Verlag, Berlin, 1953.
4. Erdélyi, A. et al., *Higher Transcendental Functions*, vol. I, McGraw Hill, New York, 1953.
5. Jeffreys, H. and Jeffreys, B. S., *Methods of Mathematical Physics*, Cambridge University Press, London, 1956.
6. Klein, F., *Vorlesungen über die hypergeometrische Funktion*, Springer–Verlag, Berlin, 1933.
7. Nörlund, N. E., *Sur les Fonctions Hypergéometriques d'Ordre Superior*, North–Holland, Copenhagen, 1956.
8. Slater, L. J., *Confluent Hypergeometric Functions*, Cambridge University Press, London, 1960.
9. Slater, L. J. *Generalized Hypergeometric Functions*, Cambridge University Press, London, 1966.
10. Snow, C., *The Hypergeometric and Legendre Functions with Applications to Integral Equations of Potential Theory*, 2nd ed., National Bureau of Standards, Washington, D.C., 1952.
11. Swanson, C. A. and Erdélyi, A., *Asymptotic Forms of Confluent Hypergeometric Functions*, Memoir 25, American Mathematical Society, Providence, Rhode Island, 1957.
12. Tricomi, F. G., *Lezioni sulla funzioni ipergeometriche confluenti*, Gheroni, Torino, 1952.

# Integral transforms

1. Bochner, S., *Vorlesungen über Fouriersche Integrale*, Akad. Verlag, Leipzig, 1932. Reprint Chelsea, New York, 1948.

2.  Bochner, S. and Chandrasekharan, K., *Fourier Transforms*, Princeton University Press, Princeton, New Jersey, 1949.

3.  Campbell, G. and Foster, R., *Fourier Integrals for Practical Applications*, Van Nostrand, New York, 1948.

4.  Carslaw, H. S. and Jaeger, J. C., *Conduction of Heat in Solids*, Oxford University Press, London, 1948.

5.  Doetsch, G., *Theorie und Anwendung der Laplace-Transformation*, Springer–Verlag, Berlin, 1937. (Reprinted by Dover Publications, New York, 1943)

6.  Doetsch, G., *Theory and Application of the Laplace-Transform*, Chelsea, New York, 1965.

7.  Doetsch, G., *Handbuch der Physik, Mathematische Methoden II*, 1st ed., Springer–Verlag, Berlin, 1955.

8.  Doetsch, G., *Guide to the Applications of the Laplace and Z-Transforms*, 2nd ed., Van Nostrand-Reinhold, London, 1971.

9.  Doetsch, G., *Handbuch der Laplace-Transformation*, Vols. I–IV, Birkhäuser Verlag, Basel, 1950–56.

10. Doetsch, G., Kniess, H., and Voelker, D., Tabellen zur Laplace- Transformation, Springer–Verlag, Berlin, 1947.

11. Erdélyi, A., *Operational Calculus and Generalized Functions*, Holt, Rinehart and Winston, New York, 1962.

12. Exton, H., *Multiple Hypergeometric Functions and Applications,* Horwood, Chichester, 1976.

13. Exton, H., *Handbook of Hypergeometric Integrals: Theory, Applications, Tables, Computer Programs*, Horwood, Chichester, 1978.

14. Hirschmann, J. J. and Widder, D. V., *The Convolution Transformation*, Princeton University Press, Princeton, New Jersey, 1955.

15. Marichev, O. I., *Handbook of Integral Transforms of Higher Transcendental Functions, Theory and Algorithmic Tables*, Ellis Horwood Ltd., Chichester (1982).

16. Oberhettinger, F., *Tabellen zur Fourier Transformation*, Springer–Verlag, Berlin (1957).

17. Oberhettinger, F., *Tables of Bessel Transforms*, Springer–Verlag, New York (1972).

18. Oberhettinger, F., *Fourier Expansions: A Collection of Formulas,* Academic Press, New York, 1973.

19. Oberhettinger, F., *Fourier Transforms of Distributions and Their Inverses*, Academic Press, New York, 1973.

20. Oberhettinger, F., *Tables of Mellin Transforms*, Springer–Verlag, Berlin, 1974.

21. Oberhettinger, F. and Badii, L., *Tables of Laplace Transforms,* Springer–Verlag, Berlin, 1973.

22. Oberhettinger, F. and Higgins, T. P., *Tables of Lebedev, Mehler and Generalized Mehler Transforms*, Math. Note No. 246, Boeing Scientific Research Laboratories, Seattle, Wash., 1961.

23. Roberts, G. E. and Kaufman, H., *Table of Laplace Transforms*, McAinsh, Toronto, 1966.

24. Sneddon, I. N., *Fourier Transforms*, McGraw Hill, New York, 1951.

25. Titchmarsh, E. C., *Introduction to the Theory of Fourier Integrals*, Oxford University Press, London, 1937.

26. Van der Pol, B. and Bremmer, H., *Operational Calculus Based on the Two Sided Laplace Transformation*, Cambridge University Press, London, 1950.

27. Widder, D. V., *The Laplace Transform*, Princeton University Press, Princeton, New Jersey, 1941.

28. Wiener, N., *The Fourier Integral and Certain of its Applications*, Dover Publications, New York, 1951.

# Jacobian and Weierstrass elliptic functions and related functions

1.  Erdélyi, A. et al., *Higher Transcendental Functions*, vol. II, McGraw Hill, New York, 1953.

2.  Byrd, P. F. and Friedman, M. D., *Handbook of Elliptic Integrals for Engineers and Physicists*, Springer–Verlag, Berlin, 1954.

3.  Graeser, E., *Einführung in die Theorie der Elliptischen Funktionen und deren Anwendungen*, Oldenbourg, Munich, 1950.

4.  Hancock, H., *Lectures on the Theory of Elliptic Functions*, vol. I, Dover Publications, New York, 1958.

5.  Neville, E. H., *Jacobian Elliptic Functions*, Oxford University Press, London, 1944 (2nd ed. 1951).
6.  Oberhettinger, F. and Magnus, W., *Anwendungen der Elliptischen Funktionen in Physik und Technik*, Springer–Verlag, Berlin, 1949.
7.  Roberts, W. R. W., *Elliptic and Hyperelliptic Integrals and Allied Theory*, Cambridge University Press, London, 1938.
8.  Tannery, J. and Molk, J., *Eléments de la Théorie des Fonctions Elliptiques*, 4 volumes. Gauthier-Villars, Paris, 1893–1902.
9.  Tricomi, F. G., *Elliptische Funktionen*, Akad. Verlag, Leipzig, 1948.

# Legendre and related functions

1.  Erdélyi, A. et al., *Higher Transcendental Functions*, vol. I, McGraw Hill, New York, 1953.
2.  Helfenstein, H., *Ueber eine Spezielle Lamésche Differentialgleichung*, Brunner and Bodmer, Zurich, 1950 (Bibliography).
3.  Hobson, E. W., *The Theory of Spherical and Ellipsoidal Harmonics*, Cambridge University Press, London, 1931. Reprinted by Chelsea, New York, 1955.
4.  Lense, J., *Kugelfunktionen*, Geest and Portig, Leipzig, 1950.
5.  MacRobert, T. M., *Spherical Harmonics: An Elementary Treatise on Harmonic Functions with Applications*, Methuen, England, 1927. (Revised ed. 1947; reprinted Dover Publications, New York, 1948).
6.  Snow, C., *The Hypergeometric and Legendre Functions with Applications to Integral Equations of Potential Theory*, 2nd ed., National Bureau of Standards, Washington, D.C., 1952.
7.  Stratton, J. A., Morse, P. M., Chu, L. J. and Hunter, R. A., *Elliptic Cylinder and Spheroidal Wave Functions Including Tables of Separation Constants and Coefficients*, Wiley, New York, 1941.

# Mathieu functions

1.  Erdélyi, A., *Higher Transcendental Functions*, vol. III, McGraw Hill, New York, 1955.
2.  McLachlan, N. W., *Theory and Application of Mathieu Functions*, Oxford University Press, London, 1947.
3.  Meixner, J. and Schäfke, F. W., *Mathieusche Funktionen und Sphäroidfunktionen mit Anwendungen auf Physikalische und Technische Probleme*, Springer–Verlag, Heidelberg, 1954.
4.  Strutt, M. J. O., *Lamésche, Mathieusche und verwandte Funktionen in Physik und Technik*, Ergeb. Math. Grenzgeb. *1*, 199–323 (1932). Reprint Edwards Bros., Ann Arbor, Michigan, 1944.

# Orthogonal polynomials and functions

1.  *Bibliography on Orthogonal Polynomials*, Bulletin of National Research Council No. 103, Washington, D.C., 1940.
2.  Courant, R. and Hilbert, D., *Methods of Mathematical Physics*, vol. I, Interscience, New York, 1953.
3.  Erdélyi, A. et al., *Higher Transcendental Functions*, vol. II, McGraw Hill, New York, 1954.
4.  Kaczmarz, St. and Steinhaus, H., *Theorie der Orthogonalreihen*, Chelsea, New York, 1951.
5.  Lorentz, G. G., *Bernstein Polynomials*, University of Toronto Press, Toronto, 1953.
6.  Sansone, G., *Orthogonal Functions*, Interscience, New York, 1959.
7.  Shohat, J. A. and Tamarkin, J. D., *The Problem of Moments*, American Mathematical Society, Providence, Rhode Island, 1943.

8.    Szegö, G., *Orthogonal Polynomials*, American Mathematical Society Colloquim Pub. No. 23, Providence, Rhode Island, 1959.

9.    Titchmarsh, E. C., *Eigenfunction Expansions Associated with Second Order Differential Equations*, Oxford University Press, London, part I (1946), part II (1958).

10.   Tricomi, F. G., *Vorlesungen über Orthogonalreihen*, Springer–Verlag, Berlin, 1955.

# Parabolic cylinder functions

1.    Buchholz, H., *Die konfluente hypergeometrische Funktion*, Springer–Verlag, Berlin, 1953.

2.    Erdélyi, A. et al., *Higher Transcendental Functions*, vol. II, McGraw Hill, New York, 1954.

# Probability function

1.    Cramer, H., *Mathematical Methods of Statistics*, Princeton University Press, Princeton, New Jersey, 1951.

2.    Erdélyi, A. et al., *Higher Transcendental Functions*, vols. I, II, and III. McGraw Hill, New York, 1953–1955.

3.    Kendall, M. G. and Stuart, A., *The Advanced Theory of Statistics,* vol. I: *Distribution Theory*, Griffin, London, 1958.

# Riemann zeta function

1.    Titchmarsh, E. C., *The Zeta Function of Riemann*, Cambridge University Press, London, 1930.

2.    Titchmarsh, E. C., *The Theory of the Riemann Zeta Function*, Oxford University Press, London, 1951.

# Struve functions

1.    Erdélyi, A. et al., *Higher Transcendental Functions*, vol. II, McGraw Hill, New York, 1954.

2.    Gray, A., Mathews, G. B. and MacRobert, T. M., *A Treatise on Bessel Functions and Their Applications to Physics*, 2nd ed., Macmillan, London, 1922.

3.    Watson, G. N., *A Treatise on the Theory of Bessel Functions*, 2nd ed., Cambridge University Press, London, 1958.

# Index of Functions and Constants

This index shows the occurrence of functions and constants used in the expressions within the text The numbers refer to pages on which the function or constant appears.

# F

# G

# M

# N

# O

# P

# Q

# T

# U

# W

# Index of Concepts

# E

# F

Printed and bound by CPI Group (UK) Ltd, Croydon, CR0 4YY

03/10/2024

01040508-0002